Virus Taxonomy

Classification and Nomenclature of Viruses

Eighth Report
of the
International Committee on the Taxonomy of Viruses

Edited by

C.M. Fauquet, M.A. Mayo, J. Maniloff, U. Desselberger and L.A. Ball

Virology Division
International Union of Microbiological Societies

ELSEVIER
ACADEMIC
PRESS

Amsterdam Boston Heidelberg London New York Oxford
Paris San Diego San Francisco Singapore Sydney Tokyo

This book is printed on acid-free paper

Elsevier Academic Press
525 B Street, Suite 1900, San Diego, California 92101-4495, USA
http://www.elsevier.com

Elsevier Academic Press
84 Theobald's Road, London WC1X 8RR, UK
http://www.elsevier.com

Library of Congress Cataloguing in Punlication Data
A catalog record for this book is available from the Library of Congress

British Library Cataloguing in Publication Data
A catalogue record for this book is available from the British Library

02 03 04 05 06 07 08 9 8 7 6 5 4 3 2 1

ISBN 0-12-249951-4

Printed and bound in Chian

Preface and Acknowledgments

The practical need to partition the world of the viruses into distinguishable, universally agreed upon entities is the ultimate justification for developing a virus classification system. The first internationally organized attempts to introduce some order in the bewildering variety of viruses took place at the International Congress of Microbiology held in Moscow in 1966. A Committee was created, later called The International Committee on Taxonomy of Viruses (ICTV) which was given the task of developing a single, universal taxonomic scheme for all the viruses infecting animals (vertebrates, invertebrates and protozoa), plants (higher plants and algae), fungi, bacteria and archaea. Since 1971 the ICTV, operating on behalf of the world community of virologists, has produced the following seven reports describing the current state of virus taxonomy:

ICTV Report	Editors	Reporting ICTV Proceedings at the International Congresses of Virology held in :
The First Report, 1971	P. Wildy	Helsinki, 1968
The Second Report, 1976	F. Fenner	Budapest, 1971 and Madrid, 1975
The Third Report, 1979	R.E.F. Mathews	The Hague, 1978
The Fourth Report, 1982	R.E.F. Mathews	Strasbourg, 1981
The Fifth Report, 1991	R.I.B. Francki, C.M. Fauquet, D.L. Knudson, F. Brown	Sendai, 1984; Edmonton, 1987 and Berlin, 1990
The Sixth Report, 1995	F.A. Murphy, C.M. Fauquet, D.H.L. Bishop, S.A. Ghabrial, A.W. Jarvis, G.P. Martell, M.A. Mayo, M.D. Summers	Glasgow, 1993
The Seventh Report, 2000	M.H.V. van Regenmortel, C.M. Fauquet, D.H.L. Bishop, E.B. Carstens, M.K. Estes, S.M. Lemon, J. Maniloff, M.A. Mayo, D.J. McGeoch, C.R. Pringle, R.B. Wickner	Jerusalem, 1996

The present Eighth Report of the ICTV builds on the accumulated taxonomic construction of its predecessors and records the proceedings of the Committee since 2000, including decisions reached at the eleventh and the twelfth International Congresses of Virology held in Sydney in 1999 and Paris in 2002 respectively, and at mid-term ICTV meetings in 2001, 2002, and 2003.

In 1991, the ICTV agreed that the hierarchical level of species would be defined and added to the categories of genus, subfamily, family and order which were already in use in the universal virus classification system. In this 8th Report, the list of recognized virus species has been further extended and the demarcation criteria used to discriminate between individual virus species in different genera have been spelled out as far as possible. This work is still incomplete and will continue to require the input of the more than 70 Study-Groups who provide the information codified in ICTV Reports. The present Report represents the work of more than 500 virologists world-wide, i.e. the members of the Study Groups, Subcommittees and the Executive Committee of the ICTV. The compilers of the Report wish to express their gratitude to all these virologists.

We are especially grateful to Dr. Claude M. Fauquet, his assistant Ms Patricia Cosgrove and his staff (Vince Abernathy, Joe Iskra, John Lewis, Ben Fofana, and Adrien Fauquet) for taking responsibility respectively for the clerical aspects, the website ICTVbook and ICTVnet, the formatting and layout of the 8th ICTV Report and for the drawing and scanning of all the diagrams and pictures.

For the Committee
L. Andrew Ball
President of the International Committee on Taxonomy of Viruses

Editors for the VIIIth ICTV Report and ICTV Officers

Dr. C.M. Fauquet (Secretary ICTV)
Director of ILTAB
Member of Danforth Plant Science Center
975 N. Warson Rd.
St. Louis, MO 63132
USA
Ph: 1 (314) 587-1241
Fax: 1 (314) 587-1956
E-mail: iltab@danforthcenter.org

Prof. J. Maniloff (Vice-President ICTV)
Univ. Rochester,
Dept. Microbiology & Imm.
School of Medicine & Dentistry,
601 Elmwood Av., P.O. Box 672
Rochester, NY 14642, USA
Tel: 585-275-3413
Fax: 585-473-9573
E-mail: jkmf@rochester.edu

Dr. M.A. Mayo (Secretary ICTV)
Scottish Crop Research Institute
Invergowrie,
Dundee DD2 5DA
Scotland,
UK
Tel: 011-44-38-256-2731
Fax: 011-44-38-256-2426
E-mail: m.mayo@tesco.net

Dr. U. Desselberger (Co-SC Vertebrate Chair)
Virologie et Immunologie Moléculaires
INRA, Domaine de Vilvert
Jouy-en-Josas,
Yvelines 78350,
France
Tel: +33 1 34 65 26 04
Fax:+33 1 34 65 26 21
E-mail: ulrich.desselberger@gv.cnrs-gif.fr

Dr. L.A. Ball (President ICTV)
Univ. of Alabama
Dept. of Microbiology
BBRB 373
1530 3rd Ave. S.
Birmingham, AL 35294-2170
USA
Tel: 1 (205) 934-0864
Fax: 1 (205) 934-1636
E-mail: andyb@uab.edu

ICTV Executive Committee Members

Subcommittee Chairs

Dr. A.A. Brunt (Plant SC Chair)
Brayton
The Thatchway
Angmering
West Suffolk, BN16 4HJ, UK
Tel: 01903 785684
Fax: 01903 785684
E-mail: brunt@hriab.u-net.com

Dr. A.J. Della-Porta (Virus Data SC Chair)
CSIRO Animal Health, Australian Animal Health Laboratory
PO Bag 24, Geelong,
Vic., 3220, Australia
Tel: +61 3 5227 5015
Fax: +61 3 5227 5250
E-mail: Antony.Della-porta@li.csiro.au

Prof. R.W. Hendrix (Prokaryote SC Chair)
Department of Biological Sciences
University of Pittsburgh
A340 Langley Hall
Pittsburgh, PA 15260,
USA
Tel: (1-412) 624-4674
Fax: (1-412) 624-4870
E-mail: rhx+@pitt.edu

Prof. B.I. Hillman (Fungal SC Chair)
Department of Plant Pathology
Foran Hall, Cook College
Rutgers, The State University of New Jersey
59 Dudley Road
New Brunswick, NJ 08901-8520, USA
Tel: (732) 932-9375 X 334
Fax: (732) 932-9377
E-mail: hillman@aesop.rutgers.edu

Dr. D.J. McGeoch (Vertebrate SC Chair)
MRC Virology Unit/Inst. of Virology
Univ. of Glasgow
Church Street
Glasgow G11 5JR
Scotland, UK
Tel: 44-(0)141-330-4645
Fax: 44-(0)141-337-2236
E-mail: d.mcgeoch@vir.gla.ac.uk

Dr. J.M. Vlak (Invertebrate SC Chair)
Laboratory of Virology
Wageningen University
Plant Sciences
Binnenhaven 11
6709 PD Wageningen, The Netherlands
Tel: 31 317 48 30 90
Fax: 31 317 48 48 20
E-mail: just.vlak@wur.nl

Elected Members

Dr. Lois Blaine
American Type Culture Collection
Bioinformatics Division
10801 University Boulevard
Manassas, VA 20110-2209, USA
Phone: (703) 365-2749
Phone: (301) 770 1541
E-mail: lblaine@atcc.org

Dr. H. Bruessow
Nestle Research Center, Nestec Ltd.,
CH-1000 Lausanne 2, Switzerland.
Tel: (+41) 217 858 676
Fax: (+41) 217 858 925
E-mail: harald.bruessow@rdls.nestle.com

Dr. E.B. Carstens
Dept. Microbiology & Imm,
Queens University
207 Stuart Street
Kingston, Ontario K7L 3N6, Canada
Tel: 613-533-2463
Fax: 613-533-6796
E-mail: carstens@post.queensu.ca

Dr. A.-L. Haenni
Laboratoire des Systemes Vegetaux
Institut Jacques Monod
2 place Jussieu, 75251 Paris Cedex 05, France
Tel: (33) 1 44.27.40.35/36
Fax: (33) 1 44.27.35.80
E-mail: haenni@ijm.jussieu.fr

Dr. A.M.Q. King (Vertebrate SC Co-Chair)
Institute for Animal health
Pirbright Laboratory
Ash Road
Woking, Surrey GU24 0NF,
UK
Tel: 01483 232441
Fax: 01483 232448
E-mail: amq.king@bbsrc.ac.uk

Prof. A.C. Palmenberg
Institute for Molecular Virology and
Dept. of Biochemistry
433 Babcock Drive
Madison, WI 53706,
USA
Ph: 608-262-7519
Fax: 608-262-6690
E-mail: acpalmen@facstaff.wisc.edu

Dr. H.J. Vetten
Biologische Bundesanstalt
Messeweg 11-12
38104 Braunschweig,
Germany
Tel: (49) 531 299 3720
Fax: (49) 531 299 3006
E-mail: h.j.vetten@bba.de

List of Contributors to the VIIIth ICTV Report – Part I

Ball, L.A.	Fauquet, C.M.	Mayo, M.A.	Palmenberg, A.C.	Sgro, J.-Y.

List of Contributors to the VIIIth ICTV Report – dsDNA Viruses

Adair B.M.	Claverie J.-M.	He, J.G.	McFadden, G.	Stasiak, K.
Ádám, É.	Davison, A.J.	Hendrix, R.W.	McGeoch, D.J.	Stedman, K.
Almeida, J.	de Jong, J.C.	Hess, M.	Mercer, A.A.	Stoltz, D.B.
Arif, B.M.	de Lamballerie X.	Hou, J.	Miller, L.K.	Strand, M.R.
Bamford, D.M.	de Villiers, E.-M.	Hyatt, A.	Minson, A.C.	Studdert, M.J.
Bath, C.	Delaroque, N.	Jehle, J.	Miyazaki, T.	Summers, M.D.
Beckage, N.E.	Delius, H.	Jensen, P.J.	Moss, B.	Suttle, C.
Benkö, M.	Dixon, L.K.	Johnson, M.	Moyer, R.W.	Theilmann, D.A.
Bernard, H.-U.	Drancourt, M.	Kajon, A.	Nagasaki, K.	Thiem, S.
Bigot, Y.	Dumbell, K.R.	Kidd, A.H.	Nagashima, K.	Thiry, E.
Black, D.N.	Dyall-Smith, M.L.	Kou, G.H.	Newton, I.	Tripathy, D.N.
Blissard, G.W.	Eberle, R.	La Scola, B.	O'Reilly, D.R.	van der Noordaa, D.
Bonami, J.R	Escribano, J.M.	Lanzrein, B.	Pellett, P.E.	van Etten, J.L.
Bonning, B.	Esposito, J.J.	Lefkowitz, E.J.	Prangishvili, D.	Vlak, J.M.
Both, G.W.	Essbauer, S.	Lehmkuhl, H.D.	Pring-Akerblom, P..	Wadell, G.
Bratbak, G.	Federici, B.A.	Li, Q.-G.	Raoult, D.	Walker, D.
Broker, T.	Flegel, T.W.	Lightner, D.V.	Rock, D.L.	Walker, P.W.
Brussaard, C.	Frisque, R.J.	Lo, C.F.	Rohrmann, G.F.	Webb, B.A.
Buck, K.W.	Ghabrial, S.A.	Loh, P.C.	Roizman, B.	Wilkinson, P.J.
Buller, R.M.	Granados, R.R.	Lowy, D.	Russell, W.C.	Williams, T.
Butel, J.S.	Hamm, J.J.	Major, E.O.	Salas, M.L.	Wilson, W.H.
Casjens, S.R.	Harrach, B.	Maniloff, J.	Schroeder, D.S.	Zillig, W.
Cheng, D.	Hayakwa, Y.	Martins, C.	Seligy, V.	zur Hausen, H.J.
Chinchar, V.G.	Hayward, G.S.	Mautner, V.	Skinner, M,A.	

List of Contributors to the VIIIth ICTV Report – ssDNA Viruses

Bendinelli, M.	Chu, P.W.G.	Hino, S.	Niel, C.	Stanley, J.
Bergoin, M.	Dale, J.L.	Hu, J.	Okamoto, H.	Stenger, D.C.
Biagini, P.	Day, L.A.	Katul, L.	Parrish, C.R.	Tattersall, P.
Bisaro, D.M.	Fane, B.	Kojima, M.	Raidal, S.	Teo, G.C.
Bloom, M.E.	Fauquet, C.M.	Linden, R.M.	Randles, J.W.	Thomas, J.E.
Briddon, R.W.	Harding, R.	Mankertz, A.	Ritchie, B.W.	Tijssen, P.
Brown, J.K.	Harrison, B.D.	Mishiro, S.	Rybicki, E.P.	Todd, D.
Brown, K.E.	Hendrix, R.W.	Muzyczka, N.	Sano, Y.	Vetten, H.J.

List of Contributors to the VIIIth ICTV Report – RT Viruses

Boeke, J.D.	Gerlich, W.H.	Lanford, R.	Newbold, J.	Sonigo, P.
Burrell, C.J.	Hahn, B.L.	Linial, M.L.	Quackenbush, S.	Stoye, J.
Casey, J.	Harper, G.	Lockhart, B.E.	Rethwilm, A.	Taylor, J.M.
Eickbush, T.	Howard, C.R.	Löwer, R.	Sandmeyer, S.B.	Tristem, M.
Fan, H.	Hull, R.	Mason, W.S.	Schaefer, S.	Voytas, D.F.
Geering, A.	Kann, M.	Neil, J.	Schoelz, J.E.	Will, H.

List of Contributors to the VIIIth ICTV Report – dsRNA Viruses

Attoui, H.
Bamford, D.H.
Bergoin, M.
Buck, K.W.
Castón, J.R.
Chappell, J.D.
Ciarlet, M.
de Lamballerie, X.
del Vas, M.
Delmas, B.
Dermody, T.S.
Duncan, R.
Fukuhara, T.
Ghabrial, S.A.
Gibbs, M.J.
Harding, R.M.
Hillman, B.I.
Jiang, D.
Kibenge, F.S.B.
Leong, J.C.
Maan, S.
Makkay, A.
Marzachì, R.
Mertens, P.P.C.
Milne, R.G.
Mohd Jaafar, F.
Mundt, E.
Nuss, D.L.
Omura, T.
Patterson, J.L.
Pfeiffer, P.
Ramig, R.F.
Rao, S.
Rigling, D.
Samal, S.K.
Samuel, A.
Stoltz, D.
Suzuki, N.
Upadhyaya, N.M.
Vakharia, V.N.
Wang, C.C.
Wei, C.
Wickner, R.B.
Wu, J.L.
Zhou, Z.H.

List of Contributors to the VIIIth ICTV Report – (-)ssRNA Viruses

Accotto, G.P.
Beaty, B.J.
Benmansour, A.
Brunt, A.A.
Buchmeier, M.J.
Burrell, C.J.
Calisher, C.
Carbone, K.
Casey, J.
Charrel, R.N.
Clegg, J.C.S.
Collins, P.L.
Cox, N.J.
de Miranda, J.R.
Dietzgen, R.G.
Elliott, R.M.
Falk, B.W.
Fang, R.-X.
Feldmann, H.
Gago-Zachert, S.
Garcia, M.L.
Garten, W.
Geisbert, T.W.
Gerlich, W.H.
Goldbach, R.
Gonzalez, J.-P.
Grau, O.
Haenni, A.-L.
Haller, O.
Hongo, S.
Howard, C.R.
Jackson, A.O.
Jahrling, P.B.
Kaverin, N.
Kawaoka, Y.
Klenk, H.-D.
Kolakofsky, D.
Kurath, G.
Lamb, R.A.
Lukashevich, I.S.
Mason, W.S.
Mayo, M.A.
McCauley, J.
Melero, J.A.
Milne, R.G.
Morikawa, T.
Nadin-Davis, S.
Nagai, Y.
Natsuaki, T.
Netesov, S.V.
Newbold, J.
Nichol, S.T.
Nowotny, N.
Oldstone, M.B.A.
Palese, P.
Peters, C.J.
Plyusnin, A.
Pringle, C.R.
Rico-Hesse, R.
Rima, B.K.
Rimstad, E.
Romanowski, V.
Salvato, M.S.
Sanchez, A.
Sasaya, T.
Schmaljohn, C.S.
Schwemmle, M.
Shirako, Y.
Swanepoel, R.
Taylor, J.M.
Tesh, R.B.
Tomonago, K.
Tordo, N.
Toriyama, S.
Torok, V.
Vaira, A.M.
Verbeek, M.
Vetten, H.J.
Volchkov, V.E.
Walker, P.J.
Walsh, J.A.
Webster, R.G.
Will, H.

List of Contributors to the VIIIth ICTV Report – (+)ssRNA Viruses

Abou Ghanem-Sabanadzovic, N.
Accotto, G.P.
Adams, M.J.
Agranovsky, A.A.
Anderson, D.
Ando, T.
Arankalle, A.
Atabekov, J.G.
Ball, L.A.
Bar-Joseph, M.
Barnett, O.W.
Berger, P.H.
Bonami, J.R.
Boonsaeng, V.
Boscia, D.
Bragg, J.N.
Brian, D.
Brinton, M.A.
Brown, F.
Brunt A.A.
Buck, K.W.
Bujarski, J.
Candresse, T.
Carstens, E.
Carter, M.J.
Cavanagh, D.
Chang, P.S.
Christian, P.
Clarke, I.N.
Collett, M.S.
Coutts, R.H.A.
Cowley, J.A.
Culley, A.I.
D'Arcy, C.J.
de Groot, R.J.
de Zoeten, G.A.
Delsert, C.
Deom, C.M.
Ding, S.W.
Dolja, V.V.
Domier, L.L.
Dreher, T.W.
Edwards, M.C.
Emerson, S.U.
Enjuanes, L.
Esteban, R.
Estes, M.K.
Faaberg, K.S.
Falk, B.W.
Fargette, D.
Flegel, T.W.
Foster, G.D.
Frey, T. K.
Gibbs, A.J.
Gibbs, M.J.
Godeny, E.K.
Gonsalves, D.
Gorbalenya, A.E.
Gordon, K.H.J.
Gould, E.A.
Green, K.Y.
Haenni, A.-L.
Hajimorad, R.
Hammond, J.
Hammond, R.W.
Hanada, K.
Hanzlik, T.N.
Heinz, F.X.
Hendry, D.A.
Herrmann, J.
Hill, J.H.
Hillman, B.I.
Holmes, K.V.
Houghton, M.
Hovi, T.
Hu, J.S.
Huang, H.V.
Hull, R.
Hyypiä, T.
Iwanami, T.
Jackson, A.O.
Jelkmann, W.
Johnson, J.E.
Johnson, K.
Jones, A.T.
Jones, T.
Jordan, R.L.
Jupin, I.
Karasev, A.V.
Kashiwazaki, S.
King, A.M.Q.
Kinney, R.M.
Knowles, N.J.
Koenig, R.
Koopmans, M.K.
Lang, A.S.
Le Gall, O.
Lehto, K.
Lemon, S.M.

Lesemann, D.-E.
Lewandowski, D.J.
Lightner, D.V.
Loh, P.C.
Lommel, S.A.
MacLachlan, N.J.
Martelli, G.P.
Masters, P.
Matson, D.O.
Mawassi, M.
Meng, X.-J.
Mengeling, W.L.
Meyers, G.
Milne, R.G.
Minafra, A.
Minor, P.D.
Mitchel, D.K.
Monroe, S.S.
Morozov, S.Yu.
Murant, A.F.
Naidu, R.A.
Nakashima, N.
Nakata, S.
Namba, S.
Neill, J.D.
Nishizawa, T.
Ohki, S.T.
Pallansch, M.A.
Palmenberg, A.C.
Plagemann, P.G.W.
Pringle, F.M.
Purcell, R.H.
Purdy, M.
Revill, P.A.
Rice, C.M.
Richards, K.E.
Robinson, D.J.
Roehrig, J.T.
Roossinck, M.J.
Rottier, P.
Rowhani A.
Rubino, L.
Russo, M.
Rybicki, E.
Sabanadzovic, S.
Sanchez-Fauquier, A.
Sanfaçon, H.
Schlauder, G.G.
Schneemann, A.
Scott, S.
Scotti, P.
Shope, R.E.
Skern, T.
Smith, A.W.
Snijder, E.J.
Solovyev, A.G.
Spaan, W.J.M.
Spence, N.
Stanway, G.
Stenger, D.C.
Strauss, E.G.
Studdert, M.J.
Suttle, C.A.
Taguchi, F.
Talbot, P.
Taliansky, M.E.
Tang, K.
Thiel, H.-J.
Torrance, L.
Tousignant, M.
Tsarev, S.A.
Uyeda, I.
Valkonen, J.
van den Worm, S.
van der Wilk, F.
van Duin, J.
van Zaayen, A.
Vetten, H.J.
Vishnichenko, V.K.
Walker, P.J.
Ward, V.K.
Waterhouse, P.M.
Weaver, S.C.
Wellink, J.
Wetzel, T.
Wisler, C.G.
Wright, P.J.
Yoshikawa, N.
Zavriev, S.K.
Zeddam, J.-L.

List of Contributors to the VIIIth ICTV Report – Unassigned and others

Bar-Joseph, M.
Baron, H.
Brunt, A.A.
Carlson, G.
Christian, P.D.
Cohen, F.E.
DeArmond, S.J.
Diener, T.O.
Desselberger, U.
Flores, R.
Gabizon, R.
Gambetti, P.
Hillman, B.I.
Hope, J.
Kitamoto, T.
Kretzschmar, H.A.
Laplanche, J.-L.
Leibowitz, M.J.
Mayo, M.A.
Owens, R.A.
Palukaitis, P.
Prusiner, S.B.
Randles, J.W.
Scholthof, K.-B.G.
Simon, A.E.
Stanley, J.
Taliansky, M.
Tateishi, J.
Telling, G.
Wickner, R.B.
Will, R.

Contents

Part I : The Universal Taxonomy of Viruses in Theory and Practice

L. Andrew Ball
President of the International Committee on Taxonomy of Viruses

Taxonomy lies at the uneasy interface between biology and logic. The processing of information follows somewhat different rules in these two systems and the role of taxonomy is to reconcile them as tidily as possible. To this end, the International Union of Microbiological Societies (IUMS) charged the International Committee on Taxonomy of Viruses (ICTV) with the task of developing, refining, and maintaining a universal virus taxonomy. The goal of this undertaking is to categorize the multitude of known viruses into a single classification scheme that reflects their evolutionary relationships, i.e. their individual phylogenies. As discussed below however, the uncertainties that surround the origins of viruses and the complexities of their evolution create unique problems for virus taxonomy. Nevertheless, this volume - the 8th in the series of ICTV taxonomic reports - documents the overall success of this 38-year effort which has created a rational, largely satisfying, and above all useful taxonomic structure that facilitates communication among virologists around the world and enriches our understanding of virus biology. The 8th ICTV Report depicts the current status of virus taxonomy in 2004; it is a tribute to the hundreds of virologists who contributed to it. Periodic updates are published in the Virology Division News (VDN) section of *Archives of Virology* and posted on the ICTV website:
http://www.danforthcenter.org/iltab/ictvnet/asp/_MainPage.asp.

Viral taxa. The 7th ICTV Report (1) formalized for the first time the concept of the virus species as the lowest taxon (group) in a branching hierarchy of viral taxa. As defined therein, '*a virus species is a polythetic class of viruses that constitute a replicating lineage and occupy a particular ecological niche*' (2). A 'polythetic class' is one whose members have several properties in common, although they do not necessarily all share a single common defining property. In other words, the members of a virus species are defined collectively by a consensus group of properties. Virus species thus differ from the higher viral taxa, which are 'universal' classes and as such are defined by properties that are necessary for membership. These issues have been presented and debated at length in the literature (see reference 3 and citations therein), and they will not be revisited here except to reiterate a few basic points:

- Viruses are real physical entities produced by biological evolution and genetics, whereas virus species and higher taxa are abstract concepts produced by rational thought and logic (3). The virus/species relationship thus represents the front line of the interface between biology and logic.
- Viruses (including virus isolates, strains, variants, types, sub-types, serotypes, etc.) should wherever possible be assigned as members of the appropriate virus species, although many viruses remain unassigned because they are inadequately characterized.
- All virus species must be represented by at least one virus isolate.
- Almost all virus species are members of recognized genera. A few species remain unassigned in their families although they have been clearly identified as new species.
- Some genera are members of recognized sub-families.
- All sub-families and most genera are members of recognized families. Some genera are not yet assigned to a family; in the future they may either join an existing family or constitute a new family with other unassigned genera. For example, the family *Flexiviridae* was recently created by grouping the following (formerly unassigned) genera: *Potexvirus, Carlavirus, Allexivirus, Vitivirus, Mandarivirus, Foveavirus, Capillovirus,* and *Trichovirus.*
- Some families are members of the following recognized orders: *Caudovirales, Nidovirales* and *Mononegavirales*.
- The hierarchy of recognized viral taxa is therefore:
  (Order)
  Family
  (Sub-family)
  Genus
  Species
- Only the aforementioned taxa are recognized by the ICTV. Other groupings (from clade to super-family), may communicate useful descriptive information in some circumstances but they have no formally recognized taxonomic meaning. Similarly, the term 'quasi-species', although it captures an important concept (4), has no recognized taxonomic meaning.

The creation or elimination, (re)naming, and (re)assignment of a virus species, genus, (sub)family, or order are all taxonomic acts that require public scrutiny and debate, leading to formal approval by the full membership of the ICTV (see below). For detailed instructions on how to initiate this process, see Article 6 of the Statutes of the ICTV and Rule 3.20 of the International Code of Virus Taxonomy and Nomenclature in this volume. In contrast, the naming of a virus isolate and its assignment to a pre-existing species are not considered taxonomic acts and therefore do not require formal ICTV approval. Instead they will typically be accomplished by publication of a paper describing the virus isolate in the peer-reviewed virology literature.

The ~1550 virus species that were introduced for the first time in the 7th ICTV Report were assigned the common (mostly English) names of representative member viruses, and this practice has been extended to the ~1950 virus species listed in the 8th Report. The only distinction is that species names are italicized whereas virus names are not (1, 5). While the decisions that led to these new conventions have been vigorously debated (6-17), it is hard to imagine that the alternative – the creation *de novo* of ~1950 species names – would have been preferable to most virologists.

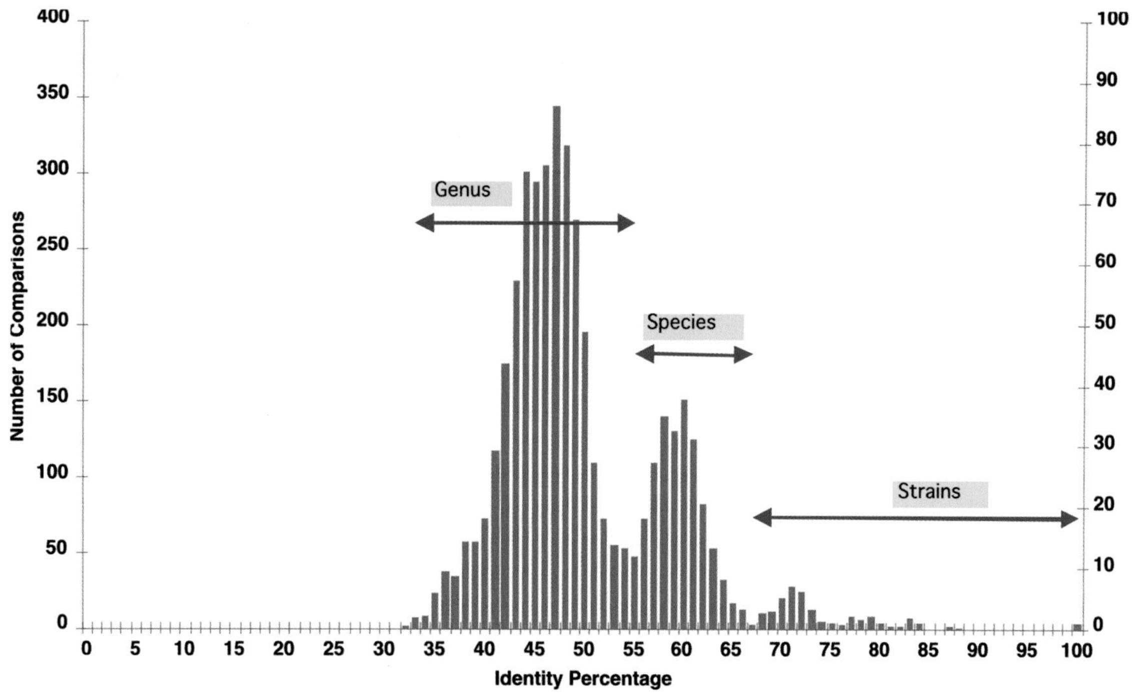

Figure 1. Distribution of pairwise identity percentages calculated for the sequences of the L1 gene of members of the family *Papillomaviridae* (Courtesy of C.M. Fauquet).

Demarcation criteria. Consistent assignment of viruses to taxa requires the specification of demarcation criteria, particularly at the species level where the differences are the smallest. Since virus species are polythetic, multiple demarcation criteria are needed to reliably delineate different species, and lists of the criteria used for each genus can be found in the corresponding descriptions. Within most genera, sequence comparisons are an increasingly dominant demarcation criterion because they provide a quantitative measure of divergence. In general, pairwise sequence identity profiles show well resolved peaks that represent typical evolutionary distances between strains, species, and genera within a given virus family (Figure 1). Different virus families may show similar overall patterns, but the absolute evolutionary distances between taxa may be different (2). Such genetic analyses provide a reassuring validation of taxonomic assignments, but they should not supersede species demarcation based on a balanced and multifaceted examination that includes phenotypic criteria. After all, virus taxonomy aims to classify organisms rather than simply to define the lineages of their genes.

Type species. Each genus contains a designated 'type species'. This is the virus species whose name typifies the name of the genus i.e. it is the 'nomenclatural type' of the genus, and its recognition creates an indissoluble link between the species and the genus. The type species is not necessarily the best characterized species in the genus and it may not even be a typical member (18). Rather, type species status is usually conferred on the species that necessitated the original creation and naming of the genus, and it should therefore seldom, if ever, be changed. Higher taxa do not have designated type species, and a proposal in the 4th ICTV Report (19) that each family should have a designated 'type genus' has since been revoked. The following definition of 'type species' has been proposed for the International Code of Virus Taxonomy and Nomenclature (18):

> "A type species is a species whose name is linked to the use of a particular genus name. The genus so typified will always contain the type species."

Abbreviations. As a general rule, there is no need to abbreviate the names of virus species because they will be used once or at most a few times in a typical paper (16). In contrast, it is convenient and appropriate to abbreviate the names of viruses, and this volume lists and uses many such abbreviations. However, there are many instances of identical abbreviations being used for very different viruses that infect plants, animals and insects. The ICTV does not formally define or endorse any particular virus name abbreviation, nor does it set rules for abbreviating virus names.

A practical guide to orthography (i.e. when to italicize). Once the distinction between a virus and a virus species is understood, when to use italics usually becomes clear: virus species names should be italicized, whereas the names of viruses themselves should not. However sentences written without the virus/species distinction in mind can be ambiguous as to whether they refer to a virus itself or to its species, and are best rewritten to resolve the ambiguity. Informal taxonomic names are widely used and should not be italicized or capitalized. For example, the informal name 'vesiculovirus' refers to a member of the genus *Vesiculovirus*; the informal name 'rhabdovirus' refers to a member of the family *Rhabdoviridae*; and the informal name 'mononegavirus' refers to a member of the order *Mononegavirales*. The International Code of Virus Taxonomy and Nomenclature in this volume gives more detailed instructions on how to correctly portray the names of viruses and their taxa.

Taxonomy and phylogeny. Since the universal scheme of virus taxonomy was derived from classical Linnaean systematics, it is well suited for classifying organisms that are related to one another via simple branched and diverging descent, with relatively long evolutionary distances between successive branch points. However, virus evolution differs from this simple paradigm in several ways. First, it seems unlikely that all viruses are descended from a single original 'protovirus'. Multiple origins appear more likely, but it is unclear how many there were or when they occurred. These uncertainties jeopardize the possibility of meaningful virus taxonomy above the level of order. Secondly, recombination and reassortment are common among some viruses, resulting in chimeric organisms with polyphyletic genomes (20, 21). It is logically impossible to accurately represent such multi-dimensional phylogeny in a monophyletic scheme. The incorporation of host genes into some viral genomes can further complicate the situation. Thirdly, viruses that can integrate into the genome of the host, such as the retroviruses and lysogenic bacteriophages, experience and respond to profoundly different selective pressures as they switch between vertical and horizontal modes of transmission by moving into and out of the host genome. Finally, viruses that infect both vertebrates and invertebrates (or other pairs of widely disparate hosts) can be expected to evolve very differently in their different host species. All these factors add complexity to virus evolution and compromise the relationship between taxonomy and phylogeny. The net result is that in any taxonomic arrangement some viruses and even some taxa will always be misfits.

Higher virus taxonomy. The uncertainties that surround the origins of viruses undermine the prospects for integrating all virus families into a single phylogenetic tree with a corresponding global taxonomy. While we can be confident that all members of a virus species and genus share common ancestors, this confidence begins to diminish at the higher taxonomic levels, which explains why there are so few recognized virus orders and no virus classes or higher taxa. However, as the abundance of sequence information and the power of methods for sequence comparisons increase it is likely that more distant phylogenetic relationships will become evident. The task of the ICTV is to recognize and reflect the collective judgment of the appropriate specialists on when to confer taxonomic status on such emergent relationships.

How to propose a taxonomic change. Any change of name or taxonomic status of a virus species, genus, sub-family, family, or order requires approval by the full membership of the ICTV before it is formally accepted by the scientific community. This includes the creation and naming of new taxa that may be deemed necessary to accommodate newly characterized viruses. For instructions on how to initiate this process, see Article 6 of the Statutes of the ICTV and Rule 3.20 of the International Code of Virus Taxonomy and Nomenclature in this volume. However most authors with new data will be ready to publish before the approval process is complete, or in some cases before it has even begun. It should be recognized by authors, referees and editors that while such publications often make valuable contributions to the taxonomic debate, the proposals they contain have no formal status, nor do they establish any sort of nomenclatural precedence (see Rules 3.10 and 2.5 in the International Code of Virus Taxonomy and Nomenclature).

Anyone wishing to propose a taxonomic change should access the ICTV website at http://www.danforthcenter.org/iltab/ictvnet/asp/ MainPage.asp and complete the appropriate taxonomic proposal template. All taxonomic proposals are first circulated to the members of the appropriate Study Group and Subcommittee for specialist review and consideration, then posted for general comments on a site accessible to all ICTV members, and then reviewed by the Executive Committee at its next annual meeting. After further rounds of consideration and review, proposals that garner the support of a majority of EC members are submitted to the full ICTV for final ratification. Approved proposals are published in the ICTV Taxonomic Reports (i.e. this volume), and in the intervening years in the Virology Division News (VDN) section of *Archives of Virology*. A current taxonomic index of approved names of virus species and other taxa is available online at http://phene.cpmc.columbia.edu/.

ICTV database (ICTVdB). Since 1991, the ICTV has been working towards the development of a comprehensive and universal database containing virus isolate data. The goals of this initiative are to provide the research community and others with online tools for precisely identifying viruses at the isolate level from constellations of characteristics, and to create reliable links to the agreed virus taxonomy on the one hand and the genome sequence databases on the other (22-24). The ICTVdB website is http://phene.cpmc.columbia.edu/ and the anticipated launch date is summer 2005.

According to authoritative estimates, the majority of virus gene sequences in both GenBank and the EMBL databases are either unassigned or misassigned with respect to the virus isolates from which they were generated. In order to prevent any further deterioration in this deplorable situation, the ICTV Executive Committee decided that future proposals for the recognition of new virus species will be considered only when supported by both a sequence accession number and data from one or more isolates entered in the ICTV database (25). It is to be hoped that the entry of virus isolate data into the ICTVdB will soon become as routine as sequence deposition in publishing the description of a new virus.

Virus taxonomy 2004. The advent of nucleotide sequence determination has revolutionized biology and largely rationalized taxonomy, including that of viruses. The universal virus taxonomy presented in this Report provides a classification scheme that is supported by verifiable data and expert consensus. It is an indispensable framework both for further study of the ~1950 currently recognized virus species and for the identification and characterization of newly emergent viruses, whether they result from natural, accidental, or deliberate dissemination. The current health of virus taxonomy is due to the efforts of hundreds of virologists from around the world, but more volunteers are always needed. Those interested in contributing their expertise are encouraged to contact the relevant Study Group Chair or any member of the ICTV Executive Committee.

References

1. van Regenmortel, M.H.V., Fauquet, C.M., Bishop, D.H.L., Carstens, E.B., Estes, M.K., Lemon, S.M., McGeogh, D.J., Maniloff, J., Mayo, M.A., Pringle, C.R. and Wickner, R.B., (eds) (2000). *Virus Taxonomy. Seventh Report of the International Committee on Taxonomy of Viruses.* Academic Press, San Diego, CA.
2. van Regenmortel, M.H.V. (2000). Introduction to the species concept in virus taxonomy. In: *Virus Taxonomy. Seventh Report of the International Committee on Taxonomy of Viruses* (van Regenmortel, M.H.V., Fauquet, C.M., Bishop, D.H.L., Carstens, E.B., Estes, M.K., Lemon, S.M., McGeogh, D.J., Maniloff, J., Mayo, M.A., Pringle, C.R. and Wickner, R.B., eds), pp. 3-16. Academic Press, San Diego, CA.
3. van Regenmortel, M.H.V. (2003). Viruses are real, virus species are man-made, taxonomic constructions. *Arch. Virol.*, **148**, 2483–2490.
4. Domingo, E., Escarmis, C., Sevilla, N., Moya, A., Elena, S.F., Quer, J., Novella, I.S. and Holland J.J. (1996). Basic concepts in RNA virus evolution. *FASEB J.*, **10**, 859-864.
5. van Regenmortel, M.H.V. (1999). How to write the names of virus species. *Arch. Virol.*, **144**, 1041–1042.
6. Bos, L. (1999). The naming of viruses: an urgent call to order. *Arch. Virol.*, **144**, 631–636.
7. Bos, L. (2000). Structure and typography of virus names. *Arch. Virol.*, **145**, 429–432.
8. Gibbs, A.J. (2000). Virus nomenclature descending into chaos. *Arch. Virol.*, **145**, 1505–1507.
9. van Regenmortel, M.H.V. (2000). On the relative merits of italics, Latin and binomial nomenclature in virus taxonomy. *Arch. Virol.*, **145**, 433–441.
10. van Regenmortel, M.H.V., Mayo, M.A., Fauquet, C.M. and Maniloff, J. (2000). Virus nomenclature: consensus versus chaos. *Arch. Virol.*, **145**, 2227–2232.
11. van Regenmortel, M.H.V. (2001). Perspectives on binomial names of virus species. *Arch. Virol.*, **146**, 1637–1640.
12. Bos, L. (2002). International naming of viruses – a digest of recent developments. *Arch. Virol.*, **147**, 1471–1477.
13. van Regenmortel, M.H.V. and Fauquet, C.M. (2002). Only italicized species names of viruses have a taxonomic meaning. *Arch. Virol.*, **147**, 2247–2250.
14. Drebot, M.A., Henchal, E., Hjelle, B., LeDuc, J.W., Repik, J.T., Roehrig, P.M., Schmaljohn, C.S., Shope, R.E., Tesh, R.B., Weaver, S.C. and Calisher, C.H. (2002). Improved clarity of meaning from the use of both formal species names and common (vernacular) virus names in virological literature. *Arch. Virol.*, **147**, 2465–2472.
15. Calisher, C.H. and Mahy, B.W.J. (2003). Taxonomy: get it right or leave it alone (editorial). *Am. J. Trop. Med. Hyg.*, **68**, 505-506.
16. van Regenmortel, M.H.V. and Mahy, B.W.J. (2004). Emerging issues in virus taxonomy. *Emerg. Infect. Dis.*, **10**, 8–13.
17. Eberhardt, M. (2004). Virus taxonomy: one step forward, two steps back. *Emerg. Infect. Dis.*, **10**, 153-154.
18. Mayo, M.A., Maniloff, J., van Regenmortel, M.H.V. and Fauquet, C.M. (2002). The Type Species in virus taxonomy. *Arch. Virol.*, **147**, 1271-1274.
19. Matthews R.E.F. (ed) (1982). Classification and nomenclature of viruses. Fourth Report of the International Committee on Taxonomy of Viruses. *Intervirology*, **17**, 1–199.
20. Pedulla, M.L., Ford, M.E., Houtz, J.M., Karthikeyan, T., Wadsworth, C., Lewis, J.A., Jacobs-Sera, D., Falbo, J., Gross, J., Pannunzio, N.R., Brucker, W., Kumar, V., Kandasamy, J., Keenan, L., Bardarov, S., Kriakov, J., Lawrence, J.G. Jacobs, W.R. Jr., Hendrix, R.W. and Hatfull, G.F. (2003). Origins of highly mosaic mycobacteriophage genomes. *Cell*, **113**, 171–182.
21. Lawrence, J.G., Hatfull, G.F. and Hendrix, R.W. (2002). Imbroglios of viral taxonomy: genetic exchange and failings of phenetic approaches. *J. Bacteriol.*, **184**, 4891-4905.
22. Büchen-Osmond C., Blaine, L. and Horzinek M.C. (2000). The universal virus database of ICTV (ICTVdB). In: *Virus Taxonomy. Seventh Report of the International Committee on Taxonomy of Viruses.* (van Regenmortel, M.H.V., Fauquet, C.M., Bishop, D.H.L., Carstens, E.B., Estes, M.K., Lemon, S.M., McGeogh, D.J., Maniloff, J., Mayo, M.A., Pringle, C.R. and Wickner, R.B., eds), pp 19-24. Academic Press, San Diego, CA.
23. Büchen-Osmond, C. (2003). Taxonomy and classification of viruses. In: Manual of Clinical Microbiology, (8th ed., Vol. 2, pp. 1217-1226), ASM Press, Washington DC.
24. Büchen-Osmond, C. (2003). The universal virus database ICTVdB. *Computing science eng.*, **5**, 16-25.
25. Ball, L.A. and Mayo, M.A. (2004). Report from the 33rd meeting of the ICTV Executive Committee. *Arch. Virol.*, **149**, 1259-1263.

Part II: The Viruses

Fauquet, C.M. and Mayo, M.A.

This report describes the taxa and viruses approved by the ICTV between 1970 and 2003. Descriptions of the most important characteristics of these taxa are provided, together with a list of species and tentative species and selected references. These descriptions represent the work of the chairpersons and members of the Subcommittees and Study-Groups of the ICTV. A glossary of abbreviations and terms is provided first, followed by a set of virus diagrams per type of host and listings of the taxa, by type of nucleic acid and size of the genome.

The different types of viruses infecting all sorts of hosts are depicted in three different ways: 1) virus diagrams are scale are represented per type of host, 2) the same diagrams are all assembled into the now famous "Virosphere" representing viruses by their biochemical structure and type of hosts, and 3) for the first time we are representing 4 pages of various virus structures at the atomic resolution level, to provide to the readers ideas about the variability and sizes of a range of viruses infecting vertebrates, invertebrates, fungi, protozoa, algae, bacteria, mycoplasma, and plants.

The names of orders, families, subfamilies, genera and species approved by the ICTV are printed in italics. Names that have not yet been approved are printed in quotation marks in standard type. Tentative species names, strain, serotype, genotype and isolate names are printed in standard type.

Throughout the Report, three categories of viruses of the various taxa have been defined: (1) *Type species:* pertains to the type species used in defining the taxon. As noted above, the choice of the type species by ICTV is not made with the kind of precision that must be used by international special groups and culture collections or when choosing substrates for vaccines, diagnostic reagents, etc. In this regard, the designation of prototype viruses and strains must be seen as a primary responsibility of international specialty groups. (2) *List of species:* other species and isolates of these species which on the basis of all present evidence definitely belong to the taxon. (3) *Tentative species:* pertains to those viruses for which there is presumptive but not conclusive evidence favoring membership of the taxon. A very limited number of species are pending ratification by the ICTV and they are marked with the sign ‡.

The ICTV has approved three orders, 73 families, 9 subfamilies, 287 genera and more than 5450 viruses belonging to more than 1950 species. Descriptions of virus satellites, viroids and the agents of spongiform encephalopathies (prions) of humans and several animal and fungal species are included. Finally a list of unassigned viruses is provided with a pertinent reference for each.

The VIIIth ICTV report is illustrated by 436 electron microscope pictures, diagrams of virus particles, diagrams of genome organization and phylogenetic trees. Most of those have been provided by the authors of the virus description but an important source of virus particle computer rendering images was the VIPER website (http://viperdb.scripps.edu/)(Reddy et al., (2001). Virus Particle Explorer (VIPER), a Website for virus capsid structures and their computational analysis. *J. Virol.* **75**:11943-11947).

Glossary of Abbreviations and Virological Terms

In addition to universally accepted abbreviations such as DNA and RNA, it is common practice to use abbreviations for virological and technical words in virology. In the Report, we have adopted commonly used abbreviations (e.g. CP for capsid protein and NC for nucleocapsid). These have been approved by the Executive Committee of ICTV for use in the ICTV Report but have no official status. The abbreviations will be used without definition throughout the book, except in a few instances where Study-Group conventions (e.g. C in place of CP for capsid protein) dictate different practice. In these instances, abbreviations are defined locally.

ABBREVIATIONS

aa	amino acid(s)
bp	base pair(s)
CF	complement fixing
CP	capsid/coat protein
CPE	cytopathic effect
D	diffusion coefficient
DI	defective interfering
DNAse	desoxyribonuclease
ds	double-stranded
gRNA	genomic RNA
HE	hemagglutination esterase
Hel	helicase
HI	hemagglutination inhibition
hr(s)	hour(s)
IRES	internal ribosome entry structure
kbp	kilobase pairs
kDa	kilodalton
min	minutes(s)
MP	movement protein
Mr	relative molar mass
mRNA	messenger RNA
Mtr	methyltransferase
N	nucleoprotein
NC	nucleocapsid
NES	nuclear export signal
NLS	nuclear localization signal
NNS	non-segmented negative strand
nt	nucleotide(s)
NTR	non-translated region
ORF	open reading frame
PAGE	poly acrylamid gel electrophoresis
PCR	polymerase chain reaction
Pol	polymerase
Pro	protease
RdRp	RNA-dependent RNA polymerase
Rep	replication associated protein
RF	replicative from
RFLP	restriction fragment length polymorphism
RI	replicative intermediate

RNAse	ribonuclease
RNP	ribonucleoprotein
RT	reverse transcriptase
sgRNA	subgenomic RNA
ss	single-stranded
T	triangulation number
UTR	untranslated region
VPg	genome-linked protein

RNA Replicases, Transcriptases and Polymerases

In the synthesis of viral RNA, the term polymerase has been replaced in general by two somewhat more specific terms: RNA replicase and RNA transcriptase. The term transcriptase has become associated with the enzyme involved in messenger RNA synthesis, most recently with those polymerases which are virion-associated. However, it should be borne in mind that for some viruses it has yet to be established whether or not the replicase and transcriptase activities reflect distinct enzymes rather than alternative activities of a single enzyme. Confusion also arises in the case of the small positive-sense RNA viruses where the term replicase (e.g., Qβ replicase) has been used for the enzyme capable both of transcribing the genome into messenger RNA via an intermediate negative-sense strand and of synthesizing the genome strand from the same template. In the text, the term replicase will be restricted as far as possible to the enzyme synthesizing progeny viral strands of either polarity. The term transcriptase is restricted to those RNA polymerases that are virion-associated and synthesize mRNA. The general term RNA polymerase (i.e., RNA-dependent RNA polymerase) is applied where no distinction between replication and transcription enzymes can be drawn (e.g., Qβ, R 17, *Poliovirus* and many plant viruses).

Other Definitions

Enveloped: possessing an outer (bounding) lipoprotein bilayer membrane.

Positive-sense (= plus strand, message strand) RNA: the strand that contains the coding triplets that are translated by ribosomes.

Positive-sense DNA: the strand that contains the same base sequence as the mRNA. However, mRNAs of some dsDNA viruses are transcribed from both strands and the transcribed regions may overlap. For such viruses this definition is inappropriate.

Negative sense (= minus strand): for RNA or DNA, the negative strand is the strand with base sequence complementary to the positive-sense strand.

Pseudotypes: enveloped virus particles in which the envelope is derived from one virus and the internal constituents from another.

Surface projections (= spikes, peplomers, knobs): morphological features, usually consisting of glycoproteins, that protrude from the lipoprotein envelope of many enveloped viruses.

Virion: morphologically complete virus particle.

Viroplasm: (= virus factory, virus inclusion, X-body): a modified region within the infected cell in which virus replication occurs, or is thought to occur.

TABLE I. Families and Genera of Viruses Listed According to the Nature of the Genome

FAMILY OR UNASSIGNED GENUS	NATURE OF THE GENOME	PRESENCE OF AN ENVELOPE	MORPHOLOGY	GENOME CONFIGURATION	GENOME SIZE kbp or kb	HOST
Myoviridae	dsDNA	-	tailed phage	1 linear	34-169	B, Ar
Siphoviridae	dsDNA	-	tailed phage	1 linear	22-121	B, Ar
Podoviridae	dsDNA	-	tailed phage	1 linear	16-70	B
Tectiviridae	dsDNA	-	isometric	1 linear	15	B
Corticoviridae	dsDNA	-	isometric	1 circular supercoiled	10	B
Plasmaviridae	dsDNA	+	pleomorphic	1 circular supercoiled	12	Ms
Lipothrixviridae	dsDNA	+	Filament- or rod-shaped	1 linear	16-42	Ar
Rudiviridae	dsDNA	-	rod-shaped	1 linear	32-35	Ar
Fuselloviridae	dsDNA	+	lemon-shaped	1 circular supercoiled	15-18	Ar
Salterprovirus	dsDNA	+	Lemon-shaped	1 linear	14.5	Ar
Guttaviridae	dsDNA	+	droplet-shaped	1 circular	20	Ar
Poxviridae	dsDNA	+	pleomorphic	1 linear	130-375	V, I
Asfarviridae	dsDNA	+	spherical	1 linear	170-190	V
Iridoviridae	dsDNA	+/-	isometric	1 linear	135-303	V, I
Phycodnaviridae	dsDNA	-	isometric	1 linear	100-560	Al
Baculoviridae	dsDNA	+	Rod-shaped	1 circular supercoiled	80-180	I
Nimaviridae	dsDNA	+	Ovoid/bacilliform	1 circular	300	I
Herpesviridae	dsDNA	+	isometric	1 linear	125-240	V
Adenoviridae	dsDNA	-	isometric	1 linear	26-45	V
Rhizidiovirus	dsDNA	-	isometric	1 linear	27	F
Polyomaviridae	dsDNA	-	isometric	1 circular	5	V
Papillomaviridae	dsDNA	-	isometric	1 circular	7-8	V
Polydnaviridae	dsDNA	+	rod, fusiform	multiple supercoiled	150-250	I
Ascoviridae	dsDNA	+	Bacilliform, ovoidal, allantoid	1 circular	120-180	I
Mimivirus	dsDNA	-	isometric	1 circular	~800	Pr
Inoviridae	ssDNA	-	Rod-shaped, filamentous	1 + circular	5-9	B, Ms
Microviridae	ssDNA	-	isometric	1 + circular	4-6	B, Sp
Geminiviridae	ssDNA	-	isometric	1 or 2 +/- circular	3-6	P
Circoviridae	ssDNA	-	isometric	1 - or +/- circular	2	V
Anellovirus	ssDNA	-	isometric	1 - circular	3-4	V
Nanovirus	ssDNA	-	isometric	6-9 + circular	6-9	P
Parvoviridae	ssDNA	-	isometric	1 +/- linear	4-6	V, I
Hepadnaviridae	dsDNA-RT	+	spherical	1 linear	3-4	V
Caulimoviridae	dsDNA-RT	-	isometric, bacilliform	1 circular	7-9	P
Pseudoviridae	ssRNA-RT	-	spherical	1 + segment	5-9	P,I,Pr
Metaviridae	ssRNA-RT	-	spherical	1 + segment	4-10	F,P,I,V
Retroviridae	ssRNA-RT	+	spherical	1 dimer + segment	7-13	V
Cystoviridae	dsRNA	+	spherical	3 segments	13	B
Reoviridae	dsRNA	-	isometric	10-12 segments	19-32	V,I,P,F
Birnaviridae	dsRNA	-	isometric	2 segments	5-6	V, I
Totiviridae	dsRNA	-	isometric	1 segment	4-7	F, Pr
Partitiviridae	dsRNA	-	isometric	2 segments	3-6	P, F
Chrysoviridae	dsRNA	-	isometric	4 segments	13	F
Hypoviridae	dsRNA	-	pleomorphic	1 segment	9-13	F
Endornavirus	dsRNA	-	none	1 segment	14-18	P

FAMILY OR UNASSIGNED GENUS	NATURE OF THE GENOME	PRESENCE OF AN ENVELOPE	MORPHOLOGY	GENOME CONFIGURATION	GENOME SIZE kbp or kb	HOST
Bornaviridae	NssRNA	+	spherical	1 - segment	9	V
Rhabdoviridae	NssRNA	+	bullet-shaped, bacilliform	1 - segment	11-15	V, I, P
Filoviridae	NssRNA	+	Bacilliform, filamentous	1 - segment	~19	V
Paramyxoviridae	NssRNA	+	pleomorphic	1 - segment	13-18	V
Varicosavirus	NssRNA	-	rod-shaped	2 - segments	13	P
Orthomyxoviridae	NssRNA	+	pleomorphic	6-8 - segments	10-15	V
Bunyaviridae	NssRNA	+	spherical	3 - or +/- segments	11-19	V, P, I
Tenuivirus	NssRNA	-	filamentous	4-6 - or +/- segments	17-18	P, I
Ophiovirus	NssRNA	-	filamentous	3/4 - segments	11-12	P
Arenaviridae	NssRNA	+	spherical	2 +/- segments	11	V
Deltavirus	NssRNA	+	spherical	1 – circular	2	V
Leviviridae	ssRNA	-	isometric	1 + segment	3-4	B
Narnaviridae	ssRNA	-	RNP complex	1 + segment	2-3	F
Picornaviridae	ssRNA	-	isometric	1 + segment	7-9	V
Iflavirus	ssRNA	-	isometric	1 + segment	9-10	I
Dicistroviridae	ssRNA	-	isometric	1 + segment	9-10	I
Marnaviridae	SsRNA	-	isometric	1 + segment	9	Al
Sequiviridae	ssRNA	-	isometric	1 + segment	10-12	P
Sadwavirus	SsRNA	-	isometric	2 + segments	11-12	P
Cheravirus	SsRNA	-	isometric	2 + segments	10	P
Comoviridae	ssRNA	-	isometric	2 + segments	9-15	P
Potyviridae	ssRNA	-	filamentous	1 / 2 + segments	8-12	P
Caliciviridae	ssRNA	-	isometric	1 + segment	7-8	V
Hepevirus	ssRNA	-	isometric	1 + segment	7	V
Astroviridae	ssRNA	-	isometric	1 + segment	6-7	V
Nodaviridae	ssRNA	-	isometric	2 + segments	4-5	V, I
Tetraviridae	ssRNA	-	isometric	1 or 2 + segment	6-8	I
Sobemovirus	ssRNA	-	isometric	1 + segment	4-5	P
Luteoviridae	ssRNA	-	isometric	1 + segment	5-6	P
Umbravirus	ssRNA	-	RNP complex	1 + segment	4	P
Tombusviridae	ssRNA	-	isometric	1 / 2 + segments	4-5	P
Coronaviridae	ssRNA	+	spherical	1 + segment	28-31	V
Arteriviridae	ssRNA	+	spherical	1 + segment	13-16	V
Roniviridae	SsRNA	+	Bacilliform	1 + segment	26	I
Flaviviridae	ssRNA	+	spherical	1 + segment	10-12	V, I
Togaviridae	ssRNA	+	spherical	1 + segment	10-12	V, I
Tobamovirus	ssRNA	-	rod-shaped	1 + segment	6-7	P
Tobravirus	ssRNA	-	rod-shaped	2 + segments	9-11	P
Hordeivirus	ssRNA	-	rod-shaped	3 + segments	9-11	P
Furovirus	ssRNA	-	rod-shaped	2 + segments	10-11	P
Pomovirus	ssRNA	-	rod-shaped	3 + segments	~12	P
Pecluvirus	ssRNA	-	rod-shaped	2 + segments	10	P
Benyvirus	ssRNA	-	rod-shaped	4/5 + segments	13-16	P
Bromoviridae	ssRNA	-	Isometric, bacilliform	3 + segments	8-9	P
Ourmiavirus	ssRNA	-	bacilliform	3 + segments	5	P
Idaeovirus	ssRNA	-	isometric	2 + segments	8	P
Tymoviridae	ssRNA	-	isometric	1 + segment	6-8	P, I
Closteroviridae	ssRNA	-	filamentous	1/2 + segments	15-19	P
Flexiviridae	SsRNA	-	filamentous	1 + segment	6-9	P
Barnaviridae	ssRNA	-	bacilliform	1 + segment	4	F

Abbreviations of the virus hosts

Algae	Al	Bacteria	B	Invertebrates	I	Protozoa	Pr
Archaea	Ar	Fungi	F	Plants	P	Vertebrates	V

Families and Genera of Viruses Infecting Vertebrates

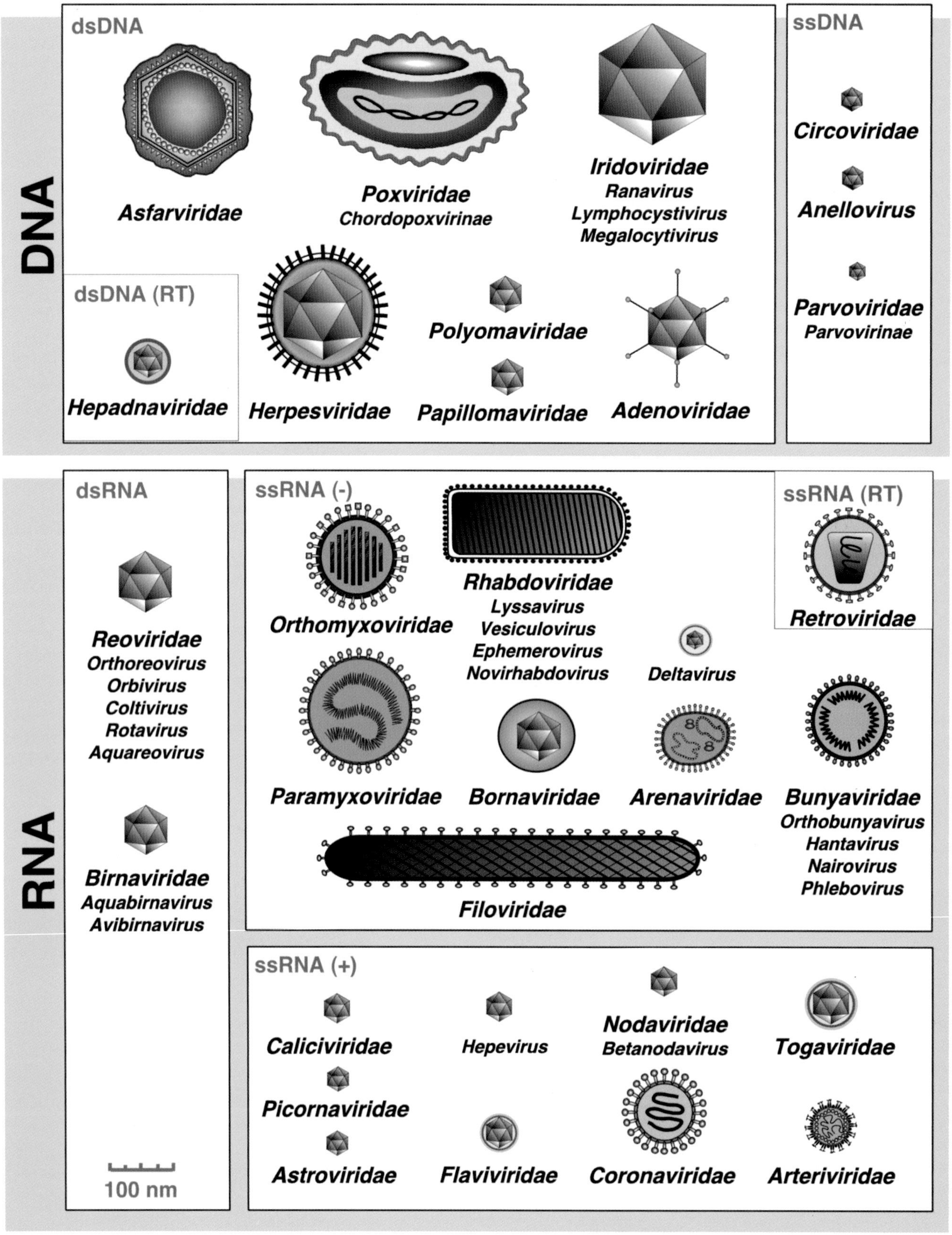
DNA
dsDNA
Asfarviridae
Poxviridae
Chordopoxvirinae
Iridoviridae
Ranavirus
Lymphocystivirus
Megalocytivirus
dsDNA (RT)
Hepadnaviridae
Herpesviridae
Polyomaviridae
Papillomaviridae
Adenoviridae
ssDNA
Circoviridae
Anellovirus
Parvoviridae
Parvovirinae
RNA
dsRNA
Reoviridae
Orthoreovirus
Orbivirus
Coltivirus
Rotavirus
Aquareovirus
Birnaviridae
Aquabirnavirus
Avibirnavirus
100 nm
ssRNA (-)
Orthomyxoviridae
Rhabdoviridae
Lyssavirus
Vesiculovirus
Ephemerovirus
Novirhabdovirus
Deltavirus
ssRNA (RT)
Retroviridae
Paramyxoviridae
Bornaviridae
Arenaviridae
Bunyaviridae
Orthobunyavirus
Hantavirus
Nairovirus
Phlebovirus
Filoviridae
ssRNA (+)
Caliciviridae
Hepevirus
Nodaviridae
Betanodavirus
Togaviridae
Picornaviridae
Astroviridae
Flaviviridae
Coronaviridae
Arteriviridae

Families and Genera of Viruses Infecting Invertebrates

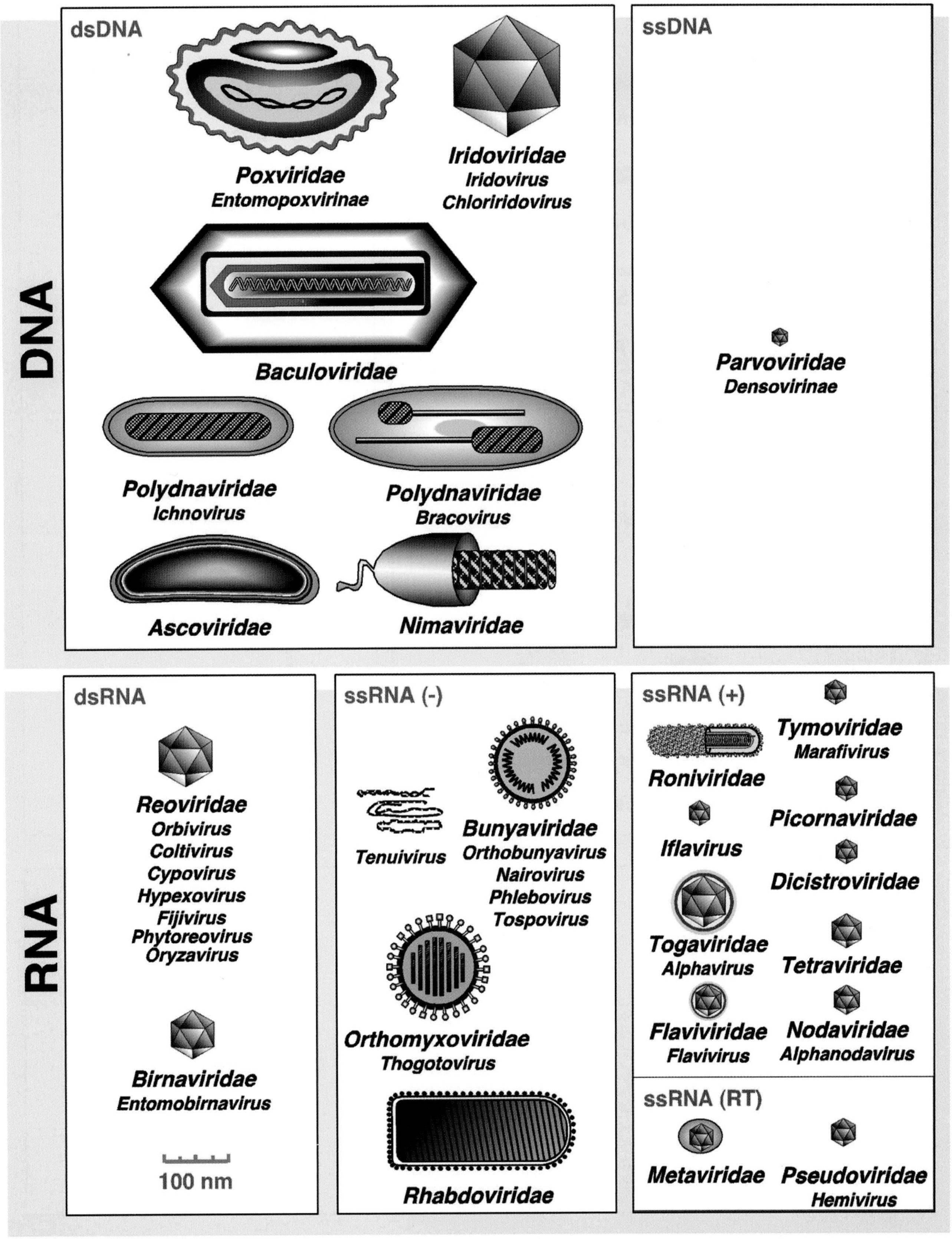
DNA
dsDNA
Poxviridae
Entomopoxvirinae
Iridoviridae
Iridovirus
Chloriridovirus
Baculoviridae
Polydnaviridae
Ichnovirus
Polydnaviridae
Bracovirus
Ascoviridae
Nimaviridae
ssDNA
Parvoviridae
Densovirinae
RNA
dsRNA
Reoviridae
Orbivirus
Coltivirus
Cypovirus
Hypexovirus
Fijivirus
Phytoreovirus
Oryzavirus
Birnaviridae
Entomobirnavirus
100 nm
ssRNA (-)
Tenuivirus
Bunyaviridae
Orthobunyavirus
Nairovirus
Phlebovirus
Tospovirus
Orthomyxoviridae
Thogotovirus
Rhabdoviridae
ssRNA (+)
Roniviridae
Tymoviridae
Marafivirus
Iflavirus
Picornaviridae
Dicistroviridae
Togaviridae
Alphavirus
Tetraviridae
Flaviviridae
Flavivirus
Nodaviridae
Alphanodavirus
ssRNA (RT)
Metaviridae
Pseudoviridae
Hemivirus

Families and Genera of Viruses Infecting Bacteria

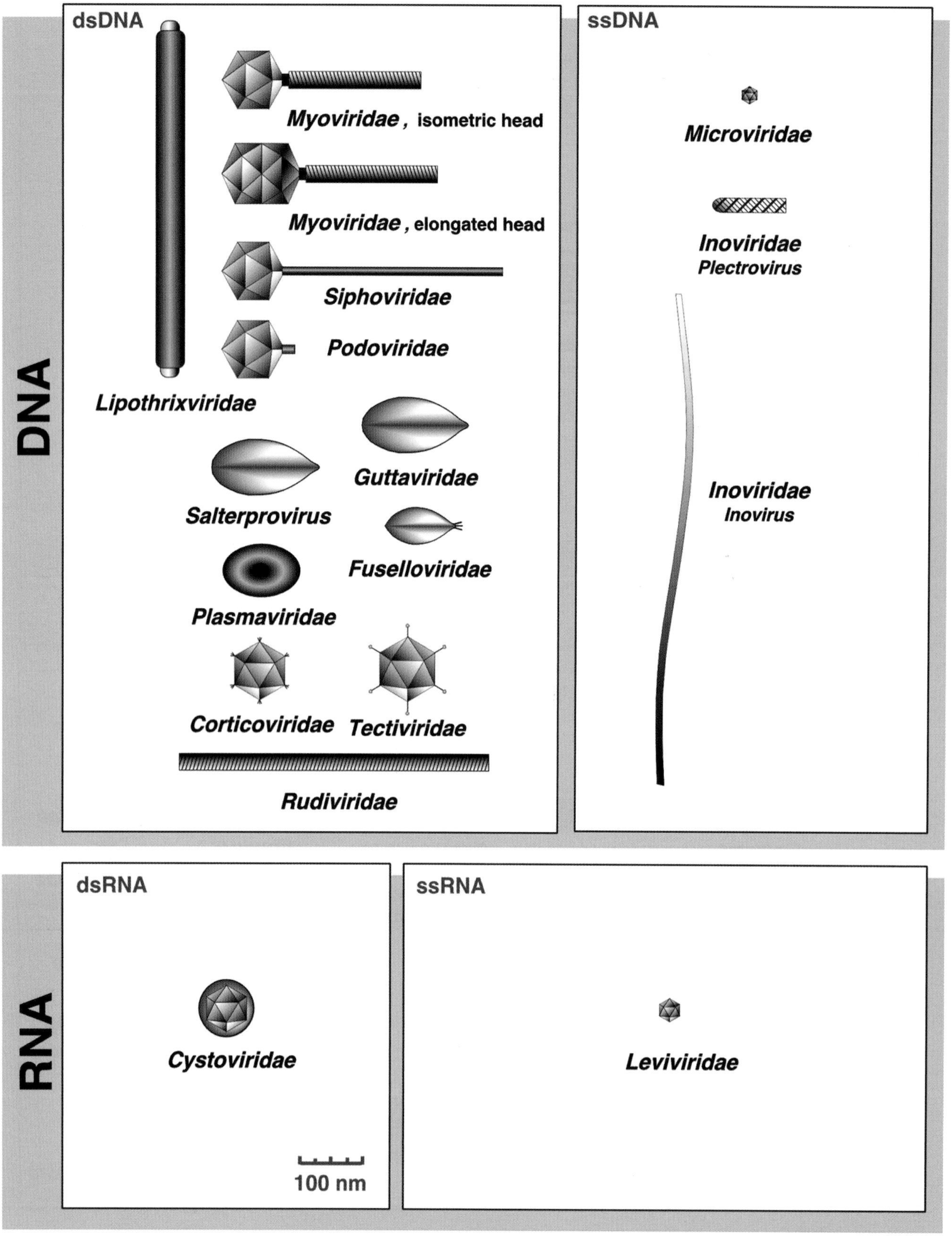

Families of Viruses Infecting Algae, Fungi, Yeast And Protozoa

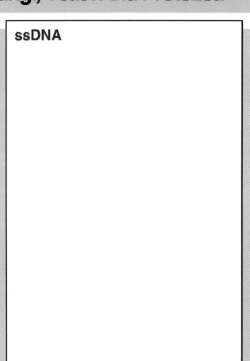

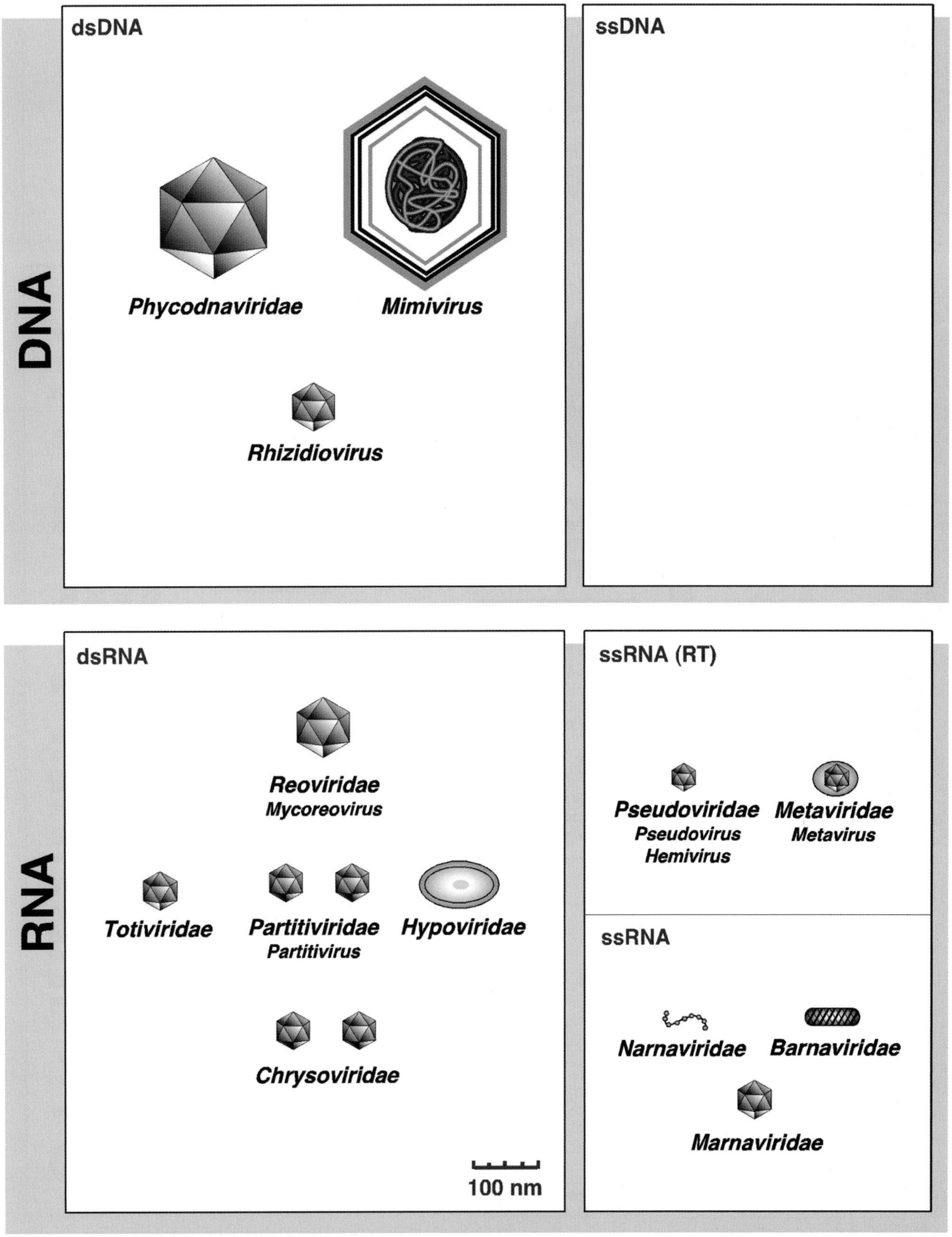

Families and Genera of Viruses Infecting Plants

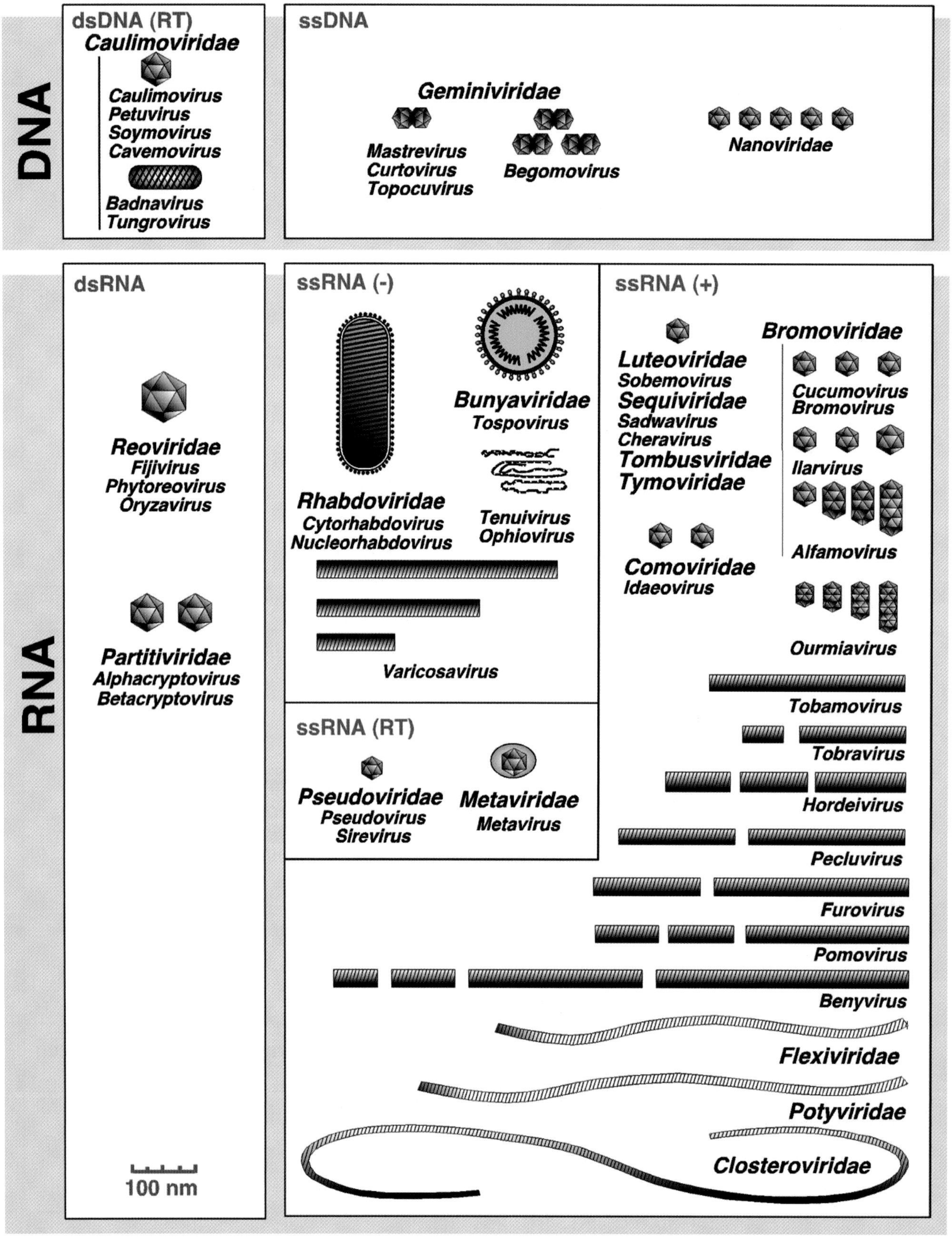
DNA
dsDNA (RT)
Caulimoviridae
Caulimovirus
Petuvirus
Soymovirus
Cavemovirus
Badnavirus
Tungrovirus
ssDNA
Geminiviridae
Mastrevirus
Curtovirus
Topocuvirus
Begomovirus
Nanoviridae
RNA
dsRNA
Reoviridae
Fijivirus
Phytoreovirus
Oryzavirus
Partitiviridae
Alphacryptovirus
Betacryptovirus
100 nm
ssRNA (-)
Rhabdoviridae
Cytorhabdovirus
Nucleorhabdovirus
Bunyaviridae
Tospovirus
Tenuivirus
Ophiovirus
Varicosavirus
ssRNA (RT)
Pseudoviridae
Pseudovirus
Sirevirus
Metaviridae
Metavirus
ssRNA (+)
Luteoviridae
Sobemovirus
Sequiviridae
Sadwavirus
Cheravirus
Tombusviridae
Tymoviridae
Comoviridae
Idaeovirus
Bromoviridae
Cucumovirus
Bromovirus
Ilarvirus
Alfamovirus
Ourmiavirus
Tobamovirus
Tobravirus
Hordeivirus
Pecluvirus
Furovirus
Pomovirus
Benyvirus
Flexiviridae
Potyviridae
Closteroviridae

Virus Particle Structures

Palmenberg, A.C. and Sgro, J.-Y.

COLOR PLATE LEGENDS

These color plates depict the relative sizes and comparative virion structures of multiple types of viruses. The renderings are based on data from published atomic coordinates as determined by X-ray crystallography. The international online repository for 3D coordinates is the Protein Databank (www.rcsb.org/pdb/), maintained by the Research Collaboratory for Structural Bioinformatics (RCSB). The VIPER web site (mmtsb.scripps.edu/viper), maintains a parallel collection of PDB coordinates for icosahedral viruses and additionally offers a version of each data file permuted into the same relative 3D orientation (Reddy, V., Natarajan, P., Okerberg, B., Li, K., Damodaran, K., Morton, R., Brooks, C. and Johnson, J. (2001). *J. Virol.*, **75**, 11943-11947). VIPER also contains an excellent repository of instructional materials pertaining to icosahedral symmetry and viral structures. All images presented here, except for the filamentous viruses, used the standard VIPER orientation along the icosahedral 2-fold axis.

With the exception of Plate 3 as described below, these images were generated from their atomic coordinates using a novel radial depth-cue colorization technique and the program Rasmol (Sayle, R.A., Milner-White, E.J. (1995). RASMOL: biomolecular graphics for all. *Trends Biochem Sci.*, **20**, 374-376). First, the Temperature Factor column for every atom in a PDB coordinate file was edited to record a measure of the radial distance from the virion center. The files were rendered using the Rasmol spacefill menu, with specular and shadow options according to the Van de Waals radius of each atom. Color was assigned on a sliding scale by individual radial distances. The composite assembly and processing used Adobe Photoshop software with attention to relative scale, visual contrast and a uniform color pallet. All graphics are copyright Dr. Jean-Yves Sgro, Institute for Molecular Virology, University of Wisconsin-Madison (E:mail: <jsgro@wisc.edu>) and are available on the VirusWorld web site (rhino.bocklabs.wisc.edu/virusworld).

PLATE 1: PICORNAVIRUSES

Bovine enterovirus 1: Picornaviridae; Enterovirus; Bovine enterovirus; strain VG-5-27.

Smyth, M., Tate, J., Hoey, E., Lyons, C., Martin, S. and Stuart, D. (1995). Implications for viral uncoating from the structure of bovine enterovirus. *Nat. Struct. Biol.*, **2**, 224-231. (PDB-ID: 1BEV)

Foot-and-mouth disease virus: Picornaviridae; Aphthovirus; Foot-and-mouth disease virus; strain disease virus.

Fry, E., Acharya, R. and Stuart, D. (1993). Methods used in the structure determination of foot-and-mouth disease virus. *Acta Crystallogr.* A, **49**, 45-55. (PDB-ID: 1BBT)

Human coxsackievirus B3: Picornaviridae; Enterovirus; Human enterovirus B; strain Nancy.

Muckelbauer, J.K., Kremer, M., Minor, I., Diana, G., Dutko, F.J., Groarke, J., Pevear, D.C., Rossmann, M.G. (1995). The structure of coxsackievirus B3 at 3.5 angstrom resolution. *Structure*, **3**, 653-667. (PDB-ID: 1COV)

Human echovirus 1: Picornaviridae; Enterovirus; Human enterovirus B; strain Farouk.

Filman, D.J., Wien, M.W., Cunningham, J.A., Bergelson, J.M. and Hogle, J.M. (1998). Structure determination of echovirus 1. *Acta Crystallogr.* D, **54**, 1261-1272. (PDB-ID: 1EV1)

Human echovirus 11: Picornaviridae; Enterovirus; Human enterovirus B; strain 207.
Stuart, A., Mckee, T., Williams, P., Harley, C., Shen, S., Stuart, D., Brown, T. and Lea, S. (2002). Determination of the structure of a decay accelerating factor-binding clinical isolate of echovirus 11 allows mapping of mutants with altered receptor requirements for infection. *J. Virol.*, **76**, 7694-7704. (PDB-ID: 1H8T)
Human poliovirus 1: Picornaviridae; Enterovirus; Poliovirus; strain Mahoney Type I.
Miller, S.T., Hogle, J.M. and Filman, D.J. (2003). Crystal structure of Mahoney strain of poliovirus at 2.2A Resolution. (PDB-ID: 1HXS)
Human rhinovirus 16: Picornaviridae; Rhinovirus; Human rhinovirus A; strain (NA).
Hadfield, A.T., Lee, W.M., Zhao, R., Oliveira, M.A., Minor, I., Rueckert, R.R. and Rossmann, M.G. (1997). The refined structure of human rhinovirus 16 at 2.15 A resolution: implications for the viral life cycle. *Structure*, **5**, 427-441. (PDB-ID: 1AYM)
Mengo virus: Picornaviridae, Cardiovirus; Encephalomyocarditis virus; strain M.
Krishnaswamy, S. and Rossmann, M.G. (1990). Structural refinement and analysis of Mengo virus. *J. Mol. Biol.*, **211**, 803-844. (PDB-ID: 2MEV)
Theiler's murine encephalomyelitis virus: Picornaviridae; Cardiovirus; Theilovirus; strain BeAn.
Luo, M., He, C., Toth, K.S., Zhang, C.X. and Lipton, H.L. (1992). Three-dimensional structure of Theiler murine encephalomyelitis virus (BeAn strain). *Proc. Natl. Acad. Sci. USA*, **89**, 2409-2413. (PDB-ID: 1TMF) A10-61.

Plate 2: Comparative Structures

Adeno-associated virus 2: *Parvoviridae; Dependovirus; Adeno-associated virus 2*; strain (recombinant).
Xie, Q., Bu, W., Bhatia, S., Hare, J., Somasundaram, T., Azzi, A. and Chapman, M.S. (2002). Atomic structure of adeno-associated virus (Aav-2), a vector for human therapy. *Proc. Nat. Acad. Sci. USA*, **99**, 10405-10410. (PDB-ID: 1LP3)
Bean pod mottle virus: *Comoviridae; Comovirus; Bean pod mottle virus*; strain Kentucky G7.
Chen, Z.G., Stauffacher, C., Li, Y., Schmidt, T., Bomu, W., Kamer, G., Shanks, M., Lomonossoff, G. and Johnson, J.E. (1989). Protein-RNA interactions in an icosahedral virus at 3.0 A resolution. *Science*, **245**, 154-159. (PDB-ID: 1BMV)
Bluetongue virus 1, Core: *Reoviridae; Orbivirus; Bluetongue virus* (VP3 core protein); serotype 1, South Africa.
Grimes, J.M., Burroughs, J.N., Gouet, P., Diprose, J.M., Malby, R., Zientara, S., Mertens, P.P. and Stuart, D.I. (1998). The atomic structure of the bluetongue virus core. *Nature*, **395**, 470-478. (PDB-ID: 2BTV)
Brome mosaic virus: *Bromoviridae; Bromovirus; Brome mosaic virus*; strain (NA).
Lucas, R.W., Larson, S.B. and McPherson, A. (2002). The crystallographic structure of brome mosaic virus. *J. Mol. Biol.*, **317**, 95-108. (PDB-ID: 1JS9)
Carnation mottle virus; *Tombusviridae; Carmovirus; Carnation mottle virus*; strain (NA).
Morgunova, E.Yu., Dauter, Z., Fry, E., Stuart, D.I., Stel'mashchuk, V.Ya., Mikhailov, A.M., Wilson, K.S. and Vainshtein, B.K. (1994). The atomic structure of carnation mottle virus capsid protein. *FEBS Lett.*, **338**, 267-271. (PDB-ID: 1OPO)
Cricket paralysis virus 1: *Dicistroviridae; Cripavirus; Cricket paralysis virus 1*; strain (NA).
Tate, J., Liljas, L., Scotti, P., Christian, P., Lin, T. and Johnson, J.E. (1999). The crystal structure of cricket paralysis virus: the first view of a new virus family. *Nat. Struct. Biol.*, **8**, 765-774. (PDB-ID: 1B35)
Canine parvovirus: *Parvoviridae; Parvovirus; Canine parvovirus*; strain D Cornell 320 (recombinant empty capsid).
Wu, H. and Rossmann, M.G. (1993). The canine parvovirus empty capsid structure. *J. Mol. Biol.*, **233**, 231-244. (PDB-ID: 2CAS)
Cucumber mosaic virus: *Bromoviridae; cucumovirus; cucumber mosaic virus*; strain Fny.
Smith, T.J., Chase, E., Schmidt, T. and Perry, K. (2000). The structure of cucumber mosaic virus and comparison to cowpea chlorotic mottle virus. *J. Virol.*, **74**, 7578-7586. (PDB-ID: 1F15)
Enterobacteria phage fd: *Inoviridae; Inovirus; Enterobacteria phage fd*;

Marvin, D.A. (1990). Model-building studies of *Inovirus*: genetic variations on a geometric theme. *Int. J. Biol. Macromol.*, **12**, 125-138. (PDB-ID: 1IFD)

Enterobacteria phage MS2: *Leviviradae; Levivirus; Enterobacteria phage MS2*, strain (NA).

Golmohammadi, R., Valegard, K., Fridborg, K. and Liljas, L. (1993). The refined structure of bacteriophage MS2 at 2.8 A resolution. *J. Mol. Biol.*, **234**, 620-639. (PDB-ID: 2MS2)

Enterobacteria phage QBeta: *Leviviridae; Allolevivirus; Enterobacteria phage Q-beta*; strain (NA).

Golmohammadi, R., Fridborg, K., Bundule, M., Valegard, K. and Liljas, L. (1996). The crystal structure of bacteriophage Q-beta at 3.5 A resolution. *Structure*, **4**, 543-554. (PDB-ID: 1QBE)

Enterobacteria phage PhiX174: *Microviridae; Microvirus; Enterobacteria phage phi-X174*, strain (NA).

McKenna, R., Xia, D., Willingmann, P., Ilag, L.L., Krishnaswamy, S., Rossmann, M.G., Olson, N.H., Baker, T.S. and Incardona, N.L. (1992). Atomic structure of single-stranded DNA bacteriophage phiX174 and its functional implications. *Nature*, **355**, 137-143. (PDB-ID: 2BPA)

Enterobacteria phage PhiX174+scaffold: *Microviridae; Microvirus; Enterobacteria phage phi-X174*; with scaffold.

Dokland, T., McKenna, R., Ilag, L.L., Bowman, B.R., Incardona, N.L., Fane, B.A. and Rossmann, M.G. (1997). Structure of a viral procapsid with molecular scaffolding. *Nature*, **389**, 308-313. (PDB-ID: 1AL0)

Galleria mellonella densovirus: *Parvoviridae; Densovirus; Galleria mellonella densovirus*; strain (NA).

Simpson, A.A., Chipman, P.R., Baker, T.S., Tijssen, P. and Rossmann, M.G. (1998). The structure of an insect parvovirus (Galleria mellonella densovirus) at 3.7 A resolution. *Structure*, **6**, 1355-1367. (PDB-ID: 1DNV)

Hepatitis B virus: *Hepadnaviridae; Orthohepadnavirus; Hepatitis B virus*; strain ayw.

Wynne, S.A., Crowther, R.A. and Leslie, A.G.W. (1999). The crystal structure of the human hepatitis B virus capsid. *Molecular Cell*, **3**, 771-780. (PDB-ID: 1QGT)

Human papillomavirus 16: *Papillomaviridae; Papillomavirus; Human papillomavirus 16*, strain (recombinant L1 protein).

Modis, Y., Trus, B.L. and Harrison, S.C. (2002). Atomic model of the papillomavirus capsid. *EMBO J.*, **21**, 4754-4762. (PDB-ID: 1L0T)

Mammalian orthoreovirus 3 - Core: *Reoviridae; Orthoreovirus; Mammalian orthoreovirus* type 3 (LMD1, LMD2, sigma2 core proteins), strain Dearing.

Reinisch, K.M., Nibert, M.L. and Harrison, S.C. (2000). Structure of the reovirus core at 3.6 A resolution. *Nature*, **404**, 960-967. (PDB-ID: 1EJ6)

Nodamura virus: *Nodaviridae; Alphanodavirus; Nodamura virus*, strain (NA).

Zlotnick, A., Natarajan, P., Munshi, S. and Johnson, J.E. (1997). Resolution of space-group ambiguity and the structure determination of Nodamura virus to 3.3 angstrom resolution from pseudo-R32 (monoclinic) crystals. *Acta Crystallogr.*, **53**, 738-746. (PDB-ID: 1NOV)

Norwalk virus: *Caliciviridae; Norovirus; Norwalk virus*; strain (recombinant capsid).

Prasad, B.V.V., Hardy, M.E., Dokland, T., Bella, J., Rossmann, M.G. and Estes, M.K. (1999). X-ray crystallographic structure of Norwalk virus capsid. *Science*, **286**, 287-290. (PDB-ID: 1IHM)

Nudaurelia capensis omega virus: *Tetraviridae; Omegatetravirus; Nudaurelia capensis omega virus*; strain (NA).

Munshi, S., Liljas, L., Cavarelli, J., Bomu, W., McKinney, B., Reddy, V. and Johnson, J.E. (1996). The 2.8 A structure of a T=4 animal virus and its implications for membrane translocation of RNA. *J. Mol. Biol.*, **261**, 1-10. (PDB-ID: NA, coordinates available from VIPER)

Rice dwarf virus: *Reoviridae; Phytoreovirus; Rice dwarf virus*; strain Akita.

Nakagawa, A., Miyazaki, N., Taka, J., Naitow, H., Ogawa, A., Fujimoto, Z., Mizuno, H., Higashi, T., Watanabe, Y., Omura, T., Cheng, R.H. and Tsukihara, T. (2003). The Atomic structure of rice dwarf virus. *Structure*, **11**, 1227-1238. (PDB-ID: 1UF2)

Simian virus 40: *Polyomaviridae; polyomavirus; simian virus 40*; strain (NA).

Stehle, T., Gamblin, S.J., Yan, Y. and Harrison, S.C. (1996). The structure of simian virus 40 refined at 3.1 A resolution. *Structure*, **4**, 165-182. (PDB-ID: 1SVA)

Southern bean mosaic virus: *Sobemovirus; Southern bean mosaic virus*; strain (NA).

Silva, A.M. and Rossmann, M.G. (1987). The refinement of southern bean mosaic virus at 2.9 A resolution. *J. Mol. Biol.*, **197**, 69-87. (PDB-ID: 4SBV)

Swine vesicular disease virus: *Picornaviridae; Enterovirus; Human enterovirus B*; strain UKG/27/72.

Fry, E.E., Knowles, N.J., Newman, J.W.I., Wilsden, G., Rao, Z., King, A.M.Q. and Stuart, D.I. (2003). Crystal structure of swine vesicular disease virus and implications for host adaptation. *J. Virol.*, **77**, 5475-5486. (PDB-ID: 1OOP)

Tobacco mosaic virus; *Tobamovirus; Tobacco mosaic virus*; strain vulgare;

Namba, K., Pattanayek, R. and Stubbs, G. (1989). Visualization of protein-nucleic acid interactions in a virus. Refined structure of intact tobacco mosaic virus at 2.9 A resolution by X-ray fiber diffraction. *J. Mol. Biol.*, **208**, 307-325. (PDB-ID: 2TMV)

Tobacco necrosis satellite virus; Satellite viruses; subgroup 2; strain (NA).

Jones, T.A. and Liljas, L. (1984). Structure of satellite tobacco necrosis virus after crystallographic refinement at 2.5 A resolution. *J. Mol. Biol.*, **177**, 735-767. (PDB-ID: 2STV)

Tobacco necrosis virus; *Tombusviridae; Necrovirus; Tobacco necrosis virus*; strain A.

Oda, Y., Saeki, K., Takahashi, Y., Maeda, T., Naitow, H., Tsukihara, T. and Fukuyama, K. (2000). Crystal structure of tobacco necrosis virus at 2.25 A resolution. *J. Mol. Biol.*, **300**, 153-169. (PDB-ID: 1C8N)

Tomato bushy stunt virus; *Tombusviridae; Tombusvirus, Tomato bushy stunt virus*; strain BS-3.

Olson, A.J., Bricogne, G. and Harrison, S.C. (1983). Structure of tomato busy stunt virus IV. The virus particle at 2.9 A resolution. *J. Mol. Biol.*, **171**, 61-93. (PDB-ID: 2TBV)

Turnip yellow mosaic virus: *Tymoviridae; Tymovirus; Turnip yellow mosaic virus*; strain (NA).

Canady, M.A., Larson, S.B., Day, J., McPherson, A. (1996). Crystal structure of turnip yellow mosaic virus. *Nat. Struct. Biol.*, **3**, 771-781. (PDB-ID: 1AUY)

PLATE 3: LARGEST AND SMALLEST VIRAL STRUCTURES. NUCLEIC ACID REVEALED WITHIN PARIACOTO VIRUS.

Both images are rendered to the same relative size scale.

Top: **Paramecium bursaria Chlorella virus 1:** *Phycodnaviridae; Chlorovirus, Paramecium bursaria Chlorella virus 1*; strain (NA), quasi-atomic model.

Nandhagopal, N., Simpson, A., Gurnon, J.R., Yan, X., Baker, T.S., Graves, M.V., van Etten, J.L. and Rossmann, M.G. (002). The structure and evolution of the major capsid protein of a large, lipid containing, DNA virus. *Proc. Nat. Acad. Sci. USA*, **99**, 14758-14763. (PDB-ID: 1M4X)

Tobacco necrosis satellite virus; Satellite viruses; subgroup 2; strain (NA).

Jones, T.A. and Liljas, L. (1984). Structure of satellite tobacco necrosis virus after crystallographic refinement at 2.5 A resolution. *J. Mol. Biol.*, **177**, 735-767. (PDB-ID: 2STV)

Inset: axial and side views of one trimeric protein unit from Chlorella. Images were created with MOLSCRIPT software (P.J. Kraulism. (1991). MOLSCRIPT: a program to produce both detailed and schematic plots of protein structures. *J. Appl. Cryst.*, **24**, 946-950) then rendered with Raster3D (Merritt, E.A. and Bacon, D.J. (1997). Raster3D: photorealistic molecular graphics. *Meth. Enzymol.*, **277**, 505-524)

Bottom: Left: radial depth-cue molecular surface of a particle of *Pariacoto virus*, split to reveal the dodecahedral arrangement of a portion of the RNA. Right: same image rotated 90°. Both images were rendered using GRASP software. (Nicholls A, Sharp K.A. and Honig B. (1991). Protein folding and association: insights from the interfacial and thermodynamic properties of hydrocarbons. *Proteins*, **11**, 281-296)

Pariacoto virus: *Nodaviridae; Alphanodavirus; Pariacoto virus*; strain (NA).

Tang, L., Johnson, K.N., Ball, L.A., Lin, T., Yeager, M. and Johnson, J.E. 2001. The structure of Pariacoto virus reveals a dodecahedral cage of duplex RNA. *Nat. Struct. Biol.*, **8**, 77-83. (PDB-ID: 1F8V)

Order of Presentation of Virus Taxonomic Descriptions

Taxonomic descriptions in this Report are organized and presented in clusters. The first level of organization is an informal grouping according to genome composition and structure (dsDNA, ssDNA, etc.) and the second level is according to taxonomic rank; i.e., order or family, if there is no order, or genus, if there is no family assignment. In general, descriptions of taxa that appear to share some level of similarity are placed close to each other.

At the end of each virus description is a list of species. The species names are in green italic script. The names of isolates within a species follow the species name and are indented and in black roman script. Isolate names are aligned with relevant accession numbers (between square brackets in the center column) and a recommended abbreviation (between parenthesis in the rightmost column). Some lists contain extra information such as details of vector and/or host. In some genera, species are clustered into groups or serogroups, while in others the isolates within a species are clustered. These clusters are not formal taxonomic groupings. When a species has been re-named recently, the former name (= synonym) is added in parentheses.

ORDER Family	Subfamily	Genus	Type Species	Host	Page
The DNA Viruses					
The dsDNA Viruses					
CAUDOVIRALES					35
Myoviridae		"T4-like viruses"	*Enterobacteria phage T4*	B	43
		"P1-like viruses"	*Enterobacteria phage P1*	B	47
		"P2-like viruses"	*Enterobacteria phage P2*	B	48
		"Mu-like viruses"	*Enterobacteria phage Mu*	B	50
		"SP01-like viruses"	*Bacillus phage SP01*	B	51
		"ϕH-like viruses"	*Halobacterium phage ϕH*	Ar	52
Siphoviridae		"λ-like viruses"	*Enterobacteria phage λ*	B	57
		"T1-like viruses"	*Enterobacteria phage T1*	B	59
		"T5-like viruses"	*Enterobacteria phage T5*	B	60
		"L5-like viruses"	*Mycobacterium phage L5*	B	61
		"c2-like viruses"	*Lactococcus phage c2*	B	63
		"ψM1-like viruses"	*Methanobacterium phage ψM1*	Ar	64
		"ϕC31-like viruses"	*Streptomyces phage ϕC31*	B	65
		"N15-like viruses"	*Enterobacteria phage N15*	B	66
Podoviridae		"T7-like viruses"	*Enterobacteria phage T7*	B	71
		"P22-like viruses"	*Enterobacteria phage P22*	B	73
		"ϕ29-like viruses"	*Bacillus phage ϕ29*	B	75
		"N4-like viruses"	*Enterobacteria phage N4*	B	76
Tectiviridae		*Tectivirus*	*Enterobacteria phage PRD1*	B	81
Corticoviridae		*Corticovirus*	*Pseudoalteromonas phage PM2*	B	87
Plasmaviridae		*Plasmavirus*	*Acholeplasma phage L2*	B	91
Lipothrixviridae		*Alphalipothrixvirus*	*Thermoproteus tenax virus 1*	Ar	95
		Betalipothrixvirus	*Sulfolobus islandicus filamentous virus*	Ar	98
		Gammalipothrixvirus	*Acidianus filamentous virus 1*	Ar	100
Rudiviridae		*Rudivirus*	*Sulfolobus islandicus rod-shaped virus 2*	Ar	103
Fuselloviridae		*Fusellovirus*	*Sulfolobus spindle-shaped virus 1*	Ar	107
		Salterprovirus	*His1 virus*	Ar	111
Guttaviridae		*Guttavirus*	*Sulfolobus newzealandicus droplet-shaped virus*	Ar	115
Poxviridae					117
	Chordopoxvirinae				122
		Orthopoxvirus	*Vaccinia virus*	V	122
		Parapoxvirus	*Orf virus*	V	123
		Avipoxvirus	*Fowlpox virus*	V	124
		Capripoxvirus	*Sheeppox virus*	V	125
		Leporipoxvirus	*Myxoma virus*	V	126
		Suipoxvirus	*Swinepox virus*	V	127
		Molluscipoxvirus	*Molluscum contagiosum virus*	V	127
		Yatapoxvirus	*Yaba monkey tumor virus*	V	128
	Entomopoxvirinae				129
		Alphaentomopoxvirus	*Melolontha melolontha entomopoxvirus*	I	129
		Betaentomopoxvirus	*Amsacta moorei entomopoxvirus 'L'*	I	130
		Gammaentomopoxvirus	*Chironomus luridus entomopoxvirus*	I	131
Asfarviridae		*Asfivirus*	*African swine fever virus*	V, I	135
Iridoviridae		*Iridovirus*	*Invertebrate iridescent virus 6*	I	145
		Chloriridovirus	*Invertebrate iridescent virus 3*	I	153
		Ranavirus	*Frog virus 3*	V	154
		Lymphocystivirus	*Lymphocystis disease virus 1*	V	156
		Megalocytivirus	*Infectious spleen and kidney necrosis virus*	V	158

ORDER Family	Subfamily	Genus	Type Species	Host	Page
Phycodnaviridae		*Chlorovirus*	*Paramecium bursaria Chlorella virus 1*	Al	163
		Coccolithovirus	*Emiliania huxleyi virus 86*	Al	168
		Prasinovirus	*Micromonas pusilla virus SP1*	Al	169
		Prymnesiovirus	*Chrysochromulina brevifilum virus PW1*	Al	170
		Phaeovirus	*Ectocarpus siliculosus virus 1*	Al	171
		Raphidovirus	*Heterosigma akashiwo virus 01*	Al	172
Baculoviridae		*Nucleopolyhedrovirus*	*Autographa californica multiple nucleopolyhedrovirus*	I	177
		Granulovirus	*Cydia pomonella granulovirus*	I	183
Nimaviridae		*Whispovirus*	*White spot syndrome virus 1*	I	187
Herpesviridae					193
	Alphaherpesvirinae				199
		Simplexvirus	*Human herpesvirus 1*	V	199
		Varicellovirus	*Human herpesvirus 3*	V	200
		Mardivirus	*Gallid herpesvirus 2*	V	201
		Iltovirus	*Gallid herpesvirus 1*	V	202
	Betaherpesvirinae				203
		Cytomegalovirus	*Human herpesvirus 5*	V	203
		Muromegalovirus	*Murid herpesvirus 1*	V	204
		Roseolovirus	*Human herpesvirus 6*	V	204
	Gammaherpesvirinae				205
		Lymphocryptovirus	*Human herpesvirus 4*	V	205
		Rhadinovirus	*Saimiriine herpesvirus 2*	V	206
		Ictalurivirus	*Ictalurid herpesvirus 1*	V	208
Adenoviridae		*Mastadenovirus*	*Human adenovirus C*	V	213
		Aviadenovirus	*Fowl adenovirus A*	V	221
		Atadenovirus	*Ovine adenovirus D*	V	223
		Siadenovirus	*Frog adenovirus*	V	225
		Rhizidiovirus	*Rhizidiomyces virus*	F	229
Polyomaviridae		*Polyomavirus*	*Simian virus 40*	V	231
Papillomaviridae		*Alphapapillomavirus*	*Human papillomavirus 32*	V	239
		Betapapillomavirus	*Human papillomavirus 5*	V	245
		Gammapapillomavirus	*Human papillomavirus 4*	V	246
		Deltapapillomavirus	*European elk papillomavirus*	V	247
		Epsilonpapillomavirus	*Bovine papillomavirus 5*	V	247
		Zetapapillomavirus	*Equine papillomavirus 1*	V	247
		Etapapillomavirus	*Fringilla coelebs papillomavirus*	V	248
		Thetapapillomavirus	*Psittacus erithacus timneh papillomavirus*	V	248
		Iotapapillomavirus	*Mastomys natalensis papillomavirus*	V	249
		Kappapapillomavirus	*Cottontail rabbit papillomavirus*	V	249
		Lambdapapillomavirus	*Canine oral papillomavirus*	V	249
		Mupapillomavirus	*Human papillomavirus 1*	V	250
		Nupapillomavirus	*Human papillomavirus 41*	V	250
		Xipapillomavirus	*Bovine papillomavirus 3*	V	251
		Omikronpapillomavirus	*Phocoena spinipinnis papillomavirus*	V	251
		Pipapillomavirus	*Hamster oral papillomavirus*	V	251
Polydnaviridae		*Bracovirus*	*Cotesia melanoscela bracovirus*	I	255
		Ichnovirus	*Campoletis sonorensis ichnovirus*	I	261
Ascoviridae		*Ascovirus*	*Spodoptera frugiperda ascovirus 1a*	I	269
		Mimivirus	*Acanthamoeba polyphaga mimivirus*	Pr, V	275

ORDER Family	Subfamily	Genus	Type Species	Host	Page
The ssDNA Viruses					
Inoviridae					279
		Inovirus	*Enterobacteria phage M13*	B	283
		Plectrovirus	*Acholeplasma phage L51*	B	285
Microviridae					289
		Microvirus	*Enterobacteria phage ϕX174*	B	292
		Chlamydiamicrovirus	*Chlamydia phage 1*	B	295
		Bdellomicrovirus	*Bdellovibrio phage MAC1*	B	296
		Spiromicrovirus	*Spiroplasma phage 4*	B	297
Geminiviridae					301
		Mastrevirus	*Maize streak virus*	P	302
		Curtovirus	*Beet curly top virus*	P	306
		Topocuvirus	*Tomato pseudo-curly top virus*	P	307
		Begomovirus	*Bean golden yellow mosaic virus*	P	309
Circoviridae					327
		Circovirus	*Porcine circovirus-1*	V	328
		Gyrovirus	*Chicken anemia virus*	V	331
		Anellovirus	*Torque teno virus*	V	335
Nanoviridae					343
		Nanovirus	*Subterranean clover stunt virus*	P	346
		Babuvirus	*Banana bunchy top virus*	P	348
Parvoviridae					353
	Parvovirinae				358
		Parvovirus	*Minute virus of mice*	V	358
		Erythrovirus	*Human parvovirus B19*	V	360
		Dependovirus	*Adeno-associated virus 2*	V	361
		Amdovirus	*Aleutian mink disease virus*	V	363
		Bocavirus	*Bovine parvovirus*	V	364
	Densovirinae				365
		Densovirus	*Junonia coenia densovirus*	I	365
		Iteravirus	*Bombyx mori densovirus*	I	366
		Brevidensovirus	*Aedes aegypti densovirus*	I	366
		Pefudensovirus	*Periplaneta fuliginosa densovirus*	I	367

ORDER Family	Subfamily	Genus	Type Species	Host	Page
The DNA and RNA Reverse Transcribing Viruses					
Hepadnaviridae					373
		Orthohepadnavirus	*Hepatitis B virus*	V	379
		Avihepadnavirus	*Duck hepatitis B virus*	V	382
Caulimoviridae					385
		Caulimovirus	*Cauliflower mosaic virus*	P	388
		Petuvirus	*Petunia vein clearing virus*	P	389
		Soymovirus	*Soybean chlorotic mottle virus*	P	390
		Cavemovirus	*Cassava vein mosaic virus*	P	391
		Badnavirus	*Commelina yellow mottle virus*	P	392
		Tungrovirus	*Rice tungro bacilliform virus*	P	394
Pseudoviridae					397
		Pseudovirus	*Saccharomyces cerevisiae Ty1 virus*	F, P	399
		Hemivirus	*Drosophila melanogaster copia virus*	F, I	402
		Sirevirus	*Glycine max SIRE1 virus*	P	403
Metaviridae					409
		Metavirus	*Saccharomyces cerevisiae Ty3 virus*	F, P, I	412
		Errantivirus	*Drosophila melanogaster Gypsy virus*	I	414
		Semotivirus	*Ascaris lumbricoides Tas virus*	I	416
Retroviridae					421
	Orthoretrovirinae				425
		Alpharetrovirus	*Avian leukosis virus*	V	425
		Betaretrovirus	*Mouse mammary tumor virus*	V	427
		Gammaretrovirus	*Murine leukemia virus*	V	428
		Deltaretrovirus	*Bovine leukemia virus*	V	431
		Epsilonretrovirus	*Walleye dermal sarcoma virus*	V	432
		Lentivirus	*Human immunodeficiency virus 1*	V	433
	Spumaretrovirinae				437
		Spumavirus	*Simian foamy virus*	V	437

ORDER Family	Subfamily	Genus	Type Species	Host	Page
The RNA Viruses					
The dsRNA Viruses					
Cystoviridae		*Cystovirus*	*Pseudomonas phage* $\phi 6$	B	443
Reoviridae					447
		Orthoreovirus	*Mammalian orthoreovirus*	V	455
		Orbivirus	*Bluetongue virus*	V, I	466
		Rotavirus	*Rotavirus A*	V	484
		Coltivirus	*Colorado tick fever virus*	V, I	497
		Seadornavirus	*Banna virus*	V	504
		Aquareovirus	*Aquareovirus A*	V	511
		Idnoreovirus	*Idnoreovirus 1*	I	517
		Cypovirus	*Cypovirus 1*	I	522
		Fijivirus	*Fiji disease virus*	P, I	534
		Phytoreovirus	*Wound tumor virus*	P, I	543
		Oryzavirus	*Rice ragged stunt virus*	P, I	550
		Mycoreovirus	*Mycoreovirus 1*	F	556
Birnaviridae					561
		Aquabirnavirus	*Infectious pancreatic necrosis virus*	V	565
		Avibirnavirus	*Infectious bursal disease virus*	V	566
		Entomobirnavirus	*Drosophila X virus*	I	567
Totiviridae					571
		Totivirus	*Saccharomyces cerevisiae virus L-A*	F	572
		Giardiavirus	*Giardia lamblia virus*	Pr	575
		Leishmaniavirus	*Leishmania RNA virus 1-1*	Pr	577
Partitiviridae					581
		Partitivirus	*Atkinsonella hypoxylon virus*	F	582
		Alphacryptovirus	*White clover cryptic virus 1*	P	585
		Betacryptovirus	*White clover cryptic virus 2*	P	587
Chrysoviridae		*Chrysovirus*	*Penicillium chrysogenum virus*	F	591
Hypoviridae		*Hypovirus*	*Cryphonectria hypovirus 1*	F	597
		Endornavirus	*Vicia faba endornavirus*	P	603

The Negative Stranded ssRNA Viruses

ORDER / Family	Subfamily	Genus	Type Species	Host	Page
MONONEGAVIRALES					609
Bornaviridae					615
		Bornavirus	*Borna disease virus*	V	615
Rhabdoviridae					623
		Vesiculovirus	*Vesicular stomatitis Indiana virus*	V, I	629
		Lyssavirus	*Rabies virus*	V	630
		Ephemerovirus	*Bovine ephemeral fever virus*	V, I	633
		Novirhabdovirus	*Infectious hematopoietic necrosis virus*	V	635
		Cytorhabdovirus	*Lettuce necrotic yellows virus*	P, I	637
		Nucleorhabdovirus	*Potato yellow dwarf virus*	P, I	638
Filoviridae					645
		Marburgvirus	*Lake Victoria marburgvirus*	V	650
		Ebolavirus	*Zaire ebolavirus*	V	651
Paramyxoviridae					651
	Paramyxovirinae				659
		Rubulavirus	*Mumps virus*	V	659
		Avulavirus	*Newcastle disease virus*	V	661
		Respirovirus	*Sendai virus*	V	662
		Henipavirus	*Hendra virus*	V	663
		Morbillivirus	*Measles virus*	V	663
	Pneumovirinae				665
		Pneumovirus	*Human respiratory syncytial virus*	V	665
		Metapneumovirus	*Avian metapneumovirus*	V	666
		Varicosavirus	*Lettuce big-vein associated virus*	P	669
		Ophiovirus	*Citrus psorosis virus*	P	673
Orthomyxoviridae					681
		Influenzavirus A	*Influenza A virus*	V	685
		Influenzavirus B	*Influenza B virus*	V	687
		Influenzavirus C	*Influenza C virus*	V	688
		Thogotovirus	*Thogoto virus*	V, I	689
		Isavirus	*Infectious salmon anemia virus*	V	691
Bunyaviridae					695
		Orthobunyavirus	*Bunyamwera virus*	V, I	699
		Hantavirus	*Hantaan virus*	V	704
		Nairovirus	*Dugbe virus*	V, I	707
		Phlebovirus	*Rift Valley fever virus*	V, I	709
		Tospovirus	*Tomato spotted wilt virus*	P, I	712
		Tenuivirus	*Rice stripe virus*	P, I	717
Arenaviridae		*Arenavirus*	*Lymphocytic choriomeningitis virus*	V	725
		Deltavirus	*Hepatitis delta virus*	V	735

ORDER Family	Subfamily	Genus	Type Species	Host	Page
The Positive Stranded ssRNA Viruses					
Leviviridae					741
		Levivirus	*Enterobacteria phage MS2*	B	743
		Allolevivirus	*Enterobacteria phage Qβ*	B	745
Narnaviridae					751
		Narnavirus	*Saccharomyces 20S narnavirus*	F	751
		Mitovirus	*Cryphonectria mitovirus 1*	F	753
Picornaviridae					757
		Enterovirus	*Poliovirus*	V	760
		Rhinovirus	*Human rhinovirus A*	V	764
		Cardiovirus	*Encephalomyocarditis virus*	V	767
		Aphthovirus	*Foot-and-mouth disease virus*	V	768
		Hepatovirus	*Hepatitis A virus*	V	770
		Parechovirus	*Human parechovirus*	V	771
		Erbovirus	*Equine rhinitis B virus*	V	772
		Kobuvirus	*Aichi virus*	V	773
		Teschovirus	*Porcine teschovirus*	V	774
		Iflavirus	*Infectious flacherie virus*	I	779
Dicistroviridae		*Cripavirus*	*Cricket paralysis virus*	I	783
Marnaviridae		*Marnavirus*	*Heterosigma akashiwo RNA virus*	F	789
Sequiviridae					793
		Sequivirus	*Parsnip yellow fleck virus*	P	794
		Waikavirus	*Rice tungro spherical virus*	P	795
		Sadwavirus	*Satsuma dwarf virus*	P	799
		Cheravirus	*Cherry rasp leaf virus*	P	803
Comoviridae					807
		Comovirus	*Cowpea mosaic virus*	P	810
		Fabavirus	*Broad bean wilt virus 1*	P	812
		Nepovirus	*Tobacco ringspot virus*	P	813
Potyviridae					819
		Potyvirus	*Potato virus Y*	P	821
		Ipomovirus	*Sweet potato mild mottle virus*	P	829
		Macluravirus	*Maclura mosaic virus*	P	831
		Rymovirus	*Ryegrass mosaic virus*	P	833
		Tritimovirus	*Wheat streak mosaic virus*	P	834
		Bymovirus	*Barley yellow mosaic virus*	P	837
Caliciviridae					843
		Lagovirus	*Rabbit hemorrhagic disease virus*	V	846
		Norovirus	*Norwalk virus*	V	847
		Sapovirus	*Sapporo virus*	V	848
		Vesivirus	*Vesicular exanthema of swine virus*	V	848
		Hepevirus	*Hepatitis E virus*	V	853
Astroviridae					859
		Avastrovirus	*Turkey astrovirus*	V	861
		Mamastrovirus	*Human astrovirus*	V	861

ORDER Family	Subfamily	Genus	Type Species	Host	Page
Nodaviridae		*Alphanodavirus*	*Nodamura virus*	I	865
		Betanodavirus	*Striped jack nervous necrosis virus*	V	869
Tetraviridae		*Betatetravirus*	*Nudaurelia capensis β virus*	I	873
		Omegatetravirus	*Nudaurelia capensis ω virus*	I	877
		Sobemovirus	*Southern bean mosaic virus*	P	885
Luteoviridae					891
		Luteovirus	*Barley yellow dwarf virus - PAV*	P	895
		Polerovirus	*Potato leafroll virus*	P	896
		Enamovirus	*Pea enation mosaic virus-1*	P	897
		Umbravirus	*Carrot mottle virus*	P	901
Tombusviridae					907
		Dianthovirus	*Carnation ringspot virus*	P	911
		Tombusvirus	*Tomato bushy stunt virus*	P	914
		Aureusvirus	*Pothos latent virus*	P	918
		Avenavirus	*Oat chlorotic stunt virus*	P	920
		Carmovirus	*Carnation mottle virus*	P	922
		Necrovirus	*Tobacco necrosis virus A*	P	926
		Panicovirus	*Panicum mosaic virus*	P	929
		Machlomovirus	*Maize chlorotic mottle virus*	P	932
NIDOVIRALES					937
Coronaviridae					947
		Coronavirus	*Infectious bronchitis virus*	V	947
		Torovirus	*Equine torovirus*	V	956
Arteriviridae					965
		Arterivirus	*Equine arteritis virus*	V	965
Roniviridae					975
		Okavirus	*Gill-associated virus*	I	975
Flaviviridae					981
		Flavivirus	*Yellow fever virus*	V, I	981
		Pestivirus	*Bovine viral diarrhea virus 1*	V	988
		Hepacivirus	*Hepatitis C virus*	V	993
Togaviridae		*Alphavirus*	*Sindbis virus*	V, I	999
		Rubivirus	*Rubella virus*	V	1006
		Tobamovirus	*Tobacco mosaic virus*	P	1009
		Tobravirus	*Tobacco rattle virus*	P	1015
		Hordeivirus	*Barley stripe mosaic virus*	P	1021
		Furovirus	*Soil-borne wheat mosaic virus*	P	1027
		Pomovirus	*Potato mop-top virus*	P	1033
		Pecluvirus	*Peanut clump virus*	P	1039
		Benyvirus	*Beet necrotic yellow vein virus*	P	1043
Bromoviridae					1049
		Alfamovirus	*Alfalfa mosaic virus*	P	1051
		Bromovirus	*Brome mosaic virus*	P	1052
		Cucumovirus	*Cucumber mosaic virus*	P	1053
		Ilarvirus	*Tobacco streak virus*	P	1055
		Oleavirus	*Olive latent virus 2*	P	1057

Order / Family	Subfamily	Genus	Type Species	Host	Page
		Ourmiavirus	*Ourmia melon virus*	P	1059
		Idaeovirus	*Rasberry bushy dwarf virus*	P	1063
Tymoviridae					1067
		Tymovirus	*Turnip yellow mosaic virus*	P	1070
		Marafivirus	*Maize rayado fino virus*	P, I	1072
		Maculavirus	*Grapevine fleck virus*	P	1073
Closteroviridae					1077
		Closterovirus	*Beet yellows virus*	P	1080
		Ampelovirus	*Grapevine leafroll-associated virus 3*	P	1082
		Crinivirus	*Lettuce infectious yellows virus*	P	1084
Flexiviridae					1089
		Potexvirus	*Potato virus X*	P	1091
		Mandarivirus	*Indian citrus ringspot virus*	P	1096
		Allexivirus	*Shallot virus X*	P	1098
		Carlavirus	*Carnation latent virus*	P	1101
		Foveavirus	*Apple stem pitting virus*	P	1107
		Capillovirus	*Apple stem grooving virus*	P	1110
		Vitivirus	*Grapevine virus A*	P	1112
		Trichovirus	*Apple chlorotic leaf spot virus*	P	1116
Barnaviridae		*Barnavirus*	*Mushroom bacilliform virus*	F	1125
Unassigned Viruses					
Unassigned Vertebrate Viruses				V	1131
Unassigned Invertebrate Viruses				I	1132
Unassigned Prokaryote Viruses				B	1139
Unassigned Fungus Viruses				F	1139
Unassigned Plant Viruses				P	1141
The Subviral Agents: Viroids, Satellites and Agents of Spongiform Encephalopathies (Prions)					
Viroids					1147
Pospiviroidae					1153
		Pospiviroid	*Potato spindle tuber viroid*	P	1153
		Hostuviroid	*Hop stunt viroid*	P	1154
		Cocadviroid	*Coconut cadang-cadang viroid*	P	1155
		Apscaviroid	*Apple scar skin viroid*	P	1156
		Coleviroid	*Coleus blumei viroid 1*	P	1157
Avsunviroidae					1158
		Avsunviroid	*Avocado sunblotch viroid*	P	1158
		Pelamoviroid	*Peach latent mosaic virus*	P	1159
Satellites					1163
Vertebrate Prions				V	1171
Fungi prions				F	1179

Abbreviations of the virus hosts

Algae	Al	Bacteria	B	Invertebrates	I	Protozoa	Pr
Archaea	Ar	Fungi	F	Plants	P	Vertebrates	V

The Double Stranded DNA Viruses

ORDER *CAUDOVIRALES*

TAXONOMIC STRUCTURE OF THE ORDER

Order	***Caudovirales***
Family	***Myoviridae***
Genus	"T4-like viruses"
Genus	"P1-like viruses"
Genus	"P2-like viruses"
Genus	"Mu-like viruses"
Genus	"SPO1-like viruses"
Genus	"φH-like viruses"
Family	***Siphoviridae***
Genus	"λ-like viruses"
Genus	"T1-like viruses"
Genus	"T5-like viruses"
Genus	"L5-like viruses"
Genus	"c2-like viruses"
Genus	"ΨM1-like viruses"
Genus	"φC31-like viruses"
Genus	"N15-like viruses"
Family	***Podoviridae***
Genus	"T7-like viruses"
Genus	"P22-like viruses"
Genus	"φ29-like viruses"
Genus	"N4-like viruses"

GENERAL

The order consists of the three families of tailed bacterial viruses infecting Bacteria and Archaea: *Myoviridae* (long contractile tails), *Siphoviridae* (long non-contractile tails), and *Podoviridae* (short non-contractile tails). Tailed bacterial viruses are an extremely large group with highly diverse virion, genome, and replication properties. Over 4,500 descriptions have been published (accounting for 96% of reported bacterial viruses): 24% in the family *Myoviridae*, 62% in the family *Siphoviridae*, and 14% in the family *Podoviridae* (as of November 2001). However, data on virion structure, genome organization, and replication properties are available for only a small number of well-studied species. Their great evolutionary age, large population sizes, and extensive horizontal gene transfer between bacterial cells and viruses have erased or obscured many phylogenetic relationships amongst the tailed viruses. However, enough common features survive to indicate their fundamental relatedness. Therefore, formal taxonomic names are used for *Caudovirales* at the order and family level, but only vernacular names at the genus level.

VIRION PROPERTIES

MORPHOLOGY

The virion has no envelope and consists of two parts, the head and the tail. The head is a protein shell and contains a single linear dsDNA molecule, and the tail is a protein tube whose distal end binds the surface receptors on susceptible bacterial cells. DNA travels through the tail tube during delivery (often called "injection") into the cell being infected. Heads have icosahedral symmetry or elongated derivatives thereof (with known triangulation numbers of T=4, 7, 13, 16 and 52). Capsomers are seldom visible: heads usually appear smooth and thin-walled (2-3 nm). When they are visible, morphological features (capsomeres) on the surface of the head commonly form 72 capsomers (T=7; 420

protein subunits), but known capsomer numbers vary from 42 to 522. Isometric heads are typically 45-170 nm in diameter. Elongated heads derive from icosahedra by addition of equatorial belts of capsomers and can be up to 230 nm long. DNA forms a tightly packed coil (without bound proteins) inside the head. Tail shafts have six-fold or (rarely) three-fold symmetry, and are helical or stacks of disks of subunits from 3 and 825 nm in length. They usually have base plates, spikes, or terminal fibers at the distal end. Some viruses have collars at the head-tail junction, head or collar appendages, transverse tail disks, or other attachments.

PHYSICOCHEMICAL AND PHYSICAL PROPERTIES

Virion Mr is 20 to 600 x 10^6; S_{20w} values are 200 to >1200S. Both upper limits may be underestimates, since these properties have not been determined for the largest tailed viruses. Buoyant density in CsCl is typically ~1.5 g/cm^3. Most tailed viruses are stable at pH 5-9; a few are stable at pH 2 or pH 11. Heat sensitivity is variable, but many virions are inactivated by heating at 55-75°C for 30 min. Tailed viruses are rather resistant to UV irradiation. Heat and UV inactivation generally follow first-order kinetics. Most tailed phages are stable to chloroform. Inactivation by nonionic detergents is variable and concentration dependent. Some virions are sensitive to osmotic shock, and many are sensitive to Mg^{++} chelators.

NUCLEIC ACID

Virions contain one molecule of linear dsDNA. Genome sizes are 18 to 500 kbp, corresponding to Mr values of 11-300 x 10^6. DNA content is 45-55% of the virions. G+C contents are 27-72% and usually resemble those of host DNA. Some viral DNAs contain modified nucleotides which partially or completely replace normal nucleotides (*e.g.*, 5-hydroxymethylcytosine instead of cytosine), and/or are glycosylated or otherwise modified.

PROTEINS

The number of different virion structural proteins ranges from 7-49. Typical head shells are made up of 60T molecules of a single main building block CP and 12 molecules of portal protein through which DNA enters and leaves, but they can also contain varied numbers of proteins that plug the portal hole, proteins to which tails bind, proteins that bind to the outside of the CP shell (decoration proteins) and other proteins whose roles are not known. Non-contractile tails are made of one major shaft or tube protein and contractile tails have a second major protein, the sheath protein, that forms a cylinder around the central tube. Tails also have small numbers of varied specific proteins at both ends. Those at the end distal from the head form a structure called the tail tip (*Siphovirus*) or baseplate (*Myovirus*) to which the tail fibers are attached. The tail fibers bind to the first-contact receptors on the surface of susceptible cells. Fibers or baseplates may include proteins with endoglycosidase or peptidoglycan hydrolase activity that aid in gaining access to the cell surface and entry of DNA into the cell. Most virions carry proteins that are injected with the DNA, such as transcription factors, RNA polymerase and others with poorly understood functions.

LIPIDS

No well-characterized virions contain lipid.

CARBOHYDRATES

Glycoproteins, glycolipids, hexosamine, and a polysaccharide have been reported in certain virions but these are not well-characterized.

GENOME ORGANIZATION AND REPLICATION

GENOME ORGANIZATION

The linear dsDNA genomes encode from 27 to over 600 genes that are highly clustered according to function and tend to be arranged in large operons. Complete functional genomic maps are very diverse and available for only a relatively small number of tailed viruses. Virion DNAs may be circularly permuted and/or terminally redundant, have single-stranded gaps, or have covalently-bound terminal proteins. The ends of these linear molecules can be blunt or have complementary protruding 5'- or 3'-ends (the "cohesive" or "sticky" ends, which can base pair to circularize the molecule). Prophages of temperate tailed viruses are either integrated into the host genome or replicate as circular or linear plasmids; these linear plasmids have covalently-closed hairpin telomeres.

REPLICATION

In typical lytic infections, after entering the host cell, viral DNA may either circularize or remain linear. All tailed viruses encode proteins that direct the replication apparatus to the replication origin, but this apparatus may be entirely host derived, partly virus encoded, or entirely virus encoded. DNA replication is semi-conservative, may be either bidirectional or unidirectional, and usually results in the formation of concatemers (multiple genomes joined head-to-tail) by recombination between phage DNAs or by rolling circle replication. Progeny viral DNA is generated during virion assembly by cleavage from this concatemeric DNA: (i) at unique sites to produce identical DNA molecules with either *cos* sites or blunt-ended, terminally redundant termini, (ii) at *pac* sites to produce circularly permuted, terminal redundant DNAs, or (iii) by a headful mechanism to produce terminally redundant, circularly permuted DNAs. A few viruses use terminal proteins to prime DNA replication and package progeny viral DNA (ϕ29 and its relatives) or replicate DNA by a duplicative transposition mechanism (Mu and its relatives). Gene expression is largely time-ordered and groups of genes are sequentially expressed. "Early genes" are expressed first and are largely involved in host cell modification and viral DNA replication. "Late genes" specify virion structural proteins and lysis proteins. The larger tailed viruses have gene expression cascades that are more complex than this simple scenario. Transcription often requires host RNA polymerase, but many tailed viruses encode RNA polymerases or transcription factors that affect the host RNA polymerase. Translational control is poorly understood and no generalizations are possible at the present state of knowledge.

VIRION ASSEMBLY AND DNA PACKAGING

Assembly of virions from newly made proteins and replicated DNA is complex and generally includes separate pathways for heads, tails and tail fibers. Coat protein shells, called procapsids or proheads, are assembled first, and DNA is inserted into these preformed proteinaceous containers. Assembly of procapsids is poorly understood, but often utilizes an internal scaffolding protein which helps CP assemble correctly and is then released from the shell after its construction. In many, but not all, tailed viruses, proteolytic cleavages (by host or virus-encoded proteases) of some proteins accompany assembly. Virus-specific DNA is recognized for packaging into procapsids by the terminase protein. One end of the DNA is then threaded through the procapsid's portal structure, and DNA is pumped into the head by an ATP hydrolysis-driven motor that is probably made up of the two terminase subunits and portal protein. Unless unit length DNA molecules are the substrate for packaging (such as with ϕ29), when the head is full of DNA a "headful sensing device" recognizes this fact and causes the terminase to cleave the DNA to release the full head from the unpackaged remainder of the DNA concatemer. The terminase subunits are usually released from the virion after DNA is packaged. Filled heads then join to tails and tail fibers to form progeny virions. Some viruses form intracellular arrays, and many produce aberrant structures (polyheads, polytails, giant, multi-tailed, or misshapen particles). Progeny viruses are liberated by lysis of the host cell. Cell lysis is caused by phage-encoded peptidoglycan hydrolases; but lysis timing is controlled by holins, phage encoded inner membrane proteins that allow the hydrolases to escape from the cytoplasm.

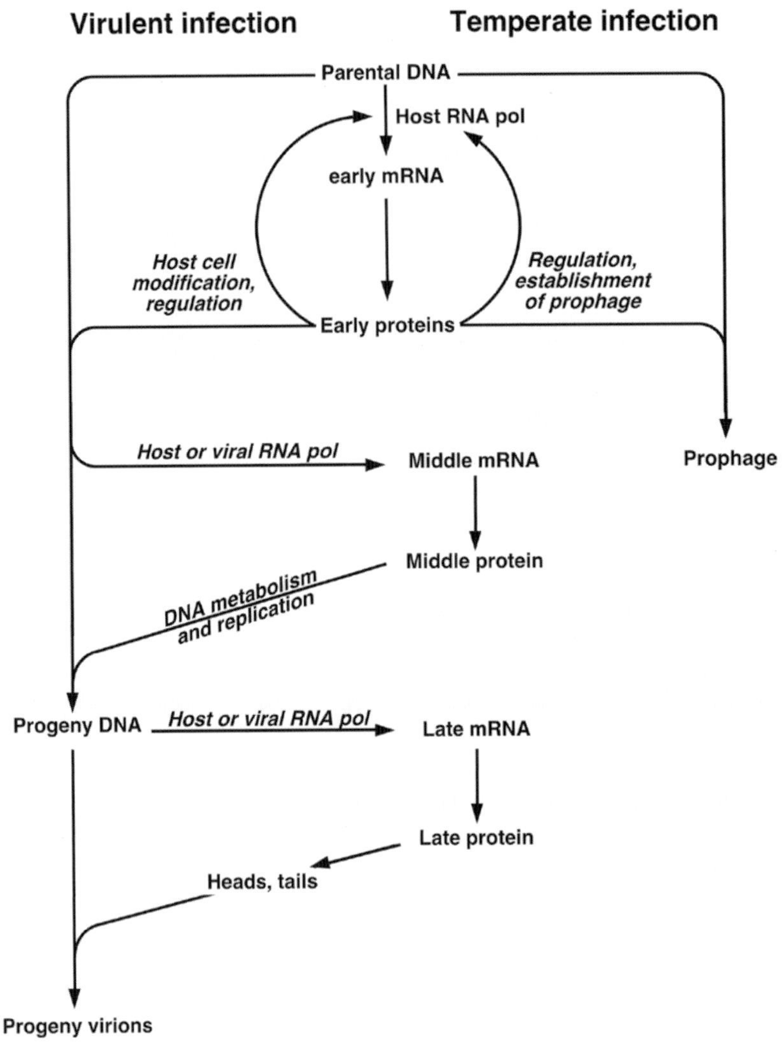

Figure 1: Flow chart of tailed phage replication. The chart depicts the replication of "typical" virulent phages such as Enterobacteria phage T4 (T4), Enterobacteria phage T7 (T7), and the temperate phages.

Antigenic Properties

Viruses are antigenically complex and efficient immunogens, inducing the formation of neutralizing and complement-fixing antigens. The existence of group antigens is likely within species or genera.

Biological Properties

Infection

Tailed-viruses are lytic or temperate. Lytic infection results in production of progeny viruses and destruction of the host. Phages adsorb tail-first to specific protein or lipoprotein host cell receptors, which are located on the outer cell surface. In a few cases, not represented by the genera described here, the primary adsorption sites are capsules, flagella, or pili. Upon adsorption to the outside of the cell, virions undergo complex and often poorly understood rearrangements which release the DNA to enter the cell through the tail. Cell walls are often locally digested by a virion-associated peptidoglycan hydrolase and viral DNA enters the cytoplasm by as yet unknown mechanisms. In some cases DNA entry is stepwise and transcription of the first DNA to enter is required for entry of the rest of the DNA. Empty virions remain outside the infected bacterium, however most viruses inject specific proteins with the DNA. Temperate viruses can, upon infection, either enter a lytic growth cycle (above) or establish a lysogenic state (below). Physiological factors in the cell can affect the decision between these two pathways.

LATENCY

All three tailed virus families include genera or species of temperate viruses. Viral genomes in lysogenized cells are called "prophages". Prophages are either integrated into host cell chromosomes or persist as extrachromosomal elements (plasmids). Integration is usually mediated by recombinases called integrases. The most common are in the tyrosine-active site class and some are in the serine-active site class. For the Mu-like viruses, integration is accomplished by transposases. Integrated prophages typically express only a very small fraction of their genes. The genes that are expressed from the prophage are called "lysogenic conversion" genes, and their products usually alter the properties of the bacterial host. Among these genes is the prophage repressor gene, whose product binds operators in the prophage to keep the lytic cascade of gene expression from initiating. Plasmid prophages typically express many of their early genes, some of which are involved in replication of the plasmid (which can be circular or linear). Prophages can often be induced to initiate a lytic growth cycle; DNA damaging agents such as ultraviolet light or mitomycin C cause many prophages to induce.

HOST RANGE

Tailed viruses have been found in over 140 prokaryote genera representing most branches of the Bacterial and Archaeal phylogenetic trees. The host specificity of these viruses can vary widely; some can infect multiple closely related genera, but perhaps more common (especially in the host family *Enterobacteriacea*, where the most varieties have been studied) are viruses that are specific for particular isolates or groups of isolates of closely related host species.

TRANSMISSION IN NATURE

Virions are typically carried and transmitted in aqueous environments, although a few are stable to drying. Virus genomes can be carried as prophages inside host bacteria. Such lysogenic bacteria can induce to release virions, either spontaneously or in response to specific environmental signals.

GEOGRAPHIC DISTRIBUTION

Tailed phages are the most abundant type of organism on Earth; the current best estimates are 10^{31} particles in our biosphere. If all these phages were laid end to end the line would extend for 2×10^{8} light years. Data from genome sequence analyses implies that these viruses can move around the globe on a time scale that is short relative to the rate at which they accumulate mutations. They have a worldwide distribution and presumably share the habitats of their hosts. An important habitat is inside lysogenic bacteria as prophages.

PHYLOGENETIC RELATIONSHIPS WITHIN THE ORDER AND THE PERILS OF MOSAICISM

The recent availability of high-throughput DNA sequencing has led to a dramatic increase in the number of complete genome sequences that are available for members of the *Caudovirales*. This has led in turn to a similarly dramatic change in our understanding of the phylogenetic relationships among members of the order. The new data substantially enrich our appreciation of the genetic structure of the global *Caudovirales* population and of the evolutionary mechanisms within that order; the new data also substantially complicate considerations of how best to represent these viruses in a taxonomy.

The hallmark of the genomes of these viruses is that they are genetic mosaics, a property that becomes apparent only when two or more genome sequences are compared. The modules of sequence that constitute the mosaic are typically individual genes, but they can also be parts of genes corresponding to protein domains, or small groups of genes such as prohead assembly genes. The mosaicism is evidently the result of non-homologous recombination during the evolution of these viruses. The novel juxtapositions of sequence produced in this way are spread through the population and reassorted with each other by means of homologous recombination. Regardless of mechanism, the overall result is as if each phage had constituted its genome by picking modules from a menu, choosing one module from each of perhaps fifty columns, each of which has alternative choices.

A consequence of the mosaic relationships among the genomes of these viruses is that if we ask how closely two viruses are related to each other — as we might do in trying to reconstruct their phylogeny or in deciding on a taxonomy — the answer will be radically different depending on which module we base our comparison on. Thus, we could look, for example, at the sequences of the major capsid proteins from a group of phages and derive a self-consistent hierarchical phylogeny, ostensibly representing the evolutionary history of the capsid genes, but if we were to construct a similar phylogeny for a different genetic module from the same group of phages, say the C-terminal domain of the integrase proteins, we would get another self-consistent phylogeny which was however completely incongruent with the phylogeny of the capsid proteins. A formal representation of such relationships is shown in figure 2.

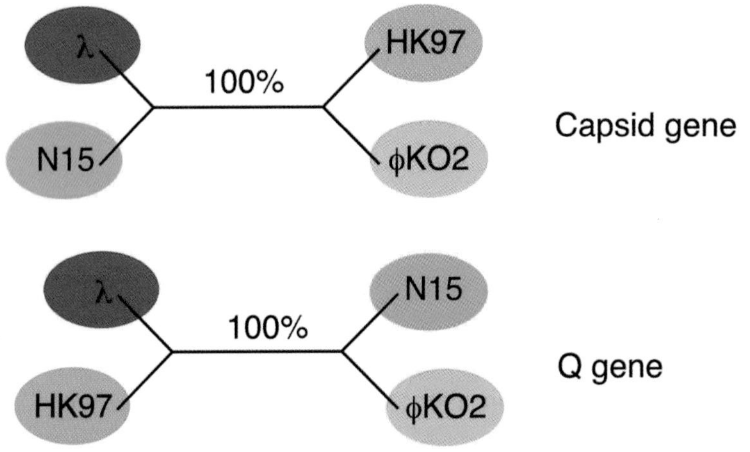

Figure 2: Phylogenetic trees showing the relationships among two different genes from four phages. Enterobacteria phage λ, Enterobacteria phage HK97 and Enterobacteria phage N15 infect *E. coli*; Klebsiella phage ϕKO2 is a phage of *Klebsiella oxytoca*. The trees are incongruent owing to mosaicism in the genomes of the phages.

A logical consequence of such mosaic relationships is that it is not possible to construct a hierarchical phylogeny for the viruses that does not misrepresent the phylogenies of some (often many) of the component genetic modules. In fact, given the degree of mosaicism in the order *Caudovirales*, any attempt at a hierarchical whole virus phylogeny will necessarily misrepresent a majority of the component modules. This is true whether the phylogeny is based on the relationships among members of a single module type (say DNA polymerases) or on some sort of average or blending of all the modules.

The question for virus taxonomists then becomes, how should we construct a taxonomy to represent these biological properties of the phage population? In the current ICTV taxonomy, represented here, the division of the order *Caudovirales* into three families is based solely on tail morphology: *Siphoviridae* have long non-contractile tails, *Myoviridae* have long contractile tails, and *Podoviridae* have short tails. As might be expected from the discussion above, this hierarchical division of phages on the basis of one character leads to many examples of inappropriate divisions of other characters. One well known and easily illustrated example of this is shown in figure 3, comparing phages λ and P22. These two phages are considered by most phage biologists to be closely related, because they share genome organization (including regulation and layout of transcription and functional order of genes), temperate lifestyle, a number of similarities of gene sequences, and they can form viable hybrids. Despite these similarities, they are classified into different families (*Siphoviridae* and *Podoviridae* for λ and P22, respectively) based on their differences in tail morphology. An argument could be made as to whether or not the similarities between these two phages are enough that they should be classified in the same family, but it is in any case clear that P22 is much closer to λ than it is to most other members of the family *Podoviridae*, such as phages T7 and N4, which have essentially no similarity to λ in sequence, genome organization, or lifestyle. The critical issue is in fact not to decide how different two phages need to be to be assigned to different families, but rather whether it is a useful exercise to try to represent a population in which the individuals are related to each other in a "reticulate" or "multi-dimensional" fashion by using a purely hierarchical taxonomy that is doomed to misrepresent the majority of those biological relationships?

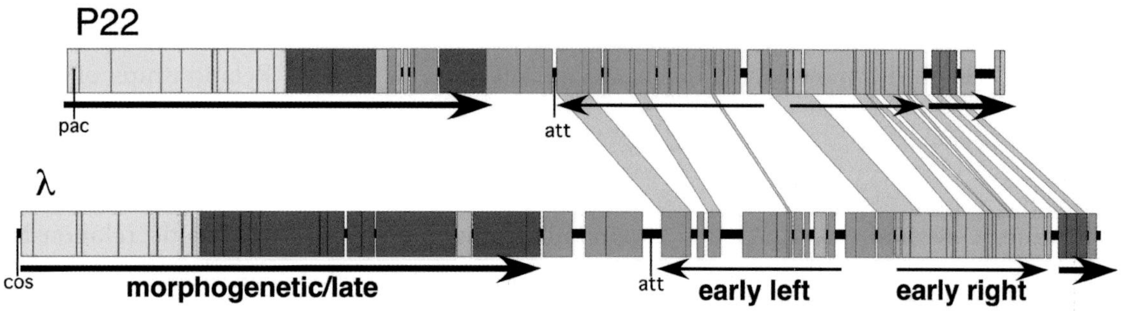

Figure 3: The mosaic relationship between the genomes of phages P22 and λ. The circular maps are opened for linear display between the lysis and head genes. The genes in each genome are represented by rectangles; white - transcribed right to left, gray - transcribed left to right. P22 genes that have sequence similarity to λ genes are connected by light gray trapezoids. The thin arrows represent transcription of the early operons and thick arrows transcription of the late operons. The circular phage genomes are opened at their attachment (att) sites for insertion of the prophage into the host chromosome in lysogens. DNA packaging initiation sites (called pac and cos in P22 and λ, respectively) are also indicated below the maps.

This discussion recapitulates a long-running controversy in the field of taxonomy over whether it is of paramount importance for a taxonomy to accurately reflect the biological relationships of the classified organisms or whether it is sufficiently useful to get organisms assigned a place in a recognized taxonomy that an occasional (or even frequent) misrepresentation of biological relationships is of little consequence. These issues are particularly sharply focused for the order *Caudovirales* due to the fact that the mosaicism is so extensive and the consequent misrepresentations so pervasive. Because of this, the ICTV considers the taxonomy of this group to be provisional, and this is the reason that the names of the genera are in a non-official vernacular format. Discussions are ongoing, both within the ICTV and in the virology community at large, and there may well be significant changes to the *Caudovirales* taxonomy in the future in response to our new understanding of the biology.

Similarity with Other Taxa

Tailed bacterial viruses resemble members of the family *Tectiviridae* by the presence of a dedicated structure for DNA injection, but differ from them by the permanent nature of their tails and lack of a lipid bilayer. Tailed viruses resemble viruses belonging to the family *Herpesviridae* in morphogenesis (use of scaffolding proteins, packaging of DNA into preformed shells, maturation of procapsids by proteolytic cleavage, and capsid conformational change) and overall strategy of replication. In addition, temperate tailed phages and members of the family *Herpesviridae* are able to establish latent infections.

Derivation of Names

Caudo: from Latin *cauda*, "tail".
Myo: from Greek *my, myos*, "muscle", referring to the contractile tail.
Sipho: from Greek *siphon*, "tube", referring to the long tail.
Podo: from Greek *pous, podos*, "foot", referring to the short tail.

References

Ackermann, H.-W. (1996). Frequency of morphological phage descriptions in 1995. *Arch. Virol.*, **141**, 209-218.

Ackermann, H.-W. (1998). Tailed bacteriophages: the order Caudovirales. *Adv. Virus Res.*, **51**, 101-168.

Ackermann, H.-W. and DuBow, M.S. (eds)(1987). *Viruses of Prokaryotes*, Vol. I and II. CRC Press, Boca Raton, Florida.

Ackermann, H.-W., DuBow, M.S., Jarvis, A.W., Jones, L.A., Krylov, V.N., Maniloff, J., Rocourt, J., Safferman, R.S., Schneider, J., Seldin, L., Sozzi, T., Stewart, P.R., Werquin, M. and Wüsche, L. (1992). The species concept and its application to tailed phages. *Arch. Virol.*, **124**, 69-82.

Ackermann, H.-W., Elzanowski, A., Fobo, G. and Stewart, G. (1995). Relationships of tailed phages: a survey of protein sequence identity. *Arch. Virol.*, **140**, 1871-1884.

Black, L.W. (1989). DNA packaging in dsDNA bacteriophages. *Annu. Rev. Microbiol.*, **43**, 267-292.

Botstein, D. and Herskowitz, I. (1974). Properties of hybrids between Salmonella phage P22 and coliphage lambda. *Nature*, **251**, 584-586.

Braithwaite, D.K. and Ito, J. (1993). Compilation, alignment and phylogenetic relationships of DNA polymerases. *Nucl. Acids Res.*, **21**, 787-802.

Calendar, R. (ed) (1988). *The Bacteriophages*, Vol. 1 and 2. Plenum Press, New York.

Canchaya, C., Proux, C., Fournous, G., Bruttin, A. and Brussow, H. (2003). Prophage Genomics. *Microbiol. Mol. Biol. Rev.*, **67**, 238-276.

Casjens, S. (2003). Prophages in sequenced bacterial genomes: what have we learned so far? *Molec. Microbiol.*, **49**, 277-300.

Casjens. S., Hatfull, G. and Hendrix, R. (1992). Evolution of dsDNA tailed-bacteriophage genomes. *Semin. Virol.*, **3**, 383-397.

Hendrix, R.W. (2002). Bacteriophages: Evolution of the majority. *Theor. Pop. Biol.*, **61**, 471-480.

Klaus, S., Krüger, D.H. and Meyer, J. (eds)(1992). *Bakterienviren*. Gustav, Fischer, Jena-Stuttgart.

Maniloff, J. and Ackermann, H.-W. (1998). Taxonomy of bacterial viruses: establishment of tailed virus genera and the order *Caudovirales*. *Arch. Virol.*, **143**, 2051-2063.

Lawrence, J.G., Hatfull, G.F. and Hendrix, R.W. (2002). Imbroglios of viral taxonomy: Genetic exchange and failings of phenetic approaches. *J. Bacteriol.*, **184**, 4891-4905.

Susskind, M.M. and Botstein, D. (1978). Molecular genetics of bacteriophage P22. *Microbiol. Rev.*, **42**, 385-413.

Contributed By

Hendrix, R.W. and Casjens, S.R.

FAMILY MYOVIRIDAE

TAXONOMIC STRUCTURE OF THE FAMILY

Family	***Myoviridae***
Genus	**"T4-like viruses"**
Genus	**"P1-like viruses"**
Genus	**"P2-like viruses"**
Genus	**"Mu-like viruses"**
Genus	**"SPO1-like viruses"**
Genus	**"ϕH-like viruses"**

DISTINGUISHING FEATURES

Tails are contractile, more or less rigid, long and relatively thick (80-455 x 16-20 nm). They consist of a central core built of stacked rings of 6 subunits and surrounded by a helical contractile sheath, which is separated from the head by a neck. During contraction, sheath subunits slide over each other and the sheath becomes shorter and thicker. This brings the tail core in contact with the bacterial plasma membrane and is an essential stage of infection. Heads and tails are assembled in separate pathways. With respect to other tailed phages, myoviruses often have larger heads and higher particle weights and DNA contents, and seem to be more sensitive to freezing and thawing and to osmotic shock. Genera are differentiated by genome organization, mechanisms of DNA replication, and packaging, and the presence or absence of unusual bases and DNA polymerases.

GENUS "T4-LIKE VIRUSES"

Type Species *Enterobacteria phage T4*

DISTINGUISHING FEATURES

Virions have elongated heads and tails with long, kinked fibers. Genomes are circularly permuted and terminally redundant, and typically code for hydroxymethylcytosine synthesizing enzymes and type B DNA polymerase. The genetic map is circular and the DNA is packaged by a headful mechanism.

VIRION PROPERTIES

MORPHOLOGY

Phage heads are prolate icosahedra (elongated pentagonal bipyramidal antiprisms), measure ~111 x 78 nm, and consist of 152 capsomers (T=13, elongated). Tails measure 113 x 16 nm and have a collar, base plate, 6 short spikes and 6 long fibers.

PHYSICOCHEMICAL AND PHYSICAL PROPERTIES

Virion Mr is ~210 x 10^6, buoyant density in CsCl is 1.50 g/cm^3, and S_{20w} ~1030S. Infectivity is ether and chloroform resistant.

NUCLEIC ACID

Genomes have a Mr ~120 x 10^6, corresponding to 48% of the particle weight. DNA contains 5-hydroxymethylcytosine (HMC) instead of cytosine (these nucleotides are glycosylated), a G+C content of 35%, and is circularly permuted and terminally redundant. The Enterobacteria phage T4 (T4) genome has been fully sequenced (168,903 bp).

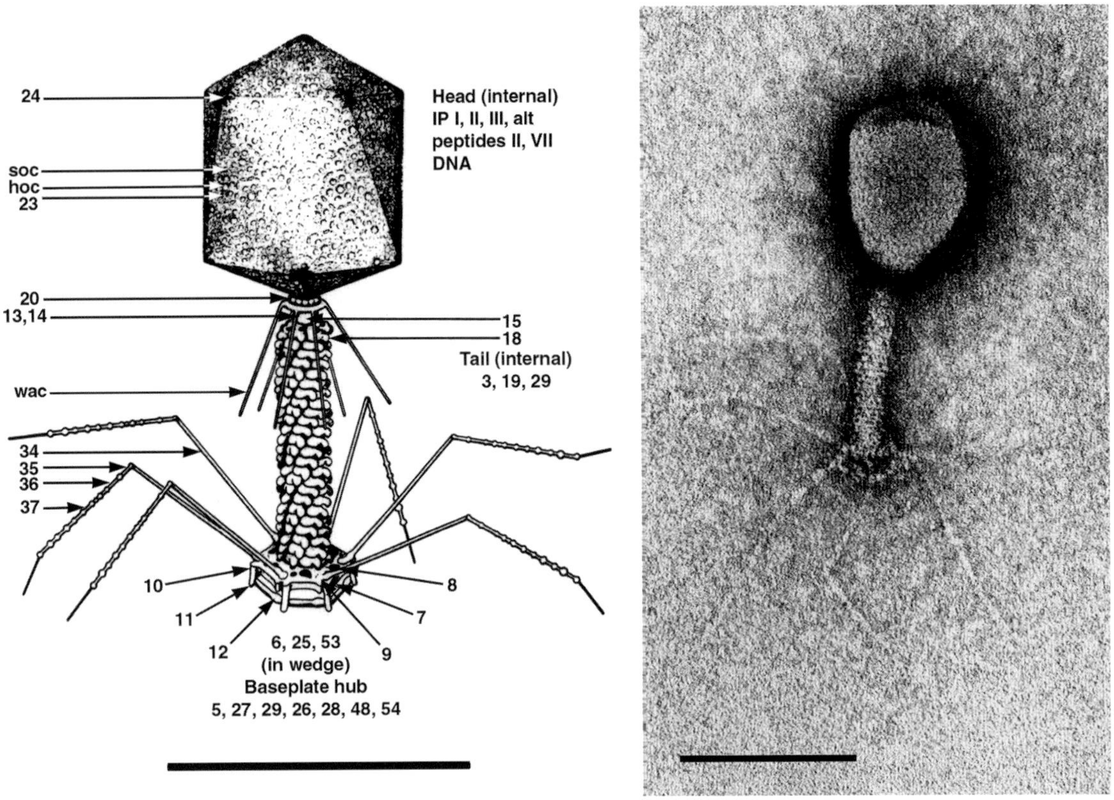

Figure 1: (Left) Diagram of Enterobacteria phage T4 (T4) showing detailed location of structural proteins. Head vertices consist of cleaved gp24. Gp20 is located at the head tail connector. Collar and whiskers appear to be made of the same protein, gp*wac*. Sheath subunits (gp18) fit into holes in the base plate and short tail proteins (gp12) are shown in the quiescent state. The complex base is assembled from a central plug and six wedges. Tail fibers consist of three proteins. (From Eiserling, F.A. (1983). *Bacteriophage T4*, (C.K., Mathews, E.M., Kutter, G., Mosig and P.B., Berget, eds). American Society for Microbiology, Washington, DC. Reproduced with permission). (Right) Negative contrast electron micrograph of T4 particle stained with uranyl acetate. The bars represent 100 nm.

PROTEINS

T4 particles contain at least 49 polypeptides (8-155 kDa), including 1,600-2,000 copies of the major CP (43 kDa) and 3 proteins located inside the head. Various enzymes are present or encoded, *e.g.* type B (*E. coli* Pol II) DNA polymerase, numerous nucleotide metabolism enzymes and lysozyme. Amino acid sequences for T4 proteins are available.

LIPIDS

None known.

CARBOHYDRATES

Glucose is covalently linked to HMC in phage DNA.

GENOME ORGANIZATION AND REPLICATION

The genetic map is circular and comprises ~300 genes. Morphopoietic genes generally cluster together, but this is not universally true, suggesting extensive translocation of genes during evolution. The genome is circularly permuted and has 1-3% terminal redundancy. After infection, the host chromosome breaks down and viral DNA replicates as a concatemer, generating forked replicative intermediates from multiple origins of replication. Transcription is regulated in part by phage-induced modification of host bacterial RNA polymerase and proceeds in three waves (early, middle, late). Heads, tails, and tail fibers are assembled in 3 separate pathways. Unique DNA molecules are packaged by a headful mechanism. Virions are assembled at the cell periphery. Aberrant head structures (polyheads and isometric heads) are frequent.

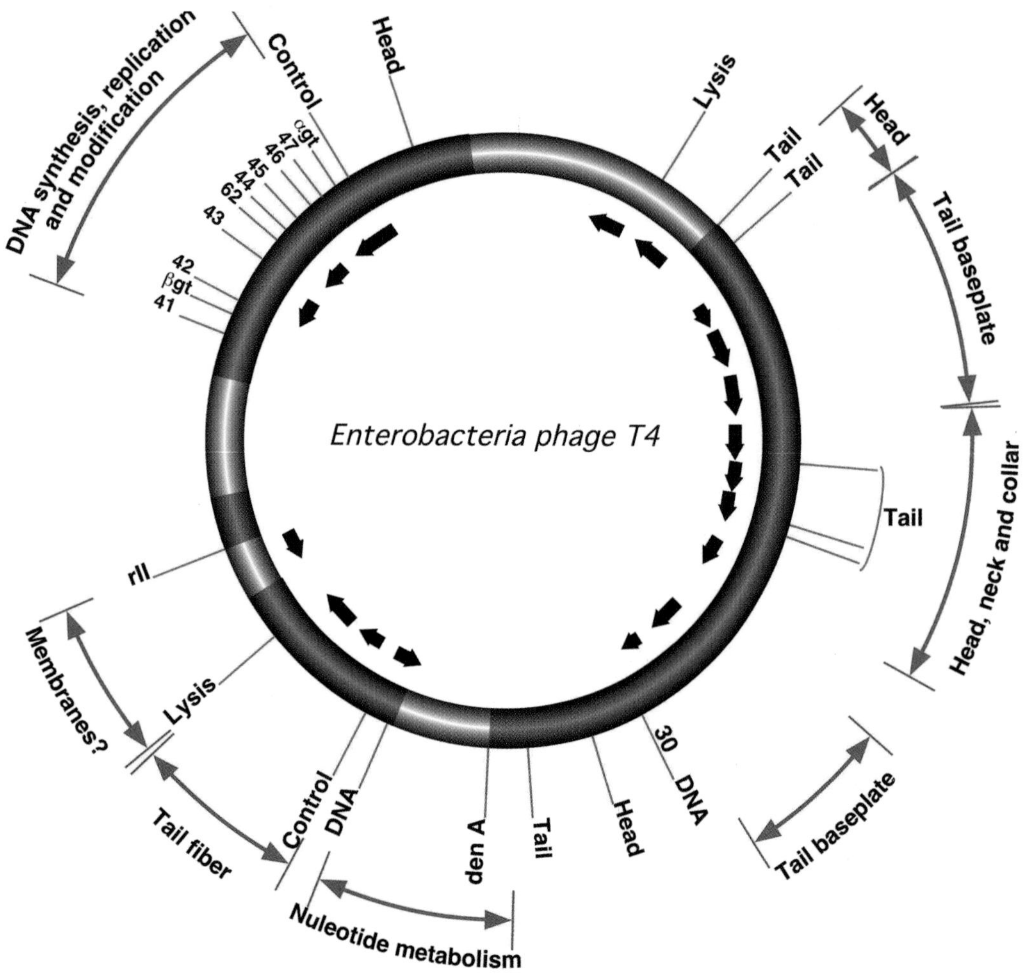

Figure 2: Simplified genetic map of Enterobacteria phage T4 (T4) showing clustering of genes with related functions, location of essential genes (solid bars), and direction and origin of transcripts (arrows). (From Freifelder, D. (ed)(1983). *Molecular Biology*. Science Books International, Boston, and Van Nostrand Reynolds, New York, p 614. With permission).

ANTIGENIC PROPERTIES

A group antigen and antigens defining 8 subgroups have been identified by complement fixation

BIOLOGICAL PROPERTIES

Phages are virulent, and infect enteric and related bacteria (γ3-subgroup of Gram-negative proteobacteria). Their distribution is worldwide.

LIST OF SPECIES DEMARCATION CRITERIA IN THE GENUS

Species differ in host range, capsid length, serological properties, and, insofar as known, DNA homology and amino acid sequences. Capsid length is 137 nm for Aeromonas phage Aeh1 (Aeh1) and Vibrio phage nt-1 (nt-1) and 111 nm for a number of other species. Phage T4 and Enterobacteria phage SV14 (Sv14) are in different hybridization groups.

LIST OF SPECIES IN THE GENUS

Species names are in green italic script; strain names and synonyms are in black roman script; tentative species names are in blue roman script. Sequence accession numbers, and assigned abbreviations ( ) are also listed.

SPECIES IN THE GENUS

Acinetobacter phage 133		
Acinetobacter phage 133		(133)
Aeromonas phage 40RR2.8t		
Aeromonas phage 40RR2.8t		(40RR2.8t)
(Aeromonas phage 40R)		(40R)
Aeromonas phage 65		
Aeromonas phage 65		(65)
Aeromonas phage Aeh1		
Aeromonas phage Aeh1		(Aeh1)
Enterobacteria phage SV14		
Enterobacteria phage D2A		(D2A)
Enterobacteria phage D8		(D8)
Enterobacteria phage SV14		(SV14)
Enterobacteria phage T4		
Enterobacteria phage C16		(C16)
Enterobacteria phage F10		(F10)
Enterobacteria phage Fsα		(Fsα)
Enterobacteria phage PST		(PST)
Enterobacteria phage SKII		(SKII)
Enterobacteria phage SKV		(SKV)
Enterobacteria phage SKX		(SKX)
Enterobacteria phage SV3		(SV3)
Enterobacteria phage T2		(T2)
Enterobacteria phage T4	[A158101]	(T4)
Enterobacteria phage T6		(T6)
Pseudomonas phage 42		
Pseudomonas phage 42		(42)
Vibrio phage nt-1		
Vibrio phage KVP20		(KVP20)
Vibrio phage KVP40		(KVP40)
Vibrio phage nt-1		(nt-1)

TENTATIVE SPECIES IN THE GENUS

Acinetobacter phage E4	(E4)
Acinetobacter phage E5	(E5)
Aeromonas phage 1	(Aer1)
Aeromonas phage 25	(25)
Aeromonas phage 31	(31)
Enterobacteria phage 1 (Phage aeI)	(aeI)
Enterobacteria phage 11F	(11F)
Enterobacteria phage 3	(3)
Enterobacteria phage 3T+	(3T+)
Enterobacteria phage 50	(50)
Enterobacteria phage 5845	(5845)
Enterobacteria phage 66F	(66F)
Enterobacteria phage 8893	(8893)
Enterobacteria phage 9/0	(9/0)
Enterobacteria phage α1	(α1)
Enterobacteria phage DdVI	(DdV1)
Enterobacteria phage F7	(F7)
Enterobacteria phage Kl3	(K13)
Enterobacteria phage RB42	(RB42)
Enterobacteria phage RB43	(RB43)

Enterobacteria phage RB49		(RB49)
Enterobacteria phage RB69	[AY303349]	(RB69)
Enterobacteria phage SMB		(SMB)
Enterobacteria phage SMP2		(SMP2)

GENUS "P1-LIKE VIRUSES"

Type Species *Enterobacteria phage P1*

DISTINGUISHING FEATURES

Virions produce head size variants. DNA is circularly permuted and terminally redundant, and is packaged from a *pac* site. The genetic map is linear, and phages can carry out generalized transduction. Prophages persist as plasmids.

VIRION PROPERTIES

MORPHOLOGY

Virions have icosahedral heads ~85 nm in diameter and produce head size variants (~47-65 nm). Tails measure 228 x 18 nm in Enterobacteria phage P1 (P1) and vary in length from 170-240 nm in other members of the genus (i.e., Enterobacteria phage P1D (P1D) and Aeromonas phage 43 (43)). Tails have base plates and six 90 nm-long kinked fibers. Particles with contracted tails aggregate side-by-side by means of exposed tail cores.

PHYSICOCHEMICAL AND PHYSICAL PROPERTIES

Phage P1 virion buoyant density is 1.48 g/cm^3.

NUCLEIC ACID

Genomes are ~100 kbp and have a G+C content of 46%.

PROTEINS

Virions contain 24-28 constitutive proteins (10-220 kDa), including a major coat protein of 44 kDa.

LIPIDS

None known.

CARBOHYDRATES

None known.

GENOME ORGANIZATION AND REPLICATION

The genetic map is linear and includes ~100 genes; related functions are often distributed over several genome regions. Prophage DNA is circular. The genome is circularly permuted and terminally redundant (8-12%), and includes a recombinational hot spot (*lox-cre*). The genome also has an invertible tail fiber segment of ~4 kbp (C-loop) that is homologous to the G-loop of Enterobacteria phage Mu (Mu). Virion DNA circularizes after injection. Replication starts at a single site and has a phase of Θ replication and then a phase of σ structures, suggesting a rolling-circle mechanism. Progeny DNA is cut from concatemers at a *pac* site.

ANTIGENIC PROPERTIES

Phages P1, P2, and Mu share tail fiber antigens.

BIOLOGICAL PROPERTIES

Phages are temperate, can carry out generalized transduction, and infect enteric and related Gram-negative bacteria. Prophages are maintained as plasmids (1-2 copies per cell) or integrate (rarely) at specific sites into the bacterial chromosome. Prophages are weakly UV inducible. The invertible C-loop codes for two sets of tail fiber genes and provides a means of extending host range.

LIST OF SPECIES DEMARCATION CRITERIA IN THE GENUS

Species differ in host range and tail length (phage P1, 228 nm; phage P1D, 240 nm; and phage 43, 170 nm).

LIST OF SPECIES IN THE GENUS

Species names are in green italic script; strain names and synonyms are in black roman script; tentative species names are in blue roman script. Sequence accession numbers, and assigned abbreviations ( ) are also listed.

SPECIES IN THE GENUS

Aeromonas phage 43	
Aeromonas phage 43	(43)
Enterobacteria phage P1	
Enterobacteria phage P1	(P1)
Enterobacteria phage P1D	(P1D)
Enterobacteria phage P7	(P7)

TENTATIVE SPECIES IN THE GENUS

Acetobacter phage pKG-2	(pKG-2)
Acetobacter phage pKG-3	(pKG-3)
Enterobacteria phage D6	(D6)
Enterobacteria phage ϕW39	(ϕW39)
Enterobacteria phage j2	(j2)
Pseudomonas phage PP8	(PP8)
Vibrio phage ϕVP25	(ϕVP253)
Vibrio phage P147	(P147)

GENUS "P2-LIKE VIRUSES"

Type Species *Enterobacteria phage P2*

DISTINGUISHING FEATURES

Virion DNA has cohesive ends. Transcription of virion structural genes is divergent.

VIRION PROPERTIES

MORPHOLOGY

Phage heads are icosahedral, measure ~60 nm in diameter, and consist of 72 capsomers (60 hexamers and 12 pentamers; T=7). Tails measure 135 x 18 nm and have a collar and 6 short kinked fibers.

PHYSICOCHEMICAL AND PHYSICAL PROPERTIES

Virion Mr is 58 x 10^6; buoyant density in CsC1 is 1.43 g/cm^3; and S_{20W} is 283S.

NUCLEIC ACID

Genomes are ~34 kbp, are ~48% of particle weight, and have a G+C content of 52%. The genomes of P2 and the related phages (HP1, HP2, 186, ϕCTX, Fels-2 and K139) have been sequenced.

PROTEINS

Virions contain at least 13 structural proteins (20-94 kDa), including 420 copies of the major CP (39 kDa). Amino acid sequences of the proteins of phages with completely sequenced genomes are available at GenBank and EMBL.

LIPIDS
None known.

CARBOHYDRATES
None known.

GENOME ORGANIZATION AND REPLICATION

The genetic map is linear and non-permuted, has *cos* sites, and includes ~40 genes. Transcription starts in the right half of the genome, has two phases (early and late), and depends on host RNA polymerase. Replication starts at a single site, is unidirectional, and follows a modified rolling-circle mechanism. DNA is cut from concatemers at specific sites during packaging into proheads.

ANTIGENIC PROPERTIES

Virions of phages P2, P1 (Genus "P1-like viruses"), and Mu (Genus "Mu-like viruses") share tail fiber antigens.

BIOLOGICAL PROPERTIES

Phages are temperate, adsorb to the cell wall, and infect enteric and related Gram-negative bacteria. Prophages may integrate at ~10 specific sites of the bacterial chromosome and are not UV-inducible. P2 acts as a "helper" for defective Enterobacteria phage P4 (P4) in by providing head and tail genes for P4 propagation.

LIST OF SPECIES DEMARCATION CRITERIA IN THE GENUS

Species differ in host range and DNA homology.

LIST OF SPECIES IN THE GENUS

Species names are in green italic script; strain names and synonyms are in black roman script; tentative species names are in blue roman script. Sequence accession numbers, and assigned abbreviations ( ) are also listed.

SPECIES IN THE GENUS

Enterobacteria phage P2		
Enterobacteria phage P2	[AF063097]	(P2)
Haemophilus phage HP1		(HP1)
Haemophilus phage HP1	[U24159]	(HP1)
Haemophilus phage S2		(S2)

TENTATIVE SPECIES IN THE GENUS

Aeromonas phage 29		(29)
Aeromonas phage 37		(37)
Agrobacterium phage PIIBNV6		(PIIBNV6)
Caulobacter phage ΦCr24		(ΦCr24)
Enterobacteria phage 186	[U32222]	(186)
Enterobacteria phage 299		(299)
Enterobacteria phage Beccles		(Beccles)
Enterobacteria phage Pk2		(Pk2)
Enterobacteria phage Wϕ		(Wϕ)
Haemophilus phage HP2	[AY027935]	(HP2)
Pasteurella phage AU		(AU)
Pseudomonas phage ϕCTX	[AB008550]	(ϕCTX)
Pseudomonas phage PsP3		(PsP3)
Rhizobium phage ϕgal-1/R		(ϕgal-1/R)
Rhizobium phage WT1		(WT1)
Salmonella phage Fels-2		(Fels-2)
Vibrio phage X29		(X29)
Vibrio phage K139	[AF125163]	(K139)

GENUS "MU-LIKE VIRUSES"

Type Species *Enterobacteria phage Mu*

DISTINGUISHING FEATURES

The viral genome contains two terminal, variable sequences of host DNA. It is able to integrate at virtually any site of the host chromosome and generate a wide range of mutations due to its unique mode of DNA replication (replicative transposition). Integration is required for establishment of lysogeny and DNA replication during lytic development.

VIRION PROPERTIES

MORPHOLOGY

Virions have icosahedral heads ~60 nm in diameter, contractile tails ~120 x 18 nm, a baseplate, and 6 short fibers.

PHYSICOCHEMICAL AND PHYSICAL PROPERTIES

Virion buoyant density in CsC1 is 1.49 g/cm^3.

NUCLEIC ACID

The phage Mu genome is ~36-40 kbp, corresponding to ~40% of particle weight, has a G+C content of 50-51%, and has been completely sequenced.

PROTEINS

Particles have 12 structural proteins (20-76 kDa), including the major coat protein (33 kDa).

LIPIDS

None known.

CARBOHYDRATES

None known

GENOME ORGANIZATION AND REPLICATION

The phage Mu genetic map is linear and includes 55 genes. Related functions cluster together. The genome is non-permuted and heterogeneous, consisting of 36,717 bp of phage-specific DNA flanked at both ends by 0.5-3 kbp of covalently bound segments of host DNA. It contains an invertible segment of ~3 kbp (the G-loop) that is homologous to the invertible C-segment of Enterobacteria phage P1 (P1) DNA. Infecting DNA undergoes either lytic or lysogenic development. Both modes require (random) integration of phage DNA into host DNA, mediated by a phage-encoded transposase. Transcription starts at the left end of the genome and depends on host RNA polymerase. Replication may start at either end of the genome, is semi-conservative, and occurs during transposition into new integration sites. Phage heads package integrated, non-concatemeric phage DNA and adjacent host DNA by an atypical headful mechanism. Progeny phage DNA is cut out of the host DNA 100-200 bp away from a phage-coded *pac* site.

ANTIGENIC RELATIONSHIPS

Enterobacteria phages Mu, D108, P1, and P2 have some common tail fiber antigens.

BIOLOGICAL PROPERTIES

Viruses are temperate and can carry out generalized transduction, and infect enteric and (possibly) other related Gram-negative bacteria. The invertible G-loop codes for two sets of tail fibers which provides a means of extending host range. Prophages are not inducible by UV light.

LIST OF SPECIES DEMARCATION CRITERIA IN THE GENUS

Not applicable.

LIST OF SPECIES IN THE GENUS

Species names are in green italic script; strain names and synonyms are in black roman script; tentative species names are in blue roman script. Sequence accession numbers, and assigned abbreviations ( ) are also listed.

SPECIES IN THE GENUS

Enterobacteria phage Mu		
Enterobacteria phage D108		(D108)
Enterobacteria phage Mu	[AF083977]	(Mu)
(Enterobacteria phage Mu-1)		

TENTATIVE SPECIES IN THE GENUS

Pseudomonas phage B3	(B3)
Pseudomonas phage B39	(B39)
Pseudomonas phage D3112	(D3112)
Pseudomonas phage PM69	(PM69)
Vibrio phage VcA3	(VcA3)

GENUS "SPO1-LIKE VIRUSES"

Type Species *Bacillus phage SPO1*

NOTE ON NOMENCLATURE

The "O" in the name SPO1 derives from Osaka, where the phage was isolated. It is therefore properly the letter "O" (oh) and not the numeral "0" (zero). However, in the published literature and earlier versions of this taxonomy, the names "SPO1" and SP01" are used interchangeably to refer to the same virus. As a consequence, database searches for SPO1 should always be done with both forms of the name.

DISTINGUISHING FEATURES

Members of this genus are large lytic phages. Heads show conspicuous capsomers. DNA is terminally redundant (but not circularly permuted), contains 5-hydroxymethyluracil, and codes for a type A (*E. coli* Pol I) DNA polymerase.

VIRION PROPERTIES

MORPHOLOGY

Virions have isometric, icosahedral heads of ~94 nm in diameter with conspicuous capsomers. Contractile tails measure 150 x 18 nm and have a small collar and a 60 nm wide baseplate.

PHYSICOCHEMICAL AND PHYSICAL PROPERTIES

SPO1 virion Mr is ~180×10^6; buoyant density in CsC1 is 1.54 g/cm^3; and S_{20w} is 794S.

NUCLEIC ACID

Genomes are ~140-160 kbp and those that have been measured have a G+C content of 42%. Thymine is replaced by 5-hydroxymethyluracil in SPO1 DNA.

PROTEINS

Virions contain ~53 proteins (16 in the head and 28 in the tail and baseplate). Type A DNA polymerase is encoded in the phage genome.

LIPIDS

None known.

CARBOHYDRATES
None known.

GENOME ORGANIZATION AND REPLICATION

The genetic map is linear and may contain as many as 200 genes. Related functions cluster together. The genome has a terminally redundancy of ~12 kbp, but is not circularly permuted. After infection, host syntheses are shut off and replication starts at two SPO1 DNA sites. Phage-encoded sigma factors are used to modify and appropriate host RNA polymerase for phage syntheses.

BIOLOGICAL PROPERTIES

Phages are virulent and so far have been characterized only from *Bacillus* and *Lactobacillus*. Distribution is worldwide.

LIST OF SPECIES DEMARCATION CRITERIA IN THE GENUS

Not applicable.

LIST OF SPECIES IN THE GENUS

Species names are in green italic script; strain names and synonyms are in black roman script; tentative species names are in blue roman script. Sequence accession numbers, and assigned abbreviations ( ) are also listed.

SPECIES IN THE GENUS

Bacillus phage SPO1	
Bacillus phage SPO1	(SPO1)
Bacillus phage SP8	(SP8)
Bacillus phage SP82	(SP82)

TENTATIVE SPECIES IN THE GENUS

Bacillus phage AR1	(AR1)
Bacillus phage GS1	(GS1)
Bacillus phage I9	(I9)
Bacillus phage NLP-1	(NLP-1)
Bacillus phage SP5	(SP5)
Bacillus phage SW	(SW)
Bacillus phage ϕe	(ϕe)
Bacillus phage ϕ25	(ϕ25)
Bacillus phage 2C	(2C)
Lactobacillus phage 222a	(222a)

GENUS "ϕH-LIKE VIRUSES"

Type Species *Halobacterium phage ϕH*

DISTINGUISHING FEATURES

The host is an archaeon. Phage DNA has a *pac* site, and is circularly permuted and terminally redundant.

VIRION PROPERTIES

MORPHOLOGY
Virions have isometric heads 64 nm in diameter, tails of 170 x 18 nm, and short tail fibers.

PHYSICOCHEMICAL AND PHYSICAL PROPERTIES
Not known.

NUCLEIC ACID
Genomes are ~59 kbp in size and have a G+C content of 64%. Cytosine is replaced by 5-methylcytosine.

PROTEINS
Virions have three major proteins (20, 45, and 70 kDa) and 10 minor components.

LIPIDS
None known.

CARBOHYDRATES
None known.

GENOME ORGANIZATION AND REPLICATION

Genomes are partially circularly permuted and ~3% terminally redundant and have a *pac* site. Halobacterium phage ϕH (ϕH) DNA is markedly variable. All DNAs harbor one or more insertion elements, and also include ordinary deletion and insertion variants. Early transcription is regulated by viral antisense mRNA. Replication results in formation of concatemers. Cutting of concatemers at *pac* sites is inaccurate and produces DNA molecules with imprecisely defined ends.

ANTIGENIC PROPERTIES

Not known.

BIOLOGICAL PROPERTIES

Phages are temperate, specific for halobacteria, and require the presence of 3.5 M NaCl. Prophages persist as plasmids and are not UV-inducible.

LIST OF SPECIES DEMARCATION CRITERIA IN THE GENUS

Not applicable.

LIST OF SPECIES IN THE GENUS

Species names are in green italic script; strain names and synonyms are in black roman script; tentative species names are in blue roman script. Sequence accession numbers, and assigned abbreviations ( ) are also listed.

SPECIES IN THE GENUS

Halobacterium phage ϕH

Halobacterium phage ϕH	(ϕH)

TENTATIVE SPECIES IN THE GENUS

Halobacterium phage Hs1	(Hs1)

LIST OF UNASSIGNED VIRUSES IN THE FAMILY

Acinetobacter phage A3/2	(A3/2)
Acinetobacter phage A10/45	(A10/45)
Acinetobacter phage BS46	(BS46)
Acinetobacter phage E14	(E14)
Actinomycetes phage SK1	(SK1)
Actinomycetes phage 108/016	(108/016)
Aeromonas phage Aeh2	(Aeh2)
Aeromonas phage 51	(51)
Aeromonas phage 59.1	(59.1)
Alcaligenes phage A6	(A6)
Bacillus phage Bace-11	(Bace-11)
Bacillus phage CP-54	(CP-54)
Bacillus phage G	(G)

Bacillus phage MP13		(MP13)
Bacillus phage PBS1		(PBS1)
Bacillus phage SP3		(SP3)
Bacillus phage SP10		(SP10)
Bacillus phage SP15		(SP15)
Bacillus phage SP50		(SP50)
Bacillus phage Spy-2		(Spy-2)
Bacillus phage Spy-3		(Spy-3)
Bacillus phage SST		(SST)
Clostridium phage HM3		(HM3)
Clostridium phage CEβ		(CEβ)
Coryneform phage A19		(A19)
Cyanobacteria phage AS-1		(AS-1)
Cyanobacteria phage N1		(N1)
Cyanobacteria phage S-6(L)		(S-6(L))
Enterobacteria phage FC3-9		(FC3-9)
Enterobacteria phage Kl9		(Kl9)
Enterobacteria phage ΦP27	[AJ298298]	(ΦP27)
Enterobacteria phage 01		(01)
Enterobacteria phage ViI		(ViI)
Enterobacteria phage ϕ92		(ϕ92)
Enterobacteria phage 121		(121)
Enterobacteria phage 16-19		(16-19)
Enterobacteria phage 9266		(9266)
Halorubrum phage HF2	[AF222060]	(HF2)
Lactobacillus phage fri		(fri)
Lactobacillus phage hv		(hv)
Lactobacillus phage hw		(hw)
Listeria phage A511		(A511)
Listeria phage 4211		(4211)
Mollicutes phage Br1		(Br1)
Mycobacterium phage I3		(I3)
Mycobacterium phage Bxz1	[AY129337]	(Bxz1)
Pseudomonas phage PB-1		(PB-1)
Pseudomonas phage PS17		(PS17)
Pseudomonas phage ϕKZ	[AF399011]	(ϕKZ)
Pseudomonas phage ϕW-14		(ϕW-14)
Pseudomonas phage 12S		(12S)
Rhizobium phage CM1		(CM1)
Rhizobium phage CT4		(CT4)
Rhizobium phage m		(m)
Shigella phage SfV	[AF339141]	(SfV)
Xanthomonas phage XP5		(XP5)
Vibrio phage kappa		(kappa)
Vibrio phage 06N-22P		(06N-22P)
Vibrio phage VP1		(VP1)
Vibrio phage II		(II)

PHYLOGENETIC RELATIONSHIPS WITHIN THE FAMILY

Not available.

SIMILARITY WITH OTHER TAXA

See *Caudovirales* chapter.

DSDNA

Derivation of Names

Myo: from Greek *my, myos,* "muscle", referring to the contractile tail.

References

Ackermann, H.-W. and DuBow, M.S. (eds)(1987). *Viruses of Prokaryotes,* Vol. II. Natural Groups of Bacteriophages. CRC Press, Boca Raton, Florida.

Ackermann, H.-W. and Krisch, H.M. (1997). A catalogue of T4-type bacteriophages. *Arch. Virol.,* **142**, 2329-2345.

Bertani, L.E. and Six, E.W. (1988). The P2-like phages and their parasite, P4. In: *The Bacteriophages,* Vol II, (R. Calendar, ed), pp 73-143. Plenum Press, New York.

Goldstein, R., Lengyel, J., Pruss, G., Barrett, K., Calendar, R. and Six, E. (1974). Head size determination and the morphogenesis of satellite phage P4. *Curr. Topics Microbiol. Immunol.,* **68**, 59-75.

Harshey, R.M. (1988). Phage Mu. In: *The Bacteriophages,* Vol. I, (R. Calendar, ed), pp 193-234. Plenum Press, New York.

Hemphill, H.E. and Whiteley, H.R. (1975). Bacteriophages of *Bacillus subtilis. Bacteriol. Rev.,* **39**, 257-315.

Karam, J.-D. (ed)(1994). *Molecular biology of Bacteriophage T4.* American Society for Microbiology, Washington DC.

Klaus, S., Krüger, D.H. and Meyer, J. (1992). *Bakterienviren.* Gustav Fischer, Jena-Stuttgart, pp 133-247.

Miller, E., Kutter, E., Mosig, G., Arisaka, F., Kunisawa, T. and Rüger, W. (2003). Bacteriophage T4. *Microbiol. Molec. Biol. Rev.,* **67**, 86-156.

Monod, C., Repoila, F., Kutateladze, M., Tart, F. and Krisch, H.M. (1997). The genome of the pseudo T-even bacteriophages, a diverse group that resembles T4. *J. Mol. Biol.,* **267**, 237-249.

Morgan, G., Hatfull, G., Casjens, S. and Hendrix, R. (2002). Bacteriophage Mu genome sequence: analysis and comparison with Mu-like prophages in *Haemophilus, Neisseria* and *Deinococcus. J. Mol. Biol.,* **317**, 337-359.

Mosig, G. and Eiserling, F. (1988). Phage T4 structure and metabolism. In: *The Bacteriophages,* Vol. II, (R., Calendar, ed), pp 521-606. Plenum Press, New York.

Stewart, C. (1988). Bacteriophage SPO1. In: *The Bacteriophages,* Vol. I, (R. Calendar, ed), pp 477-515. Plenum Press, New York.

Stolt, P. and Zillig, W. (1994). Archaebacterial bacteriophages. In: *Encyclopedia of Virology,* (R.G. Webster, A. Granoff, eds), pp 50-58. Academic Press, New York.

Symonds, N., Toussaint, A., Van de Putte, P. and Howe, M.M. (eds)(1987). *Phage Mu.* Cold Spring Harbor Laboratory, Cold Spring Harbor, New York.

Yarmolinski, M.B. and Sternberg, N. (1988). Bacteriophage P1. In: *The Bacteriophages,* Vol II, (R. Calendar, ed), pp 291-438. Plenum Press, New York.

Wang, X. and Higgins, N.P. (1994). 'Muprints' of the lac operon demonstrate physiological control over the randomness of in vivo transposition. *Molec. Microbiol.,* **12**, 665-677.

Zillig, W., Reiter, W.-D., Palm, P., Gropp, F., Neumann, H. and Rettenberger, M. (1988). Viruses of Archaebacteria. In: *The Bacteriophages,* Vol. I, (R. Calendar, ed), pp 517-558. Plenum Press, New York.

Contributed By

Hendrix, R. W. and Casjens, S. R.

FAMILY *SIPHOVIRIDAE*

TAXONOMIC STRUCTURE OF THE FAMILY

Family	***Siphoviridae***
Genus	"λ-like Viruses"
Genus	"T1-like Viruses"
Genus	"T5-like Viruses"
Genus	"L5-like Viruses"
Genus	"c2-like Viruses"
Genus	"ΨM1-like Viruses"
Genus	"ϕC31-like Viruses"
Genus	"N15-like Viruses"

DISTINGUISHING FEATURES

Virions have long, non-contractile, thin tails (65-570 x 7-10 nm) which are often flexible. Tails are built of stacked disks of 6 subunits. Heads and tails are assembled separately. Genera are differentiated by genome organization, mechanisms of DNA packaging and presence or absence of DNA polymerases.

GENUS *"λ-LIKE VIRUSES"*

Type Species *Enterobacteria phage λ*

DISTINGUISHING FEATURES

The DNA has cohesive ends and is packaged as a unit-size filament.

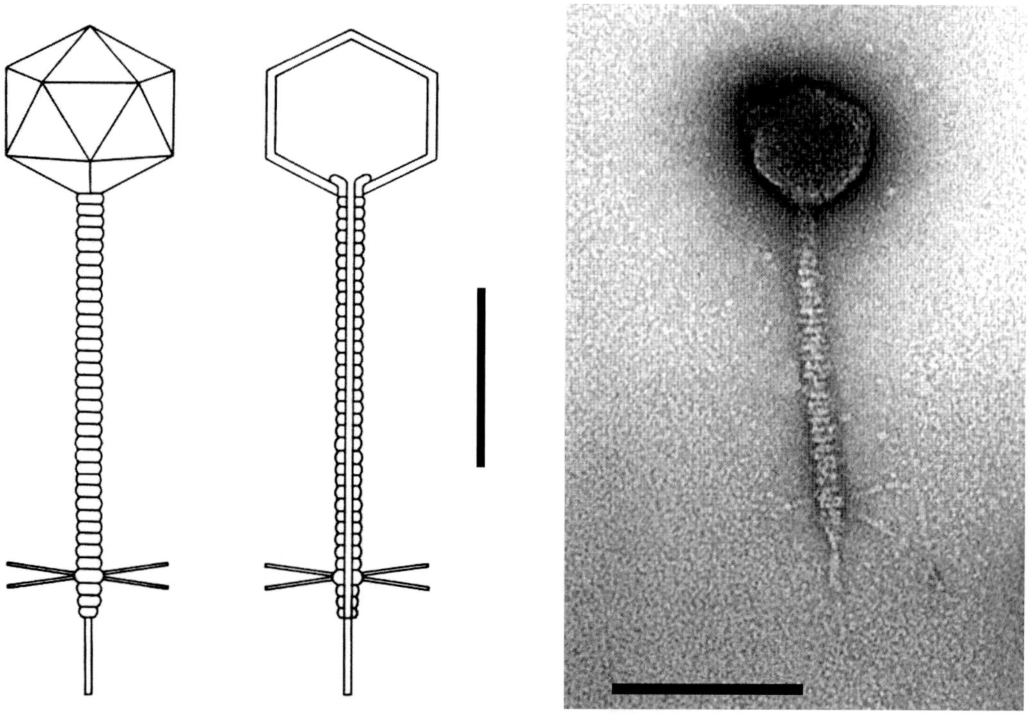

Figure 1 : Enterobacteria phage λ (λ): (Left) Representative diagram of a phage λ particle. (Right) Electron micrograph of phage λ particles with negative staining. The bar represents 100 nm..

VIRION PROPERTIES

MORPHOLOGY
Phage heads are icosahedra, ~60 nm in diameter, and consist of 72 capsomers (60 hexamers, 12 pentamers, T=7). Tails are flexible, 150 x 8 nm, and have a short terminal fiber and four long, jointed fibers attached subterminally (Fig. 1). The latter fibers are absent in most laboratory strains of the phage.

PHYSICOCHEMICAL AND PHYSICAL PROPERTIES
Virion Mr is ~60 x 10^6, buoyant density in CsCl is 1.50 g/cm^3, and the S_{20w} is ~390S. Infectivity is chloroform- and ether-resistant.

NUCLEIC ACID
The phage λ genome is 48,503 bp in size, corresponding to 54% of particle weight, has 52% G+C, cohesive ends, and are non-permuted. The genome has been sequenced.

PROTEINS
Virions contain ~14 structural proteins (11-130 kDa), including 415 copies each of major capsid proteins E and D (38 and 11 kDa, respectively).

LIPIDS
None reported.

CARBOHYDRATES
None reported.

GENOME ORGANIZATION AND REPLICATION

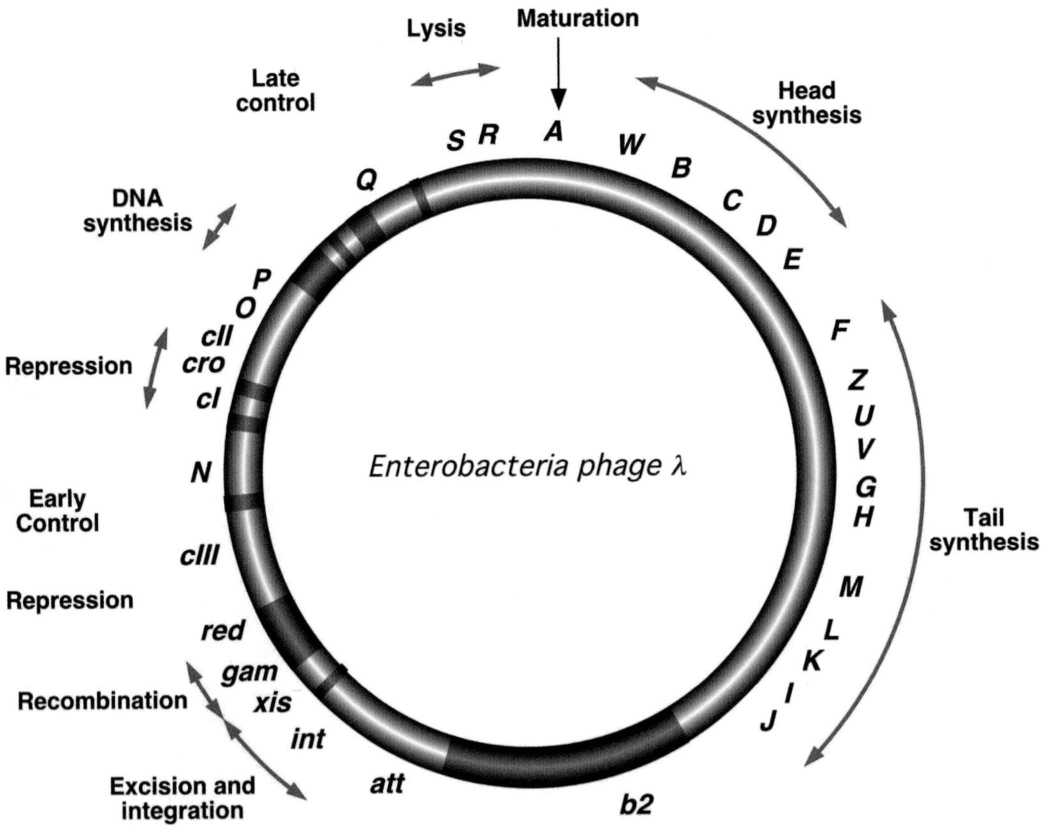

Figure 2: Simplified genetic map of Enterobacteria phage λ (λ). Dark portions of the genome indicate non-essential regions. (From Freifelder, D. (ed)(1983). *Molecular Biology*. Science Books International, Boston, and Van Nostrand Reynolds, New York, p 639. With permission).

The genome includes ~70 genes and has cohesive ends. Related functions cluster together. The infecting DNA circularizes and either replicates or integrates into the host genome. Transcription starts in the immunity region and proceeds in three waves. Bidirectional DNA replication as a Θ (theta) structure, starting from a single site, is followed by unidirectional replication via a rolling-circle mechanism. There is no breakdown of host DNA. Proheads are frequent in lysates.

BIOLOGICAL PROPERTIES

Phages are temperate and apparently specific for enterobacteria. Phages generally integrate at specific sites and are UV-inducible.

LIST OF SPECIES DEMARCATION CRITERIA IN THE GENUS

Species are distinguished by different combinations of alleles of genes encoding head proteins, homologous recombination proteins, and DNA replication proteins, in the context of very similar genome organization.

LIST OF SPECIES IN THE GENUS

Species names are in green italic script; strain names and synonyms are in black roman script; tentative species names are in blue roman script. Sequence accession numbers, and assigned abbreviations ( ) are also listed.

SPECIES IN THE GENUS

Enterobacteria phage HK022		
Enterobacteria phage HK022	[AF069308]	(HK022)
Enterobacteria phage HK97		
Enterobacteria phage HK97	[AF069529]	(HK97)
Enterobacteria phage λ		
Enterobacteria phage λ	[J02459}	(λ)

TENTATIVE SPECIES IN THE GENUS

Enterobacteria phage PA-2	(PA-2)
Enterobacteria phage FD328	(FD328)
Enterobacteria phage ϕ80	(ϕ80)
Rhizobium phage 16-6-2	(16-6-2)

GENUS "T1-LIKE VIRUSES"

Type Species *Enterobacteria phage T1*

DISTINGUISHING FEATURES

Tails are extremely flexible. Phage DNA has pac sites and is terminally redundant and circularly permuted.

VIRION PROPERTIES

MORPHOLOGY

Virions have icosahedral heads of ~60 nm and extremely flexible tails of 151 x 8 nm, with 4 short, kinked, terminal fibers. The flexible nature of the tail is best seen after phosphotungstate staining.

PHYSICOCHEMICAL AND PHYSICAL PROPERTIES

Virion buoyant density in CsC1 is 1.5 g/cm^3. Virion infectivity is stable to drying.

NUCLEIC ACID

Genomes are ~49 kbp and have a G+C content ~48%.

PROTEINS
Virions contain at least 14 proteins including two major head proteins (26 and 33 kDa).

LIPIDS
None known.

CARBOHYDRATES
None known.

GENOME ORGANIZATION AND REPLICATION
The genetic map is linear and comprises 36-41 genes; related functions cluster together. The genome is circularly permuted and terminally redundant (2.8 kbp or 6% of genome), and includes a recombinational hot spot. Host syntheses are inhibited after infection. Little is known about the mechanism of Enterobacteria phage T1 replication. Progeny DNA is cut from concatemers at *pac* sites and packaged by a headfull mechanism.

BIOLOGICAL PROPERTIES
Phages are virulent, can carry out generalized transduction, and infect enterobacteria.

LIST OF SPECIES DEMARCATION CRITERIA IN THE GENUS
Not applicable.

LIST OF SPECIES IN THE GENUS
Species names are in green italic script; strain names and synonyms are in black roman script; tentative species names are in blue roman script. Sequence accession numbers, and assigned abbreviations ( ) are also listed.

SPECIES IN THE GENUS
Enterobacteria phage T1
Enterobacteria phage T1 (T1)

TENTATIVE SPECIES IN THE GENUS
Enterobacteria phage 102 (102)
Enterobacteria phage 103 (103)
Enterobacteria phage 150 (150)
Enterobacteria phage 168 (168)
Enterobacteria phage 174 (174)
Enterobacteria phage b4 (b4)
Enterobacteria phage D20 (D20)
Enterobacteria phage fg (fg)
Enterobacteria phage Hi (Hi)
Enterobacteria phage UC-1 (UC-1)

GENUS "T5-LIKE VIRUSES"

Type Species *Enterobacteria phage T5*

DISTINGUISHING FEATURES
Virions have large heads and long, kinked tail fibers. The DNA has 5 single-stranded gaps, large terminal repetitions, codes for a type A (*E. coli* Pol I) DNA polymerase, and is injected in two steps.

VIRION PROPERTIES

MORPHOLOGY
Virions have icosahedral heads ~80 nm in diameter. Tails measure 180 x 9 nm in Enterobacteria phage T5 (T5) and 160 x 9 nm in Vibrio phage 149 (type IV), have a

subterminal disk with 3 kinked fibers ~120 nm in length, and a conical tip with a single, straight, 50-nm long fiber. Vibrio phage 149 tail fibers have terminal knobs.

PHYSICOCHEMICAL AND PHYSICAL PROPERTIES

Virion Mr is 114 x 10^6, buoyant density in CsC1 is 1.53 g/cm^3, and S_{20w} is 608S.

NUCLEIC ACID

Genomes are ~121 kbp, corresponding to 62% of particle weight, and have a G+C content of 44%.

PROTEINS

Virions contain at least 15 structural proteins (15.5-125 kDa), including ~775 copies of the major capsid protein (44 kDa). Type A (*E. coli* Pol I) DNA polymerase is encoded.

LIPIDS

None reported.

CARBOHYDRATES

None reported.

GENOME ORGANIZATION AND REPLICATION

The genetic map is linear. The genome includes at least 80 genes, is divided into 5 regions, is non-permuted, and has a large terminal repetition of ~10 kbp (8.5% of genome) and 5 single-stranded gaps at specific sites. It has neither *cos* nor *pac* sites. Pre-early, early, and late genes cluster together. Only 8% of DNA is injected immediately after adsorption; the rest follows after 3-4 minutes. Transcription involves modification of host RNA polymerase by phage gene products. DNA replication follows a bidirectional or a rolling-circle mechanism or both. Concatemers are produced.

BIOLOGICAL PROPERTIES

Infection is virulent. Known phages of the genus infect enterobacteria and vibrios.

LIST OF SPECIES DEMARCATION CRITERIA IN THE GENUS

Species differ by host range.

LIST OF SPECIES IN THE GENUS

Species names are in green italic script; strain names and synonyms are in black roman script; tentative species names are in blue roman script. Sequence accession numbers, and assigned abbreviations ( ) are also listed.

SPECIES IN THE GENUS

Enterobacteria phage T5	
Enterobacteria phage T5	(T5)
Vibrio phage 149 (type IV)	
Vibrio phage 149 (type IV)	(φ149)

TENTATIVE SPECIES IN THE GENUS

Enterobacteria phage BF23	(BF23)
Enterobacteria phage PB	(PB)
Enterobacteria phage San 2	(San2)

GENUS "L5-LIKE VIRUSES"

Type Species *Mycobacterium phage L5*

DISTINGUISHING FEATURES

Phage DNA has *cos* sites and codes for a type A (*E. coli* Pol I) DNA polymerase.

Virion Properties

Morphology

Virions have isometric heads ~60 nm in diameter and flexible tails of 135 x 8 nm with a terminal knob and a single short fiber.

Physicochemical and Physical Properties

Virion Mr is 116 x 10^6, buoyant density in CsC1 is 1.51 g/cm^3, and S_{20w} is 410 (data from Mycobacterium phage phlei). Chloroform sensitivity has been reported in possibly related phages.

Nucleic Acid

Genomes are ~52 kbp. The Mycobacterium phage L5 (L5) genome has been sequenced and has 52,297 bp. The G+C content is ~63%.

Proteins

Virions contain at least 6 structural proteins (19-22 to 250 kDa), including a major capsid protein of 35 kDa. Type A (*E. coli* Pol I) DNA polymerase is encoded. Amino acid sequences for L5 are available at GenBank and EMBL.

Lipids

None reported.

Carbohydrates

None reported.

Genome Organization and Replication

The genetic map is linear. Related genes cluster together. The genome includes 88 genes and has cohesive ends. Host syntheses are shut off during replication. Transcription of structural genes is unidirectional.

Biological Properties

Phages are temperate and specific for mycobacteria. Prophages integrate at specific sites in the bacterial genome.

List of Species Demarcation Criteria in the Genus

Species differ by relative insertions and deletions in the context of otherwise similar sequence and gene organization.

List of Species in the Genus

Species names are in green italic script; strain names and synonyms are in black roman script; tentative species names are in blue roman script. Sequence accession numbers, and assigned abbreviations ( ) are also listed.

Species in the Genus

Mycobacterium phage D29		
Mycobacterium phage D29	[NC_001900]	(D29)
Mycobacterium phage L5		
Mycobacterium phage L5	[Z18946]	(L5)

Tentative Species in the Genus

Mycobacterium phage Bxb1	[AF271693]	(Bxb1)
Mycobacterium phage Leo		(Leo)
Mycobacterium phage minetti		(minetti)
Mycobacterium phage phlei		(GS4E)

GENUS "C2-LIKE VIRUSES"

Type Species *Lactococcus phage c2*

DISTINGUISHING FEATURES

Heads are prolate; phage DNA has cos sites and codes for a putative type B DNA polymerase.

VIRION PROPERTIES

MORPHOLOGY

Virions have prolate heads ~56 x 41 nm and tails of 98 x 9 nm, with a collar (inconstant) and small base plate, and produce rare morphological aberrations (two heads joined by a bridge).

PHYSICOCHEMICAL AND PHYSICAL PROPERTIES

Virion buoyant density in CsC1 is 1.46 g/cm^3.

NUCLEIC ACID

Genomes are ~22 kbp (22,163-22,195 bp) and have a G+C content of 35-40%. The genomes of Lactococcus phage c2 (c2) and Lactococcus phage bIL67 (bIL67) have been fully sequenced.

PROTEINS

Virions have at least 6 structural proteins (19.2-175 kDa): major proteins are 29, 90, and 175 kDa. Type B (*E. coli* Pol II) DNA polymerase is apparently encoded. Amino acid sequences for c2 and bIL67 are available at GenBank and EMBL.

LIPIDS

None reported.

CARBOHYDRATES

None reported.

GENOME ORGANIZATION AND REPLICATION

The genetic map is linear. The genome includes 37-38 genes in two clusters and has cohesive ends. Early and late genes are separated by an intergenic region. Late genes are all transcribed in the same direction.

BIOLOGICAL PROPERTIES

Phages are temperate and specific for lactococci.

LIST OF SPECIES DEMARCATION CRITERIA IN THE GENUS

Not applicable.

LIST OF SPECIES IN THE GENUS

Species names are in green italic script; strain names and synonyms are in black roman script; tentative species names are in blue roman script. Sequence accession numbers, and assigned abbreviations ( ) are also listed.

SPECIES IN THE GENUS

Lactococcus phage bIL67		
Lactococcus phage bIL67	[L33769]	(bIL67)
Lactococcus phage c2		
Lactococcus phage c2	[L48605]	(c2)

TENTATIVE SPECIES IN THE GENUS

Lactococcus phage c6A	(PBc6A)
Lactococcus phage P001	(P001)

Lactococcus phage φvML3 (φvML3)
(Lactococcus phage ML3) (ML3)
(Lactococcus phage ml3) (ml3)
(Lactococcus phage 3ML) (3ML)

About 200 additional, poorly characterized lactococcal phages have been reported that are morphologically indistinguishable from c2.

GENUS "ΨM1-LIKE VIRUSES"

Type Species *Methanobacterium phage ΨM1*

DISTINGUISHING FEATURES

The host is an archaeon. Viral genomes are circularly permuted and terminally redundant.

VIRION PROPERTIES

MORPHOLOGY

Virions have isometric heads 55 nm in diameter and tails of 210 x 10 nm with a terminal knob.

PHYSICOCHEMICAL AND PHYSICAL PROPERTIES

Not known.

NUCLEIC ACID

Genomes are ~30 kbp in size.

PROTEINS

Not known.

LIPIDS

None reported.

CARBOHYDRATES

None reported.

GENOME ORGANIZATION AND REPLICATION

Genomes are circularly permuted and terminally redundant.

BIOLOGICAL PROPERTIES

Phages are lytic and infect members of the genus *Methanobacterium*.

LIST OF SPECIES DEMARCATION CRITERIA IN THE GENUS

Not applicable.

LIST OF SPECIES IN THE GENUS

Species names are in green italic script; strain names and synonyms are in black roman script; tentative species names are in blue roman script. Sequence accession numbers, and assigned abbreviations ( ) are also listed.

SPECIES IN THE GENUS

Methanobacterium phage ΨM1
Methanobacterium phage ΨM1 [AF065411, AF065412] (ΨM1)

TENTATIVE SPECIES IN THE GENUS

Methanobacterium phage FF3 (FF3)
Methanobrevibacter phage PG (PG)

DSDNA

GENUS "ϕC31-LIKE VIRUSES"

Type Species *Streptomyces phage ϕC31*

DISTINGUISHING FEATURES

Phage DNA has cos ends, codes for a type A DNA polymerase and has a serine site-specific recombinase.

VIRION PROPERTIES

MORPHOLOGY

Virions have isometric heads ~53 nm in diameter and flexible tails 100 nm long and 5 nm wide, a base plate of 15 nm and 4 tail fibres with terminal knobs ('toes').

PHYSICOCHEMICAL AND PHYSICAL PROPERTIES

Virion bouyant density is 1.493g/cm^3, and virions are chloroform sensitive.

NUCLEIC ACID

Genomes are ~43 kbp. Two genomes have been completely sequenced: phages ϕC31 and ϕBT1 are 41,491 and 41,832 bp respectively. G+C content is 63.6%.

PROTEINS

Virions contain 10 structural proteins, visible by Coomassie staining, of ~10–70 kDa.

LIPIDS

None reported.

CARBOHYDRATES

None reported.

GENOME ORGANIZATION AND REPLICATION

The genetic map is linear and related genes cluster together. The phage ϕC31 genome encodes 54 genes and has cohesive ends with 10 nt protruding at the 3'-end. Transcription of all except one gene is unidirectional. One tRNA is encoded. Mode of replication is unknown, but the genome encodes a DNA polymerase, phage P4-like primase-helicase, D29-like dCMP deaminase and T4-like nucleotide kinase. Head assembly genes most closely resemble those of *Pseudomonas* phage D3 and Enterobacteria phage HK97. The putative tail fiber gene contains a collagen motif. Lytic growth occurs via transcription from multiple conserved promoters in the early region and a single operon in the late region. A repressor gene encodes three nested N-terminally different in-frame proteins which bind to multiple highly conserved operators. The integrase belongs to the serine recombinase family of site-specific recombinases.

BIOLOGICAL PROPERTIES

Phages are temperate and specific for *Streptomyces* spp. Phages integrate at a specific site in the host genome and are not UV inducible. Phages homoimmune to ϕC31 are susceptible to a phage resistance mechanism (Pgl; phage growth limitation) in *S. coelicolor* A3(2). Lytic growth switches off host transcription.

LIST OF SPECIES DEMARCATION CRITERIA IN THE GENUS

Not applicable.

LIST OF SPECIES IN THE GENUS

Species names are in green italic script; strain names and synonyms are in black roman script; tentative species names are in blue roman script. Sequence accession numbers, and assigned abbreviations ( ) are also listed.

SPECIES IN THE GENUS

Streptomyces phage ϕC31		
Streptomyces phage ϕC31	[AJ006589]	(ϕC31)

TENTATIVE SPECIES IN THE GENUS

Streptomyces phage ϕBT1	[AJ550940]	(ϕBT1)
Streptomyces phage TG1		(TG1)
Streptomyces phage SEA		(SEA)
Streptomyces phage R4		(R4)
Streptomyces phage VP5		(VP5)
Streptomyces phage RP2		(RP2)
Streptomyces phage RP3		(RP3)

GENUS "N15-LIKE VIRUSES"

Type Species *Enterobacteria phage N15*

DISTINGUISHING FEATURES

Prophage DNA is present as a linear plasmid with covalently closed hairpin telomeres. Virion DNA has cohesive ends and is packaged as a unit-size molecule.

VIRION PROPERTIES

MORPHOLOGY

Phage heads are hexagonal in outline (probable icosahedra), ~60 nm in diameter. Tails are non-contractile, flexible, measure 140 x 8 nm, and have short brush-like terminal fibers.

PHYSICOCHEMICAL AND PHYSICAL PROPERTIES

Not characterized in detail.

NUCLEIC ACID

The genome is 46,363 bp, has a G+C content of 51.2%, 12 nt 5'-protruding cohesive ends, and is non-permuted. The genome has been sequenced.

PROTEINS

Virion proteins have not been studied, but their high level of similarity to those of phage λ heads and phage HK97 tails suggests that they are very like those phages.

LIPIDS

None reported.

CARBOHYDRATES

None reported.

GENOME ORGANIZATION AND REPLICATION

The virion genome includes ~50 genes and has cohesive ends. Related functions cluster together. The infecting DNA circularizes and replicates, or becomes established as a linear plasmid which is a circular permutation of the virion DNA. Details of DNA replication have not been studied. Organization and sequences of the late expressed virion structural protein and lysis genes are similar to lambda, but the early expressed genes are very different. An anti-repressor system and putative DNA polymerase gene (primase type) have been identified in the early left operon. Unique to this phage type is the presence of a protelomerase gene that encodes an enzyme that resolves a circular genome molecule at the *telRL* site into the linear molecule with covalently closed hairpin telomeres. An unusually large number of phage genes are expressed from the prophage

ANTIGENIC PROPERTIES

Not studied.

BIOLOGICAL PROPERTIES

Phages are temperate and mitomycin C inducible. Prophages are linear plasmids with covalently closed hairpin telomeres.

LIST OF SPECIES DEMARCATION CRITERIA IN THE GENUS

Not applicable.

LIST OF SPECIES IN THE GENUS

Species names are in green italic script; strain names and synonyms are in black roman script; tentative species names are in blue roman script. Sequence accession numbers, and assigned abbreviations ( ) are also listed.

SPECIES IN THE GENUS

Enterobacteria phage N15

Enterobacteria phage N15	[AF064539]	(N15)

TENTATIVE SPECIES IN THE GENUS

Yersinia phage PY54	(PY54)

LIST OF UNASSIGNED VIRUSES IN THE FAMILY

Acinetobacter phage 531		(531)
Acinetobacter phage E13		(E13)
Actinomycetes phage 119		(119)
Actinomycetes phage A1-Dat		(A1-Dat)
Actinomycetes phage Bir		(Bir)
Actinomycetes phage M1		(M1)
Actinomycetes phage MSP8		(MSP8)
Actinomycetes phage P-a-1		(P-a-1)
Actinomycetes phage R1		(R1)
Actinomycetes phage R2		(R2)
Actinomycetes phage SV2		(SV2)
Actinomycetes phage VP5		(VP5)
Actinomycetes phage ΦC		(ΦC)
Actinomycetes phage ϕ115-A		(ϕ115-A)
Actinomycetes phage ϕ150A		(ϕ150A)
Actinomycetes phage ϕ31C		(ϕ31C)
Actinomycetes phage ϕUW21		(ϕUW21)
Agrobacterium phage PS8		(PS8)
Agrobacterium phage PT11		(PT11)
Agrobacterium phage Ψ		(Ψ)
Alcaligenes phage 8764		(8764)
Alcaligenes phage A5/A6		(A5/A6)
Bacillus phage 1A		(1A)
Bacillus phage B1715V1		(B1715V1)
Bacillus phage BLE		(BLE)
Bacillus phage II		(II)
Bacillus phage IPy-1		(IPy-1)
Bacillus phage mor1		(mor1)
Bacillus phage MP15		(MP15)
Bacillus phage PBP1		(PBP1)
Bacillus phage SN45		(SN45)
Bacillus phage SPP1	[X97918]	(SPP1)

Bacillus phage SPβ	[AF020713]	(SPb)
Bacillus phage Tb10		(Tb10)
Bacillus phage TP-15		(TP15)
Bacillus phage type F		(type F)
Bacillus phage α		(α)
Bacillus phage ϕ105	[AB016282]	(ϕ105)
Burkholderia phage ϕE125	[AF447491]	(ϕE125)
Clostridium phage F1		(F1)
Clostridium phage HM7		(HM7)
Clostridium phage ϕ3626	[AY082069,AY082070]	(ϕ3626)
Coryneforms phage Arp		(Arp)
Coryneforms phage BL3		(BL3)
Coryneforms phage CONX		(CONX)
Coryneforms phage MT		(MT)
Coryneforms phage β		(β)
Coryneforms phage ϕA8010		(ϕA8010)
Coryneforms phage A		(A)
Cyanobacteria phage S-2L		(S-2L)
Cyanobacteria phage S-4L		(S-4L)
Enterobacteria phage H-19J		(H-19J)
Enterobacteria phage Jersey		(Jersey)
Enterobacteria phage ViII		(ViI)
Enterobacteria phage ZG/3A		(ZG/3A)
Enterobacteria phage χ		(χ)
Lactobacillus phage 223		(223)
Lactobacillus phage lb6		(lb6)
Lactobacillus phage PL-1		(PL-1)
Lactobacillus phage y5		(y5)
Lactobacillus phage ϕFSW		(ϕFSW)
Lactobacillus ϕadh	[AJ131519]	(ϕadh)
Lactococcus phage 1358		(1358)
Lactococcus phage 1483		(1483)
Lactococcus phage 936		(936)
Lactococcus phage 949		(949)
Lactococcus phage A2	[AJ251789]	(A2)
Lactococcus phage bIL170	[AF009630]	(bIL170)
Lactococcus phage bIL67	[L33769]	(bIL167)
Lactococcus phage BK5-T	[AF176025]	(BK5-T)
Lactococcus phage P107		(P107)
Lactococcus phage P335		(P335)
Lactococcus phage PO87		(PO87)
Lactococcus phage r1t	[U38906]	(r1t)
Lactococcus phage sk1	[AF011378]	(sk1)
Lactococcus phage TP901-1	[AF304433]	(TP901-1)
Lactococcus phage Tuc2009	[AF109874]	(Tuuc2009)
Lactococcus phage ul36	[AF349457]	(ul36)
Leuconostoc phage pro2		(pro2)
Listeria phage 2389		(2389)
Listeria phage 2671		(2671)
Listeria phage 2685		(2685)
Listeria phage A118	[AJ242593]	(A118)
Listeria phage H387		(H387)
Micrococcus phage N1		(N1)
Micrococcus phage N5		(N5)

Mycobacterium phage Barnyard	[AY129339]	(Barnyard)
Mycobacterium phage Bxz2	[AY129332]	(Bxz2)
Mycobacterium phage Che8	[AY129330]	(Che8)
Mycobacterium phage Che9c	[AY129333]	(Che9c)
Mycobacterium phage Che9d	[AY129336]	(Che9d)
Mycobacterium phage Cjw1	[AY129331]	(Cjw1)
Mycobacterium phage Corndog	[AY129335]	(Corndog)
Mycobacterium phage lacticola		(lacticola)
Mycobacterium phage Omega	[AY129338]	(Omega)
Mycobacterium phage R1-Myb		(R1-Myb)
Mycobacterium phage Rosebush	[AY129334]	(Rosebush)
Mycobacterium phage TM4	[AF068845]	(TM4)
Pasteurella phage 32		(32)
Pasteurella phage C-2		(C-2)
Pseudomonas phage D3	[AF165214]	(D3)
Pseudomonas phage Kf1		(Kf1)
Pseudomonas phage M6		(M6)
Pseudomonas phage PS4		(PS4)
Pseudomonas phage SD1		(SD1)
Rhizobium phage 16-2-12		(16-2-12)
Rhizobium phage 317		(317)
Rhizobium phage 5		(5)
Rhizobium phage 7-7-7		(7-7-7)
Rhizobium phage NM1		(NM1)
Rhizobium phage NT2		(NT2)
Rhizobium phage ϕ2037/1		(ϕ2037/1)
Staphylococcus phage 107		(107)
Staphylococcus phage 11	[AF424781]	(11)
(Staphylococcus phage P11)		(P11)
(Staphylococcus phage ϕ11)		(ϕ11)
(Staphylococcus phage B11-M15)		(B11-M15)
Staphylococcus phage 1139		(1139)
Staphylococcus phage 1154A		(1154A)
Staphylococcus phage 187		(187)
Staphylococcus phage 2848A		(2848A)
Staphylococcus phage 392		(392)
Staphylococcus phage 3A		(3A)
Staphylococcus phage 77		(77)
Staphylococcus phage ϕETA	[AB046707]	(ϕETA)
Staphylococcus phage ϕSLT	[AB045978]	(ϕSLT)
Streptococcus phage 24		(24)
Streptococcus phage A25		(A25)
Streptococcus phage DT1	[AF085222]	(DT1)
Streptococcus phage PE1		(PE1)
Streptococcus phage Sfi11	[AF158600]	(Sfi11)
Streptococcus phage Sfi19	[AF115102]	(Sfi19)
Streptococcus phage Sfi21	[AF115103]	(Sfi21)
Streptococcus phage VD13		(VD13)
Streptococcus phage ϕO1205	[U88974]	(ϕO1205)
Streptococcus phage ω8		(ω8)
Vibrio phage IV		(IV)
Vibrio phage OXN-52P		(OXN-52P)
Vibrio phage VP11		(VP11)
Vibrio phage VP3		(VP3)

Vibrio phage VP5	(VP5)
Vibrio phage α3α	(α3α)

Phylogenetic Relationships within the Family

Refer to the discussion of this topic under the description of Order *Caudovirales*.

Similarity with Other Taxa

Refer to the discussion of this topic under the description of Order *Caudovirales*.

Derivation of Names

Sipho: from Greek *siphon*, "tube", referring to the long tail.

References

Ackermann, H.-W. and DuBow, M.S. (eds)(1987). *Viruses of Prokaryotes*, Vol. I and II. CRC Press, Boca Raton, Florida.

Ackermann, H.-W. and Gershman, M. (1992). Morphology of phages of a general Salmonella typing set. *Res. Virol.*, **143**, 303-310.

Casjens, S., Hatfull, G. and Hendrix, R. (1992). Evolution of dsDNA-tailed bacteriophage genomes. *Semin. Virol.*, **3**, 310-383.

Clayton, T.M. and Bibb, M.J. (1990). Induction of a phi C31 prophage inhibits rRNA transcription in *Streptomyces coelicolor* A3(2). *Mol. Microbiol.*, **4**, 2179-2185.

Drexler, H. (1988). Bacteriophage T1. In: *The Bacteriophages*, Vol I, (R. Calendar, ed), pp 235-238. Plenum Press, New York.

Hatfull, G.F. and Sarkis, G.J. (1993). DNA sequence, structure and gene expression of mycobacteriophage L5: a phage system for mycobacterial genetics. *Mol. Microbiol.*, **7**, 395-405.

Hendrix, R.W., Roberts, J.W., Stahl, F.W. and Weisberg, R.A. (eds)(1983). *Lambda II*. Cold Spring Harbor Laboratory, Cold Spring Harbor, New York.

Laity, C., Chater, K.F., Lewis, C.G. and Buttner, M.J. (1993). Genetic analysis of the phic31-specific phage growth limitation (Pgl) system of *Streptomyces coelicolor* A3(2). *Mol. Microbiol.*, **7**, 329-336.

Lubbers, M.W., Waterfield, N.R., Beresford, T.P.J., Le Page, R.W.F. and Jarvis, A.W. (1995). Sequencing and analysis of the prolate-headed lactococcal bacteriophage c2 genome and identification of the structural genes. *Appl. Environ. Microbiol.*, **61**, 4348-4356.

Malanin, A., Vostrov, A., Rybchin, V. and Sverchevsky, A. (1992). The structure of the linear plasmid N15 ends. *Mol. Genet. Microb. Virol.*, **5-6**, 22-24. (in Russian)

McCorquodale, D.J. and Warner, H.R. (1988). Bacteriophage T5 and related phages. In: *The Bacteriophages*, Vol. I, (R. Calendar, ed), pp 439-475. Plenum Press, New York.

Ravin, V., Ravin, N., Casjens, S., Ford, M., Hatfull, G. and Hendrix, R. (2000). Genomic sequence and analysis of the atypical temperate bacteriophage N15. *J. Mol. Biol.*, **299**, 53-73.

Schouler, C., Ehrlich, S.D. and Chopin, M.-C. (1994). Sequence and organization of the lactococcal prolate-headed bIL67 phage genome. *Microbiology*, **140**, 3061-3069.

Sengupta, A., Ray, P. and Das, J. (1985). Characterization and physical map of choleraphage φ149 DNA. *Virology*, **140**, 217-229.

Smith, M.C.M., Burns, R.N., Wilson, S.E. and Gregory, M.A. (1999). The complete genome sequence of the *Streptomyces* temperate phage phi C31: evolutionary relationships to other viruses. *Nuc. Acids Res.*, **27**, 2145-2155.

Suarez, J.E., Caso, J.L., Rodriguez, A. and Hardisson, C. (1984). Structural characteristics of the *Streptomyces* bacteriophage phiC31. *FEMS Microbiol Lett.*, **22**, 113-117.

Thorpe, H.M. and Smith, M.C.M. (1998). *In vitro* site-specific integration of bacteriophage DNA catalyzed by a recombinase of the resolvase/invertase family. *Proc. Natl. Acad. Sci. USA.*, **95**, 5505-5510.

Wilson, S.E., Ingham, C.J., Hunter, I.S. and Smith, M.C.M. (1995). Control of lytic development in the *Streptomyces* temperate phage phi-c31. *Mol. Microbio.*, **16**, 131-143.

Contributed By

Hendrix, R.W. and Casjens, S.R.

FAMILY *PODOVIRIDAE*

TAXONOMIC STRUCTURE OF THE FAMILY

Family	***Podoviridae***
Genus	"T7-like Viruses"
Genus	"P22-like Viruses"
Genus	"ϕ29-like Viruses"
Genus	"N4-like Viruses"

DISTINGUISHING FEATURES

Virions have short, non-contractile tails ~20 x 8 nm. Heads are assembled first and tail parts are added to them sequentially. Genera are differentiated by genome organization, mechanisms of DNA packaging, and presence or absence of DNA or RNA polymerases.

GENUS *"T7-LIKE VIRUSES"*

Type Species *Enterobacteria phage T7*

DISTINGUISHING FEATURES

Medium-sized lytic phages, with nonpermuted terminally redundant DNA that codes for both DNA and RNA polymerases. Heads contain a unique 8-fold symmetric core structure. DNA is injected stepwise rather than all at once.

VIRION PROPERTIES

MORPHOLOGY

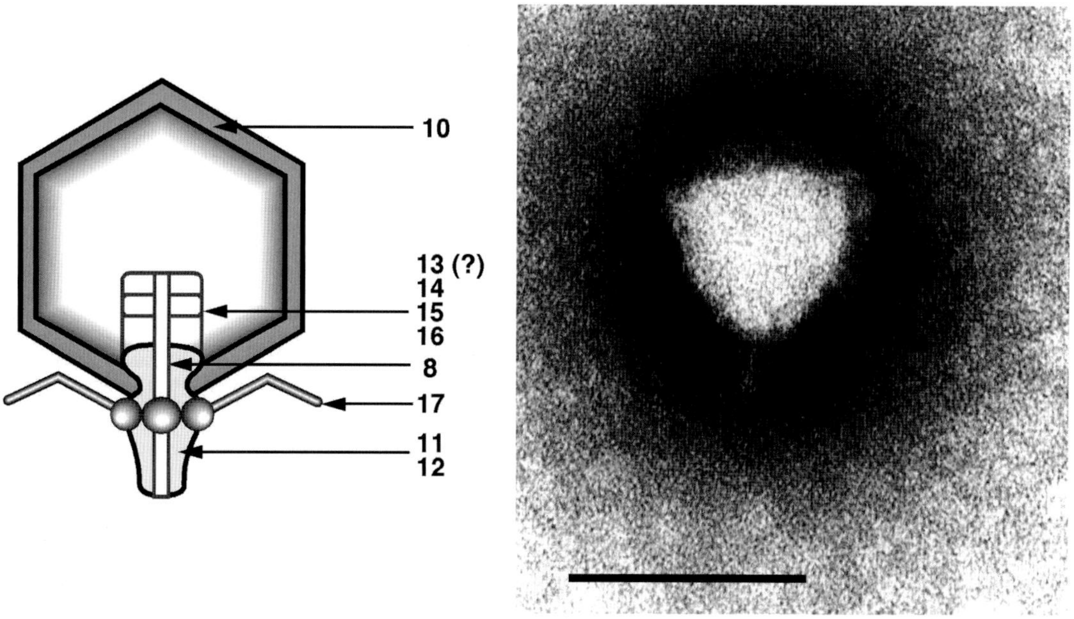

Figure 1: (Left) Diagram of particle of Enterobacteria phage T7 (T7) in section. (Modified from Eiserling, F.A. (1979). Bacteriophage structure. In: *Comprehensive Virology*, Vol 13, (H., Fraenkel-Conrat and R.R., Wagner, eds). Plenum Press, New York, p. 543. With permission). The bar represents 50 nm. (Right) Negative contrast electron micrograph of a particle of phage T7, stained with phosphotungstate. The bar represents 100 nm.

T7 phage heads are icosahedra, measure ~60 nm in diameter, and consist of 72 capsomers (60 hexamers and 12 pentamers; T=7). Tails measure 17 x 8 nm and have 6 short fibers (Fig. 1).

PHYSICOCHEMICAL AND PHYSICAL PROPERTIES

T7 virion Mr is ~48 x 10^6, buoyant density in CsCl is 1.50 g/cm^3, and S_{20w} is ~510S. Infectivity is ether and chloroform resistant.

NUCLEIC ACID

Genomes are ~40 kbp (39,936 bp for T7), corresponding to 50% of virion particle weight, have a G+C content of 50%, and are non-permuted and terminally redundant. The T7 genome has been sequenced.

PROTEINS

Particles have at least 9 structural proteins (13-150 kDa), including about 420 copies of the major capsid protein (in T7, 38 kDa). Type B (*E. coli* Pol II) DNA polymerase and RNA polymerase are encoded. Amino acid sequences for the phages whose genomes have been sequenced are available at GenBank and EMBL.

LIPIDS

None known.

CARBOHYDRATES

None known.

GENOME ORGANIZATION AND REPLICATION

The T7 genetic map is linear, non-permuted and terminally redundant, and comprises about 55 genes, several of which overlap. Related functions cluster together (Fig. 2). Infection results in shut-off of host syntheses and a breakdown of the host genome. The start of replication requires phage-encoded DNA and RNA polymerase and has multiple origins. Transcription proceeds in three waves. Only one strand is transcribed. Replication is bidirectional and produces concatemers by end-to-end joining of intermediate forms. Irregular polyheads are frequently observed. Packaged DNA is cut at fixed sites.

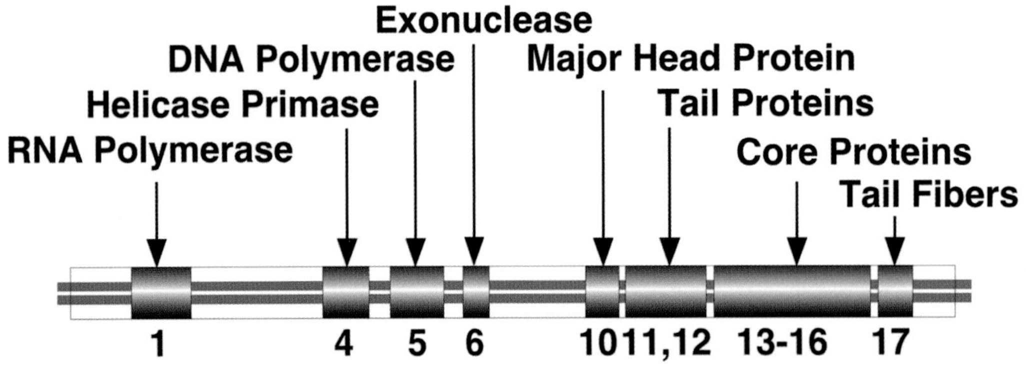

Figure 2: Simplified genetic map of Enterobacteria phage T7 (T7). (Redrawn after Freifelder, D. (ed)(1983). *Molecular Biology*. Science Books International, Boston, and Van Nostrand Reynolds, New York, p 630).

BIOLOGICAL PROPERTIES

Phages are virulent and are specific for enterics and related Gram-negative bacteria.

LIST OF SPECIES DEMARCATION CRITERIA IN THE GENUS

Species differ by host range and, insofar as known, DNA sequence similarity.

LIST OF SPECIES IN THE GENUS

Species names are in green italic script; strain names and synonyms are in black roman script; tentative species names are in blue roman script. Sequence accession numbers, and assigned abbreviations ( ) are also listed.

SPECIES IN THE GENUS

Enterobacteria phage T7		
Enterobacteria phage H		(H)
Enterobacteria phage T3	[AY318471]	(T3)
Enterobacteria phage T7	[V01156]	(T7)
Enterobacteria phage W31		(W31)
Enterobacteria phage WPK		(WPK)
Enterobacteria phage ϕI		(ϕI)
Enterobacteria phage ϕII		(ϕII)
Enterobacteria phage ϕYeO3-12	[AJ251805]	(ϕYeO3-12)
Kluyvera phage Kvp1		
Kluyvera phage Kvp1		(Kvp1)
Pseudomonad phage gh-1		
Pseudomonad phage gh-1		(gh-1)

TENTATIVE SPECIES IN THE GENUS

Caulobacter phage ϕCd1	(ϕCd1)
Enterobacteria phage BA14	(Ba14)
Enterobacteria phage ϕ1.2	(ϕ1.2)
Enterobacteria phage IV	(IV)
Enterobacteria phage K11	(K11)
Enterobacteria phage PTB	(PTB)
Enterobacteria phage R	(R)
Enterobacteria phage SP6	(SP6)
Enterobacteria phage Y	(Y)
Enterobacterial phage ViIII	(ViIII)
Pseudomonas phage ϕPLS27	(ϕPLS27)
Pseudomonas phage ϕPLS743	(ϕPLS743)
Pseudomonas phage Psy9220	(Psy9220)
Rhizobium phage 2	(2)
Rhizobium phage S	(III)
Vibrio phage III	(III)

GENUS “P22-LIKE VIRUSES”

Type Species *Enterobacteria phage P22*

DISTINGUISHING FEATURES

Virions have short tails which have 6 prominent tail spikes. DNA is circularly permuted, terminally redundant, and packaged from a *pac* site by a headful mechanism. Phages can carry out generalized transduction. The genetic map is circular.

VIRION PROPERTIES

MORPHOLOGY

P22 virions have isometric icosahedral heads with 72 faintly visible capsomers (60 hexamers, 12 pentamers; T=7) and 60-65 nm in diameter. Tails are 18 nm long and have 6 tail spikes.

PHYSICOCHEMICAL AND PHYSICAL PROPERTIES

Virion buoyant density in CsC1 is about 1.50 g/cm^3 and S_{20W} is 500S. Infectivity is ether- and chloroform-resistant.

NUCLEIC ACID

The P22 genome sequence has been determined and is 41,754 bp. The chromosome in virions ranges from 42.7-44.1 kbp, corresponds to about 55% of particle weight, and has a G+C content of 47%. Members of the genus have genome sequences of 38–42 kbp with similar organization and similar transcriptional programs.

PROTEINS

P22 virions contain 9-10 structural proteins (18-83 kDa), including ~415 copies of the major capsid protein (47 kDa). Other members of the genus are quite similar in this regard.

LIPIDS

None known.

CARBOHYDRATES

None known.

GENOME ORGANIZATION AND REPLICATION

The Enterobacteria phage P22 (P22) genetic map is circularly permuted, has terminal repeats of about 1600 bp (3.8% of the genome), and comprises at least 65 genes. The P22 genome is partially (13.5%) homologous to phage λDNA, and common sequences are scattered across the right half of the genome. Other genus members have similar but different relationships with other λ-like viruses. The integration system is dispensable for lytic growth. Transcription starts with regulatory genes and proceeds in three partly overlapping waves that are very similar to those of the λ-like viruses. Replication starts at a single site and involves replication by a Θ (theta) structure mechanism that switches at late times to a rolling-circle mechanism. Progeny DNA is cut from concatemers at a *pac* site and packaged by a headful mechanism.

ANTIGENIC PROPERTIES

No group antigens are reported.

BIOLOGICAL PROPERTIES

Phages are temperate and can carry out generalized transduction with lysogenic conversion ability. Members of the genus infect *Enterobacteria* (*Escherichia, Salmonella* and other γ-Proteobacteria) and have, under suitable conditions, very high (up to 500) burst sizes. Phages integrate at specific sites in the bacterial genome and are UV-inducible.

LIST OF SPECIES DEMARCATION CRITERIA IN THE GENUS

Not applicable

LIST OF SPECIES IN THE GENUS

Species names are in green italic script; strain names and synonyms are in black roman script; tentative species names are in blue roman script. Sequence accession numbers, and assigned abbreviations ( ) are also listed.

SPECIES IN THE GENUS

Enterobacteria phage P22		
Enterobacteria phage P22	[BK000583]	(P22)

TENTATIVE SPECIES IN THE GENUS

Aeromonas phage Aa-1		(Aa-1)
Azotobacter phage A12		(A12)
Enterobacteria phage HK620	[AF335538]	(HK620)
Enterobacteria phage L		(L)
Enterobacteria phage LP7		(LP7)

Enterobacteria phage MG40		(MG40)
Enterobacteria phage PSA78		(PSA78)
Enterobacteria phage Sf6	[AF547987]	(Sf6)
Enterobacteria phage ST64T	[AY052766]	(ST64T)
Hyphomicrobium phage Hyϕ30		(Hyϕ30)
Pseudomonas phage 525		(525)
Vibrio phage O6N-72P		(06N-72P)

DSDNA

GENUS "ϕ29-LIKE VIRUSES"

Type Species *Bacillus phage ϕ29*

VIRION PROPERTIES

MORPHOLOGY

Heads are prolate icosahedra (T=3 with 30 hexamers and 11 pentamers) and measure ~54 x 42 nm. Some members, including Bacillus phage ϕ29 (ϕ29), have ~55 fibers on the head. Tails measure 46 x 8 nm, have a distal thickening, and a collar with 12 appendages.

PHYSICOCHEMICAL AND PHYSICAL PROPERTIES

ϕ29 virion Mr is 29 x 10^7, buoyant density in CsC1 is 1.46 g/cm^3, and S_{20W} is 254S. Infectivity is chloroform-resistant.

NUCLEIC ACID

Genomes are 16-20 kbp, correspond to ~50% of particle weight, and have a G+C content of 35-38%. The genomes of phage ϕ29; the related Bacillus phages B103, GA-1 and PZA; Streptococcus phages Cp-1 and C_1; and Staphylococcus phages 44AHJD and P68 have been sequenced.

PROTEINS

Virions have 9 structural proteins (13-86 kDa), including 235 copies of the major capsid protein (49.6 kDa in ϕ29 and 42 kDa in Cp-1). Type B (*E. coli* Pol II) DNA polymerase is coded for.

LIPIDS

None known.

CARBOHYDRATES

None known

GENOME ORGANIZATION AND REPLICATION

The genetic map is linear and includes 20-29 ORFs, one of which codes for a type B (*E. coli* Pol II) DNA polymerase. Genomes are non-permuted and have inverted terminal repeats from 6-8 bp (ϕ29 species) to 230-240 bp (Cp-1 species). Both 5'-ends are covalently linked to a terminal protein. Infecting DNA does not circularize. Transcription proceeds in two waves. Early genes are transcribed from right to left on the standard map (except in Cp-1 where some are from left to right); late genes are transcribed from left to right. Replication is primed by the terminal protein and starts at either DNA end, and proceeds by strand displacement. The terminal protein is essential in DNA packaging. The packaging substrate is non-concatemeric DNA. Packaging requires phage-encoded RNA.

ANTIGENIC PROPERTIES

No group antigens are reported.

BIOLOGICAL PROPERTIES

Phages are virulent and infect Gram-positive bacteria with low G+C contents. Distribution is worldwide.

LIST OF SPECIES DEMARCATION CRITERIA IN THE GENUS

All species differ in host range. Phages φ29 and GA-1 differ in serological properties and protein molecular weights. Phage φ29, PZA, GA-1, B103, 44AHJD, P68, C_1 and Cp-1 DNAs differ in DNA sequences, the length of inverted terminal repeats (6-8 to 236-247 bp). In addition, phages φ29 and Cp-1 have opposite directions for transcription of early genes at the left end of the chromosome.

LIST OF SPECIES IN THE GENUS

Species names are in green italic script; strain names and synonyms are in black roman script; tentative species names are in blue roman script. Sequence accession numbers, and assigned abbreviations ( ) are also listed.

SPECIES IN THE GENUS

Bacillus phageφ29		
Bacillus phage GA-1	[Z47794]	(GA-1)
Bacillus phage PZA		(PZA)
Bacillus phage PZE	[X96987]	(PZE)
Bacillus phage φ15		(φ15)
Bacillus phage φ29	[M14782, V01121, V01122, J02477, J02479, M11813]	(φ29)
Kurthia phage 6		
Kurthia phage 6		(K6)
Streptococcus phage Cp-1		
Streptococcus phage Cp-1		(Cp-1)
Streptococcus phage Cp-5		(Cp-5)
Streptococcus phage Cp-7		(Cp-7)
Streptococcus phage Cp-9		(Cp-9)

TENTATIVE SPECIES IN THE GENUS

Bacillus phage AR13		(AR13)
Bacillus phage B103	[X99260]	(B103)
Bacillus phage MY2		(MY2)
Bacillus phage M2		(M2)
Bacillus phage Nφ		(Nφ)
Bacillus phage SF5		(SF5)
Kurthia phage 7		(K7)
Streptococcus C_1	[AY212251]	(C_1)
Staphylococcus 44AHJD	[AF513032]	(44AHJD)
Staphylococcus P68	[AF513033]	(P68)

About 45 additional, poorly characterized phages have been reported.

GENUS "N4-LIKE VIRUSES"

Type Species *Enterobacteria phage N4*

VIRION PROPERTIES

MORPHOLOGY

Particles have icosahedral heads ~70 nm in diameter and short tails 10 nm in length, with several short fibers originating from the junction between the head and tail.

Physicochemical and Physical Properties
Particle weight is ~84 x 10^6. Buoyant density in CsCl is 1.500 g/ml. Lipids have not been detected.

Nucleic Acid
Virions contain a single molecule of dsDNA of 70,153 bp of unique sequence plus an additional 390-440 bp (variable lengths within the population) at the right end duplicating the DNA sequence at the left end to give direct terminal repeats. There are short 3' ssDNA extensions at both ends. G+C content is 41.3%. The genome sequence is known.

DSDNA

Proteins
Particles contain at least 10 structural proteins. The major CP (~500 molecules per phage) has a subunit mass of 44.0 kDa.

Lipids
Lipids (2.4%) have been reported in phage sd, a possible member of this genus.

Carbohydrates
None reported.

Genome Organization and Replication
The physical map is linear, but a genetic linkage map has not been determined for technical reasons. There are 72 protein-coding genes identified which occupy 94% of the DNA sequence. Transcription of N4 DNA is carried out by sequential activity of three different RNA polymerases. Early transcription is done by the virion RNA polymerase, a large (3500 aa) single subunit enzyme that is present in 1-2 copies in the virion and injected with the DNA during infection. Early proteins include a two-subunit RNA polymerase responsible for middle transcription. Middle proteins include a 98 kD DNA polymerase and other DNA replication functions that replicate phage DNA in cooperation with some host functions. Late transcription is carried out by host (*E. coli)* RNA polymerase. Early and middle genes occupy a contiguous block in the left half of the genome and are transcribed rightwards. Late genes occupy the right half of the genome and are transcribed leftwards. Other noteworthy genes include three tRNA genes (Asn, Thr, Pro), a gene encoding a dCTP deaminase, homologs of the *rIIA* and *rIIB* genes originally studied in phage T4, and a homolog of gene *17* of temperate phage P22. Assembling particles form intracellular crystal-like arrays of phage heads.

Biological Properties
Phages are virulent. Infection is lytic. Host range is restricted to enterobacteria (*E. coli*).

List of Species Demarcation Criteria in the Genus
Not applicable

List of Species in the Genus
Species names are in green italic script; strain names and synonyms are in black roman script; tentative species names are in blue roman script. Sequence accession numbers, and assigned abbreviations ( ) are also listed.

Species in the Genus
Escherichia coli bacteriophage N4

Escherichia coli bacteriophage N4 (N4)

Tentative Species in the Genus
None reported

List of Unassigned Species in the Family

Acinetobacter phage A36		(A36)
Bacillus phage ϕBa1		(ϕBa1)
Brucella phage Tb		(Tb)
Clostridium phage HM2		(HM2)
Coryneforms phage 7/26		(7/26)
Coryneforms phage AN25S-1		(AN25S-1)
Cyanobacteria phage A-4(L)		(A-4(L))
Cyanobacteria phage AC-1		(AC-1)
Cyanobacteria phage LPP-1		(LPP-1)
Cyanobacteria phage SM-1		(SM-1)
Enterobacteria phage 7-11		(7-11)
Enterobacteria phage 7480b		(7480b)
Enterobacteria phage Esc-7-11		(Esc-7-11)
Enterobacteria phage sd		(sd)
Enterobacteria phage W8		(W8)
Lactococcus phage KSY1		(KSY1)
Lactococcus phage PO34		(PO34)
Mollicutes phage C3		(C3)
Mollicutes phage L3		(L3)
Mycobacterium phage ϕ17		(ϕ17)
Pasteurella phage 22		(22)
Pseudomonas phage F116		(F116)
Rhizobium phage ϕ2042		(ϕ2042)
Roseobacter phage SIO1		(SIO1)
Streptococcus phage 182		(182)
Streptococcus phage 2BV		(2BV)
Streptococcus phage Cvir	[AF33467]	(Cvir)
Streptococcus phage H39		(H39)
Synechococcus phage P60	[AF189021]	(P60)
Vibrio phage 4996		(4996)
Vibrio phage I		(I)
Vibrio phage OX N-100P		(OXN-100P)
Xanthomonas phage RR66		(RR66)

Phylogenetic Relationships within the Family

Not available.

Similarity with Other Taxa

"ϕ29-like viruses", adenoviruses, and tectiviruses have proteins linked to DNA ends and protein-primed replication, and code for type B (*E. coli* Pol II) DNA polymerase. See the paragraph "Similarity With Other Taxa" in the *Caudovirales* description for further details.

References

Casjens, S. (1989). Bacteriophage P22 DNA packaging. In: *Chromosomes, Eukaryotic, Prokaryotic and Viral DNA Packaging,* vol. III, (K. Adolph, ed), p. 241-261. CRC Press, Boca Raton, Florida.

Clark, A.J., Inwood, W., Cloutier, T. and Dhillon, T.S. (2001). Nucleotide sequence of coliphage HK620 and the evolution of lambdoid phages. *J. Mol. Biol.*, **311**, 657-679.

Dunn, J. and Studier, W. (1983). Complete nucleotide sequence of bacteriophage T7. *J. Mol. Biol.*, **166**, 477-535.

Pajunen, M.I., Elizondo, M.R., Skurnik, M., Kieleczawa, J. and Molineux, I.J. (2003). Complete nucleotide sequence and likely recombinatorial origin of bacteriophage T3. *J. Mol. Biol.*, **319**, 1115-1132.

Pajunen, M.I., Kiljunen, S.J., Söderholm, M. and Skurnik, M. (2001). Complete sequence of the lytic bacteriophage ϕY3O3-12 of *Yersinia enterocolitica* serotype O:3. *J. Bacteriol.*, **183**, 1928-1937.

Hausman, R. (1988). The T7 group. In: *The Bacteriophages*, Vol. II, (R. Calendar, ed), pp 259-289. Plenum Press, New York.

Kiino, D.R. and Rothman-Denes, L.B. (1988). Bacteriophage N4. In: *The Bacteriophages*, Vol. I, (R. Calendar, ed), pp 475-494. Plenum Press, New York.

Martín, A.C., López, R. and García, P. (1996). Analysis of the complete nucleotide sequence and functional organization of the genome of *Streptococcus pneumoniae* bacteriophage Cp-1. *J. Virol.*, **70**, 3678-3687.

Meijer, W.J., Horcajadas, J.A. and Salas, M. (2001). ϕ29 family of phages. *Microbiol. Mol. Biol. Rev.*, **65**, 261-287.

Mmolawa, P., Schmieger, H., Tucker, C. and Heuzenroeder, M. (2003). Genomic structure of the *Salmonella enterica* Serovar Typhimurium DT 64 bacteriophage ST64T: evidence for modular genetic architecture. *J. Bacteriol.*, **185**, 3473-3475.

Nelson, D., Schuch, R., Zhu. S., Tscherne, D.M. and Fischetti, V.A. (2003). Genomic sequence of C_1, the first streptococcal phage. *J. Bacteriol.*, **185**, 3325-32.

Pedulla, M.L., Ford, M.E., Karthikeyan, T., Houtz, J.M., Hendrix, R.W., Hatfull, G.F., Poteete, A.R., Gilcrease, E.B., Winn-Stapley, D.A. and Casjens, S.R. (2003). Corrected sequence of the bacteriophage P22 genome. *J. Bacteriol.*, **185**, 1475-1477.

Pencenkova, T., Benes, V., Paces, J., Vlcek, C. and Paces, V. (1997). Bacteriophage B103: complete DNA sequence of its genome and relationship to other *Bacillus* phages. *Gene*, **199**, 157-163.

Pencenkova, T. and Paces, V. (1999). Molecular phylogeny of ϕ29-like phages and their evolutionary relatedness to other protein-primed replicating phages and other phages hosted by Gram-positive bacteria. *J. Mol. Evol.*, **48**, 197-208.

Poteete, A.R. (1988). Bacteriophage P22. In: *The Bacteriophages*, Vol. II, (R. Calendar, ed.), pp 647-682. Plenum Press, New York.

Rothman-Denes, L.B. (1995). DNA supercoiling, DNA hairpins and single-stranded DNA binding proteins in bacteriophage N4 transcription regulation. *Sem. Virol.*, **6**, 15-23.

Tao, Y., Olson, N. H., Xu, W., Anderson, D.L., Rossmann, M.G. and Baker, T.S. (1998). Assembly of a tailed bacterial virus and its genome release studied in three dimensions. *Cell*, **95**, 431-437.

Vybrial, D., Takac, M., Loessner, M., Witte, A., von Ahsen, U. and Bläsi, U. (2003). Complete nucleotide sequence and molecuar characterization of two lytic *Staphylococcus aureus* phages: 44HJD and P68. *FEMS Microbiol. Lett.*, **219**, 275-283.

Zhang, Z., Greene, B., Thuman-Commike, P.A., Jakana, J. Prevelige, P.E., Jr., King, J. and Chiu, W. (2000). Visualization of the maturation transition in bacteriophage P22 by electron cryomicroscopy. *J. Mol. Biol.*, **297**, 615-626.

CONTRIBUTED BY

Hendrix, R.W., and Casjens, S.R.

FAMILY TECTIVIRIDAE

TAXONOMIC STRUCTURE OF THE FAMILY

Family ***Tectiviridae***
Genus ***Tectivirus***

Since only one genus is currently recognized, the family description corresponds to the genus description.

GENUS TECTIVIRUS

Type Species *Enterobacteria phage PRD1*

VIRION PROPERTIES

MORPHOLOGY

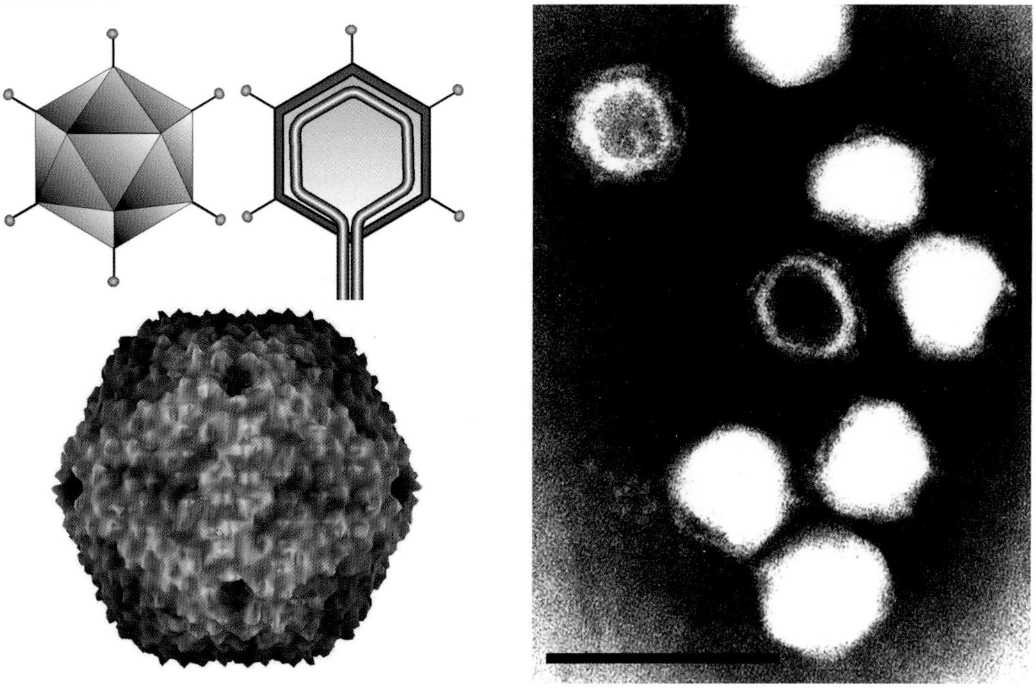

Figure 1: (Upper left) Diagram of virion surface and section of Enterobacteria phage PRD1 (PRD1). (Lower left) Pseudo atomic model of PRD1 virion (T=25)(San Martin, C. *et al.*, (2001). *Structure,* **10**, 917-30). (Right) Negative contrast electron micrograph of phage PRD1 particles. The bar represents 100 nm.

Virions are icosahedra, have no external envelope, and measure 66 nm from facet to facet (Fig. 1). Capsids have apical spikes extending ~20 nm. The capsid of Enterobacteria phage PRD1 (PRD1) is constructed of 240 major CP (P3) trimers that form a pseudo T=25 lattice. Protein P3 contains two beta barrels and forms very tight trimers. The spikes are formed by two proteins (P2 and P5) extending from the penton protein (P31) and are used for receptor recognition. The capsid encloses an inner membrane vesicle formed of approximately equal amounts of virus-encoded proteins and lipids derived from host cell plasma membrane. The DNA is coiled within this membrane. Virions are normally tail-less, but produce tail-like tubes of ~60 x 10 nm upon DNA release or after chloroform treatment (Fig. 1).

PHYSICOCHEMICAL AND PHYSICAL PROPERTIES

The virion Mr is ~66 x 10^6, the S_{20w} is 357-416S, and the buoyant density in CsCl is ~1.29 g/cm^3. Virions are usually stable at pH 5-8. Infectivity is sensitive to organic solvents and detergents.

NUCLEIC ACID

Virion contains a single molecule of linear dsDNA ~15 kbp. The DNA corresponds to 14-15% of particle weight. The complete DNA sequences of phage PRD1 and five closely related phages (Enterobacteria phage PR3; Enterobacteria phage PR4; Enterobacteria phage PR5; Enterobacteria phage PR772, and Enterobacteria phage L17) as well as Bacillus phage Bam35 have been determined. The 5'-ends of the genome have covalently linked proteins.

PROTEINS

PRD1 proteins and their functions are listed in the table below. Tectiviruses infecting members of the genus *Bacillus* (Bam35) contain approximately the same number of proteins as PRD1.

Table 1: Enterobacteria phage PRD1 (PRD1) proteins and their functions.

Phage Protein	Gene	Mutant	Mass (kDa)	Function	Location
P1	I	+	63.3	DNA polymerase	Cytoplasm
P2	II	+	63.7	Adsorption	Vertices
P3	III	+	43.1	Major capsid protein	Capsid
P5	V	+	34.3	Spike	Vertices
P6	VI	+	17.6	Packaging?	Unique vertex
P7	VII	+	27.1	DNA entry	Vertices
P8	VIII	+	29.6	Terminal protein	DNA
P9	IX	+	25.8	Packaging	Unique vertex
P10	X	+	20.6	Assembly	Cell membrane
P11	XI	+	18.2	DNA entry	Membrane surface
P12	XII	+	16.6	ssDNA binding	Cytoplasm
P14	XIV	+	17	DNA entry	Vertices
P15	XV	+	17.3	Lytic muramidase	Cytoplasm
P16	XVI	+	17	Infectivity	Viral membrane
P17	XVII	+	9.5	Assembly	Cytoplasm
P18	XVIII	+	9.8	DNA entry	Viral membrane
P19	XIX	+	10.5	ssDNA binding	Cytoplasm
P20	XX	+	4.7	Packaging	Unique vertex
P22	XXII	+	5.5	Packaging	Unique vertex
P30	XXX	+	9.2	Capsid stability	Capsid
P31	XXXI	+	13.7	Penton	Capsid
P32	XXXII	+	5.4	DNA entry	Viral membrane
P33	XXXIII	-	7.5	Assembly	Cytoplasm
P34	XXXIV	-	6.7	?	Viral membrane
P35	XXXV	+	10.6	Holin	Host plasma membrane

LIPIDS

Virions contain ~15% lipids by weight. Lipids constitute ~60% of the inner vesicle. In PRD1, lipids form a bilayer and seem to be in a liquid crystalline phase. Phospholipid content (56% phosphatidylethanolamine and 37% phosphatidyl-glycerol, 5% cardiolipin) is enriched by PG compared to the host plasma membrane lipid composition and varies somewhat depending on the host strain. The fatty acid composition is close to that of the host.

CARBOHYDRATES

Not detected.

GENOME ORGANIZATION AND REPLICATION

The linear dsDNA genome has inverted terminal repeats. Replication is protein-primed, proceeds by strand displacement, and can start at both ends of the genome as the inverted terminal repeats contain sites for the initiation of replication (Fig. 2). After DNA entry, replication and transcription, capsid proteins polymerize in the cytoplasm, whereas membrane-associated proteins are inserted into the host plasma membrane. With the help of nonstructural virion-encoded assembly factors and the coat-forming proteins, a virus-specific lipoprotein vesicle obtains an outer protein coat and is translocated to the interior of the cell where DNA packaging takes place. Mature virions are released by lysis of the host cell (Fig. 3).

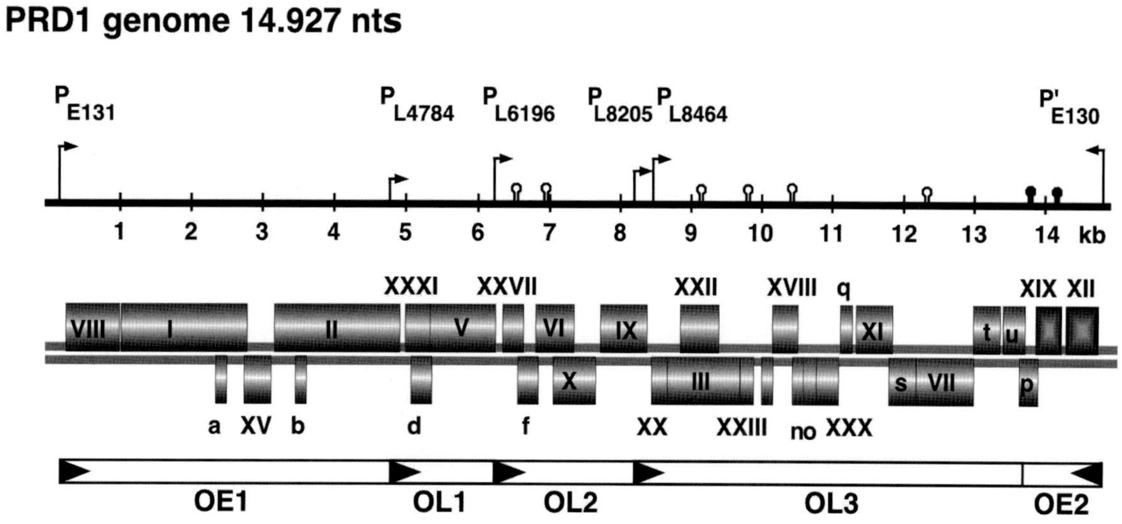

Figure 2: Representation of the genome of Enterobacteria phage PRD1 (PRD1) indicating the organization of genes and promoters (P) and their reading direction. Gene and protein nomenclature are correlated so that each protein has the same number as the gene (Roman). ORFs with no confirmed protein product are indicated by lower case letters. (Top) Promoters and terminators. (Center) Order of genes; genes XIX and XII are encoded on one strand and all other genes are on the other strand. (Bottom) Operons and their reading directions; E = early, L = late (redrawn from diagram courtesy of M. Grahn).

ANTIGENIC PROPERTIES

No information available.

BIOLOGICAL PROPERTIES

Enterobacterial phages are virulent. The tail-like tube probably acts as a DNA injection device. Phages Bam35 and AP50 and a number additional isolates are specific for *Bacillus* species. Thermus phage P37-14 and a number of similar isolates are specific for *Thermus* and are found in volcanic hot springs.

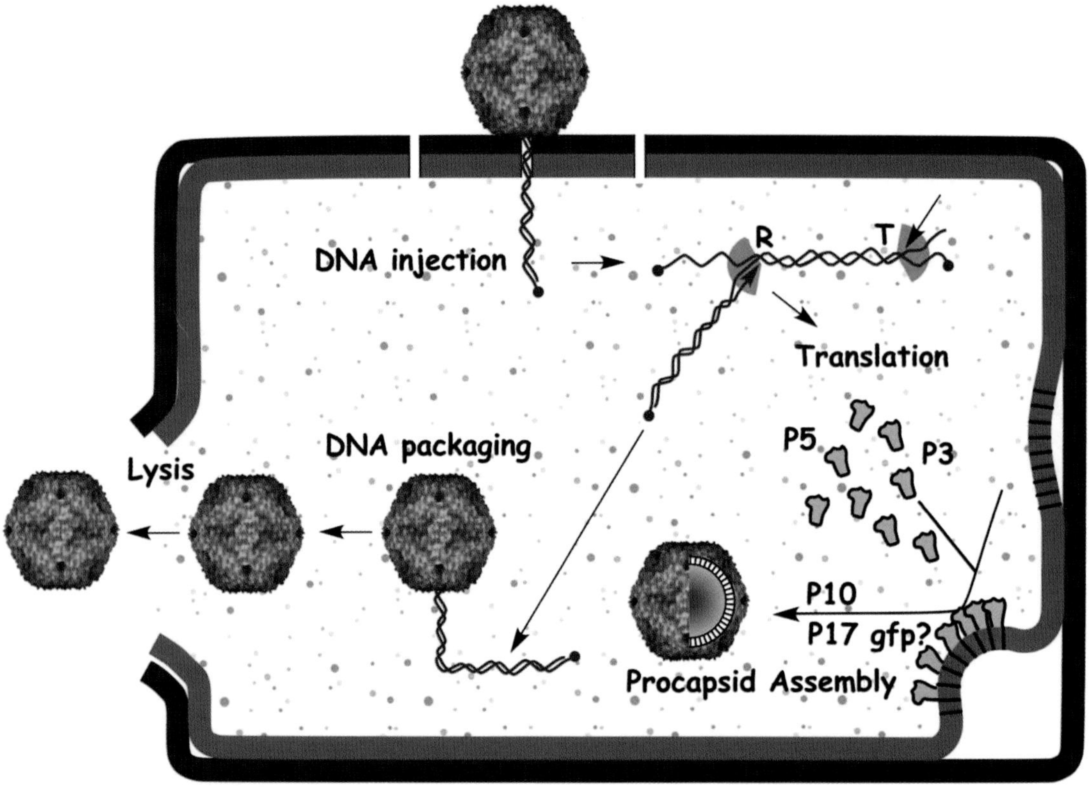

Figure 3: Diagram of the life cycle of Enterobacteria phage PRD1; for details see the text.

List of Species Demarcation Criteria in the Genus

The enterobacterial (phage PRD1) and *Bacillus* (phage Bam35) sequence comparison reveals very little identity, but genome organization and location of key genes are similar. The *Thermus* phage hosts are thermophilic and the other tectivirus hosts are mesophilic.

List of Species in the Genus

Species names are in green italic script; strain names and synonyms are in black roman script; tentative species names are in blue roman script. Sequence accession numbers, and assigned abbreviations ( ) are also listed.

Species in the Genus

Bacillus phage AP50		
Bacillus phage AP50		(AP50)
Bacillus phage Bam35 ‡		
Bacillus phage Bam35		(Bam35)
Enterobacteria phage PRD1		
Enterobacteria phage PRD1	[M69077]	(PRD1)
Enterobacteria phage PR3		(PR3)
Enterobacteria phage PR4		(PR4)
Enterobacteria phage PR5		(PR5)
Enterobacteria phage PR722		(PR772)
Enterobacteria phage L17		(L17)
Thermus phage P37-14		
Thermus phage P37-14		(P37-14)

Tentative Species in the Genus

None reported

DSDNA

LIST OF UNASSIGNED SPECIES IN THE FAMILY

Bacillus phage ϕNS11
Bacillus phage ϕNS11 (ϕNS11)

PHYLOGENETIC RELATIONSHIPS WITHIN THE FAMILY

The Enterobacteria phages sequenced show ~95% similarity. Bam35 and PRD1-type phages most probably share a common ancestor.

SIMILARITY WITH OTHER TAXA

Tectiviruses have similarities to corticoviruses. They both have an icosahedral capsid enclosing a membrane. However, the capsid and genome organizations differ considerably. PRD1, and presumably other tectiviruses, have many features in common with adenoviruses. These include: (1) a linear genome, (2) genomic inverted terminal repeats, (3) protein-primed initiation of replication, (4) type B (*E. coli* Pol II) DNA polymerase, (5) trimeric capsid proteins that contains two beta barrels and form hexagonal structures, (6) arrangement of the capsid protein in a pseudo T=25 lattice, and (7) receptor-binding spikes at capsid vertices. This principal capsid architecture and CP fold have also been described in an isolate of *Paramecium bursaria Chlorella virus 1*, belonging to the family *Phycodnaviridae.*

DERIVATION OF NAMES

Tecti from Latin tectus "covered".

REFERENCES

Ackermann, H.-W. and DuBow, M.S. (eds)(1987). *Viruses of Prokaryotes*, Vol II. Natural Groups of Bacteriophages. CRC Press, Boca Raton, Florida, pp 171 - 218.

Bamford, D.H., Caldentey, J. and Bamford, J.K.H. (1995). Bacteriophage PRD1: a broad host range dsDNA tectivirus with an internal membrane. *Adv. Vir. Res.*, **45**, 281-319.

Benson, S.D., Bamford, J.K.H., Bamford, D.H. and Burnett, R.M. (1999). Viral evolution revealed by bacteriophage PRD1 and human adenovirus coat protein structures. *Cell*, **98**, 825-833.

Caldentey, J., Bamford, J.K.H. and Bamford, D.H. (1990). Structure and assembly of bacteriophage PRD1, an Escherichia coli virus with a membrane. *J. Struct. Biol.*, **104**, 44-51.

Mindich, L. and Bamford, D.H. (1988). Lipid-containing bacteriophages. In: *The Bacteriophages*, (R. Calendar, ed), pp 145-262, Vol 2. Plenum Press, New York.

Ravantti, J.J., Gaidelyte, A., Bamford, D.H. and Bamford, J.K.H. (2003). Comparative analysis of bacterial viruses Bam35, infecting a gram-positive host, and PRD1 infecting gram-negative hosts, demonstrate a viral lineage. *Virology*, **313**, 401-14.

San Martin, C., Huiskonen, J.T., Bamford, J.K.H., Butcher, S.J., Fuller, S.D., Bamford, D.H. and Burnett, R.M. (2002). Minor proteins, mobile arms and membrane-capsid interactions in the bacteriophage PRD1 capsid. *Nature Struct. Biol.*, **9**, 756-763.

CONTRIBUTED BY

Bamford, D.H.

FAMILY *CORTICOVIRIDAE*

TAXONOMIC STRUCTURE OF THE FAMILY

Family ***Corticoviridae***
Genus ***Corticovirus***

Since only one genus is currently recognized, the family description corresponds to the genus description.

GENUS *CORTICOVIRUS*

Type Species *Pseudoalteromonas phage PM2*

VIRION PROPERTIES

MORPHOLOGY

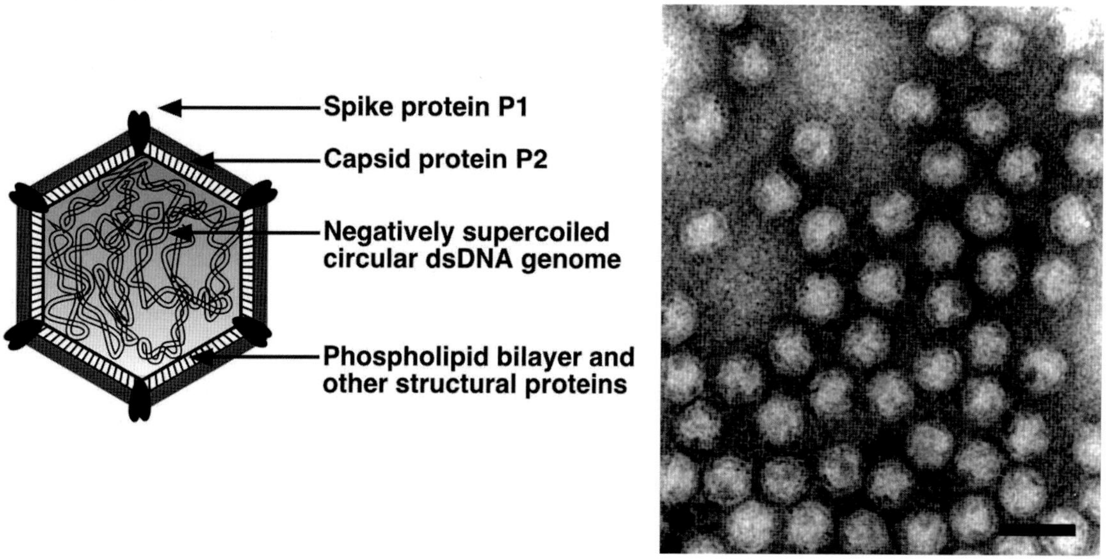

Figure 1: (Left) A schematic presentation and (right) negative stain electron micrograph of Pseudoalteromonas phage PM2 particles. (Courtesy of J.K.H. Bamford.) The bar represents 100 nm.

Icosahedral virions have a diameter ~56 nm. Spikes composed of protein P1 protrude from the five-fold vertices. The capsid, which is formed of P2 trimers, surrounds an inner lipoprotein vesicle (lipid core; Fig.1).

PHYSICOCHEMICAL AND PHYSICAL PROPERTIES

The mass of the virion is ~45 x 10^6. The buoyant density in CsCl is 1.28 g/cm^3 and in sucrose 1.26 g/cm^3, and the S_{20w} is 293S. Virions are stable at pH 6-8, and are very sensitive to ether, chloroform and detergents. The virion stability is strongly dependent on NaCl (10 mM minimum) and $CaCl_2$ (2.5 mM minimum) ions.

NUCLEIC ACID

The genome is a highly supercoiled circular dsDNA of 10,079 bp (Mr of 6.6 x 10^6). DNA comprises ~14% of the virion weight and the G+C content is 42.2%. The phage PM2 genome has been sequenced (AF155037).

PROTEINS

The genome has 21 putative genes, nine of which have been shown to code for structural proteins (P1-P9; Table 1). Five of the proteins are nonstructural (P12-P16).

Table 1: Pseudoalteromonas phage PM2 proteins.

Protein[1]	Size (kDa)	Location/function
P1	37.5	spike protein
P2	30.2	major capsid protein
P3	10.8	membrane
P4	4.4	membrane
P5	17.9	membrane
P6	14.3	membrane
P7	3.6	membrane
P8	7.3	membrane
P9	24.7	putative packaging ATPase
P12	73.4	replication initiation protein
P13	7.2	transcription factor
P14	11.0	transcription factor
P15	18.1	regulative function
P16	10.3	regulative function

1) P is for protein; arabic numeral corresponds to the Roman numeral of the gene.

LIPIDS

Particles are ~14% lipid by weight. About 90% are phospholipids, mainly phosphatidyl glycerol and phosphatidyl ethanolamine. The lipids are derived from the host plasma membrane, but their composition deviates from that of the host bacterium. Lipids form a bilayer with virus specific integral membrane proteins.

CARBOHYDRATES

Not known.

GENOME ORGANIZATION AND REPLICATION

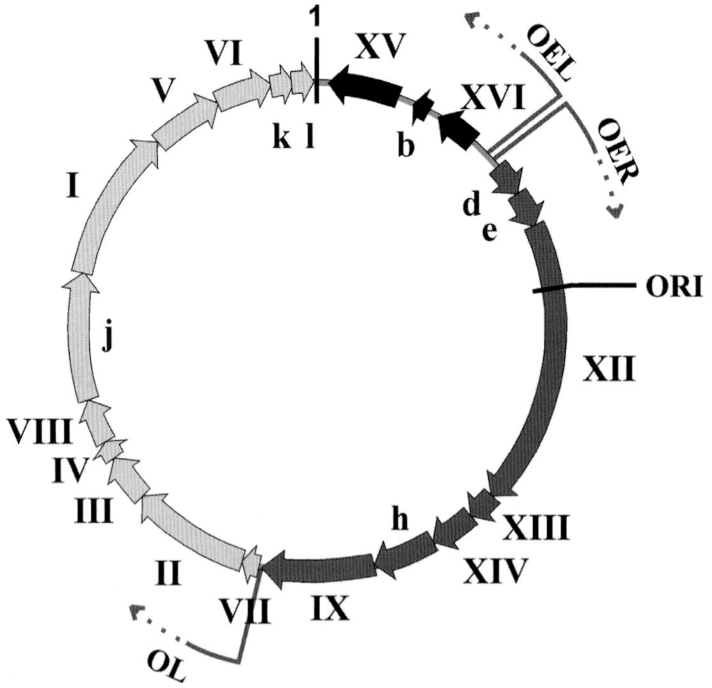

Figure 2: Map of the genome of Pseudoalteromonas phage PM2 (PM2) showing 21 ORFs. ORFs shown to code for functional proteins are classified as genes and given a Roman numeral.

The genome is organized in three operons (Fig. 2). Operons OEL and OER encode early function gene products: the replication initiation protein P12 and regulatory proteins P13, P14, P15, and P16. Expression of the genes for structural proteins is under the control of the late promoter (OL), which is activated by the phage-encoded transcription factors P13 and P14. DNA replication probably is by a rolling circle mechanism.

ANTIGENIC PROPERTIES

Not known.

BIOLOGICAL PROPERTIES

Phages are virulent and adsorb to a yet unknown receptor on the bacterial cell surface. Two strains of marine host bacteria of the genus *Pseudoalteromonas* are known: BAL-31 and ER72M2.

LIST OF SPECIES DEMARCATION CRITERIA IN THE GENUS

Not applicable.

LIST OF SPECIES IN THE GENUS

Species names are in green italic script; strain names and synonyms are in black roman script; tentative species names are in blue roman script. Sequence accession numbers, and assigned abbreviations ( ) are also listed.

SPECIES IN THE GENUS

Pseudoalteromonas phage PM2

Pseudoalteromonas phage PM2	AF155037	(PM2)

TENTATIVE SPECIES IN THE GENUS

None reported

LIST OF UNASSIGNED VIRUSES IN THE FAMILY

None reported.

PHYLOGENETIC RELATIONSHIPS WITHIN THE FAMILY

Not applicable.

SIMILARITY WITH OTHER TAXA

Corticoviruses resemble tectiviruses in having a lipid bilayer underneath the capsid. These viruses appear to differ by their capsid geometry and most probably by the infection mechanism, since no tail-like membrane tube is seen upon infection in corticoviruses.

DERIVATION OF NAMES

Cortico: from Latin *cortex, corticis*, "crust", "bark".

REFERENCES

Espejo, R.T. and Canelo, E.S. (1968). Properties of bacteriophage PM2: A lipid-containing bacterial virus. *Virology*, **34**, 738-747.

Harrison, S.C., Caspar, D.L., Camerini-Otero, R.D. and Franklin, R.M. (1971). Lipid and protein arrangement in bacteriophage PM2. *Nat. New Biol.*, **229**, 197-201.

Kivelä, H.M., Kalkkinen, N. and Bamford, D.H. (2002). Bacteriophage PM2 has a protein capsid surrounding a spherical lipid-protein core. *J. Virol.*, **76**, 8169-8178.

Kivelä, H.M., Männistö, R.H., Kalkkinen, N. and Bamford, D.H. (1999). Purification and protein composition of PM2, the first lipid-containing bacterial virus to be isolated. *Virology*, **262**, 364-374.

Männistö, R.H., Grahn, A.M., Bamford, D.H. and Bamford, J.K.H. (2003). Transcription of bacteriophage PM2 involves phage-encoded regulators of heterologous origin. *J. Bacteriol.*, **185**, 3278-3287.

Männistö, R.H., Kivelä, H.M., Paulin, L., Bamford, D.H. and Bamford, J.K.H. (1999). The complete genome sequence of PM2, the first lipid-containing bacterial virus to be isolated. *Virology*, **262**, 355-363.

Contributed By

Bamford, J.K.H.

FAMILY PLASMAVIRIDAE

TAXONOMIC STRUCTURE OF THE FAMILY

Family *Plasmaviridae*
Genus *Plasmavirus*

Since only one genus is currently recognized, the family description corresponds to the genus description.

GENUS PLASMAVIRUS

Type Species *Acholeplasma phage L2*

VIRION PROPERTIES

MORPHOLOGY

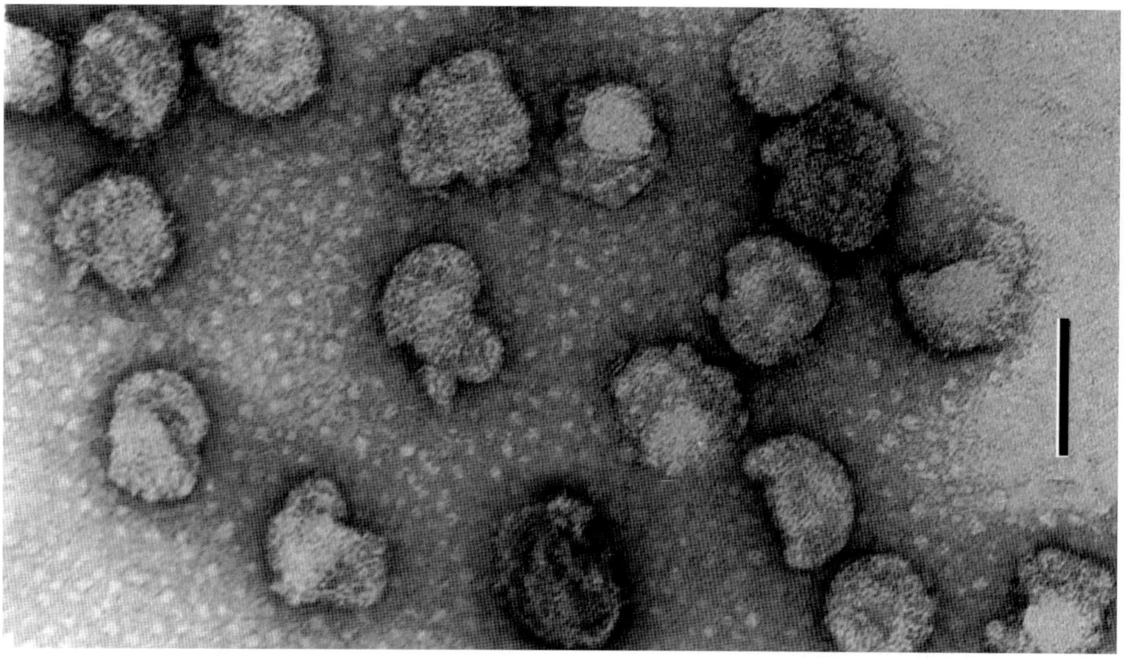

Figure 1: Negative contrast electron micrograph of Acholeplasma phage L2 (L2) virions. The pleomorphic virion appears as a core (perhaps a nucleoprotein condensation) within a baggy membrane. The bar represents 100 nm. (From Poddar, S.K., Cadden, S.P., Das, J., Maniloff, J. (1985). With permission).

Virions are quasi-spherical, slightly pleomorphic, enveloped, and ~80 nm (range 50-125 nm) in diameter (Fig. 1). Size varies due to virion heterogeneity: at least three distinct virion forms are produced during infection. Thin-sections show virions with densely stained centers, presumably containing condensed DNA, and particles with lucent centers. The absence of a regular capsid structure suggests the Acholeplasma phage L2 (L2) virion is an asymmetric nucleoprotein condensation bounded by a lipid-protein membrane.

PHYSICOCHEMICAL AND PHYSICAL PROPERTIES

Virions are extremely heat sensitive, relatively cold stable, and inactivated by nonionic detergents (Brij-58, Triton X-100, and Nonidet P-40), ether, and chloroform. Viral infectivity is resistant to DNAse I and phospholipase A, but sensitive to pronase and trypsin treatment. UV-irradiated virions can be reactivated in host cells by excision and SOS DNA repair systems. Virions are relatively resistant to photodynamic inactivation.

Nucleic Acid
Virions contain one molecule of infectious, circular, superhelical dsDNA. The phage L2 genome is 11,965 bp, with a G+C value of 32%. All ORFs are encoded in one strand. Several genes are translated from overlapping reading frames.

Proteins
Virions contain at least four major proteins: ~64, 61, 58, and 19 kDa. Several minor protein bands are also observed in virion preparations. DNA sequence analysis indicates 15 ORFs (Table 1).

Table 1: Acholeplasma phage L2 ORFs

Designation	Size	Comments
ORF 1	66,643	—
ORF 2	9,620	—
ORF 3	37,157	—
ORF 4	18,224	—
ORF 5	34,868	putative integrase, gene is upstream from *attP* site
ORF 6	9,799	—
ORF 7	14,047	—
ORF 8	7,412	—
ORF 9	9,332	—
ORF 10	16,143	—
ORF 11	25,562	—
ORF 12	17,214	basic protein, putative major virion DNA-binding protein
ORF 13	81,308	putative integral membrane protein, has 27 aa N-terminal peptidase cleavage signal sequence
ORF 13*	47,699	translation start site is 295 codons downstream from ORF13 start site and in same reading frame
ORF 14	26,105	has 26 aa N-terminal peptidase cleavage signal sequence

Lipids
Virions and host cell membranes have similar fatty acid compositions. Variation of host cell membrane fatty acid composition leads to virions with corresponding fatty acid composition variations. Data indicate viral membrane lipids are in a bilayer structure.

Carbohydrates
None reported.

Genome Organization and Replication

Acholeplasma phage L2 infection involves a noncytocidal productive infectious cycle followed by a lysogenic cycle in each infected cell. At least 11 overlapping mRNAs are transcribed from the DNA coding strand, from at least 8 promoters. In non-cytocidal infection, progeny phages are released by budding from the host cell membrane, with the host surviving as a lysogen. Lysogeny involves integration of the phage L2 genome into a unique site in the host cell chromosome. The putative phage L2 attP integration site is CATCTTCAT–7nt-CTGAAGATA. Lysogens are resistant to superinfection by homologous virus but not by heterologous virus (apparently due to a repressor), and are inducible by UV-irradiation and mitomycin C.

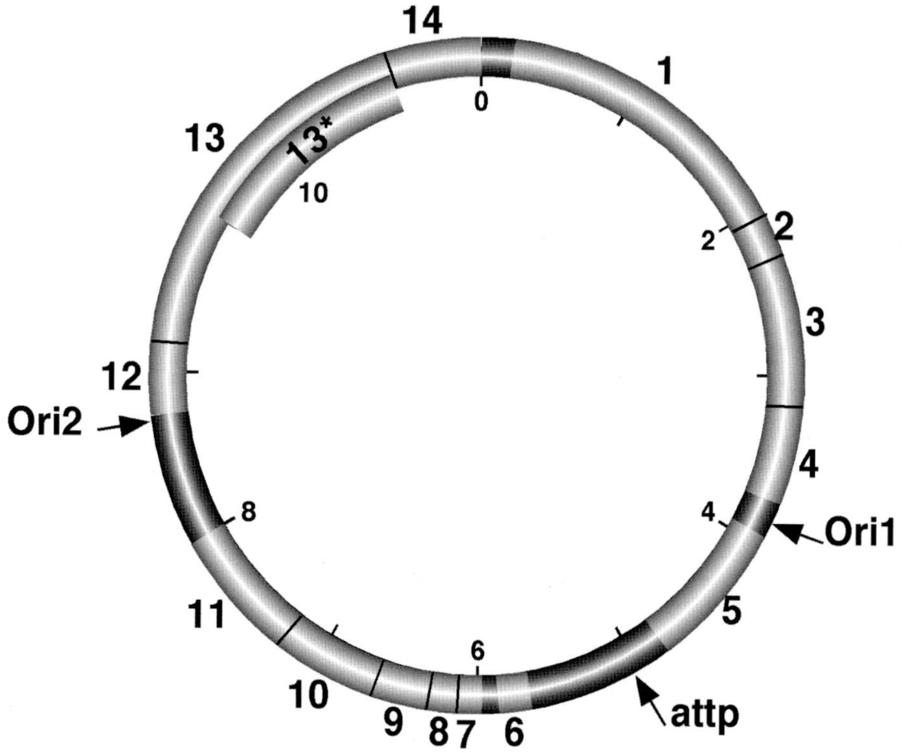

Figure 2: Map of genome organization of Acholeplasma phage L2 (L2) showing ORFs as determined from analysis of the 11,965 bp sequence. The base on the 3'-side of the single *Bst*E II cleavage site is taken as the first base of the DNA sequence. The map also shows locations of the phage L2 integration site (*attP*) and the two phage L2 DNA replication origin sites (*ori1* and *ori2*). (From J. Maniloff, G.J. Kampo, and C.C. Dascher; 1994. With permission).

Antigenic Properties

None reported.

Biological Properties

Host range: the phage L2 infects *Acholeplasma laidlawii* strains. Other putative plasmaviruses have been reported to infect *A. laidlawii (*Acholeplasma phages v1, v2, v4, v5 and v7*), A. modicum (*Acholeplasma phage M1), and *A. oculi* strains (Acholeplasma phage O1).

List of Species Demarcation Criteria in the Genus

Not applicable.

List of Species in the Genus

Species names are in green italic script; strain names and synonyms are in black roman script; tentative species names are in blue roman script. Sequence accession numbers, and assigned abbreviations ( ) are also listed.

Species in the Genus

Acholeplasma phage L2

Acholeplasma phage L2	[L13696]	(L2)

Tentative Species in the Genus

Acholeplasma phage M1	(M1)
Acholeplasma phage O1	(O1)
Acholeplasma phage v1	(v1)
Acholeplasma phage v2	(v2)

Acholeplasma phage v4 (v4)
Acholeplasma phage v5 (v5)
Acholeplasma phage v7 (v7)

List of Unassigned Viruses in the Family

None reported.

Phylogenetic Relationships within the Family

Not available.

Similarity with Other Taxa

None reported.

Derivation of Names

Plasma, from the Greek *plasma* for "shaped product" referring to the plastic virion shape.

References

Dybvig, K. and Maniloff, J. (1983). Integration and lysogeny by an enveloped mycoplasma virus. *J. Gen. Virol.*, **64**, 1781-1785.

Maniloff, J. (1992). Mycoplasma viruses. In: *Mycoplasmas: molecular biology and pathogenesis*, (J., Maniloff, R.N., McElhaney, L.R., Finch and J.B., Baseman, eds), pp 41-59. American Society for Microbiology, Washington DC.

Maniloff, J., Cadden, S.P. and Putzrath, R.M. (1981). Maturation of an enveloped budding phage: mycoplasma virus L2. In: *Bacteriophage Assembly*, (M.S. DuBow, ed), pp 503-513. AR Liss Inc, New York.

Maniloff, J., Kampo, G.K. and Dascher, C.C. (1994). Sequence analysis of a unique temperate phage: mycoplasma virus L2. *Gene*, **141,** 1-8.

Poddar, S.K., Cadden, S.P., Das, J. and Maniloff, J. (1985). Heterogeneous progeny viruses are produced by a budding enveloped phage. *Intervirology*, **23**, 208-221.

Contributed By

Maniloff, J.

FAMILY *LIPOTHRIXVIRIDAE*

TAXONOMIC STRUCTURE OF THE FAMILY

Family	***Lipothrixviridae***
Genus	***Alphalipothrixvirus***
Genus	***Betalipothrixvirus***
Genus	***Gammalipothrixvirus***

GENERAL PROPERTIES

Virions are filaments that are not flexible (*Alphalipothrixvirus*), slightly flexible (*Gammalipothrixvirus*) or very flexible (*Betalipothrixviridae*) that vary from 410-2200 nm in length and 24-38 nm in diameter. Virion ends have specific structures that vary between genera, and also between species in the genus *Betalipothrixvirus*. Virions are enveloped with host lipids and viral proteins. The envelope is non-structured. The helical nucleoprotein core contains linear dsDNA.

Virions contain one dsDNA molecule from 15.9-56 kbp, depending on the virus species. The ends of the DNA are masked from enzymatic modification in an unknown manner. The virion envelope contains host lipids in proportions different than the host membrane. Thermoproteus tenax virus 1 (TTV-1) virions contain carbohydrate in their envelope.

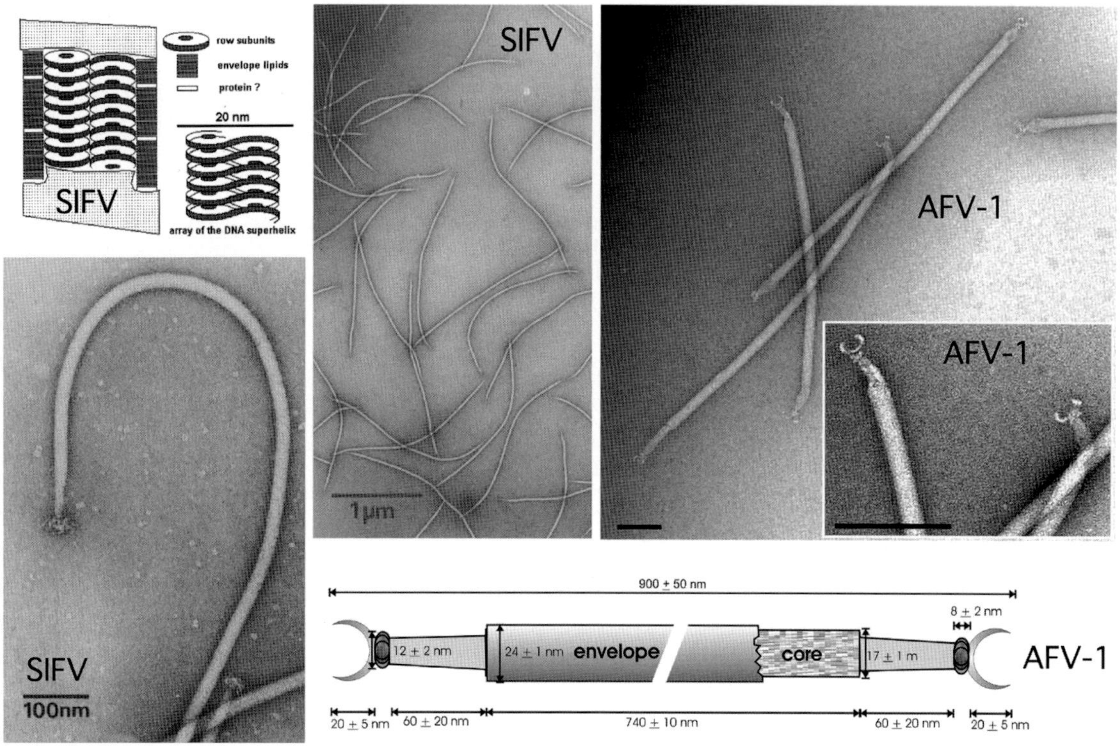

Figure 1: (Top left) A schematic model of a cross section of the Sulfolobus islandicus filamentous virus (SIFV) virion body. (Bottom left) Negative stain electron micrograph of the end of a single SIFV virion. (Center) Negative stain electron micrograph of many SIFV virions. (Top right) Negative contrast electron micrograph of Acidianus filamentous virus 1 (AFV-1) virions (bars represent 100nm). (Bottom right) Schematic model of the AFV-1 virion. (From Arnold *et al.*, 2000 and Bettstetter *et al.*, 2003 with permission).

GENUS ALPHALIPOTHRIXVIRUS

Type Species *Thermoproteus tenax virus 1*

VIRION PROPERTIES

MORPHOLOGY

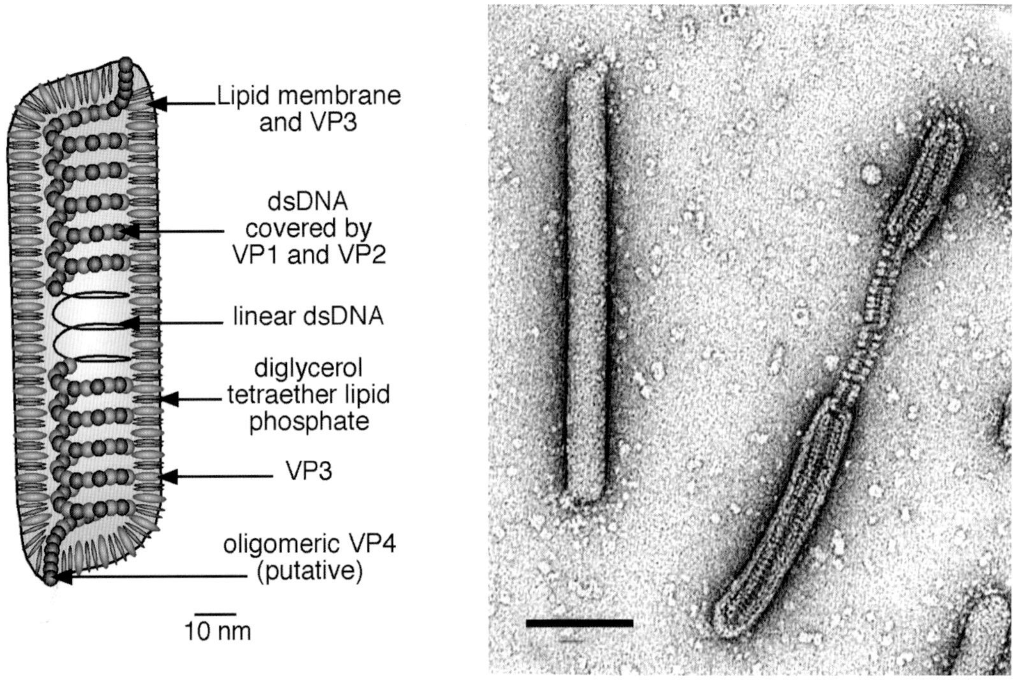

Figure 2: (Left) Diagram of a virion of an isolate of *Thermoproteus tenax virus 1*, showing a schematic representation of the composition of the coat and DNA covered by DNA-binding proteins; and superhelical DNA without covering protein molecules. (Right) Negative contrast electron micrograph of intact virus particle (on the left) and partially deteriorated particle exhibiting coat and core (on the right). The bar represents 100 nm.

Virions are rigid rods, 410 nm long and 38 nm in diameter, with asymmetric extensions at each end. The envelope has no structure in electron micrographs (Fig. 2). The envelope encloses a helical core. The DNA containing core structure and a lipid-containing coat can be released by treatment with detergent.

PHYSICOCHEMICAL AND PHYSICAL PROPERTIES

Virion Mr is 3.3 x 10^8 and buoyant density in CsCl is 1.25 g/cm^3. Virions are stable at 100°C and a fraction remains viable after autoclaving at 120°C. Particles maintain their structure in 6 M urea and 7 M guanidine hydrochloride. Detergents (e.g. Triton X-100 and octylglycoside), dissociate virions into viral cores, containing the DNA plus DNA-binding proteins, and viral envelopes, containing isoprenyl ether lipids and CP. Virions contain ~3% (w/w) DNA, ~75% protein and ~22% isoprenyl ether lipids.

NUCLEIC ACID

Virions contain one molecule of linear dsDNA of 15.9 kbp. About 85% of the total sequence has been determined (Fig. 3). The ends of the DNA molecule are masked in an unknown manner. Almost all of the ORFs greater than 150 amino acids are on one strand.

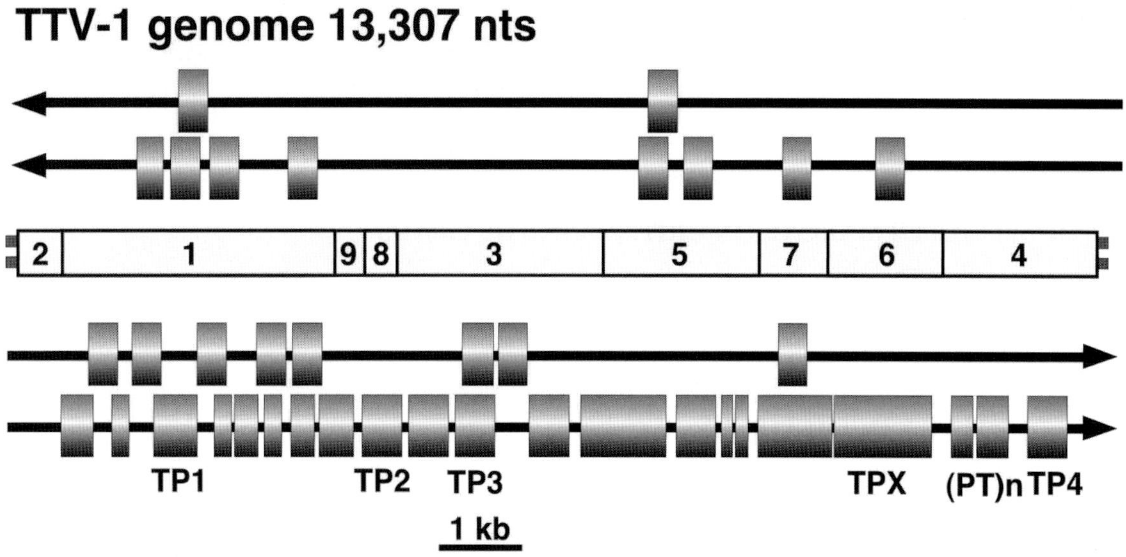

Figure 3: Partial genome of Thermoproteus tenax virus 1 (TTV-1). The *Cla*I restriction map is shown in the middle of the figure and all ORFs greater than 150 nt are shown, including the structural proteins of the virus and two regions, TPX and $(PT)_n$ containing $(thr, pro)_n$ repeats, between which recombination occurs frequently. Overlapping ORFs are shown on a separate arrow. ORFs on one strand are shown above the ClaI map and ORFs on the other strand below.

PROTEINS

Virions contain at least four proteins of the following sizes: TP1, 12.9 kDa; TP2, 16.3 kDa; TP3, 18.1 kDa; and TP4, 24.5 kDa. Proteins TP1 and TP2 are DNA associated, TP3 is the envelope protein, and the location of TP4 in the virus particle is unknown. Only TP1 is a basic protein. TP3 is highly hydrophobic and TP4 is hydrophilic. Additional minor proteins may be present. A fifth protein, TPX, carrying a C-terminal $(thr, pro)_n$ repeat, is present in infected cells in high concentration, but absent in virus particles.

LIPIDS

The virion envelope contains the same lipids as the host membrane, essentially diphytanyl tetraether lipids. The envelope has a bilayer structure. The phosphate residues of the phospholipids are oriented towards the inside and the glycosyl residues towards the outside of the particles.

CARBOHYDRATES

Virions contain glycolipids in their envelope.

GENOME ORGANIZATION AND REPLICATION

The genome contains several transcription units. So far, the function of only a few genes is known, among them those encoding the four structural proteins (TP1 to TP4) which are not linked. There are two ORFs, encoding $(TP)_n$ and $(PT)_n$, between which specific recombination occurs with high frequency. Their map positions are shown in the Cla1 restriction map of the viral genome (Fig. 3).

ANTIGENIC PROPERTIES

No information available.

BIOLOGICAL PROPERTIES

Adsorption and infection appears to proceed via interaction of the tips of host pili with the terminal protrusions of the virus. Fragments of the viral genome have sometimes been found integrated in host genomes. Complete non-integrated virus DNA exists in the cell in linear form. Virions are released by lysis. Infection may be latent.

The host range is limited to the archaeon *Thermoproteus tenax* in the kingdom *Crenarchaeota*. Other rod-shaped DNA-containing viruses with similar morphology but different dimensions have been found associated with *Thermoproteus* cultures or have been observed by electron microscopy in waters from Icelandic solfataras, but virus has not been cultivated from these sources.

LIST OF SPECIES DEMARCATION CRITERIA IN THE GENUS

Not applicable.

LIST OF SPECIES IN THE GENUS

Species names are in green italic script; strain names and synonyms are in black roman script; tentative species names are in blue roman script. Sequence accession numbers, and assigned abbreviations ( ) are also listed.

SPECIES IN THE GENUS

Thermoproteus tenax virus 1

Thermoproteus tenax virus 1	[X14855]	(TTV-1)

TENTATIVE SPECIES IN THE GENUS

None reported.

GENUS *BETALIPOTHRIXVIRUS*

Type Species *Sulfolobus islandicus filamentous virus*

VIRION PROPERTIES

MORPHOLOGY

Sulfolobus islandicus filamentous virus (SIFV) virions (1950 x 24 nm) are enveloped with a 4 nm thick lipid-containing layer and have a flexible filamentous morphology. The ends of the virion taper and end in mop-like structures. The cross-section of the virion is elliptical. The virus core consists of multiple copies of two different DNA binding proteins and a linear dsDNA that appears to wrap around the DNA binding proteins (Fig. 1).

PHYSICOCHEMICAL AND PHYSICAL PROPERTIES

The buoyant density of SIFV virions is ~1.3 g/cm^3. SIFV virions are stable when treated with up to 0.1% SDS or 0.2% Triton X-100 and in organic solvent, but are sensitive to treatment with proteinase K. SIFV virion envelopes can be dissolved by treatment with 0.3% Triton X-100: the nucleoprotein virus core thus generated has a buoyant density of 1.34 g/cm^3 and is disrupted by treatment with 0.1% SDS.

NUCLEIC ACID

The SIFV genome is linear dsDNA ~41.05 kbp with ends that are resistant to enzymatic modification. Proteins do not appear to be bound to the SIFV genome termini.

PROTEINS

SIFV virions contain two major basic proteins, one 17 kDa and the other 25 kDa, and 4 minor proteins of 40-50 kDa.

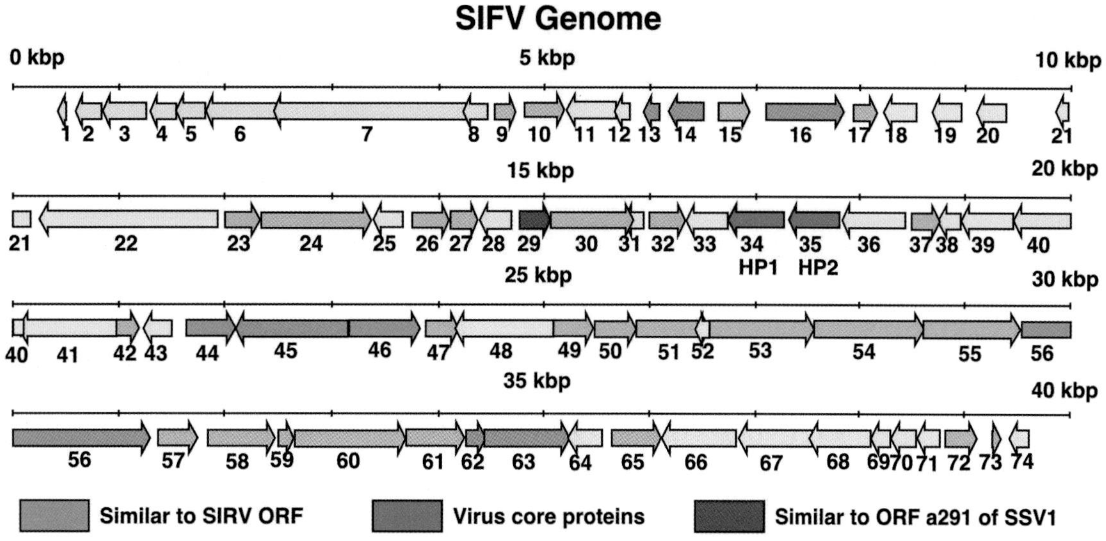

Figure 4: Linear representation of the Sulfolobus islandicus filamentous virus (SIFV) genome. The locations of each predicted protein-coding region are indicated. Arrows represent the direction of transcription for each predicted coding region. ORFs of SIFV with significant sequence similarities to ORFs in other viruses are shaded and labeled. (From Arnold *et al.*, 2000 with permission).

Lipids

Modified host lipids were detected in SIFV virions. Lipids are present in different relative amounts than in the host membrane.

Carbohydrates

None reported.

Genome Organization and Replication

About 96% of the linear dsDNA genome of SIFV was sequenced. It is tightly packed with 74 ORFs: 90% of the genome potentially encodes proteins. The ORFs are clustered in putative transcription units (Fig. 4). The major CP genes are encoded in a tandem array near the middle of the genome. Nothing is known about virus replication.

Antigenic Properties

No information available.

Biological Properties

The SIFV virus genome does not integrate into the host genome. SIFV is often lost on culture dilution. Viruses and virus-like particles with very similar morphology to betalipothrixviruses have been isolated from and detected in enrichment cultures from samples from geothermally heated environments in Iceland, Yellowstone National Park in the USA, and Pozzuoli, Italy.

List of Species Demarcation Criteria in the Genus

Species in the genus *Betalipothrixvirus* differ in virion size, genome size, stability of infection, and hosts. SIFV infects closely related strains of the extremely thermophilic archaeon *Sulfolobus islandicus*. TTV-2 and TTV-3 were found in *Thermoproteus tenax* and have differing lengths. Desulfurolobus ambivalens filamentous virus (DAFV) was observed in an *Acidanus ambivalens* (previously called *Desulfurolobus ambivalens*) culture, its host range is unclear. DAFV has a very similar protein profile to SIFV and shows weak DNA-DNA hybridization with SIFV DNA.

List of Species in the Genus

Species names are in green italic script; strain names and synonyms are in black roman script; tentative species names are in blue roman script. Sequence accession numbers, and assigned abbreviations ( ) are also listed.

Species in the Genus

Sulfolobus islandicus filamentous virus

Sulfolobus islandicus filamentous virus	[AF440571]	(SIFV)

Tentative Species in the Genus

Desulfurolobus ambivalens filamentous virus	(DAFV)
Thermoproteus tenax virus 2	(TTV-2)
Thermoproteus tenax virus 3	(TTV-3)

Genus *Gammalipothrixvirus*

Type Species *Acidianus filamentous virus 1*

Distinguishing Features

AFV-1 virions have distinctive claw-like structures at their termini. Their virions are genomes are also considerably shorter than the beta-lipothrixviruses.

Virion Properties

Morphology

Acidianus filamentous virus 1 (AFV-1) virions (900 x 24 nm) are enveloped with a 3 nm thick lipid-containing layer. The ends of the virus have 20 nm diameter claw-like structures that close on contact with host pili. Between the main virus body and the claws are tapering appendages 60 nm long and a 12 x 8 nm collar (Fig. 1).

Physicochemical and Physical Properties

The buoyant density of AFV-1 virions is ~1.3 g/cm^3. AFV-1 virion envelopes are partially degraded by treatment with 0.3% Triton X-100 and completely removed by treatment with 0.1% SDS.

Nucleic Acid

The AFV-1 genome is linear dsDNA of 20,869 bp. The ends of the AFV-1 genome contain clusters of short direct repeats of the sequence TTGTT or close variants thereof, and are reminiscent of the telomeric ends of linear eukaryotic chromosomes.

Proteins

AFV-1 virions contain 5 major proteins of 130, 100, 80, 30 and 23 kDa.

Lipids

Modified host lipids were detected in AFV-1 virions. Lipids are present in different relative amounts than in the host membrane.

Carbohydrates

None reported.

Genome Organization and Replication

The complete linear dsDNA AFV-1 genome was sequenced. It is tightly packed with 40 ORFs. These ORFs are clustered in potential transcription units (Fig. 5).

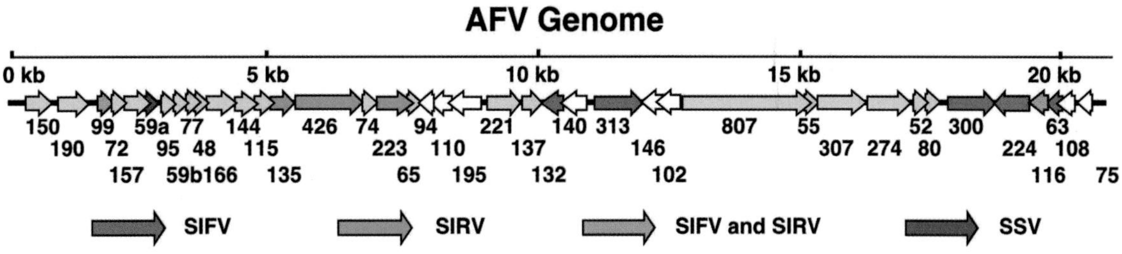

Figure 5: Linear representation of the Acidianus filamentous virus 1 (AFV-1) genome. The locations of each predicted protein-coding region are indicated. Arrows represent the direction of transcription for each predicted coding region. ORFs of AFV-1 with significant sequence similarities to ORFs in other viruses are shaded and labeled. (From Bettstetter *et al.*, 2003 with permission).

Antigenic Properties

No information available.

Biological Properties

The AFV-1 virus genome does not integrate into the host genome. In contrast to the betalipothrixvirus SIFV, AFV-1 is not lost on culture dilution. Viruses and virus-like particles with similar morphology to betalipothrixviruses and gammalipothrixviruses have been isolated from and detected in enrichment cultures from samples from geothermally heated environments in Iceland, Yellowstone National Park in the USA, and Pozzuoli, Italy.

List of Species Demarcation Criteria in the Genus

Not applicable.

List of Species in the Genus

Species names are in green italic script; strain names and synonyms are in black roman script; tentative species names are in blue roman script. Sequence accession numbers, and assigned abbreviations ( ) are also listed.

Species in the Genus

Acidianus filamentous virus 1

Acidianus filamentous virus 1	[AJ567472]	(AFV-1)

Tentative Species in the Genus

None reported.

List of Unassigned Viruses in the Family

None reported.

Phylogenetic Relationships within the Family

Genera in the family *Lipothrixviridae* differ in virion size, morphology, genome size, stability of infection, and hosts. There is no sequence similarity between the alphalipothrixvirus TTV-1 and either the betalipothrixvirus SIFV or gammalipothrixvirus AFV-1, supporting its classification into a new genus. The betalipothrixvirus SIFV infects closely related strains of the extremely thermophilic archaeon *Sulfolobus islandicus*. The betalipothrixviruses TTV-2 and TTV-3 were found in *Thermoproteus tenax* and have differing lengths. Desulfurolobus ambivalens filamentous virus (DAFV) was observed in an *Acidanus ambivalens* (previously called *Desulfurolobus ambivalens*) culture and its host range is unclear. DAFV has a very similar protein profile to SIFV and shows weak DNA-DNA hybridization with SIFV DNA. The gammalipothrixvirus AFV-1 infects closely related strains of the extremely thermophilic archaeon *Acidianus hospitalis* and *Acidianus infernus*. The AFV-1 virion is shorter than all other members of the family as is its genome. It also has unique structures on the virion termini (see Fig. 1). However, 20% of AFV-1 ORFs have homologs in the SIFV genome.

SIMILARITY WITH OTHER TAXA

Members of the families *Lipothrixviridae* and *Rudiviridae* have filamentous particles containing linear dsDNA genomes that infect extremely thermophilic *Archaea* in the kingdom *Crenarchaeota*. However, members of the family *Lipothrixviridae* are enveloped while those of the family *Rudiviridae* are not. Approximately 30% of the ORFs in the Sulfolobus islandicaus rod-shaped virus (SIRV) genomes have homologs in the SIFV genome. Similarly, 20% of AFV-1 genes have homologs in SIRV genomes. There is one ORF in the SIFV genome that shows weak similarity to a conserved ORF in the family *Fuselloviridae* (a291 in SSV-1) that is important for virus function. One ORF in the AFV-1 genome has significant similarity to ORF-a45 in the SSV-1 genome.

DERIVATION OF NAMES

Alpha: from Greek, first
Beta: from Greek, second
Gamma: from Greek, third
Lipo: from Greek, *lipos*, 'fat'.
Thrix: from Greek, *thrix*, 'hair'.

REFERENCES

Arnold, H.P., Zillig, W., Ziese, U., Holz, I., Crosby, M., Utterback, T., Weidmann, J.F., Kristjanson, J.K., Klenk, H.P., Nelson, K.E. and Fraser, C.M. (2000). A novel lipothrixvirus, SIFV, of the extremely thermophilic crenarchaeon *Sulfolobus*. *Virology*, **267**, 252-266.

Bettstetter, M., Peng, X., Garrett, R.A. and Prangishvili, D. (2003). AFV-1, a novel virus infecting hyperthermophilic archaea of the genus *Acidianus*. *Virology*, **315**, 68-79.

Neumann, H., Schwass, V., Eckerskorn, C. and Zillig, W. (1989). Identification and characterization of the genes encoding three structural proteins of the *Thermoproteus tenax* virus TTV-1. *Mol. Gen. Genet.*, **217**, 105-110.

Neumann, H. and Zillig, W. (1990). Structural variability in the genome of the *Thermoproteus tenax virus* TTV-1. *Mol. Gen. Genet.*, **222**, 435-437.

Peng, X., Blum, H., She, Q., Mallok, S., Brugger, K., Garrett, R.A., Zillig, W. and Prangishvili, D. (2001). Sequences and replication of genomes of the archaeal rudiviruses SIRV-1 and SIRV-2: relationships to the archaeal lipothrixvirus SIFV and some eukaryal viruses. *Virology*, **291**, 226-234.

Prangishvili, D., Stedman, K. and Zillig, W. (2001). Viruses of the extremely thermophilic archaeon *Sulfolobus*. *Trends Microbiol.*, **9**, 39-43.

Rachel, R., Bettstetter, M., Hedlund, B.P., Haring, M., Kessler, A., Stetter, K.O. and Prangishvili, D. (2002). Remarkable morphological diversity of viruses and virus-like particles in hot terrestrial environments Brief Report. *Arch. Virol.*, **147**, 2419-2429.

Reiter, W.-D., Zillig, W. and Palm, P. (1988). Archaebacterial viruses. In: (K. Maramorosch, F.A. Murphy, A.J. Shatkin, eds). *Adv. Virus Res.*, **34**, 143-188.

Rice, G., Stedman, K.M., Snyder, J., Wiedenheft, B., Willits, D., Brumfield, S., McDermott, T. and Young, M.J. (2001). Novel viruses from extreme thermal environments. *Proc. Natl. Acad. Sci. USA.*, **98**, 13341-13345.

Zillig, W., Kletzin, A., Schleper, C., Holz, I., Janekovic, D., Hain, J., Lanzendoerfer, M. and Kristjansson, J.K., (1994). Screening for Sulfolobales, their plasmids and their viruses in Icelandic solfataras. *Appl. Microbiol.*, **16**, 609-628.

Zillig, W., Prangishvili, D., Schleper, C., Elferink, M., Holz, I., Albers, S., Janekovic, D. and Götz, D. (1996). Viruses, plasmids and other genetic elements of thermophilic and hyperthermophilic Archaea. *FEMS Microbiol. Rev.*, **18**, 225-236.

Zillig, W., Reiter, W.D., Palm, P., Gropp, F., Neumann, H. and Rettenberger, M. (1988). Viruses of archaebacteria. In: *The Bacteriophages*, Vol 1. (R. Calendar, ed), pp. 517-558. Plenum Press, New York.

CONTRIBUTED BY

Stedman, K., Prangishvili, D. and Zillig, W.

FAMILY RUDIVIRIDAE

TAXONOMIC STRUCTURE OF THE FAMILY

Family *Rudiviridae*
Genus *Rudivirus*

Since only one genus is currently recognized, the family description corresponds to the genus description.

GENUS RUDIVIRUS

Type Species *Sulfolobus islandicus rod-shaped virus 2*

VIRION PROPERTIES

MORPHOLOGY

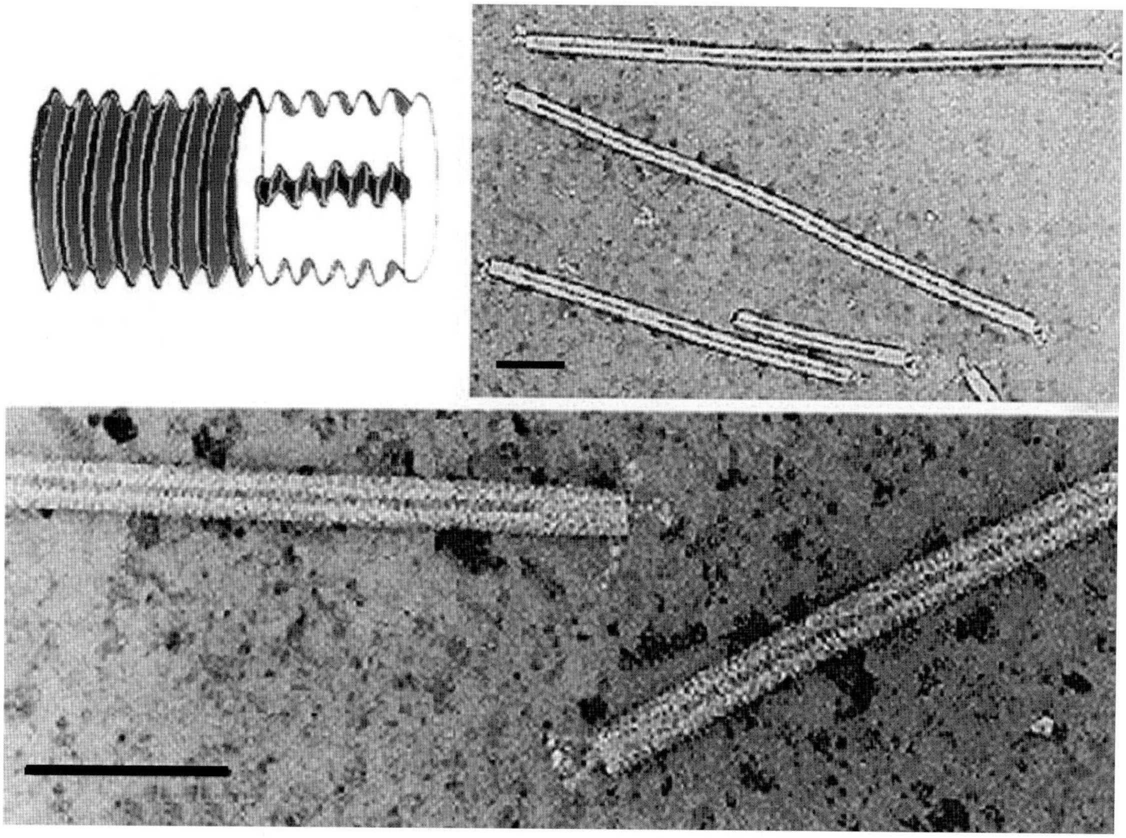

Figure 1: (Top left) Schematic model of the virion body. (Top right and Bottom) Negative contrast electron micrograph of particles of an isolate of *Sulfolobus islandicus rod-shaped virus 2*. (From Zillig *et al.*, 1998 with permission). The bars represent 100 nm.

Virions (830-900 x 23 nm) are not enveloped, have a stiff rod shape consisting of multiple copies of a single DNA binding protein and a linear dsDNA. The overall structure is helical with a repeat of 4.3 nm (Fig. 1). The tube-like superhelix formed by the nucleoprotein complex is closed at its ends by "plugs" to which tail fibers are attached.

PHYSICOCHEMICAL AND PHYSICAL PROPERTIES

The buoyant density of virus particles is 1.36 g/cm^3. Virions are completely inactivated by autoclaving at 120°C for 5 min, and are not inactivated by treatment with 6 M urea at neutral pH and 25°C, 0.1% Triton X-100, absolute ethanol, and 2-octanol. Virions are partly degraded by treatment with 0.1% SDS at 50°C.

Nucleic Acid

Genome is linear dsDNA, 32-35 kbp with covalently closed ends. There are long inverted terminal repeated sequences, 2092 bp in Sulfolobus islandicus rod-shaped virus - 1 (SIRV1) and 1628 bp in Sulfolobus islandicus rod-shaped virus - 2 (SIRV-2).

Proteins

Virions contain one major CP. A dUTPase and a Holliday junction resolvase are encoded in the virus genome and have been functionally characterized as recombinant proteins *in vitro*. The SIRV-1 and SIRV-2 genomes contain 45 and 54 ORFs respectively.

Lipids

No lipid was detected.

Carbohydrates

None reported.

Genome Organization and Replication

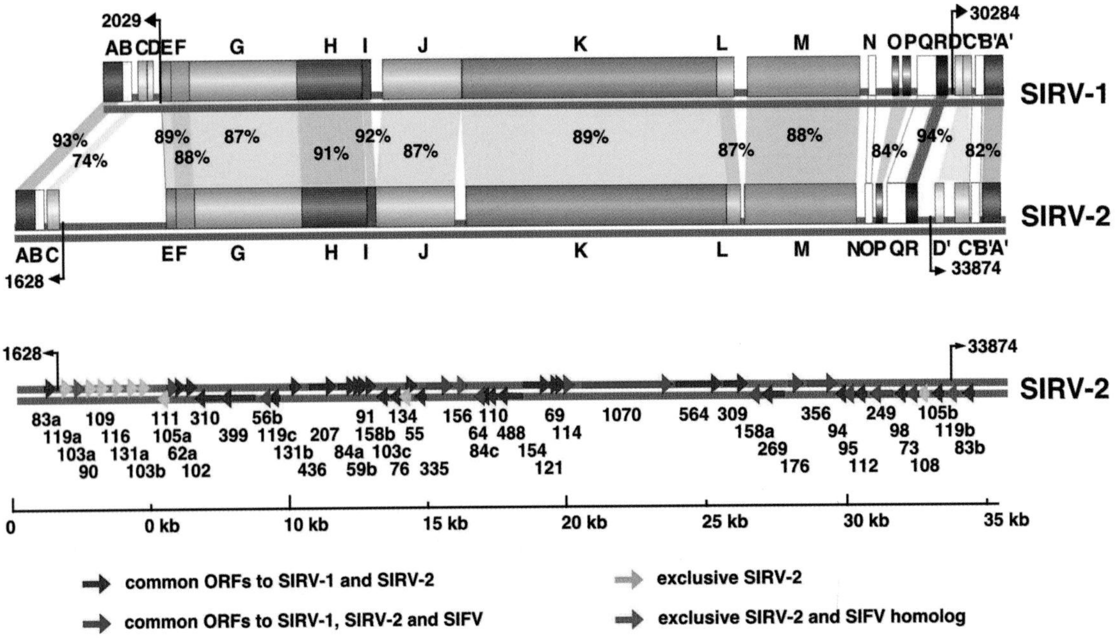

Figure 2: Comparative nucleotide sequence and predicted ORFs of the genomes of Sulfolobus islandicus rod-shaped virus – 1 (SIRV-1) and Sulfolobus islandicus rod-shaped virus – 1 (SIRV-2) genomes. (Top) Regions of similar nucleotide sequence are presented by blocks labeled A to R. The high levels of sequence similarity between blocks are indicated. Internal limits of the terminal inverted repeats (nt 2029 and 30,284 in SIRV-1, and 1602 and 33,848 in SIRV-2) are indicated. (Bottom) The ORF map of SIRV-2 is aligned with the map in (A). Locations of ORFs exclusive to SIRV-1 are indicated above the map. Red arrows represent ORFs which have homologs in SIRV-1; yellow arrows represent ORFs exclusive to SIRV-2; blue arrows represent Sulfolobus islandicus filamentous virus (SIFV) viral homologs (one of which, ORF158a, has homologs in different archaeal chromosomes); green arrows represent ORFs exclusive to SIRV-2 that are also SIFV homologs. Numbers in boldface type represent ORFs giving matches with ORFs of eukaryotic viruses. (Figure and legend from Peng *et al.*, 2001 with permission).

Antigenic Properties

No information available.

Biological Properties

Infect closely related strains of the extremely thermophilic archaeon *Sulfolobus islandicus*.

List of Species Demarcation Criteria in the Genus

The two species in the genus *Rudivirus* differ in virion size, genome size, host range, and genomic variability. Thermoproteus tenax 4 (TTV4) from *Thermoproteus tenax* was originally assigned to the family *Rudiviridae,* but it is very different in morphology and no information is available for its genome. A proposal to remove it from this family has been made and is now listed as "unassigned virus"

List of Species in the Genus

Species names are in green italic script; strain names and synonyms are in black roman script; tentative species names are in blue roman script. Sequence accession numbers, and assigned abbreviations ( ) are also listed.

Species in the Genus

Sulfolobus islandicus rod-shaped virus 1
Sulfolobus islandicus rod-shaped virus 1 (SIRV-1)
Sulfolobus islandicus rod-shaped virus 2
Sulfolobus islandicus rod-shaped virus 2 (SIRV-2)

Tentative Species in the Genus

None reported.

List of Unassigned Viruses in the Family

None reported.

Phylogenetic Relationships within the Family

SIRV-1 and SIRV-2 are very similar to each other. Their genomes contain 44 homologous ORFs and recognizable blocks of DNA sequence similarity vary from 74-96%.

Similarity with Other Taxa

Rudiviridae and *Lipothrixviridae* families have members with filamentous viruses containing linear dsDNA genomes that infect extremely thermophilic *Archaea* in the kingdom *Crenarchaeota*. However, the members of the family *Rudiviridae* are not enveloped, whereas those of the family *Lipothrixviridae* are enveloped. There is a large amount of genome conservation between members of the families *Lipothrixviridae* and *Rudiviridae.* Approximately 30% of the ORFs in the SIRV genome have homologues in the SIFV genome.

The mode of replication and topology of the genomes of the members of the family *Rudiviridae* are similar to those of poxviruses, African swine fever virus and *Chlorella* viruses.

Derivation of Names

Rudi: from the Latin '*rudis*', 'small rod'.

References

Blum, H., Zillig, W., Mallok, S., Domdey, H. and Prangishvili, D. (2001). The genome of the archaeal virus SIRV1 has features in common with genomes of eukaryal viruses. *Virology*, **281**, 6-9.

Peng, X., Blum, H., She, Q., Mallok, S., Brugger, K., Garrett, R.A., Zillig, W. and Prangishvili, D. (2001). Sequences and replication of genomes of the archaeal rudiviruses SIRV-1 and SIRV-2: relationships to the archaeal *Lipothrixvirus* SIFV and some eukaryal viruses. *Virology*, **291**, 226-234.

Prangishvili, D., Arnold, H.P., Gotz, D., Ziese, U., Holz, I., Kristjansson, J.K. and Zillig, W. (1999). A novel virus family, the *Rudiviridae*: Structure, virus-host interactions and genome variability of the sulfolobus viruses SIRV-1 and SIRV-2. *Genetics*, **152**, 1387-1396.

Prangishvili, D., Stedman, K. and Zillig, W. (2001). Viruses of the extremely thermophilic archaeon *Sulfolobus*. *Trends Microbiol.*, **9,** 39-43.

Zillig, W., Arnold, H.P., Holz, I., Prangishvili, D., Schweier, A., Stedman, K., She, Q., Phan, H., Garrett, R., Kristiansson, J.K. (1998). Genetic elements in the extremely thermophilic archaeon Sulfolobus. *Extremophiles*, **2,** 131-140.
Zillig, W., Prangishvili, D., Schleper, C., Elferink, M., Holz, I., Albers, S., Janekovic, D. and Götz, D. (1996). Viruses, plasmids and other genetic elements of thermophilic and hyperthermophilic Archaea. *FEMS Microbiol. Rev.*, **18**, 225-236.
Zillig, W., Reiter, W.D., Palm, P., Gropp, F., Neumann, H. and Rettenberger, M. (1988). Viruses of archaebacteria. In: *The bacteriophages*, Vol 1. (R. Calendar, ed), pp. 517-558. Plenum Press, New York.

CONTRIBUTED BY

Stedman, K. and Prangishvili, D.

FAMILY *FUSELLOVIRIDAE*

TAXONOMIC STRUCTURE OF THE FAMILY

Family **Fuselloviridae**
Genus **Fusellovirus**

Since only one genus is currently recognized, the family description corresponds to the genus description.

GENUS *FUSELLOVIRUS*

Type Species *Sulfolobus spindle-shaped virus 1*

VIRION PROPERTIES

MORPHOLOGY

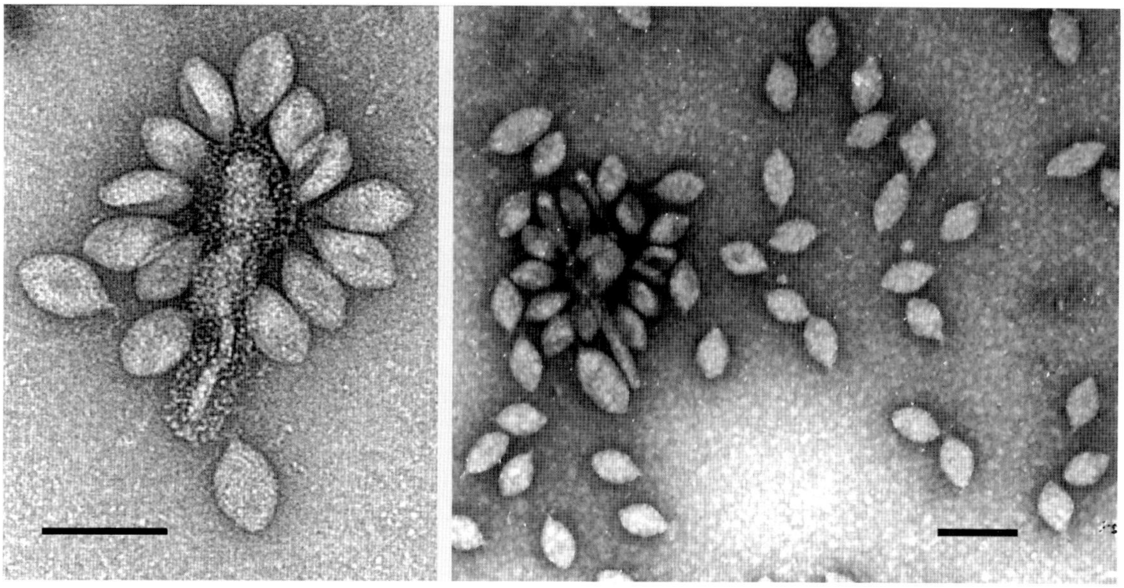

Figure 1: Negative contrast electron micrograph of Sulfolobus spindle-shaped virus 1 (SSV-1) virions, bound to a membrane vesicle (Left) or isolated (Right) (Stedman *et al.*, 1999). The bars represent 100 nm.

Virions are lemon-shaped with short tail fibers attached to one pole, and slightly heterogeneous in size. Virions are 55-60 nm in their short dimension and 80-100 nm in their long dimension. A small fraction of the Sulfolobus spindle-shaped virus 1 (SSV-1) population (up to 1%) is larger, with a particle length of ~300 nm. Some other fuselloviruses, particularly SSV-K1, have more elongated virions. Purified SSV-1 virions contain host lipids and three virus encoded proteins, two of which are associated with the coat and the other is DNA-associated.

PHYSICOCHEMICAL AND PHYSICAL PROPERTIES

Virion buoyant density in CsCl is 1.24 g/cm^3. The particles are stable at up to 97°C and are insensitive to urea, ether and pH 2. However, low pH (below 5) reduces viability due to degradation of the DNA, and virions are sensitive to pH above 11 and trichloromethane.

NUCLEIC ACID

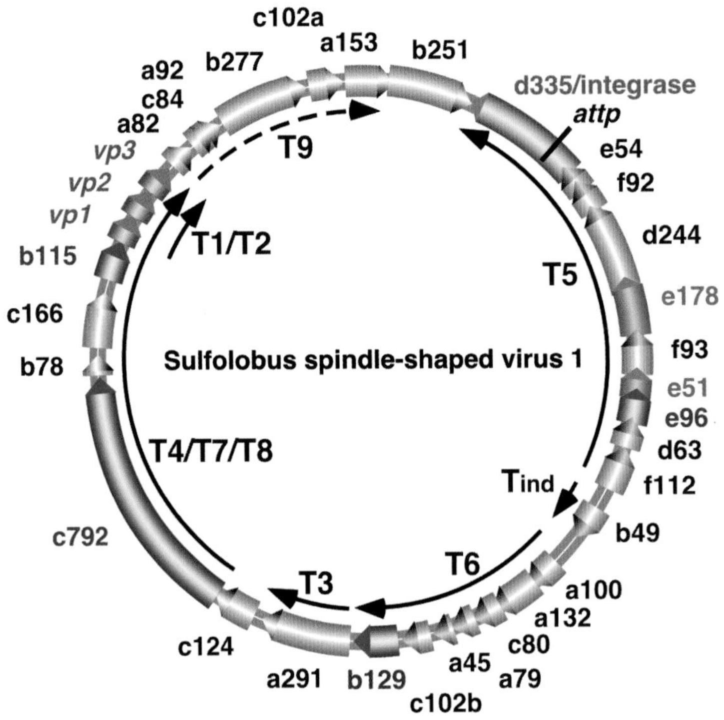

Figure 2: Genome organization of Sulfolobus spindle-shaped virus 1 (SSV-1). ORFs are shown as arrows and labeled. The major virion protein genes (VP1, VP3 and VP2) are labeled in red. The viral integrase gene, ORF d335, is labeled in green. ORFs shown to be essential for virus function are labelled in blue. ORFs shown to be non-essential for virus function are labelled in pink. The viral attachment site is labeled as *attP*. Transcripts are shown as solid arrows and labeled. Transcripts T1 and T2 start at the same promoter and overlap, and transcripts T4, T7, and T8 similarly overlap. The termination site of transcript T9 is not known.

Virions contain circular dsDNA, from 14.8-17.8 kbp. In SSV1 virions, DNA is positively supercoiled and associated with polyamines and a virus-coded basic protein. The complete DNA sequence for four fuselloviruses has been determined.

PROTEINS

The main constituents of the SSV-1 viral envelope are two basic proteins (VP1 and VP3). In SSV-1, these proteins are 73 and 92 aa, respectively, as deduced from the DNA sequence and from N-terminal protein sequencing. The VP1 protein is post-translationally proteolytically processed. In SSV-1, a very basic protein (VP2, 74 aa) is attached to the viral DNA. However, this gene is lacking in all other sequenced fusellovirus genomes. The genes encoding these structural proteins are closely linked in the fusellovirus genomes, in the order VP1, VP3, and VP2 in SSV-1 (Fig. 2). The second largest ORF of SSV-1 (ORF d335, 335 aa) is similar to the integrase family of site-specific tyrosine recombinases. This protein has been expressed in *E. coli* and recombines DNA fragments sequence-specifically *in vitro*. The gene is conserved in all sequenced fuselloviruses. Four ORFs in the virus genome have been shown to be essential for virus function and two have been shown to be non-essential for virus function in SSV-1 (see Fig. 2).

LIPIDS

It has been reported that 10% of the SSV-1 virion envelope consists of host lipids.

CARBOHYDRATES

None reported

GENOME ORGANIZATION AND REPLICATION

The virus genome is present in cells as circular dsDNA and is site-specifically integrated into a tRNA gene of the host chromosome. In SSV-1 the integrated copy is flanked by a 44 bp direct repeat (attachment core) that occurs once in the SSV-1 circular DNA genome. Upon integration, the viral integrase gene is disrupted. Eleven somewhat overlapping transcripts originating from 7 promoters have been detected and mapped. They almost completely cover the SSV-1 genome (Fig. 2). UV-irradiation stimulates virus production and progeny virions are released without host cell lysis. A small transcript (T_{ind}) is strongly induced upon ultraviolet induction. Particles appear to be assembled and are produced by extrusion at the cell membrane.

ANTIGENIC PROPERTIES

Not known.

BIOLOGICAL PROPERTIES

The host range of the fusselloviruses is limited to extremely thermophilic Archaea: *Sulfolobus shibatae, Sulfolobus solfataricus* strains P1 and P2, and *Sulfolobus islandicus* strains. Very few phage particles are produced in cultures of SSV-1 lysogens. UV-irradiation strongly induces SSV-1 production without evident lysis of the host. Fuselloviruses have been found in ~8% of *Sulfolobus* isolates from Icelandic solfataric fields. They have also been found in solfataric hot springs in Yellowstone National Park in the USA and on the Kamchatka peninsula in the Russian Federation. An infectious shuttle vector that also replicates in *E. coli* has been constructed.

LIST OF SPECIES DEMARCATION CRITERIA IN THE GENUS

In the genus *Fusellovirus*, the differentiation among species is essentially based on:

- The size of the genome,
- Genomic restriction fragment length polymorphism,
- The geographic location of virus isolation, and
- Virus genome sequence.

LIST OF SPECIES IN THE GENUS

Species names are in green italic script; strain names and synonyms are in black roman script; tentative species names are in blue roman script. Sequence accession numbers, and assigned abbreviations ( ) are also listed.

SPECIES IN THE GENUS

Sulfolobus spindle-shaped virus 1		
Sulfolobus spindle-shaped virus 1	[XO7234]	(SSV-1)
Sulfolobus shibatae virus 1		
Sulfolobus virus 1		

TENTATIVE SPECIES IN THE GENUS

Sulfolobus spindle-shaped virus 2	(SSV-2)
Sulfolobus spindle-shaped virus 3	(SSV-3)
Sulfolobus spindle-shaped virus – Yellowstone 1	(SSV-Y1)
Sulfolobus spindle-shaped virus – Kamchatka 1	(SSV-K1)

LIST OF UNASSIGNED VIRUSES IN THE FAMILY

None reported

PHYLOGENETIC RELATIONSHIPS WITHIN THE FAMILY

Phylogenetic relationships within the family *Fuselloviridae* are unclear. Different homologous genes give different phylogenetic trees.

Similarity with Other Taxa

One ORF in fusellovirus genomes is similar to a gene present in the Sulfolobus viruses of the families *Lipothrixviridae* and *Rudiviridae*. The virus of extreme halophiles, His1, originally suggested to be a fusellovirus, on further characterization, has been found to be very dissimilar and a novel genus (*Salterprovirus*) has been established for this group of viruses.

Derivation of Names

Fusello: from the Latin *fusello*, 'little spindle'.

References

Palm, P., Schleper, C., Grampp, B., Yeats, S., McWilliam, P., Reiter, W.-D. and Zillig, W. (1991). Complete nucleotide sequence of the virus SSV1 of the archaebacterium *Sulfolobus shibatae*. *Virology*, **185**, 242-250.

Prangishvili, D., Stedman, K. and Zillig, W. (2001). Viruses of the extremely thermophilic archaeon *Sulfolobus*. *Trends Microbiol.*, **9**, 39-43.

Rice, G., Stedman, K.M., Snyder, J., Wiedenheft, B., Willits, D., Brumfield, S., McDermott, T. and Young, M.J. (2001). Novel viruses from extreme thermal environments. *Proc. Natl. Acad. Sci. U.S.A.*, **98,** 13341-13345.

Stedman, K.M., Schleper, C., Rumpf, E. and Zillig, W. (1999). Genetic requirements for the function of the archaeal virus SSV1 in Sulfolobus solfataricus: Construction and testing of viral shuttle vectors. *Genetics*, **152**, 1397-1406.

Stedman, K.M., She, Q., Phan, H., Arnold, H.P., Holz, I., Garrett, R.A. and Zillig, W. (2003). Relationships between fuselloviruses infecting the extremely thermophilic archaeon *Sulfolobus*: SSV-1 and SSV-2. *Res. Microbiol.*, **154**, 295-302.

Zillig, W., Prangishvili, D., Schleper, C., Elferink, M., Holz, I., Albers, S., Janekovic, D. and Götz, D. (1996). Viruses, plasmids and other genetic elements of thermophilic and hyperthermophilic Archaea. *FEMS Microbiol. Rev.*, **18**, 225-236.

Zillig, W., Reiter, W.D., Palm, P., Gropp, F., Neumann, H. and Rettenberger, M. (1988). Viruses of archaebacteria. In: *The bacteriophages*, Vol 1. (R. Calendar, ed), pp. 517-558. Plenum Press, New York.

Contributed By

Stedman, K.

GENUS SALTERPROVIRUS

Type Species *His1 virus*

VIRION PROPERTIES

MORPHOLOGY

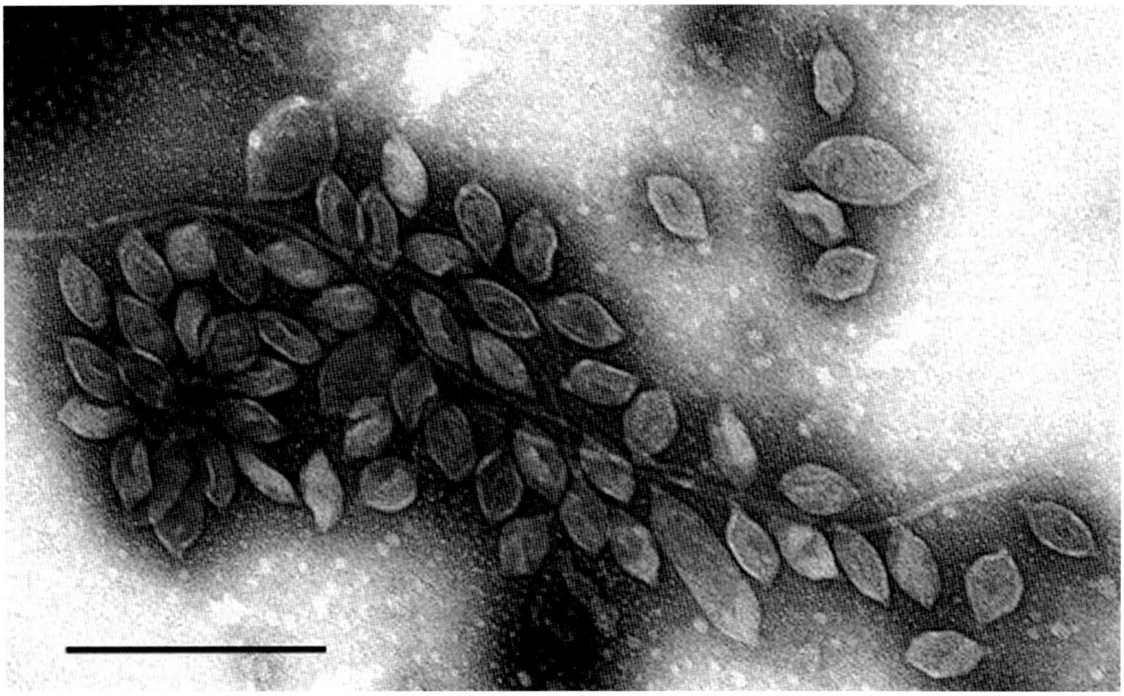

Figure 1: Negative contrast electron microscopic picture of particles of an isolate of *His1 virus*. The flagella from the host is also visible. The bar represents 200 nm.

Virus particles are lemon-shaped with a short tail at one end. The particles are 44 x 74 nm, with a 7 nm long tail. The capsid appears flexible, as the virions can vary in shape when viewed by negative stain electron microscopy. A small proportion of particles can vary in size, with some particles being much larger or more elongated (up to ~204 nm).

PHYSICOCHEMICAL AND PHYSICAL PROPERTIES

The buoyant density of His1 in CsCl is 1.28g/cm^3. The virus is sensitive to chloroform, ethanol, and detergents such as Triton X-100, and resistant to trichlorotrifluoroethane. For virus titer to be maintained, the virus needs to be stored in a high salt solution (18% salt water), as prolonged exposure to distilled water or low salt will result in destruction of the virus particles. Virus particles are stable for long periods at 37°C.

NUCLEIC ACID

The virus has a small linear dsDNA genome of 14.5kb, with imperfect inverted 105 bp terminal repeats, and terminal proteins attached at each end of the genome. The genome is 39% G+C, which is quite different from that of the host 62.7% G+C.

PROTEINS

The largest ORF of His1 virus (717 aa) appears to encode a putative protein-primed DNA-dependent DNA polymerase, based on sequence similarity with other family B and protein-primed DNA polymerases in databases. The CP(s) or the terminal protein have, as yet, not been determined.

LIPIDS

Not known.

CARBOHYDRATES

Not known.

GENOME ORGANIZATION AND REPLICATION

Given the presence of terminal proteins attached to a linear dsDNA genome, inverted terminal repeats, and a putative protein-primed DNA polymerase, it is likely that the replication of the genome occurs by the viral-encoded DNA polymerase, which would use the terminal proteins as primers for DNA replication.

As the genome does not appear to encode its own RNA polymerase, transcription must be carried out by host RNA polymerase. Virions would accumulate in the cytoplasm of the host before release. Cell lysis is seen, but viral particles may bud from the cell, like Sulfolobus spindle-shaped virus 1 (SSV-1) or the filamentous viruses.

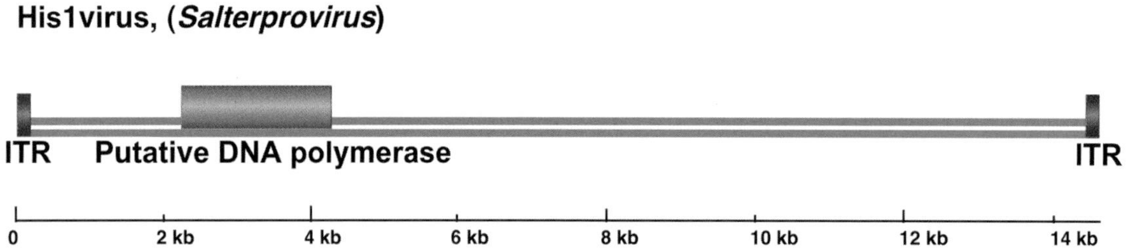

Figure 2: Genome organization of His1 virus (His1V). The inverted terminal repeats (ITR) and the putative DNA polymerase gene are shown.

ANTIGENIC PROPERTIES

Not known.

BIOLOGICAL PROPERTIES

His1 was isolated from *Haloarcula hispanica*, a halophilic archaeon, after which the virus is named. Infection is lytic, but the virus is also capable of existing in a carrier state. His1 virus was isolated from an Australian saltern pond sample, and related viruses may be possibly found in other halophilic environments.

LIST OF SPECIES DEMARCATION CRITERIA IN THE GENUS

Members of species of the *Salterprovirus* genus

- Share a lemon-shaped morphology
- Have small linear dsDNA genomes
- Have both terminal proteins and inverted terminal repeats, which therefore suggests the same replication mechanism
- Are isolated from halophilic environments, or require a high salt environment for survival.

LIST OF SPECIES IN THE GENUS

Species names are in green italic script; strain names and synonyms are in black roman script; tentative species names are in blue roman script. Sequence accession numbers, and assigned abbreviations ( ) are also listed.

FAMILY GUTTAVIRIDAE

TAXONOMIC STRUCTURE OF THE FAMILY

Family *Guttaviridae*
Genus *Guttavirus*

Since only one genus is currently recognized, the family description corresponds to the genus description

GENUS GUTTAVIRUS

Type Species *Sulfolobus newzealandicus droplet-shaped virus*

VIRION PROPERTIES

MORPHOLOGY

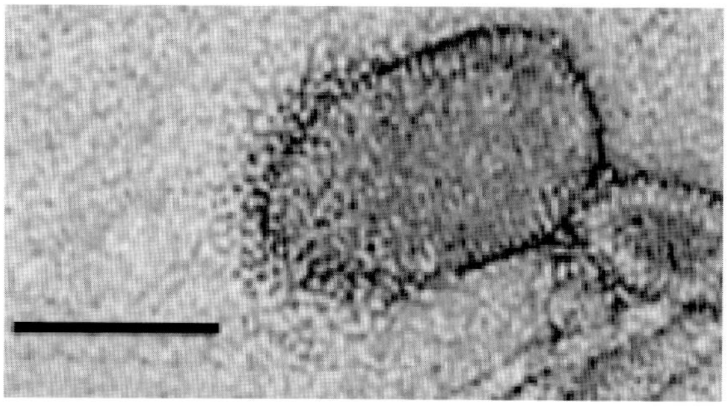

Figure 1: Negative contrast electron microscopy of particles of Sulfolobus newzealandicus droplet-shaped virus (SNDV) (Courtesy of W. Zillig). The bar represents 100 nm.

Sulfolobus newzealandicus droplet-shaped virus (SNDV) virions are somewhat pleiomorphic with a droplet shape, 75-90 x 110-185 nm with a beard of dense filaments at the pointed end.

PHYSICOCHEMICAL AND PHYSICAL PROPERTIES

Virions are unstable in CsCl and lyse.

NUCLEIC ACID

The genome is covalently-closed circular dsDNA ~20 kbp. The genome cannot be cut by many restriction endonucelases, but can be cut by the *dam*-methylation dependent restriction endonuclease *Dpn*I, indicating that it is extensively methylated by a *dam*-like methylase.

PROTEINS

By SDS-PAGE analysis, there is one major virion protein, 17.5 kDa, and two minor virion proteins, 13.5 and 13 kDa.

LIPIDS

Not known.

CARBOHYDRATES

Not known.

GENOME ORGANIZATION AND REPLICATION

Not known.

Antigenic Properties

Not known.

Biological Properties

SNDV exclusively infects *Sulfolobus* isolates from New Zealand, including the strain STH1/3. Virus production starts in the early stationary phase.

List of Species Demarcation Criteria in the Genus

Not applicable.

List of Species in the Genus

Species names are in green italic script; strain names and synonyms are in black roman script; tentative species names are in blue roman script. Sequence accession numbers, and assigned abbreviations ( ) are also listed.

Species in the Genus

Sulfolobus newzealandicus droplet-shaped virus
Sulfolobus newzealandicus droplet-shaped virus (SNDV)

Tentative Species in the Genus

None reported.

List of Unassigned Viruses in the Family

None reported.

Phylogenetic Relationships within the Family

Not applicable.

Similarity with Other Taxa

Not known.

Derivation of Names

Gutta from Latin "*gutta*" droplet.

References

Arnold, H.P., Ziese, U. and Zillig, W. (2000). SNDV, a novel virus of the extremely thermophilic and acidophilic archaeon *Sulfolobus*. *Virology*, **272,** 409-416
Prangishvili, D., Stedman, K. and Zillig, W. (2001). Viruses of the extremely thermophilic archaeon *Sulfolobus*. *Trends Microbiol.*, **9,** 39-43.
Zillig, W., Arnold, H.P., Holz, I., Prangishvili, D., Schweier, A., Stedman, K., She, Q., Phan, H., Garrett, R. and Kristjansson, J.K. (1998). Genetic elements in the extremely thermophilic archaeon *Sulfolobus*. *Extremophiles*, **2,** 131-140.

Contributed By

Stedman, K.

dsDNA

Family *Poxviridae*

Taxonomic Structure of the Family

Family	*Poxviridae*		
Subfamily		*Chordopoxvirinae*	
Genus			*Orthopoxvirus*
Genus			*Parapoxvirus*
Genus			*Avipoxvirus*
Genus			*Capripoxvirus*
Genus			*Leporipoxvirus*
Genus			*Suipoxvirus*
Genus			*Molluscipoxvirus*
Genus			*Yatapoxvirus*
Subfamily		*Entomopoxvirinae*	
Genus			*Alphaentomopoxvirus*
Genus			*Betaentomopoxvirus*
Genus			*Gammaentomopoxvirus*

Virion Properties

Morphology

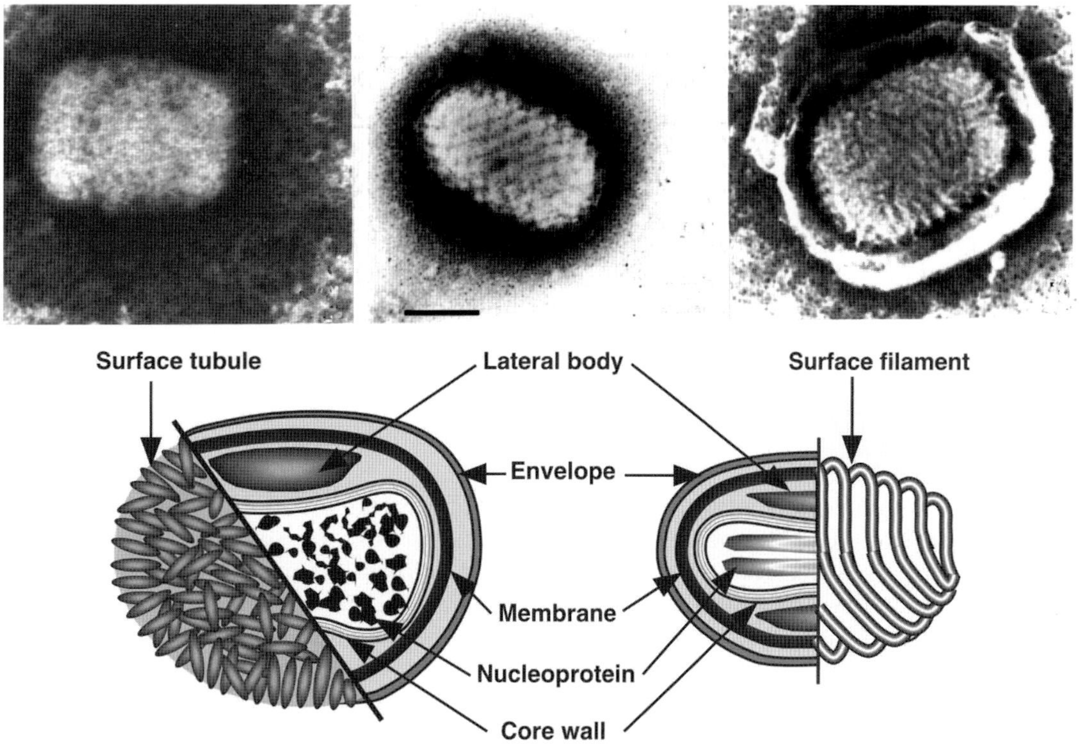

Figure 1: Negatively stained preparations of: (Upper left) an orthopoxvirus virion; (Upper center) a parapoxvirus virion and (upper right) a yatapoxvirus virion. Virions of orthopoxviruses and yatapoxviruses are morphologically similar. Yatapoxviruses are always enveloped, whereas envelopment of orthopoxviruses is relatively rare. Parapoxvirus virions are smaller and are characterized by regular surface structures. The schematic structure of the orthopoxvirus virions (Bottom left) reveals a condensed nucleoprotein organization of DNA. The core assumes a dumbell shape invaginated by the large lateral bodies, which are, in turn, enclosed within a protein shell about 12 nm thick – the outer membrane (or envelope), the surface of which appears to consist of irregularly arranged surface tubules, which in turn consist of small globular subunits.

Some evidence suggests that the "dumbell" shape of cores may be an artifact of sample preparation. Virions released from the cell by exocytosis are enclosed within an envelope, which contains host cell lipids and several additional virus-specific polypeptides not present within intracellular virus. The schematic structure of a parapoxvirus virus (Orf virus, ORFV) (bottom right) reveals an outer membrane consisting of a single, long tubule that appears to be wound around the particle. The bar represents 100 nm. (Redrawn from Esposito and Fenner (2001) with permission).

Virions are somewhat pleomorphic, generally either brick-shaped (220-450 nm long x 140-260 nm wide x 140-260 nm thick) with a lipoprotein surface membrane displaying tubular or globular units (10-40 nm). They can also be ovoid (250-300 nm long x 160-190 nm diameter) with a surface membrane possessing a regular spiral filament (10-20 nm in diameter)(Fig. 1).

Negative contrast images show that the surface membrane encloses a biconcave or cylindrical core that contains the genome DNA and proteins organized in a nucleoprotein complex. One or two lateral bodies appear to be present in the concave region between the core wall and a membrane. This virion form is known as intracellular mature virus (IMV). Some IMV is wrapped by an additional double layer of intracellular membrane to form intracellular enveloped virus (IEV). The IEV can be externalized and bound to the cell surface to form cell-associated enveloped virus (CEV) or released from the cell surface as extracellular enveloped virus (EEV). Some vertebrate viruses (e.g. isolates of *Cowpox virus* and *Ectromelia virus*) may also be sequestered within growing inclusion bodies. Others (e.g. entomopoxviruses) may be occluded into a preformed inclusion body.

Physicochemical and Physical Properties

Particle weight is about 5 x 10^{-15} g. S_{20w} is about 5000S. Buoyant density of virions is subject to osmotic influences: in dilute buffers it is about 1.16 g/cm^3, in sucrose about 1.25 g/cm^3, in CsCl and potassium tartrate about 1.30 g/cm^3. Virions tend to aggregate in high salt solution. Infectivity of some members is resistant to trypsin. Some members are insensitive to ether. Generally, virion infectivity is sensitive to common detergents, formaldehyde, oxidizing agents, and temperatures greater than 40°C. The virion surface membrane is removed by nonionic detergents and sulfhydryl reducing reagents. Virions are relatively stable in dry conditions at room temperature; they can be lyophilized with little loss of infectivity.

Nucleic Acid

Nucleic acids constitute about 3% of the particle weight. The genome is a single, linear molecule of covalently-closed, dsDNA, 130-375 kbp in length.

Proteins

Proteins constitute about 90% of the particle weight. Genomes encode 150-300 proteins depending on the species; about 100 proteins are present in virions. Virus particles contain many enzymes involved in DNA transcription or modification of proteins or nucleic acids. Enveloped virions have viral encoded polypeptides in the lipid bilayer, which surrounds the particle. Entomopoxviruses may be occluded by a virus-encoded, major structural protein, spheroidin. Similarly, orthopoxviruses may be within inclusion bodies again consisting of a single protein (the A-type inclusion [ATI] protein).

Lipids

Lipids constitute about 4% of the particle weight. Enveloped virions contain lipids, including glycolipids, that may be modified cellular lipids.

Carbohydrates

Carbohydrates constitute about 3% of the particle weight. Certain viral proteins, e.g. hemagglutinin in the envelope of orthopoxviruses, have N- and C-linked glycans.

Genome Organization and Replication

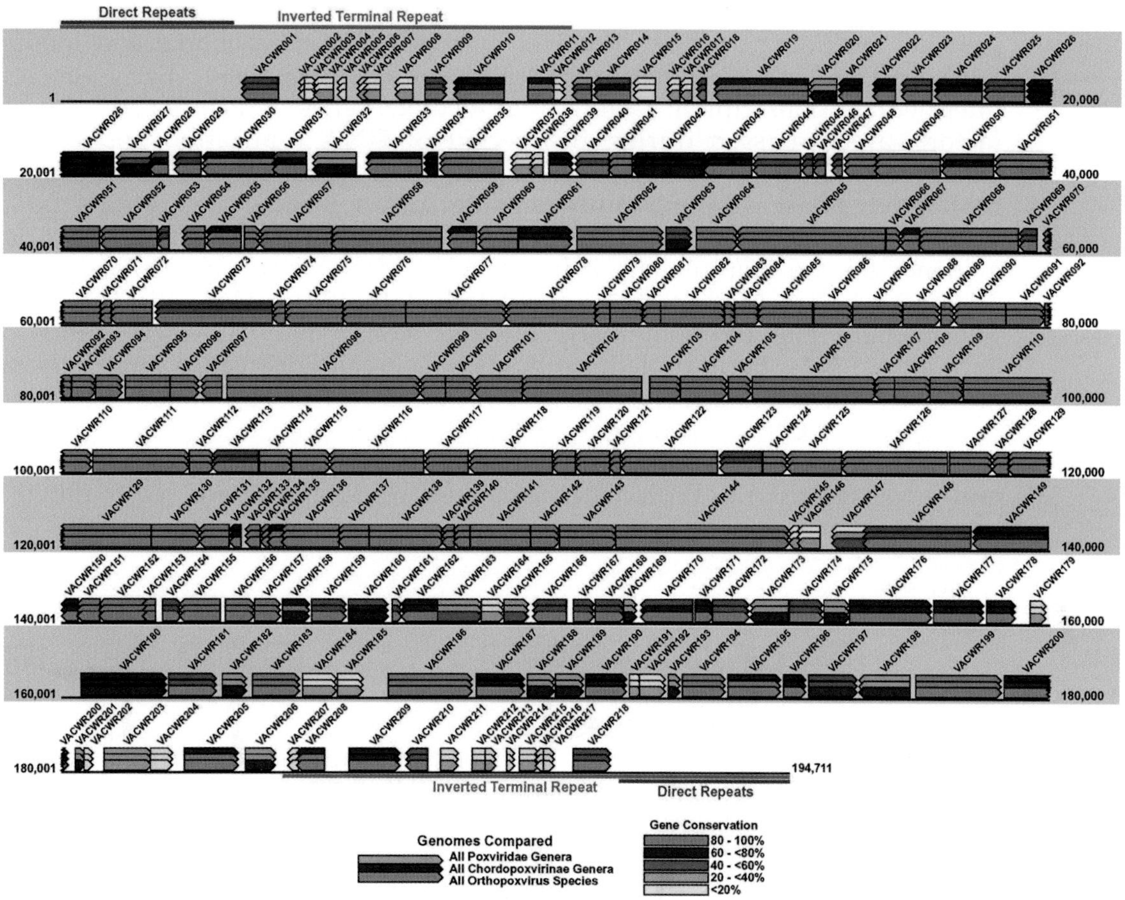

Figure 2: Schematic representation of the genome of the WR strain of *Vaccinia virus* (AY243312): The genome is a linear double-stranded molecule with terminal hairpins, inverted terminal repeats (ITR), and a series of direct repeats within the ITRs. Each overlapping bar indicates gene conservation between the WR strain and all poxviruses, vertebrate poxviruses, and orthopoviruses. The bars are color coded according to the percentage of gene conservation across the indicated taxa.

The poxvirus genome comprises a linear molecule of dsDNA with covalently closed termini; terminal hairpins constitute two isomeric, imperfectly paired, "flip-flop" DNA forms consisting of inverted complementary sequences. Variably sized, tandem repeat sequence arrays may or may not be present near the ends (Fig. 2). Replication takes place predominately if not exclusively within the cytoplasm (Fig. 3). Entry into cells of intracellular virus (IMV) and extracellular enveloped virus (EEV) is suggested to be *via* different pathways. After virion adsorption, IMV entry into the host cell is probably by fusion with the plasma membrane after which cores are released into the cytoplasm and uncoated further. EEV entry, unlike IMV, may necessitate fusion with endosomal membranes to release the core.

Polyadenylated, capped primary mRNA transcripts, representing about 50% of the genome, are initially synthesized from both DNA strands by enzymes within the core, including a virus encoded multisubunit RNA polymerase; transcripts are extruded from the core for translation by host ribosomes. During synthesis of early proteins, host macromolecular synthesis is inhibited. Virus reproduction ensues in the host cell cytoplasm, producing basophilic (B-type) inclusions termed "viroplasms" or "virus factories". The genome contains closely spaced ORFs, lacking introns, some of which may partially overlap preceded by virus-specific promoters that temporally regulate transcription of three classes of genes. One class, the early genes, are expressed from partially uncoated virions prior to DNA replication (these encode many non-structural proteins, including enzymes involved in replicating the genome and modifying DNA, RNA, and proteins designed to neutralize the host response). Early genes also encode intermediate transcription factors. Intermediate genes, which encode late transcription factors, are expressed during the period of DNA replication and are required for subsequent late gene transcription. Finally, late genes are expressed during the post-replicative phase (these mainly encode virion structural proteins but also early transcription factors). Despite a cytoplasmic site of replication, there is mounting evidence for the requirement of host nuclear proteins in post-replicative transcription. The mRNAs are capped, polyadenylated at the 3'-termini, but not spliced. Many intermediate, late and some early mRNAs have 5'-poly(A) tracts, which precede the encoded mRNA. Early protein synthesis is generally decreased during the transition to late gene expression, but some genes are expressed from promoters with both early and late activity. Certain proteins are modified post-translationally (e.g. by proteolytic cleavage, phosphorylation, glycosylation, ribosylation, sulfation, acylation, palmitylation and myristylation). Proteolytic cleavage of late proteins is required for virion morphogenesis.

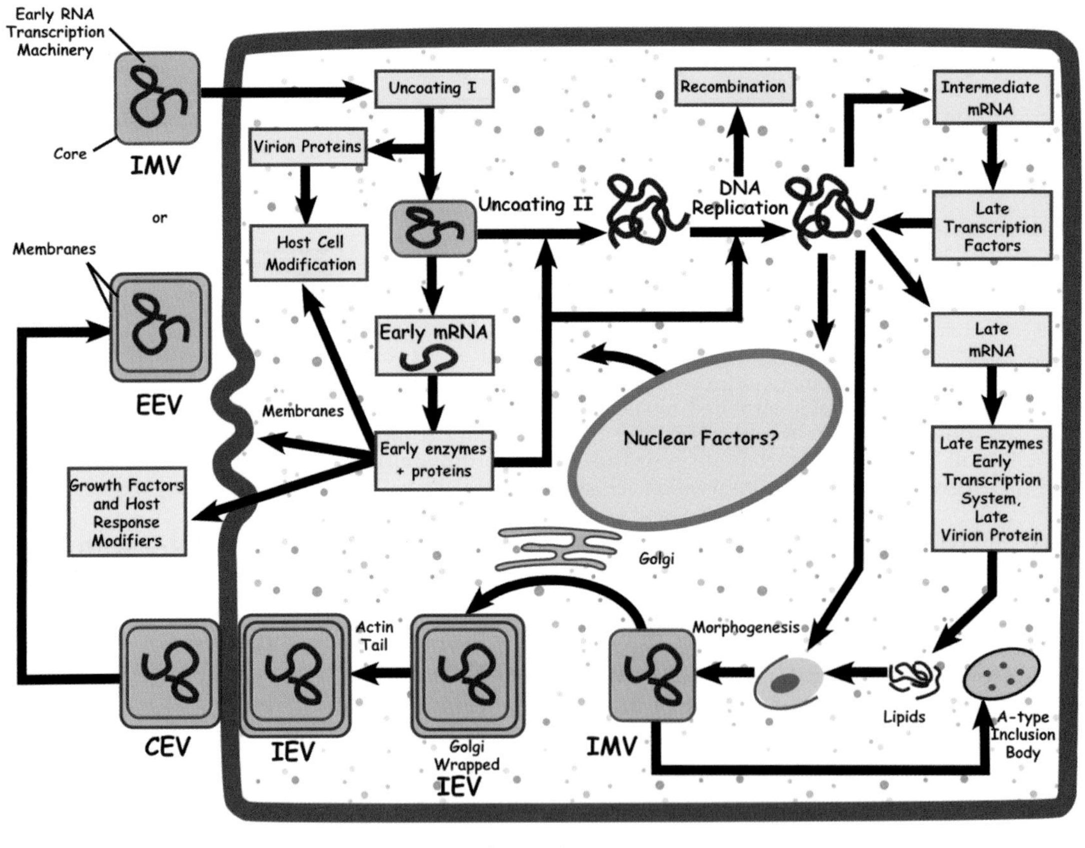

Figure 3: The infectious cycle of Vaccinia virus (VACV): IMV, intracellular mature virus; EEV, extracellular enveloped virus. See text for details.

The replication of the DNA genome appears to be mainly through the action of viral enzymes. DNA replication appears to involve a self-priming, unidirectional, strand displacement mechanism in which concatemeric replicative intermediates are generated and subsequently resolved via specific cleavages into unit length DNAs that are ultimately covalently closed. Genetic recombination within genera has been shown, and may occur between daughter molecules during replication. Non-genetic genome reactivation generating infectious virus has been shown within, and between, genera of the *Chordopoxvirinae*.

Virus morphogenesis begins following DNA replication and expression of early, intermediate and late genes. Particle assembly is initiated with the formation of crescent-shaped membrane structures in the intermediate compartment between the endoplasmic reticulum and the *trans*-Golgi network. Replicated, concatameric DNA is resolved into unit genomes and packaged, forming virion particles that mature into fully infectious IMVs. Some IMVs acquire an additional double layer of intracellular membrane derived from the early endosomes or the *trans*-Golgi network that contain unique virus proteins (IEV). These IEVs are transported to the periphery of the cell where fusion with the plasma membrane ultimately results in release of EEV or, if attached to the exterior surface of the plasma membrane, remain as CEV. While both IMVs and CEVs/EEVs are infectious, the external antigens on the two virus forms are different, and upon infection the two virion types probably bind to different cellular receptors and are likely uncoated by different mechanisms. Virus DNA and several proteins are organized as a nucleoprotein complex within the core of all infectious virions. The IMVs contain an encompassing surface membrane, lateral bodies, and the nucleoprotein core complex (see Fig. 1). For Vaccinia virus, the core wall has a regular subunit structure. Within the vaccinia virion, negative stain indicates that the core assumes a biconcave shape (Fig. 1) apparently due to the large lateral bodies. During natural infections, the virus is likely spread within an animal by extracellular virions that adhere to the cell surface (CEV) or are released from the cells (EEV) or through the movement of infected cells. Although the internal structure of vaccinia virions is revealed in thin sections, the detailed internal structure of parapoxvirus particles is less evident (Fig. 1). In negatively stained preparations of parapoxviruses, superimposition of dorsal and ventral views of the surface filament sometimes produces a distinctive "criss-cross" surface appearance.

Antigenic Properties

Within each genus of the subfamily *Chordopoxvirinae* there is considerable serologic cross-protection and cross-reactivity. Neutralizing antibodies are genus-specific. The nucleoprotein antigen, obtained by treatment of virus suspensions with 0.04 M NaOH and 56°C treatment of virus suspensions, is highly cross-reactive among members. Orthopoxviruses have hemagglutinin antigens, although this is rare in other genera.

Biological Properties

Transmission of various member viruses of the subfamily *Chordopoxvirinae* occurs by (1) aerosol, (2) direct contact, (3) arthropods (via mechanical means), or (4) indirect contact via fomites; transmission of member viruses of the subfamily *Entomopoxvirinae* occurs between arthropods by mechanical means. Host range may be broad in laboratory animals and in tissue culture; however, in nature it is generally narrow. Many poxviruses of vertebrates produce dermal maculopapular, vesicular rashes after systemic or localized infections. Poxviruses infecting humans are zoonotic except for Molluscum contagiosum virus (MOCV) and the orthopoxvirus Variola virus (VARV) (the etiologic agent of smallpox, now eradicated). Members may or may not be occluded within proteinaceous inclusions (*Chordopoxvirinae*: acidophilic (A-type) inclusion bodies, or *Entomopoxvirinae*: occlusions or spheroids). Occlusions may protect such poxviruses in environments where transmission possibilities are limited. Neutralizing antibodies and cell-mediated immunity play a major role in clearance of vertebrate poxvirus infections. Reinfection rates are generally low and usually less severe. *Molluscum contagiosum* infections may recur, especially by autoinoculation of other areas of the skin with virus derived from the original lesions (e.g., by scratching).

SUBFAMILY *CHORDOPOXVIRINAE*

TAXONOMIC STRUCTURE OF THE SUBFAMILY

Subfamily	*Chordopoxvirinae*
Genus	*Orthopoxvirus*
Genus	*Parapoxvirus*
Genus	*Avipoxvirus*
Genus	*Capripoxvirus*
Genus	*Leporipoxvirus*
Genus	*Suipoxvirus*
Genus	*Molluscipoxvirus*
Genus	*Yatapoxvirus*

DISTINGUISHING FEATURES

Includes brick-shaped or ovoid poxviruses of vertebrates with a low G+C content (30-40%), except for the parapoxviruses (64%) and MOCV (63%). Extensive serologic cross-reaction and cross-protection is observed within genera, this is less obvious among the avipoxviruses. A common, conserved co-linear signature core of genes within genera (and in the case of mammalian viruses, spanning genera) is generally maintained with most divergence amongst members occurring at the terminal extremes of the genome. The co-linear signature of core genes appears different for mammalian, avian and insect poxvirus genera. Some viruses produce pocks on the chorioallantoic membranes of embryonated chicken eggs.

GENUS *ORTHOPOXVIRUS*

Type Species *Vaccinia virus*

DISTINGUISHING FEATURES

Virions are brick-shaped, about 200 x 200 x 250 nm. Infectivity is ether-resistant. Extensive serologic cross-reactivity exists between the viruses. Virus-infected cells synthesize a hemagglutinin (HA) glycoprotein that contributes to the modification of cell membranes and enables hemadsorption and hemagglutination of certain avian erythrocytes and alteration of the envelope of extracellular enveloped viruses. Neutralization sites on enveloped viruses are distinct from those on IMVs. The host range is broad in laboratory animals and in tissue culture; in nature it may be relatively narrow. Most infections are generalized and disseminated. The genomic DNA is 170-250 kbp, and the G+C content is about 36%. The DNAs cross-hybridize extensively between members of the genus and sometimes with DNA of members of other genera. By comparison to the American species, DNA restriction maps suggest independent evolution of the Eurasian-African species.

LIST OF SPECIES DEMARCATION CRITERIA IN THE GENUS

The criteria are provisional and reflect the fact that species definitions can be rather arbitrary and reflective of attempts to define natural transmission lineages. Most orthopoxviruses contain an hemagglutinin (HA) and many contain a A-type inclusion protein, polymorphisms within these genes distinguishes species. Species can be classified by pock morphologies and by ceiling temperature for growth on the chorioallantoic membrane of embryonated chicken eggs. Ecological niche and host range are useful in some cases, but in others (Rabbitpox virus and Buffalopox virus) can be misleading. Restriction enzyme polymorphisms (RFLPs) of the terminal regions of viral

DNA outside of the core of common genes also aids the classification process. A more detailed polymerase chain reaction (PCR) polymorphism analysis throughout the entire genome and genomic DNA sequencing studies show all orthopoxviruses to be unique.

List of Species in the Genus

Species names are in green italic script; strain names and synonyms are in black roman script; tentative species names are in blue roman script. Sequence accession numbers, and assigned abbreviations ( ) are also listed.

Species in the Genus

Camelpox virus		
Camelpox virus CMS	[AY009089]	(CMLV-CMS)
Camelpox virus M-96	[AF438165]	(CMLV-M-96)
Cowpox virus		
Cowpox virus Brighton Red	[AF482758]	(CPXV-BR)
Cowpox virus GRI-90	[X94355, Y15035]	(CPXV-GRI)
Ectromelia virus		
Ectromelia virus Moscow	[AF012825]	(ECTV-MOS)
Monkeypox virus		
Monkeypox virus Zaire-96-I-16	[AF380138]	(MPXV-ZAI)
Raccoonpox virus		
Raccoonpox virus	[U08228, M94169]	(RCNV)
Taterapox virus		
Taterapox virus	[U32629]	(GBLV)
Vaccinia virus		
Buffalopox virus	[U87233]	(BPXV)
Cantagalo virus	[AF229247]	(CTGV)
Rabbitpox virus Utrecht	[M60387]	(RPXV-UTR)
Vaccinia virus Ankara	[U94848]	(VACV-ANK)
Vaccinia virus Copenhagen	[M35027]	(VACV-COP)
Vaccinia virus Tian Tan	[AF095689]	(VACV-TIA)
Vaccinia virus WR	[AY243312]	(VACV-WR)
Variola virus		
Variola major virus Bangladesh-1975	[L22579]	(VARV-BSH)
Variola major virus India-1967	[X69198]	(VARV-IND)
Variola virus minor Garcia-1966	[Y16780]	(VARV-GAR)
Volepox virus		
Volepox virus		(VPXV)

Tentative Species in the Genus

Skunkpox virus	(SKPV)
Uasin Gishu disease virus	(UGDV)

Genus *Parapoxvirus*

Type Species *Orf virus*

Distinguishing Features

Virions are ovoid, 220-300 x 140-170 nm in size, with a surface filament that may appear as a regular cross-hatched, spiral coil involving a continuous thread. Infectivity is ether-sensitive. DNA is 130-150 kbp in size; G+C content is about 64%. Most species show extensive DNA cross-hybridization and serological cross-reactivity. Cross-hybridizations and DNA maps suggest extensive sequence divergence among members, higher than seen for members of the *Orthopoxvirus* genus. Generally the member viruses come from ungulates and domesticated livestock. They exhibit a narrow cell culture host range.

List of Species Demarcation Criteria in the Genus

The most useful provisional species demarcation criterion is host range, which tends to be narrow. A coupling of host range with RFLPs and cross-hybridization analyses is suggested at the terminal regions of the genome, external to the core of conserved genes.

List of Species in the Genus

Species names are in green italic script; strain names and synonyms are in black roman script; tentative species names are in blue roman script. Sequence accession numbers, and assigned abbreviations ( ) are also listed.

Species in the Genus

Bovine papular stomatitis virus		
Bovine popular stomatitis virus		(BPSV)
Orf virus		
Orf virus	[M30023]	(ORFV)
(Contagious pustular dermatitis virus)		
(Contagious ecthyma virus)		
Parapoxvirus of red deer in New Zealand		
Parapoxvirus of red deer in New Zealand		(PVNZ)
Pseudocowpox virus		
Pseudocowpox virus		(PCPV)
(Milker's nodule virus)		
(Paravaccinia virus)		
Squirrel parapoxvirus		
Squirrel parapoxvirus		(SPPV)

Tentative Species in the Genus

Auzduk disease virus
Camel contagious ecthyma virus
Chamois contagious ecthyma virus
Sealpox virus

Genus Avipoxvirus

Type Species *Fowlpox virus*

Distinguishing Features

Virions are brick-shaped, about 330 x 280 x 200 nm. Infectivity is usually ether-resistant. Genus includes viruses of birds that usually produce proliferative skin lesions (cutaneous form) and/or upper digestive/respiratory tract lesions (diptheritic form). Cross-protection is variable. Viruses are primarily transmitted mechanically by arthropods, by direct contact or through aerosols. The genomic DNA is about 300 kbp in size. Viruses exhibit extensive serologic cross-reaction. Viruses produce A-type inclusion bodies with considerable amounts of lipid. Viruses grow productively in avian cell cultures, but abortively in mammals and mammalian cell lines that have been examined. Viruses have a worldwide distribution.

List of Species Demarcation Criteria in the Genus

Provisional species demarcation criteria include disease characteristics, nature of the host and ecological niche, growth characteristics on the chicken chorioallantoic membrane and host range in cell culture. RFLPs, cross-hybridization and genomic DNA sequencing studies are also suggested.

LIST OF SPECIES IN THE GENUS

Species names are in green italic script; strain names and synonyms are in black roman script; tentative species names are in blue roman script. Sequence accession numbers, and assigned abbreviations ( ) are also listed.

SPECIES IN THE GENUS

Canarypox virus		
Canarypox virus		(CNPV)
Fowlpox virus		
Fowlpox virus	[AF198100]	(FWPV)
Juncopox virus		
Juncopox virus		(JNPV)
Mynahpox virus		
Mynahpox virus		(MYPV)
Pigeonpox virus		
Pigeonpox virus	[M88588]	(PGPV)
Psittacinepox virus		
Psittacinepox virus		(PSPV)
Quailpox virus		
Quailpox virus		(QUPV)
Sparrowpox virus		
Sparrowpox virus		(SRPV)
Starlingpox virus		
Starlingpox virus		(SLPV)
Turkeypox virus		
Turkeypox virus		(TKPV)

TENTATIVE SPECIES IN THE GENUS

Crowpox virus	(CRPV)
Peacockpox virus	(PKPV)
Penguinpox virus	(PEPV)

GENUS *CAPRIPOXVIRUS*

Type Species *Sheeppox virus*

DISTINGUISHING FEATURES

Virions are brick-shaped, about 300 x 270 x 200 nm. Infectivity is sensitive to trypsin and ether. Genus includes viruses of sheep, goats and cattle. Viruses can be mechanically transmitted by arthropods and by direct contact, or by fomites. The genomic DNA is about 154 kbp in size. There is extensive DNA cross-hybridization between species. In addition, extensive serologic cross-reaction and cross-protection is observed among members.

LIST OF SPECIES DEMARCATION CRITERIA IN THE GENUS

Provisional species demarcation criteria include RFLPs and genomic DNA sequence analyses.

LIST OF SPECIES IN THE GENUS

Species names are in green italic script; strain names and synonyms are in black roman script; tentative species names are in blue roman script. Sequence accession numbers, and assigned abbreviations ( ) are also listed.

SPECIES IN THE GENUS

Goatpox virus		
Goatpox virus G20-LKV	[AY077836]	(GTPV-G20)

Goatpox virus Pellor	[AY077835]	(GTPV-Pell)
Lumpy skin disease virus		
Lumpy skin disease virus NI-2490	[AF325528]	(LSDV-NI)
Lumpy skin disease virus NW-LW	[AF409137]	(LSDV-NW)
Sheeppox virus		
Sheeppox virus A	[AY077833]	(SPPV-A)
Sheeppox virus NISKHI	[AY077834]	(SPPV-NIS)
Sheeppox virus 17077-99	[NC_004002]	(SPPV-17077-99)

TENTATIVE SPECIES IN THE GENUS

Not reported.

GENUS LEPORIPOXVIRUS

Type Species *Myxoma virus*

DISTINGUISHING FEATURES

Virions are brick-shaped, about 300 x 250 x 200 nm. Infectivity is ether-sensitive. Genus includes viruses of lagomorphs and squirrels with extended cell culture host range. Usually viruses are mechanically transmitted by arthropods; but they are also transmitted by direct contact and fomites. Myxoma and fibroma viruses cause localized benign tumor-like lesions in their natural hosts. Myxoma viruses cause severe generalized disease in European rabbits. The genomic DNA is about 160 kbp, and the G+C content about 40%. Extensive DNA cross-hybridization is observed between member viruses. Serologic cross-reaction and cross-protection have been demonstrated between different species.

LIST OF SPECIES DEMARCATION CRITERIA IN THE GENUS

Provisional species demarcation criteria include various serological criteria including plaque neutralization tests, cross-protection in animals and agar diffusion methods. Distribution, ecological niche, host range and disease, plaque characteristics, host range in cell culture, RFLPs, and genomic DNA sequencing studies are also suggested.

LIST OF SPECIES IN THE GENUS

Species names are in green italic script; strain names and synonyms are in black roman script; tentative species names are in blue roman script. Sequence accession numbers, and assigned abbreviations ( ) are also listed.

SPECIES IN THE GENUS

Hare fibroma virus		
Hare fibroma virus		(FIBV)
Myxoma virus		
Myxoma virus Lausanne	[AF170726]	(MYXV-LAU)
Rabbit fibroma virus		
Rabbit fibroma virus	[AF170722]	(RFV)
(Shope fibroma virus)		(SFV)
Squirrel fibroma virus		
Squirrel fibroma virus		(SQFV)

TENTATIVE SPECIES IN THE GENUS

Not reported

GENUS *SUIPOXVIRUS*

Type Species *Swinepox virus*

DISTINGUISHING FEATURES

Virions are brick-shaped, about 300 x 250 x 200 nm. The genomic DNA is about 175 kbp in size with inverted terminal repeats of about 5 kbp. Virus forms foci or plaques in pig kidney cell culture (one-step growth is about 3 days at 37°C) and plaques in swine testes cell cultures. Virus causes asymptomatic generalized skin disease in swine that appears to be localized to epithelial cells and draining lymph nodes. Virus neutralizing antibodies are not usually detected. Mechanical transmission by arthropods (probably lice) is suspected. Viruses have a worldwide distribution. Rabbits can be infected experimentally; however serial transmission in rabbits is unsuccessful.

LIST OF SPECIES DEMARCATION CRITERIA IN THE GENUS

Not applicable.

LIST OF SPECIES IN THE GENUS

Species names are in green italic script; strain names and synonyms are in black roman script; tentative species names are in blue roman script. Sequence accession numbers, and assigned abbreviations ( ) are also listed.

SPECIES IN THE GENUS

Swinepox virus

Swinepox virus	[AF410153]	(SWPV)

TENTATIVE SPECIES IN THE GENUS

Not reported.

GENUS *MOLLUSCIPOXVIRUS*

Type Species *Molluscum contagiosum virus*

DISTINGUISHING FEATURES

Virions are brick-shaped, about 320 x 250 x 200 nm. Their buoyant density in CsCl is about 1.288 g/cm^3. The genomic DNA is about 190 kbp in size, G+C content is about 60%. DNAs cross-hybridize extensively. RFLP maps suggest four sequence divergences among the isolates examined. Molluscum contagiosum virus (MOCV) has not been propagated in tissue cultures. It is transmitted mechanically by direct contact between children, or between young adults. It is often sexually transmitted. Sometimes the virus causes opportunistic infections of persons with eczema or AIDS. Virus produces localized lesions containing enlarged cells with cytoplasmic inclusions known as molluscum bodies. Infections can recur and lesions may be disfiguring when combined with bacterial infections.

LIST OF SPECIES DEMARCATION CRITERIA IN THE GENUS

There is only one species. However, RFLPs suggest that there are at least four subtypes or strains, which may require future classification.

LIST OF SPECIES IN THE GENUS

Species names are in green italic script; strain names and synonyms are in black roman script; tentative species names are in blue roman script. Sequence accession numbers, and assigned abbreviations ( ) are also listed.

Species in the Genus

Molluscum contagiosum virus

Molluscum contagiosum virus [U60315] (MOCV)

Tentative Species in the Genus

Unnamed viruses of horses, donkeys, chimpanzees.

Genus *Yatapoxvirus*

Type Species ***Yaba monkey tumor virus***

Distinguishing Features

Virions are brick-shaped, about 300 x 250 x 200 nm. The genomic DNA is about 145 kbp in size, and the G+C content is about 33%. Yaba monkey tumor virus (YMTV) in primates causes histiocytomas, tumor-like masses of mononuclear cells. Viruses have been isolated from captive monkeys, baboons, and experimentally infected rabbits. Laboratory infections of man have been reported. Although DNAs cross-hybridize extensively, DNA RFLP maps suggest major sequence divergences between Tanapox virus (TANV) and YMTV. TANV produces localized lesions in primates that likely result from the mechanical transmission by insects generally during the rainy season in African rain forests. Lesions commonly contain virions with a double-layer envelope surrounding the viral surface membrane.

List of Species Demarcation Criteria in the Genus

Species demarcation criteria include RFLPs, genomic DNA sequencing studies, serological criteria including cross-protection in animals and plaque neutralization tests, geographical distribution, ecological niche and nature of the disease.

List of Species in the Genus

Species names are in green italic script; strain names and synonyms are in black roman script; tentative species names are in blue roman script. Sequence accession numbers, and assigned abbreviations ( ) are also listed.

Species in the Genus

Tanapox virus

Tanapox virus (TANV)

Yaba monkey tumor virus

Yaba monkey tumor virus [AJ293568] (YMTV)

Tentative Species in the Genus

Not reported.

SUBFAMILY ENTOMOPOXVIRINAE

TAXONOMIC STRUCTURE OF THE SUBFAMILY

Subfamily	***Entomopoxvirinae***
Genus	***Alphaentomopoxvirus***
Genus	***Betaentomopoxvirus***
Genus	***Gammaentomopoxvirus***

DISTINGUISHING FEATURES

The viruses infect insects. The viruses include different morphologic forms, e.g., brick-shaped, or ovoid. They are about 70-250 x 350 nm in size and chemically similar to other family members. Virions contain at least 4 enzymes equivalent to those found in Vaccinia virus. Virions of several morphological types have globular surface units that give a mulberry-like appearance; some have one lateral body, others have two. The DNA G+C content is about 20%. A common co-linear signature of core genes, different from those of members of the family *Chordopoxvirinae*, is beginning to emerge, and is characteristic of the sub-family. No serologic relationships have been demonstrated between entomopoxviruses and chordopoxviruses. Entomopoxviruses replicate in the cytoplasm of insect cells (hemocytes and adipose tissue cells). Mature virions are usually occluded in spheroids comprised of a major crystalline occlusion body protein (termed "spheroidin"). The subdivision into genera is based on virion morphology, host range, and the genome sizes of a few isolates. The genetic basis for these different traits is unknown.

GENUS ALPHAENTOMOPOXVIRUS

Type Species ***Melolontha melolontha entomopoxvirus***

DISTINGUISHING FEATURES

The genus includes poxviruses of Coleoptera. Virions are ovoid, about 450 x 250 nm in size, with one lateral body and a unilateral concave core. Surface globular units are 22 nm in diameter. The genomic DNA is about 260-370 kbp in size.

LIST OF SPECIES DEMARCATION CRITERIA IN THE GENUS

The primary species demarcation criterion is currently host range recognizing that adequate molecular information is limited and available for only two members. In the future, genetic content, gene order, and RFLP between members within a defined region of the genome and cross-hybridization analysis are likely to be useful. Serological criteria including plaque and virus neutralization tests are also used.

LIST OF SPECIES IN THE GENUS

Species names are in green italic script; strain names and synonyms are in black roman script; tentative species names are in blue roman script. Sequence accession numbers, and assigned abbreviations ( ) are also listed.

SPECIES IN THE GENUS

Anomala cuprea entomopoxvirus		
Anomala cuprea entomopoxvirus	[AB005053]	(ACEV)
Aphodius tasmaniae entomopoxvirus		
Aphodius tasmaniae entomopoxvirus		(ATEV)
Demodema boranensis entomopoxvirus		
Demodema boranensis entomopoxvirus		(DBEV)
Dermolepida albohirtum entomopoxvirus		

Dermolepida albohirtum entomopoxvirus		(DAEV)
Figulus subleavis entomopoxvirus		
Figulus subleavis entomopoxvirus		(FSEV)
Geotrupes sylvaticus entomopoxvirus		
Geotrupes sylvaticus entomopoxvirus		(GSEV)
Melolontha melolontha entomopoxvirus		
Melolontha melolontha entomopoxvirus	[509284, 987084]	(MMEV)

Tentative Species in the Genus

Not reported.

Genus Betaentomopoxvirus

Type Species *Amsacta moorei entomopoxvirus 'L'*

Distinguishing Features

The genus includes poxviruses of Lepidoptera and Orthoptera. Virions are ovoid, about 350 x 250 nm in size, with a sleeve-shaped lateral body and cylindrical core. Surface globular units are 40 nm in diameter. The genomic DNA is about 225 kbp in size with covalently closed termini and inverted terminal repetitions. The G+C content is about 18.5%. Viruses produce a 115 kDa occlusion body protein encoded by the spheroidin gene.

List of Species Demarcation Criteria in the Genus

The main species demarcation criterion is currently host range, and virion morphology. Serological criteria based on plaque neutralization may also be used. Genetic content, gene order, RFLPs within specific genes or within larger selected regions of the genome and nucleic acid cross-hybridization analysis are likely to become increasingly important. The species *Melanoplus sanguinipes entomopoxvirus* was removed from the *Betaentomopoxvirus* genus based on genomic DNA sequence comparisons with the type species, and it is now listed as an unassigned member of the family *Poxviridae*. This suggests that other betaentomopoxviruses, so classified based on morphological and host range criteria, may need reclassification.

List of Species in the Genus

Species names are in green italic script; strain names and synonyms are in black roman script; tentative species names are in blue roman script. Sequence accession numbers, and assigned abbreviations ( ) are also listed.

Species in the Genus

Acrobasis zelleri entomopoxvirus 'L'		
Acrobasis zelleri entomopoxvirus 'L'		(AZEV)
Amsacta moorei entomopoxvirus 'L'		
Amsacta moorei entomopoxvirus 'L'	[AF250284]	(AMEV)
Arphia conspersa entomopoxvirus 'O'		
Arphia conspersa entomopoxvirus 'O'		(ACOEV)
Choristoneura biennis entomopoxvirus 'L'		
Choristoneura biennis entomopoxvirus 'L'	[M34140, D10680]	(CBEV)
Choristoneura conflicta entomopoxvirus 'L'		
Choristoneura conflicta entomopoxvirus 'L'		(CCEV)
Choristoneura diversuma entomopoxvirus 'L'		
Choristoneura diversuma entomopoxvirus 'L'		(CDEV)
Choristoneura fumiferana entomopoxvirus 'L'		
Choristoneura fumiferana entomopoxvirus 'L'	[D10681, U10476]	(CFEV)
Chorizagrotis auxiliars entomopoxvirus 'L'		

Chorizagrotis auxiliars entomopoxvirus 'L' (CXEV)

Heliothis armigera entomopoxvirus 'L'

Heliothis armigera entomopoxvirus 'L' [AF019224, L08077] (HAVE)

Locusta migratoria entomopoxvirus 'O'

Locusta migratoria entomopoxvirus 'O' (LMEV)

Oedaleus senigalensis entomopoxvirus 'O'

Oedaleus senigalensis entomopoxvirus 'O' (OSEV)

Operophtera brumata entomopoxvirus 'L'

Operophtera brumata entomopoxvirus 'L' (OBEV)

Schistocera gregaria entomopoxvirus 'O'

Schistocera gregaria entomopoxvirus 'O' (SGEV)

TENTATIVE SPECIES IN THE GENUS

Not reported.

GENUS GAMMAENTOMOPOXVIRUS

Type Species ***Chironomus luridus entomopoxvirus***

DISTINGUISHING FEATURES

The genus includes poxviruses of Diptera. Virions are brick-shaped, about 320 x 230 x 110 nm in size, with two lateral bodies and a biconcave core. The genomic DNA is about 250-380 kbp in size.

LIST OF SPECIES DEMARCATION CRITERIA IN THE GENUS

The major species demarcation criterion is currently host range. However, as molecular information becomes available, genetic content, gene order and RFLPs within specific genes or within larger selected regions of the genome and cross-hybridization studies are likely to become increasingly important. Again, serological criteria, such as plaque neutralization can be considered.

LIST OF SPECIES IN THE GENUS

Species names are in green italic script; strain names and synonyms are in black roman script; tentative species names are in blue roman script. Sequence accession numbers, and assigned abbreviations ( ) are also listed.

SPECIES IN THE GENUS

Aedes aegypti entomopoxvirus

Aedes aegypti entomopoxvirus (AAEV)

Camptochironomus tentans entomopoxvirus

Camptochironomus tentans entomopoxvirus (CTEV)

Chironomus attenuatus entomopoxvirus

Chironomus attenuatus entomopoxvirus (CAEV)

Chironomus luridus entomopoxvirus

Chironomus luridus entomopoxvirus (CLEV)

Chironomus plumosus entomopoxvirus

Chironomus plumosus entomopoxvirus (CPEV)

Goeldichironomus haloprasimus entomopoxvirus

Goeldichironomus haloprasimus entomopoxvirus (GHEV)

TENTATIVE SPECIES IN THE GENUS

Not reported.

List of Unassigned Species in the Subfamily

Diachasmimorpha entomopoxvirus
Diachasmimorpha entomopoxvirus (DIEV)

List of Unassigned Viruses in the Subfamily

Melanoplus sanguinipes entomopoxvirus 'O'	[AF063866]	(MSEV)

List of Unassigned Viruses in the Family

California harbor seal poxvirus		(SPV)
Cotia virus	[1060872]	(CPV)
Dolphin poxvirus		(DOV)
Embu virus		(ERV)
Grey kangaroo poxvirus		(KXV)
Marmosetpox virus		(MPV)
Molluscum-like poxvirus		(MOV)
Mule deer poxvirus		(DPV)
Nile crocodile poxvirus		(CRV)
Quokka poxvirus		(QPV)
Red kangaroo poxvirus		(KPV)
Salanga poxvirus		(SGV)
Spectacled caiman poxvirus		(SPV)
Yoka poxvirus		(YKV)

Phylogenetic Relationships within the Family

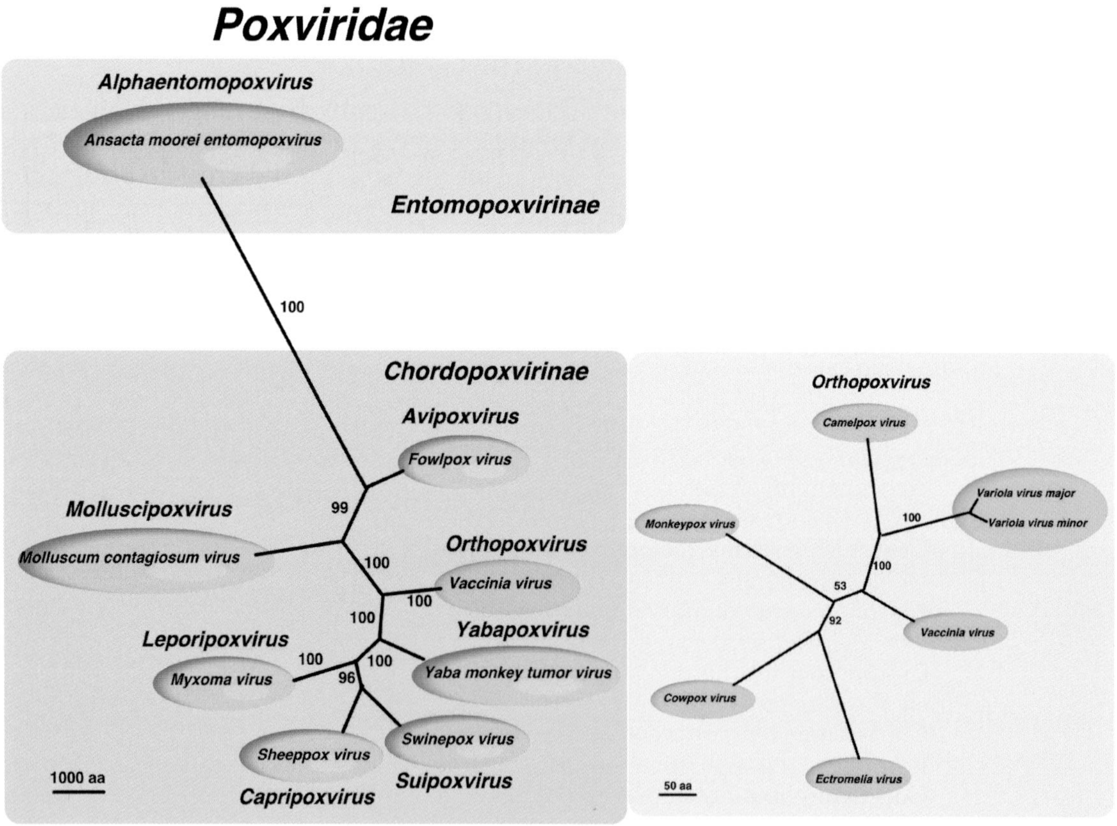

Figure 4: Phylogenetic relationships in the family *Poxviridae*. Phylogenetic predictions based upon aa alignments of 40 conserved protein coding sequences derived from the complete genomic sequences of either representative species of the family *Poxviridae* (left panel) or representative species within the *Orthopoxvirus* genus (right panel). The trees were calculated using the Branch-and-Bound search method with maximum parsimony as the optimality criterion. Bootstrap resampling confidence values (in percent) on 1,000 replicates are displayed at each branch point. Branch lengths are proportional to distance (the number of aa changes), and the distance scale for each prediction is given at the bottom of each panel (Chen *et al.*, 2004).

Similarity with Other Taxa

Not reported.

Derivation of Names

Avi: from Latin *avis*, "bird".
Capri: from Latin *caper*, "goat".
Entomo: from Greek *entomon*, "insect".
Lepori: from Latin *lepus*, "hare".
Mollusci: from Latin *molluscum*, "clam", "snail" related to appearance of lesion.
Orf: Scottish word based on Icelandic *hrufa*, "scab", "boil".
Ortho: from Greek *orthos*, "straight".
Para: from Greek *para*, "by side of".
Pox: from poc, pocc, "pustule".
Sui: from Latin *sus*, "swine".
Yata: from sigla of *Ya*ba and *ta*napox viruses.

References

Bawden, A.L., Glassberg, K.J., Diggans, J., Shaw, R., Farmerie, W. and Moyer, R.W. (2000). Complete genomic sequence of the Amsacta moorei entomopoxvirus: analysis and comparison with other poxviruses. *Virology*, **274**, 120-139.

Buller, R.M. and Palumbo, G.J. (1991). Poxvirus pathogenesis. *Microbiol. Rev.*, **55**, 80–122.

Chen, N., Danila, M.I., Feng, Z., Buller, R.M., Wang, C., Han, X., Lefkowitz, E.J. and Upton, C. (2003). The genomic sequence of ectromelia virus, the causative agent of mousepox. *Virology*, **317**, 165-186.

Esposito, J.J. and Nakano, J.H. (1991). Poxvirus infections in humans. In: *Manual of Clinical Microbiology*, 5th edn. (A. Balows, W.J. Hausler, *et al.*, eds), pp 858–867. American Society for Microbiology, Washington D.C.

Esposito, J.J. and Fenner, F. (2001). Poxviruses. In: *Fields Virology*, Fourth Edition (D.M. Knipe and P.M. Howley, eds), pp 2885-2921. Lippincott Williams & Wilkins, Philadelphia.

Fenner, F., Henderson, D.A., Arita, I., Jezek, Z. and Ladnyi, D. (1988). *Smallpox and its eradication.* World Health Organization, Geneva.

Fenner, F., Wittek, R. and Dumbell, K.R. (eds) (1989). *The Orthopoxviruses*. Academic Press, New York.

Gassmann, U., Wyler, R. and Wittek, R. (1985). Analysis of parapoxvirus genomes. *Arch.Virol.*, **83**, 17–31.

Granados, R.R. (1981). Entomopoxvirus infections in insects. In: *Pathogenesis of Invertebrate Microbial Diseases*, (E. W. Davidson, ed), pp 101-129. Allenheld, Osmun, Totowa, New Jersey.

Mercer, A.A., Lyttle, D.J., Whelan, E.M., Fleming, S.B. and Sullivan, J.T. (1995). The establishment of a genetic map of orf virus reveals a pattern of genomic organization that is highly conserved among divergent poxviruses. *Virology*, **212**, 698–704.

Meyer, H., Ropp, S.L. and Esposito, J.J. (1998). Poxviruses. In: *Methods in molecular medicine: Diagnostic virology protocols.* (A. Warnes and J. Stephenson, eds), pp 199-211. Humana Press, Totowa, New Jersey.

Moyer, R.W. and Turner, P.C. (eds) (1990). Poxviruses. *Curr. Top. Microbiol. Immunol.*, Vol. **163**. Springer-Verlag, New York.

Moss, B. (2001). *Poxviridae*: The viruses and their replication. In: *Fields Virology*, Fourth Edition (D.M. Knipe and P.M. Howley, eds), pp 2849-2883. Lippincott Williams and Wilkins, Philadelphia.

Robinson, A.J. and Lyttle, D.J. (1992). Parapoxviruses: their biology and potential as recombinant vaccine vectors. In: *Recombinant Poxviruses.* (M.M. Binns and G.L. Smith, eds), pp 285–327. CRC Press, Boca Raton, Florida.

Seet. B.T., Johnston, J. B., Brunetti, C. R., Barrett, J. W., Everett, H., Cameron, C., Sypula, J., Nazarian, S. H., Lucas, A. and McFadden, G. (2003) Poxviruses and immune evasion. *Ann. Rev Immunol.*, **21**, 377-423.

Tripathy, D.N. and Reed, W.M. (1997). Pox. In: *Diseases of Poultry*, 10th ed. (B.W. Calek and H.J. Barnes *et al.*, eds), pp 643–659. Iowa State University Press, Ames, Iowa.

Tulman, E.R., Afonso, C.L., Lu, Z., Zsak, L., Sur, J.H., Sandybaev, N.T., Kerembekova, U.Z., Zaitsev, V.L., Kutish, G.F. and Rock, D.L. (2002). The genomes of sheeppox and goatpox viruses. *J. Virol.*, **76**, 6054-6061.

Vanderplasschen, A., Hollinshead, M. and Smith, G.L. (1998). Intracellular and extracellular vaccinia virions enter cells by different mechanisms. *J. Gen. Virol.*, **79**, 877-887.

Contributed By

Buller, R.M., Arif, B.M., Black, D.N., Dumbell K.R., Esposito, J.J., Lefkowitz, E.J., McFadden, G., Moss, B., Mercer, A.A., Moyer, R.W., Skinner, M,A. and Tripathy, D.N.

FAMILY ASFARVIRIDAE

TAXONOMIC STRUCTURE OF THE FAMILY

Family *Asfarviridae*
Genus *Asfivirus*

Since only one genus is currently recognized, the family description corresponds to the genus description.

GENUS ASFIVIRUS

Type Species *African swine fever virus*

VIRION PROPERTIES

MORPHOLOGY

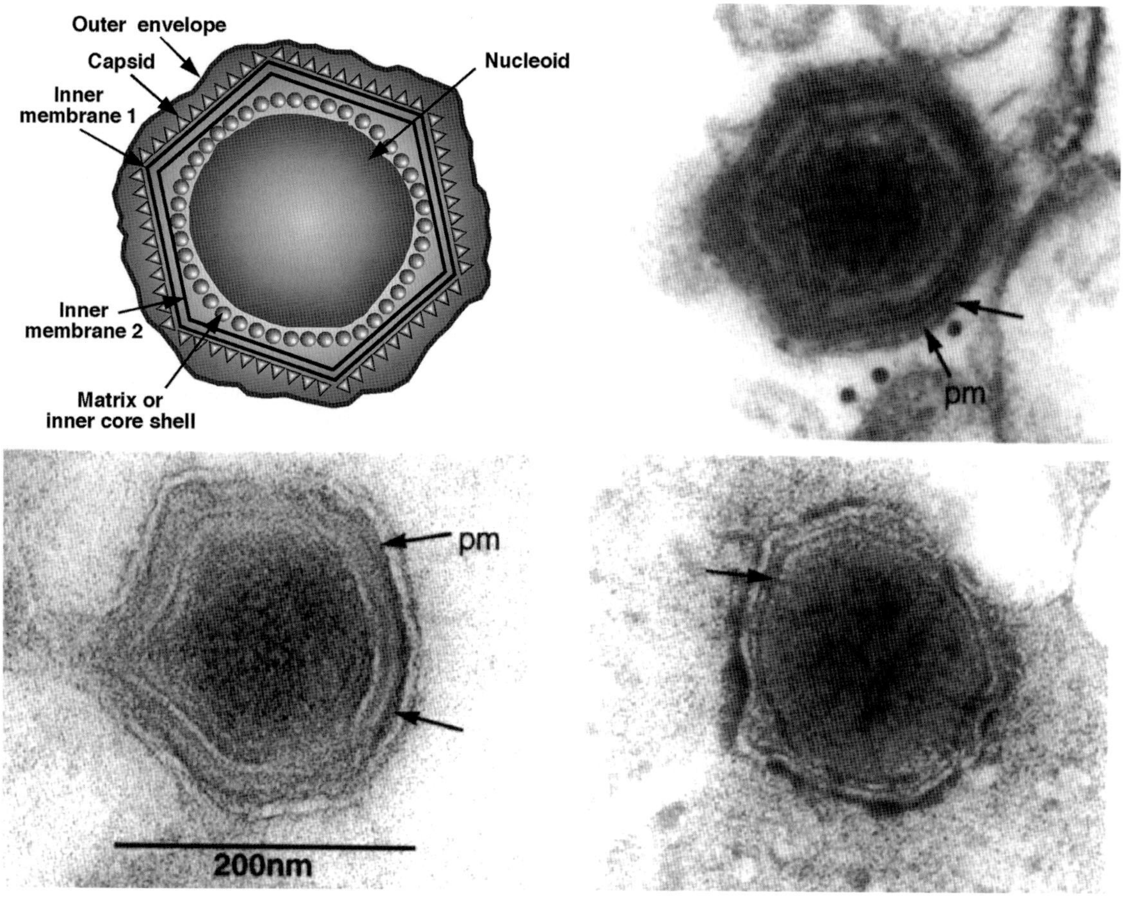

Figure 1: (Top left) Diagram of a virion section with lipid membranes, capsid and nucleoprotein core visualized of African swine fever virus (ASFV). (Top right) Thin section, (Bottom left) cryo-section and (Bottom right) a negative contrast electron micrograph of ASFV particles. The arrows indicate the membrane components of the virus; pm = plasma membrane. (Courtesy of Dr. S. Brookes, IAH Pirbright). The bar represents 200 nm.

Virions consist of a nucleoprotein core structure, 70-100 nm in diameter, surrounded by internal lipid layers and an icosahedral capsid, 170 to 190 nm in diameter, and an external lipid-containing envelope. The capsid exhibits icosahedral symmetry (T=189-217) corresponding to 1892-2172 capsomers (each capsomer is 13 nm in diameter and appears as a hexagonal prism with a central hole; intercapsomeric distance is 7.4-8.1 nm). Extracellular enveloped virions have a diameter of 175 to 215 nm (Fig. 1).

Physicochemical and Physical Properties

Virion buoyant density is 1.095 g/cm^3 in Percoll, 1.19-1.24 g/cm^3 in CsCl; S_{20W} is about 3500S. Virions are sensitive to ether, chloroform and deoxycholate and are inactivated at 60°C within 30 min., but survive for years at 20°C or 4°C. Infectivity is stable over a wide pH range. Some infectious virus may survive treatment at pH4 or pH13. Infectivity is destroyed by some disinfectants (1% formaldehyde in 6 days, 2% NaOH in 1 day); paraphenylphenolic disinfectants are very effective. Virus is sensitive to irradiation.

Nucleic Acid

The genome consists of a single molecule of linear, covalently close-ended, dsDNA 170 to 190 kbp in size (varying among isolates). The end sequences are present as two flip-flop forms that are inverted and complementary with respect to each other, adjacent to both termini are identical tandem repeat arrays about 2.1 kbp long. The complete nucleotide sequence of the tissue culture adapted Ba71V isolate of *African swine fever virus* (ASFV-Ba71V) has been published as well as about 90 kbp from the virulent ASFV-Malawi LIL20/1 isolate and 15 kbp from the ASFV-LIS 57 isolate. The genome encodes about 150 ORFs

Proteins

Virions contain more than 50 proteins including a number of enzymes and factors needed for early mRNA transcription and processing (Table 1). Enzymes packaged include RNA polymerase, poly A polymerase, guanyl transferase, protein kinase. The inhibitor of apoptosis (IAP) homolog protein is also packaged in virions. Virion structural proteins characterized include p72, p30, p12, p17, p22, p54 or j13L, p49, j18L, j5R and EP402R. The products of a 220 kDa protein which is cleaved to give four structural proteins; p150, p37, p14 and p34, and the products of a 62 kDa protein which is cleaved to give two structural proteins p35 and p15 are also present in virions. A virus-encoded protease related to the SUMO-1-specific protease family is involved in cleavage of these polyproteins. Two DNA binding proteins, p10 and p14.5 (p120R), are present in virions. Other predicted proteins encoded by the virus include enzymes involved in nucleotide metabolism (ribonucleotide reductase, thymidine kinase, thymidylate kinase, deoxyuridine triphosphatase), DNA replication and repair or transcription (DNA polymerase, DNA polymerase X, DNA ligase, topoisomerase II, guanyl transferase, three members of DNA helicase superfamily II, AP endonuclease). Deletion of the thymidine kinase and deoxyuridine triphosphatase genes does not affect virus replication in tissue culture cell lines but reduces virus replication in fully differentiated non-dividing macrophages and reduces virulence of the virus in pigs. Two enzymes involved in post-translational protein modification (a ubiquitin conjugating enzyme and a serine/threonine protein kinase) and an enzyme involved in synthesis of isoprenoid compounds (trans-prenyltransferase) are encoded by the virus. The virus-encoded nudix hydrolase preferentially degrades diphosphoinositol polyphosphates. The virus encodes two proteins involved in redox metabolism, NifS and ERV1 homologs. Deletion of the ERV1 homolog gene affects virion maturation. Five different multigene families MGF 110, MGF 360, MGF 530, MGF 300 and MGF 100 are found in genome regions close to the termini. Large length variations between genomes of different isolates are due to gain or loss of members of these multigene families. Members of families MGF 360 and 530 have been implicated as macrophage host range determinants. Virus-encoded proteins, which modulate the host response to virus infection, include homologs of the apoptosis inhibitors Bcl2 and IAP. Both of these proteins inhibit apoptosis; the IAP homolog inhibits caspase 3 activity. A bi-functional virus-encoded protein, A238L, acts both to inhibit activation of the host transcription factor NFkB and also inhibits activity of the host phosphatase calcineurin thus inhibiting calcineurin-dependent pathways, such as activation of NFAT transcription factor. This protein may therefore inhibit transcriptional activation in infected macrophages of a wide range of host immunomodulatory genes that are dependent on these factors. Virus proteins j4R and the ubiquitin-conjugating enzyme interact with host proteins involved in transcription and may also modulate host gene transcription. A virus protein, EP402R, that is similar to the host T cell adhesion protein CD2, is required for the hemadsorbtion of red blood cells around virus-infected cells and is also thought to mediate the adhesion of extracelluar virions to red blood cells. Deletion of the EP402R gene reduces virus dissemination in infected pigs and *in vitro* abrogates the ability of ASFV infected cells to inhibit proliferation of bystander lymphocytes in response to mitogens. Expression of the virus-encoded C-type lectin, EP153R, augments hemadsorbtion to infected cells. One protein, (designated NL-S, l14L or DP71L) is similar to a Herpes simplex virus-encoded neurovirulence factor (ICP34.5) and this can act as a virulence factor for pigs.

Table 1: Functions of African swine fever virus (ASFV) encoded proteins.

Nucleotide metabolism, transcription, replication and repair	**Gene Name in BA71V isolate**	**Predicted Protein size**
Thymidylate kinase	A240L	27.8
Thymidine kinase	K196R	22.4
dUTPase*	E165R	18.3
Ribonucleotide reductase (small subunit)	F334L	39.8
Ribonucleotide reductase (large subunit)	F778R	87.5
DNA polymerase a like	G1211R	139.8
DNA topoisomerase type II*	P1192R	135.5
Proliferating cell nuclear antigen (PCNA) like	E301R	35.3
DNA polymerase X like*	O174L	20.3
DNA ligase*	NP419L	48.2
AP endonuclease class II	E296R	33.5
RNA polymerase subunit 2	EP1242L	139.9
RNA polymerase subunit 6	C147L	16.7
RNA polymerase subunit 1	NP1450L	163.7
RNA polymerase subunit 3	H359L	41.3
RNA polymerase subunit 5	D205R	23.7
Helicase superfamily II	A859L	27.8
Helicase superfamily II	F1055L	123.9
Helicase superfamily II	B962L	109.6
Helicase superfamily II	D1133L	129.3
Helicase superfamily II	Q706L	80.4
Helicase superfamily II	QP509L	58.1
Transcription factor SII	I243L	28.6
Guanyl transferase*	NP868R	29.9
FTS J like methyl transferase domain	EP424R	49.3
ERCC4 nuclease domain	EP364R	40.9
Other enzymes		
Prenyl transferase*	B318L	35.9
Serine protein kinase*	R298L	35.1
Ubiquitin conjugating enzyme*	I215L	24.7
Nudix hydrolase*	D250R	29.9
Host cell interactions		
IAP apoptosis inhibitor*	A224L	26.6
Bcl 2 apoptosis inhibitor*	A179L	21.1
IkB homolog and inhibitor of calcineurin phosphatase*	A238L	28.2
C type lectin like*	EP153R	18.0
CD2 like. Causes haemadsorbtion to infected cells*	EP402R	45.3
Similar to HSV ICP34.5 neurovirulence factor	DP71L	8.5
Nif S like	QP383R	42.5
ERV 1 like. Involved in redox metabolism	B119L	14.4

Structural proteins and proteins involved in morphogenesis		
P22	KP177R	20.2
Histone-like	A104R	11.5
P11.5	A137R	21.1
P10	A78R	8.4
P72 major capsid protein. Involved in virus entry	B646L	73.2
P49	B438L	49.3
Chaperone. Involved in folding of capsid.	B602L	45.3
Sumo 1 like protease. Involved in polyprotein cleavage	S273R	31.6
P220 polyprotein precursor of p150, p37, p14, p34. Required for packaging of nucleoprotein core.	CP2475L	281.5
P32 phosphoprotein. Involved in virus entry.	CP204L	23.6
P60 polyprotein precursor of p35 and p15	CP530R	60.5
P12 attachment protein	O61R	6.7
P17	D117L	13.1
J5R	H108R	12.5
P54 (j13L) Binds to LC8 chain of dynein, involved in virus entry.	E183L	19.9
J18L	E199L	22.0
P14.5 DNA binding. Required for movement of virions to plasma membrane	E120R	13.6

Multigene family members		
Multigene family 360	KP360L	41.7
	KP362L	42.6
	L356L	41.7
	UP60L	7.0
	J319L	31.3
	A125L	14.5
	A276R	31.6
	DP363R	42.4
	DP42R	4.9
	DP311R	35.6
	DP63R	7.5
	DP148R	17.2
Multigene family 110	U104L	12.2
	XP124L	14.2
	V82L	9.4
	Y118L	13.9
Multigene family 300	J154R	17.6
	J104L	12.5
	J182L	21.7
Multigene family 505	A489R	57.7
	A280R	32.5
	A505R	59.2
	A498R	58.7
	A528R	61.8
	A505R	59.4
	A542R	63.1
	A542L	63.4
Multigene family 100	DP141L	16.8

Lipids
Enveloped virions contain lipids including glycolipids.

Carbohydrates
One virion protein is glycosylated (EP402R) and glycolipids are also incorporated into virions. The virus encodes several predicted transmembrane proteins that contain putative N-linked glycosilation sites.

Genome Organization and Replication

The 150 major ORFs are closely spaced (intergenic distances are generally less than 200 bp) and read from both DNA strands. A few intergenic regions contain short tandem repeat arrays (Fig. 2).

The primary cell types infected by the virus include those of the mononuclear-phagocytic system including fixed tissue macrophages and specific lineages of reticular cells. Virus replicates *in vitro* in macrophages, endothelial cells and several isolates have been adapted to replicate in tissue culture cell lines. Virus enters cells by receptor-mediated endocytosis and early mRNA synthesis begins in the cytoplasm immediately following entry using enzymes and factors packaged in the virus core. Virus DNA replication and assembly take place in perinuclear factory areas. At early times post-infection virus DNA is detected in the nucleus suggesting a possible role for nuclear enzymes in initial stages of DNA replication. Head to head virus DNA concatamers, which are thought to be replicative intermediates, are detected in the cytoplasm from 6 hours post-infection. The mechanism of DNA replication in the cytoplasm is similar to that of viruses in the family *Poxviridae*.

Virus transcripts are 3'-polyadenylated and 5'-capped. Genes are expressed in an ordered cascade. Early genes are expressed prior to DNA replication; expression of late genes is dependent on the onset of DNA replication. Synthesis of some early genes continues throughout infection. Intermediate genes are expressed late but their expression does not depend on the onset of DNA replication. Promoter elements are relatively short and located immediately upstream to ORFs; transcription start sites are generally a short distance from start codons. Both early and late gene transcripts are of defined length; sequences of seven or more consecutive thymidylate residues in the coding strand are signals for mRNA 3'-end formation.

Several structural proteins are expressed as polyproteins and cleaved at the sequence GlyGlyX. The polyprotein with Mr 220 x 10^3 is myristylated. Other virus encoded proteins are modified by phosphorylation (p10, p32), and N-linked glycosylation. Virus morphogenesis takes place in perinuclear virus factories. Virus factories resemble aggresomes in being surrounded by a vimentin cage and increased number of mitochondria. Aggresomes are formed in response to cell stress and function to remove misfolded proteins. Two layers of membrane derived from the endoplasmic reticulum are incorporated as internal lipid membranes in virus particles. Formation of the icosahedral capsid is thought to occur on these membranes. The virus genome and enzymes are packaged into a nucleoprotein core. Extracellular virus has a loose fitting external lipid envelope probably derived by budding through the plasma membrane. Virus is transported to and from sites of assembly on microtubules. The p54 virion protein binds to the LC8 component of the dynein motor complex and may be involved in transport of virus particles to the factory sites. The EP120R virion protein is required for virus transport from assembly sites to the plasma membrane.

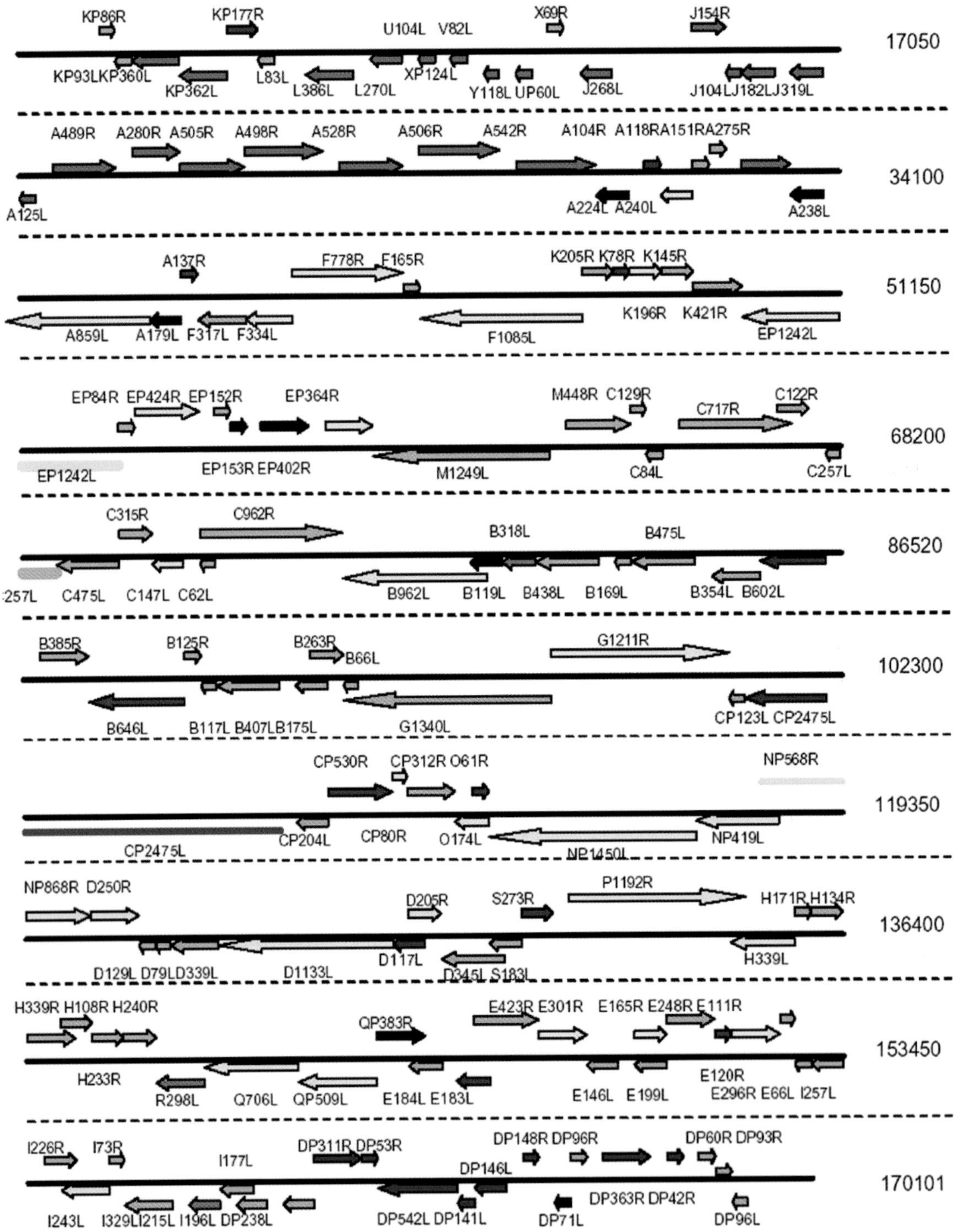

Figure 2: Genome organization map of *African swine fever virus* (ASFV) strain Ba71V. Numbers at the right end indicate the nucleotide position. ORFs are represented as arrows. The different patterns indicate the following groups of ORFs: multigene families (blue), nucleic acid metabolism (green), protein modification (magenta), structural proteins (orange), and proteins mediating virus-host interactions (black). (Modified from Yanez *et al.*, 1995)

Antigenic Properties

Infected swine mount a protective immune response against non-fatal virus strains and produce antibodies. Antibodies can neutralize virus strains with a low number of passages in tissue culture but are not as effective against virus with a high number of passages. This difference may be caused by changes in virus phospholipids. Antibodies against proteins p72, p12, p30 and p54 neutralize virus, those against p30 inhibit virus internalization rather than attachment. Monoclonal antibody analyses show that structural proteins p150, p14 and p12 vary between isolates. Several virus-encoded proteins (for example p54 or j13L and B602L or 9RL) contain tandem repeat arrays that vary in number and sequence when genomes of different isolates are compared. Several T cell epitopes have been mapped on virus-encoded proteins.

BIOLOGICAL PROPERTIES

ASFV infects domestic and wild swine (*Sus scrofa domesticus* and *S. s. ferus*), warthogs (*Phacochoerus aethiopicus*) and bushpigs (*Potamochoerus porcus*). Disease signs are only apparent in domestic and wild swine. Soft ticks of the genus *Ornithodoros* are also infected and *O. moubata* acts as a vector in parts of Africa south of the Sahara and *O. erraticus* acted as a vector in S.W. Spain and Portugal. Virus can be transmitted in ticks trans-stadially, trans-ovarially and sexually. Warthogs, bushpigs and swine can be infected by bites from infected ticks. Neither vertical nor horizontal transfer of virus between warthogs is thought to occur. However, transmission between domestic swine can occur by direct contact, or by ingestion of infected meat, or fomites, or mechanically by biting flies. Warthogs, bushpigs, wild swine and ticks act as reservoirs of virus. Disease is endemic in domestic swine in many African countries and in Europe in Sardinia. ASFV was first reported in Madagscar in 1998 and remains endemic there. Disease was first introduced into Europe in Portugal in 1957 and was endemic in parts of the Iberian peninsula from 1960 until 1995. Sporadic outbreaks have occurred in and been eradicated from Belgium, Brazil, Cuba, the Dominican Republic, France, Haiti, Holland and Malta.

ASFV causes hemorrhagic fever in domestic pigs and wild boar. Virus isolates differ in virulence and may produce a variety of disease signs ranging from acute to chronic to inapparent. Virulent isolates may cause 100% mortality in 7 to 10 days. Less virulent isolates may produce a mild disease from which a number of infected swine recover and become carriers. Viruses replicate in cells of the mononuclear phagocytic system and reticulo-endothelial cells in lymphoid tissues and organs of domestic swine. Cell surface markers expressed from intermediate stages of monocyte-macrophage differentiation are indicators of cell susceptibility to infection. Widespread cell death caused by apoptosis occurs in both T and B lymphocytes in lymphoid tissues and endothelial cells in arterioles and capillaries. This accounts for the lesions seen in acute disease. Disseminated intravascular coagulation develops during the late phase of acute infections and this may lead to the characteristic hemorrhagic syndrome.

LIST OF SPECIES DEMARCATION CRITERIA IN THE GENUS

Not applicable.

LIST OF SPECIES IN THE GENUS

Species names are in green italic script; strain names and synonyms are in black roman script; tentative species names are in blue roman script. Sequence accession numbers, and assigned abbreviations ( ) are also listed.

SPECIES IN THE GENUS

African swine fever virus

African swine fever virus - Ba71V	(ASFV- Ba71V)
African swine fever virus – LIL20/1	(ASFV- LIL20/1)
African swine fever virus – LIS57	(ASFV- LIS57)

TENTATIVE SPECIES IN THE GENUS

None reported.

List of Unassigned Viruses in the Family

None reported.

Phylogenetic Relationships within the Family

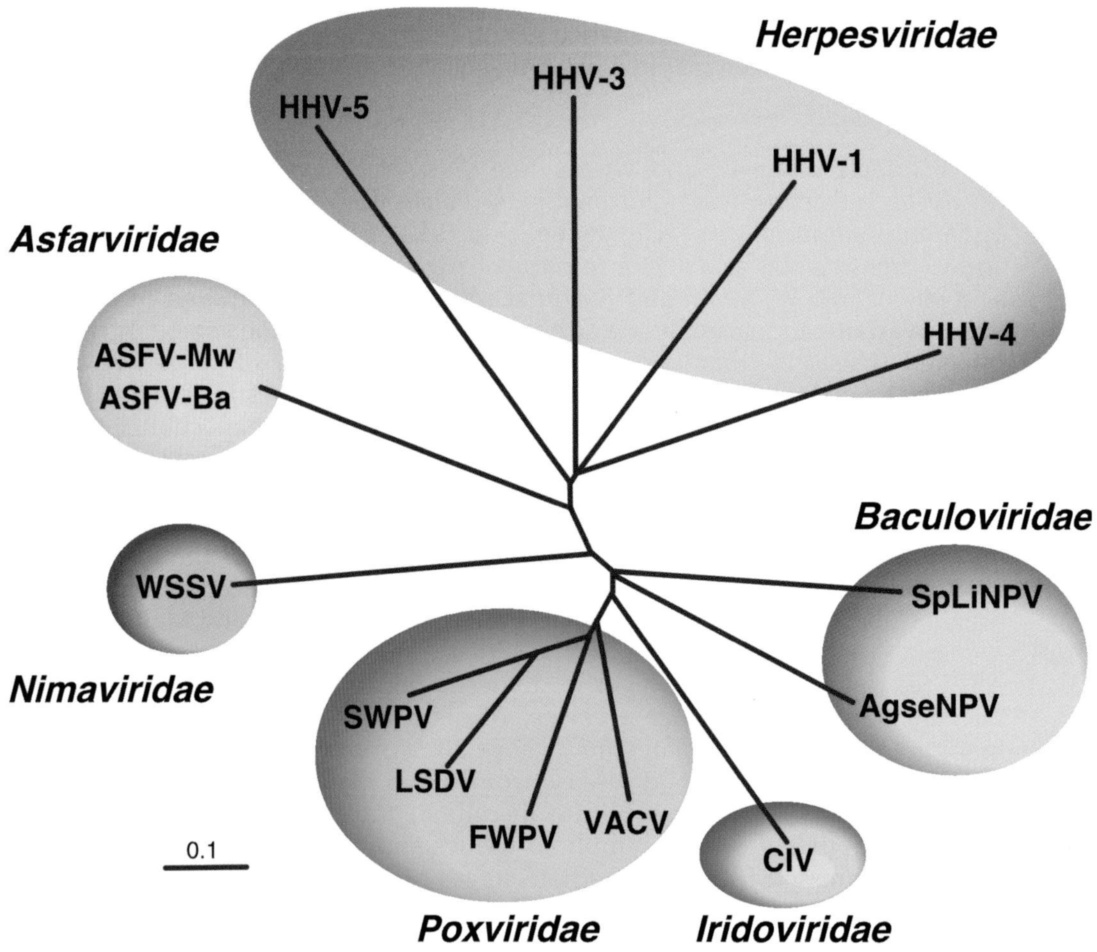

Figure 3: Phylogenetic tree comparing dUTPase proteins encoded by the African swine fever virus (ASFV) with those from other DNA viruses. Sequences were aligned using ClustalW and the tree displayed using Treeview. Sequences shown are from; ASFV-Mw and -Ba, Malawi and Ba71V isolates of African swine fever virus, WSSV White spot syndrome virus, SWPV Swinepox virus, LSDV Lumpy skin disease virus, FWPV Fowlpox virus, VACV Vaccinia virus, CIV Chilo iridescent virus, AsGV Agrotis seqetum granulosis virus, SpLiNPV Spodoptera litura nucleopolyhedron virus, HHV-1 Human herpes virus 1, HHV-3 Human herpes virus 3, HHV-4 Human herpes virus 4, HHV-5 Human herpes virus 5. (Provided by Dr D. Chapman, IAH, Pirbright).

Similarity with Other Taxa

Earlier *African swine fever virus* was listed as a member of the family *Iridoviridae*, but as more information was obtained, it was removed from this family. It exhibits some similarities in genome structure and strategy of replication to the poxviruses and phycodnaviruses, but it has a quite different virion structure from poxviruses and many other properties that distinguish it from the member viruses of the families *Poxviridae* and *Phycodnaviridae*.

Derivation of Names

Asfar sigla derived from **A**frican **s**wine **f**ever **a**nd **r**elated viruses.

References

Alonso, C., Miskin, J., Hernaez, B., Fernandez-Zapatero, P., Soto, L., Canto, C. and Rodriguez-Crespo, I. (2001). African swine fever virus protein p54 interacts with the microtubular motor complex through direct binding to light-chain dynein. *J. Virol.*, **75**, 9819-9827.

Andres, G., Alejo, A., Simon-Mateo, C. and Salas, M.L. (2001). African swine fever virus proteases, a new viral member of the SUMO-1-specific protease family. *J. Biol. Chem.*, **276**, 780-787.

Andres, G., Garcia-Escudero, R., Simon-Mateo, C. and Vinuela, E. (1998). African swine fever virus is enveloped by a two-membrane collapsed cisternae derived from the endoplasmic reticulum. *J. Virol.*, **72**, 8988-9001.

Andres, G., Garcia-Escudero, R., Vinuela, E. (2001). African swine fever virus structural protein pE120R is essential for virus transport from assembly sites to plasma membrane but not for infectivity. *J. Virol.*, **75**, 6758-6768.

Borca, M.V., Carrillo, C., Zsak, L., Laegreid, W.W., Kutish, G.F., Neilan, J.G., Burrage, T.G. and Rock, D.L. (1998). Deletion of a Cd2-like gene, 8-DR, from African swine fever virus affects viral infection in domestic swine. *J. Virol.*, **72**, 2881-2889.

Dixon, L.K., Twigg, S.R.F., Baylis, S.A., Vydelingum, S., Bristow, C., Hammond, J.M. and Smith, G.L. (1994). Nucleotide sequence of a 55 kbp region from the right end of the genome of a pathogenic African swine fever virus isolate (Malawi LIL20/1). *J. Gen. Virol.*, **75**, 1655-1684.

Goatley, L.C., Twigg, S.R.F., Miskin, J.E., Monaghan, P., St-Arnaud, R., Smith, G.L. and Dixon, L.K. (2002). The African swine fever virus protein j4R binds to the alpha chain of nascent polypeptide-associated complex. *J. Virol.*, **76**, 9991-9999.

Gomez Puertas, P., Rodriguez, F., Oviedo, J.M., Ramiro Ibanez, F., Ruiz-Gonzalez, F., Alonso, C. and Escribano, J.M. (1996). Neutralizing antibodies to different proteins of African swine fever virus inhibit both virus attachment and internalization. *J. Virol.*, **70**, 5689-5694.

Heath, C.M., Windsor, M. and Wileman, T. (2001). Aggresomes resemble sites specialized for virus assembly. *J. Cell. Biol.*, **153**, 449-455.

Powell, P., Dixon, L.K. and Parkhouse, R.M.E. (1996). An I-kappa B homolog encoded by African swine fever virus provides a novel mechanism for down regulation of proinflammatory cytokine responses in host macrophages. *J. Virol.*, **70**, 8527-8533.

Ramiro-Ibanez, F., Ortega, A., Brun, A., Escribano, J.M. and Alonso, C. (1996). Apoptosis: a mechanism of cell killing and lymphoid organ impairment during acute African swine fever virus infection. *J. Gen. Virol.*, **77**, 2209-2219.

Revilla, Y., Callejo, M., Rodriguez, J.M., Culebras, E., Nogal, M.L., Salas, M.L., Vinuela, E. and fresno, M. (1998). Inhibition of nuclear factor kappa B activation by a virus-encoded I kappa B like protein. *J. Biol. Chem.* **273**,5405-5411.

Rojo, G., Garcia-Beato, R., Vinuela, E., Salas, M.L. and Salas, J. (1999). Replication of African swine fever virus DNA in infected cells. *Virology*, **257**,524-536.

Rouiller, I., Brookes, S.M., Hyatt, A.D., Winsor, M. and Wileman, T. (1998). African swine fever is wrapped by the endoplasmic reticulum. *J. Virol.*, **72**,2373-2387.

Tulman, E.R. and Rock, D.L. (2001). Novel virulence and host range genes of African swine fever virus. *Curr. Op. Micro.*, **4**,456-461.

Yanez, R.J., Rodriguez, J.M., Nogal, M.L., Yuste, L., Enriquez, C., Rodriguez, J.F. and Vinuela, E. (1995). Analysis of the complete nucleotide sequence of African swine fever virus. *Virology*, **208**, 249-278.

Zsak, L., Lu, Z., Burrage, T.G., Neilan, J.G., Kutish, G.F., Moore, D.M. and Rock, D.L. (2001). African swine fever virus multigene family 360 and 530 genes are novel macrophage host range determinants. *J. Virol.* **75**, 3066-3076.

Contributed By

Dixon, L.K., Escribano, J.M., Martins, C., Rock, D.L., Salas, M.L. and Wilkinson, P.J.

FAMILY IRIDOVIRIDAE

TAXONOMIC STRUCTURE OF THE FAMILY

Family	***Iridoviridae***
Genus	***Iridovirus***
Genus	***Chloriridovirus***
Genus	***Ranavirus***
Genus	***Lymphocystivirus***
Genus	***Megalocytivirus***

VIRION PROPERTIES

MORPHOLOGY

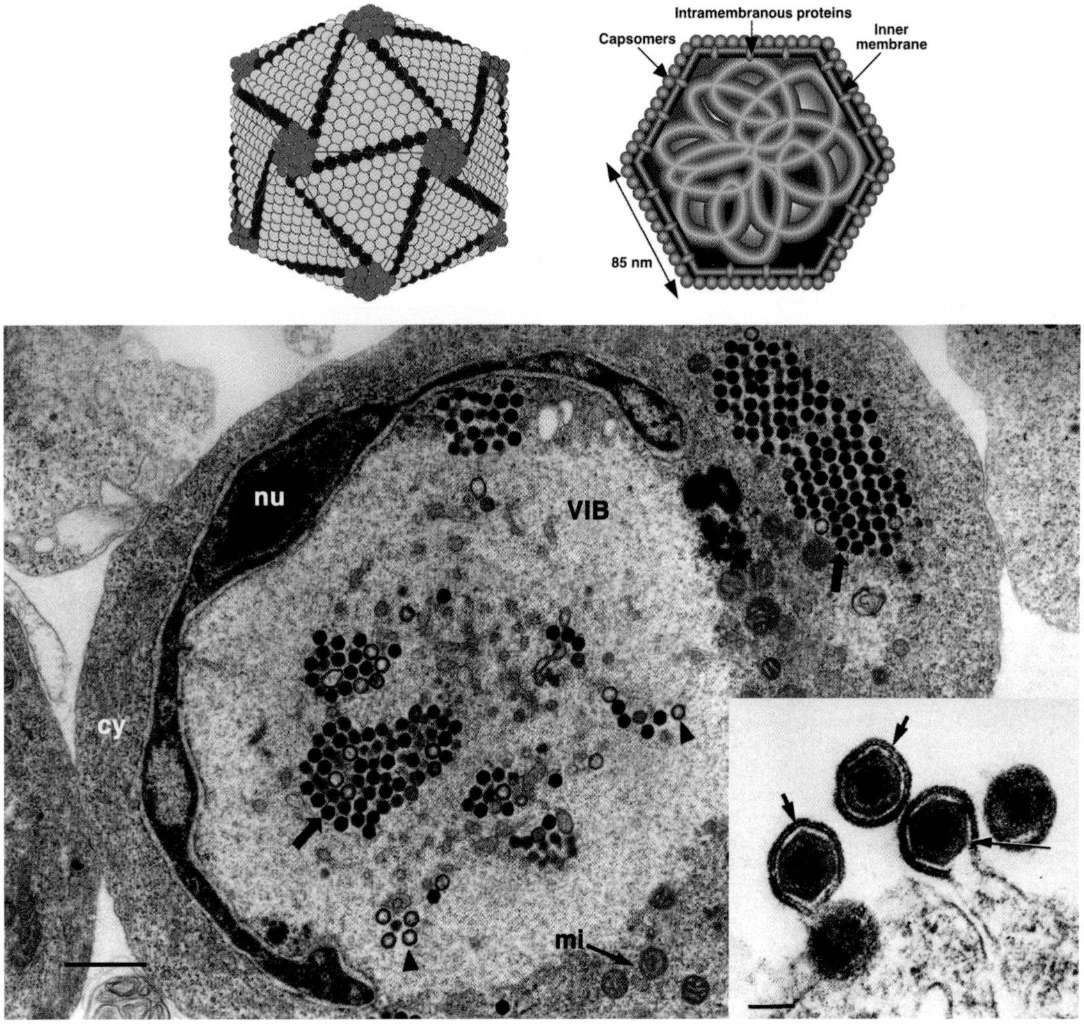

Figure 1: (Top left) Outer shell of Invertebrate iridescent virus 2 (IIV-2) (From Wrigley, *et al.* (1969). *J. Gen. Virol.,* **5,** 123. With permission). (Top right) Schematic diagram of a cross-section of an iridovirus particle, showing capsomers, transmembrane proteins within the lipid bilayer, and an internal filamentous nucleoprotein core (From Darcy-Tripier, F. *et al.* (1984). *Virology*, **138**, 287. With permission). (Bottom left) Transmission electron micrograph of a fat head minnow cell infected with an isolate of *European catfish virus*. Nucleus (Nu); virus inclusion body (VIB); paracrystalline array of non-enveloped virus particles (arrows); incomplete nucleocapsids (arrowheads); cytoplasm (cy); mitochondrion (mi). The bar represents 1 µm. (From Hyatt *et al.* (2000). *Arch. Virol.* **145**, 301, with permission). (insert) Transmission electron micrograph of particles of Frog virus 3 (FV-3), budding from the plasma membrane. Arrows and arrowheads identify the viral envelope (Devauchelle *et al.* (1985). *Curr. Topics Microbiol. Immunol.*, **116**, 1, with permission). The bar represents 200 nm.

Virions display icosahedral symmetry and are usually 120-200 nm in diameter, but may be up to 350 nm (e.g. genus *Lymphocystivirus*). The core is an electron-dense entity consisting of a nucleoprotein filament surrounded by a lipid membrane containing transmembrane proteins of unknown function. The capsid is composed of identical capsomers, the number of which depends on virion size. Capsomers are organized to form trisymmetrons and pentasymmetrons in members of the *Iridovirus* and *Chloriridovirus* genera. Capsomer dimensions are approximately 6-7 nm in diameter and 7-13 nm in height. Each capsomer is composed of an internal and external protein trimer. Fibers or short fibrils have been observed trailing from the capsid in viruses from both vertebrate and invertebrate genera. Iridoviruses may acquire an envelope by budding through the host cell membrane. The envelope increases the specific infectivity of virions, but is not required for infectivity as naked particles are also infectious (Fig. 1).

Physicochemical and Physical Properties

The Mr of virions is 1.05–2.75 x 10^9, their sedimentation coefficient ($S_{20,w}$) is 2020-4460S, and their density is 1.26-1.6 g/cm^3. Virions are stable in water at 4°C for extended periods. Sensitivity to pH varies, whereas sensitivity to ether and chloroform depends on the assay system employed. All viruses are inactivated within 30 min at >55°C. FV-3, Infectious spleen and kidney necrosis virus (ISKNV), and Invertebrate iridescent virus 6 (IIV-6) are inactivated by UV-irradiation. Some ranaviruses remain infectious after desiccation, e.g., Bohle iridovirus (BIV) survives desiccation at temperatures up to 42°C for up to 6 weeks, whereas others are sensitive to drying.

Nucleic Acid

The virion core contains a single linear dsDNA molecule of 140-303 kbp, a value which includes both unique and terminally redundant sequences. Invertebrate iridescent virus 1 (IIV-1) has been reported to have an additional genetic component of 10.8 kbp which exists as a free molecule in the particle core. DNA comprises 12-16% of the particle weight, and the G+C content ranges from ~28 to ~55%. All viruses within the family possess genomes that are circularly permuted and terminally redundant. However, the DNA of vertebrate iridoviruses (members of the genera *Ranavirus, Lymphocystivirus* and *Megalocytivirus*) is highly methylated, whereas little to no methylation is found within the genomes of the invertebrate iridoviruses (members of the genera *Iridovirus* and *Chloriridovirus*). The complete genomic sequence is known for Lymphocystis disease virus 1 (LCDV-1), IIV-6, Tiger frog virus (TFV), ISKNV, and Ambystoma tigrinum virus (ATV). Sequence analysis of Frog virus 3 (FV-3) and Epizootic haematopoietic necrosis virus (EHNV) is ongoing. Although naked genomic DNA is not infectious, non-genetic reactivation of viral DNA can be achieved in the presence of viral structural proteins.

Proteins

Iridoviruses are structurally complex, and up to 36 polypeptides, ranging from ~5–250 kDa, have been detected by two dimensional PAGE of virus particles. Sequence analysis has identified more than 100 ORFs (Tables 1 and 2, Fig. 2). The major CP (MCP), 48-55 kDa, comprises 40% of the total virion protein, and its complete aa sequence is known for several viruses. The MCP is highly conserved and shares aa sequence identity with the MCPs of African swine fever virus (ASFV, family *Asfarviridae*), several members of the family *Ascoviridae*, and Paramecium bursaria Chlorella virus 1 (PBCV-1, family *Phycodnaviridae*). At least 6 DNA associated polypeptides have been identified in the core of IIV-6, with a major species of 12.5 kDa. A virion-associated protein elicits the shutdown of host macromolecular synthesis, whereas other virion-associated proteins transactivate early viral transcription. A number of virion-associated enzymatic activities have been detected including a protein kinase, nucleotide phosphohydrolase, a ss/dsRNA-specific ribonuclease, pH 5 and pH 7.5 deoxyribonucleases, and a protein phosphatase. In addition to these polypeptides, various other proteins have been identified by BLAST analysis of recently sequenced viral genomes (Table 2; Fig 2).

Lipids
Non-enveloped particles contain 5-17% lipid, predominantly as phospholipid. The composition of the internal lipid membrane suggests that this membrane is not derived from host membranes but is produced *de novo*. Viruses released from cells by budding acquire their outer envelope from the plasma membrane.

Carbohydrates
Carbohydrates are not present in purified virions.

Genome Organization and Replication

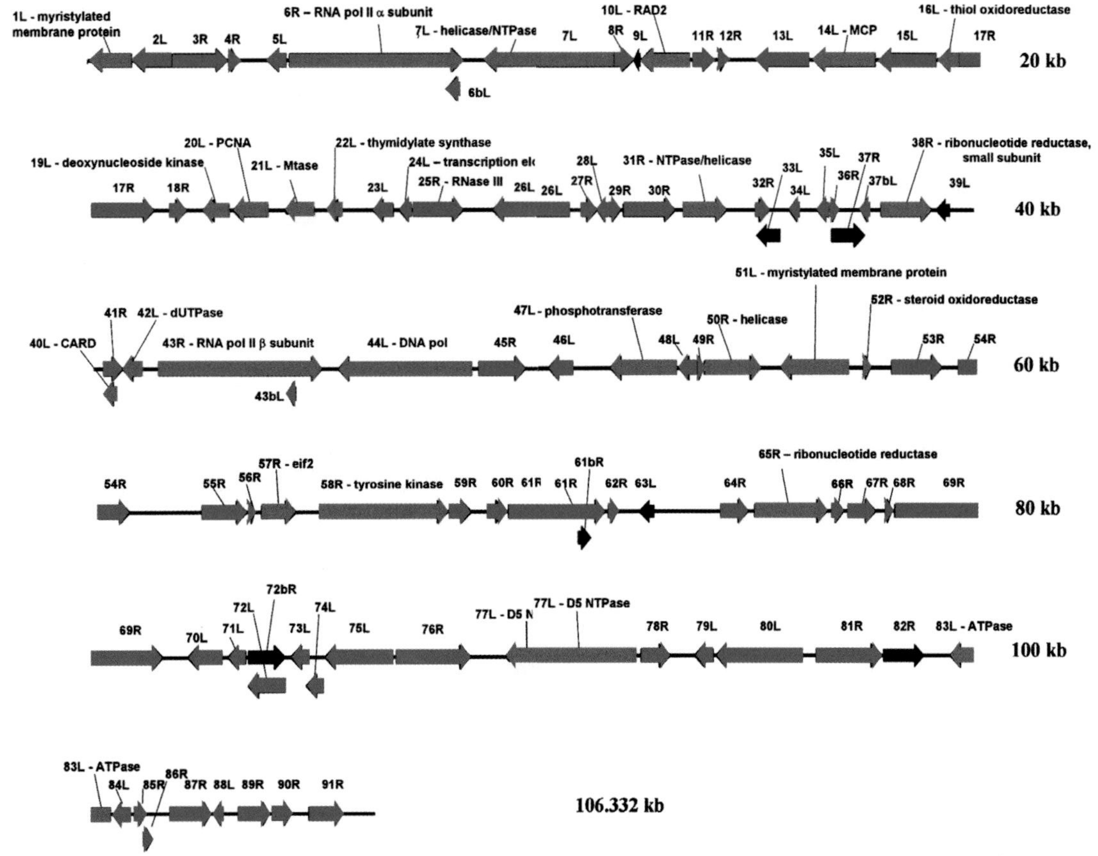

Figure 2: Genomic structure of Ambystoma tigrinum virus (ATV). Arrows represent viral ORFs with their size, position, and orientation shown. ORFs of known function are colored in red and their putative proteins identified; ORFs with known homology to Tiger frog virus (TFV) are in blue; and those of unknown function or with no homology to TFV are indicated in black. (Jancovich, and colleagues, unpublished).

The replication strategy of iridoviruses is novel and has been elucidated primarily through the study of FV-3, the type species of the genus *Ranavirus*. Virion entry occurs by either receptor mediated endocytosis (enveloped particles) or by uncoating at the plasma membrane (naked virions). Following uncoating, viral cores enter the nucleus where 1st stage DNA synthesis and the synthesis of immediate early (IE) and delayed early (DE) viral transcripts takes place. In a poorly understood process, one or more virion associated proteins act as transactivators and re-direct host RNA polymerase II to synthesize IE and DE viral mRNAs using the methylated viral genome as template. Gene products encoded by IE and DE viral transcripts include both regulatory and catalytic proteins. One of these gene products, the viral DNA polymerase, catalyzes the 1st stage of viral DNA synthesis. In this process, the parental viral genome serves as the template and progeny DNA that is genome-length, to at most twice genome length, is produced. Newly-synthesized viral DNA may serve as the template for additional rounds of DNA replication or early transcription, or it may be transported to the cytoplasm where the 2nd stage of viral DNA synthesis occurs. In the cytoplasm, viral DNA is synthesized into large, branched concatamers that serve as the template for DNA packaging.

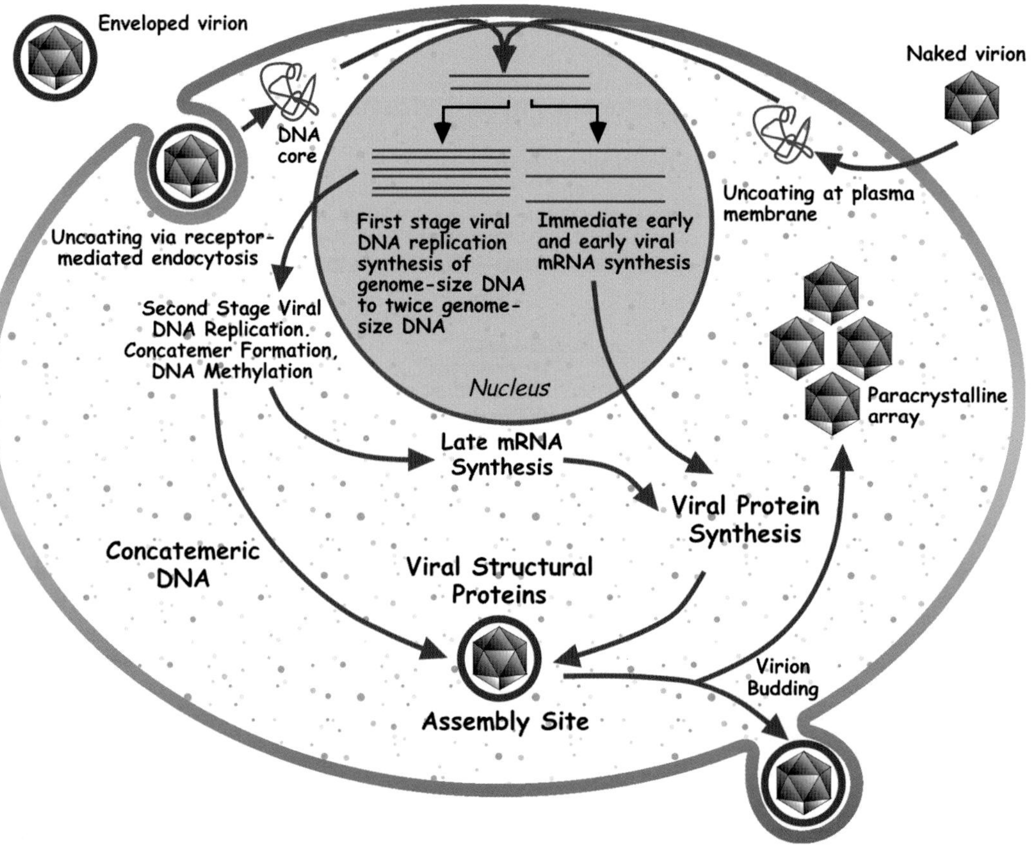

Figure 3: Replication cycle of Frog virus 3 (FV-3)(From Chinchar *et al.*, (2002). *Arch. Virol.*, **147**, 447, with permission).

Table 1. Summary of genomic sequence information for five virus species representing four genera within the family *Iridoviridae*.

Genus Virus species	*Iridovirus* IIV-6	*Ranavirus* ATV	*Ranavirus* TFV	*Lymphocysti-virus* LCDV-1	*Megalocyti-virus* ISKNV
Genome size (bp)	212,482	106,332	105,057	102, 653	111,362
G+C%	28.6%	54%	55%	29.1%	54.8%
Putative ORFs	468	102	105	195	124
ORF size (aa)	40 - 2432	40 – 1294	40 - 1294	40 - 1199	40 – 1208
Accession No.	AF303741	AY150217	NC003407	NC001824	NC003494

Viral DNA methylation also likely occurs in the cytoplasm and, although its precise role is uncertain, is thought to protect viral DNA from endonucleolytic attack. Transcription of late (L) viral transcripts occurs in the cytoplasm and full L gene transcription requires prior DNA synthesis. Homologs of the two largest subunits of RNA polymerase II are encoded by all iridoviruses. Whether these function only in the cytoplasm to transcribe L viral transcripts, or whether they also play a role in continued early transcription has not yet been determined. Virion formation takes place in the cytoplasm within morphologically distinct virus assembly sites. Within assembly sites concatameric viral DNA is packaged into virions via a "headful" mechanism that results in the generation of circularly permuted and terminally redundant genomes similar to those seen with the Enterobacteria phages T4 or P22. The degree of terminal redundancy varies from approximately 5 to 50%. Following assembly, virions accumulate in the cytoplasm within large paracrystalline arrays or acquire an envelope by budding from the plasma membrane. In the case of most vertebrate iridoviruses, the majority of virions remain cell associated (Fig 3).

Table 2. Partial listing of putative gene products encoded by viruses within the genera *Iridovirus, Ranavirus, Lymphocystivirus* and *Megalocytivirus*.

Category Gene Product	**Irido** **IIV-6**	**Rana** **ATV**	**Rana** **TFV**	**Lympho** **LCDV-1**	**Megalo** **ISKNV**
Enzymes Associated with Nucleic Acid Replication and Metabolism					
DNA polymerase	+	+	+	+	+
RNA polymerase II, α subunit	+	+	+	+	+
RNA polymerase II, β subunit	+	+	+	+	+
Transcription factor-like protein	+	+	+	+	
RAD-2, DNA repair enzyme		+	+	+	+
Cytosine DNA methyltransferase		+	+	+	+
Type II restriction enzyme *Msp* I of *Moraxella* sp.				+	
RNAse III	+	+	+	+	+
Ribonucleotide reductase, large subunit	+	+	+	+	
Ribonucleotide reductase, small subunit	+	+	+	+	+
DUTPase	+	+	+		
Thymidylate synthase	+	+	+		
Thymidine kinase	+			+	
Thymidylate kinase	+	+			
Topoisomerase II-like protein	+				
Helicase	+	+	+		
PCNA protein	+	+		+	
DNA ligase	+				
Additional Enzymatic Activities					
Tyrosine phosphatase	+				
Tyrosine kinase	+	+			
Thiol oxidoreductase		+			
Serine/threonine protein kinase	+			+	+
ATPase	+	+	+	+	+
Matrix metalloproteinase	+				
mRNA capping enzyme					+
Cathepsin B-like protein	+			+	
Putative Immune Evasion Proteins					
TNF receptor-associated factor					+
Growth factor/cytokine receptor family signature	+				
TNFR/NGFR family proteins				+	
PDGF/VEGF-like protein					+
Apoptosis inhibitor (IAP) of Cydia pomonellla granulosis virus, Bir repeat profile	+				
IAP-like protein of African swine fever virus	+				

CARD, caspase recruitment domain		+			
3' β hydroxysteroid oxidoreductase		+	+	+	
eIF-2 α homolog		+	+		
Src homology domain, suppressor of cytokine signaling					+
Structural Proteins					
Major CP	+	+	+	+	+
Myristylated membrane protein		+			

The presence of a homolog of the indicated gene is indicated by a plus sign (+).

ANTIGENIC PROPERTIES

The genera are serologically distinct from one another. In the genus *Iridovirus* there exists one main group of serologically interrelated species and others which have little sero-relatedness. Several amphibian isolates (e.g., Rana esculenta iridovirus, REIR) and piscine isolates (e.g. EHNV) show serological cross-reactivity with FV-3 (genus *Ranavirus*). Antibodies prepared against virions are often non-neutralizing.

BIOLOGICAL PROPERTIES

Iridoviruses have been isolated only from poikilothermic animals, usually associated with damp or aquatic environments, including marine habitats. Iridovirus species vary widely in their natural host range and in their virulence. Transmission mechanisms are poorly understood for the majority of these viruses. Invertebrate iridoviruses may be transmitted by endoparasitic wasps or parasitic nematodes. Viruses may be transmitted experimentally by injection or bath immersion, and naturally by co-habitation, feeding, or wounding. While many of these viruses cause serious, life-threatening infections, subclinical infections are common.

GENUS *IRIDOVIRUS*

Type Species *Invertebrate iridescent virus 6*

DISTINGUISHING FEATURES

VIRION PROPERTIES

MORPHOLOGY

Particle diameter is 120-130 nm in ultrathin section. IIV-1 and IIV-2 are assumed to contain 1472 capsomers arranged in 20 trimers and 12 pentamers.

PHYSICOCHEMICAL AND PHYSICAL PROPERTIES

Virions have an Mr of approximately 1.28 x 10^9, a buoyant density of 1.30-1.33 g/cm^3, and a sedimentation coefficient $S_{20,w}$ of 2020-2250S. IIV-6 is sensitive to chloroform, SDS, sodium deoxycholate, ethanol, pH3 and pH11, but is not sensitive to trypsin, lipase, phospholipase A2 or EDTA. The sensitivity of IIV-6 to ether differs depending on whether an *in vivo* or *in vitro* assay is used to determine residual infectivity.

NUCLEIC ACID

Genome sizes vary from 140-210 kbp. This figure includes both unique sequences and a variable amount of terminal redundancy. The unit length genome size of IIV-6 is 212,482 bp. The G+C content is typically 29-32%.

PROTEINS

Two dimensional SDS-PAGE of IIV-6 has revealed the presence of approximately 35 polypeptides ranging from 11 to 300 kDa. Sequence analysis identified 468 putative ORFs of which 234 are non-overlapping. The CP exists in two forms, a 50 kDa monomeric entity located on the external surface of the capsid and a disulphide-linked homotrimer located in the interior of the capsid.

LIPIDS

The internal lipid layer is believed to be important in the stability of these viruses in aquatic environments. Treatments with chloroform reduces the infectivity of IIV-6 and Anticarsia gemmatalis iridescent virus (AGIV) whereas IIV-2 has been reported as insensitive to treatment with ether or chloroform.

CARBOHYDRATES

Carbohydrates are not present in purified virions.

ANTIGENIC PROPERTIES

Genetic studies have indicated the presence of discrete complexes of inter-related viruses within this genus: one large complex containing 10 tentative species, and two smaller complexes. Serological relationships follow a similar pattern.

BIOLOGICAL PROPERTIES

Iridoviruses have been isolated from a wide range of arthropods, particularly insects in aquatic or damp habitats. Patently infected animals and purified viral pellets display violet, blue or turquoise iridescence. Non-apparent, non-lethal infections may be common in certain hosts. No evidence exists for transovarial transmission and where horizontal transmission has been demonstrated, it is usually by cannibalism or predation of infected invertebrate hosts. Following experimental injection, many members of the genus can replicate in a large number of insects. In nature, the host range appears to vary but there is evidence, for some viruses, of natural transmission across insect orders and even phyla. Invertebrate iridescent viruses have a global distribution.

LIST OF SPECIES DEMARCATION CRITERIA IN THE GENUS

The following species-defining characteristics and associated limits are preliminary in nature. The following definitions assume that all material being compared has been grown under near identical conditions and prepared for examination following identical protocols. It is recommended that members of both recognized virus species be included in all characterization studies of novel isolates.

- Amino acid sequence analysis of the MCP: Members of distinct species should exhibit no more than 90% aa sequence identity for the complete protein sequence. PCR primers have been designed for conserved regions of this gene. Although the complete IIV-6 genome has been determined and the sequence of a number of other proteins from different isolates has been ascertained, this information has not been used for species differentiation and quantitative limits of similarity have not been established.
- DNA-DNA dot-blot hybridization: Hybridization values should be less than 50% for members of distinct species. DNA-DNA reassociation in solution has not proven useful for species comparisons of iridoviruses.
- RFLP: Using a panel of not fewer than 4 restriction endonucleases (both rare and frequent cutters) distinct species should show completely distinct restriction endonuclease profiles.
- Serology: Antisera from members of strains within a species should exhibit high levels of cross reactivity. Within and among species, comparison by Western blot analysis using antibodies raised against disrupted virions is the preferred method. Comparisons should be performed simultaneously wherever possible and reference species should be included in each determination.

The major structural protein of IIV-1 shows 66.4% aa sequence identity to that of IIV-6 and approximately 50% or lower aa sequence identity to iridoviruses in other genera. Less than 1% DNA-DNA hybridization for genomic DNA was detected by dot-blot method between IIV-1 and IIV-6 (stringency: 26% mismatch). Restriction endonuclease profiles (*Hin*dII, *Eco*RI, *Sal*I) showed a coefficient of similarity of <66% between IIV-1 and IIV-6. These species did not share common antigens when tested by tube precipitation, infectivity neutralization, reversed single radial immunodiffusion or enzyme-linked immunosorbant assay. Genome and protein size differences are not useful in differentiating these species; genome sizes can be highly variable among strains of a species whereas the size of the MCP is well conserved among species. Little is known about the usefulness of other proteins as species demarcation criteria in the genus.

LIST OF SPECIES IN THE GENUS

Species names are in green italic script; strain names and synonyms are in black roman script; tentative species names are in blue roman script. Sequence accession numbers, and assigned abbreviations ( ) are also listed.

SPECIES IN THE GENUS

Invertebrate iridescent virus 1		
Invertebrate iridescent virus 1	[M33542, M62953]	(IIV-1)
(Tipula iridescent virus)		(TIV)
Invertebrate iridescent virus 6		
Gryllus iridovirus		(GRIV)
Invertebrate iridescent virus 6	[AF003534; M99395]	(IIV-6)
(Chilo iridescent virus)		(CIV)

TENTATIVE SPECIES IN THE GENUS

Anticarsia gemmatalis iridescent virus	[AF042343]	(AGIV)
Invertebrate iridescent virus 2	[AF042335]	(IIV-2)
Sericesthis iridescent virus		
Invertebrate iridescent virus 9	[AF025774]	(IIV-9)
Invertebrate iridescent virus 10		
Invertebrate iridescent virus 18		
Opogonia iridescent virus		
Wiseana iridescent virus		
Witlesia iridescent virus		
Invertebrate iridescent virus 16	[AF025775]	(IIV-16)
Costelytra zealandica iridescent virus		
Invertebrate iridescent virus 21		(IIV-21)
Heliothis armigera iridescent virus		
Insect iridescent virus 28		
Lethocerus columbinae iridescent virus		
Invertebrate iridescent virus 22	[AF042341; M32799]	(IIV-22)
Simulium sp. iridescent virus		
Invertebrate iridescent virus 23	[AF042342]	(IIV-23)
Black beetle iridescent virus		
Heteronychus arator iridescent virus		
Invertebrate iridescent virus 24	[AF042340]	(IIV-24)
Apis iridescent virus		
Bee iridescent virus		
Invertebrate iridescent virus 29	[AF042339]	(IIV-29)
Tenebrio molitor iridescent virus		
Invertebrate iridescent virus 30	[AF042336]	(IIV-30)
Helicoverpa zea iridescent virus		
Invertebrate iridescent virus 31	[AF042337; AJ279821]	(IIV-31)
Armadillidium vulgare iridescent virus		
Invertebrate iridescent virus 32		
Isopod iridescent virus		
Porcellio dilatatus iridescent virus		

GENUS *CHLORIRIDOVIRUS*

Type Species *Invertebrate iridescent virus 3*

DISTINGUISHING FEATURES

VIRION PROPERTIES

MORPHOLOGY

Particle diameter is approximately 180 nm in ultrathin section. The trimers and pentamers of Invertebrate iridescent virus 3 (IIV-3) are larger than the corresponding structures of the genus *Iridovirus*, with probably 14 capsomers to each edge of the trimer. Particle size has historically been used to define viruses that are members of this genus, but the validity of that characteristic is uncertain.

PHYSICOCHEMICAL AND PHYSICAL PROPERTIES

Virions have a Mr of approximately 2.49-2.75 x 10^9, a buoyant density of approximately 1.354 g/cm^3 in CsCl, and a S_{20w} of 4440-4460S. Infectivity is believed not to be sensitive to ether.

NUCLEIC ACID

The genome size is estimated to be ~135 kbp with a G+C content of 53.9%.

PROTEINS

Protein studies of chloriridoviruses are ongoing. Based on genome size, IIV-3 likely encodes ~100 proteins. Recent sequence analysis has identified a DNA polymerase delta-like protein, and SDS-PAGE detected a 55 kDa protein that is likely the MCP.

LIPIDS

The lipid content of IIV-3 is approximately 4%.

CARBOHYDRATES

None reported.

ANTIGENIC PROPERTIES

IIV-3 is serologically distinct from members of other genera.

BIOLOGICAL PROPERTIES

IIV-3 is the only virus characterized from this genus, although more than 20 host species were reported with patent infections world-wide. Chloriridovirus-like infections have only been reported from Diptera with aquatic larval stages, mainly mosquitoes. There is evidence for transovarial transmission in mosquitoes infected by IIV-3. Horizontal transmission is achieved by cannibalism or predation of infected mosquitoes of other species. Patently infected larvae and purified pellets of virus iridesce usually with a yellow-green color, although orange and red infections are known. IIV-3 appears to have a narrow host range compared to most members of the genus *Iridovirus*.

LIST OF SPECIES DEMARCATION CRITERIA IN THE GENUS

Not applicable.

LIST OF SPECIES IN THE GENUS

Species names are in green italic script; strain names and synonyms are in black roman script; tentative species names are in blue roman script. Sequence accession numbers, and assigned abbreviations ( ) are also listed.

SPECIES IN THE GENUS

Invertebrate iridescent virus 3

Invertebrate iridescent virus 3	[AJ312708]	(IIV-3)
Aedes taeniorhynchus iridescent virus		
Mosquito iridescent virus		

TENTATIVE SPECIES IN THE GENUS

None reported.

GENUS *RANAVIRUS*

Type Species ***Frog virus 3***

DISTINGUISHING FEATURES

VIRION PROPERTIES

MORPHOLOGY

Particle diameter is approximately 150 nm in ultrathin section. Enveloped virions, released by budding, measure 160-200 nm in diameter. The capsid has a skewed symmetry with a T=133 or 147. The internal lipid bilayer contains transmembrane proteins. The nucleoprotein core consists of a long coiled filament 10 nm wide.

PHYSICOCHEMICAL AND PHYSICAL PROPERTIES

Buoyant density is 1.28 g/cm^3 for enveloped particles and 1.32 g/cm^3 for unenveloped particles. Infectivity is rapidly lost at pH 2.0-3.0 and at temperatures above 50°C. Particles are inactivated by treatment with ether, chloroform, sodium deoxychlorate, and phospholipase A.

NUCLEIC ACID

Virions contain a single linear dsDNA molecule of 150-170 kbp. The genome is circularly permuted and approximately 30% terminally redundant. The unit genome size is approximately ~105 kbp with a G+C content of ~54% (Table 1). Cytosines within the dinucleotide sequence CpG are methylated by a virus-encoded cytosine DNA methyltransferase. DNA methylation likely occurs in the cytoplasm and may be important in protecting DNA from viral endonucleases.

PROTEIN

Approximately 30 structural proteins have been detected in FV-3 virions, whereas sequence analysis of TFV, and ATV predicted slightly more than 100 ORFs (Table 1) Proteins do not undergo extensive post-translational processing and no evidence for glycosylation, sulfation or cleavage from precursors has been detected. Phosphoproteins of 10-114 kDa are present in the virion core and a virion-associated protein kinase (44 kDa) has been isolated. Other virion-associated proteins include a nucleotide phosphohydrolase, pH5 and pH7.5 endodeoxyribonucleases, an endoribonuclease, a protein phosphatase, two proteins (VP44 and VP63) within the lipid layer, and one protein (VP58) within the viral envelope. All ranaviruses examined to date appear to encode a viral homolog of eIF-2α that is thought to play a role in maintaining viral protein synthesis in infected cells. Sequence analysis of key proteins among various species within the genus *Ranavirus* shows high levels (i.e., >70%) of sequence identity.

LIPIDS

Non-enveloped particles contain 9% lipid. The composition of the internal lipid membrane differs from that of the host cell. Viruses released from cells by budding acquire their envelope from the plasma membrane.

CARBOHYDRATES

None reported.

GENOME ORGANIZATION AND REPLICATION

The replication cycle of FV-3 serves as the model for iridoviruses and has been discussed above (Fig. 3). The complete genomes of two ranaviruses (e.g., TFV and ATV) have recently been sequenced and, while possessing homologous proteins, are not co-linear (Table 1, Fig. 2). Sequence analysis of FV-3 and EHNV is ongoing.

ANTIGENIC PROPERTIES

Ranaviruses such as FV-3 are serologically and genetically distinct from members of other genera. However, several piscine, reptilian and amphibian ranavirus isolates show serological and/or genetic relatedness to FV-3

BIOLOGICAL PROPERTIES

Viral transmission occurs by feeding, parenteral injection, or environmental exposure. Ranaviruses grow in a wide variety of cultured fish, amphibian, and mammalian cells, and cause marked cytopathic effect culminating in cell death, likely by apoptosis. In contrast to their marked pathogenicity *in vitro*, their effect in animals depends on the viral species, and on the identity and age of the host animal. For example, Largemouth bass virus (LMBV) shows evidence of wide-spread infection in the wild, but is only rarely linked to serious disease. Likewise, FV-3 infection leads to death in tadpoles, but often causes only non-apparent infections in adults. It is likely that environmental stress leading to immune suppression increases the pathogenicity of a given iridovirus. Ranavirus infections are often not limited to a single species or taxonomic class of animals. For example, EHNV has been reported to infect at least 13 genera of fish. In addition, BIV, a highly virulent pathogen of the burrowing frog *Lymnodynastes ornatus*, can be experimentally transmitted to fish. Therefore, isolation of a ranavirus from a new host species does not necessarily identify a new viral species. In their most severe disease manifestations, ranaviruses such as FV-3, ATV, European catfish virus (ECV) and EHNV are associated with systemic disease and show marked hemorrhagic involvement of internal organs such as the liver, spleen, kidneys, and gut.

LIST OF SPECIES DEMARCATION CRITERIA IN THE GENUS

Ranaviruses cause systemic disease in fish, amphibians, and reptiles. Members of the six viral species are differentiated from one another by multiple criteria: RFLP profiles, virus protein profiles, DNA sequence analysis, and host specificity. PCR primers have been designed to amplify 3′ and 5′ regions within the MCP gene for identification purposes. Definitive quantitative criteria based on the above features have not yet been established to delineate different viral species. Generally, if a given isolate shows a distinct RFLP profile, possesses a distinctive host range, and is markedly different from other viruses at the aa sequence level, it is considered a distinct viral species. For example, ranavirus DNA digested with *Kpn*I can be ordered into several groups based on RFLP profiles. Strains within the same species shared >70% of their bands in common and showed >95% aa sequence identity within the MCP or other key genes (e.g., ATPase, eIF-2α homolog).

LIST OF SPECIES IN THE GENUS

Species names are in green italic script; strain names and synonyms are in black roman script; tentative species names are in blue roman script. Sequence accession numbers, and assigned abbreviations ( ) are also listed.

SPECIES IN THE GENUS

Ambystoma tigrinum virus		
Ambystoma tigrinum virus	[AY150217]	(ATV)
Regina ranavirus		
Bohle iridovirus		
Bohle iridovirus	[AF157650, AF157651]	(BIV)
Epizootic haematopoietic necrosis virus		

Epizootic haematopoietic necrosis virus	[U82552, U82631, AJ007358, AJ130963, AJ130964, AJ130965]	(EHNV)
European catfish virus		
European catfish virus	[AF 157678, AF127911]	(ECV)
European sheatfish virus		(ESV)
Frog virus 3		
Box turtle virus 3		
Bufo bufo United Kingdom virus		
Bufo marinus Venezuelan iridovirus 1		
Frog virus 3	[U36913, U15575]	(FV-3)
Lucké triturus virus 1		
Rana temporaria United Kingdom virus		
Redwood Park virus		
Stickleback virus		
Tadpole edema virus		
Tadpole virus 2		
Tiger frog virus	[NC_003407]	(TFV)
Tortoise virus 5		
Santee-Cooper ranavirus		
Doctor fish virus		(DFV)
Guppy virus 6		(GV6)
Santee-Cooper ranavirus	[AF080250]	(SCRV)
(Largemouth bass virus)		(LMBV)

TENTATIVE SPECIES IN THE GENUS

Rana esculenta iridovirus	(REIR)
Singapore grouper iridovirus	(SGIV)
Testudo iridovirus	(ThIV)

GENUS *LYMPHOCYSTIVIRUS*

Type Species *Lymphocystis disease virus 1*

DISTINGUISHING FEATURES

VIRION PROPERTIES

MORPHOLOGY

Particle size varies from 198-227 nm for Lymphocystis disease virus 1 (LCDV-1) and 200 nm for LCDV-2. The capsid may show a fringe of fibril-like external protrusions ~ 2.5 nm in length and a double-layered outer envelope.

PHYSICOCHEMICAL AND PHYSICAL PROPERTIES

Virions are heat labile. Infectivity is sensitive to treatment with ether or glycerol.

NUCLEIC ACID

By restriction endonuclease analysis, the genome length is 102.6 kbp for LCDV-1 and approximately 98 kbp in LCDV-2. Contour length measurements by electron microscopy indicate the DNA molecule to be 146 kbp; the degree of terminal redundancy is approximately 50% but varies considerably among virions. The G+C content is 29.1%. Like FV-3, the genome is highly methylated. The presence of 5-methylcytosine occurs at 74% of CpG, 1% of CpC and 2-5% of CpA giving an overall level of methylation of 22%. The complete DNA sequence is known for LCDV-1.

PROTEIN

SDS-PAGE revealed the presence of 33 polypeptides, ranging from 4 to 220 kDa, in LCDV-1 virions isolated directly from flounder tumors. Analysis of LCDV-2 virions gives a discernibly different pattern of polypeptides supporting the idea that they are distinct species. The MCP is 51.4 kDa comprising 459 aa. Enzymatic activities associated with purified virions include a viral encoded adenosine triphosphate hydrolase, a protein kinase and a thymidine kinase. Genome sequence analysis indicated the presence of 38 putative proteins with significant aa sequence homologies to proteins of known function.

LIPID

Non-enveloped particles contain up to 17.1% lipid which is readily digested by treatment with phospholipase A_2, suggesting high levels of phospholipid as seen in other members of the family.

CARBOHYDRATES

None reported.

GENOME ORGANIZATION AND REPLICATION

The LCDV-1 genome contains 195 potential ORFs of which 110 are largely non-overlapping and 38 of which show significant homology to proteins of known function. These 38 ORFs represent 43% of the coding capacity of the genome. The presence of a DNA methyltransferase and a methyl-sensitive restriction endonuclease with specificity for a CCGG target site may be indicative of a restriction-modification system capable of degrading host genomic DNA while protecting viral DNA by specific methylation. LCDV-1 DNA contains numerous short direct, inverted and palindromic repetitive sequence elements. Lack of a suitable cell line has hindered studies of LCDV replication. Virus assembly occurs in and around virogenic stroma in the cytoplasm. Crescent shaped capsid precursors develop into fully formed capsids followed by condensation of the core structures.

ANTIGENIC PROPERTIES

Not known.

BIOLOGICAL PROPERTIES

LCDV-1 infects flounder and plaice while LCDV-2 infects dab. Infection results in benign, wart-like lesions comprising grossly hypertrophied cells occurring mostly in the skin and fins. The disease has been observed in over 100 teleost species although virus species other than LCDV-1 or LCDV-2 may cause a similar clinical disease. The duration of infected cell growth and viral proliferation is highly variable (5 days to 9 months) and is likely temperature dependent. Virions are released following degeneration of the lesions. Transmission is achieved by contact; external sites, including the gills are the principal portals of entry. High host population densities and external trauma are believed to enhance transmission. Implantation and injection are also effective routes of transmission. The incidence of disease may be higher in the presence of certain fish ectoparasites. LCDV is generally not considered of major economic importance. However, although infectious are usually benign and self-limiting, there many be commercial concerns among food and ornamental fish because of market rejection due to the warty appearance of infected animals. Moreover, mortalities may occur under culture conditions, especially when infections involve the gills or when there is debilitation or bacterial infection. Both LCDV-1 and LCDV-2 are difficult to culture *in vitro* although limited growth has been reported in several fish cell lines.

List of Species Demarcation Criteria in the Genus

Definitive criteria have not yet been established to delineate the viral species. LCDV-1 infects flounder and plaice, while LCDV-2 infects dab. The two species are distinguished from one another by host specificity, histopathology, and molecular criteria: viral protein profiles, DNA sequence analysis, and PCR. PCR primers targeted to regions within the MCP and ORF167L can be used to distinguish between species. While the designation of LCDV-1 as a viral species is well supported in the literature, data supporting the designation of LCDV-2 is still preliminary.

List of Species in the Genus

Species names are in green italic script; strain names and synonyms are in black roman script; tentative species names are in blue roman script. Sequence accession numbers, and assigned abbreviations ( ) are also listed.

Species in the Genus

Lymphocystis disease virus 1

Lymphocystis disease virus 1	[L63545]	(LCDV-1)
(Flounder lymphocystis disease virus)		(FLDV)
Flounder virus		

Tentative Species in the Genus

Lymphocystis disease virus 2	(LCDV-2)
Dab lymphocystis disease virus	

Genus Megalocytivirus

Type Species *Infectious spleen and kidney necrosis virus*

Distinguishing Features

Virion Properties

Morphology

Virions possess icosahedral symmetry and are ~ 140–200 nm in diameter.

Physicochemical and Physical Properties

Virions are sensitive to heat (56°C for 30 min), sodium hypochlorite, UV irradiation, chloroform and ether, and are variably inactivated by exposure to pH3 and pH11.

Nucleic Acid

The complete genomes of ISKNV and Red Sea bream iridovirus (RSIV) have been sequenced. ISKNV virions contain a single, linear dsDNA molecule of 111,362 bp with a G+C content of 54.8%. As with other members of the family, genomic DNA is circularly permuted, terminally redundant, and highly methylated.

Proteins

Sequence analysis of the ISKNV genome has identified 124 putative ORFs ranging in size from 40–1208 aa. Thirty five putative proteins with significant aa sequence homologies to proteins of known function in other species have been identified. ATPase and MCP genes (239 and 454 aa, respectively) have also been sequenced from Sea bass iridovirus (SBIV), Grouper sleepy disease iridovirus (GSDIV), Dwarf gourami iridovirus (DGIV), and African lampeye iridovirus (ALIV).

Lipids

None reported.

CARBOHYDRATES

None reported.

GENOME ORGANIZATION AND REPLICATION

The complete genome of ISKNV has been sequenced and appears similar in size and gene content to those of other iridoviruses.

ANTIGENIC PROPERTIES

No serotypes are reported and all megalocytiviruses analyzed to date appear to be members of the same viral species. Polyclonal anti-RSIV serum shows cross-reactivity with ESV- and EHNV-infected cells, whereas monoclonal anti-RSIV antibodies react only with RSIV-infected cells. Megalocytiviruses show high levels (i.e. >93%) of aa sequence identity among the proteins characterized to date.

BIOLOGICAL PROPERTIES

Iridoviruses infecting red sea bream, mandarin fish and over 20 other species of marine and tropical fish have been known since the late 1980's. Isolates from red sea bream (RSIV) and the mandarin fish (ISKNV) have been studied extensively. Viral infection is characterized by the formation of inclusion body-bearing cells (IBC). IBCs may be derived from virus-infected macrophages and enlarge by the growth of a unique inclusion body that may be sharply delineated from the host cell cytoplasm by a limiting membrane. When a limiting membrane is seen, the inclusions contain the viral assembly site and possess abundant ribosomes, rough ER and mitochondria. IBCs frequently appear in the spleen, hematopoietic tissue, gills, and digestive tract. Necrotized splenocytes are also observed. Transmission has been demonstrated by feeding, parenteral injection, and by environmental exposure. Megalocytiviruses naturally infect and cause significant mortality in freshwater and marine fish in aquaculture facilities in China, Japan, and SE Asia. A partial list of susceptible fish species includes mandarin fish (*Siniperca chuatsi*), red sea bream (*Pagrus major*), grouper (*Epinephelus* spp), yellowtail (*Seriola quinqueradiata*), striped beakperch (*Oplegnathus fasciatus*), red drum (*Sciaenops ocellata*), and African lampeye (*Aplocheilichthys normani*). The virus grows in several cultured piscine cell lines and causes a characteristic enlargement of infected cells. Outbreaks of disease caused by ISKNV only occur in fish cultured at temperatures >20° C. A vaccine targeted to RSIV has been developed.

LIST OF SPECIES DEMARCATION CRITERIA IN THE GENUS

Megalocytiviruses are distinguished from ranaviruses and lymphocystiviruses by their cytopathological presentation (i.e., inclusion body-bearing cells) and sequence analysis of key viral genes, e.g., ATPase and MCP, for which PCR primers have been developed. Megalocytiviruses show >94% sequence identity within these genes, whereas sequence identity with ranaviruses and lymphocystiviruses is <50%. Based on sequence analysis and serological studies, all megalocytiviruses isolated to date appear to be strains of the same viral species.

LIST OF SPECIES IN THE GENUS

Species names are in green italic script; strain names and synonyms are in black roman script; tentative species names are in blue roman script. Sequence accession numbers, and assigned abbreviations ( ) are also listed.

SPECIES IN THE GENUS

Infectious spleen and kidney necrosis virus		
Infectious spleen and kidney necrosis virus	[NC_003494]	(ISKNV)
Red Sea bream iridovirus	[AB006954,AB007366,AB007367,AB018418]	(RSIV)

Sea bass iridovirus	[AB043977]	(SBIV)
African lampeye iridovirus	[AB043979]	(ALIV)
Grouper sleepy disease iridovirus	[AB043978]	(GSDIV)
Dwarf gourami iridovirus		(DGIV)
Taiwan grouper iridovirus	[AAL68652]	(TGIV)

TENTATIVE SPECIES IN THE GENUS

None reported

PHYLOGENETIC RELATIONSHIPS WITHIN THE GENUS

The megalocytiviruses sequenced to date show >95% nt identity and >98% aa sequence identity within the ATPase gene, and >94% nt and >97% aa sequence identity within the MCP gene. In contrast they are distantly related to lymphocystiviruses, ranaviruses, and iridoviruses. For example, the ATPase gene of SBIV shows 54% identity and 72% similarity to that of LCDV-1; 45% identity and 62% similarity to FV-3; and 38% identity and 62% similarity to IIV-6.

LIST OF UNASSIGNED SPECIES IN THE FAMILY

The presence of large, non-enveloped virus particles in both assembly sites and paracrystalline arrays within the cytoplasm of infected cells is characteristic of iridovirus infections. Because of these distinguishing morphological features, several viruses infecting ectothermic animals have been tentatively identified as iridoviruses without further molecular or serological characterization. Furthermore, because many of these viruses have not yet been grown in culture little is known about their mode of replication and molecular organization.

Unassigned viruses within the family include:

White sturgeon iridovirus	(WSIV)
Erythrocytic necrosis virus	(ENV)

PHYLOGENETIC RELATIONSHIPS WITHIN THE FAMILY

Amino-acid sequence analysis of the MCP, DNA polymerase, ATPase, and other genes supports the existence of four genera within the family: *Iridovirus, Ranavirus, Megalocytivirus,* and *Lymphocystivirus* (Fig. 4). *Chloriridovirus* constitutes a fifth genus within the family, although little sequence information is currently available to support that assertion. Among members of the *Iridoviridae*, MCP and ATPase protein sequence identities are generally >40%.

SIMILARITY WITH OTHER TAXA

The iridovirus MCP and viral DNA polymerase genes share aa sequence similarities to *African swine fever virus* (*Asfarviridae*), species within the family *Ascoviridae*, and *Paramecium bursaria Chlorella virus 1* (*Phycodnaviridae*) (Fig. 4). Although not shown here, more distant relatedness to herpesviruses, adenoviruses, poxviruses, and baculoviruses has been noted.

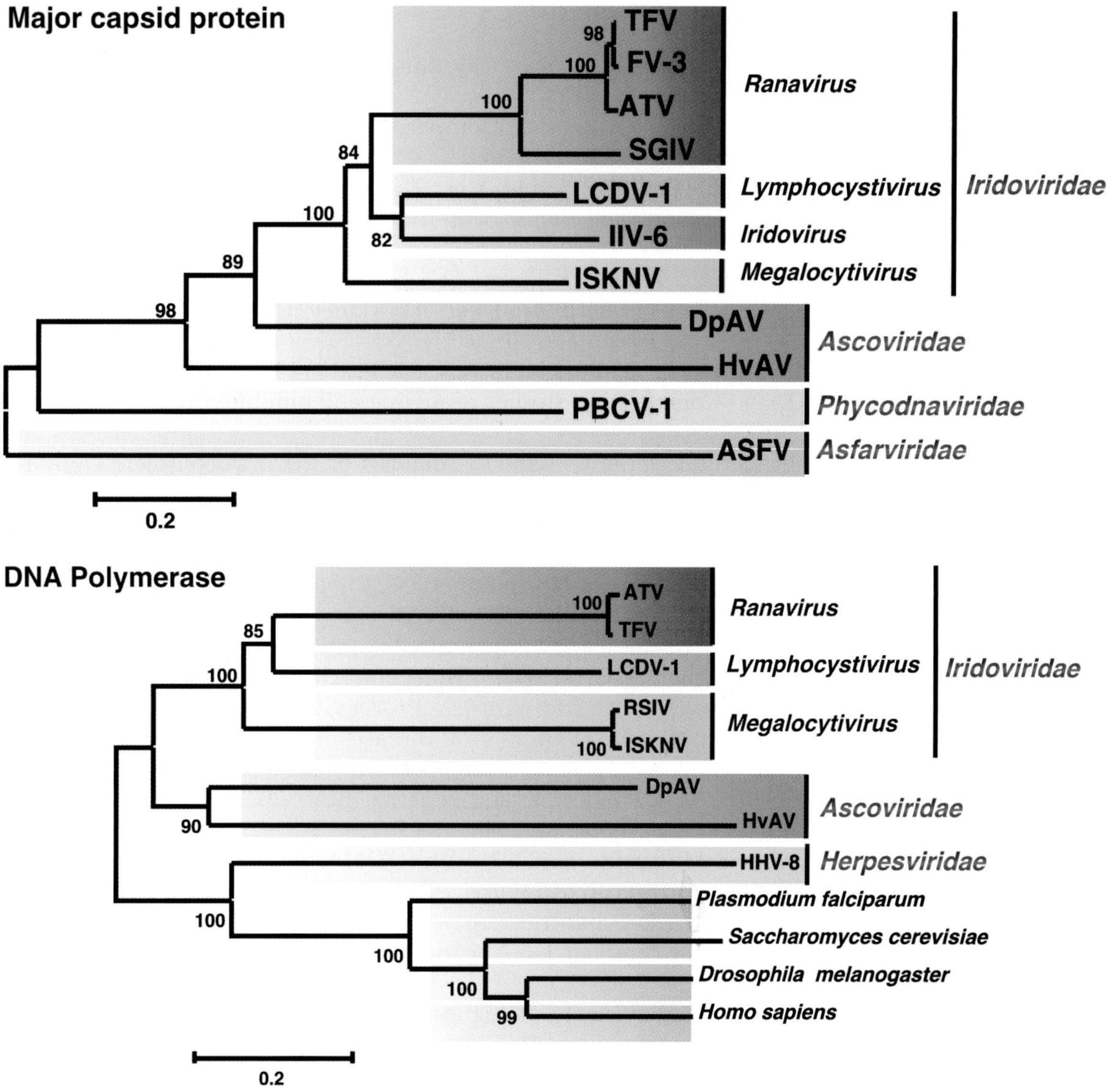

Figure 4: Phylogenetic trees depicting the relationship of selected iridoviruses to other species within the family *Iridoviridae*, viruses outside the family, and host proteins. Top panel: Comparison of the major CPs of four members of the genus *Ranavirus* (TFV, FV3, ATV, and SGIV) to LCDV (genus *Lymphocystivirus*), IIV-6 (genus *Iridovirus*), ISKNV (genus *Megalocytivirus*), two members of the family *Ascoviridae* (Diadromus pulchellus ascovirus, DpAV and Heliothis virescens ascovirus, HvAV), Paramecium bursaria Chlorella virus 1 (PBCV-1, *Phycodnaviridae*), and African swine fever virus (ASFV, *Asfarviridae*). Right panel: Comparison of the DNA polymerase gene of ATV and other iridoviruses to the DNA polymerase genes of two ascoviruses (HvAV and DpAV), Human herpesvirus 8 (HHV-8), *Plasmodium falciparum*, *Saccharomyces cerevisiae*, *Drosophilia melanogaster*, and *Homo sapiens*. (Jancovich *et al.*, (2003), *Virology* **316**, 90. With permission).

Derivation of Names

Chloro: from Greek *chloros*, meaning "green".
Cysti: from Greek *kystis* meaning "bladder/sac".
Irido: from Greek *iris, iridos*, goddess whose sign was the rainbow, hence iridescent "shining like a rainbow" from the appearance of patently infected invertebrates and centrifuged pellets of virions.
Lympho: from Latin *lympha*, meaning "water".
Megalocyti: from the Greek, meaning "enlarged cell"
Rana: from Latin *rana*, meaning "frog".

REFERENCES

Chinchar, V.G. (2002). Ranaviruses (family *Iridoviridae*): Emerging cold-blooded killers. *Arch. Virol.,* **147**, 447-470.

Essbauer, S., Bremont, M. and Ahne, W. (2001). Comparison of eIF-2∀ homologous proteins of seven ranaviruses (*Iridoviridae*). *Virus Genes,* **23**, 347-359.

He, J.G., Lu, L., Deng, M., He, H.H., Weng, S.P., Wang, X.H., Zhou, S.Y., Long, Q.X., Wang, X.Z. and Chan, S.M. (2002). Sequence analysis of the complete genome of an iridovirus isolated from the tiger frog. *Virology,* **292**, 185-197.

He, J.G., Deng, M., Weng, S.P., Li, Z., Zhou, S.Y., Long, Q.X., Wang, X.Z. and Chan, S.M. (2001). Complete genome analysis of the mandarin fish infectious spleen and kidney necrosis iridovirus. *Virology,* **291**, 126-139.

Hyatt, A.D., Gould, A.R., Zupanovic, Z., Cunningham, A.A., Hengstberger, S., Whittington, R.J., Kattenbelt, J. and Coupar, B.E.H. (2000). Comparative studies of piscine and amphibian iridoviruses. *Arch. Virol.*, **145**, 301 - 331.

Jakob, N.J., Muller, K., Bahr, U. and Darai, G. (2001). Analysis of the first complete DNA sequence of an invertebrate iridovirus: Coding strategy of the genome of Chilo iridescent virus. *Virology,* **286**, 182-196.

Jancovich J.K., Mao, J., Chinchar, V.G., Wyatt, C., Case, S.T., Kumar, S., Valente, G., Subramanian, S., Davidson, E.W., Collins, J.P. and Jacobs B.L. (2003). Genetic sequence of a ranavirus (family *Iridoviridae*) associated with salamander mortalities in North America. *Virology* **316**, 90-103.

Martinez, G., Christian, P., Marina, C. and Williams, T. (2003). Sensitivity of *Invertebrate iridescent virus 6* to organic solvents, detergents, enzymes, and temperature treatment. *Virus Res.*, **91**, 249-254.

Mao, J., Hedrick, R.P. and Chinchar, V.G. (1997). Molecular characterization, sequence analysis and taxonomic position of newly isolated fish iridoviruses. *Virology,* **229**, 212-220.

Nakajima, K, Inouye, K. and Sorimachi, M. (1998). Viral diseases in cultured marine fish in Japan. *Fish Pathol.* **33**, 181-188.

Stasiak, K., Renault, S., Demattei, M.-V., Bigot, Y. and Federici, B.A. (2003). Evidence for the evolution of ascoviruses from iridoviruses. *J. Gen. Virol.,* **84**, 2999 – 3009.

Sudthongkong, C., Miyata, M. and Miyazaki, T. (2002). Viral DNA sequences of genes encoding the ATPase and the major capsid protein of tropical iridovirus isolates which are pathogenic to fishes in Japan, South China Sea, and Southeast Asian countries. *Arch. Virol.,* **47**, 2089-2109.

Tidona, C.A. and Darai, G. (1997). The complete DNA sequence of lymphocystis disease virus. *Virology,* **230**, 207-216.

Tidona, C.A. and Darai, G. (2000). Iridovirus homologues of cellular genes: Implications for the molecular evolution of large DNA viruses. *Virus Genes,* **21**, 77-81.

Webby, R. and Kalmakoff, J. (1998). Sequence comparison of the major capsid protein gene from 18 diverse iridoviruses. *Arch. Virol.,* **143**, 1949-1966.

Williams, T. (1996). The iridoviruses. *Adv. Virus Res.,* **46**, 347-412.

Willis, D.B. (1985). *Iridoviridae. Curr. Top Microbiol. Immunol,.* **116**, 1-173.

CONTRIBUTED BY

Chinchar, V.G., Essbauer, S., He, J.G., Hyatt, A., Miyazaki, T., Seligy, V. and Williams, T.

FAMILY PHYCODNAVIRIDAE

TAXONOMIC STRUCTURE OF THE FAMILY

Family *Phycodnaviridae*
Genus ***Chlorovirus***
Genus ***Coccolithovirus***
Genus ***Prasinovirus***
Genus ***Prymnesiovirus***
Genus ***Phaeovirus***
Genus ***Raphidovirus***

VIRION PROPERTIES

MORPHOLOGY

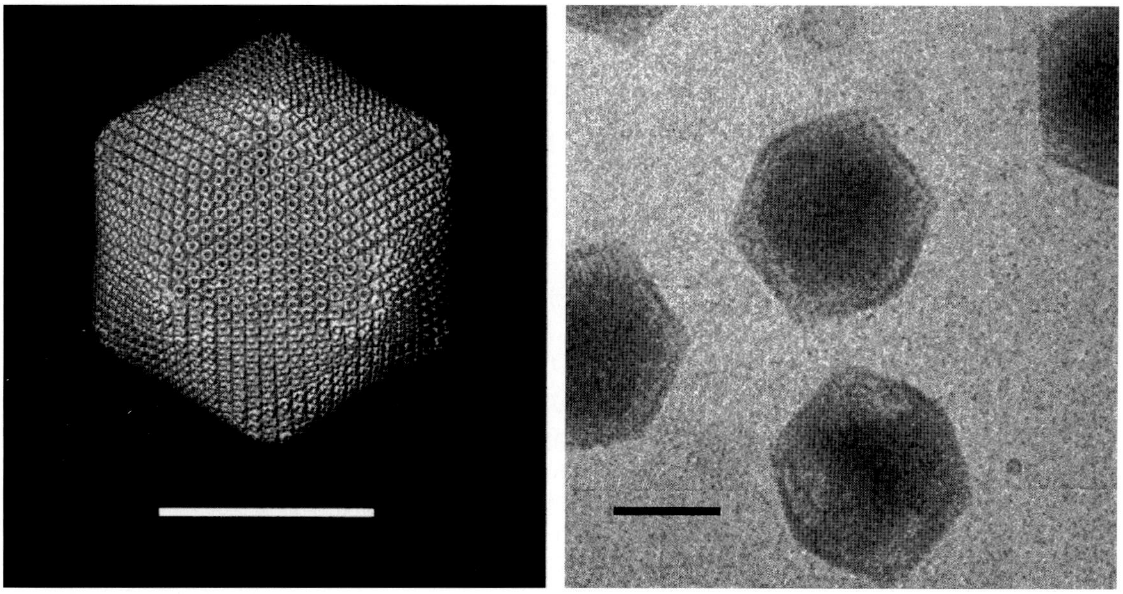

Figure 1: (Left) Shaded-surface view of 3-dimensional reconstruction of a particle of an isolate of *Paramecium bursaria Chlorella virus 1*, viewed along 3-fold axis. (Courtesy of X. Tan, N.H. Olson and T.S. Baker); (Right) Frozen hydrated virions of Paramecium bursaria Chlorella virus 1 (PBCV-1). The bars represent 100 nm.

Virions are polyhedral with a multilaminate shell surrounding an electron dense core. Virions do not have an external membrane and are 100-220 nm in diameter. Electron micrographs indicate that some virions have flexible hair-like appendages with swollen structures at their ends; these appendages may be involved in attachment to the host.

PHYSICOCHEMICAL AND PHYSICAL PROPERTIES

Virion Mr of an isolate of the type species *Paramecium bursaria Chlorella virus 1*, of the genus *Chlorovirus,* is ~1 x 10^9 and the $S_{20,w}$ is more than 2,000S. Virions in the genus *Chlorovirus* are disrupted in CsCl. The infectivity of chloroviruses is not affected by non-ionic detergents but they are inactivated by organic solvents.

NUCLEIC ACID

Virions contain large dsDNA genomes, ranging from 100 to 560 kbp. The chloroviruses have linear, non-permuted dsDNA genomes with crosslinked hairpin ends. The DNA termini contain identical inverted 1-2.2 kbp repeats. The remainder of the genome is primarily single copy DNA. Coccolithoviruses have linear dsDNA genomes. Phaeoviruses have a linear dsDNA genome with complementary ends that anneal to form a circular molecule. Phaeovirus genomes can become integrated into the genomes of their hosts. The genome structures of the prasinoviruses, prymnesioviruses and raphidoviruses are unknown.

The G + C content of the viral genomes range from 40-52%. The DNA of many, if not all, of the viruses contains methylated bases, both 5-methylcytosine (m5C) and N6-methyladenine (m6A). The percent of methylated bases in the chloroviruses ranges from no m6A and 0.1% m5C to 37% m6A and 47% m5C.

PROTEINS

Purified virions contain as many as 50 or more proteins ranging in size from <10 to >200 kDa. The chlorovirus PBCV-1 has three glycoproteins, three myristylated proteins (the major capsid protein (CP), Vp54, is both glycosylated and myristylated), and several phosphoproteins. There are 5040 copies of Vp54 per virion. The protein consists of two eight-stranded, antiparallel β-barrel, "jelly-roll" domains related by a pseudo 6-fold rotation. Four proteins, including Vp54, are located on the virus surface.

LIPIDS

The chlorovirus Paramecium bursaria Chlorella virus 1 (PBCV-1) virion contains 5-10% lipid. The lipid is in a bilayer membrane located inside the glycoprotein shell and is required for virus infectivity.

CARBOHYDRATES

At least three of the chlorovirus PBCV-1 proteins are glycosylated including the major CP Vp54. The glycan portion of Vp54 is on the external surface of the virion. Unlike other glycoprotein-containing viruses, PBCV-1 encodes most, if not all, of the components required to glycosylate its proteins.

GENOME ORGANIZATION AND REPLICATION

The 330,747 bp chlorovirus PBCV-1 genome has been sequenced (Fig. 2). The virus encodes 697 ORFs, 65 codons or longer, of which 373 are predicted to encode proteins. About 50% of these ORFs match proteins in the databases. Two of the PBCV-1 genes are interrupted by introns: a transcription factor TFIIS-like gene has a self-splicing type I intron and the DNA polymerase gene has a spliceosomal processed type of intron. The PBCV-1 genome also has 11 tRNA genes, one of which is predicted to contain a small intron.

The 335,593 bp phaeovirus Ectocarpus siliculosus virus 1 (EsV-1) genome also has been sequenced (Fig. 2). It contains tandem and dispersed repetitive elements in addition to a large number of ORFs of which 231 are currently counted as genes. Many genes can be assigned to functional groups involved in DNA synthesis, DNA integration, transposition, and polysaccharide metabolism. Furthermore, EsV-1 contains components of a surprisingly complex signal transduction system with six different hybrid histidine protein kinases and four putative serine/threonine protein kinases. Several other genes encode polypeptides with protein-protein interaction domains. However, 50% of the predicted genes have no counterparts in data banks. Only 33 of the 231 identified genes have significant sequence similarities to PBCV-1 genes.

The intracellular site of virion DNA replication is unknown. DNA packaging occurs in localized regions in the cytoplasm. PBCV-1 and EsV-1 transcription probably occurs in the nucleus because these viruses do not encode an identifiable DNA-directed RNA polymerase.

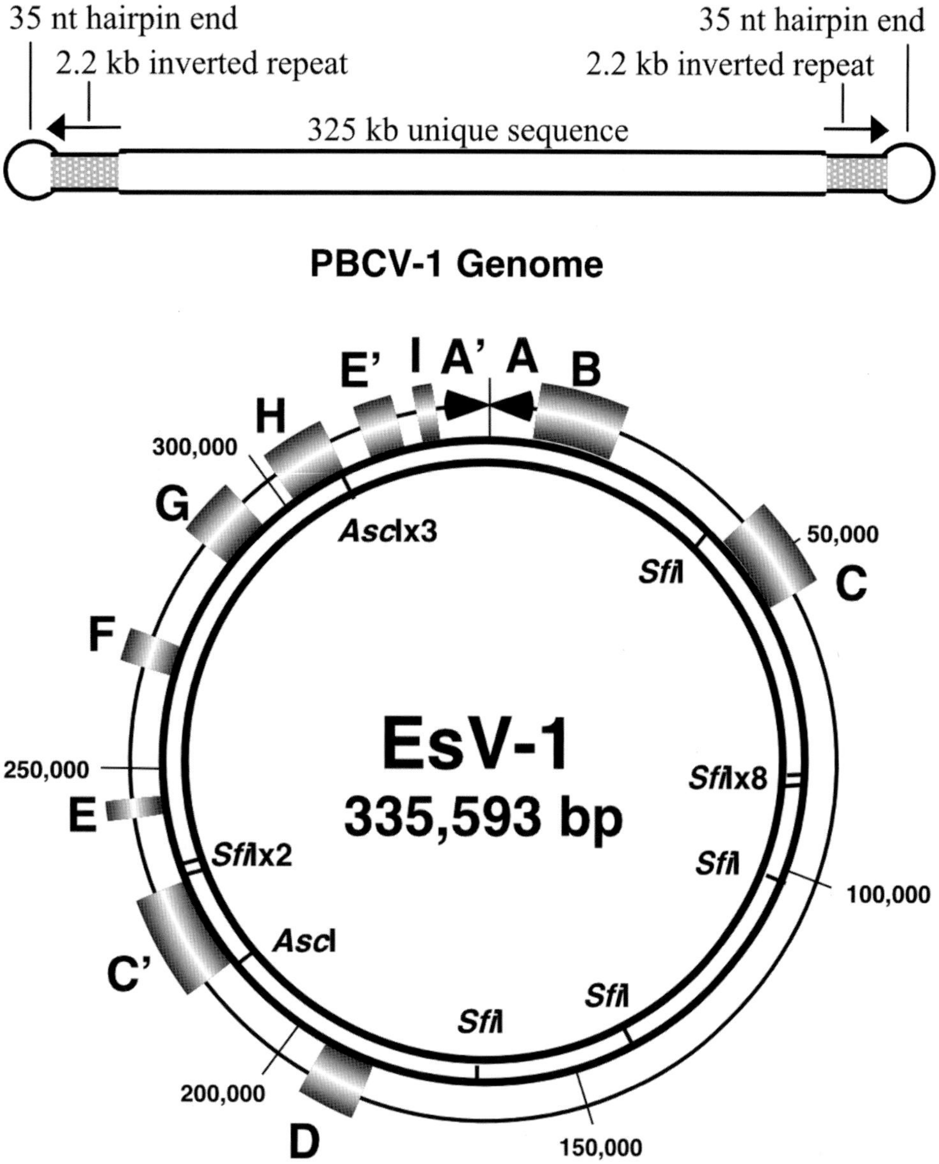

Figure 2: (TOP) Genome organization of Paramecium bursaria Chlorella virus 1 (PBCV-1). (Bottom) Circular map of the Ectocarpus siliculosus virus 1 (EsV-1) genome. Inner circle, sites for restriction endonucleases AscI and SfiI. Outer circle, nucleotide, coordinates and position of repeat regions (block rectangles: B, C, C′, etc). Triangles, the inverted terminal repeats, ITRs A and A′.

Antigenic Properties

Four distinct antigenic variants of the chlorovirus PBCV-1 can be isolated which are resistant to polyclonal antibody prepared against wildtype PBCV-1. These variants occur at a frequency of ~1×10^{-6}. The antibodies react primarily with the glycan portion of the major CP. Additional variants of these viruses can easily be isolated from natural sources.

Biological Properties

The phycodnaviruses, depending on whether they infect freshwater algae or marine algae, are ubiquitous in freshwater or seawater collected throughout the world. Some viruses are host specific and only infect single isolates or species of algae. For example, chloroviruses only attach to cell walls of certain unicellular, eukaryotic, chlorella-like green algae. Virus attachment is followed by dissolution of the host wall at the point of attachment and entry of the viral DNA and associated proteins into the cell, leaving an empty capsid on the host surface. Beginning ~2-4 hrs post-infection, progeny virions are assembled in the cytoplasm of the host. Infectious virions can be detected inside the cell ~30-40 min prior to virus release; virus release occurs by cell lysis. Coccolithoviruses, prymnesioviruses and raphidoviruses have wider host ranges, where individual viruses can infect a range of host isolates within specific algal species, however they do not cross the species barrier.

Infection of the prasinoviruses occurs when virions adhere to the wall-less host cell surface, followed by fusion of adjacent host and particle surfaces. Empty particles remain on the cell surface following the release of core contents. An eclipse period of approximately 3 hrs follows the attachment stage. The virus growth cycle is complete after approximately 14 hrs. During the replication cycle, particles appear in the cytoplasm and are associated with the production of cytoplasmic fibrils (~5-8 nm in diameter) and clusters of membrane bound vesicles that are absent in healthy cells. Particles are released into the medium via localized ruptures in the cell membrane; ruptures often appear at several locations on the same cell.

Less is known about the replication of coccolithoviruses, prymnesioviruses and raphidoviruses. Virus formation is observed in the cytoplasm and the nucleus remains intact and separate from the viroplasm that consists of a fibrillar matrix. Ultimately, viral production results in the disruption of organelles, lysis of the cell and release of the virus particles.

The phaeoviruses infect the wall-less spore or gamete stage of filamentous brown algae. These viruses appear as virus particles in sporangial or gametangial cells of the host. Depending on the virus, viral particles are formed in unilocular and plurilocular sporangia, or gametangia (EsV-1) others only form in unilocular sporangia (*Feldmannia species virus*). A few of these viruses can infect more than one species of brown algae.

The hosts for some of the chloroviruses and coccolithoviruses can easily be grown in the laboratory and the viruses can be plaque-assayed. The hosts for some of the other viruses are either cultured axenically (e.g., prymnesiovirus hosts, *P. globosa*) or non-axenically in uni-alga cultures (eg. hosts for the prasinoviruses and raphidoviruses). The brown algal viruses, which only appear in mature gametangia or sporangia cells of their hosts, can also be grown in the laboratory.

The chloroviruses, coccolithoviruses, prasinoviruses, prymnesioviruses and raphidoviruses are transmitted horizontally. The phaeoviruses are transmitted both horizontally and vertically.

GENUS *CHLOROVIRUS*

Type Species ***Paramecium bursaria Chlorella virus 1***

DISTINGUISHING FEATURES

The chloroviruses, which are ubiquitous in fresh water throughout the world, infect certain unicellular, eukaryotic, exsymbiotic chlorella-like green algae. The viruses have linear, non-permuted dsDNA genomes with cross-linked hairpin ends. The DNA termini contain identical inverted 1-2.2 kb repeats. The remainder of the genome is primarily single copy DNA. Viruses in this group can be distinguished from each other by DNA restriction digests, the levels of methylated bases in their genomes, serology, and host specificity.

LIST OF SPECIES DEMARCATION CRITERIA IN THE GENUS

Three groups of viruses are delineated based on host specificity.
Group 1. *Paramecium bursaria Chlorella* NC64A viruses (NC64A viruses)
Group 2. *Paramecium bursaria Chlorella* Pbi viruses (Pbi viruses)
Group 3. *Hydra viridis Chlorella* viruses (HVC viruses)

Chlorella strains NC64A, ATCC 30562, and N1A (originally symbionts of the protozoan *P. bursaria*), collected in the United States, are the only known host for NC64A viruses. *Chlorella* strain Pbi (originally a symbiont of a European strain of *P. bursaria*) collected in Germany, is the only known host for Pbi viruses. Pbi viruses do not infect *Chlorella* strains NC64A, ATCC 30562, and N1A. *Chlorella* strain Florida (originally a symbiont of *Hydra viridis*) is the only known host for Hydra viridis Chlorella virus (HVCV). NC64A viruses are placed in 16 species based on plaque size, serological reactivity, resistance of the genome to restriction endonucleases, virus encoded restriction endonucleases and nature and content of methylated bases.

LIST OF SPECIES IN THE GENUS

Species names are in green italic script; strain names and synonyms are in black roman script; tentative species names are in blue roman script. Sequence accession numbers, and assigned abbreviations ( ) are also listed.

SPECIES IN THE GENUS

Group 1 - *Paramecium bursaria Chlorella* NC64A virus group:

Paramecium bursaria Chlorella virus 1		
Paramecium bursaria Chlorella virus 1	[U42580]	(PBCV-1)
Paramecium bursaria Chlorella virus AL1A		
Paramecium bursaria Chlorella virus AL1A		(PBCV-AL1A)
Paramecium bursaria Chlorella virus AL2A		
Paramecium bursaria Chlorella virus AL2A		(PBCV-AL2A)
Paramecium bursaria Chlorella virus BJ2C		
Paramecium bursaria Chlorella virus BJ2C		(PBCV-BJ2C)
Paramecium bursaria Chlorella virus CA4A		
Paramecium bursaria Chlorella virus CA4A		(PBCV-CA4A)
Paramecium bursaria Chlorella virus CA4B		
Paramecium bursaria Chlorella virus CA4B		(PBCV-CA4B)
Paramecium bursaria Chlorella virus IL3A		
Paramecium bursaria Chlorella virus IL3A		(PBCV-IL3A)
Paramecium bursaria Chlorella virus NC1A		
Paramecium bursaria Chlorella virus NC1A		(PBCV-NC1A)
Paramecium bursaria Chlorella virus NE8A		
Paramecium bursaria Chlorella virus NE8A		(PBCV-NE8A)
Paramecium bursaria Chlorella virus NY2A		
Paramecium bursaria Chlorella virus NY2A		(PBCV-NY2A)
Paramecium bursaria Chlorella virus NYs1		
Paramecium bursaria Chlorella virus NYs1		(PBCV-NYs1)
Paramecium bursaria Chlorella virus SC1A		
Paramecium bursaria Chlorella virus SC1A		(PBCV-SC1A)
Paramecium bursaria Chlorella virus XY6E		
Paramecium bursaria Chlorella virus XY6E		(PBCV-XY6E)
Paramecium bursaria Chlorella virus XZ3A		
Paramecium bursaria Chlorella virus XZ3A		(PBCV-XZ3A)
Paramecium bursaria Chlorella virus XZ4A		
Paramecium bursaria Chlorella virus XZ4A		(PBCV-XZ4A)
Paramecium bursaria Chlorella virus XZ4C		
Paramecium bursaria Chlorella virus XZ4C		(PBCV-XZ4C)

Group 2 - *Paramecium bursaria Chlorella* Pbi virus group:
Paramecium bursaria Chlorella virus A1
Paramecium bursaria Chlorella virus A1 (PBCV-A1)
Group 3 - *Hydra viridis Chlorella* virus group:
Hydra viridis Chlorella virus 1
Hydra viridis Chlorella virus 1 (HVCV-1)

TENTATIVE SPECIES IN THE GENUS

Group 1 - *Paramecium bursaria Chlorella* NC64A virus group:
Paramecium bursaria Chlorella virus AL2C (PBCV-AL2C)
Paramecium bursaria Chlorella virus CA1A (PBCV-CA1A)
Paramecium bursaria Chlorella virus CA1D (PBCV-CA1D)
Paramecium bursaria Chlorella virus CA2A (PBCV-CA2A)
Paramecium bursaria Chlorella virus IL2A (PBCV-IL2A)
Paramecium bursaria Chlorella virus IL2B (PBCV-IL2B)
Paramecium bursaria Chlorella virus IL3D (PBCV-IL3D)
Paramecium bursaria Chlorella virus IL5-2s1 (PBCV-IL5-2s1)
Paramecium bursaria Chlorella virus MA1D (PBCV-MA1D)
Paramecium bursaria Chlorella virus MA1E (PBCV-MA1E)
Paramecium bursaria Chlorella virus NC1B (PBCV-NC1B)
Paramecium bursaria Chlorella virus NC1C (PBCV-NC1C)
Paramecium bursaria Chlorella virus NC1D (PBCV-NC1D)
Paramecium bursaria Chlorella virus NE8D (PBCV-NE8D)
Paramecium bursaria Chlorella virus NY2B (PBCV-NY2B)
Paramecium bursaria Chlorella virus NY2C (PBCV-NY2C)
Paramecium bursaria Chlorella virus NY2F (PBCV-NY2F)
Paramecium bursaria Chlorella virus NYb1 (PBCV-NYb1)
Paramecium bursaria Chlorella virus SC1B (PBCV-SC1B)
Paramecium bursaria Chlorella virus SH6A (PBCV-SH6A)
Paramecium bursaria Chlorella virus XZ5C (PBCV-XZ5C)
Paramecium bursaria Chlorella virus CVBII (PBCV-CVBII)
Paramecium bursaria Chlorella virus CVK2 (PBCV-CVK2)
Paramecium bursaria Chlorella virus CVU1 (PBCV-CVU1)
Group 2 - *Paramecium bursaria Chlorella* Pbi virus group:
Paramecium bursaria Chlorella virus B1 (PBCV-B1)
Paramecium bursaria Chlorella virus G1 (PBCV-G1)
Paramecium bursaria Chlorella virus M1 (PBCV-M1)
Paramecium bursaria Chlorella virus R1 (PBCV-R1)
Group 3 - *Hydra viridis Chlorella* virus group:
Hydra viridis Chlorella virus 2 (HVCV-2)
Hydra viridis Chlorella virus 3 (HVCV-3)

GENUS COCCOLITHOVIRUS

Type Species ***Emiliania huxleyi virus* 86**

DISTINGUISHING FEATURES

Viruses assigned to this genus all have large dsDNA genomes, ~410-415 kbp and probably have a common genome structure. The particles have icosohedral symmetry, are tailless and range from 150-200 nm in diameter. The latent period of these viruses is 12–14 hrs and the burst size is between 400–1000 viruses per lysed host cell (mean 620). They infect different isolates of the globally important marine coccolithophorid, *Emiliania huxleyi*, a marine alga which has a world-wide distribution and is known for forming vast coastal and mid-oceanic blooms which are easily observed by satellite imagery. The viruses described here were isolated from *E. huxleyi* blooms of the coast of Plymouth, UK, in July 1999 and July/August 2001, from an *E. huxleyi* bloom induced during a mesocosm experiment in a fjord near Bergen, Norway, during June 2000 and from *E. huxleyi* blooms off the coast of Bergen, Norway, in June 1999 and June 2000. The viruses are relatively easy to isolate and susceptible host strains usually lyse 2-7 days after the addition of filtered seawater. Clonal isolates can be obtained by plaque assay.

List of Species Demarcation Criteria in the Genus

Not applicable.

List of Species in the Genus

Species names are in green italic script; strain names and synonyms are in black roman script; tentative species names are in blue roman script. Sequence accession numbers, and assigned abbreviations ( ) are also listed.

Species in the Genus

Emiliania huxleyi virus 86

Emiliania huxleyi virus 86	(EhV-86)

Tentative Species in the Genus

Emiliania huxleyi virus 163	(EhV-163)
Emiliania huxleyi virus 201	(EhV-201)
Emiliania huxleyi virus 202	(EhV-202)
Emiliania huxleyi virus 203	(EhV-203)
Emiliania huxleyi virus 205	(EhV-205)
Emiliania huxleyi virus 207	(EhV-207)
Emiliania huxleyi virus 208	(EhV-208)
Emiliania huxleyi virus 2KB1	(EhV-2KB1)
Emiliania huxleyi virus 2KB2	(EhV-2KB2)
Emiliania huxleyi virus 84	(EhV-84)
Emiliania huxleyi virus 88	(EhV-88)
Emiliania huxleyi virus 99B1	(EhV-99B1)

Genus Prasinovirus

Type Species *Micromonas pusilla virus SP1*

Distinguishing Features

Viruses assigned to this genus probably have a common genome structure and infect the same marine alga *Micromonas pusilla* (strain UTEX 991, Plymouth 27). The particles are tailless polyhedral particles which range from 104-118 nm in diameter. The viruses are ubiquitous in sea water throughout the world, and have been isolated from the Pacific and Atlantic Oceans, and the Gulf of Mexico. The abundance of these viruses in natural samples is variable and ranges from < 20 to 1×10^4 virus particles per ml.

List of Species Demarcation Criteria in the Genus

Not applicable.

List of Species in the Genus

Species names are in green italic script; strain names and synonyms are in black roman script; tentative species names are in blue roman script. Sequence accession numbers, and assigned abbreviations ( ) are also listed.

Species in the Genus

Micromonas pusilla virus SP1

Micromonas pusilla virus SP1	(MpV-SP1)

TENTATIVE SPECIES IN THE GENUS

Micromonas pusilla virus SP2	(MpV-SP2)
Micromonas pusilla virus GM1	(MpV-GM1)
Micromonas pusilla virus PB8	(MpV-PB8)
Micromonas pusilla virus SG1	(MpV-SG1)
Micromonas pusilla virus PB7	(MpV-PB7)
Micromonas pusilla virus PB6	(MpV-PB6)
Micromonas pusilla virus PL1	(MpV-PL1)

GENUS *PRYMNESIOVIRUS*

Type Species ***Chrysochromulina brevifilum virus PW1***

DISTINGUISHING FEATURES

Viruses assigned to this genus infect hosts belonging to algal class *Haptophyceae* (also referred to as the *Prymnesiophyceae*) with the notable exception of viruses that infect *Emiliania huxleyi* (coccolithoviruses). Although viruses in this genus all have dsDNA genomes, there is a wide variety of diameters (100-170 nm) and genome sizes (120–485 kbp). Estimated burst sizes range from 320-600 viruses per infected cell. Viruses that infect members of the same marine algae, *Chysochromulina brevifilum* and *C. strobilus* were isolated from USA (Texas) coastal waters in three locations (Gulf of Mexico, Aransas Pass, and Laguna Madre). Viruses that infect *Phaeocystis globosa* were isolated from natural seawater off the coast of the Netherlands during April to July 2000 and 2001, North Sea in April 2002, from a *P. globosa* bloom induced during a mesocosm experiment during October 2000 and from a *P. globosa* bloom in the English Channel off the coast of Plymouth, UK, in April 2001. These viruses lysed susceptible host strains after 2-7 days after the addition of filtered seawater. Once isolated, susceptible host strains typically lyse within 1-2 days.

LIST OF SPECIES DEMARCATION CRITERIA IN THE GENUS

Not applicable.

LIST OF SPECIES IN THE GENUS

Species names are in green italic script; strain names and synonyms are in black roman script; tentative species names are in blue roman script. Sequence accession numbers, and assigned abbreviations ( ) are also listed.

SPECIES IN THE GENUS

Chysochromulina brevifilum virus PW1	(CbV-PW1)
Chysochromulina brevifilum virus PW1	(CbV-PW1)

TENTATIVE SPECIES IN THE GENUS

Chysochromulina brevifilum virus PW3	(CbV-PW3)
Phaeocystis globosa virus 1 (Texel)	(PgV-01T)
Phaeocystis globosa virus 2 (Texel)	(PgV-02T)
Phaeocystis globosa virus 3 (Texel)	(PgV-03T)
Phaeocystis globosa virus 4 (Texel)	(PgV-04T)
Phaeocystis globosa virus 5 (Texel)	(PgV-05T)
Phaeocystis globosa virus 6 (Texel)	(PgV-06T)
Phaeocystis globosa virus 7 (Texel)	(PgV-07T)
Phaeocystis globosa virus 9 (Texel)	(PgV-09T)
Phaeocystis globosa virus 10 (Texel)	(PgV-10T)
Phaeocystis globosa virus 11 (Texel)	(PgV-11T)
Phaeocystis globosa virus 12 (Texel)	(PgV-12T)
Phaeocystis globosa virus 13 (Texel)	(PgV-13T)

Phaeocystis globosa virus 14 (Texel)	(PgV-14T)
Phaeocystis globosa virus 15 (Texel)	(PgV-15T)
Phaeocystis globosa virus 16 (Texel)	(PgV-16T)
Phaeocystis globosa virus 17 (Texel)	(PgV-17T)
Phaeocystis globosa virus 18 (Texel)	(PgV-18T)
Phaeocystis globosa virus 102 (Plymouth)	(PgV-102P)

GENUS PHAEOVIRUS

Type Species *Ectocarpus siliculosus virus 1*

DISTINGUISHING FEATURES

Ectocarpus siliculosus virus 1 isolates infect the free-swimming, zoospore or gamete stages of filamentous brown algal hosts. The virus genome is integrated into the host genome and is inherited in a Mendelian manner. Virus particles are only formed in prospective gametangia or sporangia cells of the host. The viral genomes are circular dsDNAs.

LIST OF SPECIES DEMARCATION CRITERIA IN THE GENUS

Seven species of viruses are delineated based in part on host specificity. Field isolates of at least seven genera of the *Phaeophycae* contain 120-150 nm diameter polyhedral virus-like particles. The particles contain dsDNA genomes that vary in size from 150-350 kb, although the major CP gene and DNA polymerase sequence data indicate that they are closely related. Virus expression is variable; particles are rarely observed in vegetative cells but are common in unilocular sporangia (Feldmannia species virus, FsV) or both unilocular and plurilocular sporangia and gametangia (EsV). Some of the viruses have a narrow host range (FsV), whereas others such as Ectocarpus fasciculatus virus (EfV) and EsV infect members of more than one genus.

LIST OF SPECIES IN THE GENUS

Species names are in green italic script; strain names and synonyms are in black roman script; tentative species names are in blue roman script. Sequence accession numbers, and assigned abbreviations ( ) are also listed.

SPECIES IN THE GENUS

Ectocarpus fasciculatus virus a		
Ectocarpus fasciculatus virus a		(EfV-a)
Ectocarpus siliculosus virus 1		
Ectocarpus siliculosus virus 1	[AF204951]	(EsV-1)
Ectocarpus siliculosus virus a		
Ectocarpus siliculosus virus a		(EsV-a)
Feldmannia irregularis virus a		
Feldmannia irregularis virus a		(FiV-a)
Feldmannia species virus		
Feldmannia species virus		(FsV)
Feldmannia species virus a		
Feldmannia species virus a		(FsV-a)
Hincksia hinckiae virus a		
Hincksia hinckiae virus a		(HhV-a)
Myriotrichia clavaeformis virus a		
Myriotrichia clavaeformis virus a		(McV-a)
Pilayella littoralis virus a		
Pilayella littoralis virus a		(PlV-a)

TENTATIVE SPECIES IN THE GENUS

None reported

GENUS RAPHIDOVIRUS

Type Species ***Heterosigma akashiwo virus 01***

DISTINGUISHING FEATURES

Viruses assigned to this genus infect the harmful bloom causing raphidophyte, *Heterosigma akashiwo* (Raphidophyceae), a marine alga which has a world-wide distribution. The type species *Heterosigma akashiwo virus 01* contains members with a large dsDNA genome ~294 kbp, have icosohedral symmetry, 202 ± 6 nm in diameter. The viruses were isolated from *H. akashiwo* blooms in Unoshima Port, Fukuoka Prefecture, Japan, in June 1996, and in Nomi Bay, Kochi Prefecture, Japan, in July 1996. These viruses are relatively easy to isolate and susceptible host strains usually lyse 2-7 days after the addition of filtered seawater. The latent period and the burst size of HaV01 is 30-33 hrs and ~770 at 20°C, respectively. HaV infection is considered one of the important factors causing the disintegration of *H. akashiwo* blooms. Although these viruses rapidly degrade to lose infectivity even when kept at 4°C in the dark, they can be easily cryopreserved.

LIST OF SPECIES DEMARCATION CRITERIA IN THE GENUS

Not applicable.

LIST OF SPECIES IN THE GENUS

Species names are in green italic script; strain names and synonyms are in black roman script; tentative species names are in blue roman script. Sequence accession numbers, and assigned abbreviations ( ) are also listed.

SPECIES IN THE GENUS

Heterosigma akashiwo virus 01

Heterosigma akashiwo virus 01	(HaV-01)

TENTATIVE SPECIES IN THE GENUS

Heterosigma akashiwo virus 02	(HaV02)
Heterosigma akashiwo virus 03	(HaV03)
Heterosigma akashiwo virus 04	(HaV04)
Heterosigma akashiwo virus 05	(HaV05)
Heterosigma akashiwo virus 06	(HaV06)
Heterosigma akashiwo virus 07	(HaV07)
Heterosigma akashiwo virus 08	(HaV08)
Heterosigma akashiwo virus 09	(HaV09)
Heterosigma akashiwo virus 10	(HaV10)
Heterosigma akashiwo virus 11	(HaV11)
Heterosigma akashiwo virus 12	(HaV12)
Heterosigma akashiwo virus 13	(HaV13)
Heterosigma akashiwo virus 14	(HaV14)
Heterosigma akashiwo virus 15	(HaV15)

UNASSIGNED VIRUSES IN THE FAMILY

The following viruses have phenotypic characteristics suggesting that they belong to the family *Phycodnaviridae*. However, taxonomic assignment cannot be confirmed until phylogenetic information on their DNA polymerase genes becomes available. The following large dsDNA viruses have been reported and it is possible that after further characterization, they could be assigned either as a separate genus or fall within one of the six genera in the family *Phycodnaviridae*.

Aureococcus anophagefference virus (AaV)(Brown tide virus**):** This virus was isolated from Great South Bay New York, USA, in 1992. Viruses are 140–160 nm in diameter and cultures of *Aureococcus anophagefference* usually lyse within 24-48 hrs.

Chrysochromulina ericina virus 01B (CeV-01B): This virus was isolated from water collected off the coast of Bergen, Norway, in June 1998. Viruses have particle diameters of 160 nm and a genome size of 510 kbp. The lytic cycle is 14–19 hrs and the burst size is ~1800–4100 viruses per host cell.

Heterocapsa circularisquama viruses 01 – 10 (HcV-01 to HcV-10): These viruses were isolated from *H. circularisuqama* blooms in Wakinoura Fishing Port, Fukuoka Prefecture, Japan, in August 1999 and in Fukura Bay, Hyogo Prefecture, Japan in August 1999. HcV-01 contains a large dsDNA genome ~350 kbp, icosohedral symmetry and is 180-210 nm in diameter (197 ± 8 nm). These viruses are relatively easy to isolate and susceptible host strains usually lyse 2-7 days after the addition of filtered seawater. The latent period and the burst size of HcV-03 are 40-56 hrs and ~1800-2440 at 20-25°C, respectively. Although these viruses rapidly degrade to lose infectivity even when kept at 4°C in the dark, they can be cryopreserved by using a suitable cryoprotectant.

Phaeocystis pouchetii virus 01 (PpV-01): This virus was isolated from water collected at the end of a *P. pouchetii* bloom off the coast of Bergen, Norway, in May 1995. Viruses have particle diameters of 130-160 nm and a genome size of 485 kbp. The latent period is 12-18 hrs, complete lysis of cultures is observed after 48 hrs and the burst size is ~350-600 viruses per host cell.

Pyramimonas orientalis virus 01B (PoV-01B): This virus was isolated from water collected off the coast of Bergen, Norway, in June 1998. Viruses have particle sizes of 220 x 180 nm and a genome size of 560 kbp. The lytic cycle is 14–19 hrs and the burst size is ~800–1000 viruses per host cell.

TENTATIVE SPECIES UNASSIGNED IN THE FAMILY

Aureococcus anophagefference virus (Brown tide virus)	(AaV)
Chrysochromulina ericina virus 01B	(CeV-01B)
Heterocapsa circularisquama virus 01	(HcV-01)
Heterocapsa circularisquama virus 02	(HcV-02)
Heterocapsa circularisquama virus 03	(HcV-03)
Heterocapsa circularisquama virus 04	(HcV-04)
Heterocapsa circularisquama virus 05	(HcV-05)
Heterocapsa circularisquama virus 06	(HcV-06)
Heterocapsa circularisquama virus 07	(HcV-07)
Heterocapsa circularisquama virus 08	(HcV-08)
Heterocapsa circularisquama virus 09	(HcV-09)
Heterocapsa circularisquama virus 10	(HcV-10)
Phaeocystis pouchetii virus 01	(PpV-01)
Pyramimonas orientalis virus 01B	(PoV-01B)

PHYLOGENETIC RELATIONSHIPS WITHIN THE FAMILY

Analysis of the DNA polymerase gene from selected members of the family *Phycodnaviridae* indicate that the viruses are more closely related to each other than to other dsDNA viruses and they form a distinct phyletic group suggesting that they share a common ancestor (Fig. 2). However, the viruses fall into six clades, which correlate with their hosts. For example, the nucleotide sequence from the chlorovirus PBCV-1 is 77% identical to both Paramecium bursaria Chlorella virus NY2A (PBCV-NY2A) and

Paramecium bursaria Chlorella virus A1 (PBCV-A1), while CVA-1 and NY-2A are 64% identical. The two coccolithoviruses Emiliania huxleyi virus 86 (EhV-86) and Emiliania huxleyi virus 163 (EhV-163) are 97% identical. Among prasinoviruses, the proportion of identical nucleotides for all pairwise combinations ranged from 78-99%. Among prymnesioviruses Chrysochromulina brevifilum virus PW1 (CbV-PW1), Chrysochromulina brevifilum virus PW3 (CbV-PW3) and Phaeocystis globosa virus 102P (PgV-102P) the percent of identical nucleotides for all pairwise combinations ranged from 85-97%. Overall, sequence identities among the DNA polymerase gene fragments of all coccolithoviruses, prasinoviruses, three prymnesioviruses, three chloroviruses and one raphidovirus ranged from 29-98%. Despite the fact that these six clades of the family *Phycodnaviridae* have wide sequence variation within the family, this variation (29 to 98%) is not significant enough to belong to a branch with members of any other virus family.

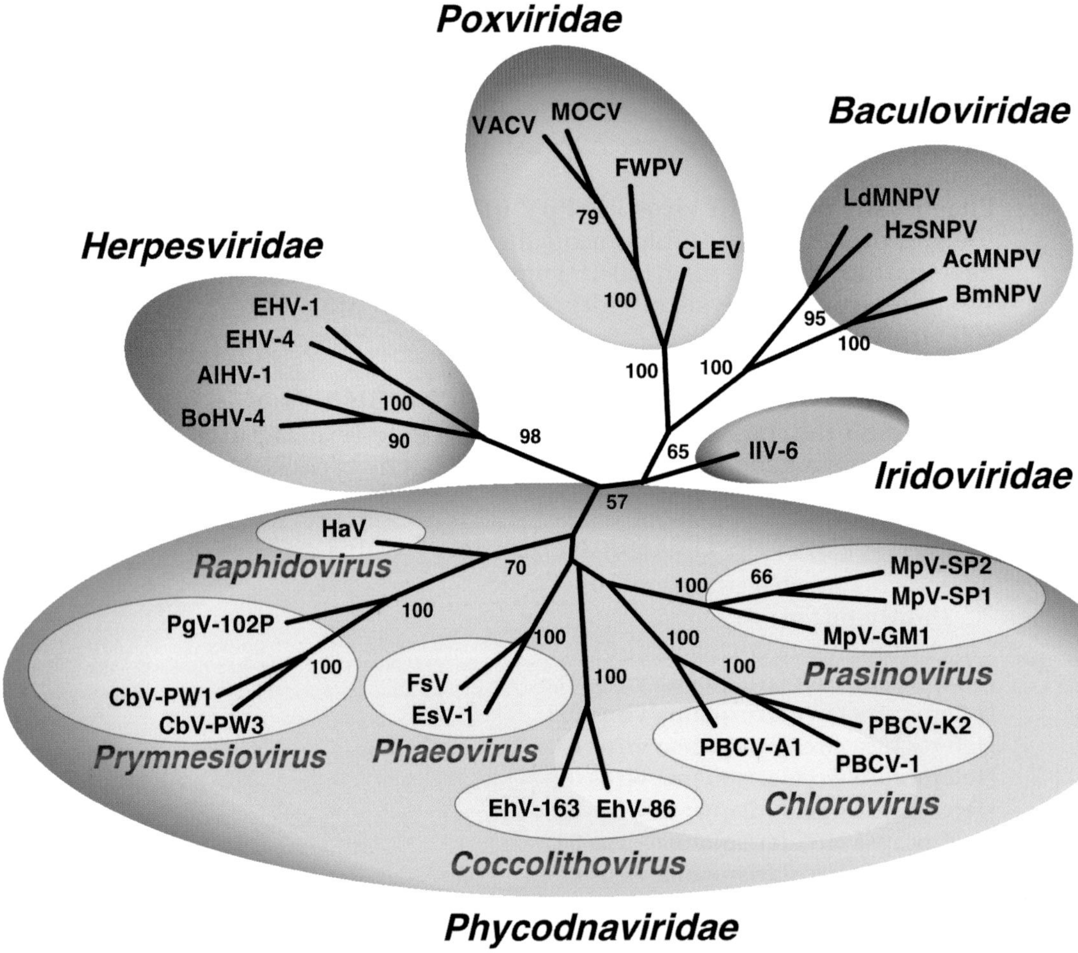

Figure 2: Phylogenetic tree of DNA polymerase gene fragments from members of the family *Phycodnaviridae* and other large dsDNA viruses generated using protein parsimony analysis of the 100 bootstrapped data sets. The numbers at the branches are the bootstrap values indicating the relative strengths of those branches. Abbreviations are: *Phycodnaviridae*: HaV, Heterosigma akashiwo virus; CbV-xx, viruses that infect Chrysochromulina brevifilum; PgV-102P, Phaeocystis globosa virus 102 (Plymouth); FsV, Feldmannia sp. virus; EsV-1, Ectocarpus siliculosus virus 1; EhV-xx, viruses that infect Emiliania huxleyi; MpV-xx, viruses that infect Micromonas pusilla; PBCV-1 and PBCV-NY2A, viruses that infect Chlorella NC64A; *Herpesviridae*: EHV-1, *Equine herpesvirus 1*; EHV-4, *Equine herpesvirus 4*; BoHV-4, *Bovine herpesvirus 4*; AlHV-1, Alcelaphine herpesvirus 1; Poxviridae: VACV, Vaccinia virus; MOCV, Molluscum contagiasum virus; FWPV, Fowl poxvirus; CLEV, Choristoneura biennis entomopoxvirus; *Baculoviridae*: AcMNPV, Autographa californica nucleopolyhedrovirus; BmNPV, Bombyx mori nucleopolyhedrovirus; HzSNPV, Helicoverpa zea nucleopolyhedrovirus; LdMNPV, Lymantria dispar nucleopolyhedrovirus; *Iridoviridae*: IIV-6, Invertebrate iridescent virus 6 (Adapted from Schroeder et al. 2002).

SIMILARITY WITH OTHER TAXA

Many large polyhedral virus-like particles have been observed in electron micrographs of eukaryotic algae. However, for the most part these particles have not been characterized.

DERIVATION OF NAMES

Chloro: from Greek *chloro,* meaning "green".
Cocco: derived from Greek *kokkis,* meaning "grain" or "berry" (referring to their shape).
dna: sigla for *d*eoxyribo*n*ucleic *a*cid.
Lith: from Greek *Lithos,* meaning "stone".
Phaeo: from Greek *phaeo,* meaning "brown".
Phyco: from Greek *phycos,* meaning "plant".
Prasino: from Latin *prasino,* meaning "green".
Prymnesio: from Greek *prymne,* meaning "stern of a ship".
Raphido: from Greek *raphido,* meaning "spine"

REFERENCES

Chen, F. and Suttle, C.A. (1996). Evolutionary relationships among large double-stranded DNA viruses that infect microalgae and other organisms as inferred from DNA polymerase genes. *Virology*, **219**, 170-178.

Delaroque N., Muller D.G., Bothe G., Knippers R. and Boland W. (2001). The complete DNA sequence of the *Ectocarpus siliculosus virus 1* (EsV-1) genome. *Virology*, **287**, 112-132.

Gastrich, M.D., Anderson, O.R., Benmayor, S.S. and Cosper, E.M. (1998). Ultrastructural analysis of viral infection in the brown-tide alga, *Aureococcus anophagefferens* (*Pelagophyceae*). *Phycologia,* **37**, 300-306.

Ivey, R.G., Henry, E.C., Lee, A.M., Klepper, L., Krueger, S.K. and Meints, R.H. (1996). A feldmannia algal virus has two genome size-classes. *Virology*, **220**, 267-273.

Jacobsen, A., Bratbak, G. and Heldal, M. (1996). Isolation and characterization of a virus infecting *Phaeocystis pouchetii* (*Prymnesiophyceae*). *J. Phycol.,* **32**, 923-927.

Kapp, M., Knippers, R. and Muller, D.G. (1997). New members to a group of DNA viruses infecting brown algae. *Phycological Res.*, **45**, 85-90.

Lee, A.M., Ivey, R.G. and Meints, R.H. (1998). The DNA polymerase gene of a brown algal virus: Structure and phylogeny. *J. Phycol.,* **34**, 608-615.

Li, Y., Lu, Z., Sun, L., Ropp, S., Kutish, G.F., Rock, D.L. and Van Etten, J.L. (1997). Analysis of 74 kb of DNA located at the right end of the 330 kb Chlorella virus PBCV-1 genome. *Virology*, **237**, 360-377.

Muller, D.G., Kapp, M. and Knippers, R. (1996). Viruses in marine brown algae. *Adv. Virus Res.*, **50**, 49-67.

Nagasaki, K., Tarutani, K. and Yamaguchi, M. (1999). Cluster analysis on algicidal activity of HaV clones and virus sensitivity of *Heterosigma akashiwo* (*Raphidophyceae*). *J. Plankt. Res.*, **21**, 2219-2226.

Nagasaki, K. and Yamaguchi, M. (1998). Intra-species host specificity of HaV (*Heterosigma akashiwo virus*) clones. *Aquat. Microb. Ecol.,* **14**, 109-112.

Nishida, K., Susuki, S., Kimura, Y., Nomura, N., Fujie, M. and Yamada, T. (1998). Group I introns found in chlorella viruses: Biological implications. *Virology*, **242**, 319-326.

Sandaa, R. A., Heldal, M., Castberg, T., Thyrhaug, R. and Bratbak, G. (2001). Isolation and characterization of two viruses with large genome size infecting *Chrysochromulina ericina* (*Prymnesiophyceae*) and *Pyramimonas orientalis* (*Prasinophyceae*). *Virology,* **290**, 272-280.

Schroeder, D. C., Oke, J., Malin, G. and Wilson, W. H. (2002). Coccolithovirus (*Phycodnaviridae*): Characterisation of a new large dsDNA algal virus that infects *Emiliania huxleyi. Arch. Virol.,* **147**, 1685-1698.

Tarutani, K., Nagasaki, K., Itakura, S. and Yamaguchi, M. (2001). Isolation of a virus infecting the novel shellfish-killing dinoflagellate *Heterocapsa circularisquama. Aquat. Microb. Ecol.,* **23**, 103-111.

Tarutani, K., Nagasaki, K. and Yamaguchi, M. (2000). Viral impacts on total abundance and clonal composition of the harmful bloom-forming phytoplankton *Heterosigma akashiwo. Appl. Environ. Microbiol.,* **66**, 4916-4920.

Van Etten, J.L. and Meints, R.H. (1999). Giant viruses infecting algae. *Ann. Rev. Microbiol.*, **53**, 447-494.

Van Etten, J.L., Graves, M.V., Muller, D.G., Boland, W. and Delaroque, N. (2002). Phycodnaviridae – large DNA algal viruses. *Arch. Virol.,* **147**,1479-1516.

Wilson, W. H., Tarran, G. A., Schroeder, D., Cox, M., Oke, J. and Malin, G. (2002). Isolation of viruses responsible for the demise of an *Emiliania huxleyi* bloom in the English Channel. *J. Mar. Biol. Ass. UK.*, **82**, 369-377.

CONTRIBUTED BY

Wilson, W.H., Van Etten, J.L., Schroeder, D.S., Nagasaki, K., Brussaard, C., Delaroque, N., Bratbak, G. and Suttle, C.

FAMILY *BACULOVIRIDAE*

TAXONOMIC STRUCTURE OF THE FAMILY

Family	*Baculoviridae*
Genus	*Nucleopolyhedrovirus*
Genus	*Granulovirus*

VIRION PROPERTIES

MORPHOLOGY

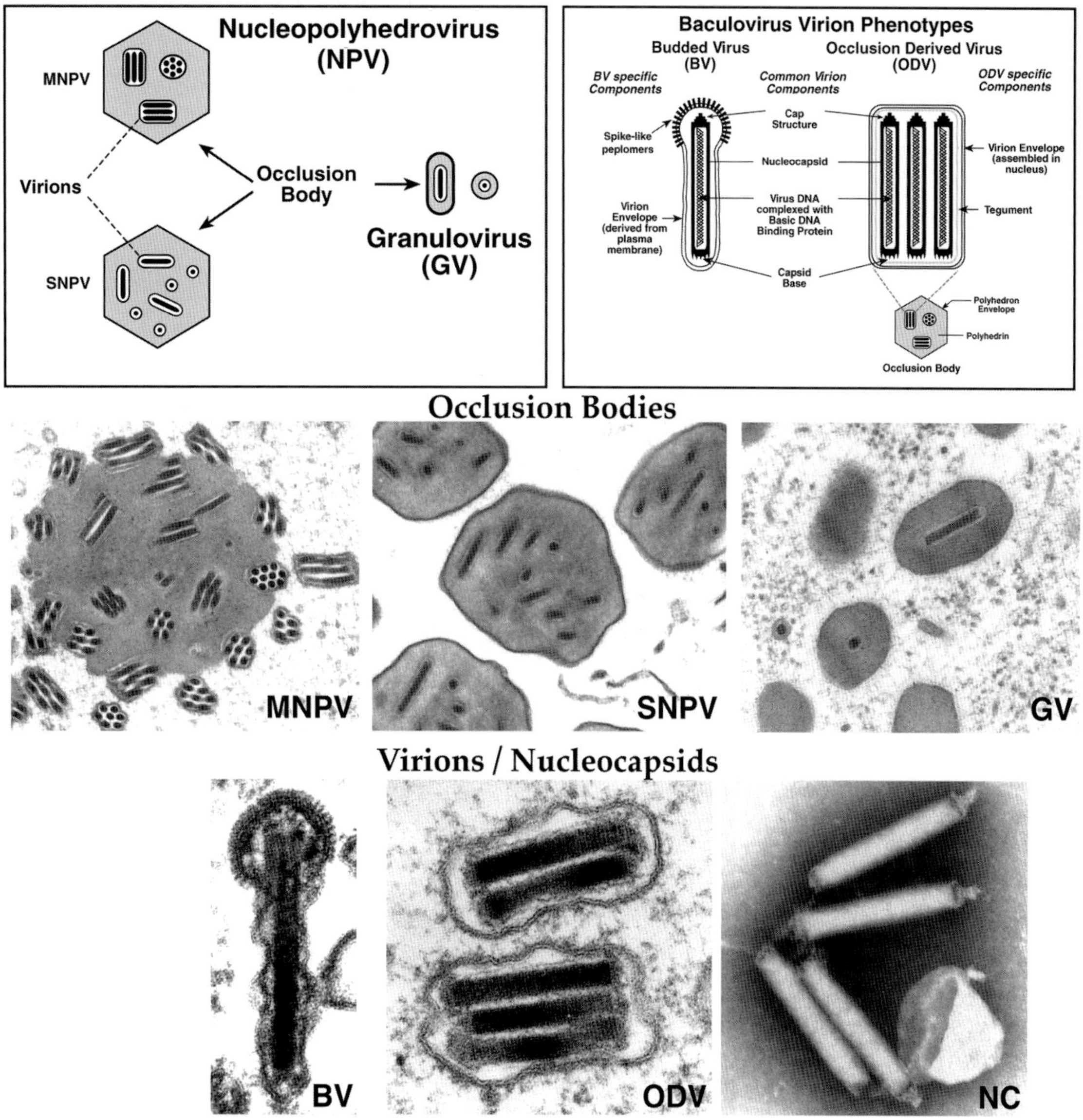

Figure 1: Baculovirus occlusion bodies, virions, and nucleocapsids. (Upper left) The structures of occlusion bodies from baculoviruses in the genera *Nucleopolyhedrovirus* (NPV) and *Granulovirus* (GV) are illustrated. Virions embedded in nucleopolyhedrovirus occlusion bodies may contain multiple nucleocapsids (MNPV) or single nucleocapsids (SNPV). (Upper right) The two baculovirus virion phenotypes are illustrated as diagrams with shared and phenotype-specific components (From Blissard, 1996). (Bottom) Transmission electron micrographs of occlusion bodies (MNPV, SNPV, and GV), virion phenotypes BV (budded virions), ODV (occlusion-derived virions), and nucleocapsids (NC). Nucleopolyhedrovirus occlusion bodies of the MNPV (Autographa californica MNPV, top left) and SNPV (Trichoplusia ni SNPV, top middle) types are compared to

granulovirus occlusion bodies (Estigmine acrea GV, top right). Transmission electron micrographs of virions of the BV (Lymantria dispar MNPV, bottom left) and ODV (Autographa californica MNPV, bottom center) phenotypes are shown beside negatively stained nucleocapsids (Autographa californica MNPV, bottom right). Electron micrographs courtesy of J.R. Adams (LdMNPV BV virion) and R. Granados (all others).

One or two virion phenotypes may be involved in baculovirus infections. The virion phenotype that initiates infection in the gut epithelium is occluded in a crystalline protein matrix which may be: a) polyhedral in shape, range in size from 0.15 to 15 μm and contain many virions (genus *Nucleopolyhedrovirus*), or b) ovicylindrical (about 0.3 x 0.5 μm) and contain only one, or rarely two or more virions (genus *Granulovirus*). Virions within occlusions consist of one or more rod-shaped nucleocapsids that have a distinct structural polarity and are enclosed within an envelope. For occluded virions, nucleocapsid envelopment occurs within the nucleus (genus *Nucleopolyhedrovirus*) or in the nuclear-cytoplasmic milieu after rupture of the nuclear membrane (genus *Granulovirus*). Nucleocapsids average 30-60 nm in diameter and 250-300 nm in length. Spike-like structures have not been reported on envelopes of the occlusion-derived virions (ODV). If infection is not restricted to the gut epithelium, a second virion phenotype is required for infection of other tissues. Virions of the second phenotype (termed budded virions or BV) are generated when nucleocapsids bud through the plasma membrane at the surface of infected cells. BV typically contain a single nucleocapsid. Envelopes of the BV are derived from the cellular plasma membrane and characteristically appear as a loose-fitting membrane that contains terminal peplomers 14-15 nm in length composed of a glycoprotein (Fig. 1).

Physicochemical and Physical Properties

ODV buoyant density in CsCl is 1.18-1.25 g/cm^3, and that of the nucleocapsid is 1.47 g/cm^3. BV buoyant density in sucrose is 1.17-1.18 g/cm^3. Virions of both phenotypes are sensitive to organic solvents and detergents. BV is marginally sensitive to heat and pH 8-12, inactivated by pH 3.0, and stable in Mg^{++} (10^{-1} M to 10^{-5} M).

Nucleic Acid

Nucleocapsids contain one molecule of circular supercoiled dsDNA, 80-180 kbp in size.

Proteins

Genomic analyses suggest that baculoviruses encode 100-200 proteins. Virions may contain from 12 to 20 or more different polypeptides. Nucleocapsids from both virion phenotypes (ODV and BV) contain a major CP, a basic DNA binding protein that is complexed with the viral genome, and at least 2-3 additional proteins. BV contain a major envelope glycoprotein (the peplomer protein) that serves as an envelope fusion protein (EFP). The EFPs identified to date include GP64, present in Autographa californica MNPV (AcMNPV) and close relatives, the F proteins including the Ld130 protein from Lymantria dispar MNPV (LdMNPV), and the Se8 protein from Spodoptera exigua MNPV (SeMNPV). Several ODV envelope proteins have been identified. The major protein of the occlusion body matrix is a virus encoded polypeptide of 25-33 kDa. This protein is called polyhedrin for nucleopolyhedroviruses and granulin for granuloviruses. The occlusion body is surrounded by an envelope that contains at least one major protein.

Lipids

Lipids are present in the envelopes of ODV and BV. Lipid composition differs between the two virion phenotypes.

Carbohydrates

Carbohydrates are present as glycoproteins and glycolipids.

GENOME ORGANIZATION AND REPLICATION

Circular genomic DNA is infectious suggesting that after cellular entry and uncoating, no virion-associated proteins are essential for infection. Genomes encode approximately 100 to 180 proteins. The Lepidopteran baculoviuses appear to share 30-60 homologs of a core group of genes. These conserved genes are involved in DNA replication, late gene transcription, virion structure, and other functions. In some cases, larger genome sizes may result from the presence of families of repeated genes. Transcription of baculovirus genes is temporally regulated, and two main classes of genes are recognized: early and late. Late genes may be further subdivided as late and very late. Gene classes (early, late, and very late) are not clustered on the baculovirus genome, and both strands of the genome are involved in coding functions. Early genes are transcribed by host RNA polymerase II, while late and very late genes are transcribed by an alpha-amanitin resistant RNA polymerase activity. RNA splicing occurs, but appears to be rare since only two instances have been identified. Transient early and late gene transcription and DNA replication studies suggest that at least three virus encoded proteins regulate early gene transcription, while approximately 19 viral encoded proteins known as Late Expression Factors (LEFs) are necessary for late gene transcription. Of the approximately 19 LEFs, 9-10 appear to be involved in DNA replication. Late gene transcription initiates within or near a highly conserved 5'-TAAG-3' sequence, which appears to be an essential core element of the baculovirus late promoter. Putative replication origins consist of repeated sequences found at multiple locations within the baculovirus genome. These sequences, termed homologous repeat (*hr*) regions, do not appear to be highly conserved between different baculovirus species. Single copy, non-*hr* putative replication origins have also been identified. DNA replication is required for late gene transcription. Most structural proteins of the virions are encoded by late genes. While transcription of late and very late genes appears to begin immediately after DNA replication, some very late genes that encode occlusion body-specific proteins are transcribed at extremely high levels at a later time. BV production occurs primarily during the late phase, and occlusion body production occurs during the very late phase (Fig. 2).

In infected animals, viral replication begins in the midgut (insects) or digestive gland epithelium (shrimp) of the arthropod host. Following ingestion, occlusions are solubilized in the gut lumen releasing the enveloped virions which are thought to enter the target epithelial cells via fusion with the plasma membrane at the cell surface. In lepidopteran insects, viral entry into midgut cells occurs in an alkaline environment, up to pH 12. Infection of the midgut is required for initiation of infection in the animal. In most cases, the virus is believed to undergo one round of replication in the midgut epithelium prior to transmission of infection to secondary tissues within the hemocoel. A mechanism for direct movement from the midgut to the hemocoel has also been proposed. DNA replication takes place in the nucleus. In granulovirus-infected cells, the integrity of the nuclear membrane is lost during the replication process. With some baculoviruses, replication is restricted to the gut epithelium and progeny virions become enveloped and occluded within these cells, and may be shed into the gut lumen with sloughed epithelium, or released upon death of the host. In other baculoviruses, the infection is transmitted to internal organs and tissues. The second virion phenotype, BV, which buds from the basolateral membrane of infected gut cells is required for transmission of the infection into the hemocoel. In secondarily infected tissues, BV is produced during the late phase and occluded virus during the very late phase of the infection. Infected fat body cells are the primary location of occluded virus production in lepidopteran insects. Occluded virus matures within nuclei of infected cells for nucleopolyhedroviruses and within the nuclear-cytoplasmic milieu for granuloviruses. Occlusion bodies containing infectious ODV virions are released upon death, and usually liquefaction, of the host.

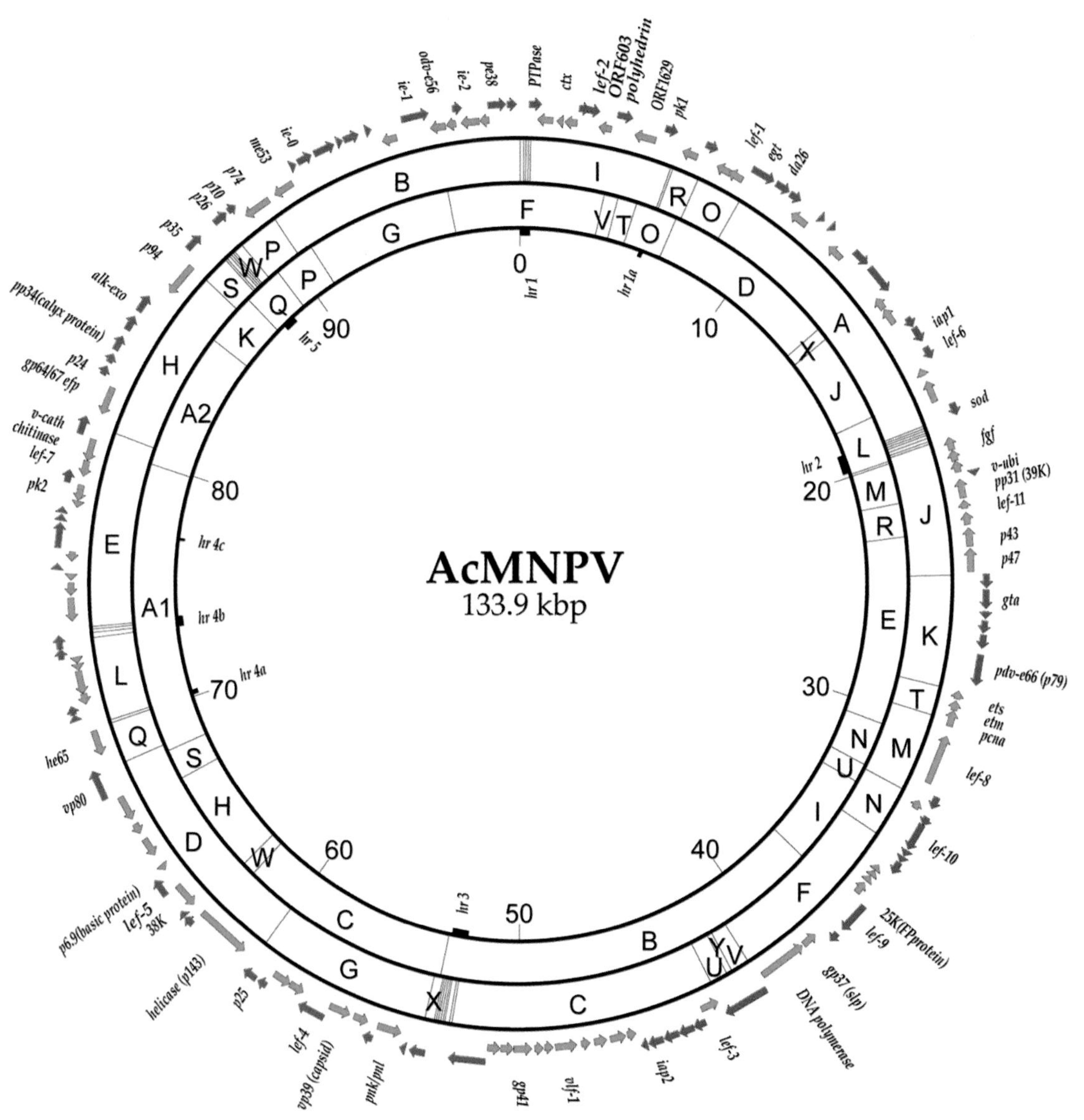

Figure 2: The covalently closed circular genome of Autographa californica multiple nucleopolyhedrovirus (AcMNPV) is illustrated with locations and orientations of known and predicted ORFs (arrows). Restriction maps for *Eco*RI and *Hind*III are indicated by letters on outer and inner rings, respectively. Locations of homologous repeat (*hr*) sequences are indicated on the inside of the circle as small filled boxes. Map units are indicated on the inside of the map (1 map unit = 1.339 kbp). (Redrawn from Ayres *et al*, 1994).

ANTIGENIC PROPERTIES

Antigenic determinants that cross react between different baculoviruses exist on virion proteins and on the major occlusion body polypeptide: polyhedrin or granulin. Neutralizing antibodies react with the major surface glycoprotein of BV.

BIOLOGICAL PROPERTIES

Baculoviruses have been isolated from arthropods only; primarily from insects of the order Lepidoptera, but also Hymenoptera, and Diptera as well as from the crustacean order Decapoda (shrimp). Transmission: (i) natural-horizontal transmission by contamination of food, egg surface, etc. with occlusion bodies; (ii) vertical transmission within the egg has been reported; (iii) experimental-by injection of intact hosts with BV or by infection or transfection of cell cultures. Typically the infection process in a permissive insect host requires approximately one week, and as an end result, the diseased insect liquifies releasing infectious occlusion bodies into the environment. Occlusion bodies represent an environmentally stable form of the virus with increased resistance to inactivation by light.

GENUS NUCLEOPOLYHEDROVIRUS

Type Species *Autographa californica multiple nucleopolyhedrovirus*

DISTINGUISHING FEATURES

Two virion phenotypes may be characteristic of a virus species. Virions of the ODV phenotype are embedded within an occlusion body composed of a crystalline matrix of a single viral protein (polyhedrin). Each occlusion body measures 0.15 to 15 μm in size, matures within nuclei of infected cells and characteristically contains many enveloped virions. The occluded virions are packaged with either single (S) or multiple (M) nucleocapsids within a single viral envelope. Factors that regulate nucleocapsid packaging are unknown and for some species, packaging may be variable. S and M designations in common usage have been retained for species where variability has not been reported and for distinct viruses that would otherwise have identical designations under the current nomenclature. Nucleocapsids are rod-shaped (30-60 nm x 250-300 nm) and contain a single molecule of circular supercoiled dsDNA of approximately 80-180 kbp in size. Nucleocapsid length appears to be proportional to genome size. Nucleocapsids are thought to be transported through the nuclear membrane and into the nucleus, where uncoating and viral replication occur. Hosts include at least three orders of insects and an order of Crustacea.

LIST OF SPECIES DEMARCATION CRITERIA IN THE GENUS

Because detailed comparative data are lacking in most cases, species parameters are not well defined. However, species distinctions indicated here are broadly based on host range and specificity, DNA restriction profiles, DNA sequences from various regions of the genome, and predicted protein sequence similarities.

LIST OF SPECIES IN THE GENUS

Species names are in green italic script; strain names and synonyms are in black roman script; tentative species names are in blue roman script. Sequence accession numbers and assigned abbreviations ( ) are also listed. The acronyms MNPV, SNPV and NPV in the following names are abbreviations for 'multiple nucleopolyhedrovirus', 'single nucleopolyhedrovirus', and 'nuclepolyhedrovirus', respectively.

SPECIES IN THE GENUS

Adoxophyes honmai NPV		
Adoxophyes honmai NPV	[NC_004690]	(AdhoNPV)
Agrotis ipsilon MNPV		
Agrotis ipsilon MNPV		(AgipMNPV)
Anticarsia gemmatalis MNPV		
Anticarsia gemmatalis MNPV		(AgMNPV)
Autographa californica MNPV		
Autographa californica MNPV	[L22858]	(AcMNPV)
Anagrapha falcifera MNPV		(AnfaMNPV)
Galleria mellonella MNPV		(GmMNPV)
Rachiplusia ou MNPV	[NC_004323]	(RoMNPV)
Spodoptera exempta MNPV		(SpexMNPV)
Trichoplusia ni MNPV		(TnMNPV)
Bombyx mori NPV		
Bombyx mori NPV	[L33180]	(BmNPV)

Buzura suppressaria NPV		
Buzura suppressaria NPV		(BuzuNPV)
Choristoneura fumiferana DEF MNPV		
Choristoneura fumiferana DEF MNPV		(CfDefNPV)
Choristoneura fumiferana MNPV		
Choristoneura fumiferana MNPV		(CfMNPV)
Choristoneura rosaceana NPV		
Choristoneura rosaceana NPV		(ChroNPV)
Culex nigripalpus NPV		
Culex nigripalpus NPV	[C_003084]	(CuniNPV)
Ecotropis obliqua NPV ‡		
Ecotropis obliqua NPV		(EcobNPV)
Epiphyas postvittana NPV		
Epiphyas postvittana NPV	[AY043265]	(EppoNPV)
Helicoverpa armigera NPV		
Helicoverpa armigera NPV	[NC_003094, NC_002654]	(HearNPV)
Helicoverpa zea SNPV		
Helicoverpa zea SNPV	[AF334030]	(HzSNPV)
Lymantria dispar MNPV		
Lymantria dispar MNPV	[NC_00002654]	(LdMNPV)
Mamestra brassicae MNPV		
Mamestra brassicae MNPV		(MbMNPV)
Mamestra configurata NPV-A ‡		
Mamestra configurata NPV-A	[NC_003529, AF539999]	(MacoNPV-A)
Mamestra configurata NPV-B ‡		
Mamestra configurata NPV-B	[NC_004117]	(MacoNPV-B)
Neodiprion lecontii NPV		
Neodiprion lecontii NPV		(NeleNPV)
Neodiprion sertifer NPV		
Neodiprion sertifer NPV		(NeseNPV)
Orgyia pseudotsugata MNPV		
Orgyia pseudotsugata MNPV	[NC_001875]	(OpMNPV)
Spodoptera exigua MNPV		
Spodoptera exigua MNPV	[NC_002169]	(SeMNPV)
Spodoptera frugiperda MNPV		
Spodoptera frugiperda MNPV		(SfMNPV)
Spodoptera littoralis NPV		
Spodoptera littoralis NPV		(SpliNPV)
Spodoptera litura NPV ‡		
Spodoptera litura NPV	[NC_003102]	(SpltNPV)
Thysanoplusia orichalcea NPV		
Thysanoplusia orichalcea NPV		(ThorNPV)
Trichoplusia ni SNPV		
Trichoplusia ni SNPV		(TnSNPV)
Wiseana signata NPV		
Wiseana signata NPV		(WisiNPV)

TENTATIVE SPECIES IN THE GENUS

Aedes sollicitans NPV	(AesoNPV)
Hyphantria cunea NPV	(HycuNPV)
Orgyia pseudotsugata SNPV	(OpSNPV)
Penaeus monodon NPV	(PemoNPV)
Panolis flammea NPV	(PaflNPV)

GENUS GRANULOVIRUS

Type Species *Cydia pomonella granulovirus*

DISTINGUISHING FEATURES

Two virion phenotypes (BV and ODV) may be characteristic of a virus species. One (ODV) is occluded within an ovicylindrical occlusion body composed mainly of a single crystallized protein (granulin). Each occlusion body measures approximately 0.13 x 0.50 μm in size and characteristically contains one virion. Each ODV virion typically contains a single nucleocapsid within a single envelope. Occluded virions may mature among nuclear-cytoplasmic cellular contents after rupture of the nuclear membrane of infected cells. Nucleocapsids are rod-shaped (30-60 nm x 250-300 nm) and contain a single molecule of circular supercoiled dsDNA approximately 80-180 kbp in size. Uncoating is thought to occur by a mechanism in which viral DNA is extruded into the nucleus through the nuclear pore while the capsid remains in the cytoplasm. Species of this genus have been isolated only from the insect Order Lepidoptera.

LIST OF SPECIES DEMARCATION CRITERIA IN THE GENUS

Because detailed comparative data are lacking in most cases, species parameters are not well defined. However, species distinctions indicated here are broadly based on host range and specificity, DNA restriction profiles, DNA sequences from various regions of the genome, and predicted protein sequence similarities.

LIST OF SPECIES IN THE GENUS

Species names are in green italic script; strain names and synonyms are in black roman script; tentative species names are in blue roman script. Sequence accession numbers, and assigned abbreviations ( ) are also listed.

SPECIES IN THE GENUS

Adoxophyes orana granulovirus		
Adoxophyes orana granulovirus		(AdorGV)
Artogeia rapae granulovirus		
Artogeia rapae granulovirus		(ArGV)
Pieris brassicae granulovirus		(PbGV)
Choristoneura fumiferana granulovirus		
Choristoneura fumiferana granulovirus		(ChfuGV)
Cryptophlebia leucotreta granulovirus		
Cryptophlebia leucotreta granulovirus		(CrleGV)
Cydia pomonella granulovirus		
Cydia pomonella granulovirus	[U53466]	(CpGV)
Harrisina brillians granulovirus		
Harrisina brillians granulovirus		(HabrGV)
Helicoverpa armigera granulovirus		
Helicoverpa armigera granulovirus		(HearGV)
Lacanobia oleracea granulovirus		
Lacanobia oleracea granulovirus		(LaolGV)
Phthorimaea operculella granulovirus ‡		
Phthorimaea operculella granulovirus	[AF499596]	(PhopGV)
Plodia interpunctella granulovirus		
Plodia interpunctella granulovirus		(PiGV)
Plutella xylostella granulovirus		
Plutella xylostella granulovirus	[NC_002593]	(PlxyGV)
Pseudalatia unipuncta granulovirus		
Pseudalatia unipuncta granulovirus		(PsunGV)
Trichoplusia ni granulovirus		

Trichoplusia ni granulovirus		(TnGV)
Xestia c-nigrum granulovirus		
Xestia c-nigrum granulovirus	[AF162221]	(XecnGV)

TENTATIVE SPECIES IN THE GENUS

None reported.

LIST OF UNASSIGNED VIRUSES IN THE FAMILY

None reported. The VIth ICTV Report listed numerous other nucleopolyhedroviruses and granuloviruses whose taxonomic status was unclear since either no isolates were available, or isolates were insufficiently characterized. These viruses were omitted from the VIIth and VIIIth Reports.

PHYLOGENETIC RELATIONSHIPS WITHIN THE FAMILY

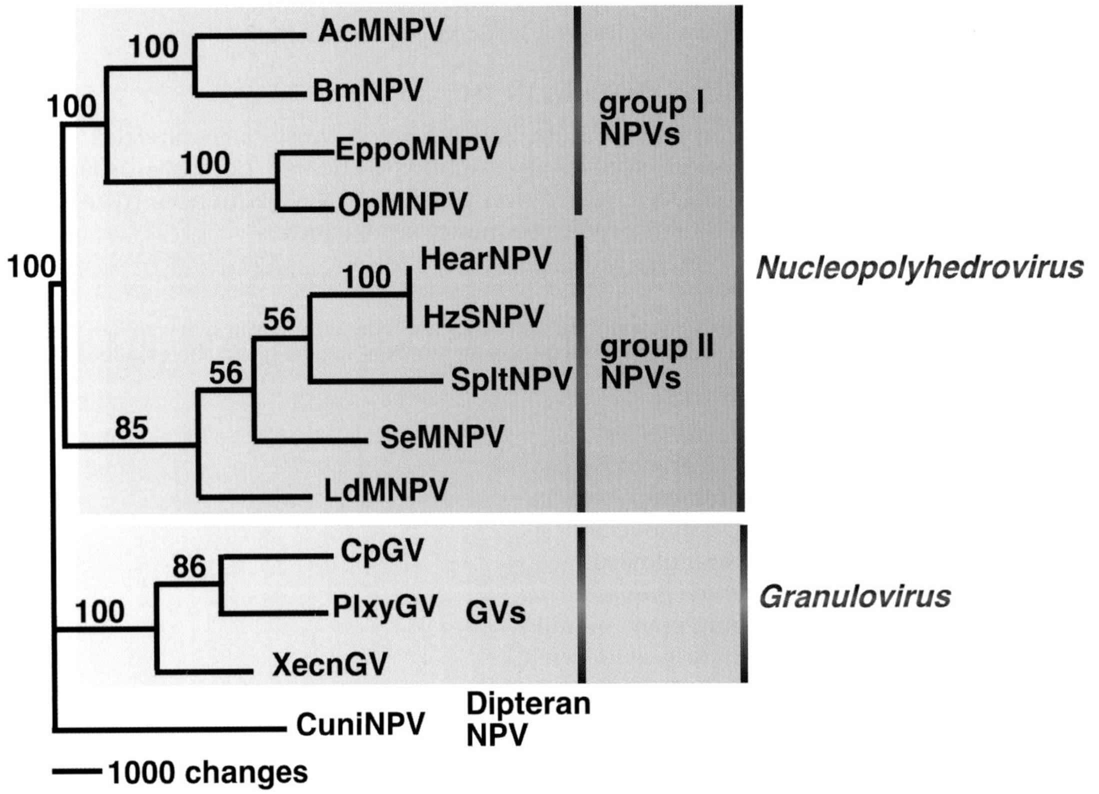

Figure 3: The phylogenetic tree obtained for nine sequenced viruses, based on an alignment of 63 genes common to all of them. This is a maximum parsimony tree with % bootstrap scores from 1000 replicates (From Herniou *et al.*, 2003).

SIMILARITY WITH OTHER TAXA

None reported.

DERIVATION OF NAMES

Baculo: from '*baculum*,' meaning 'stick', which refers to the morphology of the nucleocapsid.
Granulo: from 'granule' which refers to the relatively small size and granular appearance of GV occlusion bodies in infected cells.
Polyhedro: from 'polyhedron,' which refers to the shape of occlusion bodies.

REFERENCES

Ayres, M.D., Howard, S.C., Kuzio, J., Lopez-Ferber, M. and Possee, R.D. (1994). The complete DNA sequence of *Autographa californica* nuclear polyhedrosis virus. *Virology*, **202**, 586-605.

Ahrens, C.H., Russell, R.L.Q., Funk, C.J., Evans, J.T., Harwood, S.H. and Rohrmann, G.F. (1997). The sequence of the *Orgyia pseudotsugata* multinucleocapsid nuclear polyhedrosis virus genome. *Virology*, **229**, 381-399.

Blissard, G.W. (1996). Baculovirus-Insect Cell Interactions. *Cytotechnology*, **20**, 73-93.

Consigli, R.A., Russell, D.L. and Wilson, M.E. (1986). The biochemistry and molecular biology of the granulosis virus that infects *Plodia interpunctella*. *Cur. Topics Microbiol. Immunol.*, **131**, 69-101.

Fraser, M.J. (1986). Ultrastructural observations of virion maturation in *Autographa californica* Nuclear Polyhedrosis virus infected *Spodoptera frugiperda* cell cultures. *J. Ultrastruct. Mol. Struct. Res.*, **95**, 189-195.

Granados, R.R. and Federici, B.A. (eds)(1986). The biology of baculoviruses. CRC Press, Boca Raton, Florida.

Herniou, E.A., Olszewski, J.A., Cory, J.S. and O'Reilly D.R. (2003). The genome sequence and evolution of baculoviruses. *Ann. Rev. Ento.*, **48**, 211-234

Kool, M., Ahrens, C.H., Vlak, J.M. and Rohrmann, G.R. (1995). Replication of baculovirus DNA. *J. Gen. Virol.*, **76**, 2103-2118.

Miller, L.K. (1995). Insect Viruses. In: *Fields Virology*, (B.N. Fields, D.M. Knipe, and P. Howley, eds), pp. 533-556. Lippincott-Raven, New York.

Miller, L.K. (ed)(1997). The Baculoviruses. Plenum Press. New York.

Rohrmann, G.F. (1992). Baculovirus structural proteins. *J. Gen. Virol.*, **73**, 749-761.

Volkman, L.E. (1997). Nucleopolyhedrovirus interactions with their insect hosts. *Adv. Virus Res.*, **48**, 313-348.

CONTRIBUTED BY

Theilmann, D.A., Blissard, G.W., Bonning, B., Jehle, J., O'Reilly, D.R., Rohrmann, G.F., Thiem, S. and Vlak, J.M.

DSDNA

FAMILY NIMAVIRIDAE

TAXONOMIC STRUCTURE OF THE FAMILY

Family *Nimaviridae*
Genus *Whispovirus*

Since only one genus is currently recognized, the family description corresponds to the genus description

GENUS WHISPOVIRUS

Type Species *White spot syndrome virus 1*

VIRION PROPERTIES

MORPHOLOGY

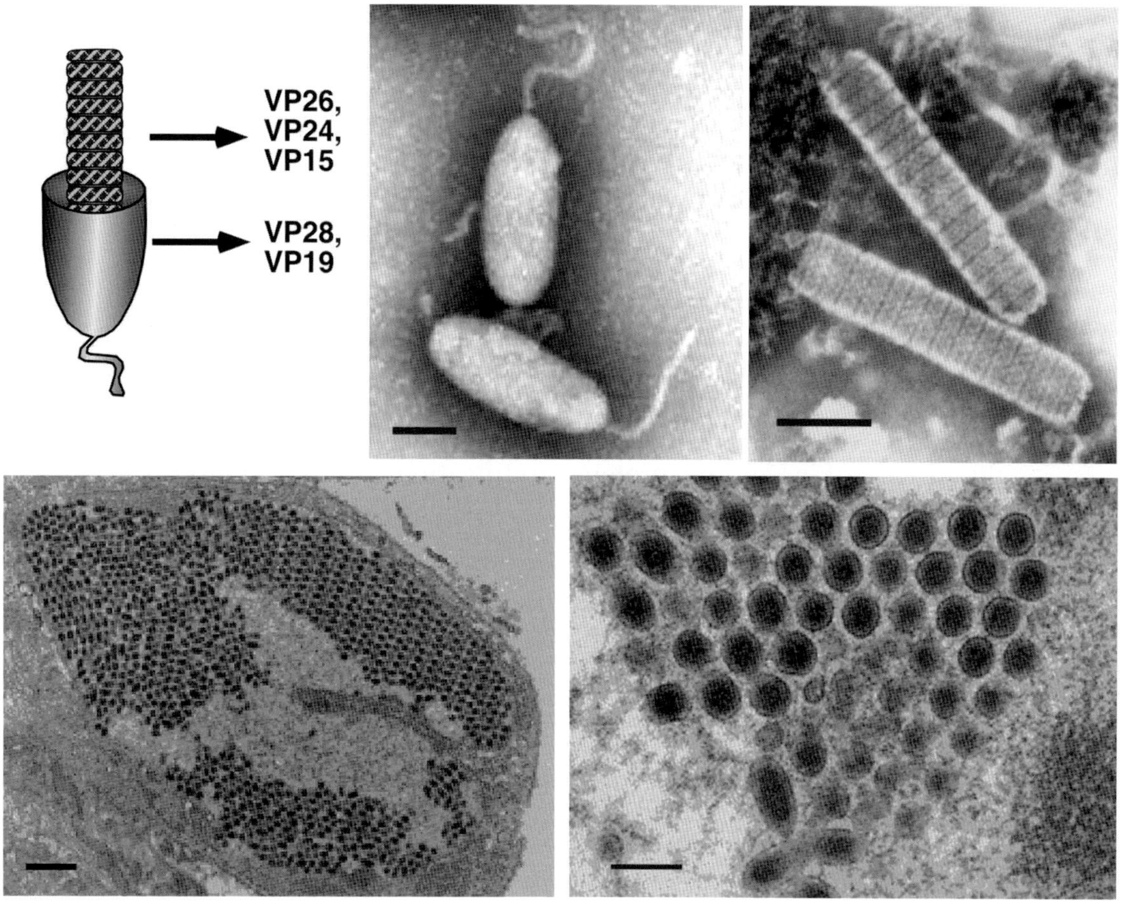

Figure 1: (Top) Morphology of virions of White spot syndrome virus (WSSV). (Left) Schematic illustration of the structure of a typical whispovirus virion. (Top center and right) Negative contrast electron micrographs of WSSV virions (center, courtesy of Marielle van Hulten) and nucleocapsids (right, courtesy of Don Lightner) from hemolymph of infected Penaeus monodon. The bars represent 100 nm. (Bottom left) Thin section of WSSV-infected stomach epithelium showing the parallel arrangement of virions in the nucleoplasm (the bar represents 1 μm) and (Bottom right) a cross-section of virions (the bar represents 100 nm)(courtesy of Don Lightner).

Whispovirus virions are ovoid or ellipsoid to bacilliform in shape, have a regular symmetry, and measure 120-150 nm in diameter and 270-290 nm in length. Most notable is the thread- or flagella-like extension (appendage) at one end of the virion. The virion consists of an inner, rod-shaped nucleocapsid with a tight-fitting capsid layer, and is surrounded by a loose-fitting outer lipid-containing trilaminar envelope. The isolated nucleocapsid typically measures 65-70 nm in diameter and 300-350 nm in length. It contains a DNA-protein core bounded by a distinctive capsid layer, giving it a cross-hatched or striated appearance (Fig. 1).

PHYSICOCHEMICAL AND PHYSICAL PROPERTIES

Virions have a buoyant density of 1.22 g/cm^3 in CsCl, whereas the nucleocapsids have a buoyant density of 1.31 g/cm^3. The virions are sensitive to detergents. Other properties are not known.

NUCLEIC ACID

The nucleocapsid contains a single molecule of circular dsDNA with an approximate size of 300 kbp. G+C ratio of WSSV is about 41%.

PROTEINS

The virions contain at least 5 major and at least 13 minor polypeptides ranging in size from 14 to 190 kDa. The envelope contains two major proteins, VP28 and VP19; the nucleocapsid contains three major proteins, VP26, VP24 and VP15. The latter protein is a very basic, histone-like DNA-binding nucleoprotein. The three major proteins VP28, VP26 and VP24 are related.

LIPIDS

Detergent sensitivity of the virion indicates the presence of lipid in the envelope. Specific lipid composition is unknown.

CARBOHYDRATES

None of the major virion structural proteins is glycosylated.

GENOME ORGANIZATION AND REPLICATION

The WSSV genome has been sequenced and 184 ORFs have been identified (WSSV-Th; Fig. 2). A few genes have been identified ranging from those encoding structural virion proteins (see below) to those involved in DNA replication (DNA polymerase, ribonucleotide reductase subunits, dUTPase, thymidylate synthase, thymidine-thymidylate kinase) and protein modification (protein kinase), but most of them are unassigned. The WSSV genome is further characterized by the presence of nine homologous repeat regions distributed along the genome. Viral transcripts are polyadenylated and probably capped. No evidence has been found for the occurrence of RNA splicing. Temporal expression studies showed classes of early and late genes. However, no WSSV specific promoter motifs involved in the expression of early and late genes have been identified yet. Consensus TATA box sequences have been found at a functional distance upstream of transcription initiation sites and late transcripts seem to start 25 nt downstream of an A/T-rich region.

Replication of whispoviruses occurs in the nucleus, where virions are assembled (Fig. 1).

Genotypic variants exist that can be discriminated by restriction fragment length polymorphism and on the basis of the complete genome sequence. Three WSSV isolates have been sequenced and they vary in size from 297,967 nt (WSSV-Th) to 305,107 nt (WSSV-Ch) and 307,287 nt (WSSV-Tw). Major differences can be attributed to insertions/deletions and are located at map position +10.6 (Fig. 2).

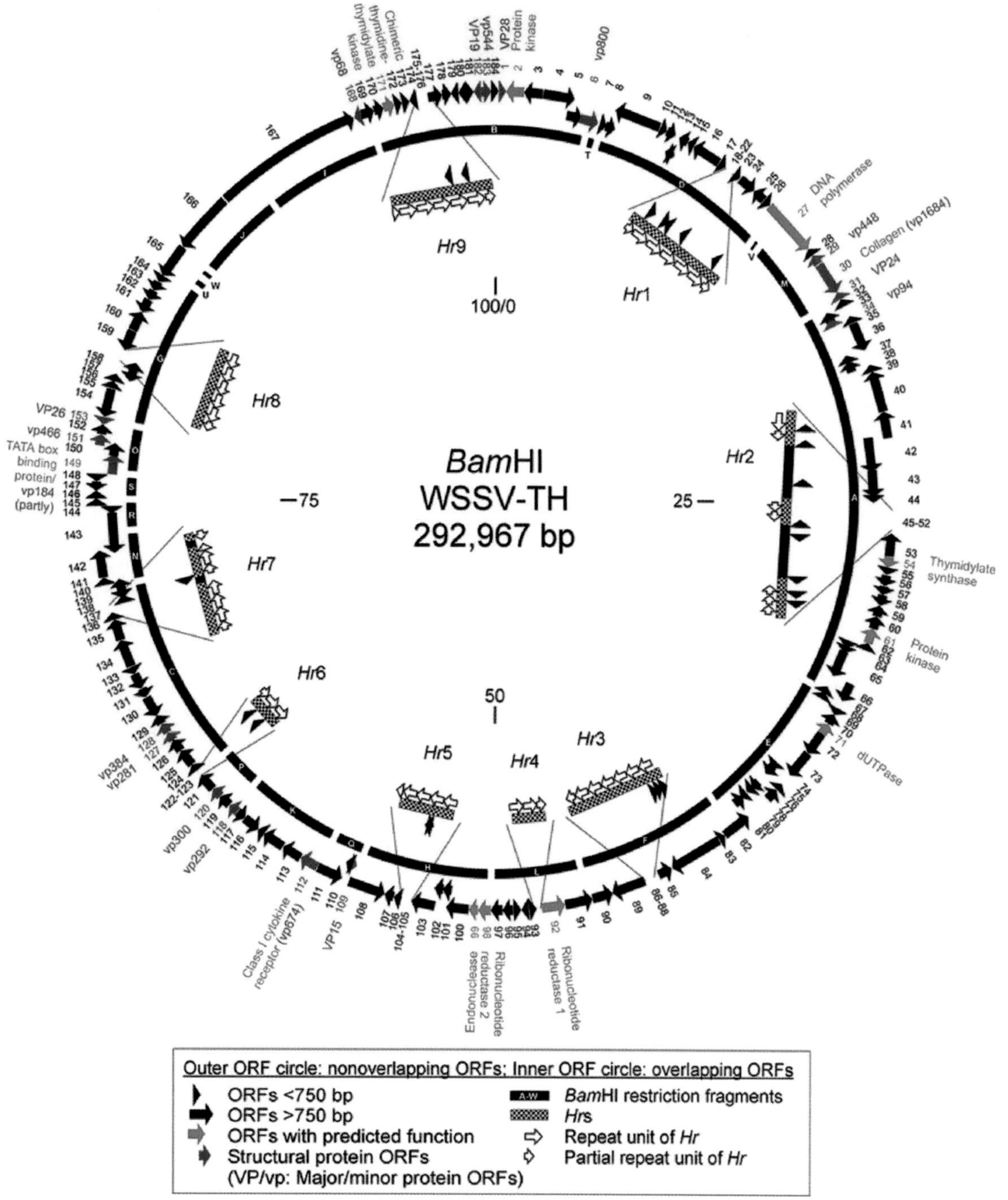

Figure 2: Circular map of the circular double-stranded White spot syndrome virus 1 - Th (WSSV-1-Th) genome showing the genomic organization. The A of the ATG initiation codon of VP28 (ORF1) has been designated position 1. *Bam*HI fragments, ORFs, and homologous repeats (*hrs*) are presented according to the figure key.

Antigenic Properties

Polyclonal and monoclonal antibodies have been raised against WSSV virions, which can be used as a diagnostic tools. Virus infection can be neutralized with antiserum against the envelope protein VP28.

BIOLOGICAL PROPERTIES

HOST RANGE

WSSV can infect a wide range of aquatic crustaceans including salt, brackish and fresh water penaeids, crabs and crayfish.

TRANSMISSION

The virus is transmitted *per os* by predation on diseased individuals or via water through the gills

GEOGRAPHICAL DISTRIBUTION

Whispovirus isolates have been identified from crustaceans in China, Japan, Southeast Asia, the Indian continent, the Mediterranean, the Middle East and the Americas. An unusual property of the virus is the extremely wide host range despite a very low level of genetic polymorphism. The latter suggests that white spot syndrome disease is a recent epizootic.

CYTOPATHIC EFFECTS

The major targets for WSSV infection are tissues of ecto-mesodermal origin, such as those of the gills, lymphoid organ, cuticular epithelium, midgut and connective tissue of the hepatopancreas. WSSV can grow in primary cultures of lymphoid origin. Infection sometimes causes disease and sometimes not, depending on factors as yet poorly understood but related to species tolerance and environmental triggers. Shrimp and other crustaceans susceptible to disease from this virus show gross signs of lethargy, such as lack of appetite and slow movement, and often a reddish coloration of the whole body. 'White spots' embedded within the exoskeleton are often seen with slow WSSV infections. These spots are the result of calcified deposits that range in size from a few mm to 1 cm or more in diameter. However, in some cases such as acute experimental infections resulting from injected viral preparations there may be no gross signs of infection other than lethargy and lack of appetite. Most affected animals die within 5-10 days after infection. With an appropriate infection dose to allow sufficient time before death, animals susceptible to disease show large numbers of virions circulating in the hemolymph but this may also occur for tolerant species that show no mortality. Thus, high viral loads *per se* do not cause disease or mortality for all susceptible species.

LIST OF SPECIES DEMARCATION CRITERIA IN THE GENUS

Only a single species within the genus has been identified to date (*White spot syndrome virus 1*). Various isolates with small genetic polymorphisms have been identified. It should be realized, however, that as the family *Nimaviridae* is newly recognized, the species organization may change once existing and new isolates are studied in more detail.

LIST OF SPECIES IN THE GENUS

Species names are in green italic script; strain names and synonyms are in black roman script; tentative species names are in blue roman script. Sequence accession numbers, and assigned abbreviations ( ) are also listed.

SPECIES IN THE GENUS

White spot syndrome virus 1

(Chinese baculo-like virus)
(Hypodermic and hematopoietic necrosis baculovirus)
(*Penaeus mondon* non-occluded baculovirus II)
(*Penaeus mondon* non-occluded baculovirus III)
(*Penaeus japonicus* rod-shaped nuclear virus)
(Systemic ectodermal and mesodermal baculovirus)
(White spot bacilliform virus)
(White spot baculovirus)

White spot syndrome virus 1 - Ch	AF369029	(WSSV-1-Ch)
White spot syndrome virus 1 - Th	AF332093	(WSSV-1-Th)
White spot syndrome virus 1 - Tw	AF440570	(WSSV-1-Tw)

TENTATIVE SPECIES IN THE GENUS

Carcinus maena B virus
Carcinus mediterraneus B virus
Carcinus mediterraneus Tau virus
Callinectes sapidus baculo-A virus
Callinectes sapidus baculo-B virus

LIST OF UNASSIGNED VIRUSES IN THE FAMILY

None reported.

PHYLOGENETIC RELATIONSHIPS WITHIN THE FAMILY

Not applicable.

SIMILARITY WITH OTHER TAXA

Whispovirus morphogenesis of the virions, including the rod-shaped nucleocapsid, bear resemblance to the budded virus particles of baculoviruses, polydnaviruses and the unassigned Oryctes virus. The presence of repeat regions dispersed over the DNA genome is a property shared with members of the families *Baculoviridae* and *Ascoviridae*. WSSV is phylogenetically distinct from other large DNA viruses.

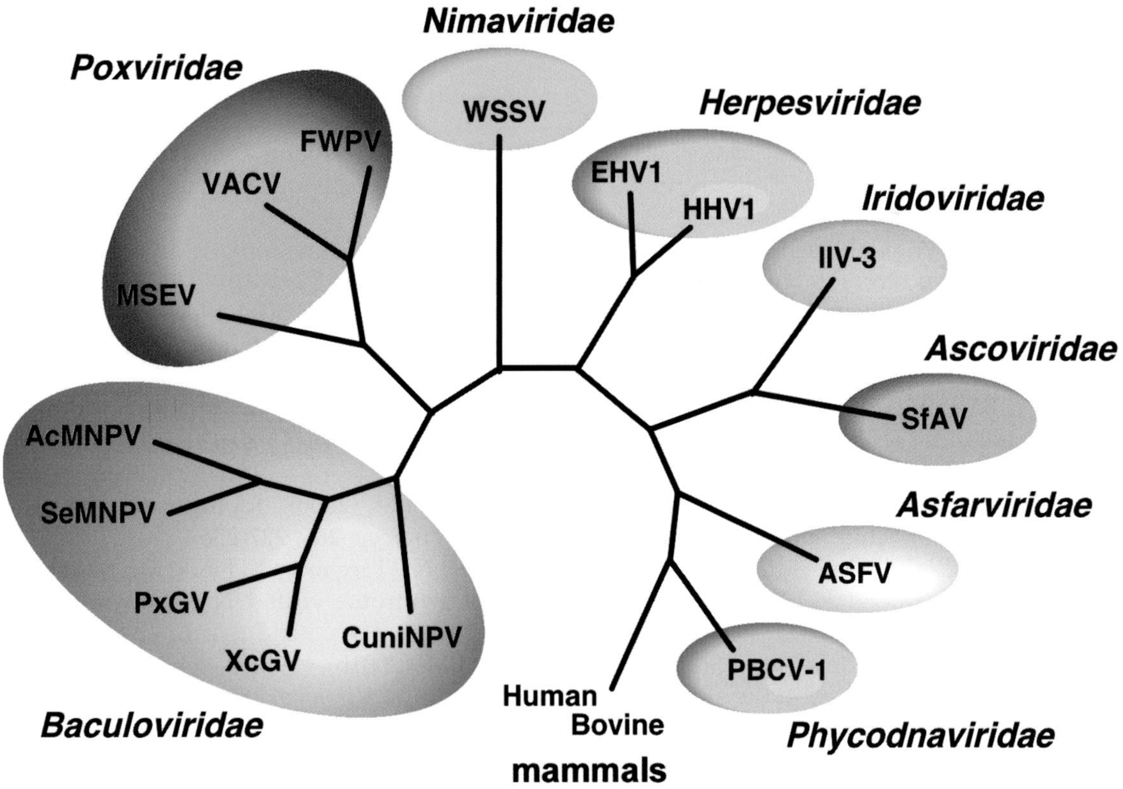

Figure 3: Cladogram based on genetic distances between aa sequences derived from the DNA polymerase gene of WSSV and a number of other large DNA viruses, as well as two mammals represenatives.

Derivation of Names

Nima: Latin for 'thread'; referring to the thread- or tail-like polar extension (appendage) on the virus particle.

References

Chou, H., Huang, C., Wang, C., Chiang, H., Lo, C., Chou, H.Y., Huang, C.Y., Wang, C.H., Chiang, H.C. and Lo, C.F. (1995). Pathogenicity of a baculovirus infection causing White spot syndrome in cultured penaeid shrimp in Taiwan. *Dis. Aquat. Org.*, **23**, 165-173.

Durand, S., Lightner, D.V., Redman, R.M. and Bonami, J.R. (1997). Ultrastructure and morphogenesis of White spot syndrome baculovirus (WSSV). *Dis. Aquat. Org.*, **29**, 205-211.

Flegel, T.W. (1997). Major viral diseases of the black tiger prawn (*Penaeus monodon*) in Thailand. *World J. Microbiol. Biotechnol.*, **13**, 433-442.

Huang, C., Zhang, X., Lin, Q., Xu, X., Hu, Z.H. and Hew, C.L. (2002). Proteomic analysis of shrimp White spot syndrome virus proteins and characterization of a novel envelope protein VP466. *Mol. Cell. Proteom.*, **1**, 223-231.

Lightner, D.V. (1996). A handbook of pathology and diagnostic procedures for diseases of penaeid shrimp. *Special publication of the World Aquaculture Society*. Baton Rouge, Louisiana.

Lo, C.F., Hsu, H.C., Tsai, M.F., Ho, C.H., Peng, S.E., Kou, G.H. and Lightner, D.V. (1999). Specific genomic fragment analysis of different geographical clinical samples of shrimp White spot syndrome virus. *Dis. Aquat. Org.*, **35**, 175-185.

Lo, C.F., Ho, C.H., Chen, C.H., Liu, K.F., Chiu, Y.L., Yeh, P.Y., Peng, S.E., Hsu, H.C., Liu, H.C., Chang, C.F., Su, M.S., Wang, C.H. and Kou, G.H. (1997). Detection and tissue tropism of White spot syndrome baculovirus (WSBV) in captured brooders of *Penaeus monodon* with a special emphasis on reproductive organs. *Dis. Aquat. Org.*, **30**, 53-72.

Loh, P.C., Tapay, L.M. and Nadala, E.C. (1997). Viral pathogensis of penaeid shrimp. *Adv. Virus Res.*, **48**, 263-312.

Nadala, E.C.B., Tapay, L.M. and Loh, P.C. (1998). Characterization of a non-occluded baculovirus-like agent pathogeneic to shrimp. *Dis. Aquat. Org.*, **33**, 221-229.

Tsai, M.F., Lo, C.F., van Hulten, M.C.W., Tzeng, H.F., Chou, C.M., Huang, C.J., Wang, C.S., Lin, J.Y., Vlak, J.M. and Kou, G.S. (2000). Transcriptional analysis of the ribonucleotide reductase genes in shrimp white spot syndrome virus. *Virology*, **277**, 92-99.

van Hulten, M.C.W., Goldbach, R.W. and Vlak, J.M. (2000a). Three functionally diverged major White spot syndrome virus structural proteins evolved by gene duplication. *J. Gen. Virol.*, **81**, 2525-2529.

van Hulten, M.C.W., Reijns, M., Vermeesch, A.M.G., Zandbergen, F. and Vlak, J.M. (2002). Identification of VP19 and VP15 of White spot syndrome virus (WSSV) and glycosylation of the WSSV major structural proteins. *J. Gen. Virol.*, **83**, 257-265.

van Hulten, M.C.W. and Vlak, J.M. (2002). Genetic evidence for a unique taxonomic position of white spot syndrome virus of shrimp: genus *Whispovirus*. In: *Diseases in Asian Aquaculture IV* (C.R. Lavilla-Pitogo and E.R. Cruz-Lacierda, eds), p. 25-35.

van Hulten, M.C.W., Westenberg, M., Goodall, S.D. and Vlak, J.M. (2000). Identification of two major virion protein genes of White spot syndrome virus of shrimp. *Virology*, **266**, 227-236.

van Hulten, M.C.W., Witteveldt, J., Peters, S., Kloosterboer, N., Tarchini, R., Fiers, M., Sandbrink, H., Klein Lankhorst, R. and Vlak, J.M. (2001a). The white spot syndrome virus DNA genome sequence. *Virology* **286**, 7-22.

van Hulten, M.C.W., Witteveldt, J., Snippe, M. and Vlak, J.M. (2001b). White spot syndrome virus envelope protein VP28 is involved in the systemic infection of shrimp. *Virology*, **285**, 228-233.

Wang, C.H., Yang, H.S., Tang, C.Y., Lu, C.H., Kou, G.H. and Lo, C.F. (2000). Ultrastructure of white spot syndrome virus development in primary lymphoid organ cell cultures. *Dis. Aquat. Org.*, **41**, 91-104.

Yang, F., He, J., Lin, X., Li, Q., Pan, D., Zhang, X. and Xu, X. (2001). Complete genome sequence of the shrimp white spot bacilliform virus. *J. Virol.*, **75**, 11811-20.

Contributed By

Vlak, J.M., Bonami, J.R., Flegel, T.W., Kou, G.H., Lightner, D.V., Lo, C.F., Loh, P.C. and Walker, P.W.

FAMILY HERPESVIRIDAE

TAXONOMIC STRUCTURE OF THE FAMILY

Family	***Herpesviridae***		
Subfamily		***Alphaherpesvirinae***	
Genus			***Simplexvirus***
Genus			***Varicellovirus***
Genus			***Mardivirus***
Genus			***Iltovirus***
Subfamily		***Betaherpesvirinae***	
Genus			***Cytomegalovirus***
Genus			***Muromegalovirus***
Genus			***Roseolovirus***
Subfamily		***Gammaherpesvirinae***	
Genus			***Lymphocryptovirus***
Genus			***Rhadinovirus***
Unassigned genus			***Ictalurivirus***

VIRION PROPERTIES

MORPHOLOGY

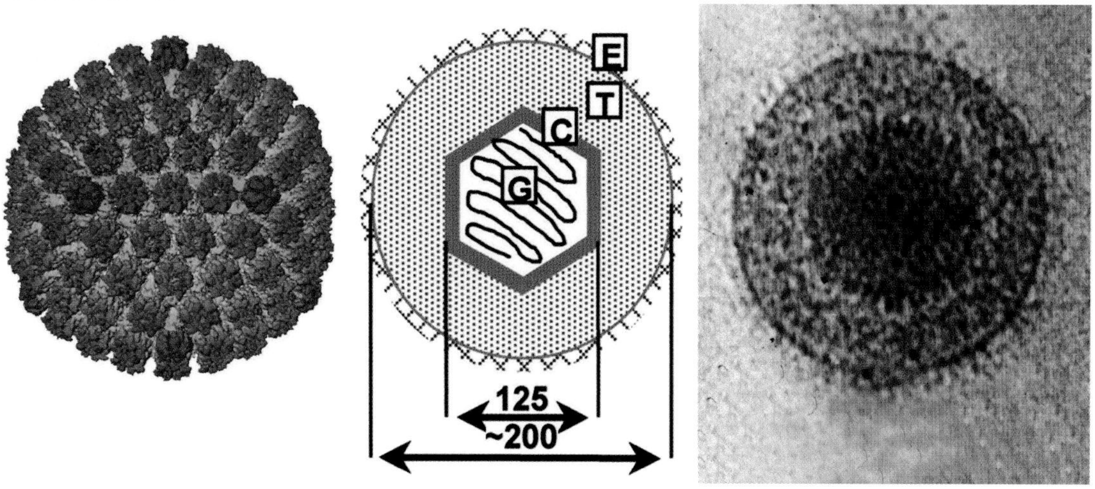

Figure 1: Herpesvirus morphology. (Left) Reconstruction of a Human herpesvirus 1 (HHV-1) capsid generated from cryo-electron microscope images, viewed along the 2-fold axis. The hexons are shown in blue, the pentons in red, and the triplexes in green. (Courtesy of W. Chiu and H. Zhou). (Center) Schematic representation of a virion with diameters in nm. (G) genome, (C) capsid, (T) tegument, (E) envelope. (Right) Cryo-electron microscope image of a HHV-1 virion. (Reproduced from Rixon (1993) with permission from Elsevier).

Virions of herpesviruses have complex and characteristic structures consisting of both symmetrical and non-symmetrical components. The spherical virion comprises core, capsid, tegument and envelope (Fig. 1). The core consists of the viral genome packaged as a single, linear, dsDNA molecule into a preformed capsid. DNA is packed in a liquid-crystalline array that fills the entire internal volume of the capsid. The mature capsid is a T = 16 icosahedron. In virions of Human herpesvirus 1 (HHV-1) the 16 nm thick protein shell has an external diameter of 125 nm. The 12 pentons and 150 hexons (a total of 162 capsomers) are composed primarily of 5 and 6 copies, respectively, of the same protein and are joined by masses, termed triplexes, which are made of two smaller proteins present in a 2:1 ratio (Fig. 1). Capsids assemble by co-condensation around a protein scaffold to form a procapsid in which the subunits are weakly connected. Proteolytic cleavage of the scaffolding protein triggers loss of scaffold and reorganisation of the shell into the characteristic capsid form. The structure of the tegument is poorly defined, with evidence of symmetry only in the region immediately adjacent to the capsid. The tegument contains many proteins, not all of which are required for the formation of virions. Some individual tegument proteins can vary markedly in abundance. Enveloped tegument structures lacking capsids can assemble and are released from cells along with virions. The envelope is a lipid bilayer that is intimately associated with the outer surface of the tegument. It contains a number (at least 10 in HHV-1) of different integral viral glycoproteins.

PHYSICOCHEMICAL AND PHYSICAL PROPERTIES

The mass of the HHV-1 virion is about 13 x 10^{-16} g, of which the DNA comprises about 10%. The mass of a full capsid is about 5 x 10^{-16} g. The buoyant density of virions in CsCl is 1.22-1.28 g/cm^3. The stability of different herpesviruses varies considerably, but they are generally unstable to desiccation and low pH. Infectivity is destroyed by lipid solvents and detergents.

NUCLEIC ACID

The genomes are composed of linear dsDNA ranging from 125 to 240 kbp in size and from 32 to 75% in G+C content. Those genomes examined in sufficient detail contain a single nucleotide extension at the 3'-ends, and no terminal protein has been identified. The arrangement of reiterated sequences, either at the termini or internally, results in a number of different genome structures (Fig. 2), and in the existence of isomers due to recombination between terminal and inverted internal reiterations.

In Fig. 2, structure 1 shows a unique sequence flanked by a direct repeat that may be larger than 10 kbp (Human herpesvirus 6; HHV-6) or as short as 30 bp (Murid herpesvirus 1; MuHV-1). Structure 2 also contains a single unique sequence but in this case it is flanked by a variable number of repeated sequences at each terminus (e.g. Human herpesvirus 8; HHV-8). Structure 3 contains different elements at each terminus, which are present internally in inverted orientation. The genome is thus divided into two unique regions (one 'long' and one 'short'), which are flanked by inverted repeats. The repeated sequence flanking the long unique sequence is very short (88 bp in Human herpesvirus 3; HHV-3). Homologous recombination in replicated concatemeric DNA results in inversion of the two regions, and cleavage largely or entirely at one of the two junction regions results in unit length genomes that are composed entirely or predominantly of two isomers differing in the orientation of the short unique sequence. Structure 4 is the most complex. Like structure 3, it contains long and short unique regions, but in this case both are flanked by large inverted repeat sequences. Homologous recombination and cleavage with equal probability at either of the two junction regions results in the formation of four isomers differing in the orientations of the unique sequences, with each isomer equimolar in virion populations (e.g. HHV-1). In addition, structure 4 genomes contain a short terminal repeat, which is present internally in inverse orientation in the junction region. The different isomers of type 3 and 4 genomes appear to be functionally equivalent. It should be stressed that Fig. 2 is a simplified depiction of herpesvirus genome structures. Some herpesviruses contain large repetitive elements within the genome that are unrelated to those found at the termini (e.g. Human herpesvirus 4; HHV-4), and a more complex set of structures can be catalogued if these are included. Particular genome structures are associated with certain herpesvirus taxa. Thus the presence of multiple repeated elements at both termini (structure 2) is associated with the subfamily *Gammaherpesvirinae* (though not all members have this structure) while structure 3 is associated with members of the genus *Varicellovirus*. Distantly related viruses may, however, have equivalent genome structures, which have presumably evolved independently (e.g. HHV-1, Human herpesvirus 5; HHV-5).

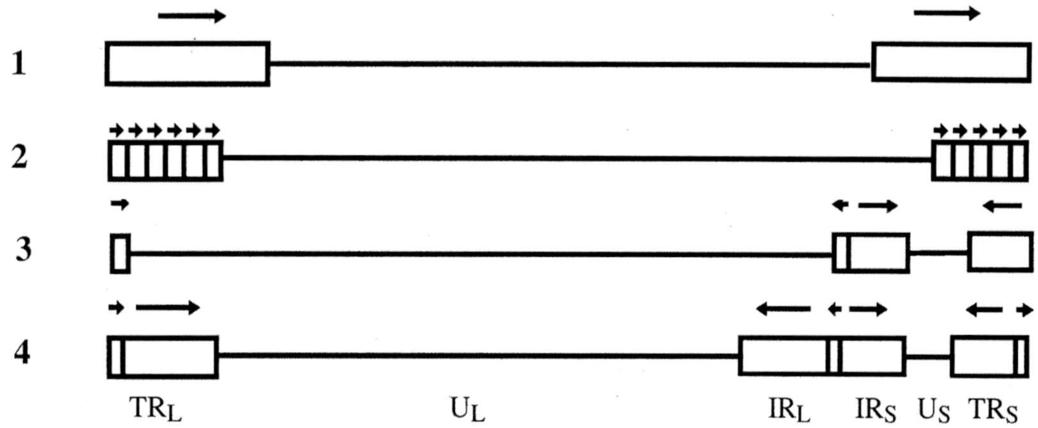

Figure 2: Simplified illustration of herpesvirus genome structures. Unique and repeated sequences are shown as solid lines and rectangles, respectively. The orientations of repeated sequences are indicated by arrows. The nomenclature used to describe regions of type 3 and 4 genomes is shown: U_L = unique long; U_S = unique short. The repeated sequences flanking the unique regions are named 'terminal repeat short' (TR_S) and 'internal repeat short' (IR_S), etc.

PROTEINS

The polypeptide composition of the mature virion varies greatly among different herpesviruses. HHV-1 is the best studied and more than 30 different polypeptides have been identified, though others doubtless remain to be found. The mature capsid is composed of 6 proteins, while the tegument contains at least 15 different polypeptides, many of which are dispensable *in vitro* and are therefore not required for virion morphogenesis. The viral envelope contains at least 10 (and in some cases many more) integral membrane proteins, a subset of which is required for adsorption and penetration of the host cell.

LIPIDS

The lipid composition of few herpesvirus envelopes has been reported. The lipid composition of the HHV-1 envelope is reported to resemble that of Golgi membranes more closely than that of other cellular membranes.

CARBOHYDRATES

The virion envelopes contain multiple proteins that carry N-linked and O-linked glycans. Mature, cell free virions contain complex glycans while a proportion of intracellular virions contain N-linked glycans of the immature high mannose type.

GENOME ORGANIZATION AND REPLICATION

The number of ORFs contained within herpesvirus genomes that potentially encode protein ranges from about 70 to more than 200. A subset of about 40 genes is conserved among the viruses of mammals and birds, arranged into 7 gene blocks (Fig. 3). These gene blocks have different orders and orientations in different herpesvirus subfamilies, but genes within a block maintain order and transcriptional polarity. The conserved genes encode capsid proteins, components of the DNA replication and packaging machinery, and to a lesser extent control proteins, nucleotide modifying enzymes, membrane proteins and tegument proteins, reinforcing the view that despite their genetic diversity these viruses share common features in many aspects of their replication strategies. The herpesviruses of fish and amphibians are very distant from those of mammals and birds and analysis of the sequences of these viruses has not revealed common coding sequences other than in a few genes that might have been captured independently. The classification of these viruses as herpesviruses is based on morphology of the virion rather than on genetic content.

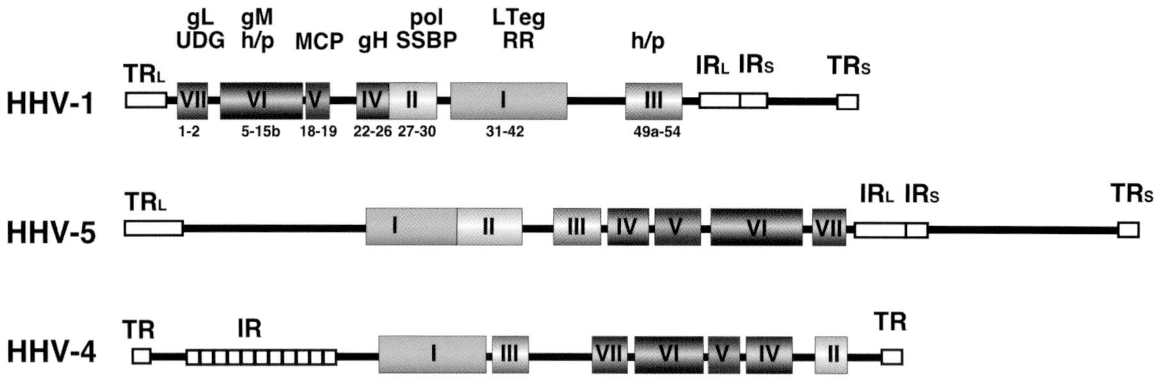

Figure 3: Organization of conserved gene blocks in members of different herpesvirus subfamilies. Prototype genome arrangements are shown for Human herpesvirus 1 (HHV-1) (*Alphaherpesvirinae*), Human herpesvirus 5 (HHV-5) (*Betaherpesvirinae*) and Human herpesvirus 4 (HHV-4) (*Gammaherpesvirinae*). Seven conserved gene blocks (I-VII) are shown and are identified by their UL ORF numbers in HHV-1. Relative to HHV-1, III is inverted in HHV–4, and I, III, VI and VII are inverted in HHV-5. Examples of genes in each block are indicated. UDG = uracil-DNA glycosylase; gL, gM, gH = glycoproteins L, M, H; h/p = subunit of the helicase-primase complex; pol = catalytic subunit of DNA polymerase; SSBP = single stranded DNA binding protein; RR = ribonucleotide reductase; MCP = major capsid protein; LTeg = large tegument protein. (Data derived from Gompels *et al.* (1995)).

Given the genetic diversity of members of the family *Herpesviridae*, it is probable that the details of their replication strategy will vary, perhaps substantially. What follows, therefore, is a brief description based on well-studied members of the family and, in particular, on HHV-1 (Figure 4). Adsorption and penetration involve the interaction of multiple virion envelope proteins with multiple cell surface receptors. Entry takes place at the cell surface by fusion of the envelope with the plasma membrane. The nucleocapsid is transported to the region of a nuclear pore by unknown mechanisms, while tegument proteins, many of whose functions are unknown, are thought to modify cellular metabolism. In HHV-1, one tegument protein (the UL41 gene product, vhs) acts in the cytoplasm to inhibit host protein synthesis while another (the UL48 gene product, VP16) is a transcription factor that enters the nucleus and activates immediate early viral genes. In permissive cells, entry of the genome into the nucleus is followed by a transcriptional cascade. Immediate early (α) genes, which are largely distinct among different subfamilies, regulate subsequent gene expression by transcriptional and post-transcriptional mechanisms. Early (β) genes encode the DNA replication complex and a variety of enzymes and proteins involved in modifying host cell metabolism, while the structural proteins of the virus are encoded primarily by late (γ) genes. Transcription is directed by host RNA polymerase II.

Viral DNA synthesis occurs from one or more origins of replication probably by a rolling circle mechanism. Replication of HHV-1 DNA requires 7 gene products: an origin-binding protein, an ssDNA-binding protein, a DNA polymerase composed of 2 subunits and a helicase-primase complex composed of 3 gene products. Homologs of all but the origin-binding protein have been identified in members of all subfamilies of herpesviruses. Newly synthesized DNA is packaged from the concatemer into pre-formed immature capsids within the nucleus by processes that involve several viral proteins. Immature capsids contain a core of scaffolding proteins, which are expelled by proteolytic cleavage during maturation. The subsequent steps in morphogenesis of the secreted enveloped virion are uncertain. Nucleocapsids are observed budding through the inner nuclear membrane into the perinuclear space and the resulting enveloped virions may then be transported by exocytosis to the cell surface. An alternative, and increasingly accepted, view is that the enveloped particles in the perinuclear space become 'de-enveloped' by fusion with the outer nuclear membrane and that the resulting nucleocapsids are re-enveloped in a Golgi or post-Golgi compartment. Only a subset of herpesvirus genes is required to achieve this basic replication cycle *in vitro*. Almost half the genes of HHV-1 are dispensable for *in vitro* culture; the products of these genes 'fine tune' the replication cycle or are required for survival *in vivo*.

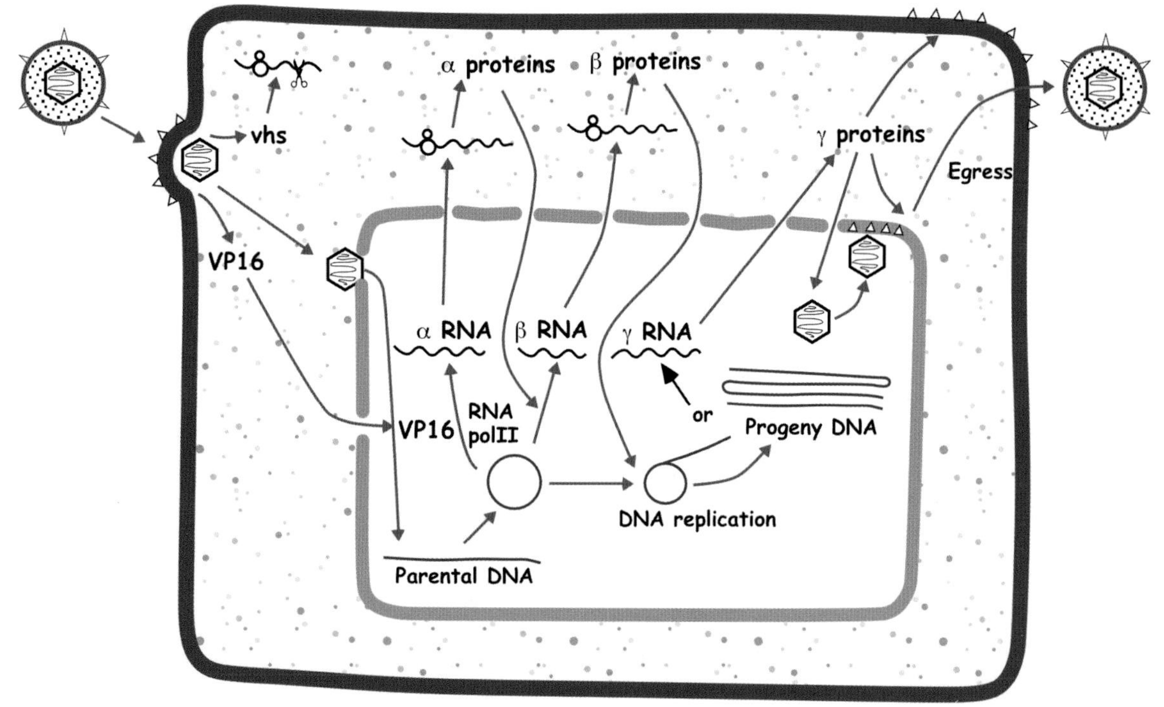

Figure 4: Schematic representation of the lytic replication cycle of Human herpesvirus 1 (HHV-1) in permissive cells. (From Roizman and Knipe (2001) with permission of Lippincott Williams and Wilkins).

The alternative to the productive cycle, and consequent cell death, is latent infection. The establishment and maintenance of the latent state is not thoroughly understood, but the weight of evidence favors a 'default' mechanism in which failure of immediate early gene expression leads to the maintenance of the input genome as a circular episomal element. Changes in the transcription factor milieu of the latently infected cell, due to external stimuli or cell differentiation, lead to immediate early gene expression and entry into the productive cycle. Like other large eukaryotic DNA viruses, herpesviruses are used as potential vectors for gene therapy.

Antigenic Properties

Infected hosts produce antibodies to a wide variety of structural and non-structural virus antigens. Some of the envelope glycoproteins are particularly immunogenic and are the targets for neutralizing antibodies. Cross-neutralization is observed only between closely related viruses within a genus.

Biological Properties

The range of host species is very wide. It is probable that all vertebrates carry multiple herpesvirus species, and a herpesvirus has also been identified in molluscs. As a general rule, the natural host range of individual viruses is highly restricted, and most herpesviruses are thought to have evolved in association with single host species, though occasional transfer to other species can occur in nature. In experimental animal systems the host range varies considerably: some members of the subfamily *Alphaherpesvirinae* can infect a wide variety of animal species while members of the subfamilies *Betaherpesvirinae* and *Gammaherpesvirinae* exhibit a very restricted experimental host range. The *in vitro* host range also varies considerably, though the same general rule holds true: members of the subfamily *Alphaherpesvirinae* will often infect a variety of cells of different species *in vitro* whereas members of the subfamilies *Betaherpesvirinae* and *Gammaherpesvirinae* exhibit greater restriction. The basis of host restriction both *in vivo* and *in vitro* is poorly understood. In a few instances (e.g. HHV-4), cell surface receptors play an important part in determining host range, but more commonly the virion is capable of entering a wide variety of cells, with intracellular factors determining susceptibility (e.g. HHV-5). Natural transmission routes range from highly contagious aerosol spread (HHV-3) to intimate oral contact (HHV-4) or sexual transmission (HHV-2). Vector-mediated transmission has not been reported.

Herpesviruses are highly adapted to their hosts, and severe infection is usually observed only in the very young, the foetus, the immunosuppressed or following infection of an alternative host. Most herpesviruses establish a systemic infection, a cell-associated viraemia being detectable during primary infection. Some members of the genus *Simplexvirus* appear to be an exception to the rule; in the normal host, infection is limited to epithelium at the inoculation site and to sensory nerves innervating the site. A variety of immune evasion mechanisms have been identified in different viruses, including the evasion of complement, antibody, MHC class I presentation and NK cell killing, but the key to survival of herpesviruses is their ability to establish life-long latent infection, a feature that is assumed to be the hallmark of all herpesviruses. The cell type responsible for harboring the latent virus has been established in relatively few instances. Nevertheless, the picture that emerges is that members of the subfamily *Alphaherpesvirinae* establish latent infection in neurones, members of the subfamily *Betaherpesvirinae* establish latent infection in cells of the monocyte series and members of the subfamily *Gammaherpesvirinae* establish latent infection in lymphocytes. It should be emphasized, however, that this general picture is based on a very limited number of examples and that there are reports of latent infection at other sites.

LIST OF SPECIES DEMARCATION CRITERIA IN THE FAMILY

Related herpesviruses are classified as distinct species if (a) their nucleotide sequences differ in a readily assayable and distinctive manner across the entire genome and (b) they occupy different ecological niches by virtue of their distinct epidemiology and pathogenesis or their distinct natural hosts. A paradigm is provided by HHV-1 and HHV-2, which differ in their sequence throughout the genome, tend to infect different epithelial surfaces and exhibit distinct epidemiological characteristics. These two viruses recombine readily in culture, but despite the fact that they can infect the same sites in the host, no recombinants have been isolated in nature, and the two viruses appear to have evolved independently for millions of years.

SUBFAMILY ALPHAHERPESVIRINAE

TAXONOMIC STRUCTURE OF THE SUBFAMILY

Subfamily	***Alphaherpesvirinae***
Genus	***Simplexvirus***
Genus	***Varicellovirus***
Genus	***Mardivirus***
Genus	***Iltovirus***

DISTINGUISHING FEATURES

The nucleotide sequences or predicted amino acid sequences of the subfamily members form a distinct lineage within the family. A region of the genome comprising the unique short sequence (U_S) and flanking inverted repeats (IR_S and TR_S) contains genes homologous to those found in HHV-1 and characteristic of the subfamily. The viruses productively infect fibroblasts in culture and epithelial cells *in vivo*. Many members cause overt, usually vesicular epithelial lesions in their natural hosts.

GENUS SIMPLEXVIRUS

Type Species *Human herpesvirus 1*

DISTINGUISHING FEATURES

Nucleotide sequences or predicted amino acid sequences of members of the genus form a distinct lineage within the subfamily. The viruses have a broad host cell range *in vitro* with a relatively rapid cytolytic productive cycle. Latent infection is established in neurones. Most members of the genus are viruses of primates. Members of the genus may have serological cross-reaction with other members.

LIST OF SPECIES DEMARCATION CRITERIA IN THE GENUS

See above under "List of Species Demarcation Criteria in the Family".

LIST OF SPECIES IN THE GENUS

Species names are in green italic script; strain names and synonyms are in black roman script; tentative species names are in blue roman script. Sequence accession numbers, and assigned abbreviations ( ) are also listed.

SPECIES IN THE GENUS

Ateline herpesvirus 1
Ateline herpesvirus 1 (AtHV-1)
(Spider monkey herpesvirus)
Bovine herpesvirus 2
Bovine herpesvirus 2 (BoHV-2)
(Bovine mamillitis virus)
Cercopithecine herpesvirus 1
Cercopithecine herpesvirus 1 [AF533768] (CeHV-1)
(B-virus)
(Herpesvirus simiae)
Cercopithecine herpesvirus 2
Cercopithecine herpesvirus 2 (CeHV-2)
(SA8)
Cercopithecine herpesvirus 16
Cercopithecine herpesvirus 16 (CeHV-16)
(Herpesvirus papio 2)

Human herpesvirus 1
Human herpesvirus 1 [X14112] (HHV-1)
(Herpes simplex virus 1)
Human herpesvirus 2
Human herpesvirus 2 [Z86099] (HHV-2)
(Herpes simplex virus 2)
Macropodid herpesvirus 1
Macropodid herpesvirus 1 (MaHV-1)
(Parma wallaby herpesvirus)
Macropodid herpesvirus 2
Macropodid herpesvirus 2 (MaHV-2)
(Dorcopsis wallaby herpesvirus)
Saimiriine herpesvirus 1
Saimiriine herpesvirus 1 (SaHV-1)
(Herpesvirus tamarinus)
(Marmoset herpesvirus)

TENTATIVE SPECIES IN THE GENUS

None reported.

GENUS VARICELLOVIRUS

Type Species ***Human herpesvirus 3***

DISTINGUISHING FEATURES

Nucleotide sequences or predicted amino acid sequences of members of the genus form a distinct lineage within the subfamily. Latent infection is established in cells of the sensory nervous system, though latent infection of other sites has also been reported. Members of the genus have been found in a wide range of mammalian hosts. Some members cross-react serologically.

LIST OF SPECIES DEMARCATION CRITERIA IN THE GENUS

See above under "List of Species Demarcation Criteria in the Family".

LIST OF SPECIES IN THE GENUS

Species names are in green italic script; strain names and synonyms are in black roman script; tentative species names are in blue roman script. Sequence accession numbers, and assigned abbreviations ( ) are also listed.

SPECIES IN THE GENUS

Bovine herpesvirus 1
Bovine herpesvirus 1 [AJ004801] (BoHV-1)
(Infectious bovine rhinotracheitis virus)
Bovine herpesvirus 5
Bovine herpesvirus 5 [AY261359] (BoHV-5)
(Bovine encephalitis virus)
Bubaline herpesvirus 1
Bubaline herpesvirus 1 (BuHV-1)
(Water buffalo herpesvirus)
Canid herpesvirus 1
Canid herpesvirus 1 (CaHV-1)
(Canine herpesvirus)
Caprine herpesvirus 1
Caprine herpesvirus 1 (CpHV-1)
(Goat herpesvirus)

Cercopithecine herpesvirus 9		
Cercopithecine herpesvirus 9	[AF275348]	(CeHV-9)
(Simian varicella virus)		
(Liverpool vervet herpesvirus)		
(Patas monkey herpesvirus)		
(Medical Lake macaque herpesvirus)		
Cervid herpesvirus 1		
Cervid herpesvirus 1		(CvHV-1)
(Red deer herpesvirus)		
Cervid herpesvirus 2		
Cervid herpesvirus 2		(CvHV-2)
(Reindeer herpesvirus)		
Equid herpesvirus 1		
Equid herpesvirus 1	[M86664]	(EHV-1)
(Equine abortion virus)		
Equid herpesvirus 3		
Equid herpesvirus 3		(EHV-3)
(Equine coital exanthema virus)		
Equid herpesvirus 4		
Equid herpesvirus 4	[AF030027]	(EHV-4)
(Equine rhinopneumonitis virus)		
Equid herpesvirus 8		
Equid herpesvirus 8		(EHV-8)
(Asinine herpesvirus 3)		
Equid herpesvirus 9		
Equid herpesvirus 9		(EHV-9)
(Gazelle herpesvirus)		
Felid herpesvirus 1		
Felid herpesvirus 1		(FeHV-1)
(Feline rhinotracheitis virus)		
Human herpesvirus 3		
Human herpesvirus 3	[X04370, AB097932-3]	(HHV-3)
(Varicella-zoster virus)		
Phocid herpesvirus 1		
Phocid herpesvirus 1		(PhoHV-1)
(Harbour seal herpesvirus)		
Suid herpesvirus 1		
Suid herpesvirus 1	[BK001744]	(SuHV-1)
(Pseudorabies virus)		

TENTATIVE SPECIES IN THE GENUS

Equid herpesvirus 6	(EHV-6)
(Asinine herpesvirus 1)	

GENUS MARDIVIRUS

Type Species *Gallid herpesvirus 2*

DISTINGUISHING FEATURES

Nucleotide sequences or predicted amino acid sequences of members of the genus form a distinct lineage within the subfamily. Members of the genus have been found only in birds and are the only members of the subfamily associated with malignancy. Production of infectious extracellular virus appears to be limited to feather-follicle epithelium. All current members cross-react serologically.

LIST OF SPECIES DEMARCATION CRITERIA IN THE GENUS

See above under "List of Species Demarcation Criteria in the Family".

LIST OF SPECIES IN THE GENUS

Species names are in green italic script; strain names and synonyms are in black roman script; tentative species names are in blue roman script. Sequence accession numbers, and assigned abbreviations ( ) are also listed.

SPECIES IN THE GENUS

Gallid herpesvirus 2		
Gallid herpesvirus 2	[AF243438]	(GaHV-2)
(Marek's disease virus type 1)		
Gallid herpesvirus 3		
Gallid herpesvirus 3	[AB049735]	(GaHV-3)
(Marek's disease virus type 2)		
Meleagrid herpesvirus 1		
Meleagrid herpesvirus 1	[AF291866]	(MeHV-1)
(Turkey herpesvirus)		

TENTATIVE SPECIES IN THE GENUS

None reported.

GENUS ILTOVIRUS

Type Species ***Gallid herpesvirus 1***

DISTINGUISHING FEATURES

Predicted amino acid sequences of the single member of this genus place it in a lineage distinct from other genera within the subfamily.

LIST OF SPECIES DEMARCATION CRITERIA IN THE GENUS

See above under "List of Species Demarcation Criteria in the Family".

LIST OF SPECIES IN THE GENUS

Species names are in green italic script; strain names and synonyms are in black roman script; tentative species names are in blue roman script. Sequence accession numbers, and assigned abbreviations ( ) are also listed.

SPECIES IN THE GENUS

Gallid herpesvirus 1	
Gallid herpesvirus 1	(GaHV-1)
(Infectious laryngotracheitis virus)	

TENTATIVE SPECIES IN THE GENUS

None reported.

UNASSIGNED SPECIES IN THE SUBFAMILY *ALPHAHERPESVIRINAE*

Psittacid herpesvirus 1		
Psittacid herpesvirus 1	[AY372243]	(PsHV-1)
(Parrot herpesvirus)		

SUBFAMILY *BETAHERPESVIRINAE*

TAXONOMIC STRUCTURE OF THE SUBFAMILY

Subfamily	***Betaherpesvirinae***
Genus	***Cytomegalovirus***
Genus	***Muromegalovirus***
Genus	***Roseolovirus***

DISTINGUISHING FEATURES

The nucleotide or predicted amino acid sequences of the subfamily members form a distinct lineage within the family. Genes corresponding to the HHV-5 US22 gene family are characteristic of the subfamily. The viruses tend to be species-specific and cell-type specific in culture. The growth cycle is slow and virus tends to remain cell-associated. Infection is often clinically non-apparent in immune-competent hosts. In some instances latent infection has been associated with cells of the monocyte series.

GENUS *CYTOMEGALOVIRUS*

Type Species *Human herpesvirus 5*

DISTINGUISHING FEATURES

The nucleotide or predicted amino acid sequences of members of the genus form a distinct lineage within the subfamily. Cytomegalovirus genomes are large (>200 kbp). Infected cells become enlarged (cytomegalic).

LIST OF SPECIES DEMARCATION CRITERIA IN THE GENUS

See above under "List of Species Demarcation Criteria in the Family".

LIST OF SPECIES IN THE GENUS

Species names are in green italic script; strain names and synonyms are in black roman script; tentative species names are in blue roman script. Sequence accession numbers, and assigned abbreviations ( ) are also listed.

SPECIES IN THE GENUS

Cercopithecine herpesvirus 5		
Cercopithecine herpesvirus 5		(CeHV-5)
(African green monkey cytomegalovirus)		
Cercopithecine herpesvirus 8		
Cercopithecine herpesvirus 8	[AY186194]	(CeHV-8)
(Rhesus monkey cytomegalovirus)		
Human herpesvirus 5		
Human herpesvirus 5	[X17403, AC146851, AC146904-7, AC146999, AY315197, AY446894, BK000394,]	(HHV-5)
(Human cytomegalovirus)		
Pongine herpesvirus 4		
Pongine herpesvirus 4	[AF480884]	(PoHV-4)
(Chimpanzee cytomegalovirus)		

TENTATIVE SPECIES IN THE GENUS

Aotine herpesvirus 1	(AoHV-1)
(Herpesvirus aotus 1)	
Aotine herpesvirus 3	(AoHV-3)
(Herpesvirus aotus 3)	

GENUS MUROMEGALOVIRUS

Type Species *Murid herpesvirus 1*

DISTINGUISHING FEATURES

The nucleotide or predicted amino acid of members of the genus form a distinct lineage within the subfamily. Muromegalovirus genomes are large (>200 kbp). Infected cells become enlarged (cytomegalic).

LIST OF SPECIES DEMARCATION CRITERIA IN THE GENUS

See above under "List of Species Demarcation Criteria in the Family".

LIST OF SPECIES IN THE GENUS

Species names are in green italic script; strain names and synonyms are in black roman script; tentative species names are in blue roman script. Sequence accession numbers, and assigned abbreviations ( ) are also listed.

SPECIES IN THE GENUS

Murid herpesvirus 1		
Murid herpesvirus 1 (Mouse cytomegalovirus)	[U68299]	(MuHV-1)
Murid herpesvirus 2		
Murid herpesvirus 2 (Rat cytomegalovirus)	[AF232689]	(MuHV-2)

TENTATIVE SPECIES IN THE GENUS

None reported.

GENUS ROSEOLOVIRUS

Type Species *Human herpesvirus 6*

DISTINGUISHING FEATURES

The nucleotide or predicted amino acid sequences form a distinct lineage within the subfamily and the genomes are smaller than those of members of the other genera within the subfamily (<200 kbp). The viruses productively infect T lymphocytes. Current members are serologically related.

LIST OF SPECIES DEMARCATION CRITERIA IN THE GENUS

See above under "List of Species Demarcation Criteria in the Family".

LIST OF SPECIES IN THE GENUS

Species names are in green italic script; strain names and synonyms are in black roman script; tentative species names are in blue roman script. Sequence accession numbers, and assigned abbreviations ( ) are also listed.

SPECIES IN THE GENUS

Human herpesvirus 6		
Human herpesvirus 6	[X83413, AB021506, AF157706]	(HHV-6)
Human herpesvirus 7		
Human herpesvirus 7	[U43400, AF037218]	(HHV-7)

TENTATIVE SPECIES IN THE GENUS

None reported.

UNASSIGNED SPECIES IN THE SUBFAMILY *BETAHERPESVIRINAE*

Caviid herpesvirus 2
Caviid herpesvirus 2 (CavHV-2)
(Guinea pig cytomegalovirus)
Tupaiid herpesvirus 1
Tupaiid herpesvirus 1 [AF281817] (TuHV-1)
(Tree shrew herpesvirus)

SUBFAMILY *GAMMAHERPESVIRINAE*

TAXONOMIC STRUCTURE OF THE SUBFAMILY

Subfamily	***Gammaherpesvirinae***
Genus	***Lymphocryptovirus***
Genus	***Rhadinovirus***

DISTINGUISHING FEATURES

The nucleotide or predicted amino acid sequences form a distinct lineage within the family. Certain genes may be unique to members of the family. These include BNRF-1, BTRF-1 and BRLF-1 of HHV-4 (the corresponding ORFs of Saimiriine herpesvirus 2 are ORFs 3, 23 and 50, respectively). Many members of the subfamily infect lymphocytes *in vitro* and 'carrier' cultures can be established in which a minority of cells is productively infected. Latent infection *in vivo* occurs in lymphocytes or lymphoid tissue. Acute infection is frequently associated with lymphoproliferative disorders and many members of the subfamily are associated with malignancies of lymphoid and non-lymphoid origin.

GENUS *LYMPHOCRYPTOVIRUS*

Type Species ***Human herpesvirus 4***

DISTINGUISHING FEATURES

The nucleotide or predicted amino acid sequences of members of the genus form a distinct lineage within the sub-family. The EBNA genes of HHV-4 and their homologs appear to be unique to members of the genus. The viruses infect B lymphocytes in culture but infection is usually non-productive and can result in immortalization. B cells or their precursors are thought to be the site of latent infection *in vivo*. Current members of the genus have been found only in primates.

LIST OF SPECIES DEMARCATION CRITERIA IN THE GENUS

See above under "List of Species Demarcation Criteria in the Family".

LIST OF SPECIES IN THE GENUS

Species names are in green italic script; strain names and synonyms are in black roman script; tentative species names are in blue roman script. Sequence accession numbers, and assigned abbreviations ( ) are also listed.

SPECIES IN THE GENUS

Callitrichine herpesvirus 3
Callitrichine herpesvirus 3 [AF319782] (CalHV-3)
(Marmoset lymphocryptovirus)
Cercopithecine herpesvirus 12
Cercopithecine herpesvirus 12 (CeHV-12)
(Herpesvirus papio)
(Baboon herpesvirus)

Cercopithecine herpesvirus 14		
Cercopithecine herpesvirus 14 (African green monkey EBV-like virus)		(CeHV-14)
Cercopithecine herpesvirus 15		
Cercopithecine herpesvirus 15 (Rhesus EBV-like herpesvirus) (Rhesus lymphocryptovirus)	[AY037858]	(CeHV-15)
Human herpesvirus 4		
Human herpesvirus 4 (Epstein-Barr virus)	[AJ507799]	(HHV-4)
Pongine herpesvirus 1		
Pongine herpesvirus 1 (Herpesvirus pan)		(PoHV-1)
Pongine herpesvirus 2		
Pongine herpesvirus 2 (Orangutan herpesvirus)		(PoHV-2)
Pongine herpesvirus 3		
Pongine herpesvirus 3 (Gorilla herpesvirus)		(PoHV-3)

TENTATIVE SPECIES IN THE GENUS

None reported.

GENUS *RHADINOVIRUS*

Type Species *Saimiriine herpesvirus 2*

DISTINGUISHING FEATURES

The nucleotide or predicted amino acid sequences of the members of the genus form a distinct lineage within the subfamily. Many members productively infect fibroblasts, often of multiple species, *in vitro*. Members have been isolated from a wide range of mammalian species. Latent infection has been reported in T or B lymphocytes.

LIST OF SPECIES DEMARCATION CRITERIA IN THE GENUS

See above under "List of Species Demarcation Criteria in the Family".

LIST OF SPECIES IN THE GENUS

Species names are in green italic script; strain names and synonyms are in black roman script; tentative species names are in blue roman script. Sequence accession numbers, and assigned abbreviations ( ) are also listed.

SPECIES IN THE GENUS

Alcelaphine herpesvirus 1		
Alcelaphine herpesvirus 1 (Malignant catarrhal fever virus)	[AF005370]	(AlHV-1)
Alcelaphine herpesvirus 2		
Alcelaphine herpesvirus 2 (Hartebeest malignant catarrhal fever virus)		(AlHV-2)
Ateline herpesvirus 2		
Ateline herpesvirus 2 (Herpesvirus ateles)		(AtHV-2)
Bovine herpesvirus 4		
Bovine herpesvirus 4 (Movar virus)	[AF318573]	(BoHV-4)
Cercopithecine herpesvirus 17		

Cercopithecine herpesvirus 17 (Rhesus rhadinovirus)	[AF083501, AF210726]	(CeHV-17)
Equid herpesvirus 2		
Equid herpesvirus 2	[U20824]	(EHV-2)
Equid herpesvirus 5		
Equid herpesvirus 5		(EHV-5)
Equid herpesvirus 7		
Equid herpesvirus 7 (Asinine herpesvirus 2)		(EHV-7)
Hippotragine herpesvirus 1		
Hippotragine herpesvirus 1 (Roan antelope herpesvirus)		(HiHV-1)
Human herpesvirus 8		
Human herpesvirus 8 (Kaposi's sarcoma-associated herpesvirus)	[U75698 ,U93872]	(HHV-8)
Murid herpesvirus 4		
Murid herpesvirus 4 (Mouse herpesvirus strain 68) (Murine gammaherpesvirus 68)	[U97553, AF105037]	(MuHV-4)
Mustelid herpesvirus 1		
Mustelid herpesvirus 1 (Badger herpesvirus)		(MusHV-1)
Ovine herpesvirus 2		
Ovine herpesvirus 2 (Sheep-associated malignant catarrhal fever virus)		(OvHV-2)
Saimiriine herpesvirus 2		
Saimiriine herpesvirus 2 (Herpesvirus saimiri)	[X64346]	(SaHV-2)

[a] Certain genome sequences lack the terminal repeat in the accession numbers listed. Separate accession numbers available for terminal repeats are not listed here.

TENTATIVE SPECIES IN THE GENUS

Leporid herpesvirus 1	(LeHV-1)
(Cottontail rabbit herpesvirus)	
Leporid herpesvirus 2	(LeHV-2)
(Herpesvirus cuniculi)	
Leporid herpesvirus 3	(LeHV-3)
(Herpesvirus sylvilagus)	
Marmodid herpesvirus 1	(MarHV-1)
(Woodchuck herpesvirus) (Herpesvirus marmota)	

UNASSIGNED SPECIES IN THE SUBFAMILY *GAMMAHERPESVIRINAE*

Callitrichine herpesvirus 1	
Callitrichine herpesvirus 1 (Herpesvirus saguinus)	(CalHV-1)

UNASSIGNED GENUS ICTALURIVIRUS

Type Species *Ictalurid herpesvirus 1*

DISTINGUISHING FEATURES

The nucleotide and predicted amino acid sequences of the single member of the genus are at best tenuously related to those of other herpesviruses and identify a distinct lineage.

LIST OF SPECIES DEMARCATION CRITERIA IN THE GENUS

Not applicable.

LIST OF SPECIES IN THE GENUS

Species names are in green italic script; strain names and synonyms are in black roman script; tentative species names are in blue roman script. Sequence accession numbers, and assigned abbreviations ( ) are also listed.

SPECIES IN THE GENUS

Ictalurid herpesvirus 1

Ictalurid herpesvirus 1 (Channel catfish virus)	[M75136]	(IcHV-1)

TENTATIVE SPECIES IN THE GENUS

None reported.

LIST OF UNASSIGNED VIRUSES IN THE FAMILY

Acipenserid herpesvirus 1 (White sturgeon herpesvirus 1)		(AciHV-1)
Acipenserid herpesvirus 2 (White sturgeon herpesvirus 2)		(AciHV-2)
Accipitrid herpesvirus 1 (Bald eagle herpesvirus)		(AcHV-1)
Anatid herpesvirus 1 (Duck plague herpesvirus)		(AnHV-1)
Anguillid herpesvirus 1 (Japanese eel herpesvirus)		(AngHV-1)
Ateline herpesvirus 3 (Herpesvirus ateles strain 73)	[AF083424]	(AtHV-3)
Boid herpesvirus 1 (Boa herpesvirus)		(BoiHV-1)
Callitrichine herpesvirus 2 (Marmoset cytomegalovirus)		(CalHV-2)
Caviid herpesvirus 1 (Guinea pig herpesvirus) (Hsiung-Kaplow herpesvirus)		(CavHV-1)
Caviid herpesvirus 3 (Guinea pig herpesvirus 3)		(CavHV-3)
Cebine herpesvirus 1 (Capuchin herpesvirus AL-5)		(CbHV-1)
Cebine herpesvirus 2 (Capuchin herpesvirus AP-18)		(CbHV-2)
Cercopithecine herpesvirus 3 (SA6 virus)		(CeHV-3)
Cercopithecine herpesvirus 4 (SA15 virus)		(CeHV-4)

Cercopithecine herpesvirus 10		(CeHV-10)
(Rhesus leukocyte-associated herpesvirus strain 1)		
Cercopithecine herpesvirus 13		(CeHV-13)
(Herpesvirus cyclopis)		
Chelonid herpesvirus 1		(ChHV-1)
(Grey patch disease of turtles)		
Chelonid herpesvirus 2		(ChHV-2)
(Pacific pond turtle herpesvirus)		
Chelonid herpesvirus 3		(ChHV-3)
(Painted turtle herpesvirus)		
Chelonid herpesvirus 4		(ChHV-4)
(Argentine turtle herpesvirus)		
Ciconiid herpesvirus 1		(CiHV-1)
(Black stork herpesvirus)		
Columbid herpesvirus 1		(CoHV-1)
(Pigeon herpesvirus)		
Cricetid herpesvirus		(CrHV-1)
(Hamster herpesvirus)		
Cyprinid herpesvirus 1		(CyHV-1)
(Carp pox herpesvirus)		
Cyprinid herpesvirus 2		(CyHV-2)
(Goldfish herpesvirus)		
(Haematopoietic necrosis herpesvirus of goldfish)		
Elapid herpesvirus 1		(EpHV-1)
(Indian cobra herpesvirus)		
(Banded krait herpesvirus)		
(Siamese cobra herpesvirus)		
Elephantid herpesvirus 1		(ElHV-1)
(Elephant [loxodontol] herpesvirus)		
Erinaceid herpesvirus 1		(ErHV-1)
(European hedgehog herpesvirus)		
Esocid herpesvirus 1		(EsHV-1)
(Northern pike herpesvirus)		
Falconid herpesvirus 1		(FaHV-1)
(Falcon inclusion body diseases)		
Gruid herpesvirus 1		(GrHV-1)
(Crane herpesvirus)		
Lacertid herpesvirus		(LaHV-1)
(Green lizard herpesvirus)		
Lorisine herpesvirus 1		(LoHV-1)
(Kinkajou herpesvirus)		
(Herpesvirus pottos)		
Murid herpesvirus 3		(MuHV-3)
(Mouse thymic herpesvirus)		
Murid herpesvirus 5		(MuHV-5)
(Field mouse herpesvirus)		
(Microtus pennsylvanicus herpesvirus)		
Murid herpesvirus 6		(MuHV-6)
(Sand rat nuclear inclusion agents)		
Ostreid herpesvirus 1	[AY509253]	(OsHV-1)
(Pacific oyster herpesvirus)		
Ovine herpesvirus 1		(OvHV-1)
(Sheep pulmonary adenomatosis associated herpesvirus)		
Percid herpesvirus 1		(PeHV-1)

(Walleye epidermal hyperplasia)	
Perdicid herpesvirus 1	(PdHV-1)
(Bobwhite quail herpesvirus)	
Phalacrocoracid herpesvirus 1	(PhHV-1)
(Cormorant herpesvirus)	
(Lake victoria cormorant herpesvirus)	
Pleuronectid herpesvirus 1	(PlHV-1)
(Herpesvirus scophthalmus)	
(Turbot herpesvirus)	
Ranid herpesvirus 1	(RaHV-1)
(Lucké frog herpesvirus)	
Ranid herpesvirus 2	(RaHV-2)
(Frog herpesvirus 4)	
Salmonid herpesvirus 1	(SalHV-1)
(Herpesvirus salmonis)	
Salmonid herpesvirus 2	(SalHV-2)
(Oncorhynchus masou herpesvirus)	
Sciurid herpesvirus 1	(ScHV-1)
(European ground squirrel cytomegalovirus)	
Sciurid herpesvirus 2	(ScHV-2)
(American ground squirrel cytomegalovirus)	
Sphenicid herpesvirus 1	(SpHV-1)
(Black footed penguin herpesvirus)	
Strigid herpesvirus 1	(StHV-1)
(Owl hepatosplenitis herpesvirus)	
Suid herpesvirus 2	(SuHV-2)
(Swine cytomegalovirus)	

PHYLOGENETIC RELATIONSHIPS WITHIN THE FAMILY

The family *Herpesviridae* is a diverse collection of viruses whose members can, nevertheless, be identified by their characteristic morphology. The diversity of the family has meant that criteria such as serology or nucleic acid hybridization have only rarely been of value in determining relationships between different viruses, and the construction of a satisfactory taxonomic structure has been a significant challenge. Historically, the herpesviruses were grouped into subfamilies on the basis of broad biological criteria, and division of the subfamilies into genera was based on molecular criteria, primarily the size and structure of the genome. The increasing volume of sequence data has led to the widespread use of sequence comparisons in assigning herpesviruses to taxa and in molecular phylogenetic analyses. Consequently, modern classification of herpesviruses is based formally on genetic content. The taxa are described as corresponding to 'distinct genetic lineages' and these lineages are defined by two criteria: (a) comparison of nucleotide or predicted amino acid sequences of conserved herpesvirus genes and (b) identification of particular genes that are unique to a virus subset. While genetic content is used as the primary criterion, some viruses have been assigned to taxa on the basis of serological cross reaction. Given the diversity of the family it is clear that serological cross reaction allows confident assignment at the genus level, even in the absence of genetic data.

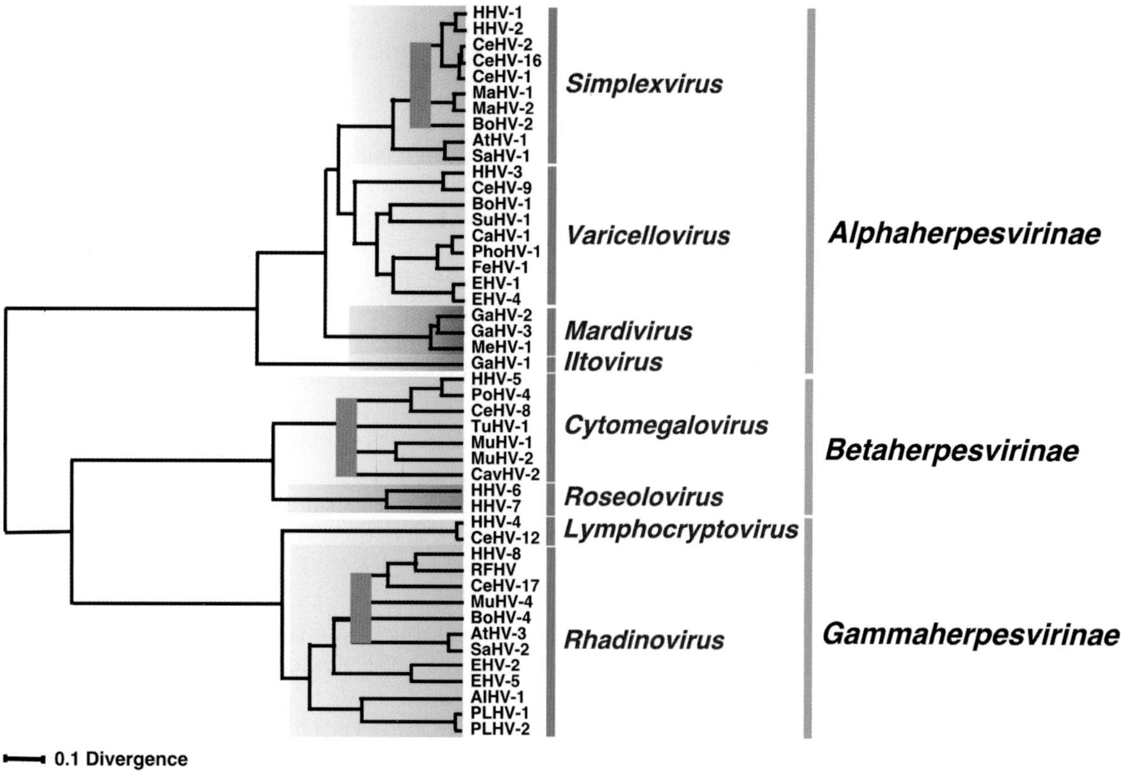

Figure 5: Composite phylogenetic tree for herpesviruses with molecular clock imposed, based on amino acid sequence alignments of eight sets of homologous genes from representatives of 46 species. Abbreviations are in the "List of Species"; three as yet unclassified viruses are included (RFHV, retroperitoneal fibromatosis herpesvirus; PLHV-1 and PLHV-2, porcine lymphotropic herpesviruses 1 and 2). The tree was constructed from maximum-likelihood analyses for subsets of the genes using the PHYLIP, MOLPHY and PAML packages. Regions of uncertain branching are shown as multifurcations (heavy lines). Genera and subfamilies are designated at the right. (From McGeoch *et al.* (2000) with permission of the American Society for Microbiology).

SIMILARITY WITH OTHER TAXA

Herpesviruses possess several genes (e.g. DNA polymerase, dUTPase) with cellular relatives that are assumed to have been gained by capture from RNA intermediates. The equivalent feature in certain other virus families results in genetic similarities that probably indicate independent capture events rather than direct evolutionary relationships. One speculative exception is the ATPase subunit of the DNA packaging terminase complex in T4 and related dsDNA bacteriophages, which has distant counterparts in all herpesviruses.

DERIVATION OF NAMES

Alpha: Greek letter α, "a".
Beta: Greek letter β, "b".
Cytomegalo: from Greek *kytos*, "cell" and *megas*, "large".
Gamma: Greek letter γ, "g".
Herpes: from Greek *herpes*, "creeping".
Ictaluri: from "ictalurid".
Ilto: from "infectious laryngotracheitis".
Lymphocrypto: from Latin *lympha*, "water" and Greek *kryptos*, "concealed".
Mardi: from "Marek's disease".
Muromegalo: from Latin *mus*, "mouse" and Greek *megas*, "great".
Rhadino: from Greek adjective *rhadinos*, "slender, taper".
Roseolo: from Latin *rose* "rose, rosy".

Simplex: from Latin *simplex*, "simple".
Varicello: derived from Latin *varius*, "spotted", and its diminutive variola, "smallpox".

REFERENCES

Davison, A.J. (2002). Evolution of the herpesviruses. *Vet. Microbiol.*, **86**, 69-88.

Davison, A.J. and Clements, J.B. (1997). Herpesviruses: General properties. In: *Topley and Wilson's Microbiology and Microbial Infections*, 9th edn (B.W.J., Mahy, and L.H., Collier, eds), pp. 309-323. Arnold, London.

Gompels, U.A., Nicholas, J., Lawrence, G., Jones, M., Thomson, B.J., Martin, M.E., Efstathiou, S., Craxton, M. and Macaulay, H.A. (1995). The DNA sequence of human herpesvirus-6: structure, coding content, and genome evolution. *Virology*, **209**, 29-51.

Homa, F.L. and Brown, J.C. (1997). Capsid assembly and DNA packaging in herpes simplex virus. *Rev. Med. Virol.*, **7**, 107-122.

McGeoch, D.J., Dolan, A. and Ralph, A.C. (2000). Toward a comprehensive phylogeny for mammalian and avian herpesviruses. *J. Virol.*, **74**, 10401-10406.

Rixon, F.J. (1993). Structure and assembly of herpesviruses. *Semin. Virol.*, **4**, 135-144.

Roizman, B. and Knipe, D.M. (2001). Herpes simplex viruses and their replication. In: *Fields Virology*, 4th ed. (D.M. Knipe, and P.M. Howley, eds), pp. 2399-2459. Lippincott Williams and Wilkins, Philadelphia.

Steven, A.C. and Spear, P.G. (1997). Herpesvirus capsid assembly and envelopment. In: *Structural Biology of Viruses* (W. Chiu, R.M. Burnett and R. Garcea, eds), pp. 312-351. Oxford University Press, New York and Oxford.

Zhou, Z.H., Dougherty, M., Jakana, J., He, J., Rixon, F.J. and Chiu, W. (2000). Seeing the herpesvirus capsid at 8.5 A. *Science*, **288**, 877-880.

CONTRIBUTED BY

Davison, A.J., Eberle, R., Hayward, G.S., McGeoch, D.J., Minson, A.C., Pellett, P.E., Roizman, B., Studdert, M.J. and Thiry, E.

FAMILY ADENOVIRIDAE

TAXONOMIC STRUCTURE OF THE FAMILY

Family *Adenoviridae*
Genus *Mastadenovirus*
Genus *Aviadenovirus*
Genus *Atadenovirus*
Genus *Siadenovirus*

VIRION PROPERTIES

MORPHOLOGY

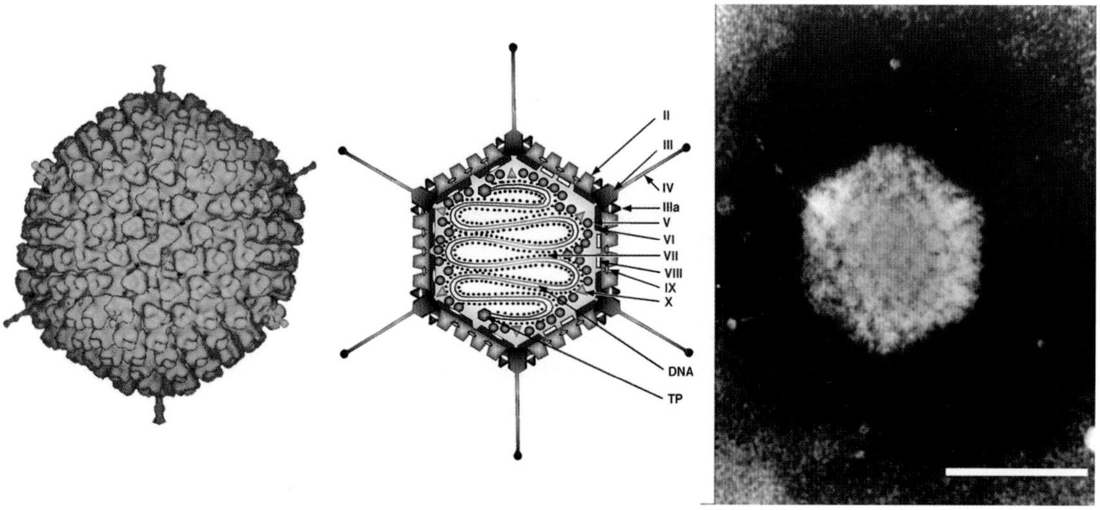

Figure 1: (Left) Cryo-electron reconstruction of a particle of an isolate of Human adenovirus 2 (Stewart et al. (1991). *Cell*, **67**:145-154). (Center) Stylized section of a mastadenovirus particle. For the description of the capsid (II, III, IIIa, IV, VI, VIII, IX) and core proteins (V, VII, X, TP), see text. As the structure of the nucleoprotein core has not been established, the polypeptides associated with the DNA are shown in hypothetical locations. (Adapted from Stewart, P.L. and Burnett, R.M. (1993). *Jpn. J. Appl. Phys*., **32**, 1342-1347). (Right) Negative contrast electron micrograph of a particle of an isolate of Human adenovirus 2 (Valentine, R.C. and Pereira, H.G. (1965). *J. Mol. Biol*., **13**, 13-20). The bar represents 100 nm.

Virions are non-enveloped, 70–90 nm in diameter. The icosahedral capsid consists of 240 non-vertex capsomers (hexons), 8–10 nm in diameter, and 12 vertex capsomers (pentons) each with a fiber protruding from the virion surface giving the characteristic morphology (Fig. 1). The length of fibers examined so far ranges between 9 and 77.5 nm. Members of the genus *Aviadenovirus* have two fiber proteins per vertex. The 240 hexons are formed by the interaction of three identical polypeptides (designated II) and consist of two distinct parts: a triangular top with three "towers", and a pseudohexagonal base with a central cavity. The hexon bases are tightly packed together and form a protein shell that protects the inner components. In the members of the genus *Mastadenovirus*, twelve copies of polypeptide IX are found between 9 hexons in the center of each facet. Polypeptide IX is not present in the other three genera. The positions of proteins IIIa, VI, and VIII are tentatively assigned. Two monomers of IIIa penetrate the hexon capsid at the edge of each facet. Multiple copies of protein VI form a ring underneath the peripentonal hexons. The 12 penton bases are each formed by the interaction of five polypeptides (III) and are tightly associated with one or two (only in aviadenoviruses) fibers each consisting of three polypeptides (IV) that interact to form a shaft of characteristic length with a distal knob. The 12 pentons (III and IV) are less tightly associated with the neighboring (peripentonal) hexons. Polypeptide VIII has been assigned to the inner surface of the hexon capsid. Other polypeptides (monomers of IIIa, trimers of IX, and multimers of VI) are in contact with hexons completing a continuous protein shell. Polypeptides VI and VIII appear to link the capsid to the virus core. The core consists of the DNA genome complexed with four polypeptides (V, VII, X, TP). Protein V was found in mastadenoviruses only.

PHYSICOCHEMICAL AND PHYSICAL PROPERTIES

Virion Mr is 150–180 x 10^6; buoyant density in CsCl is 1.30–1.37 g/cm^3. Viruses are stable on storage in the frozen state. They are stable to mild acid and insensitive to lipid solvents. Heat sensitivity varies in the different genera.

NUCLEIC ACID

The genome is a single linear molecule of dsDNA and contains an inverted terminal repetition (ITR). A virus-coded terminal protein (TP) is covalently linked to the 5'-end of each DNA strand. The size of genomes fully sequenced to date ranges between 26,163 and 45,063 bp, with ITRs of 36 to 371 bp in length. The G+C content of DNA varies between 33.7% and 63.8%. The central part of the genome is well conserved throughout the family, whereas the two ends show large variations in length and content (Fig. 2).

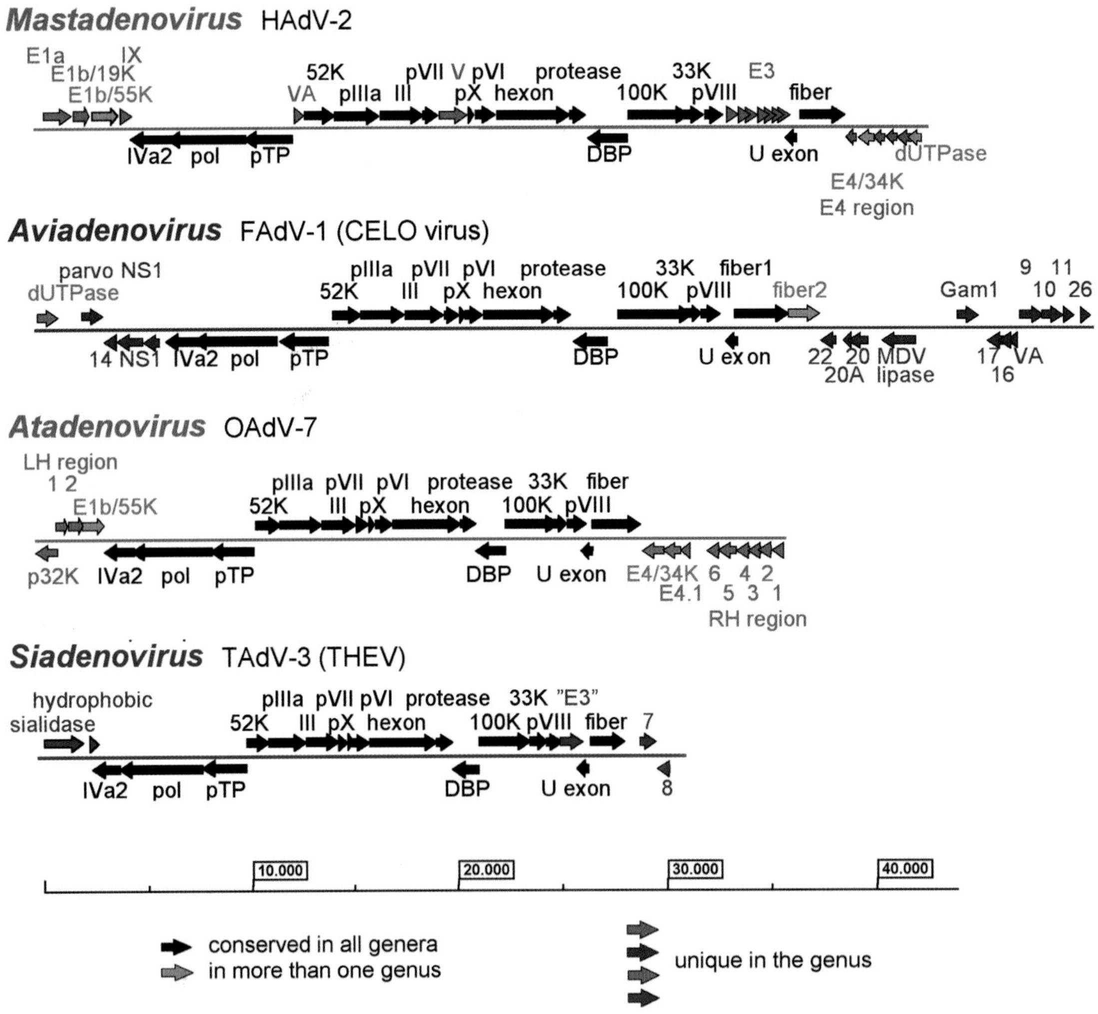

Figure 2: Schematic illustration of the different genome organizations found in members of the four genera. Black arrows depict genes conserved in every genus, gray arrows show genes present in more than one genus, colored arrows show genus-specific genes.

PROTEINS

About 40 different polypeptides are produced mostly via complex splicing mechanisms (Fig. 2, Table 1). Almost a third of these compose the virion including a virus-encoded cystein protease (23 kDa), which is necessary for the processing of some precursor proteins (marked with p). With the exception of protein V and IX, the other structural proteins are well conserved in every genus. Products of the four early regions facilitate extensive modulation of the host cell's transcriptional machinery (E1 and E4), assemble the virus DNA replication complex (E2) and provide means for subverting host defense mechanisms (E3). E2 is well conserved throughout the family, while the length and content of E1, E3 and E4 show a great variability even within the genera. Intermediate and late gene products (L1–L5) are concerned with the assembly and maturation of the virion.

Table 1: Virus proteins as deduced from gene sequence of Human adenovirus 2

kDa	Transcription class	Description	Note
13, 27, 32	E1A	NS	only in *Mastadenovirus*
16, 21	E1B	NS	only in *Mastadenovirus*
55	E1B	NS	only in *Mast-* and *Atadenovirus*
59	E2A	NS; 72kDa* DBP	
120	E2B	NS; 140kDa* DNA pol	
75	E2B	S; Term, 87kDa* pTP†	
4, 7, 8, 10, 12 13, 15, 15, 19	E3	NS	only in *Mastadenovirus*
7, 13, 13, 14	E4	NS	only in *Mastadenovirus*
15	E4	NS; 31kDa* dUTPase	only in some mast- and aviadenoviruses
17	E4	NS; 34kDa*	only in *Mast- and Atadenovirus*
47	L1	NS; scaffolding 52/55kDa*	
64	L1	S (pIIIa)†; p-protein	
63	L2	S (III); penton*	
22	L2	S (pVII)†; major core	
42	L2	S (V); minor core	only in *Mastadenovirus*
10	L2	S (pX)†; X/μ	
27	L3	S (pVI)†	
109	L3	S (II); hexon	
23	L3	S; protease	
90	L4	NS; 100kDa*	
25	L4	NS; 33kDa* p-protein	
25	L4	S (pVIII)†	
62	L5	S (IV); fiber	
14	Intermediate	S (IX)	only in *Mastadenovirus*
51	Intermediate	S (IVa2)	

Molecular masses are rounded to nearest 1000, are presented as unmodified and uncleaved gene products. NS = non-structural; S = structural; p = precursor; p-protein = phosphoprotein; DBP = DNA-binding protein, DNA pol = DNA polymerase; Term = terminal protein; * = Mr are significantly different from those obtained by SDS-PAGE; † = cleaved by viral protease; other ORFs are not yet identified.

LIPIDS

None reported.

CARBOHYDRATES

Fiber proteins and some of the nonstructural proteins are glycosylated.

GENOME ORGANIZATION AND REPLICATION

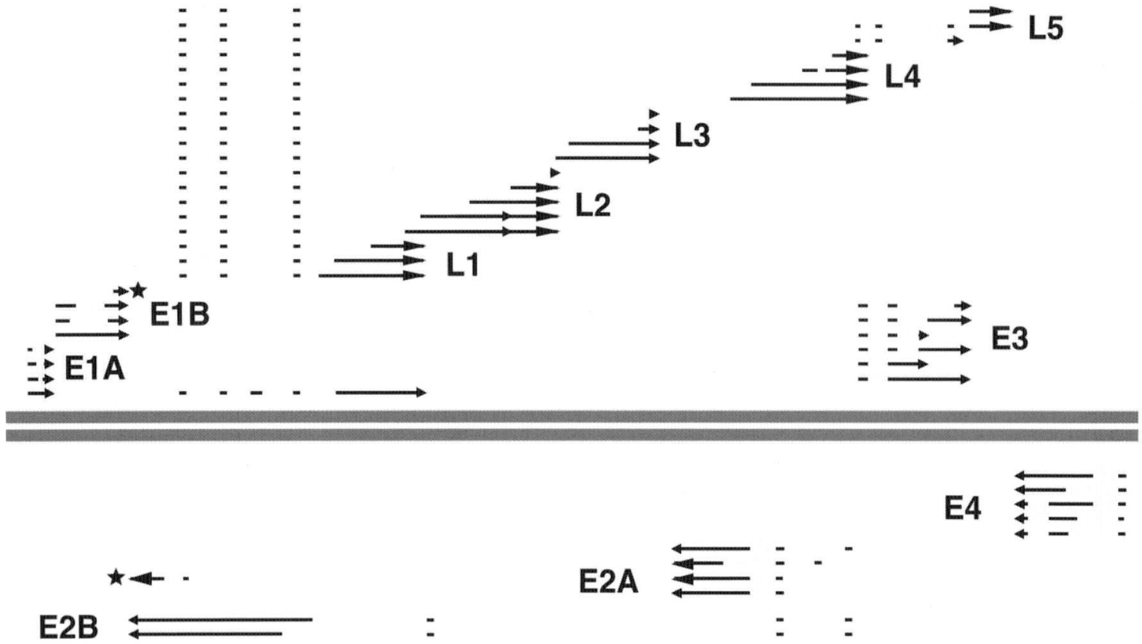

Figure 3: Schematic of the transcription pattern of Human adenovirus 2 (HAdV-2). The parallel lines indicate the linear duplex genome of 36 kbp. The dots, broken lines and split arrows indicate the spliced structures of the mRNAs. E1A, E3, etc., refer to early transcription units. Most (but not all) late genes are in the major late transcription unit which initiates at map position 16 of the indicated *r* strand (transcribed right-ward), and which includes the L1, L2, L3, L4 and L5 families of mRNAs. Other (intermediate) genes include those with a star. (Adapted from Wold, W.S. and Gooding, L.R. (1991). *Virology*, **184**, 1-8).

Virus entry is by attachment via the fiber knob to different receptors on the surface of susceptible cells, and subsequent internalization by the interaction between the penton base and cellular α_v integrins. After uncoating, the virus core is delivered to the nucleus, which is the site of mRNA transcription, virus DNA replication and assembly. Virus infection mediates the early shut-down of host DNA synthesis and, later, host RNA and protein synthesis. Transcription by the host RNA polymerase II involves both DNA strands and initiates (in Human adenovirus 2, HAdV-2) from five early (E1A, E1B, E2, E3, and E4), two intermediate, and one major late (L) promoter in a pattern as shown in Figure 3. All primary transcripts are capped and polyadenylated. There are complex splicing patterns to produce families of mRNAs. In primate adenoviruses, there are usually one or two VA RNA genes, which are transcribed by cellular RNA polymerase III and these encode RNA products, which facilitate translation of late mRNAs and the blocking of the cellular interferon response. Such VA RNA genes could not be identified in other adenoviruses. In some fowl adenoviruses the existence of one non-homologous VA RNA gene located at a different genome position has been described.

ANTIGENIC PROPERTIES

Adenovirus serotypes are differentiated on the basis of neutralization assays. A serotype is defined as one which either exhibits no cross-reaction with others, or shows a homologous/heterologous titer ratio greater than 16 (in both directions). For homologous/heterologous titer ratios of 8 or 16, a serotype assignment is made if either the viral haemagglutinins are unrelated (as shown by lack of cross-reaction in haemagglutination-inhibition tests), or if substantial biophysical, biochemical or phylogenetic differences exist. Antigens at the surface of the virion are mainly type-specific. Hexons are involved in neutralization, fibers in neutralization and haemagglutination-inhibition. Soluble antigens associated with virus infections include surplus capsid proteins, which have not been assembled. As defined with monoclonal antibodies, hexons and other soluble antigens carry numerous epitopes that can be genus-, species- or type-specific. Free hexon protein mainly reacts as a genus-specific antigen. The genus-specific antigen is located on the basal surface of the hexon, whereas serotype-specific antigens are located mainly on the tower region of the hexon.

BIOLOGICAL PROPERTIES

The natural host range of adenovirus types is usually confined to one species, or to closely related species. This also applies for cell cultures. Some human adenoviruses (mainly from members of the species *Human adenovirus C*) cause productive infection in different animal (rodent, ruminant) cells. Several viruses cause tumors in newborn hosts of heterologous species. Subclinical infections are frequent in various virus-host systems. Direct or indirect transmission occurs from throat, feces, eye, or urine, depending on the virus serotype. Certain HAdV types (in brackets) are predominantly associated with specific pathology, like adenoidal–pharyngeal conjunctivitis (3, 4, 7, 14), acute respiratory outbreaks (4, 7, 14, 21), epidemic kerato-conjunctivitis (8, 19, 37), or venereal disease (37). HAdV types 40 and 41 can be isolated in high yield from feces of young children with acute gastroenteritis and are second only to rotaviruses as a major cause of infantile viral diarrhea. HAdV-11 is associated with haemorrhagic cystitis occurring most frequently in immuno-suppressed patients after organ transplantation. The newer HAdV types (42 to 51) were all isolated from AIDS patients. In mammalian animals, mastadenovirus infection is common, but manifest disease usually appears only if predisposing factors (management problems, crowding, shipping, concurrent bacterial infections) are present. Canine adenovirus seems to be an exception, which can cause hepatitis or respiratory disease in dogs, and have caused epizootic in foxes, bears, wolves, coyotes, and skunks. Adenoviruses infecting susceptible cells cause similar gross pathology e.g., early rounding of cells and aggregation or lysis of chromatin followed by the later appearance of characteristic basophilic or eosinophilic nuclear inclusions. HAdV-5 has been engineered and is extensively used as gene vector. Other (even non-human) serotypes are being tested to overcome the problem posed by pre-existing neutralizing antibodies in the population, and also to achieve better targeting of specific organs and tissues.

GENUS MASTADENOVIRUS

Type Species *Human adenovirus C*

DISTINGUISHING FEATURES

Mastadenoviruses infect mammals only, and can be distinguished from members of other adenovirus genera serologically. Virus infectivity is inactivated after heating at 56°C for more than 10 min. The size of mastadenovirus genomes fully sequenced to date ranges between 30,288 and 36,521 bp. The G+C content of the DNA varies between 40.8 and 63.8%. The ITRs of mastadenoviruses are considerable longer (93 to 371 bp) and more complex (containing a variety of cellular factor binding sites) than in members of the other genera. HAdV-2 comprises 35,937 bp and its ITR is 103 bp long.

Unique proteins of mastadenoviruses are protein V and IX, and most of those coded by the E1A, E1B, E3, and E4 regions. Protein IX, besides cementing the hexons on the outer surface of the capsid, was demonstrated to act as a transcriptional activator, and it also takes part in the nuclear re-organization. Protein V is a core protein which, in association with cellular protein p32, seems to be involved in the transport of viral DNA into the nucleus of the infected cell. The E3 and E4 proteins are often different also in the different species of mastadenoviruses.

The genome organization and replication is best studied for isolates of the species *Human adenovirus C* (Fig. 3), but those findings seem to be generally applicable to all mastadenoviruses except the organization of the E3 and E4 regions. These early regions are different in the animal adenoviruses. In the E4 region, only a homolog of the 34K protein of HAdV-2 seems to exist in all mastadenoviruses; and is even duplicated in Bovine adenovirus 3 and Porcine adenovirus 5. The E3 region is also considerably shorter and simpler in the animal adenoviruses. The simplest E3 is in Murine adenovirus 1 containing a single 12.5 kDa homolog.

List of Species Demarcation Criteria in the Genus

The serologically distinguishable serotypes (synonym of types, designated by Arabic numbers) are being grouped into species. Species name reflects the first described host complemented by a letter if there are more than one adenovirus species bearing the same host name. The data available at present suggest the separation of at least 19 species within the genus. Several virus types listed as "Tentative Species in the Genus" need further study before being placed into the existing or additional species.

Species designation depends on several of the following characteristics:

- Calculated phylogenetic distance (more than 5-10% based primarily on the distance matrix analysis of the protease, pVIII, hexon, and DNA polymerase aa sequence comparisons)
- DNA hybridization
- RFLP analysis
- Percentage of GC in the genome
- Oncogenicity in rodents
- Growth characteristics
- Host range
- Cross-neutralization
- Possibility of recombination
- Number of VA RNA genes
- Haemagglutination
- Genetic organization of the E3 region

The lack of cross neutralization combined with a calculated phylogenetic distance of more than 10% separates two serotypes into different species. If the phylogenetic distance is less than 5%, any additional common grouping criteria from the above may classify separate serotypes into the same species even if they had been isolated from different hosts. For example, numerous properties of Bovine adenovirus 2 (BAdV-2), including results of the distance matrix analysis on protease and hexon aa sequences, RFLP analysis, GC content of the genome, DNA hybridization, ability to infect sheep, all implied its close relationship with Ovine adenovirus 2, 3, 4, and 5, and were therefore placed in a common species. Similarly, BAdV-9 proved to be almost indistinguishable from HAdV-2 and 5 based on partial sequencing, DNA hybridization, and restriction enzyme fragment profile. BAdV-1, 3, and 10, however, show considerable evolutionary distance from BAdV-2, 9, and from each other (more than 10%). In addition, they share only limited sequence homology detectable by DNA hybridization, possess very differently organized E3 region, and consequently they should be seen as separate species. The most numerous serotypes from the same host, the human adenoviruses, can be clearly separated into 6 species along the old subgenus lines supported by distance matrix analysis. HAdV-1, 2, 5, 6 may recombine with each other; HAdV-40 and 41 show similar restricted growth characteristics; the members of former subgenus A (HAdV-12, HAdV-18 and HAdV-31) share high oncogenicity in rodents and low G+C percentage in their genome. Adenoviruses isolated from chimpanzee resemble certain HAdVs in such extent that they are classified into "Human" adenovirus species. Simian adenovirus 22 to 25

(SAdV-22–25) belong to the species *Human adenovirus E*, while SAdV-21 belongs to *Human adenovirus B*.

LIST OF SPECIES IN THE GENUS

Species names are in green italic script; strain names and synonyms are in black roman script; tentative species names are in blue roman script. Sequence accession numbers, and assigned abbreviations ( ) are also listed.

SPECIES IN THE GENUS

Bovine adenovirus A		
Bovine adenovirus 1	[AF038868]	(BAdV-1)
Bovine adenovirus B		
Bovine adenovirus 3	[BK000401]	(BAdV-3)
Bovine adenovirus C		
Bovine adenovirus 10	[AF027599]	(BAdV-10)
Canine adenovirus		
Canine adenovirus 1	[BK000402]	(CAdV-1)
Canine adenovirus 2	[BK000403]	(CAdV-2)
Equine adenovirus A		
Equine adenovirus 1	[L79955]	(EAdV-1)
Equine adenovirus B		
Equine adenovirus 2	[L80007]	(EAdV-2)
Human adenovirus A		
Human adenovirus 12	[BK000405]	(HAdV-12)
Human adenovirus 18	[Y17249]	(HAdV-18)
Human adenovirus 31	[X76548]	(HAdV-31)
Human adenovirus B		
Human adenovirus 3	[M15952]	(HAdV-3)
Human adenovirus 7	[X03000]	(HAdV-7)
Human adenovirus 11	[BK001453]	(HAdV-11)
Human adenovirus 14	[AB070505]	(HAdV-14)
Human adenovirus 16	[AJ315931]	(HAdV-16)
Human adenovirus 21	[AB073222]	(HAdV-21)
Human adenovirus 34	[AB079724]	(HAdV-34)
Human adenovirus 35	[AY271307]	(HAdV-35)
Human adenovirus 50	[AJ272612]	(HAdV-50)
Simian adenovirus 21	[BK000412]	(SAdV-21)
Human adenovirus C		
Bovine adenovirus 9		(BAdV-9)
Human adenovirus 1	[AF534906]	(HAdV-1)
Human adenovirus 2	[BK000407]	(HAdV-2)
Human adenovirus 5	[BK000408]	(HAdV-5)
Human adenovirus 6	[Y16037]	(HAdV-6)
Human adenovirus D		
Human adenovirus 8	[AB110079]	(HAdV-8)
Human adenovirus 9	[AF099665]	(HAdV-9)
Human adenovirus 10	[AB023548]	(HAdV-10)
Human adenovirus 13	[AJ296009]	(HAdV-13)
Human adenovirus 15	[X74667]	(HAdV-15)
Human adenovirus 17	[BK000406]	(HAdV-17)
Human adenovirus 19	[X98359]	(HAdV-19)
Human adenovirus 20	[U52541]	(HAdV-20)
Human adenovirus 22–30	[28:Y14242]	(HAdV-22–30)
Human adenovirus 32	[U52551]	(HAdV-32)
Human adenovirus 33	[U52552]	(HAdV-33)
Human adenovirus 36-39	[37: AF217408]	(HAdV-36-39)

Human adenovirus 42–49	[48:U20821]	(HAdV-42–49)
Human adenovirus 51		(HAdV-51)
Human adenovirus E		
Human adenovirus 4	[AJ315930]	(HAdV-4)
Simian adenovirus 22		(SAdV-22)
Simian adenovirus 23	[U10684]	(SAdV-23)
Simian adenovirus 24		(SAdV-24)
Simian adenovirus 25	[BK000413]	(SAdV-25)
Human adenovirus F		
Human adenovirus 40	[NC_001454]	(HAdV-40)
Human adenovirus 41	[M21163]	(HAdV-41)
Simian adenovirus 19	[U03007]	(SAdV-19)
Murine adenovirus A		
Murine adenovirus 1	[BK000415]	(MAdV-1)
Ovine adenovirus A		
Bovine adenovirus 2	[BK000400]	(BAdV-2)
Ovine adenovirus 2		(OAdV-2)
Ovine adenovirus 3	[AF153447]	(OAdV-3)
Ovine adenovirus 4		(OAdV-4)
Ovine adenovirus 5		(OAdV-5)
Ovine adenovirus B		
Ovine adenovirus 1		(OAdV-1)
Porcine adenovirus A		
Porcine adenovirus 1	[L43364]	(PAdV-1)
Porcine adenovirus 2	[L43365]	(PAdV-2)
Porcine adenovirus 3	[BK000410]	(PAdV-3)
Porcine adenovirus B		
Porcine adenovirus 4	[U13893]	(PAdV-4)
Porcine adenovirus C		
Porcine adenovirus 5	[BK000411]	(PAdV-5)
Tree shrew adenovirus		
Tree shrew adenovirus 1	[BK001455]	(TSAdV-1)

TENTATIVE SPECIES IN THE GENUS

Goat adenovirus		
Goat adenovirus 2		(GAdV-2)
Guinea pig adenovirus		
Guinea pig adenovirus 1	[X95630]	(GPAdV-1)
Murine adenovirus B		
Murine adenovirus 2		(MAdV-2)
Ovine adenovirus C		
Ovine adenovirus 6		(OAdV-6)
Simian adenovirus*		
Simian adenovirus 1–18	[16("SAdV-7"/SA7):S70513]	(SAdV-1–18)
Simian adenovirus 20		(SAdV-20)
Squirrel adenovirus		
Squirrel adenovirus 1		(SqAdV-1)

* From simian adenoviruses, the chimpanzee types (SAdV-21–25) could unambiguously be classified into species. Based on partial DNA sequence data, the remaining 20 (old world) monkey adenovirus types cluster into several putative species. SAdV-19 (probably along with 8 others) belongs to HAdV-F. SAdVs yet unassigned to species are provisionally shown as members of a single tentative species until further evidences for their proper classification are provided.

GENUS *AVIADENOVIRUS*

Type Species *Fowl adenovirus A*

DISTINGUISHING FEATURES

Aviadenoviruses are serologically distinct from members of the other adenovirus genera and they only infect birds. The virions contain two fibers per vertex. Fowl adenovirus type 1 (FAdV-1) has two fiber genes, and two projections of considerably different lengths in each penton base. Other FAdVs also have two fibers, but apparently only one fiber gene, and the fiber shafts are of similar lengths. The long fiber of FAdV-1 uses coxsackievirus and adenovirus receptor (CAR) for attachment to the cell.

The DNA of aviadenoviruses is considerably (20 to 45%) larger in size compared to that of mastadenoviruses. Two aviadenovirus (FAdV-1 and FAdV-9) genomes, which were fully sequenced, comprise 43,804 and 45,063 bp, respectively, representing the longest adenoviral DNA molecules known to date. The G+C content of partial or full sequences of aviadenovirus genomes varies between 53.8 and 59%. The sequenced ITRs are 54 and 71 bp long.

The genomic organization of aviadenoviruses is also different (Fig. 2). The genes of protein V and IX are missing, as well as homologues of mastadenoviral early regions E1 and E3. The E4 region seems to be translocated resulting in the dUTP pyrophosphatase (dUTPase, not present in every mastadenovirus) being the first gene on the left end of the genome in aviadenoviruses studied to date. The organization of the central part of the genome containing the late genes and the E2 region is similar to that of mastadenoviruses. The right end of the genome contains several transcription units, which are unique for aviadenoviruses. The majority of genes and proteins from this region have not yet been characterized in detail. A novel protein GAM-1 of FAdV-1 was demonstrated to have an antiapoptotic effect, and to activate heat-shock response in the infected cell. GAM-1, in synergism with another novel protein, encoded by ORF22, binds the retinoblastoma protein and can activate the E2F pathway. Additional, yet uncharacterized predicted gene products were found to show homology with proteins of other viruses, like the nonstructural protein NS1 (syn. Rep protein) of parvoviruses, or a triacylglycerol lipase, a homolog of which also occurs in an avian herpesvirus (Marek's disease virus).

Aviadenoviruses possess no common complement-fixing antigen with the members of the other genera. There exist isolates where serum neutralization cannot differentiate clearly the serotypes. The introduction of 8a and 8b was deemed necessary because of the inconsistency in the type numbering scheme used in different countries and continents over the years. Newer results have not yet confirmed unambiguously the distinctness of different isolates typed as FAdV-8a and b. Avian adenoviruses have been associated with diverse disease patterns including inclusion body hepatitis, bronchitis, pulmonary congestion and oedema in different bird species. The hydropericardium syndrome is caused by FAdV-4 in chickens mainly in Asia. FAdV-1 (CELO), 9 and 10 are extensively studied for their potential feasibility as gene delivery vectors.

LIST OF SPECIES DEMARCATION CRITERIA IN THE GENUS

The serologically distinguishable avian adenovirus types can be classified at least into 6 species named by the host and additional letters if more than one species exist in the same host.

Species designation depends on several of the following characteristics:

- Calculated phylogenetic distance (more than 5-10% based primarily on the distance matrix analysis of the protease, pVIII, hexon, and DNA polymerase aa sequence comparisons)
- RFLP analysis
- Host range
- Pathogenicity
- Cross-neutralization
- Ability to recombine

For example the fowl adenovirus serotypes can be grouped into five species on the basis of RFLP profiles and the lack of significant cross-neutralization.

LIST OF SPECIES IN THE GENUS

Species names are in green italic script; strain names and synonyms are in black roman script; tentative species names are in blue roman script. Sequence accession numbers, and assigned abbreviations ( ) are also listed.

Note: A special problem that has been addressed but not resolved is the lack of consensus in the numbering of the individual serotypes. Strains deposited in American Type Culture Collection are numbered inconsistently with the majority of newer publications. For this reason, one representative strain of each serotype is also listed (in parenthesis).

SPECIES IN THE GENUS

Fowl adenovirus A		
Fowl adenovirus 1 (CELO)	[BK001452]	(FAdV-1)
Fowl adenovirus B		
Fowl adenovirus 5 (340)	[AF508952]	(FAdV-5)
Fowl adenovirus C		
Fowl adenovirus 4 (KR95)	[AJ431719]	(FAdV-4)
Fowl adenovirus 10 (CFA20)	[AF160185]	(FAdV-10)
Fowl adenovirus D		
Fowl adenovirus 2 (P7-A)	[AF339915]	(FAdV-2)
Fowl adenovirus 3 (75)	[AF508949]	(FAdV-3)
Fowl adenovirus 9 (A2-A)	[("FAdV-8"): BK001451]	(FAdV-9)
Fowl adenovirus 11 (380)	[("FAdV-12"): AF339925]	(FAdV-11)
Fowl adenovirus E		
Fowl adenovirus 6 (CR119)	[AF508954]	(FAdV-6)
Fowl adenovirus 7 (YR36)	[AF508955]	(FAdV-7)
Fowl adenovirus 8a (TR59)	[("FAdV-8"): AF508956]	(FAdV-8a)
Fowl adenovirus 8b (764)	[("FAdV-9"): AF508958]	(FAdV-8b)
Goose adenovirus		
Goose adenovirus 1		(GoAdV-1)
Goose adenovirus 2		(GoAdV-2)
Goose adenovirus 3		(GoAdV-3)

TENTATIVE SPECIES IN THE GENUS

Duck adenovirus B	
Duck adenovirus 2	(DAdV-2)
Pigeon adenovirus	
Pigeon adenovirus 1	(PiAdV)
Turkey adenovirus B	
Turkey adenovirus 1	(TAdV-1)
Turkey adenovirus 2	(TAdV-2)

GENUS *ATADENOVIRUS*

Type Species *Ovine adenovirus D*

DISTINGUISHING FEATURES

Atadenoviruses are serologically distinct from viruses within other adenovirus genera and their genomic organization and capsid protein complements are also different. Atadenoviruses have broad host range including animal species (snakes, lizards, duck, goose, chicken, possum and ruminants) from several vertebrate classes (reptilia, birds, mammals). Virions are relatively heat stable and retain substantial infectivity after treatment for 30 min at 56°C which inactivates mastadenovirions. The proteins encoded by an atadenovirus genome are summarized in Table 2. A gene for a novel structural protein p32K is present at the left end of the genome. Protein p32K is unique for atadenoviruses, and occurs in every members studied so far. In addition, genes LH1, E4.1, and RH1–6 are unique to the genus but not present within all members. Similarly, genes at the right end of DAdV-1 are so far unique to that virus and may be host-specific in function. Genes LH3 and E4.3 (and its homolog, the E4.2) show limited similarity with mastadenovirus proteins E1B 55K and E4 34K, respectively. No immunomodulatory genes such as those found in the mastadenovirus E3 region have been identified. The genome size of sequenced isolates ranges from 29,574 (OAdV-7) to 33,213 bp (DAdV-1) with ITRs of 46 (OAdV-7) to 59 (BAdV-4) nt. For ruminant, marsupial and avian atadenoviruses the GC content of the DNA is low and varies between 33.7 (OAdV-7) and 43% (DAdV-1). The corresponding high AT content was deemed to be sufficiently characteristic to justify the name of the genus. Atadenoviruses originating from reptiles seem to have a balanced G+C content.

Atadenoviruses have several unique proteins, and some that show very little similarity to their suspected counterparts in other adenovirus genera.

Table 2: Virus proteins as deduced from the gene sequence of Ovine adenovirus 7

kDa	Transcription class	Description	Note
32	timing not known	S (p32K†)	unique for *Atadenovirus*
13	LH1	NS	ORF not in DAdV-1
14.7	LH2	NS	
42.8	LH3 timing not known	NS	homology with mastadenovirus E1B 55K
43	E2	NS; DBP	
123	E2	NS; DNA pol	
67.1	E2	S; Term, pTP†	
12.9, 20.9, 19.8, 19.8	RH1, RH2, RH4, RH6 early	NS	F-box proteins unique for *Atadenovirus* homology to each other
22.6	RH5, early	NS	ORF not in DAdV-1
17.1	E4.1, early	NS	ORF not in DAdV-1
25.6, 30.8	E4.2, E4.3, early	NS	homologs of mastadenovirus E4 34K
38.2	early and late	NS; scaffolding 52/55kDa*	
58.4	late	S (pIIIa)†; p-protein	
51	late	S (III); penton*	lacks integrin binding motif
12.9	late	S (pVII)†; major core	

7.3	late	S (pX)†; X/μ	
24.5	late	S (pVI)†	
102	late	S (II); hexon	
23	late	S; protease	
72	late	NS; 100kDa*	hexon assembly protein
15.7	late	NS; 33kDa* p-protein	not found in DAdV-1
24.7	late	S (pVIII)†	
58.2	late	S (IV); fiber	cell attachment protein
37.5	timing not known	S (IVa2)	

Molecular masses are presented as unmodified and uncleaved gene products. S = structural; NS = non-structural; LH = left-hand end [genes]; RH = right-hand end [genes], p = precursor; p-protein = phosphoprotein; DBP = DNA-binding protein, DNA pol = DNA polymerase; Term = terminal protein; * = Mr are significantly different from those obtained by SDS-PAGE; † = cleaved by viral protease. All NS proteins are hypothetical until characterized.

The central part of the genome of atadenoviruses is similar to that of mastadenoviruses (except that there is no protein V and IX genes), while the extremities of the genomes are different. The first gene on the left-hand end of the genome is that of a novel protein p32K, which occurs only in atadenoviruses. The right-hand end of the genome contains several genes homologous with each other. There are two E4 34K homolog genes, and two to four RH homologs. DAdV-1 has a unique genome region at the far right-hand end with seven uncharacterized ORFs. ORF5 and 6 show homology to each other. The unique region of DAdV-1 also contains a VA RNA gene seemingly homologous with that of FAdV-1. Certain atadenoviruses can cause haemorrhagic epizooty in free-living ruminants. DAdV-1 was also associated with a specific clinical disease of hens, characterized by sharp decrease in egg production (the egg drop syndrome), all over the world. Several vectors have been constructed from OAdV-7 for human gene therapy purposes.

List of Species Demarcation Criteria in the Genus

Serologically distinguishable serotypes are grouped into species. The data available at present suggest the separation of several species within the genus, but only 4 of them have been approved yet (Fig. 4). Virus types listed as "Tentative Species in the Genus" need further study before being placed into the existing or additional species.
Species designation depends on several of the following characteristics:

- Calculated phylogenetic distance (>5-10%)
- Host range
- DNA hybridization
- Percentage of G+C in the genome
- Cross-neutralization
- Organization of the right-hand end of the genome

List of Species in the Genus

Species names are in green italic script; strain names and synonyms are in black roman script; tentative species names are in blue roman script. Sequence accession numbers, and assigned abbreviations ( ) are also listed.

Species in the Genus

Bovine adenovirus D

Bovine adenovirus 4	[NC_002685]	(BAdV-4)
Bovine adenovirus 5	[AF207658]	(BAdV-5)
Bovine adenovirus 8	[AF238233]	(BAdV-8)
Bovine adenovirus - Rus	[AF238880]	(BAdV-Rus)

Duck adenovirus A

Duck adenovirus 1 (Egg drop syndrome virus)	[BK000404]	(DAdV-1)

Ovine adenovirus D		
Goat adenovirus 1	[AF207660]	(GAdV-1)
Ovine adenovirus 7	[NC_004037]	(OAdV-7)
(Ovine adenovirus isolate 287)		
Possum adenovirus		
Possum adenovirus 1	[AF338822]	(PoAdV-1)

TENTATIVE SPECIES IN THE GENUS

Bearded dragon adenovirus		
Bearded dragon adenovirus 1		(BDAdV-1)
Bovine adenovirus E		
Bovine adenovirus 6	[AF207659]	(BAdV-6)
Bovine adenovirus F		
Bovine adenovirus 7	[AF238232]	(BAdV-7)
Cervine adenovirus		
Odocoileus adenovirus 1	[AF198354]	(OdAdV-1)
(from white-tailed deer, mule deer, black-tailed deer, moose)		
Chameleon adenovirus		
Chameleon adenovirus 1		(ChAdV-1)
Gecko adenovirus		
Gecko adenovirus 1		(GeAdV-1)
Snake adenovirus		
Snake adenovirus 1	[AY082603]	(SnAdV-1)

DSDNA

GENUS *SIADENOVIRUS*

Type Species *Frog adenovirus*

DISTINGUISHING FEATURES

Siadenoviruses are serologically distinct from members of the other adenovirus genera. This genus comprises only two known members, Frog adenovirus 1 (FrAdV-1) and Turkey adenovirus 3 (TAdV-3), which were isolated from an amphibian (frog) and from birds (turkey, pheasant, and chicken). The genomic organization of siadenoviruses is also different. The genes of protein V and IX are missing, as well as any homologs of mastadenovirus early regions E1, E3 and E4. The genome of both siadenovirus types is fully sequenced, and they represent the shortest adenovirus genomes known to date. Their length is 26,163 and 26,263 bp, with G+C contents of 38 and 35% respectively, and with ITRs of 36 and 39 bp long, in FrAdV-1 and TAdV-3, respectively.

Beside the proteins conserved in all adenoviruses, there are only five ORFs supposed to code for novel proteins. At the left-hand end of the genome, the first putative gene shows homology with sialidase genes. Adjacent to it, is another novel ORF predicted to code for a highly hydrophobic protein. The gene named "E3" solely because of its position between the pVIII and fiber genes is not homologous with any of the mastadenovirus E3 genes (or with any other known genes). The right-hand end of the genome harbors ORF 7 and 8 on the opposite (*l*) strand (transcribed left-ward). TAdV-3 has no common complement-fixing antigen with other adenoviruses isolated from birds and classified into the aviadenovirus or atadenovirus genera. FrAdV-1 is supposed to be nonpathogenic. TAdV-3 is associated with specific disease entities in different hosts (haemorrhagic enteritis in turkey, marble spleen disease in pheasants, and splenomegaly in chickens).

LIST OF SPECIES DEMARCATION CRITERIA IN THE GENUS

There are only two members of the genus, and they represent clearly two different species. Species designation depends on the following characteristics:

- Calculated phylogenetic distance (>10%)
- Host range

LIST OF SPECIES IN THE GENUS

Species names are in green italic script; strain names and synonyms are in black roman script; tentative species names are in blue roman script. Sequence accession numbers, and assigned abbreviations ( ) are also listed.

SPECIES IN THE GENUS

Frog adenovirus		
Frog adenovirus 1	[NC_002501]	(FrAdV-1)
Turkey adenovirus A		
Turkey adenovirus 3	[BK001454]	(TAdV-3)
(Avian adenovirus splenomegaly virus)		
(Marble spleen disease virus)		
(Turkey haemorrhagic enteritis virus)		

TENTATIVE SPECIES IN THE GENUS

None reported.

LIST OF UNASSIGNED VIRUSES IN THE FAMILY

White sturgeon adenovirus 1	[AJ495768]	(WSAdV-1)

PHYLOGENETIC RELATIONSHIPS WITHIN THE FAMILY

Consistently with the specific characteristics in the genome organization, the phylogenetic calculations, based either on distance matrix or parsimony analysis, on aa or nt sequence alignments of any suitable genes, always resulted in the clear separation of 5 different clusters corresponding to the four genera (*Mastadenovirus*, *Aviadenovirus*, *Atadenovirus*, *Siadenovirus*) and one proposed genus (supposedly for fish adenoviruses) (Fig. 4). The evolutionary distances among the adenoviruses seemed to be proportional to that among their hosts. There are some exceptions, where very distantly related viruses infect the same host. Adenovirus types isolated from cattle appeared on very distant branches (BAdV-10 versus BAdV-1 or BAdV-3), or even in separate clusters corresponding to different genera (BAdV-4, 6, and 7). Apparently, beside the co-evolution hypothesized for the vertebrate animals and their adenoviruses, multiple host switches might have also occurred. Reptilian adenoviruses are assumed to switch to ruminants, birds and marsupials. Similarly, siadenoviruses seem to have amphibian origin.

SIMILARITY WITH OTHER TAXA

The fibers of many adenovirus types use the same cellular receptor (CAR) for attachment as coxsackie B viruses. Adenovirus fibers were reported to show structural similarity with reovirus attachment protein sigma1 binding the JAM (junction adhesion molecule) receptor. Adenovirus may occur together with dependent parvovirus, for which it may provide helper functions. A dsDNA bacteriophage, PRD1 (member of the family *Tectiviridae*) was shown to share similar virion architecture (icosahedral capsid with fiber-like projections) with adenoviruses. The 15 kbp genome of PRD1 has ITRs, and contains the genes encoding two important proteins (terminal protein and DNA polymerase) in the same order as in adenoviruses. The terminal protein also acts as primer in the phage DNA replication. A study on the resolution of the main capsid proteins (P3 of PRD1 and hexon of HAdV-2) revealed a very similar arrangement and structure and suggested an evolutionary link between the two viruses. Fungi and plants have a linear plasmid (killer plasmid in yeast) either in the cytoplasm or within the mitochondria that shares similar features (ITR, a terminal protein gene adjacent to a DNA polymerase gene) with adenoviruses. Homology was demonstrated between certain proteins of the E3 region of HAdVs and the RL11 gene family of Human cytomegalovirus. The primary structure of the p32K protein, characteristic for atadenoviruses, was shown to have similarity with bacterial small acid soluble proteins (SASPs) commonly found in different spore-forming bacteria.

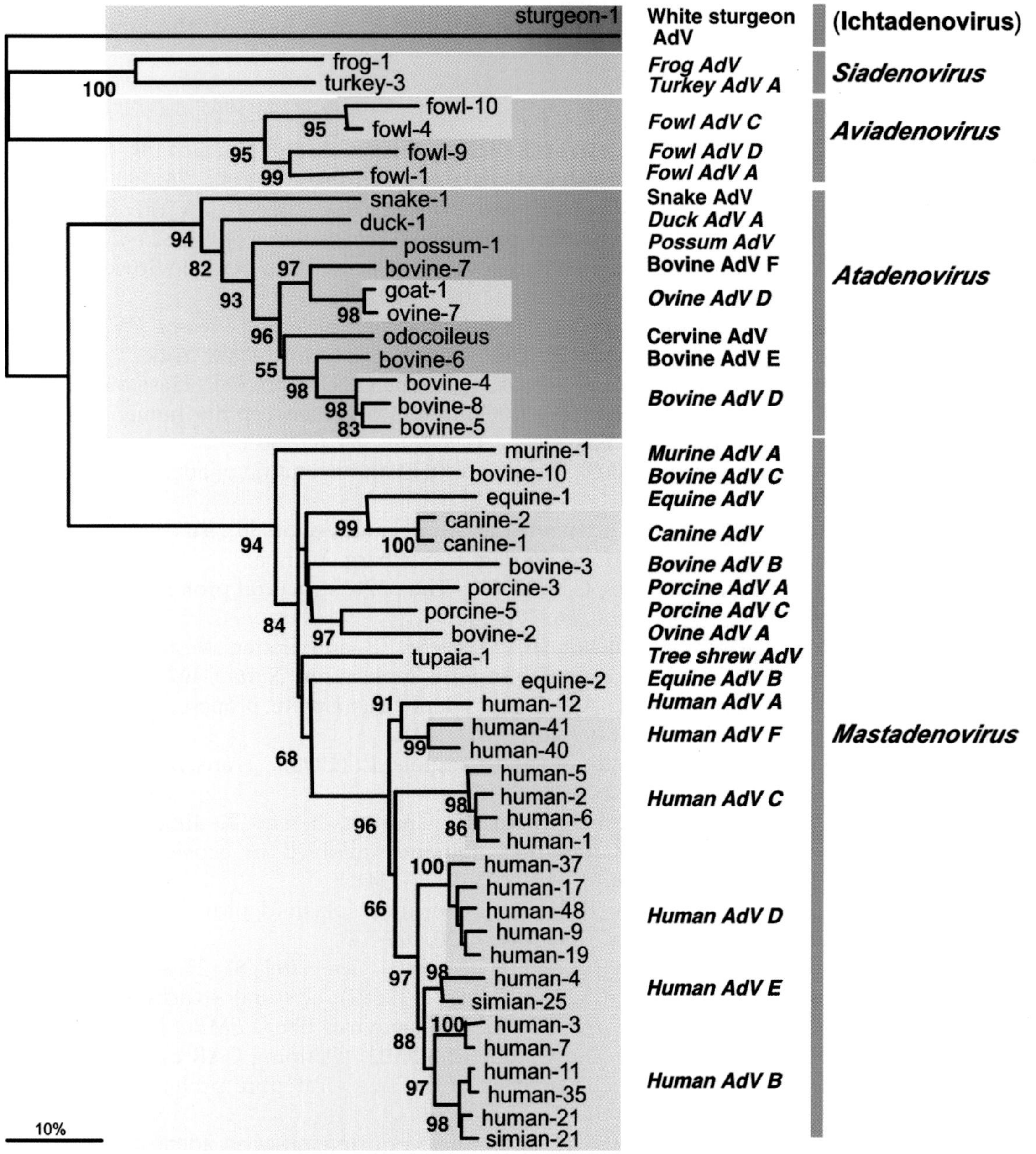

Figure 4: Phylogenetic tree of adenoviruses based on distance matrix analysis of hexon aa sequences. (Certain sequences were combined from different partial GenBank entries and may contain short unknown parts). The Seqboot (bootstrap), Protdist (Dayhoff PAM 001 matrix), Fitch (global rearrangements) programs of the PHYLIP 3.6 package were used, with the Consense program from PHYLIP 3.5c. Unrooted tree; white sturgeon adenovirus was chosen as outgroup. Adenoviruses are marked by the name of the host and the serotype number. Species names are indicated in an abbreviated mode for graphic convenience. Possible new species and genus names are not italicized. Bootstrap values higher than 50 (from 100 re-samplings) are shown for every confirmed branching.

Derivation of Names

Adeno: from Greek *aden, adenos,* "gland"; in recognition of the fact that adenoviruses were first isolated from human adenoid tissue.
At: from English *adenine* and *thymine,* in recognition that the genome of the first recognized members of the genus (from ruminant, avian and marsupial hosts) has a remarkably high AT content.
Avi: from Latin *avis,* "bird".
Mast: from Greek *mastos,* "breast".
Si: from English *sialidase,* in recognition that members of the genus have a putative sialidase homolog.

References

Benkö, M., Élö, P., Ursu, K., Ahne, W., LaPatra, E.S., Thomson, D. and Harrach, B. (2002). First molecular evidence for the existence of distinct fish and snake adenoviruses. *J. Virol.,* **76**, 10056-10059.

Benson, S.D., Bamford, J.K., Bamford, D.H. and Burnett, R.M. (1999). Viral evolution revealed by bacteriophage PRD1 and human adenovirus coat protein structures. *Cell,* **98**, 825-833.

Both, G.W. (2002). Identification of a unique family of F-box proteins in atadenoviruses. *Virology,* **304**, 425-433.

Burnett, R.M. (1997). The structure of adenovirus. In: *Structural Biology of Viruses* (W., Chiu, R.M., Burnett, and R.L., Garcea, eds), pp. 209-238. Oxford University Press, Oxford, New York.

Davison, A.J., Akter, P., Cunningham, C., Dolan, A., Addison, C., Dargan, D.J., Hassan-Walker, A.F., Emery, V.C., Griffiths, P.D. and Wilkinson, G.W. (2003). Homology between the human cytomegalovirus RL11 gene family and human adenovirus E3 genes. *J. Gen. Virol.,* **84**, 657-663.

Davison, A.J., Benkö, M., Harrach, B. (2003). Genetic content and evolution of adenoviruses. *J. Gen. Virol.,* **84,** 2895-2908.

Doerfler, W. and Böhm, P. (eds)(2003). Adenoviruses: Model and vectors in virus host interactions. *Curr. Top. Microbiol. Immunol.,* **272** (Pt 1). Springer Verlag, Heidelberg, Vienna.

Élö, P., Farkas, S. L., Dán, Á. and Kovács, G.M. (2003). The p32K structural protein of the atadenovirus might have bacterial relatives. *J. Mol. Evol.,* **56**, 175-180.

Glotzer, J.B., Saltik, M., Chiocca, S., Michou, A.I., Moseley, P. and Cotten, M. (2000). Activation of heat-shock response by an adenovirus is essential for virus replication. *Nature,* **407**, 207-211.

Lehmkuhl, H.D. and Cutlip, R.C. (1999). A new goat adenovirus isolate proposed as the prototype strain for goat adenovirus serotype 1. *Arch. Virol.,* **144**, 1611-1618.

Payet, V., Arnauld, C., Picault, J.P., Jestin, A. and Langlois, P. (1998). Transcriptional organization of the avian adenovirus CELO. *J. Virol.,* **72**, 9278-9285.

Rosa-Calatrava, M., Grave, L., Puvion-Dutilleul, F., Chatton, B. and Kedinger, C. (2001). Functional analysis of adenovirus protein IX identifies domains involved in capsid stability, transcriptional activity, and nuclear reorganization. *J. Virol.,* **75**, 7131-7141.

Rohe, M., Schrage, K. and Meinhardt, F. (1991). The linear plasmid pMC3-2 from *Morchella conica* is structurally related to adenoviruses. *Curr. Genet.,* **20**, 527-533.

Russell, W.C. (2000). Update on adenovirus and its vectors. *J. Gen. Virol.,* **81**, 2573-2604.

Chappell, J.D., Prota, A.E., Dermody, T.S. and Stehle T. (2002). Crystal structure of reovirus attachment protein sigma1 reveals evolutionary relationship to adenovirus fiber. *EMBO J.,* **21**, 1-11.

Tan, P.K., Michou, A.I., Bergelson, J.M. and Cotten, M. (2001). Defining CAR as a cellular receptor for the avian adenovirus CELO using a genetic analysis of the two viral fibre proteins. *J. Gen. Virol.,* **82**, 1465-1472.

Thomson, D., Meers, J. and Harrach, B. (2002). Molecular confirmation of an adenovirus in brushtail possums (*Trichosurus vulpecula*). *Virus Res.,* **83**, 189-195.

Contributed By

Benkö, M., Harrach, B., Both, G.W., Russell, W.C., Adair B.M, Ádám, É., de Jong, J.C., Hess, M., Johnson, M., Kajon, A., Kidd, A.H., Lehmkuhl, H.D., Li, Q.-G., Mautner, V., Pring-Akerblom, P. and Wadell, G.

GENUS RHIZIDIOVIRUS

Type Species *Rhizidiomyces virus*

VIRION PROPERTIES

MORPHOLOGY

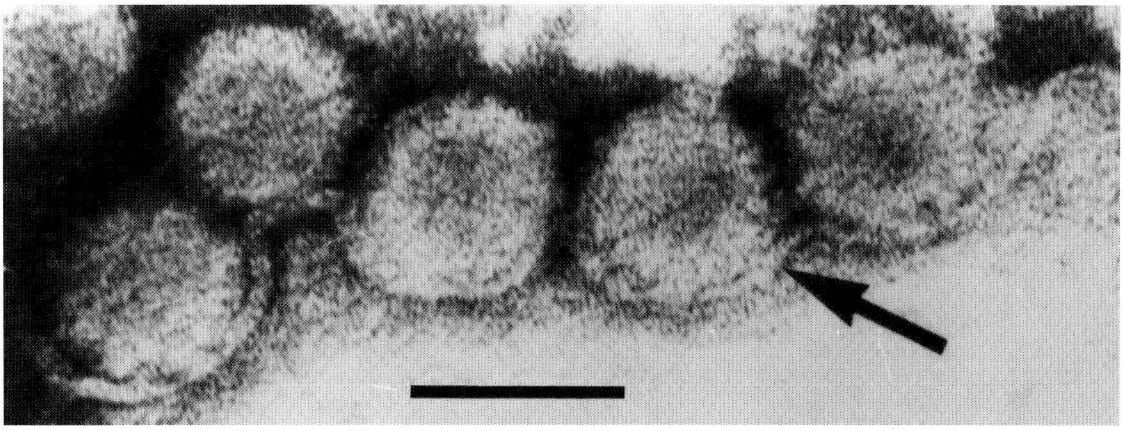

Figure 1: Negative contrast electron micrograph of *Rhizidiomyces virus* (RZV) particles which have been physically separated from the fungus are observed attached on a membrane-like structure (arrow)(From Dawe and Kuhn, *Virology* **130**, 10-20, 1983). The bar represents 50 nm.

Virions are isometric, 60 nm in diameter (Fig. 1).

PHYSICOCHEMICAL AND PHYSICAL PROPERTIES

The buoyant density of virions in CsCl is 1.31 g/cm^3; S_{20w} is 625S. Virions contain 10% nucleic acid.

NUCLEIC ACID

Virions contain a single molecule of dsDNA with a Mr of 16.8 x 10^6 and a G+C ratio of 42%.

PROTEINS

Virions contain at least 14 polypeptides with sizes in the range of 26-84.5 kDa.

LIPIDS

None reported.

CARBOHYDRATES

None reported.

GENOME ORGANIZATION AND REPLICATION

Particles appear first in the nucleus.

ANTIGENIC PROPERTIES

No information available.

BIOLOGICAL PROPERTIES

The virus appears to be transmitted in a latent form in the zoospores of the fungus. Activation of the virus, which occurs under stress conditions such as heat, poor nutrition, or aging, results in cell lysis.

List of Species Demarcation Criteria in the Genus

Not applicable.

List of Species in the Genus

Species names are in green italic script; strain names and synonyms are in black roman script; tentative species names are in blue roman script. Sequence accession numbers, and assigned abbreviations ( ) are also listed.

Species in the Genus

Rhizidiomyces virus

Rhizidiomyces virus (RhiV)

Tentative Species in the Genus

None reported.

Phylogenetic Relationships within the Genus

Not applicable.

Similarity with Other Taxa

None reported.

Derivation of Names

Rhizidio: from name of the host *Rhizidiomyces* sp.

References

Dawe, V.H. and Kuhn, C.W. (1983). Virus-like particles in the aquatic fungus, *Rhizidiomyces*. *Virology*, **130**, 10-20.

Dawe, V.H. and Kuhn, C.W. (1983). Isolation and characterization of a double-stranded DNA mycovirus infecting the aquatic fungus, *Rhizidiomyces*. *Virology*, **130**, 21-28.

Contributed By

Ghabrial, S.A. and Buck, K.W.

FAMILY POLYOMAVIRIDAE

TAXONOMIC STRUCTURE OF THE FAMILY

Family *Polyomaviridae*
Genus *Polyomavirus*

Since only one genus is currently recognized, the family description corresponds to the genus description.

GENUS POLYOMAVIRUS

Type Species *Simian virus 40*

VIRION PROPERTIES

MORPHOLOGY

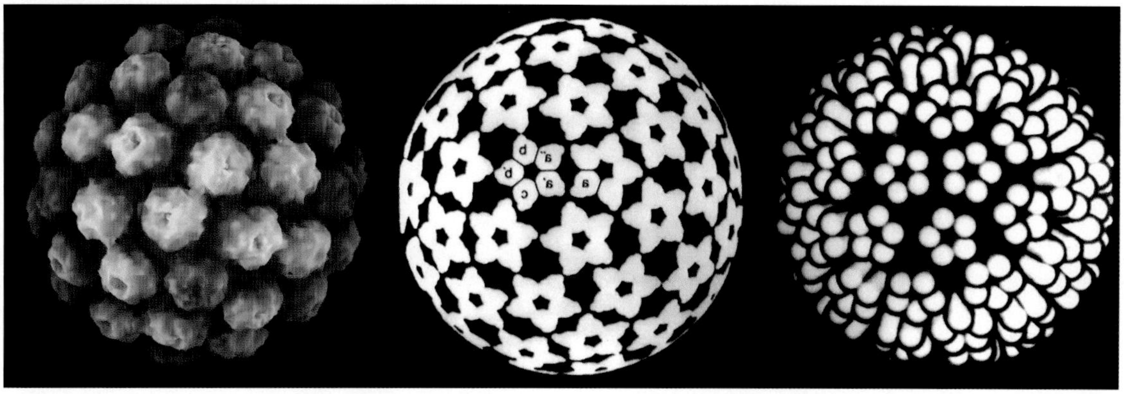

Figure 1: (Left) Computer rendering of a particle of the strain A2 of *Murine polyomavirus* (Stehle T. and Harrison S.C. (1996). *Structure*, **4**, 183-194). (Center) Capsomer bonding relations, each icosahedral asymmetric unit comprises six Vp1 subunits, including one (a) from a pentavalent pentamer. The six symmetrically different subunits are designated a, a´, a´´, b, b´ and c, corresponding to three different bonding states). (Right) Computer graphics representation of the surface of the capsid of the strain A2 of *Murine polyomavirus*. Five Vp1 subunits form the basis of a polyomavirus capsomer and 72 capsomers that link together in a 12 pentavalent/60 hexavalent arrangement, convey icosahedral capsid structure (from Eckhart 1991; adapted from Salunke *et al.*, 1986; with permission).

Virions are non-enveloped and approximately 40 to 45 nm in diameter. The icosahedral capsid is composed of 72 capsomers in a skewed (T=7d) lattice arrangement (Fig. 1). Right-handed (dextro) skew has been shown for all polyomaviruses examined in cryoelectron-microscopy tilt experiments. Aberrant structures such as empty capsids, microcapsids and tubular forms are regularly observed.

PHYSICOCHEMICAL AND PHYSICAL PROPERTIES

Virion Mr is 2.5 x 10^7. Buoyant density of virions in sucrose and CsCl gradients is 1.20 and 1.34–1.35 g/cm^3, respectively. Virion S_{20w} is 240S. Virions are resistant to ether, acid and heat treatment (50°C, 1 hr). Virions are unstable at 50°C for 1 hr in the presence of 1M $MgCl_2$. Greater than 70% of the total virion protein content is Vp1. Recombinant polyomavirus Vp1 (rVp1) expressed from baculovirus plasmid constructs self-assembles into virus-like particles (VLP) under specific chemical and physical conditions. These rVp1-VLP resemble native virions in electron microscopy and are purified by identical procedures (Fig. 2). The rVp1 independently forms pentameric assembly units (pentamers) analogous to the capsomers of virions, minus the centrally located Vp2 or Vp3 proteins. Linking of the caboxyl-termini of pentameric rVp1 creates the icosahedral lattice structure of VLP. Pentamers that form from rVp1 with truncated carboxyl-termini are stable, but do not form VLP. By modifying chemical conditions rVp1-VLP can be dissociated and subsequently reconstituted. During self-assembly or reconstitution, rVp1-VLP will non-specifically encapsidate genetic material that is present.

NUCLEIC ACID

Virions contain a single molecule of circular dsDNA. The genomic size is fairly uniform within the genus, averaging approximately 5 kbp (e.g., Simian virus 40 (SV-40) [strain 776] is 5,243 bp, JC polyomavirus (JCPyV) Mad-1 is 5,130 bp, BK polyomavirus (BKPyV) Dunlop is 5,153 bp, Murine polyomavirus (MPyV) [A2] is 5,297 bp, Baboon polyomavirus 2 (BPyV) is 4,697 bp). The DNA constitutes about 10-13% of the virion by weight. The G+C content varies between 40-50%. In the mature virion, the viral DNA is associated with the host cell histone proteins H2a, H2b, H3 and H4 in a supercoiled, chromatin-like complex.

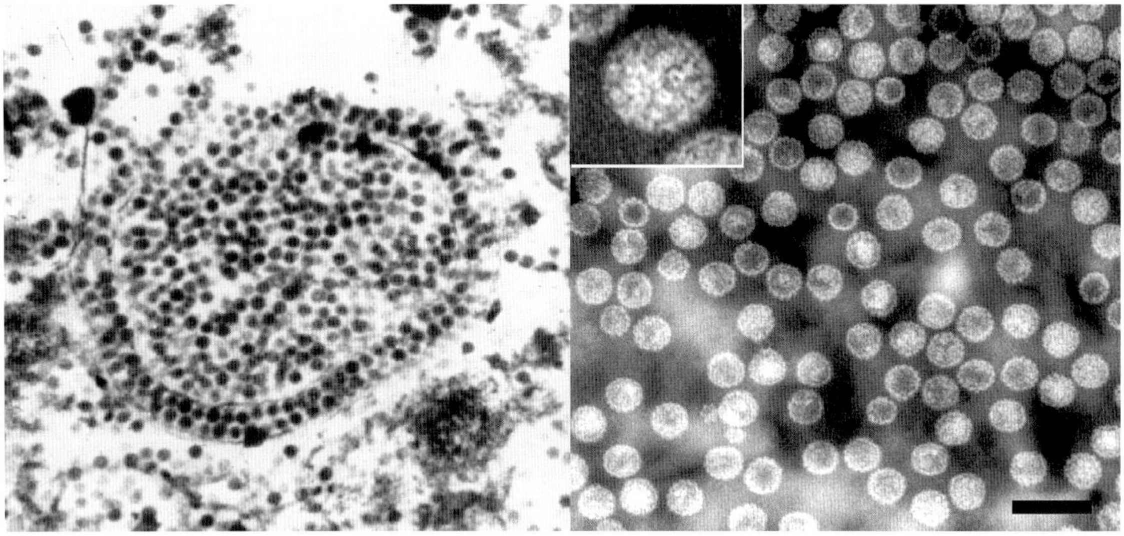

Figure 2: Electron micrographs of: (Left) brain tissue from a progressive multifocal leukoencephalopathy (PML) patient showing the assembly of JC polyomavirus (JCPyV) particles in the nucleus of an infected oligodendrocyte; and (Right) composite of virus-like particles (VLP) self-assembled from recombinant Vp1 of BK polyomavirus (BKPyV)(rBKVp1), purified by CsCl ultracentrifugation techniques from supernatants of Sf9 insect cell cultures infected with recombinant BKVp1-baculovirus. Negatively stained particles with 2% phosphotungstic acid (PTA). Composite includes an insert with enlargement of a single VLP. The bar represents 100 nm.

PROTEINS

Currently, polyomavirus genomes are known to code for between 5 and 9 proteins with predicted sizes from the nucleic acid sequences ranging from 7 to 88 kDa (Table 1). Transcription from one side of the viral origin of DNA replication (ORI) results in message for the early proteins. These non-structural proteins are referred to as T proteins because they interfere with cell cycle regulation and, in some cases, induce cellular transformation, or tumor formation. Alternative splicing appears to be responsible for the two to five related, yet distinct, proteins expressed from each polyomavirus T gene. The set of proteins expressed from a single T gene shares amino-terminus sequence. The T proteins initiate bi-directional viral genome replication, as well as the transcription of late viral message. Late message is transcribed from the strand complementary to that used for early transcription and is also initiated from the opposite side of the ORI. Late transcripts code for three structural proteins Vp1, Vp2 and Vp3 as well as another non-structural protein known as LP1, or agnoprotein. Of the three structural proteins, Vp1 makes up more than 70% of the total virion protein content and hence, is also referred to as the major structural protein. Five Vp1 proteins surround either a Vp2 or Vp3 to form stable assembly units, or capsomers; 72 random capsomers link together in icosahedral symmetry to form the capsid of each virion. The Vp2 and Vp3 molecules may be necessary to ensure specific encapsidation of replicated polyomavirus genome. The agnoprotein may have some role in facilitating capsid assembly, but it is not a component of the mature virion.

Table 1: Virus deduced size of polyomavirus proteins in kDa, ND = not detected.

Virus	MPyV	SV-40	JCPyV	BKPyV	KPyV	LPyV	BPyV
Structural proteins:							
VP1	42.4	39.9	39.6(40)	40.1(40)	41.7	40.2	40.5
VP2	34.8	38.5	37.4	38.3	37.4	39.3	39.1
VP3	22.9	27.0	25.7	26.7	25.2	27.3	26.9
Non-structural proteins:							
T	88.0	81.6(94)	79.3(94)	80.5	72.3	79.9	66.9
mT	48.6	ND	ND	ND	ND	ND	ND
T'_{135}	ND	ND	(17)	*	ND	ND	ND
T'_{136}	ND	ND	(17)	*	ND	ND	ND
T'_{165}	ND	ND	(22-23)	*	ND	ND	ND
17kT	ND	(17)	ND	ND	ND	ND	ND
tT	*	ND	(17)	ND	ND	ND	ND
t	22.8	20.4	20.2	20.5	18.8	22.2	14.0
LP1/agno	ND	7.3	8.1	7.4	ND	ND	13.1

numbers below are predicted sizes; numbers in ( ) are observed sizes of the expressed proteins. An * labels reported proteins, that lack both predicted and expressed size

LIPIDS

None present.

CARBOHYDRATES

None present.

GENOME ORGANIZATION AND REPLICATION

Virions that attach to cellular receptors are engulfed by the cell and are transported to the nucleus. During a productive infection, transcription of the viral genome is divided into an early and late stage. Transcription of the early and late coding regions is controlled by separate promoters through the binding of specific transcription factors and cis-acting elements. The sequence that codes for early transcripts is exclusive to one strand of the viral DNA and spans approximately half the genome. Late transcripts are generated in the opposite direction from the other half of the complementary strand (Fig. 3).

Precursor mRNAs undergo post-transcriptional processing that includes capping and polyadenylation of the 5′ and 3′ termini, respectively, as well as splicing. Efficient use of coding information involves differential splicing of the messages and use of overlapping ORFs. Early mRNAs encode regulatory, non-structural proteins that may exhibit cis- or trans-activating properties. These include proteins that are required for initiation of viral DNA replication and late protein production. Their expression leads to de-repression of some host cell enzymes and stimulation of cellular DNA synthesis. Prior to the start of the late events, viral DNA replication is initiated in the nucleus. Translation of most of the late transcripts produces structural proteins that are involved in capsid assembly. Post-translational modifications of some early and late viral proteins include phosphorylation, N-acetylation, fatty acid acylation, ADP-ribosylation, methylamination, adenylation, glycosylation and sulphation. Several of the viral proteins contain sequences, termed nuclear localization signals, which facilitate transport of the proteins to the host cell nucleus where virion maturation occurs. Virions are released by lysis of infected cells.

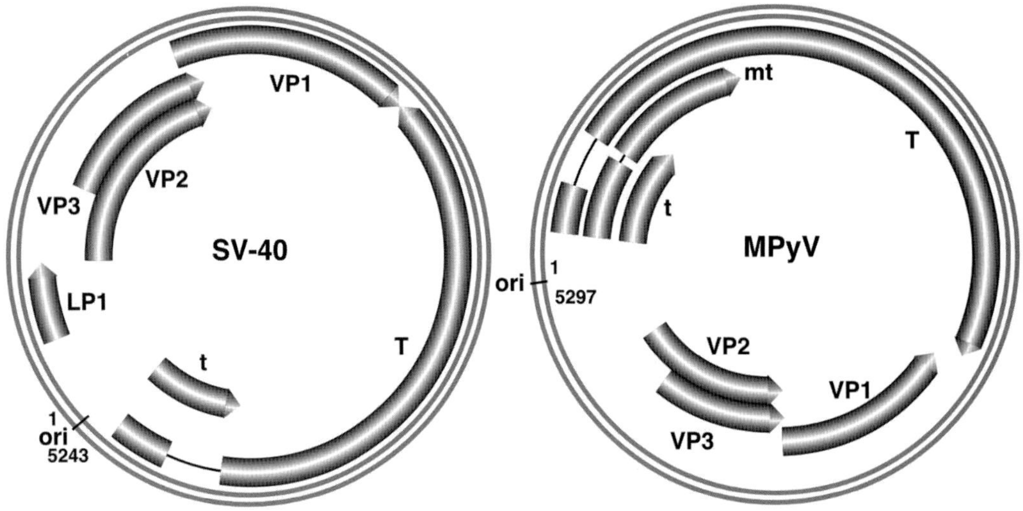

Figure 3: Diagram representative of polyomavirus genomes and encoded proteins. White and black rings represent the viral dsDNAs (origin of replication: ori), arrows indicate encoded viral proteins, or ORFs, as well as direction of transcription. Introns are denoted by solid lines. Small dotted lines represent exons. Large dotted lines represent non-coding regions including the regulatory region, RR. The hyper-variable region, v, within the JC polyomavirus (JCPyV) RR is composed of sequence sections a, b, c, d and e. Alternative splicing is a common characteristic of polyomavirus coding regions. The JCPyV Large T gene codes for Large T, small t, T'_{135}, T'_{136} and T'_{165} which all share identical amino termini.

Identified non-structural proteins include: large T, middle (m)T and small t for mouse and hamster polyomaviruses; large T, 17kT and small t for Simian virus 40 (SV-40). In addition to large T and small t, three other large T intermediates, termed T Prime (T'_{135}, T'_{136} and T'_{165}) have been described for JCPyV. Similar T′ proteins expressed from BKPyV have been identified, but the precise size of these proteins has not been reported. No mRNA encoding a protein of size comparable to the small t proteins of other polyomaviruses has been identified in BKPyV (Table 1). T proteins, first named for their involvement in tumorigenicity and transformation, play key roles in the regulation of transcription and DNA replication. The best characterized of these, the SV-40 large T protein, exhibits multiple functions that can be mapped to discrete domains.

Replication of the viral genome is initiated by the specific binding of the T antigen at a unique origin of replication and its interaction with host DNA polymerase(s). Due to the limited amount of genetic information encoded by the viral genomes, the polyomaviruses rely heavily upon host cell machinery, including nuclear transcription factors, to replicate their DNA. Replication proceeds bi-directionally via a "Cairns" structure and terminates about 180° from the origin of replication. Late in the replication cycle, rolling circle-type molecules have been identified. The viral proteins involved in initiation may also promote elongation through helicase and ATPase activities.

The non-coding regulatory region of each polyomavirus is positioned between the early and late protein-coding sequences. This sequence contains promoter/enhancer elements. Within each polyomavirus species, the nucleotide sequence of the regulatory region is hypervariable. Nucleotide sequencing studies have uncovered numerous variations of regulatory region structure. For the human polyomavirus JCPyV, as with other polyomaviruses, the nucleotide sequence of the regulatory region has been shown to control levels of viral transcription and replication. The JCPyV "archetype" regulatory region sequence, which conveys relatively inefficient viral activity, contains a single copy of all sequence sections observed in all other variant forms (Fig. 4). From the early side of archetype, the initial regulatory region sequence section contains the origin of DNA replication (ORI) followed by sequence sections designated *a, b, c, d, e* and *f*. From variant to variant, sequence sections *a* through *e* are those most likely to present deletions, replications and/or unique arrangements; in example, deletion of sequence sections *b* and *d* leaves *ace*, a 98 base-pair sequence-unit. While the *ace* sequence-unit conveys more activity than archetype, it appears to be the minimal sequence-unit required for functional viral activity. Also, tandem *ace* sequence-units, or repeats, constitute the regulatory region of the more robust "prototype" JCPyV sequence, Mad-1. Such modification to the regulatory region structure appears to alter the cellular host range and may be responsible for switching JCPyV between states of lytic and latent infection. Therefore, it has been proposed that regulatory region structure be used to distinguish JCPyV variants within the species. Arranging all known variant JCPyV regulatory regions into quadrants, according to integration of unique sequence sections and/or repetition of sequence section groups, also links variants by viral activity. This arrangement of regulatory region structures establishes a nomenclature that helps describe relationships between the JCPyV variants. Four distinct structural forms (I-S, I-R, II-S and II-R) are defined along with tissue tropisms. This design, known as the JCV Compass (Fig. 4), provides logical connections between variant regulatory regions and may be useful for elucidating crucial steps in JCPyV pathogenesis. Currently it is not known if similar arrangements of other polyomavirus species variants render logical relationships.

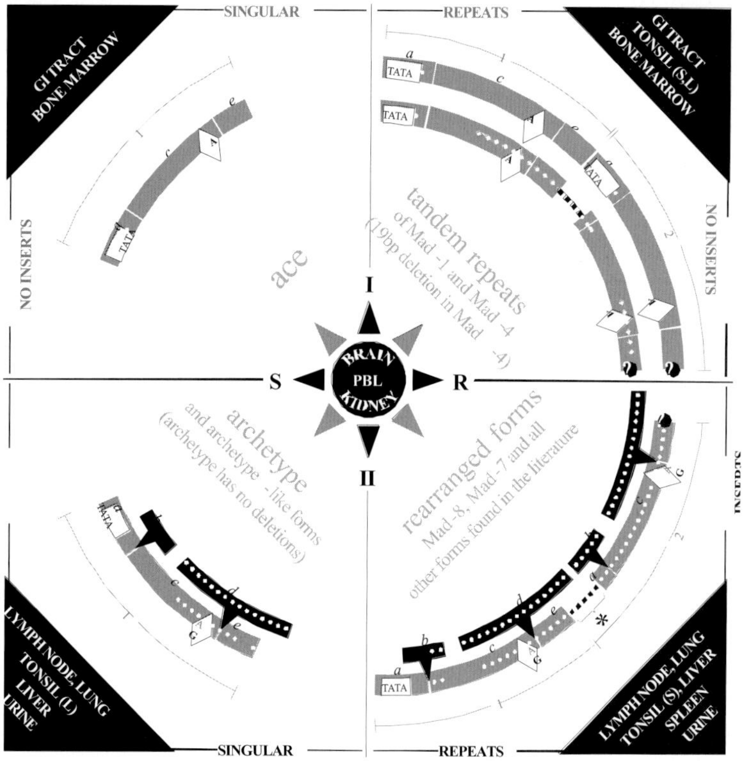

Figure 4: The Compass: A schematic diagram of the relationships between JC polyomavirus (JCPyV) regulatory region sequences published worldwide. JCPyV variant regulatory regions grouped into quadrants (I-S, I-R, II-S and II-R) with ace sequence-units lightly-shaded. Upper quadrant variant types (I) have no additional sequence integrated into the ace units (no inserts). Lower quadrant variant types (II) have dark integrated sequence sections (inserts), b (23 bp) and d (66 bp). Both types I and II are divided into singular (S) and repeat (R) forms by the left and right quadrants, respectively. Unshaded boxes are TATA boxes. Dots represent sites of possible base-pair deletions. Unshaded diamonds contain the base that occupies the 49th nt of sequence section c (nt 85 of I-S, or 108 of II-S) which is adenine (A) in type I variants, but predominantly guanine (G) in type II variants. Right quadrants (R-forms) have dark dashes where sequence is deleted and ⊕ where additional repeats may occur. The * in lower right quadrant (II-R) identifies one reported sequence that retains the second TATA box (Ciappi et al, 1999). JCPyV tropism common to all variant regulatory region forms is contained in dark central circle. Specific JCPyV tropisms are contained in dark corner triangles. Cells from tonsil are either (L) lymphocytes, or (S) stromal cells (Monaco et al, 1998). Cells in bone marrow that contain JCPyV have been identified as B-lymphocytes (Houff et al, 1988).

Antigenic Properties

The human polyomaviruses JCPyV and BKPyV can be detected by hemagglutination of human type O erythrocytes. The CPs bind to the surface of the erythrocytes resulting in a three dimensional lattice-like suspension known as hemagglutination. Using serial dilutions, a titer expressed as hemagglutination units (HA units), can be determined.

Antisera prepared against disrupted virions can also detect antigens shared with other species in the genus. Members of the genus Polyomavirus can be distinguished antigenically by neutralization, hemagglutination inhibition and immuno-electron microscopy tests. Serum levels of antibodies to JCPyV, BKPyV, and SV-40 can also be detected by enzyme linked immunosorbent assay (ELISA) by coating microtiter plates with either whole virion or recombinant virus-like particles. Polyclonal and monoclonal antibodies can be used to demonstrate cross-reactivity between the T proteins of the primate polyomaviruses. However, there are also specific antibodies currently available, which can distinguish amongst the T antigen epitopes of JCPyV, BKPyV, and SV-40.

Biological Properties

Each virus has a specific host range in nature and in cell culture. The host range is often highly restricted, although cells which fail to support viral replication may be transformed via the action of the early gene products.

Although the exact route of transmission is unclear, virus spread occurs by reactivation of persistent, latent infections during periods of immune suppression, including pregnancy. Low level shedding of virus in urine and tissue transplantation (in humans) are thought to play roles. Transmission may also involve contact and air-borne infection, as is suspected for JCPyV, where viral nucleotide sequences have been detected in human tonsillar tissue. Currently, vectors are not known to play a role in transmission. The human polyomaviruses are distributed worldwide, as demonstrated by detectable levels of circulating antibodies in the majority of the healthy human population. Persistent infections are frequently established, usually early in life, after which the virus can remain latent in several body compartments, including the tonsils, the kidneys, lymphoid tissues and bone marrow. The human polyomaviruses often demonstrate highly tissue-specific expression. Involvement of the kidney is frequently observed, with viruria noted, especially in immunodeficient hosts and patients undergoing renal transplant. BKPyV infection is an increasingly common complication in transplant recipients, resulting in nephropathy or cystitis. Infection in humans has been associated with some pathologic changes in the urinary tract. The human polyomavirus, JCPyV, can infect and destroy oligodendrocytes of the central nervous system in severely immunocompromised hosts, thereby leading to a fatal demyelinating disease termed progressive multifocal leucencephalopathy (PML). PML is a common complication in HIV-1 infection of the human central nervous system, eventually affecting 5% of the AIDS population. SV-40 may also cause a PML-like disease in rhesus monkeys. Most polyomaviruses have oncogenic potential in rodents and some primates. JCPyV can induce brain tumors in owl and squirrel monkeys. There has been intense interest in the potential association between polyomavirus infection and the development of central nervous system tumors in humans. However, recent reports suggest that such a correlation is not likely. Under some conditions mouse polyomavirus produces a wide variety of tumors in its natural host. Transformation and oncogenicity result from an expression of virus-specific early proteins and their interaction with specific cellular proteins (p53, pRB and others). In transformed and tumor cells, the polyomavirus genomes are usually integrated into chromosomes of the host cell.

List of Species Demarcation Criteria in the Genus

Until the species demarcation criteria are established the list of species in the genus is provisional.

List of Species in the Genus

Species names are in green italic script; strain names and synonyms are in black roman script; tentative species names are in blue roman script. Sequence accession numbers, and assigned abbreviations ( ) are also listed.

Species in the Genus

African green monkey polyomavirus		
African green monkey polyomavirus	[K02562]	(AGMPyV)
B-lymphotropic polyomavirus		(LPyV)
Baboon polyomavirus 2		
Baboon polyomavirus 2		(BPyV-2)
BK polyomavirus		
BK polyomavirus		(BKPyV)
Bovine polyomavirus		
Bovine polyomavirus	[D00755]	(BPyV)
Stump-tailed macaques virus		
Fetal rhesus kidney virus		
Budgerigar fledgling disease polyomavirus		
Budgerigar fledgling disease polyomavirus		(BFPyV)
Hamster polyomavirus		
Hamster papovavirus		(HapV)
Hamster polyomavirus	[X02449]	(HaPyV)
Human polyomavirus ‡		
Human polyomavirus		
JC polyomavirus		
JC polyomavirus	[J02226]	(JCPyV)
Murine pneumotropic virus		
Kilham polyomavirus (K virus)		(KPyV)
Murine pneumotropic virus	[M55904]	(MPtV)
Murine polyomavirus		
Murine polyomavirus	[J02288]	(MPyV)
Rabbit kidney vacuolating virus		
Rabbit kidney vacuolating virus		(RKV)
Simian virus 12		
Simian virus 12		(SV-12)
Simian virus 40		
Simian virus 40	[J02400]	(SV-40)

Tentative Species in the Genus

Athymic rat polyomavirus

List of Unassigned Viruses in the Family

None reported.

Phylogenetic Relationships within the Family

Not available.

Similarity with Other Taxa

Until the VIIth ICTV report, the genus *Polyomavirus* was assigned as one of two genera within the family *Papovaviridae* (the other genus being *Papillomavirus*).

Derivation of Names

Polyoma: from Greek *poly*, "many", and *-oma*, denoting "tumors".

References

Belnap, D.M., Grochulski, W.D., Olson, N.H. and Baker, T.S. (1993). Use of radial density plots to calibrate image magnification for frozen-hydrated specimens. *Ultramicroscopy*, **48**, 347-358.

Borroweic, J.A., Dean, F.B., Bullock, P.A. and Hurwitz, J. (1990). Binding and unwinding - how T antigen engages the SV40 origin of DNA replication. *Cell*, **60**, 181-184.

Ciappi, S., Azzi, A., De Santis, R., Leoncini, F., Sterrantino, G., Mazzotta, F. and Meococci, L. (1999). Archetypal and rearranged sequences of human polyomavirus JC transcription control region in peripheral blood leukocytes and in cerebrospinal fluid. *J. Gen. Virol.*, **80**, 1017-1023.

Cole, C.N. and Conzen, S.D. (2001). Polyomaviridae: The viruses and their replication. In: *Fields Virology*, 4th ed, (D.M. Knipe and P.M. Howley, eds), pp 2141-2174. Lippincott Williams and Wilkins, Philadelphia.

Fanning, E. (1992). Simian virus 40 large T antigen: the puzzle, the pieces, and the emerging picture. *J. Virol.*, **66**, 1289-1293.

Garcea, R.L., Salunke, D.M. and Caspar, D.L.D. (1987). Site directed mutation affecting polyomaviruscapsid self-assembly in vitro. *Nature*, **329**, 86-87.

Griffin, B.E., Soeda, E., Barrell, B.G. and Staden, R. (1981). Sequences and analysis of polyomavirus DNA. (J. Tooze, ed), pp 843-910. Cold Spring Harbor Laboratory, New York.

Hamilton, R.S., Gravell, M. and Major, E.O. (2000). Comparison of antibody titers determined by hemagglutination inhibition and enzyme immunoassay for JC virus and BK virus. *J. Clin. Microbiol.*, **38**, 105-109.

Houff, S.A., Major, E.O., Katz, D.A., Kufta, C.V., Sever, J.L., Pittaluga, S., Roberts, J.R., Gitt, J., Saini, N. and Lux, W. (1988). Involvement of JC Virus-infected mononuclear cells from the bone marrow and spleen in the pathogenesis of progressive multifocal leukoencephalopathy. *N. Engl. J. Med.*, **318**, 301-305.

Jensen, P.N. and Major, E.O. (2001). A classification scheme for human polyomavirus JCV variants based on the nucleotide sequence of the noncoding regulatory region. *J. Neuro. Virol.*, **7**, 280-287.

Major, E.O., Amemiya, K., Tornatore, C.S., Houff, S.A. and Berger, J.R. (1992). Pathogenesis and molecular biology of progressive multifocal leukoencephalopathy, the JC Virus-induced demyelinating disease of the human brain. *Clin. Microbiol. Rev.*, **5**, 49-73.

Major, E.O. (2001). Human Polyomavirus. In: *Fields Virology*, 4th ed (D.M. Knipe and P.M. Howley, eds), pp 2175-2196. Lippincott Williams and Wilkins, Philadelphia.

Mattern, C.F., Takemoto, K.K. and DeLeva, A.M. (1967). Electron microscopic observations on multiple polyoma virus-related particles. *Virology*, **32**, 378-392.

Monaco M.C., Jensen P.N., Hou J., Durham L.C. and Major E.O. (1998). Detection of JC virus DNA in human tonsil tissue: evidence for site of initial viral infection. *J. Virol.*, **72**, 9918-23.

Riley, M.I., Yoo, W., Mda, N.Y. and Folk, W.R. (1997). Tiny T antigen: an autonomous polyomavirus T antigen amino-terminal domain. *J. Virol.*, **71**, 6068-6074.

Salunke, D.M., Caspar, D.L.D. and Garcea, R.L. (1986). Self-assembly of purified polyomavirus capsid protein VP1. *Cell*, **46**, 895-904.

Salzman, N.P. (ed)(1986). The *Papovaviridae*, the polyomaviruses, vol 1. Plenum Press, New York.

Trowbridge, P.W. and Frisque, R.J. (1995). Identification of three new JC virus proteins generated by alternative splicing of the early viral mRNA. *J. Neuro. Virol.*, **1**, 195-206.

Zerrahn, J., Knippschild, U., Winkler, T. and Deppert, W. (1993). Independent expression of the transforming amino-terminal domain of SV40 large I antigen from an alternatively spliced third SV40 early mRNA. *EMBO J.*, **12**, 4739-4746.

zur Hausen, H. and Gissmann, L. (1979). Lymphotropic papovavirus isolated from African green monkey and human cells. *Med. Microbiol. Immunol.*, **167**, 137-153.

Contributed By

Hou, J., Jensen, P.J., Major, E.O., Zur Hausen, H.J., Almeida, J., Van Der Noordaa, D., Walker, D., Lowy, D., Bernard, U., Butel, J.S, Cheng, D., Frisque, R.J. and Nagashima, K.

FAMILY *PAPILLOMAVIRIDAE*

TAXONOMIC STRUCTURE OF THE FAMILY

Family	***Papillomaviridae***
Genus	***Alphapapillomavirus***
Genus	***Betapapillomavirus***
Genus	***Gammapapillomavirus***
Genus	***Deltapapillomavirus***
Genus	***Epsilonpapillomavirus***
Genus	***Zetapapillomavirus***
Genus	***Etapapillomavirus***
Genus	***Thetapapillomavirus***
Genus	***Iotapapillomavirus***
Genus	***Kappapapillomavirus***
Genus	***Lambdapapillomavirus***
Genus	***Mupapillomavirus***
Genus	***Nupapillomavirus***
Genus	***Xipapapillomavirus***
Genus	***Omikronpapillomavirus***
Genus	***Pipapillomavirus***

VIRION PROPERTIES

MORPHOLOGY

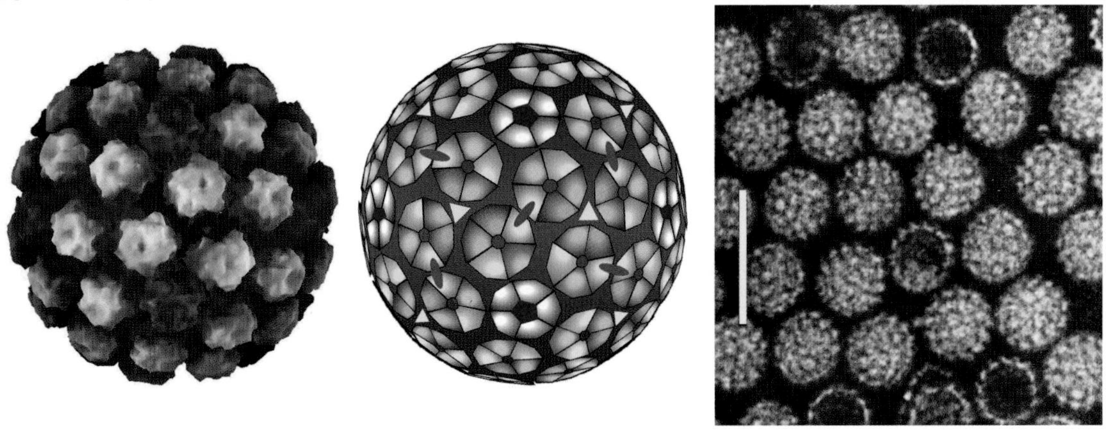

Figure 1: (Left) Atomic rendering of a papillomavirus capsid with combined image reconstructions from electron cryomicroscopy of Bovine papillomavirus (BPV) at 9 A resolution with coordinates from the crystal structure of small virus-like particles of the Human papillomavirus 16 (HPV-16) L1 protein (Modis, *et al.*, 2002). (Center) Schematic diagram representing the 72 capsomers in a T=7 arrangement of a papillomavirus capsid (the icosahedral structure includes 360 VP1 subunits arranged in 12 pentavalent and 60 hexavalent capsomers). (Right) Negative contrast electron micrograph of Human papillomavirus 1 (HPV-1) virions. The bar represents 100 nm.

Virions are non-enveloped, 55 nm in diameter. The icosahedral capsid is composed of 72 capsomers in skewed (T=7) arrangement (Fig. 1). Filamentous and tubular forms are observed as a result of aberrant maturation.

PHYSICOCHEMICAL AND PHYSICAL PROPERTIES

Virion Mr is 47 x 10^6. Buoyant density of virions in sucrose and CsCl gradients is 1.20 and 1.34-1.35 g/cm^3, respectively. Virion S_{20w} is 300S. Virions are resistant to ether, acid and heat treatment (50°C, 1 hr).

NUCLEIC ACID

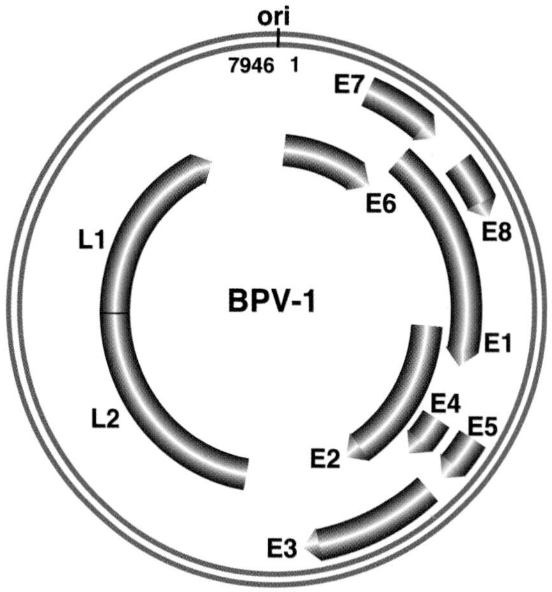

Figure 2: Diagram of the genome of the Bovine papillomavirus 1 (BPV-1). The viral dsDNA (size in bp, origin of replication: ori), the inner arrows indicate the encoded viral proteins, or ORFs, as well as the direction of transcription.

Virions contain a single molecule of circular dsDNA. The genomic size ranges between 6800 and 8400 bp. The DNA constitutes about 10-13% of the virion by weight. The G+C content is 40-60%. In the mature virion the viral DNA is associated with host cell histone proteins H2a, H2b, H3 and H4 in a chromatin-like complex.

PROTEINS

The virus genomes encode 8-10 proteins with sizes ranging from 7-73 kDa (Table 1). L1 and L2 make up the papillomavirus capsid. E1 and E2 are involved in papillomavirus replication and in intragenomic regulation (E2). E5, E6 and E7 induce cellular DNA replication. E4 may represent a late function and binds to specific cytoskeleton structures. Genetic evidence has not been presented that associates specific viral proteins with the E3 and E8 ORFs.

Table 1: Deduced papillomavirus proteins molecular ratios (kDa).

Virus:	CRPV	BPV-1	HPV-1
Structural proteins:			
L1	57.9	55.5	59.6
L2	52.8	50.1	50.7
Nonstructural proteins:			
E1	67.9	68.0	73.0
E2	44.0	48.0	41.8
E4	25.8	12.0	10.4
E5	11.3	7.0	9.4
E6	29.7	15.1	19.2
E7	10.5	14.0	11.0

LIPIDS

None present.

CARBOHYDRATES

None present.

GENOME ORGANIZATION AND REPLICATION

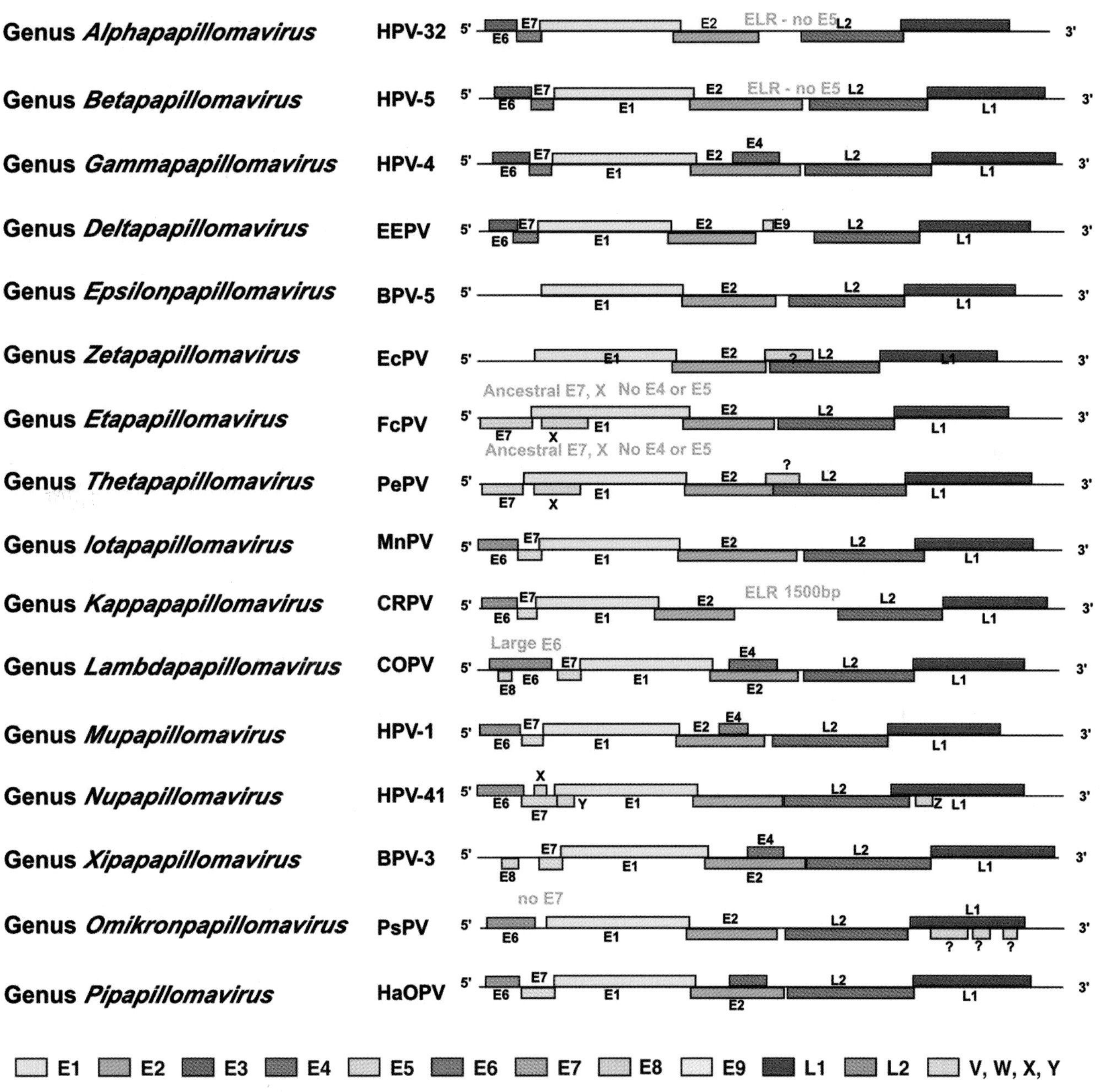

Figure 3: Comparison of the diagrams of the genome organization of the viruses corresponding to the type species of each genus in the family *Papillomaviridae*. The circular viral dsDNA genomes have been flattened for convenience and the Ori has been taken as opening site. Similar ORFs are indicated insimilar colors (see legend in figure and the nomenclature is the conventional one used in the field (see text). The meaning of the abbreviation of the virus names are listed in the description.

Virions that attach to cellular receptors are engulfed by the cell and the DNA is uncoated and transported to the nucleus. During the productive infection, transcription of the viral genome is divided into an early and late stage.

Transcription of the early and late coding regions occurs from the same strand in one direction only. Precursor mRNAs undergo post-transcriptional processing that includes capping and polyadenylation of the 5′- and 3′-termini, respectively, as well as splicing. Efficient use of coding information involves differential splicing of the messages and use of overlapping ORFs. Early mRNAs encode regulatory proteins that may exhibit trans-activating properties. These include proteins that are required for DNA replication. Their expression leads to depression of some host cell enzymes and may also stimulate host cell DNA synthesis. Prior to the start of the late events, viral DNA replication is initiated in the nucleus. Translation of the late transcripts produces structural proteins that are involved in capsid assembly. Post-translational modifications of some early and late viral proteins include phosphorylation, N-acetylation, ADP ribosylation and other events. Several of the viral proteins contain sequences, termed nuclear localization signals, which facilitate transport of the proteins to the host cell nucleus where virion maturation occurs. Virions are released by lysis of the virus-producing cells.

The genomes of most members of the family *Papillomaviridae* that have been sequenced contain 9 - 10 ORFs, labelled E1 - E8 and L1 - L2 (Fig. 3). Some members lack the E3 and E8 ORFs. Proteins encoded by the E ORFs, with the possible exception of E4, represent non-structural polypeptides involved in transcription, DNA replication and transformation, whereas those encoded by the L ORFs represent structural proteins. Replication of the viral genome is initiated bi-directionally by specific binding of the E1 and E2 proteins at a unique origin of replication.

Antigenic Properties

The L1 protein reveals type-specific domains, the L2 protein contains group-specific epitopes. The availability of papillomavirus-like particles, resulting from the expression of L1 or L1 and L2 in baculovirus, vaccinia virus or yeast systems, permits presently a detailed analysis of antigenic characteristics.

Biological Properties

Papillomaviruses are highly host-species and tissue-restricted. All known members of the Human papillomavirus (HPV) group require terminal differentiation for replication and virion production. Infection appears to occur mainly *via* microlesions of proliferating basal layer cells. Except for inefficient replication of HPVs in raft cultures of human keratinocytes or more efficiently in human skin or mucosal xenografts in immunocompromised rodents, HPV replication has not been achieved in tissue culture systems.

Virus spread occurs by virus release from the surface of warts and papillomatous lesions which frequently contain large quantities of viral particles within their superficial differentiated layers. Virus reactivation is particularly frequent under conditions of immunosuppression. The mode of viral DNA persistence and possible clearance of HPV infections by immunological interference are still poorly investigated.

Transmission of viral infections occurs by close contacts. Papillomavirus types are distributed worldwide. They cause benign tumors (warts, papillomas) in their natural host and occasionally in related species. Frequently the infection leads to microlesions, barely or not at all visible without optical aid. Papillomas are induced in the skin and in mucous membranes, often at specific sites of the body. Some papillomatous proliferations induced by specific types of papillomaviruses bear a high risk for malignant progression. Specific human cancers (e.g. cervical carcinoma, anal, vulval and penile cancers, specific squamous cell carcinomas of the skin) have been linked to certain types of HPV infection (e.g. HPV-16 and HPV-18, HPV-5 and HPV-8 and several others). The viral DNA is often, but not always, present in an integrated form, particularly in cervical cancers, whereas skin carcinomas appear to harbor the viral genome in an episomal state. Cancer-linked anogenital HPV types immortalize efficiently a wide variety of human cells in tissue culture. Immortalization results from functions of the E6 and E7 genes of these viruses which act cooperatively, although both genes are able to immortalize human cells independently at low efficiency. E6 binds and degrades the cellular p53 protein and stimulates the telomerase enzyme, whereas E7 interacts with the cellular pRB and some related proteins, and directly activates cyclins E and A. Interaction of the viral oncoproteins with cellular cyclin-dependant kinase inhibitors (p16, p21, p27) also emerge as important events in immortalization.

LIST OF SPECIES DEMARCATION CRITERIA IN THE FAMILY

This demarcation of species is extremely difficult to apply to papillomaviruses. The biological/epidemiological properties of the majority of papillomavirus types have received little or no attention and have only been reported in one, or, in exceptional cases, in a few more studies concentrating on the specific type in question. It would be very difficult to define such properties even for types which have been studied in detail. Misclassification of types in the early publications poses an additional complication. The phylogenetically related group HPV-2, HPV-27 and HPV-57, may serve as an example. HPV-2 and HPV-27 share 89,6% homology in the L1 region used for constructing the phylogenetic tree in Figure 3. HPV-57 shares 84,8% sequence homology with HPV-2 and 85,5% to HPV-27. Biological differences have been noted between HPV-2 and HPV-27 despite their close sequence homology. HPV-2 has to date not been described in malignant lesions, whereas HPV-27 has. This may be debated, because the early data on HPV-2 DNA in carcinomas based on hybridization analysis, could not distinguish between HPV-2 and HPV-27. Only in single reports was it later recognized that HPV-27 had actually been involved (identical to HPV-2c), instead of HPV-2. HPV-27 has been detected more frequently in cutaneous proliferations and more specifically in immunosuppressed patients than HPV-2. Similar misclassifications of HPV types were also earlier reported on HPV types in butcher´ s warts. The HPV-2 and HPV-3 restriction fragments demonstrated in such lesions were later recognized as actually corresponding to HPV-27, HPV-10 and HPV-28. HPV-57 frequently occurs in common warts, but was originally isolated from an inverted papilloma of the maxillary sinus and subsequently associated with nasal malignancies. Presently we have no evidence that HPV-2 and HPV-27 can be associated with inverted papillomas.

GENUS *ALPHAPAPILLOMAVIRUS*

Type Species ***Human papillomavirus 32***

DISTINGUISHING FEATURES

Members of this genus preferentially infect the oral or anogenital mucosa in humans and primates. Certain species (eg. *Human papillomavirus 2, Human papillomavirus 10*) are also found in lesions of cutaneous sites. Specific species (eg. *Human papillomavirus 16, Human papillomavirus 18)* are considered as high-risk virus in view of their regular presence in malignant tissue and their in vitro transforming activities. Other species (eg. *Human papillomavirus 53, Human papillomavirus 26, Human papillomavirus 34*) cause malignant or benign lesions, whereas the low-risk species (*Human papillomavirus 61, Human papillomavirus 7, Human papillomavirus 6, Human papillomavirus 54, Human papillomavirus cand90, Human papillomavirus 71*) mainly cause benign lesions. Genome organization: An E5 ORF is conserved between the early and late coding regions.

LIST OF SPECIES IN THE GENUS

Species names are in green italic script; strain names and synonyms are in black roman script; tentative species names are in blue roman script. Sequence accession numbers, and assigned abbreviations ( ) are also listed.

SPECIES IN THE GENUS

Human papillomavirus 2		
Human papillomavirus 2	[X55964]	(HPV-2)
Human papillomavirus 27	[X73373]	(HPV-27)
Human papillomavirus 57	[X55965]	(HPV-57)
Human papillomavirus 6		
Human papillomavirus 6	[X00203]	(HPV-6)
Human papillomavirus 11	[M14119]	(HPV-11)
Human papillomavirus 13	[X62843]	(HPV-13)
Human papillomavirus 44	[U31788; U31791]	(HPV-44)
Human papillomavirus 74	[U40822]	(HPV-74)
Pygmy champanzee papillomavirus 1	[X62844]	(PCPV-1)
Pygmy champanzee papillomavirus 1 Champanzee	[AF020905]	(PCPV-1C)
Human papillomavirus 7		
Human papillomavirus 7	[X74463]	(HPV-7)
Human papillomavirus 40	[X74478]	(HPV-40)
Human papillomavirus 43	[AJ620205]	(HPV-43)
Human papillomavirus cand91	[AF131950]	(HPV-cand91)
Human papillomavirus 10		
Human papillomavirus 3	[X74462]	(HPV-3)
Human papillomavirus 10	[X74465]	(HPV-10)
Human papillomavirus 28	[U31783]	(HPV-28)
Human papillomavirus 29	[U31784]	(HPV-29)
Human papillomavirus 77	[Y15175]	(HPV-77)
Human papillomavirus 78		(HPV-78)
Human papillomavirus 94	[AJ620021]	(HPV-94)
Human papillomavirus 16		
Human papillomavirus 16	[K02718]	(HPV-16)
Human papillomavirus 31	[J04353]	(HPV-31)
Human papillomavirus 33	[M12732]	(HPV-33)
Human papillomavirus 35	[X74476]	(HPV-35)
Human papillomavirus 52	[X74481]	(HPV-52)
Human papillomavirus 58	[D90400]	(HPV-58)
Human papillomavirus 67	[D21208]	(HPV-67)
Human papillomavirus 18		
Human papillomavirus 18	[X05015]	(HPV-18)
Human papillomavirus 39	[M62849]	(HPV-39)
Human papillomavirus 45	[X74479]	(HPV-45)
Human papillomavirus 59	[X77858]	(HPV-59)
Human papillomavirus 68	[X67161]	(HPV-68)
Human papillomavirus 70	[U21941]	(HPV-70)
Human papillomavirus cand85	[AF131950]	(HPV-cand85)
Human papillomavirus 26		
Human papillomavirus 26	[X74472]	(HPV-26)
Human papillomavirus 51	[M62877]	(HPV-51)
Human papillomavirus 69	[AB027020]	(HPV-69)
Human papillomavirus 82	[AB027021]	(HPV-82)
Human papillomavirus 32		
Human papillomavirus 32	[X74475]	(HPV-32)
Human papillomavirus 42	[M73236]	(HPV-42)
Human papillomavirus 34		
Human papillomavirus 34	[X74476]	(HPV-34)
Human papillomavirus 73	[X94165]	(HPV-73)

Human papillomavirus 53		
Human papillomavirus 30	[X74474]	(HPV-30)
Human papillomavirus 53	[X74482]	(HPV-53)
Human papillomavirus 56	[X74483]	(HPV-56)
Human papillomavirus 66	[U31794]	(HPV-66)
Human papillomavirus 54		
Human papillomavirus 54	[U37488]	(HPV-54)
Human papillomavirus 61		
Human papillomavirus 61	[U31793]	(HPV-61)
Human papillomavirus 72	[X94164]	(HPV-72)
Human papillomavirus 81	[AJ620209]	(HPV-81)
Human papillomavirus 83	[AF151983]	(HPV-83)
Human papillomavirus 84	[AF293960]	(HPV-84)
Human papillomavirus cand62	[U12499]	(HPV-cand62)
Human papillomavirus cand86	[AF349909]	(HPV-cand86)
Human papillomavirus cand87	[AJ400628]	(HPV-cand87)
Human papillomavirus cand89	[AF436128]	(HPV-cand89)
Human papillomavirus 71		
Human papillomavirus 71	[AB040456]	(HPV-71)
Human papillomavirus cand90		
Human papillomavirus cand90	[AY057438]	(HPV-cand90)
Rhesus monkey papillomavirus 1		
Rhesus monkey papillomavirus 1	[M60184]	(RhPV-1)

Tentative Species in the Genus

None reported.

Genus *Betapapillomavirus*

Type Species *Human papillomavirus 5*

Distinguishing Features

Members of this genus preferentially infect the skin of humans. These infections exist latent in the general population, but are activated under conditions of immunosuppression. Species *Human papillomavirus 5, Human papillomavirus 9 and Human papillomavirus 49* are also associated with the disease Epidermodysplasia verruciformis (EV). Genome organization: E5 ORF is absent.

List of Species in the Genus

Species names are in green italic script; strain names and synonyms are in black roman script; tentative species names are in blue roman script. Sequence accession numbers, and assigned abbreviations ( ) are also listed.

Species in the Genus

Human papillomavirus 5		
Human papillomavirus 5	[M17463]	(HPV-5)
Human papillomavirus 8	[M12737]	(HPV-8)
Human papillomavirus 12	[X74466]	(HPV-12)
Human papillomavirus 14	[X74467]	(HPV-14)
Human papillomavirus 19	[X74470]	(HPV-19)
Human papillomavirus 20	[U31778]	(HPV-20)
Human papillomavirus 21	[U31779]	(HPV-21)
Human papillomavirus 24	[U31782]	(HPV-24)
Human papillomavirus 25	[U74471]	(HPV-25)
Human papillomavirus 36	[U31785]	(HPV-36)
Human papillomavirus 47	[M32305]	(HPV-47)

Human papillomavirus 9		
Human papillomavirus 9	[X74464]	(HPV-9)
Human papillomavirus 15	[X74468]	(HPV-15)
Human papillomavirus 17	[X74469]	(HPV-17)
Human papillomavirus 22	[U31780]	(HPV-22)
Human papillomavirus 23	[U31781]	(HPV-23)
Human papillomavirus 37	[U31786]	(HPV-37)
Human papillomavirus 38	[U31787]	(HPV-38)
Human papillomavirus 80	[Y15176]	(HPV-80)
Human papillomavirus 49		
Human papillomavirus 49	[X74480]	(HPV-49)
Human papillomavirus 75	[Y15173]	(HPV-75)
Human papillomavirus 76	[Y15174]	(HPV-76)
Human papillomavirus cand92		
Human papillomavirus cand92	[AF531420]	(HPV-cand92)
Human papillomavirus cand96		
Human papillomavirus cand96	[AY382779]	(HPV-cand96)

TENTATIVE SPECIES IN THE GENUS

None reported.

GENUS GAMMAPAPILLOMAVIRUS

Type Species ***Human papillomavirus 4***

DISTINGUISHING FEATURES

Members of this genus cause cutaneous lesions in their host and are histologically distinguishable by intracytoplsmic inclusion bodies which are species specific. Genome organization: E5 ORF is absent.

LIST OF SPECIES IN THE GENUS

Species names are in green italic script; strain names and synonyms are in black roman script; tentative species names are in blue roman script. Sequence accession numbers, and assigned abbreviations ( ) are also listed.

SPECIES IN THE GENUS

Human papillomavirus 4		
Human papillomavirus 4	[X70827]	(HPV-4)
Human papillomavirus 65	[X70829]	(HPV-65)
Human papillomavirus 95	[AJ620210]	(HPV-95)
Human papillomavirus 48		
Human papillomavirus 48	[U31790]	(HPV-48)
Human papillomavirus 50		
Human papillomavirus 50	[U31790]	(HPV-50)
Human papillomavirus 60		
Human papillomavirus 60	[U31792]	(HPV-60)
Human papillomavirus 88		
Human papillomavirus 88		(HPV-88)

TENTATIVE SPECIES IN THE GENUS

None reported.

GENUS DELTAPAPILLOMAVIRUS

Type Species *European elk papillomavirus*

DISTINGUISHING FEATURES

These papillomaviruses induce fibropapillomas in their respective ungulate hosts. Trans-species transmission occurs where it induces sarcoids. Genome organization: ORFs located in the region between the early and late genes have transforming properties.

LIST OF SPECIES IN THE GENUS

Species names are in green italic script; strain names and synonyms are in black roman script; tentative species names are in blue roman script. Sequence accession numbers, and assigned abbreviations ( ) are also listed.

SPECIES IN THE GENUS

Bovine papillomavirus 1		
Bovine papillomavirus 1	[X02346]	(BPV-1)
Bovine papillomavirus 2	[M20219]	(BPV-2)
Deer papillomavirus		
Deer papillomavirus	[M11910]	(DPV)
(Deer fibroma virus)		
European elk papillomavirus		
European elk papillomavirus	[M15953]	(EEPV)
Reindeer papillomavirus	[AF443292]	(RPV)
Ovine papillomavirus 1		
Ovine papillomavirus 1	[U83594]	(OvPV-1)
Ovine papillomavirus 2	[U83595]	(OvPV-2)

TENTATIVE SPECIES IN THE GENUS

None reported.

GENUS EPSILONPAPILLOMAVIRUS

Type Species *Bovine papillomavirus 5*

DISTINGUISHING FEATURES

Infections cause cutaneous papillomas in cattle.

LIST OF SPECIES IN THE GENUS

Species names are in green italic script; strain names and synonyms are in black roman script; tentative species names are in blue roman script. Sequence accession numbers, and assigned abbreviations ( ) are also listed.

SPECIES IN THE GENUS

Bovine papillomavirus 5		
Bovine papillomavirus 5	[AF457465]	(BPV-5)

TENTATIVE SPECIES IN THE GENUS

None reported.

GENUS ZETAPAPILLOMAVIRUS

Type Species *Equine papillomavirus 1*

DISTINGUISHING FEATURES

Infections cause cutaneous lesions in horses. Genome organization: An undefined ORF overlaps with the L2 ORF.

LIST OF SPECIES IN THE GENUS

Species names are in green italic script; strain names and synonyms are in black roman script; tentative species names are in blue roman script. Sequence accession numbers, and assigned abbreviations ( ) are also listed.

SPECIES IN THE GENUS

Equine papillomavirus 1
Equus caballus papillomavirus 1 [AF498323] (EcPV)

TENTATIVE SPECIES IN THE GENUS

None reported.

GENUS *ETAPAPILLOMAVIRUS*

Type Species *Fringilla coelebs papillomavirus*

DISTINGUISHING FEATURES

Avian papillomaviruses causing cutaneous lesions in their host. Genome organization: An ancestral E7 ORF exists which have partial E6 characteristics. Typical E6 ORF is absent.

LIST OF SPECIES IN THE GENUS

Species names are in green italic script; strain names and synonyms are in black roman script; tentative species names are in blue roman script. Sequence accession numbers, and assigned abbreviations ( ) are also listed.

SPECIES IN THE GENUS

Fringilla coelebs papillomavirus
Chaffinch papillomavirus [AY957109] (FcPV)

TENTATIVE SPECIES IN THE GENUS

None reported.

GENUS *THETAPAPILLOMAVIRUS*

Type Species *Psittacus erithacus timneh papillomavirus*

DISTINGUISHING FEATURES

Avian papillomaviruses causing cutaneous lesions in their host. Genome organization: An ancestral E7 ORF exists which have partial E6 characteristics. Typical E4, E5 and E6 ORFs are absent.

LIST OF SPECIES IN THE GENUS

Species names are in green italic script; strain names and synonyms are in black roman script; tentative species names are in blue roman script. Sequence accession numbers, and assigned abbreviations ( ) are also listed.

SPECIES IN THE GENUS

Psittacus erithacus timneh papillomavirus
Psittacus erithacus timneh papillomavirus [AF420235] (PePV)

TENTATIVE SPECIES IN THE GENUS

None reported.

GENUS *IOTAPAPILLOMAVIRUS*

Type Species *Mastomys natalensis papillomavirus*

DISTINGUISHING FEATURES

Rodent papillomavirus causing cutaneous lesions in host. Genome organization: The E2 ORF is considerably larger than in other genera and the E5 ORF is absent.

LIST OF SPECIES IN THE GENUS

Species names are in green italic script; strain names and synonyms are in black roman script; tentative species names are in blue roman script. Sequence accession numbers, and assigned abbreviations ( ) are also listed.

SPECIES IN THE GENUS

Mastomys natalensis papillomavirus
Mastomys natalensis papillomavirus [U01834] (MnPV)

TENTATIVE SPECIES IN THE GENUS

None reported.

GENUS *KAPPAPAPILLOMAVIRUS*

Type Species *Cottontail rabbit papillomavirus*

DISTINGUISHING FEATURES

Members of this genus causes cutaneous and mucosal lesions in rabbits. Genome organization: The E6 ORF is larger than in other genera. An uncharacterized E8 ORF is present in the early region.

LIST OF SPECIES IN THE GENUS

Species names are in green italic script; strain names and synonyms are in black roman script; tentative species names are in blue roman script. Sequence accession numbers, and assigned abbreviations ( ) are also listed.

SPECIES IN THE GENUS

Cottontail rabbit papillomavirus
Cottontail rabbit papillomavirus [K02708] (CRPV)
Rabbitt oral papillomavirus
Rabbit oral papillomavirus [AF227240] (ROPV)

TENTATIVE SPECIES IN THE GENUS

None reported.

GENUS *LAMBDAPAPILLOMAVIRUS*

Type Species *Canine oral papillomavirus*

DISTINGUISHING FEATURES

Members of this genus infect cats and dogs, causing mucosal and cutaneous lesions. Genome organization: The region between the early and late coding regions is exceptionally large ranging between 1200-150bp.

LIST OF SPECIES IN THE GENUS

Species names are in green italic script; strain names and synonyms are in black roman script; tentative species names are in blue roman script. Sequence accession numbers, and assigned abbreviations ( ) are also listed.

SPECIES IN THE GENUS

Canine oral papillomavirus		
Canine oral papillomavirus	[L22695]	(COPV)
Feline papillomavirus		
Feline papillomavirus	[AF377865]	(FdPV)

TENTATIVE SPECIES IN THE GENUS

None reported.

GENUS *MUPAPILLOMAVIRUS*

Type Species ***Human papillomavirus 1***

DISTINGUISHING FEATURES

Human papillomaviruses causing cutaneous lesions in their host which are histologically distinguishable by intracytoplasmic inclusion bodies which are species specific. Genome organization: The control region is larger than in other genera.

LIST OF SPECIES IN THE GENUS

Species names are in green italic script; strain names and synonyms are in black roman script; tentative species names are in blue roman script. Sequence accession numbers, and assigned abbreviations ( ) are also listed.

SPECIES IN THE GENUS

Human papillomavirus 1		
Human papillomavirus 1	[V01116]	(HPV-1)
Human papillomavirus 63		
Human papillomavirus 63	[X70828]	(HPV-63)

TENTATIVE SPECIES IN THE GENUS

None reported.

GENUS *NUPAPILLOMAVIRUS*

Type Species ***Human papillomavirus 41***

DISTINGUISHING FEATURES

Human papillomaviruses causing benign and malignant cutaneous lesions in their hosts. Genome organization: Several larger ORFs are located in the L1 ORF region. The E2 binding sites in the control region are all modified.

LIST OF SPECIES IN THE GENUS

Species names are in green italic script; strain names and synonyms are in black roman script; tentative species names are in blue roman script. Sequence accession numbers, and assigned abbreviations ( ) are also listed.

SPECIES IN THE GENUS

Human papillomavirus 41		
Human papillomavirus 41	[X56147]	(HPV-41)

TENTATIVE SPECIES IN THE GENUS

None reported.

GENUS XIPAPILLOMAVIRUS

Type Species *Bovine papillomavirus 3*

DISTINGUISHING FEATURES

Infections with these papillomaviruses cause true papillomas on the cutaneous or mucosal surfaces of cattle. Genome organization: A characteristic E6 ORF is absent and the E8 ORF located in this region displays transforming properties similar to that of *Bovine papillomavirus-1*.

LIST OF SPECIES IN THE GENUS

Species names are in green italic script; strain names and synonyms are in black roman script; tentative species names are in blue roman script. Sequence accession numbers, and assigned abbreviations ( ) are also listed.

SPECIES IN THE GENUS

Bovine papillomavirus 3

Bovine papillomavirus 3	[AF486184]	(BPV-3)
Bovine papillomavirus 4	[X05817]	(BPV-4)
Bovine papillomavirus 6	[AJ620208]	(BPV-6)

TENTATIVE SPECIES IN THE GENU

None reported.

GENUS OMIKRONPAPILLOMAVIRUS

Type Species *Phocoena spinipinnis papillomavirus*

DISTINGUISHING FEATURES

These papillomaviruses have been isolated from genital warts in cetaceans. Genome organization: Several larger ORFs are located in the L1 ORF region. True E7 ORF is absent.

LIST OF SPECIES IN THE GENUS

Species names are in green italic script; strain names and synonyms are in black roman script; tentative species names are in blue roman script. Sequence accession numbers, and assigned abbreviations ( ) are also listed.

SPECIES IN THE GENUS

Phocoena spinipinnis papillomavirus

Phocoena spinipinnis papillomavirus	[AJ238272]	(PsPV)

TENTATIVE SPECIES IN THE GENUS

None reported.

GENUS PIPAPILLOMAVIRUS

Type Species *Hamster oral papillomavirus*

DISTINGUISHING FEATURES

Infections with these papillomaviruses cause mucosal lesions in hamsters. The E2 and L2 ORFs are partially overlapping.

LIST OF SPECIES IN THE GENUS

Species names are in green italic script; strain names and synonyms are in black roman script; tentative species names are in blue roman script. Sequence accession numbers, and assigned abbreviations ( ) are also listed.

SPECIES IN THE GENUS

Hamster oral papillomavirus

Hamster oral papillomavirus	[E15110]	(HaOPV)

TENTATIVE SPECIES IN THE GENUS

None reported.

LIST OF UNASSIGNED VIRUSES IN THE FAMILY

Trichosurus vulpecula papillomavirus	[AF181682]	(TvPV)
Possum papillomavirus		(PoPV)

Putative new papillomaviruses of a variety of different species have been identified by partial sequences. More than 300 such sequences are presently available in the databanks.

PHYLOGENETIC RELATIONSHIPS WITHIN THE FAMILY

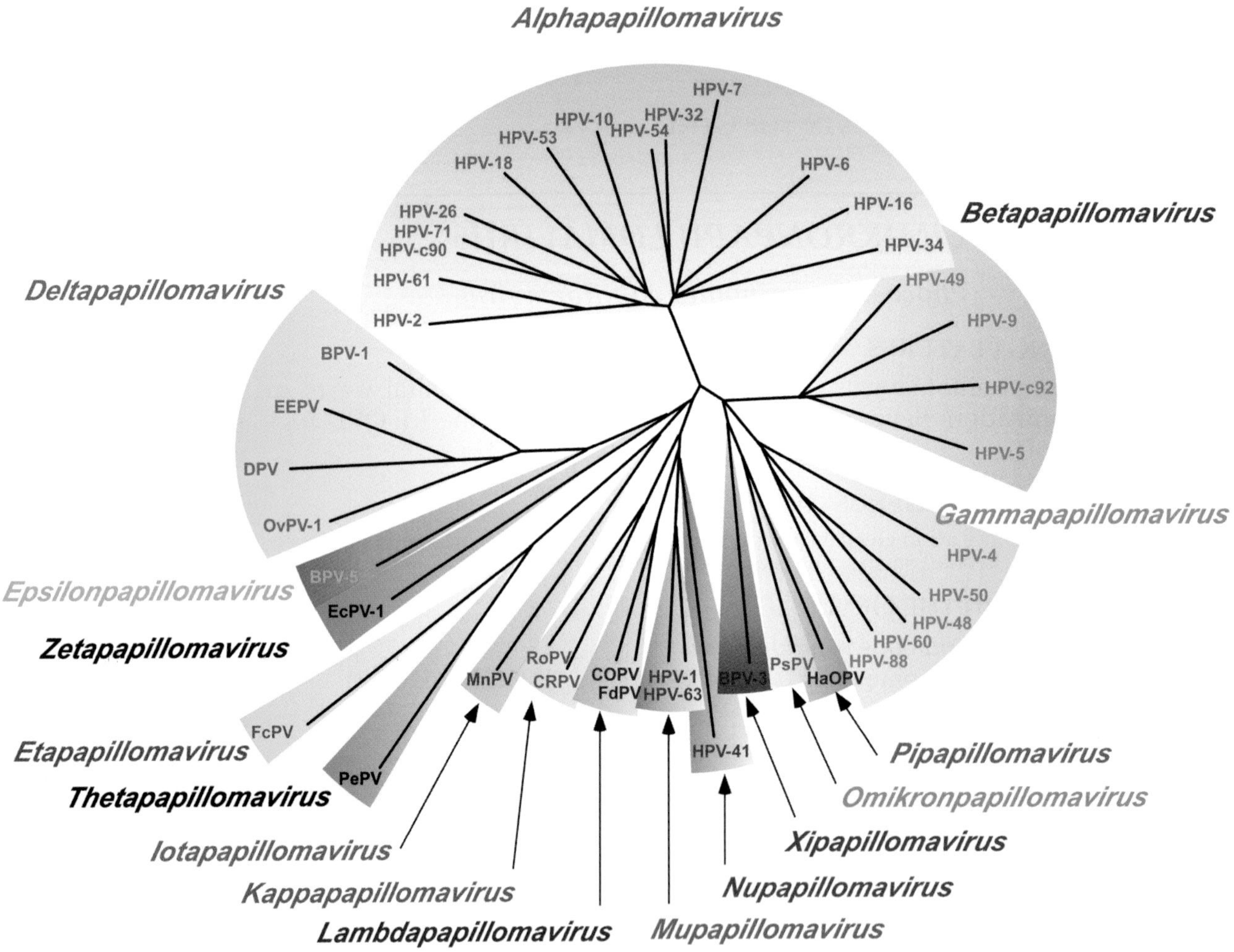

Figure 4: Phylogenetic tree containing sequences of 118 papillomaviruses. The phylogenetically informative region of the L1 ORF was used in a modified version of the Phylip version 3.572 and based on a weighted version of the neighbor-joining analysis. The accession numbers of the viruses are listed in the list of species in the genera of the family *Papillomaviridae* The tree was constructed using the Treeview program of the University of Glasgow.

Similarity with Other Taxa

The families *Papillomaviridae* and *Polyomaviridae* share some similarities in morphology, nucleic acid composition, as well as *in vitro* transforming activities of specific proteins.

Derivation of Names

Papilloma: from Latin *papilla*, "nipple, pustule", also Greek suffix *-oma*, used to form nouns denoting "tumors".

DSDNA

References

Antonsson, A., Forslund, O., Ekberg, H., Sterner, G. and Hansson, B.G. (2000). The ubiquity and impressive genomic diversity of human skin papillomaviruses suggest a commensalic nature of these viruses. *J. Virol.* **74**, 11636-11641.

Bernard, H.-U. and Chan, S.-Y. (1997). Animal Papillomaviruses. In: *Human Papillomaviruses: A compilation and analysis of nucleic acid and amino acid sequences*, (G., Myers, F., Sverdrup, C., Baker, A., McBride, K., Munger, H.-U., Bernard and J., Meissner, eds), pp II-100, III-107. Theoretical Biology and Biophysics, Los Alamos National Laboratory, Los Alamos.

Bernard, H.-U., Chan, S.-Y. and Delius, H. (1994). Evolution of papillomaviruses. *Curr. Top. Microbiol. Immunol.*, **186**, 34-53.

Chan, S.-Y., Bernard, H.-U., Ong, C.-K., Chan, S.-P., Hofmann, B. and Delius, H. (1992). Phylogenetic analysis of 48 papillomavirus types and 28 subtypes and variants: a showcase for the molecular evolution of DNA viruses. *J. Virol.*, **66**, 5714-5725.

Chan, S.-Y., Bernard, H.-U., Ratterree, M., Birkebak, T.A., Faras, A.J. and Ostrow, R.S. (1997). Genomic diversity and evolution of papillomaviruses in rhesus monkeys. *J. Virol.*, **71**, 4938-4943.

Chan, S.-Y., Chew, S.-H., Egawa, K., Gruendorf-Conen, E.-I., Honda, Y., Rübben, A., Tan, K.-C. and Bernard, H.-U. (1997). Phylogenetic analysis of the human papillomavirus type 2 (HPV-2, HPV-27 and HPV-57 group, which is associated with common warts. *Virology*, **239**, 296-302.

Chan, S.-Y., Delius, H., Halpern, A.L. and Bernard, H.-U. (1995). Analysis of genomic sequences of 95 papillomavirus types: Uniting typing, phylogeny and taxonomy. *J. Virol.*, **69**, 3074-3083.

Chen, E.Y., Howley, P.M., Levinson, A.D. and Seeburg, P.H. (1982). The primary structure and genetic organization of the bovine papillomavirus type 1 genome. *Nature*, **299**, 529-534.

de Villiers, E.-M. (1994). Human pathogenic papillomavirus types: an update. *Curr. Top. Microbiol. Immunol.*, **186**, 1-12.

de Villiers, E.-M. (1998). Human papillomaviruses infections in skin cancer. *Biomed. Pharmacother.*, **52**, 26-33.

de Villiers, E.-M., Weidauer, H., Otto, H. and zur Hausen, H. (1985). Papilllomavirus DNA in human tongue carcinomas. *Int. J. Cancer*, **36**, 575-578.

Heilmann, C.A., Law, M.-F., Israel, M.A. and Howley, P.M. (1980). Cloning of human papillomavirus genomic DNAs and analysis of homologous polynucleotide sequences. *J. Virol.*, **36**, 43-48.

Majewski, S., Jablonska, S., Favre, M. and Orth, G. (2001). Human papillomavirus type 7 and butcher´s warts. *Arch Dermatol.*, **137**, 1655-1656.

Lambert, P.F. (1991). Papillomavirus DNA replication. *J. Virol.*, **65**, 3417-3420.

Modis, Y., Trus, B.L. and Harrison, S.C. (2002). Atomic model of the papillomavirus capsid. *EMBO J.*, **21**, 4754-4762.

Myers, G., Baker, C., Munger, K., Sverdrup, F., McBride, A., Bernard, H.-U. and Meissner, J. (eds) (1997). *Human Papillomaviruses*. Theoretical Biology and Biophysics Group T-10, Mail Stop K710, Los Alamos National Laboratory, Los Alamos.

Myers, G., Lu, H., Calef, C. and Leitner, T. (1996). Heterogeneity of papillomaviruses. *Sem Cancer Biol.*, **7**, 349-358.

van Ranst, M., Kaplan, J.B. and Burk, R.D. (1992). Phylogenetic classification of human papillomaviruses: correlation with clinical manifestations. *J. Gen. Virol.*, **73**, 2653-2660.

zur Hausen, H. (2002). Papillomavirus and cancer: from basic studies to clinical application. *Nat. Rev. Cancer*, **2**, 342-350.

Contributed By

de Villiers, E.-M., Bernard, H.-U., Broker, T., Delius, H. and zur Hausen, H.

FAMILY POLYDNAVIRIDAE

TAXONOMIC STRUCTURE OF THE FAMILY

Family	***Polydnaviridae***
Genus	***Bracovirus***
Genus	***Ichnovirus***

VIRION PROPERTIES

MORPHOLOGY

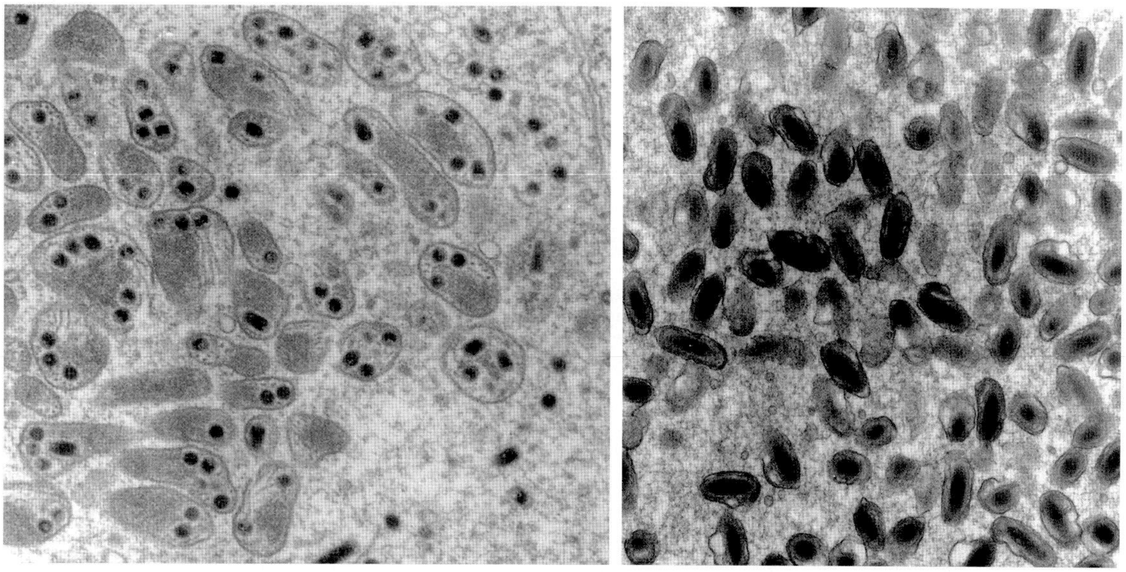

Figure 1: Bracovirus particles (Left) and ichnovirus particles (Right) illustrating morphological differences in members of the two polydnavirus genera. The bars represent 200 nm.

Polydnaviruses are enveloped DNA viruses but members of the *Ichnovirus* and *Bracovirus* genera have few other common morphological features. Morphological features relevant to each are described below in the appropriate genus description.

PHYSICOCHEMICAL AND PHYSICAL PROPERTIES

None reported.

NUCLEIC ACID

Genomes consist of multiple dsDNAs of variable size ranging from approximately 2.0 to more than 31 kbp (Fig. 2). Genome segments are present in non-equimolar amounts and there is evidence of co-migrating DNA segments. The aggregate, non-redundant, genome size of Campoletis sonorensis ichnovirus (CsIV), is 244.6 kbp while partial sequence of a representative genome, the Microplitis demolitor bracovirus (MdBV), indicates a non-redundant genome size of ~210 kbp. Estimates of polydnavirus genome size and complexity range from 150 kbp to in excess of 250 kbp. Genome size estimates are complicated by the presence of co-migrating segments and DNA sequences shared among two or more visible DNA genome segments.

PROTEINS

Virions are structurally complex and contain at least 20-30 polypeptides, with sizes ranging from 10-200 kDa.

LIPIDS

Lipids are present, but uncharacterized.

CARBOHYDRATES

Carbohydrates are present, but uncharacterized.

GENOME ORGANIZATION AND REPLICATION

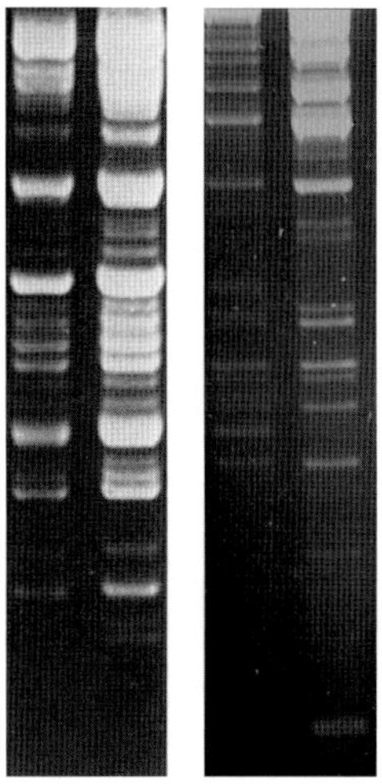

Figure 2: Two concentrations of undigested DNA genomes of members of the *Bracovirus* (Cotesia marginiventris bracovirus, CmaBV; right panel) and *Ichnovirus* (Campoletis sonorensis ichnovirus, CsIV; left panel) genera. DNA genomes were electrophoresed on a 1% agarose gel and visualized with ethidium bromide. The different amounts of DNA illustrate the non-equimolarity of polydnavirus DNA segments.

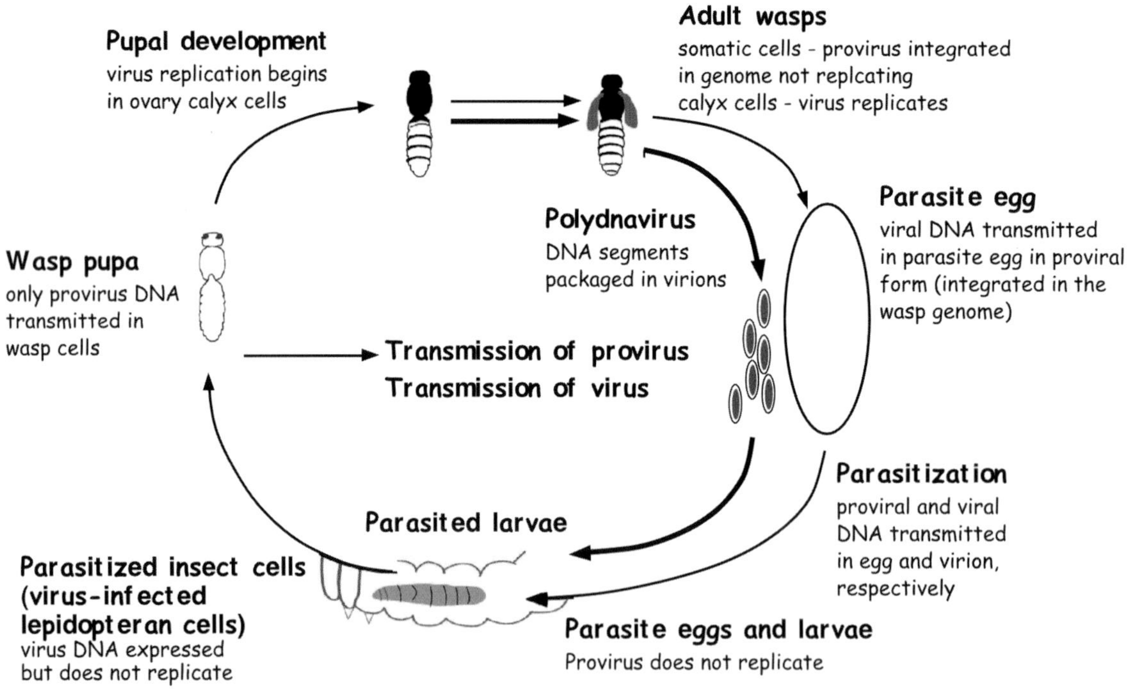

Figure 3: Polydnavirus replication and transmission cycles. The replication and transmission of polydnaviruses and the life cycle of an endoparasitic wasp are illustrated. Viral DNA is transmitted as proviral DNA in wasp cells (thin arrows) and as circular episomal DNAs within virions (thick arrows). In replicative wasp cells and in infected lepidopteran cells, viral DNA is present in an unpackaged closed circular form. In non-replicative cells (i.e. all wasp cells except female adult calyx cells) the virus does not replicate and exists predominantly in the proviral form. Polydnavirus are vertically transmitted only in the proviral DNA form (From Webb, 1998).

Unique among the dsDNA viruses, polydnaviruses have multiply segmented genomes (Fig. 2). Chromosomally integrated sequences homologous to viral DNAs are located at multiple sites within the genome of parasitic wasps (parasitoid); this proviral DNA form is responsible for the transmission of viral genomes within parasitoid species. The polydnavirus genome appears to be unusual in other respects as well. Some viral genes contain introns. Several viral gene families exist, members of which are distributed on one or more genome segments. Transcriptional activity is host-specific, in the sense that some viral genes are expressed in the wasp ovary, other genes only in the parasitized insect host and some are expressed in both hosts. Families of viral genome segments exist in both genera. Polydnavirus genomes are, at least potentially, genetically redundant (e.g., they would appear to be diploid). Polydnavirus replication is nuclear, begins during wasp pupal-adult development, and is very likely associated with changes in ecdysone titer that trigger metamorphosis of the insect. Viral DNA replication appears to involve amplification of the polydnavirus chromosomal loci and excision of viral DNA segments via site-specific recombination events. Virus morphogenesis occurs in the calyx cells of the ovaries of all female wasps belonging to all affected species. Ichnovirus particles bud directly from the calyx epithelial cells into the lumen of the oviduct while bracoviruses are released by cell lysis of calyx cells. In some species, extrachromosomal, circular DNAs are present both in male wasps and in non-ovarian female tissues but viral morphogenesis has not been demonstrated. Viral replication has not been detected in parasitized host insects.

Antigenic Properties

Cross-reacting antigenic determinants are shared by the members of a number of different *Ichnovirus* species; in some cases, viral nucleocapsids share at least one major conserved epitope. It has recently been shown that Campoletis sonorensis ichnovirus (CsIV) and *C. sonorensis* venom protein display common epitopes. Antigenic relationships among the bracoviruses are not well understood.

Biological Properties

Polydnaviruses have been isolated only from endoparasitic hymenopteran insects (wasps) belonging to the families Ichneumonidae and Brachonidae. In nature, polydnavirus genomes are transmitted vertically as proviruses and horizontally to lepidopteran insects where viral genes are expressed but the virus does not replicate and does not establish secondary infections of host cells. This defective replication has been termed genetic colonization and is similar, conceptually, to transfection of a replication-defective virus. Polydnavirus particles are injected into host animals during oviposition; virus-specific expression leads to significant changes in host physiology, some of which are required for successful parasitism of the host insect. Thus, the association between wasp and polydnavirus is often described as an obligate mutualistic symbiosis.

Genus *Bracovirus*

Type Species ***Cotesia melanoscela bracovirus***

Distinguishing Features

Bracoviruses are found only in certain species of braconid wasps that share a common evolutionary lineage. Bracovirus nucleocapsids are cylindrical, of variable length, and are surrounded by only a single unit membrane envelope. Bracoviruses encapsidate nucleocapsids either singly or multiply in a species-dependent manner.

VIRION PROPERTIES

MORPHOLOGY

Bracovirus virions consist of enveloped cylindrical electron-dense nucleocapsids of uniform diameter but of variable length (34-40 nm diameter by 8-150 nm length) and may contain one or more nucleocapsids within a single envelope; the latter appears to be assembled *de novo* within the nucleus. Bracovirus nucleocapsids in some cases possess long unipolar tail-like appendages.

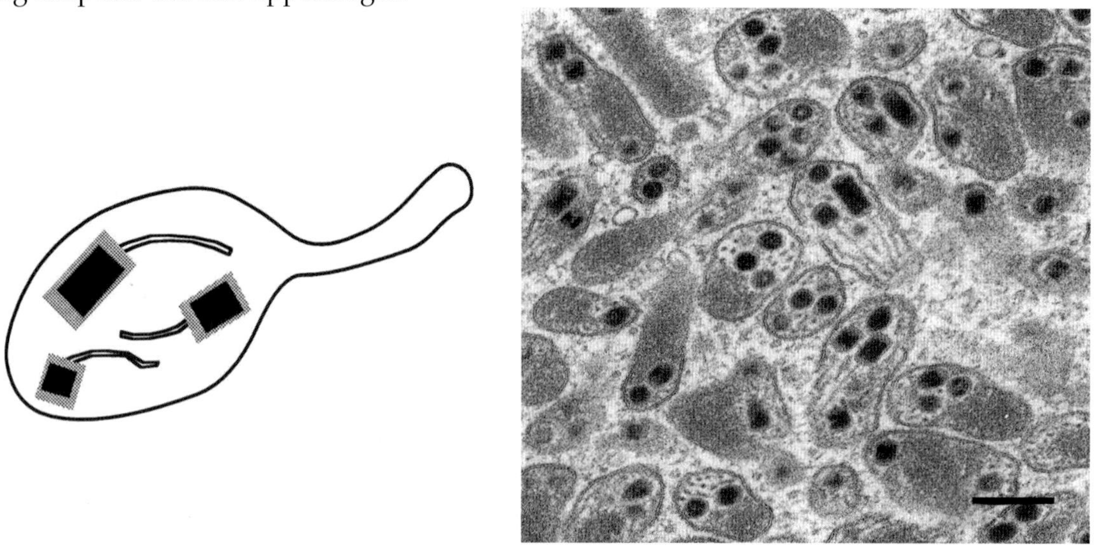

Figure 4: (Left) Sectional diagram and (Right) negative contrast electron micrograph of particles of Protapanteles paleacritae bracovirus (PpBV)(Courtesy of D. Stoltz). The bar represents 200 nm.

PHYSICOCHEMICAL AND PHYSICAL PROPERTIES

None reported.

NUCLEIC ACID

Genomes consist of multiple dsDNAs of variable size ranging from approximately 2.0 to more than 31 kbp. The smallest sequenced MdBV segment is ~3 kbp and the largest over 25 kbp. No aggregate size for a bracovirus genome has been determined by sequencing although estimates of the MdBV genome from an ongoing project is that its genome is approximately 210 kbp of unique sequence. Estimates of bracovirus genome size and complexity range from 150 kbp to in excess of 250 kbp but are complicated by the presence of related DNA sequences shared among two or more viral segments.

PROTEINS

Virions are structurally complex and contain at least 20-30 polypeptides, with sizes ranging from 10-200 kDa.

LIPIDS

Lipids are present, but uncharacterized.

CARBOHYDRATES

Carbohydrates are present, but uncharacterized.

GENOME ORGANIZATION AND REPLICATION

Bracovirus genomes exist as proviruses and are vertically transmitted in the proviral state. The proviral genome segments appear to be clustered in tandem arrays at a single locus on a wasp chromosome in Cotesia congregata bracovirus (CcBV) and Cotesia hyphantria bracovirus (ChiBV) and are likely to be similarly arranged in all bracoviruses. There is evidence that proviral DNA is amplified at the onset of replication with episomal segments then excised from the amplified cluster at repetitive sequences for packaging into virions. Some segments are hypermolar. Gene family members may be clustered on a single viral segment or distributed on multiple segments. Bracovirus DNAs appear to be largely non-coding with most, but possibly not all, viral segments encoding one or more transcribed genes. There is evidence of complex sequence homologies between segments that are partially, but not exclusively, associated with gene family members residing on different segments.

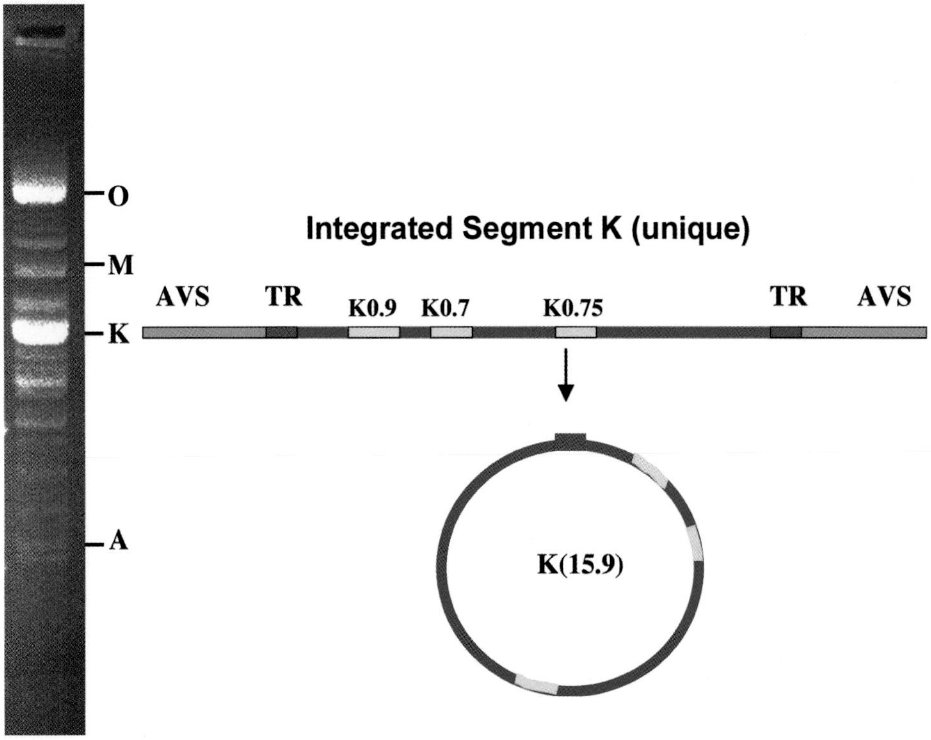

Figure 5: Bracovirus genome organization and replication. The Microplitis demolitor bracovirus (MdBV) genome is shown on left with selected viral segments identified. The right panel shows a representative bracovirus segment, MdBV segment K, in the proviral (top) and episomal form (bottom). Proviral segments may be found in tandem arrays with adjacent viral segments (AVS) flanking the integration site of proviral segments. MdBV segments encode genes but are predominantly non-coding sequence.

Antigenic Properties

Antigenic relationships among the bracoviruses have not been investigated in detail although cross-reacting epitopes have been detected.

Biological Properties

Bracoviruses have been isolated only from three braconid subfamilies of endoparasitic wasps within the microgastrine clade, the Microgastrinae, Cheloninae and Cardiochilinae. Polydnavirus particles are injected into host animals during oviposition; virus-specific expression leads to significant changes in host physiology, some of which are responsible for successful parasitism. Bracoviruses have been associated with inhibition of the immune responses of infected (parasitized) lepidopteran larvae and arrest of host development.

List of Species Demarcation Criteria in the Genus

To demonstrate experimentally that a parasitic wasp carries a bracovirus the following criteria should be met:

- Virions were isolated from reproductive tract of females of a braconid wasp species,

- Virions have bracovirus morphology having rod-shaped nucleocapsids of variable length, with a diameter of ~30 nm surrounded by a double unit membrane, singly or multiply encapsided in the virion,
- Nucleic acid isolated from virions is DNA from multiple molecules (i.e. the DNA genome is segmented),
- Reference specimens of the wasp host are identified to species by a qualified braconid systematic specialist and are deposited in an accessible insect collection.

Criteria that are thought to be of systematic significance:
- Lepidopteran host range of virus (the non-replicative host),
- DNA restriction map profiles.

LIST OF SPECIES IN THE GENUS

Species names are in green italic script; strain names and synonyms are in black roman script; tentative species names are in blue roman script. Sequence accession numbers, and assigned abbreviations ( ) are also listed.

SPECIES IN THE GENUS

Apanteles crassicornis bracovirus		
Apanteles crassicornis bracovirus		(AcBV)
Apanteles fumiferanae bracovirus		
Apanteles fumiferanae bracovirus		(AfBV)
Ascogaster argentifrons bracovirus		
Ascogaster argentifrons bracovirus		(AaBV)
Ascogaster quadridentata bracovirus		
Ascogaster quadridentata bracovirus		(AqBV)
Cardiochiles nigriceps bracovirus		
Cardiochiles nigriceps bracovirus	[Y19010, AJ440973]	(CnBV)
Chelonus altitudinis bracovirus		
Chelonus altitudinis bracovirus		(CalBV)
Chelonus blackburni bracovirus		
Chelonus blackburni bracovirus		(CbBV)
Chelonus inanitus bracovirus		
Chelonus inanitus bracovirus	[Z31378, Z58828, AJ319653-4, AJ278673-8]	(CinaBV)
Chelonus insularis bracovirus		
Chelonus insularis bracovirus		(CinsBV)
Chelonus nr. curvimaculatus bracovirus		
Chelonus nr. curvimaculatus bracovirus		(CcBV)
Chelonus texanus bracovirus		
Chelonus texanus bracovirus		(CtBV)
Cotesia congregata bracovirus		
Cotesia congregata bracovirus	[D29821, AF049876-7, AF006205-8]	(CcBV)
Cotesia flavipes bracovirus		
Cotesia flavipes bracovirus		(CfBV)
Cotesia glomerata bracovirus		
Cotesia glomerata bracovirus		(CgBV)
Cotesia hyphantriae bracovirus		
Cotesia hyphantriae bracovirus		(ChBV)
Cotesia kariyai bracovirus		
Cotesia kariyai bracovirus	[AB099714, AB086812-4, AB074136-7]	(CkBV)
Cotesia marginiventris bracovirus		
Cotesia marginiventris bracovirus		(CmaBV)
Cotesia melanoscela bracovirus		
Cotesia melanoscela bracovirus		(CmeBV)

Cotesia rubecula bracovirus		
Cotesia rubecula bracovirus	[AY234855, AF359344, U55279]	(CrBV)
Cotesia schaeferi bracovirus		
Cotesia schaeferi bracovirus		(CsBV)
Diolcogaster facetosa bracovirus		
Diolcogaster facetosa bracovirus		(DfBV)
Glyptapanteles flavicoxis bracovirus		
Glyptapanteles flavicoxis bracovirus		(GflBV)
Glyptapanteles indiensis bracovirus		
Glyptapanteles indiensis bracovirus	[AF453875, AF198385]	(GiBV)
Glyptapanteles liparidis bracovirus		
Glyptapanteles liparidis bracovirus		(GlBV)
Hypomicrogaster canadensis bracovirus		
Hypomicrogaster canadensis bracovirus		(HcBV)
Hypomicrogaster ectdytolophae bracovirus		
Hypomicrogaster ectdytolophae bracovirus		(HecBV)
Microplitis croceipes bracovirus		
Microplitis croceipes bracovirus		(McBV)
Microplitis demolitor bracovirus		
Microplitis demolitor bracovirus	[AF267174-5, AF241775, U76033-4]	(MdBV)
Phanerotoma flavitestacea bracovirus		
Phanerotoma flavitestacea bracovirus		(PfBV)
Pholetesor ornigis bracovirus		
Pholetesor ornigis bracovirus		(PoBV)
Protapanteles paleacritae bracovirus		
Protapanteles paleacritae bracovirus		(PpBV)
Tranosema rostrale bracovirus		
Tranosema rostrale bracovirus		(TrBV)

TENTATIVE SPECIES IN THE GENUS

None reported.

GENUS ICHNOVIRUS

Type Species ***Campoletis sonorensis ichnovirus***

DISTINGUISHING FEATURES

Ichnoviruses have been found only in the wasp family Ichneumonidae in the Campopleginae and Banchinae subfamilies. Ichnovirus nucleocapsids are fusiform in shape, and are enveloped by two unit membranes. Typically, virus particles each contain a single nucleocapsid. Viruses from the wasp genera Glypta and Dusona are the only known exceptions.

VIRION PROPERTIES

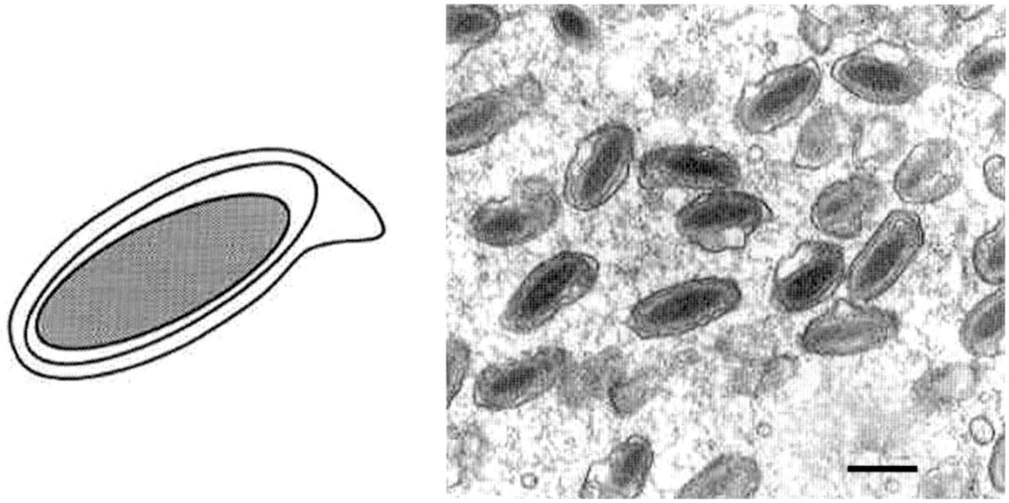

Figure 6: (Left) Sectional diagram and (Right) negative contrast electron micrograph of particles of Hyposoter exiguae ichnovirus (HeIV) (Courtesy of D. Stoltz). The bar represents 200 nm.

MORPHOLOGY

Ichnovirus virions consist of nucleocapsids of uniform size (approximately 85 x 330 nm), having the form of a prolate ellipsoid, surrounded by 2 unit membrane envelopes. The inner envelope appears to be assembled *de novo* within the nucleus of infected calyx cells, while the outer envelope is acquired by budding through the plasma membrane into the oviduct lumen.

PHYSICOCHEMICAL AND PHYSICAL PROPERTIES

None reported.

NUCLEIC ACID

Genomes consist of multiple dsDNAs of variable size ranging from approximately 2.0 to more than 20 kbp. The Campoletis sonorensis ichnovirus (CsIV) genome is 248 kbp long. Estimates of genome size and complexity in other ichnoviruses range from 150 kbp to in excess of 250 kbp but are complicated by the presence of related DNA sequences shared among two or more DNA genome segments.

PROTEINS

Virions are structurally complex and contain at least 20-30 polypeptides, with sizes ranging from 10-200 kDa.

LIPIDS

Lipids are present but uncharacterized.

CARBOHYDRATES

Carbohydrates are present but uncharacterized.

Genome Organization and Replication

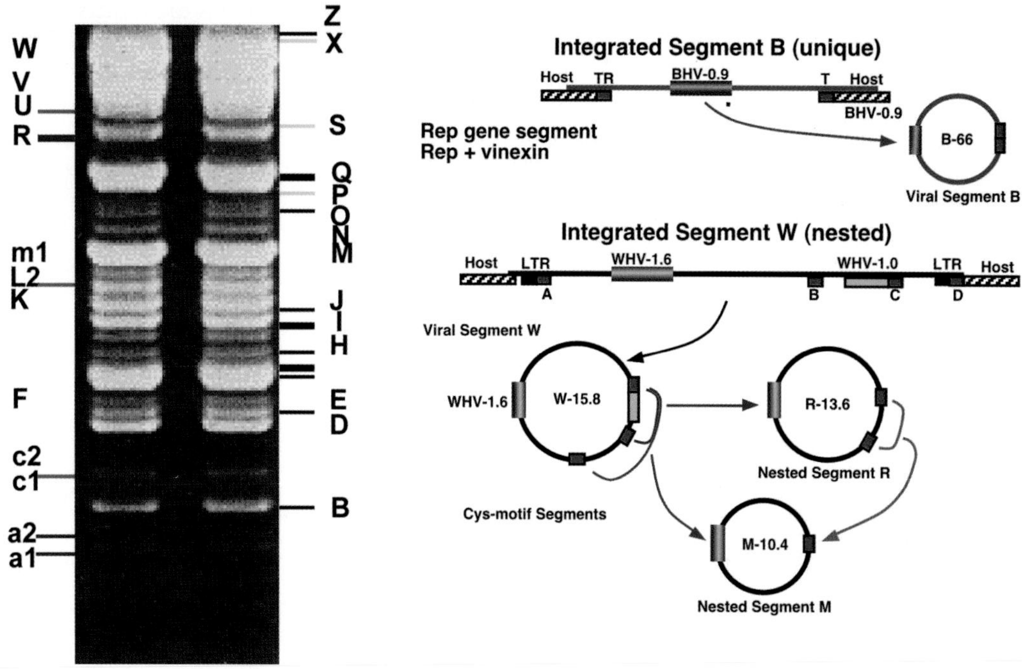

Figure 7: Undigested Campoletis sonorensis ichnovirus (CsIV) genome (Left panel) with unique segment labels on right and nested segment labels on left. Colors indicate genes known to be associated with each segment. On left all segments encode cys-motif genes. Segments indicated in red are derived from larger segments (nested). Labels on right in this panel indicate unique segments that encode rep genes (black) and 'non-rep' genes (e.g. P has similarity to NF-κβ genes). Right panel illustrates unique (Top) and nested segments (Bottom) in their proviral (integrated-linear) and episomal forms (circular).

Ichnoviruses replicate from proviral DNA. Ichnovirus proviral segments are more dispersed than bracoviruses being separated by sequence that is not encapsidated within the virion and possibly dispersed widely in the genome. Encapsidated ichnovirus genomes are non-equimolar with two recognized viral segment types, nested and unique,. Nested segments are often hypermolar and produce from 2 to 5 partially redundant segments from a single proviral locus. Nested segments may have longer and more highly conserved terminal repeats than unique segments. Unique segments excise to produce a single segment that is encapsidated. There is evidence that nested segments are preferentially associated with some gene families.

Antigenic Properties

Cross-reacting antigenic determinants are shared by a number of different ichnovirus isolates; in some cases, viral nucleocapsids share at least one major conserved epitope. CsIV and *C. sonorensis* venom protein display common epitopes.

Biological Properties

Ichnoviruses have been isolated only from two ichneumonid subfamilies of endoparasitic wasps within the Ichneumonidae clade, the Campopleginae and the Banchinae. Polydnavirus particles are injected into host animals during oviposition; virus-specific expression leads to significant changes in host physiology, some of which are responsible for successful parasitism. Ichnoviruses have been associated with inhibition of the immune responses of infected (parasitized) lepidopteran larvae and arrest of host development. There is evidence that virus infection impacts translation of some host mRNAs and this impacts the ability of the host to mount effective immune responses (i.e. melanization).

List of Species Demarcation Criteria in the Genus

To demonstrate experimentally that a parasitic wasp carries an ichnovirus the following criteria should be met:

- Virions isolated from oviduct of females of an ichneumonid wasp species,
- Virions have ichnovirus morphology having a single lenticular or fusiform nucleocapsid, a double unit membrane, and size of ~300 x 60 nm,
- Nucleic acid isolated from virions is DNA from multiple molecules (i.e. DNA genome is segmented),
- Reference specimens of the wasp host are identified to species by a qualified ichneumonid systematic specialist and are deposited in an accessible insect collection.

Criteria that are thought to be of systematic significance:

- Lepidopteran host range of virus (the non-replicative host),
- DNA restriction map profiles.

List of Species in the Genus

Species names are in green italic script; strain names and synonyms are in black roman script; tentative species names are in blue roman script. Sequence accession numbers, and assigned abbreviations ( ) are also listed.

Species in the Genus

Campoletis aprilis ichnovirus		
Campoletis aprilis ichnovirus		(CaIV)
Campoletis falvincta ichnovirus		
Campoletis flavicincta ichnovirus		(CfIV)
Campoletis sonorensis ichnovirus		
Campoletis sonorensis ichnovirus	[L08243-5, M17406, M80621-3, S47226, U41655-6, AF004366-7, AF004378, AF004557-8, AH006861, AY137756]	(CsIV)
Casinaria arjuna ichnovirus		
Casinaria arjuna ichnovirus		(CarIV)
Casinaria forcipata ichnovirus		
Casinaria forcipata ichnovirus		(CfoIV)
Casinaria infesta ichnovirus		
Casinaria infesta ichnovirus		(CiIV)
Diadegma acronyctae ichnovirus		
Diadegma acronyctae ichnovirus		(DaIV)
Diadegma interruptum ichnovirus		
Diadegma interruptum ichnovirus		(DiIV)
Diadegma terebrans ichnovirus		
Diadegma terebrans ichnovirus		(DtIV)
Eriborus terebrans ichnovirus		
Eriborus terebrans ichnovirus		(EtIV)
Enytus montanus ichnovirus		
Enytus montanus ichnovirus		(EmIV)
Glypta fumiferanae ichnovirus		
Glypta fumiferanae ichnovirus		(GfIV)
Hyposoter annulipes ichnovirus		
Hyposoter annulipes ichnovirus		(HaIV)
Hyposoter exiguae ichnovirus		
Hyposoter exiguae ichnovirus		(HeIV)
Hyposoter fugitivus ichnovirus		
Hyposoter fugitivus ichnovirus		(HfIV)

Hyposoter lymantriae ichnovirus		
Hyposoter lymantriae ichnovirus		(HlIV)
Hyposoter pilosulus ichnovirus		
Hyposoter pilosulus ichnovirus		(HpIV)
Hyposoter rivalis ichnovirus		
Hyposoter rivalis ichnovirus		(HrIV)
Olesicampe benefactor ichnovirus		
Olesicampe benefactor ichnovirus		(ObIV)
Olesicampe geniculatae ichnovirus		
Olesicampe geniculatae ichnovirus		(OgIV)
Synetaeris tenuifemur ichnovirus		
Synetaeris tenuifemur ichnovirus		(StIV)

TENTATIVE SPECIES IN THE GENUS

Campoletis sp. ichnovirus		(CspIV)
Casinaria sp. ichnovirus		(CaspIV)
Dusona sp. ichnovirus		(DspIV)
Glypta sp. ichnovirus		(GspIV)
Lissonota sp. ichnovirus		(LspIV)
Tranosema rostrales ichnovirus	[AF421353, AF217758]	(TrIV)

LIST OF UNASSIGNED VIRUSES IN THE FAMILY

None reported.

PHYLOGENETIC RELATIONSHIPS WITHIN THE FAMILY

The members of the genera *Bracovirus* and *Ichnovirus* are unrelated by serological and hybridization analyses. Nucleic acid homology between the two polydnavirus genera has not been detected even though representative bracovirus and ichnovirus genome sequences are complete or nearly so. Within the genera hybridization between closely related viral species is evident and genes are conserved although segment profiles and the strength of hybridization signals suggest that considerable divergence has occurred within these genera. As suggested by their dissimilar morphologies it is now thought that the bracoviruses and ichnoviruses do not share a common ancestor. Because of similarities in their genome structure and life cycle, ichnoviruses and bracoviruses have been classified within the family *Polydnaviridae*. However, it is recognized that these similarities may well be the result of convergent evolution.

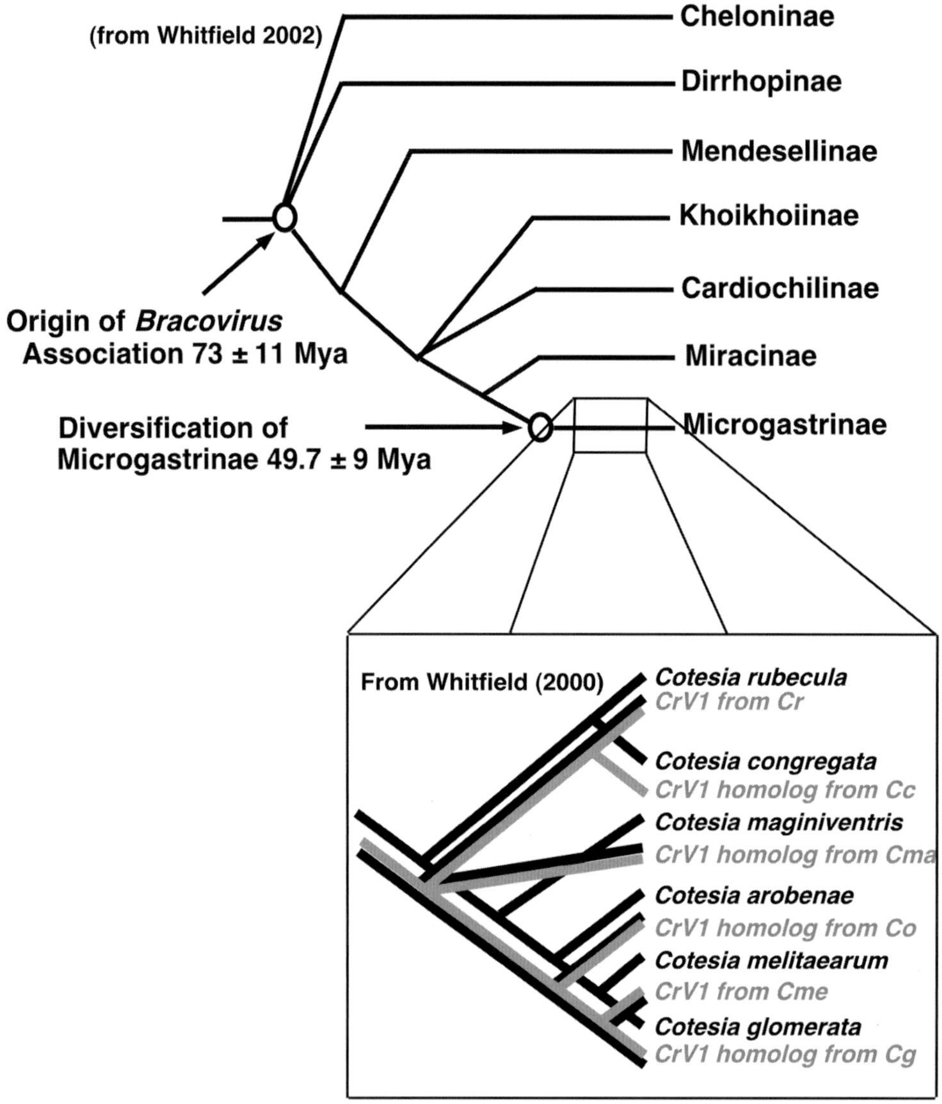

Figure 8: Bracovirus phylogenetic tree illustrating associations with braconid wasp clades and correlative phylogenies of the wasps and viruses in this evolutionary lineage (From Whitfield 2000).

Similarity with Other Taxa

Occasionally, very long bracovirus nucleocapsids are observed; at least superficially, these resemble baculovirus nucleocapsids. Ichnoviruses resemble no other known type of virus.

Derivation of Names

Braco: from *Brachonida*e, a family of wasps.
Ichno: from *Ichneumonida*e, a family of wasps.
*Polydn*a: from *poly* (meaning several) and DNA.

References

Albrecht, A., Wyler, T., Pfister-Wilhelm, R., Heiniger, P., Hurt, E., Gruber, A., Schumperli, D. and Lanzrein, B. (1994). Polydnavirus of the parasitic wasp *Chelonus inanitus (Braconidae)*: characterization, genome organization and time point of replication. *J. Gen. Virol.*, 75, 3353-3363.

Belle, E., Beckage, N.E., Rousselet, J., Poirié M., Lemeunier, F. and Drezen J.M. (2002). Visualization of polydnavirus sequences in a parasitoid wasp chromosome. *J. Virol.*, **76**, 5793-5796.

Blissard, G.W., Fleming, J.G.W., Vinson, S.B. and Summers, M.D. (1986). Campoletis sonorensis virus: expression in *Heliothis virescens* and identification of expressed sequences. *J. Insect Physiol.*, **32**, 352-359.

De Buron, I. and Beckage, N.E. (1992). Characterization of a polydnavirus (PDV) and virus-like filamentous particle (VLFP) in the braconid wasp *Cotesia congregata (Hymenoptera: Braconidae). J. Invert Path.*, **59**, 315-327.

Cui, L. and Webb, B.A. (1997). Homologous sequences in the *Campoletis sonorensis* polydnavirus genome are implicated in replication and nesting of the W segment family. *J. Virol.*, **71**, 8504-8513.

Dib-Hajj, S.D., Webb, B.A. and Summers, M.D. (1993). Structure and evolutionary implications of a "cysteine-rich" *Campoletis sonorensis* polydnavirus gene family. *Proc. Natl. Acad. Sci. USA*, **90**, 3765-3769.

Fleming, J.G.W. and Summers, M.D. (1986). *Campoletis sonorensis* endoparasitic wasps contain forms of *C. sonorensis virus* DNA suggestive of integrated and extrachromosomal polydnavirus DNAs. *J. Virol.*, **57**, 552-562.

Fleming, J.G.W. and Summers, M.D. (1991). Polydnavirus DNA is integrated in the DNA of its parasitoid wasp host. *Proc. Natl. Acad. Sci. USA*, **88**, 9770-9774.

Fleming, J.G.W. (1992). Polydnaviruses: mutualists and pathogens. *Ann. Rev. Entomol.*, **37**:401-425.

Marti, D., Grossniklaus-Bürgin, C., Wyder, S., Wyler, T. and Lanzrein, B. (2002). Ovary development and polydnavirus morphogenesis in the parasitic wasp Chelonus inanitus, part I: ovary morphogenesis, amplification of viral DNA and ecdysteroid titres. *J. Gen. Virol.*, **84**, 1141-1150.

Shelby, K.S and Webb, B.A. (1999). Polydnavirus-mediated suppression of insect immunity. *J. Insect Physiol.*, **45**, 507-514.

Stoltz, D.B. and Whitfield, J.B. (1992). Viruses and virus-like entities in the parasite Hymenoptera. *J. Hymeno. Res.*, **1**, 125-139.

Strand, M.R. and Pech, L.L. (1995). Immunological basis for compatibility in parasitoid-host relationships. *Ann. Rev. Entomol.*, **40**, 31-56.

Turnbull, M.W. and Webb B.A. (2002). Perspectives on polydnavirus origins and evolution. *Adv. Virus. Res.*, **58**, 203-254.

Webb, B.A. (1998). Polydnavirus biology genome structure and evolution. In: *The Insect Viruses* (L.K. Miller and L. A. Ball, eds), pp 105-139. Plenum Press, New York.

Whitfield, J.B. (2000). Phylogeny of microgastroid braconid wasps, and what it tells us about polydnavirus evolution. In: *The Hymenoptera: Evolution, Biodiversity and Biological Control.* (A.D. Austin & M. Dowton, eds), pp. 97-105. CSIRO Publishing, Melbourne.

Wyder, S., Tschannen, A., Hochuli, A., Gruber, A., Saladin, V., Zumbach, S. and Lanzrein, B. (2002). Characterization of Chelonus inanitus polydnavirus segments: sequences and analysis and demonstration of clustering. *J. Gen. Virol.*, **83**, 247-256.

Wyler, T. and Lanzrein, B. (2003). Ovary development and polydnavirus morphogenesis in the parasitic wasp *Chelonus inanitus*, part II: Ultrastructural analysis of calyx cell development, virion formation and release. *J. Gen. Virol.*, **84**, 1151-1163.

CONTRIBUTED BY

Webb, B.A., Beckage, N.E., Hayakwa, Y., Lanzrein, B., Stoltz, D.B., Strand, M.R. and Summers, M.D.

FAMILY ASCOVIRIDAE

TAXONOMIC STRUCTURE OF THE FAMILY

Family *Ascoviridae*
Genus *Ascovirus*

Since only one genus is currently recognized, the family description corresponds to the genus description

GENUS ASCOVIRUS

Type Species *Spodoptera frugiperda ascovirus 1a*

VIRION PROPERTIES

MORPHOLOGY

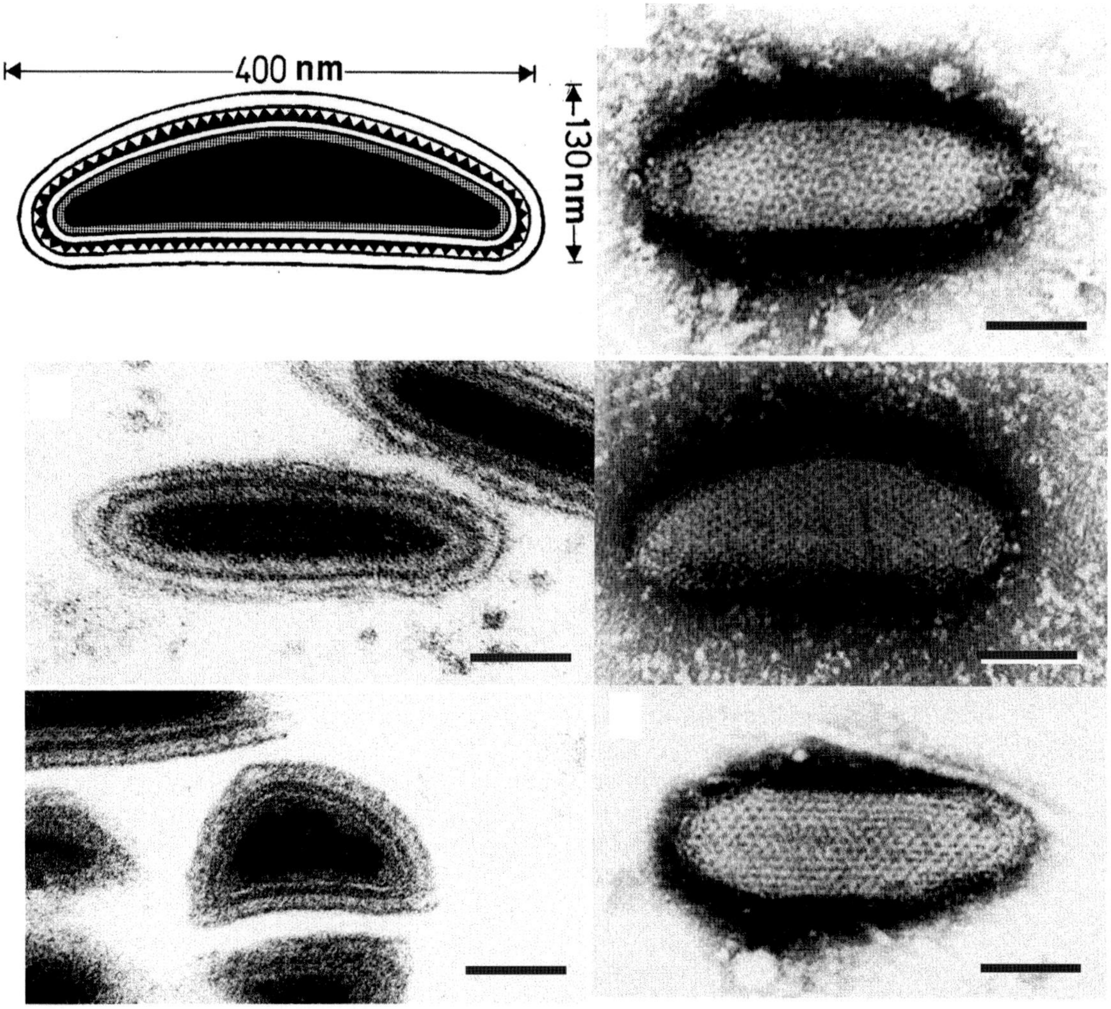

Figure 1: Morphology of ascovirus virions. (Top left) Schematic illustration of the structure of a typical ascovirus virion. The virion consists of an inner particle and an outer envelope. The inner particle is complex and contains a DNA/protein core surrounded by an apparent unit membrane, the external surface of which bears a layer of distinctive protein subunits. (Center left, Bottom left) Respectively, ultrathin longitudinal- and cross-sections through typical ascovirus virions. The dense inner layer corresponds with the distinctive layer of subunits shown in Top left. (Right) Negatively stained preparations of virions from three different ascovirus species: Top right, Spodoptera frugiperda ascovirus 1a (SfAV-1a); Center right, Trichoplusia ni ascovirus 2a (TnAV-2a), and Bottom right, Heliothis virescens ascovirus 3a (HvAV-3a). The reticulate appearance of the

virions is thought to be due to the superimposition of top and bottom layers of the inner particle and outer envelope.

Virions of ascoviruses are either bacilliform, ovoidal or allantoid in shape, depending on the species, have complex symmetry, and are large, measuring about 130 nm in diameter by 200-400 nm in length. The virion consists of an inner particle surrounded by an outer envelope. The inner particle typically measures 80 x 300 nm and contains a DNA/protein core bounded by an apparent lipid bilayer, the external surface of which bears a distinctive layer of protein subunits. The virion, therefore, appears to contain two lipid membranes, one associated with the inner particle, the other forming the envelope. In negatively stained preparations, virions have a distinctive reticulate appearance thought to result from superimposition of the protein subunits on the surface of the internal particle with those in the external envelope (Fig. 1).

PHYSICOCHEMICAL AND PHYSICAL PROPERTIES

Virions are sensitive to organic solvents and detergents. Other properties are not known.

NUCLEIC ACID

The inner particle contains a single molecule of circular dsDNA ranging in size from 120 to 180 kbp. G + C ratio ranges from 42% to 60% depending on the species.

PROTEINS

Virions contain at least 15 polypeptides ranging in size from 6 to 200 kDa. Genes for the following proteins have been cloned and sequenced: Spodoptera frugiperda ascovirus 1a (SfAV-1a): δ DNA polymerase (AJ279830) and major CP (AJ312690). Diadromus pulchellus ascovirus 4a (DpAV-4a): δ DNA polymerase (AJ279812) and major CP (AJ312705).

LIPIDS

Ultrastructural evidence and detergent sensitivity indicate the presence of lipid in both the outer envelope and inner particle of the virion. Specific lipid composition is unknown.

CARBOHYDRATES

Unknown.

GENOME ORGANIZATION AND REPLICATION

The genomes of ascoviruses are composed of circular dsDNA. Sequences of the genomes of two ascoviruses, SfAV-1a and DpAV-4a, are nearly complete and genomic organization should be soon available.

Ascoviruses initiate replication in the nucleus. The nucleus enlarges and ruptures, after which the plasmalemma invaginates forming internal membranous folds that cleave the cell into a cluster of virion-containing vesicles. Virion assembly becomes apparent after the nucleus ruptures. The first recognizable structural component of the virion to form is the multilaminar layer of the inner particle. Based on its ultrastructure, this layer consists of a unit membrane and an exterior layer of protein subunits. As the multilaminar layer forms, the dense DNA/protein core assembles along the inner surface. This process continues and the allantoid, ovoidal or bacilliform shape of the inner particle becomes apparent. After the inner particle is assembled, it is enveloped by a membrane, synthesized *de novo* or elaborated from cell membranes, within the cell or vesicle. In SfAV-1a, the virions are occluded in an occlusion body composed of minivesicles and protein.

ANTIGENIC PROPERTIES

Unknown.

Biological Properties

Host Range

Ascoviruses cause disease in lepidopterous larvae, and have been reported most commonly from species of the family Noctuidae, including *Trichoplusia ni, Heliothis virescens, Helicoverpa zea, Spodoptera frugiperda,* and *Autographa precationis.* An ascovirus disease has also been reported in the lepidopteran *Acrolepiopsis assectella* (Family Yponomeutidae). Trichoplusia ni ascovirus 2a (TnAV-2a) and Heliothis virescens ascovirus 3a (HvAV-3a) have been shown to have a broad experimental host range among larvae of the lepidopteran family Noctuidae, but the host range of SfAV-1a is restricted primarily to species of *Spodoptera.* DpAV-4a is restricted to species of the lepidopteran family Yponomeutidae, in which it replicates extensively; the virus also replicates in its ichneumonid vector, *D. pulchelles,* but replication is limited and relatively few particles are produced in comparison to the number produced in the lepidopteran host.

Transmission

Ascoviruses are difficult to transmit *per os*, and experimental studies as well as field observations indicate most are transmitted horizontally by endoparasitic wasps (Hymenoptera), many species of which belong to the families Braconidae and Ichneumonidae. The ovipositor of female wasps becomes contaminated during egg-laying with virions circulating in the blood of infected caterpillars. Wasps contaminated in this manner subsequently transmit ascovirus virions to new caterpillar hosts during oviposition. The virus that causes disease in *A. assectella* is transmitted vertically by the ichneumonid wasp, *Diadromus pulchellus,* in which the DpAV-4a genome is carried as unintegrated DNA in wasp nuclei.

Geographical Distribution

Ascoviruses are known from the United States, Europe, Australia, Indonesia, and Mexico, and likely occur worldwide, that is, wherever species of Lepidoptera and their hymenopeteran parasities occur. Ascoviruses cause a chronic, fatal disease that greatly retards larval development, but which typically exhibits little other gross pathology. This lack of easily recognizable gross pathology probably accounts for the lack of host records from many geographical regions.

Cytopathic Effects

Ascoviruses vary in tissue tropism with some attacking most host tissues, such as TnAV-2a, HvAV-3a, and DpAV-4a, whereas others, such as the type species SfAV-1a, are restricted to the fat body. Their most unique property is a novel cytopathology in which host cells cleave forming virion-containing vesicles by a developmental process resembling apoptosis. Infection results in nuclear hypertrophy followed by lysis. The anucleate cell enlarges 5-10 fold, and then cleaves into 10-30 virion-containing vesicles. Membranes delimiting vesicles form by invagination of the plasmalemma and membrane synthesis. Millions of vesicles accumulate in the blood, turning it milky white. Opaque white hemolymph containing refractile virion vesicles is diagnostic for ascovirus disease.

List of Species Demarcation Criteria in the Genus

The following list of characters is used in combination to differentiate species in the genus.

- Virion morphology
- Presence or absence of occlusion bodies
- Lack of DNA/DNA hybridization with other species at low stringency
- Restriction enzyme fragment length polymorphisms (RFLPs)
- Host of isolation and experimental host range
- Tissue tropism
- Association with specific hymenopteran parasites, if apparent

Ascoviruses can have broad host ranges among the larvae of lepidopteran species, and the fat body tissue is a major site of replication for most species. In addition, the virions of most isolates are similar in size and shape. The above characters are therefore used in combination to distinguish existing and new ascovirus species from one another. Hybridization studies have proven particularly useful, and when combined with RFLPs can also be used to distinguish variants within a species.

For example, SfAV-1a DNA does not hybridize with HvAV-3a, TnAV-2a, or DpAV-4a DNAs under conditions of low stringency, nor does DpAV-4a hybridize with the DNA of the other species. TnAV-2a DNA does hybridize to some extent with HvAV-3a DNA, but not as strongly as it does with homologous DNA. In addition, the TnAV-2a replicates in a range of larval tissues including the fat body, tracheal matrix and epidermis, but SfAV-1a and HvAV-3a appear to replicate, respectively, only or primarily in the fat body tissue of most hosts. DpAV-4a replicates primarily in the pupal stage, where the primary tissues attacked are the fat body and midgut. SfAV-1a virions are bacilliform and are occluded in vesiculate occlusion bodies, whereas TnAV-2a virions are allantoid and are not occluded in occlusion bodies. HvAV-3a virions vary from allantoid to bacilliform, and are not occluded in occlusion bodies. The DpAV genome is carried as free circular DNA in nuclei of its wasp host, *Diadromus pulchellus*, and is transmitted vertically to wasp progeny. However, vertical transmission in wasp vectors is not known to occur with the other ascoviruses.

When the genome of a new isolate cross-hybridizes with that of an existing species member, RFLPs can be used to distinguish variants. Numerous ascovirus isolates, for example, have been obtained from larvae of different noctuid species, including *Heliothis virescens*, *Helicoverpa zea*, *Autographa precationis* and *Spodoptera exigua* in the U.S., as well as from *Helicoverpa* and *Spodoptera* species in Australia and Indonesia. The DNA of many of these isolates shows strong reciprocal hybridization with HvAV-3a DNA under conditions of high stringency. RFLP profiles of these isolates, however, often show variations from HvAv-3a that range from minor to major. Because these isolates cross-hybridize strongly with HvAV-3a, they are considered variants of this viral species. Moreover, experimentally these isolates have been shown to have host ranges that overlap with HvAV-3a, providing additional evidence that they are variants of the same species. A similar situation occurs with isolates of TnAv-2a.

List of Species in the Genus

Species in the Genus

Diadromus pulchellus ascovirus 4a		
Diadromus pulchellus ascovirus 4a	[AJ279812, AJ312705, AJ312706]	(DpAV-4a)
Heliothis virescens ascovirus 3a		
Heliothis virescens ascovirus 3a		(HvAV-3a)
Heliothis virescens ascovirus 3b	[AJ312697]	(HvAV-3b)
Heliothis virescens ascovirus 3c	[AJ312696, AJ312704, AJ312698]	(HvAV-3c)
Spodoptera frugiperda ascovirus 1a		
Spodoptera frugiperda ascovirus 1a	[AJ279830, AJ312690, AJ312695]	(SfAV-1a)
Trichoplusia ni ascovirus 2a		
Trichoplusia ni ascovirus 2a	[AJ312707]	(TnAV-2a)

Tentative Species in the Genus

Helicoverpa armigera ascovirus 7a	(HaAV-7a)
Helicoverpa punctigera ascovirus 8a	(HpAV-8a)
Spodoptera exigua ascovirus 5a	(SeAV-5a)
Spodoptera exigua ascovirus 6a	(SeAV-6a)

PHYLOGENETIC RELATIONSHIPS WITHIN THE FAMILY

Recent studies of ascovirus DNA polymerase and major CP genes indicate two main groups of ascoviruses within the genus *Ascovirus*. The first group contains a single species, *Diadromus pulchellus ascovirus 4a*, a vertically transmitted ascovirus, whereas the second group consists of the three ascoviruses that are mechanically transmitted, *Spodoptera frugiperda ascovirus 1a*, *Trichoplusia ni ascovirus 2a*, and *Heliothis virescens ascovirus 3a*.

SIMILARITY WITH OTHER TAXA

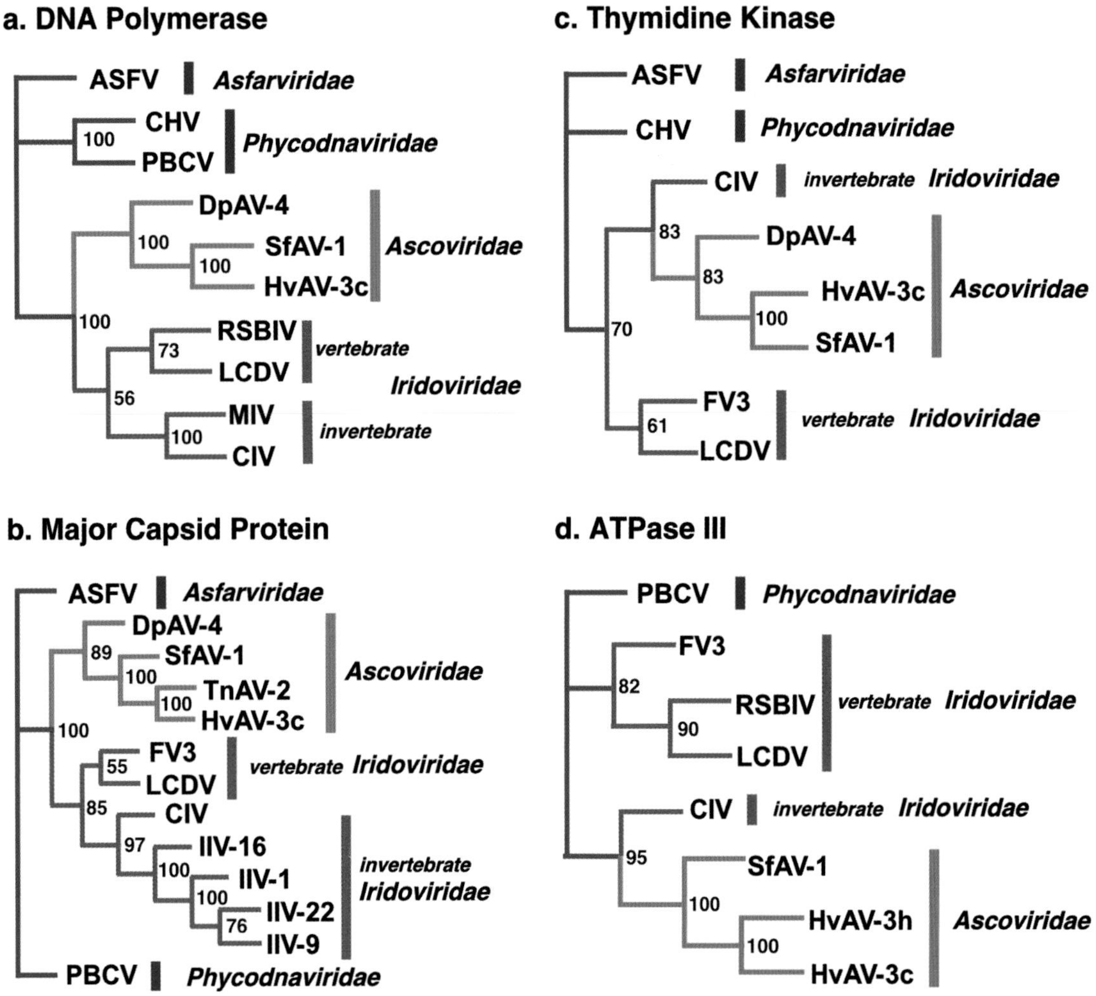

Figure 2: Consensus trees resulting from phylogenetic analyses of four ascoviruses and iridoviruses proteins. a) DNA polymerase; b) Major capsid protein; c) Thymidine kinase; and d) ATPase III. The sequences are from one asfarvirus, African swine fever virus (ASFV), two phycodnaviruses, Chorella virus (CHV) and Paramecium bursaria Chlorella virus (PBCV-1), nine iridoviruses, Invertebrate iridescent virus 1 (IIV-1), IIV-9, IIV-16, IIV-22, Chilo iridescent virus (CIV), Frog virus 3 (FV3), Lymphocystis disease virus (LCDV), Mosquito iridescent virus (MIV) and Red Sea bream iridescent virus (RSBIV), and four ascoviruses, Spodoptera frugiperda ascovirus 1a (SfAV-1a), Trichoplusia ni ascovirus 2a (TnAV-2a), Heliothis virescens ascovirus 3b and 3c (HvAV-3b and HvAV-3c), and Diadromus pulchellus ascovirus 4a (DpAV-4a). Consensus trees were rooted with the same proteins of African swine fever virus (ASFV, *Asfarviridae*) in a) b) and c), or Chlorella virus (CHV, *Phycodnaviridae*) in d). The trees were obtained using the parsimony procedure. Numbers given at each node correspond to the percent of bootstrap values (for 1000 repetitions).

The virions of ascoviruses resemble the particles produced by ichnoviruses of the viral family *Polydnaviridae*. Though the virions of members of the family *Ascoviridae* differ considerably from viruses of the family *Iridoviridae*, evidence is mounting that the ascoviruses and iridoviruses shared a common ancestor. In fact, phylogenetic analyses of several major proteins found in most enveloped dsDNA viruses provide strong evidence that ascoviruses evolved from iridoviruses (Fig. 2), despite the marked differences in the morphology of the virions characteristic of these two families, and differences in their cytopathology. The modifications in virion morphology may have evolved as a result of the transmission of the virions on the ovipositors of parasitic wasps. Similarities in virion structure may also indicate that ichnoviruses evolved from ascoviruses.

DERIVATION OF NAMES

Asco: From the Greek for "*Sac*;" referring to the virion-containing vesicles produced by cleavage of host cells, which are characteristic for all known viruses of this family.

REFERENCES

Bigot Y., Rabouille A., Doury G., Sizaret., P.Y., Delbost F., Hamelin M.H. and Periquet, G. (1997). Biological and molecular features of the relationship between Diadromus pulchellus ascovirus, a parasitoid hymenopteran wasp (*Diadromus pulchellus*) and its lepidopteran host, *Acrolepiopsis assectella*. *J. Gen. Virol.*, **78**, 1139-1147.

Bigot Y., Rabouille, A., Sizaret, P.Y., Hamelin, M.H. and Periquet, G. (1997). Particle and genomic characteristics of a new member of the *Ascoviridae*: Diadromus pulchellus ascovirus. *J. Gen. Virol.*, **78**, 1139-1147.

Cheng, X.-W., Carner, G.R. and Arif, B.M. (2000). A new ascovirus from *Spodoptera exigua* and its relatedness to an isolate from *Spodoptera frugiperda*. *J. Gen. Virol.*, **81**, 3083-3092.

Federici, B.A. (1983). Enveloped double-stranded DNA insect virus with novel structure and cytopathology. *Proc. Natl. Acad. Sci.* USA, **80**, 7664-7668.

Federici, B.A. and Bigot, Y. (2003). Origin and evolution of polydnaviruses by symbiogenesis of insect DNA viruses in endoparasitic wasps. *J. Insect Physiol.*, **49**, 419-432.

Federici, B.A. and Govindarajan, R. (1990). Comparative histopathology of three ascovirus isolates in larval noctuids. *J. Invert. Pathol.*, **56**, 300-311

Federici, B.A., Vlak, J.M. and Hamm, J.J. (1990). Comparison of virion structure, protein composition, and genomic DNA of three Ascovirus isolates. *J. Gen. Virol.*, **71**, 1661-1668.

Govindarajan, R. and Federici, B.A. (1990). Ascovirus infectivity and the effects of infection on the growth and development of Noctuid larvae. *J. Invert. Pathol.*, **56**, 291-299.

Hamm, J.J., Nordlung, D.A. and Marti, O.G. (1985). Effects of a nonoccluded virus of *Spodoptera frugiperda* (Lepidoptera: Noctuidae) on the development of a parasitoid, *Cotesia marginiventris* (Hymenoptera: Braconidae). *Environ. Entomol.*, **14**, 258-261

Hamm, J.J., Pair, S.D., and Marti, O.G. (1986). Incidence and host range of a new ascovirus isolated from fall armyworm, *Spodoptera frugiperda* (Lepidoptera: Noctuidae). *Fla. Entomol.*, **69**, 525-531.

Hamm, J.J., Styer, E.L. and Federici, B.A. (1998). Comparison of field-collected ascovirus isolates by DNA hybridization, host range, and histopathology. *J. Invert. Pathol.*, **72**, 138-146.

Pellock, B.J., Lu, A., Meagher, R.B., Weise, M.J. and Miller, L.K. (1996). Sequence, function and molecular analysis of an ascovirus DNA polymerase gene. *Virology*, **216**, 146-157.

Stasiak, K., Demattei, M.-V., Federici, B.A. and Bigot, Y. (2000). Phylogenetic position of the DpAV-4a ascovirus DNA polymerase among viruses with a large double-stranded DNA genome. *J. Gen. Virol.*, **81**, 3059-3072.

CONTRIBUTED BY

Federici, B.A., Bigot, Y., Granados, R.R., Hamm, J.J., Miller, L.K., Newton, I., Stasiak, K. and Vlak, J.M.

GENUS MIMIVIRUS

Type Species *Acanthamoeba polyphaga mimivirus*

VIRION PROPERTIES

MORPHOLOGY

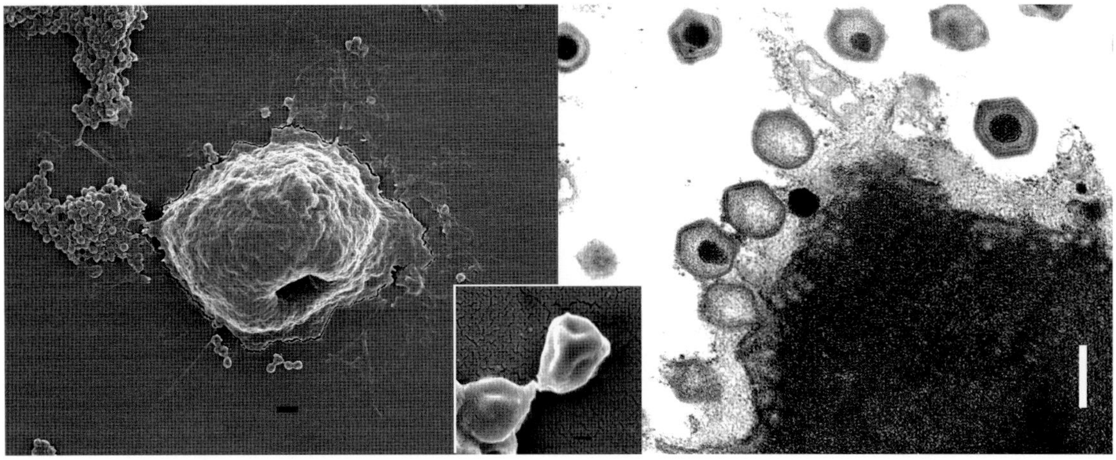

Figure 1: (Left) Clusters of particles of Acanthamoeba polyphaga mimivirus (APMV) close to an *Acanthamoeba polyphaga* amoeba. The bar represents 1 µm. (Right) Formation of particles of an isolate of *Acanthamoeba polyphaga mimivirus* at the periphery of an *Acanthamoeba polyphaga* nucleus. Notice the presence of apparently empty particles and fibrils surrounding mature particles. The bar represents 200 nm. (Insert) Isolated APMV particle showing an icosahedral morphology (Courtesy of La Scola and Raoult with permission). The bar represents 100 nm.

Virions of Acanthamoeba polyphaga mimivirus (APMV) are 400 nm in diameter, spherical in shape and surrounded by an icosahedral capsid. They are not filterable through 0.2 µm pore size filters. No envelope has been observed. Fibrils 80 nm in length can be seen attached to the capsid (Fig. 1). Virions resemble small Gram-positive cocci when Gram stained and was therefore named "mimivirus" (for mimicking microbe).

PHYSICOCHEMICAL AND PHYSICAL PROPERTIES

After suspension in a CsCI solution (0.2M Tris-HCI, pH 7.5, density 1.4) and ultracentrifugation (140,000g, 18hrs, 10°C), the density was estimated to be ~1.36 g/cm^3. Inactivation procedures have not been studied.

NUCLEIC ACID

The APMV genome is composed of circular, dsDNA demonstrated by restriction enzyme treatment and pulsed-field gel electrophoresis. Genome size is estimated of about 800 kb pairs. Partial sequence information is available.

PROTEINS

No protein has been isolated so far. Twenty-one mimivirus proteins have been identified with known functional attributes and are clear homologs to proteins highly conserved in most nucleocytoplasmic large DNA viruses. The SDS-PAGE showed the presence of proteins from 200 to 8 kDa including three major protein bands of ~75, 45 and 20 kDa.

LIPIDS

None present.

CARBOHYDRATES

None present.

Genome Organization and Replication

Unknown.

Antigenic Properties

In immunoblotting, serum from immunized mice detected several epitopes, of which the immunodominant epitope is a 35 kDa polypeptide. Sera from control mice do not react

Biological Properties

Host Range

Acanthamoeba polyphaga mimivirus was isolated from the water of a cooling tower in Brandford, England. The host was the amoeba *Acanthamoeba polyphaga*. There is indirect (serological) and direct (PCR amplification and sequencing) evidence for infection in humans.

Transmission

Unknown.

Geographical Distribution

Indirect evidences (serology) indicated human infection in England, Canada and France.

Cytopathic Effects

APMV virions are cytolytic for the amoeba *Acanthamoeba polyphaga*.

List of Species Demarcation Criteria in the Genus

No applicable.

List of Species in the Genus

Species names are in green italic script; strain names and synonyms are in black roman script; tentative species names are in blue roman script. Sequence accession numbers, and assigned abbreviations ( ) are also listed.

Species in the Genus

Acanthamoeba polyphaga mimivirus		
Acanthamoeba polyphaga mimivirus	[AABV01000000]	(APMV)

Tentative Species in the Genus

None reported.

Phylogenetic Relationships within the Genus

Not applicable.

Similarity with Other Taxa

Not available.

Derivation of Names

mimi for *mimi*cking microbe.

References

La Scola, B., Audic S., Robert, C., Jungang, L., de Lamballerie, X., Drancourt, M., Birtles, R., Claverie, L.M. and Raoult, D. (2003). A giant virus in amoebae. *Science*, **299**, 2033.

Contributed By

La Scola B., de Lamballerie X., Claverie J-M., Drancourt M. and Raoult D.

The Single Stranded DNA Viruses

FAMILY INOVIRIDAE

TAXONOMIC STRUCTURE OF THE FAMILY

Family	*Inoviridae*
Genus	*Inovirus*
Genus	*Plectrovirus*

VIRION PROPERTIES

MORPHOLOGY

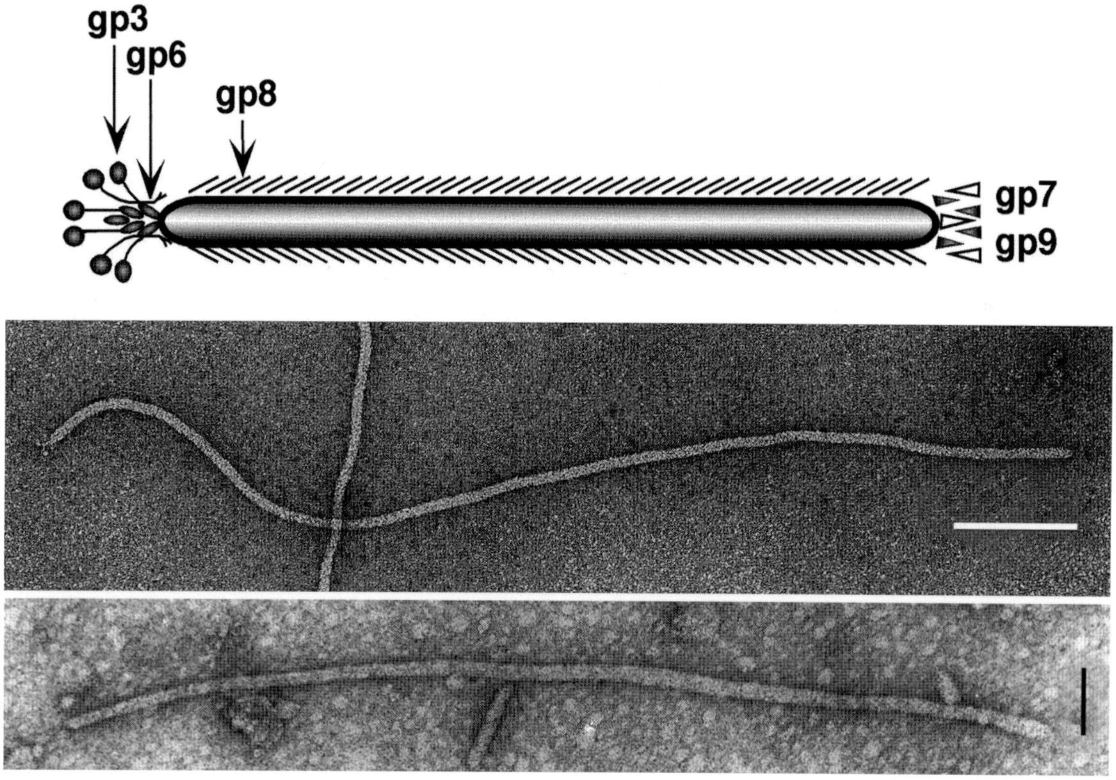

Figure 1: (Upper) Diagram of the protein shell, the loop of ssDNA, and the two ends of inovirus Enterobacteria phage fd (fd) (From Kornberg, A., Baker, T.A. (1991). *DNA replication*, 2nd (W.H. Freeman et alo. eds). New York, p 562). (Center) Negative contrast electron micrograph of inovirus phage fd, showing a tapered end with extensions due to adsorption proteins. The bar represents 100 nm. (From Gray *et al.*, 1981). *J. Mol. Biol.*, **146**, 621-627.; courtesy of Dr. Gray). (Lower) Negative contrast electron micrograph of a preparation of plectrovirus Acholeplasma phage MV-L51 (L51) showing a rod-shaped virion and an abnormally long particle. The bar represents 50 nm. (From Maniloff, J., Das, J. and Christensen, J.R. (1977). *Adv. Vir. Res.*, **21**, 343-380).

Virions in this family are rods or filaments containing a circular ssDNA genome within a cylindrical protein shell. There are no lipid components. Virion length is a function of the genome size and can change during evolution either through insertions or deletions in the genome as well as through mutations that alter the coat proteins. Inovirus virions are all ~7 nm in diameter, but lengths vary almost three fold, from 700 nm (Pseudomonas phage Pf3 (Pf3)) to 2,000 nm (Pseudomonas phage Pf1 (Pf1)). Electron micrographs reveal different structures at the virion ends, one small and blunt and the other larger and more variable. Engineered virions of Ff (the collective designation for isolates of *Enterobacteria phage M13*, *Enterobacteria phage f1* and *Enterobacteria phage fd*), in which the shell protein (gp8) or the adsorption protein (gp3) are fusions with foreign peptides are longer than wild type, due to the extra DNA, and can have other changes in shape depending on the size and properties of the fusion proteins. Wild type DNA conformations are diverse, with different spectroscopic properties and dramatically different extensions, and axial projections of the distance between neighboring nucleotides (the rise-per-nucleotide) ranging from a low of 0.23 nm for Pf3, over 0.27 nm for Ff, to a high of 0.61 nm for Pf1. X-ray fiber diffraction shows some capsids have 5-fold rotational and 2-fold screw symmetry (C_5S_2), and others have 1-fold rotational and 5.4- to 5.5-fold screw symmetry ($C_1S_{5.4}$). Plectrovirus virions are nearly straight rods with one end rounded and the other more variable. Virions of *Acholeplasma* phages are 70-90 nm long and 14-16 nm in diameter, and *Spiroplasma* phages are 230-280 nm long and 10-15 nm in diameter. Very long rods are frequently observed. Negative stained images suggest a 4 ± 2 nm hollow core. Optical diffraction of images (Acholeplasma phage L1 (L1)) suggests morphological units arranged with 2-fold rotational and 5.6-fold screw symmetry ($C_2S_{5.6}$). DNA conformations in this genus have apparent rise-per-nucleotide values in the range of 0.02-0.03 nm.

PHYSICOCHEMICAL AND PHYSICAL PROPERTIES

Virions of members of the family *Inoviridae* are sensitive to chloroform and are generally resistant to heat and a wide range of pH. Inovirus virions have buoyant densities in CsCl of 1.29 ± 0.01 and DNA contents from 6-14%. Virion Mr values are 12-34 x 10^6, almost a three-fold range, whereas S_{20w} values are in a narrow range, 41-44S: the closely similar sedimentation rates are largely determined by the closely similar mass-per-length of these virions. Translational and rotational diffusion constants are consistent with essentially rigid rods, yet inovirus virions can bend considerably without breaking. Particles of two species (*Enterobacteria phages C2* and *X*) appear to be more flexible than the others according to electron microscopy. Spectroscopic measurements reveal similar protein conformations, all highly helical, but various DNA conformations, some base-stacked and some not. For the genus *Plectrovirus*, buoyant densities are 1.39 g/cm^3 in CsCl and 1.21 g/cm^3 in metrizamide, as reported for Spiroplasma phage 1 (SpV1).

NUCLEIC ACID

Virions contain one molecule of infectious, circular, positive-sense ssDNA. Inovirus genomes range from 6-9 kb. Plectrovirus genomes are 4.5 kb for *Acholeplasma* phages and ~8 kb for *Spiroplasma* phages. DNA sequences of inoviruses fd, M13, f1, Ike (Enterobacteria phage Ike), I2-2 (Enterobacteria phage I2-2), Pf1 (Pseudomonas phage Pf1), Pf3 (Pseudomonas phage Pf3), Cf1c (Xanthomonas phage Cf1c), Cf1t (Xanthomonas phage Cf1t), and Vibrio phage fs1 (fs1), as well as plectrovirus Spiroplasma phage 1-R8A2B (SpV1-R8A2B) and Acholeplasma phage L1 (L1) are available.

PROTEINS

In genus *Inovirus*, Ff (M13, f1, fd) virions, the long shells are composed of 2700 copies of gp8 (5.2 kDa), the adsorption end has several copies (probably 5 each) of gp3 (43 kDa) and gp6 (12 kDa), and several copies (probably 5 each) of gp7 (3.5 kDa) and gp9 (3.3 kDa) form the assembly nucleation end. Six non-structural proteins have been identified: morphogenetic proteins gp1 (35 kDa), gp11 (8 kDa) and gp4 (50 kDa), and DNA replication proteins gp2 (46 kDa), gp10 (12 kDa), and gp5 (9.8 kDa). In genus *Plectrovirus* L1 and L51 virions, the major CP is probably 19 kDa, a protein with a strong tendency to aggregate, and there is at least one minor protein. The genome has only four ORFs. In genus *Plectrovirus* SpV1-R8A2B and Spiroplasma phage 1-T78 (SpV1-T78), the major CP is 7.5 kDa.

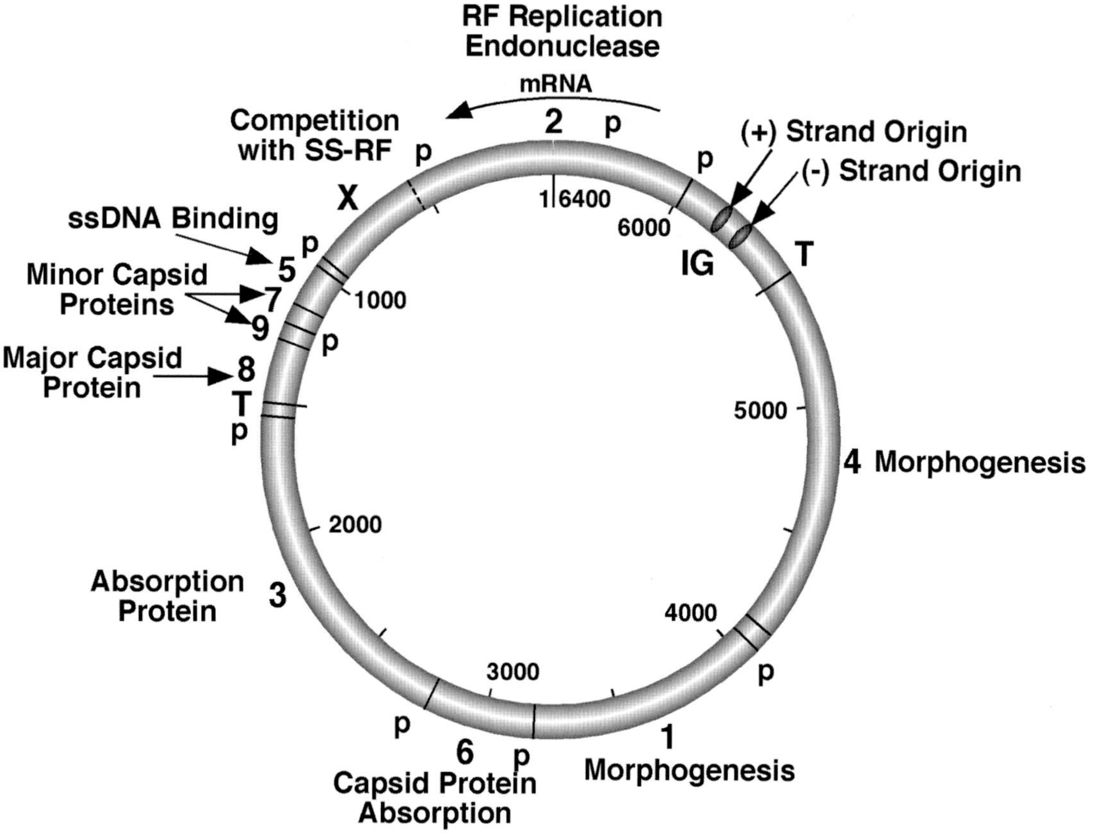

Figure 2: Genetic map of inovirus F pilus-specific coliphages showing functions of gene products. Not shown are the locations of gene *10* (DNA replication) and gene *11* (morphogenesis): these are the C-terminal thirds of gp2 and gp1, respectively, and are both derived translationally. DNA replication origins in the intergenic region (IG) are shown. P = promoter, T = transcription terminator. (From Kornberg, A., Baker, T.A. (1991). *DNA replication*, 2nd edn. WH Freeman and Co., New York, p. 561; see Hill *et al.*, (1991) for comparisons of the Ff, Pf3, and Pf1 genomes, and an alternative map numbering convention).

LIPIDS
None reported.

CARBOHYDRATES
None reported.

GENOME ORGANIZATION AND REPLICATION

Genomes replicate either independently, by a rolling circle mechanism like free plasmids or with the chromosome, if the viral genome becomes integrated. In the normal productive infectious cycle there are five steps: phage adsorption and uptake of the infecting circular ssDNA, conversion of the circular ssDNA to a parental replicative form (RF) by host cell enzymes, semi-conservative RF replication initiated by a viral endonuclease, synthesis of progeny ssDNA, which is sequestered by a ssDNA binding protein, and a membrane based assembly process that extrudes progeny virions into the medium. Genome organization reflects this sequence in that viral genes for DNA replication, virion structure, and virion morphogenesis are grouped in succession around the circle. The most detailed studies have been done on inovirus Ff. In the other phages that have been examined, closely similar or parallel processes have been observed. Genes are closely spaced and several genes are translated from overlapping reading frames or from alternate starts in the same frame. Intergenic regions contain the complementary- and viral-strand replication origins and DNA packaging signals. Phage adsorption involves specific interaction between the F pilus and a domain of gp3 on the infecting phage, and ssDNA translocation into the cytoplasm involves specific interactions between another domain of gp3 and other host membrane proteins. The ssDNA is converted to a supercoiled dsDNA replicative form (RF) by cellular enzymes. Phage DNA replication begins when the viral endonuclease gp2, expressed from parental RF, nicks this RF at a specific, high symmetry site. Progeny RF produced via ssDNA intermediates by rolling circle replication become templates for further RF replication and further mRNA synthesis. Gp10 and gp5 can down-regulate the nicking activity of gp2. When sufficient gp5 is made, complementary strand synthesis is blocked and complexes of gp5 and progeny viral ssDNA accumulate. Assembly is initiated at the membrane by concerted interactions of gp7, gp9, gp1, and gp11 and a specific packaging signal in a hairpin on the ssDNA in the gp5-ssDNA complex. Assembly proceeds at the inner membrane where ~1500 subunits of gp5 are displaced by 2700 subunits of gp8. Both gp1 and gp11 appear to be involved in this transfer of DNA from gp5-ssDNA complexes into the assembling virion. Gp1 may also function in the formation of adhesion zones between the inner and outer membranes by interacting with outer membrane pores formed by subunits of viral protein gp4. Assembly of virions is completed by addition of gp6 and gp3. There are notable exceptions to this overall pathway. Lysogenic strains encode integrases and viral sequences, partial or complete, are found integrated at several chromosomal sites. Xanthomonas phages Cf1t and Cf16, (Xanthomonas phage Cf16) variants of Cf1c, were the first inovirus lysogens to be characterized. In these Xanthomonas inoviruses, the gene for site-specific integration shows no homology with any Ff genes. Another species with a lysogenic phase is *Vibrio phage CTX*, the genome of which encodes the two subunits of cholera toxin. Upon conversion of the lysogen to the nonlytic productive cycle the toxin genes become highly expressed and toxin is released. The plectrovirus Acholeplasma phage L51 (L51) has been shown to be similar to this scheme with respect to DNA replication pathways, and virions are assembled at cell membranes and are released into the medium without cell lysis. This is presumably true for all plectrovirus species replicating as independent plasmids and producing virus by extrusion. The genomes of two viruses (L1 and SpV1) have been found integrated into the host chromosome at one or more sites.

Biological Properties

In nature, members of the family infect their hosts without causing lysis and infected cells continue to divide and produce virus indefinitely. The hosts are plant and animal pathogens. In several systems, the phage are lysogenic. Cell growth rates are slowed marginally by infection. On plates the slower growth usually allows formation of turbid plaques. Sometimes there is phage multiplication but no plaque formation. Inovirus hosts are mostly Gram-negative bacteria (i.e., *E. coli*, *Salmonella*, *Pseudomonas*, *Vibrio*, *Xanthomonas*, etc.), but two examples of inoviruses infecting Gram-positive hosts have been reported recently. Host ranges are determined primarily by host cell receptors, which are usually conjugative pili. Some pili are encoded chromosomally and some are encoded on plasmids of different incompatibility groups; e.g., phage Ff (M13, f1, fd) adsorbs to IncF pili, Pf3 to IncP pili, tf-1 to IncT pili, X to IncX pili. Transmission of these plasmids to new bacterial species usually transfers phage sensitivity. Additional host range determinants include restriction–modification systems, host periplasmic proteins involved in viral ssDNA translocation into the cytoplasm, and host protein(s) involved in membrane assembly. Transfection of non-natural hosts with naked ssDNA or dsDNA are sometimes possible. When *Vibrio cholera* lysogens colonize the human intestine, there is elevated cholera toxin expression and release and induction of progeny filamentous choleraphage extrusion. Thus, inovirus lysogeny is a critical virulence factor in cholera pathogenesis. Plectroviruses infect wall-less *Acholeplasma* and *Spiroplasma* and their receptors may contain both polysaccharide and protein components, but these are not well characterized. Indications of lysogeny in L1 and SpV1-R8A2B hosts suggest that the existence of two potential modes of carrier states, as free plasmids with virus extrusion or as lysogens, might be generally true for all members of the family *Inoviridae*.

GENUS INOVIRUS

Type Species *Enterobacteria phage M13*

DISTINGUISHING FEATURES

Infectivity is sensitive to sonication; species differ in their sensitivity to ether. Nucleic acid is 6-14% by weight of particles, and G+C is 40-60%. Sedimentation rates are 42S ± 2S. Virions have no carbohydrate. Hosts belong to certain genera in the γ-proteobacteria phylogenetic branch of Gram-negative bacteria; i.e., *Enterobacteria, Pseudomonas, Vibrio,* and *Xanthomonas.*

LIST OF SPECIES DEMARCATION CRITERIA IN THE GENUS

Demarcation criteria in the genus are:

- Particle length,
- Host range,
- Capsid symmetry,
- Antigenic properties, and
- DNA conformation and ratios of nucleotides per subunit.

An individual species in this genus is distinguishable by its host range, major CP sequence, and capsid symmetry. For example, the species Ff includes strains fd, f1, and M13 which have the same host range, capsid symmetry, and CP sequence (their DNA sequences differ less than ~1.5%). Although mutations, natural or otherwise, might dramatically affect virion length, ratios of nucleotides to subunits, DNA conformation, and antigenic properties of these strains (as in phage display work), species boundaries are not likely to be crossed. Inoviruses If1, IKe, and I2-2 have the same capsid symmetry and very similar sequences, but their host ranges differ, so they are different species.

LIST OF SPECIES IN THE GENUS

Species names are in green italic script; strain names and synonyms are in black roman script; tentative species names are in blue roman script. Hosts and pilus specificity {}, genome size, sequence accession numbers [], and assigned abbreviations ( ) are also listed.

SPECIES IN THE GENUS

1-Phages of Enterobacteriaceae:

Enterobacteria phage AE2				
Enterobacteria phage AE2	{*E. coli* IncF, Hfr}			(AE2)
Enterobacteria phage C-2				
Enterobacteria phage C-2	{E. coli, S. typhi; IncC}	8.1 kb		(C-2)
Enterobacteria phage dA				
Enterobacteria phage dA (probably all Ff)				(dA)
Enterobacteria phage Ec9				
Enterobacteria phage Ec9				(Ec9)
Enterobacteria phage f1				
Enterobacteria phage f1	{*E. coli* IncF, Hfr}	6.4 kb	[J02448]	(f1, Ff)*
Enterobacteria phage fd				
Enterobacteria phage fd	{*E. coli* IncF, Hfr}	6.4 kb	[V00602]	(fd, Ff)*
Enterobacteria phage HR				
Enterobacteria phage HR				(HR)
Enterobacteria phage I_2-2				
Enterobacteria phage I_2-2	{*E. coli*; Inc I_2}	6.7 kb	[X14336]	(I_2-2)*
Enterobacteria phage If1				
Enterobacteria phage If1	{E. coli, S. typhimurium; IncI	9 kb		(If1)*
Enterobacteria phage IKe				

Enterobacteria phage IKe	{*E. coli*; Inc I_2, N, P-1}	6.9 kb	[X02139]	(IKe)*
Enterobacteria phage M13				
Enterobacteria phage M13	{*E. coli* IncF, Hfr}	6.4 kb	[V00604]	(M13, Ff)*
Enterobacteria phage PR64FS				
Enterobacteria phage PR64FS	{*E. coli*; IncR}			(PR64FS)
Enterobacteria phage SF				
Enterobacteria phage SF	{E. coli, K. pneumoniae, S. typhi, others; IncS}			(SF)
Enterobacteria phage tf-1				
Enterobacteria phage tf-1	{E. coli, S. typhi; IncT}			(tf-1)
Enterobacteria phage X				
Enterobacteria phage X	{*E. coli, S. typhi, Sr. marc.*, others; IncX, I_2, N, P-1, others}			(X)
Enterobacteria phage X-2				
Enterobacteria phage X-2	{E. coli, S. typhi, Sr. marc.; IncX, unique R775}			(X-2)
Enterobacteria phage ZJ/2				
Enterobacteria phage ZJ/2				(Zj/2)

2- Phages of *Spirillaceae*:

Vibrio phage 493				
Vibrio phage 493	{*V. cholera* 0139-Aj27, El Tor}	9.3 kb		(493)
Vibrio phage CTX				
Vibrio phage CTX	{V. cholera 01}	7 kb	[U83795-6]	(CTX)
Vibrio phage fs1				
Vibrio phage fs1		6.3 kb	[D89074]	(fs1)
Vibrio phage fs2				
Vibrio phage fs2	{V. cholera 01, 0139}	8.5 kb		(fs2)
Vibrio phage v6				
Vibrio phage v6	{*V. cholera* 0395, Peru 15, El Tor, others; lysogenic}			(v6)
Vibrio phage Vf12				
Vibrio phage Vf12	{V. parahaemolyticus}			(Vf12)
Vibrio phage Vf33				
Vibrio phage Vf33		8.4kb		(Vf33)
Vibrio phage VSK				
Vibrio phage VSK	{*V. parahaemolyticus, V. cholera* 0139-B04, lysogenic *V cholera* 0139-P07, normal}			(VSK)

3-Phages of *Pseudomonadaceae*:

Pseudomonas phage Pf1				
Pseudomonas phage Pf1	{P. aeruginosa PAK}	7.3 kb	[X52107]	(Pf1)*
Pseudomonas phage Pf2				
Pseudomonas phage Pf2	{P. aeruginosa PAK}			(Pf2)
Pseudomonas phage Pf3				
Pseudomonas phage Pf3	{P. aeruginosa PAO; IncP-1}	5.8 kb	[M11912]	(Pf3)*

4-Phages of *Xanthomonadaceae*:

Xanthomonas phage Cf16				
Xanthomonas phage Cf16	{*X. camp. citri*: neolysogenic}			(Cf16)
Xanthomonas phage Cf1c				

Xanthomonas phage Cf1c	{X. camp. citri : normal}	7.3 kb	[M57538, U41819]	(Cf1c)
Xanthomonas phage Cf1t				
Xanthomonas phage Cf1t	{X. camp. citri : lysogenic}		[U08370]	(Cf1t)
Xanthomonas phage Cf1tv				
Xanthomonas phage Cf1tv	{*X. camp. citri* -Cf1t : lytic}			(Cf1tv)
Xanthomonas phage Lf				
Xanthomonas phage Lf	{X. campestris pv. campestris lysogenic}	6.0 kb	[U10884, U38235, X70327-31, AF018286]	(Lf)
Xanthomonas phage Xf				
Xanthomonas phage Xf	{X. campestris pv. oryzea}	7.4 kb		(Xf)*
Xanthomonas phage Xfo				
Xanthomonas phage Xfo	{X. oryzae pv. oryzea}	7.6 kb		(Xfo)
Xanthomonas phage Xfv				
Xanthomonas phage Xfv	{X. campestris pv. vesicatoria}	6.8 kb		(Xfv)

An asterisk after the abbreviation indicates that capsid symmetry has been assigned; for those in group 1 it is C_5S_2; for those in groups 2 and 4 it is $C_1S_{5.4}$.
Hosts: *E. coli: Escherichia coli, S. typhi: Salmonella typhi, Sr. marc.: Serratia marcescens, K. pneumoniae: Klebsiella pneumoniae, V. cholera: Vibrio cholera, V. parahaemolyticus: Vibrio parahaemolyticus, P. aeruginosa: Pseudomonas aeruginosa, X. camp. citri: Xanthomonas campestris citri, X. oryzae: Xanthomonas oryzae.*

TENTATIVE SPECIES IN THE GENUS

5- Phages of Gram-positive bacteria

Clostridium phage CAK1	{Clostridium beijerinickii}	6.8 kb	(CAK1)
Propionibacterium phage B5	{Propionibacterium freudenreichii}	5.8 kb	(B5)

GENUS PLECTROVIRUS

Type Species *Acholeplasma phage L51*

DISTINGUISHING FEATURES

Virions are resistant to non-ionic detergents (Nonidet P-40 and Triton X-100) and slightly sensitive to ether. Genome of Spiroplasma phage 1 (SpV-1) is 23% G+C. No data on carbohydrates have been reported. Adsorption is to cell membranes of wall-less mycoplasma host cells. Host range of Acholeplasma phage L51 (L51) is some *Acholeplasma laidlawii* strains, and of SpV-1 is some *Spiroplasma citri* strains.

LIST OF SPECIES DEMARCATION CRITERIA IN THE GENUS

Species in the genus *Plectrovirus* differ in host range.

LIST OF SPECIES IN THE GENUS

Species names are in green italic script; strain names and synonyms are in black roman script; tentative species names are in blue roman script. Hosts and pilus specificity {}, genome size, sequence accession numbers [], and assigned abbreviations ( ) are also listed.

SPECIES IN THE GENUS

1-Phages of *Acholeplasma*:

Acholeplasma phage L51			
Acholeplasma phage L51	{A. laidlawii}	4.5 kb	(L51)

2-Phages of *Spiroplasma*

Spiroplasma phage 1-aa

Spiroplasma phage 1-aa	{S. citri -SP-V3}	8.5 kb		(SpV1-aa)
Spiroplasma phage 1-C74				
Spiroplasma phage 1-C74	{S. citri Corsica}	7.8 kb		(SpV1-C74)
Spiroplasma phage 1-KC3				
Spiroplasma phage 1-KC3	{S. melliferum BC3}			(SpV1-KC3)
Spiroplasma phage 1-R8A2B				
Spiroplasma phage 1-R8A2B	{S. citri Morocco}	8.3 kb	[X51344]	(SpV1-R8A2B)
Spiroplasma phage 1-S102				
Spiroplasma phage 1-S102	{S. citri Syria}	6.9 kb		(SpV1-S102)
Spiroplasma phage 1-T78				
Spiroplasma phage 1-T78	{S. citri Turkey}	8.5 kb		(SpV1-T78)

A. laidlawii: Acholeplasma laidlawii, S. melliferum : Spiroplasma melliferum, S. citri: Spiroplasma citri.

TENTATIVE SPECIES IN THE GENUS

Acholeplasma phage L1	{*A. laidlawii*: lysogenic	4.5 kb	[X58839]	(L1)
Acholeplasma phage G51	{A. laidlawii}			(G51)
Acholeplasma phage 0c1r				(0c1r)
Acholeplasma phage 10tur				(10tur)
Spiroplasma phage C1/TS2				(C1/TS2)

LIST OF UNASSIGNED VIRUSES IN THE FAMILY

None reported.

PHYLOGENETIC RELATIONSHIPS WITHIN THE FAMILY

Not available.

SIMILARITY WITH OTHER TAXA

None reported.

DERIVATION OF NAMES

Ino: from Greek *nos*, 'muscle'.
Plectro: from Greek *plektron*, 'small stick'

REFERENCES

Ackermann, H.-W. and DuBow, M.S. (eds)(1987). *Viruses of prokaryotes*, Vol 2. CRC press, Boca Raton Florida, pp 171-218.

Baas, P. (1985). DNA replication of single-stranded Escherichia coli DNA phages. *Biochim. Biophys. Acta*, **825**, 111-139.

Bradley, D.E., Coetzee, J.N., Bothma, T. and Hedges, R.W. (1981). Phage X: a plasmid dependent, broad host range, filamentous bacterial virus. *J. Gen. Microbiol.*, **126**, 389-96.

Chopin, M.C., Rouault, A., Eherlich, S.D. and Gautier, M. (2002). Filamentous phages active on the Gram-positive bacterium *Propionibacterium freudenreichii*. *J. Bacteriol.*, **184**, 2030-2033.

Day, L.A., Marzec, C.J., Reisberg, S.A. and Casadevall, A. (1988). DNA packing in filamentous bacteriophages. *Annu. Rev. Biophys. Chem.*, **17**, 509-539.

Hill, D.F., Short, N.J., Perham, R.N. and Petersen G.B. (1991). DNA sequence of the filamentous bacteriophage Pf1. *J. Mol. Biol.*, **218**, 349-3645.

Li, Y. and Blaschek, H.P. (2002). Molecular characterization and utilization of the CAK1 filamentous virus-like particle derived from *Clostridium beijerinckii*. *J. Indust. Microbiol. Biotechnol.*, **28**, 118-126.

Kostrikis, L.G., Reisberg, S.A., Simon, M.N., Wall, J.S. and Day, L.A. (1991). Export of infectious particles by E. coli transfected with the RF DNA of Pf1, a virus of *P. aeruginosa* strain K. *Mol. Microbiol.*, **5**, 2641-2647.

Maniloff, J. (1992). Mycoplasma viruses. In: *Mycoplasmas: molecular biology and pathogenesis* (J., Maniloff, R.N., McElhaney, L.R., Finch, and J.B., Baseman, eds), pp. 41-59. Amer. Soc. for Microbiology, Washington DC.
Kuo, T.T., Lin, Y.H., Huang, C.M., Chang, S.F., Dai, H. and Feng, T.Y. (1987). The lysogenic cycle of the filamentous phage Cflt from *Xanthomonas campestris pv citri*. *Virology*, **156**, 305-312.
Liu, D.J. and Day, L.A. (1994). Pf1 virus structure : helical coat protein, and DNA with paraxial phosphates. *Science*, **265**, 671-674.
Lubkowski, J., Hennecke, F., Pluckthun, A. and Wlodawer, A. (1998). The structural basis of phage display elucidated by the crystal structure of the N-terminal domains of g3p. *Nat. Struct. Biol.*, **5**, 140-147.
Marvin, D.A. (1998). Filamentous phage structure, infection and assembly. *Curr. Op. Struc. Biol.*, **8**, 150-158.
Model, P. and Russel, M. (1988). Filamentous bacteriophage. In: *The Bacteriophages*, Vol 2., (R., Calendar, ed), pp. 375-456. Plenum Press, New York.
Rakonjac, J., Feng, Jn. and Model, P. (1999). Filamentous phage are released from the bacterial membrane by a two-step mechanism involving a short C-terminal fragment of pili. *J. Mol. Biol.*, **289**, 1253-1265.
Renaudin, J. and Bove, J.M. (1994). SpV1 and SpV4, spiroplasma viruses with circular singlestranded DNA genomes, and their contribution to the biology of spiroplasmas. *Adv. Virus Res.*, **44**, 429-463.
Russel, M., Linderoth, N.A. and Sali, A. (1997). Filamentous phage assembly: variation on a protein export theme. *Gene*, **192**, 23-32.
Stassen, A.P., Schoenmakers, E.F., Yu, M., Schoenmakers, J.G. and Konings, R.N. (1992). Nucleotide sequence of the genome of the filamentous bacteriophage I_2-2: module evolution of the filamentous phage genome. *J. Mol. Evol.*, **34**, 141-152.
Waldor, M.K. and Mekalanos, J.J. (1996). Lysogenic conversion by a filamentous phage encoding cholera toxin. *Science*, **272**, 1910-1914.
Webster, R.E. (1996). Biology of the Filamentous Bacteriophage. In: *Phage Display of Peptides and Proteins*, (B.K., Kay, J., Winter, and J., McCafferty, eds), pp. 1-20. Academic Press, San Diego, London, Boston, New York, Sydney, Tokyo, Toronto.

SSDNA

CONTRIBUTED BY

Day, L.A. and Hendrix, R.W.

FAMILY MICROVIRIDAE

TAXONOMIC STRUCTURE OF THE FAMILY

Family	***Microviridae***
Genus	***Microvirus***
Genus	***Chlamydiamicrovirus***
Genus	***Bdellomicrovirus***
Genus	***Spiromicrovirus***

VIRION PROPERTIES

MORPHOLOGY

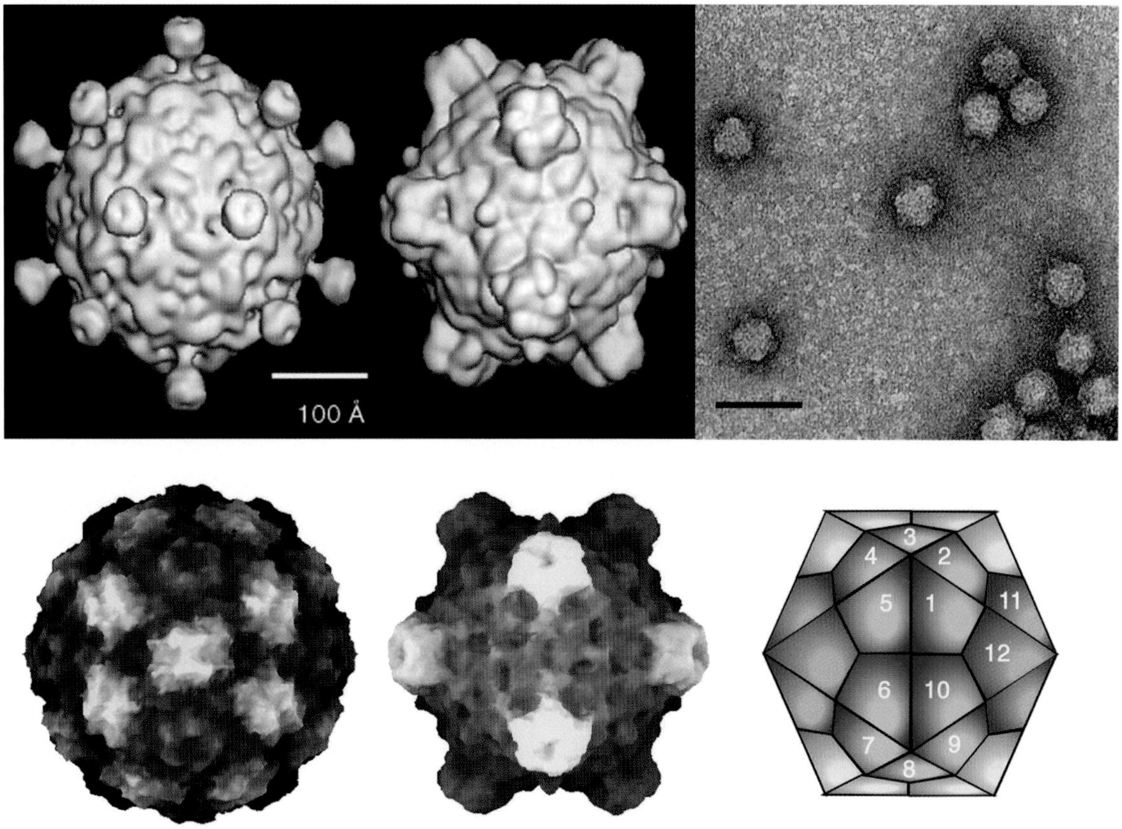

Figure 1: (Top left) Cryo-image reconstructions of the two morphologies represented within the family *Microviridae*: (Left) Spiroplasma phage 4 (SpV4)(genus *Spiromicrovirus*); (Right) Enterobacteria phage øX174 (øX174)(genus *Microvirus*). (Courtesy T. Baker, R. McKenna and M.G. Rossmann). (Top right) Negative contrast electron micrograph of øX174 particles. The bar represents 50 nm. (Bottom) Electronic rendering surface of øX174 particles (Left, scaffold; Center, procapsid) and diagram representing the T=1 lattice (Dokland *et al.,* (1997). *Nature*, **389**, 308-313).

There are two morphologies represented within the family *Microviridae*. These are non-enveloped virions with T=1 icosahedral symmetry. Members of the genus *Microvirus* infect enterobacteria, share a common morphology (typified by Enterobacteria phage øX174 (øX174) are distantly related to the phages of the other three genera. Members of the genera *Bdellomicrovirus* and *Chlamydiamicrovirus* infect obligate intracellular parasitic bacteria (*Bdellovibrio* and *Chlamydia*) and members of genus *Spiromicrovirus* infect mollicutes (*Spiroplasma*). These three genera appear closely related and share a common morphology, typified by Spiroplasma phage 4 (SpV4). Similarities between bdellomicroviruses and chlamydiamicroviruses are particularly striking. A change in the taxonomic grouping in this family is under consideration. This may mean grouping the members of the *Bdellomicrovirus, Spiromicrovirus* and *Chlamydiamicrovirus* into one genus or grouping two or three of these genera into one subfamily.

Phage øX174 typifies microvirus morphology, in which pentamers of a major spike protein decorate the five-fold axes of symmetry of the T=1 lattice. The structures of øX174, Enterobacteria phage α3 (α3) and Enterobacteria phage G4 (G4) capsids (the G4 capsids were empty particles) have been determined to at least 3.5 Å resolution. CPs can be superimposed with root mean square deviations less than or equal to 0.8 Å and exhibit the common β-barrel motif. Capsids have a diameter of 250 Å. The 70 Å-diameter spike protein pentamers rise 30 Å from the surface of the capsid. Virions of the other three genera lack major spike proteins and, hence, their five-fold axes of symmetry are not decorated. Cryo-image reconstruction of SpV4 reveals mushroom-shaped protrusions at the three-fold axes of symmetry, which rise 54 Å above the surface of the 270 Å diameter capsids. These three-fold related structures appear are composed of three interacting CPs.

PHYSICOCHEMICAL AND PHYSICAL PROPERTIES

The buoyant densities of family members range from 1.38-1.41 gm/cm^3 for microviruses and 1.30-1.31 gm/cm^3 for bdellomicroviruses and chlamydiamicroviruses. The only known spiromicrovirus, SpV4, has a reported buoyant density of 1.40 gm/cm^3. Particles of both morphologies are very stable, resistant to detergents, ether, chloroform, pH 6.0–9.0 and freezing. Microviruses have S values of ~115S, while phages in the other genera sediment with S values near 90S.

NUCLEIC ACID

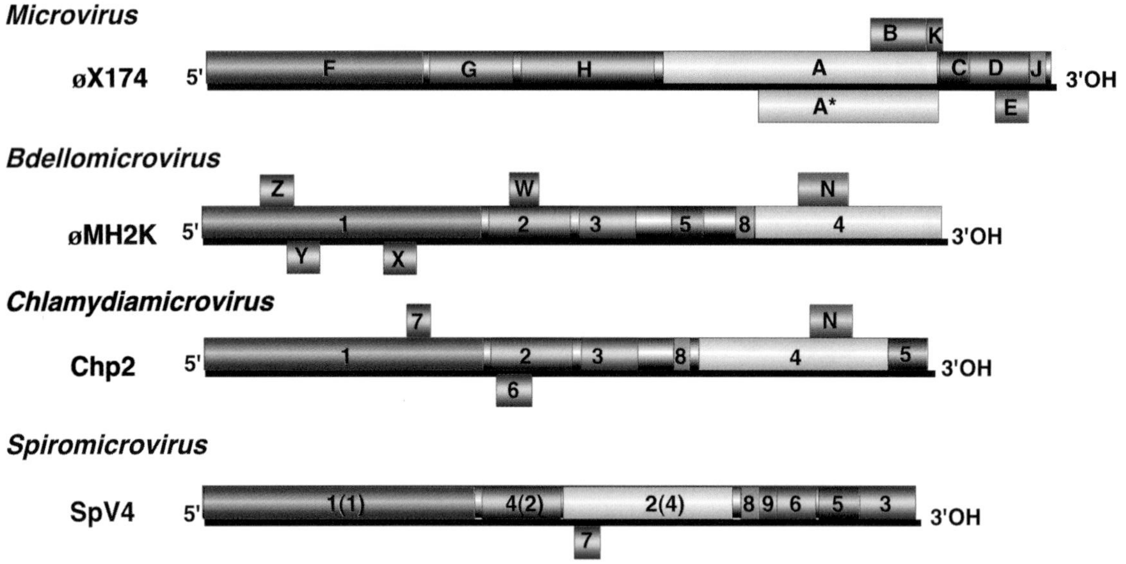

Figure 2: Genome organization of members of the family *Microviridae*. The circular genomes are presented linearly. Due to pronounced protein homologies and genome arrangements, the same gene number scheme is used for both the bdellomicroviruses (Bdellovibrio phage øMH2K; øMH2K) and chlamydiamicroviruses (Chlamydia phage 2; Chp2). Four proteins in Spiroplasma phage 4 (SpV4), (the products of genes 1, 2, 4, and 8) have homologs in the chlamydia- and bdellomicroviruses, the number in parentheses indicates the homologous gene.

Genomes are circular positive sense ssDNA molecules. As with morphological, biochemical and biophysical properties, genome sizes appear to fall into two size ranges. Microvirus genomes are 5.3-6.1 kb, while the genomes of the other three genera are 4.4-4.9

kb. The smaller genomes reflect the absence of genes encoding major spike and external scaffolding proteins.

For those genera in which multiple species have been sequenced (*Microvirus* and *Chlamydiamicrovirus*), genome arrangement is close to identical within the genus. The genome arrangement of the only sequenced member of bdellomicroviruses, Bdellovibrio phage øMH2K (øMH2K), is extremely similar to that found in chlamydiamicroviruses, with the exception of the location of gene 5. Other than this common feature, there is little or no similarity between genera.

PROTEINS

Although only the øX174-like phages have been studied in detail, sequence similarities and structurally based computational analyses have lead to reasonable hypotheses regarding some the viral proteins in the three less-studied genera (Table 1).

SSDNA

Table 1: Proteins found in members of the family *Microviridae*

Microvirus proteins	*Chlamydiamicrovirus* and *Bdellomicrovirus* proteins	*Spiromicrovirus* proteins	Protein Function
A	Vp4	Vp2	DNA replication protein
A*			Inhibition of host cell DNA synthesis, super-infection exclusion, non-essential.
B			Internal scaffolding protein
	Vp3		Internal scaffolding-like protein.
C			ssDNA synthesis, inhibitor of dsDNA synthesis
	Vp5		Hypothesized C protein homolog
D	Absent	Absent	External scaffolding protein.
E			Lysis protein.
	OrfN product (øMH2K only)		Hypothesized E protein homolog
F	Vp1	Vp1	Major capsid protein
G	Absent	Absent	Major spike protein
H			Minor spike protein, DNA pilot protein
	Vp2	Vp4	Hypothesized H protein homolog
J			DNA binding protein
	Orf8 product	Orf8 product	Hypothesized J protein homolog
K			Burst size modulation (host specific), non-essential.
L, M	6, 7 (Chp2) W, X, Y, Z (øMH2K)	3,5,6,7,8,9	Orfs of unknown coding capacity and/or proteins of unknown function

LIPIDS

None reported.

CARBOHYDRATES

None reported.

GENOME ORGANIZATION AND REPLICATION

Genome organization is summarized in Figure 2. DNA replication and capsid assembly has only been studied for microviruses. Considering the distant relationship with the three other genera, generalizations regarding common mechanisms within the family are not appropriate. Therefore, replication is discussed at the genus level.

ANTIGENIC PROPERTIES

Both neutralizing and non-neutralizing monoclonal antibodies have been produced against microviruses. These antibodies often cross react with other genus members. Non-neutralizing cross-reacting monoclonal antibodies have been produced for the chlamydiamicroviruses. Antibodies for the major CP of Chp2 recognize the CP of phage øMH2K, further demonstrating the close relationship between these two genera.

GENUS MICROVIRUS

Type Species *Enterobacteria phage øX174*

DISTINGUISHING FEATURES

All current members of the microvirus genus were isolated from Enterobacteriaceae. However, it should be noted that rigorous searches for members of the family *Microviridae* have not been conducted in other hosts. Since most isolation procedures are optimized for large dsDNA viruses, which are considerably denser than microviruses, easier to visualize by electron microscopy, and have much larger S values, special techniques must be employed for the isolation of members of the family *Microviridae*.

Although the viruses in this group are dependent on the same host cell proteins for replication, genetic studies suggest that individual phages may be particularly sensitive to host cell alleles of *rep*, *slyD*, and *mraY*. Even though the primary sequences of the microvirus proteins have diverged, some of these proteins can cross-function (e.g., DNA binding proteins and internal scaffolding proteins), indicating productive use; while other proteins cross-inhibit (e.g., external scaffolding proteins), indicating the ability to interact with other proteins across species lines. However, after interaction, other functions required for productive morphogenesis are hindered. Cross-species inhibitory protein domains have been identified with the use of chimeric proteins.

GENOME ORGANIZATION AND REPLICATION

The most distinguishing feature of microvirus genomes is the presence of overlapping reading frames. For example, 7 of the 11 genes of øX174 (genes A through E and gene K) reside in such regions (Fig. 2). Most of these genes code for non-structural proteins (Table 1). Genes found within the coding sequences of other genes (A*, B, K, and E) encode non-essential proteins (A* and K), proteins that do not affect particle formation, such as lysis proteins (E); or highly flexible proteins which tolerate substitutions, like the internal scaffolding protein (B). Genes encoding structural proteins usually do not share sequences with overlapping genes. While the use of overlapping reading frames increases the amount of genetic information encoded in a small genome, the evolution of this genetic strategy probably was not driven by fixed capsid dimensions, since packaged microvirus genomes do not form the densely condensed cores seen in most dsDNA bacteriophages. Microvirus genomes, via close association with the capsid's inner surfaces, are somewhat ordered into the icosahedral symmetry of the particle, which probably contributes to the late stages of morphogenesis and virion stability. Therefore, the acquisition of new genes (or morons) via horizontal transfer could seriously affect the structural role played by the ssDNA genome. Indeed, horizontal transfer does not appear to drive the evolution of these viruses. Instead, evolution is driven by species-jumping and the preservation of nested ORFs with weak ribosome binding sites, termed cretins.

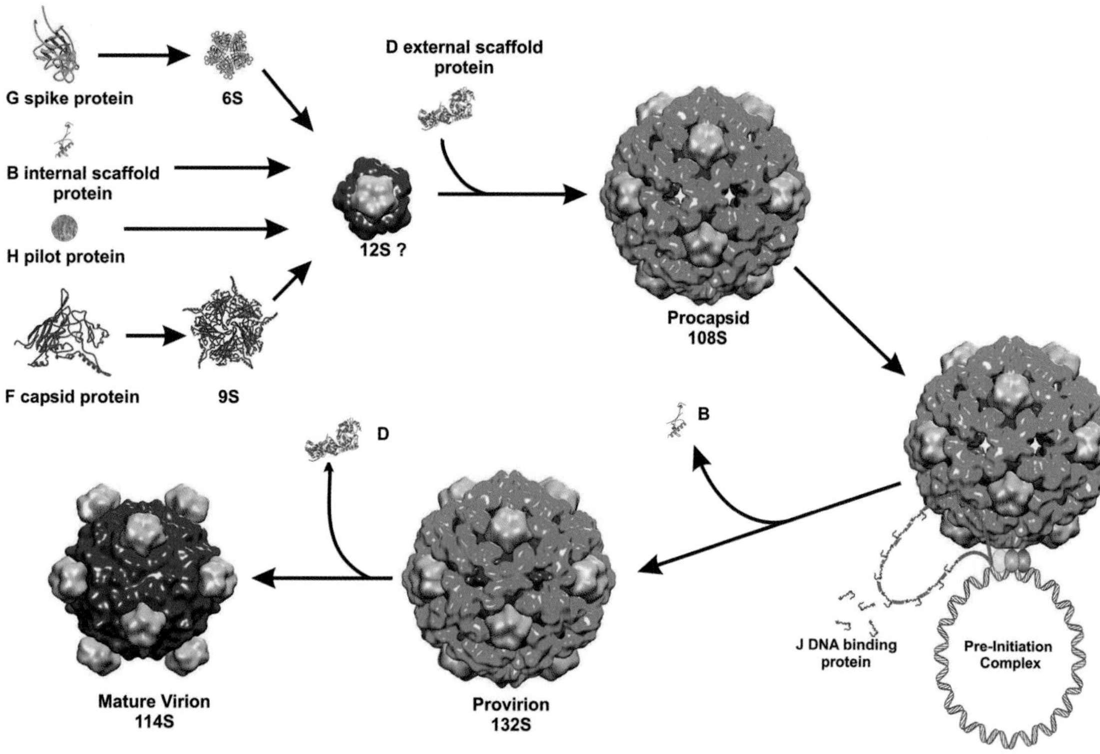

Figure 3: Microvirus capsid morphogenesis. (Courtesy R. Bernal and M.G. Rossmann).

Microvirus replication proceeds via two independent pathways, procapsid morphogenesis and DNA synthesis. After phage ssDNA enters the cell, stage I DNA replication commences. This process converts the infecting (+) single-stranded circular genome into a covalently closed double-stranded molecule called replicative form I DNA (RFI). No viral proteins are involved. With the synthesis of the (-) strand, transcription can begin. Although, *Microvirus* gene expression is not dependent on elaborate *trans*-acting mechanisms to ensure temporal regulation, the relative timing and amounts of viral proteins synthesized is controlled by highly sophisticated sets of *cis*-acting promoters, transcription terminators, and ribosome binding sites. Stage II DNA synthesis is dependent on the viral A protein, which cleaves the RFI DNA (+) strand at the origin of replication and covalently attaches itself to the DNA. This generates RF II molecules. *De novo* (+) strand replication involves a rolling circle mechanism, while *de novo* (-) strand synthesis occurs via a mechanism similar to stage I replication. Stage II DNA synthesis continues until one copy of viral protein C binds displaced ssDNA at the initiation of another round of stage II DNA synthesis, which terminates further dsDNA replication. With the binding of the C protein, the stage III DNA pre-initiation complex forms, consisting of one copy each of protein A, protein C, and the host cell rep protein, a DNA helicase which associates with protein A during stage II DNA synthesis. The pre-initiation complex then associates with the viral procapsid (see below), forming the 50S complex in which ssDNA is concurrently synthesized and packaged. The packaging mechanisms involved in members of the family *Microviridae* differ substantially from those found in other bacteriophages. Genome length is strictly governed by a single origin of replication, which determines both the initiation and termination of biosynthesis and packaging. DNA concatamers, unique translocating vertices, and head-full mechanisms are not involved. The viral A protein bound to the origin of replication in the RF II DNA is both necessary and sufficient for packaging specificity. Although any circular DNA molecule with a microvirus origin of replication can serve as a template, the secondary structure formed by the packaged genome can affect the biophysical properties of the resulting virion. After one round of rolling circle synthesis, protein A cuts the newly generated origin and acts as a ligase, generating a covalently closed circular molecule.

The first assembly intermediates in procapsid morphogenesis are 9S and 6S particles, pentamers of viral coat and spike proteins, respectively (Fig. 3). Five internal scaffolding proteins (protein B) bind to the underside of a 9S particle. This triggers conformational changes on the particle's upper surface allowing subsequent interactions with spike and external scaffolding proteins, and preventing premature 9S particle association. Protein B may facilitate the incorporation of the minor vertex protein, or DNA pilot protein, protein H. Twelve 12S particles then associate with external scaffolding proteins (protein D) to form the procapsid. In the øX174 procapsid crystal structure, four D proteins, as dimers of dimers, are associated with one CP. After particle formation, two-fold related B-B and D-D scaffolding contacts keep the CP pentamers from dissociating during DNA packaging. However, B-B interactions are only required at low temperatures.

The genome replication/packaging machinery binds to a depression along the two-fold axis of symmetry and an ssDNA genome is then concurrently synthesized and packaged, along with 60 copies of the DNA binding protein, protein J. Procapsids are probably filled through one of the 30 Å diameter pores at the 3-fold axes of symmetry. During packaging, B proteins are extruded, probably displaced by the DNA binding protein J which shares a binding cleft in the viral CP. Finally, D proteins dissociate, yielding the virion. This is accompanied by an 8.5 Å radial collapse of CPs around the packaged genome, which is tethered to the underside of the CP pentamers by F-DNA and J-DNA interactions. Genetic and biochemical data indicate that capsid-ssDNA interactions may mediate the integrity of this final stage of morphogenesis.

BIOLOGICAL PROPERTIES

The nature of the interaction of øX174-like phages with their hosts is poorly understood. Initial attachment to host cells occurs via a sugar residue, most likely glucose, in the lipopolysaccharide (LPS). A site on the surface of the capsid, near the 3-fold axes of symmetry, has been shown to bind glucose reversibly. However, host range mutations change amino acids in spike proteins G and H, suggesting that a second host cell receptor may be required for DNA ejection. The identity of this second factor is unknown. Although the members of the family *Microviridae* are tail-less, the virus may follow a pathway similar to that of the large tailed Enterobacteria phage T4 (T4). T4 reversibly interacts with LPS via its long tail fibers. The phage then "walks" along the surface of the cell until it finds a second receptor, which triggers ejection. Instead of walking, microviruses may "rock and roll" along the cell surface, until this second receptor is found.

Intracellular localization of øX174-like phage maturation is strongly dependent on the host cell *rep* allele, which must physically interact with the viral A protein and the procapsid during ssDNA synthesis and packaging. Proper interactions between the viral E protein and the gene products of host cell *slyD* and *mraY* alleles are probably required for lysis.

LIST OF SPECIES DEMARCATION CRITERIA IN THE GENUS

Currently, species demarcation criteria are temperature and host range. However, both phenotypes can be changed by single point mutations. Therefore, these criteria may not be rigorous for distinguishing between species. DNA sequencing has led to questions regarding the criteria for defining strains and species in the *Microviridae*. Species demarcation criteria should be re-examined.

LIST OF SPECIES IN THE GENUS

Species names are in green italic script; strain names and synonyms are in black roman script; tentative species names are in blue roman script. Sequence accession numbers, and assigned abbreviations ( ) are also listed.

Species in the Genus

Enterobacteria phage α3		
Enterobacteria phage α3	[X60322]	(α3)
Enterobacteria phage G4		
Enterobacteria phage G4	[J02454]	(G4)
Enterobacteria phage øX174		
Enterobacteria phage øX174	[J02482]	(ϕX174)
Enterobacteria phage S13	[M14428]	(S13)
Enterobacteria phage ϕK		
Enterobacteria phage ϕK	[X60323]	(ϕK)
Enterobacteria phage St-1		
Enterobacteria phage St-1		(St-1)

Tentative Species in the Genus

Enterobacteria phage 1ϕ1	(1ϕ1)
Enterobacteria phage 1ϕ3	(1ϕ3)
Enterobacteria phage 1ϕ7	(1ϕ7)
Enterobacteria phage 1ϕ9	(1ϕ9)
Enterobacteria phage 2D/13	(2D/13)
Enterobacteria phage α10	(α10)
Enterobacteria phage BE/1	(BE/1)
Enterobacteria phage δ1	(δ1)
Enterobacteria phage dϕ3	(dϕ3)
Enterobacteria phage dϕ4	(dϕ4)
Enterobacteria phage dϕ5	(dϕ5)
Enterobacteria phage ϕA	(ϕA)
Enterobacteria phage ϕR	(ϕR)
Enterobacteria phage G13	(G13)
Enterobacteria phage G14	(G14)
Enterobacteria phage G6	(G6)
Enterobacteria phage η8	(η8)
Enterobacteria phage M20	(M20)
Enterobacteria phage o6	(δ6)
Enterobacteria phage U3	(U3)
Enterobacteria phage WA/1	(WA/1)
Enterobacteria phage WF/1	(WF/1)
Enterobacteria phage WW/1	(WW/1)
Enterobacteria phage ζ3	(ζ3)

Note: Only stocks of S13, α3, øK, øR, øX174, and G4 are currently known to exist in laboratories or at the ATCC.

Genus *Chlamydiamicrovirus*

Type Species *Chlamydia phage 1*

Distinguishing Features

These phages infect various species of *Chlamydia.* Stocks of the type species phage no longer exist. Computational analyses indicate that chlamydiamicrovirus capsids will resemble those of SpV4. Protein sequence similarities between members of the genus *Chlamydiamicrovirus* and Bdellomicrovirus phage øMH2K are both significant and pronounced. However, locations of small ORFs, which may or may not encode proteins, are not conserved between genus members. Chlamydia phage 1 (Chp-1) appears to be more distantly related to the other *Chlamydia* phages, typified by Chlamydia phage 2

(Chp-2), than to phage øMH2K. This has lead to discussion of merging the genera *Chlamydiamicrovirus* and *Bdellomicrovirus* into one taxon.

GENOME ORGANIZATION AND REPLICATION

The genome organization of the *Chlamydiamicrovirus* Chp-2 is depicted in Figure 2. Since double stranded replicative form DNA has been isolated and homologs of microvirus proteins A and C are present, DNA replication is thought to occur via a similar mechanism. The mechanisms involved in capsid formation in the genus *Chlamydiamicrovirus* are not known, but will probably not resemble microvirus morphogenesis because chlamydiamicroviruses lack external scaffolding and major spike proteins. Vp3, believed to be an internal scaffold-like protein, may be a structural protein.

ANTIGENIC PROPERTIES

Described above.

BIOLOGICAL PROPERTIES

Unlike the microviruses, the chlamydiamicroviruses probably recognize a protein receptor.

LIST OF SPECIES DEMARCATION CRITERIA IN THE GENUS

There are no formal criteria for species demarcation.

LIST OF SPECIES IN THE GENUS

Species names are in green italic script; strain names and synonyms are in black roman script; tentative species names are in blue roman script. Sequence accession numbers, and assigned abbreviations ( ) are also listed.

SPECIES IN THE GENUS

Chlamydia phage 1		
Chlamydia phage 1	[D00624]	(Chp-1)
Chlamydia phage 2		
Chlamydia phage 2	[AJ270057]	(Chp-2)
Chlamydia pneumoniae phage CPAR39		
Chlamydia pneumoniae phage CPAR39	[AE002163]	(øCPAR39)
Guinea pig Chlamydia phage		
Guinea pig Chlamydia phage	[U41758]	(øCPG1)

TENTATIVE SPECIES IN THE GENUS

None reported.

GENUS BDELLOMICROVIRUS

Type Species ***Bdellovibrio phage MAC 1***

DISTINGUISHING FEATURES

These phages infect *Bdellovibrio* strains. DNA of isolates of the type species, Bdellovibrio phage MAC 1 (MAC-1), has not been sequenced and stocks of this phage no longer exist. All bdellovibriovirus genome and biophysical data come from the study of the tentative species øMH2K. Computational analyses suggest that the structure of øMH2K resembles SpV4.

LIST OF SPECIES DEMARCATION CRITERIA IN THE GENUS

There are no known formal criteria.

List of Species in the Genus

Species names are in green italic script; strain names and synonyms are in black roman script; tentative species names are in blue roman script. Sequence accession numbers, and assigned abbreviations ( ) are also listed.
names are in blue roman script. Sequence accession numbers, and assigned abbreviations ( ) are also listed.

Species in the Genus

Bdellovibrio phage øMH2K		
Bdellovibrio phage øMH2K	[AF306496]	(øMH2K)
Bdellovibrio phage MAC 1		
Bdellovibrio phage MAC 1		(MAC-1)

Tentative Species in the Genus

Bdellovibrio phage MAC 1'	(MAC-1')
Bdellovibrio phage MAC 2	(MAC-2)
Bdellovibrio phage MAC 4	(MAC-4)
Bdellovibrio phage MAC 4'	(MAC-4')
Bdellovibrio phage MAC 5	(MAC-5)
Bdellovibrio phage MAC 7	(MAC-7)
Bdellovibrio phage øCPG1	(øCPG1)

Genus Spiromicrovirus

Type Species *Spiroplasma phage 4*

Distinguishing Features

The one isolated phage of this group infects *Spiroplasma melliferum*.

List of Species Demarcation Criteria in the Genus

Not applicable.

List of Species in the Genus

Species names are in green italic script; strain names and synonyms are in black roman script; tentative species names are in blue roman script. Sequence accession numbers, and assigned abbreviations ( ) are also listed.

Species in the Genus

Spiroplasma phage 4		
Spiroplasma phage 4	[M17988]	(SpV4)

Tentative Species in the Genus

None reported.

List of Unassigned Viruses in the Family

None reported.

Phylogenetic Relationships within the Family

Phylogenetic trees built with 4 different proteins (CP, Rep protein, DNA pilot protein and the internal scaffolding protein) of representatives of the four genera show a relatively homogeneous clustering with the exception of Chp2 that is clustering most of the time with other chlamydiamicroviruses but forming a separate branch in the case of the Rep protein (Fig. 4). A revised classification and nomenclature for the family *Microviridae* is currently the topic of debate. (See Brentlinger *et al.*, 2002).

Similarity with Other Taxa

In some instances ssDNA genomes of viruses in the family *Microviridae* are similar in organization to those of members of the family *Inoviridae*. Structurally, viruses in the family *Microviridae* resemble those in the family *Parvoviridae.* Atomic structures of the CPs are rich in insertion loops coming off the β-barrel core. Similarities between the capsids of members of the genera *Spiromicrovirus, Bdellomicrovirus,* and *Chlamydiamicrovirus,* and those of the members of the family *Parvoviridae,* are more pronounced due to the complex interactions of CPs at the three-fold axes of symmetry.

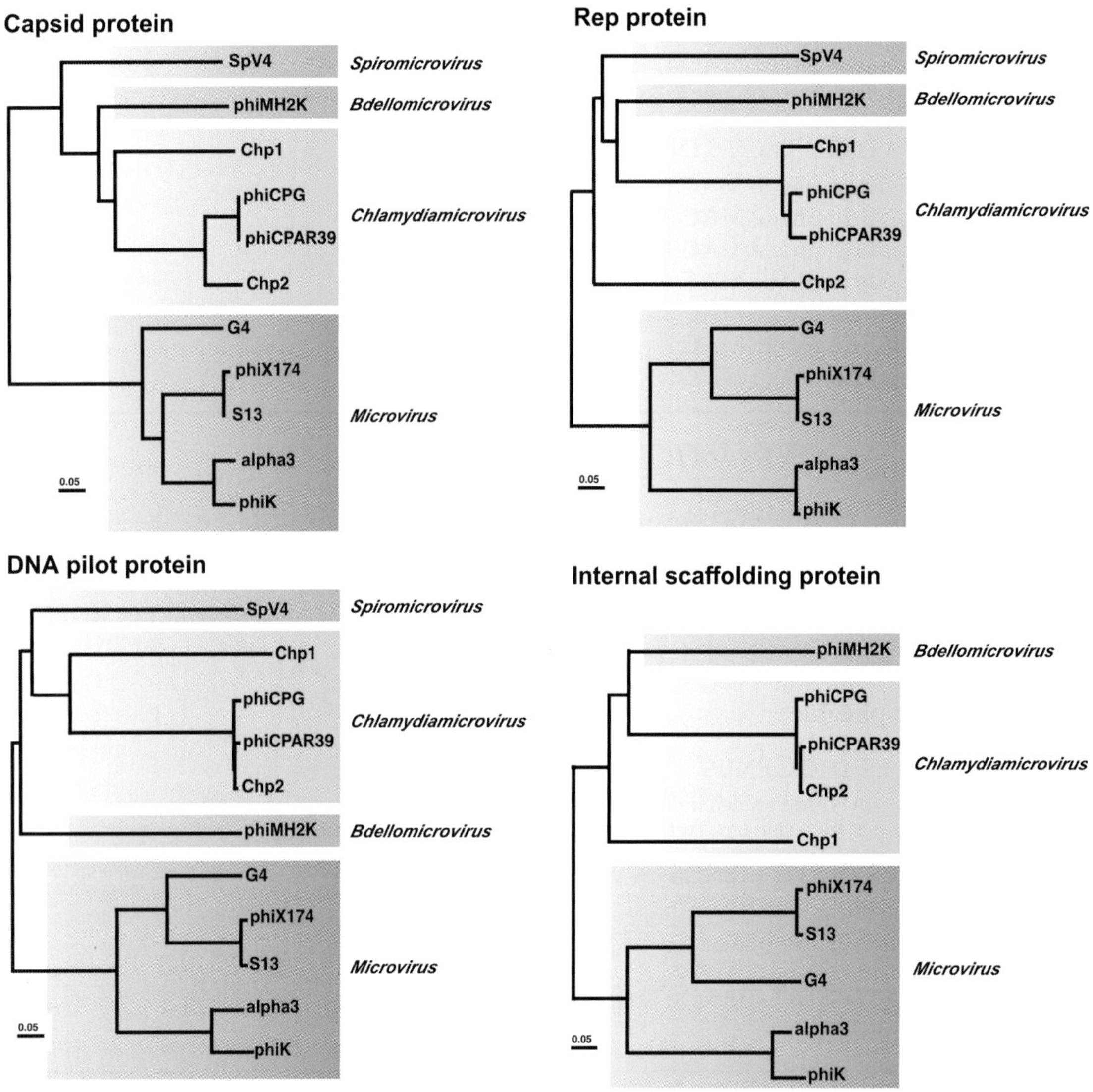

Figure 4: Phylogeny of microviruses. The aa sequences of the capsid protein, (Top left), the Rep protein (Top right), the DNA pilot protein (Bottom left) and the internal scaffolding protein (Bottom right) were used to make alignments with the CLUSTAL X software. The trees were designed with PAUP and the Bootstrap values are indicated above 50%. The abbreviations of the viruses used are listed in the List of Species of the description and accession numbers were from GeneBank.

Derivation of Names

Micro; From the Greek word *micros*, for small.

References

Bernal R.A., Hafenstein S., Olson N.H., Bowman, V.D., Chipman P.R., Baker T.S., Fane, B.A. and Rossmann, M.G. (2003). Structural studies of bacteriophage alpha3 assembly. *J Mol Biol.,* **325,** 11-24.

Bernhardt, T.G., Struck, D.K. and Young, R. (2001). The lysis protein E of øX174 is a specific inhibitor of the mraY-catalyzed step in peptidoglycan biosynthesis. *J. Biol. Chem.,* **276,** 6093-6097.

Brentlinger K., Hafenstein, S., Novak, C.R., Fane, B.A., Birgon, R., McKenna, R., and Agbandje-McKenna, M. (2002). *Microviridae*, a family divided. Isolation, characterization and genome sequence of a øMH2K, a bacteriophage of the obligate intracellular parasitic bacterium *Bdellovibrio bacteriovorus*. *J. Bacteriol.*, **184,** 1089-1094.

Bull J.J., Badgett, M.R., Wichman H.A., Huelsenbeck, J.P., Hillis, D.M., Gulati, A., Ho, C. and Molineux, I.J. (1997). Exceptional convergent evolution in a virus. *Genetics,* **147,** 1497-507.

Burch, A.D. and Fane, B.A. (2000). Foreign and chimeric external scaffolding proteins as inhibitors of *Microviridae* morphogenesis. *J. Virol.,* **74,** 9347-9352.

Chipman, P.R., Agbandje-McKenna, M., Renaudin, J., Baker, T.S. and McKenna, R. (1998). Structural analysis of the Spiroplasma virus, SpV4, implications for evolutionary variation to obtain host diversity among the Microviridae. *Structure,* **6,** 135-145.

Dokland, T., Bernal, R.A., Burch, A., Pletnev, S., Fane, B.A. and Rossmann, M.G. (1999). The role of scaffolding proteins in the assembly of the small, single-stranded DNA virus øX174. *J. Mol. Biol.,* **288,** 595-608.

Dokland, T., McKenna, R., Ilag, L.L., Bowman, B.R., Incardona, N.L., Fane, B.A. and Rossmann, M.G. (1997). Atomic structure of single-stranded DNA bacteriophage phi X174 and its functional implications. *Nature,* **389**, 308-313.

Everson, J.S., Garner, S., Fane, B.A., Liu, L., Lambden, P.R. and Clarke, I.N. (2002). Biological properties and cell tropism of Chp2, a bacteriophage of the obligate intracellular bacterium *Chlamydophila abortus*. *J. Bacteriol.*, **184,** 2748-2754.

Fane, B.A. and Prevelige, P.E. Jr. (2003). Mechanism of scaffolding-assisted viral assembly. *Adv. Prot. Chem.*, **64**, 259-299.

Hafenstein, S. and Fane B.A. (2002). øX174 genome-capsid interactions influence the biophysical properties of the virion: evidence for a scaffolding-like function for the genome during the final stages of morphogenesis. *J. Virol.*, **76**, 5350-5356.

Hayashi, M., Aoyama, A., Richardson, D.L. and Hayashi, M.N. (1988). Biology of the bacteriophage øX174. pp. 1-71. In: *The Bacteriophages*, Vol. 2, (R., Calendar, ed), pp. 1-71. Plenum Publishing, New York.

Hendrix, R.W., Lawrence, J.G., Hatfull, G.F. and Casjens, S. (2000). The origins and ongoing evolution of viruses. *Trends Microbiol.*, **8,** 504-508.

Hsia, R.C., Ting, L.M. and Bavoil, P.M. (2000). Microvirus of *Chlamydia psittaci* strain guinea pig inclusion conjunctivitis: isolation and molecular characterization. *Microbiology,* **146,** 1651-1660.

Liu, B.L., Everson, J. S., Fane, B.A., Giannikopoulou, P., Vretou, E., Lambden, P.R. and Clarke I.N. (2000). The molecular characterization of a bacteriophage (Chp2) from *Chlamydia psittaci*. *J. Virol.*, **74,** 3646 -3649.

McKenna, R., Ilag, L.L. and Rossmann, M.G. (1994). Analysis of the single-stranded DNA bacteriophage øX174 at a resolution of 3.0 A. *J. Mol. Biol.*, **237,** 517-543.

Renaudin, J., Paracel, M.C. and Bove, J.M. (1987). Spiroplasma virus 4: nucleotide sequence of the viral DNA, regulatory signals and the proposed genome organization. *J. Bacteriol.*, **169,** 4950-4961.

Storey, C.C., Lusher, M. and Richmond, S.J. (1989). Analysis of the complete nucleotide sequence of Chp1, a phage which infects *Chlamydia psittaci*. *J. Gen. Virol.*, **70,** 3381-3390.

CONTRIBUTED BY

Fane B.

FAMILY GEMINIVIRIDAE

TAXONOMIC STRUCTURE OF THE FAMILY

Family	*Geminiviridae*
Genus	*Mastrevirus*
Genus	*Curtovirus*
Genus	*Topocuvirus*
Genus	*Begomovirus*

VIRION PROPERTIES

MORPHOLOGY

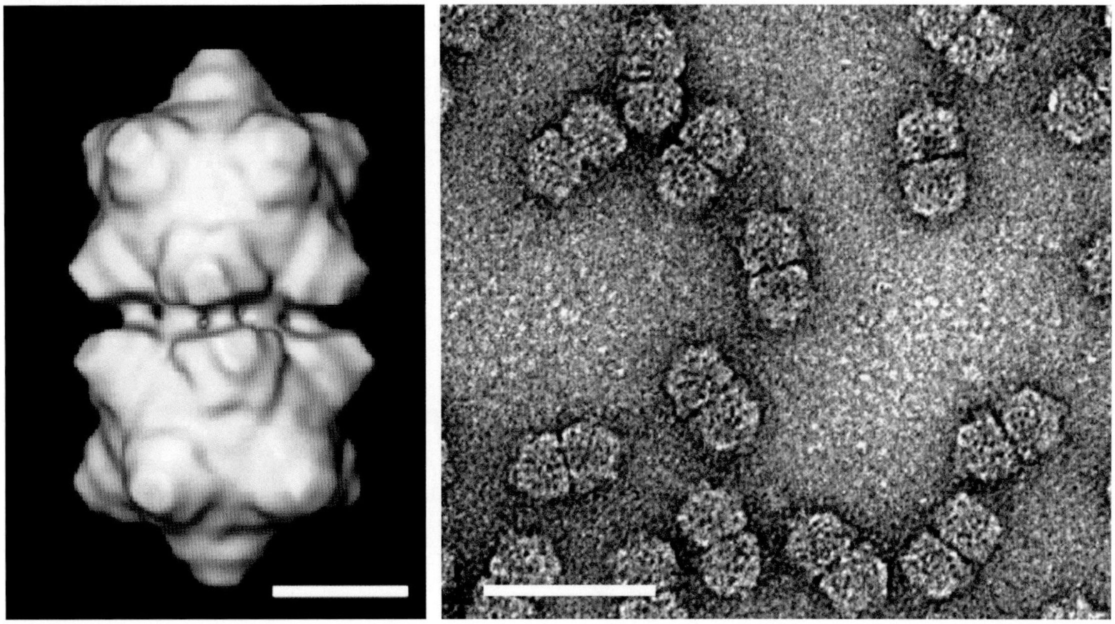

Figure 1: (Left) Cryo-electron microscopic reconstruction of Maize streak virus (MSV) viewed along a twofold axis of symmetry. The bar represents 10 nm. (From Zhang *et al.* 2001 and courtesy of R. McKenna). (Right) Purified particles of MSV stained with uranyl acetate showing typical twinned quasi-isometric subunits. The bar represents 50 nm.

Virions are typically twinned (so-called "geminate"). For Maize streak virus (MSV) particles, cryo-electron microscopy has shown that virions are about 22 x 38 nm, consisting of two incomplete icosahedra (T=1) containing a total of 110 CP subunits organized as 22 pentameric capsomers (Fig. 1).

PHYSICOCHEMICAL AND PHYSICAL PROPERTIES

Virion S_{20w} is approximately 70S.

NUCLEIC ACID

Twinned virions are presumed to contain a single copy of circular ssDNA, ranging in size from 2.5-3.0 kb. Hence, for viruses with bipartite genomes, two virions containing different genomic components will be required for infection. Half-size defective components and ssDNA satellites are also encapsidated, possibly in isometric virions.

PROTEINS

Virions contain a single structural protein (CP ~28-34 kDa). No other proteins have been found associated with virions.

LIPIDS

None reported.

CARBOHYDRATES
None reported.

GENOME ORGANIZATION AND REPLICATION
Viruses in the genera *Mastrevirus*, *Curtovirus* and *Topocuvirus* have single genomic components, those in the genus *Begomovirus* have either one or two components. Replication occurs through double-stranded replicative intermediates by a rolling circle mechanism. Complementary-sense DNA synthesis on the virion-sense (encapsidated) strand to produce dsDNA depends solely on host factors. ssDNA synthesis is initiated by cleavage of the virion-sense strand by the virus-encoded replication-associated protein (Rep) immediately downstream of the 3' thymidine residue in an absolutely conserved TAATATT/AC sequence located in the loop of a potential stem-loop structure within the intergenic region. Geminiviruses do not encode a DNA polymerase, and consequently are heavily dependent on host factors that must be recruited during early stages of replication. In all cases, coding regions in both virion-sense and complementary-sense strands diverge from an intergenic region, and transcription is bi-directional, with independently controlled transcripts initiating within the intergenic region. Viruses in the genus *Mastrevirus* use transcript splicing for gene expression, those in other genera use multiple overlapping transcripts.

GENUS *MASTREVIRUS*

Type Species *Maize streak virus*

GENOME ORGANIZATION AND REPLICATION

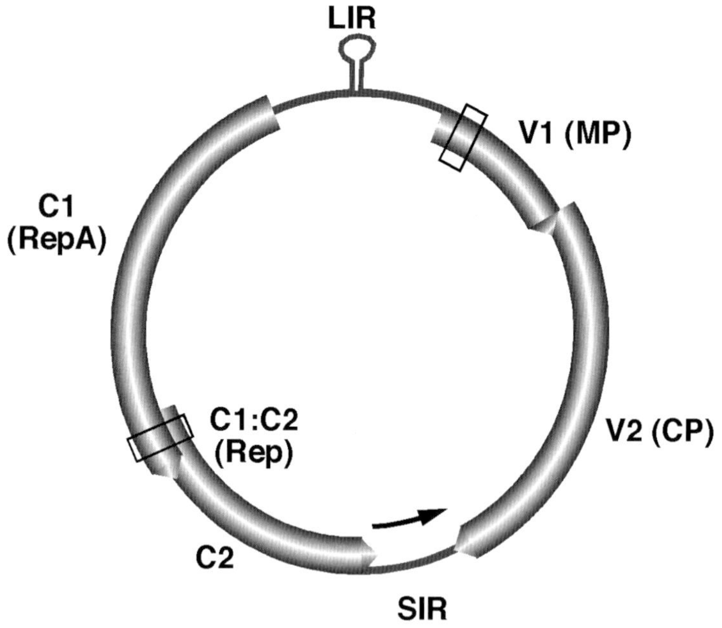

Figure 2. Typical genomic organization of mastreviruses. ORFs are denoted as either being encoded on the virion-sense (V) or complementary-sense (C) strand, and corresponding genes are indicated. The positions of the stem-loop motif containing the conserved TAATATTAC sequence in the large intergenic region (LIR) and the encapsidated complementary-sense primer-like molecule (small arrow) in the small intergenic region (SIR) are shown. Introns (open boxes) occur in ORF V1 and at the overlap between ORFs C1 and C2. MP; movement protein, CP; capsid protein, Rep; replication-associated protein.

The genomes of mastreviruses consist of a single component of circular ssDNA, 2.6-2.8 kb in size. A small complementary-sense DNA containing 5'-ribonucleotides, annealed to the genomic DNA within the small intergenic region of Chloris striate mosaic virus (CSMV), Digitaria streak virus (DSV), Maize streak virus (MSV), Tobacco yellow dwarf virus

(TYDV), and Wheat dwarf virus (WDV), may be involved in priming complementary-sense DNA synthesis. The small, annealed DNA is subsequently encapsidated with genomic ssDNA. The nt sequences of infectious genomic clones of Bean yellow dwarf virus (BeYDV), CSMV, DSV, MSV, Miscanthus streak virus (MiSV), Panicum streak virus (PanSV), Sugarcane streak virus (SSV), Sugarcane streak Egypt virus (SSEV), Sugarcane streak Reunion virus (SSREV), TYDV and WDV have been determined. Their genomes encode four proteins (Fig. 2). Two encoded on the virion-sense strand are the CP (ORF V2), that encapsidates the virion-sense ssDNA and acts as a nuclear shuttle protein for viral DNA, and the MP (ORF V1), that functions in cell-to-cell movement. The CP appears to regulate the balance of ssDNA and dsDNA accumulation. Regulation of virion-sense gene expression in MSV and grass-infecting relatives occurs by differential transcript splicing. The complementary-sense strand encodes the replication-associated protein (Rep, ORFs C1/C2), expressed from ORFs C1 and C2 by transcript splicing, that initiates rolling circle replication by introducing a nick into the conserved TAATATTAC sequence in the virion-sense strand. Rep binds to the large subunit of the replication factor C clamp loader complex, suggesting a role in the recruitment of host replication factors to the origin of replication. RepA (ORF C1), also encoded on the complementary-sense strand, binds to the plant homologue of retinoblastoma protein (Rb) to regulate cell-cycle progression, altering the environment of terminally differentiated cells to provide host factors that support viral DNA replication. Both Rep and RepA bind to the origin of replication as multimeric proteins.

ANTIGENIC PROPERTIES

Serological analyses indicate that grass-infecting geminiviruses from the same continent constitute distinct groupings: there is an African streak virus group (MSV, PanSV, SSV, SSEV and SSREV), an Australasian striate mosaic virus group (CSMV, Bromus striate mosic virus (BrSMV), Digitaria striate mosaic virus (DiSMV), Paspalum striate mosaic virus (PSMV), and the very distinct Asian MiSV and European WDV. Although DSV originates from Vanuatu, it is most closely related to the African mastreviruses. Grass geminiviruses originating from different continents are either unrelated or distantly related. Mastreviruses that infect dicotyledonous plants (TYDV and BeYDV) are not antigenically related to those that infect monocotyledonous plants.

BIOLOGICAL PROPERTIES

HOST RANGE

Mastreviruses have narrow host ranges. With the exception of TYDV, BeYDV and the tentative species Chickpea chlorotic dwarf virus (CpCDV) (which infect certain Solanaceae and Fabaceae, among others), host ranges of mastreviruses are limited to members of the Poaceae (Gramineae).

TRANSMISSION

Mastreviruses are transmitted in nature by leafhoppers (Homoptera: Cicadellidae), in most cases by a single species. Mechanism of transmission is persistent (circulative, non-propagative). Mastreviruses are normally not transmissible by mechanical inoculation, although MSV has been transmitted via a vascular puncture technique using maize seeds. Most members are transmitted experimentally to plants by *Agrobacterium*-mediated transfer (agroinoculation) from partially or tandemly repeated cloned genomic DNA.

LIST OF SPECIES DEMARCATION CRITERIA IN THE GENUS

The following criteria should be used as a guideline to establish taxonomic status:

- Nucleotide sequence identity. Full-length nt sequence identity <75% is generally indicative of a distinct species. However, decisions based on nt sequence comparisons, particularly when approaching this value, must also take into account the biological properties of the virus.

- *Trans*-replication of genomic components. The inability of Rep protein to *trans*-replicate a genomic component suggests a distinct species.
- Coat protein characteristics. Serological differences may be indicative of a distinct species.
- Different vector species.
- Natural host range and symptom phenotype. These characteristics may relate to a particular species but their commonest use will be to distinguish strains.

LIST OF SPECIES IN THE GENUS

Species names are in green italic script; strain names and synonyms are in black roman script; tentative species names are in blue roman script. Sequence accession numbers, and assigned abbreviations ( ) are also listed.

SPECIES IN THE GENUS

Bean yellow dwarf virus		
Bean yellow dwarf virus	Y11023	(BeYDV)
Chloris striate mosaic virus		
Chloris striate mosaic virus	M20021	(CSMV)
Digitaria streak virus		
Digitaria streak virus	M23022	(DSV)
Maize streak virus		
Maize streak virus - [Ethiopia]	X71956	(MSV-[ET])
Maize streak virus - [Ghana1]	X71953	(MSV-[GH1])
Maize streak virus - [Ghana2]	X71959	(MSV-[GH2])
Maize streak virus - [Malawi]		(MSV-[MW])
Maize streak virus - [Mauritius]	X71963	(MSV-[MU])
Maize streak virus - [Mozambique]	X71962	(MSV-[MZ])
Maize streak virus - [Nigeria2]	X71957	(MSV-[NG2])
Maize streak virus - [Nigeria3]	X71961	(MSV-[NG3])
Maize streak virus - [Port Elizabeth]	U20893	(MSV-[PE])
Maize streak virus - [Reunion1]	X71954	(MSV-[RE1])
Maize streak virus - [Reunion2]	X94330	(MSV-[RE2])
Maize streak virus - [Uganda]	X71958	(MSV-[UG])
Maize streak virus - [Wheat-eleusian]	U20871	(MSV-[Wel])
Maize streak virus - [Zaire]	X71964	(MSV-[ZR])
Maize streak virus - [Zimbabwe1]	X71955	(MSV-[ZW1])
Maize streak virus - [Zimbabwe2]	X71960	(MSV-[ZW2])
Maize streak virus - A[Ama]	AF329878	(MSV-A[Ama])
Maize streak virus - A[Gat]	AF329879	(MSV-A[Gat])
Maize streak virus - A[KA]	AF329885	(MSV-A[KA])
Maize streak virus - A[Kenya]	X01089	(MSV-A[KE])
Maize streak virus - A[Km]	AF395891	(MSV-A[Km])
Maize streak virus - A[Kom]	AF003952	(MSV-A[Kom])
Maize streak virus - A[MakD]	AF329884	(MSV-A[MakD])
Maize streak virus - A[MatA]	AF329881	(MSV-A[MatA])
Maize streak virus - A[MatB]	AF329882	(MSV-A[MatB])
Maize streak virus - A[MatC]	AF329883	(MSV-A[MatC])
Maize streak virus - A[Nigeria1]	X01633	(MSV-A[NG1])
Maize streak virus - A[Sag]	AF329880	(MSV-A[Sag])
Maize streak virus - A[South Africa]	Y00514	(MSV-A[ZA])
Maize streak virus - A[Vaalhart maize]	U20769, AJ012637, AF239961	(MSV-A[Vm])
Maize streak virus - B[Jam]	AF329887	(MSV-B[Jam])
Maize streak virus - B[Mom]	AF329886	(MSV-B[Mom])
Maize streak virus - B[Tas]	U20905, AJ012636, AF239962	(MSV-B[Tas])
Maize streak virus - B[Vaalhart wheat]	U20768, AJ012638, AF239960	(MSV-B[Vw])
Maize streak virus - E[Pat]	AF329888	(MSV-E[Pat])

Maize streak virus - Reunion [N2AR2]	AJ224504	(MSV-RE[N2AR2])
Maize streak virus - Reunion [N2AR3]	AJ224505	(MSV-RE[N2AR3])
Maize streak virus - Reunion [N2AR4]	AJ224506	(MSV-RE[N2AR4])
Maize streak virus - Reunion [N2AR5]	AJ224507	(MSV-RE[N2AR5])
Maize streak virus - Reunion [N2AR6]	AJ224508	(MSV-RE[N2AR6])
Maize streak virus - Reunion [N2AR8]	AJ225006	(MSV-RE[N2AR8])
Maize streak virus - Reunion [SP1]	AJ224999	(MSV-RE[SP1])
Maize streak virus - Reunion [SP1R10]	AJ225007	(MSV-RE[SP1R10])
Maize streak virus - Reunion [SP2R11]	AJ225009	(MSV-RE[SP2R11])
Maize streak virus - Reunion [SP2R12]	AJ225010	(MSV-RE[SP2R12])
Maize streak virus - Reunion [SP2R13]	AJ225011	(MSV-RE[SP2R13])
Maize streak virus - Reunion [SP2R7]	AJ225008	(MSV-RE[SP2R7])
Maize streak virus - [Raw]	AF329889	(MSV-[Raw])
Maize streak virus - [Set]	AF007881, U20870	(MSV-[Set])
Miscanthus streak virus		
Miscanthus streak virus – [91]	D01030	(MiSV-[91])
Miscanthus streak virus – [Japan 96]	E02258	(MiSV-[JP96])
Miscanthus streak virus – [Japan 98]	D00800	(MiSV-[JP98])
Panicum streak virus		
Panicum streak virus - Karino	L39638	(PanSV-Kar)
Panicum streak virus - Kenya	X60168	(PanSV-KE)
Sugarcane streak virus		
Sugarcane streak virus - [Natal]	M82918, S64567	(SSV-[Nat])
Sugarcane streak virus - [Mauritius]	D00597, AF088881	(SSV-[MU])
Sugarcane streak Egypt virus		
Sugarcane streak Egypt virus - [Aswan]	AF039528	(SSEV-[Asw])
Sugarcane streak Egypt virus - [Ben]	AF039529	(SSEV-[Ben])
Sugarcane streak Egypt virus - [Giza]	AF037752	(SSEV-[Giza])
Sugarcane streak Egypt virus - [Man]	AF039530	(SSEV-[Man])
Sugarcane streak Egypt virus - [Naga]	AF239159	(SSEV-[Naga])
Sugarcane streak Reunion virus		
(Sugarcane streak virus - [Reunion])		
Sugarcane streak Reunion virus	AF072672	(SSREV)
Tobacco yellow dwarf virus		
Tobacco yellow dwarf virus	M81103	(TYDV)
Wheat dwarf virus		
Wheat dwarf virus – [Enköping1]	AJ311031	(WDV-[Enk1])
Wheat dwarf virus – [France]	X82104	(WDV-[FR])
Wheat dwarf virus – [Sweden]	X02869	(WDV-[SE])
(Wheat dwarf virus - CJI)		

TENTATIVE SPECIES IN THE GENUS

Bajra streak virus	(BaSV)
Bromus striate mosaic virus	(BrSMV)
Chickpea chlorotic dwarf virus	(CpCDV)
Digitaria striate mosaic virus	(DiSMV)
Millet streak virus	(MlSV)
Paspalum striate mosaic virus	(PSMV)

SSDNA

GENUS CURTOVIRUS

Type Species *Beet curly top virus*

GENOME ORGANIZATION AND REPLICATION

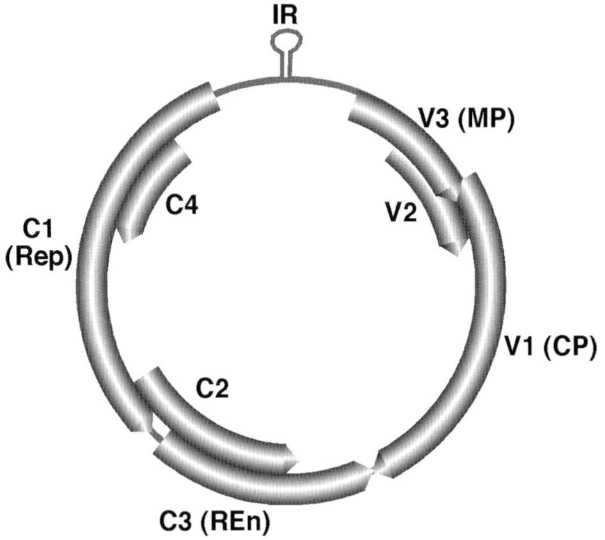

Figure 3: Typical genomic organization of curtoviruses. ORFs are denoted as either being encoded on the virion-sense (V) or complementary-sense (C) strand. Gene designations are shown where these are known. ORF C3 is not present in Horseradish curly top virus (HrCTV). The position of the stem-loop containing the conserved TAATATTAC sequence located in the intergenic region (IR), is shown. CP; capsid protein, MP; movement protein, Rep; replication-associated protein, REn; replication enhancer.

The genomes of curtoviruses consist of a single circular ssDNA component, 2.9-3.0 kb in size. The nucleotide sequences of infectious genomic clones of Beet curly top virus (BCTV), Beet mild curly top virus (BMCTV), Beet severe curly top virus (BSCTV) and Horseradish curly top virus (HrCTV) have been determined. Their genomes encode six to seven proteins (Fig. 3). Three encoded on the virion-sense strand are the CP (ORF V1), that encapsidates the virion-sense ssDNA and is involved in virus movement and insect vector transmission, V2 protein that is involved in the regulation of the relative levels of ssDNA and dsDNA, and a MP (ORF V3). The complementary-sense strand encodes the replication-associated protein (Rep, ORF C1), required for the initiation of viral DNA replication, C2 protein that acts as a pathogenicity factor in some hosts, a replication enhancer protein (REn, ORF C3), and C4 protein, an important symptom determinant that is implicated in cell-cycle control. Nucleotide sequence comparisons suggest that curtoviruses and begomoviruses diverged after a recombination event altered insect vector specificity.

ANTIGENIC PROPERTIES

Serological tests show BCTV, Tomato leaf roll virus (TLRV) and Tomato pseudo-curly top virus (TPCTV, genus *Topocuvirus*) to be relatively closely related. Distant relationships between curtoviruses and begomoviruses have been shown in serological tests.

BIOLOGICAL PROPERTIES

HOST RANGE

The type species BCTV has a very wide host range within dicotyledonous plants, including over 300 species in 44 plant families.

TRANSMISSION

Curtoviruses are transmitted in nature by leafhoppers (Homoptera: Cicadellidae) in a persistent (circulative, non-propagative) manner. BCTV may be transmitted with difficulty by mechanical inoculation. Most members are transmitted experimentally to plants by *Agrobacterium*-mediated transfer (agroinoculation) from partially or tandemly repeated cloned genomic DNA.

List of Species Demarcation Criteria in the Genus

The following criteria should be used as a guideline to establish taxonomic status:

- Nucleotide sequence identity. Full-length nt sequence identity <89% is generally indicative of a distinct species. However, decisions based on nt sequence comparisons, particularly when approaching this value, must also take into account the biological properties of the virus.
- *Trans*-replication of genomic components. The inability of Rep protein to *trans*-replicate a genomic component suggests a distinct species.
- CP characteristics. Serological differences may be indicative of a distinct species although the CP is highly conserved, suggesting that this criterion may be of limited use.
- Natural host range and symptom phenotype. These characteristics may relate to a particular species but their commonest use will be to distinguish strains.

List of Species in the Genus

Species names are in green italic script; strain names and synonyms are in black roman script; tentative species names are in blue roman script. Sequence accession numbers, and assigned abbreviations ( ) are also listed.

Species in the Genus

Beet curly top virus		
Beet curly top virus - California	X04144	(BCTV-Cal)
Beet curly top virus - California [Logan]	AF379637	(BCTV-Cal[Log])
Beet mild curly top virus		
(Beet curly top virus – Worland)		
Beet mild curly top virus – [Worland]	U56975	(BMCTV-[Wor])
Beet mild curly top virus – [Worland4]	AY134867	(BMCTV-[W4])
Beet severe curly top virus		
Beet severe curly top virus – Cfh	U02311	(BSCTV-Cfh)
(Beet curly top virus - CFH)		
Beet severe curly top virus – Cfh [Beta]	X97203	(BSCTV-Cfh[Beta])
(Beet curly top virus - CFH)		
Horseradish curly top virus		
Horseradish curly top virus	U49907	(HrCTV)

Tentative Species in the Genus

Tomato leaf roll virus	(TLRV)

Genus *Topocuvirus*

Type Species *Tomato pseudo-curly top virus*

Genome Organization and Replication

The genome of Tomato pseudo-curly top virus (TPCTV) consists of a single component of circular ssDNA, 2.8 kb in size. The genome, encoding six proteins, resembles that of members of the genus *Curtovirus* (Fig. 4). Nucleotide sequence comparisons suggest that TPCTV and begomoviruses diverged after a recombination event altered insect vector specificity. The CP is more closely related to those of the leafhopper-transmitted curtoviruses than the whitefly-transmitted begomoviruses. The V2 protein is distantly related to the curtovirus V3 protein.

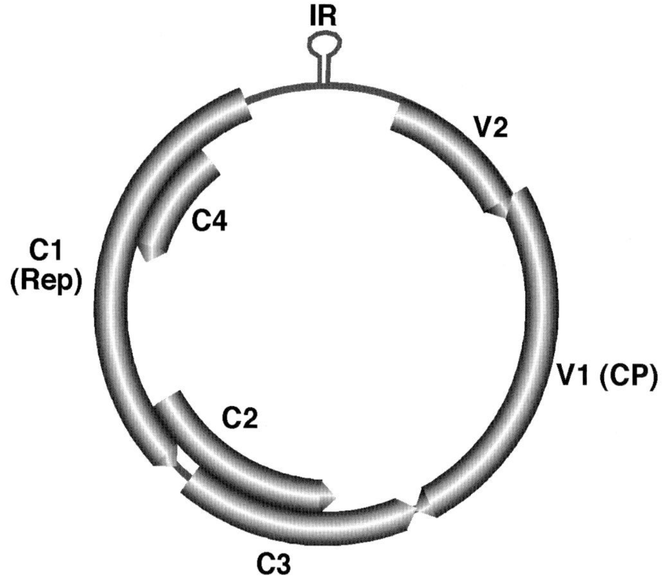

Figure 4: Genomic organization of TPCTV. ORFs are denoted as either being encoded on the virion-sense (V) or complementary-sense (C) strand. Gene designations are shown where these are known. The position of the stem-loop containing the conserved TAATATTAC sequence, located within the intergenic region, is shown. CP; capsid protein, Rep; replication-associated protein.

ANTIGENIC PROPERTIES

Serological tests show TPCTV to be relatively closely related to BCTV and TLRV in the genus *Curtovirus*.

BIOLOGICAL PROPERTIES

HOST RANGE

The host range of TPCTV is restricted to dicotyledonous plants, and includes weed species such as nightshade (*Solanum nigrum*), *Datura stramonium* and common chickweed (*Stellaria media*), crops such as tomato (*Lycopersicon esculentum*) and bean (*Phaseolus vulgaris*) and the experimental host *Nicotiana benthamiana*. TPCTV induces symptoms resembling those associated with BCTV infection in many hosts.

TRANSMISSION

TPCTV is transmitted in nature by the treehopper *Micrutalis malleifera* Fowler (Homoptera: Membracidae), and has been transmitted experimentally to plants by *Agrobacterium*-mediated transfer (agroinoculation) from a tandemly repeated cloned genomic DNA.

LIST OF SPECIES DEMARCATION CRITERIA IN THE GENUS

Currently, there is only one species in this genus. Criteria to establish taxonomic status are identical to those for the genus *Curtovirus*.

LIST OF SPECIES IN THE GENUS

Species names are in green italic script; strain names and synonyms are in black roman script; tentative species names are in blue roman script. Sequence accession numbers, and assigned abbreviations ( ) are also listed.

SPECIES IN THE GENUS

Tomato pseudo-curly top virus

Tomato pseudo-curly top virus	X84735	(TPCTV)

TENTATIVE SPECIES IN THE GENUS

None reported

GENUS BEGOMOVIRUS

Type Species *Bean golden yellow mosaic virus*

GENOME ORGANIZATION AND REPLICATION

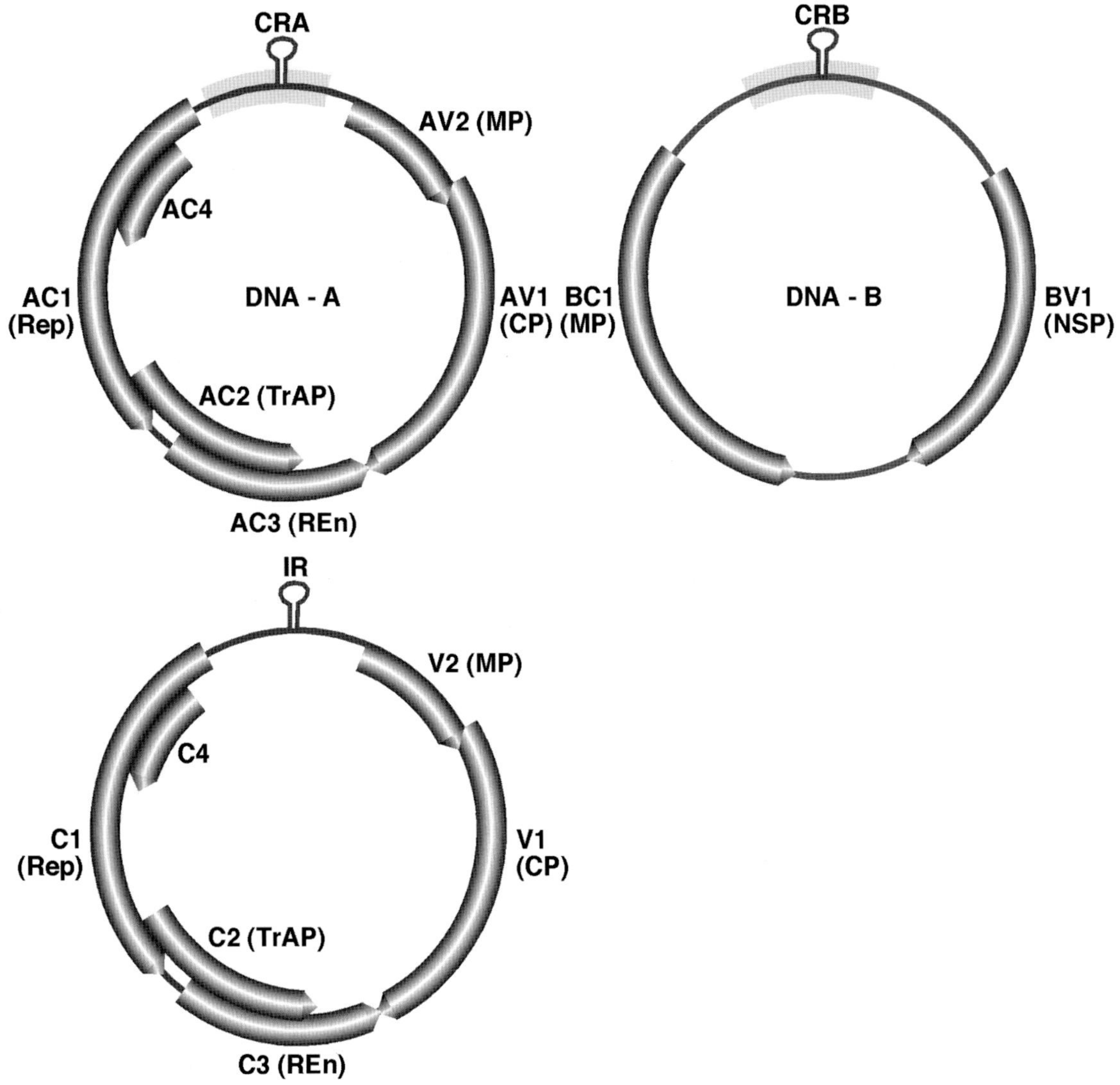

Figure 5: Genomic organization of begomoviruses. ORFs are denoted as either being encoded on the virion-sense (V) or complementary-sense (C) strand, preceded by component designation (A or B) if bipartite (top). The position of the stem-loop, containing the conserved TAATATTAC sequence, is located either within the "common region" (CRA and CRB; grey boxes representing largely intergenic sequences that are shared between the two genomic components of bipartite viruses) or the "intergenic region" (IR, bottom). CP; capsid protein, Rep; replication-associated protein, TrAP; transcriptional activator protein, REn; replication enhancer, MP; movement protein, NSP; nuclear shuttle protein.

Alteration to Figure 5: the ORFs of the monopartite begomoviruses should not have the prefix "A". Hence, change to V1 (CP), V2 (MP), C1 (Rep) and C2 (TrAP).

The genomes of most begomoviruses consist of two components, referred to as DNA A and DNA B, each 2.5-2.8 kb in size. The DNA A component of the bipartite begomoviruses can replicate autonomously and produce virions but requires DNA B for systemic infection. DNA A and DNA B share approximately 200 bp of sequence within the intergenic region, encompassing the conserved stem-loop and TAATATTAC sequence, that is termed the common region (Fig. 5). Some begomoviruses from the Old World have only a single genomic component resembling DNA A, for example Ageratum yellow vein virus (AYVV), Tomato yellow leaf curl virus (TYLCV) and Tomato leaf curl virus (ToLCV).

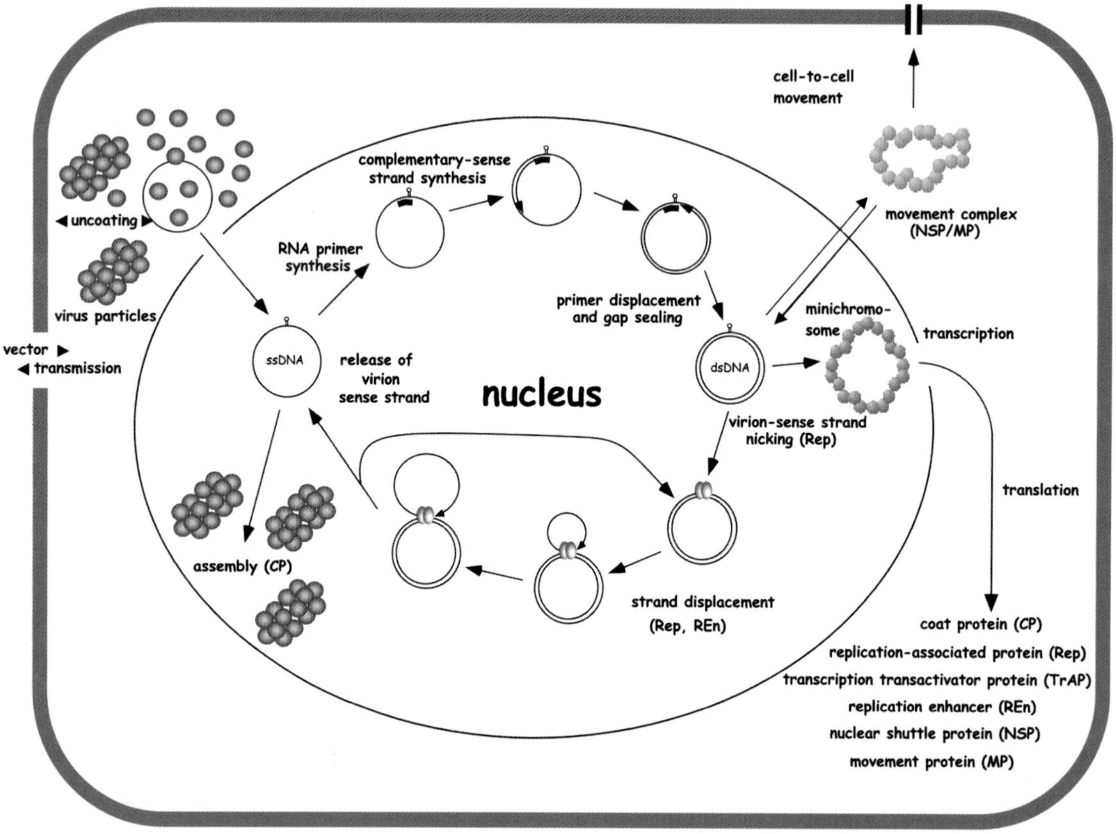

Figure 6: Diagram of the begomovirus replication strategy. Virus particles are introduced into the cell during whitefly feeding and the uncoated ssDNA is replicated in the nucleus. The mechanism of initial entry into the nucleus is not yet understood. A dsDNA intermediate is synthesised from the ssDNA template by host factors alone. ssDNA is synthesised from the dsDNA template by a rolling circle mechanism involving Rep and REn viral proteins in association with host factors, and is assembled into virus particles that accumulate in the nucleus. dsDNA adopts a transcriptionally active minichromosome form. Conflicting reports have suggested the involvement of either dsDNA (shown here) or ssDNA in nuclear transport and cell-to-cell movement, mediated by NSP and MP.

The nt sequences of full-length genomic clones of more than 100 distinct species have been established, and recombination between species occurs frequently. The DNA A virion-sense strand encodes the CP (ORF AV1/V1), that encapsidates the virion-sense ssDNA and may be involved in virus movement, and ORF AV2/V2, that has also been implicated in virus movement. The New World bipartite viruses lack an AV2 ORF. The DNA A complementary-sense strand encodes the replication-associated protein (Rep, ORF AC1/C1), a transcriptional activator protein (TrAP, ORF AC2/C2), a replication enhancer protein (REn, ORF AC3/C3), and C4 protein (ORF AC4/C4). Rep initiates viral DNA replication by binding to reiterated motifs (iterons) within the intergenic region and introducing a nick into the conserved TAATATT/AC sequence. Rep binding to the iterons is highly sequence-specific and determines the replicational competence of pseudo-recombinants. Rep also binds to the plant homologue of retinoblastoma protein (Rb) to regulate cell-cycle progression, altering the environment of terminally differentiated cells to provide host factors that support viral DNA replication. TrAP transactivates expression of virion-sense gene expression from both DNA A and DNA B, and also functions in the suppression of post-transcriptional gene silencing. REn is required for efficient viral DNA replication. C4 protein is an important symptom determinant implicated in cell-cycle control, and AC4 protein may counter a host response to Rep expression. DNA B encodes a nuclear shuttle protein (NSP, ORF BV1) on the virion-sense strand and a MP (ORF BC1) on the complementary-sense strand (Fig. 6).

Small circular ssDNA satellites, approximately 1.3 kb in size, are associated with some Old World monopartite begomoviruses (Table 1). Nothing is currently known about their evolutionary origin. The satellites, termed DNA β, have a conserved coding region (ORF βC1) that plays an important role in symptom induction, and contain the ubiquitous stem-loop and TAATATTAC sequences, but are otherwise unrelated to their helper begomovirus. The satellites depend on the begomoviruses for their replication and encapsidation, and are essential for maintenance of the disease in the field. Similar sized autonomously replicating components of unknown biological function, termed DNA 1, are frequently associated with these begomovirus/DNA β complexes (Table 1). DNA 1 components are believed to derive from nanovirus components that have become adapted to whitefly transmission by encapsidation within the begomovirus CP.

Table 1: Monopartite begomoviruses that have associated DNA β and DNA 1 components

Begomovirus species	DNA β accession number	DNA 1 accession number
Ageratum yellow vein virus	AJ252072	AJ238493, AJ416153
Bhendi yellow vein mosaic virus	AJ308425	
Chilli leaf curl virus	AJ316032	
Cotton leaf curl Multan virus	AJ292769, AJ298903	AJ132344-5
Cotton leaf curl Rajasthan virus	AY083590	
Eupatorium yellow vein virus	AJ438938-9	
Hollyhock leaf crumple virus	AF397214, AJ316044	
Honeysuckle yellow vein virus	AJ316040, AJ543430	
Malvastrum yellow vein virus	AJ421482	
Papaya leaf curl virus	AY244706	
Tobacco curly shoot virus	AJ421484-5, AJ457821-2	AJ579346-52
Tobacco leaf curl Yunnan virus	AJ536621-2, AJ536628	AJ579361
Tomato leaf curl virus	U74627	
Tomato yellow leaf curl China virus	AJ421483, AJ420313-5, AJ421619-23, AJ457818-20, AJ506791	AJ579353-8
Tomato yellow leaf curl Thailand virus	AJ566746-8	AJ579359-60

Ensure that the heading "DNA 1 accession number" does not wrap around in the final version.

ANTIGENIC PROPERTIES

Serological tests show all begomoviruses to be relatively closely related. The use of monoclonal antibodies has shown that begomoviruses may be grouped geographically based on shared epitopes.

BIOLOGICAL PROPERTIES

HOST RANGE

Collectively, begomoviruses infect a wide range of dicotyledonous plants although, individually, most have limited host ranges.

TRANSMISSION

Transmitted in nature by the whitefly *Bemisia tabaci* (Genn.), although some scientists distinguish the "B" or silverleaf biotype as *Bemisia argentifolii*. Some begomoviruses are known to be differentially adapted for efficient transmission by their local *B. tabaci* biotype. Some begomoviruses are transmissible by mechanical inoculation, although many require either *Agrobacterium*-mediated transfer (agroinoculation) from partially or

SSDNA

tandemly repeated cloned genomic DNA or biolistic delivery of cloned genomic DNA for their transmission.

LIST OF SPECIES DEMARCATION CRITERIA IN THE GENUS

The following criteria should be used as a guideline to establish taxonomic status:

- Number of genomic components. Presence or absence of a DNA B component
- Organization of the genome. Presence or absence of ORF AV2.
- Nucleotide sequence identity. Because of the growing number of recognized species, derivation of the complete nt sequence will be necessary to distinguish species. Nucleotide sequence identity <89% is generally indicative of a distinct species. However, decisions based on nt sequence comparisons, particularly when approaching this value, must also take into account the biological properties of the virus. The taxonomic status of a recombinant will depend on relatedness to the parental viruses, the frequency and extent of recombination events, and its biological properties compared with the parental viruses. Information concerning the diversity of related recombinants may be helpful to determine status.
- *Trans*-replication of genomic components. The inability of Rep protein to *trans*-replicate a genomic component suggests a distinct species. However, when considering this criterion, it should be kept in mind that small changes in the Rep binding site of otherwise identical viruses might prevent functional interaction and recombination involving a small part of the genome may confer replication competence on a distinct species.
- Production of viable pseudorecombinants. Account should be taken of the fitness of the pseudorecombinant in the natural host(s) of the parental viruses. It should be ensured that pseudorecombinant viability is not the result of inter-component recombination.
- Capsid protein characteristics. Amino acid sequence identity <90% and substantial serological differences may be indicative of a distinct species in the first instance, but derivation of the complete sequence will be necessary to confirm taxonomic status.
- Natural host range and symptom phenotype. These characteristics may relate to a particular species but their commonest use will be to distinguish strains.

LIST OF SPECIES IN THE GENUS

Species names are in green italic script; strain names and synonyms are in black roman script; tentative species names are in blue roman script. Sequence accession numbers, and assigned abbreviations ( ) are also listed.

SPECIES IN THE GENUS

Abutilon mosaic virus		
Abutilon mosaic virus	X15983-4	(AbMV)
Abutilon mosaic virus – [HW]	U51137-8	(AbMV-[HW])
African cassava mosaic virus		
(Cassava latent virus)		
African cassava mosaic virus - [Cameroon]	AF112352-3	(ACMV-[CM])
African cassava mosaic virus - [Cameroon-DO2]	AF366902	(ACMV-[CM-DO2])
African cassava mosaic virus - [Ghana]		(ACMV-[GH])
African cassava mosaic virus - [Ivory Coast]	AF259894-5	(ACMV-[CI])
African cassava mosaic virus - [Kenya]	J02057-8	(ACMV-[KE])
African cassava mosaic virus - [Nigeria]	X17095-6	(ACMV-[NG])
African cassava mosaic virus - [Nigeria-Ogo]	AJ427910-1	(ACMV-[NG-Ogo])
African cassava mosaic virus - [Uganda]	Z83252-3	(ACMV-[UG])
African cassava mosaic virus - Uganda Mild	AF126800-1	(ACMV-UGMld)
African cassava mosaic virus - Uganda Severe	AF126802-3	(ACMV-UGSvr)
Ageratum enation virus		
Ageratum enation virus	AJ437618	(AEV)
Ageratum yellow vein China virus		
Ageratum yellow vein China virus - [Hn2]	AJ495813	(AYVCNV-[Hn2])

Ageratum yellow vein China virus - [Hn2.19]	AJ564744	(AYVCNV-[Hn2.19])
Ageratum yellow vein Sri Lanka virus		
Ageratum yellow vein Sri Lanka virus	AF314144	(AYVSLV)
Ageratum yellow vein Taiwan virus		
Ageratum yellow vein Taiwan virus - [Taiwan]	AF307861,	(AYVTV-[Tai])
Ageratum yellow vein Taiwan virus - [TaiwanPD]	AF327902	(AYVTV-[TaiPD])
Ageratum yellow vein virus		
Ageratum yellow vein virus	X74516	(AYVV)
Ageratum yellow vein virus – [Tomato]	AB100305	(AYVV-[Tom])
Bean calico mosaic virus		
Bean calico mosaic virus	AF110189-90	(BcaMV)
Bean dwarf mosaic virus		
Bean dwarf mosaic virus	M88179-80	(BDMV)
Bean golden mosaic virus		
(Bean golden mosaic virus - Brazil)		
Bean golden mosaic virus - [Brazil]	M88686-7	(BGMV-[BR])
Bean golden yellow mosaic virus		
(Bean golden mosaic virus - Puerto Rico)		
Bean golden yellow mosaic virus - [Dominican Republic]	L01635-6	(BGYMV-[DO])
(Bean golden mosaic virus - Puerto Rico [Dominicar Republic])		
(Bean golden mosaic virus - Dominican Rep.)		
Bean golden yellow mosaic virus - [Guatemala]	M91604-5	(BGYMV-[GT])
(Bean golden mosaic virus - Puerto Rico [Guatemala])		
(Bean golden mosaic virus - Guatemala)		
Bean golden yellow mosaic virus - [Mexico]	AF173555-6	(BGYMV-[MX])
Bean golden yellow mosaic virus - [Puerto Rico]	M10070, M10080	(BGYMV-[PR])
(Bean golden mosaic virus - Puerto Rico)		
Bean golden yellow mosaic virus - [Puerto Rico-Japan]	D00200-1	(BGYMV-[PR-JR])
Bhendi yellow vein mosaic virus		
(Okra yellow vein mosaic virus)		
Bhendi yellow vein mosaic virus - [301]	AJ002453	(BYVMV-[301])
Bhendi yellow vein mosaic virus - [Madurai]	AF241479	(BYVMV-[Mad])
Cabbage leaf curl virus		
Cabbage leaf curl virus	U65529-30	(CaLCuV)
Chayote yellow mosaic virus		
Chayote yellow mosaic virus	AJ223191	(ChaYMV)
Chilli leaf curl virus		
Chilli leaf curl virus - [Multan]	AF336806	(ChiLCuV-[Mul])
Chino del tomate virus		
(Tomato leaf crumple virus)		
Chino del tomate virus	U57458, AF007823 L27267-8	(CdTV)
Chino del tomate virus - [B52]	AF226666	(CdTV -[B52])
Chino del tomate virus - [H6]	AF226665	(CdTV -[H6])
Chino del tomate virus - [H8]	AF226664	(CdTV -[H8])
Chino del tomate virus - [IC]	AF101476, AF101478	(CdTV -[IC])
Cotton leaf crumple virus		
Cotton leaf crumple virus	AF480940-1	(CLCrV)
Cotton leaf crumple virus – [AZ]	AY083350	(CLCrV-[AZ])

Cotton leaf crumple virus – [TX]	AY083351	(CLCrV-[TX])
Cotton leaf curl Alabad virus		
(Cotton leaf curl virus - Pakistan3)		
Cotton leaf curl Alabad virus - [802a]	AJ002455	(CLCuAV-[802a])
Cotton leaf curl Alabad virus - [804a]	AJ002452	(CLCuAV-[804a])
Cotton leaf curl Gezira virus		
(Okra enation virus)		
Cotton leaf curl Gezira virus	AF155064	(CLCuGV)
Cotton leaf curl Gezira virus - [Cotton]	AF260241	(CLCuGV-[Cot])
Cotton leaf curl Gezira virus - [Okra-Egypt]	AY036010	(CLCuGV-[Okr-EG])
Cotton leaf curl Gezira virus - [Okra-Gezira]	AY036006	(CLCuGV-[Okr-Gez])
Cotton leaf curl Gezira virus - [Okra-Shambat]	AY036008	(CLCuGV-[Okr-Sha])
Cotton leaf curl Gezira virus - [Sida]	AY036007	(CLCuGV-[Sida])
Cotton leaf curl Kokhran virus		
(Cotton leaf curl virus - Pakistan2)		
(Pakistani cotton leaf curl virus)		
Cotton leaf curl Kokhran virus - [72b]	AJ002448	(CLCuKV-[72b])
Cotton leaf curl Kokhran virus - [806b]	AJ002449	(CLCuKV-[806b])
Cotton leaf curl Kokhran virus - [Faisalabad1]	AJ496286	(CLCuKV-[Fai1])
(Cotton leaf curl virus - Pakistan2 [Faisalabad1])		
Cotton leaf curl Multan virus		
(Cotton leaf curl virus - Pakistan1)		
Cotton leaf curl Multan virus - [26]	AJ002458	(CLCuMV-[26])
Cotton leaf curl Multan virus - [62]	AJ002447	(CLCuMV-[62])
Cotton leaf curl Multan virus - [Faisalabad1]	X98995	(CLCuMV-[Fai1])
(Cotton leaf curl virus - Pakistan1 [Faisalabad1])		
Cotton leaf curl Multan virus - [Faisalabad2]	AJ496287	(CLCuMV-[Fai2])
(Cotton leaf curl virus - Pakistan1 [Faisalabad2])		
Cotton leaf curl Multan virus - [Faisalabad3]	AJ132430	(CLCuMV-[Fai3])
Cotton leaf curl Multan virus - [Multan]	AJ496461	(CLCuMV-[Mul])
(Cotton leaf curl virus - Pakistan1 [Multan])		
Cotton leaf curl Multan virus - [Okra]	AJ002459	(CLCuMV-[Ok])
(Cotton leaf curl virus - Pakistan1 [Okra])		
Cotton leaf curl Rajasthan virus		
Cotton leaf curl Rajasthan virus	AF363011	(CLCuRV)
Cowpea golden mosaic virus		
Cowpea golden mosaic virus -[Brazil]	AF188708	(CPGMV-[BR])
Cowpea golden mosaic virus -[India]	AF289058-9	(CPGMV-[IN])
Cowpea golden mosaic virus - [Nigeria]	AF029217	(CPGMV-[NG])
Croton yellow vein mosaic virus		
Croton yellow vein mosaic virus	AJ507777	(CYVMV)
Cucurbit leaf curl virus		
Cucurbit leaf curl virus	AF224760-1	(CuLCuV)
Cucurbit leaf curl virus - [Arizona]	AF256200, AF327559	(CuLCuV-[AZ])
Dicliptera yellow mottle virus		
Dicliptera yellow mottle virus	AF170101, AF139168	(DiYMoV)
Dolichos yellow mosaic virus		
Dolichos yellow mosaic virus	AY309241	(DoYMV)
East African cassava mosaic Cameroon virus		
(West African cassava mosaic virus)		
East African cassava mosaic Cameroon virus	AF112354-5	(EACMCV)
East African cassava mosaic Cameroon virus -	AF259896-7	(EACMCV-[CI])

[Ivory Coast]		
East African cassava mosaic Malawi virus		
(East African cassava mosaic virus–Malawi)		
East African cassava mosaic Malawi virus - [K]	AJ006460	(EACMMV-[K])
East African cassava mosaic Malawi virus - [MH]	AJ006459	(EACMMV-[MH])
East African cassava mosaic virus		
East African cassava mosaic virus - Uganda2 (Uganda variant)	Z83257	(EACMV-UG2)
East African cassava mosaic virus - Uganda2 Mild	AF126804	(EACMV-UG2Mld)
East African cassava mosaic virus - Uganda2 Severe	AF126806	(EACMV-UG2Svr)
East African cassava mosaic virus - Uganda3 Mild	AF126805	(EACMV-UG3Mld)
East African cassava mosaic virus - Uganda3 Severe	AF126807, AF230374	(EACMV-UG3Svr)
East African cassava mosaic virus - [Kenya – K2B]	AJ006458	(EACMV-[KE-K2B])
East African cassava mosaic virus - [Malawi]	AJ006461	(EACMV-[MW])
East African cassava mosaic virus - [Tanzania]	Z83256	(EACMV-[TZ])
East African cassava mosaic virus - [Uganda1]	AF230375	(EACMV-[UG1])
East African cassava mosaic Zanzibar virus		
East African cassava mosaic Zanzibar virus	AF422174-5	(EACMZV)
East African cassava mosaic Zanzibar virus - Kenya [Kilifi]	AJ516003	(EACMZV-KE[Kil])
Eupatorium yellow vein virus		
Eupatorium yellow vein virus	AB007990	(EpYVV)
Eupatorium yellow vein virus - [Tobacco]	E15418	(EpYVV-[Tob])
Eupatorium yellow vein virus - [MNS2]	AJ438936	(EpYVV-[MNS2])
Eupatorium yellow vein virus - [SOJ3]	AJ438937	(EpYVV-[SOJ3])
Eupatorium yellow vein virus - [Yamaguchi]	AB079766	(EpYVV-[Yam])
Euphorbia leaf curl virus		
Euphorbia leaf curl virus – [G35]	AJ558121	(EuLCV-[G35])
Hollyhock leaf crumple virus		
Hollyhock leaf crumple virus - [Giza] (Althea rosea enation virus - [Giza])	AF014881	(HoLCrV-[Giz])
Hollyhock leaf crumple virus – [Cairo]	AY036009	(HoLCrV-[Cai])
Hollyhock leaf crumple virus - [Cairo2]	AJ542539	(HolCrV-[Cai2])
Honeysuckle yellow vein mosaic virus		
Honeysuckle yellow vein mosaic virus	AB020781	(HYVMV)
Honeysuckle yellow vein mosaic virus- [Yamaguchi]	AB079765	(HYVMV-[Yam])
Honeysuckle yellow vein virus		
Honeysuckle yellow vein virus-[UK1]	AJ542540	(HYVV-[UK1])
Honeysuckle yellow vein virus-[UK2]	AJ543429	(HYVV-[UK2])
Indian cassava mosaic virus		
Indian cassava mosaic virus	Z24758-9	(ICMV)
Indian cassava mosaic virus - [Kattukuda]	AJ575821	(ICMV-[Kat])
Indian cassava mosaic virus - [Maharashstra]	AJ314739-40	(ICMV-[Mah])
Indian cassava mosaic virus - [Muvattupuzzha]	AJ575820	(ICMV-[Muv])
Ipomoea yellow vein virus		
Ipomoea yellow vein virus (Sweet potato leaf curl virus - [Ipo])	AJ132548	(IYVV)
Luffa yellow mosaic virus		
Luffa yellow mosaic virus	AF509739-40	(LYMV)
Macroptilium mosaic Puerto Rico virus		
Macroptilium mosaic Puerto Rico virus	AY044133-4	(MaMPRV)

Macroptilium mosaic Puerto Rico virus - [Bean]	AF449192-3	(MaMPRV-[Bea])
Macroptilium yellow mosaic Florida virus		
Macroptilium yellow mosaic Florida virus	AY044135-6	(MaYMFV)
Macroptilium yellow mosaic virus		
Macroptilium yellow mosaic virus - [Cuba]	AJ344452	(MaYMV-[CU])
Malvastrum yellow vein virus		
Malvastrum yellow vein virus - [Y47]	AJ457824	(MYVV-[Y47])
Melon chlorotic leaf curl virus		
Melon chlorotic leaf curl virus - [Guatemala]	AF325497	(MCLCuV-[GT])
Mungbean yellow mosaic India virus		
Mungbean yellow mosaic India virus	AF126406, AF142440	(MYMIV)
Mungbean yellow mosaic India virus - [Akola]	AY271893-4	(MYMIV-[Ako])
Mungbean yellow mosaic India virus - [Bangladesh]	AF314145	(MYMIV-[BD])
Mungbean yellow mosaic India virus - [Cowpea]	AF481865, AF503580	(MYMIV-[Cp])
Mungbean yellow mosaic India virus - [Cowpea Pakistan]	AY269990	(MYMIV-[CpPK])
Mungbean yellow mosaic India virus - [Mungbean]	AF416742, AF416741	(MYMIV-[Mg])
Mungbean yellow mosaic India virus - [Nepal]	AY271895	(MYMIV-[NP])
Mungbean yellow mosaic India virus - [Mungbean Pakistan]	AY269992	(MYMIV-[MgPK])
Mungbean yellow mosaic India virus - [Soybean]	AY049771-2	(MYMIV-[Sb])
Mungbean yellow mosaic India virus-[Soybean]	AJ416349, AJ420331	(MYMIV-[SbTN])
Mungbean yellow mosaic India virus-[Vigna 14]	AY512495	(MYMIV-[Vig14])
Mungbean yellow mosaic India virus-[Vigna 106]	AY512498	(MYMIV-[Vig106])
Mungbean yellow mosaic India virus-[Vigna 130.7]	AY512496	(MYMIV-[Vig130.7])
Mungbean yellow mosaic India virus-[Vigna 130.12]	AY512497	(MYMIV-[Vig130.12])
Mungbean yellow mosaic virus		
Mungbean yellow mosaic virus	D14703-4	(MYMV)
Mungbean yellow mosaic virus – [Aryana]	AY512496	(MYMV-[Ary])
Mungbean yellow mosaic virus - Soybean [Madurai]	AJ421642	(MYMV-Sb[Mad])
Mungbean yellow mosaic virus - Soybean [Pakistan]	AY269991	(MYMV-Sb[PK])
Mungbean yellow mosaic virus – Thailand	AB017341	(MYMV-TH)
Mungbean yellow mosaic virus – Vigna	AJ132574-5	(MYMV-Vig)
Mungbean yellow mosaic virus - Vigna [KA21]	AJ439059	(MYMV-Vig[KA21])
Mungbean yellow mosaic virus - Vigna [KA27]	AF262064	(MYMV-Vig[KA27])
Mungbean yellow mosaic virus - Vigna [KA28]	AJ439058	(MYMV-Vig[KA28])
Mungbean yellow mosaic virus - Vigna [KA34]	AJ439057	(MYMV-Vig[KA34])
Mungbean yellow mosaic virus - Vigna [Maharashstra]	AF314530	(MYMV-Vig[Mah])
Okra yellow vein mosaic virus		
Okra yellow vein mosaic virus - [201]	AJ002451	(OYVMV-[201])
Papaya leaf curl China virus		
Papaya leaf curl China virus – [G2]	AJ558123	(PaLCuCNV-[G2])
Papaya leaf curl China virus – [G8]	AJ558124	(PaLCuCNV-[G8])
Papaya leaf curl China virus – [G10]	AJ558125	(PaLCuCNV-[G10])
Papaya leaf curl China virus – [G12]	AJ558116	(PaLCuCNV-[G12])
Papaya leaf curl China virus – [G30]	AJ558117	(PaLCuCNV-[G30])

Papaya leaf curl Guandong virus		
Papaya leaf curl Guandong virus – [GD2]	AJ558122	(PaLCuGV-[GD2])
Papaya leaf curl virus		
Papaya leaf curl virus	Y15934, Y07962	(PaLCuV)
Pepper golden mosaic virus		
(Serrano golden mosaic virus)		
(Texas pepper virus)		
Pepper golden mosaic virus	U57457, AF499442 AF075591, AF077025	(PepGMV)
Pepper golden mosaic virus - [CR]	AF149227	(PepGMV-[CR])
Pepper golden mosaic virus - [GTS8]	AF136404	(PepGMV-[GTS8])
Pepper huasteco yellow vein virus		
(Pepper huasteco virus)		
Pepper huasteco yellow vein virus	X70418-9	(PHYVV)
Pepper huasteco yellow vein virus - [Sinaloa]	AY044162-3,	(PHYVV-[Sin])
Pepper leaf curl Bangladesh virus		
Pepper leaf curl Bangladesh virus	AF314531	(PepLCBV)
Pepper leaf curl virus		
Pepper leaf curl virus	AF134484	(PepLCV)
Pepper leaf curl virus - [Malaysia]	AF414287	(PepLCV-[MY])
Potato yellow mosaic Panama virus		
Potato yellow mosaic Panama virus	Y15033-4	(PYMPV)
(Potato yellow mosaic virus – Panama)		
(Tomato leaf curl virus - Panama)		
Potato yellow mosaic Trinidad virus		
Potato yellow mosaic Trinidad virus - [Trinidad & Tobago]	AF039031-2	(PYMTV-[TT])
Potato yellow mosaic virus		
Potato yellow mosaic virus - [Venezuela]	D00940-1	(PYMV-[VE])
Potato yellow mosaic virus - [Guadeloupe]	AY120882-3	(PYMV-[GP])
Rhynchosia golden mosaic virus		
Rhynchosia golden mosaic virus	AF239671	(RhGMV)
Rhynchosia golden mosaic virus - [Chiapas]	AF408199	(RhGMV-[Chi])
Sida golden mosaic Costa Rica virus		
Sida golden mosaic Costa Rica virus	X99550-1	(SiGMCRV)
Sida golden mosaic Florida virus		
Sida golden mosaic Florida virus - [A1]	U77963	(SiGMFV-[A1])
Sida golden mosaic Honduras virus		
Sida golden mosaic Honduras virus	Y11097-8	(SiGMHV)
Sida golden mosaic virus		
Sida golden mosaic virus	AF049336, AF039841	(SiGMV)
Sida golden yellow vein virus		
Sida golden yellow vein virus - [A11]	U77964	(SiGYVV-[A11])
(Sida golden mosaic Florida virus - [A11])		
Sida mottle virus		
Sida mottle virus - [Brazil]	AY090555	(SiMoV-[BR])
Sida yellow mosaic virus		
Sida yellow mosaic virus - [Brazil]	AY090558	(SiYMV-[BR])
Sida yellow vein virus		
Sida yellow vein virus	Y11099-101	(SiYVV)
(Sida golden mosaic Honduras virus - yellow vein)		

South African cassava mosaic virus		
South African cassava mosaic virus	AF155806-7	(SACMV)
South African cassava mosaic virus - [M12]	AJ422132	(SACMV-[M12])
South African cassava mosaic virus - [ZW]	AJ575560	(SACMV-[ZW])
Soybean crinkle leaf virus		
Soybean crinkle leaf virus - [Japan]	AB050781, AB020977	(SbCLV-[JR])
Squash leaf curl China virus		
Squash leaf curl China virus	AB027465	(SLCCNV)
Squash leaf curl China virus – [B]	AF509742-3	(SLCCNV-[B])
Squash leaf curl China virus – [K]	AF509741	(SLCCNV-[K])
Squash leaf curl Philippines virus		
Squash leaf curl Philippines virus	AB085793-4	(SLCPHV)
Squash leaf curl virus		
Squash leaf curl virus	M38182-3	(SLCV)
Squash leaf curl virus - Extended host	M63157-8	(SLCV-E)
Squash leaf curl virus - [Los Mochis]	L27272-3	SLCV-[Lmo])
Squash leaf curl Yunnan virus		
Squash leaf curl Yunnan virus	AJ420319	(SLCYNV)
Squash mild leaf curl virus		
Squash mild leaf curl virus - [Imperial Valley] (Squash leaf curl virus - R)	M63155-6, L20240, AF421552-3	(SMLCV-[IV])
Squash yellow mild mottle virus		
Squash yellow mild mottle virus - [CR]	AF124846, AF136447, AF440790, AY005134, AY064391	(SYMMoV-[CR])
Sri Lankan cassava mosaic virus		
Sri Lankan cassava mosaic virus - [Adivaram]	AJ579307-8	(SLCMV-[Adi])
Sri Lankan cassava mosaic virus - [Adivaram2]	AJ575819	(SLCMV-[Adi2])
Sri Lankan cassava mosaic virus - [Colombo]	AJ314737-8	(SLCMV-[Col])
Stachytarpheta leaf curl virus		
Stachytarpheta leaf curl virus - [Hn5]	AJ495814	(StaLCuV-[Hn5])
Stachytarpheta leaf curl virus - [Hn5.4]	AJ564743	(StaLCuV-[Hn5.4])
Stachytarpheta leaf curl virus - [Hn6.1]	AJ564742	(StaLCuV-[Hn6.1])
Sweet potato leaf curl Georgia virus		
Sweet potato leaf curl Georgia virus - [16]	AF326775	(SPLCGV-[16])
Sweet potato leaf curl virus		
Sweet potato leaf curl virus	AF104036	(SPLCV)
Tobacco curly shoot virus		
(Tobacco leaf curl virus - China)		
Tobacco curly shoot virus - [Y1]	AF240675	(TbCSV-[Y1])
Tobacco curly shoot virus - [Y35]	AJ420318	(TbCSV-[Y35])
Tobacco curly shoot virus - [Y41]	AJ457986	(TbCSV-[Y41])
Tobacco leaf curl Japan virus		
(Tobacco leaf curl virus - Japan)		
Tobacco leaf curl Japan virus	AB028604	(TbLCJV)
Tobacco leaf curl Japan virus - [Japan2]	AB055008	(TbLCJV-[JR2])
Tobacco leaf curl Japan virus - [Japan3]	AB079689	(TbLCJV-[JR3])
Tobacco leaf curl Kochi virus		
Tobacco leaf curl Kochi virus - [KK]	AB055009	(TbLCKoV-[KK])
Tobacco leaf curl Yunnan virus		
Tobacco leaf curl Yunnan virus - [Y3]	AF240674	(TbLCYNV-[Y3])

Tobacco leaf curl Yunnan virus - [Y136]	AJ512761	(TbLCYNV-[Y136])
Tobacco leaf curl Yunnan virus - [Y143]	AJ512762	(TbLCYNV-[Y143])
Tobacco leaf curl Yunnan virus - [Y161]	AJ566744	(TbLCYNV-[Y161])
Tobacco leaf curl Zimbabwe virus		
Tobacco leaf curl virus – [Zimbabwe]	AF350330	(TbLCZV-[ZW])
Tomato chino La Paz virus		
Tomato chino La Paz virus	AY339618	(ToChLPV)
Tomato chino La Paz virus – [Baja California Sur]	AY339619	(ToChLPV-[BCS])
Tomato chlorotic mottle virus		
Tomato chlorotic mottle virus - [Brazil]	AF490004, AF491306, AY049213	(ToCMoV-[BR])
Tomato chlorotic mottle virus - Crumple	AY090557	(ToCMoV-Cr)
Tomato curly stunt virus		
Tomato curly stunt virus - [South Africa]	AF261885	(ToCSV-[ZA])
Tomato golden mosaic virus		
Tomato golden mosaic virus - Common	M73794	(TGMV-Com)
Tomato golden mosaic virus - Yellow vein	K02029-30	(TGMV-YV)
Tomato golden mottle virus		
Tomato golden mottle virus - [GT94-R2]	AF132852, AF138298	(ToGMoV-[GT94-R2])
Tomato leaf curl Bangalore virus		
(Tomato leaf curl virus - Bangalore 1)		
(Indian tomato leaf curl virus – Bangalore 1)		
Tomato leaf curl Bangalore virus	L12738-9, Z48182	(ToLCBV)
Tomato leaf curl Bangalore virus – [Ban4]	AF165098	(ToLCBV-[Ban4])
Tomato leaf curl Bangalore virus – [Ban5]	AF295401	(ToLCBV-[Ban5])
Tomato leaf curl Bangalore virus – [Kolar]	AF428255	(ToLCBV-[Kol])
Tomato leaf curl Bangladesh virus		
Tomato leaf curl Bangladesh virus	AF188481	(ToLCBDV)
Tomato leaf curl China virus		
Tomato leaf curl China virus - [G18]	AJ558119	(ToLCCNV-[G18])
Tomato leaf curl China virus - [G32]	AJ558118	(ToLCCNV-[G32])
Tomato leaf curl Gujarat virus		
Tomato leaf curl Gujarat virus - [Mirzapur]	AF449999	(ToLCGV-[Mir])
Tomato leaf curl Gujarat virus - [Nepal]	AY234383	(ToLCGV-[NP])
Tomato leaf curl Gujarat virus - [Vadodara]	AF413671	(ToLCGV-[Vad])
Tomato leaf curl Gujarat virus - [Varanasi]	AY190290-1	(ToLCGV-[Var])
Tomato leaf curl Indonesia virus		
Tomato leaf curl Indonesia virus	AB100304	(ToLCIDV)
Tomato leaf curl Iran virus		
Tomato leaf curl Iran virus	AY297924	(ToLCIRV)
Tomato leaf curl Karnataka virus		
(Tomato leaf curl virus – Bangalore 2)		
(Indian tomato leaf curl virus – Bangalore II)		
Tomato leaf curl Karnataka virus	U38239	(ToLCKV)
Tomato leaf curl Laos virus		
Tomato leaf curl Laos virus	AF195782	(ToLCLV)
Tomato leaf curl Malaysia virus		
Tomato leaf curl Malaysia virus	AF327436	(ToLCMV)
Tomato leaf curl New Delhi virus		
(Tomato leaf curl virus - New Delhi)		
(Tomato leaf curl virus - India2)		
Tomato leaf curl New Delhi virus - Mild	U15016	(ToLCNDV-Mld)

(Tomato leaf curl virus - New Delhi [Mild])		
Tomato leaf curl New Delhi virus - Severe	U15015, U15017	(ToLCNDV-Svr)
(Tomato leaf curl virus - New Delhi [Severe])		
Tomato leaf curl New Delhi virus - [Lucknow]	X89653, X78956,	(ToLCNDV-[Luc])
(Tomato leaf curl virus - New Delhi [Lucknow])	Y16421	
Tomato leaf curl New Delhi virus - [Luffa]	AF102276	(ToLCNDV-[Luf])
(Tomato leaf curl virus - New Delhi [Luffa])		
(Angled luffa leaf curl virus)		
Tomato leaf curl New Delhi virus - [PkT1/8]	AF448059,	(ToLCNDV-[PkT1/8])
(Tomato leaf curl virus – Pakistan [T1/8])	AY150304	
Tomato leaf curl New Delhi virus - [PkT5/6]	AF448058,	(ToLCNDV-[PkT5/6])
(Tomato leaf curl virus – Pakistan [T5/6])	AY150305	
Tomato leaf curl New Delhi virus - [Potato]	AY286316,	(ToLCNDV-[Pot])
	AY158080	
Tomato leaf curl Philippines virus		
Tomato leaf curl Philippines virus	AB050597	(ToLCPV)
Tomato leaf curl Philippines virus - [LB]	AF136222	(ToLCPV-[LB])
Tomato leaf curl Sri Lanka virus		
Tomato leaf curl Sri Lanka virus	AF274349	(ToLCSLV)
Tomato leaf curl Sudan virus		
Tomato leaf curl Sudan virus – [Gezira]	AY044137	(ToLCSDV-[Gez])
Tomato leaf curl Sudan virus – [Shambat]	AY044139	(ToLCSDV-[Sha])
Tomato leaf curl Taiwan virus		
(Tomato leaf curl virus - Taiwan)		
Tomato leaf curl Taiwan virus	U88692	(ToLCTWV)
Tomato leaf curl Vietnam virus		
Tomato leaf curl Vietnam virus	AF264063	(ToLCVV)
Tomato leaf curl virus		
(Tomato leaf curl virus - Australia)		
Tomato leaf curl virus - [AU]	S53251	(ToLCV)
Tomato leaf curl virus - [Solanum species D1]	U51893, AF084006	(ToLCV-[SSpD1])
Tomato leaf curl virus - [Solanum species D2]	U51894, AF084007	(ToLCV-[SSpD2])
Tomato mosaic Havana virus		
(Havana tomato mosaic virus)		
Tomato mosaic Havana virus - [Honduras 96-H5]	AF139078	(ToMHV-[HN96-H5])
Tomato mosaic Havana virus - [Jamaica]	AF035224-5	(ToMHV-[JM])
Tomato mosaic Havana virus - [Quivican]	Y14874-5	(ToMHV-[Qui])
Tomato mottle Taino virus		
(Tomato mottle virus - Taino)		
(Taino tomato mottle virus)		
Tomato mottle Taino virus	AF012300-1	(ToMoTV)
Tomato mottle virus		
Tomato mottle virus - [Florida]	L14460-1, U65506-8	(ToMoV-[FL])
Tomato mottle virus - [Florida - B1]	M90495, L02618	(ToMoV-[FL-B1])
Tomato rugose mosaic virus		
Tomato rugose mosaic virus – [Ube]	AF291705-6	(ToRMV-[Ube])
Tomato severe leaf curl virus		
Tomato severe leaf curl virus - [Guatemala96-1]	AF130415	(ToSLCV-[GT96-1])
Tomato severe leaf curl virus - [Guatemala97-Cu1]	AF131735	(ToSLCV-[GT97-Cu1])
Tomato severe leaf curl virus - [Honduras96 - T1]	AF130416	(ToSLCV-[HN96-T1])
Tomato severe leaf curl virus - [Nicaragua]	AJ277059-61	(ToSLCV-[NI])
Tomato severe rugose virus		
Tomato severe rugose virus	AY029750	(ToSRV)

Tomato yellow leaf curl China virus		
(Tomato yellow leaf curl virus - China)		
Tomato yellow leaf curl China virus	D88773, AF186752-3, AF311734	(TYLCCNV)
Tomato yellow leaf curl China virus - [Y64]	AJ457823	(TYLCCNV-[Y64])
Tomato yellow leaf curl China virus - Tb [Y10]	AJ319675	(TYLCCNV-Tb[Y10])
Tomato yellow leaf curl China virus - Tb [Y11]	AJ319676	(TYLCCNV-Tb[Y11])
Tomato yellow leaf curl China virus - Tb [Y36]	AJ420316	(TYLCCNV-Tb[Y36])
Tomato yellow leaf curl China virus - Tb [Y38]	AJ420317	(TYLCCNV-Tb[Y38])
Tomato yellow leaf curl China virus - Tb [Y5]	AJ319674	(TYLCCNV-Tb[Y5])
Tomato yellow leaf curl China virus - Tb [Y8]	AJ319677	(TYLCCNV-Tb[Y8])
Tomato yellow leaf curl China virus - To [Y25]	AJ457985	(TYLCCNV-To[Y25])
Tomato yellow leaf curl Kanchanaburi virus		
Tomato yellow leaf curl Kanchanaburi virus – [Thailand Kan1]	AF511528, AF511529	TYLCKaV-[THKan1])
Tomato yellow leaf curl Kanchanaburi virus – [Thailand Kan2]	AF511527, AF511530	TYLCKaV-[THKan2])
Tomato yellow leaf curl Malaga virus		
Tomato yellow leaf curl Malaga virus	AF271234	(TYLCMalV)
Tomato yellow leaf curl Sardinia virus		
(Tomato yellow leaf curl virus - Sardinia)		
Tomato yellow leaf curl Sardinia virus	X61153	(TYLCSV)
(Tomato yellow leaf curl virus - Sardinia)		
Tomato yellow leaf curl Sardinia virus - [Sicily]	Z28390	(TYLCSV-[Sic])
(Tomato yellow leaf curl virus - Sardinia [Sicily])		
(Tomato yellow leaf curl virus - Sicily)		
Tomato yellow leaf curl Sardinia virus - [Spain1]	Z25751	(TYLCSV-[ES1])
(Tomato yellow leaf curl virus - Sardinia [Spain1])		
(Tomato yellow leaf curl virus - Spain)		
Tomato yellow leaf curl Sardinia virus - [Spain2]	L27708	(TYLCSV-[ES2])
(Tomato yellow leaf curl virus - Sardinia [Spain2])		
(Tomato yellow leaf curl virus - Almeria)		
Tomato yellow leaf curl Sardinia virus - [Spain3]	Z86067-8, Z92670-1	(TYLCSV-[ES3])
(Tomato yellow leaf curl virus - Sardinia [Spain3])		
(Tomato yellow leaf curl virus - European strain)		
Tomato yellow leaf curl Thailand virus		
(Tomato yellow leaf curl virus - Thailand)		
Tomato yellow leaf curl Thailand virus - [1]	X63015-6, M59838-9	(TYLCTHV-[1])
(Tomato yellow leaf curl virus - Thailand [1])		
Tomato yellow leaf curl Thailand virus - [2]	AF141897, AF141922	(TYLCTHV-[2])
(Tomato yellow leaf curl virus - Thailand [2])		
Tomato yellow leaf curl Thailand virus - [Myanmar]	AF206674	(TYLCTHV-[MM])
Tomato yellow leaf curl Thailand virus - [Y72]	AJ495812	(TYLCTHV-[Y72])
Tomato yellow leaf curl virus		
Tomato yellow leaf curl virus	X15656	(TYLCV)
(Tomato yellow leaf curl virus - Israel)		
Tomato yellow leaf curl virus - [Almeria]	AJ489258	(TYLCV-[Alm])
Tomato yellow leaf curl virus - [Cuba]	U65089	(TYLCV-[CU])
(Tomato yellow leaf curl virus - Israel [Cuba])	AJ223505	
Tomato yellow leaf curl virus - [Dominican	AF024715	(TYLCV-[DO])

SSDNA

Republic]		
(Tomato yellow leaf curl virus - Israel [DO])		
Tomato yellow leaf curl virus - [Egypt]	L12219	(TYLCV-[EG])
(Tomato yellow leaf curl virus - Israel [Egypt])		
(Tomato yellow leaf curl virus - Egypt)		
Tomato yellow leaf curl virus - [Jamaica]	U84146-7, U84397,	(TYLCV-[JM])
(Tomato yellow leaf curl virus - Israel [Jamaica])	U85782, U88889	
Tomato yellow leaf curl virus - [Lebanon]	AF160875	(TYLCV-[LB])
Tomato yellow leaf curl virus - [Puerto Rico]	AY134494	(TYLCV-[PR])
Tomato yellow leaf curl virus - [Yucatan]	AF168709	(TYLCV-[Yuc])
Tomato yellow leaf curl virus - Gezira	AY044138	(TYLCV-Gez)
Tomato yellow leaf curl virus - Iran	AJ132711	(TYLCV-IR)
(Tomato yellow leaf curl virus - Israel [Iran])		
Tomato yellow leaf curl virus - Mild	X76319	(TYLCV-Mld)
(Tomato yellow leaf curl virus - Israel [Mild])		
Tomato yellow leaf curl virus - Mild [Spain]	AJ519441	(TYLCV-Mld[ES])
Tomato yellow leaf curl virus - Mild[Aichi]	AB014347	(TYLCV-Mld[Aic])
(Tomato yellow leaf curl virus - Israel [Aichi])		
Tomato yellow leaf curl virus - Mild[Portugal]	AF105975	(TYLCV-Mld[PT])
(Tomato yellow leaf curl virus - Israel [Portugal])		
Tomato yellow leaf curl virus - Mild[Shizuokua]	AB014346	(TYLCV-Mld[Shi])
(Tomato yellow leaf curl virus - Israel [Shizuokua])		
Tomato yellow leaf curl virus - Mild[Spain7297]	AF071228	(TYLCV-Mld[ES7297])
(Tomato yellow leaf curl virus - Israel [Spain7297]		
Watermelon chlorotic stunt virus		
Watermelon chlorotic stunt virus	AJ012081-2	(WmCSV)
Watermelon chlorotic stunt virus - [IR]	AJ245652-3	(WmCSV-[IR])
Watermelon chlorotic stunt virus - [SD]	AJ245650-1	(WmCSV-[SD])

TENTATIVE SPECIES IN THE GENUS

Acalypha yellow mosaic virus		(AYMV)
Asystasia golden mosaic virus		(AGMV)
Cotton yellow mosaic virus	AF076852	(CotYMV)
Eclipta yellow vein virus		(EYVV)
Euphorbia mosaic virus		(EuMV)
Eggplant yellow mosaic virus		(EYMV)
Horsegram yellow mosaic virus		(HgYMV)
Jatropha mosaic virus		(JMV)
Leonurus mosaic virus	U92532	(LeMV)
Limabean golden mosaic virus		(LGMV)
Lupin leaf curl virus		(LLCuV)
Macroptilium golden mosaic virus		(MGMV)
Macroptilium golden mosaic virus - [Jamaica1]	AF098940	(MGMV-[JM1])
Macroptilium golden mosaic virus - [Jamaica2]	AF098939	(MGMV-[JM2])
Macroptilium golden mosaic virus - [PR]	AF176092-4	(MGMV-[PR])
Macrotyloma mosaic virus		(MaMV)
Malvaceous chlorosis virus		(MCV)
Melon leaf curl virus		(MLCuV)
Okra leaf curl India virus		(OkLCuIV)
(Okra leaf curl virus - India)		
Okra leaf curl virus		(OkLCuV)
(Okra leaf curl virus - [Ivory Coast])		
Okra mosaic Mexico virus	AF076854	(OkMMV)
Pepper mild tigré virus		(PepMTV)

Potato yellow mosaic virus - [Tomato]	AF026553	(PYMV-[Tom])
Pseuderanthemum yellow vein virus		(PYVV)
Pumpkin yellow vein mosaic virus	AY184488	(PuYVMV)
Sida golden mosaic Jamaica virus		(SiGMJV)
Sida golden mosaic Jamaica virus	U67926, U69601, U68177	(SiGMJV)
Sida golden mosaic Jamaica virus - [3]	U69157-8, U69602	(SiGMJV-[3])
Sida golden mosaic Jamaica virus - [Macroptilium 19]	U69159, U70386	(SiGMJV-[Mac19])
Solanum apical leaf curl virus		(SALCV)
Solanum yellow leaf curl virus		(SYLCV)
Tobacco apical stunt virus	AF077744, AF077746, AF076855	(TbASV)
Tobacco leaf curl India virus	AB001292-8	(TbLCIV)
(Tobacco leaf curl virus - India)	AB001301-4 AB001307-20	
Tobacco leaf rugose virus - [Cuba]	AJ488768	(TbLRV-[CU])
Tomato chlorotic vein virus - [Brazil]	AY049205	(ToCVV-[BR])
Tomato crinkle yellow leaf virus - [Brazil]	AY090556	(ToCYLV-[BR])
Tomato crinkle virus - [Brazil]	AY049218	(ToCrV-[BR])
Tomato dwarf leaf curl virus	AF035224-5	(ToDLCV)
Tomato infectious yellows virus	AY049208	(ToIYV)
Tomato leaf curl India virus	L11746	(ToLCIV)
Tomato leaf curl Nicaragua virus	AJ277057-61	(ToLCNV)
Tomato leaf curl Senegal virus	D88800, AF058028	(ToLCSV)
(Tomato leaf curl virus - Senegal)		
Tomato leaf curl Sinaloa virus	AF040635,	(ToLCSinV)
(Tomato leaf curl virus - Sinaloa)	AF131213,	
(Sinaloa tomato leaf curl virus)	AJ277052-56	
Tomato leaf curl Tanzania virus	U73498	(ToLCTZV)
(Tomato leaf curl virus - Tanzania)		
Tomato mild yellow mottle virus - [Honduras 96 - H5kw]	AF131071	(ToMYMoV-[HN96-H5kw])
Tomato mottle leaf curl virus - [Brazil]	AY049227	(ToMoLCV-[BR])
Tomato mosaic Barbados virus	AF213013-4	(ToMBV)
Tomato Uberlandia virus	AF156895	(ToUV)
Tomato yellow dwarf virus	U82829	(ToYDV)
Tomato yellow leaf curl Nigeria virus		(TYLCNV)
(Tomato yellow leaf curl virus - Nigeria)		
Tomato yellow leaf curl Kuwait virus	AF065822	(TYLCKWV)
Tomato yellow leaf curl Saudi Arabia virus		(TYLCSAV)
(Tomato yellow leaf curl virus - Saudi Arabia)		
(Tomato yellow leaf curl virus - Southern Saudi Arabia)		
Tomato yellow leaf curl Tanzania virus	U73498	(TYLCTZV)
(Tomato yellow leaf curl virus - Tanzania)		
Tomato yellow leaf curl Yemen virus	X79429	(TYLCYV)
(Tomato yellow leaf curl virus - Yemen)		
Tomato yellow mosaic virus		(ToYMV)
Tomato yellow mosaic virus – [Brazil 1]		(ToYMV-[BR1])
Tomato yellow mosaic virus – [Brazil 2]		(ToYMV-[BR2])
Tomato yellow mottle virus	AF112981	(ToYMoV)
Tomato yellow vein streak virus	U79998, U80042	(ToYVSV)
(Tomato yellow vein streak virus - Brazil)		

Watermelon curly mottle virus		(WmCMV)
Wissadula golden mosaic virus - [Jamaica1]	U69280-1, U69603-4, U69732-3	(WGMV-[JM1]
Zinnia leaf curl virus		(ZiLCV)

List of Unassigned Viruses in the Family

None reported

Phylogenetic Relationships within the Family

Phylogenetic analysis of 337 complete genomic sequences (DNA A sequences in the case of bipartite begomoviruses) shows that geminiviruses cluster according to current taxonomic classification into four genera. In addition, they cluster according to geographic distribution, at least within the begomoviruses, probably reflecting their evolutionary divergence as a consequence of isolation due to the inability of their insect vectors to fly over long distances. Despite frequent inter-species recombination events and the increasing worldwide movement of infected plants, it is remarkable that this geographical distribution is still apparent (Fig. 7)

Similarity with Other Taxa

Members of the plant virus families *Geminiviridae* and *Nanoviridae* have circular ssDNA genomes and replicate by a rolling circle mechanism. All these viruses have highly conserved sequences, either TAATATTAC (geminiviruses) or predominantly TAGTATTAC (nanoviruses) in the loop of a putative stem-loop structure within the intergenic region, in which a nick is introduced during the initiation of replication. Similar structures are found in members of the animal virus family *Circoviridae* as well as the genus *Anellovirus,* that also have small circular ssDNA genomes. It is speculated that geminiviruses derive from prokaryotic episomal replicons based on conservation of motifs in proteins that function in rolling circle replication initiation.

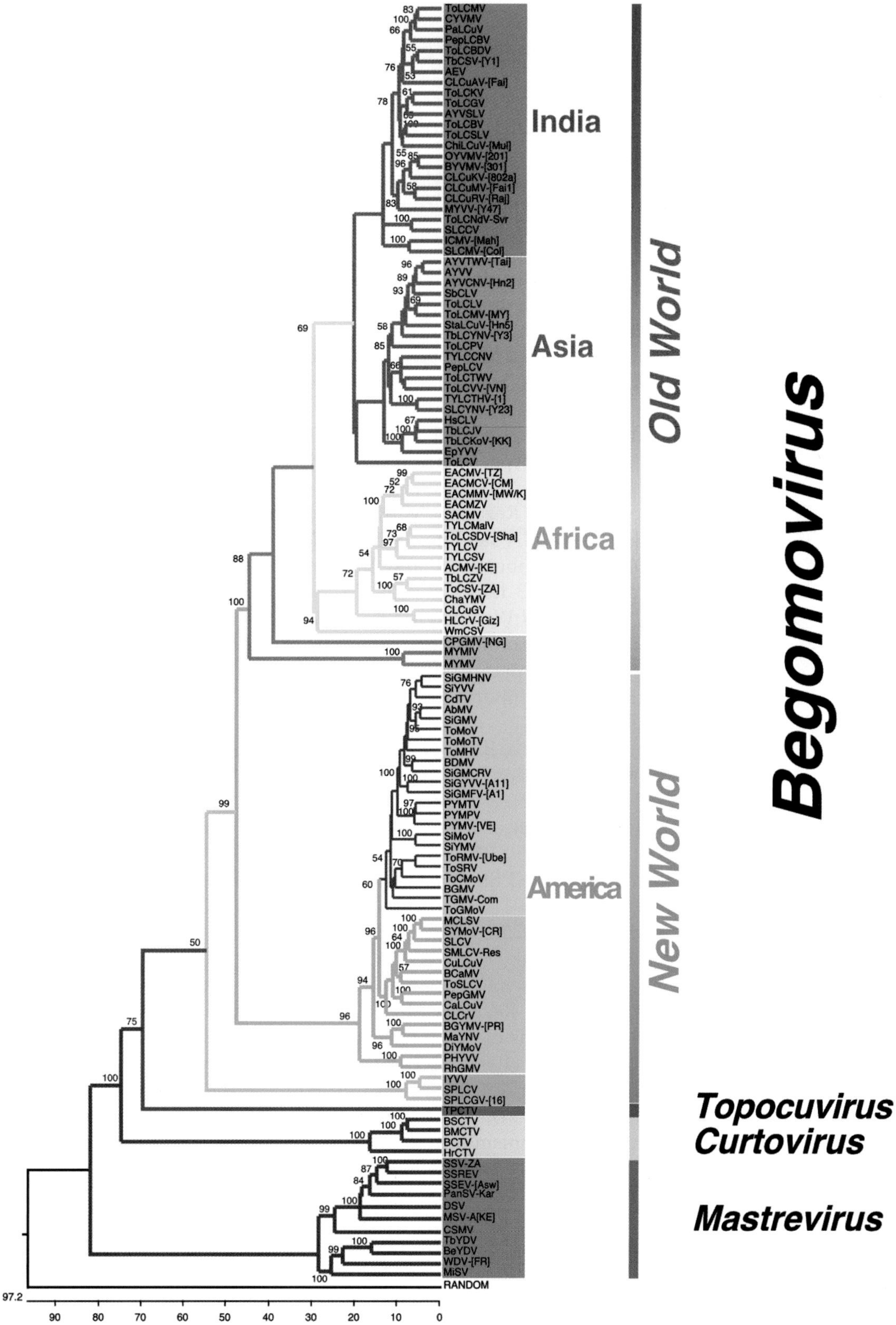

Figure 7: Dendrogram based on complete DNA A component nucleotide sequences representing 122 species of geminiviruses. Accession numbers and nucleotide sequences were obtained from GenBank. Sequences were aligned using the clustal algorithm (MegAlign 3.11, DNAstar) and the bootstrap analysis was done with PAUP 4.0. The vertical axis is arbitrary and the horizontal axis represents a distance expressed in percentage of nt substitution x100.

Derivation of Names

Begomo: sigla from the type species *Bean golden yellow mosaic virus* (previously *Bean golden mosaic virus*).

Curto: sigla from the type species *Beet curly top virus*.

Gemini: from Latin meaning "twin", describing the characteristic twinned (geminate) particle morphology.

Mastre: sigla from the type species *Maize streak virus*.

Topocu: sigla from the type species *Tomato pseudo-curly top virus*.

References

Argüello-Astorga, G.R., Guevara-González, R.G., Herrera-Estrella, L.R. and Rivera-Bustamante, R.F. (1994). Geminivirus replication origins have a group-specific organization of iterative elements: a model for replication. *Virology*, **203**, 90-100.

Briddon, R.W., Bedford, I.D., Tsai, J.H. and Markham, P.G. (1996). Analysis of the nucleotide sequence of the treehopper-transmitted geminivirus, tomato pseudo-curly top virus, suggests a recombinant origin. *Virology*, **219**, 387-394.

Campos-Olivas, R., Louis, J.M., Clérot, D., Gronenborn, B. and Gronenborn, A.M. (2002). The structure of a replication initiator unites diverse aspects of nucleic acid metabolism. *Proc. Natl. Acad. Sci. USA*, **99**, 10310-10315.

Dry, I.B., Krake, L.R., Rigden, J.E. and Rezaian, M.A. (1997). A novel subviral agent associated with a geminivirus: the first report of a DNA satellite. *Proc. Natl. Acad. Sci. USA*, **94**, 7088-7093.

Fauquet, C.M., Maxwell, D.P., Gronenborn, B. and Stanley, J. (2000). Revised proposal for naming geminiviruses. *Arch. Virol.*, **145**, 1743-1761.

Fauquet, C.M., Bisaro, D.M., Briddon, R.W, Brown, J.K., Harrison, B.D., Rybicki, E.P., Stenger, D.C. and Stanley, J. (2003). Revision of taxonomic criteria for species demarcation in the family *Geminiviridae*, and an updated list of begomovirus species. *Arch. Virol.*, **148**, 405-421.

Gutiérrez, C. (2002). Strategies for geminivirus DNA replication and cell cycle interference. Physiol. Mol. *Plant Pathol.*, **60**, 219-230.

Hanley-Bowdoin, L., Settlage, S.B., Orozco, B.M., Nagar, S. and Robertson, D. (1999). Geminiviruses: models for plant DNA replication, transcription, and cell cycle regulation. *Crit. Rev. Plant Sci.*, **18**, 71-106.

Mansoor, S., Khan, S.H., Bashir, A., Saeed, M., Zafar, Y., Malik, K.A., Briddon, R., Stanley, J. and Markham, P.G. (1999). Identification of a novel circular single-stranded DNA associated with cotton leaf curl disease in Pakistan. *Virology*, **259**, 190-199.

Mansoor, S. Briddon, R.W., Zafar, Y. and Stanley, J. (2003). Geminivirus disease complexes: an emerging threat. *Trends Plant Sci.*, **8**, 128-134.

Padidam, M., Sawyer, S. and Fauquet, C.M. (1999). Possible emergence of new geminiviruses by frequent recombination. *Virology* **265**, 218-225.

Palmer, K.E. and Rybicki, E.P. (1998). The molecular biology of mastreviruses. *Adv. Virus Res.*, **50**, 183-234.

Rigden, J.E., Dry, I.B., Krake, L.R. and Rezaian, M.A. (1996). Plant virus DNA replication processes in *Agrobacterium*: Insight into the origins of geminiviruses? *Proc. Natl. Acad. Sci. USA*, **93**, 10280-10284.

Sanderfoot, A.A. and Lazarowitz, S.G. (1996). Getting it together in plant virus movement: cooperative interactions between bipartite geminivirus movement proteins. *Trends Cell Biol.*, **6**, 353-358.

Saunders, K., Lucy, A. and Stanley, J (1991). DNA forms of the geminivirus African cassava mosaic virus consistent with a rolling circle mechanism of replication. *Nucl. Acids Res.*, **19**, 2325-2330.

Saunders, K., Bedford, I.D., Briddon, R.W., Markham, P.G., Wong, S.M. and Stanley, J. (2000). A unique virus complex causes Ageratum yellow vein disease. *Proc. Natl. Acad. Sci. USA*, **97**, 6890-6895.

Stanley, J., Markham, P.G., Callis, R.J. and Pinner, M.S. (1986). The nucleotide sequence of an infectious clone of the geminivirus beet curly top virus. *EMBO J.*, **5**, 1761-1767.

Zhang, W., Olson, N.H., Baker, T.S., Faulkner, L., Agbandje-McKenna, M., Boulton, M.I., Davies, J.W. and McKenna, R. (2001). Structure of the maize streak virus geminate particle. *Virology*, **279**, 471-477.

Contributed By

Stanley, J., Bisaro, D.M., Briddon, R.W., Brown, J.K., Fauquet, C.M., Harrison, B.D., Rybicki, E.P. and Stenger, D.C.

FAMILY CIRCOVIRIDAE

TAXONOMIC STRUCTURE OF THE FAMILY

Family	***Circoviridae***	
Genus		***Circovirus***
Genus		***Gyrovirus***

VIRION PROPERTIES

MORPHOLOGY

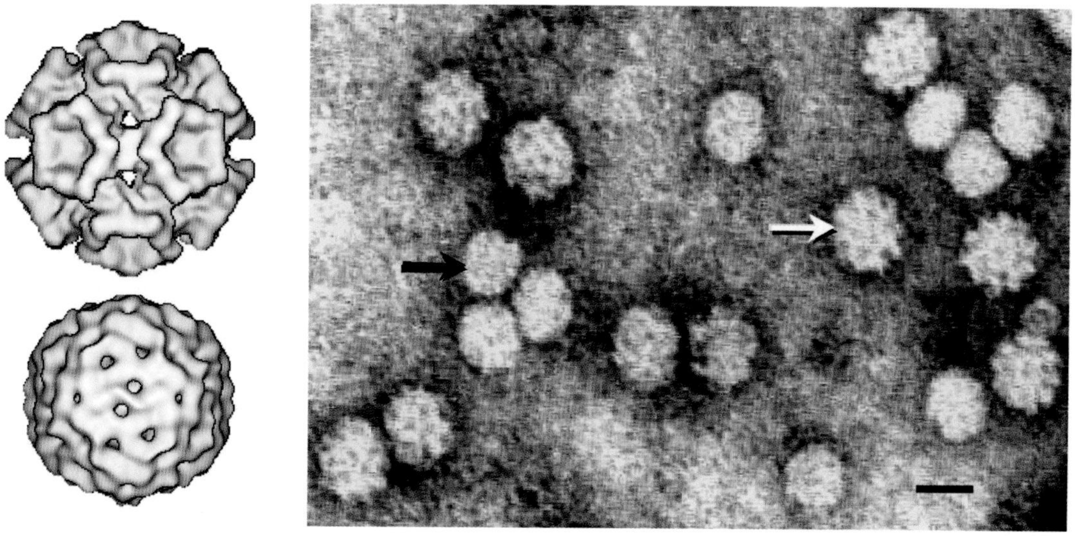

Figure 1: (Left upper) Cryo-electron microscopy image of a particle of an isolate of *Chicken anemia virus*. A structural model comprising 60 subunits (T=1) arranged in 12 protruding pentagonal trumpet-shaped units pentameric rings has been proposed. (Left lower) Cryo-electron microscopy image of a particle of an isolate of *Porcine circovirus 2*. A structural model comprising 60 subunits (T=1) arranged in 12 flat pentameric morphological units has been proposed. (Right) Negative contrast electron microscopy of particles of an isolate of *Chicken anemia virus* (black arrow) and *Beak and feather disease virus* (BFDV) (white arrow), stained with uranyl acetate. The bar represents 20 nm.

Virions exhibit icosahedral symmetry and do not possess an envelope. Ranges reported for virion diameters of Chicken anemia virus (CAV), Porcine circovirus - 1 (PCV-1), and Beak and feather disease virus (BFDV) are 19.1-26.5 nm, 17-20.7 nm, and 12-20.7 nm respectively. Comparative analysis indicated that the diameters of CAV virions are about 30% greater than those of PCV-1 and BFDV and that CAV exhibits a distinctive surface structure, that is not exhibited by PCV-1 and BFDV.

PHYSICOCHEMICAL AND PHYSICAL PROPERTIES

The buoyant densities of virions in CsCl range from 1.33 to 1.37 g/cm^3. CAV and PCV-1 are resistant to inactivation by treatment at pH 3 and both can withstand incubation at 70°C for 15 mn. Both viruses resist treatment with organic solvents such as chloroform and both show at least partial resistance to sodium dodecyl sulphate.

NUCLEIC ACID

The genomes are covalently closed, circular, ssDNAs, which range in size from 1.8 to 2.3 kb.

PROTEINS

The virions of CAV, PCV-1 and Porcine circovirus - 2 (PCV-2) are each comprised of one structural protein, for which approximate sizes of 50 kDa (CAV), 30 kDa (PCV-1) and 30 kDa (PCV-2) have been estimated respectively. BFDV is reported to contain three proteins, 26.3, 23.7 and 15.9 kDa. The protein composition of virions of Pigeon circovirus (PiCV), Goose circovirus (GoCV) and Canary circovirus (CaCV) are not known, but putative structural proteins have been identified by amino acid homology searches.

Lipids
Unknown.

Carbohydrates
Unknown.

Genome Organization and Replication

The genome organization of CAV is negative sense, whereas those of the other circoviruses are ambisense. All viruses of the family replicate their genomes using a circular, ds replicative form (RF) DNA intermediate, which is produced using host cell DNA polymerases during S phase of cell division. The RF serves as template for generation of viral ssDNA, probably using the rolling circle replication (RCR) mechanism.

Antigenic Properties

CAV is antigenically distinct from other circoviruses. PCV-1 and BFDV are antigenically different, but PCV-1 and PCV-2 share common epitopes. The antigenic relationship of avian members of the genus *Circovirus* is not known.

Biological Properties

Circoviruses are host-specific or exhibit a narrow host range and the majority of those reported infect avian species. Most infections are thought to spread by the fecal-oral route, but vertical transmission has been demonstrated. Circovirus infections are highly prevalent and have widespread geographic distributions. Although subclinical infections are common, circovirus infections are associated with a range of clinical diseases including infectious chicken anemia, psittacine beak and feather disease, circovirus disease of pigeons, and the recently described postweaning multisystemic wasting syndrome of pigs (PMWS) caused by PCV-2. Circovirus infections cause lymphoid depletion and are immunosuppressive.

Genus Circovirus

Type Species *Porcine circovirus - 1*

Distinguishing Features

Viruses have very similar ambisense genome organizations in which the genes encoding the replication-associated (Rep) and CP are divergently arranged on different strands of the ds RF. The intergenic region between the start sites of these genes contains the conserved nonanucleotide motif (TAGTATTAC), located at the apex of a potential stem-loop, at which RCR of the virus DNA initiates.

Virion Properties

Morphology
Virions are non-enveloped and show spherical morphology. The capsid structure is not completely resolved, but it follows either icosahedral or dodecahedral symmetry rules. Ranges of reported virion sizes for PCV-1 and BFDV are 17-20.7 nm and 12-20.7 nm respectively. The virion sizes reported for other circoviruses are similar and appear to depend on the type of negative stain and on whether measurement is made using negative contrast or thin section electron microscopy.

Physicochemical and Physical Properties
PCV-1 virions have a sedimentation coefficient of 57S and a buoyant density in CsCl of 1.35-1.37 g/cm^3.

Nucleic Acid
Virions contain ssDNA, which is covalently closed. The genomes of PCV-1 and PCV-2 are the smallest viruses replicating autonomously in mammalian cells (1759 and 1768 nt), while the genomes of the avian circoviruses GoCV (1821 nt), CaCV (1952 nt), BFDV (1993 nt) and PiCV (2037 nt) are slightly larger. All virus genomes carry a putative stem-loop element with the conserved nonamer (TAGTATTAC) in its apex.

Proteins
Circovirus RF DNAs contain 2 major ORFs, the *rep* gene present on the virus sense strand and the *cap* gene on the complementary sense strand. With PCV-1 and PCV-2, Rep (35.6 kDa) and Rep′ (19.2 kDa) proteins are produced by differential splicing from the *rep* gene and both proteins are essential for virus replication. The Rep protein contains 4 conserved aa motifs, three of which are associated with RCR and one of which is associated with dNTPase activity. The *cap* mRNA is also spliced, and, since splicing occurs upstream of the *cap* ORF, only one protein is produced. Some avian circoviruses use an alternative start codon for initiation of the CP protein. The N-termini of the predicted CP proteins are rich in basic aa, which suggests that this region may have a DNA-binding role within the virion. Transcription analysis of the avian circoviruses has not been reported, mainly due to the absence of cell culture growth systems. A third ORF, present on the complementary strand, exists in all circoviruses, but nothing is known about its expression and function.

Lipids
Unknown.

Carbohydrates
Unknown.

Genome Organization and Replication

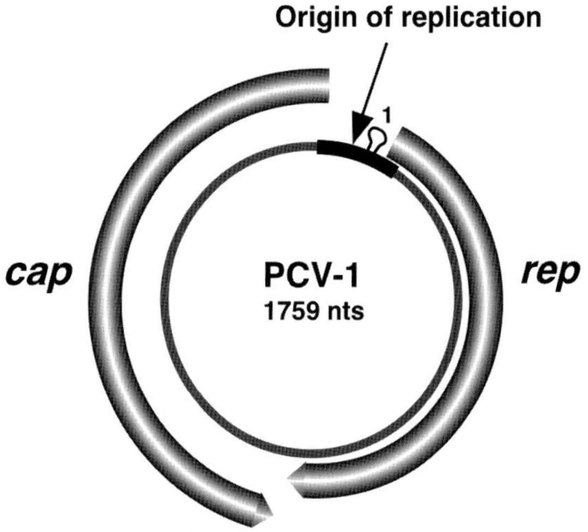

Figure 2: Genome organization of an isolate of *Porcine circovirus – 1*. The origin of replication is located between the start sites of the two major, divergently-arranged ORFs, *cap* and *rep* (thick arrows). The *cap* gene, encoding the CP, is expressed from a spliced transcript, and the *rep* gene directs the synthesis of two distinct proteins, Rep and Rep′, using differentially-spliced transcripts.

Genomes consist of one ssDNA component 1.8 to 2.0 kb in size. Coding regions are arranged divergently resulting in an ambisense organization and creating two intergenic regions, a larger one between the 5′-ends of the two major ORFs *rep* and *cap* genes and a shorter one between their 3'-ends. In the case of PCV-1 and PCV-2, the non-coding region between the ATGs of the *rep* and the *cap* genes comprises the origin of replication. After infection, the ssDNA genome is converted into dsDNA RF, presumably by host enzymes. By analogy to geminiviruses, replication is thought to follow the RCR model and is initiated at the conserved TAGTATT/AC sequence. The *rep* gene of PCV-1 and PCV-2 expresses two differentially spliced proteins, Rep and Rep'. Rep and Rep' of PCV1 bind to two genomic hexameric repeats, which are located close to the potential stem loop at the origin of replication. Both proteins and binding to the hexamers are essential for initiation of replication.

Antigenic Properties

PCV-1 and PCV-2 share antigenic epitopes and therefore show cross-reaction. A detailed serological analysis for the avian circoviruses has not yet been performed.

Biological Properties

Natural infections with PCV-1 and PCV-2 appear to be restricted to the pig, while BFDV infections are detected in over 40 species of psittacine birds. Available evidence suggests that viruses are host-specific or have narrow host ranges. The fecal-oral route of transmission is likely, although vertical transmission has been reported in cases of PCV-2, BFDV and Pigeon circovirus (PiCV).

List of Species Demarcation Criteria in the Genus

Since the *rep* genes are highly conserved, species demarcation can be best seen by phylogenetic comparison of the CP coding sequence. The suggested criteria demarcating species in the genus are (i) complete genome nt sequence less than 75%, and (ii) CP sequence identity less than 70%.

List of Species in the Genus

Species names are in green italic script; strain names and synonyms are in black roman script; tentative species names are in blue roman script. Sequence accession numbers, and assigned abbreviations ( ) are also listed.

Species in the Genus

Beak and feather disease virus		
Beak and feather disease virus	[AF080560]	(BFDV)
Canary circovirus		
Canary circovirus	[AJ301633]	(CaCV)
Goose circovirus		
Goose circovirus	[AJ304456]	(GoCV)
Pigeon circovirus		
Pigeon circovirus	[AF252610]	(PiCV)
Porcine circovirus - 1		
Porcine circovirus - 1	[Y09921]	(PCV-1)
Porcine circovirus – 2		
Porcine circovirus - 2	[AF027217]	(PCV-2)

Tentative Species in the Genus

Duck circovirus	(DuCV)
Finch circovirus	(FiCV)
Gull circovirus	(GuCV)

GENUS *GYROVIRUS*

Type Species *Chicken anemia virus*

DISTINGUISHING FEATURES

Chicken anemia virus (CAV), the only member of the genus, can be distinguished from viruses belonging to the genus *Circovirus* on the basis of its negative sense genome organization. Also, CAV virions are larger than other circoviruses and surface structure is more evident.

VIRION PROPERTIES

ssDNA

MORPHOLOGY

Virions are non-enveloped and show icosahedral morphology. The capsid structure exhibits distinctive surface structure, and a structure comprising 60 subunits (T=1) arranged in 12 pentameric rings has been proposed on the basis on cryo-electron microscopy results. The reported diameter range for the CAV virion is 19.1-26.5 nm.

PHYSICOCHEMICAL AND PHYSICAL PROPERTIES

CAV virions have a sedimentation coefficient of 91S and a buoyant density in CsCl of 1.33-1.35 g/cm^3.

NUCLEIC ACID

The genomes of *Chicken anemia virus* isolates are either 2298 or 2319 nt in size and share high nt identity. Part of the NTR of the genome is G-C rich and capable of forming stem-loop structures.

PROTEINS

Three virus proteins are synthesised. VP1 (52 kDa), encoded by ORF1, is the only structural protein. VP2 (26 kDa), encoded by ORF2, is a protein phosphatase. VP3 (14 kDa) encoded by ORF3 is capable of causing apoptosis and has been called "apoptin".

LIPIDS

Unknown.

CARBOHYDRATES

Unknown.

GENOME ORGANIZATION AND REPLICATION

CAV has a negative sense genome organization. One 1 major polycistronic mRNA (2.0 kb), containing 3 partially-overlapping ORFs, is transcribed from the circular ds RF (Fig. 3). The non-transcribed region of the genome contains transcription initiation and termination signals, and a tandemly-arranged array of either 4 or 5 nineteen nts repeats with which promoter-enhancer activity is associated. The circular ssDNA genome is thought to replicate using the RCR mechanism. However, although a semi-conserved nonanucleotide motif, at which RCR initiates in other circular ssDNA replicons, is found within the non-transcribed region of the CAV genome, it is not located at the apex of a potential stem-loop, as is the case with the other replicons. The occurrence within VP1 of 2 amino acid motifs, which have putative roles in RCR, suggests that this structural protein possesses DNA replication function.

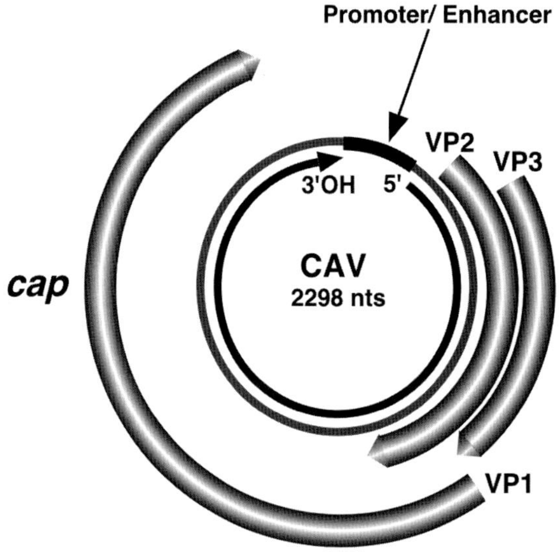

Figure 3: Genome organization of Chicken anemia virus (CAV). The unspliced CAV transcript (5`-3`) contains three partially overlapping ORFs, which are expressed in CAV-infected cells. The non-transcribed region possesses promoter-enhancer activity.

Antigenic Properties

Cell culture isolates of *Chicken anemia virus*, collected from throughout the world, belong to a single serotype but isolates can be differentiated by type-specific virus neutralising monoclonal antibodies.

Biological Properties

Although the chicken is the major host, CAV has also been isolated from turkeys, and the detection of virus-specific antibody suggests that Japanese quail may also be susceptible to infection. Vertical (egg) transmission from infected parent chickens results in clinical disease in progeny chicks at 1-2 weeks post-hatching. Clinical signs include anemia and lymphoid depletion. Sub-clinical infections with horizontally-acquired virus, probably spread by the fecal-oral route, occur when maternally-derived antibody is no longer protective.

List of Species Demarcation Criteria in the Genus

Not appropriate.

List of Species in the Genus

Species names are in green italic script; strain names and synonyms are in black roman script; tentative species names are in blue roman script. Sequence accession numbers, and assigned abbreviations ( ) are also listed.

Species in the Genus

Chicken anemia virus

Chicken anemia virus	[M55918, M81223]	(CAV)

Tentative Species in the Genus

None

List of Unassigned Viruses in the Family

None

Phylogenetic Relationships within the Family

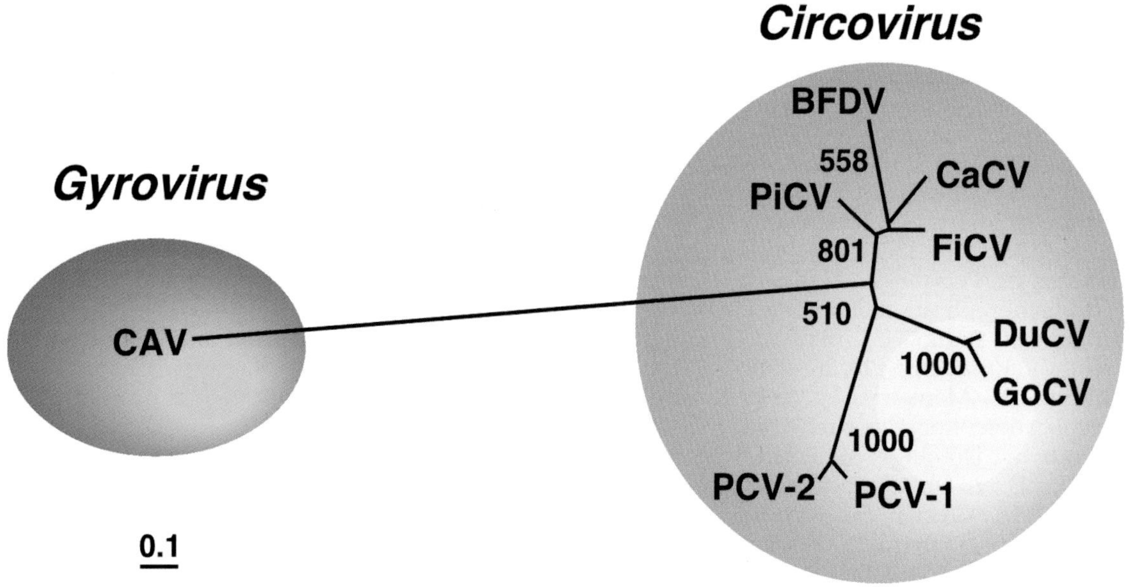

Figure 4: Phylogenetic tree of the family *Circoviridae*. The aa sequences of the *Rep* proteins of the circoviruses CAV (M81223), CaCV (AJ301633), DuCV (AY228555), FiCV(?), GoCV (AJ304456), PiCV (AF252610), BFDV (AF071878), PCV-1 (Y09921) and PCV-2 (AF201307) were compared. Analysis was performed using the bootstrap method by analyzing 1000 data sets. The bar indicates 10% aa exchange per position in the alignment. The signification of the abbreviations is in the "List of species" in the description.

Comparison of the Rep proteins of all seven members of the family *Circoviridae* revealed that CAV has no close phylogenetic relationship to the other species in the genus *Circovirus*. The cluster comprising the Rep proteins of the genus *Circovirus* shows aa identity and similarity ranging from about 37% to 85%. It can be seen that the porcine and the avian circoviruses form separate clusters. Investigation of the phylogenetic relationship of PiCV, Goose circovirus (GoCV), BFDV and CaCV, indicated that GoCV is more distantly related compared to the other three avian circoviruses. Similar observations are made when the complete viral genomes are compared.

Similarity with Other Taxa

Analysis of the genome structure, the origin of replication and the Rep proteins of viruses of the genus *Circovirus* show similarity to members of the plant virus family *Nanoviridae* and less pronounced similarity to members of the plant virus family *Geminiviridae*. CAV has features in common with viruses such as Torque Teno virus (TTV) and Torque Teno mini virus (TTMV), which are members of the genus *Anellovirus*. Members of both genera possess circular ssDNA genomes with similar but not identical negative sense genome organizations. The non-coding regions of CAV, TTV and TTMV are G-C rich and show a low level of nt homology. In addition, CAV, TTV and TTMV specify structural proteins that contain 2 aa motifs with putative roles in RCR and non-structural proteins that exhibit protein phosphatase activity.

Derivation of Names

Circo: sigla to indicate that the virus genome has a *cir*cular *co*nformation
Gyro: greek derivation from "gyrus" meaning "ring" or "circuit" to indicate that virus genome has a circular conformation

References

Allan, G.M. and Ellis, J.A. (2000). Porcine circoviruses: a review. *J. Diag. Invest.*, **12**, 3-14.
Allan, G.M., Phenix, K.V., Todd, D. and McNulty, M.S. (1994). Some biological and physico-chemical properties of porcine circovirus. *J. Vet. Med.*, **B41**, 17-26.
Bassami, M.R., Berryman D, Wilcox G.E. and Raidal S.R. (1998). Psittacine beak and feather disease virus nucleotide sequence analysis and its relationship to porcine circovirus, plant circoviruses and chicken anaemia virus. *Virology*, **249**, 453-459.
Crowther, R.A., Berriman, J.A., Curran, W.L., Allan, G.M. and Todd, D. (2003). Comparison of the structures of three circoviruses: chicken anemia virus, porcine circovirus type 2 and beak and feather disease virus. *J. Virol.*, **74**, 13036-13041.
Mankertz, A., Persson, F., Mankertz, J., Blaess, G. and Buhk, H.-J. (1997). Mapping and characterization of the origin of replication of porcine circovirus DNA. *J. Virol.*, **71**, 2562-2566.
Mankertz, J., Buhk, H.-J., Blaess, G. and Mankertz, A. (1998). Transcription analysis of porcine circovirus (PCV). *Virus Genes*, **16**, 267-276.
Mankertz, A., Mankertz, J., Wolf, K. and Buhk, H.-J. (1998). Identification of a protein essential for the replication of porcine circovirus. *J. Gen. Virol.*, **79**, 381-384.
Mankertz, A., Domingo, M., Folch, J. M., LeCann, P., Jestin, A., Segales, J., Chmielewicz, B., Plana-Duran, J. and Soike, D. (2000). Characterisation of PCV-2 isolates from Spain, Germany and France. *Virus Res.*, **66**, 65-77.
Meehan, B.M., Creelan, J.L., McNulty, M.S. and Todd, D. (1997). Sequence of porcine circovirus DNA: affinities with plant circovirus. *J.Gen. Virol.*, **78**, 221-227.
McNulty, M.S. (1991). Chicken anaemia agent: a review. *Avian Path.*, **20**, 187-203
Noteborn, M.H.M. and Koch, G. (1995). Chicken anaemia virus infection: molecular basis of pathogenicity. *Avian Path.*, **24**, 11-31.
Phenix, K.V., Weston, J.H., Ypelaar, I, Lavazza, A., Smyth, J.A., Todd, D., Wilcox, G.E. and Raidal, S.R. (2001). Nucleotide sequence analysis of a novel circovirus of canaries and its relationship to other members of the genus *Circovirus* of the family *Circoviridae*. *J. Gen. Virol.*, **82**, 2805-2809.
Ritchie, B.W., Niagro, F.D., Lukert, P.D., Steffens, W.L. and Latimer, K.S. (1989). Characterization of a new virus derived from cockatoos with psittacine beak and feather disease. *Virology*, **171**, 83-88.
Tischer, I., Gelderblom, H., Vetterman, W. and Koch, M.A. (1982). A very small porcine virus with circular single-stranded DNA. *Nature*, **295**, 64-66.
Todd, D. (2000). Circoviruses: immunosuppressive threats to avian species: a review. *Avian Path.*, **29**, 373-394
Todd, D., McNulty, M.S., Adair, B.M. and Allan, G.M. (2001). Animal circoviruses. *Adv. Virus Res.*, **57**, 1-70.
Todd, D., Weston, J.H., Soike, D. and Smyth, J.A. (2001). Genome sequence determinations and analyses of novel circoviruses from goose and pigeon. *Virology*, **286**, 354-362.
Twentyman, C.M., Alley, M.R., Meers, J., Cooke, M.M. and Duignan, P.J. (1999). Circovirus-like infection in a southern black-backed gull (*Larus dominicanus*). *Avian Path.*, **28**, 513-516.
Woods, L.W. and Latimer, K.S. (2000). Circovirus infection of nonpsittacine birds. *J. Avian Med. Surg.* **14**, 154-163
Woods, L.W., Latimer, K.S., Niagro, F.D., Riddell, C., Crowley, A.M., Anderson, M.L., Daft, B.M., Moore, J.D., Campagnoli, R.P. and Nordhausen, R.W. (1994). A retrospective study of circovirus infection in pigeons: nine cases (1986-1993). *J. Vet. Diag. Invest.*, **6**, 156-164.

Contributed By

Todd, D., Bendinelli, M., Biagini, P., Hino, S., Mankertz, A., Mishiro, S., Niel, C., Okamoto, H., Raidal, S., Ritchie, B.W. and Teo, G.C.

GENUS ANELLOVIRUS

Type Species *Torque teno virus*

VIRION PROPERTIES

MORPHOLOGY

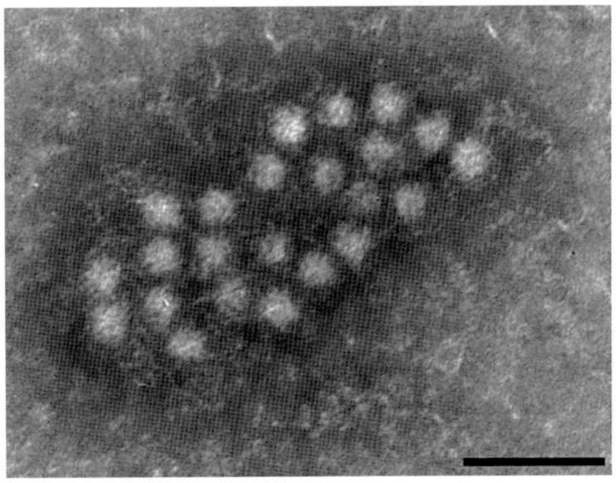

Figure 1: Negative contrast electron microscopy of particles of an isolate of *Torque teno virus*, stained with uranyl acetate (Itoh et al., 2000). The bar represents 100 nm.

Virions are non-enveloped, with Torque teno virus (TTV) and Torque teno mini virus (TTMV) having reported diameters of 30 - 32 nm and less than 30 nm respectively (Fig. 1).

PHYSICOCHEMICAL AND PHYSICAL PROPERTIES

The buoyant density of virions in CsCl is 1.31-1.33 g/cm^3 for TTV and 1.27-1.28 g/cm^3 for TTMV, both estimated using virus purified from serum.

NUCLEIC ACID

Virions contain circular ssDNAs, which range from 3.5 to 3.8 kb in size for TTV, and from 2.8 to 2.9 kb for TTMV. Both genomes are of negative sense. The putative non-coding regions of the TTV and TTMV genomes contain a sequence of about 80-110 nt with high G-C content (~90%), which is postulated to form a secondary structure composed of stems and loops. A sequence of about 120 nt, located in the putative non-coding region, is highly conserved between TTV and TTMV genomes.

PROTEINS

The proteins of TTV and TTMV have not been molecularly characterized. Two main ORFs, ORF1 and ORF2, may be deduced directly from the nt sequence (Fig. 2). These ORFs partially overlap and their estimated sizes differ slightly between isolates. ORF1 is composed of about 770 aa (TTV) or 675 aa (TTMV). The putative ORF1 proteins of TTV and TTMV possess arginine-rich, hydrophilic N-terminal sequences, and at least 2 aa sequence motifs with which rolling circle replication (RCR) of the virus DNA may be associated. On this basis, ORF1 of TTV and TTMV is believed to encode the putative nucleocapsid and replication-associated (Rep) protein of both viruses. ORF2 is composed of about 120 aa (TTV) or 100 aa (TTMV) and is considered to encode a protein with phosphatase activity.

LIPIDS

Unknown

CARBOHYDRATES

Unknown

Genome Organization and Replication

Knowledge of the genome expression and replication mechanisms of TTV and TTMV is limited mainly due to the absence of cell culture propagation systems. TTV-specific mRNAs can be detected in various tissues and organs in humans. At least 3 mRNAs of different sizes (2.9 kb, 1.2 kb and 1.0 kb) are transcribed from the negative strand of the putative circular ds replicative form (RF) TTV DNA. The presence of these mRNAs supports the view that both ORF1 and ORF2 are functional, and also suggests that two additional ORFs, namely ORF3 (~260 aa) and ORF4 (~250 aa) may be obtained by complex splicing (Fig. 2). The transcription profile of TTMV is not known, but the fact that TTV and TTMV share similar genome organizations is highly suggestive that several TTMV mRNAs may be expressed as for TTV. The presence within the ORF1 of TTV and TTMV of conserved aa sequence motifs, which occur in the Rep proteins of other animal and plant viruses with circular ssDNA genomes (genus *Circovirus*, family *Nanoviridae*), suggests that replication of anellovirus DNA may use an RCR mechanism.

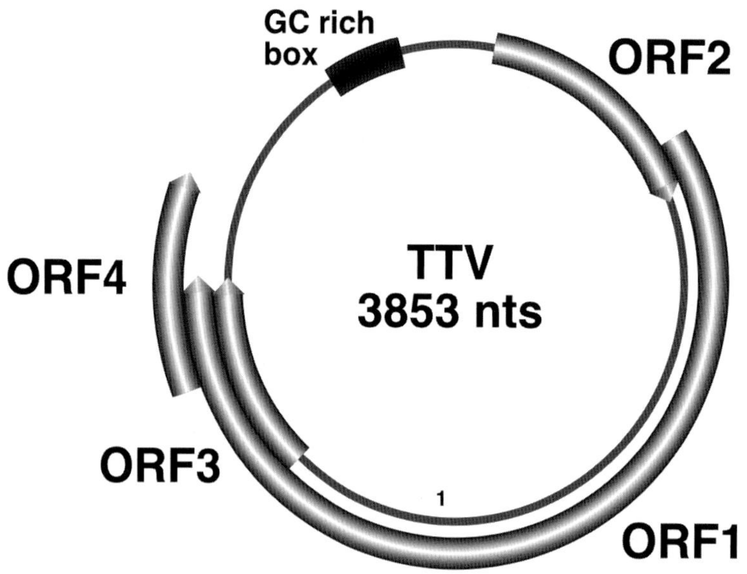

Figure 2: Genome Organization of an isolate of *Torque teno virus*.

Antigenic Properties

TTV particles in the blood are bound to immunoglobulin G (IgG), forming immune complexes.

Biological Properties

TTV and TTMV are not restricted to human hosts. So far, viruses have been detected in non-human primates (chimpanzee, macaque, tamarin, douroucouli), tupaias, cats and dogs, and in farm animals. The analysis of complete viral sequences from different animal sources reveals a high heterogeneity in the size of the viral genome (2.0 to 3.9 kb), along with a high genetic divergence when compared with human isolates. However, genomic organization and predicted transcription profiles correspond to those found in human isolates. Epidemiological studies have demonstrated the global distribution (e.g. Africa, North and South America, Asia, Europe, Oceania) of TTV in rural and urban populations. The prevalence of TTV and TTMV in the general population is high (>80%). Although initially suspected of being transmitted only by blood transfusion, the global dispersion of the viruses in populations and its detection in various biological samples (e.g. plasma, saliva, feces) suggest combined modes of diffusion, and in particular the spread by saliva droplets. Other modes of transmission, such as maternal or sexual routes have also been suggested. The link between anellovirus infection and a given pathology remains unproven, although it has been suggested that the viral load is related to the immune status of the host.

List of Species Demarcation Criteria in the Genus

The precise criteria demarcating species in the genus are currently under evaluation.

List of Species in the Genus

Species names are in green italic script; strain names and synonyms are in black roman script; tentative species names are in blue roman script. Sequence accession numbers [], and assigned abbreviations ( ) are also listed.

Species in the Genus

Torque teno virus		
Human isolates		
Group 1		
Torque teno virus - 1a	[AB017610]	(TTV-1a)
Torque teno virus – BDH1	[AF116842]	(TTV-BDH1)
Torque teno virus – CHN1	[AF079173]	(TTV-CHN1)
Torque teno virus – CHN2	[AF129887]	(TTV-CHN2)
Torque teno virus – GH1	[AF122913]	(TTV-GH1)
Torque teno virus – JA1	[AF122916]	(TTV-JA1)
Torque teno virus – JA10	[AF122919]	(TTV-JA10)
Torque teno virus – JA20	[AF122914]	(TTV-JA20)
Torque teno virus – JA2B	[AF122918]	(TTV-JA2B)
Torque teno virus – JA4	[AF122914]	(TTV-JA4)
Torque teno virus – JA9	[AF122915]	(TTV-JA9)
Torque teno virus – T3PB	[AF247138]	(TTV-T3PB)
Torque teno virus - TA278	[AB08394]	(TTV-TA278)
Torque teno virus – TK16	[AB026346]	(TTV-TK16)
Torque teno virus – TMR1	[AB026345]	(TTV-TRM1)
Torque teno virus – TP1-3	[AB026347]	(TTV-TP1-3)
Torque teno virus – US32	[AF122921]	(TTV-US32)
Torque teno virus – US35	[AF122920]	(TTV-US35)
Group 2		
Torque teno virus – KAV	[AF435014]	(TTV-KAV)
Torque teno virus – KT08F	[AB054647]	(TTV-KT08F)
Torque teno virus – KT10F	[AB054648]	(TTV-KT10F)
Torque teno virus – PMV	[AF261761]	(TTV-PMV)
Group 3		
Torque teno virus – SAa01	[AB060597]	(TTV-SAa01)
Torque teno virus – SAa10	[AB060594]	(TTV-SAa10)
Torque teno virus – SAa38	[AB060593]	(TTV-SAa38)
Torque teno virus – SAa39	[AB060592]	(TTV-SAa39)
Torque teno virus – SAf09	[AB060596]	(TTV-SAf09)
Torque teno virus – SAj30	[AB060595]	(TTV-SAj30)
Torque teno virus – SANBAN	[AB025946]	(TTV-SANBAN)
Torque teno virus – sanIR1031	[AB038619]	(TTV-sanIR1031)
Torque teno virus – sanS039	[AB038620]	(TTV-sanS039)
Torque teno virus – SENV118	[AX025761]	(TTV-SENV118)
Torque teno virus – SENV179	[AX025822]	(TTV-SENV179)
Torque teno virus – SENV187	[AX025830]	(TTV-SENV187)
Torque teno virus – SENV195	[AX025838]	(TTV-SENV195)
Torque teno virus – SENV24	[AX025667]	(TTV-SENV24)
Torque teno virus – SENV34	[AX025677]	(TTV-SENV34)
Torque teno virus – SENV75	[AX025718]	(TTV-SENV75)
Torque teno virus – SENV87	[AX025730]	(TTV-SENV87)
Torque teno virus – TJN01	[AB028668]	(TTV-TJN01)

Torque teno virus – TJN02	[AB028669]	(TTV-TJN02)
Torque teno virus – TUPB	[AF247137]	(TTV-TUPB)
Torque teno virus – TUS01	[AB017613]	(TTV-TUS01)
Torque teno virus – TYM9	[AB050448]	(TTV-TYM9)
Group 4		
Torque teno virus – CT23F	[AB064595]	(TTV-CT23F)
Torque teno virus – CT25F	[AB064596]	(TTV-CT25F)
Torque teno virus – CT30F	[AB064597]	(TTV-CT30F)
Torque teno virus – CT43F	[AB064598]	(TTV-CT43F)
Torque teno virus – JT03F	[AB064599]	(TTV-JT03F)
Torque teno virus – JT05F	[AB064600]	(TTV-JT05F)
Torque teno virus – JT14F	[AB064601]	(TTV-JT14F)
Torque teno virus – JT19F	[AB064602]	(TTV-JT19F)
Torque teno virus – JT41F	[AB064603]	(TTV-JT41F)
Torque teno virus – yonKC009	[AB038621]	(TTV-yonKC009)
Torque teno virus – yonKC186	[AB038623]	(TTV-yonKC186)
Torque teno virus – yonKC197	[AB038624]	(TTV-yonKC197)
Torque teno virus – yonLC011	[AB038622]	(TTV-yonLC011)
Group 5		
Torque teno virus – CT39F	[AB064604]	(TTV-CT39F)
Torque teno virus – CT44F	[AB064605]	(TTV-CT44F)
Torque teno virus – JT33F	[AB064606]	(TTV-JT33F)
Torque teno virus – JT34F	[AB064607]	(TTV-JT34F)
Animal isolates		
Torque teno virus – Pt-TTV6	[AB041957]	(TTV-Pt-TTV6)
Torque teno virus – Mf-TTV3	[AB041958]	(TTV-Mf-TTV3)
Torque teno virus – Mf-TTV9	[AB041959]	(TTV-Mf-TTV9)

Tentative Species in the Genus

Torque teno mini virus – CBD203	[AB026929]	(TTMV-CBD203)
Torque teno mini virus - CBD231	[AB026930]	(TTMV-CBD231)
Torque teno mini virus - CBD279	[AB026931]	(TTMV-CBD279)
Torque teno mini virus – CLC062	[AB038625]	(TTMV-CLC062)
Torque teno mini virus – CLC138	[AB038626]	(TTMV-CLC138)
Torque teno mini virus – CLC156	[AB038627]	(TTMV-CLC156)
Torque teno mini virus – CLC205	[AB038628]	(TTMV-CLC205)
Torque teno mini virus – NLC023	[AB038629]	(TTMV-NLC023)
Torque teno mini virus – NLC026	[AB038630]	(TTMV-NLC026)
Torque teno mini virus – NLC030	[AB038631]	(TTMV-NLC030)
Torque teno mini virus – PB4TL	[AF291073]	(TTMV-PB4TL)
Torque teno mini virus – Pt-TTV8-II	[AB041963]	(TTMV-Pt-TTV8-II)
Torque teno mini virus – TGP96	[AB041962]	(TTMV-TGP96)
Torque teno virus -Cat	[AB076003]	(TTV4-Fc)
Torque teno virus -Dog	[AB076002]	(TTV10-Cf)
Torque teno virus -Douroucouli	[AB041961]	(TTV3-At)
Torque teno virus -Pig	[AB076001]	(TTV31-Sd)
Torque teno virus -Tamarin	[AB041960]	(TTV2-So)
Torque teno virus -Tupaia	[AB057358]	(TTV14-Tbc)

Phylogenetic Relationships within the Genus

Comparison of complete nt sequences of TTV and TTMV, along with the corresponding aa sequences of ORF1 and ORF2, reveals very high intra-/inter-species genetic variability in the genus (Fig. 3, 4).

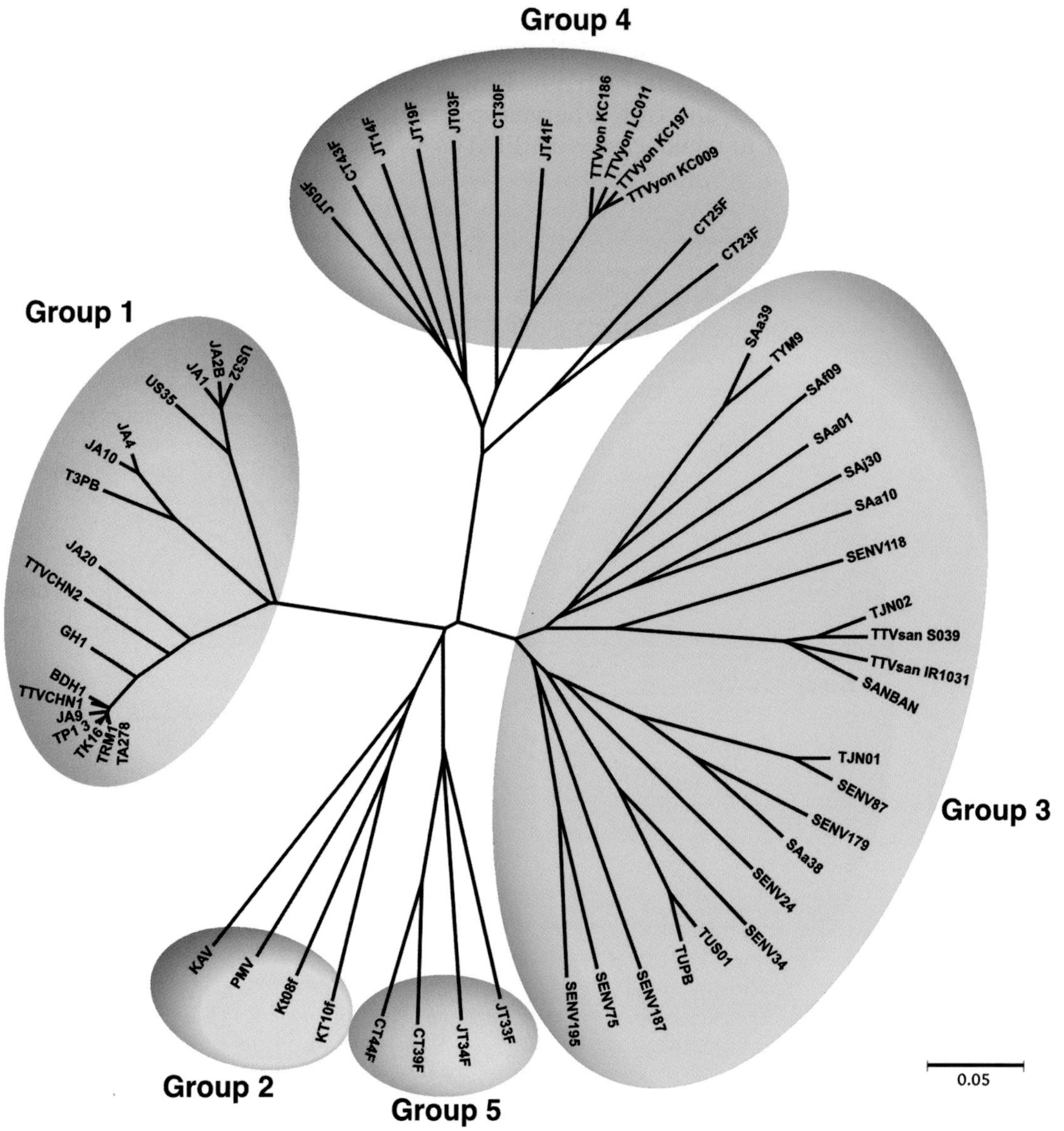

Figure 3: Neighbor-joining phylogenetic tree deduced from the comparison of nearly full-length sequences of 61 isolates of *Torque teno virus*. TTV variants are classifiable into 5 distinct clusters. Tree prepared using Clustal W and MEGA 2. Scale bar represents 5% genetic distance.

Similarity with Other Taxa

TTV and TTMV have features in common with *Chicken anaemia virus*, the type species of the genus *Gyrovirus* of the family *Circoviridae*, namely:

- All three viruses possess negative sense, circular, ssDNA genomes.
- The genome organizations of the 3 viruses are similar. Thus, the largest of the three CAV ORFs (ORF1), encoding the CP, and the largest ORF in the TTV and TTMV genomes, which encode the putative CPs, are found closest to the 3′-end of the coding region, while the ORF2s of CAV, TTV and TTMV are functionally related (see below) and are found closest to the 5′ start of the coding region.
- The CP of CAV and the putative CPs of TTV and TTMV possess aa sequence motifs that are characteristic of RCR Rep proteins. The proteins encoded by the ORF2s of CAV, TTV and TTMV contain aa sequences that are characteristic of protein tyrosine phosphatases (PTPase). ORF2s of TTV, TTMV and CAV have a common motif, $WX_7HX_3CXCX_5H$.

- The non-coding region of the CAV genome and those of the TTV and TTMV genomes contain G+C rich sequences that exhibit a low level of nt homology.
- However, CAV differs from TTV and TTMV in that it uses a single unspliced polycistronic transcript, whereas spliced transcripts have been detected with TTV and this is likely to be the case with TTMV.

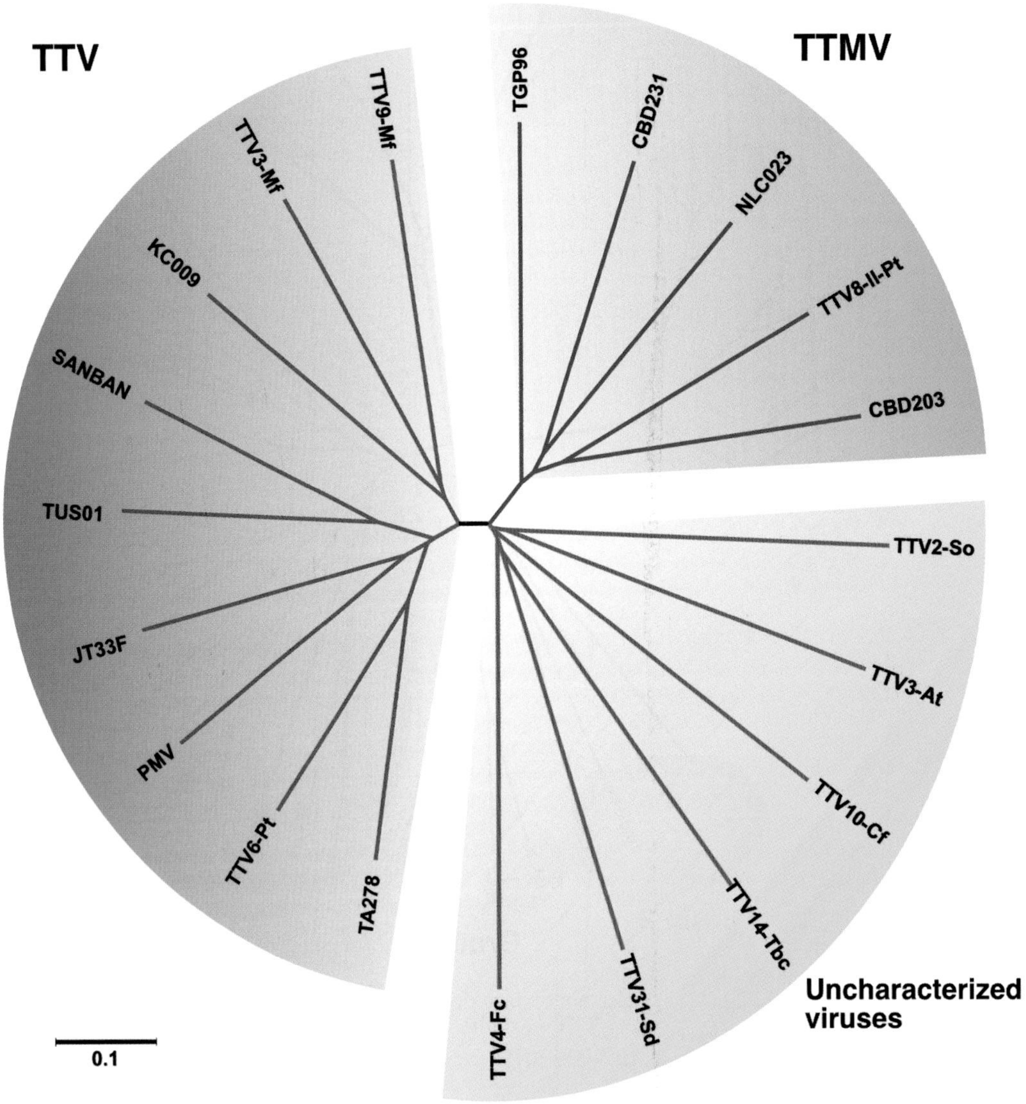

Figure 4: Neighbor-joining phylogenetic tree deduced from the comparison of ORF1 sequences of TTV, TTMV virus isolates and other unclassified viruses. Tree prepared using Clustal W and MEGA 2. Scale bar represents 10% genetic distance.

Derivation of Names

Anello is derived from latin "Anello", *the ring*, and relates to the circular nature of the DNA genome.
The name *Torque teno virus*, which replaces the name "TT virus", is derived from latin "Torques", *the necklace* and latin "Tenuis", *thin*, and relates to the circular, single-stranded nature of its DNA genome.
The name *Torque teno mini virus* replaces the name "TT virus-like mini virus".

References

Bendinelli, M., Pistello, M., Maggi, F., Fornai, C., Freer, G. and Vatteroni, M. L. (2001). Molecular properties, biology, and clinical implications of TT virus, a recently identified widespread infectious agent of humans. *Clin. Microbiol. Rev.*, **14**, 98-113.

Biagini, P., Gallian, P., Attoui, H., Touinssi, M., Cantaloube, J., de Micco, P. and de Lamballerie, X. (2001). Genetic analysis of full-length genomes and subgenomic sequences of TT virus-like mini virus human isolates. *J. Gen. Virol.*, **82**, 379-83.

Biagini, P. (2002). Les circovirus humains. *Virology*. **6**, 19-28.

Itoh, Y., Takahashi, M., Fukuda, M., Shibayama, T., Ishikawa, T., Tsuda, F., Tanaka, T., Nishizawa, T. and Okamoto, H. (2000). Visualization of TT virus particles recovered from the sera and feces of infected humans. *Biochem. Biophys. Res. Commun.*, **279**, 718-24.

Kamahora, T., Hino, S. and Miyata, H. (2000). Three spliced mRNAs of TT virus transcribed from a plasmid containing the entire genome in COS1 cells. *J. Virol.*, **74**, 9980-99.

Mankertz, A., Hattermann, K., Ehlers, B. and Soike, D. (2000). Cloning and sequencing of columbid circovirus (CoCV), a new circovirus from pigeons. *Arch. Virol.*, **145**, 2469-79.

Miyata, H., Tsunoda, H., Kazi, A., Yamada, A., Kahn, M.A., Murakami, J., Kamahora, T., Shiraki, S., and Hino, S. (1999). Identification of a novel GC-rich 113-nucleotide region to complete the circular, single-stranded DNA genome of TT virus, the first human circovirus. *J. Virol.*, **73**, 3582-86.

Mushahwar, I.K., Erker, J.C., Muerhoff, A.S., Leary, T.P., Simons, J.N., Birkenmeyr, L.G., Chalmers, M.L., Pilot-Matias, T.J., and Dexia, S. (1999). Molecular and biophysical characterization of TT virus: evidence for a new virus family infecting humans. *Proc. Natl. Acad. Sci. USA* **96**, 3177-82.

Niagro, F. D., Forsthoefel, A. N., Lawther, R. P., Kamalanathan, L., Ritchie, B. W., Latimer, K. S. and Lukert, P. D. (1998). Beak and feather disease virus and porcine circovirus genomes: intermediates between the geminiviruses and plant circoviruses. *Arch. Virol.*, **143**, 1723-44.

Okamoto, H., Fukuda, M., Tawara, A., Nishizawa, T., Itoh, Y., Hayasaka, I., Tsuda, F., Tanaka, T., Miyakawa, Y. and Mayumi, M. (2000). Species-specific TT viruses and cross-species infection in nonhuman primates. *J. Virol.*, **74**, 1132-9.

Okamoto, H. and Mayumi, M. (2001). TT virus: virological and genomic characteristics and disease associations. *J. Gastroenterol.*, **36**, 519-29.

Okamoto, H., Takahashi, M., Nishizawa, T., Tawara, A., Fukai, K., Muramatsu, U., Naito, Y. and Yoshikawa, A. (2002). Genomic characterization of TT viruses (TTVs) in pigs, cats and dogs and their relatedness with species-specific TTVs in primates and tupaias. *J. Gen. Virol.*, **83**, 1291-1297.

Peng, Y.H., Nishizawa, T., Takahashi, M., Ishikawa, T., Yoshikawa, A. and Okamoto, H. (2002). Analysis of the entire genomes of thirteen TT virus variants classifiable into the fourth and fifth genetic groups, isolated from viremic infants. *Arch. Virol.*, **147**, 21-41.

Peters, M.A., Crabb, B.S. and Browning, G.F. (2001). Chicken anaemia virus VP2 is a novel protein tyrosine phosphatase. Proceedings of the Second International Symposium on infectious bursal disease and chicken infectious anaemia. Rauischholzhausen, Germany, June 2001.

Prescott, L. E. and Simmonds, P. (1998). Global distribution of transfusion-transmitted virus. *N. Engl. J. Med.*, **339**, 776-7.

Shibayama, T., Masuda, G., Ajisawa, A., Takahashi, M., Nishizawa, T., Tsuda, F. and Okamoto, H. (2001). Inverse relationship between the titre of TT virus DNA and the CD4 cell count in patients infected with HIV. *Aids* **15**, 563-570.

Takahashi, K., Iwasa, Y., Hijikata, M. and Mishiro, S. (2000). Identification of a new human DNA virus (TTV-like mini virus, TLMV) intermediately related to TT virus and chicken anemia virus. *Arch. Virol.*, **145**, 979-93.

Todd, D., Weston, J.H., Soike, D. and Smyth, J.A. (2001). Genome sequence determinations and analyses of novel circoviruses from goose and pigeon. *Virology*, **286**, 354-62.

Contributed By

Biagini, P., Todd, D., Bendinelli, M., Hino, S., Mankertz, A., Mishiro, S., Niel, C., Okamoto, H., Raidal, S., Ritchie, B.W., and Teo, G.C.

FAMILY NANOVIRIDAE

TAXONOMIC STRUCTURE OF THE FAMILY

Family	*Nanoviridae*	
Genus		*Nanovirus*
Genus		*Babuvirus*

VIRION PROPERTIES

MORPHOLOGY

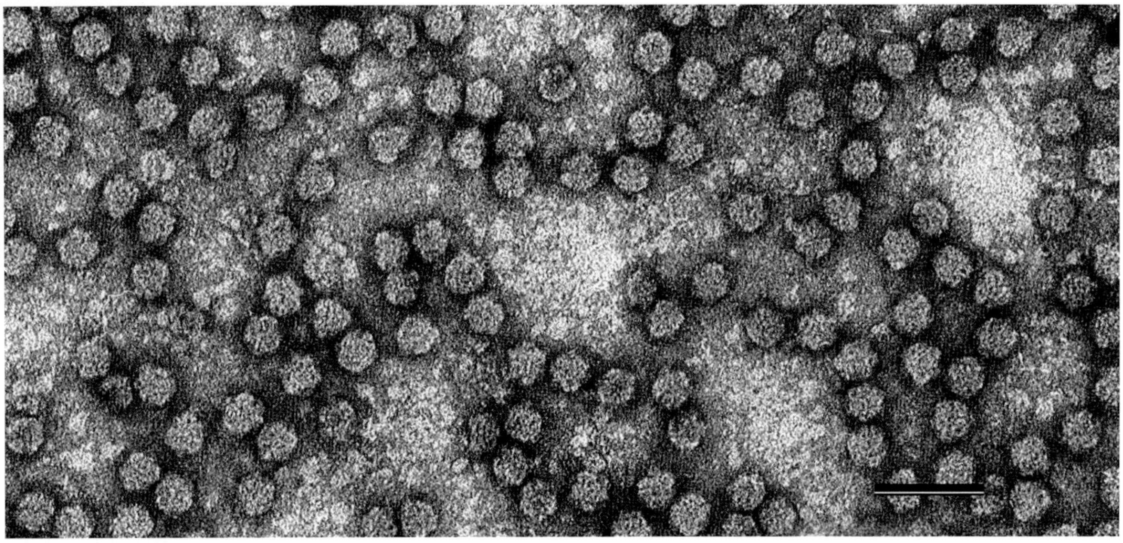

Figure 1: Negative contrast electron micrograph of particles of Faba bean necrotic yellows virus (FBNYV). The bar represents 50 nm. (Courtesy of L. Katul and D.-E. Lesemann).

Virions are 17 to 20 nm in diameter, and presumably of an icosahedral T=1 symmetry structure containing 60 subunits. They are not enveloped. Capsomeres may be evident, producing an angular or hexagonal outline (Fig. 1). Particle morphology is not affected by freezing of tissue before virion extraction.

PHYSICOCHEMICAL AND PHYSICAL PROPERTIES

Virions are stable in Cs_2SO_4 but may not be stable in CsCl. The buoyant density of virions is about 1.24 to 1.30 g/cm^3 in Cs_2SO_4, and 1.34 g/cm^3 in CsCl. They sediment as a single component in sucrose rate-zonal and Cs_2SO_4 isopycnic density gradients. The sedimentation coefficient of Banana bunchy top virus (BBTV) virions is 46S. The particle weight of Subterranean clover stunt virus (SCSV) is approximately 1.6 x 10^6. The extinction coefficient of SCSV is 3.6 at A_{260} (corrected for light scattering).

NUCLEIC ACID

Up to 12 distinct DNA components have been isolated from virion preparations of different species and their isolates. The number and types of ssDNA components constituting the integral (essential) genome parts have not been experimentally determined yet for any of the species. Available evidence suggests that the genome comprises 6-8 species of circular ssDNA ranging in size from 977 to 1111 nt. All of them seem to be structurally similar in being positive sense, transcribed in one direction, and containing a conserved stem-loop structure (and other conserved domains) in the NCR (Figs. 2 and 3). Each ssDNA component appears to be encapsidated in a separate particle.

In addition to the putative genomic DNAs, a total of 14 additional DNAs encoding replication initiator (Rep) proteins have been described from nanovirus infections: four each for BBTV [S1 (AF216221), S2 (AF216222), W1 (L32166), W2 (L32167)], Faba bean necrotic yellow virus (FBNYV) [C1 (X80879), C7 (AJ005964, AJ132185), C9 (AJ005966, AJ132187), C11 (AJ005968)], and MVDV [C1 (AB000920), C2 (AB000921), C3 (AB000922), C10 (AB009047)], and two for SCSV [C2 (U16731), C6 (U16735)]. These satellite-like DNAs are very diverse and phylogenetically distinct from the DNA-R of the nanoviruses. They are structurally similar and phylogenetically closely related to nanovirus-like *rep* DNAs that have recently been found associated with the begomoviruses *Ageratum yellow vein virus* [DNA 1 (AJ238493), DNA 2 (AJ416153)] and *Cotton leaf curl virus* [DNA 1 (AJ132345)]. However, due to the inclusion of an A-rich sequence within the intergenic region, the begomovirus-associated DNAs are larger (~1300 nt) than the nanovirus-associated DNAs (~1000 to ~1100 nt). In contrast to the genomic DNA-R which encodes the only known Rep protein essential for the replication of the multipartite genome of the nanoviruses, these additional *rep* DNAs are only capable of initiating replication of their cognate DNA but not of any heterologous genomic DNA. Since they are, moreover, only erratically associated with viral infections, they are regarded as satellite-like DNAs that depend on their helper viruses for various functions, such as encapsidation, transmission, and movement. There is no data as to whether they are of any biological significance to the helper virus.

PROTEINS

Table 1: Designation, size and functions of the proteins encoded by the various DNA components of the members of the family *Nanoviridae*

Protein*	Protein size (in kDa)	Encoded by DNA component	Identified from +				Protein function(s) §
			FBNYV	MDV	SCSV	BBTV	
M-Rep	33.1–33.6	DNA-R	+ (2)	+ (11)	+ (8)	+ (1)	**R**eplication initiator protein for all genomic DNAs
CP	18.7–19.3	DNA-S	+ (5)	+ (9)	+ (5)	+ (3)	**S**tructural protein, virion formation (encapsidation)
Clink	19.0–19.8	DNA-C	+ (10)	+ (4)	+ (3)	+ (5)	**C**ell-cycle link protein
MP	12.7–13.7	DNA-M	+ (4)	+ (8)	+ (1)	+ (4)	**M**ovement protein
NSP	17.3–17.7	DNA-N	+ (8)	+ (6)	+ (4)	+ (6)	**N**uclear shuttle protein
U1	16.9–18.0	DNA-U1	+ (3)	+ (5)	+ (7)	—	Not known (**U**nknown)
U2	14.2–15.4	DNA-U2	+ (6)	+ (7)	—	—	Not known (**U**nknown)
U3	10.3	DNA-U3	—	—	—	+ (2)	Not known (**U**nknown)
U4	10 or 12.5	DNA-U4	+ (12)	—	—	—	Not known (**U**nknown)
U5	5.0	DNA-R	—	—	—	+ (1)	Not known (**U**nknown)

* Master replication initiator protein (M-Rep), coat protein (CP), cell-cycle link protein (Clink), movement protein (MP), and nuclear shuttle protein (NSP). U1 to U5 are temporary designations until the protein function has been determined.

\+ A '+' or '–' indicates as to whether a protein has been described from a virus species or not. The former designation (number) of the encoding DNA is given in parentheses.

§ The underlined and bold letters indicate how the DNA component designations have been derived.

Virions have a single CP of about 19 kDa. No other proteins have been found associated with virions. In addition, at least 5-7 non-structural proteins are encoded by the mRNA(s) transcribed from the genomic ssDNAs (Table 1). All but one of the nanovirus DNAs encode only a single protein. A second virion-sense ORF, completely nested within the M-Rep-encoding ORF and encoding a putative 5-kDa protein of unknown function, was identified only from BBTV DNA-R but not from any of the other nanovirus DNA-R. Whereas the difference in the number and types of proteins (and encoding DNAs) between BBTV and FBNYV reflects a fundamental difference in genomic organizations between babuviruses and nanoviruses, the apparent dissimilarity in genomic organization among the nanoviruses FBNYV, SCSV and Milk vetch dwarf virus (MDV) indicates that the DNAs encoding the SCSV U2 and U4 proteins and the MDV U4 protein have not been identified yet.

LIPIDS
Not known.

CARBOHYDRATES
Not known.

GENOME ORGANIZATION AND REPLICATION

The genomic information is distributed over at least 6-8 molecules of circular ssDNA (Table 1). The fact that a typical set of 6 and 8 distinct DNAs has been consistently identified from a range of geographical isolates of BBTV and FBNYV, respectively, suggests that the babuvirus genome consists of six DNAs and the nanovirus genome of eight DNA components. The DNAs encoding a Rep, a CP, a Clink, a MP and a NSP protein have been described from all four assigned species (Table 1). DNA-U1 is shared by all three nanovirus species but is absent from the BBTV genome. The apparent absence of DNA-U2 and DNA-U4 in the SCSV genome and of DNA-U4 in MDV genome appears to be due to the fact that these genome components have not been identified yet from these nanoviruses. Hence, DNA-U1, -U2 and -U4 seem to be absent from the BBTV genome and specific components of the nanovirus genome. In contrast, DNA-U3 appears to be specific of the babuvirus genome and absent from the nanovirus genome. However, the question as to whether DNA-U3 and –U4, which potentially encode similarly sized proteins (~10 kDa), are functionally distinct, remains to be determined (Table 1).

All genomic DNAs of the assigned species contain a major virion sense ORF and appear to be transcribed unidirectionally. However, two mRNAs are transcribed from the BBTV DNA-R, whereas all the other nanovirus DNAs (incl. the other nanovirus DNA-R) encode a single protein only. Each coding region is preceded by a promoter sequence with a TATA box and followed by a poly(A) addition signal (Figs. 2 and 3). The NCRs of all genomic DNAs share a highly conserved sequence that encompasses Rep protein-binding sites and an inverted repeat sequence potentially forming a stem-loop structure and containing the origin of replication (*ori*) of the DNAs.

Since the nanovirus DNAs and some of the biochemical events determined for nanovirus replication resemble those of the geminiviruses, their replication is also thought to be completely dependent on the host cell's DNA replication enzymes and to occur in the nucleus through transcriptionally and replicationally active dsDNA intermediates by a rolling circle type of replication mechanism. Upon decapsidation of viral ssDNA, one of the first events is the synthesis of viral dsDNA with the aid of host DNA polymerase. As the virus DNAs have the ability to self-prime during dsDNA synthesis, it is likely that pre-existing primers are used for dsDNA replicative form (RF) synthesis, as has been shown for BBTV. From these dsDNA forms, host RNA polymerase then transcribes mRNAs encoding the M-Rep and other viral proteins required for virus replication. Viral DNA replication is initiated by the M-Rep protein that interacts with common sequence signals on all the genomic DNAs. Nicking and joining activity of the BBTV and FBNYV Rep proteins has been demonstrated *in vitro*. These biochemical events take place within and in close proximity to the common stem-loop region, i.e., an inverted repeat sequences that potentially forms a stem-loop structure. This stem loop including the canonical (loop-forming) nonanucleotide TAT/GTATT-AC is highly conserved in babuviruses and nanoviruses. Replication of the DNAs is by cellular enzymes, facilitated and enhanced by the action of Clink, a nanovirus-encoded cell cycle modulator protein.

ANTIGENIC PROPERTIES

Virions are strong immunogens. Most nanovirus species are serologically distinct from one another.

BIOLOGICAL PROPERTIES

HOST RANGE

Individual species have narrow host ranges. FBNYV, MDV and SCSV naturally infect a range of leguminous species, whereas BBTV has been reported only from the Musaceae. All virus species are associated with stunting of infected plants, and infected hosts may also show leaf roll, chlorosis and premature death. SCSV and FBNYV have been shown to replicate in inoculated protoplasts. All assigned species are restricted to the phloem tissue of their host plants and are not transmitted mechanically and through seeds. Apart from graft transmission, vector transmission had been the only means of experimentally infecting plants with nanoviruses for a long time, until infectivity of purified FBNYV virions by biolistic bombardment was recently demonstrated.

TRANSMISSION

Under natural conditions, all viruses are transmitted by certain aphid species in a persistent manner and do not replicate in their vectors. For FBNYV it has been demonstrated that purified virions alone are not transmissible by its aphid vector, regardless of whether they are acquired from artificial diets or directly microinjected into the aphid's hemocoel. However, faba bean seedlings biolistically inoculated with intact virions or viral DNA develop symptoms typical of FBNYV infections and are efficient sources for FBNYV transmission by aphids. These observations together with results from complementation experiments suggest that FBNYV (and other nanoviruses) require a virus-encoded helper factor for its vector transmission that is either dysfunctional or absent in purified virion preparations.

GEOGRAPHICAL DISTRIBUTION

BBTV is widely distributed in banana growing countries in the Asia-Pacific region and Africa. SCSV and MDV occur in Australia and Japan, respectively, whereas FBNYV occurs in West Asia, North and East Africa and Spain.

GENUS *NANOVIRUS*

Type Species ***Subterranean clover stunt virus***

DISTINGUISHING FEATURES

Available evidence (incl. unpublished data) suggests that the nanovirus genome consists of (at least) 8 different ssDNA components, referred to as DNA-R, -S, -M, -C, –N, and –U1, -U2, and -U4 (Fig. 2). Since they have been identified from several geographical isolates of FBNYV, they are considered integral components of the nanovirus genome. However, since information on the invariable association of DNA-U2 and -U4 with MDV and SCSV infections is lacking, the possibility that some of these DNAs play a particular role in the biology and/or replication of an individual nanovirus cannot be ruled out.

The ssDNA components of the nanoviruses range in size from 977 to 1003 nt. All of them appear to be structurally similar in being covalently closed, circular ssDNA molecules and unidirectionally transcribed in the virion sense. Moreover, by analogy to the BBTV DNA-S, -C, -M, -N, and -U3, each of which has been shown to yield only one mRNA transcript, all the nanovirus DNA components seem to contain only one major gene. In the 5 proteins shared by the 4 assigned species, FBNYV is most closely related to MDV (52–97%), and FBNYV and MDV are less closely (42–83%) and distantly (14–55%) related to SCSV and BBTV, respectively (Fig. 4).

The major differences between BBTV and nanoviruses are that the latter naturally infect legumes (dicots), are vectored by several aphid species colonizing legumes, and share aa sequence identities of 41-56% in the M-Rep and NSP, and only 18–27% in the CP, and Clink proteins with BBTV. Moreover, their DNA components are slightly smaller (by ~100 nt) than those of BBTV. Most importantly, nanoviruses have a distinct genomic organization as they appear to have a greater number (8 vs. 6) of DNA components and their DNA-R lacks an internal ORF. Consequently, nanoviruses seem to lack a homologue to the 5-kDa protein encoded by the BBTV DNA-R.

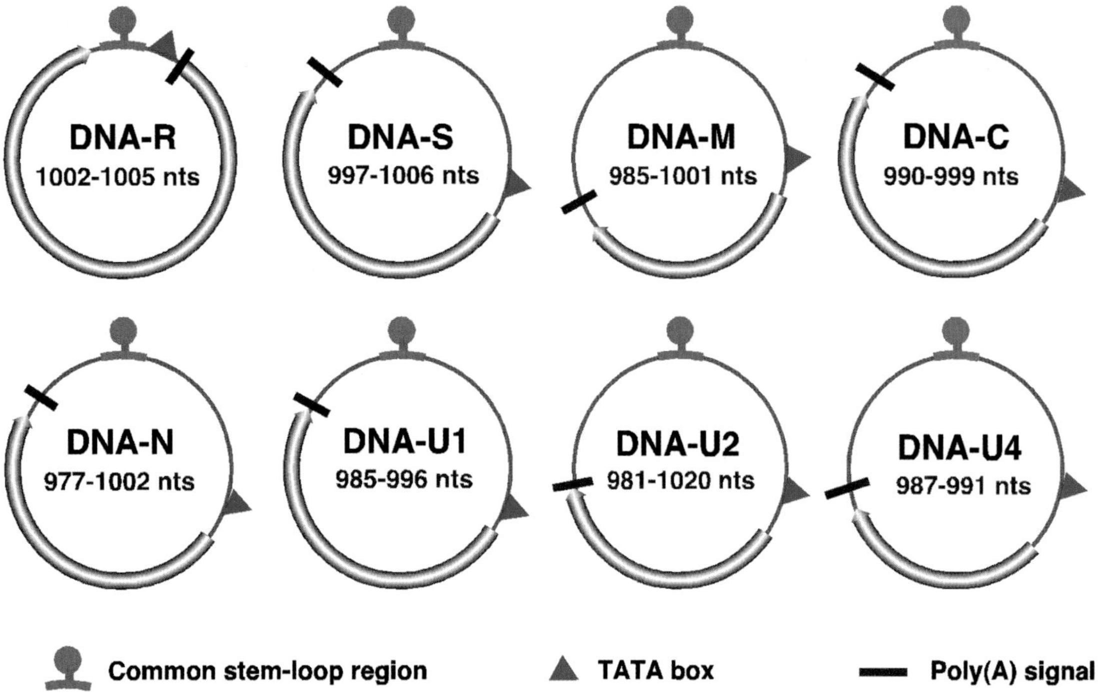

Figure 2: Diagram illustrating the putative genomic organization of the three species of the genus *Nanovirus* and depicting the structure of the eight identified viral DNA components (see also Table 1). Each DNA circle contains its designated name and its size range. Arrows refer to the location and approximate size of the ORFs and the direction of transcription. Note that DNA-U2 and U4 have not been identified from SCSV and that DNA-U4 has not been described from MDV. (Courtesy of L. Katul and H.J. Vetten)

Antisera to FBNYV and SCSV cross-react weakly with SCSV and FBNYV, respectively, in Western blots and immunoelectron microscopy but not at all in DAS-ELISA. However, MDV antigen reacts strongly not only with FBNYV antisera but also with the majority of monoclonal antibodies to FBNYV. Therefore, species-specific MAbs are required for the differentiation and specific detection of these two closely related species, which share an identity of about 83% in their CP aa sequences.

Although FBNYV infects over 50 legume species and only few non-legume species (*Stellaria media*, *Arabidopsis thaliana*, *Amaranthus* and *Malva* spp.) under experimental and natural conditions, major legume crops naturally infected by FBNYV are faba bean, lentil, chickpea, pea, French bean and cowpea. Likewise, SCSV experimentally infects numerous legume species, but its economically important natural hosts only include subterranean clover, French bean, faba bean, pea, and medics. MDV is known to cause yellowing and dwarfing in Chinese milk vetch (*Astragalus sinicus* L.), a common green manure crop in Japan, as well as in faba bean, pea, and soybean. Reports describing *Datura stramonium*,

Chenopodium quinoa, *Spinacia oleracea* and *Nicotiana tabacum* as experimental non-leguminous hosts of MDV have been recently confirmed.

Whereas only one aphid species (*Pentalonia nigronervosa*) has been reported as vector of BBTV, several aphid species transmit FBNYV, MDV, and SCSV. *Aphis craccivora* appears to be the major natural vector of these viruses as it is the most abundant aphid species on legume crops in the afflicted areas and was among the most efficient vectors under experimental conditions. Other aphid vectors of FBNYV are *Aphis fabae* and *Acyrthosiphon pisum* but in no case were *Myzus persicae* and *Aphis gossypii* able to transmit FBNYV. SCSV has been reported to be vectored also by *A. gossypii*, *M. persicae* and *Macrosiphum euphorbiae*, but some of these accounts now appear questionable. In contrast, recent data have shown that, in addition to *A. craccivora*, some other aphid species, such as *A. gossypii*, *A. fabae*, *Acyrthosiphon pisum*, and *Megoura viciae*, were able to transmit MDV at various efficiencies.

List of Species Demarcation Criteria in the Genus

The following criteria should be used as a guideline for species demarcation in the genus:

- Overall nt sequence identity of <75% is generally indicative of a distinct species,
- Different reactions to antibodies to individual species,
- Differences in CP aa sequences of >15%,
- Differences in natural host range, and
- Differences in the number and types of vector aphid species.

List of Species in the Genus

Species names are in green italic script; strain names and synonyms are in black roman script; tentative species names are in blue roman script. Sequence accession numbers, and assigned abbreviations ( ) are also listed.

Species in the Genus

Faba bean necrotic yellows virus		
Faba bean necrotic yellows virus-[Syria]	[Y11405-9, AJ005965, AJ005967]	(FBNYV-[Sy])
Faba bean necrotic yellows virus-[Egypt]	[AJ132179-84, AJ132186]	(FBNYV-[Eg])
Faba bean necrotic yellows virus-[Ethiopia]	[AF159704-5]	(FBNYV-[Et])
Milk vetch dwarf virus		
Milk vetch dwarf virus	[AB000923-7; AB009046, AB027511, AB044387]	(MDV)
Subterranean clover stunt virus		
Subterranean clover stunt virus-[F]	[U16730, U16732–4, U16736, AJ290434]	(SCSV-[F])
Subterranean clover stunt virus-[J]	[L47332]	(SCSV-[J])

Tentative Species in the Genus

None reported.

Genus *Babuvirus*

Type Species *Banana bunchy top virus*

Distinguishing Features

BBTV infects bananas (*Musa* spp.) and closely related species within the Musaceae, such as *M. textilis* Née and *Ensete ventricosum* Cheesem. There are no confirmed non-*Musa* hosts of BBTV. Symptoms of BBTV include plant stunting, foliar yellowing, and most characteristic dark green streaks on the pseudostem, petioles, and leaves.

BBTV has a genome consisting of 6 different ssDNA components, referred to as DNA-R, -S, -C, -M, -N, and -U3 (Fig. 3). Since they have been identified from all BBTV isolates tested from more than 10 countries, they are considered integral components of the BBTV genome. The major differences between nanoviruses and BBTV are that the latter naturally infects a monocot (*Musa* spp.), is vectored by the banana aphid *Pentalonia nigronervosa*, has a smaller number of DNA components (6 vs. 8), its DNA components are slightly larger (by ~100 nt), its DNA-R has an internal ORF encoding a 5-kDa protein of unknown function, and it shares aa sequence identities of < 56% in the CP and NSP and of only < 30% in the MP, and Clink proteins with the nanoviruses. In addition, the BBTV genomic DNAs share not only a highly conserved (common) stem-loop region of 69 nt, but also a short stretch (66-92 nt) of another common region, the so-called major common region (CR-M), in their non-coding sequences. An equivalent CR-M region is not found in the genomic DNAs of FBNYV, MDV, and SCSV. The DNA primers required for full-length complementary strand synthesis of BBTV DNAs map to a region within the CR-M and extend 5′ of this conserved region.

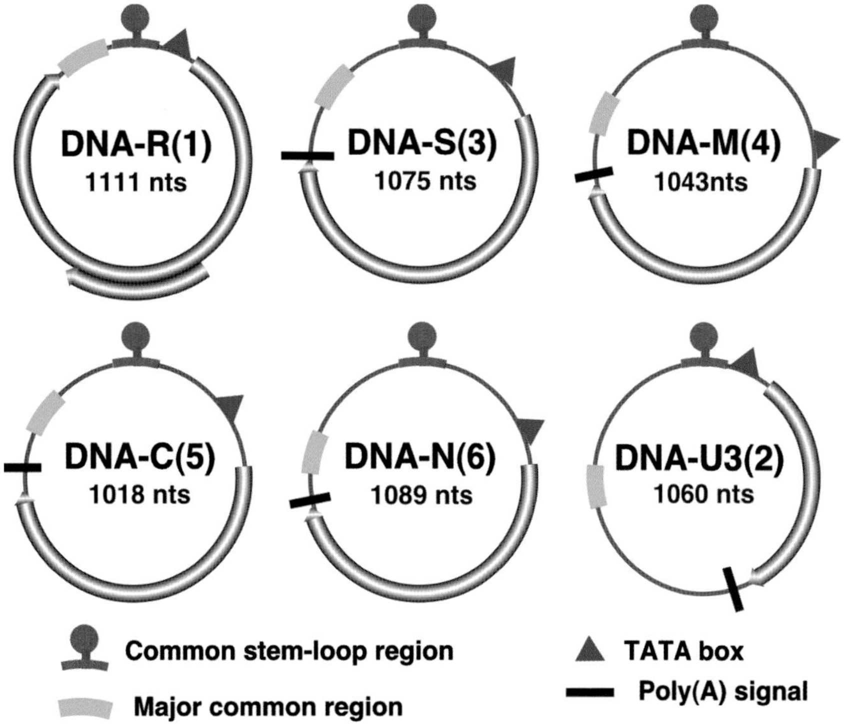

Figure 3: Diagram illustrating the putative genomic organization of Banana bunchy top virus (BBTV) and depicting the structure of the 6 identified viral DNA components (see also Table 1). Each DNA circle contains its designated name (former name in brackets) and its size (Australian isolate only). Arrows refer to the location and approximate size of the ORFs and the direction of transcription.

Two groups of BBTV isolates, designated the Asian and South Pacific groups, are distinguished based on the DNA-R, -N, and -S sequences. The nt sequences of the major gene of DNA-R, -N, and -S differ by 7.5, 8.6, and 6.3%which translate to mean differences of 5.6, 6.7, and 1.4%, respectively, in aa sequences between the two geographic groups. Particularly striking are the differences between the two groups of BBTV isolates in the CR-M of DNA-R (32%), -N (27%), and -S (39%), whereas the intra-group CR-M variation does not exceed 6%. Whereas the DNA-R, -N, and –S genes appear to be well conserved among BBTV isolates, striking differences between BBTV isolates in less conserved gene products (movement protein) have been observed (Fig. 4). Interestingly, there is far greater sequence variability between the sequences within the Asian group than those within the South Pacific group of isolates suggesting that either BBTV has been present in bananas in this region for a longer period or there has been more than one introduction of BBTV into bananas from another host in this region.

BBTV is serologically unrelated to members of the genus *Nanovirus*. In general, poly- and monoclonal antibodies raised to BBTV virions do not reveal serological variability among BBTV isolates. This is consistent with a maximum difference of only about 3% in CP aa sequences between the two geographic groups of BBTV isolates.

LIST OF SPECIES DEMARCATION CRITERIA IN THE GENUS

Not applicable.

LIST OF SPECIES IN THE GENUS

Species names are in green italic script; strain names and synonyms are in black roman script; tentative species names are in blue roman script. Sequence accession numbers, and assigned abbreviations ( ) are also listed.

SPECIES IN THE GENUS

Banana bunchy top virus		
Banana bunchy top virus-[Australia]	[S56276, L41574 to -78]	(BBTV-[Au])
Banana bunchy top virus-[Taiwan]	[AF416468, AF148942]	(BBTV-[Tw])
Banana bunchy top virus-[China-NS]	[AF238874, -6, -8; AY266417]	(BBTV-[ChNS])
Banana bunchy top virus-[China-NSP]	[AF238875, -7, -9; AY264347]	(BBTV-[ChNSP])

TENTATIVE SPECIES IN THE GENUS

None reported.

LIST OF UNASSIGNED VIRUSES IN THE FAMILY

Coconut foliar decay virus		
Coconut foliar decay virus	[M29963]	(CFDV)

This unassigned species of the family has been described only from coconut palms growing in Vanuatu. Whereas the family typically includes species that are persistently transmitted by aphids, have a CP size of about 19 kDa and possess 6-8 genomic DNAs (~1 kb) that are unidirectionally transcribed, CFDV particles have a similar morphology but differ from those of the assigned members by being transmitted by a cixiid plant hopper (*Myndus taffini*), by having a CP of about 24 kDa, and by containing a single circular ssDNA of 1291 nt which is proposed to be transcribed bidirectionally. Apart from the similarity in particle morphology, the reasons for retaining CFDV in the family *Nanoviridae* are the structural similarity of CFDV DNA with that of the nanoviruses and its capacity to encode a nanovirus-like Rep protein. In these respects, however, the CFDV DNA resembles the autonomously replicating nanovirus-like DNAs (~1300 nt) that have recently been found associated with some monopartite begomoviruses.

PHYLOGENETIC RELATIONSHIPS WITHIN THE FAMILY

Analysis of the 5 proteins identified from all 4 assigned species (Fig. 4) suggests that the most conserved nanovirus proteins are the M-Rep protein (54–97% identity) and the NSP (41–91%), followed by the CP (20–84%), the MP (14–76%), and the Clink protein (18–72%). The babuvirus BBTV shares significant levels of aa sequence similarity with the nanoviruses only in the M-Rep (54–56%) and NSP (41–45%), whereas the aa sequence similarities between the two genera are negligible in the CP (20–27%), the MP (20–23%), and the Clink protein (18–23%).

SIMILARITY WITH OTHER TAXA

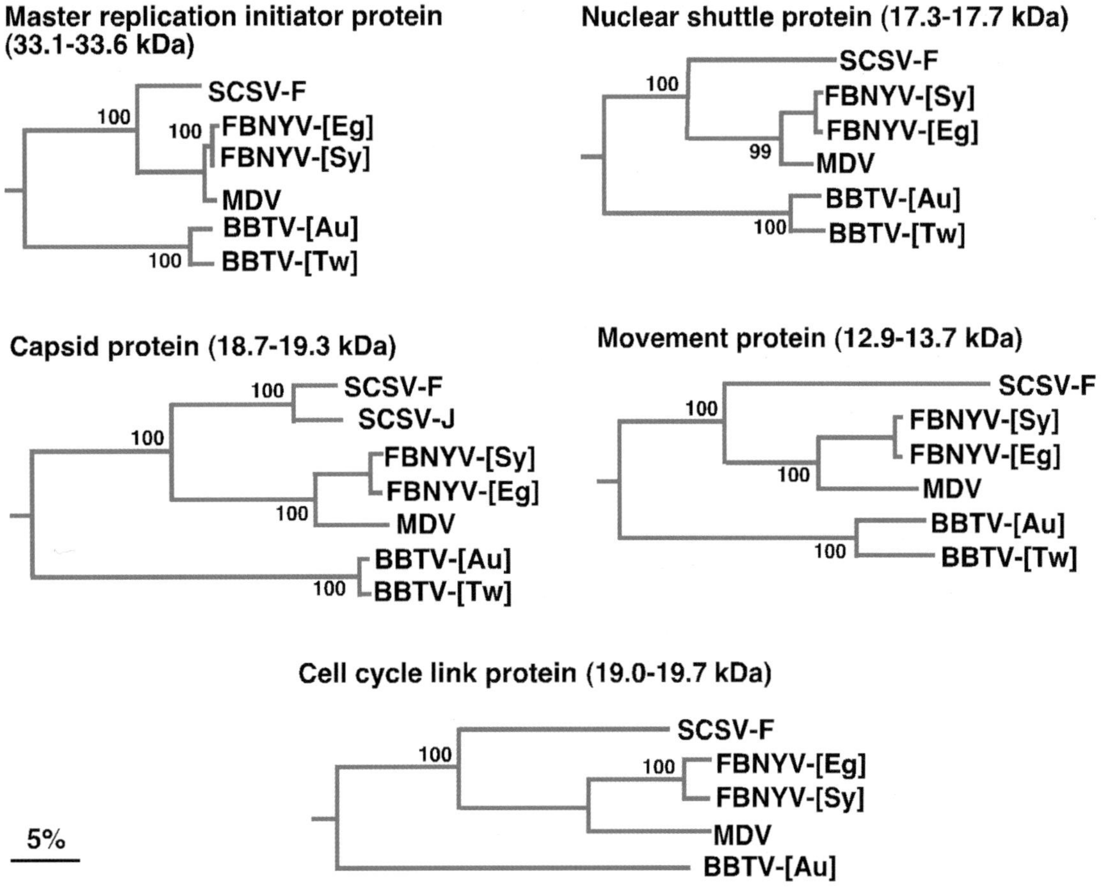

Figure 4: Neighbor-joining dendrograms illustrating aa sequence relationships among the five proteins identified from members of the family *Nanoviridae*: Banana bunchy top virus (BBTV), Faba bean necrotic yellows virus (FBNYV), Milk vetch dwarf virus (MDV), and Subterranean clover stunt virus (SCSV). Since sequence information is available for more than one isolate of each virus, the sequences of the BBTV isolates from Australia (Au), Taiwan (Tw) and China (Ch), the FBNYV isolates from Egypt (Eg) and Syria (Sy), and the SCSV strains F and J were included in the comparison. The vertical order in (top left) to (bottom) reflects decreasing identities between FBNYV and MDV as well as among all four species. Vertical branch lengths are arbitrary and horizontal distances are proportional to percent sequence differences. Sequence alignments and dendrograms were produced using DNAMAN (version 4.0 for Windows 95/98, Lynnon Biosoft, Quebec, Canada) which uses a CLUSTAL-type algorithm. The dendrograms were bootstrapped 1000 times (scores are shown at nodes). (Courtesy of L. Katul and H.J. Vetten).

All Rep proteins of the assigned species have most of the aa domains characteristic of Rep proteins of geminiviruses and other ssDNA viruses. The nanovirus Rep proteins differ from those of members of the family *Geminiviridae* in being smaller (about 33 kDa), having a slightly distinct dNTP-binding motif (GPQ/NGGEGKT), lacking the Rb-binding motif, and in sharing aa sequence identities of only 17 to 22% with them. Moreover, the assigned species are clearly distinct from the geminiviruses in particle morphology, genome size, number and size of DNA components, mode of transcription, and in vector species. Circoviruses also differ from nanoviruses in several respects. All of these viruses have a conserved nonanucleotide motif at the apex of the stem-loop sequence which is consistent with the operation of a rolling circle model for DNA replication.

DERIVATION OF NAMES

Babu: siglum from *Ba*nana *bu*nchy top virus
Nano: from the Greek *nanos*, meaning dwarf, referring to the observations that these plant viruses have the smallest known virions and genome segment sizes, and dwarf their hosts.

References

Aronson, M.N., Meyer, A.D., Gyorgyey, J., Katul, L., Vetten, H.J., Gronenborn B. and Timchenko, T. (2000). Clink, a nanovirus-encoded protein, binds both pRB and SKP1. *J. Virol.*, **74**, 2967-72.

Beetham, P.R., Hafner, G.J., Harding, R.M. and Dale, J.L. (1997). Two mRNAs are transcribed from banana bunchy top virus DNA-1. *J. Gen. Virol.*, **78**, 229-236.

Beetham, P.R., Harding, R.M. and Dale, J.L. (1999). Banana bunchy top virus DNA-2 to 6 are monocistronic. *Arch. Virol.*, **144**, 89-105.

Boevink, P., Chu, P.W.G. and Keese, P. (1995). Sequence of subterranean clover stunt virus DNA: affinities with the geminiviruses. *Virology*, **207**, 354-361.

Burns, T.M., Harding, R.M. and Dale, J.L. (1995). The genome organization of banana bunchy top virus: analysis of six ssDNA components. *J. Gen. Virol.*, **76**, 1471-1482.

Chu, P.W.G. and Helms, K. (1988). Novel virus-like particles containing circular single-stranded DNA associated with subterranean clover stunt disease. *Virology*, **167**, 38-49.

Chu, P.W.G. and Vetten, H.J. (2003). Subterranean clover stunt virus. AAB Descriptions of Plant Viruses, **396**.

Dugdale, B., Beetham, P.R., Becker, D.K., Harding, R.M. and Dale, J.L. (1998). Promoter activity associated with the intergenic regions of banana bunchy top virus DNA-1 to -6 in transgenic tobacco and banana cells. *J. Gen. Virol.*, **79**, 2301-11.

Franz, A., Katul, L., Makkouk, K.M. and Vetten, H.J. (1996). Monoclonal antibodies for the detection and differentiation of faba bean necrotic yellows virus isolates. *Ann. Appl. Biol.*, **128**, 255-268.

Franz, A.W., van der Wilk, F., Verbeek, M., Dullemans, A.M. and van den Heuvel, J.F. (1999). Faba bean necrotic yellows virus (genus *Nanovirus*) requires a helper factor for its aphid transmission. *Virology*, **262**, 210-219.

Hafner, G.J., Harding, R.M. and Dale, J.L. (1997). A DNA primer associated with banana bunchy top virus. *J. Gen. Virol.*, **78**, 479-486.

Hafner, G.J., Stafford, M.R., Wolter, L.C., Harding, R.M. and Dale, J.L. (1997). Nicking and joining activity of banana bunchy top virus replication protein in vitro. *J. Gen. Virol.*, **78**, 1795-1799.

Horser, C.L., Karan, M., Harding, R.M. and Dale, J.L. (2001). Additional Rep-encoding DNAs associated with banana bunchy top virus. *Arch. Virol.*, **146**, 71-86.

Karan, M., Harding, R.M. and Dale, J.L. (1994). Evidence for two groups of banana bunchy top virus isolates. *J. Gen. Virol.*, **75**, 3541-3546.

Katul, L., Maiss, E., Morozov, S.Y. and Vetten, H.J. (1997). Analysis of six DNA components of the faba bean necrotic yellows virus genome and their structural affinity to related plant virus genomes. *Virology*, **233**, 247-259.

Katul, L., Timchenko, T., Gronenborn, B. and Vetten, H.J. (1998). Ten distinct circular ssDNA components, four of which encode putative replication-associated proteins, are associated with the faba bean necrotic yellows virus genome. *J. Gen. Virol.*, **79**, 3101-3109.

Rohde, W., Randles, J.W., Langridge, P. and Hanold, D. (1990). Nucleotide sequence of a circular single-stranded DNA associated with coconut foliar decay virus. *Virology*, **176**, 648-651.

Sano, Y., Wada, M., Hashimoto, T. and Kojima, M. (1998). Sequences of ten circular ssDNA components associated with the milk vetch dwarf virus genome. *J. Gen. Virol.*, **79**, 3111-3118.

Timchenko, T., de Kouchkovsky, F., Katul, L., David, C., Vetten, H.J. and Gronenborn, B. (1999). A single rep protein initiates replication of multiple genome components of faba bean necrotic yellows virus, a single-stranded DNA virus of plants. *J. Virol.*, **73**, 10173-82.

Timchenko, T., Katul, L., Sano, Y., de Kouchkovsky, F., Vetten, H.J. and Gronenborn, B. (2000). The master rep concept in nanovirus replication: identification of missing genome components and potential for natural genetic reassortment. *Virology*, **274**, 189-195.

Vetten, H.J. and Katul, L. (2001). Nanoviruses. In: *Encyclopedia of Plant Pathology* (O.C., Maloy, and T.D., Murray, eds), pp. 670-674. John Wiley & Sons, Inc., New York.

Wanitchakorn, R., Hafner, G.J., Harding, R.M. and Dale, J.L. (2000). Functional analysis of proteins encoded by banana bunchy top virus DNA-4 to -6. *J. Gen. Virol,.* **81**, 299-306.

Contributed By

Vetten, H.J., Chu, P.W.G., Dale, J.L., Harding, R., Hu, J., Katul, L., Kojima, M., Randles, J.W., Sano, Y. and Thomas, J.E.

FAMILY PARVOVIRIDAE

TAXONOMIC STRUCTURE OF THE FAMILY

Family	***Parvoviridae***
Subfamily	***Parvovirinae***
Genus	***Parvovirus***
Genus	***Erythrovirus***
Genus	***Dependovirus***
Genus	***Amdovirus***
Genus	***Bocavirus***
Subfamily	***Densovirinae***
Genus	***Densovirus***
Genus	***Iteravirus***
Genus	***Brevidensovirus***
Genus	***Pefudensovirus***

VIRION PROPERTIES

MORPHOLOGY

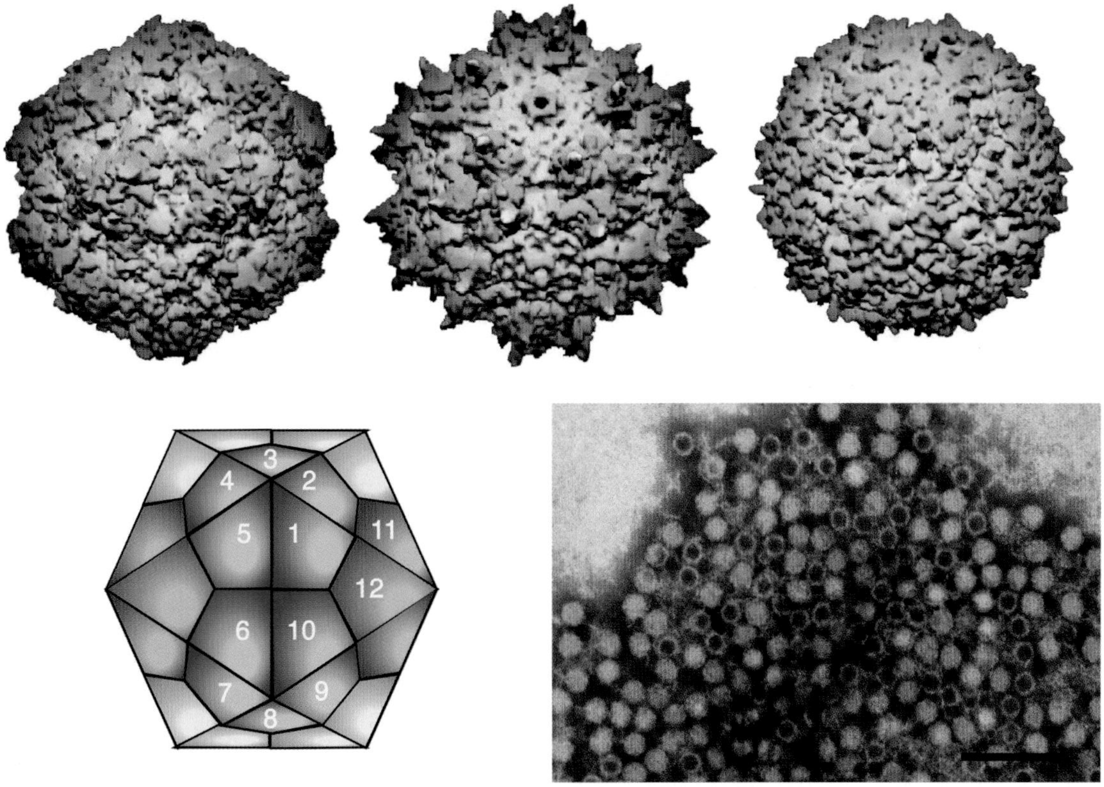

Figure 1: (Top) Space-filling models of the capsid structures of Canine parvovirus (CPV) (left); Adeno-associated virus - 2 (AAV-2) (center) and Galleria mellonella densovirus (GmDNV) (right). Each model is drawn to the same scale and is colored according to distance from the viral center. In each case, the view is down a twofold axis at the center of the virus, with threefold axes left and right of center, and fivefold axes above and below (Courtesy of M. Chapman). (Bottom left) Diagram representing a T=1 capsid structure. (Bottom right) Negative contrast electron micrograph of CPV particles. The bar represents 100 nm. (Courtesy of C.R. Parrish).

Virions are non-enveloped, 18-26 nm in diameter, and exhibit T=1 icosahedral symmetry (Fig. 1). The particles are composed of 60 copies of the region of the CP common to all forms of the structural polypeptides - usually the C-terminal three-quarters of VP1. The principal protein appears to be either VP2 or VP3, although between 9 and 12 of the copies may be VP1.

Physicochemical and Physical Properties

Virion Mr is about 5.5-6.2 x 10^6. Virion buoyant density is 1.39-1.43 g/cm^3 in CsCl. The S_{20w} is 110-122S. Infectious particles are composed of about 80% protein and about 20% DNA. Infectious particles with buoyant densities about 1.45 g/cm^3 may represent conformational or other variants, or precursors to the mature particles. Defective particles with deletions in the genome occur and exhibit lower densities. Mature virions are stable in the presence of lipid solvents, or on exposure to pH 3-9 or, for most species, incubation at 56°C for at least 60 min. Viruses can be inactivated by treatment with formalin, β-propriolactone, hydroxylamine, oxidizing agents or ultraviolet light.

Nucleic Acid

The genome is a linear, molecule of ssDNA, 4-6 kb in size (Mr 1.5 x 10^6 - 2.0 x 10^6). The G+C content is 41-53%. The 5'- and 3'-ends of the genome contain palindromic sequences, 120 to ~600 nt in length, which can be folded into hairpin structures essential for viral DNA replication. These terminal hairpins may be part of a terminal repeat, and therefore related in sequence (e.g., *Dependovirus*), while some genomes have terminal hairpins that are unrelated in sequence to one another (e.g., *Parvovirus*). Some parvoviruses preferentially encapsidate ssDNA of negative polarity (i.e., complementary to the viral mRNA species; e.g., Minute virus of mice (MVM), others may encapsidate ssDNA species of either polarity in equivalent (e.g., Adeno-associated virus, AAV), or different proportions (Bovine parvovirus, BPV). The percentage of particles encapsidating the positive strand can vary from 1 to 50% and may be influenced by the host cell in which the virus is produced (e.g., LuIII virus, LuIIIV).

Proteins

Viruses generally have 2-4 virion protein species, and up to 5 in the case of brevidensoviruses (VP1-VP5). Depending on the species, protein sizes are: VP1 80-96 kDa, VP2 64-85 kDa, VP3 60-75 kDa and VP4 49-52 kDa. The viral proteins represent alternative forms of the same gene product, predominantly differing at their N-termini. The VP1-specific regions of viruses of all genera except *Brevidensovirus* and *Amdovirus*, contains the enzymatic core of a phospholipase A2 (PLA2). The principal protein species is VP2 or VP3. Spermidine, spermine, and putrescine have been identified as components of some virus particles.

Lipids

Virions lack essential lipids.

Carbohydrates

None of the viral proteins are known to be glycosylated.

Genome Organization and Replication

Parvoviruses possess two major gene cassettes (Fig. 2). The REP ORF encodes the non-structural proteins (NS), that are required for transcription and DNA replication, and the CP ORF encodes the structural proteins of the capsid, the CAP, VP or S proteins. Both gene cassettes are present on the same DNA strand for members of the subfamily *Parvovirinae* and for members of the *Iteravirus* and *Brevidensovirus* genera of the subfamily *Densovirinae*. The REP functions and the CPs of densoviruses and pefudensoviruses are encoded on complementary strands (Fig. 2). In some viruses, other minor ORFs have been detected. For some of these, a protein product has been identified (the ORF for the amino terminus of VP1). Mutations within the REP ORF block virus replication and gene expression.

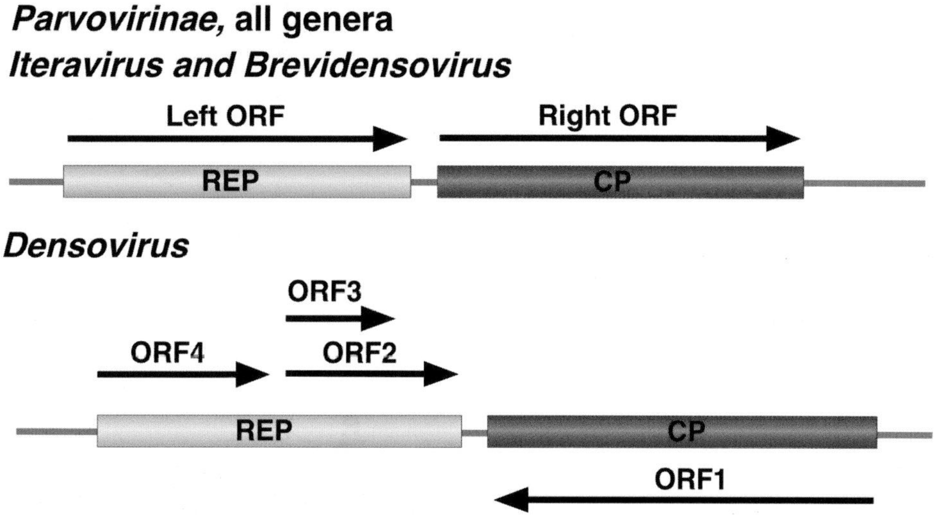

Figure 2: Gene organization for members of the monosense *Parvovirinae* sub-family, and the genera *Iteravirus* and *Brevidensovirus* (top). Gene organization for the ambisense densoviruses (bottom). REP denotes the non-structural, replicative protein genes of the virus, while CP represents the genes for the structural proteins that make up the capsid.

For some viruses, alternative splicing allows different forms of the REP gene products to be produced. The MVM REP ORF produces two major non-structural proteins, NS1, NS2P, and two minor ones, NS2Y, NS2L. A subset of the same alternative splicing strategy allows translation of the CP ORF to produce two proteins, VP1 and VP2. MVM VP3 is generated in the intact virion by proteolytic cleavage of VP2. VP1 and VP2 are identical except for their amino termini. Synthesis of VP1 derives from a minor spliced mRNA containing a methionine codon that allows translation of a small ORF, which encodes basic aa motifs, upstream (5') of the VP2-coding sequence. Parvoviruses use an alternative splice donor, while dependoviruses use an alternative splice acceptor for this purpose. Mutants in REP or CP can be complemented *in trans*. The palindromic sequences (at both termini) are required *in cis* for DNA replication to occur.

Depending on the virus there may be one (*Erythrovirus* and *Iteravirus*), two (*Parvovirus*, *Densovirus* and *Brevidensovirus*), or three (*Dependovirus*) promoters for mRNA transcription. Some of the mRNAs are spliced allowing alternative forms of the protein products to be produced. The mRNA species are capped and polyadenylated either at a common 3' site near the end of the genome (MVM, AAV), or at an alternative polyadenylation site in the centre of the genome as well as at a site near the end of the genome (Human parvovirus B19; B19V). Depending on the species, viruses may benefit from co-infection with other viruses, such as adenoviruses, or herpesviruses, or from the effects of chemical or other treatments of the host. Viral proteins accumulate in the nucleus in the form of empty capsid structures. Progeny infectious virions accumulate in the cell nucleus.

Viral entry into the cell is by receptor-mediated endocytosis and is blocked by antagonists of vacuolar ATPase. The trafficking of virus within the cell appears to vary between different members of the family, and even between different species within individual genera. The PLA2 domain probably plays a role in the trafficking or release of particles from endosomal compartments. The process of uncoating is not well understood. Virus replication takes place in the cell nucleus and appears to require the cell to go through S phase, indicating a close association between the host and virus replication processes. Autonomously replicating parvoviruses probably do not initiate gene expression until the host cell enters S-phase, whereas this transition is effected by a positive process under the control of the helper virus in the case of helper-dependent parvoviruses.

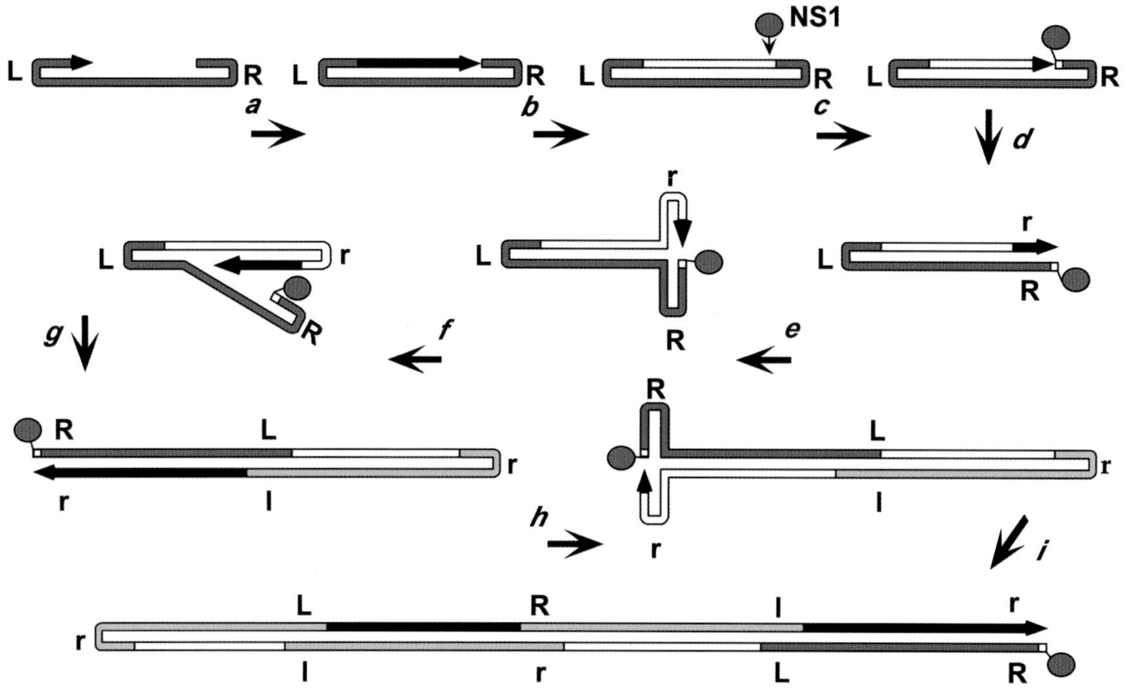

Figure 3: DNA replication model as determined for Minute virus of mice (MVM). The newly synthesized DNA of the growing 3'-end is represented by a black arrowed line, with the original genome in dark grey. Light grey denotes progeny genomes embedded within the oligomeric intermediates. Upper and lower cases of R and L represent flip and flop forms of the right and left ends, respectively. For details see Text

Replication proceeds through a series of duplex, concatemeric intermediates by the rolling hairpin mechanism (Fig. 3) and probably involves a processive host DNA polymerase(s) (probably pol δ, possibly pol ε or others). In step (a) the base-paired 3' nt of left-end hairpin is used by a host polymerase to prime conversion of virion DNA to the first duplex intermediate. This generates a monomer length duplex molecule in which the two strands are covalently continuous at the viral left-end telomere. Synthesis of this intermediate precedes viral gene expression. The 3'-end of the new DNA strand is ligated to the 5'-end of the hairpin by a host ligase, creating a covalently continuous duplex molecule (step b). Replication beyond this point requires expression of NS1, which carries out a "hairpin transfer" reaction, in which it nicks the ligated strand (step c). The replication fork now unfolds and copies the hairpin, thus replacing the original sequence of the terminus with its inverted complement (step d). The terminal sequences are imperfect palindromes, and since this inversion occurs with every round of replication, progeny genomes comprise equal numbers of each terminal orientation, dubbed "flip" and "flop". In MVM replication this hairpin transfer reaction occurs only at the right-end, because of different structural and co-factor requirements needed to activate the NS1 nickase at each terminus.

When it occurs on the first monomer formed after uncoating, it regenerates the "tether" sequence, lost during entry, now attached to a newly synthesized NS1 molecule. Extended-form right-end termini are melted out and reformed into hairpin "rabbit ear" structures in a process facilitated by the direct binding of NS1 to sequences in the terminus (step e). This allows the newly synthesized DNA to create the base paired hairpin structures needed to prime synthesis of additional linear sequences (step f). This gives rise to a palindromic duplex dimeric (step g), which can undergo the same right-end rearrangement (step h), leading to the synthesis of tetrameric concatemers (step h), in which alternating unit length genomes are fused in left-end:left-end and right-end:right-end orientations. Individual genomic monomer duplexes are then excised from these concatemers by a process called junction resolution.

ANTIGENIC PROPERTIES

Parvoviruses appear to have very stable virions that are quite simple antigenically. This has led to the use of individual serotype as a major criterion for species demarcation. Serotype has been defined by neutralization of infectivity in cell culture, hemagglutination-inhibition or specific ELISA using a capture format. Two antigenic sites, defined by mutations that confer resistance to neutralization by monoclonal antibodies, have been determined for Canine parvovirus (CPV). Some, but not all, species in a genus may be antigenically related by epitopes in the NS proteins.

BIOLOGICAL PROPERTIES

Autonomous parvoviruses require host cell passage through S-phase. Certain parvoviruses replicate efficiently only in the presence of helper viruses (e.g., adenoviruses, herpesviruses). These helper functions involve the adenovirus or herpesvirus early gene products and trans-activation of parvovirus replication. The helper functions appear to relate to effects of the helper virus upon the host cell rather than direct involvement of helper virus gene products in parvovirus replication. Association of parvoviruses with tumor cell lines appears to relate to increased DNA replication and/or the state of differentiation in such cells rather than previous involvement as an etiologic agent of oncogenesis. Co-infection involving certain parvoviruses and selected oncogenic adenoviruses (or other viruses) may reduce the oncogenic effect of those viruses, possibly by promoting cell death. In certain circumstances, parvovirus DNA may integrate into the host genome from which it may be activated by subsequent helper virus infection. The site of integration may be specific in certain hosts (e.g., the q arm of human chromosome 19 for AAV-2).

SUBFAMILY *PARVOVIRINAE*

TAXONOMIC STRUCTURE OF THE SUBFAMILY

Subfamily	*Parvovirinae*
Genus	*Parvovirus*
Genus	*Erythrovirus*
Genus	*Dependovirus*
Genus	*Amdoviruses*
Genus	*Bocavirus*

DISTINGUISHING FEATURES

Viruses assigned to the subfamily *Parvovirinae* infect vertebrates and vertebrate cell cultures, frequently in association with other viruses.

GENUS *PARVOVIRUS*

Type Species *Minute virus of mice*

DISTINGUISHING FEATURES

For some members of the genus, mature virions contain negative-strand DNA of 5 kb. In other members, positive-strand DNA occurs in variable proportions (1-50%). The linear molecule of ssDNA has hairpin structures at both the 5'- and 3'-ends. The 3'-terminal hairpin is 115-116 nt in length, the 5' structure is 200-242 nt long. There are two mRNA promoters (map units 4 and 39) and a single polyadenylation site at the 3'-end. Characteristic cytopathic effects are induced by the viruses during replication in cell culture. Many species exhibit hemagglutination with red blood cells of one or more species. Under experimental conditions the host range may be extended to a large number of vertebrate species (e.g., rodent viruses and LuIIIV replicate in Syrian hamsters). Transplacental transmission has been detected for a number of species.

GENOME ORGANIZATION AND REPLICATION

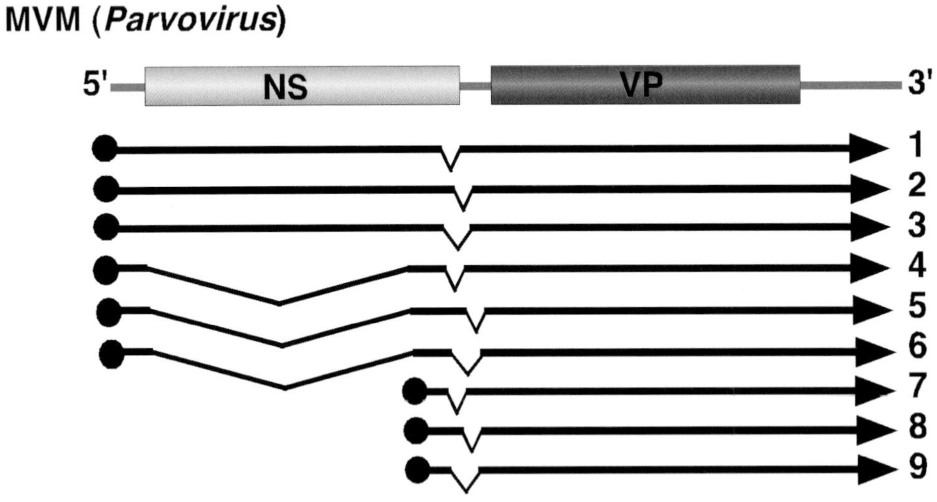

Figure 4: Gene organization and transcription scheme for members of the *Parvovirus* genus, as shown for Minute virus of mice (MVM). Genes are shown as boxes. The left ends of the mRNAs (thick lines) are the sites of the mRNA caps (filled circles), the right ends are the polyadenylation sites (arrows); introns are indicated by thin lines.

Alternate splicing controls viral gene expression (Fig. 4). For Minute virus of mice (MVM), transcripts 1-6 are made from the promoter at 4 map units. Transcripts 1-3 encode only

NS1, while transcripts 4-6, spliced into a second reading frame by the major intron, encode NS2P, NS2Y and NS2L, respectively, the C-termini of which are different due to the use of alternative donor and acceptor splice sites bordering the small intron. A promoter at 38 map units is transactivated by NS1 to drive the synthesis of transcripts 7-9. The use of the same alternative donor splice sites that are present in the P4 transcripts controls the synthesis of the CPs, such that transcripts 7 and 9 encode VP2, while transcript 8 encodes VP1.

LIST OF SPECIES DEMARCATION CRITERIA IN THE GENUS

Members of each species are antigenically distinct, as assessed by neutralization using polyclonal antisera, and natural infection is usually confined to a single host species. Generally, species are <95% related by non-structural gene DNA sequence.

SSDNA

LIST OF SPECIES IN THE GENUS

Species names are in green italic script; strain names and synonyms are in black roman script; tentative species names are in blue roman script. Sequence accession numbers, and assigned abbreviations ( ) are also listed.

SPECIES IN THE GENUS

Chicken parvovirus		
Chicken parvovirus		(ChPV)
Feline panleukopenia virus		
Canine parvovirus	[M19296]	(CPV)
Feline panleukopenia virus	[M75728]	(FPV)
Mink enteritis virus	[D00765]	(MEV)
Raccoon parvovirus	[M24005]	(RPV)
HB parvovirus		
HB parvovirus		(HBPV)
H-1 parvovirus		
H-1 parvovirus	[X01457]	(H-1PV)
Kilham rat virus		
H-3 virus		
Kilham rat virus	[AF321230]	(KRV)
Lapine parvovirus		
Lapine parvovirus		(LPV)
LuIII virus		
LuIII virus	[M81888]	(LuIIIV)
Minute virus of mice		
Minute virus of mice (Cutter)	[U34256]	(MVMc)
Minute virus of mice (immunosuppressive)	[M12032]	(MVMi)
Minute virus of mice (prototype)	[J02275]	(MVMp)
Mouse parvovirus 1		
Mouse parvovirus 1	[U12469]	(MPV-1)
Porcine parvovirus		
Porcine parvovirus Kresse	[U44978]	(PPV-Kr)
Porcine parvovirus NADL-2	[L23427]	(PPV-NADL2)
RT parvovirus		
RT parvovirus		(RTPV)
Tumor virus X		
Tumor virus X		(TVX)

TENTATIVE SPECIES IN THE GENUS

Hamster parvovirus	[U34255]	(HaPV)
Rat minute virus 1	[AF332882]	(RMV-1)
Rat parvovirus 1	[AF036710]	(RPV-1)

GENUS ERYTHROVIRUS

Type Species *Human parvovirus B19*

DISTINGUISHING FEATURES

Populations of mature virions contain equivalent numbers of positive and negative sense ssDNA, 5.5 kb in size. The DNA molecules contain inverted terminal repeats of 383 nt, the first 365 nt of which form a palindromic sequence. Upon extraction, the complementary DNA strands usually self-anneal to form dsDNA. There is a single mRNA promoter (map unit 6) and two polyadenylation signals, one near the middle of the genome, the other near the 3'-end. Efficient replication occurs in primary erythrocyte precursors. There have also been reports of productive infection in cell lines of megakaryoblastoid erythroleukemic origin.

GENOME ORGANIZATION AND REPLICATION

For Human parvovirus B19 (B19V), there is only one promoter, at 6 map units, but two alternative polyadenylation sites (Fig. 5). Transcripts 1 and 2 encode VP1, while transcript 3 encodes NS1. Two small ORFs can also be accessed by these alternatively spliced mRNAs, depending upon the relative strength of their initiation codons. Transcripts 8 and 9 encode an 11 kDa protein containing three proline-rich regions that conform to consensus Src homology 3 (SH3) ligand sequences. Transcripts 1, 4, 6 and 8 are predicted to translate a 7.5 kDa polypeptide of unknown function.

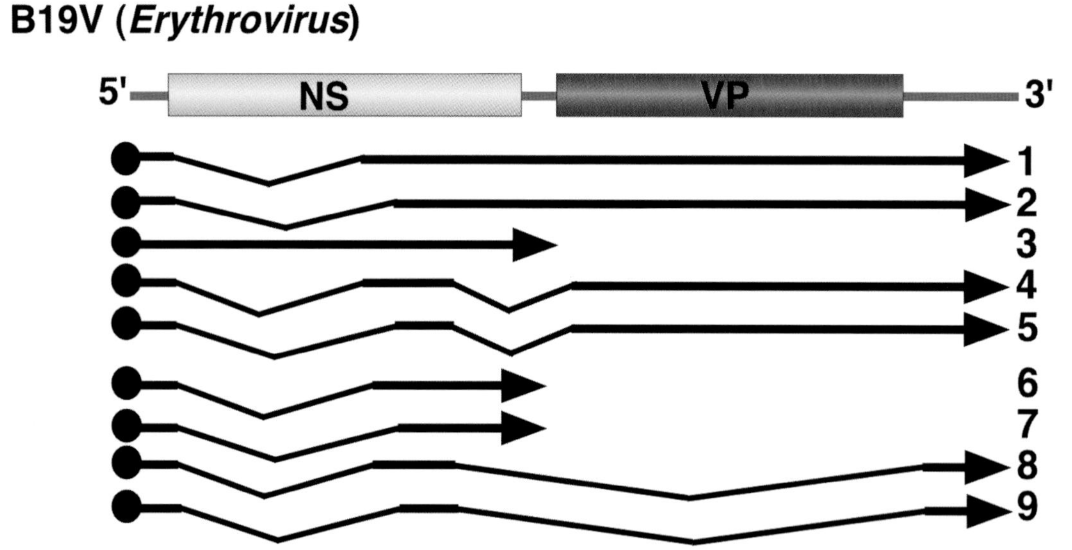

Figure 5: Gene organization and transcription scheme for members of the *Erythrovirus* genus, as shown for Human parvovirus B19 (B19V). Genes are shown as boxes. The left ends of the mRNAs (thick lines) are the sites of the mRNA caps (filled circles), the right ends are the polyadenylation sites (arrows); introns are indicated by thin lines.

BIOLOGICAL PROPERTIES

B19V causes Fifth Disease, polyarthropathia, anemic crises in children with underlying hematological diseases (eg sickle cell anemia, thalassemia) and intra-uterine infections (with hydrops fetalis in some cases).

LIST OF SPECIES DEMARCATION CRITERIA IN THE GENUS

Members of each species are probably antigenically distinct, and natural infection is confined to a single host species. Species are <95% related by non-structural gene DNA sequence.

LIST OF SPECIES IN THE GENUS

Species names are in green italic script; strain names and synonyms are in black roman script; tentative species names are in blue roman script. Sequence accession numbers, and assigned abbreviations ( ) are also listed.

SPECIES IN THE GENUS

Human parvovirus B19		
Human parvovirus B19 - A6	[AY064475, AY064476]	(B19V-A6)
Human parvovirus B19 - Au	[M13178]	(B19V-Au)
Human parvovirus B19 - LaLi	[AY044266]	(B19V-LaLi)
Human parvovirus B19 - V9	[AJ223617, AJ242810]	(B19V-V9)
Human parvovirus B19 - Wi	[M24682]	(B19V-Wi)
Pig-tailed macaque parvovirus		
Pig-tailed macaque parvovirus	[AF221123]	(PmPV)
Rhesus macaque parvovirus		
Rhesus macaque parvovirus	[AF221122]	(RmPV)
Simian parvovirus		
Simian parvovirus (cynomolgus)	[U26342]	(SPV)

TENTATIVE SPECIES IN THE GENUS

Bovine parvovirus type 3	[AF406967]	(BPV-3)
Chipmunk parvovirus	[U86868]	(ChpPV)

GENUS DEPENDOVIRUS

Type Species *Adeno-associated virus - 2*

DISTINGUISHING FEATURES

Populations of mature virions contain equivalent numbers of positive or negative strand ssDNA 4.7 kb in size. The DNA molecules contain inverted terminal repeats of 145 nt, the first 125 nt of which form a palindromic sequence. Upon extraction, the complementary DNA strands usually form dsDNA. There are three mRNA promoters (map units 5, 19, 40) (Fig. 6). For all currently accepted members of the *Dependovirus* genus except the Duck parvovirus and Goose parvovirus, efficient replication is dependent upon helper adenoviruses or herpes viruses. Under certain conditions (presence of mutagens, synchronization of cell replication with hydroxyurea), replication can also be detected in the absence of helper viruses. All isolates of *Adeno-associated virus* share a common antigen as demonstrated by fluorescent antibody staining. Transplacental transmission has been observed for AAV-1 and vertical transmission has been reported for avian AAV.

GENOME ORGANIZATION AND REPLICATION

Dependoviruses have three transcriptional promoters, at 5, 19 and 40 map units, which transcribe mRNAs in a temporally regulated fashion throughout infection (Fig. 6). P5 transcripts are the first to be expressed, followed by those from P19, then those from P40. P5 transcript 1 encodes the non-structural proteins Rep78 and transcripts 2 and 3 encode Rep68. These two forms of the Rep protein differ at their C-termini. Likewise, P19 transcripts encode two Rep forms Rep52 from transcript 4 and Rep48 from trancripts 5 and 6. P40 transcripts encode the structural proteins, transcript 8 encoding VP1 and transcript 9 encoding VP2 and VP3 by an alternate translation initiation mechanism. The unspliced P40 transcript (7) does not appear to encode a functional protein.

AAV-2 (*Dependovirus*)

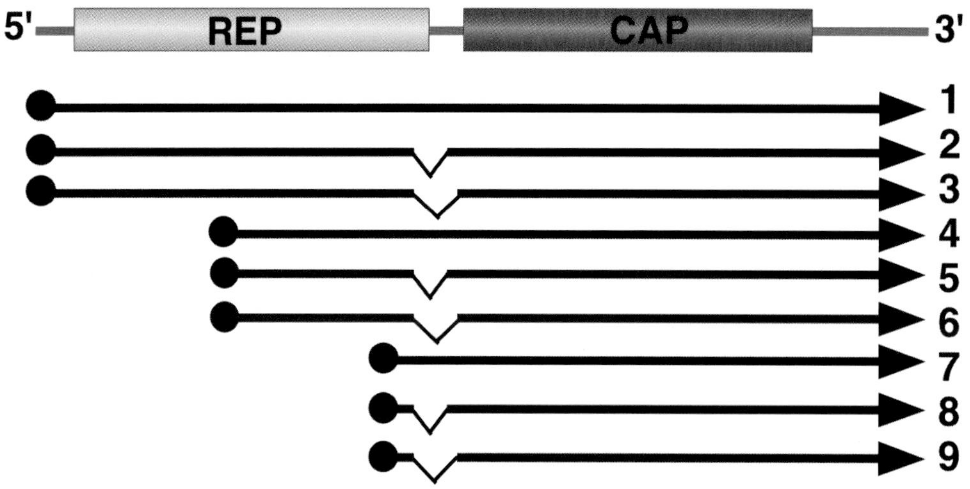

Figure 6: Gene organization and transcription scheme for members of the *Dependovirus* genus, as shown for Adeno-associated virus 2 (AAV-2). Genes are shown as boxes. The left ends of the mRNAs (thick lines) are the sites of the mRNA caps (filled circles), the right ends are the polyadenylation sites (arrows); introns are indicated by thin lines.

LIST OF SPECIES DEMARCATION CRITERIA IN THE GENUS

Members of each species are antigenically distinct, as assessed by neutralization using polyclonal antisera, and natural infection is usually confined to a single host species. Generally, species are <95% related by non-structural gene DNA sequence.

LIST OF SPECIES IN THE GENUS

Species names are in green italic script; strain names and synonyms are in black roman script; tentative species names are in blue roman script. Sequence accession numbers, and assigned abbreviations ( ) are also listed.

SPECIES IN THE GENUS

Adeno-associated virus - 1		
Adeno-associated virus - 1	[AF063497]	(AAV-1)
Adeno-associated virus - 6	[AF208704]	(AAV-6)
Adeno-associated virus - 2		
Adeno-associated virus - 2	[J01901]	(AAV-2)
Adeno-associated virus - 3		
Adeno-associated virus - 3	[AF028705]	(AAV-3)
Adeno-associated virus - 4		
Adeno-associated virus - 4	[U89790]	(AAV-4)
Adeno-associated virus - 5		
Adeno-associated virus - 5	[AF085716]	(AAV-5)
Avian adeno-associated virus		
Avian adeno-associated virus	[AY186198]	(AAAV)
Bovine adeno-associated virus		
Bovine adeno-associated virus		(BAAV)
Canine adeno-associated virus		
Canine adeno-associated virus		(CAAV)
Duck parvovirus		
Barbarie duck parvovirus	[U22967]	(BDPV)
Muscovy duck parvovirus	[X75093]	(MDPV)
Equine adeno-associated virus		
Equine adeno-associated virus		(EAAV)
Goose parvovirus		

Goose parvovirus	[U25749]	(GPV)
Ovine adeno-associated virus		
Ovine adeno-associated virus		(OAAV)

TENTATIVE SPECIES IN THE GENUS

Adeno-associated virus - 7	[AF513851]	(AAV-7)
Adeno-associated virus - 8	[AF513852]	(AAV-8)
Bovine parvovirus - 2	[AF406966]	(BPV-2)

GENUS AMDOVIRUS

Type Species *Aleutian mink disease virus*

DISTINGUISHING FEATURES

Most features are shared with the *Parvovirus* and *Bocavirus* genera. Mature virions contain negative strand DNA of 4748 nt in length. Permissive replication is observed only in Crandell feline kidney cells, although restricted replication is observed and may be antibody-dependent in cells bearing Fc receptors, e.g., macrophages. Evidence of infection has been detected in most mustelids, skunks and raccoons. Virion structure differs slightly from the *Parvovirus* and *Bocavirus* genera, and resembles that of the genus *Dependovirus*. The primary difference is the presence of three mounds elevated above capsid surface around the three-fold icosahedral axis of symmetry, similar to those observed for dependovirus virions. The VP1 N-terminus is much shorter than those found for other members of the *Parvovirinae*, and contains no evidence of a phospholipase 2A enzymatic core.

GENOME ORGANIZATION AND REPLICATION

Transcription of the Aleutian mink disease virus (AMDV) genome occurs from two promoters, at 3 and 36 map units, and the transcripts are spliced in a complex pattern, so as to access two additional small ORFs in the middle of the genome. The mRNAs generated from P3 encode two non-structural proteins, NS1 and NS2, while the structural proteins VP1 and VP2 are synthesized from P36 transcripts. In contrast to the other *Parvovirinae* members, transcription of the CP mRNAs, although upregulated by NS1, does not predominate over non-structural gene transcription. Caspase processing of the NS1 protein is required to produce a sub-species that locates to the nucleus. This processing is essential for completion of the viral life cycle.

ANTIGENIC PROPERTIES

All isolates appear to be antigenically indistinguishable.

BIOLOGICAL PROPERTIES

In susceptible adult hosts, pathogenic isolates cause a persistent, restricted infection associated with a progressive disorder of the immune system, including plasmacytosis, glomerulonephritis and hypergammaglobulinemia. Extremely high levels of antiviral antibody are directed at determinants on the virus capsid surface. In newborn animals, infection is permissive and causes a fulminant interstitial pneumonitis that is often fatal. Survivors develop adult form of disease.

LIST OF SPECIES DEMARCATION CRITERIA IN THE GENUS

Not applicable.

LIST OF SPECIES IN THE GENUS

Species names are in green italic script; strain names and synonyms are in black roman script; tentative species names are in blue roman script. Sequence accession numbers, and assigned abbreviations ( ) are also listed.

SPECIES IN THE GENUS

Aleutian mink disease virus

Aleutian mink disease virus (Aleutian disease virus)	[M20036]	(AMDV)

TENTATIVE SPECIES IN THE GENUS

None reported.

GENUS BOCAVIRUS

Type Species ***Bovine parvovirus***

DISTINGUISHING FEATURES

The genome is ssDNA of 5.5 kb, with a monosense strategy and non-identical terminal palindromes. Sequence analysis of Bovine parvovirus (BPV) and Canine minute virus (CnMV) demonstrate that, while their NS1 and VP1 genes are 34% and 41% similar to one another, the two genomes are very distinct from all other clusters of viruses in the subfamily *Parvovirinae*. In addition, the bocavirus genome encodes a 22.5 kDa nuclear phosphoprotein, NP-1, that is distinct from any other parvovirus-encoded polypeptide.

GENOME ORGANIZATION AND REPLICATION

Large ORFs within the left and right halves of the genome encode the NS and VP proteins, respectively, while a shorter ORF in the middle of the genome, overlapping the C-terminal sequence of the NS1 protein, encodes NP-1. Although there are promoter-like sequences at 4.5, 13 and 39 map units, the number of functional promoters remains to be determined.

LIST OF SPECIES DEMARCATION CRITERIA IN THE GENUS

Members of each species are probably antigenically distinct and natural infection is confined to a single host species. Species are <95% related by non-structural gene DNA sequence.

LIST OF SPECIES IN THE GENUS

Species names are in green italic script; strain names and synonyms are in black roman script; tentative species names are in blue roman script. Sequence accession numbers, and assigned abbreviations ( ) are also listed.

SPECIES IN THE GENUS

Bovine parvovirus		
Bovine parvovirus	[M14363]	(BPV)
Canine minute virus		
Canine minute virus	[AF495467]	(CnMV)

TENTATIVE SPECIES IN THE GENUS

None reported.

SUBFAMILY DENSOVIRINAE

TAXONOMIC STRUCTURE OF THE SUBFAMILY

Subfamily	***Densovirinae***
Genus	***Densovirus***
Genus	***Iteravirus***
Genus	***Brevidensovirus***
Genus	***Pefudensovirus***

DISTINGUISHING FEATURES

Viruses assigned to the subfamily *Densovirinae* infect arthropods. The ssDNA genome of virions is either ambisense (*Densovirus* and *Pefudensovirus*) or monosense (*Iteravirus* and *Brevidensovirus*). Either strand may be packaged. There are four or five structural proteins. Viruses multiply efficiently in most of the tissues of larvae, nymphs, and adult host species without the involvement of helper viruses. Cellular changes consist of hypertrophy of the nucleus with accumulation of virions therein to form dense, voluminous intranuclear masses. The known host range includes members of the orders Dictyoptera, Diptera, Hemiptera, Homoptera, Lepidoptera, Odonata and Orthoptera. Some densoviruses also infect and multiply in shrimps.

GENUS DENSOVIRUS

Type Species *Junonia coenia densovirus*

DISTINGUISHING FEATURES

The ssDNA genome is about 6 kb in size with long inverted terminal repeats and ambisense organization (Figure 2). Populations of virions encapsidate equal amounts of positive and negative strands.

GENOME ORGANIZATION AND REPLICATION

On one strand there are 3 ORFs which encode NS proteins on mRNAs transcribed from a promoter 7 map units from the left end. The four structural proteins are encoded on the complementary strand, on an mRNA transcribed from a promoter that is 9 map units from the right end. NS3 protein is produced by an unspliced mRNA, whereas NS1 and NS2 are produced by a spliced mRNA. The four VP proteins (VP1-4), and NS1 and NS2 are produced by a leaky scanning translation initiation mechanism. Junonia coenia densovirus genome has an inverted terminal repeat of 517 nt, the first 96 nt of which can fold to form a T-shaped structure of the type found in the ITR of AAV DNA.

LIST OF SPECIES DEMARCATION CRITERIA IN THE GENUS

Members of each species are probably antigenically distinct, and natural infection is confined to a single host species. Species are <95% related by non-structural gene DNA sequence.

LIST OF SPECIES IN THE GENUS

Species names are in green italic script; strain names and synonyms are in black roman script; tentative species names are in blue roman script. Sequence accession numbers, and assigned abbreviations ( ) are also listed.

SPECIES IN THE GENUS

Galleria mellonella densovirus		
Galleria mellonella densovirus	[L32896]	(GmDNV)
Junonia coenia densovirus		
Junonia coenia densovirus	[S17265]	(JcDNV)

TENTATIVE SPECIES IN THE GENUS

Diatraea saccharalis densovirus	[AF036333]	(DsDNV)
Mythimna loreyi densovirus	[AY461507]	(MlDNV)
Pseudoplusia includens densovirus		(PiDNV)
Toxorhynchites splendens densovirus	[AF395903]	(TsDNV)

GENUS *ITERAVIRUS*

Type Species ***Bombyx mori densovirus***

DISTINGUISHING FEATURES

The ssDNA genome is about 5 kb in size. Populations of virions encapsidate equal amounts of plus and minus strands. ORFs for both the structural and nonstructural proteins are located on the same strand.

GENOME ORGANIZATION AND REPLICATION

There is apparently one mRNA promoter upstream of each ORF. There is a small ORF on the complementary strand of unknown function. The DNA has an inverted terminal repeat of 230 nt, the first 159 nt are palindromic and can form a J-shaped hairpin structure when folded. There are two ORFs for non-structural proteins and one ORF for four structural proteins.

LIST OF SPECIES DEMARCATION CRITERIA IN THE GENUS

Members of each species are probably antigenically distinct, and natural infection is confined to a single host species. Species are <95% related by non-structural gene DNA sequence.

LIST OF SPECIES IN THE GENUS

Species names are in green italic script; strain names and synonyms are in black roman script; tentative species names are in blue roman script. Sequence accession numbers, and assigned abbreviations ( ) are also listed.

SPECIES IN THE GENUS

Bombyx mori densovirus

Bombyx mori densovirus	[AY033435]	(BmDNV)

TENTATIVE SPECIES IN THE GENUS

Casphalia extranea densovirus	[AF375296]	(CeDNV)
Sibine fusca densovirus		(SfDNV)

GENUS *BREVIDENSOVIRUS*

Type Species ***Aedes aegypti densovirus***

DISTINGUISHING FEATURES

The genome is about 4 kb in size with terminal hairpins, but no ITRs. ORFs for the structural and nonstructural proteins are on the same strand. The brevidensoviruses are at least as different, in sequence, from other members of the subfamily *Densovirinae*, as these are from any member of the subfamily *Parvovirinae*. Populations of virions encapsidate positive and negative strands, but a majority of strands is of negative polarity (85%).

GENOME ORGANIZATION AND REPLICATION

A palindromic sequence of 146 nt is at the 3'-end of the genome and a different palindromic sequence of 164 nt is at the 5'-end. Both terminal sequences can fold to form a T-shaped structure. The genome does not contain any sequence recognizable as a PLA2 domain. There are mRNA promoters at map units 7 and 60. There is a small ORF of unknown function on the complementary strand.

LIST OF SPECIES DEMARCATION CRITERIA IN THE GENUS

Members of each species are probably antigenically distinct, and natural infection is confined to a single host species. Species are <95% related by non-structural gene DNA sequence.

LIST OF SPECIES IN THE GENUS

Species names are in green italic script; strain names and synonyms are in black roman script; tentative species names are in blue roman script. Sequence accession numbers, and assigned abbreviations ( ) are also listed.

SPECIES IN THE GENUS

Aedes aegypti densovirus		
Aedes aegypti densovirus	[AY160976]	(AaeDNV)
Aedes albopictus densovirus		
Aedes albopictus densovirus	[AY095351]	(AalDNV)

TENTATIVE SPECIES IN THE GENUS

Aedes pseudoscutellaris densovirus		(ApDNV)
Penaeus stylirostris densovirus	[AF273215]	(PstDNV)
Simulium vittatum densovirus		(SvDNV)

GENUS *PEFUDENSOVIRUS*

Type Species *Periplaneta fuliginosa densovirus*

DISTINGUISHING FEATURES

The genome is about 5.5 kb in size, with ambisense organization (Fig. 2). The VP gene is split into a large and a small ORF (upstream) on the 5' half of the complementary strand. Unlike all other members of the family *Parvoviridae* that have proteins with a PLA2 domain, the pefudensoviruses have a PLA2 motif that is located in the C-terminal portion (centered 60-70 aa from its C-terminus) of the protein predicted to be translated from the small VP ORF.

GENOME ORGANIZATION AND REPLICATION

The genes for the three non-structural proteins of pefudensoviruses are organized in the same way as for those of the *Densovirus* genus, and are of similar sizes. The ORFs of the split VP gene are spliced in order to code for the largest VPs, although details of the splicing strategy differ between *Periplaneta fuliginosa densovirus* (PfDNV) and *Acheta domesticus densovirus* (AdDNV). Frameshifting after splicing is also required in order to generate the largest VPs. The PLA2 motif is localized in the intronand is therefore only present in the minority of transcripts that remain unspliced.

LIST OF SPECIES DEMARCATION CRITERIA IN THE GENUS

Not applicable.

LIST OF SPECIES IN THE GENUS

Species names are in green italic script; strain names and synonyms are in black roman script; tentative species names are in blue roman script. Sequence accession numbers, and assigned abbreviations ( ) are also listed.

SPECIES IN THE GENUS

Periplaneta fuliginosa densovirus		
Periplaneta fuliginosa densovirus	[AF192260]	(PfDNV)

TENTATIVE SPECIES IN THE GENUS

None reported.

UNASSIGNED VIRUSES IN THE SUBFAMILY

Acheta domesticus densovirus	[AX344110]	(AdDNV)
Blattella germanica densovirus	[AY189948]	(BgDNV)
Culex pipiens densovirus		(CpDNV)
Euxoa auxiliaris densovirus		(EaDNV)
Leucorrhinia dubia densovirus		(LduDNV)
Lymantria dispar densovirus		(LdiDNV)
Myzus persicae densovirus	[AY148187]	(MpDNV)
Pieris rapae densovirus		(PrDNV)
Planococcus citri densovirus	[AY032882]	(PcDNV)

PHYLOGENETIC RELATIONSHIPS WITHIN THE FAMILY

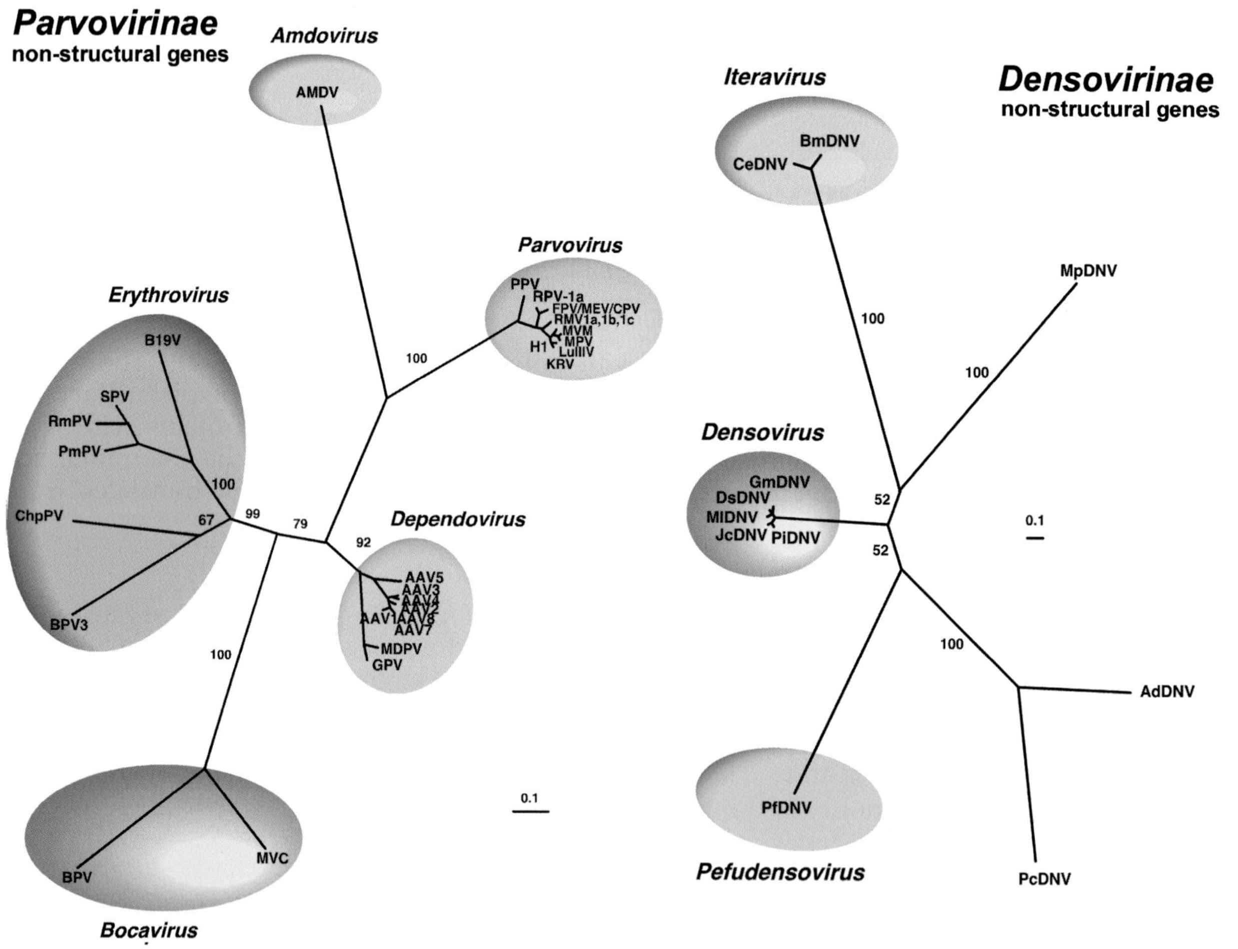

Figure 7: Phylogenetic relationship between the non-structural genes of members of the subfamily *Parvovirinae* (Left panel), and of members of the *Densovirus* and *Iteravirus* genera of the subfamily *Densovirinae* (Right panel). Not shown, because of evolutionary distance, is the genus *Brevidensovirus*, whose members are as distant from the other *Densovirinae* members shown, as these are from members of the *Parvovirinae*. Trees were constructed with the programs included in the Phylip package. For distance matrix analysis, the aligned sequences were processed first with the PROTDIST (using Dayhoff's PAM 001 scoring matrix) and then with the FITCH program (global rearrangements). For bootstrap analysis, the SEQBOOT program was run before PROTDIST and FITCH. The most probable tree was calculated with the CONSENSE program, and the resulting trees visualized using TREEVIEW. (From Z. Zadori and P. Tijssen).

SIMILARITY WITH OTHER TAXA

None reported.

DERIVATION OF NAMES

Adeno: from Greek *aden*, "gland".
Amdo: Aleutian Mink Disease.
Bocavirus: Bovine and Canine
Brevi: from Latin *brevis*, "short".
Denso: from Latin *densus*, "thick, compact".
Dependo: from Latin *dependeo*, "to hang down".
Erythro: from Greek *erythros*, "red".
Parvo: from Latin *parvus*, "small".
Pefu: sigla from *Periplaneta fuliginosa densovirus*, type species of the *Pefudensovirus* genus.

REFERENCES

Bergoin, M. and Tijssen, P. (2000). Molecular biology of *Densovirinae*. *Contrib. Microbiol.*, **4,** 12-32.

Best, S.M., Shelton, J.F., Pompey, J.M., Wolfinbarger, J.B. and Bloom, M.E. (2003). Caspase cleavage of the nonstructural protein NS1 mediates replication of Aleutian mink disease parvovirus. *J. Virol.*, **77,** 5305-5312.

Brown, K.E., Green, S.W. and Young, N.S. (1995). Goose parvovirus -- an autonomous member of the dependovirus genus? *Virology*, **210,** 283-291.

Fediere, G., Li, Y., Zadori, Z., Szelei, J. and Tijssen, P. (2002). Genome organization of Casphalia extranea densovirus, a new iteravirus. *Virology*, **292,** 299-308.

Gao, G.P., Alvira, M.R., Wang, L., Calcedo, R., Johnston, J. and Wilson, J.M. (2002). Novel adeno-associated viruses from rhesus monkeys as vectors for human gene therapy. *Proc. Natl. Acad. Sci. USA*, **99,** 11854-9.

Green, S.W., Malkovska, I., O'Sullivan, M.G. and Brown, K.E. (2000). Rhesus and pig-tailed macaque parvoviruses: identification of two new members of the *Erythrovirus* genus in monkeys. *Virology*, **269,** 105-112.

Guo, H., Zhang, J. and Hu, Y. (2000). Complete sequence and organization of Periplaneta fuliginosa densovirus genome. *Acta Virol.*, **44,** 315-322.

Heegaard, E.D. and Brown, K.E. (2002). Human parvovirus B19. *Clin. Microbiol. Rev.*, **15,** 485-505.

Li, Y., Zadori, Z., Bando, H., Dubuc, R., Fediere, G., Szelei, J. and Tijssen, P. (2001). Genome organization of the densovirus from Bombyx mori (BmDNV-1) and enzyme activity of its capsid. *J. Gen. Virol.*, **82,** 2821-2825.

Lukashov, V.V. and Goudsmit, J. (2001). Evolutionary relationships among parvoviruses: virus-host coevolution among autonomous primate parvoviruses and links between adeno-associated and avian parvoviruses. *J. Virol.*, **75,** 2729-2740.

Mouw, M.B. and Pintel, D.J. (2000). Adeno-associated virus RNAs appear in a temporal order and their splicing is stimulated during coinfection with adenovirus. *J. Virol.*, **74,** 9878-9888.

Muzyczka, N. and Berns, K.I. (2001). *Parvoviridae*: The viruses and their replication. In: *Fields Virology:* Fourth ed., (D.M. Knipe and P.M. Howley, eds), pp. 2327-2360. Lippincott, Williams and Wilkins, Philadelphia.

Nguyen, Q.T., Wong, S., Heegaard, E.D. and Brown, K.E. (2002). Identification and characterization of a second novel human erythrovirus variant, A6. *Virology*, **301,** 374-80.

Schwartz, D., Green, B., Carmichael, L.E. and Parrish, C.R. (2002). The canine minute virus (minute virus of canines) is a distinct parvovirus that is most similar to bovine parvovirus. *Virology*, **302,** 219-23.

Shike, H., Dhar, A.K., Burns, J.C., Shimizu, C., Jousset, F.X., Klimpel, K.R. and Bergoin, M. (2000). Infectious hypodermal and hematopoietic necrosis virus of shrimp is related to mosquito brevidensoviruses. *Virology*, **277,** 167-77.

Zadori, Z., Szelei, J., Lacoste, M.C., Li, Y., Gariepy, S., Raymond, P., Allaire, M., Nabi, I.R. and Tijssen, P. (2001). A viral phospholipase A2 is required for parvovirus infectivity. *Dev. Cell*, **1,** 291-302.

CONTRIBUTED BY

Tattersall, P., Bergoin, M., Bloom, M.E., Brown, K.E., Linden, R.M., Muzyczka, N., Parrish, C.R. and Tijssen, P.

The Reverse Transcribing DNA and RNA Viruses

FAMILY HEPADNAVIRIDAE

TAXONOMIC STRUCTURE OF THE FAMILY

Family	*Hepadnaviridae*
Genus	*Orthohepadnavirus*
Genus	*Avihepadnavirus*

VIRION PROPERTIES

MORPHOLOGY

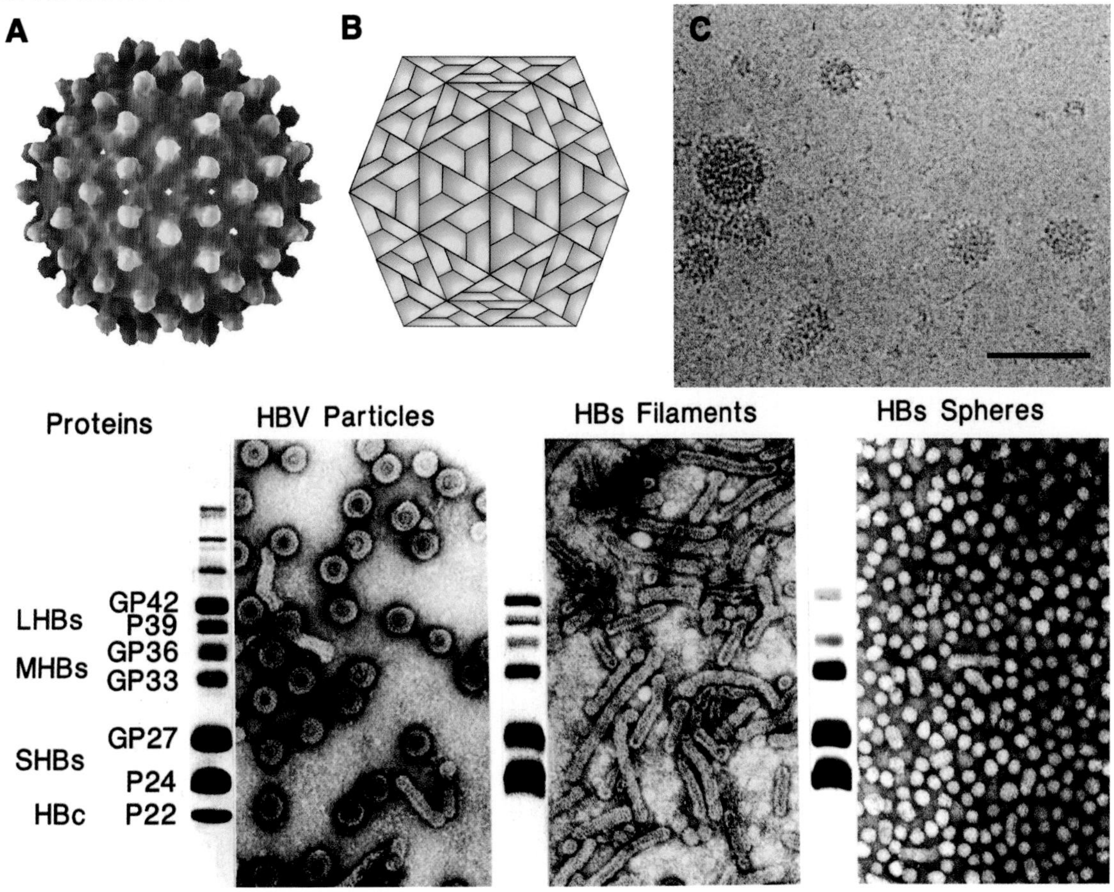

Figure 1: (Top left) Atomic resolution rendering of a particle of Hepatitis B virus (HBV)(Wynne, S.A., Crowther, R.A. and Leslie, A.G. (1999). *Mol. Cell.*, **3**, 771-80). (Top center) Diagram representing the T4 structure of a particle of HBV. (Top right). High resolution cryo-electron micrograph of normal ~42nm isometric virus with icosahedral capsid or "core" surrounded by a coat of HBsAg and of smaller ~22nm spheres and rods composed of viral envelope proteins (Courtesy of B. Boettcher, J. Monjardino, and R.A. Crowther). (Bottom). Negative contrast electron micrographs of HBV virions (Left) and virus-associated particles (Center and right) together with an SDS-PAGE protein profile of each particle form to the left of the relevant micrograph. LHBs, MHBs and SHBs refer to large, middle and small HB surface proteins, respectively. HBc, hepatitis B core proteins. GP, glycoprotein; P, protein. The identity of the slower migrating bands is unknown. (Courtesy of W. Gerlich).

Hepadnaviruses are spherical, occasionally pleomorphic, 42-50 nm in diameter and with no evident surface projections after negative staining. Projections are visible in cryo EM pictures (Fig. 1C). The outer, detergent-sensitive envelope contains the surface proteins and surrounds an icosahedral nucleocapsid core that is composed of one major protein species, the core protein. The nucleocapsid encloses the viral genome (DNA), the viral

DNA polymerase, and associated cellular protein(s), including protein kinase and chaperones that appear to play a role in the initiation of viral DNA synthesis.

In the case of Hepatitis B virus (HBV), the majority of nucleocapsid cores are around 34 nm in diameter and contain 240 core protein subunits (triangulation number T=4), while a minority are approximately 30 nm in diameter and consist of only 180 subunits (T=3). Hepadnavirus infection induces overproduction of surface proteins that are secreted as pleomorphic lipoprotein particles together with virus to the blood. In the case of HBV, these form 17-22 nm spherical particles and filaments (Fig. 1).

Physicochemical and Physical Properties

The virion S_{20w} is approximately 280S. The buoyant density of virions in CsCl is approximately 1.25 g/cm^3. Estimates of the buoyant density of particles lacking cores are 1.18-1.20 g/cm^3. Virus-derived cores (lacking envelopes) have densities of approximately 1.36 g/cm^3.

Nucleic Acid

The genome consists of a partially dsDNA that is held in a circular conformation by base pairing in a cohesive overlap between the 5'-ends of the two DNA strands. The length of the cohesive overlap is about 240 bp for the orthohepadnaviruses and 50 bp for the avihepadnaviruses. The size of the genome ranges from 3.0-3.3 kb in different family members; the viral DNA has an S_{20w} of about 14 and a G+C content of about 48%. One strand (negative sense, i.e., complementary to the viral mRNAs) is full-length, the other varies in length. For both the orthohepadnaviruses and the avihepadnaviruses, the negative strand DNA has an 8-9 nt terminal redundancy. The 5'-end of the negative strand DNA is covalently attached to the terminal-protein domain of the viral DNA polymerase, and the 5'-end of the positive sense DNA has a covalently attached 19 nt, 5'-capped oligoribonucleotide primer. The 3'-end of the positive strand terminates at a variable position in different molecules, creating a single stranded gap that may account for 60% of the HBV genome but is usually very short in avihepadnaviruses.

Proteins

Virions and empty subviral particles may contain two or three envelope proteins, with common C-termini and differing N-termini due to different sites of translation initiation. Typically, virions contain a small (S) trans-membrane protein, in the orthohepadnaviruses an intermediate sized (M) protein, and a large (L) protein which is myristylated at the N-terminus. In many cases more than one form of each of the above proteins occur due to alternative patterns of glycosylation. For HBV, virions and filaments are enriched in L proteins and empty spheres consist predominantly of S proteins, while for Duck hepatitis B virus (DHBV), L and S proteins are distributed evenly between particle types.

The core protein has a large N-terminal domain and a small RNA-binding domain at the C-terminus. Core protein above a threshold concentration can self-assemble via dimers to complete nucleocapsids in the absence of other viral components.

The polymerase protein consists of an N-terminal domain (TP) with a DNA primer function, a spacer region of variable size, a reverse transcriptase and an RNase H domain. The TP domain is covalently attached to the 5'-end of the minus strand of viral DNA via a tyrosine residue, which serves as the primer for initiation of reverse transcription.

Orthohepadnaviruses contain a fourth ORF ('X' gene) situated downstream of the S gene and partly overlapping the cohesive 5'-terminal region. This codes for a non-structural protein that can function as a promiscuous transcriptional activator and, for Woodchuck hepatitis virus (WHV), has been shown to be required for efficient *in vivo* replication. Avihepadnaviruses have an ORF in a similar location, but it remains unclear if this ORF has a role in infection. At high expression levels in cell culture systems the X proteins induce apoptosis.

Host proteins contained within nucleocapsids include heat shock protein Hsp 70 and heat shock protein Hsp 90 which, at least in the case of Duck hepatitis B virus (DHBV), appears to be part of a multicomponent chaperone complex involved in replication and nucleocapsid assembly. A protein kinase has been detected in HBV nucleocapids.

The core protein also exists in a secreted soluble form ('e' antigen) which can be translated from an additional start codon 29 codons upstream of the core start codon. This additional precore sequence functions as a signal peptide. The 'e' antigen is not essential but seems to modulate the immune response.

Lipids

Lipid constitutes 30 to 40% of the viral envelope or of the empty particles. It is derived from a host membrane compartment intermediate between the ER and Golgi, and includes phospholipids, cholesterol, cholesterol esters and triglycerides.

Carbohydrates

Demonstrated in particles and virions of orthohepadnaviruses as N-linked glycans of the complex types. Many virus isolates also contain O-linked glycans in the M surface protein.

Genome Organization and Replication

The hepadnavirus genome contains the following major ORFs; precore/core (preC/C), polymerase (P), env or surface (preS/S), and in the case of orthohepadnaviruses, a fourth ORF, the X gene (Fig. 2).

The preC/C ORF codes for two distinct products: one is the core protein forming the protein shell of the nucleocapsid, the other, made by translation of the joint preC/C ORF, is the precore protein which is targeted into the cell's secretory pathway, processed at both ends and eventually found in the serum of infected individuals as e antigen. Both products are translated from genomic, terminally redundant, polyadenylated 3.5 kb transcripts with slightly different 5'-ends. The longer precore mRNAs contain the preC initiation codon, the shorter core mRNA lacks it. The P-ORF covers some 80% of the genome and encodes the viral replication enzyme P, which is also an indispensable component in the assembly process (see below). P protein is translated from the same genomic RNA that directs synthesis of core protein by internal initiation. The env or surface gene consists of three in-phase ORFs, termed in 5'- to 3'-direction preS1, preS2 and S. S can be separately expressed to give the small or S protein; cotranslation of preS2/S yields the middle or M protein (orthohepadnaviruses), that of the entire preS1/preS2/S gene the large or L protein. Thus the S domain is common to all three forms of surface protein. As for the preC/C ORF, this is achieved by the generation of mRNAs with staggered 5'-ends in which the initiator codons of the preS1, the preS2 or the S region are the first to be encountered by translating ribosomes. L protein is translated from a 2.4 kb mRNA, M and S from a set of 2.1 kb transcripts. All viral transcripts are 3'-terminally colinear, ending after a unique polyadenylation signal located in the C gene. The X gene encodes a pleiotropic transcriptional activator that appears to be required for establishment of a normal infection with WHV, and has been implicated in one proposed mechanism for hepadnavirus carcinogenesis. The DNA sequence of HBV has 2 enhancer regions (ENHI and ENHII), 4 promoters (preC/C, preS1, preS2/S and X), two 11-base direct repeat sequences, DR1, DR2, a polyadenylation signal (TATAAA) and putative glucocorticoid-responsive elements (GRE). The 5'-end of the negative strand is located within DR1, the 5'-end of the positive strand is at the 3' boundary of DR2.

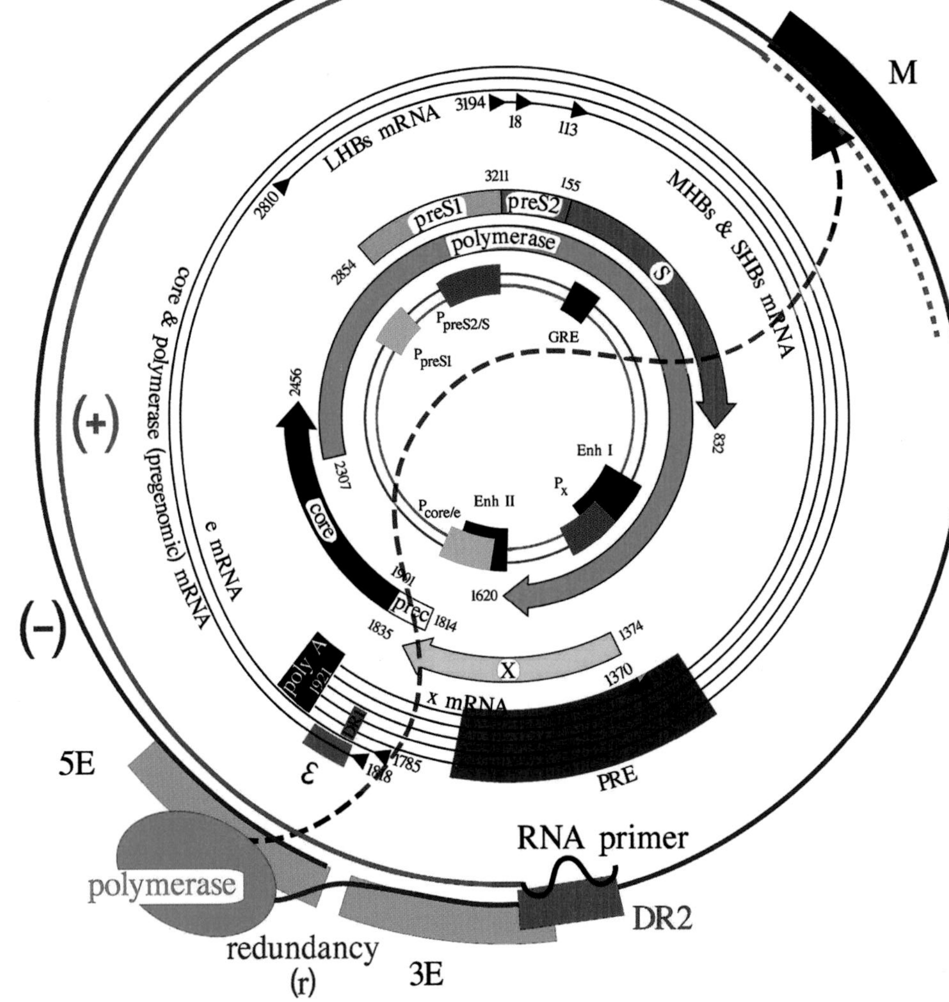

Figure 2: Genome organization and regulatory elements of orthohepadnaviruses are shown for a typical HBV isolate of genotype A. The outer circle represents the structure of relaxed circular, viral DNA found within virions, while the inner circle illustrates the structure and regulatory elements on cccDNA, the covalently closed circular DNA from which viral mRNAs are transcribed in the nucleus of the infected cell (red =positive strand; blue = negative strand). Numbering starts at the unique EcoRI restriction site located approximately at the junction of the preS1 and preS2 domains in the ORF for the viral envelope protens. The regulatory elements on the DNA are depicted at their approximate position. The promoters (P) are shown as gray boxes, the enhancers (Enh), a glucocorticoid responsive element (GRE), a negative regulatory element (NRE) and a CCAAT element (CCAAT) are depicted as black boxes. Liver-specific promoters are drawn in light gray; non-tissue specific promoters are depicted as medium gray boxes. The ORFs are drawn as arrows with their corresponding start and termination sites. The viral mRNAs are depicted as black circles in the middle region. The black triangles represent their 5´-ends; the 3´-end is common and linked to an approximately 300 nt long polyA. The regulatory elements on the RNAs are depicted as a red box (encapsidation signal ε), a black box (polyadenylation signal), in pink (DR1) and in blue (posttranscriptional regulatory element [PRE]). The genomic DNA is depicted as it is found in the virion. The minus DNA strand is drawn as a blue line with its terminal redundancy (r). The polymerase (green oval) is linked to the 5´-end of the minus strand. The plus-strand DNA is shown as a red line. The dotted red line represents the variation of the 3´-end of the plus-strand DNA. The 5´-end of the plus strand is bound to its capped RNA primer, depicted as a black wave-line. The dotted black line between the polymerase and the 3´-end of the plus-strand DNA reflects the fact that the polymerase is bound to the 5´-end of the minus-strand DNA, but interacts with the variable 3´-end of the plus-strand DNA for its elongation. The regulatory elements on the minus-strand DNA are the DR2 (red box) and the M, 5E and 3E elements, which are required for circularization of the genome. Note that their position and size are approximate since these elements are not yet completely characterized. (from Kann M. (2002). Structure and molecular virology. In: *Hepatitis B virus Human Virus Guide.* (S. Locarnini and C.L. Lai, eds) Chapter 2. International Medical Press Ltd., London. with permission).

Replication can be considered in two stages; an incoming or afferent arm in which the input viral genome enters the nucleus and is converted to covalently closed circular (ccc), supercoiled DNA (cccDNA), and an outgoing or efferent arm in which RNA transcripts from the cccDNA are encapsidated and reverse transcribed within core particles in the cytoplasm and the resulting genomic DNA is either transported to the nucleus or enveloped and secreted (Fig. 3).

There is evidence that the infectious DNA-containing virion binds to its target cell via interaction of the L protein with cellular receptor(s) that are not yet fully characterized. The nucleocapsid is presumably delivered to the cytoplasm and transported through the nuclear pore where the genome is released to the nucleoplasm. Repair of the single stranded gap is carried out, though it remains uncertain if this is achieved by the viral DNA polymerase or a host polymerase. Removal of the terminal protein and oligoribonucleotide from the negative and positive strands respectively, takes place, followed by DNA ligation, converting the viral genome to covalently closed, circular DNA (cccDNA). These steps are believed to involve the action of cellular enzymes. cccDNA, in the form of a histone-associated minichromosome, provides a stable template for transcription.

Genomic and sgRNAs are transcribed by host RNA polymerase II into a number of RNA size classes, some of which also show microheterogeneity at the 5′-end but all of which terminate at a common 3′ polyadenylation site. The RNAs of HBV and WHV contain a post-transcriptional regulatory element PRE which allows for cytoplasmic transport without splicing. The largest of these RNAs is translated to form precore protein. A slightly shorter RNA encodes the core protein and (by internal initiation) the polymerase protein and also serves as the template for reverse transcription. The polymerase protein, along with host chaperone proteins, associates with a specific encapsidation signal (ε) on pregenomic RNA, and this preassembly complex apparently triggers assembly of core protein dimers into complete nucleocapsids.

Concurrently, the different classes of sgRNA's are translated to produce the various surface gene products (L, M and S) which oligomerize and bud into the lumen of a post ER/pre-Golgi compartment to give rise to both empty particles and virions.

Reverse transcription of pregenomic RNA takes place within cytoplasmic immature cores. This process uses the terminal protein domain of polymerase as primer for first strand synthesis and a short undigested capped oligoribonucleotide derived from the 5' end of the template RNA and extending through the proximal copy of DR1, as the second strand primer. Synthesis of both strands requires transfer reactions. Reverse transcription initiates with the copying of 4 nt from a bulge in epsilon. This product is then annealed to a complementary sequence at the 3' copy of DR1, and it is from this site that synthesis of the synthesis of the full length minus strand progresses. Plus strand synthesis involves a transfer of the RNA primer from the 3'-end of the minus strand to a remote site DR2, near the 5'-end of the minus strand and identical in sequenced to DR1. It is here that plus strand synthesis normally begins. Plus strand elongation requires a second translocation, from the 5'- to the 3'-end of the minus strand template, to form an open circular genome with a less than full length (+) strand, maintained by overlapping cohesive ends. Nucleocapsids containing partly reverse transcribed DNA that have associated with cytoplasmically located pre S domains of the L envelope protein may then bud into the ER as maturing virions, or alternatively may be transported to the nucleus, thereby increasing the pool of cccDNA. While integration of viral DNA into the host genome is not required for replication and appears to be an infrequent event, integrated viral DNA, often containing deletions, inversions and duplications is found in hepatocellular carcinoma

(HCC) cells in culture and in patients as well as in apparently normal livers from chronic carriers. An aberrant linear viral DNA formed when plus strand synthesis initiates from an untranslocated primer appears to be the precursor to the majority of integration events.

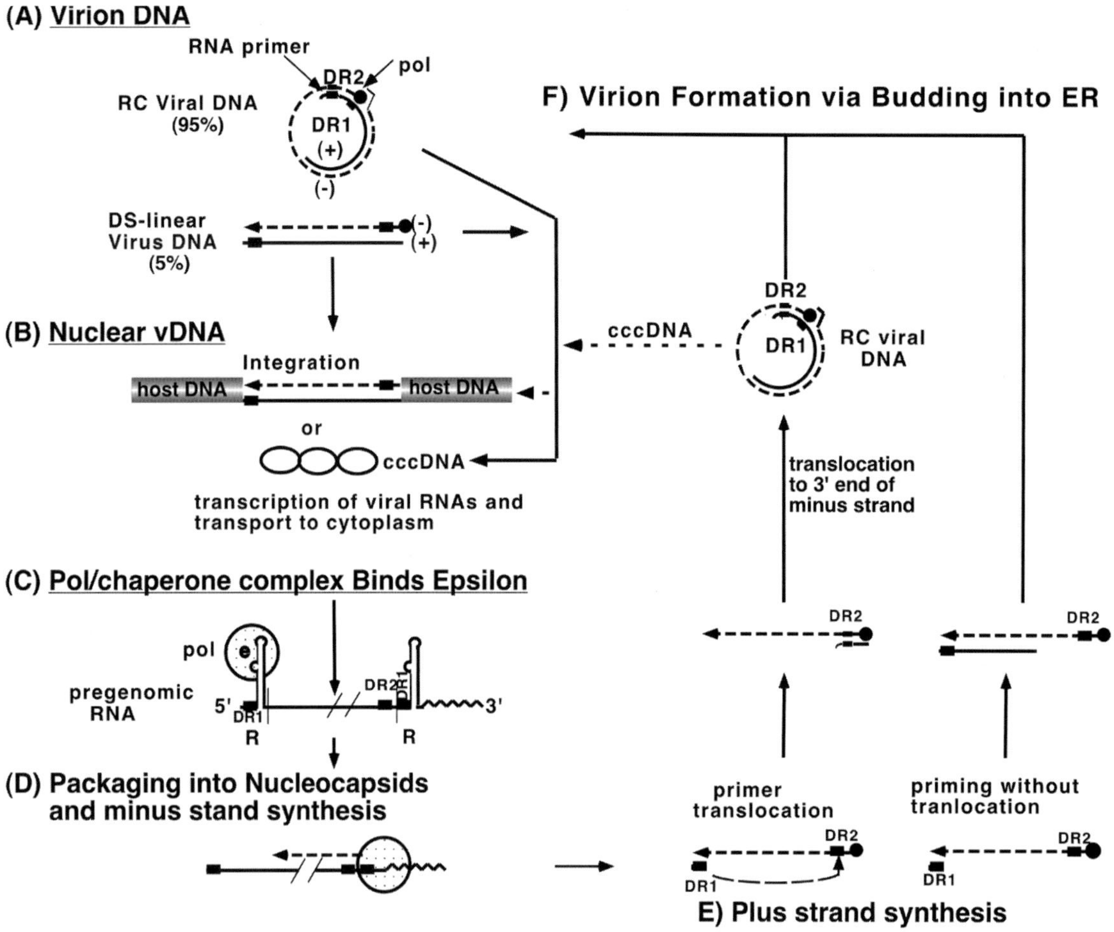

Figure 3: Hepadnavirus replication strategy. For details see Text.

Antigenic Properties

Three principal antigens have been identified for hepadnaviruses, designated surface, core and e antigen. These are abbreviated HBsAg, HBcAg, HBeAg for the HBV-related antigens, DHBsAg, DHBcAg, DHBeAg for DHBV-related antigens, etc while the corresponding antibodies are designated anti-HBs, anti-HBc, anti-DHBs, anti-DHBc etc. HBsAg is involved in neutralization. It cross-reacts to a limited extent with the analogous antigens of WHV and Ground squirrel hepatitis virus (GSHV), but not with DHBsAg. PreS antigens bear specific neutralization determinants. S proteins are sufficient to stimulate protective immunity.

HBeAg and HBcAg proteins share common sequences and epitopes but also contain epitopes which distinguish these two proteins from each other. The HBeAg is a 16 kDa truncated derivative of HBcAg. It is found as a soluble antigen in the serum of patients. HBcAg has been found to cross-react more strongly with the WHV core antigen than is seen between the corresponding surface antigens. In much of the earlier literature the term surface antigen or HBsAg is used arbitrarily to refer to either the antigenic specificity, various protein products of the preS1/preS2/S gene, or the empty 17-22 nm HBsAg-bearing particles. The term "antigen" should not be used if "protein" or "particle" is intended. Similar considerations apply to the use of "core antigen".

Biological Properties

All hepadnaviruses show narrow host specificity. *In vitro*, replication of many hepadnaviruses has only been demonstrated following transfection of tissue culture cells by cloned cDNA, resulting in the production of infectious virus. Replication of several hepadnaviruses has been achieved following inoculation of primary hepatocytes with serum that contains virus.

Hepadnavirus infections *in vivo* possess characteristic features:
They are markedly hepatotropic; although viral antigens and nucleic acids can also be detected in white blood cells (and in some extra-hepatic sites, e.g. pancreas, spleen, kidney with avihepadnaviruses).
Infection may be transient or persistent, the outcome depending on factors including host age and dose of inoculum. Persistent infection is more common in neonates and in immuno-compromised hosts. Persistent infections are typically life-long and can be accompanied by high levels of virions and subviral particles in the circulating blood.
Empty virus-like particles, composed of excess virus envelope material, are present in much greater numbers than complete virions in most individuals and at most stages of infection.
Virus replication is generally thought to be non-cytopathic, and different degrees of ongoing liver damage in different individuals are thought to be governed by different degrees of immune-mediated damage to infected hepatocytes.
In ortho-, but not avi-hepadnavirus infections, persistent virus infection confers a significantly increased rate of development of primary hepatocellular carcinoma, and a number of direct and indirect mechanisms have been described.

Genus *Orthohepadnavirus*

Type Species ***Hepatitis B virus***

Distinguishing Features

Viruses of this genus infect mammals, with a narrow host range for each virus species. The only known natural host of HBV is humans, although chimpanzees may be infected experimentally, and there are also reports of infections of uncertain origin in other great apes (gorillas, orangutans and gibbons). Virions of HBV are 40-45 nm in diameter with a 32-36 nm internal nucleocapsid, and empty envelope particles of HBV are typically spherical (16-25 nm diameter) and filamentous (20 nm diameter and variable in length). The genome of HBV is 3.2 kb with a cohesive overlap of 240 bp. The viruses have an S protein of approximately 226 aa as a major envelope protein, an M protein of ca. 271 aa (which appears unnecessary for infection in experimental situations) and an L protein of ca 400 aa.

The proteins are partially glycosylated, thus generating doublets in gel electrophoresis, eg for HBV, P24/GP27 for S, P39/GP42 for L, and, in the case of M, GP33/GP36 due to an additional glycosylation in the preS2 sequence. The HBV core protein is approximately 180 aa, and the virus encodes an HBx protein of 154 aa whose natural function in the virus life cycle in uncertain.

At least 5 antigenic specificities have been identified for HBsAg. A group determinant (a) is shared by virtually all HBsAg preparations. Mutations in this region have been found in immunized individuals who subsequently become infected, in HBV carriers and infected individuals given immunotherapy. Two pairs of subtype determinants (d,y and w,r) have been demonstrated which are generally mutually exclusive (and thus usually behave as alleles). Antigenic heterogeneity of the w determinant, and additional determinants such as q, x or g have also been described. Thus, eight major serological subtypes are found (ayw, ayw_2, ayw_3, ayw_4, ayr, adw_2, adw_4 and adr); they have distinct geographical distributions with some overlap. DNA sequence analysis has now replaced antigenic typing in defining viral genotypes and has distinguished genotypes or clades that differ between each other by 8-14% at the nucleotide level. Different genotypes also have different geographical distributions, and there is some but not complete correspondence between genotype and serological subtype. The apes each have their own HBV genotype which has been suggested to reflect co-evolution of virus and host and not horizontal transmission between the primate species.

Woolly monkey hepatitis B virus (WMHBV) is closely related in DNA sequence to HBV with 20% sequence variation but, unlike HBV, preferentially infects the Woolly monkey and can be transmitted to the spider monkey; it transmits only poorly to the chimpanzee, which is highly susceptible to HBV.

Different isolates of WHV show < 3.5% nucleotide sequence variation. A virus of arctic ground squirrels (Arctic squirrel hepatitis virus, ASHV) differs from WHV and GSHV to about the same extent (about 15% base changes) as these two latter viruses differ from each other.

Biological Properties

HBV may cause acute and chronic hepatitis, cirrhosis, hepatocellular carcinoma, and, as a consequence of the host immune response to infection, immune complex disease, polyarteritis, glomerulonephritis, infantile papular acrodermatitis and aplastic anemia. An asymptomatic carrier state with high viremia may develop particularly after perinatal infection or under immune suppression.

Horizontal transmission of HBV usually occurs by : (i) percutaneous contact with infected blood or body fluids, eg intravenous drug abuse, use of infected blood or blood products ; (ii) sexual contact ; (iii) perinatal transmission from an infected mother ; (iv) "inapparent horizontal" transmission, particularly between children in low socio economic communities and thought to be at least in part due to unrecognized exposure to open skin breaks or mucous membranes. In communities with a high prevalence of infection, routes (iii) and (iv) predominate, while in low prevalence communities infections are acquired later in life and involve particularly routes (i) and (ii).

Hepatitis occurs in woodchucks and squirrels infected with their respective viruses, and chronic infection leads to a risk of hepatocellular carcinoma even greater than that in chronic carriers of HBV. In the case of WHV, hepatocellular carcinoma frequently occurs within 2 years of infection.

Woolly monkey hepatitis B virus causes hepatitis in its host, but is not yet known to have a role in liver cancer.

List of Species Demarcation Criteria in the Genus

The species demarcation criteria in the genus are:
Nucleotide sequence diversity; WHV/HBV 40%; GHSV/WHV 15%; WMHBV/HBV 20%; WMHBV/WHV 30%
Differences in host range: HBV infection is limited to primates; GSHV infection has been experimentally transferred to chipmunks and woodchucks but not to several related ground squirrel species ; WHV also has a narrow host range, being reported not to infect ground squirrels or other rodent species. WMHBV is transmitted to the spider monkey and, inefficiently, to chimpanzees.

Oncogenicity: HBV, WHV and GSHV have been associated with primary liver cancer in infected hosts. However, proposed mechanisms are different in each case, and the incidence and typical time scales differ, being highest with WHV and lowest with HBV.

A number of related viruses that clearly belong to this genus have been isolated from subhuman primates (chimpanzees, gibbons, orangutan and gorilla) and from various rodent species (Artic ground squirrels, Richardson's ground squirrels). As illustrated in Figure 4, these isolates are quite similar to assigned isolates of HBV, WHV, or GSHV (sequencing data is not available for the Richardson's ground squirrel isolate), but nothing is known about host range divergence from existing species.

List of Species in the Genus

Species names are in green italic script; strain names and synonyms are in black roman script; tentative species names are in blue roman script. Sequence accession numbers [ ], and assigned abbreviations ( ) are also listed.

Species in the Genus

Ground squirrel hepatitis virus		
Ground squirrel hepatitis virus	[K02715]	(GSHV)
Hepatitis B virus		
Hepatitis B virus - A	[X02763]	(HBV-A)
Hepatitis B virus - B	[D00330]	(HBV-B)
Hepatitis B virus - C	[M12906]	(HBV-C)
Hepatitis B virus - D	[J02203]	(HBV-D)
Hepatitis B virus - E	[X75657]	(HBV-E)
Hepatitis B virus - F	[X69798]	(HBV-F)
Hepatitis B virus - G	[AF160501]	(HBV-G)
Hepatitis B virus - H	[AY090454]	(HBV-H)
Woodchuck hepatitis virus		
Woodchuck hepatitis virus	[J02442]	(WHV)
Woolly monkey hepatitis B virus ‡		
Woolly monkey hepatitis B virus	[AF046996]	(WMHBV)

Tentative Species in the Genus

Arctic squirrel hepatitis virus	[NC_001719]	(ASHV)

RT Viruses

Phylogenetic Relationships within the Genus

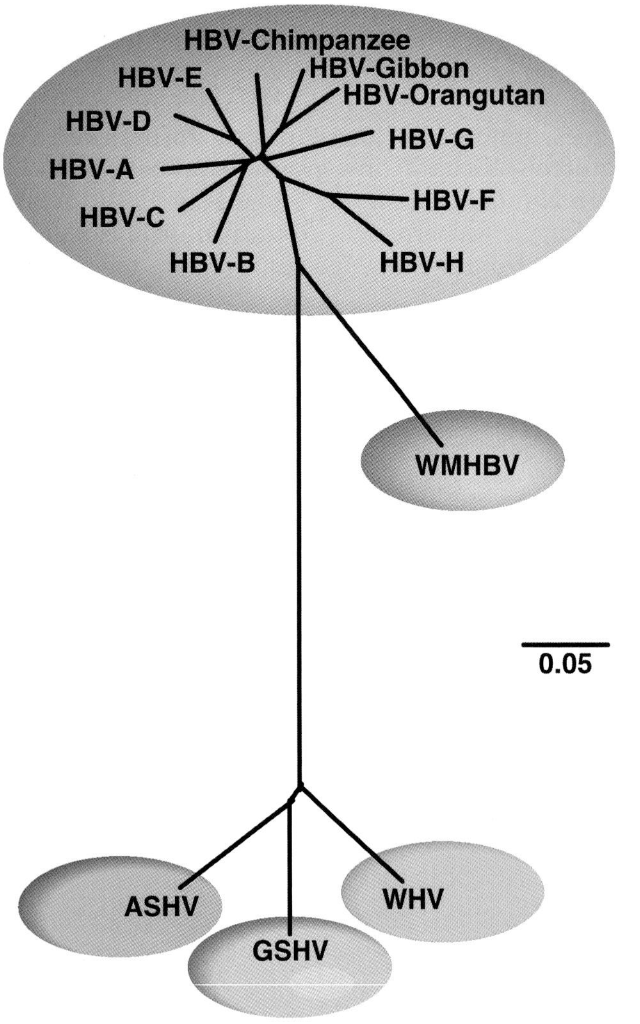

Figure 4: Phylogenetic tree of the genus orthohepadnavirus. Complete genomes of Hepatitis B virus (HBV) genotypes A (X02763), B (D00330), C (M12906), D (J02203), E (X75657), F (X69798), G (AF160501) and H (AY090454) and isolates found in chimpanzee (D00220), orangutan (NC002168), and gibbon (U46935) were aligned using clustal w with orthohepadnavirus genomes from Woolly monkey hepatitis B virus (WMHBV)(AF046996), Woodchuck hepatitis virus (WHV)(J02442), Ground squirrel hepatitis virus (GSHV)(K02715) and Arctic ground squirrel hepatitis virus (ASHV)(NC_001719). The alignment was tested with the neighbor-joining method. (Contributed by S. Schaefer). Calibration bar: substitutions per site.

Genus *Avihepadnavirus*

Type Species *Duck hepatitis B virus*

Distinguishing Features

Virions of DHBV are spherical, 46-48 nm in diameter, with a nucleocapsid that is 35 nm in diameter and exhibits projections. Empty particles composed of excess envelope material are pleomorphic and up to 60 nm diameter. The single stranded gap in the virion DNA is usually very short, 12 bases. DHBV lacks an X gene containing a conventional initiation codon but some other avihepadnaviruses may have an X gene with a regular start codon. Virus particles have only the largest (36 kDa) and smallest (17 kDa) S proteins. Transmission is predominantly vertical. Heron hepatitis B virus (HHBV) differs from DHBV in that a highly conserved ORF is present upstream of C in a position analogous to the X gene of orthohepadnaviruses, and that the S protein and not the L protein possesses a potential myristylation site.

Biological Properties

DHBV is maintained in domestic duck flocks through vertical transmission from viremic ducks. The virus infects the developing liver *in ovo* and is not recognized sufficiently by the host immune response to produce hepatitis and liver disease, or to eliminate the virus. Liver cancer has not been associated with chronic infection. Transmission may also occur to neonates leading to chronic infection. Transmission to adults generally leads to transient infection. The biology of HHBV infections in its natural host has not been studied.

List of Species Demarcation Criteria in the Genus

The species demarcation criteria in the genus are:
Nucleotide sequence diversity: HHBV/DHBV 21.6%.
HHBV can be transmitted to herons but not to ducks.

A number of other less well characterized viruses have been isolated from geese and ducks with reported sequences more closely related to that of DHBV than HHBV. A virus closely related to HHBV has recently been isolated from white storks and designated Stork hepatitis B virus (STHBV). Like HHBV, STHBV has low infectivity for duck hepatocytes. STHBV may be tentatively assigned as a variant of HHBV.

List of Species in the Genus

Species names are in green italic script; strain names and synonyms are in black roman script; tentative species names are in blue roman script. Sequence accession numbers [ ], and assigned abbreviations ( ) are also listed.

Species in the Genus

Duck hepatitis B virus		
Duck hepatitis B virus	[K01834]	(DHBV)
Heron hepatitis B virus		
Heron hepatitis B virus	[M22056]	(HHBV)
Stork hepatitis B virus	[AJ251937]	(STHBV)

Tentative Species In the Genus

None reported.

List of Unassigned Viruses in the Family

Ross's goose hepatitis B virus	[M95589]	(RGHBV)

Phylogenetic Relationships within the Family

Orthohepadnaviruses and avihepadnaviruses are distinguished by the following criteria :

- Low nucleotide sequence identity,
- Differences in genome size (≈ 3.2 kb for orthohepadnaviruses and 3.0 kb for avihepadnaviruses.
- Larger core proteins and no M surface protein for the avihepadnaviruses.
- Host range restricted to either mammals or birds respectively.

Similarity with Other Taxa

Reverse transcription as an essential step in replication is a common feature of hepadnaviruses, retroviruses and caulimoviruses. Hepadnaviruses and retroviruses also contain three major genes each with the same function and in the same order (ie core-polymerase-pre S/S and *gag-pol-env* respectively); a fundamental distinction is that, with hepadnaviruses, the form of the genome in extracellular virions is DNA and reverse transcription takes place during the efferent or outgoing arm of the replication cycle, while the reverse hold true for retroviruses (with the exception of the spumaviruses, in which some infectious particles appear to contain a DNA genome). Retroviruses use tRNAs as primer for the DNA minus strand, hepadnaviruses a tyrosine in the polymerase itself. The polymerase protein of hepadnaviruses does not contain a protease or integrase function. Many other aspects are distinctly different in both virus families, partly due to the extremely small size of the hepadnaviral genome and the need to efficiently exploit this restricted genetic space by using considerable overlap of both coding regions and regulatory elements.

Derivation of Names

Avi: from Latin *avis,* "bird",
dna: sigla for deoxyribonucleic acid,
Hepa: from Greek *hepar,* "liver",
Ortho: from Greek *orthos,* "straight".

References

Ganem, D. and Schneider, R.J. (2001). *Hepadnaviridae*: The viruses and their replication. In: *Fields Virology*, 4th ed., (D.M. Knipe, and P.M. Howley, eds), pp. 2923-2969. Lippincott, Williams & Wilkins, Philadelphia.

Hollinger, F.B. and Liang T.J. (2001). Hepatitis B Virus. In: *Fields Virology*, 4th ed., (D.M. Knipe, and P.M. Howley, eds), pp. 2971-3036. Lippincott, Williams & Wilkins, Philadelphia.

Kirschberg, O., Schuttler, C., Repp, R. and Schaefer, S. (2004). A multiplex-PCR to identify hepatitis B virus-genotypes A-F. *J. Clin. Virol.*, **29**, 39-43.

Hu, J. and Anselmo, D. (2000). *In vitro* reconstitution of a functional duck hepatitis B virus reverse transcriptase: posttranslational activation by Hsp90. *J. Virol.*, **74**, 11447-11455.

Kann, M. and Gerlich, W.H. (1998). Hepatitis B. In: *Topley and Wilson's Microbiology and Microbial Infections*, 9th ed. (L. Collier, A. Balows and M. Sussman, eds), pp 745-774. Arnold, London.

Lanford, R.E., Chavez, D., Brasky, K.M., Burns, R.B., 3rd, and Rico-Hesse, R. (1998). Isolation of a hepadnavirus from the woolly monkey, a New World primate. *Proc. Natl. Acad. Sci. USA*, **95**, 5757-5761.

Marion, P.L., Oshiro, L.S., Regnery, D.C., Scullard, G.H. and Robinson, W.S. (1980). A virus in Beechey ground squirrels which is related to hepatitis B virus of man. *Proc. Natl. Acad. Sci. USA*, **77**, 2941-2945.

Mason, W.S., Seal, G. and Summers, J. (1980). Virus of Pekin ducks with structural and biological relatedness to human hepatitis B virus. *J. Virol.*, **36**, 829-836.

Newbold, J.E., Xin, H., Tencza, M., Sherman, G., Dean, J., Bowden, S. and Locarnini, S. (1995). The covalently closed duplex form of the hepadnavirus genome exists *in situ* as a heterogeneous population of viral minichromosomes. *J. Virol.*, **69**, 3350-3357.

Netter, H.J., Chassot, S., Chang, S.F., Cova, L. and Will, H. (1997). Sequence heterogeneity of heron hepatitis B virus genomes determined by full-length DNA amplification and direct sequencing reveals novel and unique features. *J. Gen. Virol.*, **78**, 1707-1718.

Pult, I., Netter, H.J., Bruns, M., Prassolov, A., Sirma, H., Hohenberg, H., Chang, S.F., Frolich, K.,Krone, O., Kaleta, E.F. and Will, H. (2001). Identification and analysis of a new hepadnavirus in white storks. *Virology*, **289**, 114-128.

Simmonds, P. (2001). The origin and evolution of hepatitis viruses in humans. *J. Gen. Virol.*, **82**, 693-712.

Sprengel, R., Kaleta, E.F. and Will, H. (1988). Isolation and characterization of a hepatitis B virus endemic in herons. *J. Virol.*, **62**, 932-937.

Summers, J., Smolec, J. and Snyder, R. (1978). A virus similar to human hepatitis B virus associated with hepatitis and hepatoma in woodchucks. *Proc. Natl. Acad. Sci. USA*, **75**, 4533-4537.

Testut, P., Renard, C.A., Terradillas, O., Vilvitski-Trepro, L., Tekaia, F., Degott, C., Blake, J., Boyer, B. and Buendia, M.A. (1996). A new hepadnavirus endemic in arctic ground squirrels in Alaska. *J. Virol.*, **70**, 4210-4219.

Triyatni, M., Ey, P., Tran, T., Le Mire, M., Qiao, M., Burrell, C. and Jilbert, A. (2001). Sequence comparison of an Australian duck hepatitis B virus strain with other avian hepadnaviruses. *J. Gen. Virol.*, **82**, 373-378.

Yu, S.F., Baldwin, D.N., Gwynn, S.R., Yendapalli, S. and Linial, M.L. (1996). Human foamy virus replication: a pathway distinct from that of retroviruses and hepadnaviruses. *Science*, **271**, 1579-1582.

Zhou, Y.Z. (1980). A virus possibly associated with hepatitis and hepatoma in ducks. *Shanghai Med. J.*, **3**, 641-644.

Contributed By

Mason, W.S., Burrell, C.J., Casey, J., Gerlich, W.H., Howard, C.R., Kann, M., Lanford, R., Newbold, J., Schaefer, S., Taylor, J.M. and Will, H.

FAMILY CAULIMOVIRIDAE

TAXONOMIC STRUCTURE OF THE FAMILY

Family	***Caulimoviridae***
Genus	***Caulimovirus***
Genus	***Petuvirus***
Genus	***Soymovirus***
Genus	***Cavemovirus***
Genus	***Badnavirus***
Genus	***Tungrovirus***

VIRION PROPERTIES

MORPHOLOGY

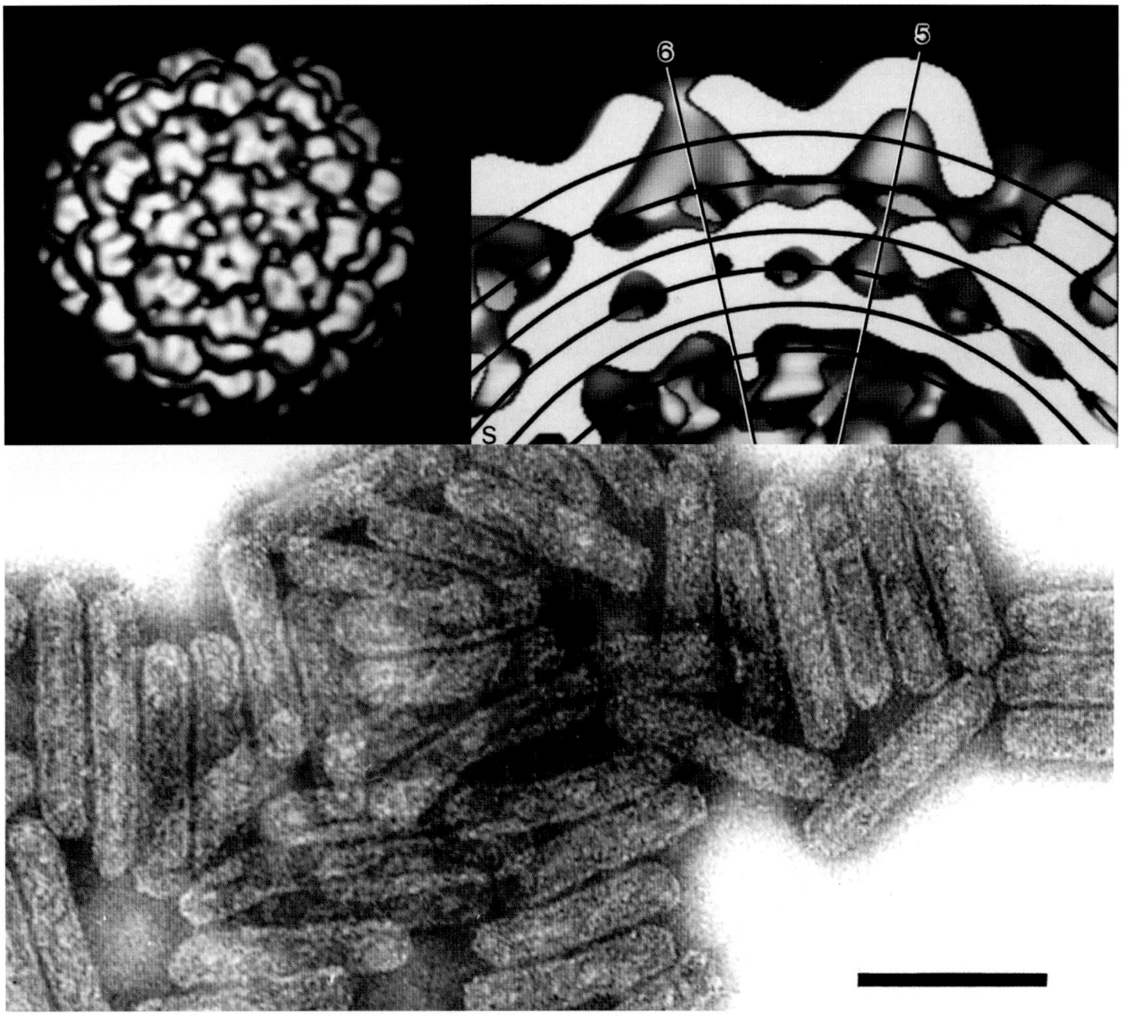

Figure 1: (Top left) Reconstruction of the surface structure of a Cauliflower mosaic virus particle showing T = 7 symmetry. (Top right) Cutaway surface reconstruction showing multilayer structure. (From Cheng *et al.* (1992). *Virology* **186**, 655-668). (Bottom) Negative contrast electron micrograph of particles of Commelina yellow mottle virus (ComYMV), stained with 2% sodium phosphotungstate, pH 7.0. The bar represents 10 nm.

Virions are either isometric or bacilliform depending on the genus (Fig. 1). There is no envelope.

PHYSICOCHEMICAL AND PHYSICAL PROPERTIES

Virions have buoyant densities in CsCl of 1.37 g/cm^3 (genera *Caulimovirus*, *Soymovirus*, *Petuvirus*, *Cavemovirus*) or in Cs_2SO_4 of 1.31 g/cm^3 (genera *Tungrovirus* and *Badnavirus*). S_{20w} is in the range of 200S to 220S. Particles are very stable between pH4 and pH9 and in high salt concentrations.

NUCLEIC ACID

Virions contain a single molecule of circular dsDNA of 7.2-8.3 kbp. Each strand of the genome has discontinuities at specific places, one strand has one discontinuity and the other has between one and three discontinuities.

PROTEINS

Genomes contain between one and seven ORFs, depending on the genus. The functions of virus-encoded proteins common to all genera are the CP, an aspartate protease, an RT and an RNAse H.

LIPIDS

None reported.

CARBOHYDRATES

The CP of Cauliflower mosaic virus (CaMV) is glycosylated. No carbohydrates have been reported for the virions of other species.

GENOME ORGANIZATION AND REPLICATION

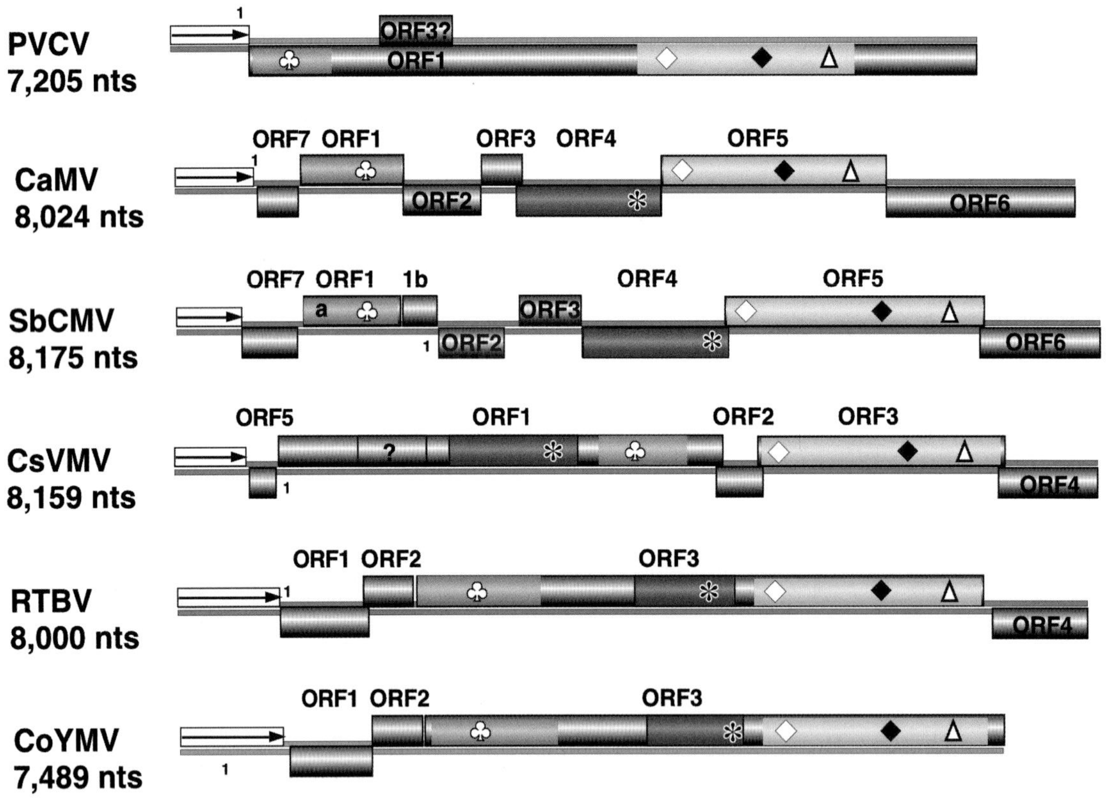

Figure 2: Comparison of genome maps characteristic of a representative of each genus for the family *Caulimoviridae*. ORFs or ORF segments encoding a protein are represented by boxes in different colors according to the different putative functions of the genes (orange MP, red CP and yellow RT protein). The symbols used are the followings: ♣ MP active site, * RNA binding site, ◊ protease active site, ♦ RT active site, Δ RNase H consensus sequence. The map starts at the intergenic region of the circular genome for convenience, the arrow shows the position of the promoter and the number 1 indicates the origin of DNA replication.

One strand of DNA is the coding sequence. The genome organization is dependent upon genus (Fig. 2) and is one of the main characteristics that distinguishes the genera from each other. Following entry into the cell, the discontinuities in the genomes are sealed to give supercoiled DNA that forms minichromosomes in the nucleus. These are transcribed asymmetrically by host DNA-dependent RNA polymerases to give a more-than-genome length transcript (35S or 34S RNA) that has a terminal redundancy of about 35 to 270 nt, dependent upon species. This transcript serves both as a template for reverse transcription to give the negative DNA strand and for expression of at least some of the ORFs. Species in genus *Caulimovirus* produce a specific mRNA (19S RNA) for ORF6; no sgRNAs have been reported for genera *Petuvirus, Soymovirus, Cavemovirus* and *Badnavirus.* ORF4 of Rice tungro bacilliform virus (RTBV) is expressed from an RNA spliced from the 35S RNA. The replication cycle, in contrast to that of retroviruses, is episomal, and does not involve an integration phase. Negative strand DNA synthesis is primed by host cytosolic $tRNA^{met}$ and synthesis of both strands is performed by the viral RT and RNAse H. The site-specific discontinuities are at the priming sites for both negative- and positive-strand DNA synthesis and are made by the oncoming strand displacing the existing strand for a short distance and not ligating to form a closed circle.

ANTIGENIC PROPERTIES

Virions range from moderate to efficient immunogens. There is pronounced antigenic variability in species in the genus *Badnavirus*. There are some serological cross-reactions among species in different genera.

BIOLOGICAL PROPERTIES

The host ranges of most species are narrow. Those of species in the genera *Petuvirus, Soymovirus* and *Cavemovirus,* are restricted to dicotyledonous plant species; species in the genus *Tungrovirus* infect monocotyledonous plant species and those of the genus *Badnavirus* infect either dicotyledonous or monocotyledonous plant species. Many virus species are spread by vegetative propagation.

The geographic range of many species is wide; most species in the genera *Tungrovirus* and *Badnavirus* are primarily tropical or subtropical with some temperate and Sub-Antarctic species whereas most of the species in the genera *Petuvirus, Caulimovirus, Soymovirus* and *Cavemovirus* are found in temperate regions.

The symptoms caused by these viruses are variable dependent on virus species, host and climatic conditions. Mosaic symptoms predominate amongst members of the genera *Petuvirus, Caulimovirus, Soymovirus* and *Cavemovirus,* whereas interveinal chlorotic mottling and streaking is the most frequent symptom of species in the genera *Tungrovirus* and *Badnavirus.*

Most species infect most cell types of their hosts although some in the genera *Tungrovirus* and *Badnavirus* are restricted to the vascular system. Virions occur in the cytoplasm and those of species in the genera *Petuvirus, Caulimovirus, Soymovirus* and *Cavemovirus,* are associated with proteinaceous inclusion bodies.

GENUS CAULIMOVIRUS

Type Species *Cauliflower mosaic virus*

DISTINGUISHING FEATURES

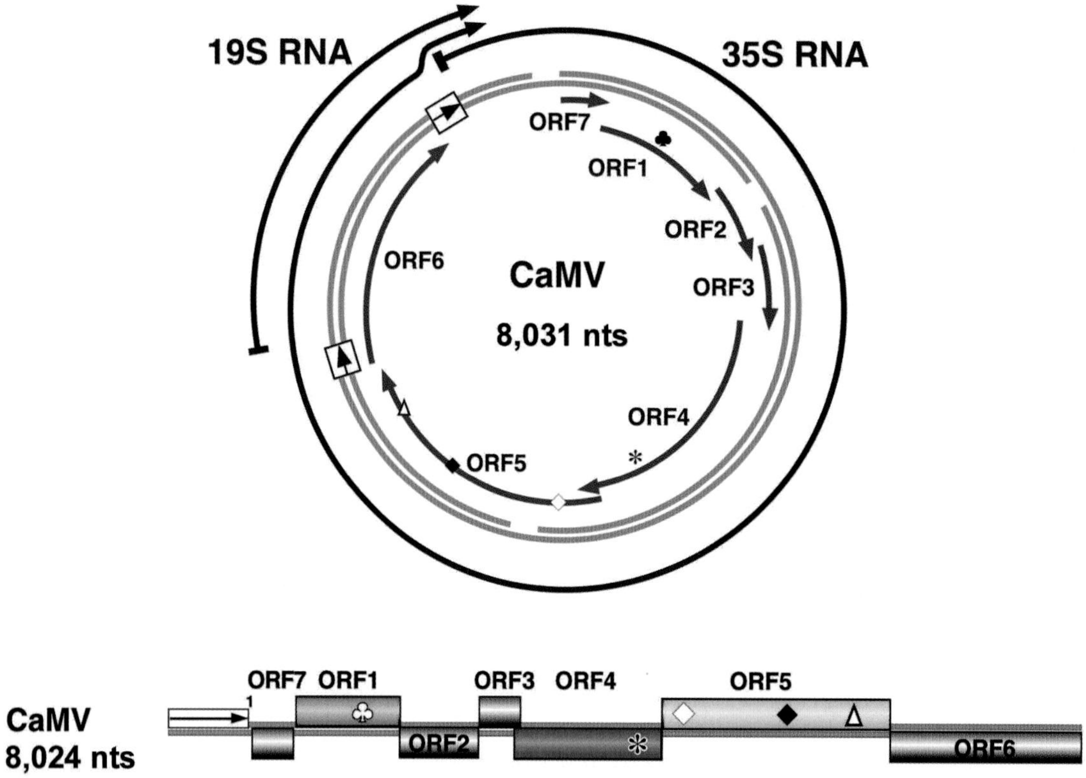

Figure 3: Circular (top) and linearized (bottom) genome maps characteristic of an isolate of *Cauliflower mosaic virus*. ORFs or ORF segments encoding a protein are represented by boxes in different colors according to the different putative functions of the genes (orange MP, red CP and yellow RT protein). The symbols used are the followings: ♣ MP active site, * RNA binding site, ◊ protease active site, ♦ RT active site, ΔRNAse H consensus sequence. The map starts at the intergenic region of the circular genome for convenience, the arrow shows the position of the promoter and the number 1 indicates the origin of DNA replication.

Members of this genus have particles and cytoplasmic inclusions similar to those of species in the genera *Soymovirus, Petuvirus* and *Cavemovirus*, but differ from them in genome organization. Isometric particles are about 50 nm in diameter with a T=7 (420 subunits) multilayered structure (Fig. 1). Most species are transmitted in a semi-persistent manner by aphids. Transmission requires a virus-encoded protein (aphid transmission factor) that is encoded by ORF2.

Virions occur in the cytoplasm, and in some cases in the nucleus. Cytoplasmic virions are associated with electron-dense proteinaceous inclusion bodies (for caulimoviruses, the product of ORF6). The product of ORF2 also forms inclusion bodies that are electron translucent. Inclusion bodies can be seen by light microscopy as well as by electron microscopy. The genome contains six ORFs that express proteins with the organization shown in Figure 3; ORF7 appears to play a regulatory role in translation.

LIST OF SPECIES DEMARCATION CRITERIA IN THE GENUS

The criteria demarcating species in the genus are:

- Differences in host ranges,
- Differences in polymerase (RT + RNAse H) nt sequences of more than 20%.
- Differences in gene product sequences.

List of Species in the Genus

Species names are in green italic script; strain names and synonyms are in black roman script; tentative species names are in blue roman script. Sequence accession numbers [ ], and assigned abbreviations ( ) are also listed.

Species in the Genus

Carnation etched ring virus		
Carnation etched ring virus	[X04658, NC003498]	(CERV)
Cauliflower mosaic virus		
Cauliflower mosaic virus - Cabb-S	[NC001497]	(CaMV-CabbS)
Cauliflower mosaic virus – B29	[X79465]	(CaMV-B29)
Cauliflower mosaic virus – G1	[V00140]	(CaMV-G1)
Cauliflower mosaic virus – G2	[V00141]	(CaMV-G2)
Cauliflower mosaic virus - Xinjiang	[AF140604]	(CaMV-Xinjiang)
Cauliflower mosaic virus - BC	[M90542]	(CaMV-BC)
Cauliflower mosaic virus – NY8153	[M90541]	(CaMV-NY8153)
Cauliflower mosaic virus – D/H	[M10376]	(CaMV-D/H)
Cauliflower mosaic virus – CMV1	[M90543]	(CaMV-CMV1)
Dahlia mosaic virus		
Dahlia mosaic virus		(DMV)
Figwort mosaic virus		
Figwort mosaic virus	[NC003554, X06166]	(FMV)
Horseradish latent virus		
Horseradish latent virus		(HRLV)
Mirabilis mosaic virus		
Mirabilis mosaic virus	[NC004036, AF454635]	(MiMV)
Strawberry vein banding virus		
Strawberry vein banding virus	[NC001725, X97304]	(SVBV)
Thistle mottle virus		
Thistle mottle virus		(ThMoV)

Tentative Species in the Genus

Aquilegia necrotic mosaic virus	(ANMV)
Cestrum yellow leaf curling virus	(CmYLCV)
Plantago virus 4	(PlV-4)
Sonchus mottle virus	(SMoV)

Genus *Petuvirus*

Type Species *Petunia vein clearing virus*

Distinguishing Features

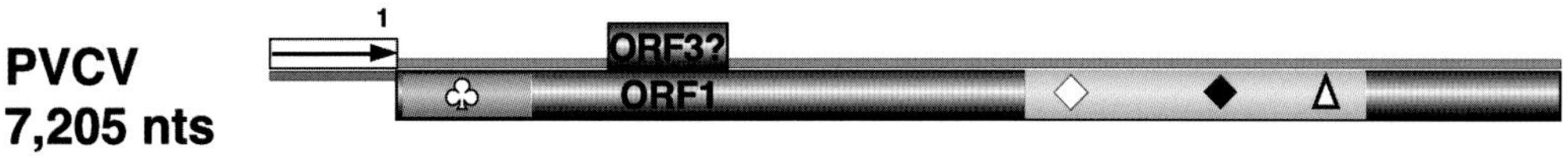

Figure 4: Linearized genome map characteristic of an isolate of *Petunia vein clearing virus*. ORFs or ORF segments encoding a protein are represented by boxes in different colors according to the different putative functions of the genes (orange MP, red CP and yellow RT protein). The symbols used are the followings: ♣ MP active site, ◊ protease active site, ♦ RT active site, Δ RNAse H consensus sequence. The map starts at the intergenic region of the circular genome for convenience, the arrow shows the position of the promoter and the number 1 indicates the origin of DNA replication.

The species in this genus has virions, and induces the formation of cytoplasmic inclusions, that are similar to those of species in the genera *Caulimovirus*, *Soymovirus* and *Cavemovirus* but differs from them in its genome organization. The genome contains one ORF organized as shown in Figure 4. There is evidence that an activatable form of the viral genome is integrated into the host genome but it is not known if this process is functional.

LIST OF SPECIES DEMARCATION CRITERIA IN THE GENUS

Not applicable.

LIST OF SPECIES IN THE GENUS

Species names are in green italic script; strain names and synonyms are in black roman script; tentative species names are in blue roman script. Sequence accession numbers [ ], and assigned abbreviations ( ) are also listed.

SPECIES IN THE GENUS

Petunia vein clearing virus

Petunia vein clearing virus	[U95208]	(PVCV)

TENTATIVE SPECIES IN THE GENUS

None reported.

GENUS *SOYMOVIRUS*

Type Species ***Soybean chlorotic mottle virus***

DISTINGUISHING FEATURES

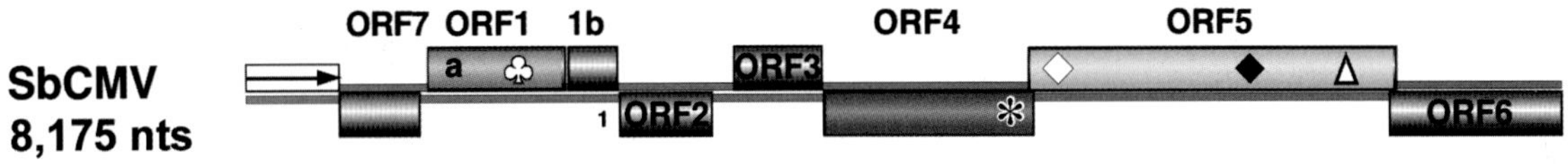

Figure 5: Linearized genome map of an isolate of *Soybean chlorotic mottle virus*. ORFs or ORF segments encoding a protein are represented by boxes in different colors according to the different putative functions of the genes (orange MP, red CP and yellow RT protein). The symbols used are the followings: ♣ MP active site, ◊ protease active site, ♦ RT active site, Δ RNAse H consensus sequence. The map starts at the intergenic region of the circular genome for convenience, the arrow shows the position of the promoter and the number 1 indicates the origin of DNA replication.

Species in this genus have virions, and induce the formation of cytoplasmic inclusions that are similar to those of species in the genera *Caulimovirus*, *Petuvirus* and *Cavemovirus*, but differ from them in their genome organization. The genome has seven (or possibly eight) ORFs that have a different distribution in relation to the negative strand DNA priming site (zero position on map) from those in genomes of species in the genus *Caulimovirus* (Fig. 5).

LIST OF SPECIES DEMARCATION CRITERIA IN THE GENUS

The criteria demarcating species in the genus are:

- Differences in host ranges,
- Differences in polymerase (RT + RNAse H) nt sequences of more than 20%,
- Differences in gene product sequences.

LIST OF SPECIES IN THE GENUS

Species names are in green italic script; strain names and synonyms are in black roman script; tentative species names are in blue roman script. Sequence accession numbers [ ], and assigned abbreviations ( ) are also listed.

SPECIES IN THE GENUS

Blueberry red ringspot virus		
Blueberry red ringspot virus	[NC003138, AF404509]	(BRRSV)
Peanut chlorotic streak virus		
Peanut chlorotic streak virus	[NC001634, U13988]	(PCSV)
Soybean chlorotic mottle virus		
Soybean chlorotic mottle virus	[NC001739, X15828]	(SbCMV)

TENTATIVE SPECIES IN THE GENUS

None reported.

GENUS *CAVEMOVIRUS*

Type Species *Cassava vein mosaic virus*

DISTINGUISHING FEATURES

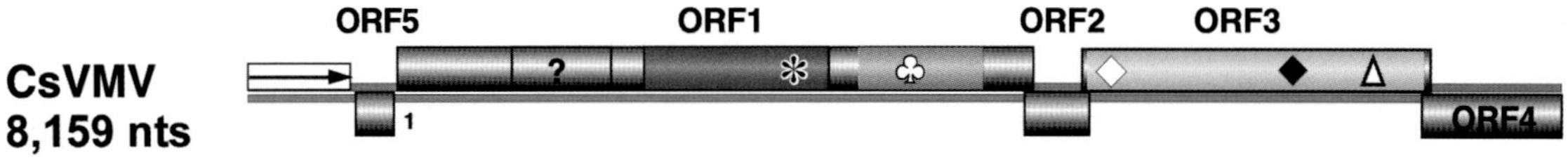

Figure 6: Linearized genome map characteristic of an isolate of *Cassava vein mosaic virus*. ORFs or ORF segments encoding a protein are represented by boxes in different colors according to the different putative functions of the genes (orange MP, red CP and yellow RT protein). The symbols used are the followings: ♣ MP active site, ◊ protease active site, ♦ RT active site, Δ RNAse H consensus sequence. The map starts at the intergenic region of the circular genome for convenience, the arrow shows the position of the promoter and the number 1 indicates the origin of DNA replication.

Cassava vein mosaic virus has virions, and induces the formation of cytoplasmic inclusions that are similar to those of species in the genera *Caulimovirus*, *Soymovirus* and *Petuvirus*, but differs from them in its genome organization. The genome contains five ORFs organized as shown in Figure 6. There is evidence that an activatable form of the genome of TVCV is integrated into the host genome but it is not known if this process is functional.

LIST OF SPECIES DEMARCATION CRITERIA IN THE GENUS

The criteria demarcating species in the genus are:

- Differences in host ranges,
- Differences in polymerase (RT + RNAse H) nt sequences of more than 20%,
- Differences in gene product sequences.

LIST OF SPECIES IN THE GENUS

Species names are in green italic script; strain names and synonyms are in black roman script; tentative species names are in blue roman script. Sequence accession numbers [ ], and assigned abbreviations ( ) are also listed.

SPECIES IN THE GENUS

Cassava vein mosaic virus		
Cassava vein mosaic virus	[NC001648, U59751,U20341]	(CsVMV)
Tobacco vein clearing virus		
Tobacco vein clearing virus	[NC003378, AF190123]	(TVCV)

TENTATIVE SPECIES IN THE GENUS

None reported.

RT VIRUSES

GENUS BADNAVIRUS

Type Species *Commelina yellow mottle virus*

DISTINGUISHING FEATURES

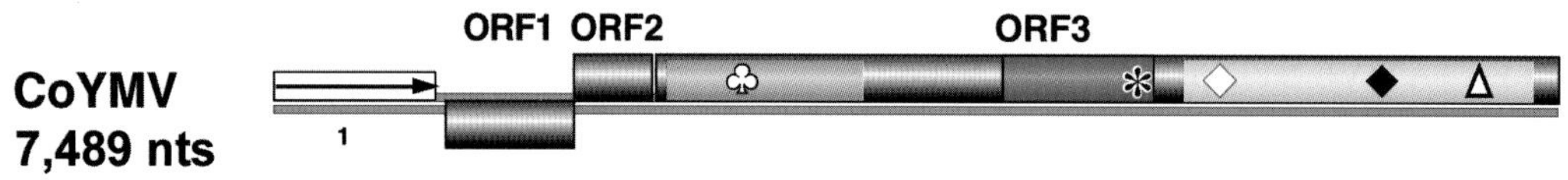

Figure 7: Linearized genome organization characteristic of an isolate of *Commelina yellow mottle virus*. ORFs or ORF segments encoding a protein are represented by boxes in different colors according to the different putative functions of the genes (orange MP, red CP and yellow RT protein). The symbols used are the followings: ♣ MP active site, ◊ protease active site, ♦ RT active site, Δ RNAse H consensus sequence. The map starts at the intergenic region of the circular genome for convenience, the arrow shows the position of the promoter and the number 1 indicates the origin of DNA replication.S

pecies in this genus resemble those in the genus *Tungrovirus* morphologically but differ from them in genome organization.

Virions are bacilliform with parallel sides and rounded ends (Fig. 1). Virions are uniformly 30 nm in width. The modal particle length is 130 nm, but particles ranging in length from 60-900 nm are commonly observed. No projections or other capsid surface features have been observed by electron microscopy. Virions have an electron-transparent central core, but there is no information on the nature of the nucleic acid-capsomer interaction. The tubular portion of the virion has a structure based on an icosahedron cut across its 3-fold axis, with a structural repeat of 10 nm and 9 rings of hexamer subunits per 130 nm length.

Within the genome, the CP, aspartate protease, RT and RNAse H are expressed in a polyprotein together with the putative cell-to-cell MP.

Transmission is in a semi-persistent manner by mealybugs and for some species also by aphids or lacebugs. The virus does not multiply in mealybug vectors and there is no transovarial transmission. All motile life stages of vectors can acquire and transmit virus. There is little information on the possible transmission of badnaviruses by other vector types.

Virions occur only in the cytoplasm either singly or in large groups, randomly distributed or arranged in palisade-like arrays. They do not occur within inclusion bodies or in membrane-bound structures.

The genomes of badnaviruses contain three ORFs organized as shown in Figure 7. There is evidence that an activatable form of the genome of some BSV species is integrated into the host genome but it is not known if this process is functional.

LIST OF SPECIES DEMARCATION CRITERIA IN THE GENUS

The criteria demarcating species in the genus are:

- Differences in host ranges,
- Differences in polymerase (RT + RNAse H) nt sequences of more than 20%,
- Differences in gene product sequences,
- Differences in vector specificities.

List of Species in the Genus

Species names are in green italic script; strain names and synonyms are in black roman script; tentative species names are in blue roman script. Sequence accession numbers [ ], and assigned abbreviations ( ) are also listed.

Species in the Genus

Aglaonema bacilliform virus		
Aglaonema bacilliform virus		(ABV)
Banana streak GF virus		
Banana streak GF virus	[AF215814*]	(BSGFV)
Banana streak Mysore virus		
Banana streak Mysore virus	[AF214005*]	(BSMyV)
Banana streak OL virus		
Banana streak OL virus	[AJ002234]	(BSOLV)
Cacao swollen shoot virus		
Cacao swollen shoot virus	[NC001374, L14546]	(CSSV)
Canna yellow mottle virus		
Canna yellow mottle virus		(CaYMV)
Citrus mosaic virus		
Citrus mosaic virus	[AF347695]	(CMBV)
Commelina yellow mottle virus		
Commelina yellow mottle virus	[X7924]	(ComYMV)
Dioscorea bacilliform virus		
Dioscorea bacilliform virus	[X94576*, X94581*]	(DBV)
Gooseberry vein banding associated virus		
Gooseberry vein banding associated virus	[AF298883*]	(GVBAV)
Kalanchoe top-spotting virus		
Kalanchoe top-spotting virus	[AY180137]	(KTSV)
Piper yellow mottle virus		
Piper yellow mottle virus		(PYMoV)
Rubus yellow net virus		
Rubus yellow net virus	[AF468454]	(RYNV)
Schefflera ringspot virus		
Schefflera ringspot virus		(SRV)
Spiraea yellow leaf spot virus		
Spiraea yellow leaf spot virus	[AF299074*]	(SYLSV)
Sugarcane bacilliform IM virus		
Sugarcane bacilliform IM virus	[AJ277091]	(SCBIMV)
Sugarcane bacilliform Mor virus		
Sugarcane bacilliform Mor virus	[M89923, NC003031]	(SCBMV)
Taro bacilliform virus		
Taro bacilliform virus	[AF357836]	(TaBV)

* indicate partial sequences

Tentative Species in the Genus

Aucuba bacilliform virus		(AuBV)
Mimosa bacilliform virus		(MBV)
Pineapple bacilliform virus	[Y12433*]	(PBV)
Stilbocarpa mosaic bacilliform virus	[AF478691*]	(SMBV)
Yucca bacilliform virus	[AF468688*]	(YBV)

* indicate partial sequences

GENUS TUNGROVIRUS

Type Species *Rice tungro bacilliform virus*

DISTINGUISHING FEATURES

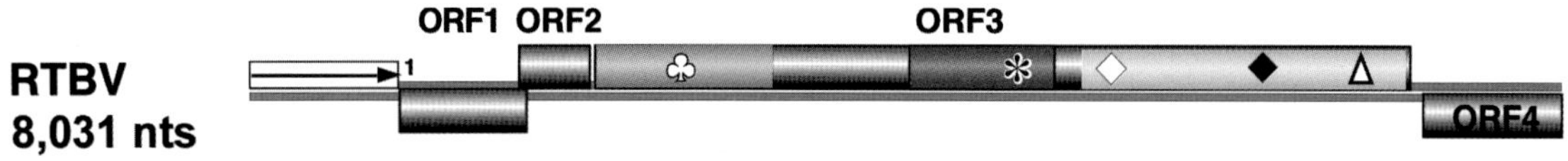

Figure 8: Linearized genome map characteristic of an isolate of *Rice tungro bacilliform virus*. ORFs or ORF segments encoding a protein are represented by boxes shaded in different patterns according to the different putative functions of the genes (vertical lines for the putative MP, slanted lines for the CP and horizontal lines for the RT protein). The symbols used are the followings: ♣ MP active site, * RNA binding site, ◊ protease active site, ♦ RT active site, Δ RNAse H consensus sequence. The map starts at the intergenic region of the circular genome for convenience, the arrow shows the position of the promoter and the number 1 indicates the origin of DNA replication.

The sole species shares many of the distinguishing features of those in the genus *Badnavirus*. The genome differs in that it contains four ORFs, organized as shown in Fig. 8. The genome has about 20-25% nt identity to the genomes of badnaviruses. RTBV is transmitted by leafhoppers with the assistance of Rice tungro spherical virus (RTSV; family *Sequiviridae*).

LIST OF SPECIES DEMARCATION CRITERIA IN THE GENUS

Not applicable.

LIST OF SPECIES IN THE GENUS

Species names are in green italic script; strain names and synonyms are in black roman script; tentative species names are in blue roman script. Sequence accession numbers [ ], and assigned abbreviations ( ) are also listed.

SPECIES IN THE GENUS

Rice tungro bacilliform virus

Rice tungro bacilliform virus	[NC001914, X57924]	(RTBV)
Rice tungro bacilliform virus – 1C	[AF113832]	(RTBV-1C)
Rice tungro bacilliform virus - Chainat	[AF220561]	(RTBV-Ch)
Rice tungro bacilliform virus – G1	[AF113830]	(RTBV-G1)
Rice tungro bacilliform virus – G2	[AF113831]	(RTBV-G2)
Rice tungro bacilliform virus – Ic	[AF113832]	(RTBV-Ic)
Rice tungro bacilliform virus - Serdang	[AF076470]	(RTBV-Ser)
Rice tungro bacilliform virus – West Bengal	[AJ314596]	(RTBV-WB)

TENTATIVE SPECIES IN THE GENUS

None reported.

Phylogenetic Relationships Within the Family

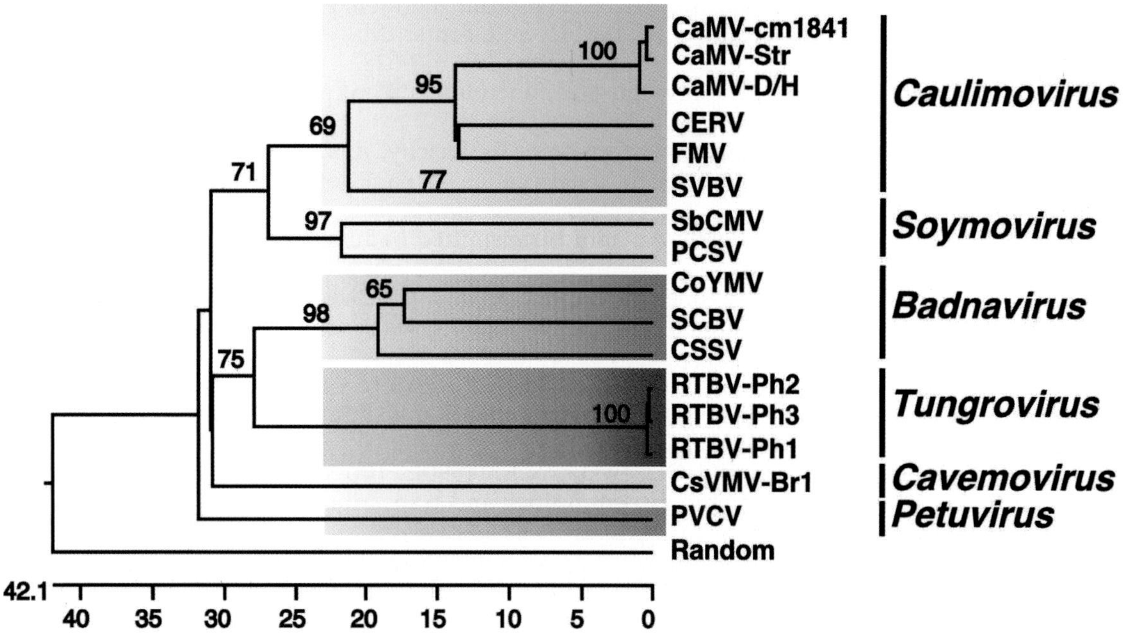

Figure 9: Neighbor-Joining dendrogram of sequence relationships among species of the different genera of the family *Caulimoviridae*. The reverse transcriptase amino-acid sequence alignment and dendrogram were produced using MegAlign (DNAStar program), which uses a CLUSTAL-type algorithm. Horizontal distances are proportional to sequence distances; vertical distances are arbitrary. The dendrogram was bootstrapped 100 times (percent scores shown at nodes), and rooted on a random sequence.

Similarities With Other Taxa

The shapes of the virions of the genera *Tungrovirus* and *Badnavirus* resemble those of viruses in *Alfalfa mosaic virus* (family *Bromoviridae*) and *Mushroom bacilliform virus* (family *Barnaviridae*) but these are smaller, being up to 60 nm long and about 18 nm diameter.

Derivation of Names

Badna: sigla from **ba**cilliform **DNA** viruses.
Caulimo: sigla from **cauli**flower **mo**saic virus.
Cavemo: sigla from **ca**ssava **ve**in **mo**ttle virus
Petu: sigla from **petu**nia
Soymo: sigla from **soy**bean chlorotic **mo**ttle virus
Tungro: sigla from rice **tungro** bacilliform virus

References

Bouhida, M., Lockhart, B.E.L. and Olszewski. N.E. (1993). An analysis of the complete sequence of a sugarcane bacilliform virus genome infectious to banana and rice. *J. Gen. Virol.*, **74**, 15-22.

Calvert, L.A., Ospina, M.D. and Shepherd, R.J. (1995). Characterization of cassava vein mosaic virus: a distinct plant pararetrovirus. *J. Gen. Virol.*, **76**, 1271-1276.

De Silva, D.P.P., Jones, P. and Shaw, M.W. (2002). Identification and transmission of Piper yellow mottle virus and Cucumber mosaic virus infecting black pepper (*Piper nigrum*) in Sri Lanka. *Plant Pathology*, **51**, 537-545.

Frank, A., Guilley, H., Jonard, G., Richards, K.E. and Hirth, L. (1980). Nucleotide sequence of cauliflower mosaic virus DNA. *Cell*, **21**, 285-294.

Geering, A.D.W., McMichael, L.A., Dietzgen, R.G. and Thomas, J.E. (2000). Genetic diversity among Banana streak virus isolates from Australia. *Phytopathology*, **90**, 921-927.

Geijskes, R.J., Braithwaite, K.S., Dale, J.L., Harding, R.M. and Smith, G.R. (2002). Sequence analysis of an Australian isolate of sugarcane bacilliform badnavirus. *Arch. Virol.*, **147**, 2393-2404.

RT Viruses

Glasheen, B.M., Polashock, J.J., Lawrence, D.M., Gillett, J.M., Ramsdell, D.C., Vorsa, N. and Hillman, B.L. (2002). Cloning, sequencing and promoter identification of blueberry red ringspot virus, a member of the family *Caulimoviridae* with similarities to the "soybean chlorotic mottle-like" genus. *Arch. Virol.*, **147**, 2169-2186

Hagen, L.S., Jacquemond, M., Lepingle, A., Lot, H. and Tepfer, M. (1993). Nucleotide sequence and genomic organization of cacao swollen shoot virus. *Virology*, **196**, 619-628.

Hohn, T. and Fütterer, J. (1997). The proteins and functions of plant pararetroviruses: knowns and unknowns. *Crit. Rev. Plant Sci.*, **16**, 133-161.

Kochko, de, A., Verdaguer, B., Taylor, N., Carcamo, R., Beachy, R.N. and Fauquet, C.M. (1998). *Cassava vein mosaic virus* (CsVMV) the type species for a new genus of plant double stranded DNA viruses? *Arch. Virol.*, **143**, 945-962.

Lockhart, B.E.L. and Lachner, J. (1999). An aphid-transmitted badnavirus associated with a yellow leafspot of spiraea. *Phytopathology*, **89**, S46

Lockhart, B.E.L., Kiratiya-Angul, K., Jones, P., Eng, L., Silva, P.D., Olszewski, N.E., Lockhart, N., Deema, N. and Sangalang, J. (1997). Identification of *Piper yellow mottle virus*, a mealybug-transmitted badnavirus infecting Piper spp. in Southeast Asia. *Eur. J. Plant Pathol.*, **103**, 303-311.

Lockhart, B.E., Menke, J., Dahal, G. and Olszewski, N.E. (2000). Characterization and genomic analysis of *Tobacco vein clearing virus*, a plant pararetrovirus that is transmitted vertically and related to sequences in the host genome. *J. Gen. Virol.*, **81**, 1579-1585.

Medberry, S.L., Lockhart, B.E.L. and Olszewski, N.E. (1990). Properties of *Commelina yellow mottle virus*'s complete DNA sequence, genomic discontinuities and transcript suggest that it is a pararetovirus. *Nuc. Acids Res.*, **18**, 5505-5513.

Qu, R., Bhattacharya, M., Laco, G.S., de Kochko, A., Subba Rao, B.L., Kaniewski, M.B., Elmer, J.S., Rochester, D.E., Smith, C.E. and Beachy, R.N. (1991). Characterization of the genome of rice tungro bacilliform virus: Comparison with Commelina yellow mottle virus and caulimoviruses. *Virology*, **185**, 354-364.

Richert-Pöggeler, K.R. and Shepherd, R.J. (1997). Petunia vein-clearing virus: a plant pararetrovirus with the core sequences for an integrase function. *Virology*, **236**, 137-146.

Skotnicki, M.L., Selkirk, P.M., Kitajima, E., McBride, T.P., Shaw, J. and Mackenzie, A. (2003). The first subantarctic plant virus report: *Stilbocarpa mosaic bacilliform badnavirus* (SMBV) from Macquarie Island. *Polar Biology*, **26**, 1-7.

Yang, I.C., Hafner, G.J., Dale, J.L. and Harding, R.M (2003). Genomic characterization of *Taro bacilliform virus*. *Arch. Virol.*, **148**, 937-949.

Contributed By

Hull, R., Geering, A., Harper G., Lockhart, B.E. and Schoelz, J.E.

FAMILY PSEUDOVIRIDAE

TAXONOMIC STRUCTURE OF THE FAMILY

Family	***Pseudoviridae***
Genus	***Pseudovirus***
Genus	***Hemivirus***
Genus	***Sirevirus***

VIRION PROPERTIES

MORPHOLOGY

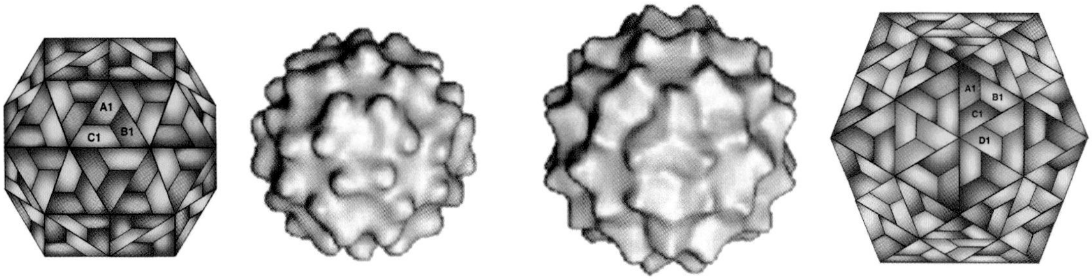

Figure 1: Saccharomyces cerevisiae Ty1 virus (SceTy1V) 1-381 virions; surface structure of two forms (T=3, left; T=4, right) determined by cryo-electron microscopy, with the corresponding diagrammatic models. (Courtesy of H. Saibil, adapted from *J. Virol.*, **71**, 6863-6868).

Pseudoviridae is a family of retrotransposable elements, primarily identified by genome sequencing. Pseudoviruses are often referred to as LTR-retrotransposons of the Ty1-copia family. They replicate via a virus-like intermediate referred to as a virus-like particle (VLP). VLPs do not display infectivity according to the traditional virological definition. However, there is good evidence that these particles are essential and direct intermediates in the life cycle of these elements. We will use the terms "virion" and "virus" here to conform to the usage in this volume.

Members of the family *Pseudoviridae* are typified by somewhat irregularly shaped VLPs that are round to ovoid, often with electron-dense centers. Although the VLPs are irregular in their native state, by expressing truncated forms of the major coat protein (Gag) icosahedral VLPs with pleasing regularity can be observed by cryo-electron microscopy (Fig. 1). Saccharomyces cerevisiae Ty1 virus (Ty1) and Drosophila melanogaster copia virus *(copia)* both make similar looking particles, but Ty1 particles are cytoplasmic, whereas those of *copia* are nuclear. The typical mean radius of the VLPs is 30-40 nm. There is no envelope, although some members encode an *env*-like gene, the function of which remains to be determined.

PHYSICOCHEMICAL AND PHYSICAL PROPERTIES

In most systems, virions are only very crudely characterized biochemically.

NUCLEIC ACID

The major virion RNA species consist of an LTR - to - LTR transcript of ~5-9 kb. In addition, most viruses package one or more host-derived primer tRNAs. The LTR - to - LTR transcript encodes one to two ORFs in most viruses, the equivalents of retroviral *gag* and *pol*; the second ORF is typically expressed at lower levels than the first. There is RNA, as well as various DNA forms (intermediates) in the virion preparation. The RNA is 5 to 9 kb, positive sense, capped and polyadenylated; the DNA is 5.5 to 10 kbp long.

RT VIRUSES

The linear ssRNA is in virions, and the linear dsDNA "provirus" is integrated into the host genome.

PROTEINS
Both Gag and Gag-Pol primary translation products are processed by the cognate protease into final products. The known *gag*-encoded proteins include analogs of retroviral capsid (CA) and nucleocapsid (NC), although the latter is only present in a subset of these viruses. The known *pol*-encoded proteins include the protease (PR), integrase (IN), and reverse transcriptase/RNase H (RT). All of these proteins appear to be required for replication.

LIPIDS
None present.

CARBOHYDRATES
None present.

GENOME ORGANIZATION AND REPLICATION

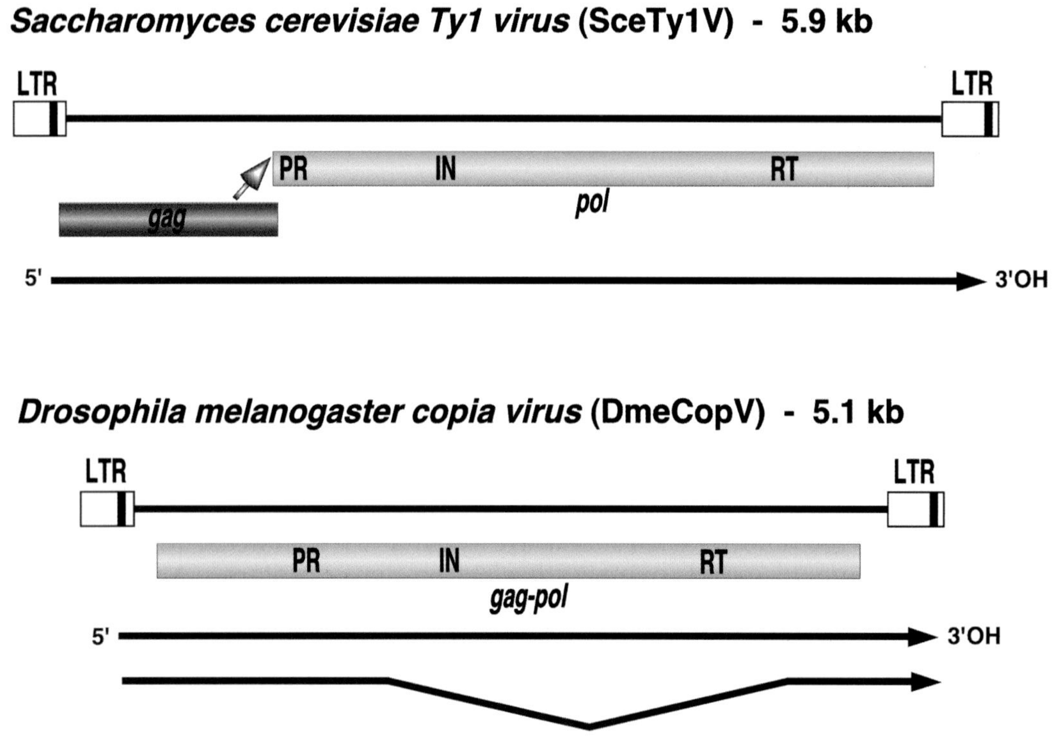

Figure 2: The genomic organization of Saccharomyces cerevisiae Ty1 virus (SceTy1V) (5.9 kb) and Drosophila melanogaster *copia* virus (DmeCopV) (5.1 kb). Black boxes within the LTRs depict sequences repeated at the 5'- and 3'-ends of the viruses transcripts (R regions); sequences 5' of R represent U3, and sequences 3' of R represent U5. Open boxes below the viruses indicate *gag* and *pol*. Conserved aa sequences in *pol* that identify protease (PR), integrase (IN) and reverse transcriptase/RNAse H (RT) are labeled. Individual mRNAs are depicted as arrows. Arrowhead indicates site of ribosomal frameshifting.

Most viruses in this genus encode a single ORF with similarity to retroviral *gag* and *pol* (Fig 2). The yeast Ty1, Ty2 and Ty4 viruses contain *pol* in the +1 frame relative to *gag*. For *copia*, *gag* is encoded on a spliced 2 kb mRNA, and differential splicing is the mechanism by which Gag and Gag-Pol stoichiometry is regulated. The mechanism(s) that regulate Gag and Gag-Pol expression for the single ORF viruses remain to be determined.

The virion-associated RT mediates the conversion of the LTR to LTR transcript into a full-length nucleic acid duplex containing full-length LTR sequences in the form of dsDNA. This DNA is then integrated into host DNA by the IN protein, where it becomes a part of the host genome and can persist there, essentially indefinitely. The integrated form (equivalent to the retroviral provirus) is then transcribed by host RNA polymerase II to generate new virus RNAs. In most viruses, the reverse transcription and integration processes closely mimic the replication of retroviral RNA, but there are some important exceptions.

ANTIGENIC PROPERTIES

No information available.

BIOLOGICAL PROPERTIES

Pseudoviruses are commonly referred to as LTR retrotransposons of the Ty1/*copia* family. They form an intrinsic and significant part of the genome of many eukaryotic species, especially plants. For most of these viruses, the virion is known to be an essential part of their multiplication cycle, but is not likely to be infectious, in the traditional virological sense, under normal conditions.

LIST OF SPECIES DEMARCATION CRITERIA IN THE GENUS

In general, viruses inhabiting different host species will be considered different species, because they will have diverged through vertical descent from a common ancestor at least as much as the host species themselves (probably more so, due to the error-prone mechanism of replication). However, there are instances in the family *Pseudoviridae* and the family *Metaviridae* in which one finds two closely related viruses inhabiting the same host species, e.g. Ty1 and Ty2 of *S. cerevisiae* and Tf1 and Tf2 of *S. pombe*. In both of these pairs of viruses, the RT aa sequences are quite similar. However, the capsid-encoding sequences are significantly diverged (e.g. < 50% aa sequence identity; their DNA sequences fail to cross hybridize). The question then arises whether these represent different species or more subtle variants. We have considered such viruses separate species if at least one of the major coding regions (e.g. capsid) is <50% identical to the reference aa sequence. for example, Ty1 and Ty2 Gag aa sequences are 49% identical.

Individual species within the genus *Sirevirus* all have greater than 50% identity in their RT aa sequences. Members of the genera *Pseudovirus*, *Hemivirus* and the Phaseolus vulgaris Tpv2-6 virus (Tpv2-6) group share less than 45% identity to sirevirus RTs.

GENUS *PSEUDOVIRUS*

Type Species *Saccharomyces cerevisiae Ty1 virus*

DISTINGUISHING FEATURES

VIRION PROPERTIES

Virions are ovoid to spheroid strictly intracellular particles, sometimes observed as paracrystalline clusters in the cytoplasm in yeast cells. Particles are quite heterodisperse structurally. However, the C-terminal deletion mutant containing residues 1-381 of the Gag protein is less heterodisperse than the wild-type, and cryo-electron micrographs of these particles reveal a surface structure with icosahedral symmetry (Fig. 1). The mean radius of the virions is 20-30 nm for wild type and 20 nm for the much better characterized and more uniform 1-381 mutant. The virions are not enveloped. The total Mr is estimated at 14 Mda and the sedimentation coefficient is estimated at ~200-300S for wild type and ~115S for 1-381 mutant. The virions are sensitive to high temperature (65°C).

The major virion RNA species consist of an LTR-to-LTR transcript of 5.6 kb. In addition Ty1 packages host-derived primer $tRNA_i^{Met}$. Ty1 particles contain two RNA molecules. This transcript encodes two overlapping ORFs, *GAG* and *POL*; the second expressed as the result of a +1 frameshift relative to the first. The major nucleic acid forms are RNA in the VLP, as well as various DNA forms (intermediates) in VLP preparations, and finally, after integration is complete, the integrated DNA "provirus". The genome size for Ty1 is RNA - 5.6 kb; DNA - 5.9 kb. There is linear ssRNA in the virion, and the provirus (integrated) consists of dsDNA.

The known *GAG*-encoded proteins are the CA and a short C-terminal peptide (the latter is only inferred to exist and has not been directly observed). No recognizable NC peptide has been identified, although the latter peptide performs similar functions. The known *POL*-encoded proteins are a protease (PR), integrase (IN), and reverse transcriptase/RNase H (RT). All of these proteins appear to be required for replication.

Genome Organization and Replication

The DNA form of the Ty1 genome consists of two 335 bp LTRs largely flanking a central coding region, although Ty1 is unusual is that *GAG* initiates within the U5 region of the 5'-LTR. The LTR sequences can be divided into three segments called U3, R and U5 (Fig. 2). The U3 region is unique to the 3'-end of the RNA, the R region is repeated in the RNA and the U5 region is unique to the 5'-end of the RNA. Ty1 encodes a 5.6 kb ssRNA with two ORFs. The first ORF is *GAG* and encodes the major CA as well as a small C-terminal peptide; these two proteins are proteolytically derived from the GAG primary translation product. The second ORF, which overlaps *GAG* in the +1 frame, is *POL* and encodes protease (PR), integrase (IN), and reverse transcriptase/RNAse H (RT). These three POL proteins are derived by proteolysis of the GAG-POL precursor by the Ty1 PR; there is evidence that cleavage at the GAG-PR boundary obligatorily precedes the other cleavages. The GAG-POL precursor is expressed by an inefficient programmed frameshift that occurs within the sequence CUU AGG C (the indicated codon boundaries represent the *GAG* frame), which is necessary and sufficient to specify the +1 frameshift. A minor shorter transcript of ~2.2 kb is reported to be 5'-coterminal with the major transcript, but has not been fully characterized. A second minor transcript is reported to be 3'-coterminal with the major transcript and is most clearly observed in yeast strains with *spt3* mutations. These *spt3* mutations eliminate or greatly reduce the abundance of the full-length transcript.

The initial step in replication is transcription of the Ty1 virus to generate the full-length RNA described above. This RNA is encapsidated in the cytoplasm in a precursor particle consisting of unprocessed GAG and GAG-POL proteins. Action of the Ty1 PR then converts this into a mature virus particle. It is thought that this somehow activates the reverse transcription process.

The first step in the reverse transcription process is the extension of the $tRNA_i^{Met}$ primer, which binds to the (-) strand primer binding site ((-)PBS) in the full-length RNA. The product of this extension is referred to as (-) strand strong stop DNA or (-) ssDNA by analogy with retroviruses. The (-) ssDNA is transferred to the 3'-end of the full-length RNA, where it can be further extended to generate a nearly full-length (-) strand DNA. Priming of the plus strand initiates at the (+) PPT1 (for polypurine tract 1) adjacent to the 3' LTR, but the mechanism of this priming and the exact nature of the primer have not yet been determined. Extension yields a product, (+)ssDNA that corresponds to the similarly named retroviral intermediate. Transfer of (+)ssDNA to the left end of the (-) strand

DNA sets up a primer-template that can be extended, in principle, to generate full-length duplex DNA. However, studies indicate that the (+) strand is not continuous because a second priming event occurring near the middle of the molecule at a site called (+) PPT2 results in a discontinuity in the middle of the plus strand. We will refer to this final product of reverse transcription as the dsDNA form, although some experiments suggest that in many of these molecules there may be stretches of RNA rather than DNA in the (+) strand. The dsDNA is imported into the cell nucleus, possibly by using a nuclear localization signal found at the C-terminus of IN.

The full length dsDNA is a substrate for the Ty1 IN, which inserts the dsDNA into a chromosomal target site, in the process generating a 5 bp target site duplication of the host target site DNA. Sequences located upstream of RNA polymerase III-transcribed genes represent strongly preferred targets for such Ty1 integration in vivo.

BIOLOGICAL PROPERTIES

Retrotransposon Ty1 is probably best thought of as a genome parasite of *Saccharomyces cerevisiae*. In typical wild strains of this yeast, 3-15 copies of Ty1 are found per haploid genome, whereas in typical laboratory isolates, there are 25-40 copies. In addition to this, the genome contains several hundred solo-LTRs (not associated with a central coding region) or fragments thereof. Ty1 appears to be restricted to this host species and to very closely related species of *Saccharomyces*, but is absent from more distant species of *Saccharomyces*. However, those species are likely to harbor related viruses. Transmission is likely to be exclusively vertical, and horizontally through conjugation. Ty1 and/or the closely related species, Ty2, has been found in virtually all isolates of *S. cerevisiae*, from all over the world. No cytopathic effects have been reported. Some strains are reported to contain very large numbers of Ty1 virus particles and be otherwise normal in every way. However, overexpression of Ty1 proteins leads to slow growth, but this phenotype is poorly characterized.

LIST OF SPECIES IN THE GENUS

Species names are in green italic script; strain names and synonyms are in black roman script; tentative species names are in blue roman script. Sequence accession numbers [ ], and assigned abbreviations ( ) are also listed.

SPECIES IN THE GENUS

Arabidopsis thaliana Art1 virus
Arabidopsis thaliana Art1 virus [Y08010] (AthArt1V)
Arabidopsis thaliana AtRE1 virus
Arabidopsis thaliana AtRE1 virus [AB021265] (AthAtRV)
Arabidopsis thaliana Evelknievel virus
Arabidopsis thaliana Evelknievel virus [AF039373] (AthEveV)
Arabidopsis thaliana Ta1 virus
Arabidopsis thaliana Ta1 virus [X13291] (AthTa1V)
Brassica oleracea Melmoth virus
Brassica oleracea Melmoth virus [Y12321] (BolMelV)
Cajanus cajan Panzee virus
Cajanus cajan Panzee virus [AJ000893] (CcaPanV)
Glycine max Tgmr virus
Glycine max Tgmr virus [U96748] (GmaTgmV)
Hordeum vulgare BARE-1 virus
Hordeum vulgare BARE-1 virus [Z17327] (HvuBV)
Nicotiana tabacum Tnt1 virus
Nicotiana tabacum Tnt1 virus [X13777] (NtaTnt1V)
Nicotiana tabacum Tto1 virus
Nicotiana tabacum Tto1 virus [D83003] (NtaTto1V)
Oryza australiensis RIRE1 virus

Oryza australiensis RIRE1 virus	[D85597]	(OauRirV)
Oryza longistaminata Retrofit virus		
Oryza longistaminata Retrofit virus	[U72726]	(OloRetV)
Physarum polycephalum Tp1 virus		
Physarum polycephalum Tp1 virus	[X53558]	(PpoTp1V)
Saccharomyces cerevisiae Ty1 virus		
Saccharomyces cerevisiae Ty1 virus	[M18706]	(SceTy1V)
Saccharomyces cerevisiae Ty2 virus		
Saccharomyces cerevisiae Ty2 virus	[M19542]	(SceTy2V)
Saccharomyces cerevisiae Ty4 virus		
Saccharomyces cerevisiae Ty4 virus	[X67284]	(SceTy4V)
Solanum tuberosum Tst1 virus		
Solanum tuberosum Tst1 virus	[X52387]	(StuTst1V)
Triticum aestivum WIS-2 virus		
Triticum aestivum WIS-2 virus	[X63184; X57168 (LTR)]	(TaeWis1V)
Zea mays Hopscotch virus		
Zea mays Hopscotch virus	[U12626]	(ZmaHopV)
Zea mays Sto-4 virus		
Zea mays Sto-4 virus	[AF082133]	(ZmaStoV)

TENTATIVE SPECIES IN THE GENUS

None reported.

GENUS *HEMIVIRUS*

Type Species *Drosophila melanogaster copia virus*

DISTINGUISHING FEATURES

For *copia*, Gag is encoded on a spliced 2 kb mRNA, and differential splicing is the mechanism by which Gag and Gag-Pol stoichiometry is regulated. As in the *Pseudovirus* genus, these viruses encode *gag* and *pol* in one or two reading frames. The mechanism(s) that regulates Gag and Gag-Pol expression for the single ORF viruses remain to be determined.

Species in this genus typically use an initiator methionine tRNA as the primer for minus-strand DNA synthesis during reverse transcription. Hemiviruses, unlike members of the families *Retroviridae, Metaviridae* and genus *Pseudovirus*, use an initiator methionine tRNA half-molecule as a primer (or an Arg tRNA half-molecule in the case of Candida albicans Tca2 virus (Tca2)). This tRNA fragment is generated by cleaving the initiator tRNA in the anticodon stem. However, relatively little is known about the detailed mechanism of this reaction. Phylogenetic analyses (Fig. 3) suggest that half-tRNA priming may be a useful character for defining this genus, since the RT sequences of this group form a distinct, albeit modestly supported cluster.

Both Gag and Gag-Pol primary translation products are processed by the cognate protease into final products. The known *gag*-encoded proteins include analogs of retroviral CA and NC. The known *pol*-encoded proteins include the protease (PR), integrase (IN), and reverse transcriptase/RNase H (RT). All of these proteins appear to be required for replication.

These viruses are found worldwide in fungi, algae and insects genomes. Their mode of transmission is unknown although presumed to be through vertical inheritance.

List of Species in the Genus

Species names are in green italic script; strain names and synonyms are in black roman script; tentative species names are in blue roman script. Sequence accession numbers [ ], and assigned abbreviations ( ) are also listed.

Species in the Genus

Aedes aegypti Mosqcopia virus		
Aedes aegypti Mosqcopia virus	[AF134899]	(AaeMosV)
Candida albicans Tca2 virus		
Candida albicans Tca2 virus	[AF050215]	(CalTca2V)
Candida albicans Tca5 virus		
Candida albicans Tca5 virus	[AF065434]	(CalTca5V)
Drosophila melanogaster 1731 virus		
Drosophila melanogaster 1731 virus	[X07656]	(Dme1731V)
Drosophila melanogaster copia virus		
Drosophila melanogaster copia virus	[M11240]	(DmeCopV)
Saccharomyces paradoxus Ty5 virus		
Saccharomyces paradoxus Ty5 virus	[U19263]	(SceTy5V)
Volvox carteri Lueckenbuesser virus		
Volvox carteri Lueckenbuesser virus	[U90320]	(VcaLeuV)
Volvox carteri Osser virus		
Volvox carteri Osser virus	[X69552]	(VcaOssV)

Tentative Species in the Genus

None reported.

Genus *Sirevirus*

Type Species *Glycine max SIRE1 virus*

Distinguishing Features

All members of this genus have thus far been identified only in plant hosts. A reverse transcriptase phylogenetic tree separates the sireviruses from all known pseudoviruses and hemiviruses (Fig. 4). Based on the sequence of putative primer binding sites, species in this genus most likely use the acceptor stem of an initiator methionine tRNA as the primer for minus-strand DNA synthesis during reverse transcription, similar to the pseudoviruses.

A unique ORF and multiple gene expression mechanisms characterize members of this genus. For example, viruses such as Glycine max SIRE1 virus (*SIRE1*), Arabidopsis thaliana Endovir virus (*Endovir*) and Lycopersicon esculentum ToRTL1 virus (ToRTL1) have an extra ORF between *pol* and the 3'-LTR (Fig. 3). This ORF has conserved transmembrane domains and is referred to as an *env*-like ORF. For *SIRE1*, *env*-like expression is regulated by stop codon suppression. However, stop codon suppression does not seem to be utilized by the other viruses (e.g. *Endovir* and ToRTL1), because the *env*-like ORF is separated from *pol* by non-coding sequence. In addition, based on ORF organization, two different Gag-Pol expression mechanisms are likely utilized by members of this genus. Some viruses such as *SIRE1* and *Endovir* encode *gag* and *pol* in a single ORF. In contrast, for viruses such as Opie-2 and Prem-2, *pol* resides in a +1 frame (Fig. 3). A unifying characteristic of this genus is that in all members, the *gag* gene is nearly twice the size of *gag* encoded by other pseudoviruses.

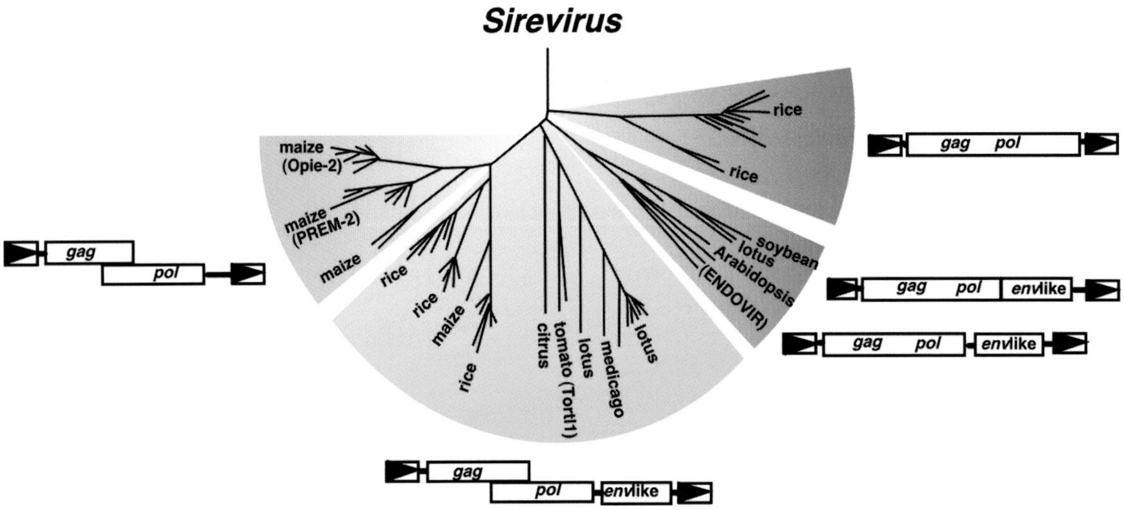

Figure 3: Genome organization for members of the genus *Serivirus*. Variability can be observed by the presence or absence of an *env*-like ORF, the organization of the *env*-like ORF relative to *pol*, and the organization of *gag* and *pol*.

List of Species in the Genus

Species names are in green italic script; strain names and synonyms are in black roman script; tentative species names are in blue roman script. Sequence accession numbers [ ], and assigned abbreviations ( ) are also listed.

Species in the Genus

Arabidopsis thaliana Endovir virus		
Arabidopsis thaliana Endovir virus	[AY016208]	(AthEndV)
Glycine max SIRE1 virus		
Glycine max SIRE1 virus	[AY205608]	(GmaSIRV)
Lycopersicon esculentum ToRTL1 virus		
Lycopersicon esculentum ToRTL1 virus	[U68072]	(LesToRV)
Zea mays Opie-2 virus		
Zea mays Opie-2 virus	[U68408]	(ZmaOp2V)
Zea mays Prem-2 virus		
Zea mays Prem-2 virus	[U41000]	(ZmaPr2V)

Tentative Species in the Genus

None reported.

List of Unassigned Viruses in the Family

Phaseolus vulgaris Tpv2-6 virus		
Phaseolus vulgaris Tpv2-6 virus	[AJ005762]	(PvuTpvV)

PHYLOGENETIC RELATIONSHIPS WITHIN THE FAMILY

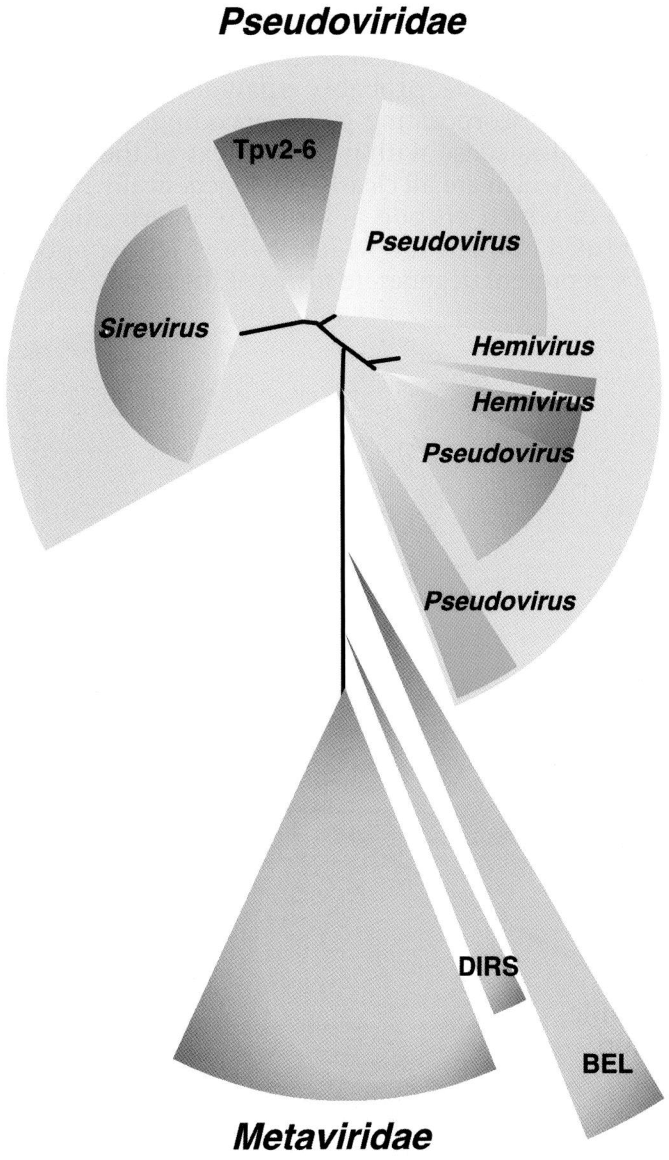

Figure 4: Phylogenetic relationships among members of the family *Pseudoviridae*, and the family *Metaviridae* as well as the BEL and DIRS groups. Based on reverse transcriptase aa sequences, the seriviruses form a distinct group from other *Pseudoviridae* (Note: Tpv2-6 and related viruses also form a distinct group, but they have not been assigned to a genus). The relationships of these viruses were determined using the neighbor-joining distance algorithm.

SIMILARITY WITH OTHER TAXA

Like the families *Hepadnaviridae* and *Metaviridae*, the *Pseudoviridae* are clearly related to the *Retroviridae* family. All four families are linked by reverse transcription and a viral core structure made up of Gag-like proteins. Pseudoviruses are different from the other two in that they have an unusual organization (PR-IN-RT-RH) of the gene *pol*. Members of the families *Pseudoviridae, Metaviridae* and *Retroviridae* also share the following: a proviral form characterized by LTRs: protease, reverse transcriptase, RNase H and integrase activities essential for multiplication; readthrough-mediated *gag-pol* gene expression (in some species); and tRNA primers (in most species).

RT VIRUSES

An important and controversial question is the extent of the relationship between the families *Pseudoviridae, Metaviridae* and the *Retroviridae*. Because the genomic structures of viruses in *Pseudoviridae-Metaviridae* are clearly related to, but typically simpler than the viruses in the family *Retroviridae*, many authors who have considered the problem have concluded that the families *Pseudoviridae* and *Metaviridae* represent more primitive groups; the family *Metaviridae* probably spawned the members of the family *Retroviridae* (presumably by incorporating genes encoding ligands for cell-surface receptors). This conclusion makes sense within the context of the enormous diversity of other types of retroelements, which are all clearly phylogenetically related by the presence of RT (Fig. 4), but not all of which encode a virus-like intermediate. An alternative viewpoint that cannot be ruled out, but for which there is less support, is that members of the family *Metaviridae* represent degenerate forms of the family *Retroviridae*.

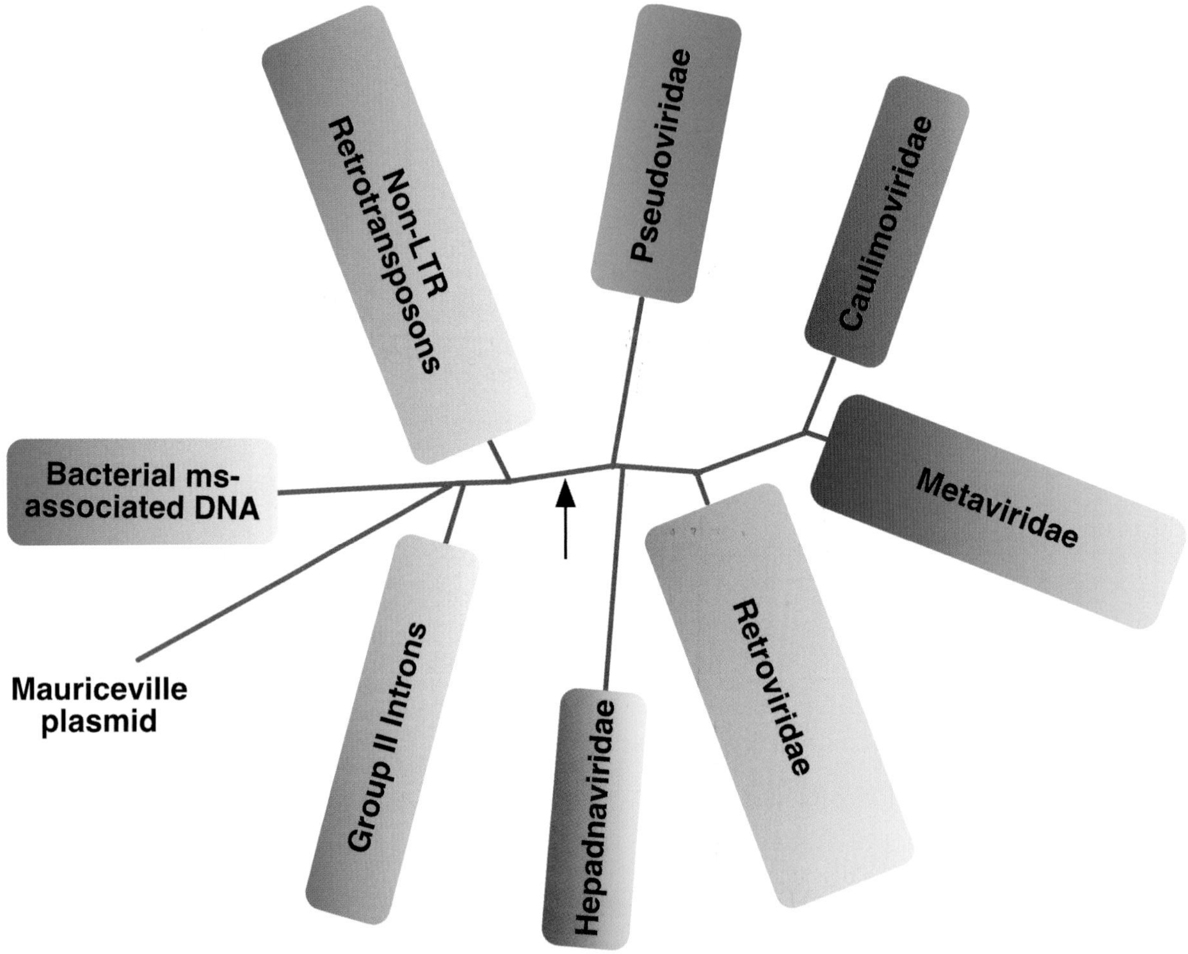

Figure 5: Unrooted phylogenetic tree of all classes of reverse transcriptase containing viruses. While over 100 reverse transcriptase sequence were used to generate this phylogeny, to simplify visual comparison of the major topologies of the tree, viruses from the same class that are located on the same branch of the tree are indicated by a box. The length of the boxes correspond to the most divergent viruses within that box. The arrow indicates a possible root of the tree using RNA polymerase sequences.

Derivation of Names

Hemi: from Greek *hemi*, "half", referring to the half-molecule of tRNA used as a primer for reverse transcription.
Pseudo: from Greek *pseudo*, "false": To connote some uncertainty as to whether these are true viruses.
Sire: from the abbreviation of the species name : *Glycine max SIRE1 virus* (SIRE).

REFERENCES

Boeke, J.D. and Sandmeyer, S.B. (1991). Yeast transposable elements. In: *The molecular and cellular biology of the yeast Saccharomyces*, vol 1, (J. Broach, E. Jones and J. Pringle, eds), pp 193-261. Cold Spring Harbor Laboratory, Cold Spring Harbor, New York.

Boeke, J.D. and Stoye, J.P. (1996). Retrotransposons, endogenous retroviruses, and the evolution of retroelements. In: *Retroviruses* (H. Varmus, S. Hughes and J. Coffin, eds), pp 343-436. Cold Spring Harbor Laboratory, Cold Spring Harbor, New York.

Burns, N.R., Saibil, H.R., White, N.S., Pardon, J.F., Timmins, P.A., Richardson, S.M.H., Richards, B.M., Adams, S.E., Kingsman, S.M. and Kingsman, A.J. (1992). Symmetry, flexibility and permeability in the structure of yeast retrotransposon virus-like particles. *EMBO J.*, **11**, 1155-1164.

Doolittle, R.F., Feng, D.F., Johnson, M.S. and McClure, M.A. (1989). Origins and evolutionary relationships of retroviruses. *Quart. Rev. Biol.*, **64**, 1-30.

Eichinger, D.J. and Boeke, J.D. (1988). The DNA intermediate in yeast Ty1 element transposition copurifies with virus-like particles: cell-free Ty1 transposition. *Cell*, **54**, 955-966.

Gao, X., Havecker, E.R., Baranov, P.V., Atkins, J.F. and Voytas, D.F. (2003). Translational recoding signals between *gag* and *pol* in diverse LTR retrotransposons. *RNA*, **9**, 1430-1422.

Garfinkel, D.J., Boeke, J.D. and Fink, G.R. (1985). Ty element transposition: reverse transcriptase and virus-like particles. *Cell*, **42**, 507-517.

Havecker, E.R. and Voytas, D.F. (2003). The soybean retroelement SIRE1 uses stop codon suppression to express its envelope-like protein. *EMBO Rep.*, **4**, 274-277.

Kikuchi, Y., Ando, Y. and Shiba, T. (1986). Unusual priming mechanism of RNA-directed DNA synthesis in *copia* retrovirus-like particles of *Drosophila*. *Nature*, **323**, 824-826.

Kim, J.M., Vanguri, S., Boeke, J.D., Gabriel, A. and Voytas, D.F. (1998). Transposable elements and genome organization: a comprehensive survey of retrotransposons revealed by the complete *Saccharomyces cerevisiae* genome sequence. *Genome Res.*, **8**, 464-478.

Kumar, A. and Bennetzen, J.L. (2000). Plant retrotransposons. *Annu. Rev. Genet.*, **33**, 479-532.

Laten, H.M., Majumdar, A. and Gaucher, E.A. (1998). SIRE-1, a copia/Ty1-like retroelement from soybean, encodes a retroviral envelope-like protein. *Proc. Natl. Acad. Sci. USA*, **95**, 6897-902.

Mellor, J., Malim, M.H., Gull, K., Tuite, M.F., McCready, S., Dibbayawan, T., Kingsman, S.M. and Kingsman, A.J. (1985). Reverse transcriptase activity and Ty RNA are associated with virus-like particles in yeast. *Nature*, **318**, 583-586.

Palmer, K.Y., Tichelaar, W., Myers, N., Burns, N.R., Butcher, S.J., Kingsman, A.J., Fuller, S.D. and Saibil, H.R. (1997). Cryo-EM structure of yeast Ty retrotransposon virus-like particles. *J. Virol.*, **71**, 6863-6868.

Peterson-Burch, B.D. and Voytas, D.F. (2002). Genes of the Pseudoviridae (Ty1/copia Retrotransposons). *Mol. Biol. Evol.*, **19**, 1832-1845.

Voytas, D.F. and Boeke, J.D. (2002). Ty1 and Ty5. In: *Mobile DNA II* (N. Craig, ed), pp 631-662. American Society for Microbiology, Washington, D.C.

Voytas, D.F., Cummings, M.P., Konieczny, A., Ausubel, F.M. and Rodermel, S.R. (1992). *Copia*-like retrotransposons are ubiquitous among plants. *Proc. Natl. Acad. Sci. USA*, **89**, 7124-7128.

Xiong, Y. and Eickbush, T.H. (1990). Origin and evolution of retroelements based on their reverse transcriptase sequences. *EMBO J.*, **9**, 3353-3362.

Zou, S., Ke, N., Kim, J.M. and Voytas, D.F. (1996). The *Saccharomyces* retrotransposon Ty5 integrates preferentially into regions of silent chromatin at the telomeres and mating loci. *Genes Dev.*, **10**, 634-645.

CONTRIBUTED BY

Boeke, J.D., Eickbush, T., Sandmeyer, S.B. and Voytas, D.F.

RT VIRUSES

FAMILY *METAVIRIDAE*

TAXONOMIC STRUCTURE OF THE FAMILY

Family	***Metaviridae***
Genus	***Metavirus***
Genus	***Errantivirus***
Genus	***Semotivirus***

VIRION PROPERTIES

MORPHOLOGY

Metaviridae is a family of retrotransposons that have been found in all studied lineages of eukaryotes. These viruses, primarily identified by their ability to induce mutations or by genome sequencing, replicate via virus-like intermediates referred to as a virus-like particles (VLPs). Members of the *Metaviridae* family are often referred to as LTR-retrotransposons of the Ty3-*gypsy* family. While there is good evidence that these particles are essential and direct intermediates in the life cycle of these viruses, only in one case, Drosophila melanogaster Gypsy virus (DmeGypV), do VLPs display infectivity according to the traditional virological definition. Viruses that generate VLPs or virions will be referred to collectively in this chapter as retrotransposons.

Morphology of particles is relatively poorly characterized and capsomeric symmetry is unknown. Members include species that produce primarily or exclusively intracellular particles (eg. Saccharomyces cerevisiae Ty3 virus (SceTy3V)) so that collections of particles are heterogeneous with respect to stage of maturation. These intracellular particles will be referred to as virus-like particles (VLPs). Extracellular particles are enveloped with ovoid cores and will be referred to as virions (e.g. DmeGypV).

PHYSICOCHEMICAL AND PHYSICAL PROPERTIES

In most systems virions are only crudely characterized biochemically.

NUCLEIC ACID

The genomes of retrotransposons in this family are positive strand RNAs. The genomic RNA is polyadenylated at the 3′-end; a cap structure has not yet been described, but may be presumed, given similarities of member elements with retroviruses. In addition to the RNA genome, some cellular RNAs may be randomly associated with particles including specific tRNAs in the case of virus replication which is primed by tRNAs. Particle fractions from cells are heterogeneous with respect to maturation and so are associated with intermediates and products of reverse transcription in addition to genomic RNA.

PROTEINS

Proteins present in characterized VLPs include a major structural protein or capsid (CA), an aspartate protease (PR), reverse transcriptase containing an RNase H domain (RT-RH), and integrase (IN). For most viruses, these proteins are not yet characterized, but are predicted based on similarity of the protein sequence with those of proteins encoded by the internal domain. In most cases, the CA is not markedly similar to retroviral CA. VLPs of some retrotransposons in this family contain proteins with the metal finger characteristic of nucleocapsid (NC) of retroviruses. All viruses of the genus *Errantivirus* are distinguished by the presence of processed envelope proteins that apparently correspond to retroviral transmembrane (TM) and surface (SU) proteins. Only a limited number of viruses of the genera *Metavirus* and *Semotivirus* contain a putative *envelope* gene (*env*).

LIPIDS

In the case of members that generate virions, the virion membrane appears to be derived from the membrane of the host cell.

CARBOHYDRATES

Carbohydrates have not been characterized, although their presence is inferred from sensitivity of the DmeGypV envelope precursor protein to digestion with endoglycosidase F.

GENOME ORGANIZATION AND REPLICATION

The integrated form of these retrotransposons is composed of long terminal repeats (LTRs) flanking a central unique domain (Fig. 1). The length of the viral genomes ranges from 4 kbp to more than 10 kbp. The LTRs are from 77 nt in the case of the Bombyx mori mag virus (BmoMagV) to greater than 2 kbp in length, in the cases of Drosophila virilis Ulysses virus (DviUlyV) and the Tribolium castaneum Woot virus (TcaWooV). Chromosomal copies of the viruses are flanked by short direct repeats of sequences derived from the insertion site. The length of the repeat is characteristic of the virus and ranges from 4 to 6 bp. The internal domain contains one to three ORFs. The 3′-end of the final ORF can extend into the downstream LTR. In all cases, the order of domains encoded in the ORFs is inferred to be: 5′-CA-(NC where present)-PR-RT-RH-IN-3′. Where characterized, envelope proteins are encoded downstream of the IN domain by spliced mRNAs. These ORFs are referred to differently for different viruses and in this discussion, will be generally referred to as *gag*, *pol* and *env*. Thus viruses may have one *gag-pol* ORF, two (*gag* and *pol*) or three (*gag*, *pol* and *env*) ORFs.

Transcription of the genomic RNA is initiated in the upstream LTR and terminates at a position downstream of that site in the downstream LTR. This divides the long terminal repeats into regions represented uniquely in the 5′-end of the genomic RNA (U5), uniquely in the 3′-end of the genomic RNA (U3) or repeated at the 5′- and 3′-ends (R). Thus, the LTRs are comprised of U3-R-U5 regions analogous to those found in integrated retroviruses. By analogy with retroviruses, these species may carry two copies of the RNA genome per virion or particle; however, this has not yet been demonstrated, and dimerization functions have not yet been characterized.

Genomic RNA is translated into proteins required for particle formation, polyprotein maturation, reverse transcription and integration. Intracellular particle preparations show that particle fractions are comprised predominantly of species derived from the upstream portion of the ORF or where two or three ORFs are present from the first ORF. Where two ORFs occur, they usually overlap, and the second ORF is translated as a fusion protein of the first and second ORF translation products. The mechanism of frameshifting is not uniform among the member viruses. In case of the Schizosaccaromyces pombe Tf1 virus (SpoTf1V), the most completely characterized virus of this family containing one ORF, it appears that a polyprotein is produced and that later proteolytic events are responsible for a high ratio of major structural proteins to catalytic proteins. Little is known about where in the cell particle assembly occurs. PR is required for maturation of viral proteins. Catalytic proteins are PR, RT-RH, and IN. Shortly after production of protein precursors, processed species are observed. Based on similarity of these metaviruses to retroviruses, it is likely that processing follows, and is dependent upon, intracellular assembly. Particle fractions are associated with genomic RNA and extrachromosomal DNA. RT activity associated with the particle fraction can be measured by exogenous assays.

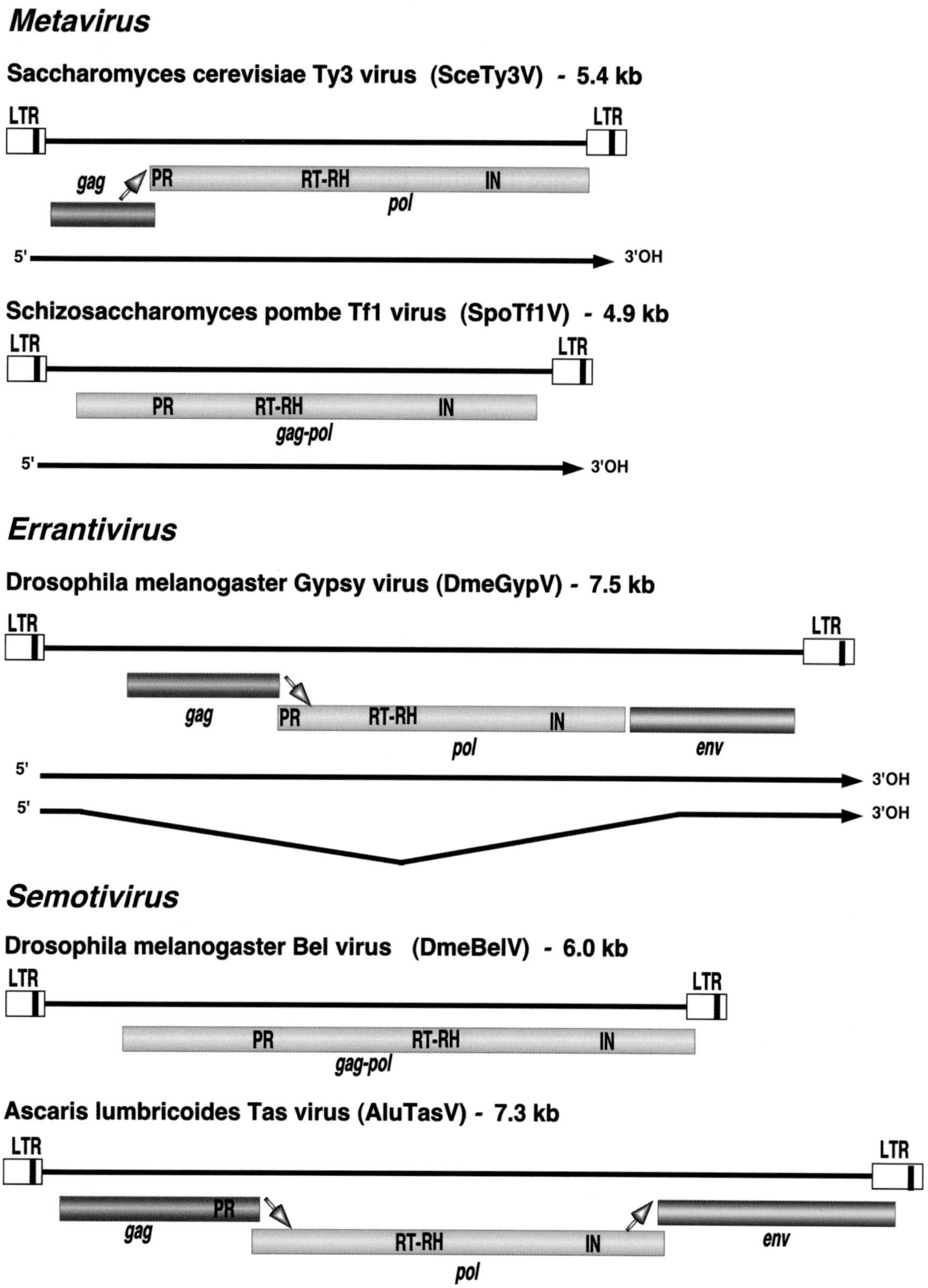

Figure 1: Genome organization of representative members of the *Metaviridae* family. The integrated genome of each virus contains Long Terminal Repeats (LTRs) flanking a central sequence. Black boxes within the LTRs depict sequences repeated at the 5′- and 3′-ends of the virus transcripts (R regions). Open boxes below the viruses indicate *gag*, *pol* and *env* ORFs. Not all viruses within the genera *Metavirus* and *Semotivirus* encode an Env-like protein. Arrows indicate sites of ribosomal frameshifting. Conserved aa sequences in *pol* that identify protease (PR), integrase (IN) and reverse transcriptase/RNAse H (RT-RH) are labeled. Individual mRNAs are depicted below the ORF diagrams as arrows. Transcription of the DmeBelV and AluTasV has not been studied.

Reverse transcription of genomic RNA of known members of this family is primed from either the 5′-end of the genomic RNA or from the 3′-end of a tRNA. In each case, the complementarity is overlapping, adjacent to, or just downstream of the U5 region of the genomic transcript. In cases in which the reverse transcription intermediates have been characterized (DmeGypV, Saccharomyces cerevisiae Ty3 virus (SceTy3V), and Schizosaccharomyces pombe Tf1 virus (SpoTf1V), data are consistent with a species representing a minus-strand copy templated from the site of priming up to the 5′-end of the genomic RNA. This is a minor species. By analogy with retroviruses, this intermediate is probably transferred to the 3′-end of the genomic RNA, where an overlap of the R region minus strand represented in the cDNA, and the R region plus strand, represented at the 3′-end of the genomic RNA, allow transfer of the minus-strand strong stop which then acts to prime copying of the template plus-strand genomic RNA. Plus-strand priming probably occurs, as in retroviruses, from a polypurine tract or related sequence overlapping, adjacent to, or just upstream of the U3 region in the genomic RNA. This is consistent with priming from a site of cleavage by RH. Plus-strand, strong-stop species have been identified for some representatives (DmeGypV and SceTy3V) which are consistent with this position of priming and copying through to the first modified base in the primer tRNA. This family is heterogeneous with respect to the presence of extra terminal nt in the extrachromosomal replicated DNA and with respect to the presence of TG-CA inverted repeats at the ends of the integrated sequence.

Biological Properties

Activation of transposition of these viruses can cause disruption of host physiology depending on the site of insertion. Several members exhibit preferential patterns of insertion. It is notable that germline activation, which is a feature of some retroviruses, also occurs for some of these viruses. For example, SceTy3V transcription is induced by mating pheromone and transposition occurs after mating. In the case of the DmeGypV and the Drosophila melanogaster Zam virus (DmeZamV), transposition occurs in germline cells.

Genus *Metavirus*

Type Species ***Saccharomyces cerevisiae Ty3 virus***

Distinguishing Features

Saccharomyces cerevisiae Ty3 virus (SceTy3V) forms generally spherical, but irregular, intracellular particles of about 50 nm in diameter. These are observed as clusters or as individual particles in the cytoplasm of cells expressing high levels of SceTy3V RNA. Particles sediment as a heterodisperse population around 156S. The major particle-associated RNA species is a 5.2 kb, polyadenylated RNA. The primer of minus-strand reverse transcription is $tRNA^{iMet}$ which is complementary to a primer binding site (pbs). A minor 3.1 kb species is also observed, but the extent to which this is associated with particles has not been characterized. The 5.4 kb RNA contains two ORFs analogous to *gag* and *pol*, *GAG3* and *POL3*, which overlap in the +1 frame.

The genomic RNA is translated into Gag3 and Gag3-Pol3 polyproteins which are processed by SceTy3V PR. Mature proteins include CA (26 kDa), NC (15 kDa), PR (15 kDa), RT-RH and IN (58 kDa and 61 kDa, respectively), and an RT-RH-IN fusion protein of approximately 115 kDa. The molecular composition of the RT is not yet known. A protein of 10 kDa is predicted to be encoded between PR and RT but has not yet been identified. The integrated form of SceTy3V is 5.4 kbp in length and consists of an internal domain flanked by two LTRs (sigma elements) 340 bp in length. Insertions of SceTy3V are flanked by 5-bp direct repeats derived from insertion site cleavage and repair. SceTy3V is transcribed into a 5.2 kb genomic RNA. The $tRNA^{iMet}$ pbs has its 5′-end two nt downstream of the junction of the 5′-LTR with the internal domain. The full-length SceTy3V DNA molecule is two bp longer at each end than the integrated molecule, consistent with predictions based on the positions of the priming sequences. Two nt are removed from each 3′-end prior to integration. SceTy3V integrates within one or two nt of the transcription initiation site of genes transcribed by RNA polymerase III. SceTy3V is found in one to five copies in typical laboratory strains of *Saccharomyces cerevisiae*. In addition, there are approximately thirty to forty copies of the isolated LTRs present. The latter presumably arose by recombination between the LTRs of complete viruses. SceTy3V transcription is induced by pheromone signal transduction. Although proteins are produced in cells undergoing signaling, DNA is not made in cells arrested in G1 of the cell cycle. Consequently, it is most likely that in natural populations, transposition only occurs after the fusion of mating cells to form diploids.

LIST OF SPECIES DEMARCATION CRITERIA IN THE GENUS

Although the members of this family encode Pol proteins which are similar to those of SceTy3V, other properties of the members are distinct from SceTy3V. Several of the elements (Drosophila melanogaster micropia virus, DmeMicV; Lilium henryi del1 virus (LheDel1V); and Schizosaccharomyces pombe Tf1 virus (SpoTf1V), Schizosaccharomyces pombe Tf2 virus (SpoTf2V) and Cladosporium fulvum T1 virus (CfuT1V)) encode a single long ORF from which major structural proteins, as well as Pol proteins, are expressed. In the most completely characterized case, SpoTf1V, the ORF has been shown to be expressed as a single polyprotein which is processed by a mechanism dependent on the SpoTf1V PR. Cells in stationary phase have the highest ratio of major structural to catalytic protein and products of reverse transcription accumulate concomitant with this transition. Members of this genus are also distinguished from SceTy3V by aspects of replication priming. SpoTf1V has been shown to form an RNA structure involving 89 bases at the 5′-end of the RNA which is processed by RNase H to cleave within the structure between nt 11 and 12 from the 5′-end. This cleavage allows priming of SpoTf1V RT from the 3′-end of the 11 nt fragment annealed immediately downstream of the 5′-LTR. Other members of the genus (SpoTf2V and CfuT1V) have sequences consistent with a similar mechanism of self-priming. Thus these viruses are distinguished by self-priming, i.e. by an apparent lack of requirement for 3′-end processing. Individual species in the genus all have less than 50% identity in their Gag protein sequences compared to all other species. For example, although Drosophila melanogaster Mdg virus (DmeMdg1V) and Drosophila melanogaster 412 virus (Dme412V), each infect *Drosophila melanogaster*, their *gag* sequences are only 39% identical. Two viruses, Drosophila buzzatti Osvaldo virus (DbuOsvaV) and Arabidopsis thaliana Athila virus (AthAthV) have an *env*-like gene. Based on this property it would be possible to place these viruses within the *Errantivirus* genus. However, the envelope-like proteins of DbuOsvV and AthAthV are unrelated in sequence to that of the errantiviruses, and based on the sequence of their RT domain, can not be placed within the *Errantivirus* genus.

LIST OF SPECIES IN THE GENUS

Species names are in green italic script; strain names and synonyms are in black roman script; tentative species names are in blue roman script. Sequence accession numbers [ ], and assigned abbreviations ( ) are also listed.

SPECIES IN THE GENUS

Arabidopsis thaliana Athila virus		
Arabidopsis thaliana Athila virus	[AC007209]	(AthAthV)
Arabidopsis thaliana Tat4 virus		
Arabidopsis thaliana Tat4	[AB005247]	(AthTat4V)
Bombyx mori Mag virus		
Bombyx mori Mag virus	[X17219]	(BmoMagV)

RT VIRUSES

Caenorhabditis elegans Cer1 virus		
Caenorhabditis elegans Cer1 virus	[U15406]	(CelCer1V)
Cladosporium fulvum T-1 virus		
Cladosporium fulvum T-1 virus	[Z11866]	(CfuT1V)
Dictyostelium discoideum Skipper virus		
Dictyostelium discoideum Skipper virus	[AF049230]	(DdiSkiV)
Drosophila buzzatii Osvaldo virus		
Drosophila buzzatii Osvaldo virus	[AJ133521]	(DbuOsvV)
Drosophila melanogaster Blastopia virus		
Drosophila melanogaster Blastopia virus	[Z27119]	(DmeBlaV)
Drosophila melanogaster Mdg1 virus		
Drosophila melanogaster Mdg1 virus	[X59545]	(DmeMdg1V)
Drosophila melanogaster Mdg3 virus		
Drosophila melanogaster Mdg3 virus	[X95908]	(DmeMdg3V)
Drosophila melanogaster Micropia virus		
Drosophila melanogaster Micropia virus	[X14037]	(DmeMicV)
Drosophila melanogaster 412 virus		
Drosophila melanogaster 412 virus	[X04132]	(Dme412V)
Drosophila virilis Ulysses virus		
Drosophila virilis Ulysses virus	[X56645]	(DviUlyV)
Fusarium oxysporum Skippy virus		
Fusarium oxysporum Skippy virus	[L34658]	(FoxSkiV)
Lilium henryi Del1 virus		
Lilium henryi Del1 virus	[X13886]	(LheDel1V)
Saccharomyces cerevisiae Ty3 virus		
Saccharomyces cerevisiae Ty3 virus	[M34549]	(SceTy3V)
Schizosaccharomyces pombe Tf1 virus		
Schizosaccharomyces pombe Tf1 virus	[M38526]	(SpoTf1V)
Schizosaccharomyces pombe Tf2 virus		
Schizosaccharomyces pombe Tf2 virus	[L10324]	(SpoTf2V)
Takifugu rubripes Sushi virus		
Takifugu rubripes Sushi virus	[AF030881]	(TruSusV)
Tribolium castaneum Woot virus		
Tribolium castaneum Woot virus	[U09586]	(TcaWooV)
Tripneustis gratilla SURL virus		
Tripneustis gratilla SURL virus	[M75723]	(TgrSurV)

TENTATIVE SPECIES IN THE GENUS

None reported.

GENUS ERRANTIVIRUS

Type Species ***Drosophila melanogaster Gypsy virus***

DISTINGUISHING FEATURES

All elements contain a similar *env*-like ORF.

VIRION PROPERTIES

Expression of DmeGypV results in production of enveloped irregular particles of approximately 100 nm in diameter and also much smaller non-enveloped particles. The *env* gene predicts a protein of 54 kDa. The actual protein has an apparent size of 66 kDa and is N-glycosylated as indicated by susceptibility to endoglycosidase F. More rapidly migrating molecules of 54 kDa and 28 kDa are also observed and are inferred to result from proteolytic processing by host enzymes, as is the case of members of the family *Retroviridae*.

GENOME ORGANIZATION AND REPLICATION

DmeGypV is 7469 bp in length including two LTRs of 482 bp (Fig. 1). It differs from most retroviruses and retrotransposons in that the termini are composed of AG...TT rather than TG...CA. The 11 nt immediately adjacent to the upstream U3 element and overlapping by one nt is complementary to $tRNA^{Lys}$. The DNA flanking the insertion includes 4 bp repeats, and the insertion site preference is for YRYRYR (where Y=purine and R=pyrimidine) sequence. The genomic RNA contains one ORF encoding major structural protein and a second ORF overlapping in the -1 frame, encoding homologues of retroviral PR, RT-RH and IN. A third ORF, *env*, encoding a 54 kDa envelope protein, occurs in a spliced 2.1 kb mRNA. This protein is apparently N-glycosylated, processed into smaller species, and is analogous to retroviral envelope proteins by virtue of hydrophobic putative membrane spanning domains, localization to the viral membrane, similarity of processing sites for cleavage into trans-membrane and surface domains, and glycosylation. An envelope protein of similar sequence to that in DmeGypV has been identified for all members of this genus. The Env protein of DmeZamV is translated from a 1.7 kb spliced message. The envelope proteins of errantiviruses have been shown to have sequence similarity to the viral *env* gene of certain baculoviruses.

DmeGypV is transcribed into a 6.5 kb genomic RNA. A minus-strand strong stop species of approximately 242 nt (with RNA removed) has been identified. Plus-strand strong-stop DNA species of 479 nt, which are similar in length to the LTR, and a species longer by 15 to 18 nt, presumed to result from copying of the tRNA primer, have been observed.

BIOLOGICAL PROPERTIES

DmeGypV transposition is repressed by the activity of the *flamenco* gene. In females homozygous for the permissive allele of *flam*, the somatic follicle cells surrounding maternal germline cells appear to accumulate DmeGypV RNA and envelope protein. Transposition, however, is observed in the maternal germ cells, and this has led to the hypothesis that transposition is attributable to infection from surrounding follicle cells. A similar path of activity has also been suggested for DmeZamV and DmeIdeV. Infection has been demonstrated to result when susceptible strains are raised in the presence of DmeGypV particles mixed into their food. Incubation with antibodies against the Env protein decreased the level of infection, implicating Env in this process.

LIST OF SPECIES DEMARCATION CRITERIA IN THE GENUS

At least one member of this group, DmeGypV, is infectious, and all members of the genus *Errantivirus* have a similar third ORF resembling *env*. While this property makes the errantiviruses candidates for inclusion into the family *Retroviridae* rather than the family *Metaviridae*, there is no sequence similarity between the *env* genes of these two groups, suggesting independent acquisition events. In addition, examination of phylogenetic relationship of these viruses based on their reverse transcriptase sequences (Fig. 2) places the errantiviruses as an independent lineage distinct from the family *Retroviridae*. Individual species in the genus all have less than 50% identity in their Gag protein sequences compared to all other species. For example DmeZamV and DmeGypV are two species of viruses in this family from *Drosophila melanogaster*, but their *gag* sequences are only 35% identical.

LIST OF SPECIES IN THE GENUS

Species names are in green italic script; strain names and synonyms are in black roman script; tentative species names are in blue roman script. Sequence accession numbers [ ], and assigned abbreviations ( ) are also listed.

RT VIRUSES

SPECIES IN THE GENUS

Ceratitis capitata Yoyo virus		
Ceratitis capitata Yoyo virus	[U60529]	(CcaYoyV)
Drosophila ananassae Tom virus		
Drosophila ananassae Tom virus	[Z24451]	(DanTomV
Drosophila melanogaster Zam virus		
Drosophila melanogaster virus	[AJ000387]	(DmeZamV)
Drosophila melanogaster Gypsy virus		
Drosophila melanogaster Gypsy virus	[M12927]	(DmeGypV)
Drosophila melanogaster Idefix virus		
Drosophila melanogaster Idefix virus	[AJ009736]	(DmeIdeV)
Drosophila melanogaster Tirant virus		
Drosophila melanogaster Tirant virus	[X93507]	(DmeTirV)
Drosophila melanogaster 17.6 virus		
Drosophila melanogaster 17.6 virus	[X01472]	(Dme176V)
Drosophila melanogaster 297 virus		
Drosophila melanogaster 297 virus	[X03431]	(Dme297V)
Drosophila virilis Tv1 virus		
Drosophila virilis Tv1 virus	[AF056940]	(DviTv1V)
Trichoplusia ni TED virus		
Trichoplusia ni TED virus	[M32662]	(TniTedV)

TENTATIVE SPECIES IN THE GENUS

None reported.

GENUS *SEMOTIVIRUS*

Type Species ***Ascaris lumbricoides Tas virus***

DISTINGUISHING FEATURES

The integrated form of Ascaris lumbricoides Tas virus (AluTasV) is 7.6 kbp in length and consists of an internal domain flanked by two LTRs 256 bp in length. Insertions of AluTasV are flanked by 5 bp direct repeats derived from the insertion sites. There are approximately 50 copies of AluTasV distributed in the genome of *A. lumbricoides*. RNA transcripts and VLPs have not been observed but can be inferred based on the similarity in structure and coding capacity to that of other members of the family *Metaviridae* or the family *Retroviridae*. Reverse transcription is primed by $tRNA^{arg}$ which anneals to the pbs 6 bp downstream of the 5′-LTR. AluTasV encodes three overlapping ORFs: the first encodes the major structural protein and PR, the second ORF overlapping in the -1 frame encoding RT-RH and IN, and the third overlapping in the +1 frame encoding the *env*. The Env-like protein encoded by AluTasV contains a transmembrane domain but exhibits no sequence similarity with the *env* gene of *Errantivirus* or *Retroviridae* suggesting its independent acquisition. The likely origin of the AluTasV third ORF is the glycoprotein gB gene of herperviruses.

LIST OF SPECIES DEMARCATION CRITERIA IN THE GENUS

Members of this group have been identified in vertebrates, insects and nematodes. A reverse transcriptase phlylogenetic tree (Fig. 2) indicates that all members of this group are well separated from members of the *Metavirus* and *Errantivirus* genera. Based on the sequence of the putative primer binding site, most viruses in this genus use either the acceptor stem of various $tRNA^{Arg}$ or $tRNA^{Gly}$ as the primer for minus-strand synthesis during reverse transcription. One continuous or two overlapping ORFs characterize the members of this group with the order of domains within the *pol* ORF (PR-RT-RH and IN) identical to that in the genera *Metavirus* and *Errantivirus*. Semotiviruses are particularly abundant in nematodes. The sequence of the *Caenorhabditis elegans* genome has revealed 13 families of viruses, but the majority of these are no longer active. An unusual feature of many of the *C. elegans* viruses is the presence of additional DNA between the 5'-LTR and the beginning of the first ORF. These additional sequences are variable within a family and completely different between families. One active group of sequences in the Caenorhabditis elegans Cer13 virus (CelCer13V) also contains a third Env-like domain between the *pol* encoded enzymatic domains and the 3'-LTR. This domain exhibits no sequence similarity with the domain in AluTasV, suggesting an independent acquisition. The likely origins of the CelCer13 Env-like domain is the G2 glycoprotein gene from phleboviruses. All individual species in the genus have less than 50% identity in their Gag protein sequences compared to all other species.

LIST OF SPECIES IN THE GENUS

Species names are in green italic script; strain names and synonyms are in black roman script; tentative species names are in blue roman script. Sequence accession numbers [ ], and assigned abbreviations ( ) are also listed.

SPECIES IN THE GENUS

Anopheles gambiae Moose virus		
Anopheles gambiae Moose virus	[AF060859]	(AgaMooV)
Ascaris lumbricoides Tas virus		
Ascaris lumbricoides Tas virus	[Z29712]	(AluTasV)
Bombyx mori Pao virus		
Bombyx mori Pao virus	[Z79443]	(BmoPaoV)
Caenorhabditis elegans Cer13 virus		
Caenorhabditis elegans Cer13 virus	[Z81510]	(CelCer13V)
Drosophila melanogaster Bel virus		
Drosophila melanogaster Bel virus	[U23420]	(DmeBelV)
Drosophila melanogaster Roo virus		
Drosophila melanogaster Roo virus	[AY180917]	(DmeRooV)
Drosophila simulans Ninja virus		
Drosophila simulans Ninja virus	[D83207]	(DsiNinV)
Fugu rubripes Suzu virus		
Fugu rubripes Suzu virus	[AF537216]	(FruSuzV)

TENTATIVE SPECIES IN THE GENUS

None reported.

LIST OF UNASSIGNED VIRUSES IN THE FAMILY

None reported.

PHYLOGENETIC RELATIONSHIPS WITHIN THE FAMILY

See Figure 2.

SIMILARITY WITH OTHER TAXA

Like viruses of the family *Pseudoviridae*, the viruses of the family *Metaviridae* are clearly related to the viruses of the family *Retroviridae*. All of these families are related by reverse transcription and a viral core structure made up of Gag-like proteins. All members of the families *Metaviridae*, *Pseudoviridae* and *Retroviridae* also share the following: a proviral form characterized by LTRs, protease, RNase H and integrase activities essential for multiplication, readthrough-mediated (Gag-Pol) *pol* gene expression and tRNA primers (in some species). An important and somewhat controversial question is therefore the extent of the relationship of the members of the family *Metaviridae* to those of the family *Retroviridae*. Reverse transcriptase aa sequences are the most conserved sequences in retroelements and hence are the best character on which to base the phylogenetic relationship of these elements. Based on the phylogeny of their RT domains (Fig. 2, see also Fig. 4 in the chapter *Pseudoviridae*) members of the families *Metaviridae* and *Pseudoviridae* probably shared a common ancestor. The only major structural change between the genomes of members of these two groups was the movement of the IN domain from upstream to downstream of the RT-RH domain. The family *Metaviridae* is a numerous and diverse group of viruses distributed throughout eukaryotes. One lineage of the family *Metaviridae* in vertebrates appears to have given rise to the family *Retroviridae*.

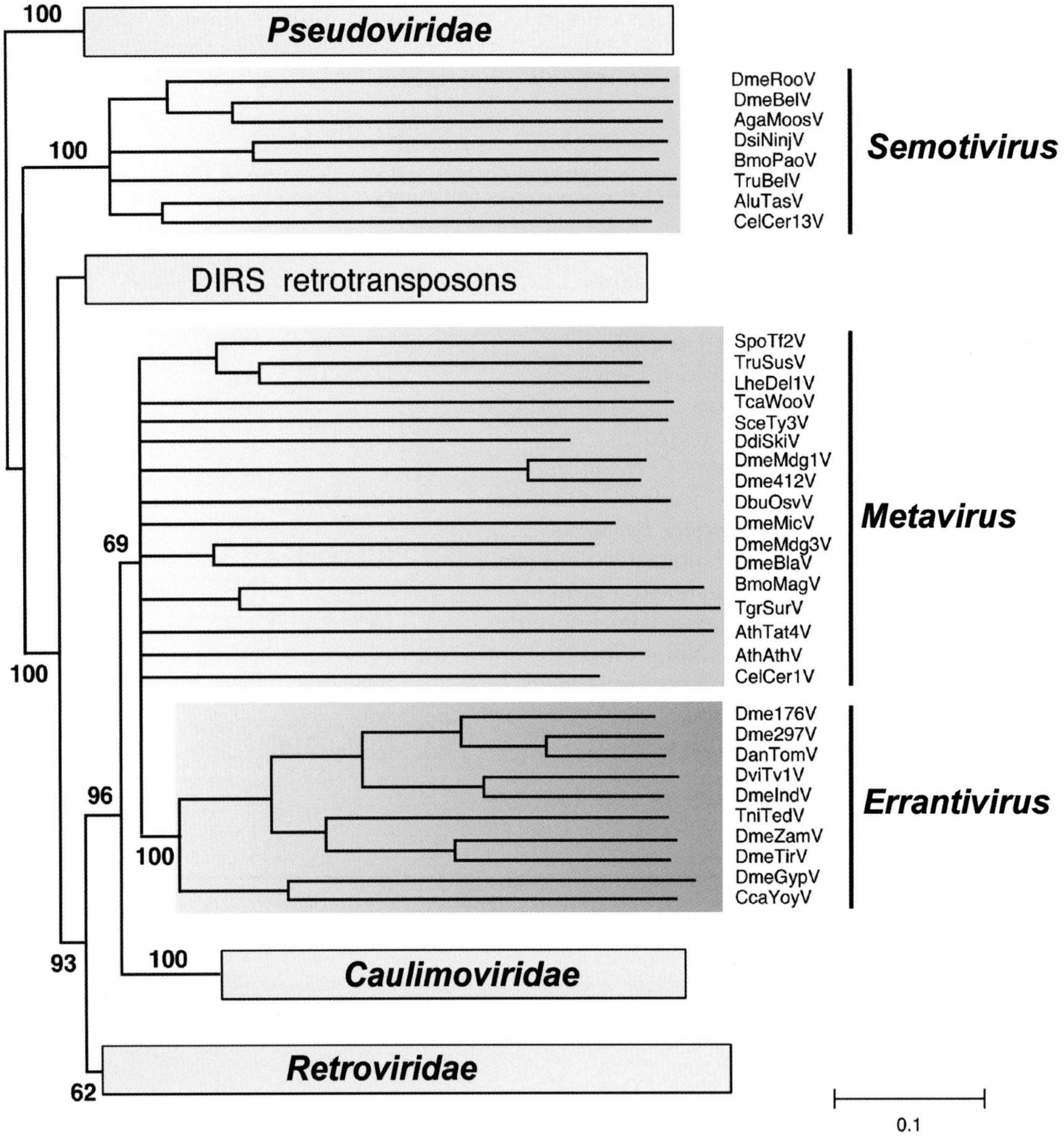

Figure 2: Phylogeny of the family *Metaviridae* and related groups based on their reverse transcriptase domain. The portion of the reverse transcriptase domain used in this analysis spans approximately 250 aa and includes the most conserved residues found in all retroelements. The phylogram is a 50% consensus tree of the viruses based on Neighbor-Joining distance algorithms and was rooted using sequences of members of the family *Pseudoviridae* (see Fig. 4 in the description *Pseudoviridae*). Bootstrap values (percentage of the time all elements are located on that branch) are shown for the major branches only. Viruses that are included in the various divisions of *Metaviridae* are indicated by the vertical lines to the right of each systematic name. Each group of viruses that are not part of the family *Metaviridae* is represented by a box with the length of the box related to the sequence diversity within that group. DIRS retrotransposons are mobile elements that utilize a reverse transcriptase closely related to that of the family *Metaviridae* but lack many structural features of this group and integrate by a different mechanism. Scale bar at bottom represents divergence per site.

The conversion of a lineage of the family *Metaviridae* into the family *Retroviridae* presumably occurred by the transduction of a gene encoding a ligand for cell-surface receptors or a cell fusion protein. The independent acquisition of a cell-surface receptor or fusion protein has occurred on at least five other occasions within the family *Metaviridae*. These include the *Errantivirus* genus in insects, Drosophila buzzatti Osvaldo virus (DbuOsvaV) and AthAthV within the genus *Metavirus* in insects and plants respectively, and AluTasV and CelCer13V within the genus *Semotivirus* in nematodes. The ease with which these viruses can gain, and presumably lose, an *env*-like gene means that this property is not always a reliable indicator of phylogenetic relationships. Finally, a second lineage in plants has become the family *Caulimoviridae* by the acquisition of a number of new genes that resulted in a number of changes to its life cycle (see description on *Caulimoviridae*).

Derivation of Names

Erranti: from Latin *errans* "to wander".
Meta: from Greek *metathesis* for "transposition". Also to connote some uncertainty as to whether these are true viruses or not.
Semoti: from Latin *semotus* meaning "distant, removed". This prefix refers to the observation that based on the sequence of their RT domain, the viruses in this genus are distantly related to the other two genera of the family *Metaviridae*.

References

Boeke, J.D. and Stoye, J.P. (1997). Retrotransposons, endogenous retroviruses, and the evolution of retroelements. In: *Retroviruses* (H. Varmus, S. Hughes and J. Coffin, eds), pp 343-435. Cold Spring Harbor Laboratory, Cold Spring Harbor, New York.

Bowen, N.J. and McDonald, J.F. (1999). Genomic analysis of *Caenorhabditis elegans* reveals ancient families of retroviral-like elements. *Genome Res.*, **9**, 924-935.

Eickbush, T.H. and Malik, H.S. (2002). Evolution of retrotransposons. In: *Mobile DNA II* (N. Craig, R. Craigie, M. Gellert, and A. Lambowitz, eds), pp 1111-1144. American Society of Microbiology Press, Washington, D.C.

Felder, H., Herzceg, A., de Chastonay, Y., Aeby, P., Tobler, H. and Müller. F. (1994). Tas, a retrotransposon from the parasitic nematode *Ascaris lumbricoides*. *Gene*, **149**, 219-225.

Frame, I.G., Cutfield, J.F. and Poulter, R.T.M. (2001). New BEL-like LTR-retrotransposons in *Fugu rubripes*, *Caenorhabditis elegans* and *Drosophila melanogaster*. *Gene*, **263**, 219-230.

Goodwin, T.J.D. and Poulter, R.T.M. (2001). The DIRS group of retrotransposons. *Mol. Biol. Evol.*, **18**, 2067-2082.

Kim, A.I., Terzian, C., Santamaria, P., Pélisson, A., Prud'homme, N. and Bucheton, A. (1994). Retroviruses in invertebrates: the *gypsy* retrotransposon is apparently an infectious retrovirus of *Drosophila melanogaster*. *Proc. Natl. Acad. Sci. USA*, **91**, 1285-1289.

Leblanc, P., Desset, S., Giorgi, F., Taddei, A.R., Fausto, A.M., Mazzini, M., Dastugue, B. and Vaury C. (2000). Life cycle of an endogenous retrovirus, ZAM, in *Drosophila melanogaster*. *J. Virol.*, **74**, 10658-10669.

Levin, H.L. (1995). A novel mechanism of self-primed reverse transcription defines a new family of retroelements. *Mol. Cell. Biol.*, **15**, 3310-3317.

Levin, H.L. (2002). Newly identified retrotransposons of the Ty3/gypsy class in fungi, plants and vertebrates. In: *Mobile DNA II*, (N. Craig, R. Craigie, M. Gellert and A. Lambowitz, eds), pp 684-701. American Society of Microbiology Press, Washington, D.C.

Malik, H.S. and Eickbush, T.H. (1999). Modular evolution of the integrase domain in the Ty3/Gypsy class of LTR-retrotransposons. *J. Virol.*, **73**, 5186-5190.

Malik, H.S., Henikoff, S. and Eickbush, T.H. (2000). Poised for contagion: evolutionary origins of the infectious abilities of insect errantiviruses and nematode retroviruses. *Genome Res.*, **10**, 1307-1318.

Marin, I. and Llorens, C. (2000). Ty3/Gypsy retrotransposons: description of new *Arabidopsis thaliana* elements and evolutionary perspectives derived from comparative genomic data. *Mol. Biol. Evol.*, **17**, 1040-1049.

Pélisson, A., Song, S.U., Prud'homme, N., Smith, P., Bucheton, A. and Corces, V.G. (1994). *Gypsy* transposition correlates with the production of a retroviral envelope-like protein under the tissue-specific control of the Drosophila *flamenco* gene. *EMBO J.*, **8**, 4401-4411.

Peterson-Burch, B.D., Wright, D.A., Laten, H.M. and Voytas, D.F. (2000). Retroviruses in plants? *Trends Genet.*, **16**, 151-152.

Sandmeyer, S.B., Aye, M. and Menees, T. (2002). Ty3, a position-specific, gypsy-like element in *Saccharomyes cerevisiae*. In: *Mobile DNA II,* (N. Craig, R. Craigie, M. Gellert, and A. Lambowitz, eds), pp 663-683. American Society of Microbiology Press, Washington, D.C.

Song, S.U., Gerasimova, T., Kurkulos, M., Boeke, J.D. and Corces, V.G. (1994). An Env-like protein encoded by a Drosophila retroelement: evidence that *gypsy* is an infectious retrovirus. *Genes Dev.*, **8**, 2046-2057.

Zolotova, L.I., Andrianov, B.V., Gorelova, T.V., Klitsunova, N.V., Reznik, N.L. and Schuppe, N.G. (1996). Polymorphism of viruslike particles of retrotransposons in *Drosophila* and yeast cells. *Genetika*, **32**, 1326-1332.

CONTRIBUTED BY

Eickbush, T., Boeke, J.D., Sandmeyer, S.B. and Voytas, D.F.

FAMILY *RETROVIRIDAE*

TAXONOMIC STRUCTURE OF THE FAMILY

Family	***Retroviridae***
Subfamily	*Orthoretrovirinae*
Genus	*Alpharetrovirus*
Genus	*Betaretrovirus*
Genus	*Gammaretrovirus*
Genus	*Deltaretrovirus*
Genus	*Epsilonretrovirus*
Genus	*Lentivirus*
Subfamily	*Spumaretrovirinae*
Genus	*Spumavirus*

VIRION PROPERTIES

MORPHOLOGY

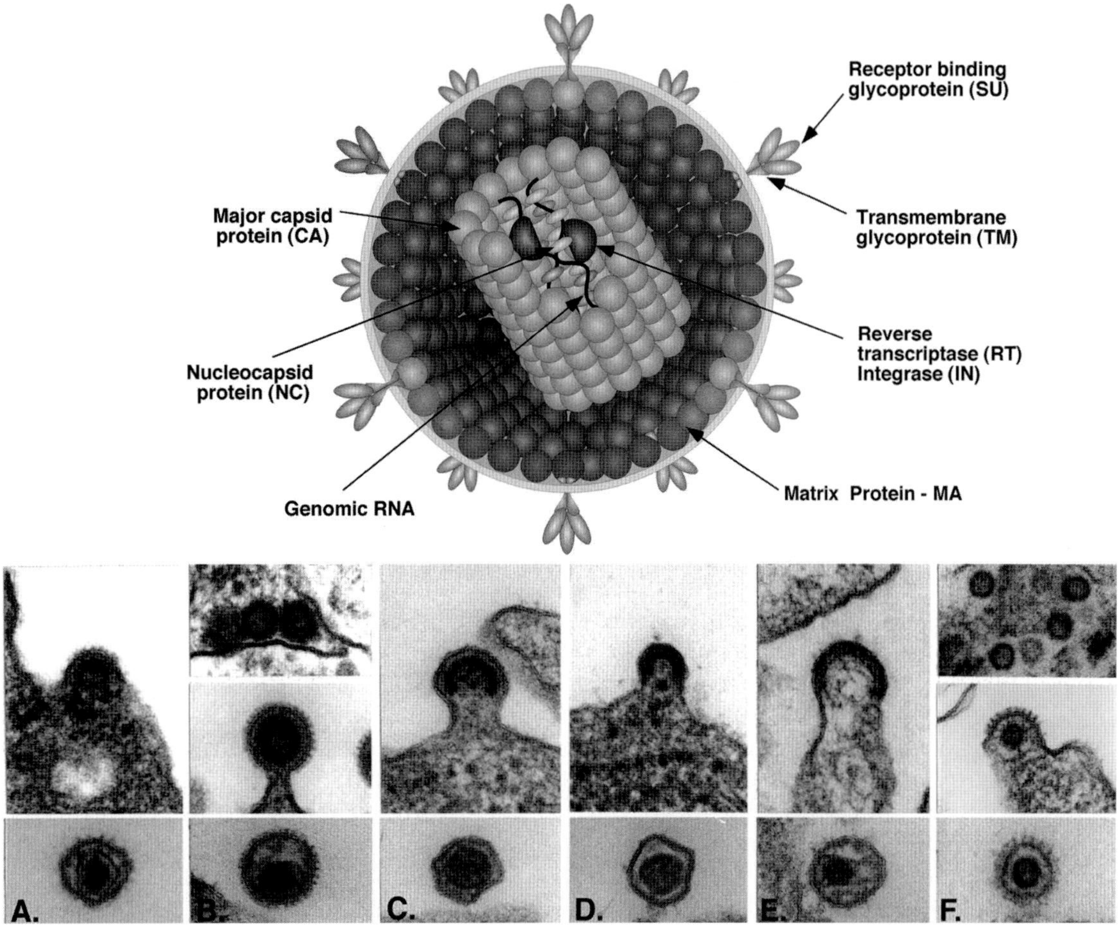

Figure 1: (Top) Schematic cartoon (not to scale) shows the inferred locations of the various structures and proteins. (Bottom) In panel (A) *Alpharetrovirus*: Avian leukosis virus (ALV); type "C" morphology; panel (B) *Betaretrovirus:* Mouse mammary tumor virus (MMTV); type "B" morphology; panel (C) *Gammaretrovirus:* Murine leukemia virus (MLV); panel (D) *Deltaretrovirus;* Bovine leukemia virus (BLV); panel (E) *Lentivirus:* Human immunodeficiency virus 1 (HIV-1); panel (F) *Spumavirus:* Simian foamy virus (SFVcpz(hu)) (formerly called HFV). (Courtesy of M. Gonda reproduced from "Retroviruses", CSH Press, with permission).

Virions are spherical, enveloped and 80-100 nm in diameter. Glycoprotein surface projections are about 8 nm in length. The internal core encapsidates the viral nucleocapsid. The apparently spherical nucleocapsid (nucleoid) is eccentric for members of the genus *Betaretrovirus*, concentric for members of the genera *Alpharetrovirus*, *Gammaretrovirus*, *Deltaretrovirus* and *Spumavirus*, and rod or truncated cone-shape for members of the genus *Lentivirus*.

Two distinct morphogenic pathways exist. Historically, a nomenclature based on electron microscopy classified members of the *Alpharetrovirus* and *Gammaretrovirus* genera, which assemble their immature capsids at the plasma membrane, as C-type viruses. Members of the *Betaretrovirus* genus in contrast were said to assemble A-type particles (immature capsids) in the cytoplasm which then budded with either a B-type (Mouse mammary tumor virus, MMTV) or D-type (Mason-Pfizer monkey virus, MPMV) morphology.

PHYSICOCHEMICAL AND PHYSICAL PROPERTIES

Virion buoyant density is 1.16-1.18 g/cm^3 in sucrose. Virion S_{20w} is approximately 600S in sucrose. Virions are sensitive to heat, detergents and formaldehyde. The surface glycoproteins may be partially removed by proteolytic enzymes. Virions are relatively resistant to UV light.

NUCLEIC ACID

The virus genome characteristic of members of the subfamily *Orthoretrovirinae* consists of a dimer of linear, positive sense, ssRNA, each monomer 7 to 11 kb in size. The RNA constitutes about 2% of the virion dry weight. The monomers are held together by hydrogen bonds. Each monomer of RNA is polyadenylated at the 3'-end and has a cap structure (type 1) at the 5'-end. The purified virion RNA is not infectious. Each monomer is associated with a specific molecule of tRNA that is base-paired to a region (termed the primer binding site) near the 5'-end of the RNA and involves about 18 nt at the 3'-end of the tRNA. Other host derived RNAs (and small DNA fragments) found in virions are believed to be incidental inclusions. The virus genome characteristic of members of the subfamily *Spumaretrovirinae* is dsDNA, as reverse transcription is a late step in the viral life cycle. The exact structure of the DNA has not been determined.

PROTEINS

Proteins constitute about 60% of the virion dry weight. There are 2 envelope proteins: SU (surface) and TM (transmembrane) encoded by the viral *env* gene. Some members of the subfamily *Spumaretrovirinae* have a third Env protein, LP (leader peptide). There are 3-6 internal, non-glycosylated structural proteins (encoded by the *gag* gene). These are, in order from the amino terminus, (1) MA (matrix), (2) in some viruses a protein of undetermined function, (3) CA (capsid protein), and (4) NC (nucleocapsid). The MA protein is often acylated with a myristyl moiety covalently linked to the amino terminal glycine. Other proteins are a protease (PR, encoded by the *pro* gene), a reverse transcriptase (RT, encoded by the *pol* gene) and an integrase (IN, encoded by the *pol* gene). In some viruses a dUTPase (DU, role unknown) is also present. Members of the *Spumaretrovirinae* encode only a single Gag protein which is cleaved once near the carboxyl-terminus in about half of the proteins. The complex retroviruses in the *Deltaretrovirus*, *Epsilonretrovirus*, *Lentivirus*, and *Spumavirus* genera also encode non-structural proteins. Many of these viruses encode transcriptional transactivators which are required for expression of the LTR promoters.

LIPIDS

Lipids constitute about 35% of the virion dry weight. They are derived from the plasma membrane of the host cell.

CARBOHYDRATES

Virions are composed of about 3% carbohydrate by weight. This value varies, depending on the virus. At least one (SU), but usually both envelope surface proteins are glycosylated. Cellular glycolipids are also found in the viral envelope

GENOME ORGANIZATION AND REPLICATION

Virions carry two copies of the genome. Infectious viruses have 4 main genes coding for the virion proteins in the order: 5'-*gag-pro-pol-env*-3'. Some retroviruses contain genes encoding non-structural proteins important for the regulation of gene expression and virus replication. Others carry cell-derived sequences that are important in pathogenesis. These cellular sequences are inserted either in a complete retrovirus genome (e.g., some strains of *Rous sarcoma virus*), or in the form of substitutions for deleted viral sequences (e.g., some isolates of *Murine sarcoma virus*). Such deletions render the virus replication-defective and dependent on non-transforming helper viruses for production of infectious progeny. In many cases the cell-derived sequences form a fused gene with a viral structural gene that is then translated into one chimeric protein (e.g. *gag-onc* protein).

Entry into the host cell is mediated by interaction between a virion glycoprotein and specific receptors at the host cell surface, resulting in fusion of the viral envelope with the plasma membrane, either directly or following endocytosis. Receptors are cell surface proteins. Several have been identified. For Human immunodeficiency virus (HIV), both the CD4 protein, which is an immunoglobulin-like molecule with a single transmembrane region and a chemokine receptor (CCR5 or CXCR4) which span the membrane seven times are required for membrane fusion. The receptors for ecotropic Murine leukemia virus (MLV), amphotropic MLV and Gibbon ape leukemia virus (GALV), are involved in the transport of small molecules and have a complex structure with multiple transmembrane domains. For the avian leukosis viruses (ALVs) two receptors have been identified: that for subgroup A viruses is a small protein with a single transmembrane domain, distantly related to a cell receptor for low-density lipoprotein while that for subgroup B viruses is related to the TNF-receptor family of proteins.

The process of intracellular uncoating of viral particles is not understood. Subsequent early events are carried out in the context of a nucleoprotein complex derived from the capsid.

For members of the subfamily *Orthoretrovirinae*, replication starts with reverse transcription (by RT) of virion RNA into cDNA using the 3'-end of the tRNA as primer for synthesis of a negative-sense cDNA transcript. The initial short product (to the 5'-end of the genome) transfers and primes further cDNA synthesis from the 3'-end of the genome by virtue of duplicated sequences at the ends of the viral RNA. cDNA synthesis involves the concomitant digestion of the viral RNA (RNAse H activity of the RT protein). The products of this hydrolysis serve to prime positive-sense cDNA synthesis on the negative-sense DNA copies. In its final form, the linear dsDNA derived from the viral ssRNA genome contains long terminal repeats (LTRs) composed of unique sequences from the 3' (U3) and 5' (U5) ends of the viral RNA flanking a repeated sequence (R) found near both ends of the RNA. The process of reverse transcription is characterized by a high frequency of recombination due to the transfer of the RT from one template RNA to the other. Reverse transcription is thought to follow the same pathway in members of the subfamily *Spumaretrovirinae*, but the timing is different as it occurs during viral assembly and/or release from the cell. The mechanism of reverse transcription allows for high rates of recombination and genetic diversity for many of the retroviruses. The high rate of genetic recombination *in vivo* can lead to formation of a quasispecies consisting of a large number of genetically diverse virions.

Retroviral DNA becomes integrated into the chromosomal DNA of the host to form a provirus by a mechanism involving the viral IN protein. The ends of the virus DNA are joined to cell DNA, involving the removal of two bases from the ends of the linear viral DNA and generating a short duplication of cell sequences at the integration site. Virus DNA can integrate at many sites in the cellular genome. However, once integrated, a sequence is apparently incapable of further transposition within the same cell. The map of the integrated provirus is co-linear with that of non-integrated viral DNA. Integration appears to be a prerequisite for virus replication.

The integrated provirus is transcribed by cellular RNA polymerase II into virion RNA and mRNA species in response to transcriptional signals in the viral LTRs. In some genera, transcription is also regulated by virally encoded transactivators. There are several classes of mRNA depending on the virus and its genetic map. An mRNA comprising the whole genome serves for the translation of the *gag*, *pro*, and *pol* genes (positioned in the 5' half of the RNA). This results in the formation of polyprotein precursors which are cleaved to yield the structural proteins, protease, RT and IN, respectively. A smaller mRNA consisting of the 5'-end of the genome spliced to sequences from the 3'-end of the genome and including the *env* gene and the U3 and R regions, is translated into the precursor of the envelope proteins. In viruses that contain additional genes, other forms of spliced mRNA are also made; however, all these spliced mRNAs share a common sequence at their 5'-ends. Members of the *Spumaretrovirinae* are unique in that they make use of an internal promoter (IP) located in the *env* gene upstream of the accessory reading frames, for transcription of these distal genes. Most primary translation products in retrovirus infections are polyproteins which require proteolytic cleavage before becoming functional. The *gag*, *pro* and *pol* gene products are generally produced from a nested set of primary translation products. For *pro* and *pol*, translation involves bypassing translational termination signals by ribosomal frameshifting or by read-through at the Gag-Pro and/or the Pro-Pol boundaries. However, members of the subfamily *Spumaretrovirinae* synthesize Pol protein from its own mRNA rather than as a Gag-Pol fusion protein.

The retroviral genomic RNA contains sequences of varying lengths, usually near the 5′ end between U3 and *gag*, which comprise a packaging signal (Ψ). Ψ is required for efficient encapsidation of the genome into particles, and is generally not present on the subgenomic mRNAs, a notably exception being the alpharetroviruses. In the case of the spumaviruses, Ψ does not appear to be a the 5′ end of the genome. In all cases, Ψ activity is not defined by the primary sequence, but by a complex structure.

Capsids assemble either at the plasma membrane (for a majority of the genera), or as intracytoplasmic particles (for members of the *Betaretrovirus* and *Spumavirus* genera) and are released from the cell by a process of budding. Virions of the spumaviruses and deltaretroviruses are highly cell-associated. Polyprotein processing of the internal proteins occurs concomitant with or just subsequent to the maturation of virions of members of the subfamily *Orthoretrovirinae*.

ANTIGENIC PROPERTIES

Virion proteins contain type-specific and group-specific determinants. Some type-specific determinants of the envelope glycoproteins are involved in antibody-mediated virus neutralization. Group-specific determinants are shared by members of a serogroup and may be shared between members of different serogroups within a particular genus. There is evidence for weak cross-reactivities between members of different genera. Epitopes that elicit T-cell responses are found on many of the structural proteins. Antigenic properties are infrequently used in classification of members of the family *Retroviridae*.

BIOLOGICAL PROPERTIES

Retroviruses are widely distributed as exogenous infectious agents of vertebrates. Endogenous proviruses that have resulted at some time from infection of germ line cells are inherited as Mendelian genes. They occur widely among vertebrates.

Retroviruses are associated with a variety of diseases. These include: malignancies including certain leukemias, lymphomas, sarcomas and other tumors of mesodermal origin; mammary carcinomas and carcinomas of liver and kidney; immunodeficiencies (such as AIDS); autoimmune diseases; lower motor neuron diseases; and several acute diseases involving tissue damage. Some retroviruses are non-pathogenic. Transmission of retroviruses is horizontal via a number of routes, including blood, saliva, sexual contact, etc., and via direct infection of the developing embryo, or via milk or perinatal routes. Endogenous retroviruses are transmitted vertically by inheritance of proviruses.

SUBFAMILY ORTHORETROVIRINAE

GENUS ALPHARETROVIRUS

Type Species *Avian leukosis virus*

DISTINGUISHING FEATURES

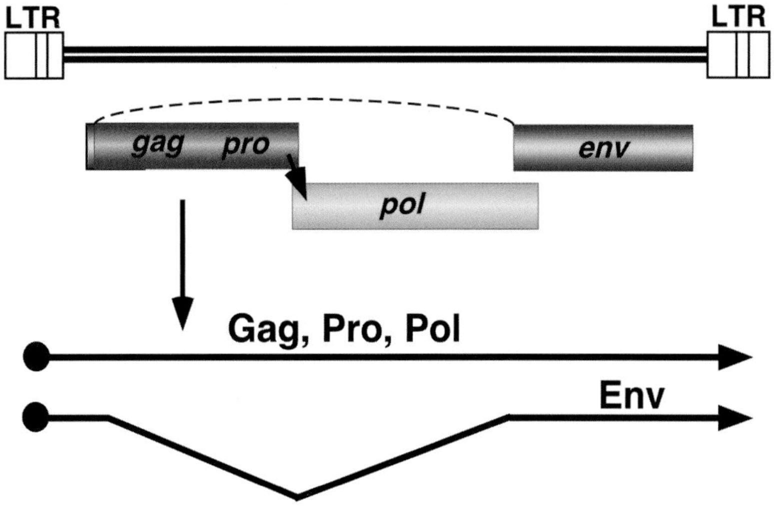

Figure 2: The 7.2 kbp Avian leukosis virus (ALV) provirus is shown, indicating the positions of the LTRs and encoded genes (*gag, pro, pol, env*), their relative reading frames (ribosomal frameshift site: arrowhead; individual mRNAs: solid line arrows with gene products).

Virus particles assemble at the plasma membrane and exhibit a 'C-type' morphology. Protein sizes are: MA ~19 kDa; p10 ~10 kDa; CA ~27 kDa; NC ~12 kDa; PR ~15 kDa; RT ~68 kDa; IN ~32 kDa; SU ~85 kDa; TM ~37 kDa. The genome is about 7.2 kb in size (one monomer); its organization is illustrated in Figure 2. There are no known additional genes other than *gag, pro, pol,* and *env*. The tRNA primer is $tRNA^{Trp}$. The LTR is about 350 nt long, of which the U3 region is 250 nt, the R sequence is 20 nt and the U5 region is some 80 nt in size. The viruses have a widespread distribution and include both exogenous (vertical and horizontal transmission) and endogenous viruses of chickens and some other birds. ALV isolates are classified into subgroups (e.g. -A, -J) by their distinct receptor usage. Distantly related endogenous sequences are found in birds and mammals. Virus infections are associated with malignancies and some other diseases such as wasting, and osteopetrosis. Many oncogene-containing members of the genus have been isolated.

List of Species Demarcation Criteria in the Genus

The list of species demarcation criteria is:

- Differences in genome sequence,
- Differences in gene product sequences,
- Differences in natural host range,
- Different oncogenes that may be incorporated.

For example, the isolates of *Avian leukosis virus* can be readily distinguished from those of *Rous sarcoma virus* because they lack oncogene sequences while encoding *gag, pol* and *env*. The replication-defective alpharetroviruses can be distinguished from Rous sarcoma virus by the variable deletion of portions of the *gag, pol* and *env* genes and the presence of a unique oncogene in each species. *Rous sarcoma virus* strains encode the *src* oncogene whereas Avian myeloblastosis virus, for example, encodes the *myb* oncogene. Host range, defined by SU interaction with a specific receptor, is generally used in defining strains within a species.

List of Species in the Genus

Species names are in green italic script; strain names and synonyms are in black roman script; tentative species names are in blue roman script. Sequence accession numbers [ ], and assigned abbreviations ( ) are also listed.

Species in the Genus

Avian leukosis virus		
Avian leukosis virus - RSA	[M37980]	(ALV-A)
Avian leukosis virus - HPRS103	[Z46390]	(ALV-J)
Oncogene containing viruses:		
Replication Competent:		
Rous sarcoma virus		
Rous sarcoma virus - Prague C	[J02342]	(RSV-Pr-C)
Rous sarcoma virus - Schmidt-Ruppin B	[AF052428]	(RSV-SR-B)
Rous sarcoma virus - Schmidt-Ruppin D	[D10652]	(RSV-SR-D)
Replication Defective:		
Avian carcinoma Mill Hill virus 2		
Avian carcinoma Mill Hill virus 2	[K02082]	(ACMHV-2)
Avian myeloblastosis virus		
Avian myeloblastosis virus	[J02013]	(AMV)
Avian myelocytomatosis virus 29		
Avian myelocytomatosis virus 29	[J02019]	(AMCV-29)
Avian sarcoma virus CT10		
Avian sarcoma virus CT10	[Y00302]	(ASV-CT10)
Fujinami sarcoma virus		
Fujinami sarcoma virus	[J02194]	(FuSV)
UR2 sarcoma virus		
UR2 sarcoma virus	[M10455]	(UR2SV)
Y73 sarcoma virus		
Y73 sarcoma virus	[J02027]	(Y73SV)

Tentative Species in the Genus

None reported.

GENUS BETARETROVIRUS

Type Species *Mouse mammary tumor virus*

DISTINGUISHING FEATURES

Virions of Mouse mammary tumor virus (MMTV) exhibit a 'B-type' morphology with prominent surface spikes and an eccentric condensed core. Other members of the genus *Betaretrovirus* have a 'D-type' morphology with less dense surface spikes and a cylindrical core. Capsid assembly occurs within the cytoplasm (to yield structures previously termed 'A-type' particles) prior to transport to, and budding from the plasma membrane. Protein sizes are: MA ~10 kDa; p21 21 kDa; p8/p12 8-12 kDa; CA ~27 kDa; NC ~14 kDa; DU ~30 kDa; PR ~15 kDa; RT ~50 kDa; IN ~? kDa; SU ~52 kDa; TM ~36 kDa. The genome is 8-10 kb in size (one monomer); its organization for MMTV is illustrated in Figure 3.

In MMTV there is an additional gene (*sag*) whose product functions as a superantigen, which is located at the 3'-end of the genome, overlapping U3. This gene is absent from other members of the genus. The tRNA primer is $tRNA^{Lys\text{-}3}$ for MMTV and $tRNA^{Lys\text{-}1,2}$ for other members of the genus. The LTR of MMTV is about 1300 nt long primarily due to the *sag* encoding U3 region of 1200 nt. The R sequence (15 nt) and the U5 region (95-120 nt) are of similar length in all members of the genus.

Viruses assigned to this genus include exogenous, (milk-transmitted) and endogenous viruses of mice, as well as exogenous, horizontally transmitted and endogenous viruses of new and old world primates and sheep. Murine viruses are associated with mammary carcinoma and T-lymphomas, while the exogenous primate viruses are associated with immuno-deficiency diseases; Jaagsiekte sheep virus is associated with pulmonary cancer of sheep. No oncogene-containing member is known.

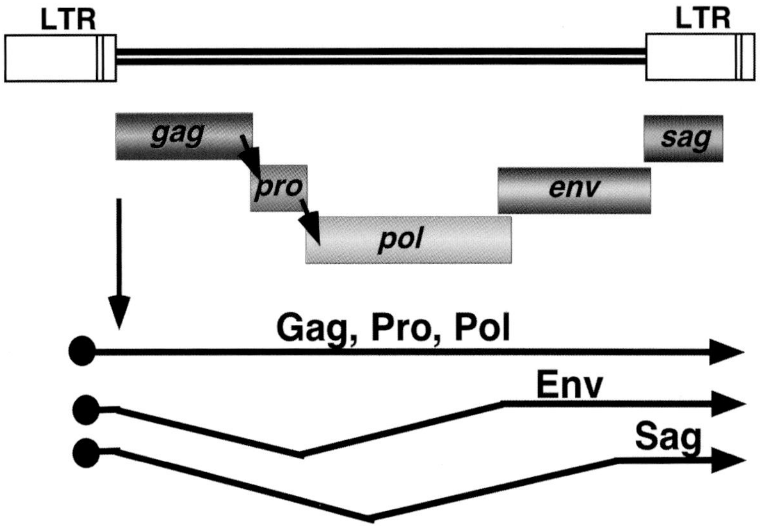

Figure 3: The 10 kbp Mouse mammary tumor virus (MMTV) provirus is shown indicating the positions of the LTRs and encoded genes (*gag, pro, pol, env, sag*), their relative reading frames (ribosomal frame-shift sites: arrow heads; individual mRNAs: solid line arrows with gene products).

LIST OF SPECIES DEMARCATION CRITERIA IN THE GENUS

The list of species demarcation criteria is:
- Differences in genome sequence,
- Differences in gene product sequences,

- Differences in natural host range,
- Different oncogenes that may be incorporated.

Several primate retroviruses have been described that appear to be divergent members of a single virus that arose from a recombination event in which the *env* gene of a primate gammaretrovirus was captured. The most divergent of these are the endogenous Squirrel monkey retrovirus (SMRV) and Langur virus (LNGV), which are unable to infect cells from the primate species of origin. Several serologically distinct strains exist within the *Mason-Pfizer monkey virus* species. The most divergent of these are the endogenous SMRV and LNGV, nevertheless, the most closely related isolates Mason-Pfizer monkey virus, Simian retrovirus-1 and Simian retrovirus-2 are serologically distinct. Mouse mammary tumor virus is assigned to a separate species because of the unique *sag* coding region and a widely divergent and distinct *env* gene. Jaagsiekte sheep retrovirus is also assigned to a separate species on the degree of nt sequence divergence. Related endogenous proviruses have been identified in other mammalian species (rodents, primates).

LIST OF SPECIES IN THE GENUS

Species names are in green italic script; strain names and synonyms are in black roman script; tentative species names are in blue roman script. Sequence accession numbers [ ], and assigned abbreviations ( ) are also listed.

SPECIES IN THE GENUS

Jaagsiekte sheep retrovirus		
Jaagsiekte sheep retrovirus	[M80216]	(JSRV)
(Ovine pulmonary adenocarcinoma virus)		
Langur virus		
Langur virus		(LNGV)
Mason-Pfizer monkey virus		
Mason-Pfizer monkey virus	[M12349]	(MPMV)
Simian retrovirus 1	[M11841]	(SRV-1)
Simian retrovirus 2	[M16605]	(SRV-2)
Mouse mammary tumor virus		
Mouse mammary tumor virus	[M15122]	(MMTV)
Squirrel monkey retrovirus		
Squirrel monkey retrovirus	[M23385]	(SMRV)

TENTATIVE SPECIES IN THE GENUS

None reported.

GENUS GAMMARETROVIRUS

Type Species ***Murine leukemia virus***

DISTINGUISHING FEATURES

Virions exhibit a 'C-type' morphology with barely visible surface spikes. They have a centrally located, condensed core. Capsid assembly occurs at the inner surface of the membrane at the same time as budding. Protein sizes are: MA ~15 kDa; p12 12 kDa; CA ~30 kDa; NC ~10 kDa; PR ~14 kDa; RT ~80 kDa; IN ~46 kDa; SU ~70 kDa; TM ~15 kDa. The genome is about 8.3 kb in size (one monomer); its organization is illustrated in Figure 4. The *pro-pol* region is translated following ribosomal readthrough at the *gag* gene termination codon. There are no known additional genes. The tRNA primer is $tRNA^{Pro}$, ($tRNA^{Glu}$ is found in a few endogenous mouse viruses). The LTR is about 600 nt long of which the U3 region is 500 nt, the R sequence 60 nt and the U5 region some 75 nt in size.

The viruses are widely distributed; exogenous (vertical and horizontal transmission) and endogenous viruses are found in many mammals. The reticuloendotheliosis viruses comprise a few isolates from birds with no known corresponding endogenous relatives. Related endogenous sequences are found in mammals. The viruses are associated with a variety of diseases including malignancies, immunosuppression, neurological disorders, and others. Many oncogene-containing members of the mammalian and reticuloendotheliosis virus groups have been isolated.

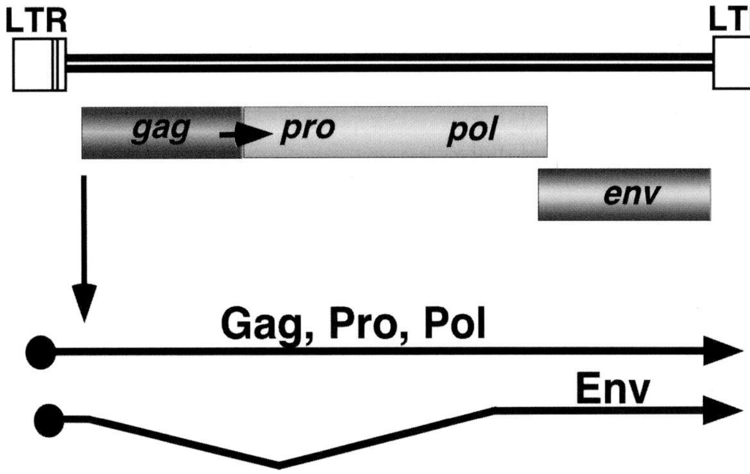

Figure 4: The 8.3 kbp Murine leukemia virus (MLV) provirus is shown indicating the positions of the LTRs and encoded genes (*gag, pro, pol, env*), their relative reading frames (ribosomal readthrough site, arrow head; individual mRNAs: solid line arrows with gene products).

LIST OF SPECIES DEMARCATION CRITERIA IN THE GENUS

The list of species demarcation criteria is:
- Differences in genome sequence and viral oncogenes,
- Differences in antigenic properties,
- Differences in natural host range,
- Differences in pathogenicity.

There are mammalian, reptilian and avian (reticuloendotheliosis) viruses. The mammalian viruses include replication competent viruses that lack cell-derived oncogenes and replication defective viruses that have acquired a variety of oncogenes from their hosts.

Members of the *Murine leukemia virus* species can be distinguished from isolates of *Gibbon ape leukemia virus,* for example, by sequence divergence, distinct receptors for virus entry and only limited antigenic cross-reactivity in ELISA assays. Murine sarcoma viruses, which are invariably replication defective, can be distinguished from the murine leukemia viruses and from one another by the presence of distinct cell-derived oncogenes (*Moloney murine sarcoma virus* (*mos*) versus *Woolly monkey sarcoma virus* (*sis*)) and the characteristic loss of portions of *gag, pol* or *env.*

LIST OF SPECIES IN THE GENUS

Species names are in green italic script; strain names and synonyms are in black roman script; tentative species names are in blue roman script. Sequence accession numbers [ ], and assigned abbreviations ( ) are also listed.

SPECIES IN THE GENUS

Mammalian virus group:

Replication competent viruses:		
Feline leukemia virus		
Feline leukemia virus	[M18247]	(FeLV)
Gibbon ape leukemia virus		
Gibbon ape leukemia virus	[M26927]	(GALV)
Guinea pig type-C oncovirus		
Guinea pig type-C oncovirus		(GPCOV)
Murine leukemia virus		
Abelson murine leukemia virus	[J02009]	(AbMLV)
AKR (endogenous) murine leukemia virus	[J01998]	(AKRMLV)
Friend murine leukemia virus	[M93134, Z11128]	(FrMLV)
Murine leukemia virus	[J02255]	(MLV)
Moloney murine leukemia virus	[J02255]	(MoMLV)
Porcine type-C oncovirus		
Porcine type-C oncovirus		(PCOV)
Replication defective viruses:		
Finkel-Biskis-Jinkins murine sarcoma virus		
Finkel-Biskis-Jinkins murine sarcoma virus	[K02712]	(FBJMSV)
Gardner-Arnstein feline sarcoma virus		
Gardner-Arnstein feline sarcoma virus		(GAFeSV)
Hardy-Zuckerman feline sarcoma virus		
Hardy-Zuckerman feline sarcoma virus		(HZFeSV)
Harvey murine sarcoma virus		
Harvey murine sarcoma virus		(HaMSV)
Kirsten murine sarcoma virus		
Kirsten murine sarcoma virus		(KiMSV)
Moloney murine sarcoma virus		
Moloney murine sarcoma virus	[J02266]	(MoMSV)
Snyder-Theilen feline sarcoma virus		
Snyder-Theilen feline sarcoma virus		(STFeSV)
Woolly monkey sarcoma virus		
Woolly monkey sarcoma virus (Simian sarcoma virus)	[J02394]	(WMSV)
Reptilian virus group:		
Viper retrovirus		(VRV)
Viper retrovirus		(VRV)
Avian (Reticuloendotheliosis) virus group:		
Chick syncytial virus		
Chick syncytial virus		(CSV)
Reticuloendotheliosis virus		
Reticuloendotheliosis virus (strain A)		(REV-A)
Reticuloendotheliosis virus (strain T)		(REV-T)
Trager duck spleen necrosis virus		
Trager duck spleen necrosis virus		(TDSNV)

TENTATIVE SPECIES IN THE GENUS

None reported.

GENUS DELTARETROVIRUS

Type Species ***Bovine leukemia virus***

DISTINGUISHING FEATURES

Virions are similar to gammaretroviruses in terms of morphology and assembly. Protein sizes are: MA ~19 kDa; CA ~24 kDa; NC ~12-15 kDa; PR ~14 kDa; RT, IN, and SU ~60 kDa; TM ~21 kDa. The genome is about 8.3 kb in size (one monomer); its organization is illustrated in Figure 5. There are non-structural genes, designated *tax* and *rex,* which are involved in regulation of synthesis and processing of virus RNA, in addition to *gag, pro, pol* and *env*. The tRNA primer is $tRNA^{Pro}$. The LTR is about 550-750 nt long, of which the U3 region is 200-300 nt, the R sequence 135-235 nt and the U5 region 100-200 nt in size.

The exogenous viruses (horizontal transmission) in this genus are found in only a few groups of mammals. No related endogenous viruses are known. Virus infections are associated with B or T cell leukemias or lymphomas as well as neurological disease (tropical spastic paraparesis, or HTLV-associated myopathy) and exhibit a long latency with an incidence of much less than 100%. No oncogene-containing members of this genus have been identified.

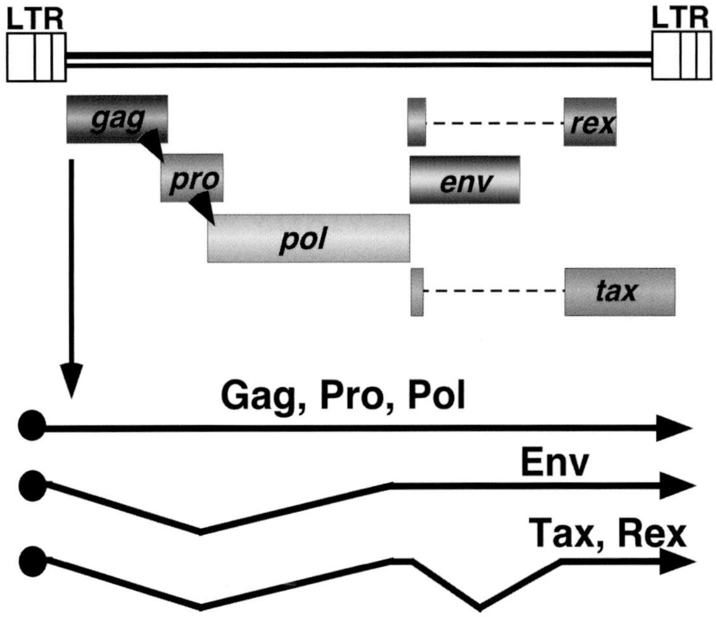

Figure 5: The 8.7 kbp Human T-lymphotropic virus 1 (HTLV-1) provirus genome is shown indicating the positions of the LTRs and encoded structural genes (*gag, pro, pol, env*) and certain other non-structural genes (*tax, rex*), their reading frames (ribosomal frameshift sites: arrow heads; individual mRNAs: solid lines with gene products).

LIST OF SPECIES DEMARCATION CRITERIA IN THE GENUS

The list of species demarcation criteria is:
- Differences in genome sequence and viral oncogenes,
- Differences in antigenic properties,
- Differences in natural host range,
- Differences in pathogenicity.

Human T-lymphotropic virus 1 (HTLV-1) and Simian T-lymphotropic virus 1 (STLV-1) are not clustered according to host species but rather according to geographic origin. All

HTLV-1 subtypes described so far have most probably originated from separate interspecies transmissions from simians to humans. *Primate T-lymphotropic virus 1* is distinguished from *Primate T-lymphotropic virus 2* for example primarily on the basis of phylogenetic differences. The two virus species have a similar coding strategy, but only the first one has been associated with human disease.

LIST OF SPECIES IN THE GENUS

Species names are in green italic script; strain names and synonyms are in black roman script; tentative species names are in blue roman script. Sequence accession numbers [ ], and assigned abbreviations ( ) are also listed.

SPECIES IN THE GENUS

Bovine leukemia virus		
Bovine leukemia virus	[K02120]	(BLV)
Primate T-lymphotropic virus 1		
Human T-lymphotropic virus 1	[D13784]	(HTLV-1)
Simian T-lymphotropic virus 1		(STLV-1)
Primate T-lymphotropic virus 2		
Human T-lymphotropic virus 2	[M10060]	(HTLV-2)
Simian T-lymphotropic virus 2		(STLV-2)
Simian T-lymphotropic virus - PP		(STLV-PP)
Primate T-lymphotropic virus 3	[Y07616]	
Simian T-lymphotropic virus 3		(STLV-3)
Simian T-lymphotropic virus - L		

TENTATIVE SPECIES IN THE GENUS

None reported.

GENUS *EPSILONRETROVIRUS*

Type Species ***Walleye dermal sarcoma virus***

DISTINGUISHING FEATURES

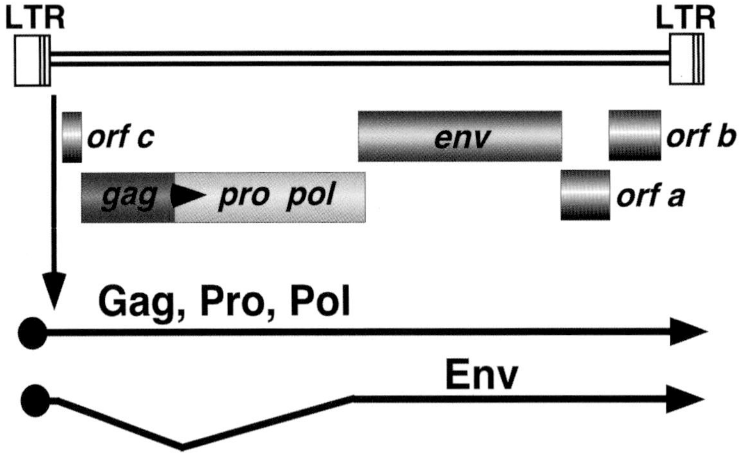

Figure 6: The 12.3 kbp Walleye dermal sarcoma virus (WDSV) provirus genome is shown indicating the positions of the LTRs and encoded structural genes (*gag*, *pro*, *pol*, *env*) and certain other non-structural genes (ORFs a, b, and c, their reading frames (ribosomal readthrough site: arrow head); individual mRNAs: solid lines with gene products).

All members of the genus are exogenous retroviruses. They are complex retroviruses in that their genomes range in size from 11.7 kb to 12.8 kb and contain from one to three

ORFs, presumably encoding accessory proteins, in addition to those encoding the structural proteins and enzymes of the virion (Fig. 6). ORFa, present in all three walleye retroviruses appears to be a viral homolog of cyclin D. The function of the two additional ORFs in the Walleye dermal sarcoma virus (WDSV) or the single ORF in the snakehead retrovirus has not been determined. The LTRs of the fish retroviruses range from about 500 to 650 nt in length, of which the U3 region is about 450 nt, the R sequence about 80 nt and the U5 region about 75 nt in size. The primer used by the walleye retroviruses is $tRNA^{His}$ while the snakehead retrovirus uses $tRNA^{Arg}$. Phylogenetic analysis comparing the polymerase region shows that the walleye and perch retroviruses cluster and have diverged significantly from the snakehead retrovirus. Nevertheless all piscine retroviruses to date appear to group with the mammalian type-C retroviruses.

LIST OF SPECIES DEMARCATION CRITERIA IN THE GENUS

The list of species demarcation criteria is:
- Differences in genome sequence and viral oncogenes,
- Differences in gene product sequence,
- Differences in natural host range,

The genus *Epsilonretrovirus* is comprised of three species of fish retroviruses, *Walleye dermal sarcoma virus*, *Walleye epidermal hyperplasia virus 1*, and *Walleye epidermal hyperplasia virus 2*. These species are distinguished from one another on the basis of phylogenetic diversity. In addition, two tentative species, Snakehead retrovirus (SnRV) and Perch hyperplasia virus (PHV) are listed. As additional fish retroviruses are identified and characterized the Snakehead retrovirus may provide a basis for an additional genus (Fig. 9). The Perch hyperplasia virus has been sequenced only within the polymerase gene and its status remains tentative until further sequence information becomes available.

LIST OF SPECIES IN THE GENUS

Species names are in green italic script; strain names and synonyms are in black roman script; tentative species names are in blue roman script. Sequence accession numbers [ ], and assigned abbreviations ( ) are also listed.

SPECIES IN THE GENUS

Walleye dermal sarcoma virus		
Walleye dermal sarcoma virus	[AF033822]	(WDSV)
Walleye epidermal hyperplasia virus 1		
Walleye epidermal hyperplasia virus 1	[AF014792]	(WEHV-1)
Walleye epidermal hyperplasia virus 2		
Walleye epidermal hyperplasia virus 2	[AF014793]	(WEHV-2)

TENTATIVE SPECIES IN THE GENUS

Perch hyperplasia virus		(PHV)
Snakehead retrovirus	[U26458]	(SnRV)

GENUS *LENTIVIRUS*

Type Species ***Human immunodeficiency virus 1***

DISTINGUISHING FEATURES

Virions have a distinctive morphology with a bar, or cone-shaped core (nucleoid). Viruses assemble at the cell membrane. Proteins sizes are: MA ~17 kDa; CA ~24 kDa; NC ~7-11 kDa; PR ~14 kDa; RT ~66 kDa; DU (in all except the primate lentiviruses), IN ~32 kDa; SU ~120 kDa; TM ~41 kDa. The genome is about 9.3 kb in size (one monomer); its organization is illustrated in Figure 7. Detailed structural data are availabe for RT, SU and PR.

In addition to the structural *gag, pro, pol,* and *env* genes, there are additional genes, depending on the virus (e.g., for Human immunodeficiency virus 1, HIV-1): *vif, vpr, vpu, tat, rev, nef*) whose products are involved in regulation of synthesis and processing of virus RNA and other replicative functions. Most are located 3' to *gag-pro-pol* and, at least in part, 5' to *env*, one (*nef* in HIV-1) is 3' to *env*. For other viruses there may be additional non-structural genes (e.g. *vpx* in HIV-2). The tRNA primer is $tRNA^{Lys1,2}$. The LTR is about 600 nt long, of which the U3 region is 450 nt, the R sequence 100 nt and the U5 region some 80 nt in size.

The viruses in the genus include exogenous viruses (horizontal and vertical transmission) of humans and many other mammals. No related endogenous viruses are known. The primate lentiviruses are distinguished by the use of a chemokine receptor and the CD4 protein as receptor and the absence of DU. Some groups have cross-reactive Gag antigens (e.g., the ovine, caprine and feline lentiviruses). Viruses related to isolates of *Feline immunodeficiency virus* have been isolated from other large felids (e.g. the *Puma lentivirus*), and antibodies to Gag antigens in lions and other large felids indicate the existence of other viruses related to FIV and the ovine/caprine lentiviruses.

The viruses are associated with a variety of diseases including immunodeficiencies, neurological disorders, arthritis, and others. No oncogene-containing member of this genus has been isolated.

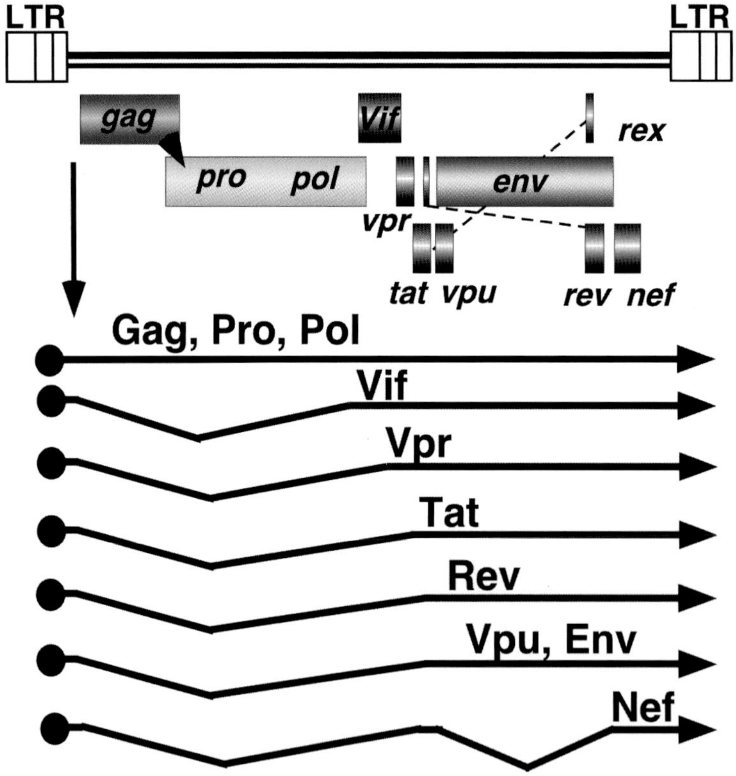

Figure 7: The 9.3 kbp Human immunodeficiency virus 1 (HIV-1) provirus is shown indicating the positions of the LTRs and encoded structural genes (*gag, pro, pol, env*) and certain non-structural genes (*vif, vpr, tat, rev, nef*), their reading frames (ribosomal frameshift site, arrow head; individual mRNAs: solid lines with gene products). The genes in other members of the genus may occupy different reading frames.

List of Species Demarcation Criteria in the Genus

The list of species demarcation criteria is:

- Differences in genome and gene product sequences,
- Differences in antigenic properties,
- Differences in natural host range,
- Differences in pathogenicity.

Five groups of lentiviruses can be clustered on the basis of the hosts they infect (primates, sheep and goats, horses, cats, and cattle). Within the primate lentivirus group, HIV-1 is distinguished from HIV-2, for example, primarily on the basis of sequence divergence that exceeds 50% and the presence in HIV-2 of the *vpx* gene. There is limited cross-reactivity in ELISA tests based on Gag components but essentially none in those based on *env* gene products.

LIST OF SPECIES IN THE GENUS

Species names are in green italic script; strain names and synonyms are in black roman script; tentative species names are in blue roman script. Sequence accession numbers [ ], and assigned abbreviations ( ) are also listed.

SPECIES IN THE GENUS

Bovine lentivirus group:

Bovine immunodeficiency virus

Bovine immunodeficiency virus	[M32690]	(BIV)

Equine lentivirus group:

Equine infectious anemia virus

Equine infectious anemia virus	[M16575]	(EIAV)

Feline lentivirus group:

Feline immunodeficiency virus

Feline immunodeficiency virus (Oma)	[FIU56928]	(FIV-O)
Feline immunodeficiency virus (Petuluma)	[M25381]	(FIV-P)

Puma lentivirus

Puma lentivirus	[PLU03982]	(PLV-14)

Ovine/caprine lentivirus group:

Caprine arthritis encephalitis virus

Caprine arthritis encephalitis virus	[M33677]	(CAEV)

Visna/maedi virus

Visna/maedi virus (strain 1514)	[M60609, M60610]	(VISNA)

Primate lentivirus group:

Human immunodeficiency virus 1

Human immunodeficiency virus 1		(HIV-1)
At least 10 genomic clades of HIV-1 are recognized. Examples include:		
Clade A		
U455	[M62320]	(HIV-1.U455)
Clade B		
ARV-2/SF-2	[K02007]	(HIV-1.ARV-2/SF-2)
BRU (LAI)	[K02013]	(HIV-1.BRU (LAI))
HXB2	[K03455]	(HIV-1.HXB2)
MN	[M17449]	(HIV-1.MN)
RF	[M17451]	(HIV-1.RF)
Clade C		
ETH2220	[U46016]	(HIV-1.ETH2220)
Clade D		
ELI	[X04414]	(HIV-1.ELI)
NDK	[M27323]	(HIV-1.NDK)

RT VIRUSES

Clade F		
93BR020	[AF005494]	(HIV-1.93BR020)
Clade H		
90CR056	[AF0055494]	(HIV-1.90CR056)
Clade O		
ANT70	[L20587]	(HIV-1.ANT70)
Human immunodeficiency virus 2		
Human immunodeficiency virus 2		(HIV-2)
At least 5 genomic clades of HIV-2 are recognized. Examples include:		
Clade A:		
BEN	[M30502]	(HIV-2.BEN)
ISY	[J04498]	(HIV-2.ISY)
ROD	[M15390]	(HIV-2.ROD)
ST	[M31113]	(HIV-2.ST)
Clade B:		
D205	[X61240]	(HIV-2.D205)
EHOA	[U27200]	(HIV-2.EHOA)
UC1	[L07625]	(HIV-2.UC1)
Simian immunodeficiency virus		
Simian immunodeficiency virus		(SIV)
African green monkey		
African green monkey TYO	[X07805]	(SIV-agm.TYO)
African green monkey 155	[M29975]	(SIV-agm.155)
African green monkey 3	[M30931]	(SIV-agm.3)
African green monkey gr-1	[M58410]	(SIV-agm.gr)
African green monkey Sab-1	[U04005]	(SIV-agm.sab)
African green monkey Tan-1	[U58991]	(SIV-agm.tan)
chimpanzee	[X52154]	(SIV-cpz)
mandrill	[M27470]	(SIV-mnd)
pig-tailed macaque*	[M32741]	(SIV-mne)
red capped mangabey	[AF028607]	(SIV-rcm)
Rhesus* (Maccaca mulatta)	[M195499]	(SIV-mac)
sooty mangabey SIV-H4	[X14307]	(SIV-sm)
stump-tailed macaque*	[M83293]	(SIV-stm)
sykes monkey	[L06042]	(SIV-syk)

*Represent cross-species transmission of SIV-sm in captivity
For HIV-1 there are now Clades A to J, forming the M group, of which groups N and O are differentiated.
For HIV-2 there are at least 5 Clades (A-E)

TENTATIVE SPECIES IN THE GENUS

None reported.

SUBFAMILY SPUMARETROVIRINAE

GENUS SPUMAVIRUS

Type Species *Simian foamy virus*

DISTINGUISHING FEATURES

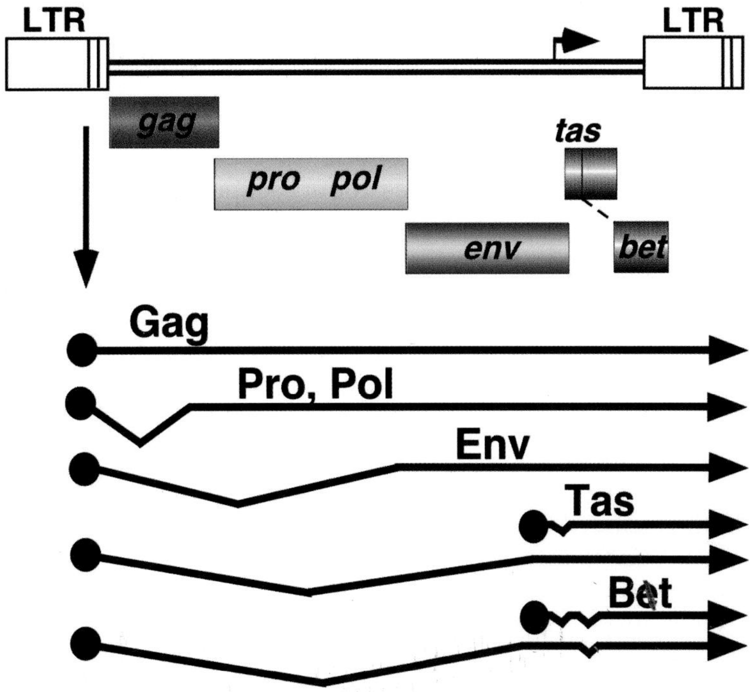

Figure 8: The 13.2 kbp Simian foamy virus (SFV) provirus is shown indicating the positions of the LTRs, the internal promoter, encoded structural genes (*gag, pro, pol, env*), the accessory reading frames (*tas* and *bet*) and individual mRNAs leading to known proteins.

Virions exhibit a distinctive morphology with prominent surface spikes and a central uncondensed core. Capsid assembly occurs in the cytoplasm prior to budding into the endoplasmic reticulum or from the plasma membrane. Capsid budding requires the presence of Env protein. No cleavage of Gag protein precursors into MA, CA, NC subunits is detectable in infectious virions. The Gag protein is cleaved once near the carboxyl-terminus. Protein sizes are: Gag precursor ~71 kDa; N-terminal Gag cleavage product ~68 kDa; Pol precursor ~127 kDa; RT ~85 kDa; IN ~40 kDa; Env precursor ~130 kDa; SU ~80 kDa; TM ~48 kDa; LP ~18 kDa; Tas ~35 kDa; Bet ~60 kDa; Env-Bet fusion protein ~170 kDa. The genome is about 11.6 kb in size; its organization is illustrated in Figure 8. The genomic organization is identical to other members of the family *Retroviridae*,, as is the mechanism of reverse transcription, which allows inclusion of these viruses into the family. There are two genes (designated *tas* and *bet*) that are expressed in cells in addition to *gag, pol,* and *env*. Tas is a DNA binding protein with transactivating function. The exact function of the other accessory protein Bet is unknown, but may be involved in viral latency. The tRNA primer is $tRNA^{lys1,2}$. The LTR of primate foamy viruses is about 1770 nt long, of which the U3 region is about 1400 nt, the R region about 200 nt and the U5 region is some 150 nt in length. In the bovine, equine and feline viruses the LTR is 950-1400 nt. Spumaviruses make use of two start sites of transcription, R in the LTR and an internal promoter (IP) located upstream of the accessory reading frames within the *env* gene. The activity of both promoters is Tas dependent. The additional major criteria distinguishing spumaviruses from members of the other genera of the family *Retroviridae* are the expression of the *Pol* protein from a spliced sgRNA and the presence of a large amount of reverse transcribed DNA in the virion which is required for infection.

Viruses have a widespread distribution and exogenous spumaviruses viruses are found in many mammals. No natural human infections are known. Human infections have been documented as a result of rare zoonotic transmissions from non-human primates. A distantly related endogenous virus has been reported. Many isolates cause characteristic "foamy" cytopathology in cell culture. No diseases have been associated with spumavirus infections. No oncogene-containing member of the genus has been found.

LIST OF SPECIES DEMARCATION CRITERIA IN THE GENUS

The list of species demarcation criteria is:
- Differences in genome and gene product sequences,
- Differences in natural host range.

LIST OF SPECIES IN THE GENUS

Species names are in green italic script; strain names and synonyms are in black roman script; tentative species names are in blue roman script. Sequence accession numbers [ ], and assigned abbreviations ( ) are also listed.

SPECIES IN THE GENUS

Simian foamy virus group:		
African green monkey simian foamy virus ‡		
African green monkey simian foamy virus	[M74895]	(SFVagm)
(Simian foamy virus 3, SFV-3)		
Macaque simian foamy virus ‡		
Macaque simian foamy virus	[X54482]	(SFVmac)
(Simian foamy virus 1, SFV-1)		
Simian foamy virus		
Simian foamy virus, human isolate	[NC001736]	(SFVcpz(hu))
(Chimpanzee foamy virus, CFV)		
(Human foamy virus, HFV)		
Simian foamy virus, chimpanzee isolate	[NC_001736]	(SFVcpz)
Bovine foamy virus group:		
Bovine foamy virus		
Bovine foamy virus	[U94514]	(BFV)
Equine foamy virus group:		
Equine foamy virus ‡		
Equine foamy virus	[AF201902]	(EFV)
Feline foamy virus group:		
Feline foamy virus		
Feline foamy virus	[Y08851]	(FFV)

TENTATIVE SPECIES IN THE GENUS

None reported.

Phylogenetic Relationships within the Family

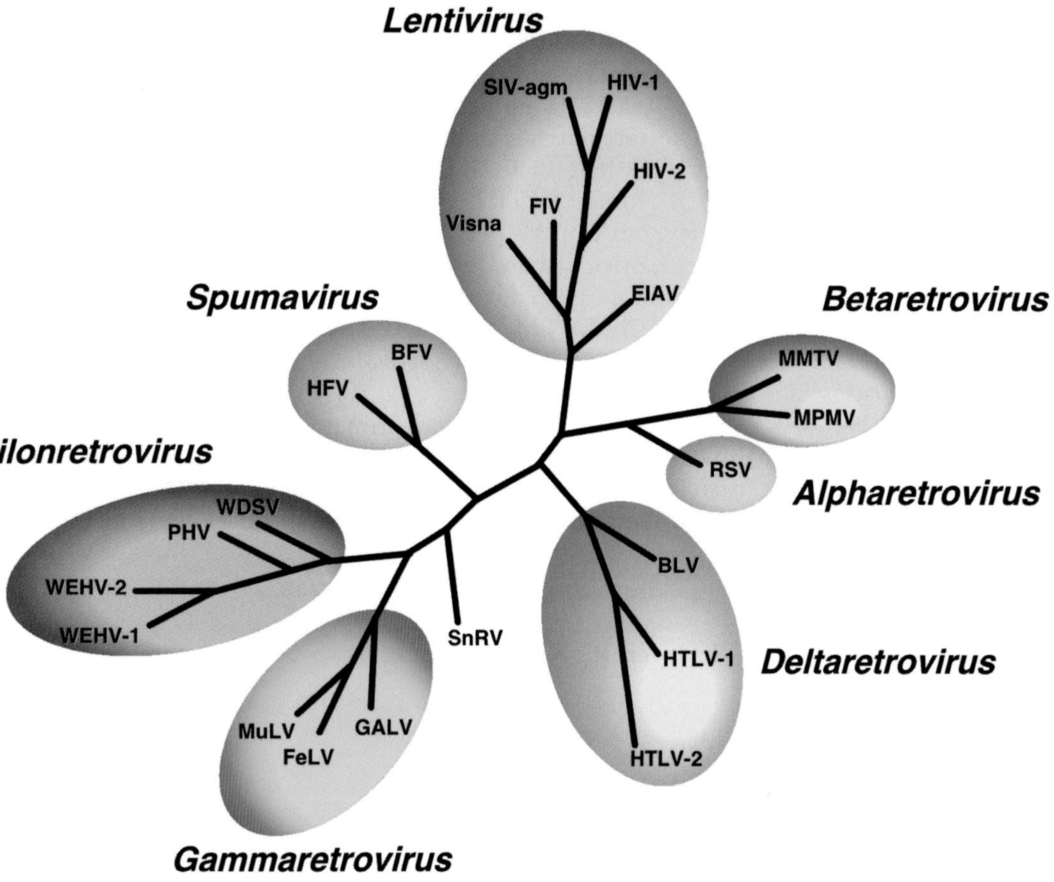

Figure 9 : Phylogenetic analysis of conserved regions of the polymerase genes of retroviruses (Courtesy of Quackenbush, S and Casey, J.). An unrooted neighbor-joining phylogenetic tree was constructed with the PHYLIP package (Felsenstein, J., 1995. "PHYLIP [Phylogeny Inference Package] Version 3.57c." University of Washington, Seattle) based on an alignment of the aa residues contained within domains 1 through 4 and part of domain 5 (Xiong, Y. and Eickbush, T.H. (1990). *EMBO J.*, **9**, 3353-3362) of reverse transcriptase genes of several retroviruses.

Similarity with Other Taxa

There is no sequence similarity between members of the families *Hepadnaviridae, Caulimoviridae, Pseudoviridae,* and *Metaviridae,* as well as with non-viral retroelements. There is however similarity through the replication strategy with members of the family *Hepadnaviridae.*

Derivation of Names

Lenti: from Latin *lentus,* "slow".
Ortho: from Greek *orthos,* "straight"
Retro: from Latin *retro,* "backwards", refers to the activity of reverse transcriptase and the transfer of genetic information from RNA to DNA.
Spuma: from Latin *spuma,* "foam".

References

Coffin, J.M. (1992). Structure and classification of retroviruses. In: *The Retroviridae,* Vol 1, (J. Levy, ed), pp 19-50. Plenum Press, New York.

Coffin, J.M., Hughes, S.H. and Varmus, H. (eds) (1997). *Retroviruses*. Cold Spring Harbor Laboratory, Cold Spring Harbor, New York.

RT VIRUSES

Desrosiers, R.C. (2001). Nonhuman lentiviruses. In: *Fields Virology*, 4th ed. (D.M. Knipe and P.M. Howley, eds), pp 2095-2122. Lippincott Williams and Wilkins, Philadelphia, Pa.

Doolittle, R.F., Feng, D.F., Johnson, M.S. and McClure, M.A. (1989). Origins and evolutionary relationships of retroviruses. *Quart. Rev. Biol.*, **64**, 1-30.

Freed, E.O., and Martin, M.A. (2001). HIVs and their replication. In: *Fields Virology*, 4th ed. (D.M. Knipe and P.M. Howley, eds), pp 1971-2042. Lippincott Williams and Wilkins, Philadelphia, Pa.

Goff, S.P. (2001). *Retroviridae*: The retroviruses and their replication. In: *Fields Virology*, 4th ed. (D.M. Knipe and P.M. Howley, eds), pp 1871-1940. Lippincott Williams and Wilkins, Philadelphia, Pa.

Green, P.L. and Chen, I.S.Y. (2001). Human T-Cell Leukemia Virus Types 1 and 2. In: *Fields Virology*, 4th ed. (D.M. Knipe and P.M. Howley, eds), pp 1941-1970. Lippincott Williams and Wilkins, Philadelphia, Pa.

Korber, B.T.M., Foley, B., Leitner, T., McCutchan, F., Hahn, B., Mellors, J.W., Myers, G. and Kuiken, K. (eds) (1997). HIV Sequence Compendium - 1997. Los Alamos National Laboratory, Los Alamos, New Mexico.

Leis, J., Baltimore, D., Bishop, J.M., Coffin, J.M., Leissner, E., Goff, S.P., Roszlan, S., Robinson, H., Skalka, A.M., Temin, H.M. and Vogt, V. (1988). Standardized and simplified nomenclature for proteins common to all retroviruses. *J. Virol.*, **62**, 1808-1809.

Linial, M.L. and Weiss, R.A. (2001). Other Human and Primate Retroviruses. In: *Fields Virology*, 4th ed. (D.M. Knipe and P.M. Howley, eds), pp 2123-2140. Lippincott Williams and Wilkins, Philadelphia, Pa.

Rethwilm, A. (ed) (2003). *Foamy viruses. Curr. Top. Microbiol. Immunol.*, **277**, Springer-Verlag, Berlin.

Varmus, H. and Brown, P. (1989). Retroviruses. In: *Mobile DNA*, (M. Howe and D. Berg, eds), pp 53-108. ASM Press, Washington.

CONTRIBUTED BY

Linial, M.L., Fan, H., Hahn, B., Lwer, R., Neil, J., Quackenbush, S., Rethwilm, A., Sonigo, P., Stoye, J. and Tristem, M.

The Double Stranded RNA Viruses

FAMILY *CYSTOVIRIDAE*

TAXONOMIC STRUCTURE OF THE FAMILY

Family *Cystoviridae*
Genus *Cystovirus*

Since only one genus is currently recognized, the family description corresponds to the genus description.

GENUS *CYSTOVIRUS*

Type Species *Pseudomonas phage ϕ6*

DISTINGUISHING FEATURES

The virion is enveloped and contains a segmented dsRNA genome (3 segments). The innermost protein capsid is a polymerase complex responsible for genome packaging, replication and transcription.

VIRION PROPERTIES

MORPHOLOGY

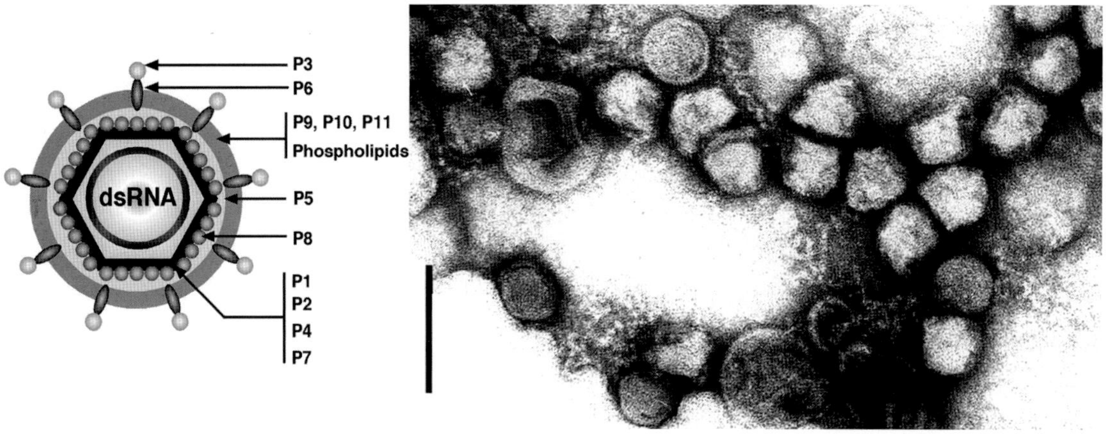

Figure 1: (Left) Schematic of cystovirus particle of an isolate of *Pseudomonas phage ϕ6*, with location of virion proteins. (Right) Negative contrast electron micrograph of Pseudomonas phage ϕ6. The bar represents 100 nm.

Virions are ~85 nm in diameter, spherical, with an envelope covered by spikes. The envelope surrounds an icosahedral nucleocapsid, ~58 nm in diameter. Removal of the nucleocapsid surface protein reveals a polymerase complex, ~43 nm in diameter.

PHYSICOCHEMICAL AND PHYSICAL PROPERTIES

Virion Mr is ~99 x 10^6 and that of the nucleocapsid is ~40 x 10^6. Virion S_{20w} is ~405S. The buoyant density of the virion is 1.27 g/cm^3 in CsCl and 1.24 g/cm^3 in sucrose. Pseudomonas phage ϕ6 virions, (ϕ6) are stable at pH 6-9 but very sensitive to ether, chloroform, and detergents.

NUCLEIC ACID

Virions contain 3 linear dsRNA segments: L (6374 bp), M (4057 bp), and S (2948 bp). The segments have a base composition of 55.2, 56.7, and 55.5% G+C, respectively. Virions contain ~10% RNA.

PROTEINS

The genome codes for 12 proteins (Fig. 2). Early proteins (P1, P2, P4, and P7) are expressed from L segment genes and form the viral polymerase complex. The association of protein P8, the NC surface protein, and P5, the viral lytic enzyme, with the polymerase complex forms the NC. These proteins are encoded in the S genome segment. Proteins P9, P10, and P13 are in the envelope. The absorption and fusion complex is formed by proteins P3 and P6. P3 is the spike protein that recognizes the host cell receptor, whereas P6 is a membrane protein with membrane fusion activity. P3 is associated with the virion through protein P6. So far, there is only one identified nonstructural protein, P12, which is needed in membrane assembly inside the host cell. Virions are ~70% protein.

Table 1: Phage ϕ6 proteins and their functions.

Segment	Protein P	Function
L	1	Structural framework (dodecahedron)
	2	RNA polymerase active site
	4	NTPase, provides the RNA packaging energy
	7	packaging factor
M	3	Spikes, host attachment
	6	Membrane, anchor for P3
	10	Membrane, lysis
	13	Membrane
S	8	Nucleocapsid surface protein
	12	Envelopment of capsid, non-structural
	9	Membrane assembly
	5	Endopeptidase, lysis and entry

LIPIDS
Virions contain ~20% phospholipids, located in the envelope. There is enough lipid to cover about one-half of the envelope surface area (the rest being protein).

CARBOHYDRATES
None reported

GENOME ORGANIZATION AND REPLICATION

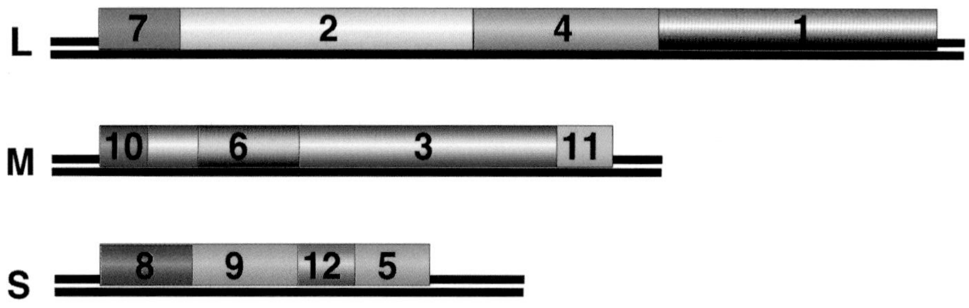

Figure 2: Genome organization of Pseudomonas phage ϕ6, (ϕ6), see text for functions of the proteins encoded by the viral genes. The gene and protein numbers are the same.

Virions adsorb to *Pseudomonas syringae* pili which retract bringing the virion into contact with the host outer membrane. The virus membrane fuses with the host outer membrane and the nucleocapsid associated lytic enzyme locally digests the peptidoglycan. The nucleocapsid enters the cell and viral polymerase is activated to produce early transcripts. Translation of L transcripts produces the early proteins, which assemble to form polymerase complexes. These package all three positive strand transcripts. Negative strand synthesis then takes place inside the polymerase complex. Transcription by these polymerase complexes produces messages for late gene synthesis. The nucleocapsid surface protein assembles on the polymerase complex and inactivates transcription. The nucleocapsid acquires membrane from the host plasma membrane with the aid of a virus specific nonstructural assembly factor. The cell eventually lyses and liberates mature progeny particles (Fig. 3).

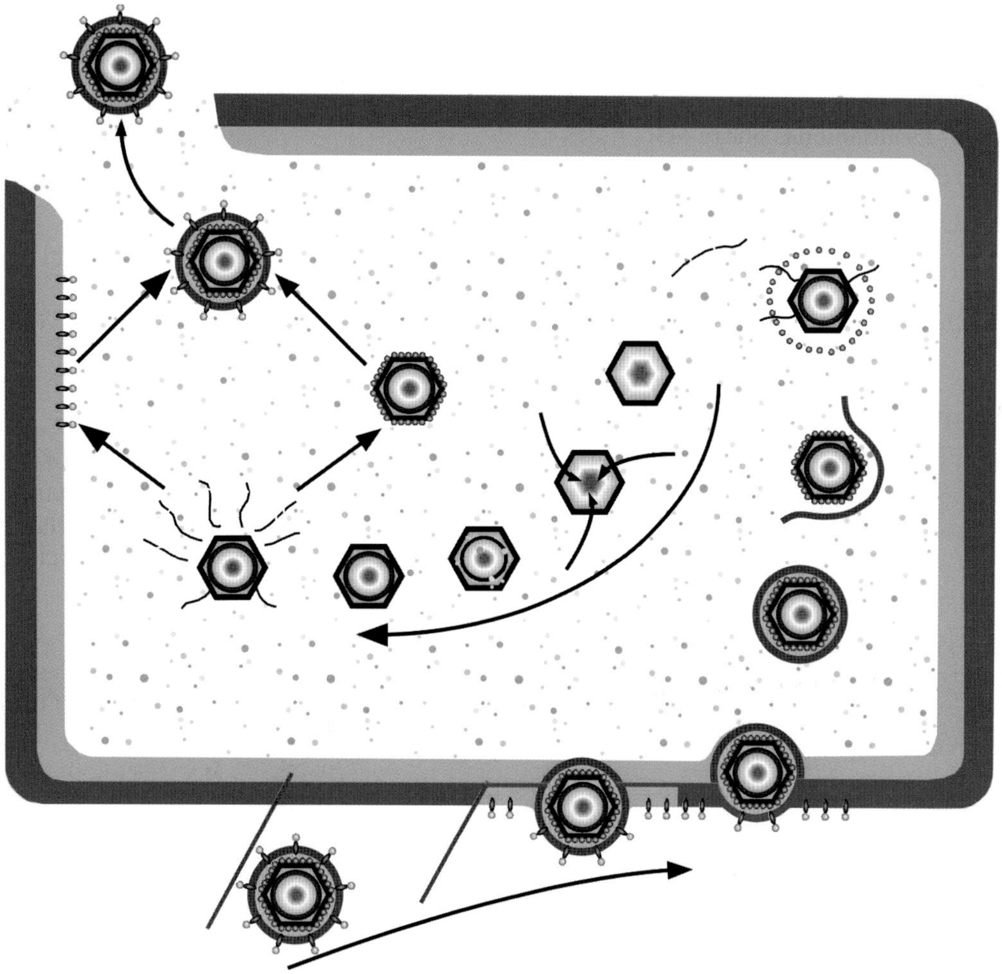

Figure 3: Schematic of the Pseudomonas phage ϕ6 (ϕ6) life cycle.

Antigenic Properties

No information available.

Biological Properties

ϕ6 infects many phytopathogenic *Pseudomonas* species. In addition, some *Pseudomonas pseudoalcaligenes* strains are sensitive to this virus.

List of Species Demarcation Criteria in the Genus

Not applicable.

List of Species in the Genus

Species names are in green italic script; strain names and synonyms are in black roman script; tentative species names are in blue roman script. Sequence accession numbers [ ], and assigned abbreviations ( ) are also listed.

Species in the Genus

Pseudomonas phage ϕ6

Pseudomonas phage ϕ6	M17461, M17462, M12921	(ϕ6)

TENTATIVE SPECIES IN THE GENUS

Pseudomonas phage ϕ7	(ϕ7)
Pseudomonas phage ϕ8	(ϕ8)
Pseudomonas phage ϕ9	(ϕ9)
Pseudomonas phage ϕ10	(ϕ10)
Pseudomonas phage ϕ11	(ϕ11)
Pseudomonas phage ϕ12	(ϕ12)
Pseudomonas phage ϕ13	(ϕ13)
Pseudomonas phage ϕ14	(ϕ14)

LIST OF UNASSIGNED VIRUSES IN THE FAMILY

None reported.

PHYLOGENETIC RELATIONSHIPS WITHIN THE FAMILY

Not applicable.

SIMILARITY WITH OTHER TAXA

In terms of genome strategy, *Pseudomonas phage ϕ6* ressembles some of the members of the family *Reoviridae*. The structure and functions of the polymerase particle containing the genome segments is the major similarity. The polymerase particle also is surrounded by two additional layers that are involved in determining host specificity as well as being crucial in the entry of the polymerase particle into the cell.

DERIVATION OF NAMES

Cysto: from Greek *kystis*, 'bladder, sack'.

REFERENCES

Ackermann, H.-W. and DuBow, M.S. (eds)(1987). Cystoviridae. In: *Viruses of Prokaryotes*, Vol II. CRC Press, Boca Raton Florida, pp 171-218.

Bamford, D.H. and Wickner, R.B. (1994). Assembly of double-stranded RNA viruses: bacteriophage ϕ6 and yeast virus L-A. *Sem. Virol.*, **5**, 61-69.

Butcher, S.J., Dokland, T., Ojala, P.M., Bamford, D.H. and Fuller, S. (1997). Intermediates in assembly pathway of the double-stranded RNA virus ϕ6. *EMBO J.*, **16**, 4477-4487.

Mindich, L. (1988). Bacteriophage ϕ6: A unique virus having a lipid-containing membrane and a genome composed of three dsRNA segments. *Adv. Virus Res.*, **35**, 137-176.

Mindich, L., Qiao, X., Qiao, J., Onodera, S., Romantschuk, M. and Hoogstraten, D. (1999). Isolation of additional bacteriophages with genomes of segmented double-stranded RNA. *J. Bacteriol.*, **181**, 4505-4508.

Olkkonen, V.M., Gottlieb, P., Strassman, J., Qiao, X., Bamford, D.H. and Mindich, L. (1990). *In vitro* assembly of infectious nucleocapsid of bacteriophage ϕ6: Formational a recombinant double-stranded RNA virus. *Proc. Natl. Acad. Sci. USA*, **87**, 9173-9177.

Onodera, S., Qiao, X., Qiao, J. and Mindich, L. (1998). Directed changes in the number of double-stranded RNA genomic segments in bacteriophage phi6. *Proc. Acad. Natl. Sci. USA*, **95**, 3920-3924.

Poranen, M.M, Paatero, A.O., Tuma, R. and Bamford, D.H. (2001). Self-assembly of a viral molecular machine from purified protein and RNA constituents. *Mol. Cell*, **7**, 845-854.

Qiao, X., Qiao, J. and Mindich, L. (1997). Stoichiometric packaging of the three genomic segments of double-stranded RNA bacteriophage ϕ6. *Proc. Acad. Natl. Sci. USA*, **94**, 4074-4079.

CONTRIBUTED BY

Bamford, D.H.

FAMILY REOVIRIDAE

TAXONOMIC STRUCTURE OF THE FAMILY

Family	*Reoviridae*
Genus	*Orthoreovirus*
Genus	*Orbivirus*
Genus	*Rotavirus*
Genus	*Coltivirus*
Genus	*Seadornavirus*
Genus	*Aquareovirus*
Genus	*Idnoreovirus*
Genus	*Cypovirus*
Genus	*Fijivirus*
Genus	*Phytoreovirus*
Genus	*Oryzavirus*
Genus	*Mycoreovirus*

VIRION PROPERTIES

MORPHOLOGY

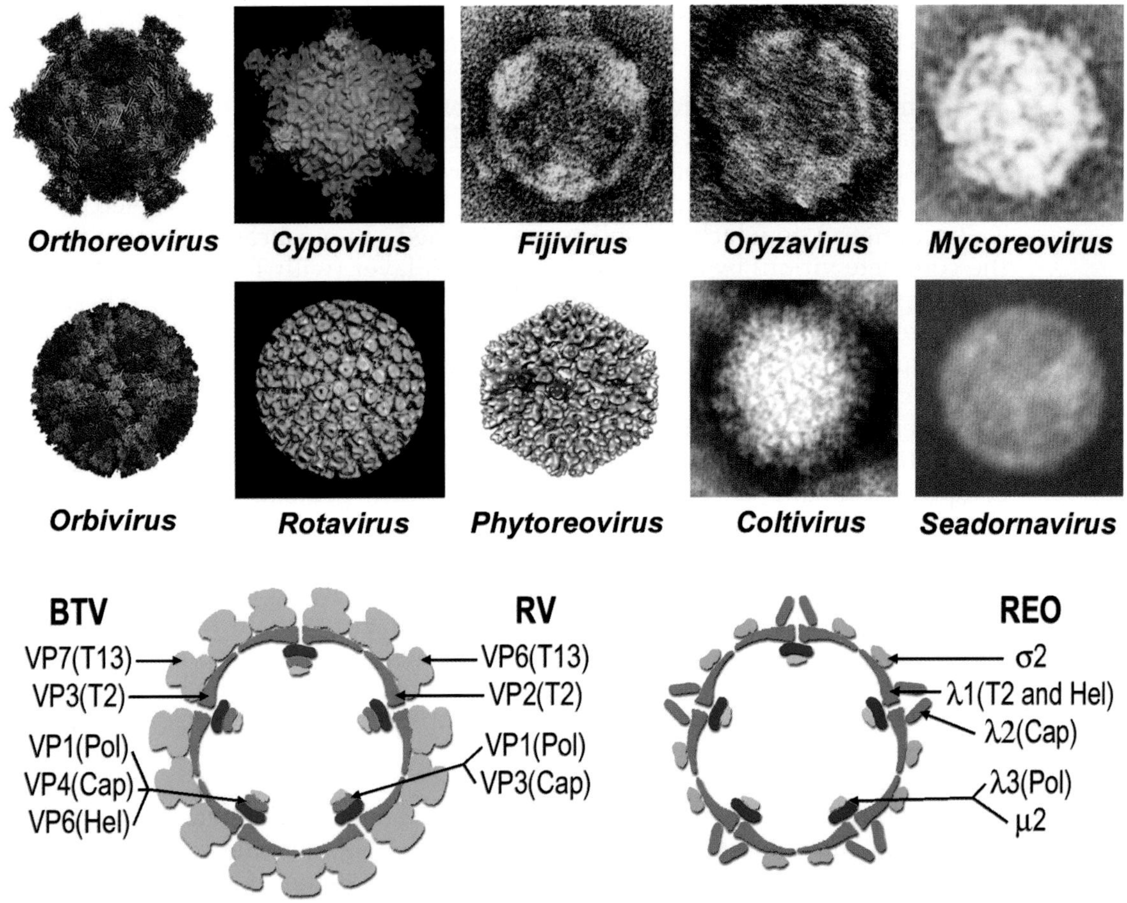

Figure 1. (Top and Center) A comparison of two distinct core particle morphologies (spiked and unspiked) present amongst members of different genera of the family *Reoviridae*. *Orbivirus*: a 3D model from x-ray crystallography of the core particle of an isolate of *Bluetongue-1*. *Orthovirus:* a 3D model from x-ray crystallography studies of a core particle of an isolate of *Reovirus 3*. *Cypovirus*: a 3D cryo-EM reconstruction of a particle of an isolate of *Cypovirus 5*, at 25 Å resolution. *Rotavirus*: a 3D cryo-EM reconstruction of a double

shelled particle of an isolate of *Rotavirus A* (SiRV-A/SA11), at 25 Å resolution. *Fijivirus*: an electron micrograph of a core particle of an isolate of *Maize rough dwarf virus*. *Phytoreovirus*: a 3D cryo-EM reconstruction of the double-shelled particle of an isolate of *Rice dwarf virus*, at 25-Å resolution (highlighted in colour are a contiguous "group of 5 trimers" found in each asymmetric unit). *Coltivirus*: an electron micrograph of a negatively stained double-shelled particle of an isolate of *Colorado tick fever virus*. *Oryzavirus*: an electron micrograph of a negatively stained core particle of an isolate of *Rice ragged stunt virus*. *Mycoreovirus*: an electron micrograph of a negatively stained core particle of *Mycoreovirus 1* (Rosallinia necatrix mycoreovirus-1). *Seadornavirus*: an electron micrograph of a negatively stained core particle of an isolate of *Banna virus*. The reconstructions and electron micrographs are not shown to exactly the same scale. The outer capsid morphologies of members of the different genera of the family *Reoviridae* are more variable and may appear smooth, or with surface projects, or may even be absent. (Bottom) shows a diagrammatic representations (on the left) of the core particles of an orbivirus (BTV), or rotavirus (RV), which have a well defined capsomeric structure but lack large surface projections at the 5 fold icosahedral axes, as compared to the 'turreted' (spiked) core particle of an orthoreovirus (Reo). (Courtesy of J. Diprose).

Particles of the family *Reoviridae* have icosahedral symmetry but may appear spherical in shape. The protein capsid is organized as one, two, or three concentric capsid layers, which surround the dsRNA segments of the viral genome, with an overall diameter of 60-80 nm. The twelve genera can be divided between two groups. One group contains viruses which have relatively large "spikes" or "turrets" situated at the 12 icosahedral vertices of either the virus or core particle, including: *Orthoreovirus*, *Cypovirus*, *Aquareovirus*, *Fijivirus*, *Oryzavirus*, *Idnoreovirus* and *Mycoreovirus*, as well as some of the unclassified or unassigned viruses from invertebrates). The second group includes those genera containing 'smooth' viruses that do not have these large surface projections on their virions or core particles, giving them an almost spherical appearance, including *Orbivirus*, *Rotavirus*, *Coltivirus*, *Seadornavirus* and *Phytoreovirus*.

The nomenclature used to describe different viral particles (with different numbers of capsid layers) varies between the genera. The current nomenclature will be explained in each case. The transcriptionally active core particle of the "spiked" viruses appears to contain only a single complete capsid layer (which has been interpreted as having T=1 or T=2 symmetry), to which the spikes are attached. In most cases the core is surrounded (in the complete virion) by an incomplete protein layer (with T=13 symmetry) that forms the outer capsid, which is penetrated by the spikes on the core surface. These virus particles are therefore usually regarded as double-shelled (= triple layered). One exception is the cypoviruses, which are distinct, having transcriptionally active virions with only a single capsid shell that are equivalent to the "core" particles of viruses from other genera. However, virus particles of most cypoviruses are characteristically occluded (either singly or multiply) within the matrix of proteinaceous crystals called polyhedra. These are composed primarily (> 90%) of the viral 'polyhedrin' protein.

In contrast, virions of the "non-spiked" viruses have an inner protein layer, which may be relatively fragile. This 'subcore' (interpreted as having T=2 symmetry) has structural similarities to the innermost shell of the spiked viruses. In the transcriptionally active 'core' particle, the subcore is surrounded and reinforced by a complete core-surface layer, which has T=13 symmetry. These double-layered cores have no surface spikes and in the intact virions are surrounded by an outer capsid shell, giving rise to three-layered virus particles.

The innermost protein layer of the virus has an internal diameter of approximately 50 to 60 nm and surrounds the 10, 11, or 12 linear dsRNA genome segments. In the 'smooth-cored' genera, the enzymatically active minor proteins of the virion are also situated within this central space, attached to the inner surface at the five-fold axes of symmetry; these include RdRp (transcriptase and replicase), NTPase, helicase, capping and transmethylase enzymes. In the "spiked" genera some of these enzyme proteins form the turrets on the surface of the core. These hollow projections also appear to act as conduits for the exit of nascent mRNA synthesized by the core-associated enzymes.

Particles of some genera can leave infected cells by budding (*Orbivirus*) or can bud into the endoplasmic reticulum (*Rotavirus*), acquiring a membrane envelope derived from cellular membranes, although in most cases this envelope appears to be transient. In some genera the protein components of the outer capsid shell can be modified by proteases (such as trypsin or chymotrypsin) forming "infectious" or "intermediate" subviral particles (ISVPs). ISVP formation may occur intracellularly (within endocytic vesicles, which represent an entry route for virus particles taken in from the cell surface), extracellularly (e.g. in the intestinal lumen following peroral inoculation, or in the host's blood stream), or *in vitro*, following treatment with proteases). The virion-to-ISVP transition is essential for the infectivity of these viruses.

Physicochemical and Physical Properties

The virion Mr is about 12 x 10^7. The buoyant density in CsCl is 1.36-1.39 g/cm^3. Virus infectivity is moderately resistant to heat, organic solvents (e.g., ether), and non-ionic detergents (variable, depending on both virus strain and detergent). The pH stability of virions varies among the genera.

Nucleic Acid

Virions contain 10, 11, or 12 segments of linear dsRNA, depending on the genus. The individual Mr of these RNAs ranges from 0.2 to 3.0 x 10^6. The total Mr of the genome is 12 to 20 x 10^6. The RNA constitutes about 15-20% of the virion dry weight. The positive strands of each duplex have a 5'-terminal Cap (type 1 structure: ^{7m}GpppN$^{2'Om}$pNp....). There are data to suggest that negative strands may have phosphorylated 5'-termini. However, in some cases (e.g., BTV, *Orbivirus*) the negative strand has been shown to be poorly labelled (with the same efficiency as the positive strand) by treatment with polynucleotide kinase and [gamma-32P] ATP, suggesting that it may also have a 'blocked' 5'-structure. Both RNA strands have a 3'OH, and viral mRNAs lack 3'-poly (A) tails. The viral dsRNA species are present within virus particles in equimolar proportions, representing one copy of each segment per virion. Intact virions of some genera also contain significant amounts of short ssRNA oligonucleotides.

Proteins

At least three internal virion structural-proteins have enzyme activities involved in RNA synthesis and capping, including a conservative RdRp (which may function as transcriptase [positive strand synthesis on a dsRNA template] or replicase [negative strand synthesis on a positive strand ssRNA template], nucleotide phosphohydrolase, guanylyltransferase, two distinct transmethylase activities, and dsRNA template-unwinding [helicase] activities). Some of the minor proteins also may play a structurally significant role as components of the virion, together with at least three major capsid proteins. The proteins range in size from 15 to 155 kDa and constitute about 80-85% of the dry weight of virions.

Lipids

Mature virions lack a lipid envelope but may be associated with cell membranes. Depending on the genus, a myristyl residue may be covalently attached to one of the virion proteins. Coltiviruses, rotaviruses, and orbiviruses have an intermediate in virus morphogenesis or release, which may have a lipid envelope, which is subsequently lost or removed.

Carbohydrates

In some genera, one of the outer CP can be glycosylated with high mannose glycans or O-linked N-acetylglucosamine. A small non-structural (NS) viral protein also may be glycosylated.

Genome Organisation and Replication

The viral RNA species are mostly monocistronic, although some segments have second functional in frame initiation codons or additional ORFs. Proteins are encoded on one strand only of each duplex (the mRNA species). The mode of entry of viruses into cells varies between genera but usually results in loss of outer capsid components. Transcriptionally active, parental-virion derived particles (cores or double layered particles) are released into the cell cytoplasm. Repetitive asymmetric transcription of full-length mRNA species from each dsRNA segment occurs, within these particles, throughout the course of infection. The mRNA products, which are produced in larger copy numbers from the smaller segments, are extruded from the icosahedral apices of these particles. Structures, termed 'viroplasms' or 'virus inclusion bodies' (VIB), occur in localized areas of the cytoplasm. They appear to be the sites of viral mRNA synthesis, genome replication and particle assembly. VIB have a granular and moderately electron dense appearance when viewed by electron microscopy and may contain nascent subviral particles. Outer capsid components appear to be added at the periphery of the VIB.

The mechanism of genome assembly and synthesis remains largely uncharacterized. Evidence has been obtained for orthoreoviruses and rotaviruses that sets of capped mRNAs and certain NS proteins are incorporated into "assortment complexes" that are considered to be the precursors of progeny virus particles. These mRNAs are then used as templates for a single round of minus strand synthesis, thereby reforming the dsRNA genome segments of a progeny virus particle. The different species of mRNAs in the cell cytoplasm are present in non-equimolar ratios. However, the dsRNA genome segments are usually packaged in exactly equimolar ratios (one copy of each genome segment per particle). The selection of viral mRNAs for packaging is therefore thought to be highly specific, involving recognition signals on each mRNA species. Genome segment reassortment, involving the selection and packaging of mRNAs from different parental strains, occurs readily in cells that are co-infected with different viruses of the same species (which presumably share the same packaging signals).

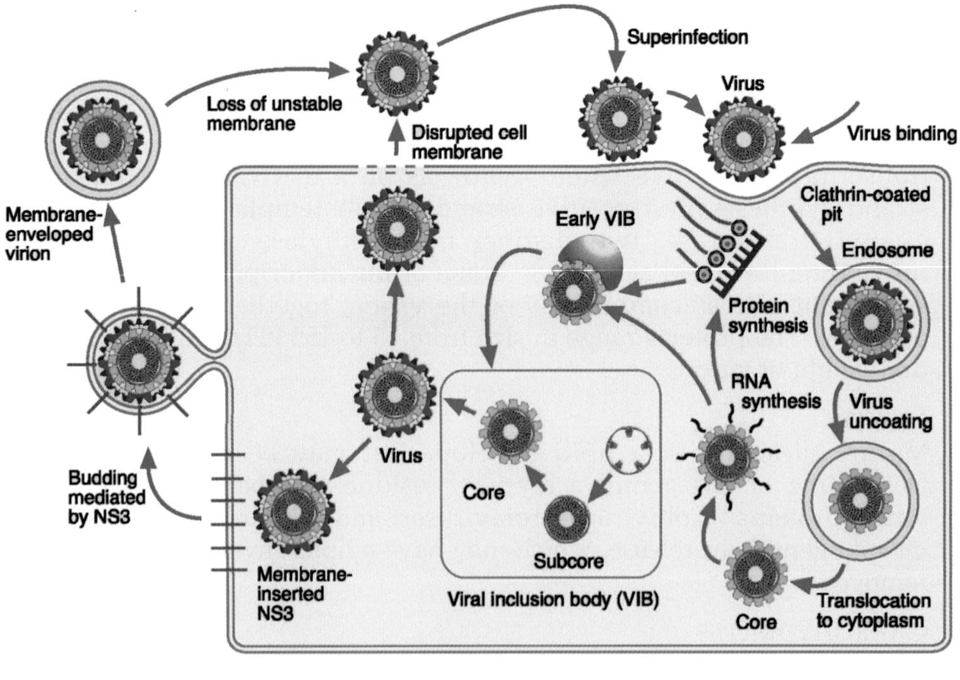

Figure 2. Typical virus replication cycle of a reovirus (presented for an orbivirus).

The RNA segments have conserved terminal sequences at both ends, which may be involved as recognition signals for the viral transcriptase and replicase functions. These sequences may also be essential for selection and incorporation of the RNAs into the nascent progeny particles. In many cases, sequences near to the 5′- and 3′-termini of the positive-sense viral RNAs share extensive complementarity, interrupted by short sequences predicted to form stem-loops and other secondary structures. These findings and mounting experimental evidence suggest that the viral mRNAs contain both primary sequences and higher-order structures that are involved in regulation of RNA function (i.e., translation, replication, or packaging). A consistent feature of the secondary structures predicted for viral positive-sense RNAs is that the conserved 5′- and 3′-terminal sequences remain non-hybridized. Indeed, non-hybridized 3′-terminal sequences has been shown to be required for efficient negative-strand synthesis. The dsRNA within assembled particles has been shown, in at least in some genera, to be packaged as a series of concentric and highly organized shells, which also have elements of icosahedral symmetry.

In addition to the parental virus derived subviral particles (parentalcores), progeny core or double layered particles also synthesize mRNAs, providing an amplification step in replication. Depending on the genus, some NS proteins are involved in translocation of virus particles within cells or virus egress by budding. Many cypoviruses also form polyhedra, which are large crystalline protein matrices that occlude virus particles and which appear to be involved in transmission between individual insect hosts. The steps involved in virion morphogenesis and virus egress from cells vary according to the genus. The only known examples of non-enveloped viruses that induce cell-cell fusion and syncytium formation in virus-infected cells, are members of the family *Reoviridae*. In the case of the fusogenic orthoreoviruses, syncytium formation promotes a rapid lytic response and release of progeny virions.

DsRNA

Antigenic Properties

The viruses that infect vertebrate hosts generally possess both "serogroup" (virus-species) specific antigens, and (within each species or serogroup) more variable "serotype-specific" antigens. The viruses that infect plants and insects only, may show a greater uniformity and less antigenic variation in their different proteins, possibly due to the lack of neutralizing antibodies in the host and therefore the absence of antibody selective pressure on neutralization-specific antigens. No antigenic relationship has been found between the viruses in different genera. Some viruses bind erythrocytes (hemagglutination).

Biological Properties

The biological properties of the viruses vary according to the genus. Some viruses replicate only in certain vertebrate species (orthoreoviruses, rotaviruses) and are transmitted between hosts by respiratory or fecal-oral routes. Other vertebrate viruses (orbiviruses, coltiviruses, and seadornaviruses) replicate in both, arthropod vectors (e.g., gnats, mosquitoes, or ticks, etc.) and vertebrate hosts. Plant viruses (phytoreoviruses, fijiviruses, oryzaviruses) replicate in both, plants and arthropod vectors (leafhoppers). Viruses that infect insects (cypoviruses) are transmitted by contact or fecal-oral routes.

Species Demarcation Criteria in the Family

The number of genome segments is in most cases characteristic for viruses within a single genus (usually 10, 11, or 12), although the genus *Mycoreovirus* currently contains viruses with both 11 and 12 genome segments. The host (and vector) range and the disease symptoms are also important indicators helping to identify viruses from different genera. Capsid structure (number of capsid layers, the presence of 'spiked' or 'unspiked' cores, symmetry and structure of outer capsid) can also be significant. The level of sequence variation, particularly in the more conserved genome segments and proteins (for example as detected by comparisons of the polymerase, or inner capsid shell proteins, and the segments from which they are translated) can be used to distinguish members of the different genera. Available data suggests that isolates from different genera will usually show greater than 74% aa variation in the polymerase, while the level of variation within a single genus is lower, usually below 67%. However the polymerase of *Rotavirus B* isolates shows a high level of aa divergence (up to 79%) from that of other rotaviruses. The parameters that are used to identify and distinguish individual virus species within each genus (see relevant sections) can also be used to distinguish different genera (since no species can belong to two genera at the same time).

Unassigned Viruses in the Family

Virus	Source or host species	Abbreviation	Characteristics
Viruses of Arachnida			
Buthus occitanus reovirus	*Buthus occitanus* (Scorpionidae: scorpion)	(BoRV)	uncharacterized
Viruses of Arthropoda			
Cimex lactularius reovirus	*Cimex lactularius* (Hemiptera: bed bug)	(ClRV)	11 segments, icosahedral double-shelled capsid ~50 nm diameter
Viruses of Crustacea			
Carcinus mediterraneus W2 virus	(Decapoda: crab)	(CcRV-W2)	12 segments
Macropipus depurator P virus	*Macropipus depurator* (Decapoda: crab)	(DpPV) (MdRV-P)	12 segments double-shelled capsid ~65nm diameter
Porcelio dilatatus reovirus	*Porcelio dilatatus* (Isopoda: terrestrial crustacean)	(PdRV)	uncharacterized

Phylogenetic Relations in the Family

Table 1. Comparison of sequences surrounding the conserved RdRp motifs of reoviruses.

Motifs	**I**	**IV,1,A**	**V,2,B**	**VI,3,C**	**D**
Consensus	.grrtRiI	D.s.wd..	SGe.aTs.a....nla	.qvqGDDtlm.ikdg	he.n.sK.s
BTV (515)	PIKATRTI 72	DYSEYDTH 119	SGENSTLIANSMHNMA 21	EQYVGDDTLFYTKLD 22	HEASPSKTM (804)
Rotavirus (455)	PGRRTRII 57	DVSQWDSS 63	SGEKQTKAANSIANLA 19	IRVDGDDNYAVLQFN 20	RMNAKVKAL (669)
RDV (643)	AWRPVRPI 73	DCTSWDQT 76	SGRLDTFFMNSVQNLI 20	FQVAGDDAIM.VYDG 24	HIINPQKTV (890)
Reovirus S3 (521)	VQRRPRSI 56	DISACDAS 89	SGSTATSTEHTANNST 31	YVCQGDDGLM.IIDG 21	GEEFGWKYD (772)
NLRV (646)	IDRRGRII 60	DMSGMDAH 90	SGLFATSGQHT.MFLV 20	NYVMGDDIFQNIKNG 24	IDGNYSKYS (894)
RRSV (500)	IGRRQRAI 62	DASVQASV 83	SGQPFTTVHHTFTLSN 1	LTVQGDDTRT.INYG 15	VSDWGFKVS (735)

The regions covering the putative polymerase module in RRSV P4 (aa 500 to 735) and other reoviruses were analyzed using the GCG program PILEUP and further aligned manually taking into account the polymerase motifs presented and aligned by Poch *et al.*, (1989) (A-D), Bruenn (1991)(1-3), and Koonin (1992) (I, IV-VI).

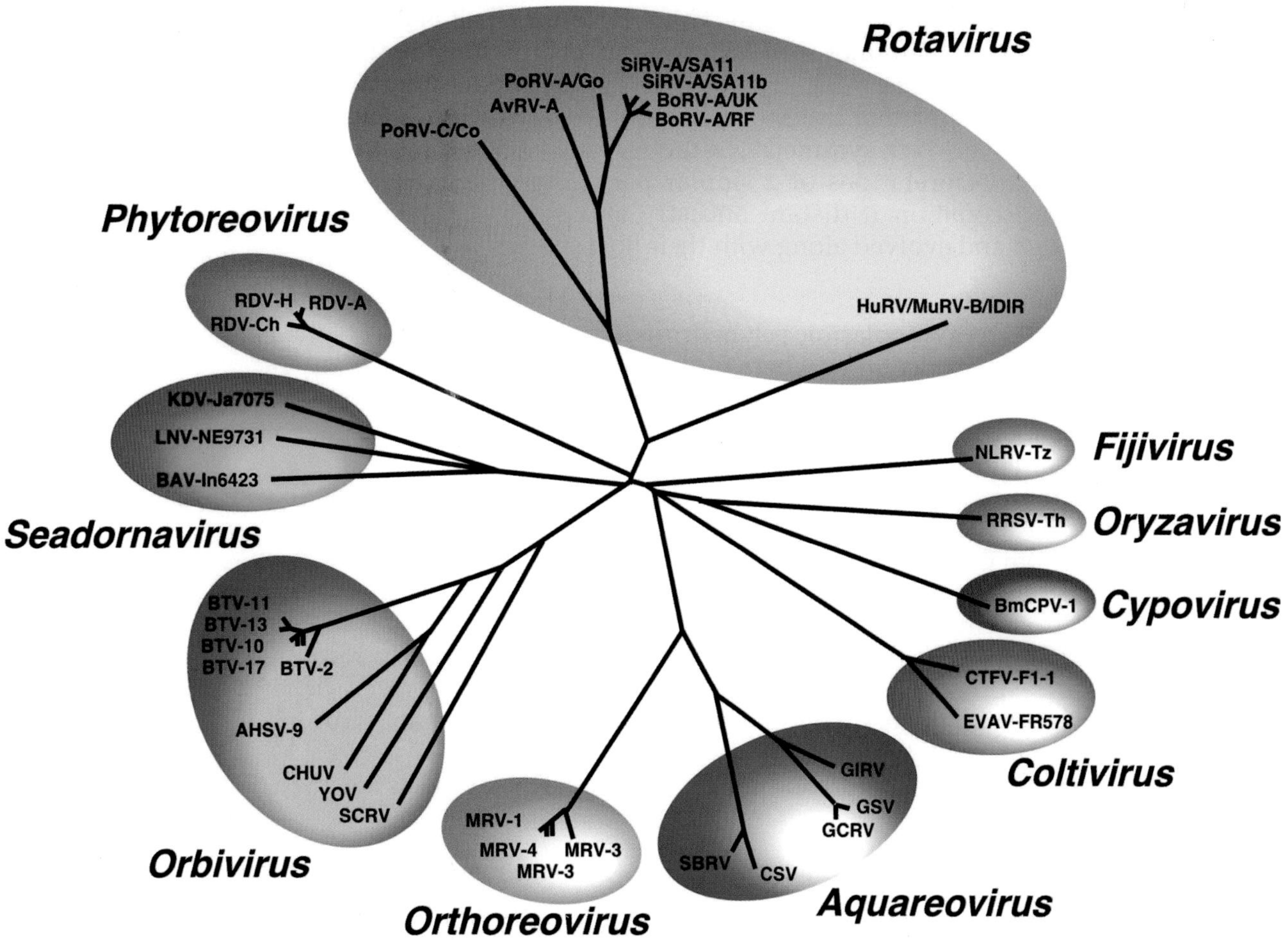

Figure 3: Neighbor joining tree constructed with the AA sequences of putative RdRp of representative viruses from the following genera of the family *Reoviridae* [accession numbers]: *Seadornavirus*, *Banna virus*: isolate BAV-In6423 [AF133430], *Kadipiro virus*: isolate KDV-Ja7075 [AF133429]. *Coltivirus*, *Colorado tick fever virus*, isolate CTFV-Fl [AF134529]. *Orthoreovirus*, *Mammalian orthoreovirus*, serotype-1 (MRV-1) [M24734], serotype-2 (MRV-2) [M31057], serotype-3 (MRV-3) [M31058], Ndelle virus (NDEV) [AF368033]. *Aquareovirus*, *Aquareovirus C*, isolate Golden shiner virus (GSV) [AF403399], Grass carp reovirus (GCRV) [AF260512], *Aquareovirus A*, isolate Striped bass reovirus (SBRV) [AF450318], isolate Chum salmon reovirus (CSRV) [AF418295], tentative species Golden ide reovirus (GIRV) [AF450323]. *Orbivirus*, *African horse sickness virus*, serotype-9 (AHSV-9) [U94887], *Bluetongue virus*, serotype-2 (BTV-2) [L20508], serotype-10 (BTV-10) [X12819], serotype-11 (BTV-11) [L20445], serotype-13 (BTV-13) [L20446], serotype-17 (BTV-17) [L20447], species *Palyam virus*, isolate CHUV [Baa76549]. *Rotavirus*, *Rotavirus A*, strain BoRV-A/RF [J04346], strain BoRV-A/UK [X55444], strain SiRV-A/SA11b [X16830], strain SiRV-A/SA11 [AF015955], strain PoRV-A/Go [M32805], strain AvRV-A [Baa24146], *Rotavirus B*, strain Hu/MuRV-B/IDIR [M97203], *Rotavirus C*, strain PoRV-C/Co [M74216], *Fijivirus Nilaparvata lugens reovirus*, strain NLRV-Iz [D49693]. *Phytoreovirus*, *Rice dwarf virus*, isolate RDV-Ch [U73201], isolate RDV-H [D10222], isolate RDV-A [D90198]. *Mycoreovirus*, *Mycoreovirus-1*, isolate CpMYRV-1 [AY277888]. *Oryzavirus*, *Rice ragged stunt virus*, strain RRSV-Th [U66714]. *Cypovirus*, *Bombyx mori cytoplasmic polyhedrosis virus*-1 strain Bm-1 CPV [AF323781].

It was found that all member-viruses of a single genus have aa identity over 30%. The only exception is the *Rotavirus B*, which is only 22% identical to other rotaviruses. Between member-viruses of genera *Aquareovirus* and *Orthoreovirus*, the aa identity ranged between 40 and 42%. This value is therefore comparable with aa identity observed between member-viruses of a single genus.

Similarities with Other Taxa

Although there is little evidence for nucleotide sequence similarities with other families of dsRNA viruses, it may be significant that some (e.g. the families *Cystoviridae* and

Totiviridae) also have particles in which the inner shell is characteristically composed of a 120 copies of a triangular protein, arranged in a manner similar to that of the members of the family *Reoviridae*. This protein provides an apparently simple yet elegant mechanism of assembling the inner icosahedral capsid shell, which has alternatively been described as having T=1 or T=2 symmetry, although it is important to note these are essentially academic interpretations of a similar particle architecture. These similarities may also indicate a common if distant ancestry and suggest that these viruses may even have diversified and evolved along with their host species.

DERIVATIONS OF NAMES

Cypo: sigla from *cy*toplasmic *po*lyhedrosis.
Fiji: from name of country where virus was first isolated.
Idno: from 'I.D.' indicating 'insect-derived'; 'N.O.' indicating that in contrast to the cypoviruses, they are 'non-occluded',
Myco: from 'Myco' from Latin fungus,
Orbi: from Latin orbis, "ring" or "circle" in recognition of the ring-like structures observed in micrographs of the surface of core particles.
Ortho: from Greek orthos "straight".
Oryza: from Latin oryza, "rice".
Phyto: from Greek phyton, "plant".
Reo: sigla from *r*espiratory *e*nteric *o*rphan, due to the early recognition that the viruses caused respiratory and enteric infections, and the (incorrect) belief that they were not associated with disease, hence they were considered "orphan" viruses.
Rota: from Latin rota, "wheel".
Seadorna: sigla from '**S**outh **e**astern **A**sia **do**deca **RNA** **virus**es'

REFERENCES

Eley, S.M., Gardner, R., Molyneux, D.H. and Moore, N.F. (1987). A reovirus from the bedbug, *Cimex lectularius*. *J. Gen. Virol.*, **68**, 195-199.

Joklik, W.K. (ed)(1983). The *Reoviridae*. Plenum Press, New York.

Juchault, P., Louis, C., Martin, G. and Noulin, G. (1991). Masculinization of female isopods (*Crustacea*) correlated with non-mendelian inheritance of cytoplasmic viruses. *Proc. Nat. Acad. Sci. USA*, **88** ,10460-10464.

Lopez-Ferber, M., Veyrunes, J.C. and Croizier, L. (1989). Drosophila S virus is a member of the *Reoviridae* family. *J. Virol.*, **63**, 1007-1009.

Montanié, H., Bossy, J.P. and Bonami, J.R (1993). Morphological and genomic characterization of two reoviruses (P and W2) pathogenic for marine crustaceans; do they constitute a novel genus of the *Reoviridae* family? *J. Gen. Virol.*, **74**, 1551-1561.

Morel, G. (1975). Un virus cytoplasmique chez le scorpion *Buthus occitanus* Amoreux. *CRAS*, Paris, **280**, 2893-2894.

Moussa, A.Y. (1978). A new virus disease in the House fly *Musca domestica* (*Diptera*). *J. Inver. Pathol.*, **31**, 204-216.

Page, R.D.M. (1996). Treeview: An application to display phylogenetic trees on personal computers. *Comp. App. Biosciences*, **12**, 357-358.

Plus, N., Gissman, L., Veyrunes, J.C., Pfister, H. and Gateff, E. (1981). Reoviruses of *Drosophila* and *Ceratitis* populations and of drosophila cell lines ; a new genus of the Reoviridae family. *Ann. Virol.*, **132E**, 261-270.

Plus, N. and Croizier, G. (1982). Further studies on the genome of *Ceratitis capitata* I virus (Reoviridae). *Ann. Virol.*, **133E**, 489-492.

Rabouille, A., Bigot, Y., Drezen, J.M., Sizaret, P.Y., Hamelin, M.H. and Periquet, G. (1994). A member of the *Reoviridae* (DpRV) has a ploidy-specific genomic segment in the wasp *Diadromus pulchellus* (*Hymenoptera*). *Virology*, **205**, 228-237.

Thompson, J.D., Gibson, T.J., Plewniak, F., Jeanmougin, F. and Higgins, D.G. (1997). The Clustal X windows interface: flexible strategies for multiple sequence alignment aided by quality analysis tools. *Nuc. Acids Res.*, **25**, 4876-4882.

CONTRIBUTORS

Mertens, P.P.C., Duncan, R., Attoui, H. and Dermody, T.S.

GENUS ORTHOREOVIRUS

Type Species *Mammalian orthoreovirus*

DISTINGUISHING FEATURES

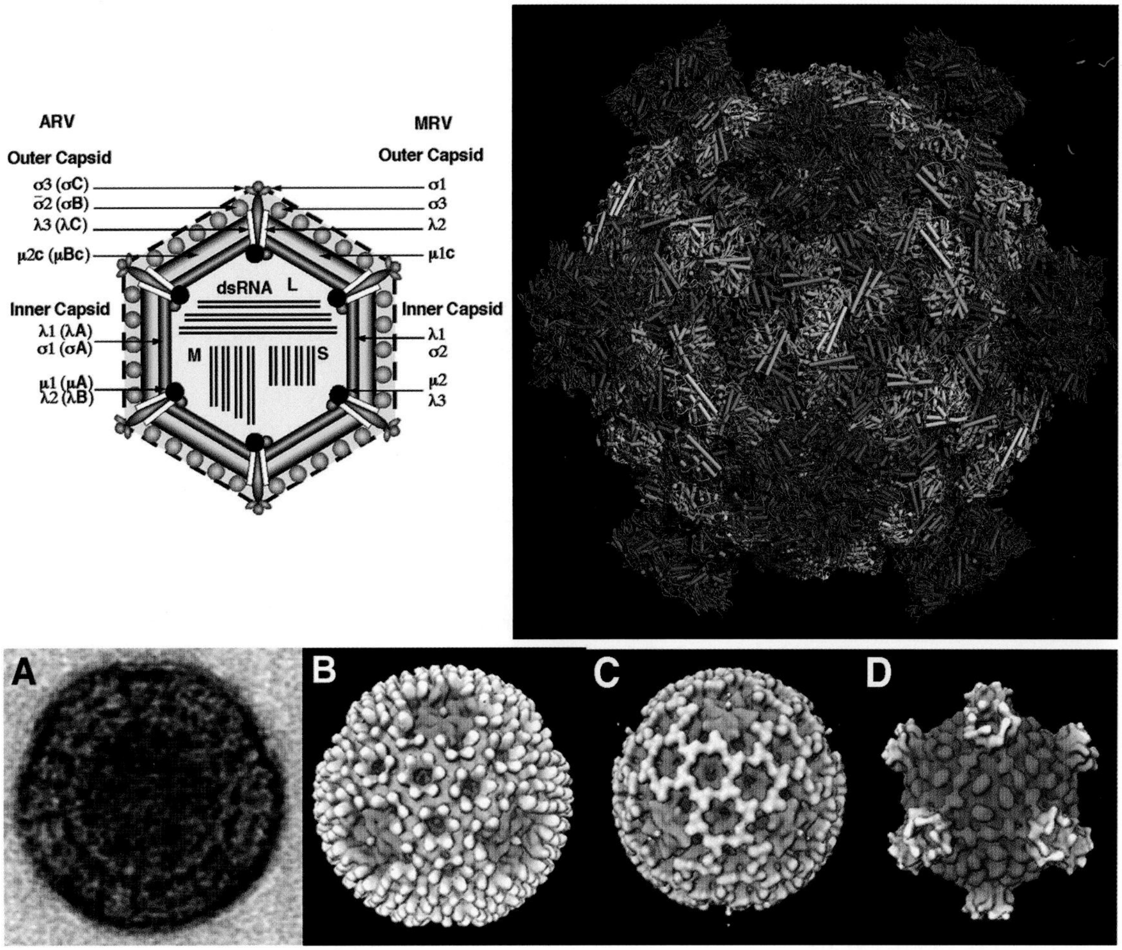

Figure 1: (Top left) Diagrammatic representation of an orthoreovirus particle in cross section. The locations and identities of the virus structural proteins are indicated using the nomenclature scheme for both Mammalian orthoreovirus (MRV) and Avian orthoreovirus (ARV). The protein components of the inner and outer capsids are indicated (from Duncan, (1999). and Martin *et al.*, (1975)). (Top right) Computer rendering of the inner capsid of Mammalian orthoreovirus 1 (MRV-1). (Bottom) Electron micrograph of a negatively stained MRV-1 particle (panel A). Image reconstructions from cryoelectron microscopy of MRV-1 virions (panel B), infectious subviral particles (ISVPs) (panel C), and cores (panel D). All particles are viewed from the three-fold axis of rotational symmetry (Courtesy of M. Nibert and T. Baker).

Orthoreoviruses infect only vertebrates and are spread by the respiratory or fecal-oral routes. All members of the genus have: (1) a well defined capsid structure, as observed by electron microscopy and negative staining, with twelve 'spikes' or 'turrets' situated on the surface of the core particle, at the icosahedral vertices; (2) ten dsRNAs including three large (L), three medium (M), and four small (S) size-class RNA genome segments; (3) a characteristic protein profile with three λ, three μ, and four σ primary translation products; (4) additional small gene products encoded by a polycistronic genome segment. Members of all of the five species, except the *Mammalian reovirus*, induce syncytium formation.

VIRION PROPERTIES

MORPHOLOGY

Virions are icosahedral with a roughly spherical appearance and possess a double-layered protein capsid discernible by negative staining (Fig. 1A). High-resolution images have been obtained by cryoelectron microscopy and image reconstruction of various Mammalian orthoreovirus (MRV) particles. A similar morphological examination of Avian orthoreovirus (ARV) virions has revealed only minor differences in the particle morphology of the two species. The virion consists of a central compartment (about 48 nm in diameter) containing the dsRNA genome segments, surrounded by an inner capsid that has T=1 symmetry (60 nm diameter: composed of 120 copies of protein λ1(Hel)) and an outer capsid (85 nm diameter) that has T=13 (laevo) symmetry. The virion surface is covered by 600 finger-like projections arranged in 60 hexameric and 60 tetrameric clusters that surround solvent channels that extend radially into the outer capsid layer (Fig. 1B). Intact virions also contain large, open depressions with a flower-shaped structure at the five-fold axes, resulting in an angular capsid profile when viewed in the three-fold orientation (Fig. 1A, 1B). Infectious subviral particles (ISVPs), generated by partial removal of the outer CP (Fig. 1C), are approximately 80 nm in diameter. The flower-shaped structures at the five-fold axes of the ISVPs may contain an extended form of the viral attachment protein, σ1, which protrudes as a 40 nm fibre from the vertices. MRV core particles generated by more extensive removal of the outer CPs (Fig. 1D) have also been examined by x-ray crystallography and have 150 ellipsoidal nodules (protein σ2) on their surface and distinctive 'turrets' located at the 5-fold axes. These projections, which are altered conformations of the flower-shaped structures observed on intact virions (composed of protein λ2(Cap), the viral capping enzyme) are about 10 nm in length, possessing central channels 5-8 nm in diameter extending into the central compartment.

PHYSICOCHEMICAL AND PHYSICAL PROPERTIES

The virion Mr is about 13 x 10^7 with a buoyant density in CsCl of 1.36 g/cm^3 (1.38 g/cm^3 for ISVPs, 1.43 g/cm^3 for core particles). The virion, ISVP, and core S_{20w} values are about 730S, 630S, and 470S, respectively. Virions are remarkably stable and withstand extremes in ionic conditions, temperatures up to 55°C, pH values between 2 and 9, lipid solvents, and detergents. Exposure to UV irradiation reduces infectivity.

NUCLEIC ACID

All orthoreoviruses have 10 linear dsRNA segments that range from 0.6 x 10^6 to 2.6 x 10^6 Mr. The total Mr of the MRV-3 (strain Dearing) genome is about 1.5 x 10^7 (23,549 bp) and constitutes approximately 11.5% of the virion mass. Based on their resolution by gel electrophoresis, the genomic dsRNAs are grouped into three size classes commonly referred to as large (L1-L3, about 3.9-3.8 kbp), medium (M1-M3, about 2.3-2.2 kbp) and small (S1-S4, about 1.6-0.9 kbp). The gel mobilities of certain genome segments are characteristic of the five distinct species of orthoreoviruses. In comparison to the type species MRV, most ARV isolates, Nelson Bay orthoreovirus (NBV), and Reptilian orthoreovirus (RRV) display retarded genome segment migration of their polycistronic S1 genome segments. Baboon orthoreovirus (BRV) and the Muscovy duck isolates of ARV (ARV-Md) have truncated polycistronic S1 genome segments that migrate as the S4 genome segment by PAGE.

Complete virus particles contain numerous oligonucleotides (2 to 20 nt in length) representing approximately 25% of the total RNA content. Three quarters of these are abortive reiterative 5'-terminal transcripts, produced by the reovirus core-associated transcriptase and capping enzymes, while the remainder are oligoadenylates. The 5'-terminus of the positive-sense RNA strand of each genome segment contains a dimethylated cap 1 structure ($^{m7}GpppG^{m2'OH}$). The genomic RNAs lack polyA tails and do not contain covalently linked proteins. The complete genomic sequences of prototype strains representing three MRV serotypes (serotype 1 Lang, [MRV-1La], serotype 2 Jones [MRV-2Jo], and serotype 3 Dearing [MRV-3De]) have been determined, along with numerous genome segments of other MRV isolates. All four S-class genome segments have been sequenced from several ARV isolates, as well as from NBV and BRV. Additional S-class genome segments have been sequenced from other avian isolates and from RRV. The M3 genome segment of ARV-Md and the L3 genome segment of Avian reovirus strain 1733 (ARV-1733) also have been sequenced. Genomic dsRNA segments contain 5'- and 3'-terminal sequences of four or five bp conserved in all ten genome segments within a particular virus species. The 3'-terminal consensus sequence (UCAUC-3') is also conserved between orthoreovirus species, at least based on the available sequences of the four S-class genome segments. The 5'-terminal conserved sequences vary and may be useful for assigning new isolates to one of the five species (Table 1).

Table 1: Conserved terminal sequences of orthoreovirus genome segments.

Virus species	Serotype or strain	Conserved RNA terminal sequences (positive strand)
Mammalian orthoreovirus		
	MRV-1La	5'- GCUA .... UCAUC -3'
	MRV-2Jo	5'- GCUA .... UCAUC -3'
	MRV-3De	5'- GCUA .... UCAUC -3'
	MRV-4Nd	5'- GCUA .... UCAUC -3'
Avian orthoreovirus		
	ARV-138	5'- GCUUUUU ......UCAUC -3'
	ARV-176	5'- GCUUUUU ..... UCAUC -3'
	ARV-Md	5'- GCUUUUU...... UCAUC -3'
Nelson Bay orthoreovirus		
	NBV	5'- GCUUUA .... UCAUC -3'
Baboon orthoreovirus		
	BRV	5'- GUAAAUUU .. UCAUC -3'
Reptilian orthoreovirus		
	RRV-Py	5'- GUUAUUUU .. UCAUC -3'

The conserved terminal sequences are shown for the positive RNA strands of members of the five *Orthoreovirus* species.

PROTEINS

The orthoreovirus structural proteins are designated in terms of their relative sizes and size classes: λ1, λ2, λ3; μ1, μ2; and σ1, σ2, σ3. In ARV, these proteins are referred to as λA, λB, λC; μA, μB; and σA, σB, σC (Table 2). The following discussion refers to the nomenclature scheme for prototype strain MRV-3.

The stabilizing lattice of the outer capsid is composed of 200 interlocking trimers of the cleaved 76 kDa μ1 protein (72 kDa μ1C and 4 kDa μ1N). The μ1 subunits also interact with monomers of the σ3 protein, which represent the fingerlike projections on the surface of the virion. Pentameric subunits of the λ2 protein make up the flower-like structures and turrets at the vertices of viral particles and cores, respectively. The λ2 structures interact with subunits of the tetrameric σ3 clusters and with the μ1C lattice and represent essential structural components of the outer capsid. This essentially outer CP remains associated with core particles, unlike the other outer CPs. The fourth component of the outer capsid, the σ1 protein, exists as 12 homotrimers associated with the vertices of virions and ISVPs. It may assume either a retracted or extended conformation. The λ1 (120 copies) and σ2 proteins (150 copies) represent the major structural proteins of the inner capsid. The final two structural proteins of the virus, λ3 and μ2, are present at ~12 copies per virion and located on the inside of the inner capsid. The λ3 protein forms 7 nm projections that extend toward the interior of the core, underlying the 12 vertices of the capsid. The μ2 protein may be associated with these λ3 structures.

Table 2: List of the dsRNA segments of Mammalian orthoreovirus – 3De (MRV-3De) with their respective size (bp) and their encoded proteins. For the proteins calculated size (in kDa), copy number per virion, location, and functions or properties are indicated.

Genome Segment	Size (bp)	Proteins (§:structure/ function)	Size (kDa)	Protein copies per particle	Location	Function
L1	3854	λ3 (Pol)	142	12	core	RNA polymerase
L2	3916	λ2 (Cap)	144	60	core spike	Guanylyl transferase, methyl transferase "turret" protein
L3	3896	λ1 (Hel)	143	120	core	Inner capsid structural protein, binds dsRNA and zinc, NTPase, helicase
M1	2304	μ2	83	12	core	NTPase
M2	2203	μ1	76	600	outer capsid	Multimerizes withσ3 and cleaved to μ1C and μ1N, which assume T=13 symmetry in the outer capsid
		μ1C (T13) δ ϕ μ1N	72 59 13 4			μ1C cleaved to δ and ϕ during the entry process, myristoylated N-terminus, membrane penetration
M3	2235	μNS	8075	0	NS	Binds ssRNA and cytoskeleton, phosphoprotein, genome packaging?
		μNSC				μ NSC from alternate translation start site, unknown function
S1	1416	σ1	49	36	outer capsid	Viral attachment protein, homotrimer, hemagglutinin, type-specific antigen
		σ1s	16	0	NS	Basic protein, blocks cell-cycle progression
S2	1331	σ2	47	150	core	Inner capsid structural protein, weak dsRNA-binding, morphogenesis?
S3	1189	σNS	41	0	NS	ssRNA-binding, genome packaging?
S4	1196	σ3	41	600	outer capsid	dsRNA-binding, multimerizes with m1, nuclear and cytoplasmic localization, translation control

§: protein structure/function: RNA **pol**ymerase = "(Pol)"; **cap**ping enzyme (guanylyltransferase and transmethylase) = "(Cap)"; Virus structural protein with **T = 13** symmetry = "(T13)". Protein with **hel**icase activity = "(Hel)".

Lipids

Mature virions lack a lipid envelope. The major outer capsid lattice protein, μ1, and its μ1N cleavage product are N-terminally myristoylated. The small NS proteins responsible for syncytium formation induced by the fusogenic orthoreoviruses are either N-terminally myristoylated or palmitoylated at internal cysteine residues. These acylations are essential for the membrane fusion activity of the proteins.

Carbohydrates

Convincing evidence that any of the orthoreovirus proteins are glycosylated has not been reported. Moreover, no carbohydrate has been observed in the structures of any of the mammalian reovirus proteins that have been determined by X-ray crystallography (λ1, λ2, λ3; μ1; and σ1, σ 2, σ3).

Genome Organization and Replication

The genome consists of ten segments of dsRNA, which are packaged in equimolar ratios (one copy of each within each virion)(Fig. 2). The segments possess terminal NTRs that are shorter at the 5'-terminus (12-32 bp for MRV-3De) than at the 3'-terminus (35-85 bp). The major ORFs vary in length from 353 to 1,298 codons. The MRV S1 segment is bicistronic, encoding the 49 kDa σ1 protein and the 14 kDa σ1s protein from a second overlapping ORF (Fig. 2). The S1 genome segments of ARV and NBV are functionally tricistronic, encoding the viral attachment protein σC, a membrane-associated protein (p17) of unknown function, and a fusion-associated small transmembrane (FAST) protein (p10) responsible for virus-induced syncytium formation. The RRV S1 genome segment is bicistronic, encoding a σC viral attachment protein homolog and a novel FAST protein (p14). The truncated S1 genome segment-equivalent (S4, 1124 bp) of Muscovy duck reovirus (ARV-Md) encodes a σC viral attachment protein homolog and a p10 protein that shares limited sequence similarity to the p10 FAST proteins of NBV and other ARV isolates. The truncated S1 genome segment-equivalent of BRV (S4, 887 bp), contains two sequential 140-141 codon ORFs, one of which encodes a third unique FAST protein (p15) and the other a novel NS protein (p16) of unknown function (Fig. 3).

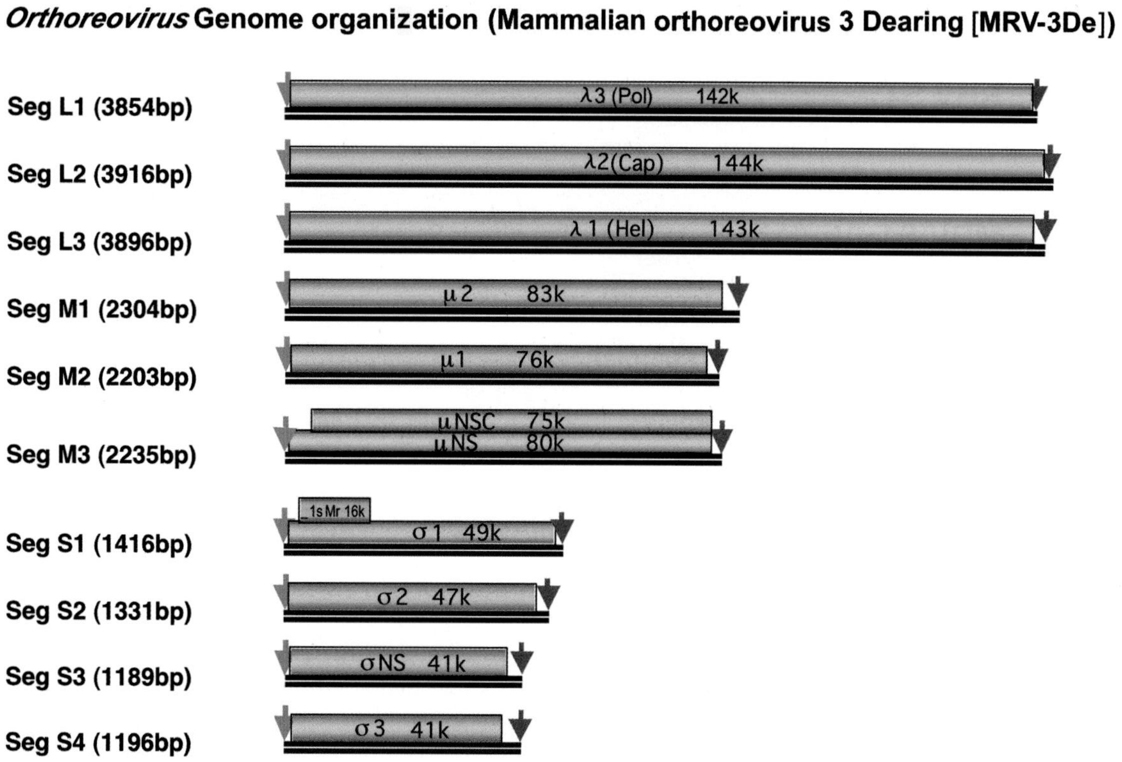

Figure 2: Genome organization of Mammalian orthoreovirus 3 Dearing (MRV-3De), containing 10 dsRNA segments, each contain an ORF, except SegM3 and SegS1, which contains two ORFs. The green arrows indicate the upstream conserved terminal sequence (+ve 5'- GCUA.....), while the red arrows indicate the downstream conserved terminal sequence (+ve.....UCAUC-3').

The overall course of infection involves adsorption, low pH-dependent penetration and uncoating to core particles, asymmetric transcription of capped, non-polyadenylated mRNAs via a fully conservative mechanism (the nascent strand is displaced), translation, assembly of positive-strands into progeny subviral particles, conversion of positive-strands to dsRNA, and further rounds of mRNA transcription and translation. The efficiency of translation of the various orthoreovirus mRNA species varies over a 100-fold range, while the proportions of the mRNA species found in infected cells vary inversely to their proportionate size. The final stage of the replication cycle involves the assembly of the outer capsid onto progeny subviral particles to form infectious virions. Based on studies of MRV replication, virion morphogenesis is thought to proceed along a pathway involving a series of assembly intermediates. Progeny particles accumulate in paracrystalline arrays in the perinuclear region of the cytoplasm and are released when infected cells lyse late in the replication cycle. The exception to the above generalized replication cycle involves the formation of multinucleated syncytia by members of the ARV, BRV, RRV, and NBV species. Syncytia formation commences 10-12 hrs post-infection resulting in a more rapid lytic response and enhanced kinetics of virus release.

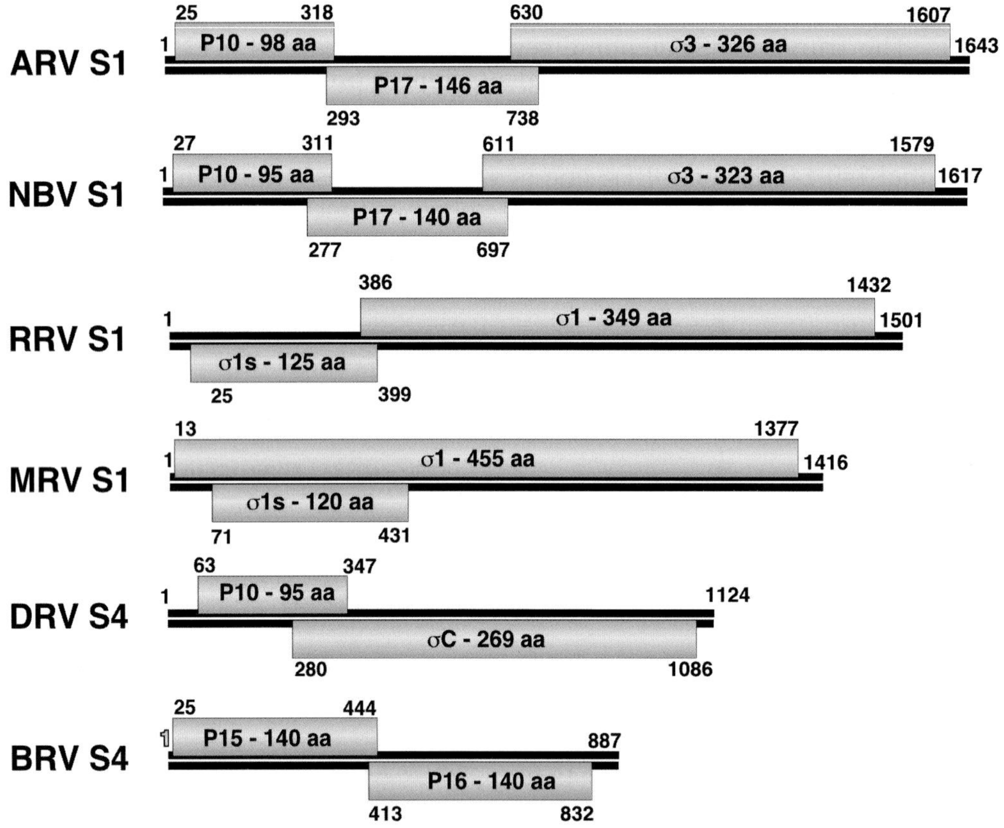

Figure 3: Gene organization of the polycistronic genome segments of the five species of orthoreoviruses. The solid line indicates the dsRNA, and the numbers refer to the first and last nucleotides of the genome segment, along with the nt positions of the various ORFs (excluding the termination codons) indicated by the rectangles. The identities of the gene products encoded by the various ORFs are indicated within the rectangles. The virus species and the genome segment are indicated on the left. The code for the abbreviations can be found in the "list of Species".

The functions and properties of specific viral proteins influence various stages of the MRV replication cycle (Table 2). The MRV σ1 viral attachment protein determines the cell and tissue tropism of the virus strain and has hemagglutination activity. The σ1 protein binds cell-surface carbohydrate and junctional adhesion molecule 1. The μ1 protein is N-terminally myristoylated and forms a complex with σ3 in solution that triggers cleavage of μ1 to μ1N and μ1C. The μ1C fragment is further proteolytically cleaved into δ and ϕ polypeptides during virus entry into cells and is responsible for membrane penetration. The μ1 protein also influences strain-specific differences in capsid stability, transcriptase activation, and neurovirulence. In the case of ARV, the μ1 homolog (μB) has been implicated in strain-specific differences in macrophage infection. In addition to interacting with μ1 and forming the outer capsid layer of the virion, the σ3 protein is a dsRNA-binding protein involved in translation regulation, altering the activity of PKR, and modulating the interferon response. The λ2 core spike is the guanylyl transferase involved in mRNA capping, while the λ1 and σ2 major inner capsid proteins both bind dsRNA. The λ1 protein also may function as a helicase and an RNA triphosphatase. The minor inner CP λ3 is the viral polymerase, while the second minor inner CP μ2, along with the major inner CP λ1, is involved in the NTPase activity associated with core particles.

There are also at least three NS proteins encoded by the MRV genome: μNS, σNS, and σ1s. The μNS and σNS proteins are produced in high abundance during infection and, together with σ3, associate with mRNA to form virus mRNA-containing complexes, which are presumed to be precursors of progeny virus assembly.

The σNS protein binds ssRNA, and the μNS protein associates with the cytoskeleton. Core protein μ2 stabilizes microtubules within viral inclusions and associates with μNS, which is an organizing center for inclusion formation. Co-expression of μNS and σNS proteins in mammalian cells from cloned viral cDNAs yields punctate structures resembling intracytoplasmic inclusions of virally infected cells. The σ1s protein is a small, basic protein expressed in cells infected by all three MRV serotypes. This protein is dispensable for growth in cell culture but is involved in cell cycle arrest at the G2/M checkpoint.

Replication strategies used by ARV, NBV, RRV, and BRV are similar to that described for MRV, with some notable exceptions. The truncated viral attachment protein of ARV, RRV, and NBV, σC (35 kDa), exists as a multimer with a coiled-coil domain similar to that of MRV but possesses no hemagglutination activity. BRV is unique in that the S-class genome segments encode no homolog of the ARV, NBV, RRV, or MRV viral attachment proteins. The dsRNA-binding domain of the MRV major sigma-class outer CP σ3 is not conserved in the homologous σB proteins of ARV, NBV, or BRV. As with the MRV σ2 protein, the major sigma-class core protein of ARV, σA, displays dsRNA-binding activity. The ARV σA core protein may function analogously to the σ3 major outer CP of MRV by regulating the PKR activity and the interferon response. Viruses in the ARV, NBV, RRV, and BRV species encode an additional FAST protein responsible for syncytium formation. The p10 FAST proteins of ARV and NBV share sequence and structural similarities, but are unrelated to the p15 and p14 FAST proteins of BRV and RRV, respectively. All of these FAST proteins are small, basic, acylated, transmembrane proteins and induce fusion in transfected cells in the absence of other viral proteins.

Antigenic Properties

The serotype-specific antigen of the orthoreoviruses is protein σ1 (σC of the avian species), which is recognized by neutralizing antibodies. Antigenic recognition of this protein is the basis for three major serotypes of MRV and 5-11 serotypes of ARV. Ndelle virus was isolated from a mouse and originally classified as an orbivirus. Recent sequence data reveals Ndelle virus is actually an orthoreovirus and closely related to MRV-1 and MRV-3. However, neutralizing antibodies against the three major MRV serotypes do not neutralize Ndelle virus indicating it represents a fourth MRV serotype. The MRV σ1 and σ1s proteins elicit strain-specific and cross-reactive cytotoxic T-cell activities. The MRV proteins λ2 and σ3 are species-specific antigens, similar to the λB and σB proteins of ARV

(Fig. 1). The considerable sequence homology that exists between different isolates in the same species, but not between species, is reflected by limited antigenic cross reactivity between species. The most extensive antigenic similarity between species subgroups occurs between ARV and NBV, which is in accordance with the increased aa identity between these species.

Biological Properties

Transmission is by the enteric or respiratory routes, no arthropod vectors are involved, and infection is restricted to a variety of vertebrate species (baboons, bats, birds, cattle, humans, monkeys, sheep, snakes, and swine). Orthoreovirus distribution is worldwide. Human orthoreoviruses are generally benign but may cause upper respiratory tract illness and possibly enteritis in infants and children (albeit rare). In mice, orthoreovirus infection can cause diarrhoea, runting, oily hair syndrome, hepatitis, jaundice, myocarditis, myositis, pneumonitis, encephalitis, and neurologic symptoms. A variety of symptoms may be associated with orthoreovirus infection of domestic animals including upper and lower respiratory illnesses and diarrhoea. In monkeys, orthoreoviruses cause hepatitis, extrahepatic biliary atresia, meningitis, and necrosis of ependymal and choroid plexus epithelial cells. The prototype baboon reovirus (BRV) isolate was obtained from baboons with meningoencephalomyelitis. Isolates from snakes were obtained from animals displaying neurological symptoms. The outcome of avian orthoreovirus infection in birds may range from inapparent to lethal depending on the virus strain and the age of the host. Systemic infection results in virus dissemination to numerous tissues. Disease presentations in chickens include feathering abnormalities, gastroenteritis, hepatitis, malabsorption, myocarditis, paling, pneumonia, stunted growth, and weight loss. In turkeys, avian orthoreoviruses cause enteritis. Birds that survive an acute systemic infection may develop obvious joint and tendon disorders (tenosynovitis) that resemble the pathology of rheumatoid arthritis in humans. Avian orthoreoviruses do not infect mammalian species.

MRV and ARV induce the biochemical and morphologic hallmarks of apoptosis in cultured cells. MRV infection leads to activation of nuclear factor kappa B (NF-κB), a family of transcription factors known to play important roles in regulating cellular stress responses, including apoptosis. The capacity of viral attachment protein σ1 to bind cell-surface receptors sialic acid and junctional adhesion molecule 1 regulates NF-κB activation and the efficiency of apoptosis induction. Additionally, the M2 genome segment encoding outer CP μ1 contributes to MRV-induced apoptosis, suggesting important roles for viral disassembly and membrane penetration in this cellular response. As with MRV, ARV-induced apoptosis requires virus disassembly but not viral transcription.

Recent studies indicate that MRV preferentially replicates in a lytic manner in transformed cells. The basis for this cell tropism relates to the effects of an activated Ras pathway in transformed cells on modulation of PKR activity and regulation of the translation machinery. These observations have led to the development of reovirus as an oncolytic agent for cancer therapy.

List of Species Demarcation Criteria in the Genus

The orthoreoviruses include five species subgroups. Conclusive species classification requires the direct demonstration of exchange of genetic material via reassortment of genome segments. To date, there is no evidence of reassortment between the five species identified, which reflects the extensive sequence divergence between species.

Members of a species in the genus *Orthoreovirus* may be identified by:

- The capacity to exchange genetic material by genome segment reassortment during mixed infections, thereby producing viable progeny virus strains;
- Identification of conserved terminal genomic RNA sequences within a species (absolute conservation of the 5'- and 3'-terminal 4-8 bp);
- Identification of extensive sequence identity between the proteins encoded by homologous genome segments (for conserved core proteins, greater than 85% aa identity within a species versus less than 65% identity between species; for the more divergent outer CPs, >55% identity within a species and <35% between species);
- Identification of extensive sequence identity between homologous genome segments (for most genome segments, greater than 75% nt sequence identity within a species versus less than 60% between species);
- Identification of virus serotype (based on cross neutralization) with a virus type already classified within a named orthoreovirus species;
- Demonstration of extensive antigenic similarity in the major structural proteins within a species, as determined by ELISA or immunoprecipitation;
- Analysis of 'electropherotype' by agarose gel electrophoresis but not by PAGE (some similarities can exist between closely related species);
- Similar organization of the polycistronic genome segment;
- Identification of host species and clinical signs.

List of Species in the Genus

Species I includes all the nonfusogenic MRV isolates, with three major serotypes (MRV-1, MRV-2, and MRV-3) representing numerous isolates, and a fourth serotype with only one isolate, Ndelle reovirus (MRV-4Nd). Amino acid identities of the sigma-class major outer CPs and core proteins of various MRV serotypes range from 90-97%. The other four species represent the fusogenic reoviruses whose members all induce syncytium formation.

Species II contains numerous ARV isolates from commercial poultry flocks, including chickens, Muscovy ducks, turkeys, and geese, and includes several different serotypes. Sequence diversity is more extensive between the various ARV isolates than between MRV isolates (54-95% in the sigma-class major outer CP).

Species III represents Nelson Bay virus (NBV), an atypical syncytium-inducing mammalian reovirus isolated from a flying fox. The sequence similarity between NBV and ARV exceeds that between NBV and the other species subgroups. ARV and NBV also share more extensive antigenic similarity than other species, possess more similar conserved terminal genome segment sequences, display a similar gene organization of the polycistronic S1 genome segment, and encode homologous p10 fusion proteins (Fig. 2). These observations indicate that NBV is more closely related to ARV isolates than to other mammalian or reptilian reovirus isolates. Although ARV and NBV clearly share a more recent evolutionary past than the other reovirus species, in view of the extent of sequence divergence (59-61% identity in the sigma-class core protein and only 29-36% identity in the sigma-class major outer CP) and the absence of evidence for reassortment between the ARV and NBV isolates, these isolates are considered as two separate species.

Species IV contains a single isolate, BRV. This atypical mammalian isolate induces syncytium formation but shares little sequence (16-32% aa identity between homologous S-class gene products) or antigenic similarity with the other fusogenic species. BRV contains a truncated, fusion-inducing, polycistronic S1 genome segment-equivalent (the S4 genome segment) with a distinct gene organization, a fusion protein (p15) with no sequence or sequence-predicted structural similarity to the fusion proteins of ARV or NBV, and a unique 5'-terminal consensus sequence. This isolate clearly represents a distinct species subgroup of the orthoreoviruses.

Species V represents the reptilian reoviruses. Sequence information is available for the polycistronic S1 genome segment and the S-class genome segment encoding the sigma-class major outer CP of a reptilian reovirus isolate from a python (RRV-Py). Several additional isolates have been obtained from other snakes and iguanas, but no sequence information is currently available. RRV contains the conserved 3′-terminal pentanucleotide sequence of the orthoreoviruses (UCAUC-3′) but possesses a unique 5′-terminal conserved sequence (5′-GUUA) (Table 1). The S1 genome segment of RRV-Py is bicistronic, encoding a viral attachment protein homolog and a novel p14 FAST protein that induces syncytium formation. Amino acid sequence identities between the RRV sigma-class major outer CP and the homologous protein of other species subgroups are 16-25%, clearly indicating RRV represents a distinct species subgroup of orthoreoviruses.

SPECIES IN THE GENUS

Species names are in green italic script; strain names and synonyms are in black roman script; tentative species names are in blue roman script. Sequence accession numbers [L1 to L3, M1 to M4, and S1 to S4 indicate RNA segment numbers], and assigned abbreviations ( ) are also listed.

Avian orthoreovirus		
Avian orthoreovirus 1733 {chicken isolate}	[L3: AF384171]	(ARV-1733)
Avian orthoreovirus 176 {chicken isolate}	[S1: AF218358, S2: AF059716, S3: AF059720, S4: AF059724]	(ARV-176)
Avian orthoreovirus 89026 {Muscovy duck isolate}	[S1: AJ278102, S2: AJ006476, S3: AJ133122, S4: AJ310525, AJ310526]	(ARV-Md89026)
Avian orthoreovirus 89330 {Muscovy duck isolate}	[M3: AJ293969]	(ARV-Md89330)
Avian orthoreovirus D15/99 {goose isolate}	[S1: AY114138]	(ARV-Go)
Avian orthoreovirus NC98 {turkey isolate}	[S3: AF465799]	(ARV-Tu)
Avian orthoreovirus S1133 {chicken isolate}	[S1: L39002, S2: AF104311, S3: U20642, S4: U95952]	(ARV-1133)
Avian orthoreovirus SK138a {chicken isolate}	[S1: AF218359, S2: AF059717, S3: AF059721, S4: AF059725]	(ARV-138)
Baboon orthoreovirus		
Baboon orthoreovirus	[S1: AF059719, S2: AF059723, S3: AF059727, S4: AF406787]	(BRV)
Mammalian orthoreovirus		
Mammalian orthoreovirus 1 Lang		(MRV-1La)
Mammalian orthoreovirus 2 D5/Jones		(MRV-2Jo)
Mammalian orthoreovirus 3 Dearing	[L1: M24734, L2: J03488, L3: M13139, M1: M27261, M2: M19408, M3: M27262, S1: M10262, S2: M25780, S3: X01627, S4: K02739]	(MRV-3De)
Mammalian orthoreovirus 4 Ndelle	[L1: AF368033, M2: AF368034, S1: AF368035, S2: AF368036, S4: AF368037]	(MRV-4Nd)
Nelson Bay orthoreovirus		
Nelson Bay orthoreovirus	[S1: AF218360, S2: AF059718, S3: AF059726, S4: AF059722]	(NBV)
Reptilian orthoreovirus		
Reptilian orthoreovirus - Python	[S1: AY238887, S3: AY238886]	(RRV-Py)

Tentative Species in the Genus

None reported.

Phylogenetic Relationships within the Genus

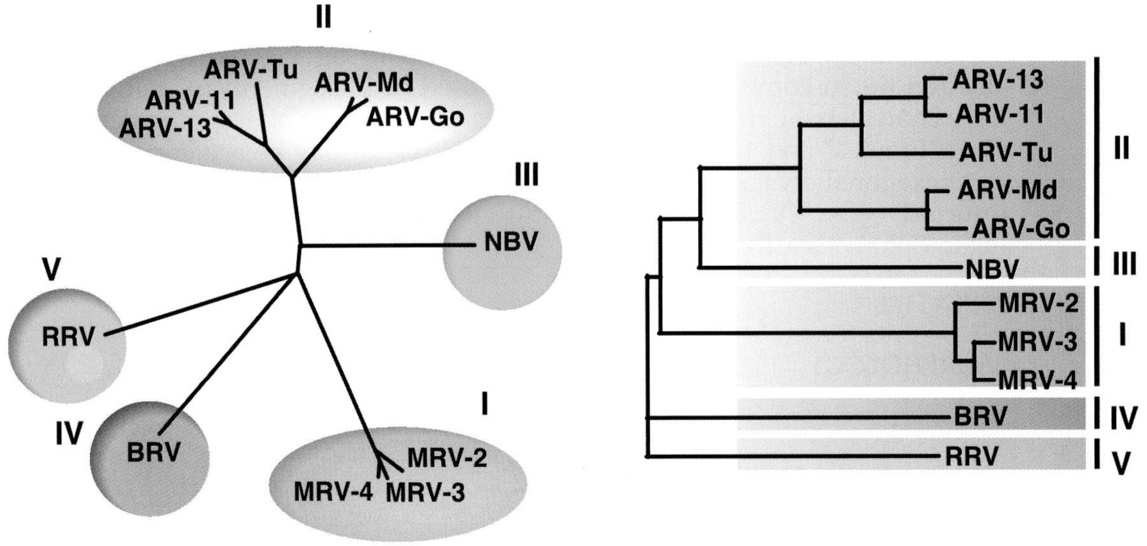

Figure 4: Phylogenetic relationships between the orthoreovirus species. (Left) Unrooted neighbor-joining tree using the aa sequences of the sigma-class major outer CPs of various *Orthoreovirus* isolates prepared using Clustal W and drawn with Phylodendron. Branch lengths are proportional to inferred evolutionary distances. (Right) Unrooted dendrogram presentation of the results from left panel. Horizontal branch lengths are proportional to inferred evolutionary distances. The designation of the five species (I-V) is indicated in both panels. In the ARV and MRV sequences analyzed included MRV-2, -3, and -4, and ARV isolates S1133 (ARV-11), 138a (ARV-13), Muscovy duck (ARV-Md), turkey (ARV-Tu), and goose (ARV-Go) (accession# and abbreviations are given in the "List of species"). (Courtesy from Duncan *et al.*, 2003).

The five species of orthoreoviruses represent five evolutionarily distinct entities, as illustrated by phylogenetic analysis using the aa sequence of the sigma-class major outer CP for which the greatest number of sequences from diverse isolates is available (Fig. 4). Identical phylogenetic relationships are generated by comparison of the more conserved sigma-class major outer CPs and NS proteins (data not shown).

References

Duncan, R. (1999). Extensive sequence divergence and phylogenetic relationships between the fusogenic and nonfusogenic orthoreoviruses: a species proposal. *Virology* **260**, 316-328.

Duncan, R., Corcoran, J., Shou, J. and Stoltz, D. (2004). Reptilian reovirus represents a new species of fusogenic orthoreovirus. *Virology,* **319**, 131-140.

Murphy, F.A., Hilliard, J. and Mirkovic, R. (1995). Characterization of a novel syncytium-inducing baboon reovirus. *Virology,* **212**, 752-756.

Nibert, M.L. (1998). Structure of mammalian orthoreovirus particles. *Curr. Top. Micro. and Imm.,* **223**, 1-30.

Roner, M. and Joklik, W.K. (1996). Molecular recognition in the assembly of the segmented reovirus genome. *Nuc. Acids Res. and Mol. Biol.,* **53**, 249-281.

Varela, R., Martinez-Costas, J., Mallo, M. and Benavente, J. (1996). Intracellular posttranslational modifications of S1133 avian reovirus proteins. *J. Virol.,* **70**, 2974-2981.

Derivations of Names

Ortho: from Greek orthos "straight".

Contributors

Chappell, J.D., Duncan, R., Mertens, P.P.C. and Dermody, T.S.

GENUS *ORBIVIRUS*

Type Species ***Bluetongue virus***

DISTINGUISHING FEATURES

Virions have a relatively featureless outer capsid as viewed by negative staining and electron microscopy and a genome composed of 10 segments of dsRNA. Core particles have characteristic ring shaped capsomers. Replication is accompanied by production of viral "tubules" and viral inclusion bodies (VIB) and may be accompanied by formation of flat hexagonal crystals of the major outer core protein (VP7 (T13)) in the cytoplasm of infected cells. Viruses are transmitted between vertebrate hosts by a variety of hematophagous arthropods.

VIRION PROPERTIES

MORPHOLOGY

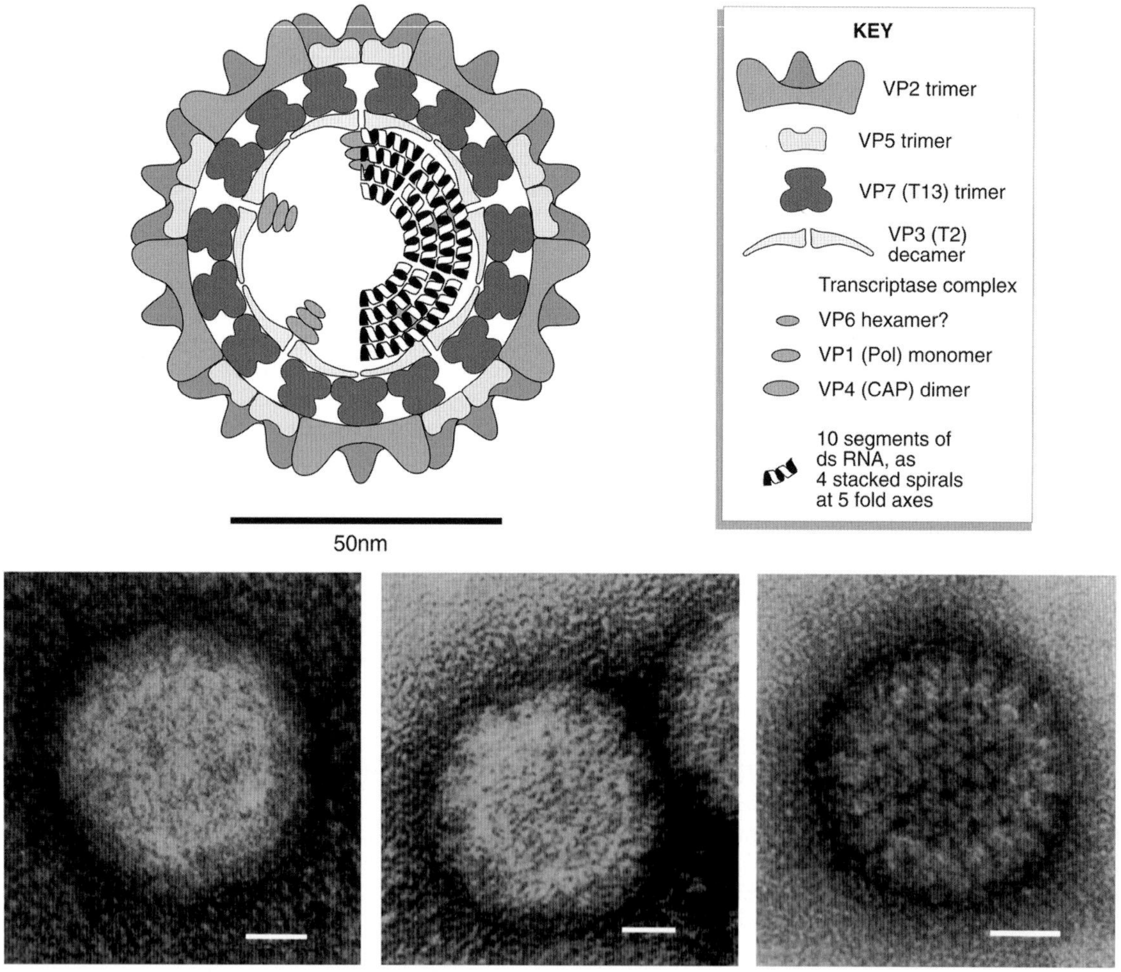

Figure 1: (Top) Diagram of the Bluetongue virus (BTV) particle structure, constructed using data from biochemical analyses, electron microscopy, cryo-electron microscopy and X-ray crystallography. (Courtesy of P.P.C. Mertens and S. Archibald). (Bottom) Electron micrographs of African horse sickness virus (AHSV) serotype 9 particles stained with 2% aqueous uranyl acetate (Left) virus particles, showing the relatively featureless surface structure; (Center) infectious subviral particles (ISVP), containing chymotrypsin cleaved outer capsid protein VP2 and showing some discontinuities in the outer capsid layer; (Right) core particles, from which the entire outer capsid has been removed, to reveal the structure of the VP7 (T13) core surface layer and showing the ring shaped capsomeres (Courtesy of P.P.C. Mertens).

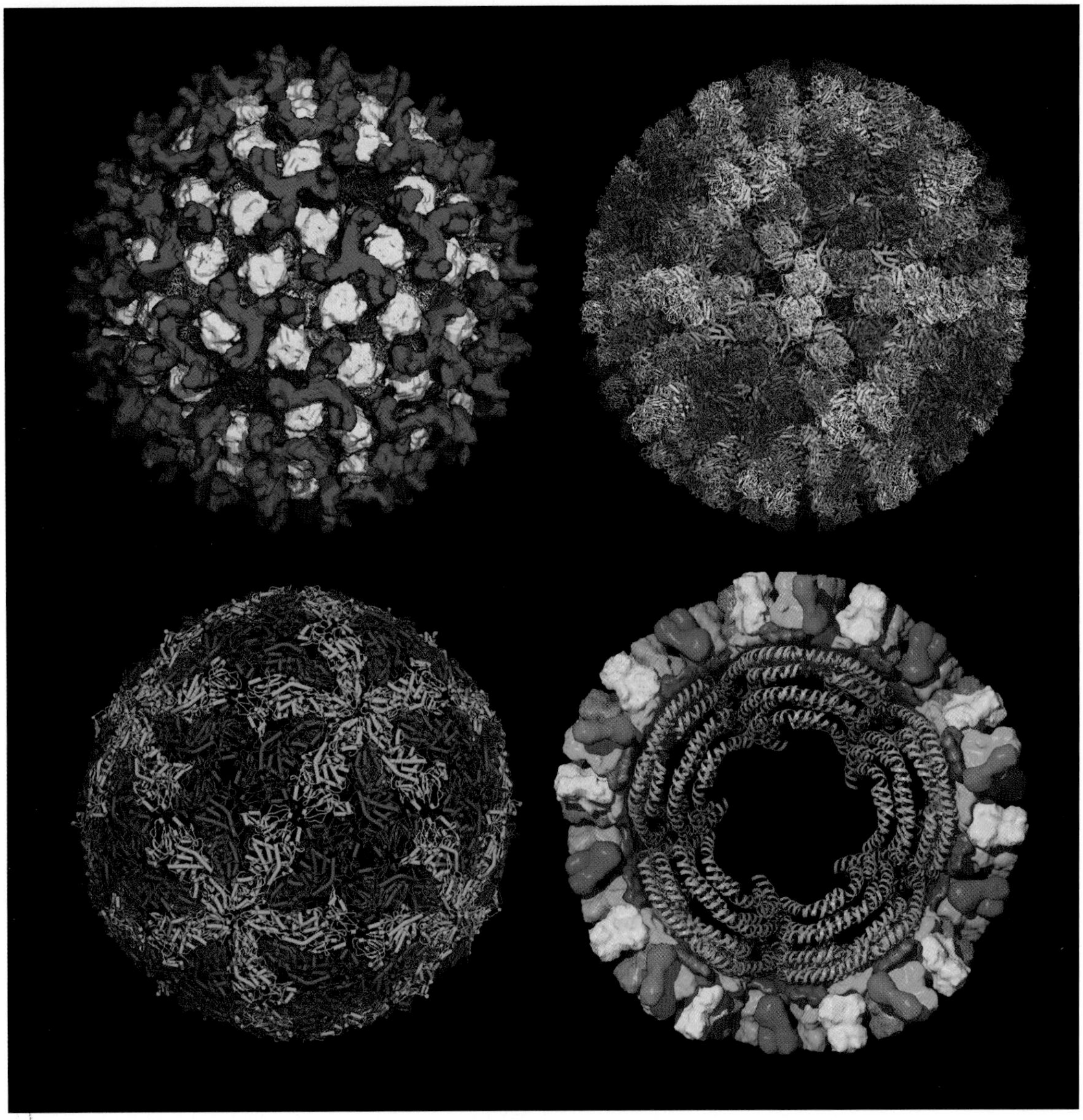

Figure 2: (Top left) The outer capsid layer of Bluetongue virus (BTV) (from cryo-electron microscopy) showing trimers of VP2 in red and trimers of VP5 in yellow, superimposed on the underlying x-ray crystallography structure for the BTV core. (Top right) the structure of the BTV core as determined by X-ray crystallography of the native core particle. The outer core surface, composed of 260 trimers of VP7 (T13) arranged with T=13 *l* symmetry. The chemically identical but structurally different trimers are named and colored in order of increasing distance from the five fold axes of symmetry (P [red], Q [orange], R [green], S [yellow] and T [blue]- situated at the three fold axes). (Bottom left) The BTV 1 subcore shell (from x-ray crystallography) is composed of 120 copies of VP3(T2), arranged with T=2 symmetry. The chemically identical but structurally different molecules are shown: 'A' (green: surrounding the five fold axis) and 'B' (red: surrounding the three fold axis). (Bottom right) Model cross section of the BTV core showing packaging of the dsRNA as four concentric shells (Courtesy of D.I. Stuart, J. Grimes, P. Gouet, J. Diprose, R. Malby, P. Roy, B.P.V. Prasad and P.P.C. Mertens).

Virions of Bluetongue virus (BTV) are approximately 90 nm in diameter, core particles have a maximum diameter of 73 nm, sub-cores have a maximum diameter of 59 nm and an internal diameter of 46 nm (Fig. 1). The virion is spherical in appearance but has icosahedral symmetry. Although no lipid envelope is present on mature virions, they can leave the host cell by budding through the cell plasma membrane. During this process they transiently acquire an unstable membrane envelope. Unpurified virus is often associated with cellular membranes. By conventional electron microscopy, the surface of intact virions is indistinct (Fig. 1). However, the outer capsid does have an ordered structure, with icosahedral symmetry and 'sail' shaped surface projections that can be observed on virions where the particle structure is maintained (e.g., using cryo-electron microscopy: Fig. 2). When the outer capsid layer is removed, it is possible to view the surface layer of the core particle, which is composed entirely of capsomeres of VP7 (T13) arranged as hexameric rings (pentameric at the 5-fold axes: Fig. 1 and 2). These rings, which are readily observed by conventional electron microscopy, give rise to the name of this genus. The core particle also contains a complete inner capsid shell (the subcore layer), which surrounds the 10 dsRNA genome segments and minor structural proteins. The minor core proteins (the transcriptase complexes) are attached to the inner surface of the subcore at the 5-fold symmetry axes (Fig. 1). Assembly of the subcore layer appears to control the overall assembly, size and symmetry of the particle.

Physicochemical and Physical Properties

The virion Mr is about 10.8 x 10^7, the core Mr is about 6.7 x 10^7. Their buoyant densities in CsCl are 1.36 g/cm^3 (virions) and 1.40 g/cm^3 (cores). The S_{20W} is 550S (virions) and 470S (cores). Virus infectivity is stable at pH 8-9 but virions exhibit a marked decrease in infectivity outside the pH range 6.5-10.2. In part, this may be related to the loss of outer coat proteins, particularly at the lower pH range. The sensitivity of the outer capsid proteins and their removal by cation treatment (e.g. by treatment with $MgCl_2$, or CsCl) varies markedly with both pH and virus strain. At low pH values (less than 5.0), virions and cores are both disrupted. Unlike orthoreoviruses, at pH 3.0 virus infectivity is abolished. In blood samples, serum, or albumin, viruses held *in vitro* at less than 15°C may remain infectious for decades. Purified BTV-1 virions held at 4°C in 0.1 M tris/HCl pH 8.0, showed no significant reduction in infectivity after 1 year. Crystals of core particles are very stable when kept at 29°C. Virus infectivity is rapidly inactivated on heating to 60°C. In general, orbiviruses are considered to be relatively resistant to treatment with solvents, or detergents, although the sensitivity to specific detergents varies with virus species. However, sodium dodecyl sulphate will disrupt the particle and destroy its infectivity. Freezing reduces virus infectivity by about 90%, possibly due to particle disruption. However, once frozen and held at -70°C, virus infectivity remains stable.

Nucleic Acid

The genomic RNA represents 12% and 19.5% of the total molecular mass of virus particles or cores, respectively. The genome is composed of 10 linear dsRNA segments that are packaged in exactly equimolar ratios, one of each segment per particle. The genomic RNA is packaged as a series of ordered concentric shells within the VP3 (T2) layer of the subcore (Fig. 2). Four layers of RNA, each of which has elements of icosahedral symmetry, can be detected by X-ray crystallography of the BTV core. Within the central space of the subcore, there appears to be an association between the dsRNA molecules and the protein density at the 5-fold axes of symmetry (vertices of the icosahedron), which is thought to represent the transcriptase complexes (TCs). From the 5-fold axis, the RNA, in the outmost layer, appears to spiral away from the 5 fold axes outward around the TC for two turns until it clashes with an icosahedrally related neighbor. At this point it is thought to move inward forming the next concentric shell of RNA. The genomic RNA contains 5'-terminal Cap 1 structures (7mGpppG$^{(2\text{-}Om)}$...).

For BTVs, the genome segments range in size from 3,954 to 822 bp (total size is 19.2 kbp, total Mr of 13.1 x 10^6). There is no evidence for short ssRNA oligonucleotides in intact virions. The genomic RNAs are named "segment 1 to 10" (Seg1 to 10), in order of increasing electrophoretic mobility in 1% agarose gels and in order of decreasing Mw. For BTVs, the segments migrate as three size classes 3 large (Seg1-3: 3.9-2.8 kbp), 3 medium (Seg4-6: 2.0-1.6 kbp) and 4 small segments (Seg6-10: 1.2-0.8 kbp). For other members of the genus, different sizes and size classes exist. For an individual virus species the dsRNA sizes from different isolates, or different serotypes, are comparable such that a uniform segment migration pattern is observed when the genomic RNAs of 'normal' isolates are analyzed by agarose gel electrophoresis. However, variations in primary sequence cause significant variations in rate and order of migration of genome segments during polyacrylamide gel electrophoresis (PAGE), particularly in high percentage gels (> 5% polyacrylamide). Earlier BTV genome segment nomenclature based on PAGE is inconsistent and the migration of Seg5 and Seg6 is usually reversed. In the genome segments that have been analyzed to date, there is only a single major ORF, which is always on the same strand (see conserved terminal sequences below). However, the ORF may have more than one functional initiation site near to the 5'-end of the RNA, resulting in production of two related proteins.

For BTV-10, the 5'-NTR range from 8 to 34 bp, while for the 3'-ends they are 31 to 116 bp in length. For other serotypes and other viruses the lengths differ. In general, however, the 5'-NTRs are shorter than the 3'-NTRs. The NTRs of almost all the orbivirus genome segments that have been sequenced (Table 1) contain two conserved bp at either terminus (+ve 5'-GU...AC-3'). The NTRs of BTV include terminal sequences of 6 bp that are identical for all 10 dsRNA segments and which are conserved between different BTV isolates. Other orbiviruses have terminal sequences comparable to those of BTVs, but which are not always identical and which may not be conserved in all 10 segments (Table 1).

Table 1: Conserved terminal sequences of orbivirus genome segments.

Virus isolate	Conserved RNA terminal sequences (positive strand)
Bluetongue virus (BTV)	5'-GUUAAA................................ACUUAC-3'
African horse sickness virus (AHSV)	5'-GUU $^{A}/_{U}$ A$^{A}/_{U}$..................AC$^{A}/_{U}$UAC-3'
Epizootic hemorrhagic disease virus (EHDV)	5'-GUUAAA............................$^{A}/_{G}$CUUAC-3'
Great Island virus (BRDV)	5'-GUAAAA..........................A$^{A}/_{G}$GAUAC-3'
Palyam virus (CHUV)	5'-GU $^{A}/_{U}$ AAA.......................$^{A}/_{G}$CUUAC-3'
Equine encephalosis virus (EEV)*	5'-GUUAAG.............................UGUUAC-3'
St Croix River virus (SCRV)	5'-$^{A}/_{G}$UAAU$^{G}/_{A/U}$......$^{G}/_{A/U}$ $^{C}/_{U}$ $^{C}/_{A}$ TAC-3′
Peruvian horse sickness virus (PHRV)	5'-GUUAAAA..................$^{A}/_{G}$$^{C}/_{G}$$^{A}/_{G}$UAC-3'
Yunan orbivirus (YUOV)	5′-GUUAAAA...........................$^{A}/_{G}$UAC-3′

* based on genome segment 10 (only) of the seven different serotypes

PROTEINS

There are 7 virus structural proteins (VP1-7: table 2). Proteins constitute 88% and 80.5% of the dry weight of virions and cores, respectively. In BTV, the outer capsid consists of 180 copies of the 111 kDa 'sail-shaped' VP2 protein arranged as 'triskellion' structures, and 360 copies of an interdispersed and underlying VP5 protein (59 kDa), which may be arranged as 120 trimers (Fig. 1 and 2). The electrophoretic migration order and nomenclature of proteins may vary in other orbivirus species. Both VP2 and VP5 of BTV are attached to VP7(T13). The surface of the core particle consists entirely of 780 copies of VP7, which are arranged with T=13 *l* symmetry, as a network of hexameric and pentameric rings (in a near perfect example of quasi-equivalence: Fig. 2). The VP7(T13) trimers of the core surface can bind dsRNA molecules, although the functional significance of this binding remains undetermined. Beneath the VP7(T13) layer, the subcore capsid shell is composed of 120 copies of VP3 arranged with T=2 symmetry, displaying "geometrical-quasi-equivalence" (Fig. 2). The VP3(T2) capsid shell encloses the 10 dsRNA segments of the genome (Fig. 2), as well as the three minor structural proteins. These include: the 150 kDa VP1 (Pol), which is the RNA polymerase; the 76 kDa VP4(Cap), which forms functional dimers and is both the guanylyl transferase, as well as Mtr 1 (forming the 7-methyl guanosine of the cap structure) and Mtr 2 (forming the 2-0

DSRNA

methyl guanosine, as the terminal nucleotide of the RNA chain); and the 36 kDa VP6/VP6a (Hel), which binds ss or dsRNA and has both helicase and NTPase activities.

X-ray diffraction studies indicate that the minor proteins are attached as a transcriptase complex (TC), to the inner surface of the subcore layer [VP3(T2)] at the 5-fold symmetry axes (at the vertices of the icosahedron). However, because there is only a single TC at each position they do not have full icosahedral symmetry and it has not yet been possible to determine their organization at the atomic level.

The VP7(T13) protein of some viruses like the African horse sickness virus (AHSV) can also form flat hexagonal crystals, typically up to 5 μm in diameter, within the cytoplasm of the infected cell. These are composed of flat sheets of hexameric rings, similar to the rings of trimers seen in the outer core surface layer.

There are three distinct non-structural viral proteins produced in cells infected with BTV or other orbiviruses. The 64 kDa NS1(TuP) protein forms tubules that vary in length, up to 4 μm, which are of unknown function but regarded as a characteristic feature of orbivirus replication. These tubules may have a ladder like structure, as observed in those of BTV and Epizootic hemorrhagic disease virus (EHDV) (68 and 52 nm in diameter) or may be finer (23 nm in diameter) with a reticular cross weave pattern, like those produced by AHSV.

The 41 kDa NS2(ViP) protein can be phosphorylated and is an important component of the matrix of VIB, which are the site of virus replication and assembly. VIBs also contain relatively large amounts of the virus core proteins. NS2(ViP) has ssRNA binding activity, suggesting it has an active role in replication. In conjunction with other virus proteins it is believed to be involved in the recruitment of viral mRNA for encapsidation. The NS3/NS3a proteins are two small non-structural membrane proteins (25 and 24 kDa), translated from different frame initiation sites on a single ORF, which are involved in the release of virus particles from cells. This function may be essential for dissemination of progeny virus, particularly from insect vector cells, which can become persistently infected and do not show CPE or high levels of cell death. In the process of particle release, the NS3 proteins are also released from the cell.

LIPIDS

Although mature virions can acquire a membrane envelope during the process of cell exit, by budding through the cell membrane and may be intimately associated with membraneous cell debris, they are usually considered to be non-enveloped.

CARBOHYDRATES

The BTV/VP5 protein may be glycosylated. NS3 and NS3a synthesized in mammalian cells can become glycosylated, forming high molecular weight products.

GENOME ORGANIZATION AND REPLICATION

The BTV genome segment coding assignments, based on the dsRNA migration in 1% agarose are: Seg1-VP1(Pol); Seg2-VP2; Seg3-VP3(T2); Seg4-VP4(Cap); Seg5-NS1(TuP); Seg6-VP5; Seg7-VP7(T13); Seg8-NS2(ViP); Seg9-VP6/VP6a(Hel); Seg10-NS3/NS3a (Fig. 3). Cognate genes of different BTV strains are similar. The S9 and S10 mRNAs are translated from either of 2 in-frame AUG codons. The significance of the 2 forms of the S9 and S10 gene products (NS3, NS3A; VP6, VP6A) is not known. In some cases other virus proteins form morphologically defined structures in infected cells [e.g. the flat hexagonal crystals formed of VP7 (T13) of AHSVs] but these are of unknown functional significance (Table 2).

Table 2: List of the dsRNA segments of Bluetongue virus serotype 10 (BTV-10) with their respective size (bp) and their encoded proteins for which the name, calculated size (kDa) and function and/or location are indicated.

dsRNA Segment number	dsRNA Segment Size (bp)	ORF (bp)	Protein name § (protein structure/ function)	Protein Size (Da)	Protein copy number / particle	Function (location)
1	3,954	12-3917	VP1(Pol)	149,588	10	RdRp
2	2,926	20-2887	VP2	111,112	180	Outer layer of the outer capsid, controls virus serotype, cell attachment protein, involved in determination of virulence, readily cleaved by proteases. Most variable protein. Reacts with neutralizing antibodies. Trimer.
3	2,770	18-2720	VP3(T2)	103,304	120	Forms the innermost protein capsid shell subcore capsid layer, T=2 symmetry, controls overall size and organization of capsid structure, RNA binding, interacts with minor internal proteins.
4	2,011	9-1970	VP4(Cap)	76,433	20	Dimer, Mtr 1 and 2, capping enzyme (guanylyl-transferase).
5	1,769	35-1690	NS1(TuP)	64,445	0	Forms tubules of unknown function in the cell cytoplasm. These are characteristic of orbivirus replication.
6	1,638	30-1607	VP5	59,163	360	Inner layer of the outer capsid, glycosylated, helps determine virus serotype, variable protein. Trimer.
7	1,156	18-1064	VP7(T13)	38,548	780	Trimer, forms outer core surface, which can bind dsRNA, T=13 symmetry, in some species (AHSV) it can form flat hexagonal crystals, involved in cell entry and core particle infectivity in adults and cells of vector insects, reacts with "core neutralizing" antibodies, immuno dominant virus species specific antigen.
8	1,124	20-1090	NS2(ViP)	40,999	0	Important viral inclusion body matrix protein, ssRNA binding, phosphorylated. Can be associated with outer capsid.
9	1,046	16-999	VP6(Hel) VP6a	35,750	60	ssRNA and dsRNA binding, Helicase, NTPase.
10	822	20-706	NS3 NS3a	25,572 24,020	0 0	Glycoproteins, membrane proteins, involved in cell exit. In some species (AHSV) these are variable proteins and are involved in determination of virulence.

§: protein structure/function: RNA **pol**ymerase = "Pol"; **cap**ping enzyme = "Cap"; inner virus structural protein with **T=13** symmetry = "T13"; Inner virus structural protein with **T=2** symmetry = "T2"; **vi**ral inclusion body or **vi**roplasm matrix **p**rotein = "ViP"; virus **tu**bule **p**rotein = "TuP"; protein with **hel**icase activity = "Hel". Other species within the genus may have proteins with significant differences in sizes.

Virus adsorption involves components of the outer capsid, although cell entry may also involve VP7(T13). VP2 (possibly also VP5) is involved in determination of virulence. VP5 may be involved in penetration of the cell membrane (release from endosomes into the cytoplasm) and the expressed protein can induce cell fusion. The outer capsid layer is lost during the early stages of replication. The transcription frequency of mRNA from individual genome segments varies, with more copies produced from the smaller segments.

Details of the process of virus replication are lacking. The viral inclusion bodies (VIB) are considered to be the sites of morphogenesis of transcriptionally active virus cores containing dsRNA. The smallest particles containing RNA that are observed in VIBs appear to represent progeny subcore particles. The outer core protein [VP7(T13)] is added within the VIB and the outer CP at the periphery of the VIB.

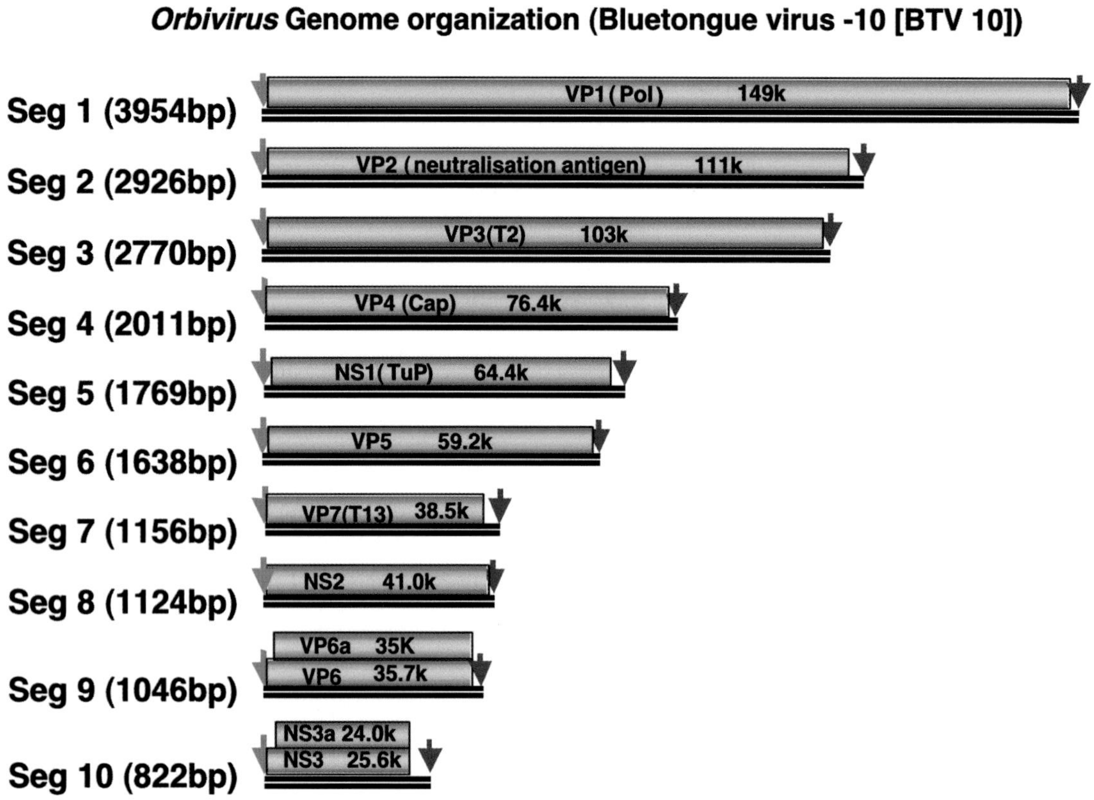

Figure 3: Genome organization of Bluetongue virus 10 (BTV-10), containing 10 dsRNA segments, each contain an ORF, except Seg9 and Seg10, which contains two ORFs. The green arrows indicate the upstream conserved terminal sequence (+ve 5'-GUUAAA...), while the red arrows indicate the downstream conserved terminal sequence (+ve....ACUUAC-3').

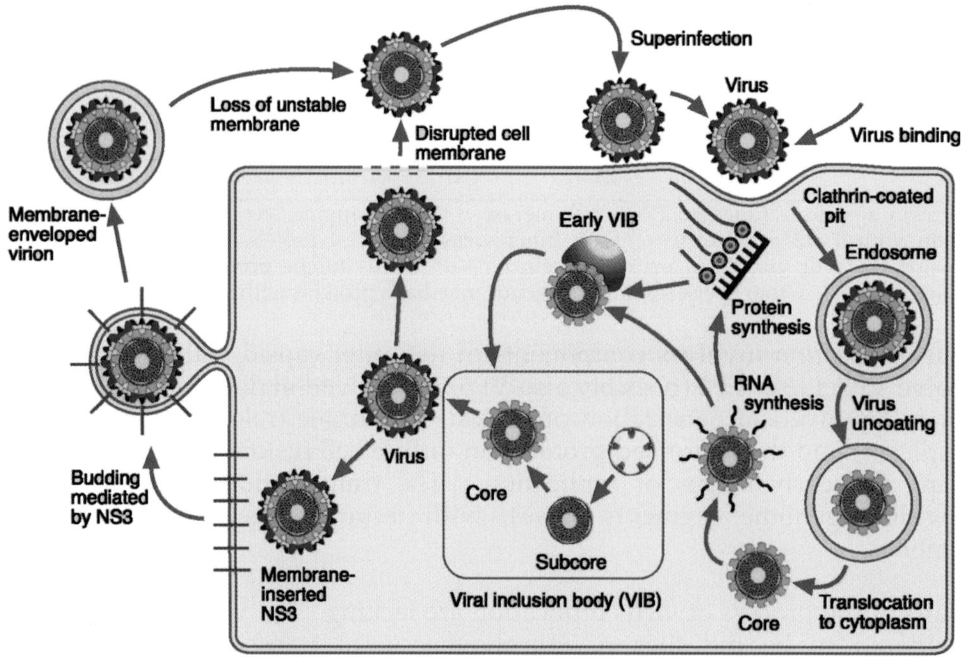

Figure 4. Diagram representing the Bluetongue virus (BTV) replication cycle.

Virus particles are transported within the cell by specific interaction with the cellular cytoskeleton and can be released from the cell prior to lysis through interaction with membrane-associated NS3 proteins (Fig. 4). There is also evidence of specific association between NS1 tubules and intact virions in the cell cytoplasm. In most mammalian cells, replication of orbiviruses leads to shut-off of host protein synthesis and usually results in cell lysis and the release of virus particles. In persistently infected insect cells (or gamma delta T cells), there is no evidence for shut-off of host protein synthesis, extensive cell lysis or CPE. In some species (AHSV), NS3 is involved in determination of virulence for the mammalian host and by controlling virus dissemination within the insect may at least partially determine their ability to transmit the virus (vector competence). Virus particles can leave viable mammalian cells by two distinct mechanisms, extrusion (involving cell membrane damage) and budding. Only budding has been observed in cells of the BTV vector *Culicoides variipennis,* resulting in particles which have a membrane envelope, although this is unstable and is rapidly lost. Continuous release of virus particles from infected cells and reinfection appears to be a feature of orbivirus replication.

ANTIGENIC PROPERTIES

The main virus serogroup (species) specific antigen is the immunodominant outer core protein VP7(T13). Monoclonal or polyclonal antibodies against VP7(T13) can neutralize core particle infectivity, but do not attach to, or neutralize undamaged virus particles or ISVP in aqueous suspension, indicating that VP7(T13) is not exposed on the intact virion surface. Other viral proteins are also conserved between virus species (in particular core proteins NS1 and NS2). Some of these antigens may also show cross-reactions with viruses in other species, particularly those regarded as closely related. These cross-reactions are usually at a significantly lower level than with other viruses from the same virus species and may be 'one way'. Such relationships between species are also demonstrated by comparisons of the RNA sequences of conserved segments (for example those of genome segment 3, coding for VP3(T2), Fig. 4). These data indicate that the different orbivirus species may be divided into at least four 'groups'. The first group (A) contains: AHSV, BTV, EHDV, Equine encephalosis virus (EEV), Eubenangee virus (EUBV), Palyam virus (PALV), Wallal virus (WALV) and Warrego virus (WARV). The second group (B) contains: Chenuda virus (CNUV), Ieri virus (IERIV), Wad Medani virus (WMV) and Great Island virus (GIV). The third group (C) contains Corriparta virus (CORV). The fourth group (D) contains Wongorr virus (WGRV). However insufficient comparisons have been made to conclusively assign all of the species of the genus *Orbivirus* to such groups.

Each species of the genus *Orbivirus* includes a number of serotypes that can be identified and distinguished by serum neutralization assays of intact virus particles (primarily via interactions of antibodies with the outer CPs). VP2 is the main neutralization antigen of BTV, while VP5 is also involved in determination of virus serotype, possibly by imposing conformational constraints on VP2. The VP2 and VP5 proteins of BTV exhibit the greatest antigenic and sequence variation (Fig. 5). In other viruses (GIV) the relative sizes of the outer CPs (VP4 and VP5) are very different and their individual roles may also be different. There is evidence that VP2 of BTV (particularly in association with VP5) and VP7 (T13) can be protective antigens.

In AHSV the small nonstructural proteins, NS3 and NS3a are also 'variable' and may be divided into three groups (α, β and γ) based on sequence analysis. Preliminary serological evidence suggests that NS3 cross-reacts poorly between these groups. NS3 can also be involved in determination of virulence (AHSV), possibly as a result of its involvement in release of virus particles from cells (budding) and its consequent effect on virus dissemination. Recent sequencing studies of BTV NS3 also indicate that it can be highly variable. NS3 variation does not correlate with virus serotype.

BIOLOGICAL PROPERTIES

The specific infectivity of purified (disaggregated) BTV virus particles is equivalent to a particle infectivity ratio of 1000: 1 in both mammalian and insect cell systems. However core particle infectivity varies from 1000 fold less than that of intact virions (BHK cells) to non-infectious (CHO cells) in mammalian systems, depending on the cell line used. However, in some insect cells (KC cells, from *Culicoides sonorensis*) and adult vector insects, core particles are only slightly less infectious than intact virions (particle infectivity ratio of 1900: 1). Treatment of virus with chymotrypsin or trypsin results in production of infectious subviral particles (ISVPs), in which VP2 is cleaved. BTV ISVPs have lost hemaglutinating activity, as well as the tendency to aggregate but have a significantly elevated infectivity for adults of insect vectors and for some insect cell lines (a particle infectivity ratio of approximately 13:1 for KC cells).

Different orbiviruses infect a wide range of vertebrate hosts including ruminants (domesticated and wild), equids (domesticated and wild), rodents, bats, marsupials, birds, sloths, and primates, including humans. Orbiviruses replicate in, and are primarily transmitted by arthropod vectors (gnats, mosquitoes, phlebotomines, or ticks, depending on the virus). Trans-stadial transmission in ticks has been demonstrated for some viruses. Infection of vertebrates *in utero* may also occur. Orbiviruses, particularly those transmitted by short-lived vectors (gnats, mosquitoes, phlebotomines), are only enzootic in areas where adults of the competent vector species persist and are present all, or most of the year. There is no evidence of trans-ovarial transmission of orbiviruses in *Culicoides*, although orbiviruses have been detected in cell lines derived from tick eggs. BTV and EHDV are distributed worldwide between about 50° North and about 30° South in the Americas and between 40° North and 35° South in the rest of the world. However, there is evidence for persistence of these viruses over winter in the absence of overt disease. Mechanisms for persistence in the vertebrate host species even at low levels may be of particular importance. Virus distribution also depends on the initial introduction into areas containing susceptible vertebrate hosts and competent vector species. For this reason not all serotypes of each species (e.g., BTV) are present at locations where some serotypes are endemic.

Orbivirus infection of arthropods has little or no evident effect. Infection in vertebrates, can vary from inapparent to fatal, depending on both the virus and the host. Some BTV strains cause death in sheep, others cause a variety of pathologies, including hemorrhagic conditions, lameness, oedema, a transitory cyanotic appearance of the tongue (giving rise to the species name), nasal and mouth lesions, etc.; still others cause no overt pathology. BTV infection of cattle may show no signs of disease but involve long-lived viraemias. AHSV, EHDV and EEV can cause severe pathology in their respective vertebrate hosts. Mortality rates in serologically naive populations can be over 98% (AHSV).

LIST OF SPECIES DEMARCATION CRITERIA IN THE GENUS

In common with the other genera within the family *Reoviridae*, the prime determinant for inclusion of virus isolates within an orbivirus species is compatibility for reassortment of genome segments during co-infection, thereby exchanging genetic information and generating viable progeny virus strains. However, data providing direct evidence of segment reassortment between isolates is limited and serological comparisons (primarily involving the immunodominant serogroup/species specific antigen VP7(T13)), form the usual basis of diagnostic assays for each of the virus species (serogroups).

Members of a species of the genus *Orbivirus* may be identified by:

- The ability to exchange genetic material by genome segment reassortment during dual infections, thereby producing viable progeny virus strains.
- High levels of serological cross reaction by ELISA, or assays such as complement fixation (CF), or agar gel immunodiffusion (AGID), using either polyclonal sera, or monoclonal antibodies against conserved antigens such as VP7(T13). For example in competition ELISA, at a test serum dilution of 1/5, a positive serum will show >50% inhibition of color formation, while a negative control serum, or serum that is specific for a different species will normally produce <25% inhibition of color compared to a no antibody control. Distinct but related species may show low level serological cross-reaction, which may be only 'one way'.
- High levels of RNA sequence similarities in "conserved" genome segments. Viruses within the same species will normally show <24% sequence variation in genome Seg3 (encoding the major subcore structural protein, VP3(T2)). Viruses in different species will normally contain >26% sequence variation in genome Seg3; these differences are also reflected in the aa sequences of the viral proteins.
- Relatively efficient cross hybridization of "conserved" genome segments (those not encoding outer capsid components, or other variable proteins) under high stringency conditions (>85% homology) (Northern or dot blots, with probes made from viral RNA or cDNA).
- PCR using primers to conserved genome regions or segments such as Seg3 or Seg7. Can be coupled with cross-hybridization analysis (Northern or dot blots);
- Identification by virus serotype with a virus type already classified within a specific orbivirus species. None of the serotypes from different species will cross-neutralize.
- Analysis of "electropherotypes" by agarose gel electrophoresis but not by PAGE. (Viruses within a single species will show a relatively uniform electropherotype. However, a major deletion / insertion event may result in two distinct electropherotypes within a single species (for example EHDV) and some similarities can exist between more closely related species;
- Identical conserved terminal regions of the genome segments (some closely related species can have identical terminal sequences on at least some segments).
- Identification of common vector or host species and the clinical signs produced. For example BTV is transmitted only by certain *Culicoides* species and will infect cattle and sheep producing clinical signs of varying severity but is not thought to infect horses. The reverse is true of AHSV.

LIST OF SPECIES IN THE GENUS

Species names are in green italic script; strain names and synonyms are in black roman script; tentative species names are in blue roman script. Sequence accession numbers [L1 to L3, M1 to M4, and S1 to S4 indicate RNA segment numbers], arthropod vector and host names { }, and assigned abbreviations ( ) are also listed.

SPECIES IN THE GENUS

African horse sickness virus

(9 serotypes)

{*Culicoides*: Equids, dogs, elephants, camels, cattle, sheep, goats, predatory carnivores and (in special circumstances) humans}

African horse sickness virus 1	[Seg2: AY163329, Seg10: U02711]	(AHSV-1)
African horse sickness virus 2	[Seg2: AY163332, Seg10: U59279]	(AHSV-2)
African horse sickness virus 3	[Seg2: U01832, Seg4: AF246225, Seg7: 545433, Seg8: AF545434, Seg9: U19881, Seg10: AJ007303]	(AHSV-3)

African horse sickness virus 4	[Seg2: U21956, Seg3: D26572, Seg4: D14402, Seg5: 11390, Seg6: M94731, Seg7: A27209, Seg10: U02712]	(AHSV-4)
African horse sickness virus 5	[Seg2: AY163331, Seg10: U60188]	(AHSV-5)
African horse sickness virus 6	[Seg2: AF021235, Seg3: AF021236, Seg5: U73658, Seg6: AF021237, Seg7: AF021238, Seg9: U33000, Seg10: U26171]	(AHSV-6)
African horse sickness virus 7	[Seg2: AY16330, Seg10: U60190]	(AHSV-7)
African horse sickness virus 8	[Seg2: AY163333, Seg10: U02713]	(AHSV-8)
African horse sickness virus 9	[Seg1: U94887, Seg2: AF043926, Seg5: U01069, Seg6: U74489, Seg7: S69829, Seg8: M69090, Seg10: D12480]	(AHSV-9)
Bluetongue virus		
(24 serotypes)		
Bluetongue virus 1	[Seg2: X06464, Seg3: AF529048, Seg5: M36713, Seg6 M63417, Seg7: X58064, Seg8: X58064, Seg9: D10905, Seg10: D00253]	(BTV-1)
Bluetongue virus 2	[Seg1: L20508, Seg2: M21946, Seg3: L19967, Seg4: L08637, Seg5: AY138895, Seg6: X62283, Seg7: M64997, Seg8: AY138896, Seg9: L08668, Seg10: AF1235224]	(BTV-2)
Bluetongue virus 3	[Seg2: X55801, Seg7: AF172827, Seg10: AF135225]	(BTV-3)
Bluetongue virus 4	[Seg2: AF135220, Seg7: AF172828, Seg9: 403423, Seg10: AF135226]	(BTV-4)
Bluetongue virus 5		(BTV-5)
Bluetongue virus 6	[Seg7: AF188653]	(BTV-6)
Bluetongue virus 7		(BTV-7)
Bluetongue virus 8	[Seg7: AF188671, Seg10: AF512924]	(BTV-8)
Bluetongue virus 9	[Seg2: L46686]	(BTV-9)
Bluetongue virus 10	[Seg1: X12819, Seg2: M11787, Seg3: M22096, Seg4: Y00421, Seg5: Y00422, Seg6: D12532, Seg7: X06463, Seg8: D00500, Seg9: D00509, Seg10: AF044372]	(BTV-10)
Bluetongue virus 11	[Seg1: L20445, Seg2: M17437, Seg3: L19968, Seg4: L08638, Seg6: M73715, Seg7: M32102, Seg9: L08670, Seg10: L08631]	(BTV-11)
Bluetongue virus 12	[Seg7: AY263377, Seg10: AF135227]	(BTV-12)

Bluetongue virus 13	[Seg1: L20446, Seg2: D00153, Seg3: L19969, Seg4: L08640, Seg5: M97762, Seg6: X54308, Seg7: J04365, Seg9: L08671, Seg10: L08629]	(BTV-13)
Bluetongue virus 14		(BTV-14)
Bluetongue virus 15	[Seg2: AF135221, Seg7: L11723, Seg10: AF135228]	(BTV-15)
Bluetongue virus 16	[Seg2: AF530067, Seg7: AF172831, Seg10: AF135229]	(BTV-16)
Bluetongue virus 17	[Seg1: L20447, Seg2: M17438, Seg3: AF017280, Seg4: L08639, Seg5: X17041, Seg6: X55359, Seg7: X53693, Seg8: P33473, Seg9: L08672, Seg10: L08630]	(BTV-17)
Bluetongue virus 18	[Seg10: AF512915]	(BTV-18)
Bluetongue virus 19		(BTV-19)
Bluetongue virus 20	[Seg5: X56735, Seg10: AF529055]	(BTV-20)
Bluetongue virus 21	[Seg2: L46684, Seg3: AF529047, Seg10: AF529053]	(BTV-21)
Bluetongue virus 22		(BTV-22)
Bluetongue virus 23	[Seg2: L46685, Seg7: AJ277802, Seg10: AF529051]	(BTV-23)
Bluetongue virus 24		(BTV-24)
Changuinola virus		
(12 serotypes)		
{phlebotomines, culicine mosquitoes: humans, rodents, sloths}		
Almeirim virus		(ALMV)
Altamira virus		(ALTV)
Caninde virus		(CANV)
Changuinola virus		(CGLV)
Gurupi virus		(GURV)
Irituia virus		(IRIV)
Jamanxi virus		(JAMV)
Jari virus		(JARIV)
Monte Dourado virus		(MDOV)
Ourem virus		(OURV)
Purus virus		(PURV)
Saraca virus		(SRAV)
Chenuda virus		
(7 serotypes)		
{ticks: seabirds}		
Baku virus		(BAKUV)
Chenuda virus		(CNUV)
Essaouira virus		(ESSV)
Huncho virus		(HUAV)
Kala Iris virus		(KIRV)
Mono Lake virus		(MLV)
Sixgun city virus		(SCV)
Chobar Gorge virus		
(2 serotypes)		
{ticks: bats}		

Chobar Gorge virus		(CGV)
Fomede virus		(FV)
Corriparta virus		
(6 serotypes/strains*)		
{culicine mosquitoes: humans, rodents}		
Acado virus		(ACDV)
Corriparta virus (CS109)		(CORV-CS109)
Corriparta virus (V654)		(CORV-V654)
Corriparta virus (V370)		(CORV-V370)
Corriparta virus MRM1	[Seg3: AF530086]	(CORV-MRM1)
Jacareacanga virus		(JACV)
Epizootic hemorrhagic disease virus		
(10 serotypes / strains*)		
{*Culicoides*: cattle, sheep, deer, camels, llamas, wild ruminants, marsupials}		
Epizootic hemorrhagic disease virus 1	[Seg2: D10767, Seg3: M76616, Seg5: X55782, Seg7: D10766, Seg8: L31764, Seg10: L29023]	(EHDV-1)
Epizootic hemorrhagic disease virus 2	[Seg2: L33818, Seg3: L33819, Seg5: L27648, Seg6: X59000, Seg7: U43560, Seg8: M69091, Seg10: L29023]	(EHDV-2)
Ibaraki virus	[Seg2: AB030735, Seg3: AB041933, Seg6: AB030736, Seg7: AB041934]	(IBAV)
Epizootic hemorrhagic disease virus 3		(EHDV-3)
Epizootic hemorrhagic disease virus 4		(EHDV-4)
Epizootic hemorrhagic disease virus 5		(EHDV-5)
Epizootic hemorrhagic disease virus 6		(EHDV-6)
Epizootic hemorrhagic disease virus 7		(EHDV-7)
Epizootic hemorrhagic disease virus 8		(EHDV-8)
Epizootic hemorrhagic disease virus 318		(EHDV-318)
Equine encephalosis virus		
(7 serotypes)		
{*Culicoides*: equids}		
Equine encephalosis virus 1	[Seg10: AY115878]	(EEV-1)
Equine encephalosis virus 2	[Seg10: AY115871]	(EEV-2)
Equine encephalosis virus 3	[Seg10: AY115874]	(EEV-3)
Equine encephalosis virus 4	[Seg10: AY115868]	(EEV-4)
Equine encephalosis virus 5	[Seg10: AY115869]	(EEV-5)
Equine encephalosis virus 6	[Seg10: AY115873]	(EEV-6)
Equine encephalosis virus 7	[Seg10: AY115870]	(EEV-7)
Eubenangee virus		
(4 serotypes)		
{*Culicoides*, anopheline and Culicine mosquitoes: unknown hosts}		
Eubenangee virus	[Seg3: AF530087]	(EUBV)
Ngoupe virus		(NGOV)
Pata virus		(PATAV)
Tilligerry virus		(TILV)
Ieri virus		
(3 serotypes)		
{mosquitoes: birds}		
Ieri virus		(IERIV)

Gomoka virus		GMKV)
Arkonam virus		(ARKV)
Great Island virus		(GIV)
(36 serotypes / strains*)		
{*Argas, Ornithodoros, Ixodes* ticks: seabirds, rodents, humans}		
Above Maiden virus		(ABMV)
Arbroath virus		(ABRV)
Bauline virus		(BAUV)
Broadhaven virus	[Seg2: M87875, Seg5: M58030, Seg6: X82599, Seg7: M87876, Seg10: M83197]	(BRDV)
Cape Wrath virus		(CWV)
Colony virus		(COYV)
Colony B North virus		(CBNV)
Ellidaey virus		(ELLV)
Foula virus		(FOUV)
Great Island virus		(GIV)
Great Saltee Island virus		(GSIV)
Grimsey virus		(GSYV)
Inner Farne virus		(INFV)
Kemerovo virus		(KEMV)
Kenai virus		(KENV)
Kharagysh virus		(KHAV)
Lipovnik virus		(LIPV)
Lundy virus		(LUNV)
Maiden virus		(MDNV)
Mill Door virus		(MDRV)
Mykines virus		(MYKV)
North Clett virus		(NCLV)
North End virus		(NEDV)
Nugget virus		(NUGV)
Okhotskiy virus		(OKHV)
Poovoot virus		(POOV)
Rost Island virus		(RSTV)
St Abb's Head virus		(SAHV)
Shiant Islands virus		(SHIV)
Thormodseyjarlettur virus		(THRV)
Tillamook virus		(TLMV)
Tindholmur virus		(TDMV)
Tribec virus		(TRBV)
Vearoy virus		(VAEV)
Wexford virus		(WEXV)
Yaquina Head virus		(YHV)
Lebombo virus		
(1 serotype)		
{culicine mosquitoes: humans, rodents}		
Lebombo virus 1		(LEBV-1)
Orungo virus		
(4 serotypes)		
{culicine mosquitoes: humans, camels, cattle, goats, sheep, monkeys}		
Orungo virus 1		(ORUV-1)
Orungo virus 2		(ORUV-2)

DSRNA

Orungo virus 3		(ORUV-3)
Orungo virus 4		(ORUV-4)
Palyam virus		
(13 serotypes / strains*)		
{*Culicoides*, culicine mosquitoes: Cattle, sheep}		
Abadina virus		(ABAV)
Bunyip creek virus	[Seg5: AB034595, Seg7: AB034671, Seg9: AB034681]	(BCV)
CSIRO village virus	[Seg5: AB034594, Seg7: AB034670, Seg9: AB034680]	(CVGV)
D'Aguilar virus	[Seg3: AF530085, Seg5: AB034593, Seg7: AB034666, Seg9: AB034676]	(DAGV)
Gweru virus	[Seg5: AB034598, Seg7: AB034674, Seg9: AB034683]	(GWV)
Kasba virus (Chuzan virus)	[Seg1: AB018086, Seg2: AB014725, Seg3: AB014728, Seg4: AB018087, Seg5: AB018089, Seg6: AB014726, Seg7: AB014727, Seg8: AB018090, Seg9: AB018088, Seg10: AB018091]	(KASV)
Kindia virus		(KINV)
Marrakai virus	[Seg5: AB034592, Seg7: AB034668, Seg9: AB034678]	(MARV)
Marondera virus	[Seg5: AB034597, Seg7: AB034673]	(MRDV)
Nyabira virus	[Seg5: AB034596, Seg7: AB034672, Seg9: AB034682]	(NYAV)
Palyam virus		(PALV)
Petevo virus		(PETV)
Vellore virus		(VELV)
Peruvian horse sickness virus		
(1 serotype)		
{ mosquitoes: horses }		
Peruvian horse sickness virus - 1		(PHSV-1)
St Croix River virus ‡		
{ticks: onknown hosts}		
St Croix River virus -1	[Seg1: AF133431, Seg2: AF133432, Seg3: AF145400, Seg4: AF145401, Seg5: AF145402, Seg6: AF145403, Seg7: AF145404, Seg8: AF145405, Seg9: AF145406, Seg10: AF145407]	(SCRV-1)
Umatilla virus		
(4 serotypes)		
{culicine mosquitoes: birds}		
Llano Seco virus		(LLSV)
Minnal virus		(MINV)
Netivot virus		(NETV)
Umatilla virus		(UMAV)

Wad Medani virus		
(2 serotypes)		
{Boophilus, Rhipicephalus, Hyalomma, Argas ticks: domesticated animals}		
Seletar virus		(SELV)
Wad Medani virus		(WMV)
Wallal virus		
(3 serotypes/strains*)		
{Culicoides: marsupials}		
Mudjinbarry virus		(MUDV)
Wallal virus	[Seg3: AF530084]	(WALV)
Wallal K virus		(WALKV)
Warrego virus		
(3 serotypes/strains*)		
{Culicoides, anopheline and culicine mosquitoes: marsupials}		
Mitchell river virus		(MRV)
Warrego virus	[Seg3: AF530083]	(WARV)
Warrego K virus		(WARKV)
Wongorr virus		(WGRV)
(8 serotypes/strains*)		
{Culicoides, mosquitoes: Cattle, macropods}		
Paroo river virus	[Seg3: U56993]	(PRV)
Picola virus	[Seg3: U56994]	(PIAV)
Wongorr virus CS131		(WGRV-CS131)
Wongorr virus MRM13443	[Seg3: U56992]	(WGRV-MRM13443)
Wongorr virus V1447	[Seg3: U56989]	(WGRV-V1447)
Wongorr virus V195	[Seg3: U56990]	(WGRV-V195)
Wongorr virus V199	[Seg3: U56991]	(WGRV-V199)
Wongorr virus V595		(WGRV-V595)

TENTATIVE SPECIES IN THE GENUS

Andasibe virus	{mosquitoes: unknown hosts}	(ANDV)
Codajas virus	{mosquitoes: rodents}	(COV)
Ife virus	{mosquitoes: rodents, birds, ruminants}	(IFEV)
Itupiranga virus	{mosquitoes: unknown hosts}	(ITUV)
Japanaut virus	{mosquitoes: unknown hosts}	(JAPV)
Kammavanpettai virus	{unknown vectors: birds}	(KMPV)
Lake Clarendon virus	{ticks: birds}	(LCV)
Matucare virus	{ticks: unknown hosts}	(MATV)
Tembe virus	{mosquitoes: unknown hosts}	(TMEV)
Tracambe virus	{mosquitoes: unknown hosts}	(TRV)
Yunnan orbivirus	{Culex tritaeniorhyncus}	(YUOV)

* In some species the serological relationships between strains has not been fully determined

Phylogenetic Relationships Within the Genus

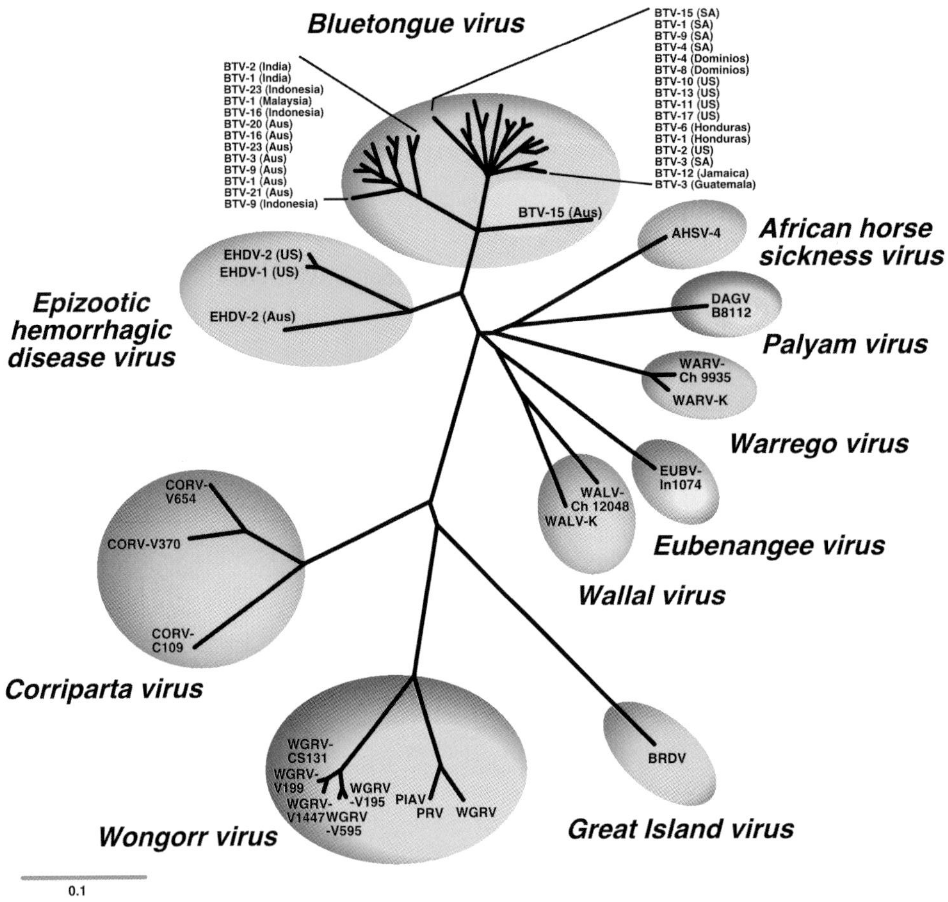

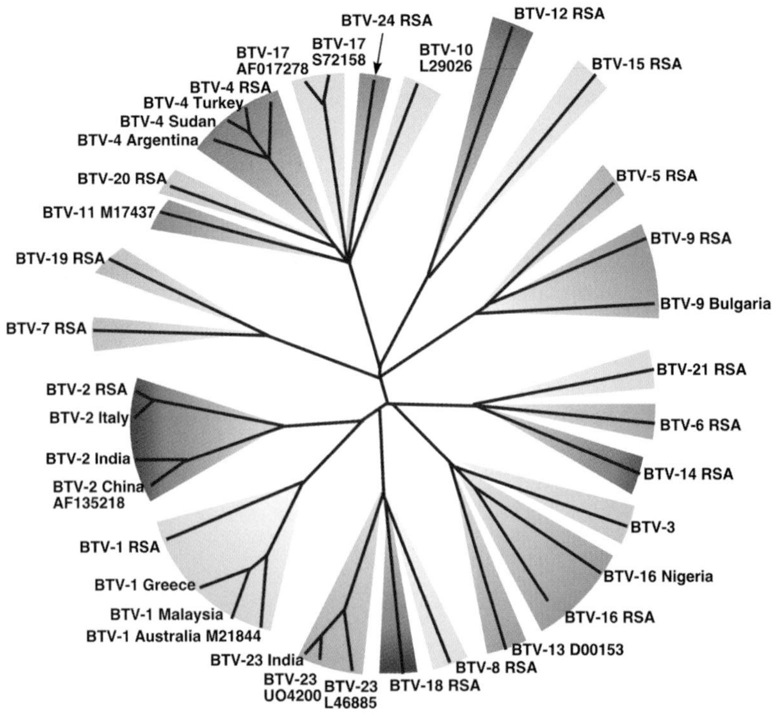

Figure 5: (Top) Phylogenetic tree for the genus *Orbivirus*, constructed using partial genome Seg3 (1193-1661 bp), or equivalent sequences (encoding the major structural protein VP3(T2) of the subcore capsid shell). (Bottom) Phylogenetic tree for Bluetongue virus (*Orbivirus* type species) constructed using complete genome Seg2 sequences. Genome Seg2 codes for VP2, the larger outer CP and major neutralization (most variable) BTV antigen. Nucleotide sequences of genome Seg2 and Seg3 were aligned in Bioedit (Hall, 1999). The tree was prepared using Clustal X) and drawn with TreeView 1.5 (Page, 1996) (Courtesy of S. Maan, A. Samuel, N. Knowles and P.P.C. Mertens). The virus indicated by (p) is a partial sequence. Names corresponding to the abbreviations are from the "List of Species".

DERIVATIONS OF NAMES

Orbi: from Latin orbis, "ring" or "circle" in recognition of the ring-like structures observed in micrographs of the surface of core particles.

REFERENCES

Attoui, H., Stirling, J.M., Munderloh, U.G., Billoir, F., Brookes, S.M., Burroughs, J.N., de Micco , P., Mertens, P.P.C. and de Lamballerie, X. (2001). Complete sequence characterization of the genome of the St. Croix River Virus, a new orbivirus isolated from *Ixodes scapularis* cells. *J. Gen. Virol.*, **82**, 795-804.

Diprose J.M., Burroughs, J.N., Sutton, G.C., Goldsmith A., Gouet, P., Malby, R., Overton, I., Zientara, S., Mertens, P.P.C., Stuart, D.I. and Grimes J.M. (2001). Translocation portals for the substrates and products of a viral transcriptase complex: the Bluetongue virus core. *EMBO J.*, **20**, 7229-7239.

Diprose, J.M., Grimes, J.M., Sutton, G.C., Burroughs, J.N., Maan, S., Meyer, A., Mertens, P.P.C. and Stuart, D.I. (2002). The core of bluetongue virus binds double-stranded RNA. *J. Gen. Virol.*, **76**, 9533-6.

Gouet, P., Grimes, J.M., Diprose, J.M., Malby, R., Burroughs, J.N., Zeintara, S., Stuart, D.I. and Mertens P.P.C. (1999). The highly ordered double-stranded RNA genome of Bluetongue virus revealed by cristallography. *Cell*, **27**, 481-490.

Gould, A.R. and Pritchard, L.I. (1991). Phylogenetic analyses of the complete nucleotide sequence of the capsid protein (VP3) of Australian epizootic hemorrhagic disease of Deer virus (serotype-2) and cognate genes from other orbiviruses. *Virus Res.*, **21**, 1-18.

Grimes, J.M., Burroughs, J.N., Gouet, P., Diprose, J.M., Malby, R., Zeintara, S., Mertens P.P.C. and Stuart, D.I. (1998). The atomic structure of the bluetongue virus core. *Nature*, **395**, 470-478.

Mellor, P.S., Baylis, M., Hamblin C., Calisher, C.H. and Mertens, P.P.C. (eds) (1998). *African horse sickness. Arch. Virol.*, **S14**.

Mertens, P.P.C., Burroughs, J.N., Walton, A., Wellby, M.P., Fu, H., O'Hara, R.S., Brookes, S.M. and Mellor, P.S. (1996). Enhanced infectivity of modified bluetongue virus particles for two insect cell lines and for two Culicoides vector species. *Virology*, **217**, 582-593.

Nuttall, P.A. and Moss, S.R. (1989) Genetic reassortment indicates a new grouping for tick borne orbiviruses. *Virology*, **171**, 156-161.

Roy, P. (1989). Bluetongue virus genetics and genome structure. *Virus Res.*, **13**, 179-206.

Roy, P. (1989). Bluetongue virus proteins. *J. Gen. Virol.*, **73**, 3051-3064.

Roy, P. and Gorman, B.M. (eds)(1990). *Bluetongue viruses. Curr. Top. Micro. Immunol.*, **162**, 1-200.

Takamatsu, H., Mellor, P.S., Mertens, P.P.C., Kirkham, P.A., Burroughs, J.N. and Parkhouse, R.M.E. (2003). A possible overwintering mechanism for bluetongue virus in the absence of the insect vector. *J. Gen. Virol.*, **84**, 227-35.

CONTRIBUTORS

Mertens, P.P.C., Maan, S., Samuel, A. and Attoui, H.

GENUS ROTAVIRUS

Type Species ***Rotavirus A***

DISTINGUISHING FEATURES

Virus particles have a 'wheel-like' appearance when viewed by negative contrast electron microscopy (Fig. 1), from which the genus derives its name. The triple layered capsid encloses a genome of 11 dsRNA segments and is formed in a unique morphogenic pathway involving the acquisition of a transient lipid envelope following budding of immature particles into the endoplasmic reticulum. Viruses infect only vertebrates and are transmitted by the fecal-oral route.

VIRION PROPERTIES

MORPHOLOGY

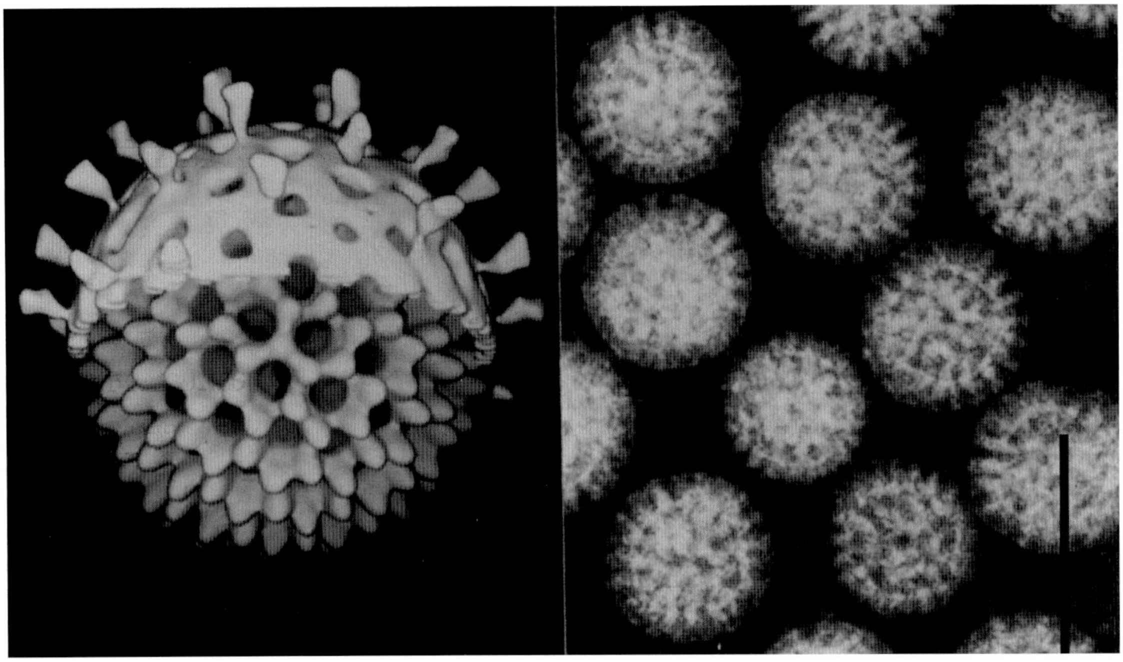

Figure 1: (Left) Cryo-electron micrograph reconstruction of a Simian rotavirus A/SA11 (SiRV-A/SA11) particle. (Right) Electron micrograph of SiRV-A/SA11 particles viewed by negative staining. (Courtesy of B.V.V. Prasad). The bar represents 100 nm.

The data from Simian rotavirus A/SA11 (SiRV-A/SA11) represent a paradigm for the other viruses within the genus. The mature infectious virion has an overall diameter of approximately 100 nm and is made up of three concentric protein layers, with no lipid-containing envelope. The detailed topology of these layers and their protein components has been revealed using a combination of cryo-electron microscopy followed by image processing (Fig. 2) and analysis of the virus like particles formed using baculovirus recombinants expressing specific rotavirus structural proteins. The innermost layer composed of VP2, is approximately 51 nm in diameter and 3.5 nm thick. This layer appears to be directly comparable to the internal capsid layer of members of some other genera of the family *Reoviridae* (for example the VP3(T2) layer of Bluetongue virus, which is referred to as the orbivirus "sub-core"). The rotavirus VP2(T2) layer surrounds the genomic dsRNAs and two structural proteins [VP1(Pol) and VP3(Cap)] that are organized as a series of up to 12 enzymatic complexes, attached to the inner surface of

VP2(T2) at the 5-fold axes of symmetry. The assembled genomic RNAs, together with the enzymatic complexes, have a collective diameter of approximately 44 nm and have been referred to as the rotavirus "subcore". However, this is not regarded as a distinct coherent particle and is not equivalent to the 'subcores' of some other genera (see *Orbivirus*). (The last two sentences of this paragraph could be omitted)

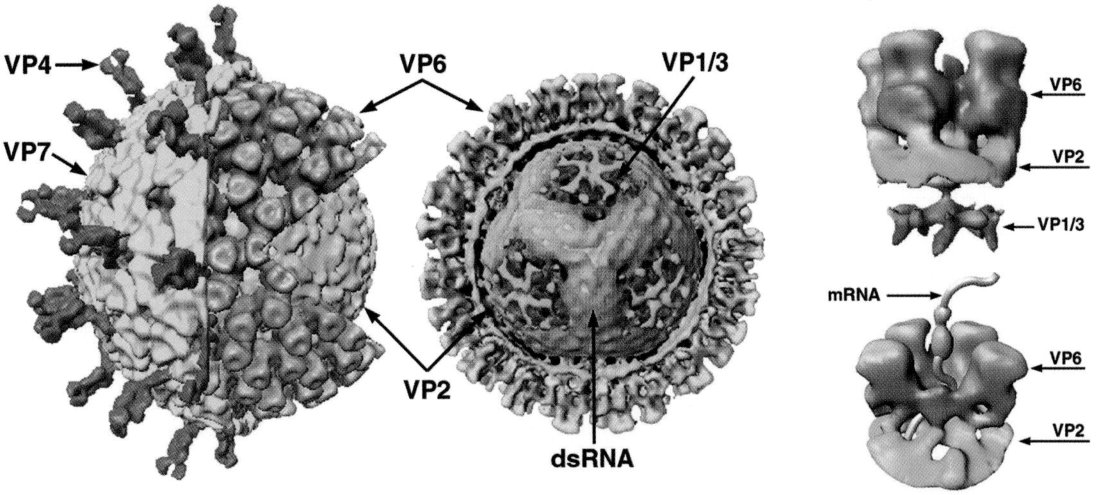

Figure 2: (Left) Cutaway view of the mature particle of Simian rotavirus A/SA11 (SiRV-A/SA11), illustrating the triple-layered capsid structure taken from cryo-EM data and following image reconstruction at 24Å resolution. (Center) Cutaway view of the transcriptionally-competent double layered particle at 19Å. (Top right) Transcription enzyme complex composed of VP1 and VP3, shown anchored to the inner surface of VP2 at the icosahedral vertex. This figure has been computationally isolated from the 22Å reconstruction of a VP1/3/2/6-VLP (virus-like particle). (Bottom right) Proposed pathway of mRNA translocation through the double-layered capsid during genome transcription. The mass of density at the extremity of the mRNA represents the structurally discernible portion of nascent mRNA visible in the 25Å structure of the actively transcribing particle (Courtesy of B.V.V. Prasad).

The two outer layers of the intact rotavirus particle both have a surface lattice with T=13 *l* (laevo) icosahedral symmetry and a uniquely characteristic set of 132 large channels that span both layers, linking the outer surface with the inner most VP2 protein layer. The middle capsid layer is composed of 780 copies of VP6(T13), arranged as 260 trimeric morphological units, positioned at all the local and strict 3-fold axes of the icosahedral lattice. This VP6(T13) layer forms the outer surface of the 'double layered' particle (approximately 70.5 nm in diameter) and is directly comparable to one of the capsid layers of viruses from some other genera within the family *Reoviridae.* For example the 'core' particle of the orbiviruses has an outer core layer composed of 780 copies of the orbivirus VP7(T13) protein, which is structurally similar to the VP6(T13) protein of the rotaviruses. The outermost layer of the rotavirus capsid, which is composed of two proteins (VP4 and VP7) and is required for infectivity, has a diameter of approximately 75 nm. The glycoprotein VP7 makes up the surface of the outermost shell and also appears to be arranged in 260 trimers, that interact with the tips of the trimers of VP6(T13). The VP4 protein associates to form 60 dimeric spikes, approximately 20 nm in length which project approximately 12 nm from the surface of the outer shell, giving a final maximum diameter of approximately 100 nm. The VP4 spikes also extend into the particle, so that they must interact with VP7 and VP6 and possibly even have contact with VP2(T2). The spike structure is stabilized by protease cleavage of VP4 and undergoes dramatic, irreversible change in conformation at high pH.

PHYSICOCHEMICAL AND PHYSICAL PROPERTIES

The triple layered infectious (complete) rotavirus particle (Fig. 1 and 2) has a density of 1.36 g/cm^3 in CsCl and sediments at 520-530S in sucrose. Virus infectivity is absolutely dependent upon the presence of the outermost protein layer, the integrity of which requires calcium. This outer layer can be removed by treatment of virions with calcium-chelating agents, such as EGTA or EDTA. Infectivity is stable within the pH range 3-9 and, when stabilized by the presence of 1.5 mM $CaCl_2$, is not significantly affected by storage for months at 4°C or even 20°C. Infectivity is also relatively thermostable at 50°C but can be destroyed by repeated cycles of freeze-thawing. Infectivity is generally resistant to fluorocarbon extraction, treatment with solvents such as ether and chloroform, or non-ionic detergents such as deoxycholate, all of which reflect the absence of a lipid containing envelope on the mature particle. However, infectivity is lost by treatment with sodium dodecyl sulphate (0.1%), or a number of disinfectants such as betapropiolactone, chlorine, formalin and phenols, with 95% ethanol, which removes the outer shell, being perhaps the most effective disinfectant against these viruses. Hemaglutinating activity of the intact particle is lost rapidly at 45°C, as a result of freeze-thawing, or through the conformational change in the VP4 spikes by treatment at pH 10. Some variation has been observed in the physicochemical properties and stability of intact virions of different rotavirus strains. For example human rotaviruses do not all exhibit hemaglutinating activity and lose the proteins of their outer layer more easily than other strains. In reassortment studies, some of this variation has been attributed to the parental origin of the VP4 present in the virus particle.

Double layered particles are non-infectious (Fig. 2), have a density of 1.38 g/cm^3 in CsCl and sediment at 380-400S in sucrose. Single layered particles can be produced by treatment of double-shelled particles with either chaotropic agents such as sodium thiocyanate or high concentrations of $CaCl_2$. Single layered particles have a density of 1.44 g/cm^3 in CsCl and sediment at 280S in sucrose.

Condensation of the genomic RNA to a radius of ~180Å occurs on treatment of virions at pH 11.5 in the presence of ammonium ions. The condensation is reversed to the native radius of 220Å by removal of ammonium ions and return to physiologic pH.

NUCLEIC ACID

The genome of Simian rotavirus A (SiRV-A) is packaged within the innermost protein shell of the particle. In each particle the genome is composed of a single copy of 11 discrete segments of dsRNA, which range in size (*Rotavirus A* isolates) from 3302 bp to 663 bp (Table 2; Fig. 3), and have an average combined size of 18,550 bp. *Rotavirus A* strains (not including the isolates from avian species), normally have a 4:2:3:2 pattern of segments following fractionation of genomic dsRNAs by PAGE. The genome segments are numbered 1-11 in order of increasing mobility during PAGE, although even within SiRV-A, the migration order of cognate segments, particularly in the Seg7-9 triplet, does vary. In other *Rotavirus A* isolates deviation from the 4:2:3:2 RNA migration pattern is indicative of a genome segment concatemerization, in which as much as an additional 1800 bp are packaged into viable virions, in the form of a partial duplication of a genomic segment. The genome RNA sequences are A+U rich (58-67%). The segments are completely base-paired, and the plus-sense strand contains a 5'-terminal cap structure ($m^7GpppG^{(m)}GPy$) but does not have a polyadenylation signal near its 3'-end. In contrast to members of the genus *Orthoreovirus*, there is no evidence for the presence of single-strand oligonucleotides in the virion. Two levels of terminal sequence conservation are evident in *Rotavirus A* isolates; firstly all genomic segments share short conserved 5'- and 3'-termini. The 5'-terminal conservation is 10 nt and has a consensus sequence of 5'-($GGC^A/_U{}^A/_UU^A/_UA^A/_U{}^A/_U$...)-3' and the 3'-terminus has an 8 nt conservation whose consensus is 5'-(...$^A/_UU^G/_U{}^U/_G{}^G/_U{}^A/_G$CC)-3' (Table 1). Immediately internal to these terminal regions at each end of the different segments, there is a second region of conservation of at least 30-40 nt which is genome segment specific. These two levels of conservation may be indicative of cis-acting signals that are important for controlling transcription, replication, and segment selection for packaging. The 5′-NTRs vary in length but are all less than 50 nt and in all segments are followed by at least one long ORF after the first AUG. Some gene segments contain additional in-frame (genes 7, 9 and 10) or 2nd frame (gene 11) ORFs but only in the case of genes 9 and 11 are these used to give more than a single primary translational product (ie: monocistronic) from each genome segment. The 3′-NTRs vary in length from 17 nt (Seg1) to 182 nt (Seg10) which in the latter case constitutes ~25% of the total segment length.

Detailed genomic information for species *Rotavirus B* and *Rotavirus C* is more sparse. The genome is made up of 11 segments and in all cases the PAGE pattern differs most significantly from *Rotavirus A* isolates in the absence of the tight triplet of migrating bands (Seg7-9). Sequence information is available for representatives of almost all (10 out of 11) genome segments of an isolate of *Rotavirus C* but relatively few in the case of *Rotavirus B* (4/11). In both cases, the segments have broadly similar properties in terms of length and presence of ORFs.

***Rotavirus* Genome organization (Simian rotavirus A / SA11)**

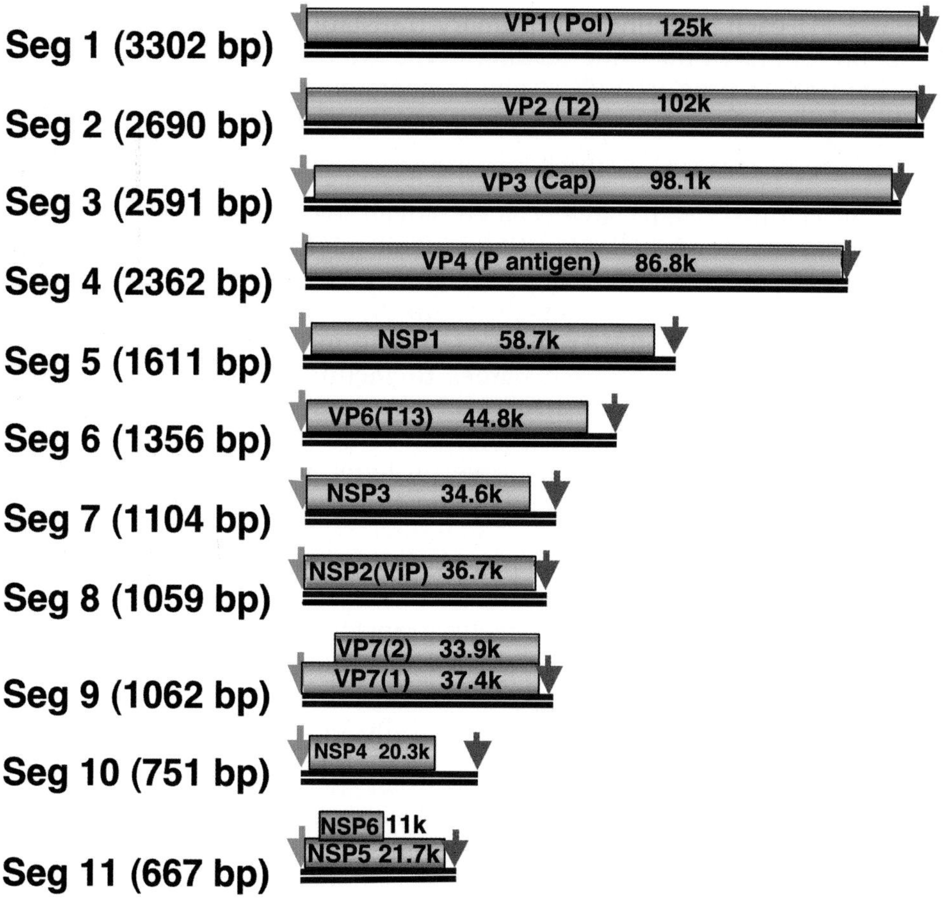

Figure 3: Genome organization of Simian rotavirus A (SV-A/SA11), containing 11 dsRNA segments, each contain an ORF, except Seg9 and Seg11, which contains two ORFs. The green arrow indicates the upstream conserved terminal sequence (+ve 5′-GGC(A/U)$_2$U(A/U)A(A/U)$_2$...), while the red arrow indicates the downstream conserved terminal sequence (+ve......A/UU(G/U)$_3$A/GCC-3′).

Table 1: Conserved terminal sequences of rotavirus genome segments.

Virus species	(strain)	Conserved RNA terminal sequences (positive strand)	
Rotavirus A	SA11	5'-GGC$(^{A}/_{U})_2$U$(^{A}/_{U})$A$(^{A}/_{U})_2$...	$^{A}/_{U}$U$(^{G}/_{U})_3$$^{A}/_{G}$CC-3'
Rotavirus B		5'-GG$(^{U}/_{C})(^{A}/_{U})$N$(^{A}/_{U})_5$...	$(^{U}/_{A})(^{A}/_{U})$AA$(^{A}/_{G})$ACCC-3'
Rotavirus C	Bristol	5'-GCC$(^{A}/_{U})_7$	UGUGGCU-3'

* The data for *Rotavirus B* are derived from 5 sequences of Mu-RV-B/IDIR and 4 sequences of Hu-RV-B/ADRV. N at position 5 of the 5'-sequence indicates it can be any base. There are many other *Rotavirus B* sequences in the database, but of those that are published the PCR cloning was based on primers designed on the IDIR or ADRV sequences, so the real terminal sequences are not known. Others were unpublished and the terminal sequence data are not available.

PROTEINS

Thirteen primary gene products have been defined in the case of SiRV-A with two viral genes (9 and 11) each encoding two primary translation products. In the case of gene 9, two initiation codons in the same reading frame are both used, giving largely overlapping forms of the protein product VP7. Gene 11 contains two long ORFs in different reading frames, translation of which results in two unrelated non-structural proteins NSP5 and NSP6 (Table 2).

Several nomenclature systems have been employed for rotavirus proteins but recently one in which proteins are numbered according to their migration rates on SDS-PAGE, starting with the slowest (i.e. highest molecular weight), with structural proteins being given the prefix VP and non-structural proteins the prefix NSP, has become accepted as logical and likely to minimize confusion. An abbreviation to indicate structure/function has been added in brackets to some protein names to facilitate comparisons between different viruses/genera (Table 2). Six structural proteins have been identified and their approximate localization within the mature virus particle defined. The viral core containing the dsRNA genome, has three proteins associated with it, two of which VP1(Pol) and VP3(Cap) are directly associated with the genome whilst the third [VP2(T2)] makes up the core shell, the innermost protein shell of the capsid. VP1(Pol), the largest viral protein at 125 kDa, possesses RdRp activity which can be assayed in purified virus preparations. VP3(Cap) is a 88 kDa protein which has been shown to carry ssRNA binding, guanylyl-transferase, methyltransferase and GTP binding activities when expressed by recombinant baculoviruses. It is therefore thought to be involved in adding the 5'-cap structure present on viral mRNAs. VP1 and VP3 form a complex that is attached to the interior of the 5-fold vertices of the VP2 capsid layer. The amount of VP1 and VP3 in the virion is known to be low (<25 molecules/particle) but has not been measured precisely. VP2(T2) (94 kDa) is the most abundant protein of the viral core with 120 molecules per virion. It has nucleic acid binding activity although this does not appear to show any sequence specificity. From its deduced aa sequence, VP2(T2) contains two leucine zipper motifs, which are thought to be characteristically involved in the dimerisation of nucleic acid binding proteins. This suggests that VP2 forms a functional dimer.

Table 2: List of the dsRNA segments of Simian rotavirus A/SA11 (SiRV-A/SA11), with their respective size (bp) and their encoded proteins, for which the name, calculated size, location, copy number per virion, and functions or properties are indicated.

dsRNA segment number	dsRNA Segment Size (bp)	ORF	Protein names (§)	Protein Size (aa)	Protein copies /particle	Function (location)
1	3302	18 - 3285	VP1 (Pol)	125,005 (1088)	<25	RdRp, complex with VP3 (subcore)
2	2690	16 - 2662	VP2 (T2)	102,431 (880)	120	Binds RNA, two leucine zipper motifs, myristoylated, cleaved, interaction with NSP5, NH2-terminus required for packaging VP1 and VP3 (innermost core shell protein)
3	2591	49 - 2557	VP3 (Cap)	98,120 (835)	<25	ssRNA binding, guanylyltransferase, Mtr, GTP-binding, complex with VP1, basic protein (subcore)
4	2362	9 -2340	VP4 VP5* VP8*	86,782 (776) 60,000 (529) 28,000 (247)	120	Surface spike protein of outer virion shell, P-type neutralization antigen. Dimer hemagglutinin, cell attachment protein, involved in virulence. Cleavage by trypsin into VP5* and VP8* enhances infectivity and stabilizes spike structure, VP8 crystal structure (galectin fold).
5	1611	30 -1518	NSP1	58,654 (495)	0	Viral protein showing greatest intra-species diversity, conserved cysteine rich zinc finger region near amino terminus, viral 5′ RNA binding. Component of pre-core RI (non-structural).
6	1356	23 - 1217	VP6 (T13)	44,816 (397)	780	Major virion protein, making up middle shell in form of trimeric units. Carries group and sub-group antigenic determinants. Myristoylated, hydrophobic, crystal structure (inner CP).
7	1104	25 - 976	NSP3	34, 600 (315)	0	Cytoskeleton associated, binds specifically to 3′-end of rotavirus mRNAs, as dimer. Interacts with cellular eIF4G eviciting cellular PABP from translation initiation complexes; shuts off host translation. Allows circularization of mRNA in translation. Crystal structures: (NH2 term + RNA) and (COOH-term + eIF4G).
8	1059	46 - 1000	NSP2 (VIP)	36,700 (317)	0	Basic protein, role in RNA replication, associates with VP1, involved in viroplasm formation with NSP5, involved in hyperphosphorylation of NSP5, exhibits non-specific ssRNA binding, functional octomer has NTPase and helix destabilizing activities. Crystal structure (HIT-like fold). (non-structural : present in viroplasms).
9	1062	48 - 1029 135 -1029	VP7 (1) VP7 (2) mature cleaved form	37,368 (326) 33,919 (297) (276)	780	Cleaved signal sequence, high mannose glycosylation, RER integral membrane glycoprotein, G type neutralization antigen, 2 hydrophobic N terminal regions, bicistronic gene (same reading frame), putative Ca^{2+} binding site, (aa 127-157), (surface glycoprotein).
10	751	41 - 569	NSP4	20,290 (175)	0	N linked high mannose, glycosylation and trimming, uncleaved signal sequence, RER trans membrane glycoprotein, 2 hydrophobic N-terminal regions, role in morphogenesis, ER budding, putative Ca^{2+} binding site. Age-dependent diarrhea inducing enterotoxin disrupting cell Ca^{2+} homeostasis, membrane destabilization (non-structural).
11	667	20 - 618 80 - 355	NSP5 NSP6	21,725 (198) 11,012 (92)	0 0	Phosphorylated, O linked glycosylation. Slightly basic, serine threonine rich, RNA binding, kinase activity, interacts with VP2, NSP2 and NSP6, multimeric (non-structural: present in viral inclusion bodies). Product of 2nd ORF (different frame). Interacts with NSP5, (non-structural: present in viral inclusion bodies)

§: protein structure/function: RNA polymerase = "Pol"; capping enzyme = "Cap"; Inner virus structural protein with T=13 symmetry = "T13"; Inner virus structural protein with T=2 symmetry = "T2"; viral inclusion body or viroplasm matrix protein = "ViP". Other species within the genus may have proteins with significant differences in sizes.

The middle protein shell of the virion is made up of 780 molecules of VP6(T13) (41 kDa) arranged in 260 trimeric units. The two remaining structural proteins of the virion, VP4 (88 kDa) and VP7 (38 kDa) of which there are 120 and 780 molecules per virion

DSRNA

respectively make up the outermost shell. The spike protein VP4 (776 aa in most animal strains and 775 aa in most human strains) contains a trypsin cleavage site approximately one third of the way along it. Cleavage of the protein by treatment with protease *in-vitro* produces two products VP5 (60 kDa) and VP8 (28 kDa) and enhances virus infectivity and stabilizes the spike structure. (In part of the literature the cleavage products are designated VP5* and VP8*, respectively). The VP8 cleavage product has hemagglutinin activity, has been crystalized and contains a novel galectin fold-like carbohydrate binding site. VP7 which makes up the surface of the outer shell is formed by N-linked gylcosylation of the primary translation products (vpr7) from genome Seg9. As indicated earlier, Seg9 contains two in-frame initiation codons, of which the first is in a weak Kozak consensus, and two bands of VP7 are seen following SDS-PAGE analysis of purified virions. Together with biochemical studies, this suggests that these bands arise from post-translational processing of VP7 initiated from each of the two initiation codons. It also implies that the two forms of VP7, with largely overlapping primary sequence, are synthesized and incorporated into virions but at present there is no formal proof of this hypothesis. (The last sentence of this paragraph could be omitted).

Six non-structural proteins are encoded by the viral genome and less is known about the functions of these. The largest, NSP1 (53 kDa), is the most variable of all the rotavirus proteins within a single rotavirus species, showing as much as 65% sequence diversity between strains of *Rotavirus A*. Lower levels of variation are seen in the aa sequences of proteins within a single rotavirus species (e.g. <25% variation in VP2(T2) and <45% variation in VP4). However variation in these proteins can also be high between virus species (>87% for VP4 and >84% for VP2). Despite this variation, NSP1 does have a conserved cysteine-rich motif near its amino terminus, which suggests a 'zinc finger' metal binding domain. Such domains are present in some nucleic acid binding proteins and NSP1 has been shown to bind both zinc and the 5′-end of viral ssRNA. NSP1 is a component of the pre-core replication intermediate (RI), found in infected cells, suggesting that it has a role in the early stages of virus assembly, which include the process of genome segment selection. However, the precise nature of this role has not yet been defined. It is worth noting that there are rotaviruses which replicate *in vitro* without a functional NSP1. NSP2(ViP) (35 kDa) is also found in early RIs. Viruses which contain temperature sensitive mutations in genome Seg8 (encoding NSP2(ViP)), have an RNA-negative phenotype at the non-permissive temperature indicating that NSP2(ViP) has a direct role in the mechanism of virus replication, although this is as yet undefined. NSP2(ViP) has non-specific RNA binding activity for both ssRNA and dsRNA, with no apparent sequence specificity. NSP2 has NTPase and helix destabiliziation activities and interacts with NSP5 to form viroplasms. NSP5 is hyperphosphorylated in this interaction. NSP2 functions as an octomer. The crystal structure of NSP2 has been solved, and the unique feature is a HIT-like fold. NSP3 (34 kDa) is also found in early RIs suggesting a role in viral morphogenesis. It self assembles into dimers and has ssRNA binding activity, which in this case appears to show sequence specificity for a conserved sequence present at the 3′-end of viral mRNAs. NSP3 interacts with the human translation initiation factor eIF4GI. Current data indicate that by taking the place of poly(A) binding protein present in eIF4GI, NSP3 is responsible for the shut-off of cellular protein synthesis and for promoting viral translation by facilitating circularization of the viral mRNA. The crystal structure of the amino-terminal portion of the NSP3 dimer complexed with viral mRNA 3′-sequences has been solved and shows that the 3-terminal GACC of the RNA is bound within an NSP3 "tunnel." The crystal structure of the carboxy-terminal portions of NSP3 dimers complexed to a fragment of translation initiation factor eIF4G has also been solved and shows that NSP3 binds the same site on eIF4G as does cellular poly(A) binding protein.

NSP4 (20 and 28 kDa) is post-translationally modified by N-linked glycosylation (like the structural protein VP7). However, in contrast to VP7 where the non-glycosylated form of the protein is only observed if inhibitors of N-linked glycosylation are used, the faster migrating non-glycosylated form of NSP4 is observed in normally infected cells. Not all of the protein can be chased into the glycosylated form, indicating that NSP4 exists in two forms that may have different functions. NSP4 has been shown to be involved in the later stages of virion maturation in the ER where it functions as a receptor for binding of double-layered particles during budding through the ER membrane. NSP4 (or NSP4-derived peptide aa 114-135) is a viral enterotoxin that leads to Ca^{2+} release from internal stores in the ER and induction of age-dependent diarrhea in mice. An NSP4 cleavage product is secreted from infected cells and retains toxin activity. Secreted NSP4 binds to cells triggering a signalling pathway activating phospholipase C and elevating inositol 1,4,5-triphosphate leading to Ca^{2+} release from internal stores. This pathway is distinct from that of endogenously expressed NSP4.

The two remaining non-structural proteins, NSP5 (26 kDa) and NSP6 (12 kDa), are encoded in two different reading frames of the same viral gene. NSP5 which is rich in serine and threonine is post-translationally modified by being both phosphorylated and O-link glycosylated. It has single-stranded and double-stranded RNA binding activity but its function is unknown. NSP5 takes multiple forms, as differentially phosphorylated proteins, some of which are autophosphorylated. The kinase domain in NSP5 has not yet been mapped. The highly phosphorylated NSP5 may play a structural role in the organization of viroplasms. NSP5 forms oligomers and can interact with VP2, NSP2(ViP) and NSP6. Interaction with NSP2 promotes hyperphosphorylation of NSP5. A carboxy-terminal domain of NSP5 is required for multimerization, hyperphosphorylation and interactions with NSP6. The role of NSP5 may be critical for RNA replication. The ORF for NSP6 is conserved in most virus isolates examined, although its function is undefined. NSP6 is also phosphorylated and localizes in the viroplasms.

Information on the proteins of isolates of species other than *Rotavirus A* is very sparse and is primarily drawn from sequence analysis of viral genes. It is clear that these viruses have homologs of the proteins characterized in *Rotavirus A* viruses but better systems for routine cultivation in tissue culture are required to facilitate detailed characterization of their proteins.

LIPIDS

None reported. Immature particles acquire a transient membrane during its passage through the ER.

CARBOHYDRATES

Three viral proteins have been shown to be glycosylated. In two cases (VP7 and NSP4) linkage of the sugar is through an amino linkage to asparagine and in the third (NSP5) through an O-linkage to serine and/or threonine.

GENOME ORGANIZATION AND REPLICATION

The complete RNA-protein coding assignments have been determined for several *Rotavirus A* isolates. The coding assignments, for the SiRVA/SA11 strain, are given in Table 2. The replication cycle which is completed in 10-12 hrs at 37°C has been studied primarily in continuous cell cultures derived from monkey kidneys. There is little conclusive information about the early steps in the replication cycle. VP4 is the viral attachment protein but the cellular receptor has not been conclusively identified. Some rotavirus strains initially attach to the sialic acid residues on the cell surface. Current data suggest that rotaviruses may enter cells by two mechanisms, either by receptor-mediated endocytosis, or by an alternative mechanism of direct virus entry. In both cases the virus entry process removes the outer virus shell and releases the transcriptionally active double-layered particle (DLP) into the cytoplasm of the infected cell. DLP-associated enzymes produce 5′-capped, non-polyadenylated mRNAs, which are full length transcripts from the minus strand of each of the virion genome segments (Transcripts are only made from minus strands). Gene expression is regulated by the level of transcription from individual genomic segments, with differences evident in both the kinetics and level of production of different mRNAs. The viral mRNAs derived from each of the genome segments serve two functions, firstly they are translated to generate the viral proteins encoded by the segment and further control of individual gene expression occurs during this process. For example there is ~250-fold difference in the level of expression between the most (NSP4) and least [VP1(Pol)] abundant protein. Secondly viral mRNAs are also the templates for genome replication. As with other members of the family *Reoviridae,* it remains unclear whether a given mRNA molecule has the potential to fulfill either role, or if there are two different forms of mRNA from individual genome segments, that can each fulfill only one of these roles. Genome segment assembly takes place by selection of the different viral mRNAs required to form the precore RI and may involve an association with the cytoskeleton. Assembly of the eleven mRNAs is followed by minus strand synthesis, which occurs in 'core-RI' and 'VP6(T13)-RI', that are present in the 'viroplasms' found within the cytoplasm of the infected cell. *Cis*-acting replication elements for minus strand synthesis have been identified using an *in vitro* replication system, but the minimal element necessary to allow synthesis is the seven 3′-terminal nt (GUGUACC-3′) on the mRNAs. The next steps in the morphogenesis of progeny virions are unique to rotaviruses and involve the double layered particle budding into the ER in a process that involves NSP4. This results in the particle transiently acquiring an envelope that is lost during the final maturation step(s) when the outer virion shell of VP4 and VP7 is added.

ANTIGENIC PROPERTIES

Three viral proteins (VP4, VP6(T13) and VP7) of SiRV-A have been subjected to detailed antigenic characterization. VP6(T13), which forms the intermediate capsid shell, is both a highly conserved and highly immunogenic protein, carrying both virus group and sub-group determinants. It does not elicit the production of neutralizing antibodies but may play a role in the induction of protective immunity. It is the major target of diagnostic assays for rotaviruses and specifies the "group" (A–E; F, G), i.e. the species. Within group A, subgroups I and II have been differentiated. The outer shell glycoprotein VP7 was the first to be recognized as eliciting a virus type specific neutralizing antibody response and hence has been subjected to extensive molecular and epidemiological analysis. Tissue culture based neutralization assays have allowed the recognition of 15 glycoprotein or "G" serotypes and where tested in animal cross protection studies, these serotype definitions have been confirmed. Sequencing of genes encoding VP7 from multiple isolates has shown that those falling into the same serotype have <10% sequence variation, whereas between serotypes variation falls in the 15-25% range. VP4 has also been shown to elicit neutralizing antibodies but in this case serotypic definition is less advanced. Cross neutralization studies have identified 14 P serotypes, with subtypes being recognized in three of these. As with the gene encoding VP7, there has been extensive epidemiological analysis of the gene encoding VP4. This has allowed the rapid recognition of 23 "P" genotypes in which members falling within a genotype show <~10% sequence divergence. On the whole there has been good correlation between genotyping and definitive serotyping.

In the case of other rotaviruses, although homologs of VP4, VP6(T13) and VP7 have been identified at least in *Rotavirus B* and *Rotavirus C,* no information is available at present on the existence or extent of antigenic diversity.

BIOLOGICAL PROPERTIES

All rotaviruses have proved difficult to cultivate *in vitro,* with growth being restricted to a few epithelial cell lines that have been derived mainly from monkey kidneys. The infection of these cells can be enhanced by pre-treatment of virus with trypsin. This restriction of virus growth *in vitro* parallels the *in vivo* situation where virus replication is normally restricted to the terminally differentiated enterocytes lining the tips of the microvilli in the small intestine.

There are several mechanisms of pathogenesis, including the destruction of enterocytes that leads to malabsorption and an osmotic diarrhea. Prior to the appearance of histologic changes, a watery diarrhea is often seen and this is thought to be secretory, possibly induced by the action of the rotavirus enterotoxin NSP4. It has been proposed that the destruction of these enterocytes causes a loss of the permeability barrier between the gut lumen and the vasculature, resulting in the osmotic pull of fluid from the circulation into the gut and the 'watery' characteristic of rotavirus infection. Rotaviruses infect a wide range of avian and mammalian species, with disease being restricted in the great majority of cases to the young. However, isolates of *Rotavirus B* have caused large epidemics in human adults.

LIST OF SPECIES DEMARCATION CRITERIA IN THE GENUS

In common with the other genera within the family *Reoviridae*, it has been agreed that the prime determinant for inclusion of virus isolates within a single virus species will be "ability to exchange (reassort) genome segments during co-infection, thereby exchanging genetic information and generating viable and novel progeny virus strains". However, data providing direct evidence of segment reassortment between isolates is very limited and serological comparisons, together with comparisons of RNA or protein sequences, represent major factors used to examine the level of similarity that exists between isolates. These and other data can also be used to predict the "compatibility" of strains for reassortment.

The rotaviruses are currently divided into five species (*Rotavirus A* to *Rotavirus E*) with two possible additional species (Rotavirus F and Rotavirus G). Viruses within different species are thought to be unable to reassort their genome segments under normal circumstances, and each species may therefore represent a separate gene pool.

Members of a single rotavirus species may be identified by:

- The ability to exchange genetic material by genome segment reassortment during dual infections, thereby producing viable progeny virus strains.
- High levels of serological cross-reaction by ELISA, using either polyclonal sera, or monoclonal antibodies against VP6(T13), or its homolog in groups other than SiRV-A.
- High levels of RNA sequence conservation in terminal and near terminal regions of the genome. Two levels of sequence conservation, namely, short (~10 nt) conserved consensus sequences at the termini of all genomic segments and, internal to these, longer (~40-50 nt), segment-specific consensus conservations have been found by terminal RNA fingerprinting. The actual conserved fingerprint patterns are different in each of the virus species and hence can be used as a diagnostic feature for species assignment.
- High levels of RNA sequence similarities in "conserved" genome segments. Viruses within the same species will normally show < 10% sequence variation in genome Seg6 (encoding the major inner structural protein, VP6(T13)). Viruses in different species will normally contain >30% sequence variation in genome Seg6. These differences are also reflected in the aa sequences of the viral proteins, such as VP2(T2) (Fig. 4).
- Identification by virus serotype with a virus type already classified within a specific rotavirus species. None of the serotypes within different species will cross neutralize.

DSRNA

- Identification of host range. For example *Rotavirus E* has to-date only been found in pigs. *Rotavirus D* and the tentative species Rotavirus F and Rotavirus G have only been isolated from avian species.

LIST OF SPECIES IN THE GENUS

Species names are in green italic script; strain names and synonyms are in black roman script; tentative species names are in blue roman script. Sequence accession numbers [Seg1 to 11 indicate RNA segment numbers], and assigned abbreviations ( ) are also listed.

SPECIES IN THE GENUS

Rotavirus A		
Simian rotavirus A/SA11	[Seg1: X16830, Seg2: X16831, Seg3: X16062, Seg4: X14204, Seg5: X14914, Seg6: X00421, Seg7: X00355, Seg8: J02353, Seg9: K02028, Seg10: KO1138, Seg11: X07831]	(SiRV-A/SA11)
Rotavirus B		
Rotavirus B	[Seg2: AB037932, Seg3: X16949, Seg4: M91434, V03556, Seg5: M55982, Seg6 M84456, Seg7: AF230974 Seg8: AF205850, Seg9: M33872, D00911, Seg10: U03557, Seg11: M34380, D00912]	(RV-B)
Rotavirus C		
Human rotavirus C/Bristol	[Seg1: AJ304859, Seg2: AJ303139, Seg3: X79442, Seg4: X96697, Seg5: X59843, Seg6: AJ132203, Seg7: AJ132204, Seg8: X77257, Seg9: AJ132205, Seg10: M81488 Seg11: X83967]	(HRV-C/Bristol)
Rotavirus D		
Chicken rotavirus D/132		(AvRV-D/132)
Rotavirus E		
Porcine rotavirus E/DC-9		(PoRV-E/DC-9)

TENTATIVE SPECIES IN THE GENUS

Rotavirus F	(RV-F)
Chicken rotavirus F/A4	(AvRV-F/A4)
Rotavirus G	(RV-G)
Chicken rotavirus G/555	(AvRV-G/555)

Note, the cognate genes do not necessarily correspond to the RNA segments with the same number (e.g., PoRV-C/Co Seg 5 - 8 correspond to SiRV-A/SA11 Seg 6, 7, 5, and 9, respectively).

PHYLOGENETIC RELATIONSHIPS WITHIN THE GENUS

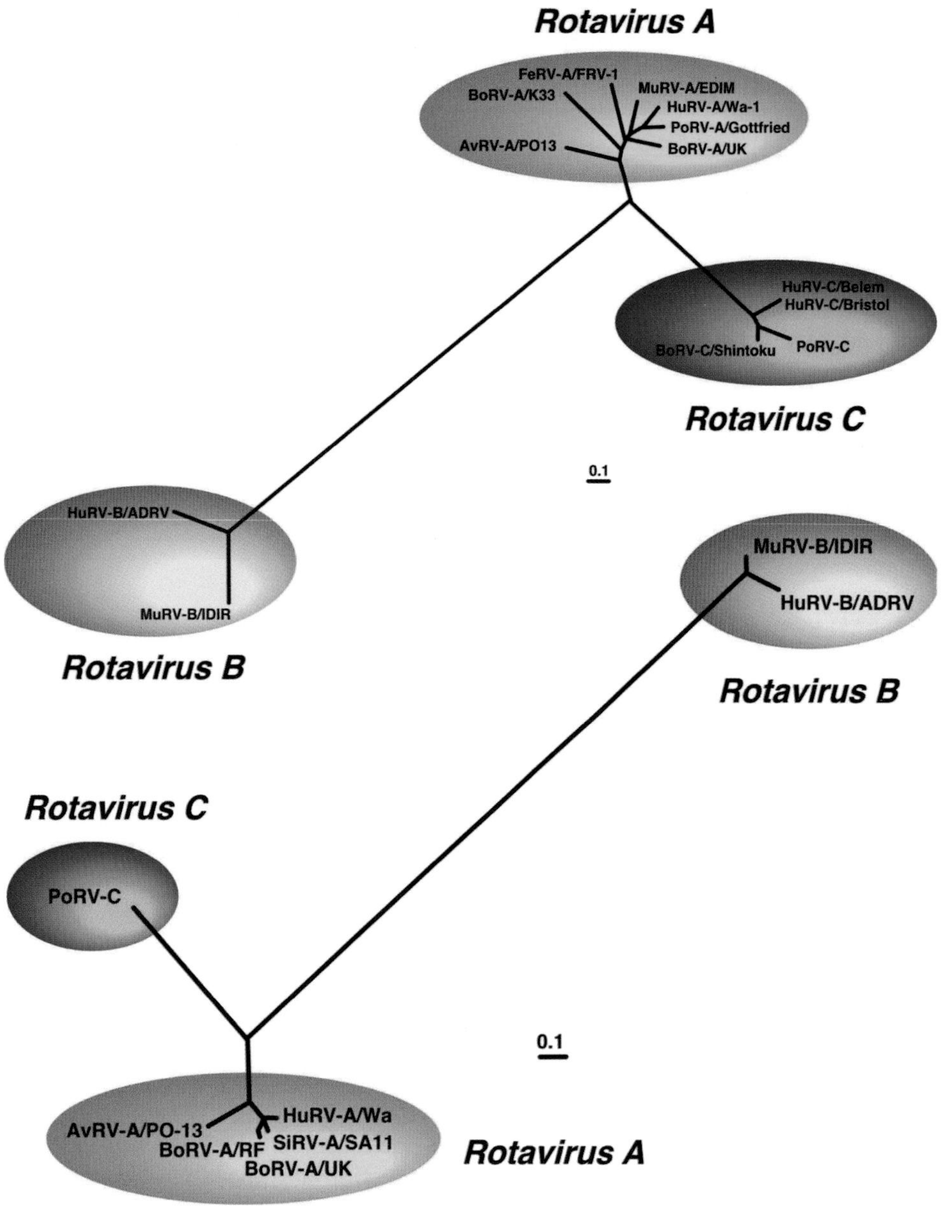

Figure 4: (Top) Phylogenetic tree of rotaviruses comparing the sequences of rotavirus VP4, the P neutralization antigen from a limited number of *Rotavirus A* isolates. (Bottom) Phylogenetic tree of rotaviruses comparing VP2(T2) (major core protein) (aa differences, corrected for multiple substitutions and excluding gaps). The isolates and accession numbers used are given in Table 6. Sequences were aligned and trees calculated (Neighbor-joining method) using ClustalX. The trees were drawn using TreeView v.1.5.2. The matrix was of aa similarities. The abbreviations used to indicate host species are: Av, avian; Bo, bovine; Ca, canine; Eq, equine; Fe, feline; Hu, human; Mu, murine; Po, porcine; Si, simian.

Table 3: Percentage aa differences between rotavirus VP2(T2) major core protein from isolates of *Rotavirus A*, *B* and *C*.

		Rotavirus-A					*Rotavirus-C*	*Rotavirus-B*	
		BoRV-A/UK	BoRV-A/RF	SiRV-A/SA11	HuRV-A/Wa	AvRV-A/PO-13	PoRV-C	MuRV-B/IDIR	HuRV-B/ADRV
	Accession #	X52589	X14057	X16831	X14942	AB00963(	M74217	U00673	M91433
Rotavirus A	BoRV-A/UK	0.0	2.2	7.4	9.1	24.9	55.2	81.6	82.4
	BoRV-A/RF		0.0	6.3	8.1	24.3	54.5	81.6	82.3
	SiRV-A/SA11			0.0	8.4	24.0	54.8	81.9	82.7
	HuRV-A/Wa				0.0	24.3	54.5	81.9	82.7
	AvRV-A/PO-13					0.0	52.9	82.9	84.0
Rotavirus C	PoRV-C						0.0	84.1	84.3
Rotavirus B	MuRV-B/IDIR							0.0	14.0
	HuRV-B/ADRV								0.0

BoRV; Bovine rotavirus, SiRV; Simian rotavirus, HuRV; Human rotavirus, MuRV; Murine rotavirus, AvRV; Avian rotavirus. PoRV; Porcine rotavirus.

Derivations of Names

Rota: from Latin rota, "wheel".

References

Conner, M.E. and Ramig, R.F. (1997). Viral enteric diseases. In: *Viral Pathogenesis*, (N. Nathanson, R. Ahmed, K.V. Holmes, F. Gonzalez-Scarano, F.A. Murphy, D.E. Griffin and H.I. Robinson, eds), pp 713-743. Lippincott-Raven, Philadelphia.

Desselberger, U. and Gray, J. (eds) (2003). *Viral Gastroenteritis*. Elsevier Science, Amsterdam.

Desselberger, U. and McCrae, M.A. (1994). The rotavirus genome. *Curr. Top. Microbiol. Immunol.*, **185**, 31-67.

Estes, M.K. (2001). Rotaviruses and their replication. In: *Fields Virology*, Fourth Ed., (D.M. Knipe and P.M. Howley, eds), pp 1747-1785. Lippincott, Williams and Wilkins, Philadelphia.

Estes, M.K., Kang, G., Zeng, C.Q.-Y., Crawford, S. and Ciarlet, M. (2001). Pathogenesis of rotavirus gastroenteritis. *Novartis Found. Symp.*, **238**, 82-100.

Hoshino, Y. and Kapikian, A.Z. (1996). Classification of rotavirus VP4 and VP7 serotypes. *Arch. Virol.*, **S12**, 99-111.

Kapikian, A.Z., Hoshino, Y. and Chanock, R.M. (2001). Rotaviruses. In: *Fields Virology*, Fourth Ed., (D.M. Knipe and P.M. Howley, eds), pp 1787-1833. Lippincott, Williams and Wilkins, Philadelphia.

Offit, P.A. (2001). Correlates of protections against rotavirus infection and disease. *Novartis Found. Symp.*, **238**, 106-124.

Pedley, S., Bridger, J.C., Chasey, D. and McCrae, M.A. (1986). Molecular definition of two new groups of atypical rotaviruses. *J. Gen. Virol.*, **67**, 131-137.

Prasad, B.V.V., Crawford, S., Lawton, J.A., Pesavento, J., Hardy. M. and Estes, M. (2001). Structural studies on gastroenteritis viruses. *Novartis Found. Symp.*, **238**,, 26-46.

Ramig, R.F. (1997). Genetics of the rotaviruses. *Annu. Rev. Microbiol.*, **51**, 225-255.

Taniguchi, K., Kojima, K. and Urasawa, S. (1996). Non-defective rotavirus mutants with an NSP1 gene which has a deletion of 500 nucleotides, including a cysteine-rich zinc finger motif-encoding region (nucleotides 156-248), or which has a non-sense codon at nucleotides 153-155. *J. Virol.* **70,** 125-130.

Contributors

Ramig, R.F., Ciarlet, M., Mertens, P.P.C. and Dermody, T.S.

GENUS COLTIVIRUS

Type Species *Colorado tick fever virus*

DISTINGUISHING FEATURES

The coltivirus genome consists of 12 segments of dsRNA. During replication, viruses are found in the cell cytoplasm, associated with granular matrices (viral inclusion bodies: VIB), arrays of filaments or "tubules" and fine kinky threads. Immunofluorescent staining reveals nucleolar fluorescence. Viruses are transmitted to vertebrate hosts by tick vectors.

VIRION PROPERTIES

MORPHOLOGY

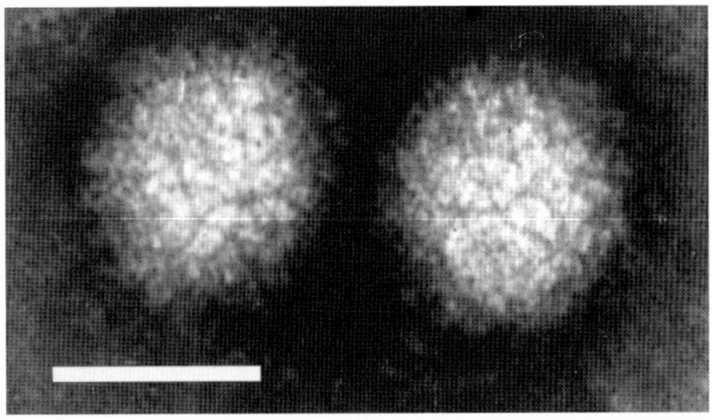

Figure 1: Negative contrast electron micrograph of particles of Colorado tick fever virus (CTFV) (Courtesy of F.A. Murphy). The bar represents 50 nm

Coltivirus particles are 60-80 nm in diameter having two concentric capsid shells with a core that is about 50 nm in diameter. Electron microscopic studies, using negative staining have shown that particles have a relatively smooth surface capsomeric structure and icosahedral symmetry (Fig. 1). Particles are found intimately associated with filamentous structures and granular matrices in the cytoplasm. The majority of the viral particles are non-enveloped, but a few acquire an envelope structure during the passage through the endoplasmic reticulum.

PHYSICOCHEMICAL AND PHYSICAL PROPERTIES

The buoyant density of the virus in CsCl is 1.38 g/cm^3. Viruses are stable between pH 7 and 8, but lose infectivity at pH 3.0. At 4°C, the virus is stable for long periods when stored in presence of 50% fetal calf serum in 0.2 M Tris-HCl pH 7.8. Heating to 55°C considerably decreases the viral infectivity. Coltiviruses are fairly stable upon treatment with non-ionic detergents, sodium lauroyl sarcosine, or freon but the viral infectivity is abolished by treatment with sodium deoxycholate or sodium dodecyl sulfate. Moderate ultrasonic oscillation treatment does not destroy infectivity and can be used in virus purification. Viruses can be stored for long periods at -80°C, and infectivity is further protected by addition of 50% foetal calf serum.

NUCLEIC ACID

The genome consists of 12 dsRNA segments that are named "genome segment 1" (Seg1) to "genome segment 12" (Seg12) in order of decreasing size, or increasing electrophoretic mobility during agarose gel electrophoresis. The genome comprises approximately 29,000 bp and the segment length range between 4350 and 675 bp. The genomic RNA of Colorado tick fever virus (CTFV) migrates in three size classes (Fig. 2) during 1% agarose gel electrophoresis (AGE): the large (long) or L-segments (Seg1-4), the medium length or M-segments (Seg5-10) and the small (short) or S-segments (Seg11 and 12). The genome of Eyach virus (EYAV) dsRNA has so far not been analyzed by AGE. The terminal 5′ and 3′ sequences of the coltivirus genome are conserved (Table 1).

DSRNA

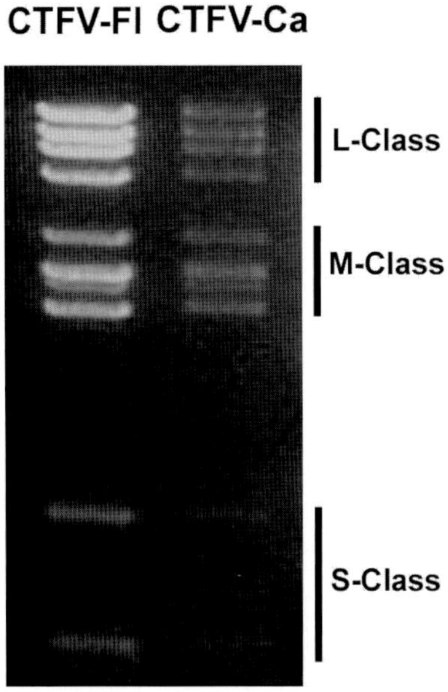

Figure 2: Agarose gel electrophoretic profiles of genome segments of Colorado tick fever virus isolate Florio (CTFV-Fl) and California hare virus (CTFV-Ca) in 1% agorose gel. The genome of the North-American, CTFV-Ca showed similar nt identities to those observed between isolates of CTFV. These migration patterns (electropherotype) are thought to be characteristic of each virus species.

RNA cross-hybridization analysis shows that CTFV isolates have remained relatively homogenous, and distinct CTFV serotypes are difficult to define (although some variation does occur, for example in genome Seg4 and Seg6). The overall similarities between nt sequences from all genome segments of different CTFV isolates ranged between 90% and 100%. The degree of similarity between nt sequences of the homologous segments from CTFV and EYAV isolates, ranged from 55% to 86%. In particular, the Seg6 of CTFV is the homologous of Seg7 of EYAV. The putative protein encoded by Seg7 of EYAV, shows similarity with a sarcolemmal-associated protein of the European rabbit *Oryctolagus cunniculis*, the suspected major host of the virus. However, the calculated degree of similarity (50%) finding suggest that there might have been a recombination between the viral RNA and the RNA encoding the sarcolemmal-associated protein of an ancestor of the European rabbit. It is thought that EYAV may have been introduced into Europe during migration of lagomorph ancestors from America to Europe during the Oligocene epoch (34-23 MYA).

Table 1: Conserved terminal sequences of coltivirus genome segments

Virus	(strain)	Conserved RNA terminal sequences (positive strand)
CTFV	(CTFV-Fl)	5'-$^{G}/_{C}$ACAUUUUGU.........................UGCAGU$^{G}/_{C}$-3'
EYAV	(EYAV-Fr578)	5'-GACA$^{A}/_{T}$UU ....$^{A}/_{T}$UG$^{C}/_{T}$AGUC-3'

Based on analysis of all segments of CTFV and EYAV, the length of the 5'-NTRs of the different coltiviruses is 11-77 nt, while that of the 3'-NTRs is 26-209 nt. The G + C content was found to range between 48 and 52%.

PROTEINS
Viral proteins of CTFV were translated from dsRNA in a cell free system. Viral proteins of the EYAV have not been characterized by translation (Table 2).

LIPIDS
None reported.

CARBOHYDRATES
None reported.

GENOME ORGANIZATION AND REPLICATION

In cells infected by CTFV, granular matrices are produced which contain virus-like particles. These structures appear similar to VIBs produced during orbivirus infections. In addition, bundles of filaments (tubules), characterized by cross-striations, are found in the cytoplasm and, in some cases, in the nucleus of infected cells. These may also be comparable to the 'tubules' found in orbivirus infected cells. There is no evidence for virus release prior to cell death and disruption, after which more than 90% of virus particles remain cell associated. Immunofluorescence shows that viral proteins accumulate in the cytoplasm and could be detected from the 12th hour post-infection. Nucleolar fluorescence was also observed. Mosquito cells infected by EYAV show syncitial foci. Electron microscopy on EYAV-infected mouse brain shows identical intracellular structure as those observed in CTFV-infected cells.

Table 2: List of the dsRNA segments of Colorado tick fever virus (CTFV) with their respective size (bp) and their encoded proteins for which the calculated size (kDa) and functions (where known) are shown.

dsRNA segment number	dsRNA size (bp)	Protein nomenclature	Protein (kDa), determined from translation products of the dsRNA ¥	Structure / Function
1	4350	VP1	125 (163)	RdRp
2	3909	VP2	117 (136)	Methyltransferase, cell-receptor
3	3586	VP3	113 (135)	RNA replication factors
4	3157	VP4	100 (112)	-
5	2432	VP5	90 (84)	Guanylyl transferase
6	2141	VP6	82 (78)	Nucleotide binding, NTPase
7	2133	VP7	75 (76)	RNA replication factors
8	2029	VP8	60 (74)	-
9	1884	VP9	42 (38 and 67)	-
10	1880	VP10	55 (69)	Kinase, helicase
11	998	VP11	34 (28.5)	-
12	675	VP12	25 (20.4)	RNA replication factors

¥: Values between parenthesis represent the size of the proteins calculated from the aa sequences deduced from RNA nt sequences.

DSRNA

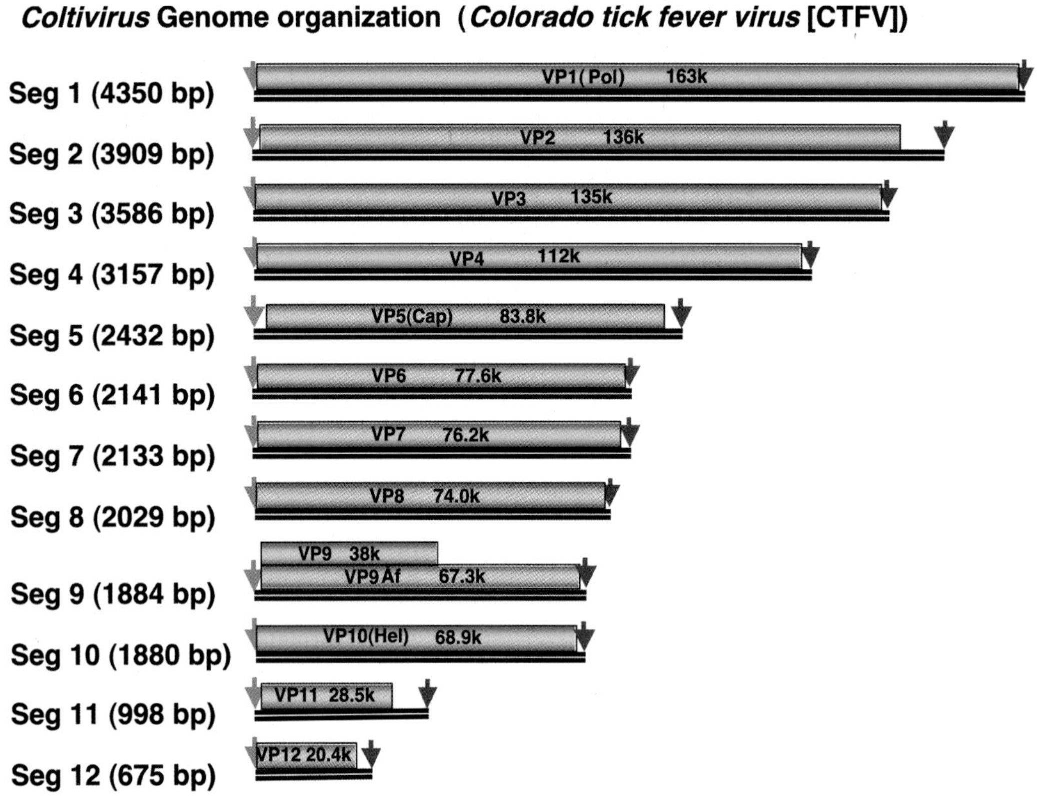

Figure 3: Genome organization of Colorado tick fever virus (CTFV), containing 12 dsRNA segments, each contains an ORF, except Seg9, which contains two ORFs. The green arrows indicate the upstream conserved terminal sequence (+ve 5'-G/CACAUUUUGU.....), while the red arrows indicate the downstream conserved terminal sequence (+ve.......UGCAGUG/C -3').

ANTIGENIC PROPERTIES

CTFV from North America and EYAV from Europe, which are classified as distinct species, show little cross-reaction in neutralization tests. An isolate, S6-14-03, obtained from a hare (*Lepus californicus*) in Northern California (CTFV-Ca) should be considered as a serotype of CTFV. Two species of coltiviruses could therefore be distinguished.

BIOLOGICAL PROPERTIES

Coltiviruses have been isolated from several mammalian species (including humans) and from ticks and mosquitoes which serve as vectors. The tick species include *Dermacentor andersoni, D. occidentales, D. albipictus, D. parumapertus, Haemaphysalis leporispalustris, Otobius lagophilus, Ixodes sculptus, I. spinipalpis, I. ricinus* and *I. ventalloi.*

Ticks become infected with CTFV on ingestion of a blood meal from an infected vertebrate host. Both adult and nymphal ticks become persistently infected and provide an overwintering mechanism for the virus. CTFV is transmitted trans-stadially but not trans-ovarially. Some rodent species have prolonged viraemia (more than 5 months) which may also facilitate overwintering and virus persistence. Humans usually become infected with CTFV when bitten by the adult wood tick *D. andersoni* but probably do not act as a source of reinfection for other ticks. Transmission from person to person has been recorded as the result of blood transfusion. The prolonged viraemia observed in humans and rodents is thought to be due to the intra-erythrocytic location of virions, protecting them from immune clearance.

CTFV is characterized in humans by an abrupt onset of fever, chills, headache, retro-orbital pains, photophobia, myalgia and generalized malaise. Abdominal pain occurs in about 20% of patients. Rashes are uncommon (less than 10%). A diphasic, or even triphasic, febrile pattern has been observed, usually lasting for 5-10 days. Severe forms of the disease, involving infection of the central nervous system, or haemorrhagic fever, or both, have been infrequently observed (nearly always in children under 12 years of age). A small number of such cases are fatal. Congenital infection with CTFV may occur, although the risk of abortion and congenital defects remains uncertain. Antibodies to EYAV have been found in patients with meningoencephalitis and polyneuritis but a causal relationship to the virus has not been established.

CTFV causes leukopaenia in adult hamsters and in about two-thirds of infected humans. Suckling mice, which usually die at 6-8 days post-infection, suffer myocardial necrosis, necrobiotic cerebellar changes, widespread focal necrosis and perivascular inflammation in the cerebral cortex, degeneration of skeletal myofibers, hepatic necrosis, acute involution of the thymus, focal necrosis in the retina and in brown fat. The pathologic changes in mice due to CTFV infection (in skeletal muscle, heart and brain), are consistent with the clinical features of human infection which may include meningitis, meningo-encephalitis, encephalitis, gastro-intestinal bleeding, pneumonia and myocarditis.

CTFV occurs in forest habitats at 4,000-10,000 ft. elevation in the Rocky Mountain region of North America. Antibodies to the virus have been detected in hares in Ontario and a virus isolate has been reported from Long Island, New York. *Eyach virus* appears to be widely distributed in Europe.

LIST OF SPECIES DEMARCATION CRITERIA IN THE GENUS

In common with the other genera within the family *Reoviridae,* the prime determinant for inclusion of virus isolates within a single coltivirus species is defined by their compatibility for reassortment of genome segments during co-infections, allowing the exchange of genetic information and generating viable progeny virus strains. However, data providing direct evidence of segment reassortment between isolates is very limited. Therefore serological comparisons, comparisons of RNA sequence, cross-hybridization, analysis of electropherotypes and analysis of conserved RNA sequences in near terminal regions of the genome segments represent major factors used to examine the level of similarity that exists between isolates.

Members of a single species of the genus *Coltivirus* may be identified by:

- An ability to exchange genetic material by genome segment reassortment during dual infections, thereby producing viable progeny virus strains.
- RNA cross-hybridization assays (Northern or dot blots, with probes made from viral RNA or cDNA). Within a single species, the amount of RNA sequence similarity is higher than 74% under hybridization conditions of Tm (RNA) –36°C.
- Serological comparisons by complement fixation and neutralisation assays. The isolates of the species *Colorado tick fever virus* show high levels of cross neutralization. Cross reactivity between CTFV and EYAV has only been detected by CF tests.
- RNA sequence analysis (Seg1 to Seg12). Within a single species, high levels of sequence similarities are observed in conserved segments, e.g. Seg12 shows less than 11% sequence variation within a single virus species.
- Comparisons of aa sequences (for example those of the translation products of genome Seg 6, 7 and 12, Fig. 4) indicate that different species will only have 55, 57 and 60% aa identity within Seg 6, 7 and 12, respectively.
- The analysis of electropherotypes by agarose gel electrophoresis. Within a single species the electropherotype is relatively uniform.
- Analysis of conserved RNA terminal sequences. These sequences show conservation within species isolates. Sequences at the 5′- and 3'-terminus are also similar between CTFV and EYAV (Table 1).

CTFV-Ca from a hare collected in California in 1976 shows some one-way cross-reaction in serum neutralization tests with EYAV, but is clearly distinguishable and has been reported as a distinct serotype. Sequence analysis has shown that CTFV-S6-14-03 should be considered as a serotype of *Colorado tick fever virus* species. Serological variants of *Eyach virus*, Eyach virus - France 577 and Eyach virus - France 578, have also been reported.

LIST OF SPECIES IN THE GENUS

Species names are in green italic script; strain names and synonyms are in black roman script; tentative species names are in blue roman script. Sequence accession numbers [S1 to S12 indicate RNA segments 1 – 12], arthropod vector and host names { }, and assigned abbreviations ( ) are also listed.

SPECIES IN THE GENUS

Colorado tick fever virus

(2 serotypes) >22 isolates		
{*Ixodidae* ticks : Rodents, humans }		
California hare virus - California s6-14-03	[Seg1: AF343051, Seg6: AF343054, Seg7: AF343057, Seg12: AF343060]	(CTFV-Ca)
Colorado tick fever virus - Florio	[Seg1: AF133428, Seg2: AF139758, Seg3: AF139759, Seg4: AF139760, Seg5: AF139761, Seg6: AF139762, Seg7: AF139763, Seg8: AF139764, Seg9: AF000720, Seg10: AF139765, Seg11: U72694, Seg12: U53227]	(CTFV-Fl)

Eyach virus

{*Ixodidae* ticks : Possibly humans }		
Eyach virus - France-577	[Seg1: AF343052, Seg6: AF343055, Seg7: AF343058, Seg12: AF343061]	(EYAV-Fr577)
Eyach virus - France-578	[Seg 1: AF282467, Seg 2: [AF282468] Seg3: AF282469, Seg4: AF282470, Seg5: AF282471, Seg6: AF282472, Seg7: AF282473, Seg8: AF282474, Seg9: AF282475, Seg10: AF282476, Seg11: AF282477, Seg12: AF282478]	(EYAV-Fr578)
Eyach virus - Germany	[Seg1: AF343053, Seg6: AF343056, Seg7: AF343059, Seg12: AF343062]	(EYAV-Gr)

TENTATIVE SPECIES IN THE GENUS

Salmon River virus* (SRV)

*has been reported as a distinct serotype from CTFV.

PHYLOGENETIC RELATIONSHIPS WITHIN THE GENUS

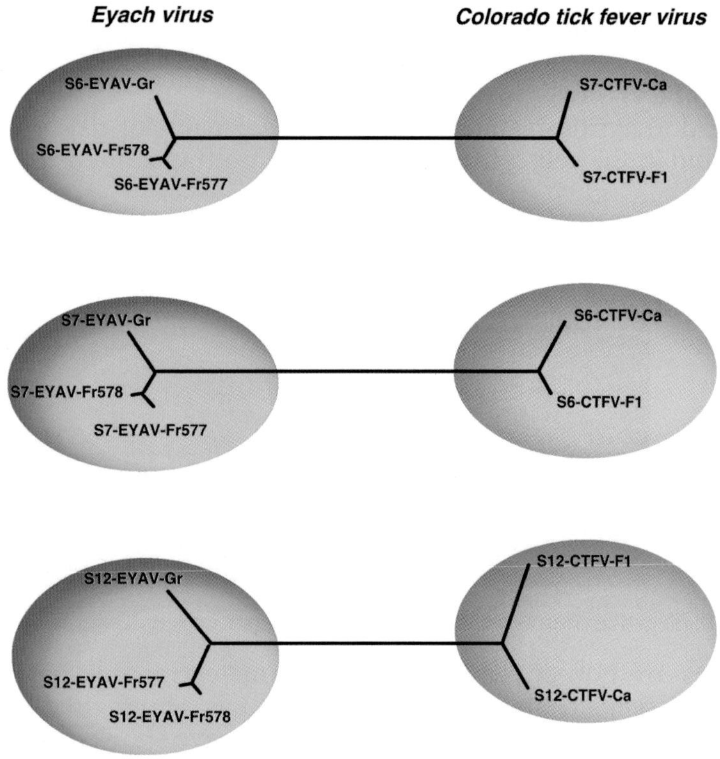

Figure 4: Phylogenetic tree for the most variable genome segments of different coltiviruses: Seg6 (Top), Seg7 (Center) and Seg12 (Bottom). Unrooted neighbor-joining tree created using MEGA program from alignments generated with the program ClustalW, depicting groupings for the deduced aa sequence of translation products from homologous genome segments of different coltiviruses. Members of the two species *Colorado tick fever virus* (CTFV) and *Eyach virus* (EYAV) that are currently recognized are labeled (from Attoui, H. *et al.*, 2002). All segments produce almost identical trees. The percentage genetic distance between species calculated from the aa sequence of Seg12 are; CTFV-Fl/EYAV-Gr: 40%; among isolates of the species *Eyach virus*: < 8%; among isolates of the species *Colorado tick fever virus*: < 8%.

DERIVATIONS OF NAMES

Colti: sigla from *Co*lorado *ti*ck fever.

REFERENCES

Attoui, H., Charrel, N., Billoir, F., Cantaloube, J.-F., de Micco P. and de Lamballerie, X. (1998). Comparative sequence analysis of American, European and Asian isolates of genus *Coltivirus*. *J. Gen. Virol.*, **79**, 2481-2489.

Attoui, H., Billoir, F., Biagini, P., Cantaloube, J.F., de Chesse, R., de Micco, P. and de Lamballerie, X. (2000). Sequence determination and analysis of the full-length genome of Colorado tick fever virus, the type species of genus. *Coltivirus. Biochem. Biophys. Res. Com.*, **273**, 1121-1125.

Attoui, H., Mohd Jaafar, F., Biagini, P., Cantaloube, J.F, de Micco, P., Murphy, F.A. and de Lamballerie, X. (2002). Genus *Coltivirus* (Family *Reoviridae*): Genomic and morphologic characterization of Old World and New World viruses. *Arch. Virol.*, **147**, 533-561.

Chastel, C., Main, A.J., Couatarmanac'h, A., Le Lay, G., Knudson, D.L., Quillien, M.C. and Beaucournu, J.C. (1984). Isolation of Eyach virus (*Reoviridae*, Colorado tick fever group) from *Ixodes ricinus* and *I. ventalloi* ticks in France. *Arch. Virol.*, **82**, 161-171.

Emmons, R.W. (1988). Ecology of Colorado tick fever. *Ann. Rev. Microbiol.*, **42**, 49-64.

CONTRIBUTORS

Attoui, H., de Lamballerie, X., and Mertens, P.P.C.

GENUS SEADORNAVIRUS

Type Species *Banna virus*

DISTINGUISHING FEATURES

The seadornavirus genome consists of 12 segments of dsRNA. During replication, viruses are found in the cell cytoplasm. Viruses are transmitted to vertebrate hosts by mosquito vectors.

VIRION PROPERTIES

MORPHOLOGY

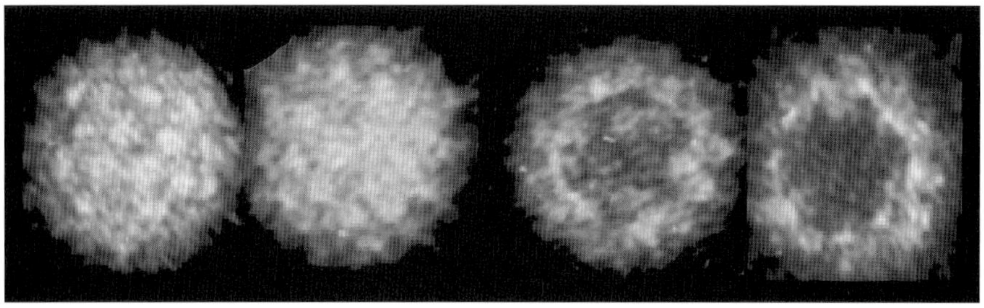

Figure 1: Negative contrast electron micrograph of particles of Kadipiro virus (KDV)(Courtesy of V.L. Popov).

Particles are non-enveloped with a diameter ranging between 60-70 nm having two concentric capsid shells with a core that is about 40-50 nm in diameter. Electron microscopic studies, using negative staining have shown that particles have well identified surface capsomeric structure and icosahedral symmetry (Fig. 1).

PHYSICOCHEMICAL AND PHYSICAL PROPERTIES

The buoyant density of the virus in CsCl is 1.36 g/cm^3. Viruses are stable around neutral pH, but lose infectivity at pH 3.0. At 4°C, the virus is stable for long periods, even non-purified in cell culture lysate, which is a convenient way for medium term storage. Heating to 55°C considerably decreases the viral infectivity. Seadornaviruses are stable upon treatment with freon, which could be used for purification of viral particles from cell lysate. The viral infectivity is abolished by treatment with sodium dodecyl sulfate. Viruses can be stored for long periods at -80°C, and infectivity is further protected by addition of 50% foetal calf serum.

NUCLEIC ACID

The genome consists of 12 dsRNA segments that are named "genome segment 1" (Seg1) to "genome segment 12" (Seg12) in order of reducing Mr, or increasing electrophoretic mobility during agarose gel electrophoresis. The genome comprises approximately 21,000 bp, and the segment lengths range between 3747 and 862 bp. The genomic RNA of BAV shows a 6-6 electrophoretic profile in 1% agarose gel electrophoresis (AGE). The genome of KDV migrates in 6-5-1 profile (Fig. 2). These patterns are thought to be characteristic of each virus species.

Sequence analysis of full-length genomes from different isolates of *Banna virus* have shown that two genotypes could be distinguished within the species: genotype A represented by BAV-Ch (China) and BAV-In6423 (Indonesia) and genotype B represented by BAV-In6969 and BAV-In7043 (Indonesia). This distinction is based on sequences of Seg7 and Seg9. Seg7 shows 72% aa identity between genotypes A and B, while Seg9 shows a lower identity of 54 % only. All other segments exhibit 83 to 98% identities. Between BAV and KDV, aa identities between homologous segments were found to be 24-42 %, the highest being within the polymerase gene.

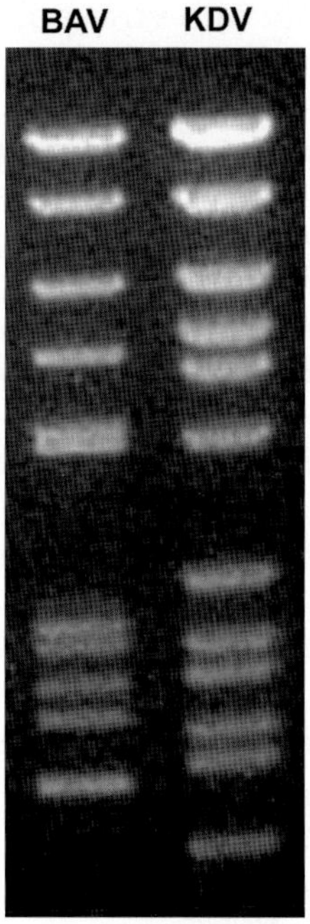

Figure 2 : Agarose gel electrophoretic profiles of genome segments of an isolate of *Banna virus* (BAV) and an isolate of *Kadipiro virus* (KDV) in 1% agarose gel. These migration patterns (electropherotype) are thought to be characteristic of each virus species.

Based on analysis of all segments of BAV and KDV, the length of the 5'-NTRs of the different coltiviruses is 17–102 nt, while that of the 3'-NTRs is 76-200 nt. The G+C content was found to range between 37 and 39%.

Table 1: Conserved terminal sequences of seadornavirus genome segments

Virus	**(strain)**	**Conserved RNA terminal sequences (positive strand)**
BAV	(CTFV-In6423)	$5'\ \mathrm{GUAU^{A}/_{U}{}^{A}/_{U}AA^{A}/_{U}{}^{A}/_{U}U\ldots{}^{A}/_{G}C^{C}/_{U}GAC\text{-}3'}$
KDV	(KDV-Ja7075)	$\mathrm{5'\text{-}GUAGAA^{A}/_{U}{}^{A}/_{U}{}^{A}/_{U}U\ldots\ldots\ldots A^{A}/_{C}{}^{C}/_{U}GAC\text{-}3'}$
LNSV	(LNSV-NE97-31)	$\mathrm{5'\text{-}GUUAU^{A}/_{U}{}^{A}/_{U}{}^{A}/_{U}\ldots\ldots\ldots\ldots{}^{A}/_{cU}/cCGAC\text{-}3'}$

PROTEINS
Native viral proteins have not been characterized. Putative protein sequences were translated from sequenced cDNA (complementary to viral genomic RNA). Their putative functions are discussed in Table 2.

LIPIDS
None reported.

CARBOHYDRATES
None reported.

GENOME ORGANISATION AND REPLICATION
Seadornavirus isolates replicate in mosquitoes cell lines, and considerable amounts of virus (over 40% of progeny) are liberated in the culture medium prior to cell death and gross CPE, which is attained by 40 hrs post-infection with BAV, and 72 hrs post-infection with KDV. Intracellular radio labelling of viral polypeptides has shown that label is

DSRNA

incorporated predominantly into viral polypeptides, even in absence of inhibitors of DNA replication, demonstrating 'shut off' of host cell protein synthesis.

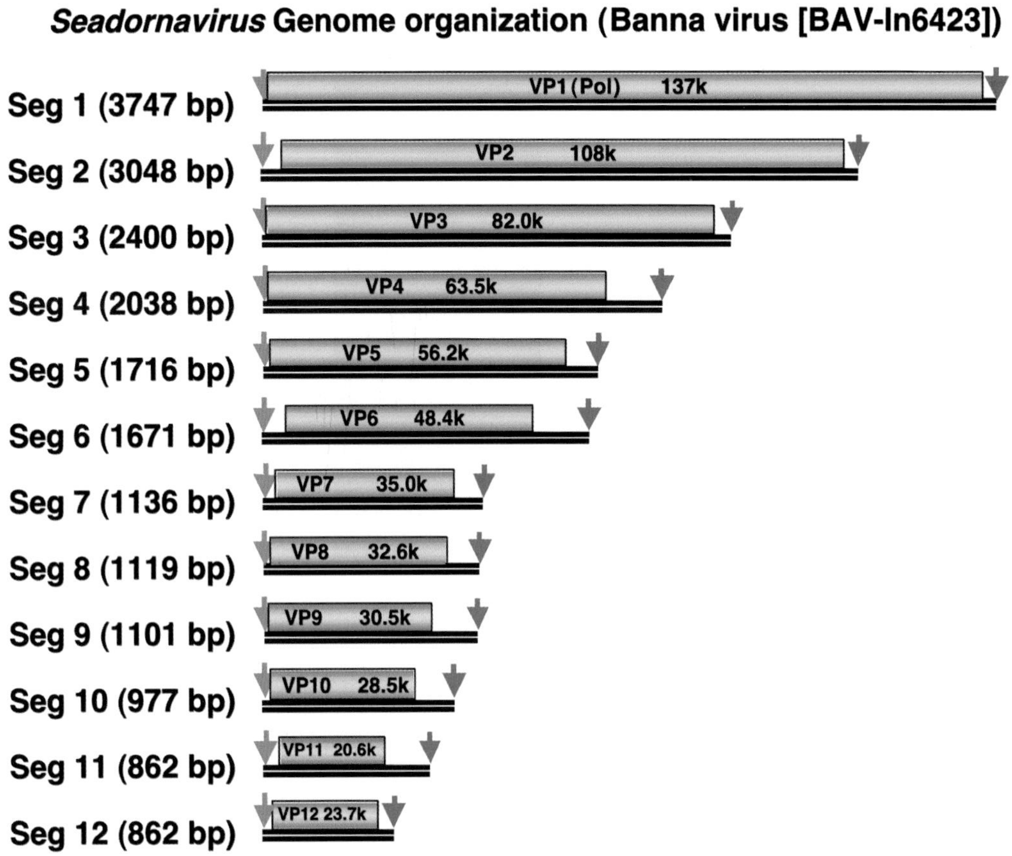

Figure 3: Genome organization of Banna virus – In6423 (BAV-In6423), containing 12 dsRNA segments, each contains an ORF. The green arrow indicates the upstream conserved terminal sequence (+ve 5'-GUAUA/UA/UAA....), while the red arrow indicates the downstream conserved terminal sequence (+ve....A/GCC/UGAC-3').

Table 2: List of the dsRNA segments of BAV-In6423, Seg1-12 with their respective size (bp) and their encoded proteins for which the name, calculated kDa and function are indicated.

dsRNA segment #	dsRNA Size (bp)	Protein nomenclature	kDa Calculated from the sequenced segments	Structure / Function
1	3747	VP1	137	RdRp
2	3048	VP2	108	Nucleotide binding, cell-receptor recognition site
3	2400	VP3	82	Mtr, helicase
4	2038	VP4	64	Mtr
5	1716	VP5	56	
6	1671	VP6	48	NTPase, contains a leucine zipper
7	1136	VP7	35	Protein kinase
8	1119	VP8	32	
9	1101	VP9	31	
10	977	VP10	29	
11	867	VP11	21	
12	862	VP12	24	dsRNA-binding

ANTIGENIC PROPERTIES

BAV from southern China and Indonesia and KDV from Indonesia which are classified in distinct species, show little cross-reaction in neutralisation tests. Recently a new virus isolate found in the Northeast of China is also regarded as belonging to a distinct species. Many other strains have been isolated in China and show cross-reactions with BAV.

BIOLOGICAL PROPERTIES

Seadornaviruses have been isolated from mosquitoes, which serve as vectors and humans. The mosquito species include *Culex vishnui, C. fuscocephalus, Anopeles vagus, Anopheles aconitus, Anopheles subpictus* and *Aedes dorsalis.*

Experimentally, the viruses were found to replicate in adult mice and were detected in infected mouse blood at 3 days post-infection.

BAV was isolated from serum and cerebrospinal fluids of patients showing neurological manifestations. The pathology provoked by this virus is only poorly described. KDV was only isolated from mosquitoes.

BAV and KDV occur in tropical and subtropical regions, where other mosquito-borne viral disease especially Japanese encephalitis and dengue have been reported as endemic. Despite the isolation of BAV from infected human patients, there have been no serious reports in literature about any detection of antibodies to the viruses in sera of humans.

LIST OF SPECIES DEMARCATION CRITERIA IN THE GENUS

In common with the other genera within the family *Reoviridae,* the prime determinant for inclusion of virus isolates within a single coltivirus species is defined by their compatibility for reassortment of genome segments during co-infections, allowing the exchange of genetic information and generating viable progeny virus strains. However, data providing direct evidence of segment reassortment between isolates is very limited. Therefore serological comparisons, comparisons of RNA sequence, cross-hybridization, analysis of conserved RNA sequences in near terminal regions of the genome segments (and analysis of electropherotypes) represent major factors used to examine the level of similarity that exists between isolates.

Members of a single species of the genus *Seadornavirus* may be identified by:

- An ability to exchange genetic material by genome segment reassortment during dual infections, thereby producing viable progeny virus strains.
- RNA cross-hybridization assays (Northern or dot blots, with probes made from viral RNA or cDNA). Within a single species, the amount of RNA sequence similarity is higher than 74% under hybridization conditions of Tm (RNA) –36°C.
- Serological comparisons by neutralization assays. Hyper immune ascitic fluids against genotype A viruses of BAV, do not cross-neutralize efficiently those of genotype B. Within a single genotype, isolates show high levels of cross neutralization. There is no cross-neutralization nor any cross reactivity between members of the species *Banna virus* and *Kadipiro virus.*
- RNA sequence analysis (e.g. Seg1 to 12). Within a single species, high levels of sequence similarities are observed in conserved segments, e.g. Seg12 shows less than 11% sequence variation within a single virus species.
- Comparisons of aa sequences (for example those of the translation products of genome Seg1 or Seg12) indicate that different species will contain less than 50% aa identity within the polymerase sequence.
- The analysis of electropherotypes by agarose gel electrophoresis. Within a single species the electropherotype is relatively uniform. However, deletions or additions can occur (as in Seg7 and Seg9 of isolates of the *Banna virus* species) resulting in variations in electropherotype.

- Analysis of conserved RNA terminal sequences. These sequences show conservation within species isolates. Sequences at the 3'-terminus may be similar on at least some segments in different species (e.g. *Banna virus* and *Kadipiro virus*) (Table 1).

LIST OF SPECIES IN THE GENUS

Species names are in green italic script; strain names and synonyms are in black roman script; tentative species names are in blue roman script. Sequence accession numbers [Seg1 to Seg12 indicate RNA segments 1 – 12], arthropod vector and host names { }, and assigned abbreviations ( ) are also listed.

SPECIES IN THE GENUS

Banna virus		
{Culex and Anopheles mosquitoes : Humans}		
Banna virus - China	[Seg1: AF134525, Seg2: AF13526, Seg6: AF13527, Seg7: AF052035, Seg8: AF052034, Seg9: AF0520333, Seg10: AF052032, Seg11: AF052031, Seg12: AF052030]	(BAV-Ch)
Banna virus - Indonesia-6423	[Seg1: AF133430, Seg2: AF134514, Seg3: AF134515, Seg4: AF134516, Seg5: AF134517, Seg6: AF134518, Seg7: AF052018, Seg8: AF052017, Seg9: AF052016, Seg10: AF052015, Seg11: AF052014, Seg12: AF019908]	(BAV-In6423)
Banna virus - Indonesia-6969	[Seg1: AF134522, Seg2: AF134523, Seg6: AF134524, Seg7: AF052013, Seg8: AF052012, Seg9: AF052011, Seg10: AF052010, Seg11: AF052009, Seg12: AF052008]	(BAV-In6969)
Banna virus - Indonesia-7043	[Seg1: AF134519, Seg2: AF134520, Seg6: AF134521, Seg7: AF052029, Seg8: AF052028, Seg9: AF052027, Seg10: AF052026, Seg11: AF052025, Seg12: AF052024]	(BAV-In7043)
Kadipiro virus		
{Culex mosquitoes}		
Kadipiro virus - Java-7075	[Seg1: AF133429, Seg2: AF134509, Seg3: AF134510, Seg4: AF134511, Seg5: AF134512, Seg6: AF134513, Seg7: AF052023, Seg8: AF052022, Seg9: AF052021, Seg10: AF052020, Seg11: AF052019, Seg12: AF019909]	(KDV-Ja7075)
Liao ning virus		
{Aedes dorsalis : mosquito}		
Liao ning virus - NE97-12		(LNV-NE97-12)
Liao ning virus - NE97-31		(LNV-NE97-31)

TENTATIVE SPECIES IN THE GENUS

Banna virus ACH*	(BAV-ACH)
Banna virus HN59	(BAV-HN59)
Banna virus HN131	(BAV-HN131)
Banna virus HN191	(BAV-HN191)
Banna virus HN295	(BAV-HN295)
Banna virus LY1	(BAV-LY1)
Banna virus LY2	(BAV-LY2)
Banna virus LY3	(BAV-LY3)

Banna virus M14	(BAV-M14)
Banna virus TRT2	(BAV-TRT2)
Banna virus TRT5	(BAV-TRT5)
Banna virus WX1	(BAV-WX1)
Banna virus WX2	(BAV-WX2)
Banna virus WX3	(BAV-WX3)
Banna virus WX8	(BAV-WX8)

*These isolates of seadornaviruses have been isolated in many provinces in China including Beijing, Gansu, Yuannan, Hainan, Henan, Shanshi, Xinjiang and recently from Liao ning. Viruses other than Banna, Kadipiro, and Liao-ning viruses, are still uncharacterized and have been temporary designated Banna virus isolates. Their serological relationship to Banna virus was not fully explored. These isolates are probably distinct from *Banna virus*, and at least some should represent new species within genus *Seadornavirus*.

PHYLOGENETIC RELATIONSHIPS WITHIN THE GENUS

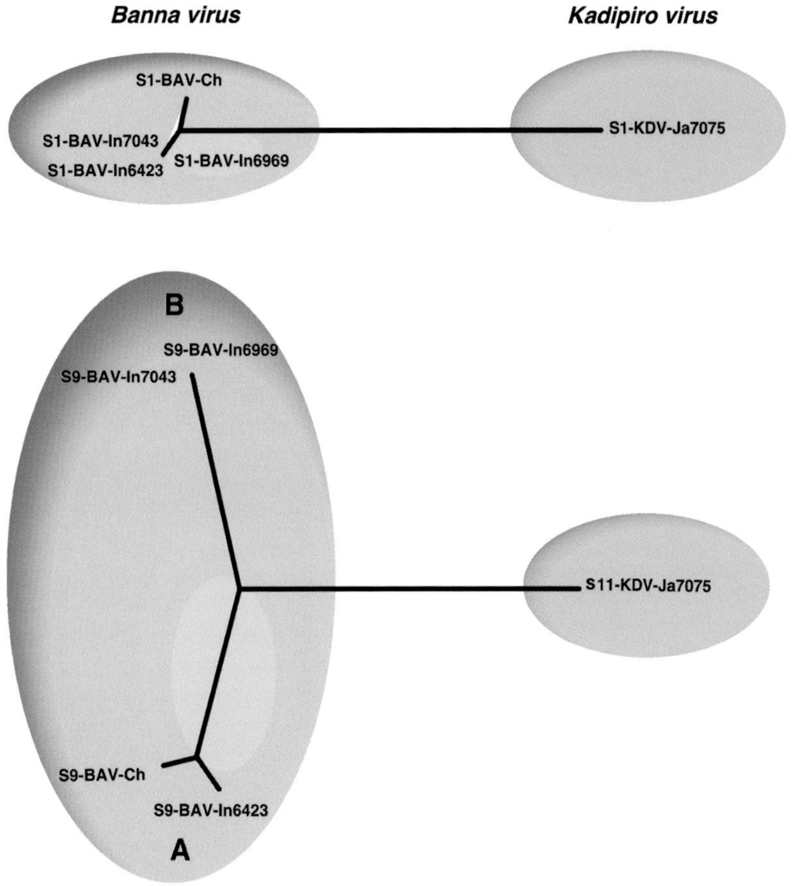

Figure 4: Phylogenetic tree for genome homologous segments of members of the species *Banna virus* and *Kadipiro virus*. (Top) Phylogenetic tree based on the polymerase aa sequence of isolates of Banna virus (BAV) and Kadipiro virus. (KDV); (Bottom) Phylogenetic tree based on aa sequences of Seg9 of BAV and its homologous Seg11 of KDV, this tree also shows the two genotypes of BAV (A and B). Genetic distance based on the polymerase gene between the two species range from 56 to 57%. Within BAV species this distance ranges between 1 and 5%. The genetic distance based on Seg9 of BAV and Seg11 of KDV ranges between 77 and 78%. Between the two genotypes of BAV, it ranges from 57 to 59%. Within a given genotype it ranges between 4 and 10%.

DERIVATIONS OF NAMES

Seadorna: sigla from '**S**outh **e**astern **A**sia **do**deca **RNA** **virus**es'

REFERENCES

Attoui, H., Mohd Jaafar, F., Biagini, P., Cantaloube, J.F., de Micco, P., Murphy, F.A. and de Lamballerie, X. (2002). Genus *Coltivirus* (family *Reoviridae*): genomic and morphologic characterization of Old World and New World viruses. *Arch. Virol.*, **147**, 533-561.

Attoui, H., Billoir, F., Biagini, P., de Micco, P. and de Lamballerie, X. (2000). Complete sequence determination and genetic analysis of Banna virus and Kadipiro virus: proposal for assignment to a new genus (*Seadornavirus*) within the family *Reoviridae*. *J. Gen. Virol.*, **81**, 1507-1515.

CONTRIBUTORS

Attoui, H., Mohd Jaafar, F., Mertens, P.P.C. and de Lamballerie, X.

GENUS AQUAREOVIRUS

Type Species *Aquareovirus A*

DISTINGUISHING FEATURES

Aquareoviruses physically resemble orthoreoviruses but possess 11 dsRNA genome segments. They infect a variety of aquatic animals, including finfish and Crustacea. Aquareoviruses replicate in cell cultures of piscine and mammalian origins, at temperatures between 15 and 25°C. Large syncytia are produced as a typical cytopathic effect of infection by a majority of aquareoviruses.

VIRION PROPERTIES

MORPHOLOGY

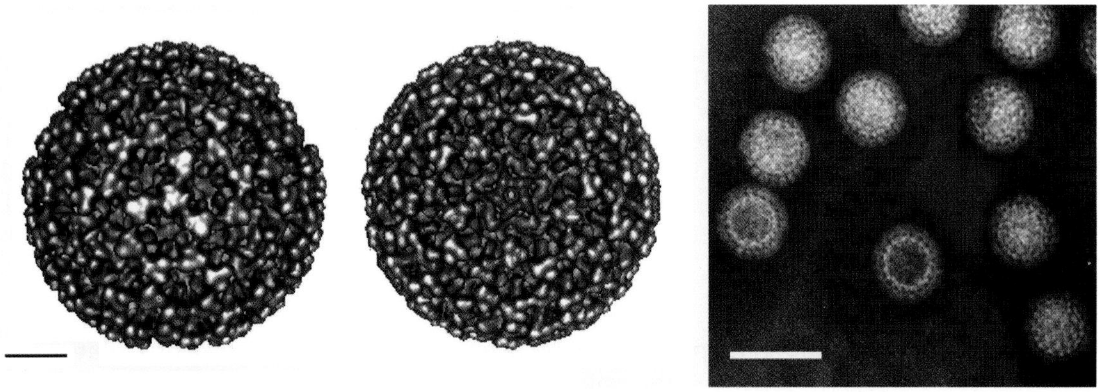

Figure 1: (Left and Center) Surface representations of the mature aquareovirus particle viewed at (Left) the icosahedral 3-fold axis and (Center) the icosahedral 5-fold axis. The scale bar represents 200Å (Courtesy of B.V.V. Prasad). (Right) Negative contrast electron micrograph of negatively stained Striped bass reovirus (SBRV) particles (Courtesy of S.K. Samal). The bar represents 100 nm.

Virus particles are spherical in appearance but have icosahedral symmetry (Fig. 1 Left and Center). The virions are approximately 80 nm in diameter, with two concentric capsid shells that appear to be made up of three layers of proteins. The outer capsid (~10 nm thick) surrounds a core ~60 nm in diameter. The boundary between the outer capsid and inner core is evident as a prominent white ring in negatively stained electron micrographs (Fig. 1 Right). Small projections on the surface of the core particle interconnect inner and outer capsid layers. The core particle also contains an inner layer, which contains the 11 dsRNA genome segments and the internal core proteins (the transcriptase complexes). The aquareovirus particle morphology is strikingly similar to that of the orthoreovirus "intermediate subviral particle" (ISVP). However, a noticeable distinction is that aquareovirus particles lack the hemagglutinin spikes observed in orthoreoviruses.

PHYSICOCHEMICAL AND PHYSICAL PROPERTIES

The virion buoyant density in CsCl is 1.36 g/cm^3. Sedimentation coefficient of the virus particles is ~550S. Virus infectivity is stable between pH 3 and pH 10. Virus infectivity is not affected by treatment with ether or chloroform. Exposure to UV irradiation reduces infectivity. None of the viral proteins is removed from the particle by treatment with 3 mM EDTA or cesium salts. Aquareoviruses held at 4, 16 or 23°C in minimal essential medium (MEM) with 5% serum showed no significant reduction in infectivity over a period of 28 days. However, all virus infectivity is lost after incubation at 45°C for 7 days. Virus infectivity is rapidly inactivated by heating to 56°C.

Nucleic Acid

The aquareovirus genome is composed of 11 segments of dsRNA that are packaged in equimolar ratios. The Mr of the dsRNA segments range from 0.4 x 10^6 - 2.6 x 10^6. The total Mr of the Golden shiner reovirus (GSRV) is about 1.5 x 10^7 (23,695 bp). The genomic RNAs are named "segment 1" (Seg1) to "segment 11" (Seg11) in order of increasing electrophoretic mobility in 1% agarose gels. The genome segments migrate as three size classes. There are three large (Seg 1-3, about 3.9–3.8 kbp), three medium (Seg 4-6, about 2.3–2.0 kbp) and five small segments (Seg 7-11, about 1.4–0.8 kbp). Six distinct species have been identified by reciprocal RNA-RNA hybridization studies (*Aquareovirus-A* to *Aquareovirus-F*). The genome segment migration pattern (electropherotype), as analyzed by electrophoresis in 1% agarose gel, is consistent within a single species but shows significant variation between species. Viruses within a single species show variations in electropherotype, when their dsRNA genome segments are analyzed by electrophoresis in high percentage polyacrylamide gels (>6%).

The G+C content of aquareoviruses ranges between 52 and 60%. The complete genomic sequences of GSRV and Grass carp reovirus (GCRV) have been determined from cloned cDNAs, along with several genome segments of other aquareovirus isolates. Genomic dsRNA segments contain 7 nt at the 5'-terminus and 6 nt at the 3'-terminus, which are conserved in all 11 genome segments within a particular virus species. The 5' and 3' conserved terminal sequences of *Aquareovirus-C* are 5'GUUAUUU/G3' and 5'A/UUCAUC3', compared to 5'GUUUUAU/G3' and 5'A/UUCAUC3' in *Aquareovirus-A*.

Proteins

Twelve primary gene products have been identified in the case of isolates of *Aquareovirus-A* (Table 1). With the exception of genome Seg11 in *Aquareovirus-A* isolates, which encodes two primary translation products, each viral genome segment encodes one primary translation product. The virions of *Aquareovirus-A* isolates contain 7 structural proteins: VP1, 130 kDa; VP2, 127 kDa; VP3, 126 kDa; VP4, 73 kDa; VP5, 71 kDa; VP6, 46 kDa; VP7, 35 kDa. VP1, VP2, VP3 and VP6 form the core of the virus particle. VP3 and VP6 are more abundant than VP1 and VP2. VP1 is present in larger copy numbers than VP2. VP6 and VP3 probably form the nodules and the spherical shell of the core, respectively. VP1 most likely forms the turret structure present at the fivefold axis. VP2 is presents in very small amounts per virion and is thought to be present beneath the fivefold axis. VP7, VP4 and VP5 are present in the outer coat of the virion. All three proteins are removed upon prolonged trypsinization, resulting in the emergence of core particles. VP7 is the most external protein. VP5 is the next most accessible protein after VP7. It is thought that the removal of VP7 by trypsin may expose some regions of VP5 critical for efficient entry into cells. Members of *Aquareovirus-A* also encodes five non-structural proteins: NS1, 97 kDa; NS2, 39 kDa; NS3, 29 kDa; NS4, 28 kDa; NS5, 15 kDa. The functions of these NS proteins are not known. The observed variations in dsRNA electropherotype suggest that viruses from different species may have proteins with significant differences in size.

Lipids

Aquareoviruses have no known lipid components.

Carbohydrates

VP7 of *Aquareovirus-A* isolates may be glycosylated.

Genome Organization and Replication

The coding assignments of Striped bass reovirus (SBRV), a strain of *Aquareovirus A*, have been determined (Fig. 2, Table 1). Genome Seg 1 - 10 of the SBRV genome encode one protein each, while Seg 11 encodes two proteins.

Aquareovirus Genome organization (Golden Shiner virus [GSRV])

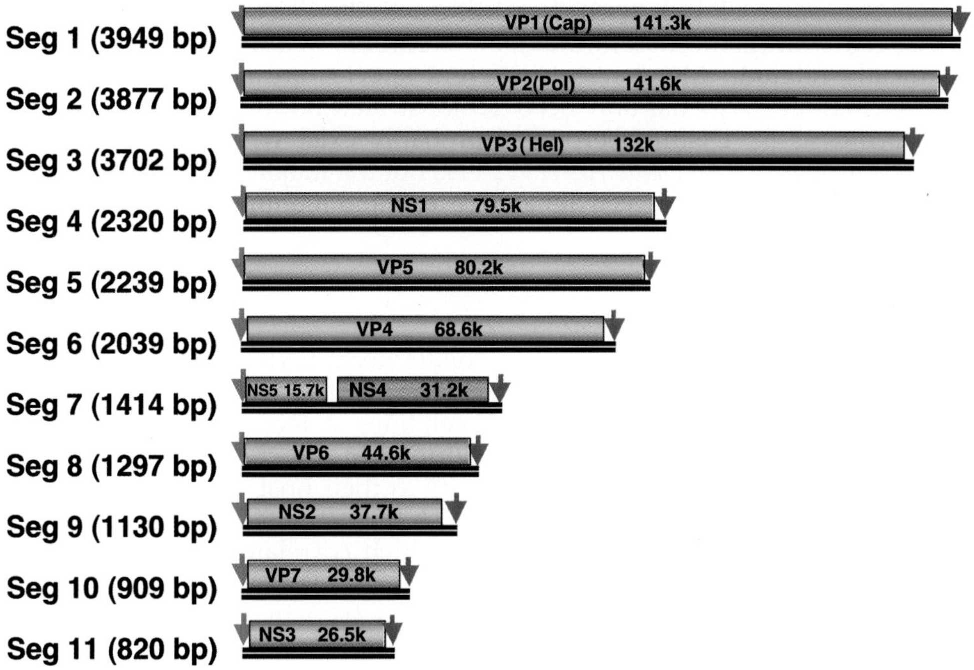

Figure 2: Genome organization of Golden shiner virus (GSRV), containing 11 dsRNA segments, each contains an ORF, except Seg7 which contains two ORFs . The green arrows indicate the upstream conserved terminal sequence (+ve 5'-GUUAUUU/G.....), while the red arrows indicate the downstream conserved terminal sequence (+ve.......A/UUCAUC-3').

Table 1: List of the dsRNA genome segments of Striped bass reovirus (SBRV) (*Aquareovirus-A* species), with their estimated size (kbp), corresponding proteins with name, size (estimated), and location.

Genome segment	dsRNA segment size (kbp)	Protein nomenclature	Protein size (kDa)	Protein location
Seg 1	3.8	VP1	130	Inner capsid (core)
Seg 2	3.6	VP2	127	Inner capsid (core)
Seg 3	3.3	VP3	126	Inner capsid (core)
Seg 4	2.5	VP4	97	Non-structural
Seg 5	2.4	VP5	71	Inner capsid (core)
Seg 6	2.2	VP4	73	Inner capsid (core)
Seg 7	1.5	NS4	28	Non-structural
Seg 8	1.4	VP6	46	Inner capsid (core)
Seg 9	1.2	NS2	39	Non-structural
Seg 10	0.9	VP7	34	Major outer capsid
Seg 11	0.8	NS3	29	Non-structural
		NS5	15	Non-structural

ANTIGENIC PROPERTIES

Aquareovirus outer CP lack hemaglutinating activity. Viruses possess type-specific and group-specific antigenic determinants. Members within a single species may be antigenically related. Members of different species are antigenically distinct. Minor antigenic cross-reactivity has only been demonstrated between members of *Aquareovirus A* and *Aquareovirus B*. Distinct serotypes probably exist within each species. The major outer CP of isolates of *Aquareovirus A* (VP7) is not the major neutralizing antigen. There is no antigenic relationship between aquareoviruses and mammalian reoviruses.

BIOLOGICAL PROPERTIES

HOST RANGE

Aquareoviruses have been isolated from poikilotherm vertebrates as well as invertebrates (hosts include fish, molluscs, etc.) obtained from both fresh and sea water. The viruses replicate efficiently in fish and mammalian cell lines at temperatures ranging from 15°C to 25°C. They produce a characteristic cytopathic effect consisting of large syncytia. Generally, the viruses are of low pathogenicity in their host species. However, grass carp reovirus is highly pathogenic in grass carp. The infectivity of aquareoviruses is enhanced by treatment with trypsin or chymotrypsin, which correlates with digestion of the outer capsid protein VP7. The most infectious stage of the virus is produced by a 5-min treatment with trypsin. However, prolonged trypsin treatment almost completely abolishes the infectivity.

LIST OF SPECIES DEMARCATION CRITERIA IN THE GENUS

Within the family *Reoviridae*, the prime determinant for inclusion of virus isolates within a single virus species is their ability to exchange genetic information during co-infection, by genome segment reassortment, thereby generating viable progeny virus strains. However, data providing direct evidence of segment reassortment between isolates of aquareoviruses are not available. RNA cross-hybridization studies and serological comparisons are therefore the methods most commonly used to examine the level of similarity that exists between isolates. The methods which form the "species parameters" can in effect be used to predict the compatibility of related viruses for genome segment reassortment.

Members of a single aquareovirus species may be identified by:

- Their ability to exchange genetic material by genome segment reassortment during dual infections, thereby producing viable progeny virus strains.
- Cross-hybridization assays (Northern or dot blot), with probes made from viral RNA or cDNA. For example, in Northern hybridization assays, conditions (stringency) that do not allow <17% mismatch will not show any hybridization between viruses from two different species, while viruses within a species will show hybridization.
- RNA sequence analysis (viruses within different species should have low levels of sequence homology among the cognate genome segments). For example, genome segment 10 that encodes the major outer CP, VP7, will show > 45% sequence variation between viruses from two different species. These nucleotide sequence differences should also be reflected in the aa sequence variation (> 64%) of the VP7 proteins.
- Serological comparisons of antigens or antibodies by neutralization (or other) assays using, either polyclonal antisera or monoclonal antibodies against conserved antigens. For example, cross-neutralization assays, using polyclonal rabbit antisera separate different aquareoviruses into the same groupings as the cross-hybridization assays.
- Analysis of "electropherotype" by agarose gel electrophoresis (AGE). For example, viruses within *Aquareovirus-A* will show a relatively uniform electropherotype, while viruses belonging to *Aquareovirus-B* will show a different, but also relatively uniform,

electropherotype. However, similarities in migration rates of some genome segments can exist between species.

- Identification of the conserved terminal regions of the genome segments (Some closely related species can also have identical terminal sequences on at least some segments).

Six species (*Aquareovirus-A* to *Aquareovirus-F*) and some unassigned viruses have been recognized on the basis of RNA-RNA hybridization.

LIST OF SPECIES IN THE GENUS

Species names are in green italic script; strain names and synonyms are in black roman script; tentative species names are in blue roman script. Sequence accession numbers [Seg1 to Seg12 indicate RNA segments 1 – 12], and assigned abbreviations ( ) are also listed.

SPECIES IN THE GENUS

Aquareovirus A		
American oyster reovirus 13p2		(13p2RV)
Angel fish reovirus		(AFRV)
Atlantic salmon reovirus AS		(ASRV)
Atlantic salmon reovirus HBR		(HBRV)
Atlantic salmon reovirus TS		(TSRV)
Chinook salmon reovirus DRC		(DRCRV)
Chum salmon reovirus CS	[Seg 1: AF418294, Seg 2: AF418295, Seg 3: AF418296, Seg 4: AF418297, Seg 5: AF418298, Seg 6: AF418299, Seg 7: AF418300, Seg 8: AF418301, Seg 9: AF418302, Seg 10: AF418303, Seg 11: AF418304]	(CSRV)
Masou salmon reovirus MS		(MSRV)
Smelt reovirus		(SRV)
Striped bass reovirus	[Seg 2: AF450318, Seg 3: AF450319, Seg 4: AF450320, Seg 8: AF450321, Seg 10: AF450322, U83396]	(SBRV)
Aquareovirus B		
Chinook salmon reovirus B		(GRCV)
Chinook salmon reovirus ICR		(ICRV)
Chinook salmon reovirus LBS		(LBSV)
Chinook salmon reovirus YRC		(YRCV)
Coho salmon reovirus CSR	[Seg 10: U90430]	(CSRV)
Coho salmon reovirus ELC		(ELCV)
Coho salmon reovirus SCS		(SCSV)
Aquareovirus C		
Golden shiner reovirus	[Seg 1: AF403398, Seg 2: AF403399, Seg 3: AF403400, Seg 4: AF403401, Seg 5: AF403402, Seg 6: AF403403, Seg 7: AF403404, Seg 8: AF403405, Seg 9: AF403406, Seg 10: AF403407, Seg 11: AF403408]	(GSRV)
Grass carp reovirus	[Seg 1: AF260511, Seg 2: AF260512, Seg 3: AF260513, Seg 4: AF403390, Seg 5: AF403391, Seg 6: AF403392, AF239175, Seg 7: AF403393, Seg 8: AF403394, Seg 9: AF403395, Seg 10: AF403396, Seg 11: AF403397]	(GCRV)
Aquareovirus D		
Channel catfish reovirus		(CCRV)

Aquareovirus E
Turbot reovirus (TRV)
Aquareovirus F
Chum salmon reovirus PSR (PSRV)
Coho salmon reovirus SSR (SSRV)

TENTATIVE SPECIES IN THE GENUS

Chub reovirus		(CHRV)
Golden ide reovirus	[Seg 2: AF450323, Seg 5: AF450324]	(GIRV)
Hard clam reovirus		(HCRV)
Landlocked salmon reovirus		(LSRV)
Tench reovirus		(TNRV)

PHYLOGENETIC RELATIONSHIPS WITHIN THE GENUS

Not available.

DERIVATIONS OF NAMES

Aqua: from Latin aqua, "water".

Reo: sigla from *r*espiratory *e*nteric *o*rphan, due to the early recognition that the viruses caused respiratory and enteric infections, and the (incorrect) belief that they were not associated with disease, hence they were considered "orphan" viruses .

REFERENCES

Attoui, H., Fang, Q., Mohd Jaafar, F., Cantaloube, J.F., Biagini, P., de Micco, P. and de Lamballerie. X. (2002). Common evolutionary origin of aquareoviruses and orthoreoviruses revealed by genome characterization of Golden shiner reovirus, Grass carp reovirus, Striped bass reovirus and golden ide reovirus (genus Aquareovirus, family *Reoviridae*). *J. Gen. Virol.*, **83**, 1941-1951.

Lupiani, B., Subramanian, K. and Samal, S.K. (1995). Aquareoviruses. *Annu. Rev. Fish Dis.*, **5**, 175-208.

Shaw, A.L., Samal, S.K., Subramanian, K. and Prasad, B.V.V. (1996). The structure of aquareovirus shows how the different geometries of the two layers of the capsid are reconciled to provide symmetrical interactions and stabilization. *Structure*, **4**, 957-967.

Subramanian, K., McPhillips, T.H. and Samal, S.K. (1994). Characterization of the polypeptides and determination of genome coding assignment of an aquareovirus. *Virology*, **205**, 75-81.

Subramanian, K., Hetrick, F.M. and Samal, S.K. (1997). Identification of a new genogroup of aquareovirus by RNA-RNA hybridisation. *J. Gen. Virol.*, **78**, 1385-1388.

Winton, J.R., Lannan, C.N., Fryer, J.L., Hetrick, R.P., Meyers T.R., Plumb, J.A. and Yamamoto, T. (1987). Morphological and biological properties of four members of a novel group of reoviruses isolated from aquatic animals. *J. Gen. Virol.*, **68**, 353-364.

CONTRIBUTED BY

Samal, S.K., Attoui, H., Mohd Jaafar, F. and Mertens, P.P.C.

GENUS IDNOREOVIRUS

Type Species *Idnoreovirus 1*

DISTINGUISHING FEATURES

Virions have a roughly spherical outer capsid which lacks prominent features as viewed by negative staining and electron microscopy, while the core particles have 12 icosahedrally arranged surface turrets. The genome is composed of 10 segments of dsRNA. All of the members of the genus that have been described infect insects.

VIRION PROPERTIES

MORPHOLOGY

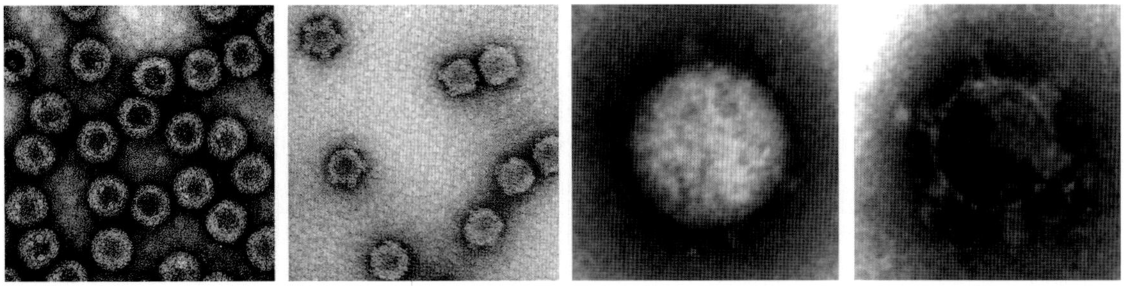

Figure 1: Electron micrographs of purified virus particles (Left) and core particles (second Left) of Hyposoter exiguae idnoreovirus 2 (HeIRV-2), stained with uranyl acetate (Courtesy of A. Makkay and D. Stoltz). Electron micrographs of a virus particle (second Right) and core particle (Right) stained with sodium phosphotungstate (from purified preparations) of Dacus oleae idnoreovirus 4 (DoIRV-4)(Courtesy of M. Bergoin). DoIRV-4 virions have small icosahedrally arranged surface projections (probably up to 12 in number). The DoIRV-4 cores show much larger 'spikes' or 'turrets', which (like those of the cypoviruses) may lose a portion near to the tip. The bars represent 50 nm.

Particles are non-enveloped. Unlike the cypoviruses, there are no polyhedra, and the virus particles have a clearly defined outer capsid layer. Electron microscopy and negative staining of virions (e.g. with aqueous uranyl acetate) shows that they are double shelled, are roughly spherical in appearance (with icosahedral symmetry), and have an estimated diameter of ~70 nm. Core particles (estimated diameter of ~60 nm) display twelve icosahedrally arranged, prominent (and at least in some cases apparently tubular) surface projections ('turrets' or 'spikes').

PHYSICOCHEMICAL AND PHYSICAL PROPERTIES

Limited studies of some viruses within the genus indicate that virus particles are resistant to freon (trichlorotrifluoroethane) and CsCl. They may also be resistant to chymotrypsin. Intact particles and cores of the prototype isolate Diadromus pulchellus idnoreovirus 1 (DpIRV-1) have densities of 1.370 g/ml and 1.385 g/ml respectively, while intact virions and empty particles of Dacus oleae idnoreovirus-4 (DoIRV-4) have a density of ~1.38 g/ml and ~1.28 g/ml respectively, as determined by CsCl gradient centrifugation. The outer capsid layer of Hyposoter exiguae idnoreovirus 2 (HeIRV-2), was disrupted by brief exposure to 0.4% sodium sarcosinate, releasing the virus core.

NUCLEIC ACID

The genome usually consists of 10 dsRNA segments that can be identified as "genome segment 1" (Seg1) to "genome segment 10" (Seg10) in order of reducing molecular weight (or increasing electrophoretic mobility during agarose gel electrophoresis; AGE). The total genome of DpIRV-1 contains an estimated 25.15 kbp of dsRNA, with the length of individual segments ranging between ~4.8 to ~0.98 kbp, showing an electrophoretic migration pattern by 1% AGE, with two groups containing 5 larger and 5 smaller segments (showing a '5, 5' pattern). However, the virions of DpIRV-1 may be unusual within the genus, since they can sometimes also contain an eleventh, 3.33kbp dsRNA segment, the presence of which is related to the sex and ploidy of the individual wasp host. This additional dsRNA (migrating between Seg3 and Seg4) contains sequences similar to and therefore possibly derived from Seg3 (3.8 kbp).

The genome segments of HeIRV-2 range in size from an estimated ~3.9 to ~1.35 kbp, with a '4, 6' electrophoretic migration pattern by 12.5 % polyacrylamide gel electrophoresis (PAGE). DoIRV-4 contains an estimated 23.4 kbp, with the length of individual segments ranging between and estimated ~3.8 to ~0.7 kbp and a '5, 3, 2' electrophoretic migration pattern by 7 % PAGE. Ceratitis capitata idnoreovirus 5 (CcIRV-5) has a '3, 3, 4' genome segment migration pattern by 6% PAGE, and has clear similarities to Drosophila melanogaster idnoreovirus-5, (DmIRV-5), as analyzed by 0.5% agarose-2% polyacrylamide gels, suggesting that despite some serological differences, they belong to the same virus species. It is unclear how closely related Drosophila S virus (which causes the 'S' phenotype in *D. simulans*) is to the other Drosophila viruses. It is therefore currently classified as a 'tentative species' within the genus.

By analogy with other members of the family *Reoviridae*, the genome segment migration patterns during AGE are likely to be characteristic for each idnoreovirus species.

Initial sequencing studies suggest that the 3' termini of DpIRV-1 genome segments are more variable than those of other virus species within the family *Reoviridae*, with little sign of conservation. However, conserved sequences were detected at the 5' termini (Table 1), which are different from those found in the other species that have been characterized within the family. No sequence data are currently available for other members of this genus.

Table 1: Conserved 5' terminal sequences of idnoreovirus genome segments*

Virus species	(strain)	Conserved RNA terminal sequences (+ve strand)
Idnoreovirus 1	DpIRV-1	(5' $^{A}/^{U}/_{G}$ CAAUUU........... 3')

PROTEINS

The purified virions of DpIRV-1 contain 11 proteins with size ranging from 21-140 kDa (as analyzed by SDS-PAGE). Three of these proteins appeared to be glycosylated (~21, 15 and 35 kDa). Native viral proteins have not been extensively characterized. However, some of the protein sequences were deduced from sequences of viral genomic RNAs (as indicated in Table 2), where they are named as VP1 to VP10 based on the molecular weight of the genome segment (segment number) from which they are translated.

LIPIDS

None reported.

CARBOHYDRATES

Three proteins from DpIRV-1 appeared to be glycosylated (~21, 15 and 35 kDa).

GENOME ORGANIZATION AND REPLICATION

On the basis of the overall similarity of idnoreoviruses to other members of the family *Reoviridae* it can be assumed that many aspects of the genome organization and replication will also be similar. On this basis it is likely that the virus core will contain transcriptase complexes that synthesise mRNA copies of the individual genome segments. These will be exported and then translated to produce viral proteins within the host cytoplasm. These positive sense RNAs are also likely to form templates for negative strand synthesis during progeny virus assembly and maturation. The Genome genome segments that have so far been characterized appear to represent single genes, with a large ORF and relatively short terminal NCRs (Fig. 2).

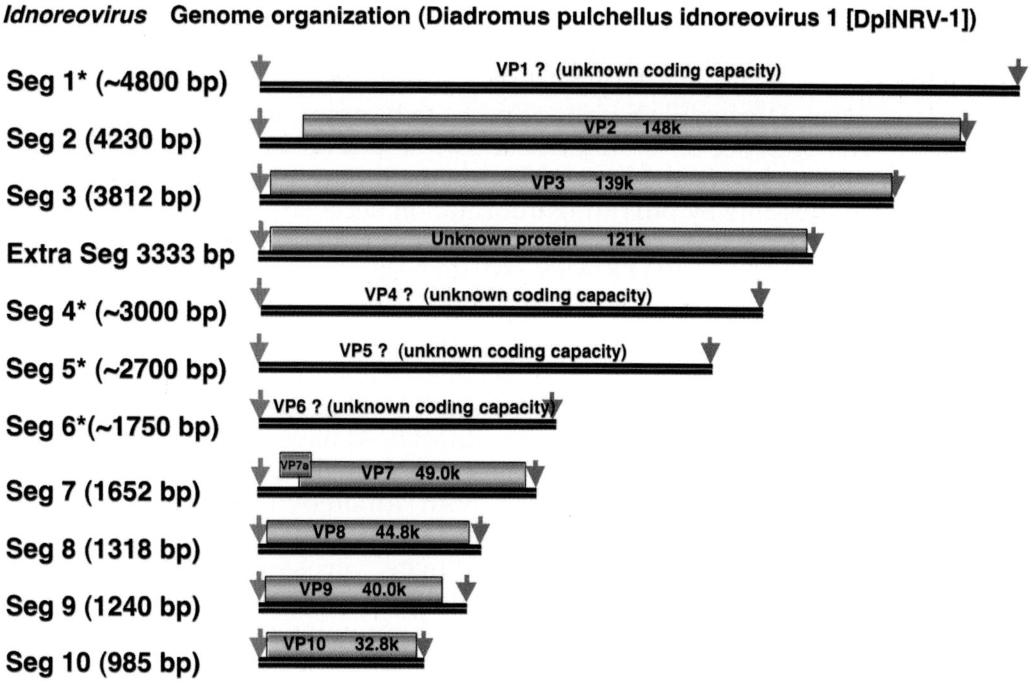

Figure 2: Genome organization of Diadromus pulchellus idnovirus 1 (DpINDRV-1), containing 10 dsRNA segments, each contains an ORF. The green arrows indicate the upstream conserved terminal sequence (+ve 5'-A/U/CCAAUUU.....), while the red arrows indicate the downstream conserved terminal sequence (under determination).

Table 2: List of Diadromus pulchellus idnoreovirus 1 (DpIRV-1) dsRNA Seg1-10, with their respective size (bp) and their encoded proteins for which the name, calculated size (kDa) and function and/or location are indicated.

dsRNA Seg #	dsRNA Size (bp)	Protein nomenclature	Protein size aa	(kDa)	Structure / Function
1	(4, 800)	VP1			
*2	*(4,230)	VP2	1312 77	(148) (???)	
* 3	*(3,812)	VP3	1241	(138,7)	Contains RdRp motifs
* additional segment	*(3,333)	??	1072	(121,9)	RNA sequence closely related to Seg3. Presence of this RNA is related to sex and ploidy of the host.
4	(3,000)	VP4			
5	(2, 700)	VP5			
6	(1,750)	VP6			
*7	*(1, 652)	VP7	436 52	(49) (???)	
*8	*(1,318)	VP8	399	(44,8)	
*9	*(1,240)	VP9	352	(40)	
*10	*(985)	VP10	294	(32,8)	

* The sizes of some dsRNA segments and their putative translation products have been determined by sequence analyses.

Antigenic Properties

Unknown.

Biological Properties

The idnoreoviruses infect insect species, where in many cases they appear to case few pathological effects. However, they may significantly alter the biological properties of the individual host. Drosophila S virus appears to be associated with the 'S' phenotype in *D. simulans*. The presence of an additional 3.33 kbp-dsRNA segment in DpIRV-1 is related to the sex and ploidy of the host. This segment may play a role in the biology of this wasp species, possibly by providing information necessary for larval development.

List of Species Demarcation Criteria in the Genus

In common with the other genera within the family *Reoviridae*, the prime determinant for inclusion of virus isolates within a single species of the genus *Idnoreovirus*, is defined by their compatibility for reassortment of genome segments during co-infections, allowing the exchange of genetic information and generating viable progeny virus strains. However, no further mycoreovirus isolates (other than RnMDV-1) have been confirmed. It may also be difficult to introduce multiple virus strains into a single fungal mycelium under laboratory conditions. Co-infection experiments with idnoreoviruses have not been reported. Therefore, as with other members of the family *Reoviridae*, serological comparisons, comparisons of RNA/protein sequences, cross-hybridization, analysis of electropherotypes, and analysis of conserved RNA sequences in near terminal regions of the genome segments, represent factors used to examine the level of similarity that exists between those species currently recognized and future isolates.

Members of a single species in the genus *Idnoreovirus* may be identified by:

- Ability to exchange genetic material by genome segment reassortment during dual infections, thereby producing viable progeny virus strains.
- RNA cross-hybridization assays (e.g. Northern or dot blots, with probes made from viral RNA or cDNA).
- Serological comparisons (e.g. by ELISA or AGID assays.
- RNA sequence analysis. Within a single species, high levels of sequence similarity would be expected in conserved segments, particularly the polymerase or internal CP genes.
- Comparisons of aa sequences (e.g. polymerase or internal CPs).
- The genome segment number and size distribution, for example as analyzed by AGE (electropherotypes). Within a single species the electropherotype is relatively uniform. However, deletions or additions could occur in individual segments resulting in unusual variations in electropherotype.
- Analysis of conserved sequences at the termini of the RNA genome segments. These sequences show conservation within individual species of many members of the family *Reoviridae*.

List of Species in the Genus

Species names are in green italic script; strain names and synonyms are in black roman script; tentative species names are in blue roman script. Sequence accession numbers, and assigned abbreviations ( ) are also listed.

Species in the Genus

Idnoreovirus 1

Diadromus pulchellus idnoreovirus 1 (Diadromus pulchellus reovirus)	[Seg2: X82049; Seg3: X80481; Seg7: X82048; Seg8: X82047; Seg9: X82046; Seg10: X82045; 'Additional segment': X80480]	(DpIRV-1)

Idnoreovirus 2

Hyposoter exiguae idnoreovirus 2 (HeIRV-2)
(Hyposoter exiguae reovirus)

Idnoreovirus 3

Musca domestica idnoreovirus 3 (MdIRV-3)
(Musca domestica reovirus)
(Housefly virus)

Idnoreovirus 4

Dacus oleae idnoreovirus 4 (DoIRV-4)
(Dacus oleae reovirus)

Idnoreovirus 5

Ceratitis capitata idnoreovirus 5 (CcIRV-5)
(Ceratitis capitata I virus)
Drosophila melanogaster idnoreovirus 5 (DmIRV-5)
(Drosophila F virus)

TENTATIVE SPECIES IN THE GENUS

Drosophila S virus (DSV)

PHYLOGENETIC RELATIONSHIPS WITHIN THE GENUS

Unknown.

SIMILARITY WITH OTHER TAXA

There is no evidence of significant sequence homology between idnoreovirus genes and those of the other members of the family *Reoviridae*.

DERIVATION OF NAMES

Idno: from 'I.D.' indicating 'insect-derived'; 'N.O.' indicating that in contrast to the cypoviruses, they are 'non-occluded',
Reo: sigla from 'reovirus' indicating membership of the family *Reoviridae.*

REFERENCES

Anagnou-Veroniki, M., Veyrunes, J.C., Kuhl, G. and Bergoin, M. (1997). A non-occluded reovirus of the olive fly, *Dacus oleae*. *J. Gen. Virol.*, **78**, 259-263.

Bigot, Y., Drezen, J.M., Sizaret, P.Y., Rabouille, A., Hamelin, M.H. and Periquet, G. (1995). The genome segments of DpRV, a commensal reovirus of the wasp Diadromus pulchellus (Hymenoptera). *Virology*, **210**, 109-119.

Lopez-Ferber, M., Veyrunes, J.C. and Croizier, L. (1989). Drosophila S virus is a member of the Reoviridae family. *J. Virol.*, **63**, 1007-1009.

Moussa, A.Y. (1978). A new virus disease in the House fly *Musca domestica* (*Diptera*). *J. Inver. Pathol.*, **31**, 204-216.

Plus, N., Gissman, L., Veyrunes, J.C., Pfister, H. and Gateff, E. (1981). Reoviruses of *Drosophila* and *Ceratitis* populations and of drosophila cell lines; a new genus of the Reoviridae family. *Ann. Virol.*, **132E**, 261-270.

Plus, N. and Croizier, G. (1982). Further studies on the genome of *Ceratitis capitata* I virus (*Reoviridae*). *Ann. Virol.*, **133E**, 489-492.

Rabouille, A., Bigot, Y., Drezen, J.M., Sizaret, P.Y., Hamelin, M.H. and Periquet, G. (1994). A member of the *Reoviridae* (DpRV) has a ploidy-specific genomic segment in the wasp *Diadromus pulchellus* (*Hymenoptera*). *Virology*, **205**, 228-237.

Stoltz, D. and Makkay, A. (2000). Co-replication of a reovirus and a polydnavirus in the ichneumonid parasitoid *Hyposoter exiguae*. *Virology*, **278**, 266-275.

CONTRIBUTED BY

Mertens, P.P.C., Makkay, A., Stoltz, D., Duncan, R., Bergoin, M. and Dermody, T.S.

GENUS *CYPOVIRUS*

Type Species *Cypovirus 1*

DISTINGUISHING FEATURES

Virus particles may be singly or multiply occluded by a virus coded polyhedrin protein, which forms "polyhedra" within the cytoplasm of infected cells. Cypoviruses only infect and are pathogenic for arthropods. Cypovirus virions have a single capsid shell with surface spikes. Virions have transcriptase and capping enzymes that are active without particle modification. They can retain RNA polymerase activity despite particle disruption into ten distinct RNA protein complexes, each representing a single genome segment and a transcriptase complex. Consequently transcriptase activity is resistant to repeated freeze-thawing, which disrupts the particle structure. The transcriptase activity may show very pronounced dependence on the presence of S-adenosyl-L-methionine or related compounds.

VIRION PROPERTIES

MORPHOLOGY

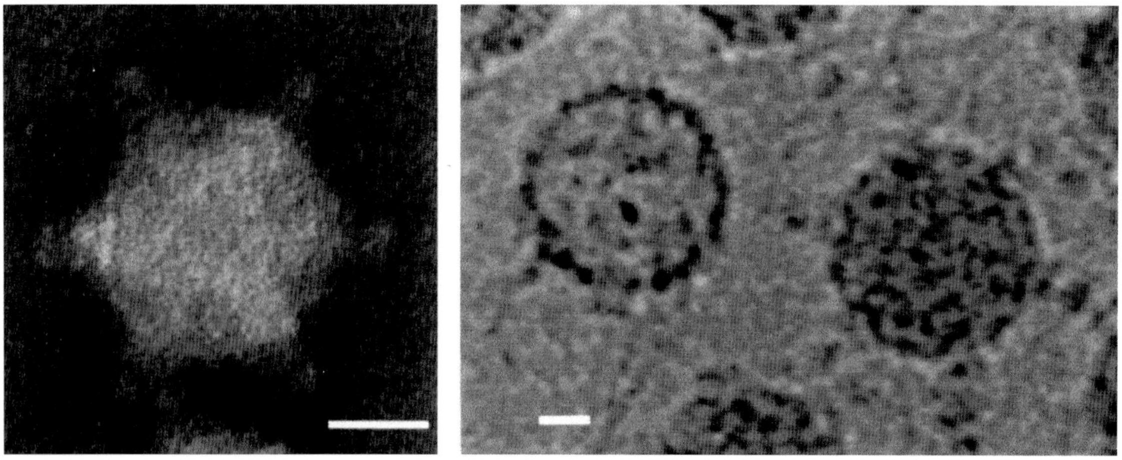

Figure 1: (Left) Negative contrast electron micrograph of non-occluded Orgyia pseudosugata cypovirus 5 (OpCPV-5) virion. (Right) Negative contrast electron micrograph of empty and full occluded virions (purified from polyhedra) of Orgyia pseudosugata cypovirus 5 (OpCPV-5), stained with uranyl acetate (Courtesy of C.L. Hill). The bars represent 20 nm.

Virus particles have a single-layered capsid, composed of a central capsid shell of 57 nm diameter, which extends to 71.5 nm (determined by cryo-electron microscopy) when the 12 'turrets' on the icosahedral 5-fold vertices are included. These surface 'spikes' are hollow and have previously been estimated to be up to 20 nm in length and 15-23 nm wide (by conventional microscopy and negative staining). The virus particle has a central compartment about 35 nm in diameter. Cypovirus virions are structurally comparable to the core particles of members of other genera within the family *Reoviridae*, particularly genera containing viruses with 'spiked' cores (*Orthoreovirus*, *Aquareovirus*, and *Oryzavirus*) (Fig. 1 and 2). The virus particles contain three major structural proteins that have been identified as: 'capsid shell protein' (CSP, 120 copies, equivalent to the VP3(T2) protein of bluetongue virus, and orthoreovirus lambda1; 'large protrusion protein' (LPP, 120 copies, comparable to orthoreovirus lambda3), and 'turret protein' (TP, 60 copies, comparable to orthoreovirus lambda2). The virion also contains transcriptase enzyme complexes attached to the inner surface of the capsid shell at the icosahedral 5-fold vertices.

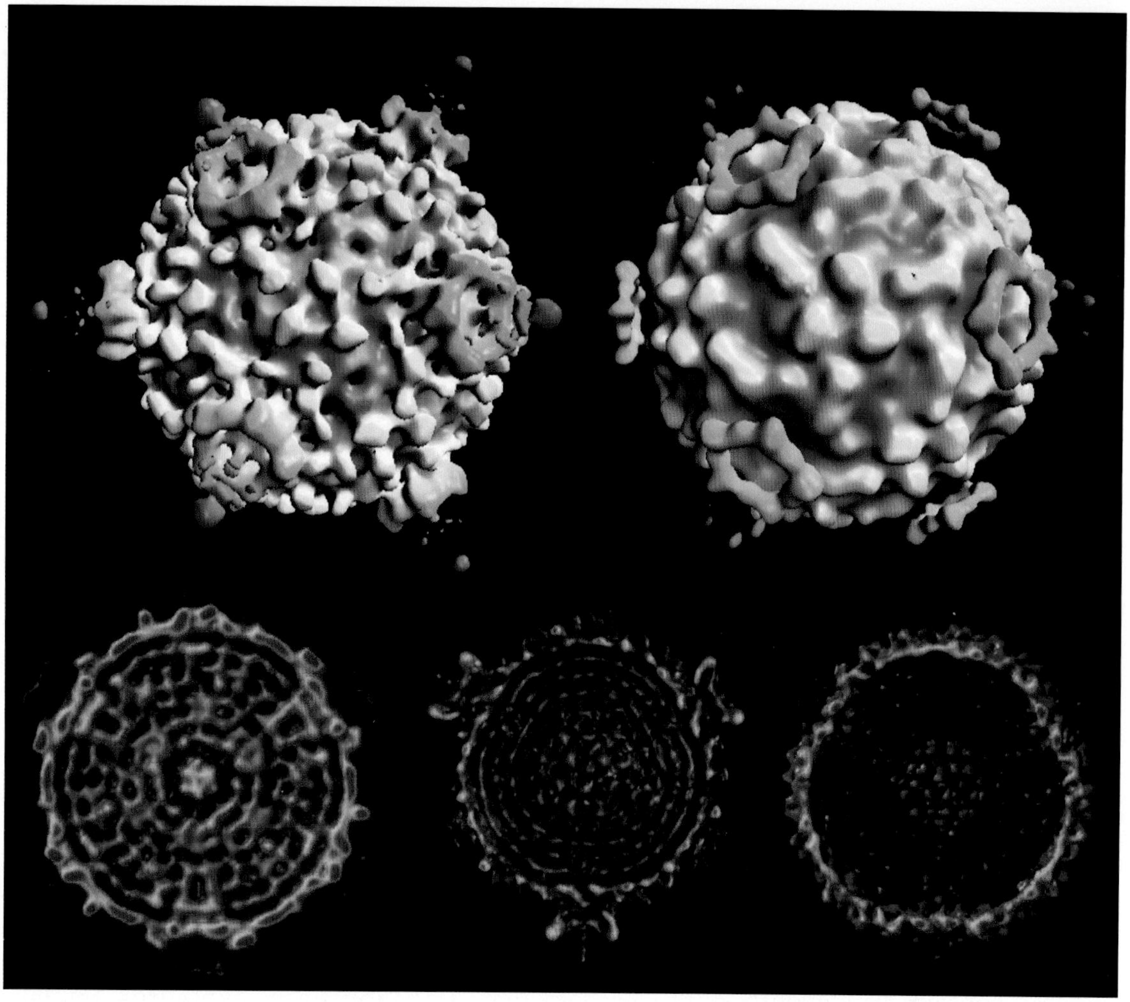

Figure 2: (Top left) cryo-electron microscopy reconstruction of a non-occluded virion of Orgyia pseudosugata cypovirus 5 (OpCPV-5), to 26 Å resolution. (Top right) Cryo-electron microscopy reconstruction of an occluded virion of OpCPV-5, to 26 Å resolution. (Bottom left) cross-section of cryo-electron microscopy reconstruction of a full occluded virion of OpCPV-5 to 26 Å resolution. (Bottom center) Cross-section of cryo-electron microscopy reconstruction of full non-occluded OpCPV-5 virion to 26 Å resolution. (Bottom right) Cross-section of cryo-electron microscopy reconstruction of empty OpCPV-5 virion to 26 Å resolution. The cross-sections show evidence of dsRNA packaged as distinct layers and suggest localization of the transcriptase complexes at the 5-fold axes of symmetry. (Courtesy of C. L. Hill).

Virus particles may be occluded by a crystalline matrix of polyhedrin protein, forming a polyhedral inclusion body. These 'polyhedra' have a symmetry (e.g., cubic, icosahedral, or irregular) that is dependent on both the virus strain (polyhedrin sequence) and the host. The polyhedrin protein appears to be arranged as a face-centered cubic lattice with center to center spacing varying between 4.1 and 7.4 nm.

PHYSICOCHEMICAL AND PHYSICAL PROPERTIES

The virion Mr is about 5.4 x 10^7. The buoyant density in CsCl is 1.44 g/cm^3 for virions, approximately 1.30 g/cm^3 for empty particles, and 1.28 g/cm^3 for polyhedra. The S_{20W} is approximately 420S for virions and 260S for empty particles. Polyhedra vary considerably in size and Mr and do not have a single characteristic S value. Polyhedra may occlude many virus particles or only single particles. Large empty polyhedra (apparently containing no virions) have also been observed.

Cypoviruses retain infectivity for several weeks at -15°C, 5°C, or 25°C. The virus retains full enzymatic activity (dsRNA-dependent ssRNA polymerase and capping activity) after repeated freeze-thawing (up to sixty cycles). However, it appears likely that this results in the breakdown of the virus particle into ten active and distinct enzyme/template complexes. Each complex contains one genome segment and a complete transcriptase complex, derived from the virion capsid and including one of the "spike" structures from the vertices of the icosahedron. Polymerase activity is therefore a poor indicator of virion integrity. Within the family *Reoviridae* the ability to retain enzyme function despite particle breakdown may be unique to the cypoviruses.

Cations have relatively little effect on the virus structure. Heat treatment of virions at 60°C for 1 hr leads to degradation and release of genomic RNA. Virus particles are relatively resistant to treatment with trypsin, chymotrypsin, ribonuclease A, deoxyribonuclease, or phospholipase. Virion enzyme functions also show some resistance to treatment with proteinase K. However, this may reflect the retention of enzyme activities despite particle disruption, particularly during the early stages of digestion. Cypovirus particles are resistant to detergents such as sodium deoxycholate (0.5-1%) but are disrupted by 0.5-1% SDS, which releases the genomic dsRNA. Treatment with triton X-100, NP40, or urea also causes diruption of the virus particle structure. One or two fluorocarbon treatments have little effect on virus infectivity, however treatment with ethanol leads to release of RNA from virions. Viruses and polyhedra are readily inactivated by UV-irradiation. It has been reported that UV also releases the dsRNA template from individual genome segment/transcriptase complexes. Polyhedra remain infectious for years at temperatures below 20°C. Virions can be released from polyhedra by treatment with carbonate buffer at pH greater than 10.5 but are disrupted below pH 5. As in permissive insects' mid-guts, high pH treatment completely dissolves the polyhedral protein matrix. This process is partly due to increased solubility of polyhedrin at high pH but is also aided by alkaline activated proteases associated with polyhedra.

Nucleic Acid

Polyhedra (but not virions) contain significant amounts of adenylate-rich oligonucleotides. In the majority of cases cypovirus particles contain 10 dsRNA genome segments. However there is evidence to indicate that in some cases the virus particles may also contain an eleventh small segment (e.g. Trichoplusia ni cypovirus 15, TnCPV-15: see Tables 2 and 4). In Bombyx mori cypovirus 1 (BmCPV-1), the genome segments vary in size from 4190 to 994 bp with a total genome size of 24,809 bp. In other isolates which have not yet been sequenced the sizes of the genome segments have been estimated by electrophoretic comparisons and have calculated Mr that vary from 0.42×10^6 to 3.7×10^6 (0.6 to 5.6 kbp) and a total genome Mr which varies from 19.3×10^6 to 22.0×10^6 (29.2 to 33.3 kbp).

The pattern of size distribution of the genome segments varies widely between different cypoviruses (e.g. the smallest dsRNA has an estimated Mr which varies between 0.42×10^6 and 0.79×10^6). These size differences have formed a basis for the recognition and classification of distinct species (electropherotypes) of cypoviruses (distinct patterns of dsRNA migration, which differ significantly in the migration of at least three genome segments, as analyzed by electrophoresis using 1% agarose or 3% SDS-PAGE). The genome segment migration patterns of types 1, 12 and 14 have some overall similarity, although in each case at least 3 segments show significant migrational differences during agarose gel electrophoresis. More recently it has been shown that members of different *Cypovirus* species can also be distinguished on the basis of RNA sequence comparisons (e.g. by comparison of genome segment 10: the polyhedrin gene, Fig. 4)

The termini of the coding strands are common or very closely related for the different genome segments members of the species *Cypovirus 1*, but differ from those reported for other cypovirus species (Table 1). Choristoneura fumiferana cypovirus 16 (CfCPV-16) shows high levels of overall sequence variation when compared to members of species

Cypovirus 1, 2, 5, 14, or *15* (Table 2) and is therefore considered to be a different species, although it has a similar 5'- but different 3'-ends to representatives of *Cypovirus 5*. These data demonstrate that different cypovirus electropherotypes are likely to have different conserved RNA terminal sequences.

Table 1: The conserved terminal sequences of cypovirus genome segments

Virus species	(isolate)	Conserved RNA terminal sequences (+ve strand)
Cypovirus 1	(BmCPV-1)	5'-AGUAA..................................GUUAGCC-3'
	(DpCPV-1)	5'-AGTAA...............................GUUAGCC-3'
	(LdCPV-1)	5′-AGU$^{A}/_{G}{}^{A}/_{G}$..................................$G^{U}/_{C}$UAGCC-3′
Cypovirus 2	(IiCPV-2)	5′-AGUUUUA...................................UAGGUC-3′
*Cypovirus 4**	(ApCPV-4)	5'-AAUCGACG..................GUCGUAUG-3'
	(AaCPV-4)	5'-AAUCGACG..................GUCGUAUG-3'
	(AmCPV-4)	5'-AAUCGACG..................GUCGUAUG-3'
Cypovirus 5	(OpCPV-5)	5'-AGUU......................................UUGC-3'
Cypovirus 14	(LdCPV-14)	5′-AGAA......................................CAGCU-3′
Cypovirus 15	(TnCPV-15)	5′-AUUAAAAA......................................GC-3′
*Cypovirus 16**	(CfCPV-16)	5'-AGUUUUU.....................UUUGUGC-3'

* Based on genome Seg 9 only of ApCPV-4, AaCPV-4, AmCPV-4 and Seg 10 only of CfCPV-16.

Table 2: Cypovirus genome segment size distribution (kbp) determined by sequence analyses or estimated from electrophoretic comparisons of the genomic dsRNA of cypovirus 1 to 16.

Genome segment number	*Cypovirus 1* to *Cypovirus 16*															
	1	**2**	**3**	**4**	**5**	**6**	**7**	**8**	**9**	**10**	**11**	**12**	**13**	**14**	**15**	**16**
Total genome	**24.8**	25.5	26.7	27.5	26.3	27.2	25.6	27.0	24.1	27.6	25.5	26.1	25.2	**25.3**	**24.9**	
1	**4.19**	4.06	4.29	4.17	4.17	4.17	4.32	4.54	4.32	4.31	4.60	4.43	4.26	**4.33**	**4.36**	
2	**3.86**	4.06	4.12	4.17	4.17	4.06	4.15	4.54	4.18	4.31	4.40	4.12	4.26	**4.06**	**4.19**	
3	**3.85**	3.83	4.12	4.17	4.17	4.00	4.02	4.40	4.07	4.02	4.40	4.12	4.03	**3.92**	**3.88**	
4	**3.26**	3.65	3.69	3.90	3.69	3.72	3.81	3.92	3.62	4.02	3.83	3.67	3.60	**3.34**	**3.31**	
5	**2.85**	2.21	3.60	2.43	3.22	2.73	2.54	3.69	2.34	2.50	1.98	3.30	3.20	**3.16**	**2.26**	
6	**1.80**	1.93	2.29	2.17	2.17	2.36	2.27	1.90	1.72	2.29	1.98	2.00	1.60	**1.78**	**1.86**	
7	**1.50**	1.79	2.15	1.95	2.06	2.23	2.02	1.30	1.72	2.29	1.35	1.44	1.40	**1.39**	**1.78**	
8	**1.33**	1.56	1.08	1.72	1.21	1.63	1.08	1.19	0.78	1.69	1.27	1.27	1.14	**1.25**	**1.23**	
9	**1.19**	1.38	0.83	**1.47**	0.88	1.40	0.85	0.88	0.69	1.21	0.98	1.13	0.98	**1.14**	**1.16**	
10	**0.99**	0.98	0.60	1.44	**0.88**	0.90	0.53	0.65	0.69	0.99	0.71	0.64	0.78	**0.96**	**0.90**	**11.71**
11															**0.20***	

Sizes are indicated in kbp. Sizes in bold are given in kilo base pairs, derived from sequence analysis of the genome segment. Previously published estimates of genome segment sizes for members of *Cypovirus 2* to *Cypovirus 13* have been adjusted in line with base pair values derived from sequencing studies of cDNA copies of genome segments from BmCPV-1.
*TnCPV-15 has been reported to contain an 11th small genome segment (200 bp).

PROTEINS

Cypovirus particles generally contain five to six distinct proteins, two to three with sizes of more than 100 kDa. For BmCPV-1 the structural proteins are 148 (VP1), 136 (VP2), 140 (VP3), 120 (VP4), 64 (VP6) and 31 kDa (VP7) (see Table 3). Polyhedra also contain a 25-37 kDa polyhedrin protein (28.5 kDa for BmCPV-1) that constitutes about 95% of the polyhedra protein dry weight. Due to the high level of variation between different cypoviruses it is unlikely that their homologous proteins can be identified simply by their migration order during PAGE.

LIPIDS

Cypoviruses are not known to contain any lipids in either virus particles or polyhedra.

CARBOHYDRATES

The polyhedrin protein is glycosylated.

GENOME ORGANIZATION AND REPLICATION

For BmCPV-1 the coding assignments are indicated in Table 3 and Figure 3. The cognate genes of other cypoviruses are not known. The large variations in the sizes of genome segments between most cypoviruses (apart from *Cypovirus 1, 12* and *14*) indicate that these assignments will not apply to other cypovirus species. Genome segment coding assignments generated by in vitro translation of individual denatured genome segment RNA have been published for members of *Cypovirus 1* and *Cypovirus 2*. These data and subsequent sequencing studies indicate that in many cases polyhedrin may be encoded by the smallest segment.

Table 3: List of dsRNA segments of Bombyx mori cypovirus 1 (BmCPV-1, strain I) and the proteins for which they code. RNA segment sizes (from sequence analyses), deduced protein sizes and functions where known are indicated.

dsRNA segment #	Size (bp)	Protein[1] nomenclature ([2])	Size (kDa)	Protein copy # per particle	Function (location)
1	4190	VP1 (VP1)	148	No information	Major core CP (Virion)
2	3854	VP2 (VP2)	136	"	RdRp (virion)
3	3846	(VP3)	140	"	(virion)
4	3262	VP3 (VP4)	120	"	Possible Mtr (virion)
5	2852	NS1 (NS5) NS2 (NS5a) NS6 (NS5b)	101 80* 23*	0	Non-structural, contains auto cleavage aa sequence, similar to FMDV $2A^{pro}$
6	1796	VP4 (VP6)	64	No information	Leucine zipper ATP/GTP binding protein (virion)
7	1501	NS3 NS4 (VP7)	50 (61*) 58* 31*	0	Non-structural, with 'structural' cleavage products
8	1328	VP5 or P44 (NSP8)	44	No information	Unknown (shows anomalous migration during PAGE, with apparent size 55 kDa)
9	1186	NS5 (NSP9)	36	0	Non-structural, dsRNA binding
10	944	Polyhedrin (Pod)	28.5	Unnown	Polyhedron matrix protein (Pod)

Size of genome segments and encoded proteins determined by sequence analysis of the genome segments.
* Sizes of some proteins estimated from electrophoretic migration.
[1]: Protein nomenclature suggested by McCrae and Mertens (1983)
[2]: Alternative nomenclature suggested by Hagiwara *et al.* (2002)

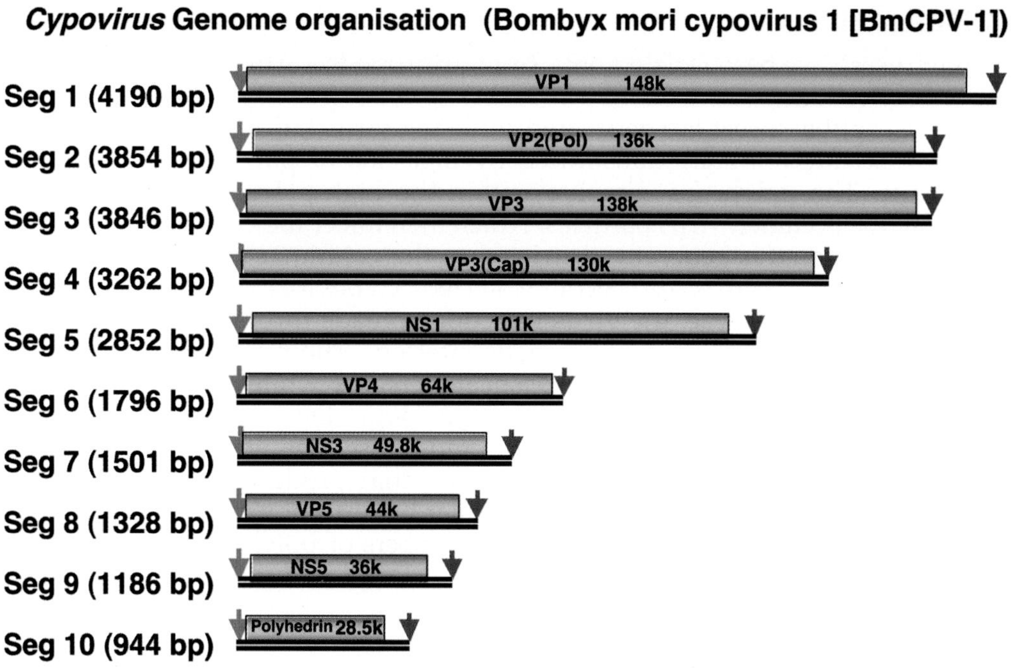

Figure 3: Genome organization of Bombyx mori cypovirus 1– In6423 (BmCPV-1), containing 10 dsRNA segments, each contains an ORF. The green arrows indicate the upstream conserved terminal sequence (+ve 5'-AGUAA.....), while the red arrows indicate the downstream conserved terminal sequence (+ve....GUUAGCC-3').

Unlike orthoreoviruses, cell entry and initiation of cypovirus replication in insect cells does not require modification of the virions for activation of the core-associated enzymes. Uptake appears to be a relatively inefficient process in cell cultures, which can be very significantly improved by the use of liposomes. Virus replication and assembly occur in the host cell cytoplasm, although there is some evidence that viral RNA synthesis can occur within the nucleus. Replication is accompanied by the formation of viroplasm (or virogenic stroma) within the cytoplasm. Viroplasms contain large amounts of virus proteins and virus particles. How genome segments are selected for packaging and assembly into progeny particles is not known. The importance of the terminal regions in this process is indicated by the packaging and transcription of a mutant Seg10 of an isolate of *Cypovirus 1* that contained only 121 bp from the 5'-end and 200 bp from the 3'-end. Particles are occluded within polyhedra apparently at the periphery of the virogenic stroma, from about 15 hr post-infection. The polyhedrin protein is produced late in infection and in large excess compared to the other viral proteins. How polyhedrin synthesis is regulated is not known.

ANTIGENIC PROPERTIES

Serological cross-comparisons of cypovirus structural and polyhedrin proteins support the use of genomic dsRNA electropherotypes as one of the species parameters for the genus *Cypovirus.* Virus isolates within a single electropherotype exhibit high levels of antigenic cross-reaction (in both polyhedrin and virion structural proteins), as well as efficient cross-hybridization of denatured genomic RNA, even under high-stringency conditions. In contrast there is evidence of little or no serological cross-reaction between viruses representing different electropherotypes. Exceptions are viruses of *Cypovirus 1* and *Cypovirus 12*, which show low level serological cross-reactions but these viruses also have some overall similarity in their electropherotype pattern and show a low level of cross-hybridization of their genome segments. *Cypovirus 14* members also show some similarity in its RNA electropherotype pattern to viruses of both *Cypovirus 1* and *Cypovirus 12.* It may therefore also show some antigenic relationship and RNA sequence homology with these viruses.

BIOLOGICAL PROPERTIES

Cypoviruses have been isolated only from arthropods. Attempts to infect vertebrates, or vertebrate cell lines, have failed. In addition, cypovirus replication is inhibited at 35°C. Even susceptible insect larvae treated with the virus fail to develop infections at 35°C. Cypoviruses are normally transmitted by ingestion of polyhedra on contaminated food materials. The polyhedra dissolve within the high pH environment of the insect gut releasing the occluded virus particles, which then infect the cells lining the gut wall. Virus infection in larvae is generally restricted to the columnar epithelial cells of the midgut, although goblet cells may also become infected. Cypovirus replication in the fat body has been reported. In larvae, the virus infection spreads throughout the midgut region. In some species the entire gut is occasionally infected. The production of very large numbers of polyhedra give the gut a characteristically creamy-white appearance. In infected cells the endoplasmic reticulum is progressively degraded, mitochondria enlarge and the cytoplasm becomes highly vacuolated. In most cases the nucleus shows few pathological changes. An exception is a cypovirus strain that produces inclusion bodies within the nucleus. In the later stages of infection cellular hypertrophy is common and microvillae are reduced or completely absent. Very large numbers of polyhedra are released by cell lysis into the gut lumen and excreted. The gut pH is lowered during infection and this prevents dissolution of progeny polyhedra in the gut fluid.

The majority of cypovirus infections produce chronic disease, often without extensive larval mortality. Consequently, many individuals reach the adult stage even though heavily diseased. However, cypovirus infections produce symptoms of starvation due to changes in the gut cell structure and reduced adsorptive capacity. Infected larvae stop feeding as early as two days post-infection. Larval body size and weight are often reduced and diarrhea is common. The larval stage of the host can be significantly increased (about by 1.5 times the normal generation time). The size of infected pupae is frequently reduced and the majority of diseased adults are malformed. They may not emerge correctly, and may be flightless. Infected females may exhibit a reduced egg laying capacity.

Virus can be transmitted on the surface of eggs, producing high levels of infection in the subsequent generation. However, no transovarial transmission has been observed provided the egg surface is disinfected. The infectious dose increases dramatically in the later larval instars. Different virus strains vary significantly in virulence. Larvae can recover from cypovirus infection, possibly because the gut epithelium has considerable regenerative capacity and because infected cells are shed at each larval moult.

LIST OF SPECIES DEMARCATION CRITERIA IN THE GENUS

Cypoviruses are currently classified within 16 species that were initially characterized by their distinctive dsRNA electropherotype patterns. Cross-hybridization analyses of the dsRNA, serological comparisons of cypovirus proteins and more recently comparison of RNA sequences have confirmed the validity of this classification and have identified new virus species. However, only relatively few cypoviruses have been characterized, suggesting that there may be many more distinct species that are as yet unidentified.

The system of nomenclature currently used to identify different cypovirus isolates takes account of both the virus species and the host species from which the virus was originally isolated (e.g. BmCPV-1). The relationships between different cypoviruses within a single electropherotype, or with other cypovirus types, are not fully understood at the molecular level. Sequence analyses of genome segments from distinct isolates have shown very high levels of identity within a single cypovirus species. For example, the different isolates of *Cypovirus 5* that have been analyzed (see Table 2) show >98% homology in genome Seg 10 (the polyhedrin gene), while isolates of *Cypovirus 1* show 80-98% nt sequence identity in this gene. In contrast, comparisons of unrelated types showed only relatively low levels of sequence identity (20-23%). Studies of the genomic RNA from different cypovirus isolates have demonstrated that although there may be slightly higher conservation in the largest genome segments (possibly as a result of functional constraints) the level of variation is relatively uniform across the whole genome.

Within the family *Reoviridae,* the prime determinant for inclusion of virus isolates within a single virus species is their ability to exchange genetic information by reassortment of their genome segments during co-infection, thereby generating viable progeny virus strains. There is no direct evidence concerning genome segment reassortment between different cypovirus isolates. However, evidence of similarity and therefore of the genetic compatibility required for reassortment, can be provided by other methods.

Members of a single species of the genus *Cypovirus* may be identified by:

- Their ability to exchange genetic material by genome segment reassortment during dual infections, thereby producing viable progeny virus strains.
- Similar electrophoretic migration of at least 7 genome segments, as analysed using either an agarose, or a low percentage (3%) polyacrylamide gel system. Viruses of different species will have significant migrational differences in at least three genome segments.
- High levels of serological cross-reaction by ELISA or agar gel immuno diffusion (e.g. using polyclonal antisera to purified virions or polyhedrin proteins). Different but more closely related species may show low levels of serological cross-reaction (eg. CPV-1 and -12).
- A high degree of sequence conservation (estimated >80%).
- Cross-hybridization of genome segments under high stringency conditions (designed to detect >90% homology) (Northern or dot blots, with probes made from viral RNA or cDNA).
- Current limited evidence suggests that the conserved terminal sequences are likely to be the same (or closely related) within a single species but likely to be different between different species. The similarity between more closely related species (e.g. CPV-1 and -12) is unknown.

DSRNA

LIST OF SPECIES IN THE GENUS

Below a list is provided of some of the lepidopteran cypoviruses for which the RNA electropherotypes have been deduced. In addition to many other lepidopteran cypoviruses that have been described (but are otherwise uncharacterized), there are dipteran and hymenopteran cypoviruses. One isolate from a freshwater daphnid has been reported. In total, more than 230 cypoviruses have been described, however the total number of species is unknown.

Species names are in green italic script; strain names and synonyms are in black roman script; tentative species names are in blue roman script. Sequence accession numbers [Seg1 to Seg11 indicate RNA segments 1 – 12], and assigned abbreviations ( ) are also listed.

SPECIES IN THE GENUS

Cypovirus 1

Bombyx mori cypovirus 1	[S1: AF323781, S2: AF323782, S3: AF323783, S4: AF323784, S5: AB035733, S6: AB030014, S7: AB030015, S8: AB016436, S9: AF061199, S10: D37768]	(BmCPV-1)
Dendrolimus punctatus cypovirus 1	[S1: AY163247, S2: AY147187, S3: AY167578, S4: AF542082, S5: AY163248, S6: AY163249,	(DpCPV-1)

	S7: AY211091, S8: AF513912, S9: AY310312, S10: AF541985]	
Dendrolimus spectabilis cypovirus 1		(DSCPV-1)
Lymantria dispar cypovirus 1	[S1: AF389462, S2: AF389463, S3: AF389464, S4: AF389465, S5: AF389466, S6: AF389467, S7: AF389468, S8: AF389469, S9: AF389470, S10: AF389471]	(LdCPV-1)
Cypovirus 2		
Aglais urticae cypovirus 2		(AuCPV-2)
Agraulis vanillae cypovirus 2		(AvaCPV-2)
Arctia caja cypovirus 2		(AcCPV-2)
Arctia villica cypovirus 2		(AviCPV-2)
Boloria dia cypovirus 2		(BdCPV-2)
Dasychira pudibunda cypovirus 2		(DpCPV-2)
Eriogaster lanestris cypovirus 2		(ElCPV-2)
Hyloicus pinastri cypovirus 2		(HpCPV-2)
Inachis io cypovirus 2		(IiCPV-2)
Lacanobia oleracea cypovirus 2		(LoCPV-2)
Malacosoma neustria cypovirus 2		(MnCPV-2)
Mamestra brassicae cypovirus 2		(MbCPV-2)
Operophtera brumata cypovirus 2		(ObCPV-2)
Papilio machaon cypovirus 2		(PmCPV-2)
Phalera bucephala cypovirus 2		(PbCPV-2)
Pieris rapae cypovirus 2		(PrCPV-2)
Cypovirus 3		
Anaitis plagiata cypovirus 3		(ApCPV-3)
Arctia caja cypovirus 3		(AcCPV-3)
Danaus plexippus cypovirus 3		(DpCPV-3)
Gonometa rufibrunnea cypovirus 3		(GrCPV-3)
Malacosoma neustria cypovirus 3		(MnCPV-3)
Operophtera brumata cypovirus 3		(ObCPV-3)
Phlogophera meticulosa cypovirus 3		(PmCPV-3)
Pieris rapae cypovirus 3		(PrCPV-3)
Spodoptera exempta cypovirus 3		(SexmCPV-3)
Cypovirus 4		
Actias selene cypovirus 4		(AsCPV-4)
Antheraea assamensis cypovirus 4	[S9: AF374299]	(AaCPV-4)
Antheraea mylitta cypovirus 4	[S9: AF374298]	(AmCPV-4)
Antheraea pernyi cypovirus 4		(ApCPV-4)
Antheraea proylei cypovirus 4	[S9: AF374300]	(AprCPV-4)
Cypovirus 5		
Euxoa scandens cypovirus 5	[S10: J04338]	(EsCPV-5)
Heliothis armigera cypovirus 5	[S10: U06196]	(HaCPV-5)
Orgyia pseudosugata cypovirus 5	[S10: U06194]	(OpCPV-5)
Spodoptera exempta cypovirus 5		(SexmCPV-5)
Trichoplusia ni cypovirus 5		(TnCPV-5)
Cypovirus 6		
Aglais urticae cypovirus 6		(AuCPV-6)
Agrochola helvolva cypovirus 6		(AhCPV-6)
Agrochola lychnidis cypovirus 6		(AlCPV-6)
Anaitis plagiata cypovirus 6		(ApCPV-6)
Anti xanthomista cypovirus 6		(AxCPV-6)
Biston betularia cypovirus 6		(BbCPV-6)

Eriogaster lanestris cypovirus 6		(E1CPV-6)
Lasiocampa quercus cypovirus 6		(LqCPV-6)
Cypovirus 7		
Mamestra brassicae cypovirus 7		(MbCPV-7)
Noctua pronuba cypovirus 7		(NpCPV-7)
Cypovirus 8		
Abraxas grossulariata cypovirus 8		(AgCPV-8)
Heliothis armigera cypovirus 8		(HaCPV-8)
Malacosoma disstria cypovirus 8		(MdCPV-8)
Nudaurelia cytherea cypovirus 8		(NcCPV-8)
Phlogophora meticulosa cypovirus 8		(PmCPV-8)
Spodoptera exempta cypovirus 8		(SexmCPV-8)
Cypovirus 9		
Agrotis segetum cypovirus 9		(AsCPV-9)
Cypovirus 10		
Aporophyla lutulenta cypovirus 10		(AlCPV-10)
Cypovirus 11		
Heliothis armigera cypovirus 11		(HaCPV-11)
Heliothis zea cypovirus 11		(HzCPV-11)
Lymantria dispar cypovirus 11		(LdCPV-11)
Mamestra brassicae cypovirus 11		(MbCPV-11)
Pectinophora gossypiella cypovirus 11		(PgCPV-11)
Pseudaletia unipuncta cypovirus 11		(PuCPV-11)
Spodoptera exempta cypovirus 11		(SexmCPV-11)
Spodoptera exigua cypovirus 11		(SexgCPV-11)
Cypovirus 12		
Autographa gamma cypovirus 12		(AgCPV-12)
Mamestra brassicae cypovirus 12		(MbCPV-12)
Pieris rapae cypovirus 12		(PrCPV-12)
Spodoptera exempta cypovirus 12		(SexmCPV-12)
Cypovirus 13		
Polistes hebraeus cypovirus 13		(PhCPV-13)
Cypovirus 14		
Heliothis armigera cypovirus 14 ('A' strain)		(HaCPV-14)
Lymantria dispar cypovirus 14	[S1: NC_003006, S2: NC_003007, S3: NC_003008, S4: NC_003009, S5: NC_003010, S6: NC_003011, S7: NC_003012, S8: NC_003013, S9: NC_003014, S10: NC_003015]	(LdCPV-14)
Cypovirus 15		
Trichoplusia ni cypovirus 15	[S1: NC_002557, S2: NC_002558, S3: NC_002559, S4: NC_002567, S5: NC_003060, S6: NC_002561, S7: NC_002562, S8: NC_002563, S9: NC_002564, S10: NC_002565, S11*: NC_002566]	(TnCPV-15)
Cypovirus 16 ‡		
Choristoneura fumiferana cypovirus 16	[S10, U95954]	(CfCPV-16)

* TnCPV-15 has been reported as having 11 distinct genome segments

DSRNA

Tentative Species in the Genus

Heliothis armigera cypovirus ('B' strain)	(HaCPV-B)
Plutella xylostella cypovirus	(PxCPV)
Maruca vitrata cypovirus (A strain)	(MvCPV-A)
Maruca vitrata cypovirus (B strain)	(MvCPV-B)

Phylogenetic Relationships within the Genus

The available sequence data for members of *Cypovirus 1, 2, 5, 14, 15* and *16* allow a comparison of some genes of these viruses, showing not only that different cypovirus species are quite distantly related but also (at least for the viruses analyzed) that there is a high level of conservation within a single species. For example a comparison of polyhedrin genes shows only 20-23% sequence identity between cypovirus species, but with 89-98% identity between different isolates within a single species (Fig. 4). Earlier cross-hybridization studies suggest that the level of nt sequence variation is also relatively uniform across the whole genome, possibly reflecting the absence of neutralizing antibodies in the insect hosts. These data indicate that sequence analyses and comparisons are effective methods for distinguishing and identifying the members of different cypovirus species.

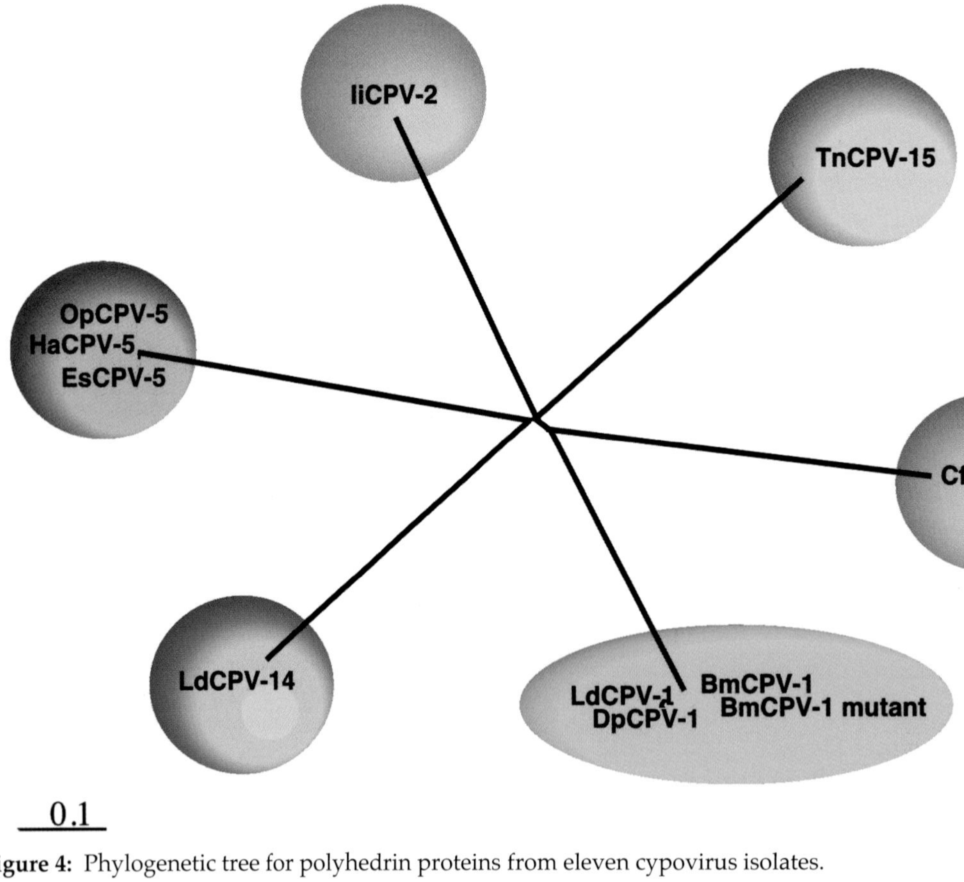

Figure 4: Phylogenetic tree for polyhedrin proteins from eleven cypovirus isolates.

Derivations of Names

Cypo: sigla from *cy*toplasmic *po*lyhedrosis.

References *Cypovirus*

Belloncik, S., Liu, J., Su, D. and Arella, M. (1996). Identification and characterisation of a new *Cypovirus* type 14, isolated from *Heliothis armigera*. *J. Invert. Pathol.*, **67**, 41-47.

Fouillaud, M. and Morel, G. (1994). Characterization of cytoplasmic and nuclear polyhedrosis viruses recovered from the nest of *Polistes hebraeus* F. (Hymenoptera; *Vespidae*). *J. Invert. Pathol.*, **64**, 89-95.

Galinsi, M.S., Yu, Y., Heminway, B.R. and Beaudreau, G.S. (1994). Analysis of the c-polyhedrin genes from different geographical isolates of a type 5 cytoplasmic polyhedrosis virus. *J. Gen. Virol.*, **75**, 1969-1974.

Hagiwara, K., Rao, S., Scott, S.W., and Carner, G.R. (2002). Nucleotide sequences of segments 1, 3 and 4 of the genome of Bombyx mori cypovirus 1 encoding putative capsid proteins VP1, VP3 and VP4, respectively. *J. Gen. Virol.*, **83**, 1477-14782.

Hagiwara, K., Kobayashi, J., Tomita, M., and Yoshimura, T. (2001). Nucleotide sequence of genome segment 5 from Bombyx mori cypovirus 1. *Arch. Virol.*, **146**, 181-187.

Hagiwara, K. and Matsumoto, T. (2000). Nucleotide sequences of genome segments 6 and 7 of Bombyx mori cypovirus 1, encoding the viral structural proteins V4 and V5, respectively. *J. Gen. Virol.*, **81**, 1143-1147.

Hagiwara, K., Tomita, M., Kobayashi, J., Miyajima, S. and Yoshimura, T. (1998). Nucleotide sequence of Bombyx mori cytoplasmic polyhedrosis virus segment 8. *Biochem. Biophys. Res. Commun.*, **247**, 549-553.

Hagiwara, K., Tomita, M., Nakai, K., Kobayashi, J., Miyajima, S. and Yoshimura, T. (1998). Determination of the nucleotide sequence of Bombyx mori cytoplasmic polyhedrosis virus segment 9 and its expression in BmN4 cells. *J. Virol.*, **72**, 5762-5768.

Hill, C.L., Booth, T.F., Prasad, B.V.V., Grimes, J.M., Mertens, P.P.C., Sutton, G.C. and Stuart, D.I. (1999). The structure of a cypovirus and the functional organization of dsRNA viruses. *Nat. Struct. Biol.*, **6**, 565-568.

McCrae, M.A. and Mertens, P.P.C. (1983). *In vitro* translation of the genome and RNA coding assignments for cytoplasmic polyhedrosis type 1. In: *Double-stranded RNA viruses* (R.W. Compans and D.H.L. Bishop, eds), pp 35-42. Elsevier Science Publishing Co., Amsterdam.

Mertens, P.P.C., Croo, N.E., Rubinstein, R., Pedley, S. and Payne, C.C. (1989). Cytoplasmic polyhedrosis virus classification by electropherotype: Validation by serological analyses and agarose gel electrophoresis. *J. Gen. Virol.*, **70**, 173-185.

Mertens, P.P.C., Pedley, S., Crook, N.E., Rubinstein, R. and Payne, C.C. (1999). A comparison of six cypovirus isolates by cross-hybridisation of their dsRNA genome segments. *Arch. Virol.*, **144**, 561-566.

Payne, C.C. and Mertens, P.P.C. (1983). Cytoplasmic polyhedrosis viruses, In: *The Reoviridae*. (W.K. Joklik, ed), pp 425-504. Plenum Press, New York.

Rao, S., Carner, G.R., Scott, S.W., Omura T. and Hagiwara, K. (2003). Comparison of the amino acid sequences of RNA-dependent RNA polymerases of cypoviruses in the family *Reoviridae*. *Arch. Virol.*, **148**, 209-219.

Wu, A.-Z. and Sun, Y.-K. (1986). Isolation and reconstitution of the RNA replicase of the cytoplasmic polyhedrosis virus of silkworm, *Bombyx mori*. *TAG*, **72**, 662-664.

Xia, Q., Jakana, J., Zhang, J.Q. and Zhou, Z.H. (2002). Structural comparisons of empty and full cytoplasmic polyhedrosis virus. Protein-RNA interactions and implications for endogenous RNA transcription mechanism. *J. Biol. Chem.*, **278**, 1094-1100.

Zhang, H., Yu, X., Lu, X.Y., Zhang, L.Q. and Ahou, Z.H. (2002). Molecular interactions and viral stability revealed by structural analysis of chemically treated cypovirus capsids. *Virology*, **29**, 45-52.

Zhou, Z.H., Zhang, H., Jaana, J., Ju, X.-Y. and Zhang, J.-Q. (2003). Cytoplasmic polyhedrosis virus structure at 8Å by electron cryo-microscopy: structural basis of capsid stability and mRNA processing regulation. *Structure*, **11**, 651-663.

CONTRIBUTED BY

Mertens, P.P.C., Rao, S. and Zhou, Z.H.

DSRNA

GENUS FIJIVIRUS

Type Species *Fiji disease virus*

DISTINGUISHING FEATURES

Fijivirus particles have a double-shelled, icosahedral structure, with a spherical rather than angular appearance and short surface spikes ('A' spikes) on each of the twelve vertices of the icosahedron. The outer shell is fragile and easily breaks down leaving the inner shell bearing twelve 'B' spikes. There are 10 genome segments. The viruses replicate in delphacid planthoppers. Nilaparvata lugens reovirus (NLRV) has the above properties but replicates only in insects, whereas other fijiviruses can also replicate in phloem cells of susceptible plants of the families Graminae (in which they induce small tumours or enations), or Liliaceae.

VIRION PROPERTIES

MORPHOLOGY

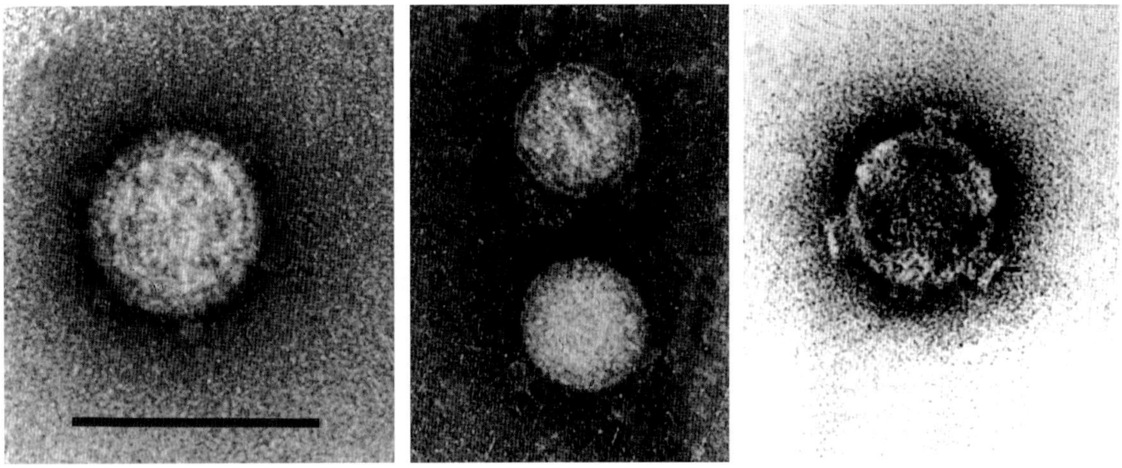

Figure 1: (Left) Negative contrast electron micrograph of Maize rough dwarf virus (MRDV) virions stained with uranyl acetate showing 'A' spikes; (Center) smooth subcores derived from MRDV on staining with neutral phosphotungstate. (Right) 'B' spikes on virus-derived MRDV cores stained with uranyl acetate; (Courtesy of R.G. Milne). The bar represents 100 nm.

Virions are double-shelled, spherical, 65-70 nm in diameter with 'A' spikes of about 11 nm in length and breadth, at the 12 vertices on the icosahedral (Fig. 1 Left). Unless pre-fixed, viruses readily break down *in vitro* to give cores, about 55 nm in diameter, with 12 'B' spikes, about 8 nm long and 12 nm in diameter (Fig. 1 Right). Some treatments (shaking with butan-1-ol or incubation with 1.9 M $MgCl_2$) produce smooth subcores (Fig. 1 Center).

PHYSICOCHEMICAL AND PHYSICAL PROPERTIES

The physicochemical properties of the virions have not been determined.

NUCLEIC ACID

Fijiviruses have 10 dsRNA segments that are identified in Table 1 as genome Segment 1 (Seg1) to Segment 10 (Seg10) in order increasing electrophoretic mobility during PAGE. However, Seg2 and Seg3 of some viruses (NLRV and Mal de Rio Cuarto virus; MRCV), Seg8 and Seg9 (Oat sterile dwarf virus, OSDV) do not migrate in order of their Mr and may migrate in a reverse order during (1%) agarose gel electrophoresis (AGE). The individual genome segments have Mr of about 1.0 to 2.9 x 10^6 (1430-4391 bp) with a total genome Mr of 19.58 x 10^6 (28,699 bp, based on the complete RNA sequence analysis of NLRV: Table 2). The coding strands of each segment of MRCV, Maize rough dwarf virus (MRDV), Rice black streaked dwarf virus (RBSDV), OSDV, and NLRV, contain terminal conserved nt sequences (Table 1). Within the genus, only the 3'-terminal sequence ...GUC-3' is conserved. Adjacent to the conserved terminal oligonucleotide sequences, each genome segment possesses inverted repeats, which are several bases long, similar to those in phytoreovirus and oryzavirus RNAs, although the sequences involved differ in these other genera. Characteristic of the genus is the low G+C content of the genomic RNAs, mostly around 34-36%. The sizes and groupings of the 10 dsRNA species are characteristic and distinctive for the five groups of fijiviruses that are recognized.

Table 1: Terminally conserved oligonucleotide sequences of some fijiviruses.

Virus	Sequence
MRDV	5'-AAGUUUUUU----------------------------UGUC-3'
RBSDV	5'-AAGUUUUU--------------AGCUNN(C/U)GUC-3'
MRCV	5′-AAGUUUUU-----------------CAGCUNNNGUC-3′
FDV	5'-AAGUUUUU----------------CAGCNNNNGUC-3'
OSDV	5'-AACGAAAAAAA----------UUUUUUUUAGUC-3'
NLRV	5'-AGU-------------------------------------GUUGUC-3'

Proteins

Six polypeptides, numbered respectively I to VI (139, 126, 123, 111, 97 and 64 kDa), can be detected by SDS PAGE of purified MRDV. The B-spiked cores contain peptides I, II and III, while the smooth core contains peptides I and II. The B spikes should therefore be composed of peptide III. Peptides IV-VI form the outer capsid. During infection of most, possibly all fijiviruses, tubules about 90 nm in diameter accumulate in the cytoplasm. Sometimes these are incompletely closed and form scrolls. They are presumably composed of a non-structural protein whose function and genome segment assignment are unknown.

Three major proteins (130, 120, and 56 kDa) and three minor ones (148, 65, and 51 kDa) can be detected by SDS PAGE of purified virions of RBSDV. The 120 kDa protein is the "B" spike protein. Smooth subcore particles consist of 148, 130, and 65 kDa proteins. The 56 kDa protein is the major component of the outer capsid shell and the 51 kDa protein is a partial degradation of it. In NLRV protein, three major proteins (140, 135, and 65 kDa), three intermediate (160, 110, and 75 kDa), and one minor protein (120 kDa) can be resolved. The 135 kDa protein is the 'B' spike. The 65 kDa protein is the major component of the outer capsid shell and the 140 kDa protein is the major core protein. In addition to the above structural proteins, there is an 'A' spike but its protein has not yet been identified.

Lipids

Not known.

Carbohydrates

Not known.

Genome Organization and Replication

Genome organizations and coding assignments of fijiviruses are summarized in Table 2. Most of the genome segments are monocistronic (Fig. 2). Some segments possess two ORFs but expression of the second ORF has not been demonstrated *in vivo* in insect or plant cells. For viruses other than NLRV, replication occurs in the cytoplasm of phloem-related cells in association with viroplasms composed partly of fine filaments.

Table 2: Genome organization of the members of the genus *Fijivirus*.

dsRNA segment	Size (bp)	G+C content (%)	ORFs	Protein size (kDa)	§, ¥ : Protein Function (location)	Equivalent genome segment from other fijiviruses
Fiji disease virus, FDV						
Seg 1	4532	31.5	39-4451	170.6	Core, RNA polymerase	RBSDV Seg1 MRCV Seg1
Seg 2	3820	31.4	50-3631	137.0	Unknown	RBSDV Seg2 MRCV Seg3
Seg 3	3623	30.5	36-3536	135.5	Unknown	RBSDV Seg4 MRCV Seg2
Seg 4	3568	31.4	14-3454	133.2	Unknown	RBSDV Seg3 MRCV Seg4
Seg 5	3150	32.8	59-3067	115.3	Unknown	RBSDV Seg5
Seg 6	2831	32.5	77-2614	96.8	Unknown	RBSDV Seg6 MRCV Seg6
Seg 7	2194	31.8	42-1136 1189-2110	41.7 36.7	Unknown Unknown	MRDV Seg6 RBSDV Seg7
Seg 8	1959	32.4	25-1809	68.9	Unknown	RBSDV Seg8 MRCV Seg8 MRDV Seg7
Seg 9	1843	32.8	50-1057 1115-1741	38.6 23.8	Structural protein? Non-structural protein?	MRDV Seg8
Seg 10	1819	34.1	24-1691	63.0	Unknown	RBSDV Seg10 MRDV Seg10
Rice black streaked dwarf virus, RBSDV (Chinese isolate)						
Seg 1	4501#	32.1	35-4430	168.8	Core, RNA polymerase	NLRV Seg1 MRCV Seg1
Seg 2	3812#	33.9	45-3726	141.5	Unknown (possible core protein)	NLRV Seg3 FDV Seg2 MRCV Seg3
Seg 3	3572#	34.2	15-3455	132.0	Unknown (¥ possible major core protein	MRCV Seg4 FDV Seg4 NLRV Seg4
Seg 4	3617#	30.8	33-3543	135.6	Unknown (possible B-spike protein)	NLRV Seg2
Seg 5	3164#	37.6	15-2829	107.1	Unknown	Not known
Seg 6	2645#	38.4	81-2460	89.9	Unknown	RBSDV Seg6 FDV Seg6
Seg 7	2193	33.9 34.3#	42-1130 1183-2112	41.2 36.4	Nonstructural Tubular structure (TuP) Non-structural	MRDV Seg6 OSDV Seg7 NLRV Seg10*
Seg 8	1927 1936#	34.5 34.8#	25-1800	68.1	Core protein (possible NTP-binding†)	MRDV Seg7 OSDV Seg9 NLRV Seg7
Seg 9	1900	34.0 33.4#	52-1095 1160-1789	39.9 24.2	Non-structural, Viroplasm (ViP) Non-structural	MRDV Seg8 OSDV Seg10 NLRV Seg9
Seg 10	1801	36.2 36.6#	22-1698	63.3	Major outer shell	MRDV Seg10 OSDV Seg8 NLRV Seg8
Mal de Rio Cuarto virus, MRCV						
Seg 1	4501	32.3	38-4432	168.4	Core, RNA polymerase	RBSDV Seg1 FDV Seg1

						NLRV Seg1
Seg 2	3617	31.4	34-3549	134.4	Unknown (¥ possible outer shell B spike)	RBSDV Seg4 FDV Seg3 NLRV Seg2
Seg 3	3826	33.4	47-3727	141.7	Unknown (¥ possible major core protein)	RBSDV Seg2 FDV Seg2 NLRV Seg3
Seg 4	3566	32.7	16-3453	131.7	Unknown	RBSDV Seg3 FDV Seg4 NLRV Seg4
Seg 6	2638	37.7	80-2446	90	Unknown	RBSDV Seg6 FDV Seg6
Seg 8	1931	34.9	25-1800	68.3	(NTP-binding†)	MRDV Seg7 RBSDV Seg8 OSDV Seg9 NLRV Seg7
Maize rough dwarf virus, MRDV						
Seg 6	2193	34.6	42-1130	41.0	Unknown (¥ possible non-structural and TuP)	RBSDV Seg7 OSDV Seg7
			1183-2112	36.3	Unknown	NLRV Seg10*
Seg 7	1936	34.6	25-1800	68.1	Unknown (¥ possible core protein and NTP-binding†)	RBSDV Seg8 OSDV Seg9 NLRV Seg7
Seg 8	1900	34.2	52-1095	40.0	Unknown (¥ possible non-structural and ViP)	RBSDV Seg9 OSDV Seg10
			1160-1789	24.2	Unknown	NLRV Seg9
Seg 10	1802	36.5	23-1699	62.9	Unknown (¥ possible Major outer shell)	RBSDV Seg10 OSDV Seg8 NLRV Seg8
Oat sterile dwarf virus, OSDV						
Seg 7	1944	35.6	42-1148	42.0	Unknown (¥ possible non-structural and TuP)	MRDV Seg6 RBSDV Seg7
			1186-1863	30.0	Unknown	NLRV Seg10*
Seg 8	1874	33.8	20-1786	66.2	Unknown (¥ possible Major outer shell)	MRDV Seg10 RBSDV Seg10 NLRV Seg8
Seg 9	1893	34.6	15-1766	68.2	Unknown (¥ possible core protein and NTP-binding†)	MRDV Seg7 RBSD Seg8 NLRV Seg7
Seg 10	1761	34.5	51-998	35.7	Unknown (¥ possible non-structural and ViP)	MRDV Seg8 RBSDV Seg9
			1028-1612	22.7	Unknown	NLRV Seg9
Nilaparvarta lugens reovirus, NLRV						
Seg 1	4391	33.5	21-4349	165.9	Core, RNA polymerase (Pol)	MRCV Seg1
Seg 2	3732	33.1	22-3621	136.6	Outer shell, B spike	MRCV Seg2
Seg 3	3753	34.2	15-3686	138.5	Major core	MRCV Seg3
Seg 4	3560	33.8	76-3474	130	Unknown	MRCV Seg4
Seg 5	3427	39	202-3120	106.4	Unknown	
Seg 6	2970	36.8	150-2642	95.1	Unknown	
Seg 7	1994	34.1	41-1930	73.5	Core protein, NTP-binding†	OSDV Seg9 MRDV Seg7 RBSDV Seg8 MRCV Seg8
Seg 8	1802	35.3	7-1695	62.4	Major outer shell	OSDV Seg8 MRDV Seg10 RBSDV Seg10

dsRNA

Seg 9	1640	33.2	53-925	33.0	Non-structural (¥ possible ViP)	OSDV Seg10 MRDV Seg8
			982-1602	23.6	Non-structural	RBSDV Seg9
Seg 10*	1430	35.2	46-1341	49.4	Non-structural (¥ possible TuP)	OSDV Seg7 MRDV Seg6 RBSDV Seg7

*genome Seg 10 of NLRV does not contain a second ORF.
† NTP binding proteins of some other genera [eg. VP4 (Cap) of Blue tongue virus {*Orbivirus*}, or P5 (Cap) of Rice dwarf virus {*Phytoreovirus*} have guanylyltransferase and/or Mtr activities involved in cap formation].
§: protein structure/function: RNA polymerase = "Pol"; Capping enzyme = "Cap"; Virus structural Viral inclusion body or viroplasm matrix protein = "ViP". Virus tubule protein = "TuP".
¥ The probable function of some of the proteins that are uncharacterized may be indicated by the equivalence of the genome segments from which they are translated, to those of other virus species.

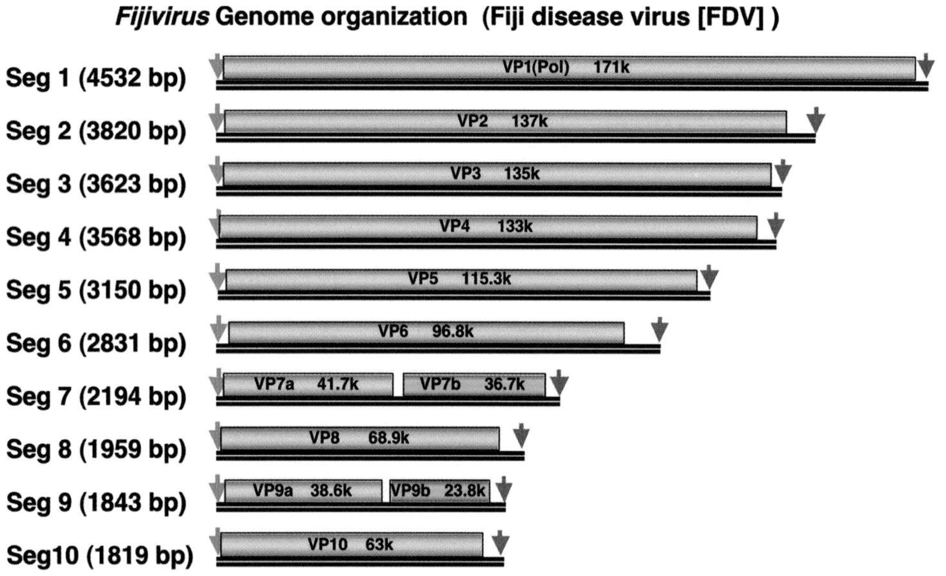

Figure 2: Genome organization of Fiji disease virus (FDV), containing 10 dsRNA segments, each contains an ORF with the exception of Seg7 and Seg9, that contain two ORFs. The green arrows indicate the upstream conserved terminal sequence (+ve 5'-AAGUUUUU.....), while the red arrows indicate the downstream conserved terminal sequence (+ve.....CGC-3').

Antigenic Properties

Proteins of some fijivirus species are serologically unrelated, but some proteins of other viruses (MRCV, MRDV, Pangola stunt virus (PaSV) and RBSDV) are distantly related.

Biological Properties

All the plant-infecting fijiviruses induce hypertrophy of the phloem (both expansion and multiplication of cells) leading to vein swellings and sometimes galls (enations or tumors) derived from phloem cells, especially on the backs of leaves. MRDV in maize induces longitudinal splitting of the roots. Other effects include the suppression of flowering, plant stunting, increased production of side shoots, and induction of a dark green coloring. In insect hosts, no particular tissue tropism or severe disease is recognized. Viruses are transmitted propagatively by delphacid planthoppers (Hemiptera, Delphacidae, e.g. *Perkinsiella, Laodelphax, Toya, Sogatella, Javesella, Ribautodelphax, Dicranotropis, Delphacodes, Sogatella, Unkanodes*). Following virus acquisition from infected plants, the latent period is about two weeks and leads to a lifelong capacity for virus transmission to plants. No transovarial transmission, or seed transmission of virus has been identified. Mechanical transmission from plant to plant can only be demonstrated with difficulty. Virus is spread by offsets in vegetatively propagated crops (e.g., pangolagrass and sugarcane). Viruses can over-winter in diapausing planthoppers, in certain weed species and in autumn-sown cereals.

Generally, fijiviruses are widespread in nature although apparently absent from North America and not reported from Africa, or confirmed from India. FDV has been reported from Australia and the Pacific islands. RBSDV occurs in Japan, Korea, and China. PaSV occurs in northern countries of South America, Oceania, Taiwan and northern Australia, and OSDV occurs in northern Europe. GDV has been found only in southern France. MRDV is found in Scandinavia and in areas bordering the northern and eastern Mediterranean. MRCV occurs in Argentina.

NLRV was found in the planthopper *Nilaparvata lugens*, which occurs in south-east Asia. Experimentally it infects a second hopper, *Laodelphax striatellus*. There is no evidence that NLRV can multiply in rice plants, a natural host of *N. lugens*, but the virus is transmitted from hopper to hopper through contaminated rice plants and moves through the phloem and/or xylem of rice plants once injected by the viruliferous hoppers.

List of Species Demarcation Criteria in the Genus

Fijivirus species can be clustered into "groups" of more closely related species. Thus, some species are distinct but others (RBDSV, MRDV, PaSV, MRCV) appear to be more closely related to each other than any is to other species in the genus *Fijivirus*. Further information about these viruses may necessitate a revision of their species status, for example MRDV and RBSDV may be considered close enough to constitute one species. However, they are retained as distinct species in accordance with previous practice.

For the family *Reoviridae* as a whole, the prime determinant for inclusion of virus isolates within a single species is "an ability to exchange (reassort) genome segments during co-infection, thereby exchanging genetic information and generating viable and novel progeny virus strains". However, data are not available that could extend this criterion to judgments about fijiviruses.

Within the genus, sequenced segments of MRDV, RBSDV, MRCV and FDV share the 5'-terminal conserved sequence 5'-AAGUUUUU... The 3'-terminal conserved nt sequence of ...CAGCUNNNGUC-3' is common among MRDV, MRCV, RBSDV, OSDV and NLRV, with the corresponding sequence of FDV differing in only one position. Values of nt and aa sequence relatedness among some species (for example MRDV and RBSDV, both in group 2) are much higher (>85% nt sequence identity among genome segments coding for major capsid proteins) than those among viruses from different groups (<55% identity) (Fig. 2). Between RBSDV or MRDV (group 2) and either OSDV or NLRV, there are detectable but low levels of aa sequence homology in some proteins. cDNA probes from some but not all of the genome segments, can cross-hybridize with corresponding segments among different viruses within group 2, whereas no cross-hybridization has been found between viruses species in different groups. NLRV does not have a counterpart to ORF2 present in the corresponding RBSDV Seg7, MRDV Seg6, and OSDV Seg7, and this may reflect its inability to replicate in plant hosts.

Members of a single species in the genus *Fijivirus* may be identified by:

- Capacity to exchange genetic material by genome segment re-assortment during dual infections, thereby producing viable progeny virus strains.
- Relatively high aa sequence homology. Members of different species have low aa sequence similarity (< 40% for counterparts corresponding to those encoded by RBSDV Seg7, Seg8, Seg9, and Seg10).
- Showing cross-hybridization of some segments using RNA or cDNA probes. Essentially, viruses within a species should have cross hybridization signals in all

genome segments as assessed by using cDNA probes under standard conditions. Members of different species will show a lack of cross-hybridization at high stringency. Some species (MRDV, RBSDV) share 94% nt sequence identity in genome Seg10, which encodes a major outer shell protein (highly conserved genome segment). The most practical and reliable criterion for distinguishing species is a lack of cross-hybridization using cDNA probes to less conserved genome segments. Hybridization using RBSDV Seg5 and Seg6 cDNA probes to detect the homologous sequences is more than 20 times more sensitive than hybridization using their counterparts from MRDV.

- Showing serological crossreactions within a species; virus species in different groups do not cross-react, those in group 2 do so to a limited extent that is dependent on the proteins being compared.
- Having differences in the near terminal conserved sequences (Table 1), although the 3'-terminal trinucleotide is identical in at least RBSDV, OSDV and NLRV.
- The identity or family of the plant host species, or the absence of a plant host, as well as the insect vector or insect host species may help to indicate virus species.

List of Species in the Genus

Species names are in green italic script; strain names and synonyms are in black roman script; tentative species names are in blue roman script. Sequence accession numbers [Seg1 to Seg10 indicate RNA segments 1 – 10], insect vector and host names { }, and assigned abbreviations ( ) are also listed.

Species in the Genus

Fijivirus group 1

Fiji disease virus

Fiji disease virus {*Perkinsiella saccharicida*, *P. vastatrix*, *P. vitiensis*: Graminae}	[Seg1: AY029520, Seg2: AF049704, Seg3: AF359556, Seg4: AF049705, Seg5: AY029521, Seg6: AF356083, Seg8: AY297693, Seg9: AF050086, Seg10: AY297694]	(FDV)

Fijivirus group 2

Maize rough dwarf virus

Maize rough dwarf virus {*Ribautodelphax notabilis*: Graminae}	[Seg6: X55701, Seg7: L76562, Seg8: L76561, Seg10: L76560]	(MRDV)

Mal de Rio Cuarto virus

Mal de Rio Cuarto virus {*Delphacodes kuscheli:* Graminae}	[Seg1: AF499925, Seg2: AF499926, Seg3: AF499928, Seg4: AF395873, Seg6: AF499927, Seg8: AF395872]	(MRCV)

Pangola stunt virus

Pangola stunt virus {*Sogatella furcifera S. kolophon*: Graminae}		(PaSV)

Rice black streaked dwarf virus

Rice black streaked dwarf virus {*Laodelphax striatellus*, *Ribautodelphax albifascia*, *Unkanodes sapporona*: Graminae}	[Seg1: AJ294757, Seg2 AJ409145, Seg3: AF432355, Seg4: AJ409146, Seg5: AJ409147, Seg6: AJ409148, Seg7: S63917, Seg8: S63914, Seg9: AB011403, Seg10: D00606]	(RBSDV)

Fijivirus group 3

Oat sterile dwarf virus

Oat sterile dwarf virus {*Javesella pellucida*, *J. discolor*, *J. dubia*, *J. obscurella*, *Dicranotropis hamata*: Graminae}	[Seg7: AB011024, Seg8: AB011025, Seg9: AB011026, Seg10: AB011027]	(OSDV)

Fijivirus group 4

Garlic dwarf virus

Garlic dwarf virus {unknown: Liliaceae}		(GDV)

Fijivirus group 5

Nilaparvata lugens reovirus

Nilaparvata lugens reovirus [Seg1: D49693, Seg2: D49694, Seg3: D49695, Seg4: D49696, Seg5: D49697, Seg6: D49698, Seg7: D49699, Seg8: D26127, Seg9: D49700, Seg10: D14691] (NLRV)
{*Nilaparvata lugens, Laodelphax striatellus*: no plant hosts}

TENTATIVE SPECIES IN THE GENUS

None reported.

PHYLOGENETIC RELATIONSHIPS WITHIN THE GENUS

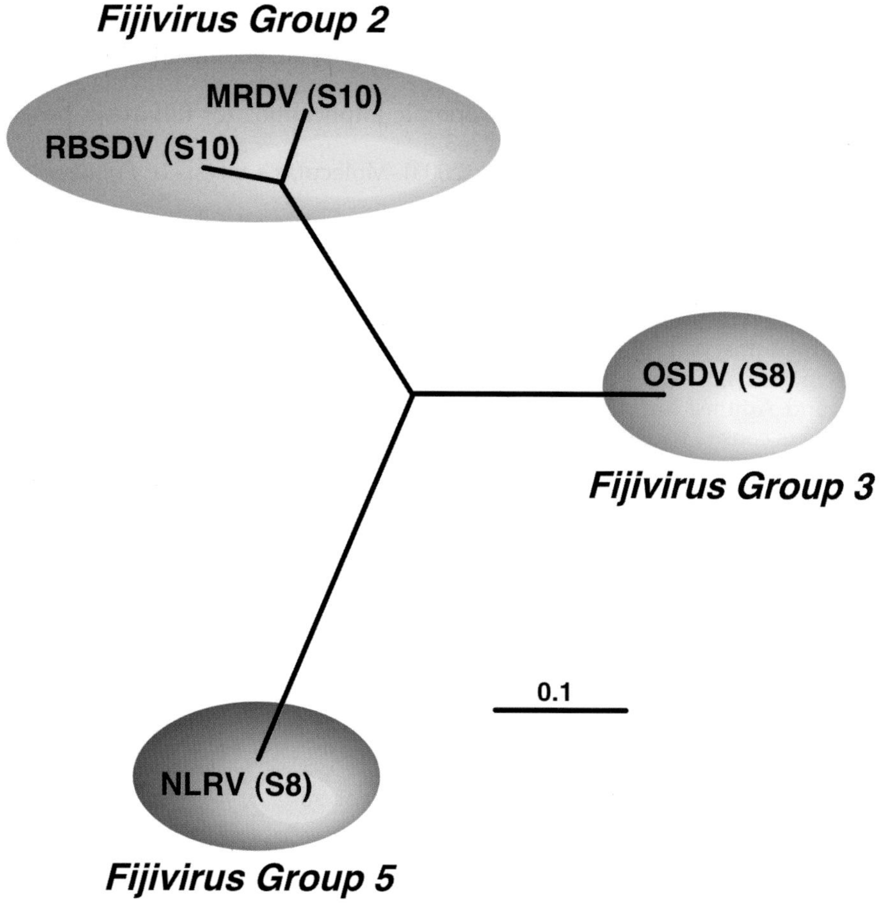

Figure 3: Phylogenetic tree produced by comparison of the nt sequences for genome segments that code for the *Fijivirus* major outer shell. (Maize rough dwarf virus, MRDV (S10) [L76560]; Rice black streaked dwarf virus, RBSDV (S10) [D00606]; Oat sterile dwarf virus, OSDV (S8) [AB011025]; Nilaparvata lugens reovirus, NLRV (S8) [D26127]). Alignment and Neighbor-joining tree produced with ClustalX. Tree drawn with TreeView. Note: Core protein sequences showed a similar tree (data not shown).

DERIVATIONS OF NAMES

Fiji: from name of country where virus was first isolated.

REFERENCES *FIJIVIRUS*

Bai, F.W., Yan, J., Qu, Z.C., Zhang, H.W., Xu, J., Ye, M.M. and Shen, D.L. (2002). Phylogenetic analysis reveals that a dwarfing disease on different cereal crops in China is due to rice black streaked dwarf virus (RBSDV). *Virus Genes*, **25**, 201-206.

Boccardo, G. and Milne, R.G. (1975). The maize rough dwarf virion I. Protein composition and distribution of RNA in different virus fractions. *Virology*, **68**, 79-85.

Distéfano, A.J., Conci, L.R., Muñoz Hidalgo, M., Guzmán, F.A., Hopp, H.E. and del Vas, M. (2002). Sequence analysis of genome segments S4 and S8 of Mal de Río Cuarto virus (MRCV): evidence that this virus should be a separate *Fijivirus* species. *Arch. Virol.*, **147**, 1699-1709.

Distéfano, A.J., Conci, L.R., Muñoz Hidalgo, M., Guzmán, F.A., Hopp, H.E. and del Vas, M. (2003). Sequence and phylogenetic analysis of genome segments S1, S2, S3 and S6 of Mal de Rio Cuarto virus, a newly accepted *Fijivirus* species. *Virus Res.*, **92**, 113-121.

Isogai, M., Uyeda, I. and Lindsten, K. (1998). Taxonomic characteristics of fijiviruses based on nucleotide sequences of the oat sterile dwarf virus genome. *J. Gen. Virol.*, **79**, 1479-1485.

Isogai, M., Uyeda, I. and Lee, B. (1998). Detection and assignment of proteins encoded by rice black streaked dwarf fijivirus S7, S8, S9 and S10. *J. Gen. Virol.*, **79**, 1487-1494.

Lot, H., Delecolle, B., Boccardo, G., Marzachì, C. and Milne, R.G. (1994). Partial characterization of reovirus-like particles associated with garlic dwarf disease. *Plant Pathol.*, **43**, 537-546.

Marzachì, C., Antoniazzi, S., d'Aquilio, M. and Boccardo, G. (1996). The double-stranded-RNA genome of maize rough dwarf virus contains both mono and dicistronic segments. *Eur. J. Plant Pathol.*, **102**, 601-605.

Marzachì, C., Boccardo, G., Milne, R.G., Isogai, M. and Uyeda, I. (1995). Genome structure and variability of fijiviruses. *Sem. Virol.*, **6**, 103-108.

McMahon, J.A., Dale, J.L. and Harding, R.M. (1999). Taxonomic implications for fijiviruses based on terminal sequences of Fiji disease virus. *Arch. Virol.*, **144**, 2259-2263.

McQualter, R.B., Smith, G.R., Dale, J.L. and Harding, R.M. (2003). Molecular analysis of *Fiji disease virus* genome segments 1 and 3. *Virus Genes*, **26**, 283-289.

Nakashima, N., Koizumi, M., Watanabe. H. and Noda, H. (1996). Complete nucleotide -sequence of the Nilaparvata-lugens reovirus - a putative member of the genus Fijivirus. *J. Gen. Virol.*, **77**, 139-146.

Soo, H.M., Handley, J.A., Maugeri M.M., Burns, P., Smith, G.R., Dale, J.L. and Harding, R.M. (1998). Molecular characterization of Fiji disease virus genome segment 9. *J. Gen. Virol.*, **79**, 3155-3161.

Zhang, H.M., Chen, J.P. and Adams, M.J. (2001). Molecular characterization of segments 1 to 6 of Rice black streaked dwarf virus from China provides the complete genome. *Arch. Virol.*, **146**, 2331-2339.

Zhang, H., Chen, J., Lei, J. and Adams, M.J. (2001). Sequence analysis shows that a dwarfing disease on rice, wheat and maize in China is caused by Rice black streaked dwarf virus. *Eur. J. Plant Pathol.*, **107**, 563-567.

Contributed By

Milne, R.G., del Vas, M., Harding, R.M., Marzachì, R. and Mertens, P.P.C.

GENUS PHYTOREOVIRUS

Type Species *Wound tumor virus*

DISTINGUISHING FEATURES

Phytoreoviruses particles have icosahedral symmetry with a distinctive angular appearance and possess 12 dsRNA species. They are transmitted by cicadellid leafhoppers to susceptible plant species, replicating in both hosts and vectors.

VIRION PROPERTIES

MORPHOLOGY

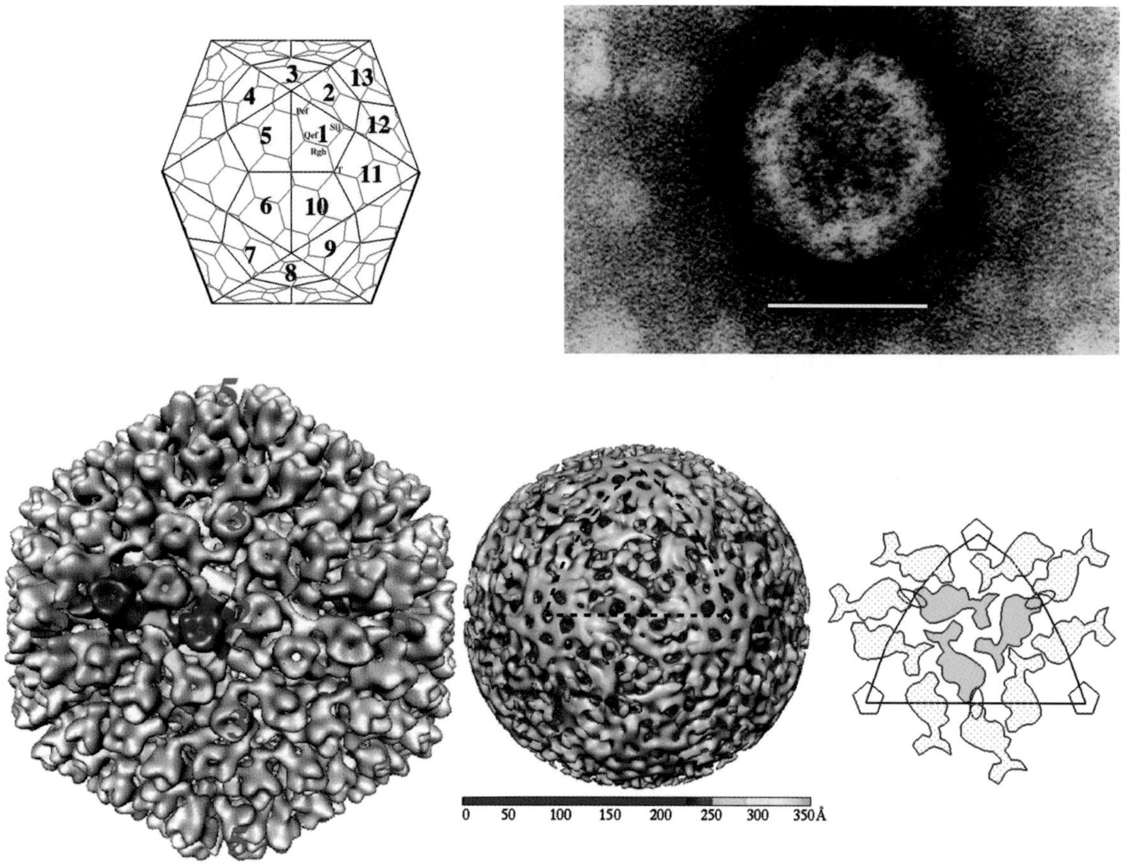

Figure 1: (Top left) Schematic diagram representing a T=13 capsid structure. (Top right) Negative contrast electron micrograph of Rice gall dwarf virus (RGDV) particles, negatively stained with phosphotunstic acid. The bar represents 50 nm. (Bottom left) Electron cryo-microscopic image and 25Å resolution three-dimensional structure of the double-shelled Rice dwarf virus (RDV). (Bottom center) Inner shell computationally extracted with 59 nm diameter. It exhibits T=1 lattice. Dashed triangle designates one triangular face of the icosahedron. (Bottom right) Schematic diagram of fish-shaped density distribution within a triangle in a T=1 lattice. (Courtesy of Hong Zhou and Wah Chiu, from Lu *et al*, 1998).

Virions of Rice dwarf virus (RDV) are icosahedral, appear to be double-shelled and ~70 nm in diameter (Fig. 1). The outer layer of RDV contains 260 trimers of P8 (46 kDa): total of 780 molecules, arranged with T=13 *l* symmetry (Fig. 1). The relative location of the neighboring capsomers on the icosahedral particle, is such that they form pentameric or hexameric rings. The inner capsid layer is reported to be a complete protein shell, composed of 60 dimers of P3 (114 kDa), a total of 120 molecules, arranged with a suggested T=1 icosahedral symmetry (Fig. 1). The outer capsid P8 trimers bind more tightly at the threefold positions of the single-layered core. The RDV particle structure appears to be comparable to that of 'core' or double layered particles of some other genera (*Orbivirus* and *Rotavirus* respectively). Ordered structures are visible in the periphery of the RNA region.

Wound tumor virus (WTV) is reported to possess three protein shells, including an outer amorphous layer, an internal layer of distinct capsomers, and a smooth core that is about 50 nm in diameter but lacking spikes.

PHYSICOCHEMICAL AND PHYSICAL PROPERTIES

The Mr of phytoreoviruses is about 75 x 10^6. The virion S_{20}, w is about 510. The optimal stability of particles is at pH 6.6. The buoyant density of RDV is 1.39-1.42 and the virion is unstable losing P8 in CsCl. CCl_4 removes P2 from the RDV virion.

NUCLEIC ACID

Phytoreoviruses have 12 genome segments of linear dsRNA. Seg1 to Seg12 are named according to their migration during PAGE. However, their relative sizes based on RNA sequence data indicate that Seg4 and Seg5, or Seg9 and Seg10, may migrate in the reverse order during agarose gel electrophoresis (Table 1). The RNA constitutes about 22% of the virion dry weight. The dsRNA Mr is in the range 0.3 to 3.0 x 10^6, with characteristic sizes for each virus. For WTV Seg4=2565bp; Seg5=2613bp; Seg6=1700bp; Seg7=1726bp; Seg8=1472bp; Seg9=1382bp; Seg10=1172bp; Seg11=1128bp; Seg12=851bp. G+C content is 38-44% and 41-48% for the genomic segments of WTV and RDV respectively. The positive strand of each genome segment, of all viruses in the genus, contains the conserved sequence; 5'GG(U/C)A---UGAU3' except for RDV Seg9 which has 5'GGUA---CGAU3'. These genus-specific terminal sequences are situated adjacent to inverted repeats, which are 6-14 bases long. These sequences differ for each RNA segment. Individual isolates of RDV can frequently be distinguished by electrophoretic profiles of at least one of the twelve genomic segments in PAGE. RDV particles encapsidate the genomic dsRNA in supercoiled form.

PROTEINS

Phytoreoviruses have 6 to 7 structural proteins in the range 45 to 160 kDa. RDV has 6 structural proteins (P1(Pol), P2, P3, P5(Cap), P7, and P8). For WTV the seven CPs are organized in three shells consisting of an amorphous outer layer of 2 CPs, an inner shell of 2 CPs and a core of three CPs. Protein constitutes about 78% of the particle dry weight. Removal of the outer shell is not required for activation of the virus transcriptase and associated enzymes. Removal of RDV P2 abolishes the ability to infect vector cell monolayers but virus particles without P2 retain viral transcriptase activity and can infect vector insects by an injection method. P1 is the transcriptase/polymerase and binds to genomic dsRNA. P7 has non-specific nucleic acid binding activity. P3 binds to P3, P7, and P8. P7 binds to P1 and P8. P5 is probably a guanylyl transferase and has GTP, ATP and UTP binding activities. P3 and P8 form virus-like particles in transgenic rice plants. Of the non-structural proteins, Pns11 has nonspecific nucleic acid binding activity and Pns 12 can be phosphorylated.

LIPIDS

None known.

CARBOHYDRATES

None known.

GENOME ORGANIZATION AND REPLICATION

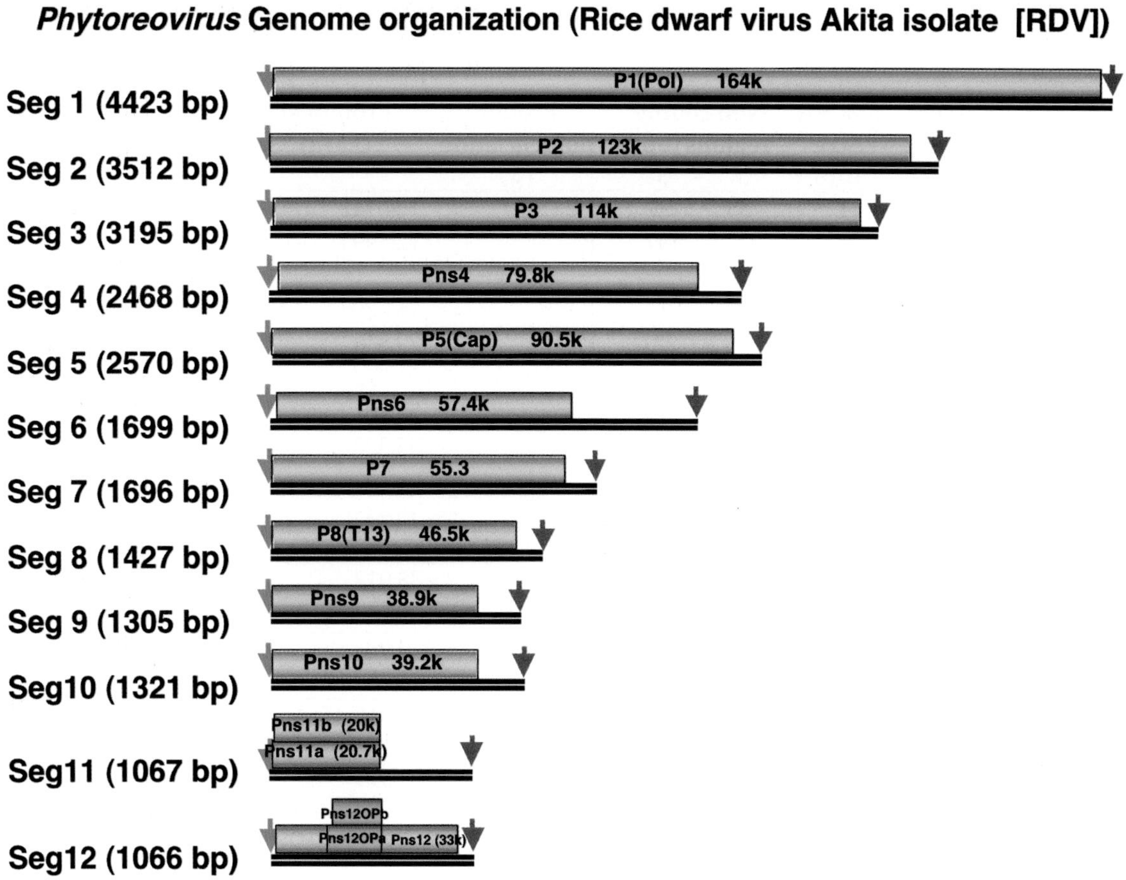

Figure 2: Genome organization of Rice dwarf virus Akita isolate (RDV), containing 12 dsRNA segments, each contains an ORF with the exception of Seg11 and Seg12, that contain two ORFs. The green arrows indicate the upstream conserved terminal sequence (+ve 5'-GGU/CA.....), while the red arrows indicate the downstream conserved terminal sequence (+ve.....U/CGAU-3').

The coding strand of each dsRNA has a single ORF, except for Seg11 and Seg12 of RDV (Fig. 2), Seg9 of RGDV and Seg9 of WTV. RDV Seg11 has two in-frame initiation codons, thus resulting in two ORFs. RDV Seg12, Rice gall dwarf virus (RGDV) Seg9 and WTV Seg9 possess a second, small out-of-frame and over-lapping ORF, downstream within the major ORF. No evidence has yet been obtained for the expression of this second ORF. Five structural and five NS WTV proteins have been assigned to their respective genome segments. RDV Seg1 encodes the putative transcriptase. Genus-specific and segment-specific sequence motifs appear to be necessary for successful replication, translation and encapsidation. Laboratory strains having internal deletions in some segments, but intact termini, replicate and compete favorably with wild-type virus, although the proteins expressed are aberrant, and the ability of the viruses to be transmitted by vectors may be lost. Virus replication occurs in the cytoplasm of infected cells in association with viroplasms. WTV and RGDV are confined to phloem tissues of the plant host, whereas RDV can also multiply elsewhere.

Table 1: Genome organization of Rice dwarf virus, Akita isolate, listing the dsRNA segments, with their size (bp), and corresponding proteins with name, size (*estimated by SDS PAGE), and function and/or location.

Genome segment (Seg#)	dsRNA Size (bp)	Non-coding regions (bp) 5' - 3'	Protein nomenclature (§)	Protein size (kDa)*.		Function and location (Number per particle)
Seg1	4423	35 - 53	P1 (Pol)	164,142	*170	Core, RNA polymerase.
Seg2	3512	14 - 147	P2	122,994	*130	"Outer capsid", essential for vector transmission
Seg3	3195	38 - 97	P3	114,298	*110	Major Core (120)
Seg4	2468	63 - 221	Pns4	79,836	*83	Non-structural
Seg5	2570	26 - 138	P5 (Cap)	90,532	*89	Core, Guanylyltransferase
Seg6	1699	48 - 121	Pns6	57,401	*56	Non-structural
Seg7	1696	25 - 150	P7	55,287	*58	Core, Nucleic acid binding protein"
Seg8	1427	23 - 138	P8 (T13)	46,483	*43	Major outer capsid (780) (trimer)
Seg9	1305	24 - 225	Pns9	38,912	*49	Non-structural
Seg10	1321	26 - 233	Pns10	39,196	*35	Non-structural
Seg11	1067	29 - 492	Pns11a	19,988	*23	Non-structural (nucleic acid binding protein)
		5 - 492	Pns11b	20,759	*24	
Seg12	1066	41 - 86	Pns12	33,916	*34	Non-structural
		312 - 475	Pns12OPa	10,551	*8	
		336 - 475	Pns12OPb	9,597	*7	

§: protein structure/function. *: size determined by SDS PAGE. RNA polymerase = "(Pol)"; capping enzyme = "(Cap)": structural protein arranged with T= 13 icosahedral symmetry = "(T13)".

ANTIGENIC PROPERTIES

The three recognized phytoreoviruses are antigenically distinct. Epitopes representing the outer surface are unrelated to each other, while the inner surface epitopes of the capsid of RDV and RGDV will cross react.

BIOLOGICAL PROPERTIES

Plant hosts are either dicotyledonous (WTV), or graminaceous (RDV and RGDV). WTV was originally identified in northeastern USA in the leafhopper *Agalliopsis novella*. The virus was recently found in New Jersey USA in a single periwinkle (*Catharanthus*) plant set out as bait for mycoplasmas in a blueberry (*Vaccinium*) field. The experimental plant host range of WTV is wide and encompasses many dicotyledonous. plants The name of this virus derives from the fact that infected plants develop phloem-derived galls (tumors) at wound sites, notably at the emergence of side roots.

RDV and RGDV have narrow and overlapping host ranges. RDV causes severe disease in rice crops in south-east Asia, China, Japan and Korea, Nepal and the Philippines. RGDV has been reported in Thailand, Malaysia and China. RDV induces white flecks and streaks on leaves, with stunting and excessive production of side shoots. RDV is the only plant reovirus that is not limited to the phloem. Plants infected with RDV are stunted and fail to bear seeds. Since the virus is widespread among rice plants in southern China and other Asian countries, it is considered likely to be the cause of a significant overall reduction in rice production. RDV does not provoke enlargement or division of infected cells and does not induce galls, enations, or tumors. RGDV was found in rice field in Thailand and induces stunting, shoot proliferation, a dark green color and enations in rice.

Phytoreoviruses induce no marked disease in the insect vectors. Virus replication occurs in the cytoplasm of infected cells in association with viroplasm. In the vector, there are no particular tissue tropisms. However, RDV induces abnormalities in fat body cells and mycetocytes. They are all transmitted propagatively by cicadellid leafhoppers (Hemiptera, Cicadellidae, e.g., *Agallia*, *Agalliopsis*, *Nephotettix*, and *Recilia*). Virus is acquired from plants shortly after feeding. The latent period in leafhoppers is about 10-20 days. Thereafter, infected insects have a lifelong ability to transmit virus to plants. Phytoreoviruses are also transmitted transovarially in their insect vectors. Experimental data suggest that, phytoreoviruses are not mechanically transmissible from plant to plant. No seed transmission has been reported.

List of Species Demarcation Criteria in the Genus

In common with the other genera within the family *Reoviridae*, it has been agreed that the prime determinant for inclusion of virus isolates within a single virus species will be "ability to exchange (reassort) genome segments during co-infection, thereby exchanging genetic information and generating viable and novel progeny virus strains". Data providing direct evidence of segment reassortment between isolates is very limited and other techniques are normally used to examine the level of similarity that exists between different isolates.

- The ability to exchange genetic material by genome segment reassortment during dual infections, thereby producing viable progeny virus strains. RDV isolates from Japan, China, and Philippines can exchange genomic segments. Exchange between RDV, WTV, RGDV has not been tested.
- Nucleotide sequence identity. So far, nt sequence identities among RDV isolates including those from different countries are more than 90%. Some isolates have deletions (Chinese strain of RDV Seg11 (U36568) compared to Japanese RDV-Akita Seg11 (D10249)) or duplications (RDV-P Seg12). The most extensive comparisons of Seg12 among 12 RDV isolates from five countries showed identities of 94.7 to 98.7% (Fig. 2 and Table 3).
- Protein sequence homology. Protein sequence homology between species is less than 56% and within species is more than 80%. These data indicate that high levels of serological cross reaction would be detected within species by ELISA complement fixation (CF), or agar gel immunodiffusion (AGID), using either polyclonal sera, or monoclonal antibodies against conserved antigens. Distinct species may show significantly lower levels of cross reaction. However, such serological methods are not routinely used to compare these viruses.
- Conserved terminal oligonucleotide-sequences. The tetranucleotides at either end of the genome segments are highly conserved within the genus. Two additional nt flanking the 5'-terminal tetranucleotide on the plus-strand are common within species.
- Cross hybridization (northern or dot blots, with probes made from viral RNA or cDNA using conditions designed to detect >80% homology). RNA-RNA hybridization detects all the segments within a species.
- Analysis of "electropherotype" by agarose gel electrophoresis but not by PAGE.
- Identification of host plant species; dicotyledons (WTV), or the family Graminae (RDV and RGDV).

List of Species in the Genus

Species names are in green italic script; strain names and synonyms are in black roman script; tentative species names are in blue roman script. Sequence accession numbers [Seg1 to Seg12 indicate RNA segments 1 – 12], insect vector and host names { }, and assigned abbreviations ( ) are also listed.

Species in the Genus

Rice dwarf virus

Rice dwarf virus [Seg1: D90198, Seg2: D00608, (RDV)

{: *Nephotettix cincticeps, N. nigropictus, Recilia dorsalis*: Graminae} — Seg3: D00607, Seg4: U36562, Seg5: D90033, Seg6: M91653, Seg7: D00639, Seg8: D00536, Seg9: D00465, Seg10: D00473, Seg11: D10249, Seg12: D90200]

Rice gall dwarf virus

Rice gall dwarf virus — [Seg2: D86439, Seg3: D13774, Seg5: D76429, Seg8: D13410, Seg9: D01047, Seg10: D13411, Seg11: AB030009] — (RGDV)
{*Nephotettix cincticeps, N. nigropictus, N. virescens, N. malayanus, Recilia dorsalis*: Gramineae}

Wound tumor virus

Wound tumor virus (34) — [Seg4: M24117, Seg5: J03020, Seg6: M24116, Seg7: X14218, Seg8: J04344, Seg9: M24115, Seg10: M24114, Seg11: X14219, Seg12: M11133] — (WTV)
{*Agallia constricta, A. quadripunctata, Agalliopsis novella:* many dicotyledons}

TENTATIVE SPECIES IN THE GENUS

Tobacco leaf enation virus* — (TLEP)
{tobacco: dicotyledon}

*first phytoreovirus isolated from Africa

PHYLOGENETIC RELATIONSHIPS WITHIN THE GENUS

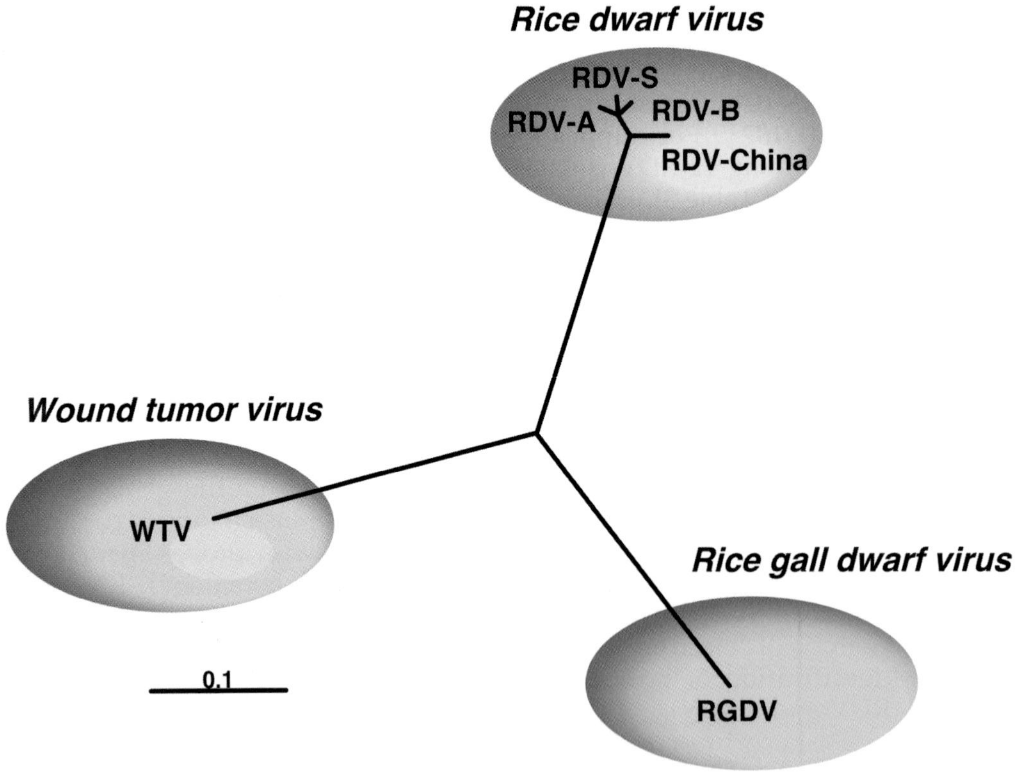

Figure 3: Phylogenetic tree of phytoreoviruses. Constructed using the RNA sequences of genome Seg8 from the following accession numbers: Rice dwarf virus A (RDV-A) (D10219), RDV-B (D00536), RDV-S (D13773), RDV-China (U36565); Wound tumor virus (J04344); Rice gall dwarf virus (D13410). Sequences were aligned and a Neighbor-joining tree was prepared using Clustal X (Thompson et al. (1997). *Nucl. Acids Res.*, **25**, 4876-4882), and drawn with TreeView 1.5 (Page (1996). *Comp. Appl. Biosci.*, **12**, 357-358).

Table 3: Percentage nucleotide differences in genome Seg8, from different phytoreoviruses.

Phytoreovirus Seg8 – percentage nucleotide differences						
Virus/virus	RDV				RGDV	WTV
Rice dwarf virus (RDV-B)	0.0	1.8	2.4	5.2	44.8	47.4
Rice dwarf virus (RDV-S)		0.0	2.7	5.2	45.0	47.1
Rice dwarf virus (RDV-A)			0.0	5.5	44.4	47.4
Rice dwarf virus (RDV-China)				0.0	45.4	46.5
Rice gall dwarf virus (RGDV)					0.0	44.9
Wound tumor virus (WTV)						0.0

DERIVATIONS OF NAMES

Phyto: from Greek phyton, "plant".

Reo: sigla from *r*espiratory *e*nteric *o*rphan, due to the early recognition that the viruses caused respiratory and enteric infections, and the (incorrect) belief that they were not associated with disease, hence they were considered "orphan" viruses.

REFERENCES

Lee, B.C., He, Y.K., Murao, K., Isogai, M., Dahal, G. and Uyeda, I. (1997). Phylogenetic relationships between rice dwarf phytoreovirus isolates from five different countries. *Europ. J. Pl. Path.*, **103**, 493-499.

Lu, G., Zhou, Z.H., Baker, M. L., Jakana, J., Cai, D., Wei, X., Chen, S., Gu, X. and Chiu, W. (1998). Structure of double-shelled Rice dwarf virus. *J. Virol.*, **72**, 8541-8549.

Murao, K., Uyeda, I., Ando, Y., Kimura, I., Cabauatan, P.Q. and Koganezawa, H. (1996). Genomic rearrangement in genome segment-12 of rice dwarf phytoreovirus. *Virology*, **216**, 238-240.

Omura, T. (1995). Genomes and primary protein structures of phytoreoviruses. *Sem. Virology*, **6**, 97-102.

Omura, T., Morinaka, T., Inoue, H. and Y. Saito. (1982). Purification and some properties of rice gall dwarf virus, a new phytoreovirus. *Phytopathology*, **72**, 1246-1249.

Rey, M.E.C., D'Andrea, E., Calver-Evers, J., Paximadis, M. and Boccardo, G. (1999). Evidence for a phytoreovirus associated with tobacco exhibiting leaf curl symptoms in South Africa. *Phytopathology*, **89**, 303-307.

Suzkuki, N. (1995). Molecular analysis of the rice dwarf virus genome. *Sem. Virology*, **6**, 89-95.

Suzuki, N., Sugawara, M. and T. Kusano. (1992). Rice dwarf phytoreovirus segment S12 transcript is tricistronic *in vitro*. *Virology*, **191**, 992-995.

Tomaru, M., Maruyama, W., Kikuchi, A., Yan, J., Zhu, Y. F., Suzuki, N., Isogai, N., Oguma, Y., Kimura, I. and Omura, T. (1997). The loss of outer capsid protein P2 results in nontransmissibility by the insect vector of rice dwarf phytoreovirus. *J. Virol.*, **71**, 8019-8023.

Uyeda, I., Ando, Y., Murao, K. and Kimura, I. (1995). High resolution genome typing and genomic reassortment events of rice dwarf Phytoreovirus. *Virology*, **212**, 724-727.

Wu, B., Hammar, L., Xing, L., Markarian, S., Yan, J., Iwasaki, K., Fujiyoshi, Y., Omura, T. and Holland, C. (2000). Phytoreovirus T=1 core plays critical roles in organizing the outer capsid of T=13 quasi-equivalence. *Virology*, **271**, 18-25.

Yan, J., Tomarj, M., Takahashi, A., Kimura, I., Hibino, H. and Omura, T. (1996). Protein encoded by genome segments 2 of rice dwarf phytoreovirus is essential for virus-infection. *Virology*, **224**, 539-541.

Zhou, Z. H., Baker, M. L., Jiang, W., Dougherty, M., Jakana, J., Dong, G., Lu, G. and Chiu W. (2001) Electron cryomicroscopy and bioinformatics suggest protein fold models for rice dwarf virus. *Nat. Struct. Biol.*, **8**, 868-873.

Zhu, Y. F., Hemmings, A. M., Iwasaki, K., Fujiyoshi, Y., Zhong, B. X., Yan, J., Isogai, M. and Omura, T. (1997) Details of the arrangement of the outer capsid of rice dwarf phytoreovirus, as visualized by two-dimensional crystallography. *J. Virol.*, **71**, 8899-8901.

CONTRIBUTED BY

Omura, T. and Mertens, P.P.C.

GENUS ORYZAVIRUS

Type Species *Rice ragged stunt virus*

VIRION PROPERTIES

MORPHOLOGY

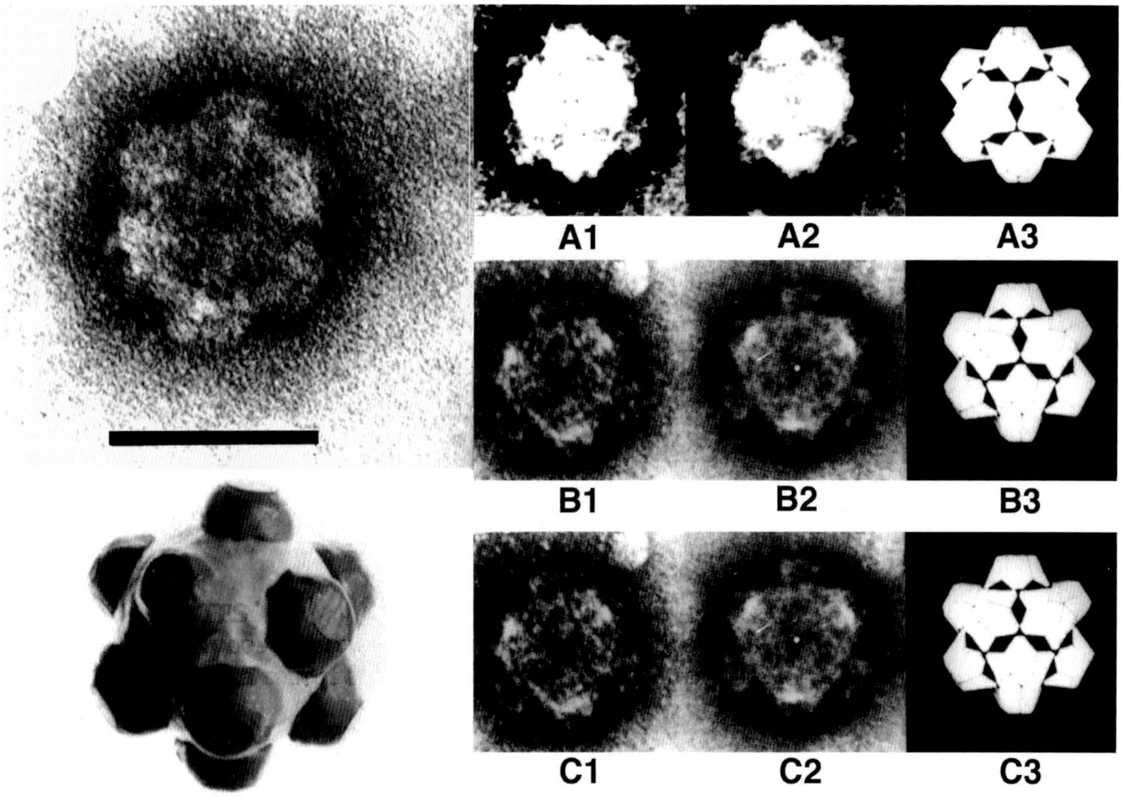

Figure 1: (Top left) Electron micrograph of Rice ragged stunt virus (RRSV) particles (Courtesy of R.G. Milne); (Bottom left) Schematic of RRSV particle; (Right panel) micrographs of the virus showing 2-, 3- and 5-fold symmetries (A1, B1 and C1, respectively) images of the same rotated by increments of 180° (A2), or 120° (B2), or 72° (C2) and proposed models of the 2-, 3- and 5-fold symmetries (A3, B3 and C3 respectively); (Courtesy of E. Shikata). The bar represents 50 nm.

Intact Rice ragged stunt virus (RRSV) particles appear to be icosahedral in symmetry and double shelled. It has a particle diameter in the range of 75-80 nm and surface 'A-spikes' (approximately 10-12 nm wide and 8 nm in length), attached to the end of 'B-spikes' situated at the 5 fold axes of the viral core. The subviral or 'core' particles have an estimated diameter of 57-65 nm (Fig. 1) and possess 12 "B"-type spikes, 8-10 nm in height, 23-26 nm wide at the base and 14-17 nm wide at the top. In negatively stained preparations of RRSV, "B-spiked subviral" particles have been seen but intact double-shelled particles are not seen without pretreatment with fixative. Echinochloa ragged stunt virus (ERSV) particles are slightly larger than RRSV particles.

PHYSICOCHEMICAL AND PHYSICAL PROPERTIES

RRSV particles sediment as one component and are stable at pH 6.0-9.0. They are stable in 0.1M $MgCl_2$. The B spikes dissociate from the core particle in 0.5M $MgCl_2$ and the entire particle is disrupted in 2M $MgCl_2$. The particles retain infectivity after 7 days at 4°C and after 10 min at 50°C but lose their infectivity after 10 min at 60°C. They retain infectivity after 3 cycles of freezing and thawing. The particles contain an RdRp.

NUCLEIC ACID

The oryzavirus genome consists of 10 linear dsRNA segments. The genomes of RRSV and ERSV have similar sizes and segment profiles (RRSV Mr 18.15 x 10^6 (26,066 bp); ERSV Mr 17.78 x 10^6) with segments ranging in size from 1,162 to 3,849 bp. The genomic dsRNAs are termed Seg1 to Seg10, in order of increasing electrophoretic mobility in 7.5% polyacrylamide gels. The entire genome of RRSV has recently been sequenced; the Seg4 and Seg10 segments are larger than they appear from migration in polyacrylamide gels, suggesting that they may migrate in the position 3 and 9 respectively during agarose gel electrophoresis (AGE). The conserved terminal sequences of the ERSV genome segments are identical to those of RRSV (5'-GAUAAA...(G)GUGC-3'). The RRSV and ERSV conserved terminal sequences differ from those of phytoreoviruses or fijiviruses. RRSV RNAs hybridize weakly with their counterparts in ERSV but not with segments of the fijivirus Rice dwarf virus (RDV).

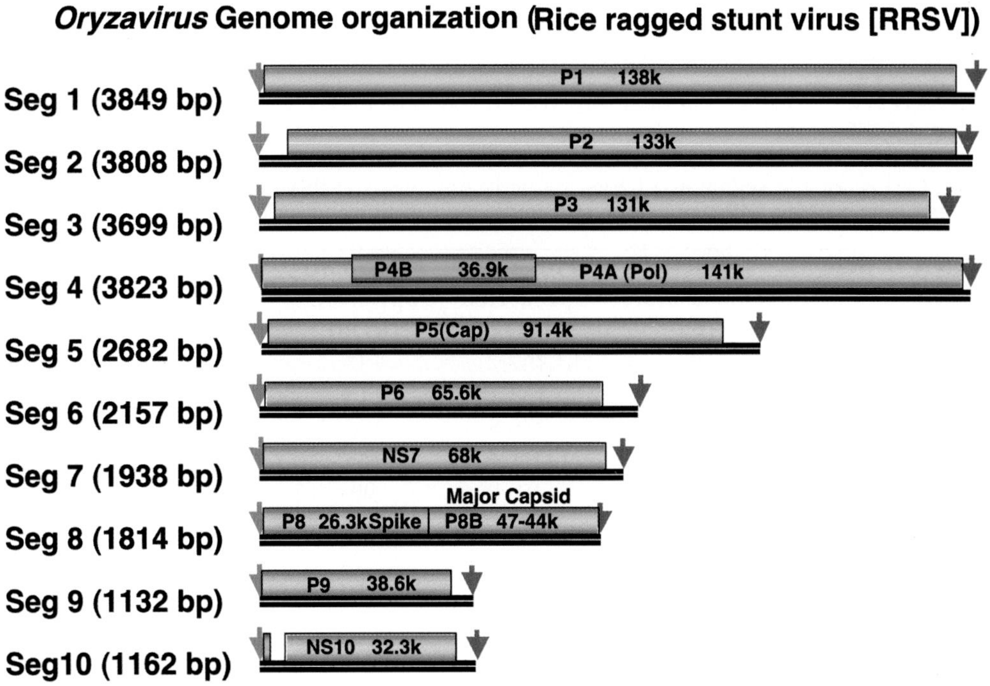

Figure 2: Genome organization of Rice ragged stunt virus (RRSV), containing 10 dsRNA segments, each contain an ORF, except Seg4, which contains two ORFs of which one encodes an RdRp. At least 6 segments encode structural proteins to make up the complex virus particle and two encode for non-structural (NS) proteins. The green arrows indicate the upstream conserved terminal sequence (+ve 5'-GAUAAA.....), while the red arrows indicate the downstream conserved terminal sequence (+ve.......(G)GUGC-3').

PROTEINS

RRSV particles are composed of five major, highly immunoreactive structural proteins, with estimated sizes of 33, 39, 43, 70 and 120 kDa, and at least five minor structural proteins (49, 60, 76, 90 and 94 kDa). Three more proteins (31, 63 and 88 kDa) have also been identified by *in vitro* translation of RRSV genomic dsRNA, and designated as non-structural proteins. RRSV S5, S8 and S9, respectively, encode a 90 kDa minor structural protein (possibly a guanylyltransferase), a 67 kDa major structural protein, which is further self-processed to 46, 43 and 26 kDa proteins, and a 38 kDa major structural protein. P9 is thought to be involved in vector transmission. RRSV segments S7 and S10 encode non-structural proteins of ~68 and 32 kDa, respectively. RRSV S4 probably encodes an RdRp and a second protein of unknown function. ERSV particles have four major structural proteins (127, 123, 63 and 34 kDa) and three minor proteins (103, 50 and

49 kDa). The reported differences in morphology of the outer capsids of RRSV and ERSV could be at least partially due to differences in the sizes of these structural proteins.

Lipids

None reported.

Carbohydrates

There are no evidence for the glycosylation of oryzavirus proteins.

Genome Organization and Replication

The genome organization is well characterized only for RRSV (Table 1). The dsRNA genome segments contain a single large ORFs (in one strand of the pair) except S4, which contains two large ORFs. The proteins encoded by S3, S8 and S9 are major components of the RRSV particle, which of those encoded by segments S7 and S10 are not found in the virion. Seg8 codes for a polyprotein that appears to autocatalytically cleave into at least two polypeptides one of which is a major structural protein. The larger protein encoded by Seg4 appears to be an RdRp. The tentative functions of the proteins encoded by the other segments are shown in Table 1. The viruses induce viroplasms in the cytoplasm of infected cells.

Table 1: List of the segments of Rice ragged stunt virus (RRSV), with their estimated size (bp), and corresponding proteins with name, Mr (estimated), and function and/or location.

dsRNA segment number	Size (bp)	Protein nomenclature (§)	Protein Mr predicted (kDa)	Protein Mr apparent (kDa)	Function (location)
Seg 1	3849	P1	137.7	137	Virus core associated (B Spike)
Seg 2	3810	P2	133.1	118	(Inner core capsid)
Seg 3	3699	P3	130.8	130	(Major core capsid)
Seg 4	3823	P4A (Pol)P4B	141.436.9	145	RDR polymerase (Unknown)
Seg 5	2682	P5 (Cap)	91.4	90	Capping enzyme / guanyltransferase
Seg 6	2157	P6	65.6		
Seg 7	1938	NS7	68	66	(Nonstructural)
Seg 8	1814	P8 P8A P8B	67.325.6 41.7	67 47/44	Precursor Protease (Major capsid)
Seg 9	1132	P9	38.6	37	Vector transmission (Spike)
Seg10	1162	NS10	32.3	32	Non-structural

§: protein structure/function: RNA polymerase = "Pol"; Capping enzyme = "Cap"

Antigenic Properties

RRSV and ERSV cross-react in serological tests. Polyclonal antisera raised against RRSV particle preparations react most strongly with P3, P8 and P9 (both the native state and resulting from *in vitro* production) suggesting that they are highly immunogenic. P5 is weakly immunogenic. Glutathione-S-transferase-NSP7 fusion protein is highly immunogenic and antibodies against this protein are useful in ELISA for the detection of RRSV in infected plants and insects

Biological Properties

Oryzaviruses infect plants in the family Graminae, causing disease in rice (RRSV) and *Echinochloa* (ERSV). They are transmitted by, and replicate in phloem-feeding, viruliferous delphacid planthoppers (RRSV: *Nilaparvata lugens*; ERSV: *Sogatella longifurcifera and S. vibix*). RRSV is ingested when the hopper feeds on rice plants, usually at the seedling stage. The minimum acquisition access period for the vector is about 3 hrs, the latent period is about 9 days, and the minimum inoculation access time is about 1 hr. Planthopper nymphs are more efficient vectors than adults but all forms of the insect can act as vectors. Any individual viruliferous hopper gives intermittent transmission. The virus is not passed though the egg.

Oryzaviruses appear to replicate in fibrillar viroplasms within the cytoplasm of phloem, or phloem-associated plant cells and in cells of the salivary glands, fat body, gut and brain of the planthopper. The phloem cells proliferate to form galls on the plant. RRSV has been reported in southeastern and far-eastern Asian countries where it affects rice yields (generally 10-20% loss, but up to 100% in severely affected areas). ERSV has been reported in Taiwan.

LIST OF SPECIES DEMARCATION CRITERIA IN THE GENUS

The prime determinant for inclusion of virus isolates within a single virus species of members of the family *Reoviridae,* is the "ability to exchange (reassort) genome segments during co-infection, thereby exchanging genetic information and generating viable and novel progeny virus strains". Data providing direct evidence of segment reassortment between isolates for oryzaviruses is very limited. Therefore serology, nucleic acid hybridization, and comparison of nucleic acid and protein sequences are the most common ways of determining the levels of similarity between isolates.

Members of a single species in the genus *Oryzavirus* may be identified by:

- Their ability to exchange genetic material by genome segment reassortment during dual infections, thereby producing viable progeny virus strains.
- RNA sequence identities;
- PCR (using primers to conserved segments coupled with cross-hybridization analysis);
- High levels of serological cross reaction by ELISA or agar gel immunodiffusion, using antibodies against the whole virus particle proteins or against specific viral proteins.
- Cross-hybridization of "conserved" genome segments under high stringency conditions (Northern or dot blots, with probes made from viral RNA or cDNA).
- Identical conserved terminal regions of the genome segments (some closely related species can also have identical terminal sequences on at least some segments).
- Analysis of "electropherotype" by agarose gel electrophoresis.
- Identification of vector species; RRSV and ERSV have different planthopper vector species.
- Identification of plant host species.

LIST OF SPECIES IN THE GENUS

Species names are in green italic script; strain names and synonyms are in black roman script; tentative species names are in blue roman script. Sequence accession numbers [Seg1 to Seg10 indicate RNA segments 1 – 10], insect vector and host names { }, and assigned abbreviations ( ) are also listed.

SPECIES IN THE GENUS

Echinochloa ragged stunt virus
{*Sogatella longifurcifera, S. vibix*}
{Graminae: *Echinochloa*}
Echinochloa ragged stunt virus (ERSV)

Rice ragged stunt virus
{*Nilaparvata lugens*}
{Graminae: Rice}

Rice ragged stunt virus - India	[Seg9: L38900]	(RRSV-Ind)
Rice ragged stunt virus - Philippines	[Seg9: 179969]	(RRSV-Phi)
Rice ragged stunt virus - Thailand	[Seg1: AF020334; Seg2: AF020335; Seg3:AF020336; Seg4: U66714; Seg5: U33633; Seg6: AF020337, Seg7: U66713; Seg8: U46682; Seg9: L38899; Seg10: U66712]	(RRSV-Tai)

DSRNA

TENTATIVE SPECIES IN THE GENUS

None reported.

PHYLOGENETIC RELATIONSHIPS WITHIN THE GENUS

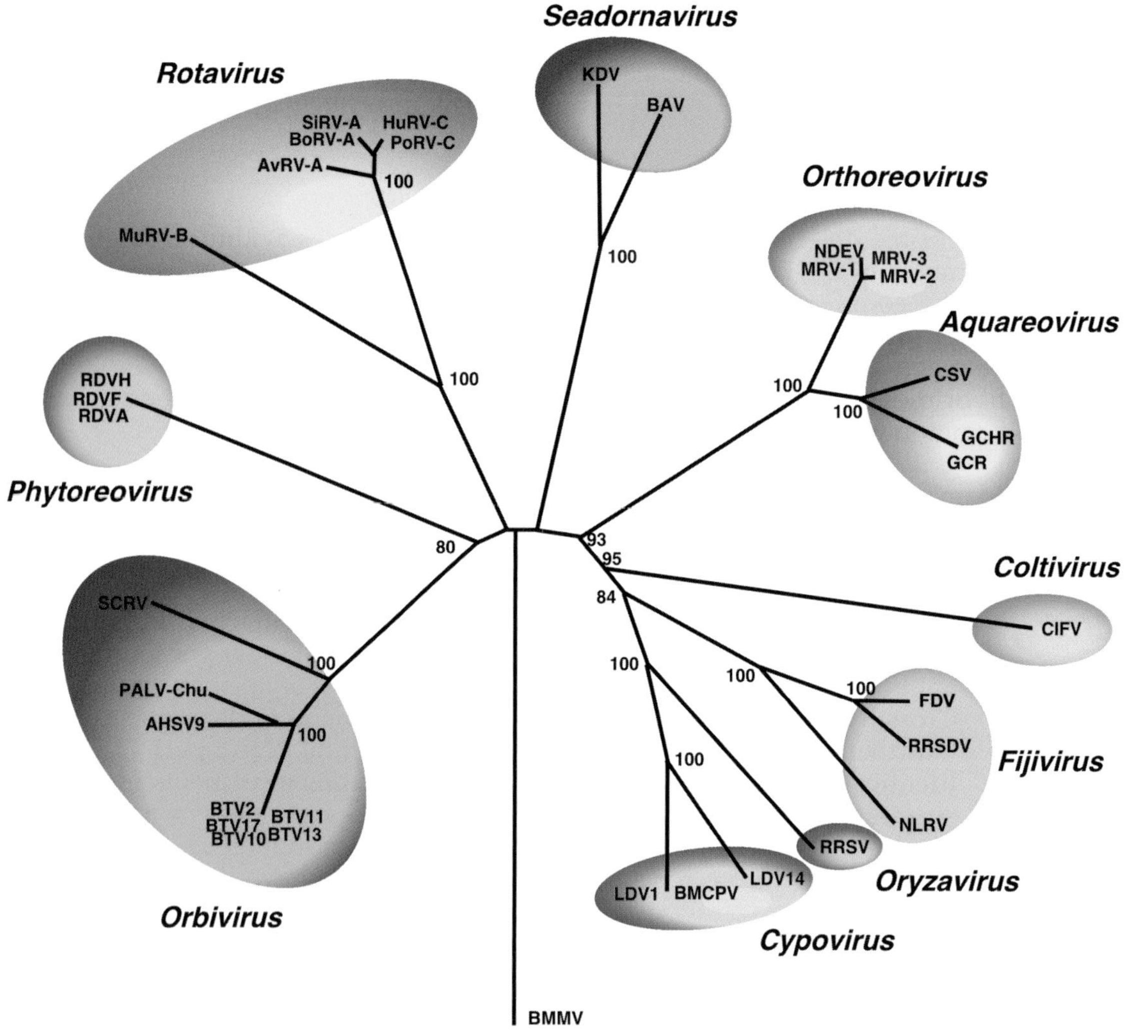

Figure 3: Phylogenetic tree of members of the *Reoviridae* family, from a Neighbor joining method, based on the aa sequences of the RdRp. A bootstrapped analysis with 100 iterations was performed and the percentages are indicated on the branches. Isolates of viruses used and their sequences are: *Aquareovirus*: Grass carp reovirus (GCRV) AAG10436; Grass carp hemorrhagic virus (GCHV) AF284502; Chum salmon virus (CSV) AAL31497. Coltivirus: Colorado tick fever virus (CTFV) AAG34362; Banna virus (BAV) AAF78849; Kadipiro virus (KDV) AAF78848. *Cypovirus*: Bombyx mori cytoplasmic polyhedrovirus 1 (BMCPV) AAK20302; Lymantria dispar cypovirus 1 (LDV-1) NP149147; Lymantria dispar cypovirus 14 (LDV-14) NP149135. *Fijivirus*: Fiji disease virus (JDV) AAK40249; Rice blacked-streaked dwarf virus (RBSDV) CAC82519; Nilaparvata lugens reovirus (NLRV) D49693. *Orbivirus*: African horse sickness virus 9 (AHSV-9); Bluetongue virus – 2 (BTV-2) L20508; BTV-10, X12819; BTV-11, L20445; BTV-13, L20446; BTV-17, L20447; Chuzan virus (PALV-Chu) BAA76549; Saint Croix river virus (SCRV) AAG34363. *Orthoreovirus*: Mammalian orthoreovirus, subgroup 1, serotype Lang 1 (MRV-1) MWXR31; serotype Jones 2 (MRV-2) MWXR32; serotype Dearing 3 (MRV-3) M31058; Ndelle virus (NDEV) AAL36027. *Oryzavirus*: Rice ragged stunt virus (RRSV) AAC36456. *Phytoreovirus*: Rice dwarf virus strain H (RDV-H) BAA01074; RDV strain A (RDV-A) Q02119; RDV strain F (RDV-F) Q98631. *Rotavirus*: *Rotavirus A* (RV-A); Avian rotavirus (AvRV-A) BAA24146; Bovine rotavirus A (BoRV-A) CAA39085; Simian rotavirus A strain SA11 (SiRV-A) CAA34732; *Rotavirus B*: Murine rotavirus B (MuRV-B) P35942; *Rotavirus C*, Porcine rotavirus C (PoRV-C) AAB00801; Human rotavirus C, strain KU (HuRV-C) BAA84962. The ssRNA foveavirus, Banana mild mosaic virus (BMMV) NP112029 was used as an outgroup. (Courtesy of R.M. Harding).

DERIVATIONS OF NAMES

Oryza: from Latin oryza, "rice".

REFERENCES *ORYZAVIRUS*

Chen, C.C., Chen, M.J., Chiu, R.J. and Hsu, H.T. (1989). Morphological comparisons of *Echinochloa* ragged stunt and rice ragged stunt viruses by electron microscopy. *Phytopathology,* **79**, 235-241.

Chen, C.C., Chen, M.J., Chiu, R.J. and Hsu, H.T. (1997). Rice ragged stunt virus (*Oryzavirus*) possesses an outer shell and A-spikes. *Plant Protection Bull.,* (Taichung) **39**, 383-388.

Chen, C.C., Hsu, Y.H., Chen, M.J. and Chiu, R.J. (1989). Comparison of proteins and nucleic acids of *Echinochloa* ragged stunt and rice ragged stunt viruses. *Intervirology,* **30**, 278-284.

Li, Z., Upadhyaya, N.M., Kositratana, W., Gibbs, A.J. and Waterhouse, P.M. (1996). Genome segment 5 of rice ragged stunt virus encodes a virion protein. *J. Gen. Virol.,* **77**, 3155-3160.

Upadhyaya, N.M., Ramm, K., Gellatly, J.A., Li, Z., Kositratana, W. and Waterhouse P.M. (1997). Rice ragged stunt oryzavirus genome segments S7 and S10 encode non-structural proteins of Mr 68,025 and Mr 32364. *Arch. Virol.,* **142**, 1719-1726.

Upadhyaya, N.M., Ramm, K., Li, Z., Kositratana, M. and Waterhouse, P.M. (1997). Rice ragged stunt oryzavirus genome segment S4 could encode an RNA dependent RNA polymerase and a second protein of unknown function. *Arch. Virol.,* **143**, 1815-1822.

Upadhyaya, N.M., Yang, M., Kositratana, W., Gosh, A. and Waterhouse, P.M. (1995). Molecular analysis of rice ragged stunt oryzavirus segment S9 and sequence conservation among isolates from Thailand and India. *Arch. Virol.,* **140**, 1945-1956.

Upadhyaya, N.M., Zinkowsky, E., Li, Z., Kositratana, W. and Waterhouse, P.M. (1996). The Mr 43K major capsid protein of rice ragged stunt oryzavirus is a post-translationally processed product of a Mr 67,348 polypeptide encoded by genome segment 8. *Arch. Virol.,* **141**, 1689-1701.

`Yan, J., Uyeda, I., Kimura, I., Shikata, E., Chen, C.C. and Chen, M.J. (1994). *Echinochloa* ragged stunt virus belongs to the same genus as rice ragged stunt virus. *Ann. Phytopath. Soc. Japan,* **60**, 613-616.

CONTRIBUTED BY

Upadhyaya, N.M. and Mertens, P.P.C.

GENUS MYCOREOVIRUS

Type Species *Mycoreovirus 1*

DISTINGUISHING FEATURES

Virions have a relatively featureless outer capsid as viewed by negative staining and electron microscopy, while the core particles have 12 icoahedrally arranged surface turrets. The genome is composed of 11 or 12 segments of dsRNA. The members of the genus that have been described all infect fungi.

VIRION PROPERTIES

MORPHOLOGY

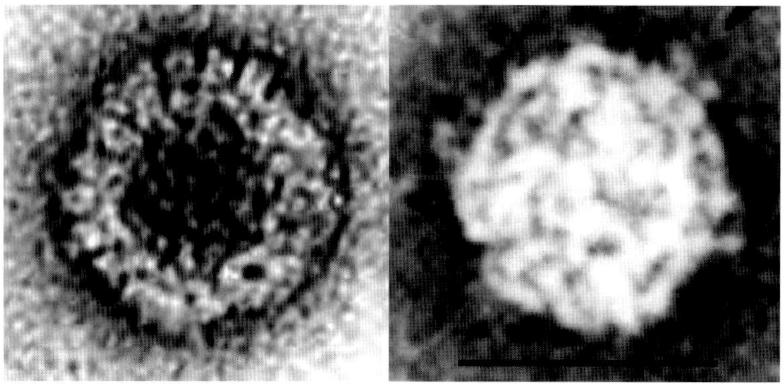

Figure 1: (Left) Electron micrograph of virus particles of Cryphonectria parasitica mycoreovirus 1 (CpMYRV-1) after purification by sucrose gradient centrifugation stained with 1% uranyl acetate (Courtesy of B. Hillman). (Right) Core particle of Rosellinia necatrix mycoreovirus 1 (RnMYRV-1) showing icosahedral arrangement surface projections or 'spikes', stained with 1% uranyl acetate (Courtesy of C. Wei). The bar represents 50 nm.

Particles are non-enveloped. Electron microscopy and negative staining of mycoreovirus virions with aqueous uranyl acetate indicates that they are double shelled, spherical in appearance (icosahedral symmetry), and approximately 80 nm in diameter. The viral core (estimated as 50 nm in diameter) has twelve icosahedrally arranged surface projections ('turrets' or 'B - spikes'). Particles were disrupted by 2% phosphotungstic acid (pH 7.0).

PHYSICOCHEMICAL AND PHYSICAL PROPERTIES

Not determined.

NUCLEIC ACID

Table 1: Conserved terminal sequences of mycoreovirus genome segments.

Virus species	(virus-strain)	Conserved RNA terminal sequences (+ve strand)
Mycoreovirus 1	(CpMYRV-1-9B21)	5′-GAUCA ....CGCAGUCA -3′
Mycoreovirus 3	(RnMYRV-1-W370)	5′-ACAAUUU.................... UGCAGAC -3′

The genome consists of 11 (group 1) or 12 (group 2) dsRNA segments that are named "genome segment 1" (Seg1) to "genome segment 11 or 12" (Seg11 or Seg12) in order of reducing molecular weight or increasing electrophoretic mobility following agarose gel electrophoresis. The total genome of Cryphonectria parasitica mycoreovirus-1 (CpMRV-1) contains 23,436 bp with the length of individual segments ranging from 732 bp to 4,127 bp, showing a 3, 3, 2, 3 electrophoretic profile following either 11% PAGE or 1% agarose gel electrophoresis (AGE). The genomic RNA of Rosellinia necatrix mycoreovirus-1 (RnMRV-

3) shows a 3, 3, 6 electrophoretic profile following 5% PAGE. Like other members of the family *Reoviridae* the genome segment migration patterns during AGE (or low percentage PAGE <5%) are likely to be characteristic of each virus species. Terminal sequences of the genome segments are shown in Table 1.

PROTEINS

Native viral proteins have not been characterized. However, protein sequences were deduced from sequences of the viral genomic RNAs. Their putative functions are shown in Table 2. Proteins are currently named as VP1 to VP11 or VP12 based on the molecular weight of the genome segment (segment number) from which they are translated.

LIPIDS

None reported.

CARBOHYDRATES

None reported.

GENOME ORGANIZATION AND REPLICATION

On the basis of the available sequence data for several of the genome segments and the overall similarity of mycoreoviruses to other members of the family *Reoviridae* it can be assumed that many aspects of the genome organization and replication also will be similar. On this basis it is likely that the viral core will contain transcriptase complexes that synthesize mRNA copies of the individual genome segments. These will be exported and translated to produce viral proteins within the host cytoplasm. These positive sense RNAs also are likely to form templates for negative strand synthesis during progeny virus assembly and maturation. Like other *Reoviridae* members, most of the mycoreovirus genome segments appear to represent single genes, with a large ORF and relatively short terminal non-coding regions (Fig. 2).

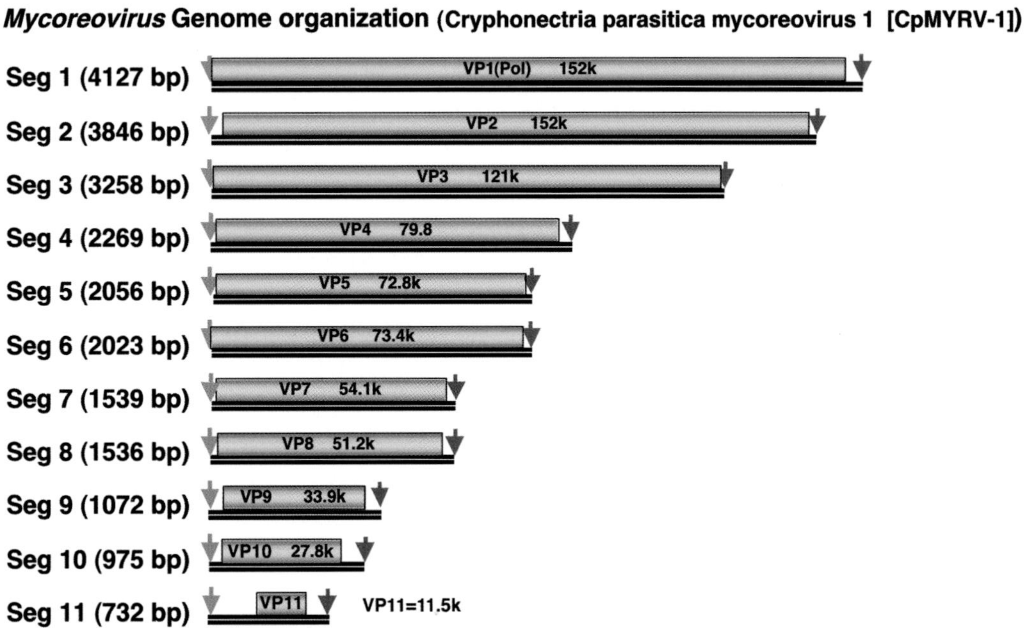

Figure 2: Genome organization of Cryphonectria parasitica mycoreovirus 1 (CpMYRV-1), containing 11 dsRNA segments, each contains an ORF. The green arrows indicate the upstream conserved terminal sequence (+ve 5′-GAUCA.....), while the red arrows indicate the downstream conserved terminal sequence (+ve.....CGCAGUCA-3').

Table 2: List of Cryphonectria parasitica mycoreovirus 1 (CpMYRV-1) dsRNA segments 1-11, with their respective size (bp) and their encoded proteins for which the name, calculated size (kDa), and function and/or location are indicated.

dsRNA segment #	dsRNA Size (bp)	Protein nomenclature	Protein size aa (kDa)	Structure / Function
Seg 1	4127	VP1	1354 (151.8)	RdRp; sequence similarity to coltivirus VP1
Seg 2	3846	VP2	1238 (138.5)	possible Mtr, sequence similarity to coltivirus VP2
Seg 3	3251	VP3	1065 (120.8)	sequence similarity to coltivirus VP3
Seg 4	2269	VP4	721 (79.8)	Sequence similarity with RnMYRV-1-W370 Seg4 and coltivirus Seg4
Seg 5	2023	VP5	648 (72.8)	
Seg 6	2056	VP6	650 (73.4)	Sequence similarity with RnMYRV-1-W370 Seg6 and coltivirus Seg10
Seg 7	1536	VP7	482 (54.1)	
Seg 8	1539	VP8	470 (51.2)	
Seg 9	1072	VP9	298 (32.9)	Sequence similarity with RnMYRV-1-W370 Seg11
Seg10	975	VP10	248 (27.8)	
Seg11	732	VP11	102 (11.5)	

Table 3: List of Rosellinia necatrix mycoreovirus 2 (RnMRV-2) dsRNA Seg1-12, with their respective size (bp) and their encoded proteins for which the name, calculated size (kDa), and function and/or location are indicated.

dsRNA segment #	dsRNA Size (bp)	Protein nomenclature	Protein size aa (kDa)	Structure / Function
Seg 1		VP1		
Seg 2	3773	VP2	1226 (138.5)	possible Mtr, shows some sequence similarity to coltivirus VP2
Seg 3		VP3		
Seg 4		VP4		
Seg 5	2089	VP5	646 (72)	
Seg 6	2030	VP6	634	
Seg 7	1509	VP7	482	
Seg 8	1299	VP8	325	
Seg 9	1226	VP9	380	
Seg10	1171	VP10	310	
Seg11	1003	VP11	282	
Seg12	943	VP12	265	

Antigenic Properties

Not available.

Biological Properties

The virus RnMYRV-1 is found in the mycelium of a strain of the white root rot fungus *Rosellinia necatrix*. The virus itself appears to make the fungus hypovirulent and may represent a useful biological control for the damage caused by the wild-type fungus. The uninfected fungus can be regenerated by hyphal tip culture. CpMYRV-1 and CpMYRV-2 are found in the mycelium of the filamentous fungus that causes chestnut blight disease (*Cryphonectria parasitica*). Purified particles of CpMYRV-1 can be used to infect protoplasts of virus free mycelium. Infection with CpMYRV-1 greatly reduced virulence of the fungal strain and may represent a useful biological control for the disease.

List of Species Demarcation Criteria in the Genus

In common with the other genera in the family *Reoviridae,* the prime determinant for inclusion of virus isolates within a single mycoreovirus species is defined by their compatibility for reassortment of genome segments during coinfections, allowing the exchange of genetic information and generation of viable progeny virus strains. Coinfection experiments with mycoreoviruses have not been reported. Therefore, as with other members of the family *Reoviridae*, serological comparisons, comparisons of RNA and protein sequences, cross-hybridization, analysis of electropherotypes, and analysis of conserved RNA sequences in near terminal regions of the genome segments represent factors used to examine the level of similarity that exists between those species currently recognized and future isolates. CpMYRV-1-9B21 and RnMYRV-1-W370 share less than 50% overall sequence identity and are therefore less closely related to each other than the two coltivirus species *Colorado tick fever virus* and *Eyach virus*.

Members of a species of the genus *Mycoreovirus* may be identified by:

- The capacity to exchange genetic material by genome segment reassortment during mixed infections, thereby producing viable progeny virus strains.
- RNA cross-hybridization assays (Northern or dot blots, with probes made from viral RNA or cDNA).
- Serological comparisons, for example by ELISA or AGID assays.
- RNA sequence analysis (e.g., Seg1 to 12). Within a single species, high levels of sequence similarities would be expected in conserved segments, e.g., genes encoding polymerase or internal capsid layer proteins.
- Comparisons of aa sequences (e.g., polymerase or internal CPs).
- The genome segment number and size distribution, for example as analyzed by electropherotypes. Within a single species the electropherotype should be relatively uniform. However, deletions or additions can occur resulting in unusual variations in electropherotype.
- Analysis of conserved RNA terminal sequences. These sequences show conservation within individual species of many members of the family *Reoviridae*.

List of Species in the Genus

Species names are in green italic script; strain names and synonyms are in black roman script; tentative species names are in blue roman script. Sequence accession numbers [Seg1 to Seg12 indicate RNA segments 1–12], and assigned abbreviations ( ) are also listed.

Species in the Genus

Group 1 (11 genome segments)

Mycoreovirus 1 ‡		
Cryphonectria parasitica mycoreovirus 1 (9B21)	[Seg1: AY277888, Seg2: AY277889, Seg3: AY277890]	(CpMYRV-1/9B21)
Mycoreovirus 2 ‡		
Cryphonectria parasitica mycoreovirus 2 (C18)		(CpMYRV-2/C18)

Group 2 (12 genome segments)

Mycoreovirus 3 ‡		
Rosellinia necatrix mycoreovirus 3 (W370) (Rosellinia anti-rot virus)	[Seg2: AB098022, Seg5: AB098023, Seg6: AB073277, Seg7: AB073278, Seg8: AB073279,	(RnMYRV-3/W370)

Seg9: AB073280,
Seg10: AB073281,
Seg11: AB073282,
Seg12: AB073283]

Tentative Species in the Genus

None reported.

Phylogenetic Relationships within the Genus

Not available.

Similarity with Other Taxa

The deduced translation product of the large ORF on genome Seg2 of RnMYRV-1 shows approximately 23% aa identity with VP2 of coltiviruses (Fig. 3). Although significant, this relatively low level of identity is consistent with Maize dwarf virus 3 (MDV-3) and the coltiviruses belonging to separate genera. Both *C. parasitica* mycoreoviruses contain 11 genome segments, the larger being homologous to RnMYRV-1 and coltivirus segments, but the smaller segments are not. The two *C. parasitica* mycoreoviruses show approximately 50% overall sequence identity, as determined by partial genome sequence analysis, confirming their identity as distinct species within the genus.

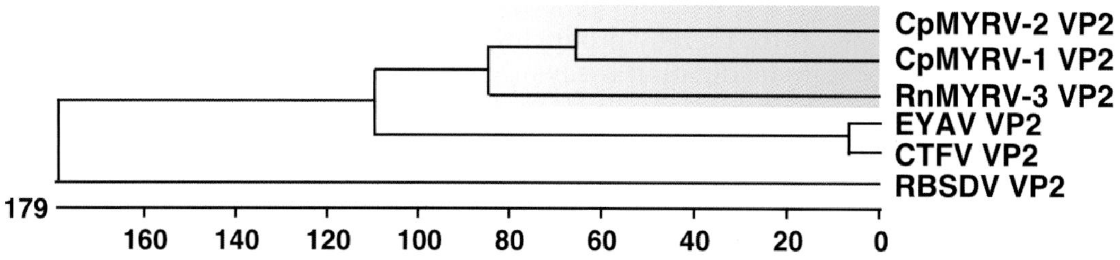

Figure 3: Phylogenetic tree showing relationships between VP2 of Cryphonectria parasitica mycoreovirus 1 (CpMYRV-1), Cryphonectria parasitica mycoreovirus 2 (CpMYRV-2), Rosellinia necatrix mycoreovirus 1 (RnMYRV-1) and selected other members of the family *Reoviridae*. The scale represents the number of nt substitutions (x100).

Derivation of Names

Mycoreo: from 'Myco' from Greek fungus; and 'reovirus' indicating membership in the family *Reoviridae*.

References

Enebak, S.A., Hillman, B.I. and MacDonald, W.L. (1994). A Hypovirulent isolate of *Cryphonectria parasitica* with multiple genetically unique dsRNA segments. *MPMI*, **5**, 590-595.

Hillman, B.I., Supyani, S., Kondo, H. and Suzuki, N. (2004). A reovirus of the fungus *Cryphonectria parasitica* that is infectious as particles and related to the *Coltivirus* genus of animal pathogens. *J. Virol.*, **78**, 892-898.

Osaki, H., Wei, C.Z., Arakawa, M., Iwanami, T., Nomura, K., Matsumoto, N. and Ohtsu, Y. (2002). Nucleotide sequences of double stranded RNA segments from a hypovirulent strain of the white root rot fungus Rosellinia necatrix: possibility of the first member of the *Reoviridae* from a fungus. *Virus Gene*, **25**, 101-107.

Wei, C.Z., Osaki, H., Iwanami, T., Matsumoto, N. and Ohtsu, Y. (2003). Molecular characterization of double–stranded segments S2 and S5 and electron microscopy of a novel reovirus from a hypovirulent isolate W370 of the plant pathogen *Rosellinia necatrix*. *J. Gen. Virol.*, **84**, 2431-2437.

Liu, Y.C., Linder-Basso, D., Hillman, B.I., Kaneko, S. and Milgroom, M.G. (2003). Evidence for interspecies transmission of viruses in natural populations of filamentous fungi in the genus Cryphonectria. *Mol. Ecol.*, **12**, 1619-1628.

Contributed By

Mertens, P.P.C., Wei, C. and Hillman, B.

FAMILY BIRNAVIRIDAE

TAXONOMIC STRUCTURE OF THE FAMILY

Family	***Birnaviridae***
Genus	***Aquabirnavirus***
Genus	***Avibirnavirus***
Genus	***Entomobirnavirus***

VIRION PROPERTIES

MORPHOLOGY

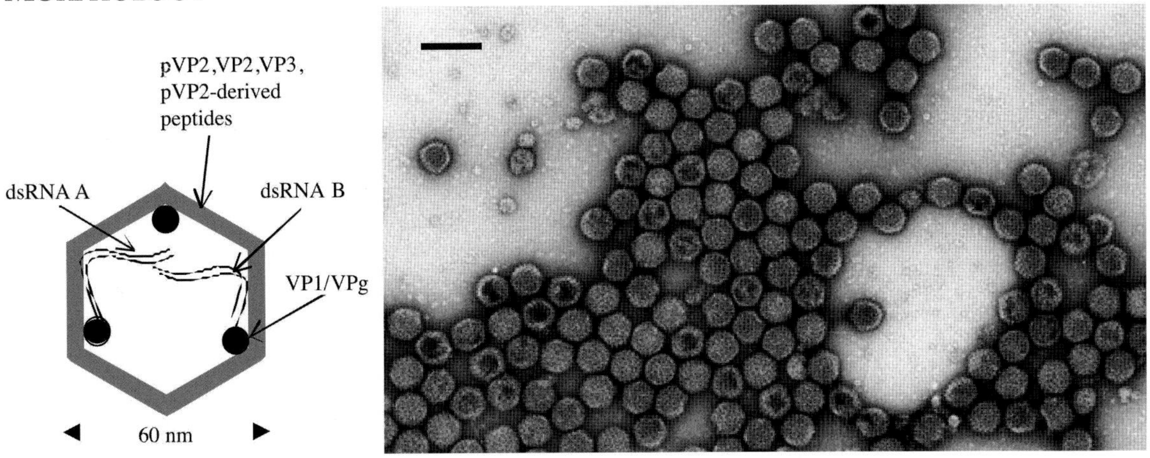

Figure 1: (Left) Diagram of a particle of Infectious bursal disease virus (IBDV); (Right) negative contrast electron micrograph of IBDV virions. The bar represents 100 nm (Courtesy of J. Lepault).

Virions are about 60 nm in diameter, single-shelled, non-enveloped icosahedra (Fig. 1). Electron micrographs of frozen, hydrated, unstained virions indicate that the capsid structure is based on a T=13 lattice. Computer image processing at about 2 nm resolution produces a map of the virion that shows the subunits to be predominantly clustered in trimers. The outer trimers correspond to the protein VP2, and the inner trimers to protein VP3.

PHYSICOCHEMICAL AND PHYSICAL PROPERTIES

Virion Mr is about 55 x 10^6, S_{20W} is 435S; buoyant density in CsCl is 1.33 g/cm^3. Defective virions with interfering activity have been demonstrated to band at 1.30 g/cm^3. Viruses are stable at pH 3-9, resistant to heat (60°C, 1 hr), ether and 1% SDS at 20°C, pH 7.5 for 30 min.

NUCLEIC ACID

Virions contain two segments (A, B) of dsRNA which constitute between 9-10% of the particle by weight. The size of the larger segment A for Infectious pancreatic necrosis virus (IPNV) ranges from 2962 bp (strain Sp) to 3097 bp (strain Jasper) and 3104 bp (strains N1 and DRT). The size of segment B for IPNV ranges from 2731 (strain DRT) to 2784 bp (strain Jasper). For Infectious bursal disease virus (IBDV), segment A ranges in size from 3063 bp (strain Farragher) to 3261 bp (strain P2). Segment B for IBDV ranges in size from 2715 bp (isolate UK661) to 2922 bp (strain QC-2). The Drosophila X virus (DXV) was found to have A and B segments containing 3365 and 3243 bp, respectively. The sizes of Blotched snakehead virus (BSNV) A and B segments were found to be 3429 and 2750 bp, respectively. Despite the differences in the sizes of the A segments among members of the same genus, the coding region of the genome produces a polyprotein of 972 aa for IPNV, 1012 aa for IBDV, 1032 aa for DXV and 1069 aa for BSNV. Similarly, the IPNV virion polymerases have a reported coding region for 844-845 aa; for IBDV it is 879 aa and there is a report of an unusually large polymerase of 891 aa for IBDV strain QC-2. For DXV and BSNV, the polymerases are predicted to be 997 and 867 aa long, respectively. Both genome segments contain a 94 kDa 5' genome-linked VPg. There are no poly(A) tracts at the 3'-ends of the RNA segments. The defective particles banding at 1.30 g/cm^3 appear to contain a truncated A segment.

PROTEINS

Virions contain five polypeptides and several small peptides: VP1 (94 kDa) which is the RdRp as well as the VPg; pre-VP2 (62 kDa, [the precursor of VP2]) and VP2 (54 kDa), the major capsid polypeptides and type-specific antigens; VP3 (30 kDa), an intemal CP and group-specific antigen. A nonstructural protein (VP4 for IBDV, DXV and BSNV or NS for IPNV; [29 kDa]) is a virus-encoded protease that autocatalytically cleaves the polyprotein to produce pre-VP2, VP2, VP4 (NS), and VP3. The four small structural peptides for IBDV are 46, 7, 7 and 11 aa long. For BSNV, they are 43, 7, 7 and 12 aa long. These peptides are derived from the C-terminus of pre-VP2. An additional nonstructural polypeptide that is positively charged has been designated as VP5 for IBDV (16.5 kDa) or the 17 kDa protein for IPNV. This polypeptide is encoded by segment A in an ORF preceding the segment A-encoded polyprotein. The VP5 and the 16.5 kDa protein have been shown to be nonessential for IBDV and IPNV rcplication, respectively. A second ORF encoding an arginine-rich protein has also been identified in genome segment A for DXV and BSNV. Although VP1 can guanylylate itself, it has no guanylyl transferase activity, and the viral mRNA made in the cell retains its 5' VPg. There is no evidence for 5' capping of any of the viral mRNAs.

LIPIDS

None present.

CARBOHYDRATES

No N-linked glycosylation of any of the virion proteins of IPNV or IBDV has been detected. There is a report of O-linked glycosylation for VP2 of IPNV.

GENOME ORGANIZATION AND REPLICATION

The genome organization of segment A of the four different birnavirus genomes is illustrated in Figure 2. The arrangements are similar although there is some variation in the sizes of the individual proteins. For all birnaviruses, genome segment A contains two ORFs: ORF2 encoding a large polyprotein of about 105 kDa and an overlapping (IPNV and IBDV) or internal (DXV and BSNV) reading frame ORF1 encoding a 17 to 27 kDa protein. The primary cleavage of the polyprotein which generate pre-VP2, VP4 and VP3 have been fully determined for either IBDV, IPNV and BSNV by N-terminal aa sequence analyses and site-directed mutagenesis. For DXV, the cleavage site between pre-VP2 and VP4 was located by N-terminal aa sequencing. Cleavage occurs at the pVP2-VP4 junction between Ala-508 and Ser-509 for IPNV, Ala-512 and Ala-513 for IBDV, Ser-500 and Ala-501 for DXV and Ala-486 and Ser-487 for BSNV. At the VP4-VP3 junction, cleavage occurs between Ala-734 and Ala-735 for IPNV, Ala-755 and Ala-756 for IBDV and Ala-791 and Ala-792 for BSNV. In IBDV and BSNV, the processing of pre-VP2 was shown to generate VP2 and four small peptides deriving from the C-terminus of pre-VP2. Identification of the peptides was carried out by mass spectrometry and N-terminal sequencing of purified virions. In IPNV and DXV, an homologous domain is present in the C-terminus of pre-VP2. Genome segment B contains one large 94 kDa product (Fig. 3, ORF3).

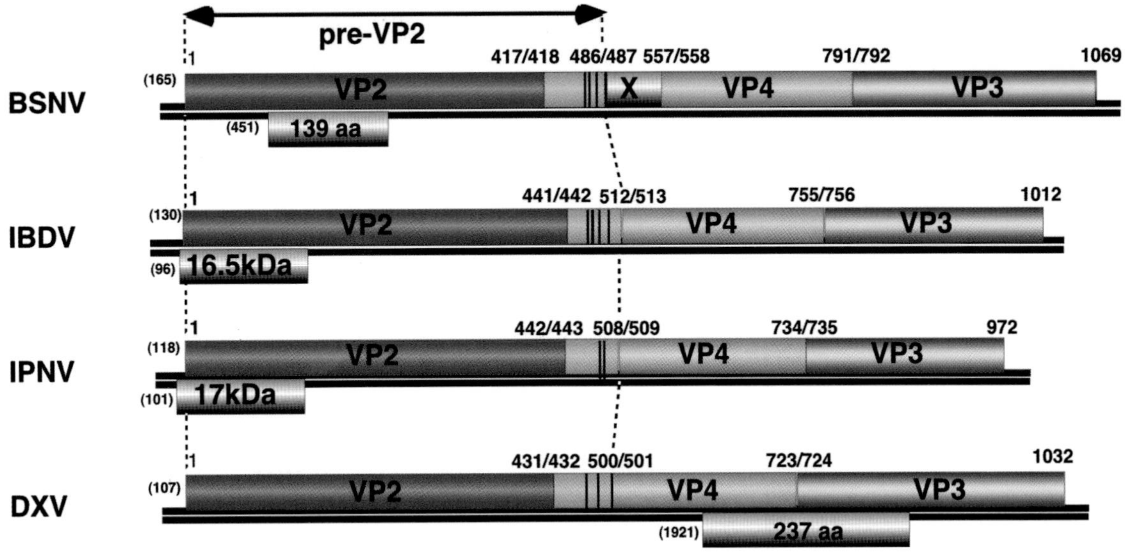

Figure 2: Schematic representation of the gene arrangement of genome segment A of Blotched snakehead virus (BSNV), Infectious bursal disease virus (IBDV), Infectious pancreatic necrosis virus (IPNV) and Drosophila X virus (DXV).

A single cycle of replication takes about 18-22 hrs for IPNV and 68 hrs for IBDV. For IBDV, virus-binding proteins have been found at the surface of various types of chicken cells. After entry into the host cell, the virion RdRp becomes activated and produces two genome length (24S) mRNA molecules from each of the 14S dsRNA genome segments. These mRNAs are not capped; rather, nascent mRNAs have a VPg attached to their 5'-ends and they lack 3' poly(A) tracts. Replicative intermediates have been identified in infected cells. Virus RNA is transcribed by a semi-conservative strand displacement mechanism *in vitro*. There is no information on minus strand RNA synthesis. The two mRNAs can be detected in infected cells by 3-4 hrs post infection and are synthesized in the same relative proportions throughout the replicative cycle (i.e., about twice as many A as B mRNA molecules). Virus-specific polypeptides can be detected at 4-5 hrs post infection and are present in the same relative proportions until the end of the replication cycle. There are no specifically early or late proteins. The segment A mRNA is translated to a 105 kDa polyprotein which contains (5' to 3') the pre-VP2, VP4 (NS) and VP3 polypeptides, with the notable exception of BSNV which contains an additional polypeptide [X] between the pre-VP2 and the VP4 domain (Fig. 2). It has been shown that the VP4 protease co-translationally cleaves the polyprotein to generate the three (or four in BSNV) polypeptides. Pre-VP2 is later processed by a slow maturation cleavage to produce VP2 and the small structural peptides deriving from the C-terminus of pVP2. This cleavage is incomplete since both pre-VP2 and VP2 are found in purified virus, although VP2 predominates. The VP4 (NS) protease is a serine-lysine protease and was shown to possess a region related to the protease domain of bacterial and organelle ATP-dependent Lon proteases. A similar catalytic dyad is reminiscent of the protease active site of bacterial leader peptidases and the bacterial LexA-like proteases. The catalytic residues were experimentally ascertained on IBDV, IPNV and BSNV. The product of the 16.5-17 kDa ORF has been detected in both IBDV and IPNV infected cells.

The mRNA from segment B is translated to a 94 kDa polypeptide which is the viral RdRp (VP1, Fig. 3). VP1 is found in virions both in a free and genome-linked form (VPg). Virus particles assemble and accumulate in the cytoplasm. Encapsidation of the RdRp VP1 is mediated by its interaction with the inner capsid VP3. For IBDV, the VP4 accumulates in the nucleus as well as in the cytoplasm of infected cells as tubules of 24 to 26 nm in diameter. The mechanism of virus release is unknown. In tissue culture about half of the progeny virions remains cell-associated. *In vitro*, DI particles are formed depending on the multiplicity of infection.

A reverse genetics system has been elaborated for IBDV and IPNV. Synthetic RNAs of segments A and B were found infectious (when transfected together), facilitating studies of birnavirus replication.

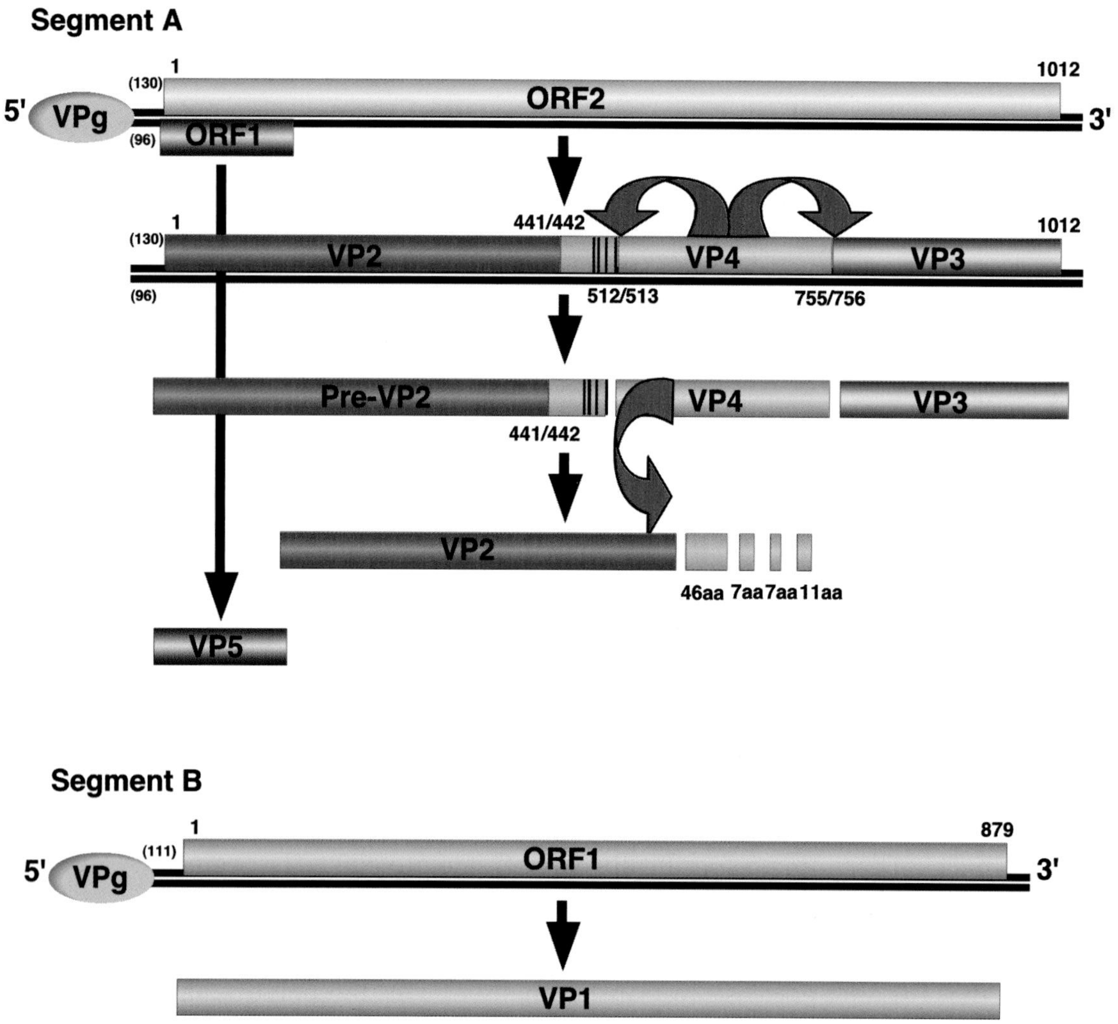

Figure 3: Schematic representation of the genome of Infectious bursal disease virus (IBDV) and of the processing of the encoded proteins.

ANTIGENIC PROPERTIES

The major CP VP2 is the type-specific antigen and contains the virus neutralizing epitopes. Anti-VP3 antibodies do not neutralize virus infectivity. There is no serological cross-reaction between the fish, avian, and insect birnaviruses nor between the aquabirnaviruses IPNV, BSNV and Tellina virus (TV-2).

BIOLOGICAL PROPERTIES

The natural hosts of IPNV are salmonid fish, although the virus has also been isolated from other freshwater and marine fishes, as well as from bivalve molluscs. The virus is transmitted both vertically and horizontally. There are no known vectors. The geographic distribution is worldwide. IPNV can cause epizootics resulting in high mortality in hatchery-reared salmonid fry and fingerlings. The virus causes necrotic lesions in the pancreas and is also found, without lesions, in other organs such as kidney, gonad, intestine, brain, etc. Infected adult fish become lifelong carriers without exhibiting overt signs of infection.

The natural hosts of IBDV are chickens and turkeys. Rarely, IBDV has been isolated from ducks and other domestic fowl. The mode of transmission is horizontal. There are no known vectors. IBDV has a worldwide distribution. The virus destroys the bursa of Fabricius of young chicks causing B-lymphocyte deficiency. Death occurs between 3 and 6 weeks of age and is associated with inflammation and necrosis of the bursa of Fabricius, formation of immune complexes, depletion of complement, and clotting abnormalities.

Drosophila melanogaster populations are the natural host of DXV. The mode of transmission is horizontal and there are no known vectors. The geographic distribution is unknown. Infected fruitflies become sensitive to CO_2. The target organs and histopathology are not known. DXV has also been isolated from populations of *Culicoides* species. BSNV has been isolated from a cell line developed from *Channa lucius*.

GENUS AQUABIRNAVIRUS

Type Species *Infectious pancreatic necrosis virus*

DISTINGUISHING FEATURES

Members of the genus infect only fish, molluscs and crustaceans.

BIOLOGICAL PROPERTIES

Aquabirnaviruses have been isolated from a variety of aquatic animals in freshwater, brackish, and seawater environments. The ubiquitous nature of these viruses and, in some cases, the lack of any association with disease has led to difficulty in nomenclature. The first reports of isolation of IPNV were limited to epizootics in cultured brook trout (*Salvalinus fontinalis*). Soon IPNV was found to be responsible for disease in a variety of salmonid fish including members of the genera Salmo, Salvalinus, and Oncorhynchus. The virus has also been associated with disease in Japanese eels (*Anguilla japonica*) where it causes a nephritis, menhaden (*Brevoortia tryrranus*) where it causes a "spinning disease," and in yellowtail fingerlings (*Seriola quinqueradiata*) where it causes an ascites and cranial hemorrhage. In salmonid fish, it causes an acute gastroenteritis and destruction of the pancreas in the very young. A birnavirus has been associated with hematopoietic necrosis causing high mortalities in turbot (*Scophthalmus maximus*) with renal necrosis, and birnaviruses have been isolated from clams exhibiting darkened gills and gill necrosis. A non-typical apoptosis has been observed in cultured cells infected by IPNV.

LIST OF SPECIES DEMARCATION CRITERIA IN THE GENUS

There are no agreed species demarcation criteria in the genus. The aquatic birnaviruses have been divided into two serogroups. These groups display no cross-reactivity by neutralization tests, but show some cross-reaction by immunofluorescence. Serogroup A contains most of the aquabirnavirus isolates. They have been divided into nine serotypes (A1-A9) based on reciprocal cross-neutralization tests. All of the archetype strains were originally isolated from North America or Europe in association with IPN disease in salmonid fishes. The exceptions are the Hecht strain (serotype A4) isolated from pike (*Esox lucius*) and serotype A5, isolated from the marine bivalve mollusc, *Tellina tenuis*. Sequence analyses of an extensive number of IPNV and other aquatic birnavirus isolates representative of all the serotypes show that the aa sequences of the VP2 proteins differ by less than 18% and no particular cluster regarding host species origin is identified. Serogroup B consists of a single serotype with fewer than 10 isolates from European episodes in fish and marine invertebrates. No sequence information is available for this serogroup.

LIST OF SPECIES IN THE GENUS

Species names are in green italic script; strain names and synonyms are in black roman script; tentative species names are in blue roman script. Sequence accession numbers [A to B indicate RNA segments A and B], and assigned abbreviations ( ) are also listed.

SPECIES IN THE GENUS

Infectious pancreatic necrosis virus		
Infectious pancreatic necrosis virus - DRT	[A:D26526, B:D26527]	(IPNV-DRT)
Infectious pancreatic necrosis virus - He	[A:AF342730]	(IPNV-He)
Infectious pancreatic necrosis virus - Jasper	[A:M18049, B:M58756]	(IPNV-Jas)
Infectious pancreatic necrosis virus - N1	[A:D00701]	(IPNV-N1)
Infectious pancreatic necrosis virus - Sp	[A:U56907, B: M58757]	(IPNV-Sp)
Infectious pancreatic necrosis virus - WB	[A:AF078668, B :AF078669]	(IPNV-WB)
Tellina virus ‡		
Tellina virus 2	[A:AF342730]	(TV-2)
Yellowtail ascites virus		
Yellowtail ascites virus - Y-6	[A:AB006783, B :AY129662]	(YTAV-Y-6)
Yellowtail ascites virus – YT-01A	[B :AY129663]	(YTAV-YT-01A)

TENTATIVE SPECIES IN THE GENUS

Marine birnavirus		
Marine birnavirus – AY-98	[B: AY123970]	(MABV-AY-98)
Marine birnavirus – H-1	[B: AY129665]	(MABV-H-1)

GENUS *AVIBIRNAVIRUS*

Type Species *Infectious bursal disease virus*

DISTINGUISHING FEATURES

Members of the genus infect only birds.

BIOLOGICAL PROPERTIES

Isolates of the type species, *Infectious bursal disease virus*, cause an immunosuppressive disease in chickens by destruction of the lymphoid cells in the bursa of Fabricius. Apoptosis has also been observed in this and other lymphoid organs. It was shown that the VP2 protein induces apoptosis in transfected mammalian cells. This finding correlates well with evidence of apoptosis and B cell death in chickens infected with IBDV. The rapid depletion of B cells in the bursa of Fabricius leads to immunosuppression and increased susceptibility to other infections and diseases. The virus is highly contagious and of major importance to the poultry industry worldwide. Two serotypes of IBDV have been identified by cross-neutralization assays, serotype 1 and serotype 2. Serotype 1 strains are pathogenic in chickens, whereas serotype 2 strains are non-pathogenic.

LIST OF SPECIES DEMARCATION CRITERIA IN THE GENUS

Not applicable.

LIST OF SPECIES IN THE GENUS

Species names are in green italic script; strain names and synonyms are in black roman script; tentative species names are in blue roman script. Sequence accession numbers [A to B indicate RNA segments A and B], and assigned abbreviations ( ) are also listed.

SPECIES IN THE GENUS

Infections bursal disease virus

Infections bursal disease virus - 00273	[A:M64738, B:M19336]	(IBDV-00273)
Infections bursal disease virus - 23/82	[A:Z21971]	(IBDV-23/82)
Infections bursal disease virus - 52/70	[A:D00869]	(IBDV-52/70)
Infections bursal disease virus - Australian 00273	[A:X03993, M24529, M27967, M64738]	(IBDV-Aus00273)
Infections bursal disease virus - Cu1	[A:D00867, X16107]	(IBDV-Cu1)
Infections bursal disease virus - Edgar	[A:A33255]	(IBDV-Edgar)
Infections bursal disease virus - Farragher	[A:A38328]	(IBDV-Far)
Infections bursal disease virus - GPF1E	[A:E05443]	(IBDV-GPF1E)
Infections bursal disease virus - KS	[A:L42284]	(IBDV-KS)
Infections bursal disease virus - OH	[A:U30818, B:U30819]	(IBDV-OH)
Infections bursal disease virus - OKYM	[A:D49707]	(IBDV-OKYM)
Infections bursal disease virus - attenuated OKYM	[A:D83985, A:D49706]	(IBDV-attOKYM)
Infections bursal disease virus - P2	[A:X84034]	(IBDV-P2)
Infections bursal disease virus - PBG98	[A:D00868]	(IBDV-PBG98)
Infections bursal disease virus - QC2	[B:U20950]	(IBDV-QC2)
Infections bursal disease virus - STC	[A:D00499]	(IBDV-STC)
Infections bursal disease virus - UK661	[A:X92761, X92760]	(IBDV-UK661)

TENTATIVE SPECIES IN THE GENUS

None reported

GENUS ENTOMOBIRNAVIRUS

Type Species *Drosophila X virus*

DISTINGUISHING FEATURES

The single member of the genus infects only insects.

LIST OF SPECIES DEMARCATION CRITERIA IN THE GENUS

Not applicable.

LIST OF SPECIES IN THE GENUS

Species names are in green italic script; strain names and synonyms are in black roman script; tentative species names are in blue roman script. Sequence accession numbers [A to B indicate RNA segments A and B], and assigned abbreviations ( ) are also listed.

SPECIES IN THE GENUS

Drosophila X virus

Drosophila X virus	[A:U60650, B :AF196645]	(DXV)

TENTATIVE SPECIES IN THE GENUS

None reported

LIST OF UNASSIGNED VIRUSES IN THE FAMILY

Blotched snakehead virus *(Channa lucius)*	[A :AJ459382, B :AJ459383]	(BSNV)
Rotifer birnavirus (*Brachiorus plicatilis*)		(RBV)

PHYLOGENETIC RELATIONSHIPS WITHIN THE FAMILY

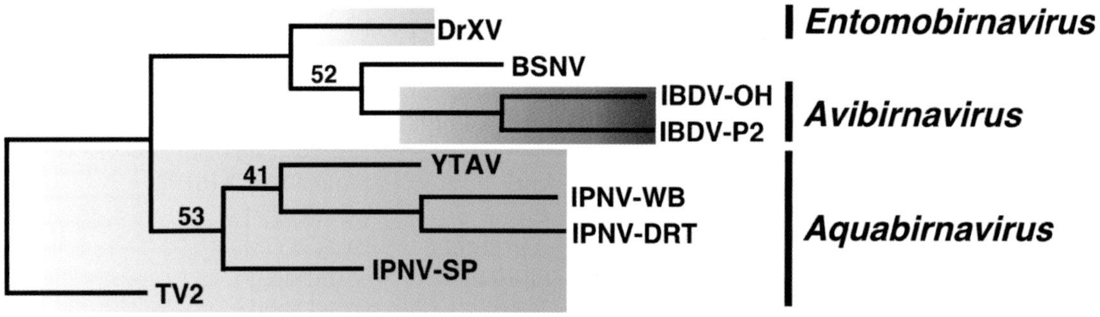

Figure 4: A phylogenetic tree representing the relationships between the different genera of this family was constructed with the complete VP2 sequences. VP2 sequences were aligned using the Clustal W Multiple Alignment Program. The tree was generated using the Protpars algorithm of the Phylogenetic Inference Package Phylip. The bootstrap values are indicated in the branches when different from value 100. Branch lengths are proportional to genetic distances. Blotched snakehead virus (BSNV) separates from the aquabirnavirus cluster. The protein accession numbers used for comparison were (top to bottom): DrXV: 1545998, BSNV: 27262817, IBDV-OH: 1469290, IBDV-P2: 854199, YTAV: 2920318, IPNV-WB: 13604261, IPNV-DRT: 561661, IPNV-SP: 13604263 and TV2: 13604269.

SIMILARITY WITH OTHER TAXA

Birnaviruses share no nt sequence similarity with other taxa. At the aa sequence level, the capsid proteins encoded by the A segment have no similarities with corresponding proteins of any other virus family. There is no antigenic similarity detectable by serological cross-reactivity. In contrast, the VP4 protease has sequence homologies with a region of the protease domain of bacterial and organelle ATP-dependent Lon proteases. This domain includes the conserved Ser-Lys catalytic dyad. The RdRp encoded by segment B displays evidence of partial conservation with other viral RdRps. Whereas the motifs A and B of the RdRp's palm subdomain were identified in the birnavirus RdRps, the motif C, which contains the Asp-Asp dipeptide critical for enzymatic activity, was not evident in IPNV, DXV and BSNV RdRps. A recent study has shown that birnavirus RdRps have sequence conservation with viral RdRps of Thosea asigna virus (TaV, [AF82930]) and Euprosterna elaeasa virus (EeV, [AF461742]), two insect viruses with a single-stranded RNA with positive polarity (*Tetraviridae*). Homologous domains include a putative RdRp with a unique C-A-B motif arrangement in the palm subdomain. Protein modeling shows that the canonical palm subdomain could accommodate the identified sequence permutation through few changes in backbone connectivity of the major structural elements of the palm subdomain. These RdRps would form a minor and deeply separated cluster in the viral RdRp tree.

Picobirnavirus, a suggested name for a possible new genus of viruses detected in humans and several species of animals, is different from the existing members of the family *Birnaviridae*. The viruses are detected in the fecal material of animals with and without diarrhea. These viruses are 30-40 nm in diameter and have icosahedral symmetry with triangulation number T=3. Their buoyant density in CsCl is 1.4 g/cm^3. Their genome is bi- or trisegmented dsRNA with segment lengths of 2.6 and 1.9 kbp for the bisegmented members of the group, and 2.9, 2.4, and 0.9 kbp for the trisegmented genomes. Genome sequence analysis of human and rabbit isolates (AF246939, AF246940, AF246941 and AJ244022) also supports the view that picobirnaviruses constitute a distinct family of viruses. Human picobirnavirus RdRp is predicted to possess the canonical A-B-C motif arrangement of the palm subdomain.

DERIVATION OF NAMES

Aqua: from Latin *aqua*, "water."
Avi: from Latin *avis*, "bird."

Birna: Bi from Latin prefix *bi*, "two," signifies the bisegmented nature of the viral genome as well as the presence of dsRNA. *Rna:* Sigla from *r*ibo *n*ucleic *a*cid, indicating the nature of the viral genome.
Entomo: from Greek *entomon*, "insect."

REFERENCES

Birghan, C., Mundt, E. and Gorbalenya, A.E. (2000). A non-canonical lon proteinase lacking the ATPase domain employs the ser-Lys catalytic dyad to exercise broad control over the life cycle of a double-stranded RNA virus. *EMBO J.*, **19**, 114-123.

Blake, S., Ma, J.-Y., Caporale, D. A., Jairath, S. and Nicholson, B. L. (2001). Phylogenetic relationships of aquatic birnaviruses based on deduced amino acid sequences of genome segment A cDNA. *Dis. Aquat. Organ.*, **45**, 89-102.

Bottcher, B., Kiselev, N.A., Ste'maschuk, V.Y., Perevoxchikova, N.A., Borisov, A.V. and Crowther, R.A. (1997). Three-dimensional structure of infectious bursal disease virus determined by electron cryomicroscopy. *J. Virol.*, **71**, 325-330.

Chung, H.K., Kordyban, S., Cameron, L. and Dobos, P. (1996). Sequence analysis of the bicistronic Drosophila X virus genome Segment A and its encoded polypeptides. *Virology*, **225**, 359-368.

Da Costa, B., Soignier, S., Chevalier, C., Henry, C., Thory, C., Huet, J.-C. and Delmas, B. (2003). Blotched snakehead virus is a new aquatic birnavirus that is slightly more related to avibirnavirus than to aquabirnavirus. *J. Virol.*, **77**, 719-725.

Da Costa, B., Chevalier, C., Henry, C., Huet, J.-C., Petit, S., Lepault, J., Boot, H. and Delmas, B. (2002). The capsid of infectious bursal disease virus contains several small peptides arising from the maturation process of pVP2. *J. Virol.*, **76**, 2393-2402.

Dobos, P. (1995). The molecular biology of Infectious pancreatic necrosis virus. *Annu. Rev. Fish Dis.*, **5**, 25-54.

Gorbalenya, A.E., Pringle, F.M., Zeddam, J.L., Luke, B.T., Cameron, C.E., Kalmakoff, J., Hanzlik, T.N., Gordon, K.H. and Ward, V.K. (2002). The palm subdomain-based active site is internally permuted in viral RNA-dependent RNA polymerases of an ancient lineage. *J. Mol. Biol.*, **324**, 47-62.

Granzow, H., Birghan, C., Mettenleiter, T.C., Beyer, J., Kollner, B. and Mundt, E. (1997). A second form of infectious bursal disease virus-associated tubule contains VP4. *J. Virol.*, **71**, 8879-8885.

Green, J., Gallimore, C. I., Clewley, J.P. and Brown, D.W.G. (1999). Genomic characterization of the large segment of a rabbit picobirnavirus and comparison with the atypical picobirnavirus of *Cryptosporidium parvum*. *Arch. Virol.*, **144**, 2457-2465.

Hong, J.-R., Lin, T.-L., Yang, J.-Y., Hsu, Y.-L. and Wu, J.-L. (1999). Dynamics of nontypical apoptotic morphological changes visualized by green fluorescent protein in living cells with infectious pancreatic necrosis virus infection. *J. Virol.*, **73**, 5056-5063.

John, K.R. and Richards R.H. (1999). Characteristics of a new birnavirus associated with a warm-water fish cell line. *J. Gen. Virol.*, **80**, 2061-2065.

Kibenge, F.S.B., Dhillon, A.S. and Russell, R.G. (1988). Biochemistry and immunology of infectious bursal disease virus. *J. Gen. Virol.*, **19**, 1757-1775.

Lombardo, E., Maraver, A., Caston, J.R., Rivera, J., Fernandez-Arias, A., Serrano, A., Carrascosa, J.L., and Rodriguez, J.F. (1999). VP1, the putative RNA-dependent RNA polymerase of infectious bursal disease virus, forms complexes with the capsid protein VP3, leading to efficient encapsidation into virus-like particles. *J. Virol.*, **73**, 6973-6983.

Mundt, E. and Vakharia, V.N. (1996). Synthetic transcripts of double-stranded birnavirus genome are infectious. *Proc. Natl. Acad. Sci., USA*, **93**, 11131-11136.

Perez, L., Chiou, P.P. and Leong, J.C. (1996). The structural proteins of infectious pancratic necrosis virus are not glycosylated. *J. Virol.*, **70**, 7247-7249.

Rosen, B.I., Fang, Z.Y., Glass, R.I. and Monroe, S.S. (2000). Cloning of human picobirnavirus genomic segments and development of an RT-PCR detection assay. *Virology*, **277**, 316-329.

Yao, K. and Vakharia, V.N. (1998). Generation of infectious pancreatic necrosis virus from cloned cDNA. *J. Virol.*, **72**, 8913-8920.

CONTRIBUTED BY

Delmas, B., Kibenge, F.S.B., Leong, J.C., Mundt, E., Vakharia, V.N. and Wu, J.L

DSRNA

Family Totiviridae

Taxonomic Structure of the Family

Family	***Totiviridae***
Genus	***Totivirus***
Genus	***Giardiavirus***
Genus	***Leishmaniavirus***

Virion Properties

Morphology

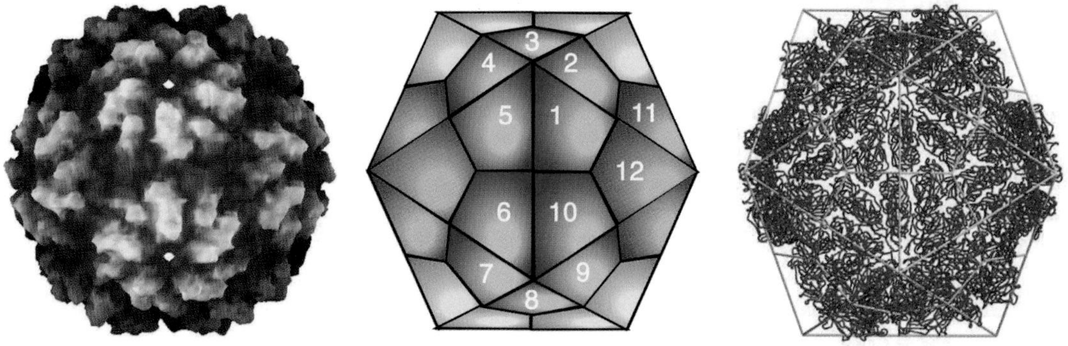

Figure 1: (Left) Reconstruction of the atomic resolution structure of the *Saccharomyces cerevisiae virus L-A* (ScV-L-A) virion. (Center) Schematic representation of a T=1 capsid structure. (Right) The ScV-L-A particle viewed down an icosahedral two-fold axis with the Cα positions traced. Red Gag molecules contact the three-fold axes, but not the two-foldor five- orfive- five-fold axes. Purple molecules surround the five-fold axes and contact the two-fold axes. Thus, two kinds of Gag molecules with identical covalent structure are in distinct environments in the viral particle.

Virions in this family are isometric with no lipid or carbohydrate content reported and no surface projections. There is a close similarity in many aspects of the virus particles (Fig. 1, 2) of members of the family *Totiviridae* and the cores of dsRNA viruses of higher organisms.

Physicochemical and Physical Properties

Virion buoyant density in CsCl is 1.33-1.43 g/cm^3. Additional components with different sedimentation coefficients are found in preparations of some viruses in the genus *Totivirus*. These consist of particles containing satellite or defective dsRNA.

Nucleic Acid

Virions contain a single molecule of linear uncapped dsRNA, 4.6-7.0 kbp in size.

Proteins

Virions contain a single major CP, of 70-100 kDa. Virion-associated RNA polymerase activity is present.

Lipids

None reported.

Carbohydrates

None reported.

Genome Organization and Replication

The genome contains two large, usually overlapping, ORFs: the 5'-proximal ORF encodes the major CP (Gag) and the 3'-proximal ORF encodes an RdRp. Virion-associated RdRp catalyzes *in vitro* end-to-end transcription of dsRNA to produce mRNA for CP and RdRp, by a conservative mechanism. When provided with viral (+) strand template, the RdRp specifically binds (+) strands and catalyzes (-) strand synthesis to form dsRNA. When provided with viral dsRNA, the RdRp carries out *in vitro* transcription in a template-specific reaction. The polymerase is usually expressed as a gag-pol-like fusion protein containing two ORFs.

Antigenic Properties

Not reported.

Biological Properties

These viruses are associated with latent infections of their fungal or protozoal hosts.

Genus Totivirus

Type Species *Saccharomyces cerevisiae virus L-A*

Distinguishing Features

Virus replicates in *Saccharomyces cerevisiae* and supports the replication of one of several satellite dsRNAs (called M dsRNAs) encoding a secreted toxin and immunity to that toxin (killer toxins).

Virion Properties

Morphology

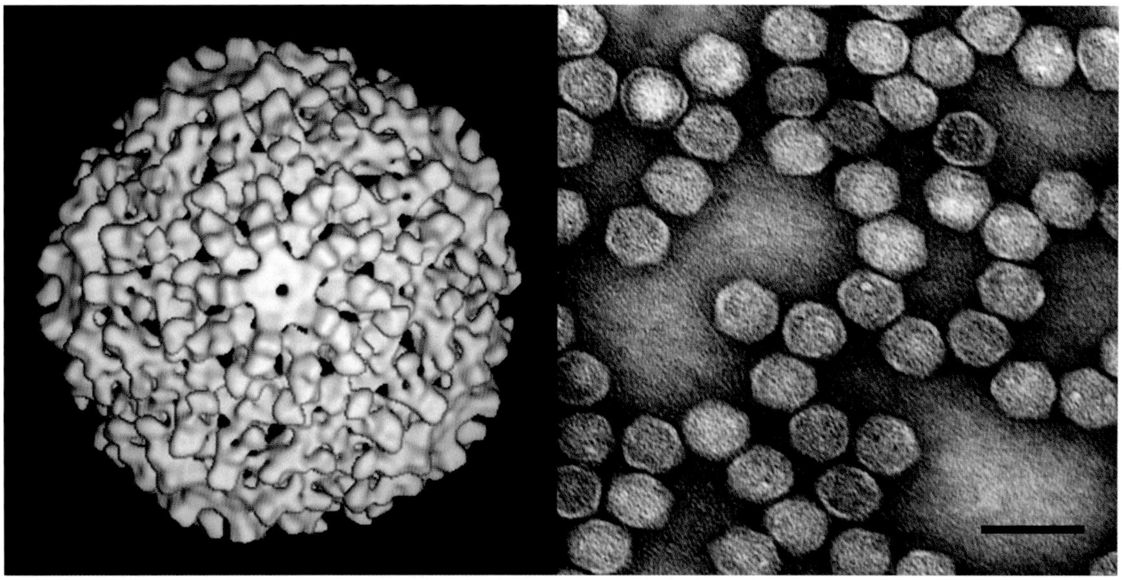

Figure 2: (Left) Cryo-electron microscopic reconstruction of Saccharomyces cerevisiae virus L-A (ScV-L-A) at 16 Å resolution (Caston *et al., 1997*). The view shown is along a 5-fold axis of the icosahedral particles. The virions show the "T=2" symmetry found in the cores of all dsRNA viruses. (Right) Negative contrast electron micrograph of particles of an isolate of Helminthosporium victoriae virus 190S, a representative species in the genus *Totivirus*. The bar represents 50 nm.

Virions are 40 nm in diameter and icosahedral with T=1 with a dimer of the major CP as the assymetric unit (Fig. 1, 2). Pores seen in the capsid structure presumably function in the uptake of nucleotides and the release of viral mRNA.

PHYSICOCHEMICAL AND PHYSICAL PROPERTIES

Virion Mr is estimated as 12.3 x 10^6. Buoyant density in CsCl is 1.40-1.43 g/cm^3 and S_{20w} is 160-190S. Additional components with different sedimentation coefficients and buoyant densities are present in virus isolates with satellite or defective RNAs. Particles lacking nucleic acid have an S_{20w} of 98-113S.

NUCLEIC ACID

Virions contain a single linear molecule of uncapped dsRNA (4.6-6.7 kbp). Some virus isolates contain additional satellite dsRNAs which encode "killer" proteins; these satellites are encapsidated separately in capsids encoded by the helper virus genome. Some virus isolates may contain (additionally or alternatively) defective dsRNAs which arise from the satellite dsRNAs; these additional dsRNAs are also encapsidated separately in capsids encoded by the helper virus genome. The complete RNA sequences of Saccharomyces cerevisiae virus L-A (ScV-L-A) (4,579 bp), Saccharomyces cerevisiae virus L-BC(La) (ScV-L-BC) and Helminthosporium victoriae virus 190S (HvV-190S) dsRNAs are available. The positive strand has two large overlapping ORFs; the length of the overlap varies from 16-130 nt. The first ORF encodes the viral major CP with a predicted size of 76-81 kDa. In the case of ScV-L-A, the two reading frames together encode, via a translational frameshift, the putative RdRp as a fusion protein (analogous to gag-pol fusion proteins of the retroviruses) with a predicted size of 170 kDa. Sites essential for encapsidation, and replication and ribosomal frameshifting have been defined.

PROTEINS

Virions contain a single major CP with of 73-88 kDa. Protein kinase activity is associated with HvV-190S virions; capsids contain phosphorylated forms of the CP. The ScV-L-A Gag removes the 5'-cap structure of host mRNA and covalently attaches it to His154, an activity necessary for expression of viral mRNA. This necessity is relieved by mutation of the host *SKI1/XRN1* 5' exoribonuclease specific for uncapped RNAs (e.g., viral (+) strands). RNA polymerase (replicase-transcriptase) is present. In ScV-L-A virions, RNA polymerase occurs as 1-2 molecules of the 170 kDa fusion protein. The Pol domain of the Gag-Pol fusion protein has three single-stranded RNA binding activities. The N-terminal 1/4 of Pol is necessary for packaging viral (+) strands and includes one of the RNA binding activities. The virion-associated RdRp of HvV-190S is present as a separate nonfused minor protein of 92 kDa that may be expressed by an internal initiation mechanism.

LIPIDS

Virions contain no lipids.

CARBOHYDRATES

None reported.

GENOME ORGANIZATION AND REPLICATION

ScV-L-A virus has a single 4.6 kbp dsRNA segment with two ORFs (Fig. 3). The 5' ORF is *Gag* and encodes the major CP that can bind and covalently remove the 5'-cap structure from mRNAs. The 3' ORF, *Pol*, encodes the RdRp, and has ssRNA binding activity. *Pol* is expressed only as a Gag-Pol fusion protein formed by a -1 frameshift in the 130 bp overlap region between the two ORFs. The -1 ribosomal frameshift is produced by a 72 bp region that has a 7 bp slippery site and an essential pseudoknot structure. The efficiency of frameshifting is critical for viral replication. The HvV-190S RdRp is a separate nonfused virion-associated polypeptide that may be expressed by an internal initiation mechanism.

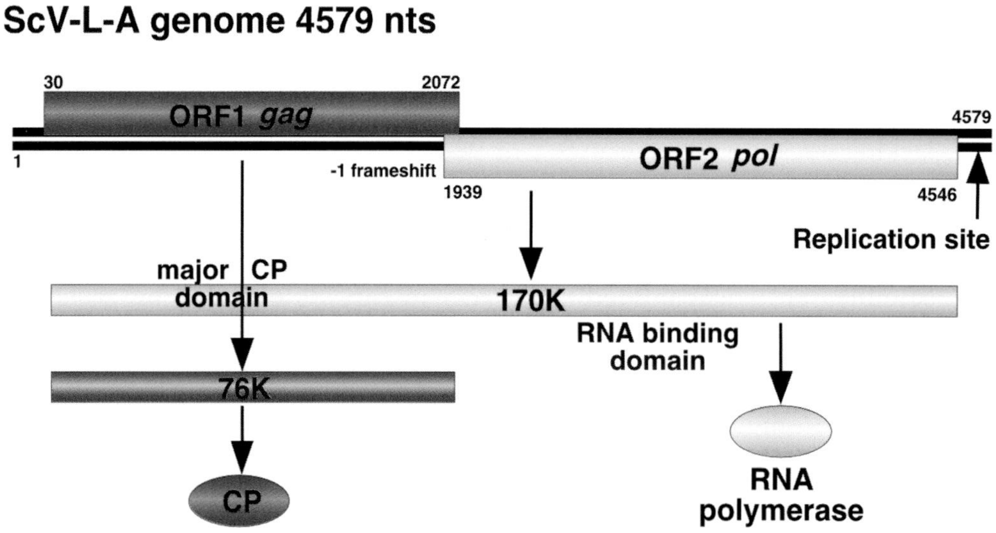

Figure 3: Genome organization of Saccharomyces cerevisiae virus L-A (ScV-L-A). The virion-associated RNA polymerase catalyzes *in vitro* end-to-end transcription of dsRNA by a conservative mechanism to produce mRNA for CP. In the case of ScV-L-A, all of the positive strand transcripts are extruded from the particles. The positive strand of satellite RNA M_1, or deletion mutants of L-A or M_1, on the other hand, often remain within the particle where they are replicated to produce two or more dsRNA molecules per particle (headful replication). The positive ssRNA of ScV-L-A is the molecule encapsidated to form progeny virus particles. The encapsidation signal on ScV-L-A or M_1 positive sense ssRNA is a 24 nt stem-loop sequence located 400 nt from the 3'-end in each case. The *Gag* protein must be acetylated (by the cellular Mak3p) for assembly and packaging to proceed. These particles have a replicase activity that synthesizes the negative strand on the positive strand template to produce dsRNA, thus completing the replication cycle. Replication requires an internal site overlapping the packaging signal, and a specific 3'-end sequence and secondary/tertiary structure. Virions accumulate in the cytoplasm.

ANTIGENIC PROPERTIES

Virions are efficient immunogens.

BIOLOGICAL PROPERTIES

TRANSMISSION

Virions remain intracellular and are transmitted during cell division, sporogenesis, and cell fusion. In some ascomycetes (e.g., *Gaeumannomyces graminis*) virus is usually eliminated during ascospore formation.

HOST RANGE

ScV-L-A depends for its multiplication on host genes *MAK3*, *MAK10*, and *MAK31*. The *MAK3* gene encodes an N-acetyltransferase that acetylates the N-terminus of the major CP and Mak10p and Mak31p are complexed with Mak3p. Reduced levels of 60S ribosomal subunits result in lower levels of ScV-L-A virus and loss of M_1 dsRNA, due to poor translation of the viral poly(A)$^-$ mRNA. The *S. cerevisiae* antiviral gene, *SKI1*, encodes a 5' to 3' exoribonuclease specific for uncapped RNAs (such as viral mRNAs). The *SKI2, 3, 6, 7,* and *8* system blocks expression of non-poly(A) (e.g., viral) mRNAs. Ski2p is an RNA helicase, homologous to nucleolar human homologs. Ski6p is a 3' to 5' exoribonuclease involved in 60S ribosomal subunit biogenesis. If an *SKI* gene is defective, ScV-L-A becomes pathogenic; but only the M dsRNA causes a cytopathogenic effect. Cells become cold sensitive and temperature sensitive for growth.

LIST OF SPECIES DEMARCATION CRITERIA IN THE GENUS

According to the virus species definition, viruses found only in distinct host species are for that reason different species. Totiviruses generally replicate stably within the cell as the cells grow. Different virus strains are expected to segregate relative to each other as the cells grow, whereas different virus species should be stably co-maintained. Viruses of the same species should be similarly affected by host chromosomal mutations. Viruses that can recombine or exchange segments with each other to give viable progeny should be considered the same species. Although these biological criteria are the prime determinants of species, sequence criteria also are used. Less than 50% sequence identity at the protein level generally reflects a species difference. None of the above criteria is absolute, but totiviruses described so far leave little doubt about species demarcation. For example, ScV-L-A and ScV-L-BC are only 30% identical in the 717 aa region of highest similarity. More important they are stably compatible with each other in the same yeast strain, and respond differently to chromosomal mutations. Mutations resulting in loss of ScV-L-BC do not result in loss of ScV-L-A and vice versa.

LIST OF SPECIES IN THE GENUS

Species names are in green italic script; strain names and synonyms are in black roman script; tentative species names are in blue roman script. Sequence accession numbers and assigned abbreviations ( ) are also listed.

SPECIES IN THE GENUS

Helminthosporium victoriae virus 190S		
Helminthosporium victoriae virus 190S	[U41345]	(HvV190S)
Saccharomyces cerevisiae virus L-A (L1)		
Saccharomyces cerevisiae virus L-A (L1)	[J04692, X13426]	(ScV-L-A, L-A)
Saccharomyces cerevisiae virus L-BC (La)		
Saccharomyces cerevisiae virus L-BC (La)	[U01060]	(ScV-L-BC, L-BC)
Ustilago maydis virus H1		
Ustilago maydis virus H1	[NC_003823]	(UmV-H1)

TENTATIVE SPECIES IN THE GENUS

Aspergillus foetidus virus S	(AfV-S)
Aspergillus niger virus S	(AnV-S)
Gaeumannomyces graminis virus 87-1-H	(GgV-87-1-H)
Mycogone perniciosa virus	(MpV)

DSRNA

GENUS GIARDIAVIRUS

Type Species *Giardia lamblia virus*

DISTINGUISHING FEATURES

Giardia lamblia virus (GlV) is a dsRNA virus that replicates in growing *Giardia lablia* and is released into the medium without lysing the host cells. It is the only member of the *Totiviridae* for which transfection has succeeded. Virions are isometric, 36 nm in diameter.

VIRION PROPERTIES

MORPHOLOGY

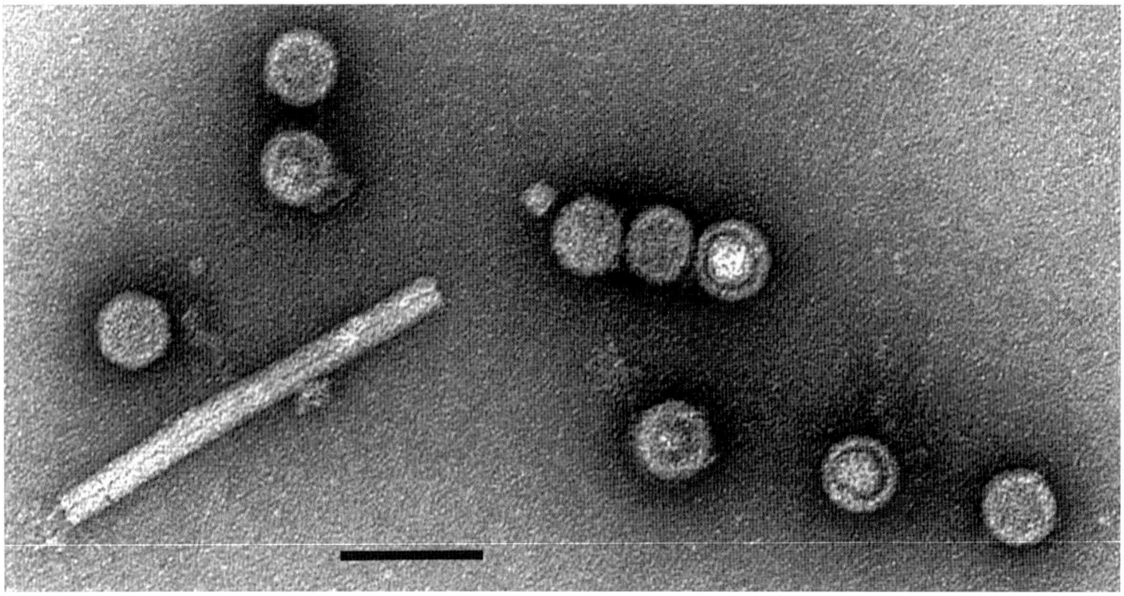

Figure 4: Negative contrast electron micrograph of particles of an isolate of *Giardia lamblia virus*. TMV (rod-shaped) is included as an internal size marker. The bar represents 100 nm.

PHYSICOCHEMICAL AND PHYSICAL PROPERTIES

Virion buoyant density in CsCl is 1.368 g/cm^3.

NUCLEIC ACID

Virions contain a single molecule of dsRNA, 6277 bp in size.

PROTEINS

Virions contain a single major CP of 98 kDa and a viral RNA polymerase of 190 kDa.

LIPIDS

None reported.

CARBOHYDRATES

None reported.

GENOME ORGANIZATION AND REPLICATION

The virus is found in the nuclei of infected *G. lamblia*. Virus replicates without inhibiting the growth of *G. lamblia* trophozoites. Virus is also extruded into the culture medium and the extruded virus can infect many virus-free isolates of the protozoan host. There are isolates of the protozoan parasite, however, that are resistant to infection by GLV. A single-stranded copy of the viral dsRNA genome is present in infected cells. The concentration of the ssRNA observed during the time course of GLV infection is consistent with a role as a viral replicative intermediate or mRNA. The ssRNA is not capped or polyadenylated.

BIOLOGICAL PROPERTIES

The virus infects many isolates of *G. lamblia*, a flagellated protozoan human parasite. The virus does not seem to be associated with the virulence of the parasite. It is not observed in the cyst form of the parasite, and it is not known whether it can be carried through the transformation between cyst and trophozoite. The virus is infectious as purified particles and can infect uninfected *G. lamblia*.

LIST OF SPECIES DEMARCATION CRITERIA IN THE GENUS

According to the virus species definition, viruses found only in distinct host species are for that reason different species. Totiviruses generally replicate stably within the cell as the cells grow. Virus strains of the same species are expected to segregate relative to each other as the cells grow, whereas those of different species should be stably co-maintained. Viruses of the same species should be similarly affected by host chromosomal mutations. Viruses that can recombine or exchange segments with each other to give viable progeny should be considered the same species. Although these biological criteria are the prime determinants of species, sequence criteria also are used. Less than 50% sequence identity at the protein level generally reflects a species difference. None of the above criteria is absolute, but totiviruses described so far leave little doubt about species demarcation.

LIST OF SPECIES IN THE GENUS

Species names are in green italic script; strain names and synonyms are in black roman script; tentative species names are in blue roman script. Sequence accession numbers and assigned abbreviations ( ) are also listed.

SPECIES IN THE GENUS

Giardia lamblia virus

Giardia lamblia virus	[L13218]	(GLV)

TENTATIVE SPECIES IN THE GENUS

Trichomonas vaginalis virus	(TVV)

GENUS LEISHMANIAVIRUS

Type Species *Leishmania RNA virus 1 - 1*

VIRION PROPERTIES

MORPHOLOGY

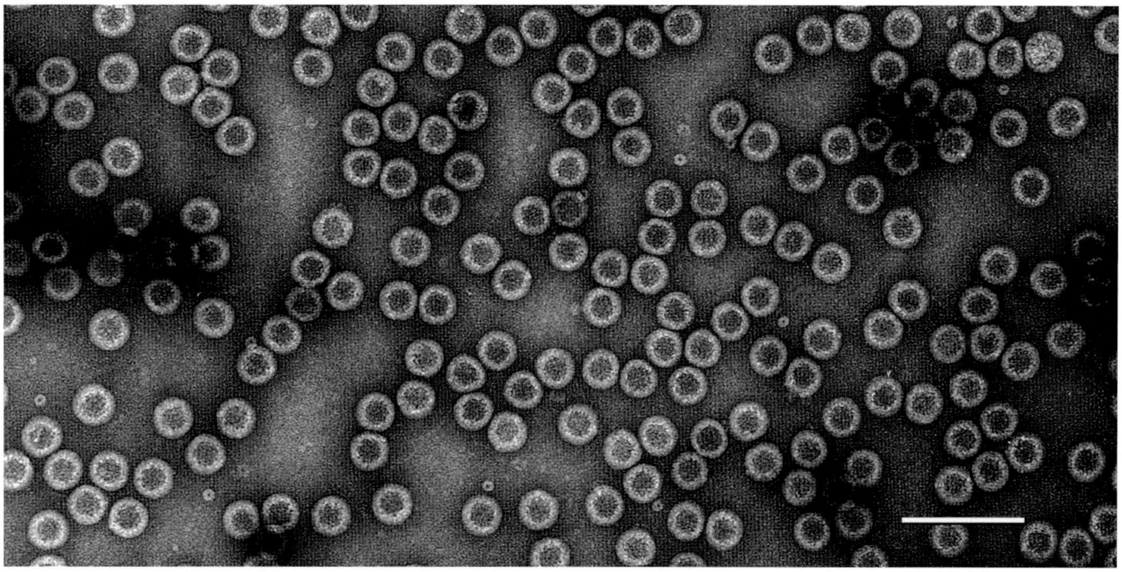

Figure 5: Negative contrast electron micrograph of particles of an isolate of *Leishmania RNA virus 1 - 1*. The bar represents 100 nm.

Virions are isometric, 33 nm in diameter, with no envelope or surface projections.

PHYSICOCHEMICAL AND PHYSICAL PROPERTIES

Virion buoyant density in CsCl is 1.33 g/cm^3.

DSRNA

NUCLEIC ACID
Virions contain a single molecule of linear uncapped dsRNA, 5.3 kbp in size. The complete 5284 nt sequence is available.

PROTEINS
Virions contain a single major CP of 82 kDa.

LIPIDS
None reported.

CARBOHYDRATES
None reported.

GENOME ORGANIZATION AND REPLICATION
The positive strand contains three ORFs. The predicted aa sequence of ORF3 has motifs characteristic of viral RdRp. ORF2 encodes the major CP and overlaps ORF3 by 71 nt, suggesting a +1 translational frameshift to produce a *gag-pol*-like fusion protein with a predicted size of 176 kDa. Sequencing data support the idea that the abundant ssRNA found in infected cells is the message sense RNA.

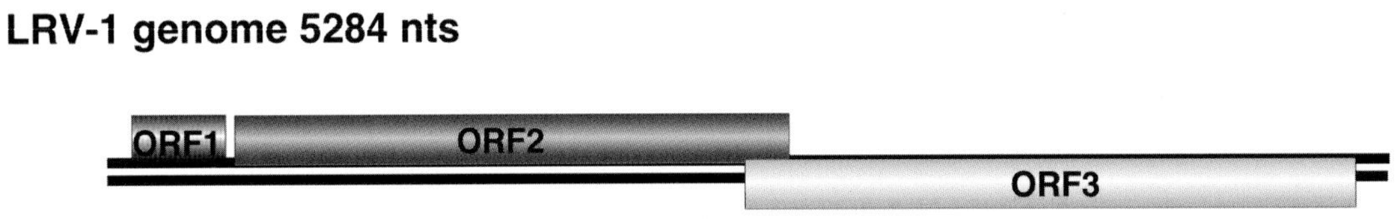

Figure 6: Genome organization of Leishmania RNA virus 1 – 1 (LRV-1-1).

BIOLOGICAL PROPERTIES
Leishmania RNA virus 1 - 1 (LRV-1-1) is found in infected *Leishmania braziliensis* strain CUMC1. Viruses infecting several other strains of *L. braziliensis* and *L. guyanensis* are possibly strains of LRV-1-1. A single strain of *L. major* is known to be infected with LRV-1-1-like virus. The latter is designated LRV-2-1 in order to distinguish it from the viruses infecting new world strains of *Leishmania.*

LIST OF SPECIES DEMARCATION CRITERIA IN THE GENUS
According to the virus species definition, viruses found only in distinct host species are for that reason different species. Totiviruses generally replicate stably within the cell as the cells grow. Virus strains of the same species are expected to segregate relative to each other as the cells grow, whereas those of different species should be stably co-maintained. Viruses of the same species should be similarly affected by host chromosomal mutations. Viruses that can recombine or exchange segments with each other to give viable progeny should be considered the same species. Although these biological criteria are the prime determinants of species, sequence criteria also are used. Less than 50% sequence identity at the protein level generally reflects a species difference. None of the above criteria is absolute, but totiviruses described so far leave little doubt about species demarcation.

LIST OF SPECIES IN THE GENUS
Species names are in green italic script; strain names and synonyms are in black roman script; tentative species names are in blue roman script. Sequence accession numbers and assigned abbreviations ( ) are also listed.

SPECIES IN THE GENUS
Leishmania RNA virus 1 - 1

Leishmania RNA virus 1 - 1	[CUMC1]	[M92355]	(LRV-1-1)

Leishmania RNA virus 1 - 2

Leishmania RNA virus 1 - 2	[CUMC3]	(LRV-1-2)
(Leishmania RNA 2)		(LR2)
Leishmania RNA virus 1 - 3		
Leishmania RNA virus 1 - 3	[M2904]	(LRV-1-3)
Leishmania RNA virus 1 – 4		
Leishmania RNA virus 1 – 4	[M4147]	(LRV-1-4)
(Leishmania B virus)	[U01899]	(LBV)
Leishmania RNA virus 1 – 5		
Leishmania RNA virus 1 – 5	[M1142]	(LRV-1-5)
Leishmania RNA virus 1 – 6		
Leishmania RNA virus 1 – 6	[M1176]	(LRV-1-6)
Leishmania RNA virus 1 – 7		
Leishmania RNA virus 1 – 7	[BOS12]	(LRV-1-7)
Leishmania RNA virus 1 – 8		
Leishmania RNA virus 1 – 8	[BOS16]	(LRV-1-8)
Leishmania RNA virus 1 – 9		
Leishmania RNA virus 1 – 9	[M6200]	(LRV-1-9)
Leishmania RNA virus 2 - 1		
Leishmania RNA virus 2 - 1		(LRV-2-1)
Leishmania RNA virus 1 – 10		
Leishmania RNA virus 1 – 10	[LC76]	(LRV-1-10)
Leishmania RNA virus 1 – 11		
Leishmania RNA virus 1 – 11	[LH77]	(LRV-1-11)
Leishmania RNA virus 1 – 12		
Leishmania RNA virus 1 – 12	[LC56]	(LRV-1-12)

TENTATIVE SPECIES IN THE GENUS

None reported.

PHYLOGENETIC RELATIONSHIPS WITHIN THE FAMILY

Not available.

SIMILARITY WITH OTHER TAXA

The RdRp of each virus has the consensus sequences typical of (+) ssRNA and dsRNA viruses. The capsid structures have the "T=2" structure with 120 monomers, typical of the cores of all dsRNA viruses but of no other viruses. The replication and transcription strategies of the L-A virus resemble those of other dsRNA viruses.

DERIVATION OF NAMES

Giardia derived from the name of the host.
Leishmania derived from the name of the host.
Toti, from *totus*, Latin for 'whole' or 'undivided'.

REFERENCES

Caston, J.R., Trus, B.L., Booy, F.P., Wickner, R.B., Wall, J.S. and Steven, A.C. (1997). Structure of L-A virus: a specialized compartment for the transcription and replication of double-stranded RNA. *J. Cell Biol.*, **138**, 975-985.

Dinman, J.D., Icho, T. and Wickner, R.B. (1991). A -1 ribosomal frameshift in a double stranded RNA virus of yeast forms a gag-pol fusion protein. *Proc. Natl. Acad. Sci. USA*, **88**, 174-178.

Fujimura, T., Esteban, R., Esteban, L.M. and Wickner, R.B. (1990). Portable encapsidation signal of the L-A double-stranded RNA virus of *S. cerevisiae*. *Cell*, **62**, 819-828.

Fujimura, T., Ribas, J.C., Makhov, A.M. and Wickner, R.B. (1992). Pol of Gag-Pol fusion protein required for encapsidation of viral RNA of yeast L-A virus. *Nature*, **359**, 746-749.

DSRNA

Huang, S. and Ghabrial, S.A. (1996). Organization and expression of the double-stranded RNA genome of Helminthosporium victoriae 190S virus, a totivirus infecting a plant pathogenic filamentous fungus. *Proc. Natl. Acad. Sci. USA,* **93**, 12541-12546.

Icho, T. and Wickner, R.B. (1989). The double-stranded RNA genome of yeast virus L-A encodes its own putative RNA polymerase by fusing two open reading frames. J. *Biol. Chem.,* **264**, 6716-6723.

Kang, J., Wu, J., Bruenn, J.A. and Park C. (2001). The H1 double-stranded RNA genome of *Ustilago maydis* virus-H1 encodes a polyprotein that contains structural motifs for capsid polypeptide, papain-like protease, and RNA-dependent RNA polymerase. *Virus Res.,* **76**, 183-9.

Koltin, Y. (1988). The killer systems of Ustilago maydis. In: *Viruses of fungi and simple eukaryotes* (Y. Koltin, and M. Leibowitz, eds), pp 209-243. Marcel Dekker, New York.

Masison, D.C., Blanc. A., Ribas, J.C., Carroll, K., Sonenberg, N. and Wickner, R.B. (1995). Decoying the cap-mRNA degradation system by a dsRNA virus and poly(A)-mRNA surveillance by a yeast antiviral system. *Mol. Cell. Biol.,* **15**, 2763-2771.

Naitow, H., Tang, J., Canady, M., Wickner, R.B. and Johnson, J.E. (2002). L-A virus at 3.4 A resolution reveals particle architecture and mRNA decapping mechanism. *Nat. Struct Biol.,* **9**, 725-728.

Ohtake, Y. and Wickner, R.B. (1995). Yeast virus propagation depends critically on free 60S ribosomal subunit concentration. *Mol. Cell. Biol.,* **15**, 2772-2781.

Park, C.M., Lopinski, J.D., Masuda, J., Tzeng, T.H. and Bruenn, J.A. (1996). A second double-stranded RNA virus from yeast. *Virology,* **216**, 451-454.

Shelbourn, S.L., Day, P.R. and Buck, K.W. (1988). Relationships and functions of virus double-stranded RNA in a P4 killer strain of *Ustilago maydis. J. Gen. Virol.,* **69**, 975-982.

Stuart, K.D., Weeks, R., Guilbride, L. and Myler, P.J. (1992). Molecular organization of Leishmania RNA virus. *Proc. Natl. Acad. Sci. USA,* **89**, 8596-8600.

Tarr, P.I., Aline, R.F., Smiley, B.L., Sholler, J., Keithly, J. and Stuart, K.D. (1988). LR1: a candidate RNA virus of Leishmania. *Proc. Natl. Acad. Sci. USA,* **85**, 9572-9575.

Tercero, J.C. and Wickner, R.B. (1992). *MAK3* encodes an N-acetyltransferase whose modification of the L-A *gag* N-terminus is necessary for virus particle assembly. *J. Biol. Chem.,* **267**, 20277-20281.

Wang, A.L. and Wang, C.C. (1991). Viruses of the protozoa. *Ann. Rev. Microbiol.,* **45**, 251-263.

White, T.C. and Wang, C.C. (1990). RNA dependent RNA polymerase activity associated with the double-stranded RNA virus of *Giardia lamblia. Nucl. Acids. Res.,* **18**, 553-559.

Widmer, G. and Patterson, J.L. (1991). Genome structure and RNA polymerase activity in Leishmania virus. *J. Virol.,* **65**, 4211-4215.

Wickner, R.B. (1996). Double-stranded RNA viruses of yeast. *Microbiol. Rev.,* **60**, 250-265.

Wickner, R.B. (2001). Viruses of yeasts, fungi and parasitic microorganisms. In: *Fields Virology,* 4th Ed., (D.M. Knipe and P.M. Howley, eds), pp 629-658. Lippincott, Williams and Wilkins, Philadelphia.

CONTRIBUTED BY

Wickner, R.B., Wang, C.C. and Patterson, J.L.

FAMILY PARTITIVIRIDAE

TAXONOMIC STRUCTURE OF THE FAMILY

Family	***Partitiviridae***
Genus	***Partitivirus***
Genus	***Alphacryptovirus***
Genus	***Betacryptovirus***

VIRION PROPERTIES

MORPHOLOGY

Virions are isometric, non-enveloped, 30-40 nm in diameter. Virion symmetry has not been determined.

PHYSICOCHEMICAL AND PHYSICAL PROPERTIES

Virion buoyant density in CsCl is in the range of 1.34-1.39 g/cm^3. Virions are stable in butanol and chloroform.

NUCLEIC ACID

Virions contain two unrelated linear dsRNA segments (1.4-3.0 kbp in size). The two segments of the individual viruses are usually of similar size.

PROTEINS

There is a single major CP. Virion-associated RNA polymerase activity is present.

LIPIDS

None reported.

CARBOHYDRATES

None reported.

GENOME ORGANIZATION AND REPLICATION

The genome consists of two linear dsRNA segments, the smaller usually codes for the CP and the larger usually codes for the virion-associated RNA polymerase. Each dsRNA is monocistronic. *In vitro* transcription/replication occurs by a semi-conservative mechanism. Virions accumulate in the cytoplasm.

ANTIGENIC PROPERTIES

Virions are efficient immunogens. A single precipitin line is formed in gel diffusion tests. Members that are serologically related may be strains of a single virus. No serological relationships between the fungal viruses and the plant viruses in the family *Partitiviridae* have been detected.

BIOLOGICAL PROPERTIES

The viruses are associated with latent infections of their fungal and plant hosts. There are no known natural vectors. The fungal partitiviruses are transmitted intracellularly during cell division and sporogenesis. In some ascomycetes (e.g., *Gaeumannomyces graminis),* virus is usually eliminated during ascospore formation. Experimental transmission of purified fungal partitiviruses has been reported by fusing virions with fungal protoplasts. The plant cryptoviruses are transmitted by ovule and by pollen to the seed embryo. There is no graft transmission and apparently no cell-to-cell transport, except at cell division; seed transmission is the only known mode for the transmission of cryptoviruses.

GENUS PARTITIVIRUS

Type Species *Atkinsonella hypoxylon virus*

VIRION PROPERTIES

MORPHOLOGY

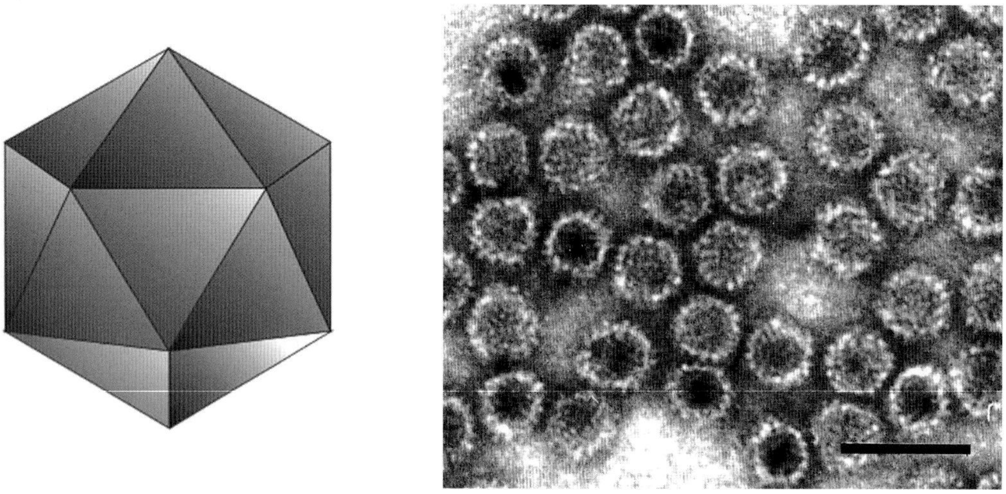

Figure 1: (Left) Diagrammatic representation of a partitivirus capsid. (Right) Negative contrast electron micrograph of virions of an isolate of *Penicillium stoloniferum virus S*, a representative species of the genus *Partitivirus*. The bar represents 50 nm. (From Ghabrial, *et al.*, (2000). Family *Partitiviridae*. In: *Virus Taxonomy: Seventh Report of the International Committee on Taxonomy of Viruses*, pp 503-513, M.H.V. van Regenmortel, *et al.*, eds. Academic Press. Reproduced with permission).

Virions are 30-35 nm in diameter. Negatively stained particles are often penetrated by stain giving the appearance of empty particles even though physical data indicate that they contain dsRNA.

PHYSICOCHEMICAL AND PHYSICAL PROPERTIES

Virion Mr is estimated to range from 6-9 x 10^6. S_{20w} values range from 101-145S. Particles lacking nucleic acid have an S_{20w} of 66-100S. Virion buoyant density in CsCl is 1.29-1.30 and 1.34-1.36 g/cm^3 for particles without and with nucleic acid, respectively. Components with other density values and sedimentation rates are found in preparations of some viruses and are believed to be replicative intermediates. These consist of particles containing ssRNA and particles with both ssRNA and dsRNA. Virus purification is usually carried out at neutral pH.

NUCLEIC ACID

Virions contain two unrelated linear dsRNA segments, 1.4-2.2 kbp in size, which are separately encapsidated. The dsRNA segments of the individual viruses are of similar size. Additional segments of dsRNA (satellite or defective) may be present.

PROTEINS

Virions contain a single major CP of 42-73 kDa. Virion-associated RNA polymerase activity is present.

GENOME ORGANIZATION AND REPLICATION

Atkinsonella hypxylon virus (AhV), an isolate of the type species of the genus *Partitivirus*, has a bipartite genome consisting of dsRNA1 and dsRNA2 (Fig. 2). Each is monocistronic: dsRNA1 codes for the RdRp and dsRNA2 codes for the major CP. The virion-associated RdRp catalyzes *in vitro* end-to-end transcription of each dsRNA to produce mRNA by a semi-conservative mechanism. Virions accumulate in the cytoplasm.

A model for the replication strategy of Penicillium stoloniferum virus S is shown in Figure 3.

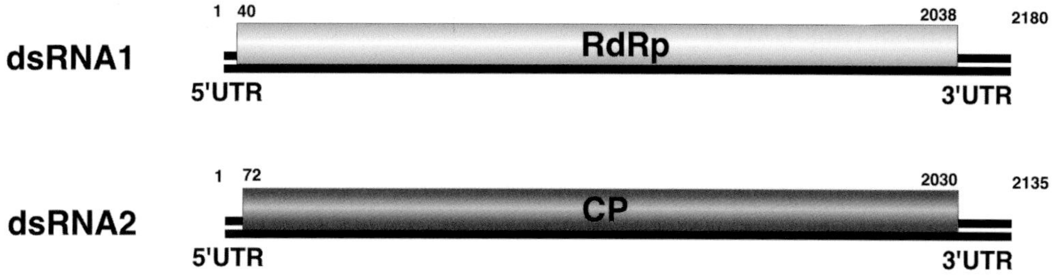

Figure 2: Genome organization of Atkinsonella hypxylon virus (AhV). The RdRp ORF (nt positions 40-2038 on dsRNA1) and the CP ORF (nt positions 72-2030 on dsRNA2) are represented by rectangular boxes.

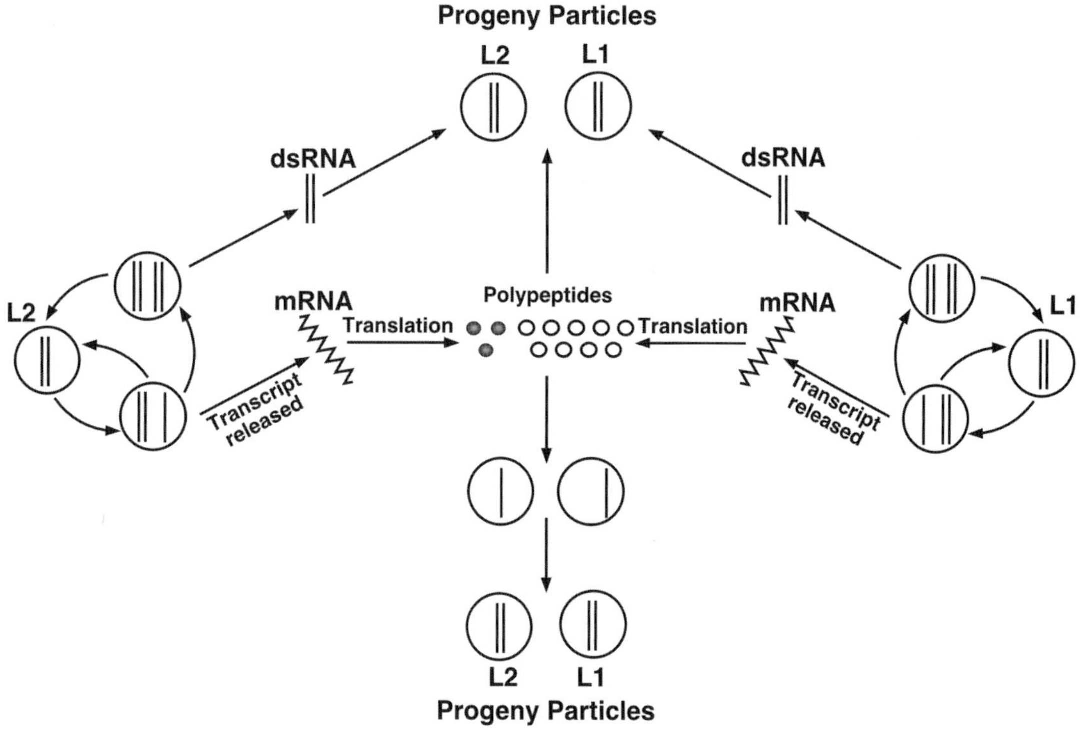

Figure 3: Model for replication of Penicillium stoloniferum S virus (PsV-S). The open circles represent CP subunits and the closed circles represent RNA polymerase subunits. Solid lines represent RNA strands whereas wavy lines represent mRNA.

LIST OF SPECIES DEMARCATION CRITERIA IN THE GENUS

The criteria to differentiate species in the genus are :

- Host species in which the viruses naturally occur; partitiviruses lack natural vectors and they are confined to the fungal host species from which they were first isolated.
- Size of dsRNA segments.
- Protein sequence similarity. Amino acid sequence similarity >40% between RdRps of viruses from different species in the same phylogenetic cluster and <40% between members of species in different clusters (see Fig. 6).
- Serological relationships.

LIST OF SPECIES IN THE GENUS

Species names are in green italic script; strain names and synonyms are in black roman script; tentative species names are in blue roman script. Sequence accession numbers and assigned abbreviations ( ) are also listed.

SPECIES IN THE GENUS

Agaricus bisporus virus 4		
Agaricus bisporus virus 4 (Mushroom virus 4, MV4)		(AbV-4)
Aspergillus ochraceous virus		
Aspergillus ochraceous virus		(AoV)
Atkinsonella hypxylon virus		
Atkinsonella hypxylon virus	[L39125, L39126, L39127]	(AhV)
Discula destructiva virus 1 ‡		
Discula destructiva virus 1	[AF316992, AF316993, AF316994, AF316995]	(DdV-1)
Discula destructiva virus 2 ‡		
Discula destructiva virus 2	[AY033436, AY033437]	(DdV-2)
Fusarium poae virus 1		
Fusarium poae virus 1	[AF015924, AF047013]	(FpV-1)
Fusarium solani virus 1		
Fusarium solani virus 1	[D55668, D55668]	(FsV-1)
Gaeumannomyces graminis virus 019/6-A		
Gaeumannomyces graminis virus 019/6-A		(GgV-019/6-A)
Gaeumannomyces graminis virus T1-A		
Gaeumannomyces graminis virus T1-A		(GgV-T1-A)
Gremmeniella abietina RNA virus MS1 ‡		
Gremmeniella abietina RNA virus MS1	[NC004018, NC004019, NC004020]	(GaRV-MS1)
Helicobasidium mompa virus ‡		
Helicobasidium mompa virus	[AB025903]	(HmV)
Heterobasidion annosum virus ‡		
Heterobasidion annosum virus	[AF473549]	(HaV)
Penicillium stoloniferum virus S		
Penicillium stoloniferum virus S		(PsV-S)
Rhizoctonia solani virus 717		
Rhizoctonia solani virus 717	[AF133290, AF133291]	(RhsV-717)

TENTATIVE SPECIES IN THE GENUS

Diplocarpon rosae virus	(DrV)
Penicillium stoloniferum virus F	(PsV-F)
Phialophora radicicola virus 2-2-A	(PrV-2-2-A)

GENUS ALPHACRYPTOVIRUS

Type Species *White clover cryptic virus 1*

VIRION PROPERTIES

MORPHOLOGY

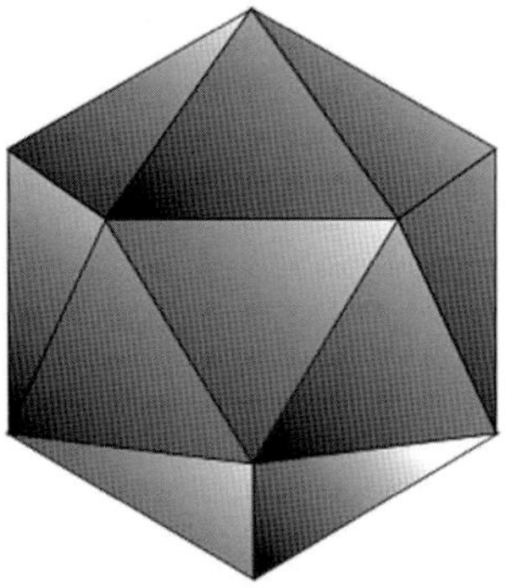

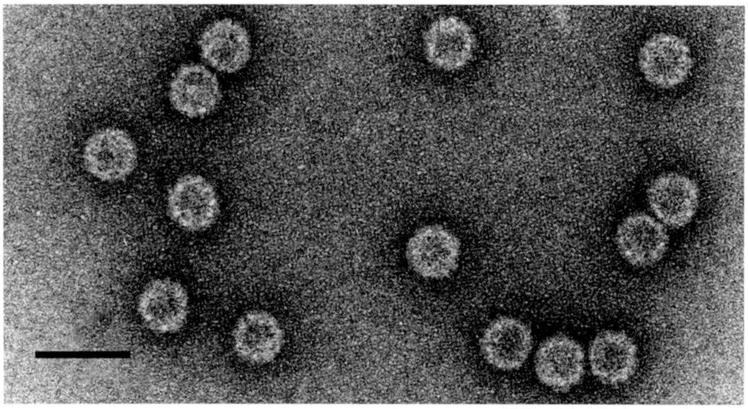

Figure 4: (Left) Diagrammatic representation of an alphacryptovirus capsid. (Right) Negative contrast electron micrograph of particles of an isolate of *White clover cryptic virus 1*, the type species of the genus *Alphacryptovirus*. The bar represents 50 nm. (From Ghabrial, *et al.*, (2000). Family *Partitiviridae*. In: *Virus Taxonomy: Seventh Report of the International Committee on Taxonomy of Viruses*, (M.H.V. van Regenmortel, et al., eds), pp 503-513. Academic Press, San Diego. Reproduced with permission).

Virions are isometric, 30 nm in diameter. Particles lack fine structural detail, appearing rounded, usually penetrated by the stain to give a ring-like appearance.

PHYSICOCHEMICAL AND PHYSICAL PROPERTIES
Density in CsCl is 1.392 g/cm^3.

NUCLEIC ACID
The vVirions typically contain two dsRNA segments, 1.7 and 2.0 kbp in size. The larger dsRNA segment codes for the virion-associated RNA polymerase and the smaller codes for the major CP. The dsRNA segments are believed to be separately packaged.

PROTEINS
Capsids are made up of single polypeptide species (55 kDa). RNA polymerase activity is present.

LIPIDS
None reported.

CARBOHYDRATES
None reported.

ANTIGENIC PROPERTIES

Some viruses in the genus are serologically related; none are related to viruses in the genus *Betacryptovirus*. There are no serological relationships with fungal viruses in the genus *Partitivirus*.

LIST OF SPECIES DEMARCATION CRITERIA IN THE GENUS

The criteria to differentiate species in the genus are :
- host range,
- size of dsRNA segments,
- serological relationships.

Species in the genus are not serologically related (serological differentiation index of 5 or greater). Electrophoretic profiles of the genomic RNAs are distinct.

LIST OF SPECIES IN THE GENUS

Species names are in green italic script; strain names and synonyms are in black roman script; tentative species names are in blue roman script. Sequence accession numbers and assigned abbreviations ( ) are also listed.

SPECIES IN THE GENUS

Alfalfa cryptic virus 1		
Alfalfa cryptic virus 1		(ACV-1)
Beet cryptic virus 1		
Beet cryptic virus 1		(BCV-1)
Beet cryptic virus 2		
Beet cryptic virus 2		(BCV-2)
Beet cryptic virus 3		
Beet cryptic virus 3	[S63913]	(BCV-3)
Carnation cryptic virus 1		
Carnation cryptic virus 1		(CCV-1)
Carrot temperate virus 1		
Carrot temperate virus 1		(CteV-1)
Carrot temperate virus 3		
Carrot temperate virus 3		(CteV-3)
Carrot temperate virus 4		
Carrot temperate virus 4		(CteV-4)
Hop trefoil cryptic virus 1		
Hop trefoil cryptic virus 1		(HTCV-1)
Hop trefoil cryptic virus 3		
Hop trefoil cryptic virus 3		(HTCV-3)
Radish yellow edge virus		
Radish yellow edge virus		(RYEV)
Ryegrass cryptic virus		
Ryegrass cryptic virus		(RGCV)
Spinach temperate virus		
Spinach temperate virus		(SpTV)
Vicia cryptic virus		
Vicia cryptic virus		(VCV)
White clover cryptic virus 1		
White clover cryptic virus 1		(WCCV-1)
White clover cryptic virus 3		
White clover cryptic virus 3		(WCCV-3)

TENTATIVE SPECIES IN THE GENUS

Carnation cryptic virus 2	(CCV-2)
Cucumber cryptic virus	(CuCV)
Fescue cryptic virus	(FCV)
Garland chrysanthemum temperate virus	(GCTV)
Mibuna temperate virus	(MTV)
Poinsettia cryptic virus	(PnCV)
Red pepper cryptic virus 1	(RPCV-1)
Red pepper cryptic virus 2	(RPCV-2)
Rhubarb temperate virus	(RTV)
Santosai temperate virus	(STV)

GENUS BETACRYPTOVIRUS

Type Species *White clover cryptic virus 2*

VIRION PROPERTIES

MORPHOLOGY

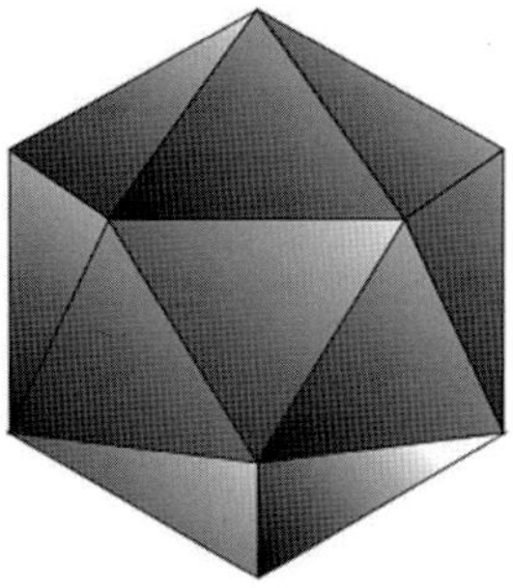

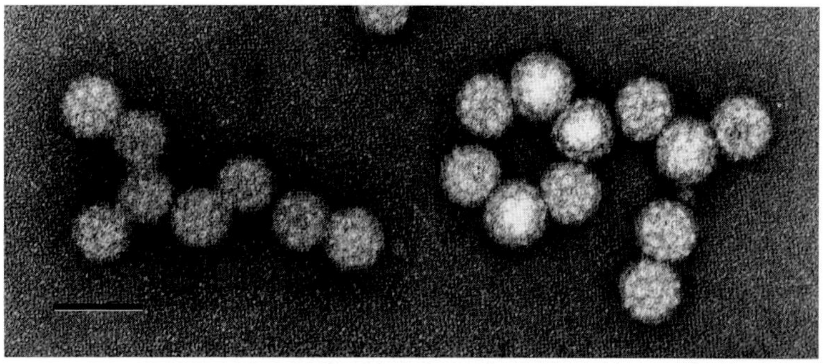

Figure 5: (Left) Diagrammatic representation of a betacryptovirus capsid. (Right) Negative contrast electron micrograph of particles of an isolate of *White clover cryptic virus 2*, the type species of the genus *Betacryptovirus*. The barrepresents 50 nm. (From Ghabrial, et al. (2000). Family *Partitiviridae*. In: *Virus Taxonomy: Seventh Report of the International Committee on Taxonomy of Viruses*, (M.H.V. van Regenmortel *et al.*, eds), pp 503-513. Academic Press, San Diego. Reproduced with permission).

Virions are isometric, 38 nm in diameter. Particles show prominent subunits, but their precise geometrical arrangement is not clear. The particles are rounded and are not penetrated by stain.

PHYSICOCHEMICAL AND PHYSICAL PROPERTIES

Virion buoyant density in CsCl is 1.375 g/cm^3.

NUCLEIC ACID

Viral nucleic acid comprises two dsRNA segments, which are about 2.1 and 2.25 kbp.

PROTEINS

Not characterized.

LIPIDS

None reported.

CARBOHYDRATES

None reported.

ANTIGENIC PROPERTIES

Some viruses in the genus are serologically related; none are related to viruses in the genus *Alphacryptovirus*.

LIST OF SPECIES DEMARCATION CRITERIA IN THE GENUS

The criteria to differentiate species in the genus are :

- host range,
- size of dsRNA segments
- serological relationships.

Species in the genus are not serologically related or are distantly related (serological differentiation index of 5 or greater). Electrophoretic profiles of the genomic RNAs are distinct.

List of Species in the Genus

Species names are in green italic script; strain names and synonyms are in black roman script; tentative species names are in blue roman script. Sequence accession numbers and assigned abbreviations ( ) are also listed.

Species in the Genus

Carrot temperate virus 2
Carrot temperate virus 2 (CTeV-2)
Hop trefoil cryptic virus 2
Hop trefoil cryptic virus 2 (HTCV-2)
Red clover cryptic virus 2
Red clover cryptic virus 2 (RCCV-2)
White clover cryptic virus 2
White clover cryptic virus 2 (WCCV-2)

Tentative Species in the Genus

Alfalfa cryptic virus 2 (ACV-2)

List of Unassigned Viruses in the Family

Pyrus pyrifolia virus [AB012616] (PPV)

Phylogenetic Relationships within the Family

The sequence of the RdRp from a number of partitiviruses has provided a database for assessing phylogenetic relationships within the family *Partitiviridae* (Fig. 6). The results indicate that the two plant members of the family *Partitiviridae* (PPV and BCV-3) are most closely related to each other but cluster with three members of the genus *Partitivirus* (DdV-1, DdV-2 and FsV). The remaining fungal partitiviruses (FpV-1, RsV, AhV, HaV and HmV) form a separate cluster.

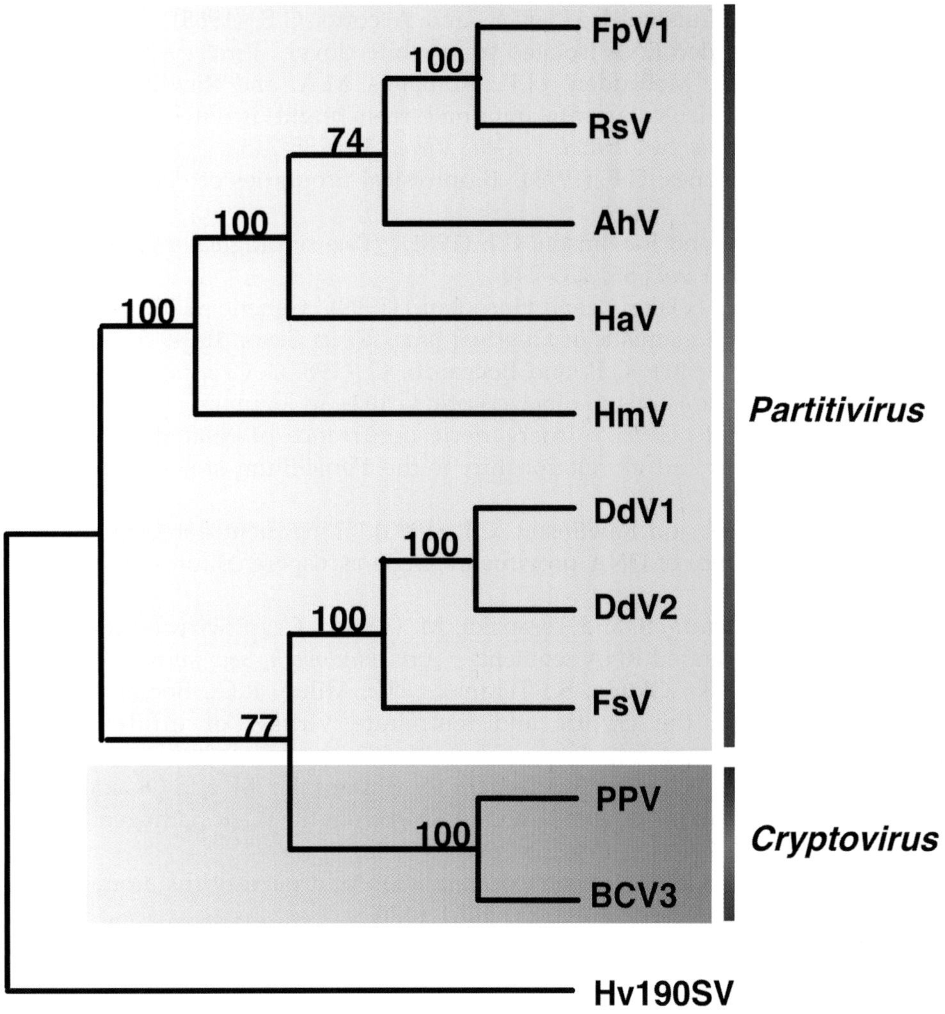

Figure 6: Phylogenetic relationships of the RdRp of partitiviruses. Amino acid sequences were aligned using the Clustal X Multiple Alignment Program and phylogenetic trees were constructed using the neighbor-joining method. The resulting consensus tree of 1000 bootstrap replicates is shown; percent scores are indicated at nodes. The tree was rooted using sequences of the totivirus Helminthosporium victoriae190S virus (Hv190SV) RNA polymerase. Abbreviations: FpV-1, Fusarium poae virus1; RsV-717; Rhizoctonia solani virus 717; AhV, Atkinsonella hypxylon virus; HaV, Heterobasidion annosum virus; HmV, Helicobasidium mompa virus; DdV-1, Discula destructiva virus 1; DdV-2, Discula destructiva virus 2; FsV-1, Fusarium solani virus 1; PPV, Pyrus pyrifolia virus; BCV-3; Beet cryptic virus 3.

Similarity with Other Taxa

Members of the family *Partitiviridae* have properties similar to members of the family *Chrysoviridae*; e.g., the genomes are segmented, genome segments are separately encapsidated, and there are sequence similarities among some of the conserved RdRp motifs.

Derivation of Names

Crypto: from Greek *crypto*, ' hidden, covered, or secret'.
Partiti: from Latin *partitius*, 'divided'.

References

Accotto, G.P., Marzachì, C., Luisoni, E. and Milne, R.G. (1990). Molecular characterization of alfalfa cryptic virus 1. *J. Gen .Virol.*, **71**, 433-437.

Boccardo, G., Milne, R.G., Luisoni, E., Lisa, V. and Accotto, G.P. (1985). Three seedborne cryptic viruses containing double-stranded RNA isolated from white clover. *Virology*, **147**, 29-40.

Buck, K.W., Almond, M.R., McFadden, J.J.P., Romanos, M.A. and Rawlinson, C.J. (1981). Properties of thirteen viruses and virus variants obtained from eight isolates of the wheat take-all fungus, *Gaeumannomyces graminis var. tritici*. *J. Gen. Virol.*, **53**, 235-245.

Buck, K.W. and Kempson-Jones, G.F. (1973). Biophysical properties of Penicillium stoloniferum virus S. *J. Gen. Virol.*, **18**, 223-235.

Buck, K.W., McGinty, R.M. and Rawlinson, C.J. (1981). Two serologically unrelated viruses isolated from a *Phialophora* sp. *J. Gen. Virol.*, **55**, 235-239.

Compel P., Papp I., Bibo M., Fekete, C. and Hornok, L. (1999). Genetic relationships and genome organization of double-stranded RNA elements of *Fusarium poae*. *Virus Genes*, **18**, 49-56.

Luisoni, E., Milne, R.G., Accotto, G.P. and Boccardo, G. (1987). Cryptic viruses in hop trefoil (*Medicago lupulina*) and their relationships to other cryptic viruses in legumes. *Intervirology*, **28**, 144-156.

Kim, J.W. and Bozarth, R.F. (1985). Intergeneric occurrence of related fungal viruses: the *Aspergillus ochraceous* virus complex and its relationship to the Penicillium stoloniferum virus S. *J. Gen. Virol.*, **66**, 1991-2002.

McFadden, J.J.P., Buck, K.W. and Rawlinson, C.J. (1983). Infrequent transmission of double-stranded RNA virus particles but absence of DNA provirus in single ascospore cultures of *Gaeumannomyces graminis*. *J. Gen. Virol.*, **64**, 927-937.

Natsuaki, T., Muroi, Y., Okuda, S. and Teranaka, M. (1990). Cryptoviruses and RNA-RNA hybridization among their double-stranded RNA segments. *Ann. Phytopath. Soc. Japan*, **56**, 354-358.

Natsuaki, T., Natsuaki, K.T., Okuda, S., Teranaka, M., Milne, R.G., Boccardo, G. and Luisoni, E. (1986). Relationships between the cryptic and temperate viruses of alfalfa, beet and white clover. *Intervirology*, **25**, 69-75.

Nogawa, M., Kageyama, T., Nakatani A., Taguchi G., Shimosaka M. and Okazaki, M. (1996). Cloning and characterization of mycovirus double-stranded RNA from the plant pathogenic fungus, *Fusarium solani* f. sp. robiniae. *Biosci. Biotechnol. Biochem.*, **60**, 784-788.

Oh, C.-S. and Hillman, B.I. (1995). Genome organization of a partitivirus from the filamentous ascomycete *Atkinsonella hypoxylon*. *J. Gen. Virol.*, **76**, 1461-1470.

Osaki, H., Nomura, K., Iwanami, T., Kanematsu, S., Okabe, I., Matsumoto, N., Sasaki, A. and Ohtsu, Y. (2002). Detection of double-stranded RNA virus from a strain of the violet root rot fungus *Helicobasidium mompa* Tanaka. *Virus Genes*, **25**, 139-145.

Rong, R., Rao, S., Scott, S.W., Carner, G.R. and Tainter, F.H. (2002). Complete sequence of the genome of two dsRNA viruses from *Discula destructive*. *Virus Res.*, **90**, 217-224.

Strauss, E.E., Lakshman, D.K. and Tavantzis, S.M. (2000). Molecular characterization of the genome of a partitivirus from the basidiomycete *Rhizoctonia solani*. *J. Gen. Virol.*, **81**, 549-555.

Xie, W.S., Antoniw, J.F. and White, R.F. (1993). Nucleotide sequence of beet cryptic virus 3 dsRNA 2 which encodes a putative RNA-dependent RNA polymerase. *J. Gen. Virol.*, **74**, 1467-1470.

CONTRIBUTED BY

Ghabrial, S.A., Buck, K.W., Hillman, B.I. and Milne, R.G.

FAMILY CHRYSOVIRIDAE

TAXONOMIC STRUCTURE OF THE FAMILY

Family *Chrysoviridae*
Genus *Chrysovirus*

Since only one genus is currently recognized, the family description corresponds to the genus description.

GENUS CHRYSOVIRUS

Type Species *Penicillium chrysogenum virus*

VIRION PROPERTIES

MORPHOLOGY

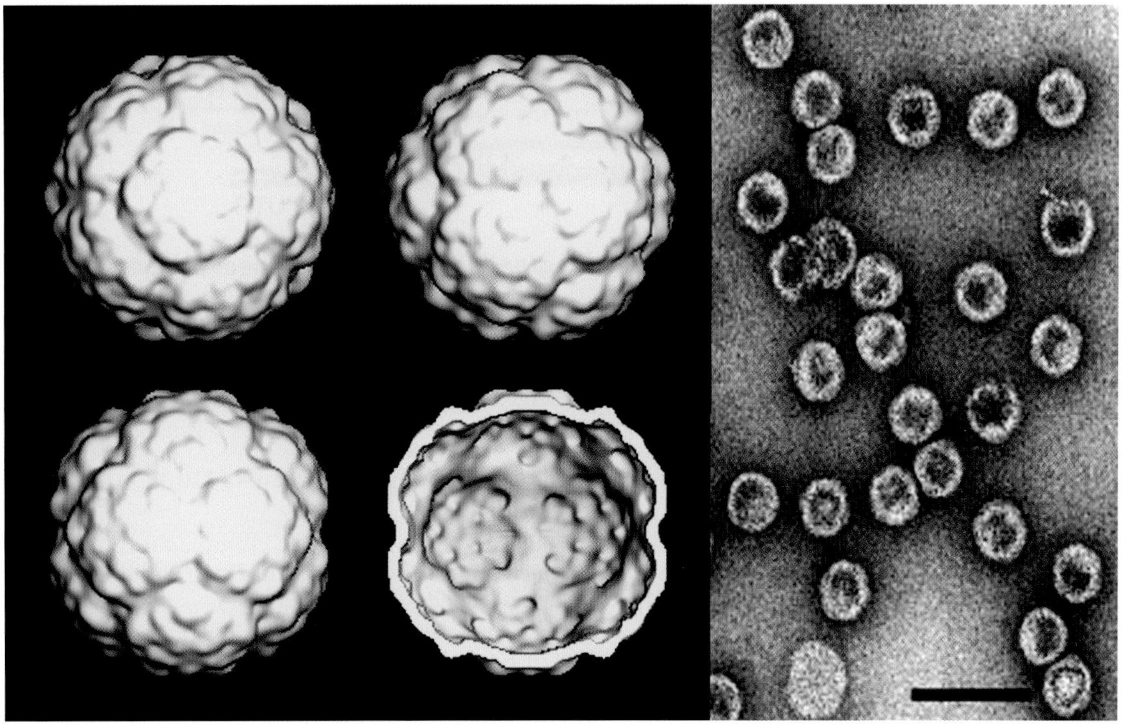

Figure 1: (Left) Cryoelectron microscopy three-dimension reconstructions of the outer surfaces of full particles of Penicillium chrysogenum virus (PcV) viewed along a five-fold (top left), three-fold (top right) and two-fold (bottom left). A model with the front half of the protein shell removed viewed along a twofold axis is shown (bottom right). (Right) Negative contrast electron micrograph of particles of an isolate of *Penicillium chrysogenum virus*, the type species of the genus *Chrysovirus*. The bar represents 100 nm. (Courtesy J.R. Castón).

Virions are isometric, non-enveloped, 35-40 nm in diameter. The capsid is composed of 60 protein subunit monomers arranged on a T=1 icosahedral lattice (Fig. 1).

PHYSICOCHEMICAL AND PHYSICAL PROPERTIES

Virion buoyant density in CsCl is in the range of 1.34-1.39 g/cm^3, and S_{20w} is 145-150S.

NUCLEIC ACID

Virions contain four unrelated linear, separately encapsidated, dsRNA segments (2.4-3.6 kbp in size, Table 1). The largest segment, dsRNA-1, codes for the virion-associated RNA polymerase and dsRNA-2 codes for the major CP. Both dsRNAs 3 and 4 code for proteins of unknown function. Sequences at the 5′- and 3′-UTRs are highly conserved among the four dsRNA segments. The 5′-UTRs are relatively long, between 140 and 400 nt in length. In addition to the absolutely conserved 5′- and 3′-termini, a 40-75 nt region with high sequence identity is present in the 5′-UTR of all four dsRNAs (box 1; Fig. 2). A second region of strong sequence similarity is present immediately downstream from Box 1 (Fig. 2). This consists of a stretch of 30-50 nt containing a reiteration of the sequence "CAA". The $(CAA)_n$ repeats are similar to the enhancer elements present at the 5′-UTRs of tobamoviruses.

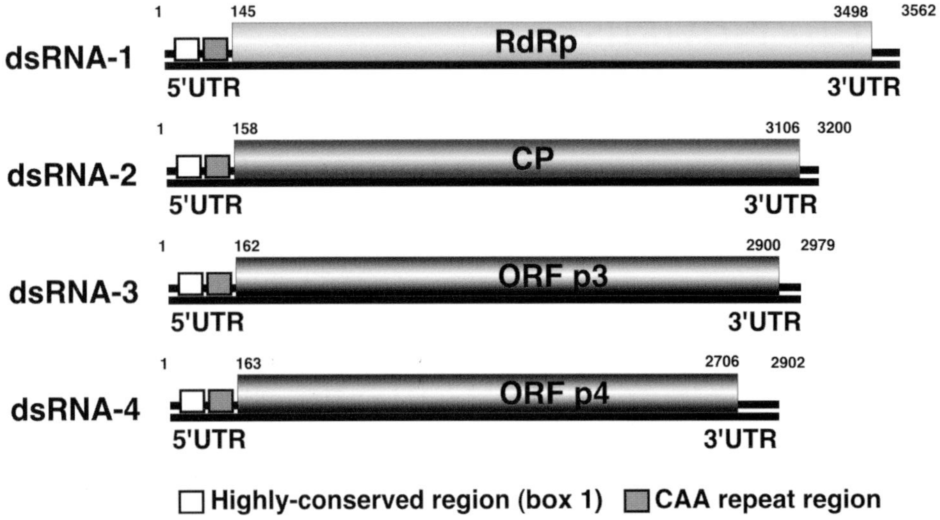

Figure 2: Genome organization of Penicillium chrysogenum virus (PcV). The genome consists of four dsRNA segments; each is monocistronic. The RdRp ORF (nt positions 145 to 3,498 on dsRNA-1), the CP ORF (nt positions 158 to 3,106 on dsRNA-2), the p3 ORF (nt positions 162 to 2900 on dsRNA-3) and the p4 ORF (nt positions 163 to 2706 on dsRNA-4) are represented by rectangular boxes.

Proteins

The capsids are made up of single major polypeptide species (110 kDa). Virion-associated RNA polymerase activity is present.

Table 1: List of the dsRNA segments of Penicillium chrysogenum virus (PcV) with their size (bp), calculated size of their encoded proteins, and function are indicated

Segment	Size (bp)[a]	Size of encoded protein (Da)[b]	Function of protein
DsRNA1	3,562	128,548 (1117 aa)	RdRp
DsRNA2	3,200	108,806 (982 aa)	CP
DsRNA3	2,976	101,458 (912 aa)	Unknown
DsRNA4	2,902	94,900 (847 aa)	Unknown

[a] Size determined by sequencing full-length cDNA clones of the indicated genome segments; bp=base pairs.
[b] Size determined from the predicted aa sequences derived from full-length cDNA clones of the indicated genome segments; The number of aa residues in the encoded viral proteins are placed in parenthesis.

Lipids

None reported.

Carbohydrates

None reported.

Genome Organization and Replication

Penicillium chrysogenum virus (PcV), has a multipartite genome consisting of four dsRNA segments (Fig. 2). Each is monocistronic; dsRNA-1 codes for the RdRp and dsRNA-2 codes for the major CP. Proteins p3 and p4, coded for by dsRNA-3 and dsRNA-4, respectively, are of unknown function (Fig. 2). The virion-associated RdRp catalyzes *in vitro* end-to-end transcription of each dsRNA to produce mRNA by a conservative mechanism. Virions accumulate in the cytoplasm.

Antigenic Properties

Virions are efficient immunogens.

Biological Properties

The viruses are associated with latent infections of their fungal hosts. There are no known natural vectors. The chrysoviruses are transmitted intracellularly during cell division and sporogenesis.

List of Species Demarcation Criteria in the Genus

The criteria to differentiate species in the genus are :

- host range,
- size of dsRNA segments,
- length of 5′-UTR,
- serological relationships.

List of Species in the Genus

Species names are in green italic script; strain names and synonyms are in black roman script; tentative species names are in blue roman script. Sequence accession numbers, and assigned abbreviations ( ) are also listed.

Species in the Genus

Helminthosporium victoriae 145S virus		
Helminthosporium victoriae 145S virus	[AF297176, AF297177, AF297178, AF297179]	(Hv145SV)
Penicillium brevicompactum virus		
Penicillium brevicompactum virus		(PbV)
Penicillium chrysogenum virus		
Penicillium chrysogenum virus	[AF296439, AF296440, AF296441, AF296442]	(PcV)
Penicillium cyaneo-fulvum virus		
Penicillium cyaneo-fulvum virus		(Pc-fV)

Tentative Species in the Genus

Agaricus bisporus virus 1	[X94361, X94362]	(AbV-1)

List of Unassigned Viruses in the Family

None reported.

Phylogenetic Relationships within the Family

The complete nt sequences of the genomes of two members in the genus *Chrysovirus*, Hv145SV and PcV, have been determined and have provided a database for sequence comparisons and for assessing phylogenetic relationships to similar virus families. As the genus *Chysovirus* was originally classified with the family *Partitiviridae*, it was important to compare its phylogenetic relationship to members of this family. The neighbor-joining dendrogram shown in Figure 3, based on multiple alignment of the conserved motifs of the RNA polymerases of chrysoviruses and partitiviruses, indicated that the two chrysoviruses Hv145SV and PcV are most closely related to each other. It is of interest that the previously unclassified AbV-1 with multipartite genome is a sister clade to the chrysoviruses. Furthermore, the chrysoviruses appear to be more closely related to the totivirus UmVH-1 than to the partitiviruses. It is quite clear from the phylogenetic tree that the chrysovirus cluster is distinct from the genetic cluster comprising the partitiviruses.

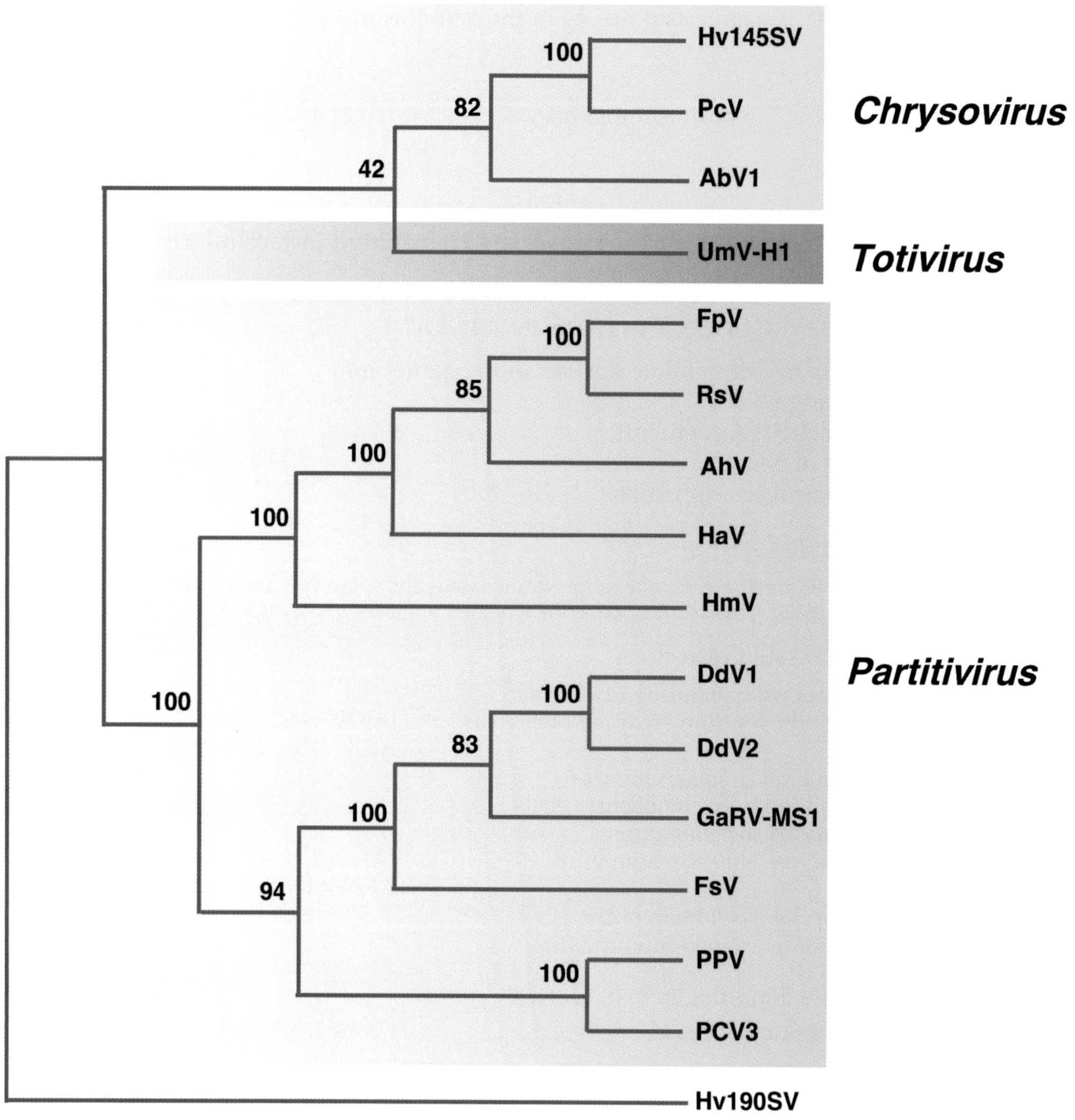

Figure 3: Phylogenetic relationships of the family *Chrysoviridae*. Neighbor-joining dendrogram of sequence relationships among species of the families *Chrysoviridae* and *Partitiviridae*. The aa sequences of the conserved motifs and flanking sequences of the RdRp were aligned using the Clustal X Multiple Alignment Program and bootstrapped 1000 times using the PAUP* program (percent scores are shown at nodes). The tree was rooted using equivalent sequences of the totivirus Helminthosporium victoriae 190S virus (Hv190SV) RNA polymerase. Abbreviations: Hv145SV, Helminthosporium victoriae 145S virus; PcV, Penicillium chrysogenum virus; AbV-1, Agaricus bisporus virus 1; UmV-H1, Ustilago maydis virus H1; FpV-1, Fusarium poae virus1; RsV; Rhizoctonia solani virus 717; AhV, Atkinsonella hypxylon virus; HaV, Heterobasidion annosum virus; HmV, Helicobasidium mompa virus; DdV-1, Discula destructiva virus 1; DdV-2, Discula destructiva virus 2; GaRV-MS1, Gremmeniella abietina RNA virus MS1; FsV-1, Fusarium solani virus 1; PPV, Pyrus pyrifolia virus; BCV-3; Beet cryptic virus 3.

Similarity with Other Taxa

Phylogenetic analysis of the RNA polymerase sequences (Fig. 3) suggests that the previously unclassified AbV1 may belong to the family *Chrysoviridae*, and justifies its designation as a tentative species in the family.

Derivation of Names

Chryso: from the specific epithet of *Penicillium chrysogenum*

References

Buck, K.W. and Grivan, R.F. (1977). Comparison of the biophysical and biochemical properties of Penicillium cyaneo-fulvum virus and Penicillium chrysogenum virus. *J. Gen. Virol.*, **34**, 145-154.

Castón, J.R., Ghabrial, S.A., Jiang, D., Rivas, G., Alfonso, C., Roca, R., Luque, D. and Carrascosa, J.L. (2003). Three-dimensional structure of *Penicillium chrysogenum virus*: a double-stranded RNA virus with a genuine T=1 capsid. *J. Mol. Biol.*, **331**, 417-431.

Edmondson, S.P., Lang, D. and Gray, D.M. (1984). Evidence for sequence heterogeneity among double-stranded RNA segments of Penicillium chrysogenum mycovirus. *J. Gen. Virol.*, **65**, 1591-1599.

Ghabrial, S.A., Soldevila, A.I. and Havens, W.M. (2002). Molecular genetics of the viruses infecting the plant pathogenic fungus *Helminthosporium victoriae.* In: *Molecular Biology of Double-stranded RNA: Concepts and Applications in Agriculture, Forestry and Medicine,* (S. Tavantzis, ed), pp 213-236. CRC Press, Boca Raton, Florida.

Jiang, D. and Ghabrial, S.A. (2004). Molecular characterization of Penicillium chrysogenum virus: reconsideration of the taxonomy of the genus *Chrysovirus. J. Gen. Virol.*, **85**, 2111-2121.

Wood, H.A. and Bozarth, R.F. (1972). Properties of virus-like particles of Penicillium chrysogenum: one double-stranded RNA molecule per particle. *Virology*, **47**, 604-609.

Contributed By

Ghabrial, S.A., Jiang, D. and Castón, J.R.

dsRNA

FAMILY HYPOVIRIDAE

TAXONOMIC STRUCTURE OF THE FAMILY

Family *Hypoviridae*
Genus *Hypovirus*

Since only one genus is currently recognized, the family description corresponds to the genus description.

GENUS HYPOVIRUS

Type Species *Cryphonectria hypovirus 1*

VIRION PROPERTIES

MORPHOLOGY

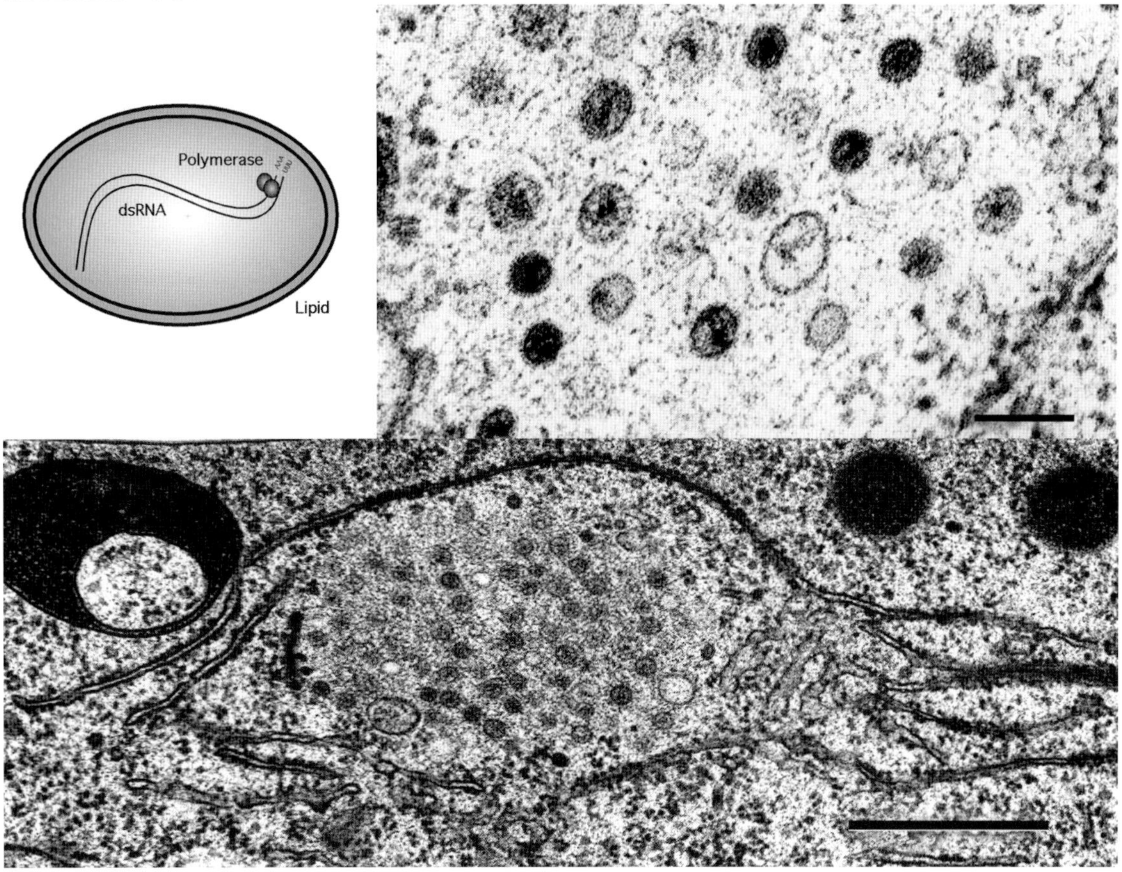

Figure 1: (Top left) Schematic diagram of a vesicle of a member of the family *Hypoviridae*; (Top right) thin section showing vesicles in fungal tissue; (Bottom) thin section showing vesicle aggregate in fungal tissue surrounded by rough ER (from Newhouse *et al*., 1983). The bar represents 100 nm.

No true virions are associated with members of this family. Pleomorphic vesicles 50-80 nm in diameter, devoid of any detectable viral structural proteins but containing dsRNA and polymerase activity are the only virus-associated particles that can be isolated from infected fungal tissue.

PHYSICOCHEMICAL AND PHYSICAL PROPERTIES

Mr of vesicles is unknown. They have a buoyant density in CsCl of approximately 1.27-1.3 g/cm^3 and sediment through sucrose as a broad component of approximately 200S.

DSRNA

Their pH stability is unknown. The vesicles can be purified in pH 5.0 buffer and resuspended in pH 7.0 buffer. pH optimum for polymerase activity *in vitro* is 8.0; the optimum Mg^{++} for polymerase activity is 5 mM. Activity decreases dramatically at pH less than 7.0 or more than 9.0. The vesicles are unstable when heated, or dispersed in lipid solvents. Optimal temperature for polymerase activity is 30°C; temperatures over 40°C inactivate polymerase activity. Deoxycholate at concentrations of more than 0.5% inactivates polymerase activity.

NUCLEIC ACID

Vesicles contain linear dsRNA, approximately 9-13 kbp in size. The genome of an isolate of the type species, Cryphonectria hypovirus 1 (CHV-1), is 12,712 nt. Apparently only one strand is employed in transcription. The coding (positive) strand contains a short 3'-poly(A) tail, which is 20-30 nt in length when analyzed as a component of the dsRNA. Synthetic RNA representing the positive strand of the full-length dsRNA segment is sufficient for infection. The presence of shorter-than-full-length, internally deleted dsRNA molecules is common among some members, and satellite-like dsRNAs are present in other members. No function has been ascribed to any ancillary dsRNA. The 5' terminus of the positive strand of dsRNA from CHV-1 is blocked, but nature of the blocking group is unknown. The 5'-terminus of the negative strand is unblocked. Both 5'-termini of dsRNA from Cryphonectria hypovirus 3-GH2 (CHV-3/GH2) are unblocked.

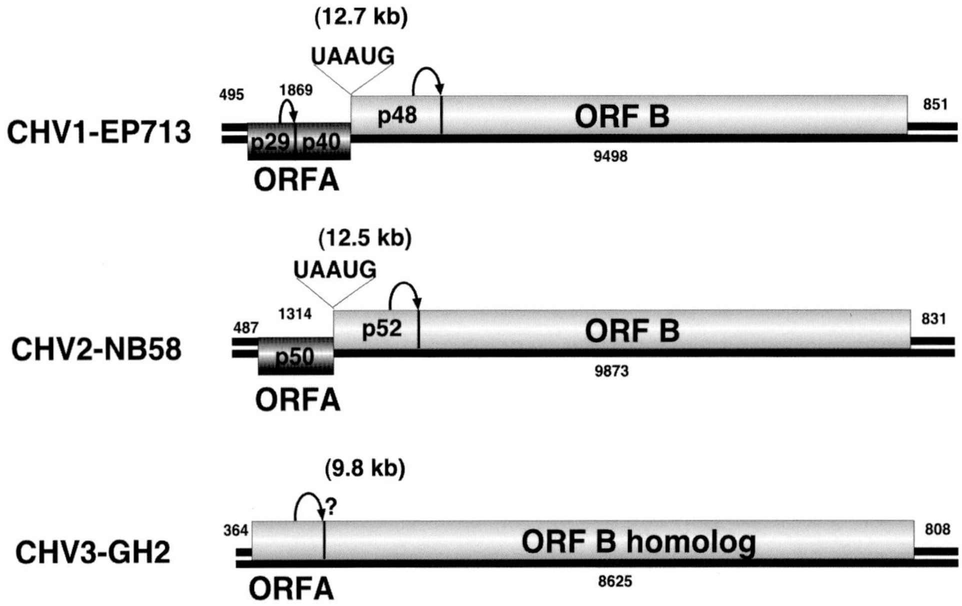

Figure 2: Genome organization of three members of the family *Hypoviridae*. Arrows represent known or suspected sites of autoproteolysis. Genome organization of Cryphonectria hypovirus 4 (CHV-4), which is 9.1 kb, is similar to Cryphonectria hypovirus 3 (CHV-3), but it is unknown whether CHV-4 undergoes autoproteolysis.

PROTEINS

No structural proteins have been described for members of this family. Functions have been assigned to several nonstructural polypeptides encoded by members of the family. The 5'-proximal coding domain, ORFA, of CHV-1/EP713 dsRNA encodes a papain-like protease, p29, and a highly basic protein, p40, derived, respectively, from the N-terminus and C-terminus of polyprotein p69, by a p29-mediated cleavage event. A presumptive NS protein identified *in vitro* and *in vivo*, p29 has been shown by DNA-mediated transformation to contribute to suppression of pigmentation, reduced sporulation, and reduced laccase accumulation. Protein p40 has been shown to enhance viral RNA accumulation. RdRp activity is associated with isolated vesicles of CHV-1/EP713. The calculated size of the CHV-1 ORFB product, which contains putative RNA polymerase and helicase domains and a second papain-like protease, p48, is approximately 250 x 10^3 based on deduced aa sequence from cDNA clones, but no protein of that size has yet been isolated from vesicles. Smaller virus-encoded proteins have been identified in the vesicle-associated polymerase complex, suggesting extensive processing of replication proteins and that ORFB processing occurs *in vivo*. There are no known external viral proteins. The polymerase transcribes ssRNA molecules *in vitro* that correspond in size to full-length dsRNA. Approximately 90% of the polymerase products *in vitro* are of positive polarity.

LIPIDS

Host-derived lipids make up the vesicles that encapsulate the viral dsRNA.

CARBOHYDRATES

Carbohydrates similar to those involved in fungal cell wall synthesis are associated with vesicles.

GENOME ORGANIZATION AND REPLICATION

A 5'-leader of approximately 300-500 nt, including several AUG triplets, precedes the AUG codon that initiates the first long ORF. The viral coding region may be expressed from a single long ORF, or may be divided into two ORFs. If two ORFs are present, the shorter, 5'-proximal ORF is designated ORFA. Its product may or may not be autocatalytically cleaved, depending on the virus. The UAA termination sequence at the end of ORFA is part of the pentanucleotide UAAUG in all members with two ORFs investigated to date. The AUG of the UAAUG pentanucleotide initiates the other long ORF, ORFB. The N-terminal product of ORFB is a papain-like cysteine protease that autocatalytically releases from the growing polypeptide chain (e.g., P48 for CHV-1/EP713 or P52 for CHV-2/NB58). No further processing *in vitro* has been demonstrated for the remaining 300 x 103 polypeptide from this ORF. Phylogenetic relatedness to members of the positive-sense, ssRNA genus *Potyvirus* has been demonstrated by comparisons of protease, polymerase, and helicase domains, although these domains are positioned differently in the two families.

ANTIGENIC PROPERTIES

No antibody has ever been raised from virus particle preparations. Anti-dsRNA antibodies have been used to confirm the genomic constituent. Chimeric β-galactosidase/EP713 ORFA fusion proteins have successfully been used to raise antiserum that is immunoreactive with a virus-specific protein in the infected fungal host, but the location of the protein in the cell is unknown. Antibodies directed against the conserved RNA polymerase domain of ORFB, expressed in bacteria, were used to identify an 87 kDa protein in a CHV-1/EP713 infected isolate.

BIOLOGICAL PROPERTIES

Confirmed members infect the chestnut blight fungus, *Cryphonectria parasitica*. Confirmed members result in reduced virulence (hypovirulence) on chestnut trees and altered fungal morphology in culture, but many possible family members have little or no discernible effect on the fungal host. Some possible members infect other filamentous fungi, e.g., *Sclerotinia sclerotiorum*. Infection of fungal mycelium is known only through fusion, or anastomosis, of infected with uninfected hyphae. The transmission rate through asexual spores (conidia) varies from a few to close to 100%. Transmission through sexual spores (ascospores) is not known to occur. Transmission via cell-free extracts has not been demonstrated but transfection of protoplasts with full-length synthetic transcripts has been successful for CHV-1/EP713. Confirmed members have been identified throughout chestnut growing areas of Europe, North America, and Asia. DsRNA-containing vesicles have been associated with abnormal Golgi apparatus in freeze-substituted thin sections. No nuclear or mitochondrial associations, nor virus-associated inclusions, have been noted.

List of Species Demarcation Criteria in the Genus

Species are differentiated based on major differences in genetic organization or genome expression, as well as by major differences in nucleic acid sequence identity. Thus, CHV-1 differs from CHV-2 in the presence or absence, respectively, of a papain-like proteinase in ORFA. CHV-1 and CHV-2 isolates share less than 60% overall sequence identity with each other. CHV-3 and CHV-4 each contain a single ORF, but isolates of the two species share less than 50% overall sequence identity. Infection by CHV-1 isolates results in a white or near-white phenotype in the fungus; CHV-2 infection results in an orange-brown phenotype; CHV-3 and CHV-4 isolates have little effect on fungal pigment. Infection by members of any of the four species may reduce fungal virulence.

List of Species in the Genus

Species names are in green italic script; strain names and synonyms are in black roman script; tentative species names are in blue roman script. Sequence accession numbers and assigned abbreviations ( ) are also listed.

Species in the Genus

Cryphonectria hypovirus 1		
Cryphonectria hypovirus 1 – EP713	[M57938]	(CHV-1/EP713)
Cryphonectria hypovirus 1 – Euro7	[AF082191]	(CHV-1/Euro7)
Cryphonectria hypovirus 2		
Cryphonectria hypovirus 2-NB58	[L29010]	(CHV-2/NB58)
Cryphonectria hypovirus 3		
Cryphonectria hypovirus 3	[AF188515]	(CHV-3/GH2)
Cryphonectria hypovirus 4		
Cryphonectria hypovirus 4 – SR2	[AY307099]	(CHV-4/SR2)

Many other virus isolates within these four species have been identified by partial sequence analysis.

Tentative Species in the Genus

None reported.

List of Unassigned Viruses in the Family

None reported.

Phylogenetic Relationships within the Family

The hierarchy of relationships among members of this family is unknown. Many strains and isolates of confirmed and tentative members of the family have been identified by RNA hybridization, RFLP, or nt sequence analysis. Among the four species, CHV-1 and CHV-2 are more closely related to each other and CHV-3 and CHV-4 are more closely related to each other.

Similarity with Other Taxa

Deduced aa sequences of polymerase, helicase, and protease motifs of members of the family *Hypoviridae* suggest that their closest relatives are bymoviruses in the family *Potyviridae*.

Derivation of Names

Hypo: from *hypo*virulence

References

Chen, B., Choi, G.H. and Nuss, D.L. (1994). Attenuation of fungal virulence by synthetic infectious hypovirus transcripts. *Science*, **264**, 1762-1764.

Chen, B. and Nuss, D.L. (1999). Infectious cDNA clone of hypovirus CHV1-Euro7: a comparative virology approach to investigate virus-mediated hypovirulence of the chestnut blight fungus *Cryphonectria parasitica*. *J. Virol.*, **73**, 985-992.

Choi, G.H. and Nuss, D.L. (1992). Hypovirulence of chestnut blight fungus conferred by an infectious viral cDNA. *Science*, **257**, 800-803.

Choi, G.H., Pawlyk, D.M. and Nuss, D.L. (1991). The autocatalytic protease p29 encoded by a hypovirulence-associated virus of the chestnut blight fungus resembles the potyvirus encoded protease HC-Pro. *Virology*, **183**, 747-752.

Fahima, T., Kazmierczak, P., Hansen, D.R., Pfeiffer, P. and Van Alfen, N.K. (1993). Membrane associated replication of an unencapsidated double-stranded RNA of the fungus, *Cryphonectria parasitica*. *Virology*, **195**, 81-89.

Fahima, T., Wu, Y., Zhang, L. and Van Alfen, N.K. (1994). Identification of the putative RNA polymerase of Cryphonectria hypovirus in a solubilized replication complex. *J. Virol.*, **68**, 6116 - 6119.

Hansen, D.R., van Alfen, N.K., Gillies, K. and Powell, W.A. (1985). Naked dsRNA associated with hypovirulence of *Endothia parasitica* is packaged in fungal vesicles. *J. Gen. Virol.*, **66**, 2605-2614.

Hillman, B.I., Foglia, R. and Yuan, W. (2000). Satellite and defective RNAs of *Cryphonectria hypovirus 3*, a virus species in the Family *Hypoviridae* with a single open reading frame. *Virology*, **276**, 181-189.

Hillman, B.I., Halpern, B.T. and Brown, M.P. (1994). A viral dsRNA element of the chestnut blight fungus with a distinct genetic organization. *Virology*, **201**, 241-250.

Hiremath, S., L'Hostis, B., Ghabrial, S.A. and Rhoads, R.E. (1986). Terminal structure of hypovirulence-associated dsRNAs in the chestnut blight fungus *Endothia parasitica*. *Nuc. Acid. Res.*, **14**, 9877-9896.

Newhouse, J.R., Hoch, H.C. and MacDonald, W.L. (1983). The ultrastructure of *Endothia parasitica*. Comparison of a virulent with a hypovirulent isolate. *Can. J. Bot.*, **61**, 389-399.

Nuss, D.L. (1996). Using hypoviruses to probe and perturb signal transduction underlying fungal pathogenesis. *Plant Cell*, **8**, 1845-1853.

Peever, T.L., Liu, Y.-C. and Milgroom, M.G. (1997). Diversity of hypoviruses and other double-stranded RNAs of *Cryphonectria parasitica*. *Phytopathology*, **87**, 1026-1033.

Shapira, R., Choi, G.H. and Nuss, D.L. (1991). Virus-like genetic organization and expression strategy for a double-stranded RNA genetic element associated with biological control of chestnut blight. *EMBO J.*, **10**, 731-739.

Shapira, R. and Nuss, D.L. (1991). Gene expression by a hypovirulence-associated virus of the chestnut blight fungus involves two papain-like protease activities. *J. Biol. Chem.*, **266**, 19419-19425.

Smart, C.D., Yuan, W., Foglia, R., Nuss, D.L., Fulbright, D.W. and Hillman, B.I. (1999). *Cryphonectria hypovirus 3*, a virus species in the Family *Hypoviridae* with a single open reading frame. *Virology*, **265**, 66-73.

Suzuki, N., Chen, B. and Nuss, D.L. (1999). Mapping of a hypovirus p29 protease symptom determinant domain with sequence similarity to potyvirus HC-Pro protease. *J. Virol.*, **73**, 9478-9484.

Suzuki, N. and Nuss, D.L. (2002). The contribution of p40 to hypovirus-mediated modulation of fungal host phenotype and viral RNA accumulation. *J. Virol.*, **76**, 7747-7759.

CONTRIBUTED BY

Nuss, D.L., Hillman, B.I., Rigling, D. and Suzuki, N.

GENUS ENDORNAVIRUS

Type Species *Vicia faba endornavirus*

VIRION PROPERTIES

MORPHOLOGY

None reported. Endornaviruses do not produce virions.

PHYSICOCHEMICAL AND PHYSICAL PROPERTIES

None reported.

NUCLEIC ACID

The linear dsRNA genomes of these viruses range in length from ~14 kbp to ~17.6 kbp. Each of the characterized genomes includes a site-specific break (nick) in the coding strand ~1.2 to 2.7 kbp from the 5'-terminus.

PROTEINS

None yet characterized. RdRp activity has been detected in cytoplasmic vesicles, which also contain the genomic dsRNA. Endornaviruses lack virion proteins.

LIPIDS

None yet characterized. A lipid membrane that is probably derived from the host bounds the cytoplasmic vesicles.

CARBOHYDRATES

None yet characterized. Carbohydrate, possibly a glycolipid, has been detected in preparations of the cytoplasmic vesicles in plants infected with endornaviruses.

GENOME ORGANIZATION AND REPLICATION

Each characterized genome encodes a single long polypeptide that crosses the break in the coding strand. These polypeptides include aa sequences typical of RdRps and viral helicases. The polypeptides of Oryza rufipogon virus (ORV) and Oryza sativa virus (OSV) are about 4600 aa residues long and that of Vicia faba virus (VFV) is 5825 aa residues long. RNA replication occurs in cytoplasmic vesicles where RdRp activity has been detected in association with the genomic dsRNA. The cytoplasmic vesicles, sometimes called "virus-like particles", are bounded by a unit membrane and are believed to be the functional equivalent of the replication complexes of the positive-strand RNA viruses. Endornavirus RNA occurs in every tissue and at every developmental stage and is maintained at an almost constant concentration (20-100 copies/cell) except in the pollen of some species. VFV probably does not spread from cell to cell.

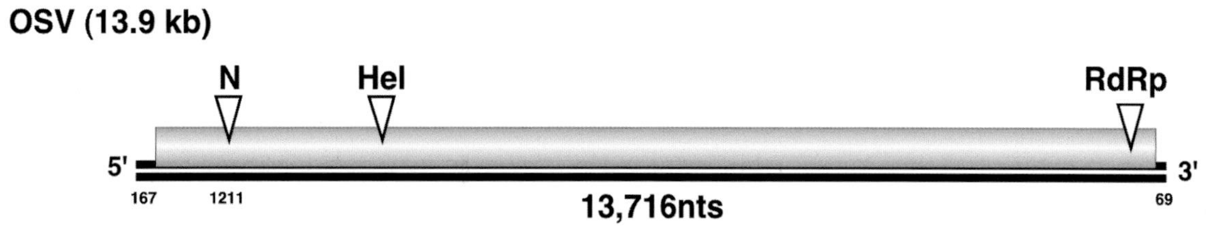

Figure 1: A genome map for Oryza sativa virus (OSV). Triangles mark the positions of the break in the coding strand, the helicase GKT motif and the RdRp GDD motif (N, Hel and RdRp respectively).

ANTIGENIC PROPERTIES

Monoclonal antibodies raised against purified cytoplasmic vesicles from plants infected with VFV allowed early detection of associated male sterility in the progeny of crosses. The antibodies recognized an epitope that contains sugars, possibly a glycolipid.

Biological Properties

Natural infections have been confirmed in some varieties of cultivated rice (*Oryza sativa*), wild rice (*Oryza rufipogon*), broad bean (*Vicia faba*) and kidney bean (*Phaselous vulgaris*). Other plants that may be infected by species from this family include alfalfa, barley, cassava and pepper. Endornaviruses are transmitted through seed via both ova and pollen. No horizontal spread has been observed in the field and no potential vectors have been identified. No attempt to transmit the viruses other than through seed has succeeded, although the deduced phylogeny suggests that inter-species transmission has occurred in the past. Endornaviruses are not mechanically transmissible. None is associated with disease symptoms, except VFV, which induces cytoplasmic male sterility.

List of Species Demarcation Criteria in the Genus

At present, species have been distinguished on the basis of their host-range and sequence differences. The nucleotide sequences of the genomes range in identity from about 30% to 75% and each species has been isolated from a different host species.

List of Species in the Genus

Species names are in green italic script; strain names and synonyms are in black roman script; tentative species names are in blue roman script. Sequence accession numbers and assigned abbreviations ( ) are also listed.

Species in the Genus

Oryza rufipogon endornavirus		
Oryza rufipogon virus	[AB014344]	(ORV)
Oryza sativa endornavirus		
Oryza sativa virus	[AB014343]	(OSV)
Phaseolus vulgaris endornavirus		
Phaseolus vulgaris virus	[X16637]	(PVuV)
Vicia faba endornavirus		
Vicia faba virus	[AJ000929]	(VFV)

Tentative Species in the Genus

None reported.

Phylogenetic Relationships Within the Genus

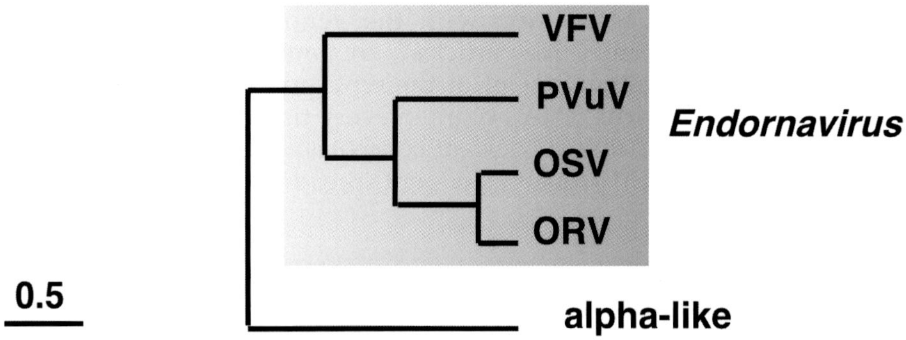

Figure 2: A phylogenetic tree constructed by combining maximum likelihood trees found for the aligned aa sequences of the RdRps and helicases of the endornaviruses and a selection of members of the "alpha-like" virus supergroup. The scale relates branch-lengths to an estimate of substitutions per site. The actual estimate of the evolutionary distance to the alpha-like viruses is about four-times that shown. Trees were found using TreePuzzle version 4 and matched those found from the equivalent nucleotide sequences by using the maximum parsimony and maximum likelihood methods. The same branching pattern was found in 1000 neighbor-joining trees inferred from bootstrap samples.

Similarity with Other Taxa

Comparisons and analyses of RdRp and helicase sequences suggest that members of the *Endornavirus* genus are related to viruses of the "alpha-like" supergroup. .

Derivation of Names

Endo, from Greek: within, and *RNA*

References

Desvoyes, B. and Dulieu, P. (1996). Purification by monoclonal antibody affinity chromatography of virus-like particles associated with the "447" cytoplasmic male sterility in *Vicia faba* and investigation of their antigenic composition. *Plant Science*, **116**, 239-246.

Duc, G., Scalla, R. and Lefebvre, A. (1984). New developments in cytoplasmic male sterility in *Vicia faba*. In: *Vicia faba: Agronomy, Physiology and Breeding*, (P.D. Hebblethwaite, T.C.K. Dawkins, M.C. Heath and G. Lockwood, eds), pp. 254-260. Martinus Nijhoff/Dr. W. Junk Publishers, The Hague.

Dulieu, P., Lemoine, A. and Dulieu, H. (1994). Production of monoclonal antibodies highly specific for the virus-like particles from cytoplasmic male sterile *Vicia faba*. Development of a double sandwich ELISA allowing quantitative analysis of VLPs. *Plant Science*, **97**, 103-108.

Fukuhara, T., Moriyama, H., Pak, J.K., Hyakutake, T. and Nitta, T. (1993). Enigmatic double-stranded RNA in Japonica rice. *Plant Mol. Biol.*, **21**, 1121-1130.

Fukuhara, T., Moriyama, H. and Nitta, T. (1995). The unusual structure of a novel RNA replicon in rice. *J. Biol. Chem.*, **270**, 18147-18149.

Gabriel, C.J., Walsh, R. and Nolt, B.L. (1987). Evidence for a latent virus-like agent in cassava. *Phytopathology* **77**, 92-95.

Gibbs, M.J., Koga, R., Moriyama, H., Pfeiffer, P., Fukuhara, T. (2000). Phylogenetic analysis of some large double-stranded RNA replicons from plants suggests they evolved from a defective single-stranded RNA virus. *J. Gen. Virol.*, **81**, 227-233.

Horiuchi, H., Udagawa, T., Koga, R., Moriyama, H. and Fukuhara T. (2001). RNA-dependent RNA polymerase activity associated with endogenous double-stranded RNA in rice. *Plant Cell Phys.*, **42**, 197 - 203.

Lefebvre, A., Scalla, R. and Pfeiffer, P. (1990). The double-stranded RNA associated with the '447' cytoplasmic male sterility in Vicia faba is packaged together with its replicase in cytoplasmic membranous vesicles. *Plant Mol. Biol.*, **14**, 477-490.

Moriyama, H., Kanaya, K., Wang, J. Z., Nitta, T. and Fukuhara, T. (1996). Stringently and developmentally regulated levels of a cytoplasmic double-stranded RNA and its high-efficiency transmission via egg and pollen in rice. *Plant Mol. Biol.*, **31**, 713-719.

Moriyama, H., Horiuchi, H., Koga, R. and Fukuhara, T. (1999). Molecular characterization of two endogenous double-stranded RNAs in rice and their inheritance by interspecific hybrids. *J. Biol. Chem.*, **274**, 6882-6888.

Pfeiffer, P. (1998). Nucleotide sequence, genetic organization and expression strategy of the double-stranded RNA associated with the '447' cytoplasmic male sterility trait in *Vicia faba*. *J. Gen. Virol.*, **79**, 2349-2358.

Pfeiffer, P. (2002). Large dsRNA genetic elements in plants and the novel dsRNA associated with the "447" cytoplasmic male sterility in *Vicia faba*. In: *dsRNA genetic elements: Concepts and Applications in Agriculture, Forestry, and Medicine*. (S.M. Tavantzis, ed), pp 259-274. CRC Press, Boca Raton, Florida.

Valverde, R.A. and Fontenot, J.F. (1990). Variation in double-stranded ribonucleic acid among pepper cultivars. *J. Amer. Soc. Hort. Sci.*, **116**, 903-905.

Valverde, R.A., Nameth, S., Abdallha, O., Al-Musa, O., Desjardins, P. and Dodds, J.A. (1990). Indigenous double-stranded RNA from pepper (*Capsicum annuum*). *Plant Science*, **67**, 195-201.

Wakarchuk, D.A. and Hamilton, R.I. (1985). Cellular double-stranded RNA in *Phaseolus vulgaris*. *Plant Mol. Biol.*, **5**, 55-63.

Wakarchuk, D.A. and Hamilton, R.I. (1990). Partial nucleotide sequence from enigmatic dsRNAs in *Phaseolus vulgaris*. *Plant Mol. Biol.*, **14**, 637-639.

Zabalgogeazcoa, I.A. and Gildow, F.E. (1992). Double-stranded ribonucleic acid in 'Barsoy' barley. *Plant Science*, **83**, 187-194.

Contributed By

Gibbs, M., Pfeiffer, P. and Fukuhara, T.

The Negative Sense Single Stranded RNA Viruses

ORDER MONONEGAVIRALES

TAXONOMIC STRUCTURE OF THE ORDER

Order	***Mononegavirales***			
Family		***Bornaviridae***		
Genus				***Bornavirus***
Family		***Rhabdoviridae***		
Genus				***Vesiculovirus***
Genus				***Lyssavirus***
Genus				***Ephemerovirus***
Genus				***Novirhabdovirus***
Genus				***Cytorhabdovirus***
Genus				***Nucleorhabdovirus***
Family		***Filoviridae***		
Genus				***Marburgvirus***
Genus				***Ebolavirus***
Family		***Paramyxoviridae***		
Subfamily			***Paramyxovirinae***	
Genus				***Rubulavirus***
Genus				***Avulavirus***
Genus				***Respirovirus***
Genus				***Henipavirus***
Genus				***Morbillivirus***
Subfamily			***Pneumovirinae***	
Genus				***Pneumovirus***
Genus				***Metapneumovirus***

VIRION PROPERTIES

GENERAL

The order comprises four families of enveloped viruses possessing linear non-segmented, negative sense, ssRNA genomes. The families *Filoviridae*, *Rhabdoviridae* and *Paramyxoviridae* have features in common, which taken together with the apparent absence of genetic recombination, suggest a phylogenetic relationship. The common features include the negative strandedness of the monopartite genome, a similar gene order (3′-UTR - core protein genes - envelope protein genes - a polymerase gene - 5′-UTR), helical nucleocapsids, initiation of transcription by a virion associated RdRp from a single 3′-promoter. The genomes of these three members of the order exhibit complementarity of the 3′- and 5′-termini, and 93-99% of the genome is protein-encoding. The ribonucleoprotein cores, but not the deproteinized RNA, are infectious. Maturation is by budding, predominately from the plasma membrane, rarely from internal membranes (rabies virus), or the inner nuclear membrane (several plant viruses). The family *Bornaviridae* exhibits a unique pattern of mRNA processing and has been included in the order on the basis of the negative strandedness of the monopartite ssRNA genome, similarity in genetic organization (Table 1), complementarity of the non-coding 5′- and 3′-termini, and homology of the transcription start and stop signal consensus sequences. The bornaviruses are distinctive because replication and transcription occurs in the nucleus and the polymerase contains a putative nuclear localization signal. Regions of conserved amino acid sequence homology are present in the L (polymerase) proteins, with the highest homology in the putative catalytic domain. The closest relationship of the L protein of an isolate of *Borna disease virus* appears to be with the L protein of an isolate of *Sonchus yellow net virus*, a member of the genus *Nucleorhabdovirus* in the family *Rhabdoviridae*. With the exception of the bornaviruses and the plant viruses classified in the genus

Nucleorhabdovirus, all other members of the Order *Mononegavirales* multiply in the cytoplasm.

MORPHOLOGY

The virions are large enveloped structures with a prominent fringe of peplomers, 5-10 nm long and spaced 7-10 nm apart, in all except the family *Bornaviridae*. The morphologies are variable, but in general distinguish the families: 90 nm diameter spherical particles with a 50 nm diameter electron-dense core and without peplomers in the family *Bornaviridae*; simple, branched, U-shaped, 6-shaped or circular filaments of uniform diameter (about 80 nm) extending up to 14,000 nm are characteristic of viruses classified in the family *Filoviridae*, although purified virions are bacilliform and of uniform length (e.g. 790 nm in the case of Marburg virus); filamentous, pleomorphic or spherical structures of variable diameter are characteristic of viruses belonging to the family *Paramyxoviridae*; and regular bullet-shaped or bacilliform particles are characteristic of the member viruses of the family *Rhabdoviridae*. The ribonucleoprotein core has a diameter of 13-20 nm, which in viruses belonging to the families *Filoviridae* and *Rhabdoviridae* is organized into a helical nucleocapsid of about 50 nm in diameter.

PHYSICOCHEMICAL AND PHYSICAL PROPERTIES

Virion Mr is 300-1,000 x 10^6. S_{20W} is 550-1,045S (plant rhabdoviruses have larger S_{20W} values). Virion buoyant density in CsCl 1.18-1.22 g/cm^3. Virus infectivity is rapidly inactivated by heat treatment at 56°C, or following UV- or X-irradiation, or exposure to lipid solvents.

NUCLEIC ACID

Virions contain one molecule of linear, non-infectious, negative sense, ssRNA, 8.9-19 kb in size, Mr of 3-5 x 10_6 which comprises about 0.5 to 2.0% of the particle weight. The viral RNA lacks a capped 5'-terminus, or a covalently associated protein. The 3'-terminus of viral RNA lacks a poly(A) tract. The 5'- and 3'-terminal regions exhibit inverse complementarity, and there are conserved motives in the terminal regions of all four families. Full-length positive sense (anti-genomic) RNAs are found in infected cells. The genome comprises a linear sequence of genes, with limited overlaps in some viruses, and with short terminal non-coding regions. There are conserved motifs in the transcription start and end signals in all families. The intergenic regions range from two to several hundred nucleotides. Exceptionally, genetic information may be encoded in all three reading frames in the P genes of respiroviruses and morbilliviruses. Splicing of some mRNA and overlapping start/stop signals are characteristic of bornaviruses. In the subfamily *Paramyxovirinae* of the family *Paramyxoviridae*, but not the subfamily *Pneumoviridae*, the number of nt in the genome is divisible by six ("the rule of six"), presumably reflecting a nucleocapsid structural constraint.

PROTEINS

There are a limited number of proteins in relation to the large particle size. The 5-7 structural proteins comprise envelope glycoprotein(s), a matrix protein, a major RNA-binding protein, other nucleocapsid-associated protein(s), plus a large molecular weight polymerase protein, and in some viruses several non-structural proteins which may be phosphorylated. The matrix protein is non-glycosylated in all except the bornaviruses. The matrix protein of *Borna disease virus* is N-glycosylated and expressed on the surface of virions. Enzymatic activities associated with the virions may include transcriptase, polyadenylate transferase, mRNA transferase, and neuraminidase.

LIPIDS

Virions are composed of about 15-25% lipids, their composition reflecting that of the host cell membrane where virions bud. Generally, phospholipids represent about 55-60%, and sterols and glycolipids about 35-40% of the total lipids. Glycoproteins may have a covalently associated fatty acid proximal to the lipid envelope.

CARBOHYDRATES

Virions are composed of about 3% carbohydrate by weight. The carbohydrates are present as N- and O-linked glycan chains on surface proteins and on glycolipids. When made in mammalian cells the oligosaccharide chains are generally of the complex type, in insect cells they are of the non-complex types.

Table 1: A diagrammatic representation of the 3′ to 5′ arrangement of the transcriptional units in the genomes of viruses classified in the four families (*Bornaviridae*, *Filoviridae*, *Paramyxoviridae* and *Rhadoviridae*) comprising the Order *Mononegavirales.* Genes encoding polypeptides of presumed homologous function are aligned vertically.

Abbreviations: viruses; BDV - Borna disease virus; BEFV - Bovine ephemeral fever virus; CDV - Canine distemper virus; HeV - Hendra virus; HMPV - Human metapneumovirus; HRSV - Human respiratory syncytial virus; MARV – Lake Victoria marburgvirus; MeV – Measles virus; MuV – Mumps virus; NDV - Newcastle disease virus; NiV - Nipah virus; PIV3 - Parainfluenza virus type 3; PVM - Pneumonia virus of mice; RABV – Rabies virus; SeV – Sendaï virus; SV5 - Simian virus 5; TRTV - Turkey rhinotracheitis virus; VSV - Vesicular stomatitis Indiana virus; ZEBOV – Zaire ebolavirus; transcriptional units; le – non-coding leader region, NS – non-structural protein gene, N - nucleoprotein gene, P - phosphoprotein gene, V and C - dispensible non-structural protein genes, sc4, 4b and (RT) - genes of unknown function, M and M1 – non-glycosylated matrix protein gene, (M) glycosylated matrix protein gene, F - fusion protein gene, SH - small hydrophobic protein gene, G (or H or HN) - glycosylated (or haemagglutinin or haemagglutinin/neuraminidase) attachment protein gene, M2 - non-glycosylated (BDV excepted) envelope protein gene, Ps - pseudogene, NV – non-virion protein gene, Gns - presumptive duplicated G sequence, L - large (polymerase) protein gene, tr – non-coding trailer region.

FAMILY SUBFAMILY	GENUS	VIRUS	3'						GENE ORDER								5'
Bornaviridae	*Bornavirus*	BDV	le			N	(P)		(M)				(G)			L	tr
Rhabdoviridae	*Vesiculovirus*	VSV	le			N	P		M				G			L	tr
	Lyssavirus	RV	le			N	P		M				G	Ps		L	tr
	Cytorhabdovirus	LNYV	le			N	P	4b	M				G			L	tr
	Nucleorhabdovirus	SYNV	le			N	P	Sc4	M				G			L	tr
	Novirhabdovirus	IHNV	le			N	P		M				G	NV		L	tr
	Ephemerovirus	BEFV	le			N	P		M				G	Gns	(α1.α2.,β.,χ)	L	tr
	"	ARV	le			N	P		M				G	Gns	(α1,α2,β)	L	tr
Filoviridae	*Ebolavirus*	ZEBOV	le			N	P		(M1)				GP/SP	(?)	(M2)	L	tr
	Marburgvirus	MARV	le			N	P		(M1)				G	(?)	(M2)	L	tr
Paramyxoviridae																	
Paramyxovirinae	*Avulavirus*	NDV	le			N	P/V		M	F			H			L	tr
	Henipavirus	HeV	le			N	P/C/V		M	F			H			L	tr
	Morbillivirus	MV	le			N	P/C/V		M	F			H			L	tr
	Respirovirus	SeV	le			N	P/C/V		M	F			H/N			L	tr
	Rubulavirus	MuV	le			N	P/V		M	F		SH	H/N			L	tr
Pneumovirinae	*Metapneumovirus*	TRTV	le			N	P		M1	F	M2	SH	G			L	tr
	Pneumovirus	HRSV	le	NS1	NS2	N	P		M1	SH	G	F	M2			L	tr

GENOME ORGANIZATION AND REPLICATION

In the families *Filoviridae*, *Paramyxoviridae* and *Rhabdoviridae*, discrete mRNAs are transcribed by sequential interrupted synthesis. Transcription is polar with step-wise attenuation. Generally genes do not overlap, the exceptions being the M2 and L genes of pneumoviruses, the VP30 and VP24 of the marburgviruses, and the VP35/VP40, GP/VP30 and VP24/L of ebolaviruses. The site of multiplication is the cytoplasm, with the exception of viruses classified in the genus *Nucleorhabdovirus*. The P genes of respiroviruses and morbilliviruses are exceptional in that additional open reading frames may be accessed via alternative non-AUG start codons, and mRNA editing occurs by insertion of non-templated nucleotides to change the reading frame for expression of P gene products. A non-templated insertion event occurs during transcription of the glycoprotein gene of ebolaviruses generating both membranes-inserted and secreted forms of the glycoprotein. Replication occurs by synthesis of a complete positive sense anti-genomic RNA. Genomic and anti-genomic RNAs are present as nucleocapsids. In the family *Bornaviridae*, the site of multiplication is the nucleus Transcription of bornavirus genomes is complex with splicing of mRNA and overlapping stop/start signals. The mRNAs of bornaviruses are capped, but synthesis is not inhibited by alpha-amanitin suggesting that a cap-snatching mechanism is not involved. In the filoviruses, paramyxoviruses and rhabdoviruses maturation of the independently assembled helical nucleocapsid occurs by budding through host membranes with investment by a host-derived lipid envelope containing transmembrane proteins. The process of assembly and maturation of bornaviruses is not known at present.

ANTIGENIC PROPERTIES

Membrane glycoproteins are involved in antibody-mediated neutralization. Virus serotypes are defined by the surface antigens. Filoviruses are an exception in that they are poorly neutralized *in vitro*. In bornaviruses, antibodies to both the glycosylated matrix protein, which may function as an attachment protein, and the gp94 envelope protein neutralize infectivity.

BIOLOGICAL PROPERTIES

The host ranges vary from restricted to unrestricted. Filoviruses have only been isolated from primates. Paramyxoviruses occur only in vertebrates and no vectors are known. Rhabdoviruses infect invertebrates, vertebrates and plants. Some rhabdoviruses multiply in both invertebrates and vertebrates, some in invertebrates and plants, but none in all three. In human hosts the pathogenic potential tends to be characteristic of the family: i.e. haemorrhagic fever (*Filoviridae*); respiratory and neurological diseases (*Paramyxoviridae*); mild febrile to fatal neurological diseases (*Rhabdoviridae*). Bornaviruses have been isolated from horses, cattle, sheep, rabbits, rats, cats, ostriches, and man. The pathology associated with virus infection is variable. Infection of animals is associated with conditions ranging from behavioral disturbances to severe non-purulent encephalomyelitis. Cytopathology varies from none (bornaviruses and filoviruses) to rapidly lytic (rhabdoviruses and paramyxoviruses); syncytium formation is common in paramyxoviruses.

PHYLOGENETIC RELATIONSHIPS WITHIN THE ORDER

Phylogenetic relationships between the families *Bornaviridae*, *Filoviridae*, *Paramyxoviridae* and *Rhabdoviridae* are illustrated in Figure 1.

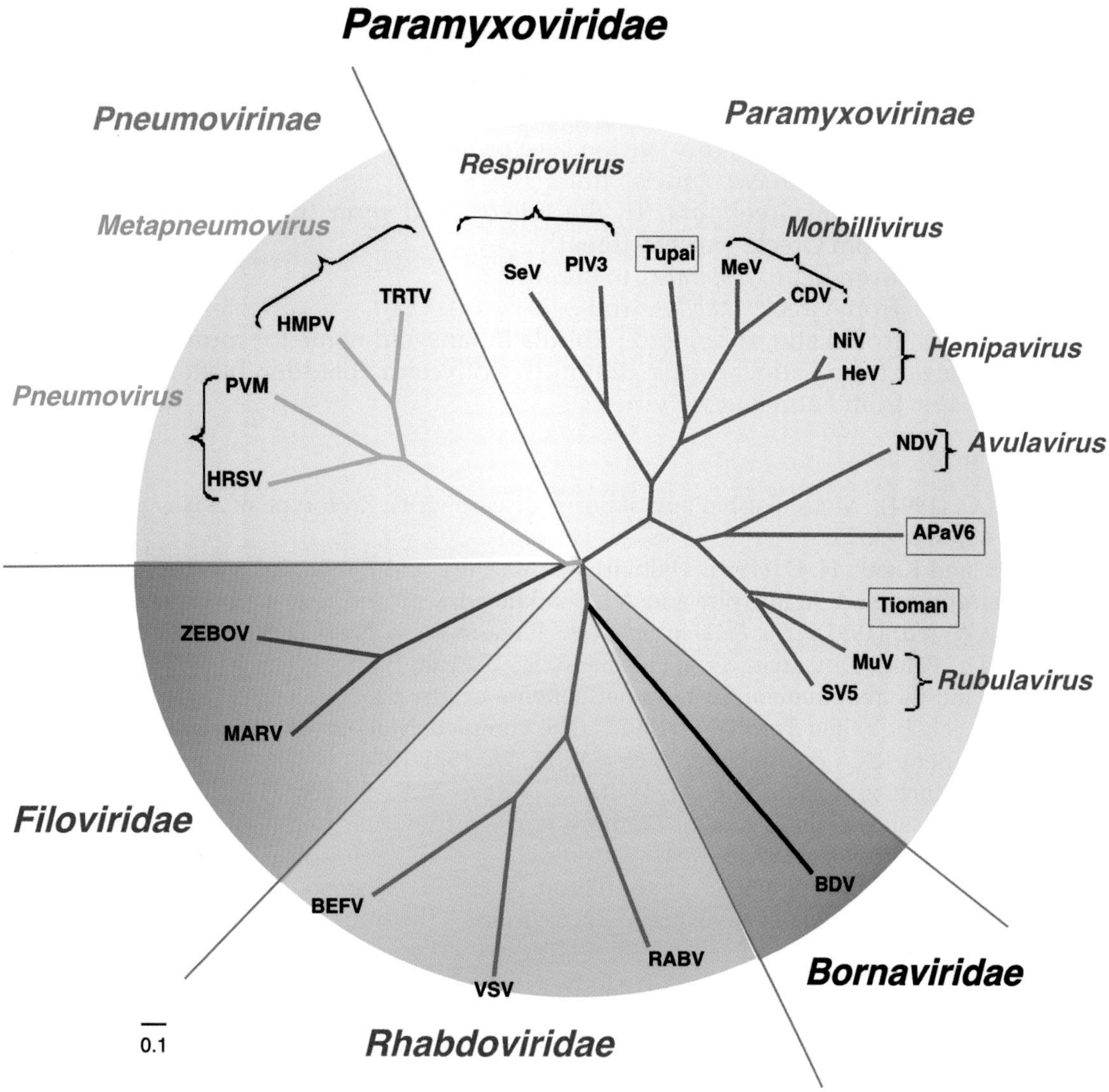

Figure 1: Unrooted phylogenetic tree of members of the Order *Mononegavirales*. The tree was constructed using the CLUSTALX program with the sequences of the conserved domain III of the polymerase proteins (Poch et al., 1989, 1990). Three unclassified paramyxoviruses are included: Tupaia paramyxovirus (Tupaia), Avian parainfluenza virus type 6 (ApaV6) and Tioman virus (Tioman). Abbreviations: BDV - Borna disease virus; BEFV - Bovine ephemeral fever virus; CDV - Canine distemper virus; HeV - Hendra virus; HMPV - Human metapneumovirus; HRSV - Human respiratory syncytial virus; MARV - Marburg virus; MeV - Measles virus; MuV - Mumps virus; NDV - Newcastle disease virus; NiV - Nipah virus; PIV3 - Parainfluenza virus type 3; PVM - Pneumonia virus of mice; RABV - Rabies virus; SeV - Sendaï virus; SV5 - Simian virus 5; TRTV - Turkey rhinotracheitis virus; VSV - Vesicular stomatitis Indiana virus; ZEBOV - Zaire Ebola virus. (Courtesy of Easton, A.J. and Pringle, C.R.).

Similarity with Other Taxa

None reported.

Derivation of Names

Borna: from Borna, a town in Saxony.
Cyto: from Greek, *kytos*, 'cell'.
Ebola: from the river Ebola, in Sudan and Zaire.
Ephemero: from Greek, *ephemeros*, 'short-lived'.
Filo: from Latin, *filo*, 'thread-like'.
Lyssa: from Greek, *lyssa*,'rage, fury, canine madness'.
Marburg: from the city of Marburg, in Germany.

Meta: from Greek, *meta*, 'after'.
Mono: from Greek, *monos*, 'single'.
Morbilli: from Latin *morbillus*, diminutive of morbus, 'disease'.
Nega: from negative sense RNA.
Novi: sigla for the non (no-) virion (vi-) protein gene characterisitic of the genus.
Nucleo: from Latin *Nux*, nucis, 'nut'.
Paramyxo: from Greek *para*, 'by the side of', and *myxa*, 'mucus'.
Pneumo: from Greek, *pneuma*, 'breath'.
Respiro: from Latin, *respirare*, 'to breathe'.
Rhabdo: from Greek, *rhabdos*, 'rod'.
Rubula: from Latin, *ruber*, 'red'; Rubula inflans, old name for mumps.
Vesiculo: from Latin, *vesicula*, diminutive of vesica, 'bladder, blister'.
Virales: from Latin *virales*, 'viruses'.

REFERENCES

De la Torre, J.C. (1994). Molecular biology of Borna disease virus: Prototype of a new group of animal viruses. *J. Virol.*, **68**, 7669-7675.

Feldmann, H. and Klenk, H.-D. (1996). Filoviruses: Marburg and Ebola. In: *Advances in Virus Research*, Vol. 47 (K. Maramorosch, F.A. Murphy and A.J. Shatkin, eds), pp. 1-52. Academic Press, San Diego.

Kingsbury, D.W. (ed) (1991). *The Paramyxoviruses*. Plenum Press, New York and London.

Kolakofsky, D., Pelet, T., Garcin, D., Hausmann, S., Curran, J. and Roux, L. (1998). Paramyxovirus RNA synthesis and the requirement for hexamer genome length: the Rule of Six revisited. *J. Virol.*, **72**, 891-899.

Le Mercier, P., Jakob, Y. and Tordo, N. (1997). The complete Mokola virus genome sequence: structure of the RNA-dependent RNA polymerase. *J. Gen. Virol.*, **78**, 1571-1576.

Mulberger, E., Sanchez, A., Randolf, A., Will, C., Kiley, M.P., Klenk, H. and Feldmann, H. (1992). The nucleotide sequence of the L gene of Marburg virus, a filovirus: homologies with paramyxoviruses and rhabdoviruses. *Virology*, **187**, 534-547.

Poch, O., Blumberg, B.M., Bougueleret, L. and Tordo, N. (1990). Sequence Comparisons of five polymerase (L-proteins) of unsegmented negative-strand RNA viruses – theoretical assignment of functional domains. *J. Gen. Virol.*, **71**, 1153-1162.

Poch, O., Sauvaget, I., delaRue, M., and Tordo N. (1989). Identification of four conserved motifs among the RNA-dependent polymerase encoding elements. *EMBO J.*, **8**, 3867-3874.

Pringle, C.R. (1991). The order *Mononegavirales*. *Arch. Virol.*, **117**, 137-140.

Pringle, C.R. (1997). The order *Mononegavirales* - current status. *Arch. Virol.*, **142**, 2321-2326.

Pringle, C.R. and Easton, A.J. (1997). Monopartite negative strand RNA genomes. *Semin. Virol.*, **8**, 49-57.

Schneemann, E., Schneider, P.A., Lamb, R.A. and Lipkind, W.I. (1995). The remarkable coding strategy of Borna disease virus: a new member of the non-segmented negative strand RNA viruses. *Virology*, **210**, 1-9.

Tordo, N., Charlton, K. and Wandeler, A. (1997). Rhabdoviruses: rabies. In: *Topley and Wilson's Principles of Bacteriology, Virology and Immunity*, 9th edition, Vol. 1, (B.W.J. Mahy and L. Collier, eds), pp. 665-692. Arnold, London.

Wagner, R.R. (ed) (1987). *The Rhabdoviruses*. Plenum Press, New York and London.

Wang, L-F., Yu, M., Hansson, E., Pritchard, L.I., Shiell, B., Michalski, W.P. and Eaton, B. (2000). The exceptionally large genome of Hendra virus: support for creation of a new genus within the family *Paramyxoviridae*. *J. Virol.*, **74**, 9972-9979.

CONTRIBUTED BY

Pringle, C.R.

FAMILY BORNAVIRIDAE

TAXONOMIC STRUCTURE OF THE FAMILY

Family *Bornaviridae*
Genus *Bornavirus*

Since only one genus is currently recognized, the family description corresponds to the genus description.

GENUS BORNAVIRUS

Type Species *Borna disease virus*

VIRION PROPERTIES

MORPHOLOGY

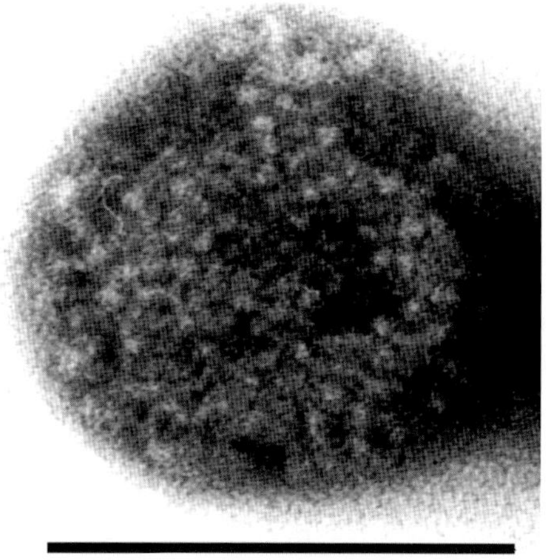

Figure 1: Negative contrast electron micrograph of a particle of *Borna disease virus*. The bar represents 100 nm. (Courtesy of Dr. M. Eickmann).

Electron microscopy studies of negatively stained infectious particles of an isolate of *Borna disease virus,* have shown that virions have a spherical morphology with a diameter of 90 ± 10 nm containing an internal electron-dense core (50 to 60 nm) (Fig. 1).

PHYSICOCHEMICAL AND PHYSICAL PROPERTIES

Virion Mr and the $S_{20,w}$ are not known. Partially purified Borna disease virus (BDV) infectious particles have a buoyant density in CsCl of 1.16-1.22 g/cm^3, in sucrose of 1.22 g/cm^3, in renografin of 1.13 g/cm^3. Virus infectivity is rapidly lost by heat treatment at 56^{0}C. Virions are relatively stable at 37^{0}C, and only minimal infectivity loss is observed after 24 hrs incubation at 37 °C in the presence of serum. Virions are inactivated below pH 5.0, as well as by treatment with organic solvents, detergents, and exposure to UV radiation. Infectivity is completely and rapidly destroyed by chlorine-containing disinfectants or formaldehyde treatment.

NUCLEIC ACIDS

The genome consists of a single molecule of a linear, NNS RNA (~8.9 kb in size and Mr of ~3 x 10^6). The RNA genome is not polyadenylated. Extracistronic sequences are found at the 3′(leader) and 5′(trailer) -ends of the BDV genome. BDV 3′-terminal genomic sequences have a high A+U content with a U/A ratio of ca 2:1. The ends of the BDV genome RNA exhibit partial inverted complementarity. Full-length plus-strand (antigenomic) RNAs are present in infected cells and in viral ribonucleoproteins. Defective RNAs have not been identified in BDV-infected cells and tissues. BDV can be classified into two subtypes based on the complete genome sequences of several BDV strains. With the unique exception of strain No/98, all isolates of BDV to date (independent of year, species and area isolation) have approximately 95% homology at the nt level (subtype 1), while BDV strain No/98 shows only 85-86% nt sequence identity compared to other BDV strains (subtype 2). The nt changes are distributed fairly evenly over the entire genome of BDV. However, all strains predict the same BDV genomic organization and differ by only one nt in absolute genome size.

PROTEINS

Six major ORFs are found in the BDV genome sequence (Fig. 2). These ORFs code for polypeptides with predicted size of 40 kDa (p40), 24 kDa (p24), 10 kDa (p10), 16 kDa (p16), 56 kDa (p56) and 180 kDa (p180), respectively. Based on their positions in the viral genome and abundance in infected cells and virion particles, together with their biochemical and sequence features, p40, p24 and p16 BDV polypeptides correspond to the viral nucleoprotein (N), the phosphoprotein (P) transcriptional activator, and matrix (M) proteins, respectively, found in other NNS RNA viruses. Two isoforms of the BDV N (p39 and p38) are found in BDV-infected cells. These two forms of the viral N appear to be encoded by two different mRNA species. Differential usage of two in-frame initiation codons present in the BDV p40 gene may also contribute to the production of BDV p39/38. BDV p39 contains both a NLS and NES, whereas p38 harbors only the NES. BDV p24 is an acidic polypeptide (predicted I.P. of 4.8), that has a high Ser-Thr content (16%), with phosphorylation at serine residues which is mediated by both PKC and casein kinase II. These features are consistent with those of the phosphoprotein (P) transcriptional activator found in other NNS RNA viruses. BDV p24 contains a bipartite NLS in the sequence. In addition to P, a 16 kDa polypeptide (P') is also translated from the second in-frame AUG codon in the P ORF. An additional ORF, p10, encodes a polypeptide of 10 kDa present in BDV-infected cells. BDV X starts within the same mRNA transcription unit, 46 nt upstream from p24 and overlaps, in a different frame, with the 71 N-terminal aa of p24. Recent study indicates that BDV X harbors a NLS in the N-terminus of the sequence.

Consistent with other NNS RNA viruses, BDV p16, the putative BDV M protein, is a non-glycosylated matrixprotein, associated at the inner surface of the viral membrane. BDV ORF4 (p56) overlaps, in a different frame, with the C-terminus of ORF p16, and is capable of encoding a 503 aa polypeptide with a predicted size of 56 kDa. Based on its sequence features, BDV p56 is the counterpart of the virus surface glycoproteins (G) found in other NNS RNA viruses. The p56 gene directs the synthesis of three glycosylated polypeptides of about 84 or 94 kDa (GP-84/94, G), 43 kDa (GP-43, GP-C) and 45 to 55 kDa (GP-N). G corresponds to the full length of the p56 gene, whereas GP-C and GP-N represent the C-terminal subunit and the N-terminal subunit of ORF p56, respectively. Both GP-C and GP-N are associated with BDV infectious particles. Antibodies to p56 have neutralizing activity, suggesting that BDV p56 gene products play an important role in the early steps of BDV infection. BDV ORF5 (p180) is capable of encoding a polypetide with a predicted size of 180 kDa, whose deduced aa sequence displays strong homology to other NNS-RNA virus polymerases, members of the L protein family. An additional ORF predicted in mRNA species generated via RNA splicing would encode a variant BDV L with a predicted size of 190 kDa (BVp190). BDVp190 corresponds to BDVp180 with 153 aa added to its N-terminus. Recent evidence suggests that p190, rather than p180, is the active BDV L. BDV L contains the NLS in the sequence.

LIPIDS

Not known.

CARBOHYDRATES

Only N-glycans, mannose-rich type and partially hybrid types.

GENOME ORGANIZATION AND REPLICATION

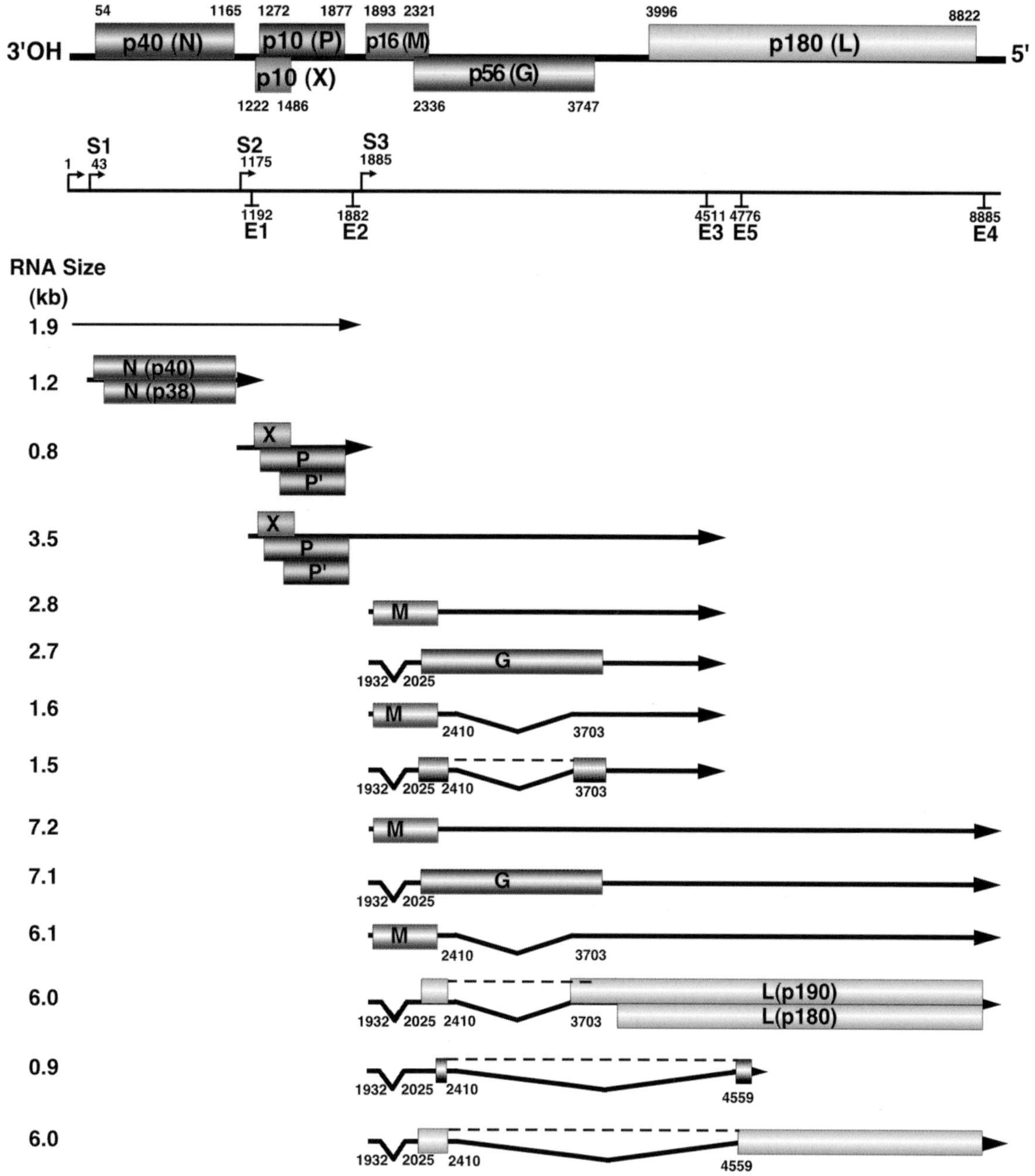

Figure 2: Genomic organization and transcriptional map of an isolate of *Borna disease virus*. ORFs are represented by boxes at the top. The location of transcription initiation and transcription termination sites are indicated by S and E, respectively. Positions of introns: I (nt 1932-2025), II (nt 2410-3703), and III (nt 2410-4559) are indicated. For details see Text.

The negative-sense BDV RNA genome codes for at least six ORFs in the order 3'-N-P/X-M-G-L-5'. The genomic polarity has a very limited coding capability, and none of its predicted ORFs has a favorable translational start signal; further they are not flanked by putative transcription start and termination/polyadenylation signals. Therefore, it seems unlikely that BDV uses an ambisense coding strategy. BDV has the property, unique among known NNS RNA animal viruses, of a nuclear site for genome transcription and replication. Full-length genome complementary RNA molecules (antigenomes) act as templates for new viral genome RNA synthesis. Genome and antigenome RNA molecules are neither capped nor polyadenylated. These RNAs exist as nucleocapsids in the nucleus of infected cells. It is unknown whether RNA species corresponding to the leader RNA are transcribed in BDV-infected cells.

BDV cell entry occurs by receptor-mediated endocytosis. The virus G protein has been implicated in entry. The identity of the BDV cellular receptor is unknown. In endosomes, low pH-dependent fusion occurs between viral and cellular membranes. This fusion event releases the BDV ribonucleoproteins (RNP) which are then transported to the cell nucleus where viral transcription and replication occur. Sequential and polar transcription results in decreasing molar quantity of BDV transcripts from the 3'- to the 5'-encoded cistrons. The viral mRNAs are polyadenylated, and their 5'-ends contain a blocking group, presumably a cap structure. Virus specific mRNA synthesis is not inhibited by α-amanitin, and sequences at the 5' of the BDV mRNAs are homogeneous and genome encoded. Thus, it is unlikely that transcription initiation of BDV mRNAs involves a cap-snatching mechanism similar to the one used by influenza viruses. Monocistronic viral mRNAs in BDV-infected cells are detected only for the N gene (Fig. 2). The BDV G and L polymerase gene products are synthesized from downstream ORFs within polycistronic mRNAs. Mapping of BDV sgRNAs present in infected cells to the viral genome revealed that the BDV genome contains three transcription initiation sites (S signals), and four transcription termination/polyadenylation sites (E signals)(Fig. 2). In addition, a putative E signal (E5) is found at nt 4776. The S signals contain a semi-conserved U-rich motif that is partially copied into the respective transcripts. A similar motif is not found within the S signals of previously described NNS RNA viruses. BDV E signals consist of six or seven U residues preceded by a single A residue, resembling the E signal motif found in other NNS RNA viruses. The BDV genome lacks the characteristic configuration of E signal / intergenic (IG) region / S signal, found at the gene boundaries of other NNS RNA viruses. Instead, BDV transcription units and transcriptive signals frequently overlap (Fig. 2). Two of the BDV primary transcripts are post-transcriptionally processed by the cellular RNA splicing machinery. Three introns (I, II and III) have been identified in the BDV genome. BDV introns I and II span nt 1932 –to 2025 and 2410 to 3703, respectively, in the BDV antigenomic sequence (Fig. 2). Splicing of intron I places the aa in position 13 of M next to a stop codon, whereas splicing of intron II, and I+II, results in a mRNA containing a predicted ORF that corresponds to the first 58 aa of G fused to a new C-terminus of 20 aa. RNA species resulting from splicing of intron II, and I+II, predicts also an additional ORF that would encode a variant BDV L protein with 153 aa added to the N-terminus. Intron III is generated by alternative 3' splice site choice and spans nt 2410 to 4559 in the BDV antigenome (Fig. 2). Splicing of introns II and III is regulated by the utilization of an alternative E signal (E5) and a putative *cis*-acting exon splicing suppressor signal located within the L gene. Transcripts lacking Intron III have the capacity to encode two new proteins with predicted size of 8.4 kDa (p8.4) and 165 kDa (p165). Whether these new predicted BDV polypeptides are synthesized in infected cells is unknown. BDV strain No/98 lacks the alternative 3' splice site and thus cannot generate transcripts lacking intron III. RNA splicing can also modulate the efficiency of termination-reinitiation of translation and leaky scanning mechanisms, thus contributing to the regulation of the expression of BDV M, G and L gene products.

BDV-infected cells exhibit a heterogeneous pattern of viral antigen expression. BDV N, P, and X polypeptides are expressed both in the nucleus and cytoplasm. N and P are the viral antigens expressed at higher levels, and they are expressed by the majority of the cells within an infected population. In contrast, only 1 to 10% of the infected cells express detectable levels of BDV G. Expression of full length BDV G (GP-84/94) is restricted to the ER and nuclear envelope. The subcellular distribution of the BDV M is cytosolic and associated with cellular membranes. G is post-translationally modified by N-glycosylation. BDV G undergoes post-translational cleavage by the cellular protease furin, with the resulting GP-N and GP-C reaching the cell surface. Cleavage of G likely occurs in the trans-Golgi compartment. G, GP-N and GP-C are partially Endo H sensitive and PNGase F sensitive. The newly exposed N-terminus of GP-C is highly hydrophobic, and BDV-infected cells form extensive syncytia upon low-pH treatment. These findings suggest that GP-C is involved in pH-dependent fusion after internalization of BDV by receptor-mediated endocytosis. GP-N is most likely responsible for receptor binding.

The mechanisms involved in nucleocytoplasmic transport of viral RNP through the nuclear pore complex remain largely unknown. However, recent studies suggest that the nuclear import activity of BDV is mediated by the NLS-containing viral antigens, such as N, P, X and L, that form complexes in infected cells. In contrast, a nuclear export activity is found only in N protein. The NES of BDV N contains the canonical leucine-rich motif, and the nuclear export activity of the protein is mediated through the chromosome region maintenance protein (CRM1) pathway.

The assembly process and site of virus maturation have not been identified and budding of BDV particles from infected cells has been documented only from the surface of BDV-infected MDCK cells after treatment with n-butyrate. BDV RNP accumulate in the nucleus and, as with other NNS RNA viruses, they are also infectious on the basis of an ability to direct synthesis of BDV macromolecules, as well as the production of BDV cell-associated infectivity upon transfection of BDV-susceptible cells. Thin sections of BDV-infected cells revealed the presence of intracytoplasmic virus-like particles with morphological characteristics similar to those described for partially purified cell-free BDV infectious particles. These particles showed no association with cisternae of the endoplasmic reticulum, the Golgi complex, or other intracytoplasmic membranes.

As for many other members of the order *Mononegavirales*, a reverse genetics system has recently been established for BDV.

Antigenic Properties

BDV possess a number of distinct antigenic determinants. The so-called soluble antigen (s-antigen) obtained from the supernatant after ultracentrifugation of ultrasonicated BDV-infected brain tissue, contains the viral N, P and M proteins. Serum antibodies from BDV-infected animals frequently recognize all the components of the s-antigen, but rarely recognize the viral G products. BDV field isolates from the same or different animal species, as well as viruses recovered from experimental infections with different histories of passages exhibit strong serological cross-reactivity. There is only one recognized serotype of BDV, but monoclonal antibodies have revealed minor antigenic differences among BDV isolates. Complement independent IgG-specific neutralizing antibodies have been documented in experimentally infected animals. Titers of neutralizing antibodies are usually very low and dependent on the infected host species. BDV G protein have been implicated in virus neutralization.

Biological Properties

Horses and sheep have been regarded as the main natural hosts of BDV. In these species BDV can cause a fatal neurologic disease, Borna disease (BD). Evidence indicates that the natural host range of BDV is wider than originally thought. Naturally occurring BDV infections have been documented in cattle, rabbits and cats. In addition, sporadic cases of natural infection with BDV have been reported in several other species including donkeys, mules, and llamas. Moreover, experimental infections have revealed a remarkable wide host range for BDV, from birds to rodents and non-human primates. BDV-induced neurobehavioral abnormalities in animals are reminiscent of some human neuropsychiatric disorders. Serological data and molecular epidemiological studies indicate that BDV can infect humans, and is possibly associated with certain neuropsychiatric disorders.

BDV is thought to be transmitted through salival, nasal, or conjunctival secretions. Infection may therefore result from direct contact with these secretions. Intranasal infection is the most likely route of natural infection, allowing BDV access to the CNS by intraaxonal migration through the olfactory nerve. Cases of BD are more frequent in some years than others and tend to occur in spring and early summer, suggesting arthropods as a potential vector. BDV has not been isolated from insects, but ticks have been implicated in the transmission of an infectious encephalomyelitis similar to BD affecting ruminants in the Middle East.

Asymptomatic naturally infected animals of different species have been documented in Europe, North America, Africa and Asia, suggesting that the prevalence and geographic distribution of BDV may have been underestimated. However, a definite natural reservoir of BDV has not been identified. Phenotypic differences have been described among different BDV field isolates, and among viruses with different histories of passages in animals and cultured cells. Despite its wide host range and phenotypic variation, molecular epidemiological data have shown a remarkable sequence conservation of BDV, not only within the same host species but also amongst sequences derived from different animal species.

BDV is highly neurotropic and has a non-cytolytic strategy of multiplication. BDV causes CNS disease in several non-human vertebrate species, which is characterized by neurobehavioral abnormalities that are often, but not always, associated with the presence of inflammatory cell infiltrates in the brain. BDV exhibits a variable period of incubation, from weeks to years, and diverse pathological manifestations that depend on the genetics, age and immune status of the host, as well as route of infection and viral determinants. Classic BD is caused by a T-cell dependent immune mechanism. Inflammatory cells are found forming perivascular cuffs and also within the brain parenchyma. Both $CD4^+$ and $CD8^+$ T-cells are present in the CNS cell infiltrates and contribute to the immune-mediated pathology associated with BD. BDV can also induce distinct deficiencies in emotional and cognitive functions that are associated with specific neurochemical disturbances in the absence of lymphoid infiltration. Heightened viral expression in limbic system structures, together with astrocytosis and neuronal structural alterations within the hippocampal formation are main histopathological hallmarks of BDV infection.

List of Species Demarcation Criteria in the Genus

Not applicable.

List of Species in the Genus

Species names are in green italic script; strain names and synonyms are in black roman script; tentative species names are in blue roman script. Sequence accession numbers and assigned abbreviations ( ) are also listed.

Species in the Genus

Borna disease virus

Borna disease virus	[L27077, U04608, AJ311524, AJ311523]	(BDV)

Tentative Species in the Genus

None reported.

List of Unassigned Viruses in the Family

None reported.

PHYLOGENETIC RELATIONSHIPS WITHIN THE FAMILY

Not applicable.

SIMILARITY WITH OTHER TAXA

BDV has a genomic organization similar to that of other NNS RNA viruses. The size of the BDV genome (ca 8.9 kb) is significantly smaller than those of the other known members of the order *Mononegavirales*: *Rhabdoviridae* (~11-15 kb), *Paramyxoviridae* (~15 kb) and *Filoviridae* (~19 kb). BDV replication and transcription take place in the nucleus. This is a unique feature among known NNS RNA animal viruses, but shared with the plant nucleorhabdoviruses. Expression of the BDV genome is regulated by an overlap of transcription units and transcriptive signals, an overlap of ORFs, readthrough of transcription termination signals and differential use of translational initiation codons. There is precedent for use of each of these strategies by other members of the order *Mononegavirales*. However, the concurrent use by BDV of such a diversity of strategies for the regulation of its gene expression is unique among known NNS RNA viruses. In addition, as with viruses belonging to the family *Orthomyxoviridae*, BDV uses RNA splicing to generate some of its mRNAs. This represents another unique feature in the order *Mononegavirales*. BDV has one single surface glycoprotein gene (G) which is responsible for viral attachment and fusion upon endocytosis and endosome acidification. This pH-dependent fusogenic activity of G requires its post-translational cleavage by the cellular protease furin. Thus, BDV G expression and function appear to be a unique feature in NNS RNA viruses, representing a combination of the strategies adopted by rhabdoviruses and paramyxoviruses.

DERIVATION OF NAME

Borna refers to the city of Borna in Saxony, Germany, where many horses died in 1885 during an epidemic of a neurological disease, designated as Borna disease (BD), caused by the infectious agent presently known as Borna disease virus (BDV).

NEG. SSRNA

REFERENCES

Carbone, K.M. (ed)(2002). *Borna disease virus* : Its role in neurobehavioral disease. ASM Press, Washington DC.

Furrer E., Bilzer T., Stitz L. and Planz O. (2001). Neutralizing antibodies in persistent borna disease virus infection: prophylactic effect of gp94-specific monoclonal antibodies in preventing encephalitis. *J. Virol.*, **75**, 943-951.

Gonzalez-Dunia, D., Cubitt, B., Grässer, F.A. and de la Torre, J.C. (1997). Characterization of Borna disease virus p56 protein, a surface glycoprotein involved in virus entry. *J. Virol.*, **71,** 3208-3218.

Gonzalez-Dunia, D., Cubitt, B. and de la Torre, J.C. (1998). Mechanism of Borna disease virus entry into cells. *J. Virol.*, **72,** 783-788.

Kiermayer S., Kraus I., Richt J.A., Garten W. and Eickmann M. (2002). Identification of the amino terminal subunit of the glycoprotein of Borna disease virus. *FEBS Lett.*, **531**, 255-258.

Kohno, T., Goto, T., Takasiki, T., Mozita, C., Nakaya, T., Ikuta, K., Kurane, I., Sano, K. and Nakai, M. (1999). Fine structure and morphogenesis of Borna disease virus. *J. Virol.*, **73**, 760-766.

Kraus I., Eickmann M., Kiermayer S., Scheffczik H., Fluess M., Richt J.A. and Garten W. (2001). Open reading frame III of borna disease virus encodes a non-glycosylated matrix protein. *J. Virol.*, **75,** 12098-12104.

Ludwig, H., Bode, L. and Gosztonyi, G. (1988). Borna disease; a persistent virus infection of the central nervous system. *Prog. Med. Virol.*, **35**, 107-151.

Nowotny, N., Kolodziejek, J., Jehle, C.O., Suchy, A., Staeheli, P. and Schwemmle, M. (2000). Isolation and characterization of a new subtype of Borna disease virus. *J. Virol.*, **74**, 5655-5668.

Pleschka, S., Staeheli, P., Kolodziejek, J., Richt, J.A., Nowotny, N. and Schwemmle, M. (2001). Conservation of coding potential and terminal sequences in four different isolates of Borna disease virus. *J. Gen. Virol.*, **82**, 2681-2690.

Schneemann, A., Schneider, P.A., Lamb, R.A. and Lipkin, W.I. (1995). The remarkable coding strategy of Borna disease virus: a new member of the nonsegmented negative strand RNA viruses. *Virology*, **210,** 1-8.
Schneider, P.A., Hatalski, C.G., Lewis, A.J. and Lipkin, W.I. (1997). Biochemical and functional analysis of Borna disease virus G protein. *J. Virol.*, **71**, 31-336.
Schwemmle, M., De, B., Shi, L., Banerjee, A. and Lipkin, W.I. (1997). Borna disease virus P-protein is phosphorylated by protein kinase Ce and casein kinase II. *J. Biol. Chem.*, **272,** 21818-21823.
Staeheli, P., Sauder, C., Hausmann, J., Ehrensperger, F. and Schwemmle, M. (2000). Epidemiology of Borna disease virus. *J. Gen. Virol.*, **81**, 2123-35.
Tomonaga, K., Kobayashi, T., Lee, B.J., Watanabe, M., Kamitani, W. and Ikuta, K. (2000). Identification of alternative splicing and negative splicing activity of a non-segmented negative-strand RNA virus, Borna disease virus. *Proc. Natl. Acad. Sci. USA*, **97**, 12788-12793.
Schneider, U., Naegele, M., Staeheli, P. and Schwemmle, M. (2003). Active borna disease virus polymerase complex requires a distinct nucleoprotein-to-phosphoprotein ratio but no viral X protein. *J. Virol.*, **77**, 11781-11789.
Perez, M., Sanchez, A., Cubitt, B., Rosario, D. and de la Torre, J.C. (2003). A reverse genetics system for Borna disease virus. *J. Gen. Virol.*, **84**, 3099-3104.
Rudolph, M.G., Kraus, I., Dickmanns, A., Eickmann, M., Garten, W. and Ficner, R. (2003). Crystal structure of the borna disease virus nucleoprotein. *Structure*, **11**, 1219-1226.

CONTRIBUTED BY

Schwemmle, M., Carbone, K.M., Tomonago, K., Nowotny, N. and Garten, W.

FAMILY RHABDOVIRIDAE

TAXONOMIC STRUCTURE OF THE FAMILY

Family	***Rhabdoviridae***
Genus	***Vesiculovirus***
Genus	***Lyssavirus***
Genus	***Ephemerovirus***
Genus	***Novirhabdovirus***
Genus	***Cytorhabdovirus***
Genus	***Nucleorhabdovirus***

VIRION PROPERTIES

MORPHOLOGY

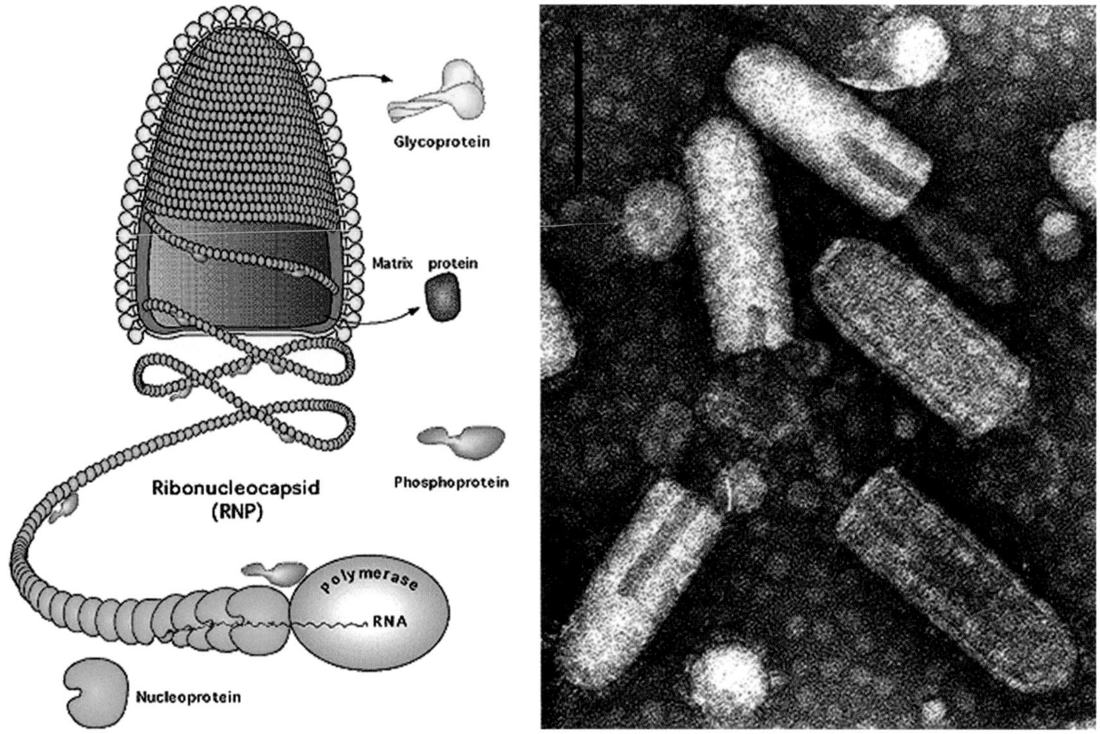

Figure 1: (Left) Diagram illustrating a rhabdovirus virion and the nucleocapsid structure (Courtesy of P. Le Mercier); (Right) Negative contrast electron micrograph of particles of an isolate of *Vesicular stomatitis Indiana virus* (Courtesy of P. Perrin). The bar represents 100 nm.

Virions are 100-430 nm long and 45-100 nm in diameter. Defective virus particles are proportionally shorter. Viruses infecting vertebrates are bullet-shaped or cone-shaped; viruses infecting plants mostly appear bacilliform when fixed prior to negative staining; in unfixed preparations they may appear bullet-shaped or pleomorphic. Some putative plant rhabdoviruses lack envelopes. The outer surface of virions (except for the quasi-planar end of bullet-shaped viruses) is covered with projections (peplomers) which are 5-10 nm long and about 3 nm in diameter. They consist of trimers of the viral glycoprotein (G). A honeycomb pattern of peplomers is observed on the surface of some viruses. Internally, the nucleocapsid, about 30-70 nm in diameter, exhibits helical symmetry and can be seen as cross-striations (spacing 4.5-5 nm) in negatively stained and thin-sectioned virus particles. The nucleocapsid consists of an RNA and N protein complex together with L and P proteins. A lipid envelope containing the G glycoprotein interacts with nucleocapsids via the M protein. The nucleocapsid contains transcriptase activity and is infectious. Uncoiled it is filamentous, about 700 nm long and 20 nm in diameter.

Physicochemical and Physical Properties

Virion Mr is 300 x 10^6 -1,000 x 10^6 and S_{20w} is 550-1,045S (plant rhabdoviruses have larger S_{20w} values). Virion buoyant density in CsCl is 1.19-1.20 g/cm^3; in sucrose it is 1.17-1.19 g/cm^3. Virus infectivity is rapidly inactivated at 56°C, or following UV- or X-irradiation, or exposure to lipid solvents.

Nucleic Acid

Viruses contain a single molecule of linear, negative-sense ssRNA (Mr 4.2 x 10^6 - 4.6 x 10^6, about 11-15 kb in size). The RNA represents about 1-2% of particle weight. The RNA has a 3′-terminal free OH group and a 5′-triphosphate and is not polyadenylated. The ends have inverted complementary sequences. Defective RNAs, usually significantly shorter than full-length RNA (less than half size), may be identified in RNA recovered from virus populations. They are usually negative sense; however, hairpin RNA forms are also found. Defective RNAs replicate only in the presence of homologous and, occasionally, certain heterologous helper rhabdoviruses which provide the functional genes. Full-length positive-strand RNA may constitute up to 5% of a viral RNA population.

Proteins

Viruses generally have 5 structural polypeptides (designated L, G, N, P and M; see Table 1 for summary of their location, sizes and functions). The functions of other proteins, including additional glycoproteins (ephemeroviruses) or C proteins (in a different ORF of the P mRNA for vesiculoviruses) are not known. The structural proteins represent 65 - 75% of the virus dry weight. For certain viruses, other nomenclature has previously been used for the P protein (NS, M1 or M2) and the M protein (M1 or M2). For *Vesicular stomatitis Indiana virus* (VSIV) the numbers of molecules per infectious virus particle is estimated as: L (20-50); G (500-1,500); N (1,000-2,000); P (100-300); and M (1,500-4,000). The enzymes identified in virions include the RdRp, a 5' capping enzyme complex (guanylyl transferase and Mtrs), a poly(A) polymerase, and possibly a protein kinase, nucleoside triphosphatase and a nucleoside diphosphate kinase. These activities are harbored by the nucleocapsid (transcription/replication complex), i.e. the genome RNA associated with N, P and L proteins. Most of the catalytic functions may be due to L.

Table 1: Location, molecular ratio and functions of rhabdovirus structural proteins

Protein	Location, size and function
L	A component of the viral nucleocapsid (~ 220–240 kDa) responsible for most of the functions required for transcription and replication: RdRp, mRNA 5'-capping, 3'-poly[A] synthesis and protein kinase activities. Observed sizes on SDS-PAGE are 150 -190 kDa.
G	Associated into trimers to form the virus surface peplomers (monomer ~ 65-90 kDa). Binds to host cell receptor(s), induces virus endocytosis then mediates fusion of viral and endosomal membranes. G is variously N-glycosylated and palmitoylated; it lacks O-linked glycans. G has hemagglutinin activity. G induces and binds virus-neutralizing antibodies and elicits cell-mediated immune responses. G is involved in tropism and pathogenicity.
N	N is a major component of the viral nucleocapsid (~ 47-62 kDa). It associates with full-length negative- and positive-sense RNAs, or defective RNAs, but not mRNAs. N is not "inert" but an active element of the template, presenting the bases to the polymerase. Newly synthesized N probably modulates the balance between genome transcription and replication by influencing the recognition of the transcription signals. N elicits cell-mediated immune responses and humoral antibodies.
P	A cofactor of the viral polymerase (~ 20-30 kDa). It is variously phosphorylated and migrates on SDS-PAGE as a 40-50 kDa protein. The P of the nucleorhabdoviruses migrates faster. It is required for transcription and replication. A soluble form is present

	in the cytoplasm of infected cells which associates to N, thus preventing self-aggregation of N protein and aiding in N encapsidation of RNA species. P elicits cell-mediated immune responses.
M	A basic protein that is an inner component of the virion (~ 20-30 kDa). It is believed to regulate genome RNA transcription. M binds to nucleocapsids and the cytoplasmic domain of G, thereby facilitating the process of budding. Sometimes M is phosphorylated or palmitoylated. M is found in the nucleus and inhibits host cell transcription. It also mediates other pathological effects (cell rounding for VSIV, etc.).

Lipids

Virions are composed of about 15-25% lipids, their composition reflecting that of the host cell membrane where virions bud. Generally phospholipids represent about 55-60%, and sterols and glycolipids about 35-40% of the total lipids. G protein has a covalently associated fatty acid proximal to the lipid envelope.

Carbohydrates

Virions are composed of about 3% carbohydrate by weight. The carbohydrates are present as N-linked glycan chains on G protein and as glycolipids. In mammalian cells, the oligosaccharide chains are generally of the complex type, in insect cells they are of the non-complex types.

Genome Organization and Replication

Viruses contain at least 5 ORFs in the negative-sense genome in the order 3'-N-P-M-G-L-5' (e.g., for VSIV), or the equivalent. The corresponding cistrons are flanked by conserved start and stop transcription signals, about 10 nt in length. For certain viruses additional genes are interposed (Fig. 2). Genes are transcribed processively (from the 3' to 5' of the template virus RNA and in decreasing molar abundances) as 5'-capped, 3'-polyadenylated and generally monocistronic mRNAs (Fig. 3). Polycistronic mRNAs have been identified for some species, they result from read-through of a stop transcription signal thereby allowing transcription extension across the adjacent 5′-cistron. A short uncapped, unpolyadenylated and untranslated "leader" RNA, corresponding to the complement of the 3'-terminus of the viral RNA (i.e., preceding the N mRNA), is also transcribed. Unlike mRNA species, it has a 5' triphosphate terminus (Fig. 3). Leader RNA has been identified in the nucleus of infected cells. The mRNAs generally have common 5'-terminal sequences (e.g. m^7Gppp(m)AmA(m)CA for vesiculoviruses and lyssaviruses) corresponding to the cap structure fused to the first nucleotides copied from the start transcription signal. The mRNAs also each contain a 3′-poly(A) tail which is produced by the viral transcriptase upon copying in a reiterative mode the 7 U residues terminating each stop transcription signal. Intergenic sequences are generally short but may be up to ~ 50 nt in some species. In certain cases the 5'-end of an mRNA overlaps the 3'-end of the preceding gene.

Except for plant rhabdoviruses that generally penetrate the cell through mechanical damage provoked by insect vectors, rhabdovirus adsorption is mediated by G protein attachment to cell surface receptors and penetration of the cell is by endocytosis via coated pits. Various candidate receptors have been postulated for Rabies virus (RABV)(nicotinic acetylcholine receptor AChR, neural cell adhesion molecule NCAM, low affinity nerve growth factor receptor p75NTR), Vesicular stomatitis virus (VSV) (phosphatidyl serine), Viral hemorrhagic septicemia virus (VHSV) (fibronectin), and others. In addition, carbohydrate moieties, phospholipids and gangliosides may play a complementary role for virus binding. After penetration by endocytosis, the pH decreases within the endosome, provoking fusion between endosomal and viral membranes. This liberates the RNP into the cytoplasm. The pH-dependent fusion depends on conformational changes of the glycoprotein, a process that is reversible upon raising the pH, a property typical of rhabdovirus G proteins. Once the nucleocapsid (RNA, N, L, P) is released into the cytoplasm, the genome RNA is repetitively transcribed (primary transcription) by the virion transcriptase. N protein removal is not required since the transcriptase recognizes the RNA-N protein complex as template. The capped and polyadenylated mRNAs are generally translated in cytoplasmic polysomes except for the G mRNA which is translated on membrane-bound polysomes. Transcription occurs in the presence of protein synthesis inhibitors indicating that it does not depend on *de novo* host protein synthesis. Following translation, RNA replication occurs in the cytoplasm (full-length positive and then full-length negative RNA synthesis).

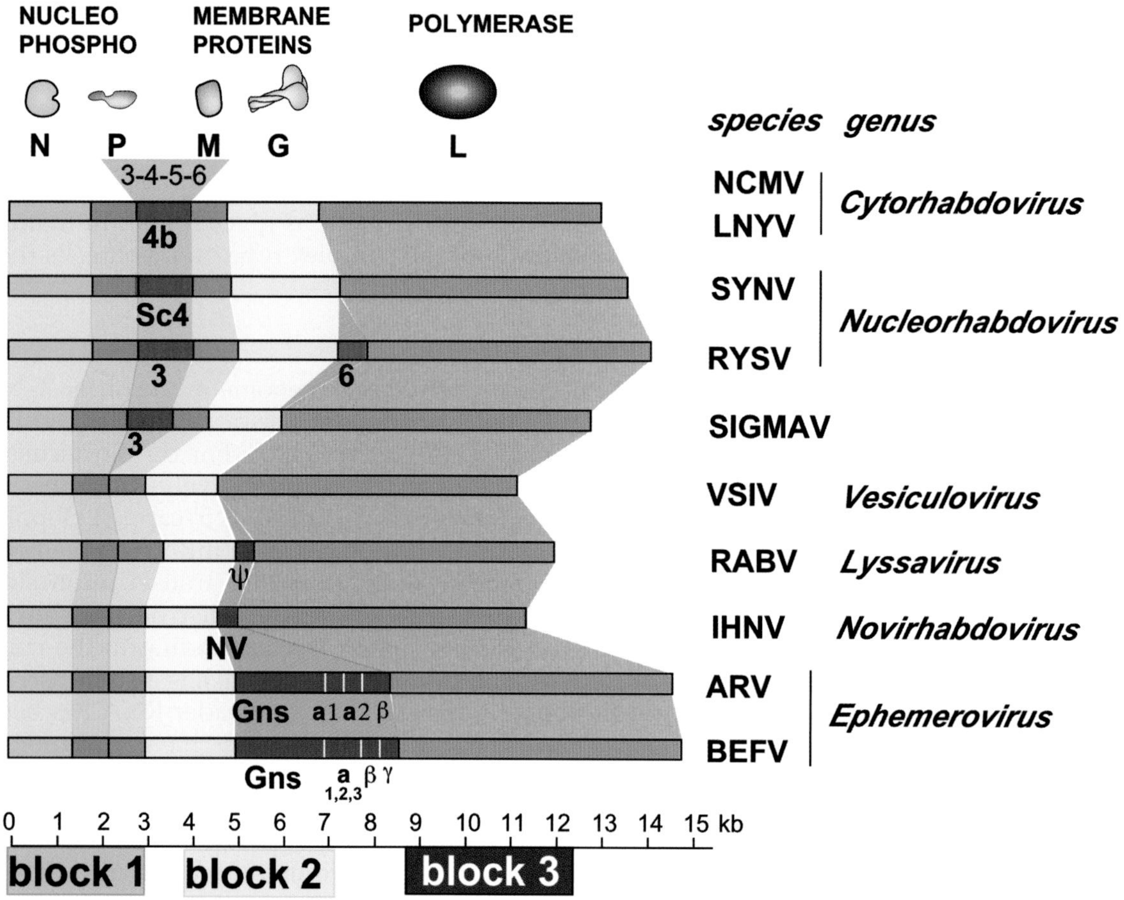

Figure 2: Comparison of genome organization of representative members from all genera of the family *Rhabdoviridae*. A modular organization into 3 blocks has been conserved: the 3′ block #1 encodes N/P proteins required in large amounts; the central block #2 encodes the membrane M/G proteins; the 5′ block #3 encodes the polymerase L required in limited amounts. Between blocks, viruses have inserted typical genes adapted to their particular biology, for example the movement protein (3, 4b) in plant rhabdoviruses. (courtesy of N. Tordo)

Certain plant rhabdoviruses may replicate RNA in the cell nucleus. Replication again occurs on the RNA-N protein complex and requires the newly synthesized N, P and L protein species to concomitantly encapsidate the nascent RNA into a nucleocapsid structure. Apart from freshly translated N-P-L proteins, replication may require host factors. However, vesiculoviruses can replicate in enucleated cells, indicating that newly synthesized host gene products are not required. It has been proposed that the concomitant binding of N protein to the nascent positive or negative-sense viral RNA species may promote replication rather than transcription, by favoring read-through of transcription termination signals. Replication leads to the synthesis of a full length positive strand antigenome RNA. This, in turn, serves as a replicative intermediate for the synthesis of negative strand genome RNAs for the progeny virions. Following replication, further rounds of transcription (secondary transcription), translation and replication ensue. A very typical feature of negative strand RNA viruses (shared by all members of the order *Mononegavirales*) is that the RNA genome (or antigenome) is never "naked" in the cell but is always encapsidated by the nucleoprotein. This RNA-N complex is the true template recognized by the viral polymerase (transcriptase or replicase).

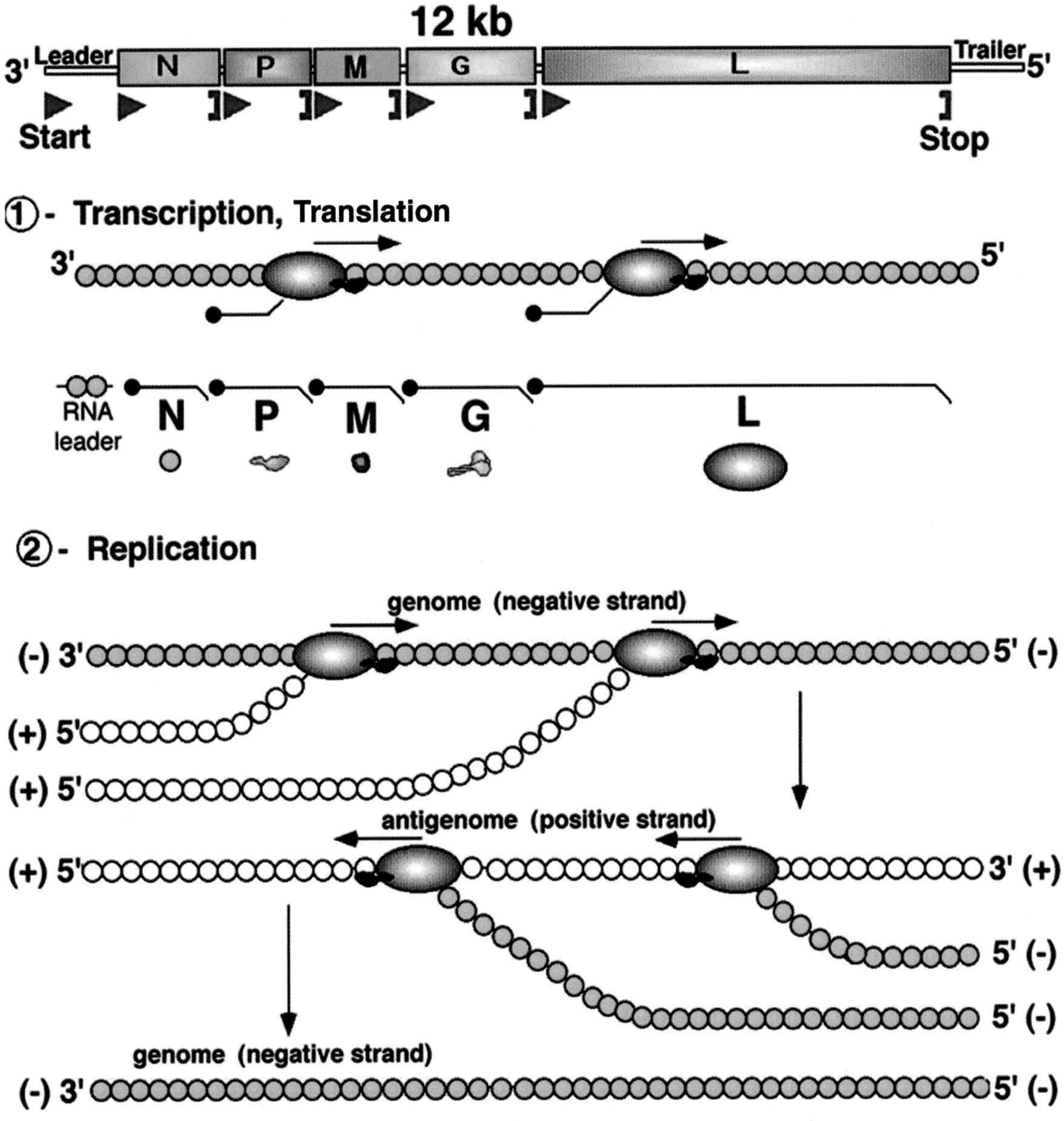

Figure 3: Panel (1) shows the genome organization of Vesicular stomatitis Indiana virus (VSIV) and the process of consecutive transcription of leader RNA and monocistronic mRNAs. Panel (2) illustrates the replication of the negative sense genome via a positive sense antigenome intermediate. The switch from transcription to replication appears to be regulated by the N protein. (Courtesy of P. Le Mercier).

Post-translational trafficking and modification of G protein involves translocation across the membrane of the endoplasmic reticulum, removal of the amino-proximal signal sequence and step-wise glycosylation in compartments of the Golgi apparatus.

Depending on the cell, the G protein may move to the plasma membrane, in particular, to the basolateral surfaces of polarized cells.

Viral nucleocapsid structures are assembled in association with M and lipid envelopes containing viral G protein. The site of formation of particles depends on the virus and host cell. For vesiculoviruses, lyssaviruses, ephemeroviruses and novirhabdoviruses, nucleocapsids are synthesized in the cytoplasm and viruses bud from the plasma membrane in most, but not all cells. Some lyssaviruses bud predominantly from intracytoplasmic membranes and in some cases prominent virus-specific cytoplasmic inclusion bodies containing N protein are observed in infected cells (RABV inclusion bodies are called Negri bodies). Cytorhabdoviruses bud from intracytoplasmic membranes associated with viroplasms. None has been observed to bud from plasma membranes. Nucleorhabdoviruses bud from the inner nuclear membrane and accumulate in the perinuclear space.

Depending on the virus and host cell type, virus infections may inhibit cellular macromolecular syntheses. The mechanisms are under investigation. Generally, 5 complementation groups of mutants have been defined by using temperature-sensitive mutants. Host range and temperature-sensitive mutants with altered polymerase functions have also been described. Complementation may occur between related viruses (e.g., between vesiculoviruses), but not between viruses representing distinct genera. Complementation is also reported to occur involving re-utilization of the structural components of UV-irradiated virus (VSIV). Recombination of genes between different virus isolates has not been demonstrated although recombination will occur during the formation of defective RNAs. Phenotypic mixing occurs between some animal rhabdoviruses and other enveloped animal viruses (e.g., paramyxoviruses, orthomyxoviruses, retroviruses, herpesviruses).

Six genera have been established, on the basis of significant differences in antigenicity, genome organization, aa sequence of structural proteins, replication site and host range. Phylogenetic relationships based on available N, G or L protein sequences support assignments of species to the identified genera (Fig. 4).

Reverse genetics systems have been established for a number of rhabdoviruses (*Vesiculovirus, Lyssavirus*), and this technique has allowed to create viruses with rearranged genomes and profoundly changed biological properties.

Antigenic Properties

G protein is involved in virus neutralization and defines the virus serotype. N protein is a cross-reacting, complement-fixing (CF) antigen. Weak serological cross-reactions may occur between viruses in different genera. Protection follows vaccination with attenuated viruses, inactivated viruses, subunits consisting of G protein alone or G protein together with the ribonucleoprotein complex, expression vectors (e.g., *Vaccinia virus*) or plasmid DNA that synthesize G and/or N.

Biological Properties

Some species multiply only in mammals, or birds, or fish, or arthropods, or other invertebrates, many have both arthropod and vertebrate hosts (arboviruses), while some species infect plants and certain plant-feeding arthropods. Some of the viruses of vertebrates have a wide experimental host range. A diverse range of vertebrate and invertebrate cells is susceptible to vertebrate rhabdoviruses *in vitro*. The viruses of plants usually have a narrow host range among higher plants; some replicate in insect vectors and grow in insect cell cultures.

Sigma virus (SIGMAV) was recognized first as a congenital infection of *Drosophila*. No rhabdovirus is transmitted vertically in vertebrates or plants. Viruses are transmitted mechanically between plants. Vector transmission may involve mosquitoes, sandflies, mites, culicoides, aphids, lacewings, leafhoppers, or planthoppers (etc.). Some viruses are transmitted mechanically in sap or from the body fluids of infected hosts. Mechanical transmission of viruses infecting vertebrates may be by contact, aerosol, bite, or venereal.

GENUS VESICULOVIRUS

Type Species *Vesicular stomatitis Indiana virus*

DISTINGUISHING FEATURES

Vesiculoviruses encode 5 major polypeptides (from 3′ to 5′ genomic end: N, P, M, G and L) as well as a minor C protein of which the precise function remains to be clarified (in a different ORF of the P mRNA). In addition, the 11.2 kb genome encodes about 50 nt of leader sequence (produced only during transcription) that precedes the N gene, and about 60 nt of a trailer region that follows the L gene. Small untranscribed intergenic sequences (generally dinucleotides) separate the polyadenylation signal terminating each cistron (usually 3'-AUACUUUUUUU) from the sequence that indicates the start of the next mRNA species (usually 3′-UUGUCNNUAG which templates the capped mRNA 5′-end: m^{7}Gppp(m)Am-A(m)CAGNNAUC...). Some viruses (e.g., Mount Elgon bat virus, MEBV, Kwatta virus, KWAV) are distinctly larger than the type species.

BIOLOGICAL PROPERTIES

Vesiculoviruses have been obtained from a variety of animals, including mammals, fish and invertebrates (insects). Vesicular stomatitis of horses, cattle and swine is one of the oldest known infectious diseases of livestock which provokes severe loss of milk production in cows. It was first recognized as distinct from foot-and-mouth disease (provoked by a picornavirus) early in the nineteenth century. The clinical manifestations are febrile illness with development of vesicles in the mouth, and excessive salivation. Epizootics periodically occur throughout the Western hemisphere. They are sudden in onset and quickly involve many animals in a herd. A number of vesiculoviruses (Chandipura virus, CHPV; Cocal virus, COCV; Isfahan virus, ISFV, Piry virus, PIRYV; Vesicular stomatitis Alagoas virus, VSAV; Vesicular stomatitis Indiana virus, VSIV; and Vesicular stomatitis New Jersey virus, VSNJV) also infect humans. Antibodies to these viruses are relatively common amongst people living in rural areas where the viruses are endemic (farmers in North and Central America). Laboratory-acquired and natural infections with these viruses have usually resulted in mild influenza-like symptoms. Many of the vesiculoviruses that infect mammals have been isolated from naturally infected arthropods, primarily phlebotomine sandflies, suggesting that they may be vector-borne. Several vesiculoviruses infect fish and are responsible for epidemics of disease. Some may be vectored by fish ectoparasites.

LIST OF SPECIES DEMARCATION CRITERIA IN THE GENUS

Members of species in the genus cross-react in CF and immunofluorescence tests and exhibit low to no cross-neutralization. Genomic sequence analyses indicate sequence similarities. Higher homologies are observed between the N genes by comparison to the G genes. N protein sequence identity between species is in the range 49% (CHPV and VSNJV) to 68.6% (VSIV and VSNJV).

List of Species in the Genus

Species names are in green italic script; strain names and synonyms are in black roman script; tentative species names are in blue roman script. Sequence accession numbers and assigned abbreviations ( ) are also listed.

Species in the Genus

Carajas virus		
Carajas virus		(CJSV)
Chandipura virus		
Chandipura virus	[J04350, V01208, M16608]	(CHPV)
Cocal virus		
Cocal virus	[V01208]	(COCV)
Isfahan virus		
Isfahan virus		(ISFV)
Maraba virus		
Maraba virus		(MARAV)
Piry virus		
Piry virus	[Z15093, D26175, M14719, M14714, V01208]	(PIRYV)
Vesicular stomatitis Alagoas virus		
Vesicular stomatitis Alagoas virus		(VSAV)
Vesicular stomatitis Indiana virus		
Vesicular stomatitis Indiana virus	[J02428]	(VSIV)
Vesicular stomatitis New Jersey virus		
Vesicular stomatitis New Jersey virus	[K02379, S61075, J02433, M20166, M14553, K02747]	(VSNJV)

Tentative Species in the Genus

BeAn 157575 virus		(BeAnV-157575)
Boteke virus		(BTKV)
Calchaqui virus		(CQIV)
Eel virus American		(EVA)
Gray Lodge virus		(GLOV)
Jurona virus		(JURV)
Klamath virus		(KLAV)
Kwatta virus		(KWAV)
La Joya virus		(LJV)
Malpais Spring virus		(MSPV)
Mount Elgon bat virus		(MEBV)
Perinet virus		(PERV)
Pike fry rhabdovirus		(PFRV)
Grass carp rhabdovirus		
Porton virus		(PORV)
Radi virus		(RADIV)
Spring viremia of carp virus	[U18101, AJ318079]	(SVCV)
Tupaia virus		(TUPV)
Ulcerative disease rhabdovirus		(UDRV)
Yug Bogdanovac virus		(YBV)

Note that only selected representative sequence accession numbers have been provided for some species.

Genus Lyssavirus

Type Species *Rabies virus*

Distinguishing Features

Lyssaviruses such as RABV have 5 major polypeptides encoded from the 3′ to the 5′ genomic ends: N (58-62 kDa), P (35-40 kDa), M (22-25 kDa), G (65-80 kDa), and L (190 kDa). The G protein of RABV is glycosylated and palmitoylated at sites that have been mapped. N and P are both phosphoproteins and in the case of RABV, phosphorylation has been shown to involve several different host protein kinases: for N casein kinase II is implicated; P is phosphorylated by certain isomers of protein kinase C as well as an additional kinase, RVPK, yet to be clearly defined. In addition a viral encoded protein kinase activity has been postulated in the L protein. The 11.9 kb RABV genome encodes a leader RNA (3′-end) of about 60 nt in length that precedes the N gene, and a trailer region of about 70 nt that follows the L gene (5′-end). Non-transcribed intergenic sequences are of variable length: di-, penta-nucleotides, and a longer G-L intergene (19 nt up to 423 nt in certain vaccine strains, i.e. PV rabies virus strain. The start and stop transcription signals are closely related to those of VSIV.

Biological Properties

Lyssaviruses are the etiological agents of rabies encephalitis, the oldest known disease caused by a rhabdovirus; it is among the most lethal of all infectious diseases since neither natural nor drug-driven recovery has virtually ever been observed after the appearance of the first symptoms. For unvaccinated individuals, only careful wound cleaning, local instillation of immunoglobulins and post-exposure vaccination performed rapidly after the exposure ensures disease prevention. Rabies is enzootic in most regions of the world except in Antarctica and several islands. It is maintained by mammalian vectors of the chiropteran (frugivorous, insectivorous and hematophagous) and carnivoran orders (skunk, mongoose, raccoon, fox, wolf, jackal, dog, etc.). These animals transmit the disease to other mammalian species including livestock, domestic animals and wildlife. Transmission usually involves the transfer of infectious saliva, although other modes of transmission have been described (aerosols, corneal transplants). Transmission from dogs to humans remains the major concern in developing countries (up to 50,000 human deaths/year). Some developed countries are considered rabies-free since they have eliminated terrestrial rabies by animal control activities as well as parenteral (dog) or oral (wild carnivores) vaccination. However, bat rabies remains present in most parts of the world and contributes a low but significant dead-end spillover infection level to terrestrial species and humans, thus causing significant public health concerns. Moreover, phylogenetic studies suggest that occasional host-switching occurs from chiropteran to carnivorous vectors to extend the virus host range.

Lyssaviruses are neurotropic. Contamination generally occurs in non-neuronal tissues (muscles) where a local multiplication is possible (myotubes). The virus then reaches the motor or sensory neurones and propagates up to the central nervous system by following neuronal connections and using retrograde axonal transport. Late in the infection, multiplication in other tissues (e.g., salivary gland) is required for transmission. Seven genotypes of lyssaviruses have been characterized so far and this genetic diversity will probably increase due to the recent isolations from bats in Central Asia. Currently, two principal phylogroups with distinct genetic, antigenic and pathogenic properties have been described. The pathogenesis is still poorly understood but apoptosis appears to play a role. Death occurs without obvious neuronal damage and is probably due to induced neuronal dysfunctions that remain to be precisely characterized. Typically, Negri bodies composed of local accumulation of nucleocapsids are observed in infected neurons. The molecular basis of neurotropism and pathogenesis are being elucidated. Three potential protein receptors (nicotinic acetylcholine receptor AChR, neural cell adhesion molecule NCAM, low affinity nerve growth factor receptor p75NTR) can attach to the G protein. The P protein in infected cells has been shown to exist as multiple N-terminally truncated forms for at least two divergent lyssaviruses (RABV and Mokola virus MOKV). The conservation of methionine at certain positions of P suggests that this may be a common feature for all lyssaviruses. P protein has been proposed to interact with dynein/dynactin motors during axonal transport. Apoptosis is a multigenic phenomenon implicating at least G and possibly M proteins. In general, the lyssavirus growth cycle is slower than that of vesiculoviruses, both *in vivo* and *in vitro*.

List of Species Demarcation Criteria in the Genus

Until 1956 and the first isolations of rabies-related viruses in Africa, RABV was believed to be antigenically unique. This warranted the creation of the genus *Lyssavirus* (Greek *lyssa*: rabies) for viruses responsible for rabies-like encephalitis. The genus was at first divided into 4 serotypes by antigenic cross-reactivity with sera and monoclonal antibodies: 1 RABV; 2, Lagos bat virus (LBV); 3, MOKV; 4, Duvenhage virus (DUVV). Further isolations of new bat lyssaviruses in Europe, then Australia and the progress in genetic characterization at several genomic regions (N, P, and G) supported the delineation of 7 genotypes, agreeing and expanding antigenic data: 1, RABV; 2, LBV; 3, MOKV; 4, DUVV; 5, European bat lyssavirus 1 (EBLV-1); 6, European bat lyssavirus 2 (EBLV-2); 7, Australian bat lyssavirus (ABLV). These genotypes super-segregate in two main phylogroups including genotypes 1-4-5-6-7 and 2-3, respectively. Within each genotype, sublineages correspond to variants circulating in specific vectors. Recent isolations of bat lyssaviruses in Central-Asia (currently tentative species in the genus) should further expand the diversity.

It is important to note that chiropters are vectors for 6 out of the 7 genotypes characterized so far (the precise vector for genotype 3, MOKV, remains to be determined). For 5 genotypes bats are the exclusive vectors, and only genotype 1 (RABV) also includes terrestrial vectors (mainly carnivores). Genotype 1 corresponds to classical RABV and is spread in domestic or wild animals world-wide. Genotypes 2–7 have a narrower geographic and host-range distribution (so far most confined in the Old World plus Australia).

Lyssaviruses show a broad antigenic cross-reactivity at the nucleocapsid level, mainly due to the N protein sequence conservation: aa identities range from 78 % (MOKV and EBLV-2) to 93% (DUVV and EBLV-1). This allows the use of similar reagents for diagnosis by immunofluorescence and complement fixation. The ectodomain of the G protein (carrying the main antigenic sites) is more variable and cross-neutralization exists among lyssaviruses of the same phylogroup (either genotypes 1-4-5-6-7 or genotypes 2-3, ectodomain conservation > 75%), but not between phylogroups (ectodomain conservation < 65%). Thus, vaccine strains (all belonging to genotype 1 within phylogroup 1) are inappropriate for protection against infection by lyssaviruses from phylogroup 2. Using a chimeric G protein approach, the molecular basis to increase the vaccine spectrum from anti-rabies to anti-lyssavirus has been established. The tentative assignment of RBUV to the genus remains to be confirmed. Kotonkon virus (KOTV) and Obodhiang virus (OBOV) were formerly assigned to the genus but are now listed as unassigned viruses as serological and molecular data link them to viruses in the genus *Ephemerovirus*.

List of Species in the Genus

Species names are in green italic script; strain names and synonyms are in black roman script; tentative species names are in blue roman script. Sequence accession numbers and assigned abbreviations ( ) are also listed.

Species in the Genus

Australian bat lyssavirus		
Australian bat lyssavirus	[AF006497, AF081020, AF418014, NC003243]	(ABLV)
Duvenhage virus		
Duvenhage virus	[U89483, U22848, AF049115, AF049120, AF298146, AF298147]	(DUVV)
European bat lyssavirus 1		
European bat lyssavirus 1	[U22845, U22844, AF049113,	(EBLV-1)

	AF049117, AF298142, AF298143]	
European bat lyssavirus 2		
European bat lyssavirus 2	[U22846, U22847, AF049121, AF298144, AF298145]	(EBLV-2)
Lagos bat virus		
Lagos bat virus	[U22842, AF049114, AF049119, AF298148, AF298149, AF298225, AF298226]	(LBV)
Mokola virus		
Mokola virus	[D00491, D00492, D13767, D13766, S59447, S59448, U22843, U17064, AF298227, Y09762]	(MOKV)
Rabies virus		
Rabies virus	[D10499, D10482, D42112, J02293, L04522-3, L20672, L40426, M13215, A14671, AF298141, AF325461 to AF325495]	(RABV)

Tentative Species in the Genus

Rochambeau virus		(RBUV)
Aravan virus	[AY262023]	(ARAV)
Khujand virus	[AY262024]	(KHUV)
Irkut virus	[AY333112]	(IRKV)
West Caucasian bat virus	[AY333113]	(WCBV)

Genus *Ephemerovirus*

Type Species ***Bovine ephemeral fever virus***

Distinguishing Features

Virions are bullet- or cone-shaped with a length of ~140-200 nm and diameter ~60-80 nm. Virions have a buoyant density in CsCl of 1.19 g/cm^3 and sedimentation coefficient of 625S. Viruses are sensitive to acid or alkali and most stable at pH 7.0-8.0.

Bovine ephemeral fever virus (BEFV) particles contain at least five structural proteins, designated: L (180 kDa); G (81 kDa); N (52 kDa); P (43 kDa); and M (29 kDa). The G protein is a virus membrane-associated glycoprotein which contains 5 potential sites for attachment of N-linked glycans. The N protein is phosphorylated. The M protein is also phosphorylated in virions. In addition to these proteins, a 90 kDa, non-virion glycoprotein (G_{NS}) has been identified in BEFV-infected mammalian cells. G_{NS} is highly glycosylated (8 potential sites for N-linked glycans). The G and G_{NS} proteins, although not identical, exhibit homologies with each other and to lesser extents with the G proteins of other animal rhabdoviruses. In Adelaide River virus (ARV), the G protein (90 kDa) contains 6 potential sites for N-linked glycans, the G_{NS} protein contains 9. Two glycoproteins have been identified in mammalian cells infected with Berrimah virus (BRMV).

The 14.9 kb negative sense viral RNA genome of BEFV includes 10 genes in the order 3'-N-P-M-G-G_{NS}-α_1-α_2-β-γ-L-5' and intergenic regions of between 26 and 53 nt. The γ and L genes overlap by 21 nt. Additional small ORFs occur in alternative frames in the P and α_2 genes. Each gene, except α_1, is initiated from a UUGUCC sequence (mRNA: 5'-cap-AACAGG...) and terminates at a putative polyadenylation site: GNAC(U_{6-7})-3'. In ARV, the 14.6 kb genome contains 9 genes in the order 3'-N-P-M-G-G_{NS}-α_1-α_2-β-L-5' and intergenic regions of 1-4 nt. The β and L genes overlap by 22 nt. ARV lacks a γ gene comparable to that of BEFV. An additional ORF occurs in an alternative frame in the P gene. Each gene is initiated from a viral 3'-UUGUC sequence (mRNA: 5'-cap-AACAG...), however the putative polyadenylation signals are more variable than those of BEFV and may account for the synthesis of polycistronic mRNAs. The products of ephemerovirus α_1, α_2, β and γ genes have not been identified. The α_1 gene product appears to be a viroporin but the functions of other products have not been established. Proteins encoded in the ARV α_1, α_2 and β genes share homology with the corresponding BEFV proteins.

The G protein of bovine ephemeral fever virus BEFV contains 4 distinct neutralization sites. G protein purified from virions or expressed from recombinant vaccinia virus protects cattle from experimental infection. The G_{NS} glycoprotein does not induce neutralizing antibodies and is not protective.

Biological Properties

Bovine ephemeral fever is an economically important enzootic disease of cattle and water buffalo in most tropical and sub-tropical regions of Africa, Australia, the Middle-East and Asia. BEFV infection causes a sudden onset of fever and other clinical signs including lameness, anorexia and ruminal stasis, followed by a sustained drop in milk production. Although the mortality rate is low (1-2%), it is highest in well-conditioned beef cattle and high producing dairy cattle. The virus is transmitted by and replicates in hematophagous arthropods and has been isolated from both culicoides and mosquitoes.

Other species in the genus are not recognized as animal pathogens, but are known to infect cattle and have been isolated from healthy sentinel cattle (ARV, BRMV) or from insects (Kimberley virus, KIMV; Malakal virus, MALV; Puchong virus, PUCV).

List of Species Demarcation Criteria in the Genus

Members of different species exhibit low to no cross-neutralization. They cross-react strongly in CF or indirect immunofluorescence tests and may show low level cross-reactions by indirect immunofluorescence with viruses of the genus *Lyssavirus*. However, sequence comparisons with other rhabdoviruses indicate that in evolutionary terms the ephemeroviruses are closer to members of the genus *Vesiculovirus* than to those of other defined genera in the family. Analyses of the aa sequences of BEFV and ARV proteins indicate highly significant sequence homologies between most of the corresponding proteins, with the higher homologies in the L and N proteins than the G proteins. BEFV and ARV N proteins share 48% aa sequence identity. Amino acid sequence data are not available for other species. Only those viruses known to encode both G and G_{NS} glycoproteins are assigned as defined species.

List of Species in the Genus

Species names are in green italic script; strain names and synonyms are in black roman script; tentative species names are in blue roman script. Sequence accession numbers and assigned abbreviations ( ) are also listed.

Species in the Genus

Adelaide River virus		
Adelaide River virus	[L09206, L09208, U05987, U10363]	(ARV)
Berrimah virus		
Berrimah virus		(BRMV)
Bovine ephemeral fever virus		
Bovine ephemeral fever virus	[M94266, U04166, U18106, U72399, AY062166]	(BEFV)

Tentative Species in the Genus

Kimberley virus	(KIMV)
Malakal virus	(MALV)
Puchong virus	(PUCV)

GENUS NOVIRHABDOVIRUS

Type Species *Infectious hematopoietic necrosis virus*

DISTINGUISHING FEATURES

This genus comprises one of the two subgroups of rhabdoviruses known to infect aquatic hosts. Members of the other subgroup, represented by Spring viremia of carp virus (SVCV), are tentatively assigned to the genus *Vesiculovirus* based on available sequence data.

The replication temperature range and thermal inactivation temperatures for these viruses are typically lower than those of other rhabdoviruses, due to the aquatic poikilotherm nature of the host species for this genus. Optimum virus replication temperatures range from 15-28°C, depending roughly on the ambient water temperature in the geographic range of each virus.

Novirhabdoviruses have five major structural proteins, designated L (150-225 kDa), G (63-80 kDa), N (38-47 kDa), P (22-26 kDa, formerly designated M1), and M (17-22 kDa, formerly designated M2). In addition to the structural proteins, novirhabdoviruses encode a small, sixth, non-virion protein designated NV (12-14 kDa), which is expressed at variable levels in infected cells but is not detectable in purified virions. The function of the NV protein is not well defined, but the open reading frame is preserved in numerous diverse virus species and strains, and modifications of infectious clones indicate a non-essential role in virus replication and pathogenicity. The NV protein aa sequences are significantly less conserved between virus species than sequences of the other structural proteins, such that there is no significant aa sequence similarity between the NV proteins of Infectious hematopoietic necrosis virus (IHNV) and Viral hemorrhagic septicemia virus (VHSV).

The genomic RNA is approximately 11.1 kb, with six genes in the order 3'-N-P-M-G-NV-L-5'. For IHNV it is known that the genome contains a leader region of approximately 60 nt preceding the transcription start of the N gene, and a trailer of about 100 nt following the transcription termination of the L gene. Genes begin with the conserved putative transcription start signal 3'-CCRWG (vRNA sense, most often 3'-CCGUG), and the signal 3'-UUGU is also found upstream of the translation initiation site in many genes. Transcription terminates at the signal 3'-UCURUC$(U)_7$, and non-transcribed intergenic regions are single nucleotides, G or A (vRNA sense).

BIOLOGICAL PROPERTIES

Novirhabdoviruses infect numerous species of fish. The natural host ranges of individual viruses are relatively broad, often infecting several species as diverse as salmonids and herring. In nature and in artificial environments novirhabdoviruses can be transmitted horizontally, from fish to fish, by waterborne virus. Egg-associated transmission has also been clearly demonstrated by several cases in which the spread of virus to new geographic regions has occurred with transport of contaminated eggs. It is increasingly apparent that wild fish can serve as reservoirs of virus. The existence of invertebrate reservoirs or vectors of virus has been postulated but their importance is uncertain. Similarly, the potential for a carrier state in survivors of IHNV infections has been demonstrated, but the frequency and significance of this phenomenon is still under investigation.

The geographic distribution of novirhabdoviruses is broad. IHNV is enzootic to western North America, but inadvertent transport of the virus with contaminated eggs and infected fish has resulted in spread and establishment of IHNV in western Europe, Korea, Taiwan, Japan, and mainland China. VHSV is enzootic to western Europe, but more recently several North American strains have been described, and the virus has been found in marine fish in the northeastern Pacific Ocean, as well as the Baltic and North Seas. Hirame rhabdovirus (HIRRV) is at present only isolated in Japan.

Species in the genus *Novirhabdovirus* cause disease in cultured fish hosts, resulting in significant economic losses to the aquaculture industry. Both IHNV and VHSV have been well documented as severe pathogens of cultured salmonids since the 1950s, often resulting in losses of 50-100%. IHNV has also been reported to cause epizootics in wild salmonid populations. IHNV and VHSV both cause hemorrhagic diseases, with petechial hemorrhages evident both externally and internally. Major degenerative changes and necrosis in the kidneys and hematopoietic tissue are evident upon histopathological examination, and are believed to be the actual cause of mortality.

List of Species Demarcation Criteria in the Genus

Members of species within the genus have been distinguished serologically on the basis of cross-neutralization with polyclonal rabbit antisera. In general, strains within a species are neutralized by a single polyclonal antiserum. Thus, IHNV and HIRRV each comprise single serotypes, and VHSV has one major serotype with a small number of associated strains. Viruses from different species do not show cross-neutralization, but in some cases there is a low level of cross-reaction with specific proteins in western blot analyses. Nucleotide sequence data are available for most genes of these viruses, and will undoubtedly contribute to the distinction of viral species in the future. For strains within a virus species the nucleotide sequence divergence values range from 3-4% for IHNV G and NV genes, to a maximum of 18% for the G genes of European and North American VHSV. N protein aa sequence identity between IHNV and VHSV is approximately 34%.

List of Species in the Genus

Species names are in green italic script; strain names and synonyms are in black roman script; tentative species names are in blue roman script. Sequence accession numbers and assigned abbreviations ( ) are also listed.

Species in the Genus

Hirame rhabdovirus		
Hirame rhabdovirus	[D45422, U24073, U47847]	(HIRRV)
Infectious hematopoietic necrosis virus		
Infectious hematopoietic necrosis virus	[L40883, M16023, U47846, U50401, U50402, X89213]	(IHNV)
Snakehead virus		
Snakehead virus	[NC 000903]	(SHRV)
Viral hemorrhagic septicemia virus		
Egtved virus		
Viral hemorrhagic septicemia virus	[D00687, U02624, U02630, U03502, U03503, U28745, U28746, X59241]	(VHSV)

Tentative Species in the Genus

Eel virus B12	(EEV-B12)
Eel virus C26	(EEV-C26)

GENUS CYTORHABDOVIRUS

Type Species *Lettuce necrotic yellows virus*

DISTINGUISHING FEATURES

Two genera of plant rhabdoviruses have been established. The viruses are primarily distinguished on the basis of the sites of virus maturation (cytoplasm: *Cytorhabdovirus*; nucleus: *Nucleorhabdovirus*). However, exceptions exist, and the significance of this property is not known. The interrelationships of the different plant viruses within or between the two genera or with the unassigned plant viruses have yet to be established at the genetic level. There is no significant sequence similarity between analogous genes of the species analyzed to date. A wide variety of plants are susceptible to plant rhabdoviruses although each virus usually has a restricted host range. Most of the plant rhabdoviruses are transmitted by leafhoppers, planthoppers, or aphids, although mite- and lacebug-transmitted viruses (one each) have also been identified. Some viruses are transmitted in contaminated sap. In all carefully examined cases, viruses have been shown to replicate in the insect vector as well as in the plant host.

Cytorhabdoviruses replicate in the cytoplasm of infected cells in association with masses of thread-like structures (viroplasms). Virus morphogenesis occurs in association with vesicles of the endoplasmic reticulum. A nuclear phase has been suggested but not proven in the replication of some cytorhabdoviruses(e.g. *Lettuce necrotic yellows virus* (LNYV). Evidence of the nuclear involvement in the replication of others is lacking (e.g. Barley yellow striate mosaic virus, BYSMV). Endogenous transcriptase activity is readily detectable in virus preparations. The genome of LNYV is about 12.8 kb and the genome organization is similar to that of Sonchus yellow net virus (SYNV) (see nucleorhabdoviruses). Preceded by a non-coding 84 nt leader sequence, the gene order is 3′-N-P-4b-M-G-L-5′. N represents a 51 kDa nucleoprotein, P, M, G and L are the putative phospho-, matrix- and glycoproteins and RNA polymerase, respectively. Protein 4b has been predicted to represent a movement protein. The intergenic regions contain highly conserved consensus sequences. The 5′-non-coding trailer sequence of 187 nt has extensive complementarity to the 3′-leader. The genome of Northern cereal mosaic virus (NCMV) is about 13.2 kb with a gene order similar to LNYV, except for the presence of four small genes of unknown function between P and M.

LIST OF SPECIES DEMARCATION CRITERIA IN THE GENUS

In the genus *Cytorhabdovirus*, species are primarily differentiated by host range and vector specificity. Nucleic acid hybridization has been used to provide confirmation of species and serological criteria have enabled verification of common species that infect different hosts. However, no virus strains have been defined unambiguously using serology. The complete nt sequence is available for only two species in the *Cytorhabdovirus* genus, LNYV and NCMV. Thus, this criterion is not presently sufficient for discrimination of species. Hybridization using cloned probes has been used to verify species within the genus and these analyses should be emphasized in future studies.

LIST OF SPECIES IN THE GENUS

Species names are in green italic script; strain names and synonyms are in black roman script; tentative species names are in blue roman script. Natural vector species { }, sequence accession numbers and assigned abbreviations ( ) are also listed.

SPECIES IN THE GENUS

Barley yellow striate mosaic virus
Barley yellow striate mosaic virus {leafhopper} (BYSMV)
Broccoli necrotic yellows virus

Broccoli necrotic yellows virus	{aphid}		(BNYV)
Festuca leaf streak virus			
Festuca leaf streak virus			(FLSV)
Lettuce necrotic yellows virus			
Lettuce necrotic yellows virus	{aphid}	[L24364-5, L30103, AF209033-35, AJ251533]	(LNYV)
Northern cereal mosaic virus			
Northern cereal mosaic virus	{leafhopper}	[AB030277]	(NCMV)
Sonchus virus			
Sonchus virus			(SonV)
Strawberry crinkle virus			
Strawberry crinkle virus	{aphid}	[AY250986, AY005146]	(SCV)
Wheat American striate mosaic virus			
Wheat American striate mosaic virus	{leafhopper}		(WASMV)

TENTATIVE SPECIES IN THE GENUS

None reported.

GENUS *NUCLEORHABDOVIRUS*

Type Species *Potato yellow dwarf virus*

DISTINGUISHING FEATURES

Nucleorhabdoviruses multiply in the nucleus of plants forming large granular inclusions that are thought to be sites of virus replication. Viral proteins are synthesized from discrete polyadenylated mRNAs and accumulate in the nucleus. Virus morphogenesis occurs at the inner nuclear envelope, and enveloped virus particles accumulate in perinuclear spaces. In protoplasts treated with tunicamycin, morphogenesis is interrupted and nucleocapsids accumulate in the nucleoplasm. The genome of SYNV is about 13.7 kb. Preceded by a non-coding 144 nt leader sequence, the gene order is 3'-N-P-SC4-M-G-L-5'. The leader sequence transcript is polyadenylated. N represents the 54 kDa viral nucleocapsid, P is a 38 kDa phosphoprotein, SC4 is probably a movement protein, M is a 32 kDa matrix protein, G a 70 kDaglycoprotein (unglycosylated form) and L the 241 kDa polymerase. The intergenic regions are similar in length and have sequence relatedness to those of other rhabdoviruses. The 5'-NCR (trailer) region is 160 nt long with extensive complementarity to the leader sequence. N and P contain NLSs, are independently imported into the nucleus, where they associate and move to a sub-nuclear location. A distinct nuclear polymerase complex composed of N, P and L is present in nuclei of infected cells. The genome of RYSV is 14 kb with a gene order similar to SYNV, except for the presence of an additional gene between G and L which encodes a virion-associated protein.

LIST OF SPECIES DEMARCATION CRITERIA IN THE GENUS

In the genus *Nucleorhabdovirus*, species are primarily differentiated by host range and vector specificity. Nucleic acid hybridization has been used to provide confirmation of species and serological criteria have enabled verification of common species that infect different hosts. However, no virus strains have been defined unambiguously using serology. The complete nt sequence is available for members of only two species in the *Nucleorhabdovirus* genus (SYNV and Rice yellow stunt virus, RYSV). Thus this criterion is not presently sufficient for discrimination of species. Hybridization using cloned probes has been used to verify species within the genus and these analyses should be emphasized in future studies.

List of Species in the Genus

Species names are in green italic script; strain names and synonyms are in black roman script; tentative species names are in blue roman script. Natural vector species { }, sequence accession numbers and assigned abbreviations ( ) are also listed.

Species in the Genus

Datura yellow vein virus			
Datura yellow vein virus			(DYVV)
Eggplant mottled dwarf virus			
Eggplant mottled dwarf virus			(EMDV)
Pittosporum vein yellowing virus			(PVYV)
Tomato vein yellowing virus			(TVYV)
Pelargonium vein clearing virus			(PVCV)
Maize mosaic virus			
Maize mosaic virus	{leafhopper}		(MMV)
Potato yellow dwarf virus			
Potato yellow dwarf virus	{leafhopper}		(PYDV)
Rice yellow stunt virus			
Rice yellow stunt virus	{leafhopper}	[X75534, U47053, D87843-4, AB002822, AB010258, AB003092, D89654, AB0011257]	(RYSV)
Rice transitory yellowing virus			(RTYV)
Sonchus yellow net virus			
Sonchus yellow net virus	{aphid}	[M13950, M17210, M23023, M35689, M73626, M87829]	(SYNV)
Sowthistle yellow vein virus			
Sowthistle yellow vein virus	{aphid}		(SYVV)

Tentative Species in the Genus

None reported.

List of Unassigned Viruses in the Family

There are at least six serogroups of rhabdoviruses infecting animals that have not been assigned to an existing genus, and there are a number of ungrouped viruses. SIGMAV is transmitted vertically through the germinal cells of *Drosophila* species and confers CO_2-sensitivity to infected insects. Both host and viral genes contribute to the maintenance of the virus in the host. SIGMAV encodes a 6th gene located between the P and M genes. The function of this gene is not known. The intergenic regions of the virus are variable (up to 36 nts in length) and one gene (M) overlaps that of the following gene (G). For most of the other listed viruses, no biochemical characterization has been reported. Their assignment to the family relies on the distinctive morphology of rhabdoviruses.

1- Bahia Grande group:	
Bahia Grande virus	(BGV)
Muir Springs virus	(MSV)
Reed Ranch virus	(RRV)
2- Hart Park group:	
Flanders virus	(FLAV)
Hart Park virus	(HPV)
Kamese virus	(KAMV)
Mosqueiro virus	(MQOV)
Mossuril virus	(MOSV)
3- Kern Canyon group:	
Barur virus	(BARV)

Fukuoka virus			(FUKAV)
Kern Canyon virus			(KCV)
Nkolbisson virus			(NKOV)
4- Le Dantec group:			
Le Dantec virus			(LDV)
Keuraliba virus			(KEUV)
5- Sawgrass group:			
Connecticut virus			(CNTV)
New Minto virus			(NMV)
Sawgrass virus			(SAWV)
6- Timbo group:			
Chaco virus			(CHOV)
Sena Madureira virus			(SMV)
Timbo virus			(TIMV)

List of Unassigned Animal Rhabdoviruses

Almpiwar virus			(ALMV)
Aruac virus			(ARUV)
Bangoran virus			(BGNV)
Bimbo virus			(BBOV)
Bivens Arm virus			(BAV)
Blue crab virus			(BCV)
Charleville virus			(CHVV)
Coastal Plains virus			(CPV)
Curionopolis virus	{Culicoides spp . midge}		(CURV)
DakArK 7292 virus			(DAKV-7292)
Entamoeba virus			(ENTV)
Farmington virus	{bird}		(FARV)
Garba virus			(GARV)
Gossas virus			(GOSV)
Harlingen virus	{Culex spp . mosquitoes}		(HARV)
Humpty Doo virus			(HDOOV)
Iriri virus	{Lutzomyia spp. Sandfly}		(IRIV)
Itacaiunas virus	{Culicoides spp . midge}		(ITAV)
Joinjakaka virus			(JOIV)
Kannamangalam virus			(KANV)
Kolongo virus			(KOLV)
Koolpinyah virus			(KOOLV)
Kotonkon virus			(KOTV)
Landjia virus			(LJAV)
Manitoba virus			(MNTBV)
Marco virus			(MCOV)
Nasoule virus			(NASV)
Navarro virus			(NAVV)
Ngaingan virus			(NGAV)
Oak-Vale virus			(OVRV)
Obodhiang virus			(OBOV)
Oita virus			(OITAV)
Ouango virus			(OUAV)
Parry Creek virus			(PCRV)
Rhode Island virus	{bird}		(RHIV)
Rio Grande cichlid virus			(RGRCV)
Sandjimba virus			(SJAV)
Sigma virus		[X91062]	(SIGMAV)
Sripur virus			(SRIV)
Sweetwater Branch virus			(SWBV)
Tibrogargan virus			(TIBV)
Xiburema virus			(XIBV)
Yata virus			(YATAV)

List of Unassigned Plant Rhabdoviruses

There are many putative plant rhabdoviruses that have not been assigned to a genus. Many of these agents have not been characterized beyond electron microscopic observations in infected plants and occasionally in their vectors. Hence, their tentative assignment to the family relies on morphological criteria. Some have been transmitted experimentally by mechanical means and via their vectors.

Species names are in green italic script; strain names and synonyms are in black roman script; tentative species names are in blue roman script. Natural vector species { }, sequence accession numbers and assigned abbreviations ( ) are also listed.

Atropa belladonna virus			(AtBV)
Beet leaf curl virus	{lacewing}		(BLCV)
Callistephus chinensis chlorosis virus			(CCCV)
Carnation bacilliform virus			(CBV)
Carrot latent virus	{aphid}		(CtLV)
Cassava symptomless virus			(CsSLV)
Cereal chlorotic mottle virus	{leafhopper}		(CCMoV)
Chrysanthemum frutescens virus			(CFV)
Chrysanthemum vein chlorosis virus			(CVCV)
Citrus leprosis virus			(CiLV)
Clover enation virus			(ClEV)
Coffee ringspot virus	{mite}		(CoRSV)
Colocasia bobone disease virus	{leafhopper}		(CBDV)
Coriander feathery red vein virus	{aphid}		(CFRVV)
Cow parsnip mosaic virus			(CPaMV)
Cynara virus			(CraV)
Dendrobium leaf streak virus			(DLSV)
Digitaria striate virus	{leafhopper}		(DiSV)
Euonymus fasciation virus			(EFV)
Finger millet mosaic virus	{leafhopper}		(FMMV)
Gerbera symptomless virus			(GeSLV)
Gomphrena virus			(GoV)
Holcus lanatus yellowing virus			(HLYV)
Iris germanica leaf stripe virus			(IGLSV)
Ivy vein clearing virus			(IVCV)
Laelia red leafspot virus			(LRLV)
Launea arborescens stunt virus			(LArSV)
Lemon scented thyme leaf chlorosis virus			(LSTCV)
Lolium ryegrass virus			(LoRV)
Lotus stem necrosis			(LoSNV)
Lucerne enation virus	{aphid}		(LEV)
Lupin yellow vein virus			(LYVV)
Malva silvestris virus			(MaSV)
Maize fine streak virus	{leafhopper}	[AF518001-2]	(MFSV)
Maize sterile stunt virus	{leafhopper}		(MSSV)
Melilotus latent virus			(MeLV)
Melon variegation virus			(MVV)
Oat striate mosaic virus	{leafhopper}		(OSMV)
Parsley virus			(PaV)
Phalaenopsis chlorotic spot virus			(PhCSV)
Pigeon pea proliferation virus			(PPPV)
Pineapple chlorotic leaf streak virus			(PCLSV)
Pisum virus			(PisV)
Plantain mottle virus			(PlMV)
Ranunculus repens symptomless virus			(RaRSV)
Raphanus virus			(RaV)
Raspberry vein chlorosis virus	{aphid}		(RVCV)
Red clover mosaic virus			(RClMV)

NEG. SSRNA

Sainpaulia leaf necrosis virus			(SLNV)
Sambucus vein clearing virus			(SVCV)
Sarracenia purpurea virus			(SPV)
Sorghum virus	{leafhopper}		(SrV)
Soursop yellow blotch virus			(SYBV)
Triticum aestivum chlorotic spot virus			(TACSV)
Vigna sinensis mosaic virus			(VSMV)
Winter wheat Russian mosaic virus	{leafhopper}		(WWMV)
Wheat chlorotic streak virus	{leafhopper}		(WCSV)
Wheat rosette stunt virus	{leafhopper}	[AF059602-04, AF059677, AF064784]	(WRSV)
Zea mays virus			(ZMV)

Phylogenetic Relationships within the Family

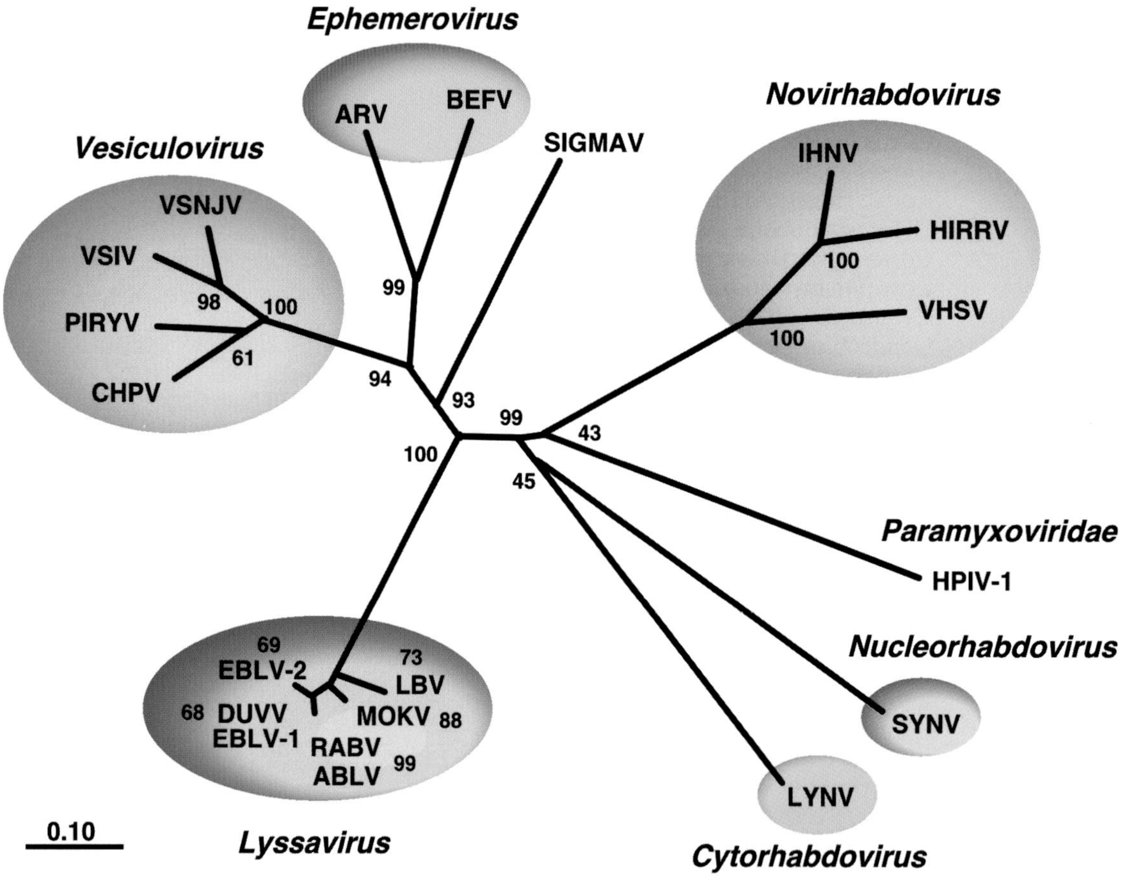

Figure 4: Phylogenetic relationships between rhabdoviruses based on a GDE alignment of a relatively conserved region of the N protein (119 aa), and using the paramyxovirus Human parainfluenza virus 1 (HPIV-1) as the outgroup. The tree was generated by the neighbor-joining method and bootstrap values (indicated in percent for each branch node) were estimated using 1000 tree replicas. Branch lengths are proportional to genetic distances. The scale bar corresponds to substitutions per aa site (Courtesy of H. Badrane and P.J. Walker).

Molecular phylogenies determined by using N, G, or L protein sequences support the integrity of the family *Rhabdoviridae* and the assignment of species within the established genera. For G proteins, relatively low sequence identities across the family prevent the construction of a universal phylogeny. L protein sequences are most highly conserved. A global phylogeny can be constructed by using representative available sequences of a relatively conserved central region (119 aa) of the N protein (Fig. 4). Similar results are obtained with the most conserved domain III of the L protein. Both phylogenetic analyses indicate that vesiculoviruses and ephemeroviruses are the most closely related of the established genera.

Similarity with Other Taxa

Rhabdoviruses share several features with viruses of the families *Filoviridae*, *Paramyxoviridae* and *Bornaviridae* in the order *Mononegavirales*. Features they have in common include the unsegmented negative-sense, single-strand, non-infectious RNA genome, the helical nucleocapsid with the genome template intimately associated with the nucleoprotein (RNAse resistant), the initiation of primary transcription by a virion-associated RdRp, similar gene order, and single 3' promoter with short terminal untranscribed regions and intergenic regions. The virions are large enveloped structures with a prominent fringe of spikes. They generally replicate in the cytoplasm (except nucleorhabdoviruses). They mature by budding, predominantly from the plasma membrane with the exception of RABV which buds occasionally from internal membranes and plant rhabdoviruses of the genus *Nucleorhabdovirus* which bud from the inner nuclear membrane. They transcribe discrete unprocessed messenger RNAs for which they ensure 5′-capping and 3′-polyadenylation.

Derivation of Names

Cyto: from Greek *kytos* "cell".
Ephemero: from Greek ephemeros, "short-lived".
Lyssa: from Greek *lyssa* "rage, fury, canine madness".
Novi: sigla from "non-virion" protein.
Nucleo: from Latin *nux*, *nucis*, "nut".
Rhabdo: from Greek *rhabdos*, "rod".
Vesiculo: from Latin *vesicula*, diminutive of *vesica*, "bladder, blister".

References

Arai, Y.T., Kuzmin, I.V., Kameoka, Y. and Botvinkin, A.D. (2003). New lyssavirus genotype from the Lesser Mouse-eared Bat *(Myotis blythi)*, Kyrghyzstan. *Emerg. Infect. Dis.*, **9**, 333-337.

Badrane, H., Bahloul, C., Perrin, P. and Tordo, N. (2001). Evidence of two lyssavirus phylogroups with distinct pathogenicity and immunogenicity. *J. Virol.*, **75**, 3268-3276.

Badrane H. and Tordo N. (2001). Host-switching in lyssavirus history from chiroptera to carnivora orders. *J. Virol*, **75**, 8096-8104.

Benmansour, A., Basurco, B., Monnier, A.F., Vende, P., Winton, J.R. and de Kinkelin, P. (1997). Sequence variation of the glycoprotein gene identifies three distinct lineages within field isolates of viral haemorrhagic septicaemia virus, a fish rhabdovirus. *J. Gen. Virol.*, **78**, 2837-2846.

Bjorklund, H.V., Higman, K.H. and Kurath, G. (1996). The glycoprotein genes and gene junctions of the fish rhabdoviruses spring viremia of carp and Hirame rhabdovirus; analysis of their relationships with other rhabdoviruses. *Virus Res.*, **42**, 65-80.

Calisher, C.H., Karabatsos, N., Zeller, H., Digoutte, J.-P., Tesh, R.B., Shope, R.E., Travassos da Rosa, A.P.A. and St. George, T.D. (1989). Antigenic relationships among rhabdoviruses from vertebrates and hematophagous arthropods. *Intervirology*, **30**, 241-257.

Frerichs, G.N. (1989). Rhabdoviruses of fishes. In: *Viruses of Lower Vertebrates*, (W. Ahne and E. Kurstak, eds), pp 316-332. Springer-Verlag, New York.

Goodin, M.M., Austin, J., Tobias, R., Fujita, M., Morales, C. and Jackson, A.O. (2001). Interactions and nuclear import of the N and P protein of sonchus yellow net virus, a plant nucleorhabdovirus. *J. Virol.*, **75**, 9393-9406.

Huang Y., Zhao H., Luo Z., Chen X. and Fang R. (2003). Novel structure of the genome of *Rice yellow stunt virus*: identification of the gene 6-encoded virion protein. *J. Gen. Virol.*, **84**, 2259-2264.

Johnson, M.C., Maxwell, J.M., Loh P.C. and Leong, J.C. (1999). Molecular characterisation of the glycoproteins from two warm water rhabdoviruses: snakehead rhabdovirus (SHRV) and rhabdovirus of penaeid shrimp (RPS)/spring viremia carp virus (SVCV). *Virus Res.*, **64**, 95-106.

NEG. SSRNA

Kurath, G., Higman, K.H. and Bjorklund, H.V. (1997). Distribution and variation of NV genes in fish rhabdoviruses. *J. Gen. Virol.*, **78,** 113-117.
McWilliam, S.M., Kongsuwan, K., Cowley, J.A., Byrne, K.A. and Walker, P.J. (1997). Genome organization and transcription strategy in the complex G_{NS}-L intergenic region of bovine ephemeral fever rhabdovirus. *J. Gen. Virol.*, **78,** 1309-1317.
Nadin-Davis, S.A., Abdel-Malik, M., Armstrong, J. and Wandeler, A. (2002). Lyssavirus P gene characterization provides insights into the phylogeny of the genus and identifies structural similarities and diversity within the encoded phosphoprotein. *Virology*, **298**, 286-305.
Redinbaugh, M.G., Seifers, D.L., Meulina, T., Abt, J.J., Anderson, R.J., Styer, W.E., Ackerman, J., Salomon, R., Houghton, W., Creamer, R., Gordon, D.T. and Hogenhout, S.A. (2002). Maize fine streak virus, a new leafhopper-transmitted rhabdovirus. *Phytopathology*, **92**, 1167-1174.
Tanno, F., Nakatsu, A., Toriyama, S. and Kojima, M. (2000). Complete nucleotide sequence of Northern cereal mosaic virus and its genome organization. *Arch. Virol.*, **145**, 1373-1384.
Travassos da Rosa, A.P.A., Mather T.N., Takeda T., Whitehouse C.A., Shope R.E., Popov V.L., Guzman H., Coffey L., Araujo T.P. and Tesh R.B. (2002). Two new rhabdoviruses (*Rhabdoviridae)* isolated from birds during surveillance for arboviral encephalitis, Northeastern United States. *Emerg. Infect. Dis.*, **8**, 614-618.
Travassos da Rosa, J.F.S., Travassos da Rosa, A.P.A., Vasconcelos, P.F.C., Pinheiro, F.P., Rodrigues, S.G., Travassos da Rosa, E.S., Dias, L.B. and Cruz, A.C.R. (1998). Arboviruses isolated in the Evandro Chagas Institute, including some described for the first time in the Brazilian Amazon region, their known hosts, and their pathology for man. In: *An overview of arbovirology in Brazil and neighboring countries.* (A.P.A. Travassos da Rosa, P.F.C. Vasconcelos and J.F.S. Travassos da Rosa, eds), pp 19-31. Instituto Evandro Chagas, Belem.
Tordo, N., Charlton, K. and Wandeler, A. (1998). Rhabdoviruses: rabies. In: *Topley and Wilson's Microbiology and Microbial Infections*, 9th Edition, (W.W.C. Topley, G.S. Wilson and L. Collier, eds), pp 665-692. Arnold, London.
Walker, P.J., Byrne, K.A., Riding, G.A., Cowley, J.A., Wang, Y. and McWilliam, S. (1992). The genome of bovine ephemeral fever rhabdovirus contains two related glycoprotein genes. *Virology*, **191,** 49-61.
Wetzel, T., Dietzgen, R.G. and Dale, J.L. (1994). Genomic organization of lettuce necrotic yellows rhabdovirus. *Virology*, **200,** 401-412.

Contributed By

Tordo, N., Benmansour, A., Calisher, C., Dietzgen, R.G., Fang, R.-X., Jackson, A.O., Kurath, G., Nadin-Davis, S., Tesh, R.B. and Walker, P.J.

FAMILY FILOVIRIDAE

TAXONOMIC STRUCTURE OF THE FAMILY

Order	*Mononegavirales*
Family	*Filoviridae*
Genus	*Marburgvirus*
Genus	*Ebolavirus*

VIRION PROPERTIES

MORPHOLOGY AND STRUCTURE

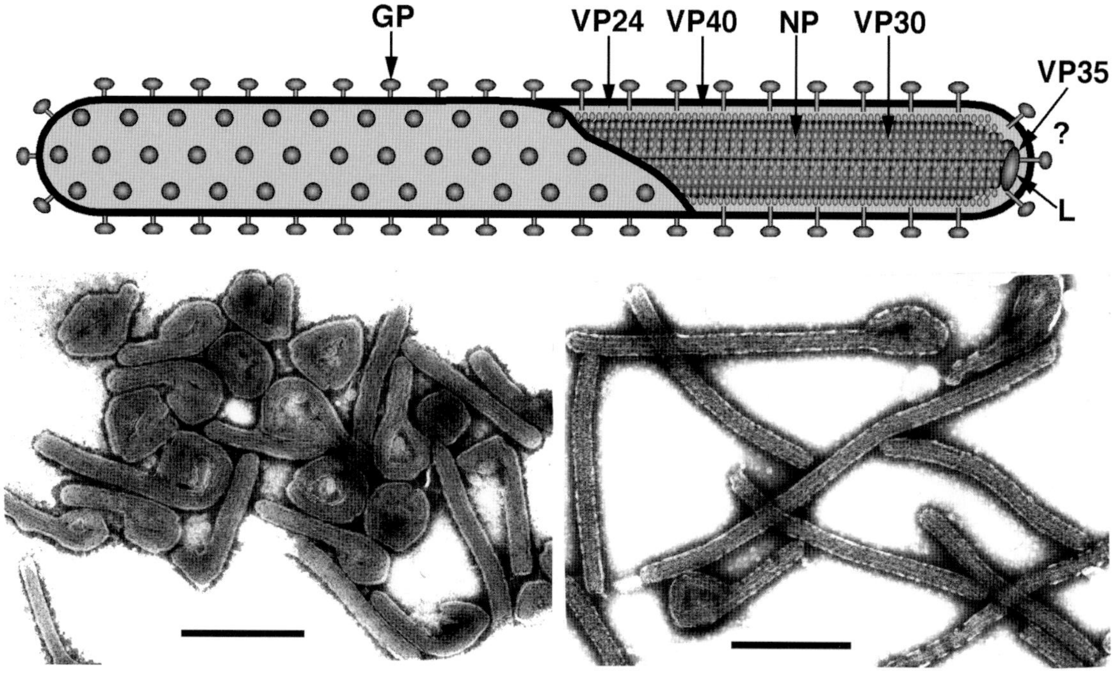

Figure 1: (Top) Diagram of virion in cross section. (Bottom left) Negative contrast electron micrograph of Lake Victoria marburgvirus (MARV) particles purified and concentrated by centrifugation from guinea pig serum, 7 days after intra-peritoneal infection, and stained with 1% phosphotungstate. (Bottom right) Negative contrast electron micrograph of Zaire ebolavirus (ZEBOV) particles, from Vero cell culture supernatant, 4 days after infection. (Courtesy of T.G. Geisbert, United States Army Medical Research Institute for Infectious Diseases, Fort Detrick, Maryland, USA). The bars represent 450 nm.

Virions are bacilliform in shape, but particles can also appear as branched, circular, U or 6-shaped and long filamentous forms (Fig. 1). This morphology is unusual for viruses and, thus, became characteristic for the family. Virions vary greatly in length but show a uniform diameter of ~80 nm. Family members differ in length of virion particles, but seem to be very similar in morphology. Peak infectivity has been associated with particles of 665 nm for Lake Victoria marburgvirus (MARV) and 805 nm for ebolaviruses.

Virions are composed of a central core formed by a nucleocapsid or ribonucleoprotein (RNP) complex, surrounded by a lipid envelope derived from the host cell plasma membrane. Electron micrographs reveal an axial channel (10-15 nm in diameter) surrounded by a central dark layer (20 nm in diameter) and an outer helical layer (50 nm in diameter) with cross-striations of 5 nm intervals (Fig. 1). Spikes about 7 nm in diameter and spaced at intervals of ~10 nm are seen as globular structures on the surface of virions.

The RNP complex is composed of a genomic RNA molecule and 4 of the 7 virion structural proteins: the nucleoprotein (NP), virion structural proteins VP30 and VP35, and the large (L) protein. Genomic RNA has a molecular weight of 4.2 x 10^6 and constitutes 1.1% of the virion mass. The 3 remaining structural proteins are membrane-associated: the glycoprotein (GP) shows a type I transmembrane protein profile, while the two non-glycosylated proteins, VP24 and VP40, are located at the inner side of the membrane (membrane-associated). VP40 functions as the matrix protein; the structure and function of VP24 is unknown.

Physicochemical and Physical Properties

Virions have a molecular weight of 3.82 x 10^8. The buoyant density of virions in potassium tartrate is 1.14 g/cm^3. The $S_{20,w}$ of bacilliform particles is 1.40, for longer particles it is very high. The nucleocapsid has a buoyant density in CsCl ~1.32 g/cm^3. The infectivity of marburgviruses and ebolaviruses is stable at less than 20°C, but drastically reduced within 30 min at 60°C. There are no data about metal cation stability. Virus infectivity is sensitive to quarternary ammonium salt, hypochlorite and phenolic disinfectants, lipid solvents, β-propiolactone, formaldehyde, and ultraviolet and gamma irradiation.

Proteins

Viral particles contain seven proteins: the L protein is an RNA-dependent RNA transcriptase-polymerase; the surface glycoprotein (GP) makes up the viral spikes in the form of homotrimers; the nucleoprotein (NP); the matrix protein VP40; VP35, which functions as the phosphoprotein-equivalent VP30, another polymerase co-factor; and VP24, a second membrane-associated protein of unknown function. The sizes and functions of the proteins are shown in Table 1.

Table 1: Filovirus structural proteins.

Protein	Encoding Gene	Function	*Lake Victoria marburgvirus* (kDa) A / B	*Zaire ebolavirus* (kDa) A / B
NP	1	Nucleoprotein	77.9 / 96.0	83.3 / 104.0
VP35	2	phosphoprotein-equivalent	31.0 / 32.0	38.8 / 35.0
VP40	3	matrix protein	31.7 / 38.0	35.3 / 40.0
GP	4	Glycoprotein	74.8 / 170.0	74.5 / 140.0
VP30	5	polymerase co-factor	31.5 / 28.0	29.7 / 30.0
VP24	6	Unknown	28.8 / 24.0	28.3 / 24.0
L	7	polymerase/replicase	267.2 / >200.0	252.7/ >200.0

A, size calculated from the deduced amino acid sequences of the corresponding ORFs; B, size estimated from SDS-PAGE analysis

Nucleic Acid

The genome is a non-segmented, negative stranded, linear RNA molecule of ~19 kb: Lake Victoria marburgvirus (MARV) is 19.1 kb and Zaire ebolavirus (ZEBOV), is 18.9 kb. The Mr of the genomic RNA is 4.2 x 10^6 and the genome represents about 1.1% of the total virion mass. Genomic RNA is not polyadenylated at the 3′-end and there is no evidence for a 5'-terminal cap structure or covalently-linked protein. Full-length nt sequences of the genomes of MARV-Mus, MARV-Pop, ZEBOV-May, Reston ebolavirus-Res (REBOV-Res), and REBOV-Phi (strains Philippines and Pennsylvania) have been determined.

Lipids

The viral envelope is derived from host cell membranes and is considered to have a lipid composition similar to that of the plasma membrane.

Carbohydrates

The glycoproteins of filoviruses are highly glycosylated with N-linked glycans of the complex, hybrid and oligomannosidic type, and O-linked glycans of the neutral mucin type. Glycans constitute > 50% of the GP total mass. ZEBOV shows a higher level of sialylation than MARV, and in certain cell lines the GP of MARV completely lacks sialic acid. The GPs of MARV and ebolaviruses are acylated and form homotrimers.

Genome Organization and Replication

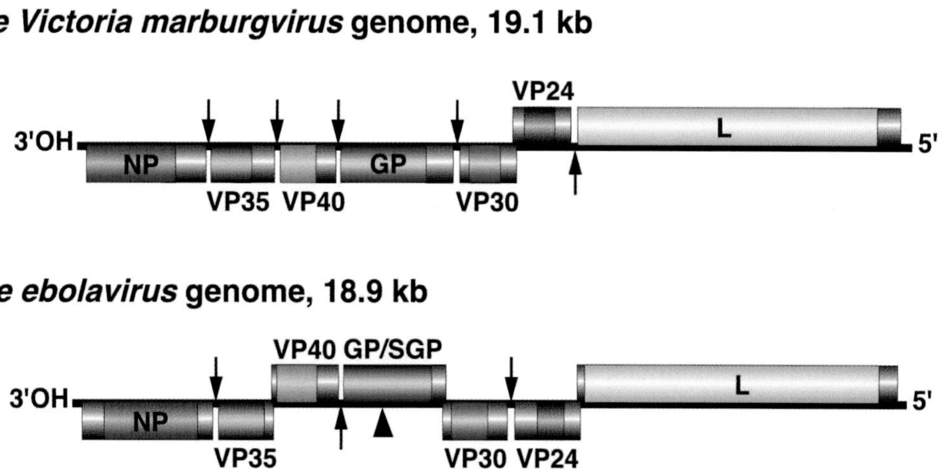

Figure 2: Diagram of filovirus genome organization. The genes which encode structural proteins are identified in the genomes and are drawn to scale. Shaded or colored squares designate coding regions and dark lines designate non-coding regions. Genes begin with a conserved transcriptional start site and end with a transcriptional stop site (polyadenylation site). Adjoining genes are either separated from one another by an intergenic region (short arrow) or overlap. The editing site on the GP gene of ebolaviruses is indicated by a black triangle. At the extreme 3′- and 5′-ends of the genomes are leader and trailer sequences, respectively, that are in part complementary

Filovirus genomes are characterized by the gene order: 3'-NP-VP35-VP40-GP-VP30-VP24-L-5' (Fig. 2). The extragenic sequences at the extreme 3′-(leader) and 5′-(trailer) ends of the genomes are conserved. They demonstrate a significantly high complementarity at their very ends. Genes are flanked by conserved transcriptional start and stop (polyadenylation) sites. Those sites contain a highly conserved pentamer 3'-UAAUU-5'. Most genes are separated by non-conserved intergenic sequences, but some genes overlap. Those overlaps are extremely short and limited to the highly conserved pentamer. There is a single gene overlap with MARV and several overlaps with ebolaviruses. The functional significance of these short overlaps is unclear, but transcriptional attenuation of the downstream gene has been postulated for other viruses of the order *Mononegavirales*. In addition, most genes possess relatively long 3'- and 5'-NCR. In contrast to MARV, the GP gene of ZEBOV consists of two overlapping ORFs which are fused through transcriptional editing (Fig. 2).

The replication strategy of filoviruses is not well-studied. Ultrastructural studies indicate an association of viral particles with coated pits for the initiation of infection, suggesting that filoviruses enter cells by endocytosis. The glycoprotein mediates receptor binding and subsequent fusion. The asialo glycoprotein receptor, folate receptor alpha, integrins, especially the beta 1 group, DC-SIGN and L-SIGN have been described as potential attachment factors to promote filovirus infection. Uncoating is presumed to occur in a manner analogous to that of other negative sense RNA viruses. Transcription and genome replication take place in the cytoplasm and, in general, follow the models of me4mbers of the families *Paramyxoviridae* and *Rhabdoviridae*. Transcription starts at the conserved start site and polyadenylation occurs at a run of uridine residues within the stop site. The 5'-terminal non-coding sequences favor hairpin-like structures for all sgRNAs (mRNAs). Replication involves the synthesis of a full-length positive-strand copy. During infection, massive amounts of nucleocapsids accumulate intracellularly and form intracytoplasmic inclusion bodies. Virions are released via budding through plasma membranes. The expression strategy of ZEBOV GP genes is unique and involves transcriptional editing (Fig. 2). The primary product of the unedited transcript (ORF1) yields a smaller non-structural glycoprotein sGP which is efficiently secreted from infected cells. Only RNA editing allows expression of full-length GP.

Antigenic and Genetic Properties

Virus infectivity is poorly neutralized *in vitro*. There is almost no antigenic cross-reactivity between the two genera. MARV and ZEBOV GP genes differ by 57%, and phylogenetic analysis clearly separates both genera (Fig. 3). The cross-reactive species of the *Ebolavirus* genus can be differentiated antigenically and genetically. All four species differ from one another by 37-41% at the nucleotide level. The genus *Marburgvirus*, however, is antigenically and genetically more homogenous. Comparative sequence analysis shows that two genetic lineages coexist with the recent isolate from Kenya (MARV-Ravn) differing from the others by 21-23% at the nt level. Among strains of individual species of filoviruses the variation in nt sequences has been shown to be extremely low, < 2% among distinct strains of the species *Zaire ebolavirus*. There seems to be less or even no genetic variability between isolates from different patients of single outbreaks (e.g., Gabon and Kikwit). All data indicate a remarkable degree of stability over time.

Biological Properties

History

MARV was first isolated from hemorrhagic fever patients in Germany and Yugoslavia in 1967 infected by contact with tissues and blood from infected, but apparently healthy, monkeys (*Cercopithecus aethiops*) imported from Uganda. A second small outbreak of Marburg hemorrhagic fever occurred in Zimbabwe in 1975, and isolated episodes have occurred subsequently in Kenya in 1980 and 1987. In 1998 MARV emerged in the northern part of the Democratic Republic of Congo. This has been the largest outbreak so far with more than 100 cases, high mortality, and several distinct introductions. MARV mortality rates in humans vary from 25-70%.

The first ebolavirus outbreaks were observed in northern DRC (former Zaire) in 1976 (ZEBOV) and in southern Sudan in 1976 and 1979 (Sudan ebolavirus; SEBOV). In 1994 the first case of ebolavirus disease occurred in West Africa, Cote d'Ivoire, when an ecologist was infected by examining a dead chimpanzee. In 1995, ebolavirus re-emerged in Kikwit, DRC, with 316 cases and 245 deaths. From 1994–1997, there were three outbreaks of ebolavirus disease in Gabon. An outbreak of Ebola disease occurred in Uganda in 2000 and there is currently an outbreak in Congo-Brazzaville that has claimed the lives of ~100 gorillas. Ebolavirus mortality rates in humans range from 50-90% depending on the species.

Reston ebolavirus (REBOV) was first isolated from *Cynomolgus* monkeys imported from the Philippines into the United States in 1989-1990, and from monkeys at an export facility located in the Philippines. Further isolates have been made from exported Asian monkeys in 1992 in Italy and in 1996 in Texas, USA. While pathogenic for naturally and experimentally infected monkeys, REBOV may be less or even nonpathogenic for humans.

Geographic Distribution

Filoviruses appear to be endemic in Central Africa in an area approximately between the 10th parallel north and south of the equator as indicated by the locations of known outbreaks and seroepidemiological studies. Additional endemic areas in Africa or on other continents may exist. In 1994, Cote d'Ivoire ebolavirus (CIEBOV) was isolated in West Africa during an outbreak of hemorrhagic fever among wild chimpanzees in the Tai Forest. Before that, the identification of REBOV suggested, for the first time, the presence of a filovirus in Asia that may be associated with wild non-human primates.

Natural Reservoir and Transmission

The natural reservoir and history of filoviruses are still unknown. There is no connection of virus spread with any vector. The usual pattern seen with large outbreaks of disease in man begins with a focus of infection that disseminates to a number of contacts. Secondary and subsequent episodes of disease occur following close contact with patients; such infections usually occur in family members or medical personnel. The major route of interhuman transmission of filoviruses requires direct contact with blood or body fluids, although droplet and aerosol infections may occur. Usage of contaminated syringes and needles are main sources for nosocomial infections. In the laboratory, monkeys, mice, guinea pigs and hamsters have been infected experimentally.

Clinical Syndrome

The onset of the disease is sudden with fever, chills, headache, myalgia, and anorexia. This may be followed by symptoms such as abdominal pain, sore throat, nausea, vomiting, cough, arthralgia, diarrhea, and pharyngeal and conjunctival infection. Patients are dehydrated, apathetic, disoriented, and may develop a characteristic, non-pruritic, maculopapular centripetal rash associated with varying degrees of erythema and desquamate by day 5 or 7 of the illness. Hemorrhagic manifestations develop during the peak of the illness; they are of prognostic value for the disease. Bleeding into the gastrointestinal tract is most prominent, in addition to petechia and hemorrhages from puncture wounds and mucous membranes. Laboratory parameters are less characteristic. Mortality is high for the African members of the family and varies from 25-90% depending on the virus strain. REBOV seems to possess a very low pathogenicity for humans or even to be non-pathogenic.

Pathology / Pathophysiology

The pathologic changes in fatal marburgvirus and ebolavirus hemorrhagic fever human cases include hemorrhagic diatheses into skin, mucous membranes, visceral organs and the lumen of the stomach and intestine. The most striking lesions are found in liver, spleen, and kidney. These lesions are characterized by focal hepatic necrosis with little inflammatory response and by follicular necrosis of lymph nodes and spleen. In late stages of the disease, hemorrhage occurs in the gastrointestinal tract, pleural, pericardial, and peritoneal spaces and into the renal tubules with deposition of fibrin. Abnormalities in coagulation parameters include fibrin-split products and prolonged prothrombin and partial thromboplastin times, suggesting that disseminated intravascular coagulation is a terminal event. There is usually also profound leukopenia in association with secondary bacteremia. Virus replication is usually extensive in fixed tissue macrophages, interstitial fibroblasts of many organs, circulating macrophages and monocytes, and less frequently in vascular endothelial cells, hepatocytes, adrenal cortical cells, and renal tubular epithelium. Macrophages seem to be the first and preferred site of replication by filoviruses.

Clinical and biochemical findings support anatomical observations of extensive liver involvement, renal damage, and changes in vascular permeability, and activation of the clotting cascade. Visceral organ necrosis is the consequence of virus replication in parenchymal cells. However, no single organ is sufficiently damaged to explain the fatal outcome. Fluid distribution problems and platelet abnormalities indicate dysfunction of endothelial cells and platelets. Virus-induced release of mediators (e.g., cytokines) may increase endothelial permeability and thus trigger the shock syndrome in severe and fatal cases. The cytotoxicity of viral proteins may contribute to cell death, especially of endothelial cells.

Filoviruses seem to induce immunosuppression in the infected host. This may be achieved by cytolytic virus replication in monocytes/macrophages, the destruction of antigen-presenting dendritic cells, bystander apoptosis of lymphocytes, and/or decoy through the secretion of non-structural glycoproteins. In addition, the ebolavirus-specific small secreted glycoprotein (sGP) has been suggested to bind and inactivate neutrophils, and an immunosuppressive function has been postulated for a domain at the carboxy-terminal end of the transmembrane glycoprotein.

DIAGNOSIS

For acute diagnosis, PCR and antigen-detection ELISA are the assays of choice. Virus isolation (Vero E6 cells, MA 104 cells), immunohistochemistry and electron microscopy (high viremia) can be used as confirmatory tests. Virus-specific antibodies can be detected by ELISA (IgM μ-capture, IgG) or indirect immunofluorescence assay (IFA). Reference centers should be contacted immediately after a presumptive diagnosis.

PREVENTION AND CONTROL

Neither vaccines for human application nor specific chemotherapeutic treatment are available at present. Patients have to be isolated and clinical personal to be protected. Human interferon, human convalescent plasma, and anticoagulation therapy have been used in the past with limited success.

At present different vaccine approaches are being tested in animal models. The glycoprotein, nucleoprotein, VP24, VP30, VP35 and VP40 have been tested as vaccine candidates using naked DNA, adenovirus, vaccinia, liposomes, and Venezuelan equine encephalitis virus replicons as delivery mechanisms. In mice and guinea pigs, many strategies, particularly using the glycoprotein and the nucleoprotein, have been successful. However, in nonhuman primates only the use of a glycoprotein-based alphavirus replicon vaccine has been successful in the protection against MARV, and a prime-boost approach using glycoprotein-based naked DNA followed by glycoprotein-based adenovirus vectors has been effective against a ZEBOV challenge. Recently, an accelerated approach was successful where non-human primates were protected against a lethal ZEBOV challenge 28 days after vaccination with adenoviral vectors that encode the ZEBOV glycoprotein and nucleoprotein.

GENUS MARBURGVIRUS

Type Species *Lake Victoria marburgvirus*

DISTINGUISHING FEATURES

- Almost no antigenic cross-reactivity with ebolaviruses,
- Virion length is ~665 nm compared to 805 nm for ebolavirus virions,
- Genome length of 19.1 kb compared to 18.9 kb for ebolaviruses,
- Single gene overlap compared to several overlaps in ebolaviruses,
- Glycoprotein expression does not involve transcriptional editing of gene four,
- Protein profile distinct from ebolaviruses; but homogenous for all strains,
- Glycoprotein gene nt sequence difference of 57% in comparison with ebolaviruses,

- Fatality rate in humans varies from 25-70%.

List of Species Demarcation Criteria in the Genus

Not applicable.

List of Species in the Genus

Species names are in green italic script; strain names and synonyms are in black roman script; tentative species names are in blue roman script. Sequence accession numbers, and assigned abbreviations ( ) are also listed.

Species in the Genus

Lake Victoria marburgvirus		
Lake Victoria marburgvirus - Musoke (Kenya, 1980)	[Z12132]	(MARV-Mus)
Lake Victoria marburgvirus - Ozolin (Zimbabwe, 1975)		(MARV-Ozo)
Lake Victoria marburgvirus - Popp (West Germany, 1967)	[Z29337]	(MARV-Pop)
Lake Victoria marburgvirus - Ratayczak (West Germany, 1967)		(MARV-Rat)
Lake Victoria marburgvirus - Ravn (Kenya, 1987)		(MARV-Ravn)
Lake Victoria marburgvirus - Voege (Yugoslavia, 1967)		(MARV-Voe)

Tentative Species in the Genus

None reported.

Genus *Ebolavirus*

Type Species *Zaire ebolavirus*

Distinguishing Features

- Almost no antigenic cross-reactivity with marburgviruses,
- Virion length is ~805 nm compared to 665 nm for marburgviruses,
- Genome length of 18.9 kb compared to 19.1 kb for marburgviruses,
- Several gene overlaps compared to a single overlap in marburgviruses,
- Glycoprotein expression involves transcriptional editing,
- Transcription of only the first ORF of gene four yields a soluble small glycoprotein not observed with marburgviruses,
- Protein profile distinct from marburgviruses; but are species-specific,
- Glycoprotein gene nt sequence difference of 57% compared to marburgviruses
- Fatality rates for humans 50-90% with species *Sudan ebolavirus* and *Zaire ebolavirus*,. Species *Reston ebolavirus* appears to be non-pathogenic for humans.

List of Species Demarcation Criteria in the Genus

Virus species in the genus may be distinguished on the basis of glycoprotein gene sequence differences (more than 30% aa difference), cross-protection data (where available) and differences in geographic origins.

List of Species in the Genus

Species names are in green italic script; strain names and synonyms are in black roman script; tentative species names are in blue roman script. Sequence accession numbers, and assigned abbreviations ( ) are also listed.

Species in the Genus

Cote d'Ivoire ebolavirus		
Cote d'Ivoire ebolavirus - Cote d'Ivoire (Tai Forest, 1994)		(CIEBOV-CI)
Reston ebolavirus		
Reston ebolavirus - Philippines (1989)	[AB050936]	(REBOV-Phi)
Reston ebolavirus - Reston (1989)	[AF522874]	(REBOV-Res)
Reston ebolavirus - Siena (1992)		(REBOV-Sie)

Reston ebolavirus - Texas (1996) (REBOV-Tex)
Sudan ebolavirus (SEBOV-Gul)
Sudan ebolavirus - Boniface (1976) (SEBOV-Bon)
Sudan ebolavirus – Gulu (2000)
Sudan ebolavirus - Maleo (1979) (SEBOV-Mal)
Zaire ebolavirus
Zaire ebolavirus - Eckron (Zaire, 1976) (ZEBOV-Eck)
Zaire ebolavirus - Gabon (1994-1997) (ZEBOV-Gab)
Zaire ebolavirus - Kikwit (1995) (ZEBOV-Kik)
Zaire ebolavirus - Mayinga (Zaire, 1976) [AF272001] (ZEBOV-May)
Zaire ebolavirus - Tandala (1977) (ZEBOV-Tan)
Zaire ebolavirus - Zaire (Zaire, 1976) (ZEBOV-Zai)

Tentative Species in the Genus

None reported.

Unassigned Viruses in the Family

None reported.

Phylogenetic Relationships within the Family

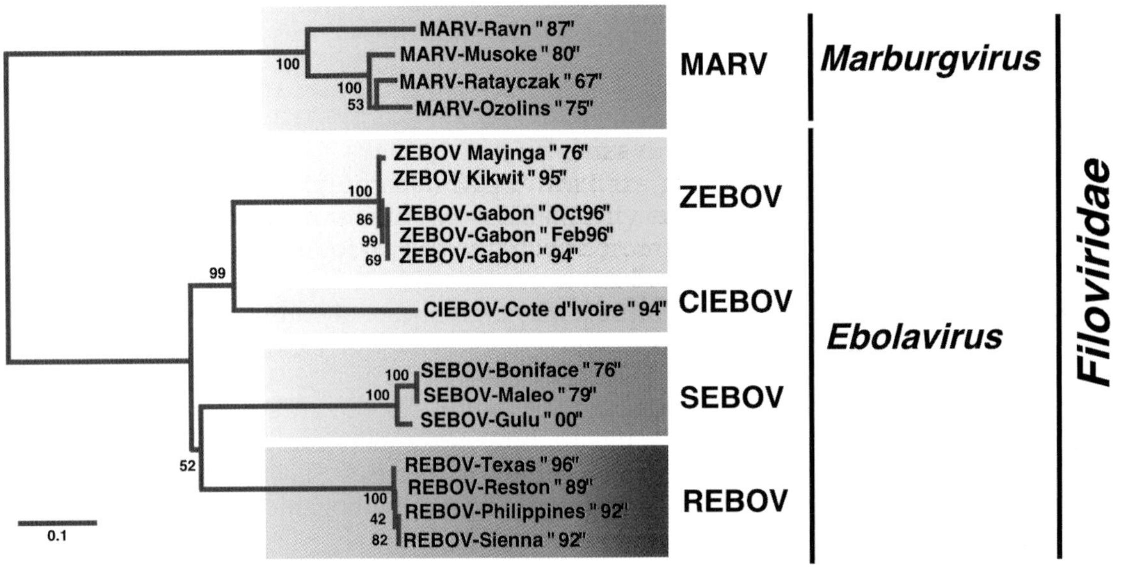

Figure 3: Phylogenetic tree showing the relationships between the glycoproteins genes of filoviruses. The entire coding region for the glycoprotein gene of the viruses shown was produced using MEGA ver. 2.1 (Kimar S, Tamura T, Jakobsen IB and Nei M, 2001, MEGA2: Molecular Evolutionary Genetics Analysis Software, Arizona State University, Tempe, Arizona, USA). A neighbor-joining tree using Tajima-Nei distance method and 1000 bootstrap replicates for branch points values was prepared. The analysis strongly supports a common evolutionary origin for the viruses, early separation of evolution of marburgviruses and ebolaviruses, and substantial differences between four species of ebolaviruses. (Courtesy of A. Sanchez, Centers for Disease Control and Prevention, Atlanta, USA)

Similarity with Other Taxa

Comparison of filovirus genomes with other *Mononegavirales* demonstrates a similar structure and suggests comparable mechanisms of transcription and replication. Comparative sequence analyses of single genes indicate that filoviruses are phylogenetically quite distinct from other families of the order *Mononegavirales*. Limited homology exists between the carboxy-terminal part of filovirus GP and the transmembrane p15E-related glycoproteins of oncogenic retroviruses.

Derivation of Names

Ebola: from Ebola river in Zaire where one of the first registered outbreaks of the disease occurred.

Filo: from Latin *filum*, "thread", to indicate the morphology of virus particles.

Marburg: from Marburg, town in Germany, where the first known outbreak of filovirus disease occurred.

References

Feldmann, H., Jones, S., Klenk, H.D. and Schnittler, H.J. (2003). Ebola virus: from discovery to vaccine. *Nat. Rev. Immunol.*, **3**, 677-685.

Feldmann, H., Volchkov, V.E., Volchkova, V.A., Stroeher, U. and Klenk, H.D. (2001). Biosynthesis and role of filoviral glycoproteins. *J. Gen. Virol.*, **82**, 2839-2848.

Geisbert, T.W. and Jahrling, P.B. (1995). Differentiation of filoviruses by electron microscopy. *Virus Res.*, **39**, 129-150.

Geisbert, T.W., Pushko, P., Anderson, K., Smith, J., Davis, K.J. and Jahrl;ing, P.B. (2002). Evaluation in nonhuman primates of vaccines against Ebola virus. *Emerg. Infect. Dis.*, **8**, 503-507.

Hevey, M., Negley, D., Pushko, P., Smith, J., Schmaljohn, A. (1998). Marburg virus vaccines based upon aplhavirus replicons protect guinea pigs and nonhuman primates. *Virology*, **251**, 28-37.

Kiley, M.P., Bowen, E.T.W., Eddy, G.A., Isaäcson, M., Johnson, K.M., McCormick, J.B., Murphy, F.A., Pattyn, S.R., Peters, D., Prozesky, O.W., Regnery, R.L., Simpson, D.I.H., Slenczka, W., Sureau, P., van der Groen, G., Webb, P.A. and Wulff, H. (1982). *Filoviridae*: a taxonomic home for Marburg and Ebolaviruses? *Intervirology*, **18**, 24-32.

Klenk, H.D. (ed)(1999). Marburg and Ebolaviruses. *Curr. Top. Microbiol. Immunol.*, **235**, 1-230.

Martini, G.A. and Siegert, R. (eds) (1971). *Marburg Disease*. Springer Verlag, New York.

Neumann, G., Feldmann, H., Watanabe, S., Lukashevich, L., and Kawaoka, Y. (2002). Reverse genetics demonstrates that proteolytic processing of the Ebola virus glycoprotein is not essential for replication in cell culture. *J Virol.*, **76**, 406-410.

Pattyn, S.R. (ed)(1978). *Ebola Virus Hemorrhagic Fever.* Amsterdam: Elsevier/North-Holland Biomedical Press.

Peters, C.J. and LeDuc, J.W. (1999). Ebola: The virus and the disease. *J. Infect. Dis.*, **179**, S1-S288.

Sanchez, A., Kiley, M.P., Holloway, B.P. and Auperin, D.D. (1993). Sequence analysis of the Ebola virus genome: organization, genetic elements, and comparison with the genome of Marburgvirus. *Virus Res.*, **29**, 215-240.

Sanchez, A., Trappier, S.G., Mahy, B.W.J., Peters, C.J. and Nichol, S.T. (1996). The virion glycoprotein of Ebolaviruses are encoded in two reading frames and are expressed through transcriptional editing. *Proc. Natl. Acad. Sci. USA*, **93**, 3602-3607.

Sanchez, A., Trappier, S.G., Ströher, U., Nichol, S.T., Bowen, M. and Feldmann, H. (1998). Variation in the glycoprotein and VP35 genes of Marburg virus strains. *Virology*, **240**, 138-146.

Sanchez, A. et al. (2001). *Filoviridae*: Marburg and Ebolaviruses. In: *Fields Virology*, 4th ed., (D.M. Knipe and P.M. Howley, eds), pp 1279-1304. Lippincott, Williams and Wilkins, New York.

Sullivan, N.J., Sanchez, A., Rollin, P.E., Yang, Z.Y. and Nabel, G.J. (2000). Development of a preventive vaccine for Ebola virus infection in primates. *Nature*, **408**, 605-609.

Sullivan, N.J., Geisbert, T.W., Geisbert, J.B., Xu, L., Yang, Z.Y., Roederer, M., Koup, R.A., Jahrling, P.B. and Nabel, G.J. (2003). Accelerated vaccination for Ebola virus haemorrhagic fever in non-human primates. *Nature*, **424**, 681-684.

Volchkov, V.E., Becker, S., Volchkova, V.A., Ternovoi, V.A., Kotov, A.N., Netesov, S.V. and Klenk, H.-D. (1995). GP mRNA of Ebola virus is edited by the Ebola virus polymerase and by T7 and vaccinia virus polymerases. *Virology*, **214**, 421-430.

Volchkov, V.E., Volchkova, V.A., Mühlberger, E., Kolesnikova, L.V., Weik, M., Dolnik, O. and Klenk, H.-D. (2001). Recovery of infectious Ebola virus from cDNA: transcriptional RNA editing of the GP gene controls viral cytotoxicity. *Science*, **291**, 1965-1969.

Contributed By

Feldmann, H., Geisbert, T.W., Jahrling, P.B., Klenk, H.-D., Netesov, S.V., Peters, C.J., Sanchez, A., Swanepoel, R., and Volchkov, V.E.

NEG. SSRNA

FAMILY *PARAMYXOVIRIDAE*

TAXONOMIC STRUCTURE OF THE FAMILY

Family	***Paramyxoviridae***
Subfamily	***Paramyxovirinae***
Genus	***Rubulavirus***
Genus	***Avulavirus***
Genus	***Respirovirus***
Genus	***Henipavirus***
Genus	***Morbillivirus***
Subfamily	***Pneumovirinae***
Genus	***Pneumovirus***
Genus	***Metapneumovirus***

VIRION PROPERTIES

MORPHOLOGY

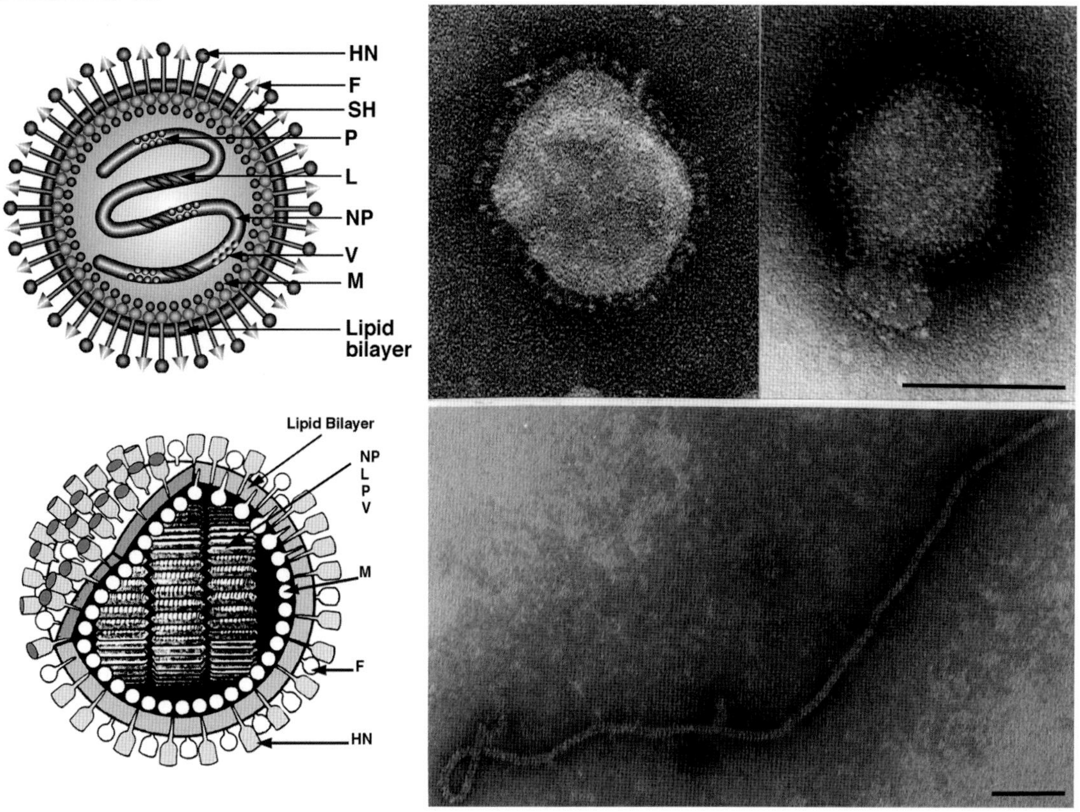

Figure 1: (Right) Negative contrast electron micrographs of intact Simian virus-5 (SV-5) particles (genus *Rubulavirus*) (Top) and the SV-5 nucleocapsid after detergent lysis of virions (Bottom)(Courtesy of G.P. Leser and R.A. Lamb). The bars represent 100 nm. (Left top and bottom) Schematic diagrams of SV-5 particles in cross section (N) (formerly NP): nucleocapsid, P: phosphoprotein, L: large polymerase protein, V: cysteine rich protein that shares its N-terminus with P sequence and for SV-5 is found in virions, M: matrix or membrane protein, F: fusion protein, HN: hemagglutinin-neuraminidase, SH: small hydrophobic protein). Adapted from Kingsbury, D.W. (1990). *Paramyxoviridae*: the viruses and their replication. In: *Virology*, 2nd Edn (B.N. Fields and D.M. Knipe, eds). Raven Press, New York, and from Scheid, H. (1987). Animal Virus Structure, (M.V. Nermut, and A.C. Steven, eds). Elsevier, Amsterdam. With permission).

Virions are 150 nm or more in diameter, pleomorphic, but usually spherical in shape, although filamentous and other forms are common. Virions consist of a lipid envelope surrounding a nucleocapsid. The envelope is derived directly from the host cell plasma membrane by budding and contains 2 or 3 transmembrane glycoproteins. These are present as homo-oligomers and form spike-like projections, 8-12 nm in length, spaced 7-10 nm apart (depending on the genus). One non-glycosylated membrane or matrix protein is associated with the inner face of the envelope. The viral nucleocapsid consists of a single species of viral RNA and associated proteins. It has helical symmetry and is 13-18 nm in diameter with a 5.5-7 nm pitch (depending on the subfamily); its length can be up to 1,000 nm in some genera. Multiploid virions are found, although the vast majority of virions contain a single functional genome. The viral polymerase is packaged in the virion.

PHYSICOCHEMICAL AND PHYSICAL PROPERTIES

Virion Mr is about 500 x 10^6, and much greater for multiploid virions. Virion buoyant density in sucrose is 1.18-1.20 g/cm^3. Virion S_{20w} is at least 1000S. Virions are very sensitive to heat, lipid solvents, ionic and non-ionic detergents, formaldehyde and oxidizing agents.

NUCLEIC ACID

Virions contain a single molecule of linear, negative sense, ssRNA that is not infectious alone but is infectious in the form of the nucleocapsid. The RNA genome size is fairly uniform: 15,384 nt for Sendai virus, (SeV); 15,462 nt for Human parainfluenza virus 3, (HPIV-3); 15,384 nt for Mumps virus, (MuV); 15,246 nt for Simian virus 5, (SV-5); 15,450 nt for Simian virus 41, (SV-41); 15,156 nt for Newcastle disease virus, (NDV); 15,654 nt for Human parainfluenza virus 2, (HPIV-2); 18,234 nt for Hendra virus, (HeV); 18,246 nt for Nipah virus, (NiV); 15,894 nt for Measles virus, (MeV); 15,690 nt for Canine distemper virus, (CDV); 15,882 nt for Rinderpest virus, (RPV); 15,702 nt for Cetacean morbillivirus, (CeMV); 15,190-15,225 nt for Human respiratory syncytial virus, (HRSV), and 13,280-13,378 nt for Human metapneumovirus (HMPV). All the above genome lengths are multiples of 6, which is a requirement for the efficient replication of the members of the subfamily *Paramyxovirinae*, but does not apply to the members of the subfamily *Pneumovirinae*. Some virions may contain positive sense RNA. Thus, partial self-annealing of extracted RNA may occur. The Mr of the genome is 5 x 10^6 and this constitutes about 0.5% of the virion by weight. Intracellularly, or in virions, genome-size RNA is found exclusively as nucleocapsids. The genome RNA does not contain a 5′-cap, nor a covalently linked protein. The genome 3′-end is not polyadenylated.

PROTEINS

Members of the subfamily *Paramyxovirinae* encode 7-9 proteins (5–250 kDa) of which 2-4 (or more) are derived from the 2-3 overlapping ORFs in the P locus (Fig. 2). Pneumoviruses encode 9-11 proteins of 4.8–250 kDa. Virion proteins common to all genera include: three nucleocapsid-associated proteins, i.e., an RNA-binding protein (N) (formerly NP), a phosphoprotein (P), and a large polymerase protein (L); three membrane associated proteins, i.e., an unglycosylated inner membrane or matrix protein (M), and two glycosylated envelope proteins, comprising a fusion protein (F) and an attachment protein (G, or H, or HN). The F protein is synthesized within an infected cell as a precursor (F_0) which is activated following cleavage by cellular protease(s) to produce the virion disulfide-linked F_1 and F_2 subunits (order: amino F_2-S-S-F_1 carboxyl). Variable proteins include putative non-structural proteins (C, NS1, NS2), a cysteine-rich protein that binds zinc (V) (in the subfamily *Paramyxovirinae* only) that can be structural or non-structural depending on the virus, a small integral membrane protein (SH), a transcription processivity factor (M2-1, formerly called 22K protein) which previously was thought to be a second M-like protein, and a non-abundant protein (M2-2) involved in the balance between genome replication and transcription. Virion enzyme activities (variously represented among the genera) include a RNA-dependent RNA transcriptase, mRNA guanylyl and methyl transferases, and a neuraminidase. A protein kinase is associated with many members but it is probably of cellular origin.

Lipids
Cell plasma membrane.

Carbohydrates
Virions are composed of 6% carbohydrate by weight; composition is dependent on the host cell. Fusion and attachment proteins are glycosylated by N-linked carbohydrate side chains. In the subfamily *Pneumovirinae* the attachment protein (G) is heavily glycosylated by O-linked as well as N-linked carbohydrate side chains. The SH protein of respiratory syncytial virus contains polylactosaminoglycan.

Genome Organization and Replication

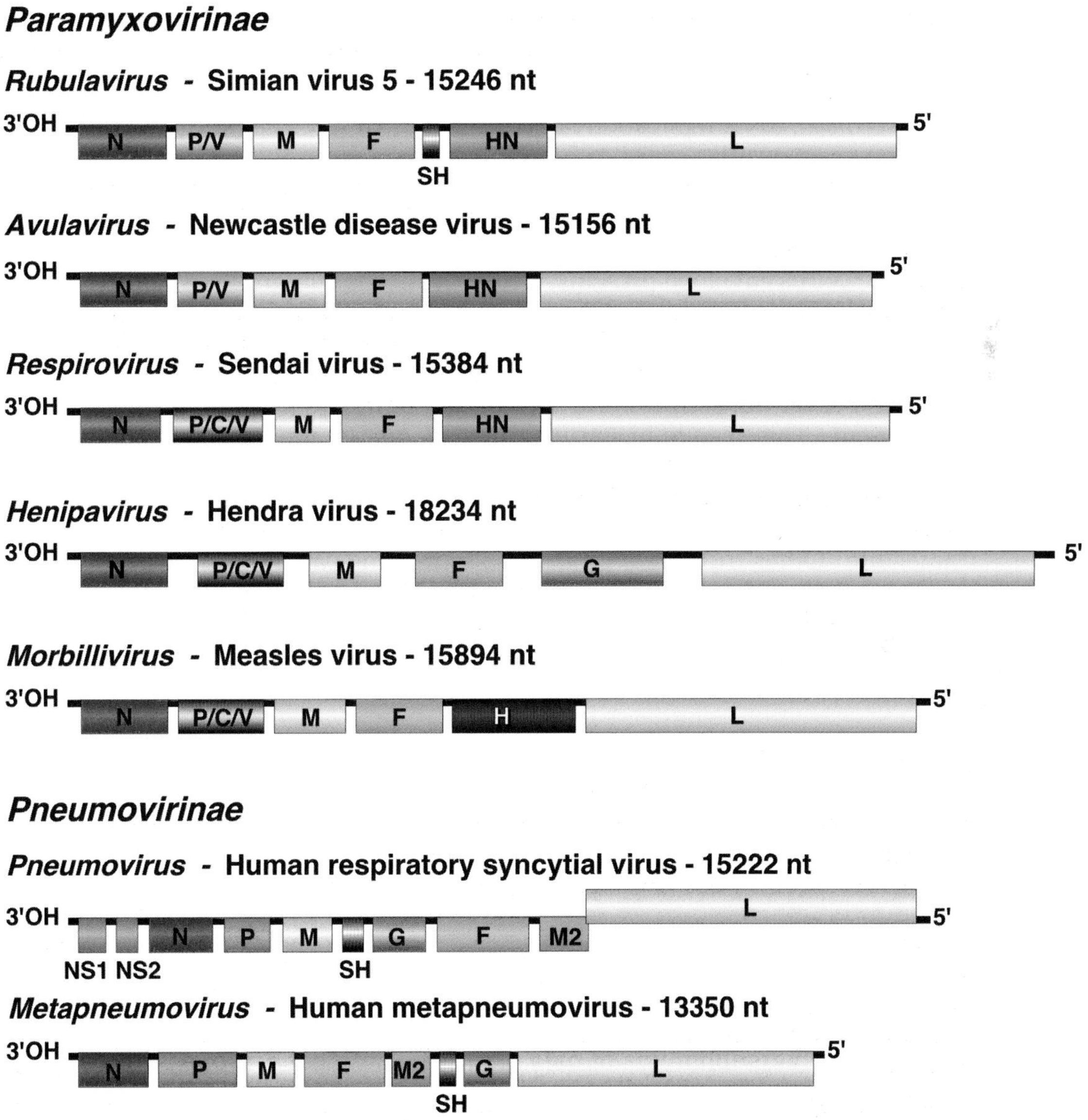

Figure 2: Maps of genomic RNAs (3′-to-5′) of viruses belonging to the seven genera of the family *Paramyxoviridae*. Each box represents a separately encoded mRNA; multiple distinct ORFs within a single mRNA are indicated by slashes. The M2 mRNA of members of subfamily *Pneumovirinae* has two overlapping ORFs, M2-1 and M2-2, (not shown). The lengths of the boxes are approximately to scale although the intervening or preceding sequences are not to scale. The D ORF present in some respiroviruses is not shown. Certain viruses give rise to additional proteins by the utilization of secondary translational start sites within some of the ORFs: these are not shown. In some viruses of the genus *Respirovirus* the V ORF may be a non-expressed relic. In the genus *Rubulavirus* some species lack the SH gene. In the genus *Pneumovirus*, human respiratory syncytial virus (HRSV) has a transcriptional overlap at M2 and L (staggered boxes). There are

conserved trinucleotides that serve as intergenic sequences for the respiroviruses, henipaviruses and morbilliviruses. For rubulaviruses, avulaviruses, pneumoviruses and metapneumoviruses, the intergenic sequences are variable (1-190 nt long).

The genome organization is illustrated in Figure 2 for viruses representing the 7 genera of the family. After attachment to cell receptors, virus entry is achieved by fusion of the virus envelope with the cell surface membrane. This can occur at neutral pH. Virus replication occurs in the cell cytoplasm and is thought to be independent of host nuclear functions. The genome is transcribed processively from the 3'-end by virion-associated enzymes into 6-10 separate, subgenomic, positive sense mRNAs. Transcription is guided by short (10-13 nt) conserved transcription start and termination /polyadenylation signals flanking each transcriptional element. The mRNAs are capped and possess 3'-poly(A) tracts synthesized by reiterative copying of the polyadenylation site. Intergenic regions are either highly conserved in sequence and length (*Respirovirus, Henipavirus, Morbillivirus*) or are not conserved in sequence and length (*Rubulavirus, Avulavirus, Pneumovirinae*). RNA replication occurs through an intermediate, the antigenome, that is a complete exact positive sense copy of the genome.

Nucleocapsid assembly occurs in the cytoplasm and is tightly linked to RNA synthesis. Nucleocapsids are enveloped by budding at the cell surface plasma membrane at sites containing virus envelope proteins. Members of the subfamily *Paramyxovirinae* contain 6-7 transcriptional elements that encode 7-11 proteins. Each element encodes a single mRNA with the sole exception of the P/V element. This element is transcribed into an exact-copy mRNA (P or V mRNA, depending on the genus) as well as into an alternative version in which the RNA transcriptase stutters on the template at an editing motif midway down the element. This results in the insertion of one of more pseudo-templated nucleotides ("RNA editing") and shifts the reading frame to access an alternative ORF. The exact-copy and edited mRNAs synthesize two alternative proteins, P and V, which have identical amino-terminal domains but have different carboxy-terminal domains due to the frameshift. Other truncated, or chimeric, proteins can be produced by shifting into the third reading frame. The C ORF present in respiroviruses, henipaviruses, and morbilliviruses overlaps the P ORF and can initiate synthesis at a non-AUG codon that is accessed by ribosomal choice or at AUG codons in poor with translation initiation at alternative start codons in the same ORF.

Members of the subfamily *Pneumovirinae* have 8 (*Metapneumovirus*) or 10 (*Pneumovirus*) transcriptional elements, each of which encodes one mRNA. Each mRNA has a single ORF, except for the M2 mRNA which encodes two proteins from separate ORFs. There is overlap between the M2 and L transcriptional elements in some pneumoviruses (Fig. 2), but these elements nonetheless give rise to separate mRNAs.

ANTIGENIC PROPERTIES

The attachment (HN, or H, or G) and fusion (F) proteins are of primary importance in inducing virus-neutralizing antibodies and immunity against reinfection. Antibodies to N and, variably, to other viral proteins also are induced by infection. Various proteins of members of the subfamily *Paramyxovirinae* have been reported broken into specific peptides that when complexed to major histocompatibility glycoproteins serve as recognition molecules for cytotoxic or helper T cells.

BIOLOGICAL PROPERTIES

Paramyxoviruses have been conclusively identified only in vertebrates and almost exclusively in mammals and birds, although there are reports of paramyxoviruses of reptiles and fish. Most viruses have a narrow specific host range in nature, but in cultured cells they display a broad host range. Infection of cultured cells generally is lytic, but temperate or persistent infections *in vitro* are common. Other features of infection include the formation of inclusion bodies and syncytia. Cell surface molecules reported to serve as receptors for the attachment of respiroviruses and rubulaviruses include sialoglycoproteins and glycolipids. The cell surface proteins CD46 and SLAM 150 are major receptors for measles virus. Nucleocapsids associate with viral membrane proteins at the plasma membrane and are enveloped by budding. Transmission is horizontal, mainly through airborne routes; no vectors are known. Paramyxovirus infection typically begins in the respiratory tract and may remain at that site (e.g., HRSV and HPIV) or may spread to secondary sites (e.g., lymphoid and endothelial tissues for MeV, or the parotid gland, CNS and endothelial tissues for MuV). In general, paramyxovirus infections are limited by, and eliminated by, host immunity. However, virus sometimes can be shed for periods of weeks or months in normal and, especially, immunocompromised individuals. Latent infection is unknown, and long term persistent infection is known only for subacute sclerosing panencephalitis, a rare complication that involves defective measles virus, and old dog distemper, which can involve persistence of defective or fully infectious virus for weeks or months in normal and, especially, immuno-compromised individuals.

SUBFAMILY *PARAMYXOVIRINAE*

TAXONOMIC STRUCTURE OF THE SUBFAMILY

Subfamily	***Paramyxovirinae***
Genus	***Rubulavirus***
Genus	***Avulavirus***
Genus	***Respirovirus***
Genus	***Henipavirus***
Genus	***Morbillivirus***

DISTINGUISHING FEATURES

Member species of the subfamily *Paramyxovirinae* have six to seven transcriptional elements. Amino acid sequence relatedness is much greater within the subfamily than between subfamilies. Within the subfamily *Paramyxovirinae,* sequence relatedness between corresponding proteins is greater for N, P, M and L, with F and HN being somewhat less conserved and C and P being poorly conserved although the unique region of V that is not shared with P is highly conserved. The division of this subfamily into the five genera is consistent with phylogenetic grouping based on amino acid sequence relationships. Their nucleocapsids have diameters of 18 nm and a pitch of 5.5 nm, the length of the surface F and H/HN spikes is 8 nm. The genome length must be a multiple of six nt for efficient genome replication (the "rule of six"), perhaps reflecting the precise packing of nt by a nucleocapsid protein subunit. RNA editing of the P/V transcriptional element occurs for all members except Human parainfluenza virus 1 (HPIV-1). Some genera (*Morbillivirus)* lack a neuraminidase activity and some genera (Henipaviruses) and viruses (Canine distemper virus, Phocine distemper virus and Rinderperst virus) lack a detectable hemagglutinating activity.

GENUS *RUBULAVIRUS*

Type Species *Mumps virus*

DISTINGUISHING FEATURES

All species of the genus *Rubulavirus* have hemagglutination and neuraminidase activities. They share greater sequence relatedness within the genus than with members of other genera. For example, the N protein of HPIV-2 is 39-74% identical with that of MuV, SV-5, HPIV-4 and SV-41, compared to 18% and 24% identical with HPIV-1 and MeV. Some members (SV-5 and MuV) contain an extra transcriptional element (SH) between the F and HN loci (Fig. 2). The unedited and edited versions of the mRNA from the P locus encode the V and P proteins, respectively. The intergenic sequences are of variable length. All members lack a C protein ORF. The rubulavirus P protein is substantially smaller than that of the respiroviruses or morbilliviruses. MuV and HIPV2 are significant human pathogens.

LIST OF SPECIES DEMARCATION CRITERIA IN THE GENUS

HPIV-2 and Human parainfluenza virus 4 (HPIV-4) represent distinct serotypes that lack significant cross-neutralization and cross-protection. HPIV-2, SV-5, and SV-41 exhibit considerable sequence relatedness and some antigenic relatedness, but these viruses can be distinguished on either basis (for example, the N protein of HPIV-2 is 57% or 74% identical to those of SV-5 or SV-41, respectively) as well as by host range: HPIV-2, HPIV-4a and HPIV-4b infect humans, SV-5 dogs, monkeys and humans, and SV-41 monkeys. They also have differences in their gene maps: SV-5 and MuV have an additional gene, SH, and SV-41 lacks a functional transcription termination signal for the M gene and thus does not express a monocistronic M mRNA. HPIV-4 contains two antigenic subgroups (a and b) that are distinguished by differences in reactivity with monoclonal antibodies but are highly related by sequence – 84% and 95% identity for the HN and F protein, respectively - and should not be considered distinct species. MuV also does not exhibit significant cross-neutralization and cross-protection with other paramyxoviruses, and it is distinguished by its gene map (it contains an SH gene, found within this group only in SV-5), by sequence divergence (the MuV N protein shares 44% or less aa sequence identity with other rubulaviruses), and by its disease.

LIST OF SPECIES IN THE GENUS

Species names are in green italic script; strain names and synonyms are in black roman script; tentative species names are in blue roman script. Sequence accession, and assigned abbreviations ( ) are also listed.

SPECIES IN THE GENUS

Human parainfluenza virus 2		
Human parainfluenza virus 2	[AF533010-12]	(HPIV-2)
Human parainfluenza virus 4		
Human parainfluenza virus 4a	[M32982, M55975, D10241, M34033, D49821]	(HPIV-4a)
Human parainfluenza virus 4b	[AB006958, M32983, M55976, D10242, D49822,]	(HPIV-4b)
Mapuera virus		
Mapuera virus	[X85128]	(MPRV)
Mumps virus		
Mumps virus	[AB040874]	(MuV)
Porcine rubulavirus		
Porcine rubulavirus (La-Piedad-Michoacan-Mexico virus)	[AF416650]	(PoRV)
Simian virus 5		
Simian virus 5	[AF052755]	(SV-5)
Simian virus 41		
Simian virus 41	[X64275]	(SV-41)

TENTATIVE SPECIES IN THE GENUS

Tioman virus	[AF298895]
Menangle virus	[AF326114]

GENUS AVULAVIRUS

Type Species *Newcastle disease virus*

DISTINGUISHING FEATURES

All species of the genus *Avulavirus* have hemagglutinin and neuraminidase activities. They share greater sequence relatedness within the genus than with members of other genera but are closely related to the rubulaviruses. The major distinguishing feature between avulaviruses and rubulaviruses is that for avulaviruses the exact copy of the P/V mRNA encodes P and the edited form encodes V, and none of the avulaviruses have an SH gene. The intergenic sequences are of variable length. All members lack a C protein ORF. The avulavirus P protein is substantially smaller than that of the respiroviruses or morbilliviruses. These viruses, especially NDV, include significant avian pathogens.

LIST OF SPECIES DEMARCATION CRITERIA IN THE GENUS

There are many strains of NDV (virulent and avirulent) for chickens that have been extensively analyzed. The many other avian paramyxoviruses are antigenically related and are not well studied. These are serotypes defined by hemagglutination-inhibition tests, although weak interactions occur between the types. Also, each serotype has a distinct pattern of electrophoretically-separated polypeptides. The genome encodes the P protein and the edited mRNA is the V protein-encoding mRNA.

LIST OF SPECIES IN THE GENUS

Species names are in green italic script; strain names and synonyms are in black roman script; tentative species names are in blue roman script. Sequence accession, and assigned abbreviations ( ) are also listed.

SPECIES IN THE GENUS

Avian paramyxovirus 2		
Avian paramyxovirus 2 (Yucaipa)	[D14040]	(APMV-2)
Avian paramyxovirus 3		
Avian paramyxovirus 3		(APMV-3)
Avian paramyxovirus 4		
Avian paramyxovirus 4	[D14031]	(APMV-4)
Avian paramyxovirus 5		
Avian paramyxovirus 5 (Kunitachi)		(APMV-5)
Avian paramyxovirus 6		
Avian paramyxovirus 6		(APMV-6)
Avian paramyxovirus 7		
Avian paramyxovirus 7		(APMV-7)
Avian paramyxovirus 8		
Avian paramyxovirus 8		(APMV-8)
Avian paramyxovirus 9		
Avian paramyxovirus 9		(APMV-9)
Newcastle disease virus		
Newcastle disease virus	[AF077761, AF309418, AF375823]	(NDV)
(Avian parainfluenza virus 1)		(APMV-1)

TENTATIVE SPECIES IN THE GENUS

None reported.

NEG. SSRNA

GENUS RESPIROVIRUS

Type Species *Sendai virus*

DISTINGUISHING FEATURES

Member viruses of the genus *Respirovirus* possess a hemagglutinin and a neuraminidase. These viruses have six transcriptional elements. All members encode a C protein. Unedited P mRNA encodes P and C, whereas insertion of a G nucleotide in P mRNA transcripts accesses the V ORF. Amino acid sequence relatedness within the genus ranges from low to high, depending on the protein, and always is higher than in comparisons with other genera. For example, within respiroviruses the N protein is 88% identical comparing HPIV-1 to SeV, its "murine counterpart", and 63% identical compared to HPIV-3. Between genera, the N protein is 18% identical between HPIV-1 (*Respirovirus*) and the N protein of MuV or HPIV-2 or HPIV-4 (*Rubulavirus*), and 21% compared with that of MeV (*Morbillivirus*). HPIV-1 and HPIV-3 are significant agents of respiratory tract disease.

LIST OF SPECIES DEMARCATION CRITERIA IN THE GENUS

HPIV-1 and HPIV-3 represent distinct serotypes defined by a lack of significant cross-neutralization and cross protection. Their gene maps are similar but not identical (HPIV-3 edits to make a D but not V protein, whereas HPIV-1 lacks editing). They share low to high sequence relatedness among the various proteins (see above). SeV, which resembles HPIV-1, is now predominantly a pathogen of laboratory mice. However, this virus was isolated from the lungs of infants during a fatal epidemic of newborn pneumonitis in Japan in 1952 and this virus is not found in wild mice either in Japan or the USA. SeV is distinguished from HPIV-1 by host range: specifically, HPIV-1 is permissive and pathogenic in humans whereas in mice it grows poorly or not at all and is nonpathogenic. Conversely, SeV is highly permissive, transmissible and pathogenic for mice. The two viruses have considerable sequence relatedness (see above) and antigenic similarity, but also can be clearly distinguished on either basis; also, HPIV-1 lacks editing and a V protein. Bovine parainfluenza virus 3 (BPIV-3) differs from HPIV-3 by their host ranges, which overlap but exhibit specificity. For example, in humans HPIV-3 replicates efficiently, is easily transmissible and causes disease, whereas BPIV-3 is highly attenuated, nonpathogenic, and poorly transmissible. Furthermore, compared to HPIV-3, BPIV-3 is restricted 100 to 1000-fold in Old World primates. HPIV-3 and BPIV-3 exhibit considerable genetic and antigenic similarity, but also can clearly be distinguished on either basis. For example, HPIV-3 and BPIV-3 are 25% related antigenically by reciprocal cross neutralization and hemagglutination inhibition studies. Also, BPIV-3 makes a V protein whereas HPIV-3 does not. With the exception of Simian virus 10, each species represents significant pathogens in its respective host.

LIST OF SPECIES IN THE GENUS

Species names are in green italic script; strain names and synonyms are in black roman script; tentative species names are in blue roman script. Sequence accession, and assigned abbreviations ( ) are also listed.

SPECIES IN THE GENUS

Bovine parainfluenza virus 3		
Bovine parainfluenza virus 3	[AF178654]	(BPIV-3)
Human parainfluenza virus 1		
Human parainfluenza virus 1	[AF457102]	(HPIV-1)
Human parainfluenza virus 3		
Human parainfluenza virus 3	[AB012132]	(HPIV-3)
Sendai virus		
Sendai virus	[AB005795]	(SeV)
(Murine parainfluenza virus 1)		
Simian virus 10		

Simian virus 10		(SV-10)

TENTATIVE SPECIES IN THE GENUS

None reported.

GENUS HENIPAVIRUS

Type Species *Hendra virus*

DISTINGUISHING FEATURES

The two species of the genus *Henipavirus* have an attachment protein (G) that lacks hemagglutinating and neuraminidase activities. They share greater sequence relatedness within the genus than with members of other genera. The major distinguishing features between henipaviruses and other paramyxoviruses are (i) the long 5′- and 3′-UTRs in the mRNAs and (ii) a genome that as a result is approximately 3,000 nt or more longer than other members of the family *Paramyxoviridae*. The unedited P mRNA encodes P protein. The edited P mRNA encodes V protein. The intergenic sequences are three nt at each gene junction. Both members encode a C protein ORF. Both HeV and NiV are indigenous to fruit bats. Each species has been associated with limited outbreaks with high mortality in domesticated animals and humans.

LIST OF SPECIES DEMARCATION CRITERIA IN THE GENUS

HeV and NiV are antigenically distinct and aredistinct by genome sequence.

LIST OF SPECIES IN THE GENUS

Species names are in green italic script; strain names and synonyms are in black roman script; tentative species names are in blue roman script. Sequence accession, and assigned abbreviations ( ) are also listed.

SPECIES IN THE GENUS

Hendra virus		
Hendra virus	[AF017149]	(HeV)
Nipah virus		
Nipah virus	[AF212302]	(NiV)

TENTATIVE SPECIES IN THE GENUS

None reported.

GENUS MORBILLIVIRUS

Type Species *Measles virus*

DISTINGUISHING FEATURES

Members of the genus *Morbillivirus,* lack a neuraminidase activity. Member viruses exhibit greater amino acid sequence relatedness within the genus than with other genera. They have an identical gene order, number of transcriptional elements and size of intergenic sequences with members of the genus *Respirovirus* (Fig. 2). All morbilliviruses have a P/C/V transcription unit with RNA editing like respiroviruses, namely the templated exact-copy mRNA encodes P and the predominant edited mRNA form (1 G added) encodes V. All morbilliviruses produce both intracytoplasmic and intranuclear inclusion bodies containing nucleocapsid-like structures. Viruses cross-react in serological tests. Sialic acid does not appear to be a receptor for morbilliviruses. Narrow host-range distribution of receptor defines susceptibility of organisms to infection. For MeV one receptor is CD46 and another CD150. CD150 also appears to be a receptor for CDV and RPV, which have a preference for canine and bovine CD150 respectively. Each species is a significant cause of disease in its respective host.

LIST OF SPECIES DEMARCATION CRITERIA IN THE GENUS

The morbilliviruses are distinguished by host range, genetic (sequence) and antigenic differences. There is a low to moderate degree of sequence relatedness between members, depending on the protein (for example, the N protein of MeV is 65% related to that of CDV, viruses that represent two branches of the genus. Cross-neutralization and cross-protection also occurs between members of the genus, although members can also be distinguished on that basis. MeV infects primates, CDV infects members of the Order *Carnivora*, and RPV and Peste-des-petits-ruminants virus (PPRV) infect members of the Order *Artiodactyla* (even-toed ungulates), especially ruminants and swine. PPRV is distinguished from RPV by sequence analysis (the N protein of PPRV shares 68-72% identity with that of MeV, RPV or CDV), and because it does not readily infect cattle. Phocine distemper virus (PDV) is most closely related to CDV and is distinguished by host range and sequence divergence. Two cetacean morbilliviruses have been described, Dolphin morbillivirus (DMV) and Porpoise morbillivirus (PMV), but these are closely related and now are considered to be members of a single species, now named *Cetacean morbillivirus*. Members of this species are most closely related to RPV and MeV: these viruses are distinguished by host range and sequence divergence.

LIST OF SPECIES IN THE GENUS

Species names are in green italic script; strain names and synonyms are in black roman script; tentative species names are in blue roman script. Sequence accession, and assigned abbreviations ( ) are also listed.

SPECIES IN THE GENUS

Canine distemper virus		
Canine distemper virus	[AF014953]	(CDV)
Cetacean morbillivirus virus		
Cetacean morbillivirus virus	[AJ608288]	(CeMV)
Measles virus		
Measles virus	[K01711, AB016162, AF266288]	(MeV)
Peste-des-petits-ruminants virus		
Peste-des-petits-ruminants virus	[L39878, Z37017, Z81358]	(PPRV)
Phocine distemper virus		
Phocine distemper virus	[D10371, Y09630]	(PDV)
Seal distemper virus		
Rinderpest virus		
Rinderpest virus	[Z30697]	(RPV)

TENTATIVE SPECIES IN THE GENUS

None reported.

SUBFAMILY PNEUMOVIRINAE

TAXONOMIC STRUCTURE OF THE SUBFAMILY

Subfamily	***Pneumovirinae***	
Genus		***Pneumovirus***
Genus		***Metapneumovirus***

DISTINGUISHING FEATURES

Member species differ from those of the subfamily *Paramyxovirinae* in several features: (a) possession of 8-10 separate transcriptional elements; (b) smaller average ORF size; (c) possession of an additional nucleocapsid-associated protein (M2-1, formerly called 22K) and an RNA regulatory protein (M2-2); (d) extensive O-linked glycosylation of the G protein; (e) the P mRNA has a single ORF and does not have RNA editing; (f) a lack of amino acid sequence relatedness with members of the subfamily *Paramyxovirinae* except for a low level in the F and L proteins, and (g) differences in nucleocapsid diameter (13-14 nm compared with 18 nm in the subfamily *Paramyxovirinae*), nucleocapsid pitch (7 nm), and length of glycoprotein spikes (10-14 nm), and (h) lack of a "rule of six" governing the nucleotide length. Species also lack a neuraminidase and a hemagglutinin except in the case of Murine pneumonia virus (MPV) (formerly Pneumonia virus of mice [PVM]), which has a hemagglutinin. The G attachment protein is structurally unrelated to the HN or H proteins of the other genera of the family *Paramyxoviridae* and exhibits a high level of inter-strain diversity: only 53% identity among HRSV isolates, 21-30% identity between human and non-human respiratory syncytial viruses, and 37% identity between HMPV strains

GENUS PNEUMOVIRUS

Type Species ***Human respiratory syncytial virus***

DISTINGUISHING FEATURES

Pneumoviruses are distinguished from metapneumoviruses by (i) the presence of the NS1 and NS2 genes, (ii) the SH, G, F and M2 genes being in the order SH-G-F-M2 as opposed to F-M2-SH-G for metapneumoviruses, (iii) a greater genome length (15,190-15,225 nt compared to 13,280-13,378), and (iv) a higher degree of nucleotide and amino acid sequence relatedness within the genus than between genera.

LIST OF SPECIES DEMARCATION CRITERIA IN THE GENUS

HRSV and Murine pneumonia virus (MPV) (formerly pneumonia virus of mice [PVM]) are distinguished by host range (humans versus mice) and a lack of cross-neutralization. Amino acid sequence relatedness between these two viruses varies from undetectable to intermediate, depending on the protein (for example, the NS1 and NS2 proteins lack demonstrable relatedness, whereas the N or F proteins share 60% or 40% identity, respectively). Their gene maps differ only in the absence of the M2/L gene overlap in MPV. Bovine respiratory syncytial virus (BRSV) differs from HRSV in host range, specifically cattle versus humans, but the difference is not absolute. For example, both viruses grow efficiently in cultured human or bovine cells, although some specificity may be evident. In chimpanzees, HRSV replicates efficiently, is transmissible and is pathogenic, whereas BRSV is very attenuated and non-pathogenic. The two viruses share considerable sequence and antigenic relatedness, but also can clearly be distinguished on either basis. For example, the N or F proteins are each 81% identical between BRSV and HRSV, compared to 96% or 89% identical, respectively, among different HRSV strains. Antiserum against one virus will cross-neutralize the other with a 6 to 64-fold reduction in efficiency. There are two antigenic subgroups of HRSV, called A and B, which exhibit aa sequence identity ranging from 96% (N) to 53% (G), and which are approximately 50% or

5% related antigenically in the F or G protein, respectively, with the overall difference in reciprocal cross-neutralization being up to four-fold. Comparable antigenic dimorphism also may exist for BRSV.

LIST OF SPECIES IN THE GENUS

Species names are in green italic script; strain names and synonyms are in black roman script; tentative species names are in blue roman script. Sequence accession, and assigned abbreviations ( ) are also listed.

SPECIES IN THE GENUS

Bovine respiratory syncytial virus		
Bovine respiratory syncytial virus	[AF295543, AF092942]	(BRSV)
Human respiratory syncytial virus		
Human respiratory syncytial virus A2	[M74568]	(HRSV-A2)
Human respiratory syncytial virus B1	[AF013254]	(HRSV-B1)
Human respiratory syncytial virus S2	[U39662]	(HRSV-S2)
Murine pneumonia virus		
Murine pneumonia virus (formerly Pneumonia virus of mice [PVM])	[D11128, D11130, D10331, U66893, U09649]	(MPV)

TENTATIVE SPECIES IN THE GENUS

Caprine and ovine strains of BRSV also have been described but might represent, with BRSV, a subgroup of ruminant strains rather than different species.

GENUS *METAPNEUMOVIRUS*

Type Species *Avian metapneumovirus*

DISTINGUISHING FEATURES

The relative placements of SH-G versus F-M2 in the gene order are reversed as compared to pneumoviruses. NS1 and NS2 genes are absent in pneumoviruses, and the genome is nearly 2,000 nt shorter. The intergenic regions are longer (up to 190 nt compared with 57nt). The extent of sequence relatedness is greater within than between genera.

LIST OF SPECIES DEMARCATION CRITERIA IN THE GENUS

Metapneumovirus species are distinguished on the basis of having an avian or human host. Interestingly, the sequence diversity between avian and human isolates is less than that between certain avian isolates, and thus sequence relatedness is not a reliable distinguishing feature.

LIST OF SPECIES IN THE GENUS

Species names are in green italic script; strain names and synonyms are in black roman script; tentative species names are in blue roman script. Sequence accession, and assigned abbreviations ( ) are also listed.

SPECIES IN THE GENUS

Avian metapneumovirus ‡		
Avian metapneumovirus (Avian pneumovirus) (Turkey rhinotracheitis virus)	[U22110, U65312, Y14294, U39295, U37586, L34032, S40185]	(AMPV)
Human metapneumovirus ‡		
Human metapneumovirus	[AF371337]	(HMPV)

TENTATIVE SPECIES IN THE GENUS

None reported.

LIST OF UNASSIGNED VIRUSES IN THE FAMILY

Fer-de-Lance virus	[AY141760]	(FDLV)

Nariva virus		(NarPV)
Salem virus	[AF237881]	(SaPV)
Tupaia paramyxovirus	[AF079780]	(TuPV)

In addition there are several viruses from penguins which are known to be distinct from *Avian paramyxoviruses 1-9*.

PHYLOGENETIC RELATIONSHIPS WITHIN THE FAMILY

The literature on the relationships of members of the subfamily *Paramyxovirinae* is consistent with the phylogeny.

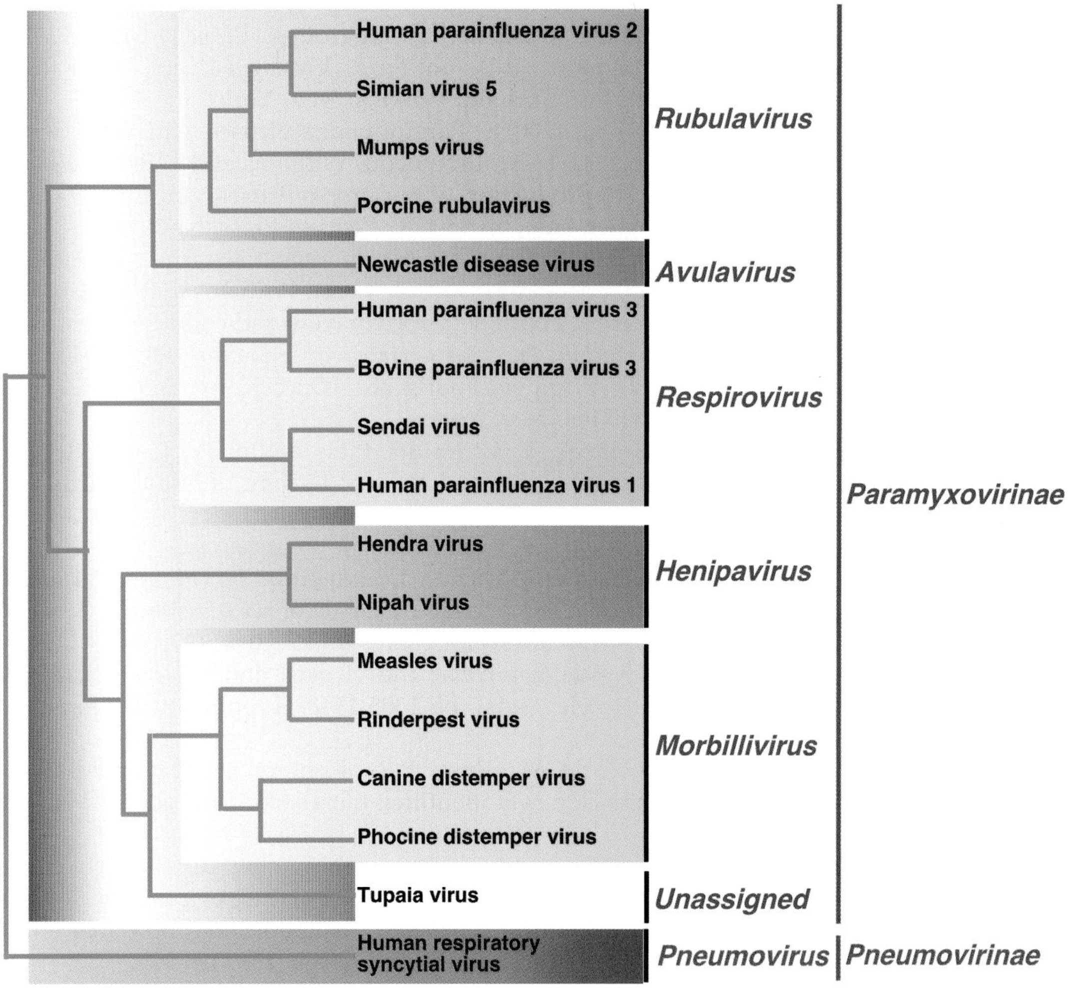

Figure 3: Phylogenetic analysis of the L proteins of members of the family *Paramyxoviridae*. Phylogenetic analysis using PAUP 4.02 was performed on the aa sequence of L proteins from various members of the family *Paramyxoviridae*. The tree shown was based on maximum parsimony; however, analysis of the same data using maximun likelihood produced a tree with nearly identical topology (data not shown). The genus *Metapneumovirus* is not shown, but would cluster with the genus *Pneumovirus* (Adapted from Harcourt *et al.*, (2001) *Virology* **287**, 192-201).

SIMILARITY WITH OTHER TAXA

The member viruses of the family *Paramyxoviridae* have a similar strategy of gene expression and replication and gene order to those of other families in the order *Mononegavirales*, specifically the families *Rhabdoviridae* and *Filoviridae*.

DERIVATION OF NAMES

Avula: from Avian Rubula virus

Henipa: from Hendra and Nipah viruses
Meta: from Greek *meta* for "after"
Morbilli: from Latin *morbillus*, diminutive of *morbus*, "disease".
Ortho: from Greek *orthos* "straight".
Paramyxo: from Greek *para*, "by the side of", and *myxa* 'mucus'.
Pneumo: from Greek *pneuma*, "breath".
Respiro from Latin *respirare*, "respire, breath".
Rubula: *Rubula inflans* - old name for mumps.

REFERENCES

Biacchesi, S., Skiadopoulos, M.H., Boivin, G., Hanson, C.T., Murphy, B.R., Collins, P.L., Buchholz, U.J. (2003). Genetic diversity between human metapneumovirus subgroups. *Virology*, **315**, 1-9.

Carbone, K.M. and Wolinsky, J.S. (2001). Mumps virus. In: *Fields Virology*, 4th ed., (D.M. Knipe and P.M. Howley, eds), pp 1381-1441. Lippincott, Williams and Wilkins, New York.

Chanock, R.M., Murphy, B.R. and Collins, P.L., (2001). Parainfluenza viruses. In: *Fields Virology*, 4th ed., (D.M. Knipe and P.M. Howley, eds), pp 1341-1379. Lippincott, Williams and Wilkins, New York.

Choppin P.W. and Compans, R.W. (1975). Reproduction of paramyxoviruses. In: *Comprehensive Virology*, Vol 4, (H. Fraenkel-Conrat and R.R. Wagner, eds), pp 95-178. Plenum Press, New York.

Collins, P.L., Chanock, R.M. and Murphy, B.R. (2001). Respiratory syncytial virus. In: *Fields Virology*, 4th ed., (D.M. Knipe and P.M. Howley, eds), pp 1443-1485. Lippincott, Williams and Wilkins, New York.

De Leeuw, O and Peeters B. (1999). Complete nucleotide sequence of Newcastle disease virus: evidence for the existence of a new genus within the subfamily *Paramyxovirinae*. *J. Gen. Virol.*, **80**, 131-136.

Griffin, D.E (2001). Measles virus. In: *Fields Virology*, Fourth ed., (D.M. Knipe and P.M. Howley, eds), pp 1401-1441. Lippincott, Williams and Wilkins, New York.

Harcourt, B.H., Tamin, A., Halpin, K., Ksiazek, T.G., Rollin, P.E., Bellini, W.J. and Rota, P.A. (2001). Molecular characterization of the polymerase gene and genomic termini of Nipah virus. *Virology*, **287**, 192-201.

Kingsbury, D.W. (ed) (1991). *The paramyxoviruses.* Plenum Press, New York.

Kolakofsky, D., Pelet, T., Garcin, D., Hausmann, S., Curran J. and Roux. L. (1998). Paramyxovirus RNA synthesis and the requirement for hexamer genome length: the rule of six revisited. *J. Virol.*, **72**, 891-899.

Lamb, R.A. (1993). Paramyxovirus fusion: A hypothesis for changes. *Virology*, **197**, 1-11.

Lamb, R.A. and Kolakofsky, D. (2001). *Paramyxoviridae*: the viruses and their replication. In: *Fields Virology*, 4th ed., (D.M. Knipe and P.M. Howley, eds), pp 1305-1340. Lippincott, Williams and Wilkins, New York.

Osterhaus, A.D.M.E., de Swart, R.K., Vos, H.W., Ross, P.S., Kenter, M.J.H., and Barrett T. (1991). Morbillivirus infections of aquatic mammals: newly identified members of the genus. *Vet. Microbiol.*, **44**, 161-165.

Thomas, S.M., Lamb, R.A. and Paterson, R.G. (1988). Two mRNAs that differ by two non-templated nucleotides encode the amino co-terminal proteins P and V of the paramyxovirus SV5. *Cell*, **54**, 891-902.

Tidona, C.A., Kurz, H.W., Gelderblom, H.R. and Darai, G. (1999). Isolation and molecular characterization of a novel cytopathogenic paramyxovirus from tree shrews. *Virology*, **258**, 425-434.

Van den Hoogen, B.G., De Jong, J.C., Groen, J., Kuiken, T., De Groot, R., Fouchier, R.A.M. and Osterhaus, D.M.E. (2001). A newly discovered human pneumovirus isolated from young children with respiratory tract disease. *Nat. Med.*, **7**, 719-724.

Wang, L.-F., Harcourt, B.H., Yu, M., Tamin, A., Rota, P.A., Bellini, W.J. and Eaton, B.T. (2001). Molecular biology of Hendra and Nipah viruses. *Microb. Inf.*, **3**, 279-287.

Wang, L.-F., Yi, M., Hansson, E., Pritchard, L.I., Shiell, B., Michalski, W.P. and Eaton, B.T. (2000). The exceptionally large genome of Hendra virus: support for creation of a new genus within the family *Paramyxoviridae*. *J. Virol.*, **74**, 9972-9979.

CONTRIBUTED BY

Lamb, R.A., Collins, P.L., Kolakofsky, D., Melero, J.A., Nagai, Y., Oldstone, M.B.A., Pringle, C.R. and Rima, B.K.

GENUS VARICOSAVIRUS

Type Species *Lettuce big-vein associated virus*

VIRION PROPERTIES

MORPHOLOGY

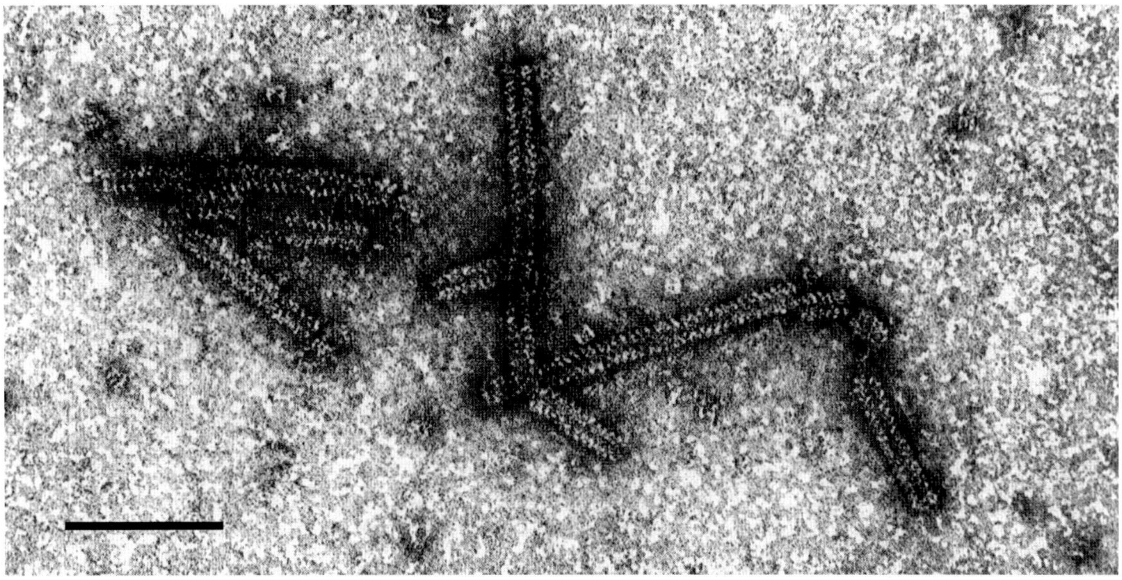

Figure 1: Negative contrast electron micrograph of virions of an isolate of *Lettuce big-vein associated virus*. (Courtesy J.A. Walsh and C.M. Clay). The bar represents 100 nm.

Virions are fragile non-enveloped rods mostly measuring 320-360 x c. 18 nm; each has a central canal c. 3 nm in diameter and an obvious helix with a pitch of c. 5 nm (Fig. 1). As virions are very unstable *in vitro*, their detection and visualization may be facilitated by prior fixation with glutaraldehyde. The helix of particles, especially those in purified preparations, tends to loosen and particles are then seen as partially uncoiled filaments.

PHYSICOCHEMICAL AND PHYSICAL PROPERTIES

Virus particles have a density in Cs_2SO_4 of c. 1.27 g/cm^3.

NUCLEIC ACID

The genome of Lettuce big-vein associated virus (LBVaV) consists of ssRNA, but after deproteinization of purified virions both ds- and ssRNA are detected. The genome is 12.9 kb in size, divided into two components of 6.8 kb (RNA-1) and 6.1 kb (RNA-2); their 3′ termini have no poly(A) tracts. The 3′- and 5′-terminal sequences of the two RNAs are similar, but do not exhibit inverse complementarities. Conserved motifs in the transcription start and end signals resemble those of viruses in the order *Mononegavirales*.

PROTEINS

The size of the CP, when estimated by PAGE, is ~48 kDa. The CP gene, located on RNA-2, encodes a protein of 397 aa with a predicted size of 44.5 kDa.

LIPIDS

None reported.

CARBOHYDRATES

None reported.

GENOME ORGANIZATION AND REPLICATION

The first genomic segment (RNA-1) contains one small ORF, and one large ORF that encodes a protein (designated L protein by analogy with the L polymerase of rhabdoviruses). The second genomic segment (RNA-2) contains five ORFs; the first ORF encodes the CP, and the second to the fifth proteins with unknown function; these four ORFs have coding capacities of 36, 32, 19 and 41 kDa, respectively. Although the genome organization of LBVaV is similar to that of rhabdoviruses, LBVaV does not have a non-coding leader sequence. The aa sequence of the L protein is homologous with the L polymerases of some negative-sense RNA viruses, especially those of rhabdoviruses.

ANTIGENIC PROPERTIES

LBVaV and the tentative species, Tobacco stunt virus (TStV), are serologically closely related.

BIOLOGICAL PROPERTIES

HOST RANGE

Varicosaviruses occur naturally in species in two families (Compositae and Solanaceae). LBVaV and TStV are serologically closely related and infect some common experimental host species.

TRANSMISSION

Both LBVaV and TStV are transmitted in soil and in hydroponic systems by zoospores of the chytrid fungus *Olpidium brassicae*. The viruses are also transmitted experimentally, sometimes with difficulty, by mechanical inoculation. Neither of the viruses is reported to be seed–transmitted.

CYTOPATHIC EFFECTS

None reported.

LIST OF SPECIES DEMARCATION CRITERIA IN THE GENUS

Not yet definable; the molecular properties of TStV have yet to be determined.

LIST OF SPECIES IN THE GENUS

Species names are in green italic script; strain names and synonyms are in black roman script; tentative species names are in blue roman script. Sequence accession, and assigned abbreviations ( ) are also listed.

SPECIES IN THE GENUS

Lettuce big-vein associated virus

Lettuce big-vein associated virus	[AB050272; AB075039]	(LBVaV)

TENTATIVE SPECIES IN THE GENUS

Tobacco stunt virus	(TStV)

PHYLOGENETIC RELATIONSHIPS WITHIN THE GENUS

As the molecular properties of TStV have yet to be determined, detailed comparisons with LBVaV are not possible.

SIMILARITY WITH OTHER TAXA

The genome structure and probable transcription mechanism of LBVaV indicates that it has a close relationship with rhabdoviruses. The aa sequences of both the CP and the L protein of LBVaV have significant similarities with those of rhabdoviruses. LBVaV also resembles rhabdoviruses in possessing conserved transcription termination/ polyadenylation signal-like poly(U) tracts and in transcribing monocistronic RNAs. The presence of poly(U) tracts in the NCRs of LBVaV RNA-1 and RNA-2 suggest that transcription of LBVaV is regulated by a mechanism similar to that of rhabdoviruses (order, *Mononegavirales*; family *Rhabdoviridae*). However, whereas rhabdoviruses contain a single negative-sense ssRNA, LBVaV has two such RNAs.

DERIVATION OF NAMES

Vari from Latin *varix*, meaning abnormal dilation or enlargement of a vein or artery and referring to the symptom previously thought to be induced by the type species. However, lettuce big vein disease, although long thought to be induced by a virus previously designated Lettuce big-vein virus, is now considered to be caused by an *Ophiovirus* which, although tentatively designated *Mirafiori lettuce virus*, has now been renamed *Mirafiori lettuce big-vein virus*. Viruses of this species are soil-borne and often occur in lettuce together with isolates of *Lettuce big-vein associated virus*.

REFERENCES

Brunt, A.A., Crabtree, K., Dallwitz, M.J., Gibbs, A.J. and Watson, L. (eds)(1996). *Viruses of Plants*. CAB International, Wallingford, U.K.

Huijberts, N., Blystad, D.R. and Bos, L. (1990). Lettuce big-vein virus: mechanical transmission and relationships to tobacco stunt virus. *Ann. App. Biol.*, **116**, 463-475.

Kuwata, S. and Kubo, S. (1986). Tobacco stunt virus. *AAB Descriptions of Plant Viruses* No. 313, 4 pp.

Kuwata, S., Kubo, S., Yamashita S. and Doi, Y. (1983). Rod-shaped particles, a probable entity of lettuce big vein virus. *Ann. Phytopath. Soc. Japan*, **49**, 246-251.

Roggero, P., Ciuffo, M., Vaira, A.M., Accotto, G.P., Masenga, V. and Milne, R.G. (2000). An *Ophiovirus* isolated from lettuce with big-vein symptoms. *Arch. Virol.*, **145**, 2629-2642.

Sasaya, T., Ishikawa, K. and Koganezawa, H. (2001). Nucleotide sequence of the coat protein gene of *Lettuce big vein virus*. *J. Gen. Virol.*, **82**, 1509-1515.

Sasaya, T., Ishikawa, K. and Koganezawa, H. (2002). The nucleotide sequence of RNA-1 of *Lettuce big-vein virus*, genus *Varicosavirus*, reveals its relation to nonsegmented negative strand RNA viruses. *Virology*, **297**, 289-297.

Vetten, H.J., Lesemann, D.-E. and Dalchow, J. (1987). Electron microscopical and serological detection of virus-like particles associated with lettuce big-vein disease. *J. Phytopathol.*, **120**, 53-59.

Walsh, J.A. (1994). Effects of some biotic and abiotic factors on symptom expression of lettuce big-vein virus in lettuce (*Lactuca sativa*). *J. Hort. Sci.*, **69**, 21-28.

CONTRIBUTED BY

Brunt, A.A., Milne, R.G., Sasaya, T., Verbeek, M., Vetten, H.J. and Walsh, J.A.

GENUS OPHIOVIRUS

Type Species *Citrus psorosis virus*

VIRION PROPERTIES

MORPHOLOGY

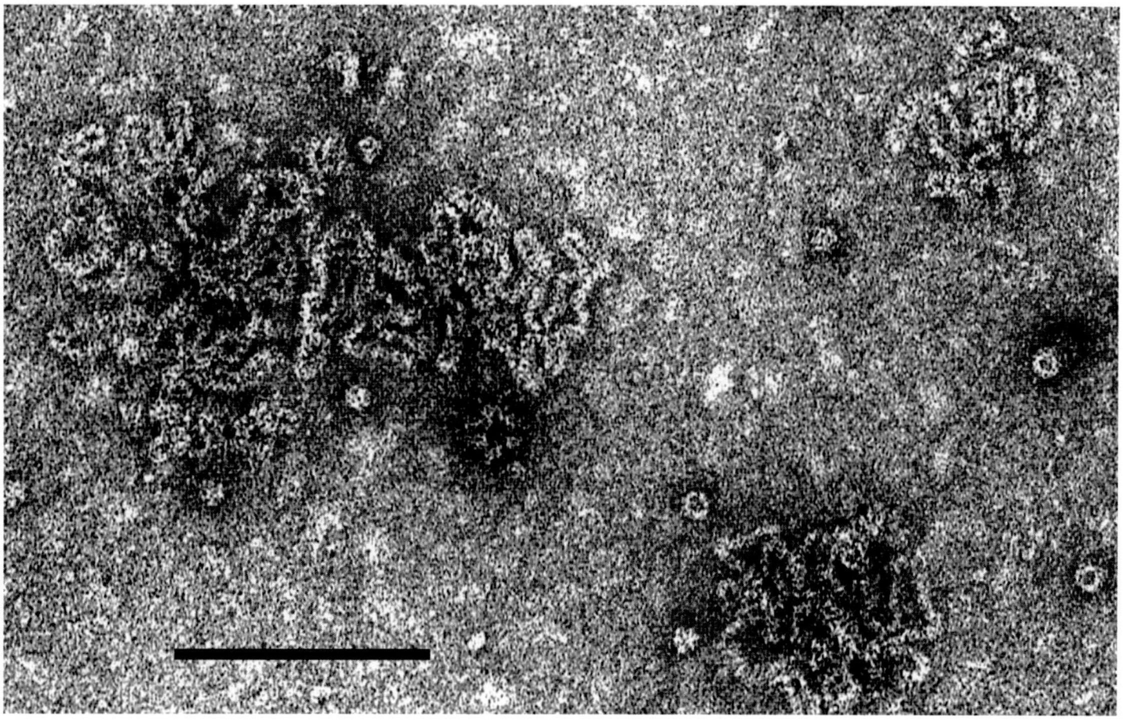

Figure 1: Negative contrast electron micrograph (uranyl acetate) of virions of an isolate of *Mirafiori lettuce virus*. The bar represents 100 nm. (Courtesy of R.G. Milne).

The virions are naked filamentous nucleocapsids about 3 nm in diameter (Fig. 1), forming kinked (probably internally coiled) circles of at least two different contour lengths, the shortest length about 760 nm. The circles can collapse to form pseudolinear duplex structures about 9-10 nm in diameter.

PHYSICOCHEMICAL AND PHYSICAL PROPERTIES

The Mr and sedimentation coefficients of virions are not known. The particles are unstable in CsCl but the buoyant density in cesium sulfate (for both Ranunculus white mottle (RWMV), and Mirafiori lettuce virus (MiLV)) is 1.22 g/cm^3. The particles have limited stability between pH 6 and 8. Infectivity does not survive in crude sap held at 50°C for 10 min (40-45°C for Tulip mild mottle mosaic virus, TMMMV). Particle structure survives limited treatment with organic solvents and nonionic or zwitterionic detergents.

NUCLEIC ACID

The ssRNA genome is 11.3-12.5 kb in size and consists of three or four segments. Viral RNA of both polarities is present in purified virus preparations. As virions appear circularized, the presence of panhandle structures is suggested; however, significant complementation between the 5′ and the 3′ terminal sequences of genomic RNAs has not been demonstrated either for CPsV or MiLV. At the 5′- and 3′-ends of MiLV genomic RNAs there are orthomyxovirus-like palindromic sequences that could fold into a symmetrically hooked conformation. These structures could not be predicted for Citrus psorosis virus (CPsV). The size of RNA-1 is 8.2 kb for CPsV, about 7.5 kb for RWMV, 7.8

kb for MiLV and 7.6 kb for Lettuce ring necrosis virus (LRNV). RNA-2 is about 1.8 kb for RWMV, MiLV and LNRV, and 1.6 kb for CPsV. RNA-3 is about 1.5 kb. For MiLV and LRNV a fourth genomic RNA of about 1.4 kb has been reported.

PROTEINS

There is one CP, varying in size according to species; 48.6 kDa for CPsV, ~43 kDa for RWMV, ~47 kDa for TMMMV, 48.5 kDa for MiLV and 48 kDa for LRNV. RNA1-derived protein sequences of RWMV, CPsV and MiLV, encoded by the largest ORF, contain the core polymerase module with the five conserved motifs proposed to be part of the RdRp active site. A paramyxovirus-RdRp domain (pfam 00946) is present. LRNV RNA-1 shows sequence similarity with these sequences. The SDD sequence, a signature for segmented negative-stranded RNA viruses (*Orthomyxoviridae, Arenaviridae* and *Bunyaviridae*), is also present in motif C. NLS are present in the CPsV, MiLV and RWMV polymerases and in the RNA2-encoded proteins of CPsV and MiLV.

LIPIDS

None reported.

CARBOHYDRATES

None reported.

GENOME ORGANIZATION AND REPLICATION

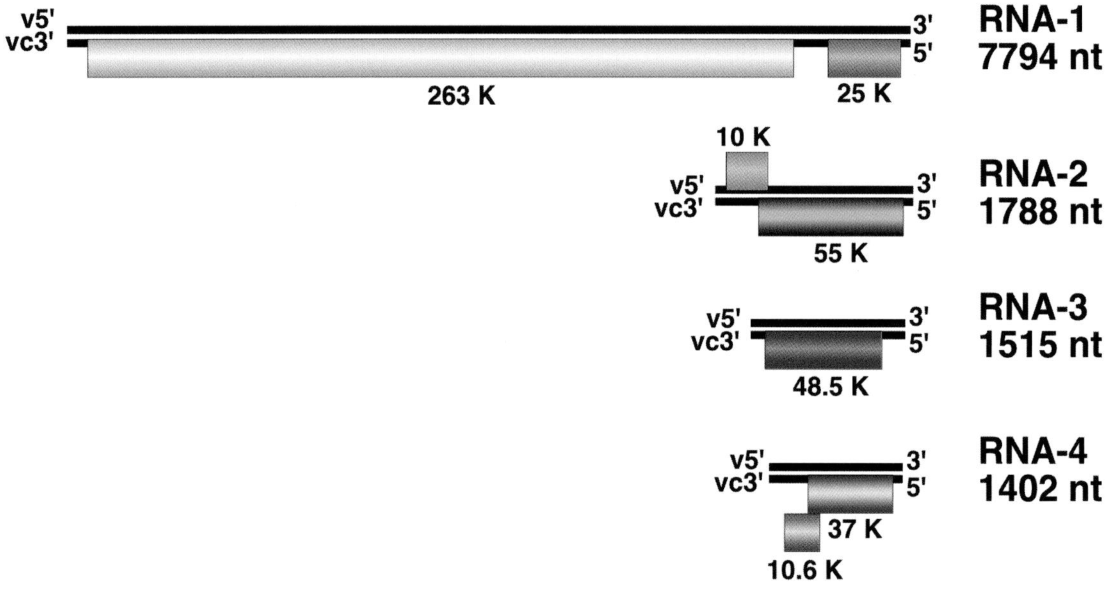

Figure 2: Genome organization of Mirafiori lettuce virus (MiLV). Boxes represents ORFs. The length of the RNA segments and the predicted sizes of the ORF products are indicated. V, viral RNA; vc, viral complementary RNA. RNA-4 is not reported for all ophioviruses. (Modified from van der Wilk *et al.*, 2002).

The ophiovirus genome appears to be of negative-sense (Fig. 2). RNA-1 is of negative polarity and contains a large ORF coding for the RdRp; a further small ORF has also been reported. RNA-2 contains a single ORF in the complementary strand for CPsV; for MiLV two ORFs are reported: the second small one, present in the virion sense strand, has not yet been confirmed as functional. RNA-3 codes for the CP and is of negative polarity. RWMV CP antigen accumulates in the cytoplasm of parenchyma cells. RNA-4, reported only for MiLV and LRNV, is about 1.4 kb. For MiLV, it contains two ORFs overlapping by 38 nt. The second ORF lacks an initiation codon and it is probably expressed by a +1 translational frameshift. For CPsV the 5′-ends of the complementary strand of RNA-1, RNA-2 and RNA-3 are identical, GATAC(T)$_7$, but the 3′-ends are all different. For MiLV, the respective 5′-ends and 3′-ends are conserved among the four viral RNA segments.

Antigenic Properties

The CP is a relatively poor antigen and is the only significant antigenic element. In Western blots, CPs of RWMV, TMMMV, MiLV and LRNV appear to be slightly to moderately serologically related. CPsV CP is serologically unrelated to the others.

Biological Properties

The viruses can be mechanically transmitted to a limited range of test plants, inducing local lesions and systemic mottles. The natural hosts for CPsV, RWMV, MiLV and LRNV are dicotyledonous plants of widely differing taxonomy, and tulip (*Tulipa gesneriana* L., hybrids, Liliaceae) is the natural host of TMMMV. CPsV is commonly transmitted by vegetative propagation of the host and no natural vectors have been identified. No vector is known for RWMV. The zoospores of *Olpidium brassicae* transmit TMMMV, MiLV and LRNV.

CPsV has a wide geographical distribution in citrus, it has been reported in the Americas and the Mediterranean; chlorotic flecking and ringspots may be produced in the young leaves of infected trees and later bark scaling reduces productivity and may eventually kill the plants. RWMV has been reported in two species, ranunculus (*Ranunculus asiaticus* hyb.) and the anemone (*Anemone coronaria*), in Northern Italy; symptomatology in the field is not clear as mixed virus infections are always present; infection of ranunculus seedlings by mechanical inoculation results in limited necrosis and deformation of stems and leaves. TMMMV has been reported in tulips in Japan since 1979; symptoms include veinal chlorotic mottle mosaic on leaves and color break in flowers. MiLV is the causal agent of big-vein symptoms (zones cleared of chlorophyll, parallel to the veins) in lettuce; the widespread and damaging disease probably occurs world-wide (reported in California, France, Germany, Italy, Norway, The Netherlands, UK and Japan). LRNV is closely associated with lettuce ring necrosis disease, first described in The Netherlands and in Belgium as "kring necrosis" and observed in France where it was called "maladie des taches orangées".

List of Species Demarcation Criteria in the Genus

The different criteria considered for species demarcation in the genus are:

- Differing sizes of CP.
- No or distant serological relationship between CPs of different species.
- Natural host range.
- Different number, organization and/or size of genome segments.

Preliminary information is available about genome sequences of ophioviruses. Alignments between CPsV, MiLV and LRNV CP amino acid sequences show 31-52% identity (53-70% similarity) for isolates belonging to different species, while showing almost 100% identity for isolates of the same species. A preliminary result indicates that homology percentage between N-terminal halves of the CP of MiLV and TMMMV is about 80%.

The non-overlapping natural hosts and the slightly different CP sizes, together with the almost complete lack of information regarding TMMMV genome sequences/organization, suggest that MiLV and TMMMV are different species.
FOV particle morphology is ophiovirus-like and the virus is amplified in RT-PCR using ophiovirus-specific primers. The CP is reported to be ~ 31 kDa in size.

List of Species in the Genus

Species names are in green italic script; strain names and synonyms are in black roman script; tentative species names are in blue roman script. Sequence accession, and assigned abbreviations ( ) are also listed.

Species in the Genus

Citrus psorosis virus		
Citrus psorosis virus (Citrus ringspot virus)	[AY224663, AF060855, AY204675, AF218572, AF149101-7, AF036338]	(CPsV)
Lettuce ring necrosis virus		
Lettuce ring necrosis virus		(LRNV)
Mirafiori lettuce virus		
Mirafiori lettuce virus (Mirafiori lettuce big-vein virus)	[AY204672, AH012103, AF525933-6, AY204674, AF532872]	(MiLV)
Ranunculus white mottle virus		
Ranunculus white mottle virus	[AF335430, AF335429, AY204671]	(RWMV)
Tulip mild mottle mosaic virus		
Tulip mild mottle mosaic virus	[AY204673]	(TMMMV)

Tentative Species in the Genus

Freesia ophiovirus	[AY204676]	(FOV)

Phylogenetic Relationships within the Genus

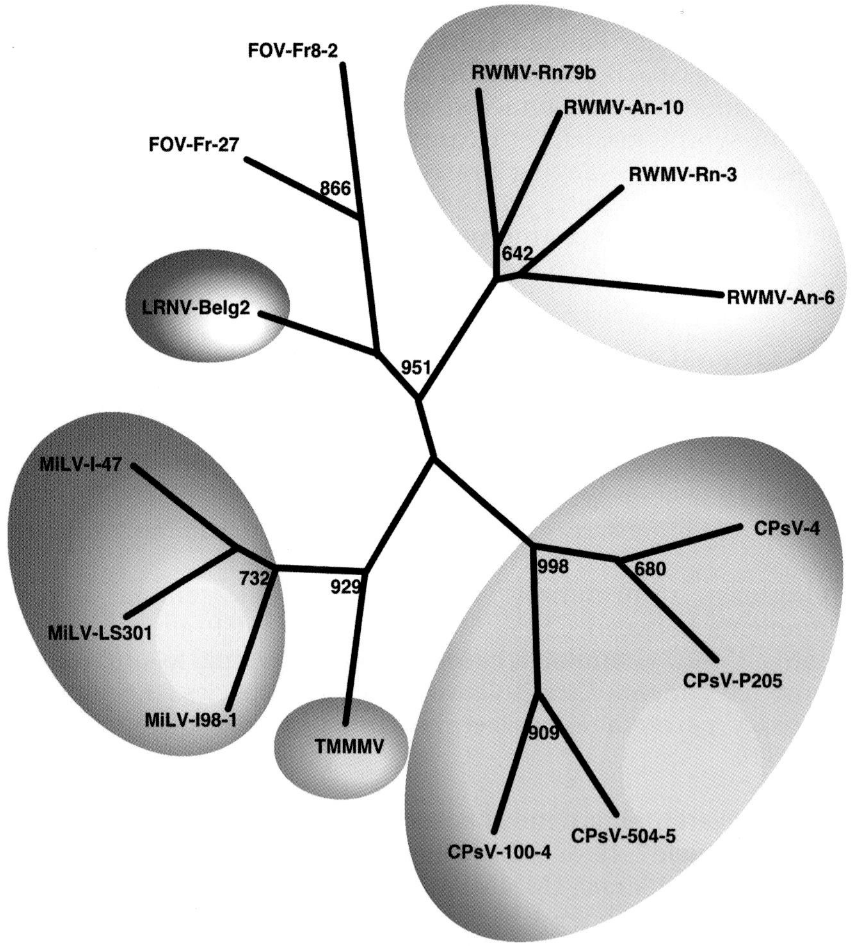

Figure 3: Unrooted phylogenetic tree of 15 available official and tentative ophiovirus isolates. The tree is based on the predicted amino acid sequences (45 aa) encoded by the conserved 136 bp fragment of the RdRp gene. Multiple alignments were done using ClustalX, and analyzed by the neighbor-joining method, with 1000 bootstrap replications. The bootstrapping values below 500 are not included. (Courtesy of A.M. Vaira).

Recently, ophiovirus-specific primers, based on a highly conserved sequence of RNA-1, have been tested in RT-PCR with all ophiovirus species and with FOV. In all cases, a 136 bp fragment was amplified. The 45 aa strings derived from the amplified sequences of at least two isolates for each ophiovirus were subjected to phylogenetic analysis. The results supported the positions of CPsV, RWMV and LRNV as distinct species and a closer relationship between MiLV and TMMMV, already suggested on serological grounds (Fig. 3). The FOV isolates formed a distinct clade.

SIMILARITY WITH OTHER TAXA

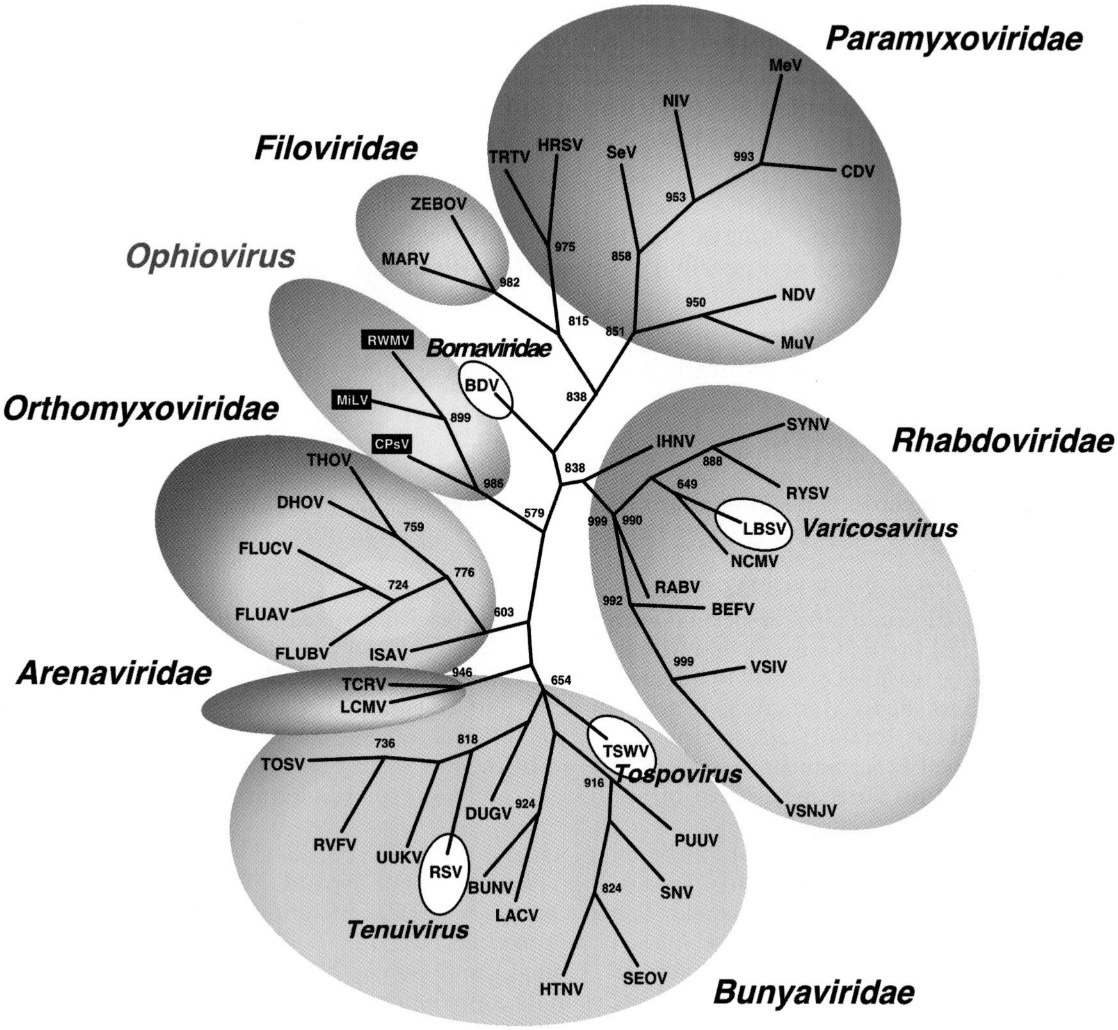

Figure 4: Unrooted phylogenetic tree of members of genus *Ophiovirus* and other negative-stranded RNA viruses based on their conserved RdRp modules. The tree was generated by the neighbor-joining method and bootstrap values (indicated for each branch node) were estimated using 1000 replicates. The bootstrapping values below 500 are not included. Viruses included in the analysis, accession numbers [ ] are: BDV [L27077], BEFV [AF234533], BUNV [X14383], CDV [NC_001921], CPsV [AY224663], DHOV [M65866], DUGV [U15018], FLUAV [J02151], FLUBV [M20170], FLUCV [M28060], HRSV [NC_001781], HTNV [X55901], IHNV [L40883], ISAV [AJ002475], LACV [U12396], LBVV [AB075039], LCMV [J04331], MARV [M92834], MEV [K01711], MiLV [AF525933], MuV [D10575], NCMV [NC_002251], NDV [X05399], NIV [AF212302], PUUV [M63194], RABV [M13215], RSV [D31879], RVFV [X56464], RWMV [AF335429], RYSV [NC_003746], SEOV [X56492], SeV [M19661], SNV [L37901], SYNV [L32603], TCRV [J04340], THOV [AF004985], TOSV [X68414], TRTV [U65312], TSWV [D10066], UUKV [D10759], VSIV [J02428], VSNJV [M20166], and ZEBOV [AF499101]. (Courtesy of Gago-Zachert S.).

Ophiovirus virion morphology resembles that of the tenuiviruses and the internal nucleocapsid component of members of the family *Bunyaviridae*. However, members of the genus *Ophiovirus* do not, like the tenuiviruses, infect plants in the Graminae, and do not, like members of the family *Bunyaviridae*, have enveloped virions. The conservation of identical nucleotides at the genomic RNA termini, observed for the tenui- and phleboviruses (family *Bunyaviridae*) appears to be absent in ophioviruses, in which, as proposed for MiLV, a "corkscrew"-like conformation (as in *Orthomyxoviridae*) or other not yet identified structures may be present. RdRp aa sequences show ophioviruses to be similarto members of the *Paramyxoviridae*, *Rhabdoviridae*, *Bornaviridae* and *Filoviridae, and Varicosavirus* (Fig. 4). The RdRp aa sequence also contains a motif typical of viruses in the families *Orthomyxoviridae*, *Arenaviridae* and *Bunyaviridae*. However, phylogenetic reconstructions using the sequences of the conserved RdRp motifs of representative negative-stranded RNA viruses, reinforce the taxonomic relatedness of the ophioviruses studied (CPsV, MiLV, RWMV and LRNV), and suggest their separation as a monophyletic group.

The virus-like particles recently reported for Pigeonpea sterility mosaic virus (Unassigned viruses), suggest a possible relationship with ophioviruses.

Derivation of Names

Ophio is derived from the Greek "ophis", a serpent, and refers to the snaky appearance of the virions.

References

Barthe, G.A., Ceccardi, T.L., Manjunath, K.L. and Derrick, K.S. (1998). Citrus psorosis virus: nucleotide sequencing of the coat protein gene and detection by hybridization and RT-PCR. *J. Gen. Virol.*, **79**, 1531-1537.

Derrick, K.S., Brlansky, R.H., Da Graça, J.V., Lee, R.F., Timmer. L.W. and Nguyen, T.K. (1988). Partial characterization of a virus associated with citrus ringspot. *Phytopathology*, **78**, 1298-1301.

Garcia, M.L., Dal Bo, E., Grau, O. and Milne, R.G. (1994). The closely related citrus ringspot and citrus psorosis viruses have particles of novel filamentous morphology. *J. Gen. Virol.*, **75**, 3585-3590.

Kawazu, Y., Sasaya, T., Morikawa, T., Sugiyama, K. and Natsuaki, T. (2003). Nucleotide sequence of the coat protein gene of Mirafiori lettuce virus. *J. Gen. Plant Pathol.*, **69**, 55-60.

Lot H., Campbell R.N., Souche S., Milne R.G. and Roggero P. (2002). Transmission by *Olpidium brassicae* of Mirafiori lettuce virus and Lettuce big-vein virus, and their roles in lettuce big-vein etiology. *Phytopathology*, **92**, 288-293.

Morikawa, T., Nomura, Y., Yamamoto, T. and Natsuaki, T. (1995). Partial characterization of virus-like particles associated with tulip mild mottle mosaic. *Ann. Phytopathol. Soc. Jpn.*, **61**, 578-581.

Natsuaki, K.T., Morikawa, T., Natsuaki, T. and Okuda, S. (2002). Mirafiori lettuce virus detected from lettuce with big vein symptoms in Japan. *Jpn. J. Phytopathol.*, **68**, 309-312.

Naum-Onganìa, G., Gago-Zachert, S., Pena, E., Grau, O. and Garcia, M.L. (2003). Citrus psorosis virus RNA 1 is of negative polarity and potentially encodes in its complementary strand a 24K protein of unknown function and 280K putative RNA dependent RNA polymerase. *Virus Res.*, **96**, 49-61.

Navas-Castillo, J. and Moreno, P. (1995). Filamentous flexuous particles and serologically related proteins of variable size associated with citrus psorosis and ringspot diseases. *Europ. J. Plant Pathol.*, **101**, 343-348.

Roggero P., Ciuffo M., Vaira A.M., Accotto G.P., Masenga V. and Milne R.G. (2000). An Ophiovirus isolated from lettuce with big-vein symptoms. *Arch. Virol.*, **145**, 2629-2624.

Sánchez de la Torre, M.E., Lopez, C., Grau, O. and Garcia, M.L. (2002). RNA 2 of Citrus psorosis virus is of negative polarity and has a single open reading frame in its complementary strand. *J. Gen. Virol.*, **83**, 1777-1781.

Sanchez de la Torre, M.E., Riva, O., Zandomeni, R., Grau, O. and Garcia, M.L. (1998). The top component of citrus psorosis virus contains two ssRNAs, the smaller encodes the coat protein. *Mol., Plant Pathol., On-Line*, http://www.bspp.org.uk/mppol/1998/1019sanchez.

Torok, V.A. and Vetten, H.J. (2002). Characterisation of an ophiovirus associated with lettuce ring necrosis. *Joint Conf Int Working Groups on Legume and Vegetable Viruses*, Bonn 4-9 August 2002. Abstract p. 4

Vaira, A.M., Accotto, G.P., Costantini, A. and Milne R.G. (2003). The partial sequence of RNA1 of the ophiovirus Ranunculus white mottle virus indicates its relationship to rhabdoviruses and provides candidate primers for an ophiovirus-specific RT-PCR test. *Arch. Virol.*, **148**, 1037-1050.

Vaira, A.M., Milne, R.G., Accotto, G.P., Luisoni, E., Masenga, V. and Lisa, V. (1997). Partial characterization of a new virus from ranunculus with a divided RNA genome and circular supercoiled thread-like particles. *Arch. Virol.*, **142**, 2131-2146.

Vaira, A.M., Accotto, G.P., Lisa, V., Vecchiati, M., Masenga V. and Milne, R.G. (2002). Molecular diagnosis of ranunculus white mottle virus in two ornamental species. *Acta Hort.*, **568**, 29-33.

Van der Wilk, F., Dullemans, A.M., Verbeek, M. and van den Heuvel, J.F.J.M. (2002). Nucleotide sequence and genomic organization of an ophiovirus associated with lettuce big-vein disease. *J. Gen. Virol.*, **83**, 2869-2877.

Contributed By

Vaira, A.M., Accotto, G.P., Gago-Zachert, S., Garcia, M.L., Grau, O., Milne, R.G., Morikawa, T., Natsuaki, T., Torok, V., Verbeek, M. and Vetten, H.J.

FAMILY ORTHOMYXOVIRIDAE

TAXONOMIC STRUCTURE OF THE FAMILY

Family	*Orthomyxoviridae*
Genus	*Influenzavirus A*
Genus	*Influenzavirus B*
Genus	*Influenzavirus C*
Genus	*Thogotovirus*
Genus	*Isavirus*

VIRION PROPERTIES

MORPHOLOGY

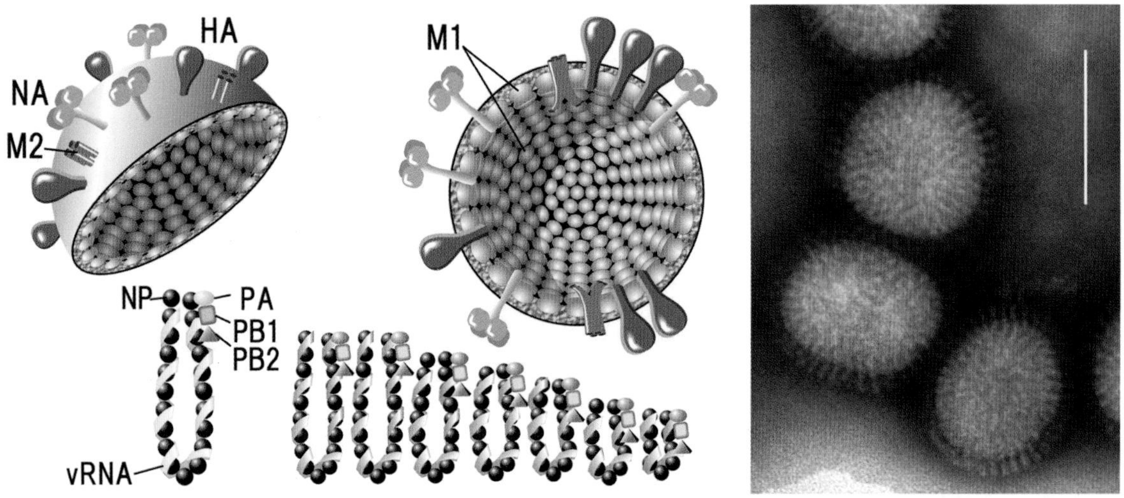

Figure 1: (Left) Diagram of an Influenza A virus (FLUA) virion in section. The indicated glycoproteins embedded in the lipid membrane are the trimeric hemagglutinin (HA), which predominates, and the tetrameric neuraminidase (NA). The envelope also contains a small number of M_2 membrane ion channel proteins. The internal components are the M_1 membrane (matrix) protein and the viral ribonucleoprotein (RNP) consisting of RNA segments, associated nucleocapsid protein (NP), and the PA, PB_1 and PB_2 polymerase proteins. NS_2 (NEP), also a virion protein, is not shown. (Right) Negative contrast electron micrograph of particles of FLUAV (Courtesy of N. Takeshi). The bar represents 100 nm.

Virions are spherical or pleomorphic, 80-120 nm in diameter (Fig. 1). Filamentous forms several micrometers in length also occur. The virion envelope is derived from cell membrane lipids, incorporating variable numbers of virus glycoproteins (1-3) and non-glycosylated proteins (1-2). Virion surface glycoprotein projections are 10-14 nm in length and 4-6 nm in diameter. The viral nucleocapsid is segmented, has helical symmetry, and consists of different size classes, 50-150 nm in length.

PHYSICOCHEMICAL AND PHYSICAL PROPERTIES

Virion Mr is 250 x 10^6. Virion buoyant density in aqueous sucrose is 1.19 g/cm^3. S_{20w} of non-filamentous particles is 700-800S. Virions are sensitive to heat, lipid solvents, non-ionic detergents, formaldehyde, irradiation, and oxidizing agents.

NUCLEIC ACID

Depending on the genus, virions contain different numbers of segments of linear, negative sense ssRNA: 8 segments: Influenza A virus (FLUAV), Influenza B virus (FLUBV), and Infectious salmon anemia virus (ISAV); 7 segments: Influenza C virus (FLUCV) and Dhori virus (DHOV); 6 segments: Thogoto virus (THOV). Segment lengths range from 874-2396 nt. Genome size ranges from 10.0-14.6 kb. RNA segments possess conserved and partially complementary 5'- and 3'-end sequences with promoter activity. Defective (shorter) viral RNAs may occur.

PROTEINS

Structural proteins common to all genera include: 3 polypeptides that form the viral RdRp (e.g., PA, PB_1, PB_2 in FLUAV); a nucleocapsid protein (NP), which is a group-specific protein associated with each genome ssRNA segment in the form of ribonucleoprotein; a hemagglutinin (HA, HEF or GP), which is an integral, type I membrane glycoprotein involved in virus attachment, envelope fusion and neutralization; and a non-glycosylated membrane or matrix protein (M_1 or M). The HA of FLUAV is acylated at the membrane-spanning region and has N-linked glycans at a number of sites. In addition to its hemagglutinating and fusion properties, the HEF protein of FLUCV has esterase activity that functions as a receptor-destroying enzyme. In contrast, the GP of THOV is unrelated to influenzavirus proteins, but shows sequence homology to a baculovirus surface glycoprotein. Members of the genera *Influenzavirus A* and *Influenzavirus B* have an integral, type II envelope glycoprotein (neuraminidase, NA), which contains sialidase activity. Depending on the genus, viruses possess small integral membrane proteins (M_2, NB, BM_2, or CM_2) that may be glycosylated. M_2 and BM_2 function as proton-selective ion channels in mammalian cells, acidifying the virion interior during uncoating and equilibrating the intralumenal pH of the Golgi apparatus with that of the cytoplasm. The channel activity of only the former is inhibited by the anti-influenza A drug amantadine. In addition to the structural proteins and depending on the genus, viruses may code for 2 nonstructural proteins (NS_1, NS_2 [NEP]). Virion enzymes (variously represented and reported among genera) include a transcriptase (PB_1 in influenzaviruses A, B, C and thogotoviruses), an endonuclease (PB_1 in influenzaviruses A, B, C), and a receptor-destroying enzyme (neuraminidase for FLUAV and FLUBV or 9-0-acetyl-neuraminyl esterase in the case of the FLUCV HEF protein).

LIPIDS

Lipids in the virion envelope constitute ~18-37% of the particle weight. They resemble lipids of the host cell plasma membrane.

CARBOHYDRATES

Carbohydrates in the form of glycoproteins and glycolipids constitute about 5% of the particle weight. They are present as N-glycosidic side chains of glycoproteins, as glycolipids, and as mucopolysaccharides. Their composition is host- and virus-dependent.

GENOME ORGANIZATION AND REPLICATION

The genome codes for up to 11 proteins of 14-96 kDa. The 5 largest genome segments encode 1 protein each (with the exception of the PB1 segment, which for most (but not all) strains, encodes PB1 and PB1-F2 proteins). By contrast, smaller segments often code for additional proteins from spliced or bicistronic mRNAs. Generally the three largest RNAs encode the P proteins, and the 4th and 5th the viral HA (HEF, GP) and NP proteins. Depending on the virus, the smaller RNA species encode the NA protein (FLUAV NA and FLUBV NA, NB: 6th RNA), the membrane proteins (FLUAV M_1, M_2 and FLUBV M_1, BM_2: 7th RNA; FLUCV M_1, CM_2 and THOV M, ML and DHOV M_1: 6th RNA) and NS proteins (FLUAV and FLUBV NS_1, NS_2 (NEP): 8th RNA; FLUCV NS_1, NS_2 (NEP): 7th RNA; putative DHOV 7th RNA: unknown). Gene reassortment occurs during mixed infections involving viruses of the same genus, but not between viruses of different genera (e.g., FLUAV and FLUBV).

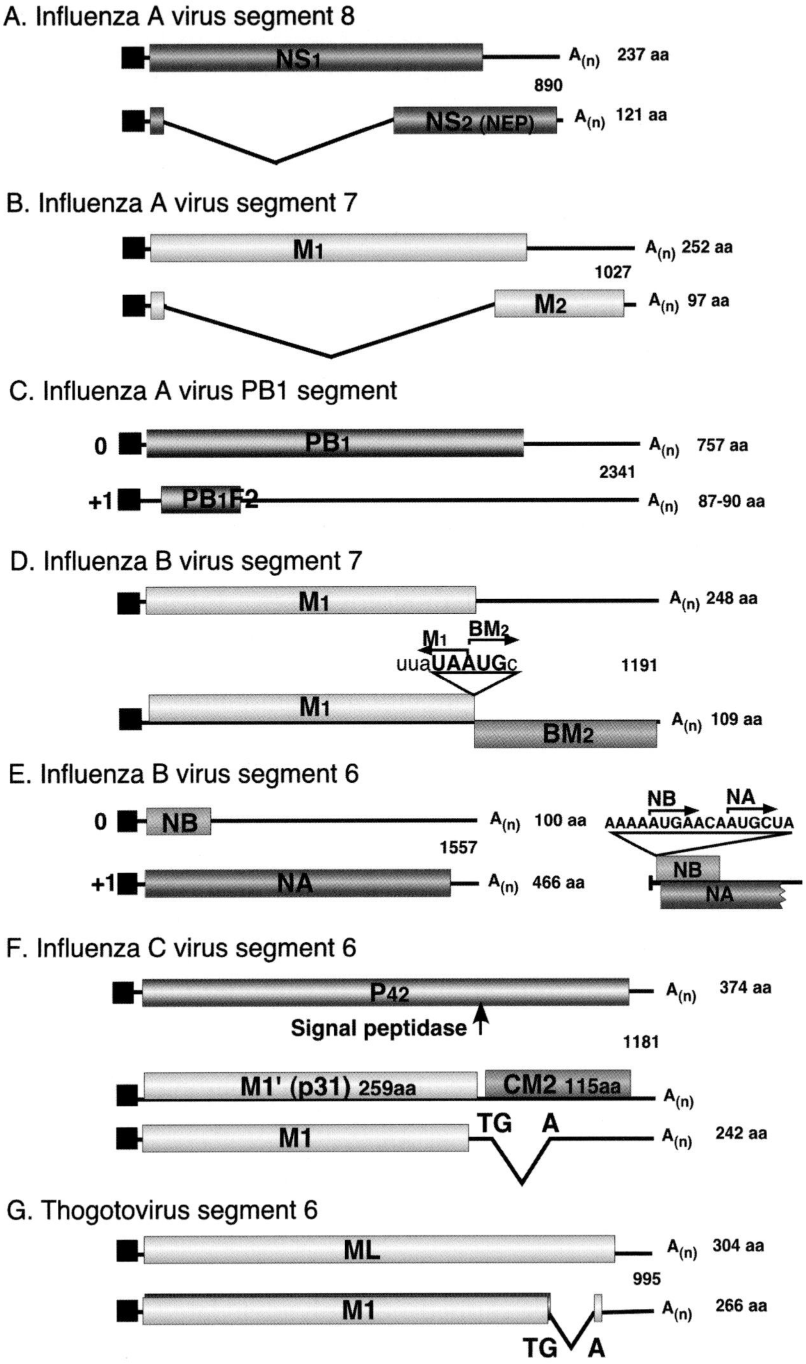

Figure 2: Orthomyxovirus genome organization. The genomic organization and ORFs are shown for genes that encode multiple proteins. Segments encoding the polymerase, hemagglutinin, and nucleoprotein genes are not depicted as each encodes a single protein. (A) Influenza A virus segment 8 showing NS_1 and NS_2 (NEP) mRNAs and their coding regions. NS_1 and NS_2 (NEP) share 10 amino-terminal residues, including the initiating methionine. The ORF of NS_2 (NEP) mRNA (nt 529-861) differs from that of NS_1. (B) Influenza A virus segment 7 showing M_1 and M_2 mRNAs and their coding regions. M_1 and M_2 share 9 amino-terminal residues, including the initiating methionine; however, the ORF of M_2 mRNA (nt 740-1004) differs from that of M_1. A peptide that could be translated from mRNA3 has not been found *in vivo*. (C) Influenza A virus PB1 segment ORFs. Initiation of PB1 translation is thought to be relatively inefficient based on Kozak's rule, likely allowing initiation of PB1-F2 translation by ribosomal scanning. (D) Influenza B virus RNA segment 7 ORFs and the organization of the ORFs used to translate the M_1 and BM_2 proteins. A stop-start pentanucleotide, thought to couple translation between the two ORFs, is illustrated. (E) ORFs in Influenza B virus RNA segment 6, illustrating the overlapping reading frames of NB and NA. Nucleotide sequence surrounding the 2

AUG initiation codons, in the mRNA sense, is shown. Dark lines at the 5'- and 3'-termini of the mRNAs represent untranslated regions. (F) Influenza C virus mRNAs derived from RNA segment 6. The unspliced and spliced mRNAs encode P42 and M1, respectively. The cleavage of P42 by a signal peptidase produces M1'(p31) and CM_2. (G) Thogoto virus segment 6 showing M and ML. M is translated from a spliced mRNA with a stop codon that is generated by the splicing process itself, as in Influenza C virus M1 mRNA. ML is translated from the unspliced transcript and represents an elongated form of M with a C-terminal extension of 38 aa. The boxes represent different coding regions. Introns in the mRNAs are shown by the V-shaped lines; filled rectangles at the 5'-ends of mRNAs represent heterogeneous nucleotides derived from cellular RNAs that are covalently linked to viral sequences. (Modified from Cox and Kawaoka, 1998 and Lamb and Horvath, 1991).

Virus entry involves the HA (HEF, GP) and occurs by receptor-mediated endocytosis. The receptor determinant of influenzaviruses consists of sialic acid bound to glycoproteins or glycolipids. In endosomes, low pH-dependent fusion occurs between viral and cell membranes. For influenzaviruses, fusion (and infectivity) depends on cleaved virion HA (FLUAV and FLUBV: HA_1, HA_2; FLUCV: HEF_1, HEF_2). In addition to the activity of signal peptidases, the HA of the influenzaviruses must undergo post-translational cleavage by cellular proteases to acquire infectivity and fusion activity. Cleavability depends, among other factors, on the number of basic amino acids at the cleavage site. It produces a hydrophobic amino terminal HA_2 molecule. No requirement for glycoprotein cleavage has been demonstrated for the GP species of thogotoviruses. Integral membrane proteins migrate through the Golgi apparatus to localized regions of the plasma membrane. New virions form by budding, thereby incorporating matrix protein and the viral nucleocapsids which align below regions of the plasma membrane containing viral envelope proteins. Budding is from the apical surface in polarized cells.

Viral nucleocapsids are transported to the cell nucleus where the virion transcriptase complex synthesizes mRNA species. For influenzaviruses, mRNA synthesis is primed by capped RNA fragments 10-13 nt in length that are generated from host heterogeneous nuclear RNA species by viral endonuclease activity associated with the PB_1 protein, after cap recognition by PB_2. Thogotoviruses differ from influenzaviruses in having capped viral mRNA without host-derived sequences at the 5'-end. Virus-specific mRNA synthesis is inhibited by actinomycin D or α-amanitin due to inhibition of host DNA-dependent RNA transcription and a (presumed) lack of newly synthesized substrates that allow the viral endonuclease to generate the required primers. Virus-specific mRNA species are polyadenylated at the 3'-termini, and lack sequences corresponding to the 5'-terminal (~16) nucleotides of the viral RNA segment. Certain mRNAs are spliced to provide alternative products (Fig. 2). Protein synthesis occurs in the cytoplasm. However, NP, M_1, and NS_1 proteins accumulate in the cell nucleus during the first few hours of replication, then migrate to the cytoplasm. Cytoplasmic inclusions of NS_1 may be formed.

Complementary RNA molecules which act as templates for new viral RNA synthesis are full-length transcripts and are neither capped nor polyadenylated. These RNAs exist as nucleocapsids in the nucleus of infected cells.

Reverse genetics systems (technologies that allow one to genetically engineer viruses) have been established for FLUAV, FLUBV, and THOV, allowing their generation entirely from cloned cDNA.

Antigenic Properties

The best studied antigens are the NP, HA, NA, M_1, and NS_1 proteins of FLUAV and FLUBV. NP and M_1 are genus specific for the influenzaviruses. Considerable variation occurs among the FLUAV HA and NA antigens, less for FLUBV or the HEF surface antigens of FLUCV. THOV and DHOV do not cross-react in standard serologic tests, while DHOV and Batken virus do. Antibody to HA, HEF, and GP neutralizes virus infectivity. Two major antigenic groups based on the properties of the HA have been identified in ISAV.

Influenzaviruses agglutinate erythrocytes of many species. Serotype-specific antibodies may block agglutination. The NA or HEF of attached influenza virions may destroy sialic acid containing virus receptors of erythrocytes, resulting in elution of virus. Hemolysis of erythrocytes may be produced by HA at acid pH. By comparison to the influenzaviruses, thogotoviruses and isaviruses exhibit limited hemagglutination with certain erythrocyte species

BIOLOGICAL PROPERTIES

Certain influenzaviruses A naturally infect humans and cause respiratory disease. Particular influenzaviruses A infect other mammalian species and a variety of avian species. Interspecies transmission, though rare, is well documented. Influenza B virus strains appear to naturally infect mainly humans and cause epidemics every few years. Influenzaviruses C cause more limited outbreaks in humans and may also infect pigs. Influenzaviruses A and B replicate in the amniotic cavity of embryonated hen eggs, and after adaptation they can also be propagated in the allantoic cavity. Influenzaviruses C replicate only in the amniotic cavity. Primary kidney cells from monkeys, humans, calves, pigs, and chickens support replication of many FLUAV and FLUBV strains. The host range of these viruses may be extended by addition of trypsin to the growth medium, so that multiple cycles of replication can occur in some continuous cell lines. Clinical specimens from influenza-infected hosts often contain sub-populations of virus with minor sequence differences in HA proteins. These subpopulations may differ in their receptor specificity or their propensity for growth in different host cells.

Natural transmission of influenzaviruses is by aerosol (human and most non-aquatic hosts) or is water-borne (waterfowl). Thogoto and Dhori viruses are transmitted by ticks and replicate in both ticks and a variety of tissues and organs in mammalian species as well as in mammalian cell cultures. In some laboratory species (e.g., hamsters for THOV) these infections have a fatal outcome. Unlike influenzaviruses, these viruses do not cause respiratory disease and do not replicate in embryonated hens' eggs. Orthomyxoviruses have an Mx1-sensitive step in their multiplication cycle.

GENUS *INFLUENZAVIRUS A*

Type Species *Influenza A virus*

DISTINGUISHING FEATURES

Member viruses of the genus *Influenzavirus A* all have 8 genome segments. The hemagglutinin and neuraminidase receptor-destroying enzyme are different glycoproteins. The conserved end sequences of the viral RNAs are 5'-AGUAGAAACAAGG and 3'-UCG(U/C)UUUCGUCC. The exact order of electrophoretic migration of the RNA segments varies with strain and electrophoretic conditions. On the basis of the gene sequences, for *Influenza A virus* the segments 1-3 encoded PB_1, PB_2 and PA proteins are estimated to be 87 kDa (observed: 96 kDa), 84 kDa (observed: 87 kDa) and 83 kDa (observed: 85 kDa), respectively. The segment 4 encoded (unglycosylated) HA is 63 kDa (glycosylated HA_1 is 48 kDa, HA_2 is 29 kDa). The segment 5 encoded NP is 56 kDa (observed: 50-60 kDa). The segment 6 encoded NA is 50 kDa (observed: 48-63 kDa). The segment 7 encoded M_1 and M_2 proteins are 28 kDa (observed: 25 kDa) and 11 kDa (observed: 15 kDa), respectively. The segment 8 encoded NS_1 and NS_2 (NEP) are 27 kDa (observed: 25 kDa) and 14 kDa (observed: 12 kDa), respectively.

Antigenic Properties

Antigenic variation occurring within the HA and NA antigens of influenzaviruses A has been analyzed in detail. Based on antigenicity, 15 subtypes of HA and nine subtypes of NA are recognized for influenzaviruses A. Additional variation occurs within subtypes. By convention, new isolates are designated by their serotype / host species / site of origin / strain designation / and year of origin and (HA [H] and NA [N] subtype); e.g., A/tern/South Africa/1/61 (H5N3). In humans, continual evolution of new strains occurs, and older strains apparently disappear from circulation. Antibody to HA neutralizes infectivity. If NA antibody is present during multicycle replication, it inhibits virus release and thus reduces virus yield. Antibody to the amino terminus of M_2 reduces virus yield in tissue culture.

Biological Properties

Epidemics of respiratory disease in humans during the 20th-21st century have been caused by influenzaviruses A having the antigenic composition H1N1, H2N2 and H3N2. H1N2 reassortants between H1N1 and H3N2 human viruses appeared in 2001 and seem to have established themselves in humans. Limited outbreaks of respiratory disease in humans caused by antigenically novel viruses occurred in 1976 in Fort Dix, New Jersey, when classical swine H1N1 viruses infected military recruits; in 1997 and 2003 in Hong Kong when H5N1 viruses caused outbreaks in poultry and contemporaneous illnesses and deaths in humans; and in 1998 and 1999 when H9N2 viruses present in poultry caused illness in humans in China. Influenzaviruses A of subtype H7N7 and H3N8 (previously designated equine 1 and equine 2 viruses, respectively) cause outbreaks of respiratory disease in horses; but H7N7 virus has not been isolated from horses since the late 1970s. Influenzaviruses A (H1N1), and (H3N2) have been isolated frequently from swine. The H1N1 viruses isolated from swine in recent years appear to be of three general categories: those closely related to classical "swine influenza" and which cause occasional human cases; those first recognized in avian specimens, but which have caused outbreaks and infections among swine in Europe and China; and those resembling viruses isolated from epidemics in humans since 1977. H3N2 viruses from swine appear to contain HA and NA genes closely related to those from human epidemic strains. Triple gene reassortant viruses possessing the H3 HA and N2 NA from a recent human virus and other genes from a swine and/or avian virus have been circulating in the US pig population since 1998. In addition, H1N2 viruses, distinct from those in humans, have been isolated from pigs in France, Japan, and the US. Influenzaviruses A (H7N7 and H4N5) have caused outbreaks in seals, with virus spread to non-respiratory tissues in this host. In two separate cases, H7N7 viruses were isolated from conjunctival infections of a laboratory worker and a farm worker in 1980 and 1996, respectively. In addition, in 2003, a human was fatally infected with a pathogenic avian H7N7 virus. Pacific Ocean whales have reportedly been infected with type A (H1N1) virus. Other influenza subtypes have also been isolated from lungs of Atlantic Ocean whales off North America. FLUAV (H10N4) has caused outbreaks in mink. All subtypes of HA and NA, in many different combinations, have been identified in isolates from avian species, particularly wild aquatic birds, chickens, turkeys, and ducks. Pathology in avian species varies from non-apparent infection (often involving replication in, and probable transmission via, the intestinal tract), to more severe infections (observed with subtypes H5 and H7) with spread to many tissues and high mortality rates. The structure of the HA protein, in particular the specificity of its receptor binding site and its cleavability by host protease(s), appears to be critical in determining the host range and organ tropisms of influenza viruses. The NS1 also contributes to the outcome of infection by mitigating host defense mechanisms; e.g., through anti-interferon activity. In addition, interactions between gene products determine the outcome of infection. Interspecies transmission apparently occurs in some instances without genetic reassortment (e.g., H1N1 virus from swine to humans and vice versa, H3N2 virus from humans to swine, and the recent transmissions of H5N1 and H9N2 viruses from poultry to humans). In other cases, interspecies transmission may involve RNA segment reassortment in hosts infected with more than one strain of virus, each with distinct host ranges, or epidemic properties (e.g., 1968 isolates of H3N2 viruses apparently were derived by reassortment between a human H2N2 virus and a virus containing an H3 HA). Laboratory animals that may be infected with influenzaviruses A include ferrets, mice, hamsters, and guinea pigs as well as some small primates such as squirrel monkeys.

List of Species Demarcation Criteria in the Genus

No separate species are currently recognized in the genus *Influenzavirus A*. This genus is comprised of a cluster of strains that replicate as a continuous lineage and can genetically reassort with each other. Therefore, although 15 different HA subtypes and 9 different NA subtypes are recognized among influenzaviruses A replicating in birds, separate species designations have not been accorded to these subtypes. All isolates are capable of exchanging of RNA segments (reassortment).

List of Species in the Genus

Species names are in green italic script; strain names and synonyms are in black roman script; tentative species names are in blue roman script. Sequence accession, and assigned abbreviations ( ) are also listed.

Species in the Genus

Influenza A virus

Influenza A virus A/PR/8/34 (H1N1)	[J02144, J02146, J02148, J02151, V00603, V01099, V01104, V01106]	(FLUAV-A/PR/8/34 (H1N1)

No attempt has been made here to cite the accession numbers of the many sequences deposited in the EMBL (http://www.ebi.ac.uk/embl); Genbank (http://www.ncbi.nlm.nih.gov/Genbank) and LANL (http://www.flu.lanl.gov) databases.

Tentative Species in the Genus

None reported.

Genus *Influenzavirus B*

Type Species *Influenza B virus*

Distinguishing Features

Member viruses of the genus *Influenzavirus B* all have 8 genome segments. As for members of the genus *Influenzavirus A*, hemagglutinin and the neuraminidase receptor-destroying enzyme are different glycoproteins. The conserved end sequences of the viral RNAs of the influenzaviruses B are 5'-AGUAG(A/U)AACAA and 3'-UCGUCUUCGC. Influenza B virus proteins have sizes similar to those for Influenza A virus A. NB, the second product of FLUBV segment 6, 11 kDa (glycosylated 18 kDa).

Antigenic Properties

Antigenic variation within the HA and NA antigens of influenzaviruses B has also been analyzed in some detail. In contrast to influenzaviruses A, no distinct antigenic subtypes are recognized for members of the species *Influenza B virus*, however, viruses with antigenically and genetically distinguishable lineages of HA and NA (e.g., the B/Victoria/2/87-like and the B/Yamagata/16/88-like viruses) have been shown to co-circulate in humans for over a decade. Influenzaviruses B are designated by their serotype/ site of origin/ strain designation/ and year of origin. Antibody to HA neutralizes infectivity

Biological Properties

Influenzaviruses B, first isolated in 1940, have been circulating continuously in humansand causing recurrent epidemics of respiratory disease. Although antigenic change (antigenic drift) occurs more slowly among influenzaviruses B than influenzaviruses A, influenzaviruses B cause epidemics characterized by high attack rates in student populations every 3-years.

List of Species Demarcation Criteria in the Genus

No separate species are currently recognized in the genus *Influenzavirus B*. This genus is comprised of a cluster of strains that replicate as a continuous lineage and can reassort genetically with each other. Although considerable antigenic and sequence differences exist among viruses in this genus, these differences are not sufficient for designation of separate species.

List of Species in the Genus

Species names are in green italic script; strain names and synonyms are in black roman script; tentative species names are in blue roman script. Sequence accession, and assigned abbreviations ( ) are also listed.

Species in the Genus

Influenza B virus

Influenza B virus B/Lee/40	[J02094, J02095, J02096, K00423, K01395, M20168, M20170, M20172]	(FLUBV-B/Lee/40)

No attempt has been made here to cite the accession numbers of the many sequences deposited in the EMBL (http://www.ebi.ac.uk/embl), Genbank (http://www.ncbi.nlm.nih.gov/Genbank) and LANL (http://www.flu.lanl.gov) databases.

Tentative Species in the Genus

None reported.

Genus *Influenzavirus C*

Type Species *Influenza C virus*

Distinguishing Features

Member viruses of the genus *Influenzavirus C* naturally infect humans. Viruses have 7 genome segments. They lack neuraminidase. The HEF protein contains the receptor binding and fusion activities and also functions as the receptor-destroying enzyme, 9-0-acetylneuraminyl esterase. The conserved end sequences of the viral RNAs of the influenzaviruses C are 5'-AGCAGUAGCAA... and 3'-UCGU(U/C)UUCGUCC. RNA segments 1-3 encode the P proteins (87.8 kDa, 86.0 kDa, and 81.9 kDa, respectively). Segment 4 encodes HEF (unglycosylated: 72.1 kDa), segment 5 NP (63.5 kDa), segment 6 M1 and P42 (27.0 kDa and 42.0 kDa, respectively) and segment 7 NS_1 (27.7 kDa) and NS_2 (NEP) (21.0 kDa). Proteolytic cleavage of P42 at an internal signal peptidase cleavage site gives rise to M1' (p31) and CM2 proteins (31.0 kDa and 18.0 kDa, respectively).

Antigenic Properties

Antigenic drift characterized by the emergence of successive antigenic variants which have descended from those that circulated previously apparently does not occur among influenzaviruses C; however, antigenic variation between distinct co-circulating lineages has been detected in HI tests with both anti-HEF Mabs and polyclonal antisera. Viruses exhibit no cross-reactivity with influenzaviruses A and B, although homologies of HEF to influenzavirus A and B HA were identified near the amino and carboxy termini and several of the cysteines co-aligned in the sequences. Antibody to HEF neutralizes infectivity.

BIOLOGICAL PROPERTIES

Infection in humans is common in childhood. Occasional outbreaks, but not epidemics, have been detected. Swine in China have been reported to be infected by viruses similar to human *Influenza C virus* strains.

LIST OF SPECIES DEMARCATION CRITERIA IN THE GENUS

No separate species are currently recognized in the genus *Influenzavirus C*. This genus is comprised of a cluster of strains that replicate as a continuous lineage and can reassort genetically with each other. Although detectable antigenic and sequence differences exist among this genus, these differences are not sufficient for separate species designation.

LIST OF SPECIES IN THE GENUS

Species names are in green italic script; strain names and synonyms are in black roman script; tentative species names are in blue roman script. Sequence accession, and assigned abbreviations ( ) are also listed.

SPECIES IN THE GENUS

Influenza C virus

Influenza C virus C/California/78	[K01689, M10087, M17700]	(FLUCV-C/California/78)

No attempt has been made here to cite the accession numbers of the many sequences deposited in the EMBL (http://www.ebi.ac.uk/embl), Genbank (http://www.ncbi.nlm.nih.gov/Genbank) and LANL (http://www.flu.lanl.gov) databases.

TENTATIVE SPECIES IN THE GENUS

None reported.

GENUS THOGOTOVIRUS

Type Species *Thogoto virus*

NEG. SSRNA

DISTINGUISHING FEATURES

Morphology and morphogenesis of these viruses show similarities with the influenzaviruses. Virions are reported to contain 6 (THOV) or 7 (DHOV) segments of linear, negative sense ssRNA. Total genomic size is ~10 kb. Sequences of the ends of vRNA are partially complementary and resemble those of influenzaviruses. The conserved end sequences of THOV viral RNAs are 5'-AGAGA(U/A)AUCAA(G/A)GC and 3'-UCGUUUUUGU(C/U)CG (segments 1-5) or 3'-UCACCUUUGUCCG (segment 6). Intrastrand base-pairings are favored above interstrand base-pairings, leading to a "hook-like" or cork-screw structure. THOV RNA segments 1-3 encode PB2, PB1, and PA proteins (88, 81, and 71.5 kDa, respectively) that exhibit homology to the respective influenzavirus proteins. The single glycoprotein GP (THOV: 75 kDa; DHOV: 65 kDa) is encoded by the fourth segment. It is unrelated to any influenzavirus protein but shows aa sequence similarity with the glycoprotein (gp64) of baculoviruses. The fifth segment encodes the NP (THOV: 52 kDa; DHOV: 54 kDa), which is related to influenzavirus NP. The sixth segment of THOV encodes the matrix protein M (29 kDa, translated from a spliced mRNA) and a second protein ML (32 kDa, translated from the unspliced mRNA). ML represents an elongated version of M with a C-terminal extension of 38 aa. The sixth segment of DHOV encodes the M1 protein (30 kDa) and may encode another protein, M2 (15 kDa, but not detected in virions) of unknown function. The coding of the DHOV putative seventh segment is not known.

Antigenic Properties

Antigenic relationships between THOV and DHOV viruses are not apparent and none of the virus proteins are related antigenically to those of influenzaviruses; however, serological cross reactivity between DHOV and Batken virus has been demonstrated. For THOV and DHOV, several viruses have been isolated; however, the relationships of these isolates to the prototype viruses are not known.

Biological Properties

THOV and DHOV are transmitted between vertebrates by ticks. Comparatively low levels of hemagglutination occur at acidic pH and not at physiological pH. No receptor-destroying enzyme has been observed. Fusion of infected cells occurs at acidic pH and is inhibited by neutralizing monoclonal antibodies directed against GP, indicating that cell entry is via the endocytic pathway as for the influenzaviruses. Replication is inhibited by actinomycin-D or α-amanitin. Nucleoprotein accumulates early in replication within the nucleus. M of THOV is required for the generation of virus-like particles and infectious recombinant viruses by reverse genetics. In contrast, ML is dispensable for virus growth in cell culture, but appears to be a virulence factor with interferon antagonistic function. THOV is inhibited by the interferon-induced Mx GTPases at an early step in the virus multiplication cycle. Reassortment between THOV temperature sensitive mutants has been demonstrated experimentally in co-infected ticks and in vertebrates.

List of Species Demarcation Criteria in the Genus

THOV has been isolated from *Boophilis* sp. and *Rhipicephalus* sp. ticks in Kenya and Sicily, from *Amblyomma variegatum* ticks in Nigeria, and from *Hyalomma* sp. ticks in Nigeria and Egypt. THOV is known to infect humans in natural settings, and serological evidence suggests that other animals (including cattle, sheep, donkeys, camels, buffaloes and rats) are also susceptible to this virus. THOV has been isolated in the central African Republic, Cameroon, Uganda, and Ethiopia as well as in southern Europe. DHOV has a somewhat different, but overlapping geographic distribution that includes India, eastern Russia, Egypt, and southern Portugal. DHOV has been isolated from *Hyalomma* sp. ticks. As demonstrated by the accidental infection of laboratory workers, DHOV is able to infect humans, causing a febrile illness and encephalitis. Serologic evidence suggests that cattle, goats, camel, and waterfowl are also susceptible to this virus. There is no detectable serological reactivity between THOV and DHOV and the structural differences (THOV has 6 RNA segments and DHOV has 7) and sequence diversity of 37% and 31% in the nucleoprotein and the envelope protein, respectively, argues for separate species status. Batken virus isolated from mosquitoes and ticks from Russia cross reacts serologically with DHOV and shares 98% identity in a portion of the nucleoprotein and 90% identity in a portion of the envelope protein. These data suggest that Batken virus, although isolated from mosquitoes and ticks, is closely related to DHOV.

List of Species in the Genus

Species names are in green italic script; strain names and synonyms are in black roman script; tentative species names are in blue roman script. Sequence accession, and assigned abbreviations ( ) are also listed.

Species in the Genus

Dhori virus		
Dhori virus	[M65866, M34002, M17435, M95567]	(DHOV)
Batken virus	[X97338, X97339]	(BATV)
Thogoto virus		
Thogoto virus	[D00540, M77280]	(THOV)

Tentative Species in the Genus

None reported.

GENUS ISAVIRUS

Type Species *Infectious salmon anemia virus*

DISTINGUISHING FEATURES

Isaviruses are similar in morphology to influenzaviruses. Surface projections are 10 nm in length. Virion surface glycoproteins have both hemagglutinating and receptor destroying activities, the latter being an acetylesterase activity. It is not known if these activities are encoded by different glycoproteins. Isaviruses have 8 linear genome segments of negative sense ssRNA. The vRNA 5′-AGUAAAAA(A/U) and 3′-UCG(U/A)UUCUA end sequences are conserved among isaviruses and partially complementary, with some sequence resemblance to those of influenzaviruses. Total genome size is ~13.5 kb. The two smallest segments each encode two proteins. Segment 7 gives a spliced mRNA product. mRNA synthesis is primed by RNA fragments 8-18 nt in length. The genes of segments 1, 2 and 4 are thought to encode the P proteins based on limited homologies to other RdRp; estimated sizes are 79.9 kDa, 80.5 kDa, and 65.3 kDa, respectively. Segment 3 is thought to encode NP, which is phosphorylated (68 kDa). The segment 5-encoded protein is of unknown function, but is assumed to be a type 1 membrane protein (48.8 kDa). Segment 6 encodes HA, a type 1 membrane protein (42.7 kDa). Segment 7 encodes proteins of unknown function, with estimated sizes of 34.2 kDa and 17.6 kDa. Segment 8 encodes proteins that have estimated sizes of 27.6 kDa and 22 kDa, the latter being a major structural protein.

ANTIGENIC PROPERTIES

There is no known antigenic relationship between isavirus proteins and those of influenzaviruses. Antibodies against ISAV HA neutralize infectivity. The humoral immune response of the host recognizes mainly the HA and NP. There are many isolates of ISAV, and they have tentatively been divided into two major antigenic groups based on properties of the HA.

BIOLOGICAL PROPERTIES

ISAV is transmitted through water. It agglutinates erythrocytes of many fish species, but not avian or mammalian erythrocytes. Fusion of the virus with infected cells occurs at low pH, suggesting endocytic cell entry. The maximum rate of virus replication in the salmon head kidney cell line (SHK-1) is in the temperature range 10-15°C; at 20°C the production of infectious virus is reduced by more than 99% and no replication is observed at 25°C. Replication is inhibited by actinomycin D or α-amanitin. NP accumulates in the nucleus early in the replication cycle. The HA is synthesized in the cytoplasm and accumulates on the cell surface.

LIST OF SPECIES DEMARCATION CRITERIA IN THE GENUS

Isavirus has been isolated from Atlantic salmon (*Salmo salar*) in salmon farming areas in the Atlantic coasts of Northern Europe and North America and from Coho salmon (*Oncorhynchus kisutch*) in Chile. Under experimental conditions, ISAV may be transmitted to several other fish species. Disease caused by this virus has mainly been found in Atlantic salmon.

LIST OF SPECIES IN THE GENUS

Species names are in green italic script; strain names and synonyms are in black roman script; tentative species names are in blue roman script. Sequence accession, and assigned abbreviations ( ) are also listed.

NEG. SSRNA

Species in the Genus

Infectious salmon anemia virus

Infectious salmon anemia virus	[AF220607, AY168787, AJ002475]	(ISAV)

Tentative Species in the Genus

None reported.

List of Unassigned Viruses in the Family

None reported.

Phylogenetic Relationships within the Family

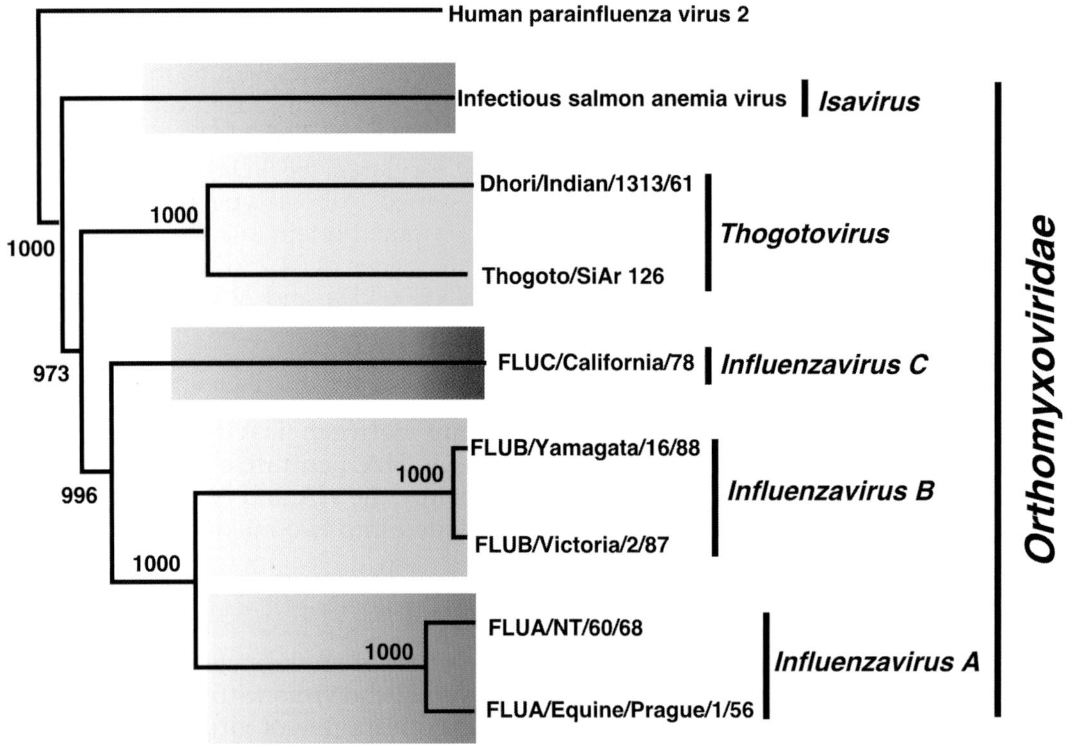

Figure 3. Phylogenetic relationships within the family *Orthomyxoviridae* (Courtesy of Hideki Ebihara). Predicted aa sequences of the nucleocapisd proteins (NP) were aligned, and their phylogenetic relationships were determined using the CLUSTAL X Multiple Sequence Alignment Program (version 1.8, June 1999, Thompson et al., 1997). Human parainfluenza virus 2 was used as an outgroup. The bootstrap values (out of 1000 times) are indicated on the branches. The GenBank accession numbers for the sequences used for comparison were (top to bottom) X57559, AF306549, M17435, X96872, M17700, L49385, AF100359, J02137, and M63748.

Similarity with Other Taxa

Not reported.

Derivation of Names

Influenza: Italian form of Latin *influentia*, "epidemic", originally used because epidemics were thought to be due to astrological or other occult "influences".
Isavirus: sigla from *I*nfection *s*almon *a*nemia *virus*
Myxo: from Greek *myxa*, "mucus".
Ortho: from Greek *orthos*, "straight".
Thogoto: from Thogoto forest near Nairobi, Kenya, where Thogoto virus was first isolated from ticks.

REFERENCES

Albo, C., Martin J. and Portela, A. (1996). The 5′-ends of Thogoto virus (*Orthomyxoviridae*) mRNAs are homogeneous in both length and sequence. *J. Virol.*, **70**, 9013-9017.

Chen, W., Calvo, P.A., Malide, D., Gibbs, J., Schubert, U., Bacik, I., Basta, S., O'Neill, R., Schickli, J., Palese, P., Henklein, P., Bennink, J.R. and Yewdell, J.W. (2001). A novel influenza A virus mitochondrial protein that induces cell death. *Nat. Med.*, **12**, 1306-1312.

Clerx, J., Fuller, F. and Bishop, D. (1983). Tick-borne viruses structurally similar to orthomyxoviruses. *Virology*, **127**, 205-219.

Falk, K., Namork, E., Rimstad, E., Mjaaland, S. and Dannevig, B.H. (1997). Characterization of infectious salmon anemia virus, an orthomyxo-like virus isolated from Atlantic salmon (Salmo salar L.). *J. Virol.*, **71**, 9016-9023.

Fodor, E., Devenish, L., Engelhardt, O.G., Palese, P., Brownlee, G.G. and Garcia-Sastre, A. (1999). Rescue of influenza A virus from recombinant DNA. *J. Virol.*, **73**, 9679-9682.

Frese, M., Weeber, M., Weber, F., Speth, V. and Haller, O. (1997). MX1 sensitivity: Batken virus is an orthomyxovirus closely related to Dhori virus. *J. Gen. Virol.*, **78**, 2453-2458.

Fuller, F., Freedman-Faulstich, E. and Barnes, J. (1987). Complete nucleotide sequence of the tick-borne, orthomyxo-like Dhori/Indian/1313/61 virus nucleoprotein gene. *Virology*, **160**, 81-87.

Hagmaier, K., Jennings, S., Buse, J., Weber, F. and Kochs, G. (2003). Novel gene product of Thogoto virus segment 6 codes for an interferon antagonist. *J. Virol.*, **77**, 2747-2752.

Hoffmann, E., Mahmood, K., Yang, C.F., Webster, R.G., Greenberg, H.B. and Kemble, G. (2002). Rescue of influenza B virus from eight plasmids. *Proc. Natl. Acad. Sci. USA.*, **99**, 11411-11416.

Kochs, G. and Haller, O. (1999). Interferon-induced human MxA GTPase blocks nuclear import of Thogoto virus nucleocapsids. *Proc. Natl. Acad. Sci. USA.*, **96**, 2082-2086.

Krossoy, B., Hordvik, I., Nilsen, F., Nylund, A. and Endresen, C. (1999). The putative polymerase sequence of infectious salmon anemia virus suggests a new genus within the Orthomyxoviridae. *J. Virol.*, **73**, 2136-2142

Lamb, R.A. and Horvath, C.M. (1991). Diversity of coding strategies in influenza viruses. *Trends Genet.*, **7**, 261-266.

Lamb, R.A. and Krug, R.M. (2001). *Orthomyxoviridae*: The Viruses and Their Replication, In: *Fields Virology*, 4rd Edition, (D.M. Knipe and P.M. Howley, eds), pp 1487-1531. Lippincott Williams & Wilkins, Philadelphia.

Mjaaland, S., Rimstad, E., Falk, K. and Dannevig, B.H. (1997). Genomic characterization of the virus causing infectious salmon anemia in Atlantic salmon (Salmo salar L.): an orthomyxo-like virus in a teleost. *J. Virol.*, **71**, 7681-7686.

Neumann, G., Watanabe, T., Ito, H., Watanabe, S., Goto, H., Gao, P., Hughes, M., Perez, D.R., Donis, R., Hoffmann, E., Hobom, G. and Kawaoka, Y. (1999). Generation of influenza A viruses entirely from cloned cDNAs. *Proc. Natl. Acad. Sci. USA.*, **96**, 9345-9350.

Paterson, R.G., Takeda, M., Ohigashi, Y., Pinto, L.H. and Lamb, R.A. (2003). Influenza B virus BM2 protein is an oligomeric integral membrane protein expressed at the cell surface. *Virology*, **306**, 7-17.

Rimstad, E., Mjaaland, S., Snow, M., Mikalsen, A.B. and Cunningham, C.O. (2001). Characterisation of the genomic segment of infectious salmon anemia virus that encodes the putative hemagglutinin. *J. Virol.*, **75**, 5352-5356.

Sandvik, T., Rimstad, E. and Mjaaland, S. (2000). The viral RNA 3'- and 5'-end structure and mRNA transcription of infectious salmon anaemia virus resemble those of influenza viruses. *Arch. Virol.*, **145**, 1659-1669.

Thompson, J.D., Higgins, D.G. and Gibson, T.J. (1994). Clustal W: improving the sensitivity of progressive multiple sequence alignment through sequence weighting, position-specific gap penalties and weight matrix choice. *Nuc. Acids Res.*, **22**, 4673-4680.

Wagner, E., Engelhardt, O.G., Gruber, S., Haller, O. and Kochs, G. (2001). Rescue of recombinant Thogoto virus from cloned cDNA. *J. Virol.*, **75**, 9282-9286.

CONTRIBUTED BY

Kawaoka Y., Cox, N.J., Haller, O., Hongo, S., Kaverin, N., Klenk, H.-D., Lamb, R.A., McCauley, J., Palese, P., Rimstad E. and Webster, R.G.

FAMILY BUNYAVIRIDAE

TAXONOMIC STRUCTURE OF THE FAMILY

Family	*Bunyaviridae*
Genus	*Orthobunyavirus*
Genus	*Hantavirus*
Genus	*Nairovirus*
Genus	*Phlebovirus*
Genus	*Tospovirus*

VIRION PROPERTIES

MORPHOLOGY

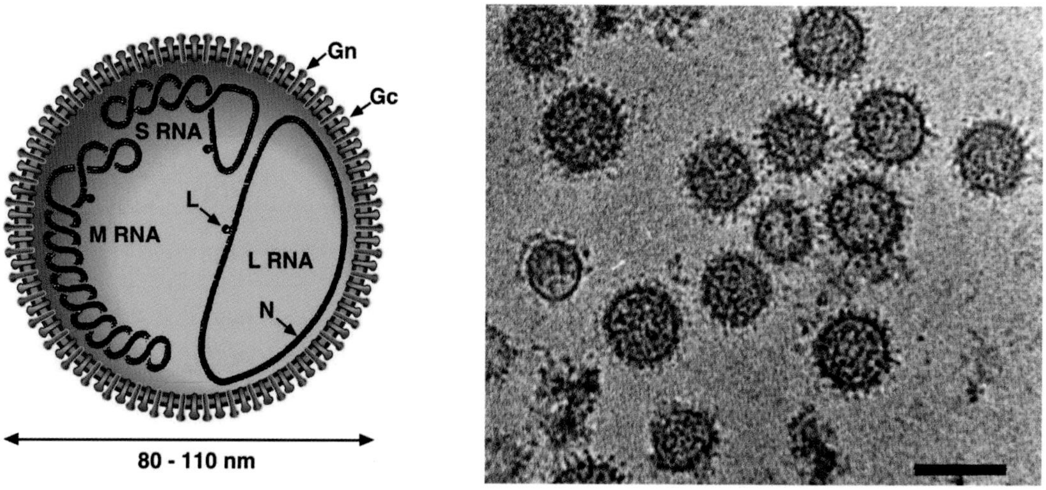

Figure 1: (Left) Diagrammatic representation of a bunyavirus virion in cross-section. The surface spikes comprise two glycoproteins termed Gn and Gc (previously referred to as G1 and G2). The three helical nucleocapsids are circular and comprise one each of the unique ssRNA segments (L, large; M, medium; S, small) encapsidated by N protein and associated with the L protein (Courtesy of R. Pettersson). (Right). Cryo-electron micrograph of particles of California encephalitis virus strain La Crosse virus, taken with large defocus value which demonstrates the glycoprotein spikes (Courtesy of B.V.V. Prasad; see Elliott, 1996).

Morphological properties vary among viruses in each of the five genera; however, virions generally are spherical or pleomorphic, 80-120 nm in diameter, and display surface glycoprotein projections of 5-10 nm which are embedded in a lipid bilayered envelope approximately 5 nm thick. Virion envelopes are usually derived from cellular Golgi membranes, or on occasion, from cell surface membranes. Viral ribonucleocapsids are 2-2.5 nm in diameter, 200-3,000 nm in length, and display helical symmetry.

PHYSICOCHEMICAL AND PHYSICAL PROPERTIES

The virion Mr is 300 x 10^6 - 400 x 10^6 and has an S_{20W} of 350-500S. Virion buoyant densities in sucrose and CsCl are 1.16-1.18 and 1.20-1.21 g/cm^3, respectively. Virions are sensitive to heat, lipid solvents, detergents, and formaldehyde.

NUCLEIC ACID

The viral genome comprises 3 unique molecules of negative or ambisense ssRNA, designated L (large), M (medium) and S (small), which total 11-19 kb (Table 1). The terminal nucleotides of each genome RNA segment are base-paired forming non-covalently closed, circular RNAs (and ribonucleocapsids). The terminal nt sequences of genome segments are conserved among viruses in each genus but are different from those of viruses in other genera. The genomic RNAs are not modified at their 5′-ends. The Mr of the genome ranges from 4.8 x 10^6 - 8 x 10^6 and this constitutes 1-2% of the virion by weight. Viral mRNAs are not polyadenylated and are truncated relative to the genomic RNAs at the 3'-termini. mRNAs have 5'-methylated caps and 10-18 non-templated nt at the 5′-end which are derived from host-cell mRNAs.

Table 1: Nucleotide lengths of selected completely sequenced genomes.

Genus Virus	RNA Segment L	M	S
Orthobunyavirus			
Bunyamwera virus	6875	4458	961
California encephalitis virus - La Crosse virus	6980	4526	980
Hantavirus			
Hantaan virus - 76-118	6533	3616	1696
Seoul virus - HR80-39	6530	3651	1796
Puumala virus - Cg18-20	6550	3682	1784
Sin Nombre virus - NMH10	6562	3696	2059
Nairovirus			
Dugbe virus	12255	4888	1712
Phlebovirus			
Rift Valley fever virus	6404	3884	1690
Sandfly fever Naples virus - Toscana virus	6404	4215	1869
Uukuniemi virus	6423	3231	1720
Tospovirus			
Tomato spotted wilt virus	8897	4821	2916
Impatiens necrotic spot virus	8776	4972	2992

Proteins

All viruses have four structural proteins, 2 external glycoproteins (Gn, Gc, named in accordance with their relative proximity to the amino or carboxy terminus of the polyprotein encoded by the M segment), a nucleocapsid protein (N), and a large (L) transcriptase protein. Non-structural proteins of generally unknown function are expressed from the S segments of bunyaviruses, phleboviruses, and tospoviruses, and from the M segments of bunyaviruses, nairoviruses, tospoviruses and some phleboviruses. Hantaviruses apparently do not encode nonstructural proteins. Proteins encoded by each of the genome segments of viruses in each genus of the family are listed in Table 2.

Table 2: Deduced protein sizes (kDa).

RNA Protein	*Orthobunyavirus*	*Hantavirus*	*Genus* *Nairovirus*	*Phlebovirus*	*Tospovirus*
L segment					
L	259-263	246-247	459	238-241	330-332
M segment					
Gn	29-41	68-76	30-45	50-72	46-58
Gc	108-120	52-58	72-84	55-75	72-78
NSm	15-18	none	78-85, 92-115	none or 78	34-37
S segment					
N	19-26	48-54	48-54	24-30	29
NSs	10-13	none	none	29-32	52

Lipids

Virions contain 20-30% lipids by weight. Lipids are derived from the membranes where viruses mature and include phopholipids, sterols, fatty acids and glycolipids.

Carbohydrates

Virions contain 2-7% carbohydrate by weight. Asparagine-linked sugars on the Gn and Gc proteins are largely of the high mannose type when viruses are grown in vertebrate cells.

Genome Organization and Replication

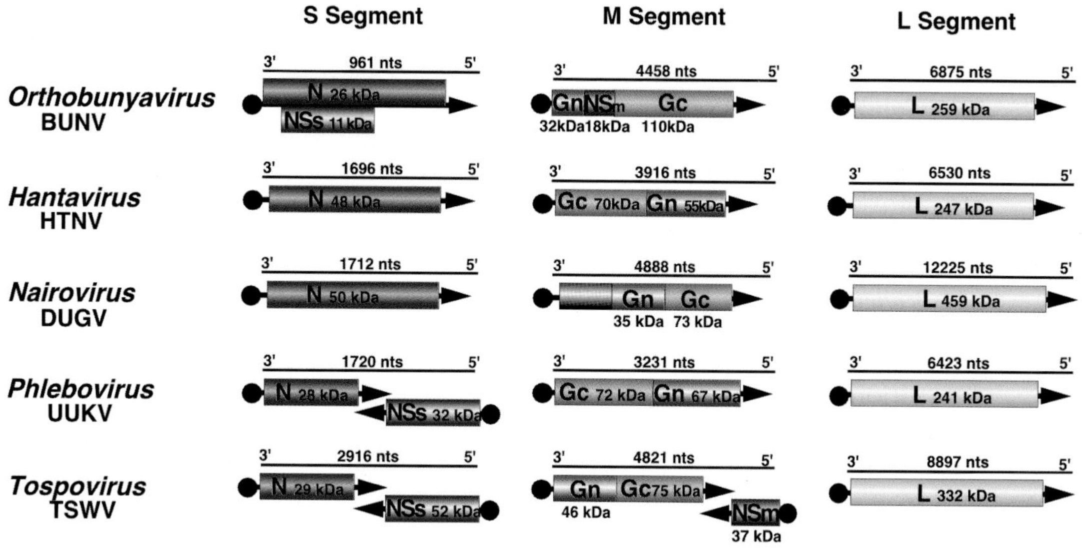

Figure 2: Coding strategies of genome segments of members of the family *Bunyaviridae*. Genomic RNAs are represented by thin lines (the number of nucleotides is given above the line) and mRNAs are shown as arrows (• indicates host derived primer sequence at 5′-end). Gene products, with their size (in kDa), are represented by solid rectangles. (Modified from Elliott, 1996).

The genome organization of the different genera is shown in Figure 2. For all viruses, the L, M and S genome segments encode, respectively, the viral RNA polymerase (L protein), envelope glycoproteins (Gn and Gc) and nucleocapsid protein (N) in the virus-complementary sense RNA. The L protein is encoded in the complementary mRNA. A single, continuous ORF in the M RNA encodes the glycoproteins, and the primary gene product is co-translationally cleaved (except for nairoviruses) to give mature Gn and Gc. Hantaviruses and Uukuniemi-like phleboviruses encode no additional proteins in their M genome segments. Orthobunyaviruses and other phleboviruses encode a nonstructural protein (NSm) in the virion-complementary sense RNA. Nairoviruses encode at least two nonstructural proteins which are precursors to the viral glycoproteins. Tospoviruses encode a NSm protein in an ambisense ORF at the 5'-end of virion-sense RNA. Orthobunyaviruses encode a nonstructural protein (NSs) in an overlapping ORF to that encoding N in the 3'-half of the virion-sense S RNA. There is no direct evidence that hantaviruses or nairoviruses encode any additional proteins in their S genome segments. However, all hantaviruses, with the exception of those associated with *Murinae* subfamily rodents, contain an overlapping ORF with potential to encode an NSs. Phleboviruses and tospoviruses encode a NSs protein in an ambisense ORF in the 5'-half of virion-sense S RNA. The NSs proteins of orthobunyaviruses and phleboviruses have been shown to act as interferon antagonists.

All stages of replication occur in the cytoplasm. The principal stages of replication are:

- Attachment, mediated by an interaction of one or both of the integral viral envelope proteins with, as yet unidentified, host receptors.
- Entry and uncoating, by endocytosis of virions and fusion of viral membranes with endosomal membranes.

- Primary transcription; *i.e.*, the synthesis of mRNA species complementary to the genome templates by the virion-associated polymerase using host cell-derived capped primers (Fig. 3).
- Translation of primary L and S segment mRNAs by free ribosomes; translation of M segment mRNAs by membrane-bound ribosomes and primary glycosylation of nascent envelope proteins. Co-translational cleavage of a precursor to yield Gn and Gc, and for some viruses, NSm.
- Synthesis and encapsidation of antigenome RNA to serve as templates for genomic RNA or, in some cases, for sgRNA.
- Genome replication (Fig. 3).
- Secondary transcription; *i.e.*, the amplified synthesis of the mRNA species and ambisense transcription.
- Morphogenesis, including accumulation of Gn and Gc in the Golgi, terminal glycosylation, acquisition of modified host membranes, generally by budding into the Golgi cisternae; budding at the cell surface has been observed with isolates of *Rift Valley fever virus* (*Phlebovirus*) in rat hepatocytes and *Sin Nombre virus* (*Hantavirus*) in polarized epithelial cells.
- Fusion of cytoplasmic vesicles with the plasma membrane and release of mature virions.

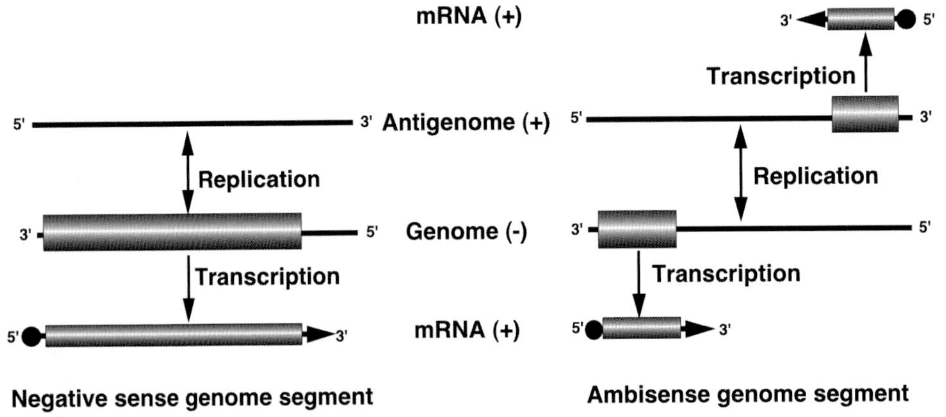

Figure 3: Transcription and replication scheme of genome segments of members of the family *Bunyaviridae* for a negative-strand segment (Left) and for an ambisense segment (Right). The genome RNA and the positive-sense viral complementary RNA, known as anti-genome RNA, are only found as ribonucleoprotein complexes and are encapsidated by N protein. The mRNA species contain host derived primer sequences at their 5′-ends (•) and are truncated at the 3′-end relative to the vRNA template; the mRNAs are not polyadenylated.

Antigenic Properties

One or both of the envelope glycoproteins display hemagglutinating and neutralizing antigenic determinants. Complement fixing antigenic determinants are principally associated with nucleocapsid protein.

Biological Properties

Viruses in the genera *Orthobunyavirus*, *Nairovirus* and *Phlebovirus* are capable of alternately replicating in vertebrates and arthropods, and generally are cytolytic for their vertebrate hosts, but cause little or no cytopathogenicity in their invertebrate hosts. Different viruses are transmitted by mosquitoes, ticks, phlebotomine flies, and other arthropod vectors. Some viruses display a very narrow host range, especially for arthropod vectors. No arthropod vector has been demonstrated for hantaviruses. Tospoviruses can be transmitted by thrips between plants and are capable of replicating in both thrips and plants. Transovarial and venereal transmission have been demonstrated for some mosquito-borne viruses. Aerosol infection occurs in certain situations or is the principal means of transmission for some viruses, particularly hantaviruses. In some instances, avian host and/or vector movements may result in virus dissemination. Some viruses cause a reduction in host-cell protein synthesis in vertebrate cells. Hantaviruses cause no detectable reduction in host macromolecular synthesis and routinely establish persistent, non-cytolytic infections in susceptible mammalian host cells, a finding consistent with their non-pathogenic persistence in their natural rodent hosts. In natural infections of mammals, viruses are often targeted to a particular organ or cell type. Some viruses induce cell fusion at low pH. Some members have ion-dependent hemagglutinating activity. Genetic reassortment has been demonstrated for certain members both *in vitro* and *in vivo*.

GENUS ORTHOBUNYAVIRUS

Type Species *Bunyamwera virus*

DISTINGUISHING FEATURES

The consensus terminal nucleotide sequences of the L, M and S genome segments are UCAUCACAUGA... at the 3'-end and AGUAGUGUGCU... at the 5'-end. The N and NSs proteins are encoded in overlapping reading frames by the S RNA and are translated from the same complementary mRNA as the result of alternate AUG initiation codon usage. Both glycoproteins and a NSm protein of 15-18 kDa are encoded as a precursor polyprotein by the M RNA. Genetic reassortment has been demonstrated between viruses, belonging to the same species but not between viruses from different species.

Viruses are serologically unrelated to members of other genera. Most viruses are mosquito-transmitted though some (e.g. the Tete group) are tick-transmitted. Occasionally alternate arthropods such as culicoid flies and phlebotomines transmit bunyaviruses. Some viruses are transmitted transovarially and venereally in arthropods.

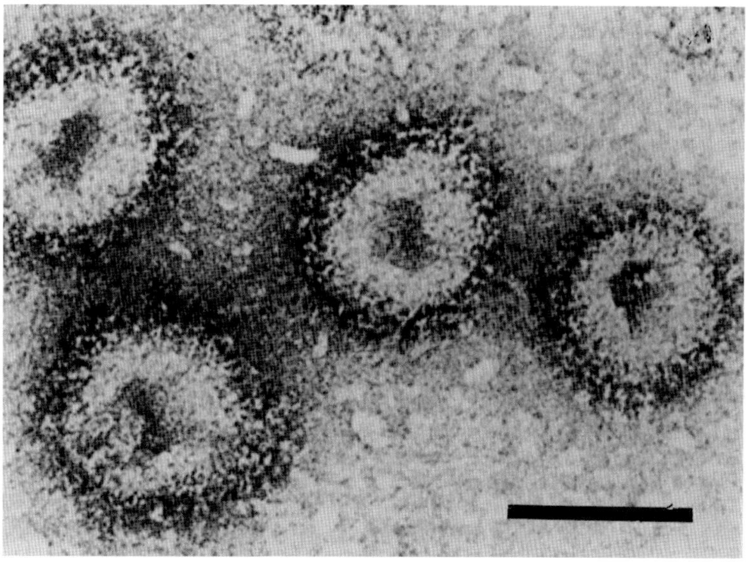

Figure 4: Electron micrograph of negatively stained particles of California encephalitis virus strain La Crosse virus. The bar represents 100 nm. (Courtesy of D.H.L. Bishop).

LIST OF SPECIES DEMARCATION CRITERIA IN THE GENUS

The demarcation of orthobunyavirus species has proven difficult due to the lack of biochemical characterization of most of the named virus isolates. Species are thus primarily defined by serological criteria (cross-neutralization and cross-hemagglutination-inhibition tests). The limited available data indicate that one bunyavirus species is unable to form a reassortant with another species. Where known the aa sequences of the N proteins differ by more than 10%.

List of Species in the Genus

Species names are in green italic script; strain names and synonyms are in black roman script; tentative species names are in blue roman script. Vector type, genome sequence accession numbers [L, M, S, designating RNA segments], and assigned abbreviations ( ) are also listed.

Species in the Genus

Acara virus			
Acara virus –BeAn27639	mosquitoes		(ACAV)
Moriche virus–TRVL57896	mosquitoes		(MORV)
Akabane virus			
Akabane virus – JaGAr39	mosquitoes, culicoid flies	[S: M22011]	(AKAV)
Sabo virus – AN9398	culicoid flies		(SABOV)
Tinaroo virus – CSIRO153	culicoid flies	[S: AB000819]	(TINV)
Yaba-7 virus	N.D.		(Y7V)
Alajuela virus			
Alajuela virus - 75V 2374	mosquitoes		(ALJV)
Alajuela virus - 78V 2441	mosquitoes		(ALJV)
San Juan virus – 75V 446	mosquitoes		(SJV)
Anopheles A virus			
Anopheles A virus – 1940 prototype	mosquitoes		(ANAV)
Anopheles A virus - CoAr3624	mosquitoes		(ANAV)
Anopheles A virus - ColAn57389	mosquitoes		(ANAV)
Las Maloyas virus – AG8-24	mosquitoes		(LMV)
Lukuni virus – TRVL 10076	mosquitoes		(LUKV)
Trombetas virus	mosquitoes		(TRMV)
Anopheles B virus			
Anopheles B virus – 1940 prototype	mosquitoes		(ANBV)
Boraceia virus – SPAr395	mosquitoes		(BORV)
Bakau virus			
Ketapang virus – MM2549	mosquitoes		(KETV)
Bakau virus - MM-2325	mosquitoes		(BAKV)
Nola virus – DakArB 2882	mosquitoes		(NOLAV)
Tanjong Rabok virus – P9-87	N.D.		(TRV)
Telok Forest virus – P72-4	N.D.		(TFV)
Batama virus			
Batama virus – AnB 1292a	N.D.		(BMAV)
Benevides virus			
Benevides virus – BeAn 153564	mosquitoes		(BENV)
Bertioga virus			
Bertioga virus – SPAn 1098	N.D.		(BERV)
Cananeia virus – SPAn 64962	mosquitoes		(CNAV)
Guaratuba virus – APAn 12252	mosquitoes		(GTBV)
Itimirim virus – SPAn47817	N.D.		(ITIV)
Mirim virus – BeAn7722	mosquitoes		(MIRV)
Bimiti virus			
Bimiti virus – TRVL 8362	mosquitoes		(BIMV)

Botambi virus			
Botambi virus – DakArB 937	mosquitoes		(BOTV)
Bunyamwera virus			
Bunyamwera virus - AG83-1746	mosquitoes		(BUNV)
Batai virus – MM2222	mosquitoes	[S: X73464]	(BATV)
Birao virus – DakArB 2198	mosquitoes		(BIRV)
Bozo virus – ArB7343	mosquitoes		(BOZOV)
Bunyamwera virus – 1943 prototype	mosquitoes	[L: X14383; M: M11852; S: X73465]	(BUNV)
Bunyamwera virus - CbaAr 426	mosquitoes		(BUNV)
Cache Valley virus 6V633	mosquitoes	[S: X73465]	(CVV)
Fort Sherman virus – 86MSP18	mosquitoes	[M: M21951; S: M19420]	(FSV)
Germiston virus – Ar1050	mosquitoes		(GERV)
Iaco virus – BeAn 314206	mosquitoes		(IACOV)
Ilesha virus – KO/2	mosquitoes, culicoid flies		(ILEV)
Lokern virus – FMS 4332	mosquitoes	[S: D00354]	(LOKV)
Maguari virus – BeAr7272	mosquitoes		(MAGV)
Mboke virus – DakArY 357	mosquitoes		(MBOV)
Ngari virus – DAKArD 28542	mosquitoes	[S: X73470]	(NRIV)
Northway virus - 0234	mosquitoes		(NORV)
Playas virus – 75V3066	mosquitoes		(PLAV)
Potosi virus	mosquitoes		(POTV)
Santa Rosa virus – M2-1493	mosquitoes		(SARV)
Shokwe virus – SAAr4042	mosquitoes		(SHOV)
Tensaw virus – A9-171b	mosquitoes		(TENV)
Tlacotalpan virus – 61D240	mosquitoes		(TLAV)
Tucunduba virus	mosquitoes		(TUCV)
Xingu virus	mosquitoes		(XINV)
Bushbush virus			
Benfica virus – BeAn 84381	mosquitoes		(BENV)
Bushbush virus – TRVL 26668	mosquitoes		(BSBV)
Bushbush virus - GU71U 344	mosquitoes		(BSBV)
Juan Diaz virus – MARU 8563	N.D.		(JDV)
Bwamba virus			
Bwamba virus - M459	mosquitoes		(BWAV)
Pongola virus – SAAr 1	mosquitoes		(PGAV)
California encephalitis virus			
California encephalitis virus - BFS-283	mosquitoes	[S: U12797]	(CEV)
California encephalitis virus - AG83 497 virus	mosquitoes		(CEV)
Inkoo virus KN3641	mosquitoes	[S: Z68496; M: U88059]	(INKV)
Jamestown Canyon virus 61V-2235	mosquitoes	[S: U12796; M: U88058]	(JCV)
Keystone virus C14031-33	mosquitoes	[S: U12801]	(KEYV)
La Crosse virus	mosquitoes	[L: U12396; M: D00202; S: K00610]	(LACV)
Lumbo virus	mosquitoes	[S: X73468]	(LUMV)
Melao virus	mosquitoes	[S: U12802; M: 88057]	(MELV)

San Angelo virus	mosquitoes	[S: U47139]	(SAV)
Serra do Navio virus	mosquitoes	[S: U47140]	(SDNV)
Snowshoe hare virus	mosquitoes	[M: K02539; S: J02390]	(SSHV)
South River virus	mosquitoes	[S: U47141]	(SORV)
Tahyna virus	mosquitoes	[S: Z68497]	(TAHV)
Trivittatus virus	mosquitoes	[S: U12803]	(TVTV)
Capim virus			
Capim virus - BeAn 8582	mosquitoes		(CAPV)
Caraparu virus			
Apeu virus – BeAn 848	mosquitoes		(APEUV)
Bruconha virus	mosquitoes		(BRUV)
Caraparu virus – BeAn 3994	mosquitoes		(CARV)
Ossa virus – BT 1820	mosquitoes		(OSSAV)
Vinces virus – 75V-807	mosquitoes		(VINV)
Catu virus			
Catu virus – BeH 151	mosquitoes		(CATUV)
Estero Real virus			
Estero Real virus – K329	ticks		(ERV)
Gamboa virus			
Gamboa virus - 75V 2621	mosquitoes		(GAMV)
Gamboa virus - MARU 10962	mosquitoes		(GAMV)
Pueblo Viejo virus – E4-816	mosquitoes		(PVV)
Guajara virus			
Guajara virus – BeAn10615	mosquitoes		(GJAV)
Guajara virus - GU71U 350	mosquitoes		(GJAV)
Guama virus			
Guama virus - BeAn 277	mosquitoes		(GMAV)
Ananindeua virus – BeAn 109303	mosquitoes		(ANUV)
Moju virus – BeAr 12590	mosquitoes		(MOJUV)
Mahogany Hammock virus – FE4-2s	N.D.		(MHV)
Guaroa virus			
Guaroa virus - 352111	mosquitoes	[S: X73466]	(GROV)
Kairi virus			
Kairi virus	mosquitoes	[S: X73467]	(KRIV)
Kaeng Khoi virus			
Kaeng Khoi virus	nest bugs		(KKV)
Koongol virus			
Koongol virus - MRM31	mosquitoes		(KOOV)
Wongal virus	mosquitoes		(WONV)
Madrid virus			
Madrid virus	mosquitoes		(MADV)
Main Drain virus			
Main Drain virus	mosquitoes, culicoid flies	[S: X73469]	(MDV)
Manzanilla virus			
Buttonwillow virus	culicoid flies		(BUTV)
Ingwavuma virus	mosquitoes		(INGV)
Inini virus	N.D.		(INIV)
Manzanilla virus	N.D.		(MANV)
Mermet virus	mosquitoes		(MERV)
Marituba virus			
Gumbo Limbo virus	mosquitoes		(GLV)

Marituba virus	mosquitoes		(MTBV)
Marituba virus - 63U-11	mosquitoes		(MTBV)
Murutucu virus	mosquitoes		(MURV)
Nepuyo virus	mosquitoes		(NEPV)
Restan virus	mosquitoes		(RESV)
Minatitlan virus			
Minatitlan virus M67U5	N.D.		(MNTV)
Palestina virus	mosquitoes		(PLSV)
M'Poko virus			
M'Poko virus	mosquitoes		(MPOV)
Yaba-1 virus	mosquitoes		(Y1V)
Nyando virus			
Nyando virus MP 401	mosquitoes		(NDV)
Eret virus - 147	mosquitoes		(ERETV)
Olifantsvlei virus			
Bobia virus	mosquitoes		(BIAV)
Dabakala virus	mosquitoes		(DABV)
Olifantsvlei virus - SAAAr 5133	mosquitoes		(OLIV)
Oubi virus	mosquitoes		(OUBIV)
Oriboca virus			
Itaqui virus	mosquitoes		(ITQV)
Oriboca virus	mosquitoes		(ORIV)
Oropouche virus			
Facey's Paddock virus	N.D.		(FPV)
Oropouche virus	mosquitoes, culicoid flies	[M: AF312381; L: AF484424]	(OROV)
Utinga virus	N.D.		(UTIV)
Utive virus	N.D.		(UVV)
Patois virus			
Abras virus	mosquitoes		(ABRV)
Babahoya virus	mosquitoes		(BABV)
Pahayokee virus	mosquitoes		(PAHV)
Patois virus - BT 4971	mosquitoes		(PATV)
Shark River virus	mosquitoes		(SRV)
Sathuperi virus			
Douglas virus	culicoid flies		(DOUV)
Sathuperi virus	mosquitoes, culicoid flies		(SATV)
Simbu virus			
Simbu virus - SAAr 53	mosquitoes, culicoid flies		(SIMV)
Shamonda virus			
Peaton virus	culicoid flies		(PEAV)
Sango virus	mosquitoes, culicoid flies		(SANV)
Shamonda virus	culicoid flies		(SHAV)
Shuni virus			
Aino virus	mosquitoes, culicoid flies	[S: M22011]	(AINOV)
Kaikalur virus	mosquitoes		(KAIV)
Shuni virus	mosquitoes, culicoid flies		(SHUV)
Tacaiuma virus			
CoAr 1071 virus	mosquitoes		(CA1071V)
CoAr 3627 virus	mosquitoes		(CA3627V)
Tacaiuma virus - BeAn73	mosquitoes		(TCMV)
Tacaiuma virus - H-32580	mosquitoes		(TCMV)
Virgin River - SPAr 2317	mosquitoes		(VRV)

Virgin River virus – 743-366	mosquitoes	(VRV)
Tete virus		
Bahig virus	ticks	(BAHV)
Matruh virus	ticks	(MTRV)
Tete virus - SAAn 3518	N.D.	(TETEV)
Tsuruse virus	N.D.	(TSUV)
Weldona virus	culicoid flies	(WELV)
Thimiri virus		
Thimiri virus	N.D.	(THIV)
Timboteua virus		
Timboteua virus	mosquitoes	(TBTV)
Turlock virus		
Lednice virus	mosquitoes	(LEDV)
Turlock virus - S 1954-847-32	mosquitoes	(TURV)
Umbre virus	mosquitoes	(UMBV)
Wyeomyia virus		
Anhembi virus	mosquitoes	(AMBV)
BeAr 328208 virus	mosquitoes	(BAV)
Macaua virus	mosquitoes	(MCAV)
Sororoca virus	mosquitoes	(SORV)
Taiassui virus	mosquitoes	(TAIAV)
Wyeomyia virus	mosquitoes	(WYOV)
Zegla virus		
Zegla virus	N.D.	(ZEGV)

Tentative Species in the Genus

Leanyer virus	Mosquitoes	(LEAV)
Mojui Dos Campos virus	N.D.	(MDCV)
Termeil virus	Mosquitoes	(TERV)

Genus Hantavirus

Type Species *Hantaan virus*

Distinguishing Features

The consensus terminal nt sequences of the L, M and S genomic segments are AUCAUCAUCUG... at the 3′-end and UAGUAGUA... at the 5′-end. Non-structural polypeptides are not encoded in any of the 3 genomic segments. Viruses are serologically unrelated to members of other genera. Certain hantaviruses are etiologic agents of hemorrhagic fever with renal syndrome or hantavirus pulmonary syndrome (HPS). The host range of hantaviruses is primarily rodents, and genetically distinct hantaviruses are usually associated with a single rodent species. Human infection is incidental to viral maintenance and is almost always a dead end in the infection chain, with the exception of a recently reported human-to-human transmission of Andes virus. In contrast to other viruses in the family, hantaviruses are not transmitted by arthropods, and both rodent and human infections are acquired by aerosol exposure to infectious virus in rodent urine, feces or saliva, and less frequently by rodent bite. Hantaviruses cause no detectable cytopathology in vertebrate cell cultures and cause persistent, non-pathogenic infections of rodents.

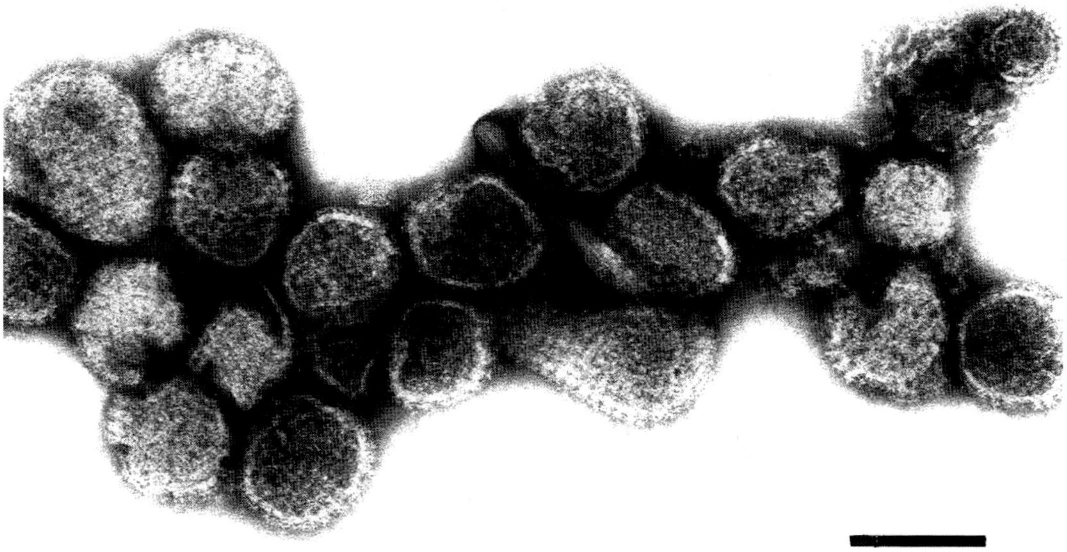

Figure 5: Electron micrograph of negatively stained particles of an isolate of *Hantaan virus*. The bar represents 100 nm. (Courtesy of C.S. Schmaljohn).

List of Species Demarcation Criteria in the Genus

Species are found in unique ecological niches, i.e. in different primary rodent reservoir species or subspecies. Species exhibit at least 7% difference in aa identity on comparison of the complete glycoprotein precursor and nucleocapsid protein sequences. Species show at least 4-fold difference in two-way cross neutralization tests, and species do not naturally form reassortants with other species. However, since reassortants have been generated in the laboratory between some of the listed viruses, the issue of whether these are distinct virus species is uncertain.

List of Species in the Genus

Species names are in green italic script; strain names and synonyms are in black roman script; tentative species names are in blue roman script. Vector type, genome sequence accession numbers [L, M, S, designating RNA segments], and assigned abbreviations ( ) are also listed.

Species in the Genus

Andes virus			
Andes virus	*Oligoryzomys longicaudatus*		(ANDV)
Bermejo virus	*Oligoryzomys chacoensis*		(BMJV)
Lechiguanas virus	*Oligoryzomys flavescens*	[M: AF028022]	(LECV)
Maciel virus	*Bolomys obscurus*		(MCLV)
Oran virus	*Oligoryzomys longicaudatus* (?)	[M: AF028024]	(ORNV)
Pergamino virus	*Akadon azarae*		(PRGV)
Bayou virus			
Bayou virus	*Oryzomys palustris*	[M: L36930; S: L36929]	(BAYV)
Black Creek Canal virus			
Black Creek Canal virus	*Sigmodon hispidus*	[M: L39950]	(BCCV)
Cano Delgadito virus			
Cano Delgadito virus	*Sigmodon alstoni*		(CADV)
Dobrava-Belgrade virus			
Dobrava-Belgrade virus	*Apodemus flavicollis*	[L: AJ410619; M: L33685; S: L41916]	(DOBV)
Saaremaa virus	*Apodemus agrarius*	[L: AJ410618; M: AJ009774; S:	(SAAV)

		AJ009773]	
El Moro Canyon virus			
El Moro Canyon virus - RM-97	*Reithrodontomys megalotis*	[M: U26828; S: U11427]	(ELMCV)
Hantaan virus			
Amur virus	*Apodemus peninsulae*		(AMRV)
Da Bie Shan virus	*Niviventer confucianus*		(DBSV)
Hantaan virus - 76-118	*Apodemus agrarius*	[L: X55901; M: M14627; S: M14626]	(HTNV)
Hantaan virus - HV114		[M: L08753]	(HTNV)
Isla Vista virus			
Isla Vista virus	*Microtus californicus*	[S: U19302]	(ISLAV)
Khabarovsk virus			
Khabarovsk virus	*Microtus fortis*	[S: U35255]	(KHAV)
Laguna Negra virus			
Laguna Negra virus	*Calomys laucha*	[M: AF005728; S: AF005727]	(LANV)
Muleshoe virus			
Muleshoe virus	*Sigmodon hispidus*		(MULV)
New York virus			
New York virus - RI-1	*Peromyscus leucopus*	[M: U36801; S: U09488]	(NYV)
Prospect Hill virus			
Bloodland Lake virus	*Microtus ochrogaster*	[S: U19303]	(BLLV)
Prospect Hill virus	*Microtus pennsylvanicus*	[M: X55129; S: X55128]	(PHV)
Puumala virus			
Hokkaido virus - Kamiiso-8Cr-95	*Clethrionomys rufocanus*	[S: AB010730]	(HOKV)
Muju virus	*Eothenomys regulus*		(MUJV)
Puumala virus - Sotkamo	*Clethrionomys glareolus*	[L: Z66548; M: X61034; S: X61035]	(PUUV)
Puumala virus - Urmurtia/338Cg/92		[S: Z30708]	(PUUV)
Rio Mamore virus			
Rio Mamore virus	*Oligoryzomys microtis*		(RIOMV)
Rio Segundo virus			
Rio Segundo virus	*Reithrodontomys mexicanus*	[S: U18100]	(RIOS)
Seoul virus			
Seoul virus - HR80-39	*Rattus norvegicus, Rattus rattus*	[L: X56492; M: S47716]	(SEOV)
Seoul virus - L99	*Rattus losea*	[M: AF035833]	(SEOV)
Seoul virus - SR-11 virus		[M: M34482; S: M34881]	(SEOV)
Sin Nombre virus			
Blue River virus - Indiana	*Peromyscus leucopus*	[M: AF030551]	(BRV)
Blue River virus - Oklahoma	*Peromyscus leucopus*	[M: AF030552]	(BRV)
Monongahela virus	*Peromyscus maniculatus*	[S: U32591]	(MGLV)
Sin Nombre virus - Convict Creek 107		[L: L35008; M: L33474; S: L33683]	(SNV)

Sin Nombre virus - NMH10 virus	*Peromyscus maniculatus*	[L: L37901; M: L25783; S: L25784]	(SNV)
Thailand virus			
Thailand virus	*Bandicota indica*	[M: L08756]	(THAIV)
Thottapalayam virus			
Thottapalayam virus	*Suncus murinus*		(TPMV)
Topografov virus			
Topografov virus	*Lemmus sibiricus*		(TOPV)
Tula virus			
Tula virus -Tula/Ma76/87	*Micotus arvalis, M. rossiaemeridionalis*	[S: Z30941]	(TULV)
Tula virus - Moravia/Ma5302V	*Micotus arvalis, M. rossiaemeridionalis*	[L: AJ005637; M: Z69993; S: Z69991]	(TULV)

TENTATIVE SPECIES IN THE GENUS

None reported.

GENUS NAIROVIRUS

Type Species *Dugbe virus*

DISTINGUISHING FEATURES

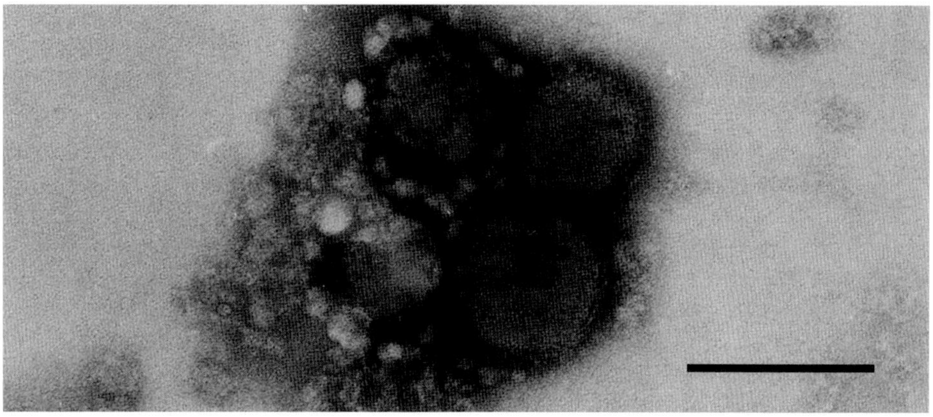

Figure 6. Electron micrograph of negatively stained particles of Crimean-Congo hemorrhagic fever virus (CCHFV). The bar represents 100 nm. (Courtesy of C.S. Schmaljohn)

Virions are morphologically similar to other members of the family with very small surface units which appear as a peripheral fringe 7 nm in length (Fig. 6). The L RNA segment (12.2 kb) is considerably larger than the L segments of other members of the family. The consensus terminal nt sequences of the L, M, and S segments are AGAGUUUCU... at the 3′-end and UCUCAAAGA... at the 5′-end. The S segment does not encode a nonstructural protein. The M segment encodes a single gene product which is processed in a complex and poorly defined manner to yield the structural glycoproteins; at least 3 non-structural proteins have been observed, 2 of which are precursors of the glycoproteins. The L protein is predicted to be much larger than those of other members of the family but has yet to be identified. Viruses are serologically unrelated to members of other genera. Most viruses are transmitted by ticks: members of the CCHF, NSD, and SAK groups mainly by ixodid ticks and DGK, HUG and QYB groups mainly by argasid ticks. Some viruses are transmitted transovarially in arthropods.

List of Species Demarcation Criteria in the Genus

The paucity of biochemical data dictates that nairovirus species are defined by serological reactivities. There are 7 species recognized in the genus *Nairovirus* containing 34 virus strains.

List of Species in the Genus

Species names are in green italic script; strain names and synonyms are in black roman script; tentative species names are in blue roman script. Under each virus species name is listed a number of other viruses of unknown exact taxonomic status. They differ from the listed virus species antigenically as well as in terms of host range vector species, geographical origins and pathogenic properties. Vector type, genome sequence accession numbers [L, M, S, designating RNA segments], and assigned abbreviations ( ) are also listed.

Species in the Genus

Crimean-Congo hemorrhagic fever virus			
Crimean-Congo hemorrhagic fever virus - AP92	culicoid flies	[S:U04958]	(CCHFV)
Crimean-Congo hemorrhagic fever virus - C68031	ticks,	[S:M86625]	(CCHFV)
Hazara virus	ticks	[S: M86624]	(HAZV)
Khasan virus	ticks		(KHAV)
Dera Ghazi Khan virus			
Abu Hammad virus	ticks		(AHV)
Abu Mina virus	N.D.		(ABMV)
Dera Ghazi Khan virus - JD254	ticks		(DGKV)
Kao Shuan virus	ticks		(KSV)
Pathum Thani virus	ticks		(PTHV)
Pretoria virus	ticks		(PREV)
Dugbe virus			
Dugbe virus	ticks	[S: M25150; M: M94133; L: U15018]	(DUGV)
Nairobi sheep disease virus (Ganjam virus)	ticks, culicoid flies, mosquitoes		(NSDV)
Hughes virus			
Farallon virus	ticks		(FARV)
Fraser Point virus	ticks		(FPV)
Great Saltee virus	ticks		(GRSV)
Hughes virus	ticks		(HUGV)
Puffin Island virus	ticks		(PIV)
Punta Salinas virus	ticks		(PSV)
Raza virus	ticks		(RAZAV)
Sapphire II virus	ticks		(SAPV)
Soldado virus	ticks		(SOLV)
Zirqa virus	ticks		(ZIRV)
Qalyub virus			
Bakel virus	ticks		(BAKV)
Bandia virus	ticks		(BDAV)
Omo virus	N.D.		(OMOV)
Qalyub virus	ticks		(QYBV)
Sakhalin virus			
Avalon virus (Paramushir virus)	ticks		(AVAV)
Clo Mor virus	ticks		(CMV)
Kachemak Bay virus	ticks		(KBV)
Sakhalin virus	ticks		(SAKV)
Taggert virus	ticks		(TAGV)

Tillamook virus	ticks	(TILLV)
Thiafora virus		
Erve virus	N.D.	(ERVEV)
Thiafora virus	N.D.	(TFAV)

TENTATIVE SPECIES IN THE GENUS

None reported.

GENUS PHLEBOVIRUS

Type Species *Rift Valley fever virus*

DISTINGUISHING FEATURES

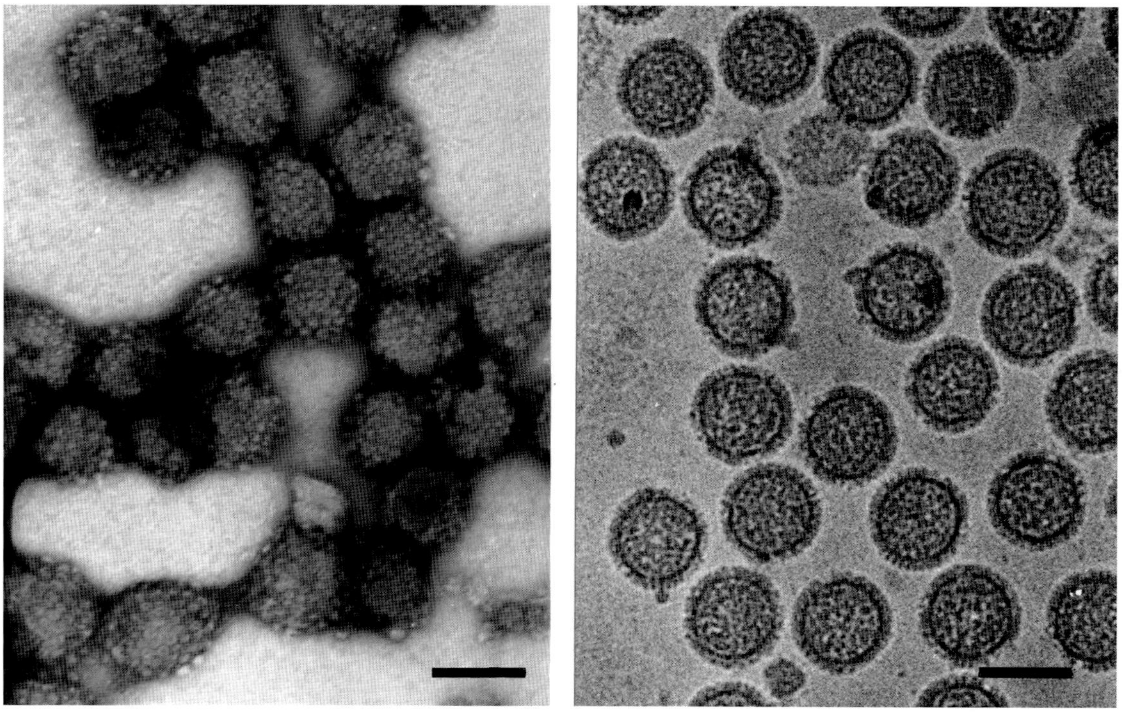

Figure 7. (Left). Electron micrograph of negatively stained particles of Uukuniemi virus (UUKV). The bar represents 100 nm. (Right) Cryo-electron micrograph of purified UUKV particles (Both courtesy of C-H. von Bornsdorff).

The surface morphology of phleboviruses is distinct in having small round subunits with a central hole (Fig. 7). The consensus terminal nucleotide sequences of the L, M and S segments are UGUGUUUC... at the 3′-end and ACACAAAG... at the 5′-end. The S RNA exhibits an ambisense coding strategy, i.e. it is transcribed by the virion RNA polymerase to a subgenomic virus-complementary sense mRNA that encodes the N protein and, from a full-length antigenome S RNA, to a subgenomic virus-sense mRNA that encodes a nonstructural (NSs) protein. The M segment of viruses in the Sandfly fever group but not viruses in the Uukuniemi group have a preglycoprotein coding region that codes for a nonstructural protein(s) (NSm). The Gn and Gc glycoproteins were earlier referred to as G1 and G2 based on apparent size on gel electrophoresis. However, the similar sizes of the G1 and G2 proteins resulted in the different G1:G2 order in the M segments of different viruses. The further adoption of the Gn/Gc nomenclature is strongly encouraged so as to achieve more consistency across the *Bunyaviridae* family. Phleboviruses are antigenically unrelated to members of other genera, but cross-react serologically among themselves to different degrees. Sandfly fever group viruses are transmitted by phlebotomines, mosquitoes or ceratopogonids of the genus *Culicoides*; Uukuniemi group viruses are transmitted by ticks.

List of Species Demarcation Criteria in the Genus

The lack of biochemical data for most phleboviruses dictates that the species are defined by the serological relationships, and are distinguishable by 4-fold differences in 2-way neutralization tests.

List of Species in the Genus

Species names are in green italic script; strain names and synonyms are in black roman script; tentative species names are in blue roman script. Under each virus species name is listed a number of other viruses of unknown exact taxonomic status. They differ from the listed virus species antigenically as well as in terms of host range vector species, geographical origins and pathogenic properties. Vector type, genome sequence accession numbers [L, M, S, designating RNA segments], and assigned abbreviations ( ) are also listed.

Species in the Genus

Bujaru virus			
Bujaru virus - BeAn 47693	N.D.		(BUJV)
Munguba virus	phlebotomines		(MUNV)
Chandiru virus			
Alenquer virus	N.D.		(ALEV)
Chandiru virus - BeH 2251	N.D.		(CDUV)
Itaituba virus	N.D.		(ITAV)
Nique virus	phlebotomines		(NIQV)
Oriximina virus	phlebotomines		(ORXV)
Turuna virus	phlebotomines		(TUAV)
Chilibre virus			
Cacao virus	phlebotomines		(CACV)
Chilibre virus VP-118D	phlebotomines		(CHIV)
Frijoles virus			
Frijoles virus VP-161A	phlebotomines		(FRIV)
Joa virus	N.D.		(JOAV)
Punta Toro virus			
Buenaventura virus	phlebotomines		(BUEV)
Punta Toro virus - D-4021A	phlebotomines	[M: M11156; S: K02736]	(PTV)
Rift Valley fever virus			
Belterra virus	N.D.		(BELTV)
Icoaraci virus	phlebotomines, mosquitoes		(ICOV)
Rift Valley fever virus	mosquitoes	[L: X56464; M: M11157; S: X53771]	(RVFV)
Salehebad virus			
Arbia virus	phlebotomines		(ARBV)
Salehebad virus - I-81	phlebotomines		(SALV)
Sandfly fever Naples virus			
Karimabad virus	phlebotomines		(KARV)
Sandfly fever Naples virus - Sabin	phlebotomines		(SFNV)
Tehran virus	phlebotomines		(THEV)
Toscana virus	phlebotomines	[L: X68414; M: X89628; S: X53794]	(TOSV)
Uukuniemi virus			
EgAN 1825-61 virus	N.D.		(EGAV)
Fin V 707 virus	N.D.		(FINV)
Grand Arbaud virus	ticks		(GAV)

Manawa virus	ticks		(MWAV)
Murre virus	N.D.		(MURV)
Oceanside virus	ticks		(OCV)
Ponteves virus	ticks		(PTVV)
Precarious Point virus	ticks		(PPV)
RML 105355 virus	ticks		(RMLV)
St. Abbs Head virus	ticks		(SAHV)
Tunis virus	N.D.		(TUNV)
Uukuniemi virus - S 23	ticks	[L: D10759; M: M17417; S: M33551]	(UUKV)
Zaliv Terpeniya virus	ticks		(ZTV)

TENTATIVE SPECIES IN THE GENUS

Aguacate virus	phlebotomines		(AGUV)
Anhanga virus	N.D.		(ANHV)
Arboledas virus	phlebotomines		(ADSV)
Arumowot virus	mosquitoes		(AMTV)
Caimito virus	phlebotomines		(CAIV)
Chagres virus	phlebotomines, mosquitoes		(CHGV)
Corfou virus	phlebotomines		(CFUV)
Gabek Forest virus	N.D.		(GFV)
Gordil virus	N.D.		(GORV)
Itaporanga virus	mosquitoes		(ITPV)
Odrenisrou virus	mosquitoes		(ODRV)
Pacui virus	phlebotomines		(PACV)
Rio Grande virus	N.D.		(RGV)
Sandfly fever Sicilian virus	phlebotomines	[S: J04418]	(SFSV)
Saint-Floris virus	N.D.		(SAFV)
Urucuri virus	N.D.		(URUV)

GENUS TOSPOVIRUS

Type Species *Tomato spotted wilt virus*

DISTINGUISHING FEATURES

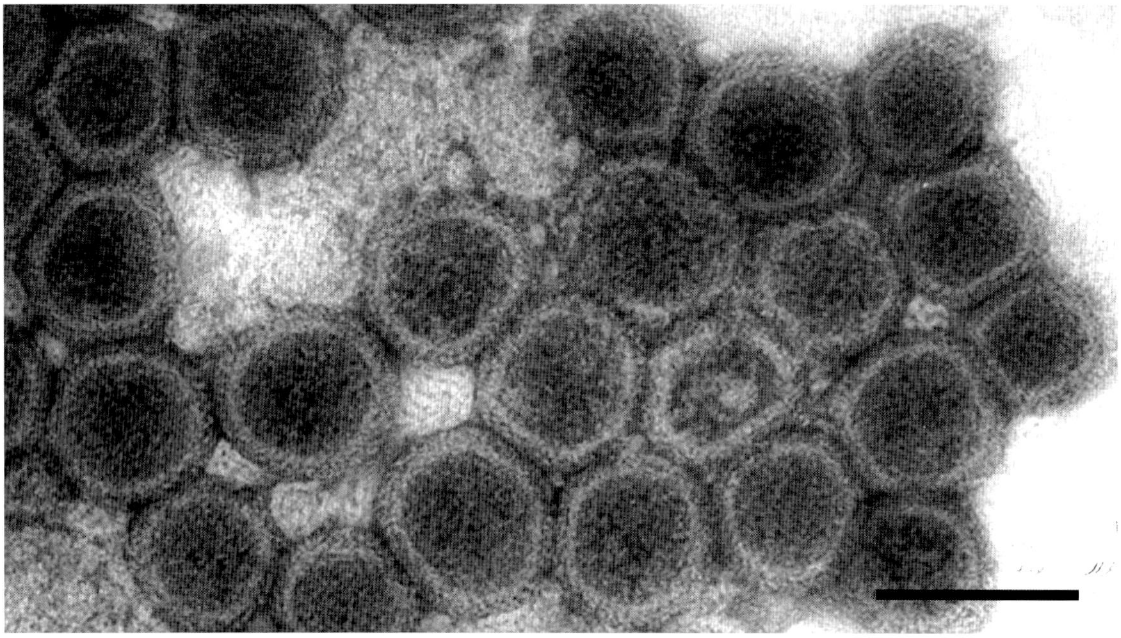

Figure 8. Electron micrograph of negatively stained particles of Tomato spotted wilt virus (TSWV). The bar represents 100 nm. (Courtesy of J. van Lent).

Morphogenesis occurs in clusters in the cisternae of the endoplasmic reticulum of host cells. Nucleocapsid material may accumulate in the cytoplasm in dense masses; these masses may be composed of defective particles. The morphology of a tospovirus is shown in Figure 8. The consensus terminal sequences of the L, M and S genomic segments are UCUCGUUA... at the 3′-end and AGAGCAAU... at the 5′-end. Both the M and S segment RNAs of tospoviruses utilize an ambisense coding strategy. The virion glycoproteins Gn and Gc are encoded in the complementary-sense RNA of the M segment, and a nonstructural protein, NSm, is encoded in the genome-sense RNA. The S segment encodes the nucleocapsid protein in the complementary-sense RNA and a nonstructural protein, NSs, in the genome-sense RNA. The NSm protein plays a role in cell-to-cell movement of the virus during systemic infection of plants; NSs may form paracrystalline or filamentous inclusions in infected plant cells or when expressed in insect cells from recombinant baculovirus.

At least 8 species of thrips in the genera Frankliniella (5) and Thrips (3) have been reported to transmit tospoviruses, and the Gn and/or Gc glycoproteins are involved in virus-vector interactions. Transmission can also be achieved through infected plant sap, and within an infected plant tospoviruses are transported mainly as free nucleocapsids. For isolates of the type species *Tomato spotted wilt virus*, more than 925 plant species belonging to 70 botanical families are known to be susceptible whereas the other tospoviruses have much narrower host ranges.

LIST OF SPECIES DEMARCATION CRITERIA IN THE GENUS

Species are defined on the basis of their vector specificity, their plant host range, serological relationships of the N protein and on the criterion that their N protein sequence should show less than 90% aa identity with that of any other described tospovirus species.

List of Species in the Genus

Species names are in green italic script; strain names and synonyms are in black roman script; tentative species names are in blue roman script. Vector type, genome sequence accession numbers [L, M, S, designating RNA segments], and assigned abbreviations ( ) are also listed.

Species in the Genus

Groundnut bud necrosis virus			
Groundnut bud necrosis virus (Peanut bud necrosis virus)	*T. palmi*	[L: AF025538; M: U42555; S: U27809]	(GBNV)
Groundnut ringspot virus			
Groundnut ringspot virus	*F. occidentalis, F. schultzei*	[S(N): S54327]	(GRSV)
Groundnut yellow spot virus			
Groundnut yellow spot virus (Peanut yellow spot virus)	N.D.	[S: AF013994]	(GYSV)
Impatiens necrotic spot virus			
Impatiens necrotic spot virus	*F. occidentalis*	[L: X93218; M: M74904; S: S40057]	(INSV)
Tomato chlorotic spot virus			
Tomato chlorotic spot virus	*F. occidentalis, F. schultzei, F. intonsa*	[S(N): S54325]	(TCSV)
Tomato spotted wilt virus			
Tomato spotted wilt virus	*F. occidentalis, F. schultzei, F. intonsa, F. fusca, F. bispinosa, T. tabaci, T. setosus, T. palmi*	[L: D10066; M: S48091; S: D00645; S: (B) L12048; S: (BL) L20953; S: (L3) D13926]	(TSWV)
Watermelon silver mottle virus			
Watermelon silver mottle virus	*T. palmi*	[M: U75379; S: Z46419]	(WSMoV)
Zucchini lethal chlorosis virus			
Zucchini lethal chlorosis virus	N.D.	[S(N): AF067069]	(ZLCV)

Tentative Species in the Genus

Capsicum chlorosis virus (Gloxinia tospovirus) (Thailand tomato tospovirus)	N.D.	[M: AF023172 S: AY036058 (N), AF059578 (N), AF059577 (NSs) S (N): AF134400]	(CACV)
Chrysanthemum stem necrosis virus	*F. schultzei*	[S(N): AF067068]	(CSNV)
Iris yellow spot virus	N.D.	[S: AF001387; M: AF214014]]	(IYSV)
Groundnut chlorotic fan-spot virus	N.D.		(GCFSV)
Physalis severe mottle virus	N.D.	[S: AF067151]	(PhySMV)
Watermelon bud necrosis virus	N.D.	[S(N): AF045067]	(WBNV)

Unassigned Species in the Family

There are 7 groups (19 viruses) and 21 ungrouped viruses which have not been assigned to a recognized genera in the family *Bunyaviridae*. For most, no biochemical characterization of the viruses has been reported to determine their taxonomic status.

The groups are:

Bhanja virus	(BHAV)
Forecariah virus	(FORV)
Kismayo virus	(KISV)
Kaisodi virus	(KSOV)
Lanjan virus	(LJNV)
Silverwater virus	(SILV)
Mapputta virus	(MAPV)
Gan Gan virus	(GGV)
Maprik virus	(MPKV)
Trubanaman virus	(TRUV)
Okola virus	(OKOV)
Tanga virus	(TANV)
Resistencia virus	(RTAV)
Antequera virus	(ANTV)
Barranqueras virus	(BQSV)
Upolu virus	(UPOV)
Aransas Bay virus	(ABV)
Yogue virus	(YOGV)
Kasokero virus	(KASV)

Ungrouped viruses:

Bangui virus	(BGIV)
Belem virus	(BLMV)
Belmont virus	(BELV)
Bobaya virus	(BOBV)
Caddo Canyon virus	(CDCV)
Chim virus	(CHIMV)
Enseada virus	(ENSV)
Issyk-Kul virus	(ISKV)
Keterah virus	(KTRV)
Kowanyama virus	(KOWV)
Lone Star virus	(LSV)
Pacora virus	(PCAV)
Razdan virus	(RAZV)
Salanga virus	(SGAV)
Santarem virus	(STMV)
Sunday Canyon virus	(SCAV)
Tai virus	(TAIV)
Tamdy virus	(TDYV)
Tataguine virus	(TATV)
Wanowrie virus	(WANV)
Witwatersrand virus	(WITV)
Yacaaba virus	(YACV)

Phylogenetic Relationships within the Family

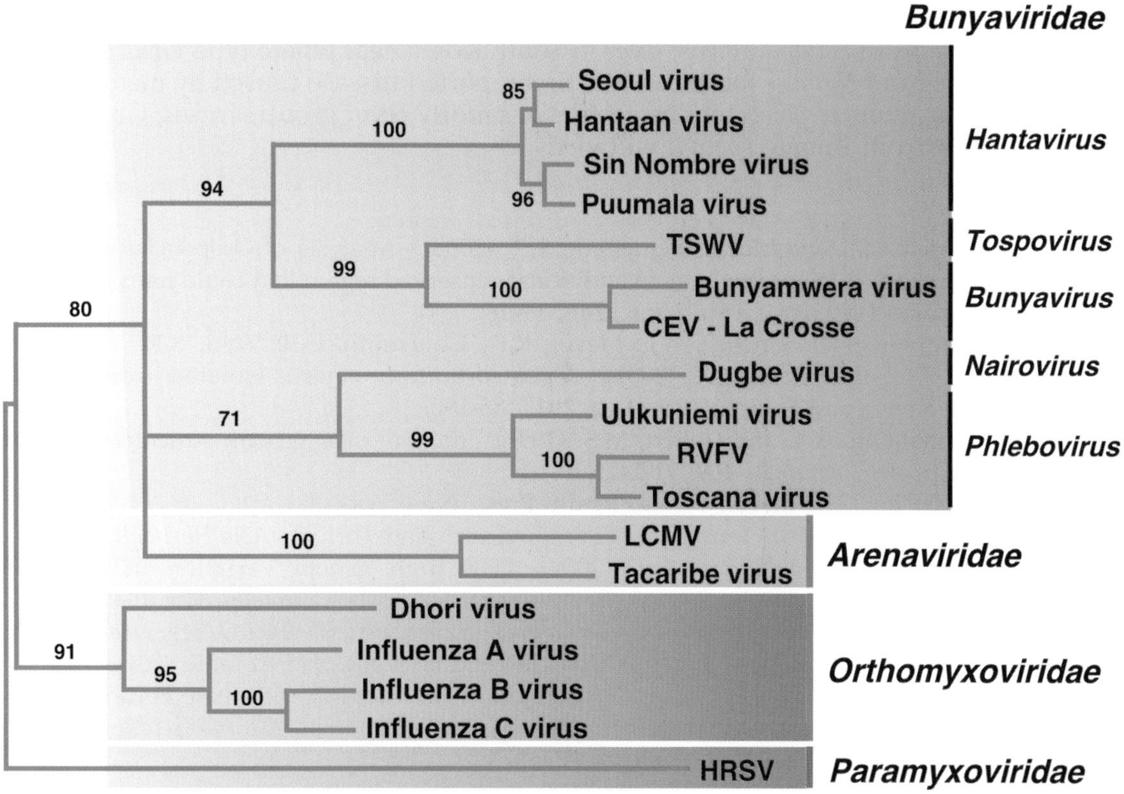

Figure 9: Phylogenetic tree of aligned polymerase domains from the L proteins of members of the family *Bunyaviridae* and from analogous proteins of other segmented (*Arenaviridae, Orthomyxoviridae*) and non-segmented (*Paramyxoviridae*) negative-strand RNA viruses. The tree was constructed by the neighbor-joining method. Figures on the branches represent percentages of trees containing each cluster out of 500 bootstrap replicates. Branch lengths are proportional to the genetic distance. The polymerase domain from Human respiratory syncytial virus (HRSV) is used as an out-group. (Redrawn from Marriott and Nuttall, 1996).

As documented above the analogous genes and gene products of viruses in the different genera vary widely in size, and there is little obvious global similarity at either nt or aa level. Attempts to produce convincing alignments of either genome segments or structural proteins from which to generate phylogenetic trees have so far proved unsuccessful, with the exception of the putative polymerase domain of the L proteins. Such analysis suggests that viruses in the family *Bunyaviridae* fall into two major lineages, comprising bunyaviruses, hantaviruses and tospoviruses on one, and nairoviruses and phleboviruses on the other. The significant point to note is that L protein phylogeny does not segregate with the use of an ambisense coding strategy.

Similarity of this Family with Other Taxa

The plant-infecting tenuiviruses show some similarities to members of the family *Bunyaviridae*, particularly the genus *Phlebovirus*. Tenuiviruses have a ssRNA genome comprising 4 or 5 segments which encode proteins using a negative or ambisense coding strategy. The tenuivirus RNA terminal sequences are conserved and the 3′ and 5′-sequences exhibit inverted complementarity; the conserved 3′-sequence, UGUGUUUCAG..., is similar to the consensus phlebovirus sequence. Tenuiviruses employ a cap-snatching mechanism to prime viral mRNA synthesis, similar to members of the family *Bunyaviridae*. Weak sequence homology has been noted between the Rice stripe virus 94 kDa protein and phlebovirus glycoproteins, and between tenuivirus nucleocapsid proteins and those of phleboviruses.

Derivation of Names

Bunya: from *Bunya*mwera; place in Uganda, where type virus was isolated.
Hanta: from *Hanta*an virus; river in South Korea near where type virus was isolated.
Nairo: from *Nairo*bi sheep disease; first reported disease caused by member virus.
Phlebo: refers to *phlebo*tomine vectors of sandfly fever group viruses; Greek *phlebos*, "vein".
Tospo: from *To*mato *spo*tted wilt virus.

References:

Aquino, V.H., Moreli, M.L. and Moraes Figueiredo, L.T. (2003). Analysis of Oropouche virus L protein amino acid sequence showed the presence of an additional conserved region that could harbour an important role for the polymerase activity. *Arch. Virol.*, **148**, 19-28.

Bowen, M.D., Trappier, S.G., Sanchez, A.J., Meyer, R.F., Goldsmith, C.S., Zaki, S.R., Dunster, L.M., Peters, C.J., Ksiazek, T.G. and Nichol, S.T. (2001). A reassortant bunyavirus isolated from acute hemorrhagic fever cases in Kenya and Somalia. *Virology*, **291**, 185-190.

Elliott, R.M., Schmaljohn, C.S. and Collett, M.S. (1991). *Bunyaviridae* genome structure and gene expression. *Curr. Top. Microbiol. Immunol.*, **169**, 91-142.

Elliott, R.M. (ed) (1996). The *Bunyaviridae*. Plenum Press, New York.

Goldbach, R. and Kuo, G. (1996). Introduction (Tospoviruses and Thrips). *Acta Horticult.*, **431**, 21-26.

Honig, J.E., Osborne, J.C. and Nichol, S.T. (2004). The high genetic variation of viruses of the genus *Nairovirus* reflects the diversity of their predominant tick hosts. *Virology*, **318**, 10-16.

Karabatsos, N. (ed) (1985). *International catalogue of arboviruses including certain other viruses of vertebrates.* American Society of Tropical Medicine and Hygiene, San Antonio, Texas.

Liu, D.Y., Tesh, R.B., Travassos Da Rosa, A.P., Peters, C.J., Yang, Z., Guzman, H. and Xiao, S.Y. (2003). Phylogenetic relationships among members of the genus *Phlebovirus* (*Bunyaviridae*) based on partial M segment sequence analyses. *J. Gen. Virol.*, 84, 465-473.

Marriott, A.C. and Nuttall, P.A. (1996). Large RNA segment of Dugbe nairovirus encodes the putative RNA polymerase. *J. Gen Virol.*, **77**, 1775-1780.

Osborne, J.C., Rupprecht, C. E., Olson, J. G., Ksiazek, T.G., Rollin, P.E., Niezgoda, M., Goldsmith, C.S., An, U.S. and Nichol, S.T. (2003). Isolation of Kaeng Khoi virus from dead *Chaerephon plicata* bats in Cambodia. *J. Gen. Virol.*, **84**, 2685-2689.

Nichol, S.T. (1999). Genetic analysis of hantaviruses and their host relationships. In: *Emergence and control of rodent-borne viral diseases*, (J.F. Saluzzo and B. Dodet, eds), pp. 99-109. Elsevier, Paris.

Nichol, S.T. (2001). *Bunyaviridae*. In: *Fields Virology*, 4th Edn, (D.M. Knipe and P. Howley, eds), pp 1603-1633. Lippincott, Williams and Wilkins, Philadelphia.

Plyusnin, A., Vapalahti, O. and Vaheri, A. (1996). Hantaviruses: genome structure, expression and evolution. *J. Gen. Virol.*, **77**, 2677-2687.

Plyusnin A. (2002). Genetics of hantaviruses: implications to taxonomy. *Arch. Virol.*, **147**, 665-682.

Ramirez, B.C. and Haenni, A.L. (1994). Molecular biology of tenuiviruses, a remarkable group of plant viruses. *J. Gen. Virol.*, **75**, 467-475.

Saeed, M.F., Li, L., Wang, H., Weaver, S.C. and Barrett, A.D. (2001). Phylogeny of the Simbu serogroup of the genus *Bunyavirus*. *J. Gen. Virol.*, **82**, 2173-2181.

Saeed, M.F., Wang, H., Suderman, M., Beasley, D.W., Travassos da Rosa, A., Li, L., Shope, R.E., Tesh, R.B. and Barrett, A.D. (2001). Jatobal virus is a reassortant containing the small RNA of Oropouche virus. *Virus Res.*, **77**, 25-30.

Schmaljohn C.S. and Hooper, J.W. (2001). *Bunyaviridae*: The viruses and their replication. In: *Fields Virology*, 4th Edn, (D.M. Knipe and P. Howley, eds), pp 1581-1602. Lippincott, Williams and Wilkins, Philadelphia.

Schmaljohn, C.S. and Nichol, S.T. (eds) (2001). Hantaviruses. *Curr. Top. Microbiol. Immunol.*, volume **256**.

Contributed By:

Nichol, S.T., Beaty, B.J., Elliott, R.M., Goldbach, R., Plyusnin, A., Schmaljohn, C.S. and Tesh, R.B.

GENUS *TENUIVIRUS*

Type Species *Rice stripe virus*

VIRION PROPERTIES

MORPHOLOGY

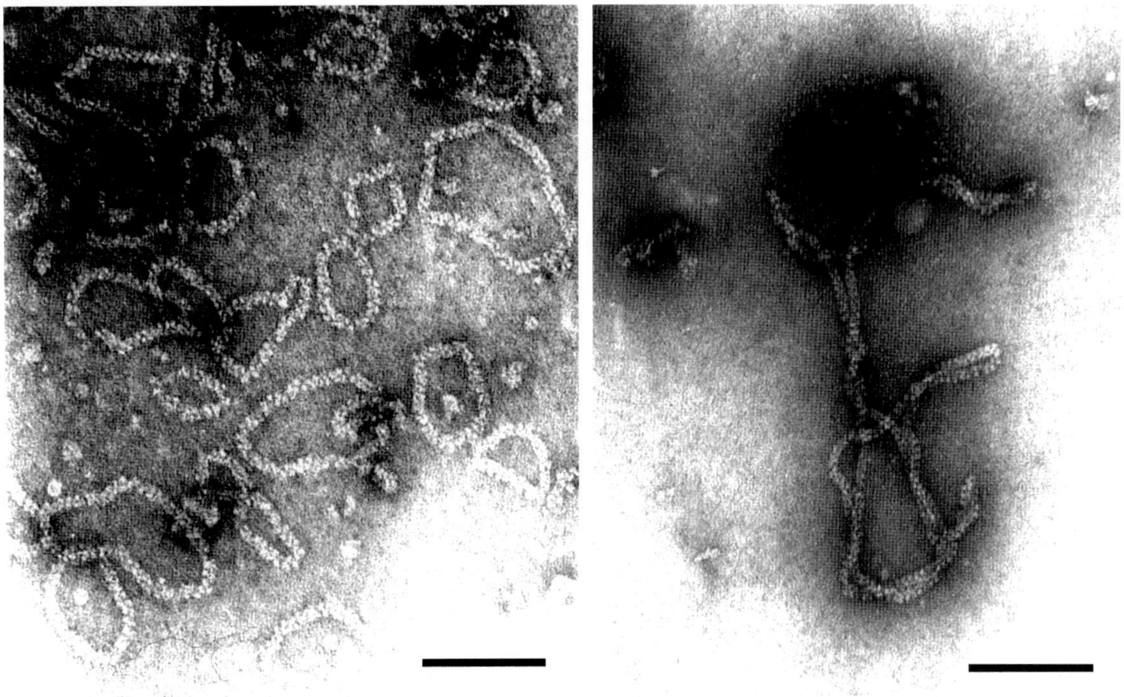

Figure 1: Electron micrographs of sucrose density gradient purified RNPs of Rice hoja blanca virus (RHBV). (Left) Small circular RNPs from the slowest sedimenting RHBV RNP. (Right) Larger, circular RNPs from the fastest sedimenting RHBV RNP. The bar represents 100 nm. (Courtesy of A.M. Espinoza).

The RNPs have a thin filamentous shape; they consist of nucleocapsids, 3-10 nm in diameter, with lengths proportional to the sizes of the RNAs they contain. The filamentous particles may appear to be spiral-shaped, branched or circular (Fig. 1). No envelope has been observed.

PHYSICOCHEMICAL AND PHYSICAL PROPERTIES

RNP preparations can be separated into four or five components by sucrose density gradient centrifugation, but form one component with a buoyant density 1.282-1.288 g/cm^3 when centrifuged to equilibrium in CsCl solutions.

NUCLEIC ACID

The ssRNA genome consists of four or more segments. The sizes are ~9 kb (RNA-1, generally of negative polarity), 3.3-3.6 kb (RNA-2, ambisense), 2.2-2.5 kb (RNA-3, ambisense), and 1.9-2.2 kb (RNA-4, ambisense). Virion preparations of Maize stripe virus (MSpV) and Echinochloa hoja blanca virus (EHBV) contain a fifth RNA of negative polarity and with a size of 1.3 kb. A fifth RNA segment has also been reported for some isolates of *Rice stripe virus*. Rice grassy stunt virus (RGSV) preparations contain six segments, all of which are ambisense. RGSV RNAs-1, -2, -5 and -6 are homologous to RNA-1, -2, -3 and –4 respectively of other tenuiviruses, whereas RNA-3 (3.1 kb) and RNA-4 (2.9 kb) are unique to RGSV.

PROTEINS
The nucleocapsid proteins are of 34-35 kDa. Small amounts of a minor 230 kDa protein co-purify with RNPs of Rice stripe virus (RSV), RHBV and RGSV. This protein may be an RdRp polymerase, as this activity is associated with filamentous nucleoprotein particles.

LIPIDS
None reported.

CARBOHYDRATES
None reported.

GENOME ORGANIZATION AND REPLICATION

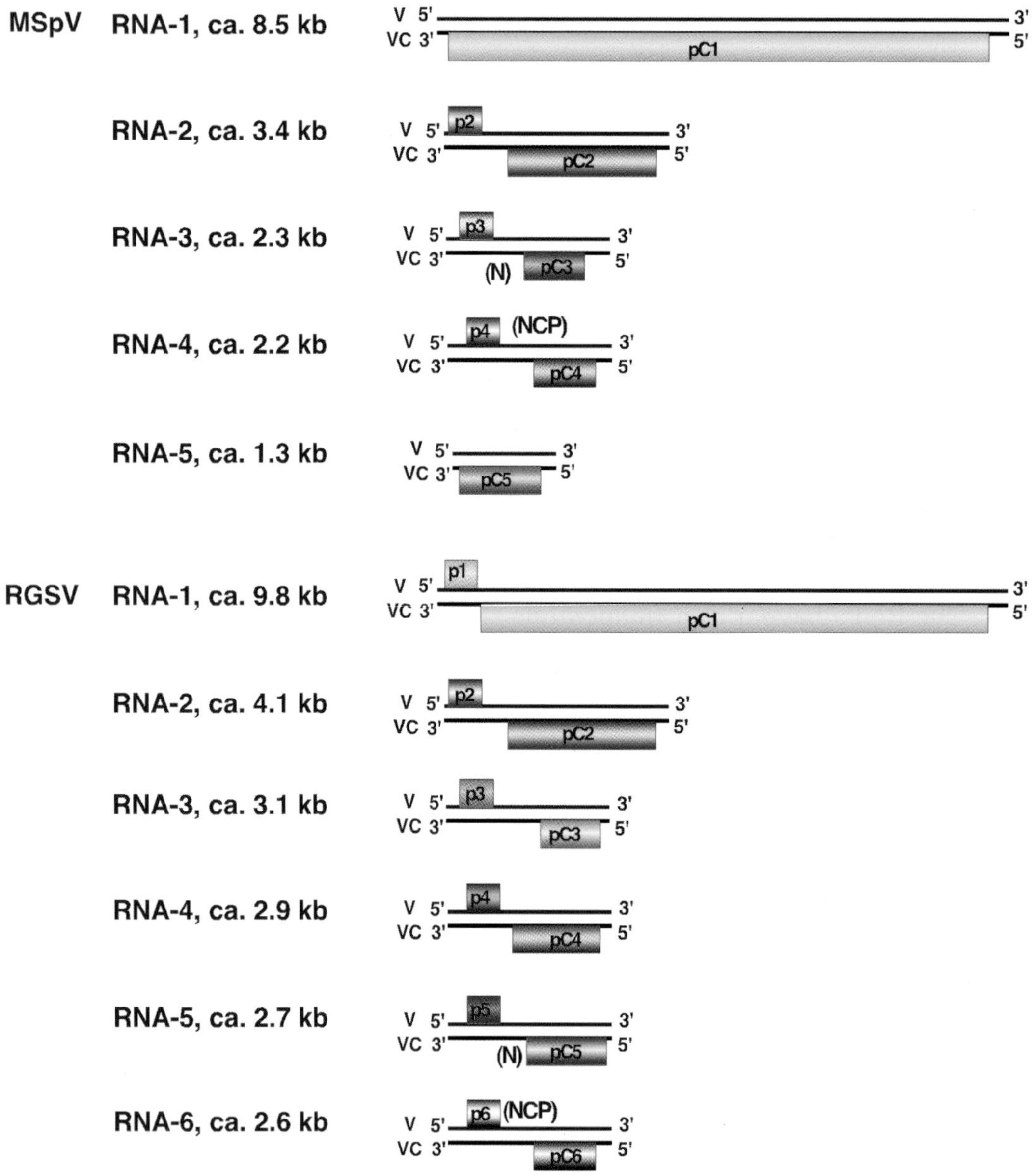

Figure 2: Genome organization characteristic of (Top) Maize stripe virus (MSpV), and (Bottom) Rice grassy stunt virus (RGSV). Boxes indicate the positions and designations of the ORF translation products. V signifies virion-sense RNA and VC signifies virion-complementary sense RNA.

The 3'- and 5'-terminal sequences of each ssRNA are almost complementary for about 20 bases. Several RNA segments encode two proteins in an ambisense arrangement (Fig. 2). In most tenuiviruses, the nucleocapsid protein (pC3; N) is encoded by the 5′-proximal region of the virion-complementary sense strand of RNA-3. Virion-sense RNA-4 encodes in its 5′-proximal region a major non-structural protein (p4; NCP) that accumulates in infected plants. Some of the intergenic NCR between the ORFs can adopt hairpin structures. Some segments (e.g. RNA-1 of RSV and RNA-5 of MSpV) are of negative polarity. RNA-1 encodes the RdRp (pC1; RdRp). Some proteins are translated from sgRNAs (Fig. 3). For MSpV, RHBV and RSV mRNA, the production of mRNAs involves a cap-snatching mechanism. An RNA polymerase has been found associated with purified preparations of RSV, RHBV and RGSV. The RNA polymerase activity of RHBV is capable of replicating and transcribing the RNA segments *in vitro*. In RHBV, p3 is a suppressor of RNA silencing.

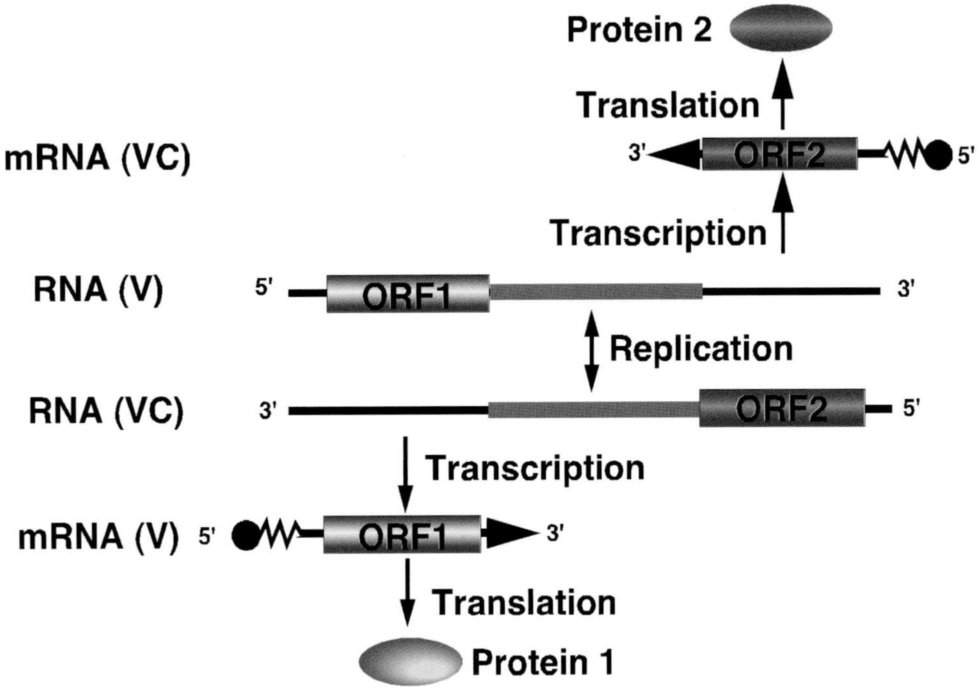

Figure 3: Diagram of the expression of the ambisense RNA of tenuiviruses. The black circle signifies the Cap, the broken line signifies non-viral nucleotides, V signifies virion-sense RNA and VC signifies virion-complementary sense RNA.

Antigenic Properties

The N proteins of RSV and MSpV are serologically related, and both the N and NCP proteins of RSV and RGSV are related. Likewise, the N proteins of RHBV, EHBV and Urochloa hoja blanca virus (UHBV) are serologically related. The RSV N protein reacts weakly with antibodies made to virion preparations of RGSV or RHBV.

Biological Properties

Host Range

Plant hosts of tenuiviruses are all in the family Graminae.

Transmission

Each species is transmitted by a particular species of planthopper in a circulative, propagative manner. The major vectors are *Laodelphax striatellus* (RSV), *Peregrinus maidis* (MSpV), *Tagosodes orizicolus* (RHBV), *T. cubanus* (EHBV), *Nilaparvata lugens* (RGSV),

Caenodelphax teapae (UHBV), *Ukanodes tanasijevici* (Iranian wheat stripe virus; IWSV), *Javesella pellucida* (European wheat striate mosaic virus; EWSMV) and *Sogatella kolophon* (Brazilian wheat spike virus; BWSV).

Tenuiviruses can be transmitted transovarially by viruliferous female planthoppers to their offspring, and through sperm from viruliferous males. Mechanical transmission using sap extracts is difficult.

CYTOPATHOLOGY

Characteristic inclusion bodies, consisting almost entirely of NCP are formed in cells of infected plants. The protein p5 of RGSV accumulates in large amounts in both infected plants and vector insects.

LIST OF SPECIES DEMARCATION CRITERIA IN THE GENUS

The criteria demarcating species in the genus are:

- Vector specificity, i.e. transmission by different species of vector
- Host range, i.e. different abilities to infect key plant species
- Different sizes and/or numbers of RNA components
- <85% aa sequence identity between any corresponding gene products
- <60% nt sequence identity between corresponding non-coding intergenic regions

An example of species discrimination is that between RSV and MSpV. RSV is transmitted by *Laodelphax striatellus* and infects 37 species in the Graminae including wheat and rice. MSpV is transmitted by *Peregrinus maidis* and infects maize, occasionally sorghum and a few other graminaceous plants but not wheat or rice. RSV isolates have genomes of four RNA segments of 9090, 3514, 2475 to 2504 and 2137 to 2157 nt; the MSpV genome has five segments (9121, 3575, 2357, 2227 and 1317 nt. Also, the differences in sequence among the components of these viruses all fall outside the limits set by the Species Demarcation Criteria.

It is difficult to decide if the hoja blanca viruses are one or several species. They have different vectors, different hosts, different sizes and numbers of RNA segments and the nt identity of their intergenic regions is less than 60%. However, the aa sequences of the 4 proteins on RNA-3 and RNA-4 are about 90% identical between RHBV, EHBV and UHBV (although the nt identities of these same coding regions are about 81% identical among them). So 4 out of 5 criteria are met, therefore they could be considered distinct species, possibly only recently separated and now diverging, with little contact in the field between them.

LIST OF SPECIES IN THE GENUS

Species names are in green italic script; strain names and synonyms are in black roman script; tentative species names are in blue roman script. Sequence accession, and assigned abbreviations ( ) are also listed.

SPECIES IN THE GENUS

Echinochloa hoja blanca virus

Echinochloa hoja blanca virus	[RNA3, L75930; RNA4, L48441; RNA5, L47430]	(EHBV)

Maize stripe virus

Maize stripe virus	[RNA2, U53224; RNA3, M57426; RNA4, L13438; RNA5, L13446]	(MSpV)

Rice grassy stunt virus

Rice grassy stunt virus – cn	[RNA1, AF509470; RNA2, AF511072; RNA3, AF397468; RNA4, AF290946; RNA5, AF290947; RNA6, AF287949]	(RGSV-cn)

Rice grassy stunt virus – ph(Laguna)	[RNA1, AB009656; RNA2, AB010376; RNA3, AB010377, RNA4, AB010378; RNA5, AB000403; RNA6, AB000404]	(RGSV-ph(l))
Rice grassy stunt virus – ph(South Cotabato)	[RNA1, AB032180; RNA2, AB023777; RNA3, AB029894; RNA4, AB023778; RNA5, AB023779; RNA6, AB023780]	(RGSV-ph(sc))
Rice hoja blanca virus		
Rice hoja blanca virus – co	[RNA3, AF004658; RNA4, L14952]	(RHBV-co)
Rice hoja blanca virus – cr	[RNA1, AF009569; RNA2, L54073; RNA3, L07940; RNA4, AF004657]	(RHBV-cr)
Rice stripe virus		
Rice stripe virus – cn	[RNA3, Y11095; RNA4, Y11096]	(RSV-cn)
Rice stripe virus – cn (bs)	[RNA3, AF220103; RNA4, AF221830]	(RSV-cn(bs))
Rice stripe virus – cn (cx)	[RNA1, AY186787; RNA2, AY186790; RNA4, AY185501]	(RSV-cn(cx))
Rice stripe virus – cn (dl)	[RNA4, AY185502]	(RSV-cn(dl))
Rice stripe virus – cn (dw)	[RNA3, AF509500]	(RSV-cn(dw))
Rice stripe virus – cn (hz)	[RNA1, AY186788; RNA2, AY186789; RNA3, AF508865; RNA4, AF513505]	(RSV-cn(hz))
Rice stripe virus – cn (jd)	[RNA3, AF220104; RNA4, AF221831]	(RSV-cn(jd))
Rice stripe virus – cn (jn)	[RNA3, AF220105; RNA4, AF221832]	(RSV-cn(jn))
Rice stripe virus – cn (km)	[RNA3, AF508912; RNA4, AY185499]	(RSV-cn(km))
Rice stripe virus – cn (ly)	[RNA3, AF220106; RNA4, AF221833]	(RSV-cn(ly))
Rice stripe virus – cn (pj)	[RNA3, AF220107; RNA4, AF221834]	(RSV-cn(pj))
Rice stripe virus – cn (sq)	[RNA3, AF220108; RNA4, AF221835]	(RSV-cn(sq))
Rice stripe virus – cn (yl)	[RNA3, AF508913; RNA4, AF221836]	(RSV-cn(yl))
Rice stripe virus – cn (yr)	[RNA4, AY185500]	(RSV-cn(yr))
Rice stripe virus – jp (m)	[RNA3, D01094; RNA4, D01039]	(RSV-jp(m))
Rice stripe virus – jp (o)	[RNA2, D13787]	(RSV-jp(o))
Rice stripe virus – jp (t)	[RNA1, D31879; RNA2, D13176; RNA3, X53563; RNA4, D10979]	(RSV-jp(t))
Urochloa hoja blanca virus		
Urochloa hoja blanca virus	[RNA1, U82448; RNA3, U82447; RNA4, U82446]	(UHBV)

TENTATIVE SPECIES IN THE GENUS

Brazilian wheat spike virus	(BWSpV)
European wheat striate mosaic virus	(EWSMV)
Iranian wheat stripe virus	(IWSV)
Rice wilted stunt virus	(RWSV)
Winter wheat mosaic virus	(WWMV)

PHYLOGENETIC RELATIONSHIPS WITHIN THE GENUS

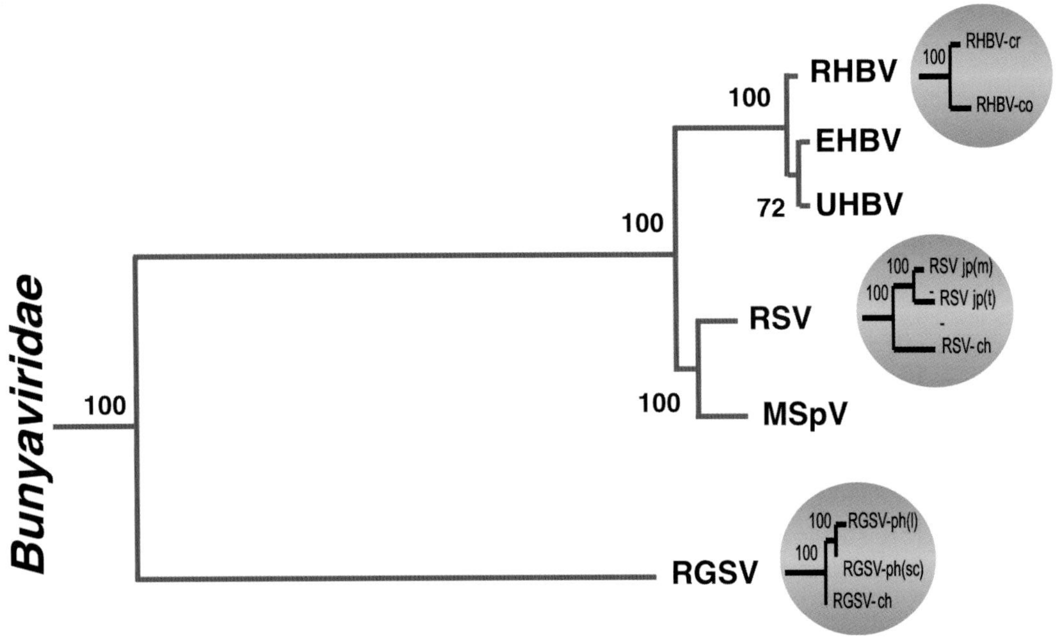

Figure 4: Phylogenetic tree showing the relationships between tenuiviruses. Input data were nt sequences of the coding regions of RNA-3 and -4 (RNA-5 and -6 respectively for RGSV) as aligned by PileUp (GCG-Wisconsin 9.0). A total of 3020 characters with clear positional homology across the alignment were used for the phylogenetic analysis. The phylogenetic tree was generated using Maximum Likelihood criteria as implemented by PAUP 4.0(b10), allowing for variable nucleotide substitution rates, rate heterogeneity between characters and rate heterogeneity between lineages. The tree was significantly superior to alternative trees, as determined by likelihood analysis. Bootstrap probabilities were calculated separately for the main tree and for the resolution among the strains of RHBV, RSV and RGSV. The following sequences were used: RHBV-co (AF004658, L14952); RHBV-cr (L07940, AF004657); EHBV (L75930, L48441); UHBV (U82447, U82446); RSV-jp(t) (X53563, D10979); RSV-jp(m) (D01094, D01039); RSV-cn (Y11095,Y11096); MSpV (M57426, L13438); RGSV-ph(l) (AB000403, AB000404); RGSV-ph(sc) (AB023779, AB023780); RGSV-cn (AF290947, AF287949).

Table 1. Comparisons between the tenuiviruses for the combined RNA-3 and -4 (RNA-5 and –6 of RGSV). Shown are the percent nt identity for the intergenic regions (upper diagonal), and sthe percent aa identity for the coding regions (lower diagonal).

	RSV	MSpV	RHBV	EHBV	UHBV	RGSV
RSV		59.3	48.3	49.3	49.6	45.6
MSpV	69.6		46.0	47.2	45.5	44.3
RHBV	52.8	53.1		59.8	59.9	43.9
EHBV	52.6	52.5	90.8		64.0	45.5
UHBV	53.2	53.6	92.2	92.6		41.9
RGSV	20.8	21.6	21.1	21.8	20.9	

SIMILARITY WITH OTHER TAXA

Tenuiviruses have some similarities with viruses classified in the family *Bunyaviridae*, particularly those in the genus *Phlebovirus*. The multipartite genomes of tenuiviruses contain negative sense and ambisense components. RNPs containing the genomic RNAs can be purified from infected plants. The genomic RNA 5′- and 3′-ends can base-pair, and probably give rise to circular RNPs. Generation of mRNA involves a cap-snatching mechanism. Like viruses in most genera in the family *Bunyaviridae*, tenuiviruses infect their insect vectors as well as their primary hosts, plants. The number of genome components (four to six) and the apparent lack of a membrane-bound virus particle distinguish tenuiviruses from viruses in the family *Bunyaviridae*. Recent data raise the possibility that RGSV be classified in a separate genus. It has six RNA segments that in total encode four or five proteins in addition to those characteristic of the expression of a tenuivirus genome. Moreover, the sequence relatedness of the RGSV gene products with those of other tenuiviruses are all unusually low.

DERIVATION OF NAMES

Tenui: from Latin *tenuis*, "thin, fine, weak".

REFERENCES

Bucher, E., Sijen, T., de Haan, P., Goldbach, R. and Prins, M. (2003). Negative-strand tospoviruses and tenuiviruses carry a gene for a suppressor of gene silencing at analogous genomic positions. *J. Virol.*, **77**, 1329-1336.

Chomchan, P., Miranda, G.J. and Shirako, Y. (2002). Detection of rice grassy stunt tenuivirus nonstructural proteins p2, p5 and p6 from infected rice plants and from viruliferous brown planthoppers. *Arch. Virol.*, 147, 2291-2300.

Chomchan, P., Li, S.F. and Shirako, Y. (2003). *Rice grassy stunt tenuivirus* nonstructural protein p5 interacts with itself to form oligomeric complexes in vitro and in vivo. *J. Virol.*, **77**, 769-775.

de Miranda, J.R., Munoz, M., Wu, R. and Espinoza, A.M. (2001). Phylogenetic position of a novel tenuivirus from the grass *Urochloa plantaginea*. *Virus Genes*, **22**, 329-333.

Falk, B.W. and Tsai, J. H. (1984). Identification of single- and double-stranded RNAs associated with maize stripe virus. *Phytopathology*, **74**, 909-915.

Estabrook, E.M., Suyenaga, K., Tsai, J.H. and Falk, B.W. (1996). Maize stripe tenuivirus RNA 2 transcripts in plant and insect hosts and analysis of pvc2, a protein similar to the Phlebovirus virion membrane glycoproteins. *Virus Genes*, **12**, 239-247.

Huiet, L., Feldstein, P.A., Tsai, J.H. and Falk, B.W. (1993). The maize stripe virus major noncapsid protein messenger RNA transcripts contain heterogeneous leader sequences at their 5′ termini. *Virology*, **197**, 808-812.

Madriz, J., de Miranda, J.R., Cabezas, E., Oliva, M., Hernandez, M. and Espinoza, A.M. (1998). Echinochloa hoja blanca virus and rice hoja blanca virus occupy distinct ecological niches. *J. Phytopath.*, **146**, 305-308.

Nguyen, M., Ramirez, B.C., Goldbach, R. and Haenni, A.-L. (1997). Characterization of the *in vitro* activity of the RNA-dependent RNA polymerase associated with the ribonucleoproteins of rice hoja blanca tenuivirus. *J. Virol.*, **71**, 2621-2627.

Ramirez, B.-C. and Haenni, A.-L. (1994). Molecular biology of tenuiviruses, a remarkable group of plant viruses. *J. Gen. Virol.*, **75**, 467-475.

Shimizu, T., Toriyama, S., Takahashi, M., Akutsu, K. and Yoneyama, K. (1996). Non-viral sequences at the 5′-termini of mRNAs derived from virus-sense and virus-complementary sequences of the ambisense RNA segments of rice stripe tenuivirus. *J. Gen. Virol.*, **77**, 541-546.

Swofford, D.L. (1998). PAUP* Phylogenetic analysis using parsimony (*and other methods). Version 4.0. Sinnauer Associates, Sunderland, Massachusetts.

Toriyama, S., Takahashi, M., Sano, Y., Shimizu, K. and Ishihama, A. (1994). Nucleotide sequence of RNA 1, the largest genomic segment of rice stripe virus, the prototype of the tenuiviruses. *J. Gen. Virol.*, **75**, 3569-3579.

Toriyama, S., Kimishima, T. and Takahashi, M. (1997). The proteins encoded by rice grassy stunt virus RNA5 and RNA6 are only distantly related to the corresponding proteins of other members of the genus *Tenuivirus*. *J. Gen. Virol.*, **78**, 2355-2363.

Toriyama, S., Kimishima, T., Takahashi, M., Shimizu, T., Minaka, N. and Akutsu, K. (1998). The complete nucleotide sequence of the rice grassy stunt virus genome and genomic comparisons with viruses of the genus *Tenuivirus*. *J. Gen. Virol.*, **79**, 2051-2058.

CONTRIBUTED BY

Haenni, A.-L., de Miranda, J.R., Falk, B.W., Goldbach, R., Mayo, M.A., Shirako, Y. and Toriyama, S.

NEG. SSRNA

FAMILY *ARENAVIRIDAE*

TAXONOMIC STRUCTURE OF THE FAMILY

Family **_Arenaviridae_**
Genus **_Arenavirus_**

Since only one genus is currently recognized, the family description corresponds to the genus description.

GENUS *ARENAVIRUS*

Type Species *Lymphocytic choriomeningitis virus*

VIRION PROPERTIES

MORPHOLOGY

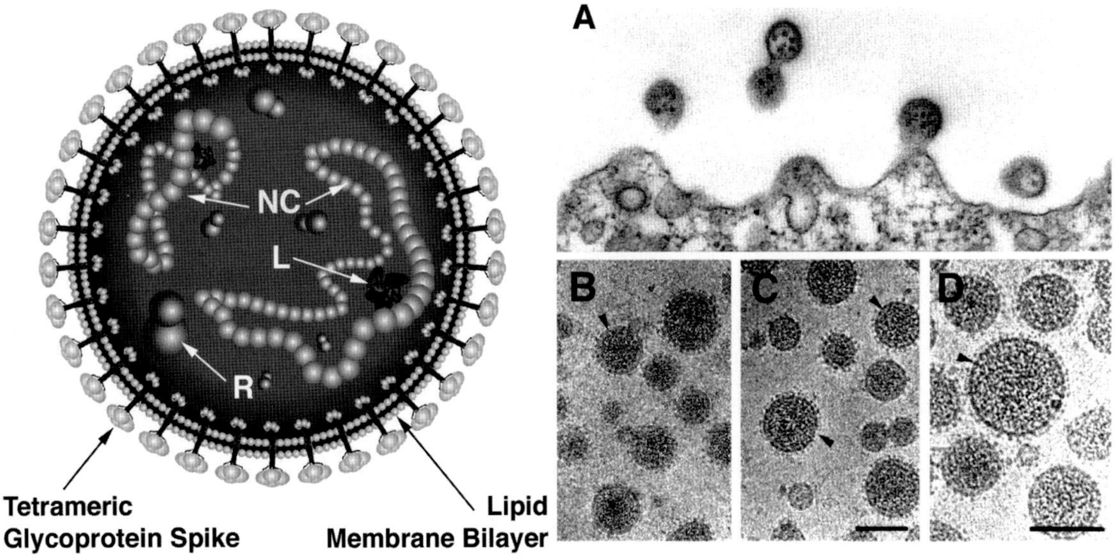

Figure 1: (Left) Diagrammatic representation of virion structure. L, L protein (RNA polymerase); NC, nucleocapsid; R, ribosome. (Courtesy A. Featherstone and C. Clegg). (Right) Electron microscopic images of Lymphocytic choriomeningitis virus (LCMV). A. Thin section showing several virions budding from the surface of an infected BHK-21 cell. (B–D). Cryo-electron microscopic images of purified unstained virions frozen in vitreous ice, taken at –1.5, –3 and –4 microns defocus. Arrowheads indicate glycoprotein spikes which are composed of a trans-membrane GP2 and globular GP1 head arranged in a tetrameric configuration. Bars indicate 100 nm. (Courtesy R. Milligan, J. Burns and M. Buchmeier).

Virions are spherical to pleomorphic, 50-300 nm in diameter (mean 110-130 nm), with a dense lipid envelope and a surface layer covered by club-shaped projections, 8-10 nm in length. A variable number of 20-25 nm ribosomes are generally present within virus particles. Isolated nucleocapsids, free of contaminating host ribosomes, are organized in closed circles of varying length (450-1300 nm), which have been shown to assume supercoiled forms, and display a linear array of nucleosomal subunits.

PHYSICOCHEMICAL AND PHYSICAL PROPERTIES

Virion Mr has not been determined. The S_{20w} is 325-500S. The buoyant density in sucrose is about 1.17-1.18 g/cm^3, in CsCl it is about 1.19-1.20 g/cm^3, in amidotrizoate compounds it is about 1.14 g/cm^3. Virions are relatively unstable *in vitro*, and are rapidly inactivated below pH 5.5 and above pH 8.5. Virus infectivity is inactivated at 56°C, by treatment with organic solvents, or by exposure to UV- and gamma-irradiation.

NEG. SSRNA

Nucleic Acid

The genome consists of two single stranded, ambisense RNA molecules, L and S, of lengths of about 7.5 kb and 3.5 kb, respectively. There are no poly(A) tracts at the 3'-termini. The 3'-terminal sequences (19-30 nt) are similar in the two RNAs and among different arenaviruses. Overall, they are largely complementary to the 5'-end sequences. Although the RNA genomic species are thought to be present in virions in the form of circular nucleocapsids, the genomic RNA is not covalently closed. Variable amounts of full-length viral-complementary RNAs (predominantly S) and viral subgenomic mRNA species have been reported in virus preparations. Preparations of purified virus may also contain RNAs of cellular origin with sedimentation coefficients of 28S, 18S and 4-6S. These include ribosomal RNAs. The viral mRNA species are presumably associated with encapsidated ribosomes. The RNA species are not present in equimolar amounts, apparently due to the packaging of multiple RNA species per virion.

Proteins

The most abundant structural protein is the nucleoprotein (N or NP), a non-glycosylated polypeptide (~63 kDa) found tightly associated with the virus genomic RNA in the form of a ribonucleoprotein complex or nucleocapsid structure. A minor component is the L protein, an RNA polymerase (~200 kDa). A putative zinc binding protein (Z or p11; 10-14 kDa) is also a structural component of the virus. Two glycosylated proteins (GP1 or G1, GP2 or G2; 34-44 kDa) are found in all members of the family and are derived by posttranslational cleavage from an intracellular precursor, GPC (~75-76 kDa). Other minor proteins and enzymatic activities have been described associated with virions including poly (U) and poly (A) polymerases, and a protein kinase that can phosphorylate N. It is thought unlikely that these are virally encoded.

Lipids

Lipids represent about 20% of virion dry weight and are similar in composition to those of the host plasma membrane.

Carbohydrates

Carbohydrates in the form of complex glycans on GP1 (5 or 6 sites in LCMV) and GP2 (2 sites in LCMV) represent about 8% of virion dry weight.

Genome Organization and Replication

The L and S RNAs of arenaviruses each have an ambisense coding arrangement (Fig. 2). The L RNA encodes in its viral-complementary sequence the L protein, and in the viral-sense 5′-end sequence the Z protein. The Z mRNA is small (<0.5 kb). The N protein is encoded in the viral-complementary sequence corresponding to the 3′-half of the S RNA, while the viral glycoprotein precursor (GPC) is encoded in the viral-sense sequence corresponding to the 5′-half of the S RNA. The 2 proteins are made from sg mRNA species transcribed from the viral (for N mRNA) or full-length viral-complementary S RNA species (for GPC mRNA). The intergenic regions of both S and L RNAs contain nt sequences with the potential of forming one or more hairpin configurations. These secondary structural features may function to terminate mRNA transcription from the viral and viral-complementary S RNAs. The mRNAs are capped and contain 1-5 non-templated nt of heterogeneous sequence at their 5′-ends. The mRNAs are not polyadenylated. The transcription mechanism is not fully elucidated. Initiation of transcription may involve cap-snatching. The 3′ termini of the mRNAs have been mapped to locations in the intergenic regions.

The process of infection involves attachment to cell receptors, entry via the endosomal route, uncoating and mRNA transcription in the cytoplasm of infected cells. Because of the ambisense coding arrangement, only N and L mRNAs can be synthesized from the genomic RNAs by the virion polymerase prior to translation. The products of these mRNAs are presumed to be involved in the synthesis of full-length viral complementary species which serve as templates for the synthesis of GPC and Z mRNAs and the synthesis of full-length viral RNAs. The process of RNA replication, which may involve a slippage mechanism during initiation, and read-through of transcription termination signals, has not been fully elucidated. However, the presence of full-length viral-complementary genomic RNAs and viral sgRNA species in virus preparations may affect this perceived temporal order of RNA and protein synthesis.

The viral envelope glycoproteins are synthesized in cells as a single mannose-rich precursor molecule that is proteolytically cleaved and processed to contain complex glycans during transport to the plasma membrane. Virions mature by budding at sites on the surface of cells. Ribosomes are also observed at such sites. Interstrain reassortant progeny can be formed, including diploid (or multiploid) species with respect to the genomic RNA segments. Some evidence for interspecies reassortment between *Lassa virus* and Mopeia virus has been obtained. The replication *in vitro* of a number of arenaviruses is inhibited by a variety of antiviral agents, including amantadine, alpha-amanitin, glucosamine, and thiosemicarbazones. Ribavirin inhibits the replication of several arenaviruses in vitro and is effective in the therapy of humans and primates infected with Lassa virus.

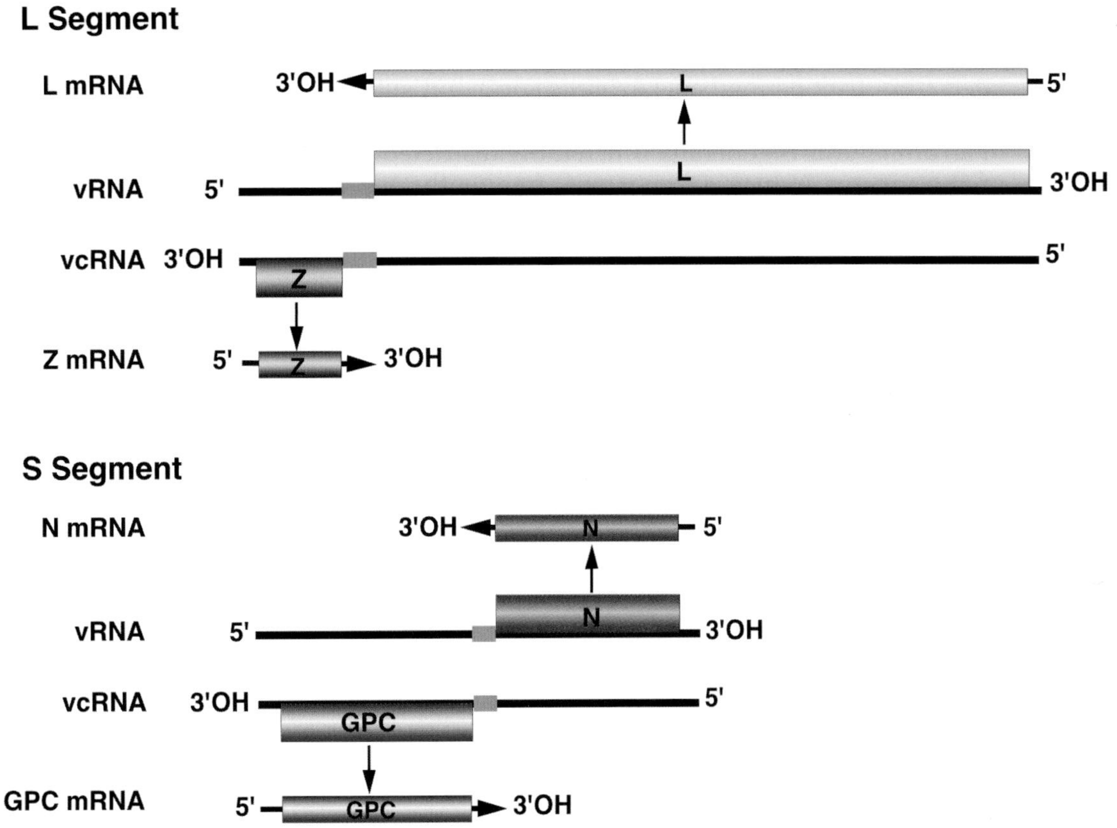

Figure 2: Organization, transcription and replication of the arenavirus L and S RNAs. Regions encoding the L, Z, GPC and N proteins are shown as boxes with arrowheads indicating the notional direction of translation. The intergenic regions separating the ORFs are indicated by gray boxes. The sgRNAs which function as messengers are shaded grey. RNA transcription processes are indicated by solid arrows. (from V. Romanowski).

ANTIGENIC PROPERTIES

Viruses possess a number of distinct antigenic determinants as shown by monoclonal and polyclonal antibody analyses. Antigens on the 44 kDa GP1 of Lymphocytic choriomeningitis virus (LCMV) are involved in virus neutralization. These are type-specific, although cross-neutralization tests have demonstrated partially shared antigens between Tacaribe virus and Junín virus. Cross-protection has also been demonstrated against Junín virus following prior infection by Tacaribe virus, or against Lassa virus following infection by Mopeia virus. Major complement-fixing antigens are associated with the viral N proteins, which were used to define the Tacaribe complex of arenaviruses. Monoclonal antibodies react with common epitopes on the N proteins of all arenaviruses and a single highly conserved epitope has also been described in the transmembrane GP2 glycoprotein.

By analyses using monoclonal and polyclonal antibody, the African arenaviruses are distinguishable from the New World arenaviruses. Fluorescent antibody studies show that antisera against New World viruses, as well as those against African viruses, react with LCMV. Cytotoxic T-lymphocyte epitopes have been identified on the nucleoprotein and glycoproteins of LCMV. The number and location of epitopes varies depending on the virus strain and host MHC class I molecules. No hemagglutinin has been identified.

Biological Properties

The reservoir hosts of almost all the arenaviruses are species of rodents. LCMV is found in mouse and the African viruses mainly in the rodents *Mastomys* and *Praomys*, in the subfamily *Murinae*. The New World viruses are mostly found in the Sigmodontine rodents *Calomys*, *Neacomys*, *Neotoma*, *Oryzomys* and *Sigmodon*. Exceptionally, Tacaribe virus was isolated from fruit-eating bats (*Artibeus* spp.), but subsequent attempts to recover it from bats or from other potential hosts have been unsuccessful. It is notable that the geographic range of an arenavirus is generally much more restricted than that of its cognate rodent host. Most of the viruses induce a persistent, frequently asymptomatic infection in their reservoir hosts, in which chronic viremia and viruria occur. Such infections are known or suspected to be caused by a slow and/or insufficient host immune response. Most arenaviruses do not normally infect other mammals or humans. However, Lassa virus is the cause of widespread human infection (Lassa fever) in West Africa (Nigeria, Sierra Leone, Liberia, Guinea), and Junín virus causes Argentine hemorrhagic fever in agricultural workers in an increasingly large area of that country. Machupo virus has caused isolated outbreaks of similar disease in Bolivia, and Guanarito virus is associated with human disease in Venezuela. Sabiá virus was isolated from a fatal human case in Brazil. Human infection with LCMV may occur in some rural and urban areas with high rodent populations, and has been acquired from pet hamsters. LCMV acquired from mice has also caused a highly fatal hepatitis in captive Callitrichid primates. Severe laboratory-acquired infections have occurred with LCM, Lassa, Junín, Machupo, Sabiá and Flexal viruses. Asymptomatic infections with Pichinde virus have been reported.

Success of experimental infection in laboratory animals (mouse, hamster, guinea pig, rhesus monkey, marmoset, rat) varies with the animal species and the virus. In general, New World viruses are pathogenic for suckling but not weaned mice; LCMV and Lassa virus produce the opposite effect. Viruses grow moderately well in many mammalian cells. LCMV can grow in murine T-lymphocytes.

Vertical and horizontal (including venereal) transmissions occur in the natural hosts. These include transuterine, transovarian and post-partum transmission and can be via milk-, saliva- or urine-borne routes. Horizontal transmission within and between host species occurs by contamination and aerosol routes. No arthropod vectors are thought to be involved in the normal transmission process.

List of Species Demarcation Criteria in the Genus

The parameters used to define a species in the genus are :

- an association with a specific host species or group of species;
- presence in a defined geographical area;
- etiological agent (or not) of disease in humans;
- significant differences in antigenic cross-reactivity, including lack of cross-neutralization activity where applicable;
- significant aa sequence difference from other species in the genus.

For example, although both Pirital virus and Guanarito virus circulate in the same region of Venezuela, they are distinguished by their isolation from different rodent hosts (*Sigmodon alstoni* and *Zygodontomys brevicauda*, respectively). In addition, in ELISA with hyperimmune mouse ascitic fluids, titers differ by at least 64-fold, and sequence analysis shows less than 55% aa identity between partial nucleocapsid protein sequences. In another example, both Lassa virus and Mopeia virus share a common rodent host (*Mastomys*) at the genus level. However, they are distinguished by their different geographical range, different profiles of reactivity with panels of monoclonal antibodies, and by N protein aa sequence divergencies of about 26%. Also, Lassa virus is the cause of hemorrhagic fever in humans and other primates, while Mopeia virus is not associated with human disease and does not cause disease in experimentally infected primates.

List of Species in the Genus

Species names are in green italic script; strain names and synonyms are in black roman script; tentative species names are in blue roman script. Natural hosts, geographic location, sequence accessions, and assigned abbreviations ( ) are also listed.

Species in the Genus

Old World Arenaviruses

Ippy virus		
Ippy virus - Dak AN B 188d *Arvicanthis* sp., Central African Republic	[N gene (partial): U80003]	(IPPYV)
Lassa virus		
Lassa virus - GA391 *Mastomys* sp., West Africa	[S segment: X52400]	(LASV-GA391)
Lassa virus - Josiah *Mastomys* sp., West Africa	[S segment: J04324 L segment: U73034]	(LASV-Jo)
Lassa virus - LP *Mastomys* sp., West Africa	[N gene (partial): U80004]	(LASV-LP)
Lymphocytic choriomeningitis virus		
Lymphocytic choriomeningitis virus - Armstrong *Mus musculus*, Europe, Americas	[S segment: M20869 : L segment: J04331, M27693]	(LCMV-Ar)
Lymphocytic choriomeningitis virus - WE *Mus musculus*, Europe, Americas	[S segment: M22138]	(LCMV-WE)
Mobala virus		
Mobala virus - 3076 *Praomys* sp., Central African Republic	[N gene (partial): AF012530]	(MOBV-3076)
Mobala virus - 3099 *Praomys* sp., Central African Republic	[N gene (partial): U80007, U80008]	(MOBV-3099)
Mopeia virus		
Mopeia virus - AN 20410 *Mastomys natalensis*, Mozambique, Zimbabwe	[N gene (partial): U80005]	(MOPV-AN20410)
Mopeia virus - AN 21366 (or 800150) *Mastomys natalensis*, Mozambique, Zimbabwe	[S segment: M33879]	(MOPV-AN21366)

New World Arenaviruses

Allpaahuayo virus		
Allpaahuayo virus - CLHP-2472 *Oecomys bicolor, Oe. paricola*	[S segment: AY012687]	(ALLV-CLHP2472)
Amapari virus		
Amapari virus - BeAn 70563 *Oryzomys capito, Neacomys guianae*, Brazil	[N gene (partial): U43685]	(AMAV-BeAn70563)
Bear Canyon virus		
Bear Canyon virus - A0060209 *Peromyscus californicus*	[S segment: AF512833]	(BCNV-A0060209)
Cupixi virus		
Cupixi virus - BeAn 119303 *Oryzomys sp.*	[S segment: AF512832]	(CPXV- BeAn 119303)
Flexal virus		
Flexal virus - BeAn 293022 *Oryzomys* spp., Brazil	[N gene (partial): U43687]	(FLEV-BeAn293022)
Guanarito virus		
Guanarito virus - INH-95551 *Zygodontomys brevicauda*, Venezuela	[N gene (partial): L42001 N gene (partial): U43686]	(GTOV-INH95551)
Junín virus		
Junín virus - MC2 *Calomys musculinus*, Argentina	[S segment: D10072]	(JUNV-MC2)
Junín virus - XJ *Calomys musculinus*, Argentina	[GPC gene : U70799 N gene :U70802]	(JUNV-XJ)
Latino virus		
Latino virus - 10924 *Calomys callosus*, Bolivia	[N gene (partial): U43688]	(LATV-10924)
Machupo virus		
Machupo virus - AA288-77 *Calomys callosus*, Bolivia	[N gene: X62616]	(MACV-10924)
Oliveros virus		
Oliveros virus - RIID 3229 *Bolomys obscurus*, Argentina	[S segment: U34248]	(OLVV-RIID3229)
Paraná virus		
Paraná virus - 12056 *Oryzomys buccinatus*, Paraguay	[N gene (partial): U43689]	(PARV-12056)
Pichinde virus		
Pichinde virus - 3739 *Oryzomys albigularis*, Colombia	[S segment: K02734]	(PICV-3739)
Pirital virus		
Pirital virus - VAV488 *Sigmodon alstoni*, Venezuela	[N gene (partial): U62561]	(PIRV-VAV488)
Pirital virus - VAV-499 *Sigmodon alstoni*, Venezuela	[N gene (partial): U62562]	(PIRV-VAV499)
Sabiá virus		
Sabiá virus - SPH114202 Natural host unknown, Brazil	[S segment: U41071]	(SABV-SPH114202)
Tacaribe virus		
Tacaribe virus - p2b2 *Artibeus* spp., Trinidad	[S segment: M20304]	(TCRV-p2b2)
Tacaribe virus - TRVLII573 *Artibeus* spp., Trinidad	[L segment: M65834, J04340, M33513]	(TCRV-TRVLII573)

Tamiami virus

Tamiami virus - W10777 *Sigmodon hispidus*, Florida, U.S.A.	[N gene (partial): U43690]	(TAMV-W10777)

Whitewater Arroyo virus

Whitewater Arroyo virus - AV 9310135 *Neotoma albigula*, New Mexico, U.S.A.	[N gene (partial): U52180]	(WWAV-AV9310135)

Tentative Species in the Genus

Rio Cacarana virus ? / Argentina	[GPC and N partial]	(RCAV)
Pampa virus *Bolomys* sp./ Argentina	[GPC and NP]	(PAMV)

Pampa virus should be considered as a genotype of Oliveros virus because it has only 6.3 % aa divergence between GPC genes and 4.2% divergence between N genes. By way of comparison, aa distances observed between Amapari and Cupixi viruses (GPC gene = 31.4%, NP gene = 14.6%) and between Junín and Machupo viruses (GPC gene = 30.0%, NP gene = 14.4%) are the lowest observed between distinct arenaviruses species.

List of Unassigned Viruses in the Family

None reported.

Phylogenetic Relationships within the Family

Nucleic acid sequences from the N genes of all the known arenaviruses have provided the basis for phylogenetic analysis that supports previously defined antigenic groupings and further defines virus relationships within them (Fig. 3). Sequence data derived from other regions of the genome, where available, are largely consistent with this analysis. Among the Old World viruses, *Lassa virus, Mopeia virus* and *Mobala virus* are monophyletic, while *Ippy virus* and *Lymphocytic choriomeningitis virus* are more distantly related. The New World viruses can be divided into three groups on the basis of the sequence data. In group A are *Pirital virus, Pichinde virus, Paraná virus, Flexal virus,* and newly added *Allpahuayo (Peru) virus* from South America, together with *Tamiami virus, Whitewater Arroyo virus* and newly added *Bear canyon virus* from North America. Group B contains the human pathogenic viruses *Machupo virus, Junín virus, Guanarito virus,* and *Sabiá virus* and the non-pathogenic *Tacaribe virus, Amapari virus,* and the recently added *Cupixi virus* (from Brazil). *Latino virus* and *Oliveros virus* form a small separate group (group C). The division of the arenaviruses into Old World and New World groups, as well as the subdivision of New World arenaviruses into three groups, is strongly supported by bootstrap resampling analysis. It is important to note that the trait of human pathogenicity appears to have arisen on at least two independent occasions during arenavirus evolution.

It is apparent that recombination has influenced the evolution of several RNA viruses, including arenaviruses. The nucleocapsid and glycoprotein genes of *Whitewater Arroyo virus, Tamiami virus,* and the *Bear Canyon virus* have divergent phylogenetic histories. Separate analysis of full-length aa sequences using maximum parsimony or neighbor-joining methods show that the nucleocapsid protein genes of these three viruses are related to those of *Pichinde virus* and *Pirital virus* (New World lineage A), while the glycoprotein genes are more closely related to those of *Junín virus, Tacaribe virus, and Sabia virus* (New World lineage B). Recombination seems to have played a role in the genome of Río Carcarañá virus, but sequence data are incomplete. There is published evidence that the Whitewater Arroyo virus, Tamiami virus, and Bear canyon virus S segments are the product of recombination between ancestral arenaviruses from different lineages.

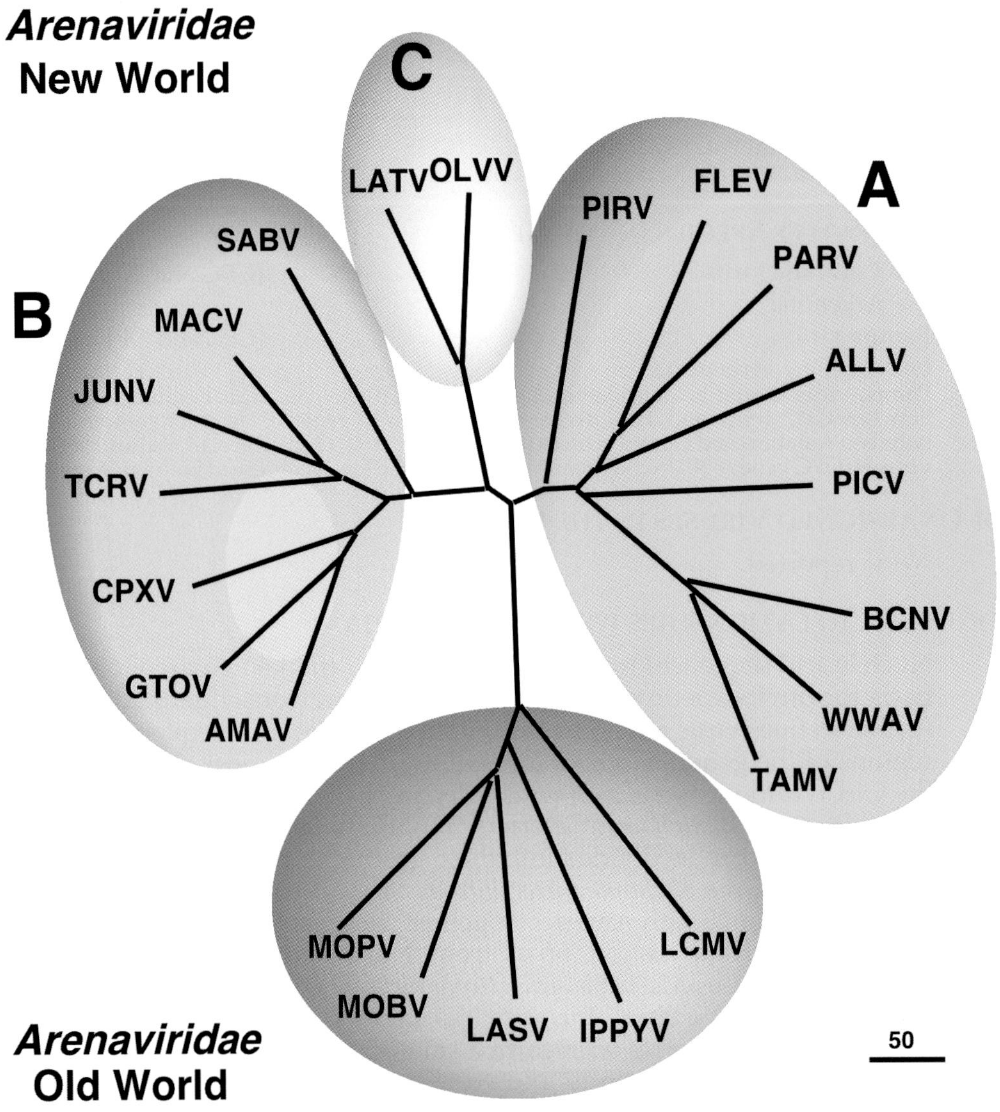

Figure 3: Phylogenetic relationships among the *Arenaviridae*. Partial N gene nt sequences corresponding to nt 1770-2418 of Tacaribe virus S RNA sequence (GenBank accession no. M20304) were aligned (PILEUP, adjusted manually) and analyzed by maximum parsimony using PAUP (Michael Bowen, C.J. Peters, Stuart Nichol, CDC). BCN, and ALL recently joined group A, and CPX joined group B.

Similarity with Other Taxa

None reported.

Derivation of Names

arena: from Latin *arenosus*, "sandy" and *arena*, "sand", in recognition of the sand-like ribosomal contents of particles in thin section. The name originally proposed was arenovirus, but was subsequently changed to avoid possible confusion with adenovirus.

References

Archer, A.M. and Rico-Hesse, R. (2002). High genetic divergence and recombination in Arenaviruses. *Virology*, **304**, 274-281.

Bowen, M.D., Peters, C.J. and Nichol, S.T. (1997). Phylogenetic analysis of the *Arenaviridae*: patterns of virus evolution and evidence for cospeciation between arenaviruses and their rodent hosts. *Mol. Phylogenet. Evol.*, **8**, 301-316.

Charrel, R.N., de Lamballerie, X. and Fullhorst, C.F. (2001). The Whitewater Arroyo virus: natural evidence for genetic recombination among Tacaribe serocomplex viruses (family *Arenaviridae*). *Virology*, **283**, 161-166.

Charrel, R.N., Feldmann, H., Fulhorst, C.F., Khelifa, R., de Chesse, R. and de Lamballerie, X. (2002). Phylogeny of New World arenaviruses based on the complete coding sequences of the small genomic segment identified an evolutionary lineage produced by intra-segmental recombination. *Biochem. Biophys. Res. Commun.*, **296,** 1118-1124.

Fulhorst, C.F., Bennett, S.G., Milazzo, M.L., Murray, Jr. H.L., Webb, Jr. J.P., Cajimat, M.N.B. and Bradley, R.B. (2002). Bear canyon virus: an arenavirus naturally associated with the California mouse (*Peromyscus californicus*). *Emerg. Infect. Dis.*, **8,** 717-721.

Moncayo, A.C., Hice, C.L., Watts, D.W., Travassos da Rosa, A.P., Guzman, H., Russell, K.L., Calampa, C., Gozalo, A., Popov, V.L., Weaver, S.C., Tesh, R.B. (2001). Allpahuayo virus: a newly recognized arenavirus (*Arenaviridae*) from arboreal rice rats (*Oecomys bicolor* and *Oecomys paricola*) in northeastern Peru. *Virology*, **284,** 277-286.

Oldstone, M.B.A. (ed.) (2002). Arenaviruses I. *Curr. Top. Microbiol. Immunol.*, **262**, 1-197.

Oldstone, M.B.A. (ed.) (2002). Arenaviruses II. *Curr. Top. Microbiol. Immunol.*, **263**, 1-268.

Peters, C.J., Buchmeier, M., Rollin, P.E. and Ksiazek, T.G. (1996). Arenaviruses. In: *Fields Virology*, 3rd ed., (B.N. Fields, D.M. Knipe and P.M. Howley, eds), pp 1521-1552. Lippincott-Raven, Philadelphia.

Salvato, M.S. (ed.) (1993). The *Arenaviridae*. Plenum Press, New York.

Southern, P.J. (1996). *Arenaviridae*: the viruses and their replication. In: *Fields Virology*, 3rd ed., (B.N. Fields, D.M. Knipe and P.M. Howley, eds), pp 1505-1520. Lippincott-Raven, Philadelphia.

Contributed By

Salvato, M.S., Clegg, J.C.S., Buchmeier, M.J., Charrel, R.N., Gonzalez, J.-P., Lukashevich, I.S., Peters, C.J., Rico-Hesse, R. and Romanowski, V.

GENUS *DELTAVIRUS*

Type Species *Hepatitis delta virus*

VIRION PROPERTIES

MORPHOLOGY

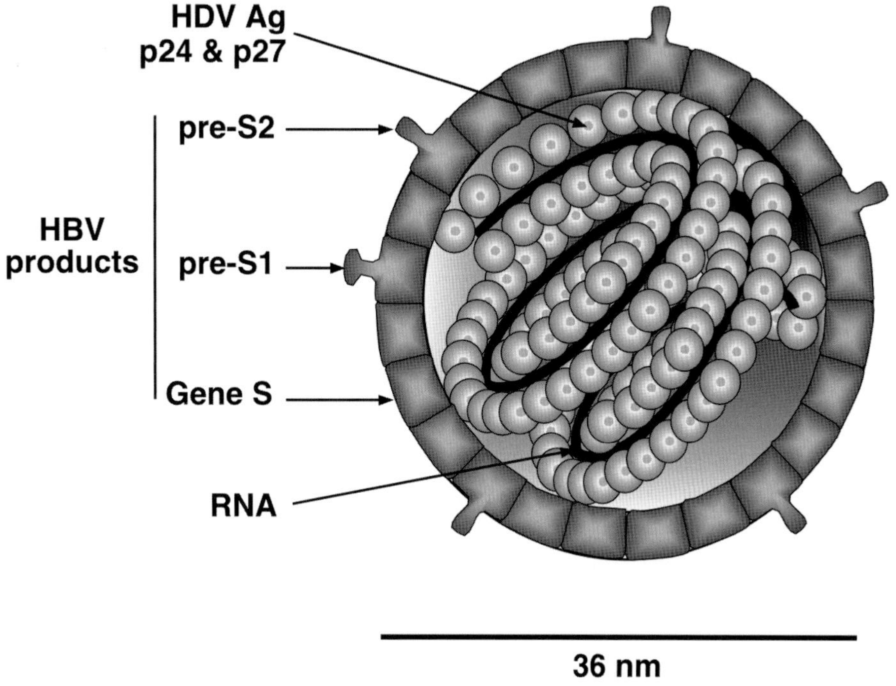

Figure 1: Schematic representation of a particle of Hepatitis delta virus (HDV).

Virions of Hepatitis delta virus (HDV) are roughly spherical with an average diameter of 36 to 43 nm and no identified surface projections (Fig. 1). They consist of an outer envelope containing lipid and all three envelope proteins of the co-infecting helper hepadnavirus (Human hepatitis B virus in nature, though Woodchuck hepatitis virus can also act as a helper) (see below), and an inner nucleocapsid of 19 nm comprising the RNA genome of HDV and approximately 70 copies of the only HDV-encoded protein, known as delta antigen (HDAg). HDAg exists in two forms (large HDAg; L-HDAg, p27 and small HDAg; S-HDAg, p24) which differ only by a 19 aa C-terminal extension. Virions contain variable amounts of L-HDAg and S-HDAg in close association with virion RNA. Nucleocapsid symmetry has not been confirmed. Nucleocapsids can be released by treatment of virions with non-ionic detergent and dithiothreitol.

PHYSICOCHEMICAL AND PHYSICAL PROPERTIES

None reported.

NUCLEIC ACID

The genome consists of a single molecule of circular negative sense ssRNA about 1.7 kb in length. With a high degree (~70%) of intramolecular base pairing, it has the potential to fold on itself forming an unbranched rod-like structure. Both genomic and antigenomic RNA species can function as a ribozyme to carry out self-cleavage and possibly self-ligation. The above properties make this genome unique and distinct from all other known animal viruses.

PROTEINS

HDV RNA encodes one known protein S-HDAg (see above). A second species, L-HDAg, arises via an RNA editing event mediated by the cellular enzyme dsRNA adenosine deaminase, in which the UAG stop codon for S-HDAg is converted to UGG, thereby allowing read-through translation giving rise to L-HDAg. Both HDAg species are multi-functional with domains identified that result in (from the N-terminus) (i) dimerisation via a coiled coil structure; (ii) nuclear localisation via a bipartite signal; (iii) RNA binding via two arginine-rich motifs. In addition, L-HDAg has a domain, which includes a prenylation site required for packaging. The two HDAg species play distinct roles in replication; S-HDAg is essential for HDV replication while L-HDAg is essential for packaging and in some situations, may inhibit replication. HDAg can be phosphorylated at serine residues.

The remaining structural proteins of the HDV virion consist of the surface proteins and glycoproteins of the helper hepadnavirus located in the HDV envelope.

LIPIDS

Not characterized.

CARBOHYDRATES

Not characterized.

GENOME ORGANIZATION AND REPLICATION

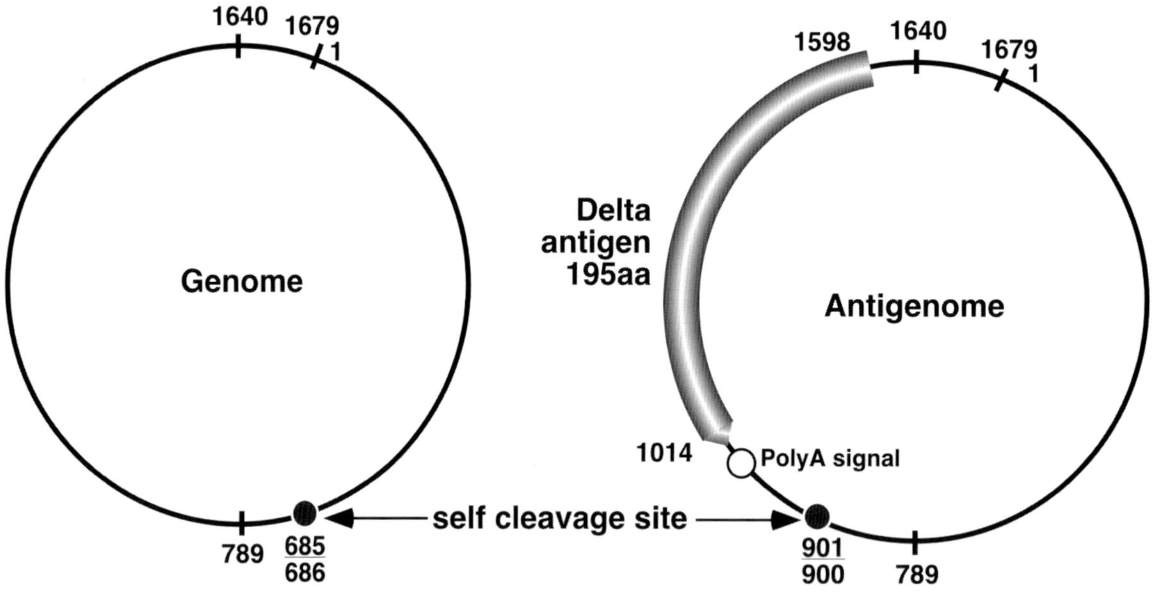

Figure 2: Organization of the genome and antigenome of Hepatitis delta virus (HDV) (Adapted from Taylor *et al.*, 1996).

Attachment, entry and uncoating of HDV may be similar to the steps that occur with the helper hepadnavirus. Both viruses require the preS1 domain of the Hepatitis B virus large envelope protein for attachment. Genome replication involves RNA-directed RNA synthesis carried out by host cell RNA polymerase II in the nucleus. It is thought to occur by a double rolling circle mechanism that generates oligomeric forms of each complementarity, which then undergo site specific autocatalytic cleavage and ligation to generate circular genomic and antigenomic monomers (Fig. 2).

Only one HDV mRNA species has been identified, coding for HDAg. In transfected cells only S-HDAg is made initially and L-HDAg appears subsequently as a result of the RNA editing event described above. Such editing occurs during replication in tissue culture as well as in infected chimpanzees and woodchucks.

As HDV assembly requires the envelope proteins of a helper hepadnavirus, its assembly pathway is likely to overlap with that of the helper virus. In doubly transfected cells, L-HDAg must be present for delta antigen-containing particles to be released, while S-HDAg is packaged if present in the cell, but is not essential for particle formation. Full size, or deleted, HDV RNA molecules are incorporated if present in the cell, as long as they are capable of folding into rod-like structures and binding with HDAg's. In cells undergoing HDV RNA replication this process is highly specific for genomic RNA, while in cells expressing but not replicating HDV RNA, either sense can be assembled.

ANTIGENIC PROPERTIES

HDAg's have a unique antigenicity, and antibodies to these epitopes are diagnostic of current or past infections.

BIOLOGICAL PROPERTIES

Full replication of HDV requires the presence of a helper hepadnavirus to provide envelope proteins, and it can therefore be considered as a subviral satellite virus. Natural HDV infection is found only in humans with HBV as helper virus. However, it can be transmitted to chimpanzees if accompanied by HBV, and experimental transmission of HDV to woodchucks has also been achieved using Woodchuck hepatitis virus as helper virus.

Helper-independent HDV infection has also been seen in human liver transplant recipients and in experimental woodchuck infection. In this situation, "latent" persistent HDV infection with little or no virus release can be rescued by subsequent HBV reactivation or reinfection. Transmission of HDV to laboratory mice has been reported leading to a single round of HDV genome replication in hepatocytes but no further replication presumably due to the absence of helper virus.

Transmission of HDV in man occurs by similar routes to HBV although in many parts of the world transmission by parenteral contact (e.g. sharing of intravenous needles) is more prominent than sexual or vertical routes. If transmission occurs to an individual with chronic HBV infection, this situation is termed superinfection and HDV infection usually then persists. On the other hand, if both HDV and HBV are simultaneously transmitted to a naive host the situation is termed co-infection and both infections are usually transient. HDV distribution is world-wide, but the proportion of HBV carriers who also have chronic HDV infection varies greatly between 0% and 60% in different geographical areas.

Clinical sequelae of acute and chronic HDV infection are variable and cover a similar spectrum to HBV alone, including acute hepatitis, chronic active hepatitis, cirrhosis, fulminant acute hepatitis and hepatocellular carcinoma. However, the frequency of severe sequelae and their rates of progression are significantly higher in chronic HDV infection than in chronic HBV infection alone. A subacute rapidly progressive form of HDV superinfection has been seen in HBV carriers in Venezuela, and other forms of severe acute and chronic hepatitis D, often fatal, occur in indigenous populations of Venezuela, Colombia, Brazil and Peru.

List of Species Demarcation Criteria in the Genus

Not applicable. There is only one species in the genus. A study of 14 independent isolates of HDV revealed up to 40% variation in nucleotide sequence with some geographical clustering. Studies of nucleotide homologies have distinguished genotype 1 (USA, Europe, China), genotype 2 (Japan) and genotype 3 (S. America).

List of Species in the Genus

Species names are in green italic script; strain names and synonyms are in black roman script; tentative species names are in blue roman script. Sequence accession, and assigned abbreviations ( ) are also listed.

Species in the Genus

Hepatitis delta virus

Hepatitis delta virus - 1 (USA, Europe, China)	(HDV-1)
Hepatitis delta virus - 2 (Japan)	(HDV-2)
Hepatitis delta virus - 3 (S. America)	(HDV-3)

Tentative Species in the Genus

None reported.

Phylogenetic Relationships within the Genus

Not applicable.

Similarity with Other Taxa

Several features of HDV (genome structure, RNA-RNA transcription using RNA polymerase II, and autocatalytic RNA sites) are similar to properties of some viroids. However, unlike viroids, HDV possesses a larger genome, encodes a functional protein and requires a specific hepadnavirus helper function.

HDV also possesses some features in common with ssRNA satellites of plants, including B type mRNA satellites and circular single-stranded satellite RNAs and also (in terms of the satellite-helper relationship) with ssRNA satellite viruses such as Chronic bee-paralysis virus associated satellite virus and Tobacco necrosis satellite virus.

However, on the basis of genome size and structure, mode of replication, protein coding strategy, structure of virion and satellite-helper virus relationship, none of the above examples warrant inclusion in a distinct family together with HDV.

Derivation of Names

Delta. A novel antigen in HBV infected tissue, unrelated to previously described HBV antigens, was named delta antigen (δAg).

References

Rizzetto, M., Canese, M.G., Arico, S., Crivelli, O., Bonino, F., Trepo, C.G. and Verme, G. (1977). Immunofluorescence detection of a new antigen-antibody system (δ/anti-δ) associated to hepatitis B virus in liver and serum of HBsAg carriers. *Gut*, **18**, 977 - 1003.

Taylor, J.M. (1996). Hepatitis delta virus and its replication. In: *Fields Virology*, 3rd ed., (B.N. Fields, D.M. Knipe and P.M. Howley, eds), pp 2809–2818. Lippincott-Raven, Philadelphia.

Contributed By

Mason, W.S., Burrell, C.J., Casey, J., Gerlich, W.H., Howard, C.R., Newbold, J., Taylor, J.M. and Will, H.

The Positive Sense Single Stranded RNA Viruses

FAMILY LEVIVIRIDAE

TAXONOMIC STRUCTURE OF THE FAMILY

Family	***Leviviridae***
Genus	***Levivirus***
Genus	***Allolevivirus***

VIRION PROPERTIES

MORPHOLOGY

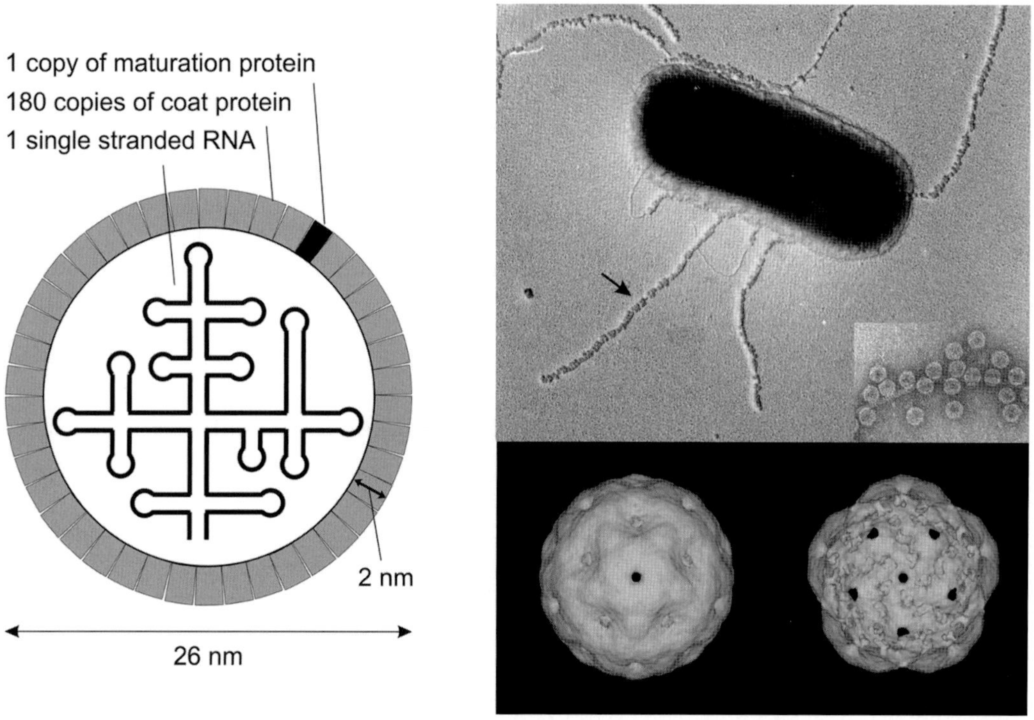

Figure 1: (Left) Schematic representation of a levivirus: the RNA inside the virion is highly structured. (Upper right) *Escherichia coli* bacterium with Enterobacteria phage MS2 (MS2) particles attached its F-pili (Courtesy A.B. Jacobson). The inset is a -pilus with phage- enlargement. (Courtesy R.I. Koning and H.K. Koerten). (Lower right) Reconstruction of images obtained from cryo-electron microscopy of infectious MS2. View from outside (left) and inside (right). (Courtesy R.I. Koning and H.K. Koerten).

Virions are spherical and exhibit icosahedral symmetry (T=3) with a diameter of about 26 nm. There is no envelope (Fig. 1).

PHYSICOCHEMICAL AND PHYSICAL PROPERTIES

Virion Mr varies from 3.6-4.2 x 10^6 depending on the genus. The S_{20w} value is 80-84S and buoyant density in CsCl is 1.46 g/cm^3. Infectivity is ether-, chloroform-, and low-pH-resistant, but is sensitive to RNase and detergents. Inactivation by UV light and chemicals is comparable to that of other icosahedral viruses containing ssRNA.

NUCLEIC ACID

Virions contain one molecule of positive sense ssRNA of 3466-4276 nt: size and gene arrangement vary with genus. RNA makes up 39% of the virion weight. The 5′ nucleotide carries a triphosphate, while at the 3′ terminus a non-templated A residue is added by the replicase (Figs. 2, 3).

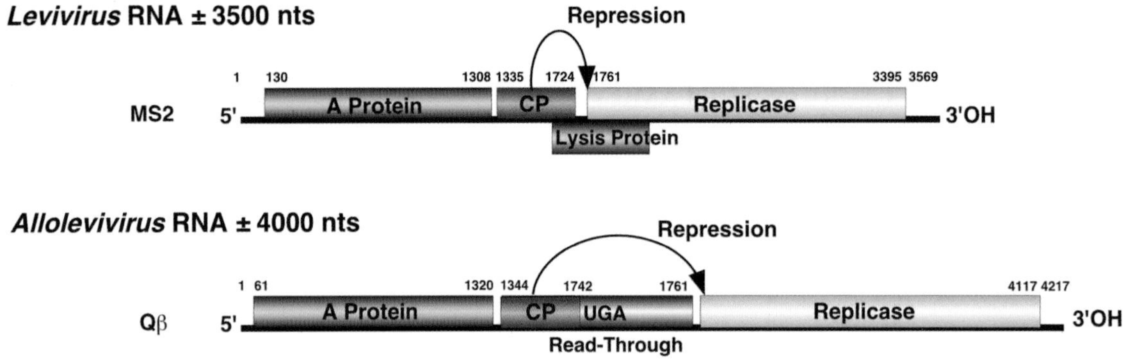

Figure 2: General genetic map of a representative levivirus; Enterobacteria phage MS2 (MS2) and an allolevivirus; Enterobacteria phage Qβ (Qβ). The maturation protein is also called A-protein. The lysis gene overlaps the replicase gene in a +1 frameshift. Arrows indicate repression of replicase translation by capsid protein binding to an RNA hairpin structure (the operator) present at the start of the gene. The UGA nonsense codon (nt 1742) is occasionally (~6%) misread as tryptophan to produce the read-through protein.

PROTEINS
The capsid contains 180 copies of the CP (14 kDa) arranged in an icosahedron. The structure of the protein shell of several ssRNA phages has been solved by X-ray crystallography, and shows 60 quasi-symmetric AB- and 30 symmetric CC'-dimers. The A and C subunits are situated around the 3-fold axes, and the B subunits around the 5-fold axes of the icosahedron. The CP has no structural similarity to those of eukaryote icosahedral RNA viruses. The X-ray structure of the capsid in a complex with the 19 nt operator shows interaction of the dimers with this hairpin. Each virion contains one copy of the A-protein (35-61 kDa), which is required for maturation of the virion and for pilus attachment (Fig. 1). Alloleviviruses also contain several copies of the read-through protein in their capsid. Virions lacking the A-protein are RNase-sensitive.

LIPIDS
None reported.

CARBOHYDRATES
None reported.

GENOME ORGANIZATION AND REPLICATION

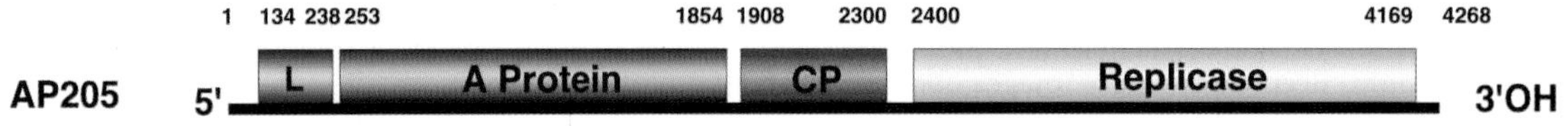

Figure 3: Genetic map of Acinetobacter phage AP205 (AP205). Note the location of the tentative lysis gene at the 5'-terminus. AP205 is unusually long for a levivirus. This map corrects the one previously published (Klovins *et al*, 2002). A: A-protein CP: capsid protein R: replicase L: lysis.

Members of the family *Leviviridae* that propagate in *E. coli* infect by adsorption to the sides of F(ertility) pili (Non-coliphages such as Pseudomonas phage PP7 (PP7) and Acinetobacter phage AP205 (AP205) bind to other pili). This event leads to cleavage of the A-protein and release of the RNA from the virion into the bacterium. The infecting RNA encodes a replicase, which assembles with three host proteins (ribosomal protein S1 and translation elongation factors EF-Tu and EF-Ts) to form the active RNA polymerase. A fourth protein, called Host Factor, not associated with the polymerase complex but acting directly on the RNA, is needed for synthesis of the minus strand. Members of the two genera use different Host Factors. Plus-strand synthesis requires, besides the virus-coded replicase, only EF-Tu and EF-Ts as cofactors. Late in infection coat-protein dimers act as translational repressors of the replicase gene by binding to an RNA hairpin, the operator, that contains the start site of this gene. This protein-RNA complex is considered to also be the nucleation site for encapsidation. Virions assemble in the cytoplasm around phage RNA. It is unknown at which point the A-protein (and read-through protein) is assembled in the virion but it is assumed to be an early step since the A-protein can not be incorporated into preformed virions lacking the protein. Infection usually results in cell lysis releasing thousands of phages per cell. The lysis protein short-circuits the membrane potential and somehow activates the bacterial autolysins leading to degradation of the peptidoglycan network.

ANTIGENIC PROPERTIES

Members of the family *Leviviridae* are highly antigenic.

BIOLOGICAL PROPERTIES

Members of the *Leviviridae* family occur worldwide and are abundantly present in sewage, waste water, animal and human faeces. In Asia a particular geographic distribution has been noticed with respect to the four levivirus species. It has also been proposed that the various species have a preference for particular hosts *e.g.* Members of *Enterobacteria phage Qβ* are found predominantly in human waste. The evidence is not conclusive. RNA bacteriophages are harmless for humans. Members of the family *Leviviridae* not only infect enterobacteria but also species of the genera *Caulobacter*, *Pseudomonas*, and *Acinetobacter* and probably many other Gram-negative bacteria, provided they express the appropriate pili on their surface. RNA coliphages are often used as indicators for the presence of enteroviruses in waste and surface water. There is renewed interest in phage therapy to combat bacterial infections.

GENUS *LEVIVIRUS*

Type Species *Enterobacteria phage MS2*

DISTINGUISHING FEATURES

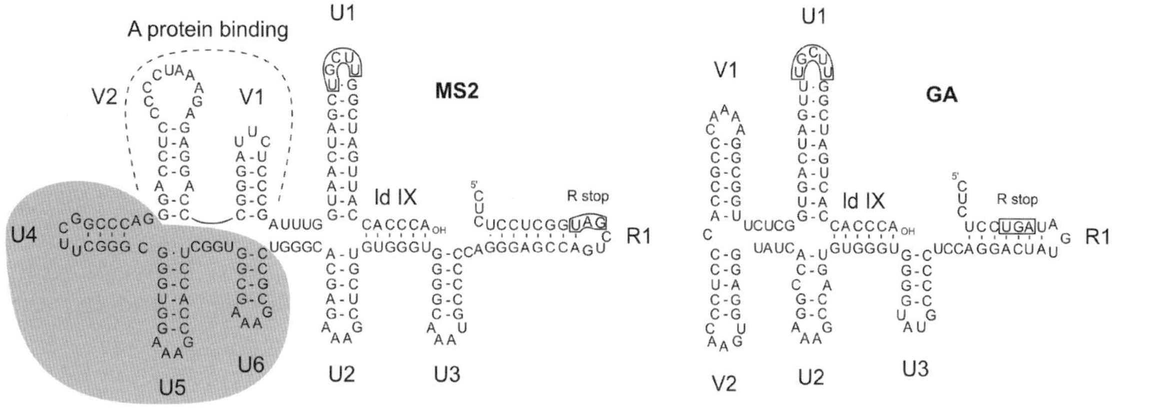

Figure 4: Comparison of the RNA folding in the 3'UTR of Enterobacteria phage MS2 (MS2) and Enterobacteria phage GA (GA). GA lacks the three stem-loops U4, U5 and U6. In MS2 stem-loops V1 and V2 are part of the A-protein binding site. The other part of the protein's binding site is located around nt 400.

Leviviruses contain the short version of the genome and have a separate gene for cell lysis, which partly overlaps the replicase coding region in the +1 reading frame (Fig. 2). Overlap with the CP gene is variable. Genome size ranges from 3466 for GA (*Enterobacteria phage BZ13*) to 3577 for fr (*Enterobacteria phage MS2*) (Fig. 2). Leviviruses and alloleviviruses use different Host Factors for their polymerase holoenzyme. The levivirus Host Factor has been isolated but has not been genetically identified. Generally, the replicases from leviviruses poorly replicate allolevivirus RNA and *vice versa.*

Recently, the sequence of RNA of AP205, an RNA phage growing on *Acinetobacter* tentatively identified its lysis gene in the unusual location of the 5'-end. The absence of a read-through protein was taken as criterion to classify AP205 as a levivirus (Fig. 3).

GENOME ORGANIZATION AND REPLICATION

Figure 2 shows the map of the levivirus genome. Lysis and replicase synthesis are dependent on translation of the CP gene: early CP nonsense mutants are deficient in replicase and lysis protein synthesis. Translational starts at the lysis gene were shown to be reinitiations by ribosomes that had completed CP-gene reading but had not yet detached themselves from the message. A small fraction of these ribosomes manages to back up to the lysis start. Part of the replicase ribosome binding site is base-paired to an upstream sequence located in the coat coding region. A ribosome translating the CP cistron disrupts this interaction, thereby exposing the replicase start site (when not blocked by a CP dimer, which is the case late in infection). The CP gene is freely accessible to ribosomes.

Maturation or A-protein is translated from an RNA folding intermediate which has an accessible ribosome-binding site. This intermediate exists for a short time on nascent strands. Full-length RNA reaches an equilibrium folding in which the start site of the A-protein gene is inaccessible. It is believed that the purpose of these control mechanisms is to facilitate the switch from translation of the viral RNA to its replication. One of the binding sites of the replicase holoenzyme is the start of the CP gene. Binding of the enzyme to this site squeezes out ribosomes from CP, lysis and replicase genes. At this stage the A-protein gene is folded in its ribosome-inaccessible state and replication can proceed without interference from translation.

The polymerase of GA has been purified, that of MS2 may be unstable. Except for the Host Factor the polymerases of leviviruses and alloleviviruses contain the same subunits.

ANTIGENIC PROPERTIES

Antigenic specificity is distinct from that of members of the genus *Allolevivirus*.

LIST OF SPECIES DEMARCATION CRITERIA IN THE GENUS

A major difference between members of the species *Enterobacteria phage MS2* and *Enterobacteria phage BZ13* (formerly called subgroups I and II) is a ~60 nt deletion in the 3'-UTR of members of *Enterobacteria phage BZ13*, comprising three small RNA hairpins (Fig. 4). There is also a 35 nt deletion in the replicase gene of members of *Enterobacteria phage BZ13* producing a shorter hairpin stem. Furthermore, the percentage of aa or nt sequence identity is dramatically lower between the two species than between strains within a species. Species can also be distinguished by serological means and by species-specific antisense DNA probes.

LIST OF SPECIES IN THE GENUS

Species names are in green italic script; strain names and synonyms are in black roman script; tentative species names are in blue roman script. Sequence accession numbers, and assigned abbreviations ( ) are also listed.

SPECIES IN THE GENUS

Enterobacteria phage MS2

Enterobacteria phage f2		(f2)
Enterobacteria phage fr	[X15031]	(fr)
Enterobacteria phage JP501	[AF227251]	(JP501)
Enterobacteria phage M12	[AF195778]	(M12)
Enterobacteria phage MS2	[GB-PH:MS2CG]	(MS2)
Enterobacteria phage R17		(R17)

Enterobacteria phage BZ13		
Enterobacteria phage BZ13		(BZ13)
Enterobacteria phage GA	[NC_001426]	(GA)
Enterobacteria phage JP34	[J04343]	(JP34)
Enterobacteria phage KU1	[AF227250]	(KU1)
Enterobacteria phage TH1		(TH1)

TENTATIVE SPECIES IN THE GENUS

Acinetobacter phage AP205	[AF334111]	(AP205)
Pseudomonas phage PP7	[X80191]	(PP7)

GENUS *ALLOLEVIVIRUS*

Type Species *Enterobacteria phage Qβ*

DISTINGUISHING FEATURES

Alloleviviruses contain the longer version of the genome (Fig. 2). The extra RNA encodes a C-terminal extension of the CP arising by occasional suppression of the CP gene termination codon. The read-through protein, is present at ~12 copies per virion, together with the A-protein, is necessary for infection. Its precise role is not known. There is no separate lysis gene. Cell lysis is a secondary function of the A-protein. Genome length varies between 4217 nt for Qβ (*Enterobacteria phage Qβ*) and 4276 nt for SP (*Enterobacteria phage F1*) (Fig. 2).

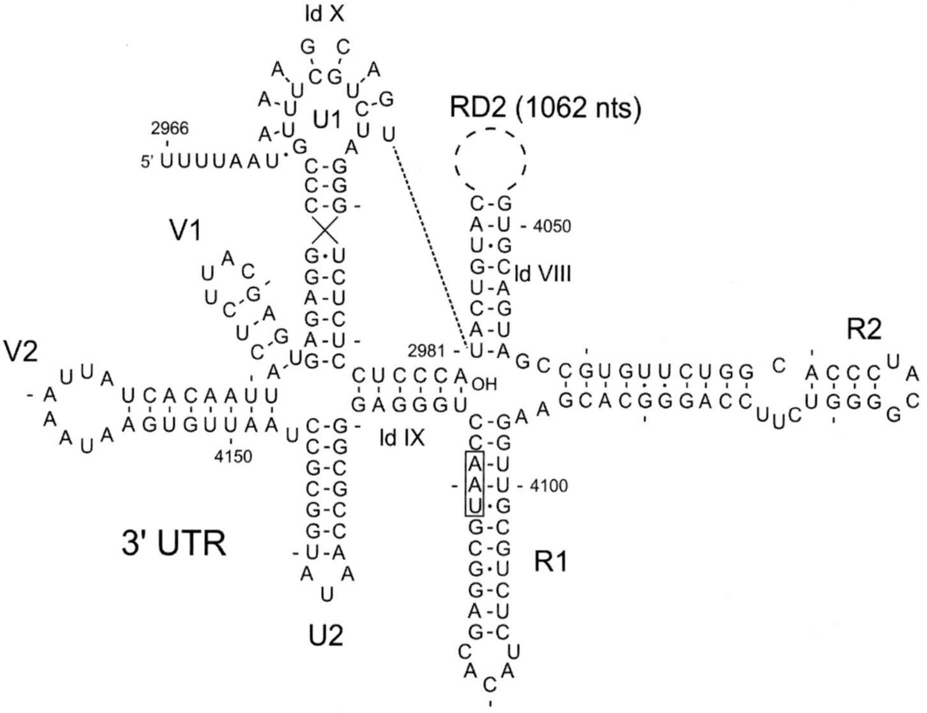

Figure 5: RNA secondary structure for Enterobacteria phage Qβ (Qβ) RNA from nt 2966 to the 3′-end (nt 4217) marked as A_{OH}. The UAA stop codon (nt 4119) of the replicase gene is boxed. Replicase Domain 2 (RD2) containing 1062 nt has been replaced by a dotted circle. Breaking two or three basepairs in the central pseudoknot (IdX) or IdVIII abolishes replication. However, breaking the pairs in Id IX, which buries the 3′ terminal nucleotides, stimulates replication. Production of minus strand is also inhibited by deletion of stem-loops U1, V1, V2 or U2. (R1 and R2 were not tested).

Genome Organization and Replication

Genome organization is shown Figure 2. The RNA polymerase of Qβ has been purified and the enzyme can amplify Qβ RNA *in vitro*. The Host Factor has been purified and genetically characterized. It is the product of the *hfq* gene. In the uninfected cell the protein functions in the transition to stationary phase. In particular, it stimulates translation of the mRNA encoding the σ^{38} protein involved in transcription of stationary phase genes. Hfq is a sequence non-specific ssRNA binding protein with some preference for A-residues. It is heat resistant and acts as a pentamer. The protein helps the polymerase to get access to the 3'-end of the plus strand, which exists in a base-paired and therefore inactive state. In Fig. 5 the secondary structure of the 3'UTR of Qβ RNA is shown; the 3'-terminal 6 nt are taken up in long-distance interaction with ld IX.

Although the polymerases are specific for their own RNA, the interaction with RNA involves host-encoded subunits (EF-Tu, S1 and Hfq) that have no sequence specificity. An important contribution to template activity is provided by the higher order structure of Qβ RNA (Fig. 5). For instance, destroying 2 out of the 8 base pairs that make up the central pseudoknot in Qβ RNA, here indicated as ld X, lowers replication 100-fold. The higher order structures of the RNAs of phages PP7 (tentative) and SP (*Enterobacteria phage F1*) are shown in Fig. 6.

The switch from translation to replication is as in leviviruses and was first formulated for Qβ. Control of the maturation protein is slightly different. The time window for producing the A-protein is not set by the lifetime of a folding intermediate, as for MS2, but by the time it takes the polymerase to move from position ~60, the start of the A-protein gene, to position ~470 where the complement to the Shine-Dalgarno sequence of the A-protein gene is located. Once this complement is synthesized pairing between the two regions blocks further translation.

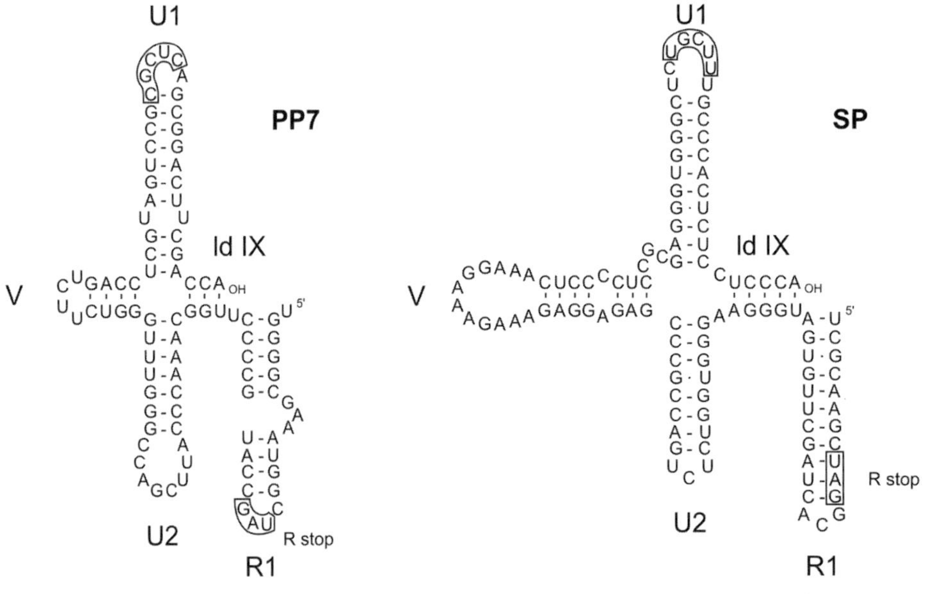

Figure 6: RNA secondary structure in the 3'UTR of Pseudomonas phage PP7 (PP7) and Enterobacteria phage SP (SP). The folding of PP7 RNA is much more like that of SP RNA than that of either MS2 or GA (Fig. 4). Compared to MS2 the stem-loops U3, U4, U5, U6 and one of the two V-loops are missing. The boxed sequence in the loop of hairpin U1 is conserved in all viruses of the family *Leviviridae*. The sequence is part of the central pseudoknot in Qβ. The pseudoknot is believed to exist also in the other phages.

Antigenic Properties

Antigenic specificity is distinct from that of members of the genus *Levivirus*.

List of Species Demarcation Criteria in the Genus

The major difference between *Enterobacteria phage Qβ* and *Enterobacteria phage F1* (formerly called subgroups III and IV respectively) is a ~90-nt deletion in the maturation-protein gene of Qβ, corresponding to a bifurcated hairpin. There is also the extra stem-loop (V1) in the 3'UTR of members of *Enterobacteria phage Qβ* that is lacking in members of *Enterobacteria phage F1*. Species can also be differentiated by serological criteria and by species-specific antisense DNA probes. Finally, the percentage of aa or nt sequence identity is dramatically lower between the two species than between strains within a species.

List of Species in the Genus

Species names are in green italic script; strain names and synonyms are in black roman script; tentative species names are in blue roman script. Sequence accession numbers, and assigned abbreviations ( ) are also listed.

Species in the Genus

Enterobacteria phage F1		
Enterobacteria phage F1		(F1)
Enterobacteria phage ID2		(ID2)
Enterobacteria phage NL95	[AF059243]	(NL95)
Enterobacteria phage SP	[NC_004301]	(SP)
Enterobacteria phage TW28		(TW28)
Enterobacteria phage Qβ		
Enterobacteria phage Qβ	[AY099114]	(Qβ)
Enterobacteria phage M11	[NC_004304]	(M11)
Enterobacteria phage MX1	[NC_001890]	(MX1)
Enterobacteria phage ST		(ST)
Enterobacteria phage TW18		(TW18)
Enterobacteria phage VK		(VK)

Tentative Species in the Genus

None reported.

List of Unassigned Viruses in the Family

Caulobacter phage φCb12r	(φCb12r)
Caulobacter phage φCb2	(φCb2)
Caulobacter phage φCb23r	(φCb23r)
Caulobacter phage φCb4	(φCb4)
Caulobacter phage φCb5	(φCb5)
Caulobacter phage φCb8r	(φCb8r)
Caulobacter phage φCb9	(φCb9)
Caulobacter phage φCP18	(φCP18)
Caulobacter phage φCP2	(φCP2)
Caulobacter phage φCr14	(φCr14)
Caulobacter phage φCr28	(φCr28)
Enterobacteria phage β	(β)
Enterobacteria phage τ	(τ)
Enterobacteria phage α15	(α15)
Enterobacteria phage μ2	(μ2)
Enterobacteria phage B6	(B6)
Enterobacteria phage B7	(B7)
Enterobacteria phage C-1	(C-1)
Enterobacteria phage C2	(C2)
Enterobacteria phage fcan	(fcan)
Enterobacteria phage Folac	(Folac)

Enterobacteria phage Iα	(Iα)
Enterobacteria phage M	(M)
Enterobacteria phage pilHα	(pilHα)
Enterobacteria phage R23	(R23)
Enterobacteria phage R34	(R34)
Enterobacteria phage ZG/1	(ZG/1)
Enterobacteria phage ZIK/1	(ZIK/1)
Enterobacteria phage ZJ/1	(ZJ/1)
Enterobacteria phage ZL/3	(ZL/3)
Enterobacteria phage ZS/3	(ZS/3)

(other enterobacteriophages, with many plasmid specificities, have been reported).

Pseudomonas phage 7s	(7s)
Pseudomonas phage PRR1	(PRR1)

PHYLOGENETIC RELATIONSHIPS WITHIN THE FAMILY

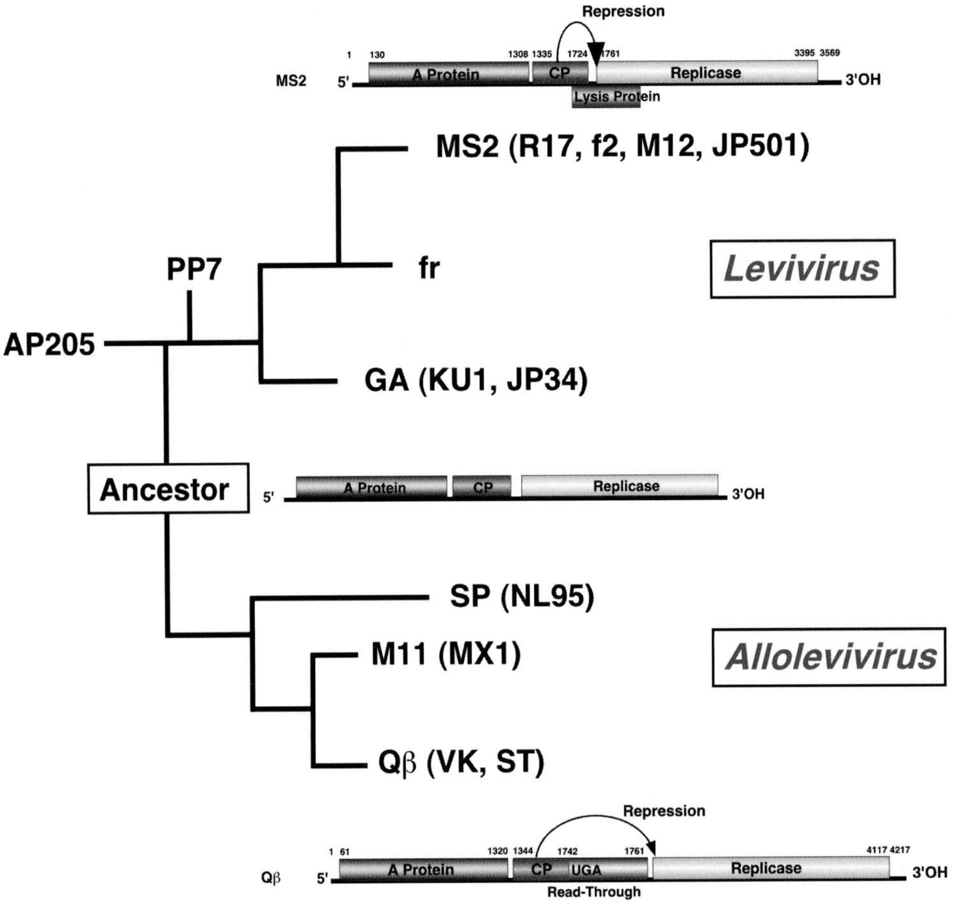

Figure 7: Proposed phylogenetic tree for the family *Leviviridae*. Distances are arbitrary. The ancestor only has the three basic genes. Lysis is effected by the A-protein as it still is today in Qβ. Presumably, fitness of the ancestor was restricted by the double function of the A-protein (Bollback and Huelsenbeck, 2001). The leviviruses solved the problem by evolving a separate lysis protein either encoded on a vacant region of the genome (AP205) or resulting from a ribosomal restart following translation termination at the end of the capsid gene (other leviviruses). Once restrictions on the A-protein were relaxed the gene could evolve in various directions to better fulfill its remaining function: virion maturation and infection. Two features of leviviruses can be explained by this scenario: first, lysis genes have variable startpoints (even between MS2 and fr or between GA and KU1) and secondly, of the three "old" genes, the A-protein gene shows the lowest sequence conservation. The alloleviviruses solved the dual-function problem by transfering part of the maturation and infection function to a new protein, read-through, which arose by an insertion between coat and replicase genes. Presumably, this allowed the A-protein to improve its lysis function. Such a scenario would provide a

different reason why also in the alloleviviruses the A-protein is least conserved of the "old" genes. Signification of the abbreviations of virus names are to be found in "List of Species in the genus".

A tentative phylogenetic tree of the family *Leviviridae* is given in Figure 7. Relationships have been based first on deeply rooted features such as the genetic map and second on similarity in RNA folding, in particular the one present at the 3'UTR which is conserved in its outline. As a result there is a fundamental split between leviviruses and alloleviviruses because they have different maps. The two non-coli leviviruses AP205 and PP7 have been placed closer to the ancestor than the coli leviviruses because they have the same folding of their 3'UTR as the alloleviviruses (Fig. 6). As a result MS2 and GA are closer to the non-coliphages than to coliphage Qβ. PP7 is placed closer to MS2 than AP205 because AP205 has its lysis gene in a different position. PP7 has it in the same position as MS2, fr and GA.

In this scheme, the ancestor contains only the three basic genes and the A-protein has the double function of lysis and maturation (infection). We assume that its 3'-UTR is folded in the simple way of PP7 (AP205) and Qβ (SP) (Fig. 6).

The subdivision of each genus in two species is based on criteria explained above. Based on the sequence it is possible to make subtle distinctions between strains within a species. For example, MS2, R17, f2, M12 and JP501 are extremely close (~95% identity) whereas fr is much further away (~80% identity), has some features of members of *Enterobacteria phage BZ13*, but is still clearly a member of *Enterobacteria phage MS2*.

Similarity with Other Taxa

Not reported.

Derivation of Names

Levi: from Latin *levis*, *"light"*.
Allo: from Greek *allov*, *"other"*.

References

Ackermann, H.W. and Dubow, M.S. (eds) (1987). *Viruses of Prokaryotes*, Vol. 2. CRC Press, Boca Raton, Florida.

Bollback, J.P. and Huelsenbeck, J.P. (2001). Phylogeny, genome evolution and host specificity of single-stranded RNA bacteriophage (Family *Leviviridae*). *J. Mol. Evol.*, **52**, 117-128.

Brown, D. and Gold, L. (1996). RNA replication by Qβ replicase: A working model. *Proc. Natl. Acad. Sci. USA*, **93**, 11558-11562.

Convery, M.A., Rowsell, S., Stonehouse, N.J., Ellington, A.D., Hirao, I., Murray, J.B., Peabody, D.S., Phillips, S.E.V. and Stockley, P.G. (1998). Crystal structure of an RNA aptamer-protein complex at 2.8 Å resolution. *Nature Struct. Biol.*, **5**, 133-139.

Fiers, W. (1979). Structure and function of RNA bacteriophages. In: *Comprehensive Virology*, (H. Fraenkel-Conrat and R.R. Wagner, eds), Vol. 13, pp 69-204. Plenum Press, New York.

Furuse, K. (1987). Distribution of the coliphages in the environment. In: *Phage Ecology* (S.M. Goyal, C.P. Gerber and G. Bitton, eds), pp 87-124. John Wiley and Sons, New York.

Havelaar, A.H. (IAWPRC Study group on health related water microbiology). (1991). Bacteriophages as model viruses in water quality control. *Wat. Res.*, **25**, 529-545.

Jacobson, A.B., Arora, R., Zuker, M., Priano, C., Liu, C.H. and Mills, D.R. (1998). Structural plasticity in RNA and its role in the regulation of translation in Qβ. *J. Mol. Biol.*, **275**, 589-600.

Klovins, J., Overbeek, G.P., van den Worm, S.H.E., Ackermann, H.-W. and van Duin, J. (2002). Nucleotide sequence of a ssRNA phage from *Acinetobacter*: kinship to coliphages. *J. Gen. Virol.*, **83**, 1523-1533.

Miranda, G., Schuppli, D., Barrera, I., Hausherr, C., Sogo, J.M. and Weber, H. (1997). Recognition of Qβ plusstrand RNA as a template by Qβ replicase: role of RNA interactions mediated by ribosomal protein S1 and Host Factor. *J. Mol. Biol.*, **267**, 1089 -1103.

Rohde, N., Daum, H. and Biebricher, K. (1995). The mutant distribution of an RNA species replicated by Qβ replicase. *J. Mol. Biol.*, **294**, 754-762.

Schuppli, D., Georgijevic, J. and Weber, H. (2000). Synergism of mutations in Qβ RNA affecting host factor dependence of Qβ replicase. *J. Mol. Biol.*, **295**, 149-154.

Sledjeski, D., Whitman, C. and Zhang, A. (2001). Hfq is necessary for regulation by the untranslated RNA DsrA. *J. Bacteriol.*, **183**, 1997-2005.

van Duin, J. (1988). Single stranded RNA bacteriophages. In: *The Bacteriophages* (R. Calendar, ed), pp 117-167. Plenum Press, New York.

Valegård, K., Liljas, L., Fridborg, K. and Unge, T. (1990). The three-dimensional structure of the bacterial virus MS2. *Nature*, **345**, 36-41.

Valegård, K., Murray, J.B., Stockley, P.G., Stonehouse, N.J. and Liljas, L. (1994). Crystal structure of an RNA bacteriophage coat protein-operator complex. *Nature*, **371**, 623-626.

Zinder, N.D. (ed) (1975). *RNA phages*. Cold Spring Harbor Laboratory Press. Monograph Series. Cold Spring Harbor, New York.

Contributed By

van Duin, J. and van den Worm, S.

FAMILY NARNAVIRIDAE

TAXONOMIC STRUCTURE OF THE FAMILY

Family	***Narnaviridae***
Genus	***Narnavirus***
Genus	***Mitovirus***

Viruses in the family *Narnaviridae* consist of a single molecule of non-encapsidated positive-strand RNA of 2.3-2.9 kb, which encodes a single protein of 80-104 kDa with amino acid sequence motifs characteristic of RdRps.

GENUS NARNAVIRUS

Type Species *Saccharomyces 20S RNA narnavirus*

VIRION PROPERTIES

MORPHOLOGY

No true virions are found associated with members of this genus. The genomes, however, are associated with their RdRps forming ribonucleoprotein complexes in 1:1 stoichiometry. Genetic and biochemical evidence show that they are cytoplasmically-located.

PHYSICOCHEMICAL AND PHYSICAL PROPERTIES

The ribonucleoprotein complex sediments through a sucrose gradient with a sedimentation coefficient ~20S. These complexes are quite stable at pH 9.0 and have *in vitro* RNA polymerase activity that synthesizes mainly 20S RNA, and a minor amount of complementary strands.

NUCLEIC ACID

The Saccharomyces 20S RNA narnavirus (ScNV-20S) genome is a linear ssRNA of 2.5 kb in size with a high G+C content (~60%). There is no poly(A) tail at the 3'-end and it is not known whether the 5'-end is capped. It is present in a high copy number under stress conditions, such as growth under nitrogen starvation, reaching up to 100,000 copies/cell.

PROTEINS

No structural proteins have been described for members of this family. ScNV-20S has coding capacity for a protein of 91 kDa (p91), with sequences conserved among RdRps. The conserved sequences are more similar to those of replicases of ssRNA enterobacteria phages than polymerases of members of the family *Totiviridae* in the same host. This protein is quite basic (estimated pI of 11) and has ssRNA binding activity. Protein p91 is essential for replication and responsible for the *in vitro* RdRp activity that synthesizes 20S RNA. P91 does not undergo proteolytic processing after translation. Studies using antibodies against this protein show that it is expressed in yeast cells grown exponentially or under induction conditions.

LIPIDS

No lipids have been described associated to ScNV-20S.

CARBOHYDRATES

None reported.

GENOMIC ORGANIZATION AND REPLICATION

ScNV-20S has only one ORF that encodes p91, and there are no ORFs with coding capacity larger than 100 aa in the complementary strand. The ORF for p91 spans almost the entire sequence of 20S RNA, with a short untranslated leader sequence at the 5'-end (12 nt) and an UTR at the 3'-end of 12 nt. Two replication models for 20S RNA have been proposed based on the similarity of p91 to the replicases of RNA enterobacteria phages and the replication intermediates obtained in the *in vitro* RNA polymerase reaction. One model is similar to the replication cycle of ssRNA enterobacteria phages such as Qβ; that is, ScNV-20S is copied into the complementary strands and these copies serve as templates for 20S RNA synthesis. Annealing of 20S RNA and its complementary strand gives a double-stranded form of ScNV-20S. This dsRNA called W can be easily isolated from all ScNV-20S-containing yeast strains. The other model hypothesizes that W dsRNA is the replicative form of ScNV-20S. At present, available data support the first model. Recently, a reverse genetics system for ScNV-20S has been established. Like native viruses, viruses generated from cDNA vectors can be transmitted to daughter cells indefinitely without the vector or any selection.

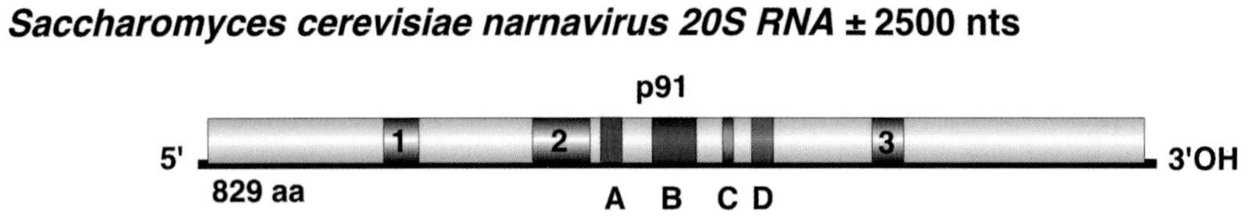

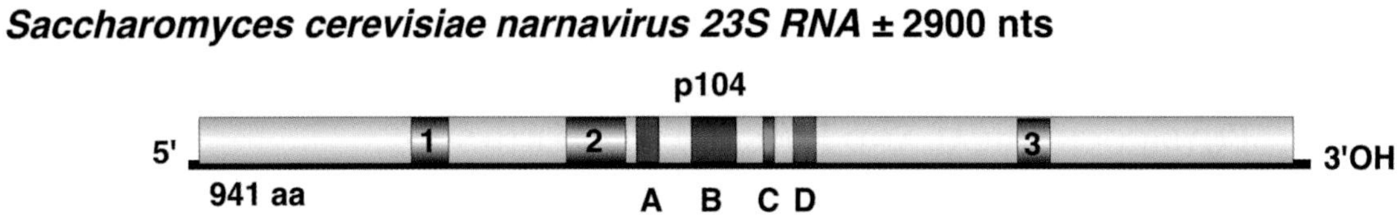

Figure 1. Genomic organization of Saccharomyces 20S RNA narnavirus (ScNV-20S) and Saccharomyces 23S RNA narnavirus (ScNV-23S) and the proteins encoded on them (p91 and p104, respectively). Sequence motifs (A to D) conserved in RdRp are boxed and shaded. Motifs 1, 2 and 3 are present only in p91 and p104.

Biological Properties

ScNV-20S infects more than 90% of laboratory strains of the baker's yeast *Saccharomyces cerevisiae*. Some strains isolated from the brewery industry also have been found to carry ScNV-20S. There is no phenotype associated with the presence of this RNA. Like other viruses of fungi, there is no extracellular stage in the ScNV-20S life cycle. Transmission takes place through mating or cytoplasmic mixing. These viruses are very stable. Known curing procedures that eliminate members of the family *Totiviridae* in the same host, such us growth at high temperature, or with cycloheximide, acridine orange, or guanidine HCl, do not eliminate ScNV-20S.

List of Species Demarcation Criteria in the Genus

Narnaviruses generally replicate stably within the cell as the cells grow. Virus strains of the same species are expected to segregate relative to each other as the cells grow, whereas those of different species should be stably co-maintained. Viruses of the same species should be similarly affected by host chromosomal mutations. Viruses that can recombine or exchange segments with each other to give viable progeny should be considered the same species. Although these biological criteria are the prime determinants of species, sequence criteria also are used. Less than 50% sequence identity at the protein level generally reflects a species difference. None of the above criteria is absolute, but narnaviruses described so far leave little doubt about species demarcation. For example, ScNV-20S and ScNV-23S are only 30% identical in the 439 aa region of highest similarity. More important, they are stably compatible with each other in the same yeast strain.

List of Species in the Genus

Species names are in green italic script; strain names and synonyms are in black roman script; tentative species names are in blue roman script. Sequence accession numbers, and assigned abbreviations ( ) are also listed.

Species in the Genus

Saccharomyces 20S RNA narnavirus		
Saccharomyces 20S RNA narnavirus	[M63893]	(ScNV-20S)
Saccharomyces 23S RNA narnavirus		
Saccharomyces 23S RNA narnavirus	[M86595]	(ScNV-23S)

Tentative Species in the Genus

None reported.

Genus *Mitovirus*

Type Species *Cryphonectria mitovirus 1*

Virion properties

No virions have been reported for members of this genus.

Nucleic Acid

The virus genome consists of a single molecule of RNA of 2.3-2.7 kb. Double-stranded RNAs in this size range can be isolated from mitochondria of infected isolates. Single-stranded RNA of the same size, and corresponding to the coding strand of the dsRNA, is present in infected tissue in greater molar amount than the dsRNA. The 5' and 3' sequences can be folded into stable stem-loop structures. For some mitoviruses, the 5' and 3' sequences are complementary. The coding strand has 62-73% A+U residues, but no poly(A) tail is associated with the 3'-end.

Proteins

No structural proteins are known to be associated with the virus ssRNA or dsRNA.

Antigenic Properties

None reported.

Genome Organization and Replication

The putative coding strand is predicted to be translatable only in mitochondria, not in cytoplasm. When mitochondrial codon usage is invoked (UGA coding for tryptophan), the deduced translation product is a protein of 80-97 kDa, containing RdRp motifs. RdRp activity and an 80 kDa RdRp protein have been detected in mitochondria from an infected *Ophiostoma novo-ulmi* isolate. No large polypeptide is predicted from the complementary strand of any mitovirus.

Biological Properties

Mitoviruses have been found in isolates of the chestnut blight fungus, *Cryphonectria parasitica*, Dutch elm disease fungi, *Ophiostoma novo-ulmi* and *O. ulmi*, and *Sclerotinia homoeocarpa*, the cause of dollar spot of turf grass. Fungal isolates may contain one or several mitoviruses. Some, but not all, member viruses reduce virulence of the fungus (i.e., cause "hypovirulence"). Mitoviruses are localized in mitochondria. They can be transmitted to uninfected strains by hyphal fusion (anastomosis). The transmission rate through asexual spores (conidia) is virus-specific and varies from 10-100%. In *C. parasitica*, transmission through sexual spores (ascospores) occurs at 20-50% when the infected parent is the female in matings, but does not occur when the infected parent is male in matings. In *O. novo-ulmi*, viruses are usually excluded from ascospores, even when both parents are infected. Identical mitoviruses have been found in *O. novo-ulmi* and *O. ulmi*, and a strain of *Ophiostoma mitovirus 3a* has been reported in *Sclerotinia homoeocarpa*, suggesting that both interspecies and intergenus virus transmission occurs in nature.

List of Species Demarcation Criteria in the Genus

Species demarcation criteria have not been precisely defined. However, amino acid sequence identities of putative RdRp proteins between the different mitovirus species so far defined are less than 40%. Amino acid sequence identities of putative RdRp proteins between strains of the same mitovirus species are greater than 90%.

List of Species in the Genus

Species names are in green italic script; strain names and synonyms are in black roman script; tentative species names are in blue roman script. Sequence accession numbers, and assigned abbreviations ( ) are also listed.

Species in the Genus

Cryphonectria mitovirus 1		
Cryphonectria mitovirus 1	[L31849]	(CMV1)
Ophiostoma mitovirus 3a		
Ophiostoma mitovirus 3a	[AJ004930; AY172454]	(OMV3a)
Ophiostoma mitovirus 4		
Ophiostoma mitovirus 4	[AJ132754]	(OMV4)
Ophiostoma mitovirus 5		
Ophiostoma mitovirus 5	[AJ132755]	(OMV5)
Ophiostoma mitovirus 6		
Ophiostoma mitovirus 6	[AJ132756]	(OMV6)

Tentative Species in the Genus

Gremmeniella mitovirus S1	[AF534641]	(GMVS1)
Ophiostoma mitovirus 1a		(OMV1a)
Ophiostoma mitovirus 1b		(OMV1b)
Ophiostoma mitovirus 2		(OMV2)
Ophiostoma mitovirus 3b		(OMV3b)

List of Unassigned Viruses in the Family

Rhizoctonia virus M2	[U51331]	(RVM2)

Phylogenetic Relationships Within the Family

In a neighbor-joining phylogenetic tree based on aa sequences of the putative RdRp proteins, the mitovirus and narnavirus genera are clearly distinguished, but nevertheless form a significant cluster (Fig. 2). The putative RdRp protein of the unassigned virus, Rhizoctonia virus M2 (RVM2), clusters with those of the mitoviruses (Fig. 2). However, since only a small proportion of RVM2 copurifies with mitochondria with most being found in the cytoplasm, RVM2 does not use the mitochondrial code, and there is evidence for a DNA copy in the host genome, this suggests significant differences from the mitoviruses.

Similarity with Other Taxa

The putative RdRp proteins of narnaviruses and mitoviruses are distantly related to those of bacteriophages in the family *Leviviridae* (Fig. 2). Furthermore, the 3'-end secondary structures of members of the genus *Narnavirus* resemble those of coliphages in the family *Leviviridae*. In a neighbor-joining phylogenetic tree of families of fungus viruses and related viruses in other taxa, based on aa sequences of the putative RdRp proteins, the families *Narnaviridae* and *Leviviridae* form a cluster with 69.2% bootstrap support (Fig. 2).

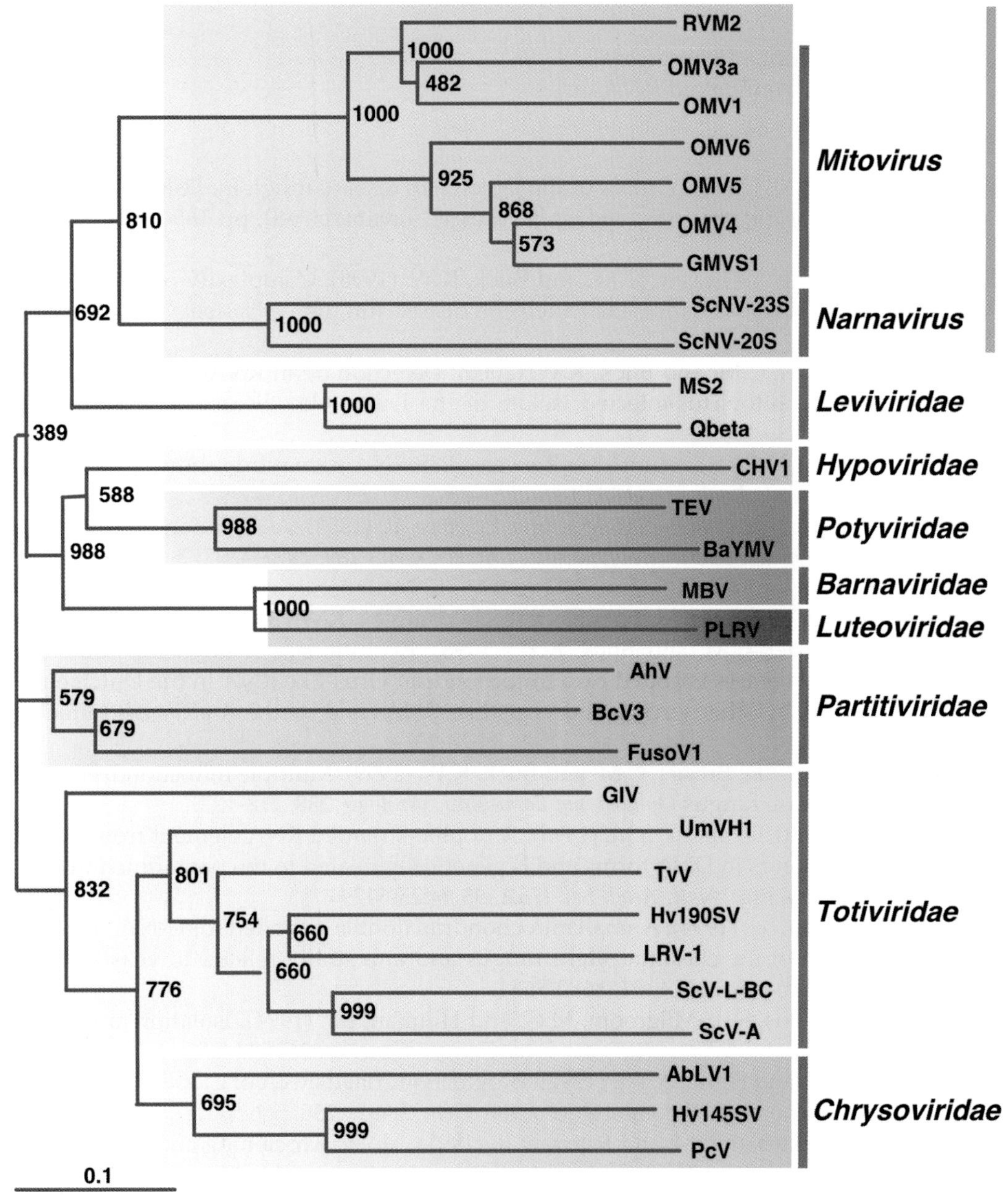

Figure 2. Phylogenetic tree based on aa sequences of motifs A to E (Hong *et al.* 1998) of the putative RdRp proteins of members of the family *Narnaviridae*, other families of RNA viruses of fungi and related viruses in other host taxa, and the family *Leviviridae* of RNA bacteriophages. Sequence alignments and the neighbor-joining tree were made using the Clustal X program. Bootstrap numbers (1000 replicates) are shown on the nodes. Abbreviations and sequence acquisition numbers [ ] are: AbVL1, Agaricus bisporus virus L1 [X94361]; AhV, Atkinsonella hypoxylon virus [L39126]; BaYMV, Barley yellow mosaic virus [D01091]; BcV3, Beet cryptic virus 3 [S63913]; CHV1, Cryphonectria hypovirus 1 [M57938]; CMV1, Cryphonectria mitovirus 1 [L31849]; FusoV1, Fusarium solani virus 1 [D55668]; GlV, Giardia lamblia virus [L13218]; GMVS1, Gremmeniella mitovirus S1 [AF534641]; Hv145SV, Helminthosporium victoriae 145S virus [AF297176]; Hv190SV, Helminthosporium victoriae 190S virus [U41345]; LRV1, Leishmania RNA virus 1-1 [M92355]; MBV, Mushroom bacilliform virus [U07551]; MS2, Enterobacteria phage MS2 [GB-PH:MS2CG]; OMV3a, Ophiostoma mitovirus 3a [AJ004930]; OMV4, Ophiostoma mitovirus 4 [AJ132754]; OMV5, Ophiostoma mitovirus 5 [AJ132755]; OMV6, Ophiostoma mitovirus 6 [AJ132756]; PcV, Penicillium chrysogenum virus [AF296439]; PLRV, Potato leafroll virus [X14600]; Qbeta, Enterobacteria phage Qβ [AY099114]; RVM2, Rhizoctonia virus M2 [U51331]; ScV-L-A, Saccharomyces cerevisiae virus L-A [J04692]; ScV-L-BC, Saccharomyces cerevisiae virus L-BC [U01060]; ScNV-20S, Saccharomyces 20S RNA narnavirus [M63893]; ScNV-23S, Saccharomyces 23S RNA narnavirus [M86595]; TEV, Tobacco etch virus [M15239]; TvV, Trichomonas vaginalis virus [U08999]; UmVH1, Ustilago maydis virus H1 [NC_003823].

Derivation of Names

Mito: sigla from *mito*chondrial.
Narna: sigla from *na*ked *RNA* virus.

References

Buck, K.W. and Brasier, C.M. (2002). Viruses of the Dutch elm disease fungi. In: *DsRNA genetic elements: concepts and applications in agriculture, forestry and medicine* (S.M. Tavantzis, ed), pp 165-190. CRC Press, Boca Raton, Florida.

Cole, T.E., Müller, B., Hong, Y., Brasier, C.M. and Buck, K.W. (1998). Complexity of virus-like double-stranded RNA elements in a diseased isolate of the Dutch elm disease fungus, *Ophiostoma novo-ulmi*. *J. Phytopathol.*, **146**, 593-598.

Cole, T.E., Hong, Y., Brasier, C.M. and Buck, K.W. (2000). Detection of an RNA-dependent RNA polymerase in mitochondria from a mitovirtus-infected isolate of the Dutch elm disease fungus, *Ophiostoma novo-ulmi*. *Virology*, **268,** 239-243.

Esteban, R., and Fujimura,T. (2003). Launching the yeast 23S RNA narnavirus shows 5′ and 3′ *cis*-acting signals for replication. *Proc. Natl. Acad. Sci. USA*, **100**, 2568-2573.

Esteban, L.M., Fujimura, T., García-Cuéllar, M.P. and Esteban, R. (1994). Association of yeast viral 23S RNA with its putative RNA-dependent RNA polymerase. *J. Biol. Chem.*, **269**, 29771-29777.

García-Cuéllar, M.P., Esteban, R. and Fujimura, T. (1997). RNA-dependent RNA polymerase activity associated with the yeast viral p91/20S RNA ribonucleoprotein complex. *RNA*, **3**, 27-36.

Hong. Y., Cole, T.E., Brasier, C.M. and Buck, K.W. (1998). Evolutionary relationships among putative RNA-dependent RNA polymerases encoded by a mitochondrial virus-like RNA in the Dutch elm disease fungus, *Ophiostoma novo-ulmi*, by other viruses and virus-like RNAs and by the *Arabidopsis* mitochondrial genome. *Virology*, **246,** 158-169.

Hong, Y., Dover, S.L., Cole, T.E., Brasier, C.M. and Buck, K.W. (1999). Multiple mitochondrial viruses in an isolate of the Dutch elm disease fungus *Ophiostoma novo-ulmi*. *Virology*, **258**, 118-127.

Lakshman, D.K., Jian, J. and Tavantzis, S.M. (1998). A double-stranded RNA element from a hypovirulent strain of *Rhizoctonia solani* occurs in DNA forms and is genetically related to the pentafunctional AROM protein of the shikimate pathway. *Proc. Natl. Acad. Sci. USA*, **95,** 6425-6429.

Polashock, J.J. and Hillman, B.I. (1994). A small mitochondrial double-stranded (ds) RNA element associated with a hypovirulent strain of the chestnut blight fungus and ancestrally related to yeast cytoplasmic T and W dsRNAs. *Proc. Natl. Acad. Sci. USA*, **91**, 8680-8684.

Polashock, J.J., Anagnostakis, S.L., Milgroom, M.G. and Hillman, B.I. (1994). Isolation and characterization of a virus-resistant mutant of *Cryphonectria parasitica*. *Current Genetics*, **26**, 528-534.

Polashock, J.J., Bedker, P.J. and Hillman, B.I. (1997). A mitochondrial dsRNA of *Cryphonectria parasitica*: Ascospore inheritance and mitochondrial recombination. *Mol. Gen. Genet.*, **256**, 566-571.

Rodríguez-Cousiño, N., Esteban, L.M. and Esteban, R. (1991). Molecular cloning and characterization of W double-stranded RNA, a linear molecule present in *Saccharomyces cerevisiae*: identification of its single-stranded RNA form as 20S RNA. *J. Biol. Chem.*, **266**, 12772-12778.

Rodriguez-Cousiño, N., Solórzano, A., Fujimura, T. and Esteban, R. (1998). Yeast positive-strand virus-like RNA replicons: 20S and 23S RNA terminal nucleotide sequences and 3′-end secondary structures resemble those of RNA coliphages. *J. Biol. Chem.*, **273**, 20363-20371.

Solórzano, A. Rodríguez-Cousiño, N., Esteban, R. and Fujimura, T. (2000). Persistent yeast single-stranded RNA viruses exist in vivo as genomic RNA:RNA polymerase complexes in 1:1 stoichiometry. *J. Biol. Chem.*, **275**, 26428-26435.

Tavantzis, S.M., Lakshman, and Liu, C. (2002). Double-stranded RNA elements modulating virulence in *Rhizoctonia solani*. In: *DsRNA genetic elements: concepts and applications in agriculture, forestry and medicine* (S.M. Tavantzis, ed), pp 191-211. CRC Press, Boca Raton, Florida.

Thompson, J.D., Gibson, T.J., Plewniak, F., Jeanmougin, F. and Higgins, D.G. (1997). The ClustalX windows interface: flexible strategies for multiple sequence alignment aided by quality analysis tools. *Nuc. Acids Res.*, **24,** 4876-4882.

Widner, W.R., Matsumoto, Y. and Wickner, R.B. (1991). Is 20S RNA naked? *Mol. Cell. Biol.*, **11**, 2905-2908.

Contributed By:

Buck, K.W., Esteban, R. and Hillman, B.I.

FAMILY PICORNAVIRIDAE

TAXONOMIC STRUCTURE OF THE FAMILY

Family	***Picornaviridae***
Genus	***Enterovirus***
Genus	***Rhinovirus***
Genus	***Cardiovirus***
Genus	***Aphthovirus***
Genus	***Hepatovirus***
Genus	***Parechovirus***
Genus	***Erbovirus***
Genus	***Kobuvirus***
Genus	***Teschovirus***

VIRION PROPERTIES

MORPHOLOGY

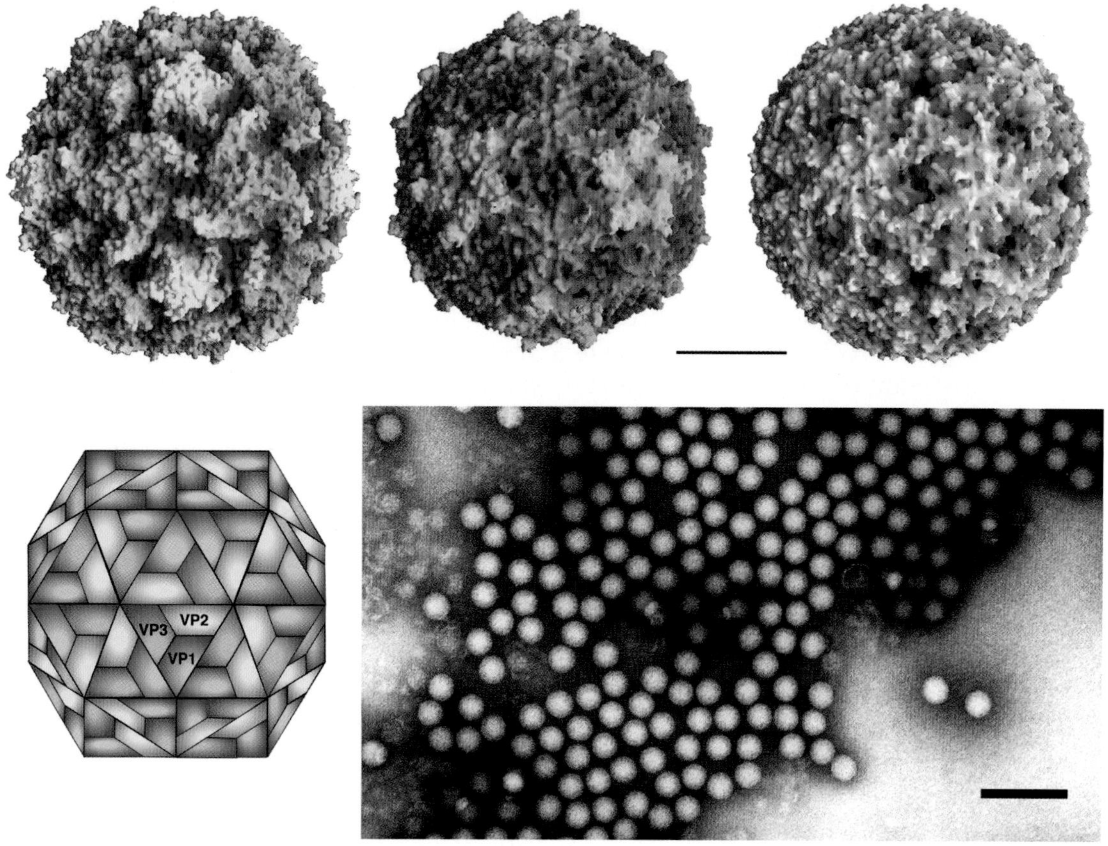

Figure 1: (Top) Pictures of picornavirus structures; Poliovirus type 1 (PV-1) (Left), Mengo virus (Center) and Foot-and-mouth disease virus serotype O (FMDV-O) (Right), PDB entries: 2PLV, 2MEV and 1FOD respectively. The bar represents 10 nm. (Images courtesy of J.Y. Sgro, with permission). (Bottom left) Diagram of a picornavirus particle. The surface shows proteins VP1, VP2 and VP3. The fourth capsid protein, VP4, is located about the internal surface of the pentameric apex of the icosahedron. (Right) Negative contrast electron micrograph of Poliovirus (PV) particles. The bar represents 100 nm. (Courtesy of Ann C. Palmenberg).

Virions consist of a capsid, with no envelope, surrounding a core of ssRNA. Hydrated native particles are 30 nm in diameter, but vary from 22-30 nm in electron micrographs due to drying and flattening during preparation. Electron micrographs reveal no projections, the virion appearing as an almost featureless sphere (Fig. 1). The capsid is composed of 60 identical units (protomers), each consisting of three surface proteins, 1B, 1C and 1D, of 24-41 kDa, and, in most picornaviruses, an internal protein, 1A of 5.5-13.5 kDa. Total protomer is 80–97 kDa. Proteins 1A, 1B, 1C and 1D are also commonly named VP4, VP2, VP3, and VP1, respectively. Proteins 1B, 1C and 1D each possess a core structure comprising an eight-stranded β-sandwich ("β-barrel"). The β-barrels pack together in the capsid with T=l, pseudo T=3, icosahedral symmetry. (These structural features are shared by certain plant viruses that exhibit T=3, or pseudo T=3, symmetry, e.g. *Sobemovirus* and *Comoviridae,* respectively). Genera differ in the external loops that interconnect the β strands. These loops account for differences in surface relief of each genus (Fig. 1) and in thickness of the capsid wall. Assembly occurs via pentameric intermediates (pentamer=five protomers). Proteins within each pentamer are held together by an internal network formed from the N-termini of the three major CPs, the C-termini lying on the outer capsid surface. Empty capsids, which are produced by some picornaviruses, are very similar to virions, except that 1A and 1B are normally replaced by the uncleaved precursor, 1AB.

PHYSICOCHEMICAL AND PHYSICAL PROPERTIES

Virion Mr is 8 x 10^6 - 9 x 10^6, S_{20w} is 140-165S (empty particle S_{20w} is 70-80S). Buoyant density in CsCl is 1.33-1.45 g/cm^3, depending on the genus. Some species are unstable below pH 7; many are less stable at low ionic strength than at high ionic strength. Virions are insensitive to ether, chloroform, or non-ionic detergents. Viruses are inactivated by light when grown with, or in the presence of photodynamic dyes such as neutral red or proflavin. Thermal stability varies with viruses as does stabilization by divalent cations.

NUCLEIC ACID

Virions contain one molecule of positive sense, ssRNA, 7-8.8 kb in size, and possessing a single long ORF. A poly(A) tail, heterogeneous in length, is located after the 3'-terminal heteropolymeric sequence. A small protein, VPg (~2.4 kDa), is linked covalently to the 5'-terminus. The NTRs at both termini contain regions of secondary structure which are essential to genome function. The very long 5'-NTR (0.5-1.5 kb) includes a 5'-terminal domain involved in replication (e.g. the poliovirus "clover-leaf") and an IRES of 400-450 nt upstream of the translational start site; most picornaviral IRES elements can be assigned to one of two types, according to their secondary structure. Between the 5'-terminal domain and the IRES there may be one, or more, pseudoknots and/or a poly(C) tract (Fig. 2). The 3'-NTR, which may also contain a pseudoknot, ranges from 40 to 165 nt in length. The overall sequence identity between the genomes of viruses of different genera is typically less than 40%.

PROTEINS

In addition to the major CPs, 1A, 1B, 1C and 1D, and 3B (Vpg), described above, small amounts of 1AB (VP0) are commonly seen in lieu of one or more copies of 1A and 1B. Protein 1A is small in hepatoviruses, and 1AB is uncleaved in parechoviruses and kobuviruses. Traces of other proteins, including the viral RdRp, $3D^{pol}$, may also be present in purified virus preparations.

LIPIDS

Some picornaviruses carry a sphingosine-like molecule ("pocket factor") in a cavity ("pocket") located inside 1D. Protein 1A, where present, has a molecule of myristic acid covalently attached to the amino terminal glycine.

CARBOHYDRATES

None of the viral proteins are glycosylated.

Genome Organization and Replication

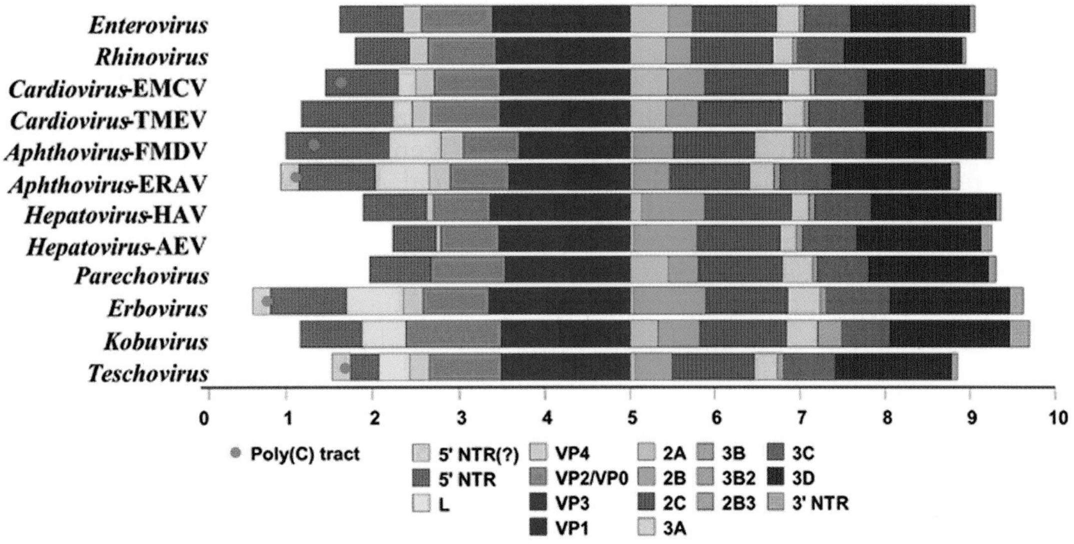

Figure 2: Genome structure and gene organization of members of the family *Picornaviridae*. Each of the 9 genera is represented, as are species where there is a significant difference within a genus. Circles within the 5′-NTR indicate poly(C) tracts that are present in some members. The 1A gene products of many members are myristylated at the amino terminal glycine. The 5-′NTR is followed by a long ORF encoding the polyprotein, that is in turn followed by the 3′-NTR and a poly(A) tail. The eventual cleavage products of the polyprotein are indicated by vertical lines and different shading. The nomenclature of the polypeptides follows an L:4:3:4 scheme corresponding to the genes (numbers) encoded by the L, P1, P2, P3 regions. The P1 region encodes the structural proteins 1A, 1B, 1C and 1D, also referred to as VP4, VP2, VP3 and VP1, respectively. VP0 (1AB) is the intermediate precursor for VP4 and VP2 and in parechoviruses and kobuviruses it remains uncleaved. In all viruses 3C is a protease, in enteroviruses and rhinoviruses 2A is a protease, while in all viruses 3D is considered to be a component of the RNA replicase. Only Foot-and-mouth disease virus encodes 3 VPg proteins that map in tandem.

The virion RNA is infectious and serves as both the genome and the viral mRNA. Gene maps are shown in Figure 2. Initiation of protein synthesis is stimulated by the IRES. Translation of the single ORF produces the polyprotein precursor 240–250 kDa) to the structural proteins (derived from the P1 region of the genome) and the nonstructural proteins (from the P2 and P3 regions). In some viruses P1 is preceded by a leader protein (L). The polyprotein is cleaved to functional proteins by specific proteases contained within it. Intermediates are denoted by letter combinations (e.g. 3CD, the uncleaved precursor of 3C and 3D). The viral proteases are as follows: Protease $3C^{pro}$, a serine-like cysteine protease encoded by all picornaviruses, performs most of the cleavages. In most genera, 2A is also associated with proteolytic activity; the $2A^{pro}$ of cardioviruses and aphthoviruses acts only *in cis*. The leader protein of aphthoviruses has proteolytic activity (L^{pro}). Some intermediates are stable and serve functions distinct from those of their cleavage products (e.g. cleavage of poliovirus P1 by $3CD^{pro}$, not by $3C^{pro}$). The cleavage of 1AB, which accompanies RNA encapsidation, is thought to be autocatalytic.

Replication of viral RNA occurs in complexes associated with cytoplasmic membranes. These complexes contain proteins derived from the whole of the 2BC-P3 region of the polyprotein, including the polymerase ($3D^{pol}$, an RNA chain-elongating enzyme), and 2C (an ATPase containing a nucleotide binding sequence motif). The poliovirus $3C^{pro}$ component has been shown to be required for binding to the 5'-terminal RNA cloverleaf. Many compounds that specifically inhibit replication have been described. Mutants resistant to, or dependent on drugs have been reported. Genetic recombination, complementation, and phenotypic mixing occur. Defective particles, carrying deletions in the CPs or L, have been produced experimentally but have not been observed in natural virus populations.

ANTIGENIC PROPERTIES

Serotypes are classified by cross-protection, neutralization of infectivity, complement-fixation, specific ELISA using a capture format or immunodiffusion. Some serotypes can be identified by hemagglutination. Antigenic sites, defined by mutations that confer resistance to neutralization by monoclonal antibodies, typically number 3 or 4 per protomer. Neutralization by antibody follows first-order inactivation kinetics.

BIOLOGICAL PROPERTIES

Most picornaviruses are specific for one, or a very few host species (exceptions are Foot-and-mouth disease virus (FMDV) and Encephalomyocarditis virus (EMCV)). Members of most species can be grown in cell culture. Resistant host cells (e.g., mouse cells in the case of the primate-specific polioviruses) can often be infected (for a single round) by transfection with naked, infectious RNA. Transmission is horizontal, mainly by fecal-oral, fomite or airborne routes. Transmission by arthropod vectors is not known, although EMCV has been isolated from mosquitoes and ticks.

Infection is generally cytolytic, but persistent infections are common with some species and reported with others. Poliovirus infected cells undergo extensive vacuolation as membranes are reorganized into viral replication complexes. Infection may be accompanied by rapid inhibition of cap-dependent translation of cellular mRNAs ($2A^{pro}$ of poliovirus and L^{pro} of aphthovirus are each powerful inhibitors), mRNA synthesis, and the cellular secretary pathway (poliovirus 2B and 3A have been implicated).

SPECIES DEMARCATION CRITERIA IN THE FAMILY

A picornavirus species is a polythetic class of phylogenetically related serotypes or strains which would normally be expected to share (i) a limited range of hosts and cellular receptors, (ii) a significant degree of compatibility in proteolytic processing, replication, encapsidation and genetic recombination, and (iii) essentially identical genome maps.

GENUS *ENTEROVIRUS*

Type Species *Poliovirus*

DISTINGUISHING FEATURES

VIRION PROPERTIES

MORPHOLOGY

CPs 1B, 1C and 1D of the human enteroviruses are among the largest in the family (VP1-3 chain lengths, 238-302 aa), and this is reflected in the typically long inter-β-strand loops, the larger than average thickness of the capsid wall (46 Å), and a surface relief that is strongly pronounced compared to most other picornaviruses. Encircling a raised area at the 5-fold axis is a 25 Å deep groove, or "canyon", into which the cellular receptor for poliovirus binds. The binding site for the pocket factor lies beneath the floor of this canyon within the 1D β-barrel. Virions can be converted by a variety of treatments (gentle heating, binding to receptor, or some neutralizing antibodies) to altered ('A') particles of 135S which lack VP4 and possess altered antigenicity.

PHYSICOCHEMICAL AND PHYSICAL PROPERTIES

Typically virions are stable at acid pH. Buoyant density in CsCl is 1.30-1.34 g/cm^3. Empty capsids are often observed in virus preparations. Sometimes a small proportion (about 1% of the population) of heavy particles (density: 1.43 g/cm^3) is observed.

NUCLEIC ACID

The genome contains a type-1 IRES and no poly(C) tract. Sequence identities for different enteroviruses, or between enteroviruses and rhinoviruses are more than 50% over the genome as a whole.

GENOME ORGANIZATION AND REPLICATION

Genomes encode a single VPg and no L protein. Protease 2A^{pro}, which is related to the family of small bacterial serine proteases, cleaves the polyprotein at its own N-terminus. Certain hydrophobic molecules that bind to the capsid in competition with pocket factor exert a powerful antiviral action by interfering with receptor binding and/or uncoating.

ANTIGENIC PROPERTIES

Native virions are antigenically serotype-specific (designated "N" or "D" for poliovirus), whereas 'A' particles exhibit group specificity (designated "H" or "C" for poliovirus).

BIOLOGICAL PROPERTIES

Viruses multiply primarily in the gastrointestinal tract, but they can also multiply in other tissues, e.g., respiratory mucosa, nerve, muscle, etc. Infection may frequently be asymptomatic. Clinical manifestations include mild meningitis, encephalitis, myelitis, myocarditis and conjunctivitis. Cap-dependent translation of host mRNA is inhibited by 2A^{pro}, which cleaves the host eukaryotic initiation factor 4G (eIF-4G). Many different cell surface molecules, many of them uncharacterized, serve as viral receptors.

LIST OF SPECIES DEMARCATION CRITERIA IN THE GENUS

Members of a species of the genus *Enterovirus*:

- share greater than 70% aa identity in P1,
- share greater than 70% aa identity in the non-structural proteins 2C + 3CD,
- share a limited range of host cell receptors,
- share a limited natural host range,
- have a genome base composition (G+C) which varies by no more than 2.5%,
- share a significant degree of compatibility in proteolytic processing, replication, encapsidation and genetic recombination.

LIST OF SPECIES IN THE GENUS

Swine vesicular disease virus is a porcine variant of Human coxsackievirus B5 (CV-B5). Certain viruses initially reported as novel echoviruses were later shown to have been misidentified. Thus E-8 is the same serotype as E-1, E-10 is now Reovirus 1, E-28 is now Human rhinovirus 1A, E-22 is now Human parechovirus 1, E-23 is now Human parechovirus 2. Similarly CV-A23 is the same serotype as E-9, and CV-A15 is the same serotype as CV-A11 and CV-A18. Porcine enteroviruses belonging to CPE group I have been moved to the genus *Teschovirus* and renamed Porcine teschovirus 1-10.

Species names are in green italic script; strain names and synonyms are in black roman script; tentative species names are in blue roman script. Sequence accession numbers, and assigned abbreviations ( ) are also listed.

SPECIES IN THE GENUS

Bovine enterovirus

Bovine enterovirus 1	[D00214]	(BEV-1)
Bovine enterovirus 2	[X79369]	(BEV-2)

POS. SSRNA

Human enterovirus A		
Human coxsackievirus A2	[L28146, X87585]	(CV-A2)*
Human coxsackievirus A3	[X87586]	(CV-A3)
Human coxsackievirus A4	[AF081295]	(CV-A4)
Human coxsackievirus A5	[X87588]	(CV-A5)
Human coxsackievirus A6	[AF081297]	(CV-A6)
Human coxsackievirus A7	[X87589]	(CV-A7)
Human coxsackievirus A8	[X87590]	(CV-A8)
Human coxsackievirus A10	[X87591]	(CV-A10)
Human coxsackievirus A12	[X87593]	(CV-A12)
Human coxsackievirus A14	[X87595]	(CV-A14)
Human coxsackievirus A16	[U05876]	(CV-A16)
Human enterovirus 71	[U22521]	(EV-71)
Human enterovirus 76		(EV-76)
Human enterovirus B		
Human coxsackievirus B1	[M16560]	(CV-B1)
Human coxsackievirus B2	[AF081485]	(CV-B2)
Human coxsackievirus B3	[M88483]	(CV-B3)
Human coxsackievirus B4	[X05690]	(CV-B4)
Human coxsackievirus B5 (including Swine vesicular disease virus)	[X67706, D00435]	(CV-B5)
Human coxsackievirus B6	[AF039205]	(CV-B6)
Human coxsackievirus A9	[D00627]	(CV-A9)
Human echovirus 1	[X89531, AF029859]	(E-1)
Human echovirus 2	[X89532, AY302545]	(E-2)
Human echovirus 3	[X89533, AY302553]	(E-3)
Human echovirus 4	[X89534, AY302557]	(E-4)
Human echovirus 5	[X89535, AF083069]	(E-5)
Human echovirus 6	[U16283, AY302558]	(E-6)
Human echovirus 7	[X89538, AY036579, AY036578]	(E-7)
Human echovirus 9	[X84981, X92886]	(E-9)
Human echovirus 11	[X80059]	(E-11)
Human echovirus 12	[X79047]	(E-12)
Human echovirus 13	[X89542, AY302539]	(E-13)
Human echovirus 14	[X89543, AY302540]	(E-14)
Human echovirus 15	[X89544, AY302541]	(E-15)
Human echovirus 16	[X89545, AY302542]	(E-16)
Human echovirus 17	[X89546, AY302543]	(E-17)
Human echovirus 18	[X89547, AF317694]	(E-18)
Human echovirus 19	[X89548, AY302544]	(E-19)
Human echovirus 20	[X89549, AY302546]	(E-20)
Human echovirus 21	[X89550, AY302547]	(E-21)
Human echovirus 24	[X89551, AY302548]	(E-24)
Human echovirus 25	[X90722, X89552, AY302549]	(E-25)
Human echovirus 26	[X89553, AY302550]	(E-26)
Human echovirus 27	[X89554, AY302551]	(E-27)
Human echovirus 29	[X89555, AY302552]	(E-29)
Human echovirus 30	[X89556, AF102711]	(E-30)
Human echovirus 31	[X89557, AY302554]	(E-31)
Human echovirus 32	[X89558, AY302555]	(E-32)
Human echovirus 33	[X89559, AY302556]	(E-33)
Human enterovirus 69	[X87605, AY302560]	(EV-69)
Human enterovirus 73	[AF241359]	(EV-73)
Human enterovirus 74	[AY208118]	(EV-74)

Human enterovirus 75	[AF152280]	(EV-75)
Human enterovirus 77	[AY208119]	(EV-77)
Human enterovirus 78	[AY208120]	(EV-78)
Human enterovirus C		
Human coxsackievirus A1	[X87584, AF499035]	(CV-A1)
Human coxsackievirus A11	[X87592, AF499636, AF499638]	(CV-A11)
Human coxsackievirus A13	[X87594, AF499637, AF499640]	(CV-A13)
Human coxsackievirus A17	[X87597, AF499039]	(CV-A17)
Human coxsackievirus A19	[X87599, AF499641]	(CV-A19)
Human coxsackievirus A20	[X87600, AF499642]	(CV-A20)
Human coxsackievirus A21	[D00538]	(CV-A21)
Human coxsackievirus A22	[X87603, AF499643]	(CV-A22)
Human coxsackievirus A24	[D90457]	(CV-A24)
Human enterovirus D		
Human enterovirus 68	[X87604]	(EV-68)
Human enterovirus 70	[D00820]	(EV-70)
Poliovirus		
Human poliovirus 1	[J02281]	(PV-1)
Human poliovirus 2	[M12197]	(PV-2)
Human poliovirus 3	[K01392]	(PV-3)
Porcine enterovirus A		
Porcine enterovirus 8	[AF406813]	(PEV-8)
Porcine enterovirus B		
Porcine enterovirus 9	[Y14459]	(PEV-9)
Porcine enterovirus 10	[AF363455]	(PEV-10)
Simian enterovirus A		
Simian enterovirus A1		(SEV-A1)
Simian enterovirus A2-plaque virus**	[AF201894]	(SEV-A2)
Simian enterovirus SV4**	[AF326759]	(SEV-SV4)
Simian enterovirus SV28**	[AF326757]	(SEV-SV28)
Simian enterovirus SA4**		(SEV-SA4)

* Note: The alternative abbreviations, CAV-2, etc, are widely used.
** the 4 isolates are closely related and probably constitute a single serotype.

TENTATIVE SPECIES IN THE GENUS

Simian enterovirus A13	[AF326750]	(A13)
Simian enterovirus N125	[AF414372]	(N125)
Simian enterovirus N203	[AF414373]	(N203)
Simian enterovirus SA5	[AF326751]	(SA5)
Simian enterovirus SV16	[AY064715; AF326752; AY064709]	(SV16)
Simian enterovirus SV18	[AY064716; AF326753; AY064710]	(SV18)
Simian enterovirus SV19	[AF326754]	(SV19)
Simian enterovirus SV2	[AY064708]	(SV2)
Simian enterovirus SV26	[AF326756]	(SV26)
Simian enterovirus SV35	[AF326758]	(SV35)
Simian enterovirus SV42	[AY064717; AF326760; AY064711]	(SV42)
Simian enterovirus SV43	[AF326761]	(SV43)
Simian enterovirus SV44	[AY064718; AF326762; AY064712]	(SV44)
Simian enterovirus SV45	[AY064719; AF326763; AY064713]	(SV45)
Simian enterovirus SV47		(SV47)
Simian enterovirus SV49	[AY064720; AF326765; AY064714]	(SV49)
Simian enterovirus SV6	[AF326766]	(SV6)

GENUS RHINOVIRUS

Type Species ***Human rhinovirus A***

DISTINGUISHING FEATURES

VIRION PROPERTIES

MORPHOLOGY

Rhinoviruses share with human enteroviruses the same, comparatively uneven, surface, with its characteristic canyon around the 5-fold axis (attachment site for the intercellular adhesion molecule-1 (ICAM-1) receptor), thick-walled capsid, and pocket factor-binding cavity.

PHYSICOCHEMICAL AND PHYSICAL PROPERTIES

Typically virions are unstable below pH 5-6. They can also be distinguished from enteroviruses by their porosity to CsCl, which gives rise to a buoyant density in the range 1.38-1.42 g/cm^3.

NUCLEIC ACID

The 5'-NTR of 650 nt is shorter than that of enteroviruses, owing to a deletion of approximately 100 nt between the IRES and the translation start site. The IRES is of type 1 and there is no poly(C) tract. Nucleotide sequence identity over the entire genome for different species of the genus *Rhinovirus*, or between enteroviruses and rhinoviruses is more than 50%, although it may be greater or less than this for particular genomic regions.

PROTEINS

Virion proteins are very similar in size to those of human enteroviruses.

GENOME ORGANIZATION AND REPLICATION

These are similar to human enteroviruses. Antiviral, pocket-binding drugs, analogous to those used against enteroviruses, have been described.

ANTIGENIC PROPERTIES

Antigenic properties, including the N-D conversion, are as for human enteroviruses.

BIOLOGICAL PROPERTIES

Human rhinoviruses can be divided into major and minor receptor group viruses. Eighty-nine serotypes (major group) use ICAM-l as receptor, 10 serotypes (minor group) bind members of the low-density lipoprotein receptor (LDLR) family. Clinical manifestations include the common cold and other upper and lower respiratory tract illnesses of humans. Cap-dependent translation of host mRNA is inhibited by $2A^{pro}$, which cleaves the host eIF-4G.

LIST OF SPECIES DEMARCATION CRITERIA IN THE GENUS

Members of a species in the genus *Rhinovirus* share:

- greater than 70% aa identity in P1,
- greater than 70% aa identity in 2C + 3CD,
- similar susceptibility of receptor attachment to inhibition by pocket-binding antiviral agents ("inhibitor group" A or B).

LIST OF SPECIES IN THE GENUS

Human rhinovirus 87 is now considered to be the same serotype as Human enterovirus 68.

Species names are in green italic script; strain names and synonyms are in black roman script; tentative species names are in blue roman script. Sequence accession numbers, and assigned abbreviations ( ) are also listed.

SPECIES IN THE GENUS

Human rhinovirus A

Human rhinovirus 1†	[D00239]	(HRV-1)
Human rhinovirus 2	[X02316]	(HRV-2)
Human rhinovirus 7	[Z47564]	(HRV-7)
Human rhinovirus 8	[AF343594]	(HRV-8)
Human rhinovirus 9	[AF343605]	(HRV-9)
Human rhinovirus 10	[AF343609]	(HRV-10)
Human rhinovirus 11	[Z47565]	(HRV-11)
Human rhinovirus 12	[AY016405]	(HRV-12)
Human rhinovirus 13	[AF343599]	(HRV-13)
Human rhinovirus 15	[AF343630]	(HRV-15)
Human rhinovirus 16	[L24917]	(HRV-16)
Human rhinovirus 18	[AY016407]	(HRV-18)
Human rhinovirus 19	[AF343632]	(HRV-19)
Human rhinovirus 20	[AF343644]	(HRV-20)
Human rhinovirus 21	[Z47566]	(HRV-21)
Human rhinovirus 22	[AF343628]	(HRV-22)
Human rhinovirus 23	[AF343597]	(HRV-23)
Human rhinovirus 24	[AF343619]	(HRV-24)
Human rhinovirus 25	[AF343617]	(HRV-25)
Human rhinovirus 28	[AY016406]	(HRV-28)
Human rhinovirus 29	[Z47567]	(HRV-29)
Human rhinovirus 30	[AF343596]	(HRV-30)
Human rhinovirus 31	[AF343583]	(HRV-31)
Human rhinovirus 32	[AF343584]	(HRV-32)
Human rhinovirus 33	[AF343625]	(HRV-33)
Human rhinovirus 34	[AF343634]	(HRV-34)
Human rhinovirus 36	[Z49123]	(HRV-36)
Human rhinovirus 38	[AF343614]	(HRV-38)
Human rhinovirus 39	[AF343637]	(HRV-39)
Human rhinovirus 40	[AF343641]	(HRV-40)
Human rhinovirus 41	[AF343600]	(HRV-41)
Human rhinovirus 43	[AY040232]	(HRV-43)
Human rhinovirus 44	[AF343616]	(HRV-44)
Human rhinovirus 45	[AY016409]	(HRV-45)
Human rhinovirus 46	[AY040235]	(HRV-46)
Human rhinovirus 47	[AF343607]	(HRV-47)
Human rhinovirus 49	[Z47568]	(HRV-49)
Human rhinovirus 50	[Z47569]	(HRV-50)
Human rhinovirus 51	[AF343585]	(HRV-51)
Human rhinovirus 53	[AF343592]	(HRV-53)
Human rhinovirus 54	[AF343612]	(HRV-54)
Human rhinovirus 55	[AF343621]	(HRV-55)
Human rhinovirus 56	[AF343610]	(HRV-56)
Human rhinovirus 57	[AF343622]	(HRV-57)
Human rhinovirus 58	[Z47570]	(HRV-58)
Human rhinovirus 59	[AF343611]	(HRV-59)
Human rhinovirus 60	[AF343627]	(HRV-60)
Human rhinovirus 61	[AF343601]	(HRV-61)
Human rhinovirus 62	[Z47571]	(HRV-62)
Human rhinovirus 63	[AF343636]	(HRV-63)
Human rhinovirus 64	[AF343629]	(HRV-64)
Human rhinovirus 65	[Z47572]	(HRV-65)

Human rhinovirus 66	[AF343640]	(HRV-66)
Human rhinovirus 67	[AF343603]	(HRV-67)
Human rhinovirus 68	[AF343591]	(HRV-68)
Human rhinovirus 71	[AF343587]	(HRV-71)
Human rhinovirus 73	[AF343602]	(HRV-73)
Human rhinovirus 74	[AF343631]	(HRV-74)
Human rhinovirus 75	[AF343639]	(HRV-75)
Human rhinovirus 76	[AF343624]	(HRV-76)
Human rhinovirus 77	[AF343608]	(HRV-77)
Human rhinovirus 78	[AY016408]	(HRV-78)
Human rhinovirus 80	[AF343593]	(HRV-80)
Human rhinovirus 81	[AF343606]	(HRV-81)
Human rhinovirus 82	[AY040233]	(HRV-82)
Human rhinovirus 85	[AF343642]	(HRV-85)
Human rhinovirus 88	[AF343590]	(HRV-88)
Human rhinovirus 89	[M16248]	(HRV-89)
Human rhinovirus 90	[AF343620]	(HRV-90)
Human rhinovirus 94	[AF343638]	(HRV-94)
Human rhinovirus 95	[AF343595]	(HRV-95)
Human rhinovirus 96	[AF343604]	(HRV-96)
Human rhinovirus 98	[AF343613]	(HRV-98)
Human rhinovirus 100	[AF343643]	(HRV-100)
Human rhinovirus Hanks	[AY040234]	(HRV-Hanks)
Human rhinovirus B		
Human rhinovirus 3	[U60874]	(HRV-3)
Human rhinovirus 4	[AF343655]	(HRV-4)
Human rhinovirus 5	[AF343651]	(HRV-5)
Human rhinovirus 6	[AY016402]	(HRV-6)
Human rhinovirus 14	[K02121, K01087, L05355]	(HRV-14)
Human rhinovirus 17	[AF343645]	(HRV-17)
Human rhinovirus 26	[AF343653]	(HRV-26)
Human rhinovirus 27	[AF343654]	(HRV-27)
Human rhinovirus 35	[AY040241]	(HRV-35)
Human rhinovirus 37	[AY016401]	(HRV-37)
Human rhinovirus 42	[AY016404]	(HRV-42)
Human rhinovirus 48	[AY016400]	(HRV-48)
Human rhinovirus 52	[AY016398]	(HRV-52)
Human rhinovirus 69	[AY016399]	(HRV-69)
Human rhinovirus 70	[AF343646]	(HRV-70)
Human rhinovirus 72	[Z47574]	(HRV-72)
Human rhinovirus 79	[AF343649]	(HRV-79)
Human rhinovirus 83	[AF343647]	(HRV-83)
Human rhinovirus 84	[AY040240]	(HRV-84)
Human rhinovirus 86	[AF343648]	(HRV-86)
Human rhinovirus 91	[AY040237]	(HRV-91)
Human rhinovirus 92	[AY040238]	(HRV-92)
Human rhinovirus 93	[AY040239]	(HRV-93)
Human rhinovirus 97	[AY040242]	(HRV-97)
Human rhinovirus 99	[AF343652]	(HRV-93)

† HRV-1 is divided into two antigenic subtypes referred to as HRV-1A and HRV-1B.

TENTATIVE SPECIES IN THE GENUS

Bovine rhinovirus 1	(BRV-1)
Bovine rhinovirus 2	(BRV-2)
Bovine rhinovirus 3	(BRV-3)

GENUS CARDIOVIRUS

Type Species *Encephalomyocarditis virus*

DISTINGUISHING FEATURES

VIRION PROPERTIES

MORPHOLOGY

Empty capsids are seen only rarely, if ever. When compared by mean wall thickness, surface unevenness, and chain length of the major proteins, the cardiovirus capsid is intermediate between the enteroviruses and aphthoviruses. In place of a continuous, circular, canyon, seen in enteroviruses, is a five-fold repeated pit. There is no pocket factor.

PHYSICOCHEMICAL AND PHYSICAL PROPERTIES

Virion buoyant density in CsCl is 1.33-1.34 g/cm^3. Virions are moderately stable to acidic pH.

NUCLEIC ACID

EMCV has a poly(C) tract of variable length (usually 80-250 nt) about 150 nt from the 5'-terminus of the viral RNA, while Theilovirus isolates lack this feature. All EMCV members have two pseudoknots 5' to their poly(C) tracts. The IRES is of type 2. The nt sequence identity over the entire genome for different species of the genus *Cardiovirus* is more than 50% (e.g. TMEV has 54% nt sequence identity to EMCV).

GENOME ORGANIZATION AND REPLICATION

The viral genome encodes a leader (L) protein which lacks proteolytic activity, unlike the L of aphthoviruses; thus L is cleaved from P1 by the virus encoded protease 3C. The 1D/2A junction is also cleaved by $3C^{pro}$, rather than by 2A. The 2A protein causes cleavage, or polypeptide chain interruption, between P1-2A and downstream sequences at an essential sequence, --NPG/P--.

ANTIGENIC PROPERTIES

Four independent antigenic sites have been described. There is no evidence of an N-D conversion, nor of 'A' particles.

BIOLOGICAL PROPERTIES

Encephalomyocarditis viruses have been isolated from over 30 host species including mammals, birds and insects. Clinical manifestations include encephalitis and myocarditis in mice and many other animals. TMEV can be divided into two biological subgroups which both infect mice; one causes an acute and fatal polioencephalomyelitis and the other causes a chronic persistent demyelinating infection of the white matter. Vilyuisk human encephalomyelitis virus (VHEV) is thought to be the cause of a degenerative neurological disease in man which has been reported in the Vilyuy valley in Siberia. Cardiovirus infection does not cause cleavage of the host eIF-4G. The cellular receptor used by EMCV to attach to murine vascular endothelial cells has been identified as VCAM-1. However, in human cell lines an as yet unidentified sialoglycoprotein(s) has been found. EMCV binds to human erythrocytes via glycophorin A.

LIST OF SPECIES DEMARCATION CRITERIA IN THE GENUS

Members of a species of the genus *Cardiovirus*:

- share greater than 70% aa identity in P1,
- share greater than 70% aa identity in 2C + 3CD,
- share a natural host range,

- share a common genome organization.

LIST OF SPECIES IN THE GENUS

Mengovirus, Columbia SK virus and Maus Elberfeld virus are strains of EMCV, based on serological cross-reaction and sequence identity. The rat encephalomyelitis virus MHG appears to be a strain of TMEV; however, the serological relationship of a genetically divergent Theiler-like virus (TLV) of rats to TMEV is not presently clear.

Species names are in green italic script; strain names and synonyms are in black roman script; tentative species names are in blue roman script. Sequence accession numbers, and assigned abbreviations ( ) are also listed.

SPECIES IN THE GENUS

Encephalomyocarditis virus		
Columbia SK virus		
Encephalomyocarditis virus*	[M81861]	(EMCV)
Maus Elberfeld virus		
Mengovirus	[L22089]	
Theilovirus		
Theiler's murine encephalomyelitis virus	[M20562]	(TMEV)
Theiler-like virus of rats	[AB090161]	(TLV)
Vilyuisk Human encephalomyelitis virus	[M80888, M94868]	(VHEV)

*The significance of the reported serological cross-reaction between Cricket paralysis virus, a member of the family *Dicistroviridae,* and EMCV is not presently understood.

TENTATIVE SPECIES IN THE GENUS

None reported.

GENUS *APHTHOVIRUS*

Type Species *Foot-and-mouth disease virus*

DISTINGUISHING FEATURES

VIRION PROPERTIES

MORPHOLOGY

The capsid of FMDV is thin-walled (mean thickness ~33 Å), and has an unusually smooth surface. A long (17-23 aa), mobile loop, the G-H loop, projects from the surface of 1D. There is a pore at the 5-fold axis, where part of the underlying 1C is exposed. Some serotypes of FMDV accumulate empty capsids.

PHYSICOCHEMICAL AND PHYSICAL PROPERTIES

Virions are acid labile; FMDV particles are unstable below pH 6.8; Equine rhinitis A virus (ERAV) particles are unstable below pH 5.5. The buoyant density in CsCl is 1.43-1.45g/cm^3. Virions of FMDV sediment at 146S, empty capsids at 75S.

NUCLEIC ACID

There is a poly(C) tract close to the 5'-terminus of the genome. In FMDV it is located about 360 nt from the end, and varies in length from 100 to more than 400 nt. Current data suggest that the poly(C) tract in ERAV is shorter (~40 nt) and closer to the 5'-end. In the RNA of members of both species there is a series of pseudoknots on the 3'-side of the poly(C); the total 5'-NTR is thus extremely long (1.1-1.5 kb). ERAV and FMDV differ by approximately 50% in nt sequence across the entire genome.

PROTEINS

The major CPs of FMDV have the shortest chain lengths of any picornavirus (208-220 aa); those of ERAV are only slightly longer. At the tip of the 1D G-H loop of FMDV is the conserved integrin recognition motif, RGD.

GENOME ORGANIZATION AND REPLICATION

Translation starts at two alternative in-frame initiation sites, resulting in two forms of the L protein (Lab and Lb). L is a papain-like cysteine protease which cleaves itself from the virus polyprotein. The 2A polypeptide is very short (chain length = 18 aa in FMDV), and is involved in NPGP-dependent polypeptide chain interruption at its C-terminus as in cardioviruses. The genome of FMDV encodes 3 species of VPg while that of ERAV encodes only one.

ANTIGENIC PROPERTIES

Five independent antigenic sites have been reported in FMDV type O, two of which have determinants in the G-H loop of 1D. There is no evidence of N-D conversion, nor 'A' particles.

BIOLOGICAL PROPERTIES

FMDV infects mainly cloven-hooved animals, but has been isolated from at least 70 species of mammals. Clinical manifestations of FMDV infections include foot-and-mouth disease (vesicular lesions), sometimes with associated acute fatal myocarditis in young animals; of ERAV, upper respiratory tract infections of horses. Both species may produce persistent upper respiratory tract infections. FMDV infects cells by binding to integral membrane proteins of the integrin family through its 1D G-H loop; heparan sulfate proteoglycans may also serve as receptors. Cap-dependent translation of host mRNA is inhibited by L^{pro}, which cleaves the host eIF-4G.

LIST OF SPECIES DEMARCATION CRITERIA IN THE GENUS

Members of a species of the genus *Aphthovirus*:

- share greater than 50% aa identity in P1,
- share greater than 70% aa identity in 2C + 3CD,
- share a natural host range,
- have a genome base composition which varies by no more than 1%,
- share a common genome organization.

LIST OF SPECIES IN THE GENUS

Species names are in green italic script; strain names and synonyms are in black roman script; tentative species names are in blue roman script. Sequence accession numbers, and assigned abbreviations ( ) are also listed.

SPECIES IN THE GENUS

Equine rhinitis A virus		
Equine rhinitis A virus	[L43052, X96870]	(ERAV)
(Equine rhinovirus 1)		
Foot-and-mouth disease virus		
Foot-and-mouth disease virus - A	[L11360, M10975]	(FMDV-A)
Foot-and-mouth disease virus - Asia 1	[U01207]	(FMDV-Asia1)
Foot-and-mouth disease virus - C	[X00130, J02191]	(FMDV-C)
Foot-and-mouth disease virus - O	[M35873, X00871]	(FMDV-O)
Foot-and-mouth disease virus - SAT 1	[Z98203]	(FMDV-SAT1)
Foot-and-mouth disease virus - SAT 2	[AJ251473]	(FMDV-SAT2)
Foot-and-mouth disease virus - SAT 3	[M28719]	(FMDV-SAT3)

TENTATIVE SPECIES IN THE GENUS

None reported.

POS. SSRNA

GENUS *HEPATOVIRUS*

Type Species *Hepatitis A virus*

DISTINGUISHING FEATURES

VIRION PROPERTIES

PHYSICOCHEMICAL AND PHYSICAL PROPERTIES

Viruses are very stable, resistant to acid pH and elevated temperatures (60°C for 10 min). Buoyant density in CsCl is 1.32-1.34 g/cm^3.

NUCLEIC ACID

There is little similarity between the genome sequences of hepatoviruses and those of other picornaviruses, although the IRES is distantly related to the type 2 IRES. The 5′-NTR contains a 5′-terminal hairpin, two putative pseudoknots, and a short (~40 nt) pyrimidine-rich (i.e. not pure polyC) tract upstream of the IRES. Nucleotide sequence identity between different Hepatitis A virus (HAV) strains is generally greater than 80%. Avian encephalomyelitis virus (AEV) RNA contains the shortest of all picornavirus 5'-NTRs, at 494 nt.

PROTEINS

In contrast to those of other picornaviruses, protein lA of hepatoviruses is extremely small, does not appear to be myristylated at its N-terminus, and may not be a component of the mature virus particle. Immature HAV may contain uncleaved 1D2A (PX) precursor protein.

GENOME ORGANIZATION AND REPLICATION

The polyprotein contains only a single protease (3C^{pro}). There is no clearly defined L protein, and 2A has no proteolytic activity. The primary cleavage of the polyprotein occurs at the 2A/2B junction, and is catalyzed by 3C^{pro}. The 1D/2A cleavage may be directed by an unknown cellular protease, or the VP1 protein may be subject to C-terminal trimming as in cardioviruses. Replication in cell culture occurs slowly, with little CPE, and with low yields of virus compared to most other picornaviruses. The IRES differs from those of other picornaviruses in that its activity is dependent on intact eIF-4G. The 2A protein of hHepatitis A virus (which is unique among picornaviruses) is distinct from that in the tentative species Avian encephalomyelitis-like virus (which is distantly related to the 2A of parechoviruses and kobuviruses).

ANTIGENIC PROPERTIES

Hepatitis A viruses are strongly conserved in their antigenic properties. Most antibodies are directed against a single, conformationally defined immunodominant antigenic site that is comprised of aa residues of the VP3 and VP1 proteins on the surface of the virion.

BIOLOGICAL PROPERTIES

HAV infects epithelial cells of the small intestine and hepatocytes of primates. Virus is predominantly replicated within the liver, excreted via the bile and present in feces in high titer. Viral shedding is maximal shortly before the onset of clinical signs of hepatitis, which probably represents immunopathologically mediated liver injury. Clinical manifestations are fever, jaundice, light stools, abdominal pain, and occasionally diarrhea. Hepatoviruses generally establish persistent infection when inoculated on to any of a wide range of primate cells in vitro, but persistent infection does not occur *in vivo,* and the viruses are not associated with chronic hepatitis. HAVs can be divided into two distinct biotypes that are phylogenetically distinct and have different preferred hosts (all species of primates: humans, chimpanzees, owl monkeys and marmosets, for one biotype, vs.

green monkeys and cynomolgus monkeys for the other). These two biotypes share cross-reacting antigens, but have biotype-specific epitopes that can be distinguished by monoclonal antibodies. AEV causes encephalomyelitis in young chickens, pheasants, quail and turkeys. It can be transmitted both vertically and by the fecal-oral route; field strains are enterotropic.

LIST OF SPECIES DEMARCATION CRITERIA IN THE GENUS

Members of a species of the genus *Hepatovirus* have:

- greater than 70% aa identity in P1,
- greater than 70% aa identity in 2C + 3CD,
- greater than 75% nt sequence identity over the genome as a whole,
- cross-protective antigens,
- a defined tissue tropism and host range,
- a similar genome base composition which varies by no more than 1%,
- a common genome organization.

LIST OF SPECIES IN THE GENUS

Species names are in green italic script; strain names and synonyms are in black roman script; tentative species names are in blue roman script. Sequence accession numbers, and assigned abbreviations ( ) are also listed.

SPECIES IN THE GENUS

Hepatitis A virus		
Human hepatitis A virus	[M14707]	(HHAV)
Simian hepatitis A virus	[D00924]	(SHAV)

TENTATIVE SPECIES IN THE GENUS

Avian encephalomyelitis-like virus	[AJ225173]	(AEV)

GENUS PARECHOVIRUS

Type Species *Human parechovirus*

DISTINGUISHING FEATURES

VIRION PROPERTIES

PHYSICOCHEMICAL AND PHYSICAL PROPERTIES

Virions are acid stable. The buoyant density in CsCl is 1.36g/cm^3.

NUCLEIC ACID

The 5′-NTR is 710 (Human parechovirus) - 730 (Ljungan virus) nt and contains a typical type 2 IRES. The ORF is 2180/2250 codons and the 3′-NTR 87 and 111 nt in Human parechovirus and Ljungan virus, respectively.

PROTEINS

Predicted protein sequences of parechoviruses are highly divergent, no protein having a greater than 30% level of identity when compared with corresponding proteins of any other picornavirus. In contrast to most other picornaviruses, protein 1AB of parechoviruses appears not to be cleaved, and its N-terminus, also unusually, is not myristylated. The mature capsid therefore appears to comprise only three proteins, 1AB, 1C and 1D.

GENOME ORGANIZATION AND REPLICATION

The polyprotein contains only a single protease (3C^{pro}). The 2A protein is believed to lack protease activity and is related distantly to a family of cellular proteins involved in the control of cell proliferation, as well as to that of Kobuvirus and AEV. Ljungan virus isolates additionally contain sequences resembling the aphthovirus 2A at this locus.

BIOLOGICAL PROPERTIES

Human parechoviruses replicate in the respiratory and gastrointestinal tract. Infection is particularly prevalent in young children but it is probably often asymptomatic. In addition to respiratory infections and diarrhea, infections of the central nervous system have occasionally been reported. The cytopathology may be unusual in including changes in granularity and chromatin distribution in the nucleus, when viewed in the electron microscope. Isolates of ljungan viruses appear to infect predominantly rodents.

LIST OF SPECIES DEMARCATION CRITERIA IN THE GENUS

Members of a species of the genus *Parechovirus*:

- share greater than 70% aa identity in P1,
- share greater than 70% aa identity in 2C + 3CD,
- share a natural host range,
- share a common genome organization,
- have a similar genome base composition which varies by no more than 1%.

LIST OF SPECIES IN THE GENUS

Species names are in green italic script; strain names and synonyms are in black roman script; tentative species names are in blue roman script. Sequence accession numbers, and assigned abbreviations ( ) are also listed.

SPECIES IN THE GENUS

Human parechovirus		
Human parechovirus 1	[L02971]	(HPeV-1)
(Human echovirus 22)		
Human parechovirus 2	[AF055846, AJ005695]	(HPeV-2)
(Human echovirus 23)		
Human parechovirus 3	[AB084913]	(HPeV-3)
Ljungan virus		
Ljungan virus*	[AF327920, AF327921, AF327922, AF538689]	(LV)

*The American isolate M1146 is relatively divergent from the Swedish isolates 87-012, 174F and 145SL, but it is not known if they are distinct serotypes.

TENTATIVE SPECIES IN THE GENUS

None reported.

GENUS ERBOVIRUS

Type Species *Equine rhinitis B virus*

DISTINGUISHING FEATURES

VIRION PROPERTIES

PHYSICOCHEMICAL AND PHYSICAL PROPERTIES

Equine rhinitis B virus (ERBV) has a buoyant density in CsCl of 1.41-1.45 g/cm^3. The virus is unstable below pH 5.

NUCLEIC ACID

ERBV possesses an unusually long 3'-NTR of 167 nt. The IRES is of type 2, and a poly(C) tract is thought to be present. No pseudoknots have been identified.

PROTEINS
The CPs have between 25% and 47% aa sequence identity to those of ERAV, FMDV and EMCV, though protein modelling studies indicate that they more closely resemble those of EMCV.

GENOME ORGANIZATION AND REPLICATION
No evidence for alternative sites of initiation of protein synthesis is available. The L protein appears to be a protease, but has only 23% and 18% aa sequence identity to the L proteins of FMDV and ERAV, respectively. The 2B and 3C proteins have exceptionally large chain lengths (283 and 251 aa). The 2A protein has a chain length of 18 aa, ending in NPGP, and there is only one VPg.

BIOLOGICAL PROPERTIES
ERBV causes upper respiratory tract disease in horses, with a viremia and fecal shedding. Infections may be persistent.

LIST OF SPECIES DEMARCATION CRITERIA IN THE GENUS
Not applicable

LIST OF SPECIES IN THE GENUS
Species names are in green italic script; strain names and synonyms are in black roman script; tentative species names are in blue roman script. Sequence accession numbers, and assigned abbreviations ( ) are also listed.

SPECIES IN THE GENUS
Equine rhinitis B virus

Equine rhinitis B virus 1		
(Equine rhinovirus 2)	[X96871]	(ERBV-1)
Equine rhinitis B virus 2		
(Equine rhinovirus 3)	[AF361253]	(ERBV-2)

TENTATIVE SPECIES IN THE GENUS
None reported.

GENUS KOBUVIRUS

Type Species *Aichi virus*

DISTINGUISHING FEATURES

VIRION PROPERTIES

PHYSICOCHEMICAL AND PHYSICAL PROPERTIES
Unlike other picornaviruses, kobuviruses exhibit icosahedral surface structure under the electron microscope. Virions are stable at pH 3.5.

NUCLEIC ACID
The genome of Aichi virus (AiV) has a high G+C base composition (59%), and a very long 3'-NTR (240 nt). There is a 5′-proximal stem-loop involved in RNA replication and encapsidation.

PROTEINS
Protein 1AB appears not to be cleaved.

POS. SSRNA

Genome Organization and Replication

There is a leader polypeptide of unknown function, and distinctive length (170 aa rather than 67 aa or 217 aa in EMCV and FMDV, respectively). The 2A protein is distantly related to that of parechoviruses.

Biological Properties

AiV grows in cell cultures (BSC-1, Vero). AiV is thought to be a cause of human gastroenteritis associated with eating shellfish.

List of Species Demarcation Criteria in the Genus

Members of a species of the genus *Kobuvirus*:

- share greater than 70% aa identity in P1,
- share greater than 70% aa identity in 2C + 3CD,
- share a common genome organization.

List of Species in the Genus

Species names are in green italic script; strain names and synonyms are in black roman script; tentative species names are in blue roman script. Sequence accession numbers, and assigned abbreviations ( ) are also listed.

Species in the Genus

Aichi virus		
Aichi virus	[AB040749]	(AiV)
Bovine kobuvirus		
Bovine kobuvirus	[AB084788]	(BKV)

Tentative Species in the Genus

None reported.

Genus *Teschovirus*

Type Species *Porcine teschovirus*

Distinguishing Features

Virion Properties

Physicochemical and Physical Properties

Virions are stable at acid pH. Buoyant density in CsCl is 1.33 g/cm^3. Empty capsids are often observed in virus preparations.

Nucleic Acid

Teschoviruses have an IRES which is unlike that of other picornaviruses in being shorter (290 nt) and functional in the absence of eIF-4G. In both these properties the IRES resembles that of Hepatitis C virus (family *Flaviviridae*) and sequence similarity has also been observed.

Proteins

Genomes encode a single VPg and a leader (L) protein. The 2A polypeptide is very short and ends in NPGP, indicative of an aphthovirus 2A-like molecule.

Biological Properties

Clinical manifestations may include a polioencephalomyelitis ("Teschen disease"), which may vary in severity.

List of Species Demarcation Criteria in the Genus

Not applicable.

List of Species in the Genus

Species names are in green italic script; strain names and synonyms are in black roman script; tentative species names are in blue roman script. Sequence accession numbers, and assigned abbreviations ( ) are also listed.

Species in the Genus

Porcine teschovirus

Porcine teschovirus 1 (Porcine enterovirus 1)	[AF231769, AB038528]	(PTV-1)
Porcine teschovirus 2 (Porcine enterovirus 2)	[AF296087]	(PTV-2)
Porcine teschovirus 3 (Porcine enterovirus 3)	[AF296088]	(PTV-3)
Porcine teschovirus 4 (Porcine enterovirus 4)	[AF296089]	(PTV-4)
Porcine teschovirus 5 (Porcine enterovirus 5)	[AF296090]	(PTV-5)
Porcine teschovirus 6 (Porcine enterovirus 6)	[AF296091]	(PTV-6)
Porcine teschovirus 7 (Porcine enterovirus 7)	[AF296092]	(PTV-7)
Porcine teschovirus 8 (Porcine enterovirus 11)	[AF296093]	(PTV-8)
Porcine teschovirus 9 (Porcine enterovirus 12)	[AF296094]	(PTV-9)
Porcine teschovirus 10 (Porcine enterovirus 13)	[AF296119, AF296095]	(PTV-10)
Porcine teschovirus 11	[AF296096]	(PTV-11)

Tentative Species in the Genus

None reported.

List of Unassigned Viruses in the Family

Acid-stable equine picornaviruses	(EqPV)
Avian entero-like virus 2	(AELV-2)
Avian entero-like virus 3	(AELV-3)
Avian entero-like virus 4	(AELV-4)
Avian nephritis virus 3*	(ANV-3)
Barramundi virus-1†	(BaV)
Cockatoo entero-like virus	(CELV)
Duck hepatitis virus 1	(DHV-1)
Duck hepatitis virus 3	(DHV-3)
Guineafowl transmissible enteritis virus	(GTEV)
Harbour seals picorna-like virus	(SPLV)
Sea-bass virus-1†	(SBV)
Sikhote-Alyn virus	(SAV)
Smelt virus-1†	(SmV-1)
Smelt virus-2†	(SmV-2)
Syr-Daria Valley fever virus	(SDFV)
Turbot virus-1	(TuV-1)
Turkey entero-like virus	(TELV)
Turkey hepatitis virus	(THV)
Turkey pseudo enterovirus 1	(TPEV-1)
Turkey pseudo enterovirus 2	(TPEV-2)

*Avian nephritis virus 1 (ANV-1) and Avian nephritis virus 2 (ANV-2) have been shown to be members of the family *Astroviridae*. Taura syndrome virus (TSV) of marine penaeid shrimp has been shown to be a member of the family *Dicistroviridae*.

POS. SSRNA

PHYLOGNETIC RELATIONSHIPS WITHIN THE FAMILY

Figure 3: Phylogenetic trees showing the relationships between the species and genera of the family *Picornaviridae*. (Top) Protein P1 and (Bottom) Proteins 2C+3CD. The Neighbor-joining trees were produced and bootstrapped (1000 replicates) using CLUSTALX and an aa weight matrix (BLOSUM). The trees were drawn using TreeView v1.5.2. Only bootstrap values of >90% are indicated.

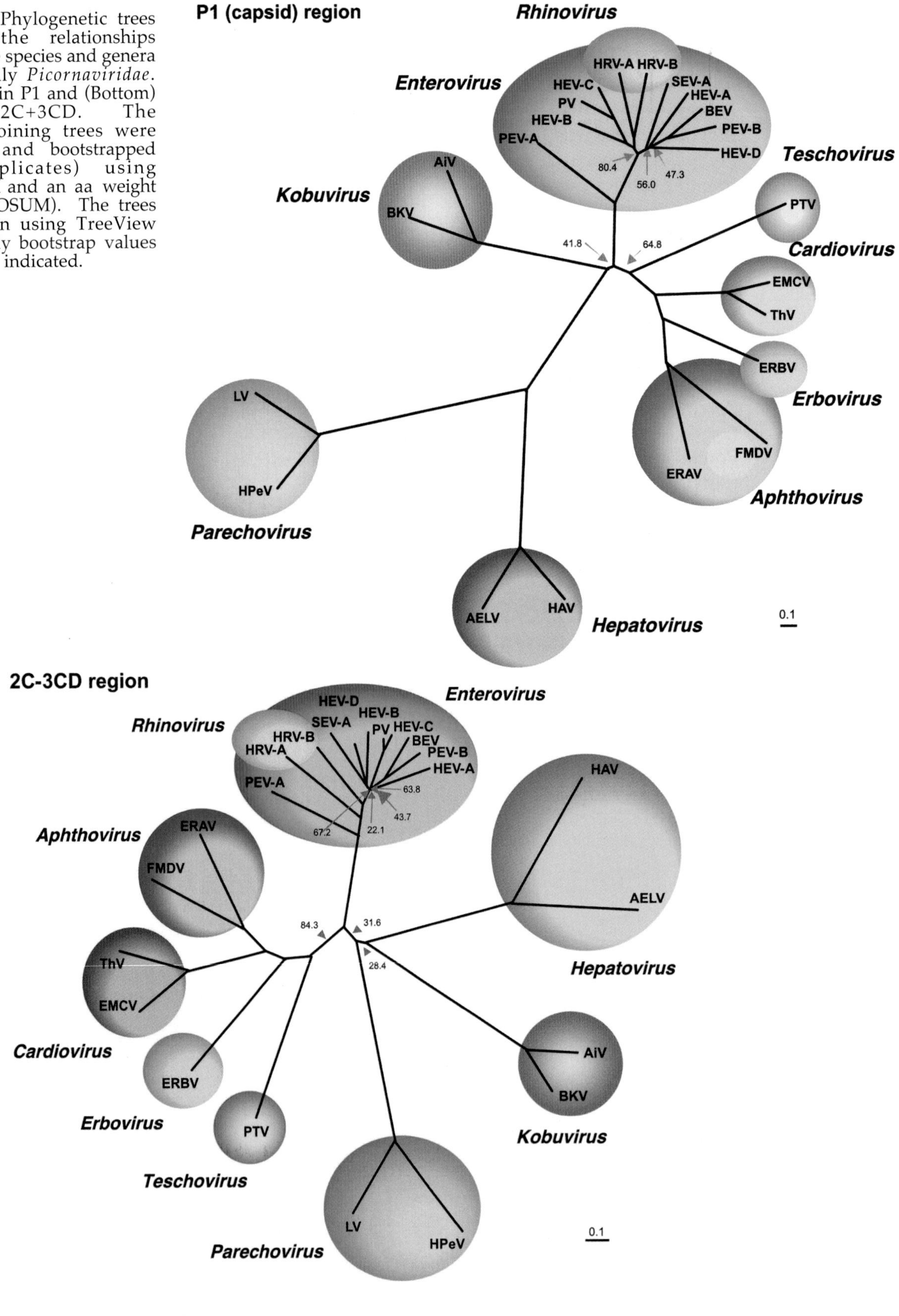

SIMILARITY WITH OTHER TAXA

A "picornavirus-like superfamily" has previously been proposed to include the families *Picornaviridae, Sequiviridae, Comoviridae* and *Potyviridae,* which share the following features:
(i) Genome: This consists of one, or two (in the case of the bipartite plant viruses), molecules (segments) of ssRNA of positive sense. Each segment acts as the exclusive mRNA for the genes it carries (i.e. there are no sgRNAs), contains a single ORF, and is linked at its 5′-end to a tyrosine or serine residue of a genome-linked protein (VPg) via an O-phosphate ester bond.
(ii) Gene expression: The single polyprotein encoded by each genome segment is processed proteolytically to functional proteins by proteases, and these proteases are exclusively virus encoded.
(iii) Genes: All members of the "superfamily" encode, in addition to VPg, a 2C-like protein having a nucleotide binding sequence motif, a $3C^{pro}$-like protease, and a $3D^{pol}$-like polymerase.
(iv) Gene maps: The gene order, 2C-VPg-$3C^{pro}$-$3D^{pol}$, is common to all members of the "superfamily", the $3D^{pol}$ gene always being located at the 3'-terminus of the relevant ORF. Similarly, where CP gene(s) are present in a single ORF, they are always 5'-proximal.

In addition to the families already mentioned above, there are two groups of insect viruses, those in the unassigned genus *Iflavirus* and those comprising the *Dicistroviridae,* that show many of the above characteristics and clear affinities with the putative "picornavirus-like superfamily". At a structural level the "superfamily" is polythetic and contains both icosahedral and rod-shaped viruses. Where the crystal structure is known for icosahedral representatives, of members of the *Picornaviridae* and *Comoviridae,* for example, they exhibit the same, pseudo T=3, arrangement of three major protein subunits, which themselves share the same, β-barrel tertiary structure. This structure is also shared with members of the *Dicistroviridae.*

Despite these notable affinities, the "picornavirus-like superfamily" has no formal taxonomic status.

DERIVATION OF NAMES

Aphtho: from Greek *aphthae,* "vesicles in the mouth"; English: *aphtha,* "thrush"; French: fièvre aphteuse.
Cardio: from Greek *kardia,* "heart".
Entero: from Greek *enteron,* "intestine".
Erbo: sigla for *e*quine *r*hinitis *B* virus.
Hepato: from Greek *hepatos,* "liver".
Kobu: from Japanese *kobu,* "knuckle" (reference to surface structure of virus particle).
Parecho: from *par*(a)*echo* (*echo,* the former name of the type species, a sigla for "enteric cytopathic human orphan").
Picorna: from the prefix "*pico*" (='micro-micro') and RNA.
Rhino: from Greek *rhis, rhinos,* "nose".
Tescho: from *Teschen* disease.

REFERENCES

Agol, V.I. (2002). Picornavirus genome: an overview. In: *Molecular biology of picornaviruses,* (B.L. Semler and E. Wimmer, eds), pp 127-148. ASM Press, Washington DC.

Blomqvist, S., Savolainen, C., Raman, L., Roivainen, M. and Hovi, T. (2002). Human rhinovirus 87 and enterovirus 68 represent a unique serotype with rhinovirus and enterovirus features. *J. Clin. Miocrobiol.* **40**, 4218-422.

Calnek, B.W. (1993). Avian encephalomyelitis. In: *Virus infections of vertebrates, virus infections of birds,* (J.B. McFerran and M.S. McNulty, eds), Vol. 4, pp 469-478. Elsevier, Amsterdam.

Calnek, B.W. (1993). Duck virus hepatitis. In: *Virus infections of vertebrates, virus infections of birds,* (J.B. McFerran and M.S. McNulty, eds), Vol. 4, pp 485-495. Elsevier, Amsterdam.

Hamparian, V.V., Colonno, R.J., Dick, E.C., Gwaltney, J.M., Hughes, J.H., Jordan, W.S., Kapikian, A.Z., Mogabgab, W.J., Mores, A., Phillips, C.A., Rueckert, R.R., Scheble, J.H., Stott, E.J. and Tyrrell, D.A.J. (1987). A collaborative report: rhinoviruses - extension of the numbering system from 89 to 100. *Virology,* **159**, 191-192.

Huang, J.A., Ficorilli, N., Hartley, C.A., Wilcox, R.S., Weiss, M. and Studdert, M.J. (2001). Equine rhinitis B virus: a new serotype. *J. Gen. Virol.,* **82**, 2641-2645.

Hyypiä, T., Hovi, T., Knowles, N.J. and Stanway, G. (1997). Classification of enteroviruses based on molecular and biological properties. *J. Gen. Virol.,* **78**, 1-11.

Hyypiä, T., Horsnell, C., Maaronen, M., Khan, M., Kalkinnen, N., Auvinen, P., Kinnuren, L. and Stanway, G. (1992). A novel picornavirus group identified by sequence analysis. *Proc. Natl. Acad. Sci. USA,* **89**, 8847-8851.

Imada, T. (1993). Avian nephritis virus infection. In: *Virus infections of vertebrates, virus infections of birds,* (J.B. McFerran and M.S. McNulty, eds), Vol. 4, pp 479-483. Elsevier, Amsterdam.

McFerran, J.B. (1993). Other avian enterovirus infections. In: *Virus infections of vertebrates, virus infections of birds,* (J.B. McFerran and M.S. McNulty, eds), Vol. 4, pp 497-503. Elsevier, Amsterdam.

Oberste, M.S., Maher, K., Kilpatrick, D.R. and Pallansch, M.A. (1999). Molecular evolution of the human enteroviruses: correlation of serotype with VP1 sequence and application to picornavirus classification. *J. Virol.,* **73**, 1941-1948.

Oberste, M.S., Maher, K. and Pallansch, M.A. (2002). Molecular phylogeny and proposed classification of the simian picornaviruses. *J. Virol.,* **76**, 1244-1251.

Rueckert, R.R. and Wimmer, E. (1984). Systematic nomenclature of picornavirus proteins. *J. Virol.,* **50**, 957-959.

Savolainen, C., Blomqvist, S., Mulders, M.N. and Hovi, T. (2002). Genetic clustering of all 102 human rhinovirus prototype strains: serotype 87 is close to human enterovirus 70. *J. Gen. Virol.,* **83**, 333-340.

Stanway, G., Hovi, T., Knowles, N.J. and Hyypiä, T. (2002). Molecular and biological basis of picornavirus taxonomy. In: *Molecular Biology of Picornaviruses,* (B.L. Semler and E. Wimmer, eds), pp 17-24. ASM Press, Washington DC

Wutz, G., Auer, H., Nowotny, N., Grosse, B., Skern, T. and Kuechler, E. (1996). Equine rhinovirus serotypes 1 and 2: relationship to each other and to aphthoviruses and cardioviruses. *J. Gen. Virol.,* **77**, 1719-1730.

Yamashita, T., Sakae, K., Tsuzuki, H., Suzuki, Y., Ishikawa, N., Takeda, N., Miyamura, T. and Yamazaki, S. (1998). Complete nucleotide sequence and genetic organization of Aichi virus, a distinct member of the Picornaviridae associated with acute gastroenteritis in humans. *J. Virol.,* **72**, 8408-8412.

Zell, R., Dauber, M., Krumbholz, A., Henke, A., Birch-Hirschfeld, E., Stelzner, A., Prager, D. and Wurm, R. (2001). Porcine teschoviruses comprise at least eleven distinct serotypes: molecular and evolutionary aspects. *J. Virol.,* **75**, 1620-1631.

CONTRIBUTED BY

Stanway, G., Brown, F., Christian, P., Hovi, T., Hyypiä, T., King, A.M.Q., Knowles, N.J., Lemon, S.M., Minor, P.D., Pallansch, M.A., Palmenberg, A.C. and Skern, T.

GENUS *IFLAVIRUS*

Type Species *Infectious flacherie virus*

VIRION PROPERTIES

MORPHOLOGY

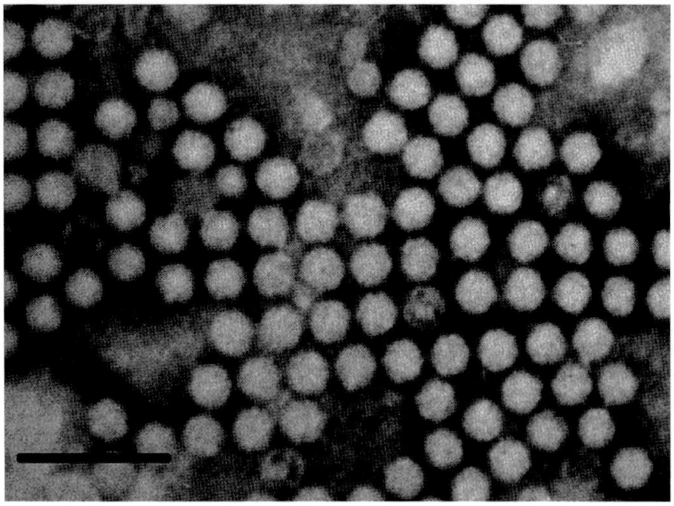

Figure 1: Negative contrast electron micrograph of isometric particles of an isolate of *Infectious flacherie virus*. The bar represents 100nm (Courtesy of H. Bando).

Virions are roughly spherical with a particle diameter of approximately 30 nm and no envelope.

PHYSICOCHEMICAL AND PHYSICAL PROPERTIES

Virions have a buoyant density of between 1.33 g/ml and 1.38 g/ml.

NUCLEIC ACID

Particles contain a single molecule of linear, positive sense, ssRNA of between ~8,800-9,700 nt in size. The 3′-end of the viral RNA is polyadenylated and in Infectious flacherie virus (IFV) there is a protein, VPg, covalently linked to the 5′-end of the genome. The 5′-UTR is quite small and ranges from 156 nt (IFV) to 473 nt (Perina Nuda virus - PnV). The 3′-UTR ranges from 45 nt (PnV) to 239 nt (IFV).

PROTEINS

Mature virions contain three major structural proteins with size generally between 28-35 kDa. In IFV and PnV a fourth smaller structural protein (4-12 kDa) protein has also been reported which may be analogous to the VP4 of picornaviruses and dicistroviruses. There are no reports of other minor structural proteins in the capsid. The structural proteins have been termed VP2, VP3, and VP1 (N-C) to reflect the sequence and deduced structural homology with the equivalent proteins in picornaviruses and dicistroviruses.

LIPIDS

None reported.

CARBOHYDRATES

None reported.

GENOME ORGANIZATION AND REPLICATION

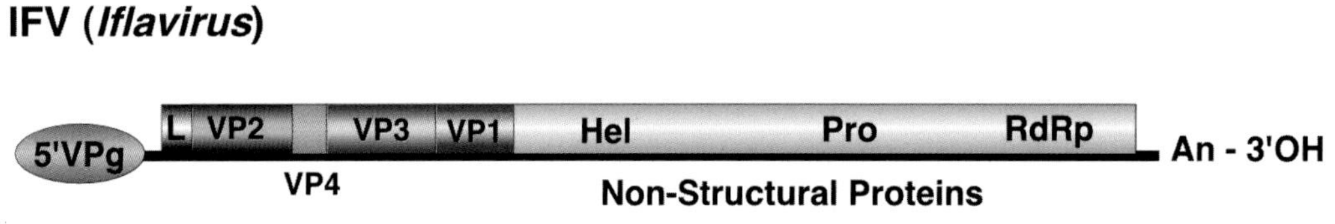

Figure 2: Genome structure of Infectious flacherie virus (IFV). The genome encodes a single polyprotein that is processed to produce the three major structural proteins (VP2, VP3 and VP1) and the non-structural proteins. The structural proteins are encoded at the 5′-end of the polyprotein and the non-structural proteins at the 3′-end. The polyprotein is preceded by a leader sequence (L) of unknown function that is removed from VP2 before capsid assembly. The VP4 is in analogous to the VP4 present in some dicistroviruses and in the case of IFV is present as a minor structural component of the capsid. The approximate positions of the helicase (Hel), protease (Pro) and replicase (RdRp) domains in the non-structural protein are shown.

The genome consists of a ssRNA with a 5′-UTR of 156-473 nt followed by a single ORF of 8,400-9,500 nt and a 3′-UTR of 45-239 nt. The genome is arranged with the structural proteins at the 5′-end of the genome and the non-structural proteins at the 3′-end. The CPs are preceded by a leader sequence of greater than 140 aa which has no known function. Sequence analysis reveals the presence of a small protein between VP2 and VP3 that is analogous to the VP4 of dicistroviruses. In IFV this protein appears to be present as a minor CP, however, its structural role is yet to be elucidated.

The mechanism of virus entry into cells is unknown. *In vitro* translation studies with genomic RNA of IFV have shown that the viral polyprotein is processed to form an array of smaller polypeptides including the CPs. There is no evidence to indicate that translation of the polyprotein is mediated by an IRES. Mechanisms of polyprotein processing and the effects on host cell macromolecular synthesis during infection have not been well studied as suitable cell culture systems have not been available.

ANTIGENIC PROPERTIES

No reported antigenic relationships between species.

BIOLOGICAL PROPERTIES

All members appear to have restricted host ranges. IFV is known only from the lepidopteran species *Bombyx mori* and *Glyphodes pyloalis* and is not know to replicate in any cultured cell line. *Sacbrood virus* (SBV) is a common infection of larvae of the honeybee, *Apis mellifera*. No other hosts or permissive cell lines are known. PnV is known only from the lepidopteran *Perina nuda*. Unlike the other two viruses in the genus, PnV replicates in a homologous cell line established from *Perina nuda*.

LIST OF SPECIES DEMARCATION CRITERIA IN THE GENUS

The list of species demarcation criteria is:

- Natural host range: species can be differentiated on the basis of their natural host range
- Sequence identity between the CPs of isolates and strains of a species is above 90%.

LIST OF SPECIES IN THE GENUS

Species names are in green italic script; strain names and synonyms are in black roman script; tentative species names are in blue roman script. Sequence accession numbers, and assigned abbreviations ( ) are also listed.

SPECIES IN THE GENUS

Infectious flacherie virus

Infectious flacherie virus	[AB000906]	(IFV)

Perina nuda virus		
Perina nuda virus	[AF323747]	(PnV)
(Perina nuda picorna-like virus)		(PnPV)
Sacbrood virus		
Sacbrood virus	[AF469603]	(SBV)

TENTATIVE SPECIES IN THE GENUS

None reported.

PHYLOGENETIC RELATIONSHIPS WITHIN THE GENUS

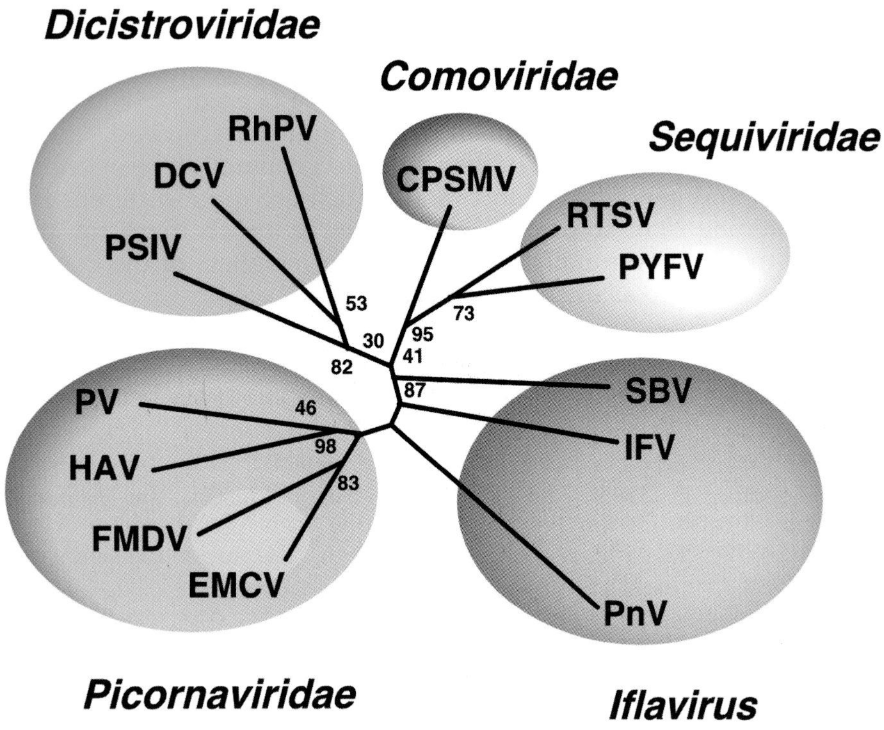

Figure 3: Unrooted phenogram derived from the RdRp domain of the viral non-structural proteins showing the relationships of representative picornaviruses, sequiviruses, dicistroviruses and the three members of the genus *Iflavirus* (IFV, PnV and SBV). Taxa used (with sequence accession numbers shown in brackets [ ]) were Cowpea severe mosaic virus (CPSMV) [M83830], Drosophila C virus (DCV) [AF014388], Encephalomyocarditis virus (EMCV) [M81861], Foot-and-mouth disease virus (FMDV) [X00871], Hepatitis A virus (HAV) [M14707], Poliovirus (PV) [J02281], Parsnip yellow fleck virus (PYFV) [D14066], Plautia stali intestine virus (PSIV), Rhopalosiphum padi virus (RhPV) [AF022937], and Rice tungro spherical virus (RTSV) [M95497]. A bootstrap analysis was performed and values obtained are shown next to the branching points. Branch lengths are proportional to distance.

The current members of the genus are related only distantly to each other. Phenetic analyses of the CP coding regions show that the VP2 and VP3 of the iflaviruses share between 17-22% and 18-24% aa identity in pairwise comparisons. These levels of variation are below those seen between members of the genus *Cripavirus* in the *Dicistroviridae* and even below the levels of variation seen between genera within the *Picornaviridae*.

SIMILARITY WITH OTHER TAXA

Members of the genus *Iflavirus* have similarities to other viruses with positive sense ssRNA genomes within the picornavirus "superfamily" *(Comoviridae, Dicistroviridae, Picornaviridae, Potyviridae* and *Sequiviridae).* For instance the gene order of the non-structural proteins is the same for all groups within this assemblage i.e. Hel-Pro-Rep.

However, in many respects the iflaviruses superficially appear to be entomogenous picornaviruses. The genome organization (with the structural proteins located at the 5′-end of the genome), three CPs, a single ORF and no sgRNAs, all point to strong affinities with the picornaviruses. However, there are some important differences between the iflaviruses and the picornaviruses. First, the position of VP4 in the CPs is very different from the picornaviruses and suggests an affinity with the dicistroviruses. Second the very small size of the 5′-UTR - which is much smaller than that of either the picornaviruses or the dicistroviruses - may well indicate the lack of an IRES-like element in this region which would make this group unique amongst the animal-infecting members of the "picornavirus-like superfamily".

In addition to the known members of this genus and the *Dicistroviridae* there are a large number of viruses of insects/invertebrates with icosahedral/spherical particles, 3 major CPs and ssRNA genomes. While many have been described as either picornaviruses or picorna-like viruses many remain relatively uncharacterized. In the case of *Acyrthosiphon pisum* virus (APV) [AF14514], the complete genome has been sequenced. This virus has a structural and genomic organization, which are quite different from any other. Among the remaining 20 or so picorna-like viruses of insects there are undoubtedly a number of viruses that will eventually be aligned with members of the genus *Iflavirus* but which are currently classified as unassigned viruses.

Derivation of Names

Ifla: a sigla from the type virus of the genus **I**nfectious **fla**cherie virus

References

Christian, P.D. and Scotti, P.D. (1998). The picorna-like viruses of insects. In: *The Insect Viruses* (L.K. Miller and L.A. Ball, eds), pp 301-336. Plenum Publishing Company, New York.

Ghosh, R., Ball, B.V., Willcocks, M.M. and Carter, M.J. (1999). The nucleotide sequence of sacbrood virus of the honey bee: an insect picorna-like virus. *J. Gen. Virol.*, **80,** 1541-9.

Isawa, H., Asano, S., Sahara, K., Iizuka, T. and Bando, H. (1998). Analysis of the genetic information of an insect picorna-like virus, infectious flacherie of silkworm: evidence for evolutionary relationships among insect, mammalian and plant picorna(-like) viruses. *Arch. Virol.*, **143**, 127-143.

Wu, C.-Y., Lo, C.-F., Huang, C.-J., Yu, H.-T. and Wang, C.-H. (2002). The complete genome sequence of *Perina nuda* picorna-like virus, an insect-infecting RNA virus with a genome organization similar to that of the mammalian picornaviruses. *Virology*, **294**, 312-323.

Contributed By

Christian, P., Carstens, E., Domier, L., Johnson, J., Johnson, K., Nakashima, N., Scotti, P. and van der Wilk, F.

Family *Dicistroviridae*

Taxonomic Structure of the Family

Family	***Dicistroviridae***
Genus	***Cripavirus***

Since only one genus is currently recognized, the family description corresponds to the genus description.

Genus *Cripavirus*

Type Species *Cricket paralysis virus*

Virion Properties

Morphology

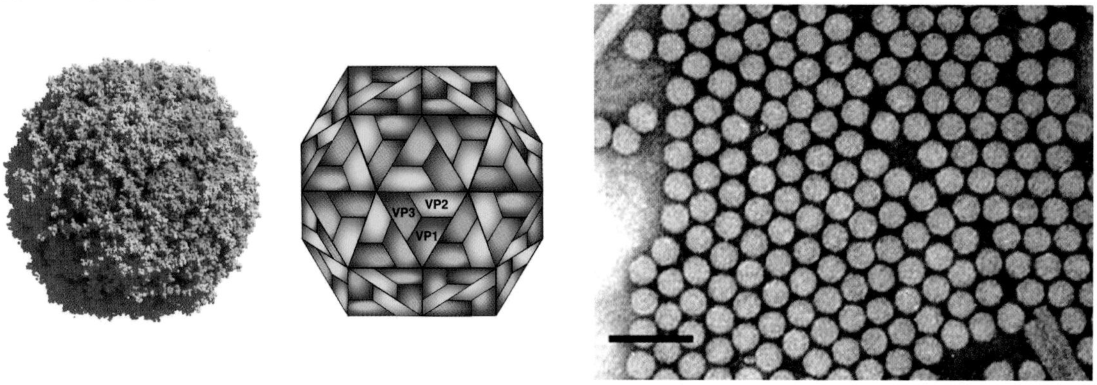

Figure 1: (Left) Rendering of a particle of an isolate of *Cricket paralysis virus* at 2.4Å resolution (Courtesy of Reddy et al., 2001). (Center) Diagram showing the packing of surface proteins of Cricket paralysis virus (CrPV). (Right) Negative contrast electron micrograph of isometric particles of CrPV. The bar represents 100nm (Courtesy of C. Reinganum).

Virions are roughly spherical with a particle diameter of approximately 30 nm and no envelope (Fig. 1). The virions exhibit icosahedral, pseudo T=3 symmetry and are composed of 60 protomers, each comprised of a single molecule of each of VP2, VP3 and VP1 (Fig. 1). A smaller protein, VP4, is also present in the virions of some members and is located on the internal surface of the 5-fold axis below VP1.

Physicochemical and Physical Properties

Virions have a buoyant density in CsCl of between 1.34 and 1.39 g/cm^3 and sedimentation coefficients of between 153 and 167S. For those viruses where physicochemical stability has been assessed like Cricket paralysis virus (CrPV) the virions are stable at pH 3.0.

Nucleic Acid

Particles contain a single molecule of linear, positive sense, ssRNA of approximately 9,000-10,000 nt in size. The 3′-end of the viral RNA is polyadenylated and in most species there is a protein, VPg, covalently linked to the 5′-end of the genome. The 500-800 nt 5′-UTR and the untranslated intergenic region (IGR) between the two ORFS can both initiate translation as IRES. In the case of the IGR the predicted secondary structure is found to be highly conserved across all members of the family and to have a characteristic series of stem-loop structures and a pseudo-knot immediately upstream of the initiation codon. Structural conservation in the putative 5′-IRES is much less conserved between viruses than the IGR IRES and there are no clear structural homologies between the 5′- and IGR IRES.

CrPV (*Dicistrovirus*)

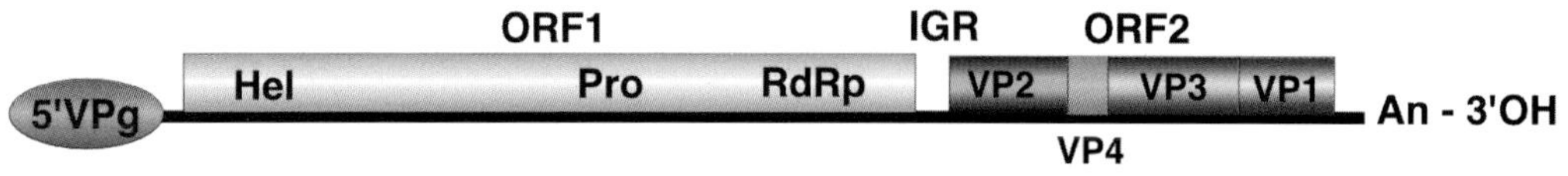

Figure 2: Genome structure of Cricket paralysis virus (CrPV). The approximate positions of the helicase (Hel), protease (Pro) and replicase (RdRp) domains in the non-structural protein encoded by the 5′ ORF are shown. The structural proteins are encoded by ORF 2 and are expressed as a polyprotein. This is processed to produce the three major structural proteins (VP2, VP3 and VP1). VP4 is presumed to be cleaved from a precursor comprising VP4-VP3 and is a minor structural component of the virion. VP4 is not produced by all members of the family.

PROTEINS

Mature virions contain three major structural proteins with size generally between 28-37 kDa. In the case of Taura syndrome virus (TSV) one of the structural proteins, VP3, is much larger than this with a deduced size (from nt sequence analysis) of 56 kDa. In some species a fourth smaller structural protein (4.5-9 kDa) protein has also been reported. In most species a minor structural component - larger than the major CPs - has also been reported and is presumed to be the precursor of one of the major structural proteins (VP3) and the minor structural proteinVP4.

LIPIDS

None reported.

CARBOHYDRATES

None reported.

GENOME ORGANIZATION AND REPLICATION

The genome consists of ssRNA with a 5′-UTR of 500-800 nt followed by two ORFs of around 5,500 nt and 2,600 nt separated by an untranslated region of approximately 190 nt known as the IGR (Fig. 2). The CPs have been shown by direct sequence analysis to be encoded by the ORF proximal to the 3′-end. The 5′ ORF encodes protein(s) with sequence motifs related to the helicase, protease and replicase domains of other positive sense RNA viruses of plants and animals e.g. picornaviruses, comoviruses, sequiviruses and iflaviruses.

The mechanism of virus entry into cells in unknown. Initiation of protein synthesis co-incides with shutdown or down-regulation of host cell protein synthesis. Large precursor proteins are produced in infected cells which are then cleaved to produce an array of smaller polypeptides.

In the case of CrPV the structural proteins are observed to be synthesized in supramolar excess relative to the non-structural proteins. No sgRNA is produced during infection and translation of ORF 2 is presumed to be initiated by an IRES-like element. Studies have shown that the 5′-UTR and IGR of several members can direct initiation of translation in *in vitro* translation systems and in the case of CrPV, both regions function as IRES elements and direct translation from bi-cistronic messages transfected into cultured cells. Initiation of translation from the IGR does not require the presence of a methionine residue and in all known cases is initiated from alanine or glutamine codons.

Virions are assembled in the cytoplasm of infected cells and tend to form large paracrystalline arrays with most species.

ANTIGENIC PROPERTIES

All species are serologically distinct. However, there is some serological relatedness between CrPV and Drosophila C virus (DCV) which show a reaction of partial identity in double diffusion in agar with some sera raised against CrPV.

Biological Properties

All member viruses have been isolated from invertebrate species. CrPV has been isolated from species of Orthoptera, Hymenoptera, Lepidoptera, Hemiptera and Diptera. DCV has only been isolated from dipteran species while Himetobi P virus (HiPV), Plautia stali intestine virus (PSIV), Rhopalosiphon padi virus (RhPV) and Triatoma virus (TrV) have only come from hemipteran species. Acute bee paralysis virus (ABPV) and Black queen cell virus (BQCV) are known only from honeybees (*Apis mellifera*) and TSV only from penaeid shrimps. No vectors are known to be involved in transmission.

CrPV is known to have a wide host range and to be widely distributed in nature. DCV is commonly associated with *Drosophila* species both in nature and in laboratory cultures. Infections with most viruses are usually not associated with a noticeable disease state although they commonly lead to reduced life expectancy of infected individuals. Under some circumstances CrPV shows increased tropism for neural cells and can lead to an obvious paralytic disease state.

CrPV and DCV replicate readily in several established *Drosophila* cell lines. CrPV has also been found to replicate in a number of other established insect cell lines. There are no cell lines known that are able to support replication of the other members.

List of Species Demarcation Criteria in the Genus

The list of species demarcation criteria is:

- Natural host range: species can be differentiated on the basis of their natural host range and their relative ability to replicate in a range of cultured insect cells.
- Serology: all species are serologically distinct.
- Sequence identity between the CPs of isolates and strains of a species is above 90%.

List of Species in the Genus

Species names are in green italic script; strain names and synonyms are in black roman script; tentative species names are in blue roman script. Sequence accession numbers, and assigned abbreviations ( ) are also listed.

Species in the Genus

Aphid lethal paralysis virus		
Aphid lethal paralysis virus	[AF536531]	(ALPV)
Black queen cell virus		
Black queen cell virus	[AF183905]	(BQCV)
Cricket paralysis virus		
Cricket paralysis virus	[AF218039]	(CrPV)
Drosophila C virus		
Drosophila C virus	[AF014388]	(DCV)
Himetobi P virus		
Himetobi P virus	[AB017037]	(HiPV)
Plautia stali intestine virus		
Plautia stali intestine virus	[AB006531]	(PSIV)
Rhopalosiphum padi virus		
Rhopalosiphum padi virus	[AF022937]	(RhPV)
Triatoma virus		
Triatoma virus	[AF178440]	(TrV)

Tentative Species in the Genus

None reported.

LIST OF UNASSIGNED SPECIES IN THE FAMILY

Acute bee paralysis virus		
Acute bee paralysis virus	[AF150629]	(ABPV)
Taura syndrome virus		
Taura syndrome virus	[AF277675]	(TSV)

PHYLOGENETIC RELATIONSHIPS WITHIN THE FAMILY

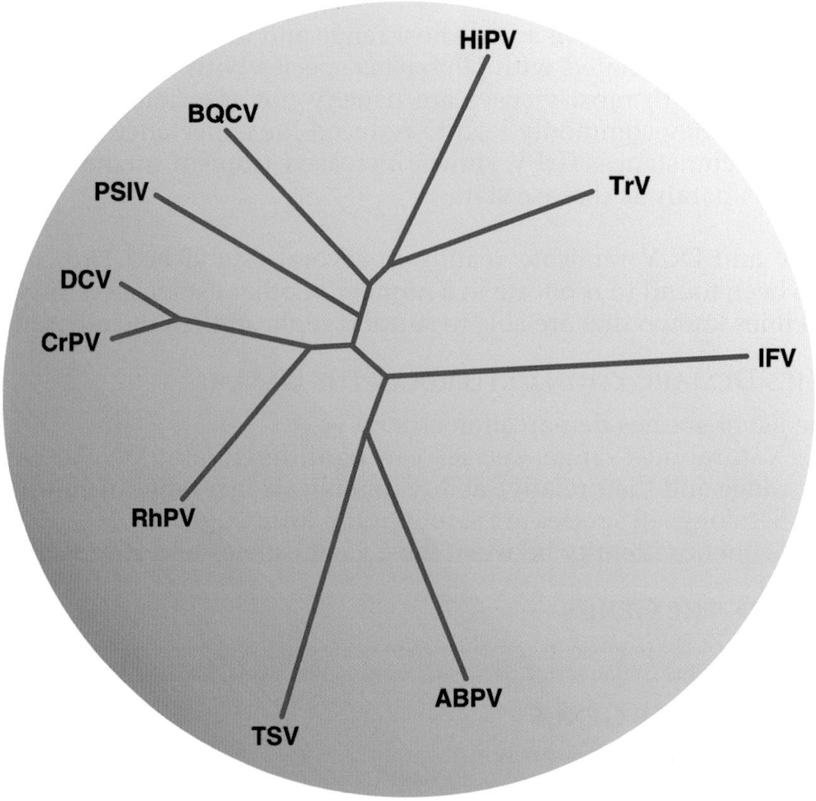

Figure 3: Phenogram showing the relationships among members of the family *Dicistroviridae,* constructed from the aa similarity of VP2 using the neighbor-joining algorithm of the MEGA software. The sequence from Infectious flacherie virus (IFV) [AB000906] was used as an outgroup for the analysis. The abbreviations refer to the virus names above. Branch lengths are drawn to scale.

SIMILARITY WITH OTHER TAXA

Members of the *Dicistroviridae* have similarities to other viruses with positive sense ssRNA genomes within the "picornavirus-like superfamily" *(Comoviridae, Iflavirus, Picornaviridae, Potyviridae* and *Sequiviridae).* For instance the gene order of the non-structural proteins is the same for all groups within this assemblage i.e. Hel-Pro-RdRp. Like other isometric members of this "superfamily" for which the structure is known, CrPV has a pseudo T=3 symmetry. However, the members of the family *Dicistroviridae* can be distinguished from the members of the taxa *Iflavirus, Picornaviridae* and *Sequiviridae* by the organization of the genome i.e. having the structural proteins at the 3′-end of the genome rather than the 5′-end and by the presence of the IGR, and they can be separated from members of the family *Comoviridae* by having only a single rather than two genomic segments.

There are a large number of RNA-containing viruses of approximately 30 nm in diameter that have been described from insects for which the taxonomic status is not known. Many of these have characteristics that are superficially similar to members of the family *Dicistroviridae* and have been described in the literature as either picornaviruses or picorna-like viruses. While many of these viruses remain relatively uncharacterized, for Acyrthosiphon pisum virus (APV) [AF14514], the complete genome has been sequenced.

This virus has a structural and organization which is quite different from either the members of the family *Dicistroviridae* or of the recently established unassigned genus, *Iflavirus*. Among the remaining 20 or so picorna-like viruses of insects there are undoubtedly a number of viruses that will eventually be members of the family *Dicistroviridae* but they are currently classified as unassigned viruses.

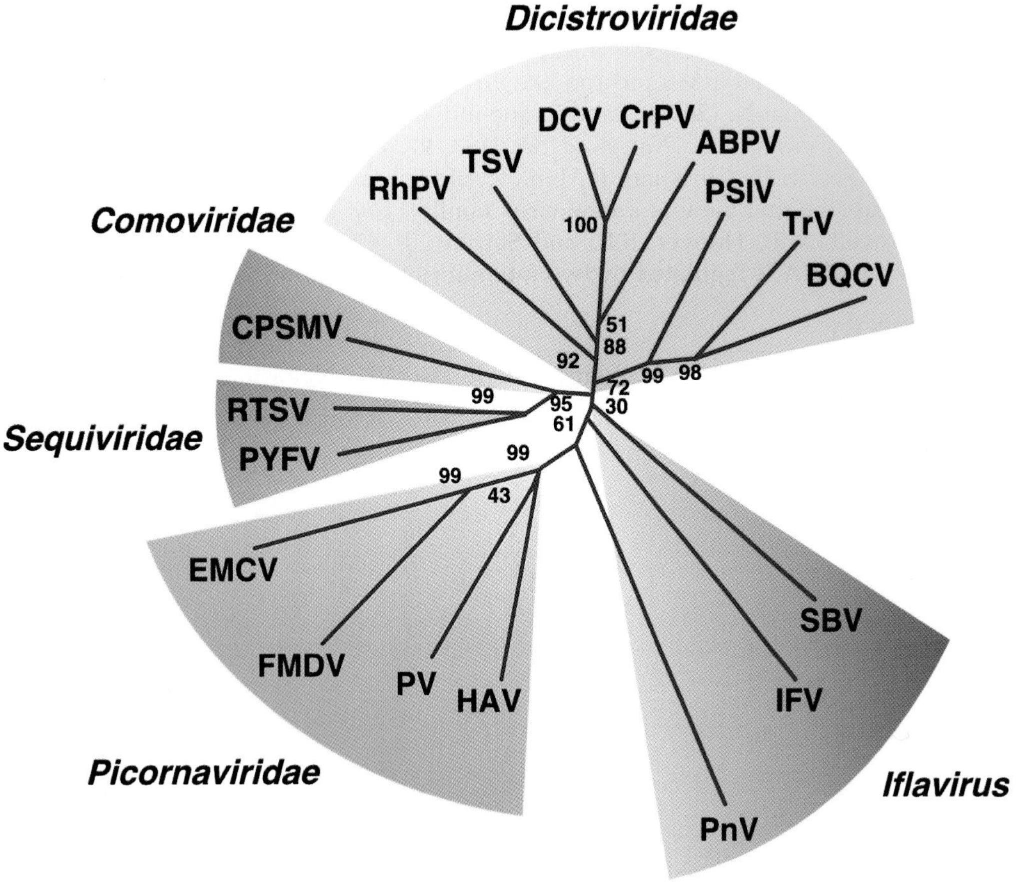

Figure 4. Unrooted phenogram showing the relationships of members of the family *Dicistroviridae* to representatives of the families *Picornaviridae*, *Comoviridae*, *Sequiviridae* and the unassigned genus *Iflavirus*. The phenogram was constructed from an aa similarity matrix of the replicase (RdRp) region of the non-structural proteins using the neighbor-joining method. A bootstrap analysis was performed and the percentage values are indicated at the branching points. Viruses other than dicistroviruses included in the analysis, abbreviation ( ) and accession numbers [ ] are; Cowpea severe mosaic virus (CPSMV) [M83830], Encephalomyocarditis virus (EMCV) [M81861], Foot-and-mouth disease virus (FMDV) [X00871], Hepatitis A virus (HAV) [M14707], Infectious flacherie virus (IFV) [AB000906], Parsnip yellow fleck virus (PYFV) [D14066], Perina nuda virus (PnV) [AF323747], Poliovirus (PV) [J02281], Rice tungro spherical virus (RTSV) [M95497], Sacbrood virus (SBV) [AF469603]. Branch lengths are drawn to scale.

Derivation of Names

Cripa: sigla from the name of the type member of the genus, ***Cri**cket **pa**ralysis virus*
Dicistro: sigla from the characteristic **di-cistro**nic arrangement of the genome.

References

Christian, P.D. and Scotti, P.D. (1998). The picorna-like viruses of insects. In: *The insect viruses*, (L.K. Miller and A. Ball, eds), pp 301-336. Plenum Press, New York.

POS. SSRNA

Domier, l.L., McCoppin, N.K. and D'Arcy, C.J. (2000). Sequence requirements for translation initiation of *Rhopalosiphum padi* virus ORF2. *Virology*, **268**, 264-71.
Kanamori, Y. and Nakashima, N. (2001). A tertiary structure model of the internal ribosome entry site (IRES) for methionine-independent initiation of translation. *RNA*, **7**, 266-274.
Liljas, L., Tate, J., Christian, P. and Johnson, J.E. (2002). Evolutionary and taxonomic implications of conserved structural; motifs between picornaviruses and insect picorna-like viruses. *Arch. Virol.*, **147**, 59-84.
Mari, J., Poulos, B.T., Lightner, D.V. and Bonami, J.-R. (2002). Shrimp Taura syndrome virus: genomic characterization and similarity with members of the genus *Cricket paralysis-like viruses*. *J. Gen. Virol.*, **83**, 915-926.
Sasaki, J. and Nakashima, N. (1999). Translation initiation at the CUU codon is mediated by the internal ribosome entry site of an insect picorna-like virus *in vitro*. *J. Virol.*, **73**, 1219-1226.
Sasaki, J. and Nakashima, N. (2000). Methionine-independent initiation of translation in the capsid protein of an insect RNA virus. *Proc. Nat. Acad. Sci. USA*, **97**, 1512-1515.
Tate, J. Liljas, L., Scotti, P., Christian, P., Lin, T. and Johnson, J.E. (1999). The crystal structure of cricket paralysis virus: the first view of a new virus family. *Nature Struct. Biol.*, **6**, 765-774.
Wilson, J.E., Powell, M.J., Hoover, S.E. and Sarnow, P. (2000). Naturally occurring dicistronic Cricket Paralysis Virus RNA is regulated by two internal ribosome entry sites. *Mol. Cell. Biol.*, **20**, 4990-4999.

CONTRIBUTED BY

Christian, P., Carstens, E., Domier, L., Johnson, J., Johnson, K., Nakashima, N., Scotti, P. and van der Wilk, F.

FAMILY MARNAVIRIDAE

TAXONOMIC STRUCTURE OF THE FAMILY

Family ***Marnaviridae***
Genus ***Marnavirus***

Since only one genus is currently recognized, the family description corresponds to the genus description.

GENUS MARNAVIRUS

Type Species ***Heterosigma akashiwo RNA virus***

VIRION PROPERTIES

MORPHOLOGY

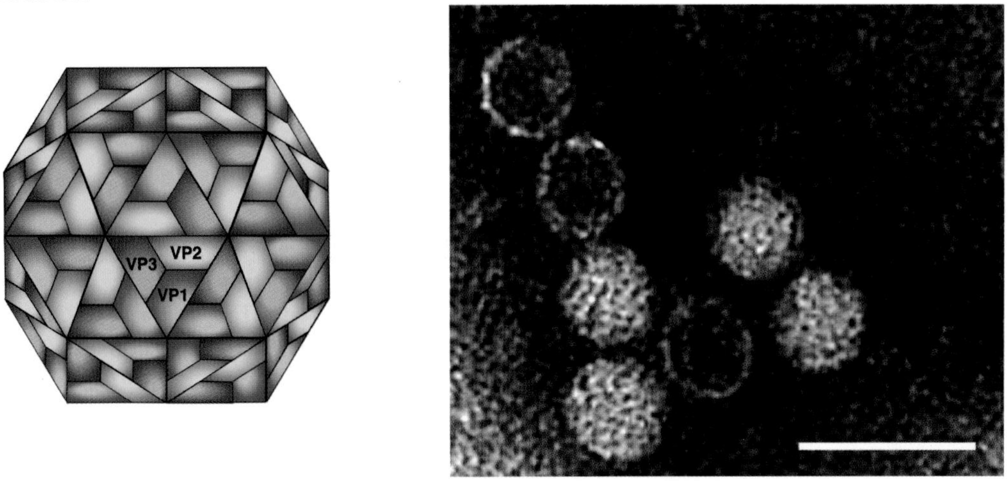

Figure 1: (Left) Diagrammatic representation of the possible structure of Heterosigma akashiwo RNA virus (HaRNAV) particles. (Right) Electron micrograph of HaRNAV particles stained with phosphotungstic acid. The bar represents 50 nm.

At present, the only characterized representative of the family *Marnaviridae* is Heterosigma akashiwo RNA virus (HaRNAV). Based on electron micrographs, HaRNAV virions are approximately 25 nm in diameter, polyhedral in shape, do not appear to have an envelope, and have no discernable projections (Fig. 1).

PHYSICOCHEMICAL AND PHYSICAL PROPERTIES

HaRNAV is not sensitive to chloroform.

NUCLEIC ACID

HaRNAV has a 8.6-kb ssRNA genome containing a single ORF. The genome has a poly(A) tail at the 3′-terminus. The 5′- and 3′-UTRs are 483 and 361 nt long, respectively, accounting for a total of 9.8% of the genome. Computer predictions (*mfold* 3.0) of secondary structure of the 5′-UTR and a notable pyrimidine-rich stretch of sequence upstream of the predicted start codon, suggest the presence of an IRES, a feature observed in many picorna-like viruses. The genome sequence has 2 large pseudo repeats in the 5′- and 3′-UTRs. A 136 nt sequence in the 5′-UTR shares 123 exact bases with a 137 nt sequence in the 3′-end. These repeated sequences may have some function in replication or translation.

POS. SSRNA

PROTEINS

The major structural proteins of HaRNAV are characterized in Table 1. The 33 and 29 kDa protein sequence revealed similarities to the VP3 proteins from the *Dicistroviridae* and *Picornaviridae* and the VP1 proteins from the *Dicstroviridae*, respectively. The lack of a recognizable pattern at the protein cleavage site suggests there may be more than one protease involved in polyprotein processing.

Table 1. Characteristics of major HaRNAV structural proteins

Protein (kDa)[a]	Position of N-teminus in polyprotein[b]	Putative sequence at cleavage site[b]
39	1990	PTST-SEIV
33	2318	FVST-SEII
29	2060	LFGY-SRPP
26	1776	EKLL-TETL
24	1810	RPGE-VDGD

a. Based on SDS-PAGE
b. Based on N-terminal sequencing and genome sequence analysis

LIPIDS

Undetermined.

CARBOHYDRATES

Undetermined.

GENOME ORGANIZATION AND REPLICATION

The map of protein domains within the predicted HaRNAV polyprotein sequence is shown in Figure 2. Domains were identified on the basis of similarities with conserved domains for picorna-like helicases, RdRps, and CPs. HaRNAV encodes an amino acid sequence that matches the chymotrypsin-related serine protease catalytic domain. A VPg-like protein, characteristic of most picorna-like viruses, has not been identified.

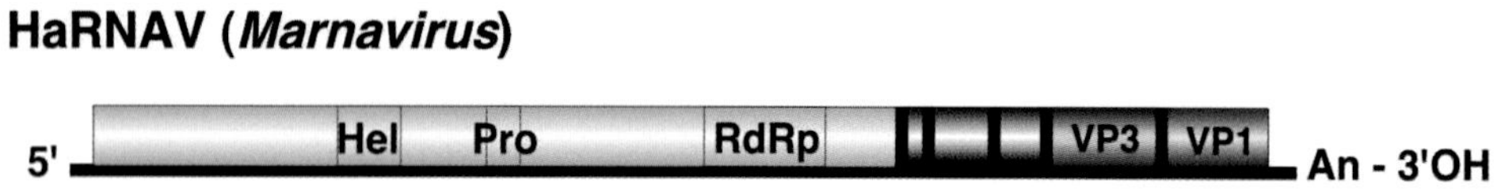

Figure 2. Representation of the genome organization of Heterosigma akashiwo RNA virus (HaRNAV). The location of conserved picorna-like protein domains are indicated within the polyprotein box: Hel, helicase; Pro, protease; RdRp, RNA-dependent RNA polymerase; VP3 and VP1, structural proteins. The locations of N-termini found by sequencing the HaRNAV structural proteins are shown by black lines in the box.

ANTIGENIC PROPERTIES

Undetermined.

BIOLOGICAL PROPERTIES

The host range of HaRNAV is restricted to specific strains of *Heterosigma akashiwo*. Of 15 host strains isolated from the Northwest Pacific, Western Atlantic Ocean, and Japanese coastal waters, five were permissive to HaRNAV infection. HaRNAV replication appears to be cytolytic. Cytopathic effects begin approximately 48 hrs after infection. Ultrastructural changes include swelling of the endoplasmic reticulum, vacuolation and disintegration of the cytoplasm, and the appearance of fibrous material in vacuolated areas. Particles of HaRNAV are distributed in the cytoplasm in crystalline arrays or as individuals.

LIST OF SPECIES DEMARCATION CRITERIA IN THE GENUS

Not applicable.

List of Species in the Genus

Species names are in green italic script; strain names and synonyms are in black roman script; tentative species names are in blue roman script. Sequence accession numbers, and assigned abbreviations ( ) are also listed.

Species in the Genus

Heterosigma akashiwo RNA virus

Heterosigma akashiwo RNA virus (HaRNAV)

Tentative Species in the Genus

None reported.

List of Unassigned Viruses in the Family

None reported.

Phylogenetic Relationships within the Family

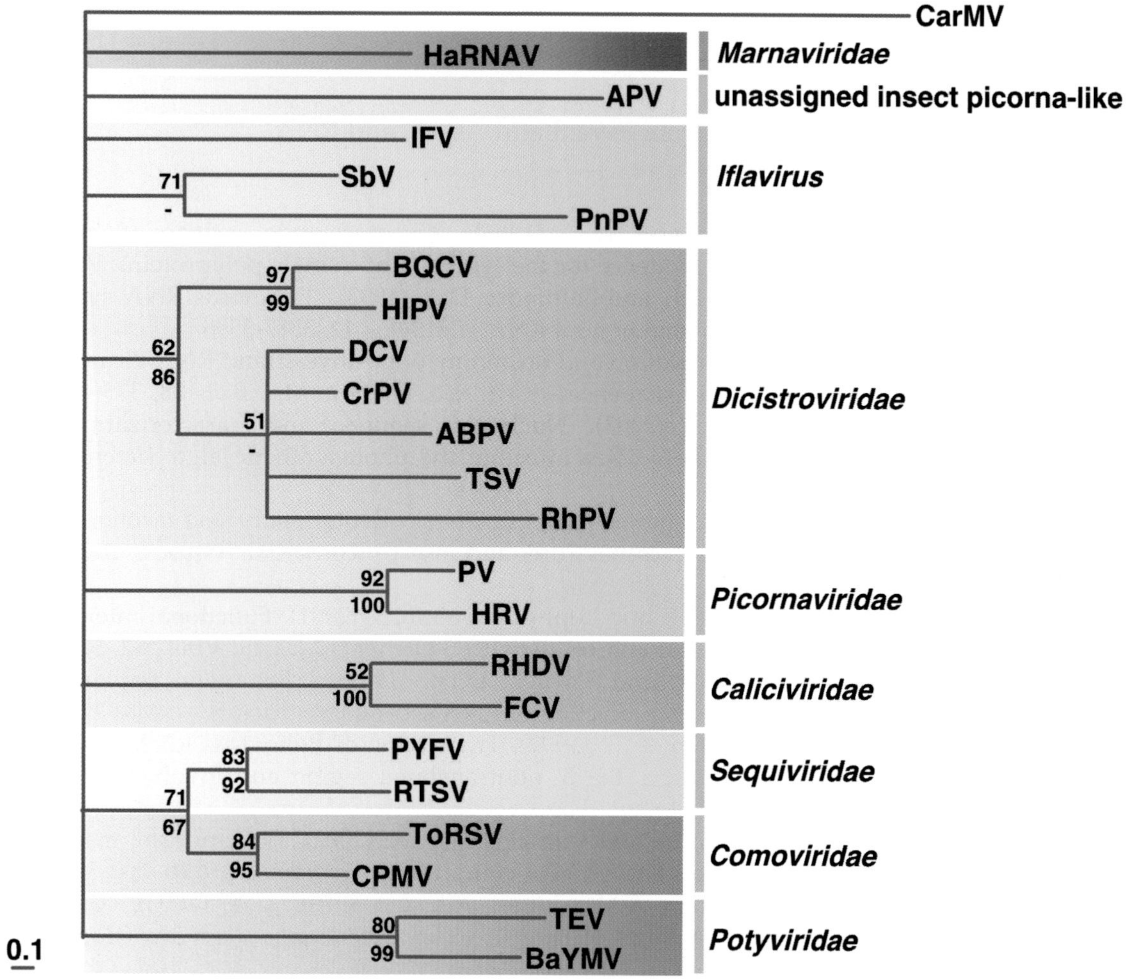

Figure 3: Phylogenetic analysis of picorna-like RdRp domain protein sequences. CLUSTAL_X alignments were done with residues 1362-1619 of the HaRNAV polyprotein that represent the conserved regions I-VIII (defined in Koonin and Dolja (1993)) and the corresponding regions from the other viruses included. The tree is based on maximum likelihood distances generated with TREE-PUZZLE. The sequence from the Carnation mottle virus (CarMV) was used as an outgroup. Support values based on 10,000 puzzling steps are shown above the branches. Bootstrap values (based on 1,000 replicates) for branches that are supported by >50% by neighbor-joining analysis are labeled below the branches (a dash indicates there was no corresponding branch in the neighbor-joining tree). The maximum likelihood scale bar is shown.

POS. SSRNA

SIMILARITY WITH OTHER TAXA

There is strong evidence placing the family *Marnaviridae* in the "picorna-like superfamily". The HaRNAV genome is composed of one molecule of ssRNA of positive sense that exhibits the 2C-3C^{pro}-3D^{pol} gene order and particles are icosahedral with a diameter ~25 nm, a size and structure consistent with picorna-like viruses. However, the structure of the viral genome and the patterns of sequence relationships of HaRNAV proteins to other known viral picorna-like proteins clearly show that it does not belong in any of the currently established picorna-like families. The HaRNAV genome structure is mostly like the potyviruses (e.g., Tobacco etch virus), in that the non-structural protein domains are located at the N-terminus and the structural proteins are at the C-terminus in a single large polyprotein encoded on a monopartite genome. However, potyvirus capsids are filamentous and phylogenetic analyses demonstrated no significant homology with this family (Fig. 3). Moreover, a phylogenetic analysis of picorna-like RdRps does not place the HaRNAV sequence within any established family of picorna-like viruses (Fig. 3). It is not surprising that HaRNAV represents a new family as it is the first picorna-like virus that has been described that infects a protist.

DERIVATION OF NAMES

Marna is a sigla derived from *mare*: Latin, "sea", and RNA

REFERENCES

Allison, R.F., Johnson, R.E. and Dougherty, W.G. (1986). The nucleotide sequence of the coding region of Tobacco etch virus genomic RNA: Evidence for the synthesis of a single polyprotein. *Virology*, **154**, 9-20.

Andino, R., Rieckhof, G.E., Achacoso, P.L. and Baltimore, D.A. (1993). Poliovirus RNA synthesis utilizes an RNP complex formed around the 5′-end of viral RNA. *EMBO J.*, **12**, 3587-3598.

Koonin, E.V. and Dolja, V.V. (1993). Evolution and taxonomy of positive-strand RNA viruses: Implications of comparative analysis of amino acid sequences. *Crit. Rev. Biochem. Mol. Biol.*, **28**, 375-430.

Lang, A.S., Culley, A.I. and Suttle, C.A. (2003). Nucleotide sequence and characterization of HaRNAV: a marine virus related to picorna-like viruses infecting the photosynthetic alga *Heterosigma akashiwo*. *Virology*, **310**, 359-371.

Liljas, L., Tate, J., Lin, T., Christian, P. and Johnson, J.E. (2002). Evolutionary and taxonomic implications of conserved structural motifs between picornaviruses and insect picorna-like viruses. *Arch. Virol.*, **147**, 59-84.

Martinez-Salas, E., Ramos, R., Lafuente, E. and Lopex de Quinto, S. (2001). Functional interactions in internal translation initiation directed by viral and cellular IRES elements. *J. Gen. Virol.*, **82**, 973-984.

Mathews, D.H., Sabina, J., Zuker, M. and Turner, D.H. (1999). Expanded sequence dependence of thermodynamic parameters improves prediction of RNA secondary structure. *J. Mol. Biol.*, **288**, 911-940.

Pestova, T.V., Hellen, C.U.T. and Wimmer, E. (1991). Translation of Poliovirus RNA: Role of essential cis-acting oligopyrimidine element within the 5′ nontranslated region and involvement of a cellular 57-kilodalton protein. *J. Virol.*, **65**, 6194-6204.

Schmidt, H.A., Strimmer, K., Vingron, M. and Von Haeseler, A. (2002). Tree-puzzle: maximum likelihood phylogenetic analysis using quartets and parallel computing. *Bioinformatics*, **18**, 502-504.

Tai, V., Lawrence, J.E., Lang, A.S., Chan, A.M., Culley, A.I. and Suttle, C.A. (2003). Characterization of HaRNAV, a single-stranded RNA virus causing lysis of *Heterosigma akashiwo* (Raphidophyceae). *J. Phycol.*, **39**, 343-352.

CONTRIBUTED BY

Culley, A.I., Lang, A.S. and Suttle, C.A.

FAMILY SEQUIVIRIDAE

TAXONOMIC STRUCTURE OF THE FAMILY

Family *Sequiviridae*
Genus *Sequivirus*
Genus *Waikavirus*

VIRION PROPERTIES

MORPHOLOGY

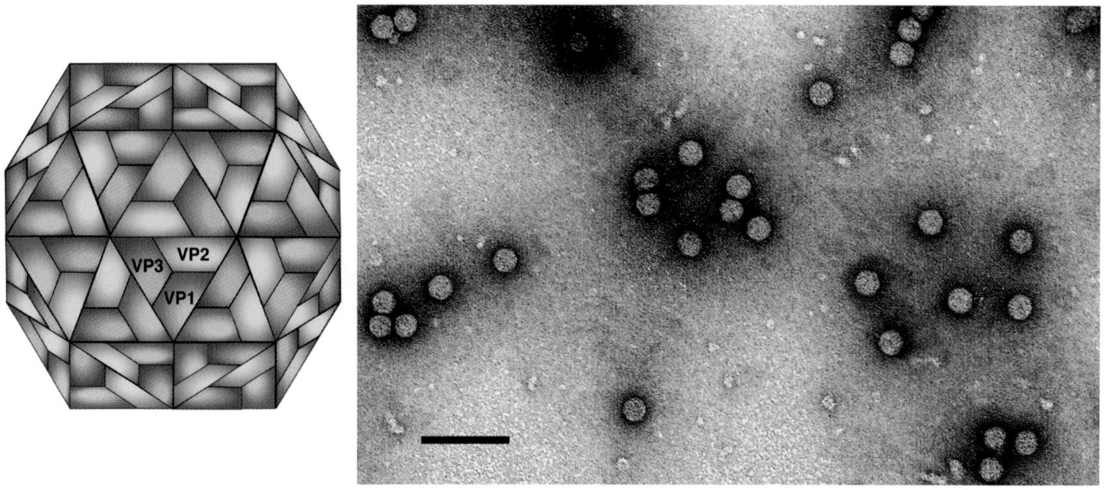

Figure 1: (Left) Putative diagram representation of the capsid structure of sequiviruses. (Right) Negative contrast electron micrograph of particles of an isolate of *Parsnip yellow fleck virus*, stained in 1% uranyl acetate. The bar represents 100 nm. (Courtesy I.M. Roberts).

Particles are icosahedral, about 25-30 nm in diameter (Fig. 1).

PHYSICOCHEMICAL AND PHYSICAL PROPERTIES

For a given virus, two classes of virions are distinguished according to their buoyant densities: the main virion component contains RNA and sediments at 150-190S, and some preparation also contain empty shells that sediment at about 60S.

NUCLEIC ACID

The genome is a single positive-sense ssRNA, encoding a polyprotein. Infectivity is protease-sensitive and a 5'-linked VPg molecule is probably present.

PROTEINS

Virions contain three major CP of about 32-34, 22-26 and 22-24 kDa. Virion and non-structural proteins arise by proteolytic cleavage of polyproteins.

LIPIDS

None reported.

CARBOHYDRATES

None reported.

GENOME ORGANIZATION AND REPLICATION

Genetic information encoded by the RNA genome is organized as a single ORF (Fig. 2). The genomic organizations are essentially similar between each genus, and similar to that of other "picorna-like" viruses of plants and animals, with domains characteristic of proteins with NTP binding, proteinase and RdRp. The domain encoding the structural proteins is located upstream of this "replication block", as in picornavirus genomes but unlike in those of other plant "picorna-like" viruses with monopartite genomes.

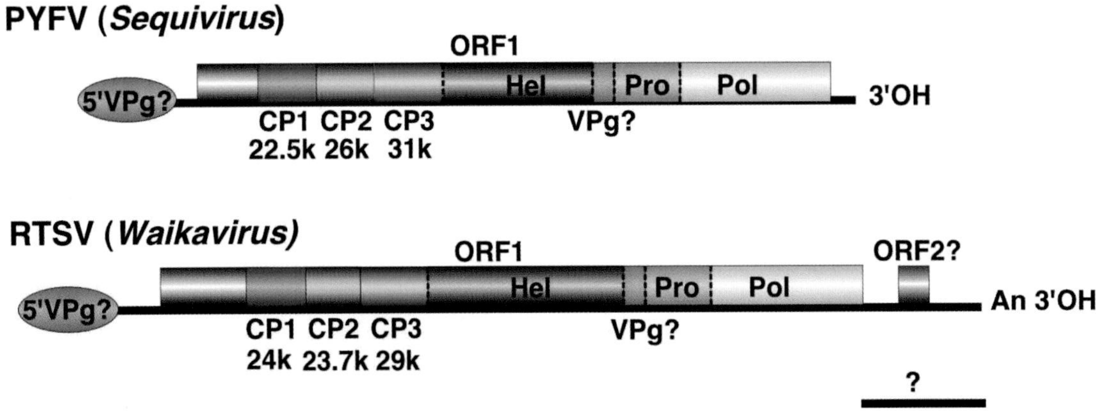

Figure 2: Genome organizations characteristic of Parsnip yellow fleck virus (PYFV; *Sequivirus*) and Rice tungro spherical virus (RTSV; *Waikavirus*). The boxes represent the polyproteins. The vertical solid lines show where cleavages are known to occur in the polyproteins and the dashed lines show where cleavages are presumed to occur. The approximate positions of NTP-binding (Hel), proteinase (Pro) and polymerase (Pol) are shown. Other putative proteins include the movement protein (MP). Ovals represent the putative VPg, and An the 3′ poly-A in RTSV.

ANTIGENIC PROPERTIES

Polyclonal sera contain antibodies to all virion proteins.

BIOLOGICAL PROPERTIES

Natural host ranges are usually restricted. Transmission is in the semi-persistent mannerby aphids, or by leafhoppers. Co-infection with a "helper" waikavirus seems to be required for aphid transmission of sequiviruses. RTSV also serves as a helper virus for leafhopper transmission of Rice tungro bacilliform virus, a member of the family *Caulimoviridae*. The viruses are graft-transmissible. Sequiviruses, but not waikaviruses, are also mechanically transmitted in standard laboratory conditions.

GENUS SEQUIVIRUS

Type Species *Parsnip yellow fleck virus*

DISTINGUISHING FEATURES

The main virion component sediments around 150S, contains about 40% RNA and has a correspondingly high equilibrium density in cesium salts (1.49-1.52 g/cm^3). Some preparations also contain less dense particles (about 60S) that contain no RNA. Virions of PYFV contain three major CP of about 32, 26 and 23 kDa.

The RNA is about 10 kb. PYFV RNA is not polyadenylated and has no small ORF near the 3'-end. There are about 400 aa upstream of the structural proteins in the large polyprotein. Aphid transmission of PYFV depends on simultaneous or prior access to plants infected by a helper virus, Anthriscus yellows virus (genus *Waikavirus*).

GENOME ORGANIZATION AND REPLICATION

The genomic RNA contains one major ORF that encodes a polyprotein of about 3,000 to 3,500 aa. The structural proteins are in the N-terminal half of the polyprotein but are separated from the N-terminus by polypeptide(s) of about 40-60 kDa. Sequences downstream of the structural proteins contain domains characteristic of proteins with NTP binding, protease and RdRp activities.

BIOLOGICAL PROPERTIES

The natural host range includes several species in several families. Transmission is in the semi-persistent manner by aphids. However, it is dependent on the presence of a helper virus in the genus *Waikavirus*.

LIST OF SPECIES DEMARCATION CRITERIA IN THE GENUS

Members of the species *Parsnip yellow fleck virus* and the species *Dandelion yellow mosaic virus* are distinguished because :

- they do not cross-react with heterologous antibodies; PYFV can be divided into two serotypes (formerly regarded as separate species) that differ by serological differentiation index of 4 to 5,
- their principal hosts belong to different families; PYFV infects umbelliferous plants (although the serotypes differ in host range) and Dandelion yellow mosaqic virus infects plants in the Compositae,
- they are transmitted by different species of vector aphid; PYFV is transmitted by *Cavariella* spp. and DaYMV is transmitted by *Aulacorthum solani* and some *Myzus* spp.

LIST OF SPECIES IN THE GENUS

Species names are in green italic script; strain names and synonyms are in black roman script; tentative species names are in blue roman script. Sequence accession numbers, and assigned abbreviations ( ) are also listed.

SPECIES IN THE GENUS

Dandelion yellow mosaic virus		
Dandelion yellow mosaic virus		(DaYMV)
Parsnip yellow fleck virus		
Celery yellow net virus		(CeYNV)
Parsnip yellow fleck virus	[NC_003628]	(PYFV)

TENTATIVE SPECIES IN THE GENUS

Lettuce mottle virus	(LeMoV)

GENUS *WAIKAVIRUS*

Type Species *Rice tungro spherical virus*

DISTINGUISHING FEATURES

The main virion component sediments at 180-190S, contains about 40% RNA and has a correspondingly high equilibrium density in cesium salts (1.5 g/cm^3). Virions contain three major CPs of about 33-34, 22-24 and 22-25 kDa. Particles of some waikaviruses are thought to contain other proteins that may be derived from one of the three major proteins.

The RNA is about 12 kb and has a poly(A) tail. Genomes of Rice tungro spherical virus (RTSV) and Maize chlorotic dwarf virus (MCDV) RNA contain a small ORF near the 3'-end and have about 600 to 700 aa upstream of the structural proteins in the large polyproteins. Transmission by aphids or leafhoppers is thought to depend on a self-encoded helper protein. The helper protein of some species can assist insect transmission of other viruses, e.g. sequiviruses, when they are present in co-infection.

GENOME ORGANIZATION AND REPLICATION

The genomic RNA contains one major ORF that encodes a polyprotein. A smaller ORF is present downstream of the region encoding the polyprotein. The structural proteins are in the N-terminal half of the polyprotein but are separated from the N-terminus by polypeptide(s) of about 40-60 kDa. Sequences downstream of the structural proteins contain domains characteristic of proteins with NTP-binding, protease and RdRp activities.

BIOLOGICAL PROPERTIES

Natural host ranges are restricted to few species within few families. Waikaviruses are not sap-transmitted. Field transmission is in the semi-persistent manner by aphids or leafhoppers. A virus-encoded helper protein is probably needed. Some waikaviruses are helper viruses for the insect transmission of other viruses: *Parsnip yellow fleck virus* (*Sequivirus*) in the case of *Anthriscus yellows virus* and *Rice tungro bacilliform virus* (*Caulimoviridae*) in the case of *Rice tungro spherical virus* (this association being responsible for the very damaging rice tungro disease).

LIST OF SPECIES DEMARCATION CRITERIA IN THE GENUS

Isolates belong to distinct species if:

- gene products differ in aa sequence; from limited comparisons values of <70% homology over the entire polyprotein and <80% between NTP-binding domains or proteinase domains or polymerase domains would suggest distinct species (the extent is not possible to define with certainty as too few sequences are at hand),
- they differ serologically; at most, there is a very weak cross-reaction between RTSV and MCDV in immunoblots,
- they differ in host range; RTSV infects rice and some other graminaceous hosts, MCDV infects maize and some other graminaceous hosts, but not rice and AYV infects umbelliferous (dicotyledonous) hosts,
- they are transmitted by different vector species; MCDV and RTSV are transmitted by leafhoppers (*Graminella* spp. and *Nephotettix* spp. respectively) and AYV is transmitted by aphids.

LIST OF SPECIES IN THE GENUS

Species names are in green italic script; strain names and synonyms are in black roman script; tentative species names are in blue roman script. Sequence accession numbers, and assigned abbreviations ( ) are also listed.

SPECIES IN THE GENUS

Anthriscus yellows virus		
Anthriscus yellows virus		(AYV)
Maize chlorotic dwarf virus		
Maize chlorotic dwarf virus	[NC_003626]	(MCDV)
Rice tungro spherical virus		
Rice tungro spherical virus	[NC_001632]	(RTSV)
Rice tungro spherical virus - Vt6	[AB064963]	(RTSV-Vt6)

TENTATIVE SPECIES IN THE GENUS

None reported.

LIST OF UNASSIGNED VIRUSES IN THE FAMILY

None reported.

Phylogenetic Relationships within the Family

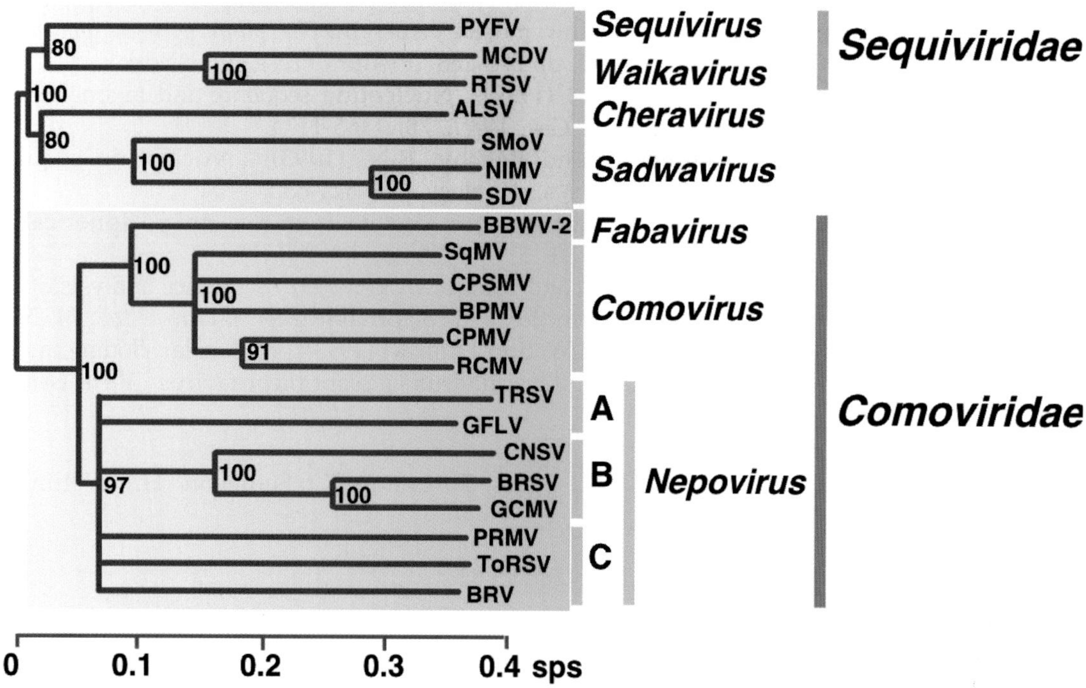

Figure 3: Unrooted dendrogram derived from the alignment of the sequences of the proteinase-polymerase domains of virus species belonging to the families *Comoviridae* and *Sequiviridae*, and the unassigned genera *Cheravirus* and *Sadwavirus*. Branches supported by bootstrap values lower than 75% were merged. The horizontal scale is proportional to the level of divergence while the vertical scale is arbitrary.

Similarity with Other Taxa

The amino acid sequences in the conserved NTP-binding and RNA polymerase domains of the polyproteins resemble those in the polyproteins encoded by RNA of viruses in the unassigned genera *Cheravirus* and *Sadwavirus*, and in the families *Comoviridae*, *Picornaviridae* and *Dicistroviridae*. The number and sizes of the CPs resemble those of viruses in the family *Picornaviridae* although the protein(s) upstream of the CPs is larger than the L protein of aphthoviruses. The properties of the particles and the genomes of these viruses have sometimes prompted their description as 'plant picornaviruses'.

Derivation of Names

Sequi: sigla from latin *sequi*, to follow, accompany, attend (in reference to the dependent aphid transmission of PYFV).
Waika: from Japanese, describing the symptoms induced in rice by infection with RTSV alone (i.e. in the absence of Rice tungro bacilliform virus, RTBV).

References

Bos, L., Huijberts, N., Huttinga, H. and Maat, D.Z. (1983). Further characterization of dandelion yellow mosaic virus from lettuce and dandelion. *Neth. J. Pl. Path.*, **89**, 207-222 .

Elnagar, S. and Murant, A.F. (1976). The role of the helper virus, anthriscus yellows, in the transmission of parsnip yellow fleck virus by the aphid *Cavariella aegopodii*. *Ann. Appl. Biol.*, **84**, 169-181.

Ge, X., Gordon, D.T. and Gingery, R.E. (1989). Occurrence of a small RNA in maize chlorotic dwarf virus-like particles. *Phytopathology*, **79**, 1195.

Gingery, R.E. (1988). Maize chlorotic dwarf and related viruses. In: *The plant viruses; polyhedral virions with monopartite RNA* (R. Koenig, ed), Vol 3, pp 259-272. Plenum Press, New York.

Hemida, S.K., Murant, A.F. and Duncan, G.H. (1989). Purification and some particle properties of anthriscus yellows virus, a phloem-limited, semi-persistent, aphid-borne virus. *Ann. Appl. Biol.*, **114**, 71-86.

Hunt, R.E., Nault, L.R. and Gingery, R.E. (1988). Evidence for infectivity of maize chlorotic dwarf virus and for a helper component in its leafhopper transmission. *Phytopathology*, **78**, 499-504 .
Murant, A.F. (1988). Parsnip yellow fleck virus, type member of a proposed new plant virus group, and a possible second member, dandelion yellow mosaic virus. In: *The plant viruses; polyhedral virions with monopartite RNA*, (R. Koenig, ed), Vol 3, pp 273-288. Plenum Press, New York..
Reddick, B.B., Habera, L.F. and Law, M.D. (1997). Nucleotide sequence and taxonomy of maize chlorotic dwarf virus within the *Sequiviridae*. *J. Gen. Virol.*, **78**, 1165-1174.
Shen, P., Kaniewska, M.B., Smith, C. and Beachy, R.N. (1993). Nucleotide sequence and genomic organization of rice tungro spherical virus. *Virology*, **193**, 621-630.
Thole, V. and Hull, R. (2002). Characterization of a protein from rice tungro spherical virus with serine proteinase-like activity. *J. Gen. Virol.*, **83**, 3179-3186
Turnbull-Ross, A.D., Mayo, M.A., Reavy, B. and Murant, A.F. (1993). Sequence analysis of the parsnip yellow fleck virus polyprotein: evidence of affinities with picornaviruses. *J. Gen. Virol.*, **74**, 555-561.
Zhang, S., Jones, M.C., Barker, P., Davies, J.W. and Hull, R. (1993). Molecular cloning and sequencing of coat protein-encoding cDNA of rice tungro spherical virus - a plant picornavirus. *Virus Genes*, **7**, 121-132

CONTRIBUTED BY

Le Gall, O., Iwanami, T., Karasev, A.V., Jones, T., Lehto, K., Sanfaçon, H., Wellink, J., Wetzel, T. and Yoshikawa, N.

GENUS *SADWAVIRUS*

Type Species *Satsuma dwarf virus*

VIRION PROPERTIES

MORPHOLOGY

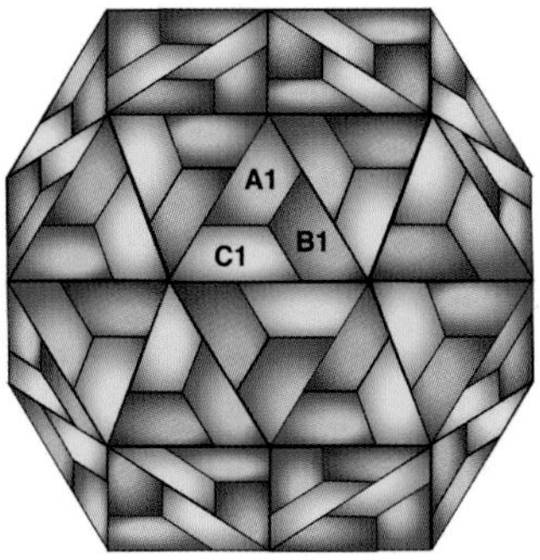

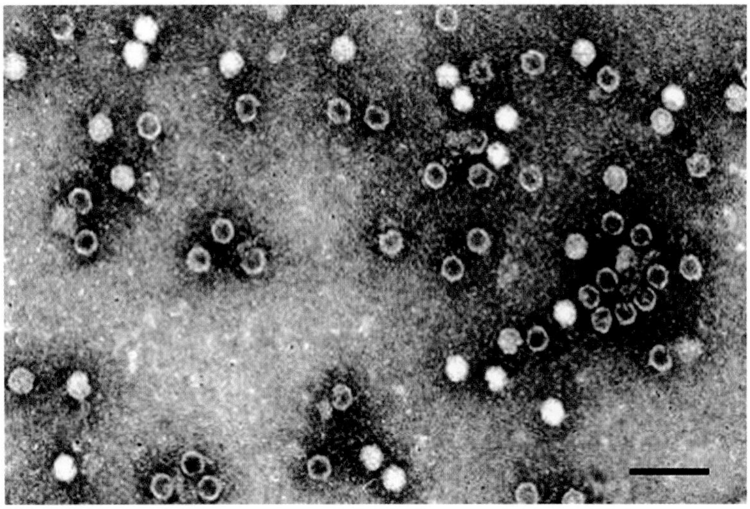

Figure 1: Negative contrast electron micrograph of particles of isolates of *Satsuma dwarf virus*. The bar represents 100 nm. (Photograph T. Iwanami).

Particles are icosahedral, about 25-30 nm in diameter (Fig. 1).

PHYSICOCHEMICAL AND PHYSICAL PROPERTIES

Three types of virions are found, differing in their buoyant densities. Top (T) particles, found in small amounts in purified virus preparations, are empty and 50-60S. Middle (M) and bottom (B) particles contain genomic RNA. Their buoyant densities in CsCl are about 1.43 and 1.46 g/cm^3 respectively. Strawberry latent ringspot virus (SLRSV) has only B particles that contain either one RNA with Mr of ~ 2.6×10^6 or two RNAs of 1.6×10^6, while Satsuma dwarf virus (SDV) has M and B particles that contain either of the different RNA molecules.

NUCLEIC ACID

The genome consists of two species of linear positive-sense ssRNA. Both RNAs are polyadenylated at their 3'-end and encode a single polyprotein that is processed to yield the mature proteins. Both RNAs are necessary for systemic infection. RNA-1 is about 7,000 nt in length, and RNA-2 is 4,600-5,400 nt long. RNA-2 of SLRSV, with 3,800 nt, is somewhat shorter. SLRSV RNAs have a VPg attached at the 5'-end of their genome. Some isolates of *Strawberry latent ringspot virus* are associated with a large ss satellite RNA that encodes a protein.

PROTEINS

Sadwaviruses have two CP subunits (Large subunit: 40-45 kDa and Small subunit: 21-29 kDa). Virion and non-structural proteins arise by proteolytic cleavage of polyproteins.

LIPIDS

None reported.

CARBOHYDRATES

None reported.

GENOME ORGANIZATION AND REPLICATION

Genetic information encoded by the RNA genome is organized as a single ORF for each RNA molecule (Fig. 2). The general genetic organization is similar to that of other "picorna-like" viruses of plants and animals, with domains characteristic of proteins with NTP-binding, proteinase and RdRp. As in the family *Comoviridae*, the polyprotein encoded by RNA-1 contains the domains for proteins likely to be involved in the replication of the virus genome, while RNA-2 encodes the CPs and the putative MP. Extensive sequence identity between RNA-1 and RNA-2 are found in the 5' and 3' UTR as well as in the 5'-end of the putative coding region.

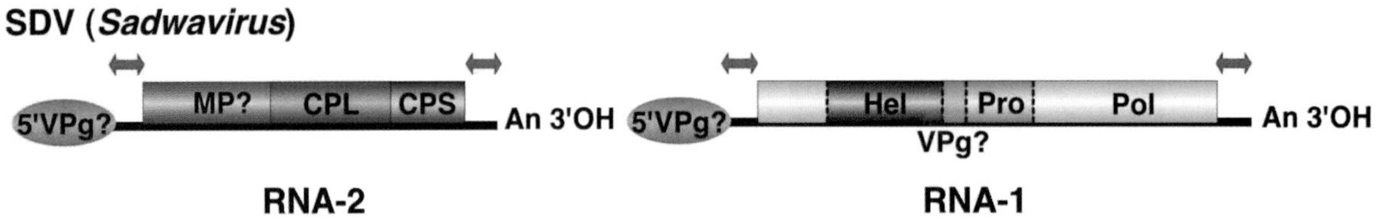

Figure 2: Genome organization characteristic of Satsuma dwarf virus (SDV). The boxes represent the polyproteins. The vertical solid lines show where cleavages are known to occur in the polyproteins and the dashed lines show where cleavages are presumed to occur. The approximate positions of NTP-binding (Hel), proteinase (Pro) and polymerase (Pol) are shown. Other putative proteins include the MP. Arrows show the portions of the RNAs that are similar in both genomic RNAs. Ovals represent the putative VPg, and An the 3′ poly-A.

ANTIGENIC PROPERTIES

Purified virions are moderate to good immunogens.

BIOLOGICAL PROPERTIES

SLRSV is transmitted by nematodes of several species within the genus *Xiphinema*, while SDV has no known vector and Strawberry mottle virus (SMoV) is transmitted in a semi-persistent manner by *Chaetosiphon* spp. and other aphids. In several hosts, SLRSV is transmitted through seed at a relatively high rate.

LIST OF SPECIES DEMARCATION CRITERIA IN THE GENUS

- Type of biological vector
- Host range
- Absence of serological cross-reaction
- Absence of cross-protection
- Sequence similarity:
  - less than 75% aa sequence identity in the Large CP within a species.
  - less than 75% aa sequence identity in the proteinase-polymerase region.

Citrus mosaic virus (CiMV), Natsudaidai dwarf virus (NDV) and Navel orange infectious mottling virus (NIMV) are considered to be distantly related strains of *Satsuma dwarf virus* because:

- Their host ranges are similar to that of "type" SDV.
- They are serologically related although they form three different serogroups.
- SDV protects Satsuma orange against infection (except in the case of NDV).
- The aa sequence identities in the Large CP sequence is 81-85% between groups.
- The aa sequence identity between NIMV and SDV is 82% in the proteinase-polymerase region.

Natural isolates of SMoV have more than 90% sequence identity with each other in the large CP aa sequence and more than 94% in the RdRp region.

LIST OF SPECIES IN THE GENUS

Species names are in green italic script; strain names and synonyms are in black roman script; tentative species names are in blue roman script. Sequence accession numbers, and assigned abbreviations ( ) are also listed.

SPECIES IN THE GENUS

	RNA-1 Accession#	RNA-2 Accession #	
Satsuma dwarf virus			
Citrus mosaic virus		[AB032751 (part.)]	(CiMV)
Natsudaidai dwarf virus		[AB032750]	(NDV)
Navel orange infectious mottling virus	[AB022887 (part.)]	[AB000282 (part.)]	(NIMV)
Satsuma dwarf virus	[NC_003785]	[NC_003786]	(SDV)
Strawberry latent ringspot virus			
Rhubarb virus 5			(RhuV5)
Strawberry latent ringspot virus		[X77466]	(SLRSV)
Strawberry mottle virus			
Strawberry mottle virus	[NC_003445]	[NC_003446]	(SMoV)

TENTATIVE SPECIES IN THE GENUS

Lucerne Australian symptomless virus	(LASV)
Rubus Chinese seed-borne virus	(RCSV)

PHYLOGENETIC RELATIONSHIPS WITHIN THE GENUS

Not available.

SIMILARITY WITH OTHER TAXA

The aa sequences in the conserved NTP-binding and RNA polymerase domains of the polyproteins resemble those in the polyproteins encoded by RNA of viruses in the unassigned genera *Cheravirus*, in the families *Sequiviridae*, *Comoviridae*, *Picornaviridae* and *Dicistroviridae*. Like fabaviruses and comoviruses (family *Comoviridae*), sadwaviruses have 2 CPs of significantly different sizes. Sadwaviruses were previously considered as atypical but tentative members of the genus *Nepovirus* (family *Comoviridae*), but were distinguished on the basis of their genomic organization, in particular the number of CP species, as well as sequence homologies and, for some of them, natural transmission by insects.

DERIVATION OF NAMES

Sadwa: from *Satsuma dwarf virus*, the type member.

REFERENCES

Barbara, D.J., Ashby, S.C. and McNamara, D.G. (1985). Host range, purification and some properties of Rubus Chinese seed-borne virus. *Ann. Appl. Biol.*, **107**, 45-55.

Bellardi, M.G. and Bertaccini, A. (1991). Parsley seeds infected by strawberry latent ringspot virus (SLRV). *Phytopath. Medit.*, **30**, 198-199.

Everett, K.R., Milne, K.S. and Forster, R.L. (1994). Nucleotide sequence of the coat protein genes of strawberry latent ringspot virus: lack of homology to the nepoviruses and comoviruses. *J. Gen. Virol.*, **75**, 1821-1825.

Hellen, C.U.T., Yuanyi, L. and Cooper, J.I. (1991). Synthesis and proteolytic processing of arabis mosaic nepovirus, cherry leaf roll nepovirus, and strawberry latent ringspot nepovirus proteins in reticulocyte lysate. *Arch. Virol.*, **120**, 19-31.

POS. SSRNA

Iwanami, T., Kondo, Y. and Karasev, A.V. (1999). Nucleotide sequences and taxonomy of satsuma dwarf virus. *J. Gen. Virol.*, **80**, 793-797.

Iwanami, T., Kondo, Y., Kobayashi, M., Han, S.S. and Karasev, A.V. (2001). Sequence diversity and interrelationships among isolates of satsuma dwarf-related viruses. *Arch. Virol.*, **146**, 807-813.

Karasev, A.V., Han, S.S. and Iwanami, T. (2001). Satsuma dwarf and related viruses belong to a new lineage of plant picorna-like viruses. *Virus Genes*, **23**, 45-52.

Kreiah, S., Strunk, G. and Cooper, J.I. (1994). Sequence analysis and location of the capsid proteins within RNA 2 of strawberry latent ringspot virus. *J. Gen. Virol.*, **75**, 2527-2532.

Mayo, M.A., Barker, H. and Harrison, B.D. (1979). Polyadenylate in the RNA of five nepoviruses. *J. Gen. Virol.*, **43**, 603-610.

Mayo, M.A., Barker, H. and Harrison, B.D. (1982). Specificity and properties of the genome-linked proteins of nepoviruses. *J. Gen. Virol.*, **59**, 149-162.

Mayo, M.A., Barker, H. and Robinson, D.J. (1982). Satellite RNA in particles of strawberry latent ringspot virus. *J. Gen. Virol.*, **63**, 417-423.

Mayo, M.A., Murant, A.F., Harrison, B.D. and Goold, R.A. (1974). Two protein and two RNA species in particles of strawberry latent ringspot virus. *J. Gen. Virol.*, **24**, 29-37.

Remah, A., Jones, A.T. and Mitchell, M.J. (1986). Purification and properties of lucerne Australian symptomless virus, a new virus infecting lucerne in Australia. *Ann. Appl. Biol.*, **109**, 307-315.

Roberts, I.M. and Harrison, B.D. (1970). Inclusion bodies and tubular structures in *Chenopodium amaranticolor* plants infected with strawberry latent ringspot virus. *J. Gen. Virol.*, **7**, 47-54.

Thompson, J.R., Leone, G., Lindner, J.L., Jelkmann, W. and Schoen, C.D. (2002). Characterization and complete nucleotide sequence of Strawberry mottle virus: a tentative member of a new family of bipartite plant picorna-like viruses. *J. Gen. Virol.*, **83**, 229-239.

CONTRIBUTED BY

Le Gall, O., Iwanami, T., Karasev, A.V., Jones, T., Lehto, K., Sanfaçon, H., Wellink, J., Wetzel, T. and Yoshikawa, N.

GENUS CHERAVIRUS

Type Species *Cherry rasp leaf virus*

VIRION PROPERTIES

MORPHOLOGY

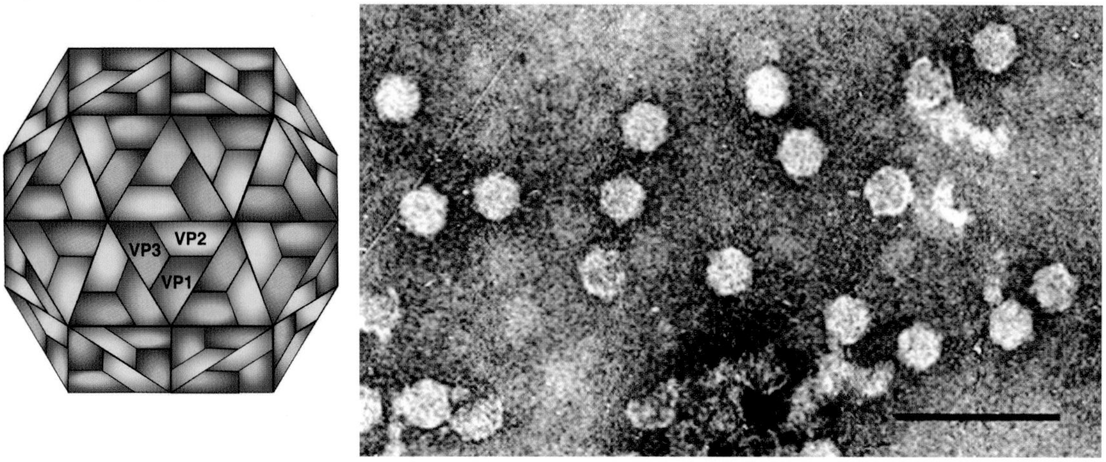

Figure 1: Negative contrast electron micrograph of particles of an isolate of *Cherry rasp leaf virus,* stained in 2% ammonium molybdate, pH 5. The bar represents 100 nm. (Photograph A.T. Jones).

Particles are icosahedral, about 25-30 nm in diameter (Fig. 1).

PHYSICOCHEMICAL AND PHYSICAL PROPERTIES

Three types of virions are found in purified Cherry rasp leaf virus (CRLV) preparations that differ in their buoyant densities. Top (T) particles, found in small amounts in purified virus preparations, are empty and sediment at 56S. Middle (M) and bottom (B) particles contain genomic RNA and sediment at 96S and 120S respectively for CRLV. Apple latent spherical virus (ALSV) has two components (M and B) with densities in CsCl of 1.41 and 1.45 g/cm^3, respectively. M and B particles probably contain two molecules of RNA-2 and a single molecule of RNA-1, respectively.

NUCLEIC ACID

The genome consists of two species of linear positive-sense ssRNA. RNA-1 is about 7,000 nt in length, and RNA-2 is about 3,300 nt long. Both RNAs are polyadenylated at their 3′-end and encode a single polyprotein that is processed to yield the mature proteins (Fig. 2). Proteinase treatment abolishes the infectivity of purified RNA, which suggests the presence of a VPg.

PROTEINS

Cheraviruses have three CP subunits of similar sizes (24, 22 and 20 kDa for CRLV, and 25, 24 and 20 kDa for ALSV). In some cases, these proteins are not fully or reproducibly resolved from each other by electrophoresis. Virion and non-structural proteins arise by proteolytic cleavage of polyproteins.

LIPIDS

None reported.

CARBOHYDRATES

None reported.

GENOME ORGANIZATION AND REPLICATION

Genetic information encoded by the RNA genome is organized as a single ORF for each RNA molecule (Fig. 2). The general genetic organization is similar to that of other picorna-like viruses of plants and animals, with domains characteristic of proteins with NTP-binding, proteinase and RdRp. As in the family *Comoviridae*, the polyprotein encoded by RNA-1 contains the domains for proteins likely to be involved in the replication of the virus genome, while RNA-2 encodes the CPs and the putative movement protein.

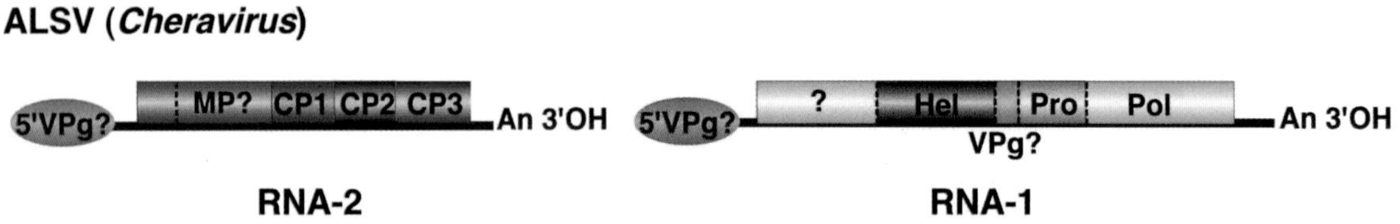

Figure 2: Genome organization characteristic of Apple latent spherical virus (ALSV). The boxes represent the polyproteins. The vertical solid lines show where cleavages are known to occur in the polyproteins and the dashed lines show where cleavages are presumed to occur. The approximate positions of NTP-binding (Hel), proteinase (Pro) and polymerase (Pol) are shown. Other putative proteins include the movement protein (MP). Ovals represent the putative VPg, and An the 3′ poly-A.

ANTIGENIC PROPERTIES

Purified virions are moderate immunogens.

BIOLOGICAL PROPERTIES

The host range is broad or narrow, depending on viruses, and includes weed plants found in the vicinity of infected crops. Symptoms are usually mild or absent. CRLV is transmitted by nematodes (*Xiphinema americanum*) in the field, and is readily seed-transmitted. No information is available for other cheraviruses.

LIST OF SPECIES DEMARCATION CRITERIA IN THE GENUS

- Type of biological vector
- Host range
- Absence of serological cross-reaction
- Absence of cross-protection
- Sequence similarity:
  - less than 75% aa sequence identity in the CPs within a species.
  - less than 75% aa sequence identity in the proteinase-polymerase region.

Members of CRLV and ALSV were distinguished because:

- The host ranges differ, although some hosts including apple are common.
- Purified ALSV particles did not react with polyclonal antibodies against CRLV and AVBV in either gel diffusion tests or immunoblot analysis.
- The aa sequence identities between three CPs, Vp25, Vp20, and Vp24, are 54%, 59%, and 65%, respectively.

LIST OF SPECIES IN THE GENUS

Species names are in green italic script; strain names and synonyms are in black roman script; tentative species names are in blue roman script. Sequence accession numbers, and assigned abbreviations ( ) are also listed.

SPECIES IN THE GENUS

	RNA-1 Accession#	RNA-2 Accession #	
Apple latent spherical virus			
Apple latent spherical virus	[NC_003787]	[NC_003788]	(ALSV)
Cherry rasp leaf virus			
Cherry rasp leaf virus			(CRLV)

Flat apple virus [AY122330]

TENTATIVE SPECIES IN THE GENUS

Arracacha virus B (AVB)
Artichoke vein banding virus (AVBV)

PHYLOGENETIC RELATIONSHIPS WITHIN THE GENUS

Not available.

SIMILARITY WITH OTHER TAXA

The aa sequences in the conserved NTP-binding and RdRp domains of the polyproteins resemble those in the polyproteins encoded by RNA of viruses in the unassigned genus *Sadwavirus*, in the families *Sequiviridae*, *Comoviridae*, *Picornaviridae* and *Dicistroviridae*. The number and sizes of the CPs resemble those of viruses in the families *Picornaviridae* and *Sequiviridae*.

Like sadwaviruses, cheraviruses were previously considered as atypical but tentative members of the genus *Nepovirus* (family *Comoviridae*), but were distinguished on the basis of their genomic organization, in particular the number of CP species, as well as sequence homologies and, for some of them, natural transmission by insects.

DERIVATION OF NAMES

Chera: from *Cherry rasp leaf virus*, the type member. CRLV was preferred as a type member to ALSV despite the lack of a complete sequence because more biological data were available.

REFERENCES

Brown, D.J.F., Halbrendt, J.M., Jones, A.T., Vrain, T.C. and Robbins, R.T. (1994). Transmission of three North American nepoviruses by populations of four distinct species of the *Xiphinema americanum* group. *Phytopathology*, **84**, 646-649.

Hansen, A.J., Nyland, G., McElroy, F.D. and Stace-Smith, R. (1974). Origin, cause, host range and spread of cherry rasp leaf disease in North America. *Phytopathology*, **64**, 721-727.

James, D. and Upton, C. (2002). Nucleotide sequence analysis of RNA-2 of a flat apple isolate of Cherry rasp leaf virus with regions showing greater identity to animal picornaviruses than to related plant viruses. *Arch. Virol.*, **147**, 1631-1641.

Jones, A.T., Mayo, M.A. and Henderson, S.J. (1985). Biological and biochemical properties of an isolate of cherry rasp leaf virus from red raspberry. *Ann. Appl. Biol.*, **106**, 101-110.

Li, C., Yoshikawa, N., Takahashi, T., Ito, T., Yoshida, K. and Koganezawa, H. (2000). Nucleotide sequence and genome organization of apple latent spherical virus: a new virus classified into the family *Comoviridae*. *J. Gen. Virol.*, **81**, 541-547.

Parish, C.L. (1977). A relationship between flat apple disease and cherry rasp leaf disease. *Phytopathology*, **67**, 982-984.

Stace-Smith, R. and Hansen, A.J. (1976). Cherry rasp leaf virus. *CMI/AAB Description of Plant Viruses*, n°159.

CONTRIBUTED BY

Le Gall, O., Iwanami, T., Karasev, A.V., Jones, T., Lehto, K., Sanfaçon, H., Wellink, J., Wetzel, T. and Yoshikawa, N.

FAMILY *COMOVIRIDAE*

TAXONOMIC STRUCTURE OF THE FAMILY

Family	***Comoviridae***
Genus	***Comovirus***
Genus	***Fabavirus***
Genus	***Nepovirus***

VIRION PROPERTIES

MORPHOLOGY

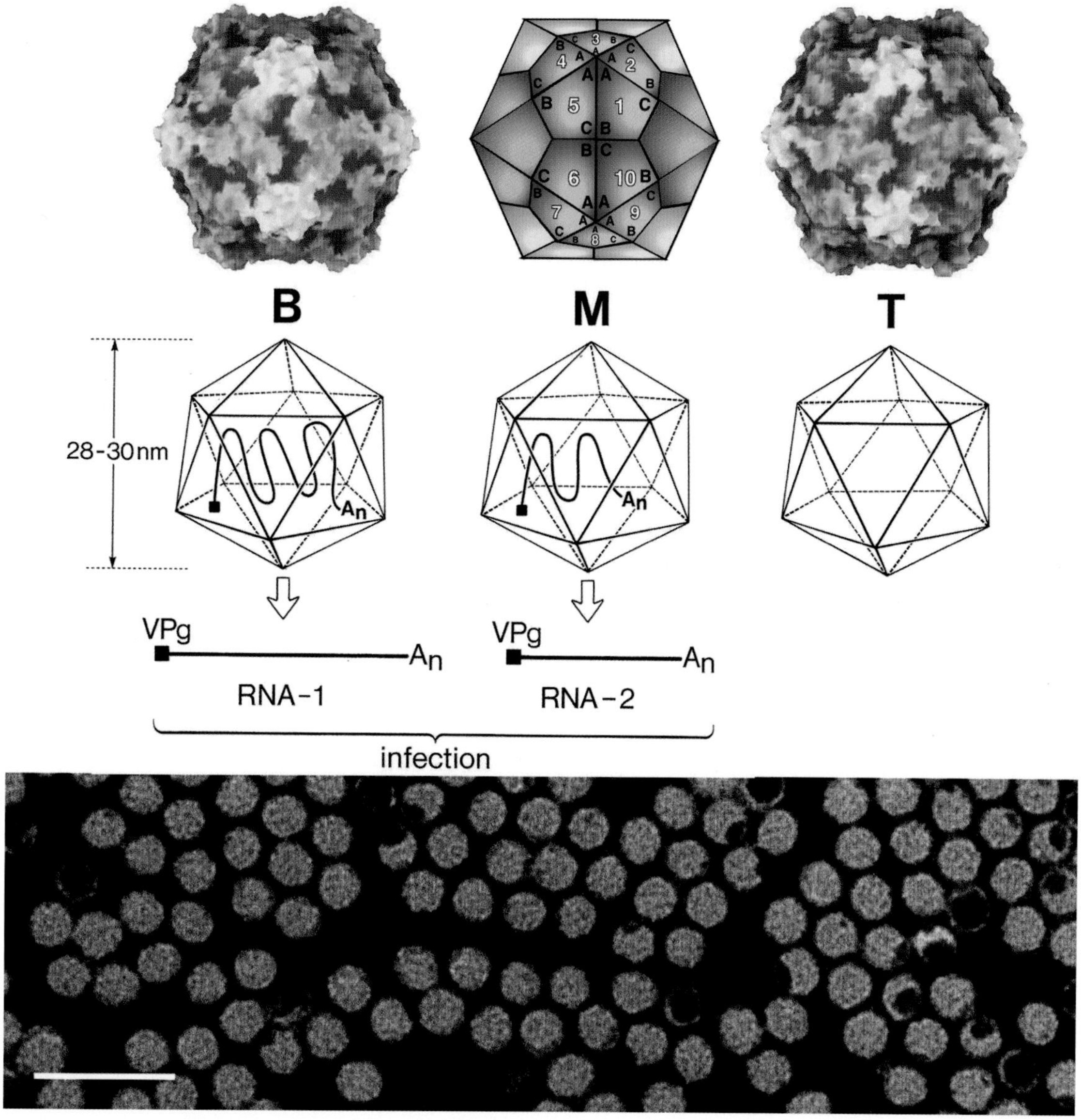

Figure 1: (Top left) Molecular rendering of the Cowpea mosaic virus (CPMV) particle (Lin *et al.*, 1999, with permission). (Top central) Diagrammatic representation of a T=1 lattice. A= Small capsid protein, B= C-terminal & C= N-terminal domains of the Large capsid protein. (Top right) Molecular rendering of the Red clover mosaic virus (RCMV) particle (Lin *et al.*, 2000, with permission). (Center) Diagram of the three types of comovirus particles. (Bottom) Negative contrast electron micrograph of particles of CPMV. The bar represents 100 nm.

Virions are non-enveloped 28-30 nm in diameter and exhibit icosahedral symmetry (T=1, pseudo T=3). They contain two types of positive sense ssRNA molecules, RNA-1 and RNA-2. Virus preparations contain three types of components differing in their buoyant properties: T ("Top", empty particles), M ("Middle", particles usually containing a single molecule of RNA-2) and B ("Bottom", particles containing a single molecule of RNA-1 or, in some nepoviruses, two molecules of RNA-2).

Physicochemical and Physical Properties

Virions are heat-stable (thermal inactivation is usually above 60°C), and most are insensitive to organic solvents. Particles sediment as three components, T, M and B, with S_{20w} values of 49-63S, 84-128S and 111-134S, respectively, (values vary within each genus). Mr of particles are $3.2\text{-}3.8 \times 10^6$ (T), $4.6\text{-}5.8 \times 10^6$ (M) and $6.0\text{-}6.2 \times 10^6$ (B). Buoyant densities in CsCl are 1.28-1.30 (T), 1.41-1.48 (M) and 1.44-1.53 (B) g/cm^3 (density values refer only to comoviruses and nepoviruses).

Nucleic Acid

The genome consists of two species of linear positive-sense ssRNA. Both RNAs are necessary for systemic infection. Sizes of RNAs differ among genera; nepovirus RNA-1 (7.2-8.4 kb) and RNA-2 (3.9-7.2 kb) are larger than fabavirus and comovirus RNA-1 (5.9-7.2 kb) and RNA-2 (3.5-4.5 kb). For the genera *Comovirus* and *Nepovirus* the genomic RNAs contain a 3'-terminal poly(A) tract of variable length, and a polypeptide, designated VPg (2 - 4 kDa), covalently bound at the 5'-end (this has not been confirmed for fabaviruses).

Table 1: Sizes of the genomes of viruses in the family *Comoviridae*, in nucleotides.

Genus **(virus)**	**RNA-1**	**RNA-2**
Comovirus	5,900 – 7,200	3,300 – 3,800
Cowpea mosaic virus (CPMV)	(5,889)	(3,810)
Fabavirus	5,900 – 6,300	3,100 - 4,500
Broad bean wilt virus 2 (BBWV2)	(5,951)	(3,607)
Nepovirus	7,200 - 8,400	3,700 - 7,300
Grapevine fanleaf virus (GFLV)	(7,342)	(3,774)
Beet ringspot virus (BRSV)	(7,356)	(4,662)
Tomato ringspot virus (ToRSV)	(8,214)	(7,273)

Proteins

Comoviruses and fabaviruses have two CPs of 40-45 kDa and 21-27 kDa; nepoviruses normally have a single CP of 52-60 kDa. Virions have 60 copies of each CP per particle. For three comoviruses (CPMV, BPMV and RCMV) and one nepovirus (TRSV) the atomic structure has been solved and found to be very similar (pseudo T=3) to that of viruses belonging to the family *Picornaviridae*. Each capsid subunit is made of three beta-barrels that are present in two CPs (*Comovirus*) or a single CP (*Nepovirus*).

Lipids

None reported.

Carbohydrates

None reported. No carbohydrate in comoviruses (a report that the CPs contain about 1.9% carbohydrates covalently linked has recently been shown to be mistaken).

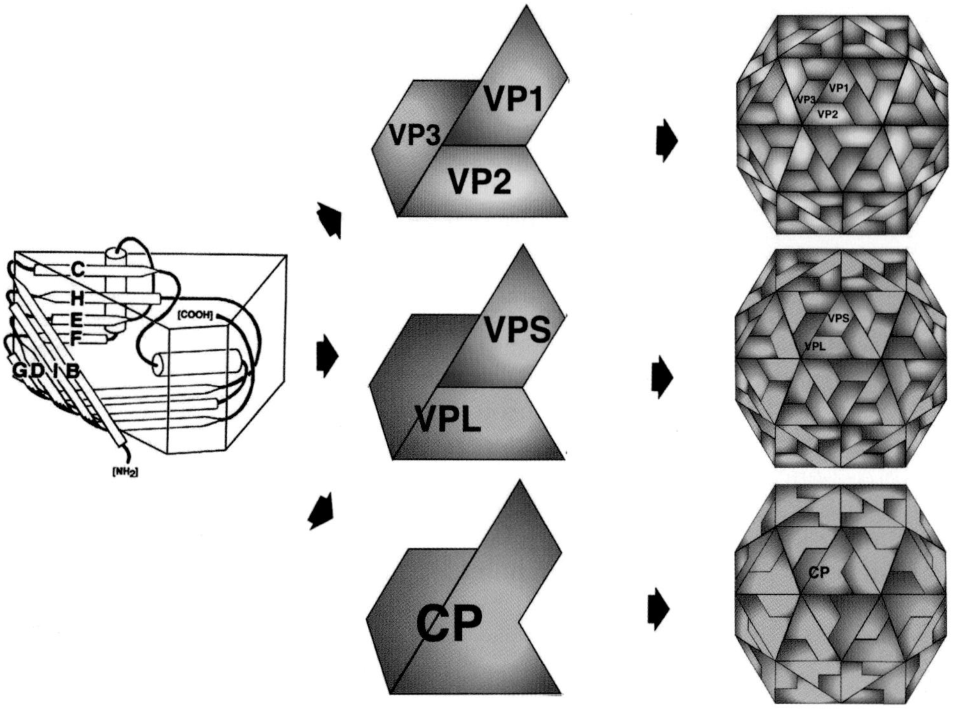

Figure 2: Capsid architecture of a picornavirus (top; 3 beta-barrels in 3 CPs), a comovirus (middle; 3 beta-barrels in 2 CPs) and a nepovirus (bottom; 3 beta-barrels in 1 CP).

Genome Organization and Replication

Unfractionated RNA is highly infective but neither RNA species alone can infect plants systemically. Cytoplasm of infected cells contains conspicuous inclusions consisting primarily of membranous elements and electron-dense material that are sites of virus genome replication. The following information only refers to como- and nepoviruses (fabaviruses have not been studied): RNA-1 can replicate in protoplasts but in the absence of RNA-2 (encoding the CPs) no virus particles are produced. RNA-1 carries all the information for RNA replication, including the polymerase. Both RNA species are translated into polyproteins that are cleaved by a viral proteinase (encoded by RNA-1) to give several intermediate and final processing products. Virions assemble and accumulate in the cytoplasm, often in crystalline or paracrystalline arrays. They are also found within infection-specific tubules, which contain the viral MP and cross cell walls, and which have been implicated in cell-to-cell transport.

Antigenic Properties

Virus preparations are good immunogens. Species belonging to the same genus are serologically interrelated (especially comoviruses), but often distantly.

Biological Properties

Comoviruses have narrow host ranges; nepoviruses and fabaviruses have wide host ranges. Symptoms vary widely within each genus. Viruses in the family *Comoviridae* all have biological vectors, comoviruses are transmitted by beetles (especially members of the family *Chrysomelidae*), fabaviruses are transmitted by aphids and some nepoviruses are transmitted by nematodes. All are readily transmissible experimentally by mechanical inoculation. Seed and/or pollen transmission is very common among nepoviruses, but is rare for comoviruses and fabaviruses.

GENUS COMOVIRUS

Type Species *Cowpea mosaic virus*

DISTINGUISHING FEATURES

The comovirus genome is made of two genomic ssRNAs with a 5′-bound polypeptide (VPg) and a 3′ poly-A (see section on *Comoviridae*). The comovirus capsid is made of two types of polypeptides (Large subunit: 40-45 kDa and Small subunit: 21-27 kDa).

RNA-2 is translated into two largely overlapping polyproteins that are processed into three domains. P2A is involved in RNA-2 replication. P2B is the MP, with a typical "LPL" motif. The CP domains are encoded at the C-terminus of the polyprotein. RNA-1 is translated into a single polyprotein that is processed into five domains. P1A limits the processing of the RNA-1-encoded polyprotein *in cis* and assists the processing of the RNA-2-encoded polyprotein. P1B has sequence motifs characteristic of an NTP-binding helicase, P1C is the VPg, P1D is the proteinase and P1E has sequence motifs characteristic of an RdRp. P1A and P1B are involved in inducing the cytopathic structure. The 5′ and 3′ NTRs of RNA-1 and RNA-2 are homologous.

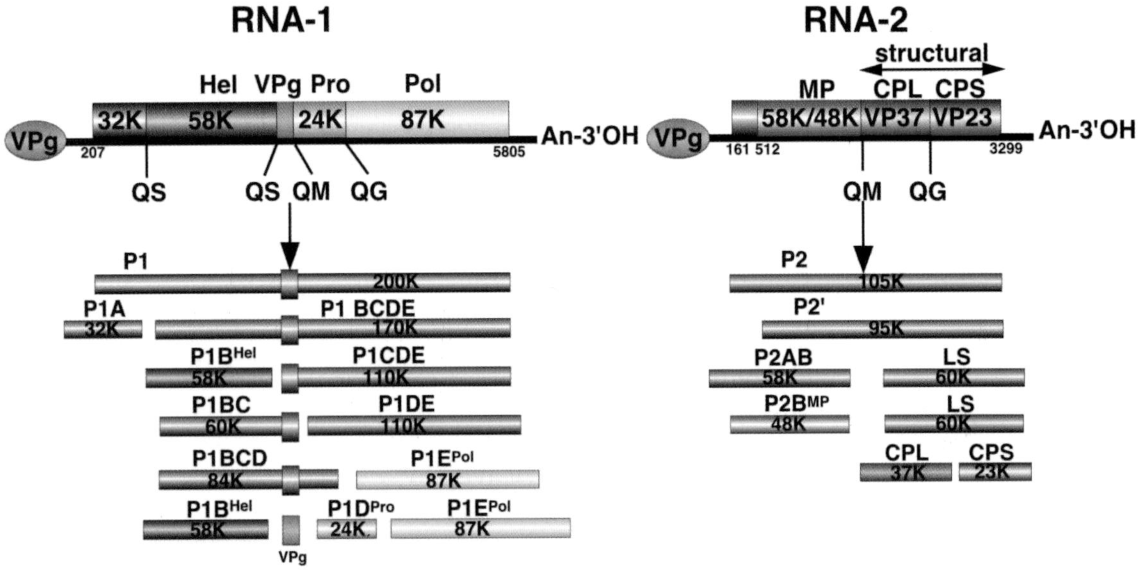

Figure 3: Genome organization characteristic of Cowpea mosaic virus (CPMV). The ORFs are boxed, and functions of the proteins are indicated. MP, movement protein; CPL and CPS, large and small capsid proteins; Hel, helicase; Pro, proteinase; Pol, polymerase. Proteolytic cleavage sites are indicated on the polyproteins. All intermediate and final cleavage products have been detected in infected cells. The ovals at the 5′-end of the RNAs represents the VPg, and An at the 3′-end the poly-A.

BIOLOGICAL PROPERTIES

Comoviruses have narrow host ranges, 11 of the 15 species being restricted to a few species of the family Leguminosae. Mosaic and mottle symptoms are characteristic, but usually not ringspots. Transmission in nature is exclusively by beetles, especially members of the family *Chrysomelidae*. Beetles retain their ability to transmit virus for days or weeks.

LIST OF SPECIES DEMARCATION CRITERIA IN THE GENUS

The criteria demarcating species in the genus are:

- Large CP aa sequence less than 75% homologous,

- Polymerase aa sequence less than 75% homologous,
- No pseudo-recombination between components possible,
- Differences in antigenic reactions.

Using polyclonal rabbit antisera in agar gel diffusion tests, isolates of *Cowpea mosaic virus* and *Cowpea severe mosaic virus* differ by four two-fold steps. Their CPs have less than 54% identity and 61% homology between the aligned respective nt and aa sequences. Their Pol proteins have less than 55% identity and 62% homology between the aligned respective nt and aa sequences. No component reassortment has been demonstrated between these viruses.

List of Species in the Genus

Species names are in green italic script; strain names and synonyms are in black roman script; tentative species names are in blue roman script. Sequence accession numbers, and assigned abbreviations ( ) are also listed.

Species in the Genus

	RNA-1 Acc. #	RNA-2 Acc. #	
Andean potato mottle virus			
Andean potato mottle virus		[L19239]	(APMoV)
Bean pod mottle virus			
Bean pod mottle virus - KentuckyG7	[NC_003496]	[NC_003495]	(BPMV-KenG7)
Bean pod mottle virus - K-Hopkins1	[AF394608]	[AF394609]	(BPMV-KHop)
Bean pod mottle virus - K-Hancock1	[AF394606]	[AF394607]	(BPMV-KHan)
Bean rugose mosaic virus			
Bean rugose mosaic virus		[AF263548]	(BRMV)
Broad bean stain virus			
Broad bean stain virus			(BBSV)
Broad bean true mosaic virus			
Broad bean true mosaic virus			(BBTMV)
Echtes Ackerbohnemosaik virus			(EABV)
Vicia virus 1			(VV1)
Cowpea mosaic virus			
Cowpea mosaic virus	[NC_003549]	[NC_003550]	(CPMV)
Cowpea yellow mosaic virus			(CPYMV)
Cowpea severe mosaic virus			
Arkansas cowpea mosaic virus			(CPSMV-Ark)
Cowpea mosaic virus - severe			(CPSMV-Svr)
Cowpea severe mosaic virus - DG	[NC_003545]	[NC_003544]	(CPSMV-DG)
Trinidad cowpea mosaic virus			(CPSMV-Tri)
Glycine mosaic virus			
Glycine mosaic virus			(GMV)
Pea green mottle virus			
Pea green mottle virus			(PGMV)
Pea mild mosaic virus			
Pea mild mosaic virus			(PMiMV)
Quail pea mosaic virus			
Bean curly dwarf mosaic virus			(BCDMV)
Quail pea mosaic virus			(QPMV)
Radish mosaic virus			
Radish enation mosaic virus			(RaEMV)
Radish mosaic virus			(RaMV)
Red clover mottle virus			
Red clover mottle virus - S	[NC_003741]	[NC_003738]	(RCMV-S)
Squash mosaic virus			
Cucurbit ring mosaic virus			(CuRMV)
Muskmelon mosaic virus			(MuMV)
Squash mosaic virus - Arizona		[AF059533]	(SqMV-Ari)
Squash mosaic virus - Kimble		[AF059532]	(SqMV-Kim)

Squash mosaic virus - Y	[NC_003799]	[NC_003800]	(SqMV-Y)
Ullucus virus C			
Ullucus virus C			(UVC)

TENTATIVE SPECIES IN THE GENUS

None reported.

GENUS *FABAVIRUS*

Type Species *Broad bean wilt virus 1*

DISTINGUISHING FEATURES

Capsids contain two polypeptide species (Large and Small). road bean wilt virus 2 (BBWV-2) RNA-2 is translated into a polyprotein that is cleaved into the two CPs and a 52 kDa protein that is possibly involved in movement (Fig. 4). Fabaviruses have wide host ranges among dicotyledonous plants and some families of monocotyledonous plants. Symptoms are ringspots, mottling, mosaic, distortion, wilting and apical necrosis. In nature, fabaviruses are transmitted by aphids in a non-persistent manner. In other respects, fabaviruses are similar to comoviruses.

RNA-2 is translated into a polyprotein that is processed into three domains. The occurrence of two overlapping reading frames as in comoviruses is not known. P2B is the MP, with a typical "LPL" motif. The CP domains are encoded at the C-terminus of the polyprotein. RNA-1 is translated into a single polyprotein. A cleavage site between putative P1A and P1B domains has not been investigated. P1B has sequence motifs characteristic of a NTP-binding helicase, P1C is the (putative) VPg, P1D is the proteinase and P1E has sequence motifs characteristic of an RdRp. The 5′ and 3′ non-translated regions (NTRs) are homologous between RNA-1 and RNA-2.

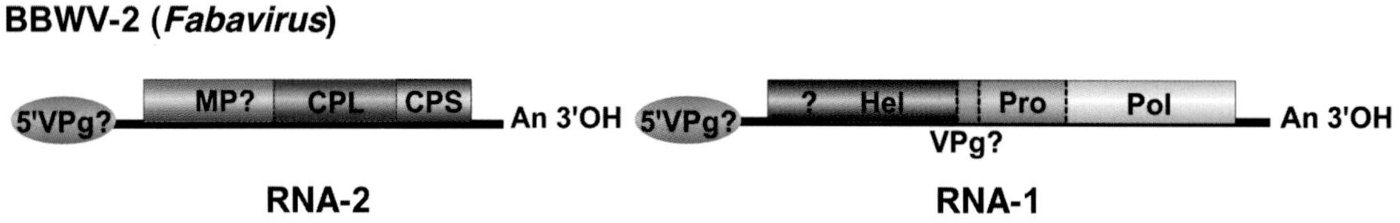

Figure 4: Genome organization characteristic of Broad bean wilt virus 2 (BBWV-2). The ORFs are boxed, and the putative functions of the proteins are indicated. MP, movement protein; CPL and CPS, large and small capsid proteins; Hel, helicase; Pro, proteinase; Pol, polymerase. Proteolytic cleavage sites are indicated on the polyproteins. The ovals at the 5′-end of the RNAs represents the putative VPg, question marks some uncertainty regarding the presence of a cleavage site in the polyprotein, and An at the 3′-end the poly-A.

BIOLOGICAL PROPERTIES

Fabaviruses have wide host ranges among dicotyledonous plants and some families of monocotyledonous plants. Symptoms are ringspots, mottling, mosaic, distortion, wilting and apical necrosis. In nature, fabaviruses are transmitted by aphids in a non-persistent manner.

LIST OF SPECIES DEMARCATION CRITERIA IN THE GENUS

The criteria demarcating species in the genus are:

- No pseudo-recombination between components possible,
- Differences in antigenic reactions.
- Sequence identity lower than 75% in the CP and polymerase coding regions.

Using polyclonal rabbit antisera in agar gel diffusion tests Broad bean wilt virus 1 (BBWV-1) and Lamium mild mosaic virus (LMMV) differ by eight twofold steps. No reassortment has been demonstrated between these viruses.
On the other hand, Patchouli mild mosaic virus is now believed to be a strain of BBWV-2 because these two viruses differ only by 3-21% in sequence (21% in the Large CP).

List of Species in the Genus

Species names are in green italic script; strain names and synonyms are in black roman script; tentative species names are in blue roman script. Sequence accession numbers, and assigned abbreviations ( ) are also listed.

Species in the Genus

	RNA-1 Acc. #	RNA-2 Acc. #	
Broad bean wilt virus 1			
Broad bean wilt virus 1		[AF225955]	(BBWV-1)
Broad bean wilt virus 2			
Broad bean wilt virus		[E31398]	(BBWV)
Broad bean wilt virus 2	[AF144234]	[E31397]	(BBWV-2)
Broad bean wilt virus 2 - B935	[AF149425]		(BBWV2-B935)
Broad bean wilt virus 2 - Chinese		[AJ132844]	(BBWV2-Chi)
Broad bean wilt virus 2 - IA	[AB051386]	[AB032403]	(BBWV2-IA)
Broad bean wilt virus 2 - IP	[AB023484]	[AB018698]	(BBWV2-IP)
Broad bean wilt virus 2 - Korean		[AF104335]	(BBWV2-Kor)
Broad bean wilt virus 2 - MB7	[AB013615]	[AB013616]	(BBWV2-MB7)
Broad bean wilt virus 2 - ME	[NC_003003]	[NC_003004]	(BBWV2-ME)
Broad bean wilt virus 2 - P158		[AF228423]	(BBWV2-P158)
Broad bean wilt virus 2 - PV131		[U65985]	(BBWV2-PV131)
Nasturtium ringspot virus			(NaRSV)
Parsley virus 3			(PaV-3)
Patchouli mild mosaic virus ‡	[NC_003975]	[NC_003974]	(PatMMV)
Petunia ringspot virus			(PeRSV)
Plantago II virus			(PlIIV)
Lamium mild mosaic virus			
Lamium mild mosaic virus			(LMMV)

Tentative Species in the Genus

None reported.

Genus Nepovirus

Type Species Tobacco ringspot virus

Distinguishing Features

The capsid of nepoviruses contains a single polypeptide species. Genome organization and expression are similar to those of comoviruses, except that RNA-2 specifies a single primary translation product of 105-207 kDa (Fig. 5). Definitive nepoviruses can be divided in three subgroups. Subgroup A has an RNA-2 with Mr ± 1.3-1.5 x 10^6, present in both M and B components. Subgroup B has an RNA-2 with Mr ± 1.4-1.6 x 10^6, present only in M component. Subgroup C has an RNA-2 with Mr ± 1.9-2.2 x 10^6, present in M component particles that are sometimes barely separable from those of B component.
The nepovirus capsid is typically made of 60 subunits of a 52–60 kDa polypeptide. Additional linear or circular satellite RNAs, which sometimes modulate symptoms, are found associated with several nepoviruses of all three subgroups. They are either linear (1100-1800 nt) with a VPg, a poly-A and encoding a 36–48 kDa polypeptide, or circular (300-460 nt) and apparently non-coding, are present in some natural isolates but are not necessary for virus accumulation.

Pos. ssRNA

Genome Organization and Replication

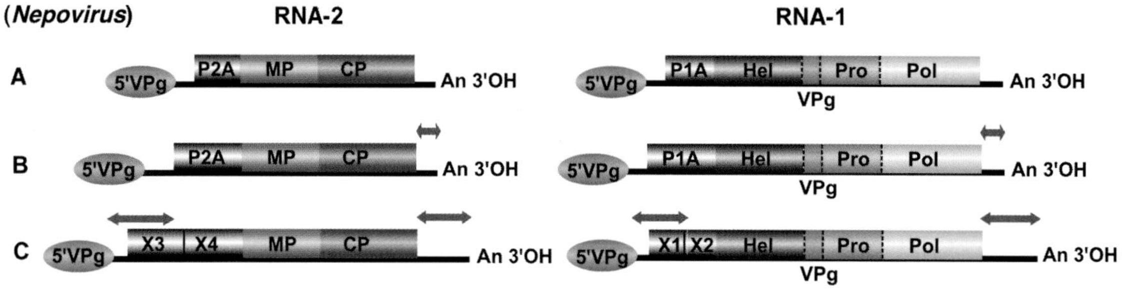

Figure 5: Genome organization characteristic of Grapevine fanleaf virus (GFLV) (subgroup A; top), BRSV; Beet ringspot virus (subgroup B; middle) and ToRSV; Tomato ringspot virus (subgroup C; bottom). The ORFs are boxed, and functions of the proteins released by proteolysis are indicated. MP, movement protein; CP, capsid protein; Hel, helicase; Pro, proteinase; Pol, polymerase. Proteolytic cleavage sites are indicated on the polyproteins. The ovals at the 5′-end of the RNAs represents the VPg, and An at the 3′-end the poly-A. Braces show the portions of the RNAs that are highly homologous or identical in both RNAs in a subgroup. Diamonds indicate typical sequence motifs.

RNA-2 is translated into a single polyprotein that is processed into three domains. In Grapevine fanleaf virus (GFLV), P2A is involved in RNA-2 replication. P2B is the MP, with a typical "LPL" motif. P2C is the single CP, with two conserved motifs. In Tomato ringspot virus (ToRSV) (subgroup C), a third cleavage site occurs within the 5′-most domain.

RNA-1 is translated into a single polyprotein that is processed into five domains. The function of P1A is unknown, P1B has sequence motifs characteristic of an NTP-binding helicase, P1C is the VPg, P1D is the proteinase and P1E has sequence motifs characteristic of an RdRp. In ToRSV (subgroup C), a fifth cleavage site is present in the 5′-terminal region of the polyprotein. The 5′ and 3′ NTRs are homologous but different between RNA-1 and RNA-2 in subgroup A. The 5′-NTRs of RNA-1 and RNA-2 are also homologous in subgroup B, while the 3′-NTRs are identical. Both NTRs are identical or nearly identical in subgroup C, and include in part the coding region of the polyproteins in ToRSV,but not in Blackcurrant reversion virus (BRV).

Biological Properties

Nepoviruses are widely distributed in temperate regions. Natural host ranges vary from wide to restricted to a single plant species, depending on the virus. Ringspot symptoms are characteristic, but mottling and spotting are equally frequent. Twelve species are acquired and transmitted persistently by longidorid nematodes (*Xiphinema, Longidorus* or *Paralongidorus* spp), three are transmitted by pollen, one (BRV) is transmitted by mites (Blackcurrant reversion virus, BRV) and the others have no known biological vector. Seed and/or pollen transmission is very common. In herbaceous plants, the symptoms induced by nepoviruses are often transient, with newly emerging leaves appearing symptomless a few weeks after infection (the so-called "recovery" phenomenon) due to inefficient plant defense inhibition by the virus.

List of Species Demarcation Criteria in the Genus

The criteria demarcating species in the genus are:

- CP aa sequence less than 75% homologous,
- Polymerase aa sequence less than 75% homologous,
- No pseudo-recombination between components possible,
- Differences in antigenic reactions,
- Different vector species.

Using polyclonal rabbit antisera in agar gel diffusion tests, Grapevine chrome mosaic virus (GCMV) and Grapevine fanleaf virus (GFLV) differ by more than 10 two-fold dilution steps. Their CP sequences have less than 25% identity and 32% homology between the aligned nt and aa sequences, respectively. Their RdRp sequences have less than 40% identity and 51% homology between the aligned nt and aa sequences, respectively. No reassortment has been demonstrated between these viruses. GFLV is vectored by *Xiphinema index* nematodes while the vector of GCMV is not known (but is not *Xiphinema index*).

List of Species in the Genus

Species names are in green italic script; strain names and synonyms are in black roman script; tentative species names are in blue roman script. Sequence accession numbers, and assigned abbreviations ( ) are also listed.

Species in the Genus

Three clusters of species in the genus, designated as Subgroups A, B and C, are based on length and packaging of RNA-2, sequence similarities and serological relationships.

Subgroup A

	RNA-1 Acc. #	RNA-2 Acc. #	
Arabis mosaic virus			
Arabis mosaic virus - NW	[AY303786]	[AY017339]	(ArMV-NW)
Arabis mosaic virus - P2		[X81814, X81815]	(ArMV-P2)
Raspberry yellow dwarf virus			(RYDV)
Rhubarb mosaic virus			(RhuMV)
Arracacha virus A			
Arracacha virus A			(AVA)
Artichoke Aegean ringspot virus			
Artichoke Aegean ringspot virus			(AARSV)
Cassava American latent virus			
Cassava American latent virus			(CsALV)
Grapevine fanleaf virus			
Grapevine fanleaf virus - F13	[NC_003615]	[NC_003623]	(GFLV-F13)
Grapevine fanleaf virus - NW		[AY017338]	(GLFV-NW)
Grapevine infectious degeneration virus			(GIDV)
Potato black ringspot virus			
Potato black ringspot virus			(PBRSV)
Tobacco ringspot virus - potato calico			(TobRSV-PC)
Raspberry ringspot virus			
Raspberry leaf curl virus			(RpLCV)
Raspberry ringspot virus		[S46011]	(RpRSV)
Raspberry ringspot virus - Cherry	[AY303787]	[AY303788]	(RpRSV-Che)
Raspberry ringspot virus - Grapevine	[AY310444]	[AY310445]	(RpRSV-Gra)
Redcurrant ringspot virus			(RcRSV)
Tobacco ringspot virus			
Tobacco ringspot virus	[U50869]		(TRSV)
Tobacco ringspot virus n°1			(TRSV-1)

Subgroup B

	RNA-1 Acc. #	RNA-2 Acc. #	
Artichoke Italian latent virus			
Artichoke Italian latent virus		[X87254 (part.)]	(AILV)
Beet ringspot virus			
Beet ringspot virus	[NC_003693]	[NC_003694]	(BRSV)
Tomato black ring virus - Scottish			(TBRV-S)

Pos. ssRNA

Cocoa necrosis virus			
Cocoa necrosis virus			(CoNV)
Cocoa swollen shoot virus - S			(CSSV-S)
Crimson clover latent virus			
Crimson clover latent virus			(CCLV)
Cycas necrotic stunt virus			
Cycas necrotic stunt virus	[NC_003791]	[NC_003792]	(CNSV)
Grapevine chrome mosaic virus			
Grapevine chrome mosaic virus	[NC_003622]	[NC_003621]	(GCMV)
Mulberry ringspot virus			
Mulberry ringspot virus			(MRSV)
Olive latent ringspot virus			
Olive latent ringspot virus		[AJ277435]	(OLRSV)
Tomato black ring virus			
Bean ringspot virus			(BRSV)
Lettuce ringspot virus			(LRSV)
Potato bouquet virus			(PBV)
Tomato black ring virus - ED		[X80831]	(TBRV-ED)
Tomato black ring virus - MJ	[NC_004439]	[NC_004440]	(TBRV-MJ)

Subgroup C

	RNA-1 Acc. #	RNA-2 Acc. #	
Apricot latent ringspot virus			
Apricot latent ringspot virus		[AJ278875 (part.)]	(ALRSV)
Artichoke yellow ringspot virus			
Artichoke yellow ringspot virus			(AYRSV)
Blackcurrant reversion virus			
Blackcurrant reversion virus	[NC_003509]	[NC_003502]	(BRV)
Blueberry leaf mottle virus			
Blueberry leaf mottle virus	[U20622 (part.)]	[U20621 (part.)]	(BLMoV)
Cassava green mottle virus			
Cassava green mottle virus			(CsGMV)
Cherry leaf roll virus			
Cherry leaf roll virus	[Z34265 (part.)]	[U24694 (part.)]	(CLRV)
Elm mosaic virus			(ElMV)
Golden elderberry virus			(GEBV)
Walnut black line virus			(WBLV)
Chicory yellow mottle virus			
Chicory yellow mottle virus			(ChYMV)
Parsley carrot leaf virus			(PaCLV)
Grapevine Bulgarian latent virus			
Grapevine Bulgarian latent virus			(GBLV)
Grapevine Tunisian ringspot virus			
Grapevine Tunisian ringspot virus			(GTRSV)
Hibiscus latent ringspot virus			
Hibiscus latent ringspot virus			(HLRSV)
Lucerne Australian latent virus			
Lucerne Australian latent virus			(LALV)
Lucerne latent virus			(LLV)
Myrobalan latent ringspot virus			
Myrobalan latent ringspot virus			(MLRSV)
Peach rosette mosaic virus			
Grape decline virus			(GrDV)
Grapevine degeneration virus			(GraDV)
Peach rosette mosaic virus	[AF016626]		(PRMV)
Potato virus U			
Potato virus U			(PVU)
Tomato ringspot virus			
Grape yellow vein virus			(GraYVV)

Nicotiana virus 13			(NV13)
Peach yellow bud mosaic virus			(PYBMV)
Tobacco ringspot virus n°2			(TbRSV-2)
Tomato ringspot virus	[NC_003840]	[NC_003839]	(ToRSV)

TENTATIVE SPECIES IN THE GENUS

None reported.

LIST OF UNASSIGNED VIRUSES IN THE FAMILY

None reported.

PHYLOGENETIC RELATIONSHIPS WITHIN THE FAMILY

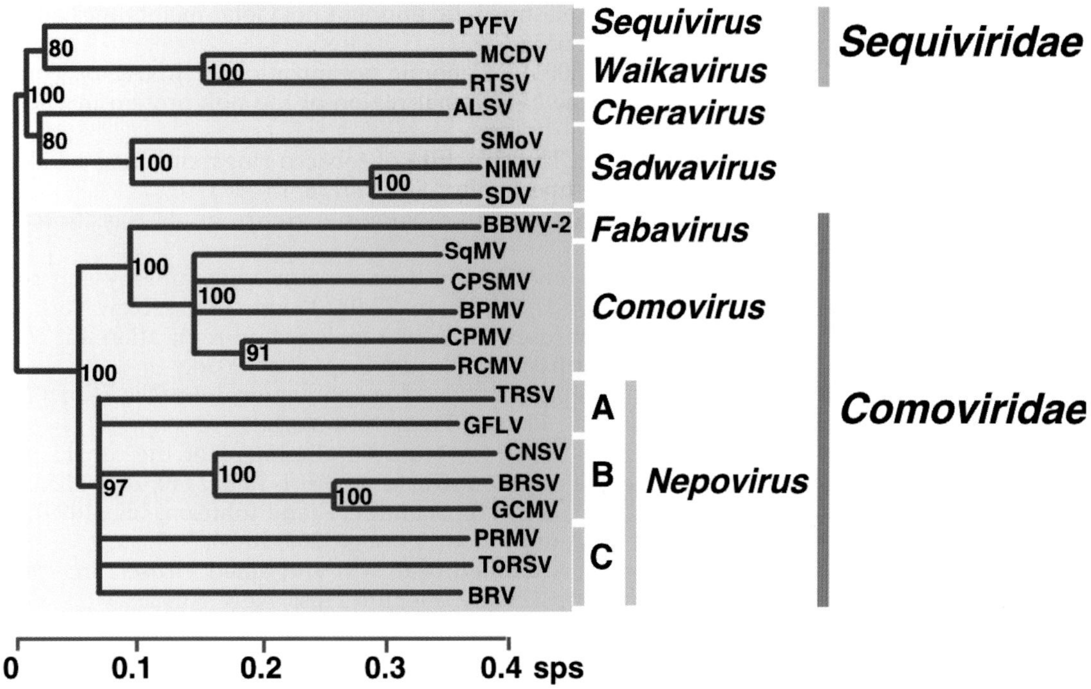

Figure 6: Unrooted dendrogram derived from the alignment of the sequences of the proteinase-polymerase domains of virus species belonging to the families *Comoviridae* and *Sequiviridae*, and the unassigned genera *Cheravirus* and *Sadwavirus*. Branches supported by bootstrap values lower than 75% were merged. The horizontal scale is proportional to the level of divergence while the vertical scale is arbitrary.

The dendrograms derived from multiple sequence alignments from various regions of the genome agree with each other to indicate that the genera *Comovirus* and *Fabavirus* are more closely related to each other than they are from the genus *Nepovirus*. This is supported by genetic organization data. Within the genus *Nepovirus*, subgroup B forms a distinct entity according to the dendrograms. Some viruses considered previously as tentative members of the genus *Nepovirus* are now considered to form two unassigned genera, *Cheravirus* and *Sadwavirus*.

SIMILARITY WITH OTHER TAXA

Several features of the family *Comoviridae* are similar to those of the families and genera *Sequiviridae*, *Cheravirus*, *Sadwavirus*, *Picornaviridae*, *Dicistroviridae*, *Caliciviridae* and *Potyviridae* e.g. genome organization, VPg at 5'-end and poly(A) tract at 3'-end of genomes, post-translational processing of polyproteins and sequence similarities among nonstructural proteins (Fig. 6). Moreover all these virus families except the family *Potyviridae* have very similar capsid morphologies.

Derivation of Names

Como: sigla from *Co*wpea *mo*saic
Faba: Latin *Faba*, bean; also *Vicia faba*, broad bean
Nepo: sigla from *Ne*matode-transmitted, *po*lyhedral particles (to distinguish them from the tobraviruses, also nematode-transmitted but with elongated particles).

References

Altmann, F. and Lomonossoff, G.P. (2000). Glycosylation of the capsid proteins of cowpea mosaic virus: a reinvestigation shows the absence of sugar residues. *J. Gen. Virol.*, **81**, 1111-1114.

Carette, J.E., Stuiver, M., Van Lent, J., Wellink, J. and Van Kammen, A. (2000). Cowpea mosaic virus infection induces a massive proliferation of endoplasmic reticulum but not Golgi membranes and is dependent on de novo membrane synthesis. *J. Virol.*, **714**, 6556-6563.

Carrier, K., Xiang, Y. and Sanfacon, H. (2001). Genomic organization of RNA2 of Tomato ringspot virus: processing at a third cleavage site in the N-terminal region of the polyprotein in vitro. *J. Gen. Virol.*, **82**, 1785-1790.

Chandrasekar, V. and Johnson, J.E. (1998). The structure of tobacco ringspot virus: a link in the evolution of icosahedral capsids in the picornavirus superfamily. *Structure*, **6**, 157-171.

Francki, R.I.B., Milne, R.G. and Hatta, T. (eds) (1985). Comovirus group. In: *Atlas of Plant Viruses*, Vol II, pp 1-22. CRC Press, Boca Raton, Florida.

Gergerich, R.C. and Scott, H.A. (1996). Comoviruses: transmission epidemiology and control. In: *The Plant Viruses* (B.D. Harrison and A.F. Murant, eds), 5th ed., pp 77-98. Plenum Press, New York.

Goldbach, R. and Wellink, J. (1996). Comoviruses: molecular biology and replication. In: *The Plant Viruses* (B.D. Harrison and A.F. Murant, eds), 5th ed., pp 35-76. Plenum Press, New York.

Harrison, B.D. and Murant, A.F. (1996). Nepoviruses: ecology and control. In: *The Plant Viruses* (B.D. Harrison and A.F. Murant, eds), 5th ed., pp 211-228. Plenum Press, New York.

Le Gall, O., Candresse, T. and Dunez, J. (1995). A multiple alignment of the capsid protein sequences of nepoviruses and comoviruses suggests a common structure. *Arch. Virol.*, **140**, 2041-2053.

Lin, T., Chen, Z., Usha, R., Stauffacher, C.V., Dai, J.B., Schmidt, T. and Johnson, J.E. (1999). The refined crystal structure of cowpea mosaic virus at 2.8 A resolution. *Virology*, **265**, 20-34.

Lisa, V. and Boccardo, G. (1996). Fabaviruses: broad bean wilt and allied viruses. In: *The Plant Viruses* (B.D. Harrison and A.F. Murant, eds), 5th ed., pp 229-250. Plenum Press, New York.

Mayo, M.A. and Robinson, D.J. (1996). Nepoviruses: molecular biology and replication. In: *The Plant Viruses* (B.D. Harrison and A.F. Murant, eds), 5th ed., pp 139-186. Plenum Press, New York.

Murant, A.F., Jones, A.T., Martelli, G.P. and Stace-Smith, R. (1996). Nepoviruses: general properties, diseases and virus identification. In: *The Plant Viruses* (B.D. Harrison and A.F. Murant, eds), 5th ed., pp 99-138. Plenum Press, New York.

Pouwels, J., Carette, J.E., Van Lent, J. and Wellink, J. (2002). Cowpea mosaic virus: effects on host cell processes. *Mol. Plant Pathol.*, **3**, 411-418.

Rochon, D. and Sanfaçon, H. (2001). Nepoviruses. In: *Encyclopedia of Plant Pathology* (O.C. Maloy and T.D. Murray, eds), pp 704-708. Academic Press, San Diego.

Ritzenthaler, C., Laporte, C., Gaire, F., Dunoyer, P., Schmitt, C., Duval, S., Piequet, A., Loudes, A.M., Rohfritsch, O., Stussi-Garaud, C. and Pfeiffer, P. (2002). Grapevine fanleaf virus replication occurs on endoplasmic reticulum-derived membranes. *J. Virol.*, **76**, 8808-8819.

Valverde, R.A. and Fulton, J.P. (1996). Comoviruses: identification and diseases caused. In: *The Plant Viruses* (B.D. Harrison and A.F. Murant, eds), 5th ed., pp 17-34. Plenum Press, New York.

Contributed By

Le Gall, O., Iwanami, T., Karasev, A.V., Jones, T., Lehto, K., Sanfaçon, H., Wellink, J., Wetzel, T. and Yoshikawa, N.

FAMILY *POTYVIRIDAE*

TAXONOMIC STRUCTURE OF THE FAMILY

Family	***Potyviridae***
Genus	***Potyvirus***
Genus	***Ipomovirus***
Genus	***Macluravirus***
Genus	***Rymovirus***
Genus	***Tritimovirus***
Genus	***Bymovirus***

VIRION PROPERTIES

MORPHOLOGY

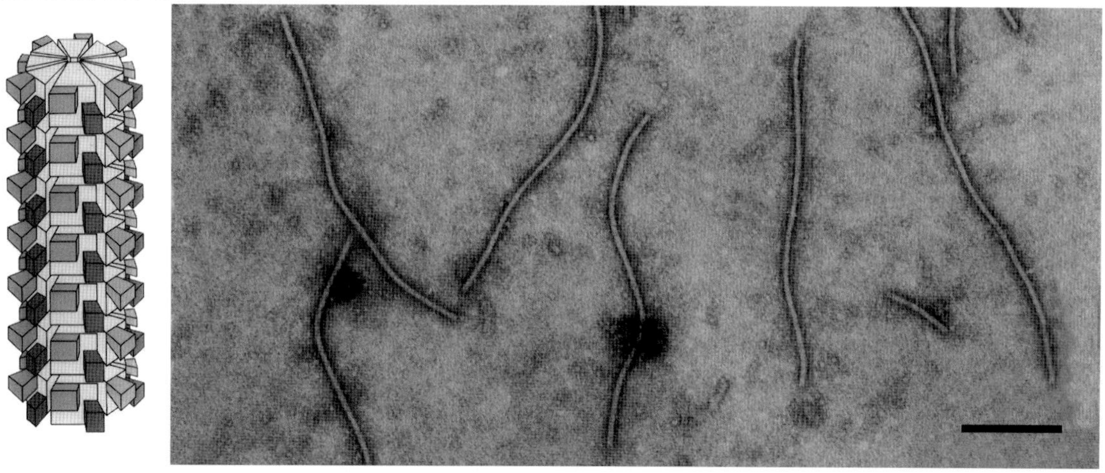

Figure 1: (Left) Schematic diagram of a potyvirus particle. The N-terminal ~30 aa (large rectangle) and C-terminal ~19 aa (small rectangle) of the CP molecules are exposed on the surface of the intact virus particle (from Shukla and Ward, 1989). (Right) Negative contrast electron micrograph of particles of an isolate of *Plum pox virus*, stained with 1% PTA, pH 6.0. The bar represents 200 nm (Courtesy of I.M. Roberts).

Virions are flexuous filaments with no envelope and are 11-15 nm in diameter, with a helical pitch of about 3.4 nm (Fig. 1). Particle lengths of members of some of the six genera differ. Members of the genera *Potyvirus, Ipomovirus, Macluravirus, Rymovirus, Tritimovirus* and the unassigned viruses are monopartite with particle modal lengths of 650-900 nm; members of the genus *Bymovirus* are bipartite with particles of two modal lengths of 250-300 and 500-600 nm.

PHYSICOCHEMICAL AND PHYSICAL PROPERTIES

Virions of viruses in the genera *Potyvirus* and *Rymovirus* have densities in CsCl of about 1.31 g/cm^3 and S_{20w} of 137-160S. Those of viruses of the genus *Bymovirus* have a density in CsCl of about 1.29 g/cm^3.

NUCLEIC ACID

Viruses in all genera except *Bymovirus* have a single molecule of positive sense, ssRNA, 9.3-10.8 kb in size. Virions are infectious. A VPg of about 24 kDais covalently linked to the 5'-terminal nt. A polyadenylate tract (20 to 160 adenosines) is present at the 3' terminus. The complete nt sequence is known for over 30 potyviruses, one ipomovirus, one rymovirus, two tritimovirus, and four bymoviruses. Bymoviruses have two positive sense, ssRNA molecules; RNA-1 is 7.3-7.6 kb in size and RNA-2 is 3.5-3.7 kb in size. Both RNAs have 3' terminal polyadenylate tracts and probably a VPg at the 5'-termini.

PROTEINS

The genome-derived polyprotein is cleaved into several proteins, some of which form inclusion bodies in the cell. Virions contain one type of CP of 28.5-47 kDa. N- and C-terminal residues are positioned on the exterior of the virion. Mild trypsin treatment removes N- and C-terminal segments, leaving a trypsin-resistant core of about 24 kDa. Plant proteases may degrade the CP *in vivo,* as happens *in vitro* during purification using some procedures or from certain hosts. All potyvirus CPs display significant aa sequence identity in the trypsin-resistant core, but little identity in their N and C-terminal segments.

LIPIDS

None reported.

CARBOHYDRATES

None reported.

GENOME ORGANIZATION AND REPLICATION

Genetic information encoded by the RNA genome is organized as a single ORF. Genetic maps for members of the genus *Potyvirus,* and BaYMV (genus *Bymovirus*) are presented in genera descriptions. For potyviruses, the genome is expressed initially as a polyprotein, which then undergoes co- and post-translational proteolytic processing by three viral-encoded proteinases to form individual gene products. Genomic RNA replicates via the production of a full-length negative sense RNA.

ANTIGENIC PROPERTIES

The viral proteins are moderately immunogenic; there are serological relationships among members. An epitope of the CP in the conserved internal trypsin-resistant core has been identified that is similar in most members of the family.

BIOLOGICAL PROPERTIES

INCLUSION BODY FORMATION

All members of the family *Potyviridae* form cytoplasmic cylindrical inclusion (CI) bodies during infection. The CI is an array of a 70 kDa viral protein that possesses ATPase and helicase activities. Some potyviruses induce nuclear inclusion bodies that are co-crystals of two viral-encoded proteins – NIa and NIb – present in equimolar amounts. The small nuclear inclusion (NIa) protein (49 kDa) is a polyprotein consisting of the VPg and proteinase. The large nuclear inclusion (NIb) protein has aa motifs of RdRps. NIa and NIb are also found in the cytoplasm. Amorphous inclusion bodies are also evident in the cytoplasm during certain potyvirus infections and represent aggregations of the protein HC-Pro and perhaps other non-structural proteins. HC-Pro has a helper component activity and a proteolytic activity associated with it. Bymoviruses do not encode a protein similar in length to the helper component, but a 28 kDa protein from RNA-2 of BaYMV has aa domains with sequence similarities to the potyvirus protein HC-Pro.

HOST RANGE

Some members have a narrow host range, most members infect an intermediate number of plants, and a few members infect species in up to 30 families. Transmission to most hosts is readily accomplished by mechanical inoculation. Many viruses are widely distributed. Distribution is aided by seed transmission in some cases.

TRANSMISSION

Potyviruses are vectored by a variety of organisms. Members of the genera *Potyvirus* and *Macluravirus* have aphid vectors that transmit in a non-persistent, non-circulative manner. A helper component and a particular CP aa triplet (i.e., DAG for some potyviruses) are required for aphid transmission. Rymoviruses and tritimoviruses are transmitted by eriophyid mites, in a semi-persistent manner. Bymoviruses are transmitted persistently by a fungus vector. Ipomoviruses appear to be transmitted by whiteflies.

GENUS *POTYVIRUS*

Type Species *Potato virus Y*

VIRION PROPERTIES

MORPHOLOGY

Virions are flexuous filaments, 680-900 nm long and 11-13 nm wide, with helical symmetry and a pitch of about 3.4 nm. Particles of some viruses are longer in the presence of divalent cations than in the presence of EDTA.

PHYSICOCHEMICAL AND PHYSICAL PROPERTIES

Virion S_{20w} is 137-160S; density in CsCl is 1.31 g/cm^3; $E^{0.1\%}{}_{1\ cm,\ 260\ nm}$ = 2.4-2.7.

NUCLEIC ACID

Virions contain a single molecule of linear, positive sense ssRNA, about 9.7 kb in size; virions contain 5% RNA by weight. RNA molecules have poly (A) tracts at their 3'-ends. A 24 kDa VPg is covalently linked at or near the 5' terminus (Fig. 2).

PROTEINS

Virions contain a single CP, 30 to 47 kDa in size. The CP of most isolates of the type species, PVY, contains 267 aa.

LIPIDS

None reported.

CARBOHYDRATES

None reported.

GENOME ORGANIZATION AND REPLICATION

TEV (*Potyvirus*) 9496 nts

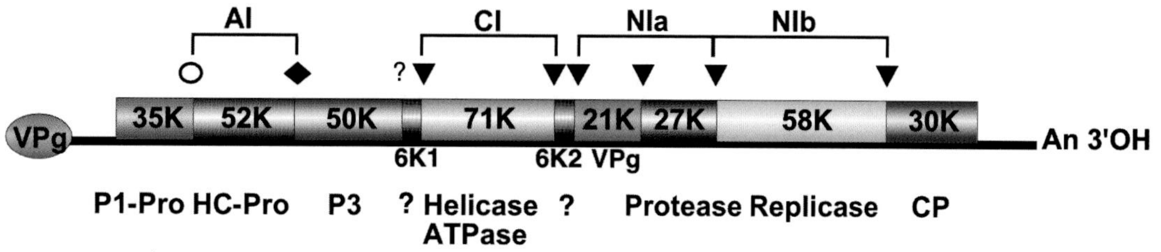

Figure 2: Genomic map of a member of the genus *Potyvirus,* using a strain of *Tobacco etch virus* as an example. The RNA genome is represented by a thin line and an open box which represents translated segments of the ssRNA. Functions associated with these products are shown. VPg, genome-linked viral protein covalently attached to the 5' terminal nt (represented by the oval at the 5'-end); P1-Pro, a protein with a proteolytic activity responsible for cleavage at typically Tyr/Phe-Ser (O); HC-Pro, a protein with aphid transmission helper component activity and proteolytic activity responsible for cleavage at typically Gly-Gly (◆); Pro, serine-like proteolytic activity responsible for cleavage at Gln/Glu-(Ser/Gly/Ala) (▼). Some of these proteins of particular viruses of the family *Potyviridae* aggregate to form inclusion bodies during infection. The protein involved and the particular type of inclusion body is shown above the genetic map; AI, amorphous inclusion; CI, cylindrical-shaped inclusion body found in the cytoplasm; NIa and NIb, small and large nuclear inclusion proteins, respectively, which aggregate in the nucleus to form a nuclear inclusion body.

Potyvirus genomes are ~9.7 kb in size (Fig. 2).

ANTIGENIC PROPERTIES

Virions are moderately immunogenic; there are serological relationships among many members. One monoclonal antibody reacts with most aphid-transmitted potyviruses. The CP aa sequence identity among aphid-transmitted viruses is 40-70%. Some viruses are serologically related to viruses in the genera *Rymovirus* and *Bymovirus*.

BIOLOGICAL PROPERTIES

Many individual viruses have a narrow host range, but a few infect plant species in up to 30 host families. The viruses are transmitted by aphids in a non-persistent manner and are transmissible experimentally by mechanical inoculation. Some isolates are inefficiently transmitted by aphids and others are not transmissible by aphids at all. This is apparently due to mutations within the helper component and/or CP cistrons. Some viruses are seed-transmitted.

LIST OF SPECIES DEMARCATION CRITERIA IN THE GENUS

- Genome sequence relatedness.
  - CP aa sequence identity less than ca. 80%,
  - nt sequence identity of less than 85% over whole genome,
  - different polyprotein cleavage sites.
- Natural host range.
  - host range may be related to species but usually not helpful in identifying species; may delineate strains.
- Pathogenicity and cytopathology.
  - different inclusion body morphology,
  - lack of cross protection,
  - seed transmissibility, or lack thereof,
  - some aspects of host reaction may be useful (e.g., different responses in key host species, and particular genetic interactions).
- Mode of transmission.
  - different primary vectors, but vector species not use in identification to virus species.
- Antigenic properties.
  - serological differences.

LIST OF SPECIES IN THE GENUS

Species names are in green italic script; strain names and synonyms are in black roman script; tentative species names are in blue roman script. Sequence accession numbers, and assigned abbreviations ( ) are also listed.

SPECIES IN THE GENUS

Alpinia mosaic virus		
Alpinia mosaic virus		(AlpMV)
Alstroemeria mosaic virus		
Alstroemeria mosaic virus		(AlMV)
Alstroemeria streak virus		
Amaranthus leaf mottle virus		
Amaranthus leaf mottle virus		(AmLMV)
Apium virus Y ‡		
Apium virus Y		(ApVY)
Parsley virus Y		
Araujia mosaic virus		
Araujia mosaic virus		(ArjMV)
Artichoke latent virus		
Artichoke latent virus		(ArLV)
Asparagus virus 1		
Asparagus virus 1		(AV-1)
Banana bract mosaic virus		
Banana bract mosaic virus		(BBrMV)
Bean common mosaic necrosis virus		
Bean common mosaic necrosis virus	[AY138897, AY282577, U19287]	(BCMNV)
Bean common mosaic virus serotype A		

Bean common mosaic virus		
Azuki bean mosaic virus		
Bean common mosaic virus	[AY112735, U34972]	(BCMV)
Blackeye cowpea mosaic virus	[AJ312437-8]	
Dendrobium mosaic virus		
Guar green sterile virus		
Peanut chlorotic ring mottle virus		
Peanut mild mottle virus		
Peanut stripe virus	[U05771-2]	
Bean yellow mosaic virus		
Bean yellow mosaic virus	[AY192568, D83749, U47033]	(BYMV)
Crocus tomasinianus virus		
Pea mosaic virus		
White lupin mosaic virus		
Beet mosaic virus		
Beet mosaic virus	[AY206394]	(BtMV)
Dioscorea alata ring mottle virus		
Bidens mottle virus		
Bidens mottle virus		(BiMoV)
Calanthe mild mosaic virus		
Calanthe mild mosaic virus		(CalMMV)
Carnation vein mottle virus		
Carnation vein mottle virus		(CVMoV)
Carrot thin leaf virus		
Carrot thin leaf virus		(CTLV)
Carrot virus Y		
Carrot virus Y		(CtVY)
Celery mosaic virus		
Celery mosaic virus		(CeMV)
Ceratobium mosaic virus		
Ceratobium mosaic virus		(CerMV)
Chilli veinal mottle virus		
Chilli veinal mottle virus	[AJ237843]	(ChiVMV)
Indian pepper mottle virus		
Pepper vein banding mosaic virus		
Clitoria virus Y		
Clitoria virus Y		(ClVY)
Clover yellow vein virus		
Clover yellow vein virus	[AB011819]	(ClYVV)
Pea necrosis virus		
Statice virus Y		
Cocksfoot streak virus		
Cocksfoot streak virus	[AF499738]	(CSV)
Colombian datura virus		
Colombian datura virus		(CDV)
Petunia flower mottle virus ‡		
Commelina mosaic virus		
Commelina mosaic virus		(ComMV)
Cowpea aphid-borne mosaic virus		
Cowpea aphid-borne mosaic virus	[AF348210]	(CABMV)
Sesame mosaic virus		
South African passiflora virus		
Cowpea green vein banding virus		
Cowpea green vein banding virus		(CGVBV)

POS. SSRNA

Cypripedium virus Y		
Cypripedium virus Y		(CypVY))
Dasheen mosaic virus		
Dasheen mosaic virus	[AJ298033]	(DsMV)
Datura shoestring virus		
Datura shoestring virus		(DSSV)
Diuris virus Y		
Diuris virus Y		(DiVY)
Endive necrotic mosaic virus		
Endive necrotic mosaic virus		(ENMV)
Freesia mosaic virus		
Freesia mosaic virus		(FreMV)
Gloriosa stripe mosaic virus		
Gloriosa stripe mosaic virus		(GSMV)
Groundnut eyespot virus		
Groundnut eyespot virus		(GEV)
Guinea grass mosaic virus		
Guinea grass mosaic virus		(GGMV)
Helenium virus Y		
Helenium virus Y		(HVY)
Henbane mosaic virus		
Henbane mosaic virus		(HMV)
Hibbertia virus Y ‡		
Hibbertia virus Y		(HiVY)
Hippeastrum mosaic virus		
Hippeastrum mosaic virus		(HiMV)
Hyacinth mosaic virus		
Hyacinth mosaic virus		(HyaMV)
Iris fulva mosaic virus		
Iris fulva mosaic virus		(IFMV)
Iris mild mosaic virus		
Iris mild mosaic virus		(IMMV)
Iris severe mosaic virus		
Bearded iris mosaic virus		
Iris severe mosaic virus		(ISMV)
Japanese yam mosaic virus		
Japanese yam mosaic virus	[AB016500, AB027007]	(JYMV)
Johnsongrass mosaic virus		
Johnsongrass mosaic virus	[Z26920]	(JGMV)
Kalanchoë mosaic virus		
Kalanchoë mosaic virus		(KMV)
Konjac mosaic virus		
Konjac mosaic virus		(KoMV)
Leek yellow stripe virus		
Garlic mosaic virus		
Garlic virus		
Garlic virus 2		
Leek yellow stripe virus	[AJ307057]	(LYSV)
Lettuce mosaic virus		
Lettuce mosaic virus	[AJ278854, AJ306288, X97704-5]	(LMV)
Lily mottle virus		
Lily mild mottle virus		
Lily mottle virus	[AJ564636]	(LMoV)
Tulip band breaking virus		

Lycoris mild mottle virus		
Lycoris mild mottle virus		(LyMMoV)
Maize dwarf mosaic virus		
Maize dwarf mosaic virus	[AJ001691]	(MDMV)
Moroccan watermelon mosaic virus		
Moroccan watermelon mosaic virus		(MWMV)
Narcissus degeneration virus		
Narcissus degeneration virus		(NDV)
Narcissus late season yellows virus		
Jonquil mild mosaic virus		
Narcissus late season yellows virus		(NLSYV)
Narcissus yellow stripe virus		
Narcissus yellow stripe virus		(NYSV)
Nerine yellow stripe virus		
Nerine yellow stripe virus		(NeYSV)
Nothoscordum mosaic virus		
Nothoscordum mosaic virus		(NoMV)
Onion yellow dwarf virus		
Onion yellow dwarf virus	[AJ510223]	(OYDV)
Ornithogalum mosaic virus		
Ornithogalum mosaic virus		(OrMV)
Pterostylis virus Y		
Ornithogalum virus 2		
Ornithogalum virus 2		(OrV2)
Ornithogalum virus 3		
Ornithogalum virus 3		(OrV3)
Papaya leaf distortion mosaic virus		
Papaya leaf distortion mosaic virus	[AB088221]	(PLDMV)
Papaya ringspot virus		
Papaya ringspot virus	[S46722, X97251, AY010722, AY027810, AY162218, AY231130]	(PRSV)
Watermelon mosaic virus 1		
Parsnip mosaic virus		
Parsnip mosaic virus		(ParMV)
Passion fruit woodiness virus		
Passion fruit woodiness virus		(PWV)
Pea seed-borne mosaic virus		
Pea seed-borne mosaic virus	[AJ252242, D10930, X89997]	(PSbMV)
Peanut mottle virus		
Peanut mottle virus	[AF023848]	(PeMoV)
Pepper mottle virus		
Pepper mottle virus	[AF501591, M96425]	(PepMoV)
Pepper severe mosaic virus		
Pepper severe mosaic virus		(PepSMV)
Pepper veinal mottle virus		
Pepper veinal mottle virus		(PVMV)
Pepper yellow mosaic virus		
Pepper yellow mosaic virus		(PepYMV)
Peru tomato mosaic virus		
Peru tomato mosaic virus		(PTV)
Pleione virus Y		
Pleione virus Y		(PlVY)
Plum pox virus		
Plum pox virus	[AF401295-6, AJ243957, AY184478,	(PPV)

	D13751, M92280, X16415, X81083, Y09851]	
Pokeweed mosaic virus		
Pokeweed mosaic virus		(PkMV)
Potato virus A		
Potato virus A	[AF543212, AF543709, AJ131400-2, AJ296311]	(PVA)
Tamarillo mosaic virus	[AJ131403]	
Potato virus V		
Potato virus V	[AJ243766]	(PVV)
Potato virus Y		
Potato virus Y	[AF237963, AF463399, AF522296, AJ439544-5, AY166866-7, D00441, M95491, U09509, X12456, X97895]	(PVY)
Sunflower chlorotic mottle virus		
Rhopalanthe virus Y		
Rhopalanthe virus Y		(RhVY)
Sarcochilus virus Y		
Sarcochilus virus Y		(SaVY)
Scallion mosaic virus		
Scallion mosaic virus	[AJ316084]	(ScMV)
Shallot yellow stripe virus		
Shallot yellow stripe virus		(SYSV)
Welsh onion yellow stripe virus		
Sorghum mosaic virus		
Sorghum mosaic virus	[AJ310197-8, U57358]	(SrMV)
Soybean mosaic virus		
Soybean mosaic virus	[AB100442-3, AF241739, AJ310200, AJ312439, AJ507388, AJ628750, AY216010, AY216987, AY294044-5, D00507, S42280]	(SMV)
Sugarcane mosaic virus		
Sugarcane mosaic virus	[AF494510, AJ278405, AJ297628, AJ310102-5, AY042184, AY149118]	(SCMV)
Sunflower mosaic virus		
Sunflower mosaic virus		(SuMV)
Sweet potato feathery mottle virus		
Sweet potato chlorotic leafspot virus		
Sweet potato feathery mottle virus	[D86371]	(SPFMV)
Sweet potato internal cork virus		
Sweet potato russet crack virus		
Sweet potato virus A		
Sweet potato latent virus		
Sweet potato latent virus		(SPLV)
Sweet potato mild speckling virus ‡		
Sweet potato mild speckling virus		(SPMSV)
Sweet potato virus G ‡		
Sweet potato virus G		(SPVG)
Telfairia mosaic virus		
Telfairia mosaic virus		(TeMV)
Tobacco etch virus		
Tobacco etch virus	[L38714, M11458, M15239]	(TEV)
Tobacco vein banding mosaic virus		
Tobacco vein banding mosaic virus		(TVBMV)

Tobacco vein mottling virus		
Tobacco vein mottling virus	[U38621, X04083]	(TVMV)
Tropaeolum mosaic virus		
Nasturtium mosaic virus		
Tropaeolum mosaic virus		(TrMV)
Tuberose mild mosaic virus		
Tuberose mild mosaic virus		(TuMMV)
Tulip breaking virus		
Tulip breaking virus		(TBV)
Tulip mosaic virus ‡		
Tulip mosaic virus		(TulMV)
Turnip mosaic virus		
Tulip chlorotic blotch virus		
Tulip top breaking virus		
Turnip mosaic virus	[AB093596-627, AF169561, AF394601-2, AY090660, AY227024, D10927, D83184]	(TuMV)
Watermelon leaf mottle virus ‡		
Watermelon leaf mottle virus		(WLMV)
Watermelon mosaic virus		
Vanilla necrosis virus		
Watermelon mosaic virus		(WMV)
Watermelon mosaic virus 2		
Wild potato mosaic virus		
Wild potato mosaic virus	[AJ437279]	(WPMV)
Wisteria vein mosaic virus		
Wisteria vein mosaic virus		(WVMV)
Yam mild mosaic virus ‡		
Dioscorea alata virus		
Yam mild mosaic virus		(YMMV)
Yam mosaic virus		
Dioscorea green banding virus		
Yam mosaic virus	[U42596]	(YMV)
Zantedeschia mosaic virus ‡		
Japanese hornwort mosaic virus		
Zantedeschia mosaic virus		(ZaMV)
Zea mosaic virus		
Iranian johnsongrass mosaic virus		
Zea mosaic virus		(ZeMV)
Zucchini yellow fleck virus		
Zucchini yellow fleck virus		(ZYFV)
Zucchini yellow mosaic virus		
Zucchini yellow mosaic virus	[AF014811, AF127929, AJ307036, AJ316228-9, AJ515911, AY278998-9, AY279000, L29569, L31350]	(ZYMV)

TENTATIVE SPECIES IN THE GENUS

Aphid-borne (*aphid transmission not confirmed; + denotes plant species with a report of a potyvirus infection)

Alstroemeria flower banding virus	(AlFBV)
Amazon lily mosaic virus	(ALiMV)
Aneilema mosaic virus	(AneMV)
Anthoxanthum mosaic virus*	(AntMV)
Aquilegia necrotic ringspot virus*	(AqNRSV)

POS. SSRNA

Arracacha virus Y	(AVY)
Asystasia gangetica mottle virus*	(AGMoV)
Bidens mosaic virus	(BiMV)
Bramble yellow mosaic virus	(BrmYMV)
Bryonia mottle virus	(BryMoV)
Canary reed mosaic virus	(CRMV)
Canavalia maritima mosaic virus	(CnMMV)
Carrot mosaic virus	(CtMV)
Cassia yellow spot virus	(CasYSV)
Celery yellow mosaic virus	(CeYMV)
Chickpea bushy dwarf virus	(CpBDV)
Chickpea filiform virus	(CpFV)
Chrysanthemum spot virus	(ChSV)
Clitoria yellow mosaic virus	(CtYMV)
Cowpea rugose mosaic virus	(CPRMV)
Crinum mosaic virus*	(CriMV)
Croatian clover virus+	(CroCV)
Cypripedium chlorotic streak virus*	(CypCSV)
Daphne virus Y	(DVY)
Datura virus 437	(DV-437)
Datura distortion mosaic virus	(DDMV)
Datura mosaic virus*	(DTMV)
Datura necrosis virus	(DNV)
Desmodium mosaic virus	(DesMV)
Dioscorea dumentorum virus	(DDV)
Dioscorea trifida virus+	(DTV)
Dipladenia mosaic virus	(DipMV)
Dock mottling mosaic virus	(DMMV)
Eggplant green mosaic virus	(EGMV)
Eggplant severe mottle virus	(ESMoV)
Euphorbia ringspot virus	(EuRSV)
Fig leaf chlorosis virus	(FigLCV)
Guar symptomless virus*	(GSLV)
Habenaria mosaic virus	(HaMV)
Holcus streak virus*	(HSV)
Hungarian datura innoxia virus*	(HDIV)
Isachne mosaic virus*	(IsaMV)
Kennedya virus Y	(KVY)
Malva vein clearing virus	(MVCV)
Marigold mottle virus	(MaMoV)
Melilotus mosaic virus	(MeMV)
Melon vein-banding mosaic virus	(MVBMV)
Melothria mottle virus	(MeMoV)
Mungbean mosaic virus*	(MbMV)
Mungbean mottle virus	(MMoV)
Palm mosaic virus*	(PalMV)
Passion fruit mottle virus	(PFMoV)
Passion fruit ringspot virus	(PFRSV)
Patchouli mottle virus	(PatMoV)
Peanut chlorotic blotch virus	(PeClBlV)
Peanut green mottle virus	(PeGMoV)
Peanut top paralysis virus	(PeTPV)
Pecteilis mosaic virus	(PcMV)
Pepper mild mosaic virus	(PMMV)

Perilla mottle virus	(PerMoV
Plantain virus 7	(PlV-7)
Pleioblastus mosaic virus	(PleMV)
Poplar decline virus*	(PopDV)
Primula mosaic virus	(PrMV)
Primula mottle virus	(PrMoV)
Radish vein clearing virus	(RaVCV)
Ranunculus mottle virus	(RanMoV)
Rembrandt tulip breaking virus	(ReTBV)
Rudbeckia mosaic virus	(RuMV)
Sri Lankan passion fruit mottle virus	(SLPMoV)
Sweet potato vein mosaic virus	(SPVMV)
Sweet potato mild speckling virus	(SPMSV)
Sword bean distortion mosaic virus	(SBDMV)
Taro feathery mottle virus	(TFMoV)
Teasel mosaic virus	(TeaMV)
Tobacco wilt virus	(TWV)
Tongan vanilla virus	(TVV)
Tradescantia mosaic virus	(TraMV)
Trichosanthes mottle virus	(TrMoV)
Tropaeolum virus 1	(TV-1)
Tropaeolum virus 2	(TV-2)
Ullucus mosaic virus	(UMV)
Vallota mosaic virus	(ValMV)
Vanilla mosaic virus	(VanMV)
White bryony virus	(WBV)
Zoysia mosaic virus	(ZoMV)

GENUS IPOMOVIRUS

Type Species *Sweet potato mild mottle virus*

VIRION PROPERTIES

MORPHOLOGY

Virions are flexuous filaments 800-950 nm long.

PHYSICOCHEMICAL AND PHYSICAL PROPERTIES

Virion S_{20w} is 155S for Sweet potato mild mottle virus (SPMMV).

NUCLEIC ACID

Virions contain a positive sense ssRNA, ~10.8 kb in size, with a 3′-poly(A) terminus.

PROTEINS

The viral CP of is a single peptide of 270-275 aa (37.7 kDa).

LIPIDS

None reported.

CARBOHYDRATES

None reported.

GENOME ORGANIZATION AND REPLICATION

The SPMMV genome consists of 10818 nt excluding the 3′-terminal poly(A) tail. Sequence analysis reveals an ORF of 3456 aa (Fig. 3). The structure and organization of the SPMMV genome appear to be similar to those of other members of the family *Potyviridae* except the bymoviruses. Almost all known potyvirus motifs are present in the polyprotein

POS. SSRNA

of SPMMV. However, motifs in the putative HC-Pro and CP of SPMMV are incomplete or missing, which may account for its vector relations. Comparative sequence analyses show only limited similarities between the nine mature proteins and those of other species in other genera of the family *Potyviridae*.

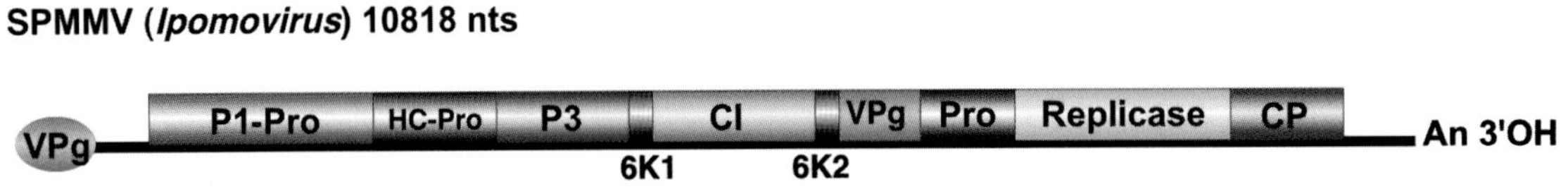

Figure 3: Genomic map of an isolate of *Sweet potato mild mottle virus*. The RNA genome is represented by thin lines and an open box which represent translated segments of the ssRNA. Conventions are as for the potyvirus genome organization map (Fig. 2). Activities of most gene products are postulated by analogy with genus *Potyvirus*.

ANTIGENIC PROPERTIES

Moderately immunogenic. No serological relationships with other members of the family *Potyviridae* have been found.

BIOLOGICAL PROPERTIES

INCLUSION BODY FORMATION

Characteristic cytoplasmic cylindrical ("pinwheel") inclusions are present in infected cells.

HOST RANGE

The natural host range of SPMMV is wide, with more than nine families susceptible.

TRANSMISSION

SPMMV is transmitted by the whitefly *Bemisia tabaci* in a non-persistent manner and is transmissible experimentally by mechanical inoculation and by grafting.

LIST OF SPECIES DEMARCATION CRITERIA IN THE GENUS

- Genome sequence relatedness.
  - CP aa sequence identity less than ca. 80%,
  - nt sequence identity of less than 85% over whole genome,
  - different polyprotein cleavage sites.
- Natural host range.
  - host range may be related to species but usually not helpful in identifying species. May delineate strains.
- Pathogenicity and cytopathology.
  - different inclusion body morphology,
  - lack of cross protection,
  - seed transmissibility, or lack thereof,
  - some aspects of host reaction may be useful (e.g., resistance genes, different responses in key host species).
- Antigenic properties.
  - serological differences.

LIST OF SPECIES IN THE GENUS

Species names are in green italic script; strain names and synonyms are in black roman script; tentative species names are in blue roman script. Sequence accession numbers, and assigned abbreviations ( ) are also listed.

SPECIES IN THE GENUS

Cassava brown streak virus
Cassava brown streak virus (CBSV)
Cucumber vein yellowing virus
Cucumber vein yellowing virus (CVYV)

Sweet potato mild mottle virus
Sweet potato mild mottle virus [Z73124] (SPMMV)

TENTATIVE SPECIES IN THE GENUS
Sweet potato yellow dwarf virus (SPYDV)

GENUS *MACLURAVIRUS*

Type Species *Maclura mosaic virus*

VIRION PROPERTIES

MORPHOLOGY
Virions are flexuous filaments mostly 650-675 nm x 13-16 nm.

PHYSICOCHEMICAL AND PHYSICAL PROPERTIES
Virion S_{20w} is 155-158S; density in CsCl is 1.31-1.33 g/cm^3.

NUCLEIC ACID
Virions contain one molecule of linear positive sense, ssRNA. RNA is ~8.0 kb.

PROTEINS
Macluraviruses have a single CP species of 33-34 kDa.

LIPIDS
None reported.

CARBOHYDRATES
None reported

GENOME ORGANIZATION AND REPLICATION

The aa sequences of macluravirus CPs show limited (14-23%) identity with CP sequences of some aphid-transmitted potyviruses. Macluraviruses show significant aa sequence identity in portions of the replicase protein with viruses in other genera of the family *Potyviridae*. Characteristic cytoplasmic cylindrical ("pinwheel") inclusions are present in infected cells. The macluraviruses seem to have a genome organization and replication strategy typical of viruses in the family *Potyviridae*.

ANTIGENIC PROPERTIES

Moderately immunogenic. No serological relationships to members of the genus *Potyvirus* have been found except for a weak reactions between Maclura mosaic virus (MacMV) and Narcissus latent virus (NLV) and between MacMV and Bean yellow mosaic virus (BYMV).

BIOLOGICAL PROPERTIES

HOST RANGE
Both MacMV and NLV have a narrow host range, infecting species in up to 9 host families. MacMV has only been reported from the former Yugoslavia while NLV is likely to occur wherever narcissus, gladiolus and bulbous iris are grown.

TRANSMISSION
The viruses are transmitted by aphids in a non-persistent manner and experimentally by mechanical inoculation.

LIST OF SPECIES DEMARCATION CRITERIA IN THE GENUS

- Genome sequence relatedness.

- CP aa sequence identity less than ca. 80%,
- nt sequence identity less than 85% over whole genome,
- different polyprotein cleavage sites.

• Natural host range.
- host range may be correlated with species but usually not helpful in identifying species; may delineate strains.

• Pathogenicity and cytopathology.
- different inclusion body morphology,
- lack of cross protection,
- seed transmissibility, or lack thereof,
- some aspects of host reaction may be useful (e.g., different responses in key host species, and particular genetic interactions).

• Mode of transmission.
- different primary vectors, but vector species are not of use in demarcating virus species.

• Antigenic properties.
- serological differences.

LIST OF SPECIES IN THE GENUS

Species names are in green italic script; strain names and synonyms are in black roman script; tentative species names are in blue roman script. Sequence accession numbers, and assigned abbreviations ( ) are also listed.

SPECIES IN THE GENUS

Cardamom mosaic virus
Cardamom mosaic virus (CdMV)
Indian cardamom mosaic virus
Maclura mosaic virus
Maclura mosaic virus (MacMV)
Narcissus latent virus
Narcissus latent virus (NLV)

TENTATIVE SPECIES IN THE GENUS

Chinese yam necrotic mosaic virus (CYNMV)

GENUS *RYMOVIRUS*

Type Species *Ryegrass mosaic virus*

VIRION PROPERTIES

MORPHOLOGY

Virions are flexuous filaments 690-720 x 11-15 nm in size.

PHYSICOCHEMICAL AND PHYSICAL PROPERTIES

Virion density in CsCl is 1.325 g/cm^3 (for Ryegrass mosaic virus, RGMV). Virion S_{20w} is 165-166S for most members.

NUCLEIC ACID

Virions contain a single molecule of linear positive sense ssRNA with a 3'-poly(A) terminus. Virion RNA is about 9-10 kb in size.

PROTEINS

Rymoviruses have one type of CP, of 29.2 kDa for RGMV.

LIPIDS

None reported

CARBOHYDRATES

None reported

GENOME ORGANIZATION AND REPLICATION

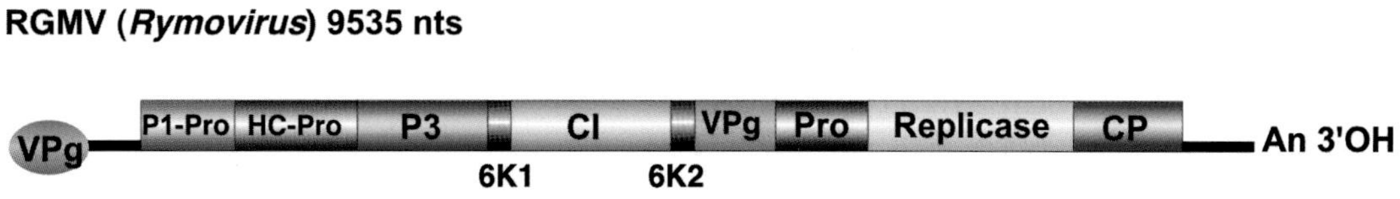

Figure 4: Genomic map of an isolate of *Ryegrass mosaic virus*. The RNA genome is represented by thin lines and an open box which represent translated segments of the ssRNA. Conventions are as for the potyvirus genome organization map (Fig. 2). Activities of most gene products are postulated by analogy with genus *Potyvirus*.

The rymoviruses presumably have a genome organization (Fig. 4) and replication strategy typical of viruses in the family *Potyviridae*.

ANTIGENIC PROPERTIES

Particles of most rymoviruses are moderately immunogenic. No serological relationships have been found among member viruses.

BIOLOGICAL PROPERTIES

HOST RANGE

Most rymoviruses have limited but widespread host ranges within the family Graminae but some have relatively narrow host ranges.

TRANSMISSION

Transmission by eriophyid mites and mechanical transmission have been reported for most members.

LIST OF SPECIES DEMARCATION CRITERIA IN THE GENUS

- Genome sequence relatedness.
  - CP aa sequence identity less than ca. 80%,

  - nt sequence identity less than 85% over whole genome,
  - different polyprotein cleavage sites.
- Natural host range.
  - host range may be related to species but usually not helpful in identifying species; may delineate strains.
- Pathogenicity and cytopathology.
  - different inclusion body morphology,
  - lack of cross protection,
  - seed transmissibility, or lack thereof,
  - some aspects of host reaction may be useful (e.g., different responses in key host species, and particular genetic interactions).
- Mode of transmission.
  - different primary vectors, but vector species not of use in demarcating virus species.
- Antigenic properties.
  - serological differences.

LIST OF SPECIES IN THE GENUS

Species names are in green italic script; strain names and synonyms are in black roman script; tentative species names are in blue roman script. Sequence accession numbers, and assigned abbreviations ( ) are also listed.

SPECIES IN THE GENUS

Agropyron mosaic virus		
Agropyron mosaic virus		(AgMV)
Hordeum mosaic virus		
Hordeum mosaic virus		(HoMV)
Ryegrass mosaic virus		
Ryegrass mosaic virus	[AF035818, Y09854]	(RGMV)

TENTATIVE SPECIES IN THE GENUS

None reported

GENUS *TRITIMOVIRUS*

Type Species ***Wheat streak mosaic virus***

VIRION PROPERTIES

MORPHOLOGY

Virions are flexuous filaments 690-700 nm long.

PHYSICOCHEMICAL AND PHYSICAL PROPERTIES

Virion S_{20w} is 166S for Wheat streak mosaic virus (WSMV).

NUCLEIC ACID

Virions contain a positive sense ssRNA, ~9.4-9.6 kb in size, with a 3′-poly(A) terminus.

PROTEINS

The viral CP is a single peptide of ~349 aa for WSMV and 320 aa for Brome streak mosaic virus (BrSMV). The Mr estimated by electrophoresis is ~42 kDa.

LIPIDS

None reported

CARBOHYDRATES

None reported

GENOME ORGANIZATION AND REPLICATION

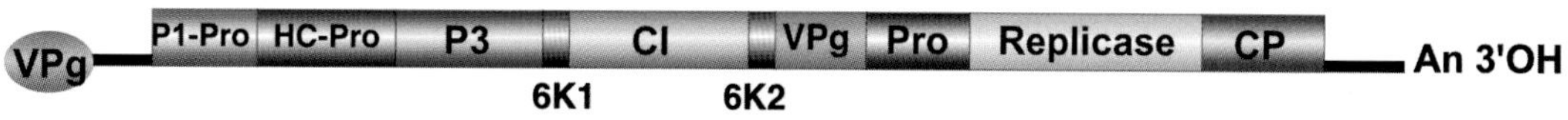

Figure 5: Genomic map of an isolate of *Wheat steak mosaic virus*. The RNA genome is represented by thin lines and an open box which represent translated segments of the ssRNA. Conventions are as for the potyvirus genome organization map (Fig. 2). Activities of most gene products are postulated by analogy with genus *Potyvirus*.

The WSMV genome consists of 9384 nt excluding the 3′-terminal poly(A) tail. Sequence analysis reveals an ORF of 3035 aa (Fig. 5). The structure and organization of the WSMV genome is similar to those of other members of the family *Potyviridae* except the bymoviruses. Most known potyvirus motifs are present in the polyprotein of WSMV. However, motifs in the putative helper-component and CP of BrSMV are incomplete or missing, which may account for different vector relations of the tritimoviruses. The WSMV CP sequence shows limited (22-25%) identity with CP sequences of some aphid-transmitted potyviruses. WSMV shows significant aa sequence identity with aphid-transmitted potyviruses in the cylindrical inclusion protein and portions of the nuclear inclusion proteins. WSMV RNA has been translated *in vitro* into several large proteins immunoprecipitable with WSMV CP antiserum, suggesting that WSMV uses a proteolytic processing strategy to express functional proteins such as the CP. Antiserum to TEV 58 kDa nuclear inclusion protein also reacts with *in vitro* translation products of WSMV. An *in vitro* translation product is precipitated with antiserum to HC-Pro helper component of an isolate of *Tobacco vein mottling virus*. Comparative sequence analyses show similarities with other members of the family *Potyviridae*, but these are limited to the nine mature proteins. WSMV has CP molecules of 42, 36 and 32 kDa; the two smaller proteins are parts of the 42 kDa protein.

Antigenic Properties

Moderately immunogenic. The WSMV and ONMV are serologically related to each other, but not to the other members of the family *Potyviridae*.

Biological Properties

Cytology

Characteristic cytoplasmic cylindrical ("pinwheel") inclusions are present in infected cells.

Host Range

WSMV has a wide host range within the Graminae. BrSMV and ONMV have a narrow host range, also restricted to the Graminae.

Transmission

WSMV and BrSMV are transmitted by eriophyid mites in a semi-persistent manner. All definitive tritimoviruses are transmissible experimentally by mechanical inoculation.

List of Species Demarcation Criteria in the Genus

- Genome sequence relatedness.
  - CP aa sequence identity less than ca. 80%,
  - nt sequence identity less than 85% over whole genome,
  - different polyprotein cleavage sites.
- Natural host range.
  - host range may be related to species but usually not helpful in identifying species. May delineate strains.
- Pathogenicity and cytopathology.
  - different inclusion body morphology,
  - lack of cross protection,
  - seed transmissibility, or lack thereof,
  - some aspects of host reaction may be useful (e.g., resistance genes, different responses in key host species).
- Mode of transmission.
  - different primary vectors, but vector species not of use in demarcating virus species.
- Antigenic properties.
  - serological differences.

List of Species in the Genus

Species names are in green italic script; strain names and synonyms are in black roman script; tentative species names are in blue roman script. Sequence accession numbers, and assigned abbreviations ( ) are also listed.

Species in the Genus

Brome streak mosaic virus		
Brome streak mosaic virus	[Z48506]	(BrSMV)
Oat necrotic mottle virus		
Oat necrotic mottle virus	[AY377938]	(ONMV)
Wheat streak mosaic virus		
Wheat streak mosaic virus	[AF057533, AF285169-70, AF454454-5]	(WSMV)

Tentative Species in the Genus

None reported

GENUS *BYMOVIRUS*

Type Species *Barley yellow mosaic virus*

VIRION PROPERTIES

MORPHOLOGY

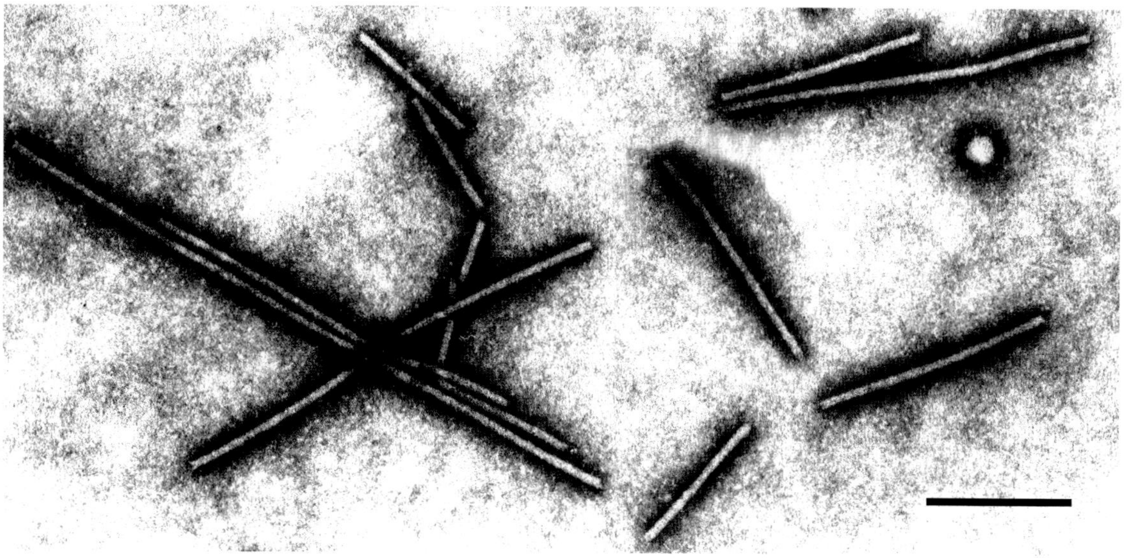

Figure 6: Virions of an isolate of *Barley yellow mosaic virus*, stained with 1% PTA, pH 7.0. The bar represents 200 nm (from D. Lesemann).

Virions are flexuous filaments of two modal lengths, 250-300 and 500-600 nm; both are 13 nm in width (Fig. 6).

PHYSICOCHEMICAL AND PHYSICAL PROPERTIES

Virion buoyant density in CsCl is 1.28-1.30 g/cm^3.

NUCLEIC ACID

Virions contain two molecules of linear positive sense, ssRNA. RNA-1 is 7.5-8.0 kb and RNA-2 is 3.5-4.0 kb; RNA makes up 5% by weight of particles. Both RNA molecules have 3'-terminal poly(A) tracts. There is little base sequence homology between the two RNAs except in the 5' NTR. The CP gene is located in the 3'-proximal region of RNA-1

PROTEINS

Virions have a single CP of 28.5-33 kDa. The CP of Barley yellow mosaic virus (BaYMV) has 297 aa.

LIPIDS

None reported

CARBOHYDRATES

None reported

GENOME ORGANIZATION AND REPLICATION

The two RNA molecules appear to be translated initially into precursor polypeptides from which functional proteins are derived by proteolytic processing (Fig. 4).

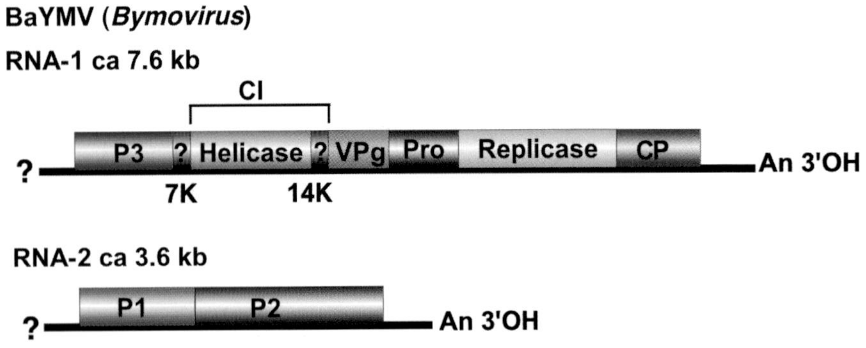

Figure 7: Genomic map of the bymovirus bipartite genome, using, as an example, an isolate of *Barley yellow mosaic virus*. Conventions are as for potyvirus genome organization map (Fig. 2). Function of most gene products are postulated by analogy with genus *Potyvirus*. P1 corresponds to the C-terminal protease of HC-Pro

ANTIGENIC PROPERTIES

The viral proteins are moderately immunogenic; serological relationships exist among members except Barley mild mosaic virus (BaMMV). The CP aa sequence identity among members is 35-74%.

BIOLOGICAL PROPERTIES

CYTOLOGY

There are characteristic pinwheel-like inclusions and membranous network structures are formed in the cytoplasm of infected plant cells. No nuclear inclusions are found.

HOST RANGE

The host range of member viruses is narrow, restricted to the host family Graminae.

TRANSMISSION

These viruses are transmitted by *Polymyxa graminis* in a persistent manner, surviving in resting spores as long as these remain viable; transmissible experimentally by mechanical inoculation.

LIST OF SPECIES DEMARCATION CRITERIA IN THE GENUS

- Genome sequence relatedness.
  - CP aa sequence identity less than ca. 80%,
  - nt sequence identity less than 85% over whole genome,
  - different polyprotein cleavage sites.
- Natural host range.
  - host range may be related to species but usually not helpful in identifying species.
- Mode of transmission.
  - transmitted by *Polymyxa graminis.*
- Antigenic properties.
  - serological differences.

LIST OF SPECIES IN THE GENUS

Species names are in green italic script; strain names and synonyms are in black roman script; tentative species names are in blue roman script. Sequence accession numbers, and assigned abbreviations ( ) are also listed.

SPECIES IN THE GENUS

Barley mild mosaic virus		
Barley mild mosaic virus	[AF536942, AJ242725, L49381, D83408-9, X75933, X82625, X84802, X90904, Y10973-4,]	(BaMMV)
Barley yellow mosaic virus		

Barley yellow mosaic virus	[AF536958, AJ132268-9, AJ515479-85, D01091-2, D01099, X69757]	(BaYMV)
Oat mosaic virus		
Oat mosaic virus	[AJ306718-9]	(OMV)
Rice necrosis mosaic virus		
Rice necrosis mosaic virus		(RNMV)
Wheat spindle streak mosaic virus		
Wheat spindle streak mosaic virus		(WSSMV)
Wheat yellow mosaic virus		
Wheat yellow mosaic virus	[AF041041, AF067124, AJ131981-2, AJ239039, AJ242490, D86634-5]	(WYMV)

TENTATIVE SPECIES IN THE GENUS

None reported.

LIST OF UNASSIGNED SPECIES IN THE FAMILY

Species names are in green italic script; strain names and synonyms are in black roman script; tentative species names are in blue roman script. Sequence accession numbers, and assigned abbreviations ( ) are also listed.

Spartina mottle virus		
Spartina mottle virus	[AF491351-2]	(SpMoV)
Sugarcane streak mosaic virus		
Sugarcane streak mosaic virus	[AY189681, AY193783-4, U75456, Y17738]	(SCSMV)
Tomato mild mottle virus		
Tomato mild mottle virus	[AF359575]	(TomMMoV)

Three viruses have been assigned to the family *Potyviridae* on the basis of similarity to recognized members of the family. TomMMoV, SpMoV and SCSMV have CP aa sequence similarity to other members of the family *Potyviridae*, but are sufficiently different from other members of the family to suggest that they may represent new genera. TomMMoV is most closely related to members of the genus *Ipomovirus*, but differs from other members of this genus on the basis of vector relationships; ipomoviruses are transmitted by the whitefly *Bemisia tabaci* while TomMMoV is transmitted in a non-circulative manner by aphids. No vectors have been identified for SpMoV or SCSMV.

PHYLOGENETIC RELATIONSHIPS WITHIN THE FAMILY

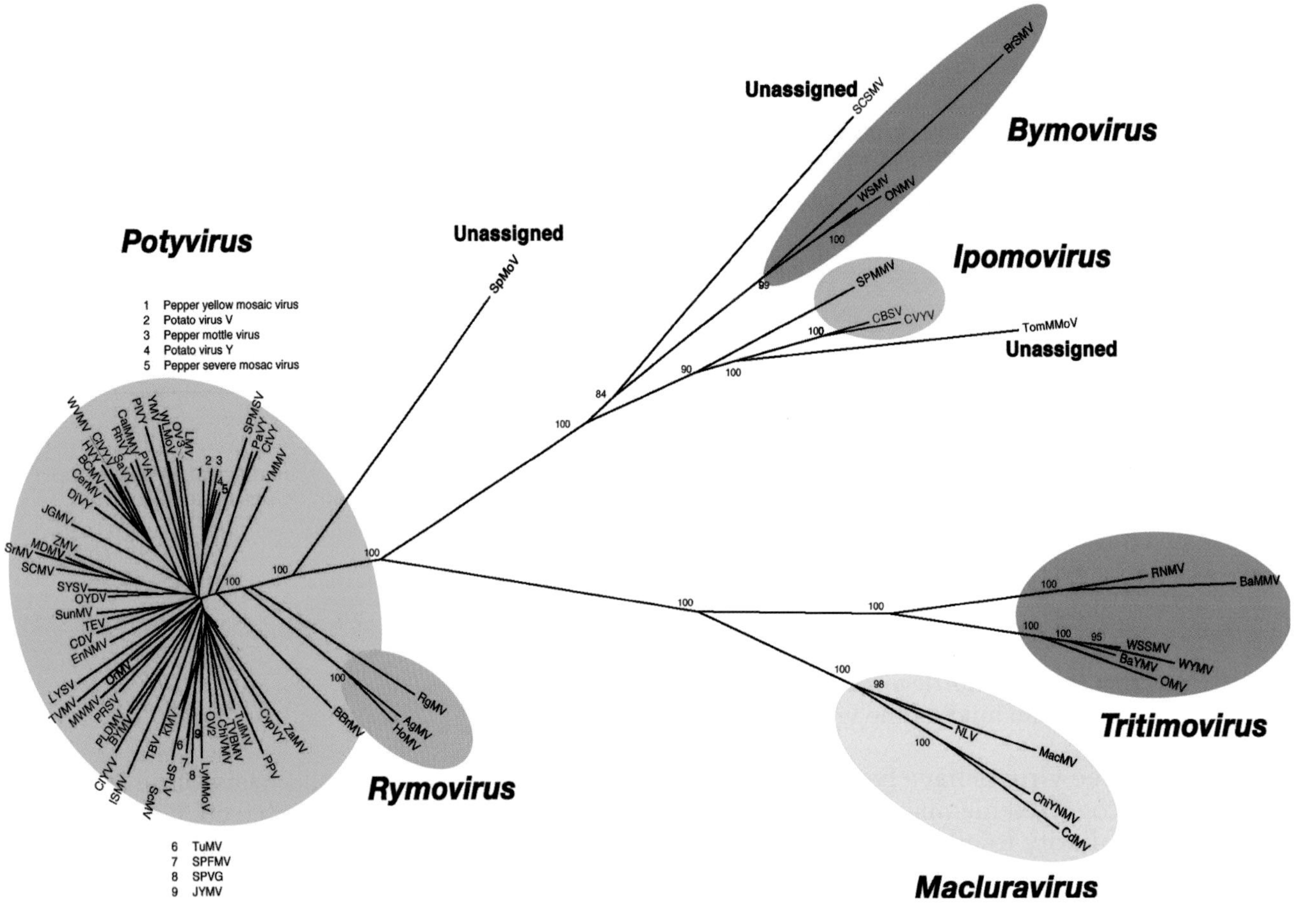

Figure 8: Phylogenetic tree for the family *Potyviridae* using aa sequence relationships of the CP. Inference based on the Fitch-Margoliash (1967) least squares method. The sequences were aligned using PILEUP (Devereux *et al.*, 1984). Branch lengths are proportional to sequence distances. The dendrogram was bootstrapped 1000 times (percent scores shown at nodes; bootstrap values for genus *Potyvirus* not shown), and the tree is not rooted. Many branches are condensed and may contain multiple viruses, e.g., BCMV, for example, includes BCMNV, CABMV, DMV, PWV, SAPV, SMV, WMV and ZYMV, AzMV, BlCMV, DeMV, and PStV. Signification of abbreviations are in the "List of Species".

SIMILARITY WITH OTHER TAXA

Viruses in the family *Potyviridae* have similarity to members of the families *Comoviridae*, *Picornaviridae*, and *Hypoviridae*. Genomes of member viruses of these taxa are positive sense ssRNAs, except those of the family *Hypoviridae*, which have dsRNA genomes. Most have a VPg at their 5'-termini and a poly(A) tract at their 3'-termini. Their genomes are expressed initially as high molecular weight polyprotein precursors which are processed by virus-encoded proteases. Gene products involved in replication are conserved in gene order and gene sequence.

DERIVATION OF NAMES

Bymo: sigla from *B*arley *y*ellow *mo*saic.
Ipomo: sigla from *Ipom*ea and *mo*saic.
Maclura: sigla from host genus name "*Maclura*".
Poty: sigla from *Pot*ato virus *Y*.
Rymo: siglafrom *Ry*egrass *mo*saic.
Tritimo: sigla from *Triti*cum and *mo*saic.

REFERENCES

Berger, P.H., Wyatt, S.D., Shiel, P.J., Silbernagel, M.J., Druffel, K. and Mink, G.I. (1997). Phylogenetic analysis of the *Potyviridae* with emphasis on legume-infecting potyviruses. *Arch. Virol.,* **142**, 1979-1999.

Badge, J., Robinson, D.J., Brunt, A.A. and Foster, G.D. (1997). 3'-terminal sequences of the RNA genomes of narcissus latent and Maclura mosaic viruses suggest that they represent a new genus of the *Potyviridae*. *J. Gen. Virol.,* **78**, 253-257.

Barnett, O.W. (ed)(1992). *Potyvirus taxonomy*. Springer-Verlag, Vienna.

Colinet, D., Kummert, J. and Lepoivre, P. (1998). The nucleotide sequence and genome organization of the whitefly transmitted sweet potato mild mottle virus: A close relationship with members of the family *Potyviridae*. *Virus Res.,* **53**, 187-196.

Dougherty, W.G. and Carrington, J.C. (1988). Expression and function of potyviral gene products. *Ann. Rev. Phytopathol.,* **26**, 123-143.

Dougherty, W.G. and Parks, T.D. (1991). Post-translational processing of the tobacco etch virus 49-kDa small nuclear inclusion polyprotein: identification of an internal cleavage site and delimitation of VPg and proteinase domains. *Virology,* **183**, 449-456.

Edwardson, J.R. and Christie, R.G. (eds)(1991). *The potyvirus group*. Vol I-IV. Univ Florida, Agric Exp Stn. Mono 16, Gainesville Florida.

Götz, R., Huth, W., Lesemann, D.-E. and Maiss, E. (2002). Molecular and serological relationships of Spartina mottle virus (SpMV) strains from *Spartina* spec. and from *Cynodon dactylon* to other members of the *Potyviridae*. *Arch. Virol.,* **147**, 379–391

Götz, R. and Maiss, E. (1995). The complete nucleotide sequence and genome organization of the mite-transmitted brome streak mosaic rymovirus in comparison with those of potyviruses. *J. Gen. Virol.,* **76**, 2035-2042

Hall, J.S., Adams, B., Parsons, T.J., French, R., Lane, L.C. and Jensen, S.G. (1998). Molecular cloning, sequencing, and phylogenetic relationships of a new potyvirus: sugarcane streak mosaic virus, and a reevaluation of the classification of the *Potyviridae*. *Mol. Phylogenet. Evol.,* **10**, 323-332.

Jordan, R. and Hammond, J. (1991). Comparison and differentiation of potyvirus isolates and identification of strain-, virus-, subgroup-specific and potyvirus group-common epitopes using monoclonal antibodies. *J. Gen. Virol.,* **72**, 25-36.

Kashiwazaki, S., Minobe, Y. and Hibino, H. (1991). Nucleotide sequence of barley yellow mosaic virus RNA 2. *J. Gen. Virol.,* **72**, 995-999.

Kashiwazaki, S., Minobe, Y., Omura, T. and Hibino, H. (1990). Nucleotide sequence of barley yellow mosaic virus RNA 1: a close evolutionary relationship with potyviruses. *J. Gen. Virol.,* **71**, 2781-2790.

Monger, W.A., Spence, N.J. and Foster, G.D. (2002). Molecular evidence that the aphid-transmitted *Tomato mild mottle virus* belongs to the *Potyviridae* family but not the *Potyvirus* genus. *Arch. Virol.,* **146,** 2345-2441.

Revers, F., Le Gall, O., Candresse, T. and Maule, A. J. (1999). New advances in understanding the molecular biology of plant/potyvirus interactions. *MPMI,* **12**, 367-76.

Salm, S.N., Rey, M.E.C. and Rybicki, E.P. (1996). Phylogenetic justification for splitting the *Rymovirus* genus of the taxonomic family *Potyviridae*. *Arch. Virol.,* **141**, 2237-2242.

Shukla, D.D., Ward, C.W. and Brunt, A.A. (eds)(1994). *The Potyviridae*. CAB International, Wallingford, UK.

Stenger, D.C., Hall, J.S., Choi, I.R. and French, R. (1998). Phylogenetic relationships within the family *Potyviridae*: Wheat streak mosaic virus and brome streak mosaic virus are not members of the genus *Rymovirus*. *Phytopathology,* **88**, 782-787.

CONTRIBUTED BY

Berger, P.H., Adams, M.J., Barnett, O.W., Brunt, A.A., Hammond, J., Hill, J.H., Jordan, R.L., Kashiwazaki, S., Rybicki, E., Spence, N., Stenger, D.C., Ohki, S.T., Uyeda, I., van Zaayen, A. Valkonen, J. and Vetten, H.J.

FAMILY CALICIVIRIDAE

TAXONOMIC STRUCTURE OF THE FAMILY

Family	***Caliciviridae***
Genus	***Lagovirus***
Genus	***Norovirus***
Genus	***Sapovirus***
Genus	***Vesivirus***

VIRION PROPERTIES

MORPHOLOGY

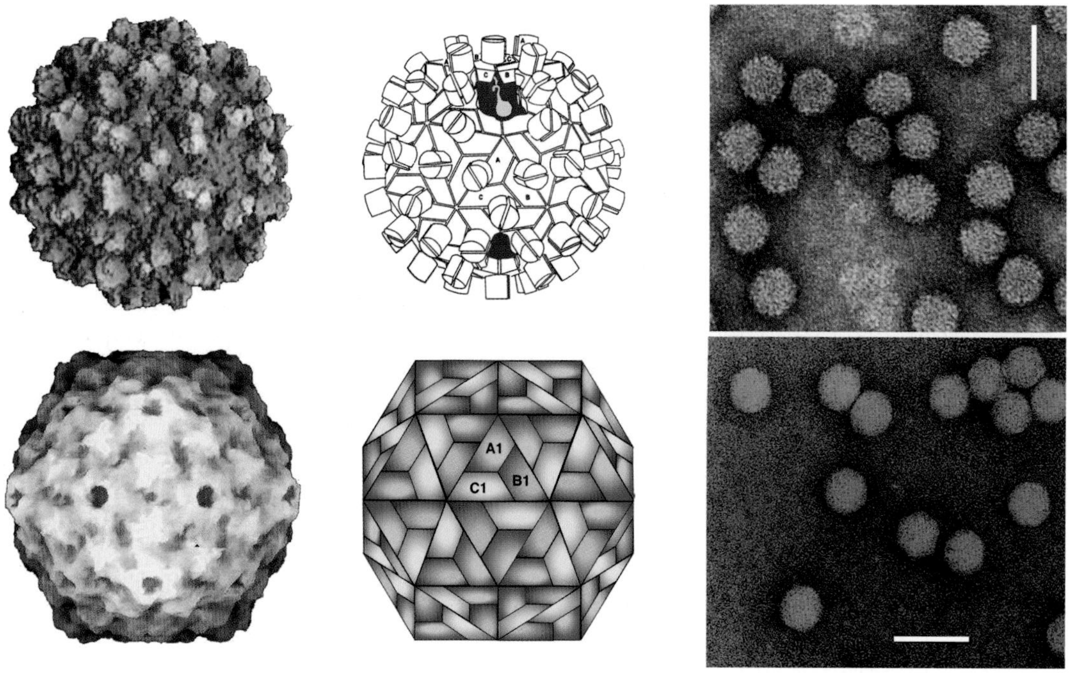

Figure 1: (Top left) Cryo-image reconstruction of recombinant Norwalk virus (NV)-like particles (rNV VLPs). (Top central) Cryo-image reconstruction of Primate calicivirus. A set of icosahedral 5- and 3-fold axes is marked. (Courtesy of Prasad, B.V.V.). (Top right) Central cross-section of rNV VLPs. (Bottom left) Electronic rendering of Norwalk virus (Prasad et al.,1999). (Bottom center) Diagram representing a T=3 icosahedral structure. (Bottom right) Negative stain electron micrographs of Bovine calicivirus particles (Courtesy of S. McNulty). The bar represents 100 nm.

Virions are nonenveloped with icosahedral symmetry. They are 27-40 nm in diameter by negative stain electron microscopy and 35-40 nm by electron cryo-microscopy. The capsid is composed of 90 dimers of the major structural protein arranged on a T=3 icosahedral lattice. A characteristic feature of the capsid architecture is the 32 cup-shaped depressions at each of the icosahedral 5-fold and 3-fold axes. In some negative stain virus preparations, the cup-shaped depressions appear distinct and well-defined, while in others, these depressions are less prominent (Fig. 1).

PHYSICOCHEMICAL AND PHYSICAL PROPERTIES

Virion Mr is ~15×10^6. Virion buoyant density is 1.33-1.41 g/cm^3 in CsCl and 1.29 g/cm^3 in glycerol-potassium tartrate gradients. Virion S_{20w} is 170-187S. Physicochemical properties have not been fully established for all members of the family. Rabbit hemorrhagic disease virus (RHDV) in the genus *Lagovirus* has been reported as stable over a wide range of pH values (4.-10.5). The genus *Norovirus* has been shown in studies with one of its members (Norwalk virus, NV) to be acid, ether, and relatively heat stable. For strains examined in the genus *Vesivirus*, inactivation occurs at pH 3-5, thermal inactivation is accelerated in high concentrations of Mg^{++} ions, and virions are insensitive to treatment with ether, chloroform, or mild detergents.

Nucleic Acid

The genome consists of a linear, positive sense, ssRNA molecule of 7.4–8.3 kb. A protein (VPg, 10-15 kDa) is covalently attached to the 5'-end of the genomic RNA and the 3′-end is polyadenylated. SgRNA (2.2-2.4 kb) is synthesized intracellularly and is VPg-linked in RHDV and Feline calicivirus (FCV). The FCV sgRNA can be packaged into viral particles with lower density than the particles with the full-length genome. The gene order for RHDV and FCV was determined by *in vitro* translation studies and cleavage mapping, respectively, as 5′-p16-p23-p37(helicase)-p30-VPg-protease-polymerase-VP60 (major CP)-VP10-3′ (RHDV), and 5′-p5.6-p32-p39 (helicase)- p30-p13 (VPg)-p76 (Pro-Pol)-VP62 (major CP)-VP8.5 (minor CP).

Proteins

Virions are constructed predominantly from one major species of CP (58-60 kDa). A second minor structural protein (8.5-23 kDa) has been found in association with FCV, NV, and RHDV virions. Nonstructural proteins have homology with those of the family *Picornaviridae* replicative enzymes and include 2C helicase, 3C cysteine protease, and 3D RdRp domains. The calicivirus VPg (10-15 kDa) is covalently linked to the viral RNA and maps to the region of the calicivirus genome analogous to the 3B region of picornaviruses, but has no apparent amino acid homology with those of picornaviruses. Mapping studies are in progress to establish precursor and product relationships of the calicivirus nonstructural proteins.

Lipids

None reported.

Carbohydrates

None reported.

Genome Organization and Replication

The positive strand genomic RNA is organized into either two or three major ORFs. The nonstructural proteins are encoded in the 5′-end of the genome and the structural proteins in the 3′-end. Replication occurs in the cytoplasm and two major positive-sense RNA species are found in infected cells. The genome-sized positive-sense RNA serves as the template for translation of a large polyprotein that undergoes cleavage by a virus-encoded protease to form the mature nonstructural proteins. A subgenomic-sized positive strand RNA, co-terminal with the 3′-end of the genome, is the template for translation of the major viral CP as well as the 3′-terminal ORF product. A dsRNA corresponding in size to full-length genomic RNA has been identified in FCV and San Miguel sea lion virus (SMSV)-infected cells, indicating that replication occurs via a negative strand intermediate (Fig. 2).

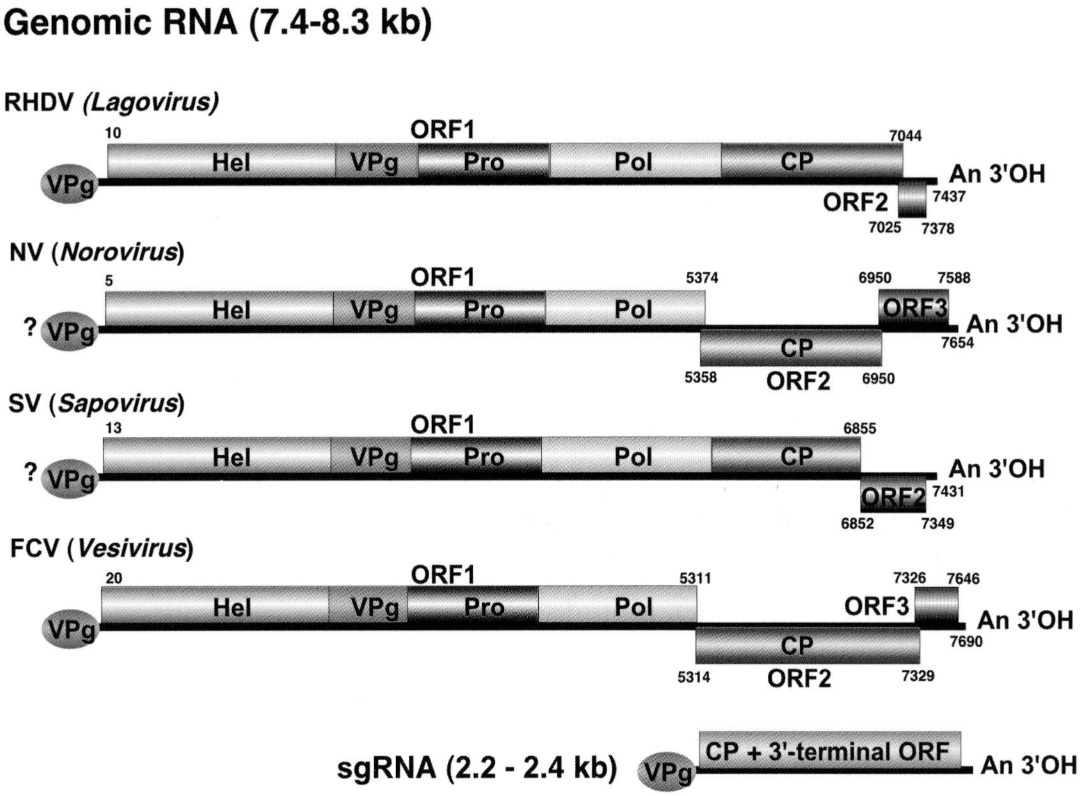

Figure 2: Genome organizations of viruses of the family *Caliciviridae*. The genomic organization and ORF usage are shown for representative species (with strain indication shown in brackets) in the following genera: *Lagovirus:* Rabbit hemorrhagic disease virus (RHDV) [Ra/LV/RHDV/GH/1988/GE]; *Norovirus*: Norwalk virus (NV) [Hu/NV/Nor/1968/US]; *Sapovirus*: Sapporo virus (SV) [Hu/SV/Man/1993/UK]; and *Vesivirus*: Feline calicivirus (FCV) [Fe/VV/FCV/F9/1958/US. Viruses in two genera (*Lagovirus* and *Sapovirus)* contain a large ORF1 in which the nonstructural polyprotein gene is continuous and in frame with the CP coding sequence. Some strains in the genus *Sapovirus* encode a third predicted ORF that overlaps ORF1 (not shown). Viruses in the other two genera (*Norovirus* and *Vesivirus*) encode the major structural CP in a separate reading frame (ORF2). The RNA helicase (HEL), protease (PRO), and polymerase (POL) regions of the genome are indicated. The linkage of VPg to the RNA of viruses in the genera *Norovirus* and *Sapovirus* has not been confirmed. The designated VPg region of the genomes of representative viruses from the genera *Norovirus*, *Sapovirus*, and *Vesivirus* is shown by homology with the mapped VPg of RHDV. The shaded region of the ORF2 of the representative member of the genus *Vesivirus* illustrates the leader sequence (approximately 125 aa in length) of the precursor CP. Studies of FCV and RHDV have identified two major positive sense RNA molecules in infected cells. One RNA molecule corresponds in size to the full-length genome and the other, a subgenomic-sized RNA, is co-terminal with the 3′-end of the genome. The sgRNA is the template for translation of the major viral CP and the 3′-terminal ORF product that has been identified as a minor structural protein in FCV.

Antigenic Properties

Cross-challenge studies in the natural host and experiments with monoclonal antibodies indicate that RHDV and European brown hare syndrome virus (EBHSV) are antigenically distinct. Antigenic types have been defined by cross-challenge studies, immune electron microscopy or solid phase immune electron microscopy for noncultivatable strains in the genera *Norovirus* and *Sapovirus*. Numerous serotypes have been established by neutralization for Vesicular exanthema of swine virus (VESV) and SMSV strains. One serotype has been described for FCV strains, but considerable antigenic variation within this serotype has been reported. Recombinant virus-like particles (rVLPs) have been generated by expression of the major calicivirus structural CP in baculovirus and plant expression systems. These VLPs are highly immunogenic and similar in antigenicity to native virions.

Biological Properties

Caliciviruses infect a broad range of animals that includes hares, rabbits, pigs, cats, pinnipeds, mice, cattle, reptiles, skunks, cetaceans, chimpanzees, and humans. Although individual calicivirus species generally exhibit a natural host restriction, the VESV species of the genus *Vesivirus* is an exception, showing a broad host range. For example, VESV-like viruses have been isolated from several marine animal species (including fish), birds, reptiles, and land mammals. The geographic distribution of each calicivirus species usually reflects the host distribution.

Transmission is via contaminated food, water, fomites, and on occasion via aerosolization of fecal material, vomitus or respiratory secretions. In general, no vectors appear to be involved in transmission; however, mechanical arthropod vector transmission of RHDV has been described.

Caliciviruses are associated with a number of disease syndromes. RHDV is associated with a generalized viremic infection in which there is massive liver necrosis that triggers a disseminated intravascular coagulation and rapid death in rabbits greater than three months of age. A nonvirulent RHDV strain has been described. EBHSV is similar to RHDV but appears to be less virulent. Human caliciviruses in the genera *Norovirus* and *Sapovirus* induce a generally self-limited gastroenteritis with symptoms that may include nausea, diarrhea, vomiting, abdominal cramping, fever, and malaise. VESV produces clinical signs in swine that are sometimes indistinguishable from foot-and-mouth disease, including vesicles in the mouth, tongue, lips, snout and feet between the digits. In addition, the virus may cause encephalitis, myocarditis, fever, diarrhea, abortion and failure of infected animals to thrive. SMSV is similar to VESV, although there have been limited studies of natural infection in the marine host. Primate calicivirus causes mucosal vesiculation and persistent infection. FCV is associated in cats with conjunctivitis, rhinitis, pneumonia, mucosal vesiculation, diarrhea, urinary tract infection, and paresis and can produce a persistent infection with virus latent in the tonsils.

Genus Lagovirus

Type Species *Rabbit hemorrhagic disease virus*

Distinguishing Features

The strains in this genus form a distinct phylogenetic clade within the family. The genome is organized into two major ORFs. ORF1 encodes the nonstructural polyprotein, with the major structural CP gene (VP60) in frame with the nonstructural polyprotein coding sequence. ORF2 overlaps ORF1 by 17 nt in the RHDV genome and 5 nt in the EBHSV genome. The ORF2 encodes a small protein (VP10) of unknown function that has been identified as a minor structural component in the RHDV virion. These viruses have characteristically been associated with infection in rabbits and hares (lagomorphs), and can cause epidemics with high mortality in these animals.

List of Species Demarcation Criteria in the Genus

Members of a *Lagovirus* species share :

- a major phylogenetic branch within the genus,
- a common genome layout,
- greater than 80% aa identity in the CP,
- a natural host range,
- cross-protection antigens.

List of Species in the Genus

Species names are in green italic script; strain names and synonyms are in black roman script; tentative species names are in blue roman script. Sequence accession numbers, and assigned abbreviations ( ) are also listed.

Species in the Genus

European brown hare syndrome virus		
EBHSV-BS89	[X98002]	(Ha/LV/EBHSV/BS89/1989/IT)
EBHSV-FRG	[U09199]	(Ha/LV/EBHSV/FRG/1989/GE)
EBHSV-GD	[Z69620]	(Ha/LV/EBHSV/GD/1989/FR)
EBHSV-UK91	[U65372]	(Ha/LV/EBHSV/UK91/1991/UK)
Rabbit hemorrhagic disease virus		
Rabbit calicivirus	[X96868]	(Ra/LV/RHDV/RCV/1995/IT)
RHDV-AST89	[Z49271]	(Ra/LV/RHDV/AST89/1989/SP)
RHDV-BS89	[X87607]	(Ra/LV/RHDV/BS89/1989/IT)
RHDV-FRG	[M67473]	(Ra/LV/RHDV/GH/1988/GE)
RHDV-SD	[Z29514]	(Ra/LV/RHDV/SD/1989/FR)
RHDV-V351	[U54983]	(Ra/LV/RHDV/V351/1987/CK)

Tentative Species in the Genus

None reported.

Genus *Norovirus*

Type Species *Norwalk virus*

Distinguishing Features

The strains in this genus form a distinct phylogenetic clade within the family. The genome is organized into three major ORFs. ORF1 encodes the nonstructural polyprotein. ORF2 encodes the major structural CP and overlaps by 14 nt with ORF1 in the Norwalk and Southampton virus strains and by 17 nt in the Lordsdale virus strain, resulting in a -2 frameshift of ORF2 in all three viruses. ORF3 overlaps by one nt with ORF2 in a -1 frameshift and encodes a small virion-associated protein. Members of the genus often (but not always) have a less-defined surface structure when observed by negative stain electron microscopy, leading to the now historic descriptive term "small round structured viruses" for the viruses in this genus associated with epidemic gastroenteritis in humans.

List of Species Demarcation Criteria in the Genus

Not applicable. Additional characterization of the viruses in this genus will be required in order to delineate species criteria.

List of Species in the Genus

Species names are in green italic script; strain names and synonyms are in black roman script; tentative species names are in blue roman script. Sequence accession numbers, and assigned abbreviations ( ) are also listed.

Species in the Genus

Norwalk virus		
Desert Shield virus	[U04469]	(Hu/NV/DSV395/1990/SR)
Hawaii virus	[U07611]	(Hu/NV/HV/1971/US)
Lordsdale virus	[X86557]	(Hu/NV/LD/1993/UK)
Mexico virus	[U22498]	(Hu/NV/MX/1989/MX)
Norwalk virus	[M87661]	(Hu/NV/NV/1968/US)
Snow Mountain virus	[L23831]	(Hu/NV/SMV/1976/US)
Southampton virus	[L07418]	(Hu/NV/SHV/1991/UK)

TENTATIVE SPECIES IN THE GENUS

Bovine norovirus – CH126	[AF320625]	(Bo/NV/CH126/1980/DE)
Bovine norovirus - Jena	[AJ011099]	(Bo/NV/JV/1980/DE)
Human norovirus - Alphatron	[AF195847]	(Hu/NV/Alphatron/98-2/1998/NET)
Murine norovirus 1	[AY228235]	(Mu/NV/mouse1/2002/US)
Swine norovirus	[AB009412]	(Sw/NV/Sw43/1997/JA)

GENUS SAPOVIRUS

Type Species *Sapporo virus*

DISTINGUISHING FEATURES

The strains in this genus form a distinct phylogenetic clade within the family. The full-length genomic sequence is available for the Manchester virus, and for a porcine enteric calicivirus (PEC strain Cowden). The Manchester virus genome is organized into three major ORFs. ORF1 encodes the non-structural polyprotein, with the major structural CP gene in frame with the nonstructural polyprotein coding sequence. ORF2 overlaps ORF1 in a -1 frameshift and encodes a predicted small protein of unknown function. ORF3 begins 11 nt downstream from the predicted start codon of the CP in a +1 frameshift and encodes a predicted protein of ~160 aa. However, in certain strains of this genus, and in the PEC Cowden strain, ORF3 is absent. The viruses in this genus have characteristically been associated with sporadic outbreaks and cases of gastroenteritis in humans and diarrhea in pigs, and often (but not always) have distinct calicivirus cup-like morphology when observed by negative stain electron microscopy.

LIST OF SPECIES DEMARCATION CRITERIA IN THE GENUS

Not applicable. Additional characterization of the viruses in this genus will be required in order to delineate species criteria.

LIST OF SPECIES IN THE GENUS

Species names are in green italic script; strain names and synonyms are in black roman script; tentative species names are in blue roman script. Sequence accession numbers, and assigned abbreviations ( ) are also listed.

SPECIES IN THE GENUS

Sapporo virus

Sapporo virus	[U65427]	(Hu/SV/SV/1982/JA)
Sapporo virus Houston/86	[U95643]	(Hu/SV/Hou/1986/US)
Sapporo virus Houston/90	[U95644]	(Hu/SV/Hou 27/1990/US)
Sapporo virus London 29845	[U95645]	(Hu/SV/Lon 29845/1992/UK)
Sapporo virus Manchester virus	[X86560]	(Hu/SV/Man/1993/UK)
Sapporo virus Parkville virus	[U73124]	(Hu/SV/Park/1994/US)

TENTATIVE SPECIES IN THE GENUS

Porcine enteric sapovirus - Cowden	[AF182760]	(Sw/SV/Cowden/1980/US)
Mink enteric sapovirus	[AF338404]	(Mi/SV/MEC/1999/US)

GENUS VESIVIRUS

Type Species *Vesicular exanthema of swine virus*

DISTINGUISHING FEATURES

The species in this genus form a distinct phylogenetic clade within the family. The genome is organized into three major ORFs. ORF1 encodes the nonstructural polyprotein.

ORF2 encodes the major structural CP that is translated as a larger precursor protein before cleavage into the mature CP, a feature that appears unique to this genus. ORF1 and ORF2 of the viruses in this genus are separated by either 2 nt (GC for FCV strains) or 5 nt (CCACT/C for SMSV and VESV strains). A third ORF (ORF3) encodes a small, basic protein of unknown function and overlaps by one nt with ORF2 in a -1 frameshift. The ORF3 product has been detected in FCV-infected cells. Most members of this genus can be readily propagated in cell culture, which contrasts with viruses in the other three genera, none of which has been cultured in conventional cell culture systems. FCV grows most efficiently in cells of feline origin; *in vivo*, the primary site of replication is the upper respiratory tract. VESV isolates grow in a number of cell lines and infect a broad range of hosts, with vesicles of the skin a prevalent disease symptom with the possible exception of the canine caliciviruses. The RNA genomes of VESV and FCV are infectious, as is synthetic RNA derived from a full-length cDNA clone of the FCV genome.

List of Species Demarcation Criteria in the Genus

Members of a *Vesivirus* species share :

- a major phylogenetic branch within the genus,
- a common genome layout,
- greater than 60% aa identity in the CP.

List of Species in the Genus

Species names are in green italic script; strain names and synonyms are in black roman script; tentative species names are in blue roman script. Sequence accession numbers, and assigned abbreviations ( ) are also listed.

Species in the Genus

Feline calicivirus		
Feline calicivirus CFI/68	[U13992]	(Fe/VV/FCV/CFI/1968/US)
Feline calicivirus F9	[M86379]	(Fe/VV/FCV/F9/1958/US)
Vesicular exanthema of swine virus		
Bovine calicivirus	[U18741]	(Bo/VV/VESV/Bos-1/1981/US)
Cetacean calicivirus	[U52091]	(Ce/VV/VESV/Tur-1/1977/US)
Primate calicivirus	[U52086]	(Pr/VV/VESV/Pan-1/1979/US)
Reptile calicivirus	[U52092]	(Re/VV/VESV/Cro-1/1978/US)
San Miguel sea lion virus, serotype 1	[M87481]	(Pi/VV/VESV/SMSV-1/1972/US)
San Miguel sea lion virus, serotype 4	[M87482]	(Pi/VV/VESV/SMSV-4/1973/US)
San Miguel sea lion virus, serotype 17	[U52005]	(Pi/VV/VESV/SMSV-17/1991/US)
Skunk calicivirus	[U14667]	(Pi/VV/VESV/SCV/1992/US)
Vesicular exanthema of swine virus-A48	[U76874]	(Sw/VV/VESV/A48/1948/US)

Tentative Species in the Genus

Mink calicivirus	[AF338407]	(Mi/VV/MCV 20/1980/US)

List of Unassigned Viruses in the Family

Amyelosis chronic stunt virus (insects)	(ACSV)
Bovine enteric calicivirus strain NB	(BEC-NB)
Canine calicivirus	(CaCV)
Fowl calicivirus	(FCV)
Walrus calicivirus	(WCV)

Pos. ssRNA

Phylogenetic Relationships within the Family

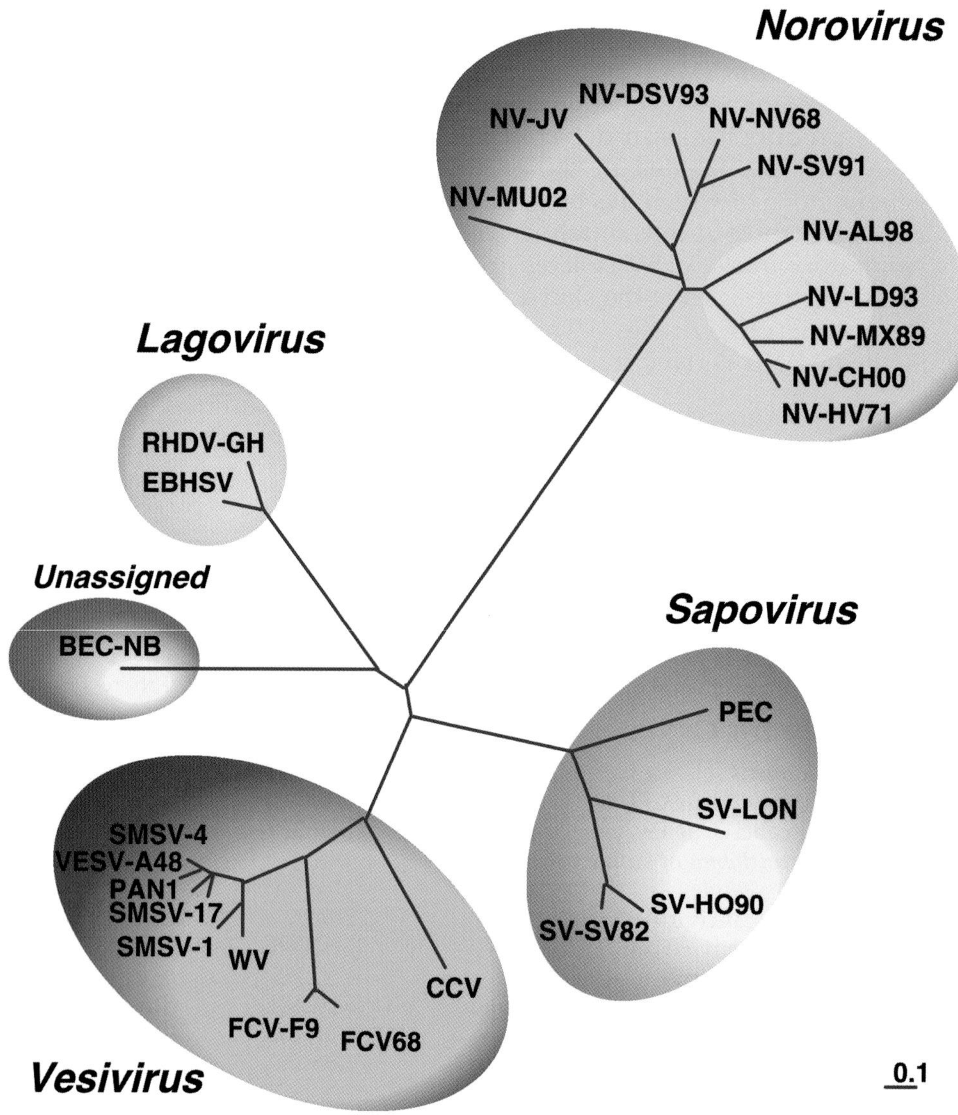

Figure 3: Phylogenetic relationships within the family *Caliciviridae* and comparison with the family *Picornaviridae* (Courtesy of E. van Strien and H. Vennema). Full-length capsid aa sequences were used for the phylogenetic analysis and included representative strains from each genus in the family *Caliciviridae*. Clustal W analysis was used to create a multiple alignment for the aa sequences, which were verified by alignment of known motifs in the region (e.g., PPG/N). The PHYLIP v3.5c package was used for the phylogenetic analyses of the aligned nt sequences. The multiple alignment data were bootstrapped (N = 1000) and submitted for the distance method. The unrooted phylogenetic tree is presented. Scale bar represents units for expected number of substitutions per site. The following numbers represent the phylogenetic distances, in units of expected number of substitutions per site, among the calicivirus genera (representing distinct clades) as measured by the distances between the first branchpoints of: *Norovirus* and *Sapovirus*, 1.59; *Norovirus* and *Lagovirus*, 1.48; *Norovirus* and *Vesivirus*, 1.46; *Sapovirus* and *Lagovirus*, 1.21; *Sapovirus* and *Vesivirus*, 0.84; and *Lagovirus* and *Vesivirus*, 1.07. A fifth branch is constituted by the Bovine enteric calicivirus strain NB, which has not yet been assigned. The BEC-NB is at present the sole representative of this potential new genus. Genbank accession numbers for the strains in this analysis were: AB032758, AF053720, AF091736, AF182760, AF195847, AJ011099, AY082890, AY228235, L07418, M67473, M86379, M87481, M87482, M87661, NC_004541, U04469, U07611, U13992, U22498, U52005, U65427, U76874, U95644, U95645, X86557, Z69620.

Similarity with Other Taxa

Caliciviruses have some properties similar to the viruses of the families *Picornaviridae*, *Potyviridae* and *Comoviridae* relative to the presence of a VPg at the 5′-end and a poly(A) tract at the 3′-end of the positive sense ssRNA genome. The putative viral replicase of caliciviruses shares sequence homology with that of picornaviruses.

Derivation of Names

Calici: from Latin calix, "cup" or "goblet", from cup-shaped depressions on the virion surface observed by electron microscopy.
Lago: from *Lago*morpha, the mammalian host order for the prototype strain rabbit hemorrhagic disease virus.
Noro: modified from the type species name, Norwalk virus
Sapo: modified from the type species name, Sapporo virus
Vesi: from the type species name, *vesi*cular exanthema of swine virus.

References

Berke, T., Golding, B., Jiang, X., Cubitt, D.W., Wolfaardt, M., Smith, A.W. and Matson, D.O. (1997). Phylogenetic analysis of the caliciviruses. *J. Med. Virol.*, **52**, 419-424.

Carter, M.J., Milton, I.D., Meanger, J., Bennett, M., Gaskell, R.M. and Turner, P.C. (1992). The complete nucleotide sequence of a feline calicivirus. *Virology*, **190**, 443-448.

Green K.Y., Chancock R.M. and Kapikian A.Z. (2001). Human caliciviruses. In: *Fields Virology*, 4th ed., (D.M. Knipe and P.M. Howley, eds), pp 841-874. Lippincott Williams and Wilkins, Philadelphia.

Guo, M., Chang, K.O., Hardy, M.E., Zhang, Q., Parwani, A.V. and Saif, L.J. (1999). Molecular characterization of a porcine enteric calicivirus genetically related to Sapporo-like human caliciviruses. *J. Virol.*, **73**, 9625-9531.

Guo, M., Evermann, J.F. and Saif, L.J. (2001). Detection and molecular characterization of cultivable caliciviruses from clinically normal mink and enteric caliciviruses associated with diarrhea in mink. *Arch Virol.*, **146,** 479-493.

Jiang, X., Wang, M., Wang, K. and Estes, M.K. (1993). Sequence and genome organization of Norwalk virus. *Virology*, **195**, 51-61.

Lambden, P.R., Caul, E.O., Ashley, C.R. and Clarke, I.N. (1993). Sequence and genome organization of a human small round-structured (Norwalk-like) virus. *Science*, **259**, 516-519.

le Gall, G., Huguet, S., Vende, P., Vautherot, J.-F. and Rasschaert, D. (1996). European brown hare syndrome virus: molecular cloning and sequencing of the genome. *J. Gen. Virol.*, **77**, 1693-1697.

Liu, B.L., Clarke, I.N., Caul, E.O. and Lambden, P.R. (1995). Human enteric caliciviruses have a unique genome structure and are distinct from the Norwalk-like viruses. *Arch. Virol.*, **140**, 1345-1356.

Liu, B.L., Lambden, P.R., Gunther, H., Otto, P., Elschner, M. and Clarke, I.N. (1999). Molecular characterization of a bovine enteric calicivirus: relationship to the Norwalk-like viruses. *J. Virol.*, **73**, 819-825.

Meyers, G., Wirblich, C. and Thiel, H.-J. (1991). Rabbit hemorrhagic disease virus-molecular cloning and nucleotide sequence of a calicivirus genome. *Virology*, **184**, 664-676.

Neill, J.D., Meyer, R.F. and Seal, B.S. (1995). Genetic relatedness of the caliciviruses: San Miguel sea lion and vesicular exanthema of swine viruses constitute a single genotype within the *Caliciviridae*. *J. Virol.*, **69**, 4484-4488.

Neill, J.D. (2002). The subgenomic RNA of feline calicivirus is packaged into viral particles during infection. *Virus Res.*, **87**, 89-93.

Noel, J.S., Liu, B.L., Humphrey, C.D., Rodriguez, E.M., Lambden, P.R., Clarke, I.N., Dwyer, D.M., Ando, T., Glass, R.I. and Monroe, S.S. (1997). Parkville virus: a novel genetic variant of human calicivirus in the Sapporo virus clade, associated with an outbreak of gastroenteritis in adults. *J. Med. Virol.*, **52**, 173-178.

Numata, K., Hardy, M.E., Nakata, S., Chiba, S. and Estes, M.K. (1997). Molecular characterization of morphologically typical human calicivirus Sapporo. *Arch. Virol.*, **142**, 1537-1552.

Prasad, B.V., Hardy, M.E., Dokland, T., Bella, J., Rossmann, M.G. and Estes, M.K. (1999). X-ray crystallographic structure of the Norwalk virus capsid. *Science*, **286**, 287-290.

Wirblich, C., Meyers, G., Ohlinger V.F., Capucci, L., Eskens, U., Haas, B. and Thiel, H.-J. (1994). European brown hare syndrome virus: relationship to rabbit hemorrhagic disease virus and other caliciviruses. *J. Virol.*, **68**, 5164-5173.

Contributed By

Koopmans, M.K., Green, K.Y., Ando, T., Clarke, I.N., Estes, M.K., Matson, D.O., Nakata, S., Neill, J.D., Smith, A.W., Studdert, M.J. and Thiel, H.-J.

GENUS *HEPEVIRUS*

Type Species *Hepatitis E virus*

VIRION PROPERTIES

MORPHOLOGY

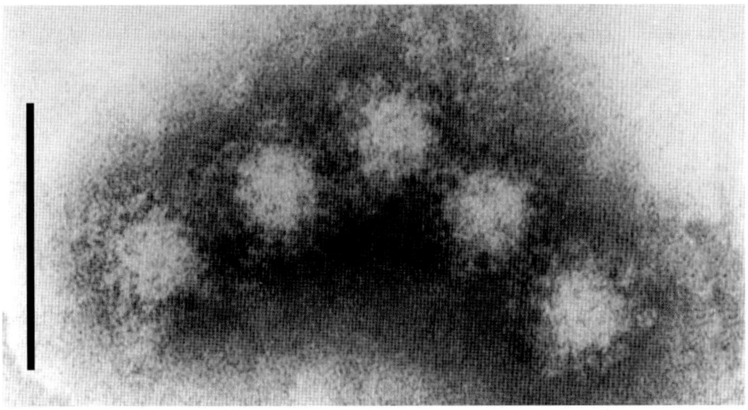

Figure 1: Negative contrast electron micrograph of virions of an isolate of *Hepatitis E virus*, in the bile fluid from a monkey challenged with the Mexico strain of human *Hepatitis E virus*. The bar represents 100 nm. (From Ticehurst *et al.*, 1992, with permission).

Virions (27-34 nm) are icosahedral and non-enveloped.

PHYSICOCHEMICAL AND PHYSICAL PROPERTIES

Virion buoyant density is 1.35 g/cm^3 in CsCl and 1.29 g/cm^3 in glycerol potassium tartrate gradients. Virion S_{20w} is 183S.

NUCLEIC ACID

The genome is a positive sense ssRNA of approximately 7.2 kb, with a 5' m^7G cap and a 3'- poly(A) tail.

PROTEINS

Virions are constructed from a major CP (72 kDa) which may be proteolytically processed: its size in the virion is unknown. A small immunoreactive protein (1.5 kDa) of unknown function has been identified. Non-structural proteins have limited similarity with the "alpha-like supergroup" of viruses and include domains consistent with a Mtr, RNA helicase, papain-like cysteine protease and RdRp. RdRp and Mtr/guanlytransferase activities have been demonstrated for ORF1 recombinant proteins.

LIPIDS

None reported.

CARBOHYDRATES

Evidence for glycosylation of the major CP in mammalian expression studies has been reported, but the biological significance is unknown.

GENOME ORGANIZATION AND REPLICATION

The RNA genome of Hepatitis E virus (HEV) is organized into three ORFs, with the non-structural proteins encoded toward the 5'-end of the genome and the structural protein(s) toward the 3'-end. Capped genomic RNA is infectious for rhesus monkeys and chimpanzees. The 5'-NTR is only ~26 nt long. The 3'-NTR contains a *cis*-reactive element. ORF1 encodes the non-structural polyprotein. ORF2 encodes the major CP. A third ORF overlaps ORF1 and ORF2 and encodes a small phosphoprotein (123 aa) of unknown function (Fig. 2). ORF2 and ORF3 are thought to be translated from subgenomic messenger RNAs.

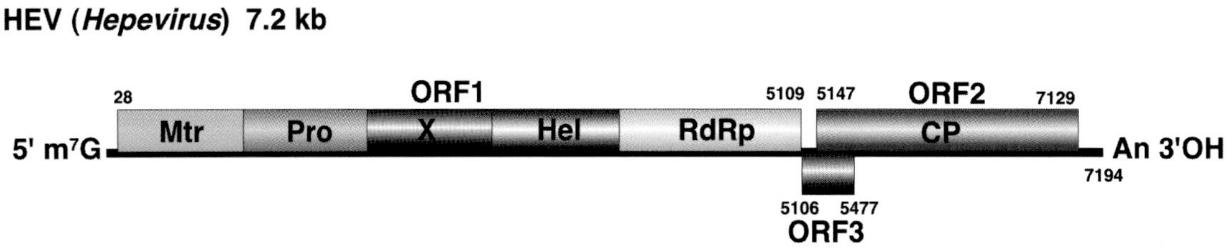

Figure 2: Genome organization of Hepatitis E virus (HEV) (human strain Burma, M73218). The putative MTR, Pro, "X", Hel, and RdRp domains are indicated.

Antigenic Properties

A single serotype has been described, with extensive cross-reactivity among circulating human and swine strains. Antibodies cross-reactive with ORF2 epitopes of human strains have been identified in numerous species of rodents and other mammals but the putative viruses have not been characterized. A distantly related virus that infects, and causes hepatitis in, chickens also cross-reacts serologically.

Biological Properties

HEV is associated in humans with outbreaks and sporadic cases of enterically transmitted acute hepatitis. The virus is considered endemic in tropical and subtropical countries of Asia and Africa, as well as in Mexico, but antibody prevalence studies suggest global distribution of this virus, perhaps in a non-pathogenic form. Antibody prevalence studies have found evidence for HEV or a related agent in animals and there is speculation that the virus may be zoonotic. HEV strains have been identified in swine and are closely related antigenically and genetically to human HEV genotype 3 and 4 strains. Interspecies transmission of genotype 3 HEV between swine and primates has been experimentally demonstrated but direct evidence for natural interspecies transmission has not been obtained.

List of Species Demarcation Criteria in the Genus

Not applicable.

List of Species in the Genus

Species names are in green italic script; strain names and synonyms are in black roman script; tentative species names are in blue roman script. Sequence accession numbers, and assigned abbreviations ( ) are also listed.

Species in the Genus

Hepatitis E virus		
Hepatitis E virus 1 (Burma)	[M73218]	(HEV-1)
Hepatitis E virus 2 (Mexico)	[M74506]	(HEV-2)
Hepatitis E virus 3 (Meng)	[AF082843]	(HEV-3)
Hepatitis E virus 4	[AJ272108]	(HEV-4)

Tentative Species in the Genus

Avian hepatitis E virus	[AY043166]	(AHEV)

PHYLOGENETIC RELATIONSHIPS WITHIN THE GENUS

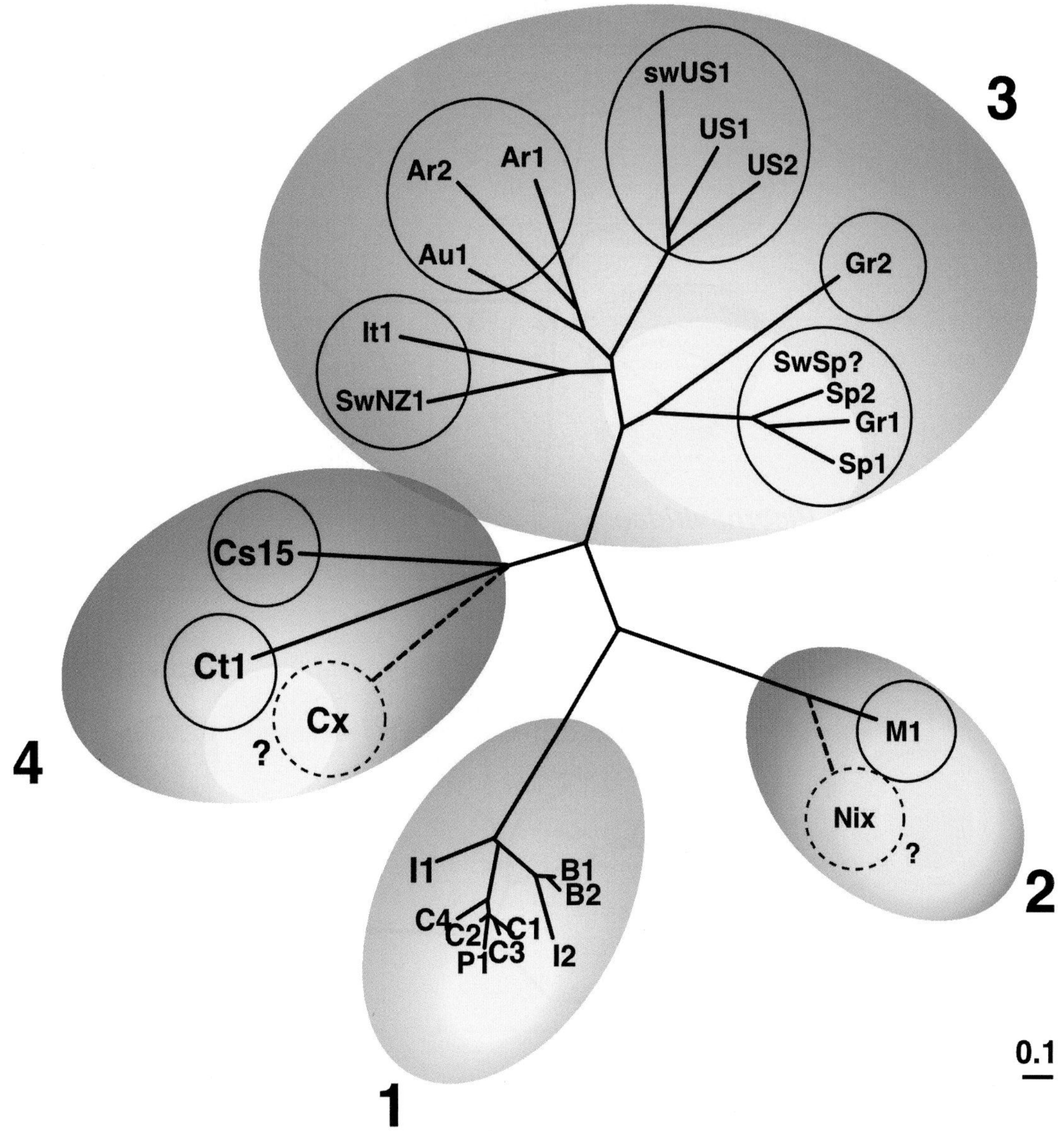

Figure 4: Unrooted phylogenetic tree depicting the relationship of nt sequences over a 287 nt fragment of ORF1. Isolates represented are Burma (B1, B2), China/Taiwan (C1, C2, C3, C4, Cs15, Ct1), Pakistan, (P1), India (I1, I2), Mexico (M1), United States (US1, US2), Greece (Gr1, Gr2), Spain (Sp1, Sp2), Italy (It1), Argentina (Ar1, Ar2), Austria (Au1), swine US (swUS1), swine New Zealand (swNZ1). The Arabic number adjacent to the shaded circles indicates genotypic designations. The dashed lines and circles indicate the putative positions of additional isolates from China/Taiwan (Cx), Nigeria (Nix) and swine sewage from Spain (swSp?) based on overlap of ORF 2 sequence. (From Schlauder and Mushahwar, 2001, with permission.)

SIMILARITY WITH OTHER TAXA

HEV is similar to members of the family *Caliciviridae* in its structural morphology, as assessed by electron microscopy, and in its genome organization. HEV shows highest, but limited, aa similarity in its replicative enzymes with *Rubella virus* and alphaviruses of the family *Togaviridae* and with plant furoviruses. The HEV capping enzyme has properties very similar to those of viruses within the "alphavirus-like supergroup".

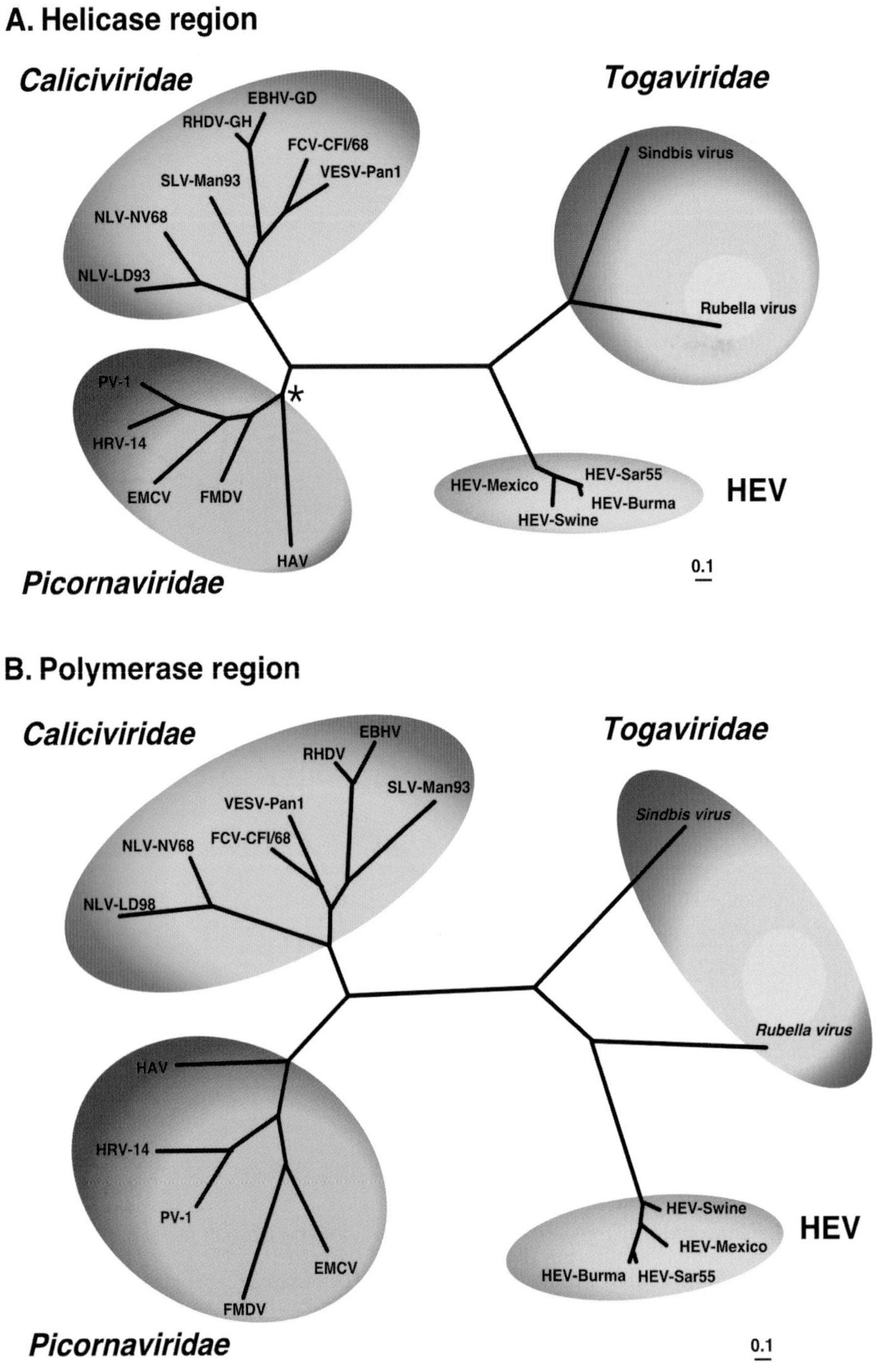

Figure 3: Phylogenetic relationships of *Hepatitis E virus* (HEV) with members of the families *Picornaviridae*, *Caliciviridae*, and *Togaviridae*. The helicase (Hel) and polymerase (Pol) regions of the genome were analyzed (Courtesy of Berke, T. and Matson, D.O.).
A. Partial gene sequences (200 aa) from the proposed helicase region were used for the phylogenetic analysis and included representative strains from each family. Clustal W v1.7 was used to create a multiple alignment for the aa sequences, which was verified by alignment of known motifs in the region (e.g. GxGKS/T). The nt sequences were added and aligned by hand using the corresponding aa sequences as template resulting in a consensus length of 608 nt. A phylogenetic tree was constructed from the nt sequence alignment using the maximum likelihood algorithm in the program DNAML from the PHYLIP 3.52c package within UNIX environment. For the algorithm, the global rearrangement option was invoked and the order of sequence input was randomized ten times. Other menu options were left as default. The resultant tree is unrooted and the phylogenetic distances are in the unit of expected number of substitutions per site. Branch points of the resulting tree had a confidence level of $P<0.01$ ($P<0.05$*). Genbank accession numbers for the strains in this analysis

FAMILY ASTROVIRIDAE

TAXONOMIC STRUCTURE OF THE FAMILY

Family	*Astroviridae*
Genus	*Avastrovirus*
Genus	*Mamastrovirus*

VIRION PROPERTIES

MORPHOLOGY

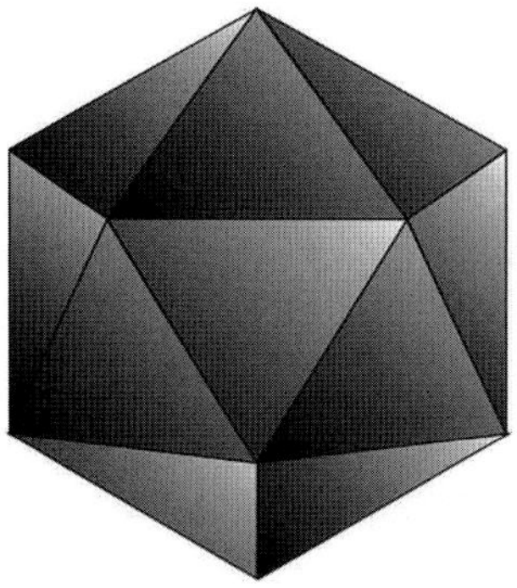

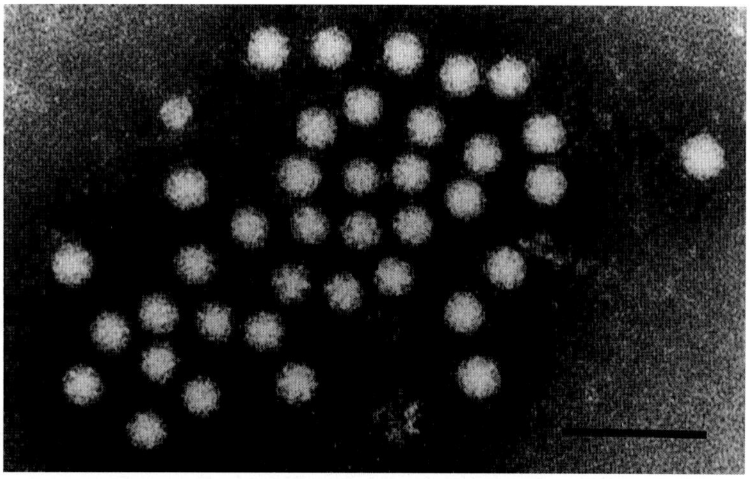

Figure 1: Negative contrast electron micrograph of virions of Human astrovirus (HAstV), from a stool specimen. The bar represents 100 nm (Courtesy of Dr. C. Humphrey).

Virions shed in feces samples are 28-30 nm in diameter, spherical in shape and non-enveloped. A distinctive five- or six-pointed star is discernible on the surface of about 10% of virions. Virions derived from cell culture are up to 41 nm in diameter, with well –defined surface spikes.

PHYSICOCHEMICAL AND PHYSICAL PROPERTIES

Virion M_r is about 8 x 10^6, S_{20w} is about 160S. Virion buoyant density in CsCl is 1.36-1.39 gm/cm^3. Virions are resistant to pH 3, 50°C for 1 hr, 60°C for 5 min, chloroform, lipid solvents and non-ionic, anionic, and zwitterionic detergents.

NUCLEIC ACID

Virions contain one molecule of infectious, positive sense, ssRNA, 6.4–7.4 kb in size. A poly(A) tract is located after the 3-terminal heteropolymeric sequence. The structure of the 5′-end of the genome in not known. A small protein with similarity to picornavirus VPg has been identified by sequence comparisons, but its presence on viral RNA has not been experimentally verified. The lengths of the non-translated regions at both ends of the genome vary by genus.

PROTEINS

Virion protein composition remains unclear; however, all isolates have at least two, usually three, major proteins of between 24 and 39 kDa.

LIPIDS

Virions do not contain a lipid envelope.

CARBOHYDRATES

None of the viral proteins is glycosylated.

Genome Organization and Replication

The virion RNA is infectious and serves as a messenger RNA for the non-structural polyproteins, p1A and p1AB. A polyadenylated, sgRNA (~2.8 kb) is detected in the cytoplasm of infected cells. Viral RNA replication is resistant to actinomycin D. A precursor of CP of 86-90 kDa has been detected in the cytoplasm of infected cells. Virions cultivated in a trypsin-free environment are composed of a single 79 kDa protein (VP79). N-terminal microsequencing indicates that VP79 results from the cleavage of the N-terminal 70 aa of the full-length ORF2 product. When VP79-containing particles are exposed to trypsin, capsid subunits of 26, 29 and 34 kDa are detected, and infectivity is enhanced.

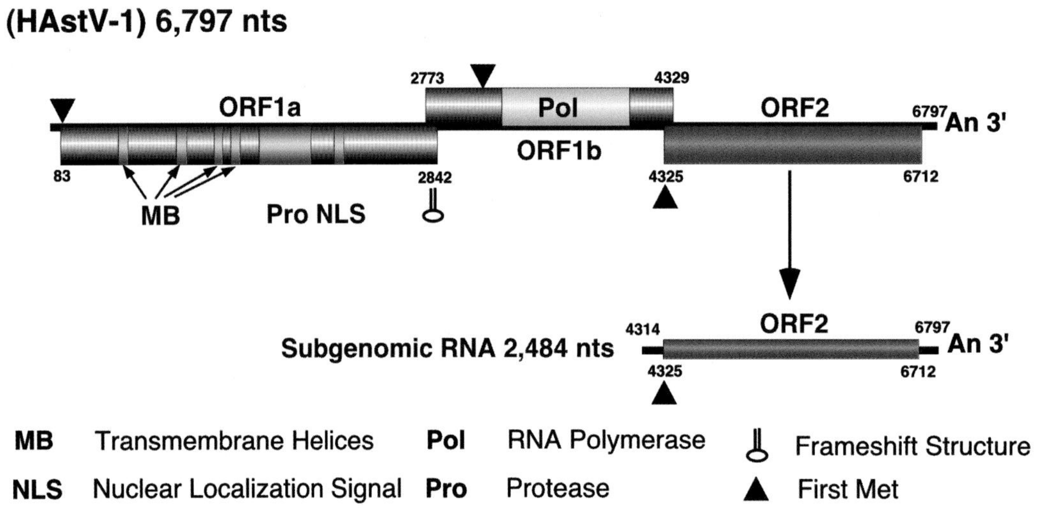

Figure 2: Genome organization and replication strategy of Human astrovirus - 1 (HAstV-1). ORF 2 is expressed from a sgRNA detected in the cytoplasm of infected cells.

Antigenic Properties

Eight serotypes of *Human astrovirus* have been defined by immune electron microscopy and neutralization tests and been confirmed by sequence comparisons. They share at least one common epitope recognized by monoclonal antibody. Neutralization epitopes have been mapped to the VP26 and VP29 proteins of Human astrovirus 1 (HAstV-1) and HAstV-2. Two distinct serotypes of bovine astrovirus have been defined by neutralization.

Biological Properties

Astroviruses appear to have very limited host ranges. They have been detected in stool samples from humans, cats, cattle, deer, dogs, ducks, mice, pigs, sheep, mink, turkeys, and chickens. Transmission is by the fecal-oral route and no intermediate vectors have been described. Astrovirus pathogenesis appears to vary by genus.

GENUS *AVASTROVIRUS*

Type Species *Turkey astrovirus*

DISTINGUISHING FEATURES

Members of the genus *Avastrovirus* infect avian species.

BIOLOGICAL PROPERTIES

Infection with Avastrovirus species often involves extra-intestinal manifestations (e.g. damage to liver, kidney, or the immune system). Duck astrovirus (DastV) causes an often fatal hepatitis in ducklings. Astroviruses infecting turkeys (TAstV) and chickens (ANV) affect multiple organs, including the kidney and thymus.

Duck astrovirus grows in embryonated hen's eggs following blind passage in the amniotic sac. Few infected embryos die in less than 7 days. Infected embryos appear stunted and have greenish, necrotic livers in which astrovirus particles have been identified.

LIST OF SPECIES DEMARCATION CRITERIA IN THE GENUS

Species are defined on the basis of host of origin. Serotypes are defined on the basis of twenty-fold, or greater, two-way cross-neutralization titers. Serotypes assigned to the species are given consecutive numbers.

LIST OF SPECIES IN THE GENUS

Species names are in green italic script; strain names and synonyms are in black roman script; tentative species names are in blue roman script. Sequence accession numbers, and assigned abbreviations ( ) are also listed.

SPECIES IN THE GENUS

Chicken astrovirus ‡		
Avian nephritis virus 1	[AB033998]	(ANV-1)
Avian nephritis virus 2	[AB046864]	(ANV-2)
Chicken astrovirus		(CAstV)
Duck astrovirus		
Duck astrovirus 1		(DAstV-1)
Turkey astrovirus		
Turkey astrovirus 1	[Y15936]	(TAstV-1)
Turkey astrovirus 2	[AF206663]	(TAstV-2)

TENTATIVE SPECIES IN THE GENUS

None reported.

POS. SSRNA

GENUS *MAMASTROVIRUS*

Type Species *Human astrovirus*

DISTINGUISHING FEATURES

Members of the genus *Mamastrovirus* infect mammalian species.

BIOLOGICAL PROPERTIES

The predominant feature of infection with mamastroviruses is gastroenteritis. In humans, astrovirus has been detected in duodenal biopsies in epithelial cells located in the lower part of villi. In experimentally infected sheep, astrovirus was found in the small intestine in the apical two-thirds of villi. In calves, astrovirus infection was localized to specialized M cells overlying the Peyer's patches. Human astroviruses are distributed worldwide and have been associated with 2–8% of acute, non-bacterial gastroenteritis in children. Astroviruses have also been associated with gastroenteritis outbreaks and with gastroenteritis in immunocompromised children and adults.

LIST OF SPECIES DEMARCATION CRITERIA IN THE GENUS

Species are defined on the basis of host of origin. Serotypes are defined on the basis of twenty-fold, or greater, two-way cross-neutralization titers. Serotypes assigned to the species are given consecutive numbers.

LIST OF SPECIES IN THE GENUS

Species names are in green italic script; strain names and synonyms are in black roman script; tentative species names are in blue roman script. Sequence accession numbers, and assigned abbreviations ( ) are also listed.

SPECIES IN THE GENUS

Bovine astrovirus		
Bovine astrovirus 1		(BAstV-1)
Bovine astrovirus 2		(BAstV-2)
Feline astrovirus		
Feline astrovirus 1		(FAstV-1)
Human astrovirus		
Human astrovirus 1	[Z25771; L13745]	(HAstV-1)
Human astrovirus 2	[L12745]	(HAstV-2)
Human astrovirus 3	[AF141381]	(HAstV-3)
Human astrovirus 4	[Z33883]	(HAstV-4)
Human astrovirus 5	[U15136]	(HAstV-5)
Human astrovirus 6	[Z46658]	(HAstV-6)
Human astrovirus 7	[Y08632]	(HAstV-7)
Human astrovirus 8	[AF260508]	(HAstV-8)
Mink astrovirus ‡		
Mink astrovirus 1	[AY179509]	(MastV-1)
Ovine astrovirus		
Ovine astrovirus 1	[Y15937]	(OAstV-1)
Porcine astrovirus		
Porcine astrovirus 1	[Y15938]	(PAstV-1)

TENTATIVE SPECIES IN THE GENUS

None reported.

LIST OF UNASSIGNED VIRUSES IN THE FAMILY

None reported.

Phylogenetic Relationships within the Family

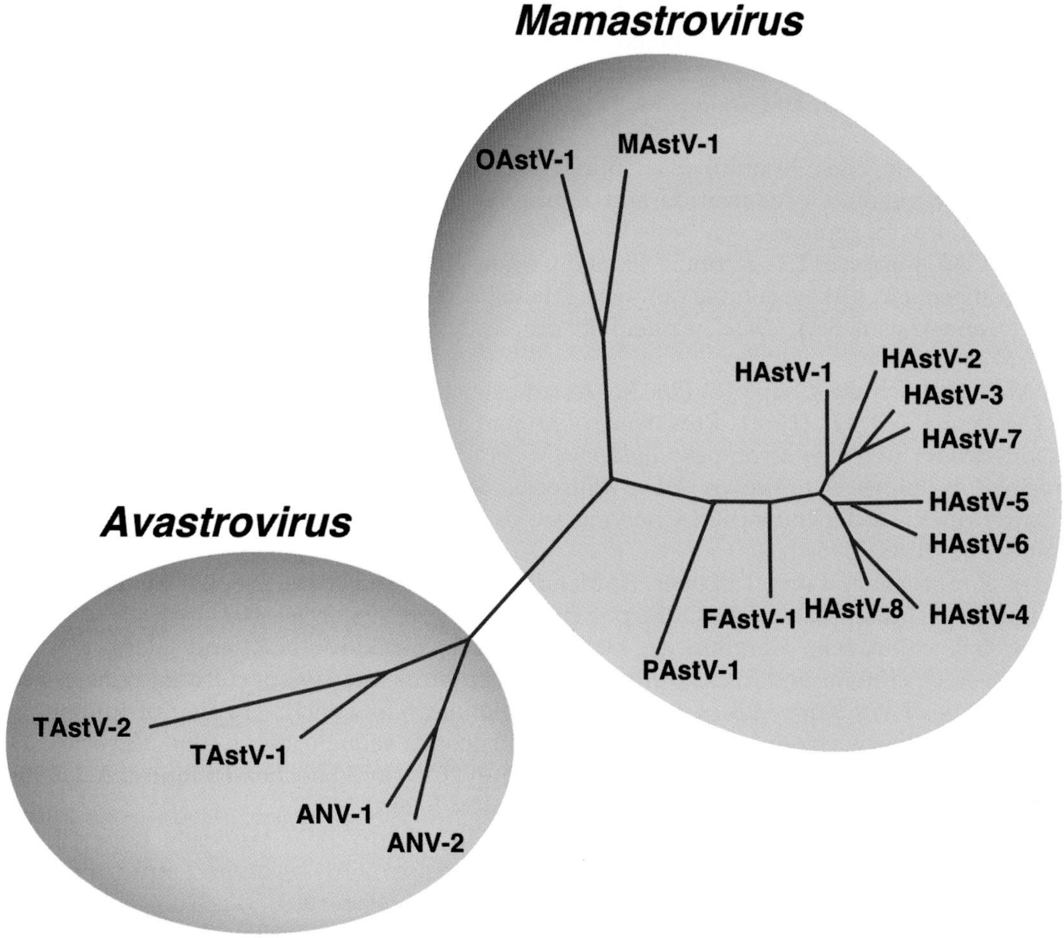

Figure 3: Phylogenetic relationships between the species and serotypes of the family *Astroviridae*. The predicted aa sequences of the entire capsid polyprotein were aligned using CLUSTALX (v1.83) and the nt sequences were aligned on the basis of the protein alignments using DAMBE (v4.1.19). The phylogenetic tree was generated using the maximum likelihood algorithm as implemented in the DAMBE program. Sequences used for the comparison are indicated in the list of species.

Similarity with Other Taxa

None reported.

Derivation of Names

Astro: from Greek astron, "star", representing the star-like surface structure on virions.
Av: from Latin avis, "bird", representing avian host species
Mam: from Latin mamma, "breast", representing mammalian host species

References

Aroonprasert, D., Fagerland, J.A., Kelso, N.E., Zheng, S. and Woode, G.N. (1989). Cultivation and partial characterization of bovine astrovirus. *Vet. Microbiol.*, **19**, 113-125.

Bass, D.M. and Qiu, S. (2000). Proteolytic processing of the astrovirus capsid. *J. Virol.*, **74**, 1810-1814.

Behling-Kelly, E., Schultz-Cherry, S., Koci, M., Kelley, L., Larsen, D. and Brown, C. (2002). Localization of astrovirus in experimentally infected turkeys as determined by *in situ* hybridization. *Vet. Pathol.*, **39**, 595-598.

Englund, L., Chriel, M., Dietz, H.H. and Hedlund, K.O. (2002). Astrovirus epidemiologically linked to pre-weaning diarrhoea in mink. *Vet. Microbiol.*, **85**, 1-11.

Geigenmueller, U., Ginzton, N.H. and Matsui, S.M. (1997). Construction of a genome-length cDNA clone for human astrovirus serotype 1 and synthesis of infectious RNA transcripts. *J. Virol.*, **71**, 1713-1717.

Geigenmueller, U., Chew, T., Ginzton, N. and Matsui, S.M. (2002). Processing of nonstructural protein 1a of human astrovirus. *J. Virol.*, **76**, 2003-2008.

Herring, A.J., Gray, E.W. and Snodgrass, D.R. (1981). Purification and characterization of ovine astrovirus. *J. Gen. Virol.*, **53**, 47-55.

Jiang, B., Monroe, S.S., Koonin, E.V., Stine, S.E. and Glass, R.I. (1993). RNA sequence of astrovirus: distinctive genomic organization and a putative retrovirus-like ribosomal frameshifting signal that directs the viral replicase synthesis. *Proc. Natl. Acad. Sci. USA*, **90**, 10539-10543.

Jonassen, C.M., Jonassen, T.T., Sveen, T.M. and Grinde, B. (2003). Complete genomic sequences of astroviruses from sheep and turkey: comparison with related viruses. *Virus Res.*, **91**, 195-201.

Kiang, D. and Matsui, S.M. (2002). Proteolytic processing of a human astrovirus nonstructural protein. *J. Gen. Virol.*, **83**, 25-34.

Koci, M.D. and Schultz-Cherry, S. (2002). Avian astroviruses. *Avian Pathol.*, **31**, 213-227.

Lee, T.W. and Kurtz, J.B. (1994). Prevalence of human astrovirus serotypes in the Oxford region 1976-92, with evidence for two new serotypes. *Epidemiol. Infect.*, **112**, 187-93.

Marczinke, B., Bloys, A.J., Brown, T.D., Willcocks, M.M., Carter, M.J. and Brierley, I. (1994). The human astrovirus RNA-dependent RNA polymerase coding region is expressed by ribosomal frameshifting. *J. Virol.*, **68**, 5588-5595.

Mendez, E., Fernandez-Luna, T., Lopez, S., Mendez-Toss, M. and Arias, C.F. (2002). Proteolytic processing of a serotype 8 Human Astrovirus ORF2 polyprotein. *J. Virol.*, **76**, 7996-8002.

Monroe, S.S., Stine, S.E., Gorelkin, L., Herrmann, J.E., Blacklow, N.R. and Glass, R.I. (1991). Temporal synthesis of proteins and RNAs during human astrovirus infection of cultured cells. *J. Virol.*, **65**, 641-648.

Noel, J.S., Lee, T.W., Kurtz, J.B., Glass, R.I. and Monroe, S.S. (1995). Typing of human astroviruses from clinical isolates by enzyme immunoassay and nucleotide sequencing. *J. Clin. Microbiol.*, **33**, 797-801.

Risco, C., Carrascosa, J.L., Pedregosa, A.M., Humphrey, C.D. and Sanchez-Fauquier, A. (1995). Ultrastructure of human astrovirus serotype 2. *J. Gen. Virol.*, **76**, 2075-2080.

Sanchez-Fauquier, A., Carrascosa, A.L., Carrascosa, J.L., Otero, A., Glass, R.I., Lopez, J.A., San Martin, C. and Melero, J.A. (1994). Characterization of a human astrovirus serotype 2 structural protein (VP26) that contains an epitope involved in virus neutralization. *Virology*, **201**, 312-320.

Willcocks, M.M., Carter, M.J., Laidler, F.R. and Madeley, C.R. (1990). Growth and characterisation of human faecal astrovirus in a continuous cell line. *Arch. Virol.*, **113**, 73-81.

CONTRIBUTED BY

Monroe, S.S., Carter, M.J., Herrmann, J., Mitchel, D.K. and Sanchez-Fauquier, A.

FAMILY *NODAVIRIDAE*

TAXONOMIC STRUCTURE OF THE FAMILY

Family ***Nodaviridae***
Genus ***Alphanodavirus***
Genus ***Betanodavirus***

GENUS *ALPHANODAVIRUS*

Type Species *Nodamura virus*

VIRION PROPERTIES

MORPHOLOGY

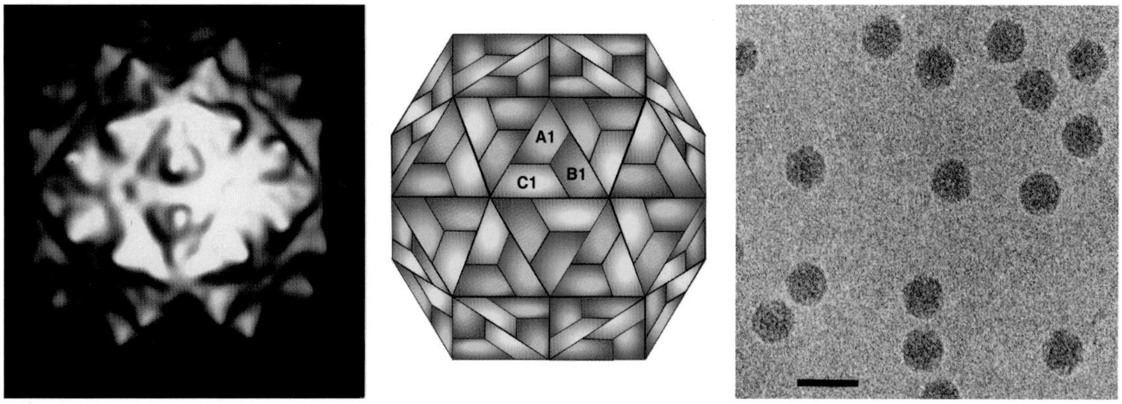

Figure 1: (Left) Image reconstruction of a particle of Flock House virus (FHV). (Center) Schematic representation of a T=3 icosahedral lattice. (Right) Cryo-electron micrograph of particles of FHV; the bar represents 50 nm. (Courtesy of N. Olson and T. Baker).

Virions are non-enveloped, roughly spherical in shape, 32-33 nm in diameter and have icosahedral symmetry (T=3). No distinct surface structure is seen by electron microscopy of negatively stained preparations. Empty shells are seldom seen in virus preparations.

PHYSICOCHEMICAL AND PHYSICAL PROPERTIES

Virion Mr is about 9×10^6; S_{20w} is 135-145S. Virion buoyant density in CsCl is 1.30-1.34 g/cm^3 (varies with species). Infectivity of aqueous suspensions is stable to extraction with chloroform. Infectivity of Nodamura virus (NoV), Black beetle virus (BBV), or Flock House virus (FHV) is stable at room temperature in 1% sodium dodecyl sulfate but Boolarra virus (BoV) is inactivated. Virions are stable at acid pH. The RNA content of the virion is about 16%.

NUCLEIC ACID

The genome consists of two molecules of positive sense ssRNA: RNA1 (Mr 1.1×10^6, 3.1 kb) and RNA2 (Mr 0.48×10^6, 1.4 kb). Both molecules are required for infectivity, and both are encapsidated in the same virus particle. Both RNA molecules are capped at their 5′-ends with cap zero structures and lack poly(A) tails at their 3′-ends. RNA 3′-ends cannot be chemically derivatized even after treatment with denaturing solvents, indicating that the expected 3′-terminal-OH groups are unreactive.

PROTEINS

The capsid consists of 180 protein subunits (protomers) arranged on a T=3 surface lattice. Each protomer is composed of a single CP (protein α) or the two products of its cleavage (proteins β and γ). Mass spectrometry of the FHV CP indicates that the initiating methionine is removed. Thus, for FHV, the capsid proteins are: protein α: (44 kDa), aa 2-407; protein β: (39 kDa), aa 2–363; protein γ: (4 kDa), aa 364–407. Morphogenesis involves the formation of a non-infectious provirion which acquires infectivity by autocatalytic cleavage of protein α to form proteins β and γ. Maturation cleavage is often incomplete and virions typically contain residual uncleaved protein α.

LIPIDS
None.

CARBOHYDRATES
None.

GENOME ORGANIZATION AND REPLICATION

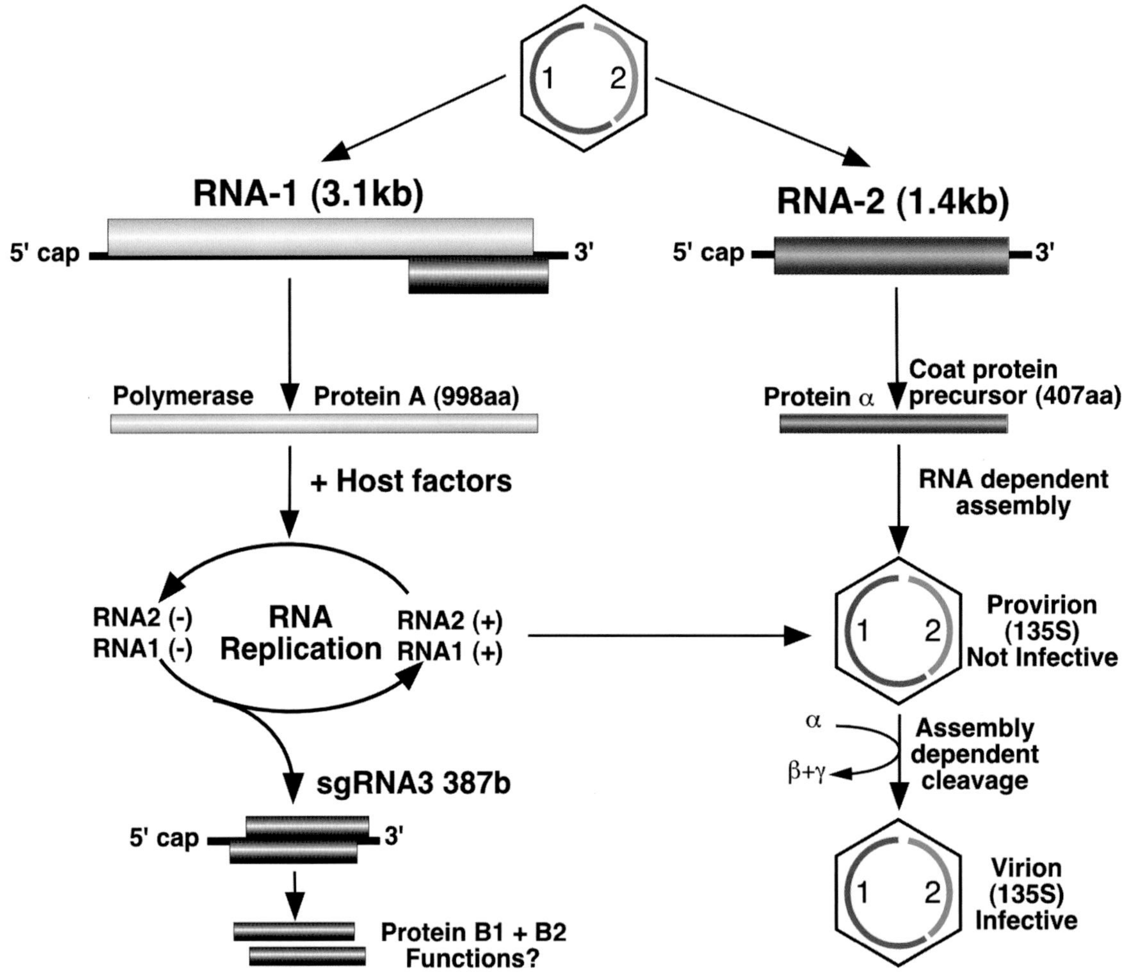

Figure 2: *Alphanodavirus* (Flock House virus; FHV) genome organization and strategy of replication. (Adapted from L.A. Ball and K.L. Johnson).

Alphanodaviruses replicate in the cytoplasm of infected cells (Fig.2). RNA synthesis is resistant to actinomycin D. Infected cells contain three ssRNAs: RNA1 (Mr 1.1 x 10^6; 3.1 kb); RNA2 (Mr 0.48 x 10^6; 1.4 kb) and a sgRNA3 (Mr 1.10.13 x 10^6; 0.39 kb), whose nt sequence corresponds to the 3′-end of RNA1 (387 nt in the case of FHV). RNA3 is not packaged into virions; its 3′-end is chemically unreactive like those of RNAs 1 and 2. RNA1 encodes protein A (112 kDa), which is the catalytic subunit of the viral RdRp. RNA2 encodes protein α, the CP precursor (44 kDa). Depending on virus species, RNA3 encodes one or two small proteins (proteins B1 and B2, 11 kDa). B1 is encoded in the same ORF as protein A. Protein B2 is encoded in an overlapping ORF. BoV RNA3 does not encode protein B1 but all known alphanodavirus RNA3 molecules encode protein B2. Protein B2 of FHV has been shown to function as a suppressor of RNA silencing in cultured *Drosophila melanogaster* cells (Schneider's line 2) and tobacco plants (*Nicotiana benthamiana*). The function of protein B1 is unknown. Cells transfected with isolated RNA1 synthesize RNA1 and overproduce RNA3, but do not make RNA2. RNA2 replication strongly inhibits synthesis of RNA3 and the translation of RNA2 suppresses the translation of RNA1.

ANTIGENIC PROPERTIES

NoV, BBV, FHV and BoV are cross-reactive by double-diffusion immunoprecipitation tests, but all four members represent different serotypes (neutralization titer of each antiserum less than 0.5% in heterotypic crosses).

BIOLOGICAL PROPERTIES

HOST RANGE

All species of alphanodaviruses were isolated in nature from insects, although serological data suggest that NoV also naturally infects pigs and perhaps herons. NoV seems to be unique among the nodaviruses in its ability to infect both vertebrates and invertebrates. It is also very unusual in being able to kill both insect and mammalian hosts. The other alphanodaviruses do not show strict specificity for particular insect hosts.

In the laboratory, most alphanodaviruses can be propagated in larvae of the common wax moth, *Galleria mellonella,* where they cause paralysis and death. NoV, isolated from mosquitoes, also grows in suckling mice but not in cultured cells of *Drosophila melanogaster*. FHV, BBV, and BoV grow well in cultured *Drosophila melanogaster* cells and form plaques on monolayers of these cells. Defective-interfering particles are readily formed unless the viruses are passaged at low multiplicity of infection. Persistent infections, with subsequent resistance to superinfection, occur readily in cultured *Drosophila melanogaster* cells. NoV multiplies poorly in most cultured cells but can be propagated by transfecting insect or vertebrate cell cultures with virion RNA at temperatures below about 34°C.

TRANSMISSION

NoV is transmissible to suckling mice by *Aedes aegypti* mosquitoes. It causes paralysis and death when injected into suckling mice, but no disease in adult animals. In their insect hosts, alphanodaviruses typically cause stunting, paralysis, and death.

LIST OF SPECIES DEMARCATION CRITERIA IN THE GENUS

The following criteria can be applied to the demarcation of species within the *Alphanodavirus* genus:

- **Biological properties (host range, vectors, mode of transmission).** Since the natural host ranges of the nodaviruses have generally not been examined in detail but may in some cases be broad, virus isolation from a new host is not, in itself, evidence of a new nodavirus species.
- **Antigenic properties.** Antisera raised against different isolates or strains of a single nodavirus species should exhibit high levels of cross-reactivity in Western blot and/or neutralization analyses. Lower levels of cross-reactivity in these assays using antisera against all previously recognized nodavirus species can provide evidence of a novel nodavirus.
- **Virion physical/physicochemical characteristics.**
    - **Virion electrophoretic mobility.** Intact virus particles migrate with characteristic electrophoretic mobilities in non-denaturing agarose gel, so virion mobility should be compared with those of other nodavirus species.

- **Sedimentation coefficient, buoyant density.** Virion sedimentation coefficient and buoyant density should be compared with those of other nodavirus species.

- **Structural protein characteristics.** The electrophoretic mobilities in SDS-PAGE of the CP precursor or its cleavage products should be compared with those of other nodavirus species.
- **Genome molecular characteristics.**
  - **RNA electrophoretic mobilities.** In the absence of sequence information, the electrophoretic mobilities of the viral genomic RNAs should be compared with those of other nodaviruses.
  - **RNA hybridization properties.** In the absence of differences in RNA electrophoretic mobilities, the molecular hybridization properties of the viral genomic RNAs should be compared with those of other nodaviruses.
  - **Genome sequence characteristics.** The nt sequence of the two genomic RNAs should be compared with those of other nodaviruses. Because the nodavirus genome is segmented, reassortment can occur and the two genome segments may have distinct evolutionary lineages.

Application of these criteria. In practice, while the five criteria above may be suggestive of a new species, definitive demarcation is based on the nt sequence of the viral CP gene. The two closest recognized species are BBV and FHV, whose RNA2 sequences show 80% identity at the nt level and 87% identity at the aa sequence level. Their RNA1 sequences, however, are 99% identical.

LIST OF SPECIES IN THE GENUS

Species names are in green italic script; strain names and synonyms are in black roman script; tentative species names are in blue roman script. Sequence accession numbers, and assigned abbreviations ( ) are also listed.

SPECIES IN THE GENUS

Black beetle virus		
Black beetle virus	RNA1[X02396] RNA2[X00956]	(BBV)
Boolarra virus		
Boolarra virus	RNA1[AF329080] RNA2[X15960]	(BoV)
Flock House virus		
Flock House virus	RNA1[X77156] RNA2[X15959]	(FHV)
Nodamura virus		
Nodamura virus	RNA1[AF174533] RNA2[X15961]	(NoV)
Pariacoto virus		
Pariacoto virus	RNA1[AF171942] RNA2[AF171943]	(PaV)

TENTATIVE SPECIES IN THE GENUS

Gypsy moth virus	(GMV)
Lymantria ninayi virus Greenwood	(LNV)
Manawatu virus	(MwV)
New Zealand virus	(NZV)
Drosophila line 1 virus	(DLV)

Genus *Betanodavirus*

Type Species *Striped jack nervous necrosis virus*

Virion Properties

Morphology

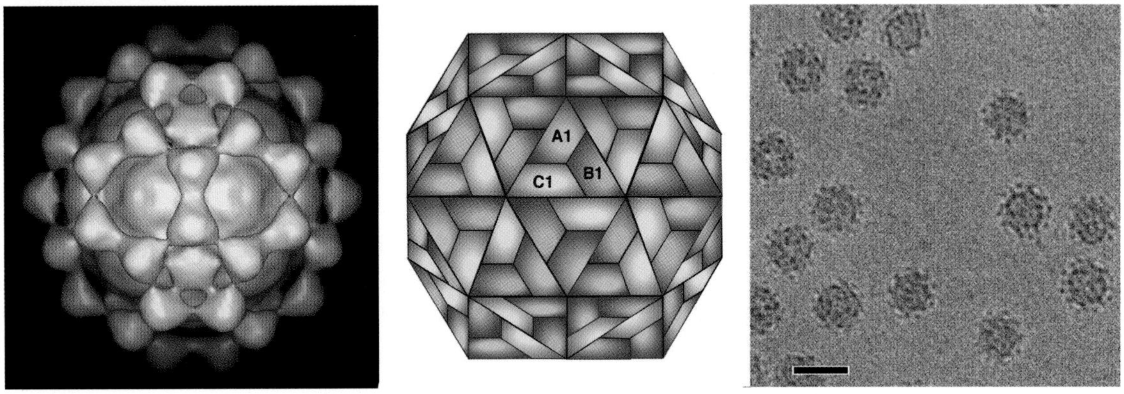

Figure 3: (Left) Image reconstruction of virus-like particles of Malabaricus grouper nervous necrosis virus (MGNNV) generated in *Spodoptera frugiperda* cells from a recombinant baculovirus expressing the MGNNV coat protein gene. (Center) Schematic representation of a T=3 icosahedral lattice. (Right) Cryo-electron micrograph of virus-like particles of MGNNV; the bar represents 40 nm. (Courtesy of L. Tang and J.E. Johnson).

Virions are non-enveloped, spherical in shape, and have icosahedral symmetry (T=3). Distinct surface protrusions are observed by electron microscopy of negatively stained preparations (Fig. 3). Image reconstruction of virus-like particles of Malabaricus grouper nervous necrosis virus (MGNNV) indicates that the CP of betanodaviruses has a two domain structure compared to the single domain structure of the CP of alphanodaviruses. The average diameter of the particle is 37 nm. In contrast with most alphanodaviruses, empty particles have been seen by electron microscopy of some preparations of betanodaviruses.

Physicochemical and Physical Properties

Virion buoyant density in CsCl of Striped jack nervous necrosis virus (SJNNV) has not been reported but that of Dicentrarchus labrax encephalitis virus (DlEV) is about 1.31–1.36 g/cm^3. Virions of DlEV are stable between pH 2–9 and resistant to heating at 56°C for 30 min. Infectivity is resistant to extraction of virions with chloroform.

Nucleic Acid

The genome consists of two molecules of positive sense ssRNA: RNA1 (Mr 1.01 x 10^6) and RNA2 (Mr 0.49 x 10^6). Both RNA molecules lack poly(A) tails at their 3′-ends.

Proteins

Betanodavirus capsids contain 180 copies of a single structural protein of 42 kDa. In contrast to alphanodaviruses, maturation cleavage of this protein is not observed. The appearance of a protein doublet on denaturing polyacrylamide gels is an artefact of incomplete reduction of a disulfide bond.

Lipids

None reported.

Carbohydrates

None reported.

Genome Organization and Replication

The betanodaviruses replicate in the cytoplasm. Infected cells contain three ssRNAs: RNA1 (Mr 1.01 x 10^6; 3.1 kb); RNA2 (Mr 0.49 x 10^6; 1.4 kb) and a sgRNA3 (Mr about 0.13 x 10^6; 0.4 kb) derived from RNA1. RNA3 is not packaged into virions. RNA1 encodes protein 1a (110 kDa), theRdRp. RNA2 encodes protein 2a (42 kDa), the CP. The protein(s) encoded by RNA3 have not yet been identified.

Antigenic Properties

Betanodaviruses are cross-reactive by immunoblot analysis using polyclonal antisera but differential reactivity is observed with monoclonal antibodies.

Biological Properties

Host Range

Nature: All species of the betanodaviruses were isolated from juvenile marine fish, in which they cause a vacuolating encephalopathy and retinopathy associated with behavioral abnormalities and high mortalities. These diseases have been detected particularly in commercial fish hatcheries, where they cause significant problems for the marine aquaculture industry.
Laboratory: Betanodaviruses replicate in cultured cells from striped snakehead fish (SNN-1) and sea bass larvae (SBL). A low level of virus replication was observed in mammalian (COS-1 and HeLa) cells at 28°C.

Transmission

Antibodies to SJNNV were found in 65% of plasma samples collected from wild and domestic brood stocks of striped jack, suggesting that the virus is very prevalent. Viral antigens were detected in eggs, larvae, and ovaries of hatchery-reared and wild spawner fish, suggesting both horizontal and vertical modes of transmission of the virus.

List of Species Demarcation Criteria in the Genus

The species demarcation criteria applied above for the alphanodaviruses also apply to betanodaviruses.

List of Species in the Genus

Species names are in green italic script; strain names and synonyms are in black roman script; tentative species names are in blue roman script. Sequence accession numbers, and assigned abbreviations ( ) are also listed.

Species in the Genus

Barfin flounder nervous necrosis virus		
Barfin flounder nervous necrosis virus	RNA2[D38635]	(BFNNV)
Redspotted grouper nervous necrosis virus		
Redspotted grouper nervous necrosis virus	RNA2[D38636]	(RGNNV)
Striped jack nervous necrosis virus		
Striped jack nervous necrosis virus	RNA1[AB025018, AB056571] RNA2[D30814, AB056572]	(SJNNV)
Tiger puffer nervous necrosis virus		
Tiger puffer nervous necrosis virus	RNA2[D38637]	(TPNNV)

Tentative Species in the Genus

Atlantic cod nervous necrosis virus	RNA2[AF445800]	(ACNNV)
Atlantic halibut nodavirus	RNA2[AJ245641]	(AHNV)
Dicentrarchus labrax encephalitis virus	RNA2[U39876, Y08700, AJ277803-10, AF175509-20]	(DIEV)
Dragon grouper nervous necrosis virus	RNA2[AF245004]	(DGNNV)
Greasy grouper nervous necrosis virus	RNA1[AF319555]	(GGNNV)

	RNA2[AF318942]	
Grouper nervous necrosis virus		(GNNV)
Halibut nervous necrosis virus		(HNNV)
Japanese flounder nervous necrosis virus	RNA2[D38527]	(JFNNV)
Lates calcarifer encephalitis virus		(LcEV)
Malabaricus grouper nervous necrosis virus	RNA2[AF245003]	(MGNNV)
Seabass nervous necrosis virus		(SBNNV)
Umbrina cirrosa nodavirus		(UCNV)

List of Unassigned Viruses in the Family

None reported

Phylogenetic Relationships within the Family

Within the alphanodaviruses, CP sequences are 44–87% identical to one another at the aa level, whereas within the betanodaviruses, CP aa sequence identities are 80% or greater. The phylogenetic relationship between betanodavirus species and tentative species has not yet been rigorously defined. However, the sequences of RNA2 of RGNNV, GGNNV, MGNNV and DGNNV are >95% identical at both the nt and aa level suggesting that future reclassification of some tentative species as strains of current betanodavirus species may be possible. The CP aa sequences of the alphanodaviruses are only about 10% identical to those of the betanodaviruses, insufficient to indicate common ancestry.

Similarity with Other Taxa

The omegatetraviruses such as Nudaurelia capensis ω virus (NωV) and Helicoverpa armigera stunt virus (HaSV) contain bipartite ssRNA genomes, but their RNAs are about twice the size of nodavirus RNAs and they have no 3'-terminal blockage. Tetraviruses also have larger capsids with T=4 icosahedral symmetry.

Derivation of Names

Noda is from *Noda*mura, the name of a village (now a city; Nodashi) in the vicinity of the site where NoV was isolated in Japan. Other nodaviruses are similarly named after the place of isolation or after the host name from which they were isolated.

References

Ball, L.A. and Johnson, K.L. (1998). Nodaviruses of insects. In: *The Insect Viruses,* (L.K. Miller and L.A. Ball, eds), pp 225-267. Plenum Publishing Company, New York.

Ball, L.A., Amann, J.M. and Garrett, B.K. (1992). Replication of nodamura virus after transfection of viral RNA into mammalian cells in culture. *J. Virol.,* **66,** 2326-2334.

Comps, M., Pepin, J.F. and Bonami, J.R. (1994). Purification and characterization of two fish encephalitis viruses (FEV) infecting *Lates calcarifer* and *Dicentrarchus labrax. Aquaculture,* **123,** 1-10.

Dearing, S.C., Scotti, P.D., Wigley, P.J. and Dhana, S.D. (1980). A small RNA virus isolated from the grass grub, *Costelytra zealandica* (Coleoptera: *Scarabaeidae*). *N. Z. J. Zool.,* **7,** 267-269.

Delsert, C., Morin, N. and Comps, M. (1997). Fish nodavirus lytic cycle and semipermissive expression in mammalian and fish cell cultures. *J. Virol.,* **71,** 5673-5677.

Frerichs, G.N., Rodger, H.D. and Peric, Z. (1996). Cell culture isolation of piscine neuropathy nodavirus from juvenile sea bass, *Dicentrarchus labrax. J. Gen. Virol.,* **77,** 2067-2071.

Garzon, S. and Charpentier, G. (1992). *Nodaviridae.* In: *Atlas of Invertebrate Viruses,* (J.R. Adams and J.R. Bonami, eds), pp 351-370. CRC Press, Boca Raton. Florida.

Hendry, D.A. (1991). *Nodaviridae* of Invertebrates. In: *Viruses of Invertebrates.* (E. Kurstak, ed), pp. 227-276. Marcel Dekker, New York.

Johnson, J.E. and Reddy, V. (1998). Structural studies of noda and tetraviruses. In: *The Insect Viruses,* (L.K. Miller and L.A. Ball, eds), pp 171-223. Plenum Publishing Company, New York.

Krondiris, J.V. and Sideris, D.C. (2002). Intramolecular disulfide bonding is essential for betanodavirus coat protein conformation. *J. Gen. Virol.* **83,** 2211-2214.

Li, H., Li, W.X. and Ding, S.W. (2002). Induction and suppression of RNA silencing by an animal virus. *Science,* **296,** 1319-1321.

Mori, K.I., Nakai, T., Muroga, K., Arimoto, M., Musiake, K. and Firusawa, I. (1992). Properties of a new virus belonging to the *Nodaviridae* found in larval striped jack (*Pseudocaranx dentex*) with nervous necrosis. *Virology,* **187,** 368-385.

Munday, B.L., Nakai, T. and Nguyen, H.D. (1994). Antigenic relationship of the picorna-like virus of larval barramundi, *Lates calcarifer* Bloch to the nodavirus of larval striped jack, *Pseudocaranx dentex* (Bloch and Schneider). *Aust. Vet. J.,* **71,** 384-385.

Nishizawa, T., Mori, K., Furuhashi, M., Nakai, T., Furusawa, I. and Muroga, K. (1995). Comparison of the coat protein genes of five fish nodaviruses, the causative agents of viral nervous necrosis in marine fish. *J. Gen. Virol.,* **76,** 1563-1569.

Nishizawa, T., Furuhashi, M., Nagai, T., Nakai, T. and Muroga, K. (1997). Genomic classification of fish nodaviruses by molecular phylogenetic analysis of the coat protein gene. *Appl. Environ. Microbiol.,* **63,** 1633-1636.

Nishizawa, T., Takano, R. and Muroga, K. (1999). Mapping a neutralizing epitope on the coat protein of striped jack nervous necrosis virus (SJNNV). *J. Gen. Virol.,* **80,** 3023-3028.

Reinganum, C., Bashirrudin, J.B. and Cross, G.F. (1985). Boolarra virus: a member of the *Nodaviridae* isolated from *Oncopera intricoides (Lepidoptera: Hapealidae). Intervirology,* **24,** 10-17.

Schneemann, A., Reddy, V. and Johnson, J.E. (1998). The structure and function of nodavirus particles: a paradigm for understanding chemical biology. *Adv. Virus Res.,* **50,** 381-446.

Schneemann, A., Zhong, W., Gallagher, T.M. and Rueckert, R. R. (1992). Maturation cleavage required for infectivity of a nodavirus. *J. Virol.,* **66,** 6728-6734.

Scotti, P.D. and Fredericksen, S. (1987). Manawatu virus: a nodavirus isolated from *Costelytra zealandica* (White) (Coleoptera: *Scarabaeidae*). *Arch. Virol.,* **97,** 85-92.

Tang, L., Lin, C.S., Krishna, N.K., Yeager, M., Schneemann, A. and Johnson, J.E. (2002). Virus-like particles of a fish nodavirus display a capsid subunit domain organization different from insect nodaviruses. *J. Virol.,* **76,** 6370-6375.

CONTRIBUTED BY

Schneemann, A., Ball, L.A., Delsert, C., Johnson, J.E. and Nishizawa, T.

FAMILY *TETRAVIRIDAE*

TAXONOMIC STRUCTURE OF THE FAMILY

Family *Tetraviridae*
Genus *Betatetravirus*
Genus *Omegatetravirus*

GENUS *BETATETRAVIRUS*

Type Species *Nudaurelia capensis β virus*

VIRION PROPERTIES

MORPHOLOGY

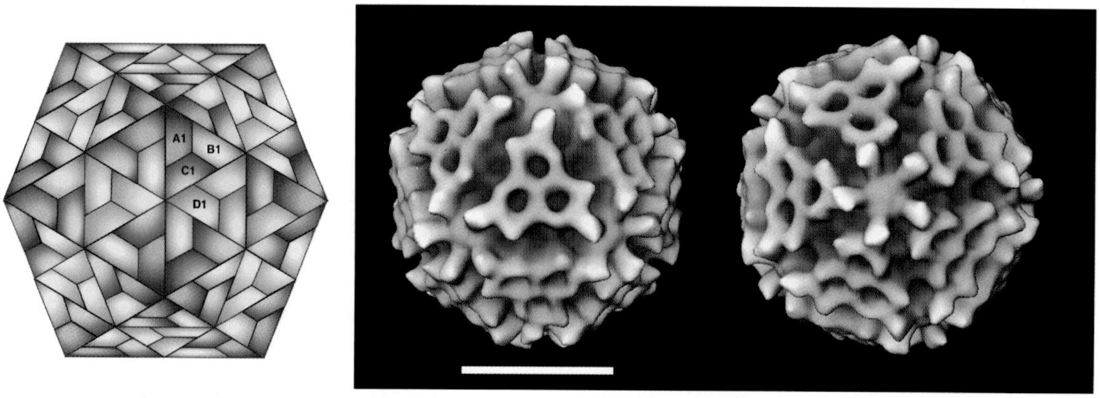

Figure 1: (Left) Schematic representation of a T=4 icosahedral lattice. (Center and Right) Cryo-electron image reconstruction of a particle of Nudaurelia capensis β virus (NβV) on the symmetry axis 3 and 5; the bar represents 20 nm. (Courtesy of H.R. Cheng, N. Olson and T. Baker).

Virions are non-enveloped, roughly spherical, about 40 nm in diameter and exhibit T=4 icosahedral shell *quasi*-symmetry. Distinct capsomers have been resolved by cryo-electron microscopy and image reconstruction (Fig. 1). The genome consists of ssRNA. Viruses in the genus *Betatetravirus* have monopartite genomes, whereas those in the genus *Omegatetravirus* have bipartite genomes.

PHYSICOCHEMICAL AND PHYSICAL PROPERTIES

Virion Mr is about 18 x 10^6. Virion S_{20w} is 194-217S. Virion buoyant density in CsCl is usually 1.28-1.30 g/cm^3 but occasionally as high as 1.33 g/cm^3 (varies with species). Virions are stable over a broad range of pH and their infectivity can resist desiccation and protease treatment.

NUCLEIC ACID

Virions of the type species, *Nudaurelia capensis β virus,* contain a single, positive sense, ssRNA segment of about 6.5 kb (Mr 1.8 x 10^6) which represents about 10% of the particle mass. This genomic RNA is not polyadenylated at its 3'-end, nor blocked like nodaviral RNAs, but terminates instead with a distinctive tRNA-like structure. A subgenomic message for the CPs, which is derived from the 3'-end of the genomic RNA, can also be encapsidated in some species.

PROTEINS

Capsids consist of 240 protein subunits (protomers) arranged on a T=4 surface lattice. Each protomer consists of the two cleavage products (β, 58.4 kDa and γ, 8 kDa), of a single

CP precursor (α, 66.4 kDa). Minor amounts of the uncleaved precursor may be found in virions.

Lipids
None reported.

Carbohydrates
None reported.

Genome Organization and Replication

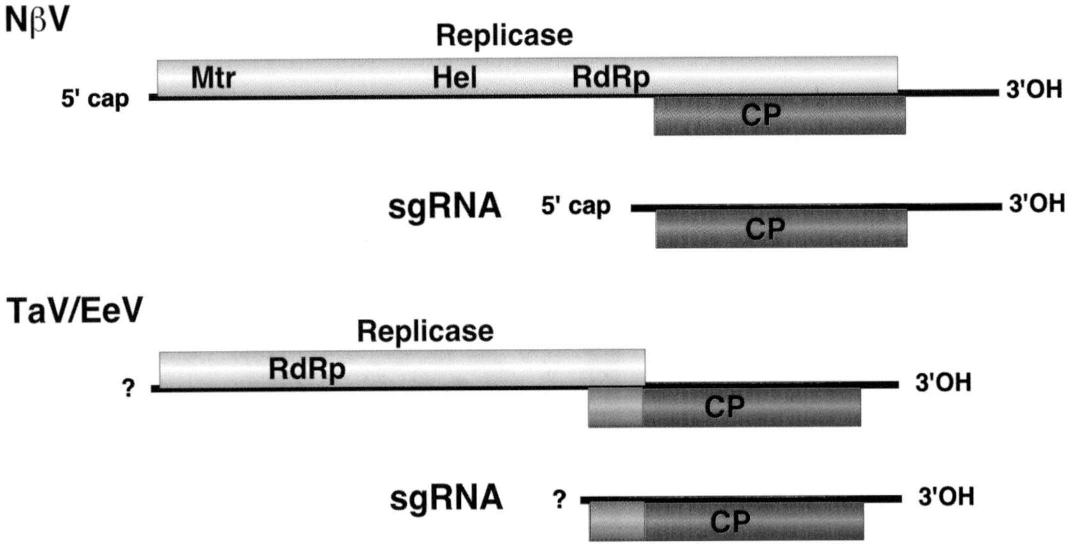

Figure 2. Genome organization of betatetraviruses. The locations of the methyltransferase (Mtr), RNA helicase (Hel) and RdRp motifs within the replicase ORFs are indicated. The amino-terminal portion of the TaV and EeV CP precursor removed by processing is boxed. The 5′-ends of the TaV/EeV genomic and sgRNAs have not been characterized and are indicated by "?".

The viruses replicate in the cytoplasm. The 6,625 nt genomic RNA of Nudaurelia capensis β virus (NβV), contains two ORFs that overlap for 1517 nt: that nearer the 5′-end contains 1821 codons and encodes the multidomain RNA replicase (204 kDa), whereas that nearer the 3′-end contains 612 codons and encodes a precursor (66.4 kDa) to the two Cps (Fig. 2). The RNA replicase includes three highly conserved enzymatic domains, Mtr, RNA Hel and RdRp that are yet to be characterized experimentally.

The genomic RNA of the recently sequenced Thosea asigna virus (TaV) is 5,715 nt. Like that of the closely related Euprosterna elaeasa virus (EeV)(5,698 nt), it carries a shorter ORF (1257 codons) encoding the RdRp-containing replicase, resulting in a much shorter overlap between the replicase and CP ORFs. The TaV CP ORF is longer than that of NβV, yielding a putative precursor of 757 aa in length that is processed by an additional step to remove an amino terminal portion of 17 kDa (Fig. 2).

During RNA replication of NβV and TaV, a sgRNA, which represents the 3′ 2.5 kb of the genome is synthesized, and this serves as the mRNA for the CP precursor (Fig. 2). In some betatetravirus species, including both NβV and TaV, the sgRNA can be encapsidated in virus particles, which complicates the distinction between the monopartite genome organization of the betatetraviruses and the bipartite genome organization of the omegatetraviruses.

Antigenic Properties

Most of the members of the group are serologically interrelated but distinguishable. The majority of the isolates were identified on the basis of their serological reaction with antiserum raised against NβV.

Biological Properties

Host Range

Nature: All virus species were isolated from *Lepidoptera* species (moths and butterflies), principally from Saturniid, Limacodid and Noctuid moths and no replication in other animals has been detected. In larvae, virus replication is restricted predominantly to the cells of the midgut.

Laboratory: With the exception of Providence virus (PrV), no infections by members of the *Betatetravirus* genus have been achieved in cultured cells, even when genomic RNA was transfected directly into cells.

Transmission

Oral transmission of NβV to *Antherea eucalypti* (the emperor gum moth) has been demonstrated experimentally. Oral transmission is implied by the midgut site of viral replication and by reports of some tetraviruses being used as sprayed insecticides in Malaysia (e.g. DtV; Darna trima virus). At high host densities, horizontal spread appears to be the major route of infection. Suggestive evidence exists for vertical transmission, which could be responsible for the observed persistence of tetraviruses within insect populations.

Cytopathic Effects

The viruses replicate primarily in the cytoplasm of gut cells of several *Lepidoptera* species. Crystalline arrays of virus particles are often seen within cytoplasmic vesicles. Different isolates vary considerably in pathogenicity; symptoms can vary from inapparent to acutely lethal infections.

List of Species Demarcation Criteria in the Genus

The following criteria can be applied to the demarcation of species within the betatetravirus genus:

- **Biological properties (host range, vectors, mode of transmission).** Since the natural host-ranges of individual recognized tetravirus species appear to be narrow, virus isolation from a new host can provide suggestive evidence of a new tetravirus species.
- **Antigenic properties.** Antisera raised against different isolates or strains of a single tetravirus species should exhibit high levels of cross-reactivity in Western blot and/or neutralization analyses. Lower levels of cross-reactivity in these assays using antisera against previously recognized tetraviruses can provide evidence of a new tetravirus species.
- **Virion physical/physicochemical characteristics.** In the absence of more definitive criteria, significant (>5%) differences in virion sedimentation coefficient or buoyant density from those of all previously recognized tetravirus species can provide evidence of a new virus species.
- **Structural protein characteristics.** The electrophoretic mobilities in SDS-PAGE of the CP precursor or its cleavage products should be compared with those of other tetravirus species.
- **Genome molecular characteristics.**
    - **RNA electrophoretic mobilities.** In the absence of sequence information, the electrophoretic mobilities of the viral genomic RNAs should be compared with those of other tetraviruses.

 - **RNA hybridization properties.** In the absence of differences in RNA electrophoretic mobilities, the molecular hybridization properties of the viral genomic RNAs should be compared with those of other tetraviruses.
- **Genome sequence characteristics.** The nt sequences of the genomic RNA(s) should be compared with those of other tetraviruses.

Application of these criteria. In practice, while criteria 1 - 5 above may be suggestive of a new species, definitive demarcation is based on the nt sequence of the viral CP gene.

List of Species in the Genus

Species names are in green italic script; strain names and synonyms are in black roman script; tentative species names are in blue roman script. Sequence accession numbers, and assigned abbreviations ( ) are also listed.

Species in the Genus

Antheraea eucalypti virus		
Antheraea eucalypti virus		(AeV)
Darna trima virus		
Darna trima virus		(DtV)
Dasychira pudibunda virus		
Dasychira pudibunda virus		(DpV)
(Calliteara pudibunda virus)		(CpV)
Euprosterna elaeasa virus		
Euprosterna elaeasa virus		(EeV)
Nudaurelia capensis β virus		
Nudaurelia capensis β virus	[AF102884]	(NβV)
Antheraea eucalypti virus		(AeV)
Philosamia cynthia x ricini virus		
Philosamia cynthia x ricini virus		(PxV)
Providence virus		
Providence virus		(PrV)
Pseudoplusia includens virus		
Pseudoplusia includens virus		(PiV)
Thosea asigna virus		
Thosea asigna virus	[AF82930, AF062037]	(TaV)
(Setothosea asigna virus)		(SaV)
Trichoplusia ni virus		
Trichoplusia ni virus		(TnV)

Tentative Species in the Genus

None reported.

GENUS *OMEGATETRAVIRUS*

Type Species *Nudaurelia capensis ω virus*

VIRION PROPERTIES

MORPHOLOGY

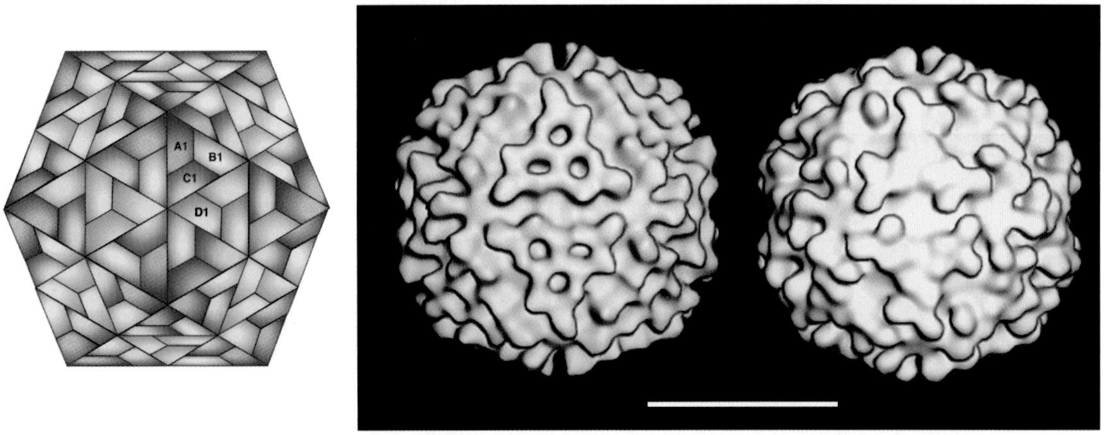

Figure 3: (Left) Schematic representation of a T=4 icosahedral lattice. (Center and Right) Cryo-electron image reconstruction of a particle of Nudaurelia capensis β virus (NβV)(Center) and of Nudaurelia capensis ω virus (NωV)(Right). The bar represents 25 nm. (From Johnson *et al.* (1994)).

Virions are non-enveloped, roughly spherical, about 40 nm in diameter and exhibit T=4 icosahedral shell *quasi*-symmetry. Distinct capsomers have been resolved by cryo-electron microscopy and image reconstruction (Fig. 3). The genome consists of ssRNA. Viruses in the genus *Omegatetravirus* have bipartite genomes, whereas those in the genus *Betatetravirus* have monopartite genomes.

PHYSICOCHEMICAL AND PHYSICAL PROPERTIES

Virion Mr is about 16 x 10^6. Virion S_{20w} is 194-217S. Virion buoyant density in CsCl is 1.28-1.30 g/cm^3 (varies with species). Virions of Helicoverpa armigera stunt virus (HaSV) are stable between pH 3.0-11.0, and at temperatures up to 55°C.

NUCLEIC ACID

Virions of the type species, *Nudaurelia capensis ω virus*, contain two positive-sense, ssRNA segments of 5.3 kb (RNA1, Mr 1.75 x 10^6) and 2.45 kb (RNA2, Mr 0.8 x 10^6). These genomic RNAs are capped at their 5′-ends but their 3′-ends are not polyadenylated, nor blocked like nodaviral RNAs, but instead terminate with a distinctive tRNA-like structure. It is likely that omegatetraviruses encapsidate both genomic RNAs within a single particle, in which case the RNAs would represent about 10% of the particle mass.

PROTEINS

Capsids consist of 240 protein subunits (protomers) arranged on a T=4 surface lattice. Each protomer consists of the two cleavage products (β, 62 kDa and γ, 7.8 kDa) of a single CP precursor (α, 69.8 kDa). Overall, the CPs of Nudaurelia capensis ω virus (NωV) and NβV share less than 20% aa sequence homology, indicating that the beta and omega genera of the tetraviruses have substantially diverged.

LIPIDS

None reported.

CARBOHYDRATES

None reported.

Genome Organization and Replication

Omegatetraviruses replicate in the cytoplasm. The viral genome consists of two unique molecules of RNA, probably encapsidated in the same virus particle (Fig. 4). A single ORF of 1704 codons on Helicoverpa armigera stunt virus (HaSV) RNA1 (Mr 1.75 x 10^6; 5.3 kb) encodes the RNA replicase with a domain organization similar to that encoded by the betatetravirus NβV. RNA2 (Mr 0.8 x10^6; 2.45 kb) encodes the CP precursor (70 kDa). A 157-codon ORF precedes and partially overlaps that for the CP; it encodes a non-structural protein (p17) of unknown function. Separate dsRNAs that correspond to the two genome segments are observed.

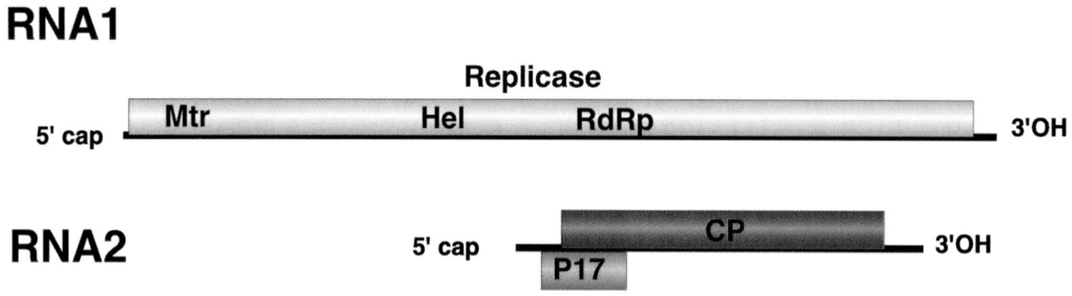

Figure 4: Genome organization of omegatetraviruses. The locations of the Mtr, RNA Hel and RdRp motifs within the replicase ORF are indicated.

Antigenic Properties

The two recognized omegatetraviruses (NωV and HaSV) display no serological relationship although their CPs show a high degree of aa sequence homology.

Biological Properties

Host Range

Nature: All species were isolated from *Lepidoptera* species, principally from Saturniid, Limacodid and Noctuid moths.

Laboratory: No infections by members of the genus *Omegatetravirus* have yet been achieved in cultured cells, but infectious HaSV particles were produced by plant protoplasts transfected with plasmids carrying full-length cDNAs that corresponded to the viral genome segments.

Transmission

As with the betatetraviruses, oral transmission is implied by the midgut site of viral replication. At high host densities, horizontal spread appears to be the major route of infection, but evidence exists for vertical transmission which might be responsible for the observed persistence of tetraviruses within insect populations.

Cytopathic Effects

The viruses replicate primarily in the cytoplasm of midgut cells of the larvae of several *Lepidoptera* species. Crystalline arrays of virus particles are often seen within cytoplasmic vesicles. There is a considerable range of pathogenicity with different isolates, and symptoms can vary from inapparent to acutely lethal infections.

List of Species Demarcation Criteria in the Genus

The species demarcation criteria of betatetraviruses also apply to omegatetraviruses. Also, because the genome of omegatetraviruses is segmented, reassortment is possible and the two genome segments may have different evolutionary lineages. However, no chimeric omegatetraviruses have yet been detected.

List of Species in the Genus

Species names are in green italic script; strain names and synonyms are in black roman script; tentative species names are in blue roman script. Sequence accession numbers, and assigned abbreviations ( ) are also listed.

Species in the Genus

Helicoverpa armigera stunt virus		
Helicoverpa armigera stunt virus	RNA1 [U18246] RNA2 [L37299]	(HaSV)
Nudaurelia capensis ω virus		
Nudaurelia capensis ω virus	RNA2 [S43937]	(NωV)

Tentative Species in the Genus

None reported.

List of Unassigned Viruses in the Family

Acherontia atropas virus	(AaV)
Agraulis vanillae virus	(AvV)
Callimorpha quadripuntata virus	(CqV)
Eucocytis meeki virus	(EmV)
Euploea corea virus	(EcV)
Hyalophora cecropia virus	(HcV)
Hypocritae jacobeae virus	(HjV)
Lymantria ninayi virus	(LnV)
Saturnia pavonia virus	(SpV)
Setora nitens virus	(SnV)
Nudaurelia capensis ε virus[a]	(NεV)

[a] NεV resembles the tetraviruses in appearance but is serologically unrelated to any known species.

Phylogenetic Relationships within the Family

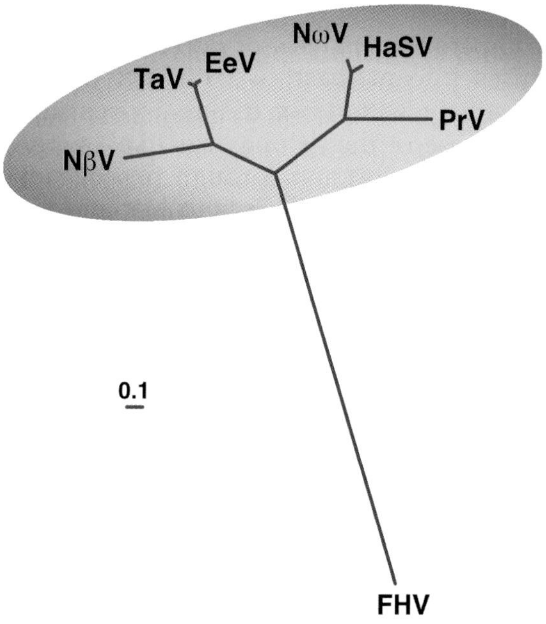

Figure 5: Unrooted phenogram showing the relationships amongst CPs of members of the family *Tetraviridae*. The phenogram was constructed from an aa similarity matrix of CP alignments using the neighbor-joining method. Branch lengths are drawn to scale. Viruses included in the analysis, abbreviation ( ) and accession numbers [ ] are: Nudaurelia capensis ω virus (NωV) [S43937], Helicoverpa armigera stunt virus (HaSV) [L37299], Nudaurelia capensis β virus (NβV) [AF102884], Providence virus (PrV) [AF548354], Thosea asigna virus (TaV) [AF062037] and Euprosterna elaeasa virus (EeV) (AF461742). The nodavirus Flock House virus (FHV) [X15959] is included as an outgroup.

Pos. ssRNA

Capsid proteins of all sequenced tetraviruses form a compact group (Fig. 5). Two major clusters are evident in the tetravirus capsid phylogram, with one comprising NβV, TaV and EeV, and the other HaSV, NωV and PrV. It is striking that in each of these clusters, both CP processing strategies (i.e. that employed by NβV, HaSV and NωV and that identified for TaV, EeV, and PrV are present.

In contrast to the CPs, the replicases of the tetraviruses for which genomic sequences have been determined fall into at least two distinct phylogenetic groups that do not reflect the taxonomic demarcation. The first of these groups includes the betatetravirus NβV and the omegatetravirus HaSV. (Although no complete sequence has been published for the replicase of NωV, unpublished data show it to be very closely related to that of HaSV). Replicases of these viruses include three conserved domains, Mtr, Hel and RdRp, and cluster together with the similarly organized replicases of a dozen other ssRNA+ virus families (Fig. 6; see also below). The second group includes the betatetraviruses TaV and EeV. The replicases of these viruses are also multi-domain proteins with RdRp being the only domain with provisionally assigned function. The phylogenetic neighborhood of the TaV, EeV RdRps includes diverse viruses that are discussed below but not those of the NβV/HaSV group (Fig. 7). That there appears to be a third distinct RdRp lineage within the tetraviruses is indicated by available (unpublished) data on PrV, whose RdRp clusters with viruses other than those that form either of the above two groups.

Similarity with Other Taxa

The tetravirus CPs form a monophyletic group with the jelly-roll fold β subunits being distantly related to the CPs of nodaviruses having the T=3 capsids. It has been speculated that the tetravirus capsid might have evolved from a nodavirus-like ancestor through a process that included insertion of an immunoglobulin-like protein domain coding sequence (either acquired or evolved through sequence duplication) within the CP gene.

In contrast, comparative analysis of currently available non-structural protein sequences split tetraviruses into at least two distinct lineages, prototyped by NβV/HaSV and TaV/EeV respectively, within two different virus superclusters. The replicases of NβV and HaSV resemble those of the "alphavirus-like" supercluster, having the distinct Mtr-Hel-RdRp domain organization and through phylogenetic clustering of these three domains. The replicases of TaV and EeV lack both Mtr and Hel domains. Furthermore, their RdRp domain has a unique C-A-B motif arrangement in the palm subdomain of the active site that differs from the canonical A-B-C arrangement found in the other tetraviruses, all "alphavirus-like" viruses and indeed almost all known template-dependent polynucleotide polymerases (viral and cellular) carrying the palm sub-domain. Interestingly, the same C-A-B permutation of the motif arrangement is also found in replicases of all dsRNA birnaviruses. This motif rearrangement is a result of migration of ~22 aa residues encompassing motif C between two internal positions, separated by ~110 aa, in a conserved region of ~400 aa. The permuted TaV, EeV and birnavirus enzymes form a minor, deeply separated cluster in the RdRp tree that also includes viruses of the "picornavirus-like supercluster" and order *Nidovirales*. Thus, TaV/EeV and birnaviruses may represent their own virus supercluster. The RdRp of PrV clusters with viruses of other families and therefore appears to belong to yet another lineage.

These complex and incongruent relationships of CP and replicase proteins imply that viruses currently classified as tetraviruses on the basis of their CP and other properties form a polyphyletic group. It is likely that immediate ancestors of TaV/EeV and PrV have independently acquired CP genes from ancestral tetraviruses resembling NβV and HaSV, respectively. Future revision of tetravirus taxonomy will need to address these complexities.

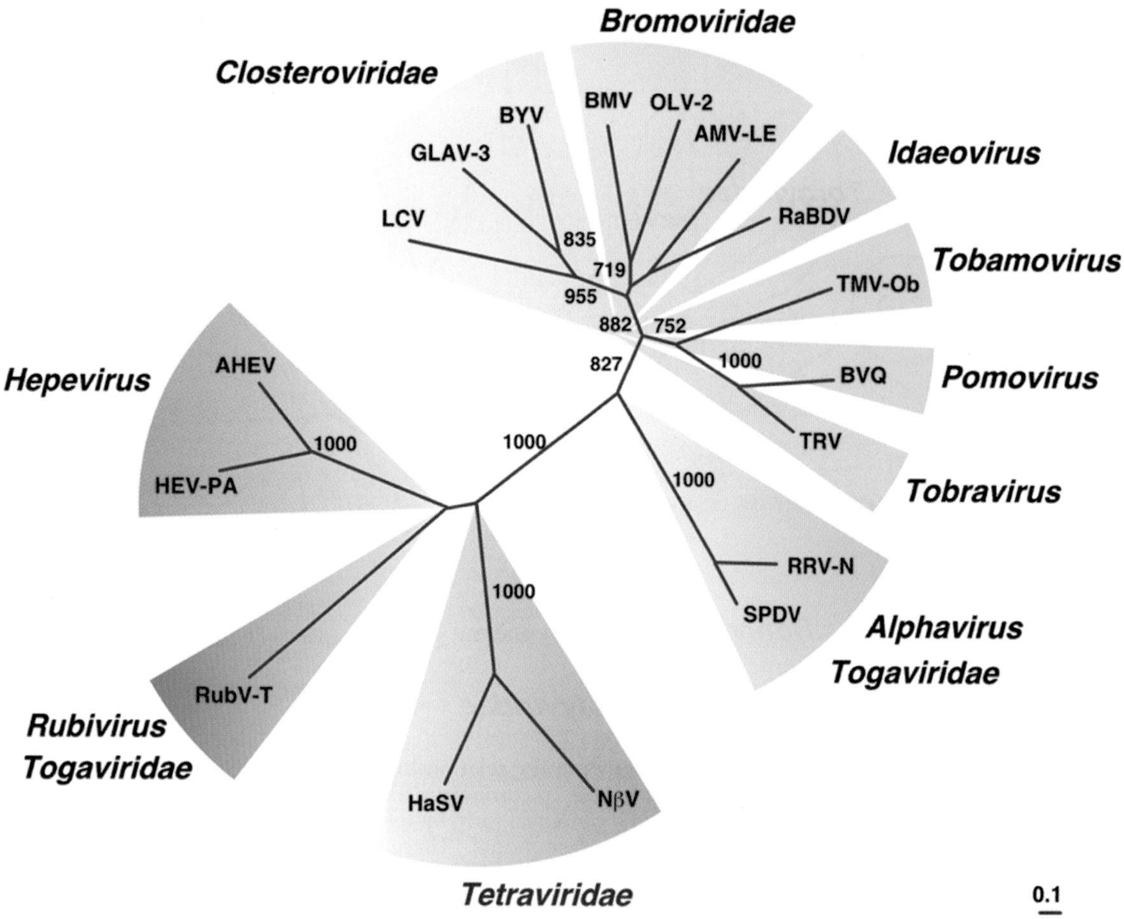

Figure 6: Unrooted phenogram showing the relationships of the RdRps of the tetraviruses NβV and HaSV to representatives of other virus families in the "alphavirus-like supercluster". From the alignment, an unrooted neighbor-joining tree was inferred by the ClustalX1.82 program. Columns containing gaps were removed from the alignment and the Kimura correction for multiple substitutions was on. Branch lengths are drawn to scale. A bootstrap analysis was performed and the values of all bifurcations with support in > 700 out of 1000 bootstraps are indicated at the branching points.
Viruses included in the analysis, abbreviation ( ) and accession numbers [ ] are: *Bromoviridae*: Alfalfa mosaic virus (AMV-LE) [L00163, K02792, K02730], Olive latent virus 2 (OLV-2) [X94346, X94327, X76933, X77115], Brome mosaic virus (BMV) [V00099, J02042, J02043, K02706, K02707, X01678, X02380, M25172]; *Closteroviridae*: Beet yellows virus (BYV) [X73476], Grapevine leafroll-associated virus 3 (GLAV-3) [U82937], Little cherry virus (LCV) [Y10237]; *Tetraviridae*: Helicoverpa armigera stunt virus (HaSV) [U18246], Nudaurelia capensis beta virus (NβV) [AF102884]; *Togaviridae*: Ross River virus strain NB5092 (RRV-N) [M20162], Salmon pancreas disease virus (SPDV) [AJ316244], Rubella virus strain Therien (RUBV-T) [M15240, X05259, X72393, L78917]; *Hepevirus*: Hepatitis E virus strain Pakistan (HeV-PA) [AF185822], Avian hepatitis E virus (AHEV) [AY043166]; *Idaeovirus*: Raspberry bushy dwarf virus (RBDV) [S51557, S55890, D01052]; *Pomovirus*: Beet virus Q (BVQ) [AJ223596 to AJ223598]; *Tobamovirus*: Tobacco mosaic virus strain Ob (TMV-Ob) [L11665]; *Tobravirus*: Tobacco rattle virus (TRV) [AF034622].

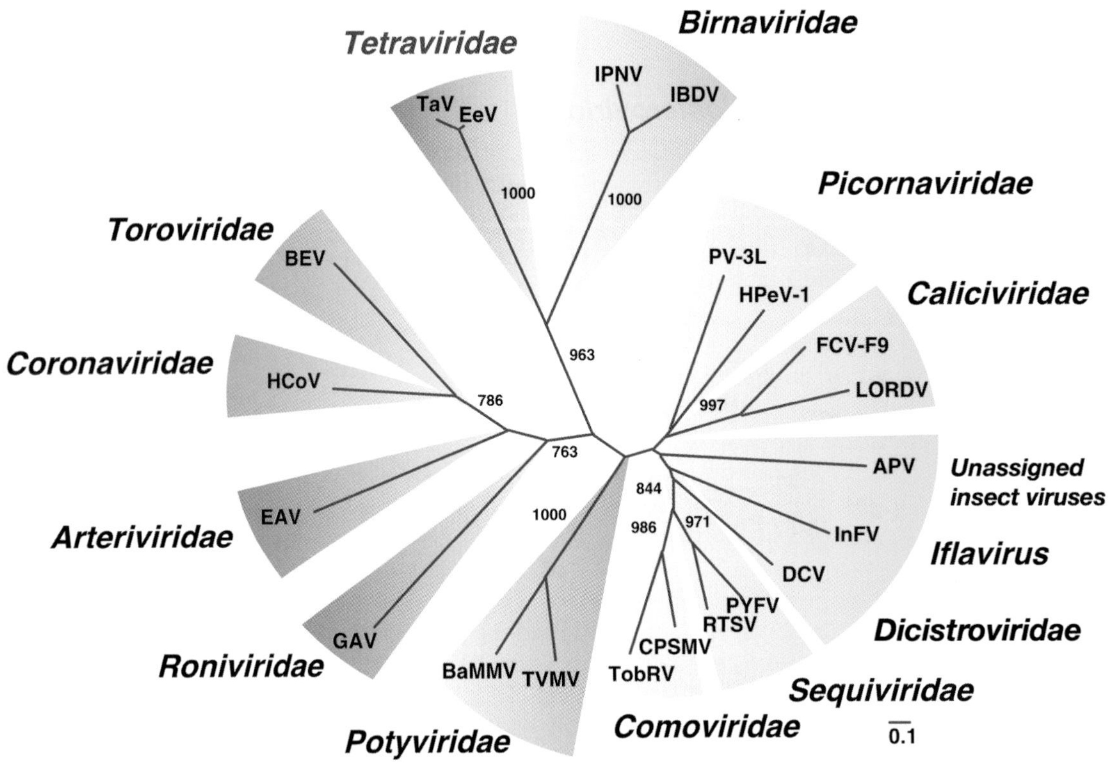

Figure 7: Unrooted phenogram showing the relationships of the RdRps of the tetraviruses TaV and EeV to other virus families and viruses in the "picornavirus-like supercluster". The RdRps of TaV, EeV and the birnaviruses were converted into the canonical form by relocating the motif C sequence (18-20 aa) downstream of the motif B, as in canonical polymerase motifs. These sequences were aligned with those of polymerases from representative viruses in the *Picornaviridae, Dicistroviridae, Sequiviridae, Comoviridae, Caliciviridae, Potyviridae, Coronaviridae, Roniviridae, Arteriviridae,* the genus *Iflavirus* and unclassified insect viruses. Using an extended, gap-free version of the alignment containing 332 informative characters, an unrooted neighbor-joining tree was inferred by the ClustalX1.81 program. All bifurcations with support in > 700 out of 1000 bootstraps are indicated. Different groups of viruses are highlighted.
Virus families and groups, viruses included in the analysis, abbreviations ( ) and the NCBI protein (unless other specified) IDs [ ] are as follows: *Picornaviridae*: Human poliovirus type 3 Leon strain (PV-3L) [130503] and Human parechovirus 1 (HpeV-1) [6174922]; *Iflavirus*: Infectious flacherie virus (InFV) [3025415]; unclassified insect virus Acyrthosiphon pisum virus (APV) [7520835]; *Dicistroviridae*: Drosophila C virus (DCV) [2388673]; *Sequiviridae*: Rice tungro spherical virus (RTSV) [9627951] and Parsnip yellow fleck virus (PYFV) [464431]; *Comoviridae*: Cowpea severe mosaic virus (CPSMV) [549316] and Tobacco ringspot virus (TobRV) [1255221]; *Caliciviridae*: Feline calicivirus F9 (FCV-F9) [130538] and Lordsdale virus (LORDV) [1709710]; *Potyviridae*: Tobacco vein mottling virus (TVMV) [8247947] and Barley mild mosaic virus (BaMMV) [1905770]; *Coronaviridae*: Human coronavirus 229E (HCoV) [12175747] and Berne torovirus (BEV) [94017]; *Arteriviridae*: Equine arteritis virus (EAV) [14583262]; *Roniviridae*: Gill-associated virus (GAV) [9082018]; *Tetraviridae*: Thosea asigna virus (TaV) [AF82930; nt sequence] and Euprosterna elaeasa virus (EeV) [AF461742; nt sequence]; *Birnaviridae*: Infectious pancreatic necrosis virus (IPNV) [133634] and Infectious bursal disease virus (IBDV) [1296811]. *Coronaviridae, Arteriviridae* and *Roniviridae* belong to the order *Nidovirales*.

Derivation of Names

Nudaurelia capensis is the emperor pine moth.
Tetra: from Greek *'tettares'* meaning four, as $T = 4$.

References

Agrawal, D.K. and Johnson, J.E. (1992). Sequence and analysis of the capsid protein of Nudaurelia capensis ω virus, an insect virus with $T = 4$ icosahedral symmetry. *Virology*, **190**, 806-814.

Agrawal, D.K. and Johnson, J.E. (1995). Assembly of the $T = 4$ Nudaurelia capensis ω virus capsid protein, post-translational cleavage, and specific encapsidation of its mRNA in a baculovirus expression system. *Virology*, **207**, 89-97.

Brooks, E.M., Gordon, K.H.J., Dorrian, S.J., Hines, E.R. and Hanzlik, T.N. (2002). Infection of its lepidopteran host by the *Helicoverpa armigera* stunt virus (Tetraviridae). *J. Invertebrate Pathol.*, **80**, 97-111.

Gorbalenya, A.E., Pringle, F.M., Zeddam, J.-L., Luke, B.T., Cameron, C.E., Kalmakoff, J., Hanzlik, T.N., Gordon, K.H.J. and Ward, V.K. (2002). The palm subdomain-based active site is internally permuted in viral RNA-dependent RNA polymerases of an ancient lineage. *J. Mol. Biol.*, **324**, 47-62.

Gordon, K.H.J. and Hanzlik, T.N. (1998). Tetraviruses. In: *The Insect Viruses*, (L.K. Miller and L.A. Ball, eds), pp 269-299. Plenum Publishing Company, New York.

Gordon, K.H.J., Johnson, K.N. and Hanzlik, T.N. (1995). The larger genomic RNA of Helicoverpa armigera stunt tetravirus encodes the viral RNA polymerase and has a novel 3´-terminal tRNA-like structure. *Virology*, **208**, 84-98.

Gordon, K.H.J., Williams, M.R., Hendry, D.A. and Hanzlik, T.N. (1999). Sequence of the genomic RNA of Nudaurelia β virus (*Tetraviridae*) defines a novel virus genome organization. *Virology*, **258**, 42-53.

Gordon, K.H.J., Williams, M.R., Baker, J.S., Gibson, J., Bawden, A.L., Millgate, A., Larkin P.J. and Hanzlik, T.N. (2001). Replication-independent assembly of an insect virus (*Tetraviridae*) in plant cells. *Virology*, **288**, 36-50.

Hanzlik, T.N., Dorrian, S.J., Gordon, K.H.J. and Christian, P.D. (1993). A novel small RNA virus isolated from the cotton bollworm, *Helicoverpa armigera*. *J. Gen. Virol.*, **74**, 1105-1110.

Hanzlik, T.N., Johnson, K.N. and Gordon, K.H.J. (1995). Sequence of RNA2 of the *Helicoverpa armigera* stunt virus (*Tetraviridae*) and bacterial expression of its genes. *J. Gen. Virol.*, **76**, 799-811.

Hanzlik, T.N. and Gordon, K.H.J. (1997). The *Tetraviridae*. *Adv. Virus Res.*, **48**, 101-168.

Johnson, J.E., Munshi, S., Liljas, L., Agrawal, D., Olson, N.H., Reddy, V., Fisher, A., McKinney, B., Schmidt, T. and Baker, T.S. (1994). Comparative studies of $T = 3$ and $T = 4$ icosahedral RNA insect viruses, *Arch. Virol.*, [suppl.] **9**, 497-512.

Moore, N.F. (1991). The Nudaurelia β family of insect viruses. In: *Viruses of Invertebrates*, (E. Kurstak, ed), pp 277-285. Marcel Dekker, New York.

Munshi, S., Liljas, L., Cavarelli, J., Bomu, W., McKinney, B., Reddy, V. and Johnson, J.E. (1996). The 2.8 Å structure of a $T = 4$ animal virus and its implications for membrane translocation of RNA. *J. Mol. Biol.*, **261**, 1-10.

Olson, N.H., Baker, T.S., Johnson, J.E. and Hendry, D.A. (1990). The three-dimensional structure of frozen-hydrated Nudaurelia capensis β virus, a $T = 4$ insect virus. *J. Structural Biol.*, **105**, 111-122.

Pringle, F.M., Gordon, K.H.J., Hanzlik, T.N., Kalmakoff, J., Scotti, P.D. and Ward, V.K (1999). A novel capsid expression strategy for *Thosea asigna* virus, a member of the *Tetraviridae*. *J. Gen. Virol.*, **80**, 1855-63.

Pringle, F.M., Johnson, K.N., Goodman, C.L., McIntosh, A.H. and Ball, L.A. (2003). Providence virus: a new member of the *Tetraviridae* that infects cultured insect cells. *Virology*, **306**, 359-370.

Reinganum, C. (1991). *Tetraviridae*. In: *Atlas of Invertebrate Viruses*, (J.R. Adams and J.R. Bonami, eds), pp 553-592. CRC Press, Boca. Raton, Florida.

Contributed By

Hanzlik, T.N., Gordon, K.H.J., Gorbalenya, A.E., Hendry, D.A., Pringle, F.M., Ward, V.K. and Zeddam, J.-L.

GENUS SOBEMOVIRUS

Type Species *Southern bean mosaic virus*

VIRION PROPERTIES

MORPHOLOGY

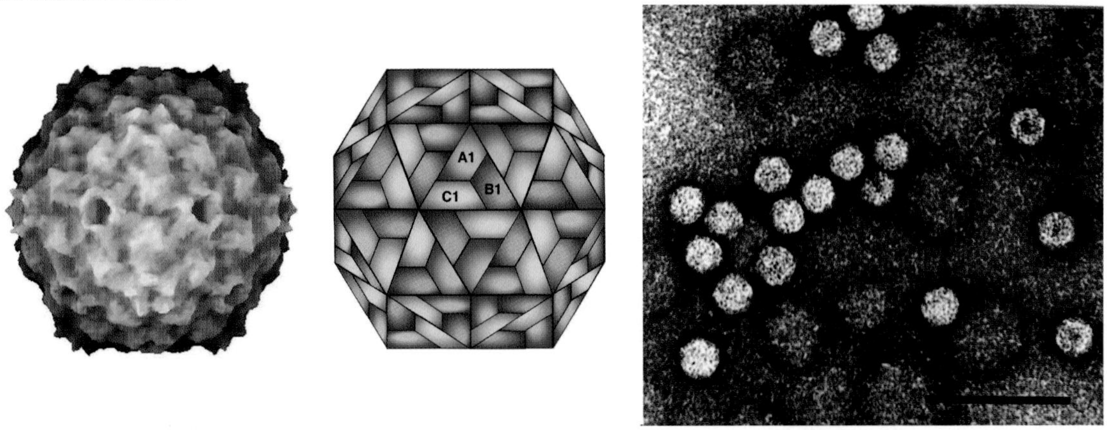

Figure 1: (Left) Electronic rendering of particles of Southern bean mosaic virus (SBMV) (T=3); (Center) Diagrammatic representation of a T=3 structure capsid; (Right) Negative contrast electron micrograph of Rice yellow mottle virus (RYMV) particles stained in uranyl acetate. The bar represents 100 nm.

Virions are about 30 nm in diameter. They have a single tightly packed capsid layer with 180 subunits of about 26-34 kDa assembled on a T = 3 icosahedral lattice. Sobemoviruses are stabilized by divalent cations, pH-dependent protein-protein interactions and salt bridges between protein and RNA. Upon alkali treatment in the presence of divalent chelators, the capsid swells and become sensitive to enzymes and denaturants.

PHYSICOCHEMICAL AND PHYSICAL PROPERTIES

The virion Mr is about 6.6 x 10^6; S_{20w} is about 115S; density is about 1.36 g/cm^3 in CsCl (but virus forms two or more bands in Cs_2SO_4); particles swell reversibly in EDTA and higher pH with concomitant changes in capsid conformation and partial loss of stability.

NUCLEIC ACID

Particles contain a single molecule of positive sense ssRNA, ~ 4.0-4.5 kb in size. A sgRNA molecule, co-terminal with the 3'-end of the genomic RNA, with an Mr of 0.3-0.4 x 10^6, is synthesized in the virus infected cell. Both genomic and subgenomic RNAs have a viral protein (VPg) covalently bound to their 5'-end. The 3' terminus is non-polyadenylated and does not contain a tRNA-like structure. Several sobemoviruses encapsidate a circular viroid-like satellite RNA (220-390 nt).

PROTEINS

The CP subunits are chemically identical but structurally not equivalent. Three types of CP subunits termed A, B, C are related by quasi three-fold axes of symmetry and are involved in different inter-subunit contacts. The CP has two distinct domains. The N-terminal R (random) domain, partially ordered, is localized to the interior of the particle. The S domain which forms the surface of the particle displays a canonical ß-barrel motif. The arrangement of the N-terminal part (in particular the ßA arm) of the sub-unit plays a crucial role in determining the capsid size.

The CP of RYMV is required for cell-to-cell movement as well as for long-distance movement. The protein P1, coded by ORF1, has been associated with cell-to-cell movement for SBMV and RYMV and also for suppression of RNA silencing in RYMV-infected plants. Sequence similarities, suggest that the ORF2 protein has replication functions. No function has been attributed to the protein encoded by ORF3.

LIPIDS

None reported.

CARBOHYDRATES

None reported.

GENOME ORGANIZATION AND REPLICATION

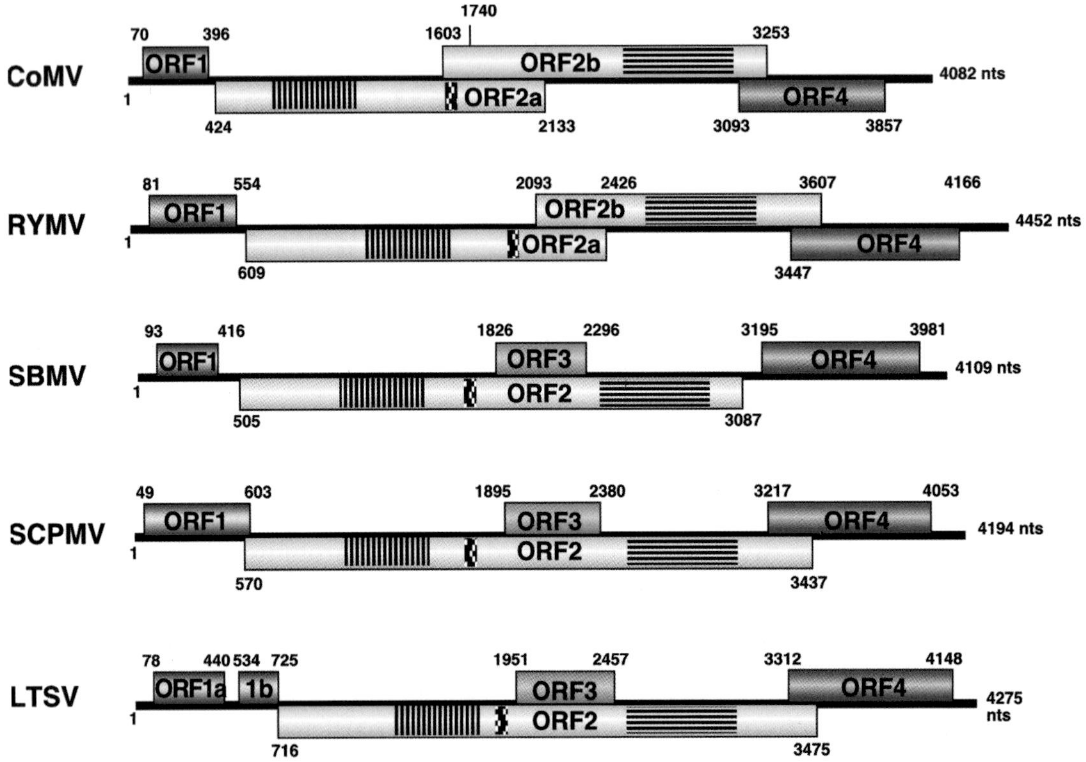

Figure 2: Genomic organization for representatives of four species of sobemoviruses. The lines represent the viral RNA genomes, the boxes indicate the ORFs and the numbers correspond to the nt numbers for the beginning and the ending of the ORFs. The vertical lines indicate the position of serine and cysteine protease motifs and the horizontal lines indicate a polymerase conserved domain of luteoviruses. The ORF frameshift signal: UUUAAAC is positioned by a striped bar and the presence of a second ATG in ORF2b of CoMV is indicated at nt 1740.

The genome of sobemoviruses harbors four ORFs. Based on organizational differences in the central part of the genome (encoding the viral polyprotein), the sobemoviruses are subdivided into Southern cowpea mosaic virus (SCPMV)-like and Cocksfoot mottle virus (CoMV)-like types. The polyprotein of SCPMV is encoded by a large continuous ORF2. The genome of SCPMV also contains an internal coding region, ORF3, situated in the -1 reading frame within ORF2. Similar organization has been reported for LTSV, RGMoV, SBMV and SeMV. In contrast, CoMV lacks the continuous ORF2 and the nested coding region of the SCPMV. Instead, CoMV has two overlapping ORFs, ORF2a and ORF2b. ORF2b is expressed as a fusion protein through a -1 ribosome frameshift mechanism. A similar genomic organization is characteristic of SCMoV and of all strains of RYMV.

Translation initiation of sobemovirus ORF1 and ORF2 proteins occurred via a leaky ribosomal scanning mechanism. ORF1 encodes a small protein involved in virus movement and in suppressing gene silencing. ORF2 codes for a polyprotein having the putative serine-protease, VPg and RdRp domains. The polyprotein is cleaved by the N-terminal serine protease. The CP gene (ORF4) is encoded by a sgRNA at the 3′-end of the genome. The coat protein is required for cell-to-cell and long-distance movement.

ANTIGENIC PROPERTIES

Viral proteins and virions are efficient immunogens. A single precipitin line is formed in gel diffusion tests. There are serological relationships between SBMV, SCPMV and SeMV. SCMoV and LTSV virions are serologically distantly related. Several serotypes with different geographical origins have been identified in some species.

BIOLOGICAL PROPERTIES

HOST RANGE

Sobemoviruses infect both monocotyledonous and dicotyledonous plants, but the natural host range of each virus species is relatively narrow. Disease symptoms are mainly mosaics and mottles. Systemic infections are caused in most natural hosts with most cell types being infected.

TRANSMISSION

Seed transmission occurs in several host plants for some sobemoviruses (SBMV, Southern cowpea mosaic virus; SCPMV). The viruses are transmitted by beetles or, for Velvet tobacco mottle virus (VTMoV), a myrid. All sobemoviruses are readily transmitted mechanically.

GEOGRAPHICAL DISTRIBUTION

Most members have limited distribution but some species are found worldwide.

CYTOPATHIC EFFECTS

Virions are found in both the cytoplasm and nuclei, and late in infection occur as large crystalline aggregates in the cytoplasm and the vacuoles. Infected cells show extensive cytoplasmic vacuolation. Some members invade phloem and xylem cells (SBMV, RYMV), and virions are also found in pit membranes of primary cell walls (RYMV).

LIST OF SPECIES DEMARCATION CRITERIA IN THE GENUS

The criteria demarcating species in the genus are:

- Different host ranges in certain plant species.
- Antigenic differences, and
- Values of 40% or more sequence difference as assessed by hybridization tests or by comparisons of sequence data.

LIST OF SPECIES IN THE GENUS

Species names are in green italic script; strain names and synonyms are in black roman script; tentative species names are in blue roman script. Sequence accession numbers, and assigned abbreviations ( ) are also listed.

SPECIES IN THE GENUS

Blueberry shoestring virus		
Blueberry shoestring virus	[NC003138]	(BSSV)
Cocksfoot mottle virus		
Cocksfoot mottle virus	[Z48630, NC002618, L40905]	(CoMV)
Lucerne transient streak virus		
Lucerne transient streak virus	[NC001696]	(LTSV)
Rice yellow mottle virus		
Rice yellow mottle virus – Cote d'Ivoire	[NC001575]	(RYMV-CI)
Rice yellow mottle virus - Nigeria	[U23142]	(RYMV-Ni)
Ryegrass mottle virus		
Ryegrass mottle virus	[NC003747]	(RGMoV)

POS. SSRNA

Sesbania mosaic virus		
Sesbania mosaic virus	[NC802568]	(SeMV)
Solanum nodiflorum mottle virus		
Solanum nodiflorum mottle virus		(SNMoV)
Southern bean mosaic virus		
Southern bean mosaic virus	[L34672]	(SBMV)
Southern bean mosaic virus – B/Ark	[NC004060]	(SBMV-BAr)
Southern cowpea mosaic virus		
Southern cowpea mosaic virus (previously cowpea strain of SBMV)	[NC001625]	(SCPMV)
Sowbane mosaic virus		
Sowbane mosaic virus		(SoMV)
Subterranean clover mottle virus	[NC004346]	
Subterranean clover mottle virus		(SCMoV)
Turnip rosette virus		
Turnip rosette virus	[NC004553]	(TRoV)
Velvet tobacco mottle virus		
Velvet tobacco mottle virus		(VTMoV)

TENTATIVE SPECIES IN THE GENUS

Cocksfoot mild mosaic virus	(CMMV)
Cynosurus mottle virus	(CnMoV)
Ginger chlorotic fleck virus	(GCFV)
Rottboellia mottle virus	(RoMoV)

PHYLOGENETIC RELATIONSHIPS WITHIN THE GENUS

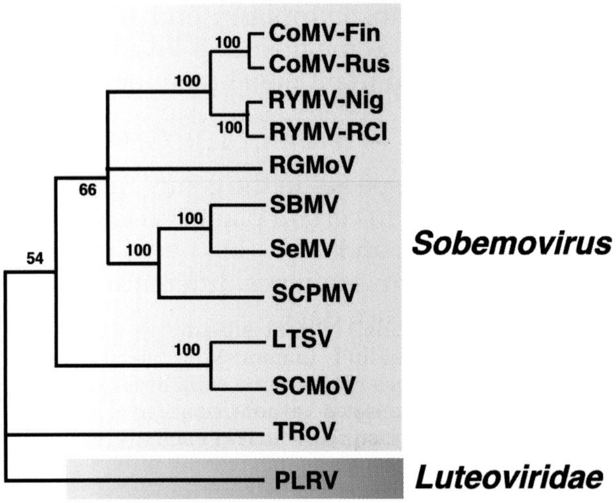

Figure 3: Phylogenetic tree reconstructed by maximum parsimony of representatives of the nine species of sobemoviruses. It includes Cocksfoot mottle virus (CoMV; NC 002618, L40905), Lucerne transient streak virus (LTSV; NC 001696), Rice yellow mottle virus (RYMV; NC 001575, AJ608219), Ryegrass mottle virus (RGMoV; NC 003747), Sesbania mosaic virus (SeMV; NC 002568), Southern bean mosaic virus (SBMV; NC 004060, L34672), Southern cowpea mosaic virus (SCPMV; NC 001625), Subterranean clover mottle virus (SCMoV; NC 004346) and Turnip rosette virus (TRoV; NC 004553). The sequences of the Potato leaf roll virus (*Luteoviridae*)(PLRV ; AF453394) and of the Mushroom bacilliform virus (*Barnaviridae*)(MBV ; NC001633) were used as out-groups. The numbers at each node indicate the percentage of bootstrap support after 1000 replicates.

From their genomic organization, sobemoviruses can be divided into two distinct subgroups: those that express the RdRp from a single in-frame polyprotein, and those that express it via a -1 translational frameshifting mechanism. However, the sobemoviruses with either of the two genomic organizations do not cluster into two discrete sub-groups in phylogenetic analyses.

SIMILARITY WITH OTHER TAXA

The 5′-terminal half of the sobemovirus genome resembles that of the poleroviruses and enamoviruses with the successive functional domains of a serine protease-like, a VPg and an RdRp. Moreover, polymerase sequences of the sobemoviruses are phylogenetically related to those of the poleroviruses and enamoviruses. By contrast, the CP of the sobemoviruses - encoded by the 3′terminal half of the genome - shows sequences and structural similarities with the CP of members of the genus *Necrovirus* of the family *Tombusviridae*. A member of the species *Mushroom bacilliform virus*, the unique species of the family *Barnaviridae*, has a genomic organization similar to that of the sobemoviruses and showed sequence identities with them, both in the polymerase and in the CP genes. Genomic organization of animal viruses of the family *Astroviridae* also shows similarities to that of sobemoviruses, but the sequence identities are more remote.

DERIVATION OF NAMES

Sobemo: sigla derived from the name of type species *so*uthern *be*an *mo*saic.

REFERENCES

Bonneau, C., Brugidou, C., Chen, L., Beachy, R.N. and Fauquet, C.M. (1998). Expression of the Rice yellow mottle virus P1 protein in vitro and in vivo and its involvement in virus spread. *Virology*, **244**, 79-96.

Dwyer, G.I., Njeru, R., Williamson, S., Fosu-Nyarko, J., Hopkins, R., Jones, R.A.C., Waterhouse, P.M. and Jones, M.G.K. (2003). The complete nucleotide sequence of Subterranean clover mottle virus. *Arch. Virol.*, **148**, 2237-2247.

Fargette, D., Pinel, A., Abubakar, Z., Traoré, O., Brugidou, C., Fatogoma, S., Hébrard, E., Choisy, M., Yacouba, S., Fauquet, C. and Konaté, G. (2004). Inferring the evolutionary history of Rice yellow mottle virus from genomic, phylogenetic and phylogeographic studies. *J. Virol.*, **7**, 3252-3261.

Hacker, D.L. and Sivahumaran, K. (1997). Mapping and expression of Southern bean mosaic virus genomic and subgenomic RNAs. *Virology*, **234**, 317-327.

Jones, A.T. and Mayo, M.A. (1984). Satellite nature of the viroid-like RNA-2 of Solanum nodiflorum mottle virus and the ability of other plant viruses to support the replication of viroid-like RNA molecules. *J. Gen. Virol.*, **65**, 1713-1721.

Konate, G., Sarra, S. and Traore, R. (2001). Rice yellow mottle virus is seed-borne but not seed transmitted in rice seeds. *Eur. J. Plant Pathol.*, **107**, 361-364.

Lee, L. and Anderson, E. (1998). Nucleotide sequence of a resistance breaking mutant of southern bean mosaic virus. *Arch. Virol.*, **143**, 2189-2201.

Lokesh, G.L., Gopinath, K., Satheshkumar, P.S. and Savithri, H.S. (2001). Complete nucleotide sequence of Sesbania mosaic virus: a new virus of the genus *Sobemovirus*. *Arch. Virol.*, **146**, 209-223

Mäkinen, K., Tamm, T., Næss, V., Truve, E., Puurand, U., Munthe, T. and Saarma, M. (1995). Characterization of Cocksfoot mottle virus genomic RNA and sequence comparisons with related viruses. *J. Gen. Virol.*, **76**, 2817-2825.

Ngon A Yassi, M., Ritzenthaler, C., Brugidou, C., Fauquet, C.M. and Beachy, R.N. (1994). Nucleotide sequence and genome characterization of Rice yellow mottle virus RNA. *J. Gen. Virol.*, **75**, 249-257.

Opalka, N., Brugidou, C., Bonneau, C., Nicole, M., Beachy, R., Yeager, M. and Fauquet, C.M. (1998). Movement of rice yellow mottle virus between xylem cells through pit membranes. *Proc. Natl. Acad. Sci. USA*, **95**, 3323-3328.

Othman, Y. and Hull, R. (1995). Nucleotide sequence of the bean strain of Southern bean mosaic virus. *Virology*, **206**, 287-297.

Qu, C., Liljas, L., Opalka, N., Brugidou, C., Yeager, M., Beachy, R., Fauquet, C., Johnson, J. and Lin, T. (2000). 3D domain swapping modulates the stability of members of an icosahedral virus group. *Structure*, **8**, 1095-1103.

Sara, S. and Peters, D. (2003). Rice yellow mottle virus is transmitted by cows, donkeys, and grass rats in irrigated rice crops. *Plant Disease*, **87**, 804-808.

Satheshkumar, P., Lokesh, G. and Savithri, H. (2004). Polyprotein processing: *cis* and *trans* proteolytic activities of Sesbania mosaic virus serine protease. *Virology*, **318**, 429-438.
Sivakumaran, K. and Hacker, D. (1998). The 105-k-Da polyprotein of Southern bean mosaic virus is translated by scanning ribosomes. *Virology*, **246**, 34-44.
Tars, K., Zeltins, A. and Liljas, L. (2003). The three-dimensional structure of cocksfoot mottle virus at 2.7 A resolution. *Virology*, **310**, 287-297.
Tamm, T. and Truve, E. (2000). Sobemoviruses. *J. Virol.*, **74**, 6231-6241.
Voinnet, O., Pinto, M. and Baulcombe, D. (1999). Suppression of gene silencing: a general strategy used by diverse DNA and RNA viruses of plants. *Proc. Natl. Acad. Sci. USA*, **96**, 14147-14152.
Wu, S., Rinehart, C.A. and Kaesberg, P. (1987). Sequence and organization of Southern bean mosaic virus genomic RNA. *Virology*, **161**, 73-80.
Zhang, F.Y., Toriyama, S. and Takahashi, M. (2001). Complete nucleotide sequence of Ryegrass mottle virus: a new species of the genus *Sobemovirus*. *J. Gen. Plant Pathol.*, **67**, 63-68.

Contributed By

Hull, R. and Fargette, D.

FAMILY LUTEOVIRIDAE

TAXONOMIC STRUCTURE OF THE FAMILY

Family	*Luteoviridae*	
Genus		*Luteovirus*
Genus		*Polerovirus*
Genus		*Enamovirus*

VIRION PROPERTIES

MORPHOLOGY

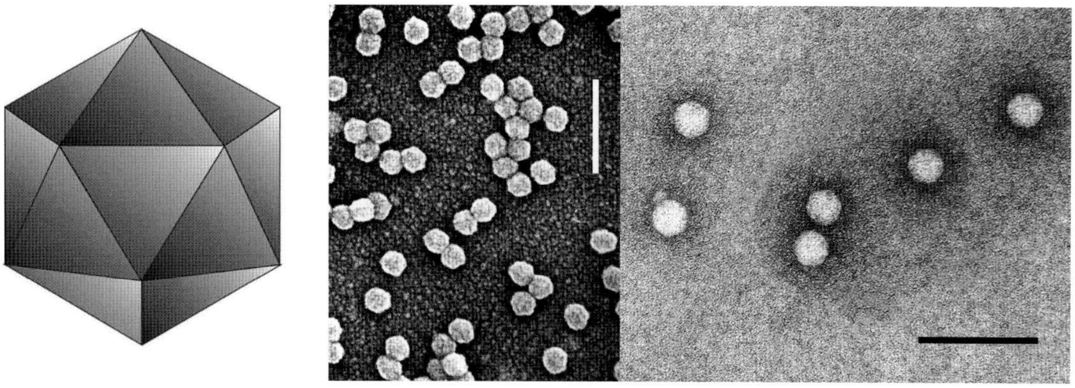

Figure 1: (Left) Diagram of the proposed structure of luteovirus particles. (Center) Negative contrast electron micrograph of particles of Barley yellow dwarf virus-PAV (BYDV-PAV) and (Right) Pea enation mosaic virus-1 (PEMV-1), isolated by means of sucrose density gradient centrifugation and stained with uranyl acetate. The bars represent 100 nm.

Virions are 25 to 30 nm in diameter, hexagonal in outline and have no envelope (Fig. 1). They exhibit icosahedral symmetry (T=3). Particles are composed of two proteins and a core of genomic ssRNA. A small protein (VPg) has been reported to be covalently linked to the 5'-end of the genomic RNA of two poleroviruses and the one enamovirus.

PHYSICOCHEMICAL AND PHYSICAL PROPERTIES

For luteoviruses and poleroviruses, virion Mr is about 6 x 10^6; buoyant density in CsCl is 1.40 g/cm^3; S_{20w} is 106-127S. For enamoviruses (B component), the Mr is about 5.6 x 10^6; buoyant density in CsCl is 1.42 g/cm^3; S_{20w} is 107-122S. Virions are moderately stable and are insensitive to treatment with chloroform or non-ionic detergents, but are disrupted by prolonged treatment with high concentrations of salts. Luteovirus and polerovirus particles are insensitive to freezing.

NUCLEIC ACID

Virions contain a single molecule of infectious, linear, positive-sense ssRNA. The genome size is fairly uniform: 5,677 nt for BYDV-PAV, 5,697 nt for BYDV-PAS, 5,964 nt for Bean leafroll virus(BLRV), 5,708-5,853 nt for Soybean dwarf virus (SbDV), 5,882 nt for Potato leafroll virus (PLRV), 5,776 nt for Beet chlorosis virus (BChV), 5,722 nt for Beet mild yellowing virus (BMYV), 5,641 nt for Beet western yellows virus (BWYV), 5,600 nt for Cereal yellow dwarf virus (CYDV-RPV) (partial sequence), 5,669 nt for Cucurbit aphid borne yellows virus (CABYV), 5,899 nt for Sugarcane yellow leaf virus (ScYLV) and 5,705 nt for (PEMV-1). The RNAs do not have a 3'-terminal poly(A) tract. A VPg is linked to the genome RNA of the poleroviruses PLRV and BYDV-RPV and the enamovirus PEMV-1.

PROTEINS

The five or six proteins encoded by genome RNA are between 4 and 84 kDa (Table 1). The CP gene has been assigned to ORF3, which is followed in frame by ORF5, and the polerovirus VPg has been assigned to ORF1.

Virion structural proteins are CP and a "readthrough" protein, which is a fusion of the products of the CP gene and the contiguous ORF5. The readthrough protein may be associated with aphid transmission or virus particle stability. The product of ORF4 has been shown to be required for long distance movement of some luteoviruses and poleroviruses. ORF4 is absent from enamovirus RNA. The region containing ORF6 in the luteoviruses, but not the ORF6 translation product, acts as a translational enhancer of the expression of BYDV-PAV RNA.

Table 1: Proteins of the different ORFs with sizes (kDa) and possible function(s).

ORF	*Luteovirus*	*Polerovirus*	*Enamovirus*	Function of product
0	NA	28-30	34	possible membrane-linked replication factor
1	39-42	66-72	84	helicase motifs in luteoviruses; protease and VPg in poleroviruses
2	60-62	65-72	67	probable RdRp
3	22	22-23	21	CP gene
4	16-21	17-21	NA	probable MP
5	43-59	50-56	29	possible aphid transmission or virus particle stability factor
6	4-7	7-9	NA	unknown

LIPIDS

None reported.

CARBOHYDRATES

None reported.

GENOME ORGANIZATION AND REPLICATION

The genomes contain 5 or 6 ORFs (Fig. 2). The genera can be distinguished on the basis of the arrangements and sizes of the ORFs. The ORFs encoding the replication-related proteins (1 and 2) of the luteoviruses are not homologous to the corresponding ORFs of the poleroviruses. The products of ORFs 1 and 2 of the luteoviruses are most similar to those of viruses in the family *Tombusviridae*, while the products of ORFs 1 and 2 of the poleroviruses and enamoviruses are related to those of sobemoviruses. Also, ORF0 is present in the genomes of poleroviruses and enamoviruses, ORF4 is present in the genomes of luteoviruses and poleroviruses, and ORF6 is present in the genomes of some luteoviruses and poleroviruses. ORF0 overlaps ORF1 (poleroviruses and enamoviruses), which overlaps ORF2. ORF4 is contained completely within ORF3 (luteoviruses and poleroviruses). Finally, ORF5 is positioned directly downstream of, and contiguous with, ORF3.

The differences among luteoviruses, poleroviruses and enamoviruses are principally in the 5'-end of the genome. ORFs 0, 1 and 2 are translated from the genomic RNA. ORF2 is translated by frameshift from ORF1 and thus shares an amino terminus with the product of ORF1. ORFs 3, 4 (in luteoviruses and poleroviruses) and 5 are expressed from a sgRNA. ORF5 is probably translated *via* a readthrough following translation of ORF3. Luteoviruses produce two additional sgRNAs, the larger of which contains ORF6. Some poleroviruses produce a second sgRNA.

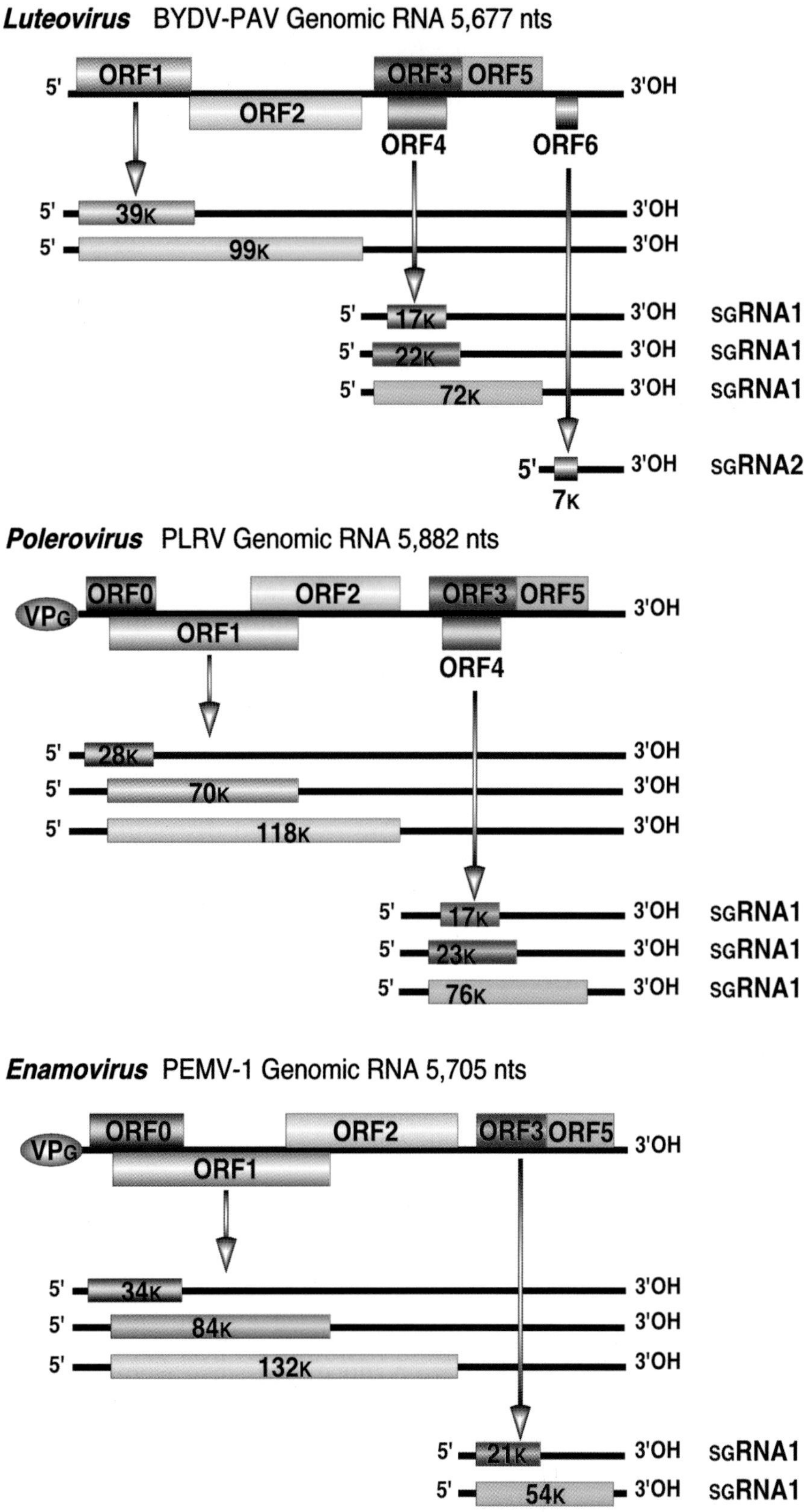

Figure 2: Diagram of the genome organization and map of the translation products typical of viruses in each genus of the family *Luteoviridae*. Solid lines represent RNA; boxes represent ORFs; thinner boxes represent translation products; grey circles represent VPgs.

There are no data on post-translational modification. Particles of some strains of CYDV-RPV contain 322 nt satellite RNAs and virions of some isolates that consist of PEMV-1 together with the umbravirus Pea enation mosaic virus-2 (PEMV-2) contain 717 nt satellite RNAs in addition to genomic RNAs.

Antigenic Properties

Luteovirus and polerovirus particles are strongly immunogenic. Species within a genus are more closely related serologically than are species in different genera. Serological relationships may be detected when comparing disrupted virus particles that are not detectable when intact virions are tested. Virions produced in plants infected with PEMV-1 together with PEMV-2 (*Umbravirus*) are moderately antigenic. In gel diffusion assays, aphid-transmissible isolates display an antigenic determinant that is absent from aphid-non-transmissible isolates. No serological relationships have been reported between enamoviruses and either luteoviruses or poleroviruses.

Biological Properties

Host Range

Several members of the family *Luteoviridae* have host ranges largely restricted to one plant family. For example, BYDV and CYDV infect many grasses, BLRV infects mainly legumes, and Carrot red leaf virus infects mainly plants in the family Umbelliferae. Other members of the family *Luteoviridae* infect plants in several or many different families. For example, BWYV infects more than 150 species of plants in over 20 families.

Geographic Distribution

Members of the family *Luteoviridae* have been reported from arctic, temperate, sub-tropical, and tropical regions. Some of the viruses are found worldwide, such as BYDV, BWYV and PLRV. Others have more restricted distributions, such as Tobacco necrotic dwarf virus, which has been reported only from Japan, and Groundnut rosette assistor virus, which has been reported from south Saharan countries in Africa.

Transmission

Transmission is in a circulative, non-propagative manner by specific aphid vectors. Virus is acquired by phloem feeding, enters the hemocoel of the aphid via the hindgut (e.g., BYDV-PAV) or posterior midgut (e.g., PLRV), circulates in hemolymph and enters the accessory salivary gland. Inoculation probably results from transport of virus into the salivary duct and introduction of saliva into the plant during feeding. PEMV-1 is readily transmitted mechanically, a property dependent on its multiplication in cells co-infected with PEMV-2 (*Umbravirus*), but aphid transmissibility can be lost after several mechanical passages.

Cytopathology

Luteovirus and polerovirus particles are largely confined to phloem cells; PEMV-1, with PEMV-2, is found in phloem and mesophyll tissue. Virus particles are found in both the nuclei and cytoplasm of infected cells. Luteoviruses and poleroviruses often cause phloem necrosis that spreads from inoculated sieve elements and causes symptoms by inhibiting translocation, slowing plant growth and inducing the loss of chlorophyll. The genome of PEMV-1 is capable of autonomous replication in protoplasts, but is dependent on PEMV-2 to support systemic invasion, which induces mosaic and enation symptoms.

List of Species Demarcation Criteria in the Family

Criteria used to demarcate species of the family *Luteoviridae* include:
- Differences in breadth and specificity of host range;
- Failure of cross protection in either one-way or two-way relationships;

- Differences in serological specificity with discriminatory polyclonal or monoclonal antibodies;
- Differences in aa sequences of any gene product of greater than 10%.

The nt sequences of BLRV and SbDV (genus *Luteovirus*) lack ORF0, like those of luteoviruses, and the predicted aa sequences of their replication proteins are similar to those of the luteoviruses. However, BLRV and SbDV structural proteins are more closely related to those of the poleroviruses. The genome of the polerovirus ScYLV contains an ORF0. ScYLV ORFs 1 and 2 are most closely related to those of the polerovirses, ORFs 3 and 4 are most closely related to those of the luteoviruses and ORF5 is most closely related to the read-through protein gene of the enamovirus. These viruses may represent hybrids of sequences from these three genera.

GENUS LUTEOVIRUS

Type Species *Barley yellow dwarf virus – PAV*

DISTINGUISHING FEATURES

Virion buoyant density in CsCl is 1.39-1.40 g/cm^3; S_{20w} is 106-118S. Genome sizes are 5273 nt (BYDV-MAV) (partial sequence), 5,677 nt (BYDV-PAV), and 5,697 nt (BYDV-PAS). The genome RNA does not have a VPg. Genome properties are the key features. There is no ORF0 and frameshift from ORF1 into ORF2 occurs at the termination codon of ORF1. The translation products of ORF1 and ORF2 form replication-related proteins, which are most similar to those of viruses in the family *Tombusviridae*. The length of the non-coding sequence between ORF2 and ORF3 is about 100 nt. There is no evidence for the presence of a genome-linked protein and translation is by a cap-independent mechanism. ORF4 is present and contained within ORF3.

LIST OF SPECIES DEMARCATION CRITERIA IN THE GENUS

See above "List of Species Demarcation Criteria in the Family".

LIST OF SPECIES IN THE GENUS

Species names are in green italic script; strain names and synonyms are in black roman script; tentative species names are in blue roman script. Sequence accession numbers, and assigned abbreviations ( ) are also listed.

SPECIES IN THE GENUS

Barley yellow dwarf virus-MAV		
Barley yellow dwarf virus-MAV	[D01213]	(BYDV-MAV)
Barley yellow dwarf virus-PAS ‡		
Barley yellow dwarf virus-PAS	[AF218798, U29604]	(BYDV-PAS)
Barley yellow dwarf virus-PAV		
Barley yellow dwarf virus-PAV	[D01214, L25299, D85783, AF235167]	(BYDV-PAV)
(Barley yellow dwarf virus-RGV(		
(Rice giallume virus)		
Bean leafroll virus		
Bean leafroll virus	[AF441393]	(BLRV)
(Legume yellows virus)		
(Michigan alfalfa virus)		
(Pea leafroll virus)		
Soybean dwarf virus		
Soybean dwarf virus	[L24049]	(SbDV)
(Subterranean clover red leaf virus)		

TENTATIVE SPECIES IN THE GENUS

None reported.

GENUS POLEROVIRUS

Type Species *Potato leafroll virus*

DISTINGUISHING FEATURES

Particles are thought to have 180 subunits arranged in a T=3 icosahedron. Virion buoyant density in CsCl is 1.39-1.42; S_{20w} is 115-127S. Genome sizes range from 5,641 nt for BWYV to 5,882 nt for PLRV. Poleroviruses and enamoviruses are distinguished from luteoviruses by genome features. The polerovirus genome has a VPg linked to the 5'-end of the genome RNA. The presence of a VPg has been confirmed for PLRV and CYDV-RPV. Poleroviruses possess an ORF0 and a non-coding region between ORF2 and ORF3 of about 200 nt. The translation products of ORF1 and ORF2 form replication-related proteins, which are most similar to those of sobemoviruses. Frameshift from ORF1 into ORF2 occurs upstream of the termination of ORF1. Polerovirus genomes differ from those of enamoviruses in that ORF4 is present within ORF3 and ORF5 is about 1400 nt.

LIST OF SPECIES DEMARCATION CRITERIA IN THE GENUS

See above "List of Species Demarcation Criteria in the Family".

LIST OF SPECIES IN THE GENUS

Species names are in green italic script; strain names and synonyms are in black roman script; tentative species names are in blue roman script. Sequence accession numbers, and assigned abbreviations ( ) are also listed.

SPECIES IN THE GENUS

Beet chlorosis virus		
Beet chlorosis virus	[NC_002766]	(BChV)
Beet mild yellowing virus		
Beet mild yellowing virus	[X83110]	(BMYV)
Beet western yellows virus		
Beet western yellows virus	[X13062, X13063]	(BWYV)
(Malva yellows virus)		
(Turnip mild yellows virus)		
Cereal yellow dwarf virus-RPS		
Cereal yellow dwarf virus-RPS	[AF235168]	(CYDV-RPS)
Cereal yellow dwarf virus-RPV		
Cereal yellow dwarf virus-RPV	[Y07496]	(CYDV-RPV)
Cucurbit aphid-borne yellows virus		
Cucurbit aphid-borne yellows virus	[X76931]	(CABYV)
Potato leafroll virus		
Potato leafroll virus	[X14600, X74789, D13954,	(PLRV)
(Solanum yellows virus)	D00734, D13953, D00733,	
(Tomato yellow top virus)	D00530, X14600]	
Sugarcane yellow leaf virus		
Sugarcane yellow leaf virus	[AF157029, AJ249447, AY236971]	(ScYLV)
Turnip yellows virus ‡		
Turnip yellows virus	[AF168608, AF168606]	(TuYV)

TENTATIVE SPECIES IN THE GENUS

None reported.

GENUS ENAMOVIRUS

Type Species *Pea enation mosaic virus-1*

DISTINGUISHING FEATURES

PEMV-1 occurs as part of a complex with PEMV-2 (*Umbravirus*). Unlike other members of the family *Luteoviridae*, PEMV-1 is readily transmitted mechanically, a property dependent on its multiplication in cells co-infected with PEMV-2, but aphid transmissibility can be lost after several mechanical passages. Virions are found in mesophyll tissue as well as in vascular tissue. PEMV-1 will multiply when inoculated to isolated leaf protoplasts, but there is no evidence that it can spread in plants.

Enamovirus particles (B component) are 25-28 nm in diameter. A 180 subunit arrangement in a T=3 icosahedron has been proposed. The virions have Mr of about 5.6 x 10^6, buoyant densities in CsCl of 1.42 g/cm^3 and S_{20w} of 107-122S.

Genome size is 5,706 nt (PEMV-1). A VPg is associated with virion RNA of PEMV-1. The PEMV-1 genome contains an ORF0, but does not contain an ORF4 (present in luteoviruses and poleroviruses). The non-coding intergenic region between ORF2 and ORF3 is about 200 nt in length. The translation products of ORF1 and ORF2 form replication-related proteins, which are most similar to those of sobemoviruses. Frameshift from ORF1 into ORF2 occurs upstream of the termination of ORF1. The PEMV-1 genome contains an ORF5 of about 730 nt.

LIST OF SPECIES DEMARCATION CRITERIA IN THE GENUS

Not applicable.

LIST OF SPECIES IN THE GENUS

Species names are in green italic script; strain names and synonyms are in black roman script; tentative species names are in blue roman script. Sequence accession numbers, and assigned abbreviations ( ) are also listed.

SPECIES IN THE GENUS

Pea enation mosaic virus-1
Pea enation mosaic virus-1 [L04573] (PEMV-1)

TENTATIVE SPECIES IN THE GENUS

None reported.

LIST OF UNASSIGNED SPECIES IN THE FAMILY

Barley yellow dwarf virus-GPV
Barley yellow dwarf virus-GPV [L10356] (BYDV-GPV)
Barley yellow dwarf virus-RMV
Barley yellow dwarf virus-RMV [Z14123, L12757-9] (BYDV-RMV)
Barley yellow dwarf virus-SGV
Barley yellow dwarf virus-SGV [U06865] (BYDV-SGV)
Carrot red leaf virus
Carrot red leaf virus (CtRLV)
Chickpea stunt disease associated virus
Chickpea stunt disease associated virus [Y11530] (CpSDaV)
Groundnut rosette assistor virus
Groundnut rosette assistor virus [Z68894] (GRAV)
Indonesian soybean dwarf virus
Indonesian soybean dwarf virus (ISDV)
Strawberry mild yellow edge associated virus
Strawberry mild yellow edge associated (SMYEaV)

virus

Sweet potato leaf speckling virus

Sweet potato leaf speckling virus (SPLSV)

Tobacco necrotic dwarf virus

Tobacco necrotic dwarf virus (TNDV)

Tobacco vein distorting virus ‡

Tobacco vein distorting virus [AJ459320, AJ457176, AF402621] (TVDV)

PHYLOGENETIC RELATIONSHIPS WITHIN THE FAMILY

The three genera within the family *Luteoviridae* share very similar structural protein genes (ORFs 3 and 5) whose products show varying levels of serological relatedness. Phylogenetic analysis of the predicted aa sequences of the polymerases (ORF2) most clearly separate the members of the family *Luteoviridae* into the three genera (Fig. 3). ORFs 1 and 2 of the luteoviruses are most closely related to the polymerase genes of viruses in the family *Tombusviridae*, while ORFs 1 and 2 of the poleroviruses and enamoviruses are related to those of viruses in the genus *Sobemovirus*. This is manifest in differences in gene number and placement and the presence or absence of a VPg. BLRV, SbDV and ScYLV, unassigned species, group with different genera depending on which sequences are analyzed. The CP sequences of BYDV-GPV, BYDV-RMV, CpSDaV, GRAV and SPLSV group with the poleroviruses; BYDV-SGV groups with the luteoviruses.

SIMILARITY WITH OTHER TAXA

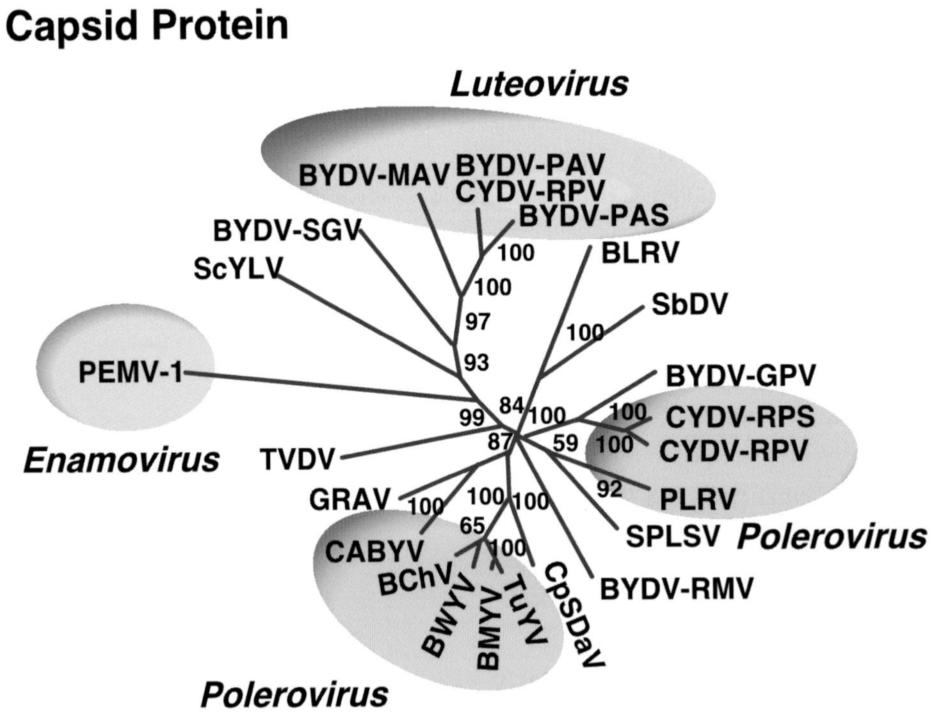

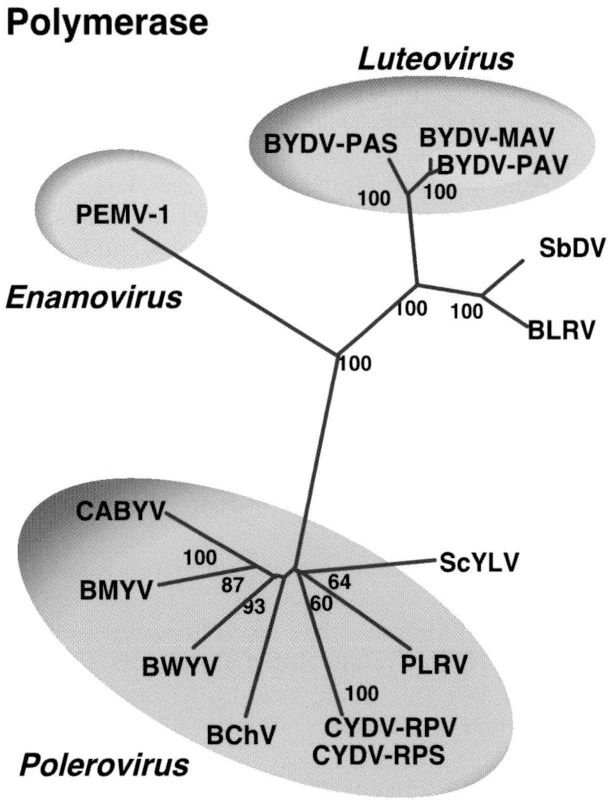

Figure 3: Phylogenetic analyses of the (Top) Capsid Protein, and (Bottom) Polymerase sequences of representatives of species in the family *Luteoviridae*. Amino acid sequences were aligned with CLUSTALX and trees constructed with PAUP. Bootstraps values above 50% are indicated.

Viruses in the family *Luteoviridae* have replication-related proteins which are sufficiently similar to those in other genera to suggest evolutionary relationships. The putative luteovirus polymerases resemble those of members of the family *Tombusviridae*. In contrast, polymerases of poleroviruses and enamoviruses resemble those of viruses in the genus *Sobemovirus*. These polymerase types are thought to be very distant in evolutionary terms and it has been suggested that the origin of these genomes was recombination between ancestral genomes containing the CP genes characteristic of the family *Luteoviridae* and genomes containing either of the two polymerase types. The CP sequences of PLRV and Rice yellow mottle virus, a sobemovirus, share 33% similarity, which has been used to predict the structure of PLRV and other members of the family *Luteoviridae*.

Pos. ssRNA

Derivation of Names

Enamo: sigla from Pea *ena*tion *mo*saic virus
Luteo: from Latin luteus, "yellow"
Polero: sigla from *Po*tato *le*af *ro*ll virus

References

Ashoub, A., Rohde, W. and Prufer, D. (1998). *In planta* transcription of a second subgenomic RNA increases the complexity of the subgroup 2 luteovirus genome. *Nuc. Acids Res.*, **26,** 420-426.

Bencharki, B., Mutterer, J., El Yamani, M., Ziegler-Graff, V., Zaoui, D. and Jonard, G. (1999). Severity of infection of Moroccan barley yellow dwarf virus PAV isolates correlates with variability in their coat protein sequences. *Ann. Appl. Biol.*, **134**, 89-99.

Brault, V., van Den Heuvel, J.F.J.M., Verbeek, M., Ziegler-Graff, V., Reutenauer, A., Herrbach, E., Garaud, J.-C., Guilley, H., Richards, K. and Jonard, G. (1995). Aphid transmission of beet western yellows luteovirus requires the minor capsid read-through protein P74. *EMBO J.*, **14**, 650-659.

Demler, S.A. and De Zoeten, G.A. (1991). The nucleotide sequence and luteovirus-like nature of RNA 1 of an aphid non-transmissible strain of pea enation mosaic virus. *J. Gen. Virol.*, **72**, 1819-1834.

Demler, S.A., De Zoeten, G.A., Adam, G. and Harris, K.F. (1995). Pea enation mosaic enamovirus: properties and aphid transmission. In: *The Plant Viruses 5*: polyhedral virions and bipartite RNA genomes (B.D. Harrison and A.F. Murant, eds), pp 303-344. Plenum, New York and London.

Domier, L.L., McCoppin, N.K., Larsen, R.C. and D'Arcy, C.J. (2002). Nucleotide sequence shows that *Bean leafroll virus* has a Luteovirus-like genome organization. *J. Gen. Virol.*, **83**, 1791-1798.

Fuentes, S., Mayo, M.A., Jolly, C.A., Nakano, M., Querci, M. and Salazar, L.F. (1996). A novel luteovirus from sweet potato, sweet potato leaf speckling virus. *Ann. Appl. Biol.*, **128**, 491-504.

Hauser, S., Stevens, M., Beuve, M. and Lemaire, O. (2002). Biological properties and molecular characterization of *Beet chlorosis virus* (BChV). *Arch. Virol.*, **147**, 745-762.

Hauser, S., Stevens, M., Mougel, C., Smith, H.G., Fritsch, C., Herrbach, E. and Lemaire, O. (2000). Biological, serological, and molecular variability suggest three distinct polerovirus species infecting beet or rape. *Phytopathology*, **90**, 460-466.

Martin, R.R. and D'Arcy, C.J. (1990). Relationships among luteoviruses based on nucleic acid hybridization and serological studies. *Intervirology*, **31**, 23-30.

Mayo, M.A. and Ziegler-Graff, V. (1996). Molecular biology of luteoviruses. *Adv. Virus Res.*, **46**, 413-460.

Miller, W.A. and Rasochova, L. (1997). Barley yellow dwarf viruses. *Ann. Rev. Phytopath.*, **35**, 167-190.

Mo, X.H., Qin, X.Y., Wu, J., Yang, C., Wu, J.Y., Duan, Y.Q., Li, T.F. and Chen, H.R. (2003). Complete nucleotide sequence and genome organization of a Chinese isolate of tobacco bushy top virus. *Arch. Virol.*, **148,** 389-397.

Moonan, F., Molina, J. and Mirkov, T.E. (2000). Sugarcane yellow leaf virus: An emerging virus that has evolved by recombination between luteoviral and poleroviral ancestors. *Virology*, **269**, 156-171.

Naidu, R.A., Mayo, M.A., Reddy, S.V., Jolly, C.A. and Torrance, L. (1997). Diversity among the coat proteins of luteoviruses associated with chickpea stunt disease in India. *Ann. Appl. Biol.*, **130**, 37-47.

Sadowy, E., Maasen, A., Juszczuk, M., David, C., Zagorski-Ostoja, W., Gronenborn, B. and Hulanicka M.D. (2001). The ORF0 product of Potato leafroll virus is indispensable for virus accumulation. *J. Gen. Virol.*, **82**, 1529-1532.

Terauchi, H., Kanematsu, S., Honda, K., Mikoshiba, Y., Ishiguro, K. and Hidaka, S. (2001). Comparison of complete nucleotide sequences of genomic RNAs of four Soybean dwarf virus strains that differ in their vector specificity and symptom production. *Arch. Virol.*, **146**, 1885-1898.

Terradot, L., Souchet, M., Tran, V. and Giblot Ducray-Bourdin, D. (2001). Analysis of a three-dimensional structure of Potato leafroll virus coat protein obtained by homology modeling. *Virology*, **286**, 72-82.

van Der Wilk, F., Verbeek, M., Dullemans, A.M. and van Den Heuvel, J.F.J.M. (1997). The genome-linked protein of potato leafroll virus is located downstream of the putative protease domain of the ORF1 product. *Virology*, **234**, 300-303.

Contributed By

D'Arcy, C.J. and Domier, L.L.

GENUS UMBRAVIRUS

Type Species *Carrot mottle virus*

VIRION PROPERTIES

MORPHOLOGY

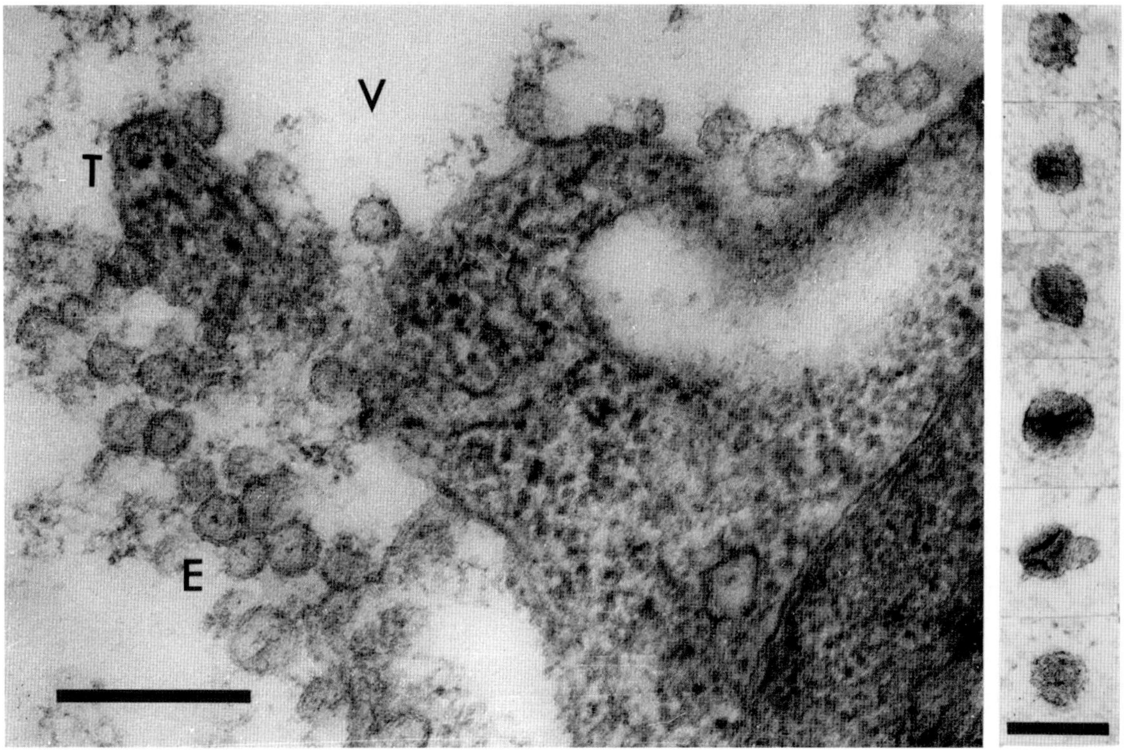

Figure 1: (Left) Section of palisade mesophyll cell from a leaf of *Nicotiana clevelandii* systemically infected with Carrot mottle virus (CMoV), showing enveloped structures (E) ~52 nm in diameter in the cell vacuole (V) in association with the tonoplast (T). The bar represents 250 nm. (Right) Enveloped structures ~52 nm in diameter in a partially purified preparation from CMoV-infected *N. clevelandii*, stained with 2% uranyl acetate. The bar represents 100 nm.

Umbraviruses do not form conventional virus particles, and the four genomes whose complete sequences are known lack plausible ORFs for capsid CPs. Umbraviruses rely on the CP of a helper virus, characteristically from a virus in the family *Luteoviridae*, for encapsidation and for transmission by the vector of the helper virus. However, in single infections by umbraviruses, the infectivity in buffer extracts of leaves is surprisingly stable, though very sensitive to treatment with organic solvents, suggesting that the infective RNA is protected in lipid-containing structures. In plants infected with Carrot mottle virus (CMoV), enveloped structures ~52 nm in diameter (Fig. 1) occur in the vacuoles of infected cells and in partially purified preparations. These structures may be involved in virus replication and/or serve to protect the RNA. Similar structures occur in plants infected with the bean yellow vein-banding strain of *Pea enation mosaic virus-2* (BYVBV), Groundnut rosette virus (GRV) and Lettuce speckles mottle virus (LSMV), but no information is available for other umbraviruses.

PHYSICOCHEMICAL AND PHYSICAL PROPERTIES

Infectivity in leaf extracts is stable for several hours at room temperature or several days at 5°C, but is abolished by treatment with organic solvents. Partially purified preparations of CMoV consist predominantly of cell membranes but contain infective components which, because they have a sedimentation coefficient of ~270S and a buoyant density of ~1.15 g/cm^3 in CsCl, are probably the 52 nm-diameter enveloped structures observed in these preparations. An infective fraction from GRV-infected tissue contained complexes with a buoyant density of 1.34–1.45 g/cm^3 consisting of filamentous ribonucleoprotein particles, composed of the ORF3 protein and virus RNA, embedded in a matrix. The relationship between these two kinds of structure is unclear.

Nucleic Acid

Nucleic acid preparations made by extracting leaves with phenol are often much more infective than buffer extracts. The infective agent in these preparations is a ssRNA, but the preparations also contain abundant dsRNA. The genome consists of one linear segment of positive-sense ssRNA. Nucleotide sequences have been determined for five umbraviruses: Carrot mottle mimic virus, (CMoMV) (4,201 nt), GRV (4,019 nt), Pea enation mosaic virus (PEMV-2) (4,253 nt), Tobacco bushy top virus (TBTV) (4,152 nt) and Tobacco mottle virus (TMoV) (incomplete). These genomes are probably not polyadenylated at their 3'-ends; there is no information about the structures at their 5'-ends.

Proteins

No structural proteins are reported. The nucleotide sequences lack plausible ORFs for CPs but possess ORFs for four potential non-structural protein products.

Lipids

Although no conventional virus particles are formed, the sensitivity to organic solvents, and low buoyant density, of the infective components in partially purified preparations of CMoV suggests that this infectivity is associated with lipid, probably of plant origin. The infective components probably correspond to the enveloped structures seen in sections of infected leaves.

Carbohydrates

None reported.

Genome Organization and Replication

Figure 2 shows the genome organization of GRV; those of other umbraviruses are very similar. For each RNA, there is at the 5'-end a very short non-coding region preceding ORF1, which encodes a putative product of 31-37 kDa. ORF2, which slightly overlaps the end of ORF1, could encode a product of 63-65 kDa but lacks an AUG initiation codon near its 5'-end. However, immediately before the stop codon of ORF1 there is a 7-nt sequence that is associated with frameshifting in several plant and animal viruses, and it seems probable that ORF1 and ORF2 are translated as a single polypeptide of 94-98 kDa by a mechanism involving a -1 frameshift. The predicted product contains, in the ORF2 region, sequence motifs characteristic of viral RdRp. A short untranslated region separates ORF2 from ORF3 and ORF4, which overlap each other almost completely in different reading frames and each yield a putative product of 26-29 kDa. The ORF4 product contains sequences characteristic of plant virus MPs. The ORF3 product of different umbraviruses studied have up to 50% similarity to each other but no significant similarity to any other viral or non-viral proteins; their function is to protect viral RNA and enable its transport through the phloem.

Umbravirus-infected leaf tissue contains abundant dsRNA that is not itself infective but that becomes so when heat-denatured. Two dsRNA species are common to all umbraviruses: dsRNA-1 (~4.2-4.8 kbp) and dsRNA-2 (~1.1-1.5 kbp). cDNA copies of dsRNA-1 hybridize with dsRNA-2 and these molecules are thought to represent double-stranded forms of, respectively, genomic and subgenomic ssRNA species. ORFs 3 and 4 are probably expressed from sgRNA. There is evidence for the presence in GRV-infected plants of two less-than-full-length RNA species of very similar size, close to that expected for such sgRNAs, and corresponding to that of dsRNA-2. The dsRNA-2 of CMoMV has been shown to include the sequences of ORFs 3 and 4, and the 3' UTR.

Some umbraviruses possess one or more additional dsRNA species, associated in at least one instance (GRV) with the presence of a satellite RNA. PEMV-2 too has a satellite RNA, and each of these satellites can be supported by the helper virus of the other.

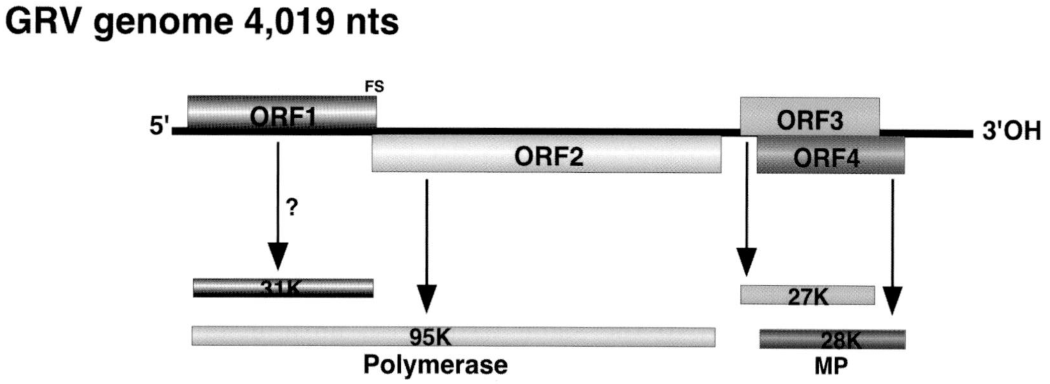

Figure 2: Diagram showing the genomic organization of Groundnut rosette virus (GRV). The continuous horizontal line represents the genome RNA, and the numbered blocks the correspondingly numbered ORFs. The lower part of the diagram shows the predicted translation products, with their size. The potential product of ORF1 has not been shown to exist. The single product produced from ORFs 1 and 2, probably as a result of a -1 frameshift event (FS), is thought to be a polymerase because it contains, in the ORF2 region, sequences characteristic of viral RdRp. The ORF3 product functions to protect viral RNA and enable its transport through the phloem. The ORF4 product, marked MP, has a cell-to-cell movement function.

Antigenic Properties

None reported.

Biological Properties

Host Range

Individual umbraviruses are confined in nature to one or a few host plant species. Their experimental host ranges are broader but still restricted. The symptoms induced in infected plants are usually mottles or mosaics. Symptoms of GRV are greatly influenced by the associated satellite RNA.

Transmission

Umbraviruses are transmissible, sometimes with difficulty, by mechanical inoculation. However, in nature each is dependent on a specific helper virus for transmission in a persistent (circulative, non-propagative) manner by aphids. All helper viruses that have been characterized are members of the family *Luteoviridae*. The mechanism of this dependence is encapsidation of the dependent virus RNA in the CP of the helper. In GRV, the satellite RNA plays an essential role in mediating this luteovirus-dependent aphid transmission. There is no evidence for multiplication of umbraviruses in the insect vector. Seed transmission has not been reported.

Geographical Distribution

CMoV and/or CMoMV, and PEMV-2, apparently occur worldwide wherever their crop hosts are grown; other umbraviruses have a restricted distribution. Several umbraviruses, notably GRV, occur only in Africa.

Pathogenicity

Although all umbraviruses depend on a helper virus for transmission by vector insects, several of them are as important or more important than their helpers in the causation of disease symptoms. The umbravirus of greatest economic importance is GRV, which causes the most devastating virus disease of groundnut (peanut) in Africa. However, in this case it is a GRV-dependent satellite RNA that is the actual cause of the symptoms. In most instances umbraviruses have not been shown to contribute functions essential for the biological success of their associated helper viruses. However, a notable exception is PEMV-2, which is essential for the systemic spread of PEMV-1 in plants, and even allows it, unlike typical members of the family *Luteoviridae*, to spread out of the phloem into mesophyll tissue and thereby to become transmissible by manual inoculation. Another member of the *Luteoviridae*, Beet western yellows virus, has been reported to show limited manual transmissibility when in the presence of Lettuce speckles mottle virus, a member of the genus *Umbravirus*.

Cytopathic Effects

Umbraviruses, even in the absence of their helper viruses, exhibit rapid systemic spread in plants. They infect cells throughout the leaf, though presumably the aphid transmissible particles are restricted to the same tissues (in most instances the phloem) as the luteoviruses that provide their CP. In mesophyll cells infected with CMoV there is extensive development of cell wall outgrowths sheathing elongated plasmodesmatal tubules.

List of Species Demarcation Criteria in the Genus

The criteria demarcating species in the genus are:

- Natural host range,
- dsRNA band pattern (bearing in mind that some bands may represent satellite RNA species),
- Nucleotide sequence identity less than 55%, and
- Little or no hybridization with cDNA probes representing most parts of the genome.

List of Species in the Genus

Species names are in green italic script; strain names and synonyms are in black roman script; tentative species names are in blue roman script. Sequence accession numbers, and assigned abbreviations ( ) are also listed.

Species in the Genus

Carrot mottle mimic virus		
Carrot mottle mimic virus	[U57305]	(CMoMV)
Carrot mottle virus		
Carrot mottle virus		(CMoV)
Groundnut rosette virus		
Groundnut rosette virus	[Z69910; satellite RNA: Z29702-Z29711]	(GRV)
Lettuce speckles mottle virus		
Lettuce speckles mottle virus		(LSMV)
Pea enation mosaic virus-2		
Bean yellow vein-banding virus		(BYVBV)
Pea enation mosaic virus-2	[U03563; satellite RNA: U03564]	(PEMV-2)
Tobacco bushy top virus		
Tobacco bushy top virus	[AF431890]	(TBTV)
Tobacco mottle virus		
Tobacco mottle virus	[AY007231]	(TMoV)

Tentative Species in the Genus

Sunflower crinkle virus	(SuCV)

(Sunflower rugose mosaic virus)
Sunflower yellow blotch virus (SuYBV)
(Sunflower yellow ringspot virus)
Tobacco yellow vein virus (TYVV)

Phylogenetic Relationships within the Genus

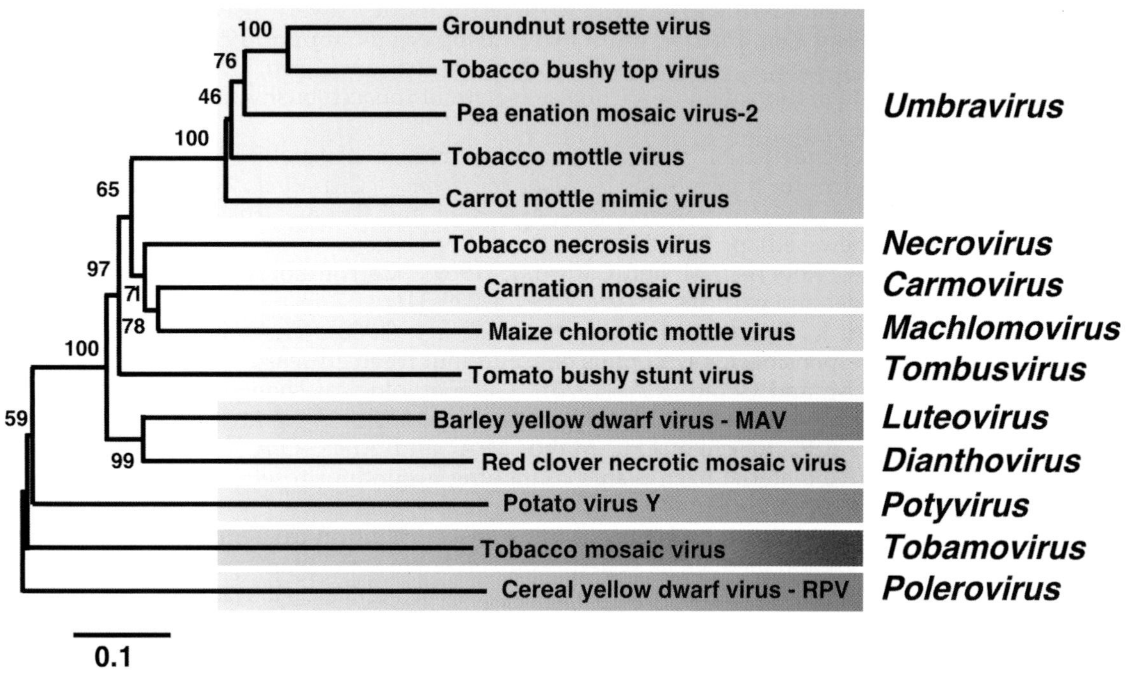

Figure 3: Phylogenetic relationships of the RdRps of umbraviruses and some other plant viruses. Amino acid sequences were aligned using the "Clustal" method and the phylogenetic tree was constructed with MEGA software using the following parameters: P-distance; Neighbor-joining; pairwise deletion; 2000 bootstraps.

Similarity with Other Taxa

Amino acid sequence comparisons show that the putative RdRp encoded by the genomic RNA molecules of CMoMV, GRV, PEMV-2 and TBTV belong to the so-called supergroup 2 of RNA polymerases, as do those of viruses in the genera *Carmovirus*, *Dianthovirus*, *Luteovirus*, *Machlomovirus*, *Necrovirus*, and *Tombusvirus* (Fig. 3). Since these enzymes are the only universally conserved proteins of positive-strand RNA viruses, the genus *Umbravirus* might be considered to be in or close to the family *Tombusviridae*.

Derivation of Names

Umbra: From Latin, a shadow. In English, a shadow, an uninvited guest that comes with an invited one.

References

Adams, A.N. and Hull, R. (1972). Tobacco yellow vein, a virus dependent on assistor viruses for its transmission by aphids. *Ann. Appl. Biol.*, **71**, 135-140.

Demler, S.A., de Zoeten, G.A., Adam, G. and Harris, K.F. (1996). Pea enation mosaic enamovirus: properties and aphid transmission. In: *The Plant Viruses*, polyhedral virions and bipartite RNA genomes, (B.D. Harrison and A.F. Murant, eds), Vol. 5, pp 303-344. Plenum Press, New York.

Demler, S.A., Rucker, D.G., de Zoeten, G.A., Ziegler, A., Robinson, D.J. and Murant, A.F. (1996). The satellite RNAs associated with the groundnut rosette disease complex and pea enation mosaic virus: sequence similarities and ability of each other's helper virus to support their replication. *J. Gen. Virol.*, **77**, 2847-2855.

Falk, B.W., Duffus, J.E. and Morris, T.J. (1979). Transmission, host range, and serological properties of the viruses that cause lettuce speckles disease. *Phytopathology*, **69**, 612-617.

Falk, B.W., Morris, T.J. and Duffus, J.E. (1979). Unstable infectivity and sedimentable ds-RNA associated with lettuce speckles mottle virus. *Virology*, **96**, 239-248.

Gibbs, M.J., Cooper, J.I. and Waterhouse, P.M. (1996). The genome organization and affinities of an Australian isolate of carrot mottle umbravirus. *Virology*, **224**, 310-313.

Gibbs, M.J., Ziegler, A., Robinson, D.J., Waterhouse, P.M. and Cooper, J.I. (1996). Carrot mottle mimic virus (CMoMV): a second umbravirus associated with carrot motley dwarf disease recognised by nucleic acid hybridisation. *Mol. Plant Path. On-Line*, http://www.bspp.org.uk/mppol/1996/1111gibbs.

Mo, X.H., Gin, X.Y., Wu, J, Yang, C., Wu, J.Y., Duan, Y.G., Li, T.F. and Chen, H.R. (2003). Complete nucleotide sequence and genome organization of a Chinese isolate of tobacco bushy top virus. *Arch. Virol.*, **148**, 389-397

Murant, A.F. (1990). Dependence of groundnut rosette virus on its satellite RNA as well as on groundnut rosette assistor luteovirus for transmission by *Aphis craccivora*. *J. Gen. Virol.*, **71**, 2163-2166.

Murant, A.F. (1993). Complexes of transmission-dependent and helper viruses. In: *Diagnosis of Plant Virus Diseases* (R.E.F. Matthews, ed), pp 333-357. CRC Press, Boca Raton.

Murant, A.F., Goold, R.A., Roberts, I.M. and Cathro, J. (1969). Carrot mottle - a persistent aphid-borne virus with unusual properties and particles. *J. Gen. Virol.*, **4**, 329-341.

Murant, A.F., Rajeshwari, R., Robinson, D.J. and Raschké, J.H. (1988). A satellite RNA of groundnut rosette virus that is largely responsible for symptoms of groundnut rosette disease. *J. Gen. Virol.*, **69**, 1479-1486.

Murant, A.F., Roberts, I.M. and Goold, R.A. (1973). Cytopathological changes and extractable infectivity in *Nicotiana clevelandii* leaves infected with carrot mottle virus. *J. Gen. Virol.*, **21**, 269-283.

Reddy, D.V.R., Murant, A.F., Raschké, J.H., Mayo, M.A. and Ansa, O.A. (1985). Properties and partial purification of infective material from plants containing groundnut rosette virus. *Ann. Appl. Biol.*, **107**, 65-78.

Ryabov, E.V., Robinson, D.J. and Taliansky, M.E. (2001). Umbravirus-encoded proteins both stabilize heterologous viral RNA and mediate its systemic movement in some plant species. *Virology*, **288**, 391-400.

Smith, K.M. (1946). The transmission of a plant virus complex by aphides. *Parasitology*, **37**, 131-134.

Taliansky, M.E., Robinson, D.J. and Murant, A.F. (1996). Complete nucleotide sequence and organization of the RNA genome of groundnut rosette umbravirus. *J. Gen. Virol.*, **77**, 2335-2345.

Taliansky, M.E., Roberts, I.M., Kalinina, N.O., Ryabov, E.V., Raj, S.K., Robinson D.J. and Oparka K.J. (2003). An umbraviral protein, involved in long-distance RNA movement, binds viral RNA and forms unique, protective ribonucleoprotein complexes. *J. Virol.*, **77**, 3031-3040.

Theuri, J.M., Bock, K.R. and Woods, R.D. (1987). Distribution, host range and some properties of a virus disease of sunflower in Kenya. *Trop. Pest Manag.*, **33**, 202-207.

Contributed By

Taliansky, M.E., Robinson, D.J., Waterhouse, P.M., Murant, A.F., de Zoeten, G.A., Falk, B.W. and Gibbs, M.J.

FAMILY TOMBUSVIRIDAE

TAXONOMIC STRUCTURE OF THE FAMILY

Family	*Tombusviridae*
Genus	*Dianthovirus*
Genus	*Tombusvirus*
Genus	*Aureusvirus*
Genus	*Avenavirus*
Genus	*Carmovirus*
Genus	*Necrovirus*
Genus	*Panicovirus*
Genus	*Machlomovirus*

VIRION PROPERTIES

MORPHOLOGY

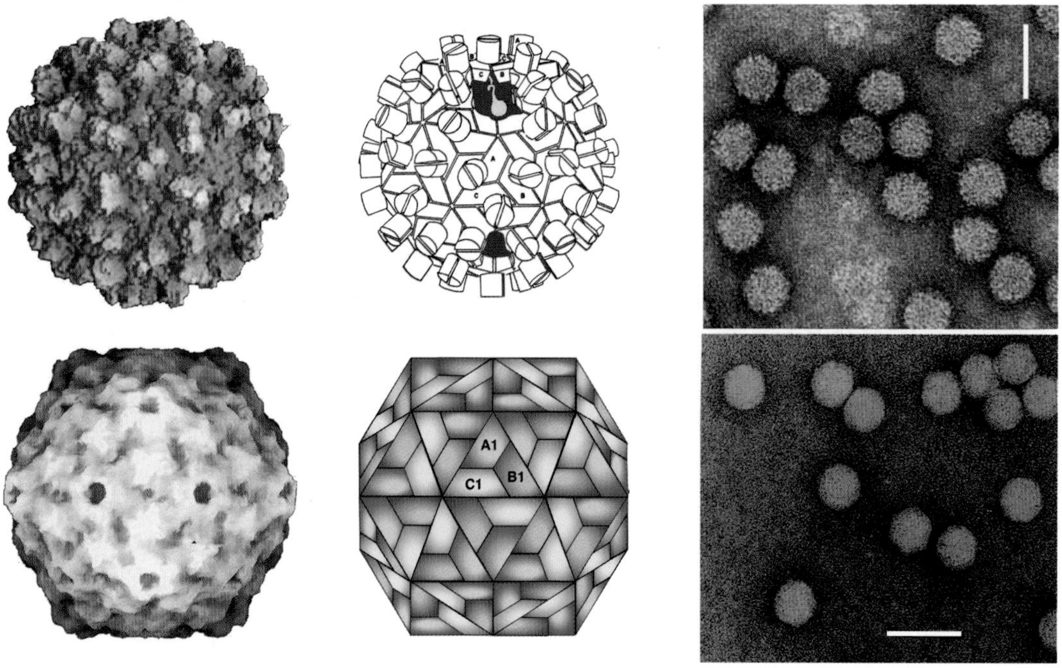

Figure 1: (Top row, left) Computer reconstruction of a Tomato bushy stunt virus (TBSV) particle based on X-ray crystallography at 2.9Å resolution (J.Y. Sgro, Univ. Wisconsin-Madison; Olson *et al.*, 1983). (Top row center) Diagrammatic representation of T=3 TBSV particles (from Hopper *et al.*, 1984, with permission). (Top row, right) Negative contrast electron micrograph of TBSV particles. (Bottom row, left) Computer reconstruction of a Tobacco necrosis virus A (TNV-A) particle, based on X-ray crystallography at 2.25Å resolution (from Oda *et al.*, 2000; Reddy V. *et al.*, 2001). (Bottom row center) Schematic representation of the T=3 structure of TNV particles. (Bottom row, right) Negative contrast electron micrograph of TNV particles. The bars represent 50 nm.

Capsids exhibit a T=3 icosahedral symmetry and are composed of 180 protein subunits (Fig. 1). Capsids are formed with CP having one of two distinct phylogenetic origins. The virions from the genera *Aureusvirus, Avenavirus, Carmovirus, Dianthovirus,* and *Tombusvirus,* have a rounded outline, a granular surface, and a diameter of about 32-35 nm. Each subunit folds into three distinct structural domains: R, the N-terminal internal domain interacting with RNA; S, the shell domain constituting the capsid backbone; and P, the protruding C-terminal domain, which gives the virus its granular appearance. P domains are clustered in pairs to form 90 projections. These dimeric contacts are important in the assembly and stabilization of the capsid structure. The R domain, which contains many positively charged residues, binds RNA. The S domain forms a barrel structure made up of β-strands. Two Ca^{2+} binding sites stabilize contacts between adjacent S domains. The capsids of viruses in the genera *Machlomovirus, Necrovirus,* and *Panicovirus* are composed of CPs that lack the protruding domain. Consequently, the surfaces of the virions have a smooth appearance. They range in diameter between 30-32 nm, and the shell domain is related to the CPs of sobemoviruses.

PHYSICOCHEMICAL AND PHYSICAL PROPERTIES

Virions sediment as one component with an S_{20w} of 118-140S, have a buoyant density ranging from 1.34-1.36 g/cm^3 in CsCl, and a virion Mr of 8.2-8.9 x 10^6. Virions are stable at acidic pH, but expand above pH 7 and in the presence of EDTA. Lowering pH or adding Ca^{++} recompacts the particles. Virions are resistant to elevated temperatures (thermal inactivation usually occurs above 80°C) and are insensitive to organic solvents and non-ionic detergents.

NUCLEIC ACID

With the exception of those of the genus *Dianthovirus,* virions contain a single molecule of positive sense, linear ssRNA, that constitutes about 17% of the particle weight, and have a size ranging from 3.7 to 4.8 kb, depending on the genus. Dianthovirus virions contain two genomic RNAs. The large RNA1 is ~3.9 kb and the smaller RNA2 is 1.5 kb. The 3'-ends are not polyadenylated. The 5'-termini are probably not protected however the presence of a cap was demonstrated for Carnation mottle virus (CarMV), Red clover necrotic mosaic virus (RCNMV) and Maize chlorotic mottle virus (MCMV). Addition of a cap analogue to *in vitro* RNA transcripts enhances infectivity little or not at all. With the possible exception of MCMV, all species express the CP from a sgRNA. DI RNAs occur in some genera. In addition some members have satellite RNAs or satellite viruses associated with them.

PROTEINS

All capsids are composed of 180 copies of a single CP type. Capsids are composed of one of two phylogenetically distinct groups of CP. Those CPs form one phylogenetically conserved group, containing a protruding domain, possess a single major CP of 37-48 kDa. In those genera that have a CP from the second phylogenetically distinct group, lacking a protruding domain, the CPs range in size from 25-29 kDa.

LIPIDS

None reported.

CARBOHYDRATES

None reported.

GENOME ORGANIZATION AND REPLICATION

Even though variability exists in the number and location of genes within members of the family, a number of organizational features are highly conserved (Fig. 2). The unifying feature of the family is that each member species possesses a highly conserved polymerase that is interrupted by an in-frame termination codon that is periodically suppressed to express catalytic quantities of the core polymerase containing the canonical "GDD" motif. Dianthoviruses utilize a –1 ribosomal frameshifting mechanism to accomplish the same result. The polymerase is further characterized by possessing no obvious helicase motif. In this description, as well as those for each genus, the polymerase is labeled as a single ORF with the read through portion labeled ORF1-RT or ORF1-FS. With the exception of MCMV, the polymerase is the first 5'-proximal ORF encountered when translated from the genomic RNA.

Genomes of members of the genera *Dianthovirus* and *Avenavirus* encode 3 ORFs while all others encode 5 ORFs. The genera *Machlomovirus* and *Panicovirus* have additional terminator readthrough and ribosomal frameshifting events to extend putative MP and accessory ORFs. Products of the 5'-proximal ORFs 1 and 1RT or 1FS are expressed by translation directly from the genomic RNA, whereas translation products of the internal and 3'-proximal ORFs 2, 3 and 4, are expressed from sgRNAs. dsRNAs corresponding in size to virus-related RNAs (genomic and subgenomic) are present in infected tissues. For all genera, the CP ORF is either internal or 3′-proximal and requires the synthesis of a sgRNA for expression *in vivo*.

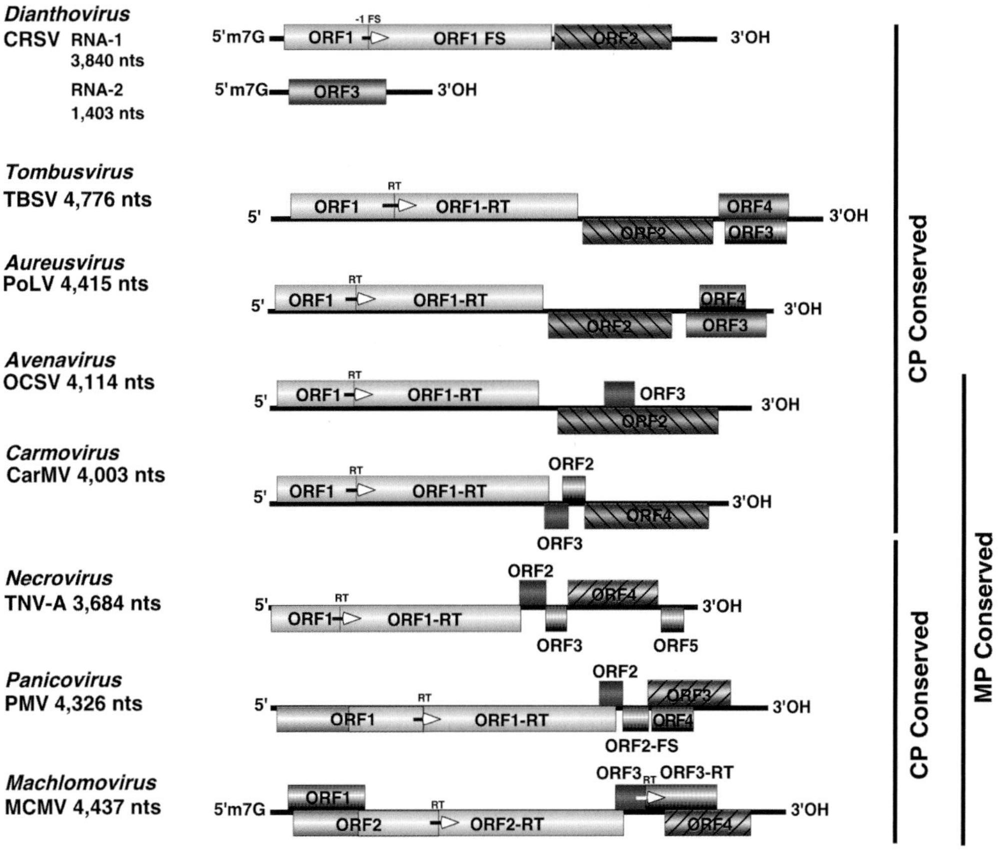

Figure 2: Genome organization of the type species for each genus in the family *Tombusviridae*. Boxes represent known and predicted ORFs. Similarly shaded boxes represent proteins with extensive sequence conservation. The yellow boxes represent ORFs encoding the phylogenetically conserved polymerase. Red hatched boxes represent CP encoding ORFs. Right-hatched boxes identify CPs lacking a protruding domain that are related to those of the genus *Sobemovirus* while those which are left-hatched represent tombusvirus-like CPs that contain protruding domains. Dark blue boxes represent viral MPs that have a high degree of sequence conservation at the carboxyl-terminus. Light blue boxes identify MPs (*Dianthovirus* ORF3, *Tombusvirus* ORF4, *Aureusvirus* ORF3) involved in movement with no sequence similarity, and shaded boxes identify other unrelated ORFs whose proteins encode an accessory function. RT: translational readthrough of termination codon. -1FS: -1 ribosomal frameshifting event. CP: capsid protein.

Non-structural proteins include the phylogenetically conserved polymerase proteins of 22-50 kDa and its 82-112 kDa readthrough product. Viruses in the family utilize at least three phylogenetically distinct MPs. The avenaviruses, carmoviruses, machlomoviruses, necroviruses, and panicoviruses encode a 7-9 kDa MP that in all these genera, excluding *Avenavirus,* is associated with another small ORF encoding an 8-9 kDa polypeptide. Genomes of the genera *Tombusvirus* and *Aureusvirus* encode a 22-27 kDa MP and that of the genus *Dianthovirus* utilizes the third type of MP of around 35 kDa. The genera *Tombusvirus* and *Aureusvirus* encode a 14-19 kDa accessory protein whose function is associated with the suppression of virus-induced gene silencing. The genera *Panicovirus* and *Machlomovirus* also have several additional accessory ORFs whose functions have not been determined.

Replication occurs in the cytoplasm, possibly in membranous vesicles that may be associated with endoplasmic reticulum, or modified organelles such as peroxisomes, mitochondria and, more rarely, chloroplasts. Virions are assembled in the cytoplasm and occasionally in mitochondria and nuclei. Virions accumulate, sometimes in crystalline form, in the cytoplasm and in vacuoles.

ANTIGENIC PROPERTIES

Virions are efficient immunogens. Antisera yield single precipitin lines in immunodiffusion tests. Depending on the genus, serological cross-reactivity among species ranges from nil to near-homologous titers. Many serologically related strains have been identified in several species.

BIOLOGICAL PROPERTIES

HOST RANGE

The natural host range of individual virus species is relatively narrow. Members can either infect monocotyledonous or dicotyledonous plants, but no species can infect both. The experimental host range is wide. Infection is often limited to the root system, but when hosts are invaded systemically, viruses enter all tissues. Many members induce a necrosis symptom in the foliar parts of the plant. Diseases are characterized by mottling, crinkling, necrosis, and deformation of foliage. Some virus species infections are symptomless in their natural hosts.

TRANSMISSION

All species are readily transmitted by mechanical inoculation and through plant material used for propagation. Some may be transmitted by contact and through seeds. Viruses are often found in natural environments, i.e. surface waters and soils from which they can be acquired without assistance of vectors. Transmission by the chytrid fungi in the genus *Olpidium* and beetles has also been reported for members of several genera. Most, if not all, members can be transmitted through the soil either dependent on, or independent of, a biological vector.

GEOGRAPHICAL DISTRIBUTION

Geographical distribution of particular species varies from wide to restricted. The majority of the species occur in temperate regions. Legume-infecting carmoviruses and one tentative member of the genus *Dianthovirus* have been recorded from tropical areas.

CYTOPATHIC EFFECTS

Distinctive cytopathological features occur in association with exceedingly high accumulations of virus particles in cells and "multivesicular bodies", i.e. cytoplasmic membranous inclusions originated from profoundly modified mitochondria and/or peroxisomes.

LIST OF GENUS DEMARCATION CRITERIA IN THE FAMILY

The list of criteria demarcating genera in the family are:

- Structural criteria: T=3 icosahedral virions 28-35 nm in diameter that are insensitive to organic solvents, composed of 180 copies of a single subunit.
- Genomic criteria: genome composed of positive polarity ssRNA on one or two segments with the total genome size being less that 5.5 kb.
- Polymerase criteria: gene interrupted by either a termination codon or a -1 ribosomal frameshifting element that is periodically readthrough. Polymerase with at least 25% aa sequence identity. Polymerase located at or near the 5′-end of the genomic RNA. Polymerase lacking obvious helicase motifs.
- Capsid protein criteria: one of two phylogenetic origins; a CP with a protruding domain having 25% aa sequence identity with other CPs in the family having a

protruding domain, a CP lacking a protruding domain with 20% or higher sequence identity with the sobemovirus CP. CP expressed from a sgRNA *in vivo*.

- Transmission criteria: virions that are mechanically transmissible. Soil transmission either with or without the aid of a biological vector.

GENUS DIANTHOVIRUS

Type Species *Carnation ringspot virus*

DISTINGUISHING FEATURES

Virions sediment in sucrose gradients as a single band of S_{20w} 126–135S. The genome is in two genomic RNAs of 3.9 and 1.5 kb. The first ORF1 encoding the polymerase is interrupted by a ribosomal frameshifting event yielding a pre-frameshift 27 kDa protein and a 88 kDa frameshift polypeptide. The CP possessing a protruding domain, is encoded by the 3′-proximal ORF on RNA1 and is expressed from a 1.5 kb sgRNA *in vivo*. The ORF for the phylogenetically distinct 34-35 kDa MP is in the monocistronic RNA2. Species are transmitted through the soil without the aid of a biological vector.

VIRION PROPERTIES

MORPHOLOGY

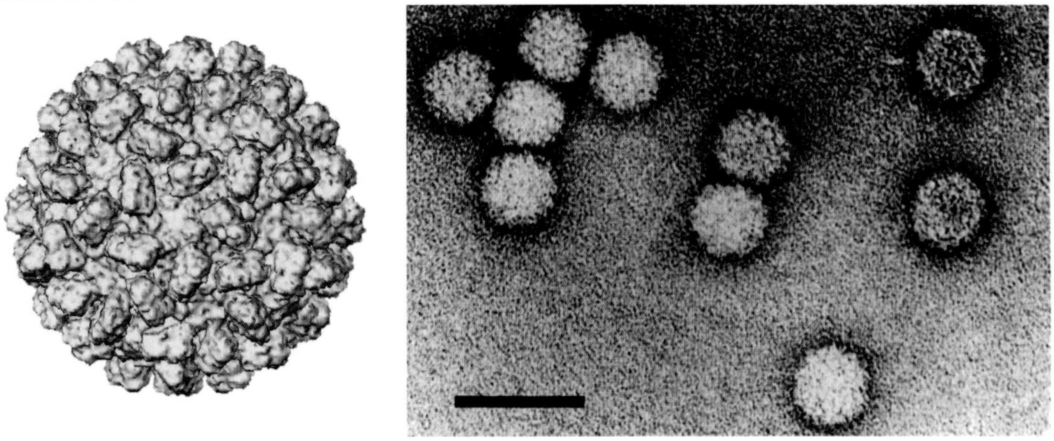

Figure 3: (Left) Cryo-reconstruction image at 10Å resolution of a particle of Red clover necrotic mosaic virus (RCNMV) (from Baker, Sherman and Lommel, with permission). (Right) Negative contrast electron micrograph of RCNMV particles (from S. A. Lommel, with permission). The bar represents 50 nm.

Virions are not enveloped. Capsids are 32-35 nm in diameter and have a T=3 icosahedral symmetry (Fig. 3). The isometric nucleocapsids have an obvious regular surface structure giving a granular appearance in the electron microscope. The surface capsomer arrangement is not obvious. Capsids are composed of 180 protein subunits, but the detailed structure is not known. However, based on similarity of CP sequence, it is predicted that the capsid is similar in structure to those of species in the genera *Carmovirus* and *Tombusvirus*. Each subunit is predicted to fold into three distinct structural domains: R, the N-terminal internal domain interacting with RNA; S, the shell domain constituting the capsid backbone; and P, the protruding C-terminal domain. P domains are clustered in pairs to form 90 projections.

PHYSICOCHEMICAL AND PHYSICAL PROPERTIES

Virions sediment as one component in sucrose with S_{20w} of 126-135S. The buoyant density in CsCl is 1.363-1.366 g/cm^3, and the virion Mr is 8.6 x 10^6. Particles exhibit an A_{260}/A_{280} ratio of 1.67 and a thermal inactivation point between 80-90°C. A longevity *in vitro* of around 10 weeks has been reported for most species. Virions are insensitive to ether, chloroform and non-ionic detergents. Virions are stable at pH 6 and lower; alkaline conditions (pH 7-8) induce particle swelling. Virions are stabilized by divalent cations.

Nucleic Acid

Virions contain two molecules of infectious linear positive sense ssRNA. Genomic RNA1 is 3,876-3,940 nt and RNA2 is between 1,412-1,449 nt in size. The genomic RNAs are not capped however there is a report that the 5'-end of each RNA is capped with m^7GpppA. The RNAs do not contain a 3'-terminal poly(A) tract or a tRNA-like structure. It is assumed that the virion packages only one copy of each genomic RNA. There is no evidence that the sgRNAs are packaged into virions. Three virus-specific dsRNA species are found in infected cells. The largest and smallest dsRNAs correspond to the genomic RNA1 and RNA2, respectively. The intermediate sized dsRNA corresponds to a sgRNA of 1.5 kb representing the 3' portion of genomic RNA1.

Proteins

Capsids are composed of 180 copies of a single CP species of 339 aa (37-38 kDa).

Lipids

None reported.

Carbohydrates

None reported.

Genome Organization and Replication

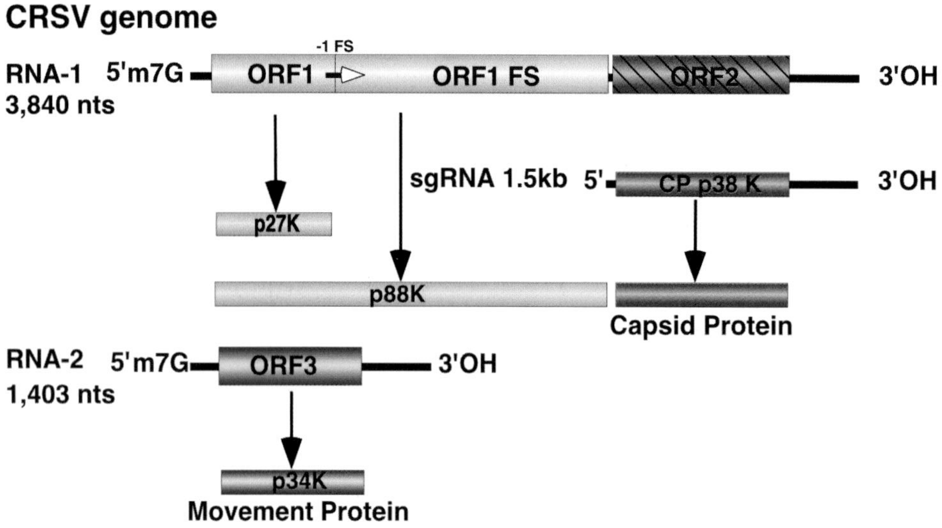

Figure 4: Genome organization and replication strategy of Carnation ringspot virus (CRSV). Boxes represent known ORFs with the sizes of the respective proteins (or readthrough products) indicated within. Yellow ORFs on RNA1 indicate polymerase proteins that have a high degree of sequence conservation within the family *Tombusviridae*. Left-hatched red box on RNA1 identifies the CP that is highly conserved among other genera within the family *Tombusviridae* that have a protruding domain. The blue box on RNA2 identifies the ORF that encodes the MP. This protein exhibits a small region of sequence conservation with MPs in the family *Bromoviridae*. The line under RNA1 depicts the 1.5 kb CP sgRNA. -1 FS = site of -1 ribosomal frameshifting.

Only the 5'-terminal 13 nt and 3'-terminal 27 nt are identical in RNA1 and RNA2 (Fig. 4). The 3' 27 nt are predicted to form a stem-loop structure. RNA1 contains two ORFs. ORF1 is capable of encoding a 27 kDa protein. A -1 ribosomal frameshift event at the canonical shifty heptanucleotide allows translation to continue into ORF1-FS in about 5% of the times the RNA is translated to yield an 88 kDa protein. Both the 27 kDa and 88 kDa proteins are observed *in vivo* and are made by translation of virion RNA *in vitro*. The ORF1 and ORF1-FS encoded proteins form the viral polymerase. ORF2 encodes the 37-38 kDa CP. This ORF is expressed *in vivo* from the 1.5 kb sgRNA. ORF3 on RNA2 encodes the 34-35 kDa MP. For the sgRNA to be expressed, the loop of a stem-loop in RNA2 must base pair within the RNA1 sgRNA promoter.

ANTIGENIC PROPERTIES

Virus particles are moderately to highly immunogenic. Various serologically distinct strains have been identified. Antisera yield a single precipitin line in agar gel-diffusion assays. Monoclonal antibodies have been identified that cross-react between species.

BIOLOGICAL PROPERTIES

HOST RANGE

In nature, dianthoviruses have moderately broad natural host ranges restricted to dicotyledonous plants. In the laboratory, the experimental host range is much broader, and includes a wide range of herbaceous species in the families Solanaceae, Leguminosae, Cucurbitaceae, and Compositae. Species infect a larger number of plants locally (non-systemically).

TRANSMISSION

The viruses are readily transmitted by mechanical inoculation; they are not known to be seed-transmitted. The viruses are not transmitted by insects, nematodes, or soil-inhabiting fungi. However, viruses are readily transmitted through the soil without the aid of a biological vector.

GEOGRAPHIC DISTRIBUTION

Dianthoviruses, with the possible exception of FNSV, which appears to be tropical in range, are widespread throughout the temperate regions of the world.

CYTOPATHIC EFFECTS

None reported.

LIST OF SPECIES DEMARCATION CRITERIA IN THE GENUS

The list of species demarcation criteria in the genus is:

- Extent of serological relationship as determined by immunodiffusion and/or ELISA,
- Extent of sequence identity between relevant gene products,
  - Less than 79% aa sequence identity of the CP,
  - Less than 54% aa sequence identity of the polymerase,
- Ability to form pseudorecombinants with the two RNA components,
- Transmission through the soil,
- Natural host range,
- Artificial host range reactions.

LIST OF SPECIES IN THE GENUS

Species names are in green italic script; strain names and synonyms are in black roman script; tentative species names are in blue roman script. Sequence accession numbers, and assigned abbreviations ( ) are also listed.

SPECIES IN THE GENUS

Carnation ringspot virus		
Carnation ringspot virus	[M88589, L18870]	(CRSV)
Red clover necrotic mosaic virus		
Red clover necrotic mosaic virus	[J04357, X08021]	(RCNMV)
Sweet clover necrotic mosaic virus		
Sweet clover necrotic mosaic virus	[L07884, S46027, S46028]	(SCNMV)

TENTATIVE SPECIES IN THE GENUS

Furcraea necrotic streak virus		(FNSV)
Rice virus X	[AB033715]	(RVX)
Sesame necrotic mosaic virus		(SNMV)

GENUS TOMBUSVIRUS

Type Species ***Tomato bushy stunt virus***

DISTINGUISHING FEATURES

The genome is approximately 4.8 kb and contains four ORFs. The CP ORF is located internally on the genomic RNA and is expressed *in vivo* from a 2.1 kb sgRNA. ORFs 3 and 4 are 3' proximally located and ORF4 is contained within ORF3 in a different reading frame. Both ORFs are expressed from a second 0.9 kb sgRNA. The genome organization and expression strategy are identical to that of PoLV (genus *Aureusvirus*). The tombusvirus ORF3 is significantly smaller and ORF4 significantly larger than that in PoLV. All members elicit formation of multivesicular inclusion bodies. Diseases caused by tombusviruses prevail in temperate climates. All species are soil-borne, but only one, Cucumber necrosis virus (CNV), has a recognized fungus vector (*Olpidium bornovanus*).

VIRION PROPERTIES

MORPHOLOGY

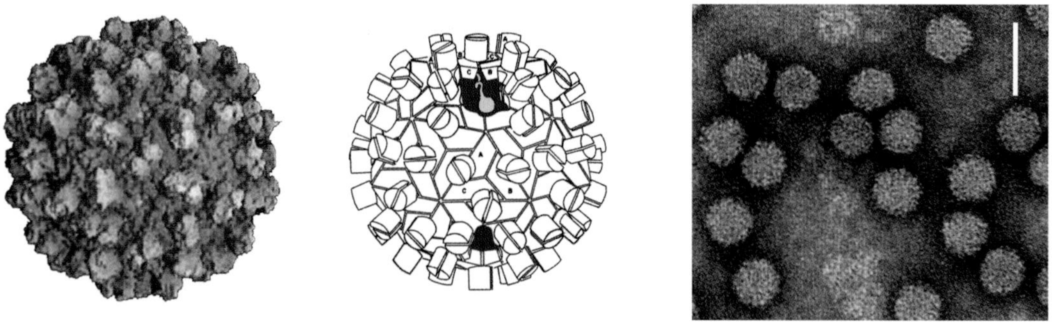

Figure 5: (Left) Computer reconstruction of a Tomato bushy stunt virus (TBSV) particle based on X-ray crystallography at 2.9Å resolution (Courtesy J.Y. Sgro, Univ. Wisconsin-Madison; Olson *et al.*, 1983). (Center) Diagrammatic representation of a T=3 TBSV particles (from Hopper *et al.*, 1984, with permission). (Right) Negative contrast electron micrograph of TBSV particles. The bar represents 50 nm.

Capsids are 32-35 nm in diameter and have a T=3 icosahedral symmetry (Fig. 5). The isometric nucleocapsids have a regular surface structure giving a granular appearance under the electron microscope. The surface capsomer arrangement is not obvious. Capsids comprise 32 capsomers composed of 180 protein subunits. Each subunit folds into three distinct structural domains: R, the N-terminal internal domain interacting with RNA; S, the shell domain constituting the capsid backbone; and P, the protruding C-terminal domain. P domains are clustered in pairs to form 90 projections. These dimeric contacts are important in the assembly and stabilization of the virion structure. The R domain, which contains many positively charged residues, binds RNA. The S domain forms a barrel structure made up of β-strands. Two Ca^{++} binding sites stabilize contacts between S domains.

PHYSICOCHEMICAL AND PHYSICAL PROPERTIES

The virus sediments as one component with an S_{20w} of 132-140S, has a buoyant density of 1.34-1.36 g/cm^3 in CsCl, and a virion Mr of 8.9 x 10^6. The virion isoelectric point is pH 4.1. Particles exhibit an A_{260}/A_{280} ratio of 1.64 and a thermal inactivation point of 80-90°C. Longevity *in vitro* of 130-150 days has been reported. Virions have a dilution end point in excess of 10^{-6}. Virions are insensitive to ether, chloroform and non-ionic detergents. Virions are stabilized by divalent cations.

NUCLEIC ACID

Nucleic acid represents 17% of the virion, and consists of one molecule of linear positive-sense ssRNA. Total genome length averages around 4.8 kb. The 5'-end of the genome lacks a cap structure and the 3′-terminus has neither a poly(A) tract nor a terminal tRNA-like structure. A 3'-proximal segment is involved in facilitating cap-independent translation. In addition to genomic RNA, virions of some species harbor and package DI and/or satellite RNAs. SgRNAs may also be packaged into virions at various levels. SgRNAs are generated by premature termination during genome minus strand synthesis, followed by sgRNA production using the truncated minus strand RNA as template. Three virus-specific dsRNA species are found in infected cells. The size of largest virus specific dsRNA corresponds to the genomic RNA. The second largest, 2.1 kbp, and the smallest, 0.9 kbp, dsRNAs correspond to sgRNAs 1 and 2, respectively.

PROTEINS

Virions contain 83% protein. One species of structural protein found in virions. The CP is about 41 kDa and is not glycosylated nor phosphorylated.

LIPIDS

None reported.

CARBOHYDRATES

None reported.

GENOME ORGANIZATION AND REPLICATION

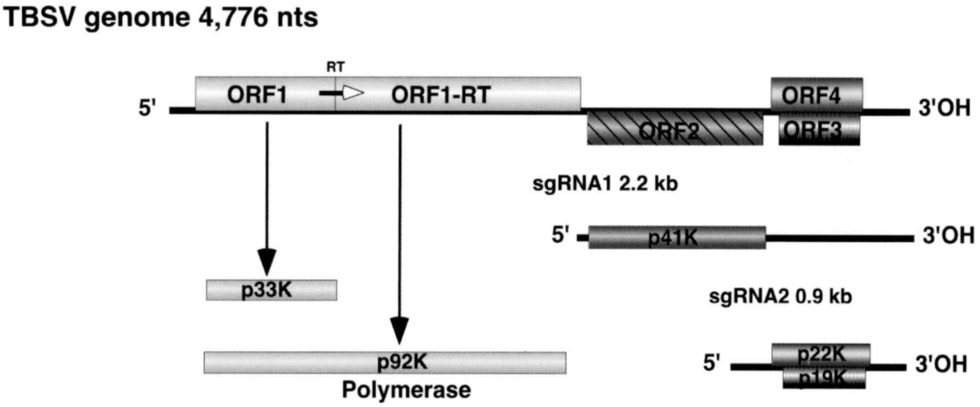

Figure 6: Genome organization and replication strategy of Tomato bushy stunt virus (TBSV). Boxes represent known and predicted ORFs with the sizes of the respective proteins (or readthrough products) indicated within. Shaded ORFs indicate polymerase proteins that have a high degree of sequence conservation within the family *Tombusviridae*. Left-hatched box identifies the CP that is highly conserved among other genera within the family *Tombusviridae* that have a protruding domain. The grey boxes identify ORFs whose proteins are unique to the genus. Lines underneath depict the two sgRNAs that are synthesized in infected cells and allow for the expression of ORF2 and ORFs 3 and 4 from sgRNAs 1 and 2, respectively. RT = Amber termination codon that is read through.

The genomic RNA contains four ORFs (Fig. 6). ORF1 encodes a 32-36 kDa protein. Readthrough of the ORF1 amber termination codon allows the expression of a 92-95 kDa protein (ORF1-RT). Both the 32-36 and 92-95 kDa proteins are produced by translation of virion RNA. The ORF1 and ORF1-RT-encoded proteins are the viral polymerase. The internal ORF2 encodes the CP, which is expressed from the 2.1 kb sgRNA1. Two nested ORFs (ORF3 and ORF4) located at the 3' terminus of the genome, encode 22 (p22) and 19 (p19) kDa proteins, respectively. ORF3 and ORF4 initiation codons are in a sub-optimal and optimal translational context, respectively. Ribosome scanning occurs to allow for translation of ORF4. p22 has a role in symptom induction and is required for cell-to-cell movement, interacting with a host homeodomain leucine-zipper protein. p19 is a suppressor of post-transcriptional gene silencing. It has a role in the systemic spread of the virus, and is involved in the development of necrotic host response to infection.

Necrotic response may be determined by the interaction of p19 with the protein encoded by ORF1, which, in turn, is mediated by DI RNAs.

Genome replication is carried out by the virus-coded RdRp, probably in conjunction with host factors. The replication process begins with the synthesis of a minus-strand RNA from a plus-strand template. The minus-strand is then used as template for the synthesis of the progeny genomes. Replication requires the presence of several *cis*-acting elements in the 5' and 3' UTRs and in internal positions. DI RNAs are generated during replication of some species following multiple and progressive deletions of genomic RNA templates. Polymerase proteins of CymRSV and CIRV are expressed in cells of the yeast *Saccharomyces cerevisiae* and the sequences involved in targeting and anchoring to peroxisomal and mitochondrial membranes were identified. In addition, yeast cells expressing CIRV polymerase replicate molecules of DI RNA following the same basic mechanisms as in plant cells.

ANTIGENIC PROPERTIES

Most species are serologically interrelated, though to a variable extent. Species with serologically related virions are: Artichoke mottled crinkle virus (AMCV), Pelargonium leaf curl virus (PLCV), and Petunia asteroid mosaic virus (PAMV), which are closely related; Morrocan pepper virus (MPV) which is distantly related; Eggplant mottled crinkle virus (EMCV), Carnation Italian ringspot virus (CIRV), Lato River virus (LRV) and Neckar River virus (NRV) which are very distantly related. Species with serologically unrelated virions are Cymbidium ringspot virus (CymRSV) and Cucumber necrosis virus (CNV).

BIOLOGICAL PROPERTIES

HOST RANGE

Most species have a narrow natural host range. However, most also have a wide experimental host range. Even though the host range of an individual species is restricted in nature, tombusviruses are present in a wide range of both monocotyledonous and dicotyledonous plants. Viruses tend to remain localized, forming a necrosis in artificially infected hosts.

TRANSMISSION

Members are easily transmitted mechanically in the field and experimentally. Most, if not all, species are soil-borne without the aid of a biological vector. They appear to be directly transmitted through the soil. CNV is transmitted by *Olpidium bornovanus*. Some species may be transmitted through the seed at a very low level.

GEOGRAPHICAL DISTRIBUTION

TBSV strains are present throughout North and South America, Europe, and the Mediterranean. Other tombusviruses are present wherever the primary hosts exist.

CYTOPATHIC EFFECTS

Virions are found in all parts of the host plant cells including cytoplasm, nuclei, nucleoli, mitochondria, and cell vacuoles. Virus crystals are present in the cytoplasm of infected cells. Multivesicular bodies (MVBs), i.e. cytopathic structures made up of a main body surrounded by spherical to ovoid vesicles 80-150 nm in diameter, are consistently present in infected cells. MVBs originate from the proliferation of the limiting membrane of peroxisomes (e.g. CymRSV) or mitochondria (CIRV), their formation being determined by an ORF1 sequence as short as *c.* 600 nt. MVBs do not contain virions which, however, are present within seemingly intact mitochondria in cells infected by EMCV, CymRSV, GALV, MPV, NRV, TBSV, and, occasionally, PAMV.

LIST OF SPECIES DEMARCATION CRITERIA IN THE GENUS

The list of species demarcation criteria in the genus is:

- Extent of serological relationship as determined by immunodiffusion usually not below 3, and/or ELISA,
- Extent of sequence identity between relevant gene products,
  - Less than 87% aa sequence identity of the CP,
  - Less than 96% aa sequence identity of the polymerase,
- Size of the CP,
- Differential cytopathological features; organelles from which multivesicular bodies arise,
- Natural host range,
- Artificial host range reactions.

LIST OF SPECIES IN THE GENUS

Species names are in green italic script; strain names and synonyms are in black roman script; tentative species names are in blue roman script. Sequence accession numbers, and assigned abbreviations ( ) are also listed.

SPECIES IN THE GENUS

Artichoke mottled crinkle virus		
Artichoke mottled crinkle virus	[X62493]	(AMCV)
Carnation Italian ringspot virus		
Carnation Italian ringspot virus	[X85215]	(CIRV)
Cucumber Bulgarian latent virus		
Cucumber Bulgarian latent virus	[AY163842]	(CBLV)
Cucumber necrosis virus		
Cucumber necrosis virus	[M25270]	(CNV)
Cymbidium ringspot virus		
Cymbidium ringspot virus	[X15511]	(CymRSV)
Eggplant mottled crinkle virus		
Eggplant mottled crinkle virus		(EMCV)
Grapevine Algerian latent virus		
Grapevine Algerian latent virus	[AF540885]	(GALV)
Lato river virus		
Lato river virus		(LRV)
Moroccan pepper virus		
Moroccan pepper virus	[AF540886]	(MPV)
Neckar river virus		
Neckar river virus		(NRV)
Pear latent virus		
Pear latent virus	[AY100482]	(PeLV)
Pelargonium leaf curl virus		
Pelargonium leaf curl virus	[AF290026]	(PLCV)
Petunia asteroid mosaic virus		
Petunia asteroid mosaic virus		(PAMV)
Sikte waterborne virus		
Sikte waterborne virus		(SWBV)
Tomato bushy stunt virus		
Tomato bushy stunt virus	[M21958, U80935, AJ249740]	(TBSV)

TENTATIVE SPECIES IN THE GENUS

Maize necrotic streak virus	[AF266518]	(MNeSV)

Genus *Aureusvirus*

Type Species *Pothos latent virus*

Distinguishing Features

The virion is a 30 nm icosahedron that packages the 4.4 kb genomic RNA. The RNA contains four ORFs. The CP ORF is located internally in the genomic RNA and is expressed *in vivo* from a 2 kb sgRNA. ORF3 and ORF4 are 3'-proximal and ORF4 is contained within ORF3, in a different reading frame. Both ORFs are expressed from a second 0.8 kb sgRNA. The genome organization and expression strategy are identical to those of viruses in genus *Tombusvirus*. While conserved, the polymerases of the two genera are no more closely related than either is with even the least conserved genus in the family *Tombusviridae*. The aureusvirus ORF3 is significantly larger and ORF4 significantly smaller than those in the genus *Tombusvirus*. Transmission is through the soil without (PoLV) or with (CLSV) the aid of a biological vector.

Virion Properties

Morphology

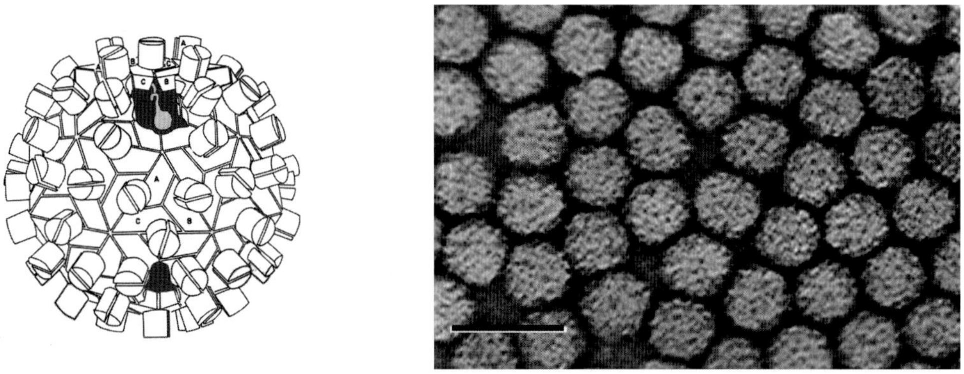

Figure 7: (Left) Diagrammatic representation of a T=3 TBSV particles (from Hopper *et al.*, 1984, with permission). (Right) Negative contrast electron micrograph of Pothos latent virus (PoLV) particles (G. P. Martelli, with permission). The bar represents 50 nm.

Virions are isometric with a rounded outline, a knobby surface and a diameter of c. 30 nm (Fig. 7). Based on comparative aa sequence alignments, the CP subunits of Pothos latent virus (PoLV) appear to be made up of three structural domains, i.e. the N-terminal internal domain, the shell domain and the C-terminal protruding domain.

Physicochemical and Physical Properties

Preparations of purified virus sediment as a single component in sucrose density gradients, and to equilibrium in solutions of CsCl and Cs_2SO_4. Buoyant density in CsCl and Cs_2SO_4 is 1.34-1.36 and 1.37 g/cm^3, respectively. The thermal inactivation point is above 80°C. Virus particles resist organic solvents but are readily disrupted by SDS.

Nucleic Acid

Virions contain a single molecule of linear, uncapped, non-polyadenylated, positive sense ssRNA of 4,415 nt, constituting 17% of the particle weight. Virions can contain two sgRNAs 2.0 and 0.8 kb in size. Three dsRNA species corresponding to the full-size genomic RNA and the two sgRNAs can be recovered from infected plants. Satellite or DI RNAs do not occur naturally, nor does PoLV genomic RNA support the replication of tombusvirus satellite or DI RNAs.

Proteins

Capsids possess 180 copies of a single CP species of 40-41 kDa.

Lipids
None reported.

Carbohydrates
None reported.

Genome Organization and Replication

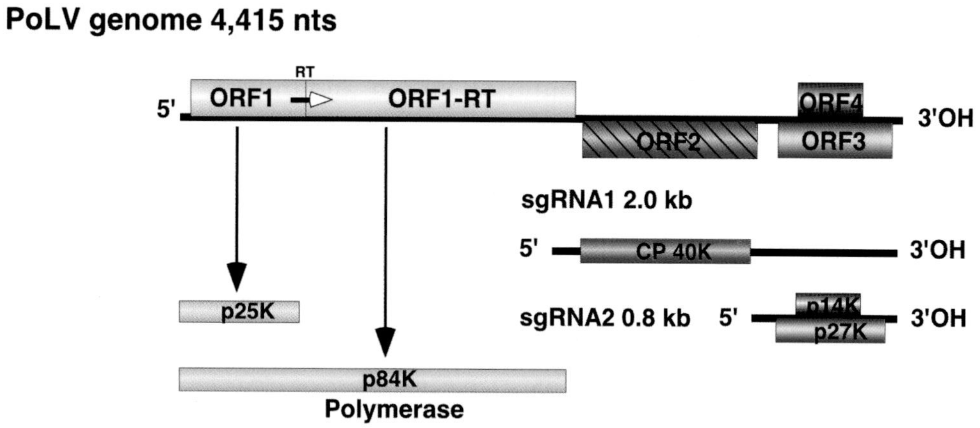

Figure 8: Genome organization and replication strategy of Pothos latent virus (PoLV). Boxes represent known and predicted ORFs with the sizes of the respective proteins (or readthrough products) indicated within. Yellow ORFs indicate polymerase proteins that have a high degree of sequence conservation within the family *Tombusviridae*. Left-hatched red box identifies the CP that is highly conserved among other genera within the family *Tombusviridae* that share a protruding domain. The grey boxes identify ORFs whose proteins are unique to the genus. Lines underneath depict the genomic RNA and two sgRNAs that are synthesized in infected cells and allow for the expression of ORF2 and ORFs 3 and 4 from sgRNAs 1 and 2, respectively. RT = termination codon that is read through.

The viral genome contains four ORFs (Fig. 8). ORF1 encodes a 25 kDa protein. The readthrough of its amber stop codon results in translation into ORF1-RT yielding an 84 kDa protein possessing the conserved motifs of RdRps. ORF2 encodes the 40 kDa CP. ORF3 and 4 are nested in different reading frames. The 27 kDa product of ORF3 is the MP and the 14 kDa product of ORF4 is responsible for symptom severity. CP is important in regulating the synthesis of the 14 kDa protein, the excess production of which is lethal to infected plants. *In vitro* translation of genome-length RNA transcribed from an infectious full-length cDNA clone, yields only one 25 kDa protein. Translation of the 2.0 kb and 0.8 kb sgRNAs gives rise to the 40 kDa CP and the 27 kDa and 14 kDa proteins, respectively. Replication may occur in the cytoplasm, possibly in association with nucleus-derived vesicles and vesiculated bodies, i.e. globose aggregates of vesicular elements surrounded by a unit membrane. The strategy of replication includes readthrough and sgRNA production. Virus particles assemble and accumulate in the cytoplasm.

Antigenic Properties

Virions are efficient immunogens. Polyclonal antisera yield a single precipitin line in immunodiffusion tests and uniformly decorate virus particles. Distant relationships were found with members of the genera *Tombusvirus* and *Carmovirus*.

Biological Properties

Host Range

Pothos (*Scindapsus aureus*) and pigeonpea (*Cajanus cajan*) are the only known natural hosts of PoLV, and cucumber (*Cucumis sativus*) is the natural host of Cucumber leaf spot virus (CLSV). The experimental host range is moderately wide. Localized infections are induced in most hosts, except for *Nicotiana benthamiana* and *N. clevelandii*, which are systemically invaded.

TRANSMISSION

PoLV is readily transmitted by mechanical inoculation. Natural transmission occurs through the soil or the circulating solution in hydroponics, apparently without the intervention of a vector. CLSV is transmitted by the soil-inhabiting fungus *Olpidium bornovanus* and, to a low rate (c. 1%), through seeds.

GEOGRAPHICAL DISTRIBUTION

Reported from several European countries, Jordan, and India.

CYTOPATHIC EFFECTS

PoLV is very invasive in systemically infected plants and is found in parenchyma and conducting tissues. Virus particles often form intracellular crystalline aggregates. Distinctive cytopathological features are the extensive vesiculation of the nuclear envelope and the single-membrane vesiculated bodies in the cytoplasm. Cytopathology of CLSV infections is characterized by occasional peripheral vesiculation of mitochondria and by the presence of membranous cytoplasmic vesicles with fibrillar content.

LIST OF SPECIES DEMARCATION CRITERIA IN THE GENUS

The list of species demarcation criteria in the genus is:

- Serological specificity (known species are serologically unrelated),
- Extent of sequence identity between relevant gene products;
  - Less than 45% aa sequence identity of the CP,
  - Less than 90% aa sequence identity of the polymerase,
- Differential cytopathological features,
- Transmission by a fungal vector,
- Natural host range,
- Artificial host range reactions.

LIST OF SPECIES IN THE GENUS

Species names are in green italic script; strain names and synonyms are in black roman script; tentative species names are in blue roman script. Sequence accession numbers, and assigned abbreviations ( ) are also listed.

SPECIES IN THE GENUS

Cucumber leaf spot virus		
Cucumber leaf spot virus	[AY038365]	(CLSV)
Pothos latent virus		
Pothos latent virus	[X87115]	(PoLV)

TENTATIVE SPECIES IN THE GENUS

None reported.

GENUS *AVENAVIRUS*

Type Species ***Oat chlorotic stunt virus***

DISTINGUISHING FEATURES

This monotypic genus is distinguished from other genera in the family *Tombusviridae* because the CP is significantly larger than other members with a protruding domain. The genome organization is intermediate between those of the genera *Carmovirus* and *Tombusvirus*. The MP of Oat chlorotic stunt virus (OCSV) is related to the MPs of viruses in the genera *Carmovirus, Machlomovirus* and *Necrovirus*, but it lacks the second MP ORF associated with these viruses.

Virion Properties

Morphology

Particles are isometric and approximately 35 nm in diameter. Based on sequence similarity with the CPs of members of the family *Tombusviridae* that contain a protruding domain, it is assumed that the particle has T=3 icosahedral symmetry and are composed of 180 protein subunits.

Physicochemical and Physical Properties

No information exists on physicochemical and physical properties.

Nucleic Acid

The genome is composed of a single positive-sense ssRNA molecule of 4,114 nt. It is not known if the 5'-terminus is capped. The 3'-end does not possess a poly(A) tail. A single sgRNA of 1,772 nt is expressed in infected tissues and is encapsidated at low concentrations within the virions.

Proteins

The capsid is probably composed of 180 copies of the single 48 kDa CP.

Lipids

None reported.

Carbohydrates

None reported.

Genome Organization and Replication

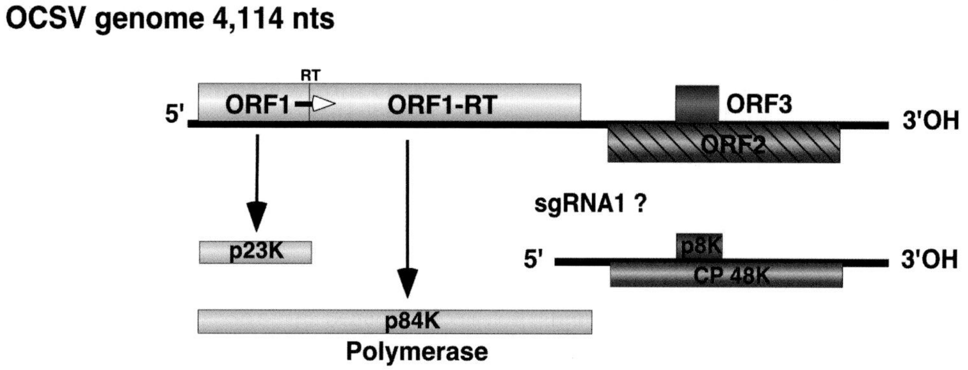

Figure 9: Genome organization of Oat chlorotic stunt virus (OCSV). Boxes represent known and predicted ORFs with the sizes of the respective proteins (or readthrough products) indicated within. Yellow ORFs indicate polymerase proteins that have a high degree of sequence conservation within the family *Tombusviridae*. Left-hatched red box identifies the CP that is highly conserved among other genera within the family *Tombusviridae* that share a protruding domain. The blue box identifies the putative cell-to-cell MP that exhibits sequence conservation with similar proteins in the genera *Carmovirus, Machlomovirus, Necrovirus*, and *Panicovirus*. A single 1.8 kb sgRNA is expressed *in vivo* for the expression of the ORF2 CP and ORF. RT = termination codon that is read through.

The genomic RNA contains 3 ORFs (Fig. 9). The genome organization is intermediate between those of the genera *Carmovirus* and *Tombusvirus*. ORF1 encodes a 23 kDa protein. Readthrough of the ORF1 amber termination codon allows the expression of a 84 kDa protein. ORF2 encodes the 48 kDa CP. ORF3 is within ORF2, in a different reading frame. This ORF encodes an 8 kDa polypeptide. A single sgRNA is formed to allow for the expression of the 3'-proximal CP ORF as well as ORF3. The ORF1 product and its amber terminator readthrough product are thought to form the viral polymerase. ORF2 encodes the virus CP. The ORF3 gene product is believed to encode the MP.

POS. SSRNA

ANTIGENIC PROPERTIES

The virus is a moderate immunogen. Antibodies do not cross-react with other unrelated icosahedral viruses of oats or with representative members of the genera *Carmovirus* and *Machlomovirus*.

BIOLOGICAL PROPERTIES

HOST RANGE

The virus has only been identified and studied in oats (*Avena sativa*).

TRANSMISSION

The virus is easily mechanically transmitted from oat plant to oat plant. Infection patterns in winter oat fields are consistent with the virus being soil-borne, and possibly transmitted by zoosporic fungi.

GEOGRAPHICAL DISTRIBUTION

This virus has only been reported in the United Kingdom.

CYTOPATHIC EFFECTS

None reported.

LIST OF SPECIES DEMARCATION CRITERIA IN THE GENUS

Not applicable.

LIST OF SPECIES IN THE GENUS

Species names are in green italic script; strain names and synonyms are in black roman script; tentative species names are in blue roman script. Sequence accession numbers, and assigned abbreviations ( ) are also listed.

SPECIES IN THE GENUS

Oat chlorotic stunt virus

Oat chlorotic stunt virus	[X89864]	(OCSV)

TENTATIVE SPECIES IN THE GENUS

None reported.

GENUS *CARMOVIRUS*

Type Species *Carnation mottle virus*

DISTINGUISHING FEATURES

Virion Mr is 8.2 x 10^6 and S_{20w} is 118–130S. Some viruses sediment as two entities in Cs_2SO_4 gradients. The genomic RNA is 4.0 kb in size and contains four ORFs. Translation of the genome yields a 28 kDa polypeptide encoded by ORF1 and an 88 kDa polypeptide (ORF1RT) originating from readthrough of the amber terminator of ORF1. ORFs 2 and 3 code for two small polypeptides of 7-8 kDa and 8-9 kDa, respectively, depending on the virus. CP contains a protruding domain and is encoded by ORF4, which is 3' co-terminal. The ORFs 2, 3 and 4 polypeptides are translated from two sgRNAs with sizes of ~1.7 and 1.5 kb, respectively. Viral species are not serologically related. Multivesicular bodies are formed only by some viruses. Most species are found in temperate regions. Those infecting legumes are reported from tropical areas. Several species are soil-borne. Melon necrotic spot virus (MNSV) is transmitted by *Olpidium bornovanus*.

VIRION PROPERTIES

MORPHOLOGY

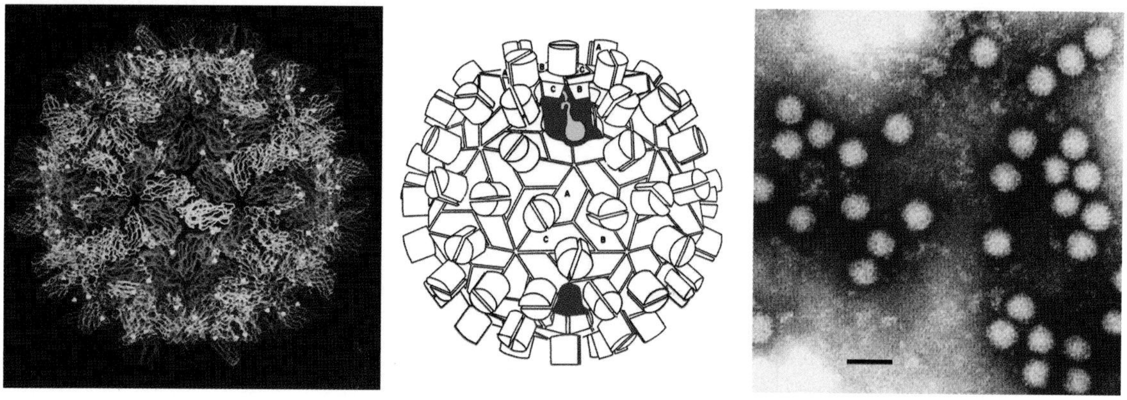

Figure 10: (Left) Computer reconstruction of a Carnation mottle virus (CarMV) particle based on X-ray crystallography at 3.2Å resolution. (from Morgunova *et al.*, 1994 , with permission). (Center) Diagrammatic representation of a carmovirus particle (from Hopper *et al.*, 1984, with permission). (Right) Negative contrast electron micrograph of CarMV particles (from Morgunova *et al.*, 1994, with permission).. (Right) Negative contrast electron micrograph of CarMV particles. The bar represents 50 nm.

Virions are 32-35 nm in diameter and have a T=3 icosahedral symmetry (Fig. 10). The isometric nucleocapsids have an obvious regular surface structure giving them a granular appearance in the electron microscope. Surface capsomer arrangement not obvious, there are 32 capsomers per nucleocapsid. Virions are composed of 180 protein subunits. Each subunit is folded into three distinct structural domains: R, the N-terminal internal domain interacting with RNA; S, the shell domain constituting the capsid backbone; and P, the protruding C-terminal domain. P domains are clustered in pairs to form 90 projections. These dimeric contacts are important in the assembly and stabilization of the virion structure. The R domain, which contains many positively charged residues, binds RNA. The S domain forms a barrel structure made up of β-strands. Two Ca^{++} binding sites stabilize contacts between S domains. X-ray crystallography analysis indicates large similarities to the TBSV structure except that CarMV lacks a β-annulus and thus may assemble by a different mechanism to that proposed for the tombusviruses.

PHYSICOCHEMICAL AND PHYSICAL PROPERTIES

Virions sediment as one component with an S_{20w} of 118-130S. The buoyant density of virions is 1.33-1.36 g/cm^3 in CsCl, and the Mr is 8.2 x 10^6. Carnation mottle virus (CarMV) has an isoelectric point of pH 5.2. Particles exhibit an A_{260}/A_{280} ratio of 1.48-1.66 and a thermal inactivation point of 95°C. Longevity *in vitro* of 395 days has been reported. Virions have dilution end points often in excess of 10^{-6}. Virions are insensitive to ether, chloroform and non-ionic detergents, and are stabilized by divalent cations.

NUCLEIC ACID

Nucleic acid comprises 14% of the virion. Virions contain one molecule of linear positive-sense ssRNA. Total genome length varies between 3,879 and 4,450 nt. For CarMV, the 5'-end of the genome is probably capped with a m^{7}GpppG or A whereas Turnip crinkle virus (TCV) genomic RNA appears not to be capped. The 3'-terminus lacks either a poly(A) tract or a terminal tRNA-like structure. Generally only the genomic RNA is encapsidated. Some species also harbor and package DI and/or satellite RNAs. SgRNAs may also be packaged into virions at a very low level. Three virus-specific dsRNA species are found in infected cells. The size of the largest virus-specific dsRNA corresponds to that of the genomic RNA. The second largest 1.5-1.9 kbp and the smallest 1.1-1.6 kbp dsRNA correspond to sgRNAs1 and 2 respectively. Complete nucleotide sequences are available for most of the species. Partial sequences are available for Pelargonium flower break virus (PFBV) and Elderberry latent virus (ELV).

Proteins
Virions contain 86% protein. The capsids are composed of 180 copies of a single structural CP ranging in size from 36.4 to 40.6 kDa. The CPs are not glycosylated or phosphorylated.

Lipids
None reported.

Carbohydrates
None reported.

Genome Organization and Replication
The genomic RNA contains four ORFs (Fig. 11). ORF1 can encode a 28 kDa protein. Readthrough of the ORF1 amber termination codon (ORF1-RT) yields an 86 kDa protein. Both the 27 and 86 kDa proteins are made *in vivo* and by translation of virion RNA *in vitro*. The proteins encoded by ORF1 and ORF1-RT make the viral polymerase. ORF2 encodes the 7 kDa MP. ORF3 encodes the 9 kDa polypeptide that also has been implicated in facilitating movement of the infection throughout the plant. Both ORFs 2 and 3 are thought to be expressed *in vivo* from the larger 1.7 kb sgRNA1 synthesized in infected cells.

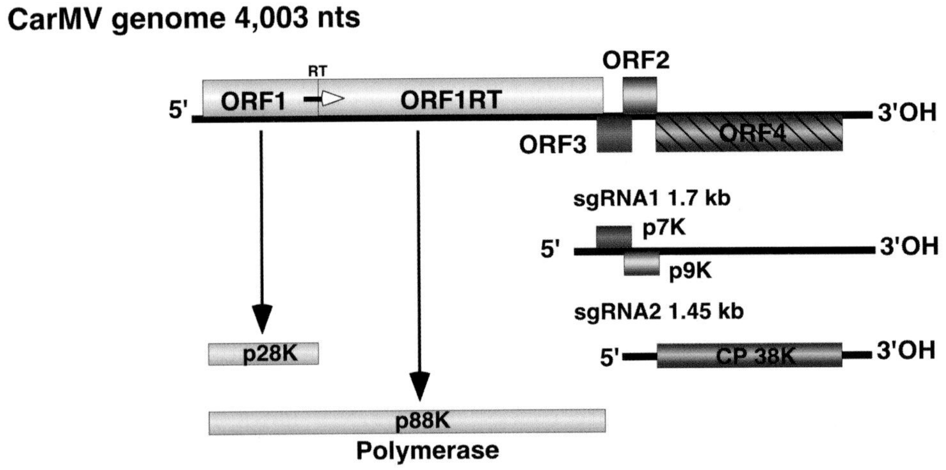

Figure 11: Genome organization and replication strategy of Carnation mottle virus (CarMV). Boxes represent known and predicted ORFs with the sizes of the respective proteins (or readthrough products) indicated within or to the right or beside. Yellow ORFs indicate polymerase proteins that have a high degree of sequence conservation within the family *Tombusviridae*. Left-hatched red box identifies the CP that is conserved among other genera in the family *Tombusviridae* that have a protruding domain. The dark blue box identifies one of the two proteins involved in cell-to-cell movement that exhibits carboxyl-terminal sequence conservation with like proteins in the genera *Avenavirus, Machlomovirus, Necrovirus,* and *Panicovirus*. An ORF for a 9kDa protein (p9K), is also involved in movement, shares no sequence similarity with RNAs of other viruses in the family *Tombusviridae*. Lines underneath the gene map depict the genomic RNA and the two sgRNAs that are synthesized in infected cells and allow for the expression of ORFs 2 and 3, and ORF4 from sgRNAs 1 and 2, respectively. RT = amber termination codon that is read through.

The ORF2 initiation codon is in a sub-optimal translational context and the ORF3 initiation codon is in an optimal translational context. Ribosome scanning allows translation of ORF3 from the 1.7 kb sgRNA1. ORF4 encodes the 38 kDa CP and is expressed *in vivo* from the 1.5 kb sgRNA2. For TCV, the CP has also been shown to be a suppressor of virus-induced gene silencing.

Antigenic Properties
Virions are efficient immunogens. Polyclonal antisera yield a single precipitin line in immunodiffusion tests. Virus species are not serologically related.

Biological Properties

Host Range

Most species have a narrow natural host range. However, most also have a wide experimental host range. Even though the host range of an individual species is restricted in nature, species infect a wide range of both monocotyledonous and dicotyledonous plants. Viruses tend to remain localized, forming a necrosis in artificially infected hosts.

Transmission

Species are easily mechanically transmitted experimentally and in nature. CarMV has spread worldwide by the dispersal of infected carnation cuttings. Some species may be transmitted through seed at a low level. Several viruses are soil-borne, but only MNSV is transmitted by *Olpidium bornovanus*. Cowpea mottle virus (CPMoV), Bean mild mosaic virus (BMMV), Blackgram mottle virus (BmoV) and TCV are transmitted by beetles (Coleoptera).

Geographical Distribution

Probably distributed worldwide. Most species are found in temperate regions of the world. Those infecting legumes are reported from tropical areas.

Cytopathic Effects

In systemically infected plants virus particles are found in parenchyma and conducting tissues, sometimes forming intracellular crystalline aggregates. Membranous vesicles are produced from the endoplasmic reticulum. Multivesicular bodies are formed only in cells infected by some viruses.

List of Species Demarcation Criteria in the Genus

The list of species demarcation criteria in the genus is:

- Extent of serological relationship as determined by immunodiffusion and/or ELISA,
- Extent of sequence identity between relevant gene products,
  - Less than 41% aa sequence identity of the CP,
  - Less than 52% aa sequence identity of the polymerase,
- Cytopathological features. Presence or absence of multivesicular bodies,
- Transmission by a fungal vector,
- Natural host range,
- Artificial host range reactions.

List of Species in the Genus

Species names are in green italic script; strain names and synonyms are in black roman script; tentative species names are in blue roman script. Sequence accession numbers, and assigned abbreviations ( ) are also listed.

Species in the Genus

Ahlum waterborne virus		
Ahlum waterborne virus		(AWBV)
Bean mild mosaic virus		
Bean mild mosaic virus		(BMMV)
Cardamine chlorotic fleck virus		
Cardamine chlorotic fleck virus	[L16015]	(CCFV)
Carnation mottle virus		
Carnation mottle virus	[X02986]	(CarMV)
Cowpea mottle virus		
Cowpea mottle virus	[U20976, U07227]	(CPMoV)
Cucumber soil-borne virus		
Cucumber soil-borne virus		(CuSBV)
Galinsoga mosaic virus		
Galinsoga mosaic virus	[Y13463, NC001818]	(GaMV)
Hibiscus chlorotic ringspot virus		

Hibiscus chlorotic ringspot virus	[X86448]	(HCRSV)
Japanese iris necrotic ring virus		
Japanese iris necrotic ring virus	[D86123]	(JINRV)
Melon necrotic spot virus		
Melon necrotic spot virus	[M29671]	(MNSV)
Pelargonium flower break virus		
Pelargonium flower break virus	[AJ003153, Z28395]	(PFBV)
Saguaro cactus virus		
Saguaro cactus virus	[U72332]	(SgCV)
Turnip crinkle virus		
Turnip crinkle virus	[M22445]	(TCV)
Weddel waterborne virus		
Weddel waterborne virus		(WWBV)

Tentative Species in the Genus

Blackgram mottle virus		(BMoV)
Elderberry latent virus	[AY038066]	(ElLDV)
Glycine mottle virus		(GMoV)
Narcissus tip necrosis virus		(NTNV)
Pea stem necrosis virus		(PSNV)
Plantain virus 6		(PlV-6)
Squash necrosis virus		(SqNV)
Tephrosia symptomless virus		(TeSV)

Genus Necrovirus

Type Species *Tobacco necrosis virus A*

Distinguishing Features

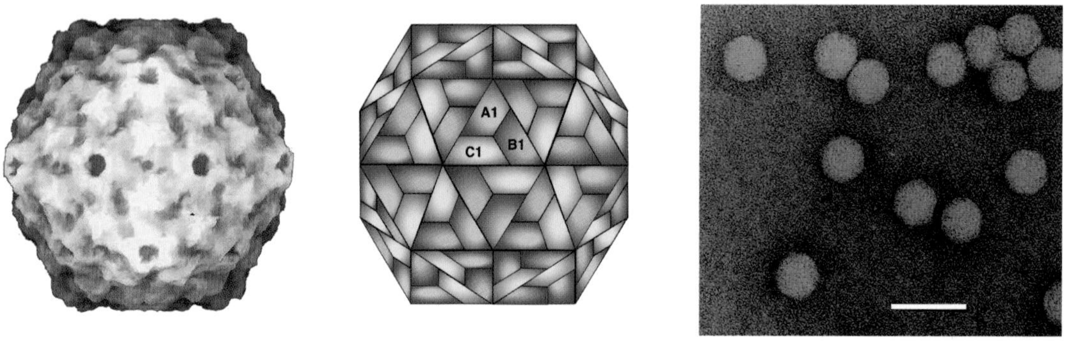

Figure 12: (Left) Computer reconstruction of a Tobacco necrosis virus A (TNV-A) particle based on X-ray crystallography at 2.25Å resolution (from Oda *et al.*, 2000; Reddy V. *et al.*, (2001). (Center) Diagrammatic representation of a necrovirus particle. (Right) Negative contrast electron micrograph of TNV-A virions. The bar represents 50 nm.

Virions sediment as a single component with an S_{20w} of 118S. The genomic RNA is about 3.8 kb in size and contains four ORFs. A fifth potential smaller ORF is predicted within the 3′-leader of the type species TNV-A. The genome organization and expression strategy are similar to members of the genus *Carmovirus*. Necroviruses have a small CP that forms smooth virions and is phylogenetically related to the sobemovirus CP. This feature distinguishes the necroviruses from carmoviruses, which have a larger CP with a protruding domain

VIRION PROPERTIES

MORPHOLOGY

Virions are approximately 28 nm in diameter and exhibit a T=3 icosahedral symmetry. The shell has a smooth appearance (Fig. 12).

PHYSICOCHEMICAL AND PHYSICAL PROPERTIES

The virus sediments as one component with S_{20w} of 118S, and has a buoyant density of 1.40 g/cm^3 in CsCl. The particle has a Mr of 7.6 x 10^6. The thermal inactivation point of TNV is between 85 and 95°C. Virion isoelectric point is pH 4.5. Virions are insensitive to ether, chloroform and non-ionic detergents.

NUCLEIC ACID

Virions contain one molecule of infectious linear positive sense ssRNA. The type species RNA is 3684 nt. The 5'-end of the RNA does not have a covalently linked virion protein and is uncapped, possessing a ppA... terminus. The RNA does not contain a 3'-terminal poly(A) tract. The virion packages exclusively the genomic RNA. Three virus specific dsRNA species are found in infected cells. The size of the largest virus specific dsRNA corresponds to that of the genomic RNA. The second largest 1.6 kbp and the smallest 1.3 kbp dsRNAs correspond to sgRNA1 and 2 respectively. Infectious RNA transcripts have been synthesized from a full-length cDNA clone of the Olive latent virus 1 (OLV-1) and TNV-D genomes.

PROTEINS

The capsid is composed of 180 copies of a single CP species. This protein has 268-275 aa and a size of 29-30 kDa.

LIPIDS

None reported.

CARBOHYDRATES

None reported.

GENOME ORGANIZATION AND REPLICATION

The genomic RNA contains four ORFs (Fig. 13). However, TNV-A also contains a small 3' proximal ORF5. ORF1 is capable of encoding a 22-23 kDa polypeptide. Readthrough of the ORF1 amber termination codon allows translation to continue into ORF1-RT for the expression of an 82 kDa polypeptide. The 82 kDa protein is predicted to be the RdRp. ORF2 can encode a 7-8 kDa polypeptide implicated in cell-to-cell movement. ORF3 encodes a 6-7 kDa polypeptide that also may be involved in movement. ORF4 encodes the 29-30 kDa CP. ORF5, present only in the TNV-A, encodes a 6.7 kDa protein. Two sgRNAs of 1.6 and 1.3 kb are synthesized in infected cells. The smaller sgRNA is the translational template for CP and the larger is the translational template for the ORF2 and possibly ORF3 products. The function of the ORF5 product is not known. The 8K protein encoded by the OLV-1 genome was detected in close proximity to plasmodesmata or within plasmodesma channels and accumulated in the cytoplasm of infected cells as bundles of thin filaments.

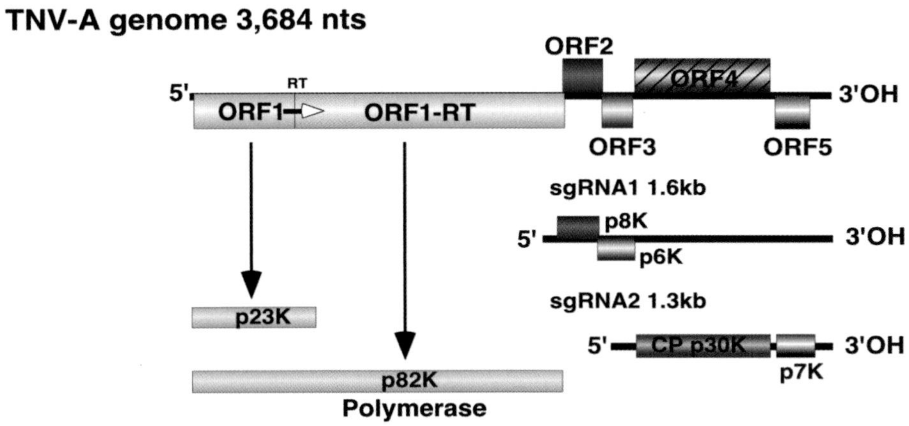

Figure 13: Genome organization and replication strategy of Tobacco necrosis virus A (TNV-A). Boxes represent known and predicted ORFs with the sizes of the respective proteins (or readthrough products) indicated within. Shaded ORFs indicate polymerase proteins that have a high degree of sequence conservation within the family *Tombusviridae*. Right-hatched red box identifies the CP that is highly conserved among other genera within the family *Tombusviridae* that lack a protruding domain. The blue box identifies one of the two proteins involved in cell-to-cell movement that exhibits sequence conservation with a like protein in the genera *Avenavirus, Carmovirus, Machlomovirus,* and *Panicovirus*. The grey boxes identify ORFs whose proteins share no sequence similarity with other viruses in the family *Tombusviridae*. Lines underneath the gene map depict the two sgRNAs that are synthesized in infected cells; these allow for the expression of ORFs 2 and 3 and ORF4 from sgRNA1 and 2, respectively. RT = amber termination codon that can be read through.

ANTIGENIC PROPERTIES

Particles of necroviruses are moderately immunogenic. Species can be distinguished serologically. Antisera yield a single precipitin line in agar gel-diffusion assays.

BIOLOGICAL PROPERTIES

HOST RANGE

Necroviruses have wide host ranges that include monocotyledonous and dicotyledonous plants. In nature, infections are typically restricted to roots. Experimental inoculations usually cause necrotic lesions on the inoculated leaves, but rarely result in systemic infection.

TRANSMISSION

Virions are readily transmitted by mechanical inoculation. Member viruses are soil-borne. Some (TNV-A, TNV-D, Beet black scorch virus; BBSV) are naturally transmitted by the chytrid fungus *Olpidium brassicae*, while others (OLV-1) are transmitted through the soil without the apparent intervention of a vector.

GEOGRAPHICAL DISTRIBUTION

TNV-A and TNV-D are ubiquitous, OLV-1 was reported from several Mediterranean countries, Leek white stripe virus (LWSV) from France, and BBSV from China.

CYTOPATHIC EFFECTS

Virus particles occur, often in prominent crystalline arrays, in infected cells in all tissue types, including vessels. Clumps of electron-dense amorphous material resembling accumulations of excess coat protein are present in the cytoplasm of cells infected by TNV or OLV-1. Membranous vesicles with fibrillar material, lining the tonoplast, or derived from the endoplasmic reticulum and accumulating in the cytoplasm are elicited by LWSV or OLV-1.

LIST OF SPECIES DEMARCATION CRITERIA IN THE GENUS

- Extent of serological relationship as determined by immunodiffusion and/or ELISA,
- Extent of sequence identity between relevant gene products,

  - o Less than 62% aa sequence identity of the CP,
  - o Less than 76% aa sequence identity of the polymerase,
- Transmission by a fungal vector,
- Natural host range,
- Artificial host range reactions.

LIST OF SPECIES IN THE GENUS

Species names are in green italic script; strain names and synonyms are in black roman script; tentative species names are in blue roman script. Sequence accession numbers, and assigned abbreviations ( ) are also listed.

SPECIES IN THE GENUS

Beet black scorch virus		
Beet black scorch virus	[NC004452]	(BBSV)
Chenopodium necrosis virus		
Chenopodium necrosis virus		(ChNV)
Leek white stripe virus		
Leek white stripe virus	[X94560]	(LWSV)
Olive latent virus 1		
Olive latent virus 1	[X85989]	(OLV-1)
Tobacco necrosis virus A		
Tobacco necrosis virus A	[M33002]	(TNV-A)
Tobacco necrosis virus D		
Tobacco necrosis virus D	[D00942, U62546]	(TNV-D)

TENTATIVE SPECIES IN THE GENUS

Carnation yellow stripe virus	(CYSV)
Lisianthus necrosis virus	(LNV)

GENUS PANICOVIRUS

Type Species *Panicum mosaic virus*

DISTINGUISHING FEATURES

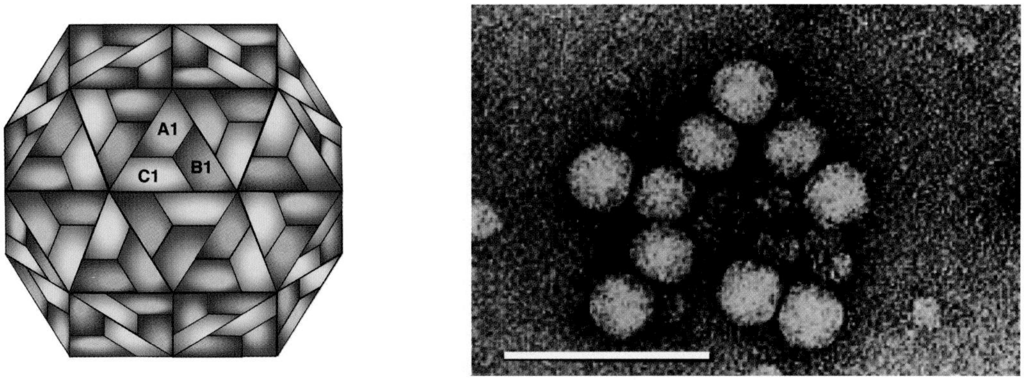

Figure 14: (Left) Diagrammatic representation of a particle of Panicum mosaic virus (PMV). (Right) Negative contrast electron micrograph of PMV particles (K. B. Scholthof, with permission).. The bar represents 100 nm.

Virions sediment at S_{20w} 109S. The genomic RNA is 4.3 kb and contains four ORFs. The polymerase is larger than those encoded by members of the family *Tombusviridae*. Like that of the machlomoviruses, the panicovirus polymerase has an amino terminal extension fused to the rest of the polymerase that is phylogenetically conserved among the family *Tombusviridae*. The virus produces only a single 1.5 kb sgRNA that is a template for the expression of ORFs 2 through 5. A second smaller sgRNA that could be used for the expression of the CP ORF4 and ORF5 has not been identified. The overall size and organization of the genome is similar to that of the genus *Machlomovirus*. However, viruses in the genus *Panicovirus* lack the additional 5′-proximally located ORF encoding a 32 kDa protein of unknown function. The virus is restricted to monocotyledonous hosts.

VIRION PROPERTIES

MORPHOLOGY

Virions are approximately 30 nm in diameter and exhibit icosahedral symmetry (Fig. 14). Detailed capsid structure is not known. Based on CP sequence similarity, it is predicted that the capsid is structurally similar to the T=3 capsid of *Southern bean mosaic virus* (genus *Sobemovirus*).

PHYSICOCHEMICAL AND PHYSICAL PROPERTIES

Virions sediment as one sedimenting component with an S_{20w} of 109S, and have a buoyant density of 1.365 g/cm^3 in CsCl, and a virion Mr of 6.1 x 10^6. Virus stored in desiccated tissue retained infectivity after twelve years. The thermal inactivation point of the type strain is 85°C, and 60°C for the St. Augustine grass decline virus strain. Virions are stable at pH 6 and lower. Virions are insensitive to ether, chloroform and non-ionic detergents. Virions are stabilized by divalent cations.

NUCLEIC ACID

Virions contain a single molecule of infectious linear positive sense ssRNA. The RNA is 4326 nt in length. The 5'-end of the RNA appears not to be capped. The RNA does not contain a 3'-terminal poly(A) tract. A 1475 nt sgRNA is also produced *in vivo* that appears not to be packaged into the virions.

PROTEINS

The virion is probably composed of 180 copies of a single 26 kDa CP.

LIPIDS

None reported.

CARBOHYDRATES

None reported.

GENOME ORGANIZATION AND REPLICATION

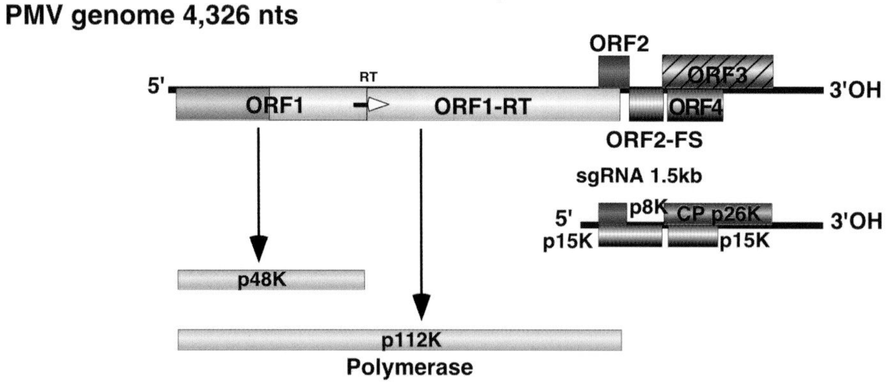

Figure 15: Genome organization and replication strategy of Panicum mosaic virus (PMV). Boxes represent known and predicted ORFs with the sizes of the respective proteins (or readthrough products) indicated within, or beside. Yellow shaded ORFs indicate polymerase proteins that have a high degree of sequence conservation within the family *Tombusviridae*. The red right-hatched box identifies the CP that is highly conserved among other genera within the family *Tombusviridae* that lack a protruding domain. The black box identifies the putative cell-to-cell movement protein that exhibits sequence conservation with similar proteins in the genera *Avenavirus, Carmovirus, Machlomovirus,* and *Necrovirus*. The gray boxes identify ORFs not having significant sequence similarity with a known viral protein. The 1.5 kb sgRNA is illustrated as a line below the genomic RNA. RT = amber termination codon that can be read through.

The genomic RNA contains five ORFs (Fig. 15). ORF1 encodes a 48 kDa protein. Readthrough of the ORF1 amber termination codon allows the production of a 112 kDa protein. Both the 48 kDa and 112 kDa polypeptides are produced by translation of virion RNA *in vitro*. ORF2 encodes an 8 kDa protein that is produced by *in vitro* translation of a transcript representing the subgenomic RNA. ORF 3 is expressed from a noncanonical start codon (GUG) and encodes a 6.6 kDa protein. ORF 4 encodes the 26 kDa CP. ORF5 is nested within ORF4 in a different reading frame and encodes a 15 kDa polypeptide. This ORF is likely expressed by leaky ribosome scanning. From a single subgenomic RNA, ORFs 2-5 are expressed, most likely through a combination of translational strategies including leaky scanning and internal ribosome entry (IRES). ORFs 2, 3, and 5 likely encode proteins involved in virus movement.

Antigenic Properties

PMV particles are highly immunogenic. Antisera yield a single precipitin line in agar gel-diffusion assays. There are several serological strains of the type strain as well as the serologically distinct St. Augustine grass decline virus strain.

Biological Properties

Host Range

In nature, PMV is restricted to grass species in the Paniceae tribe of the Poaceae. It is known to cause diseases of note in switch grass (*Panicum virgatum*), St. Augustinegrass (*Stenotaphrum secundatum*), and centipede grass (*Eremochloa ophiuroides*). In the laboratory, a number of additional species in the *Graminae* can be symptomless hosts. *Zea mays* is used as a propagation host for the type strain. St. Augustinegrass must be used as a propagation host for the St. Augustinegrass decline virus strain.

Transmission

The virus is readily transmitted by mechanical inoculation. The virus is typically transmitted by the transport and replanting of infected sod. There is one report that the St. Augustinegrass decline virus strain was seed-transmitted through *Setaria italica*.

Geographic Distribution

The virus has been reported in the USA and Mexico. The type strain is widely distributed in a number of turf grasses throughout the central United States whereas the St. Augustinegrass decline virus strain is widely distributed throughout the southern US.

Pathogenicity, Association with Disease

The virus typically forms a systemic mosaic. More severe symptoms, including chlorotic mottling, stunting and seed yield reduction, occur in forage grasses when PMV is in a mixed infection with Panicum mosaic satellite virus. The St. Augustinegrass decline virus strain can cause a severe disease on St. Augustinegrass with symptoms being more severe in the hot summer months.

List of Species Demarcation Criteria in the Genus

The list of species demarcation criteria in the genus is:

- Extent of serological relationship as determined by immunodiffusion and/or ELISA,
- Extent of sequence identity between relevant gene products,
- Soil transmission with the aid of a biological vector,
- Natural host range,
- Artificial host range reactions.

List of Species in the Genus

Species names are in green italic script; strain names and synonyms are in black roman script; tentative species names are in blue roman script. Sequence accession numbers, and assigned abbreviations ( ) are also listed.

SPECIES IN THE GENUS

Panicum mosaic virus

Panicum mosaic virus	[U55002]	(PMV)

TENTATIVE SPECIES IN THE GENUS

Molinia streak virus	(MoSV)

GENUS *MACHLOMOVIRUS*

Type Species *Maize chlorotic mottle virus*

DISTINGUISHING FEATURES

Virions sediment at an S_{20w} of 109S. The genomic RNA is about 4.4 kb and contains four ORFs. The polymerase is larger than those encoded by other species in the family *Tombusviridae*. Like the panicoviruses, the machlomovirus polymerase has an amino terminal extension fused to the rest of the polymerase that exhibits sequence conservation with the family *Tombusviridae*. It is not clear how the 3′ proximally located CP ORF is expressed *in vivo*. There are conflicting reports of either a 1.1 kb or a 1.47 kb sgRNA being made *in vivo*. Apparently another 0.34 kb sgRNA is made *in vivo* that would not act as a mRNA for any viral ORF. The overall size and organization of the genome is quite similar to that of the genus *Panicovirus*. However, genomes of machlomoviruses encode an additional 5′ proximally located ORF encoding a 32 kDa protein of unknown function. The virus is restricted to monocotyledonous hosts.

VIRION PROPERTIES

MORPHOLOGY

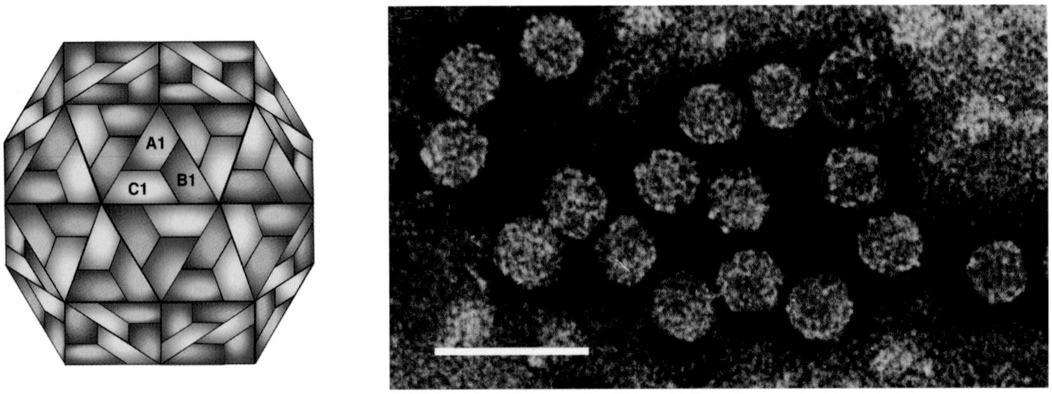

Figure 16: (Left) Diagrammatic representation of a particle of Maize chlorotic mottle virus (MCMV). (Right) Negative contrast electron micrograph of MCMV virions (S.A. Lommel, with permission).. The bar represents 100 nm.

Virions are approximately 30 nm in diameter and exhibit icosahedral symmetry (Fig. 16). Detailed structure of virions is not known. Based on CP sequence similarity, it is predicted that the capsid is structurally similar to the T=3 capsids of *Southern bean mosaic virus* (genus *Sobemovirus*).

PHYSICOCHEMICAL AND PHYSICAL PROPERTIES

Mr of virions is 6.1 x 10^6; S_{20w} is 109S; buoyant density in CsCl is 1.365 g/cm^3. Virions are insensitive to ether, chloroform and non-ionic detergents. Virions are stable *in vitro* for up to 33 days and the thermal inactivation point of virions is between 80-85°C. Virions are stable at pH 6 and lower. Virions are stabilized by divalent cations.

NUCLEIC ACID

Virions contain a single molecule of infectious linear positive-sense ssRNA. The RNA is 4437 nt in length. The 5'-end of the RNA has been reported to be capped with m^7GpppA, however the absence of a cap analog from full length transcripts does not reduce infectivity. The RNA does not contain a 3'-terminal poly(A) tract. Either a 1470 or an 1100 nt sgRNA is also packaged into virions at a very low frequency.

PROTEINS

Capsids are composed of 180 copies of a single CP of 238 aa (25.1 kDa).

LIPIDS

None reported.

CARBOHYDRATES

None reported.

GENOME ORGANIZATION AND REPLICATION

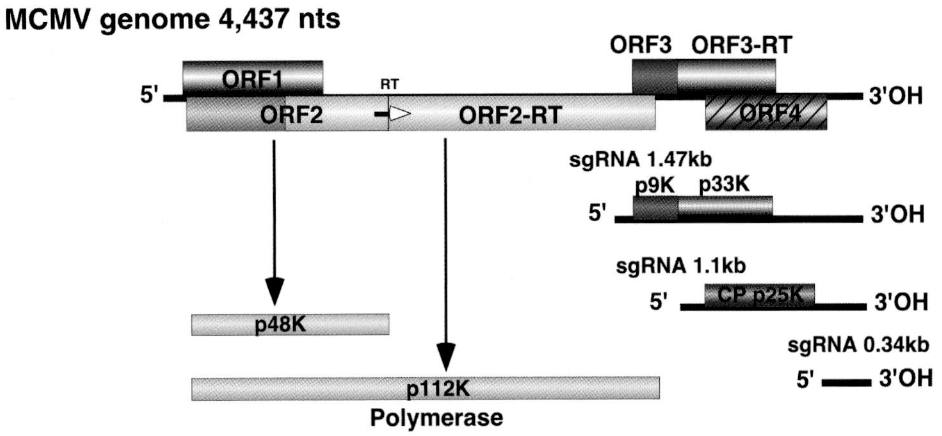

Figure 17: Genome organization and replication strategy of Maize chlorotic mottle virus (MCMV). Boxes represent known and predicted ORFs with the sizes of the respective proteins (or readthrough products) indicated within. Shaded ORFs indicate polymerase proteins that have a high degree of sequence conservation within the family *Tombusviridae*. Right-hatched red box identifies the CP that is highly conserved among those genera within the family *Tombusviridae* that lack a protruding domain. The blue box identifies the putative cell-to-cell MP that exhibits sequence conservation with similar proteins in the genera *Avenavirus, Carmovirus, Necrovirus,* and *Panicovirus*. The gray boxes identify ORFs not having significant sequence similarity with a known viral protein. The 1.47 kb, the 1.1 kb CP and the 0.34 kb sgRNAs are illustrated as lines below the genomic RNA. RT = termination codon that is read through.

The genomic RNA contains four ORFs (Fig. 17). ORF1 is capable of encoding a 32 kDa protein. ORF2 can encode a 50 kDa protein. Readthrough of the ORF2 amber termination codon allows for translation to continue into ORF2-RT, yielding a 111 kDa protein. A 111 kDa protein is produced by translation of virion RNA *in vitro*. ORF3 encodes a 9 kDa protein whose carboxyl terminus is like those of proteins encoded by similarly located small ORFs in the genomes of the genera *Avenavirus, Carmovirus, Necrovirus,* and *Panicovirus*. Assuming readthrough of the ORF3 opal termination codon, a 33 kDa protein would be produced. ORF4 encodes the 25.1 kDa CP. It is not clear how the internal ORFs are expressed *in vivo*. A major sgRNA is synthesized *in vivo* with a size of either 1.1 kb or 1.47 kb. If the sgRNA is 1.1 kb it serves as the translational template for the CP ORF. However, a 1.1 kb sgRNA does not explain the expression of the small internally located ORF. Conversely, if the sgRNA is 1.47 kb it does explain the expression of the two internal ORFs but does not explain how CP would be expressed. A second 0.34 kb sgRNA is made *in vivo* which does not act as a mRNA for any viral ORF. The functions of proteins encoded by ORF1 and ORF3 and the ORF3 readthrough products are not known. The ORF2-encoded protein and its readthrough product are thought to be the viral polymerase.

ANTIGENIC PROPERTIES

MCMV particles are moderately to highly immunogenic. Serological variants have been identified. Antisera yield a single precipitin line in agar gel-diffusion assays

BIOLOGICAL PROPERTIES

HOST RANGE

In nature, the virus systemically infects varieties of maize (*Zea mays*). In the laboratory, the virus is restricted to members of the family Graminae.

TRANSMISSION

The virus is readily transmitted by mechanical inoculation. The virus is also seed-transmitted. Kansas and Nebraska isolates can be transmitted by six species of chrysomelid beetles in the laboratory. A Hawaiian isolate is transmitted by thrips.

GEOGRAPHIC DISTRIBUTION

The virus has been reported in Argentina, Mexico, Peru, and the United States. Within the United States, the virus is restricted to the Republican River valley of Kansas and Nebraska, and to Kauai, Hawaii.

PATHOGENICITY, ASSOCIATION WITH DISEASE

(MCMV causes a mild mosaic on maize in nature. When plants are also infected with one of several Graminae-specific potyviruses, a severe necrotic disease results, termed corn lethal necrosis.

LIST OF SPECIES DEMARCATION CRITERIA IN THE GENUS

Not applicable.

LIST OF SPECIES IN THE GENUS

Species names are in green italic script; strain names and synonyms are in black roman script; tentative species names are in blue roman script. Sequence accession numbers, and assigned abbreviations ( ) are also listed.

SPECIES IN THE GENUS

Maize chlorotic mottle virus

Maize chlorotic mottle virus	[X14736]	(MCMV)

TENTATIVE SPECIES IN THE GENUS

None reported.

LIST OF UNASSIGNED VIRUSES IN THE FAMILY

None reported.

SIMILARITY WITH OTHER TAXA

The polymerases of members of the family *Tombusviridae* are related to the polymerase utilized by Barley yellow dwarf virus-PAV and similar species in the genus *Luteovirus* (Fig 18). The CPs of viruses in the genera *Machlomovirus*, *Necrovirus*, and *Panicovirus* are similar in sequence, and presumably in structure, to those of sobemoviruses (Fig. 18). The dianthovirus MP has limited sequence similarity over a limited region with MPs of viruses in the family *Bromoviridae*.

PHYLOGENETIC RELATIONSHIPS WITHIN THE FAMILY

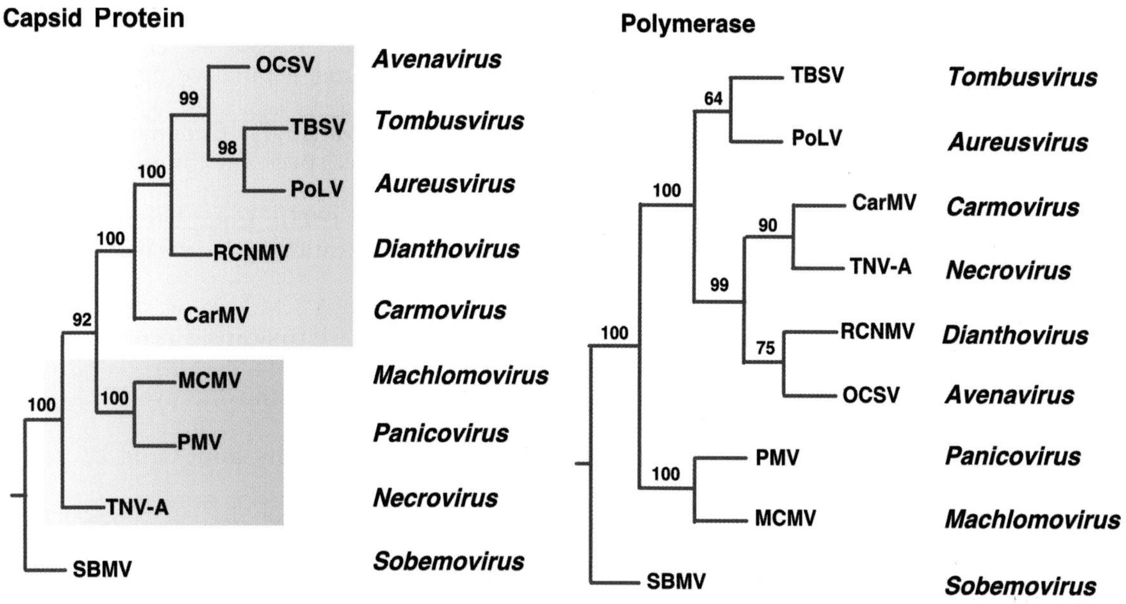

Figure 18: Phylogenetic analysis of CP (left) and polymerase (right) proteins of genera of the family *Tombusviridae*. The two different morphological types of virions, smooth versus those with a protruding domain are outlined with colored boxes. Protein sequences were aligned with CLUSTAL W and phylogenetic trees constructed with the SEQBOOT, PROTPARS, and CONSENSE programs of the PHYLIP package. SBMV CP and polymerase proteins were used as out-groups.

DERIVATION OF NAMES

Aureus: from the specific epithet of *Scindapsus aureus* (pothos), natural host of the virus.
Avena: from *Avena*, the generic name for oats.
Carmo: sigla from *car*nation *mo*ttle.
Diantho: from *Dianthus*, the generic name of carnation.
Machlomo: sigla from *ma*ize *chlo*rotic *mo*ttle.
Necro: from Greek *nekros*, "dead body".
Panico: sigla from *panic*um mosaic.
Tombus: sigla from *tom*ato *bus*hy stunt.

REFERENCES

Booonham, N., Henry, C.M. and Wood, R.R. (1998). The characterization of a subgenomic RNA and *in vitro* translation products of oat chlorotic stunt virus. *Virus Genes,* **16**, 141-145.

Campbell, R.N., Tim, S.T. and Lecoq, H. (1995). Virus transmission of host-specific strains of *Olpidium bornovanus* and *Olpidium brassicae*. *Euro. J. Plant Pathol.*, **101**, 273-282.

Cao, Y., Cai, Z., Ding, Q., Li, D., Han, C., Yu, J. and Liu, Y. (2002). The complete nucleotide sequence of beet black scorch virus (BBSV), a new member of the genus necrovirus. *Arch. Virol.*, **147**, 2431-2435.

Ciuffreda, P., Rubino, L. and Russo, M. (1998). Molecular cloning and complete nucleotide sequence of galinsoga mosaic virus genomic RNA. *Arch. Virol.*, **143**, 173-180.

Coutts, R.H.A., Rigden, J.E., Slabas, A.R., Lomonossoff, G.P. and Wise, P.J. (1991). The complete nucleotide sequence of tobacco necrosis virus strain D. *J. Gen. Virol.*, **72**, 1521-1529.

Giesman-Cookmeyer, D., Kim, K.H. and Lommel, S.A. (1995). Dianthoviruses. In: *Pathogenesis and Host Specificity in Plant Diseases; histopathological, biochemical, genetic and molecular basis,* (R.P., Singh, U.S., Singh, and K., Kohomoto, eds), Vol 3, pp 157-176. Pergamon Press, Oxford.

Grieco, F., Savino, V. and Martelli, G.P. (1996). Nucleotide sequence of a citrus isolate of olive latent virus 1. *Arch. Virol.*, **141**, 825-838.

Hamilton, R.I. and Tremaine, J.H. (1996). Dianthoviruses: Properties, Molecular Biology, Ecology, and Control. In: *The Plant Viruses*, Vol. 5, (B.D., Harrison and A.F., Murant, eds), pp. 251-282. Plenum Press, New York,

POS. SSRNA

Kumar P.L., Jones, A.T., Sreenivasulu, P., Fenton, B. and Reddy, V.R. (2001). Characterization of a virus from pigeonpea with affinities to species of the genus *Aureusvirus*, family *Tombusviridae*. *Plant Dis.* **85**, 208-215.

Lesnaw, J.A. and Reichmann, M.E. (1969). The structure of tobacco necrosis virus I. The protein subunit and the nature of the nucleic acid. *Virology*, **39**, 729-737.

Lot, H., Rubino, L., Delecolle, B., Jaquemond, M., Turturo, C. and Russo, M. (1996). Characterization, nucleotide sequence and genome organization of leek stripe virus, a putative new species of the genus *Necrovirus*. *Arch. Virol.*, **141**, 2375-2386.

Miller, J.S., Damude, H., Robbins, M.A., Reade, R.D. and Rochon D.M. (1997). Genome structure of cucumber leaf spot virus: sequence analysis suggests it belongs to a distinct species within the *Tombusviridae*. *Virus Res.*, **52**, 51-60

Molnar, A., Havelda, Z., Dalmay, T., Szutorisz, H. and Burgyan, J. (1997). Complete nucleotide sequence of tobacco necrosis virus strain D^H and genes required for RNA replication and virus movement. *J. Gen. Virol.*, **78**, 1235-1239.

Morgunova, E.Y.,Dauter, Z., Fry, E., Stuart, D.I., Stel'maschchuk, V.Y., Mikhailov, A.M., Wilson, K.S. and Vainshtein, B.K. (1994). The atomic structure of carnation mottle virus capsid protein. *FEBS Letters*, **338**, 267-271.

Oda, Y., Saeki, K., Takahashi, Y., Maeda, T., Naitow, H., Tsukihara, T. and Fukuyama, K. (2000). Crystal structure of tobacco necrosis virus at 2.25Å resolution. *J. Mol. Biol.*, **300**, 153-179.

Olson, A.J., Bricogne, G. and Harrison, S.C. (1983). Structure of tomato bushy stunt virus. IV. The virus particle at 2.9 Å resolution. *J. Mol. Biol.*, **171**, 61-93.

Pantaleo, V., Rubino, L. and Russo, M. (2003). Replication of Carnation Italian ringspot virus defective interfering RNA in *Saccharomyces cerevisiae*. *J. Virol.*, **77**, 2116-2123.

Rubino, L. and Russo, M. (1997). Molecular analysis of the pothos latent virus genome. *J. Gen. Virol.*, **78**, 1219-1226.

Rubino, L., Di Franco, A. and Russo, M. (2000). Expression of a plant virus non-structural protein in *Saccharomyces cerevisiae* causes membrane proliferation and altered mitochondrial morphology. *J. Gen. Virol.*, **81**, 279-286.

Russo, M., Burgyan, J. and Martelli, G.P. (1994). Molecular biology of *Tombusviridae*. *Adv. Virus Res.*, **44**, 381-428.

Sabanadzovic, S., Boscia, D., Saldarelli, P., Martelli, G.P., Lafortezza, R. and Koenig, R. (1995). Characterization of a pothos (*Scindapsus aureus*) virus with unusual properties. *Eur. J. Plant Pathol.*, **101**, 171-182.

Scheets, K. (2000). Maize chlorotic mottle machlomovirus expresses its coat protein from a 1.47 kb subgenomic RNA and makes a 0.34 kb subgenomic RNA. *Virology*, **267**, 90-101.

Takemoto, Y., Kanehira, T., Shinohara, M., Yamashita, S. and Hibi, T. (2000). The nucleotide sequence and genome organization of Japanese iris necrotic ringspot virus, a new species in the genus carmovirus. *Arch. Virol.*, **145**, 651-657.

Contributed By

Lommel, S.A., Martelli, G.P., Rubino, L. and Russo, M.

ORDER NIDOVIRALES

TAXONOMIC STRUCTURE OF THE ORDER

Order	***Nidovirales***
Family	***Coronaviridae***
Genus	***Coronavirus***
Genus	***Torovirus***
Family	***Arteriviridae***
Genus	***Arterivirus***
Family	***Roniviridae***
Genus	***Okavirus***

VIRION PROPERTIES

GENERAL

The order comprises three families of viruses. Currently recognized unique molecular markers that distinguish this order from all other RNA viruses are part of replicase:
1. An ORF1a-encoded cysteine or serine protease with chymotrypsin-like fold and the substrate specificity resembling that of picornavirus 3C proteases (3C-like or main protease) that is flanked by two hydrophobic trans-membrane domains (M-3CL-M);
2. An ORF1b-encoded putative multinuclear Zn-finger-like domain associated with nucleoside triphosphate (NTP)-binding/5′-to-3′-helicase domain (Zn-HEL);
3. An ORF1b-encoded putative poly(U)-specific endoribonuclease (NendoU);
4. The replicase gene constellation separated by a ribosomal frameshifting signal (FS): M-3CL-M_FS_RdRp_Zn-HEL_NendoU.

The characteristics which are common for the members of the order *Nidovirales* and not listed above are:

- Linear, non-segmented, positive-sense, ssRNA genomes.
- The general genome organization 5′-UTR-replicase gene-structural protein genes-UTR-3′.
- A 3′ co-terminal nested set of two or more sg mRNAs.
- The genomic RNA functions as the mRNA for the replicase gene.
- Only the 5′-proximal one or two ORFs of mRNAs are translationally active.
- Presence of a virion envelope.
- Genome and sg mRNAs contain a 3′-poly[A] tail.
- Two large overlapping open reading frames (ORF1a and ORF1b) encoding replicase subunits that are derived from the translation products of ORF1a and ORF1a plus ORF1b (ORF1ab), the latter generated by ribosomal frameshifting. .

MORPHOLOGY

The members of the order *Nidovirales* are enveloped viruses with an architecture that shows various similarities and differences, depending on whether the external appearance or the nucleocapsid of the virions is studied (Fig. 1). The two genera of the family *Coronaviridae* (*Coronavirus* and *Torovirus*) and members of the family *Roniviridae* show large projections protruding from the envelope (peplomers) that are formed, at least in coronaviruses, by trimers of the spike protein. These oligomeric structures provide them with the characteristic "crown" observed by electron microscopy that inspired the name of the coronavirus family. Coronaviruses have an internal core shell that protects a nucleocapsid having helical symmetry. Vitrified coronavirus particles have a diameter of 145 nm (including the extended peplomers) and an internal core shell of 65 nm. The torovirus nucleocapsid shows an unusual morphology resembling a toroid, which inspired their name. Toroviruses have a virion size similar to coronaviruses, while Roniviruses (for ro̲d-shaped n̲idoviruses) are bacilliform in shape and have a 150-200 nm long nucleocapsid with helical symmetry and a diameter of 20-30 nm. Also ronivirus envelopes are studded with prominent peplomers projecting approximately 11 nm from the surface. Arterivirus virions are spherical and significantly smaller than other nidoviruses with a complete particle of 50-70 nm in diameter, a nucleocapsid of 25-35 nm in diameter. Arteriviruses probably have an icosahedral core shell that contains the genome. No spikes are obvious on the arterivirus surface, but a surface pattern of relatively small and indistinct projections has been observed.

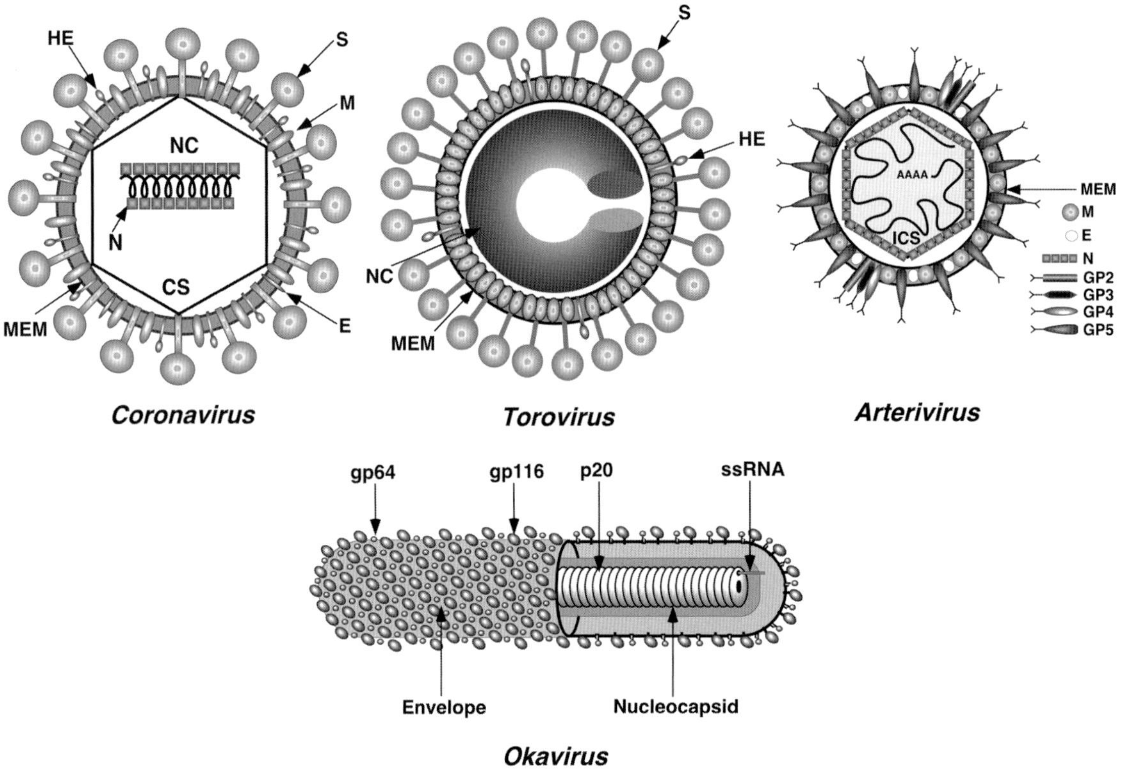

Figure 1: Schematic structure of particles of members of the order *Nidovirales*. MEM, lipid membrane; CS, core shell; NC, nucleocapsid; N, nucleocapsid protein; S, spike protein; M, membrane protein; E, envelope protein; HE, hemagglutinin-esterase.

PHYSICOCHEMICAL AND PHYSICAL PROPERTIES

The coronavirus virion Mr is 400 x 10^6, the buoyant density in sucrose is 1.15 to 1.20 g/ml, the density in CsCl is 1.23-1.24 g/ml, and viron S_{20W} is 300 to 500S. Toroviruses have bouyant densities of 1.14-1.18 g/ml in sucrose. Arterivirus virion buoyant density is about 1.13-1.17 g/ml in sucrose and 1.17-1.20 g/ml in CsCl. Virion S_{20W} is 200 to 300S. Ronivirus virion buoyant density in sucrose is 1.18-1.20 g/ml. Nidovirus virions are sensitive to heat, lipid solvents, non-ionic detergents, formaldehyde, oxidizing agents and UV irradiation.

NUCLEIC ACID

The nidoviruses genome is an infectious, single-stranded linear, positive-sense RNA molecule, which is polyadenylated and, at least for arteri- and coronaviruses, carries a 5′ cap. The size of the genomes of members of the order *Nidovirales* are 27.6 to 31 kb (*Coronavirus*), 25 to 30 kb (*Torovirus*), ~28 kb (*Okavirus*) and 12.7 to 15.7 kb (*Arterivirus*). The coronavirus genome is the largest known non-segmented viral RNA genome. Complete genome sequences have been determined for several coronaviruses (Mouse hepatitis virus (MHV), Transmissible gastroenteritis virus (TGEV), Infectious bronchitis virus (IBV), Bovine coronavirus (BCoV), Porcine epidemic diarrhea virus (PEDV), Severe acute respiratory syndrome coronavirus (SARS-CoV), Human coronavirus OC43 (HCoV-OC43) and Human coronavirus 229E (HCoV-229E), the ronivirus Gill-associated virus (GAV), and all known arteriviruses (Equine arteritis virus (EAV), Lactate dehydrogenase-elevating virus (LDV), Porcine reproductive and respiratory syndrome virus (PRRSV), Simian hemorrhagic fever virus (SHFV)). A sequence of the (estimated) 3'-proximal half of the genome is available for the Equine torovirus (EToV).

PROTEINS

No virion protein common to all nidoviruses has so far been recognized. The virion proteins typical for the various nidovirus families are summarized in Table 2.

Table 2. Nidovirus virion-associated proteins (kDa)

Protein		*Coronavirus*[a]	*Torovirus*[a]	*Arterivirus*[a]	*Okavirus*[a]
Spike glycoprotein	S	180-220	200	-	-
Major surface glycoprotein	GP_5	-	-	30-45	-
Minor surface glycoprotein	GP_2	-	-	25	-
	GP_3			36-42	-
	GP_4	-	-	15-28	-
Large spike glycoprotein	Gp116	-	-	-	110-135
Small spike glycoprotein	Gp64	-	-	-	60-65
Membrane protein	M	23-35	27	16	
Nucleocapsid protein	N	50-60	19	12	20-22
Small envelope protein	E	9-12	-	9	-
Hemagglutinin-esterase protein	HE	65	65	-	-

[a] apparent molecular weight estimated by SDS-PAGE electrophoresis (kDa)

The envelope of coronaviruses contains three or four proteins. The spike proteins (S) of coronaviruses and toroviruses have a highly exposed globular domain and a stem portion containing heptad repeats, indicative of a coiled-coil structure. The membrane (M) proteins of coronaviruses and toroviruses are different in sequence but alike in size, structure and function. The M proteins have a similar triple- or quadruple-spanning membrane topology. In addition, coronaviruses have a small structural protein (E) within the envelope (around 20 copies per virion). Toroviruses seem to lack a homolog for the E protein. Some coronaviruses (MHV, HCoV-OC43, BCV) and toroviruses contain an additional membrane protein with hemagglutinin-esterase activity (HE).

The structural proteins of arteriviruses are apparently unrelated to those of the other members of the family *Coronaviridae*. There are six envelope proteins that have been identified in EAV and PRRSV virions and may be common to all arteriviruses: a 16-20 kDa non-glycosylated membrane protein (M) is thought to transverse the membrane three times and thus structurally resembles the M protein of corona- and toroviruses. The heterogeneously N-glycosylated, putative triple-spanning major glycoprotein (GP_5 for EAV, LDV, and PRRSV) of variable size forms a disulfide-linked heterodimer with the M protein. Recently, a trimeric complex consisting of the three remaining viral glycoproteins (GP_2, GP_3, and GP_4), which are all a minor virion components, was described for EAV. The final structural protein of arteriviruses is a small, non-glycosylated, hydrophobic protein designated E (for envelope).

Ronivirus structural proteins have been studied only for Yellow head virus (YHV). YHV virions contain three major structural proteins (110-135 kDa, 63-67 kDa and 20-22 kDa). The 110-135 kDa protein (gp116) and the 63-67 kDa protein (gp64) are glycosylated and appear to be envelope proteins that form the prominent peplomers on the virion surface. Mature gp116 and gp64 are generated by post-translational processing of a precursor glycopolyprotein. Gp116 and gp64 are not linked by intramolecular disulfide bonds but each is anchored in the virion by C-terminal hydrophobic transmembrane domains. The 20-22 kDa protein (p20) is associated with nucleocapsids and appears to function as the nucleoprotein.

LIPIDS

Nidoviruses have lipid envelopes. The S protein (MHV, BCV) and the E protein (MHV) of coronaviruses are acylated (palmitic acid).

CARBOHYDRATES

Coronavirus S and HE proteins contain N-linked glycans, the S protein being heavily glycosylated (about 20-35 glycans). The M protein of coronaviruses contains a small number of either N- or O-linked glycans, depending on the species. These side chains are located near the amino-terminus, but the M protein of TGEV also has a potential glycosylation site in the carboxy-terminus. At present there is no evidence to suggest that the E protein is glycosylated or phosphorylated.

Torovirus S protein has 18 potential N-glycosylation sites. Also their HE protein (Bovine torovirus (BToV)) is N-glycosylated and binds 9-*O*-acetylated receptors, but their M protein is not glycosylated.

In the arteriviruses the GP_2, GP_3, GP_4 and GP_5 contain N-linked glycans. GP_5 of EAV, LDV, and PRRSV are modified by heterogeneous N-acetyl lactosamine addition. The M and E proteins are not glycosylated.

GENOME ORGANIZATION AND REPLICATION

Despite the differences in genetic complexity and gene composition, the genome organization of corona-, toro, roni-, and arteriviruses are remarkably similar (Fig. 2). Two thirds of each genome contain two large ORFs, designated ORF1a and ORF1b. Downstream of ORF1b there are three (*Okavirus*) to ten (BCV, MHV, SHFV), or 12 (SARS-CoV) ORFs that encode the structural proteins and, at least for coronaviruses, a number of non-structural "accessory" proteins.

Nidovirus replication proceeds through the synthesis of the full-length negative antigenome in the cytoplasm of infected cells. It is catalyzed by a poorly characterized membrane-bound replicative complex. The products of the ORF1a and ORF1ab are sufficient to maintain genomic and sgRNA synthesis in arteriviruses and sgRNA synthesis in coronaviruses. These ORFs encode a variety of (putative) enzymes (see section on *replicase*) that may be part of the replicative complex *per se* or control its composition and functioning.

The genome is expressed by diverse mechanisms including replication, transcription, translation and co- and post-translational regulation. The virion RNA functions as the mRNA (mRNA1) for the two 5′-most ORFs (1a and 1b) encoding replicase components. The translation of ORF1a yields the pp1a polyprotein. In approximately 20-30% of cases, ribosomes slip at the overlap between the ORF1a and ORF1b regions, containing a specific seven-nucleotide "slippery" sequence and a downstream pseudoknot structure (ribosomal frameshifting signal), to continue translation into ORF1b, which is in the –1 frame relative to ORF1a. Translation of ORF1ab yields the pp1ab polyprotein. The pp1a and pp1ab polyproteins have not been described in infected cells; they are believed to be co- and post-translationally processed by several virus-encoded proteases to more than a dozen mature and intermediate replicase subunits. There is no nidovirus-wide nomenclature of replicase subunits; they may be listed in the order in which they are encoded in polyproteins, or according to their molecular size or function. In arteriviruses and coronaviruses, from one to three (and possibly four) papain-like proteases (PLpro) control processing of the N-terminal part of pp1a/pp1ab polyproteins. A chymotrypsin-like protease (known as 3CLpro or 'main' protease - Mpro) is responsible for processing of the rest part of pp1a/pp1ab polyproteins at 8-11 conserved cleavage sites.

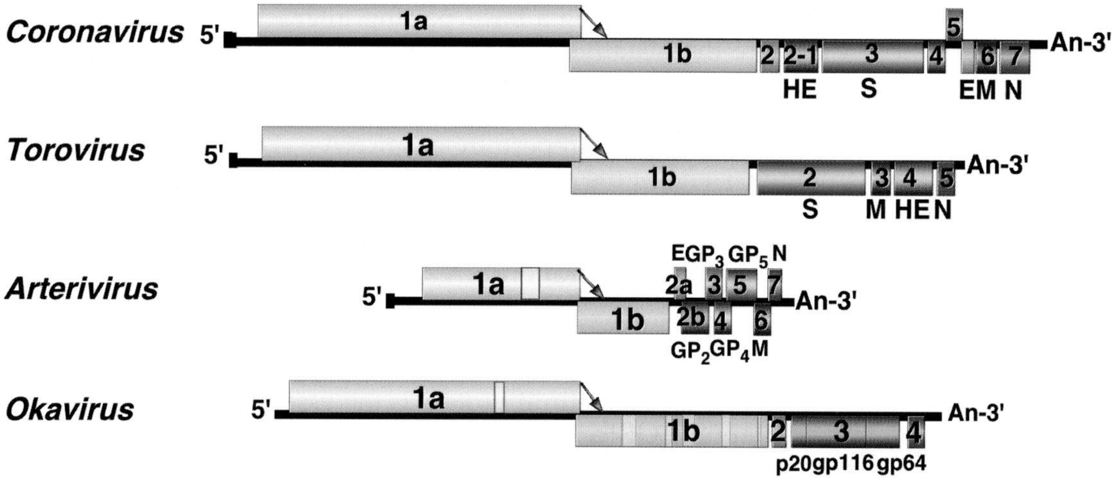

Figure 2: Schematic representation of the genome structure of members of the order *Nidovirales* (from top to bottom: MHV, BEV, EAV, GAV) and the 3' genome organization can differ between members of each genus. ORFs are represented by boxes. Untranslated sequences are indicated by solid lines. The ribosomal frameshift sites in ORF1 are aligned and represented b y arrows. The proteins encoded by the ORFs are indicated. The 5' leader sequences are depicted by a small black box. Poly(A) tails are indicated by An. S, spike protein; M, membrane protein; N, nucleocapsid protein, HE, hemagglutinin-esterase protein; I, internal ORF. PolyA, is indicated by An.

All ORFs downstream of ORF1b are expressed from a 3'-coterminal nested set of specialized sgRNAs whose number varies between 2 and 8. Except for the smallest mRNA, all of the mRNAs are structurally polycistronic. As a rule, only the 5'-most one or two ORFs of each mRNA are translated while downstream ORF(s) remain translationally silent.

The sg mRNAs of corona- and arteriviruses carry a 5' leader sequence of 55 to 92 and 170 to 210 nt, respectively, which is derived from the 5'-end of the viral genome. The sgRNA synthesis thus involves a mechanism of discontinuous RNA synthesis, which is currently presumed to occur during minus strand synthesis. The torovirus mRNAs lack a common 5' leader sequence, with the exception of mRNA-2. The ronivirus sg mRNAs also lack a common 5' leader sequence. Both coronaviruses and arteriviruses contain conserved AU-rich sequences at the fusion sites of leader and mRNA bodies. These sequences are most commonly termed transcription-regulating sequences (TRS), but were previously also referred to as intergenic sequences (IS), subgenomic promoters, or leader-to-body junction sites. Cells infected by arteriviruses or coronaviruses contain negative-stranded sgRNAs that correspond to each sg mRNA. The combined biochemical and reverse genetics data obtained for coronaviruses and arteriviruses strongly suggests that these subgenomic minus strands serve as template for sg mRNA synthesis. In agreement with this notion, replicative intermediates (RI)/replicative forms (RF) with sizes corresponding to the different sg mRNAs have been shown to be present and actively involved in RNA synthesis.

Several models have been proposed to explain the wealth of experimental data and, in particular, how the common leader sequence is 'fused' to different mRNA bodies. These models are not necessary mutually exclusive, as components of each model may operate at different stages of the viral replication cycle. The model compatible with most of the experimental data is *discontinuous extension of nascent negative strand RNA*. This model proposes that during negative strand RNA synthesis the discontinuous transcription step occurs, generating negative-stranded sgRNAs, which then serve as templates for the (uninterrupted) synthesis of sg mRNAs. In this model, the body TRSs on the genomic RNA serve as termination or pausing signals for negative strand synthesis, and the nascent negative-stranded sgRNA then jumps to the leader TRS sequence at the 5'-end of the genomic RNA where, following a base pairing interaction, minus strand synthesis is resumed to complete the negative-stranded sgRNA. The alternative model, the *leader-primed transcription* model, proposes that the genomic RNA is first transcribed into a genome-length, negative-stranded RNA, which, in turn, becomes the template for subsequent synthesis of the entire repertoire of sgRNAs. The leader of nascent sgRNAs is transcribed from the 3'-end of the negative-strand RNA and dissociates from the template to subsequently associate with the template RNA at one of the (body) TRSs to prime the synthesis of the particular sg mRNA. In this mechanism, the discontinuous transcription step takes place during positive strand RNA synthesis.

The replicase gene

A large variety of virus proteins known as non-structural proteins are not incorporated into virions. The largest non-structural protein is the essential replicase, which is encoded by the slightly overlapping ORFs 1a and 1b (gene 1) and accounts for approximately two-thirds of the genome size. The replicase gene encodes two proteins, one of which is encoded by ORF1a (pp1a), whereas the other (pp1ab) is the fused product of ORF1a and ORF1b translation and is expressed through –1 ribosomal frameshifting. Neither of these giant proteins (from the approximately 2000-aa pp1a of arteriviruses to the >7000-aa pp1ab of coronaviruses) has been observed in infected cells, and they appear to be co- and post-translationally processed at conserved junctions by viral proteases, yielding dozens of mature and intermediate replicase products. Replicases of all nidoviruses, irrespective of their more than two-fold size differences, have a common backbone of conserved domains. Sequence alignments and phylogenetic analysis imply that this replicase conservation is likely due to continuous evolution from a common nidovirus ancestor. A number of activities and functions have been provisionally assigned to many of the conserved replicase subunits. These assignments largely derived from bioinformatics analyses, although an increasing number of these predictions is being verified experimentally. Replicase subunits conserved among nidoviruses include (from the N-terminus to C-terminus): a chymotrypsin-like protease with a substrate specificity resembling that of picornavirus 3C proteases (3C-like or main protease) that is flanked by two hydrophobic trans-membrane domains (M-3CL-M), a large RdRp, a protein including a putative multinuclear Zn-finger-like domain and a nucleoside triphosphate (NTP)-binding/5′-to-3′-helicase domain (Zn-HEL), a putative poly(U)-specific endoribonuclease (NendoU).

Unfortunately, no complete genome sequence is available for any torovirus. For the best characterized torovirus (EToV), a large part of ORF1a between the known 5′- and 3′-terminal sequences is yet to be determined. The N-terminal half of the ORF1a protein is quite variable in other nidoviruses. This variability is largely responsible for the size differences in coronaviruses, or arterivirus genomes. A comparison between the coronavirus and arterivirus N-terminal ORF1a protein sequences does not yield any significant similarities beyond the presence of one or more papain-like proteases.

ANTIGENIC PROPERTIES

Serological interfamily cross-reactivity has not been demonstrated. Coronaviruses have four structural proteins: (i) the spike (S) protein forms trimers and is the major inducer of virus-neutralizing antibodies which are elicited by several domains located in the amino terminal half of the molecule; (ii) the membrane (M) protein has three or four transmembrane domains, with either the amino-terminus alone or both the amino-terminus and the carboxy-terminus being exposed at the virus surface. Most of the antibodies elicited by the M protein are directed to the carboxy-terminus. In general, polyvalent or monovalent antibodies to the amino-terminus weakly neutralize virus infectivity, but in the presence of complement they reduce infectivity around 100-fold; (iii) nucleoprotein (N) is a dominant antigen during virus infection; a N peptide is presented on the surface of infected cells and induces protective T cell responses; (iv) the envelope (E) protein is also exposed at the surface of virus and virus-infected cells. In the case of MHV, E protein-specific antiserum neutralizes viral infectivity in the presence of complement

The four toroviruses described (EToV, BToV, Porcine torovirus, Human torovirus) are serologically related. Toroviruses have four structural proteins: (i) the spike (S) protein of 180 kDa that forms large 17-20-nm spikes and induces virus-neutralizing and hemagglutination-inhibiting antibodies, (ii) the 26-kDa triple-spanning integral membrane protein (M), (iii) a 65 kDa class I membrane protein (HE) exhibiting acetylesterase activity; this protein in BToV virions forms short surface projections of 6 nm on average that act as a prominent antigen during infection, and (iv) the 19-kDa nucleocapsid protein.

Antibodies against the known arteriviruses infecting different species (EAV, LDV, PRRSV, SHFV) do not cross-react. Arteriviruses have six structural proteins: (i) a major glycoprotein (GP_5) which, due to heterogeneous glycosylation, has a size between 30 and 42 kDa and is the main determinant of virus-neutralization; (ii) a trimer of GP2, GP_3, and GP_4, which are all a minor virion components and of which (at least) GP_4 can play a role in humoral immunity; (ii) (iii) an unglycosylated transmembrane (M) protein; (iv) a 12 kDa nucleocapsid (N) protein involved in the induction of protection; and (v) a small nonglycosylated hydrophobic protein E.

BIOLOGICAL PROPERTIES

Coronaviruses infect many mammals, including humans. The main targets are epithelial cells, and consequently respiratory and gastrointestinal organ disorders result. Biological vectors are not known. Respiratory, fecal-oral and mechanical transmission are common. Swine and domestic fowl may become persistently infected with TGEV and IBV, respectively, and shed virus from the enteric tract. Hepatitis and neurological (MHV), heart and eye (RbCoV) infections have also been described.

Toroviruses infect ungulates: horses (EToV, Berne virus), bovines (BToV, Breda virus) and swine (PToV). Humans (HToV) and probably carnivores (mustellids) are also hosts for toroviruses. The transmission is probably by the fecal-oral route.

Arteriviruses infect horses (EAV), mice (LDV), monkeys (SHFV) and swine (PRRSV). EAV causes inflammation of small arteries and EAV infection can lead to extremely variable clinical signs. A fatal outcome of the disease has been reported in both natural and experimental infections, but most natural infections are either mild or subclinical. Primary host cells for all arteriviruses are macrophages. Persistent infections are frequently established. Spread is in general horizontal (respiratory, biting), by venereal routes and semen (EAV and PRRSV). Males may become persistently infected and shed virus from their reproductive tract. In pregnant animals arteriviruses (PRRSV and EAV) can cause abortions or *in utero* fetal death (PRRSV).

Roniviruses are the only known invertebrate nidoviruses and have been detected only in crustaceans. The black tiger prawn (*Penaeus monodon*) appears to be the natural host of GAV but other prawn species are susceptible to experimental infection. Infections may be chronic or acute and transmission can occur horizontally and vertically. During acute infections, mortalitiy is usually high and virus occurs in most tissues of ectodermal and mesodermal origin, and particularly in the 'Oka' or lymphoid organ. Necrotic cells display intensely basophilic cytoplasmic inclusions. The geographic range of infection presently appears to be restricted to Asia and Australia where the prevalence of sub-clinical chronic infections in *P. monodon* is commonly high. There are no known prophylactic or curative treatments.

PHYLOGENETIC RELATIONSHIPS WITHIN THE ORDER

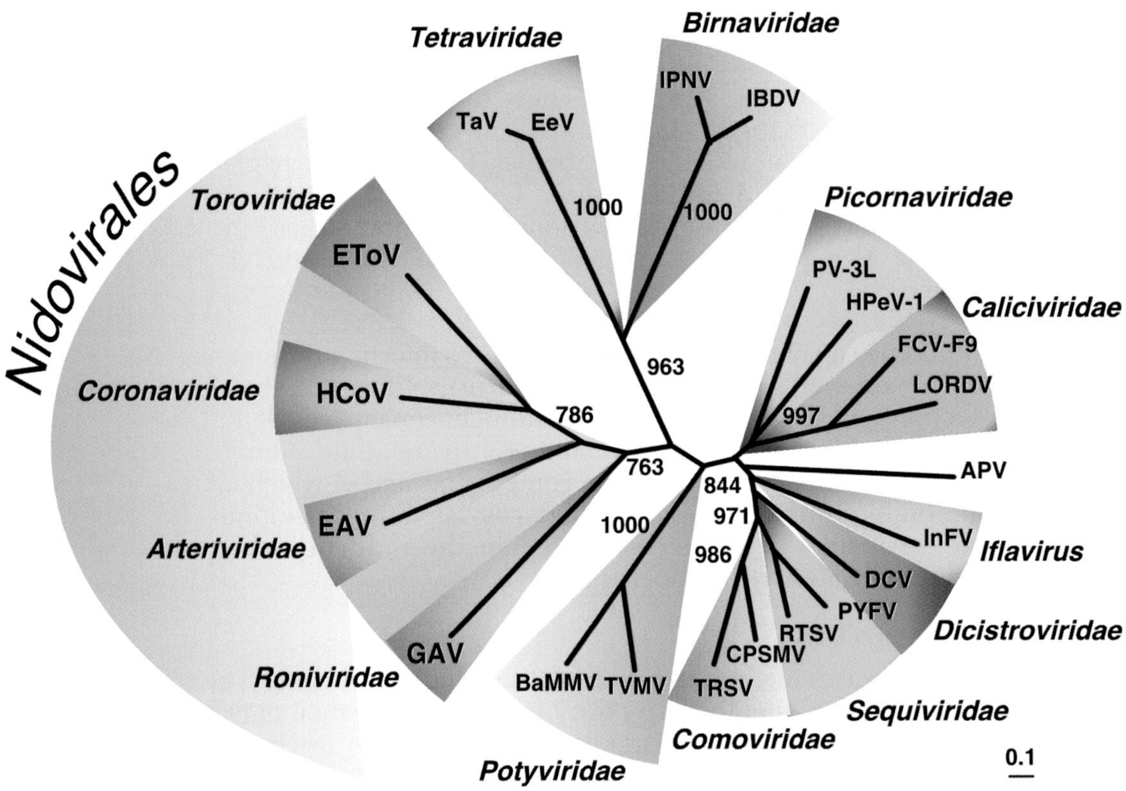

Figure 3: Unrooted phenogram showing the relationships of the RdRps of the *Nidovirales* lineages with virus of the families of the "Picornavirus-like" supergroup, *Tetraviridae* and *Birnaviridae*. The most conserved part of RdRps from representative viruses in the *Picornaviridae, Dicistroviridae, Sequiviridae, Comoviridae, Caliciviridae, Potyviridae, Coronaviridae, Roniviridae, Arteriviridae, Birnaviridae, Tetraviridae* and unclassified insect viruses was aligned. The RdRps of Thosea asigna virus (TaV), Euprosterna elaeasa virus (EeV) and the birnaviruses were converted into the canonical ABC motif form before the analysis. Using an extended, gap-free version of the alignment containing 332 informative characters, an unrooted neighbor-joining tree was inferred by the ClustalX1.81 program. All bifurcations with support in > 700 out of 1000 bootstraps are indicated. Different groups of viruses are highlighted. Note that the relative positions of members of the families *Arteriviridae* and *Roniviridae* within the *Nidovirales* are not resolved in this tree.
Virus families and groups, viruses included in the analysis, abbreviations ( ) and the NCBI protein (unless other specified) IDs [ ] are as follows: *Picornaviridae*: Human poliovirus type 3 Leon strain (PV-3L) [130503] and parechovirus 1 (HpeV-1) [6174922]; *Iflavirus*: Infectious flacherie virus (InFV) [3025415] and Unclassified insect viruses: Acyrthosiphon pisum virus (APV) [7520835]; *Dicistroviridae*: Drosophila C virus (DCV) [2388673]; *Sequiviridae*: Rice tungro spherical virus (RTSV) [9627951] and Parsnip yellow fleck virus (PYFV) [464431]; *Comoviridae*: Cowpea severe mosaic virus (CPSMV) [549316] and Tobacco ringspot virus (TRSV) [1255221]; *Caliciviridae*: Feline calicivirus F9 (FCV-F9) [130538] and Lordsdale virus (LORDV) [1709710]; *Potyviridae*: Tobacco vein mottling virus (TVMV) [8247947] and Barley mild mosaic virus (BaMMV) [1905770]; *Coronaviridae*: Human coronavirus 229E (HCoV) [12175747] and Equine torovirus (EToV) [94017]; *Arteriviridae*: Equine arteritis virus (EAV) [14583262]; *Roniviridae*: Gill-associated virus (GAV) [9082018]; *Tetraviridae*: Thosea asigna virus (TaV) [AF82930; nt sequence] and Euprosterna elaeasa virus (EeV) [AF461742; nt sequence]; *Birnaviridae*: Infectious pancreatic necrosis virus (IPNV) [133634] and Infectious bursal disease virus (IBDV) [1296811]. (Tree was modified from Gorbalenya *et al.*, (2002), with permission).

SIMILARITY WITH OTHER TAXA

Homologs of several (putative) enzymes encoded by viruses of the order *Nidovirales* have been found in non-nidoviruses. The proteolytic enzymes and RdRps cluster together with homologs of viruses of the "Picornavirus-like" supergroup, double-stranded RNA *Birnaviridae* family members and a subset of members of the family *Tetraviridae* (Fig. 3).

The HEL enzyme has a counterpart in viruses of the "Alphavirus-like" supergroup. The organization of the replicase ORFs, including the 3CLpro_FS_RdRp constellation, is also conserved in the family *Astroviridae and some Sobemoviruses and related viruses*. Some parallels in the genome organization and expression strategy are evident between members of the order *Nidovirales* and the family *Closteroviridae*.

Derivation of Names

Arteri, from equine *arteri*tis, the disease caused by the reference virus.
Corona, derived from Latin *corona*, meaning *crown*, representing the appearance of surface projections.
Nido, from Latin *nidus*, meaning nest, representing the nested set of mRNAs.
Roni refers to *ro*d-shaped *ni*doviruses and is derived from the morphology of the virus.
Toro, from Latin *torus*, the lowest convex moulding in the base of a column, refers to the nucleocapsid morphology.

References

Cavanagh, D., Brian, D.A., Briton, P., Enjuanes, L., Horzinek, M.C., Lai, M.M.C., Laude, H., Plagemann, P.G. W., Siddell, S., Spaan, W. and Talbot, P.J. (1997). *Nidovirales*: a new order comprising *Coronaviridae* and *Arteriviridae*. *Arch. Virol.*, **142,** 629-635.

De Vries, A.A.F., Horzinek, M.C., Rottier, P.J.M. and De Groot, R.J. (1997). The genome organization of the *Nidovirales*: similarities and differences between arteri-, toro-, and coronaviruses. *Sem. Virol.*, **8,** 33-47.

Enjuanes, L., Siddell, S.G. and Spaan, W.J. (eds)(1998). *Coronaviruses* and *Arteriviruses*. Plenum Press, New York.

González' J.M., Gomez-Puertas' P., Cavanagh, D., Gorbalenya, A.E. and L. Enjuanes (2003). A comparative sequence analysis to revise the current taxonomy of the family *Coronaviridae*. *Arch. Virol.*, 148, 2207-2235.

Gorbalenya, A.E. (2001). Big nidovirus genome: when count and order of domains matter. *Adv. Exp. Med. Biol.*, **494**, 1-17.

Gorbalenya, A.E., Pringle, F.M., Zeddam, J.-L., Luke, B.T., Cameron, C.E., Kalmakoff, J., Hanzlik, T.N., Gordon, K.H.J. and Ward, V.K. (2002). The palm subdomain-based active site is internally permuted in viral RNA-dependent RNA polymerases of an ancient lineage. *J. Mol. Biol.*, **324**, 47-62.

Lai, M.M.C., and Cavanagh, D. (1997). The molecular biology of coronaviruses. *Adv. Vir. Res.*, **48,** 1-100.

Siddell, S.G. (ed)(1995). "The *Coronaviridae*". In: *The Viruses*, (H. Fraenkel-Conrat and R.R. Wagner, eds), Plenum Press, New York.

Snijder, E.J., Bredenbeek, P.J., Dobbe, J.C., Thiel, V., Ziebuhr, J., Poon, L.L.M., Guan, Y., Rozanov, M., Spaan, W.J.M. and A.E. Gorbalenya (2003). Unique and conserved features of genome and proteome of SARS-Coronavirus, an early split-off from the coronavirus group 2 lineage. *J. Mol. Biol.*, **331**, 991-1004.

Snijder, E.J. and Horzinek, M.C. (1995). The molecular biology of toroviruses. In: *The Coronaviridae*, (S.G. Siddell, ed), pp 219-238. Plenum Press, New York.

Snijder, E.J. and Meulenberg, J.J.M. (1998). The molecular biology of arteriviruses. *J. Gen. Virol.*, **79,** 961-979.

Snijder, E.J. and Spaan, W.J.M. (1995). The coronavirus-like superfamily. In: *The Coronaviridae* (S.G. Siddell, ed), pp 239-252. Plenum press, New York.

Ziebuhr, J., Snijder, E.J. and Gorbalenya, A.E. (2000). Virus-encoded proteinases and proteolytic processing in the *Nidovirales*. *J. Gen. Virol.*, **81**, 853-879.

Ziebuhr, J., Bayer, S., Cowley, J.A. and Gorbalenya, A.E. (2003). The 3C-like proteinase of an invertebrate nidovirus links coronavirus and potyvirus homologs. *J. Virol.*, **77,** 1415-1426.

Contributed By

Spaan, W.J.M., Cavanagh, D., de Groot, R.J., Enjuanes, L., Gorbalenya, A.E., Snijder, E.J. and Walker, P.J.

FAMILY CORONAVIRIDAE

TAXONOMIC STRUCTURE OF THE FAMILY

Order	*Nidovirales*
Family	*Coronaviridae*
Genus	*Coronavirus*
Genus	*Torovirus*

GENUS CORONAVIRUS

Type Species *Infectious bronchitis virus*

VIRION PROPERTIES

MORPHOLOGY

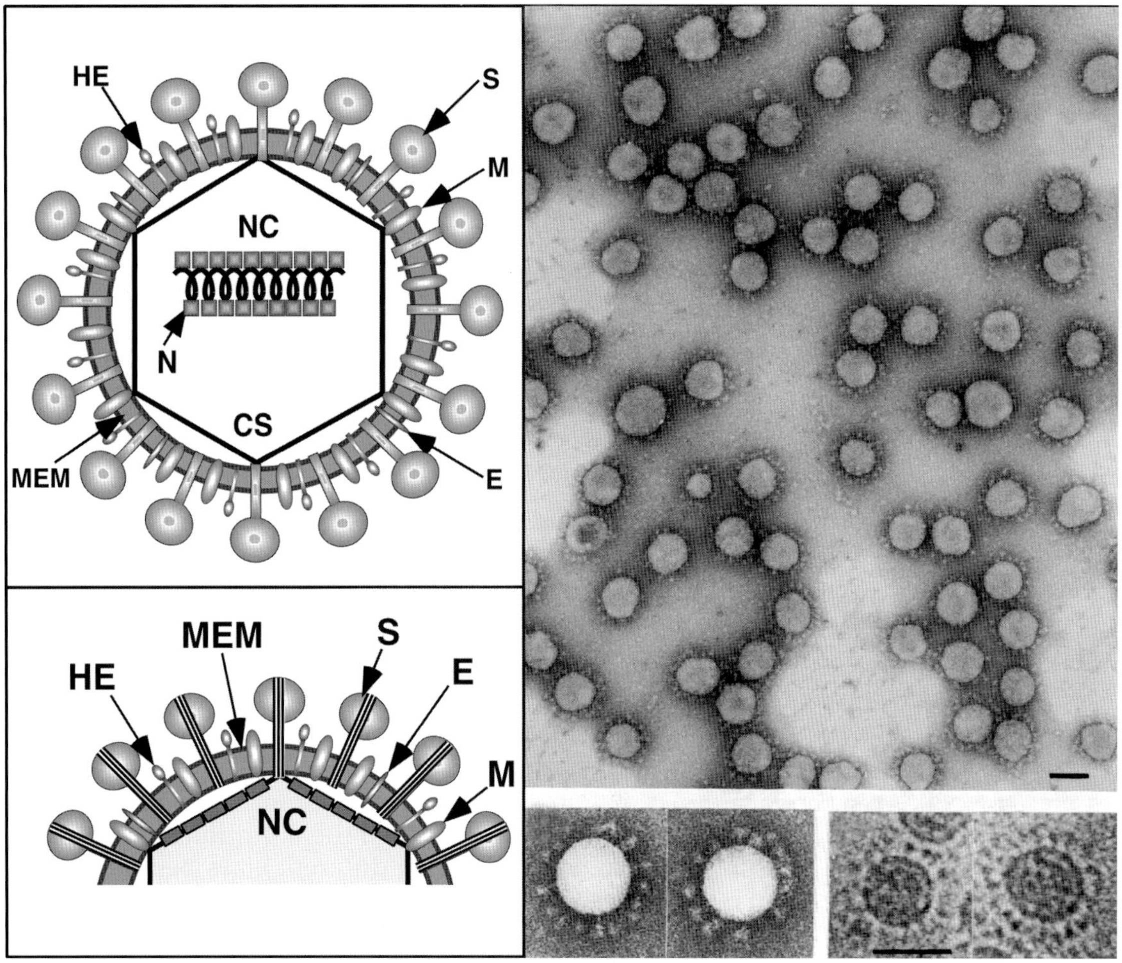

Figure 1: Structure of coronavirus virions. (Top left) Schematic diagram of virus structure; (Bottom left) Diagram of virion surface. (Top right) Electron micrograph of virus particles of Transmissible gastroenteritis virus (TGEV) stained with uranyl acetate (top right) or sodium phosphotungstate (insert bottom left) showing the surface of the virus particles. The peplomers are better defined using sodium phosphotungstate. (insert bottom right) Cryo-electron microscopic visualization of unstained TGEV in vitreous ice. The particles contain an internal structure inside the viral envelope and well extended peplomers. MEM, lipid membrane; S, spike protein; M, large membrane protein, E, small envelope protein; HE, hemagglutinin-esterase; N, nucleocapsid protein; CS, core-shell; NC, nucleocapsid. The bars represents 100 nm.

Virions are enveloped and spherical in shape; those of members of the genus *Coronavirus* being commonly 120-160 nm in diameter, with an internal, possibly icosahedral, core shell of around 65 nm, and a helical nucleocapsid comprising the nucleocapsid protein (N) and RNA genome (Fig. 1). Coronaviruses have large surface projections formed by glycoproteins (peplomers; trimers of the spike protein) with a globular and a stem portion. The peplomers (trimers of the spike protein) are about 20 nm in length. In some coronaviruses such as Bovine coronavirus (BCoV) and some strains of Mouse hepatitis virus (MHV), a second layer of peplomers formed by the hemagglutinin esterase (HE) glycoprotein is also observed. A gap separating the internal core from the envelope has been observed in coronaviruses using cryoelectron microscopy. The core can be released after treatment with detergents. Disruption of these cores releases N-protein-containing helical nucleocapsids.

Physicochemical and Physical Properties

Virion Mr has been estimated at 400 x 10^6 for coronaviruses. Virion buoyant density in sucrose is 1.15-1.20 g/ml; density in CsCl is 1.23-1.24 g/ml. Virion $S_{20,w}$ is 300-500. Virions are inactivated by heat, lipid solvents, non-ionic detergents, formaldehyde, oxidizing agents and UV irradiation. After incubating 24 hrs at 37°C an at least 10-fold decrease in virus infectivity has been observed with certain strains in tissue culture. Magnesium ions (1M) reduce the extent of heat inactivation in MHV. Some viruses in both genera are stable at pH 3.0.

Nucleic Acid

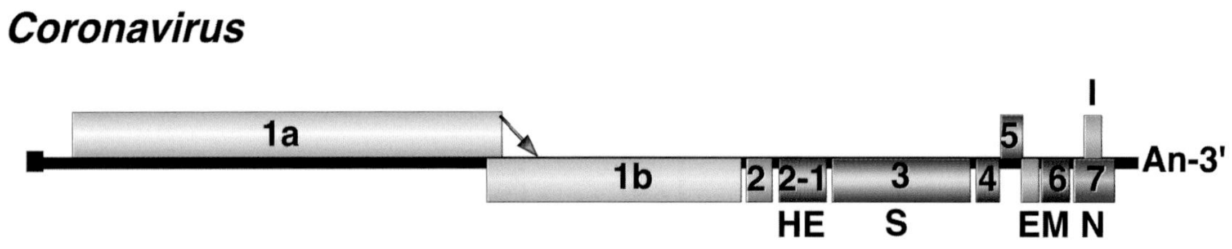

Figure 2: Representation of the genome of MHV as a coronavirus genome example. ORFs are represented by boxes. The proteins encoded by the ORFs are indicated; ORF1a encodes pp1a and, together with ORF1b, pp1ab polyproteins. The 5′ leader sequence is depicted by a small black box; poly(A) tail is indicated by An; M, membrane protein; N, nucleocapsid protein; S, spike protein; HE, hemagglutinin-esterase glycoprotein; I, internal ORF; no designated boxes, nonstructural proteins; the leader is indicated by a small black box; The arrow between ORF 1a and 1b represents the ribosomal frameshifting site.

Virions contain a single molecule of linear, positive-sense, ssRNA which functions as an mRNA and is infectious. The genomic RNA is the largest viral RNA genome known ranging from 27.6 to 31 kb in size. The cap structure at the 5′-end of genome is followed by an UTR of 200-400 nt that includes a so-called leader sequence of 65-98 nt at the very 5′-end. At the 3′ end of the genome is another UTR of 200-500 nt followed by a poly(A) tail. RNA secondary structures, including a bulged stem-loop, a pseudoknot, or both (depending on the coronavirus group), are found in the 3' untranslated regions of genomic and sgRNA. There may be from 6 to 14 ORFs of different sizes between 5′- and 3′-UTRs. The two 5′-most and overlapping ORFs comprise the replicase gene and each other ORF is a separate gene. The genes are arranged in the order 5′-replicase-(HE)-S-E-M-N-3′, with a variable number of other genes dispersed downstream of the replicase in different coronaviruses. The products of these genes (known as non-structural, ns) are believed not to be part of virions and they are largely non-essential, at least in tissue culture. A genome structure of a coronavirus is shown in Figure 2.

The complete sequences of several coronavirus genomes have been determined and are accessible from international databases: BCoV, Human coronavirus strains 229E and OC43 (HCoV-229E and HCoV-OC43, respectively), IBV, MHV, Porcine epidemic diarrhea virus (PEDV), Severe acute respiratory syndrome coronavirus (SARS-CoV), TGEV.

PROTEINS

Virions contain a large surface glycoprotein (or spike, S), an integral membrane protein (M) that has three or four transmembrane segments, a small membrane protein (E) and a nucleocapsid protein (N) (Table 1). The ratios of S:E:M:N proteins vary in different reports. For TGEV, these ratios have been estimated to be 20:1:300:140, respectively. The S protein is large, ranging from 1160 to 1452 aa, and in some coronaviruses is cleaved into S1 and S2 subunits. The S protein is responsible for attachment to cells, hemagglutination, membrane fusion, and induction of neutralizing antibodies. Immunization with S alone can induce protection from challenge with some coronaviruses (IBV, MHV, TGEV). The S protein has a carboxy-terminal half with a coiled-coil structure and belongs to the class I virus fusion proteins. The M protein contains 221 to 260 aa and can induce α-interferon. Envelopes of most coronaviruses in group 2 (see below) also contain a hemagglutinin-esterase glycoprotein (HE) that forms short surface projections. This apparently non-essential protein has a receptor binding domain for 9-O-acetylated neuraminic acid, hemagglutination activity, as well as receptor destroying activities (neuraminate-*O*-acetylesterase). The E protein (76 to 109 aa), together with the M protein, plays an essential role in coronavirus particle assembly. No E protein (gene) has been identified in toroviruses. The N protein (377 to 455 aa) is a highly basic phosphoprotein that modulates viral RNA synthesis, binds to the viral RNA and forms a helical nucleocapsid.

Table 1: Virion-associated proteins of coronaviruses.

Protein		kDa
Spike glycoprotein	S	180 - 220
Membrane protein	M	23 - 35
Nucleocapsid protein	N	50 - 60
Small-envelope protein	E	9 - 12
Hemagglutinin-esterase protein	HE	65

a, apparent molecular mass estimated by electrophoresis (kDa)

A large variety of virus proteins known as non-structural proteins are not incorporated into virions. The largest non-structural protein is the essential replicase encoded by slightly overlapping ORF1a and ORF1b (gene 1), which accounts for two-thirds of the genome (from 18 to 22 kb). The replicase gene is predicted to encode two proteins of approximately 450 kDa (encoded by ORF1a) and 740-800 kDa (a fused product of ORF1a and ORF1b expressed through –1 frameshifting). Neither of these giant proteins has been observed in infected cells, and they appear to be co- and post-translationaly processed at the conserved junctions by viral proteases to form 15 to 16 mature replicase products and an unknown number of intermediates. A number of activities and functions has been provisionally assigned to most of the conserved replicase subunits by bioinformatics methods and some of these predictions were verified experimentally. Thus, replicase of viruses of *Coronavirus* genus includes (from the N-terminus to C-terminus): putative adenosine diphosphate-ribose 1″-phosphatase (ADRP), Zn-ribbon-dependent papain-like cysteine proteinase, a chymotrypsin-like cysteine protease with the substrate specificity resembling that of picornavirus 3C proteases (3C-like or main protease) that is flanked by two hydrophobic trans-membrane domains, a small cysteine-rich protein previously known as the growth-factor-related protein, a large RdRp, a protein including putative multinuclear Zn-finger-like domain and nucleoside triphosphate (NTP)-binding/5′-to-3′-helicase domain (HEL), putative Zn-finger-containing 3′-to-5′ exonuclease, putative poly(U)-specific endoribonuclease, and S-adenosylmethionine-dependent ribose 2′-*O*-methyltransferase (MTR). Coronaviruses also differ in respect to a

POS. SSRNA

small number of proteins and domains, which are derived from the N-terminus of the replicase. They include a distant copy of a Zn-ribbon-dependent papain-like cysteine proteinase and SARS-CoV unique domain (SUD) located in the largest replicase subunit nsp3 (also known as p195/p210).

The other non-structural proteins that are encoded downstream of the replicase gene vary in type and location among coronaviruses. The locations of the genes encoding these non-structural proteins are indicated in Figure 3. These non-structural proteins are generally not essential for virus replication in tissue culture or *in vivo*, although (provisional) functional assignment was made only for two proteins including HE and putative cyclic phosphodiesterase (CPD).

LIPIDS

Virions have lipid-containing envelopes derived from the host cell. The S protein (BCoV, MHV) and E protein (MHV) of coronaviruses are acylated.

CARBOHYDRATES

The S and HE proteins contain N-linked glycans, the S protein being extensively glycosylated. The M protein of coronaviruses contains a small number of either N- or O-linked glycans, depending on the virus or strain.

GENOME ORGANIZATION AND REPLICATION

Coronavirus replication proceeds through the synthesis of the full-length negative antigenome in the cytoplasm of infected cells. It is catalyzed by a poorly characterized membrane-bound replicative complex. The products of the ORF1a and ORF1ab are sufficient to maintain the coronavirus RNA synthesis. These ORFs encode a variety of (putative) enzymes (see section on *proteins*) that may be part of the replicative complex *per se* or control its composition and functioning.

During the synthesis of positive- and negative-strand RNAs MHV undergoes recombination at very high frequency. A lower recombination frequency has been described for IBV and TGEV. Interspecies recombination limited to viruses of the same group, e.g. between Feline coronavirus (FCoV) and Canine coronavirus (CCoV), has also been documented.

The genome is expressed by different mechanisms that includes replication, transcription, translation and co- and post-translation regulation. The virion RNA functions as the mRNA1 for the two 5′-most ORF1a and ORF1b encoding replicase components. The translation of ORF1a yields the pp1a polyprotein. In approximately 20% of cases, ribosomes slip at the overlap between the ORF1a and ORF1b regions, containing a specific seven-nucleotide "slippery" sequence and a pseudoknot structure (ribosomal frameshifting signal), to continue translation into ORF1b, which is in the –1 frame relative to ORF1a. Translation of ORF1ab yields the pp1ab polyprotein. The pp1a and pp1ab polyproteins have not been described in infected cells; they are believed to be co- and post-translationally processed by two or three virus encoded proteases to 15 or 16 mature replicase subunits and an unknown number of functional intermediates. There is no universally accepted nomenclature of replicase subunits; they may be listed in the order of encoding in polyproteins, or according to the molecular size or function. For SARS-CoV a nomenclature is proposed as indicated in Figure 3. One or two papain-like proteases (PLpro) control processing of the N-terminal part of pp1a/pp1ab polyproteins at two or three conserved cleavage sites. A chymotrypsin-like protease (3CLpro or Mpro) is responsible for processing of the rest part of pp1a/pp1ab polyproteins at 11 conserved cleavage sites.

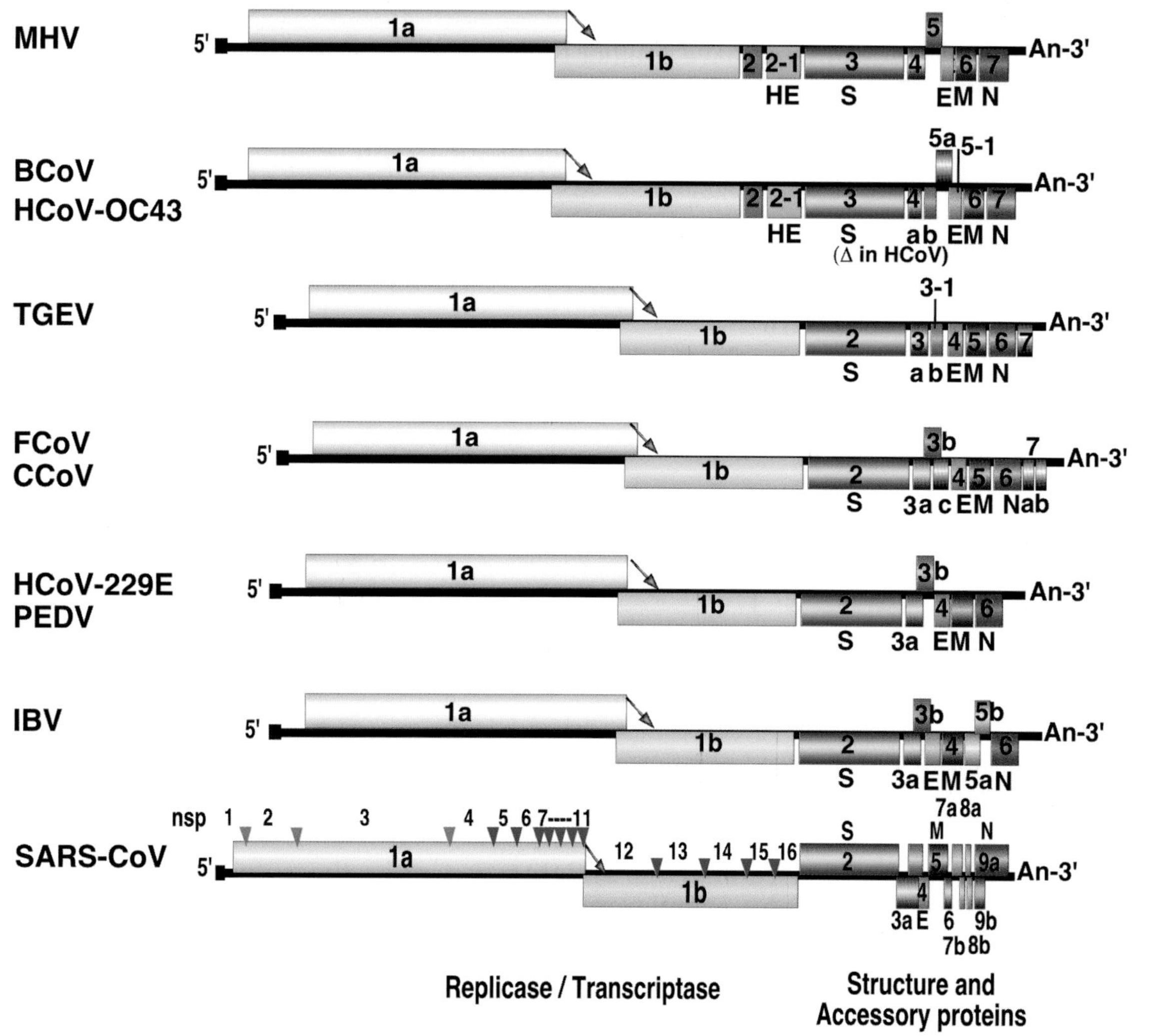

Figure 3: Genome organization in coronaviruses. ORFs are represented by boxes and indicated by numbers in, below or above the box except for the SARS CoV ORF 1a and 1b (see below) . The names below or on top of the bars indicate the protein encoded by the corresponding ORF. The structural protein genes are marked by various symbols while the proteins of the nonstructuralgenesare not designated. The 5' leader sequence is depicted by a small black box. Protein acronyms are as in figure 2. Δ, deleted. nsp, non-structural protein. The arrow between ORF 1a and 1b represents the ribosomal frameshifting site.HCoV-OC43 does not have the two non-structural proteins encoded by gene 4. The translation products of ORFs MHV 5a, BCoV 4a,b, FCoV and CCoV 3b have not been detected. ORFs TGEV 3b, FCoV and CCoV 3c, and HCV229E 3a,b are homologous. The 3a ORFs of TGEV, FCoV and CCoV are homologous. ORFs TGEV 7 and the 7a of FCoV and CCoV are homologous. Genetic structure of SARS-CoV: the numbers above ORF 1a and 1b indicate the putative proteolytic cleavage products; the red arrowheads depict the papain-like cysteine protease cleavage sites and the blue arrowheads represent the 3C-like cysteine protease cleavage sites.

All ORFs downstream of ORF1b are expressed from a 3'-coterminal nested set of specialized sgRNAs whose number varies between 5 and 7 among coronaviruses (Fig. 4). Subgenomic mRNAs were originally designated with numbers from 2 to 7, in order of decreasing size. Some later discovered mRNAs have been given a hyphenated name, e.g. mRNA2-1, to show that they are variants of previously recognized mRNAs. Except for the smallest mRNA, all of the mRNAs are structurally polycistronic. As a rule, only the 5'-most ORF of each mRNA is translated while downstream ORF(s) remain translationally silent. However, there are exceptions which, in addition to the expression of ORFs 1a and 1b from the same RNA detailed above, include 2nd and 3rd ORFs of some mRNAs, e.g. mRNA5 of MHV, mRNA3 of IBV and mRNA encoding nucleocapsid of BCoV. These ORFs may be translated by internal, as well as 5'-terminal, initiation to produce from two to three proteins from a single sg mRNA.

Coronavirus mRNAs have another unique structural feature: their 5'-ends have a leader sequence of approximately 65 to 98 nt, which is identical to the 5'-end of the genomic RNA. At the mRNA start sites on the viral genomic RNA, there is a short stretch of sequence that is an imperfect repeat of the 3'-end of the leader RNA. This sequence constitutes part of the signal for sgRNA transcription.

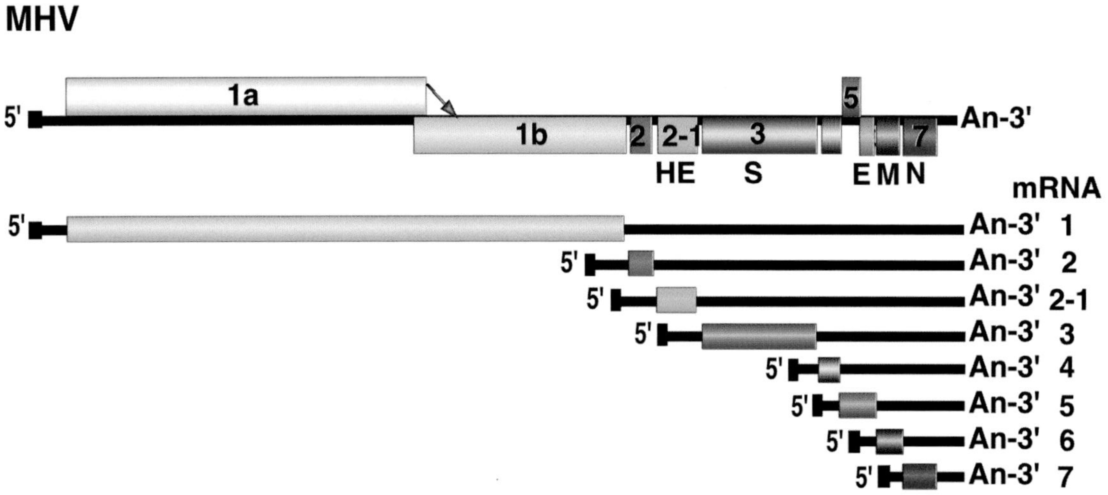

Figure 4: Structural relationship between mRNAs and the genomic RNA of coronaviruses. Thick lines represent the translated sequences. Thinner lines, untranslated sequences. The names below the boxes indicate the proteins encoded by the corresponding genes. Acronyms are explained in the legend to Figure 2.

The synthesis of coronavirus sgRNAs occurs in the cytoplasm at negative-stranded template(s) through a complex mechanism for which essential details remain a matter of debate. Both genome-size and negative-strand sgRNAs, which correspond in number of species and size to those of the virus-specific mRNAs, have been detected in infected cells. The 5'-end of the negative-strand RNA contains short stretches of oligo(U). The negative-strand sgRNAs appear to be complementary copies of the positive-strand sgRNAs. An almost perfect repeat of a sequence stretch known as the consensus or core sequence, which is UCUAAAC in MHV or a related sequence for other coronaviruses, was found immediately upstream of gene (ORFs) in the coronavirus genome. These sequences represent crucial signals for the synthesis of sgRNAs and they are also called transcription regulating sequences (TRS) (previously designated as intergenic sequences, IGSs). The 5′-most TRS is at the 3′-border of the leader sequence (leader TRS) that is common for all mRNAs including genomic RNA. Several models have been proposed to explain a wealth of experimental data and, particularly, how the common leader sequence is fused to different bodies in the variety of mRNAs. These models are not mutually exclusive, as components of each model may operate at different stages of the viral replication cycle. The model compatible with most of the experimental data is *discontinuous transcription during negative-strand RNA synthesis*. This model proposes that the discontinuous transcription step occurs during negative-strand RNA synthesis, generating negative-strand sgRNAs, which then serve as templates for uninterrupted synthesis of sg mRNAs. In this model, the body TRSs on the genomic RNA serve as termination or pausing signals for negative-strand synthesis, and the nascent negative-strand sgRNA then jumps to the leader TRS sequence at the 5'-end of the genomic RNA to complete the synthesis of the negative copy of sg mRNA. In the alternative model, the *leader-primed transcription* model proposes that the virion genomic RNA is first transcribed into a genomic-length, negative-strand RNA, which, in turn, becomes the template for subsequent synthesis of the entire repertoire of sg mRNAs. The leader of nascent sg mRNA is transcribed from the 3'-end of the negative-strand RNA and dissociates from the template to subsequently associate with the template RNA at one of the (body) TRSs to prime the synthesis of the particular sg mRNA. In this mechanism, the discontinuous transcription step takes place during positive-strand RNA synthesis. The composition of the machinery responsible for the synthesis of sgRNAs (transcriptase) is unknown, although many replicase subunits described above must be involved in transcription.

The packaging signal for MHV RNA, as determined using defective minigenomes, is localized near the 3′-end of gene 1. This packaging signal forms a stem loop and is sufficient for the packaging of DI RNA or heterologous RNAs into virions. In contrast, the packaging signal of TGEV and IBV has been localized within the first 650 nt of gene 1.

The assembly of virus particles probably starts with the formation of an RNP that interacts with components of the core shell. Virions mature in the cytoplasm by budding through the endoplasmic reticulum and other pre-Golgi membranes. The interaction between the M and E proteins appears to be a key event for virus particle assembly. The S and HE proteins are not necessary for virus particle formation though the S protein is essential for infectivity.

Coronavirus reverse genetics has been facilitated by targeted recombination, the construction of infectious cDNA clones using different strategies: as bacterial artificial chromosomes, using poxviruses as cloning vectors, and as independent fragments that are assembled *in vitro*.

ANTIGENIC PROPERTIES

Strong humoral immune responses are elicited by the structural proteins S, M, N, and, when present, HE. The S and HE proteins are the predominant antigens involved in virus neutralization. Reduction of infectivity with anti-M antibodies also has been shown, but generally in the presence of complement. Protection against coronavirus infections (IBV, MHV, TGEV) is provided by S protein that has been affinity-purified or expressed by recombinant adenovirus. N- and M-specific antibodies also give some protection *in vivo*. The most efficient induction of virus neutralizing antibodies has been achieved with a combination of S and N proteins. The globular portion of the S protein contains many dominant antigenic sites targeted by the humoral immune response and also by cytotoxic T lymphocytes. Other important antibody epitopes are also found in the stem portion, at least for MHV. Both the amino- and carboxy- termini of the M protein elicit strong immune responses. While neutralizing antibodies can prevent disease if present prior to infection, cytotoxic T cell responses are important in virus clearance. Hypervariable domains in the S1 portion of the S protein facilitate the selection of virus escape mutants that evade both humoral and cellular immune responses. N protein also elicits a protective cellular immune response.

The immune system is known to play a major role in the pathogenesis of some coronavirus-induced diseases, such as the experimental induction of demyelination by MHV, and the antibody-dependent enhancement of disease (feline infectious peritonitis) caused by FCoV.

BIOLOGICAL PROPERTIES

Coronaviruses infect birds and many mammals, including humans. The respiratory tract, gastrointestinal tractand neurological tissues are the most frequent targets of coronaviruses, but other organs including liver, kidney, heart, and eye can also be affected. Epithelial cells are the main target of coronaviruses, plus, with some coronavirus species, widely distributed cells such as macrophages. Coronaviruses have a wider host range *in vivo* than would be expected from *in vitro* studies. In experimental infections BCoV caused enteric disease in turkeys. FCoV and CCoV replicated in experimentally infected pigs, clinical disease being caused by virulent FCoV. Turkey coronavirus has been demonstrated to replicate in chickens asymptomatically. Type II FCoV has arisen by natural recombination of type I FCoV with CCoV. The cause of severe acute respiratory syndrome (SARS) in man is a coronavirus that is believed to have jumped from another animal species. BCoV and some isolates of HCoV-OC43 have >99% aa identity in their S and HE proteins, raising the possibility of a shared host range *in vivo*. Leader-switching between a defective RNA of BCoV and HCoV-OC43 virus has been demonstrated, and the BCoV defective RNA could be replicated by the replicative machinery of HCoV-OC43 and other members of coronavirus Group 2, suggesting a potential for recombination between members of Group 2. Biological vectors are not known. Respiratory, fecal-oral and mechanical transmission are common.

Although coronaviruses may bind cells through ubiquitous acetylated forms of glycoproteins and lipids, a more specific binding between the virus and a cellular receptor is required for the establishment of viral infection. Coronaviruses are divided into three groups called 1, 2 and 3 (see below). MHV in Group 2 (see below), uses as receptors CEACAM1 glycoproteins in the immunoglobulin superfamily. SARS-CoV uses angiotensin converting enzyme 2 (ACE2) as a receptor. Members of the group 1 (see below) coronaviruses (including TGEV, FCoV and HCoV-229E) use aminopeptidase N (APN or CD13) as a receptor for cell entry. Sialic acid (N-acetyl-9-O-acetylneuraminic acid)-containing glycoproteins are probably a component of the cell surface molecules required for BCoV and HCoV-OC43 infectivity, and α2,3-linked sialic acid in the case of the Group 3 coronavirus IBV. Nevertheless, binding to CEACAM1 and APN receptors does not explain the differences in coronavirus tropism. In addition to the binding to the aforementioned receptors, a second factor mapping in the S protein (possibly, a second receptor binding site) has been implicated in coronavirus tropism.

Infections of man and animals by coronaviruses seem to be ubiquitous, as evidence of infection has been obtained in every country where serological or virological studies have been done.

List of Species Demarcation Criteria in the Genus

Practically, species in the *Coronavirus* genus have been differentiated using four parameters: (i) the position and variety of the non-essential non-structural proteins in the 3′-part of genome, (ii) antigenic properties, (iii) the processing of the S protein into two halves (S1 and S2), and (iv) the host range. No rigorous guidelines have been developed to apply the above criteria for defining new coronavirus species or differentiating between existing species. A brief posteriori analysis of the coronavirus species diversity in respect to these criteria is summarized below.

The number, type and location of the non-essential genes (encoding non-structural proteins) significantly vary. Some of these genes are common for a set of closely related viruses, e.g. CPD and HE genes in MHV, BCoV and HCoV-O43, and others may be specific for a virus species, e.g. all unique ORFs in SARS-CoV. Importantly, different isolates of the same coronavirus may also differ in respect to certain 3′-located ORFs, e.g. ORF8a and ORF8b are fused in a single ORF in some human vs. animal isolates of SARS-CoV. Serological characterization allows placing a virus within an antigenic cluster and nucleic (amino) acid similarity within a genetic cluster or group; the characterized coronaviruses form at least 3 genetic clusters or groups (see below). All coronaviruses fall in one of two divisions depending whether the S protein is processed into S1 and S2 subunits or not. Those with uncleaved S protein include Group 1 coronaviruses (CCoV, FCoV, HCoV229E, PEDV and TGEV) and Group 2, SARS-CoV, and those with the cleaved S protein include species of Groups 2 (BCoV, MHV and HCoV-OC43) and 3 (IBV). Coronaviruses infect a wide range of mammals and distantly related coronaviruses may infect the same host, e.g. HCoV-229E, HCoV-O43 and SARS-CoV, from different groups of coronaviruses, all infect humans. The group 3 consists exclusively of avian coronaviruses. Future revision of the taxonomy of the family *Coronaviridae* may propose a quantitative measure of genome similarity or other criterion to discriminate coronavirus species rigorously.

LIST OF SPECIES IN THE GENUS

Species names are in green italic script; strain names and synonyms are in black roman script; tentative species names are in blue roman script. Sequence accession numbers, and assigned abbreviations ( ) are also listed.

SPECIES IN THE GENUS

Group 1 species		
Canine coronavirus		
Canine coronavirus	[DL3096]	(CCoV)
Feline coronavirus		
Feline coronavirus		(FCoV)
Feline infectious peritonitis virus		(FIPV)
Human coronavirus 229E		
Human coronavirus 229E	[X69721]	(HCoV-229E)
Porcine epidemic diarrhea virus		
Porcine epidemic diarrhea virus	[Z35758]	(PEDV)
Transmissible gastroenteritis virus		
Transmissible gastroenteritis virus	[Z24675, Z34093, D00118, X06371]	(TGEV)
Porcine respiratory coronavirus		(PRCoV)
Group 2 species		
Bovine coronavirus		
Bovine coronavirus		(BCoV)
Human coronavirus OC43		
Human coronavirus OC43		(HCoV-OC43)
Human enteric coronavirus ‡		
Human enteric coronavirus		(HECoV)
Murine hepatitis virus		
Murine hepatitis virus	[AF029248]	(MHV)
Porcine hemagglutinating encephalomyelitis virus		
Porcine hemagglutinating encephalomyelitis virus		(HEV)
Puffinosis coronavirus ‡		
Puffinosis coronavirus		(PCoV)
Rat coronavirus		
Rat coronavirus		(RtCoV)
Sialodacryoadenitis virus,		(SDAV)
Severe acute respiratory syndrome coronavirus		
Severe acute respiratory syndrome coronavirus		(SARS-CoV)
Group 3 species		
Infectious bronchitis virus		
Infectious bronchitis virus	[M95169]	(IBV)
Pheasant coronavirus ‡		
Pheasant coronavirus		(PhCoV)
Turkey coronavirus		
Turkey coronavirus		(TCoV)

TENTATIVE SPECIES IN THE GENUS

Rabbit coronavirus	(RbCoV)

GENUS TOROVIRUS

Type Species *Equine torovirus*

DISTINGUISHING FEATURES

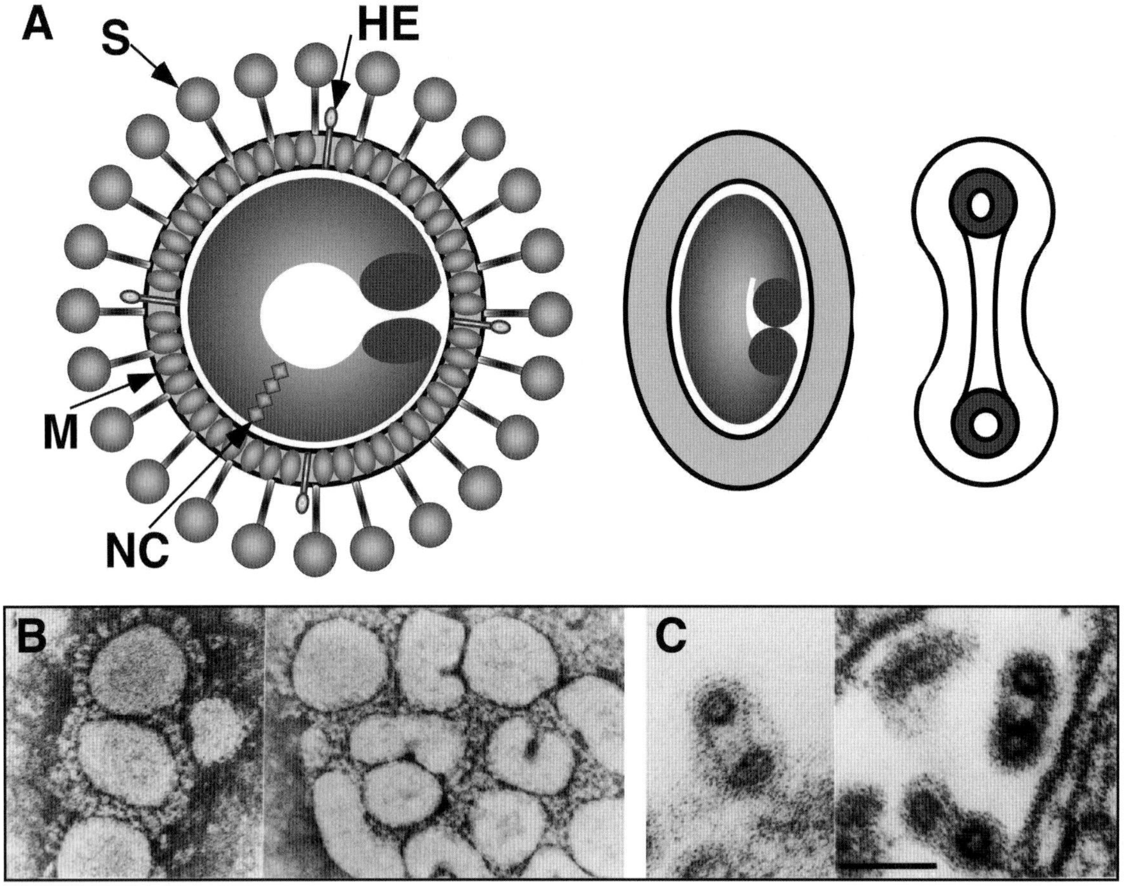

Figure 5: Torovirus particle structure. (Top) Schematic representation of the architecture of a particle of Equine torovirus (EToV). The localization of the structural proteins and genome are indicated. S, spike protein; M, large membrane protein; HE, hemagglutinin-esterase; N, nucleocapsid protein. (Bottom left) Negative contrast electron micrograph of EToV Berne strain particles. (Bottom right) Different forms of EToV particles in ultrathin sections of EoTV-infected equine dermis cells. The bar represents 100 nm.

RNAs and ORFs are summarized in Figure 6.

The nucleocapsid has a tubular appearance and virions are disc-, kidney- or rod-shaped (Fig. 5). There are 4 structural proteins: N, M, S and HE. The later protein is dispensable for replication *in vitro*; in Berne virus (BEV) most of the HE gene has been deleted and only the 3′ 426 nt are present (Table 2). The toroviral and coronaviral HE proteins share 30% sequence identity and display a similar degree of sequence relatedness with subunit 1 of the HE fusion protein of Influenza C virus. Although the nature of the presumed gene acquisition is uncertain, toroviruses and coronaviruses appear to have acquired their HE gene through independent heterologous RNA recombination events. Toroviruses seem to combine discontinuous and non-discontinuous transcription to produce their set of sgRNAs. In the case of EToV, sgRNAs 3 through 5 lack a leader; mRNA-2, however, carries a short 15-18 nt extension at its 5′-end, which is derived from the very 5′-end of the genome.

VIRION PROPERTIES

MORPHOLOGY

Toroviruses are pleomorphic and measure 120 to 140 nm at their largest diameter. Spherical, oval elongated and kidney-shaped particles are observed. The two most conspicuous features of toroviruses are the double fringe of small and large spikes on the envelope, the latter of which resemble the peplomers of coronaviruses, and the tubular nucleocapsid of helical symmetry, which appears to determine the shape of the virion. Toroviruses are enveloped viruses; an isometric core shell has not been identified, in contrast to coronaviruses.

PHYSICOCHEMICAL AND PHYSICAL PROPERTIES

Buoyant densities of 1.16, 1.18, and 1.14 g/cm^3 in sucrose were determined for EToV, BToV serotype 2 and HToV, respectively.

NUCLEIC ACID

The EToV genome is 28 kb in length, capped and polyadenylated. It contains six ORFs. The 5′-UTR includes 820 nt and 3′-UTR encompasses 200 nt, excluding the poly(A) tail. The characteristics of EToV genome RNA and ORFs are summarized in Figure 6. The full genome sequence has not yet been reported for any torovirus.

PROTEINS

In EToV strain Berne, four structural proteins have been identified: a 180 kDa spike (S) protein of 1581 aa (which is post-translationaly cleaved into two subunits, S1 and S2), a 26 kDa integral membrane (M) protein of 233 aa, a HE of 65 kDa and a 19 kDa N protein of 160 aa. The S protein has an N-terminal signal sequence, a putative C-terminal transmembrane anchor, two putative heptad-repeat domains, and a possible cleavage site for a "trypsin-like" protease. The M protein is unglycosylated, accounts for about 13% of the virion protein mass and does not contain an N-terminal signal sequence. The HE is a class I membrane protein displaying 30% sequence identity with the HE of coronaviruses and influenza C viruses. The N protein is the most abundant structural protein of the EToV particle, accounting for about 80% of its protein mass. It is a phosphorylated protein with RNA-binding properties.

Table 2: Virus-associated proteins of toroviruses.

Protein		kDa
Spike glycoprotein	S	200
Membrane protein	M	27
Nucleocapsid protein	N	19
Small-envelope protein	E	nk
Hemagglutinin-esterase protein	HE	65

[a]apparent molecular weight estimated by electrophoresis (kDa)
nk, presence not known.

LIPIDS

Virions contain lipid bilayer envelopes.

CARBOHYDRATES

The torovirus S and HE proteins are N-glycosylated, carrying up to 24 and 11 potential N-glycosylation sites, respectively. The M protein is not glycosylated.

POS. SSRNA

GENOME ORGANIZATION AND REPLICATION

Torovirus

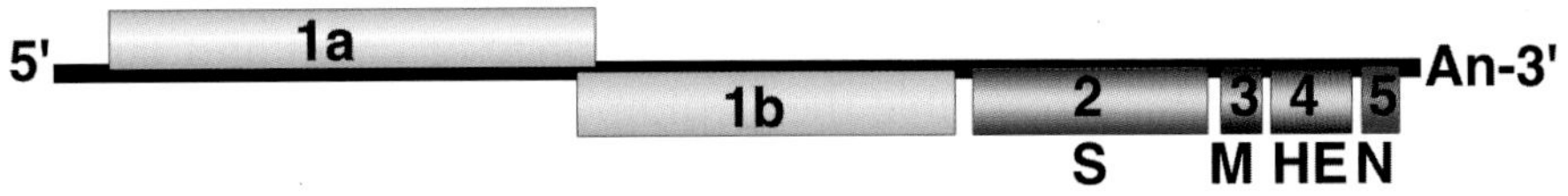

Figure 6: Genome organization structure in toroviruses. ORFs are represented by boxes. The proteins encoded by the ORFs are indicated. ORF1a encodes pp1a and, together with ORF1b, pp1ab polyproteins. Poly(A) tails are indicated by An; M, membrane protein; N, nucleocapsid protein; S, spike protein; HE, hemagglutinin-esterase glycoprotein.

The first two ORFs 1a and 1b from the 5′-end are translated from genomic RNA, to yield two large polyproteins, pp1a and pp1ab. These are believed to be proteolytically cleaved to yield the mature proteins from which the viral replicase/transcriptase is composed. The ORF1a initiation codon is located at nt position 821-823 in the genome. The four remaining ORFs 2, 3, 4, and 5 have been identified as structural genes (Fig. 6), and are expressed by the generation of a 3′-coterminal nested set of 4 sg mRNAs (Figure 7).

Sequence analysis of EToV and BToV has shown that the genes for M, HE and N are preceded by a conserved AU-rich sequence (5′AC-$N_{2\text{-}3}$-UCUUUAGA3′), which is also present at the very 5′-end of the genome. Although this sequence resembles the transcription-regulating sequences (TRSs) in coronaviruses, there is no evidence for fusion of a common leader to mRNAs 3, 4 and 5. These mRNAs are produced via a non-discontinuous transcription mechanism with transcription-initiation occurring at the AC-dinucleotide preceeding the UCUUUAGA core sequence. This appears to be an important difference between toroviruses and coronaviruses. In terms of transcription, however, the consequences of this dissimilarity between toro- and coronaviruses may be limited: whereas in coronaviruses the TRSs may act as a site for homology-assisted RNA recombination during subgenomic negative strand synthesis to add an anti-leader, the conserved intergenic sequences in toroviruses may act as a termination signal. Subgenomic mRNA synthesis would then occur via direct binding of the polymerase to the torovirus "core promoters" at the 3′-ends of the subgenomic negative strand templates. It is of note, that one subgenomic EToV mRNA, mRNA2 apparently is produced via discontinuous transcription during which a 15-18 nt "leader" derived from the 5′-end of the genome is added. Hence, toroviruses seem to use a combination of discontinuous and non-discontinuous RNA synthesis to produce their set of sgRNAs.

Evidence for two independent non-homologous recombination events during torovirus evolution have been obtained. The first putative recombination involves ORF 4, the HE gene. The second putative recombination site involves the C-terminus of the EToV ORF 1a, which contains 31 to 36% identical aa residues compared with the N-terminal 190 aa of the 30-32 kDa non-structural 2A protein (CPD) of coronavirus.

In addition to the products of ORFs 2, 3, 4, and 5, which are assumed to be synthesized by monocistronic translation of a nested set of structurally polycistronic mRNAs, the 3′ part of the EToV genome may encode one more protein in ORF5, which completely overlaps 264 nt with N gene and potentially encodes a hydrophobic 10 kDa protein. Although no such protein has been observed in virions or EToV-infected cells, it is interesting to note that a similar situation, a small hydrophobic protein expressed from an ORF that completely overlaps with the N protein gene, has been reported for the BCV.

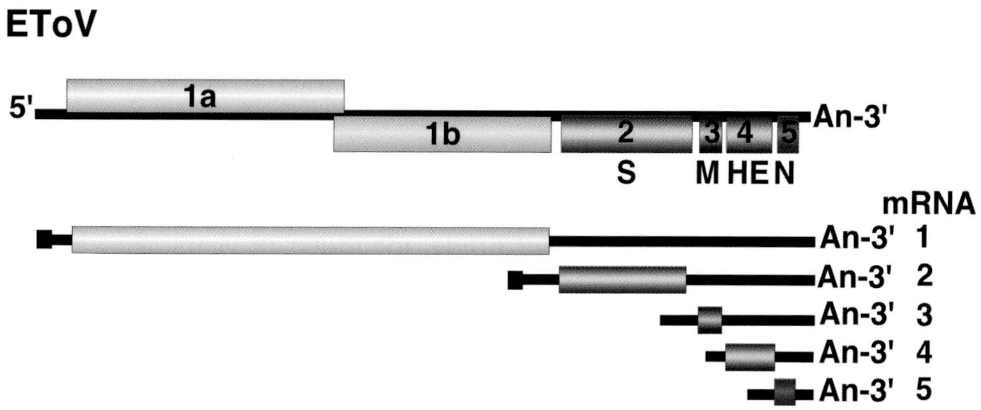

Figure 7: Structural relationship between sgRNAs and the genomic RNA of toroviruses. Thick lines represent the translated sequences, thinner lines, untranslated sequences. The names below the boxes indicate the proteins encoded by the corresponding genes. The 5′ leader sequence is depicted by a small black box; poly(A) tail is indicated by An; M, membrane protein; N, nucleocapsid protein; S, spike protein; HE, hemagglutinin-esterase glycoprotein; the leader is indicated by a small black box.

The composition of a 1 kb DI genome in a replication competent virus suggests that the minimal sequences required for EToV RNA replication (and probably also for packaging) are located in two small domains present at the termini of the genomic RNA. This suggests a difference with members of the genus coronavirus.

Extensive N-glycosylation and proteolytic cleavage of the precursor are part of the posttranslational processing of the torovirus S protein. The EToV M protein accumulates in intracellular membranes and is thought to play a role in budding through intracellular membranes.

Antigenic Properties

The S protein is recognized by neutralizing and hemagglutination-inhibiting monoclonal antibodies.

Biological Properties

The BToV has been identified as a pathogen causing gastroenteritis in calves and possibly pneumonia in older cattle. BToV infections are usually limited to the gut, although the respiratory system may be sporadically involved. No disease has been associated to EToV. Serological evidence indicates that it infects ungulates (horses, cattle, sheep, goats, pigs), rats, rabbits, and some species of feral mice. Torovirus-like particles have been detected by electron microscopy in humans, dogs and cats. The torovirus-like particles found in humans cross-reacts antigenically with BToV and sequence similarity. No antibody to toroviruses has been found in the sera of cats. Infections by BToV are also quite common in dairy cattle. The presence of maternal antibodies in calves does not prevent infection, but may modify the outcome of the disease. Infections by BToV appear to be ubiquitous, as evidence of infection has been obtained in every country where serological or virological studies have been done: Western Europe, North America, India, South Africa, and New Zealand. Torovirus infects the epithelial cells lining the small and large intestine, with progression from areas of the midje junum down to the ileum and colon. Within the small intestine, cells of the upper third of the crypt and the epithelium overlying the Peyer's patches, including M cells, are also infected. Chronic torovirus infections may also occur.

List of Species Demarcation Criteria in the Genus

So far, only limited molecular genetic information is available for toroviruses. Comparative sequence analysis of the structural protein genes of a set torovirus field variants identified three distinct genotypes, displaying 20 to 40% divergence. These are exemplified by BToV Breda strain, PToV Markelo strain and EToV Berne strain. Human torovirus, for which only the HE gene has been characterized, may represent a fourth genotype.

The bovine and porcine toroviruses apparently display host species preference. In phylogenetic analyses, all PToV variants cluster, while the extant European BToVs mostly resemble the New World BToV variant Breda, identified 19 years ago. However, there is evidence for recurring intergenotypic recombination. All newly characterized European BToV variants seem to have arisen from a genetic exchange, during which the 3'-end of the HE gene, the N gene, and the 3' -UTR of a Breda virus-like parent had been swapped for those of PToV. Moreover, some PToV and BToV variants carried chimeric HE genes, which apparently resulted from recombination events involving hitherto unknown toroviruses. From these observations, the existence of two additional torovirus genotypes can be inferred.

Sequencing of C-terminus of the N gene and the 3′-UTR have shown > 93% identity between HToV, BToV and EToV. Nevertheless, small but consistent sequence differences were noted among five HToV isolated and EToV. The published HToV HE sequence is most related to that of BToV variant Breda (83%), and more distantly related to those of the European BToV strains (~73%) and PToV strain (~56%).

BToV, PToV and HToV cause gastroenteritis and the BToV sporadically infects the respiratory system, in contrast to EToV that has remained as a virus in search of a disease.

List of Species in the Genus

Species names are in green italic script; strain names and synonyms are in black roman script; tentative species names are in blue roman script. Sequence accession numbers, and assigned abbreviations ( ) are also listed.

Species in the Genus

Bovine torovirus		
Bovine torovirus	[Y10866]	(BToV)
Breda virus		(BRV)
Equine torovirus		
Berne virus		(BEV)
Equine torovirus		(EToV)
Human torovirus		
Human torovirus		(HToV)
Porcine torovirus		
Porcine torovirus		(PToV)

Tentative Species in the Genus

Not reported.

List of Unassigned Viruses in the Family

None reported.

Distinguishing Features Between Coronaviruses and Toroviruses

Coronavirus mRNAs contain a 5' leader sequence, which is acquired via discontinuous RNA transcription, most likely during (-)strand RNA synthesis. In contrast, the torovirus mRNAs seem to be produced mainly via a non-discontinuous transcription mechanism: subgenomic mRNAs 3 through 5 of EToV strain Berne do not possess a leader. EToV mRNA2, however, contains a short 15-18 nt sequence at its 5′end, which apparently is derived from the 5′-end of the genome. Coronaviruses have a loosely wound helical nucleocapsid protected by a core shell while the toroviruses have a more rigid tubular nucleocapsid. All torovirus field strains studied so far express an HE protein; of the coronaviruses only those belonging to group 2 encode an HE protein. The N protein is much larger in coronaviruses than in toroviruses. The M protein is glycosylated only in coronaviruses. Toroviruses do not have a counterpart to the E protein and direct the synthesis of fewer subgenomic mRNAs than any coronavirus. A putative CPD was mapped to the very C-terminus of pp1a in a torovirus but it is encoded immediately downstream of ORF1b in some coronaviruses. Upon comparison of every pair of homologous protein, toroviruses and coronaviruses do not interleaved and form separate groups.

Table 3: Features of coronaviruses and toroviruses.

Feature	*Coronavirus*	*Torovirus*
Enveloped virions	+	+
Core shell	+	-
Nucleocapsid architecture	helical	tubular
Prominent spikes	+	+
Coiled-coil structure in spikes	+	+
M protein with three membrane-spanning sequences	+	+
Intracellular budding	+	+
Linear positive-sense ssRNA genome with poly (A) tail	+	+
Genome size (kb)	27-31	~25
The genome organization includes 5′-UTR-replicase gene-structural protein genes-UTR 3′	+	+
An –1 ribosomal frameshifting in the replicase gene	+	+
Transcriptome includes 3′ co-terminal nested set of ≥ 4 sgRNAs	+	+
Only 5′-most one or two ORFs of mRNAs are translationally active	+	+
5′ leader sequence in mRNAs	+	+/-

PHYLOGENETIC RELATIONSHIPS WITHIN THE FAMILY

On the basis of antigenic cross-reactivity, coronaviruses were classified into three groups, known as groups 1, 2 and 3. This classification was corroborated by phylogenetic analysis of virion proteins as well as replicase proteins that do not contribute to the virion antigenicity. Subsequent expansion of the coronavirus groups with new species was made using a phylogenetic criterion rather than results of serological characterization. Consequently, Group 1 coronaviruses include three distinct antigenic clusters, one formed by TGEV, CCoV and FCoV, and two others formed by HCoV229E and PEDV, respectively. The phylogenetic position of the recently identified SARS-CoV was subject of several studies all of which found it not to be closely related to any of previously established coronaviruses. This prompted placing SARS-CoV in a separate 4th group. However, in an analysis dealing with the most conserved pp1b part of replicase and involving a torovirus outgroup, SARS-CoV confidently clusters together with group 2 species as the most distant member. This clustering was also observed in studies of several other proteins including pp1a replicase domains and virion S, M, and N components.

Most closely related coronaviruses (members of the same antigenic group) share also a common pattern of the accessory non-structural genes in the 3′-end of genome that distinguish them from other coronaviruses. Such unique ORFs, encoding CPD and HE, were described for closely related MHV, BCoV and HCoV-O43 of group 2. They are absent in SARS-CoV, which has 5 (4) unique ORFs. Likewise, in diverse group 1, different patterns of accessory genes have been described for distant viruses. These observations argue that the 3′-located accessory ORFs may be reliable markers for antigenic clusters rather than for the evolutionary groups. In contrast, group-specific genetic markers were recently identified in the N-terminal cleavage products nsp1 and nsp3 of pp1a/pp1ab replicase. According to this criterion, SARS-CoV belongs to the coronavirus group 2.

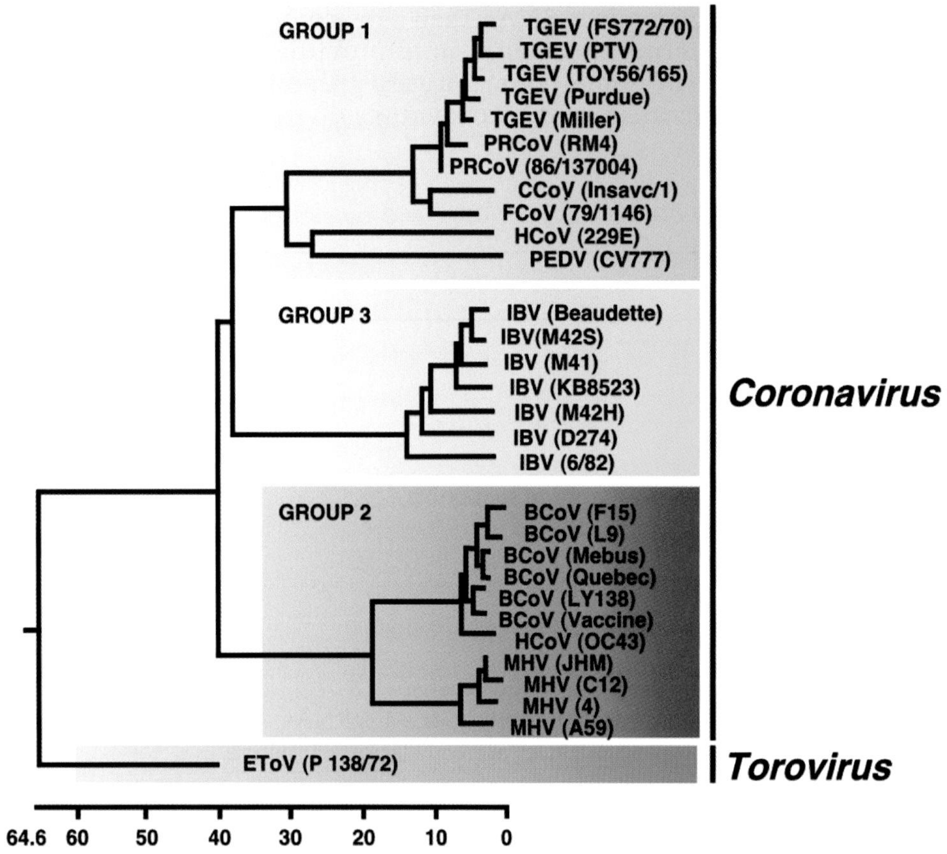

Figure 8: Phylogenetic relationship of the large-surface glycoprotein of coronaviruses and toroviruses. Amino acid sequences were aligned using the "clustal" method and phylogenetic trees were constructed using the neighborhood-joining method. The analyses were done using the MegAlign module of the Lasergene software suite (DNASTAR). The phylogenies are rooted assuming a biological clock (Siddell, 1995).

The intra-group diversity in coronaviruses is comparable to that found between viruses of different genera in other RNA virus families, e.g. *Picornaviridae*. A proposal was put forward to accept the evolutionary coronavirus groups as the basis for the three new genera in the family *Coronaviridae*.

Immunological evidence has shown that equine and bovine torovirus are antigenically related to each other, and to torovirus-like particles found in human fecal specimens but not to other animal viruses, including coronaviruses. The available sequences also show that all known toroviruses are closely related, although the actual phylogenetic diversity of toroviruses remains unknown. In a broad comparative sequence analysis aimed at a revision of the taxonomy of the family *Coronaviridae*, toroviruses proved to be well separated from coronaviruses with which they share some characteristics not found in two other families of the order *Nidovirales*. Among these unique shared characteristics are distantly related S and M proteins. If coronavirus groups are to be elevated to the genera ranks, a special distant torovirus-coronavirus relationship could be recognized through assigning the subfamily ranks to the coronaviruses and toroviruses within the family

Coronaviridae. Future revision of the *Coronaviridae* family taxonomy must address this issue.

Similarity with Other Taxa

The family *Coronaviridae* together with the *Arteriviridae* and the *Roniviridae* families form the order *Nidovirales*. The viruses of these families share the genome organization, expression strategy and have a similarly organized replicase gene. Non-nidovirus homologs of several (putative) enzymes encoded by viruses of the family *Coronaviridae* have been found in other RNA viruses. The proteolytic enzymes and RdRps cluster together with homologs of viruses of "picorna-like" supergroup. The ADRP and Hel enzymes have counterparts in viruses of the "alphavirus-like" supergroup. Mtr has homologs in alphaviruses, flaviviruses, and mononegavirales. CPD has homologs among some rotaviruses. HE has homologs in the Influenza C viruses. Also some parallels in the genome organization and expression strategy are evident between members of the *Coronaviridae* and *Closteroviridae* families.

Derivation of Names

Corona, from the Latin *corona* for "crown", representing the appearance of surface projections in negatively-stained electron micrographs of members of the *Coronavirus* genus.

Toro, from the Latin *torus*, "lowest convex moulding in the base of a column", representing the toroviruses nucleocapsid shape.

References

Cavanagh, D., Brian, D.A., Briton, P., Enjuanes, L., Horzinek, M.C., Lai, M.M.C., Laude, H., Plagemann, P.G. W., Siddell, S., Spaan, W. and Talbot, P.J. (1997). *Nidovirales*: a new order comprising *Coronaviridae* and *Arteriviridae*. *Arch. Virol.*, **145**, 629-635.

Cavanagh, D., Mawditt, K., Welchman, D., de B. Britton, P. and Gough, R.E. (2002). Coronaviruses from pheasants (*Phasianus colchicus*) are genetically closely related to coronaviruses of domestic fowl (infectious bronchitis virus) and turkeys. *Avian Pathology, **31**, 81-93.*

Cornelissen, L.A.H.M., Wierda, C.M.H., Van Der Meer, F.J., Herrewegh, A.P.M., Horzinek, M.C., Egberonk, H. F. and Groot, R.J. (1997). Hemagglutinin-esterase, a novel structural protein of torovirus. *J. Virol.*, **71,** 5277-5286.

González' J.M., Gomez-Puertas' P., Cavanagh, D., Gorbalenya, A.E. and L. Enjuanes (2003). A comparative sequence analysis to revise the current taxonomy of the family *Coronaviridae*. *Arch. Virol.*, **148,** 2207-2235.

Gorbalenya, A.E. (2001). Big nidovirus genome: when count and order of domains matter. *Adv. Exp. Med. Biol.*, **494**, 1-17.

De Vries, A.A.F., Horzinek, M.C., Rottier, P.J.M. and De Groot, R.J. (1997). The genome organization of the *Nidovirales*: similarities and differences between arteri-, toro-, and coronaviruses. *Sem. Virol.*, **8**, 33-47.

Escors, D., Capiscol, M.C. and Enjuanes, L. (2003). Location of transmissible gastroenteritis coronavirus encapsidation signal (Ψ). *J. Virol.*, **77**, 7890-7902

Enjuanes, L., Siddell, S.G. and Spaan, W.J. (eds) (1998). *Coronaviruses and Arteriviruses*. Plenum Press, New York.

Ismail, M.M., Tang, Y. and Saif, Y.M. (2003). Pathogenicity of turkey coronavirus in turkeys and chickens. *Avian Diseases*, **47**, 515-522.

Lai, M. M. C. and Cavanagh, D. (1997). The molecular biology of coronaviruses. *Adv. Vir. Res.*, **48,** 1-100.

Lai, M.M.C. and Holmes, K.V. (2001). Coronaviruses. In: *Fields Virology*, (D.M. Knipe and P.M. Howley, eds), pp 1163-1185. Lippincott, Williams & Wilkins, Philadelphia, Pa.

Sawicki, S.G. and Sawicki, D.L. (1995). Coronaviruses use discontinuous extension for synthesis of subgenome-length negative strands. *Adv. Exp. Biol. Med.*, **380**, 499-506.

Sethna, P.B., Hung, S.-L. and Brian, D.A. (1989). Coronavirus subgenomic minus-strand RNAs and the potential for mRNA replicons. *Proc. Natl. Acad. Sci. USA*, **86,** 5626-5630.

Siddell, S.G., (ed) (1995). *The Coronaviridae*. Series: The Viruses, (H. Fraenkel-Conrat and R.R. Wagner, eds), Plenum Press, New York.

Smits, S.L., Lavazza, A., Matiz, K., Horzinek, M.C., Koopmans, M.P. and de Groot, R.J. (2003). Phylogenetic and evolutionary relationships among torovirus field variants: evidence for multiple intertypic recombination events. *J. Virol.*, **77**, 9567-77.

Snijder, E.J. and Horzinek, M.C. (1995). The molecular biology of toroviruses. In: *The Coronaviridae,* (S.G. Siddell, ed), pp 219-238. Plenum Press, New York.

Snijder, E.J., Bredenbeek, P.J., Dobbe, J.C., Thiel, V., Ziebuhr, J., Poon, L.L.M., Guan, Y., Rozanov, M., Spaan, W.J.M. and Gorbalenya, A.E. (2003). Unique and conserved features of genome and proteome of SARS-Coronavirus, an early split-off from the coronavirus group 2 lineage. *J. Mol. Biol.,* **331**, 991-1004.

van Vliet, A.L., Smits, S.L., Rottier, P.J. and de Groot, R.J. (2002). Discontinuous and non-discontinuous subgenomic RNA transcription in a nidovirus. *EMBO J.*, **21**, 6571-80.

CONTRIBUTED BY

Spaan, W.J.M., Brian, D., Cavanagh, D., de Groot, R.J., Enjuanes, L., Gorbalenya, A.E., Holmes, K.V., Masters, P., Rottier, P., Taguchi, F. and Talbot, P.

FAMILY ARTERIVIRIDAE

TAXONOMIC STRUCTURE OF THE FAMILY

Order *Nidovirales*
Family *Arteriviridae*
Genus *Arterivirus*

Since only one genus is currently recognized, the family description corresponds to the genus description

GENUS ARTERIVIRUS

Type Species *Equine arteritis virus*

VIRION PROPERTIES

MORPHOLOGY

Arteriviruses are spherical particles with a diameter of 45 to 60 nm. Virions consist of an isometric nucleocapsid of about 25 to 35 nm in diameter, surrounded by a lipid envelope (Fig. 1). No spikes are obvious on the virion surface, but a surface pattern of relatively small and indistinct projections has been observed.

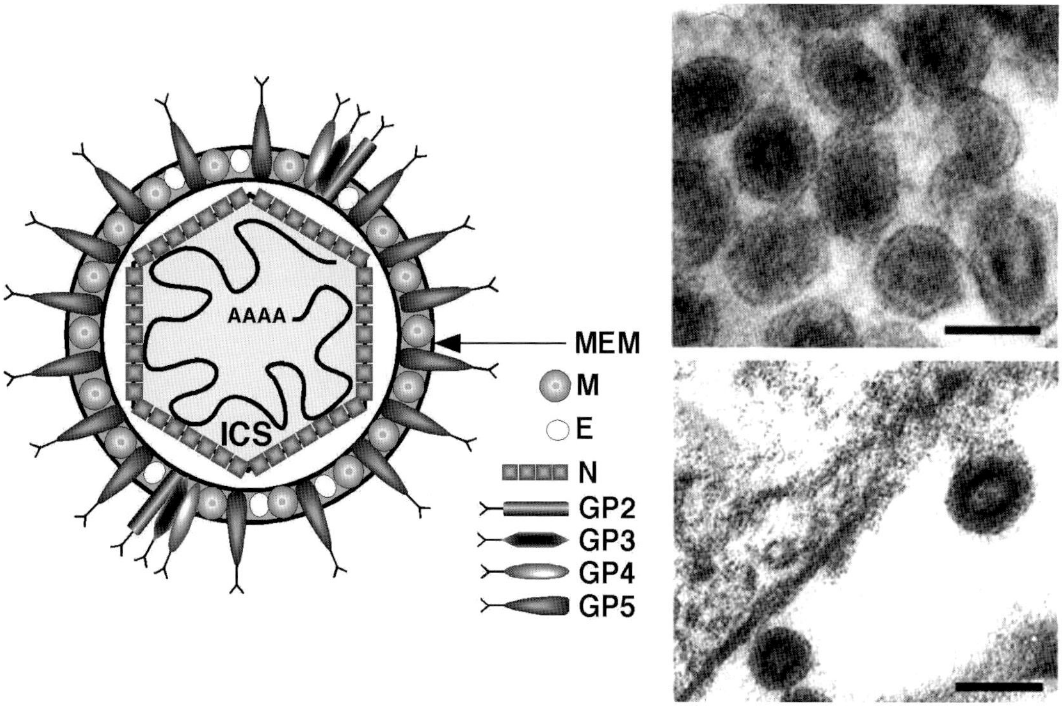

Figure 1: Structure of arterivirus virions. (Upper right) Electron micrograph of negatively stained particles of Porcine reproductive and respiratory syndrome virus (PRRSV). Bar is 50 nm. (Lower right) Electron micrograph of Equine arteritis virus (EAV) particles, budding from smooth intracellular membranes (BHK-21 cells). The bar represents 50 nm. (Left) Schematic representation of an arterivirus particle. Seven virion-associated proteins have been identified in EAV and PRRSV virions. In addition to the nucleocapsid protein (N), there are two major (GP_5 and M) and four minor (GP_2, GP_3, GP_4, E) envelope proteins. By reverse genetics (EAV), each of these proteins was shown to be required for the production of infectious progeny. The proteins encoded by ORFs 2a, 3 and 4 have not been identified as structural components of Lactate dehydrogenase-elevating virus (LDV). The virion proteins of Simian hemorrhagic fever virus (SHFV) remain largely uncharacterized. The major virion glycoprotein of arteriviruses (GP_5 in EAV, PRRSV, and LDV) forms a disulfide-linked heterodimer with M that is essential for virus infectivity. GP_2, GP_3, and GP_4 were recently described to form heterotrimers in the virus particle (EAV).

PHYSICOCHEMICAL AND PHYSICAL PROPERTIES

The buoyant density of arteriviruses has been estimated to be 1.13 to 1.17 g/cm^3 in sucrose. Reported sedimentation coefficients for arteriviruses range from 214S to 230S. Virion stability is affected by temperature and pH. Virions are stable when stored at -70°C. The half-life of arteriviruses decreases progressively with increasing temperature. Virions are relatively stable between pH 6.0 and 7.5, but are inactivated at higher or lower pHs. Arteriviruses are inactivated by lipid solvents, such as ether, butanol, and chloroform and are extremely sensitive to detergents. For example, a brief incubation with a nonionic detergent such as 0.01% NP40 or Triton X-100, efficiently disrupts the viral envelope (Lactate dehydrogenase-elevating virus, LDV; Simian hemorrhagic fever virus; SHFV).

NUCLEIC ACID

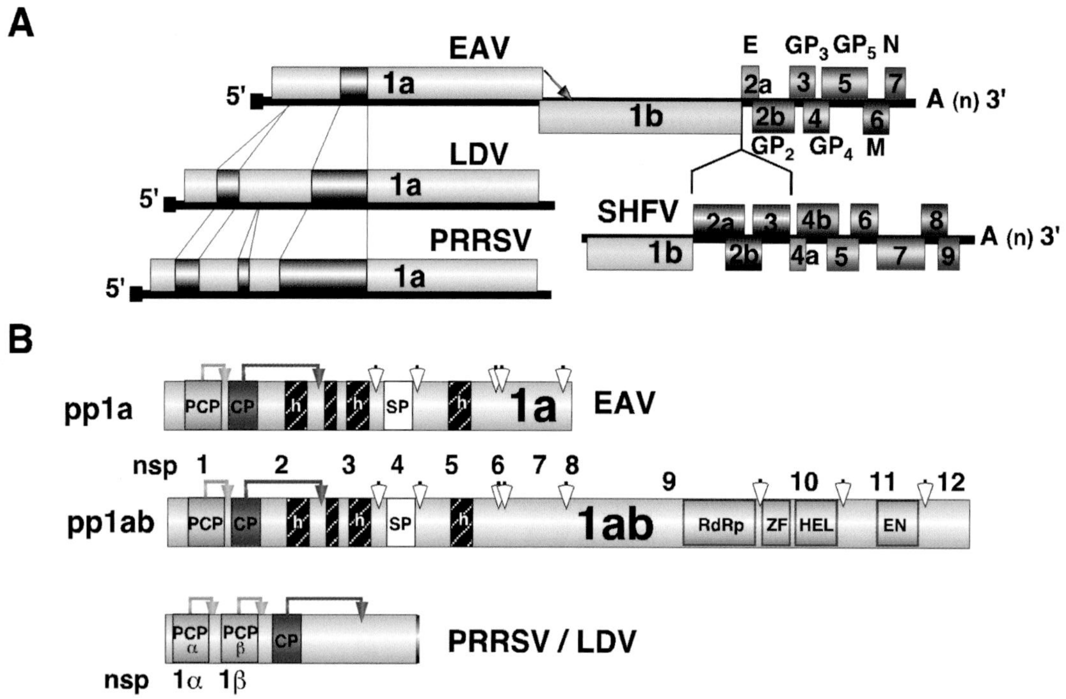

Figure 2: Overview of arterivirus genome organization and replicase polyproteins. (A) General genome organization. ORFs are represented by boxes. The proteins encoded by the EAV ORFs are indicated. The 5′ leader sequence is depicted by a small black box; 3′ poly(A) tails are not shown. The arrow between ORF1a and ORF1b represents the ribosomal frameshift site. The grey boxes represent regions where PRRSV, LDV, and SHFV contain major insertions compared to Equine arteritis virus (EAV). (B) Overview of proteolytic processing and domain organization of the EAV replicase polyproteins, pp1a and pp1ab, with differences in PRRSV and LDV indicated. Polyprotein cleavage sites are depicted with arrowheads matching the color of the proteinase involved. Abbreviations: PCP, papain-like cysteine proteinase; CP, nsp2 cysteine proteinase; SP, nsp4 chymotrypsin-like serine proteinase; h, hydrophobic domain; RdRp, RNA-dependent RNA polymerase; ZF, zinc finger; HEL, NTPase/helicase; EN, putative endoribonuclease.

Virions contain a single molecule of linear, positive-sense, ssRNA that ranges in length from 12.7 to 15.7 kb. The genome RNA has a 5' type I cap structure (SHFV) and a 3'-terminal poly(A) tract. Full-length sequences are available in the GenBank database for all currently known arteriviruses. Arterivirus genomes contain 5′- and 3′-UTR of 156-224 and 59-117 nt, respectively. The 5′ 156-211 nt are used as a common 5′ "leader sequence" on viral sgRNAs (see below). For Equine arteritis virus (EAV), Porcine reproductive and respiratory syndrome virus (PRRSV), and LDV, the genome contains 9 functional genes, whereas the single reported SHFV sequence contains 12 ORFs, due to a postulated 3-gene duplication. The genes are arranged in the order 5′-replicase-E/GP_2-GP_3-GP_4-GP_5-M-N-3′ (EAV) (Fig. 2A). The virion RNA functions as the mRNA for replicase gene translation.

Non-Structural Proteins

Arteriviruses encode two large non-structural polyproteins, the ORF1a-encoded pp1a (187-260 kDa) and the ORF1ab-encoded pp1ab (345-421 kDa), whose synthesis involves ribosomal frameshifting. The EAV replicase polyproteins are cleaved into 12 mature non-structural proteins by three ORF1a-encoded proteinases (PCP, CP, and SP; Fig. 2B), whereas PRRSV and LDV produce an additional cleavage product due to the fact that their nsp1 equivalent contains an additional internal autoproteinase that cleaves the nsp1 region into nsp1α and nsp1β. In EAV, this second papain-like proteinase (PCP β) in the nsp1 region has become inactivated, although its remnants were detected in the EAV nsp1 αequivalent. A third papain-like cysteine protease (CP) with some unique properties is located in nsp2. A chymotrypsin-like serine protease (SP), related to the members of the 3C-like cysteine proteinase family, is located in nsp4 and constitutes the main proteinase of arteriviruses. EAV nsp4 is the first arterivirus protein for which a crystal structure was determined. The nsp4 SP is responsible for 8 proteolytic cleavages that occur in the C-terminal half of the ORF1a-encoded polyprotein (5 sites) and in the ORF1b-encoded part of pp1ab (3 sites).

Table 1: Non-structural proteins of Equine arteritis virus (EAV)[a]

Protein	size[b]	mode of expression[c]	(putative) function(s)
nsp1[d]	260	TI + nsp1 PCP	proteinase (PCP), mononuclear zinc finger, role in sgRNA synthesis
nsp2	571	TI + nsp1 PCP + nsp2 CP	proteinase (CP), integral membrane protein, replication complex formation
nsp3	233	TI + nsp2 CP + nsp4 SP	integral membrane protein, replication complex formation
nsp4	204	TI + nsp4 SP	main proteinase (SP)
nsp5	162	TI + nsp4 SP	integral membrane protein
nsp6	22	TI + nsp4 SP	?
nsp7	225	TI + nsp4 SP	?
nsp8[e]	50	TI + nsp4 SP + TT	?
nsp9	693	TI + RFS + nsp4 SP	RdRp
nsp10	467	TI + RFS + nsp4 SP	RNA helicase/NTPase, polynuclear zinc finger, role in sgRNA synthesis
nsp11	219	TI + RFS + nsp4 SP	nidovirus-specific endoribonuclease
nsp12	119	TI + RFS + nsp4 SP + TT	?

[a] Based on the currently known replicase processing scheme of EAV
[b] in aa
[c] TI, translation initiation; RFS, ORF1a/ORF1b ribosomal frameshifting; TT, translation termination; PCP, nsp1 cysteine proteinase; CP, nsp2 cysteine proteinase; SP, nsp4 serine proteinase
[d] Nsp1 of LDV and PRRSV is cleaved internally by an additional papain-like proteinase to yield nsp1α and nsp1β
[e] Due to ribosomal frameshifting, nsp8 is identical to the N-terminal 50 aa of nsp9

In EAV, nsp1 was found to be fully dispensable for genome replication, but essential for the synthesis of sgRNAs. Nsp2 and nsp3 (and also nsp5) contain hydrophobic domains and have been shown to interact, a step that is essential for the formation of ER-derived double membrane vesicles that were shown to carry the viral RNA replication complex. The core replication complex is formed by nsp9, the putative RdRp, and nsp10, which was shown to possess NTPase and RNA helicase activities. In addition, nsp10 contains an N-terminal putative zinc finger region, which is also conserved in other nidoviruses. This region was implicated to have a function specific to sgRNA synthesis (EAV), but also appears to be essential for genome replication. By comparative sequence analysis, an nsp11 domain that is conserved in all nidoviruses was recently predicted to be an RNA endoribonuclease on the basis of a distant relationship with XendoU, a poly(U)-specific endoribonuclease of cellular origin. Functions or putative functions for the other arterivirus non-structural proteins remain to be identified.

Virion-associated Proteins

The arterivirus nucleocapsid is composed of a single basic nucleocapsid protein (N) that is 12-15 kDa in size. In all arteriviruses, this protein is encoded by the 3'-proximal ORF. For the two best studied arteriviruses, EAV and PRRSV, six envelope proteins (encoded by ORFs 2a to 6) have now been identified in virus particles, an unusually large number for a positive strand RNA virus. It can be assumed that, given the right reagents, the same number of virion-associated proteins may also be found in LDV and SHFV particles. So far, three envelope proteins were identified in LDV and an attempt to detect the E protein was unsuccessful. By reverse genetics (EAV), each of the seven virion-associated proteins was shown to be required for the production of infectious progeny. The arterivirus envelope contains a heterodimer and a heterotrimer, which are probably both essential for virus infectivity. The heterodimer is formed between the integral membrane protein M and the major virion glycoprotein GP_5 (Fig. 1), both of which are probably triple-spanning. The two proteins are linked by a single disulfide bond between conserved cysteine residues in their respective ectodomains. The recently described heterotrimer (EAV) is composed of the remaining three glycoproteins, GP_2, GP_3, and GP_4. Homodimers of the GP_2 protein have also been described (EAV). A soluble, non-virion-associated form of the ORF3 glycoprotein has been reported to be released from infected cells (LDV and PRRSV). SHFV encodes three additional 3' ORFs, which may be duplications of ORFs 2 to 4 (Fig. 2).

Table 2: Virion-associated proteins of arteriviruses[a]

Protein	size[b]	ORF	mRNA	(putative) function(s)
E[d]	67-70	2a/2b[c]	2[d]	small integral envelope protein
GP_2	227-249	2b/2a[c]	2[d]	minor glycoprotein, part of $GP_2/GP_3/GP_4$ heterotrimer
GP_3	163-265	3	3	minor glycoprotein, part of $GP_2/GP_3/GP_4$ heterotrimer
GP_4	152-183	4	4	minor glycoprotein, part of $GP_2/GP_3/GP_4$ heterotrimer
GP_5	199-255	5	5	major glycoprotein, carries main determinants for neutralization, part of GP_5/M heterodimer
M	162-173	6	6	integral (triple-spanning) membrane protein, part of GP_5/M heterodimer
N	110-128	7	7	nucleocapsid protein, partially localizes to the nucleus of infected cells

[a] Based on the genome organization of EAV, PRRSV and LDV
[b] in aa
[c] in PRRSV
[d] sgRNA2 is assumed to be functionally bicistronic

Lipids

Virions have lipid-containing envelopes derived from the host cell. Budding occurs from smooth intracellular membranes, probably including those of the endoplasmic reticulum and the Golgi complex.

CARBOHYDRATES

The four glycoproteins of arteriviruses all contain one or more putative N-linked glycosylation signals. Glycosylation of GP_5 (EAV, LDV, PRRSV) occurs by the addition of variable numbers of lactosamine repeats.

GENOME ORGANIZATION AND REPLICATION

About three-quarters of the arterivirus genome is occupied by the replicase gene (ORF1a and ORF1b), which is expressed directly from the genomic RNA. The very short region in which ORF1a and ORF1b overlap contains a specific "slippery" sequence. Together with a downstream pseudoknot structure, this sequence directs a -1 ribosomal frameshift which is required for the translation of ORF1b and results in the synthesis of an ORF1ab polyprotein (pp1ab), in addition to the ORF1a polyprotein (pp1a) that results from normal translation termination.

The arterivirus replicase polyproteins are co- and post-translationally processed by three or four viral proteinases (Fig. 2), which reside within the ORF1a polyprotein. Replicase subunits containing hydrophobic domains (see above) are assumed to target the replication/transcription complex to modified intracellular (double) membranes, which are probably derived from the endoplasmic reticulum. Four host cell proteins have been identified that bind to a cis-acting region required for plus-strand RNA synthesis from the minus-strand template.

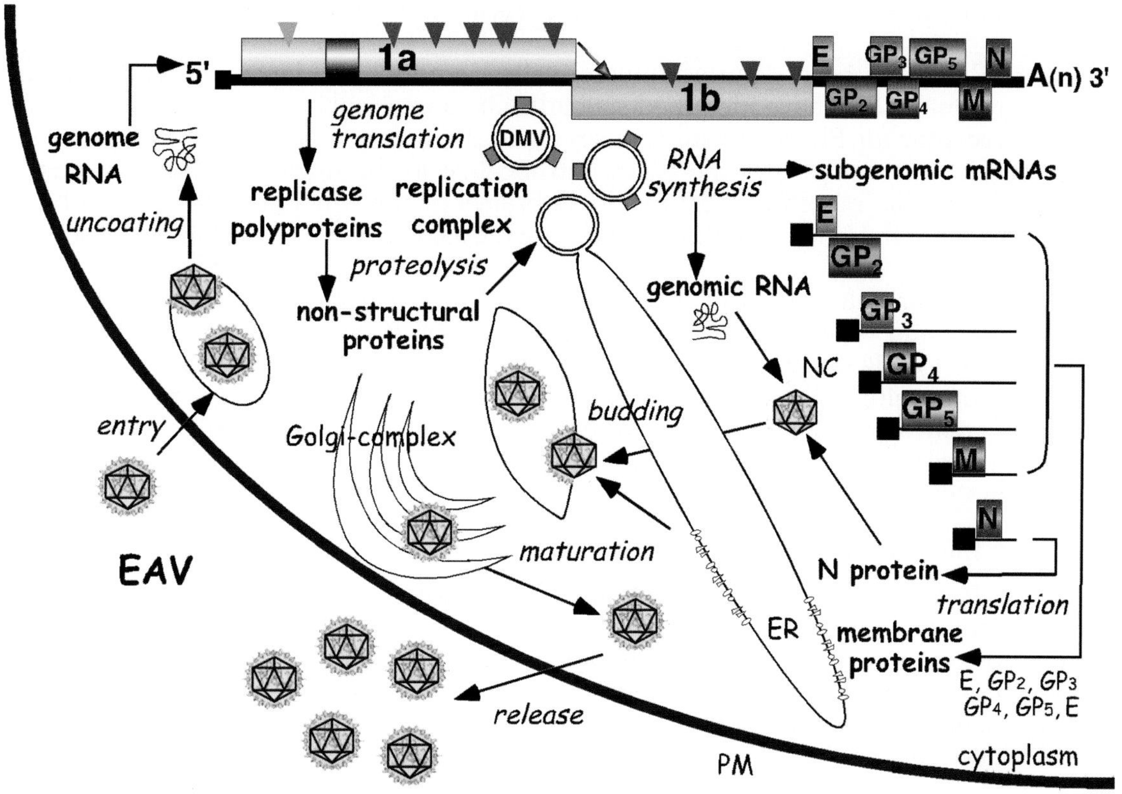

Figure 3: Overview of the life cycle of the arterivirus prototype (EAV). The genome organization, including replicase cleavage sites (arrowheads; see also Fig. 2), is shown at the top of the figure. Abbreviations: ER, endoplasmic reticulum; PM, plasma membrane; DMV, double membrane vesicle; NC, nucleocapsid.

The cytoplasmic, membrane-associated replication/transcription complex of arteriviruses engages in RNA-dependent RNA synthesis (Fig. 3), leading to the synthesis of genomic and subgenomic plus- and minus-stranded RNAs. Subgenomic mRNAs are used to express the virion-associated proteins, which are encoded by overlapping ORFs in the 3'-proximal quarter of the genome. Subgenomic mRNAs contain both 5' and 3' co-terminal sgRNAs ("nested set" structure). Whereas the coding region of the sgRNA (the mRNA "body") is colinear with a 3'-proximal portion of the genome, the 5'-proximal 156-211 nt of the sgRNAs ("leader") are identical to the 5'-end of the genome. In addition to a full-length minus strand, infected cells also contain a nested set of minus strand sgRNAs, which are the complements of the sgRNAs and are believed to function as templates for their synthesis. The production of the minus strands sgRNA, which involves the "fusion" of sequences that are noncontiguous in the genome, is currently thought to occur by discontinuous extension of minus strand synthesis. According to this model, transcription-regulating sequences (TRSs) would direct attenuation of minus strand synthesis, a step that would be followed by translocation of the nascent strand to the leader region in the 5'-end of the genomic template. Guided by a base pairing interaction, minus strand synthesis would resume to add the leader complement to the nascent minus strand sgRNA.

Arteriviruses produce 6 (EAV, LDV, PRRSV) to 8 (SHFV) major sgRNA species. In addition, the production of multiple, alternative sgRNAs (from alternative TRSs) has been reported for various arteriviruses. Despite their polycistronic structure, it is assumed that - as a rule - only the 5' proximal ORF of each sgRNA is translated. However, one (EAV, PRRSV and LDV) or two (SHFV) sgRNAs are thought to be functionally bicistronic.

PRRSV has been reported to enter cells via a low pH-dependent endocytic pathway and recent studies have identified sialoadhesin, a macrophage-restricted surface molecule, as a receptor for PRRSV. Arteriviruses replicate in the cytoplasm of infected cells, although a fraction of nsp1 and the N protein localize to the nucleus. Nucleocapids bud into the lumen of smooth intracellular membranes of the exocytic pathway, probably including those of the Golgi complex. Virions are released from infected cells via exocytosis.

Thus far, infectious molecular clones (full-length cDNA clones) have been constructed for EAV and both European and North American prototype strains of PRRSV. Using these tools, knowledge on the details of various aspects of arterivirus molecular biology and immunology has been expanded considerably. Among the topics studied by reverse genetics are nonstructural protein processing and function, sgRNA synthesis, virus assembly, virus immunogenicity, and the development of arterivirus-based vector systems.

Antigenic Properties

GP_5-specific neutralizing antibodies have been described for EAV, LDV and PRRSV. Also monoclonal antibodies against the GP5 equivalent of SHFV can neutralize virus infectivity. The neutralization domain has been mapped to various epitopes/regions in the ectodomain of this protein. Monoclonal antibodies specific for the GP_4 can also be neutralizing (PRRSV) and a neutralizing epitope in PRRSV GP_4 was mapped to the region between aa 39 and 79. No antigenic cross reactivity between different arterivirus members has been found. LDV- and PRRSV-specific T-cell responses have been described.

Biological Properties

Infections with arteriviruses can cause acute or persistent asymptomatic infections or respiratory disease (EAV and PRRSV), in utero fetal death (PRRSV) and abortion (EAV and PRRSV), fatal age-dependent poliomyelitis (LDV) or fatal hemorrhagic fever (SHFV).

The host range of arteriviruses is restricted. EAV infects horses and donkeys, LDV infects mice, PRRSV infects swine, and SHFV infects some species of African (patas monkeys, African green monkeys and baboons) and Asian (macaque) monkeys. Macrophages are the primary target cells for all arteriviruses in their respective hosts. Laboratory mutants of LDV can also replicate in ventral motor neurons in some strains of inbred mice. In tissue culture, LDV replicates only in primary mouse macrophages, SHFV and PRRSV replicate in primary macrophages (rhesus and porcine alveolar lung macrophages, respectively) as well as in the MA-104 established cell line (an African green monkey kidney cell line), and EAV replicates in primary equine macrophages and kidney cells as well as in BHK-21, RK-13, Vero and MA-104 cells.

Arteriviruses are spread horizontally and vertically. Horizontal transmission can occur via the respiratory route (EAV and PRRSV), via the sexual route in semen (EAV and PRRSV) and via infected blood or body fluids (LDV, PRRSV, and SHFV). Congenital infection is common for PRRSV. Arterivirus replication is characterized by the formation of double-membrane vesicles (see above) in infected cells. One-step growth experiments have shown that the replication cycle of arteriviruses (in cell culture) is relatively short, maximum progeny virus titers being released by 10 to 15 hrs post infection. The maximum titers obtained in cell culture are 10^6-10^7 $TCID_{50}$/ml for PRRSV, but may exceed 10^8 PFU/ml for EAV and SHFV. The infection of macrophages and cell lines is highly cytocidal, resulting in rounding of the cells and detachment from the culture plate surface. PRRSV, SHFV, and EAV are titrated by endpoint dilution or plaque assays. However, LDV is titrated in mice, because the percentage of susceptible cells in primary mouse macrophage cultures is too low to detect their destruction. Apoptosis has been observed in PRRSV-infected porcine alveolar macrophages, MA-104 cells, and testicular germ cells. Expression of PRRSV GP_5 from a Vaccinia virus recombinant also induced apoptosis.

List of Species Demarcation Criteria in the Genus

The members of the genus *Arterivirus* form a distinct phylogenic group. Their genomes are polycistronic positive-stranded RNA molecules ranging in size from 12 to 16 kb, which is considerably smaller than the genome sizes of other members of the order *Nidovirales* (corona-, toro- and ronivirus genomes are all larger than 25 kb). The genome encodes two large 5' ORFs, ORFs 1a and 1b, that are expressed from the genomic RNA. ORF1b is only expressed after a -1 frameshift has occurred. The 3' ORFs encode virion-associated proteins and are expressed from a set of 3'- and 5'-co-terminal sgRNAs. The arterivirus nucleocapsid is isometric and is composed of the genome RNA and a single nucleocapsid protein N. Virions contain 2 major and 4 minor envelope proteins, 4 of which are glycoproteins. The virion surface projections are relatively small and indistinct, and none of the virion glycoproteins contain a coiled-coil structure. With the exception of the triple-spanning character of the membrane protein M, there are no obvious similarities between the virion-associated proteins of *Arteriviridae* and those of other members of the order *Nidovirales*. Members of the four species in the genus *Arterivirus* (EAV, LDV, PRRSV, and SHFV) each constitute a phylogenetic branch within the genus (Fig. 4), with each species represented by a cluster of strains. LDV and PRRSV are most closely related to each other. There are two 'subspecies' of PRRSV, a European (type I) and an American (type II). Members of the four virus species are antigenically distinct. The viruses of each species have a restricted host range but infects different hosts. The variability in the length of the

N-terminal region of ORF1a among arteriviruses suggests that this region may contain species-specific functions.

List of Species in the Genus

Species names are in green italic script; strain names and synonyms are in black roman script; tentative species names are in blue roman script. Sequence accession numbers, and assigned abbreviations ( ) are also listed.

Species in the Genus

Equine arteritis virus		
Equine arteritis virus - Bucyrus (North American)	[X53459]	(EAV-Buc)
Equine arteritis virus - CW (European)	[AY349167]	(EAV-CW)
Lactate dehydrogenase-elevating virus		
Lactate dehydrogenase-elevating virus - C	[L13298]	(LDV-C)
Lactate dehydrogenase-elevating virus - P	[U15146]	(LDV-P)
Porcine reproductive and respiratory syndrome virus		
Porcine reproductive and respiratory syndrome virus - Lelystad (type I (European) prototype)	[M96262]	(PRRSV-L)
Porcine reproductive and respiratory syndrome virus - VR-2332 (type II (North American) prototype)	[U87392]	(PRRSV-VR)
Simian hemorrhagic fever virus		
Simian hemorrhagic fever virus - LVR42-0	[AF180391]	(SHFV-LVR)

Tentative Species in the Genus

None reported.

List of Unassigned Viruses in the Family

None reported.

Phylogenetic Relationships within the Family

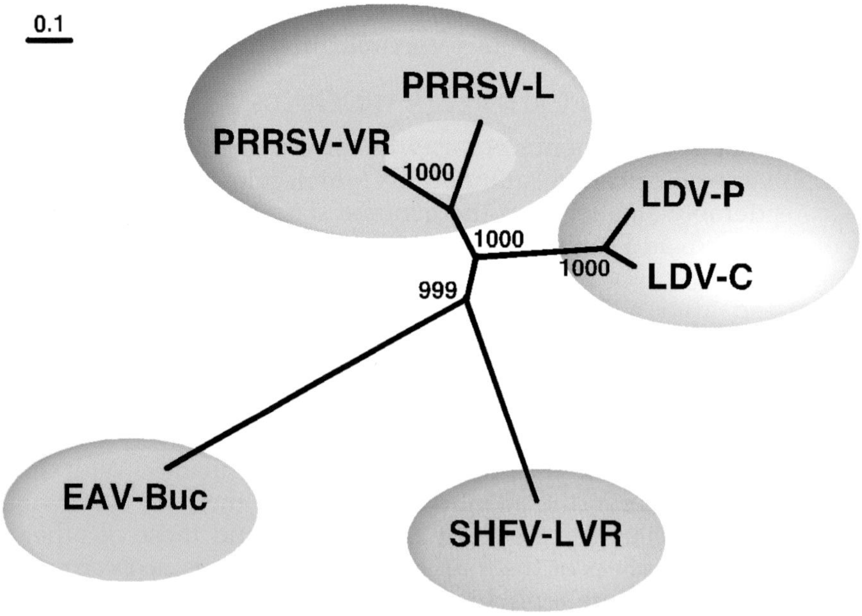

Figure 4: Unrooted phylogenetic analysis of the full-length ORF1b replicase polyprotein of the members of the family *Arteriviridae*. The tree was generated using the neighbor-joining algorithm implemented in the ClustalX program (with the Kimura correction on and using an alignment from which all columns containing gaps had been deleted). One thousand bootstrap replicates were done. Branch lengths indicate number of substitutions per residue. (A.E. Gorbalenya; unpublished data).

Similarity with Other Taxa

The family *Arteriviridae* together with the family *Coronaviridae* and the family *Roniviridae* form the order *Nidovirales*. These viruses have important common features at the level of genome organization, genome expression strategy, and phylogeny and internal organization of their large replicase gene. Despite these overall similarities, arterivirus genomes are substantially smaller, and the size, structure, and composition of their virions do not resemble those of other members of the order *Nidovirales*. Various non-structural proteins contain arterivirus- and/or nidovirus-specific domains or signatures, including an SDD signature (instead of the canonical GDD) in the RdRp (nsp9), a complex, N-terminal (putative) zinc binding domain in the helicase (nsp10), and a nidovirus-specific putative endonuclease domain in nsp11. Non-nidovirus homologs of several (putative) enzymes encoded by viruses of the family *Arteriviridae* have been found in other RNA viruses. The main proteinase and RdRp cluster together with homologs of viruses in the "picornavirus-like" and "sobemovirus-like" supergroups. The helicase has counterparts in viruses of the "alphavirus-like" supergroup. Also, some parallels are evident between the genome organization and expression strategy of members of the family *Arteriviridae* (and other members of the order *Nidovirales*) and those of the plant family *Closteroviridae*.

Derivation of Names

Arteri: from equine *arteri*tis, the disease caused by the reference virus.

References

Barrette-Ng, I.H., Ng, K.K., Mark, B.L., van Aken, D., Cherney, M.M., Garen, C., Kolodenko, Y., Gorbalenya, A.E., Snijder, E.J. and James, M.N. (2002). Structure of arterivirus nsp4: The smallest chymotrypsin-like proteinase with an alpha/beta C-terminal extension and alternate conformations of the oxyanion hole. *J. Biol. Chem.*, **277**, 39960-39966.

Chen, Z., Li, K. and Plagemann, P.G. (2000). Neuropathogenicity and sensitivity to antibody neutralization of lactate dehydrogenase-elevating virus are determined by polylactosaminoglycan chains on the primary envelope glycoprotein. *Virology*, **266**, 88-98.

Godeny, E.K., de Vries, A.A.F., Wang, X.C., Smith, S.L. and de Groot, R.J. (1998). Identification of the leader-body junctions for the viral subgenomic mRNAs and organization of the simian hemorrhagic fever virus genome: Evidence for gene duplication during arterivirus evolution. *J. Virol.*, **72**, 862-867.

Gorbalenya, A.E. (2001). Big nidovirus genome: when count and order of domains matter. *Adv. Exp. Med. Biol.*, **494**, 1-17.

Hwang, Y.-K. and Brinton, M.A. (1998). A 68-nucleotide sequence within the 3' non-coding region of simian hemorrhagic fever virus negative-strand RNA binds to four MA104 cell proteins. *J. Virol.*, **72**, 4341-4351.

Kreutz, L.C. and Ackermann, M.R. (1996). Porcine reproductive and respiratory syndrome virus enters cells through a low pH dependent endocytic pathway. *Virus Res.*, **42**, 137-147.

Meulenberg, J.J.M., Petersen-den Besten, A., de Kluyver, E.P., Moormann, R.J.M., Schaaper, W.M.M. and Wensvoort, G. (1995). Characterization of proteins encoded by ORFs 2 to 7 of Lelystad virus. *Virology*, **206**, 155-163.

Molenkamp, R., van Tol, H., Rozier, B.C.D., van der Meer, Y., Spaan, W.J.M. and Snijder, E.J. (2000). The arterivirus replicase is the only viral protein required for genome replication and subgenomic mRNA transcription. *J. Gen. Virol.* **81**, 2491-2496,

Pasternak, A.O., van den Born, E., Spaan, W.J.M. and Snijder, E.J. (2001). Sequence requirements for RNA strand transfer during nidovirus discontinuous subgenomic RNA synthesis. *EMBO J.*, **20**, 7220-7228.

Pedersen, K.W., van der Meer, Y., Roos, N. and Snijder, E.J. (1999). Open reading frame 1a-encoded subunits of the arterivirus replicase induce endoplasmic reticulum-derived double-membrane vesicles which carry the viral replication complex. *J. Virol.*, **73**, 2016-2026.

Snijder, E.J., Bredenbeek, P.J., Dobbe, J.C., Thiel, V., Ziebuhr, J., Poon, L.L., Guan, Y., Rozanov, M., Spaan, W.J.M. and Gorbalenya, A.E (2003). Unique and conserved features of genome and proteome of SARS-coronavirus, an early split-off from the coronavirus group 2 lineage. *J. Mol. Biol.*, **331**, 991-1004.

Snijder, E.J., Dobbe, J.C. and Spaan, W.J.M. (2003). Heterodimerization of the two major envelope proteins is essential for arterivirus infectivity. *J. Virol.*, **77**, 97-104.

Snijder, E.J. and Meulenberg, J.J.M. (2001). Arteriviruses. In: *Fields Virology*, 4th ed., (D.M. Knipe and P.M. Howley, eds), pp 1205-1220. Lippincott, Williams, and Wilkins, Philadelphia.

Tijms, M.A., van Dinten, L.C., Gorbalenya, A.E. and Snijder, E.J. (2001). A zinc finger-containing papain-like protease couples subgenomic mRNA synthesis to genome translation in a positive-stranded RNA virus. *Proc. Nat. Acad. Sci. USA*, **98**, 1889-1894.

Vanderheijden, N., Delputte, P.L., Favoreel, H.W., Vandekerckhove, J., Van Damme, J., van Woensel, P.A. and Nauwynck, H.J. (2003). Involvement of sialoadhesin in entry of porcine reproductive and respiratory syndrome virus into porcine alveolar macrophages. *J. Virol.*, **77**, 8207-8215.

Wieringa, R., de Vries, A.A.F. and Rottier, P.J.M. (2003). Formation of disulfide-linked complexes between the three minor envelope glycoproteins (GP2b, GP3, and GP4) of equine arteritis virus. *J. Virol.*, **77**, 6216-6226.

Ziebuhr, J., Snijder, E.J. and Gorbalenya, A.E. (2000). Virus-encoded proteinases and proteolytic processing in the *Nidovirales*. *J. Gen. Virol.*, **81**, 853-879.

Contributed By

Snijder, E.J., Brinton, M.A., Faaberg, K.S., Godeny, E.K., Gorbalenya, A.E., MacLachlan, N.J., Mengeling, W.L. and Plagemann, P.G.W.

FAMILY *RONIVIRIDAE*

TAXONOMIC STRUCTURE OF THE FAMILY

Order **Nidovirales**
Family **Roniviridae**
Genus **Okavirus**

Since only one genus is currently recognized, the family description corresponds to the genus description.

GENUS *OKAVIRUS*

Type Species *Gill-associated virus*

VIRION PROPERTIES

MORPHOLOGY

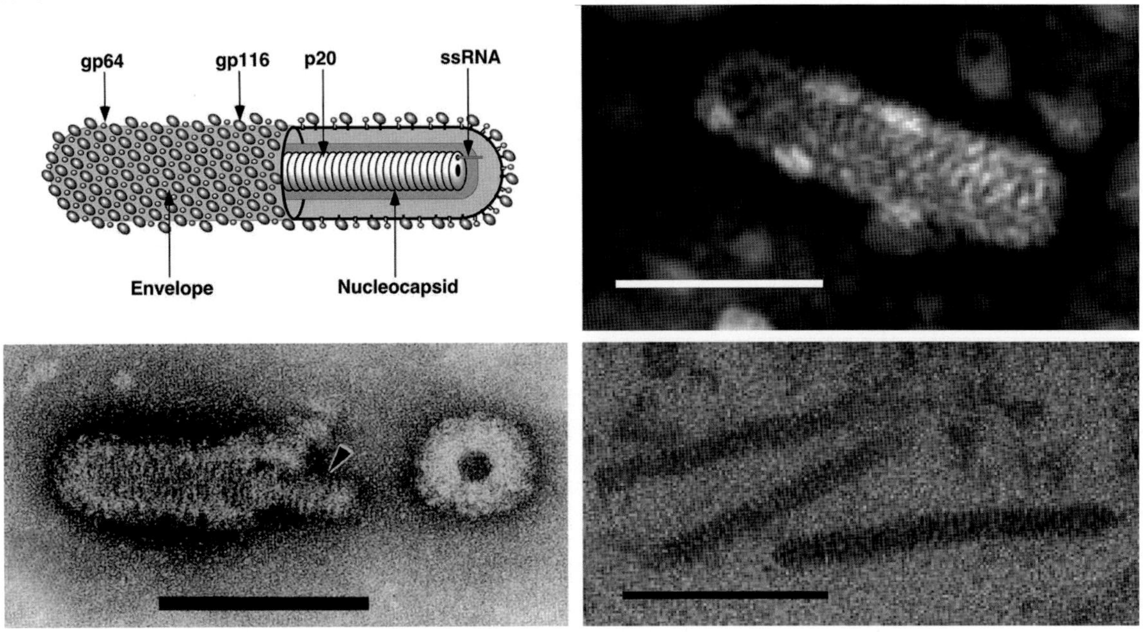

Figure 1: (Top left) Schematic illustration of an okavirus virion. (Top right) Transmission electron micrograph of negative-stained particles of Gill-associated virus (GAV). (Bottom left). Transmission electron micrograph of partially disrupted Yellow head virus (YHV) virion displaying the internal nucleocapsid and a ring-like structure which appears to be a disrupted virion in cross-section. (Bottom right) Transmission electron micrograph of cytoplasmic unenveloped nucleocapsids in a thin section of GAV-infected lymphoid organ cells. The bars represent 100 nm. (Courtesy of K. Spann, P. Loh, J. Cowley and R.J McCulloch and reproduced with permission).

Virions are enveloped and bacilliform in shape, with rounded ends and dimensions of 150-200 nm x 40-60 nm. Envelopes are studded with prominent peplomers projecting approximately 11 nm from the surface. Nucleocapsids have helical symmetry with diameter of 20-30 nm, apparently consisting of coiled structures with a periodicity of 5-7 nm. Long filamentous nucleocapsid precursors (approximately 15 x 80-450 nm) occur in the cytoplasm of infected cells and appear to acquire envelopes by budding into vesicles at the endoplasmic reticulum, generating intracytoplasmic paracrystalline arrays of enveloped virions.

PHYSICOCHEMICAL AND PHYSICAL PROPERTIES

Virion buoyant density in sucrose is 1.18-1.20 g/ml. Yellow head virus is inactivated by heating at 60°C for 15-30 min but has been reported to survive in seawater at 25-28°C for at least 4 days. Virions are sensitive to calcium hypochlorite and SDS but sensitivity to other treatments is not known.

NUCLEIC ACID

Virions contain a single segment of linear, positive-sense ssRNA. The genome of Gill-associated virus comprises 26,235 nt. A 68 nt 5'-UTR is followed by 5 long ORFs (5'-ORF1a-ORF1b-ORF2-ORF3-ORF4-3'), a 129 nt UTR and a 3'-poly[A] tail. ORF4 is significantly truncated in Yellow head virus (YHV) and other known genotypes and it may not always be expressed. Untranslated regions upstream of ORF2 (93 nt), ORF3 (57 nt) and ORF4 (256 nt) each contain a core of relatively conserved sequence. The genome structure of GAV is shown in Figure 2. The complete genome sequence of GAV and partial sequences of YHV are available.

PROTEINS

Virion structural proteins have been identified only for YHV. YHV virions contain three major structural proteins (110-135 kDa, 63-67 kDa and 20-22 kDa). The 110-135 kDa protein (gp116) and the 63-67 kDa protein (gp64) are glycosylated and appear to be envelope proteins that form the prominent peplomers on the virion surface. Mature gp116 and gp64 are generated by post-translational processing of a precursor polyglycoprotein. Gp116 and gp64 are not linked by intramolecular disulfide bonds but each is anchored in the virion by C-terminal hydrophobic transmembrane domains. The 20-22 kDa protein (p20) is associated with nucleocapsids and appears to function as the nucleoprotein.

Table 1: Virus-associated proteins of okaviruses

Protein			*Okavirus*[a]
Large spike glycoprotein	S1	gp116	110-135
Small spike glycoprotein	S2	gp64	60-65
Nucleocapsid protein	N	p20	20-22

[a] Apparent sizes by electrophoresis (kDa)

LIPIDS

Viruses have tri-laminar envelopes derived from the host cell.

CARBOHYDRATES

Gp116 and gp64 are extensively glycosylated. N-linked glycosylation sites are present in both gp116 (7-8 sites) and gp64 (4 sites) and the size estimate for each protein is consistent with glycosylation at these sites. Multiple O-linked glycosylation sites are also present but it is not known if they are utilized.

GENOME ORGANIZATION AND REPLICATION

The genome comprises a large replicase gene (ORF1a/ORF1b) followed by the nucleoprotein gene (ORF2), a glycoprotein gene (ORF3), a small gene or pseudogene of unknown function (ORF4), and a 3'-poly[A] tail. At the overlap between ORF1a and ORF1b there is a pseudoknot structure and a slippery sequence (AAAUUUU) that allow expression of ORF1b through a -1 ribosomal frame-shift. The pseudoknot structure and slippery sequence are not of the 'H-type' that occurs commonly in other nidoviruses but resemble the complex pseudoknot at the gag/pol junction of retroviruses. ORF1a contains a 3C-like protease flanked by hydrophobic domains that shares common sequence motifs with coronavirus homologues and appears to be involved in autolytic processing of the polyprotein pp1a. ORF1b contains multiple conserved sequence motifs including helicase and metal-ion binding domains, and a polymerase domain with the 'SDD' active site motif that is characteristic of nidoviruses.

Okaviruses are unique amongst members of the *Nidovirales* in that the nucleoprotein gene (ORF2) is located upstream of the glycoprotein gene (ORF3). ORF3 encodes a polyprotein that is processed by proteolysis at two predicted signal cleavage sites to generate virion envelope glycoproteins gp116 and gp64. The N-terminal fragment of the ORF3 polyprotein is predicted to be a triple-membrane-spanning glycoprotein of similar size to the M proteins of other nidoviruses. However, this product does not appear to be a major structural protein and has not yet been detected in infected cells. The 3'-terminal ORF4 varies significantly in size in different viruses. The 83 aa GAV ORF4 product is poorly characterized but has been detected in infected cells.

Okavirus transcription occurs via a nested set of 3'-coterminal polyadenylated mRNAs. Two sgRNAs are transcribed from highly conserved sequences in intergenic regions preceding ORF2 and ORF3. As for most torovirus sgRNAs, but unlike coronaviruses or arteriviruses, okavirus sgRNAs lack a common 5'-leader sequence acquired from the genomic RNA and initiate by templated transcription at common 5'-AC...sites. A less conserved but similar sequence also occurs in the long intergenic region preceding ORF4.

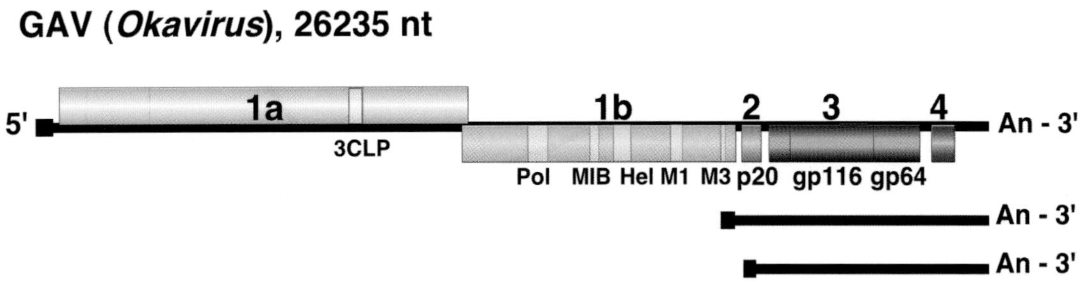

Figure 2: Schematic representation of the okavirus genome which comprises 5 long ORFs (ORF1a, ORF1b, ORF2, ORF3, ORF4). ORF 1a and ORF1b overlap at a ribosomal frame-shift site and contain the conserved sequence motifs (3CLP, Pol, MIB, Hel, M1 and M3). ORF2 encodes the nucleoprotein (p20). ORF3 encodes the virion transmembrane glycoproteins gp116 and gp64. The products of ORF4 have not yet been identified. Two sub-genomic mRNAs initiate at conserved sequences in the intergenic regions upstream of ORF2 and ORF3.

ANTIGENIC PROPERTIES

Not known.

BIOLOGICAL PROPERTIES

Okaviruses are the only known invertebrate nidoviruses and have been detected only in crustaceans. The black tiger prawn (*Penaeus monodon*) appears to be the natural host of GAV but other prawn species are susceptible to experimental infection. Infections may be chronic or acute and transmission can occur horizontally and vertically. During acute infections, mortalities are usually high and virus occurs in most tissues of ectodermal and mesodermal origin, and particularly in the 'Oka' or lymphoid organ. Necrotic cells display intensely basophilic cytoplasmic inclusions. The known geographic range of infection presently appears to be restricted to Asia and Australia where the prevalence of sub-clinical chronic infections in *P. monodon* is commonly high. There are no known prophylactic or curative treatments.

LIST OF SPECIES DEMARCATION CRITERIA IN THE GENUS

Not applicable.

LIST OF SPECIES IN THE GENUS

Species names are in green italic script; strain names and synonyms are in black roman script; tentative species names are in blue roman script. Sequence accession numbers, and assigned abbreviations ( ) are also listed.

SPECIES IN THE GENUS

Gill-associated virus

Gill-associated virus	[AF227196, AY039647]	(GAV)
Yellow head virus	[AY052786, AF540644]	(YHV)

POS. SSRNA

TENTATIVE SPECIES IN THE GENUS

Genotype 3
Genotype 4

PHYLOGENETIC RELATIONSHIPS WITHIN THE GENUS

Phylogenetic analysis of sequences in the ORF1b gene has clustered the viruses into four distinct genotypes: GAV, YHV, genotype 3 and genotype 4. GAV has been detected in Australia, Vietnam and Thailand; YHV has been detected in Thailand and Taiwan; genotype 3 has been detected in Thailand, Vietnam and Malaysia; genotype 4 has been detected only in India.

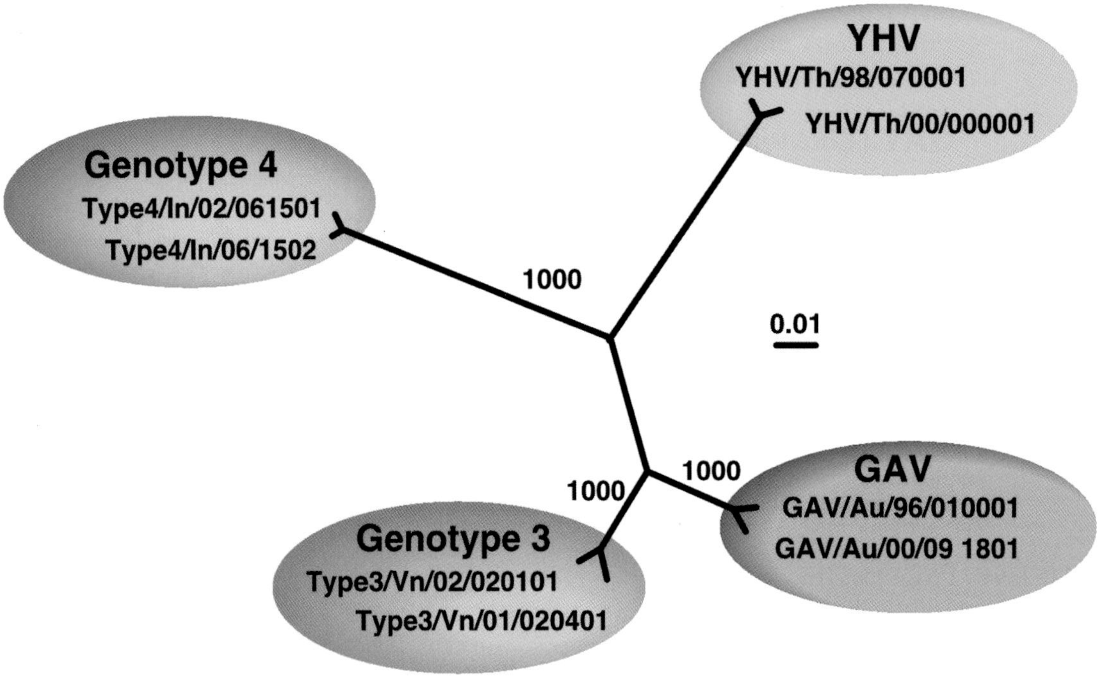

Figure 3: Unrooted phylogenetic tree of okaviruses based on a 781 nt region of ORF1b. The sequences were aligned using Clustal W, phylogenetic inference was determined by the neighbor-joining method and the tree was constructed using NJ-plot and Treeview software. Bootstrap analysis was conducted on 1000 replicates. Branch lengths are proportional to phylogenetic distance. The bar represents a sequence divergence of 1.0%.

SIMILARITY WITH OTHER TAXA

Various structural and genetic similarities with other viruses in the order *Nidovirales*. Virion morphology distantly resembles that of plant rhabdoviruses. The pseudoknot structure in the ORF1a/1b overlap resembles the gag/pol pseudoknots of some retroviruses. The structure and substrate specificity of the ORF1a protease bridges the gap between the chymotrypsin-like cysteine proteases of coronaviruses and plant potyviruses.

DERIVATION OF NAMES

Roni is a sigla derived from <u>ro</u>d-shaped <u>ni</u>dovirus referring to the unique virion morphology of viruses in the family.
Oka refers to the 'Oka' or lymphoid organ in which the viruses are commonly detected and in which pathology occurs during acute infections. Lymphoid organs are anatomical structures common to penaeid shrimp.

REFERENCES

Boonyaratpalin, S., Supamattaya, K., Kasornchandra, J., Direkbusaracom, S., Aekpanithanpong, U. and Chantanachookin, C. (1993). Non-occluded baculo-like virus, the causative agent of yellow head disease in the black tiger shrimp (*Penaeus monodon*). *Gyobyo Kenkyu*, **28,** 103-109.

Chantanachookin, C., Boonyaratpalin, S., Kasornchandra, J., Direkbusarakom, S., Ekpanithanpong, U., Supamataya, K., Siurairatana, S. and Flegel, T.W. (1993). Histology and ultrastructure reveal a new granulosis-like virus in *Penaeus monodon* affected by "yellow-head" disease. *Dis. Aquat. Org.*, **17,** 145-157.

Cowley, J.A., Dimmock, C.M., Spann, K.M. and Walker, P.J. (2000). Gill-associated virus of *Penaeus monodon* shrimp: an invertebrate virus with ORF1a and ORF1b genes related to arteri- and coronaviruses. *J. Gen. Virol.*, **81,** 1473-1484.

Cowley, J.A., Dimmock, C.M. and Walker, P.J. (2001). Gill-associated nidovirus of *Penaeus monodon* prawns transcribes 3'-coterminal subgenomic RNAs that do not possess 5'-leader sequences. *J. Gen. Virol.*, **83,** 927-935.

Cowley, J.A., Dimmock, C.M., Wongteerasupaya, C., Boonsaeng, V., Panyim, S. and Walker, P.J. (1999). Yellow head virus from Thailand and gill-associated virus from Australia are closely related but distinct prawn viruses. *Dis. Aquat. Org.*, **36,** 153-157.

Cowley, J.A., Hall, M.R., Cadogan, L.C., Spann, K.M. and Walker, P.J. (2002). Vertical transmission of covert gill-associated virus (GAV) infections in *Penaeus monodon*. *Dis. Aquat. Org.*, **50,** 95-104.

Cowley, J.A. and Walker, P.J. (2002). The complete genome sequence of gill-associated virus of *Penaeus monodon* prawns indicates a gene organisation unique among nidoviruses. *Arch. Virol.*, **147,** 1977-1987.

Jitrapakdee, S., Unajak, S., Sittidilokratna, N., Hodgson, R.A.J., Cowley, J.A., Walker, P.J., Panyim, S. and Boonsaeng, V. (2003). Identification and analysis of gp116 and gp64 structural glycoproteins of yellow head nidovirus of *Penaeus monodon* shrimp. *J. Gen Virol.*, **84,** 863-873.

Loh, P.C., Tapay, L.M., Lu, Y. and Nadala, E.C.B. (1997). Viral pathogens of penaeid shrimp. *Adv. Virus Res.*, **48,** 263-312.

Nadala, E.C.B., Tapay, L.M. and Loh, P.C. (1997). Yellow-head virus: a rhabdovirus-like pathogen of penaeid shrimp. *Dis. Aquat. Org.*, **31,** 141-146.

Sittidilokratna, N., Hodgson, R.A.J., Cowley, J.A. Jitrapakdee, S., Boonsaeng, V., Panyim, S. and Walker, P.J. (2002). Complete ORF1b-gene sequence indicates yellow head virus is an invertebrate nidovirus. *Dis. Aquat. Org.*, **50,** 87-93.

Spann, K.M., Cowley, J.A., Walker, P.J. and Lester, R.J.G. (1997). Gill-associated virus (GAV), a yellow head-like virus from *Penaeus monodon* cultured in Australia. *Dis. Aquat. Org.*, **31,** 169-179.

Spann, K.M., McCulloch, R.J., Cowley, J.A. East, I.J. and Walker, P.J. (2003). Detection of gill-associated virus (GAV) by *in situ* hybridization during acute and chronic infections of *Penaeus monodon* and *P. esculentus*. *Dis. Aquat. Org.*, *56, 1-10.*

Spann, K.M., Vickers, J.E. and Lester, R.J.G. (1995). Lymphoid organ virus of *Penaeus monodon* from Australia. *Dis. Aquat. Org.*, **23,** 127-134.

Walker, P.J., Cowley, J.A., Spann, K.M., Hodgson, R.A.J., Hall, M.R. and Withychumnarnkul, B. (2001). Yellow head complex viruses: transmission cycles and topographical distribution in the Asia-Pacific region. *In: The New Wave*: Proceedings of the Special Session on Sustainable Shrimp Culture, Aquaculture 2001, (C.L. Browdy and D.E. Jory, eds), pp 227-237. The World Aquaculture Society, Baton Rouge, LA.

Wang, Y.C. and Chang, P.S. (2000). Yellow head virus infection in the giant tiger prawn *Penaeus monodon* cultured in Taiwan. *Fish Pathol.*, **35,** 1-10.

Wongteerasupaya, C., Sriurairatana, S., Vickers, J.E., Akrajamorn, A., Boonsaeng, V., Panyim, S., Tassanakajon, A., Withyachumnarnjul, B. and Flegel, T.W. (1995). Yellow-head virus of *Penaeus monodon* is an RNA virus. *Dis. Aquat. Org.*, **22,** 45-50.

Ziebuhr, J., Bayer, S., Cowley, J.A. and Gorbalenya, A.E. (2003). The 3C-like proteinase of an invertebrate nidovirus links coronavirus and potyvirus homologs. *J. Virol.*, **77,** 1415-1426.

CONTRIBUTED BY

Walker, P.J., Bonami, J.R., Boonsaeng, V., Chang, P.S., Cowley, J.A., Enjuanes, L., Flegel, T.W., Lightner, D.V., Loh, P.C., Snijder, E.J. and Tang, K.

POS. SSRNA

FAMILY *FLAVIVIRIDAE*

TAXONOMIC STRUCTURE OF THE FAMILY

Family	***Flaviviridae***
Genus	***Flavivirus***
Genus	***Pestivirus***
Genus	***Hepacivirus***

VIRION PROPERTIES

MORPHOLOGY

Virions are 40-60 nm in diameter, spherical in shape and contain a lipid envelope. The capsid is composed of a single capsid protein (C) and the envelope contains two or three virus-encoded membrane proteins. The behavior of hepaciviruses during filtration and their susceptibility to chemical and physical treatments suggest that its overall structural properties are similar to those of flaviviruses and pestiviruses. Specific descriptions of the three individual genera are given in the corresponding sections.

PHYSICOCHEMICAL AND PHYSICAL PROPERTIES

The virion Mr, buoyant density, sedimentation coefficient and other physicochemical properties differ among the members of the three genera and are described separately in the corresponding sections.

NUCLEIC ACID

The genome RNA of all three genera is a positive sense ssRNA of approximately 11, 12.3, and 9.6 kb for flavi-, pesti-, and hepaciviruses, respectively. All members of the family lack a 3'-terminal poly(A) tract. Flaviviruses contain a 5'-terminal type I cap structure, while pestiviruses and hepaciviruses do not.

PROTEINS

The virions of all members of the family have a single, small basic C and two (*Flavivirus* and *Hepacivirus*) or three (*Pestivirus*) membrane-associated proteins. The non-structural proteins contain sequence motifs characteristic of a serine proteinase, RNA helicase, and RdRp that are encoded at similar locations along the genome in all three genera. Further details of specific functional properties are given in the corresponding sections of the individual genera.

LIPIDS

Lipids present in virions are derived from host cell membranes and make up 17% of the total virion weight in the case of flaviviruses. The lipid content of pesti- and hepaciviruses has not been determined.

CARBOHYDRATES

Virions contain carbohydrates in the form of glycolipids and glycoproteins.

GENOME ORGANIZATION AND REPLICATION

The genome RNA of all three genera has a similar organization and is the only viral mRNA found in infected cells. It contains a single long ORF flanked by 5'- and 3'-terminal NCRs that form specific secondary structures required for genome replication and translation. Translation-initiation is cap-dependent in the case of flaviviruses, whereas IRES have been demonstrated for pestiviruses and hepaciviruses. Viral proteins are synthesized as part of a polyprotein of more than 3000 aa that is co- and post-translationally cleaved by viral and cellular proteinases. The structural proteins are contained in the N-terminal portion of this polyprotein and the non-structural proteins in the remainder. The latter include a serine proteinase, an RNA helicase, and an RdRp.

RNA synthesis occurs in the cytoplasm in association with the endoplasmic reticulum via synthesis of a full-length negative-strand intermediate. Virion assembly and envelopment are thought to take place at intracellular membranes. Viral particles are transported in cytoplasmic vesicles through the secretory pathway before they are released by exocytosis, as shown for members of the genus *Flavivirus* and assumed for the pestiviruses and hepaciviruses.

Antigenic Properties

The three genera are antigenically unrelated, but serological cross-reactivities exist among members within the genera *Flavivirus* and *Pestivirus*. Hepaciviruses have so far not been amenable to antigenic analysis.

Biological Properties

The biological properties of the three genera exhibit different characteristics and are described in the corresponding sections.

Genus *Flavivirus*

Type Species *Yellow fever virus*

Distinguishing Features

The 5′-end of the genome possesses a type I cap (m-7GpppAmp) not seen in the other genera. Most flaviviruses are transmitted to vertebrate hosts by arthropod vectors, mosquitos or ticks, in which they actively replicate. Some flaviviruses are zoonotic agents transmitted between rodents or bats without known arthropod vectors.

Virion Properties

Morphology

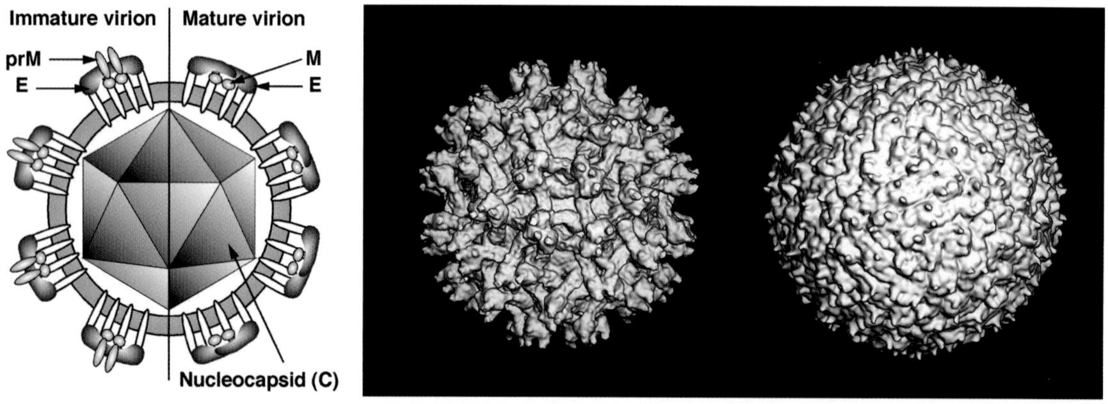

Figure 1: (Left) Schematic of immature and mature virion. (Center and right) Three-dimensional cryo-electron microscopic reconstructions of immature and mature particles of an isolate of *Dengue virus* (courtesy of R.J. Kuhn).

Virions are 50 nm in diameter and spherical in shape (Fig. 1). Two virus forms can be distinguished. Mature virions contain two virus encoded membrane-associated proteins, E and M. Intracellular, immature virions contain the precursor prM instead of M, which is proteolytically cleaved in the course of maturation. The atomic structure of the major envelope protein E from Tick-borne encephalitis virus (TBEV) and Dengue virus (DENV) has been determined by X-ray crystallography. It is a dimeric, rod-shaped molecule that is oriented parallel to the membrane and does not form spike-like projections in its neutral pH conformation. Image reconstructions from cryo-electron micrographs have shown that the virion envelope has icosahedral symmetry.

Physicochemical and Physical Properties

Virion Mr has not been precisely determined but can be estimated from the virus composition to be ~6 x 10^7. Mature virions sediment at ~200S and have a buoyant density of ~1.19 g/cm^3 in sucrose. Viruses are stable at slightly alkaline pH 8.0 but are readily inactivated at acidic pH, temperatures above 40°C, organic solvents, detergents, ultraviolet light, and gamma-irradiation.

Nucleic Acid

The virion RNA of flaviviruses is a positive-sense infectious ssRNA of ~11 kb. The 5′-end of the genome possesses a type I cap (m-7GpppAmp) followed by the conserved dinucleotide AG. The 3′-ends lack a terminal poly(A) tract and terminate with the conserved dinucleotide CU.

Proteins

Virions contain three structural proteins: C (11 kDa), E (50 kDa), the major envelope protein; and either prM (26 kDa), in immature virions, or M (8 kDa), in mature virions. The E protein is the viral hemagglutinin and is believed to mediate both receptor binding and acid pH-dependent fusion activity after uptake by receptor-mediated endocytosis. Seven nonstructural proteins are synthesized in infected cells: NS1 (46 kDa), NS2A (22 kDa), NS2B (14 kDa), NS3 (70 kDa), NS4A (16 kDa), NS4B (27 kDa) and NS5 (103 kDa). NS3 is a multi-functional protein. The N-terminal one-third of the protein forms the viral serine proteinase complex together with NS2B that is involved in processing the polyprotein. The C-terminal portion of NS3 contains an RNA helicase domain involved in RNA replication, as well as an RNA triphosphatase activity that is probably involved in formation of the 5′-terminal cap structure of the viral RNA. NS5 is the largest and most highly conserved flavivirus protein. NS5 is the flavivirus RdRp and also possesses motifs suggesting that it encodes the methyltransferase activity necessary for methylation of the 5′-cap structure.

Lipids

Virions contain about 17% lipid by weight; lipids are derived from host cell membranes.

Carbohydrates

Virions contain about 9% carbohydrate by weight (glycolipids, glycoproteins); their composition and structure are dependent on the host cell (vertebrate or arthropod). N-glycosylation sites are present in the proteins prM (1 to 3 sites), E (0 to 2 sites), and NS1 (1 to 3 sites).

Genome Organization and Replication

The genome RNA represents the only viral messenger RNA in flavivirus-infected cells. It consists of a single long ORF of more than 10,000 nt that codes for all structural and nonstructural proteins and is flanked by short NCRs at the 5′- and 3′-terminal ends (Fig. 2).

While nt sequences are divergent, the predicted secondary structures within the 5′- and 3′-NCRs are conserved among different flaviviruses. The NCRs contain stretches of conserved RNA sequences that are distinct in mosquito- and tick-borne flaviviruses. The length of the 3′-NCR of Tick-borne encephalitis virus can vary significantly, from 450 to almost 800 nt, and in some cases may contain an internal poly(A) tract. RNA synthesis appears to occur on the membranes of the perinuclear endoplasmic reticulum. After translation of the incoming genomic RNA, RNA replication begins with synthesis of complementary negative strands, which are then used as templates to produce additional genome-length positive-stranded molecules. These are synthesized by a semi-conservative mechanism involving replicative intermediates (containing double-stranded regions as well as nascent single-stranded molecules) and replicative forms (duplex RNA molecules). Negative strand synthesis in flavivirus-infected cells continues throughout the replication cycle. Translation usually starts at the first AUG of the ORF but may also occur at a second in-frame AUG located 12 to 14 codons downstream in mosquito-borne flaviviruses. The polyprotein is processed by cellular proteinases and the viral NS2B-NS3 serine proteinase to give rise to the mature structural and nonstructural proteins. Protein topology with respect to the ER and cytoplasm is determined by internal signal and stop-transfer sequences. Proliferation and hypertrophy of intracellular membranes is a characteristic feature of flavivirus-infected cells. The replication complex sediments with membranous fractions of extracts from infected cells. Virus particles can first be observed in the rough endoplasmic reticulum, which is believed to be the site of virus assembly. These immature virions are then transported through the membrane systems of the host secretory pathway to the cell surface where exocytosis occurs. Shortly before virion release, the prM protein is cleaved by furin or a furin-like cellular proteinase to generate mature virions. Flavivirus-infected cells also release a noninfectious subviral particle that has a lower sedimentation coefficient than whole virus (70S vs. 200S) and exhibits hemagglutination activity (slowly sedimenting hemagglutinin; SHA).

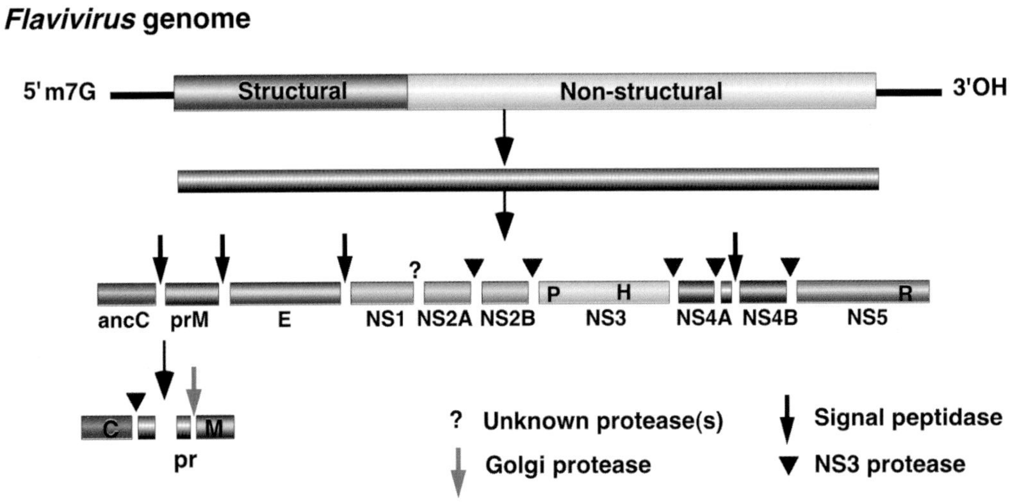

Figure 2: Flavivirus genome organization (not to scale) and polyprotein processing. At the top is the viral genome with the structural and non-structural protein coding regions and the 5'- and 3'-NCRs indicated. Boxes below the genome indicate viral proteins generated by the proteolytic processing cascade. AncC: anchored CP (with C-terminal hydrophobic domain). The viral structural proteins are shown in black. P, H, and R symbols indicate the localization of the NS3 proteinase, the NS3 RNA helicase, and the NS5 RdRp, respectively.

Antigenic Properties

All flaviviruses are serologically related, which can be demonstrated by binding assays such as ELISA and by hemagglutination-inhibition using polyclonal and monoclonal antibodies. Neutralization assays are more discriminating and have been used to define several serocomplexes of more closely related flaviviruses (see List of Species in the Genus). The envelope protein E is the major target for neutralizing antibodies and induces protective immunity. The E protein also induces flavivirus cross-reactive non-neutralizing antibodies. Antigenic sites involved in neutralization have been mapped to each of the three structural domains of the E protein. Antibodies to prM can also mediate immunity, probably by neutralizing viruses with partially uncleaved prM.

Biological Properties

Host Range

Flaviviruses can infect a variety of vertebrate species and in many cases arthropods. Some viruses have a limited vertebrate host range (e.g., only primates), others can infect and replicate in a wide variety of species (mammals, birds, etc.). Arthropods are usually infected when they feed on a vertebrate host during viremia, but non-viremic transmission has also been described for tick-borne flaviviruses.

Transmission

Most flaviviruses are arthropod-borne viruses that are maintained in nature by transmission from hematophagous arthropod vectors to vertebrate hosts. About 50% of known flaviviruses are mosquito-borne, 28% are tick-borne, and the rest are zoonotic agents transmitted between rodents or bats without known arthropod vectors. In some instances, the transmission cycle has not yet been identified. In the arthropod vectors, the viruses may also be passed on trans-ovarially (mosquitoes, ticks) and trans-stadially (ticks).

Geographical Distribution

Flaviviruses have a world-wide distribution but individual species are restricted to specific endemic or epidemic areas (e.g., *Yellow fever virus* in tropical and subtropical regions of Africa and South America; *Dengue virus* in tropical areas of Asia, Oceania, Africa, Australia, and the Americas; *Japanese encephalitis virus* in South-East Asia; *Tick-borne encephalitis virus* in Europe and Northern Asia).

Pathogenicity

More than 50% of known flaviviruses have been associated with human disease, including the most important human pathogens: *Yellow fever virus, Dengue virus, Japanese encephalitis virus, West Nile virus* and *Tick-borne encephalitis virus*. Flavivirus-induced diseases may be associated with symptoms of the central nervous system (e.g., meningitis, encephalitis), fever, arthralgia, rash, and hemorrhagic fever. Several flaviviruses are pathogenic for domestic or wild animals (turkey, pig, horse, sheep, dog, grouse, muskrat) and cause economically important diseases.

List of Species Demarcation Criteria in the Genus

Species demarcation criteria in the genus include:

- Nucleotide and deduced amino acid sequence data,
- Antigenic characteristics,
- Geographic association,
- Vector association,
- Host association,
- Disease association,
- Ecological characteristics.

Other defined members of individual species, which do not constitute a species on their own, are shown below the species. Those viruses for which insufficient information is available are listed as 'tentative' within the group of viruses to which they are most closely related by sequence analysis.

List of Species in the Genus

Virus species in the Genus can be grouped serologically and in terms of their vector preferences as shown in the list provided.

Species names are in green italic script; strain names and synonyms are in black roman script; tentative species names are in blue roman script. Sequence accession numbers, and assigned abbreviations ( ) are also listed.

Species in the Genus

1. Tick-borne viruses

Mammalian tick-borne virus group

Gadgets Gully virus		
Gadgets Gully virus	[AF013374]	(GGYV)
Kyasanur Forest disease virus		
Kyasanur Forest disease virus	[X74111]	(KFDV)
Langat virus		
Langat virus	[M73835]	(LGTV)
Louping ill virus		
Louping ill virus	[Y07863]	(LIV)
British subtype	[D12937]	(LIV-Brit)
Irish subtype	[X86784]	(LIV-Ir)
Spanish subtype	[X77470]	(LIV-Span)
Turkish subtype	[X69125]	(LIV-Turk)
Omsk hemorrhagic fever virus		
Omsk hemorrhagic fever virus	[X66694]	(OHFV)
Powassan virus		
Powassan virus	[L06436]	(POWV)
Royal Farm virus		
Karshi virus	[AF013381]	(KSIV)
Royal Farm virus	[AF013398]	(RFV)
Tick-borne encephalitis virus		
Tick-borne encephalitis virus		(TBEV)
European subtype	[M27157, M33668]	(TBEV-Eu)
Far Eastern subtype	[X07755]	(TBEV-FE)
Siberian subtype	[L40361]	(TBEV-Sib)

Seabird tick-borne virus group

Kadam virus		
Kadam virus	[AF013380]	(KADV)
Meaban virus		
Meaban virus	[AF013386]	(MEAV)
Saumarez Reef virus		
Saumarez Reef virus	[X80589]	(SREV)
Tyuleniy virus		
Tyuleniy virus	[X80588]	(TYUV)

2. Mosquito-borne viruses

Aroa virus group

Aroa virus		
Aroa virus	[AF013362]	(AROAV)
Bussuquara virus	[AF013366]	(BSQV)
Iguape virus	[AF013375]	(IGUV)
Naranjal virus	[AF013390]	(NJLV)

Dengue virus group

Dengue virus		
Dengue virus 1	[23027]	(DENV-1)
Dengue virus 2	[M19197]	(DENV-2)
Dengue virus 3	[A34774]	(DENV-3)
Dengue virus 4	[M14931]	(DENV-4)
Kedougou virus		
Kedougou virus	[AF013382]	(KEDV)

Japanese encephalitis virus group

Cacipacore virus		
Cacipacore virus	[AF013367]	(CPCV)
Japanese encephalitis virus		
Japanese encephalitis virus	[M18370]	(JEV)

Koutango virus		
Koutango virus	[AF013384]	(KOUV)
Murray Valley encephalitis virus		
Alfuy virus	[AF013360]	(ALFV)
Murray Valley encephalitis virus	[X03467]	(MVEV)
St. Louis encephalitis virus		
St. Louis encephalitis virus	[M1661]	(SLEV)
Usutu virus		
Usutu virus	[AF013412]	(USUV)
West Nile virus		
Kunjin virus	[D00246]	(KUNV)
West Nile virus	[M12294]	(WNV)
Yaounde virus		
Yaounde virus	[AF013413]	(YAOV)
Kokobera virus group		
Kokobera virus		
Kokobera virus	[AF013383]	(KOKV)
Stratford virus	[AF013407]	(STRV)
Ntaya virus group		
Bagaza virus		
Bagaza virus	[AF013363]	(BAGV)
Ilheus virus		
Ilheus virus	[AF013376]	(ILHV)
Rocio virus	[AF013397]	(ROCV)
Israel turkey meningoencephalomyelitis virus		
Israel turkey meningoencephalomyelitis virus	[AF013377]	(ITV)
Ntaya virus		
Ntaya virus	[AF013392]	(NTAV)
Tembusu virus		
Tembusu virus	[AF013408]	(TMUV)
Spondweni virus group		
Zika virus		
Spondweni virus	[AF013406]	(SPOV)
Zika virus	[AF013415]	(ZIKV)
Yellow fever virus group		
Banzi virus		
Banzi virus	[L40951]	(BANV)
Bouboui virus		
Bouboui virus	[AF013364]	(BOUV)
Edge Hill virus		
Edge Hill virus	[AF013372]	(EHV)
Jugra virus		
Jugra virus	[AF013378]	(JUGV)
Saboya virus		
Potiskum virus	[AF013395]	(POTV)
Saboya virus	[AF013400]	(SABV)
Sepik virus		
Sepik virus	[AF013404]	(SEPV)
Uganda S virus		
Uganda S virus		(UGSV)
Wesselsbron virus		
Wesselsbron virus		(WESSV)
Yellow fever virus		
Yellow fever virus	[X03700]	(YFV)

3. Viruses with no known arthropod vector

Entebbe bat virus group

Entebbe bat virus		
Entebbe bat virus	[AF013373]	(ENTV)

Sokoluk virus	[AF013405]	(SOKV)
Yokose virus		
Yokose virus	[AF013414]	(YOKV)
Modoc virus group		
Apoi virus		
Apoi virus	[AF013361]	(APOIV)
Cowbone Ridge virus		
Cowbone Ridge virus	[AF013370]	(CRV)
Jutiapa virus		
Jutiapa virus	[AF013379]	(JUTV)
Modoc virus		
Modoc virus	[AF013387]	(MODV)
Sal Vieja virus		
Sal Vieja virus	[AF013401]	(SVV)
San Perlita virus		
San Perlita virus	[AF013402]	(SPV)
Rio Bravo virus group		
Bukalasa bat virus		
Bukalasa bat virus	[AF013365]	(BBV)
Carey Island virus		
Carey Island virus	[AF013368]	(CIV)
Dakar bat virus		
Dakar bat virus	[AF013371]	(DBV)
Montana myotis leukoencephalitis virus		
Montana myotis leukoencephalitis virus	[AF013388]	(MMLV)
Phnom Penh bat virus		
Batu Cave virus	[AF013369]	(BCV)
Phnom Penh bat virus	[AF013394]	(PPBV)
Rio Bravo virus		
Rio Bravo virus	[AF013396]	(RBV)

TENTATIVE SPECIES IN THE GENUS

Cell fusing agent virus	[M91671]	(CFAV)
Tamana bat virus		(TABV)

GENUS PESTIVIRUS

Type Species *Bovine viral diarrhea virus 1*

DISTINGUISHING FEATURES

Relative to the other genera, pestiviruses encode two unique gene products, namely N^{pro} and E^{rns}. The first protein of the ORF, nonstructural protein N^{pro}, which possesses an autoproteolytic activity and is responsible for its release from the nascent polyprotein, is not essential for virus replication in cell culture. One of the three viral envelope glycoproteins, E^{rns}, possesses an intrinsic RNAse activity. Two biotypes of pestiviruses, cytopathogenic (cp) and non-cytopathogenic (noncp) viruses, are distinguished by their ability to cause cytopathic effects in cell culture.

VIRION PROPERTIES

MORPHOLOGY

Virions are 40-60 nm in diameter and spherical in shape (Fig. 3). The virion envelope has 10-12 nm ring-like subunits on its surface. Structure and symmetry of the core have not been characterized.

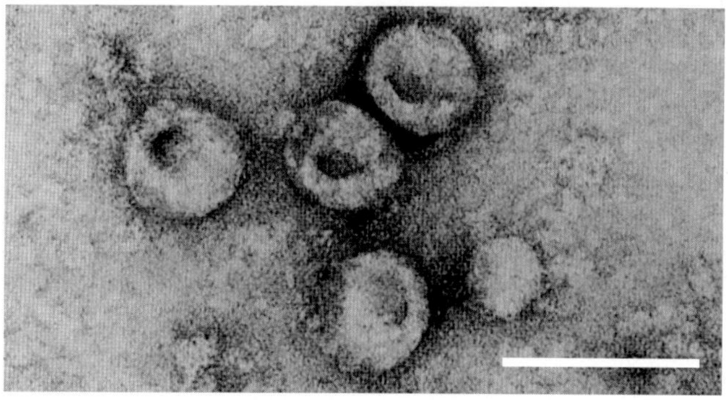

Figure 3: Negative contrast electron micrograph of particles of an isolate of *Bovine viral diarrhea virus 1*. The bar represents 100 nm. (From M. König, with permission).

PHYSICOCHEMICAL AND PHYSICAL PROPERTIES

Virion Mr has not been determined precisely, but can be estimated from the virus composition to be ~6 x 10^7. Buoyant density in sucrose is 1.10-1.15 g/cm^3; S_{20W} is 140-150S. Virion infectivity is stable over a relatively broad pH range, but unstable at temperatures above 40°C. Organic solvents and detergents rapidly inactivate the viruses.

NUCLEIC ACID

The virion RNA is a positive sense, infectious molecule of ssRNA ~12.3 kb in size. The 5'-NCR contains an IRES and is ~370-385 nt in length. The 3'-NCR with ~185-273 nt in length is complex and contains a region with variable sequences and a highly conserved terminal region. Genomic RNA contains a single ORF spanning the viral genome. For some cp pestivirus strains, a small and variable segment of host cell or viral nucleic acid is integrated into particular regions (often within NS2 or at the junction between NS2 and NS3) of the viral genome, sometimes accompanied by viral gene duplications or deletions. Other cp pestiviruses contain only viral gene duplications involving all or part of the N^{pro} and NS3 protein-coding regions, resulting in genomic RNA of up to ~16.5 kb. In all cases, the single large ORF is maintained. Finally, cp viruses may also arise by deletion of large portions of their genomes. Such defective genomes may be rescued by intact helper viruses.

PROTEINS

Virions are composed of 4 structural proteins: a basic nucleocapsid core protein, C (14 kDa), and 3 envelope glycoproteins, E^{rns} (gp44/48), E1 (gp33) and E2 (gp55). All three glycoproteins exist as intermolecular disulfide-linked complexes: E^{rns} homodimers, E1-E2 heterodimers, and E2 homodimers. E^{rns} possesses an intrinsic RNase activity. Pestiviruses encode 7-8 non-structural (NS) proteins among which N^{pro} (23 kDa), p7 (7 kDa) and NS2 (40 kDa) are not necessary for RNA replication. N^{pro} is a proteinase that autocatalytically releases itself from the nascent polyprotein. Nonstructural protein p7 is presumed to have a role in virus maturation. NS2-3 (120 kDa) is a multifunctional protein. The N-terminal 40% (NS2) is hydrophobic and contains a zinc finger motif suggesting divalent metal ion binding. The C-terminal 60% (NS3, 80 kDa) acts as both a serine proteinase involved in polyprotein processing and an RNA helicase/NTPase likely involved in RNA replication. NS2-3 is found after infection with all pestiviruses. In cells infected with cp pestiviruses, large amounts of NS3 can be detected. For some noncp BDV and CSFV strains, a minor part of NS2-3 is cleaved into NS3 and NS2. The NS4A (7 kDa) protein acts as a cofactor to the NS3 proteinase activity. The role of NS4B (33 kDa) is unknown. NS5A (58 kDa) represents a phosphorylated protein and presumably also plays a yet to be identified role in RNA replication. NS5B (75 kDa) possesses RdRp activity.

LIPIDS

The viruses are enveloped, but no reports have described the lipid composition.

CARBOHYDRATES

All virus envelope glycoproteins contain N-linked glycans.

GENOME ORGANIZATION AND REPLICATION

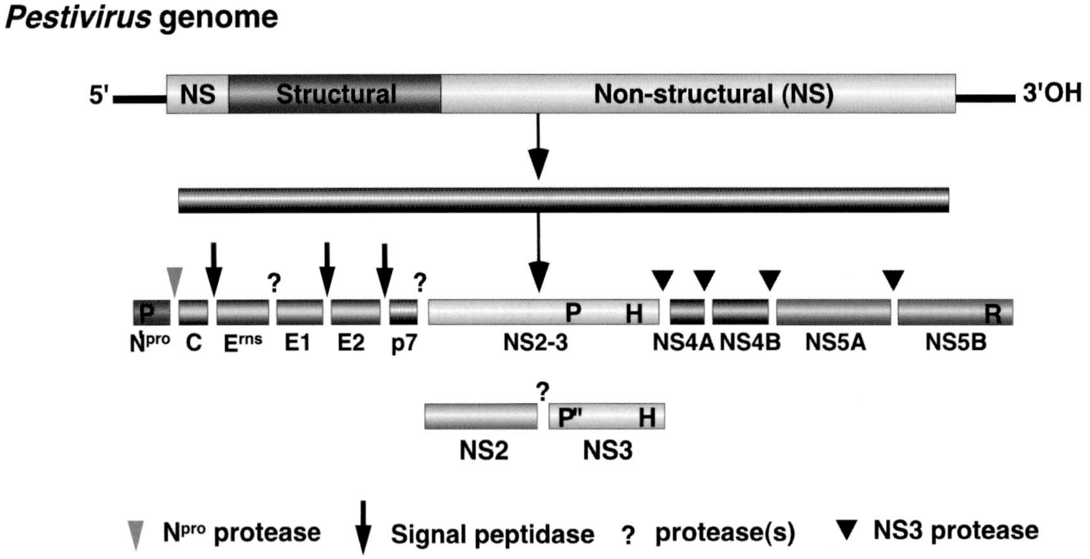

Figure 4: Pestivirus genome organization (not to scale) and polyprotein processing. The RNA is usually ~12.3 kb in size (depending on the virus). The 5'-NCR is ~370-385 nt, the ORF ~11.7 kb and the 3'-NCR is 185-273 nt. P′, P′′, H, and R symbols indicate the localization of the N^{pro} proteinase, the NS3 proteinase, the NS3 RNA helicase, and the NS5B RdRp, respectively. The viral non-structural proteins are indicated as NS. The proteinases (where known) and proteolytic steps involved in the generation of individual proteins are indicated. In noncp BVD viruses, NS2-3 is not cleaved. In cp BVD viruses, NS3 is produced in addition to NS2-3. If NS3 production is a result of NS3 gene duplication, NS2 is not necessarily produced.

The genomic RNA contains a single large ORF encoding a polyprotein of ~3,900 aa that is preceded by a 5′-NCR of ~370-385 nt and followed by a 3′-NCR of ~185-273 nt. The gene order is 5'-N^{pro}-C-E^{rns}-E1-E2-p7-NS2-3(NS2-NS3)-NS4A-NS4B-NS5A-NS5B-3' (Fig. 4).

Pestivirus replication is probably initiated by receptor-mediated endocytosis involving one or more cell surface molecules and viral glycoproteins E^{rns} and E2. After endocytosis and uncoating, the genome RNA serves as mRNA; there are no subgenomic mRNA molecules. Translation initiation occurs by a cap-independent internal initiation mechanism involving an IRES within the 5′-NCR of the RNA. Polyprotein processing occurs co- and post-translationally by both cellular and viral proteinases. Nonstructural protein N^{pro}, the first protein of the ORF, autoproteolytically removes itself from the nascent polyprotein by cleavage at the N^{pro}/C site. Downstream cleavages that produce structural proteins E^{rns}, E1 and E2 as well as p7 are probably mediated by cellular signal peptidase(s). Glycoprotein translocation to the endoplasmic reticulum probably occurs by an internal signal sequence, perhaps within the C protein. Cleavage between E2 and p7 is not complete, leading to two intracellular forms of E2 with different C-termini. Depending on the particular pestivirus and its biotype, NS2-3 remains intact or is found together with its N- and C-terminal products NS2 and NS3. The generation of NS3 in cp pestiviruses is in most cases due to RNA recombination. Most cp pestiviruses have gene insertions, deletions, duplications or rearrangements that result in (enhanced) NS3 production. The NS3/NS2-3 serine proteinase activity is responsible for all processing events downstream of NS3. NS4A facilitates cleavages by NS3 of sites 4B/5A and 5A/5B.

RNA replication occurs most likely in association with intracytoplasmic membranes, presumably in a replication complex composed of viral RNA and viral nonstructural proteins. Nonstructural proteins NS3, 4A, 4B, 5A and 5B are necessary for RNA replication; only NS5A can be provided *in trans.* Replicative forms of pestiviral RNA have been detected. The ratio of positive-to-negative sense RNA in cells 12 hr post-infection is about 10. RNA synthesis is resistant to actinomycin D. Virus maturation is poorly understood. However, viral proteins are not found on the cell surface, suggesting that viruses mature in intracellular vesicles and are released by exocytosis. Considerable amounts of infectious virus remain cell-associated. Host cell RNA and protein synthesis continue throughout infection.

Antigenic Properties

Pestiviruses are antigenically related, and cross-reactive epitopes are present on all pestiviruses. Separate antigenic determinants defined by monoclonal antibodies (Mabs) have also been identified. Antigenic variation is particularly pronounced among BVDV and BDV isolates. The N-terminal portion of E2 contains an antigenically hypervariable region. Mab binding patterns are generally consistent with the genetic relatedness of viruses.

Infected animals mount potent antibody responses to two structural glycoproteins (E^{rns}, E2) and to the NS2-3/NS3 protein, while antibody responses are weak or nonexistent to other virus-encoded polypeptides. E^{rns} and E2 are able to induce protection independently. Mabs reactive with these proteins can neutralize virus infectivity. Antibodies to NS2-3/NS3 are not able to neutralize virus infectivity.

Biological Properties

Host Range

Pestiviruses infect pigs and ruminants, including cattle, sheep, goats and wild ruminants. There are no invertebrate hosts.

Transmission

Transmission occurs by direct and indirect contact (e.g., nasal or urine secretion, contaminated food, etc.). Transplacental and congenital transmission occur in all host species.

Pathogenicity

Pestivirus infections may be subclinical or produce a range of clinical conditions including acute diarrhea, acute hemorrhagic syndrome, acute fatal disease, and a wasting disease. Diplacental infection can result in fetal death, congenital abnormalities, or lifelong persistent infection. Fatal mucosal disease can occur in cattle persistently infected with noncp viruses when a cp virus is generated by mutation or introduced by superinfection. Pestivirus infections of livestock are economically important worldwide.

Experimental Hosts

Experimental infection models have not been established for bovine or ovine pestiviruses outside their natural mammalian hosts. However, CSFV can be adapted to propagate in rabbits.

Cell Culture

Cells derived from natural host species (bovine, porcine, ovine) support virus replication. Most virus isolates are noncp and can establish persistent infections in cell culture. Infectious noncp BVDV is often present in bovine serum products used for cell culture. Cp pestiviruses induce extensive cytopathology and form virus plaques under appropriate conditions. Death of cp pestivirus infected cells is due to apoptosis. It is not known whether apoptosis is induced by cp viruses or actively blocked by noncp viruses.

Hemagglutination

No hemagglutinating activity has been found associated with pestiviruses.

List of Species Demarcation Criteria in the Genus

Species demarcation criteria in the genus includes:

- Nucleotide sequence relatedness,
- Serological relatedness,
- Host of origin.

Pestivirus species demarcation considers several parameters and their relationship to the type viruses of the currently recognized species (BVDV-1-NADL; BVDV-2-890; BDV-BD31; and CSFV-A187). Nucleotide sequence relatedness is an important criterion for pestivirus species demarcation. For example, the 5'-NCR sequences among the four currently recognized species are over 15% divergent. In most cases, the degree of homology within the 5'-NCR will allow pestivirus species demarcation. However, in some cases the nt sequence relatedness may be ambiguous and must be complemented with additional comparative analyses. Convalescent animal sera generated against members of a given pestivirus species (e.g., *Bovine viral diarrhea virus 1*) generally show a several-fold higher neutralization titer against viruses of the same same species than against viruses from the other species. Finally, differences in host of origin and disease can assist in species identification.

For example, *Bovine viral diarrhea virus 1* and *Classical swine fever virus* are considered different species because their members differ from each other by: (i) at least 25% at the sequence level (complete genomes), (ii) at least 10-fold difference in neutralization titer in cross-neutralization tests with polyclonal immune sera, and (iii) host range, in that under natural conditions CSFV infects only pigs while BVDV-1 infects ruminants as well as pigs.

List of Species in the Genus

Species names are in green italic script; strain names and synonyms are in black roman script; tentative species names are in blue roman script. Sequence accession numbers, and assigned abbreviations ( ) are also listed.

Species in the Genus

Border disease virus		
Border disease virus - BD31	[U70263]	(BDV-BD31)
Border disease virus - X818	[AF037405]	(BDV-X818)
Bovine viral diarrhea virus 1		
Bovine viral diarrhea virus 1-CP7	[U63479]	(BVDV-1-CP7)
Bovine viral diarrhea virus 1-NADL	[M31182]	(BVDV-1-NADL)
Bovine viral diarrhea virus 1-Osloss	[M96687]	(BVDV-1-O)
Bovine viral diarrhea virus 1-SD1	[M96751]	(BVDV-1-SD1)
Bovine viral diarrhea virus 2		
Bovine viral diarrhea virus 2-C413	[AF002227]	(BVDV-2-C413)
Bovine viral diarrhea virus 2-NewYork'93	[AF502399]	(BVDV-2-NY93)
Bovine viral diarrhea virus 2-strain 890	[U18059]	(BVDV-2-890)
Classical swine fever virus		
(Hog cholera virus)		
Classical swine fever virus - Alfort/187	[X87939]	(CSFV-A187)
Classical swine fever virus - Alfort-Tübingen	[J04358]	(CSFV-ATub)
Classical swine fever virus - Brescia	[M31768]	(CSFV-Bre)
Classical swine fever virus - C	[Z46258]	(CSFV-C)

Tentative Species in the Genus

Pestivirus of giraffe	[AF144617]

GENUS HEPACIVIRUS

Type Species *Hepatitis C virus*

DISTINGUISHING FEATURES

Hepaciviruses are transmitted between humans, principally via exposure to contaminated blood or blood products. There is no known invertebrate vector. Hepaciviruses differ from the other *Flaviviridae* genera by their inability to be propagated efficiently in cell culture. In the hepacivirus precursor protein, the NS2-3 junction is autocatalytically cleaved by a Zn-dependent NS2-3 proteinase activity.

VIRION PROPERTIES

MORPHOLOGY

Virions are ~50 nm in diameter, as determined by filtration and electron microscopy. They are spherical in shape and contain a lipid envelope, as determined by electron microscopy and inactivation by chloroform. The viral core is spherical and ~30 nm in diameter. Detailed structural properties have not been determined.

PHYSICOCHEMICAL AND PHYSICAL PROPERTIES

Virion Mr has not been determined. Buoyant density in sucrose is ~1.06 g/cm^3 when the virus is recovered from the serum of an acute infection and ~1.15-1.18 g/cm^3 when recovered from the serum of a chronically infected patient. The lower density results from physical association of the virion with serum very-low-density lipoproteins. The higher density results from the binding of serum antibodies to the virion. A buoyant density in sucrose of 1.12 g/cm^3 has been measured for Hepatitis C virus (HCV) recovered from cell culture. The S_{20w} is equal to or greater than 150S. The virus is stable in buffer at pH 8.0-8.7. Virions are sensitive to heat, organic solvents and detergents.

NUCLEIC ACID

Virions contain a single positive-sense, infectious ssRNA (Fig. 5). The genome length is ~9.6 kb. The 5'-NCR contains an IRES and is 341 nt in length. The 3'-NCR contains a sequence-variable region of ~50 nt, a polypyrimidine-rich region averaging ~100 nt (but variable in length) and a highly conserved terminal region (98 nt).

PROTEINS

The virion consists of at least 3 proteins: the nucleocapsid core protein C (p19), and two envelope proteins, E1 (gp31) and E2 (gp70). One or more additional proteins, resulting from ribosomal frame-shifting within the C gene, have been reported although it is not known if these are part of the virion. An additional protein, p7 (believed to have properties of an ion channel protein), is incompletely cleaved from a precursor of E2 to yield E2-p7 and p7, but it is not known whether these are virion structural components. The two envelope proteins form heterodimers (probably non-covalently linked) in virions. The recognized nonstructural proteins include NS2, (21 kDa protein that, before cleavage, is part of a Zn-dependent proteinase that bridges NS2 and NS3 and mediates autocatalytic cleavage of the NS2/NS3 junction), NS3 (70 kDa protein with additional serine proteinase, helicase and NTPase acitivities; the NS3 proteinase cleaves the remaining junctions between nonstructural proteins), NS4A (6 kDa cofactor essential for NS3 serine proteinase activity), NS4B (27 kDa protein that induces a membranous replication complex at the endoplasmic reticulum), NS5A (a serine phosphoprotein of unknown function that exists in 56 and 58 kDa forms, depending on the degree of phosphorylation) and NS5B (68 kDa protein with RdRp activity).

Lipids

Lipids have not been demonstrated directly. However, based on observed removal of the viral envelope and loss of infectivity following exposure to solvents or detergents, the presence of lipids is inferred.

Carbohydrates

Carbohydrates have not been demonstrated directly but the presence of glycosylation sites in the predicted coding sequences of the E1 and E2 genes, and the demonstration of carbohydrate associated with the products of these two genes expressed as recombinant proteins is consistent with the presence of carbohydrates in virions. When expressed in transfected cells, the E1 and E2 gene products normally remain tightly anchored within the lumen of the endoplasmic reticulum and contain high-mannose chains lacking complex carbohydrate.

Genome Organization and Replication

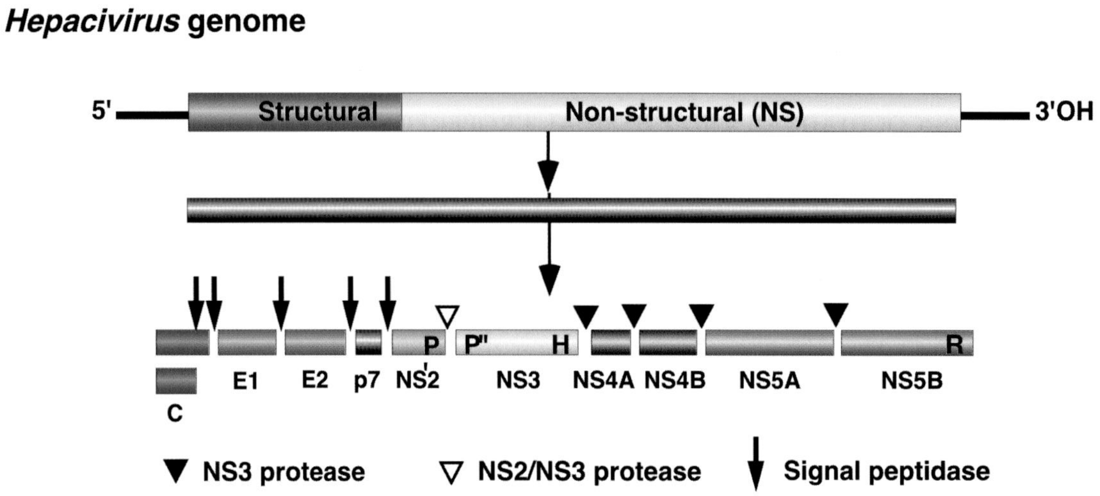

Figure 5: Hepacivirus genome organization (not to scale) and polyprotein processing. The RNA is ~9.6 kb in size. The 5'-NCR is 341 nt, the 3'-NCR is ~250 nt, and the ORF is ~9 kb. The proteinases involved in cleavage of the polyprotein are indicated. The structural proteins are shown in dark. The locations of the proteinases, helicase and RdRp are indicated by P', P'', H and R, respectively.

The genome contains a single large ORF encoding a polyprotein of ~3000 aa (Fig. 5). The gene order is 5′-C-E1-E2-p7-NS2-NS3-NS4A-NS4B-NS5A-NS5B-3′. All 3 structural proteins (C, E1, E2), are encoded within the amino-terminal portion of the large ORF. Immediately downstream is the small protein, p7. The non-structural proteins are encoded in the 3′ portion of the ORF. Replication is poorly understood but is believed to occur in association with intracytoplasmic membranes. Replicative forms of viral RNA have been detected in liver tissue. The genomic RNA is translated into a polyprotein that is rapidly processed both co- and post-translationally. Translation initiation occurs via an IRES within the 5′-NCR, which also contains several closely spaced AUGs. Translocation of the structural glycoproteins to the endoplasmic reticulum probably occurs via an internal signal sequence. Cleavage of the structural proteins is effected by host cell signal peptidases, and signal peptide peptidase. Viral proteinases cleave all non-structural protein junctions. Virus assembly is believed to occur by budding into vesicles from the endoplasmic reticulum.

Antigenic Properties

Virus-specific antibodies to recombinant-expressed structural proteins (C, E1 and E2) and non-structural proteins (principally NS3, NS4 and NS5) have been detected in individuals infected with HCV. Both linear and conformational epitopes are believed to be involved in the humoral immune response of the host to infection. Significant genetic heterogeneity throughout the genome is reflected in some serologic heterogeneity in the humoral immune response, especially to the product of the NS4 gene. The most extensive heterogeneity of HCV is found in the N-terminal 27 aa of E2 (hypervariable region 1; HVR-1). There is some evidence that HVR-1 is a neutralization epitope of HCV and that neutralization-escape variants of HVR-1 are positively selected by the humoral immune response of the host. Other neutralization epitopes may exist but have not been defined. Cell-mediated immune responses to all HCV proteins have been detected; it is believed that such responses are associated with amelioration or resolution of infection. Because there is no efficient or standardized cell culture system for the propagation of HCV, it has not been possible to carry out *in vitro* virus neutralization assays.

Biological Properties

Host Range

Humans are the natural host and apparent reservoir of HCV. The virus can be transmitted experimentally to chimpanzees. No other natural host has been identified.

Transmission

Hepatitis C virus is transmitted almost exclusively by parenteral exposure to blood, blood products and objects contaminated with blood. Effective screening of blood donors and implementation of inactivation procedures have virtually eliminated the transmission of HCV via blood and blood products, but other routes of exposure, principally via blood-contaminated syringes, are now the most important recognized risk factors in developed countries. Sexual and perinatal transmission have been reported but are relatively uncommon.

Geographical Distribution

HCV has a worldwide distribution. Antibody-based epidemiological studies suggest that ~0.1-2% of the populations of developed countries may be infected with HCV, but antibody prevalence as high as 20% has been detected in some developing countries. The high prevalence of antibody to HCV is thought to be the result of using contaminated needles and syringes in such countries. Overall, it has been estimated that about 3% of the world population has been infected with HCV, resulting in ~170 million chronically infected individuals.

Pathogenicity

Infections range from subclinical to clinical acute and chronic hepatitis, liver cirrhosis and hepatocellular carcinoma. Persistence of the virus occurs in ~80% of HCV infections. Of these, ~20% progress to chronic active hepatitis and cirrhosis, usually over the course of many years. Persistent HCV infection has been epidemiologically linked to primary liver cancer, cryptogenic cirrhosis, and some forms of auto-immune hepatitis. Extra-hepatic manifestations of HCV infection include mixed cryoglobulinemia with associated membranoproliferative glomerulonephritis and, possibly, porphyria cutanea tarda, Sjögren's-like syndromes and other autoimmune conditions.

Cell Tropism

HCV has been reported to replicate in several cell lines derived from hepatocytes and lymphocytes, but virus growth has not been sufficient for practical application of these systems. *In vivo*, HCV replicates in hepatocytes and possibly in lymphocytes.

List of Species Demarcation Criteria in the Genus

Hepatitis C virus can be classified into 6 genetic groups, based upon the genome-wide heterogeneity of isolates recovered throughout the world. These have been called HCV clades 1-6; clades differ from each other by ~25-35% at the nt level. Genotypes 7-11 have been described, but more extensive genetic analysis has placed genotypes 7, 8, 9 and 11 within clade 6 and genotype 10 within clade 3. The 6 clades have been further subdivided into over 100 subtypes. These differ from each other by ~15-25% at the nt level. Although the clades are more or less distinct, discrimination of subtypes is less clear, owing to overlap in the degree of heterogeneity. Because serotyping of HCV isolates is not possible at present, and because major genotypes do not have any other taxonomic characteristics except, in some cases, geographic distribution, the 6 genetic groups of HCV currently comprise one species.

List of Species in the Genus

A number of clades are recognized for *Hepatitis C virus*. Examples are listed below.

Species names are in green italic script; strain names and synonyms are in black roman script; tentative species names are in blue roman script. Sequence accession numbers, and assigned abbreviations ( ) are also listed.

Species in the Genus

Hepatitis C virus		
HCV clade 1		
HCV genotype 1a	[M62321]	(HCV-1)
HCV genotype 1b	[D90208]	(HCV-J)
HCV clade 2		
HCV genotype 2a	[D00944]	(HCV-J6)
HCV genotype 2b	[D01221]	(HCV-J8)
HCV clade 3		
HCV genotype 3a	[D17763]	(HCV-NZL1)
HCV genotype 10	[D63821]	(HCV-JK049)
HCV clade 4		
HCV genotype 4a	[Y11604]	(HCV-ED43)
HCV clade 5		
HCV genotype 5a	[Y13184]	(HCV-EVH1480)
HCV clade 6		
HCV genotype 6a	[Y12083]	(HCV-EUHK2)
HCV genotype 11	[D63822]	(HCV-JK046)

Tentative Species in the Genus

GB virus B	[U22304; AF179612]	(GBV-B)

Phylogenetically, GBV-B is related to but distinct from the hepatitis C viruses. Like the latter, it causes hepatitis and replicates in the liver. However, it only infects tamarins and owl monkeys, not humans or chimpanzees. Only one isolate of GBV-B has been identified to date, in contrast to the several hundred HCV isolates. While clearly related to HCV at the level of both nucleotide sequence and genetic organization, there is more sequence divergence evident between GBV-B and HCV (28% aa identity between the encoded polyproteins) than within the HCV species itself (> 60% aa identity). GBV-B causes self-limiting hepatitis in tamarins and owl monkeys, whereas HCV typically causes chronic hepatitis in man and chimpanzees.

List of Unassigned Viruses in the Family

Two distinct groups of viruses have been assigned tentatively to the family *Flaviviridae*, based upon their genomic organization and genetic similarity to recognized members of the family.

GB virus A	[U22303]	(GBV-A)
GB virus A-like agents	[U94421]	

DISTINGUISHING FEATURES

GBV-A and GBV-A-like agents are a group of related viruses that have been identified in at least 6 species of New World monkeys. They do not cause hepatitis in the unique host species of each virus nor in other susceptible species. Their organ site of replication has not been identified and, although the viruses are transmissible via blood, their natural route of transmission is unknown. These viruses share an overall genomic organization and distant genetic similarity with hepaciviruses, but differ in that they appear to lack a complete nucleocapsid protein gene and the organization of their 3'-NCR is less complex than that of the hepaciviruses.

GB virus C	[U36380, AF070476]	(GBV-C)
Hepatitis G virus - 1	[O44402]	(HGV-1)

DISTINGUISHING FEATURES

GBV-C is a genetically heterogeneous virus of human and chimpanzee origin. It is transmitted via blood and blood products and possibly sexually, but other routes of transmission may exist. Although originally described as a hepatitis virus, it rarely, if ever, causes hepatitis, and its pathogenicity and organ site of replication remain controversial. Lymphocytes may be its primary site of replication. Although distinct, GBV-C is most closely related to the GBV-A group of viruses, both in genomic organization and genetic relatedness.

PHYLOGENETIC RELATIONSHIPS WITHIN THE FAMILY

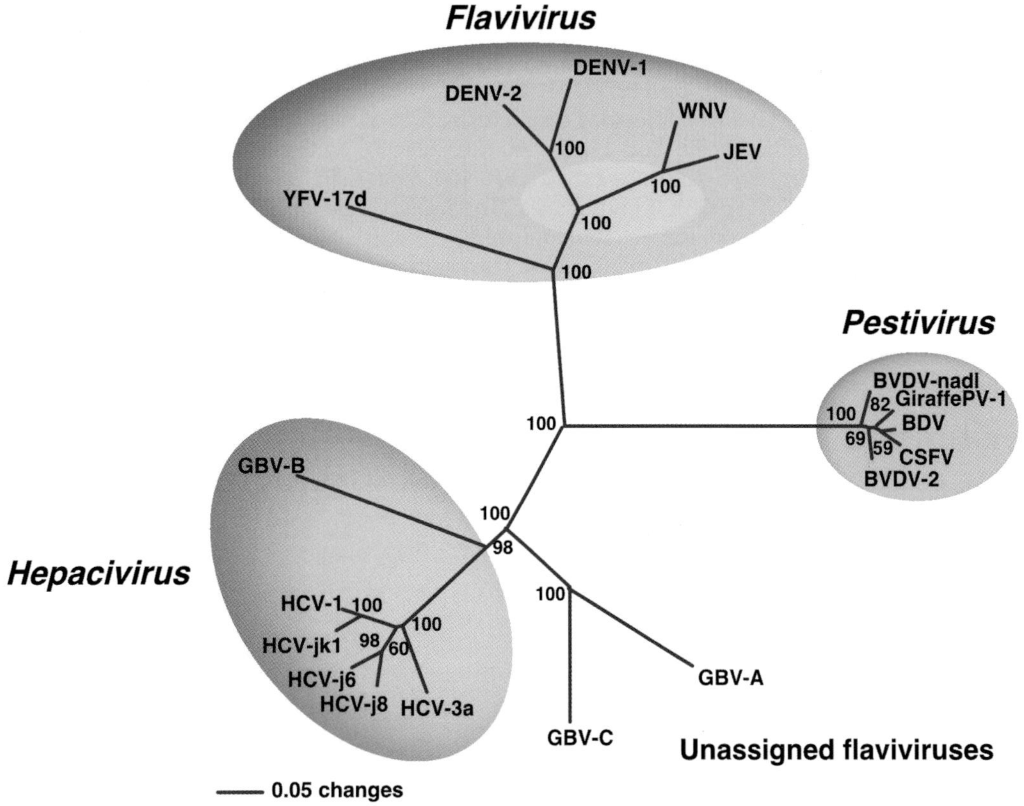

Figure 6: Phylogenetic relationship of the helicase region of members of the family *Flaviviridae*. Partial gene sequences (~280 aa) from the proposed helicase region were used for the phylogenetic analysis and included representative strains from each genus. CLUSTALX v1.81 was used to create a multiple alignment for the aa sequences which was verified by alignment of the known motifs in the region (e.g., GxGKS/T). An unrooted phylogenetic tree was constructed from the sequence alignment using the distance method, Neighbor-joining from the PAUP 4b10 package within a Macintosh environment. The virus names corresponding to the

abbreviations can be found in the "List of Species" in each genus and the Genbank accession numbers are AF037405, U18059, M31182, J04358, M87512, M29095, U22303, U22304, NC_002348, AF144617, M62321, D17763, D00944, D10988, X61596, M18370, M12294, NC_002031. Bootstrap replicates of 1000 trees was also examined using PAUP and the percentages are drawn at the branch points of the tree. (Courtesy of L. McMullan).

Similarity with Other Taxa

None reported.

Derivation of Names

Flavi: from Latin *flavus*, "yellow".
Hepaci: from Greek *hepar, hepatos*, "liver".
Pesti: from Latin *pestis*, "plague".

References

Avalos-Ramirez, R., Orlich,. M., Thiel., H.-J. and Becher, P. (2001). Evidence for the presence of two novel pestivirus species. *Virology*, **286**, 456-465.

Gould, E.A., de Lamballerie, X. de, Zanotto, P.M. and Holmes, E.C. (2001). Evolution, epidemiology and dispersal of flaviviruses revealed by molecular phylogenies. *Adv. Virus Res.*, **57,** 71-103.

Kuhn, R.J., Zhang, W., Rossmann, M.G., Pletnev, S.V., Corver, J., Lenches, E., Jones, C.T., Mukhopadhyay, S., Chipman, P.R., Strauss, E.G., Baker, T.S. and Strauss, J.H. (2002). Structure of Dengue virus: Implications for flavivirus organization, maturation, and fusion. *Cell*, **108**, 717-725.

Kuno, G., Chang, G.-J.J., Tsuchiya, K.R., Karabatsos, N. and Cropp, C.B. (1998). Phylogeny of the genus *Flavivirus*. *J. Virol.*, **72**, 73-83.

Lindenbach, B.D. and Rice, C.M. (2001). *Flaviviridae*: The viruses and their replication. In: *Fields Virology*, 4th ed., (D.M. Knipe, and P.M. Howley, eds), pp 991-1041. Lippincott Williams and Wilkins, Philadelphia.

Major, M.E., Rehermann, B. and Feinstone, S.M. (2001). Hepatitis C viruses. In: *Fields Virology*, 4th ed., (D.M. Knipe and P.M. Howley, eds), pp 1127-1161. Lippincott Williams and Wilkins, Philadephia.

Meyers, G. and Thiel, H.-J. (1996). Molecular characterization of pestiviruses. *Adv. Virus Res.*, **47**, 53-118.

Modis, Y., Ogata, S., Clements, D. and Harrison, S.C. (2003). A ligand-binding pocket in the dengue virus envelope glycoprotein. *Proc. Natl. Acad. Sci. USA*, **100**, 6986-6991.

Monath, T.P. and Heinz, F.X. (1996). Flaviviruses. In: *Fields Virology* (B.N. Fields, B.M. Knipe and P.M. Howley, eds), pp 961-1034. Lippincott-Raven, Philadelphia.

Rey, F.A., Heinz, F.X., Mandl, C., Kunz, C. and Harrison, S.C. (1995). The envelope glycoprotein from tick-borne encephalitis virus at 2 A resolution. *Nature*, **375**, 291-298.

Robertson, B. (2001). Viral hepatitis and primates: historical and molecular analysis of human and nonhuman primate hepatitis A, B, and the GB-related viruses. *J. Viral Hepatitis*, **8**, 233-242.

Simmonds, P. (2001). The origin and evolution of hepatitis viruses in humans. *J. Gen. Virol.*, **82**, 693-712.

Contributed By

Thiel, H.-J., Collett, M.S., Gould, E.A., Heinz, F.X., Houghton, M., Meyers, G., Purcell, R.H. and Rice, C.M.

FAMILY TOGAVIRIDAE

TAXONOMIC STRUCTURE OF THE FAMILY

Family *Togaviridae*
Genus *Alphavirus*
Genus *Rubivirus*

VIRION PROPERTIES

MORPHOLOGY

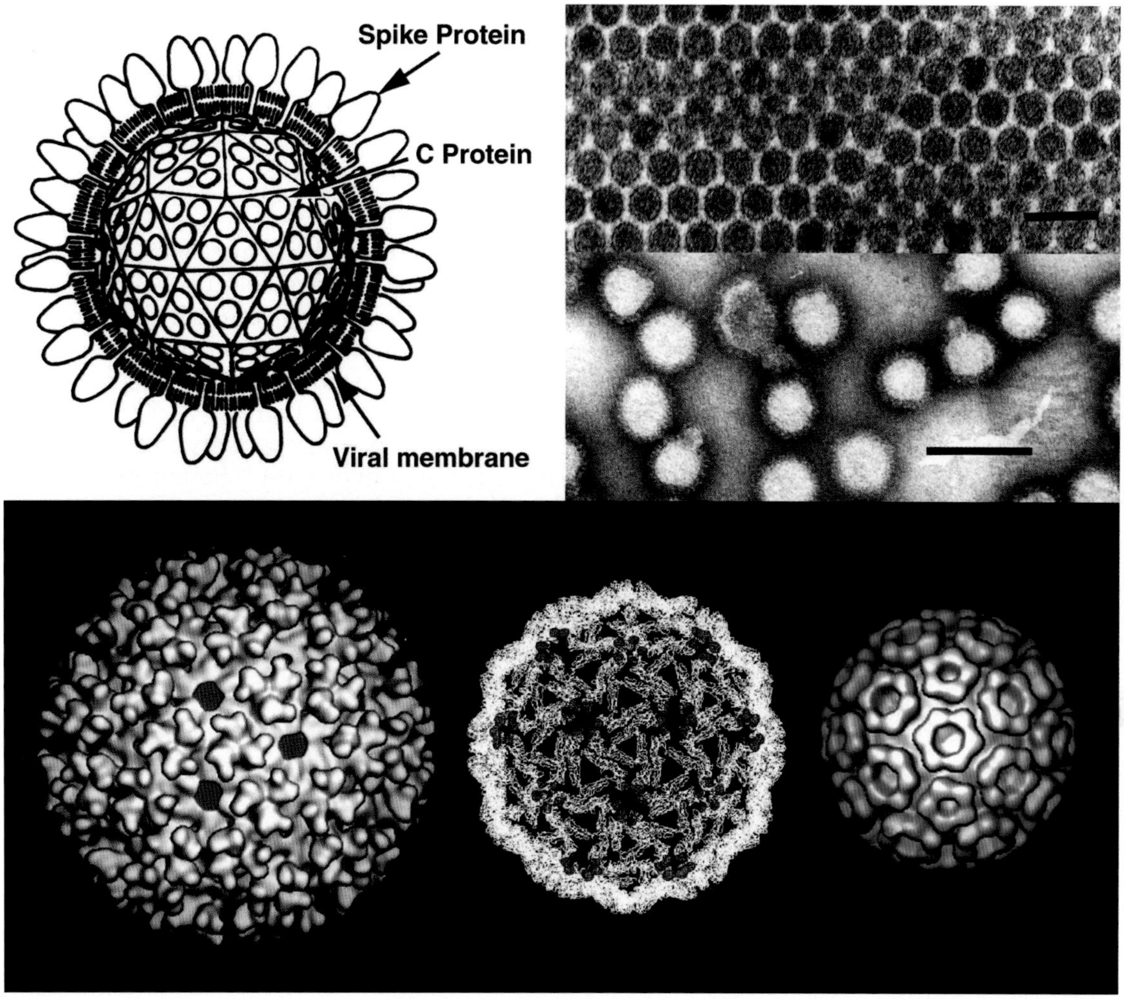

Figure 1: (Top panel) On the left a diagrammatic representation of Sindbis virus (SINV) particle. The knobs on the surface represent the external portions of the E1+E2 heterodimers. The heterodimers associate to form trimers. The 240 heterodimers and 240 copies of SINV capsid (C) proteins are arranged in an icosahedral lattice with a T=4 symmetry (from Harrison, 1990). On the upper right; thin section of pelletted particles of Semliki forest virus (SFV) (Courtesy of B.V.V. Prasad). On the lower right, negative contrast electron micrograph of particles of SFV (Courtesy of C.H. von Bonsdorff). The bars represent 100 nm. (Bottom panel) Structure of Sindbis virus particle. Left: Surface shaded view as determined by cryo-electronmicroscopy and image reconstruction. The view is looking down the icosahedral 3-fold axis. The image is calculated to 20Å resolution. Center: Surface view of SINV particle showing the organization of the E1 glycoprotein on the surface of the virus particle. The C-alpha backbone of E1 is shown in white. The view is identical to that shown in the left panel. Right: The image represents the nucleocapsid core showing the pentameric and hexameric capsomeres. 240 copies of the nucleocapsid protein together with the genome RNA form a T=4 icosahedron. (Courtesy of R. Kuhn).

Virions are 70 nm in diameter, spherical, with a lipid envelope containing heterodimeric glycoprotein spikes composed of two virus glycoproteins. For alphaviruses, the heterodimers are organized in a T=4 icosahedral lattice consisting of 80 trimers (Fig. 1). The envelope is tightly organized around an icosahedral nucleocapsid that is 40 nm in diameter. The nucleocapsid is composed of 240 copies of the CP, organized in a T=4 icosahedral symmetry, and the genomic RNA. Some virions such as those of Aura virus also encapsidate the 26S sgRNA. The Rubella virus (RUBV) core is reported to have T=3 symmetry, and is composed of genomic RNA and multiple copies of the CP in the form of homodimers. For alphaviruses, the one-to-one relation between glycoprotein heterodimers and nucleocapsid proteins is believed to be important in virus assembly. The E1 envelope glycoprotein, which as been solved structurally by crystallography, is the fusion protein for entry into the cytoplasm from acidic endosomes. The E2 envelope glycoprotein predominates in the viral spikes that extend outward from the envelope, and forms the petals of the spike that cover the underlying E1 protein fusion peptide at neutral pH. The three dimensional structure of RUBV virions has not been determined. RUBV virions are 60-70 nm in diameter and pleomorphic in nature, indicating that the capsid-glycoprotein interaction is not as tight as it is in virions of alphaviruses.

PHYSICOCHEMICAL AND PHYSICAL PROPERTIES

Virion Mr is about 52 x 10^6. Alphaviruses have a buoyant density in sucrose of 1.22 g/cm^3 and an S_{20w} of 280S. RUBV has a buoyant density of 1.18-1.19 g/cm^3 and a similar S value. Alphaviruses are stable between pH 7 and 8, but are rapidly inactivated by very acidic pH. Virions have a half-life at 37°C of about 7 hr in culture medium. Most alphaviruses are rapidly inactivated at 58°C with a half-life measured in minutes. RUBV virions are more heat labile than alphaviruses, with a half-life at 37°C of 1-2 hrs and a half-life at 58°C of 5-20 min. Generally, togaviruses are sensitive to organic solvents and detergents which solubilize their lipoprotein envelopes. Sensitivity to irradiation is directly proportional to the size of the viral genome.

NUCLEIC ACID

The genome consists of a linear, positive sense, ssRNA molecule 9.7-11.8 kb in size. The viral RNA is capped (7-methylguanosine) at the 5'-terminus and polyadenylated at the 3'-terminus.

PROTEINS

The structural proteins of togaviruses include a basic CP (30-33 kDa) and two envelope glycoproteins, E1 and E2 (45-58 kDa). Some alphaviruses may contain a third envelope protein, E3 (10 kDa). The four non-structural proteins, which are present in infected cells but not found in virions, are called nsP1-nsP4. Their functions are described below.

LIPIDS

Lipids comprise about 30% of the dry weight of virions. They are derived from the host-cell membrane from which budding occurs: the plasma membrane for alphaviruses, both intracellular membranes and the plasma membrane for RUBV. Their composition depends upon the cells in which the virus was grown. Phospholipids (including phosphatidyl ethanolamine, phosphatidyl choline, phosphatidyl serine, and sphingomyelin) and cholesterol are present in a molar ratio of about 2:1 for alphaviruses, 4:1 for RUBV, presumably because the latter matures primarily at intracellular membranes.

CARBOHYDRATES

Both high mannose and complex N-linked glycans are found on the envelope glycoproteins. In addition, RUBV E2 protein contains O-linked glycans.

GENOME ORGANIZATION AND REPLICATION

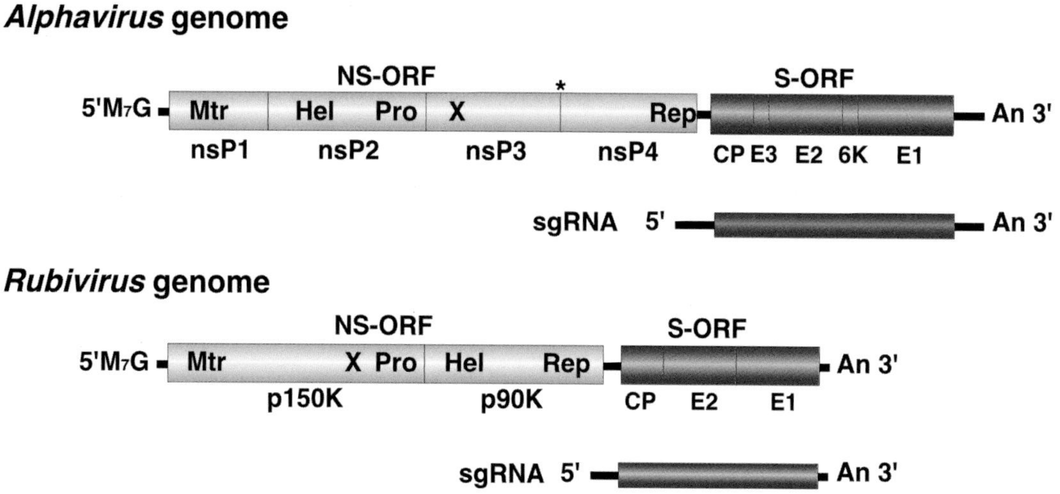

Figure 2: Togavirus genomic coding strategies. Shown are comparative schematic representations of the alphavirus and rubivirus genomic RNAs with untranslated regions represented as solid black lines and ORFs as open boxes (NS-ORF = non-structural protein ORF; S-ORF = structural protein ORF). Within each ORF, the coding sequences for the proteins processed from the translation product of the ORF are delineated. The asterisk between nsP3 and nsP4 in the alphavirus NS-ORF indicates the stop codon present in some alphaviruses that must be translationally read through to produce a precursor containing nsP4. Additionally, within the NS-ORFs, the locations of motifs associated with the following activities are indicated: (Mtr) methyl transferase, (Pro) protease, (Hel) helicase, (X) unknown function, and (Rep) replicase. The sequences encompassed by the sgRNA are also shown. -(Courtesy of T. K. Frey).

The genomic RNA serves as the mRNA for the non-structural proteins of the virus (Fig. 2). In alphaviruses, the polyprotein precursor is cleaved by a viral-encoded protease in nsP2 to produce four final products, nsP1, nsP2, nsP3 and nsP4 (Fig. 3). In eight of ten alphaviruses sequenced, there is a termination codon (UGA) between nsP3 and nsP4 genes which is read-through with moderate efficiency (5-20%), whereas in the two other alphaviruses this codon has been replaced by a codon for arginine (CGA). Polyproteins containing nsP2 are enzymes and function primarily *in trans* to produce the cleaved non-structural proteins. In RUBV the polyprotein precursor is cleaved into two products, P150 and P90. The protease mediating this cleavage is located near the C-terminus of P150, and can function either *in cis* or *in trans*.

The non-structural proteins, as individual entities and as polyproteins, are required to replicate viral RNA and probably act in association with cellular proteins. The alphavirus nsP1 protein is thought to be involved in capping of viral RNAs and in initiation of negative-strand RNA synthesis. The nsP2 functions as a protease to process the non-structural proteins and is believed to be a helicase required for RNA replication. Protein nsP4 is believed to be the viral RNA polymerase. Protein nsP3 is also required for RNA replication; P123 (the uncleaved precursor of nsP1, nsP2 and nsP3) and nsP4 form the replicase complex for minus strand synthesis, whereas efficient plus-strand synthesis requires cleavage of P123. In RUBV, P150 contains a MTR motif of unknown function that is also present in nsP3 of alphaviruses. RUBV P90 contains both helicase and replicase motifs. These motifs are in a different order than in the alphavirus genome. The difference in processing and order of non-structural protein motifs between alphaviruses and RUBV suggests that evolution of these two genera was more complicated than simple divergence from a common ancestor. In RNA replication, a negative-strand copy is produced that is used as template in the synthesis of both genome-sized RNA as well as a subgenomic 26S mRNA that corresponds to the 3' third of the viral genome and encodes the viral structural proteins.

POS. SSRNA

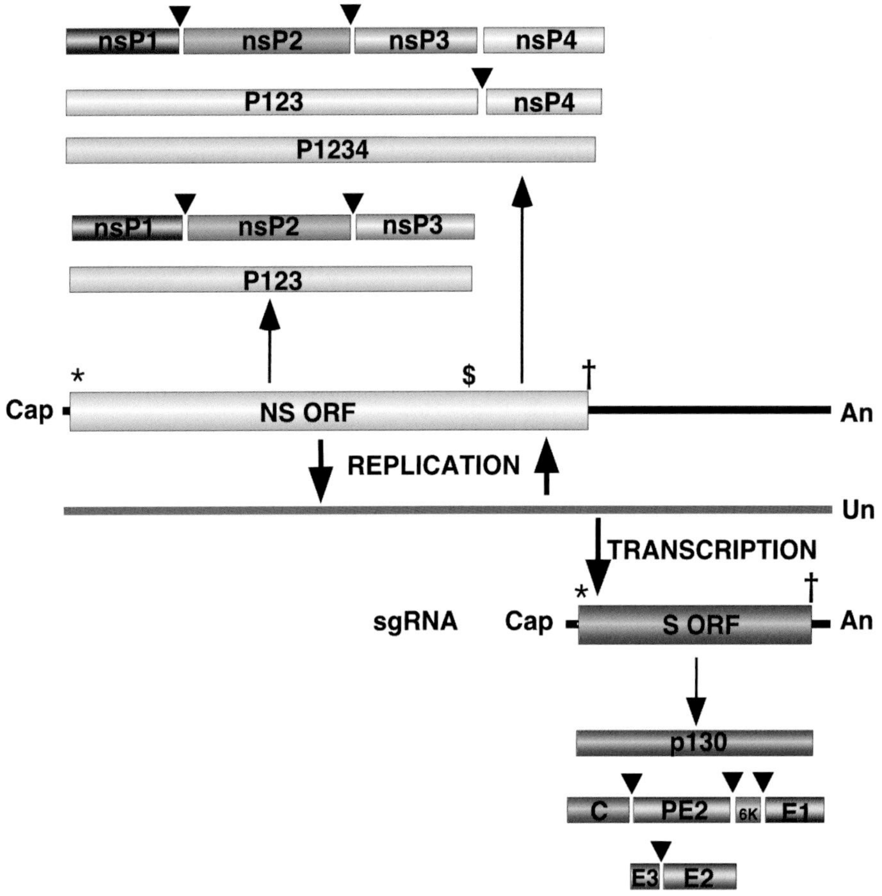

Figure 3: Genome organization, translation, transcription and replication strategies of Sindbis virus (SINV). The regions of the 11.7 kb genomic RNA and 26S sgRNA (dark lines) that code respectively for the non-structural (nsP) and structural proteins (colored boxes) are shown. Replication and transcription are indicated by thick arrows. The grey line is the replicative intermediate that is also the template for the sgRNA. E3 is a structural protein in some alphaviruses (not present in Rubella virus (RUBV). Initiation codons are indicated by (*), termination codons by (†) and ($)(the latter is readthrough to produce P1234, hence nsP4 is cleaved off). Dark triangles represent nsP2 protease activity. (From Strauss and Strauss, 1994).

The mRNA is capped and polyadenylated. It is translated as a polyprotein, which is processed in alphaviruses by a combination of an autoprotease activity present in the CP and cellular organelle-bound proteases, to produce the viral structural proteins. The RUBV CP lacks autoprotease activity, and all of the cleavages of this precursor are mediated by a cellular signal endopeptidase.

Cis-acting regulatory elements in the 5′- and 3′-non-translated regions of the genomic RNA are required to produce alphavirus minus strands and to copy the minus strand into plus strands. There are believed to be other *cis*-acting regulatory elements within the viral RNA as well. For alphaviruses, the promoter for the production of the 26S sgRNA is a stretch of 24 nt that span the start point of the sgRNA. This minimal 24 nt sequence element is upregulated by upstream sequences. The RUBV subgenomic promoter is 50 nt upstream from the sgRNA start site. RUBV also contains *cis*-acting untranslated sequences preceding each ORF that are believed to form stem-loops and to regulate translation and RNA replication. RUBV and alphaviruses share homology in the *cis*-acting elements at the 5′-end of the genome and subgenomic promoter region.

The non-structural proteins function in the cytoplasm of infected cells in association with the surface of membranes, and attachment appears to be mediated by nsP1 palmitoylation. Some alphavirus nsP2 is translocated into the nucleus. The CP assembles with the viral RNA to form the viral nucleocapsids in the cytosol. Glycoproteins inserted into the endoplasmic reticulum during translation are translocated via the Golgi apparatus to the plasma membrane for alphaviruses; for RUBV they are also found at intracellular membranes. Assembled nucleocapsids bud through these membranes and acquire a lipid envelope containing the two integral membrane glycoproteins. For RUBV, the glycoproteins are retained in the Golgi apparatus, the preferred site of budding. Unlike in alphaviruses, rubellavirus capsids are not pre-assembled in the cytosol and only form during the budding process. Late in infection, the rubellavirus glycoproteins also accumulate at the plasma membrane, and budding also occurs at this site.

ANTIGENIC PROPERTIES

Member viruses of the genus *Alphavirus* were originally defined on the basis of serological cross-reactions. Thus, all alphaviruses are antigenically related to each other. They share a minimum aa sequence identity of about 40% in the more divergent structural proteins and about 60% in the non-structural proteins. Alphaviruses can be grouped into 8 antigenic complexes based on serologic cross-reactivity: the eastern, Venezuelan and western equine encephalitis, Trocara, Middelburg, Ndumu, Semliki Forest and Barmah Forest complexes. The non-arthropod-borne alphaviruses, Salmon pancretic disease and southern elephant seal viruses, are also antigenically distinct from the remaining members of the genus. RUBV is serologically distinct from alphaviruses and no structural protein aa sequence homology can be detected between RUBV and the alphaviruses.

BIOLOGICAL PROPERTIES

Most alphaviruses are transmitted biologically between vertebrates by mosquitoes or other hematophagous arthropods. However, salmon pancreatic disease virus is not known to have an arthropod vector. Alphaviruses have a wide host range and nearly worldwide distribution. The infection of cells of vertebrate origin is generally cytolytic and involves the shutdown of hostcell macromolecular synthesis. In mosquito cells, most alphaviruses usually establish a non-cytolytic infection in which the cells survive and become persistently infected. Cytopathology has been described in the midguts of mosquitoes infected with Eastern equine encephalitis virus (EEEV) and Western equine encephalitis virus (WEEV). In contrast, humans are the only known host for RUBV, which is spread via the respiratory route. RUBV replicates in a number of mammalian cell culture lines, including lines from humans, monkeys, rabbits and hamsters. The virus is not cytopathic in most of these lines and has a propensity to initiate persistent infections.

GENUS ALPHAVIRUS

Type Species *Sindbis virus*

DISTINGUISHING FEATURES

Genomes are 11-12 kb in size, exclusive of the 3'-terminal poly(A) tract: Sindbis virus (SINV), 11,703 nt; O'nyong-nyong virus (ONNV), 11,835 nt; Ross River virus (RRV), 11,851 nt; VEEV, 11,444 nt; Semliki Forest virus (SFV) 11,442 nt; RNA S_{20w} about 49S. The order of the genes for the non-structural proteins in the genomic RNA is nsP1, nsP2, nsP3, nsP4 (Fig. 2). These are made as polyprotein precursors and processed by the nsP2 protease (Fig. 3). The gene order in the 26S mRNA is CP-E3-E2-6K-E1. The derived polyprotein is processed by an auto-proteolytic activity in the CP, by cellular signal peptidase, and by an enzyme thought to be a component of the Golgi apparatus (Fig. 3). Glycoprotein E2 is produced as a precursor, PE2 (otherwise called p62), that is cleaved during virus maturation. For some viruses the N-terminal cleavage product of PE2, referred to as E3 (~10 kDa), remains associated with the virion. Carbohydrates comprise about 14% of the mass of the envelope glycoproteins and about 5% of the mass of the alphavirus virion.

Alphaviruses possess the ability to replicate in and be transmitted horizontally by mosquitoes. Some viruses have a preferred mosquito vector; however, as a group these viruses use a wide range of mosquitoes. Fort Morgan virus (FMV) is transmitted by arthropods of the family Cimicidae (Order Hemiptera) associated with birds. Most alphaviruses can infect a wide range of vertebrates. Many alphaviruses have different species of birds as their primary vertebrate reservoir host, but most are able to replicate in mammals as well. A number of alphaviruses have mammals as their primary vertebrate reservoir host. Some of these, such as RRV, replicate poorly in birds. Alphavirus isolations from reptiles and amphibians have also been reported. Southern elephant seal virus replicates in several mammalian cell lines, and salmon pancreatic disease virus has only been propagated in salmonid cell lines. As a group, the alphaviruses are found on all continents and on many islands. However, most viruses have a more limited distribution. Sindbis virus (SINV), the type species virus, has been isolated from many regions of Europe, Africa, Asia, the Philippines and Australia. WEEV is distributed discontinuously from Canada to Argentina. At the other extreme, ONNV has been isolated only from East Africa where it caused epidemics in the years 1959-60 and 1996-97. Many Old World alphaviruses cause serious, but not life threatening illnesses that are characterized by fever, rash and a painful arthralgia. RRV, Mayaro virus (MAYV), and the Ockelbo subtype of SINV cause epidemic polyarthritis in humans with symptoms (in a minority of cases) that may persist for months, or years. The New World alphaviruses, EEEV, VEEV and WEEV, regularly cause fatal encephalitis in humans, although the fraction of infections that leads to clinical disease is small. EEEV, WEEV, and VEEV also cause encephalitis in horses, and EEEV causes encephalitis in pheasants, emus, pigs and other domestic animals. Highlands J virus (HJV) is generally not believed to be pathogenic for humans or horses, but is recognized as an important pathogen of turkeys, pheasants, chukar partridges, ducks, emus, and whooping cranes.

LIST OF SPECIES DEMARCATION CRITERIA IN THE GENUS

Not available.

LIST OF SPECIES IN THE GENUS

Species names are in green italic script; strain names and synonyms are in black roman script; tentative species names are in blue roman script. Sequence accession numbers, and assigned abbreviations ( ) are also listed.

SPECIES IN THE GENUS

Aura virus		
Aura virus	[S78478]	(AURAV)
Barmah Forest virus		
Barmah Forest virus	[U73745]	(BFV)
Bebaru virus		
Bebaru virus	[U94595]	(BEBV)
Cabassou virus		
Cabassou virus	[AF075259, U94611]	(CABV)
Chikungunya virus		
Chikungunya virus	[L37661, U94597]	(CHIKV)
Eastern equine encephalitis virus		
Eastern equine encephalitis virus	[D01034]	(EEEV)
Everglades virus		
Everglades virus	[AF075251, U94608]	(EVEV)
Fort Morgan virus		
Buggy Creek virus	[U94607, U60403, U60395]	
Fort Morgan virus	[U60399, U60404]	(FMV)

Getah virus		
Getah virus	[U94568]	(GETV)
Highlands J virus		
Highlands J virus	[J02206, U60401, U94609]	(HJV)
Mayaro virus		
Mayaro virus	[U94602]	(MAYV)
Middelburg virus		
Middelburg virus	[J02246, U94599]	(MIDV)
Mosso das Pedras virus ‡		
Mosso das Pedras virus (78V3531)	[AF075257]	(MDPV)
Mucambo virus		
Mucambo virus	[AF075253, U94615]	(MUCV)
Ndumu virus		
Ndumu virus	[U94600]	(NDUV)
O'nyong-nyong virus		
O'nyong-nyong virus	[M33999]	(ONNV)
Pixuna virus		
Pixuna virus	[AF075256, U94613]	(PIXV)
Rio Negro virus		
Rio Negro virus (strain Ag80-663)	[AF075258, U94610]	(RNV)
Ross River virus		
Ross River virus	[M20162]	(RRV)
Sagiyama virus	[U94601]	
Salmon pancreas disease virus		
Salmon pancreas disease virus	[AJ012631]	(SPDV)
Sleeping disease virus	[AJ238578]	
Semliki Forest virus		
Semliki Forest virus	[X04129]	(SFV)
Sindbis virus		
Babanki virus	[U94604, U60394, U60400]	
Kyzylagach virus	[U94605, U60396, U60402]	
Ockelbo virus	[M69205]	
Sindbis virus	[V00073]	(SINV)
Southern elephant seal virus ‡		
Southern elephant seal virus	[AF315122]	(SESV)
Tonate virus ‡		
Tonate virus	[AF075254]	(TONV)
Trocara virus		
Trocara virus	[AF252265]	(TROV)
Una virus		
Una virus	[U94603]	(UNAV)
Venezuelan equine encephalitis virus		
Venezuelan equine encephalitis virus	[X04368]	(VEEV)
Western equine encephalitis virus		
Western equine encephalitis virus	[J03854, U01065, AF109297]	(WEEV)
Whataroa virus		
Whataroa virus	[U94606, U60398, U60408]	(WHAV)

TENTATIVE SPECIES IN THE GENUS

None reported.

GENUS *RUBIVIRUS*

Type Species *Rubella virus*

DISTINGUISHING FEATURES

The genome is 9,757 nt in size exclusive of the 3'-terminal poly(A) tract (RNA S_{20W} of 40S). The order of the genes in the nonstructural protein ORF is NH_2-P150-P90-COOH. The non-structural protein precursor is cleaved by a papain-like cysteine protease located near the C-terminus of P150. The virion has a CP (34 kDa) and two envelope glycoproteins (E1, 59 kDa; E2, 44-50 kDa), but no equivalent of E3 or the 6K protein of the alphaviruses. The order of the proteins in the structural polyprotein precursor is NH_2-CP-E2-E1-COOH. The two cleavages that separate these three structural proteins are effected by signal peptidase (the E2 signal sequence remains attached to CP). Carbohydrates make up 10% of the mass of E1 and 30-40% of E2. E2 is heterogeneous in size due to differential processing of glycans (N- and O-linked). RUBV is transmitted by aerosol. The illness, rubella or German measles, is generally benign in nature, but complications such as arthritis, thrombocytopenia purpura, and encephalitis can occur. Intrauterine transmission is the most serious consequence of rubella as viral infection of the fetus during the first trimester of pregnancy leads to a constellation of serious birth defects known as congenital rubella syndrome (CRS). In addition, CRS patients suffer a variety of autoimmune and psychiatric disorders in later life, including a fatal neurodegenerative disease known as progressive rubella panencephalitis (which has also been described in post-natally infected individuals). CRS infants shed virus for up to six months. RUBV is endemic worldwide, although it is controlled by vaccination in most developed countries. Only one serotype has been described, although two genotypes have been detected.

LIST OF SPECIES DEMARCATION CRITERIA IN THE GENUS

Not appropriate.

LIST OF SPECIES IN THE GENUS

Species names are in green italic script; strain names and synonyms are in black roman script; tentative species names are in blue roman script. Sequence accession numbers, and assigned abbreviations ( ) are also listed.

SPECIES IN THE GENUS

Rubella virus

Rubella virus	[M15240, X05259, X72393, L78917]	(RUBV)

TENTATIVE SPECIES IN THE GENUS

None reported.

LIST OF UNASSIGNED VIRUSES IN THE FAMILY

Triniti virus	(TRIV)

PHYLOGENETIC RELATIONSHIPS WITHIN THE FAMILY

Amino acid sequence homology between RUBV and the alphaviruses has been detected only within the nonstructural proteins (see Fig. 2). Nucleotide and aa sequence homology has been demonstrated among all alphaviruses sequenced, and phylogenetic relationships have been estimated using the nsP1, nsP4 and E1 genes. The 8 antigenic complexes (see above) are supported as distinct monophyletic groups by these analyses (Fig. 4).

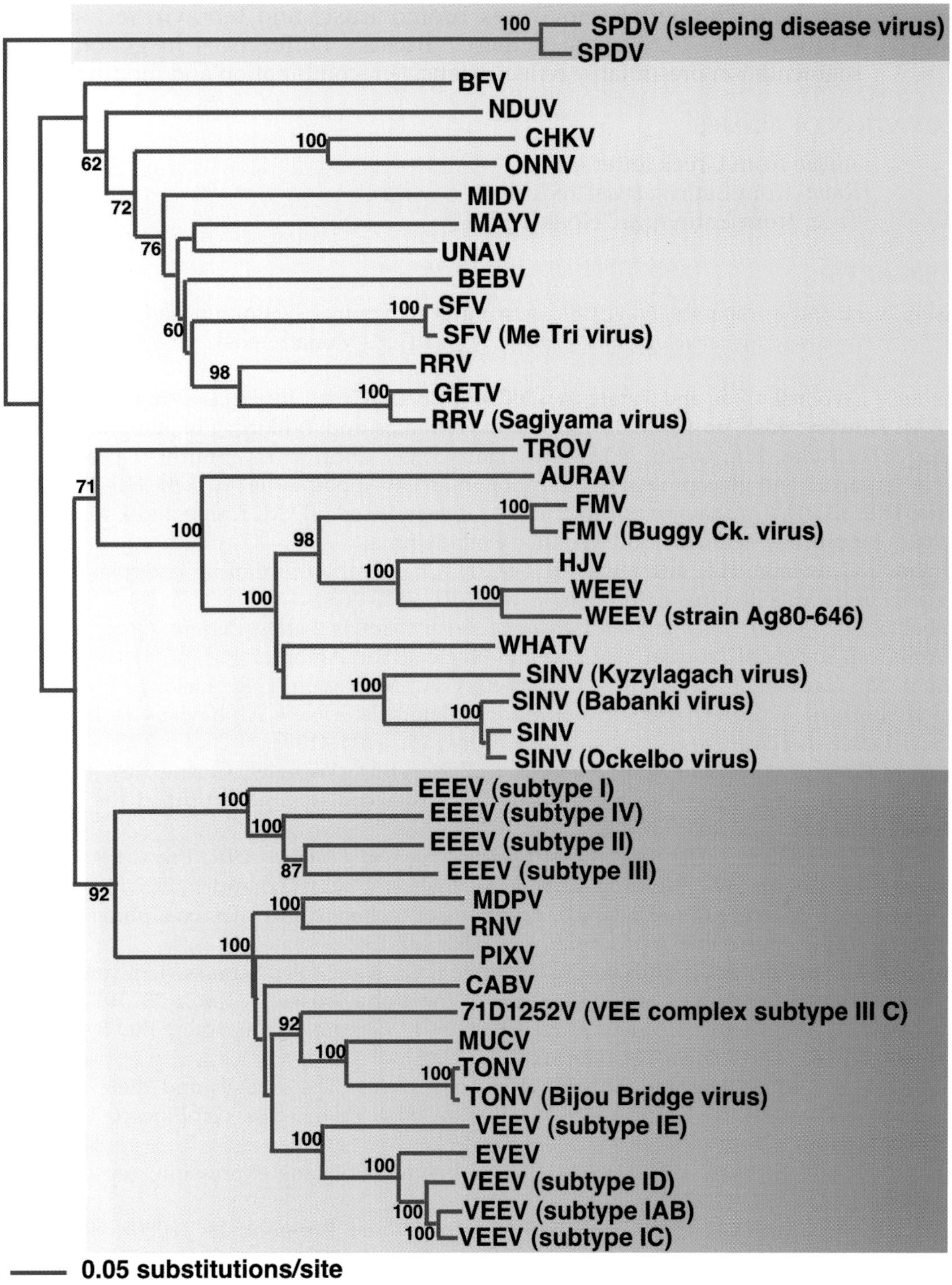

Figure 4: Phylogenetic tree of all alphavirus species except *Southern elephant seal virus* (the homologous sequence region is not available), and selected subtypes and variants, generated from partial E1 envelope glycoprotein gene sequences using the neighbor joining program with the F84 distance formula. Rubella virus cannot be included in this analysis because there is no detectable primary sequence homology in comparisons with alphavirus structural protein sequences. Bootstrap values for 100 replicates are indicated.

Similarity with Other Taxa

Togavirus non-structural proteins (alphavirus nsP1, nsP2, and nsP4 and the methyl transferase, helicase and replicase regions of RUBV) share some sequence homology with the non-structural proteins of Hepatitis E virus (*Hepevirus*) and several groups of plant viruses, including tobamoviruses, bromoviruses and tobraviruses, suggesting a common origin for the replicases of these viruses. Differences in genome organization and segmentation presumably reflect extensive recombination and modular evolution.

Derivation of Names

Alpha: from Greek letter α.
Rubi: from Latin *rubeus* "reddish".
Toga: from Latin *toga* "cloak".

References

Calisher, C.H. and Karabatsos, N. (1988). Arbovirus serogroups: Definition and geographic distribution. In: *The Arboviruses: Epidemiology and Ecology,* Vol. I (T.P. Monath, eds), pp 19-57. CRC Press, Boca Raton, Florida.

Chantler, J., Wolinsky, J.S. and Tingle, A. (2001). Rubella virus. In: *Fields' Virology*, 4th ed., (D.M. Knipe and P.M. Howley, eds), pp 963-990. Lippincott, Williams and Wilkins, Philadelphia.

Cheng, R.H., Kuhn, R.J., Olson, N.H., Rossmann, M.G., Choi, H.K., Smith, T.J. and Baker, T.S. (1995). Nucleocapsid and glycoprotein organization in an enveloped virus. *Cell*, **80**, 621-30.

Griffin, D.E. (2001). Alphaviruses. In: *Fields' Virology*, 4th ed., (D.M. Knipe and P.M. Howley, eds), pp 917-962. Lippincott, Williams and Wilkins, Philadelphia.

Hobman, T.C., Lemon, H.F. and Jewell, K. (1997). Characterization of an endoplasmic reticulum retention signal in the rubella virus E1 protein. *J. Virol.*, **71**, 7670-7680.

Karabatsos, N. (1985). *International Catalog of Arboviruses Including Certain Other Viruses of Vertebrates.* American Society of Tropical Medicine and Hygiene, San Antonio.

La Linn, M., Gardner, J., Warrilow, D., Darnell, G.A., McMahon, C.R., Field, I., Hyatt, A.D., Slade, R.W. and Suhrbier, A. (2001). Arbovirus of Marine Mammals: a New Alphavirus Isolated from the Elephant Seal Louse, *Lepidophthirus macrorhini*. *J. Virol.*, **75,** 4103-4109.

Lescar, J., Roussel, A., Wien, M.W., Navaza, J., Fuller, S.D., Wengler, G. and Rey, F.A. (2001). The fusion glycoprotein shell of Semliki Forest virus: an icosahedral assembly primed for fusogenic activation at endosomal pH. *Cell*, **105,** 137-148.

Monath, T.P. (ed)(1988). *The Arboviruses: Epidemiology and Ecology.* CRC Press, Boca Raton, Florida.

Pletnev, S.V., Zhang, W., Mukhopadhyay, S., Fisher, B.R., Hernandez, R., Brown, D.T., Baker, T.S., Rossmann, M.G. and Kuhn, R.J. (2001). Locations of carbohydrate sites on alphavirus glycoproteins show that E1 forms an icosahedral scaffold. *Cell*, **105,** 127-136.

Powers, A.M., Brault, A.C., Shirako, Y., Strauss, E.G., Kang, W., Strauss, J.H. and Weaver, S.C. (2001). Evolutionary relationships and systematics of the alphaviruses. *J. Virol*., **75,** 10118-10131.

Pugachev, K.V., Abernathy, E.S. and Frey, T.K. (1997). Genomic sequence of the RA27/3 vaccine strain of rubella virus. *Arch. Virol.*, **142**, 1165-1180.

Schlesinger, S. and Schlesinger, M.J. (2001). *Togaviridae*: The viruses and their replication. In: *Fields' Virology*, 4th ed., (D.M. Knipe and P.M. Howley, eds), pp 895-916. Lippincott, Williams and Wilkins, Philadelphia.

Strauss, J.H. and Strauss, E.G. (1994). The alphaviruses: gene expression, replication, and evolution. *Microbiol. Rev.*, **58**, 491-562.

Villoing, S., Bearzotti, M., Chilmonczyk, S., Castric, J. and Bremont, M. (2000). Rainbow trout sleeping disease virus is an atypical alphavirus. *J. Virol.*, **74,** 173-183.

Weaver, S.C., Kang, W., Shirako, Y., Rumenapf, T., Strauss, E.G. and Strauss, J.H. (1997). Recombinational history and molecular evolution of western equine encephalomyelitis complex alphaviruses. *J. Virol.*, **71**, 613-623.

Zhang, W., Mukhopadhyay, S., Pletnev, S. V., Baker, T.S., Kuhn, R.J. and Rossmann, M.G. (2002). Placement of the structural proteins in Sindbis virus. *J. Virol*., **76**, 11645-11658.

Contributed By

Weaver, S.C., Frey, T. K., Huang, H. V., Kinney, R.M., Rice, C.M., Roehrig, J.T., Shope, R.E. and Strauss, E.G.

GENUS *TOBAMOVIRUS*

Type Species *Tobacco mosaic virus*

VIRION PROPERTIES

MORPHOLOGY

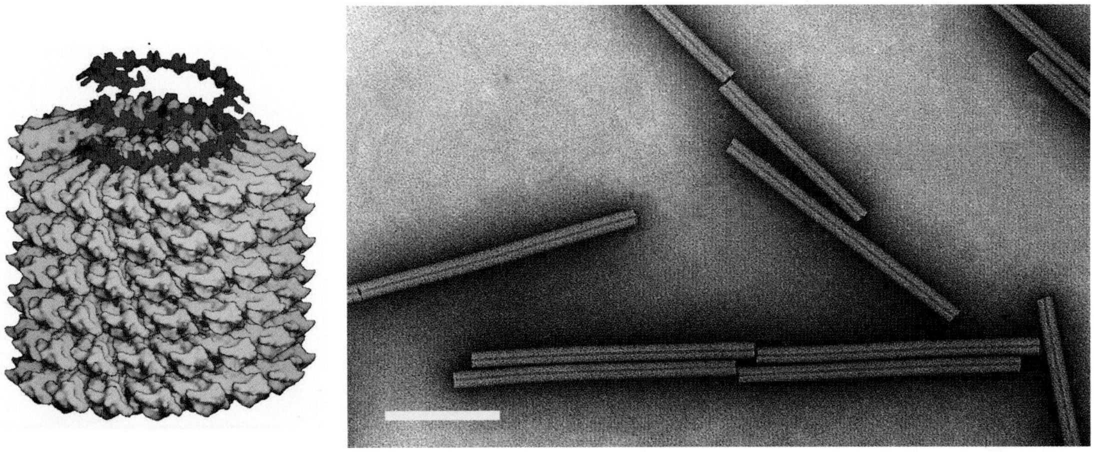

Figure 1: (Left) Model of particle of Tobacco mosaic virus (TMV). Also shown is the RNA as it is thought to participate in the assembly process. (Right) Negative contrast electron micrograph of TMV particle stained with uranyl acetate. The bar represents 100 nm.

Virions are elongated rigid cylinders, approximately 18 nm in diameter with a central hollow cavity and helical symmetry (pitch 2.3 nm) (Fig. 1). The predominant virion has a length of 300-310 nm, and contains the genomic RNA. Shorter virions produced by the encapsidation of sgRNA are a minor component of the virion population, although at least two species produce an abundant short virion 32-34 nm in length.

PHYSICOCHEMICAL AND PHYSICAL PROPERTIES

Virion Mr is 40 x 10^6. Buoyant density in CsCl is 1.325 g/cm^3. S_{20w} is 194S. Virions are very stable.

NUCLEIC ACID

The genome consists of one molecule of linear positive sense ssRNA 6.3-6.6 kb in size. An m^7GpppGp cap structure is found at the 5'-terminus of the genomic RNA, followed by an approximately 70 nt long 5'-untranslated sequence, containing many AAC repeats and few or no G residues. The 0.2-0.4 kb 3'-untranslated region contains sequences that can be folded into pseudoknots followed by 3'-terminal sequences that can be folded into a tRNA-like, amino acid-accepting structure. SgRNAs also contain a 5′-terminal cap and 3′-tRNA-like structure. The origin of assembly for encapsidation is located within the ORF for the MP in most species, but within the ORF for the CP in at least two species: *Cucumber green mottle mosaic virus* and *Sunn hemp mosaic virus*.

PROTEINS

Virions contain a single structural protein (17-18 kDa). Two nonstructural proteins are expressed directly from the genomic RNA: a 124-132 kDa protein terminated by an amber stop codon and a 181-189 kDa protein produced by readthrough of this stop codon, both of which are required for efficient replication. A third nonstructural protein (28-31 kDa) is required for cell-to-cell and long-distance movement. The MP is associated with plasmodesmata and has single-stranded nucleic acid binding activity *in vitro*. The CP is not required for cell-to-cell movement, but has a role in vascular tissue dependent virus accumulation. The replication proteins have also been implicated in virus movement. The MP and CP are expressed from individual 3′-co-terminal sgRNAs. The MP is expressed early during infection, whereas the CP is expressed later, and at higher levels. The MP and CP are not required for replication in single cells. The N-terminal one third of the 124-132 kDa protein has similarity with methyltransferase / guanylyl transferases whereas the C-terminal one third of the 124-132 kDa protein has similarity with RNA helicases (including an NTP-binding motif). The readthrough domain of the 181-189 kDa protein has motifs common to RdRps.

Lipids

Virions contain no lipids.

Carbohydrates

Virions contain no carbohydrates.

Genome Organization and Replication

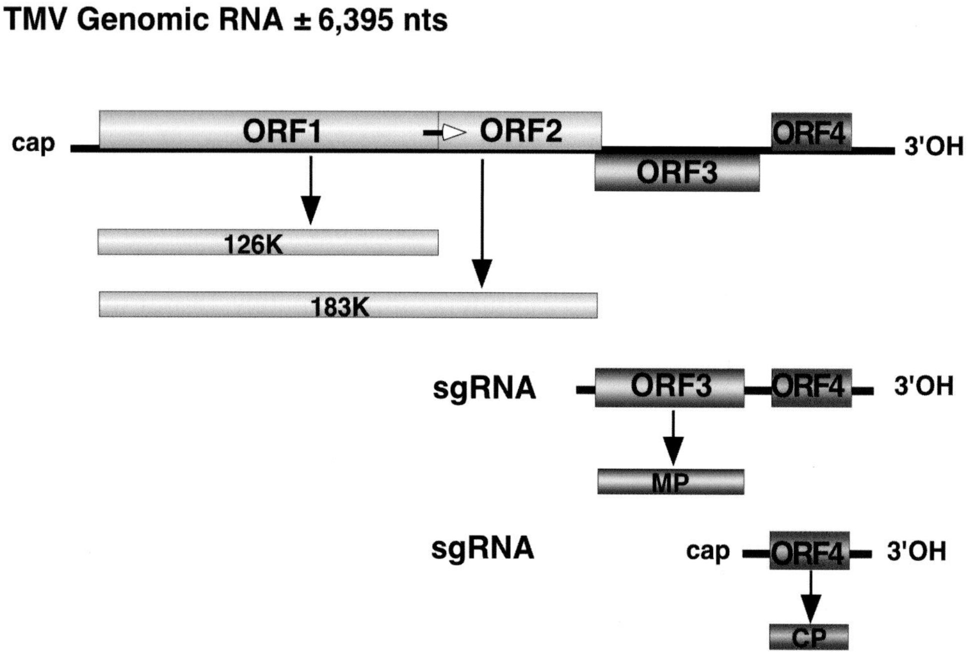

Figure 2: Genome organization of Tobacco mosaic virus (TMV). Conserved replicase domains are indicated as shaded boxes. Genomic RNA is capped and is template for expression of the 126 and 183 kDa proteins. The 3' distal movement and CP ORFs are expressed from individual 3' co-terminal sgRNAs. CP = coat protein; MP = movement protein.

The single genomic RNA encodes at least 4 proteins. The 124-132 kDa and 181-189 kDa replication proteins are translated directly from the genomic RNA. The 124-132 kDa replication protein contains the Mtr and Hel domains. The 181-189 kDa replication protein that also contains the polymerase domain is synthesized by occasional readthrough of the leaky termination codon of the 124-132 kDa ORF. The 181-189 kDa replication protein is the only protein required for replication in single cells, although the 124-132 kDa replication protein is also required for efficient replication. The next ORFs encode the 28-31 kDa MP and 17-18 kDa CP in 5' to 3' order. MP and CP are translated from their respective 3' co-terminal sgRNAs, both of which contain a 5'-cap (Fig. 2). In some species, the MP ORF overlaps both of the 181-189 kDa protein and CP ORFs, and in others does not overlap either ORF or overlaps one of the ORFs. Replication is cytoplasmic where the positive-sense genomic RNA is copied into a negative-sense RNA, which is used as template to produce positive sense genomic and sgRNAs.

ANTIGENIC PROPERTIES

The virions act as strong immunogens. Different species can be identified by intragel cross-absorption immunodiffusion tests using polyclonal antisera or by ELISA using monoclonal antibodies. Antigenic distances between individual species expressed as serological differentiation indices are correlated with the degree of sequence difference in their CPs.

BIOLOGICAL PROPERTIES

Most species have moderate to wide host ranges under experimental conditions, although in nature host ranges are usually quite narrow. Transmission occurs without the help of vectors by contact between plants and sometimes by seed, although this occurs in the absence of infection of the embryo. Geographic distribution is worldwide. The viruses are found in all parts of host plants. Virions often form large crystalline arrays visible by light microscopy.

LIST OF SPECIES DEMARCATION CRITERIA IN THE GENUS

Tobamoviruses have historically been designated as strains of TMV, but many of these are now defined as separate species based on nucleotide sequence data.
The criteria demarcating species in the genus are:

- Sequence similarity; less than 10% overall nt sequence difference is considered to characterize strains of the same species, although most of the sequenced species have considerably less than 90% sequence identity,
- Host range; however many of these viruses have wider and more overlapping host ranges in experimental rather than natural situations,
- Antigenic relationships between the CPs.

LIST OF SPECIES IN THE GENUS

Species names are in green italic script; strain names and synonyms are in black roman script; tentative species names are in blue roman script. Sequence accession numbers, and assigned abbreviations ( ) are also listed.

SPECIES IN THE GENUS

Cucumber fruit mottle mosaic virus		
Cucumber fruit mottle mosaic virus	[NC_002633]	(CFMMV)
Cucumber green mottle mosaic virus		
Cucumber green mottle mosaic virus-SH	[NC_001801]	(CGMMV-SH)
Cucumber green mottle mosaic virus-W	[AB015146]	(CGMMV-W)
Frangipani mosaic virus		
Frangipani mosaic virus	[AF165884]	(FrMV)
Hibiscus latent Fort Pierce virus		
Hibiscus latent Fort Pierce virus	[AY250831]	(HLFPV)
Hibiscus latent Singapore virus		
Hibiscus latent Singapore virus	[AF395898, AF395899]	(HLSV)
Kyuri green mottle mosaic virus		
Kyuri green mottle mosaic-C1	[NC_003610]	(KGMMV-C1)
Kyuri green mottle mosaic-Yodo	[AB015145]	(KGMMV-Y)
Obuda pepper virus		
Obuda pepper virus	[NC_003852, L11665]	(ObPV)
Odontoglossum ringspot virus		
Odontoglossum ringspot virus	[U89894, S83257, U34586]	(ORSV)
Paprika mild mottle virus		
Paprika mild mottle virus-J	[NC_004106]	(PaMMV-J)
Pepper mild mottle virus		
Pepper mild mottle virus-S	[NC_003630]	(PMMoV-S)
Pepper mild mottle virus-Ia	[AJ308228]	(PMMoV-Ia)

POS. SSRNA

Ribgrass mosaic virus		
Ribgrass mosaic virus	[U69271]	(RMV)
Sammons's Opuntia virus		
Sammons's Opuntia virus		(SOV)
Sunn-hemp mosaic virus		
Sunn-hemp mosaic virus	[U47034, J02413]	(SHMV)
Tobacco latent virus		
Tobacco latent virus	[AY137775]	(TLV)
Tobacco mild green mosaic virus		
Tobacco mild green mosaic virus-Japan	[AB078435]	(TMGMV-J)
Tobacco mild green mosaic virus-U2	[NC_001556]	(TMGMV-U2)
Tobacco mosaic virus		
Tobacco mosaic virus	[NC_001367, AF395127, AF165190, AJ011933, X68110]	(TMV)
Tobacco mosaic virus-Rakkyo	[D63809]	(TMV-Rakkyo)
Tomato mosaic virus		
Tomato mosaic virus	[X02144, NC_002692, AJ132845, AJ243571, AJ417701, Z92909]	(ToMV)
Turnip vein-clearing virus		
Turnip vein clearing virus	[NC_001873]	(TVCV)
Turnip vein clearing virus-cr	[Z29370]	(TVCV-cr)
Ullucus mild mottle virus		
Ullucus mild mottle virus		(UMMV)
Wasabi mottle virus		
Wasabi mottle virus	[NC_003355]	(WMoV)
Youcai mosaic virus		
Youcai mosaic virus	[NC_004422]	(YoMV)
Youcai mosaic virus-Cg	[D38444]	(YoMV-Cg)
Zucchini green mottle mosaic virus		
Zucchini green mottle mosaic virus	[NC_003878]	(ZGMMV)

TENTATIVE SPECIES IN THE GENUS

Maracuja mosaic virus	(MarMV)

PHYLOGENETIC RELATIONSHIPS WITHIN THE GENUS

Fully sequenced tobamovirus genomes share 43-83% identity. However, tobamoviruses grouped based on host range share a much higher degree of sequence similarity. Nucleotide sequence of viruses within the groups of tobamoviruses that infect cruciferous (CFMMV, CGMMV, KGMMV, ZGMMV), brassicaceous (RMV, TVCV, WMoV, YoMV) or solanaceous plants (ObPV, PaMMV, PMMoV, TMGMV, TMV, ToMV) share 60-83%, 81-82%, and 56-79% identity, respectively. ORSV, an orchid infecting tobamovirus, shares 56-58% and 55-56% nt sequence identity with the groups infecting solanaceous plants and cucurbitaceous, respectively. SHMV, a legume infecting tobamovirus, shares 44-47% identity with other tobamoviruses.

Relationships between deduced aa sequences of each of the ORFs are similar to the relationship based on nt sequence: aa sequence identity is higher within groups based on host range, than between groups (Fig. 3). The CP, MP and large replication protein of tobamoviruses that infect brassicas share 84-92%, 81-87% and 97% identity, respectively, but share only 30-49%, 21-36% and 42-68% identity, respectively, with other tobamoviruses. The CP, MP and large replication protein of tobamoviruses that infect cucurbits share 44-81%, 57-88% and 63-93% identity within the group, and 30-43%, 20-42% and 40-45% identity outside the group, respectively. Within the group of cucurbit-infecting tobamoviruses, CGMMV is the most distinct, sharing 44-46%, 57-61% and 63-64% identity for the CP, MP and the large replication protein, respectively. SHMV CP and large replication protein share 41-43% and 41-42% identity, respectively, with the cucurbit-infecting tobamoviruses, although the SHMV MP shares only 18-22% identity with other tobamoviruses. FrMV CP and MP share only 33-39% and 20-33% identity with other tobamovirus species. The CP and MP of tobamoviruses that infect malavaceous hosts (HLFPV, HLSV) are 73% and 70% identical, respectively, but share 36-47% and 19-28% identity to the remainder of the genus. Although the ORSV CP and MP are more similar to tobamoviruses that infect solanaceous hosts (63-72% and 50-62% identity, respectively), the large replication protein is more similar to tobamoviruses that infect brassicas (67-68% identity).

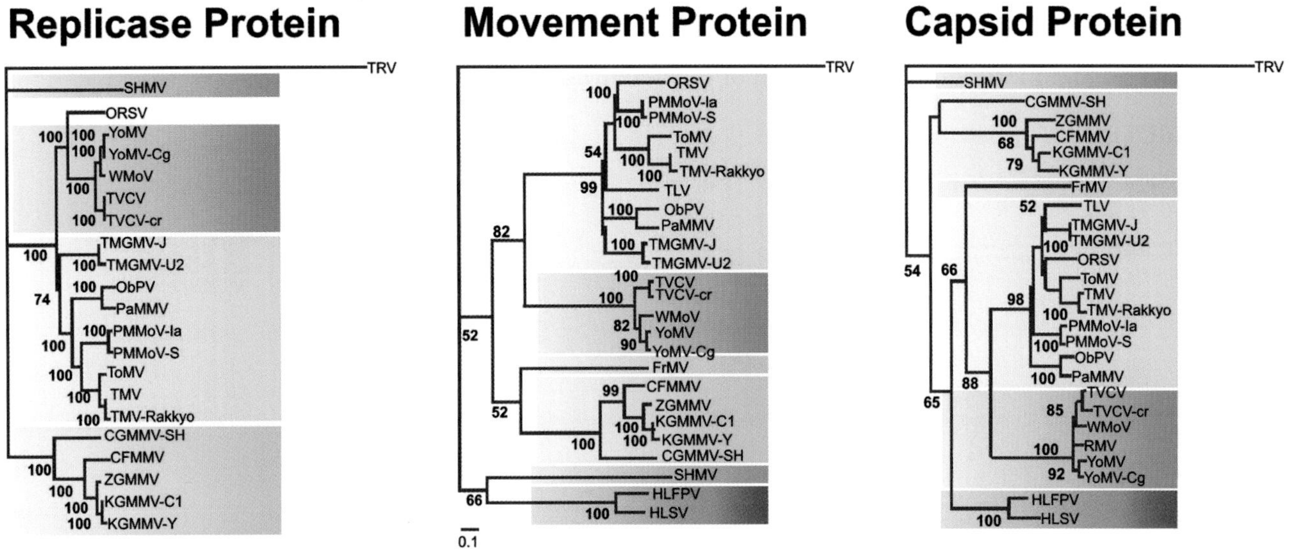

Figure 3: Dendrograms based on the deduced aa sequences of the 181-189 kDa replication protein (left), the MP (center) and the CP (right) of selected strains of tobamovirus species. The corresponding proteins from an isolate of *Tobacco rattle virus*, the type species of the *Tobravirus* genus, were used as the outgroup. Sequences were aligned using CLUSTALW, and the trees were constructed by a distance method using the PHYLIP programs PROTDIST and NEIGHBOR. The scale represents a distance value of 0.1.

Similarity with Other Taxa

The 124-132 kDa and 181-189 kDa non-structural proteins contain conserved motifs common to the RdRp of many viruses, and are most similar to the corresponding proteins of *Tobacco rattle virus*. The 28-31 kDa MP is the prototypical plant virus MP and resembles MPs of a wide variety of plant viruses, including *Tobacco rattle virus*. The 17-18 kDa structural protein share similarities with those of other plant viruses that form rod-shaped virions.

Derivation of Names

Tobamo: sigla from *toba*cco *mo*saic virus.

References

Buck, K. (1999). Replication of tobacco mosaic virus RNA. *Phil. Trans. R. Soc. Lond. B Biol. Sci.*, **354**, 613-627.

Citovsky, V. (1999). Tobacco mosaic virus: a pioneer of cell-to-cell movement. *Phil. Trans. R. Soc. Lond. B Biol. Sci.*, **354**, 637-643.

Dawson, W.O. and Lehto, K.M. (1990). Regulation of tobamovirus gene expression. *Adv. Virus Res.*, **38**, 307-342.

Deom, C.M., Lapidot, M. and Beachy, R.N. (1992). Plant virus movement proteins. *Cell*, **69**, 221-224.

Gibbs, A.J. (1977). Tobamovirus group. CMI/AAB Descriptions of Plant Viruses No **184**.

Gibbs, A. (1999). Origin and evolution of tobamoviruses. *Phil. Trans. R. Soc. Lond. B Biol. Sci.,* **354**, 593-602.
Klug, A. (1999). The tobacco mosaic virus particle: structure and assembly. *Phil. Trans. R. Soc. Lond. B Biol. Sci.,* **354**, 531-535.
Lartey, R.T., Voss, T.C. and Melcher, U. (1996). Tobamovirus evolution: gene overlaps, recombination, and taxonomic implications. *Mol. Biol. Evol.,* **13**, 1327-1338.
Nelson, R.S. and van Bel, A.J.E. (1998). The mystery of virus trafficking into, through and out of vascular tissue. In: *Progress in Botany,* (U. Luttge, ed), Vol. 59, pp 476-553. Springer-Verlag, Berlin.
Okada, Y. (1999). Historical overview of research on the tobacco mosaic virus genome: genome organization, infectivity and gene manipulation. *Phil. Trans. R. Soc. Lond. B Biol. Sci.,* **354**, 569-582.
Scholthof, K.-B., Shaw, J.G. and Zaitlin, M. (eds) (1999). *Tobacco mosaic virus*: One hundred years of contributions to virology. APS Press. St. Paul, Minnesota.
Van Regenmortel, M.H.V. and Fraenkel-Conrat, H. (eds) (1986). *The Plant Viruses.* The Rod-Shaped Plant Viruses. Plenum Press, New York.
Wang, H. and Stubbs, G. (1994). Structure determination of cucumber green mottle mosaic virus by X-ray fiber diffraction. Significance for the evolution of tobamoviruses. *J. Mol. Biol.,* **239**, 371-384.

CONTRIBUTED BY

Lewandowski, D.J.

GENUS *TOBRAVIRUS*

Type Species *Tobacco rattle virus*

VIRION PROPERTIES

MORPHOLOGY

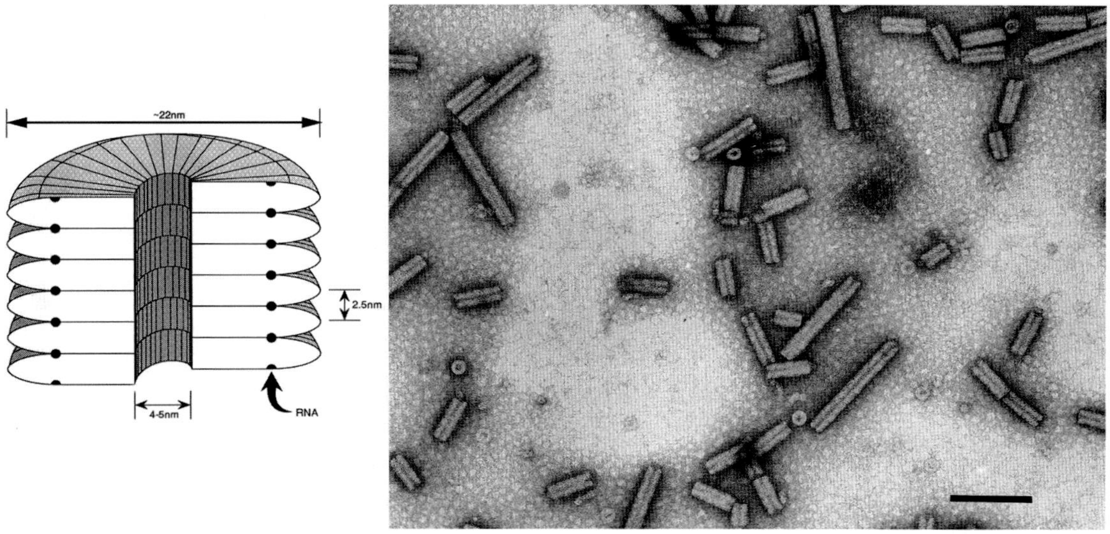

Figure 1: (Left) Diagram of a virion of Tobacco rattle virus (TRV), in section. (Right) Negative contrast electron micrograph of particles of TRV. The bar represents 100 nm.

Virions are tubular particles with no envelope. They are of two predominant lengths, (L) 180-215 nm and (S) ranging from 46 to 115 nm, depending on the isolate. Many strains produce in addition small amounts of shorter particles. The particle diameter is 21.3-23.1 nm by electron microscopy or 20.5-22.5 nm by X-ray diffraction, and there is a central canal 4-5 nm in diameter. Virions have helical symmetry with a pitch of 2.5 nm; the number of subunits per turn has been variously estimated as 25 or 32.

PHYSICOCHEMICAL AND PHYSICAL PROPERTIES

Virion Mr is 48-50 x 10^6 (L particles) and 11-29 x 10^6 (S Particles). Buoyant density in CsCl is 1.306-1.324 g/cm^3. $S_{20,W}$ is 286-306S (L particles) and 155–245S (S particles). Virions are stable over a wide range of pH and ionic conditions and are resistant to many organic solvents, but are sensitive to treatment with EDTA.

NUCLEIC ACID

The genome consists of two molecules of linear positive sense ssRNA; RNA-1 is about 6.8 kb and RNA-2 ranges from 1.8 kb to about 4.5 kb in size (varying in different isolates). The 5'-terminus is capped with the structure $m^7G^{5'}ppp^{5'}Ap$. There is no genome-linked protein or poly(A) tract.

PROTEINS

Virions contain a single structural protein of 22-24 kDa. RNA-1 codes for four non-structural proteins: a 134-141 kDa protein terminated by an opal stop codon and a 194-201 kDa protein produced by readthrough of this stop codon, both of which are probably involved in RNA replication; a 29-30 kDa protein (P1a) involved in intercellular transport of the virus; and a 12-16 kDa protein (P1b), which in Pea early-browning virus (PEBV) is involved in transmission of the virus in pea seeds. In addition to the virion structural protein, RNA-2 codes for two non-structural proteins, P2b and P2c. The size of P2b ranges from 27 to 40 kDa in different isolates, and that of P2c from 18 to 33 kDa. P2b is absolutely required for transmission by nematodes, whereas mutation of the P2c gene affects nematode transmission in some strains but not in others. The genes for P2b and P2c are missing from some laboratory strains that have been maintained by mechanical transmission. RNA-2 of some tobravirus isolates contains an additional small ORF between the CP and P2b genes, which codes for a potential 9 kDa protein.

LIPIDS
Virions contain no lipids.

CARBOHYDRATES
Virions contain no carbohydrates.

GENOME ORGANIZATION AND REPLICATION

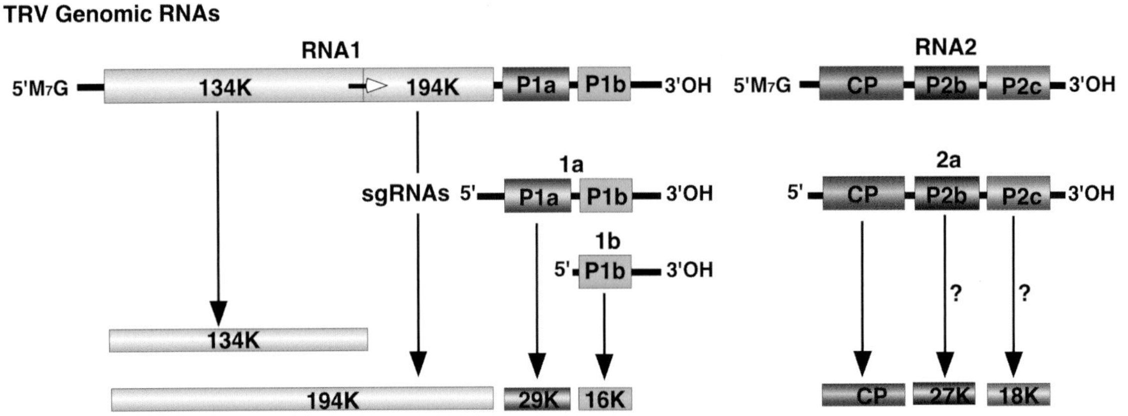

Figure 2: Genome organization and strategy of expression of Tobacco rattle virus (TRV). The means by which P2b and P2c are expressed is unknown.

RNA-1 is capable of independent replication and systemic spread in plants. The 134-141 kDa and 194-201 kDa replication proteins are translated directly from it, whereas P1a and P1b are translated from sgRNA species 1a and 1b, respectively. RNA-2 does not itself have messenger activity; the CP is translated from sgRNA-2a. The means by which the other RNA-2 encoded proteins are expressed is unknown (Fig. 2). There is sequence homology between RNA-1 and RNA-2 at both ends, but the extent of the homology varies between strains. In some strains, the homologous region at the 3′-end is large enough to include some or all of the P1a and P1b genes of RNA-1, but it is not known if these genes are expressed from RNA-2. Accumulation of virus particles is sensitive to cycloheximide but not to chloramphenicol, suggesting that cytoplasmic ribosomes are involved in viral protein synthesis. Virions accumulate in the cytoplasm. L particles of Pepper ringspot virus (PepRSV) become radially arranged around mitochondria, which are often distorted, and in cells infected with some *Tobacco rattle virus* isolates, 'X-bodies' largely composed of abnormal mitochondria and containing small aggregates of virus particles may be produced.

ANTIGENIC PROPERTIES

Tobravirus particles are moderately immunogenic. There is little or no serological relationship between members of the genus, and considerable antigenic heterogeneity among different isolates of the same virus.

BIOLOGICAL PROPERTIES

The host ranges are wide, including members of more than 50 monocotyledonous and dicotyledonous plant families. The natural vectors are nematodes in the genera *Trichodorus* and *Paratrichodorus* (Trichodoridae), different species being specific for particular virus strains. Adults and juveniles can transmit, but virus is probably not retained through the moult. Ingested virus particles become attached to the esophageal wall of the nematodes, and are thought to be released by salivary gland secretions and introduced into susceptible root cells during exploratory feeding probes. Virus can be retained for many months by non-feeding nematodes. There is no evidence for multiplication of virus in the vector and it is probably not transmitted through nematode eggs. The viruses are transmitted through seed of many host species. Tobacco rattle virus (TRV) occurs in Europe (including Russia), Japan, New Zealand and North America; PEBV occurs in Europe and North Africa, and PepRSV occurs in South America. TRV causes diseases in a wide variety of crop plants as well as weeds and other wild plants, including spraing (corky ringspot) and stem mottle in potato, rattle in tobacco, streaky mottle in narcissus and tulip, ringspot in aster, notched leaf in gladiolus, malaria in hyacinth and yellow blotch in sugar beet. PEBV is the cause of diseases in several legumes, including broad bean yellow band, distorting mosaic of bean and pea early-browning. PepRSV causes diseases in artichoke, pepper and tomato.

Most tissues of systemically invaded plants can become infected, but in many species virus remains localized at the initial infection site in the roots. In some virus-host combinations, notably TRV in some potato cultivars, limited systemic invasion occurs, and virus may not be passed on to all the vegetative progeny of infected mother plants.

Normal particle-producing isolates (called M-type) are readily transmitted by inoculation with sap and by nematodes. Other isolates (called NM-type) have only RNA-1, do not produce particles, are transmitted with difficulty by inoculation with sap, and are probably not transmitted by nematodes. NM-type isolates are obtained from M-type isolates by using inocula containing only L particles, and are also found in naturally infected plants. They often cause more necrosis in plants than do their parent M-type cultures.

List of Species Demarcation Criteria in the Genus

The criteria demarcating species in the genus are:

- Nucleotide sequences of RNA-1 show <75% identity.
- Interspecific pseudo-recombinant isolates cannot be made.
- Host ranges differ in specific hosts (e.g. legumes).
- RNA-2 sequences and serological relationships are of limited value.

List of Species in the Genus

Species names are in green italic script; strain names and synonyms are in black roman script; tentative species names are in blue roman script. Sequence accession numbers, and assigned abbreviations ( ) are also listed.

Species in the Genus

Pea early-browning virus		
Pea early-browning virus – E116	[RNA-2: AJ006500]	(PEBV-E116)
Pea early-browning virus – SP5	[RNA-1: X14006; RNA-2: X51828]	(PEBV-SP5)
Pea early-browning virus – TpA56	[RNA-2: X78455]	(PEBV-TpA56)
Pepper ringspot virus		
Pepper ringspot virus	[RNA-1: L23972; RNA-2: X03241]	(PepRSV)
Tobacco rattle virus		
Tobacco rattle virus-ON	[RNA-2: Z97357]	(TRV-ON)
Tobacco rattle virus-ORY	[RNA-1: AF034622; RNA-2: AF034621]	(TRV-ORY)
Tobacco rattle virus-PaY4	[RNA-2: AJ250488]	(TRV-PaY4)
Tobacco rattle virus-PLB	[RNA-2: J04347]	(TRV-PLB)
Tobacco rattle virus-PpK20	[RNA-1: AF314165; RNA-2: Z36974]	(TRV-PpK20)
Tobacco rattle virus-PSG	[RNA-2: X03686]	(TRV-PSG)
Tobacco rattle virus-Rostock	[RNA-2: AJ272198]	(TRV-Rostock)

Tobacco rattle virus-SP	[RNA-2: AJ007293]	(TRV-SP)
Tobacco rattle virus-SYM	[RNA-1: D00155]	(TRV-SYM)
Tobacco rattle virus-TCM	[RNA-2: X03955]	(TRV-TCM)
Tobacco rattle virus-TpO1	[RNA-2: AJ009833]	(TRV-TpO1)

TENTATIVE SPECIES IN THE GENUS

None reported.

PHYLOGENETIC RELATIONSHIPS WITHIN THE GENUS

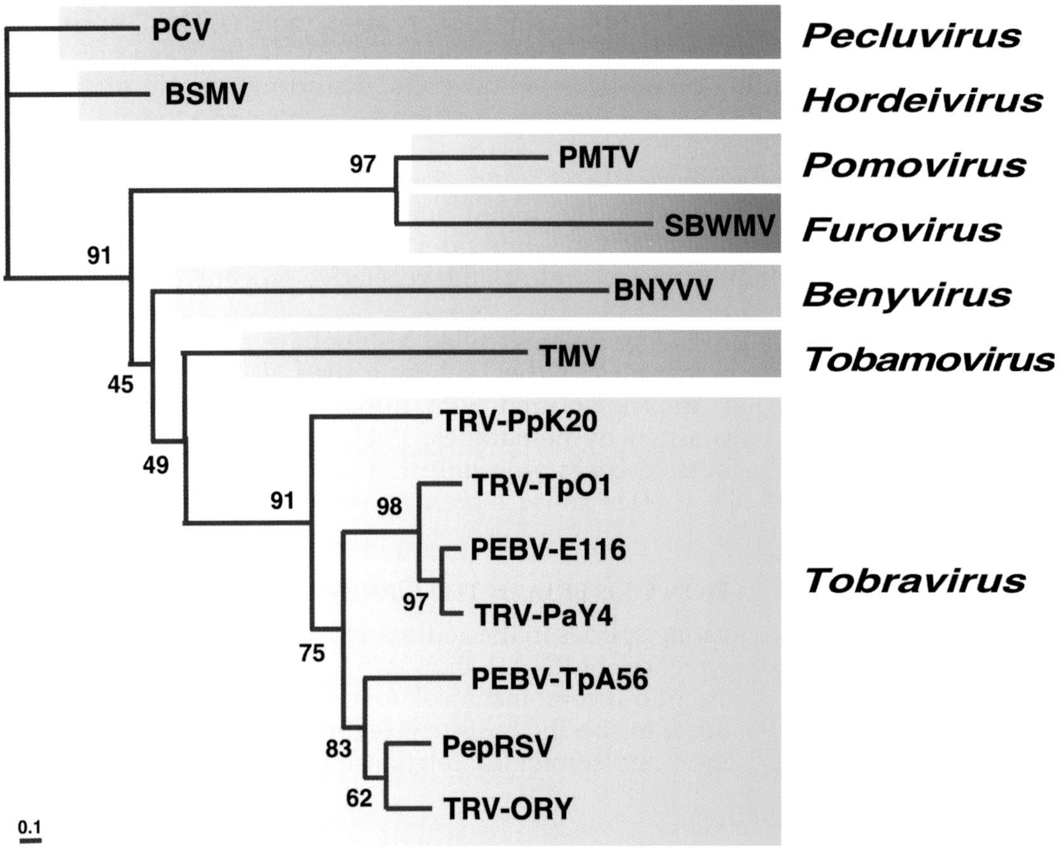

Figure 3: Dendrogram based on the coat protein amino acid sequences of representative tobravirus strains and a representative of the type species of the other genera of plant viruses with rod-shaped particles. Sequences were aligned using CLUSTAL, and the tree was constructed by a distance method using the PHYLIP programs PROTDIST and NEIGHBOR-NJ. The scale represents a distance value of 0.1.

Nucleotide sequences of RNA-1 of PEBV, PepRSV and TRV share only 40–45% identity, whereas those of three strains of TRV are more than 99% identical. However, relationships assessed using RNA-2 nt sequences or CP aa sequences are more complex. In particular, the CPs of several TRV strains are similar to those of typical PEBV strains, although at the extremities of RNA-2 they have nt sequences typical of TRV. Such strains are believed to have arisen through recombination.

SIMILARITY WITH OTHER TAXA

The 134-141 kDa and 194-201 kDa non-structural proteins contain conserved sequence motifs common to RdRp of many viruses, and are most closely related to the analogous proteins of Tobacco mosaic virus. The 29-30 kDa P1a protein also shares sequence similarities with the analogous 30 kDa protein of Tobacco mosaic virus and, to a lesser extent, with non-structural proteins of some other plant viruses. The structural proteins share sequence motifs with those of other plant viruses with rod-shaped particles. Hypochoeris mosaic virus, previously listed as a tentative species in the genus *Furovirus*, is now believed to have been an isolate of TRV.

DERIVATION OF NAMES

Tobra : sigla from *tob*acco *ra*ttle virus.

REFERENCES

Harrison, B.D. and Robinson, D.J. (1978). The tobraviruses. *Adv. Virus Res.*, **23**, 25-77.

Harrison, B.D. and Robinson, D.J. (1981). Tobraviruses. In: *Handbook of plant virus infections and comparative diagnosis* (E. Kurstak, ed), pp 515-540. Elsevier/North-Holland, Amsterdam.

Harrison, B.D. and Robinson, D.J. (1986). Tobraviruses. In: *The plant viruses*, Vol 2, (M.H.V. van Regenmortel, and H. Fraenkel-Conrat, eds), pp 339-369. Plenum Press, New York.

MacFarlane, S.A. (1999). Molecular biology of the tobraviruses. *J. Gen. Virol.*, **80**, 2799-2807.

Uhde, K., Koenig, R. and Lesemann, D.-E. (1998). An onion isolate of tobacco rattle virus: reactivity with an antiserum to Hypochoeris mosaic virus, a putative furovirus, and molecular analysis of its RNA 2. *Arch. Virol.*, **143**, 1041-1053.

CONTRIBUTED BY

Robinson, D.J.

POS. SSRNA

GENUS *HORDEIVIRUS*

Type Species *Barley stripe mosaic virus*

VIRION PROPERTIES

MORPHOLOGY

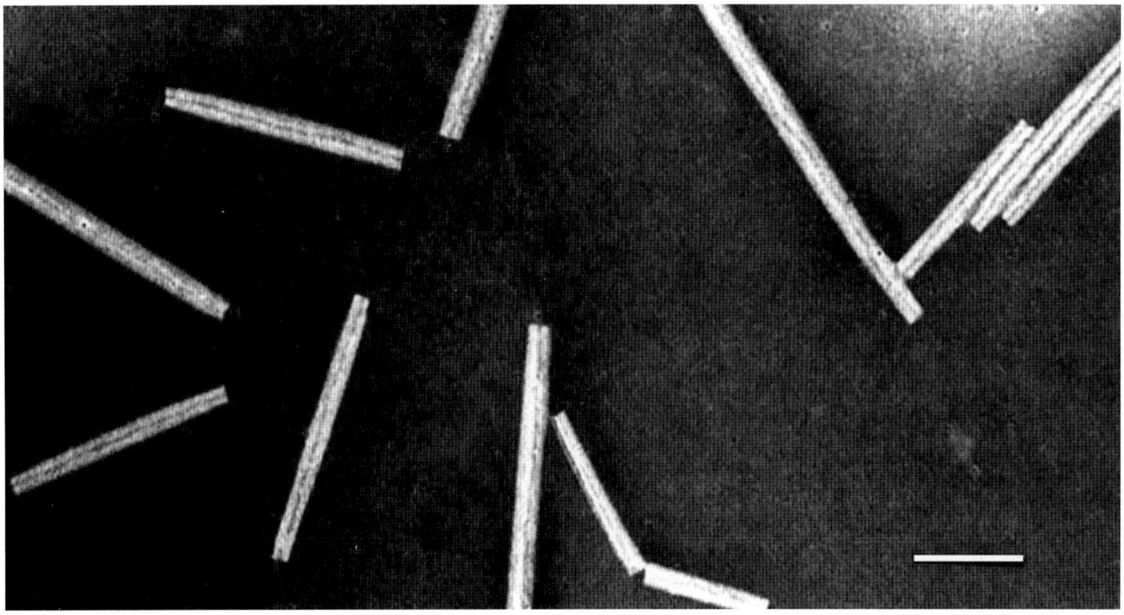

Figure 1: Electron micrograph of purified Barley stripe mosaic virus (BSMV) particles stained with 2% uranyl acetate. The particles are approximately 20 nm wide and have a length that varies depending on the size of the encapsidated RNA. The field was selected to represent monomers, but often a range of heterodisperse end-to-end aggregates up to 1000 nm in length predominate in purified preparations. The particles in the top left, bottom center, and upper left side of the micrograph are end-to-end aggregates that occur during purification. The bar represents 150 nm.

Virions are non-enveloped, elongated and rigid, about 20 x 110-150 nm in size; they are helically symmetrical with a pitch of 2.5 nm.

PHYSICOCHEMICAL AND PHYSICAL PROPERTIES

BSMV virions occur as heterodisperse sedimenting species with an S_{20w} of about 182-193S; other species have an S_{20w} of about 165-200S, depending on the virus. The BSMV isoelectric point is pH 4.5. Anionic detergents, added to purification buffers, increase virus yield by preventing particle aggregation. Thermal inactivation of infectivity occurs at 63-70°C. Virions are stable and their survival in sap ranges from a few days to several weeks.

NUCLEIC ACID

Virions normally contain three positive sense ssRNAs. The RNAs are designated α, β, and γ, and their respective sizes are 3.8, 3.2, and 2.8 kb (BSMV-ND18 strain), 3.7, 3.1, and 2.6 kb (Lychnis ringspot virus, LRSV), and 3.9, 3.6, and 3.2 kb (Poa semilatent virus, PSLV). The sizes of the α and β RNAs are similar between different strains of BSMV, whilst RNAγ varies in size. The ND18 RNAγ is 2.8 kb, that of the type strain is 3.2 kb. This difference is due to a 266 nt duplication near the 5′-end of the RNA that produces a γa protein of 87 kDa in the type strain compared to a 74 kDa γa ND18 protein. The Argentine mild strain contains mixtures of RNAγ species of 3.2, 2.8, and 2.6 kb. The 3.2 kb molecule contains a duplication similar to that of the type strain and the 2.6 kb RNA

POS. SSRNA

encodes a defective polymerase. No extensive hybridization can be detected between RNAs of BSMV, LRSV, and PSLV. Each RNA has m^7GpppGUA at its 5′-end, and a highly conserved 238 nt (BSMV), 148 nt (LRSV), or 330 nt (PSLV) tRNA-like structure at the 3′-end. In the case of BSMV, this structure can be charged with tyrosine. In the BSMV and LRSV genomes, a poly(A) sequence that is variable in length separates the coding region from the tRNA-like structure; however, this sequence is not present in the PSLV genome. A close sequence similarity between the first 70 nt of RNAα and RNAγ of the CV17 strain of BSMV suggests that a natural recombination event has occurred between RNAα and RNAγ of this strain. A similar recombination appears to have occurred between the 5′-untranslated leaders of RNAα and RNAβ of LRSV. These results plus sequence duplications in RNAγ provide persuasive evidence that RNA recombination has had a substantial role in the evolution of hordeiviruses.

Proteins

The virion capsid is constructed from subunits of a single protein. The CP of all species is 22 kDa in size, yet the proteins differ in electrophoretic mobility.

Lipids

None present.

Carbohydrates

The BSMV virion CP has been reported to be glycosylated, based on results of a staining procedure. However, glycosylation sites are not present in the deduced protein sequence and independent experiments failed to substantiate the report.

Genome Organization and Replication

All three BSMV genomic RNAs are required for systemic infection of plants, but RNAs α and γ alone can infect protoplasts. The 5'- and 3'-NCR of each BSMV RNA are required for replication. The hordeivirus genome encodes seven proteins as illustrated for BSMV in Figure 2. RNAα is monocistronic and encodes the αa protein (130 kDa in BSMV, 129 kDa in LRSV, and 131 kDa in PSLV) that functions as the helicase subunit of the viral replicase. The αa protein has two conserved sequence domains, an amino-terminal Mtr and a carboxy-terminal NTPase/Hel. The 5'-terminal RNAβ ORFs (βa) of all three viruses encode a 22 kDa CP. The BSMV CP, which is dispensable for systemic movement of the virus, is more closely related to the PSLV CP (55.2% identity) than to the LRSV CP (41.5% identity). An intergenic region separates a "triple gene block" (TGB) that encodes three non-structural proteins, βb (TGB1), βc (TGB3) and βd (TGB2), in which the βd protein overlaps the other two genes. In BSMV, The βb protein is expressed from a 2,450 nt sgRNA, and the βc and βd proteins are expressed from a second bicistronic 960 nt sgRNA with βc being expressed via a leaky scanning mechanism. In BSMV, a minor 23 kDa translational readthrough extension of the βd protein, designated βd', is present in plants. However, genetic experiments have not identified a function for βd', so it appears to be dispensable for infection in all local lesion and systemic hosts tested. The BSMV sgRNAβ1 and sgRNAβ2 promoters reside between positions –29 to –2 and –32 to –17 relative to the transcription initiation sites, respectively, and the nt sequences preceding the transcription initiation sites of these sgRNAs are conserved in LRSV and PSLV. The βb protein (58 kDa in BSMV, 50 kDa in LRSV, and 63 kDa in PSLV) contains a conserved NTPase/Hel domain. The BSMV βb protein binds RNA, NTPs, and exhibits ATPase and helicase activity *in vitro*. The βc (17 kDa in BSMV, and 18 kDa in LRSV and PSLV) and βd (14 kDa in BSMV and LRSV, and 18 kDa in PSLV) proteins are hydrophobic and membrane-associated. Each of the BSMV TGB proteins (βb, βc, and βd) is required for virus cell-to-cell movement in plants. RNAγ is bicistronic and encodes the γa polymerase subunit of the viral replicase (74 kDa in the BSMV-ND18 strain, 71 kDa in LRSV, and 84 kDa in PSLV), and the cysteine-rich γb protein (17 kDa in BSMV, 16 kDa in LRSV, and 20 kDa in PSLV).

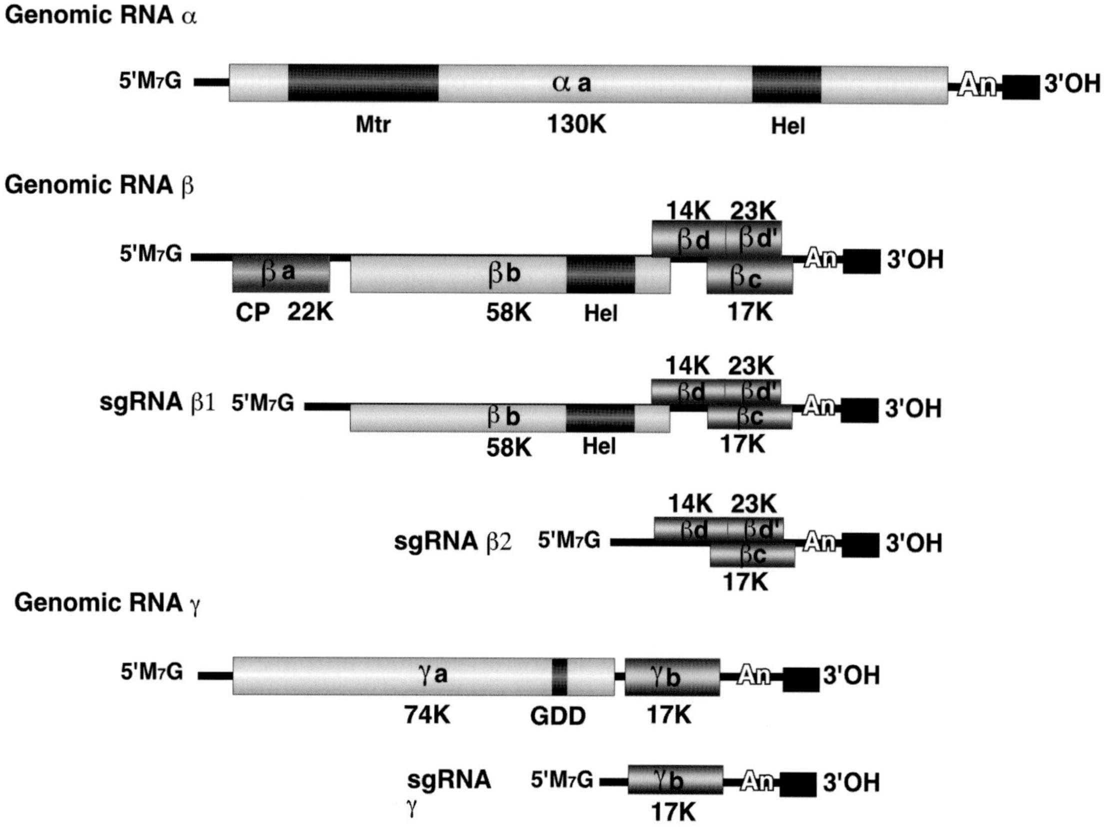

Figure 2: Genome organization of Barley stripe mosaic virus (BSMV). The filled circle, open rectangles and solid rectangles represent the 5'-cap structure, the ORFs, and the 3'-terminal tRNA-like structure, respectively. The 3'-proximal ORFs on each RNA terminate with an UAA that initiates the short poly (A) tract that directly precedes the 238 nt tRNA-like terminus. RNAα encodes a single protein, αa, with an amino-terminal Mtr domain (Mtr) and a carboxy-terminal helicase domain (Hel). This protein is referred to as the helicase subunit of the replicase (RdRp). RNAβ encodes five proteins: βa, the CP is translated from the genomic RNA; βb, a 58 kDa protein that contains a helicase domain (Hel). βb is translated from sgRNAβ1, whose promoter resides between positions –29 to –2 relative to the transcription start site; βc, a 17 kDa protein that is separated from the βb ORF by 173 nt; βd, a 14 kDa protein which overlaps the βb and the βc ORFs; and βd', a 23 kDa translational extension product of unknown function. The βd, βd', and βc proteins are translated from sgRNAβ2. The sgRNAβ2 promoter is located between nt –32 to –17 relative to its transcription start site. RNAγ encodes two proteins. The γa protein contains the GDD domain and is the polymerase subunit of the replicase. The cysteine-rich, 17 kDa γb protein has RNA binding ability, and is translated from a sgRNAγ, Note: what is this? whose promoter lies between positions –21 to +2 relative to the transcription start site.

The γa protein is variable in size because of the approximately 250 nt RNAγ repeated sequence present in different strains of BSMV. The BSMV γb protein is expressed from a 737 nt sgRNA and is a pathogenicity determinant that is involved in regulating expression of genes encoded by RNAβ. The sgRNAγ promoter is between nt –21 to +2 relative to its transcription start site, and this sequence has similarity to sequences upstream of the γb proteins in PSLV and LRSV. The BSMV γb protein has both RNA binding and zinc binding ability, participates in homologous interactions, and may act as a suppressor of post-transcriptional gene silencing. Translation of a functional αa protein is required for replication of RNAα *in cis,* whilst replication of RNAβ is dependent on the presence of the βa and βb intergenic region, and RNAγ requires approximately 600 nt of the 5'-terminal region. The TGB proteins on RNA β (b, c, d) are required for cell-to-cell and systemic movement in plants, but the CP and βd' are dispensable. The γb protein is also dispensable in some genetic backgrounds. A mutation in the 5'-leader sequence of the γa

ORF interfered with systemic infection of *Nicotiana benthamiana*, suggesting that modulation of γa expression can affect movement. Full-length dsRNAs corresponding to all viral genomic ssRNAs can be isolated from infected plants. Virus particles accumulate predominantly in the cytoplasm and also in nuclei. Infected barley plants develop pronounced enlargements of the plasmodesmata that contain the βb protein, and prominent peripheral vesicles appear in proplastids and chloroplasts. These vesicles may be the sites of replication because antibodies raised against poly(I):poly(C) have detected dsRNA in proplastids from infected barley root tips.

ANTIGENIC PROPERTIES

Hordeivirus particles are efficient immunogens. Member species are distantly related serologically with BSMV being more closely related to PSLV than to LRSV, which is in agreement with sequence analyses.

BIOLOGICAL PROPERTIES

HOST RANGE

The native hosts of three viruses (ALBV, BSMV, PSLV) are grasses (family Gramineae); strains of LRSV occur naturally in dicotyledonous plants of the families Caryophyllaceae and Labiatae. Various strains of these viruses elicit local lesions in *Chenopodium* species and are able to establish systemic infections in a common host, *Nicotiana benthamiana*.

TRANSMISSION

BSMV and LRSV are efficiently seed-transmitted, and are transmitted less efficiently by pollen. Field spread from primary infection foci occurs efficiently by direct leaf contact. There are no known vectors for any members of the family.

GEOGRAPHIC DISTRIBUTION

ALBV has been reported only from Great Britain; BSMV occurs world-wide wherever barley is grown; LRSV (mentha strain) has been isolated in Hungary, and the type strain which is highly seed-transmissible in the family Caryophyllaceae, was initially discovered in California from seed of *Lychnis divaricata* introduced from Europe. PSLV has been recovered from *Poa palustris* isolated from two locations in Western Canada.

LIST OF SPECIES DEMARCATION CRITERIA IN THE GENUS

Not available.

LIST OF SPECIES IN THE GENUS

Species names are in green italic script; strain names and synonyms are in black roman script; tentative species names are in blue roman script. Sequence accession numbers, and assigned abbreviations ( ) are also listed.

SPECIES IN THE GENUS

Anthoxanthum latent blanching virus		
Anthoxanthum latent blanching virus		(ALBV)
Barley stripe mosaic virus		
Barley stripe mosaic virus	[U35768, X03854, X52774]	(BSMV)
Lychnis ringspot virus		
Lychnis ringspot virus	[Z46630, Z46351, Z46353]	(LRSV)
Poa semilatent virus		
Poa semilatent virus	[Z46352, M81486, M81487]	(PSLV)

TENTATIVE SPECIES IN THE GENUS

None reported.

Phylogenetic Relations within the Genus

Not available.

Similarity with Other Taxa

The hordeivirus replicase sequences are most similar to the pecluvirus, Peanut clump virus (PCV; 65-66 % identity in the RdRp domain), the furovirus Soil borne wheat mosaic virus (SBWMV; 63-67 %), and to the tobravirus, Tobacco rattle virus (TRV; 52-54 %). Besides the hordeiviruses, the TGB MPs are found in allexiviruses, benyviruses, carlaviruses, foveaviruses, furoviruses, pecluviruses, pomoviruses, and potexviruses. The hordeivirus TGB proteins are most similar to PCV proteins, and to those of Potato mop-top virus and Beet soil-borne virus. However, SBWMV and TRV lack the TGB, and encode single MPs unrelated to the TGB proteins. Moreover, in the other TGB-containing viruses, the replicase sequences are only distantly related to those of hordeiviruses. Apparently, reassortment events during evolution have resulted in different phylogenetic trees of hordeivirus TGB MPs and replication-associated proteins. The hordeivirus γb proteins show significant similarity to the respective proteins of SBWMV and PCV, and have a marginal similarity to the cysteine-rich proteins of tobraviruses. The hordeivirus CP sequences are most closely related to CP of PCV. Taking into account the high degree of sequence similarity of the encoded proteins and some common features of genome organization, the hordeiviruses are most closely related to PCV.

Derivation of Names

hordei from *hordeus*, latin name of the primary host of the type species virus of the genus.

References

Beczner, L., Hamilton, R.I. and Rochon, D.M. (1992). Properties of the mentha strain of lychnis ringspot virus. *Intervirology*, **33**: 49-56.

Carroll, T.W. (1986). Hordeiviruses: biology and pathology. In: *The Plant Viruses, The Rod-shaped Plant Viruses*, Vol. 2, (M.H.V. van Regenmortel and H. Fraenkel-Conrat, eds), pp 373-395. Plenum Press, New York.

Donald, R.G.K. and Jackson, A.O. (1994). The barley stripe mosaic virus γb protein encodes a multifunctional cysteine-rich protein that affects pathogenesis. *Plant Cell*, **6**, 1593-1606.

Donald, R.G.K. and Jackson, A.O. (1996). RNA-binding activities of barley stripe mosaic virus γb fusion proteins. *J. Gen. Virol.*, **77**, 879-888.

Donald, R.G.K., Lawrence, D.M. and Jackson, A.O. (1997). The BSMV 58 kDa βb protein is a multifunctional RNA binding protein. *J. Virol.*, **71**, 1538-1546.

Edwards, M.C., Petty, I.T.D. and Jackson, A.O. (1992). RNA recombination in the genome of barley stripe mosaic virus. *Virology*, **189**, 389-392.

Edwards, M.L., Kelley, S.E., Arnold, M.K. and Cooper, J.I. (1989). Properties of a hordeivirus from *Anthoxanthum odoratum*. *Plant Pathol.*, **38**, 209-21.

Jackson, A.O., Hunter, B.G. and Gustafson, G.D. (1989). Hordeivirus relationships and genome organization. *Ann. Rev. Phytopathol.*, **27**, 95-121.

Jackson, A.O., Petty, I.T.D., Jones, R.W., Edwards, M.C. and French, R. (1991). Analysis of Barley stripe mosaic virus pathogenicity. *Semin. Virol.*, **2**, 107-119.

Johnson, J.A., Bragg, J.N., Lawrence, D.M. and Jackson, A.O. (2003). Sequences controlling expression of Barley stripe mosaic virus subgenomic RNAs. *Virology*, **313**, 66-80.

Kalinina, N.O., Rakitina, D.V., Soloyev, A.G., Schiemann, J. and Morozov, S. Yu (2002). RNA helicase activity of the plant movement proteins encoded by the first gene of the triple gene block. *Virology*, **296**, 321-329.

Lawrence, D.M. and Jackson, A.O. (2001). Interactions of the TGB1 protein during cell-to-cell movement of Barley stripe mosaic virus. *J. Virol.*, **75**, 8712-8723.

Lawrence, D.M. and Jackson, A.O. (2001). Requirements for cell-to-cell movement of barley stripe mosaic virus in monocot and dicot hosts. *Mol. Plant Path.*, **2**, 65-75.

Morozov, S. Yu and Solovyev, A.G. (2003). Triple gene block: modular design of a multifunctional machine for plant virus movement. *J. Gen. Virol.*, **84**,1351-1366.
Solovyev, A.G., Savenkov, E., Agranovsky, A.A. and Morozov, S. Yu (1996). Comparisons of the genomic *cis*-elements and coding regions in RNA β components of the hordeiviruses Barley stripe mosaic virus, Lychnis ringspot virus, and Poa semilatent virus. *Virology*, **219**, 9-18.
Savenkov, E., Solovyev, A.G. and Morozov, S. Yu (1997). Genome sequences of Poa semilatent and Lychnis ringspot hordeiviruses. *Arch. Virol.*, **143,** 1379-1393.
Yelina, N.E., Savenkov, E.I., Solovyev, A.G., Morozov, S. Yu and Valkonen, J.P.T. (2002). Long-distance movement, virulence, and RNA silencing suppression controlled by a single protein in hordeivirus and potyvirus: complementary functions between virus families. *J. Virol.*, **76**, 12981-12991.

CONTRIBUTED BY

Bragg, J.N., Solovyev, A.G., Morozov, S. Yu, Atabekov, J.G. Jackson, A.O.

GENUS FUROVIRUS

Type Species *Soil-borne wheat mosaic virus*

VIRION PROPERTIES

MORPHOLOGY

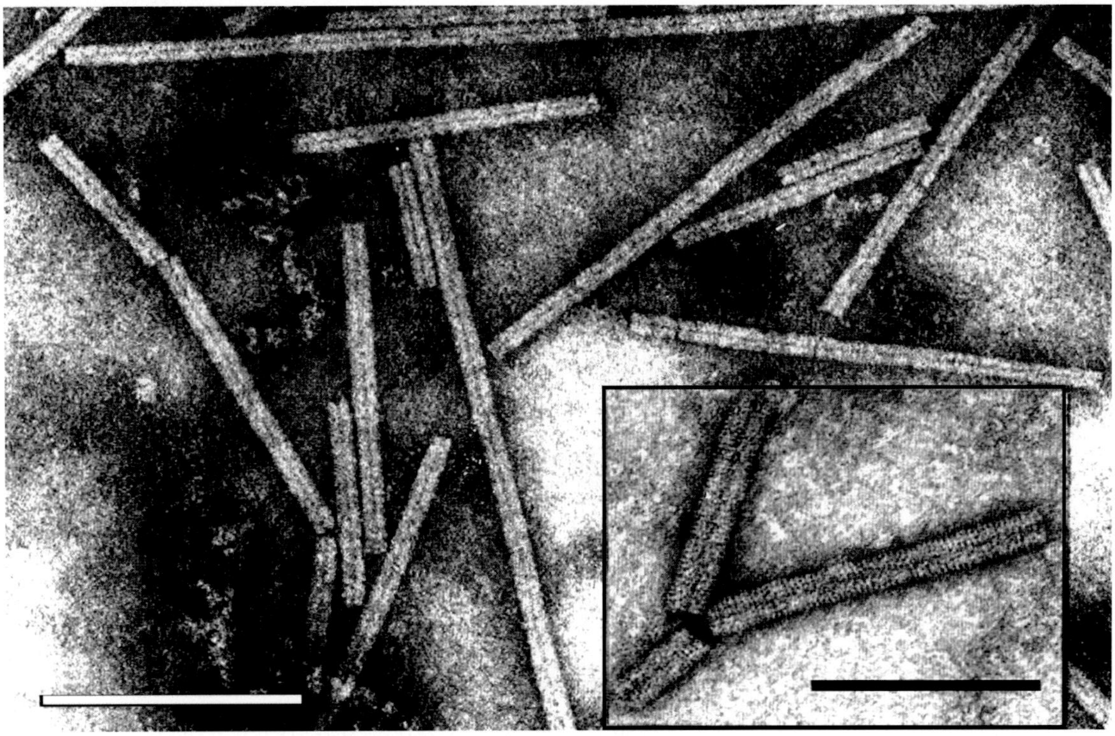

Figure 1: Negative contrast electron micrograph of stained (ammonium molybdate pH 7.0) particles of Soil-borne wheat mosaic virus (SBWMV). The bar represents 200 nm. Inset: Negative contrast electron micrograph of particles SBWMV stained with 1% uranyl acetate. The bar represents 100 nm.

Virions are non-enveloped hollow rods, which have helical symmetry. Virions are about 20 nm in diameter, with predominant lengths of 140-160 nm and 260-300 nm. The length distribution of the Soil-borne wheat mosaic virus (SBWMV) short particles is broad, 80-160 nm, due to the presence of deletion mutants in some cultures (Fig. 1).

PHYSICOCHEMICAL AND PHYSICAL PROPERTIES

Virions sediment as two (or three) components; for SBWMV the S_{20w} values are 220-230S (long particles) and 170-225S (short particles), and 126-177S (deletion mutants). SBWMV loses infectivity in extracts of wheat kept at 60-65°C for 10 min.

NUCLEIC ACID

Complete or almost complete nt sequences are available for all five species in the genus. The genome is bipartite, linear, positive-sense ssRNA. RNA-1 is c. 6-7 kb and RNA-2 c. 3.5-3.6 kb. The RNA molecules of SBWMV have a 5′-cap (m^7GpppG) and in all of the species where the complete sequences have been determined there is a 3′-terminal tRNA-like structure with a putative anticodon for valine. The 3′-terminus of SBWMV RNA was shown experimentally to accept valine.

PROTEINS

The capsid comprises multiple copies of a single polypeptide of ~19-20.5 kDa. The CPs of SBWMV, Chinese wheat mosaic virus (CWMV), Soil-borne cereal mosaic virus (SBCMV) and Oat golden stripe virus (OGSV) comprise 176 aa with 76-82% aa homologies, they share only 46% homology with SrCSV. The CP gene terminates in a leaky (UGA) stop codon that can be suppressed to produce a read-through protein (~85 kDa), which is thought to be involved in natural transmission by the plasmodiophorid vector. In addition to replicase proteins the furoviruses encode a single MP (~37 kDa) and a cysteine rich protein (~18 kDa) of unknown function.

LIPIDS
None reported.

CARBOHYDRATES
None reported.

GENOME ORGANIZATION AND REPLICATION

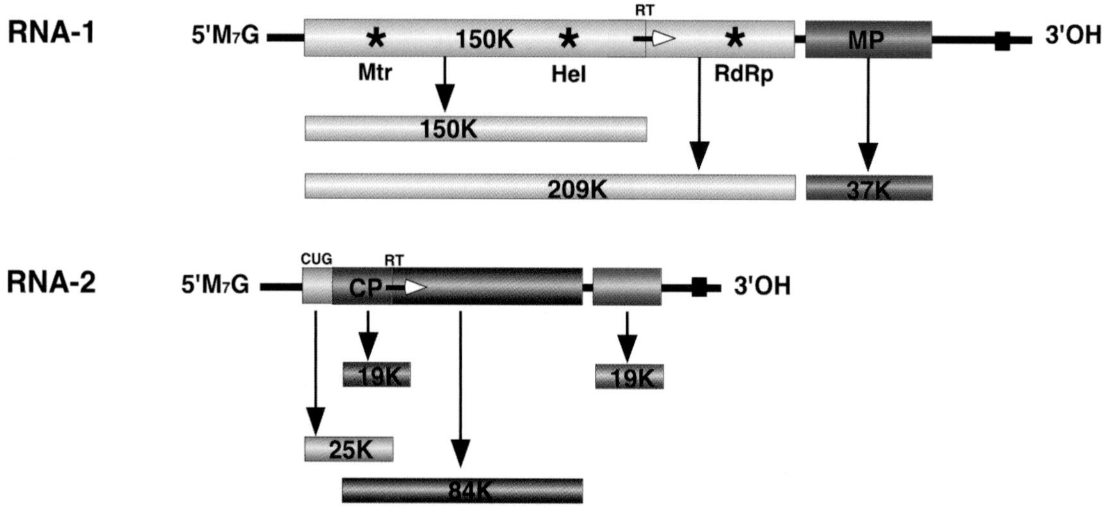

Figure 2: Genome organization of Soil-borne wheat mosaic virus (SBWMV). The tRNA structure motifs at the 3′-ends of the RNAs are represented by a dark square, the Met, Hel and RdRp motifs of the polymerase by asterisks and the readthrough of the polymerase and coat protein ORFs by RT and an arrow.

Genome organization and structure is conserved between species but there are substantial differences in the nt sequences. SBWMV RNA-1 encodes a 150 kDa protein, a 209 kDa readthrough product and a 37 kDa protein (Fig. 2). The 150 kDa protein contains Mtr and NTP-binding Hel motifs and the readthrough protein, in addition, contains RNA polymerase motifs, indicating that these proteins are involved in replication. The 37 kDa protein is thought to be involved in virus movement as it shares partial sequence similarity to the MPs of dianthoviruses. RNA-2 encodes the CP (19 kDa), the sequence of which terminates in a UGA codon that can be suppressed to give a readthrough product of 84 kDa. A 25 kDa polypeptide is initiated from a CUG codon upstream of the CP AUG. An ORF towards the 3′-end of the RNA-2 encodes a 19 kDa protein that contains 7 conserved cysteine residues. Products corresponding to the 37 kDa protein and the cysteine-rich 19 kDa protein were not found in *in vitro* transcription/translation experiments, and these proteins are thought to be expressed from sgRNAs. Spontaneous deletions in the CP readthrough domain occur on successive passage by manual inoculation, and in field isolates in older infected plants.

ANTIGENIC PROPERTIES

Virions are immunogenic and the five virus species can be distinguished serologically.

Biological Properties

Host Range

The natural host ranges of furoviruses are narrow and confined to species within the Graminae. SBWMV induces green or yellow mosaic and stunting in winter wheat (*Triticum aestivum*) causing up to 80% yield loss in severely infected crops. It also may infect barley and rye. SBCMV infects mainly wheat and triticale in Western and Southern Europe and mainly rye in Central and North-Eastern Europe. Both viruses are (not readily) mechanically transmissible to *Chenopodium quinoa*. OGSV infects oats (*Avena sativa*) but failed to infect wheat when plants were grown in viruliferous soil. Mechanically it can be transmitted to some *Nicotiana* and *Chenopodium* species. SrCSV infects *Sorghum bicolor*. Mechanically it can be transmitted to a range of species including *Chenopodium quinoa, C. amaranticolor, Nicotiana clevelandii, Arachis hypogaea, Zea mays* and *T. aestivum*.

Transmission

The viruses are soil-borne, and *Polymyxa graminis* has been identified as a vector for SBWMV. Virions are thought to be carried within the motile zoospores. Soil containing the resting spores remains infectious for many years.

Geographical Distribution

Furoviruses are found in temperate regions worldwide including the United States of America, Europe, China, Japan.

Cytopathic Effects

Virions are found scattered, or in aggregates and inclusion bodies in the cytoplasm and vacuole. Inclusion bodies can be crystalline inclusions or comprise loose clusters of virus particles in association with masses of microtubules. Amorphous inclusion bodies can be seen in tissue sections by light microscopy.

List of Species Demarcation Criteria in the Genus

The species within the genus *Furovirus* are presently mainly differentiated on the basis of the nt sequences of their RNAs and the deduced aa sequences of their putative gene products. On these bases, SBWMV (Japan) may represent an additional species. RNA-1 sequences of these viruses share 58-74% identity and RNA-2 46-80% (Fig. 3). SBWMV, SBCMV, CWMV and OGSV can be discriminated also by reactivity with selected monoclonal and polyclonal antibodies. OGSV and SrCSV differ in host range to SBWMV, SBCMV and CWMV. Especially the latter three viruses have similar biological properties and genetic reassortants can be formed with RNA-1 and RNA-2 of SBWMV (Nebraska), SBWMV (Japan) and SBCMV. With the other viruses this possibility has not yet been checked. The relative merits of using genetic reassortment in furovirus species demarcation in the future are currently being debated in the furovirus study group.

List of Species in the Genus

Species names are in green italic script; strain names and synonyms are in black roman script; tentative species names are in blue roman script. Sequence accession numbers, and assigned abbreviations ( ) are also listed.

Species in the Genus

Chinese wheat mosaic virus		
Chinese wheat mosaic virus	[RNA1: AJ012005; RNA2: AJ012006]	(CWMV)
Oat golden stripe virus		
Oat golden stripe virus	[RNA1: AJ132578; RNA2: J132579]	(OGSV)
Soil-borne cereal mosaic virus		
Soil-borne cereal mosaic virus	[RNA1: AF146278-80, AJ252151;	(SBCMV)
(European wheat mosaic virus)	RNA2: AF146281-83, AJ252152]	

Pos. ssRNA

(Soil-borne rye mosaic virus)
Soil-borne wheat mosaic virus
Soil-borne wheat mosaic virus - Japan [RNA1: AB033689; RNA2: AB033690] (SBWMV-JP)
Soil-borne wheat mosaic virus - Nebraska [RNA1: L07937; RNA2: L07938] (SBWMV-Ne)
Soil-borne wheat mosaic virus - New York [RNA1: AY016007, AF361641; RNA2: AY016008, AF361642] (SBWMV-NY)
Sorghum chlorotic spot virus
Sorghum chlorotic spot virus [RNA1: AB033691; RNA2: AB033692] (SrCSV)

TENTATIVE SPECIES IN THE GENUS

None reported.

PHYLOGENETIC RELATIONSHIPS WITHIN THE GENUS

The genome organization is identical for all furoviruses. However, there are considerable differences in percent nt sequence identity between the species with OGSV and SrCSV being the most distantly related ones (Fig 3).

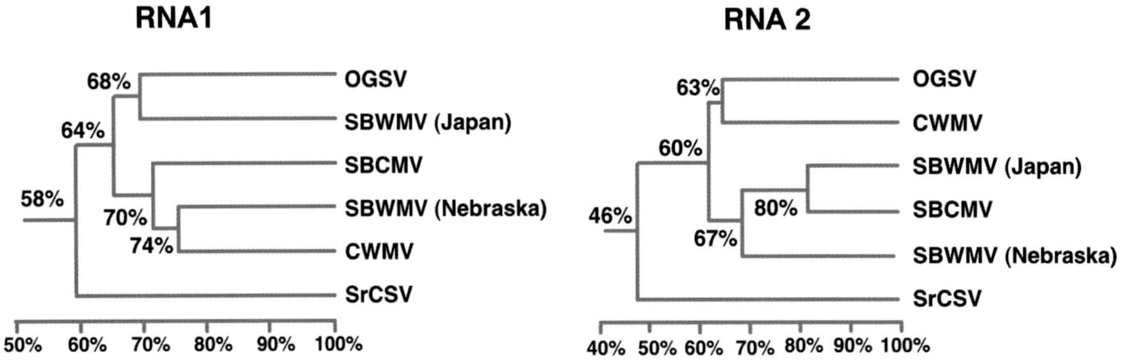

Figure 3: Percentage sequence identities of total RNAs of furoviruses.

SIMILARITY WITH OTHER TAXA

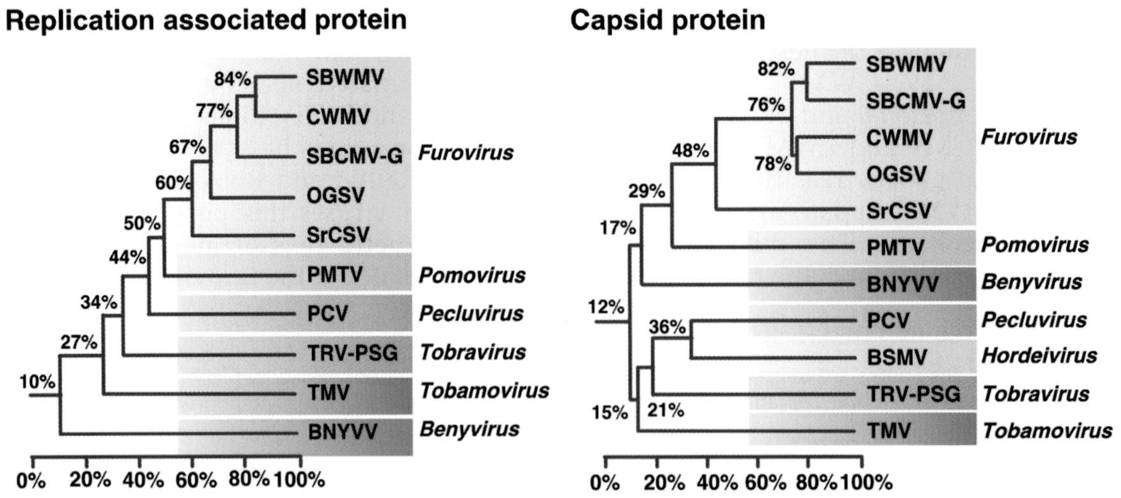

Figure 4: Percentage identities of the deduced aa sequences of replication-associated and CPs of furoviruses and other viruses with tubular particles

Furoviruses have similar particle morphology to members of the genera *Pecluvirus*, *Pomovirus*, *Benyvirus*, *Tobamovirus*, *Tobravirus* and *Hordeivirus*. The CPs of these viruses show various degrees of aa sequence homologies (Fig. 4) and have a number of conserved residues, e.g. RF and FE in their central and C-terminal parts, respectively, which are thought to be involved in the formation of salt bridges. The aa sequences of the putative replicase proteins form a cluster with those of the pomo-, peclu-, hordei-, tobra- and tobamoviruses (Fig. 4). The furoviruses also resemble the pomoviruses and surprisingly also the tymoviruses in having valine-specific tRNA-like structures at the 3'-ends of their genomic RNAs. The furoviruses differ from the pomo-, beny-, hordei- and tobamoviruses by their bipartite genomes and from peclu-, pomo-, beny- and hordeiviruses in that they have a single movement protein gene rather a triple block of genes encoding movement proteins. SBWMV has distant serological relationships with Potato mop-top virus and Tobacco mosaic virus.

Derivation of Names

Furo: sigla from *fu*ngus-borne, *ro*d-shaped virus.

References

Diao, A., Chen, J., Ye, R., Zheng, T., Yu, S., Antoniw, J.F. and Adams, M.J. (1999). Complete sequence and genome properties of Chinese wheat mosaic virus, a new furovirus from China. *J. Gen. Virol.*, **80**, 1141-1145.

Diao, A., Chen, J., Gitton, F., Antoniw, J.F., Mullins, J., Hall, A.M. and Adams, M.J. (1999). Sequences of European wheat mosaic virus and Oat golden stripe virus and genome analysis of the genus *Furovirus*. *Virology*, **261**, 331-339.

Chen, J., MacFarlane, S.A. and Wilson, T.M.A. (1994). Detection and sequence analysis of a spontaneous deletion mutant of Soil-borne wheat mosaic virus RNA2 associated with increased symptom severity. *Virology*, **202**, 921-929.

Chen, J., Torrance, L., Cowan, G.H., MacFarlane, S.A., Stubbs, G. and Wilson, T.M.A. (1997). Monoclonal antibodies detect a single amino acid difference between the coat proteins of Soil-borne wheat mosaic virus isolates: Implications for virus structure. *Phytopathology*, **87**, 295-301.

Goodwin, J.B. and Dreher, T.W. (1998). Transfer RNA mimicry in a new group of positive-strand RNA plant viruses, the furoviruses: differential aminoacylation between the RNA components of one genome. *Virology*, **246**, 170-178.

Kendall, T.L., Langenberg, W.G. and Lommel, S. (1988). Molecular characterization of Sorghum chlorotic spot virus, a proposed furovirus. *J. Gen. Virol.*, **69**, 2335-2345.

Koenig, R., Bergstrom, G.C., Gray, S.M. and Loss, S. (2002). A New York isolate of Soil-borne wheat mosaic virus differs considerably from the Nebraska type strain in the nucleotide sequences of various coding regions but not in the deduced amino acid sequences. *Arch. Virol.*, **147**, 617-265.

Koenig, R. and Huth, W. (2000). Soil-borne rye mosaic and European wheat mosaic virus: two names for a furovirus with variable genome properties which is widely distributed in several cereal crops in Europe. *Arch. Virol.*, **145**, 689-697.

Koenig, R., Pleij, C.W.A. and Huth, H. (1999). Molecular characterization of a new furovirus mainly infecting rye. *Arch. Virol.*, **144**, 2125-2140.

Miyanishi, M., Roh, S.H., Yamamiya, A., Ohsato, S. and Shirako, Y. (2002). Reassortment between genetically distinct Japanese and US strains of Soil-borne wheat mosaic virus: RNA1 from a Japanese strain and RNA2 from a US strain make a pseudorecombinant virus. *Arch. Virol.*, **147**, 1141-1153.

Rao, A.S. and Brakke, M.K. (1969). Relation of soil-borne wheat mosaic virus and its fungal vector. *Phytopathology*, **59**, 581-587.

Shirako, Y. (1998). Non-AUG translation initiation in a plant RNA virus: a forty-amino-acid extension is added to the N terminus of soil-borne wheat mosaic virus capsid protein. *J. Virol.*, **72**, 1677-1682.

Shirako, Y. and Wilson, T.M.A. (1993). Complete nucleotide sequence and organization of the bipartite genome of Soil-borne wheat mosaic virus. *Virology*, **195**, 16-32.

Pos. ssRNA

Shirako, Y., Suzuki, N. and French, R.C. (2000). Similarity and divergence among viruses in the genus *Furovirus*. *Virology*, **270**, 201-207.
Ye, R., Xu, L., Gao, Z.Z., Yang, J.P., Chen, J.P., Chen, M.J., Adams, M.J. and Yu, S.Q. (2000). Use of monoclonal antibodies for the serological differentiation of wheat and oat furoviruses. *J. Phytopathology*, **148**, 257-262.

Contributed By

Torrance, L. and Koenig, R.

GENUS POMOVIRUS

Type Species *Potato mop-top virus*

VIRION PROPERTIES

MORPHOLOGY

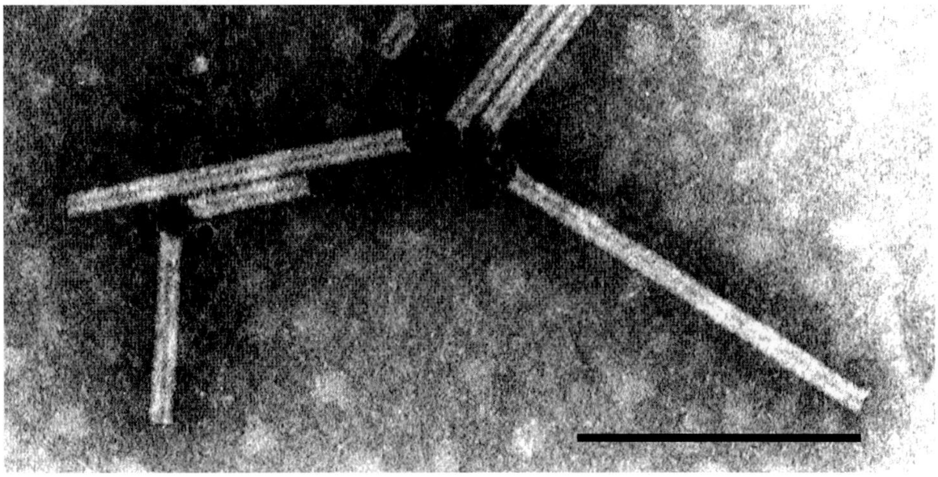

Figure 1: Negative contrast electron micrograph of particles of Potato mop-top virus (PMTV). The gold-labelling shows the binding of monoclonal antibody SCR 68 to one extremity of the particles. The bar represents 100 nm. (Courtesy I. M. Roberts).

The non-enveloped, rod-shaped particles are helically constructed with a pitch of 2.4 to 2.5 nm and an axial canal. They have predominant lengths of ~ 65-80, 150-160 and 290-310 nm and diameters of 18-20 nm. Crude extracts of plants infected with Beet soil-borne virus (BSBV), Beet virus Q (BVQ) and Potato mop-top virus (PMTV) contain characteristic small bundles of a few side-by-side aggregated particles in addition to singly dispersed particles.

PHYSICOCHEMICAL AND PHYSICAL PROPERTIES

Virions sediment as three components with S_{20w} of c. 125S, 170S and 230S, respectively. In sap at room temperature, most of the infectivity is lost within a few hours.

NUCLEIC ACID

Virions contain three molecules of linear positive-sense ssRNA of ~ 6, 3-3.5 and 2.5-3 kb, respectively. The sequence has been determined for all three RNA species of BSBV, BVQ, PMTV and Broad bean necrosis virus (BBNV). The RNAs are probably capped at the 5'-end; their 3'-ends can be folded into tRNA-like structures that are preceded by a long hairpin-like structure and an upstream pseudoknot domain. The tRNA-like structures of pomoviruses like those of tymoviruses contain an anticodon for valine and are capable of high-efficiency valylation.

PROTEINS

The major capsid protein (CP) species is 20 kDa in size. It is not needed for systemic infection. The CP readthrough protein, which may be detected in some PMTV particles near one extremity by means of immunogold labeling, apparently initiates virus assembly. Sequences in the CP readthrough protein are also necessary for the transmission of PMTV by *Spongospora subterranea*. PMTV triple gene block protein 1 and 2 (TGBp2 and TGBp3) are membrane-associated and bind ssRNA in a sequence nonspecific manner. It has been suggested that they may form a complex with PMTV RNA that is translocated and localized to the plasmodesmata by TGBp3.

POS. SSRNA

LIPIDS
None reported.

CARBOHYDRATES
None reported.

GENOME ORGANIZATION AND REPLICATION

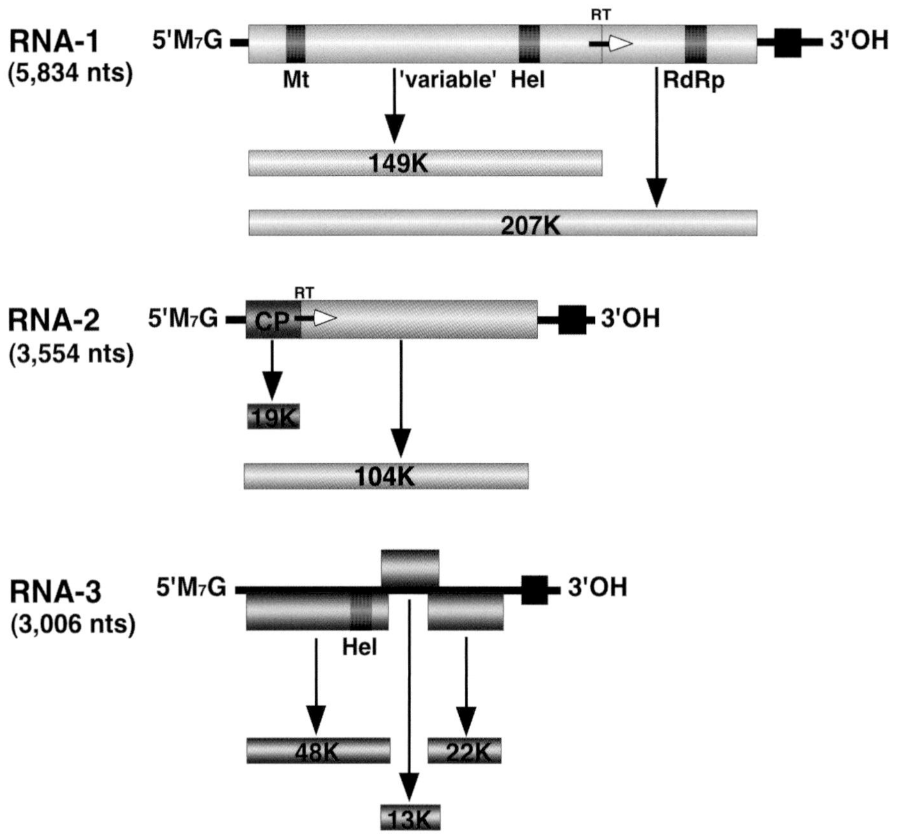

Figure 2: Genome organization of Beet soil-borne virus (BSBV). Areas in the putative translation products indicate respectively the UAA and UAG stop codons that are thought to be suppressible, and solid squares indicate a 3'-terminal tRNA-like structure. Dark areas indicate conserved motifs. Hel; helicase, Mt; methyltransferase, RdRp; RNA dependent RNA polymerase, RT; readthrough.

RNA-1 of BSBV has ORF for a 149 kDa protein and a 207 kDa readthrough protein that are presumably involved in replication. The shorter ORF is terminated by an apparently suppressible UAA stop codon (Fig. 2). Proteins of similar sizes are encoded on RNA-1 of BVQ, BBNV and PMTV. The smaller protein contains in its N-terminal part Mtr (Mt) and in its C-terminal part Hel motifs; the motifs for RdRp are found in the C-terminal part of the readthrough protein (Fig. 2). The two proteins contain also other highly conserved domains of unknown function in their N- and C-terminal parts, but their central regions (designated as 'variable' in Fig. 2) are highly specific for each virus. RNAs-2 contain the CP gene, which terminates with a suppressible UAG stop codon and then continues in frame to form a CP readthrough protein gene that varies considerably in size between different pomoviruses, possibly because it readily undergoes internal deletions. PMTV RNA-2 has therefore originally been referred to as RNA-3 and *vice versa*. PMTV RNA-2 contains a gene for a cysteine-rich protein that is not found on the RNAs of BSBV and BVQ. A triple gene block (TGB) coding for proteins involved in viral movement is found on RNAs-3. TGBp1 also contains Hel motifs. The sequences of the C-terminal part of

TGBp1, of the entire TGBp2 and of the N-terminal part of TGBp3 are highly conserved among pomoviruses. The replication mechanisms are unknown.

ANTIGENIC PROPERTIES

Virions are moderately antigenic. Distant serological relationships have been found between the particles of BSBV and BVQ but not between those of the two beet viruses and PMTV. This is probably due to the fact that PMTV CP has ten extra amino acids on its immunodominant N-terminus that are missing in the two beet viruses. A conserved sequence EDSALNVAHQL is found in the CPs of PMTV, BSBV and BVQ. It contains an epitope for which the monoclonal antibody SCR 70 is specific and which is only detectable by Western blotting after disruption of the particles. Other epitopes are either exposed along the entire particle length, e.g. the immunodominant N-terminus, or are accessible only on one extremity (Fig. 1). PMTV and BBNV show distant serological relationships to tobamoviruses.

BIOLOGICAL PROPERTIES

HOST RANGE

The natural host range of pomoviruses is very narrow; only dicotyledoneous hosts have been described.

TRANSMISSION

Pomoviruses are transmitted by soil. *Spongospora subterranea* and *Polymyxa betae* have been identified as vectors for PMTV and BSBV, respectively. The viruses are also transmissible mechanically.

GEOGRAPHICAL DISTRIBUTION

Countries with temperate climate.

CYTOPATHIC EFFECTS

PMTV-infected cells contain in the cytoplasm virions aggregated in sheaves. Infections by BSBV and BVQ induce voluminous cytoplasmic inclusions which consist of hypertrophied endoplasmic reticulum, convoluted membrane accumulations, numerous small virion bundles and rarely compact virus aggregates.

LIST OF SPECIES DEMARCATION CRITERIA IN THE GENUS

The criteria demarcating species in the genus are:

- Differences in host range,
- Effects in infected tissue: different inclusion body morphology,
- Transmission: different vector species,
- Serology: virions are distantly related serologically,
- Genome: different numbers of genome components (presence or absence of a gene for a cysteine-rich protein).
- Sequence: less than ~80% identical over whole sequence,
- Sequence: less than ~90% identical in CP amino acid sequence.

LIST OF SPECIES IN THE GENUS

Species names are in green italic script; strain names and synonyms are in black roman script; tentative species names are in blue roman script. Sequence accession numbers, and assigned abbreviations ( ) are also listed.

SPECIES IN THE GENUS

Beet soil-borne virus

Beet soil-borne virus - Ahlum	[RNA-1: NC_003020; RNA-2: NC_003518; RNA-3: NC_003519]	(BSBV-Ahl)

Beet virus Q

Beet virus Q - Wierthe	[RNA-1: NC_003510; RNA-2:	(BVQ-Wie)

POS. SSRNA

NC_003511; RNA-3: NC_003512]

Broad bean necrosis virus

Broad bean necrosis virus [RNA-1: NC_004423; RNA-2: NC_004424; RNA-3: NC_004425] (BBNV)

Potato mop-top virus

Potato mop-top virus – SW [RNA-1: NC_003723; RNA-2*: NC_003724; RNA-3*: NC_003725] (PMTV-SW)

Potato mop-top virus - T [RNA-2*: D16193; RNA-3*: D30753] (PMTV-T)

*RNA2 of PMTV has originally been referred to as RNA 3 and vice versa

TENTATIVE SPECIES IN THE GENUS

None reported.

PHYLOGENETIC RELATIONSHIPS WITHIN THE GENUS

The replication-associated proteins (RNA-1-encoded readthrough proteins) and the CPs of the pomoviruses described so far share less than 65% and 55% aa sequence identity, respectively (Fig. 3). The highest percentages of aa sequence identities are found in TGBp2 (75% for BSBV and BVQ – Fig. 3).

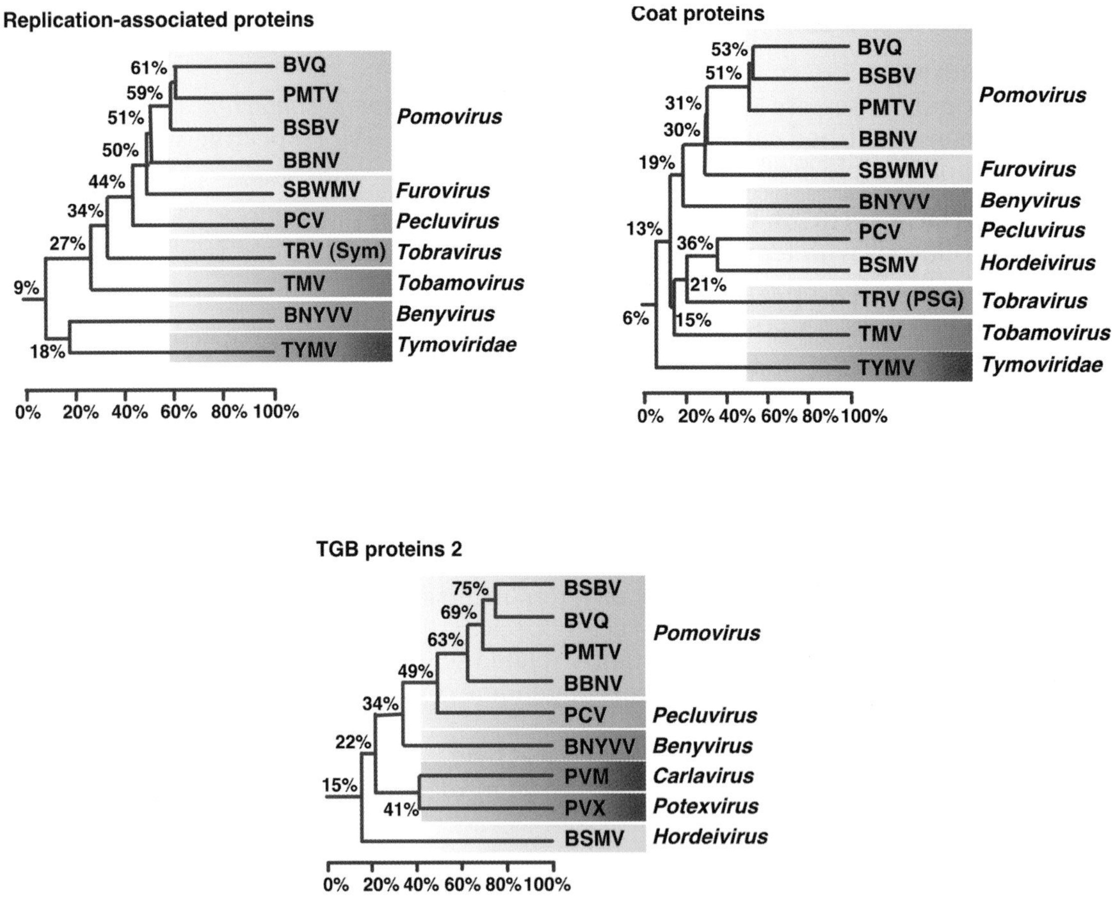

Figure 3: Phylogenetic trees showing the percentages of aa sequence identities among the replication-associated proteins, the CPs and the TGBp2 of pomoviruses, of other plant viruses with rod-shaped particles and of Turnip yellow mosaic virus (as an unrelated plant virus, only for the replication-associated and CPs). The alignments were made by the GCG programs LINEUP and PILEUP and the trees were generated by the program DNAMAN (Lynnon Bio/Soft).

Similarity with Other Taxa

Pomoviruses are morphologically similar to other rod-shaped viruses, i.e. furoviruses, pecluviruses, hordeiviruses, tobraviruses, tobamoviruses and benyviruses. The coat proteins of these viruses have a number of conserved residues, e.g. RF and FE in their central and C-terminal parts, respectively, which are thought to be involved in the formation of salt bridges. The derived aa sequences for the putative RNA-1-encoded proteins also suggest relationships to furoviruses, pecluviruses, tobraviruses, hordeiviruses and tobamoviruses (Fig. 3). Affinities not only to pecluviruses, benyviruses and hordeiviruses, but also to potexviruses and carlaviruses are suggested by the derived aa sequences of their triple gene block-encoded proteins (Fig. 3). The folding properties of the 5-UTRs of their genomic RNAs suggest affinities to the tymoviruses, those of the 3′-UTRs to tymoviruses, tobamoviruses and hordeiviruses. Pomoviruses differ from furoviruses, pecluviruses and tobamoviruses by having a tripartite genome and from hordeiviruses by having the proteins presumably involved in replication encoded on one rather than two RNA molecules.

Derivation of Names

Pomo: sigla from *Po*tato *mo*p-top virus.

References

Cowan, G.H., Lioliopoulou, F., Ziegler, A. and Torrance, L. (2002). Subcellular localisation, protein interactions, and RNA binding of Potato mop-top virus triple block proteins. *Virology,* **298**, 106-115.

Cowan, G.H., Torrance, L. and Reavy, B. (1997). Detection of potato mop-top virus capsid readthrough protein in virus particles. *J. Gen. Virol.,* **78**, 1779-1783.

Goodwin, J.B. and Dreher, T.W. (1998). Transfer RNA mimicry in a new group of positive-strand RNA plant viruses, the furoviruses: differential aminoacylation between the RNA components of one genome. *Virology,* **246,** 170-178.

Kashiwazaki, S., Scott, K.P., Reavy, B. and Harrison, B.D. (1995). Sequence analysis and gene content of potato mop-top virus RNA 3: further evidence of heterogeneity in the genome organization of furoviruses. *Virology,* **206**, 701-706.

Koenig, R., Beier, C., Commandeur, U., Lüth, U., Kaufmann, A. and Lüddecke, P. (1996). Beet soil-borne virus RNA 3 - a further example for the heterogeneity of the gene content of furovirus genomes and of triple gene block-carrying RNAs. *Virology,* **216**, 202-207.

Koenig, R., Commandeur, U., Loss, S., Beier, C., Kaufmann, A. and Lesemann, D.-E. (1997). Beet soil-borne virus RNA-2: similarities and dissimilarities to the coat protein gene-carrying RNAs of other furoviruses. *J. Gen. Virol.,* **78**, 469-477.

Koenig, R. and Loss, S. (1997). Beet soil-borne virus RNA 1: genetic analysis enabled by a starting sequence generated with primers to highly conserved helicase-encoding domains. *J. Gen. Virol.,* **78**, 3161-3165.

Koenig, R., Pleij, C.W.A., Beier, C. and Commandeur, U. (1998). Genome properties of beet virus Q, a new furo-like virus from sugarbeet, determined from unpurified virus. *J. Gen. Virol.,* **79**, 2027-2036.

Koonin, E.V. and Dolja, V.V. (1993). Evolution and taxonomy of positive-strand RNA viruses: Implications of comparative analysis of amino acid sequences. *Crit. Rev. Biochem. Mol. Biol.,* **28**, 375-430.

Lu, X., Yamamoto, S., Tanaka, M., Hibi, T. and Namaba, S. (1998). The genome organization of the broad bean necrosis virus (BBNV). *Arch. Virol.,* **143**, 1335-1348.

McGeachy, K.D. and Barker, H. (2000). Potato mop-top virus RNA can move long distance in the absence of coat protein: evidence from resistant, transgenic plants. *Mol. Plant Microbe Interact.,* **13**, 125-128.

Pereira, L.G., Torrance, L., Roberts, I.M. and Harrison, B.D. (1994). Antigenic structure of the coat protein of potato mop-top virus. *Virology,* **203**, 277-285.

Reavy, B., Arif, M., Cowan, G.H. and Torrance, L. (1998). Association of sequences in the coat protein/readthrough domain of potato mop-top virus with transmission by *Spongospora subterranea. J. Gen. Virol.,* **79**, 2343-2347.

Sandgren, M., Savenkov, E.I. and Valkonen, J.P. (2001). The readthrough region of Potato mop-top virus (PMTV) coat protein encoding RNA, the second largest RNA of PMTV genome, undergoes structural changes in naturally infected and experimentally inoculated plants. *Arch. Virol.,* **146**, 467-477.

Savenkov, E.I., Sandgren, M. and Valkonen, J.P. (1999). Complete sequence of RNA 1 and the presence of tRNA-like structures in all RNAs of Potato mop-top virus, genus Pomovirus. *J. Gen. Virol.*, **80**, 2779-2784.
Torrance, L. and Mayo, M.A. (1997). Proposed re-classification of furoviruses. *Arch. Virol.*, **142**, 435-439.

CONTRIBUTED BY

Koenig, R. and Lesemann D.-E.

GENUS *PECLUVIRUS*

Type Species *Peanut clump virus*

VIRION PROPERTIES

MORPHOLOGY

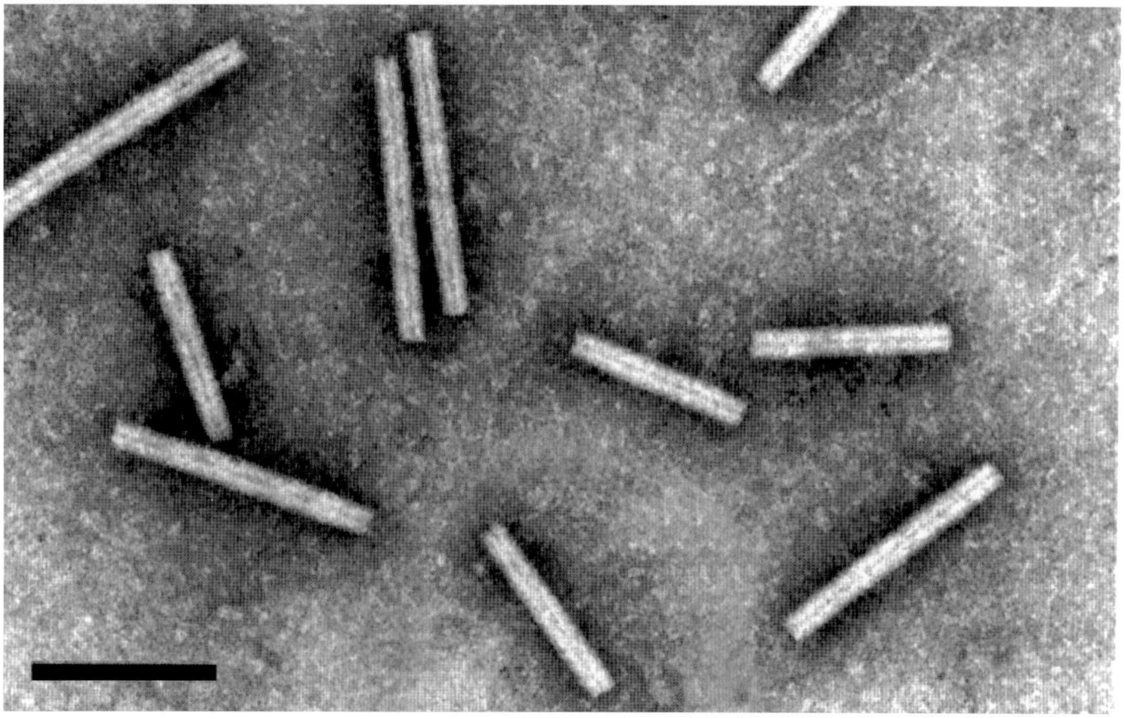

Figure 1: Negative contrast electron micrograph of virions of Indian peanut clump virus (IPCV) (L serotype) negatively stained with 2% phosphotungstic acid, pH 6. The bar represents 150 nm. (Courtesy, G.H. Duncan).

Virions are rod-shaped of about 21 nm in diameter and of two predominant lengths of 190 and 245 nm (Fig. 1). The length distribution of the short particles is broad and in some preparations an additional class of 160 nm is recognizable. Virions have helical symmetry with a pitch of 2.6 nm.

PHYSICOCHEMICAL AND PHYSICAL PROPERTIES

Virions sediment as two major components with S_{20w} of 183S and 224S. Buoyant density in CsCl is 1.32 g/cm^3. Virion isoelectric point is pH 6.45. Thermal inactivation of infectivity occurs at 64°C. Virions are stable in frozen leaves.

NUCLEIC ACID

The genome consists of two molecules of linear positive sense ssRNA; RNA-1 of ~5,900 nt and RNA-2 of ~4,500 nt. RNAs are thought to have a 5'-cap structure but this has not been confirmed. The 3'-ends of the RNAs are not polyadenylated.

PROTEINS

The virion CP subunits are 23 kDa.

LIPIDS

None reported.

CARBOHYDRATES

None reported.

POS. SSRNA

GENOME ORGANIZATION AND REPLICATION

RNA-1 contains two ORFs (Fig. 2). The 5' ORF encodes a 131 kDa protein, and suppression of a termination codon results in the synthesis of a readthrough protein of 191 kDa. The 3' ORF encodes a 15 kDa protein. The proteins of 131 and 191 kDa contain NTP-binding, Hel and RNA polymerase motifs that make a putative replication complex. For Peanut clump virus (PCV), these proteins are respectively 88%, 95% and 75% similar to the products of an isolate of one serotype of Indian peanut clump virus (IPCV). The 15 kDa protein is translated from a sgRNA. It is a suppressor of post-transcriptional gene silencing and is targeted to peroxisomes or related punctate bodies during infection. RNA-2 contains 5 ORFs: the ORF near the 5'-end encodes the CP, the second ORF which, in PCV RNA-1, overlaps the first ORF by 2 nts encodes a 39 kDa protein. This protein is expressed by leaky scanning and is thought to be involved in the transmission of PCV by its fungus vector. Further downstream, separated by a 135 nt intergenic region, is a triple gene block sequence that codes for proteins of 51, 14 and 17 kDa that are thought to be involved in the movement of virus from cell to cell. The proteins are probably expressed via sgRNAs, but these have not been clearly identified. The 3'-NCR for PCV are 298 nt for RNA-1 and of 275 nt for RNA-2; the last 96 nt are identical in both RNAs. The NCRs differ in size among isolates from the different serotypes of IPCV.

The two RNAs are required for systemic invasion of plants but RNA-1 is able to replicate in absence of RNA-2 in protoplasts. The virus is found in the cells of roots, stems and leaves of systemically infected plants.

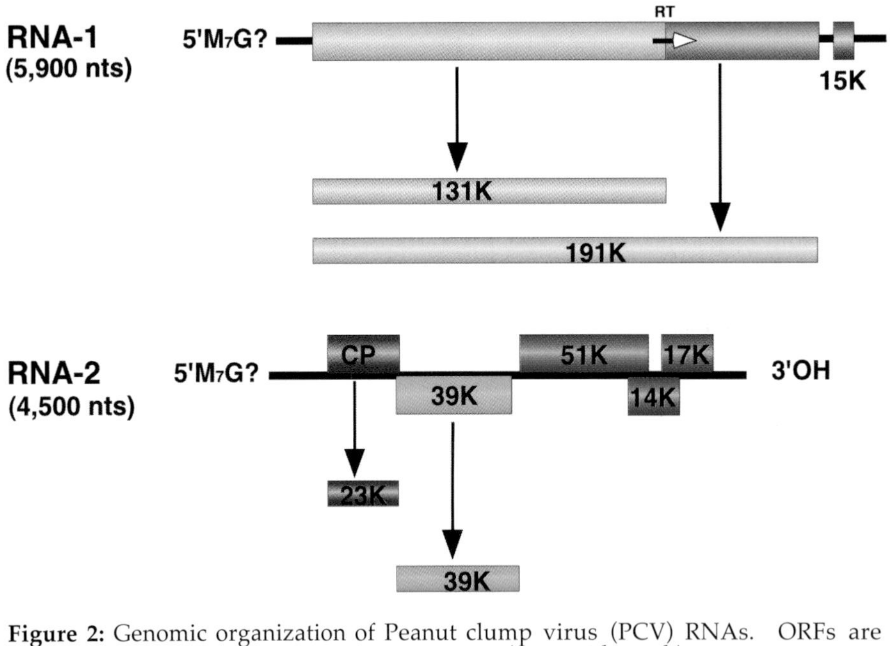

Figure 2: Genomic organization of Peanut clump virus (PCV) RNAs. ORFs are indicated by rectangles and suppressible termination codon by an arrow (RT readthrough).

ANTIGENIC PROPERTIES

The virus is highly immunogenic. There is a great serological variability among isolates of PCV. IPCV isolates fall into one of three very distinct serotypes; IPCV-H, IPCV-L, IPCV-T. All are serologically distinct from PCV.

Biological Properties

Host Range

The natural host first reported was *Arachis hypogea* (groundnut, Leguminosae). Disease symptoms are stunting - mottle - mosaic - chlorotic ringspot. PCV infects *Sorghum arundinaceum*, usually symptomlessly. IPCV infects a number of cereal crops and graminaceous weeds, some symptomlessly and others to induce stunting. The experimental host range is wide and includes species of Aizoaceae, Amaranthaceae, Chenopodiaceae, Cucurbitaceae, Graminae, Leguminosae, Scrophulariaceae, and Solanaceae. *Nicotiana benthamiana* and *Phaseolus vulgaris* are experimental propagation hosts, *Chenopodium amaranticolor*, and *Chenopodium quinoa* are local lesions hosts.

Transmission

The virus is transmitted naturally by a plasmodiophorid fungus (*Polymyxa graminis*) or by seed (in groundnuts). It is mechanically transmissible.

Geographical Distribution

PCV spreads in West Africa (Bénin, Burkina Faso, Congo, Côte d'Ivoire, Mali, Niger, Senegal and Pakistan). IPCV is widely distributed in India and Pakistan. A soil type favorable to the vector is a prerequisite for virus to cause disease.

List of Species Demarcation Criteria in the Genus

The species are distinguished by different reactions with particular antisera (heterologous reactions are weak or undetectable). Also, PCV occurs only in Africa whereas IPCV occurs in the Indian subcontinent. However, isolates of IPCV can be readily assigned to one of three serotypes as protein preparations made from particles of each serotype barely react with heterologous antisera in immunoblotting tests. Isolates of PCV are also heterogeneous in their reactions with a panel of monoclonal antibodies. Moreover, several of the proteins encoded by genes in RNA of the different serotypes of IPCV differ in sequence from corresponding proteins of other IPCV serotypes by about as much as each differs from the corresponding protein of one isolate of PCV.

List of Species in the Genus

Species names are in green italic script; strain names and synonyms are in black roman script; tentative species names are in blue roman script. Sequence accession numbers, and assigned abbreviations ( ) are also listed.

Species in the Genus

Indian peanut clump virus		
Indian peanut clump virus - H	[X99149, AF447397]	(IPCV-H)
Indian peanut clump virus - D	[- , AF447396]	(IPCV-D)
Indian peanut clump virus - L	[- , AF239729]	(IPCV-L)
Peanut clump virus		
Peanut clump virus - S	[X78602, L07269]	(PCV-S)
Peanut clump virus - Ni	[- , AF447398]	(PCV-Ni)
Peanut clump virus - N	[- , AF447399]	(PCV-N)
Peanut clump virus - M	[- , AF447400]	(PCV-M)
Peanut clump virus - B	[- , AF447401]	(PCV-B)

Tentative Species in the Genus

None reported.

Phylogenetic Relationships within the Genus

Not available.

SIMILARITY WITH OTHER TAXA

There are similarities among the putative replicase proteins of PCV and IPCV with those of *Soil-borne wheat mosaic virus* (SBWMV), the type species of the genus *Furovirus*. These proteins are more closely related to those of *Barley stripe mosaic virus* (BSMV; genus *Hordeivirus*) than to those of *Beet necrotic yellow vein virus* (BNYVV: genus *Benyvirus*, previously classified in the genus *Furovirus*). The CP sequences are 37% (IPCV) and 36% (PCV) identical to that of the CP of BSMV.

REFERENCES

Bouzoubaa, S. (1998). Furovirus isolation and RNA extraction. In: *Plant Virology Protocols. From virus isolation to transgenic resistance.* (G.D. Foster and S. Taylor, eds), Vol. 81, pp 107-114. Humana Press Inc, Totowa, NJ.

Dunoyer, P., Pfeffer, S., Fritsch, C., Hemmer, O., Voinnet, O. and Richards, K.E. (2002). Identification, subcellular localization and some properties of a cysteine-rich suppressor of gene silencing encoded by peanut clump virus. *Plant J.*, **29**, 555-567.

Herzog, E., Guilley, H. and Fritsch, C. (1995). Translation of the second gene of peanut clump virus RNA-2 occurs by leaky scanning *in vitro*. *Virology*, **208**, 215-225.

Herzog, E., Guilley, H., Manohar, S.K., Dollet, M., Richards, K., Fritsch, C. and Jonard, G. (1994). Complete nucleotide sequence of peanut clump virus RNA 1 and relationships with other fungus-transmitted rod-shaped viruses. *J. Gen. Virol.*, **75**, 3147-3155.

Legrève, A., Vanpee, B., Risopoulos, J., Ward, E. and Maraite, H. (1996). Characterization of *Polymyxa* sp. associated with the transmission of indian peanut clump virus. Third symposium of the international working group on plant viruses with fungal vectors. (J.L. Sherwood and C.M. Rush, eds), pp 157-160.

Manohar, S.K., Guilley, H., Dollet, M., Richards, K. and Jonard, G. (1993). Nucleotide sequence and genetic organization of peanut clump virus RNA-2 and partial characterization of deleted forms. *Virology*, **195**, 33-41.

Mayo, M.A., and Reddy, D.V.R. (1985). Translation products of RNA from Indian peanut clump virus. *J. Gen. Virol.*, **66**, 1347-1351.

Miller, J.S., Wesley, S.V., Naidu, R.A., Reddy, D.V.R. and Mayo, M.A. (1996). The nucleotide sequence of RNA-1 of Indian peanut clump furovirus. *Arch. Virol.*, **141**, 2301-2312.

Naidu, R.A., Sawyer, S. and Deom, C.M. (2003). Molecular diversity of RNA-2 genome segments in pecluviruses causing peanut clump disease in West Africa and India. *Arch. Virol.*, **148,** 83-98.

Nolt, B.L., Rajeshwari, R., Reddy, D.V.R., Bharathan, N. and Manohar, S.K. (1988). Indian peanut clump virus isolates: host range, symptomatology, serological relationships, and some physical properties. *Phytopathology*, **78**, 310-313.

Reddy, D.V.R., Rajeshwari, R., Iisuka, W., Lesemann, D.E., Nolt, B.L. and Goto, T. (1983). The occurrence of Indian peanut clump, a soil-borne virus disease of groundnuts (*Arachis hypogaea*) in India. *Ann. Appl. Biol.*, **102**, 305-310.

Reddy, D.V.R., Robinson, D.J., Roberts, I.M. and Harrison, B.D. (1985). Genome properties and relationships of Indian peanut clump virus. *J. Gen. Virol.*, **66**, 2011-2016.

Thouvenel, J.-C. and Fauquet, C. (1981). Further properties of peanut clump virus and studies on its natural transmission. *Ann. Appl. Biol.*, **97**, 99-107.

Thouvenel, J. C., Dollet, M. and Fauquet, C. (1976). Some properties of peanut clump, a newly discovered virus. *Ann. Appl. Biol.*, **84**, 311-320.

Torrance, L. and Mayo, M.A. (1997). Proposed re-classification of furoviruses. *Arch. Virol.*, **142**, 435-439.

Wesley, S.V., Mayo, M.A., Jolly, C.A., Naidu, R.A., Reddy, D.V.R., Jana, M.K. and Parnaik, V.K. (1994). The coat protein of Indian peanut clump virus: relationships with other furoviruses and with barley stripe mosaic virus. *Arch. Virol.*, **134**, 271-278.

CONTRIBUTED BY

Richards, K. E., Deom, C. M. and Naidu, R. A.

GENUS BENYVIRUS

Type Species *Beet necrotic yellow vein virus*

VIRION PROPERTIES

MORPHOLOGY

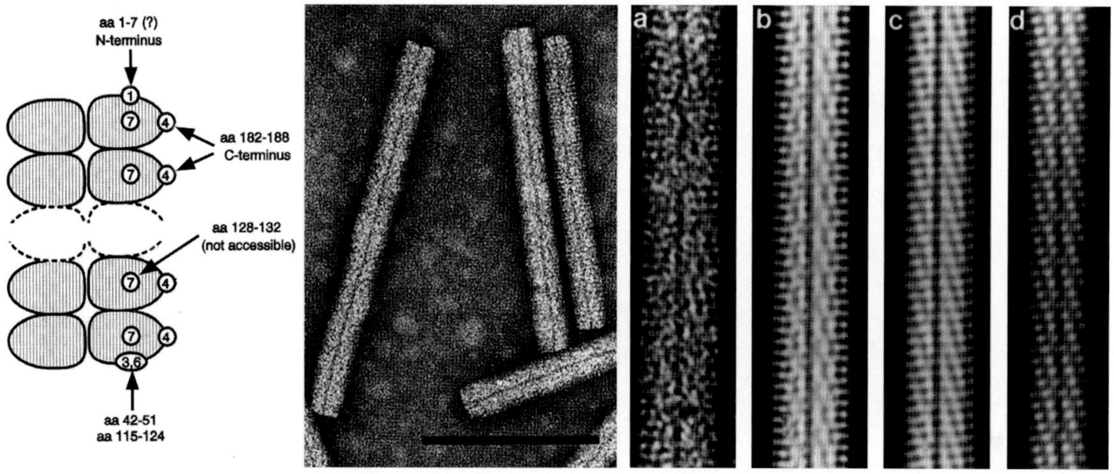

Figure 1: (Left) Scheme showing the accessibility to antibodies of various parts of the coat protein amino acid (aa) sequence in particles of Beet necrotic yellow vein virus (BNYVV). Encircled numbers designate different epitopes. (Center) Negative contrast electron micrograph of stained purified particles of BNYVV. (Right) From left (a) negative contrast electron micrograph of a BNYVV particle and (b, c, d) computer-filtered micrographs of BNYVV particles (Courtesy of A.C. Steven, from *Virology* **113**, 428, (1981), with permission). The bar represents 100 nm.

The non-enveloped, rod-shaped particles are helically constructed with an axial canal (Fig. 1). They have predominant lengths of c. 85, 100, 265 and 390 nm and diameters of 20 nm. The right-handed helix with a pitch of 2.6 nm has an axial repeat of four turns, involving 49 CP subunits. Each CP subunit occupies four nucleotides on the RNA.

PHYSICOCHEMICAL AND PHYSICAL PROPERTIES

The viruses are rather unstable in sap. At room temperature, most of the infectivity is lost within one or a few days.

NUCLEIC ACID

In naturally infected plants, virions contain four - or with some BNYVV isolates five – molecules of linear positive-sense ssRNA of c. 6.7, 4.6, 1.8, 1.4 and 1.3 kb, respectively. After mechanical transmission to test plants, RNAs 3, 4 and 5 of BNYVV may become partially deleted or may be lost entirely. The RNAs are capped at the 5'-end and - unlike the RNAs of all other plant viruses with rod-shaped particles - they are 3'-polyadenylated. The complete sequence has been determined for all five RNAs of different isolates of BNYVV and for the four RNAs of an isolate of Beet soil-borne mosaic virus (BSBMV). Partial sequences which have not yet been released have been determined for RNA-1 of Rice stripe necrosis virus (RSNV) as well as for RNAs 1 and 2 of Burdock mottle virus (BdMV).

PROTEIN

The major CP species is 21-23 kDa in size. The CP readthrough protein which may be detected in some BNYVV particles near one extremity by means of immunogold labelling apparently initiates virus assembly. A KTER motif in the C terminal part of the BNYVV CP readthrough protein is necessary for the transmission of the virus by *Polymyxa betae*. The three triple gene block (TBG)-encoded proteins (TGBp1, TGBp2 and TGBp3) are necessary for cell to cell movement. BNYVV TGBp1 labeled with green fluorescent protein (GFP) on its N-terminus is targeted by TGBp2 and TGBp3 to punctate bodies associated with plasmodesmata. The three proteins can be functionally substituted by the MP of Tobacco mosaic virus or the three TGB-encoded proteins of Peanut clump virus. The latter have to be supplied together, they are unable to substitute their BNYVV counterparts one by one. The N-terminal part of BNYVV TGBp1 has nucleic acid binding activity, its C-terminal part contains consensus sequence motifs characteristic of an ATP/GTP-dependent helicase.

Lipid

None reported.

Carbohydrates

None reported.

Genome Organization and Replication

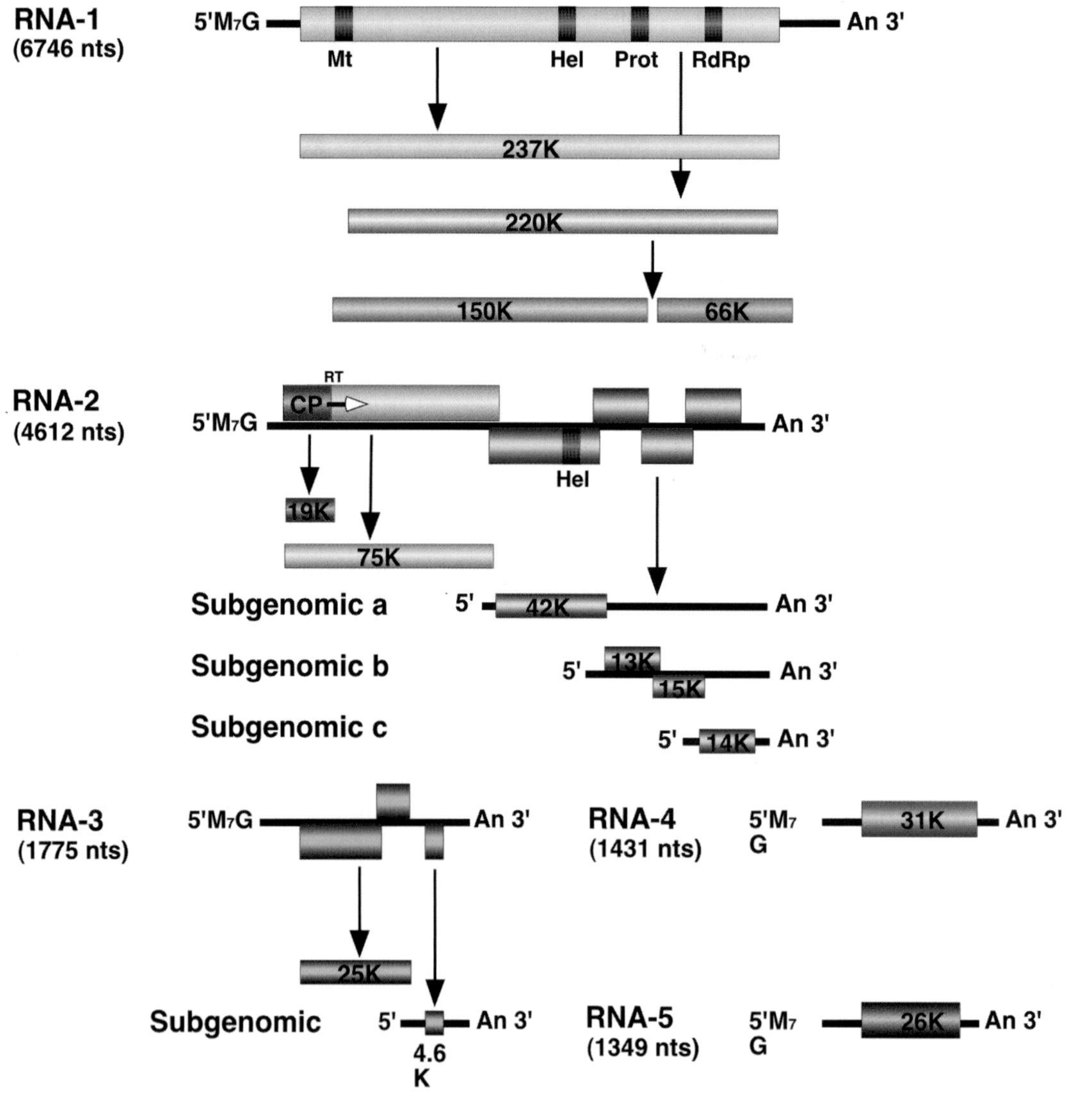

Figure 2: Genome organization and translation strategies of Beet necrotic yellow-vein virus (BNYVV). The scheme indicates a suppressible UAG stop codon (arrow), and (An) the 3′ poly (A)-tails. Dark areas indicate conserved motifs. Hel; helicase, Mtr; methyltransferase, Pro: protease, RdRp; RNA dependent RNA polymerase, RT; readthrough.

RNA-1 contains one large ORF for a putative replication-associated protein which is cleaved post-translationally. This distinguishes the benyviruses from all other viruses with rod-shaped particles, which have their replication-associated proteins encoded on two ORFs. *In vitro* translation of BNYVV RNA-1 may initiate at two sites: at the first AUG in the sequence at position 154 or at a downstream AUG at position 496. The resulting proteins of 237 and 220 kDa, respectively, both contain in their N-terminal part Mtr motifs, in their central part Hel and papain-like protease motifs (Prot) and in their C-terminal part RdRp motifs (Fig. 2). RNA-2 of BNYVV contains six ORFs, i.e. the CP gene which is terminated by a suppressible UAG stop codon, the CP readthrough protein gene, the TGB coding for TGBp1, TGBp2 and TGBp3 (42, 13 and 15 kDa respectively, with BNYVV) and a gene coding for a 14 kDa cysteine-rich protein. The TGB ORFs and the P14 gene are expressed by means of subgenomic RNAs (Fig. 2). RNAs 1 and 2 are sufficient for replication of BNYVV in the local lesion host *Chenopodium quinoa*. The typical rhizomania symptoms in beet are produced only in the presence of RNA-3; RNA-4 greatly increases the transmission rate by *Polymyxa betae* and RNA-5 may modulate the type of symptoms formed. RNAs-3 and -4 are always present in natural BNYVV infections.

Antigenic Properties

Virions are moderately to strongly antigenic. BNYVV and BSBMV are very distantly related serologically. Epitope mapping has revealed portions of the aa sequence of BNYVV CP which are either exposed along the entire particle length, e.g. the immunodominant C-terminus, or are accessible only on one extremity or after disruption of the particles (Fig. 1).

Biological Properties

Host Range

The natural host ranges of benyviruses are very narrow. Species of *Chenopodium* are infected experimentally, often only locally.

Transmission

In nature BNYVV and BSBMV are transmitted by *Polymyxa betae* and RNSV by *P. graminis*. The viruses are also mechanically transmissible.

Geographical Distribution

BNYVV has spread to most sugarbeet-growing areas world-wide. Different variants (A type, B type, P type) occur in different geographical areas. BSBMV is widely distributed in the USA. RSNV occurs in Africa, South and Central America. BdMV has been identified in a restricted area in Japan.

Cytopathic effects

At early times post infection BNYVV has been found at the cytoplasmic surface of mitochondria. Later virions of most BNYVV isolates are scattered throughout the cytoplasm of infected cells or occur in aggregates. More or less dense masses of particles arranged in parallel or angle-layer arrays may be formed. Depending on the isolate only one or both types of aggregates occur. Membraneous accumulations of endoplasmic reticulum may also be found.

List of Species Demarcation Criteria in the Genus

The criteria demarcating species in the genus are:

- Distantly related serologically,
- Less than 90% identity in their CP aa sequence

List of Species in the Genus

Species names are in green italic script; strain names and synonyms are in black roman script; tentative species names are in blue roman script. Sequence accession numbers, and assigned abbreviations ( ) are also listed.

Species in the Genus

Beet necrotic yellow vein virus		
Beet necrotic yellow vein virus - C, China	[RNA-3: AJ239200; RNA-4: AJ239199; RNA-5: AJ236894-5]	(BNYVV-C)
Beet necrotic yellow vein virus – F2, France	[RNA-1: X05147; RNA-2: X04197; RNA-3: M36894; RNA-4: M36896]	(BNYVV-B-Eur)
Beet necrotic yellow vein virus – F72, France	[RNA-2: AF197547*; RNA-4: AF197546*; RNA-5: U78293*]	(BNYVV-P-Eur)
Beet necrotic yellow vein virus – N7, The Netherlands	[RNA-3: AF197558*; RNA-4: AF197559*]	(BNYVV-A-Eur)
Beet necrotic yellow vein virus - S, Japan	[RNA-1: D84410; RNA-2: D84411; RNA-3: D84412; RNA-4: D84413; RNA-5: AB018607-10*]	(BNYVV-A-Jap)
Beet necrotic yellow vein virus, Kazakhstan	[RNA-2: AF197556; RNA-4: AF197554*; RNA 5: AF197555*]	(BNYVV-P-Kaz)
Beet soil-borne mosaic virus		
Beet soil-borne mosaic virus - EA	[RNA-1: NC_003506; RNA-2: NC_003503; RNA-3: NC_003507; RNA-4: NC_003508]	(BSBMV)

- Sequences lack short stretches at the 5' and 3'-ends

Tentative Species in the Genus

Burdock mottle virus	(BdMV)
Rice stripe necrosis virus	(RSNV)

Phylogenetic Relationships Within the Genus

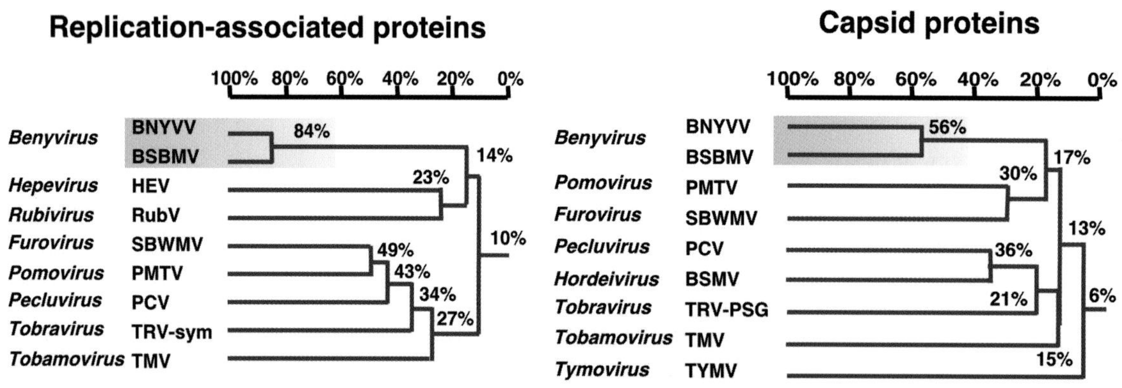

Figure 3: Phylogenetic trees showing the percentages of aa sequence identities of the replication-associated proteins and the CPs of benyviruses, of other plant viruses with rod-shaped particles, of Hepatitis virus E and Rubella virus (replication-associated proteins only) and of Turnip yellow mosaic virus as an unrelated plant virus (CP only). The alignments were made by the GCG programs LINEUP and PILEUP and the trees were generated by the program DNAMAN (Lynnon Bio/Soft).

The CP aa sequences of BNYVV and BSBMV share less than 60% homology (Fig. 3). The percentages of identity between various non-structural proteins of the two viruses range between 38% (cysteine-rich protein) and 84% (replication-associated protein). The CP of BdMV shares 38% of sequence identity with that of BNYVV. TGBp1 of BdMV shares c. 50% identity with the corresponding proteins of BNYVV and BSBMV. A 2,239 nt fragment of RSNV RNA-1 shares about 50% sequence identity with the corresponding regions of BNYVV and BSBMV RNA-1.

Similarities with Other Taxa

Benyviruses are morphologically similar to other rod-shaped viruses, i.e. furo-, peclu-, pomo-, hordei-, tobra- and tobamoviruses. The CPs of these viruses have a number of conserved residues, e.g. RF and FE in their central and C-terminal parts, respectively, which are presumably involved in the formation of salt bridges. Like pomo-, peclu- and hordeiviruses, but unlike furo-, tobamo- and tobraviruses, the benyviruses have their movement function encoded on a triple gene block. Sequence identities in the first and second triple gene block-encoded proteins reveal affinities not only to pomo- and hordei-, but also to potex- and carlaviruses. The Mtr, Hel and RdRp motifs in the putative replication-associated proteins show a higher degree of similarity to those of the viruses of the family *Togaviridae* (Rubella virus) and of Hepatitis virus E than to those of other rod-shaped plant viruses.

Derivation of Name

Beny: siglaw from *Be*et *n*ecrotic *y*ellow vein virus.

References

Commandeur, U., Koenig, R., Manteuffel, R., Torrance, L., Lüddecke, P. and Frank, R. (1994). Location, size and complexity of epitopes on the coat protein of beet necrotic yellow vein virus studied by means of synthetic overlapping peptides. *Virology*, **198**, 282-287.

Erhardt, M., Morant, M., Ritzenthaler, C., Stussi-Garaud, C., Guilley, H., Richards, K, Jonard, G., Bouzoubaa, S. and Gilmer, D. (2000). P42 movement protein of Beet necrotic yellow vein virus is targeted by the movement proteins P13 and P15 to punctate bodies associated with plasmodesmata. *Mol. Plant Microbe Interact.* **13,** 520-528.

Erhardt, M., Dunoyer, P., Guilley, H., Richards, K., Jonard, G. and Bouzoubaa, S. (2001). Beet necrotic yellow vein virus particles localize to mitochondria during infection. *Virology*, **286**, 256-62.

Haeberlé, A.M., Stussi-Garaud, C., Schmitt, C., Garaud, J.C., Richards, K.E., Guilley, H. and Jonard, G. (1994). Detection by immunogold labelling of P75 readthrough protein near an extremity of Beet necrotic yellow vein virus particles. *Arch. Virol.*, **134**, 195-203.

Hehn, A., Fritsch, C., Richards, K.E., Guilley, H. and Jonard, G. (1997). Evidence for *in vitro* and *in vivo* autocatalytic processing of the primary translation product of Beet necrotic yellow vein virus RNA 1 by a papain-like proteinase. *Arch. Virol.*, **142**, 1051-1058.

Hirano, S., Kondo, H., Maeda, T. and Tamada, T. (1999). Burdock mottle virus has a high genome similarity to beet necrotic yellow vein virus. *Proceedings of the Fourth Symposium of the International Working Group on Plant Viruses with Fungal Vectors, Asilomar, October 5-8,*1999, (J.L. Sherwood and C.M. Rush, eds), pp 33-36.

Kiguchi, T., Saito, M. and Tamada, T. (1996). Nucleotide sequence analysis of RNA-5 of five isolates of Beet necrotic yellow vein virus and the identity of a deletion mutant. *J. Gen. Virol.*, **77**, 575-580.

Koenig, R., Haeberlé, A.M. and Commandeur, U. (1997). Detection and characterization of a distinct type of beet necrotic yellow vein virus RNA 5 in a sugarbeet growing area in Europe. *Arch. Virol.*, **142**, 1499-1504.

Koenig, R., Jarausch, W., Li, Y., Commandeur, U., Burgermeister, W., Gehrke, M. and Lüddecke, P. (1991). Effect of recombinant Beet necrotic yellow vein virus with different RNA compositions on mechanically inoculated sugarbeets. *J. Gen. Virol.*, **72**, 2243-2246.

Koenig, R., Lüddecke, P. and Haeberlé, A.M. (1995). Detection of Beet necrotic yellow vein virus strains, variants and mixed infections by examining single-strand conformation polymorphisms of immunocapture RT-PCR products. *J. Gen. Virol.*, **76**, 2051-2055.

Lauber, E., Bleykasten-Grosshans, C., Erhardt, M., Bouzoubaa, S., Jonard, G., Richards. K.E. and Guilley, H. (1998). Cell-to-cell movement of Beet necrotic yellow vein virus: I. Heterologous complementation experiments provide evidence for specific interactions among the triple gene block proteins. *Mol Plant Microbe Interact.*, **11**, 618-25.

Lee, L., Telford, E.B., Batten, J.S., Scholthof, K.B. and Rush, C.M. (2001). Complete nucleotide sequence and genome organization of Beet soilborne mosaic virus, a proposed member of the genus *Benyvirus*. *Arch. Virol.*, **146**, 2443-2453.

Richards, K. and Tamada, T. (1992). Mapping functions on the multipartite genome of beet necrotic yellow vein virus. *Ann. Rev. Phytopathol.*, **30**, 291-313.

Tamada, T. and Abe, H. (1989). Evidence that beet necrotic yellow vein virus RNA-4 is essential for efficient transmission by the fungus *Polymyxa betae*. *J. Gen. Virol.*, **70**, 3391-3398.
Tamada, T., Schmitt, C., Saito, M., Guilley, H., Richards, K. and Jonard, G. (1996). High resolution analysis of the readthrough domain of Beet necrotic yellow vein virus readthrough protein: a KTER motif is important for efficient transmission of the virus by *Polymyxa betae*. *J. Gen. Virol.*, **77**, 1359-1367.

Contributed by

Koenig, R. and Lesemann, D.-E.

FAMILY BROMOVIRIDAE

TAXONOMIC STRUCTURE OF THE FAMILY

Family	***Bromoviridae***
Genus	***Alfamovirus***
Genus	***Bromovirus***
Genus	***Cucumovirus***
Genus	***Ilarvirus***
Genus	***Oleavirus***

VIRION PROPERTIES

MORPHOLOGY

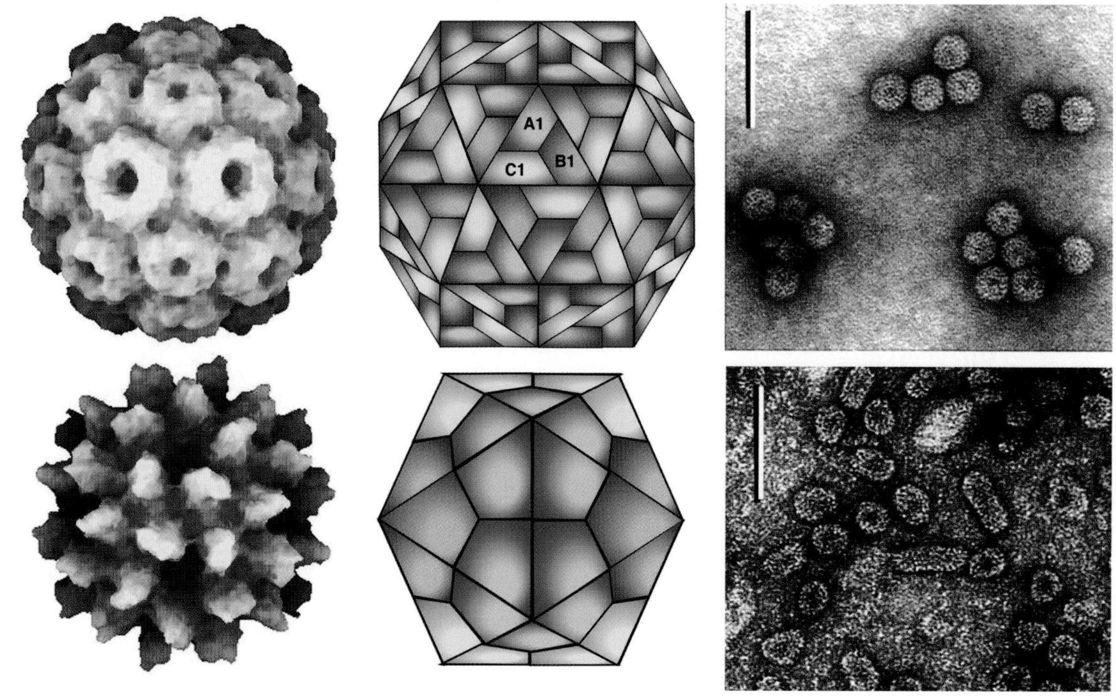

Figure 1: (Top left) Image reconstruction of a particle of Cowpea chlorotic mottle virus (CCMV)(genus *Bromovirus*), showing pentamer and hexamer clustering in a T=3 quasi-icosahedron (Lucas *et al.*, 2001); (Top central) Schematic representation of a T=3 particle; (Top right) Negative contrast electron micrograph of particles of Cucumber mosaic virus (CMV)(genus *Cucumovirus*), (Courtesy of G. Kasdorf); (Bottom left) Electron density representation of a Ta particle of Alfalfa mosaic virus (AMV)(genus *Alfamovirus*), showing T=1 structure (Kumar *et al.*, 1997); (Bottom Center) Schematic representation of a T=1 particle; (Bottom right) Negative contrast electron micrograph of particles of Prune dwarf virus (PDV)(genus *Ilarvirus*). (Courtesy of G. Kasdorf). The bars represents 100 nm.

Virions are either spherical (Fig. 1), having T=3 icosahedral symmetry, and a diameter of 26-35 nm (genera *Bromovirus*, *Cucumovirus*, and *Ilarvirus*) or bacilliform (genera *Alfamovirus*, *Ilarvirus*, and *Oleavirus*) having within a species constant diameters of 18-26 nm and lengths from 30 to 85 nm. Genomic RNAs are packaged in separate virions that may also contain sgRNAs, defective RNAs or satellite RNAs.

PHYSICOCHEMICAL AND PHYSICAL PROPERTIES

The Mr of the virions varies from 3.5-6.9 x 10^6, depending on the nucleic acid content. Virion Mr is constant among members of the genera *Bromovirus*, *Cucumovirus*, and some *Ilarvirus* members, and varies among the remaining members of the family. The buoyant density of formaldehyde-fixed virions ranges from 1.35-1.37 g/cm^3 in CsCl. The S_{20w} varies from 63S to 99S. Virion integrity is dependent on RNA-protein interactions and virion RNA is susceptible to RNAse degradation *in situ* at neutral pH. Heat inactivation occurs at 60°C in some genera, others have not been tested. In most cases, virions are unstable in the presence of divalent cations. Virions are generally stable in the presence of chloroform, but not in the presence of phenol. Virions are unstable in the presence of strong anionic detergents such as SDS, but can tolerate low concentrations of mild detergents such as Triton X-100.

NUCLEIC ACID

Total genome length is approximately 8 kb. Genomes consist of three linear, positive sense ssRNAs with 5'-terminal cap structures. The 3'-termini are not polyadenylated, but generally are highly conserved within a species or isolate, and form strong secondary structures. They are either tRNA-like and can be aminoacylated (genera *Bromovirus* and *Cucumovirus*) or form other structures that are not aminoacylated (genera *Alfamovirus*, *Ilarvirus* and *Oleavirus*) (Table 1).

Table 1: Genomic RNA sizes in nucleotide number for typical members of each genus.

Genus	species	strain	RNA-1	RNA-2	RNA-3	3' term.	DIs/sat RNAs
Alfamovirus	AMV	425	3,644	2,593	2,037	complex[a]	-/-
Bromovirus	BMV	Russian	3,234	2,865	2,117	tRNA[b]	+/-
Cucumovirus	CMV	Fny	3,357	3,050	2,216	tRNA	+/+
Ilarvirus	TSV	n.a	3,491	2,926	2,205	complex	-/-
Oleavirus	OLV-2	n.a	3,126	2,734	2,438	complex	?/?

[a]complex secondary structure. [b]aminoacylatable, with pseudoknot folding. n.a.: not applicable, only 1 strain reported

PROTEINS

A single 20-24 kDa CP is expressed from a sgRNA. The CP is generally required for systemic movement and may be required for cell to cell spread in some cases. There are at least three ORFs for non-structural proteins: the replicase (consisting of the 1a and 2a proteins, along with host factors) and the MP (3a) (Table 2). Cucumovirus RNA-2 encodes 2b proteins that are involved in post-transcriptional gene silencing. TSV RNA-2 also encodes an ORF for a 2b protein.

Table 2: Virus proteins, sizes and functional activities.

Protein	size (kDa)	mRNA	Function[1]
1a	102.7-125.8	RNA-1	Mtr, helicase
2a	89.8-96.7	RNA-2	RdRp
3a	30.5-36.5	RNA-3	cell to cell movement
CP	19.8-26.2	sgRNA-4[2]	encapsidation, movement

[1] Functions of 1a and 2a are putative in most cases, by analogy to related viruses,
[2] The sgRNA for the CP derived from RNA-3 is encapsidated in all but the genus *Oleavirus*.

LIPIDS

There are no lipids associated with the virions.

CARBOHYDRATES

There are no carbohydrates associated with the virions.

GENOME ORGANIZATION AND REPLICATION

RNA-1, -2, and -3, can act as mRNAs. The 2b ORF is found in cucumovirus RNA-2 the genus *Cucumovirus*, and is expressed from an additional sgRNA that may or may not be encapsidated. The CP ORF is expressed from a sgRNA that is usually encapsidated (RNA-4). The genomic RNA-1 and -2 each encode a single large ORF, and in some genera RNA-2 also encodes a second ORF that is translated from a sgRNA. RNA-3 encodes a 5' protein, the MP, and the CP, which is translated from a sgRNA. There is no clear evidence of proteolytic or other post-translational processing. Virus replication occurs on cytoplasmic membranes via full length minus (-) strand synthesis and subsequent plus (+) strand synthesis. The sgRNAs are synthesized from the (-) template, and may or may not be found in the virions. The CP accumulates to high levels in infected cells, whereas the nonstructural proteins accumulate to much lower levels. Virions accumulate in the cytoplasm. The life cycle of the virus takes place predominantly in the cytoplasm.

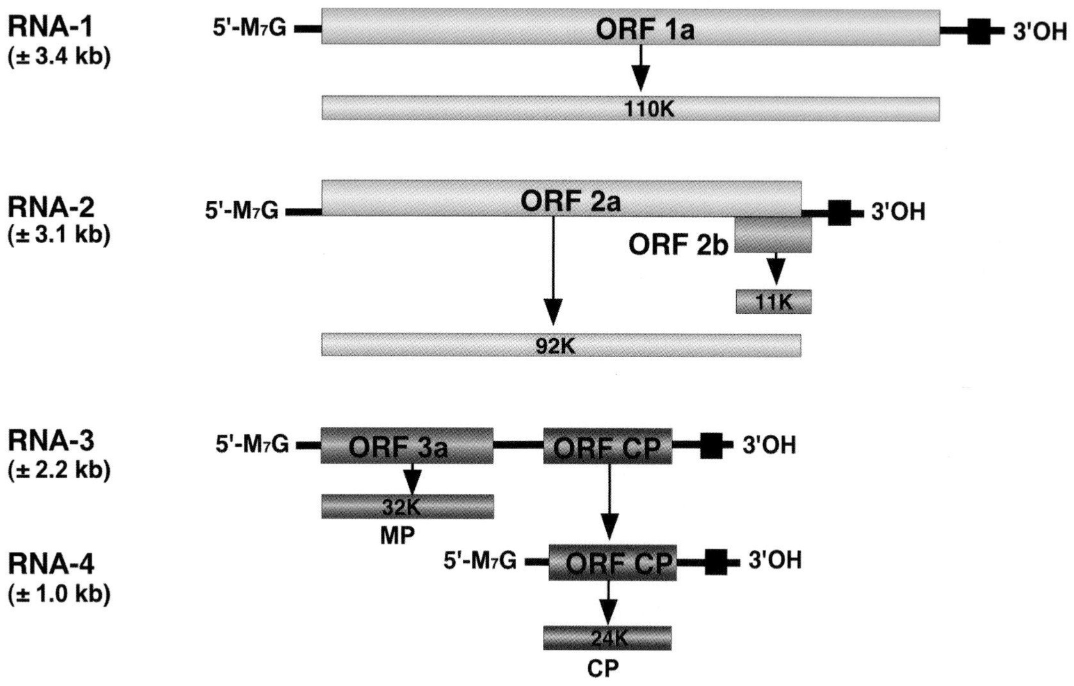

Figure 2: Schematic genome organization for members of the family *Bromoviridae*.

ANTIGENIC PROPERTIES

Native virions are generally poor immunogens and require stabilization with formaldehyde prior to use as antigens. There are little or no serological relationships between the genera, and weak relationships between species of the same genus.

BIOLOGICAL PROPERTIES

The natural host range of the viruses ranges from very narrow (genus *Bromovirus*) to extremely broad (genus *Cucumovirus*). They are predominantly transmitted by insects, in a non-persistent manner, or mechanically. Vectors have not been identified for some of the members of the family. They are distributed worldwide, and several are responsible for major disease epidemics in crop plants.

GENUS ALFAMOVIRUS

Type Species *Alfalfa mosaic virus*

DISTINGUISHING FEATURES

VIRION PROPERTIES

Virions are generally bacilliform, having a constant diameter of 18 nm, and varying from 30-57 nm in length, depending on the nucleic acid species encapsidated. The Mr are from 3.5-6.9 x 10^6. Virions are readily separated into components by sucrose density.

GENOME ORGANIZATION AND REPLICATION

Replication is activated by CP binding to the complex 3'-end structure. CP from members of the genus *Ilarvirus* can also activate replication.

BIOLOGICAL PROPERTIES

The host range is very broad. Viruses are transmitted by aphids in a non-persistent manner.

LIST OF SPECIES DEMARCATION CRITERIA IN THE GENUS

Not applicable.

LIST OF SPECIES IN THE GENUS

Species names are in green italic script; strain names and synonyms are in black roman script; tentative species names are in blue roman script. Sequence accession numbers, and assigned abbreviations ( ) are also listed.

SPECIES IN THE GENUS

Alfalfa mosaic virus

Alfalfa mosaic virus	[strain 425 Leiden; RNA1: L00163, RNA2: L00161, RNA3: L00162,]	(AMV)

TENTATIVE SPECIES IN THE GENUS

None reported.

GENUS *BROMOVIRUS*

Type Species *Brome mosaic virus*

DISTINGUISHING FEATURES

VIRION PROPERTIES

MORPHOLOGY

Virions are polyhedral, and all the same size, with a diameter of 27 nm.

PHYSICOCHEMICAL AND PHYSICAL PROPERTIES

Virions prepared below pH 6.0 have S_{20w} of 88S, are stable to high salt and low detergent concentrations, and are nuclease- and protease-resistant. At pH 7.0 and above, virions swell to a diameter of 31 nm, S_{20w} decreases to 78S, salt and detergent stability decreases dramatically, and protein and RNA are susceptible to hydrolytic enzymes. This swelling is accompanied by conformational changes of the capsid that are detectable by physical and serological means.

NUCLEIC ACID

RNA 3'-termini are tRNA-like, are very similar in all viruses sequenced so far, and can be aminoacylated with tyrosine.

PROTEINS

Coat protein size is 20 kDa, unlike the 24-26 kDa of other members of the family *Bromoviridae*. The 3a proteins and CPs of bromoviruses share sequence similarities with one another, and more distantly with cucumoviruses, but not with ilarviruses or alfamoviruses.

ANTIGENIC PROPERTIES

All members are serologically related, with large antigenic differences between species.

Biological Properties

The natural host range is narrow, and is limited to a few plant hosts for each species. All species are thought to be beetle-transmitted, although BMV is inefficiently transmitted by aphids in a non-persistent manner.

List of Species Demarcation Criteria in the Genus

The criteria demarcating species in the genus are:

- Host range,
- Serological relationships,
- Compatible replicase proteins (i.e. 1a and 2a proteins),
- Nucleotide sequence similarity between species ranges from 50 to 80% depending on the gene used for comparison.

List of Species in the Genus

Species names are in green italic script; strain names and synonyms are in black roman script; tentative species names are in blue roman script. Sequence accession numbers, and assigned abbreviations ( ) are also listed.

Species in the Genus

Broad bean mottle virus		
Broad bean mottle virus	[RNA1: M65138, RNA2: M64713, RNA3: M60291,]	(BBMV)
Brome mosaic virus		
Brome mosaic virus	Russian strain [RNA1: X02380, RNA2: X01678, RNA3: V00099]	(BMV)
Cassia yellow blotch virus		
Cassia yellow blotch virus		(CYBV)
Cowpea chlorotic mottle virus		
Cowpea chlorotic mottle virus	[RNA1: M65139, RNA2: M28817, RNA3: M28818,]	(CCMV)
Melandrium yellow fleck virus		
Melandrium yellow fleck virus		(MYFV)
Spring beauty latent virus		
Spring beauty latent virus	[RNA1: AB080598, RNA2: AB080599, RNA3: AB080600]	(SBLV)

Tentative Species in the Genus

None reported.

Genus *Cucumovirus*

Type Species *Cucumber mosaic virus*

Distinguishing Features

Virion Properties

Morphology

Virions are icosahedral, of uniform size and sedimentaion properties. In electron micrographs they appear to have electron dense centers.

Physicochemical and Physical Properties

Purified virions are labile, and are especially susceptible to anionic detergents and high ionic strength buffers that disrupt the RNA-protein interactions required for particle integrity. Most strains are unstable in the presence of Mg^{2+}, but at least one strain of CMV requires Mg^{2+} for stability.

NUCLEIC ACID

The 3'-termini of all RNAs within a species are highly similar and can form tRNA-like structures that are aminoacylatable with tyrosine. Within a species, the 5'-NTR of RNA-1 and RNA-2 are also very similar. At least one strain of CMV can form defective RNAs that arise by deletions in the 3a ORF of RNA-3. Subgroup II strains of CMV encapsidate the sgRNA for the 2b ORF, called RNA-4a, and an additional small RNA of about 300 nt, called RNA-5, that is co-terminal with the 3'-ends of RNAs-3 and 4. In addition, CMV and Peanut stunt virus (PSV) may harbor satellite RNAs of about 330 to 400 nt. The satellite RNAs are more common under experimental conditions than in field conditions, and may dramatically alter the symptoms of infection by the helper virus.

PROTEINS

Although the CPs are estimated to be about 24 kDa by sequence analysis, most migrate in polyacrylamide gels with a slower than predicted mobility.

GENOME ORGANIZATION AND REPLICATION

An additional ORF, the 2b ORF is found in all cucumoviruses and has been shown to be active in CMV and Tomato aspermy virus (TAV).

ANTIGENIC PROPERTIES

CMV has been divided into two subgroups, based on serology. PSV also has more than one serological group. Sequence analysis has upheld the divisions, although CMV subgroup I can be further divided into two groups by phylogenetic analyses.

BIOLOGICAL PROPERTIES

CMV has an extremely broad host range, infecting 85 distinct plant families, and up to 1000 species experimentally. The other cucumovirus species have narrower host ranges; PSV is largely limited to legumes and solanaceous hosts, and TAV predominantly infects composites and solanaceous plants. All species are transmitted by aphids in a non-persistent manner.

LIST OF SPECIES DEMARCATION CRITERIA IN THE GENUS

The criteria demarcating species in the genus are:

- Serological relatedness,
- Compatibility of replicase proteins (1a and 2a proteins), but these distinctions may break down in the case of naturally occurring reassortants,
- Sequence similarity.

Serology and nucleotide sequence similarity is used to distinguish subgroups within a species. Subgroups generally have at least 65% sequence similarity.

LIST OF SPECIES IN THE GENUS

Species names are in green italic script; strain names and synonyms are in black roman script; tentative species names are in blue roman script. Sequence accession numbers, and assigned abbreviations ( ) are also listed.

SPECIES IN THE GENUS

Cucumber mosaic virus		
Cucumber mosaic virus	[Fny strain; RNA1: D00356, RNA2: D00355, RNA3: D10538, Q strain; RNA1: X02733, RNA2: X00985, RNA3: M21463]	(CMV)
Peanut stunt virus		
Peanut stunt virus	[ER strain; RNA1: U15728, RNA2: U15729, RNA3: U15730, J strain; RNA1: D11126, RNA2: D11127, RNA3: D00668]	(PSV)
(Robinia mosaic virus)		
Tomato aspermy virus		
Tomato aspermy virus	[RNA1: D10663, RNA2: D10044, RNA3: D01015]	(TAV)

(Chrysanthemum aspermy virus)

TENTATIVE SPECIES IN THE GENUS

None reported.

GENUS ILARVIRUS

Type Species: *Tobacco streak virus*

DISTINGUISHING FEATURES

VIRION PROPERTIES

MORPHOLOGY

Virions are quasi-isometric or occasionally bacilliform, and are about 30 nm in diameter.

NUCLEIC ACID

There is a short region of sequence similarity at the 3'-ends of the RNAs.

PROTEINS

The CP is required for activation of replication, but may be substituted with CP from AMV (genus *Alfamovirus*).

ANTIGENIC PROPERTIES

Viruses in each subgroup are all serologically related, and there are also some serological cross-reactions between certain subgroups; however, there are no cross-reactions between subgroup 1 viruses and any other viruses.

BIOLOGICAL PROPERTIES

The viruses infect mainly woody plants.

LIST OF SPECIES DEMARCATION CRITERIA IN THE GENUS

The criteria demarcating species in the genus are:

- Serology,
- Host range,
- Sequence similarity (Specific levels of sequence similarity have not been defined).

Clusters of viruses (Subgroup 1-7) that are antigenically related are indicated in the list.

LIST OF SPECIES IN THE GENUS

Species names are in green italic script; strain names and synonyms are in black roman script; tentative species names are in blue roman script. Sequence accession numbers, and assigned abbreviations ( ) are also listed.

SPECIES IN THE GENUS

Subgroup 1

Parietaria mottle virus

Parietaria mottle virus	[RNA1: AY496068, RNA2: AY496069]	(PMoV)

Tobacco streak virus

Tobacco streak virus	[RNA1: U80934, RNA2: U75538, RNA3: X00435,]	(TSV)

Subgroup 2

Asparagus virus 2

Asparagus virus 2	[RNA1: U93601, RNA1: U93602, RNA2: U93603, RNA3: X86352]	(AV-2)

Citrus leaf rugose virus

Citrus leaf rugose virus	[RNA1: U23715, RNA2: U17726,	(CiLRV)

	RNA3: U17390]	
Citrus variegation virus		
Citrus variegation virus	[RNA1: U93604, RNA2: U93605, RNA3: U17389]	(CVV)
Elm mottle virus		
Elm mottle virus	[RNA1: U57047, RNA2: U34050, RNA3: U57048]	(EMoV)
Hydrangea mosaic virus	[RNA1: AF172968, RNA2: AF172967, RNA3: AF192965]	(HdMV)
Spinach latent virus		
Spinach latent virus	[RNA1: U93192, RNA2: U93193, RNA3: U93194]	(SpLV)
Tulare apple mosaic virus		
Tulare apple mosaic virus	[RNA1: AF226160, RNA2: AF226161, RNA3: AF226162]	(TAMV)
Subgroup 3		
Apple mosaic virus		
Apple mosaic virus	[RNA1: AF174584, RNA2: AF174585, RNA3: U15608]	(ApMV)
Blueberry shock virus		
Blueberry shock virus		(BlShV)
Humulus japonicus latent virus		
Humulus japonicus latent virus		(HJLV)
Prunus necrotic ringspot virus		
Prunus necrotic ringspot virus	[RNA1: AF278534, RNA2: AF278535, RNA3: L38823]	(PNRSV)
Subgroup 4		
Prune dwarf virus		
Prune dwarf virus	[RNA1: U57648, RNA2: AF277662, RNA3: L28145]	(PDV)
Subgroup 5		
American plum line pattern virus		
American plum line pattern virus	[RNA1: AF235033, RNA2: AF235165, RNA3: AF235166]	(APLPV)
Subgroup 6		
Fragaria chiloensis latent virus		
Fragaria chiloensis latent virus		(FClLV)
Lilac ring mottle virus		
Lilac ring mottle virus	[RNA3: U17391]	(LiRMoV)

TENTATIVE SPECIES IN THE GENUS

None reported.

GENUS OLEAVIRUS

Type Species: *Olive latent virus 2*

DISTINGUISHING FEATURES

VIRION PROPERTIES

MORPHOLOGY

Virions have different shape and size, ranging from quasispherical with a diamter of 26 nm, to bacilliform with lengths of 37, 43, 38, and 55 nm, and diameters of 18 nm. Particles up to 85 nm in length occasionally are present.

PHYSICOCHEMICAL AND PHYSICAL PROPERTIES

In sucrose gradients, virions sediment as 5 or 6 components.

NUCLEIC ACID

Virion RNA differs in size from that of other members of the family, encapsidating the three genomic RNAs and a sgRNA of about 2 kb, that is apparently not an mRNA. The sgRNA for the CP ORF is not encapsidated. Three additional RNAs of 200 to 550 nt are also present in virions. The 5'-termini of the genomic RNAs are capped, but not the 5'-terminus of the encapsidated sgRNA. The 3'-termini of the RNAs are similar to those of the genera *Alfamovirus* and *Ilarvirus*, but do not interact with CP to activate replication.

ANTIGENIC PROPERTIES

Virions are efficient immunogens.

BIOLOGICAL PROPERTIES

The only known host is olive (*Olea europaea*), which is infected asymptomatically. The virus is transmitted by inoculations, but no insect vector is known.

LIST OF SPECIES DEMARCATION CRITERIA IN THE GENUS

Not applicable.

LIST OF SPECIES IN THE GENUS

Species names are in green italic script; strain names and synonyms are in black roman script; tentative species names are in blue roman script. Sequence accession numbers, and assigned abbreviations ( ) are also listed.

SPECIES IN THE GENUS

Olive latent virus 2

Olive latent virus 2	[X94346, X94327, X76933, X77115]	(OLV-2)

TENTATIVE SPECIES IN THE GENUS

None reported.

LIST OF UNASSIGNED SPECIES IN THE FAMILY

None reported.

PHYLOGENETIC RELATIONSHIPS WITHIN THE FAMILY

Phylogenetic relationships have not been thoroughly characterized in this family. The CP and Mps are too distant to use for phylogenetic analyses. The replicase genes reveal discrepancies between the accepted taxonomy and the probable phylogenetic relationships. However, the genera *Bromovirus* and *Cucumovirus* form a distinct clade from genera *Alfamovirus* and *Ilarvirus*, and genus *Oleavirus* is the most distant from all the others.

Similarity with Other Taxa

The viruses are members of the "alpha-like" supergroup, sharing sequence similarity in the 1a protein domains for Mtr and Hel activities, and in the 2a protein polymerase domain with members of the genera *Tobravirus*, *Hordeivirus*, *Tobamovirus*, *Potexvirus*, *Carlavirus*, and *Tymovirus*, and animal viruses in the family *Togaviridae*. The 3a proteins of bromoviruses and the 35 kDa protein of the members of the genus *Dianthovirus* (RCNMV) form a distinct "family" of movement-associated proteins. *Raspberry bushy dwarf virus* (genus *Idaeovirus*) is similar to bromoviruses in genome organization and in the sequence of certain genes.

Derivation of Names

Alfamo: derived from *Alf*alfa *mo*saic virus.
Bromo: derived from *Bro*me *mo*saic, also, from *Bromus* (host of *Brome mosaic virus*).
Cucumo: derived from *Cucu*mber *mo*saic virus.
Ilar: derived from *i*sometric *la*bile *r*ingspot.
Olea: derived from the genus name of the host, olive (*Olea*).

References

Ahlquist, P. (1992). Bromovirus RNA replication and transcription. *Curr. Opin. Gen. Dev.*, **2**, 71-76.

Bernal, J.J., Moriones, E. and García-Arenal, F. (1991). Evolutionary relationships in the cucumoviruses: nucleotide sequence of tomato aspermy virus RNA-1. *J. Gen. Virol.*, **72**, 2191-2195.

Bol, J.F. (1999). Alfalfa mosaic virus and ilarviruses: involvement of coat protein in multiple steps of the replication cycle. *J. Gen. Virol.*, **80**, 1089-1102.

Ding, S.-W., Anderson, B.J., Haase, H.R. and Symons, R.H. (1994). New overlapping gene encoded by the cucumber mosaic virus genome. *Virology*, **198**, 593-601.

Kim, S.H., Palukaitis, P. and Park, Y.I. (2002). Phosphorylation of cucumber mosaic virus RNA polymerase 2a protein inhibits formation of replicase comples. *EMBO J.*, **21**, 2292-2300.

Kumar, A., Reddy, V.S., Yusibov, V., Chipman, P.R., Hata, Y., Fita, I., Fukuyama, K., Rossmann, M.G., Loesch-Fries, L.S., Baker, T.S. and Johnson, J.E. (1997). The structure of Alfalfa mosaic virus capsid protein assembled as a T=1 icosahedral particle at 4.0-A resolution. *J. Virol.*, **71**, 7911-7916.

Lucas, R.W., Larson, S.B. and McPherson, A. (2001). The crystallographic structure of Brome mosaic virus. *J. Mol. Biol.*, **317**, 95-108.

Martelli, G.P. and Grieco, F. (1997). *Oleavirus*, a new genus in the family *Bromoviridae*. *Arch. Virol.*, **142**, 1933-1936.

Naidu, R.A., Hu, C.-C., Pennington, R.E. and Ghabrial, S.A. (1995). Differentiation of eastern and western strains of Peanut stunt cucumovirus based on satellite RNA support and nucleotide sequence homology. *Phytopathology*, **85**, 502-507.

Noueiry, A.O. and Ahlquist, P. (2003). Brome mosaic virus RNA replication: revealing the role of the host in RNA virus replication. *Ann. Rev. Phytopathol.*, **41**, 77-98.

O'Reilly, D., Thomas, C.J.R. and Coutts, R.H.A. (1991). Tomato aspermy virus has an evolutionary relationship with other tripartite RNA plant viruses. *J. Gen. Virol.*, **72**, 1-7.

Palukaitis, P., Roossinck, M.J., Dietzgen, R.G. and Francki, R.I.B. (1992). Cucumber mosaic virus. *Adv. Virus Res.*, **41**, 281-348.

Roossinck, M.J. (2002). Evolutionary history of cucumber mosaic virus deduced by phylogenetic analyses. *J. Virol.*, **76**, 3382-3387.

Sánchez-Navarro, J.A. and Pállas, V. (1997). Evolutionary relationships in the ilarviruses: nucleotide sequence of Prunus necrotic ringspot virus RNA-3. *Arch. Virol.*, **142**, 749-763.

Scott, S.W., Zimmerman, M.T. and Ge, X. (2003). Viruses in subgroup 2 of the genus Ilarvirus share both serological relationships and characteristics at the molecular level. *Arch. Virol.*, **148**, 2063-2075.

Suzuki, M., Yoshida, M., Yoshinuma, T. and Hibi, T. (2003). Interaction of replicase components between *Cucumber mosaic virus* and *Peanut stunt virus*. *J. Gen. Virol.*, **84**, 1931-1939.

Contributed By

Roossinck, M.J., Bujarski, J., Ding, S.W., Hajimorad, R., Hanada, K., Scott, S. and Tousignant, M.

GENUS *OURMIAVIRUS*

Type Species *Ourmia melon virus*

VIRION PROPERTIES

MORPHOLOGY

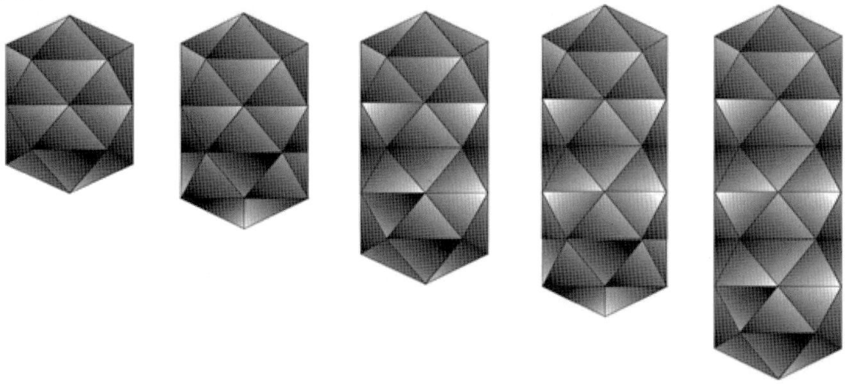

Figure 1: Diagram of virion surface of a member of the genus *Ourmiavirus*, showing arrangement of double disks and conical ends in particles of different length. Each row of 5 triangles represents a double disk.

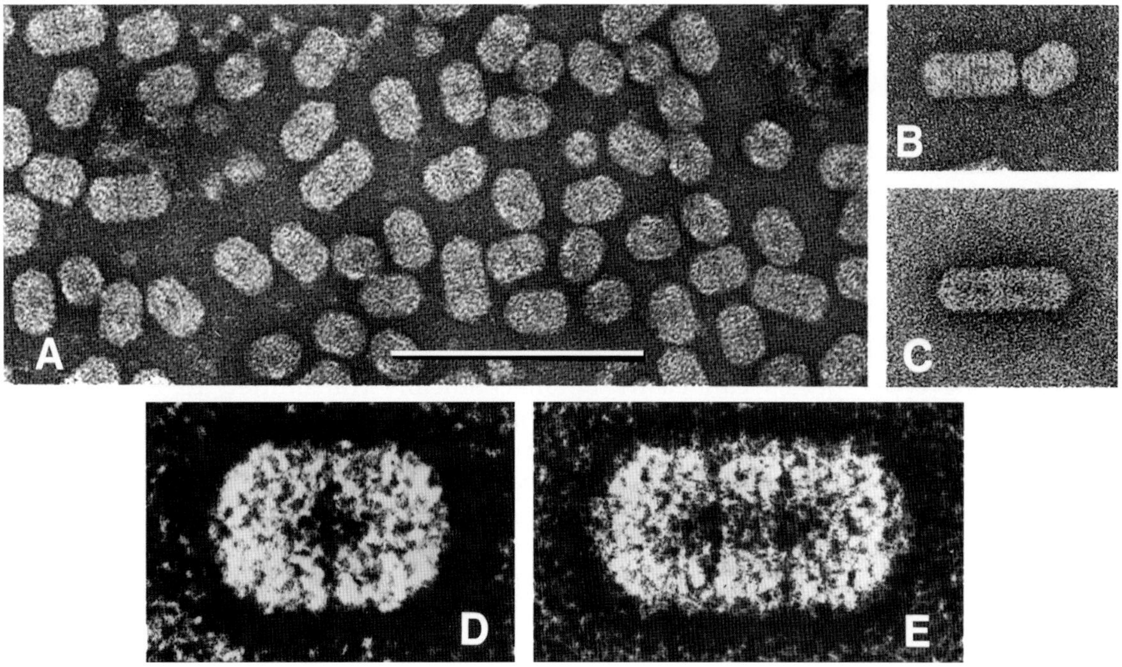

Figure 2: (A, B, C) Negative contrast electron micrographs (uranyl acetate) of purified particles of Ourmia melon virus (OuMV). The bar represents 100 nm. D and E, features of the two commonest particle types, enhanced by photographic superimposition.

The bacilliform virions constitute a series of particles with conical ends (apparently hemi-icosahedra) and cylindrical bodies 18 nm in diameter. The bodies of the particles are composed of a series of double disks, the commonest particles having 2 disks (particle length 30 nm), a second common particle having 3 disks (particle length 37 nm) with rarer particles having 4 disks (particle length 45.5 nm) and 6 disks (particle length 62 nm). There is no envelope (Fig. 1 and 2).

Physicochemical and Physical Properties

The Mr of virions and their sedimentation coefficients are not known. The buoyant density in CsCl of all particle sizes is 1.375 g/cm^3. The particles are stable near pH 7. Thermally, they are relatively stable; infectivity survives in crude sap after heating for 10 min at 70°C but not 80°C, and is retained after at least one freeze-thaw cycle. Cation stability is not known, except that the particles survive in CsCl density gradients. The particles survive treatment with chloroform but not n-butanol, and survive treatment with 1% Triton X-100 detergent.

Nucleic Acid

The nucleic acid is linear positive-sense ssRNA with an estimated Mr of 1.58 x 10^6, divided into 3 segments with estimated Mr of 0.91, 0.35 and 0.32 x 10^6. It is not known if there is a 5′-terminal cap, a 5′-terminal covalently-linked polypeptide or a 3′-terminal poly(A) tract. No nucleotide sequences are available.

Proteins

There is one CP of 25.2 kDa. In addition, one nonstructural protein of unknown function accumulates abundantly in the cytoplasm of infected plants, forming fibers or tubules.

Lipids

None reported.

Carbohydrates

None reported.

Genome Organization and Replication

The CP is encoded by RNA-3. The fibrous nonstructural protein is encoded by RNA-1 or RNA-2. Particles accumulate to high levels in the cytoplasm of parenchyma cells and become associated with the fibrous protein.

Antigenic Properties

The native virions and the nonstructural protein are good immunogens. They do not cross-react.

Biologic Properties

The type species can easily be mechanically transmitted to a rather wide range of dicotyledonous plants (34 species in 14 families reported), usually inducing systemic ringspots, mosaic and necrosis, with local lesions on some hosts. There is no particular tissue tropism. No vector has been identified. No experimental transmission was obtained with several aphid species, the whiteflies *Trialeurodes* and *Bemisia* or a *Tetranychus* mite. Experimental seed-transmission rates are 1 - 2% in *Nicotiana benthamiana* and *N. megalosiphon*. Different species in the genus are reported to occur in geographically diverse areas and on widely differing hosts, though there are experimental hosts in common.

List of Species Demarcation Criteria in the Genus

The species demarcation criteria are:

- No evident sequence homologies between the respective RNA-2s and RNA-3s (by hybridization).
- CPs of distinctly different size (25 or 21 kDa).
- No serological relationships between CPs (by gel-diffusion or EM decoration tests) or distant serological (in Western blots).
- Natural host ranges non-overlapping (cucurbits; cherry; cassava).

List of Species in the Genus

Species names are in green italic script; strain names and synonyms are in black roman script; tentative species names are in blue roman script. Sequence accession numbers, and assigned abbreviations ( ) are also listed.

Species in the Genus

Cassava virus C
Cassava virus C (CsVC)
Epirus cherry virus
Epirus cherry virus (EpCV)
Ourmia melon virus
Ourmia melon virus (OuMV)

Tentative Species in the Genus

None reported.

Phylogenetic Relationships Within the Genus

Not available.

Similarity With Other Taxa

The virion morphology, in particular the architecture of double disks and conical ends, forming particles of different lengths, is unique, differing clearly from that of other known "bacilliform" viruses of similar size. The virions are notably resistant to treatment with solvents, detergent and heat. There are 3 ssRNAs of Mr about 0.9, 0.35 and 0.32 x 10^6 combined with a single CP of 21-25 kDa. A fibrous nonstructural protein accumulates conspicuously in the cytoplasm. The genus may have affinities with other ssRNA positive-strand multipartite viruses such as the members of the family *Bromoviridae*.

Derivation of Names

Ourmia: derived from the district of *Ourmia* in Northern Iran where the type species was found.

References

Accotto, G.P., Riccioni, L., Barba, M. and Boccardo, G. (1997). Comparison of some molecular properties of Ourmia melon and Epirus cherry viruses, two representatives of a proposed new virus group. *J. Plant Pathol.*, **78**, 87-91.

Aiton, M.M., Roberts, I.M. and Harrison, B.D. (1988). Two new cassava viruses from Africa. Abstracts 5th Inter. Congress Plant Pathol., Kyoto, 1988, p. 43.

Avgelis, A., Barba, M. and Rumbos, I. (1989). Epirus cherry virus, an unusual virus isolated from cherry with rasp-leaf symptoms in Greece. *J. Phytopathol.*, **126**, 51-58.

Lisa, V., Milne, R.G., Accotto, G.P., Boccardo, G., Caciagli, P. and Parvizy, R. (1988). Ourmia melon virus, a virus from Iran with novel properties. *Ann. Appl. Biol.*, **112**, 291-302.

Contributed By

Milne, R.G.

GENUS *IDAEOVIRUS*

Type Species *Raspberry bushy dwarf virus*

VIRION PROPERTIES

MORPHOLOGY

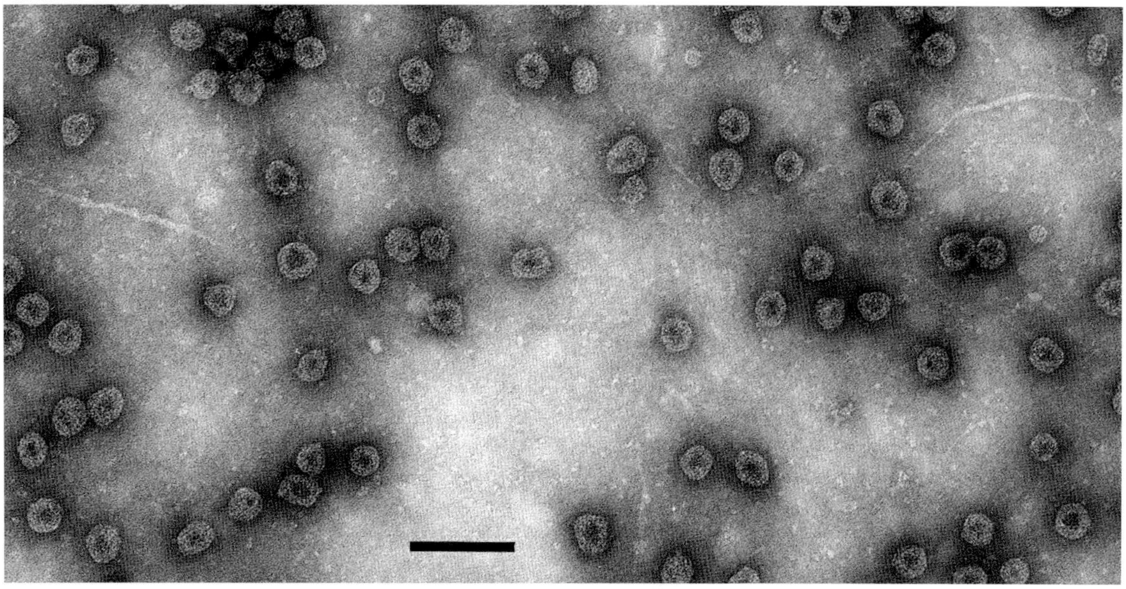

Figure 1: Negative contrast electron micrograph of particles of an isolate of *Raspberry bushy dwarf virus*, stained with uranyl formate/sodium hydroxide. The bar represents 100 nm.

Virions are isometric, about 33 nm in diameter and are not enveloped. They appear flattened in electron micrographs of preparations negatively stained with uranyl salts (Fig. 1).

PHYSICOCHEMICAL AND PHYSICAL PROPERTIES

Virion Mr is about 7.5 x 10^6 (calculated from the S_{20w} of 115S). The buoyant density of aldehyde-fixed particles in CsCl is 1.37 g/cm^3. Particles are readily disrupted in neutral chloride salts and by SDS.

NUCLEIC ACID

Virion preparations contain three species of linear, positive sense, ssRNA of about 5.5 kb (RNA-1), 2.2 kb (RNA-2) and 1 kb (RNA-3). These RNA molecules are not polyadenylated.

PROTEINS

Virions possess one major CP species of ~30 kDa. Sequence data indicate that there are two non-structural proteins of 188 and 39 kDa.

LIPIDS

Not known.

CARBOHYDRATES

Not known.

GENOME ORGANIZATION AND REPLICATION

The genome is bipartite. RNA-1 has a major ORF encoding a 188 kDa protein, and also in a different frame and overlapping the 188 kDa protein, a putative small ORF that encodes a 12 kDa protein that is not essential for infectivity. The 188 kDa protein contains sequence motifs characteristic of helicases and polymerases. RNA-2 has two in-frame ORFs: that in the 5'-terminal half encodes a ~39 kDa protein which has some slight sequence similarities with proteins of other viruses that are thought to have roles in virus transport; that in the 3'-terminal half encodes the CP. RNA-2 is probably a template for the production of RNA-3 that comprises the 3'-most 946 nt of RNA-2 and is a sgRNA for CP. The 3'-terminal non-coding 18 nt of RNA-1 and RNA-2 (and hence of RNA-3) are the same and the 3'-terminal 70 nt can be arranged in similar extensively base-paired structures. Infected leaves contain dsRNA corresponding in size to double-stranded forms of RNA-1 and RNA-2. *In vitro* translation yields three major proteins, of ~190, 44 and 31 kDa (CP), which are the translation products, respectively, of RNA-1, RNA-2 and RNA-3.

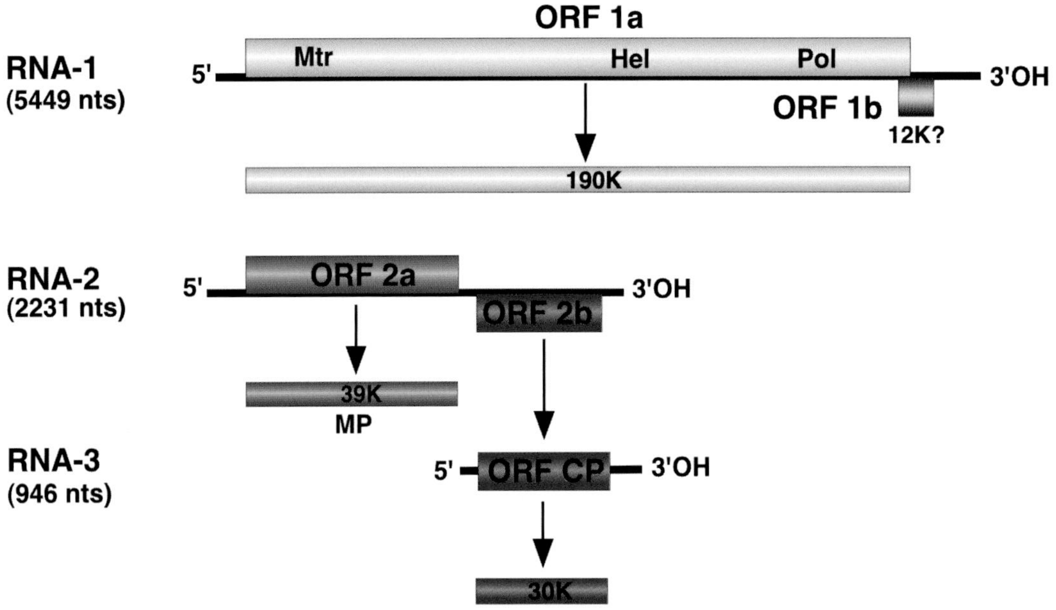

Figure 2: Scale diagram of RNA species found in particles of Raspberry bushy dwarf virus (RBDV). The open boxes represent the ORFs. The positions of putative domains are indicated as 'Mtr', a Mtr motif, 'Hel', an NTP-binding motif, 'Pol', an RdRp motif. MP signifies the putative movement protein role of the 39 kDa protein and CP the coat protein.

Antigenic Properties

Particles are moderate immunogens.

Biological Properties

In nature, the host range is confined to *Rubus* species, all but one in the subgenus *Idaeobatus*; the experimental host range is fairly wide. The virus occurs in all tissues of the plant, including seed and pollen, and Raspberry bushy dwarf virus (RBDV) is transmitted in association with pollen, both vertically to the seed and horizontally to the pollinated plant. This is the only known method of natural spread. Experimentally, the virus can be transmitted by mechanical inoculation. The virus occurs throughout the world wherever raspberry is grown. Infection of raspberry and blackberry is often symptomless but in some cultivars may be associated with 'yellows disease' and/or 'crumbly fruit'. Confusingly, Black raspberry necrosis virus is the main cause of bushy dwarf disease in Lloyd George raspberry. However, because the additional presence of RBDV in plants contributes significantly to the intensity of the disease symptoms usually observed in the field, it can also be regarded as an integral component of the disease syndrome.

List of Species Demarcation Criteria in the Genus

Not applicable.

LIST OF SPECIES IN THE GENUS

Species names are in green italic script; strain names and synonyms are in black roman script; tentative species names are in blue roman script. Sequence accession numbers, and assigned abbreviations ( ) are also listed.

SPECIES IN THE GENUS

Raspberry bushy dwarf virus

Raspberry bushy dwarf virus	[S51557, S55890, D01052]	(RBDV)

TENTATIVE SPECIES IN THE GENUS

None reported.

PHYLOGENETIC RELATIONSHIPS WITHIN THE GENUS

Not applicable.

SIMILARITY WITH OTHER TAXA

RBDV resembles viruses of the genus *Ilarvirus*, family *Bromoviridae*, in having easily deformable particles that are transmitted in association with pollen. RNA-2 resembles RNA-3 of viruses in family *Bromoviridae* in the arrangement and sizes of its encoded gene products, the generation of a 3'-terminal sgRNA and in the structured nature of the 3'-ends of the molecules. The sequence of the translation product of RBDV RNA-1 resembles, in different parts, sequences in the translation products of viruses in the family *Bromoviridae* and to a lesser extent the sequence of the helicase + polymerase protein (~183 kDa) of tobamoviruses. Idaeoviruses, therefore, belong to the "alphavirus-like" supergroup.

DERIVATION OF NAMES

Idaeo: from *idaeus*, specific name of raspberry, *Rubus idaeus*.

REFERENCES

Barnett, O.W. and Murant, A.F. (1970). Host range, properties and purification of raspberry bushy dwarf virus. *Ann. Appl. Biol.*, **65**, 435-449.

Jones, A.T., Mayo, M.A. and Murant, A.F. (1996). Raspberry bushy dwarf idaeovirus. In: *The Plant Viruses*, Vol. 5: Polyhedral Virions and Bipartite RNA Genomes, (B.D. Harrison and A.F. Murant, eds), pp 283-301. Plenum Press, New York.

Jones, A.T., Murant, A.F., Jennings, D.L. and Wood, G.A. (1982). Association of raspberry bushy dwarf virus with raspberry yellows disease; reaction of *Rubus* species and cultivars, and the inheritance of resistance. *Ann. Appl. Biol.*, **100**, 135-147.

Mayo, M.A., Jolly, C.A., Murant, A.F. and Raschke, J.H. (1991). Nucleotide sequence of raspberry bushy dwarf virus RNA-3. *J. Gen. Virol.*, **72**, 469-472.

Murant, A.F. (1975). Some properties of the particles of raspberry bushy dwarf virus. *Proc. Am. Phytopath. Soc.*, **2**, 116-117.

Murant, A.F. (1987). Raspberry bushy dwarf. In: *Virus Diseases of Small Fruits*, (R.H. Converse, ed), pp 229-234. USDA Agriculture Handbook N° 631.

Murant, A.F., Chambers, J. and Jones, A.T. (1974). Spread of raspberry bushy dwarf virus by pollination, its association with crumbly fruit, and problems of control. *Ann. Appl. Biol.*, **77**, 271-281.

Murant, A.F., Mayo, M.A. and Raschke, J.H. (1986). Some biochemical properties of raspberry bushy dwarf virus. *Acta Hort.*, **186**, 23-30.

Natsuaki, T., Mayo, M.A., Jolly, C.A. and Murant, A.F. (1991). Nucleotide sequence of raspberry bushy dwarf virus RNA-2: a bicistronic component of a bipartite genome. *J. Gen. Virol.*, **72**, 2183-2189.

Ziegler, A., Natsuaki, T., Mayo, M.A., Jolly, C.A. and Murant, A.F. (1992). Nucleotide sequence of raspberry bushy dwarf virus RNA-1. *J. Gen. Virol.*, **73**, 3213-3218.

CONTRIBUTED BY

Jones, A.T.

FAMILY TYMOVIRIDAE

TAXONOMIC STRUCTURE OF THE FAMILY

Family *Tymoviridae*
Genus *Tymovirus*
Genus *Marafivirus*
Genus *Maculavirus*

VIRION PROPERTIES

MORPHOLOGY

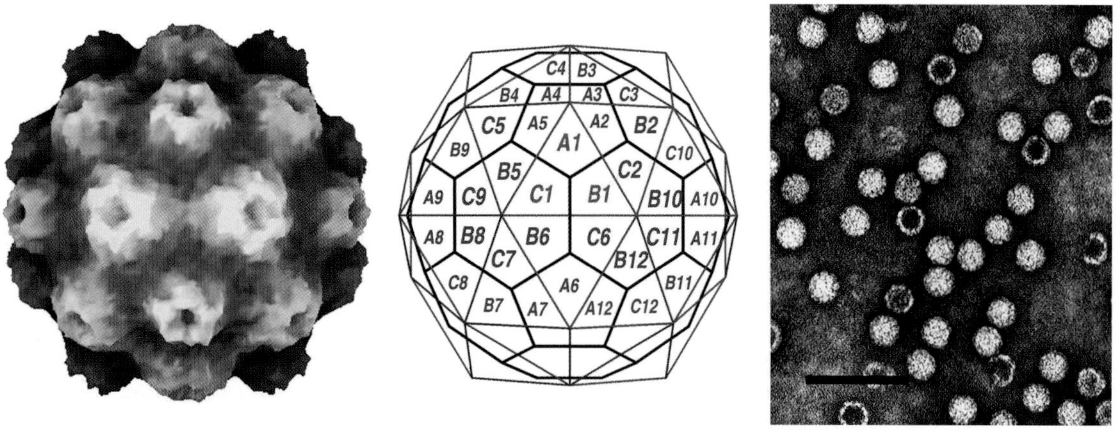

Figure 1: (Left) Atomic rendering of a virion of Turnip yellow mosaic virus (TYMV) (Canady et al., 1996). (Center) Diagram of a TYMV virion with CP clusters in hexa- and pentamers. (Right) Negative contrast electron micrograph of virions and 'empty particles' of Belladonna mottle virus (BeMV), a representative of the genus Tymovirus. Particles of members of the genera Marafivirus and Maculavirus have the same morphology. The bar represents 100 nm.

Virions are isometric, non-enveloped, ~ 30 nm in diameter, with a rounded contour and prominent surface structure, with clustering of CP subunits in pentamers and hexamers. The capsids of tymoviruses are made up of 20 hexameric and 12 pentameric subunits arranged in a T=3 icosahedron and the RNA appears to be at least partially ordered in an icosahedral arrangement in the center of the protein shell.

PHYSICOCHEMICAL AND PHYSICAL PROPERTIES

Virus particles sediment as two centrifugal components: T, made up of non-infectious protein shells that contain a small amount of RNA (primarily subgenomic CP mRNA) and B, composed of intact nucleoprotein particles. Sedimentation coefficients (S_{w20}) of component T and B range from 42 to 55 and from 109 to 125, respectively. Buoyant densities in CsCl of components T and B are 1.26-1.28 g/cm^3 and 1.40-1.46 g/cm^3 respectively. Virions resist high temperatures (thermal inactivation point is 60-65°C up to above 80°C with some tymoviruses) and organic solvents, but are disrupted by SDS.

NUCLEIC ACID

Virions contain a single molecule of positive sense, ssRNA constituting 25 to 35% of the particle weight. The RNA has a very high cytidine content (from 32 to about 50%) and ranges from 6.0 to 7.5 kb in length. The complete genomic sequences have been determined for representative of seven species of the genus *Tymovirus* (Chayote mosaic virus; ChMV, Eggplant mosaic virus; EMV, Erysimum latent virus; ErLV, Kennedya yellow mosaic virus, KYMV, Ononis yellow mosaic virus; OYMV, Physalis mottle virus; PhyMV and Turnip yellow mosaic virus; TYMV), two of the genus *Marafivirus* (Maize rayado fino virus; MRFV, Oat blue dwarf virus; OBDV), one of the genus *Maculavirus* (Grapevine fleck virus; GFkV), and for the unassigned Poinsettia mosaic virus (PnMV). The genomes of several other viruses have been partially sequenced (see lists of species). Sequenced tymovirus genomes are capped at the 5'-terminus and have a tRNA-like structure at the 3'-end, which for TYMV and several other species, accepts valine. The genomes of GFkV, PnMV and OBDV are polyadenylated at the 3'-terminus and are thought to be capped at the 5'-end. The genome of the marafivirus MRFV also is thought to be 5'-capped, but it lacks a poly(A) tail at the 3'-terminus. Infectious RNA transcripts have been generated from cDNA clones of the tymoviruses TYMV, EMV, and OYMV.

Proteins

The CP of virus particles contains either a single protein species with molecular mass of 20 kDa (tymoviruses), 24.5-25 kDa (maculaviruses), or a major protein of 21-21.5 kDa and a minor protein of 24.4-25 kDa (marafiviruses OBDV and MRFV). The capsids of PnMV and a strain of the *Bermuda grass etched line virus* are made up of a single protein of ~21 kDa.

Lipids

None reported.

Carbohydrates

None reported.

Genome Organization and Replication

The genomes of tymoviruses contain two extensively overlapping ORFs that begin 7 nt apart, and a third ORF (for CP) that is expressed from a 3′-coterminal sgRNA (Fig. 2). The longest ORF encodes a ~220 kDa replication polyprotein. The genomes of the marafiviruses, maculaviruses and PnMV possess a similar ~220 kDa-coding ORF, and a CP ORF in the same reading frame that is either fused to the long ORF (marafiviruses and PnMV) or separated from it by two adjacent stop codons (GFkV). MRFV is the sole marafivirus with an extensive ORF near the 5′-end of the genome that entirely overlaps the replication protein ORF (Fig. 3). The genome of GFkV possesses two additional ORFs towards the 3′-end. Viral RNA of tymoviruses replicates in the cytoplasm in association with the double-membraned invaginations that line the periphery of the chloroplasts. Comparable vesicles occurring at the peripheries of mitochondria or chloroplasts of cells infected with maculaviruses, the tentative marafivirus Grapevine asteroid mosaic-associated virus (GAMaV), and PnMV may have the same function. Genome expression of the tymoviruses includes post-translational autocatalytic cleavage of the ~220 kDa protein by a virus-encoded papain-like protease, and synthesis and translation of a 3'-coterminal sgRNA for CP expression.

Antigenic Properties

Virions are moderately to highly antigenic. Monoclonal antibodies have been produced to GFkV. Tymoviruses cluster roughly into two serological groups with cross-reactivity ranging from strong to weak within each group. Cross-reactivities between group members are weak to undetectable. Serological relationships occur between some marafiviruses, but not between individual maculaviruses.

Biological Properties

Host Range

Tymoviruses and maculaviruses infect dicotyledonous plants, whereas some marafiviruses primarily infect plants in the Graminae. Natural and experimental host ranges of individual virus species are narrow, sometimes restricted to a single type of host (e.g., GFkV, GRGV, GAMaV and Grapevine rupestris vein feathering virus (GRVFV) infect only *Vitis;* MRFV only *Zea*). Disease symptoms are bright yellow mosaic or mottling (tymoviruses and PnMV), chlorotic stripes, vein clearing, etched lines or dwarfing (marafiviruses), and flecking of the leaves (maculaviruses).

TRANSMISSION

Tymoviruses and PnMV, but not other members of the family, are readily transmissible by mechanical inoculation to leaf surfaces. They replicate to high titres and invade all main tissues of the host. Marafi- and maculaviruses are phloem-limited. Marafiviruses, except for tentative members GAMaV and GRVFV, are distinguished by being vectored in a persistent manner by leafhoppers, in which they replicate. Some tymoviruses are weakly seed-transmissible, and also are spread by beetles, which serve as low-efficiency local vectors. Maculaviruses do not have recognized vectors, and the grapevine-associated marafi- and maculaviruses are disseminated primarily through infected propagating material.

GEOGRAPHICAL DISTRIBUTION

Members of the family have been recorded from most parts of the world. Geographical distribution of individual species varies from restricted to widespread.

CYTOPATHIC EFFECTS

Most species elicit derangement of the internal structure and alteration of the shape of chloroplasts and/or mitochondria, which also show rows of peripheral vesicles derived from localized invaginations of the limiting membrane.

LIST OF GENUS DEMARCATION CRITERIA IN THE FAMILY

The criteria demarcating genera in the family are:

- *Biological criteria*: Tymoviruses are mechanically transmissible, invade parenchyma tissues, and infect dicotyledonous plants. Marafiviruses and maculaviruses are both phloem-limited and are not mechanically transmissible to leaf surfaces. Marafiviruses infect primarily monocotyledonous plants, maculaviruses infect dicotyledonous plants.
- *Epidemiological criteria*: Tymoviruses are transmitted by beetles, most marafiviruses by leafhoppers. Maculaviruses have no known vector.
- *Cytopathological criteria*: Tymoviruses elicit the formation of double-membraned vesicles at the periphery of chloroplasts, maculaviruses elicit them at the periphery of mitochondria. Chloroplast vesicles induced by *Poinsettia mosaic virus* are single-membraned.
- *Physicochemical criteria*: (a) Genome size: 6.0-6.7 kb (tymoviruses), 6.3-6.5 kb (marafiviruses), 7.5 kb (maculaviruses); (b) number and molecular mass of the CP: one subunit type of 20 kDa (tymoviruses), 24 kDa (maculaviruses) or 21 kDa (PnMV), or two subunit types (21 and 24 kDa) (marafiviruses).
- *Molecular criteria*: Tymovirus genomes have three ORFs and a tRNA-like structure at the 3'-terminus. Maculavirus genomes have four ORFs, a 3'-terminal poly(A) tail. Two genome types have been reported for marafiviruses: (i) a single large ORF and a 3′-terminal poly(A) tail; (ii) two ORFs and no poly(A) tract. Tymovirus genomes possess a conserved sequence that serves as a sgRNA promoter known as the "tymobox"; the genomes from marafiviruses and PnMV contain a variant termed the “marafibox”, while no such sequence is evident in maculavirus genomes.

GENUS *TYMOVIRUS*

Type Species *Turnip yellow mosaic virus*

DISTINGUISHING FEATURES

The genomic RNA (6.0-6.7 kb in size) contains three ORFs (Fig. 2). ORF1 encodes a 206 kDa protein with the conserved sequence motifs characteristic of Mtr, papain-like proteases, Hel, and RdRp. The 3′-ends of these large ORFs are highly conserved and contain a 16 nt sequence known as the "tymobox", which functions as subgenomic RNA promoter. ORF2 initiates 7 nt upstream of ORF1, almost entirely overlaps ORF1, and encodes a 50-80 kDa proline-rich protein that for TYMV is dispensable for replication but is required for cell-to-cell movement. ORF3 codes for the viral CP (20 kDa), which is expressed via a sgRNA. The narrow host range of tymoviruses makes host susceptibility an important property in distinguishing among species. All species induce vesicles at the periphery of chloroplasts and to a lesser degree of mitochondria, and they induce characteristic aggregates of swollen and modified chloroplasts. All main tissues of the hosts are invaded. Empty virion shells sometimes accumulate in the nuclei. Beetles of the families Chrysomelidae and Curculionidae serve as local vectors, transmitting the virus in a semi-persistent manner. All members of the genus are mechanically transmissible and a few (TYMV, EMV, DuMV) are transmitted through seeds.

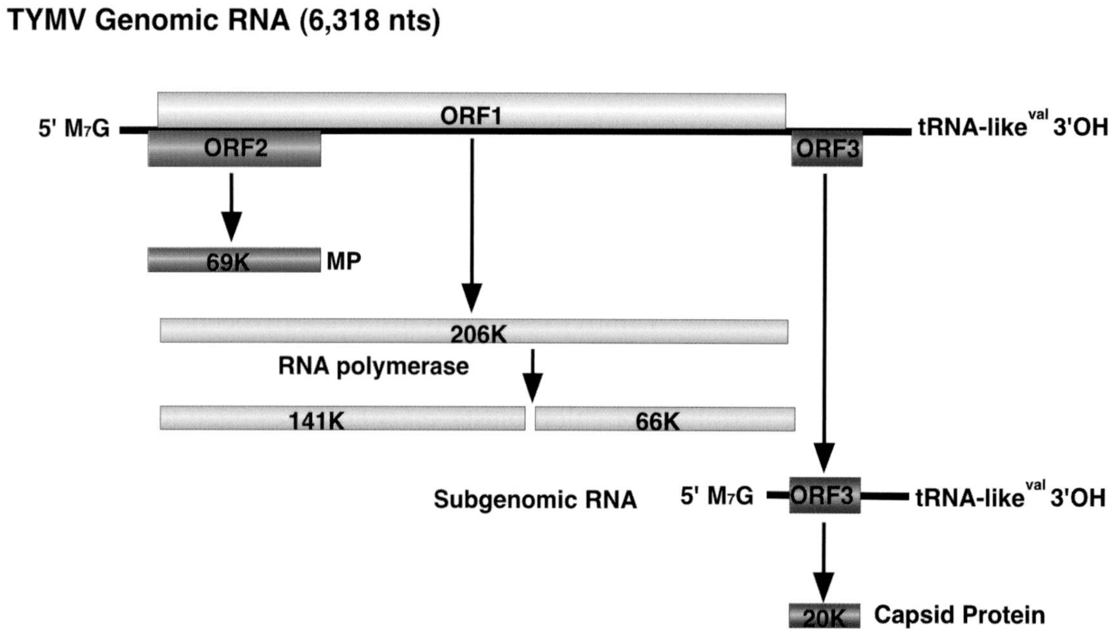

Figure 2. Organization and expression of the genome of Turnip yellow mosaic virus (TYMV).

LIST OF SPECIES DEMARCATION CRITERIA IN THE GENUS

The criteria demarcating species in the genus are:
- Overall sequence identity of less than 80%
- Capsid protein sequences less than 90% identical
- Differential host range
- Differences in the 3'-terminal structure
- Serological specificity

List of Species in the Genus

Species names are in green italic script; strain names and synonyms are in black roman script; tentative species names are in blue roman script. Sequence accession numbers, and assigned abbreviations ( ) are also listed.

Species in the Genus

Andean potato latent virus		
Andean potato latent virus	[AF035402]	(APLV)
Belladonna mottle virus		
Belladonna mottle virus	[X54529]	(BeMV)
Cacao yellow mosaic virus		
Cacao yellow mosaic virus	[X54354]	(CYMV)
Calopogonium yellow vein virus		
Calopogonium yellow vein virus	[U91413]	(CalYVV)
Chayote mosaic virus		
Chayote mosaic virus	[AF195000]	(ChMV)
Clitoria yellow vein virus		
Clitoria yellow vein virus	[M15963]	(CYVV)
Desmodium yellow mottle virus		
Desmodium yellow mottle virus	[AF035201]	(DYMoV)
Dulcamara mottle virus		
Dulcamara mottle virus	[AF035634]	(DuMV)
Eggplant mosaic virus		
Eggplant mosaic virus	[J04374]	(EMV)
Erysimum latent virus		
Erysimum latent virus	[AF098523]	(ErLV)
Kennedya yellow mosaic virus		
Kennedya yellow mosaic virus	[D00637]	(KYMV)
Melon rugose mosaic virus		
Melon rugose mosaic virus		(MRMV)
Okra mosaic virus		
Okra mosaic virus	[AF035202]	(OkMV)
Ononis yellow mosaic virus		
Ononis yellow mosaic virus	[J04375]	(OYMV)
Passion fruit yellow mosaic virus		
Passion fruit yellow mosaic virus	[AF47107]	(PFYMV)
Peanut yellow mosaic virus		
Peanut yellow mosaic virus		(PeYMV)
Petunia vein banding virus		
Petunia vein banding virus	[AF210709]	(PetVBV)
Physalis mottle virus		
Physalis mottle virus	[Y16104]	(PhyMV)
Plantago mottle virus		
Plantago mottle virus		(PlMoV)
Scrophularia mottle virus		
Scrophularia mottle virus		(SrMV)
(Anagyris vein yellowing virus)		
Turnip yellow mosaic virus		
Turnip yellow mosaic virus	[J04373, X16378, X07441]	(TYMV)
Voandzeia necrotic mosaic virus		
Voandzeia necrotic mosaic virus		(VNMV)
Wild cucumber mosaic virus		
Wild cucumber mosaic virus	[AF035633]	(WCMV)

Tentative Species in the Genus

None reported.

Pos. ssRNA

GENUS *MARAFIVIRUS*

Type Species *Maize rayado fino virus*

DISTINGUISHING FEATURES

The distinct feature of the marafivirus genome (6.3-6.5 kb in size) is a large ORF encoding a precursor polypeptide consisting of the replication-associated proteins and the larger (~25 kDa) of the two forms of CP found in the virions. The ~225 kDa replication protein domains (Fig. 3) contain the conserved signature motifs of the replication-associated proteins (Mtr, Hel, RdRp), a papain-like protease domain, and the "marafibox", a conserved 16 nt stretch comparable to the "tymobox", from which it differs by three residue changes. The ~25 kDa CP is produced initially as a C-terminal fusion of the replication protein, while the ~21 kDa CP is produced from a 3'-co-terminal sgRNA. Viruses presently classified as marafiviruses exhibit some diversity in genome design: the MRFV and OBDV genomes differ by the presence or absence of an extensive ORF overlapping the 5' regions of the replication protein ORF, and by the presence or absence of a poly(A) tail at the 3'-terminus (MRFV, OBDV, Fig. 3). Host ranges are very narrow, making host susceptibility an important identifying criterion. Marafiviruses are strictly confined to the phloem of infected hosts and are not readily transmissible by sap inoculation, although MRFV has been transmitted by vascular puncture. Except for GAMaV and GRVFV, which have no known vector, transmission is by leafhoppers in a persistent-propagative manner, with virus replication occurring in the insect host. Species-specific virus-vector associations occur: i.e., MRFV is transmitted by *Dalbulus*, OBDV by *Macrosteles*, and BELV by *Aconurella*. The insect-transmitted species infect primarily plants in the Graminae, although a notable exception is the infection of flax by OBDV. None of the species is transmitted through seeds. The tentative species GAMaV induces peripheral vesiculation of chloroplasts.

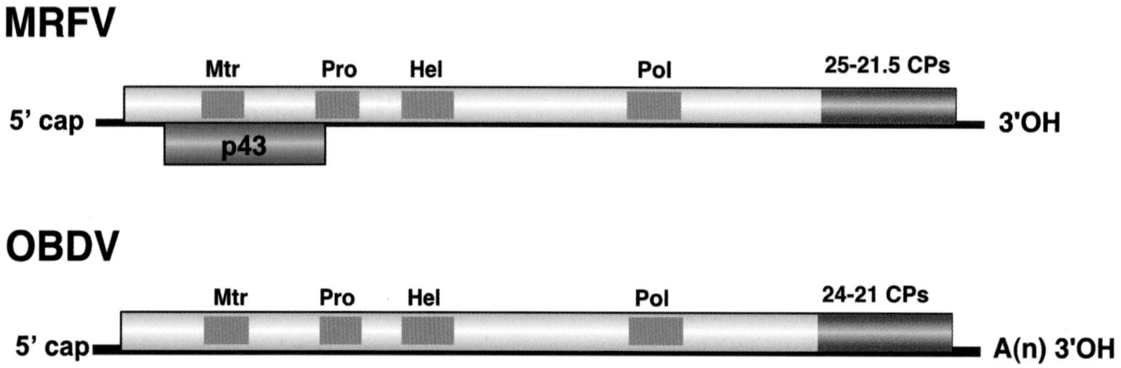

Figure 3: The two known types of genome structure in the genus *Marafivirus*, exemplified by Maize rayado fino virus (MRFV) and Oat blue dwarf virus (OBDV), showing the relative position of the ORFs and their expression products. Mtr, methyltransferase; Pro, papain-like protease; Hel, helicase; Pol, polymerase (RdRp); CPs, capsid proteins; p43, proline-rich protein.

LIST OF SPECIES DEMARCATION CRITERIA IN THE GENUS

The criteria demarcating species in the genus are:

- Overall sequence identity of less than 80%
- Capsid protein sequences less than 90% identical
- Differences in the 3'-terminal structure and in the number of ORFs
- Differential host range
- Vector specificity
- Serological specificity
- Different effects on cell ultrastructure

- Marafibox that is distinct in sequence from the tymobox

List of Species in the Genus

Species names are in green italic script; strain names and synonyms are in black roman script; tentative species names are in blue roman script. Sequence accession numbers, and assigned abbreviations ( ) are also listed.

Species in the Genus

Bermuda grass etched-line virus		
Bermuda grass etched-line virus	[AY040531]	(BELV)
Maize rayado fino virus		
Maize rayado fino virus	[AF265566]	(MRFV)
Oat blue dwarf virus		
Oat blue dwarf virus	[U87832]	(OBDV)

Tentative Species in the Genus

Grapevine asteroid mosaic-associated virus	[AJ249357, AJ249358]	(GAMaV)
Grapevine rupestris vein feathering virus	[AY128949]	(GRVFV)

Genus Maculavirus

Type Species *Grapevine fleck virus*

Distinguishing Features

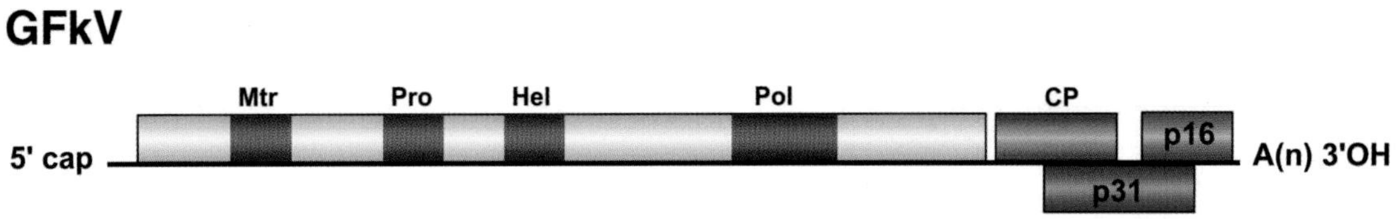

Figure 4: Genome structure of an isolate of *Grapevine fleck virus*, the type species of the genus *Maculavirus* showing the relative position of the ORFs and their expression products. Mtr, methyltransferase; Pro, papain-like protease; Hel, helicase; Pol, polymerase (RdRp); CP, capsid protein; p31 and p16, proline-rich proteins.

The genomic RNA of GFkV (7.5 kb in size) is the largest in the family, contains four ORFs and is 3'-polyadenylated (Fig.4). ORF1 encodes a polypeptide of 215 kDa that contains the conserved signature motifs of the replication-associated proteins (Mtr, Hel, RdRp) and a papain-like protease domain, but does not seem to have a conserved sequence comparable to the "tymobox" or the "marafibox". ORF2 codes for the 24 kDa CP. ORF3 and ORF4, which are located at the extreme 3'-end of the genome, code for proline-rich proteins of 31 and 16 kDa, respectively, which show a distant relationship with the putative MPs of tymoviruses. There are two known species in the genus (GFkV and GRGV), both of which are restricted to *Vitis* species, which are infected latently, with the exception of *V. rupestris,* which reacts to GFkV with translucent spots (flecks) on the leaves. Both species are strictly confined to the phloem of infected hosts and are not transmissible by sap inoculation. The cytopathology of GFkV infections is characterized by a severe modification of mitochondria into structures called "multivesiculate bodies". Field spread of GFkV has been reported, but the vector is unknown. GFkV is not transmitted through seeds, thus virus dissemination is primarily through distribution of infected propagative material.

List of Species Demarcation Criteria in the Genus

The criteria demarcating species in the genus are:

- Overall sequence identity of less than 80%
- Capsid protein sequences identity less than 90%

Pos. ssRNA

- Serological specificity
- Lack of tymobox/marafibox sequence

List of Species in the Genus

Species names are in green italic script; strain names and synonyms are in black roman script; tentative species names are in blue roman script. Sequence accession numbers, and assigned abbreviations ( ) are also listed.

Species in the Genus

Grapevine fleck virus		
Grapevine fleck virus	[AJ309022]	(GFkV)

Tentative Species in the Genus

Grapevine red globe virus	[AJ249361,AF521977]	(GRGV)

List of Unassigned Viruses in the Family

Poinsettia mosaic virus		
Poinsettia mosaic virus	[AJ271595]	(PnMV)

PnMV has characteristics of both the tymoviruses and marafiviruses, and some distinct characteristics. Like tymoviruses, PnMV is mechanically transmissible, infects all main tissues of the host, and amplifies to high titre. Like that of the marafivirus OBDV, the genome contains a single long ORF comprising fused replication and CP domains, a poly(A) tail at the 3′-end and a "marafibox" variant of the "tymobox". Unlike marafiviruses, PnMV is not insect-transmitted, and possesses only a single CP. PnMV infection induces invaginations of the chloroplast membrane, but these are bounded by a single membrane, rather than the double membrane characteristic of tymoviruses. Sequence relationships (Fig. 6) support classification distinct from the existing genera of the family *Tymoviridae*.

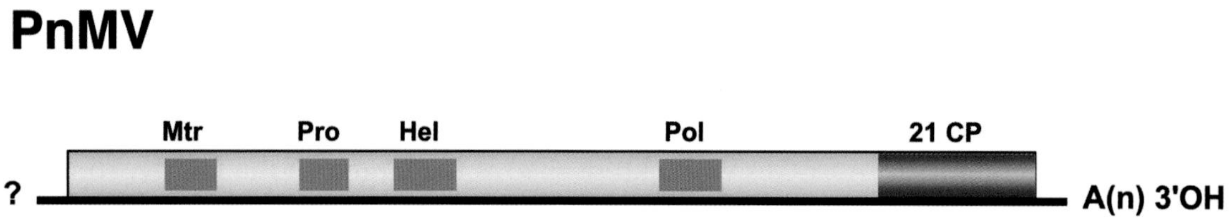

Figure 5: Genome structure of Poinsettia mosaic virus (PnMV), unassigned within the family *Tymoviridae*, showing the single ORF encoding domains characteristic of Mtr, papain-like protease (Pro), helicase (Hel), and RdRp (Pol), and the 21-kDa CP. It is unknown whether the 5′-end of the genome is capped.

Phylogenetic Relationships within the Family

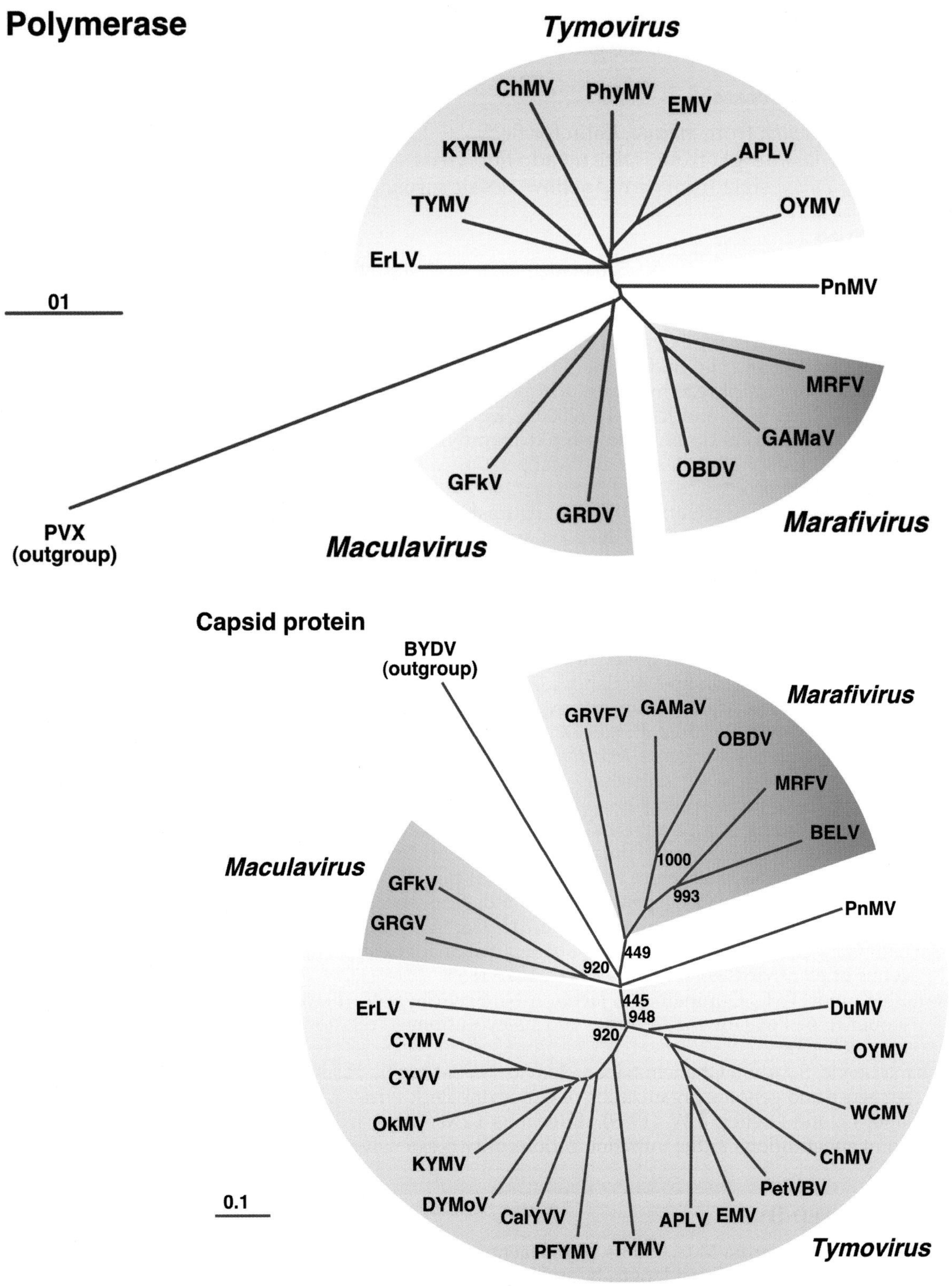

Figure 6: Phylogenetic tree showing the relationship between the species and genera of the family *Tymoviridae* based on the RdRp (Top) and CP (Bottom) sequences. The meaning of the abbreviations and the sequence accession # are indicated in the list of species in the genera descriptions. The tree was produced using CLUSTAL W. Branch lengths are proportional to sequence distances.

Similarity with Other Taxa

Replication-associated proteins (RdRp, Mtr, and Hel) contain signature sequences homologous to those of other taxa of the "alpha-like" supergroup of ssRNA viruses, especially those of the genera *Carlavirus* and *Potexvirus*.

Derivation of Names

Macula: from *macula*, Latin for fleck.
Marafi: sigla from *ma*ize *ra*yado *fi*no virus,
Tymo: sigla from *t*urnip *y*ellow *mo*saic virus,

References

Abou Ghanem-Sabanadzovic, N., Sabanadzovic, S. and Martelli, G.P. (2003). Sequence analysis of the 3'-end of three *Grapevine fleck virus*-like viruses from grapevine. *Virus Genes*, **27**, 11-16.

Boscia, D., Sabanadzovic, S., Savino, V., Kyriakopoulou, P.E., Martelli, G.P. and Lafortezza, R. (1994). A non mechanically-transmissibile virus associated with asteroid mosaic of the grapevine. *Vitis*, **33**, 101-102.

Bozarth, C.S., Weiland, J.J. and Dreher, T.W. (1992). Expression of ORF-69 of turnip yellow mosaic virus is necessary for viral spread in plants. *Virology*, **187**, 124-130.

Bradel, B.G., Preil, W.X. and Jeske, H. (2000). Sequence analysis and genome organisation of Poinsettia mosaic virus (PnMV) reveal closer relationship to marafiviruses than to tymoviruses. *Virology*, **271**, 289-297.

Canady, M.A., Larson, S.B., Day, J. and McPherson, A. (1996). Crystal structure of turnip yellow mosaic virus. *Nature Struct. Biol.*, **3**, 771-781.

Castellano, M.A. and Martelli, G.P. (1984). Ultrastructure and nature of the vesiculated bodies associated with isometric virus-like particles in diseased grapevines. *J. Ultrastruct. Res.*, **89**, 56-64.

Ding, S.W., Howe, J., Keese, P., Mackenzie, A., Meek, D., Osorio-Keese, M., Skotnicki, M., Pattana, S., Torronen, M. and Gibbs, A. (1990). The tymobox, a sequence shared by most tymoviruses: its use in molecular studies of tymoviruses. *Nucl. Acids Res.*, **18**, 1181-1187.

Dreher, T.W. and Goodwin, J.B. (1998). Transfer RNA mimicry among tymoviral genomic RNAs ranges from highly efficient to vestigial. *Nucl. Acids Res.*, **26**, 4356-4364.

Edwards, M.C., Zhang, Z. and Weiland, J.J. (1997). Oat blue dwarf marafivirus resembles the tymoviruses in sequence, genome organization, and expression strategy. *Virology*, **232**, 217-229.

Hammond, R.W. and Ramirez, P. (2001). Molecular characterization of the genome of *Maize rayado fino virus*, the type member of the genus *Marafivirus*. *Virology*, **282**, 338-347.

Hellendoorn, K., Mat, A.W., Gultyaev, A.P. and Pleij, C.W.A. (1996). Secondary structure model of the coat protein gene of turnip yellow mosaic virus RNA: long C-rich, single-stranded regions. *Virology*, **224**, 43-54.

Izadpanah, K., Zhang, Y.P., Daubert, S. and Rowhani, A. (2002). Sequence of the coat protein gene of Bermuda grass etched-line virus and of the adjacent "marafibox" motif. *Virus Genes*, **24**, 131-134.

Koenig, R. (1976). A loop structure in the serological classification system of tymoviruses. *Virology*, **72**, 1-5.

Lesemann, D.E. (1977). Virus group-specific and virus-specific cytological alterations induced by members of the tymovirus group. *Phytopathol. Z.*, **90**, 315-336.

Martelli, G.P., Sabanadzovic, S., Abou Ghanem-Sabanadzovic, N. and Saldarelli, P. (2002). *Maculavirus*, a new genus of plant viruses. *Arch. Virol.*, **147**, 1847-1853.

Ranjith-Kumar, C.T., Gopinath, K., Jacob, A.N., Srividhya, V., Elango, P. and Savithri, H.S. (1998). Genomic sequence of Physalis mottle virus and its evolutionary relationship with other tymoviruses. *Arch. Virol.*, **143**, 1489-1500.

Sabanadzovic, S., Abou Ghanem-Sabanadzovic, N., Saldarelli, P. and Martelli, G.P. (2001). Complete nucleotide sequence and genome organization of Grapevine fleck virus. *J. Gen. Virol.*, **82**, 2009-2015.

Weiland, J.J. and Dreher, T.W. (1989). Infectious TYMV RNA from cloned cDNA. Effects *in vitro* and *in vivo* of point substitutions in the initiation codons of two extensively overlapping ORFs. *Nucl. Acids Res.*, **17**, 4675-4687.

Contributed By

Dreher, T.W., Edwards, M.C., Gibbs, A.J., Haenni, A.-L., Hammond, R.W., Jupin, I., Koenig, R., Sabanadzovic, S., Abou Ghanem-Sabanadzovic, N. and Martelli, G.P.

FAMILY CLOSTEROVIRIDAE

TAXONOMIC STRUCTURE OF THE FAMILY

Family	*Closteroviridae*
Genus	*Closterovirus*
Genus	*Ampelovirus*
Genus	*Crinivirus*

VIRION PROPERTIES

MORPHOLOGY

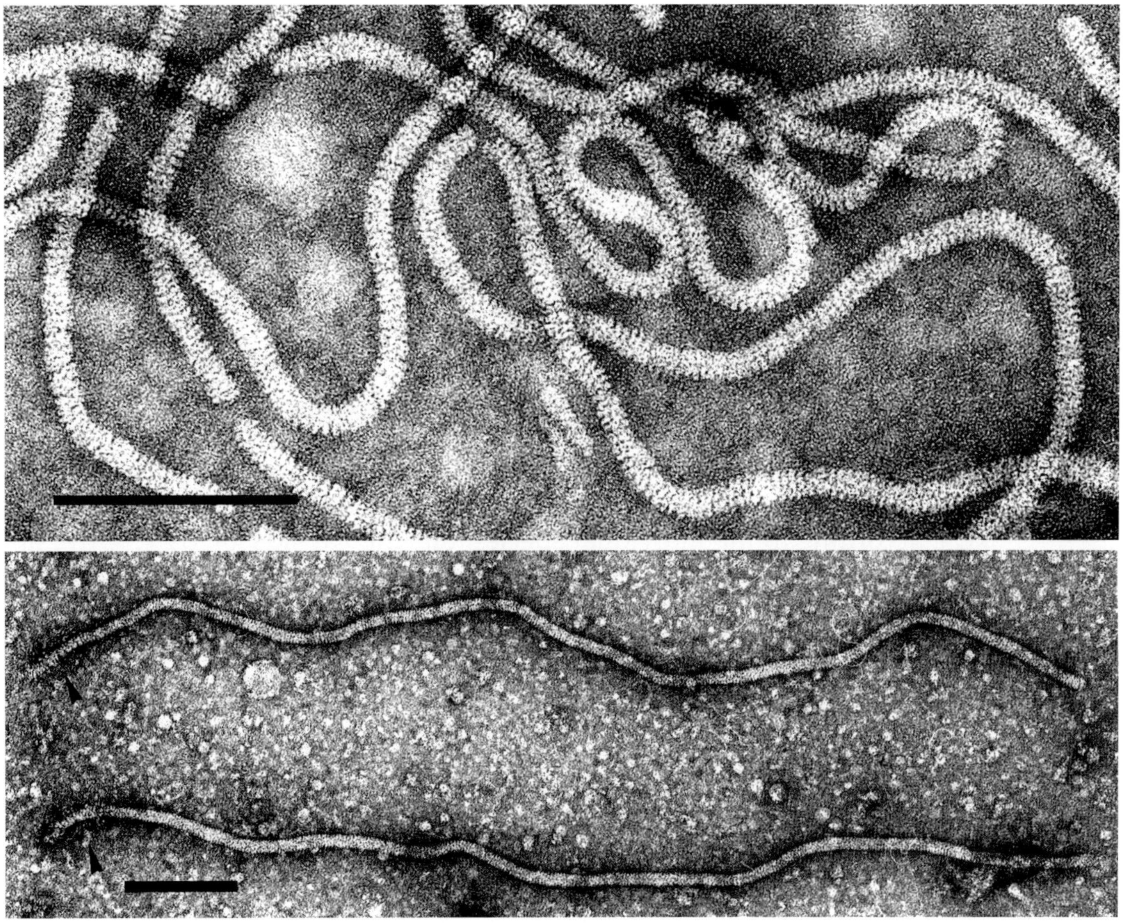

Figure 1: (Top) Negative contrast electron micrograph of virions of Citrus tristeza virus (CTV) (genus *Closterovirus*). Particles of all members of the genera *Ampelovirus* and *Crinivirus* have a similar morphology. The bar represents 100 nm (Courtesy of R.G. Milne). (Bottom) Beet yellows virus (BYV) particles showing a decorated extremity (arrow heads) following exposure to an antiserum to the N-terminal peptide expressed from the minor CP gene. The bar represents 100 nm (Courtsey of D.E. Lesemann).

PHYSICOCHEMICAL AND PHYSICAL PROPERTIES

Virions of all genera usually sediment as a single band in sucrose or Cs_2SO_4 gradients. S_{20w} range from 96 to 140, buoyant densities are 1.30-1.34 g/cm^3 in CsCl and 1.30-1.34 g/cm^3 in Cs_2SO_4. Virions of most species are degraded by CsCl and are unstable in high salt concentration. Virions resist moderately high temperatures (thermal inactivation is around 45-55°C) and organic solvents, but are sensitive to RNAse and chelation.

NUCLEIC ACID

Regardless of the genome type, monopartite or bipartite, virions contain a single molecule of linear, positive sense, ssRNA, constituting 5-6% of the particle weight. Genome size is related to particle length. In monopartite genome species (genera *Closterovirus* and *Ampelovirus*), genome sizes range from 15.5 kb (BYV) to 19.3 kb (CTV), the largest genome among positive-strand RNA plant viruses. Genome sizes of the bipartite genome criniviruses Sweet potato chlorotic stunt virus (SPCSV), Lettuce infectious yellows virus (LIYV) and Cucurbit yellow stunting disorder (CYSDV) are 17.6 kb, 15.3 kb and 16.4 kb, respectively. The 5'-end of the genome is likely to be capped. The 3'-end is not polyadenylated and does not possess a tRNA-like structure, but may have several hairpin structures. The genomic RNAs of isolates of two species of the genus *Closterovirus*, i.e., BYV and CTV (four isolates from California, Florida, Spain, and Israel), three species of the genus *Ampelovirus*, i.e. Little cherry virus 2 (LChV-2), Grapevine leafroll-associated virus 2 (GLRaV-2), Grapevine leafroll-associated virus 3 (GLRaV-3), and three viruses of the genus *Crinivirus*, i.e., LIYV, CYSDV, SPCSV, and Beet pseudo yellows virus (BPYV) have been completely sequenced. Partial to almost complete sequences are available for Beet yellow stunt virus (BYSV) and Carnation necrotic fleck virus (CNFV) from the genus *Closterovirus*; Grapevine leafroll-associated virus 1 (GLRaV-1), Grapevine leafroll-associated virus 5 (GLRaV-5), Pineapple mealybug wilt associated virus 1 (PMWaV-1), Pineapple mealybug wilt associated virus 2 (PMWaV-2), Grapevine leafroll-associated virus 4 (GLRaV-4), Grapevine leafroll-associated virus 6 (GLRaV-6), Plum bark necrosis and stem pitting associated virus (PBNSPaV) from the genus *Ampelovirus*; Tomato infectious chlorosis virus (TICV), Tomato chlorosis virus (ToCV), from the genus *Crinivirus*, and for the unassigned species in the family, Grapevine leafroll-associated virus 7 (GLRaV-7), Olive leaf yellowing associated virus (OLYaV) and Little cherry virus 1 (LChV-1).

PROTEINS

Virions of all members of the family are composed of a major CP ranging from 22 to 46 kDa, according to the individual species, but the capsids of BYV, CTV, and LIYV also incorporate the minor CP (CPm) at one end of the particle. Additional evidence suggests that HSP70h and CPm are tightly associated with the virions and that CPm is required for the assembly of virion tails. The CP of some isolates (BYV, CNFV and Lilac chlorotic leafspot virus) lack tryptophan, resulting in a high A_{260}/A_{280} ratio (1.4-1.8) for the virions. The reverse is true for the CP of CTV, which contains a significant amount of tryptophan (A_{260}/A_{280} =1.21-1.22). Non-structural proteins common to all members of the family are: (i) a large polypeptide (295 to 349 kDa) containing the conserved domains of papain-like protease (P-Pro), Mtr, and Hel; (ii) a 48 to 57 kDa protein with all sequence motifs of viral RdRp of the "alpha-like" supergroup of positive-strand RNA viruses; (iii) a 6 kDa hydrophobic protein with membrane-binding properties; (iv) a 59 to 70 kDa homolog of the cellular HSP70 heat-shock proteins (HSP70h) implicated in the cell-to-cell movement of the viruses; (v) a 55 to 64 kDa product. At least five different virus-coded proteins are required for BYV cell-to-cell movement, i.e. CP, CPm, three MPs, namely, the 6 kDa protein (p6), the HSP70h, and the 64 kDa protein (p64) (Fig. 2). The L-Pro domain of ORF1a and the p20 protein coded for by ORF7 of BYV are involved in systemic transport. The BYV HSP70h localizes near to plasmodesmata and inside plasmodesma channels and interacts with p20, providing a site for its attachment to virions. Both proteins associate with intracellular accumulations of virus particles, forming a physical complex with them. p20 is dispensable for assembly and cell-to-cell movement of BYV, but is required for long-distance transport of BYV through the phloem.

LIPIDS

None present.

Carbohydrates

None present.

Genome Organization and Replication

Closteroviruses have the largest genomes among positive-strand RNA plant viruses. The genome is monopartite for the members of the genera *Closterovirus* and *Ampelovirus* and bipartite for most of those in the genus *Crinivirus*. The genomes of viruses in all genera are characterized by the presence of unique genes coding for HSP70h proteins and for an analogue of the CP. The genome organizations, the numbers and relative positions of the ORFs vary with the genus and/or individual virus species. In species of the genus *Closterovirus* (BYV, CTV, BYSV, GLRaV-2), the CPm is upstream of the CP, whereas the reverse is true for members of the genus *Ampelovirus* (GLRaV-1, GLRaV-3, LChV-2, PMWaV-1, PMWaV-2) and *Crinivirus* (LIYV, SPSVV, CCSV, CYSDV, BPYV). The ORFs coding for the 6 kDa small hydrophic protein, the HSP70h, the 55-64 kDa product, the CP, and CPm, form a five-gene module that is conserved among members of the family (Fig. 2 and 3). The genome expression strategy is based on: (i) proteolytic processing of the polyprotein encoded by ORF1a; (ii) +1 ribosomal frameshift for the expression of the RdRp domain encoded by ORF1b; (iii) expression of the downstream ORFs via the formation of a nested set of 3′ co-terminal sgRNAs (Fig. 2 and 3). The dsRNA patterns are very complex and variable among species, reflecting the different numbers and sizes of the ORFs present in individual genomes. Replication occurs in the cytoplasm, possibly in association with endoplasmic reticulum-derived membranous vesicles and vesiculated mitochondria.

Antigenic Properties

Virion proteins are moderately antigenic. Most of the species are serologically unrelated or distantly related to one another. Monoclonal antibodies have been produced to proteins of a number of members, i.e., CTV, GLRaV-1 to -6, PMWaV-2, and SPCSV. Polyclonal antisera to CTV, LChV-2, GLRaV-2, GLRaV-3, and SPCSV have been raised from fusion proteins produced in bacterial expression systems.

Biological Properties

Host Range

The natural and experimental host ranges of individual virus species are restricted. Disease symptoms are of the yellowing type (i.e. stunting, rolling, yellowing or reddening of the leaves, small and late ripening fruits), or pitting and/or grooving of the woody cylinder. Infection is systemic, but usually limited to the phloem, which may necrotize to a varying extent

Transmission

Few species of the genus *Closterovirus* are transmissible by mechanical inoculation, though with difficulty, but none of those in the genera *Ampelovirus* and *Crinivirus* can be. In vegetatively propagated crops, long distance virus dissemination is primarily through infected propagating material. Transmission through seeds is very rare. According to the genus, natural vectors are aphids, whiteflies (*Bemisia* and *Trialeurodes*), pseudococcid (*Pseudococcus, Planococcus; Phenacoccus, Saccharicoccus,* and *Dysmicoccus),* and coccid (*Pulvinaria, Neopulvinaria,* and *Parthenolecanium*) mealybugs. Transmission is semi-persistent regardless of the type of vector.

Geographical Distribution

Geographical distribution varies from restricted to widespread, depending on the virus species, most of which occur in temperate or subtropical regions.

Cytopathic Effects

Virions are usually found in the phloem (sieve tubes, companion cells, phloem parenchyma), occasionally in the mesophyll and epidermis. Ultrastructural modifications arise by membrane proliferation, vesiculation of chloroplasts, degeneration and vesiculation of mitochondria, and formation of inclusion bodies. These are made up of aggregates of virions or membranous vesicles, or a combination of the two. Virions accumulate in conspicuous cross-banded fibrous masses or, more typically, in more or less loose bundles intermingled with single or clustered membranous vesicles. Inclusions of this type are one of the hallmarks of the family. The vesicles contain a fibrillar network and derive either from the endoplasmic reticulum, or from peripheral vesiculation of mitochondria.

Genus *Closterovirus*

Type Species *Beet yellows virus*

Distinguishing Features

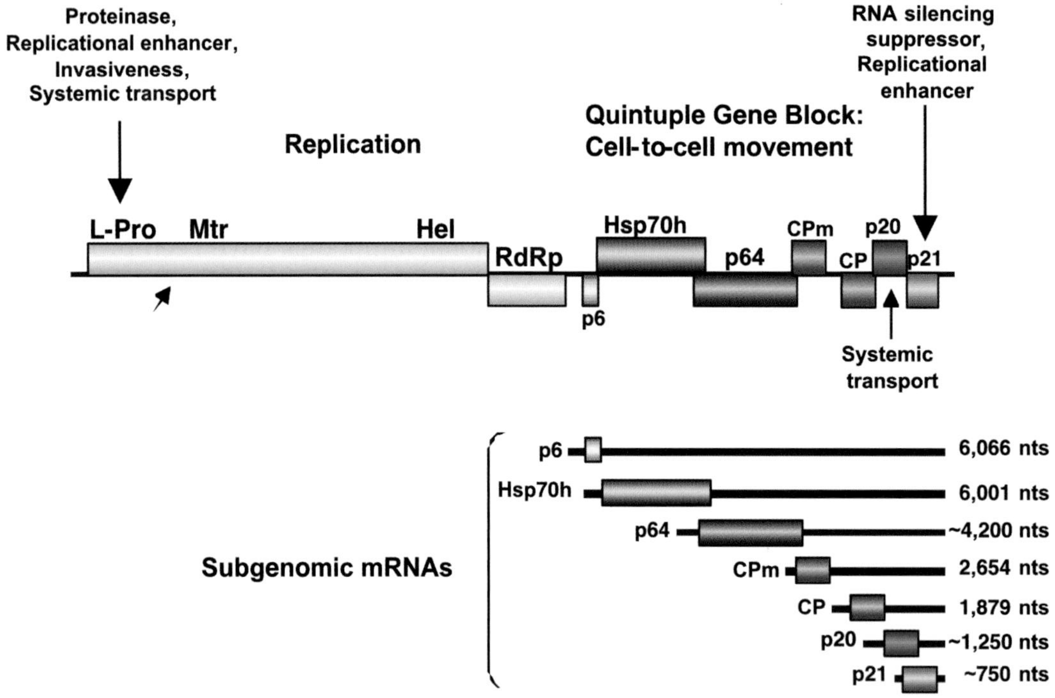

Figure 2: Genome organization and strategy of replication characteristic of Beet yellows virus (BYV) showing the relative position of the ORFs, their expression products, and the 3' nested set of sgRNAs. L-Pro, leader proteinase; Mtr, methyltransferase;Hel, helicase; RdRp, RNA polymerase; HSP70h, heat shock protein homologue; CP coat protein; CPm, minor capsid protein. The five boxes under "cell-to-cell movement" represent the five gene block conserved among closteroviruses (from Dolja, 2003).

Virions are of one size (>1000 nm, usually 1250 to 2200 nm in length) and contain a single molecule of linear, positive sense, ssRNA from 15.5 to 19.3 kb in size. CTV has also smaller than full-length particles that may encapsidate subgenomic or multiple species of defective RNAs (D-RNA) containing all of the *cis*-acting sequences required for replication. SgRNAs may be involved in the construction of recombinant D-RNAs. The presence of D-RNA makes the dsRNA pattern of CTV isolates more complex than that of other members of the genus. The major CP subunit (22-25 kDa) coats most of the virion length and CPm coats a short segment (~75 nm) at one end of the particle. There are three types of genome organization among the sequenced species of the genus, typified by BYV (Fig. 2), CTV (Fig. 3) and BYSV. The BYV genome contains nine ORFs flanked by 5'- and 3'-UTR of 107 and 181 nt, respectively. CTV has twelve ORFs and UTRs of 107 nt at the 5'-end and 275 nt at the 3'-end. Differences with BYV are the presence of two papain-like protease domains in ORF 1a, of an extra 5'-proximal ORF (ORF2) encoding a 33 kDa product with no similarity to any other protein in databases, and of two extra 3'-proximal ORFs (ORF9 and ORF11). The BYSV genome has ten ORFs and a 3' UTR of 241 nt, a length intermediate between that of the BYSV and CTV UTRs. A further difference with the BYV genome is the presence of an extra ORF (ORF2) encoding a 30 kDa polypetide with no similarity to any other protein in the databases. This ORF is located downstream of ORF1b, i.e. in the same position as the unrelated CTV ORF2. Thus, the organization of BYSV genomes is intermediate between that of BYV and CTV, suggesting that these three viruses might represent three distinct stages in closterovirus evolution. In contrast to the other two genera in the family (*Ampelovirus* and *Crinivirus*), the closterovirus gene encoding the CPm is upstream of the CP gene. The dsRNA-containing cytoplasmic vesicles that accumulate in the cytoplasm of infected cells arise primarily by proliferation of the endoplasmic reticulum. The genus contains only viruses that are transmitted by aphids in a semi-persistent manner and that infect primarily dicotyledonous hosts. Several of the members are transmissible by sap inoculation, though with some difficulty.

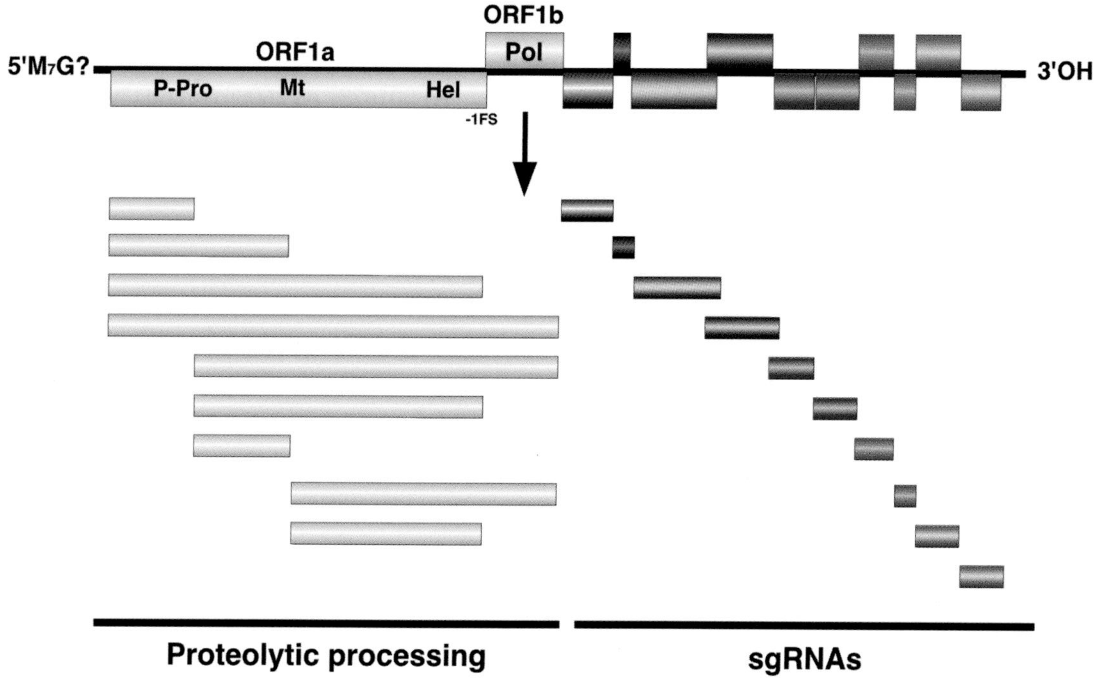

Figure 3: Organization and expression strategy of genomes characteristic of Citrus tristeza virus. Dashed lines represent putative protein products corresponding to the respective ORFs. +1 FS the putative ribosomal frameshift. Solid lines define the two genomic regions expressed through proteolytic processing of the polyprotein precursor(s) and through the formation of a nested set of 3' co-terminal sgRNAs (from Karasev and Hilf, 1997).

List of Species Demarcation Criteria in the Genus

The criteria demarcating species in the genus are:

- Particle size,
- Size of CP, as determined by deduced amino acid sequence data,
- Serological specificity using discriminatory monoclonal or polyclonal antibodies,
- Genome structure and organization (number and relative location of the ORFs),
- Amino acid sequence of relevant gene products (CP, CPm, HSP70h) differing by more than 10%,

POS. SSRNA

- Vector species and specificity,
- Magnitude and specificity of natural and experimental host range,
- Cytopathological features (i.e., aspect of inclusion bodies and origin of cytoplasmic vesicles).

LIST OF SPECIES IN THE GENUS

Species names are in green italic script; strain names and synonyms are in black roman script; tentative species names are in blue roman script. Sequence accession numbers, and assigned abbreviations ( ) are also listed.

SPECIES IN THE GENUS

1- Aphid-transmitted

Beet yellows virus		
Beet yellows virus	[X73476]	(BYV)
Beet yellow stunt virus		
Beet yellow stunt virus	[U51931]	(BYSV)
Burdock yellows virus		
Burdock yellows virus		(BuYV)
Carnation necrotic fleck virus		
Carnation necrotic fleck virus		(CNFV)
Carrot yellow leaf virus		
Carrot yellow leaf virus		(CYLV)
Citrus tristeza virus		
Citrus tristeza virus	[U56902, U163304, AY170468, Y18420, AF001623, AF260651, AB04398]	(CTV)
Wheat yellow leaf virus		
Wheat yellow leaf virus		(WYLV)

2- Vector unknown

Grapevine leafroll-associated virus 2		
Grapevine leafroll-associated virus 2	[Y14131]	(GLRAV-2)

TENTATIVE SPECIES IN THE GENUS

Clover yellows virus	(CYV)
Dendrobium vein necrosis virus	(DVNV)
Festuca necrosis virus	(FNV)
Heracleum virus 6	(HV-6)

GENUS *AMPELOVIRUS*

Type Species *Grapevine leafroll-associated virus 3*

DISTINGUISHING FEATURES

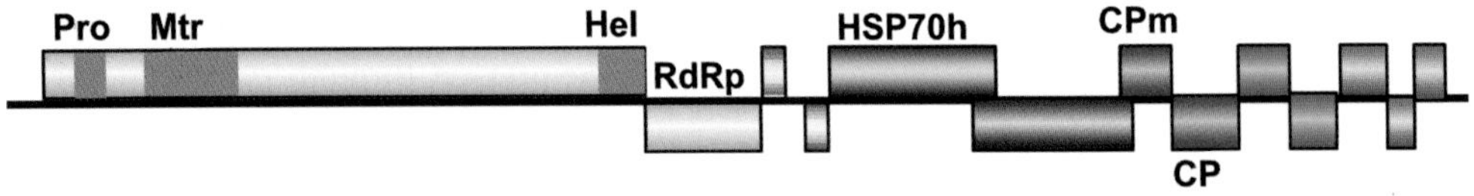

Figure 4: Genome structure characteristic of a representative of *Grapevine leafroll-associated virus 3*, the type species of the genus *Ampelovirus*, showing the relative position of the ORFs and their expression products. Pro, papain-like protease; Mtr; Hel, helicase; RdRp, RNA polymerase; HSP70h, heat shock-related protein homologue; CP, capsid protein; CPm, minor capsid protein.

Virions are of one size (>1000 nm, usually 1400-2200 nm in length) and contain a single molecule or linear, positive sense, single stranded RNA 16.9-17.9 kb in size. CP subunits have a high molecular mass (35-46 kDa). There are two types of genome structure in the genus typified by GLRaV-3 (Fig. 4) and LChV-2. The GLRaV-3 genome contains 12 ORFs with a 3' UTR of 277 nt. The LChV-2 genome (16.9 kb), the third largest monopartite closterovirus sequenced to date, contains nine ORFs and UTRs of 76 nt at the 5'-end and 210 nt at the 3'-end, respectively. LChV-2 has the largest HSP70h (70 kDa), CP (46 kDa) and CPm (76 kDa) among ampeloviruses. With GLRaV-3 and several other sequenced members of the genus, the CPm gene is downstream of the CP gene, whereas with LChV-2, the CPm gene is five ORFs upstream of the CP cistron. GLRaV-1 shows a further peculiarity in that its CPm gene is duplicated. The dsRNA-containing vesicles that accumulate in the cytoplasm of infected cells may arise either by proliferation of the endoplasmic reticulum or from vesiculation and fragmentation of mitochondria. The genus contains viruses that infect only dicotyledonous hosts, and that are transmitted semi-persistently by coccid (*Parthenolecanium, Pulvinaria, Neopulvinaria*) or pseudococcid (*Pseudococcus, Planococcus, Saccharicoccus, Dysmiococcus, Phenacoccus, Heliococcus*) mealybugs. None of the members is transmissible by sap inoculation.

List of Species Demarcation Criteria in the Genus

The criteria demarcating species in the genus are:

- Particle size,
- Size of CP, as determined by deduced aa sequence data,
- Serological specificity using discriminatory monoclonal or polyclonal antibodies,
- Genome structure and organization (number and relative location of the ORFs),
- Amino acid sequence of relevant gene products (CP, CPm, HSP70h) differing by more than 10%,
- Vector species and specificity,
- Magnitude and specificity of natural and experimental host range,
- Cytopathological features (i.e., aspect of inclusion bodies and origin of cytoplasmic vesicles).

List of Species in the Genus

Species names are in green italic script; strain names and synonyms are in black roman script; tentative species names are in blue roman script. Sequence accession numbers, and assigned abbreviations ( ) are also listed.

Species in the Genus

Grapevine leafroll-associated virus 1		
Grapevine leafroll-associated virus 1	[AF195822]	(GLRaV-1)
Grapevine leafroll-associated virus 3		
Grapevine leafroll-associated virus 3	[U82937]	(GLRaV-3)
Grapevine leafroll-associated virus 5		
Grapevine leafroll-associated virus 5	[AF039552]	(GLRaV-5)
Little cherry virus 2 ‡		
Little cherry virus 2	[AF 531505, AF333237, AF416335]	(LChV-2)
Pineapple mealybug wilt-associated virus 1		
Pineapple mealybug wilt-associated virus 1	[AF414119]	(PMWaV-1)
Pineapple mealybug wilt-associated virus 2		
Pineapple mealybug wilt-associated virus 2	[AF283103]	(PMWaV-2)

Tentative Species in the Genus

1- Mealybug-transmitted		
Sugarcane mild mosaic virus		(SMMV)
2 -Vector unknown		
Grapevine leafroll-associated virus 4	[AF039553]	(GLRaV-4)
Grapevine leafroll-associated virus 6	[AJ496796]	(GLRaV-6)
Grapevine leafroll-associated virus 8		(GLRaV-8)
Plum bark necrosis and stem pitting-associated virus	[AF195501]	(PBNSTaV)

POS. SSRNA

GENUS CRINIVIRUS

Type Species *Lettuce infectious yellows virus*

DISTINGUISHING FEATURES

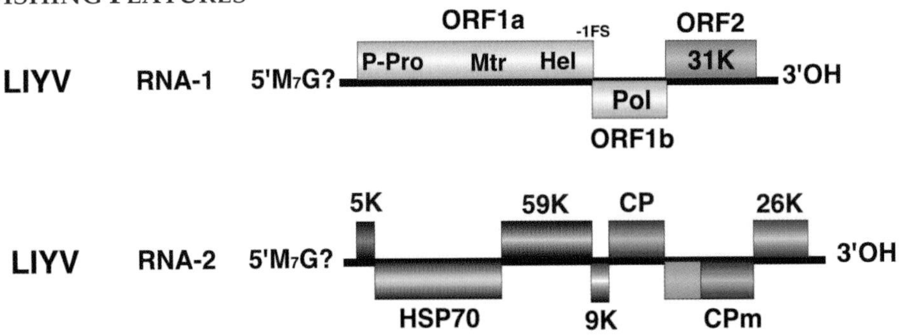

Figure 5: Genome organization of Lettuce infectious yellows virus (LIYV) showing the relative position of the ORFs and their expression products. P-Pro, papain-like protease; Mtr, methyltranferase; Hel, helicase; POL, RNA polymerase; HSP70h, heat shock-related protein homologue; CP capsid protein; CPm, minor capsid protein.

Virions are shorter that 1000 nm and have two modal lengths (650-850 and 700-900 nm). The genome is a linear, positive sense, ssRNA 15.3 kb (LIYV), 16.4 kb (CYSDV), and 17.6 kb (SPCSV) in size, divided into two molecules, both needed for infectivity and separately encapsidated (Fig. 4). CPs of known members range from 28 to 33 kDa, but the size of CPm can be as large as *c.* 80 kDa. There are three types of genome structure in the genus, typified by LIYV, SPCSV and BPYV. LIYV RNA-1 is a bicistronic molecule comprising a 5'- and 3'-UTR of 97 and 219 nt, respectively. By contrast, the SPCSV 3'-UTR is longer (208 nt). LIYV RNA-1 encodes replication-related proteins (ORF1), including RdRp, which is expressed via a putative ribosomal frameshift, and a *trans*-enhancer for RNA-2 accumulation (ORF2). SPCSV RNA-1 has two extra ORFs coding for an RNAse III-like protein (ORF2) and a putative small hydrophobic protein (p7) (ORF3). This protein could be homologous to a hydrophobic protein of similar size encoded by ORF1 of LIYV RNA-2. BPYV RNA-1 is a monocistronic molecule comprising only ORF1a and ORF1b. RNA-2 of all sequenced members of the genus (LIYV, CYSDV, SPCSV, BPYV) has seven ORFs. It contains the five-gene module typical of the family, which, however, differs from that of the genera *Closterovirus* and *Ampelovirus* because of the insertion of an extra small gene (ORF4) upstream of the CP gene. SPCSV and ToCV have notably large ORFs for the CPm (75-79 kDa), unlike LIYV (52 kDa). In all members of the genus the position of the CPm ORF is downstream of the CP gene. The dsRNA-containing vesicles that accumulate in the cytoplasm of infected cells may arise by proliferation of the endoplamic reticulum. None of the members can be transmitted by sap inoculation. Natural vectors are whiteflies (*Bemisia, Trialeurodes*) that transmit in a semi-persistent manner. PYVV and DVCV are transmitted by *Trialeurodes* seem to have undivided genomes and arrangements of genes similar to those of bipartite viruses. They have been assigned to this genus as tentative species because the type of vector transmission is a key biological character recognized as a major classification feature of closteroviruses.

LIST OF SPECIES DEMARCATION CRITERIA IN THE GENUS

The criteria demarcating species in the genus are:

- Particle size,
- Size of CP, as determined by deduced amino acid sequence data,
- Serological specificity using discriminatory monoclonal or polyclonal antibodies,
- Genome structure and organization (number and relative location of the ORFs),
- Amino acid sequence of relevant gene products (CP, CPm, HSP70h) differing by more than 10%,

- Vector species and specificity,
- Magnitude and specificity of natural and experimental host range,
- Cytopathological features (i.e., aspect of inclusion bodies and origin of cytoplasmic vesicles).

LIST OF SPECIES IN THE GENUS

Species names are in green italic script; strain names and synonyms are in black roman script; tentative species names are in blue roman script. Sequence accession numbers, and assigned abbreviations ( ) are also listed.

SPECIES IN THE GENUS

Abutilon yellows virus		
Abutilon yellows virus		(AbYV)
Beet pseudoyellows virus		
Beet pseudoyellows virus	[Y15568, U67447]	(BPYV)
(Cucumber chlorotic spot virus)		
(Cucumber yellows virus)	[AB085612, AB085613]	
Cucurbit yellow stunting disorder virus		
Cucurbit yellow stunting disorder virus	[AJ439690, AJ537493]	(CYSDV)
Lettuce chlorosis virus		
Lettuce chlorosis virus		(LCV)
Lettuce infectious yellows virus		
Lettuce infectious yellows virus	[U15440, U15441]	(LIYV)
Sweet potato chlorotic stunt virus		
Sweet potato chlorotic stunt virus	[AJ428554, AJ428555]	(SPCSV)
(Sweet potato sunken vein virus)		
Tomato chlorosis virus		
Tomato chlorosis virus		(ToCV)
Tomato infectious chlorosis virus		
Tomato infectious chlorosis virus		(TICV)

TENTATIVE SPECIES IN THE GENUS

Diodea vein chlorosis virus		(DVCV)
Potato yellow vein virus	[AF150984, AJ508757]	(PYVV)

LIST OF UNASSIGNED VIRUSES IN THE FAMILY

The species that follow could not be allocated to any of the genera either because the available information is scanty (Megakepasma mosaic virus; MegMV and Alligatorweed stunting virus; AWSV), or because more molecular (Olive leaf yellowing-associated virus; OLYaV and Grapevine leafroll-associated virus 7; GLRaV-7) or biological data (Little cherry virus 1; LChV-1) are needed for unequivocal classification.

Alligatorweed stunting virus		(AWSV)
Little cherry virus 1 ‡	[Y10237, NC001836]	(LChV-1)
Grapevine leafroll-associated virus 7	[Y15897]	(GLRaV-7)
Megakepasma mosaic virus		(MegMV)
Olive leaf yellowing-associated virus	[AJ440010]	(OLYaV)

PHYLOGENETIC RELATIONSHIPS WITHIN THE FAMILY

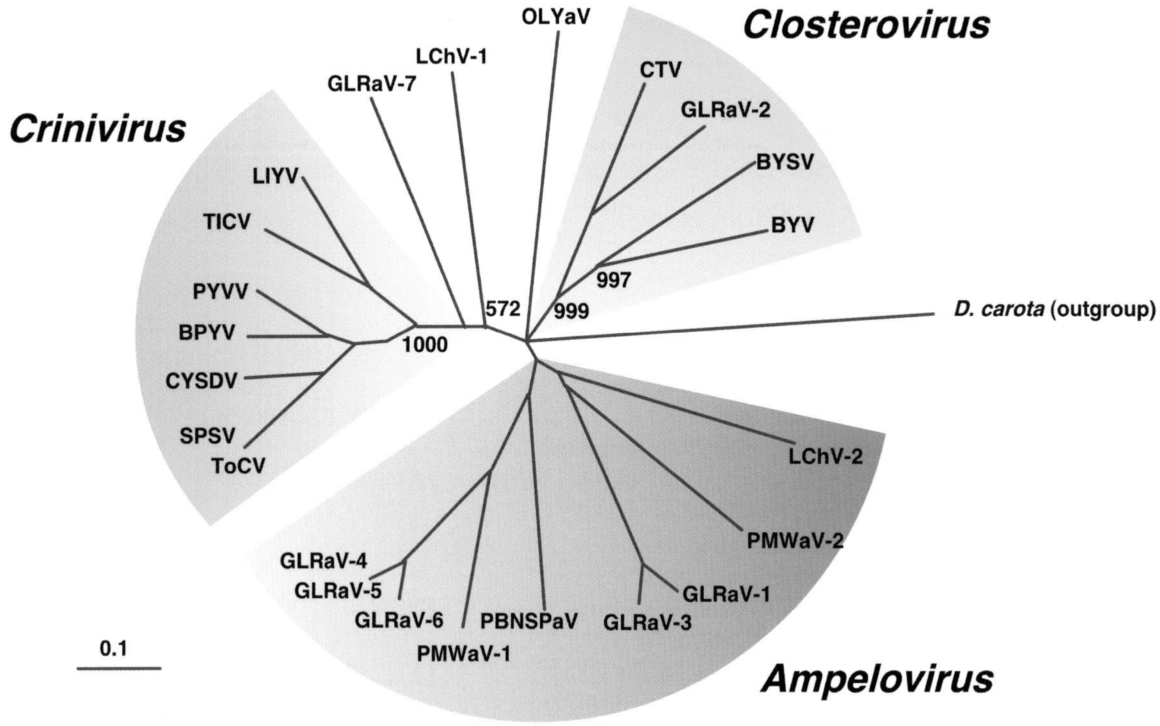

Figure 6: Phylogenetic tree showing the relationships between the species and genera of the family *Closterovoridae* based on the sequence of the HSP70h gene. The neighbor-joining tree was produced and bootsrapped using CLUSTAL W. Branch lengths are proportional to sequence distances. All the abbreviations signification can be found in the "List of Soecies" in the description. GLRaV-7, LChV-1, and OLYaV are unassigned species in the family.

SIMILARITY WITH OTHER TAXA

Virions of the genera *Capillovirus, Trichovirus,* and *Vitivirus* have the same particle morphology as those of the family *Closteroviridae*. However, the sequence of the CP of members of this family has little homology with that of CPs of viruses in the three above genera, and major differences exist in genome size and organization, and strategy of expression. Replication-associated proteins (RdRp, Mtr and Hel) contain signature sequences homologous to those of other taxa of the "alpha-like" supergroup of ssRNA viruses, the closest affinity being with the family *Bromoviridae* and the genera *Tobravirus, Tobamovirus,* and *Hordeivirus*. The replication strategy, based on polyprotein processing, translational frameshifting, and multiple sgRNA generation, closely resembles that of viruses in the families *Coronaviridae* and *Arteriviridae*. However, unlike closteroviruses, the RdRp of coronaviruses and arteriviruses belong to the "picorna-like" supergroup of polymerases. Hence, the transcriptional strategy of members of the family *Closteroviridae* follows the mechanism of other "alpha-like" viruses, and is dissimilar from the discontinuous, leader-primed transcription of coronaviruses and arteriviruses.

DERIVATION OF NAMES

ampelo: from *ampelos*, Greek for grapevine, the host of the type species of the genus
clostero: from Greek *kloster*, 'spindle, thread'.
crini from *crinis*, Latin for hair, from the appearance of the very long thread-like particles.

REFERENCES

Agranovsky, A.A. (1996). Principles of molecular organization, expression and evolution of closteroviruses: over the barriers. *Adv. Virus Res.*, **47**, 119-158.

Agranovsky, A.A., Lesemann, D.E., Maiss, E., Hull, R. and Atabekov, J.G. (1995). Rattlesnake structure of a filamentous plant RNA virus built of two capsid proteins. *Proc. Natl. Acad. Sci. USA*, **92**, 2470-2473.

Boyko, V.P., Karasev, A.A., Agranovsky, A.A., Koonin, E.V. and Dolja, V.V. (1992). Coat protein gene duplication in a filamentous RNA virus of plants. *Proc. Natl. Acad. Sci. USA*, **89**, 9156-9160.

Coffin, R.S. and Coutts, R.H.A. (1995). Relationships among *Trialeurodes vaporariorum*-transmitted yellowing viruses from Europe and North America. *J. Phytopathol.*, **143**, 375-380.

Dawson, W.O. (1995). Complete sequence of the citrus tristeza virus RNA genome. *Virology*, **208**, 511-520.

Dolja, V.V. (2003). *Beet yellows virus*: the importance of being different. *Mol. Plant Pathol.*, **4**, 91-98

Dolja, V.V., Karasev, A.V. and Koonin, E.V. (1994). Molecular biology and evolution of closteroviruses: sophisticated build up of large RNA genomes. *Annu. Rev. Phytopathol.*, **32**, 261-285.

Faoro, F. (1997). Cytopathology of closteroviruses and trichoviruses infecting grapevines. In: *Filamentous Viruses of Woody Plants*, (P.L. Monette, ed), pp 29-47. Research Signpost, Trivandrum.

Karasev, A.V. (2000). Genetic diversity and evolution of closteroviruses. *Annu. Rev. Phytopathol.*, **38**, 293-324.

Karasev, A.V. and Hilf, M.E. (1997). Molecular biology of the citrus tristeza virus. In: *Filamentous Viruses of Woody Plants*, (P.L. Monette, ed), pp 121-131. Research Signpost, Trivandrum.

Klaassen, V.A., Boeshore, M.L., Koonin, E.V., Tian, T. and Falk B.W. (1995). Genome structure and phylogenetic analysis of lettuce infectious yellows virus, a whithefly-transmitted, bipartite closterovirus. *Virology*, **208**, 99-110.

Mawassi, M., Mietkiewska, E., Hilf, M.E., Ashoulin, L., Karasev, A.V., Gafny, R., Lee, R.F., Garnsey, S.E., Dawson, W.O. and Bar-Joseph, M. (1995). Multiple species of defective RNAs in plants infected with citrus tristeza virus. *Virology*, **214**, 264-268.

Medina, V., Peremyslov, V.V., Hagiwara, Y. and Dolja, V.V. (1999). Subcellular localization of theHsp70-homolog encoded by beet yellows closterovirus. *Virology*, **260,** 173-181.

Napuli, A.J., Alzhanova D.V., Doneanu, C.E., Barofsky, D.F., Koonin, E.V. and Dolja, V.V. (2003). The 64-kilodaltons capsid protein homolog of *Beet yellows virus* is required for assembly of virion tails. *J. Virol.*, **77**, 2377-2384.

Prokhnevsky, A.I., Peremyslov, V.V., Napuli, A.J. and Dolja, V.V. (2002). Interaction between long-distance transport factor and Hsp70-related movement protein of *Beet yellows virus*. *J. Virol.*, **76**, 11003-11011.

Tian, T., Klaasssen, V.A., Soong, J., Wisler, G., Duffus, J.E. and B.W. Falk (1996). Generation of cDNAs specific to lettuce infectious yellows closterovirus and other whitefly-transmitted viruses by RT-PCR and degenerate oligonucleotide primers corresponding to the closterovirus gene encoding the heat shock protein 70 homolog. *Phytopathology*, **86**, 1167-1173.

Wisler, G.C., Duffus, J.E., Liu, H.Y. and Li, R.H. (1998). Ecology and epidemiology of whitefly-transmitted closteroviruses. *Plant Dis.*, **82**, 270-280.

Yeh, H.H., Tian, T., Rubio, L., Crawford, B. and Falk, B.W. (2000). Asynchronous accumulation of *Lettuce infectious yellows virus* RNAs 1 and 2 and identification of an RNA 1 *trans* enhancer of RNA 2 accumulation. *J. Virol.*, **74**, 5762-5768

CONTRIBUTED BY

Martelli, G.P., Agranovsky, A.A., Bar-Joseph, M., Boscia, D., Candresse, T., Coutts, R.H.A., Dolja, V.V., Falk, B.W., Gonsalves, D., Hu, J.S., Jelkmann, W., Karasev, A.V., Minafra, A., Namba, S., Vetten, H.J., Wisler, C.G. and Yoshikawa, N.

FAMILY *FLEXIVIRIDAE*

TAXONOMIC STRUCTURE OF THE FAMILY

Family	***Flexiviridae***
Genus	***Potexvirus***
Genus	***Mandarivirus***
Genus	***Allexivirus***
Genus	***Carlavirus***
Genus	***Foveavirus***
Genus	***Capillovirus***
Genus	***Vitivirus***
Genus	***Trichovirus***

VIRION PROPERTIES

MORPHOLOGY

Virions are flexuous filaments, usually 12-13 nm in diameter (range 10-15 nm) and from 470 to over 1000 nm in length, depending on the genus. They have helical symmetry with a pitch of about 3.4 nm (range 3.3-3.7 nm) and in some genera there is clearly visible cross-banding.

PHYSICOCHEMICAL AND PHYSICAL PROPERTIES

Virions sediment as single (or occasionally two very close) bands with an S_{20w} of 92-176*S*, depending on the genus.

NUCLEIC ACID

Virions contain a single molecule of linear ssRNA of ~ 5.9-9.0 kb which is 5-6% by weight of the virion. The RNA is capped (or probably capped) at the 5'-terminus and has a polyadenylated tract at the 3'-terminus. Smaller 3'-co-terminal sgRNAs are encapsidated in some, but not all, members of the genus *Potexvirus,* while in the genus *Carlavirus* some viruses have two sgRNAs of 2.1-3.3 kb and 1.3-1.6 kb, which are possibly encapsidated in shorter particles.

PROTEINS

The viral capsid of all species is composed of a single polypeptide ranging in size from 18-44 kDa. In allexiviruses a 42 kDa polypeptide was also detected as a minor component of virions.

GENOME ORGANIZATION AND REPLICATION

The number of genes is between 3 and 6 depending upon the genus (Fig. 1) but, in all species, the ORF1-encoded product, which follows a short 5'-UTR sequence, has homologies with putative polymerase proteins of the "alphavirus-like" supergroup of RNA viruses. This protein (150-250 kDa) contains conserved Mtr, Hel and RdRp motifs. Smaller ORFs encode the proteins involved in cell-to-cell movement, either a single MP of the '30K' superfamily (*Capillovirus, Trichovirus, Vitivirus, Citrus leaf blotch virus*) or a 'triple gene block' (TGB) (remaining genera and viruses). These are usually located following (3'-proximal to) the polymerase but in capillovirus genomes the MP ORF2 is nested within the ORF1. The CP gene always follows the MP(s) and in some genera (*Mandarivirus, Allexivirus, Carlavirus, Vitivirus*) a final ORF encodes a protein with a zinc binding finger motif and the ability to bind nucleic acids. ORFs downstream of the polymerase are translated from 3'-terminal sgRNAs that can often be found in infected tissue. In some viruses, notably in the genera *Vitivirus* and *Trichovirus* and in *Citrus leaf blotch virus*, nested sets of 5'-terminal sgRNAs and their associated dsRNAs can also be detected. Replication is (or is presumed to be) cytoplasmic and the product of ORF1 is the only virus-encoded protein known to be involved.

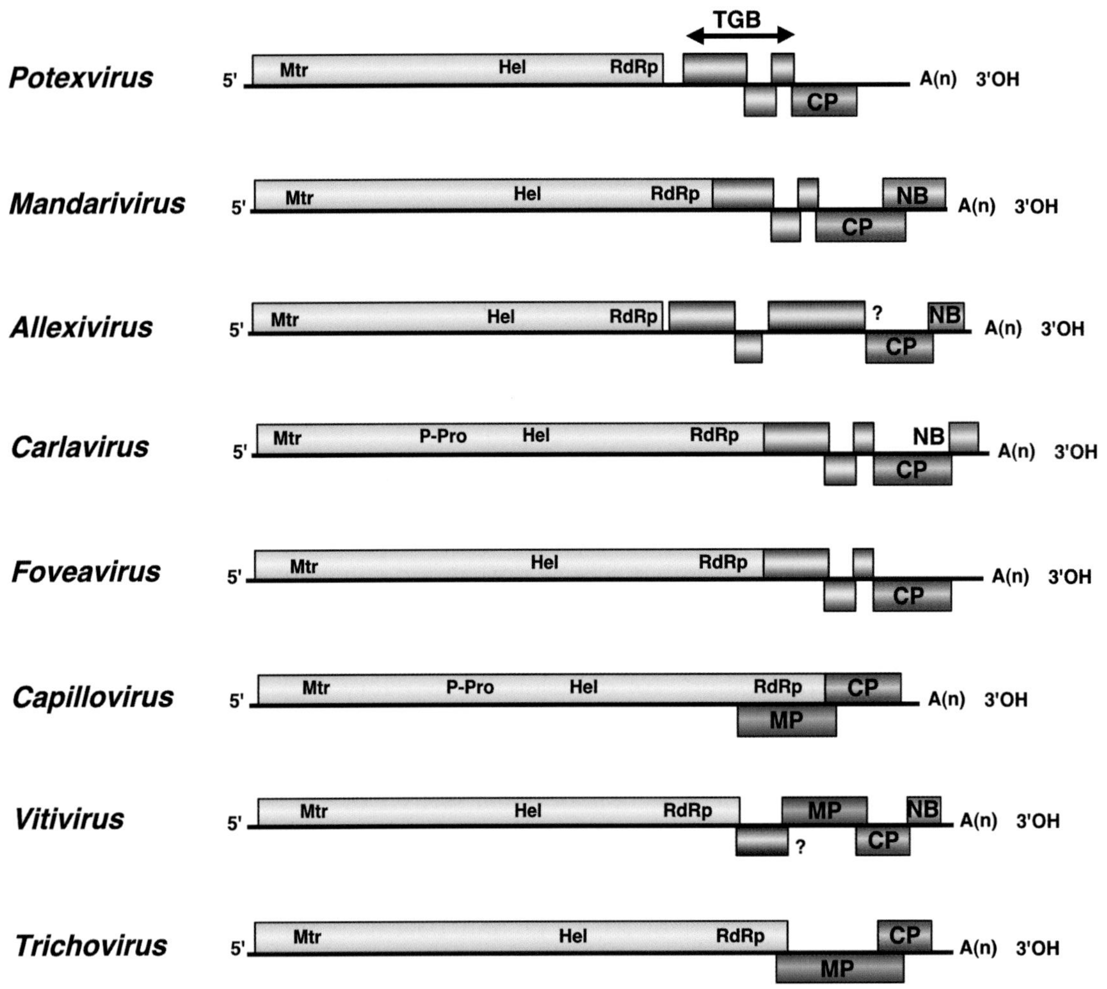

Figure 1: Genome organization of viruses in the family *Flexiviridae*. Motifs in the polymerase protein ORF1 are Methyltransferase (Mtr), Helicase (Hel), Papain-like protease (P-Pro) and RdRp (Pol). Triple gene block (TGB) proteins; capsid protein (CP) genes; movement protein of the '30K' superfamily (MP); NB, nucleic acid-binding protein.

Antigenic Properties

Virions are highly immunogenic in some genera (*Potexvirus*, *Mandarivirus*, *Allexivirus*, *Carlavirus*) but those of others are only moderate to poor antigens. Within (but not usually between) genera, some viruses are serologically related.

Biological Properties

Flexiviruses have been reported from a wide range of herbaceous and woody mono- and dicotyledonous plant species but the host range of individual members is usually limited. Natural infections by members of the genera *Mandarivirus*, *Foveavirus*, *Capillovirus*, *Vitivirus* and *Trichovirus* are mostly or exclusively of woody hosts. Many of the viruses have relatively mild effects on their host. All species can be transmitted by mechanical inoculation, often readily. Many of the viruses have no known invertebrate or fungus vectors; however, allexiviruses and some trichoviruses are thought to be mite-borne, most carlaviruses are transmitted naturally by aphids in the non-persistent manner and a range of vectors have been reported for different vitiviruses. Aggregates of virus particles accumulate in the cytoplasm. Many carlaviruses induce the formation of ovoid or irregularly shaped inclusions but otherwise there are usually no specific cytopathic structures.

SPECIES AND GENUS DEMARCATION CRITERIA IN THE FAMILY

Throughout the family, distinct species have less than ~ 72% identical nt or 80% identical aa between their CP or polymerase genes. Viruses from different genera usually have less than ~ 45% identical nt in these genes.

GENUS POTEXVIRUS

Type Species *Potato virus X*

VIRION PROPERTIES

MORPHOLOGY

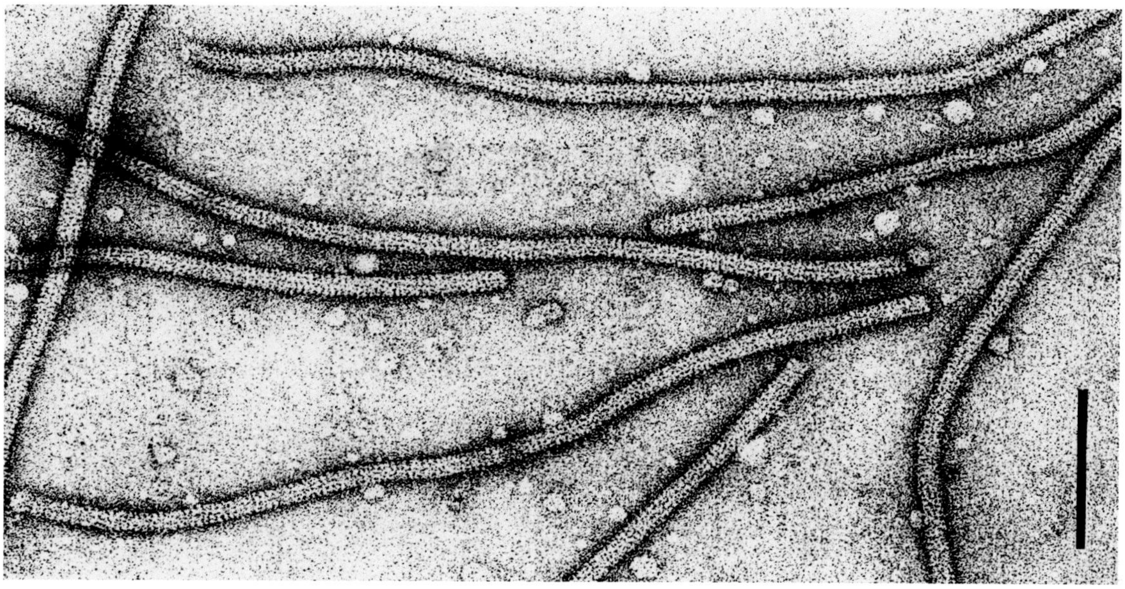

Figure 2: Negative contrast electron micrograph of particles of an isolate of *Potato virus X*. The bar represents 100 nm. (Courtesy of D.-E. Lesemann).

Virions are flexuous filaments, 470-580 nm in length and 13 nm in diameter, with helical symmetry and a pitch of 3.3-3.7 nm (Fig. 2). A central axial canal, ~ 3 nm in diameter, is discernible only in best preparations. The number of protein subunits per turn of the primary helix is slightly less than 9.0. The RNA backbone is at a radial position of 3.3 nm.

PHYSICOCHEMICAL AND PHYSICAL PROPERTIES

Virion Mr is ~ 3.5 x 10^6; S_{20w} is 115 - 130S; buoyant density in CsCl is 1.31 g/cm^3.

NUCLEIC ACID

Virions contain a single linear molecule of positive sense ssRNA of ~ 5.9-7.0 kb which is ~ 6% by weight of the virion. The RNA is capped at the 5'-terminus and has a polyadenylated tract at the 3'-terminus. The genomic RNA sequences are available for isolates of many species: Potato virus X (PVX, 6,435 nt), White clover mosaic virus (WCMV, 5,845 nt), Clover yellow mosaic virus (ClYMV, 7,015 nt), Papaya mosaic virus (PapMV, 6,656 nt), Bamboo mosaic virus (BaMV, 6,366 nt), Cactus virus X (CVX, 6,689 nt) Cymbidium mosaic virus (CymMV, 6,227 nt), Foxtail mosaic virus (FoMV, 6,151 nt), Pepino mosaic virus (PepMV, 6,450 nt), Plantago asiatica mosaic virus (PlAMV, 6,128 nt), Potato aucuba mosaic virus (7,059 nt) Strawberry mild yellow edge virus (SMYEV, 5,966 nt) Tulip virus X (PVX, 6,056 nt) and Narcissus mosaic virus (NMV, 6,955 nt). Many potexviruses have 5 ORFs; however, some (e.g., Cassava common mosaic virus, CsCMV; NMV and SMYEV have a sixth smaller ORF located completely within ORF5, although it has no known protein product and is of unknown function.

PROTEINS

The virus capsid consists of 1,000-1,500 protein subunits of a single 18-27 kDa polypeptide; the CP of some strains of *Potato virus X* is glycosylated. Partial proteolytic cleavage of the CP subunits can occur during storage of purified virus. Four non-structural proteins are coded by the PVX genome including an RNA polymerase (166 kDa) and three proteins (25, 12 and 8 kDa) involved in cell-to-cell movement of the virus (Fig. 3).

LIPIDS

None reported.

CARBOHYDRATES

The CP of some strains of PVX is glycosylated.

GENOME ORGANIZATION AND REPLICATION

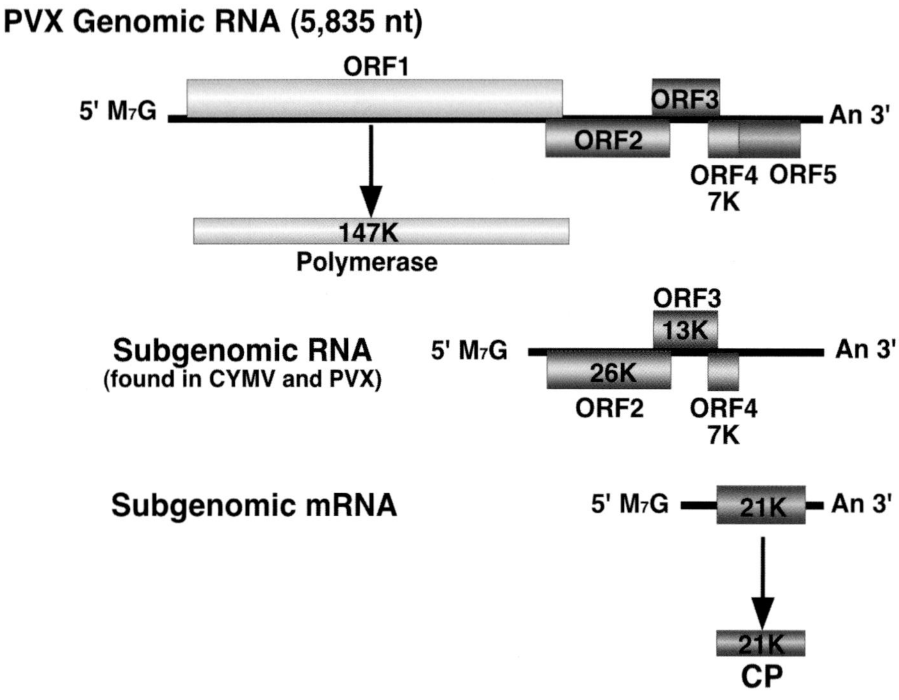

Figure 3: Potato virus X (PVX) genome structure and expression.

Virions of PVX contain only genomic RNA, but other potexviruses also encapsidate the sgRNA for the CP. Genomic RNA is translated as a functionally monocistronic message; only the 5'-proximal RNA-polymerase gene is translated directly by ribosomes, producing the RNA polymerase (150-181 kDa). The 5'-UTR leader sequence of PVX RNA (ab - leader) consists of 83 nt (excluding the cap-structure) and efficiently enhances translation. In infected plants, some potexviruses produce sgRNAs including one that acts as messenger RNA for the CP (Fig. 3).

The genomic RNA of potexviruses typically has five ORFs; however, some (including CsCMV, NMV, White clover mosaic virus (WClMV) and SMYEV) have a sixth smaller

ORF located completely within ORF5. ORF1, at the 5'-terminus, is the polymerase gene and ORF5, located at the 3'-terminus, is the CP gene. Between ORF1 and ORF5 is the TGB of three overlapping ORFs, the products of which (25, 12, and 8 kDa) are involved in cell-to-cell movement of viral RNA. The 25 kDa protein (as well as the 166 kDa replicase) contains an NTPase-helicase domain, but is not involved in RNA replication. The 12 and 8 kDa proteins contain large blocks of uncharged aa and are membrane-bound. ORFs 2 to 5 are expressed via the production (and subsequent translation) of appropriate sgRNAs. Two or three 3'-co-terminal sgRNAs can be isolated from plants infected with potexviruses (2.1; 1.2 and 1.0 kb); the double-stranded counterparts of these sgRNAs have also been detected. The medium-size sgRNA (1.2 kb) is probably functionally bicistronic, its translation yielding the 12 and 8 kDa proteins.

Antigenic Properties

Virions are highly immunogenic; some species are antigenically related, but others are serologically distinct.

Biological Properties

Host Range

Some of the viruses are moderately pathogenic, causing mosaic or ringspot symptoms in a wide range of mono- and dicotyledonous plant species, but others alone cause little damage to infected plants. The host range of individual members is limited.

Transmission

The viruses are mechanically transmissible, but none has known invertebrate or fungal vectors. The viruses are transmitted in nature by mechanical contact.

Geographical Distribution

The viruses have world-wide distribution.

Cytopathic Effects

The cytoplasm of infected cells contains fibrous, banded or irregular aggregates of virus particles, and often membrane accumulations. There is no cytopathology specific to potexviruses, although some viruses induce unique structures such as the beaded sheets found in PVX-infected cells.

List of Species Demarcation Criteria in the Genus

The list of species demarcation criteria in the genus is:

- Host range: the natural host range is usually specific to different species.
- Distinct species fail to cross-protect in infected plants.
- Serology; species and strains of some species are also readily distinguishable in differential reactions with monoclonal antibodies.
- Sequence: Distinct species have less than ~ 72% identical nt or 80% identical aa between their CP or polymerase genes.

List of Species in the Genus

Species names are in green italic script; strain names and synonyms are in black roman script; tentative species names are in blue roman script. Sequence accession numbers, and assigned abbreviations ( ) are also listed.

Species in the Genus

Alternanthera mosaic virus		
Alternanthera mosaic virus	[AF080448]	(AltMV)
Asparagus virus 3		
Asparagus virus 3		(AV-3)
Bamboo mosaic virus		
Bamboo mosaic virus	[D26017, L77962, AF018156]	(BaMV)

Cactus virus X		
Cactus virus X		(CVX)
Cassava common mosaic virus		
Cassava common mosaic virus	[U23414]	(CsCMV)
Cassava virus X		
Cassava virus X		(CsVX)
Clover yellow mosaic virus		
Clover yellow mosaic virus	[M63511, M63512, M63512, M63514, D00485, D29630]	(ClYMV)
Commelina virus X		
Commelina virus X		(ComVX)
Cymbidium mosaic virus		
Cymbidium mosaic virus	[U62963, X62663, X62664, X62665, X62133, X81051, AF016914]	(CymMV)
Daphne virus X		
Daphne virus X		(DVX)
Foxtail mosaic virus		
Foxtail mosaic virus	[AY121833, M62730]	(FoMV)
Hosta virus X		
Hosta virus X	[AY181252]	(HVX)
Hydrangea ringspot virus		
Hydrangea ringspot virus	[AJ270987, AJ550524]	(HdRSV)
Lily virus X		
Lily virus X	[X15342]]	(LVX)
Narcissus mosaic virus		
Narcissus mosaic virus	[D13747]	(NMV)
Nerine virus X		
Nerine virus X		(NVX)
Papaya mosaic virus		
Papaya mosaic virus	[D13957, AY017186-88, D00240]	(PapMV)
Pepino mosaic virus		
Pepino mosaic virus	[AF484251, AJ438767, AF340024]	(PepMV)
Plantago asiatica mosaic virus		
Plantago asiatica mosaic virus	[Z21647]	(PlAMV)
Plantago severe mottle virus		
Plantago severe mottle virus		(PlSMoV)
Plantain virus X		
Plantain virus X		(PlVX)
Potato aucuba mosaic virus		
Potato aucuba mosaic virus	[S73580]	(PAMV)
Potato virus X		
Potato virus X	[M38655, M95516, U19790, X55802, X72214, X88782, X88784-88, Z29333]	(PVX)
Scallion virus X ‡		
Scallion virus X	[AJ316085]	(ScaVX)
Strawberry mild yellow edge virus		
Strawberry mild yellow edge virus	[D12517, D12515, D01227, D00866]	(SMYEV)
Tamus red mosaic virus		
Tamus red mosaic virus		(TRMV)
Tulip virus X		
Tulip virus X	[AB066288]	(TVX)
White clover mosaic virus		
White clover mosaic virus	[X06728, X16636]	(WClMV)

Tentative Species in the Genus

Artichoke curly dwarf virus	(ACDV)
Barley virus B1	(BarV-B1)
Boletus virus X	(BolVX)
Centrosema mosaic virus	(CenMV)

Dioscorea latent virus (DLV)
Lychnis symptomless virus (LycSLV)
Malva veinal necrosis virus (MVNV)
Nandina mosaic virus (NaMV)
Negro coffee mosaic virus (NeCMV)
Parsley virus 5 (PaV-5)
Parsnip virus 3 (ParV-3)
Parsnip virus 5 (ParV-5)
Patchouli virus X (PatVX)
Rhododendron necrotic ringspot virus (RoNRSV)
Rhubarb virus 1 (RV-1)
Smithiantha latent virus (SmiLV)
Viola mottle virus (VMoV)
Zygocactus symptomless virus (ZSLV)

Phylogenetic Relationships within the Genus

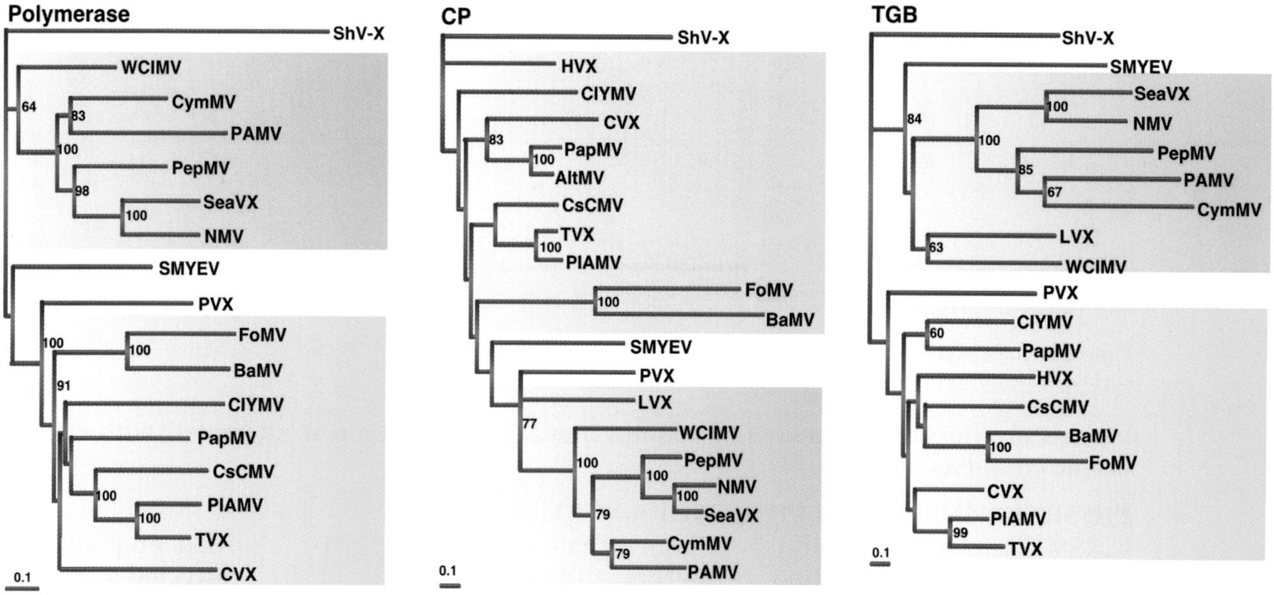

Figure 4: Phylogenetic analysis of potexviruses using the aa sequences of the polymerase (Left), the CP (Center) and the TGBp1 protein (Right). Sequences were aligned using GCG PILEUP and genetic distances estimated by PROTDIST (Dayhoff PAM method). Trees were displayed in TreeView. Bootstrap values based on 100 replicates are shown where >60%. Shallot virus X (ShV-X), genus *Allexivirus* is included for comparison. The abbreviations of the other virus names are indicated in the list of species.

GENUS MANDARIVIRUS

Type Species *Indian citrus ringspot virus*

VIRION PROPERTIES

MORPHOLOGY

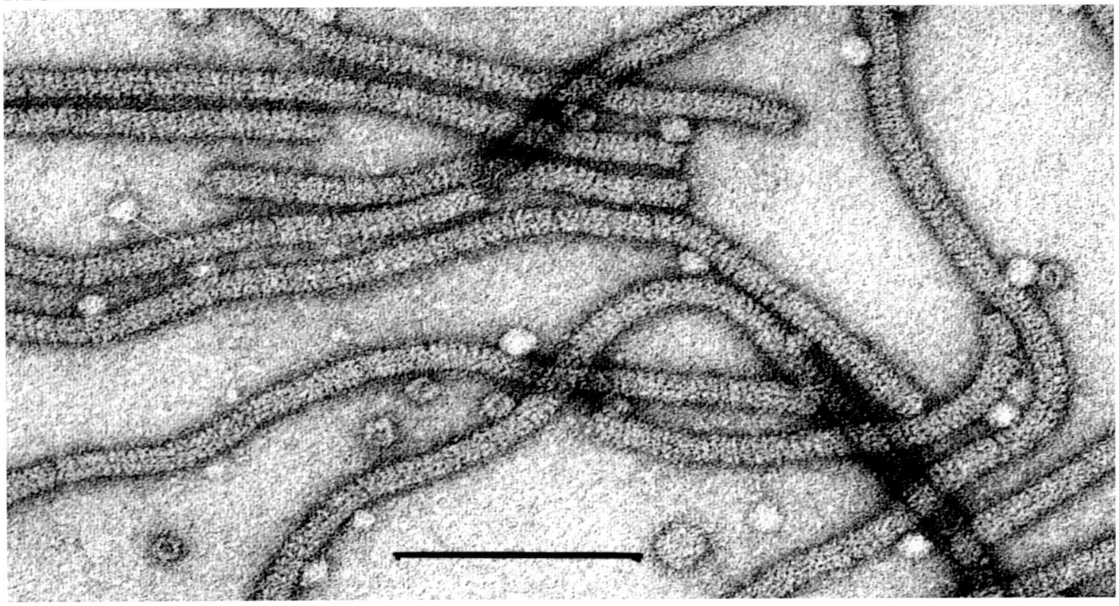

Figure 5: Electron micrograph of particles of an isolate of *Indian citrus ringspot virus*, stained in 1% uranyl acetate. The bar represents 100 nm.

Virions are flexuous filaments of 650 nm modal length, 13 nm in diameter, with clearly visible cross-banding (Fig. 5).

PHYSICOCHEMICAL AND PHYSICAL PROPERTIES

ICRSV forms a single band in cesium sulfate density gradients. Purified preparations show maximum absorption at 260 nm with a $A_{260/280}$ ratio of 1.1 (corrected for light scattering).

NUCLEIC ACID

Indian citrus ringspot virus (ICRSV) virions contain a single molecule of linear ssRNA, 7,560 nt in length, excluding the 3'-poly(A) tail.

PROTEINS

The only structural protein of Indian citrus ringspot virus (ICRSV) is the CP composed of 325 aa (34 kDa).

LIPIDS

Not reported.

CARBOHYDRATES

Not reported.

GENOME ORGANIZATION AND REPLICATION

The genomic RNA of ICRSV comprises six ORFs on the positive strand, a 5'-UTR of 78 nt and a 3'-UTR of 40 nt, followed by a poly(A) tail (Fig. 6). No significant ORFs are in the negative strand. The putative polypeptides encoded by the different ORFs are: ORF1,

1658 aa, 187.3 kDa; ORF2, 225 aa, 25 kDa; ORF3, 109 aa, 12 kDa; ORF4, 60 aa, 6.4 kDa; ORF5, 325 aa, 34 kDa, and ORF6, 222 aa, 23 kDa.

ORF1 encodes the viral polymerase. ORFs 2, 3 and 4 form the TGBck, as in potexvirus genomes. ORF5 encodes the CP. ORF6 encodes a putative protein of unknown function that shows limited similarity with nucleic acid-binding regulatory proteins encoded by ORF6 of allexi- and carlaviruses.

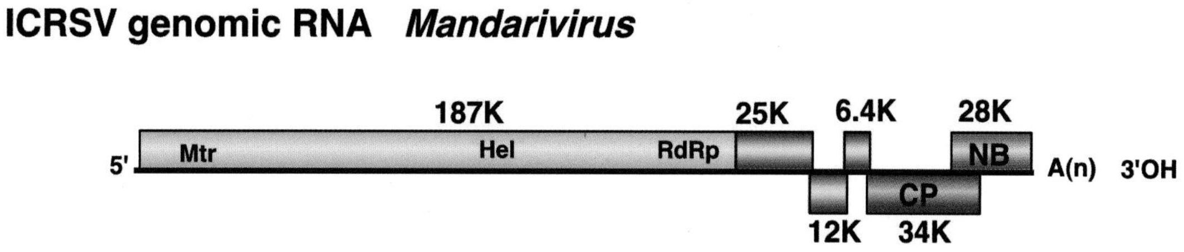

Figure 6: Genome organization of Indian citrus ringspot virus (ICRSV).

Antigenic Properties

ICRSV particles are good immunogens, rabbit antisera can have titers of 1/128 and 1/2048 in gel diffusion and EM decoration, respectively.

Biological Properties

ICRSV causes a serious disease of citrus, especially Kinnow mandarin, in India, with bright yellow ringspots on mature leaves, followed by rapid decline of the tree. Experimentally the virus can be mechanically inoculated to leaves of *Chenopodium quinoa, ~ amaranticolor, Glycine max, Vigna unguiculata* and *Phaseolus vulgaris*, giving local lesions, but systemic infection only in *P. vulgaris*. No natural vector is known, but ICRSV is transmitted by grafting and persists in the host.

List of Species Demarcation Criteria in the Genus

Not applicable.

List of Species in the Genus

Species names are in green italic script; strain names and synonyms are in black roman script; tentative species names are in blue roman script. Sequence accession numbers, and assigned abbreviations ( ) are also listed.

Species in the Genus

Indian citrus ringspot virus

Indian citrus ringspot virus	[AF406744]	(ICRSV)

Tentative Species in the Genus

None reported.

Phylogenetic Relationships within the Genus

Not applicable.

GENUS ALLEXIVIRUS

Type Species *Shallot virus X*

VIRION PROPERTIES

MORPHOLOGY

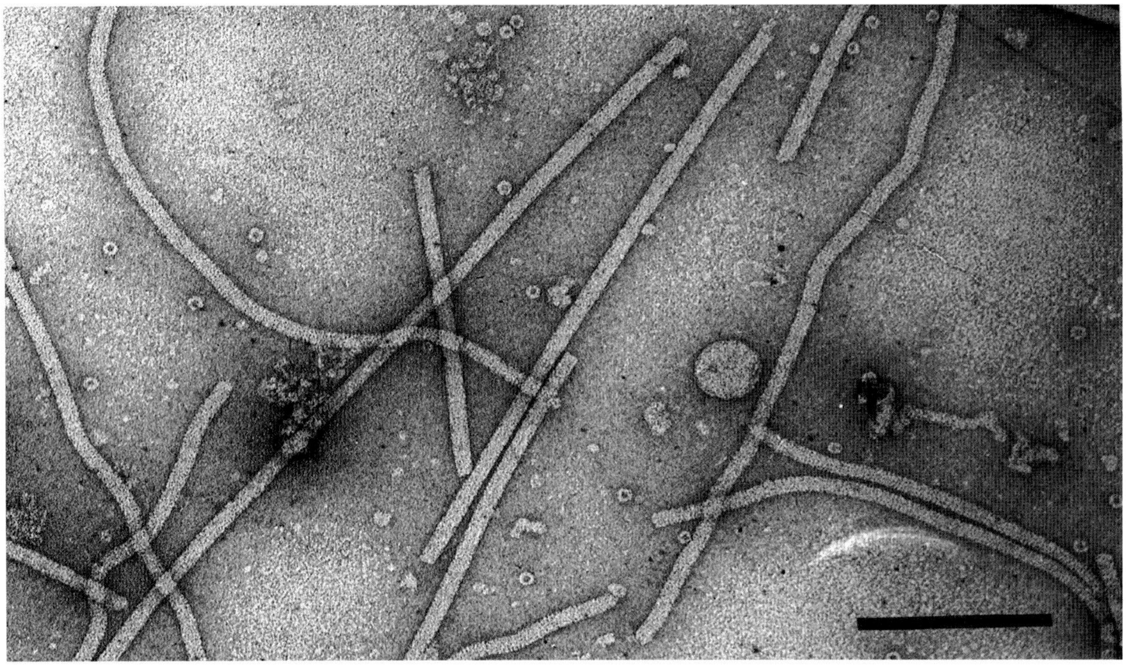

Figure 7: Negative staining electron micrograph of virions of an isolate of *Shallot virus X*. The bar represents 200 nm.

Virions are highly flexible filamentous particles, about 800 nm in length and 12 nm in diameter. They resemble potyviruses in their length, but closteroviruses in their flexibility and cross-banded substructure (Fig. 7).

PHYSICOCHEMICAL AND PHYSICAL PROPERTIES

Virions of Shallot virus X (ShVX) virions sediment with an S_{20w} of about 170S in 0.1 M tris-HC1, pH 7.5 at 20°C and have abuoyant density in CsCl of 1.33 g/cm^3.

NUCLEIC ACID

Virions contain a single molecule of linear ssRNA, about 9.0 kb in size, with a 3'poly(A) tract. RNA preparations contain genomic ssRNA and molecules of dsRNA, 1.5 kb in length, whose genesis and function(s) are unknown. The complete nt sequences of the genomic RNA of ShVX and Garlic virus X (GVX), and the partial sequences of the RNA of four allexiviruses have been determined.

PROTEINS

Virions are composed of a 28-36 kDa polypeptide as a major CP. A 42 kDa polypeptide is a minor component of virions.

LIPIDS

None reported.

CARBOHYDRATES

None reported.

Genome Organization and Replication

The genomic RNA contains six large ORFs and UTRs of 98 nt at the 5′ terminus, and 112 nt followed by poly (A) tail at the 3′ terminus (Fig. 9). The ORFs code for polypeptides of 195, 26, 11, 42, 28, and 15 kDa, respectively from 5′-end to 3′-end. The gene arrangement of other incompletely sequenced allexiviruses is similar. The 195 kDa polypeptide is probably the RdRp. In comparisons among the aa sequences of Mtr, Hel or RdRp domains, those of allexiviruses were most similar to those of potexviruses. The 26 and 11 kDa proteins are similar to the first two proteins encoded by the TGB of potexviruses and carlaviruses and are probably involved in cell-to-cell movement of the virus. There is a coding sequence for a small (7-8 kDa) TGB protein but it lacks the initiation AUG-codon. The 42 kDa polypeptide has no significant homology with any proteins known but has been shown to be expressed in plants infected with ShVX in relatively large amounts and was shown to be involved in the virion assembly. The 28 kDa polypeptide is the CP. In PAGE it migrates as an apparently 32-36 kDa protein, which could be due to its high hydrophilicity. The 15 kDa protein is similar to the 11-14 kDa proteins encoded by the 3′ ORFs of carlaviruses, has a zinc binding finger motif and an ability to bind nucleic acids. The function of this polypeptide is not known.

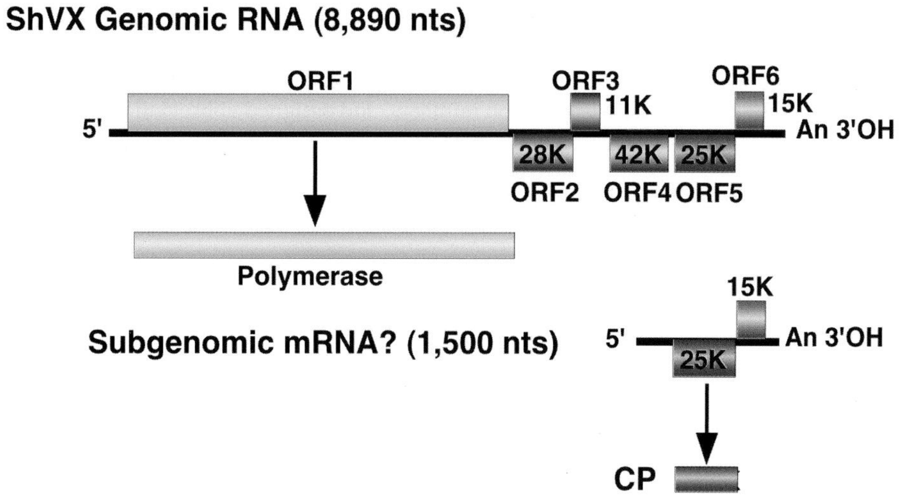

Figure 8: Genome organization of Shallot virus X (ShVX).

Antigenic Properties

Allexivirus particles are good immunogens. Some members of the genus are serologically interrelated. Specific antisera and monoclonal antibodies against pure virus particles as well as antisera against recombinant CPs have been used for differentiation purposes.

Biological Properties

Host range

Host range is extremely restricted. Some isolates from shallot, onion, garlic and sand leek have been experimentally transmitted to *Chenopodium murale*, in which they induced local lesions.

Transmission

Allexiviruses are thought to be mite-borne. Garlic virus C (GarV-C) and Garlic virus D (GarV-D) have been shown to be transmitted by the eriophyd mite, *Aceria tulipae*. All are manually transmissible by sap inoculation of healthy host plants. None could be transmitted by aphids.

Geographical Distribution

Allexiviruses have been identified in Russia, Japan, France, Germany, UK, The Netherlands, Korea, Taiwan, Thailand and Argentina.

Cytopathic Effects

Most induce no visible or only very mild symptoms in many species, although certain isolates can cause severe damage to crops. In infected tissue allexiviruses can induce formation of granular inclusion bodies and small bundles of flexible particles.

List of Species Demarcation Criteria in the Genus

The criteria demarcating species in the genus are:

- Members of distinct species have less than ~ 72% identical nt or 80% identical aa between their CP or polymerase genes,
- Different reactions with antisera.

List of Species in the Genus

The available information about the identity of the members in the genus *Allexivirus* is still fragmentary. There are almost always found in mixed infections of vegetatively propagated species and it is difficult or impossible to isolate and/or separate the viruses.

Species names are in green italic script; strain names and synonyms are in black roman script; tentative species names are in blue roman script. Sequence accession numbers, and assigned abbreviations ( ) are also listed.

Species in the Genus

Garlic mite-borne filamentous virus		
Garlic mite-borne filamentous virus		(GarMbFV)
Garlic virus A		
Garlic virus A	[AB010300, AF478197, X98991]	(GarV-A)
Garlic virus B		
Garlic virus B	[AB010301, AF543829]	(GarV-B)
Garlic virus C		
Garlic virus C	[AB010302, D49443]	(GarV-C)
Garlic virus D		
Garlic virus D	[AB010303, AF519572, L38892]	(GarV-D)
Garlic virus E ‡		
Garlic virus E	[AJ292230]	(GarV-E)
Garlic virus X		
Garlic virus X	[AJ292229, U89243]	(GarV-X)
Shallot virus X		
Shallot virus X	[M97264, L76292]	(ShVX)

Tentative Species in the Genus

Garlic mite-borne latent virus	(GarMbLV)
Onion mite-borne latent virus	(OMbLV)
Shallot mite-borne latent virus	(ShMbLV)

Phylogenetic Relationships within the Genus

A phylogenetic tree based on the coat protein aa sequence alignment shows that ShVX is closer to GarV-A, GarV-D and GarV-E.

GENUS *CARLAVIRUS*

Type Species *Carnation latent virus*

VIRION PROPERTIES

MORPHOLOGY

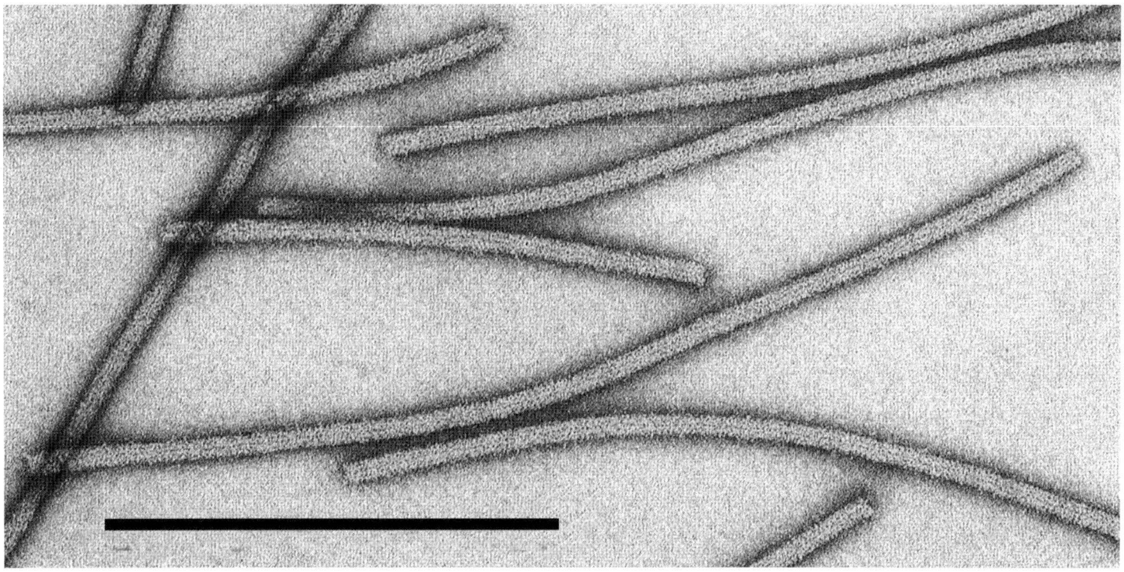

Figure 9: Filamentous particles of an isolate of *Carnation latent virus*. The bar represents 100 nm. (Courtesy R.G. Milne).

Virions are slightly flexuous filaments, 610-700 nm in length and 12-15 nm in diameter (Fig. 9). They have helical symmetry with a pitch of about 3.4 nm.

PHYSICOCHEMICAL AND PHYSICAL PROPERTIES

Virion Mr is about 60 x 10^6, with a nucleic acid content of ~ 6%. Virion $S_{20,w}$ is 147-176*S*, and the buoyant density in CsCl solutions is 1.3 g/cm^3.

NUCLEIC ACID

Virions contain a single molecule of linear ssRNA that has a size range of 7.4-7.7 kb when estimated by agarose gel analysis, although full-length sequence analysis suggests that genome sizes are in the 8.3-8.6 kb range. Some species also have two sgRNAs of 2.1-3.3 kb and 1.3-1.6 kb, which are possibly encapsidated in shorter particles. The genomic RNAs have a 3'-poly(A) tract and a 5'-cap. They contain six ORFs. The nt sequence or partial sequence of a large number of carlavirus RNAs has been determined.

Pos. ssRNA

PROTEINS

Virions contain a single 31-36 kDa polypeptide.

LIPIDS

None reported.

CARBOHYDRATES

None reported.

GENOME ORGANIZATION AND REPLICATION

The genomic RNAs of Potato virus M (PVM)(Fig. 10), (PCMV, 8,530 nt) and Blueberry scorch virus (BlScV, 8,514 nt), the better characterized viruses, each contain six ORFs; PVM RNA has UTRs of 75 nt at the 5'-terminus, 70 nt followed by a poly(A) tail at the 3'-

terminus and 38 and 21 nt between the three large blocks of coding sequences. Full-length genomic sequences have recently been obtained for Garlic common latent virus (GarCLV, 8,353 nt) and Aconitum latent virus (AcLV, 8,657 nt).

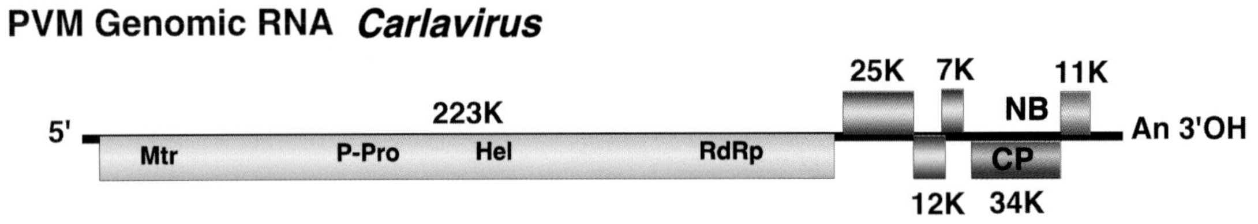

Figure 10: Genome organization of Potato virus M (PVM).

ORF1 encodes a polypeptide of 223 kDa that is the viral replicase; ORFs 2, 3 and 4 form the triple gene block and encode polypeptides of 25, 12 and 7 kDa which facilitate virus movement. ORF5 encodes the 34 kDa CP and overlaps ORF5, which encodes a cysteine-rich protein of 11-16 kDa. The gene arrangement of other incompletely sequenced carlaviruses is similar. The proteins encoded by the triple gene block facilitate cell-to-cell movement of virus. The 34 kDa polypeptide is the CP. The function of the 11-16 kDa polypeptide has yet to be determined, but its ability to bind nucleic acid indicates that it may facilitate aphid transmission or be involved in host gene transcription/gene silencing and/or viral RNA replication.

Only the 223 kDa replicase ORF is translated from the full length genomic RNA. With BlScV and probably other carlaviruses the 223 kDa protein is proteolytically processed by a papain-like proteinase activity, with ~ 30-40 kDa being removed. The 3'-terminal ORFs appear to be translated from two sgRNAs that can be found in infected tissue, and, for some viruses, can be detected in purified virus preparations. The 5'-untranslated leader sequence of the genomic RNA and the sgRNA for the CP of Potato virus S (PVS) have both been shown to act as efficient enhancers of translation.

Antigenic Properties

Carlavirus virions are good immunogens. Some species are serologically interrelated, but others are apparently distinct.

Biological Properties

Host Range

Individual viruses have restricted natural host ranges, but some can infect a wide range of experimental hosts.

Transmission

Most species are transmitted naturally by aphids in the non-persistent manner; Cowpea mild mottle virus CPMMV) is transmitted by whiteflies (*Bemisia tabaci*). Pea streak virus, PeSV; Red clover vein mosaic virus, RCVMV and Cowpea mild mottle virus CPMMV) are seedborne in their leguminous hosts. All are mechanically transmissible; some (e.g. Carnation latent virus, CLV and PVS) are sufficiently infectious to be so transmitted this way in the field.

Geographical Distribution

The geographical distribution of many species is restricted, but those infecting vegetatively-propagated crops are usually widely distributed, presumably due to inadvertent dissemination in vegetative propagules. Most species commonly occur in temperate climates, but that transmitted by whiteflies is restricted to tropical and sub-tropical regions.

Cytopathic Effects

Virions of aphid-borne species are scattered throughout the cytoplasm or occur in membrane-associated bundle-like or plate-like aggregates. Many species also induce the formation of ovoid or irregularly shaped inclusions that appear in the light microscope as vacuolate bodies; these consist of aggregates of virus particles, mitochondria, endoplasmic reticulum and lipid globules. The particles of CPMMV, the whitefly-transmitted carlavirus, also occur in aggregates in cytoplasm; those of most, but not all, strains of CPMMV form brush-like inclusions.

List of Species Demarcation Criteria in the Genus

Each distinct species usually has a specific natural host range. Distinct species do not cross-protect in infected common host plant species. Distinct species are readily differentiated by serological procedures; strains of individual species are often distinguishable in reactions with polyclonal antisera, but more readily so with monoclonal antibodies. Distinct species have less than ~72% identical nt or 80% identical aa between their CP or polymerase genes.

List of Species in the Genus

Species names are in green italic script; strain names and synonyms are in black roman script; tentative species names are in blue roman script. Sequence accession numbers, and assigned abbreviations ( ) are also listed.

Species in the Genus

Aphid-borne species:

American hop latent virus		
American hop latent virus		(AHLV)
Blueberry scorch virus		
Blueberry scorch virus	[L25658]	(BlScV)
Cactus virus 2		
Cactus virus 2		(CV-2)
Caper latent virus		
Caper latent virus		(CapLV)
Carnation latent virus		
Carnation latent virus	[AJ010697, X55331, X55897, U43905]	(CLV)
Chrysanthemum virus B		
Chrysanthemum virus B	[AJ564854, S60150]	(CVB)
Cole latent virus		
Cole latent virus	[AY340584]	(CoLV)
Dandelion latent virus		
Dandelion latent virus		(DaLV)
Elderberry symptomless virus		
Elderberry symptomless virus		(ESLV)
Elderberry virus A		
Garlic common latent virus		
Garlic common latent virus	[AB004566, AB004804-05, AF228416, AF538951, X81138, X81139]	(GarCLV)
Helenium virus S		
Helenium virus S	[D10454, S71594]	(HVS)
Honeysuckle latent virus		
Honeysuckle latent virus		(HnLV)
Hop latent virus		
Hop latent virus	[AB032469]	(HpLV)
Hop mosaic virus		
Hop mosaic virus	[AB051109]	(HpMV)
Hydrangea latent virus		
Hydrangea latent virus		(HdLV)
Kalanchoe latent virus		
Kalanchoe latent virus	[AJ293570, AJ293571, AY238136-	(KLV)

Pos. ssRNA

	AY238143]	
Lilac mottle virus		
Lilac mottle virus		(LiMoV)
Lily symptomless virus		
Alstroemeria carlavirus		
Lily symptomless virus	[AJ564638, AJ516059, AJ131812, AF103784, AF015286, X15343, D43801, U43905]	(LSV)
Mulberry latent virus		
Mulberry latent virus		(MLV)
Muskmelon vein necrosis virus		
Muskmelon vein necrosis virus		(MuVNV)
Nerine latent virus		
Hippeastrum latent virus		
Nerine latent virus		(NeLV)
Passiflora latent virus		
Passiflora latent virus	[AF354652]	(PLV)
Pea streak virus		
Alfalfa latent virus	[AY037925]	
Pea streak virus		(PeSV)
Potato latent virus		
Potato latent virus	[AY007728]	(PotLV)
Potato virus M		
Potato virus M	[X53062, X57440, D144449, AF23877]	(PVM)
Potato virus S		
Pepino latent virus		
Potato virus S	[D00461, S45593]	(PVS)
Red clover vein mosaic virus		
Red clover vein mosaic virus		(RCVMV)
Shallot latent virus		
Garlic latent virus	[AB004458, AB004565-67, AB004684-86, NC_003557]	(SLV)
Shallot latent virus	[AB004456-57, AB004544, AB004802-03]	(GarLV)
Sint-Jan's onion latent virus		
Sint-Jan's onion latent virus		(SJOLV)
Strawberry pseudo mild yellow edge virus		
Strawberry pseudo mild yellow edge virus		(SPMYEV)

Whitefly-transmitted species:

Cowpea mild mottle virus		
Bean angular mosaic virus		
Cowpea mild mottle virus	[AF024628-29]	(CPMMV)
Groundnut crinkle virus		
Psophocarpus necrotic mosaic virus		
Tomato pale chlorosis virus		
Voandzeia mosaic virus		

Unknown vector species:

Aconitum latent virus		
Aconitum latent virus	[AB051848]	(AcLV)
Narcissus common latent virus		
Narcissus common latent virus	[AJ311375, AJ311376]	(NCLV)
Poplar mosaic virus		
Poplar mosaic virus	[X65102, D13364, X97683, X97765]	(PopMV)
Verbena latent virus		
Verbena latent virus	[AF271218]	(VeLV)

Tentative Species in the Genus

Anthriscus latent virus		(AntLV)
Arracacha latent virus		(ALV)
Artichoke latent virus M		(ArLVM)
Artichoke latent virus S		(ArLVS)
Butterbur mosaic virus		(ButMV)
Caraway latent virus		(CawLV)
Cardamine latent virus		(CaLV)
Cassia mild mosaic virus		(CasMMV)
Chicory yellow blotch virus		(ChYBV)
Cynodon mosaic virus		(CynMV)
Daphne virus S	[AJ535084]	(DVS)
Dulcamara virus A		(DuVA)
Dulcamara virus B		(DuVB)
Eggplant mild mottle virus (Eggplant virus)		(EMMV)
Euonymus mosaic virus		(EuoMV)
Fig virus S		(FVS)
Fuchsia latent virus		(FLV)
Garlic mosaic virus		(GarMV)
Gentiana latent virus		(GenLV)
Gynura latent virus (Chrysanthemum virus B)		(GyLV)
Helleborus mosaic virus		(HeMV)
Impatiens latent virus		(ILV)
Lilac ringspot virus		(LiRSV)
Plantain virus 8		(PlV-8)
Potato rough dwarf virus	[AJ250314]	(PRDV)
Potato virus P		(PVP)
Prunus virus S		(PruVS)
Southern potato latent virus		(SoPLV)
White bryony mosaic virus		(WBMV)

Phylogenetic Relationships within the Genus

The phylogenetic analysis is presented for the aa sequences of the polymerase, TGBp1 and the CP are presented in Figure 11. ASPV, genus *Foveavirus* has been included as a nearest neighbor.

Virions of Narcissus latent virus (NLV) are filamentous and ~650 nm long. It was previously considered to be a carlavirus. However, it differs from carlaviruses in inducing the formation of intracellular inclusions ("pinwheels") and in having a CP of 46 kDa; it is thus now better placed in the genus *Macluravirus* (family *Potyviridae*), together with *Maclura mosaic virus*, to which it is serologically related. Chinese yam necrotic mosaic virus, (ChYNMV) which was also listed as a tentative carlavirus, has now been sequenced and is also placed in the genus *Macluravirus* (family *Potyviridae*).

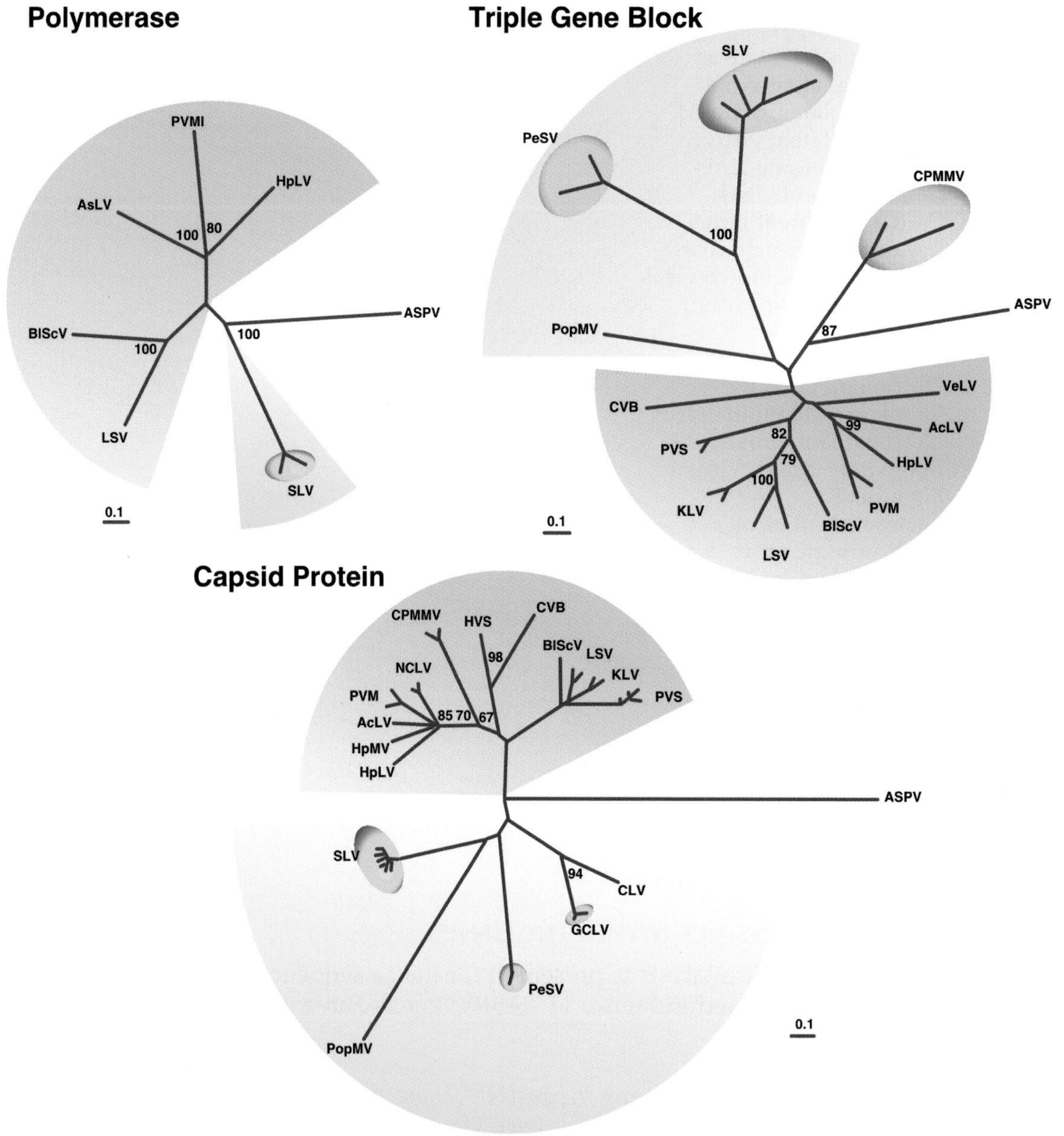

Figure 11: Phylogenetic (NEIGHBOR) analysis of carlaviruses using the aa sequences of the products of three ORFs: (Top left) Polymerase, (Top right) Triple gene block 1 (TGBp1), (Bottom) Capsid protein. Sequences were aligned using GCG PILEUP and genetic distances estimated by PROTDIST (Dayhoff PAM method). Trees were displayed in TreeView. Bootstrap values based on 100 replicates are shown where >60%. Apple stem pitting virus sequences (ASPV), genus *Foveavirus* is included for comparison. The abbreviations of the other viruses are indicated in the list of species and all the available Genbank accessions were used.

GENUS FOVEAVIRUS

Type Species *Apple stem pitting virus*

VIRION PROPERTIES

MORPHOLOGY

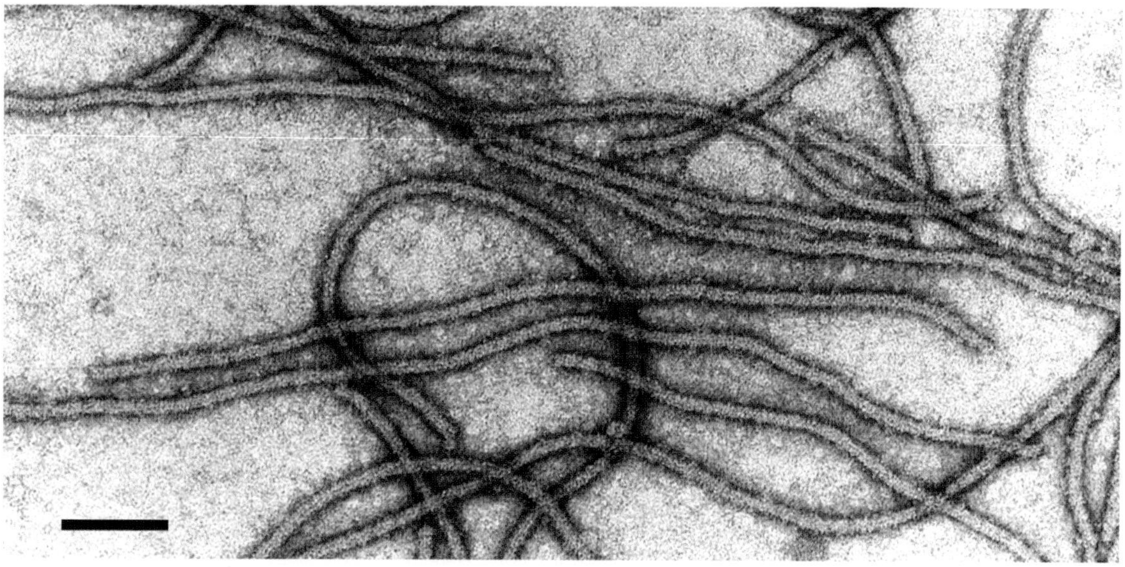

Figure 12: Negative contrast electron micrograph of particles of an isolate of *Apple stem pitting virus*. The bar represents 100 nm. (Courtesy of H. Koganezawa).

Virions are flexuous filaments, ~800 to over 1000 nm in length and 12-15 nm in diameter with helical symmetry exhibiting a surface pattern with cross-banding and longitudinal lines (Fig. 12). Particles of some viruses such as Apple stem pitting virus (ASPV) and Cherry green ring mottle virus (CGRMV) show a tendency to end-to-end aggregation.

PHYSICOCHEMICAL AND PHYSICAL PROPERTIES

ASPV virions sediment as two or three bands in sucrose density gradients but yield a single band at equilibrium in Omnipaque 350 density gradients. They resist moderately high temperatures (thermal inactivation is around 60°C) but not organic solvents, and are unstable in cesium chloride and sulfate. CGRMV virions resist organic solvents and are stable in cesium sulfate.

NUCLEIC ACID

Virions contain a single molecule of positive sense ssRNA, polyadenylated at the 3'-terminus. The genome of the tentative species CGRMV is capped at the 5'-terminus. The complete nt sequences are known for Apple stem pitting virus (ASPV, 9.3 kb), Rupestris stem pitting-associated virus (RSPaV, 8.7 kb), Cherry green ring mottle virus (CGRMV, 8.4 kb) and Cherry necrotic rusty mottle (CNRMV, 8.4 kb) and the a partial sequence of Apricot latent virus (ApLV) is available.

PROTEINS

The viral capsid of all species is composed of a single polypeptide with a size ranging from 23 kDa (SCSMaV), to 30 kDa (CGRMV), to 44 kDa (ASPV and ApLV). Non structural proteins consist of a large polypeptide (205-247 kDa) containing the Mtr, Hel, and viral RdRp signatures, a 23-25 kDa polypeptide with the NTP-binding helicase domain, and two small polypetides (12-13 kDa and 7-8 kDa) with membrane-binding functions.

Lipids
None reported.

Carbohydrates
None reported.

Genome Organization and Replication

The genomes of all fully sequenced members, except for CGRMV contain 5 ORFs (Fig. 13). The 5' region initiates with a UTR of 33-72 nt, ORF1 codes for the replication-related proteins with conserved motifs of the "alphavirus-like" supergroup of positive-strand ssRNA viruses (i.e. Mtr and RdRp); ORF2, ORF3, and ORF4 constitute the TGB thought to be involved in cell-to-cell spread of the virus; ORF5 is the CP cistron. A non-coding sequence of 176-312 nt followed by a poly(A) tail terminate the genome. The 5'-end of CGRMV RNA is capped and two additional ORFs (ORF2a and ORF4a) nested in ORF2 and ORF5, respectively, which potentially encode 14 and 18 kDa polypeptides, have been reported from computer analysis only. No similarity was found with other proteins in databases. ASPV virions accumulate in the cytoplasm, where multiplication is likely to occur following a strategy comparable to that of potexviruses, based on direct expression of the 5'-proximal ORF, and expression of downstream ORFs from sgRNAs. Multiple dsRNAs are found in infected hosts.

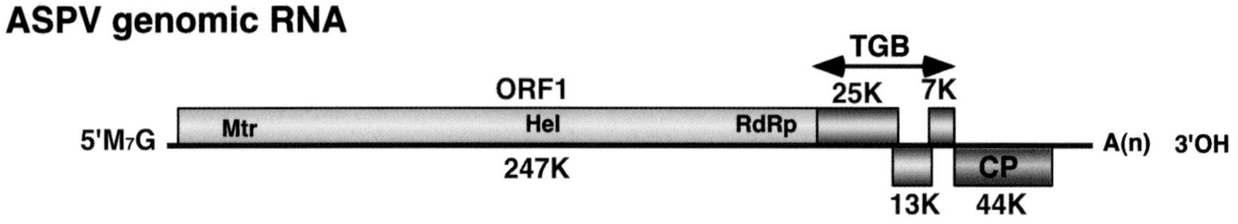

Figure 13: Genome organization of Apple stem pitting virus (ASPV) showing the relative position of the ORFs and their expression products. Mtr, methytransferase; Hel, helicase; Pol, polymerase; TGB, triple gene block; CP, capsid protein.

Antigenic Properties

Antisera to ASPV that can be used for serological detection tests have been raised from purified virions or chimeric fusion CPs expressed in *E. coli*. A CGRMV particle-decorating antiserum was also produced by using fusion proteins. ASPV and ApLV are serologically related, but there is no recognized serological relationships among other virus species.

Biological Properties

Host Range

The natural host range of individual species is restricted to a single (RSPaV) or a few hosts (ASPV, ApLV, CGRMV). ASPV infects primarily pome fruits, causing diseases of apple (topworking disease) when grafted on susceptible rootstocks, and of pear (vein yellows and necrotic spot). ApLV is the putative agent of peach asteroid spot and peach sooty ringspot diseases. RSPaV is a pathogen of grapevine. Stone fruits (sweet, sour and flowering cherries, peach, and apricot) are the natural hosts of CGRMV, ApLV, and CNRMV. Experimental host ranges are restricted.

Transmission

No vector is known for any of the viruses. ASPV, CGRMV, and CNRMV are transmitted by grafting and persist in the host propagative material. ASPV is mechanically transmissible, with some difficulty, to *Nicotiana occidentalis* and its subspecies *obliqua.* CGRMV was transmitted by sap inoculation from cherry to cherry but not to a range of herbaceous hosts.

Geographical Distribution

All species have a rather wide geographical distribution.

Cytopathic Effects

ASPV elicits a severe derangement of the cytology of infected cells but no specific cytopathic structures or inclusion bodies. Virus particles accumulate in bundles in the cytoplasm. Massive accumulations of virus-like particles were also observed in mesophyll cells of CGRMV-infected Kwazan flowering cherry.

List of Species Demarcation Criteria in the Genus

The criteria demarcating species in the genus are:

- Natural host range,
- Mechanical transmissibility,
- Serological specificity,
- CP size,
- Less than ~ 72% identical nt or 80% homologous aa between CP or polymerase genes.

List of Species in the Genus

Species names are in green italic script; strain names and synonyms are in black roman script; tentative species names are in blue roman script. Sequence accession numbers, and assigned abbreviations ( ) are also listed.

Species in the Genus

Apple stem pitting virus		
Apple stem pitting virus	[D21829, AB045731]	(ASPV)
Pear vein yellows virus	[D21828]	
Apricot latent virus		
Apricot latent virus	[AF057035]	(ApLV)
Rupestris stem pitting-associated virus		
Rupestris stem pitting-associated virus	[AF026278, AF057136]	(RSPaV)

Tentative Species in the Genus

None reported.

Phylogenetic Relationships within the Genus

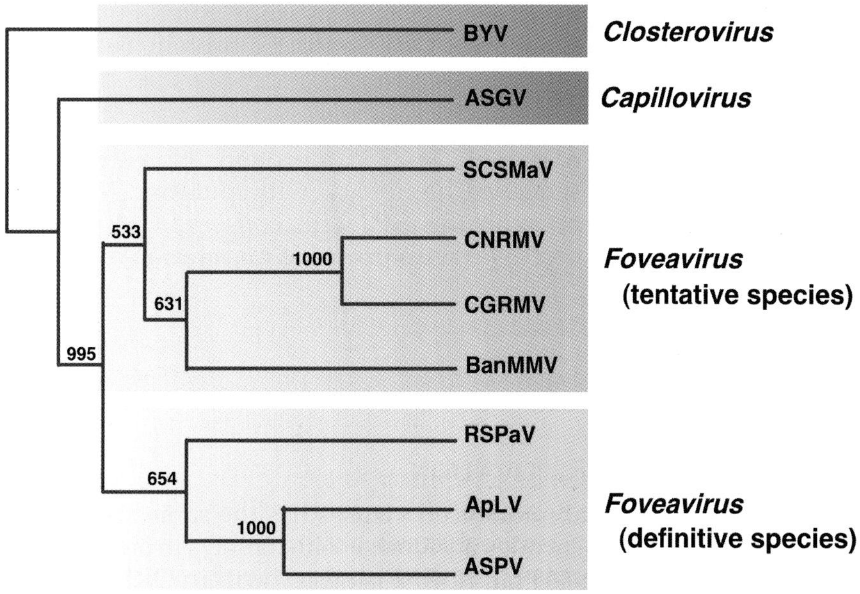

Figure 14: Phylogenetic tree showing the relationships among definitive and tentative species of the genus *Foveavirus* based on the CP gene sequence. The tree was produced and bootstrapped using CLUSTAL W.

GENUS CAPILLOVIRUS

Type Species *Apple stem grooving virus*

VIRION PROPERTIES

MORPHOLOGY

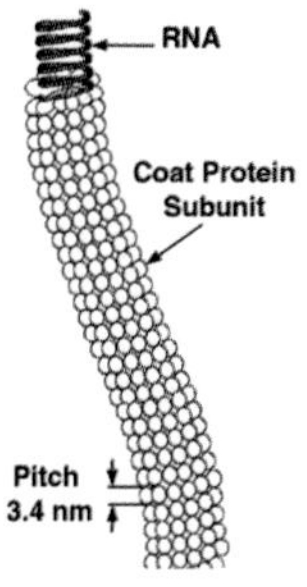

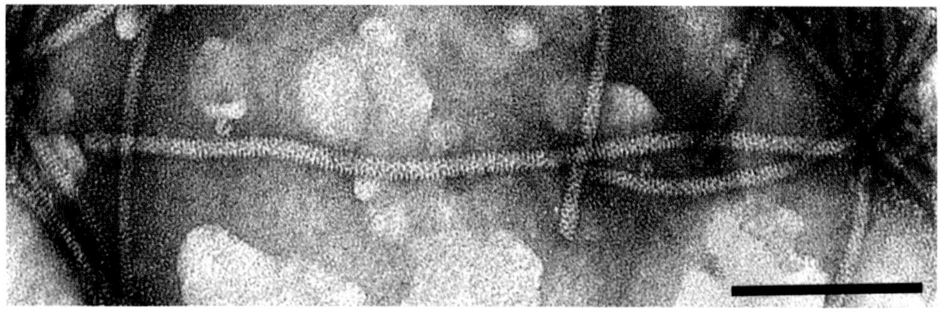

Figure 15: (Left) Schematic representation of a fragment of a particle of a capillovirus. (Right) Negative contrast electron micrograph of particles of an isolate of *Apple stem grooving virus*. The bar represents 100 nm. (Courtesy of N. Yoshikawa).

Virions are flexuous filaments, 640-700 x 12 nm, constructed from helically arranged protein subunits in a primary helix with a pitch of 3.4 nm and between 9 and 10 subunits per turn (Fig. 15).

PHYSICOCHEMICAL AND PHYSICAL PROPERTIES

S_{20w} of particles of Apple stem grooving virus (ASGV) is ~112S, isoelectric point is about pH 4.3 at ionic strength 0.1 M, and electrophoretic mobility is 10.3 and 6.5 x 10^{-5} cm^2/sec/volt, at pH 7.0 and 6.0 respectively (ionic strength 0.1 M).

NUCLEIC ACID

Virions contain linear positive sense ssRNA, 6.5-7.4 kb in size, constituting about 5%, by weight, of virions. The RNA is polyadenylated at its 3'-end. The complete nt sequences are known for genome RNA of three isolates of *Apple stem grooving virus* and one of *Cherry virus A*. Isolates of *Apple stem grooving virus* from different hosts show wide variations in the sequence of a 284 aa region of ORF1-encoded protein, between the polymerase and CP domains.

PROTEINS

Virions are composed of a single 24-27 kDa protein. Non-structural proteins include a 36-52 kDa protein with sequence homology with putative MPs, and a large replication-associated protein fused with the CP, with conserved Mtr, Hel and RNA polymerase motifs, expressed as a precursor with probable maturation by proteolysis.

LIPIDS

None reported.

CARBOHYDRATES

None reported.

GENOME ORGANIZATION AND REPLICATION

The genomic RNA of all sequenced viruses has the same structural organization, and two ORFs (Fig. 16). ORF1 encodes a putative 240-266 kDa protein followed by a UTR of 142 nt upstream of the 3'-poly(A) tail. ORF2 is nested within ORF1 near its 3'-end, and encodes a 36-52 kDa protein. The ORF1-encoded product has homologies with putative polymerase proteins of the "alphavirus-like" supergroup of RNA viruses. Although the CP cistron is located in the C-terminal end of ORF1, and ORF2 is nested within ORF1, the strategy of expression of both CP and putative MP may be based on sgRNA production, as suggested by the analysis of dsRNA patterns from infected tissues. dsRNAs of ASGV consist of five major bands with sizes of approximately 6.5, 5.5, 4.5, 2.0 and 1.0 kbp. The 6.5 kbp species probably represents the double-stranded form of the full-length genome, and the 2.0 and the 1.0 kbp species may be the double-stranded forms of sgRNAs that code for the putative MP and the CP, respectively. Replication is likely to occur in the cytoplasm, in which virus particles accumulate in discrete bundles.

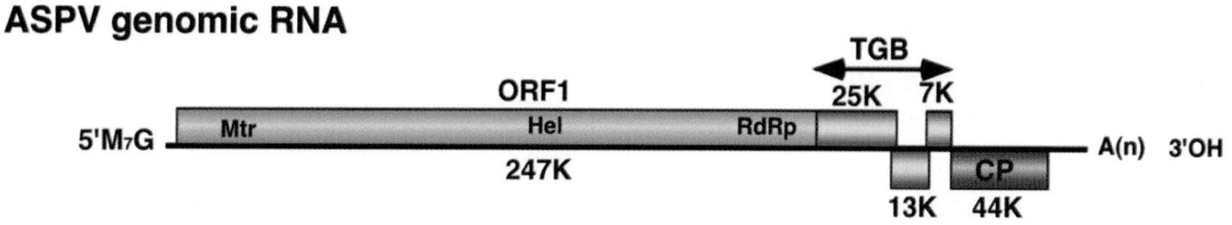

Figure 16: Genome organization of Apple stem grooving virus (ASGV) showing the relative positions of the ORFs and their expression products. Mtr, methyltransferase; P-Pro, papain-like protease; Hel, helicase; Pol, polymerase; MP, putative movement protein; CP, capsid protein.

Antigenic Properties

Virions are moderately antigenic. There are no serological relationship between species.

Biological Properties

Host Range

ASGV is pathogenic to pome fruits and citrus and induces stock/scion incompatibility, i.e. top-working disease of apple and bud union crease syndrome of citrus. It also infects lily. CVA is found in cherry but no disease has been associated with it.

Transmission

No vectors are known. ASGV was transmitted through seed to progeny seedlings of *Chenopodium quinoa,* and lily. ASGV, Cherry virus A (CVA), and Nandina stem pitting virus (NSPV) are transmitted by grafting. NSPV has not been transmitted by sap inoculation, but by slashing stems with a partially purified virus preparation.

Geographical Distribution

Geographical distribution ranges from wide to restricted according to the virus. ASGV has been recorded from most areas where apples are grown, and is widespread in citrus in China, Japan, United States, Australia, and South Africa. Lilac chlorotic leafspot virus (LiCLV) occurs in England and possibly in continental Europe and the United States. NSPV is found only in the United States and CVA in Germany.

Cytopathic Effects

No distinct cytological alterations have been observed in infected cells. Virus particles occur in bundles in mesophyll and phloem parenchyma cells, but not in the epidermis and sieve elements.

List of Species Demarcation Criteria in the Genus

The criteria demarcating species in the genus are:

- Natural host range,
- Serological specificity (all known species are serologically unrelated),
- Less than ~ 72% identical nt or 80% identical aa between the CP or polymerase genes.

List of Species in the Genus

Species names are in green italic script; strain names and synonyms are in black roman script; tentative species names are in blue roman script. Sequence accession numbers, and assigned abbreviations ( ) are also listed.

SPECIES IN THE GENUS

Apple stem grooving virus		
Apple stem grooving virus	[D14995, D16681, D16368, D14455, AB004063]	(ASGV)
Citrus tatter leaf virus		
Cherry virus A		
Cherry virus A	[X82547]	(CVA)
Lilac chlorotic leafspot virus		
Lilac chlorotic leafspot virus		(LiCLV)

TENTATIVE SPECIES IN THE GENUS

Nandina stem pitting virus (NSPV)

PHYLOGENETIC RELATIONSHIPS WITHIN THE GENUS

Not available.

GENUS *VITIVIRUS*

Type Species *Grapevine virus A*

VIRION PROPERTIES

MORPHOLOGY

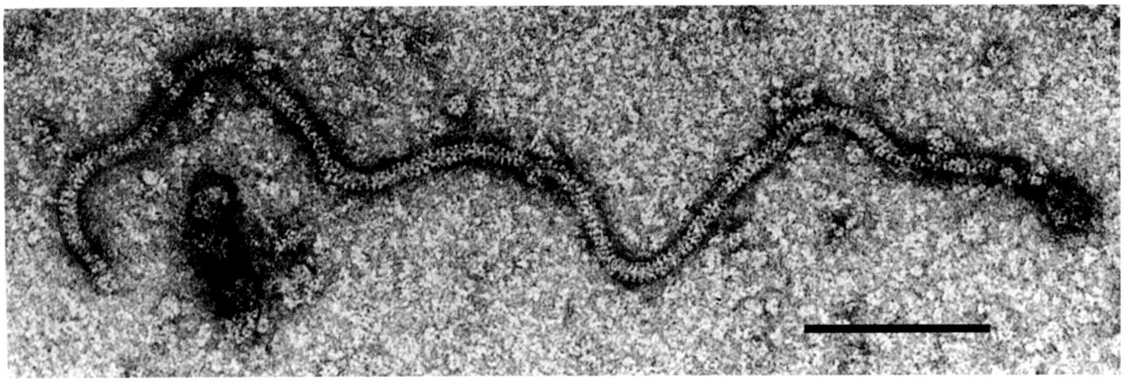

Figure 17: Negative contrast electron micrograph of particles of an isolate of *Grapevine virus A*. The bar represents 100 nm. (Courtesy of A.A. Castellano).

Virions are flexuous filaments 725-825 x 12 nm in size, showing distinct cross banding, helically constructed with a pitch of 3.3-3.5 nm and about 10 subunits per turn of the helix (Fig. 17).

PHYSICOCHEMICAL AND PHYSICAL PROPERTIES

Virions sediment as a single or two very close bands in sucrose or Cs_2SO_4 gradients, with an S_{20w} of ~92S. Virions of Heracleum latent virus (HLV), are sensitive to ribonucleases. Virions of all species resist moderately high temperatures (thermal inactivation is around 60°C) and are moderately resistant to organic solvents.

NUCLEIC ACID

Virions contain a single molecule of positive sense ssRNA, ~ 7.6 kb in size, capped at the 5'-terminus and polyadenylated at the 3'-terminus. The RNA accounts for ~ 5% of the particle weight. The complete nt sequences are available for Grapevine virus A (GVA) and Grapevine virus B (GVB). The genomes of Grapevine virus D (GVD) and HLV have been sequenced in part. Infectious cDNA clones have been produced for GVA and GVB.

PROTEINS

The CPs are composed of a single 18-21.5 kDa polypeptide. Non-structural proteins are: (i) a 194 kDa polypeptide with conserved motifs of replication-related proteins of the "alphavirus-like" supergroup of positive-strand ssRNA viruses (i.e., Mtr, Hel and RdRp, in that order from the N- to the C-terminus); (ii) a 19-20 kDa polypeptide of unknown function with no significant sequence homology to known proteins, which, in GVB infections does not accumulate in phase with MPs (iii) a 31-36.5 kDa polypeptide that possesses the G/D motif of the "30K superfamily" MP and which is associated with cell walls and plasmodesmata (GVA, GVB) and/or with cytoplasmic viral aggregates (GVA)(iv) a 10-14 kDa polypeptide with weak homologies to proteins with RNA-binding properties.

LIPIDS

None reported.

CARBOHYDRATES

None reported.

GENOME ORGANIZATION AND REPLICATION

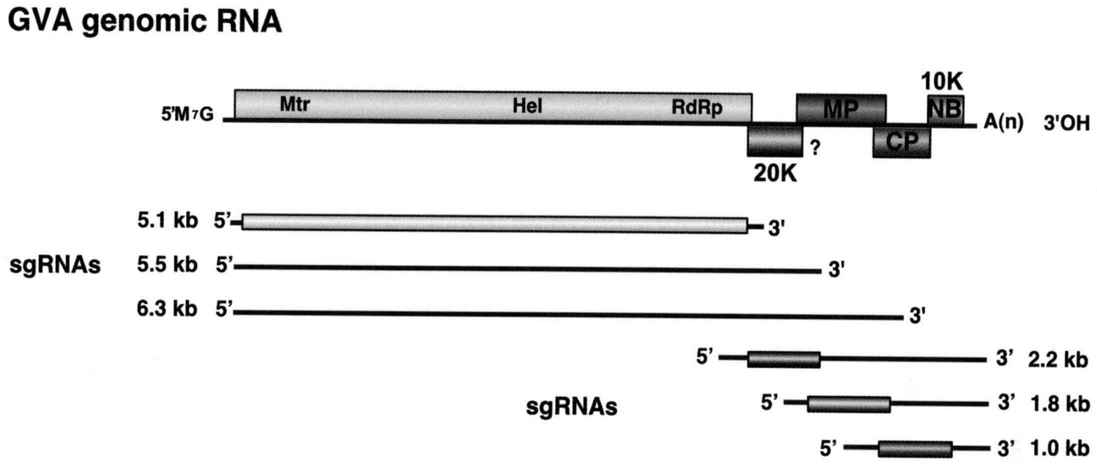

Figure 18: Organization and expression the genome of Grapevine virus A (GVA) showing the relative position of the ORFs, their expression products, and the nested sets of 5'- and 3'-terminal sgRNAs. Mtr, methyltransferase; Hel, helicase; Pol, polymerase; MP, putative movement protein; CP, capsid protein. (adapted from Galiakparov *et al.*, 2003).

The genomes of GVA, GVB and GVD contain five slightly overlapping ORFs (Fig. 19). The 5' region of GVA and GVB initiates with an A/T-rich (60-68%) UTR of 47-86 nt. In addition to the 21.5 kDa CP encoded by ORF4, virus genomes express the non-structural proteins. The strategy of expression is based on sgRNA production, as suggested by the analysis of dsRNA patterns from infected tissues. The four dsRNAs have sizes of 7.6, 6.48, 5.68 and 5.1 kbp for GVA and GVD, and 7.6, 6.25, 5.03 and 1.97 kbp for GVB. Replication occurs in the cytoplasm, possibly in association with membranous vesicles. The replication strategy of GVA encompasses the production of nested sets of 5'-terminal sgRNAs 5.1, 5.5, and 6.0 kb in size and of 3'-terminal sgRNAs 1.0, 1.8, and 2.2 kb that serve for the expression of all ORFs, except for ORF5, which may be expressed via a bi- or polycistronic mRNA. The generation of these 5'- and 3'-terminal sgRNAs appears to be controlled by internal *cis*-acting elements.

ANTIGENIC PROPERTIES

Virions are moderate or poor antigens. Most species are very distantly serologically related. Monoclonal antibodies to GVA, GVB, and GVD and recombinant protein antibodies to the putative MP of GVA have been produced. The relationship between GVA, GVB and GVD is due to a few common internal antigenic determinants (cryptotopes). GVA particles carry a highly structured epitope centered in a common peptide region of the CP sequence.

BIOLOGICAL PROPERTIES

HOST RANGE

The natural host range of individual species is restricted to a single host. Infections induce either no symptoms (HLV) or severe diseases characterized by pitting and grooving of the wood (GVA, GVB, GVD). The experimental host range varies from wide (HLV) to restricted (GVA, GVB, GVC, GVD)

TRANSMISSION

All species are transmitted by mechanical inoculation, those infecting grapevines with greater difficulty. Transmission by grafting and dispersal through propagating material is common with grapevine-infecting species. GVA and GVB are transmitted in a semi-persistent manner by different species of pseudococcid mealybugs of the genera *Pseudococcus* and *Planococcus*. GVA is also transmitted by the scale insect *Neopulvinaria innumerabilis*. HLV is transmitted semi-persistently by aphids, in association with a helper virus.

GEOGRAPHICAL DISTRIBUTION

Geographical distribution varies from very wide (GVA, GVB, GVD) to restricted (HLV).

CYTOPATHIC EFFECTS

Infected cells are damaged to a varying extent. All species elicit the formation of vesicular evaginations of the tonoplast containing finely fibrillar material, possibly representing replicative forms of viral RNA. Virions of grapevine-infecting species are strictly phloem-limited, but in herbaceous hosts they also invadethe parenchyma. Virus particles accumulate in the cytoplasm in bundles or paracrystalline aggregates.

LIST OF SPECIES DEMARCATION CRITERIA IN THE GENUS

The criteria demarcating species in the genus are:

- The natural host range,
- Serological specificity using discriminatory polyclonal and monoclonal antibodies,
- Epidemiology: individual species or groups of species are transmitted by different types and species of vectors,
- Differences in dsRNA pattern,
- Less than ~ 72% identical nt or 80% identical aa between the CP or polymerase genes.

LIST OF SPECIES IN THE GENUS

Species names are in green italic script; strain names and synonyms are in black roman script; tentative species names are in blue roman script. Sequence accession numbers, and assigned abbreviations ( ) are also listed.

SPECIES IN THE GENUS

Grapevine virus A		
Grapevine virus A	[X75433, AF007415]	(GVA)
Grapevine virus B		
Grapevine virus B	[X75448]	(GVB)
Grapevine virus D		
Grapevine virus D	[Y07764]	(GVD)
Heracleum latent virus		
Heracleum latent virus	[X79270]	(HLV)

Tentative Species in the Genus

Grapevine virus C (GVC)

Phylogenetic Relationships within the Genus

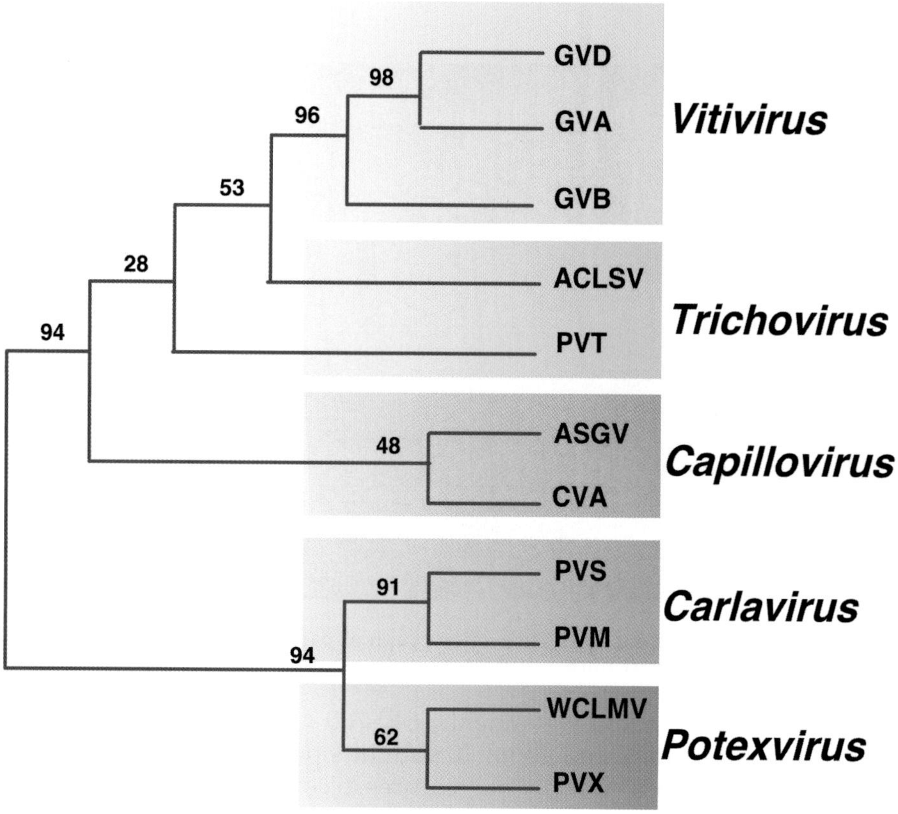

Figure 19: Phylogenetic tree based on CP sequences, showing the relationships among species of the genus *Vitivirus* and between these and some members of related genera. The tree was produced using the Neighbor, Seqboot, Protdist and Consense programs of the PHYLIP package (adapted from Abou-Ghanem *et al.*, 1997).

GENUS TRICHOVIRUS

Type Species *Apple chlorotic leaf spot virus*

VIRION PROPERTIES

MORPHOLOGY

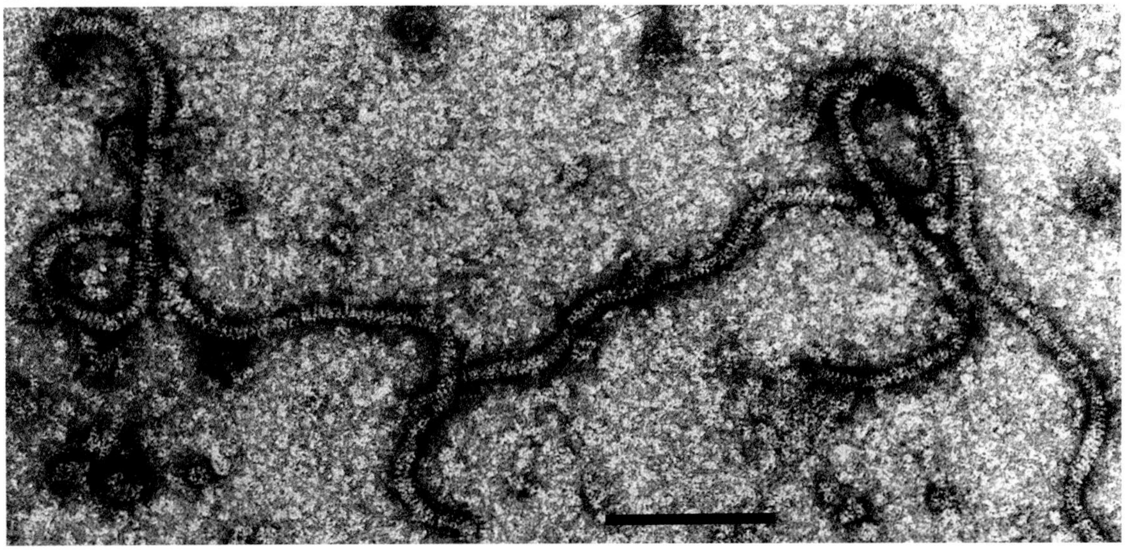

Figure 20: Negative contrast electron micrograph of particles of an isolate of *Apple chlorotic leaf spot virus*. The bar represents 100 nm. (Courtesy of M.A. Castellano).

Virions are very flexuous filaments, 640-760 x 10-12 nm in size, helically constructed with a pitch of 3.3-3.5 nm, and about 10 subunits per turn of the helix. Virions may show cross banding, criss-cross or rope-like features according to the negative contrast material used (Fig. 20).

PHYSICOCHEMICAL AND PHYSICAL PROPERTIES

Virions sediment as single or as two very close bands with an S_{20w} of ~100S. Virions of Apple chlorotic leaf spot virus (ACLSV) are sensitive to ribonucleases. Virions of all species resist moderately high temperatures (thermal inactivation is around 55-60°C) and are moderately resistant to organic solvents.

NUCLEIC ACID

Virions contain a single molecule of linear, positive sense, ssRNA ~ 7.5 to 8.0 kb in size, with a polyadenylated 3'-terminus, accounting for ~ 5% of the particle weight. Indirect evidence suggests that the genome RNA of ACLSV is capped at its 5'-end. The complete nt sequences are available for four isolates of ACLSV and for Cherry mottle leaf virus (CMLV). Partial sequences are known for RNAs of Potato virus T (PVT), Grapevine berry inner necrosis virus (GINV), and Peach mosaic virus (PcMV). An infectious cDNA clone of ACLSV has been produced. ACLSV isolates show a high variability in their nt sequence with an overall homology between 76 and 82%. The CP is the most conserved protein (87-93% homology), whilst the putative MP is the most divergent (77-85% homology)

PROTEINS

Virions of all species are composed of a single 20.5-27 kDa polypeptide. Non-structural proteins of sequenced members are: (i) a protein of about 180-220 kDa containing RdRp, Hel and Mtr signature sequences typical of replication-associated proteins of the "alphavirus-like" supergroup of ssRNA viruses; (ii) a 40-50 kDa polypeptide with weak homologies to some plant virus MP.

Lipids

None reported.

Carbohydrates

None reported.

Genome Organization and Replication

The genomes of ACLSV, PVT, and GINV contain three slightly overlapping ORFs (Fig. 21). An additional ORF is present at the 3'-terminus of CMLV genome. The large 5' ORF of ACLSV is directly expressed from genomic RNA, whereas the two smaller downstream ORFs that code, respectively, for the putative MP and CP, are expressed via sgRNAs. ACLSV-infected tissues contain six dsRNA species of approximately 7.5, 6.4, 5.4, 2.2, 1.1, and 1.0 kbp. The 7.5 kbp species represents the double-stranded form of the full-length genome, whereas the 2.2 and the 1.1 kbp species are the double-stranded forms of sgRNAs coding for the putative MP and the CP, respectively. The most abundant dsRNA species, the function of which are unknown, are 5′ co-terminal with genomic RNA, and have sizes of 6.4 and 5.4 kbp, respectively. Replication is presumed to be cytoplasmic and to involve the translation product of ORF1.

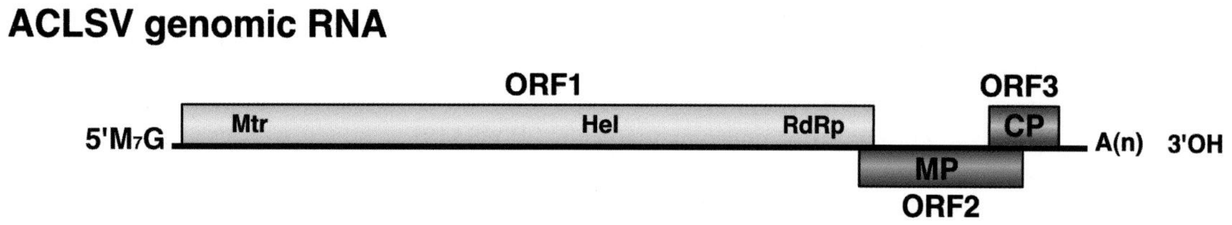

Figure 21: Genome organization of Apple chlorotic leaf spot virus (ACLSV), showing the relative positions of the ORFs and their expression products. Mtr, methyltransferase; Hel, helicase; Pol, polymerase; MP, putative movement protein; CP, capsid protein.

Antigenic Properties

Virions are moderate to poor antigens. CMLV and PcMV are serologically related with one another but with none of the other species in the genus.

Biological Properties

Host Range

The natural host range of individual species is relatively narrow (ACLSV, PcMV), or restricted to a single host (PVT, GINV, CMLV). The experimental host range is somewhat wider, but still limited to a few herbaceous species. In the natural hosts, infections induce few or no symptoms (PVT, ACLSV in certain hosts), or mottling, rings, line patterns, and fruit injuries (i.e. pseudosharka) (ACLSV), mottling with stunting and internal necrosis of shoots and berries (GINV), mottling and severe distortion of the leaves (CMLV), mottling and deformation of leaves and fruits and color break in the petals (PcMV).

Transmission

The viruses are readily transmitted by mechanical inoculation, by grafting (ACLSV, GINV, CMLV, PcMV) and through propagating material. PVT is seed-transmitted in several hosts, including *Solanum* spp. GINV is transmitted by the grape erineum mite *Colomerus vitis*, CMLV by the scale mite *Eriophyes inequalis*, and PcMV by the peach bud mite *Eriophyes insidiosus*.

Geographical Distribution

Geographical distribution varies from wide to restricted, according to the virus species. ACLSV is ubiquitous, whereas PVT is reported only from the Andean region of South America, GINV from Japan, and CMLV and PcMV from North America.

Cytopathic Effects

Infected cells are damaged by ACLSV to varying extents. Virions are found in phloem and parenchyma cells of leaves and roots and accumulate in the cytoplasm, sometimes in the nucleus, in bundles or paracrystalline aggregates. No inclusion bodies are formed.

List of Species Demarcation Criteria in the Genus

The criteria demarcating species in the genus are:

- Natural and experimental host range,
- Serological specificity,
- Less than ~ 72% identical nt or 80% identical aa between the CP or polymerase genes,
- Transmission by a vector,
- Vector specificity.

List of Species in the Genus

Species names are in green italic script; strain names and synonyms are in black roman script; tentative species names are in blue roman script. Sequence accession numbers, and assigned abbreviations ( ) are also listed.

Species in the Genus

Apple chlorotic leaf spot virus		
Apple chlorotic leaf spot virus	[M58152, D14996, X99752, AJ243438]	(ACLSV)
Cherry mottle leaf virus		
Cherry mottle leaf virus	[AF170028]	(CMLV)
Grapevine berry inner necrosis virus		
Grapevine berry inner necrosis virus	[D88448]	(GINV)
Peach mosaic virus		
Peach mosaic virus		(PcMV)

Tentative Species in the Genus

None reported.

Phylogenetic Relationships within the Genus

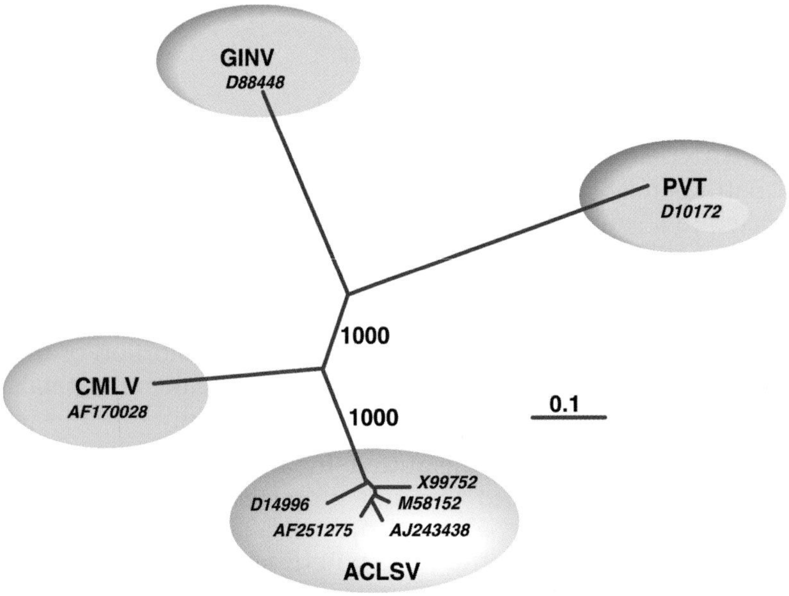

Figure 22: Phylogenetic tree showing the relationships between the species in the genus *Trichovirus*, based on the sequence of the CP gene. The tree was produced and bootstrapped using CLUSTAL W and drawn using Treeview. The figures are protein database accession numbers.

List of Unassigned Viruses in the Family

Species names are in green italic script; strain names and synonyms are in black roman script; tentative species names are in blue roman script. Sequence accession numbers, and assigned abbreviations ( ) are also listed.

Species in the Family

Banana mild mosaic virus ‡

Banana mild mosaic virus	[AF314662]	(BanMMV)

Cherry green ring mottle virus

Cherry green ring mottle virus	[AF017780, AJ291761]	(CGRMV)

Cherry necrotic rusty mottle virus

Cherry necrotic rusty mottle virus	[AF237816]	(CNRMV)

Citrus leaf blotch virus ‡

Citrus leaf blotch virus	[AJ318061]	(CLBV)

Potato virus T

Potato virus T	[D10172]	(PVT)

Sugarcane striate mosaic-associated virus ‡

Sugarcane striate mosaic-associated virus	[AF315308]	(SCSMaV)

Tentative Species in the Family

None reported.

Phylogenetic Relationships within the Family

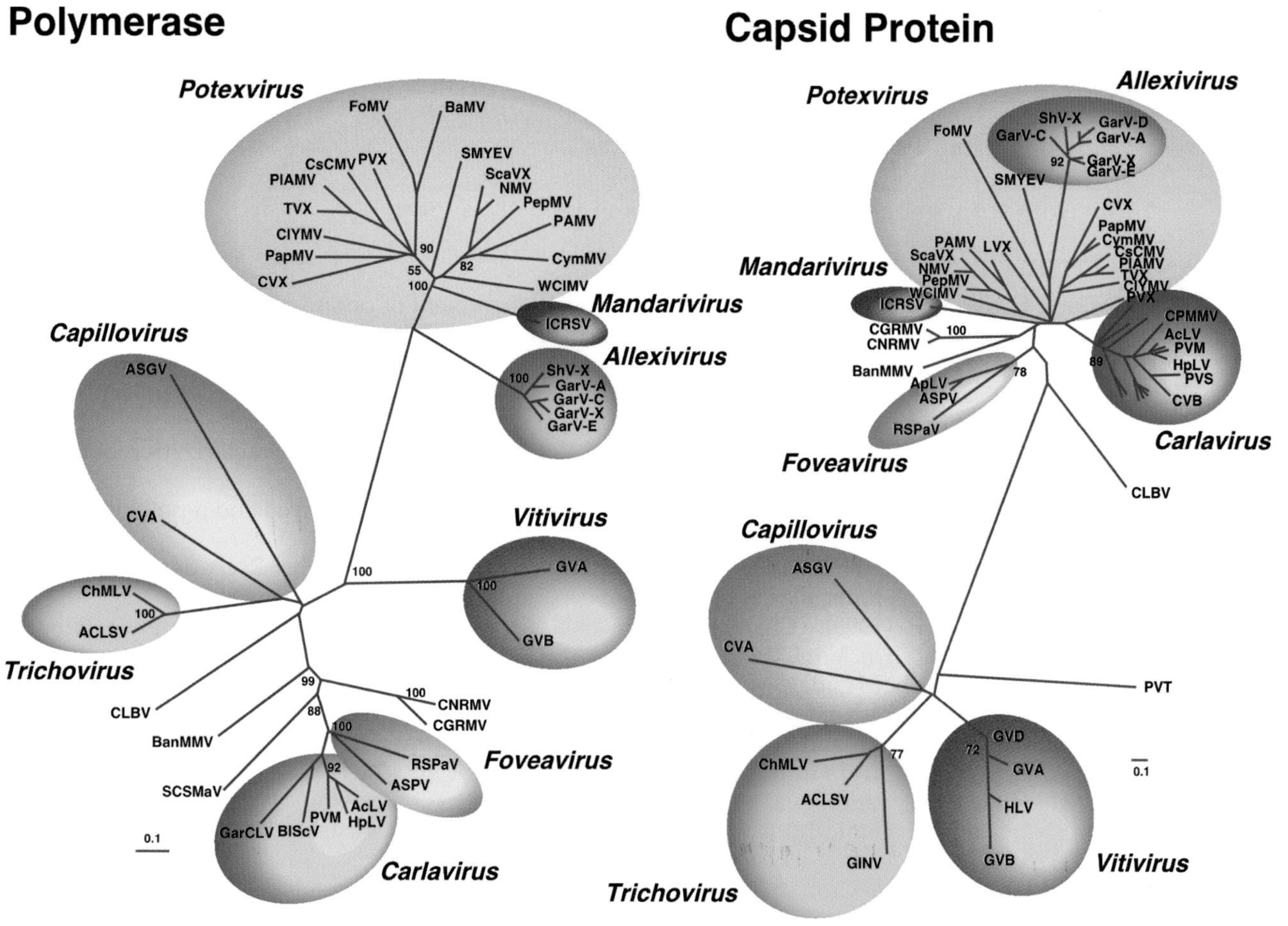

Figure 23: Phylogenetic (Neighbor) analysis of viruses in the family *Flexiviridae* using the aa sequences of the Polymerase (left) and coat protein (right). Sequences were aligned using GCG PileUp and genetic distances estimated by ProtDist (Dayhoff PAM method). Trees were displayed in TreeView. Bootstrap values based on 100 replicates are shown where >60%. The abbreviations and the accession numbers of the viruses are indicated in the lists of species in each genus.

Genera fall into two groups but these are different depending upon the part of the genome examined (Fig. 23). The polymerase analysis places the genera *Potexvirus, Allexivirus* and *Mandarivirus* in one group and the remaining genera and viruses in a second cluster. This grouping may reflect some differences in replication strategy as 5'-terminal sgRNAs are present in at least some members of the second cluster, but experimental details are lacking for some genera. In the CP analysis, the genera *Capillovirus, Trichovirus* and *Vitivirus* form one group and the remaining viruses and genera (most of which have a triple gene block) are in a distantly related cluster.

SIMILARITY WITH OTHER TAXA

The polymerase proteins are members of the "alphavirus-like" supergroup of RNA viruses and therefore are likely to have some relationship to those of other virus genera, including the tymoviruses. The TGB proteins of members of the family *Flexiviridae* are related to those of rod-shaped viruses in the genera *Benyvirus, Hordeivirus, Pecluvirus* and *Pomovirus,* but form a distinct subgroup.

DERIVATION OF NAMES

Allexi: sigla from ***Al**lium* (the genus name for the principal host, shallot) + **X**
Capillo: from Latin *capillus*, a hair.
Carla: sigla from ***Car**nation **la**tent virus.*
Flexi: from *flexus,* Latin for bent.
Fovea: from *fovea,* Latin for pit, hole, type of symptom induced by the type species.
Mandari: from mandarin (*Citrus reticulata*), the host of the type species, *Indian citrus ringspot virus*.
Potex: sigla from ***Pot**ato virus* **X**.
Tricho: from *thrix*, Greek for hair.
Viti: from *Vitis*, generic name of the grapevine, *Vitis vinifera.*

REFERENCES *POTEXVIRUS*

Angell, S.M., Davies, C. and Baulcombe, D.C. (1996). Cell-to-cell movement of potato virus X is associated with a change in the size-exclusion limit of plasmodesmata in trichome cells of *Nicotiana clevelandii. Virology,* **216**, 197-201.

Batten, J.S., Yoshinari, Y and Hemenway, C. (2003). Potato virus X: a model system for virus replication, movement and gene expression. *Mol. Plant Pathol.*, **4**, 125-131.

Calvert, L.A., Cuervo, M.I., Ospino, M.D., Fauquet, C.M. and Ramirez, B.-C. (1996). Characterization of cassava common mosaic virus and a defective RNA species. *J. Gen. Virol.*, **77**, 525-530.

Chen, J., Zheng, H.Y., Chen, J.P. and Adams, M.J. (2002). Characterisation of a potyvirus and a potexvirus from Chinese scallion. *Arch. Virol.*, **147**, 683-693.

Davenport, W.F. and Baulcombe, D.C. (1997). Mutation of the GKS motif of the RNA-dependent RNA polymerase from potato virus X disables or eliminates virus replication. *J. Gen. Virol.*, **78**, 1247-1251.

Geering, A.D. and Thomas, J.E. (1999). Characterisation of a virus from Australia that is closely related to papaya mosaic potexvirus. *Arch. Virol.*, **144**, 577-592.

Hefferon, K.L., Doyle, S. and AbouHaidar, M.G. (1997). Immunological detection of the 8K protein of potato virus X (PVX) in cell walls of PVX-infected tobacco and transgenic potato. *Arch. Virol.*, **142**, 425-433.

Hefferon, K.L., Khalilian, H., Xu, H. and AbouHaidar, M.G. (1997). Expression of the coat protein of Potato virus X from a dicistronic mRNA in transgenic potato plants. *J Gen. Virol.*, **78**, 3051-3059.

Kalinina, N.O., Fedorkin, O.N., Samuilova, O.V., Maiss, E., Korpela, T., Morozov, S.Y. and Atebekov, J.G. (1996). Expression and biochemical analysis of the recombinant potato virus X 25K movement protein. *FEBS Lett.*, **397**, 75-78.

Kavanagh, T., Goulden, M., Santa Cruz, S., Chapman, S., Barker, I. and Baulcombe, D. (1992). Molecular analysis of a resistance-breaking strain of potato virus X. *Virology*, **189**, 609-617.

Kim, K.-H. and Hemenway, C. (1997). Mutations that alter a conserved element upstream of the potato virus X triple block and coat protein genes affect subgenomic RNA accumulation. *Virology*, **232**, 187-197.

Lambrecht, S. and Jelkmann, W. (1997). Infectious cDNA clone used to identify strawberry mild yellow edge-associated potexvirus as causal agent of the disease. *J. Gen. Virol.*, **78**, 2347-2353.

Lee, Y.S., Lin, B.Y., Hsu, Y.H., Chang, B.Y. and Lin, N.S. (1998). Subgenomic RNAs of bamboo mosaic potexvirus-V isolate are packaged into virions. *J. Gen. Virol.*, **79**, 1825-1832.

Park, M.H. and Ryu, K.H. (2003). Molecular evidence supporting the classification of Hosta virus X as a distinct species of the genus *Potexvirus*. *Arch. Virol.*, **148**, 2039-2045.

Solovyev, S.G., Novikov, V.K., Merits, A., Savenkov, E.I., Zeleninu, D.A., Tyulkina, L.G. and Morozov, S.Y. (1994). Genome characterization and taxonomy of Plantago asiatica mosaic potexvirus. *J. Gen. Virol.*, **75**, 259-267.

Tomashevskaya, O.L., Solovyev, A.G., Karpova, O.V., Fedorkin, O.N., Rodionova, NP, Morozov, S.Y. and Atabekov, J.G. (1993). Effects of sequence elements in the potato virus X RNA non- translated -leader on its tramslation enhancing activity. *J. Gen. Virol.*, **74**, 2717-2724.

REFERENCES *MANDARIVIRUS*

Byadgi, A.S., Ahlawat, Y.S., Chakraborty, N.K., Varma, A., Srivastava, M. and Milne, R.G. (1993). Characterization of a filamentous virus associated with citrus ringspot in India. In: *Proc. 12th Conf. International Organization of Citrus Virologists*, (P. Moreno *et al.*, eds), pp. 155-162. Riverside, California.

Rustici, G., Accotto, G.P., Noris, E., Masenga, V., Luisoni, E. and Milne, R.G. (2000). Indian citrus ringspot virus: a proposed new species with some affinities to potex-, carla-, fovea- and allexivirusers. *Arch. Virol.*, **145**, 1895-1908.

Rustici, G., Milne, R.G. and Accotto, G.P. (2002). Nucleotide sequence, genome organization and phylogenetic analysis of Indian citrus ringspot virus. *Arch. Virol.*, **147**, 2215-2224.

REFERENCES *ALLEXIVIRUS*

Arshava, N.V., Konareva, T.N., Ryabov, E.V. and Zavriev, S.K. (1995). The 42K protein of shallot virus X is expressed in the infected *Allium* plants. *Mol. Biol.*, (Russia) **29**, 192-198.

Barg, E., Lesemann, D.E., Vetten, H.J. and Green, S.K. (1994). Identification, partial characterization and distribution of viruses infecting crops in south and south-east Asian countries. *Acta Horti.*, **358**, 251-258.

Chen J., Chen J. and Adams MJ. (2001). Molecular characterization of a complex mixture of viruses in garlic with mosaic symptoms in China. *Arch. Virol.*, **146,** 1841-1853.

Kanyuka, K.V., Vishnichenko, V.K., Levay, K.E., Kondrikov, D.Yu, Ryabov, E.V. and Zavriev, S.K. (1992). Nucleotide sequence of shallot virus X RNA reveals a 5′-proximal cistron closely related to those of potexviruses and a unique arrangement of the 3′-proximal cistrons. *J. Gen. Virol.*, **73**, 2553-2560.

Ryabov, E.V., Generozov, E.V., Vetten, H.J. and Zavriev, S.K. (1996). Analysis of the 3′-region of the mite born filamentous virus genome testifies its relation to the shallot virus X group. *Mol. Biol.* (Russia), **30**, 103-110.

Song, S.I., Song, J.T., Kim, C.H., Lee, J.S. and Choi, Y.D. (1998). Molecular characterization of the garlic virus X genome. *J. Gen. Virol.*, **79**, 155-159.

Sumi, S., Tsuneyoshi, T. and Furuntani, H. (1993). Novel rod-shaped viruses isolated from garlic, possessing a unique genome organization. *J. Gen. Virol.*, **74**, 1879-1885.

Vishnichenko, V.K., Konareva, T.N. and Zavriev, S.K. (1993). A new filamentous virus in shallot. *Pl. Path.*, **42**, 121-126.

Vishnichenko, V.K., Stelmashchuk, V.Y. and Zavriev S.K. (2002). The 42K protein of the Shallot virus X participates in Formation of virus particles. *Mol. Biol.*, (Russia) **36,** 1080-1084.

Yamashita, K., Sakai, J. and Hanada, K. (1996). Characterization of a new virus from garlic (*Allium sativum* L.), garlic mite-borne virus. *Ann. Phytopath. Soc. Japan.*, **62**, 483-489.

REFERENCES *CARLAVIRUS*

Belintani, P., Gaspar, J.O., Targon, M. and Machado, M.A. (2002). Evidence supporting the recognition of Cole latent virus as a distinct carlavirus. *J. Phytopath.*, **150**, 330-333.

Brattey, C., Badge, J.L., Burns, R., Foster, G.D., George, E., Goodfellow, H.A., Mulholland, V., McDonald, J.G. and Jeffries, C.J. (2002). Potato latent virus: a new species in the genus *Carlavirus*. *Plant Pathol.*, **51,** 495-505.

Chen, J., Chen, J.P., Langeveld, S.A., Derks, A.F.L.M. and Adams, M.J. (2003). Molecular characterisation of carla- and potyviruses from *Narcissus* in China. *J. Phytopath.*, **151**, 26-29.

Chen, J., Chen, J. and Adams, M.J. (2001). Molecular characterisation of a complex mixture of viruses in garlic with mosaic symptoms in China. *Arch. Virol.*, **146**, 1841-1853.

Cohen, J., Zeidan, M., Feigelson, L., Maslenin, L., Rosner, A. and Gera, A. (2003). Characterization of a distinct carlavirus isolated from Verbena. *Arch. Virol.*, **148**, 1007-1015

Foster, G.D. (1992). The structure and expression of the genome of carlaviruses. *Res. Virol.*, **143**, 103-112.

Foster, G.D. (1998). Carlavirus isolation and RNA extraction. Plant Virology Protocols: From virus isolation to transgenic resistance. *Methods Mol. Biol.*, **81**, 145-151.

Fuji, S., Yamamoto, H., Inoue, M., Yamashita, K., Fukui, Y., Furuya, H. and Naito, H. (2002). Complete nucleotide sequence of the genomic RNA of Aconitum latent virus (genus Carlavirus) isolated from *Delphinium* sp. *Arch. Virol.*, **147**, 865-870.

Hataya, T., Uchino, K., Arimoto, R., Suda, N., Sano, T., Shikata, E. and Uyeda, I. (2000). Molecular characterization of Hop latent virus and phylogenetic relationships among viruses closely related to carlaviruses. *Arch. Virol.*, **145**, 2503-2524.

Hataya, T., Arimoto, R., Suda, N. and Uyeda, I. (2001). Molecular characterization of Hop mosaic virus: its serological and molecular relationships to Hop latent virus. *Arch. Virol.*, **146**, 1935-1948.

Lawrence, D.M., Rozanov, M.N. and Hillman, B.I. (1995). Autocatalytic processing of the 223-kDa protein of blueberry scorch carlavirus by a papain-like proteinase. *Virology*, **207**, 127-135.

Monger, W., Seal, S., Issacs, A. and Foster, G.D. (2001). The molecular characterisation of the cassava brown streak virus coat protein. *Plant Pathol.*, **50**, 527-534.

Nicolaisen, M., Lykke, and Nielsen, S. (2001). Analysis of the triple gene block and coat protein sequences of two strains of Kalanchoe latent carlavirus. *Virus Genes*, **22**, 265-270.

Song, S.I., Choi, J.N., Song, J.T., Ahn, J.H., Lee, J.S., Kim, M., Cheong, J.J. and Choi, Y.D. (2002). Complete genome sequence of garlic latent virus, a member of the carlavirus family. *Mol. Cells*, **14**, 205-213.

Tsuneyoshi, T., Matsumi, T., Deng, T., Sako, I. and Sumi, S. (1998). Differentiation of Allium carlaviruses isolated from different parts of the world based on the viral coat protein sequence. *Arch. Virol.*, **143**, 1093-1107.

REFERENCES *FOVEAVIRUS*

Giunchedi, L, and Poggi Pollini, C. (1992). Cytopathological, negative staining and serological electron microscopy of a clostero-like virus associated with pear vein yellows disease. *J. Phytopatol.*, **134**, 329-335.

Jelkmann, W. (1994). Nucleotide sequence of apple stem pitting virus and of the coat protein gene of a similar virus from pear associated with vein yellows disease and their relationship with potex- and carlaviruses. *J. Gen. Virol.*, **75**, 1535-1542.

Jelkmann, W. and Keim-Konrad, R. (1997). An immunocapture-polymerase chain reaction and plate trapped ELISA for the detection of apple stem pitting virus. *J. Phytopathol.*, **145**, 499-504.

Jelkmann, W., Kunze, L., Vetten, H.J. and Lesemann, D.E. (1992). cDNA cloning of dsRNA associated with apple stem pitting disease and evidence of the relationship of the virus-like agents associated with apple stem pitting and pear vein yellows. *Acta Hort.*, **309**, 55-62.

Martelli, G.P. and Jelkmann, W. (1998). Foveavirus, a new plant genus. *Arch. Virol.*, **143**, 1245-1249.

Meng, B., Pang, S., Forsline, P.L., McPherson, J.R. and Gonzalves, D. (1998). Nucleotide sequence and genome structure of grapevine rupestris stem pitting associated virus 1 reveal similarities to apple stem pitting virus. *J. Gen. Virol.*, **79**, 2059-2069.

Nemchinov, L. Shamloul, A.M., Zemtchik, E.Z., Verdervskaya, T.D. and Hadidi, A. (2000). *Apricot latent virus*: a new species in the genus *Foveavirus*. *Arch, Virol.*, **145**, 1801-1813.

Rott, M.E. and Jelkmann, W. (2001). Complete nucleotide sequence of cherry necrotic rusty mottle virus. *Arch. Virol.*, **146**, 395-401.

Zagula, K., Aref, N.M. and Ramsdell, D.C. (1989). Purification serology, and some properties of a mechanically transmissible virus associated with green ring mottle disease in peach and cherry. *Phytopathology*, **79**, 451-456.

Zhang, Y.P., Kirkpatrick, B.C., Smart, C.D. and Uyemoto, J.K., (1998). cDNA cloning and molecular characterization of sour cherry green ring mottle virus. *J. Gen. Virol.*, **79**, 2275-2281.

Zhang, Y.P., Uyemoto, J.K., Golino, D.A. and Rowhani, A. (1998). Nucleotide sequence and RT-PCR detection of a virus associated with grapevine rupestris stem-pitting disease. *Phytopathology*, **88**, 1231-1237.

REFERENCES *CAPILLOVIRUS*

Magome, H., Yoshikawa, N., Takahashi T., Ito T. and Miyakawa, T. (1997). Molecular variability of the genome of capilloviruses from apple, Japanese pear, European pear, and citrus trees. *Phytopathology*, **87**, 389-396.

Magome, H., Terauchi, H., Yoshikawa, N. and Takahashi, T. (1997). Analysis of double-stranded RNA in tissues infected with apple stem grooving capillovirus. *Ann. Phytopath. Soc. Japan*, **63**, 450-454.

Ohira, K., Namba. S., Rozanov, M., Kusumi, T. and Tsuchizaki, T. (1995). Complete sequence of and infectious full-length cDNA clone of citrus tatter leaf capillovirus: comparative sequence analysis of capillovirus genomes. *J. Gen. Virol.*, **76**, 2305-2309.

Ohki, S., Yoshikawa, N., Inouye, N. and Inouye, T. (1989). Comparative electron microscopy of *Chenopodium quinoa* leaves infected with apple chlorotic leaf spot, apple stem grooving, or citrus tatter leaf virus. *Ann. Phytopath. Soc. Japan*, **55**, 245-249.

Terauchi, H., Magome, H., Yoshikawa, N., Takahashi, T. and Inouye, N. (1997). Construction of an infectious cDNA clone of the apple stem pitting capillovirus (isolate Li-23) genome containing a cauliflower mosaic virus 35S RNA promoter. *Ann. Phytopath. Soc. Japan*, **63**, 432-436.

Yoshikawa. N. and Takahashi, T. (1988). Properties of RNAs and proteins of apple stem grooving and apple chlorotic leaf spot viruses. *J. Gen. Virol.*, **69**, 241-245.

Yoshikawa, N. and Takahashi, T. (1992). Evidence for genomic translation of apple stem grooving capillovirus RNA. *J. Gen. Virol.*, **73**, 1313-1315.

REFERENCES *VITIVIRUS*

Abou-Ghanem, N., Saldarelli, P., Minafra, A., Castellano, M.A. and Martelli, G.P. (1997). Properties of grapevine virus D, a novel putative trichovirus. *J. Plant Pathol.*, **79**, 1-11.

Bonavia, M., Digiaro, M., Boscia, D., Boari, A., Bottalico, G., Savino, V. and Martelli, G.P. (1996). Studies on "corky rugose wood" of grapevine and on the diagnosis of grapevine virus B. *Vitis*, **35**, 53-58.

Boscia, D., Digiaro, M., Safi, M., Garau, R., Zhou, Z., Minafra, A., Abou Ghanem-Sabanadzovic, N., Bottalico, G. and Potere, O. (2001). Production of monoclonal antibodies to Grapevine virus D and contribution to the study of its aetiologial role in grapevine diseases. *Vitis*, **40**, 69-74.

Choueiri, E., Abou-Ghanem, N. and Boscia, D. (1997). Grapevine virus A and grapevine virus D are serologically distantly related. *Vitis*, **36**, 39-41.

Dell'Orco, M., Saldarelli, P., Minafra, A., Boscia, D. and Gallitelli D. (2002). Epitope mapping of Grapevine virus A capsid protein. *Arch. Virol.*, **147**, 627-634.

Galiakparov, N., Tanne, E., Sela, I. and Gafny, R. (1999). Infectious RNA transcripts from a grapevine virus A cDNA clone. *Virus Genes*, **19**, 235-242.

Galiakparov, N., Tanne, E., Sela, I. and Gafny, R. (2003). Functional analysis of the grapevine virus A genome. *Virology*, **306**, 42-50.

Galiakparov, N., Goszczynski, D., Che X., Batuman O., Bar-Joseph, M. and Mawassi, M. (2003). Two classes of subgenomic RNA of grapevine virus A produced by internal controller elements. *Virology*, ***312***, 434-448.

Garau, R., Prota, V.A., Boscia, D., Fiori, M. and Prota, U. (1995). Pseudococcus affinis Mask., new vector of grapevine trichovirus A and B. *Vitis*, **34**, 67-68.

Goszczynski, D., G.G.F. and Pietersen, G. (1996). Western blot reveals that grapevine viruses A and B are serologically related. *J. Phytopathol.*, **144**, 581-583.

La Notte, P., Buzkan, N., Choueiri, E., Minafra, A, and Martelli, G.P. (1997). Acquisition and transmission of grapevine virus A by the mealybug Pseudococcus longispinus. *J. Plant Pathol.*, **79**, 79-85.

Martelli, G.P., Minafra, A. and Saldarelli, P. (1997). Vitivirus, a new genus of plant viruses. *Arch. Virol.*, **142**, 1929-1932.

Rubinson, E., Galiakparov, N., Rasian, S., Sela, I., Tanne, E. and Gafny, R. (1997). Serological detection of grapevine virus A using an antiserum to a non structiural protein, the putative movement protein. *Phytopathology*, **87**, 1041-1045.

Saldarelli, P. and Minafra, A. (2000). Immunodetection of the 20kDa protein encoded by ORF 2 of Grapevine virus B. *J. Plant Pathol.*, **82**, 157-158.

Saldarelli P., Dell'Orco, M. and Minafra, A. (2000). Infectious cDNA clones of two grapevine viruses. *Arch. Virol.*, **145**, 397-405.

Saldarelli P., Minafra, A., Castellano, M.A. and Martelli, G.P. (2000). Immunodetection and subcellular localization of proteins encoded by ORF 3 of grapevine viruses A and B. *Arch. Virol.*, **145**, 1535-1542.

REFERENCES *TRICHOVIRUS*

German, S., Candresse, T., Le Gall, O., Lanneau, M. and Dunez, J. (1992). Analysis of dsRNAs of apple chlorotic leaf spot virus. *J. Gen. Virol.*, **73**, 767-773.

German-Retana, S., Bergey, B., Delbos, R.P., Candresse, T. and Dunez, J. (1997). Complete nucleotide sequence of the genome of a severe cherry isolate of apple chlorotic leafspot trichovirus (ACLSV). *Arch. Virol.*, **142**, 833-841.

James D. and Upton, C. (1999). Single primer pair designs that facilitate simultaneous detection and differentiation of peach mosaic virus and cherry mottle leaf virus. *J. Virol. Meth.*, **83**, 103-111.

James, D. (1998). Isolation and partial characterization of a filamentous virus associated with peach mosaic disease. *Plant Dis.*, **82**, 909-913.

James, D. and Mukerji, S. (1993). Mechanical transmission, identification and characterization of a virus associated with mottle leaf in cherry. *Plant Dis.*, ***77***, 271-275

James, D., Jelkmann, W. and Upton, C. (2000). Nucleotide sequence and genome organisation of cherry mottle leaf virus and its relationship to members of the Trichovirus genus. *Arch. Virol.*, ***145***, 995-1007.

Kunugi, Y., Asari, S., Terai, Y. and Shinkau, A. (2000). Studies on the grapevine berry inner necrosis virus disease. 2. Transmission of *Grapevine berry inner mecrosis virus* by the grape erineum mite *Colomerus vitis* in Yamanashi. *Bull. Yamanashi Fruit Tree Exp. Sta.*, **10**, 57-64.

Ohki, S.T., Yoshikawa, N., Inouye, N. and Inouye, T. (1989). Comparative electron microscopy of *Chenopodium quinoa* leaves infected with apple chlorotic leaf spot, apple stem grooving or citrus tatter leaf virus. *Ann. Phytopath. Soc. Japan*, **55**, 245-249.

Satoh, H., Yoshikawa N and Takahashi T. (1999). Construction and biolistic inoculation of an infectious cDNA clone of apple chlorotic leafspot trichovirus. *Ann. Phythopathol. Soc. Japan*, **65**, 301-304.

Yoshikawa, N., Iida, H., Goto, S., Magome, H., Takahashi, T. and Terai, Y. (1997). Grapevine berry inner necrosis, a new trichovirus: comparative studies with several trichoviruses. *Arch. Virol.*, **142**, 1351-1363.

References *Flexiviridae*

Gambley, C.F. and Thomas, J.E. (2001). Molecular characterisation of Banana mild mottle virus, a new filamentous virus in *Musa* spp. *Arch. Virol.*, **146**, 1369-1379.

Hataya, T., Uchino, K., Arimoto, R., Suda, N., Sano, T., Shikata, E. and Uyeda, I. (2000). Molecular characterization of Hop latent virus and phylogenetic relationships among viruses closely related to carlaviruses. *Arch. Virol.*, **145,** 2503-2524.

Melcher, U. (2000). The '30K' superfamily of viral movement proteins. *J. Gen. Virol.*, **81,** 257-266.

Morozov, S.Y. and Solovyev, A.G. (2003). Triple gene block: modular design of a multifunctional machine for plant virus movement. *J. Gen. Virol.*, **84,** 1351-1366.

Thompson, N. and Randles, J.W. (2001). The genome organization and taxonomy of *Sugarcane striate mosaic associated virus*. *Arch. Virol.,* **146,** 1441-1451.

Vives, M.C., Galipienso, L., Navarro, L., Moreno, P. and Guerri, J. (2001). The nucleotide sequence and genomic organization of Citrus leaf blotch virus: Candidate type species for a new virus genus. *Virology,* **287**, 225-233.

Vives, M.C., Galipienso, L., Navarro, L., Moreno, P. and Guerri, J. (2002). Characterization of two kinds of subgenomic RNAs produced by citrus leaf blotch virus. *Virology*, **295**, 328-336.

Wong, S.-M., Lee, K.-C, Yu, H.-H. and Leong, W.-F. (1998). Phylogenetic analysis of triple gene block viruses based on the TGB 1 homolog gene indicate a convergent evolution. *Virus Genes,* **16,** 295-302.

Contributed By

Adams, M.J., Accotto, G.P., Agranovsky, A.A., Bar-Joseph, M., Boscia, D., Brunt A.A., Candresse, T., Coutts, R.H.A., Dolja, V.V., Falk, B.W., Foster, G.D., Gonsalves, D., Jelkmann, W., Karasev, A., Martelli G.P., Mawassi, M., Milne, R.G., Minafra, A., Namba, S., Rowhani A., Vetten, H.J., Vishnichenko, V.K., Wisler, G.C., Yoshikawa, N. and Zavriev, S.K.

FAMILY BARNAVIRIDAE

TAXONOMIC STRUCTURE OF THE FAMILY

Family *Barnaviridae*
Genus *Barnavirus*

Since only one genus is currently recognized, the family description corresponds to the genus description.

GENUS BARNAVIRUS

Type Species *Mushroom bacilliform virus*

VIRION PROPERTIES

MORPHOLOGY

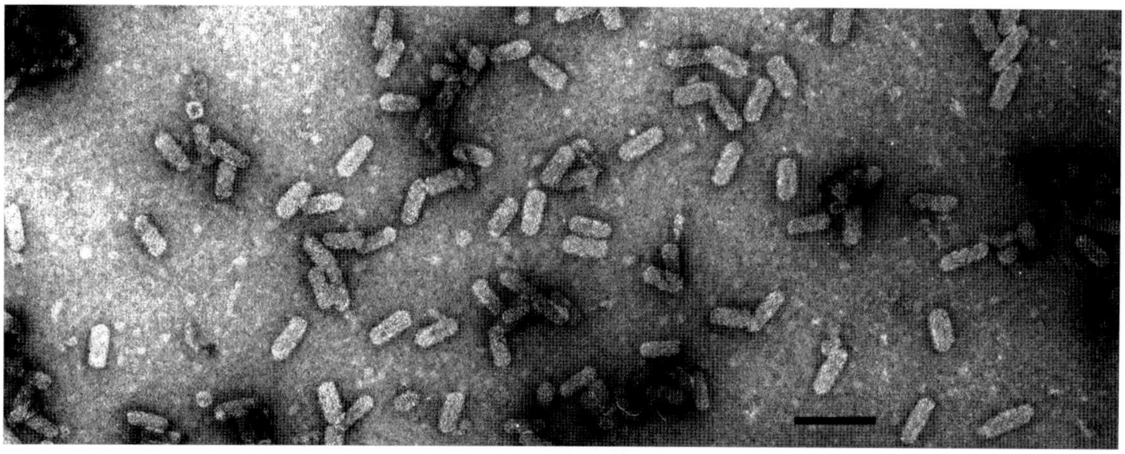

Figure 1: Negative contrast electron micrograph of particles of an isolate of *Mushroom bacilliform virus*. The bar represents 100 nm.

Virions are bacilliform, non-enveloped and lack prominent surface projections. Typically, virions are 19 x 50 nm, but range between 18-20 nm in width and 48-53 nm in length (Fig. 1). Optical diffraction patterns of the virions resemble those of virions of Alfalfa mosaic virus, suggesting a morphological subunit diameter of about 10 nm and a T=1 icosahedral symmetry.

PHYSICOCHEMICAL AND PHYSICAL PROPERTIES

Virion Mr is 7.1 x 10^6, buoyant density in Cs_2SO_4 is 1.32 g/cm^3. Virions are stable between pH 6 and 8 and ionic strength of 0.01 to 0.1M phosphate, and are insensitive to chloroform.

NUCLEIC ACID

Virions contain a single linear molecule of a positive sense ssRNA, 4.0 kb in size. The complete 4,009 nt sequence is available. The RNA has a linked VPg and appears to lack a poly(A) tail. RNA constitutes about 20% of virion weight.

PROTEINS

Virions contain a single major CP of 21.9 kDa. There are probably 240 molecules in each capsid.

LIPIDS

None reported.

CARBOHYDRATES
None reported.

GENOME ORGANIZATION AND REPLICATION
The RNA genome (4,009 nt) contains four major and three minor ORFs and has 5'- and 3'-UTRs of 60 nt and 250 nt, respectively. ORFs 1 to 4 encode polypeptides of 20, 73, 47, and 22 kDa, respectively. The deduced aa sequence of ORF2 contains putative serine protease motifs related to chymotrypsin. ORF3 encodes a putative RdRp and ORF4 encodes the CP. ORFs 5 to 7 encode 8, 6.5, and 6 kDa polypeptides, respectively. The polypeptides potentially encoded by ORFs 1, 5, 6 and 7 show no homology to known polypeptides (Fig. 2).

In a cell-free system, genomic length RNA directs the synthesis of a 21 kDA and a 77 kDa polypeptide and several minor polypeptides of 18-60 kDa. The full-length genomic RNA and a sgRNA (0.9 kb) encoding ORF4 (CP) are found in infected cells. Virions accumulate singly or as aggregates in the cytoplasm.

MBV genomic RNA (4,009 nts)

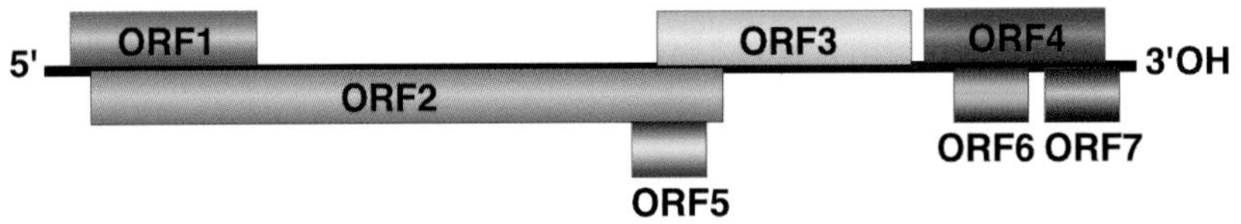

Figure 2: Genome organization of Mushroom bacilliform virus (MBV).

ANTIGENIC PROPERTIES
Virions are highly immunogenic.

BIOLOGICAL PROPERTIES
The virus infects the common cultivated button mushroom (*Agaricus bisporus*). Bacilliform particles, which are morphologically similar to MBV, have been observed in the field mushroom *A. campestris*. Transmission is horizontal via mycelium and possibly basidiospores. Distribution of MBV coincides with that of the commercial cultivation of *A. bisporus*; the virus has been reported to occur in most major mushroom-growing countries. MBV is capable of autonomous replication, but commonly occurs as a double infection with a dsRNA virus (LaFrance isometric virus, LFIV) in mushrooms afflicted with La France disease. MBV is not required in pathogenesis involving LFIV, but it remains to be determined if it is a second, minor causal agent of LaFrance disease, the etiologic agent of an unrecognized pathology or benign. MBV RNA and LFIV dsRNA do not share extensive sequence homology.

LIST OF SPECIES DEMARCATION CRITERIA IN THE GENUS
Not applicable.

LIST OF SPECIES IN THE GENUS
Species names are in green italic script; strain names and synonyms are in black roman script; tentative species names are in blue roman script. Sequence accession, and assigned abbreviations ( ) are also listed.

SPECIES IN THE GENUS
Mushroom bacilliform virus

Mushroom bacilliform virus	U07551	(MBV)

TENTATIVE SPECIES IN THE GENUS

None reported.

PHYLOGENETIC RELATIONSHIPS WITHIN THE GENUS

None reported.

SIMILARITY WITH OTHER TAXA

The aa sequences of the putative chymotrypsin-related serine protease and RdRp suggest an evolutionary relationship with some ssRNA positive-sense plant viruses, particularly poleroviruses, sobemoviruses and enamoviruses (Fig. 3).

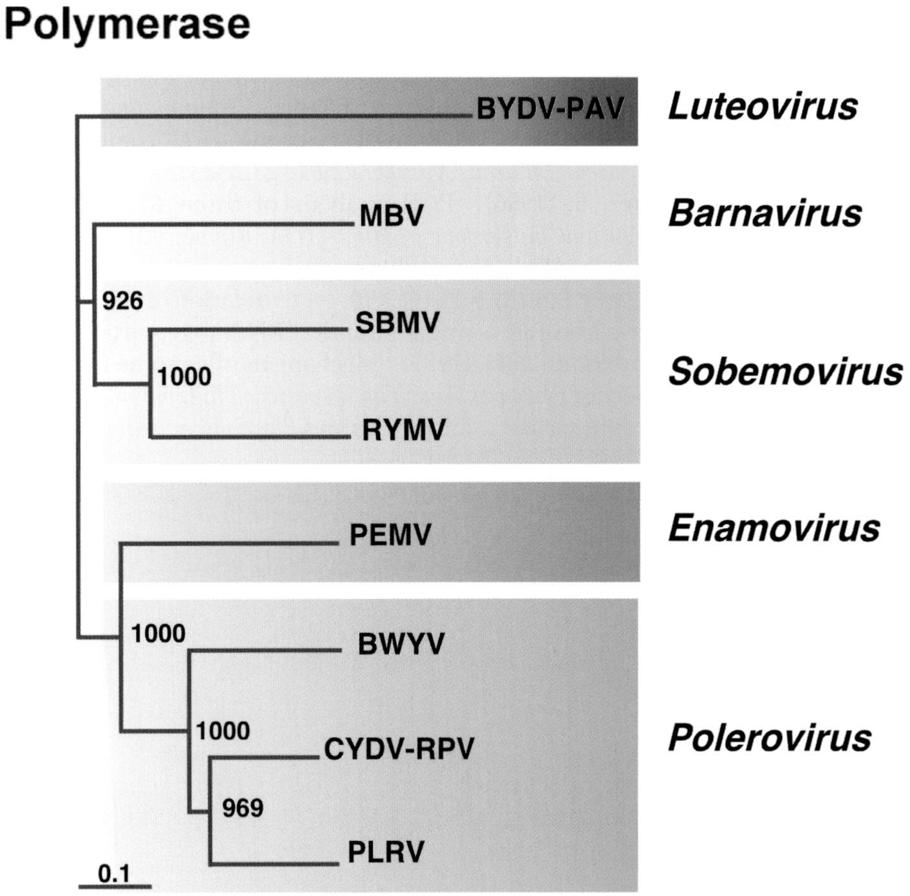

Figure 3: An unrooted neighbour-joining dendrogram showing comparison of the aa sequence of the putative Mushroom bacilliform virus RdRp with those of selected sobemoviruses and poleroviruses, and an enamovirus. The luteovirus BYDV-PAV was included as an outgroup. Amino acid sequences were aligned with Clustal X and the tree was generated with Treeview. Bootstrap values (1000 replicates) are indicated. CYDV-RPV = Cereal yellow dwarf virus-RPV; SBMV = Southern bean mosaic virus; RYMV = Rice yellow mottle virus; PEMV = Pea enation mosaic virus; BWYV = Beet western yellows virus; PLRV = Potato leafroll virus. Accession numbers are AF218798 (BYDV-PAV), U07551 (MBV), AF055887 (SBMV), L20893 (RYMV), L04573 (PEMV) X13063 (BWYV), AF020090 (CYDV-RPV), AY138970 (PLRV).

DERIVATION OF NAMES

Barna: sigla from *ba*cilliform-shaped *RNA* viruses.

REFERENCES

Buck, K.W. (1986). Fungal virology: an overview. In: *Fungal Virology* (K.W. Buck, ed), pp 1-84. CRC Press, Boca Raton, Florida.

Ghabrial, S.A. (1994). New developments in fungal virology. *Adv. Virus Res.*, **43**, 303-388.

Goodin, M.M., Schlagnhaufer, B. and Romaine, C.P. (1992). Encapsidation of the La France disease specific double-stranded RNAs in 36 nm isometric virus-like particles. *Phytopathology*, **82**, 285-290.
Moyer, J.W. and Smith, S.H. (1976). Partial purification and antiserum production to the 19 x 50 nm mushroom virus particle. *Phytopathology*, **66**, 1260-1261.
Moyer, J.W. and Smith, S.H. (1977). Purification and serological detection of mushroom virus-like particles. *Phytopathology*, **67**, 1207-1210.
Revill, P.A., Davidson, A.D. and Wright, P.J. (1994). The nucleotide sequence and genome organization of mushroom bacilliform virus: a single-stranded RNA virus of *Agaricus bisporus* (Lange) Imbach. *Virology*, **202**, 904-911.
Revill, P.A., Davidson, A.D. and Wright, P.J. (1998). Mushroom bacilliform virus RNA: the initiation of translation at the 5′-end of the genome and identification of the VPg. *Virology*, **249**, 231-237.
Revill, P.A., Davidson, A.D. and Wright, P.J. (1999). Identification of a subgenomic mRNA encoding the capsid protein of mushroom bacilliform virus, a single-stranded RNA mycovirus. *Virology*, **260**, 273-276.
Romaine, C.P. and Schlagnhaufer, B. (1991). Hybridization analysis of the single-stranded RNA bacilliform virus associated with La France disease of *Agaricus bisporus*. *Phytopathology*, **81**, 1336-1340.
Romaine, C.P. and Schlagnhaufer, B. (1995). PCR analysis of the viral complex associated with La France disease of *Agaricus bisporus*. *Appl. Environ. Microbiol.*, **61**, 2322-2325.
Romaine, C.P. and Schlagnhaufer, B. (1996). PCR analysis of three RNA genetic elements in *Agaricus bisporus*. In: *Mushroom biology and mushroom products* (D.J. Royse, ed). pp 449-458. The Pennsylvania State University Press, University Park, Pennsylvania.
Tavantzis, S.M., Romaine, C.P. and Smith, S.H. (1980). Purification and partial characterization of a bacilliform virus from *Agaricus bisporus*: a single-stranded RNA mycovirus. *Virology*, **105**, 94-102.
Tavantzis, S.M., Romaine, C.P. and Smith, S.H. (1983). Mechanism of genome expression in a single-stranded RNA virus from the cultivated mushroom *Agaricus bisporus*. *Phytopathology*, **106**, 45-50.
van Zaayen, A.M. (1979). Mushroom viruses. In: *Viruses and Plasmids in Fungi* (P.A. Lemke, ed), pp 239-324. Marcel Dekker, New York.

Contributed By

Wright, P.J. and Revill, P.A.

The Unassigned Viruses

UNASSIGNED VIRUSES

Although many viruses have been classified into genera in this Report, a significant number have not yet been assigned to a recognized genus, or been sufficiently distinguished from viruses in recognized genera so as to form a new genus. Some examples are listed here. These are viruses for which certain key characteristics are known but which are as yet unplaced; poorly characterized viruses are excluded. Some unassigned viruses listed in the 7th ICTV Report are now classified in recognized taxa but others remain unassigned. A few viruses are new to this list and these emphasize the point that the task of devising a universally applicable virus taxonomy is a continuous and complex one, and is unlikely ever to be completed.

VERTEBRATE VIRUSES

ARAGUARI VIRUS (ARAV)

Araguari virus was isolated from a *Philander opossum* collected in 1969 at Serra do Navio, T.F. Amapa, Brazil. Virions are enveloped and contained 8 RNA species and 3 major polypeptides (67, 58 and 30 kDa; 1:2:3), 2 minor polypeptides (43.5 and 35 kDa). The 67 kDa and 30 kDa proteins appear to be glycoproteins (i.e., not similar to arenaviruses). This virus is pathogenic for certain laboratory vertebrates and cell cultures.

Karabatsos, N. (ed) (1985). International Catalogue of Arboviruses Including Certain Other Virus of Vertebrates. 3rd ed., San Antonio, Texas. Am. Soc. Trop. Med. Hyg.

Zeller, H.G., Karabatsos, N., Calisher, C.H., Digoutte, J.P., Murphy, F.A. and Shope, R.E. (1989). Electron microscopic and antigenic studies of uncharacterized viruses. I. Evidence suggesting the placement of viruses in families *Arenaviridae, Paramyxoviridae*, or *Poxviridae*, determined the morphology of Araguari virus was "arenavirus-like". *Arch. Virol.*, **108**, 191-209.

JOHNSTON ATOLL VIRUS (JAV)

Johnston Atoll virus [related to Quaranfil virus, *vide infra*] was isolated from *Ornithodoros capensis* ticks collected in 1969 at Sand Island, Johnston Atoll, Pacific Ocean; all subsequent isolates have been obtained from *O. capensis* from Australasia. It is pathogenic for certain laboratory vertebrates and cell cultures. Virions are enveloped and contain three major polypeptides (67, 58 and 30 kDa; 1:2:3), and 2 minor polypeptides of 43.5 kDa and 35 kDa). The 30kDa and 67kDa proteins appear to be glycoproteins (i.e., not similar to arenaviruses).

Austin, F.J. (1978). Johnston Atoll virus (Quaranfil group) from *Ornithodoros capensis* (Ixodoidea: Argasidae) infesting a gannet colony in New Zealand. *Am. J. Trop. Med. Hyg.*, **27**, 1045-1048.

Karabatsos, N. (ed) (1985). International Catalogue of Arboviruses Including Certain Other Virus of Vertebrates, 3rd ed., San Antonio, Texas. Am. Soc. Trop. Med. Hyg.

Zeller, H.G., Karabatsos, N., Calisher, C.H., Digoutte, J.P., Murphy, F.A. and Shope, R.E. (1989). Electron microscopic and antigenic studies of uncharacterized viruses. I. Evidence suggesting the placement of viruses in families *Arenaviridae, Paramyxoviridae*, or *Poxviridae*. determined the morphology of Araguari virus was "arenavirus-like". *Arch. Virol.*, **108**, 191-209.

NYAMANINI VIRUS (NYMV)

The virus was first isolated from *Bubulcus ibis* (cattle egret) collected in 1957 in South Africa and since has been isolated from cattle egrets and ticks in Africa. It is pathogenic for certain laboratory vertebrates and cell cultures. Virions are enveloped and contain RNA.

Karabatsos, N. (ed) (1985). International catalogue of arboviruses including certain other virus of vertebrates, 3rd ed., San Antonio, Texas. Am. Soc. Trop. Med. Hyg.

Kemp, G.E., Lee, V.H. and Moore, D.L. (1975). Isolation of nyaminin and quaranfil viruses from *Argas* (*Persicargas*) *arboreus* ticks in Nigeria. *J. Med. Entomol.*, **12**, 535-537.

Quaranfil virus (QRFV)

Quaranfil virus [related to Johnston Atoll virus, *vide supra*] was isolated from *Argas* (*Persicargus*) *arboreus* ticks collected in 1953 near Cairo, Egypt. Subsequent isolates have been obtained from humans (febrile illnesses), ibis and pigeons, and ticks in Egypt, South Africa, Afghanistan, Nigeria, and Iran. Electron microscopy (numerous attempts) has been inconclusive. Virions are enveloped. QRFV is pathogenic for a wide variety of laboratory hosts.

Karabatsos, N. (ed) (1985). International catalogue of arboviruses including certain other virus of vertebrates, 3rd ed., San Antonio, Texas. Am. Soc. Trop. Med. Hyg.

Zeller, H.G., Karabatsos, N., Calisher, C.H., Digoutte, J.P., Murphy, F.A. and Shope, R.E. (1989). Electron microscopic and antigenic studies of uncharacterized viruses. I. Evidence suggesting the placement of viruses in families *Arenaviridae*, *Paramyxoviridae*, or *Poxviridae*. determined the morphology of Araguari virus was "arenavirus-like". *Arch. Virol.*, **108**, 191-209.

Invertebrate Viruses

Acyrthosiphon pisum virus (APV)

Acyrthosiphon pisum virus particles are icosahedral, 31 nm in diameter. The virus was isolated from *Acyrthosiphon pisum* (Hemiptera: Aphididae) but is able to infect many other aphid species. The virus capsid is composed of a major protein of 34 kDa and three minor proteins of 23, 24, and 66 kDa. The genome has been completely sequenced and consists of a single stranded polyadenylated ssRNA molecule of approximately 10 kb. The genome contains two large ORFs with ORF2 overlapping ORF1. The latter is thought to be expressed by a -1 translational frameshift. The CPs are encoded by the 3'-end of ORF1 and the 5'-end of ORF2. The ORF1 product contains motifs characteristic for RdRp, Hel and cysteine proteases.

GenBank Accession Number: AF024514.

Van der Wilk, F., Dullemans, A.M., Verbeek, M. and van den Heuvel, J.F.J.M. (1997). Nucleotide sequence and genomic organization of Acyrthosiphon pisum virus. *Virology*, **238**, 353-362.

Arkansas bee virus (ABV)

The virus was originally isolated from pollen gathered from hives of the honey bee, *Apis mellifera* (Hymenoptera: Apidae) in Arkansas, USA. The virus particle has a diameter of around 30 nm and contains a single segment of ssRNA of Mr 1.8×10^6. The buoyant density of the virions is 1.37g/ml and the virus contains only one major CP of 41 kDa.

Bailey, L. and Woods, R.D. (1974). Three previously undescribed viruses form the honey bee. *J. Gen. Virol.*, **25**, 175-186.

Lommel, S.A., Morris, T.J. and Pinnock, D.E. (1985). Characterization of nucleic acids associated with Arkansas bee virus. *Intervirology*, **23**, 199-207.

Bee virus X (BXV)

The virus was first isolated from honey bees in the United Kingdom. The virions are around 35 nm in diameter and contain an RNA genome. The single major CP is 52 kDa and the virus shows a distant serological relationship to BYV. BXV and BYV may be the same closely related. The virus is only transmissible *per os* to adult honey bees but it is found commonly in Great Britain; it is but also present in mainland Europe.

Bailey, L. and Woods, R.D. (1974). Three previously undescribed viruses form the honey bee. *J. Gen. Virol.*, **25**, 175-186.

Bee Virus Y (BYV)

The virus was originally isolated from honey bees, *Apis mellifera* (Hymenoptera: Apidae) in the United Kingdom. The virus shows a distant serological relationship to BXV but has a major CP of 50 kDa and an RNA genome. Like BXV, BYV has virions of 35 nm in diameter, an RNA genome and is only transmissible *per os* to adult honey bees. The virus has been found more widely in honey bee colonies (including North America and Australasia) than BXV.

Bailey, L., Carpenter, J.M., Govier, D.A. and Woods, R.D. (1980). Bee virus Y. *J. Gen. Virol.*, **51**, 405-407.

Berkeley bee virus (BBV)

The virus was isolated from honey bees, *Apis mellifera* (Hymenoptera: Apidae) in California during studies on Arkansas bee virus. The 30 nm virions contain a single species of ssRNA of Mr ~2.8×10^6. The virus has three major CPs of 37, 35 and 32.5 kDa.

Lommel, S.A., Morris, T.J. and Pinnock, D.E. (1985). Characterisation of nucleic acids associated with Arkansas bee virus. *Intervirology*, **23**, 199-207.

Ceratitis V virus (CVV)

Originally isolated from a laboratory colony of the Mediterranean fruit fly, *Ceratitis capitata* in France. The isometric virus particles are 24 nm in diameter and contain three major CPs of 39, 34 and 27 kDa. It has no serological relationship to Drosophila A, C or P viruses.

Plus, N., Veyrunes, J.C. and Cavalloro, R. (1981). Endogenous viruses of *Ceratitis capitata* Vied. "J.R.C. Ispra" strain, and of *C. capitata* permanent cell lines. *Ann. Virol.*, **132E**, 91-100.

Chronic bee paralysis virus (CBPV)

The virus was first isolated from honey bees, *Apis mellifera* (Hymenoptera: Apidae), in the United Kingdom. The virions have an unusual anisometric shape and are heterogeneous in size, usually of ~60nm in length but ranging in diameter from 20 to 30 nm. Purified virion preparations contain several ssRNA species ranging is size from 1.35 to 0.35 kb. Virions contain a single structural protein of 23.5 kDa. The virus is readily transmitted to adult honey bees *per os* and has been identified in colonies almost everywhere that honey bees are maintained.

Bailey, L., and Ball, B.V. (eds)(1991). *Honey bee pathology,* 2nd ed. Harcourt Brace Jovanovich, Sidcup, UK.

Cloudy wing virus (CWV)

The virus was originally isolated from honeybees, *Apis mellifera* (Hymenoptera: Apidae), during studies on other honeybee viruses. The virions are around 17 nm in diameter and contain a single segment of RNA of Mr ~0.45×10^6. The virus has a single 19 kDa CP and a buoyant density of 1.38 g/ml.

Bailey. L., Ball, B.V., Carpenter, J.M. and Woods, R.D. (1980). Small virus-like particles in honeybees associated with chronic bee paralysis virus and with a previously undescribed disease. *J. Gen. Virol.*, **46**, 149-155.

Deformed wing virus (DWV)

The virus was originally isolated from diseased adult honeybees *Apis mellifera* (Hymenoptera: Apidae), collected in Japan but is now known from other locations. The virions are 30 nm in diameter and contain an RNA genome that has been completely sequenced. The 10,131 nt genome is organized like those of members of the genus *Iflavirus* with the structural proteins encoded at the 5′-end of the single ORF and the non-structural polypeptides at the 3′-end. The virus shows a weak reaction with sera raised against Egypt bee virus.

GenBank Accession Number: AY292384

Bailey, L. and Ball, B.V. (eds)(1991). *Honey bee pathology*, 2nd ed. Harcourt Brace Jovanovich, Sidcup, UK.

Drosophilia A virus (DAV)

Originally isolated from a laboratory colony of *Drosophila melanogaster* (Diptera: Drosophilidae) in France, it has subsequently been found widely distributed in laboratory and natural populations of *Drosophila* spp. from around the world. The virions are 30 nm in diameter and contain a ssRNA genome polyadenylated at the 3′-end. There are three coat proteins reported of 73, 44 and 32 kDa - although they are not present in equimolar amounts. DAV is serologically distinct from Drosophila C and P viruses. It is readily transmitted both horizontally and vertically (transovarially).

Christian, P.D. and Scotti, P.D. (1998). The picorna-like viruses of insects. In: *The Insect Viruses,* (L. Miller and L.A. Ball, eds), pp 301-336. Plenum Publishing Corporation, New York.

Drosophila P virus (DPV)

Isolated from *Drosophila melanogaster* (Diptera: Drosophilidae) the virions of DPV are approximately 30nm in diameter. The virions contain a single species of ssRNA and three major proteins of 48, 30 and 26 kDa in non-equimolar amounts. DPV is serologically related to Iota virus of *Drosophila immigrans* (Diptera: Drosophilidae) and is readily transmitted both horizontally and vertically (transovarially).

Christian, P.D. and Scotti, P.D. (1998). The picorna-like viruses of insects. In: *The Insect Viruses,* (L. Miller and L.A. Ball, eds), pp 301-336. Plenum Publishing Corporation, New York.

Egypt bee virus (EBV)

The virus was isolated from European honeybees, *Apis mellifera* (Hymenoptera: Apidae), collected in Egypt in the spring of 1977. The virions contain RNA, are 30 nm in diameter and compromise three major CPs of 41, 30 and 25 kDa. The virus is serologically distinct from most other viruses of domesticated bees but does show a weak relationship to deformed wing virus.

Bailey, L., Carpenter, J.M. and Woods, R.D. (1979). Egypt bee virus and Australian isolates of Kashmir bee virus. *J. Gen. Virol.,* **43**, 641-647.

Epicerura pergrisea virus (EpV)

The virus was isolated from larvae of the lepidopteran forest pest, *Epicerura perigrisea* (Lepidoptera: Notodontidae). The virions are icosahedral, 30 nm in diameter and contain RNA. Four virion polypeptides have been reported. This virus has been successfully used as a control agent for *E. perigrisea* on *Terminalia spp* in the Ivory Coast.

Fedière, G, Lery, X., Dauthuille, D. and Kanga, L. (1987). Evidence of a new picorna-like virus *Epicerura pergrisea* (Lepidoptera Notodontidae) a pest of *Terminalia ivorensis* in Ivory Coast. In: *Symposium on Integrated Pest Management and Environmental Conservation*, Dakar, Senegal 7-10 December 1987.

Galleria cell line virus (GmCLV)

Originally isolated from a cell line derived from the lepidopteran, *Galleria mellonella* after challenging with the unassigned RNA-containing Maize stem borer virus (MSBV). The virions are 29 nm in diameter and contain a ssRNA of Mr 2.9 x 10^6 that is polyadenylated at its 3'-end. The capsid contains two major polypeptides of 34.5 and 32.5 kDa in equimolar amounts. The virus replicates in cultured *Galleria mellonella* cells and after injection into *G. mellonella* larvae.

Lery, X., Fediere, G., Taha, A., Salah, M. and Giannotti, J. (1997). A new small RNA virus persistently infecting an established cell line of *Galleria mellonella*, induced by a heterologous infection. *J. Invert. Pathol.*, **69**, 7-31.

GONOMETA PODOCARPI VIRUS (GPV)

The virus was isolated from insects collected during an epizootic in populations of the lepidopteran, *Gonometa podocarpi* (Lepidoptera: Lasiocampidae) in pine plantations in Uganda in September of 1963. The isometric 32 nm diameter virions contain a single segment of ssRNA and four major polypeptides of 37, 32, 20 and 12 kDa. Minor amounts of a 48 kDa protein are also present in the virion. A similar virus was also found to be present in another lepidopteran present at the epizootic site (*Pacymetana* sp.).

Harrap, K.A., Longworth, J.F., Tinsley, T.W. and Brown, K.W. (1966). A non-inclusion virus of *Gonometa podocarpi* (Lepidoptera: Lasiocampidae). *J. Invert. Pathol.*, **8**, 270-272.
Longworth, J.F., Payne, C.C. and MacLeod, R. (1973). Studies on a virus isolated from *Gonometa podocarpi* (Lepidoptera: Lasiocampidae). *J. Gen. Virol.*, **18**, 119-125.

GRYLLUS RUBENS VIRUS (GRV)

The virus was isolated from dead and dying nymphs of the field crickets *Gryllus rubens* and *G. firmis* (Orthoptera: Gryllidae). The enveloped virions are approximately 150 x 90 nm and contain a closed circular dsDNA genome of around 87 kbp. The virion has six major polypeptides and 11 minor polypeptides ranging in size between 132 and 19 kDa. The virus is transmissible *per os* to first instar nymphs of both *Gryllus rubens* and *G. firmis,* albeit at low frequency. Previously classified as a member of the family *Baculoviridae*, the affinities of this virus to the morphologically similar Oryctes virus and Hz-1V are not clear. DNA-DNA hybridization studies indicated little nt homology with the former.

Boucias, D.G., Maruniak, J.E. and Pendland, J. (1989). Characterization of a non-occluded baculovirus (subgroup C) from the field ricket, *Gryllus rubens*. *Arch. Virol.*, **106**, 93-102.

HELIOTHIS ZEA VIRUS 1 (HZV-1)

Heliothis zea virus 1 was isolated as a persistent virus of an insect cell line derived from *Heliothis zea* (Lepidoptera: Noctuidae). Although the virus can infect a number of insect (lepidopteran) cell lines, infection of an insect has not been observed. The virion of HzV-1 is composed of an enveloped rod-shaped nucleocapsid. Virions are released from infected cells by cell lysis. HzV-1 genome consists of a single molecule of circular dsDNA, approximately 228 kbp in length. HzV-1 was previously classified as a non-occluded member of the family *Baculoviridae,* but sequence analysis of selected genes has shown only a very distant relationship to the baculoviruses.

GenBank Acccession Number: AF451898.

Burand, J. (1998). Nudiviruses. In: *The Insect Viruses,* (L. Miller and L.A. Ball, eds), pp 69-90. Plenum Press, New York.

HELICOVERPA ZEA VIRUS- 2 (HZV-2)

(synonymous with Helicoverpa zea reproductive virus and Gonad specific virus (GSV)). Helicoverpa zea reproductive virus was first observed in the gonads of adult *Helicoverpa zea* (Lepidoptera: Noctuidae) from a rearing facility in Stoneville, Mississippi. The virus causes deformities of the reproductive tract in both sexes that in turn leads to sterility. The virus grows in several cultured insect cell lines and is infectious in insects. The virion

UNASSIGNED

of Hz-2V is composed of an enveloped rod-shaped nucleocapsid containing a genome of a single molecule of circular dsDNA, approximately 225 kbp in length. Sequence analysis of a small region of the genome (approximately 2.5 kbp) shows very high homology with Heliothis zea virus 1 (Hz-1V).

Raina, A.K., Adams, R.R., Lupiani, B., Lynn, D.E., Kim, W., Burand, J.P. and Dougherty, E.M. (2000). Further characterization of the gonad-specific virus of corn earworm, *Helicoverpa zea*. *J. Invert. Pathol.*, **76**, 6-12.

Kashmir bee virus (KBV)

Originally isolated from the Asiatic hive bee, *Apis cerana* (Hymenoptera: Apidae), KBV has subsequently been found in both the European honeybee, *Apis mellifera* and the European Wasp, *Vespa germinica*. The virions are approximately 30 nm in diameter, contain single-stranded RNA of 9,524 nt in length and have four major structural polypeptides of 41, 37, 25 and 6 kDa. The virus is transmissible *per os* to larvae, infected individuals giving rise to persistently infected adults. The virus is not usually associated with disease symptoms, but in association with other parasites/pathogens a disease state may be induced. Several serologically distinct strains of KBV have been reported. The genome has recently been fully sequenced and shows a genomic organization similar to that of the dicistroviruses. Sequence analysis indicates a close relationship with the dicistrovirus *Acute bee paralysis virus*.

GenBank Accession: AY275710
Bailey, L. and Ball, B.V. (eds)(1991). *Honey bee pathology*. 2nd ed. Harcourt Brace Jovanovich, Sidcup, UK.

Kawino virus (KaV)

Kawino virus was first isolated from the mosquito *Mansonia uniformis* (Diptera: Culicidae) collected near the village of Kawino in Kenya in 1973. The virus particles are 28 nm in diameter and contain four major structural polypeptides of 33, 30, 27 and 7 kDa with a minor component of 36 kDa. The genome comprises a single segment of ssRNA of Mr 2.6 x 10^6. The virus replicates in several mosquito cell lines but was not found to replicate in a tick cell line or several mammalian cell lines.

Pudney, M., Newman, J.F.E. and Brown, F. (1978). Characterization of Kawino virus, an entero-like virus isolated from the mosquito *Mansonia uniformis*, Diptera: *Culicidae*. *J. Gen. Virol.*, **40**, 433-441.

Kelp fly virus (KFV)

The virus was originally isolated from the kelp fly, *Chaetocoelopa sydneyensis* (Diptera: Coelopidae) collected in New South Wales, Australia. The virus has isometric particles of 29 nm in diameter with surface projections that give the particles the appearance of small reovirus particles. The virions contain ssRNA of around Mr 3.5 x 10^6 and two major CPs of 73 and 29 kDa. The virions have a buoyant density of 1.43 g/ml in neutral CsCl.

Scotti, P.D., Gibbs, A.J. and Wrigley, N.G. (1976). Kelp fly virus. *J. Gen. Virol.*, **30**, 1-9.

Latoia viridissima virus (LvV)

The virus was isolated from the lepidopteran palm pest *Latoia viridissima* (Lepidoptera, Limacodidae) in the Ivory Coast and its virions are approximately 30 nm in diameter. The genome comprises a single segment of ssRNA of Mr 2.9 x 10^6 and the virions appear to contain two major structural polypeptides of 31 and 30 kDa. Stoichiometry of the structural proteins suggests that there may actually be two 30 kDa proteins. The virus is serologically related to a virus isolated from the limacodid banana pest, *Teinorhynchha umbra*. (Cf Galleria cell line virus).

Zeddam, J.L., Phillippe, R., Veyrunes, J.C., Fediere, G., Mariau, D. and Bergoin, M. (1990). Etude du ribovirus de *Latoia viridissima* Holland, ravageur de *Palmacae* en Afrique de l'Ouest. *Oleagineux*, **45**, 493-498.

LYMANTRIA NINAYI VIRUS (LNV)

The virus was isolated from the lepidopteran pine pest *Lymantria ninayi* (Lepidoptera: Lymantriidae) collected during an epizootic in pine plantations in Papua New Guinea. The virions contain a polyadenylated ssRNA of around Mr 2.8×10^6, three major CPs of 38, 33 and 32 kDa and a minor capsid component of 43 kDa. The virus shows some serological cross-reaction with antisera to an isolate of Drosophila C virus.

Pullin, J.S.K., Black, F., King, L.A., Entwistle, P.F. and Moore, N.F. (1984). Characterization of a small RNA-containing virus in field-collected larvae of the tussock moth, *Lymantria ninayi*, from Papua New Guinea. *Appl. Envir. Microbiol.*, **48**, 504-507.

MAIZE STEM BORER VIRUS (MSBV)

The virus was isolated from larvae of the maize stem borer *Sesamia cretica* (Lepidoptera: Noctuidae) in Egypt. The virions are isometric with a diameter of 30 nm and contain one segment of ssRNA of 9.4 kb that is polyadenylated at the 3'-end. The virus has three major polypeptides of 60, 54 and 28 kDa in non-equimolar amounts and a minor component of 58 kDa. The virus is highly pathogenic *per os* for larvae of *S. cretica.*

Fediere, G., Taha, A.A., Lery, X., Giannotti, J.L., Monsarrat, A. and Abol-Ela, S. (1991). A small RNA virus isolated from the maize stem borer *Sesamia cretica* Led. (Lepidoptera: *Noctuidae*) in Egypt. *Egypt J. Biol. P. Cont.*, **1**, 115-119.

NEZARA VIRIDULA VIRUS-1 (NVV-1)

The virus was originally isolated from a diseased laboratory colony of the green stinkbug *Nezara viridula* (Hemiptera: Pentatomidae) and has isometric virions of 29 nm in diameter. The genome comprises a single segment of ssRNA of approximately Mr 3.1×10^6 and the virions contain three major structural proteins of 32, 32 and 31 kDa. The virus is transmitted both vertically and horizontally and has significant effects on the longevity and fecundity of infected individuals.

Williamson, C. and von Wechmar, M.B. (1992). Two novel viruses associated with severe disease symptoms of the green stinkbug *Nezara viridula*. *J. Gen. Virol.*, **73**, 2467-2471.

ORYCTES RHINOCEROS VIRUS (ORV)

The Oryctes rhinoceros virus infects a number of coleopteran insects in the family Scarabaeidae. The mature virion of Oryctes rhinoceros virus consists of an enveloped, rod-shaped nucleocapsid and contains a unique tail-like structure protruding from one end. The mature virion is produced by virus budding from the plasma membrane and contains two unit membranes. The genome is a single supercoiled circular dsDNA of approximately 130 kbp. The virus was previously classified as a member of the family *Baculoviridae*, but differs in several respects including virion morphology and the lack of an occlusion body. The limited amount of sequence data available indicates no clear relationship with either the baculoviruses or morphologically similar Hz-1V.

Burand, J. (1998). Nudiviruses. In: *The Insect Viruses*, (L. Miller and L.A. Ball, eds), pp 69-90. Plenum Press, New York.

Pectinophora gossypiella virus (PgV)

The virus was isolated from naturally occurring dead larvae of the pink bollworm *Pectinophora gossypiella* (Lepidoptera: Noctuidae). The icosahedral virions are 27 nm in diameter and contain RNA. Three CPs of size 47, 33 and 32 kDa have been identified but are present in non-equimolar amounts. The virus is infectious but not highly pathogenic to larvae of *P. gossypiella* exposed per os and can be vertically transmitted from infected adults to their progency.

Monsarrat, A., Abol-Ela, S., Abdel-Hamid, I., Fediere, G., El Husseini, M. and Gianotti, J. (1995). A new RNA picorna-like virus in the cotton pink bollworm *Pectinophora gossypiella*, (lepidopteran: *Gelechiidae*) in Egypt. *Entomophaga*, **40**, 47-54

Pteroteinon laufella virus-2 (PlV-2)

Isolated from the lepidopteran pest of oil palms *Pteroteinon laufella* (Lepidoptera Hesperidae) in the Ivory Coast. Two viruses were isolated one of 40 nm (Pteroteinon laufella virus-1) and one of 30 nm (Pteroteinon laufella virus-2). The virions of Pteroteinon laufella virus-2 contain RNA and show no serological relationship to CrPV or other small RNA-containing viruses from *Latoia viridissima* or *Teinorhyncha umbra.*

Herder, S., Fediere, G., Kouassi, K.N., Lery, X., Kouevidjin, R., Phillippe, R. and Mariau, D. (1994). Mise en évidence de nouveaux virus à ARN chez *Pteroteinon laufella* Hewitson (lepidopteran *Hesperidae*), défoliateur du palmier à huile en Côte-d'Ivoire. *Oleagineux*, **49**, 43-47.

Queensland fruit fly virus (QFFV)

The virus was originally isolated from a laboratory culture of the Queensland fruitfly *Bactrocera* (*Dacus*) *tryoni (Diptera: Tephritidae).* The isometric virions have a diameter of 30 nm and contain three major structural proteins of 42, 37 and 31 kDa. The genome comprises one segment of ssRNA of approximately 2.9 MDa and is polyadenylated at its 3'-end. The virus is serologically unrelated to other small isometric viruses of insects including Ceriatitis V virus.

Bashiruddin, J., Martin, J.K. and Reinganum, C. (1988). Queensland fruit fly virus, a probable member of the *Picornaviridae*. *Arch. Virol.*, **100**, 61-74.

Sitobion avenae virus (SaV)

The virus was isolated from laboratory cultures of the aphids *Sitibion avenae* and *Metopolophium dirhodum* maintained in the United Kingdom. The virions are 30 nm in diameter and comprised of a ssRNA genome and a single 29 kDa CP. The virus is readily transmitted horizontally between individuals and is serologically unrelated to *Rhopalosiphum padi virus* and *Aphid lethal paralysis virus*.

Allen, M.F. and Ball, B.V. (1990). Purification, characterization and some properties of a virus from the aphid *Sitiobion avenae*. *J. Invert. Pathol.*, **55**, 162-168.

Slow bee paralysis virus (SBPV)

The virus was first isolated from adult honey bees (Hymenoptera: Apidae) in Britain. Although capable of causing paralysis after injection into adult honey bees, it does not appear to be associated with any natural disease symptoms. The virions are 30 nm in diameter and contain ssRNA. The virus has three major CPs of 46, 29 and 27 kDa.

Bailey, L. and Ball, B.V. (eds)(1991). *Honey bee pathology*, 2nd ed. Harcourt Brace Jovanovich, Sidcup, UK.

TURNACA RUFISQUAMATA VIRUS (TRV)

The virus was isolated from larvae of the palm defoliating insect *Turnaca rufisquamata (Lepidoptera: Notodontidae)* collected during a disease epizootic in the Ivory Coast. The 30 nm diameter virions are reported to contain four proteins of 41, 33, 30, and 12 kDa. The genome comprises a single segment of ssRNA of 9.4 kb.

Fédière, G., Lery, X., Kouassi, K.N., Herder, S., Veyrunes, J.C. and Mariau, D. (1992). Etude d'un nouveau virus à ARN isolé de *Turnaca rufisquamata* (Lepidoptera, *Notodontidae*) defoliateur du palmier à huile en Côte d'Ivoire. *Oleagineux*, **47**, 107-112.

PROKARYOTE VIRUSES

THERMOPROTEUS TENAX VIRUS 4 (TTV4)

The virus TTV4, was found in a fresh sample from the Krafla volcano in Iceland. It lysed all enrichment cultures, which otherwise contained *Thermoproteus tenax*. TTV4 virions are stiff rods with thin tail fibers parallel to the virion. TTV4 appears to be one of the most resistant viruses known, as it is stable and infectious after 1 hour of autoclaving at 120 °C. TTV4 is also particularly virulent and spreads easily, complicating its study.

Arnold, H.P., Stedman, K.M. and Zillig, W. (1999). Archaeal Phages, in: *Encyclopedia of Virology*, 2nd edn, (A. Granoff and R.G. Webster, eds), 76-89. Academic Press, London.

Zillig, W., Reiter, W.D., Palm, P., Gropp, F., Neumann, H. and Rettenberger, M. (1988). Viruses of Archaebacteria, in: *The Bacteriophages*, (R. Calendar, ed), 517-558. Plenum Press, New York.

FUNGI VIRUSES

AGARICUS BISPORUS VIRUS 1 (ABV-1)

Particles are isometric, about 25 nm in diameter and sediment at 90-100 S. The single 25 kDa CP encapsidates two dsRNA species of about 2 kb.

Barton, R.J. and Hollings M (1979). Purification and some properties of two viruses infecting the cultivated mushroom *Agaricus bisporus*. *J. Gen. Virol.*, **42**, 231-240.

ALLOMYCES ARBUSCULA VIRUS (AAV)

Particles are isometric, about 40 nm in diameter and sediment as 67S and 75S components. Particles consist of proteins, with size of 38, 34, 28 and 21 kDa and dsRNA of 3.6 kbp, 2 kbp and 1.6 kbp.

Khandjian, E.W., Turian, G. and Eisen, H. (1977). Characterization of the RNA mycovirus infecting *Allomyces arbuscula*. *J. Gen. Virol.*, **35**, 415-424.

ASPERGILLUS FOETIDUS VIRUS F (AFV-F)

Particles are isometric, 40-42 nm in diameter and sediment as 164S and 145S components. Particles contain a major 87 kDa protein and minor species of 125 and 100 kDa. The dsRNA are 3.8 kbp, 2.7 kbp, 2.5 kbp, 2.1 kbp and 1.8 kbp.

Buck, K.W. and Ratti, G. (1975). Biophysical and biochemical properties of two viruses isolated from *Aspergillus foetidus*. *J. Gen. Virol.*, **27**, 211-224.

BOTRYTIS CINEREA VIRUS-CVG25 (BCV-CVG25)

Particles are isometric and approximately 40 nm in diameter. A single dsRNA of 8.3 kb is contained in the particles.

Vilches, S. and Castillo, A. (1997). A double-stranded RNA virus in *Botrytis cinerea*. *FEMS Microbiol. Let.*, **155**, 125-130.

Botrytis cinerea virus F (BCV-F)

Flexuous rod-shaped particles with modal length of 720 nm are composed of a CP of 32 kDa and one ssRNA of 6,827 nt. Two ORFs were identified from deduced translations: one encodes the putative replicase, the other the CP. Both showed considerable similarity to the genus *Potexvirus* and related genera of flexuous-rod-shaped plant-infecting viruses.

Howitt, R.L., Beever, R.E., Pearson, M.N. and Forster, R.L. (2001). Genome characterization of Botrytis virus F, a flexuous rod-shaped mycovirus resembling plant 'potex-like' viruses. *J. Gen. Virol.*, **82**, 67-78.

Colletotrichum lindemuthianum virus (CLV)

Particles are isometric, 30 nm in diameter and sediment as 110S and 85S components. Particles contain a major 52 kDa protein and a minor species of 45 kDa. The dsRNAs are 3.6 kbp, 1.6 kbp and 1.5 kbp.

Rawlinson, C.J., Carpenter, J.M. and Muthyalu, G. (1975). Double-stranded RNA virus in *Colletotrichum lindemuthianum*. *Trans. Brit. Mycol. Soc.*, **65**, 305-341.

Diaporthe RNA virus (DRV)

No particles are known to be associated with this virus, a property that is common among fungal viruses. A genome size of 4,113 nt was deduced from the sequence of cDNA clones made from the replicative form of the RNA. Two ORFs were deduced from the sense strand. The larger of the two ORFs encodes a putative RdRp that shows greatest similarity to those of viruses in the family *Tombusviridae*. Transcripts representing the coding strand from full-length cDNA clones were infectious when inoculated to fungal protoplasts and reproduced the natural disease in the fungus upon regeneration of colonies.

Moleleki, N., van Heerden, S.W., Wingfield, M.J., Wingfield, B.D. and Preisig, O. (2003). Transfection of *Diaporthe perjuncta* with Diaporthe RNA Virus. *Appl. Environ. Microbiol.*, **69**, 3952-3956.
Preisig, O., Moleleki, N., Smit, W.A., Wingfield, B.D. and Wingfield, M.J. (2000). A novel RNA mycovirus in a hypovirulent isolate of the plant pathogen *Diaporthe ambigua*. *J. Gen. Virol.*, **81**, 3107-3114.

Fusarium graminearum virus-DK21 (FGV-DK21)

No particles are associated with this virus. DsRNA of approximately 7.5 kb was consistently associated with altered fungal morphology in culture and reduced virulence. The dsRNA was transmissible by anastomosis, resulting in transmission of phenotypic traits. Partial nt sequence of the dsRNA reveals similarity with members of the fungal virus family *Hypoviridae*, known members of which have genome sizes of 9-13 kb.

Chu, Y-L., Jeon, J.-J., Yea, S.-J., Kim, Y-H., Yun, S.-H., Lee, Y.-W. and Kim, K.-H. (2002). Double-stranded RNA mycovirus from *Fusarium graminearum*. *Appl. Environ. Microbiol.*, **68**, 2529-2534.

Gaeumannomyces graminis virus 45/101-C (GGV-45/101C)

Particles are isometric, 29 nm in diameter and sediment at 127S. They consist of a 66 kDa protein and a ds RNA of 1.8 kbp.

Buck, K.W. (1984). A new dsRNA virus from *Gaeumannomyces graminis*. *J. Gen. Virol.*, **65**, 987-990.

Helminthosporium maydis virus (HMV)

Particles are isometric, 48 nm in diameter and sediment at 283S. Particles consist of a 121 kDa protein and ds RNA of 8.3 kbp.

Bozarth, R.F. (1977). Biophysical and biochemical characterization of virus-like particles containing a high molecular weight dsRNA from *Helminthosporium maydis*. *Virology*, **80**, 149-157.

LaFrance isometric virus (LFIV)

Particles are isometric, 36 nm in diameter and contain dsRNA species of 3.6 kbp, 3 kbp, 2.8 kbp, 2.7 kbp, 2.5 kbp, 1.6 kbp, 1.4 kbp, 0.9 kbp, and 0.8 kbp.

Goodin, M.M., Schlagnhaufer, B. and Romaine, C.P. (1992). Encapsidation of the LaFrance disease-specific dsRNAs in 36 nm isometric virus-like particles. *Phytopathol.*, **82**, 285-290.

Lentinus edodes virus (LEV)

Particles are isometric, 39 nm in diameter and contain 1 dsRNA of 6.5 kbp.

Ushiyama, R. and Nakai, Y. (1982). Ultrastructural features of virus-like particles from *Lentinus edodes*. *Virology*, **123**, 93-101.

Perconia circinata virus (PeCV)

Particles are isometric, 32 nm in diameter and sediment as 150S and 140S components. Particles contain dsRNA of 2.5 kbp, 2 kbp, 1.8 kbp, 1.6 kbp, 0.7 kbp and 0.6 kbp.

Dunkle, L.D. (1974). Double-stranded RNA mycovirus in *Perconia circinata*. *Physiol. Plant Pathol.*, **4**, 107-116.

Plant viruses

Black raspberry necrosis virus (BRNV)

Particles are isometric and c. 28 nm in diameter. They sediment as a 50S component that is not infective and probably does not contain nucleic acid, or a 130S component that is infective. The genome is thought to be ssRNA. BRNV is transmitted by raspberry aphids in nature and by other species experimentally. It is not seed-transmitted.

Jones, A.T. (1988). Black raspberry necrosis virus. CMI/AAB Descriptions of Plant Viruses, **333**.

Brachypodium yellow streak virus (BYSV)

Particles are isometric, c. 32 nm in diameter, sediment at about 122S and have a density in CsCl solutions of about 1.35g/cm^3. They comprise a single ssRNA component with an Mr of about 1.67 x 10^6 and a CP of c. 38 kDa.

Edwards, M.-L., Cooper, J.I., Massalskì, P.R. and Green, B. (1985). Some properties of a virus-like agent found in *Brachypodium sylvaticum* in the United Kingdom. *Plant Pathology* **34**, 95-104.

Cassava Ivorian bacilliform virus (CsIBV)

Particles are bacilliform, 18 nm in diameter and 42 nm, 49 nm and 76 nm in length, and sediment as three components. Particles contain three RNA components that, if ssRNA, would be c. 2.7 kb, 3 kb and 4 kb in size, and protein with a size of 22 kDa. No vector is known. A tentative, though distant, relationship has been proposed between CsIBV and the genus *Alfamovirus*.

Fargette, D., Roberts, I.M. and Harrison, B.D. (1991). Particle purification and properties of cassava Ivorian bacilliform virus. *Ann. Appl. Biol.*, **119**, 303 - 312.
Fargette, D. and Harrison, B.D. (1998). Cassava Ivorian Bacilliform Virus. AAB Descriptions of Plant Viruses **361**.

Chara australis virus (CAV; formerly Chara corallina virus)

The virions of CAV, like those of tobamoviruses, are rod-shaped and 17nm in diameter, but are 532 nm long. The ssRNA genome is of about 10,000 nt. The virion protein has a size of 18,5 kDa, is serologically distantly related to those of tobamoviruses, and its gene is clearly homologous with the virion protein genes of tobamoviruses. Some other parts of the genome show clear sequence homology with parts of the *Beet necrotic yellow vein virus* and *Hepatitis virus E* genomes (Keese, P., Mackenzie, A., Torronen, M. and Gibbs, A.J., pers. com.). The vector has not been identified.

Skotnicki, A., Gibbs, A. and Wrigley, N.G. (1976). Further studies on Chara corallina virus. *Virology* **75**, 457-468.

Flame chlorosis virus (FlCV)

No particles have been detected in diseased tissue. The genome is thought to be dsRNA and the probable vector is *Pythium* sp.

Haber, S., Kim, W., Gillespie, R. and Tekauz, A. (1990). Flame chlorosis: a new, soil-transmitted virus-like disease in barley in Manitoba. *J. Phytopath.*, **129**, 245-256

Haber, S., Barr, D.J.S., Chong, J., Tekauz, A. and Kim, W. (1993). Flame chlorosis, a novel, soil-transmited, viruslike disease of spring cereals and wild grasses is linked to *Pythium* spp. Proceedings of the 2nd International working group on plant viruses with fungal vectors (C. Hiruki, ed), pp.115-118.

Hart's tongue fern mottle virus (HTFMoV)

Particles are rod-shaped, 22 nm in diameter and either 135 nm or 320 nm in length. The genome is probably of ssRNA. The virus is probably soil-transmitted.

Hull, R. (1968). A virus disease of Hart's tongue fern. *Virology*, **35**, 333-335.

Hawaiian rubus leaf curl virus (HRLCV)

Phloem sieve elements in leaves of symptomless Himalaya Giant blackberry and of symptomatic *Rubus macraei* graft inoculated with scions from this blackberry, contained highly flexuous virus-like particles c. 8–10nm in diameter of indeterminate length. They had a beaded structure with an apparent periodicity of 10-11nm unlike that of known plant virus particles. The virus is probably not mechanically transmissible, and is thought to be transmitted in a persistent or semi-persistent manner. HRLCV was recently recognized as a species but is not assigned to any existing genus.

Jones, A.T., Roberts, I.M., McGavin, W.J. (2004). Unusual filamentous virus-like particles in symptomless blackberry associated with severe leaf curling in the virus indicator *Rubus macraei*, and anomalous data using a degenerate badnavirus primer in PCR of Rubus species. *Ann. Appl. Biol.*, **144**, 87-94.

Maize white line mosaic virus (MWLMV)

Particles are isometric about 35 nm in diameter and contain ssRNA of about 4 kb. CP has an size of about 33 kDa. Virus is soil-borne; transmission may be by fungi but is not possible mechanically.

de Zoeten, G.A. and Reddick, B.B. (1984). Maize white line mosaic virus. CMI/AAB Descriptions of Plant Viruses, **283**, 4pp.

Nicotiana velutina mosaic virus (NVMV)

Particles are rod-shaped, c. 18 nm in diameter and have a modal length of about 125 to 150 nm. Particles contain ssRNA molecules of either 3 kb or 8 kb. The smaller RNA contains a triple gene block that encodes proteins somewhat similar to the translation products of analogous genes in some definitively classifed viruses. The virus is mechanically transmissible, but no vector is known. NVMV has been linked to furoviruses in the past but may be no more like these viruses than viruses in other genera.

Randles, J.R. (1978). Nicotiana velutina mosaic virus. CMI/AAB Descriptions of Plant Viruses, **189**, 4pp.
Randles, J.R. and Rohde, W. (1990). Nicotiana velutina mosaic virus: evidence for a bipartite genome comprising 3 kb and 8 kb RNAs. *J. Gen. Virol.*, **71**, 1019-1027.

Orchid fleck virus (OFV)

Preparations of purified virus contain bacilliform and non-enveloped particles of 150 x 40 nm diameter. Enveloped particles are sometimes seen in infected plant tissues. The genome comprises 2 molecules of 6,413 nt (RNA-1) and 6,001 nt (RNA-2). The RNA have conserved and complementary terminal sequences. RNA-1 contains 5 ORFs, and RNA-2 encodes a single ORF. Some of the encoded proteins have sequences similar to those of proteins of rhabdoviruses. OFV was previously classified as a tentative rhabdovirus, but differs from viruses in the family *Rhabdoviridae* in having a bipartite genome.

Blanchfield, A.L., Mackenzie, A.M., Gibbs, A., Kondo, H., Tamada, T. and Wilson, C.R. (2001). Identification of Orchid fleck virus by reverse transcription-polymerase chain reaction and analysis of isolate relationships. *J. Phytopathol.*, **149**, 713-718.

Parsley latent virus (PaLV)

Particles are isometric and are c.27 nm in diameter. They sediment at about 128S and have a buoyant density in CsCl solutions of about 1.45 g/cm^3. Particles contain about 36% RNA and have a CP with a size of about 22 kDa. The virus is mechanically transmissible.

Bos, L., Huttinga, H. and Maat, D.Z. (1979). Parsley latent virus, a new and prevalent seed-transmitted, possibly harmless virus of *Petroselinum crispum*. *Neth. J. Pl. Pathol.*, **85**, 125-136.

Pelargonium zonate spot virus (PZSV)

Particles are quasi-isometric 25-35 nm in diameter, and sediment as three components. The genome comprises RNA-1 of 3,383 nt, RNA-2 of 2,435 nt and RNA-3 of 2,659 nt. RNA-1 encodes a 108,4 kDa polypeptide (1a) that contains the conserved sequence motifs I-III of type I Mtr and the seven consensus motifs of the Hel of superfamily 1. RNA-2 encodes a 78,9 kDa polypeptide (2a), which resembles an RdRp and RNA-3 contains ORFs for a putative MP and a 44 kDa CP, which is expressed via a sgRNA, RNA-4. The virus is readily transmitted by sap inoculation. Natural transmission is by thrips. The organization of the PZSV genome strongly suggests that this virus belongs to the family *Bromoviridae* and a taxonomic proposal for this classification is being prepared.

Finetti-Sialer, M. and Gallitelli, D. (2003). Complete nucleotide sequence of Pelargonium zonate spot virus and its relationship with the family *Bromoviridae*. *J. Gen. Virol.*, **84**, 3143-51.

Pigeon pea sterility mosaic virus (PPSMV)

Particles are highly flexuous, 3 to 10 nm in diameter and of indeterminate length. The CP is about 32 kDa and encapsidates 5 to 7 RNAs of between 1.1 and 6.8 kb. Sequence of cDNA is unlike any in databases. The virus is transmitted by eriophyid mites. Recently PPSMV was recognized as a species but it is not assigned to any existing genus.

Kulkani, N.K., Kumar, P.L., Muniyappa, V., Jones, A.T. and Reddy, D.V.R. (2002). Tranmission of Pigeonpea sterility mosaic virus by the Eriophyid mite *Aceria cajani* (Acari: Arthropoda). *Ann. Appl. Biol.*, **140**, 87-96.

Kumar, P.L., Duncan, G.H., Roberts, I.M., Jones, A.T. and Reddy, D.V.R. (2002). Cytology of pigeonpea sterility mosaic virus in pigeonpea and *Nicotiana benthamiana*: similarities with those of eriophyid mite-borne agents of undefined etiology. *Phytopathology*, **93**, 71-81.

TULIP STREAK VIRUS (TSTV)

Particles are filamentous 300-1600 x 7-11 nm and resemble those of tenuiviruses rather than ophioviruses. The virus is sap-transmissible and soil-borne, possibly being transmitted by an *Olpidium* sp. Purification is very difficult. Partial sequence data for the presumed CP show no particular similarity with known CP sequences.

Morikawa, T., Taga, Y. and Natsuaki, T. (2004). Tulip streak virus, a novel virus detected from tulip with streaking. *Jap. J. Phytopathol.*, **170**, (in press).

WATERCRESS YELLOW SPOT VIRUS (WYSV)

Particles are isometric and c. 37 nm in diameter. They comprise a 43.5 kDa protein and ssRNA with an Mr of about 1.6 x 10^6. Particles are unstable and show no serological relationship to carmoviruses, dianthoviruses or tombusviruses. A possible vector is *Spongospora subterranea* f. sp. *nasturtii*.

Walsh, J.A. and Clay, C.M. (1993). A summary of research on watercress yellow spot and its fungal vector *Spongospora subterranea* f. sp. *nasturtii*. *Proceedings of the 2nd International working group on plant viruses with fungal vectors* (C. Hiruki, ed), pp 111-114.

WHITE CLOVER VIRUS L (WCLVL)

The virus is mechanically transmissible but no virus particles are detectable. Infective RNA is single-stranded and lacks detectable poly(A). dsRNA of 6.8, 3.5 and 3.3 kbp were detected in infected plants and the RNA was infective after heating.

Gardiner T.J., Pearson M.N., Hopcroft D.H. and Forster R.L.S. (1995). Characterization of a labile RNA virus-like agent from white clover. *Ann. Appl. Biol.*, **126**, 91-104.

CONTRIBUTED BY

Mayo, M.A., Christian, P.D., Hillman, B.I., Brunt, A.A. and Desselberger, U.

The Subviral Agents

Subviral Agents: Viroids

Definition

Viroids are non-encapsidated, small, circular, single-stranded RNAs that replicate autonomously when inoculated into host plants. Some are pathogenic, but others replicate without eliciting symptoms.

Taxonomic Structure of Viroids

Family	*Pospiviroidae*	
Genus		*Pospiviroid*
Genus		*Hostuviroid*
Genus		*Cocadviroid*
Genus		*Apscaviroid*
Genus		*Coleviroid*
Family	*Avsunviroidae*	
Genus		*Avsunviroid*
Genus		*Pelamoviroid*

Physical Chemical and Physical Properties

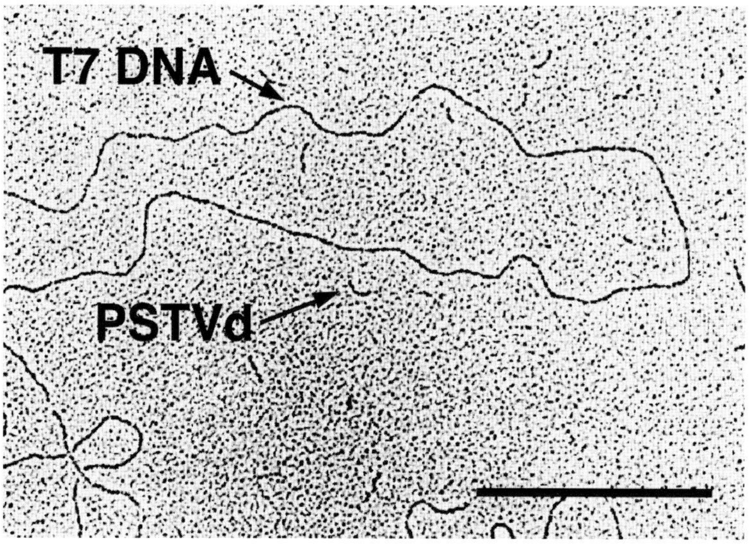

Figure 1: Electron micrograph of Potato spindle tuber viroid (PSTVd) and viral DNA (Coliphage T7 DNA) illustrating the comparative sizes and the rod-like structure of the viroid. The bar represents 0.5 μm (Reproduced from Diener, 1979).

Viroid molecules, with a Mr of 80-125 x 10^3, display extensive internal base pairing to give in most cases, *in vitro* rod-like or quasi-rod-like conformations ~50 nm long for Potato spindle tuber viroid (PSTVd) and most other viroids (Fig. 1). There is indirect *in vivo* evidence supporting this type of structure for some viroids.

These structures denature by cooperative melting (T_m in 10 mM Na^+ ~50°C) to single-stranded circles of ~100 nm contour length. Viroids can also form metastable structures with hairpins that may be functionally important. However, at least two viroids, Peach latent mosaic viroid (PLMVd) and Chrysanthemum chlorotic mottle viroid (CChMVd), do not follow this rule and their RNAs clearly appear to adopt branched conformations. There is experimental support for the contention that PLMVd and CChMVd have unique conformations because unlike other viroids, they are insoluble in 2M LiCl. Viroids also contain elements of tertiary structure. Sequences vary from 246 to 399 nt in length and are rich in G+C (53-60%), except Avocado sunblotch viroid (ASBVd) which has 38% G+C.

Genome Organization and Replication

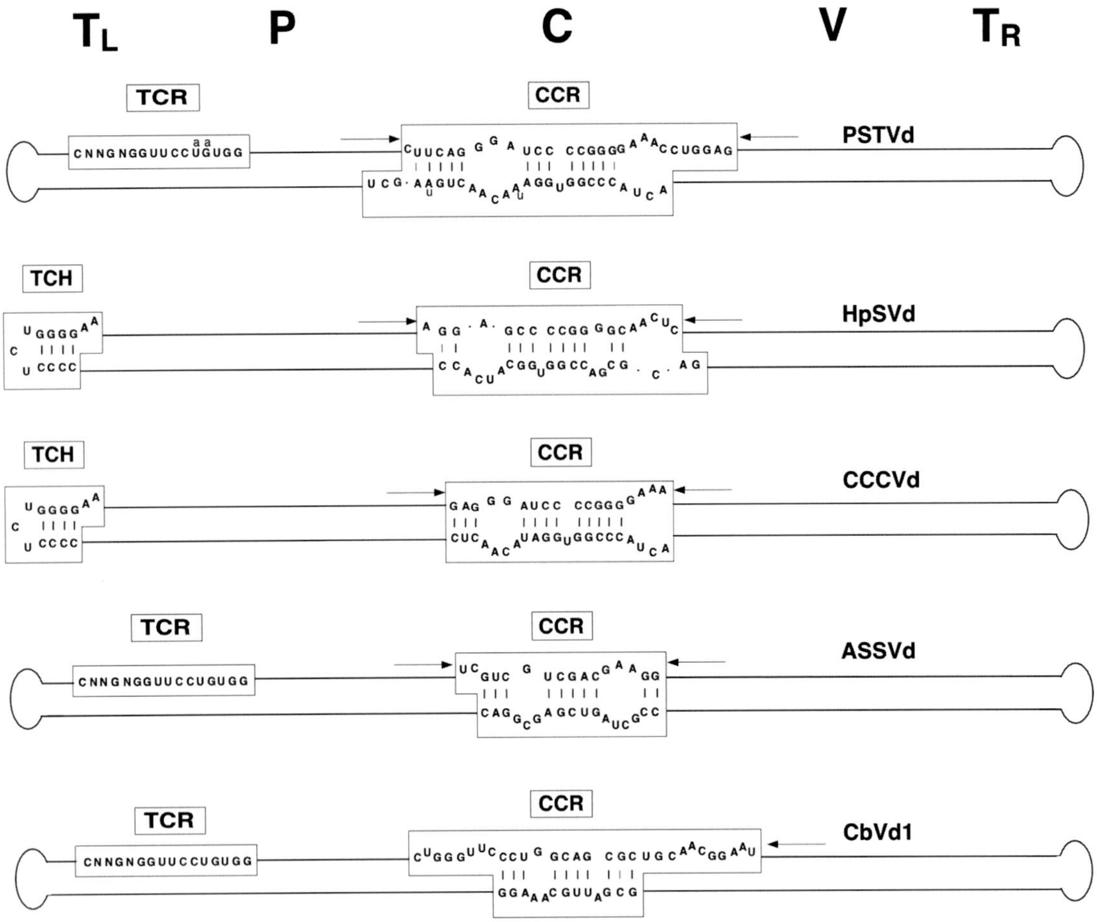

Figure 2: Rod-like structure models for viroids in the different genera of the family *Pospiviroidae*. A representative of the type species of each genus is indicated on the right and the approximate location of the five domains C (central), P (pathogenic), V (variable) and T_L and T_R (terminal left and right), on top of the figure. The core nt of the central conserved region (CCR), terminal conserved region (TCR) and terminal conserved hairpin (TCH), are shown. Arrows indicate flanking sequences which form, together with the core nt of the CCR upper strand, imperfect inverted repeats. The core nt of the Hop stunt viroid (HpSVd) CCR have been defined by comparison with Columnea latent viroid (CLVd), which on the basis of other criteria (presence of the TCR, absence of the TCH, and overall sequence similarity) is included in the genus *Pospiviroid*. Substitutions found in the CCR and TCR of Iresine viroid 1 (IrVd-1) a new member of the genus *Pospiviroid*, are indicated in lowercase. In the genus *Coleviroid* the TCR only exists in the two largest members Coleus blumei viroid 2 (CbVd-2) and Coleus blumei viroid 2 (CbVd-3) (Adapted from Flores *et al.*, 1997).

All viroids, except ASBVd, PLMVd and CChMVd, share a model of five structural functional domains within the proposed rod-like secondary structure of minimal free energy: C (central), P (pathogenic), V (variable), and T_R and T_L (terminal right and left). The C domain contains a central conserved region (CCR) formed by two sets of conserved nt located in the upper and lower strands, with those of the upper strand being flanked by an inverted repeat (Fig. 2).

The upper strand of the CCR can form either a hairpin or, in oligomers, a palindromic structure possibly relevant in replication. Two other conserved sequence motifs are the terminal conserved region (TCR), found in all members of the genera *Pospiviroid* and *Apscaviroid* and in the two largest members of the genus *Coleviroid*, and a terminal conserved hairpin (TCH) present in all members of the genera *Hostuviroid* and *Cocadviroid* (Fig. 2). The conservation of these two motifs in sequence and location in the left terminal domain suggests their involvement in some critical functions. Coconut cadang-cadang viroid (CCCVd) is unusual in occurring as RNAs of different sizes, the larger ones having sequence repetitions of the smallest one; a similar situation has been found in Citrus exocortis viroid (CEVd). Some viroids, the most striking examples being Columnea latent viroid (CLVd) and Australian grapevine viroid (AGV), appear to have arisen by intermolecular RNA recombination events since they seem to consist of a more or less complex mosaic of sequences present in other viroids. Direct evidence in support of an ongoing recombination process has come from studies on the dynamic distribution over time of the three *Coleus* viroids coinfecting individual plants.

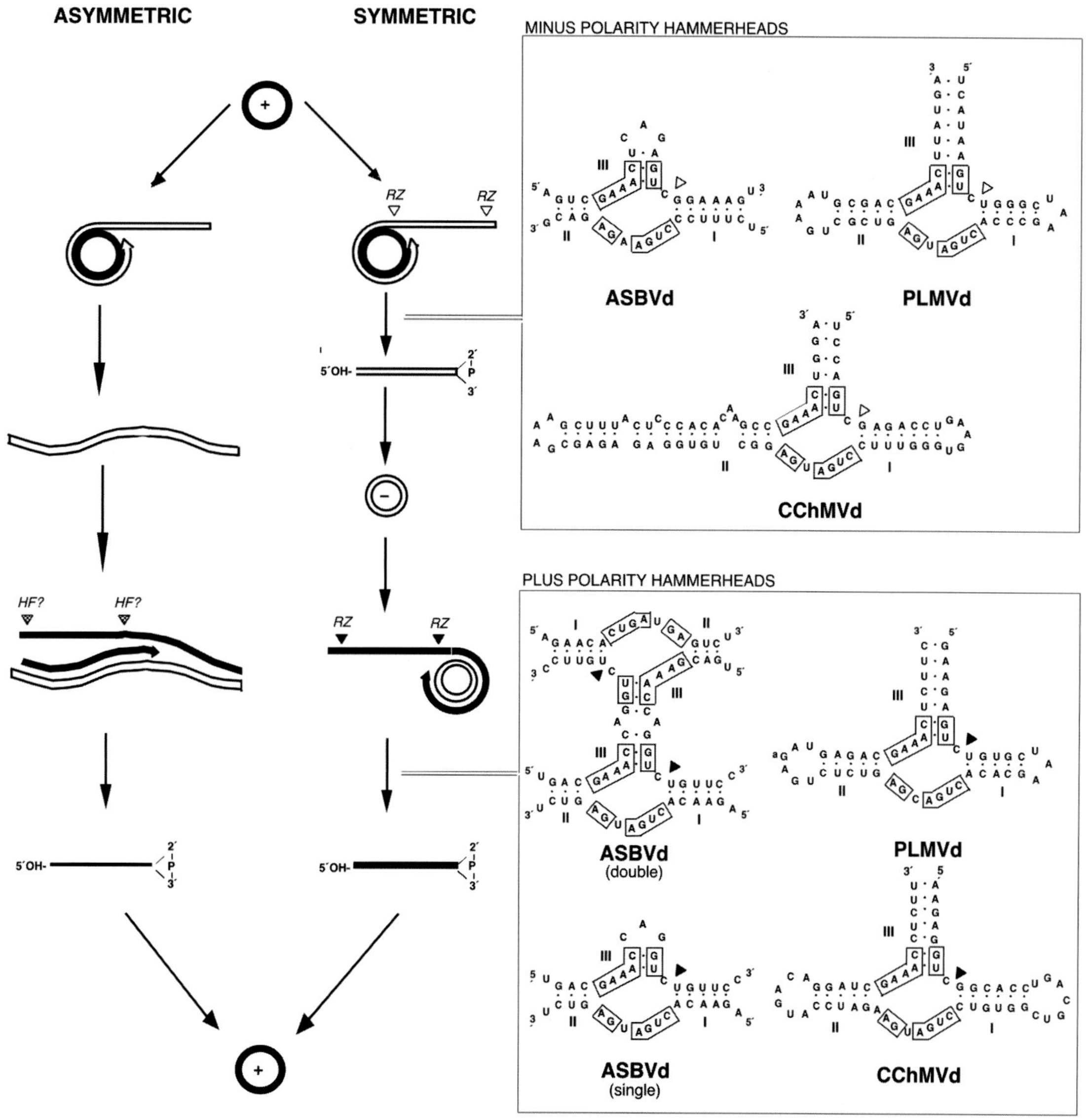

Figure 3: Models for viroid replication. The plus polarity (solid lines) is assigned by convention to the most abundant infectious RNA and the minus polarity (open lines) to its complementary strand. The alternative asymmetric and symmetric pathways involve one and two rolling circles, respectively. In the symmetric variant, cleavage of plus and minus multimeric strands is mediated by hammerhead ribozymes (RZ), which lead to linear monomeric RNAs with 5'-hydroxyl and 2'-3'-cyclic phosphate termini. Arrowheads denote the

cleavage sites. The hammerhead structures that can be formed by Avocado sunblotch viroid (ASBVd), Peach latent mosaic viroid (PLMVd) and Chrysanthemum chlorotic mottle viroid (CChMVd) RNAs are shown on the right; conserved nucleotides are boxed. In the asymmetric variant, cleavage of plus multimeric strands may rely on a host factor (HF), which generates linear monomeric RNA containing probably 5'-hydroxyl and 2'-3'-cyclic phosphate termini (Adapted from Flores *et al.*, 1997).

There is no evidence that viroids encode proteins and unlike viruses that primarily parasitize the host translation machinery, viroids mainly parasitize host transcription, possibly by using the nuclear RNA polymerase II (family *Pospiviroidae*), or a chloroplastic RNA polymerase (family *Avsunviroidae*), which are subverted to accept RNA templates. Oligomers isolated from infected tissue may be replicative intermediates produced by a rolling circle mechanism with two variants (asymmetric and symmetric) and three steps (RNA polymerization, cleavage and ligation). Initiation of RNA synthesis in PSTVd and ASBVd appears to occur at specific sites, suggesting that it is promoter-driven. ASBVd, PLMVd and CChMVd self-cleave *in vitro* and very probably *in vivo* through hammerhead structures to produce unit length strands but others do not, and may rely on host factors for cleavage, although self-cleavage through a different class of ribozymes can not be excluded (Fig. 3). Ligation appears to be mediated by a nuclear RNA ligase in PSTVd, whereas this reaction has been proposed to be autocatalytic for PLMVd leading to unusual 2′, 5′-phosphodiester bonds at the ligation sites; alternatively, a chloroplastic RNA ligase might mediate this step.

Antigenic Properties

No antigenicity demonstrated.

Biological Properties

Host Range

Some viroids have wide host ranges in the angiosperms but others, particularly members of the family *Avsunviroidae*, have narrow host ranges. CCCVd and Coconut tinangaja viroid (CTiVd) infect monocotyledons. Old cultivars of grapevine and Citrus can harbor at least five different viroids. A single nucleotide substitution converts PSTVd from a non-infectious RNA to one that is infectious for *Nicotiana tabacum*.

Symptoms

Some viroids have devastating effects, for example, CCCVd has killed millions of coconut palms in the Philippines, while others cause epinasty, rugosity, chlorosis and necrosis on leaves, internode shortening of stems leading to stunted plants, bark cracking, deformation and color alterations of fruits and storage organs, and delays in foliation, flowering and ripening. A few viroids induce only mild or no symptoms. Symptom expression is generally more pronounced at high temperature and light intensities. Small sequence changes can alter a severe strain into a symptomless strain, or *vice versa*.

Transmission

Viroids of horticultural species are transmitted mainly by vegetative propagation. In plants propagated via seeds, they may be transmitted mechanically or through seed or pollen. Only Tomato planta macho viroid (TPMVd) is known to be efficiently transmitted by aphids.

Movement

To invade plants systematically, viroids must be able to move from the initially infected cells to the surrounding ones and then to the vascular system. Three different types of movement can be considered: intracellular, cell-to-cell and long-distance. The nuclear import of PSTVd is a cytoskeleton-independent process mediated by a specific and saturable receptor, but how members of the family *Avsunviroidae* are transported into chloroplasts is unknown. PSTVd seems to possess a specific sequence or structural motif for rapid cell-to-cell movement through plasmodesmata. Long-distance movement occurs via the phloem, a process in which PSTVd levels appears to be maintained by replication in the phloem parenchyma cells and governed by plant developmental and cellular factors. *In vitro*, Hop stunt viroid (HpSVd) can form a ribonucleoprotein complex with the phloem protein 2, a dimeric widespread lectin able to move from cell-to-cell through plasmodesmata and toward sink tissues in the assimilate stream. These properties, together with its RNA-binding ability, suggest that this lectin facilitates the systemic movement of viroids.

Cross Protection

Interactions at the level of symptom expression and viroid accumulation have been detected in plants co-infected by two strains of a viroid or even by two different viroids sharing extensive sequence similarities. Interactions of this class have been observed in the case of members belonging to both viroid families (see below), suggesting the possible existence of more than one mechanism of cross protection between viroids.

Phylogenetic Relationships

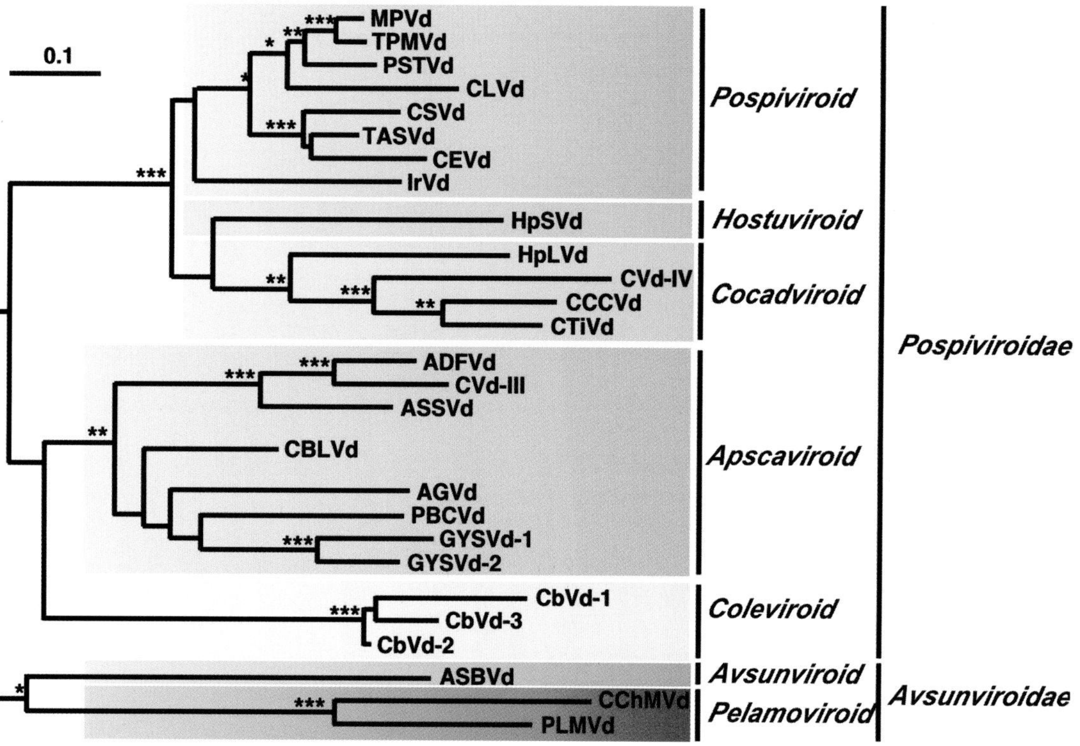

Figure 4: Consensus phylogenetic tree (based on 1000 replicates) obtained for viroids. ***, **, and *, group monophyletic in more than 95%, 85% and 75% of the replicates, respectively. Numbering of sequences was based on the similarities found between the upper strands of the central conserved regions of members of the family *Pospiviroidae* and the upper-left-hand portions of the hammerhead structures of members of the family *Avsunviroidae* (Diener, 1991). The alignment was performed with the Clustal X program, version 1.64 (gap opening and gap extension penalties 15 and 1, respectively) and minor manual adjustments between the conserved regions. Evolutionary distances were estimated according to the model of Jukes and Cantor and the phylogenetic tree was constructed by the Neighbor-Joining method using the MEGA program (version 1.01). (De la Peña and Flores, unpublished data).

Criteria such as the nuclear site of replication and accumulation, the presence and type of CCR, and the presence or absence of the two other conserved regions TCR and TCH, are used for classifying most viroids in family *Pospiviroidae*, with the two latter criteria serving for their allocation into the genera. ASBVd, PLMVd and CChMVd form a second family of viroids *Avsunviroidae* without CCR but endowed with self-cleavage through hammerhead ribozymes (Fig. 3). The type of hammerhead structure, together with the G+C content and the solubility in 2 M LiCl are used to demarcate species into genera. Essentially the same grouping is obtained by using phylogenetic trees derived from the whole sequences (Fig. 4). An arbitrary level of less than 90% sequence similarity and distinct biological properties, particularly host range, currently separates species within genera. Due to the heterogeneous nature of viroid populations it should be expected that a spectrum of closely related variants, having more than 90% sequence similarity, will co-exist within an infected plant, although one or a limited number of them may represent the bulk of the population.

FAMILY *POSPIVIROIDAE*

TAXONOMIC STRUCTURE OF THE FAMILY

Family *Pospiviroidae*
Genus *Pospiviroid*
Genus *Hostuviroid*
Genus *Cocadviroid*
Genus *Apscaviroid*
Genus *Coleviroid*

DISTINGUISHING FEATURES

Presence of a central conserved region (CCR) and lack of RNA self-cleavage mediated by hammerhead ribozymes. According to the conserved residues there are basically three types of CCRs exemplified by those of PSTVd, Apple scar skin viroid (ASSVd) and Coleus blumei viroid 1 (CbVd-1). Genera *Pospiviroid, Hostuviroid* and *Cocadviroid* share an identical subset of nucleotides within their CCRs, which are, therefore, more closely related between them than with the CCRs of members of the genera *Apscaviroid* and *Coleviroid* (Fig. 2). Replication studies have been done mostly with PSTVd and the inferred mechanism (see below) is presumed to operate in the other members of the family as well.

GENUS *POSPIVIROID*

Type Species ***Potato spindle tuber viroid***

DISTINGUISHING FEATURES

Circular ssRNAs between 356 and 375 nt (excluding duplications) depending on species and sequence variants. The most stable secondary structure is a rod-like or quasi-rod-like conformation with five domains, a central conserved region (CCR) and a terminal conserved region (TCR) (Fig. 2 and 5). Replication, which takes place in the nucleus where the viroid also accumulates, occurs by an asymmetric rolling-circle mechanism since longer-than-unit minus strands have been found in infected tissue (Fig. 3).

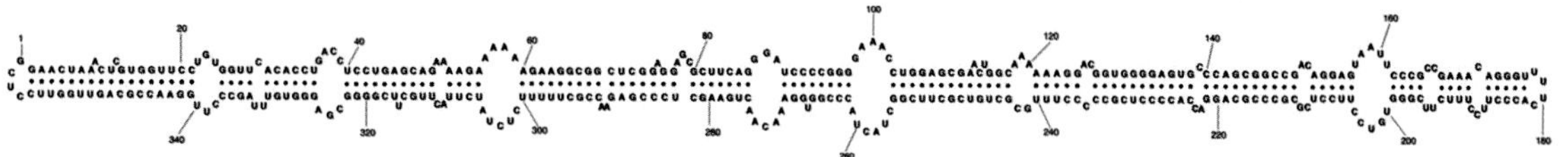

Figure 5: Primary and proposed rod-like secondary structure of lowest free energy for the reference sequence variant of Potato spindle tuber viroid (PSTVd) (Adapted with modifications from Gross *et al.*, 1978).

BIOLOGICAL PROPERTIES

Pospiviroids are found world-wide and in a wide range of plants, mostly solanaceous species, but also in other hosts including citrus, avocado, and some ornamentals. They can be transmitted experimentally to many other hosts in which symptom expression differs from symptomless to almost lethal. Some members, particularly TPMVd, can be transmitted by aphids under specific ecological conditions. PSTVd is also aphid-transmissible when encapsidated by Potato leafroll virus particles.

LIST OF SPECIES DEMARCATION CRITERIA IN THE GENUS

Having less than 90% sequence similarity and distinct biological properties, particularly host range and symptom expression, currently demarcate species within the genus.

List of Species in the Genus

Species names are in green italic script; strain names and synonyms are in black roman script; tentative species names are in blue roman script. Sequence accession numbers [ ], length in nt { }, and assigned abbreviations ( ) are also listed.

Species in the Genus

Chrysanthemum stunt viroid			
Chrysanthemum stunt viroid	[V01107]	{356}	(CSVd)
Chrysanthemum stunt viroid – Petunia	[U82445]	{355}	(CSVd-Pet)
Citrus exocortis viroid			
Citrus exocortis viroid	[M34917]	{371}	(CEVd)
Citrus exocortis viroid – broad bean	[S79831]	{373}	(CEVd-bro)
Citrus exocortis viroid - carrot			(CEVd-car)
Citrus exocortis viroid - eggplant			(CEVd-egg)
Citrus exocortis viroid - grapevine	[Y00328]	{369}	(CEVd-gra)
Citrus exocortis viroid – tomato (Indian tomato bunchy top viroid)	[X53716]	{372}	(CEVd-tom)
Citrus exocortis viroid - turnip			(CEVd-tur)
Columnea latent viroid			
Columnea latent viroid	[X15663]	{370}	(CLVd)
Columnea latent viroid – *Brunfelsia undulata*	[X95292]	{373}	(CLVd-bru)
Columnea latent viroid – *Nemathantus wettsteinii*	[M93686]	{372}	(CLVd-nem)
Iresine viroid 1			
Iresine viroid 1	[X95734]	{370}	(IrVd-1)
Mexican papita viroid			
Mexican papita viroid	[L78454]	{360}	(MPVd)
Potato spindle tuber viroid			
Potato spindle tuber viroid	[V01465]	{359}	(PSTVd)
Potato spindle tuber viroid - avocado			(PSTVd-avo)
Potato spindle tuber viroid - pepino			(PSTVd-pep)
Potato spindle tuber viroid – *Solanum spp*	[U51895]	{356}	(PSTVd-sol)
Tomato apical stunt viroid			
Tomato apical stunt viroid	[K00818]	{360}	(TASVd)
Tomato apical stunt viroid – *Solanum pseudocapsicum*	[X95293]	{360}	(TASVd-sol)
Tomato chlorotic dwarf viroid			
Tomato planta macho viroid	[AF162131]	{360}	(TCDVd)
Tomato planta macho viroid			
Tomato planta macho viroid	[K00817]	{360}	(TPMVd)

Tentative Species in the Genus

None reported.

Genus *Hostuviroid*

Type Species *Hop stunt viroid*

Distinguishing Features

A circular ssRNA of 295-303 nt depending on isolates and sequence variants. The most stable secondary structure is a rod-like or quasi-rod-like conformation with five domains, a central conserved region (CCR) similar to that of pospiviroids and cocadviroids, and a terminal conserved hairpin (TCH) (Fig. 2). Replication occurs through an asymmetric rolling-circle model since longer-than-unit minus strands have been found in infected tissue (Fig. 3).

BIOLOGICAL PROPERTIES

HpSVd infects a very broad range of natural hosts and has been reported to be the causal agent of five different diseases. It is distributed worldwide.

LIST OF SPECIES DEMARCATION CRITERIA IN THE GENUS

Not applicable.

LIST OF SPECIES IN THE GENUS

Species names are in green italic script; strain names and synonyms are in black roman script; tentative species names are in blue roman script. Sequence accession numbers [ ], length in nt { }, and assigned abbreviations ( ) are also listed.

SPECIES IN THE GENUS

Hop stunt viroid			
Hop stunt viroid	[X00009]	{297}	(HpSVd)
Hop stunt viroid - almond	[AJ011813]	{296}	(HpSVd-alm)
Hop stunt viroid - apricot	[AJ297840]	{297}	(HpSVd-apr)
Hop stunt viroid – citrus	[AF131249]	{299}	(HpSVd-cit)
(Citrus cachexia viroid)			
Hop stunt viroid – cucumber	[X00524]	{303}	(HpSVd-cuc)
(Cucumber pale fruit viroid)			
Hop stunt viroid - grapevine	[M35717]	{296}	(HpSVd-gra)
Hop stunt viroid – peach	[D13765]	{297}	(HpSVd-pea)
(Peach dapple viroid)			
Hop stunt viroid - pear			(HpSVd-pear)
Hop stunt viroid – plum	[D13764]	{297}	(HpSVd-plu)
(Plum dapple viroid)			

TENTATIVE SPECIES IN THE GENUS

None reported.

GENUS COCADVIROID

Type Species *Coconut cadang-cadang viroid*

DISTINGUISHING FEATURES

Circular ssRNAs of 246-301 nt depending on species and sequence variants. The most stable secondary structure is a rod-like or quasi-rod-like conformation with five domains, a central conserved region (CCR) similar to that of pospiviroids and hostuviroids, and a terminal conserved hairpin (TCH) (Fig. 2). The T_R domain is variable as a result of 41, 50, or 55 nt being duplications that generate variants between 287 and 301 nt. Dimeric forms of CCCVd occur with their corresponding monomeric forms. Replication most probably occurs through an asymmetric rolling-circle model by analogy with PSTVd (Fig. 3). At the subnuclear level, CCCVd is concentrated in the nucleolus with the remainder distributed throughout the nucleoplasm.

BIOLOGICAL PROPERTIES

CCCVd and CTiVd have a host range restricted to members of the family *Palmae* and a geographic distribution in the Philippines and some parts of Micronesia. As the disease progresses from the early to the medium stages the small forms are modified to the large ones that also have single or double cytosine residues at a specific locus and CVd-IV have broader geographic distributions.

List of Species Demarcation Criteria in the Genus

Having less than 90% sequence similarity and distinct biological properties, particularly host range and symptom expression, currently demarcate species within the genus.

List of Species in the Genus

Species names are in green italic script; strain names and synonyms are in black roman script; tentative species names are in blue roman script. Sequence accession numbers [ ], length in nt { }, and assigned abbreviations ( ) are also listed.

Species in the Genus

Citrus viroid IV			
Citrus viroid IV	[X14638]	{284}	(CVd-IV)
Coconut cadang-cadang viroid			
Coconut cadang-cadang viroid	[J02049]	{246}	(CCCVd)
Coconut tinangaja viroid			
Coconut tinangaja viroid	[M20731]	{254}	(CTiVd)
Hop latent viroid			
Hop latent viroid	[X07397]	{256}	(HpLVd)

Tentative Species in the Genus

None reported.

Genus Apscaviroid

Type Species ***Apple scar skin viroid***

Distinguishing Features

Circular ssRNAs between 306 and 369 nt depending on species and sequence variants. The most stable secondary structure is a rod-like or quasi-rod-like conformation with five domains, a central conserved region (CCR) and a terminal conserved region (TCR) (Fig. 2). Replication most probably occurs through an asymmetric rolling-circle model by analogy with PSTVd (Fig. 3).

Biological properties

Known hosts are confined to genera *Malus, Pyrus, Citrus* and *Vitis*. Reported worldwide, particularly those affecting the genera *Citrus* and *Vitis.*

List of Species Demarcation Criteria in the Genus

Having less than 90% sequence similarity and distinct biological properties, particularly host range and symptom expression, currently demarcate species within the genus.

List of Species in the Genus

Species names are in green italic script; strain names and synonyms are in black roman script; tentative species names are in blue roman script. Sequence accession numbers [ ], length in nt { }, and assigned abbreviations ( ) are also listed.

Species in the Genus

Apple dimple fruit viroid			
Apple dimple fruit viroid	[X99487]	{306}	(ADFVd)
Apple scar skin viroid			
Apple scar skin viroid	[M36646]	{329}	(ASSVd)
Apple scar skin viroid – dapple (Dapple apple viroid)	[X71599]	{331}	(ASSVd-dap)
Apple scar skin viroid – Japanase pear (Japanese pear fruit dimple viroid)			(ASSVd-jpf)

Apple scar skin viroid – pear rusty skin (Pear rusty skin viroid)			(ASSVd-prs)
Australian grapevine viroid			
Australian grapevine viroid	[X17101]	{369}	(AGVd)
Citrus bent leaf viroid			
Citrus bent leaf viroid	[M74065]	{318}	(CBLVd)
Citrus viroid III			
Citrus viroid III	[AF184147]	{294}	(CDVd)
Grapevine yellow speckle viroid 1			
Grapevine yellow speckle viroid 1	[X06904]	{367}	(GYSVd-1)
Grapevine yellow speckle viroid 2			
Grapevine yellow speckle viroid 2	[J04348]	{363}	(GYSVd-2)
Pear blister canker viroid			
Pear blister canker viroid	[D12823]	{315}	(PBCVd)

TENTATIVE SPECIES IN THE GENUS

Apple fruit crinkle viroid	[E29032]	{371}	(AFCVd)
Citrus viroid original source	[NC004359]	{330}	(CVd-OS)

GENUS COLEVIROID

Type Species ***Coleus blumei viroid 1***

DISTINGUISHING FEATURES

Circular ssRNAs between 248 and 361 nt depending on species and sequence variants. The most stable secondary structure is a rod-like or quasi-rod-like conformation with five domains and a central conserved region (CCR) different from that of other viroid genera, and a terminal conserved region (TCR) in the two largest members of the genus (Fig. 2). Replication most probably occurs through an asymmetric rolling-circle model by analogy with PSTVd (Fig. 3).

BIOLOGICAL PROPERTIES

Known hosts are confined to the *Coleus* genus. Identified in Brazil, Canada and Germany. Seed transmission in CbVd-1 (and presumably in CbVd-2 and CbVd-3) is particularly efficient.

LIST OF SPECIES DEMARCATION CRITERIA IN THE GENUS

Having less than 90% sequence similarity and distinct biological properties, particularly host range and symptom expression, currently demarcate species within the genus.

LIST OF SPECIES IN THE GENUS

Species names are in green italic script; strain names and synonyms are in black roman script; tentative species names are in blue roman script. Sequence accession numbers [ ], length in nt { }, and assigned abbreviations ( ) are also listed.

SPECIES IN THE GENUS

Coleus blumei viroid 1			
Coleus blumei viroid 1	[X52960]	{248, 250-251}	(CbVd-1)
Coleus blumei viroid 2			
Coleus blumei viroid 2	[X95365]	{301-302}	(CbVd-2)
Coleus blumei viroid 3			
Coleus blumei viroid 3	[X95364]	{361-362, 364}	(CbVd-3)

TENTATIVE SPECIES IN THE GENUS

None reported.

FAMILY *AVSUNVIROIDAE*

TAXONOMIC STRUCTURE OF THE FAMILY

Family	***Avsunviroidae***
Genus	***Avsunviroid***
Genus	***Pelamoviroid***

DISTINGUISHING FEATURES

Lack of a central conserved region (CCR). RNA self-cleavage is mediated by hammerhead ribozymes in strands of either polarity. Replication studies have been carried out with ASBVd and PLMVd, and the inferred mechanism (see below) is presumed to operate in the other members of the family.

GENUS *AVSUNVIROID*

Type Species *Avocado sunblotch viroid*

DISTINGUISHING FEATURES

A circular ssRNA between 246 and 250 nt depending on isolates and sequence variants. It is unique in having a base composition rich in A+U (62%) in contrast to the other viroids which are rich in G+C (53-60%). The most stable secondary structure is a rod-like or quasi-rod-like conformation in which neither five domains nor a central conserved region (CCR) can be distinguished. It is soluble in 2 M LiCl as are typical viroids like Potato spindle tuber viroid (PSTVd), Apple scar skin viroid (ASSVd) and Coleus blumei viroid 1 (CbVd-1) having also rod-like or quasi-rod-like most stable secondary structures. Plus and minus strands of Avocado sunblotch viroid (ASBVd) can form hammerhead structures. Because both single hammerhead structures of ASBVd are thermodynamically unstable, double hammerhead structures have been proposed to operate in the self-cleavage reactions, especially in that of the plus polarity RNA (Fig. 3). Replication occurs through a symmetric rolling-circle model since the minus circular monomer has been found in infected tissue (Fig. 3). ASBVd accumulates and replicates in chloroplasts.

BIOLOGICAL PROPERTIES

Found naturally only in avocado but it can be experimentally transmitted into other members of the Lauraceae family. Reported in USA, Australia, Israel, Spain, South Africa and South America.

LIST OF SPECIES DEMARCATION CRITERIA IN THE GENUS

Not applicable.

LIST OF SPECIES IN THE GENUS

Species names are in green italic script; strain names and synonyms are in black roman script; tentative species names are in blue roman script. Sequence accession numbers [ ], length in nt { }, and assigned abbreviations ( ) are also listed.

SPECIES IN THE GENUS

Avocado sunblotch viroid

Avocado sunblotch viroid	[J02020]	{247}	(ASBVd)

TENTATIVE SPECIES IN THE GENUS

None reported.

GENUS PELAMOVIROID

Type Species ***Peach latent mosaic viroid***

DISTINGUISHING FEATURES

Circular ssRNAs between 337 and 399 nt depending on isolates and sequence variants. The most stable secondary structure is a branched conformation in which neither five domains nor a central conserved region (CCR) can be distinguished (Fig. 6). PLMVd RNA is insoluble in 2 M LiCl. Thermodynamically stable single hammerhead structures can be formed in both strands, and plus and minus monomeric RNAs self-cleave *in vitro* as predicted by these structures (Fig. 3). Replication most probably occurs by a symmetric rolling-circle mechanism since the PLMVd minus circular monomers have been found in infected tissue (Fig. 3). PLMVd and most likely Chrysanthemum chlorotic mottle viroid (CChMVd), replicate and accumulate in the chloroplast.

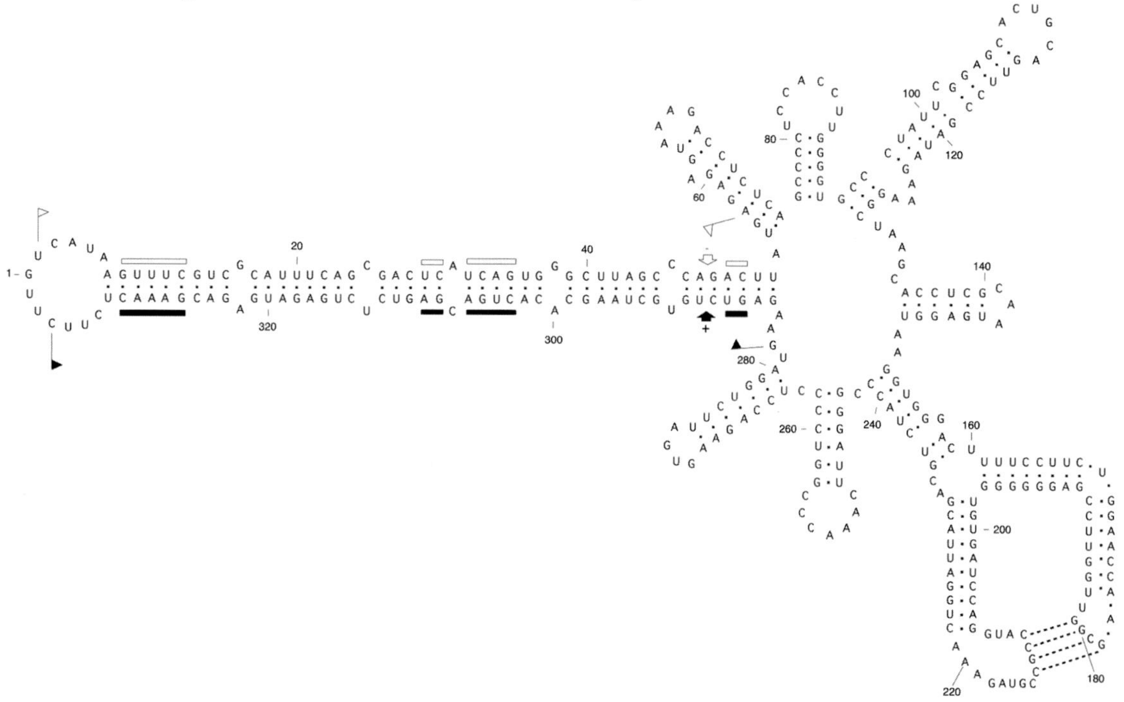

Figure 6: Primary and proposed branched secondary structure of lowest free energy for the reference sequence variant of an isolate of *Peach latent mosaic viroid*. Plus and minus self-cleavage domains are delimited by flags, the 13 conserved residues present in most of the hammerhead structures are indicated by bars, and the self-cleavage sites are shown by arrows. Solid and open symbols refer to plus and minus polarities respectively. *In vitro* chemical and enzymatic probing has revealed the probable existence of a pseudoknot between residues GCGG and CCGC (position 178-181 and 211-214 respectively) (Bussière *et al.*, 2000). (Adapted from Hernández and Flores (1992). *PNAS*, **89**, 3711-3715).

BIOLOGICAL PROPERTIES

PLMVd and CChMVd appear to be restricted to their natural hosts and very closely related species, but other reports indicate that PLMVd naturally infects additional species within and outside the genus *Prunus*. Recorded in many peach and chrysanthemum-growing areas.

LIST OF SPECIES DEMARCATION CRITERIA IN THE GENUS

An arbitrary level of less than 90% sequence similarity and distinct biological properties, particularly host range, currently separates species within genera.

List of Species in the Genus

Species names are in green italic script; strain names and synonyms are in black roman script; tentative species names are in blue roman script. Sequence accession numbers [ ], length in nt { }, and assigned abbreviations ( ) are also listed.

Species in the Genus

Chrysanthemum chlorotic mottle viroid			
Chrysanthemum chlorotic mottle viroid	[Y14700]	{399}	(CChMVd)
Peach latent mosaic viroid			
Peach latent mosaic viroid	[M83545]	{337}	(PLMVd)
Peach latent mosaic viroid - apple			(PLMVd-app)
Peach latent mosaic viroid - apricot			(PLMVd-apr)
Peach latent mosaic viroid - cherry	[AF339741]	{337}	(PLMVd-che)
Peach latent mosaic viroid - pear	[AF339740]	{338}	(PLMVd-pea)
Peach latent mosaic viroid - plum	[AF339742]	{338}	(PLMVd-plu)

Tentative Species in the Genus

None reported.

List of Unassigned Viroids in the Family

Blueberry mosaic viroid-like RNA	(BluMVd-RNA)
Burdock stunt viroid	(BuSVd)
Eggplant latent viroid	(ELVd)
Nicotiana glutinosa stunt viroid	(NGSVd)
Pigeon pea mosaic mottle viroid	(PPMMoVd)
Tomato bunchy top viroid	(ToBTVd)

Derivation of Names

Apsca: sigla from *ap*ple *sca*r skin viroid.
Avsun: sigla *av*ocado *sun*blotch viroid.
Cocad: sigla from *co*conut *cad*ang-cadang viroid.
Cole: sigla from *cole*us blumei viroid 1.
Hostu: sigla from *ho*p *stu*nt viroid.
Pelamo: sigla from *pe*ach *la*tent *mo*saic viroid.
Pospi: sigla from *po*tato *spi*ndle tuber viroid.
Viroid: from the name given to the subviral agent of potato spindle tuber disease.

References

Ben-Shaul, A., Guang, N., Mogilner, N., Hadas, M., Mawassi, M., Gafny, R. and Bar-Joseph, M. (1995). Genomic diversity among populations of two citrus viroids from different graft-transmissible dwarfing complexes in Israel. *Phytopathology*, **85**, 359-364.

Bonfiglioli, R.G., Webb, D.R. and Symons, R.H. (1996). Tissue and intra-cellular distribution of coconut cadang-cadang viroid and citrus exocortis viroid determined by in situ hybridization and confocal laser scanning and transmission electron microscopy. *Plant J.*, **9**, 457-465

Branch, A.D. and Robertson, H.D. (1984). A replication cycle for viroids and other small infectious RNAs. *Science*, **223**, 450-454.

Bussière, F., Ouellet, J., Côté, F., Lévesque, D. and Perreault, J.P. (2000). Mapping in solution shows the peach latent mosaic viroid to possess a new pseudoknot in a complex, branched secondary structure. *J. Virol.*, **74**, 2647-2654.

Diener, T.O. (1971). Potato spindle tuber "virus" IV. A replicating low molecular weight RNA. *Virology*, **45**, 411-428.

Ding, B., Kwon, M.O., Hammond, R. and Owens, R. (1997). Cell-to-cell movement of potato spindle tuber viroid. *Plant J.*, **12**, 931-936

Durán-Vila, N., Roistacher, C.N., Rivera-Bustamante, R. and Semancik, J.S. (1988). A definition of citrus viroid groups and their relationship to the citrus exocortis disease. *J. Gen. Virol.*, **69**, 3069-3080.

Elena, S.F., Dopazo, J., De la Peña, M., Flores, R., Diener, T.O. and Moya, A. (2001). Phylogenetic analysis of viroid and viroid-like satellite RNAs from plants: a reassessment. *J. Mol. Evol.*, **53**, 155-159.
Flores, R., Di Serio, F. and Hernández, C. (1997). Viroids: the non-coding genomes. *Semin. Virol.*, **8,** 65-73.
Flores, R., Daròs, J.A. and Hernández, C. (2000). The *Avsunviroidae* family: viroids containing hammerhead ribozymes. *Adv. Virus Res.*, **55**, 271-323.
Garnsey, S.M. and Randles, J.W. (1987). Biological interactions and agricultural implications of viroids; In: *Viroids and Viroid-like Pathogens,* (J.S. Semancik, ed), pp 127-160. CRC Press, Boca Raton, Florida.
Gross, H.J., Domdey, H., Lossow, C., Jank, P., Raba, M., Alberty, H. and Sänger, H.L. (1978). Nucleotide sequence and secondary structure of potato spindle tuber viroid. *Nature*, **273**, 203-208.
Hadidi, A., Flores, R., Randles, J.W. and Semancik, J.S. (eds)(2003). *Viroids*. CSIRO Publishing, Collingwood, Australia.
Hammond, R., Smith, D.R. and Diener, T.O. (1989). Nucleotide sequence and proposed secondary structure of Columnea latent viroid: a natural mosaic of viroid sequences. *Nuc. Acids Res.*, **17**, 10083-10094.
Hanold, D. and Randles, J.W. (1991). Coconut cadang-cadang disease and its viroid agent. *Plant Dis.*, **75**, 330-335.
Keese, P. and Symons, R.H. (1985). Domains in viroids: Evidence of intermolecular RNA rearrangements and their contribution to viroid evolution. *Proc. Natl. Acad. Sci. USA,* **82**, 4582-4586.
Koltunow, A.M. and Rezaian, M.A. (1989). A scheme for viroid classification. *Intervirology*, **30**, 194-201.
Pelchat, M., Rocheleau, L., Perreault, J. and Perreault, J.P. (2003). Subviral RNAs: a database of the smallest known auto-replicable RNA species. *Nuc Acids Res.*, **31**, 444-445.

Contributed By

Flores, R., Randles, J.W., Owens, R.A., Bar-Joseph M. and Diener, T.O.

SUBVIRAL

SATELLITES

DEFINITION

Satellites are sub-viral agents which lack genes that could encode functions needed for replication. Thus they depend for their multiplication on the co-infection of a host cell with a helper virus. Satellite genomes have a substantial portion or all of their nucleotide sequences distinct from those of the genomes of their helper virus.

There are two major classes of satellites: satellite viruses and satellite nucleic acids. Satellite viruses are characterized by having distinct nucleoprotein components that are present in preparations of particles of helper viruses. Satellite viruses have nucleic acid genomes that encode a structural protein that encapsidates the genome. Satellite nucleic acids either encode non-structural proteins, or no proteins at all, and are encapsidated by the CP of helper viruses.

A related class of agents is that of RNAs that resemble satellites but which encode a function necessary for the biological success of the associated virus. These may be considered as components that remedy a deficiency in a defective virus. These are denoted as satellite-like RNAs, that is they do not encode a replicase, but are classified as being part of the genome of the virus they assist. Examples are the RNAs associated with some umbraviruses, or with *Beet necrotic yellow vein virus*, that contribute to vector transmissibility. The distinction between satellite RNAs and satellite-like RNAs can in some instances be relatively slight.

DISTINGUISHING FEATURES

Satellites are genetically distinct from their helper virus by virtue of having a nucleotide sequence substantially different from that of their helper virus. However, the genomes of some satellites have short sequences, often at the termini, that are the same as those in the genome of the helper. This is presumably linked to the dependence of nucleic acids of both satellite and helper virus on the same viral enzymes and host-encoded proteins for replication. Satellites are thus distinct from defective interfering (DI) particles or defective RNAs because such RNAs are derived from their 'helper' virus genomes. An example of an intermediate state is the chimeric molecules formed from part of a satellite RNA associated with *Turnip crinkle virus* and part of the virus genome.

Satellites do not constitute a homogeneous taxonomic group. The descriptions in this section are meant only to provide a classification framework and a nomenclature to assist in the description and identification of satellites. The arrangement adopted is based largely on features of the genetic material of the satellites. The physicochemical and biological features of the helper virus and of the helper virus host are important secondary characters.

There appears to be no taxonomic correlation between the viruses that are associated with satellites. Satellites would appear to have arisen independently a number of times during virus evolution. A further complication is that some viruses are associated with more than one satellite and some satellites are supported by more than one species of helper virus. Satellites can even depend on both a second satellite and a helper virus for multiplication.

Most known satellites are ssRNA satellites, with ssRNA plant viruses as helpers. It can be very difficult to distinguish between satellite RNA and genome RNA (e.g., with the dsRNA satellites of fungus viruses) and it is very likely that other satellites, some with novel combinations of characters, remain to be discovered.

CATEGORIES OF SATELLITES

Satellite viruses
- *Chronic bee-paralysis virus-associated satellite virus*
- *Satellites that resemble Tobacco necrosis satellite virus*

Satellite nucleic acids
- *Single-stranded satellite DNAs*
- *Double-stranded satellite RNAs*
- *Single-stranded satellite RNAs*
  - *Large single-stranded satellite RNAs*
  - *Small linear single-stranded satellite RNAs*
  - *Circular single-stranded satellite RNAs*

SATELLITE VIRUSES

DISTINGUISHING FEATURES

This category comprises the satellites with ssRNA genomes encapsidated in satellite-encoded protein structures. Several types are known. In all cases, the satellite virus particles are antigenically, and usually morphologically, distinct from those of the helper virus.

Two subgroups of satellite viruses are distinguished: Chronic bee-paralysis virus associated satellite virus and satellites that resemble Tobacco necrosis satellite virus.

1: CHRONIC BEE-PARALYSIS ASSOCIATED SATELLITE VIRUS

DISTINGUISHING FEATURES

Satellite virus particles are found in bees infected with the helper, *Chronic bee-paralysis virus* (CBPV). Particles are about 12 nm in diameter and serologically unrelated to those of CBPV. Satellite RNA is also found encapsidated in CBPV capsid protein. The satellite RNA consists of 3 species, about 1 kb in size. These are distinct from CBPV RNA although T1 RNase digestion yields some oligonucleotides that appear to be the same in products from CBPV RNA and from satellite RNA. The satellite interferes with CBPV replication.

LIST OF MEMBERS

Chronic bee-paralysis satellite virus

REFERENCES

Overton, H.A., Buck, K.W., Bailey, L. and Ball, B.V. (1982). Relationships between the RNA components of Chronic bee-paralysis virus and those of Chronic bee-paralysis virus associate. *J. Gen. Virol.*, **63**, 171-179.

2: TOBACCO NECROSIS SATELLITE VIRUS-LIKE

DISTINGUISHING FEATURES

Satellite virus particles are found in plant hosts in association with taxonomically diverse helper viruses. Particles are isometric and about 17 nm in diameter. They consist of 60 copies of a single protein of 17 to 24 kDa and one molecule of RNA. The satellite virus RNA contains an ORF that encodes satellite CP; some satellite virus RNAs contain a second ORF.

LIST OF MEMBERS

Panicum mosaic satellite virus

Tobacco mosaic satellite virus
Tobacco necrosis satellite virus
Maize white line mosaic satellite virus

LIST OF TENTATIVE MEMBERS

Cereal yellow dwarf - RPV satellite virus

REFERENCES

Ban, N., Larson, S.B. and McPherson, A. (1995). Structural comparison of the plant satellite viruses. *Virology*, **214**, 571-583.

Bringloe, D.H., Gultyaev, A.P., Pelpel, M., Pleij, C.W. and Coutts, R.H. (1998). The nucleotide sequence of satellite tobacco necrosis virus strain C and helper-assisted replication of wild-type and mutant clones of the virus. *J. Gen. Virol.*, **79**, 1539-1546

Dodds, J.A. (1999). Satellite tobacco mosaic virus. *Curr. Top. Microbiol. Immunol.* **239**, 145-147.

Forde, S.M.D. (1993). Effects of temperature on BYDV concentration in oat plants and virus purification efficiency, and detection of a putative associated satellite virus. *J. Phytopathol.*, **139**, 347-356.

Masuta, C., Zuidema, D., Hunter, B.G., Heaton, L.A., Sopher, D.S. and Jackson, A.O. (1987). Analysis of the genome of satellite panicum mosaic virus. *Virology*, **159**, 329-338.

Mirkov, T.E., Mathews, D.M., du Plessis, D.H. and Dodds, J.A. (1989). Nucleotide sequence and translation of satellite tobacco mosaic virus RNA. *Virology*, **170**, 139-146.

Qiu, W.P. and Scholthof, K.-B.G. (2001). Genetic identification of multiple biological roles associated with the capsid protein of satellite panicum mosaic virus. *Mol. Plant-Microbe Interact.*, **14**, 21-31.

Scholthof, K.-B.G. (1999). A synergism induced by satellite panicum mosaic virus. *Mol. Plant-Microbe Interact.* **12**, 163-166.

Scholthof, K.-B.G., Jones, R.W. and Jackson, A.O. (1999). Biology and structure of plant satellite viruses activated by icosahedral helper viruses. *Curr. Top. Microbiol. Immunol.*, **239**, 123-143.

Zhang, L., Zitter, T.A. and Palukaitis, P. (1991). Helper virus-dependent replication, nucleotide sequence and genome organization of the satellite virus of maize white line mosaic virus. *Virology*, **180**, 467-473.

SATELLITE NUCLEIC ACIDS

1: SINGLE STRANDED DNA SATELLITES

DISTINGUISHING FEATURES

These satellites have a ssDNA genome that does not encode a satellite CP. Most are circular DNAs about 1.3 kb in size and have helper viruses among the monopartite genome viruses in the genus *Begomovirus*. These are also known as DNA β components. Tomato leaf curl virus satellite DNA is a 682 nt DNA (probably a defective β component); its replication is supported by geminiviruses in the genera *Begomovirus* and *Curtovirus*. All DNA β components have a stem-loop region containing the ubiquitous nonanucleotide TAATATTAC and an associated highly conserved sequence located immediately upstream (presumably these features are for replication), a conserved ORF (termed βC1) of unknown function, and an A-rich region that may reflect size adaptation for maintenance of the component by the helper begomovirus. The satellite DNAs readily recombine with the helper begomovirus genome DNA, and such recombinants may retain a biological activity similar to the parental DNA β.

LIST OF MEMBERS

Ageratum enation virus satellite DNA β
Ageratum yellow vein virus satellite DNA β
Bhendi yellow vein mosaic virus satellite DNA β

Chilli leaf curl virus satellite DNA β
Cotton leaf curl Alabad virus satellite DNA β
Cotton leaf curl Gezira virus satellite DNA β
Cotton leaf curl Kokhran virus satellite DNA β
Cotton leaf curl Multan virus satellite DNA β
Cotton leaf curl Rajasthan virus satellite DNA β
Eupatorium yellow vein virus satellite DNA β
Hollyhock leaf crumple virus satellite DNA β
Honeysuckle yellow vein virus satellite DNA β
Malvastrum yellow vein virus satellite DNA β
Papaya leaf curl virus satellite DNA β
Tobacco curly shoot virus satellite DNA β
Tomato leaf curl virus satellite DNA
Tomato yellow leaf curl China virus satellite DNA β

REFERENCES

Briddon, R.W., Mansoor, S., Bedford, I.D., Pinner, M.S., Saunders, K., Stanley, J., Zafar, Y., Malik, K.A. and Markham, P.G. (2001). Identification of DNA components required for induction of cotton leaf curl disease. *Virology*, **285**, 234-243.

Dry, I.B., Krake, L.R., Rigden, J.E. and Rezaian, M.A. (1997). A novel subviral agent associated with a geminivirus: The first report of a DNA satellite. *Proc. Natl. Acad. Sci. USA*, **94**, 7088-7093.

Mansoor, S. Briddon, R.W., Zafar, Y. and Stanley, J. (2003). Geminivirus disease complexes: an emerging threat. *Trends Plant Science*, **8**, 128-134.

Saunders, K., Bedford, I.D., Briddon, R.W., Markham, P.G., Wong, S.M. and Stanley, J. (2000). A unique virus complex causes Ageratum yellow vein disease. *Proc. Natl. Acad. Sci. USA*, **97**, 6890-6895.

2: DOUBLE STRANDED RNA SATELLITES

DISTINGUISHING FEATURES

The only examples in this category are satellites found in association with viruses in the family *Totiviridae*. The dsRNAs range in size from 0.5 to 1.8 kbp and are encapsidated in helper virus capsid protein; these particles often also contain a positive-sense single-stranded copy of the dsRNA. The presence of satellites in helper virus cultures can markedly affect the virulence of the helper virus.

LIST OF MEMBERS

M satellites of Saccharomyces cerevisiae L-A virus
Satellites of Trichomonas vaginalis T1 virus

LIST OF TENTATIVE MEMBERS

Satellite of Ustilago maydis killer M virus

REFERENCES

Shelbourn, S.L., Day, P.R. and Buck, K.W. (1988). Relationships and functions of virus double-stranded RNA in a P4 killer strain of *Ustilago maydis*. *J. Gen. Virol.*, **69**, 975-982.

Tai, J.-H., Chang, S.-C., Ip, C.-F. and Ong, S.-J. (1995). Identification of a satellite double-stranded RNA in the parasitic protozoan *Trichomonas vaginalis* infected with T. vaginalis virus T1. *Virology*, **208**, 189-196.

Wickner, R.B. (1996). Double-stranded RNA viruses of *Saccharomyces cerevisiae*. *Microbiol. Rev.*, **60**, 250-265.

3: SINGLE STRANDED RNA SATELLITES

DISTINGUISHING FEATURES

These satellites have ssRNA genomes that do not encode a capsid protein. Particles containing satellite RNA are antigenically identical to those of the helper virus but can sometimes be distinguished by physical features such as sedimentation rates. Three different subgroups of these satellites are distinguished: Large satellite RNAs that are mRNA, small linear satellite RNAs and circular satellite RNAs.

LARGE SINGLE STRANDED RNA SATELLITES

DISTINGUISHING FEATURES

This category comprises satellites with genomes that are 0.8 to 1.5 kb in size and encode a non-structural protein that, at least in some cases, is essential for satellite RNA multiplication. Little sequence homology exists between the satellites and their helpers, some satellites can be exchanged among different helper viruses. These satellites rarely modify the disease induced in host plants by the helper virus.

LIST OF MEMBERS

Arabis mosaic virus large satellite RNA
Bamboo mosaic virus satellite RNA
Beet ringspot virus satellite RNA
Chicory yellow mottle virus large satellite RNA
Grapevine Bulgarian latent virus satellite RNA
Grapevine fanleaf virus satellite RNA
Myrobalan latent ringspot virus satellite RNA
Strawberry latent ringspot virus satellite RNA
Tomato black ring virus satellite RNA
  (TBRV-G serotype satellite RNA)
  (TBRV-S serotype satellite RNA)

LIST OF TENTATIVE MEMBERS

Beet necrotic yellow vein virus satellite-like RNA5

REFERENCES

Fritsch, C., Mayo, M.A. and Hemmer, O. (1993). Properties of satellite RNA of nepoviruses. *Biochimie*, **75**, 561-567.

Hans, F., Pinck, M. and Pinck, L. (1993). Location of the replication determinants of the satellite RNA associated with grapevine fanleaf nepovirus (strain F13). *Biochimie*, **75**, 597-603.

Hemmer, O., Meyer, M., Greif, C. and Fritsch, C. (1987). Comparison of the nucleotide sequences of five tomato black ring virus satellite RNAs. *J. Gen. Virol.*, **68**, 1823-1833.

Kigachi, T., Saito, M. and Tamada, T. (1996). Nucleotide sequence analysis of RNA-5 of five isolates of beet necrotic yellow vein virus and the identity of a deletion mutant. *J. Gen. Virol.*, **77**, 575-580.

Kreiah, S., Cooper, J.I. and Strunk, G. (1993). The nucleotide sequence of a satellite RNA associated with strawberry latent ringspot virus. *J. Gen. Virol.*, **74**, 1163-1165.

Lin, N.S., Lee, Y.S., Lin, B.Y., Lee, C.W. and Hsu, Y.H. (1996). The open reading frame of bamboo mosaic potexvirus satellite RNA is not essential for its replication and can be replaced with a bacterial gene. *Proc. Natl. Acad. Sci. USA*, **93**, 3138-3142.

Liu, Y.Y. and Cooper, J.I. (1993). The multiplication in plants of arabis mosaic virus satellite RNA requires the encoded protein. *J. Gen. Virol.*, **74**, 1471-1474.

Mayo, M.A., Taliansky, M.E. and Fritsch, C. (1999). Large satellite RNA: Molecular parasitism or molecular symbiosis. *Curr. Top. Microbiol. Immunol.*, **239**, 65-79.

SUBVIRAL

Rubino, L., Tousignant, M.E., Steger, G. and Kaper, J.M. (1990). Nucleotide sequence and structural analysis of two satellite RNAs associated with chicory yellow mottle virus. *J. Gen. Virol.*, **71**, 1897-1903.

SMALL LINEAR SINGLE STRANDED RNA SATELLITES

DISTINGUISHING FEATURES

These satellites have genomes of less than 0.7 kb that do not encode functional proteins. No circular molecules are present in infected cells. Some satellites can attenuate the symptoms induced by helper virus infection, whereas other satellites can exacerbate the symptoms.

LIST OF MEMBERS

Cucumber mosaic virus satellite RNA
(several types)
Cymbidium ringspot virus satellite RNA
Groundnut rosette virus satellite RNA
Panicum mosaic virus satellite RNA
Pea enation mosaic virus satellite RNA
Peanut stunt virus satellite RNA
Tomato bushy stunt virus satellite RNA
(several types)
Turnip crinkle virus satellite RNA

LIST OF TENTATIVE MEMBERS

Robinia mosaic virus satellite RNA
Tobacco necrosis virus small satellite RNA

REFERENCES

Cabrera, O. and Scholthof, K.-B.G. (1999). The complex viral etiology of St. Augustine decline. *Plant Dis.*, **83,** 902-904.

Celix, A., Rodriguez-Cerezo, E. and Garcia-Arenal, F. (1997). New satellite RNAs, but no DI RNAs, are found in natural populations of tomato bushy stunt tombusvirus. *Virology,* **239**, 277-284.

Dalmay, T. and Rubino, L. (1995). Replication of cymbidium ringspot virus satellite RNA mutants. *Virology,* **206**, 1092-1098.

Demler, S.A. and de Zoeten, G.A. (1989). Characterization of a small satellite RNA associated with pea enation mosaic virus. *J. Gen. Virol.*, **70**, 1075-1084.

Francki, R.I.B. (1985). Plant virus satellites. *Ann. Rev. Microbiol.*, **39**, 151-174.

Gallitelli, D. and Hull, R. (1985). Characterization of satellite RNAs associated with tomato bushy stunt virus and five other definitive tombusviruses. *J. Gen. Virol.*, **66**, 1533-1543.

García-Arenal, F. and Palukaitis, P. (1999). Structure and functional relationships of satellite RNAs of Cucumber mosaic virus. *Curr. Top. Microbiol. Immunol.*, **239**, 37-63.

Naidu, R.A, Collins, G.B. and Ghabrial, S.A. (1992). Peanut stunt virus satellite RNA: analysis of sequences that affect symptom attenuation in tobacco. *Virology,* **189**, 668-677.

Simon, A.E. (1999). Replication, recombination, and symptom-modulation properties of the satellite RNAs of Turnip crinkle virus. *Curr. Top. Microbiol. Immunol.*, **239,** 19-36.

Simon, A.E. and Howell, S.H. (1986). The virulent satellite RNA of turnip crinkle virus has a major domain homologous to the 3′ end of the helper virus genome. *EMBO J.*, **5**, 3423-3428.

Skoric, D., Krajacic, M. and Stefanac, Z. (1997). Cucumovirus with a satellite-like RNA isolated from Robinia pseudoacacia L. *Periodicum Biologorum,* **99**, 125-128.

CIRCULAR SINGLE STRANDED RNA SATELLITES

DISTINGUISHING FEATURES

These satellites have genomes that are about 350 nt long that occur as circular as well as linear molecules. Replication of some has been shown to involve self-cleavage of circular progeny molecules by an RNA-catalyzed reaction.

LIST OF MEMBERS

Arabis mosaic virus small satellite RNA
Cereal yellow dwarf virus-RPV satellite RNA
Chicory yellow mottle virus satellite RNA
Lucerne transient streak virus satellite RNA
Solanum nodiflorum mottle virus satellite RNA
Subterranean clover mottle virus satellite RNA
(2 types)
Tobacco ringspot virus satellite RNA
Velvet tobacco mottle virus satellite RNA

LIST OF TENTATIVE MEMBERS

Rice yellow mottle virus satellite

REFERENCES

AbouHaidar, M.G. and Paliwal, Y.C. (1988). Comparison of the nucleotide sequences of viroid-like satellite RNA of the Canadian and Australasian strains of lucerne transient streak virus. *J. Gen. Virol.*, **69**, 2369-2373.

Davies, C., Haseloff, J. and Symons, R.H. (1990). Structure, self-cleavage, and replication of two viroid-like satellite RNAs (virusoids) of subterranean clover mottle virus. *Virology*, **177**, 216-224.

Etscheid, M., Tousignant, M.E. and Kaper, J.M. (1995). Small satellite of arabis mosaic virus autolytic processing of in vitro transcripts of (+) and (-) polarity and infectivity of (+) strand transcripts. *J. Gen. Virol.*, **76**, 271-282.

Passmore, B.K. and Bruening, G. (1993). Similar structure and reactivity of satellite tobacco ringspot virus RNA obtained from infected tissue and by *in vitro* transcription. *Virology*, **197**, 108-115.

Rasochova, L. and Miller, W.A. (1996). Satellite RNA of barley yellow dwarf-RPV virus reduces accumulation of RPV helper virus RNA and attenuates RPV symptoms in oats. *Mol. Plant-Microbe Interact.*, **9**, 646-650.

Rubino, L., Tousignant, M.E., Steger, G. and Kaper, J.M. (1990). Nucleotide sequence and structural analysis of two satellite RNAs associated with chicory yellow mottle virus. *J. Gen. Virol.*, **71**, 1897-1903.

Sehgal, O.P., AbouHaidar, M.G., Gellatly, D.L., Ivanov, I. and Thottapilly, G. (1993). An associated small RNA in rice yellow mottle sobemovirus homologous to the satellite RNA of lucerne transient streak sobemovirus. *Phytopathology*, **83**, 1309-1311.

Symons, R.H. (1997). Plant pathogenic RNAs and RNA catalysis. *Nuc. Acids Res.*, **25**, 2683-2689.

Symons, R.H. and Randles, J.W. (1999). Encapsidated circular viroid-like satellite RNAs. *Curr. Top. Microbiol. Immunol.*, **239**, 81-105.

CONTRIBUTED BY

Mayo, M.A., Leibowitz, M.J., Palukaitis, P., Scholthof, K.-B.G., Simon, A.E., Stanley, J. and Taliansky, M.

SUBVIRAL

PRIONS

Prions are *pro*teinaceous *in*fectious particles that lack nucleic acids. Prions are composed largely, if not entirely, of an abnormal isoform of a normal cellular protein.

In mammals, prions are composed of an abnormal, pathogenic isoform of the prion protein (PrP). A chromosomal gene encodes PrP and no PrP genes were found in purified preparations of prions. Neither prion specific nucleic acids nor virus-like particles have been detected in purified, infectious preparations.

In fungi, evidence for three different prions has been accumulated. Those studies are summarized in the section on Fungal Prions.

MAMMALIAN PRIONS

Unfortunately it has not been possible to update this description since the VIIth ICTV Report, therefore we reproduce here the description of the previous report.

GENERAL DESCRIPTION OF THE MAMMALIAN PRIONS

The mammalian prions cause scrapie and other related neurodegenerative diseases of animals and humans (Table 1). The prion diseases are also referred to as the transmissible spongiform encephalopathies (TSEs).

Table 1: The prion diseases

Disease (abbreviation)	Natural host	Prion	Pathogenic PrP Isoforms
Scrapie	sheep & goats	Scrapie prion	$OvPrP^{Sc}$
Transmissible mink encephalopathy (TME)	mink	TME prion	$MkPrP^{Sc}$
Chronic wasting disease (CWD)	mule deer & elk	CWD prion	$MDePrP^{Sc}$
Bovine spongiform encephalopathy (BSE)	cattle	BSE prion	$BoPrP^{Sc}$
Feline spongiform encephalopathy (FSE)	cats	FSE prion	$FePrP^{Sc}$
exotic ungulate encephalopathy (EUE)	nyala & greater kudu	EUE prion	$UngPrP^{Sc}$
Kuru	humans	Kuru prion	$HuPrP^{Sc}$
Creutzfeldt-Jakob disease (CJD)	humans	CJD prion	$HuPrP^{Sc}$
Gerstmann-Sträussler Sheinker syndrome (GSS)	humans	GSS prion	$HuPrP^{Sc}$
Fatal familial insomnia (FF1)	humans	FF1 prion	$HuPrP^{Sc}$

Prions are composed of an abnormal, pathogenic PrP isoform denoted PrP^{Sc}. The "Sc" superscript was initially derived from the term scrapie because scrapie was the prototypic prion disease. Since all of the known prion diseases (Table 1) of mammals involve aberrant metabolism of PrP similar to that observed in scrapie, we suggest using the "Sc" superscript for all abnormal, pathogenic PrP isoforms. In this context, the "Sc" superscript is used to designate the scrapie-like isoform of PrP; for those who desire a more general derivation "Sc" can equally well be derived from the term "prion sickness" (see Table 2).

A post-translational process converts the normal cellular isoform of the protein, designated PrP^{C}, into PrP^{Sc}. Attempts to identify a chemical difference that distinguishes PrP^{Sc} from PrP^{C} have not been successful; moreover, PrP^{C} and PrP^{Sc} are encoded by the same single copy chromosomal gene. The conformations of the two PrP isoforms are profoundly different; PrP^{C} has little, if any, β-sheet while PrP^{Sc} has a high β-sheet content.

Like viruses, prions are infectious because they stimulate a process by which more of the pathogen is produced. As prions or viruses accumulate in an infected host, they eventually cause disease. Both prions and viruses exist in different varieties or subtypes that are called strains. But many features of prion structure and replication distinguish them from viruses and all other known infectious pathogens.

Prions differ from viruses and viroids since they lack a nucleic acid genome that directs the synthesis of their progeny. Prions are composed of an abnormal isoform of a cellular protein whereas, most viral proteins are encoded by the viral genome and viroids are devoid of protein. Prions can exist in multiple molecular forms, whereas viruses exist in a single form with a distinct ultrastructural morphology. Prions are non-immunogenic, in contrast to viruses, which almost always provoke an immune response. An enlarging body of evidence argues that strains of prions are enciphered in the conformation of PrP^{Sc}; in contrast, strains of viruses and viroids have distinct nucleic acid sequences that produce pathogens with different properties.

NOMENCLATURE OF THE MAMMALIAN PRIONS

The terminology for prions is still evolving (Table 2). While some terms are borrowed from infectious diseases caused by viruses, others are taken from genetics and still others from the biology of protein structure as well as neuropathology. This new area of biological investigation, which has such diverse roots, creates some unique problems with terminology that are discussed below in the Nomenclature section.

Table 2: Glossary of Prion Terminology

Term	Description
Prion	A proteinaceous infectious particle that lacks nucleic acid. Prions are composed largely, if not entirely, of PrP^{Sc} molecules.
PrP^{Sc}	Abnormal, pathogenic isoform of the prion protein that causes sickness. This protein is the only identifiable macromolecule in purified preparations of prions.
PrP^{C}	Cellular isoform of the prion protein.
PrP 27-30	Digestion of PrP^{Sc} with proteinase K generates PrP 27-30 by hydrolysis of the N-terminus
PRNP	Human PrP gene located on chromosome 20.
Prnp	Mouse PrP gene located on syntenic chromosome 2. *Prnp* controls the length of the prion incubation time and is congruent with the incubation time genes *Sinc* and *Prn-i*. PrP-deficient ($Prnp^{o/o}$) mice are resistant to prions.
PrP amyloid	Fibril of PrP fragments derived from Prp^{Sc} by proteolysis. Plaques containing PrP amyloid are found in the brains of some mammals with prion disease.
Prion rod	An amyloid polymer composed of Prp 27 - 30 molecules. Created by detergent extraction and limited proteolysis of PrP^{Sc}.
protein X	A hypothetical macromolecule that is thought to act like a molecular chaperone in facilitating the conversion of PrP^{C} into PrP^{Sc}.

PRION PROPERTIES

MORPHOLOGY

Microsomal fractions from infected tissues enriched for prion infectivity contain numerous membrane vesicles (Fig. 1 Left); detergent extraction and limited proteolysis of brain microsomes generate rod-shaped particles (Fig. 1 Center). Most are of uniform diameter (11 nm) with mean lengths of 165 nm (range 25-550 nm). The rods are smooth, almost ribbon-like, and infrequently are twisted. The rods resemble purified amyloid, both ultrastructurally and histochemically (Fig. 1 Right). The rods are not considered the infectious entity since large PrP 27-30 polymers are not required for infectivity (Fig. 1C).

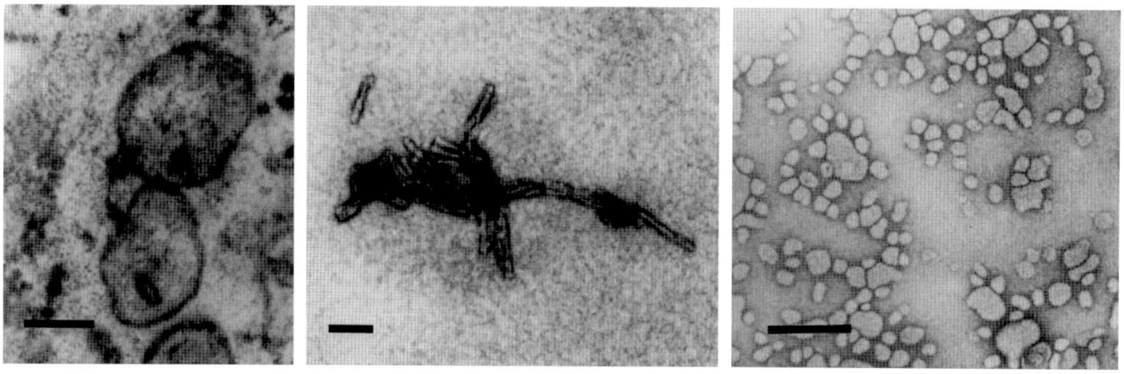

Figure 1: Multiple forms of prions isolated from infected Syrian hamster brains. (Left) microsomal membranes containing submicroscopic, infectious prion particles; (Center) purified prion rods representing a polymeric form of the infectious prion particle and generated by limited proteolysis in the presence of detergent; (Right) prion liposomes generated by sonication of infectious prion rods isolated from infected Syrian hamster (SHa) brains using limited proteolysis, detergent extraction and sedimentation through a discontinuous sucrose gradient. All three forms contain high levels of prion infectivity (>10^7 ID_{50} units/ml). Bars represent 100 nm.

PHYSICOCHEMICAL AND PHYSICAL PROPERTIES

The size of PrP^{Sc} is 33-35 kDa. The smallest infectious prion particle is probably a dimer of PrP^{Sc}; this estimate is consistent with an ionizing radiation target size in daltons of 55 ± 9 kDa. Prions aggregate into particles of non-uniform size and cannot be solubilized by detergents, except under denaturing conditions where infectivity is lost. However, solubilization of PrP^{Sc} and prions can be achieved with phospholipids (Fig. 1C). Prions resist inactivation by nucleases, UV-irradiation at 254 nm, treatment with psoralens, divalent cations, metal ion chelators, acid (between pH 3 and 7), hydroxylamine, formalin, boiling, and proteases. Prion infectivity is diminished by prolonged digestion with proteases, or by treatments such as urea, boiling in SDS, alkali (> pH 10), autoclaving at 132^{0}C for more than 2 h, denaturing organic solvents (e.g., phenol), or chaotropic agents such as guanidine isocyanate.

NUCLEIC ACIDS

No prion-specific nucleic acid has been detected.

PROTEINS

PrP^C is likely to be a three-helix bundle protein based on NMR structural studies of recombinant PrPs expressed in *Escherichia coli.* When PrP^{Sc} is formed from PrP^C, the region from residue 90 to 125 seems to undergo a profound structural change where antibody epitopes once exposed become cryptic. Optical spectroscopy has shown that concurrent with the conversion of PrP^C into PrP^{Sc}, there is a modest decrease in the α-helical content of PrP and a large increase in β-sheet. PrP^{Sc} is generally distinguished from PrP^C by its different biochemical and biophysical properties. Limited proteolysis of PrP^{Sc} produces a smaller, protease-resistant molecule of about 142 aa, designated PrP 27-30. Under the

SUBVIRAL

same conditions, PrP^{C} is completely hydrolyzed. The aa sequence of PrP^{Sc} that has been established by protein sequencing and mass spectrometry is identical to that deduced from the genomic DNA sequence. No proteins other than PrP^{Sc} have been consistently found in fractions enriched for prion infectivity.

LIPIDS

PrP^{Sc} contains a glycosylinositol phospholipid (GPI) attached to amino acid residue 231 (serine) of the Syrian hamster (SHa) PrP. The lipids of the diradylglycerol moiety of the GPI anchor are not well characterized.

CARBOHYDRATES

In addition to the GPI anchor which contains sialic acid, PrP^{Sc} has two consensus sites where it can undergo N-linked glycosylation (residues 181 and 197 of the Syrian hamster PrP). Bi-, tri- and tetra-antennary structures have been reported for the N-linked, complex type glycans of PrP^{Sc}. Some of these complex-type oligosaccharides have branched fucose residues, some have terminal sialic acid residues. Six different GPI glycans have been found, two of which are sialylated.

GENOME ORGANIZATION AND REPLICATION

The entire ORF of all known mammalian PrP genes is contained within a single exon. In general, PrP genes are composed of three exons, as clearly demonstrated for mouse and sheep. The PrP genes of humans and Syrian hamsters (SHa) appear to have three exons but most HuPrP and SHaPrP mRNAs are spliced from only two exons that are separated by ~10 kb. Exon-1 of the SHaPrP gene encodes a portion of the 5′ untranslated leader sequence while exon-3 encodes the ORF and the 3′-UTR. The ORF of the HuPrP gene encodes a protein of 253 amino acids while the mouse and SHa PrP genes encode proteins of 254 residues. The promoters of both the PrP genes of both animals contain 3 or 2 repeats, respectively, of G-C nonamers, but are devoid of TATA boxes. These nonamers represent a motif which may function as a canonical binding site for transcription factor Sp1.

The multiplication of prion infectivity involves the post-translational conversion of the precursor PrP^{C} into PrP^{Sc}. Studies with mice expressing a foreign SHa, Hu, or BoPrP transgene argue that prion formation results from infecting PrP^{C} molecules combining with homologous host-encoded PrP^{C} molecules giving rise to new PrP^{Sc} molecules.

Additional evidence to support this proposed model for prion replication comes from studies of transgenic mice expressing chimeric or mutant PrPs where the prions produced from these transgene products have an artificial host range. In the absence of any candidate post-translational chemical modification that differentiates PrP^{C} from PrP^{Sc}, these two isoforms remain distinguishable only by their conformations.

ANTIGENIC PROPERTIES

PrP^{Sc} is a weak antigen. The antigenicity of PrP^{Sc} is enhanced by immunization of PrP--deficient ($Prnp^{o/o}$) mice which are not tolerant to PrP^{C}. The immunoreactivity of PrP^{Sc} is significantly enhanced by denaturation. Some antibodies and recombinant Fabs bind to non-denatured PrP^{Sc}; moreover, anti-PrP antibodies have been used to neutralize prion infectivity that is dispersed into liposomes.

BIOLOGICAL PROPERTIES

The prion diseases are a group of neurodegenerative disorders afflicting mammals (Table 1). The diseases are transmissible under some circumstances but, unlike other transmissible disorders, the prion diseases can also be caused by mutations in the host PrP gene. The mechanism of prion spread among sheep and goats developing natural scrapie is unknown. CWD, TME, BSE, FSE, and EUE are all thought to occur after the consumption of prion-infected materials. Similarly, kuru of the New Guinea Fore people is thought to have resulted from the consumption of brains during ritualistic cannibalism. Familial CJD, GSS, and FFI are all dominantly inherited prion diseases; five different mutations of the PrP gene have been shown to be genetically linked to development of inherited prion disease. While iotragenic CJD cases can be traced to inoculation of prions through human pituitary-derived growth hormone, cornea transplants, dura mater grafts, or cerebral electrode implants, the number of cases recorded to date is small. New variant CJD is thought to be due to the consumption of BSE tainted beef products.

Most cases of CJD are sporadic, probably the result of somatic mutation of the PrP gene or the spontaneous conversion of PrP^{C} into PrP^{Sc}. About 10-15% of CJD cases and virtually all cases of GSS and FF1 appear to be caused by germline mutations in the PrP gene. Twenty different mutations of the PrP gene have been shown to segregate with the heritable human prion diseases (Table 3).

Table 3: Examples of human PrP gene mutations found in the inherited prion diseases

Inherited Prion Disease	PrP Gene Mutation
Gerstinann-Straussler-Scheinker disease	PrP P1O2L*
Gerstmann-Straussler-Scheinker disease	PrP AI17V
Familial Creutzfeldt-Jakob disease	PrP D178N, V129
Fatal familial insomnia	PrP D178N, M129*
Gerstmann-Straussler-Scheinker disease	PrP F198S*
Familial Creutzfeldt-Jakob disease	PrP E2OOK*
Gerstmann-Straussler-Scheinker disease	PrP Q217R
Familial Creutzfeldt-Jakob disease	PrP octarepeat insert*

*signifies genetic linkage between the mutation and the inherited prion disease

Prion Strains

Strains or isolates of prions are defined by specific incubation times, distribution of vacuolar lesions, and patterns of PrP^{Sc} accumulation. Strain-specific information in prions appears to reside in the conformation of PrP^{Sc}. Prions derived from fCJD(E200K) and FF1 (Table 3) patients have PrP^{Sc} molecules with different conformations as judged by the size of the protease resistant fragments after deglycosylation. These different conformations of PrP^{Sc} were replicated in mice expressing a chimeric mouse:human PrP transgene. The origin of the inoculum determined the incubation time, the distribution of neuropathologic lesions and the pattern of PrP^{Sc} deposition in the Tg mice. Additional support for strain specific information being enciphered in the conformation of PrP^{Sc} comes from two prion strains isolated from mink dying of TME and passaged in Syrian hamsters where PrP^{Sc} from each strain exhibited different sensitivities to proteolytic digestion.

Mutant PrP Genes

Humans carrying point mutations or inserts in their PrP genes produce mutant PrP^{C} molecules that are believed to spontaneously convert into PrP^{Sc}. While the initial stochastic event could be inefficient, once it happens the process may become autocatalytic. The proposed mechanism is consistent with individuals harboring germline mutations who develop CNS dysfunction only in middle or old age as is seen with all of the inherited prion diseases of humans. The level of mutant PrP transgene overexpression in mice is inversely proportional to the age at which neurologic deficits appear.

POLYMORPHISMS IN PRP GENES

The M/V polymorphism at codon 129 in HuPrP influences the incidence of sporadic and infectious CJD as well as the age of onset of inherited prion disease. Homozygosity at codon 129 is more frequent in both sporadic and infectious CJD; it also heralds an earlier age of clinical disease in people carrying pathologic mutations. Those individuals who carry the DI78N mutation develop FFI manifest by insomnia and dysautonomia if they encode M129 on the mutant allele while those who encode V129 present with fCID characterized by dementia (Table 3).

The E/K polymorphism at codon 219 in HuPrP seems to influence resistance to CJD. Individuals with wild-type (wt) PrP sequences who are heterozygous for K219 appear to be resistant to CJD. This dominant negative effect is likely to be the result of increased affinity of PrP^{C} (K219) for protein X which is thought to function like a molecular chaperone in facilitating the conversion of PrP^{C} into PrP^{Sc}.

The Q/R polymorphism at codon 171 in OvPrP seems to govern the resistance of sheep to natural scrapie. Sheep with wild-type (wt) PrP sequences that are heterozygous for R171 appear to be resistant to natural scrapie but scrapie can be transmitted to some of these animals by inoculation of prions. This dominant negative effect is also likely to be the result of increased affinity of PrP^{C} (R171) for protein X. A polymorphism at codon 136 is also thought to influence the susceptibility of some breeds of sheep to scrapie.

TAXONOMIC CONSIDERATION OF MAMMALIAN PRIONS

A listing of the different animal prions is given in Table 1. Although the prions that cause TME and BSE are referred to as TME prions and BSE prions this may be unjustified, because both are thought to originate from the oral consumption of scrapie prions in sheep-derived foodstuffs and because many lines of evidence argue that the only difference among the various prions is the sequence of PrP which is dictated by the host and not the prion itself. The human prions present a similar semantic conundrum. Transmission of human prions to laboratory animals produces prions carrying PrP molecules with sequences dictated by the PrP gene of the host, not that of the inoculum.

To simplify the terminology, the generic term PrP^{Sc} is suggested in place of such terms as PrP^{CJD}, PrP^{BSE}, and PrP^{res}. To distinguish PrP^{Sc} found in humans or cattle from that found in other animals, we suggest $HuPrP^{Sc}$ or $BoPrP^{Sc}$ instead of PrP^{CJD} or PrP^{BSE}, respectively (Table 1). Once human prions and thus, $HuPrP^{Sc}$ molecules have been passaged into animals, then the prions and PrP^{Sc} are no longer of the human species unless they were formed in an animal expressing a HuPrP transgene. In the case of mutant PrPs, the mutation and any important polymorphism can be denoted in parentheses following the particular PrP isoform. For example in FFI, the pathogenic PrP isoform would be referred to as PrP^{Sc} or $HuPrP^{Sc}$; alternatively, if it were important to identify the mutation, then it would be written as $HuPrP^{Sc}$ (D178N, M129) (Table 3). The term PrP^{res} or PrP-res is derived from the protease-*res*istance of PrP^{Sc} but protease-resistance, insolubility, and high β-sheet content should be only considered as surrogate markers of PrP^{Sc} since one or more of these may not always be present. Whether Prp^{res} is useful in denoting PrP molecules that have been subjected to procedures that modify their resistance to proteolysis but have not been demonstrated to convey infectivity or to cause disease remains debatable.

The term PrP* has been used in two different ways. First, it has been used to identify a fraction of PrP^{Sc} molecules that are infectious. Such a designation is thought to be useful since there are ~ 10^5 PrP^{Sc} molecules per infectious unit. Second, PrP* has been used to designate a metastable intermediate of PrP^{C} that is bound to protein X. It is noteworthy that neither a subset of biologically active PrP^{Sc} molecules nor a metastable intermediate of PrP^{C} has been identified, to date.

In mice, the PrP gene denoted *Prnp* is now known to be identical with two genes denoted *Sinc* and *Prn-i* that were known to control the length of the incubation time in mice inoculated with prions. These findings permit a welcome simplification. A gene designated *Pid-1* on mouse chromosome 17 also appears to influence experimental CJD and scrapie incubation times but information on this locus is limited.

Distinguishing among CJD, GSS, and FFI has grown increasingly difficult with the recognition that CJD, GSS, and FFI are autosomal dominant diseases caused by mutations in the PRNP gene. Initially, it was thought that a specific PrP mutation was associated with a particular clinico-neuropathologic phenotype but an increasing number of exceptions are being recognized. Multiple examples of variations in the clinico-neuropathologic phenotype within a single family where all affected members carry the same PrP mutation have been recorded. Most patients with a PrP mutation at codon 102 present with ataxia and have PrP amyloid plaques; such patients are generally given the diagnosis of GSS, but some individuals within these families present with dementia, a clinical characteristic that is usually associated with CJD. One suggestion is to label these inherited disorders as "prion disease" followed by the mutation in parenthesis while another is the use the terms FCJD and GSS followed by the mutation. In the case of FFI, describing the Dl78N mutation and M129 polymorphism seems unnecessary since this is the only known mutation-polymorphism combination that gives the FFI phenotype.

DERIVATION OF NAMES

Prion: singla for *pro*teinaceous and *in*fectious particle.

REFERENCES

Aiken, J.M. and Marsh, R.F. (1990). The search for scrapie agent nucleic acid. *Microbiol Rev.*, **54**, 242-246.

Brown, P., Goldfarb, L.G. and Gajdusek, D.C. (1991). The new biology of spongiform encephalopathy: infectious amyloidoses with a genetic twist. *Lancet*, **337**, 1019-1022.

Carp, R.I., Kascsak, R.J., Wisniewski, H.M., Mertz, P.A., Rubenstein, R., Bendheim, P. and Bolton, D. (1989). The nature of the unconventional slow infection agents remains a puzzle. *Alzheimer Dis. Assoc. Disord.*, **3**, 79-99.

Caughey, B. and Chesebro, B. (1997). Prion protein and the transmissible spongiform encephalopathies. *Trends Cell Biol.*, **7**, 5-62

Court, L. and Dodet, B. (eds) (1996). *Transmissible subacute spongiform encephalopathies: prion diseases.* Elsevier, Amsterdam.

Dickinson, A.G. and Outram, G.W. (1988). Genetic aspects of unconventional virus infections: the basis of the virino hypothesis. In: *Novel infectious agents and the central nervous system*, (G. Bock, and J. Marsh, eds), pp 63-83. Ciba Foundation Symposium 135. John Wiley and Sons, Chichester, UK.

Gajdusek, D.C. (1977). Unconventional viruses and the origin and disappearance of kuru. *Science*, **197**, 943-960.

Ironside, J.W. (1997). The new variant form of Creutzfeldt-Jakob disease: a novel prion protein amyloid disorder (Editorial). *Amyloid: Int. J. Exp. Clin. Invest.*, **4**, 66-69.

Prusiner, S.B., Collinge, J., Powell, J. and Anderton, B. (eds) (1992). *Prion diseases of humans and animals.* Ellis Horwood, London.

Prusiner, S.B. (1997). Prion diseases and the BSE crisis. *Science*, **278**, 245-251.

Weissmann, C. (1996). Molecular biology of transmissible spongiform encephalopathies. *FEBS Lett.*, **389**, 3-11.

CONTRIBUTED BY

Prusiner, S.B., Baron, H., Carlson, G., Cohen, F.E., DeArmond, S.J., Gabizon. R, Gambetti, P., Hope, J., Kitamoto, T., Kretzschmar, H.A., Laplanche, J.-L., Tateishi, J., Telling, G. and Will, R.

FUNGAL PRIONS

TAXONOMIC STRUCTURE OF THE FUNGAL PRIONS

Prions
- [URE3] prion
- [PSI] prion
- [Het-s] prion
- [PIN] prion
- [BETA] prion

DISTINGUISHING FEATURES

The word "prion" means infectious protein. In yeast and filamentous fungi, infection by viruses occurs exclusively by cell-cell fusion with transmission of cytoplasm. Viruses are transmitted as non-chromosomal genetic elements. Likewise, one expects prions of these organisms to show similar behavior. There are now five well documented prions in yeast and filamentous fungi. Four of these are due to self-propagating amyloids, while the fifth is a auto-activating protease.

PRION PROPERTIES

MORPHOLOGY

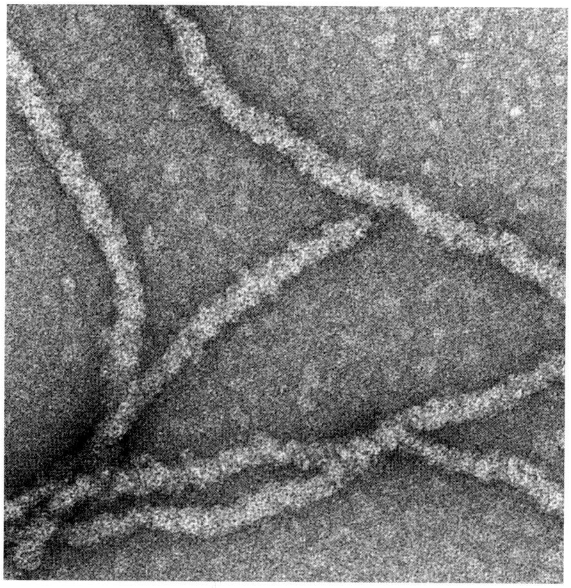

Figure 1: Amyloid filaments of the Ure2 protein (photo courtesy of Dr. Ulrich Baxa, NIH).

Prions are infectious proteins: altered forms of a normal cellular protein that may have lost their normal function, but have acquired the ability to change the normal form of the protein into the same abnormal form as themselves. In some cases they form filaments or filamentous aggregates, but this is not part of the definition of a prion, which in some cases could be based on covalent modification without substantial morphological effects.

PHYSICOCHEMICAL AND PHYSICAL PROPERTIES

The Ure2 protein (Ure2p) is relatively proteinase K resistant in [URE3] strains compared to wild-type strains, and forms filaments in [URE3]-carrying cells. The Sup35 protein (Sup35p) is slightly proteinase K-resistant in [PSI] strains compared to wild-type strains and forms large aggregates in these cells. The het-s protein (Hetsp) is relatively proteinase K-resistant in strains with the [Het-s] prion compared to wild-type cells. Rnq1p is

SUBVIRAL

aggregated in cells carrying the [PIN] prion. The vacuolar protease B is processed into its mature active form in cells carrying the [BETA] prion; the [BETA] prion is simply the mature form of protease B.

NUCLEIC ACID

None. The definition of a prion is that it is an infectious protein without the need for any accompanying nucleic acid.

PROTEINS

Prion	Protein	Function of the protein
[URE3]	Ure2p	regulator of nitrogen catabolism
[PSI]	Sup35p	translation termination
[Het-s]	Hetsp	heterokaryon incompatibility
[PIN]	Rnq1	unknown; detected by [PSI]-inducibility
[BETA]	Prb1p	Protease B: protein processing & degradation

LIPIDS

None.

CARBOHYDRATES

None.

GENOME ORGANIZATION AND REPLICATION

[URE3], [PSI], [PIN], [BETA] and [Het-s] are all inherited as non-Mendelian genetic elements (also called cytoplasmic genetic elements or non-chromosomal genetic elements). The 'genetic material' in this case is the altered protein which is self-replicating by changing the normal form of the protein (which nonetheless has the same primary sequence) into the altered form. Thus altered Ure2p, Sup35p, Rnq1p, Prb1p and Hetsp are the 'genomes' of [URE3], [PSI], [PIN], [BETA] and [Het-s].

Each amyloid forming prion protein has a prion domain. Deletions of the remainder of the molecule increase the frequency with which the prion form arises. Chaperones play an important role in all amyloid-related prions of *S. cerevisiae*, but different prions show different effects of altered chaperone activities.

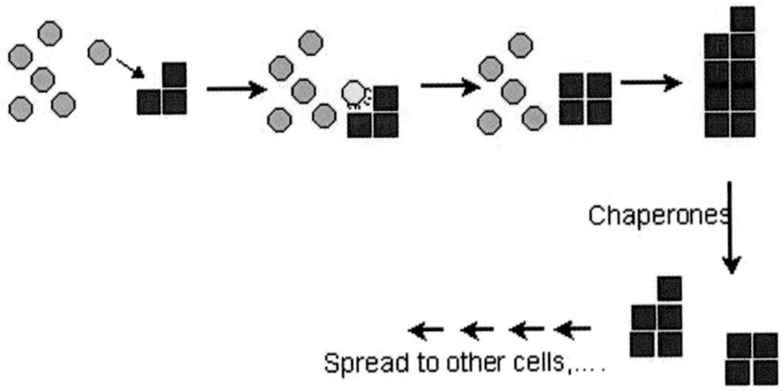

Figure 2: Amyloid prions replicate by elongation of amyloid filaments and scission of filaments by the action of chaperones to produce new filaments.

BIOLOGICAL PROPERTIES

The yeast and fungal prions are passed from cell to cell during mating and hyphal anastomosis. There are no known natural vectors or extracellular transmission. Thus, fungal prions appear first as non-Mendelian genetic elements. Three biologic (genetic) criteria have been proposed to determine in a preliminary way whether a given non-Mendelian genetic element is a prion.

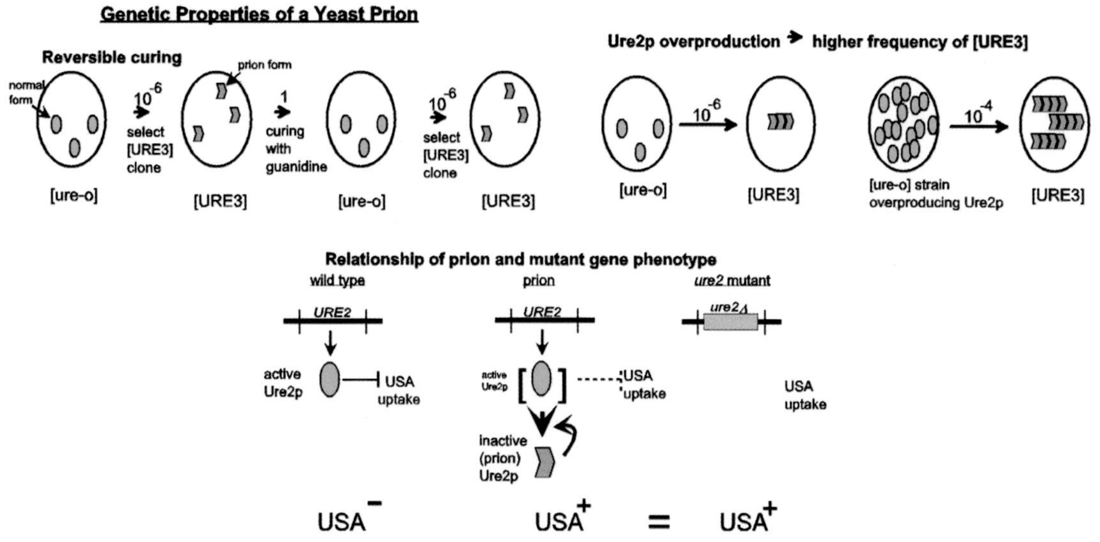

Figure 3: Genetic criteria for identification of prions.

- Prions are "reversibly curable", meaning that even if a prion can be cured from a strain, it can arise again spontaneously at some low frequency in the cured strain.
- Overexpression of the normal protein increases the frequency with which the prion form arises.
- The phenotype of the presence of the prion is the same as the phenotype of a recessive mutant in the gene for the normal form, which gene first appears as a chromosomal gene necessary for the propagation of the prion. In some cases, the prion produces a phenotype by having an action beyond inactivation of the normal form. In this case, the phenotypic relation does not provide evidence that the non-Mendelian genetic element is a prion, but, of course, it is also not evidence against it being a prion. In all cases, the prion depends for its propagation on the gene for the protein.

Further, physical or chemical evidence for an alteration of the protein in cells carrying the prion should be demonstrated.
The [Het-s] and [BETA] chaperones are advantageous for their hosts and are present in essentially all of the respective wild-type strains.

Species *[URE3-1]*

HOST

Saccharomyces cerevisiae.

DISTINGUISHING FEATURES

[URE3] is a self-propagating amyloid form of the Ure2 protein, a regulator of nitrogen catabolism. In the prion form of the Ure2 protein, its nitrogen regulation activity is largely lost. This results in a slow growth phenotype as well as derepression of activities normally repressed by a good nitrogen source.

PRION PROPERTIES

MORPHOLOGY

The synthetic prion domain of Ure2p, Ure2p$^{1-65}$, forms amyloid *in vitro*, and induces a self-propagating amyloid formation by full-length native Ure2p purified from yeast. Ure2p amyloid filaments have a 4 nm core composed of the prion domain, surrounded by globular appendages composed of the C-terminal nitrogen regulation domain. Cells carrying the [URE3] prion contain a network of 25 nm diameter filaments containing the Ure2 protein.

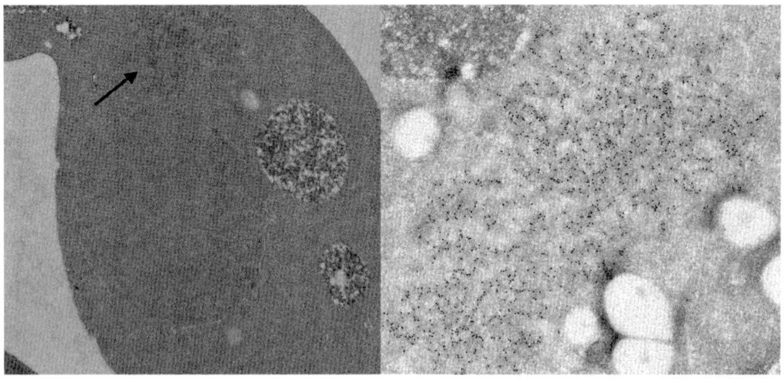

Figure 4: Amyloid filament in cells of *S. cerevisiae* infected with the [URE3] prion (Speransky *et al.* 2001).

PHYSICOCHEMICAL AND PHYSICAL PROPERTIES

Ure2p is more proteinase K-resistant in extracts of [URE3] strains than in wild-type strains, with the N-terminal prion domain the most protease resistant part. Ure2p amyloid formed *in* vitro likewise has a core composed of the prion domain which is the most protease-resistant part of the aggregate. The amyloid core of the filaments has beta-sheet structure.

NUCLEIC ACID

None.

PROTEINS

Ure2p, in its altered form, is the primary component of [URE3]. The Mks1 protein is necessary for generation of the [URE3] prion, but not for its propagation. The presence of the [PIN} prion also stimulates the rate of [URE3] prion generation.

LIPIDS

None.

CARBOHYDRATES

None.

GENOME ORGANIZATION AND REPLICATION

The *URE2* gene has an ORF coding for a 354 aa polypeptide, of which the C-terminal part, including residues 81 to 354, is capable of carrying out the nitrogen regulation function of Ure2p. The N-terminal 80 residues is necessary for the initial formation of [URE3] or for its propagation or for a molecule to be affected by the presence of [URE3]. Propagation of [URE3] requires the chaperone Hsp104 and is blocked by overproduction of the Hsp40-group chaperone Ydj1p. Generation *de novo* of [URE3] requires Mks1p, and is negatively regulated by the Ras-cAMP pathway.

Antigenic Properties

Not studied.

Biological Properties

[URE3] makes cells able to take up ureidosuccinate from the media containing a good nitrogen source such as ammonia or glutamine. The *ure2* mutants have the same phenotype, and the propagation of [URE3] requires the *URE2* gene. Yeast cells turn off transcription of genes for enzymes involved in utilization of poor nitrogen sources when a good nitrogen source is available (nitrogen catabolite regulation). *DAL5* encodes the permease for allantoate, a poor, but usable, nitrogen source for yeast. *DAL5* is thus subject to nitrogen catabolite regulation. Allantoate is structurally similar to ureidosuccinate, and so Dal5p can take up ureidosuccinate. Ureidosuccinate is an intermediate in uracil biosynthesis, the product of the first step in the pathway, aspartate transcarbamylase (*URA2*). Thus, *ura2* mutants can grow on ureidosuccinate in place of uracil if the cell has either the [URE3] prion or a mutation in *ure2*. [URE3] can be cured by growth in the presence of 5 mM guanidine HCl, but from the cured strain can again be isolated subclones which have again acquired [URE3]. Overproduction of Ure2p induces the *de novo* formation of [URE3] at 100 fold the normal frequency.

List of Species Demarcation Criteria

Prions are divided into species based on the identity of the protein that makes up the infectious element. However, even a single protein can form prions of distinct 'strains', or 'variants', apparently due to different self-propagating amyloid structures, much as a given protein can assume more than one crystal form.

List of Species

Species

[URE3] prion

URE2	[URE3]
URE3	[URE3]

Tentative Species

None reported.

Type Species *[PSI+]*

Host

Saccharomyces cerevisiae.

Distinguishing Features

[PSI] is a self-propagating amyloid form (a prion) of the Sup35 protein, a subunit of the translation termination factor of the yeast *Saccharomyces cerevisiae*. The presence of the [PSI] prion results in a relative shortage of the soluble form of Sup35p, and a resulting decrease in the efficiency of translation termination. This makes read-through of termination codons more frequent. The phenotype of [PSI] is ability to grow in the absence of adenine in spite of a premature stop mutation in *ADE1* or *ADE2*.

PRION PROPERTIES

MORPHOLOGY

Sup35p forms typical amyloid filaments *in vitro,* but their morphology *in vivo* has not been described.

PHYSICOCHEMICAL AND PHYSICAL PROPERTIES

Amyloid filaments of the Sup35 protein are high in beta sheet structure, but are only slightly protease resistant.

NUCLEIC ACID

None.

PROTEINS

Sup35p (79 kDa) is a subunit of the translation termination factor, which also includes Sup45p.

LIPIDS

None.

CARBOHYDRATES

None.

GENOME ORGANIZATION AND REPLICATION

The N-terminal 154 residues of Sup35p are sufficient to propagate the [PSI] prion. This part of Sup35p is very high in glutamine and asparagine residues, and these are critical for prion formation and propagation. Overexpression of this prion domain of Sup35p also induces the *de novo* appearance of [PSI] at a much higher frequency than does similar overexpression of the full length Sup35p.

The level of the Hsp104 chaperone is critical for [PSI] propagation with either overexpression or under-expression leading to the loss of the prion. The Hsp70s in the Ssa group are also essential for [PSI] propagation.

ANTIGENIC PROPERTIES

Not studied.

BIOLOGICAL PROPERTIES

[PSI] increases the efficiency with which ribosomes read through termination codons. The Sup35 protein is a subunit of the translation release factor that recognizes terination codons, and releases the peptidyl tRNA and cleaves the nascent peptide from the tRNA. Thus, *sup35* mutants have the same phenotype as do [PSI] strains. Furthermore, the *SUP35* gene is necessary for the propagation of [PSI]. [PSI] can be cured by growth in the presence of 5 mM guanidine HCl, but from the cured strains, can again be isolated subclones carrying [PSI]. Overproduction of Sup35p results in a 100-fold increase in the frequency with which [PSI] arises de novo.

Strains of [PSI], like the strains known for the scrapie agent and other transmissible spongiform encephalopathies, have been found for [PSI]. They are associated with different stabilities, curability by guanidine, and efficiency of suppression of nonsense mutations.

[PSI], [URE3] and [PIN], can be cured by growth of cells on media containing 1-3 mM guanidine HCl. This curing is due to specific inhibition of Hsp104, a disaggregating chaperone necessary for the propagation of each of these prions. By breaking up amyloid filaments into smaller filaments, Hsp104 increases the number of 'seeds' and so insures inheritance of the amyloid by each of the daughter cells.

List of Species Demarcation Criteria

Prions are divided into species based on the identity of the protein that makes up the infectious element. However, even a single protein can form prions of distinct 'strains', or 'variants', apparently due to different self-propagating amyloid structures, much as a given protein can assume more than one crystal form.

List of Species

Species names are in green italic script; strain names and synonyms are in black roman script; tentative species names are in blue roman script. Sequence accession numbers, and assigned abbreviations ( ) are also listed.

Species

[PSI+]

[PSI+] [PSI+]

Tentative Species

None reported.

Type Species *[PIN+]*

Distinguishing Features

[PIN+] is a self-propagating amyloid form of the Rnq1 protein, whose normal function is unknown. [PIN+] has the ability to dramatically increase the frequency with which the [PSI+] prion arises. [PIN+] also has a smaller effect on [URE3] prion generation.

Host

Saccharomyces cerevisiae.

Prion Properties

Morphology

Rnq1p is aggregated in [PIN+] strains and recombinant Rnq1p forms amyloid filaments *in vitro.*

Physicochemical and Physical Properties

Filaments of the Rnq1 protein formed *in vitro* are 11 nm in diameter and stain with Thioflavin T, suggesting amyloid structure.

Nucleic Acid

None

Proteins

Rnq1p is a 405 aa residue protein with a C-terminal domain rich in asparagine and glutamine residues (like the prion domains of Ure2p and Sup35p). This domain can act as a prion domain when fused to the C-terminal part of Sup35p.

Lipids

None

CARBOHYDRATES

None.

GENOME ORGANIZATION AND REPLICATION

The Hsp104 chaperone is essential for [PIN] propagation. The Hsp40 Sis1p is also essential for [PIN] propagation.

ANTIGENIC PROPERTIES

Not studied.

BIOLOGICAL PROPERTIES

[PIN] dramatically increases the frequency with which the [PSI] prion arises *de novo,* so much so that it is nearly essential for [PSI] generation. [PIN] also increases the frequency of [URE3] *de novo* generation, but only 10 to 100 fold. The evidence suggests that the amyloid formed by one asparagine-glutamine rich protein can prime the formation of amyloid by another such protein.

LIST OF SPECIES DEMARCATION CRITERIA

Prions are divided into species based on the identity of the protein that makes up the infectious element. However, even a single protein can form prions of distinct 'strains', or 'variants', apparently due to different self-propagating amyloid structures, much as a given protein can assume more than one crystal form.

LIST OF SPECIES

Species names are in green italic script; strain names and synonyms are in black roman script; tentative species names are in blue roman script. Sequence accession numbers, and assigned abbreviations ( ) are also listed.

SPECIES

[PIN]

[PIN]	[PIN]

TENTATIVE SPECIES

None reported.

Type Species *[Het-s]*

HOST

Podospora anserina, a filamentous fungus.

DISTINGUISHING FEATURES

[Het-s] was found as a non-chromosomal gene necessary for heterokaryon incompatibility based on the chromosomal *het-s/S* locus. [Het-s] is a self-propagating amyloid of the HET-s protein. Heterokaryon incompatibility is a normal fungal function, and so the [Het-s] prion is the first prion known to be responsible for a normal function, a conclusion further confirmed by the fact that nearly all wild strains carry this prion. Unique among all prions, amyloid formed *in vitro* from recombinant HET-s protein has been shown to be infectious to normal *Podospora.*

PRION PROPERTIES

MORPHOLOGY

The HET-s protein shows self-seeding formation of amyloid filaments 15 to 20 nm in diameter.

PHYSICOCHEMICAL AND PHYSICAL PROPERTIES

The amyloid filaments are high in beta sheet content as judged by circular dichroism and FTIR measurements. They show the yellow-green birefringence on staining with Congo Red that is typical of amyloid.

NUCLEIC ACID

None

PROTEINS

The HET-s protein is the product of the *het-s* gene. The prion domain sufficient for amyloid formation *in vitro* and prion propagation *in vivo* is located at the C-terminal part of the molecule. Deletions of the N-terminal part destabilize the prion domain.

LIPIDS

None.

CARBOHYDRATES

None.

GENOME ORGANIZATION AND REPLICATION

The prion domain of the HET-s protein is the C-terminal 72 aa residues. This region is very protease sensitive in the soluble form and protease resistant in the amyloid form. Unlike the proteins that form the [URE3], [PSI] and [PIN] prions, the HET-s protein prion domain is not rich in asparagine or glutamine residues. Residues 23 and 33 are particularly critical in determining whether the protein can undergo the prion change or not.

ANTIGENIC PROPERTIES

Not studied.

BIOLOGICAL PROPERTIES

[Het-s] makes *Podospora anserina* strains differing at the *het-s* locus able to carry out heterokaryon incompatibility. It is the first prions described that is responsible for a normal cellular function, rather than a disease of the organism. When two fungal colonies grow together, the hyphae of the two colonies fuse with each other. This process facilitates cooperation between the colonies in acquisition of nutrients, and perhaps other functions. One risk of this process is that viruses present in one colony will spread into the other colony. Apparently to prevent this, most fungi only will form heterokaryons with closely related strains, identical for alleles at each of 6 or more loci. In *Podospora anserina*, there are 8 such loci called *Het* loci. The *Het-s* locus has alleles *het-s* and *het-S*. The *het-s* strains show incompatibillity for heterokaryon formation with *het-S* strains only if the protein encoded by *het-s* is in a prion form. The presence of this prion is found as a non-Mendelian genetic element, called [Het-s]. [Het-s] is reversibly curable, the frequency of its appearance *de novo* is enhanced by overproduction of the protein encoded by the *het-s* gene, and the *het-s* gene is necessary for the propagation of [Het-s]. The protein encoded by the *het-s* gene is more protease-resistant in strains carrying [Het-s].

[Het-s] is also the only prion for which the role of amyloid has been established by an external infection experiment using recombinant protein. It has been found that amyloid from recombinant HET-s protein formed *in vitro* infects 99% of colonies when introduced by the biolistic method. In contrast, the soluble form of the protein, or aggregates produced by either heat denaturation or acid denaturation have no substantial effect.

List of Species Demarcation Criteria

Prions are divided into species based on the identity of the protein that makes up the infectious element. However, even a single protein can form prions of distinct 'strains', or 'variants', apparently due to different self-propagating amyloid structures, much as a given protein can assume more than one crystal form.

List of Species

Species names are in green italic script; strain names and synonyms are in black roman script; tentative species names are in blue roman script. Sequence accession numbers, and assigned abbreviations ( ) are also listed.

Species

[Het-s]
[Het-s] [Het-s]

Tentative Species

None reported.

Type Species *[BETA]*

Host

Saccharomyces cerevisiae.

Distinguishing Features

An enzyme that is necessary for its own activation (or maturation) will remain inactive if not initially activated or remain active in a self-propagating manner if initially active. This can be the basis for a prion (infectious protein) since transmission of the active enzyme from one individual to another lacking the active enzyme will change the self-propagating status of the recipient individual. In the absence of protease A, which normally can activate protease B, protease B acts as a prion of *Saccharomyces cerevisiae.*

Prion Properties

Morphology

The beta prion is identical with the mature active form of the vacuolar protease B of yeast.

Physicochemical and Physical Properties

Active protease B is a soluble Protease B is synthesized as a precursor protein encoded by *PRB1.* Its activation requires both N- and C-terminal cleavage, which either protease A or mature protease B can carry out.

Nucleic Acid

None

Proteins

The mature active form of Prb1p (vacuolar protease B) is a 284 aa residues protein which is derived by N- and C-terminal cleavages of a 635 residue precursor protein. Prb1p is a serine protease of the subtilisin group with roles in maturation of other vacuolar proteins, and in protein degradation by the vacuole which is critical for meiosis and spore formation and for survival of starvation.

Lipids

None.

Carbohydrates

None.

Genome Organization and Replication

Normally, protease A, as well as mature protease B, can carry out the cleavage-maturation of proprotease B. In a *pep4* deletion mutant lacking protease A, protease B acts as a prion. Cells initially lacking mature active protease B give rise to progeny, nearly all of which lack the active enzyme. Cells initially carrying active mature protease B give rise to offspring, nearly all of which also have active protease B. Transmission of active protease B to a cell lacking it, converts the pro-protease B in the recipient to the active form. This activity is then passed on to progeny by continued auto-activation. About 1 in 10^5 cells spontaneously develops protease B activity.

The protease B precursor protein requires removal of both N-terminal and C-terminal extension for maturation.

Antigenic Properties

Not studied.

Biological Properties

Like the amyloid-related prions, [BETA] is reversibly curable. Activity is lost by culturing cells on media that repress expression of Prb1p, but from a cured strain, [BETA] arises spontaneously in about 1 in 10^5 progeny cells. Overexpression of pro-Prb1p increases the frequency with which [BETA] arises about 100 to 1000 fold. The propagation of [BETA] requires the *PRB1* gene, as expected. [BETA] is present in all wild-type cells, but is inapparent because protease A can activate proprotease B. Only in the absence of protease A does the prion character become evident.

[BETA] allows cells to go through meiosis and spore formation and to better survive starvation, process which require protein breakdown in the vacuole.

List of Species Demarcation Criteria

Species demarcation criteria will be devised once there are more than one prion in this genus to classify.

List of Species

Species names are in green italic script; strain names and synonyms are in black roman script; tentative species names are in blue roman script. Sequence accession numbers, and assigned abbreviations ( ) are also listed.

Species

[BETA]

[BETA] [BETA]

Tentative Species

None reported.

Similarity with Other Taxa

The N-terminal prion domain of Sup35p includes repeat aa sequences similar to the octapeptide repeats in PrP.

Derivation of Names

[BETA] from protease B and the greed letter beta.
[Het-s] and het-s from heterokaryon incompatibility.

[PIN] from [PSI] inducibility
[PSI] from the greek letter Psi
[URE3] and URE2 from ureidosuccinate.
Prion: from infectious protein.

REFERENCES

Chernoff, Y.O., Lindquist, S.L., Ono, B.-I., Inge-Vechtomov, S.G. and Liebman, S.W. (1995). Role of the chaperone protein Hsp104 in propagation of the yeast prion-like factor [psi+]. *Science*, **268**, 880-884.

Coschigano, P.W. and Magasanik, B. (1991). The URE2 gene product of *Saccharomyces cerevisiae* plays an important role in the cellular response to the nitrogen source and has homology to glutathione S-transferases. *Mol. Cell. Biol.*, **11**, 822-832.

Coustou, V., Deleu, C., Saupe, S. and Begueret, J. (1997). The protein product of the het-s heterokaryon incopatibility gene of the fungus *Podospora anserina* behaves as a prion analog. *Proc. Natl. Acad. Sci. USA*, **94**, 9773-9778.

Cox, B.S. (1965). PSI, a cytoplasmic suppressor of super-suppressor in yeast. *Heredity*, **20**, 505-521.

DePace, A.H., Santoso, A., Hillner, P. and Weissman, J.S. (1998). A critical role for amino-terminal glutamine/asparagine repeats in the formation and propagation of a yeast prion. *Cell*, **93**, 1241 - 1252.

Derkatch, I.L., Bradley, M.E., Hong, J.Y. and Liebman, S.W. (2001). Prions affect the appearance of other prions: the story of *[PIN]*. *Cell*, **106**, 171 - 182.

Jung, G., Jones, G. and Masison, D.C. (2002). Amino acid residue 184 of yeast Hsp104 chaperone is critical for prion-curing by guanidine, prion propagation, and thermotolerance. *Proc. Natl. Acad. Sci. USA*, **99**, 9936-9941.

King, C.-Y., Tittmann, P., Gross, H., Gebert, R., Aebi, M. and Wuthrich, K. (1997). Prion-inducing domain 2-114 of yeast Sup35 protein transforms *in vitro* into amyloid-like filaments. *Proc. Natl. Acad. Sci. USA*, **94**, 6618-6622.

Maddelein, M.L., Dos Reis, S., Duvezin-Caubet, S., Coulary-Salin, B. and Saupe, S.J. (2002). Amyloid aggregates of the HET-s prion protein are infectious. *Proc. Natl. Acad. Sci. USA*, **99**, 7402-7407.

Masison, D.C., Maddelein, M.-L. and Wickner, R.B. (1997). The prion model for [URE3] of yeast: spontaneous generation and requirements for propagation. *Proc. Natl. Acad. Sci. USA*, **94**, 12503-12508.

Masison, D.C. and Wickner R.B. (1995). Prion-inducing domain of yeast Ure2p and protease resistance of Ure2p in prion-containing cells. *Science*, **270**, 93-95.

Paushkin, S.V., Kushnirov, V.V., Smirnov, V.N. and Ter-Avanesyan, M.D. (1997). *In vitro* propagation of the prion-like state of yeast Sup35 protein. *Science*, **277**, 381-383.

Roberts, B.T. and Wickner, R.B. (2003). A class of prions that propagate via covalent auto-activation. *Genes Dev.*, **17**, 2083-2087.

Sondheimer, N. and Lindquist, S. (2000). Rnq1: an epigenetic modifier of protein function in yeast. *Molec. Cell*, **5**, 163-172.

Speransky, V., Taylor, K. L., Edskes, H. K., Wickner, R. B. and Steven, A. (2001). Prion filament networks in [URE3] cells of *Saccharomyces cerevisiae*. *J. Cell. Biol.*, **153**, 1327 - 1335.

Taylor, K.L., Cheng, N., Williams, R.W., Steven, A.C. and Wickner, R.B. (1999). Prion domain initiation of amyloid formation *in vitro* from native Ure2p. *Science*, **283**, 1339 - 1343.

TerAvanesyan, A., Dagkesamanskaya, A.R. Kushnirov, V.V. and Smirnov, V.N. (1994). The SUP35 omnipotent suppressor gene is involved in the maintenance of the non-Mendelian determinant [psi+] in the yeast *Saccharomyces cerevisiae*. *Genetics*, **137**, 671-676.

Wickner, R.B. (1994). Evidence for a prion analog in *Saccharomyces cerevisiae*: the [URE3] non-Mendelian genetic element as an altered URE2 protein. *Science*, **264**, 566-569.

Wickner, R.B., Liebman, S.W. and Saupe, S.J. (2004). Prions of yeast and filamentous fungi: [URE3], [PSI^+], [PIN^+] and [Het-s]. In: *Prion Biology and Diseases*, (S.B. Prusiner, ed), pp 305-377. Cold Spring Harbor Laboratory Press, Cold Spring Harbor, New York.

CONTRIBUTED BY

Wickner, R.B.

Part III: The ICTV

Members and Officers of the ICTV 2002 - 2005

Executive Committee ICTV 2002-2005

Dr. L. Andrew Ball (President)
Dept of Microbiology
University Alabama
Birmingham
Alabama 35294
USA
Ph: 1 (205) 934 0864
Fax: 1 (205) 934 1636
E-mail: andyb@uab.edu

Prof. Jack Maniloff (Vice-President)
University of Rochester,
School of Medicine & Dentistry,
Dept. of Microbiology & Immunology
P.O. BOX 672
601 Elmwood Av.
Rochester, NY 14642, USA
Ph: 1 (716)-275-3413
Fax: 1 (716)-473-9573
E-mail: jkmf@mail.rochester.edu

Dr. Claude M. Fauquet (Secretary)
ILTAB/Danforth Plant Science Center
675 N. Warson Rd.
St Louis, MO 63141
USA
Ph: 1 (314) 587-1241
Fax: 1 (314) 587-1956
E-mail: cmf@danforthcenter.org

Dr. Mike A. Mayo (Secretary)
Scottish Crop Research Institute
Invergowrie
Dundee, DD2 5DA
Scotland, UK
Ph: (44) 38-256-2731
Fax: (44) 38-256-2426
E-mail: mmayo@scri.sari.ac.uk

Dr. Brad I. Hillman (Chair Fungus virus Subcommittee)
The State Univ. of New Jersey Rutgers
Cook College
Dept. of Plant Pathology, Foran Hall
59 Dudley Rd.
New Brunswick, NJ 08901-8520, USA
Ph: 1 (732)-932-9375 x 334
Fax: 1 (732)-932-9377
E-mail: hillman@aesop.rutgers.edu

Dr. Justus M. Vlak (Chair Invertebrate virus Subcommittee)
Laboratory of Virology
Wageningen University, Plant Sciences
Binnenhaven 11
6709 PD Wageningen
The Netherlands
Ph: (31) 317 48 30 90
Fax: (31) 317 48 48 20
E-mail: just.vlak@wur.nl

Dr. Alan A. Brunt (Chair Plant virus Subcommittee)
Brayton
The Thatchway
Angmering
W. Sussex, BN16 4HJ, UK
Ph: (44) 1903 785684
Fax: (44) 1903 785684
E-mail: brunt@hriab.u-net.com

Prof. Roger W. Hendrix (Chair Prokaryote virus Subcommittee)
University of Pittsburgh
Dept. of Biological Sciences
A340 Langley Hall
Pittsburg, PA 15260,
USA
Ph: 1 (412) 624-4674
Fax: 1 (412) 624-4870
E-mail: rhx@vms.cis.pitt.edu

Prof. Duncan J. McGeoch (Chair Vertebrate virus Subcommittee)
MRC Virology Unit
Institute of Virology
University of Glasgow
Church Street
Glasgow, G11 5JR, Scotland, UK
Ph: (44) 0 141-330-4645
Fax: (44) 0 141-337-2236
E-mail: d.mcgeoch@vir.gla.ac.uk

Dr. Tony Della-Porta (Chair Virus Data Subcommittee)
Biosecurity and Biocontainment International Consultants Pty Ltd
PO Box 531,
Geelong, Vic., 3220, Australia
Ph: (61) 3 5243 0433
Fax: (61) 3 5243 0052
E mail: Tony.dellaporta@bigpond.com

Ms Lois Blaine
American Type Culture Collection
Bioinformatics Division10801 University
Boulevard, Manassas,
VA 20110-2209, USA
Ph: 1 (703) 365-2749
Fax: 1 (703) 365-2740
E-mail: lblaine@atcc.org

Dr. Harald Brʃssow
Nestle Research Center, Nestec Ltd.,
Vers-chez-les-Blanc,
CH-1000 Lausanne 26,
Switzerland.
Tel: (41) 217 858 676
Fax: (41) 217 858 925
E-mail: harald.bruessow@rdls.nestle.com

Dr. Eric Carstens
Queen's University
Dept. of Microbiology & Immunology,
207 Stuart Street
Kingston, Ontario KTL-3N6, Canada
Ph: 1 (613) 533-2463
Fax: 1 (613) 533-6796
E-mail: carstens@post.queensu.ca

Dr. Ulrich Desselberger
Virologie Moléculaire et Structurale
UMR 2472, CNRS
1 Avenue de la Terrasse, Bât. 14B
91198 Gif-sur-Yvette Cedex
France
Ph: (33) 1 69 82 38 54
Fax: (33) 1 69 82 43 08
Email: Ulrich.Desselberger@gv.cnrs-gif.fr

Dr. Anne-Lise Haenni
Institut Jacques Monod
2 place Jussieu - Tour 43
Paris 75251 Cedex 05
France
Ph: (33) 6 44.27.47.39
Fax: (33) 6 44.27.82.10
E-mail: haenni@ijm.jussieu.fr

Dr. Andrew M.Q. King
Institute for Animal Health
Pirbright Laboratory
Ash Road, Woking
Surrey GU24 0NF, UK
Ph: (44) 01 483 232441
Fax: (44) 01 483 232448
E-mail: amq.king@bbsrc.ac.uk

Prof. Ann Palmenberg
University of Wisconsin
Institute for Molecular Virology
1525 Linden Drive
Madison, WI 53706, USA
Ph: 1 (608) 262-7519
Fax: 1 (608) 262-6690
E-mail: acpalmen@wisc.edu

Dr. H. Josef Vetten
Biologische Bundesanstalt fur Land- u.
Forstwirtschaft (BBA)
Institut fur Pflanzenvirologie,
Mikrobiologie und biologische
Sicherheit (PS), Messeweg 11-12
38104 Braunschweig, Germany
Ph: (49) 531 299 3720
Fax: (49) 531 299 3006
E-mail: h.j.vetten@bba.de

Life-Members ICTV 2002-2005

Prof. H. W. Ackermann
Dr. D. H. L. Bishop
Dr. F. Brown (to 2004)
Prof. J. Casals (to 2004)
Dr. F. J. Fenner
Prof. A. J. Gibbs
Prof. B. D. Harrison
Prof. G. P. Martelli
Dr. F. A. Murphy
Prof. C. R. Pringle
Prof. M. H. V. van Regenmortel (from 2004)

Prof. B. Falk
Dr. M. Gibbs
Dr. A. F. Murant
Dr. D. J. Robinson
Dr. E. Ryabov
Dr. P. Waterhouse

Unassigned Viruses Study Group
Dr. R. G. Milne (Chair)
Dr. A. A. Brunt
Dr. M. A. Mayo

***Varicosavirus* Study Group**
Dr. A. A. Brunt (Chair)
Dr. R. G. Milne
Dr. T. Sasaya
Dr H. J. Vetten
Dr. J. Walsh

Viroid Study Group
Dr. R. Flores (Chair)
Dr. M. Bar-Joseph
Dr. T. O. Diener
Dr. R. A. Owens
Dr. J. W. Randles

Vertebrate virus Subcommittee

Study Groups

Adenoviridae
Arenaviridae
Arteriviridae
Asfarviridae
Astroviridae
Birnaviridae
Bornaviridae
Bunyaviridae
Caliciviridae
Circoviridae
Coronaviridae
Filoviridae
Flaviviridae
Hepadnaviridae
Herpesviridae
Hepeviridae
Orthomyxoviridae
Papillomaviridae
Paramyxoviridae
Parvoviridae
Picornaviridae
Polyomaviridae
Poxviridae
Reoviridae
Retroviridae
Rhabdoviridae
Togaviridae

Subcommittee Members

Prof. D. J. McGeoch (Chair, Vertebrate virus SC)
Dr. M. Benko (Chair, *Adenoviridae* Study Group)
Dr. M. Salvato (Chair, *Arenaviridae* Study Group)
Dr. E. J. Snijder (Chair, *Arteriviridae* Study Group)
Dr. L. Dixon (Chair, *Asfarviridae* Study Group)
Dr. S. Matsui (Chair, *Astroviridae* Study Group)
Dr. J.-A. C. Leong (Chair, *Birnaviridae* Study Group)
Prof. K. M. Carbone (Chair, *Bornaviridae* Study Group)
Prof. R. Elliot (Chair, *Bunyaviridae* Study Group)
Dr. M. P.G. Koopmans (Chair, *Caliciviridae* Study Group)
Dr. D. Todd (Chair, *Circoviridae* Study Group)
Dr. L. Enjuanes (Chair, *Coronaviridae* Study Group)
Dr. S. V. Netesov (Chair, *Filoviridae* Study Group)
Dr. H.-J. Thiel (Chair, *Flaviviridae* Study Group)
Dr. W. S. Mason (Chair, *Hepadnaviridae* Study Group)
Dr. A. J. Davison (Chair, *Herpesviridae* Study Group)
Dr. S. U. Emerson (Chair, *Hepeviridae* Study Group)
Dr. Y. Kawaoka (Chair, *Orthomyxoviridae* Study Group)
Prof. Dr. E.-M. DeVilliers (Chair, *Papillomaviridae* Study Group)

Prof. R. A. Lamb (Chair, *Paramyxoviridae* Study Group)
Dr. P. Tattersall (Chair, *Parvoviridae* Study Group)
Dr. G. Stanway (Chair, *Picornaviridae* Study Group)
Dr. E. O. Major (Chair, *Polyomaviridae* Study Group)
Dr. R. M. L. Buller (Chair, *Poxviridae* Study Group)
Dr. T. Dermody (Chair, *Reoviridae* Study Group)
Dr. M. Linial (Chair, *Retroviridae* Study Group)
Dr. N. Tordo (Chair, *Rhabdoviridae* Study Group)
Dr. S. Weaver (Chair, *Togaviridae* Study Group)

***Adenoviridae* Study Group**
Dr. M. Benko (Chair)
Dr. E. Adam
Dr. B. M. Adair
Dr. G. W. Both
Dr. J. C. Dejong
Dr. M. Hess
Dr. M. Johnson
Dr. A. Kajon
Dr. A. H. Kidd
Dr. H. D. Lehmkuhl
Dr. Q.-G. Li
Dr. V. Mautner
Dr. P. Pring-Akerblom
Dr. G. Wadell

***Asfarviridae* Study Group**
Dr. L. K. Dixon (Chair)
Dr. J. Escribano
Dr. D. Rock
Dr. M. L Salas
Prof. C. Martins
Dr. P. Wilkinson,

***Arenaviridae* Study Group**
Dr. M. Salvato (Chair)
Dr. M. J. Buchmeier
Dr. J. C. S. Clegg
Dr. J.-P. Gonzalez
Dr. I. S. Lukashevich
Dr. C.J. Peters
Dr. R. Rico-Hesse
Dr. V. Romanowski

***Arteriviridae* Study Group**
Dr. E. J. Snijder (Chair)
Dr. M. A. Brinton
Dr. K. S. Faaberg
Dr. E. K. Godeny
Dr. A. E. Gorbalenya
Dr. N. J. Maclachlan
Dr. W. L. Mengeling
Dr. P. G. W. Plagemann

***Astroviridae* Study Group**
Dr. S. M. Matsui (Chair)
Dr. D. M. Bass
Dr. M. J. Carter
Dr. J. E. Herrmann
Dr. D. K. Mitchell
Dr. S. S. Monroe
Dr. A. Sanchez-Fauquier

***Birnaviridae* Study Group**
Prof. J.-A. Leong (Chair)
Dr. B. Delmas
Dr. E. Everitt
Dr. F. Kibenge
Dr. E. Mundt
Dr. V. N. Vakharia
Dr. J.-L Wu

***Bornaviridae* Study Group**
Prof. K. M. Carbone (Chair)
Dr. L. Bode
Dr. J. C. de la Torre
Dr. J. A. Richt
Dr. H. Ludwig
Dr. N. Nowotny
Dr. K. Ikuta
Prof. Dr. L. Stitz
Dr. W. I. Lipkin
Dr. P. Staeheli

***Bunyaviridae* Study Group**
Prof R. M. Elliott (Chair)
Dr. B. J. Beaty
Dr. C. H. Calisher
Dr. R. W. Goldbach
Dr. S. T. Nichol
Dr. A. Plyusnin
Dr. C. Schmaljohn

***Caliciviridae* Study Group**
Dr. M.P.G. Koopmans (Chair)
Dr. K.Y. Green
Dr. S. Nakata
Dr. J. D. Neill
Dr. I. Clarke
Dr. D. O. Matson
Dr. H.-J. Thiel
Dr. T. Ando
Dr. M. K. Estes
Prof. M. J. Studdert
Dr. A. W. Smith

Circoviridae **Study Group**
Dr. D. Todd (Chair)
Dr. S. Hino
Dr. A. Mankertz
Dr. S. Mishiro
Dr. S. Raidal
Prof. B. W. Ritchie
Dr. P. Biagini
Dr. H. Okamoto
Dr. C. Niel
Dr. M. Bendinelli
Dr. C. G. Teo

Coronaviridae **Study Group**
Dr. D. A. Brian (Chair)
Dr. D. Cavanagh
Dr. L. Enjuanes
Dr. K. V. Holmes
Dr. M. M. C. Lai
Dr. H. Laude
Dr. P. S. Masters
Dr. P. J. M. Rottier
Dr. W. J. M. Spaan
Dr. F. Taguchi
Dr. P. J. Talbot

Filoviridae **Study Group**
Dr. H. Feldmann (Chair)
Dr. T. W. Geisbert
Dr. N. Jaax
Dr. P. Jahrling
Prof. Dr. H.-D. Klenk
Dr. T. G. Ksiazek
Dr. S. V. Netesov
Dr. C. J. Peters
Dr. A. Sanchez
Dr. V. E.Volchkov

Flaviviridae **Study Group**
Dr. H.-J. Thiel (Chair)
Dr. M. S. Collett
Dr. E. A. Gould
Dr. F. X. Heinz
Dr. G. Meyers
Dr. R. H. Purcell
Dr. C. M. Rice
Dr. M. Houghton

Hepadnaviridae **Study Group**
Dr. W. S. Mason (Chair)
Dr. C. Burrell
Dr. J. Casey
Dr. W. Gerlich
Dr. C. Howard
Dr. M. Kann
Dr. R. Lanford
Dr. J. Newbold
Dr. J. M. Taylor
Dr. S. Schaefer
Dr. H. Will

Hepeviridae **Study Group**
Dr. S. U. Emerson (Chair)
Dr. D. Anderson,
Dr. M. S. Balayan
Dr. V. A. Arankalle
Dr. M. Purdy
Dr. S. A. Tsarev
Dr. I. Mushahwar

Herpesviridae **Study Group**
Dr. A. J. Davison (Chair)
Dr. R. Eberle
Dr. G. S. Hayward
Dr. D. J. McGeoch
Prof. A. Minson
Dr. P. E. Pellett
Dr. B. Roizman
Prof. M. J. Studdert
Dr. E. Thiry

Orthomyxoviridae **Study Group**
Dr. Y. Kawaoka (Chair)
Dr. N. J. Cox
Dr. N. Kaverin
Prof. Dr. H.-D. Klenk
Prof. R. A. Lamb
Dr. J. Mccauley
Dr. K. Nakamura
Dr. P. Palese
Dr. R. G. Webster

Papillomaviridae **Study Group**
Prof. Dr. E.-M. DeVilliers (Chair)
Dr. H.-U. Bernard
Dr. T. Broker
Dr. H. Delius

Polyomaviridae **Study Group**
Dr. E. O. Major (Chair)
Prof. J. Butel
Prof. D. Chang
Prof. R. Frisque
Prof. K. Nagashima
Prof. M. Tognon

Paramyxoviridae **Study Group**
Prof. R. A. Lamb (Chair)
Dr. W. J. Bellini
Dr. P. L. Collins
Dr. A. J. Easton
Dr. D. Kolakofsky
Dr. B. Eaton
Dr. B. K. Rima

Parvoviridae **Study Group**
Dr. P. Tattersall (Chair)
Dr. M, Bergoin
Dr. M. E. Bloom
Dr. K. E. Brown
Dr. R. M. Linden

Dr. N. Muzyczka
Dr. C. R. Parrish
Dr. P. Tijssen

Picornaviridae **Study Group**
Dr. G. Stanway (Chair)
Dr. F. Brown
Dr. P. Christian
Dr. T. Hovi
Dr. T. Hyypia
Dr. A. M. Q. King
Dr. N. J. Knowles
Dr. S. M. Lemon
Dr. P. D. Minor
Dr. M. A. Pallansch
Prof. A. C. Palmenberg
Dr. T. Skern

Poxviridae **Study Group**
Dr. R. M. L. Buller (Chair)
Dr. R. W. Moyer
Dr. J. J. Esposito
Dr. B. M. Arif
Dr. D. N. Black
Dr. K. Dumbell
Dr. E. J. Lefkowitz
Dr. G. McFadden
Dr. A. Mercer
Dr. B. Moss
Dr. M. A. Skinner
Prof. D. N Tripathy
Dr. R. Wittek

Reoviridae **Study Group**
Dr. T. Dermody (Chair) (*Orthoreovirus*)
Dr. H. Attoui (*Coltivirus*)
Dr. G. Boccardo (*Fijivirus*)
Dr. R. Duncan (*Orthoreovirus*)
Dr. I Holmes (*Reoviridae*)
Dr. Y. Hoshino (*Rotavirus*)
Dr. X. De Lamballerie (*Coltivirus*)
Dr. P. P. C. Mertens (*Orbivirus*)
Dr. R. G. Milne (*Fijivirus, Phytoreovirus, Oryzavirus*)
Dr. H. Mori (*Cypovirus*)
Dr. O. Nakagomi (*Rotavirus*)
Dr. R. F. Ramig (*Rotavirus*)
Dr. S. K. Samal (*Orbivirus, Aquareovirus*)
Dr. N. Upadhyaya (*Oryzavirus*)
Dr. T. Omura (*Phytoreovirus*)

Retroviridae **Study Group**
Dr. Maxine Linial (Chair)
Dr. H. Fan
Dr. B. H. Hahn
Dr. R. Löwer
Dr. J. Neill
Dr. S. Quackenbush
Dr. A. Rethwilm
Dr. P. Sonigo
Dr. J. Stoye
Dr. M. Tristem

Rhabdoviridae **Study Group**
Dr. N. Tordo (Chair)
Dr. S. Nadin-Davis
Prof. R. B. Tesh
Prof. P. Walker
Dr. R. G. Dietzgen
Dr. A. O. Jackson
Dr. R.-X. Fang
Dr. G. Kurath
Dr. A. Benmansour

Togaviridae **Study group**
Dr. S. C. Weaver (Chair)
Dr. T. K. Frey
Dr. H. V. Huang
Dr. R. M. Kinney
Dr. C. M. Rice
Dr. J. T. Roehrig
Dr. R. E. Shope
Dr. E. G. Strauss

The Statutes of ICTV

August 2002

Article 1
Official name

1.1 The official name is the International Committee on Taxonomy of Viruses (ICTV).

Article 2
Status

2.1 The ICTV is a Committee of the Virology Division of the International Union of Microbiology Societies (IUMS).

Article 3
Objectives

The objects of the Committee shall be for the public benefit and in particular to advance education in the taxonomy of viruses and in furtherance thereof. The objectives are –

3.1 To develop an internationally agreed taxonomy for viruses.
3.2 To establish internationally agreed names for virus taxa, virus species and subviral agents.
3.3 To communicate the decisions reached concerning the classification and nomenclature of viruses and subviral agents to virologists by holding meetings and publishing reports.
3.4 To maintain an Official Index of virus names.

Article 4
Membership

4.1 Membership of the ICTV shall be :-
  A. President and Vice-President
  B. Secretaries
  C. Other members of the Executive Committee
  D. National Members
  E. Life Members
  F. Members of the Prokaryote Virus, Fungal Virus, Invertebrate Virus, Plant Virus, Vertebrate Virus, and Virus Data Subcommittees.

4.2 The composition of the ICTV Executive Committee shall be:-
  (i) The Officers of ICTV, namely President, Vice-president and the Secretaries
  (ii) The Chairs of the Executive Committee Subcommittees, namely one each for the Subcommittees on viruses of prokayotes (including Archaea), fungi (including algae), invertebrates, plants and vertebrates, and one for the Virus Data Subcommittee
  (iii) eight elected Members.

4.3 Election or appointment procedures for positions on ICTV and its Executive Committee
  A. President and Vice-President shall be nominated and seconded by any members of the ICTV and elected at a plenary meeting of the full ICTV membership. They shall be elected for a term of three years and may not serve for more than two consecutive terms of three years.
  B. Two Secretaries shall be nominated by the Executive Committee and elected at a plenary meeting of the full ICTV membership. The Secretaries shall be elected for a term of six years and may be re-elected.
  C. The Chairs of the Subcommittees shall be elected by the Executive Committee, normally at an interim meeting preceding a Plenary Meeting of the full ICTV membership. Nominations shall be made to the President at least one week before the election and must be accompanied by indications from each of the nominees that they are willing to serve. The

CODE...

term of office shall be for three years and may not exceed two consecutive terms of three years each.

D. No person shall serve on the Executive Committee for more than four successive complete terms other than as Officers subject to the limitations set out in 4.3.A and 4.3.B

E. Nominees for each vacant Elected Member seat shall be proposed to an ICTV Plenary Meeting by a Nominations Subcommittee comprising the President, Vice-President, both Secretaries, and the Chairs of the Subcommittees. The Nominations Subcommittee will be chaired by the President and a quorum will consist of 5 members. The Nominations Subcommittee will seek to ensure the widest possible ranges of virological expertise and geographical representation on the EC.

In addition, nominations for an Elected Member vacant seat may be submitted to the Nominations Subcommittee by any virologist and will be allowed up to 2 days prior to the Plenary Meeting at which elections will be held. All nominations must be accompanied by a brief biographical sketch and an indication that the nominee is willing to stand for election.

Elected Members shall be nominated, seconded, and elected at a Plenary Meeting of the ICTV for a term of three years and may not serve for more than two consecutive terms of three years each. Generally, three or four of the eight elected Members shall be replaced every three years.

F. National Members shall be nominated by Member Societies of the Virology Division of the IUMS. Societies belonging to the IUMS are considered to be Member Societies of the Division if they have members actively interested in virology. Wherever practicable, each country shall be represented by at least one National Member and no country shall be represented by more than five National Members. Nominations of virologists as National Members shall not require approval by the ICTV. However, it is the responsibility of a Member Society to inform one of the ICTV Secretaries in writing of the selection of their National Member before the start of the Plenary Meeting at which that National Member is to participate.

G. Life Members shall be nominated by the Executive Committee, normally in recognition of outstanding services to virus taxonomy. Currently serving members of the Executive Committee and virologists within 6 months of their retirement from the Executive Committee will not be eligible for election. Life members shall be elected by the full ICTV.

H. Virologists shall be appointed to Subcommittees by the Chairs of the Subcommittees and the appointments will not require approval by the ICTV EC.

I. The EC has the authority to appoint virologists to fill EC positions that are vacant because of resignation or some failure in the normal channels of appointment (4.3 A, B, C, E). Such an interim appointment ends at the Plenary Session following their appointment and time spent in such a role does not count towards formal service on the EC as defined by Statute 4.3 D.

4.4 EC Finance Sub-Committee

The Finance Sub-Committee will consist of the Officers of ICTV.

4.5 Study Groups

Study Groups shall be formed by Subcommittee Chairs to examine the classification and nomenclature of particular groups of viruses. A Chair of a Study Group shall be appointed by the Chair of the appropriate Subcommittee and shall be a member of that Subcommittee *ex officio*. Study Group Chairs are non-voting members of the ICTV.

Chairs of Study Groups shall appoint the members of their Study Groups. Members of Study Groups, other than Chairs, shall not be members of the ICTV, but their names shall be published in the minutes and reports of the ICTV to recognize their valuable contribution to the taxonomy of viruses. Study Group Chairs and Study Group members will normally be appointed immediately after an ICTV Plenary Session and will serve until the following Plenary Session, which is normally a period of three years. Except in unusual circumstances, the term of office of Study Group Chairs will be limited to two consecutive periods of three years.

Article 5
Meetings

5.1 Plenary meetings of the full ICTV membership shall be held in conjunction with the International Congresses of Virology. Meetings of the ICTV Executive Committee shall be held in conjunction with the International Congresses of Virology as well as at least once between International Congresses.

Article 6
Taxonomic Proposals

6.1 Taxonomic proposals may be initiated by an individual member of the ICTV, by a Study Group or by a Subcommittee member. In addition, any virologist may submit a taxonomic proposal or suggestion to the appropriate ICTV Subcommittee Chair following the procedures described below.

Proposals will be sent to the Chair of the appropriate Subcommittee for consideration by that Subcommittee and/or its Study Groups. Proposals will also be sent for consideration to all Study Groups and Subcommittees whose interests might be affected by the taxonomic or nomenclatural changes proposed. Taxonomic proposals approved by any Subcommittee, usually in consultation with a Study Group, shall be submitted to the Executive Committee by the Subcommittee Chair. Proposals approved by the Executive Committee shall be presented for ratification to the full ICTV membership either at the subsequent Plenary meeting or by circulation of proposals by mail followed by a ballot.

Separate proposals shall be required (1) to establish a new taxon, (2) to name a taxon, (3) to designate the type species of a genus and (4) to establish a list of members of a taxon.

Article 7
Voting

7.1 Decisions will be made by a simple majority of either those eligible to vote and present at a meeting or that reply to a call for a vote within 1 month of a proposition being circulated. A quorum for the ICTV Plenary Session decisions will consist of the President or Vice-President together with 15 voting members. In the event of a tied vote the proposal will not be approved.

7.2 The members of the full ICTV who are entitled to vote are the Members of the EC, the National Members and the Life Members.

7.3 Proposals for changes to taxonomy, nomenclature or the ICTV constitution will be voted on in two stages: (1) the EC will vote either at a meeting or by circulation of a ballot if a proposal should be presented to the ICTV for a decision, and (2) the ICTV will decide either at a Plenary meeting or by circulation of a ballot.

Decisions of the EC concerned with changes to taxonomy, nomenclature or the constitution will be by simple majority of members of the EC voting. The quorum shall be the President or Vice-President together with 1 Secretary and **7** members of the EC. Only EC members are entitled to vote on such matters.

7.4 Matters of EC business not directly concerned with changes to taxonomy, nomenclature or the constitution may be decided by consensus under the President's chairmanship. Any EC member may call for a vote on such matters.

Article 8
The Code of Virus Nomenclature

8.1 Classification and nomenclature of viruses and related agents will be subject to Rules formalised into a Code. The Code, and substantive modifications to it, are subject to the approval of ICTV Executive Committee and ICTV as under Article 7.

Article 9
Duties of Officers

9.1 Duties of the President shall be:

A. To preside at meetings of the Executive Committee and plenary meetings of the full ICTV membership.

B. To prepare with the Secretaries the agendas for meetings of the Executive Committee and the plenary meetings of the full ICTV membership.
C. To act as editor-in-chief for ICTV Reports.

9.2 Duties of the Vice-President shall be:
A. To carry out the duties of the President in the absence of the President.
B. To attend meetings of the Executive Committee and plenary meetings of the ICTV.

9.3 Duties of the Secretaries shall be:
A. To attend meetings of the Executive Committee and plenary meetings of the ICTV.
B. To prepare with the President the agendas for meetings of the EC and the plenary meetings of the ICTV.
C. To prepare the Minutes of meetings of the Executive Committee and plenary meetings of the ICTV and circulate them to all ICTV members.
D. To act as Treasurer of the ICTV. To handle any funds that may be allocated to the ICTV by the Virology Division of the IUMS or other sources.
E. To keep an up-to-date record of ICTV membership.

9.4 Duties of the Subcommittee Chairs shall be:
A. To attend meetings of the Executive Committee.
B. To appoint members of the Subcommittee.
C. To appoint Study Group Chairs.
D. To organize the Subcommittee and Study Groups to study taxonomic problems and to bring forward proposals.
E. To present taxonomic proposals to the Executive Committee for voting.
F. To co-ordinate the preparation of up-dates of the ICTV Reports.
G. To report to the plenary meetings of ICTV on taxonomic changes since the preceding ICTV Report and to present any current proposals for further change.

Article 10
Publications

10.1 Changes to taxonomy, nomenclature or the International Code approved by ICTV will be communicated to the virological community by publication in rapid short form, for example in Virology Division News, and as part of the formal triennial Report of ICTV.

10.2 Whenever feasible, taxonomic information will be published in a database form under the guidance of the Virus Data Subcommittee.

10.3 No publication of the ICTV shall bear any indication of sponsorship by a commercial agency, or institution connected in any way with a commercial company, except as an acceptable acknowledgement of financial assistance.

10.4 Publications may bear the name of the ICTV only if all the material contained has been authorized, prepared, or edited by the ICTV, or a committee or subcommittee of the ICTV.

10.5 Publications containing translations of ICTV-approved material may only bear the name of ICTV if they are approved by the ICTV Executive Committee.

Article 11
ICTV Statutes

11.1 The Statutes of the ICTV, and any subsequent changes, shall be approved by votes of the ICTV EC and the full ICTV membership as under Statute 7 and then approved by the Virology Division of the IUMS.

Article 12
Disposition of Funds

12.1 In the event of dissolution of the ICTV, any remaining funds shall be turned back to the Secretary-Treasurer of the Virology Division of the IUMS.

12.2 Any surplus assets/funds must be used for the charitable purposes set out at Article 3 of this constitution or for purposes that are charitable within the context of Section 505 of the Income and Corporation Taxes Act 1988 (or statutory re-enactment thereof).

The International Code of Virus Classification and Nomenclature of ICTV

1. Statutory basis for the International Committee on Taxonomy of Viruses (ICTV)

1.1 The International Committee on Taxonomy of Viruses (ICTV) is a committee of the Virology Division of the International Union of Microbiological Societies. ICTV activities are governed by Statutes agreed with the Virology Division.

1.2 The Statutes define the objectives of the ICTV. These are:-

(i) to develop an internationally agreed taxonomy for viruses

(ii) to develop internationally agreed names for virus taxa, including species and subviral agents.

(iii) to communicate taxonomic decisions to the international community of virologists.

(iv) to maintain an Index of virus names.

1.3 The Statutes also state that classification and nomenclature will be subject to Rules set out in an International Code.

Comment: Ratified changes will be published in Virology Division News in Archives of Virology, and in subsequent ICTV Reports.

2. Principles of Nomenclature

2.1 The essential principles of virus nomenclature are:-

(i) to aim for stability.

(ii) to avoid or reject the use of names which might cause error or confusion.

(iii) to avoid the unnecessary creation of names.

2.2 Nomenclature of viruses and sub-viral agents is independent of other biological nomenclature. Virus and virus taxon nomenclature are recognised to have the status of exceptions in the proposed International Code of Bionomenclature (*BioCode*).

2.3 The primary purpose of naming a taxon is to supply a means of referring to the taxon, rather than to indicate the characters or history of the taxon.

2.4 The application of names of taxa is determined, explicitly or implicitly, by means of nomenclatural types.

2.5 The name of a taxon has no official status until it has been approved by ICTV.

***Comment**: see section 3.8*

3. Rules of Classification and Nomenclature

I - General Rules

The universal scheme

3.1 Virus classification and nomenclature shall be international and shall be universally applied to all viruses.

3.2 The universal virus classification system shall employ the hierarchical levels of Order, Family, Subfamily, Genus, and Species.

Comments: It is not obligatory to use all levels of the taxonomic hierarchy. The primary classification is of viruses into species. Most species are classified into genera and most genera are classified into families. Species not assigned to a genus will be "unassigned" in a family (see Rule 3.6) and genera not classified in families have the status of "unassigned" (sometimes referred to as "floating"). Some families are classified together into Orders, but for many, the family is the highest level taxon in use. Also, families are not necessarily divided into subfamilies. This taxon is to be used only when it is needed to solve a complex hierarchical problem (see Rule 3.29).

Contrasting examples of full classifications of some negative strand RNA viruses are: (1) species *Mumps virus*; genus *Rubulavirus*; subfamily *Paramyxovirinae*; family *Paramyxoviridae*; order *Mononegavirales*, and (2) species *Rice stripe virus*; genus *Tenuivirus* (see also Rule 3.41).

Scope of the classification

3.3 The ICTV is not responsible for classification and nomenclature of virus taxa below the rank of species. The classification and naming of serotypes, genotypes, strains, variants and isolates of virus species is the responsibility of acknowledged international specialist groups.

Comments: Particular virus isolates may be regarded as strains, variants, clusters or other subspecific entities that, together with other entities, constitute a species. Classification of such isolates is not the responsibility of the ICTV but is the responsibility of international specialty groups. It is the responsibility of ICTV Study Groups to decide if an isolate or a group of isolates should constitute a species.

Deciding the names of serotypes, genotypes, strains, variants or isolates of virus species is not the responsibility of the ICTV. However, it is recommended that new names not be the same as, or closely similar to, names already in use (Rule 3.14 for taxa). When a particular virus isolate is designated to represent a species, the decision as to which name will be adopted for the species for formal taxonomic purposes will be the responsibility of the ICTV, usually based on recommendations of a particular Study Group working on behalf of the ICTV. The Study Group will be expected to consult widely so as to ensure the acceptability of names, subject to the Rules in the Code. The policy of the ICTV is that as far as is possible, decisions on questions of taxonomy and nomenclature should reflect the majority view of the appropriate virological constituency.

3.4 Artificially created viruses and laboratory hybrid viruses will not be given taxonomic consideration. Their classification will be the responsibility of acknowledged international specialist groups.

Comments: Naturally occurring isolates that have genomes formed from parts of the genomes of different strains of a virus, either by recombination between genome nucleic acids or by re-assortment of separate genome parts, will be classified either as species or subspecific entities in the same way that other isolates are classified. Neither artificial variants made by recombination or re-assortment nor mutant viruses are subject to the Rules in the Code.

Limitations

3.5 Taxa will be established only when representative member viruses are sufficiently well characterized and described in the published literature so as to allow them to be identified unambiguously and the taxon to be distinguished from other similar taxa.

3.6 When it is uncertain how to classify a species into a genus but its classification in a family is clear, it will be classified as an unassigned species of that family.

Comments: A species can be classified as an unassigned member of a family when no genus has been devised. For example, *Bimbo virus* is a rhabdovirus of vertebrates but is not a member of any of the currently recognised genera in the family *Rhabdoviridae*. Likewise, *Groundnut rosette assistor virus* resembles viruses in the family *Luteoviridae* but is not classified in any of the genera in that family. These viruses are each classified as an unassigned member of their respective families.

3.7 Names will only be accepted if they are linked to taxa at the hierarchical levels described in Rule 3.2 and which have been approved by the ICTV.

Comments: Taxa above the rank of species must be approved before a name is assigned to them. Proposals for the creation of taxa shall be accompanied by proposals for names. A decision to create a taxon can thus be followed immediately by a decision about the name for the taxon. Species will be approved together with their names as a single taxonomic act.

The following example is of a proposal concerning an imaginary virus with the vernacular name of "beta gamma virus" that is related to another virus, "alpha beta virus".

Proposal 1. Approve *Beta gamma virus* as a species containing strains known as "beta gamma virus" and "alpha beta virus".

Proposal 2. Create a genus to contain species resembling *Beta gamma virus*.
Proposal 3. Name the genus created by Proposal 2, *Betavirus*.
Proposal 4. Nominate *Beta gamma virus* as the type species of the genus *Betavirus*.
Proposal 5. Create a family to contain genus *Betavirus* and similar genera.
Proposal 6. Name the family created by Proposal 5, *Betaviridae*.
Proposal 7. Assign species X, Y and Z to genus *Betavirus* (such a proposal should include a listing of the parameters for discriminating between species in the genus *Betavirus*).

II - Rules about naming Taxa

Status of Names

3.8 Names proposed for taxa are "valid names" if they conform to the Rules set out in the Code and they pertain to established taxa. Valid names are "accepted names" if they are recorded as approved International Names in the 6th ICTV Report or have subsequently become "accepted names" by an ICTV vote of approval for a taxonomic proposal.
Comments: A valid name is one that has been published, one that is associated with descriptive material, and one that is acceptable in that it conforms to the Rules in the Code. Accepted names will be kept in an "Index" by the ICTV.

3.9 Existing names of taxa and viruses shall be retained whenever feasible.
Comment: A stable nomenclature is one of the principal aims of taxonomy and therefore changes to names that have been accepted will only be considered in exceptional circumstances, and then only because of serious conflict with the Rules.

3.10 The rule of priority in naming taxa and viruses shall not be observed.
Comments: The earlier of candidate names for a taxon may be chosen as a convenience to virologists, but the Rule ensures that it is not possible to invalidate a name in current use by claiming priority for an older name that has been superceded.

3.11 No person's name shall be used when devising names for new taxa.
Comments: New taxon names shall not be made by adopting a person's name, by adding a formal ending to a person's name or by using part of a person's name to create a stem for a name. When existing names of species incorporate a person's name (for example, *Shope papilloma virus*) continued usage of this name, in agreement with Rule 2.3 and 3.9, is in general preferable to the creation of a new name.

3.12 Names for taxa shall be easy to use and easy to remember. Euphonious names are preferred.
Comments: In general, short names are desirable and the number of syllables should be kept to a minimum.

3.13 Subscripts, superscripts, hyphens, oblique bars and Greek letters may not be used in devising new names.
Comments: The Rule is intended to make text unambiguous and easy to manipulate and its application should often make names more pronouncable, in agreement with Rule 3.12. Existing names of some species violate this Rule (e.g. *Enterobacteria phage λ*), but international names of genera and families do not.

3.14 New names shall not duplicate approved names. New names shall be chosen such that they are not closely similar to names that are in use currently or have been in use in the recent past.
Comments: The name selected for a new taxon should not sound indistinguishable from the name of another taxon at any rank or from any taxon. For example, the existence of the genus *Iridovirus* means that forms of new name such as "irodovirus" or "iridivirus" are unacceptable as they are too easily confused with an approved name. Confusion can also be between species and genus names as both end in "virus". Thus, for example, the name selected for a genus typified by a species "Omega virus" would not be named "Omegavirus" because species and genus would then be too readily confused.

3.15 Sigla may be accepted as names of taxa, provided that they are meaningful to virologists in the field, normally as represented by Study Groups.

Comments: Sigla are names comprising letters and/or letter combinations taken from words in a compound term. The name of the genus *Comovirus* has the sigla stem "Co-" from cowpea and "-mo-" from mosaic; the name of the genus *Reovirus* has the sigla stem "R" from "Respiratory, "e" from "enteric" and "o" from "orphan".

Decision making

3.16 In the event of more than one candidate name being proposed, the relevant Subcommittee will make a recommendation to the Executive Committee of the ICTV, which will then decide among the candidates as to which to recommend to ICTV for acceptance.
Comments: When there is more than one candidate name for the same taxon, the decision as to which will be accepted shall be made on the basis of the Code and, if necessary, thereafter on the basis of likely acceptability to the majority of virologists.

3.17 If no suitable name is proposed for a taxon, the taxon may be approved and the name will be left undecided until the adoption of an acceptable international name when one is proposed to and accepted by ICTV.
Comments: Occasionally the names proposed for taxa are judged unsuitable and revisions are requested. To facilitate the classification into appropriate taxa, it is preferable to accept the creation of a taxon with a temporary name rather than delay. Temporary names are indicated by quotation marks and the type species name is used. For example, Soybean chlorotic mottle virus is the type species in an as yet un-named genus in the family Caulimoviridae. Until the genus is named, it is designated as "soybean chlorotic mottle-like viruses". This designation is regarded as temporary.

3.18 New names shall be selected such that they, or parts of them, do not convey a meaning for the taxon which would either (1) seem to exclude viruses which lack the character described by the name but which are members of the taxon being named, or (2) seem to exclude viruses which are as yet undescribed but which might belong to the taxon being named, or (3) appear to include within the taxon viruses which are members of different taxa.

3.19 New names shall be chosen with due regard to national and/or local sensitivities. When names are universally used by virologists in published work, these or derivatives shall be the preferred basis for creating names, irrespective of national origin.

Procedures for naming taxa

3.20 Proposals for new names, name changes, establishment of taxa and taxonomic placement of taxa shall be submitted to the Executive Committee of the ICTV in the form of taxonomic proposals. All relevant ICTV subcommittees and study groups will be consulted prior to a decision being taken.
Comments: Proposals concerning a family containing genera of viruses that infect diverse types of host (eg plants and vertebrates, fungi and plants, and so on) must be considered by the Subcommittees responsible for viruses of each host type (ie Plant viruses, Vertebrate viruses, and so on). For example, taxonomic proposals concerned with the family Partitiviridae would be considered by the Fungal Virus Subcommittee and one of its Study Groups but because some genera in the family contain viruses of plants, proposals affecting the family would also be considered by the Plant Virus Subcommittee.

III - Rules about Species

Definition of a virus species

3.21 A virus species is defined as a polythetic class of viruses that constitutes a replicating lineage and occupies a particular ecological niche.

Definition of "tentative" status

3.22 When an ICTV Subcommittee is uncertain about the taxonomic status of a new species or about assignment of a new species to an established genus, the new species will be listed as a tentative species in the appropriate genus or family. Names of tentative species, as of taxa

generally (Rule 3.14), shall not duplicate approved names and shall be chosen such that they are not closely similar to names that are in use currently, names that have been in use in the recent past, or names of definitive species.
Comments: Species classified as tentative are candidates for taxonomic decision by the appropriate Study Groups to resolve their tentative status.

Construction of a name

3.23 A species name shall consist of as few words as practicable but be distinct from names of other taxa. Species names shall not consist only of a host name and the word "virus".
Comments: Species names normally comprise more than one word. The styles used when virus names are devised differ according to the traditions of the particular fields of virology. For example, plant virus names are usually constructed as host + symptom + "virus" (e.g. tobacco necrosis virus) whereas, in contrast, viruses in the family *Bunyaviridae* are usually named after the location at which the virus was found + "virus" (e.g. Bunyamwera virus).

3.24 A species name must provide an appropriately unambiguous identification of the species.
Comments: Species names should be distinctive. They should not be in a form that could be easily confused with the names of other taxa.

3.25 Numbers, letters, or combinations thereof may be used as species epithets where such numbers and letters are already widely used. However, newly designated serial numbers, letters or combinations thereof are not acceptable alone as species epithets. If a number or letter series is in existence it may be continued.
Comments: The existence of species names such as Bovine adenovirus A, Bovine adenovirus B, and so on, and Human herpesvirus 1, Human herpesvirus 2, Human herpesvirus 3, and so on justify the use of the next number in the series. A name such as "22" or "A7" is not acceptable.

IV - Rules about Genera

3.26 A genus is a group of species sharing certain common characters.
3.27 A genus name shall be a single word ending in *...virus*.
3.28 Approval of a new genus must be accompanied by the approval of a type species.

V - Rules about Subfamilies

3.29 A subfamily is a group of genera sharing certain common characters. The taxon shall be used only when it is needed to solve a complex hierarchical problem.
3.30 A subfamily name shall be a single word ending in *...virinae*.

VI - Rules about Families

3.31 A family is a group of genera (whether or not these are organized into subfamilies) sharing certain common characters.
3.32 A family name shall be a single word ending in *...viridae*.

VII - Rules about Orders

3.33 An order is a group of families sharing certain common characters.
3.34 An order name shall be a single word ending in *...virales*.

VIII - Rules about Sub-viral Agents

Viroids

3.35 Rules concerned with the classification of viruses shall also apply to the classification of viroids.
3.36 The formal endings for taxa of viroids are the word "viroid" for species, the suffix "-viroid" for genera, the suffix "-viroinae" for sub-families (should this taxon be needed) and "-viroidae" for families.
Comments: For example, the species *Potato spindle tuber viroid* is classified in genus *Pospiviroid*, and the family *Pospiviroidae*.

Other sub-viral Agents

3.37 Retrotransposons are considered to be viruses in classification and nomenclature

3.38 Satellites and prions are not classified as viruses but are assigned an arbitrary classification as seems useful to workers in the particular fields.

IX - Rules for Orthography

3.39 In formal taxonomic usage, the accepted names of virus Orders, Families, Subfamilies, and Genera are printed in italics and the first letters of the names are capitalized.
Comments: See Rule 3.8 for the definition of an "accepted" name.

3.40 Species names are printed in italics and have the first letter of the first word capitalized. Other words are not capitalized unless they are proper nouns, or parts of proper nouns.
Comments: When used formally, as labels for taxonomic entities, the names "*Tobacco mosaic virus*" and "*Murray River encephalitis virus*" are in the correct form and typographical style. Examples of incorrect forms are Aspergillus niger virus S (not italic), *Murray river encephalitis virus* (River is a proper noun) or tobacco mosaic virus (not capitalized or italic).

Taxa are abstractions and thus when their names are used formally, these are written distinctively using italicization and capitalization. In other senses, such as an adjectival form (e.g. the tobacco mosaic virus polymerase) italics and capital initial letters are not needed. Equally, these are not needed when referring to physical entities such as virions (e, g. a preparation or a micrograph of tobacco mosaic virus).

This Rule was introduced in 1998 and is in contradistinction to Rules in the Code published in the 6th ICTV Report.

3.41 In formal usage, the name of the taxon shall precede the term for the taxonomic unit.
Comments: For example, the correct formal descriptions of various taxa are ... the family *Herpesviridae* ... the genus *Morbillivirus*, ...the genus *Rhinovirus*, ...the species *Tobacco necrosis virus*, and so on.

August 2002

Part IV: The Indexes

Virus Index

A

INDEXES

B

C

D

INDEXES

E

INDEXES

F

G

H

I

J

K

L

INDEXES

M

N

O

P

Q

R

S

R

S

T

U

INDEXES

V

W

X

Y

Z

Taxonomic Index